British Isles

Outer Isles

100km

Scottish Highlands

Great Britain

Lake District

R.Tees

Ireland

Yorkshire Dales

Peak District

Connemara

Isle of Man

R.Humber

East Anglia

Snowdonia

R.Thames

R.Severn

Isles of Scilly
(Scillies)

Alderney

France

Guernsey

Channel Islands

Jersey

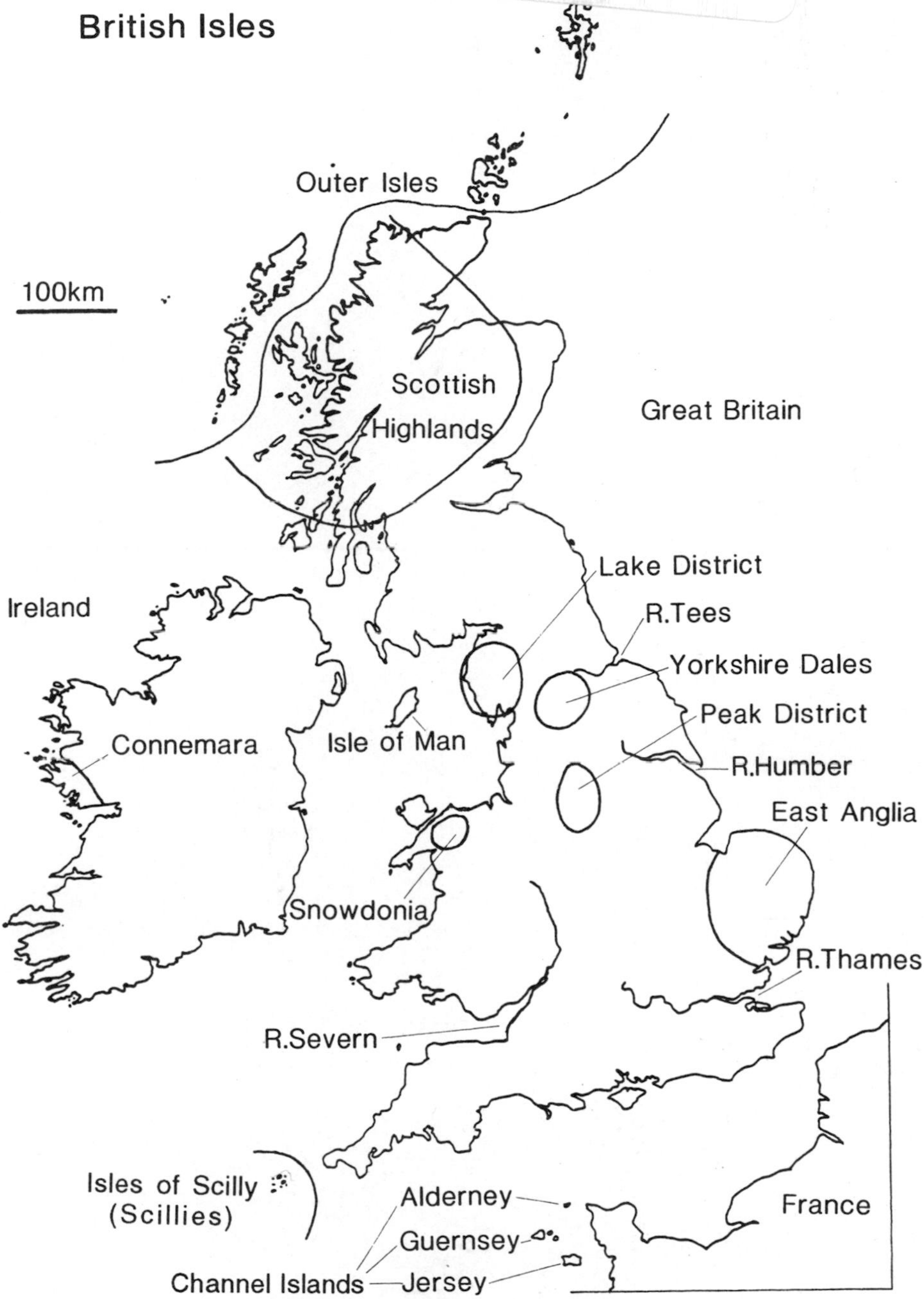

Since its publication in 1991, *New Flora of the British Isles* has become the standard work on British plant identification. It is designed to be user-friendly throughout and to serve as a practical database for taxonomists, ecologists, conservationists, plant geographers, teachers and students, as well as for amateur botanists and plant hunters. The descriptions have been drawn up using living plant material as well as the literature, and the keys have all been written anew, with the aims of simplicity, unambiguity and accuracy borne in mind. The plants included cover all natives, all naturalized plants, all crop plants and all recurrent casuals. The format of the book ensures ease of use both in the laboratory or home and in the field. Technical terms are used wherever necessary but are kept to an essential minimum. Over 160 pages of specially prepared illustrations are provided, including line diagrams or photographs of diagnostic parts of critical groups, and plant portraits of less familiar, alien species for which illustrations are not readily available.

In this new edition the text has been revised thoroughly throughout and adjustments made to many of the illustrations to ensure that the work is fully up-to-date. Over 200 species and subspecies have been added, together with numerous extra hybrids, bringing the total number of taxa covered to over 4500.

NEW FLORA OF THE BRITISH ISLES

To MARGARET, RICHARD and MARTIN

with love and gratitude

NEW FLORA

OF THE

BRITISH ISLES

SECOND EDITION

CLIVE STACE

with illustrations mainly
by
HILLI THOMPSON

CAMBRIDGE
UNIVERSITY PRESS

PUBLISHED BY THE PRESS SYNDICATE OF THE UNIVERSITY OF CAMBRIDGE
The Pitt Building, Trumpington street, Cambridge CB2 1RP, United Kingdom

CAMBRIDGE UNIVERSITY PRESS
The Edinburgh Building, Cambridge CB2 2RU, United Kingdom
40 West 20th Street, New York, NY 10011-4211, USA
10 Stamford Road, Oakleigh, Melbourn 3166, Australia

First published 1991
Reprinted with corrections 1992
Reprinted with corrections 1995
Second edition 1997

Printed in the United Kingdom by The Bath Press, Bath

A catalogue record for this book is available from the British Library

ISBN 0 521 58933 5 plastic covers

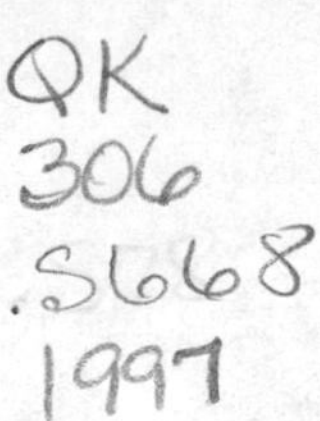

CONTENTS

PREFACE TO FIRST EDITION

In writing this book I have attempted to produce exactly the kind of Flora that for twenty to thirty years I have wanted for my own use. Such a Flora would be as complete, up-to-date and user-friendly as possible, would be selectively illustrated, and would be available at a reasonable price.

No doubt many people will find camera-ready-copy of the sort utilized here less attractive than traditional type-setting, but the costs of the latter would have increased the price of the book very steeply without, in my opinion, increasing its utility in any way. Others might well decry the consistent use of English names (albeit completely subsidiary to the scientific Latin names), but I strongly believe that the study of wild plants by many more people with very diverse backgrounds is important if we are to convince the politicians that we must effectively conserve our native plant genetic diversity. For the same reason I have used fewer technical terms and fewer abbreviations of them than is usual in Floras; indeed, with hindsight I believe I should have used even fewer.

None of the above, however, is to be seen in any way as compromising or diminishing the need for absolute accuracy. When it is necessary to use a greater magnification than a hand-lens, or to cut sections of an organ, in order to see the diagnostic features, I have never pretended otherwise. The lack of a means of magnifying objects above x20 in good illumination, or of the ability to measure accurately to within 0.1mm, not only prevents one from obtaining certain data but, more seriously, is a frequent cause of misinterpretation or mismeasurement of plants. The remedy is obvious, and no more expensive than are the essential tools of a photographer, ornithologist or golfer.

The Flora is designed to enable field-botanists and those working with herbarium specimens to identify plants that are found **in the wild** in the British Isles. This is, I believe, a new criterion; more usually the origin and performance of plants are given higher significance when deciding where to draw the line between those to be included and those to be excluded. However, when one encounters a plant it is often not possible to know whether it is native or alien, or whether it has arrived accidentally or been planted, and one cannot know whether it will still be there next year! Hence a pragmatic approach has been adopted. The list of species included, as well as the data provided for each of them (especially their nomenclature and distribution), are as up-to-date as the information I possessed in October 1990 would allow.

The decision to write this Flora was made in 1983, and the actual preparation of the text took almost seven years. This decision was made due to the collapse of plans by a group of taxonomists to write a multi-volume definitive or critical Flora of Great Britain and Ireland, which had been discussed over the period December 1973 to January 1985. In the early 1980s it became clear that this project would founder (just as a similar one did in the 1930s), and it is regrettably the case that the British Isles still lack a truly complete Flora. Much information (especially concerning aliens) used in writing the present book was originally obtained by me for the abandoned floristic project in the late 1970s. However, this Flora in no way replaces the latter.

In the Acknowledgements I have tried to convey my indebtedness to all those who have helped me over the past 40 or more years and during the writing of this book. Despite all their help I am aware that many imperfections remain, and doubtless errors will be uncovered as well. I should be very grateful to hear of those encountered by readers. If this Flora helps others to achieve anything approaching the degree of enjoyment and satisfaction that I have gained from the study of our wild plants, then it will have succeeded in its main aim.

Ullesthorpe, Leicestershire CLIVE A. STACE
March 1991

ACKNOWLEDGEMENTS

This book certainly could not have been written without the assistance so readily given me by over 200 friends and correspondents over the past twelve years, quite apart from the help and encouragement I received for many years prior to that. Rather than make lame reference to people 'too numerous to mention', I have attempted to list all of those who have directly helped me. Their participation has ranged from essentially one-word answers to questions such as 'Is the *Phyteuma spicatum* naturalized in your area blue- or yellow-flowered?', to detailed advice on taxonomic problems, the provision of specimens, or careful proof-reading or factual checking of draft accounts. Others, mentioned elsewhere, have assisted in more specific ways, such as by preparing drawings.

As well as thanking those who have directly helped in the preparation of this Flora, I wish to pay tribute to a number of people who have been instrumental in guiding me along the road to becoming a botanist. From the age of about 8 my parents actively encouraged my interest in natural history and helped to develop contacts with the local museum and societies in Tunbridge Wells, Kent. In my teens I received a tremendous amount of tuition and stimulation from Aline Grasemann (Tunbridge Wells Natural History Society) and Dik Shaw (my biology master). Our 'regional expert', Francis Rose, was also very influential, especially after the formation of the Kent Field Club in 1955. My first post-card from Miss Grasemann (dated 13th September 1953) reads: "If you are not doing anything else next Saturday what about coming over to the Dykes from Tonbridge Castle and seeing the Orange Balsam? And if you want to come and look anything up in 'Clapham' afterwards do." 'Clapham' (Clapham, Tutin & Warburg 1952) was to be a Christmas present from her three months later!

In all my years at University, as both student and lecturer, I have been fortunate in working with many very clever and helpful people. I should especially mention my postgraduate supervisor, Arthur Exell, and one of my professors at Manchester, David Valentine. The other great source of inspiration to me has been the Botanical Society of the British Isles. In the fifties and sixties those of its members who were of particular help to me, both on excursions and by post, were Joan and Peter Hall, Douglas Kent, Ted Lousley, David McClintock and Ted Bangerter.

Finally, my wife Margaret and sons Richard and Martin deserve special thanks. To someone who is totally addicted to field botany encouragement is both unnecessary and inappropriate, but their understanding and support have been crucial.

Of all those who have assisted me during the writing of this book, a few demand particular mention. This project has been very much a joint project with my wife and it has absorbed much of her spare time over the past five years. She played a major role in the planning of the format and in proof-reading, and carried out all the inputting and the preparation of camera-ready copy. Douglas Kent, who has been preparing a new checklist of vascular plants of the British Isles in parallel with my work, has given me the benefit of his vast knowledge of alien as well as native plants and of nomenclatural matters. He has corrected many nomenclatural errors in my drafts. Hilli Thompson has skilfully prepared the great majority of the line-drawings, patiently taking account of all my demanding and pernickety criticisms and requests for alterations. Peter Hall has painstakingly proof-read all the text, ensuring that there are many fewer mistakes and inconsistencies than would otherwise have been the case, and has helped in numerous places by drawing on his (and his wife Joan's) long field experience. Many people have advised me on alien plants, particularly on those that should and those that should not be treated by me, but by far the most help has been received from David McClintock and Eric Clement. Their knowledge of British alien plants is unparalleled, and they shared it freely with me. I am more indebted than I can adequately express to all the above for their generous co-operation.

The long list of others to whom I offer my sincere thanks follows: Kenneth Adams, John Akeroyd, Abdul-Karim Al-Bermani, David Allen, Mark Atkinson, John Bailey, Peter Ball, G.H. Ballantyne, Bernard Baum, E.P. Beattie, Stan Beasley, Peter Benoit, E.M. Booth, Humphry Bowen, Paul Bowman, John Bowra, Eileen Bray, Anne Brewis, Mary Briggs, Dick Brummitt, Elaine Bullard, John Burlison, Rodney Burton, Andrew Byfield, Mary Caddick, Douglas Chalk, Arthur Chater, Eric Chicken, Tony Church, Peter Clough, P.R. Colegate, Ann Conolly, Tom Cope, Pam Copson, Roderick Corner, Eva Crackles, Diana Crichton, Gigi Crompton, David Curry, Tom Curtis, Kery Dalby, Dick David, John Day, David Dow, Ursula Duncan, Trevor Elkington, Gwynn Ellis, Trevor Evans, Lynne Farrell, F. Fincher, Bryan Fowler, Christopher Fraser-Jenkins, Jeanette Fryer, Gill Gent, Joan Gibbons, Vera Gordon, Richard Gornall, Gordon Graham, Florence Gravestock, Peter Green, Eric Greenwood, Adrian Grenfell, Paul Hackney, G. Haldimann, Joan Hall, Geoffrey Halliday, Gordon Hanson, Ray Harley, Gerald Harrison, John Harron, Clare Harvey, Chris Haworth, Stan Heyward, Sonia Holland, Kathleen Hollick, Florence Houseman, Enid Hyde, Bertil Hylmö, Ruth Ingram, Martin Ingrouille, Charlie Jarvis, Charles Jeffrey, Clive Jermy, John Jobling, Bengt Jonsell, Stephen Jury, Joachim Kadereit, John Kelcey, Archie Kenneth, Michel Kerguélen, Mary Kertland, Mohammed Khalaf, Peter Knipe, Doreen Lambert, Jacques Lambinon, Meredith Lane, David Lang, Peter Langley, Ailsa Lee, Alan Leslie, Frances Le Sueur, Richard Libbey, David Long, David Mabberley, Hugh McAllister, David McCosh, Len Margetts, Mary Martin, John Mason, Brian Mathew, C.T.F. Medd, Desmond Meikle, Ronald Melville, Guy Messenger, J.M. Milner, Alan Mitchell, M. Morris, Mike Mullin, A.R.G. Mundell, Michael Nelhams, Charles Nelson, Alan Newton, Hans den Nijs, Tycho Norlindh, Elizabeth Norman, John Palmer, Richard Palmer, Richard Pankhurst, Rosemary Parslow, James Partridge, Ron Payne, Franklyn Perring, Ted Phenna, Eric Philp, Ann Powell, Chris Preston, Tony Primavesi, Cecil Prime, Jim Ratter, John Richards, Dick Roberts, A.W. Robson, R.G.B. Roe, Francis Rose, Krzysztof Rostanski, Fred Rumsey, Jocelyn Russell, Alison Rutherford, Patience Ryan, Bruno Ryves, Margaret Sanderson, Maura Scannell, H.D. Schotsman, Walter Scott, Peter Sell, Ted Shaw, B. Shepard, Alan Silverside, B. Simpson, David Simpson, W.A. Sledge, Philip Smith, Sven Snogerup, Keith Spurgin, Rod Stern, Olga Stewart, Allan Stirling, Lawrence Storer, David Streeter, Barbara Sturdy, George Swan, Joan Swanborough, Eric Swann, Pierre Taschereau, Michael Taylor, Nigel Taylor, David Tennant, Richard Thomas, Stephanie Thomson, Göran Thor, John Trist, Ian Trueman, Bill Tucker, Maureen Turner, Tom Tutin, David Valentine, Ted Wallace, Geoffrey Watts, David Webb, Mary McCallum Webster, Sarah Webster, David Welch, Derek Wells, Terry Wells, Christopher Westall, Ann Westcott, Mike Wilkinson, Arthur Willis, Chris Wilmot-Dear, Stan Woodell, Brian Wurzell, Goronwy Wynne, Peter Yeo, Jerzy Zygmunt.

PREFACE TO SECOND EDITION

This second edition of *New Flora of the British Isles* incorporates the new information that has arrived in the six years (to September 1996) since the first edition was completed, corrects errors and ambiguities detected in the latter, and introduces a number of improved features. The most important changes are

- inclusion of c.320 additional taxa, including 129 fully treated at species level, so that now over 4600 taxa are covered in one way or another
- use of superior computer ware and laser-printing to produce higher quality camera-ready-copy, resulting in fewer pages despite the increase in text
- citation of chromosome numbers for all taxa where known
- standardization of authority abbreviations with those listed in Brummitt & Powell (1992)
- provision of a full index down to subspecies level, including author citations and English names.

In addition, nomenclatural changes have been made where they have been shown to be necessary, but have been kept to a minimum; a slightly fuller treatment of *Rubus, Euphrasia, Taraxacum* and *Hieracium* has been provided; a number of extra figures have been added and others modified; the R/RR/RRR classification of rare plants has been updated; and the number of technical terms used throughout the book has been slightly reduced.

Of the 129 extra species treated, one is a suspected native (*Sorbus domestica*), one results from a new monograph raising a subspecies to species rank (*Arctium nemorosum*), and the other 127 are aliens, both those now considered to be naturalized and those that now seem frequent enough as casuals. Many of these extra aliens were formerly included as 'other spp.' in the first edition, but their removal from that category has been more than compensated for by the addition of others. The reappraisal of the status of alien plants has been made possible by two recent publications (Clement & Foster 1994; Ryves, Clement & Foster 1996). As a result of this I believe that all currently naturalized plants are included in the *Flora*, and that at least a mention is made of all other aliens that have been recorded from more than 4 localities since 1930.

I have received the same level of help and support in preparing the second edition that I enjoyed when writing the first. Duggie Kent, Peter Hall and Hilli Thompson deserve particular mention for their continuing roles, and I should also like to thank Gwynn Ellis for preparing the index to the second edition. Additions to the many correspondents that I previously listed and to whom I offer my sincere thanks are Phyllis Abbott, Dick Barrett, Richard Bateman, Gillian Beckett, John Benson, Jim Bevan, John Blame, Brian Bonnard, Chris Boon, Michael Braithwaite, John Bruinsma, A.L. Bull, John Bullock, David Coombe, Allen Coombes, Jim Dickson, Philip Drazin, Andrew Dudman, Leni Duistermaat, J. Edelsten, Ian Evans, S.B. Evans, Larch Garrad, Peter Gateley, John Gibson, Jean Green, Paul Green, Chris Hall, David Hambler, Eric Hardy, Paul Harmes, Edith Hesselgreaves, Ellen Heywood-Waddington, Nick Hodgetts, George Hutchinson, Ann Hutchison, J. Iliff, Helen Jackson, Michael Wyse Jackson, Peter Wyse Jackson, Sylvia Jay, Nejc Jogan, Graeme Kay, Quentin Kay, John Killick, Mark Kitchen, Geoffrey Kitchener, Adrian Knowles, Richard Lester, Len Livermore, Douglas McKean, Peter Macpherson, Steve Madge, David Mardon, Marcia Marsden, Roy Maycock, Ruud van der Meijden, Christoph Metelmann, J.D.A. Miller, Rose Murphy, Jackie Muscott, Philip Nethercott, Joy Newton, Jack Oliver, Margaret Palmer, John Parnell, Alison Paul,

David Pearman, Donald Piggott, Rachel Rabey, Alan Radcliffe-Smith, Syliva Reynolds, Tim Rich, Hilary Rose, Lesley Rose, Martin Sanford, Julian Shaw, Graeme Smith, Joyce Smith, Ros Smith, Roy Smith, Tony Smith, Laurie Spalton, Chris Stapleton, Paul Tabbush, Ray Takagi-Arigho, David Taylor, Bernard Thompson, John Timson, Bill Tucker, Roger Veall, Jaap Vermeulen, Max Walters, M.A. Walton, Keith Watson, John White, Geoffrey Wilmore, Brenda Woodliff.

The aims of this second edition remain exactly the same as those expressed in the first, and again I would be glad to hear the comments and criticism of readers.

Ullesthorpe, Leicestershire CLIVE A. STACE
September 1996

INTRODUCTION

The following paragraphs are intended to explain the contents and arrangement of the Flora and the reasons for the various conventions adopted and decisions taken.

TAXONOMIC SCOPE

All vascular plants (pteridophytes, gymnosperms and angiosperms) are included, as is traditional in British Floras. These are placed in classes (Lycopodiopsida, etc.), subclasses (Magnoliidae, etc.), families (Lauraceae, etc.), genera, species and subspecies. In addition the Synopsis lists within the angiosperms superorders (Magnoliiflorae, etc.) and orders (Laurales, etc.). Below the family level, subfamilies or tribes (both in the case of Asteraceae and Poaceae) are defined only for those families with 20 or more genera. Below the genus level, subgenera or sections are defined only for those genera with at least 20 species.

Apomictic microspecies are covered in full in most genera, but not for the three notorious genera *Rubus*, *Taraxacum* and *Hieracium*, for which specialist accounts already exist or are in preparation by experts. In these genera a separation into relatively easily recognized groups of microspecies (here called sections) is provided instead. A full account of these genera would have greatly exceeded my own abilities and the scope of one volume. In addition to the above 3 problem genera, *Ranunculus auricomus* is an apomictic complex in which probably over 100 microspecies could be segregated; however, this complex has not been sufficiently researched in BI to permit such a detailed treatment.

The coverage of alien taxa has been as thorough and consistent as possible. Many more aliens are included than in any other British Flora, yet a considerable number of aliens traditionally to be found in other Floras have been omitted. To merit inclusion, an alien must be either naturalized (i.e. permanent and competing with other vegetation, or self-perpetuating) or, if a casual, frequently recurrent so that it can be found in most years. All this applies as much to garden-escapes or throwouts as to unintentionally introduced plants. Rarity, or the requirement of a highly specialized habitat, have not been taken into consideration (any more than is the case with natives). Cultivated species have been included if they are field-crops or forestry-crops or, in the case of trees only, ornamentals planted on a large scale. Exclusively garden plants, however abundant, whether crops or ornamentals, have not been covered, but most of the commoner taxa are included anyway because of their occurrence as escapes or throwouts. Also excluded are non-tree ornamentals planted en masse on new roadsides or in parks, etc. The aim of this re-vamped and expanded set of criteria is to include **all taxa that the plant-hunter might reasonably be able to find 'in the wild' in any one year.** Any such plant, whether native, accidentally introduced or planted, affects wild habitats and is part of the ecosystem, and botanists and others might be expected to need or want to identify it. Ornamental trees (but not shrubs or herbs) have been included because they are long-lived and frequently persist decades after all other signs of planting have disappeared from the area, so that the finder could not be expected to know that they were once planted. Doubtless, some additions to and removals from the list finally adopted are justified, but the selection of taxa is as judicious as it is largely due to the enormous help I have received from many correspondents (but especially Eric Clement, Douglas Kent and David McClintock), who have made alien plants their special study and have generously given me the benefit of their advice.

As well as the taxa treated 'fully' (i.e. keyed out, and provided with a numbered entry), other taxa that narrowly miss qualification, or fall far short of qualification despite their frequent inclusion in other Floras, or have been erroneously included in the past, are also briefly mentioned. These are covered under families with a number followed by a letter (e.g. 11A), or under the headings 'Other genera' or 'Other spp.' that follow the keys to genera or species.

All interspecific and intergeneric hybrids are included, but the level of treatment varies. Hybrids that have attained distributions no longer tied to those of their parents (i.e. those that occur at least sometimes in the absence of both parents) are treated exactly like species, except that the multiplication sign is inserted between the generic name and the specific epithet (e.g. *Salix x rubens*), and the parental formula is given (e.g. *S. alba* x *S. fragilis*). This has been normal procedure for some genera (e.g. *Circaea, Mentha*) in the past, but a consistent application of the criterion has resulted in many more such taxa being similarly treated. Other hybrids are placed in their appropriate systematic position, but are not keyed and are not provided with their own number; they always occur with at least one parent and their identity can usually be deduced because of this. They are provided with as much information as their situation appears to warrant. The only exception to the above is that some highly fertile hybrids that can occur in the absence of both parents (e.g. *Geum, Hyacinthoides*) are not treated as separate entities, since they form a spectrum of variation linking that of their parents.

GEOGRAPHICAL SCOPE

This Flora deals with the British Isles, comprising Great Britain (England, Scotland and Wales), Ireland (Northern Ireland and Eire), the Isle of Man, and the Channel Islands (Bailiwicks of Jersey and Guernsey). The Bailiwick of Guernsey includes Guernsey, Alderney, Sark, Herm and various lesser islands. The above are always referred to in their correct, strict senses, rather than loosely, except that a distinction between Great Britain and the Isle of Man is made only where necessitated by particular patterns of distribution. The United Kingdom (Great Britain and Northern Ireland) is not referred to in the text.

The smallest geographical unit utilized is usually the vice-county. There are 111 of these in Great Britain and 40 in Ireland, with the Isle of Man and Channel Islands representing two others. All 153 are mapped and listed on the end-papers. The Isles of Scilly (Scillies) are part of W Cornwall, but they have a distinctive flora and are therefore frequently referred to separately.

CLASSIFICATION AND NOMENCLATURE

I have attempted to use the most up-to-date and accurate classification and nomenclature, but in the interests of stability I have adopted names that represent departures from normal usage in the British literature only where the evidence is unequivocal or overwhelming. The writing of this book coincided with the preparation by Douglas Kent (Kent 1992) of a new standard list of vascular plants of the British Isles (replacing that of Dandy (1958)). Similarly, preparation of the second edition coincided with that of the first supplement to Kent's *List* (Kent 1996). We have collaborated closely in our respective works and have adopted the same classification, sequence and nomenclature; I have benefited greatly from my colleague's expertise and from the results of his research, which he has generously put at my disposal.

Where a recent authoritative revision of a group is available I have generally followed it, even where I am inclined to disagree with some aspects. The family sequence and circumscription followed is that of Cronquist (1981) for the angiosperms, and Derrick *et al.* (1987) for the pteridophytes. Many sources have been followed for generic and specific sequences, but in general the less specialized taxa precede the more specialized ones.

A limited list of synonyms is provided. These are the accepted names used in

Dandy (1958), Clapham *et al.* (1962, but not 1952), and Tutin *et al.* (1964-1980, 1993), and those used in the post-1970 British literature, in Greuter *et al.* (1984-) and in Walters *et al.* (1984-). This selection should enable the reader to equate the names used in this work with those in virtually any standard modern reference.

The abbreviations of authors used are those in Brummitt & Powell (1992), a standard now universally adopted and replacing that used in the first edition.

English names are given for all the species except those in certain complex genera. I am quite convinced that the provision of English names is important in increasing the numbers of people with an interest in and knowledge of wild plants. Despite the fact that the Latin names are scientifically more meaningful and in my view always preferable, English names, if consistently and logically applied, can be no less accurate and their use by those who find them easier to remember should not be too strongly disparaged. In all but a few cases I have used the names adopted in Dony *et al.* (1986), but about 1100 species that I have included are not listed in that work. Many sources have been consulted in order to find suitable English names for these other species; the American, Australasian, South African and horticultural literature was especially helpful. However, about 350 names have been, of necessity, coined anew. My list of names not in Dony *et al.* (1986) was scrutinized by Franklyn Perring and Philip Oswald, who kindly commented upon them, and in most cases their suggestions have been adopted.

DESCRIPTIONS

The descriptions of all the taxa are brief diagnoses providing what I consider to be the most important characters, and they have been made as consistent and comparable as possible. The data provided in the family descriptions cover all the genera treated as well as 'Other genera', and the data in the generic descriptions likewise cover 'Other spp.' as well as the fully treated species. However, it is important to note that no account of variation outside the British Isles is taken; indeed, it is specifically excluded, and the reader must beware of using the descriptions as definitions of the taxa on a world-wide or even European scale.

In order to compile a description of a genus, the generic diagnosis should be read in conjunction with the family diagnosis. Species and subspecies descriptions should likewise be supplemented by the family and the generic diagnoses and the key to species (where provided). For reasons stated later on, however, it could be misleading to use the family and generic keys in order to compile descriptions.

No generic description is given if there is only one genus included in the family, subfamily or tribe, as in these cases the description of the latter would be the same as that of the former.

In compiling the descriptions I have naturally made use of a very wide range of literature. I have attempted to avoid the repetition of errors thereby encountered by checking most of the measurements and other characters on actual specimens. I have examined material of virtually all the species covered in this work, most of it in the fresh state, over the past six or seven years. I have grown (or allowed to grow, or failed to prevent from growing) about a quarter of the species in my own garden; these, and others in other gardens (including the University of Leicester Botanic Garden) and locally in the wild, I have been able to observe closely over the changing seasons.

Many measurements given, especially those describing plant heights or lengths, should be prefixed 'normally', 'usually' or 'mostly'. It is often misleading to give ranges including the extremes that have been encountered (e.g. a grass species 2-153cm high); usually the normal range is much more useful. More exceptional measurements are often given in brackets, e.g. 3-6(9)cm, but even these do not always represent the extremes. In the case of trees, however, the maximum heights known in the British Isles are given, taken from Mitchell (1982). Measurements given without qualification are lengths; those separated by a multiplication sign (e.g. 3-6 x 1-2cm) are lengths and widths respectively.

Certain conventions in terminology will become apparent after usage, especially if the *Glossary* is consulted. For example, 'above' and 'below' are used only to imply the upper and lower parts of a plant; upper and lower surfaces of an organ are referred to as 'upperside' and 'lowerside', or sometimes more specifically as adaxial and abaxial sides. The term 'leaf', unless otherwise stated, refers to the leaf-blade, excluding the petiole; this fact is especially important in the case of leaf length/breadth ratios.

Sporophytic chromosome numbers (written as 2n=) have been included for most species. The primary source for these has been *The cytological catalogue of the British and Irish flora*, an inventory of chromosome numbers of mainly native plants based on material of known wild origin in the British Isles, compiled by Richard Gornall and John Bailey and now incorporated in the BSBI Database at Leicester. I am grateful to the two compilers for kindly making their catalogue available to me. I have supplemented those data where no counts from BI are available by including counts of non-BI material (clearly distinguished - see *Signs and Abbreviations*) compiled from the standard published chromosome indices. I have usually not added any of the many counts from abroad differing from those from our islands. Sometimes the data available have been simplified by omitting reference to rare variations or to the existence of B-chromosomes.

IDENTIFICATION KEYS

The primary means of identification is the keys to families, genera, species and subspecies. The great majority of these are dichotomous keys. In order to save space no line-gap is left between couplets, but alternate couplets are slightly indented to effect visual separation. Despite this appearance all the dichotomous keys are of the bracketed version, which I consider to be generally superior to (i.e. easier to use than) the indented type followed by some Floras.

In constructing the keys I have attempted to avoid as many as possible of the pitfalls that I have personally encountered over the years. Keys are a vital part of a Flora, yet are one of the most difficult aspects to master and they provide a frequent barrier for the beginner. Long keys are particularly daunting, so I have subdivided keys wherever necessary by providing a general key to a series (A, B, etc) of supplementary keys. Hence few keys contain more than 20 couplets and very few more than 30.

In a small number of genera multi-access keys are used instead of dichotomous keys, i.e. where I consider them to be more reliable (e.g. *Epilobium*, *Cotoneaster*). These are usually cases where some important characters are difficult to observe or are likely to be misinterpreted, so that it is hazardous to rely upon them in isolation (as often encountered in a couplet). In other cases 'difficult' characters are allowed for by providing two or more routes in a key. For example the (superior) ovaries of *Rosa* are liable to be wrongly scored as inferior, the (five) leaflets of *Lotus* are often mistaken for three plus two stipules, and the (white) petals of *Berteroa* often fade to yellow when dried. In these and in many other cases both alternatives are allowed for. A consequence of this is that the 'information' given in a number of keys to families and genera is sometimes taxonomically inaccurate. These keys are provided solely for the purposes of identification, and should not be used to compile descriptions of taxa. The keys to species and subspecies, however, should be free of any such misleading data, and can be considered as part of the description of the species and subspecies. Some notoriously difficult characters (e.g. aerial stems present/absent in *Viola*; inflorescences axillary/terminal in *Trifolium*; structure of the throat of the corolla-tube in Boraginaceae) have been deliberately more or less ignored in the keys.

No species key is provided if the genus includes only two species, and no subspecies key if a species contains only two subspecies; in both cases the two taxa are immediately adjacent and no key is needed. The keys to families include all the families not fully treated (e.g. 47A. Basellaceae) and take full account of all the genera mentioned in certain families under 'Other genera'; likewise, the keys to genera take account of all the species mentioned in certain genera under 'Other spp.'.

I have assumed that the reader is familiar with the use of dichotomous and multi-access keys, but I provide here some hints that I have found very valuable in the past. The keys are intended for use both indoors and in the field, and with both fresh and dried material. However, certain characters are not suited to field observation and, where special dissection or high magnification is absolutely necessary, no pretence is made that less satisfactory characters will suffice. The use of insufficient magnification is a frequent cause of misidentification. *Before* starting on the keys it is important to examine in detail the structure of the flowers, making sure that the number, shape and arrangement of the various parts are fully ascertained. If the flowers are not all bisexual then the distribution of the sexes must be understood. The structure of the gynoecium usually presents the greatest problems; sectioning with a razor-blade vertically and transversely is often required. If fresh material is being collected, observations on underground parts, woodiness of stems and range of leaf-shape should be recorded. If possible, flowers (and fruits) of varying ages should be gathered. Mistakes are often made in distinguishing between a compound leaf (no buds in axils of leaflets) and a group of simple leaves (with buds, often very rudimentary, in axils).

In general, flowers are needed for identification by means of the keys, but there are some exceptions such as near the start of the General Key and in Keys A and B of the *Key to Families of Magnoliopsida*. Apart from non-flowering material, it is usually not possible to key out a range of abnormalities such as extreme horticultural variants (e.g. *flore pleno* or otherwise with more floral parts than usual, extremely dissected leaves or petals, unusual colour variants), abnormally tall or dwarfed plants, monstrosities such as many-headed *Plantago* and leafy-stemmed *Taraxacum*, plants with petaloid or leafy bracts, gall-induced variations, and various odd mutants (e.g. *Fraxinus* leaves with 1 leaflet). In the wild such plants usually occur with normal ones.

Finally, four tips. Firstly, before using a key to genera read carefully the family description and any notes that follow it, and before using a key to species read carefully the genus description and any subsequent notes. These descriptions and notes always contain useful data and sometimes vital ones, since special terms and conventions (e.g. 'spikelet length' in *Festuca* is not actually the total spikelet length) are often defined. Secondly, read the *whole* of *both* alternatives of each couplet before attempting to choose between them. Thirdly, if there is genuine doubt about which alternative to choose, follow both, as one will usually soon show itself to be unsuitable. Fourthly, if a nonsensical answer is obtained, check back to ensure that the frequent error of choosing the correct alternative but following the wrong subsequent route has not been committed.

ILLUSTRATIONS

Some sort of illustration is provided for over half the taxa treated. The page on which each occurs is indicated in bold in the right-hand margin of the text. The purpose is not to picture a representative sample of the taxa, but (1) to provide drawings of (mostly alien) species for which ready sources are not available in the literature; and (2) to illustrate diagnostic parts (e.g. seeds, leaves, flowers) of more critical groups of taxa on a comparative basis.

The illustrations are either line drawings or photographs. The former have mostly been executed by Hilli Thompson, to whom I am greatly indebted for the tremendous trouble she has taken to capture accurately the minute detail of the specimens. However, the choice of subject-matter, the supply of material, and the checking of accuracy, was carried out by me, and if there are faults in those respects they are my responsibility. The few drawings not made by Hilli Thompson are all attributed and acknowledged in the appropriate caption. Other artists who made drawings especially for this Flora and to whom I offer my sincere thanks are Kery Dalby (149), Jerzy Zygmunt (450), Olga Stewart (614, 616, 617, 619, 620), Fred Rumsey (629), Dick Roberts (37, 977) and Sue Ogden (989-992).

The photographs, most of them taken via the light microscope or scanning electron microscope, have been prepared by me or by various colleagues in the School of Biological Sciences, University of Leicester, on my behalf, except in the few cases specifically acknowledged. I am extremely grateful for the help in this respect given me by Abdul-Karim Al-Bermani, John Bailey, Jenny Haywood, George McTurk, Ian Riddell and Andrew Scott.

CONSERVATION AND RARITY

By far the greatest threat to our wild flora is the destruction of habitats, still continuing at a most alarming rate in the name of everything from 'economic development' to 'leisure activity'. When populations of plants (or animals) are decimated they become highly vulnerable to secondary pressures, of which collecting is one. There can be little objection to the accumulation of a reference collection of plants, providing uncommon species are excluded and populations of even common ones are not significantly reduced. Indeed, a collection of accurately determined plants is the best way of learning them and of enabling identification of extra species encountered later. Often, however, only a small part of the plant (e.g. a basal leaf or a single flower) is needed for diagnostic purposes, and rarely are underground parts essential. Usually, even where they are, they can be adequately substituted by notes made in the field. It should be noted that it is actually illegal in Great Britain to uproot any wild plant, even common weeds, without the land-owner's permission, and there are more specific regulations governing Nature Reserves and rare species (see below).

Since it is only botanists who have a good knowledge of our wild flora, it is vital that they consider themselves under a special obligation to protect it by example and by persuasion.

Rare species are referred to in this Flora under three categories, marked by **R**, **RR** and **RRR** in the right-hand margin; no plants in any of these categories should be collected, damaged or disturbed (e.g. by trampling, or by 'arranging' the immediate surroundings during photography). Even species not so marked are frequently rare in some areas (e.g. montane species in the south, southern species in Scotland, or Continental species in Ireland); where they are rare they should be respected as much as species that are rare throughout the British Isles. The 'R' signs of rarity are given for only native taxa; hybrids not treated fully and aliens are not designated. Many species that are rare as natives are much commoner as aliens; only the native distribution is considered for present purposes. The precise meanings of these signs are as follows:

R - Scarce, found in not more than 100 different 10 x 10km grid-squares in the British Isles (there are over 3500 of these grid squares in total). Such species have recently been monitored in Britain by the Scarce Plant Project of the Institute of Terrestrial Ecology (Stewart, Pearman & Preston 1994). The above scheme does not cover Ireland or the Channel Islands, but I have extended it to do this. Occasionally the abundance of a species in Ireland has changed the status of a taxon that is rarer in Britain.

RR - Rare, found in not more than 15 different 10 x 10km grid-squares in the British Isles. Such plants are listed in the British Red Data Book, but the current edition (Perring & Farrell 1983) is now out of date and I have attempted to update the list by additions or removals. The Irish Red Data Book (Curtis & McGough 1988) is also useful in this respect, but most of the species in it are not rare in a British context; sometimes the reverse is true. Lynne Farrell, of the Nature Conservancy Council, has helped to interpret the Red Data Book entries.

RRR - Rare and endangered or vulnerable. These species are all listed in the Schedule of Protected Plants of the Wildlife and Countryside Act 1981, updated after the first and second quinquennial reviews to include 107 vascular plant species. This Act covers only Great Britain. Further Acts cover Eire and Northern Ireland, and species not found in Great Britain are also given the **'RRR'** symbol where their status is comparable. Under the British Act it is an offence to pick, remove or destroy any part (including seeds) of the species in the Schedule, to attempt to do so, or to trade in these species.

Endemic or extinct taxa are indicated in the text.

BIBLIOGRAPHY

Hundreds of books and thousands of articles in journals have been used in writing this book. In addition to those listed below, which were the ones most frequently used, special mention must be made of the numerous local Floras together covering most of the counties of the British Isles. These are packed with valuable information and were freely consulted, especially those dealing with rich areas or with regions at the extremities of the British Isles.

Bailey, L.H. & Bailey, E.Z. (1976). *Hortus third* (revised by staff of L.H. Bailey Hortorium). MacMillan, New York.

Bean, W.J. (1970-88). *Trees and shrubs hardy in the British Isles*, 8th ed. (revised by Clarke, D.L.), **1-4** + Supplement. John Murray, London.

Brummitt, R.K. & Powell, C.E. (1992). *Authors of plant names.* Royal Botanic Gardens, Kew.

Clapham, A.R., Tutin, T.G. & Warburg, E.F. (1952). *Flora of the British Isles.* Cambridge University Press, Cambridge. 2nd ed. (1962); 3rd ed. (by Clapham, A.R., Tutin, T.G. & Moore, D.M.) (1987).

Clapham, A.R., Tutin, T.G. & Warburg, E.F. (1959). *Excursion Flora of the British Isles.* Cambridge University Press, Cambridge. 2nd ed. (1968); 3rd ed. (1981).

Clayton, W.D. & Renvoize, S.A. (1986). *Genera graminum.* H.M.S.O., London.

Clement, E.J. & Foster, M.C. (1994). *Alien plants of the British Isles.* Botanical Society of the British Isles, London.

Cronquist, A. (1981). *An integrated system of classification of flowering plants.* Columbia University Press, New York.

Curtis, T.G.F. & McGough, H.N. (1988). *The Irish Red Data Book*, **1**. *Vascular plants.* Stationery Office, Dublin.

Dandy, J.E. (1958). *List of British vascular plants.* British Museum, London.

Dandy, J.E. (1969). *Watsonian vice-counties of Great Britain.* Ray Society, London.

Derrick, L.N., Jermy, A.C. & Paul, A.M. (1987). Checklist of European pteridophytes. *Sommerfeltia*, **6**. Oslo.

Dony, J.G., Jury, S.L. & Perring, F.H. (1986). *English names of wild flowers*, 2nd ed. Botanical Society of the British Isles, London.

Edees, E.S. & Newton, A.N. (1988). *Brambles of the British Isles.* Ray Society, London.

Ellis, R.G. (1983). *Flowering plants of Wales.* National Museum of Wales, Cardiff.

Graham, G.G. & Primavesi, A.L. (1993). *Roses of Great Britain and Ireland.* Botanical Society of the British Isles, London.

Greuter, W., Burdet, H.M. & Long, G. (1984-). *Med-Checklist*, **1**-. Conservatoire et jardin botaniques, Geneva.

Hubbard, C.E. (1954). *Grasses.* Penguin Books, Harmondsworth. 2nd ed. (1968); 3rd ed. (revised by Hubbard, J.C.E.) (1984).

Jermy, A.C., Arnold, H.R., Farrell, L. & Perring, F.H. (1978). *Atlas of ferns of the British Isles.* Botanical Society of the British Isles, London.

Jermy, A.C., Chater, A.O. & David, R.W. (1982). *Sedges of the British Isles.* Botanical Society of the British Isles, London.

Kent, D.H. (1992). *List of vascular plants of the British Isles.* Botanical Society of the British Isles, London.

Kent, D.H. (1996). *Supplement to List of vascular plants of the British Isles.*

Botanical Society of the British Isles, London.

Lousley, J.E. & Kent, D.H. (1981). *Docks and knotweeds of the British Isles.* Botanical Society of the British Isles, London.

Mabberley, D.J. (1987). *The plant-book.* Cambridge University Press, Cambridge.

Meikle, R.D. (1984). *Willows and poplars of Great Britain and Ireland.* Botanical Society of the British Isles, London.

Mitchell, A. (1982). *The trees of Britain and northern Europe.* Collins, London.

Page, C.N. (1982). *The ferns of Britain and Ireland.* Cambridge University Press, Cambridge.

Perring, F.H. (1968). *Critical supplement to the Atlas of the British flora.* Thomas Nelson, London.

Perring, F.H. & Farrell, L. (1983). *British Red Data Books,* **1**. *Vascular plants.* Royal Society for Nature Conservation, Lincoln.

Perring, F.H. & Walters, S.M. (1962). *Atlas of the British flora.* Thomas Nelson, London.

Preston, C.D. (1995). *Pondweeds of Great Britain and Ireland.* Botanical Society of the British Isles, London.

Rich, T.C.G. (1991). *Crucifers of Great Britain and Ireland.* Botanical Society of the British Isles, London.

Ryves, T.B., Clement, E.J. & Foster, M.C. (1996). *Alien grasses of the British Isles.* Botanical Society of the British Isles, London.

Scannell, M.J.P. & Synnott, D.M. (1987). *Census catalogue of the flora of Ireland,* 2nd ed. Stationery Office, Dublin.

Stace, C.A. (1975). *Hybridization and the flora of the British Isles.* Academic Press, London.

Stearn, W.T. (1983). *Botanical Latin,* 3rd ed. Thomas Nelson, London.

Stewart, A., Pearman, D.A. & Preston, C.D. (1994). *Scarce plants in Britain.* Joint Nature Conservation Committee, Peterborough.

Tutin, T.G. et al. (1964-80). *Flora Europaea,* **1-5**. Cambridge University Press, Cambridge. 2nd ed., 1- (1993-).

Tutin, T.G. (1980). *Umbellifers of the British Isles.* Botanical Society of the British Isles, London.

Walters, S.M. et al. (1984-). *The European garden flora,* **1**-. Cambridge University Press, Cambridge

Willis, J.C. (1973). *A dictionary of the flowering plants and ferns,* 8th ed. (revised by Shaw, H.K.A.). Cambridge University Press, Cambridge.

SYNOPSIS OF FAMILIES

LYCOPODIOPSIDA
1. Lycopodiaceae
2. Selaginellaceae
3. Isoetaceae

EQUISETOPSIDA
4. Equisetaceae

PTEROPSIDA
5. Ophioglossaceae
6. Osmundaceae
7. Adiantaceae
8. Pteridaceae
9. Marsileaceae
10. Hymenophyllaceae
11. Polypodiaceae
11A. Cyatheaceae
12. Dicksoniaceae
13. Dennstaedtiaceae
14. Thelypteridaceae
15. Aspleniaceae
16. Woodsiaceae
16A. Davalliaceae
17. Dryopteridaceae
18. Blechnaceae
19. Azollaceae

PINOPSIDA
20. Pinaceae
21. Taxodiaceae
22. Cupressaceae
23. Araucariaceae
24. Taxaceae

MAGNOLIOPSIDA

MAGNOLIIDAE (Dicotyledons)

MAGNOLIIFLORAE
Laurales
25. Lauraceae
Aristolochiales
26. Aristolochiaceae
Nymphaeales
27. Nymphaeaceae
28. Cabombaceae
29. Ceratophyllaceae
Ranunculales
30. Ranunculaceae
31. Berberidaceae
Papaverales
32. Papaveraceae
33. Fumariaceae

HAMAMELIFLORAE
Hamamelidales
34. Platanaceae
Urticales
35. Ulmaceae
36. Cannabaceae
37. Moraceae
38. Urticaceae
Juglandales
39. Juglandaceae
Myricales
40. Myricaceae
Fagales
41. Fagaceae
42. Betulaceae

CARYOPHYLLIFLORAE
Caryophyllales
43. Phytolaccaceae
43A. Nyctaginaceae
44. Aizoaceae
45. Chenopodiaceae
46. Amaranthaceae
47. Portulacaceae
47A. Basellaceae
48. Caryophyllaceae
Polygonales
49. Polygonaceae
Plumbaginales
50. Plumbaginaceae

DILLENIIFLORAE
Dilleniales
51. Paeoniaceae
Theales
52. Elatinaceae
53. Clusiaceae
Malvales
54. Tiliaceae
55. Malvaceae
Nepenthales
56. Sarraceniaceae
57. Droseraceae
Violales
58. Cistaceae
59. Violaceae
60. Tamaricaceae
61. Frankeniaceae
62. Cucurbitaceae
Salicales
63. Salicaceae
Capparales
63A. Capparaceae
64. Brassicaceae
65. Resedaceae
Ericales
65A. Clethraceae
66. Empetraceae
67. Ericaceae
68. Pyrolaceae
69. Monotropaceae
Diapensiales
70. Diapensiaceae
Primulales
70A. Myrsinaceae
71. Primulaceae

ROSIFLORAE
Rosales
72. Pittosporaceae
73. Hydrangeaceae
74. Grossulariaceae
75. Crassulaceae
76. Saxifragaceae
77. Rosaceae
Fabales
78. Mimosaceae
78A. Caesalpiniaceae
79. Fabaceae
Proteales
80. Elaeagnaceae
Haloragales
81. Haloragaceae
82. Gunneraceae
Myrtales
83. Lythraceae
84. Thymelaeaceae
85. Myrtaceae
86. Onagraceae
Cornales
87. Cornaceae
Santalales
88. Santalaceae
89. Viscaceae
Celastrales
90. Celastraceae
91. Aquifoliaceae
Euphorbiales
92. Buxaceae
93. Euphorbiaceae
Rhamnales
94. Rhamnaceae
95. Vitaceae
Linales
96. Linaceae
Polygalales
97. Polygalaceae
Sapindales
98. Staphyleaceae
99. Sapindaceae
100. Hippocastanaceae
101. Aceraceae
102. Anacardiaceae
103. Simaroubaceae
104. Rutaceae
Geraniales
105. Oxalidaceae
106. Geraniaceae
107. Limnanthaceae
108. Tropaeolaceae
109. Balsaminaceae
Apiales
110. Araliaceae
111. Apiaceae

ASTERIFLORAE
Gentianales
112. Gentianaceae
113. Apocynaceae
Solanales
114. Solanaceae
115. Convolvulaceae
116. Cuscutaceae
117. Menyanthaceae
118. Polemoniaceae
119. Hydrophyllaceae
Lamiales
120. Boraginaceae
121. Verbenaceae
122. Lamiaceae
Callitrichales
123. Hippuridaceae
124. Callitrichaceae
Plantaginales
125. Plantaginaceae
Scrophulariales
126. Buddlejaceae
127. Oleaceae
128. Scrophulariaceae
129. Orobanchaceae
130. Gesneriaceae
131. Acanthaceae
132. Lentibulariaceae
Campanulales
133. Campanulaceae
Rubiales
134. Rubiaceae
Dipsacales
135. Caprifoliaceae
136. Adoxaceae
137. Valerianaceae
138. Dipsacaceae
Asterales
139. Asteraceae

LILIIDAE (Monocotyledons)

ALISMATIFLORAE
Alismatales
140. Butomaceae
141. Alismataceae
Hydrocharitales
142. Hydrocharitaceae
Najadales
143. Aponogetonaceae
144. Scheuchzeriaceae
145. Juncaginaceae
146. Potamogetonaceae
147. Ruppiaceae
148. Najadaceae
149. Zannichelliaceae
150. Zosteraceae

ARECIFLORAE
Arecales
150A. Arecaceae
Arales
151. Araceae
152. Lemnaceae

COMMELINIFLORAE
Commelinales
153. Commelinaceae
Eriocaulales
154. Eriocaulaceae
Juncales
155. Juncaceae
Cyperales
156. Cyperaceae
157. Poaceae
Typhales
158. Sparganiaceae
159. Typhaceae

ZINGIBERIFLORAE
Bromeliales
160. Bromeliaceae

LILIIFLORAE
Liliales
161. Pontederiaceae
162. Liliaceae
163. Iridaceae
164. Agavaceae
165. Dioscoreaceae
Orchidales
166. Orchidaceae

HOW TO USE THIS BOOK

It is strongly recommended that, before consulting the information in this book, the contents page, the introductory chapter and this page (including the reverse of it) be read carefully.

The Index at the end of the book should be used in order to look up a family, genus, species, subspecies or hybrid; both Latin and English names are indexed.

In order to identify a plant it is necessary first to decide whether it is a pteridophyte, gymnosperm or angiosperm. Many Floras purport to do this by means of keys, but in reality the questions posed (e.g. plant reproducing by spores; ovules enclosed in a carpel) amount to the same as the decision called for here. In practice it is best to become familiar with the range of form and structure found in the relatively few pteridophytes and gymnosperms, all other vascular plants being angiosperms. In the case of pteridophytes, the few that do not have divided fern-like leaves can easily be learnt, and in the case of gymnosperms, all have simple narrow leaves (except *Araucaria*) and woody female cones (except *Taxus* and *Juniperus*). It is especially crucial to distinguish between superficially similar but unrelated plants that provide pitfalls for the unwary. Well-known examples are mosses and Lycopodiopsida; *Equisetum* and *Hippuris*; *Lemna* and *Azolla*; *Isoetes* and *Littorella*; *Pilularia* and *Juncus*; and *Alnus* and conifers. If flowers, spore-bearing sporangia or woody cones are evident, then the task is an easy one. If not, familiarity and experience will soon prevent one from falling into traps such as the above. According to the decision, follow the generic or family keys starting on pages 1, 38 and 52, the positions of which are marked by black-edged pages. These will lead to a family or genus, which will provide further keys as necessary, enabling one to arrive at the genus, species and subspecies. Where relevant, keys to genera are given under each family, to species under each genus, and to subspecies under each species. Before using the keys the appropriate part of the introductory chapter should be studied.

The following spoof entry indicates the arrangement of the information given in each species account:

1. Accepted Latin Name Author(s) (*Synonyms* Authors) - *English Name*. Brief description to give habit and comparative diagnostic features, not always repeating those in species key. Status; most characteristic habitats; distribution in BI; area of most likely origin if not native.

Illustrations are numbered according to the page on which they appear, not in a sequence from 1 onwards. References to illustrations are given in the right-hand margin adjacent to the relevant taxon by sole means of a bold number.

Rarity and conservation status are similarly referred to in the margin by means of **R, RR** and **RRR**. For the precise meaning of these symbols, see pages xx-xxi.

A glossary is placed after the systematic accounts (marked by a black-edged page).

Signs and abbreviations are listed on the next page.

Maps and a ruler are provided on the end covers.

SIGNS AND ABBREVIATIONS

BI	- British Isles
CI	- Channel Islands
Br	- Great Britain
En	- England
Ir	- Ireland
Sc	- Scotland
Wa	- Wales
N, E, S, W, NE, etc.	- points of compass
C, M, MW, etc.	- central, Mid-, Mid-West, etc.
Leics, W Kent, etc.	- (vice-counties) see end papers
Jan, Feb, Mar, etc.	- months of year
agg.	- aggregate
auct.	- of various authors but not the original one
c.	- about
cv.	- cultivar
excl.	- excluding
FIG	- Figure (number following is the page number)
incl.	- including
ined.	- as yet not published
intrd	- introduced
natd	- naturalized
nom. illeg.	- illegitimate (but valid) name
nom. nud.	- name invalid since without description
nom. inval.	- name invalid for some other reason
R.	- River
sp., spp.	- species (singular and plural)
ssp., sspp.	- subspecies (singular and plural)
var., vars	- variety, varieties
±	- more or less
>, <	- more than, less than
≥	- over and including; at least; not less than
≤	- up to and including; at most; not more than
0	- absent
x	- times (2x, etc. = twice, etc.); or indicating a hybrid
2n=	- sporophytic chromosome number based on wild material from BI
(2n=)	- sporophytic chromosome number based on other material

KEYS TO GENERA OF PTERIDOPHYTES (LYCOPODIOPSIDA, EQUISETOPSIDA, PTEROPSIDA)

General key

1 Leaves scale-like, in whorls fused into sheath at each node; stems jointed **4/1. EQUISETUM**

1 Leaves not in a fused whorl at each node; stems not jointed 2

2 Plants free-floating on water, with 2-lobed leaves on short stem **19/1. AZOLLA**

2 Plant rooted in solid substratum 3

3 Leaves simple but lobed almost to midrib or just pinnate, when mature covered on lowerside by dense reddish-brown scales **15/3. CETERACH**

3 Leaves variously divided, not covered on lowerside by reddish-brown scales 4

4 Leaves simple, not lobed or lobed <1/2 way to midrib *Key A*

4 Leaves compound, or simple and lobed >1/2 way to midrib (rarely a few ± simple) 5

5 Sporangia borne on leaves or parts of leaves or special branches distinctly different from vegetative leaves *Key B*

5 Sporangia borne on normal foliage leaves 6

6 Sori on margins of leaves either in protruding indusia or at least partly covered by indusium-like folded-over leaf-margin *Key C*

6 Sori on underside of leaves, sometimes near margin but then not covered by folded-over leaf-margin *Key D*

Key A - Leaves simple, not lobed or lobed <1/2 way to midrib

1 Stem a rhizome or stolon, or very short and leaves single or tufted from ground; leaves usually >1cm 2

1 Stem elongated and aerial; leaves <1cm 6

2 Leaves filiform, ≤5mm wide 3

2 Leaves linear to ovate-elliptic, >5mm wide 5

3 Plant rhizomatous; leaves borne singly (often close together) and rolled in flat spiral when young **9/1. PILULARIA**

3 Plant with very short corm-like stem; leaves 1-2 or in a rosette, not rolled in flat spiral when young 4

4 Leaves borne in rosette, with sporangia at base on upperside **3/1. ISOETES**

4 Leaves 1-2; sporangia borne on spike-like special branches **5/1. OPHIOGLOSSUM**

5 Leaves cordate at base, with sporangia borne in linear sori on lowerside **15/1. PHYLLITIS**

5 Leaves cuneate at base; sporangia borne on spike-like special branches **5/1. OPHIOGLOSSUM**

6 Leaves distinctly serrate along most of margin (x10 lens), the youngest ones with minute ligule near base on upperside; heterosporous **2/1. SELAGINELLA**

6 Leaves entire, serrate only at base, or obscurely serrate along margin, without ligule; homosporous 7

7 Stems all ascending to erect, dividing into equal branches; sporangium-bearing leaves not in differentiated cones **1/1. HUPERZIA**

7 Main stems procumbent, with shorter branches; sporangium-bearing leaves in differentiated cones 8
  8 Branches flattened, with leaves in 2 alternating, opposite pairs **1/4. DIPHASIASTRUM**
  8 Branches not flattened, with leaves borne in whorls, alternately or spirally 9
9 Sterile and sporangium-bearing leaves similar, without either hair-points or scarious margins **1/2. LYCOPODIELLA**
9 Either sterile leaves with hair-points or sporangium-bearing leaves with scarious, toothed margins **1/3. LYCOPODIUM**

Key B - Leaves compound, or simple but lobed >1/2 way to midrib; sporangia borne on leaves or branches that are different from foliage leaves

1 Leaves simple and deeply lobed or 1-pinnate, the lobes or leaflets not or scarcely lobed 2
1 Leaves ≥2-pinnate or 1-pinnate with deeply lobed leaflets 5
  2 Stalk from ground bearing 1 pinnate vegetative branch and 1 sporangium-bearing branch **5/2. BOTRYCHIUM**
  2 Stalks from ground either a vegetative leaf or a sporangium-bearing leaf 3
3 Sorus-bearing pinnae with distinct flat, green central region, the sori clearly marginal **8/1. PTERIS**
3 Sorus-bearing pinnae without green flat region, or if with one then sori clearly on its lowerside 4
  4 Sterile leaves triangular-ovate in outline, <2x as long as wide **16/2. ONOCLEA**
  4 Sterile leaves oblanceolate to lanceolate in outline, >3x as long as wide **18/1. BLECHNUM**
5 Stalks from ground each bearing very different vegetative and fertile branches **6/1. OSMUNDA**
5 Stalks from ground either a vegetative leaf or a sporangium-bearing leaf 6
  6 Sterile leaves >2-pinnate, finely divided, ± parsley-like 7
  6 Sterile leaves regularly 2-pinnate, or 1-pinnate with deeply lobed pinnae 8
7 Perennial with densely scaly rhizome; sori near leaf margin which is folded over to cover it **7/1. CRYPTOGRAMMA**
7 Annual with very short sparsely scaly rhizome; sori on leaf lowerside, not covered **7/2. ANOGRAMMA**
  8 Lowest pinna on each side bearing another pinna near its base **8/1. PTERIS**
  8 Lowest pinna on each side ± like upper ones, not bearing another pinna 9
9 Leaves borne singly spaced out along rhizome; fertile leaves green on upperside **14/1. THELYPTERIS**
9 Leaves borne in tufts from apices of branches of rhizome; fertile leaves brown at maturity **16/1. MATTEUCCIA**

Key C - Leaves compound, or simple but lobed >1/2 way to midrib; sporangia borne on edge of normal vegetative leaves

1 Sori a continuous line round margins of pinnules 2
1 Sori few-many discrete groups of sporangia, sometimes close together 3
  2 Leaves 1-2-pinnate, tufted, ≤75cm excl. petiole; rhizomes short, scaly **8/1. PTERIS**
  2 Leaves (2)3-pinnate, borne singly, ≤2(5)m excl. petiole; rhizomes long, pubescent **13/1. PTERIDIUM**
3 Rhizome trunk-like, >20cm thick, covered with old leaf-bases; some

leaves >1m incl. petioles **12/1. DICKSONIA**

3 Rhizome horizontal, <1cm thick, not covered with leaf-bases; leaves <50cm incl. petioles 4

4 Ultimate leaf-segments >5mm wide; indusia formed from folded-under flap of pinnule **7/3. ADIANTUM**

4 Ultimate leaf-segments <5mm wide; indusia formed from tubular or 2-valved protruding outgrowth from pinnule 5

5 Distal part of petiole winged; rhizomes pubescent; mature indusia with protruding bristle, tubular **10/2. TRICHOMANES**

5 Petiole not winged; rhizomes glabrous; indusia without protruding bristle, of 2 valves **10/1. HYMENOPHYLLUM**

Key D - Leaves compound, or simple but lobed >1/2 way to midrib; sporangia borne on lowerside of normal vegetative leaves

1 Leaves simple, or 1-pinnate with the pinnae entire to toothed <1/2 way to midrib 2

1 Leaves 1-pinnate with the pinnae divided >1/2 way to midrib, or 2- or more-pinnate 6

2 Sori narrowly elliptic to linear **15/2. ASPLENIUM**

2 Sori circular to very broadly elliptic 3

3 Indusium 0 4

3 Indusium present 5

4 Leaves regularly pinnate or nearly so **11/1. POLYPODIUM**

4 Leaves (on 1 plant) very variably and irregularly pinnately lobed **11/2. PHYMATOSORUS**

5 Pinnae <1.5cm wide, with sori in one row either side of midrib **17/1. POLYSTICHUM**

5 Pinnae >1.5cm wide, with sori distributed ± evenly all over lowerside **17/2. CYRTOMIUM**

6 Sori linear to oblong or C- to V-shaped, >1.5x as long as wide 7

6 Sori orbicular to broadly elliptic-oblong, <1.5x as long as wide 8

7 Sori linear to oblong, with the margin next to midrib straight **15/2. ASPLENIUM**

7 Sori oblong to C- or V-shaped, with the margin next to midrib curved or bent **16/3. ATHYRIUM**

8 Leaves borne singly spaced out along rhizome 9

8 Leaves borne in tufts from apices of branches of rhizome 12

9 Leaves 2-pinnate, or 1-pinnate with the pinnae deeply lobed 10

9 Leaves 3-pinnate, at least at base 11

10 Pinnae all ± parallel, the longest ones several removed from the basal one; indusium present **14/1. THELYPTERIS**

10 Lowest pair of pinnae bent back away from plane of others, the longest one the basal or the next to basal; indusium 0 **14/2. PHEGOPTERIS**

11 Indusium 0 **16/4. GYMNOCARPIUM**

11 Indusium present, persistent **16/5. CYSTOPTERIS**

12 Indusium consisting of ring of hairs or narrow scales arching over sorus when young; petiole with joint c.1/3 way from base **16/6. WOODSIA**

12 Indusium 0, vestigial or well developed and membranous; petiole not jointed 13

13 Indusium a flap-like hood; leaves slender, with few or 0 scales on petiole **16/5. CYSTOPTERIS**

13 Indusium 0, vestigial or reniform or peltate; leaves often large and with many scales on petiole 14

14 Pinnules with teeth contracted into very fine acuminate apices;

indusium peltate **17/1. POLYSTICHUM**

14 Pinnules untoothed or with rounded to acute teeth; indusium 0, vestigial or reniform 15

15 Sori in rows on pinnules distinctly nearer margin than midrib; fresh fronds with faint lemon scent when crushed **14/3. OREOPTERIS**

15 Sori either rather scattered on pinnules or in rows no nearer margin than midrib; fresh fronds without lemon scent 16

16 Indusium reniform, very obvious; widespread **17/3. DRYOPTERIS**

16 Indusium 0 or vestigial; mountains of Sc **16/3. ATHYRIUM**

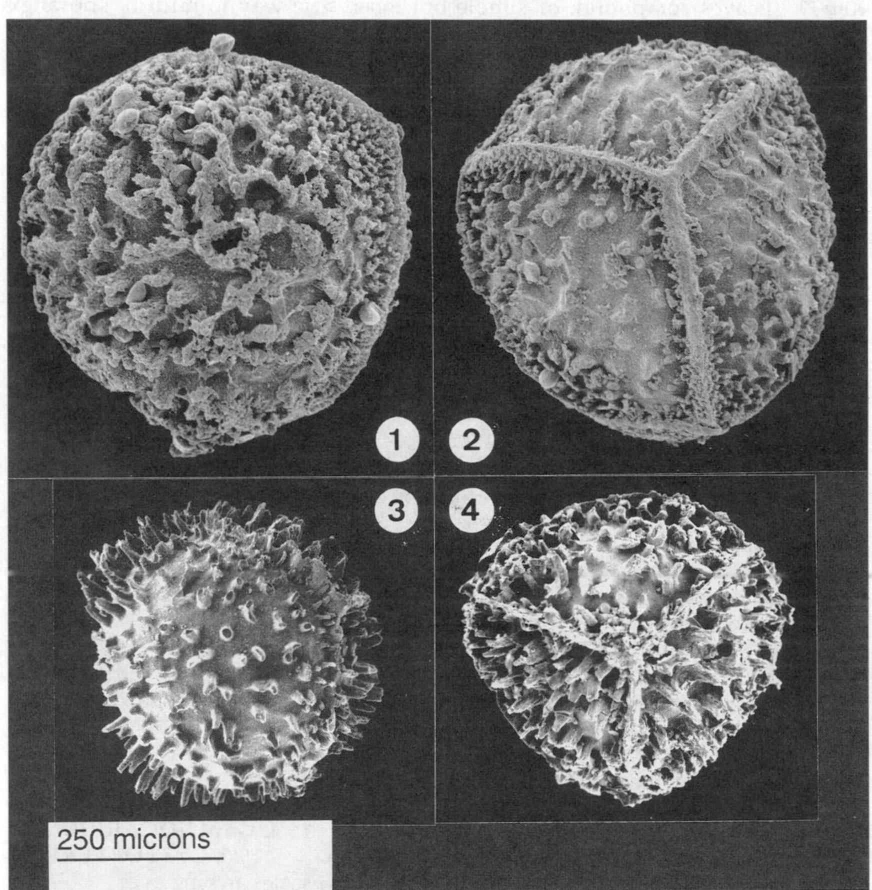

FIG 4 - ***Isoetes*** megaspores. 1-2, ***I. lacustris,*** outer and inner faces. 3-4, ***I. echinospora,*** outer and inner faces. Courtesy of A.C. Jermy and Natural History Museum, London.

LYCOPODIOPSIDA - CLUBMOSSES AND QUILLWORTS

(Lepidophyta, Lycopodineae, Lycopsida)

Herbaceous plants with simple or sparingly branched stems and simple leaves with 1 vein. Sporangia homosporous or heterosporous, borne singly in leaf axils or on upperside of leaf near its base, the sporangium-bearing leaves often aggregated into cones. Gametophyte of homosporous species free-living, subterranean, mycorrhizal and saprophytic; gametophytes of heterosporous species much reduced and retained within spore, which lies on the ground.

1. LYCOPODIACEAE - *Clubmoss family*

Stems elongated, not, little or considerably branched, bearing roots and leaves without ligules. Homosporous; sporangia in leaf-axils, the sporangium-bearing leaves often differentiated into cones.

Moss-like plants whose leaves have true midribs and stomata.

1. HUPERZIA Bernh. - *Fir Clubmoss*
Stems all ascending to erect, dividing into equal, non-flattened branches; leaves spirally arranged, often with bud-like outgrowths in their axils (these effect vegetative propagation); sporangium-bearing leaves not differentiated into cones, similar to sterile leaves.

1. H. selago (L.) Bernh. ex Schrank & Mart. (*Lycopodium selago* L.) - *Fir Clubmoss.* Stems to 30cm; leaves 4-8mm, patent to appressed, linear-lanceolate to narrowly ovate, entire or nearly so; 2n=>260. Native; heaths, moors, grassy or rocky places on mountains.

a. Ssp. selago. Stems to 30cm, 6-12mm thick; leaves linear-lanceolate, ± patent, green when healthy. Common in NW Br S to Wa, rather scattered in Ir, rare and very scattered in lowland Br, formerly locally frequent there.

b. Ssp. arctica (Grossh. ex Tolm.) Á. & D. Löve. Stems to 10cm, 5-6mm thick; leaves ovate-lanceolate to narrowly ovate, appressed to stem, yellowish-green even when healthy. Rare in parts of Sc, exact distribution not worked out.

2. LYCOPODIELLA Holub - *Marsh Clubmoss*
Stems procumbent, with non-flattened branches, giving rise to erect, fertile lateral stems; leaves spirally arranged; sporangium-bearing leaves weakly differentiated into apical cones.

1. L. inundata (L.) Holub (*Lepidotis inundata* (L.) P. Beauv., *Lycopodium inundatum* L.) - *Marsh Clubmoss.* Procumbent stems dying back quickly behind, to c.20cm; erect stems to 8(10)cm; leaves 4-6mm, erecto-patent, linear to narrowly ovate, entire; sporangium-bearing leaves broader at base; 2n=156. Native; wet heaths, often on bare peaty soil, sometimes submerged; formerly scattered almost throughout Br and Ir, now very local, extinct in C & E En. R

3. LYCOPODIUM L. - *Clubmosses*
Stems procumbent, with non-flattened branches, with erect sterile and fertile lateral stems; leaves spirally arranged or in whorls; sporangium-bearing leaves well

differentiated into apical cones.

1. L. clavatum L. - *Stag's-horn Clubmoss.* Procumbent stems to 1m or more; erect stems to 25cm, leaves 3-5mm, erect to erecto-patent, linear-lanceolate, with long white apical point, minutely toothed; sporangium-bearing leaves ovate to broadly ovate with long white apical point and scarious toothed margin; cones (1-3) borne at apex of distinct peduncle 1.5-20cm with very sparse leaves; 2n=68. Native; heaths, moors, mountains, mostly in grassy places; formerly throughout Br and Ir, now absent from much of lowlands.

2. L. annotinum L. - *Interrupted Clubmoss.* Procumbent stems to 60cm; erect stems R
to 25cm; leaves 4-10mm, patent to erecto-patent, linear-lanceolate, acute, ± entire; sporangium-bearing leaves ovate, acuminate, with scarious toothed margin; cones (1) borne at apex of leafy stems (distinct peduncle 0); 2n=68. Native; moors and mountains on thin soil over rocks, often among *Calluna*; local in C & N (± entirely mainland) Sc, extinct in S Sc, N En and N Wa except Westmorland (1 site).

4. DIPHASIASTRUM Holub - *Alpine Clubmosses*
Stems procumbent, often ± subterranean, with flattened erect branches arising in fan-like groups; leaves in alternating opposite pairs; sporangium-bearing leaves well differentiated into apical cones.

1. D. alpinum (L.) Holub (*D. complanatum* ssp. *alpinum* (L.) Jermy, *Diphasium alpinum* (L.) Rothm., *Lycopodium alpinum* L.) - *Alpine Clubmoss.* Procumbent stems to 50(100)cm; erect branches to 10cm, slightly flattened, glaucous; leaves on erect branches and upperside of procumbent stems 2-4 x c.1mm, entire, appressed, sessile; ventral leaves petiolate, c.0.5mm wide, with >1mm free from stem; lateral leaves fused to stem for c.1/2 their length; cones at apices of normal leafy shoots; 2n=c.48. Native; moors and mountains among grass and *Calluna*, often very exposed; locally common in N & W Br S to Derbys and S Wa (formerly to S Devon), N, E & W Ir.

2. D. complanatum (L.) Holub (*Diphasium complanatum* (L.) Rothm., *Lycopodium* RR
complanatum L.) - *Issler's Clubmoss.* Differs from *D. alpinum* in more robust habit; erect branches strongly flattened, scarcely glaucous; ventral leaves sessile, c.1mm wide, with <1mm free from stem; lateral leaves fused to stem for c.2/3 their length; cones at apices of sparsely-leafed peduncles. Native; heaths and lowland moors; formerly very sparsely scattered in C & N Sc and W En, extinct except in few sites in S Northumb, S Aberdeen and W Sutherland. Our plant may be ssp. **issleri** (Rouy) Jermy (*D. issleri* (Rouy) Holub, *Diphasium issleri* (Rouy) Holub), and is probably derived from hybrids with *D. alpinum.*

2. SELAGINELLACEAE - *Lesser Clubmoss family*

Stems elongated, little or considerably branched, bearing roots on end of special leaf-less branches or on small corm-like swelling at base of stem; leaves serrate, with microscopic outgrowths (ligules) on upperside near base. Heterosporous; sporangia in leaf-axils, the sporangium-bearing leaves in ill- to well-defined cones with megasporangia at base and microsporangia at apex.

Distinguished from Lycopodiaceae in presence of ligule, heterospory, and roots being borne on specialized leaf-less stem-like outgrowths or small corm-like swellings.

1. SELAGINELLA P. Beauv. - *Lesser Clubmosses*

1. S. selaginoides (L.) P. Beauv. - *Lesser Clubmoss.* Stems decumbent to procumbent, to 15cm, the branches not flattened, bearing erect fertile branches to 6(10)cm with terminal rather ill-defined cones; roots borne from small corm-like swelling at base of stem; leaves all of 1 sort, 1-3mm, those in cones similar but larger; 2n=18. Native; damp places among moss and short grass on mountains; locally common in BI S to Co Limerick, Merioneth and SE Yorks, formerly to W Cork, Derbys and S Lincs.

2. S. kraussiana (Kunze) A. Braun - *Krauss's Clubmoss.* Stems procumbent, to 1m, the branches dorsiventrally flattened, bearing well-defined cones increasing in length with age and apparently not terminal; roots borne at ends of special leafless branches; leaves of 2 sorts, 2 rows on upperside of stems c.1-2mm, 2 rows on sides of stems c.2-4mm; cones with 4 closely overlapping rows of leaves; (2n=20). Intrd; grown as ground cover in mild damp regions, ± natd in shrubberies and damp shady places; scattered in S & W Br, E, W & S Ir and CI N to Herts, Man and W Mayo, Co Durham; Africa.

3. ISOETACEAE - *Quillwort family*

Stems short and corm-like, bearing roots at base and a rosette of long, erect, ± subulate leaves with minute ligule on upperside near base. Heterosporous; sporangia ± embedded in leaf-base below ligule; megasporangia produced each year on older leaves, microsporangia on younger ones, the youngest leaves not bearing sporangia.

Similar only to certain angiosperms (notably *Lobelia, Littorella, Subularia, Juncus*); in absence of sporangia the leaves with 4 air-cavities seen in transverse section (only 1 in *I. histrix)* and the peculiar corm-like 2-3-lobed stem are diagnostic.

1. ISOETES L. - *Quillworts*

1 Plant only seasonally submerged, with leaves Oct-Jun; leaf-bases dark, shiny, horny, persistent **3. I. histrix**
1 Plant submerged for all or most of year, with leaves Jan-Dec; leaf-bases not dark, shiny and horny, not persistent 2
  2 Megaspores 530-700 microns across, with blunt, anastomosing tubercles on outer face; leaves stiff, remaining apart when plant removed from water **1. I. lacustris**
  2 Megaspores 440-550 microns across, with acute spines on all faces; leaves flaccid, falling together when plant removed from water **2. I. echinospora**

Other spp. - The amphidiploid derivative of *I. lacustris* x *I. echinospora* (**I. brochonii** Motelay) might also occur, but needs careful research, as does the relation to it of the Irish **I. morei** Moore.

1. I. lacustris L. - *Quillwort.* Leaves 8-25(40)cm x 2-5mm, usually ± erect, ± stiff, 4
with 4 longitudinal, septate air canals, parallel-sided for most of length then tapered to acute, often asymmetric point; megaspores 530-700 microns across, with blunt anastomosing tubercles; 2n=110. Native; in clear upland lakes, mostly on stony substrata, down to 6m depth; locally frequent in Ir and N & W Br, but absent in En except S Devon and Lake District.

1 x 2. I. lacustris x I. echinospora occurs by a reservoir in Cumberland with both parents; it is intermediate and sterile; 2n=66.

2. I. echinospora Durieu (*I. setacea* auct. non Lam.) - *Spring Quillwort.* Leaves 4- R
15cm x 2-3mm, usually spreading to ± erect, ± flaccid, with 4 longitudinal, septate 4
air canals, tapered to very acute apex from low down; megaspores 440-550 microns across, with acute spines; 2n=22. Native; in similar places to *I. lacustris* but rarely with it, mostly on silty substrata; similar distribution to *I. lacustris*, but rare in N, commoner in SW En E to Dorset, only in W Ir.

3. I. histrix Bory - *Land Quillwort.* Leaves 1-4(10)cm x c.1mm, variously spreading, with 1 longitudinal non-septate air canal, tapered to very acute apex from low down; megaspores 400-560 microns across, with blunt tubercles; 2n=20. Native; sandy or peaty hollows on cliff-tops near sea, where water lies in winter; extremely local in Guernsey, Alderney and Lizard Peninsula (W Cornwall). **RR**

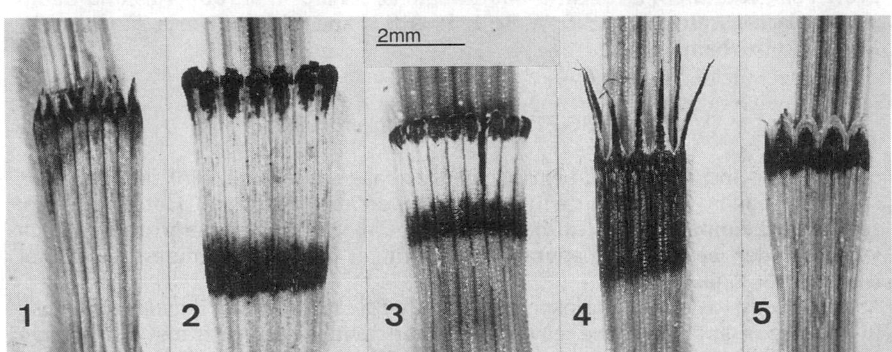

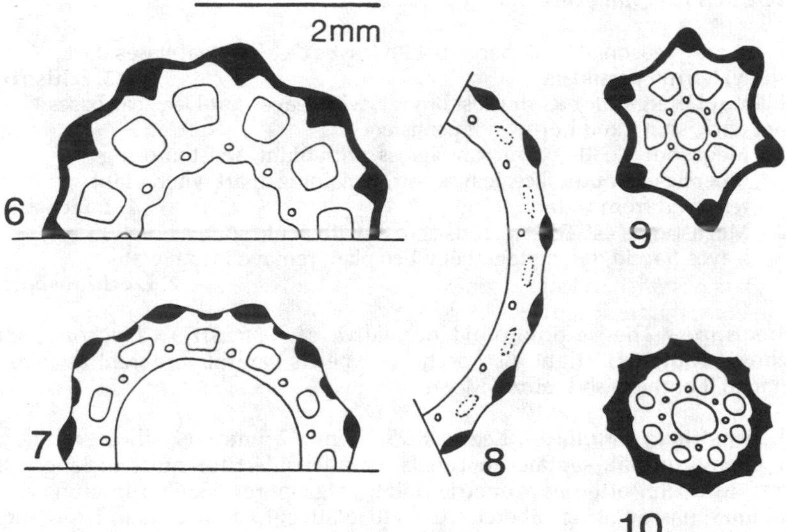

FIG 8 - ***Equisetum***. 1-5, leaf-sheaths. 1, *E.* ***ramosissimum***. 2, *E.* ***x moorei***. 3, *E.* ***hyemale***. 4, *E.* ***x trachyodon***. 5, *E.* ***variegatum***. 6-10, leaf-internode sections. 6, *E.* ***arvense***. 7, *E.* ***x litorale***. 8, *E.* ***fluviatile***. 9, *E.* ***palustre***. 10, *E.* ***variegatum***. Drawings by C.A. Stace.

EQUISETOPSIDA - HORSETAILS

(Calamophyta, Equisetinae, Sphenopsida)

Herbaceous rhizomatous perennials; aerial stems elongated, jointed, simple or bearing whorls of branches at nodes; leaves simple, with 1 vein, borne in whorls and fused into sheath round stem. Sporangia homosporous, borne in clusters under peltate specialized branches which are packed into well defined terminal cones. Gametophyte free living, green and photosynthetic.

4. EQUISETACEAE - *Horsetail family*

The jointed ridged aerial stems without branches or with whorls of branches at each node, and with whorls of leaves forming a fused sheath at each node, are unmistakable. The cones are terminal either on the normal green vegetative stems or on special unbranched brownish or whitish stems produced earlier than the vegetative ones. The common name *Mare's-tail* is often used but is a misapplication; true *Mare's-tail* is *Hippuris* (Hippuridaceae), whose whorled 'branches' at each node are actually leaves.

1. EQUISETUM L. - *Horsetails*

1 Stems brown or whitish, simple, with cone at apex — 2
1 Stems green, simple or branched, with or without cone at apex — 5
  2 Leaf-sheaths with (15)20-30(40) teeth; cones (2)4-8cm — **12. E. telmateia**
  2 Leaf-sheaths with 3-20 teeth; cones 1-4cm — 3
3 Leaf-sheaths with teeth united into 3-6 obtuse lobes at least at some nodes — **10. E. sylvaticum**
3 Leaf-sheaths with (3)6-20 separate (sometimes slightly adherent) acute teeth — 4
  4 Base of stem and leaf-sheaths usually tinged green; green branches very soon produced — **9. E. pratense**
  4 Green colour absent from stem and leaf-sheaths; branches normally not produced — **8. E. arvense**
5 Leaf-sheaths normally with conspicuous black bands near top and bottom; teeth falling off before or as soon as shoots fully expanded — 6
5 Leaf-sheaths with 0-1 conspicuous black bands, or black ± all over; teeth present on fully expanded shoots — 7
  6 Leaf-sheaths c. as long as wide, without teeth from very early on; stems perennial — **1. E. hyemale**
  6 Leaf-sheaths distinctly longer (usually c.1.5-2x) than wide, with teeth until shoots fully expanded; stems wholly or largely dying down in winter — **2. E. x moorei**
7 Stems perennial with previous year's cones persisting; cones obtuse to apiculate at apex — 8
7 Stems annual; previous year's cones not persisting; cones rounded at apex — 10
  8 Teeth of leaf-sheaths with broad scarious margins each much wider than black centre, obtuse at maturity; stem-ridges 4-10 — **5. E. variegatum**

8 Teeth of leaf-sheaths at least near tip with narrow scarious margins no wider than black centre, tapering to fine point; stem-ridges 8-20 9

9 Stems usually ± well branched; spores fertile; central hollow of stem mostly ≥1/2 as wide as stem **4. E. ramosissimum**

9 Stems not or sparsely branched; spores sterile; central hollow of stem mostly <1/2 as wide as stem **3. E. x trachyodon**

10 Stem internodes white, often ≥1cm wide, with 18-40 ridges; stems with whorls of branches ± to top **12. E. telmateia**

10 Stem internodes green, mostly <1cm wide, with 4-30 ridges, if >18 then at least top part of stem without whorls of branches 11

11 Branches regularly branched again; teeth of leaf-sheaths united into 3-6 lobes (fewer than stem-ridges) **10. E. sylvaticum**

11 Branches 0, or present but not or sparsely and irregularly branched again; teeth of leaf-sheaths not fused, as many as stem-ridges 12

12 Stem-internodes with central hollow >3/4 as wide as stem, with 10-30 ridges (usually >20 in stems >8mm wide) **6. E. fluviatile**

12 Stem-internodes with central hollow <3/4 (usually c.1/2 or less) as wide as stem, with 4-20 ridges (stems rarely >8mm wide) 13

13 Stem with peripheral hollows c. same size as central hollow; stem-internodes with 4-9(12) ridges **11. E. palustre**

13 Stem with peripheral hollows <1/2 size of central hollow; stem-internodes with 6-20 ridges 14

14 Internodes of branches mostly 3-angled, the lowest shorter than adjacent leaf-sheath on main stem **9. E. pratense**

14 Internodes of branches mostly 4(or more)-angled, the lowest as long as to longer than adjacent leaf-sheath on main stem 15

15 Cones always produced on green stems; stem-internodes with central hollow c.1/2 as wide as stem **7. E. x litorale**

15 Cones only exceptionally produced on green stems; stem-internodes with central hollow distinctly <1/2 as wide as stem **8. E. arvense**

1. E. hyemale L. - *Rough Horsetail.* Stems to 1m, evergreen, simple, rough to touch, 8
with 10-30 2-angled ridges; cones on normal vegetative shoots, apiculate at apex; 2n=216. Native; in ditches and on river- or stream-banks, often in dense vegetation; scattered through most of Br and Ir, decreasing.

2. E. x moorei Newman (*E. hyemale* x *E. ramosissimum*) - *Moore's Horsetail.* Stems RR
to 60cm, deciduous to semi-evergreen, simple, slightly rough to touch, with 10-15 2- 8
angled ridges; cones on normal vegetative shoots, apiculate, with sterile spores. Native; dunes and banks by sea in Co Wexford and Co Wicklow in absence of *E. ramosissimum* and usually of *E. hyemale*

3. E. x trachyodon A. Braun (*E. hyemale* x *E. variegatum*) - *Mackay's Horsetail.* RR
Stems to 1m, evergreen, simple to sparsely branched, slightly rough to touch, with 8
6-14 ridges; cones on normal vegetative shoots, apiculate, with sterile spores; 2n=216. Native; sandy lake-shores and river-banks and damp places in dunes; very scattered in Ir and Sc, Cheshire, often far from 1 or both parents.

4. E. ramosissimum Desf. - *Branched Horsetail.* Stems to 1m, ± evergreen, usually RRR
well branched in lower part, slightly rough to touch, with 7-20 rounded ridges; 8
cones on normal vegetative shoots, obtuse to apiculate; (2n=216). Probably intrd; in long grass by river at 1 site in S Lincs since 1947, in rough grass near sea at 1 site in N Somerset since c.1963; Europe.

5. E. variegatum Schleich. ex F. Weber & D. Mohr - *Variegated Horsetail.* Stems to R
40(80)cm (often <20cm), evergreen, usually rough to touch, simple or branched at 8
base, with 4-10 2-angled ridges; cones on normal vegetative shoots, bluntly apiculate; 2n=216. Native; dune slacks, river-banks, lake-shores, wet stony mountain sites; scattered in Ir and W & N Br, very rare in C & S En.

6. E. fluviatile L. - *Water Horsetail.* Stems to 1.5m, deciduous, smooth to touch, 8
simple or with whorls of branches in middle region, with 10-30 very low ridges; cones on vegetative but shorter and less branched shoots, rounded at apex; 2n=216. Native; ponds, ditches, marshes, backwaters, in or by water; common throughout BI.

6 x 11. E. fluviatile x E. palustre = E. x dycei C.N. Page occurs in W Sc and W & SW Ir; it is intermediate, resembling a weak plant of *E. x litorale* but with fewer whorled branches.

6 x 12. E. fluviatile x E. telmateia = E. x willmotii C.N. Page was found in Co Cavan in 1984; endemic.

7. E. x litorale Kühlew. ex Rupr. (*E. fluviatile* x *E. arvense*) - *Shore Horsetail.* Stems 8
to 1m, deciduous, smooth to touch, with dense whorls of branches except at base and extreme apex, with 6-20 low ridges; cones on normal vegetative shoots, rounded at apex. Native; wet places by rivers and lakes, in ditches and dune-slacks; scattered throughout Br and Ir, probably overlooked, sometimes in absence of 1 or both parents.

8. E. arvense L. - *Field Horsetail.* Vegetative stems to 80cm, deciduous, smooth to 8
touch, with dense whorls of long branches, with 6-20 rounded ridges; cones on special, unbranched brown stems to 20(30)cm appearing before vegetative stems, rounded at apex; 2n=216. Native; grassy places, damp places, dune-slacks and rough, waste and cultivated ground, often a very pervasive weed; abundant throughout BI.

8 x 11. E. arvense x E. palustre = E. x rothmaleri C.N. Page was found in N Ebudes in 1972, Herts in 1987 and Kintyre in 1988; it is intermediate between the parents (cones on normal vegetative shoots) and sterile; endemic.

9. E. pratense Ehrh. - *Shady Horsetail.* Vegetative stems to 50cm, deciduous, R
rough to touch, with whorls of thin branches often slightly swept to 1 side, with (8)12-20 rounded ridges; cones on shorter, pale shoots at first unbranched but then resembling others, rounded at apex; 2n=216. Native; banks of rivers and streams, often on open soil in shade, and flushed grassy areas, mostly upland; local in BI S to NW Yorks and Fermanagh. Often not coning.

9 x 10. E. pratense x E. sylvaticum = E. x mildeanum Rothm. occurs in 3 sites in M & E Perth and in Outer Hebrides; it is intermediate in all respects and no cones have been found.

10. E. sylvaticum L. - *Wood Horsetail.* Vegetative stems to 50(80)cm, deciduous, slightly rough to touch, with whorls of thin delicate branched branches, with 10-18 rather flat-topped ridges; cones on shorter, pale shoots at first unbranched but then resembling others, rounded at apex; 2n=216. Native; damp woods, hedgerows and stream-banks in lowlands, open moorland in uplands; throughout Br and Ir, common in N & W, rare in most of C & E En.

10 x 12. E. sylvaticum x E. telmateia = E. x bowmanii C.N. Page was found in S Hants in 1986; it is intermediate in all characters (cone-bearing stems with only sparse green branches) and has sterile spores; endemic.

11. E. palustre L. - *Marsh Horsetail.* Stems to 60cm, deciduous, smooth to touch, 8
simple or with usually rather sparse whorled branches, with 4-9(12) rounded ridges; cones on vegetative (but often shorter and less branched) shoots, rounded at apex; 2n=216. Native; all kinds of wet or marshy ground; very common throughout BI. Unbranched stems usually differ from *E. variegatum* in rounded, smooth stem-ridges; branched stems differ from *E. arvense* in branches with teeth appressed (not spreading) and in lowest branch-internode shorter (not longer) than main stem leaf-sheath.

11 x 12. E. palustre x E. telmateia = E. x font-queri Rothm. occurs in scattered places in W Br from S Hants to N Ebudes; it is intermediate in all characters (but with cones on vegetative shoots) and has sterile spores.

12. E. telmateia Ehrh. - *Great Horsetail.* Vegetative stems to 1.5(2)m, deciduous, smooth to touch, white, with very dense whorls of branches, with 18-40 very low ridges; cones on special, unbranched, brown stems to 25(30)cm appearing before vegetative stems, rounded at apex; 2n=216. Native; damp shady places, woods and wayside banks; throughout BI, but rare in C, E & N Sc.

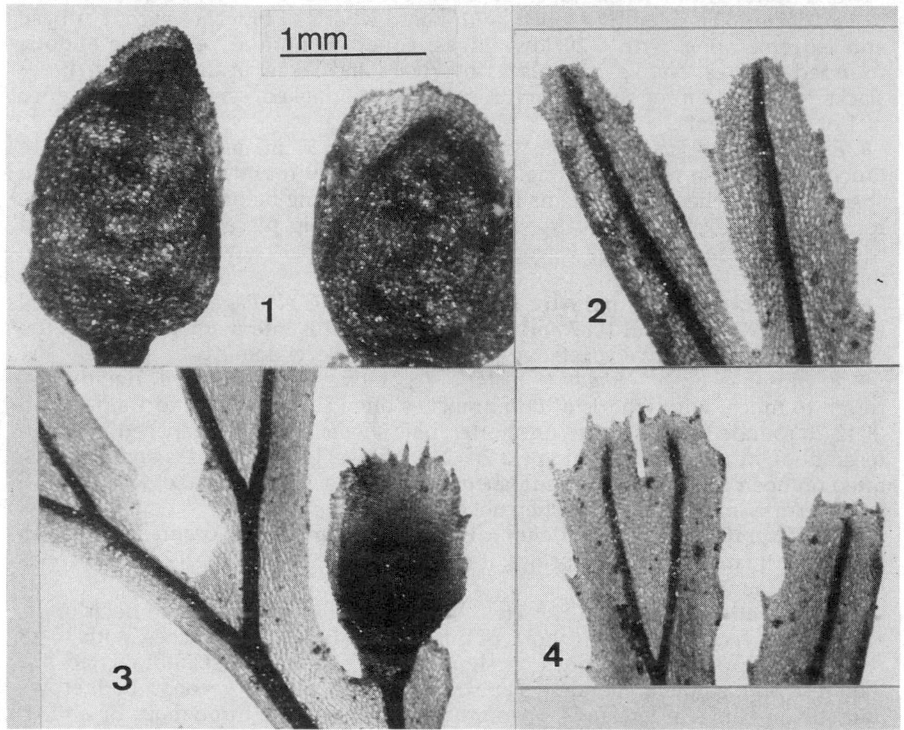

FIG 12 - ***Hymenophyllum***. 1-2, indusia and leaf-apices of ***H. wilsonii***. 3-4, indusium and leaf-apices of ***H. tunbrigense***.

PTEROPSIDA - FERNS

(Filicopsida, Filicineae)

Herbaceous plants with simple or variously branched, frequently subterranean stems (or rhizomes) and usually much divided (rarely simple) leaves with several to many veins, the veins usually much branched and often anastomosing. Sporangia homosporous or rarely heterosporous, often grouped in specialized regions (*sori*) on lowerside of leaves or on specialized leaves or parts of leaves, never in cones, often covered by special flaps of tissue (*indusia*). Gametophyte of homosporous species free living, either non-green, subterranean and mycorrhizal, or green and photosynthetic; gametophytes of heterosporous species much reduced and retained within spore, which lies on or under water. All except *Azolla*, *Ophioglossum* and *Botrychium* have the young leaves flatly spirally coiled in 1 plane. Only *Azolla* and *Pilularia* are heterosporous and only these 2 have simple non-branching veins in leaves. Only *Phyllitis*, *Pilularia* and *Ophioglossum* have undivided leaves; in *Azolla* they are only 2-lobed. Primary divisions of a leaf are *pinnae*, ultimate divisions of a leaf at least 2-pinnate are *pinnules*. In most ferns the stalk arising from ground level appears to be a stem, bearing leaves, but in fact it is the petiole of a usually much divided leaf. *Pteridium* and *Osmunda* are perhaps the spp. most likely to be misinterpreted. In all our ferns the true stem is a rhizome, subterranean or on the surface, but often it is oblique (as in *Dryopteris* etc.) and partly above ground, and in *Dicksonia* it becomes the 'trunk'.

5. OPHIOGLOSSACEAE - *Adder's-tongue family*

Rhizome short or corm-like, without scales; leaves borne singly, with erect stem-like petiole and sterile blade often plus 1 fertile blade; sterile blade simple and entire or 1-pinnate, not spirally coiled when young; fertile blade a simple spike or a panicle, the spike or panicle-branches bearing sporangia in a row either side of axis; homosporous; gametophytes non-green, subterranean.

The leaves, divided into 2 parts, are unique.

1. OPHIOGLOSSUM L. - *Adder's-tongues*

Sterile blade simple, entire; fertile blade a simple spike of sunken sporangia.

1 Sterile blade rarely >2cm, linear to narrowly elliptic, the vein-islets without free vein endings; spores ripe Jan-Mar **3. O. lusitanicum**

1 Sterile blade rarely <2cm, oblong-elliptic to broadly so, the vein-islets with minute free vein-endings within them; spores ripe Apr-Aug 2

 2 Sterile blade mostly 3-3.5cm; sporangia 6-14 either side of spike **2. O. azoricum**

 2 Sterile blade mostly 4-15cm; sporangia 10-40 either side of spike **1. O. vulgatum**

1. O. vulgatum L. - *Adder's-tongue*. Leaves to 30(45)cm; sterile blade (3)4-15(30)cm, rounded to cuneate at base; fertile blade 1.5-5(7)cm; spores 26-41 microns across; 2n=500-520, c.540 (highest chromosome number of any British plant). Native; grassland, dune-slacks, ditches, open woods, mostly in lowlands;

frequent throughout most of BI.

2. O. azoricum C. Presl (*O. vulgatum* ssp. *ambiguum* (Coss. & Germ.) E.F. Warb.) - *Small Adder's-tongue*. Leaves to 10cm; sterile blade (1.5)3-3.5cm, strongly narrowed to ± stalked at base; fertile blade 0.8-2cm; spores 38-47 microns across; (2n=c.480). Native; barish or grassy places on sandy or peaty damp soils near sea; very scattered round coasts of BI, but not in E Ir, E Sc or E En except Cheviot. Possibly derived from *O. vulgatum* x *O. lusitanicum*. **R**

3. O. lusitanicum L. - *Least Adder's-tongue*. Leaves to 2cm; sterile blade 0.6-3cm; fertile blade 0.3-1.5cm, with 3-8 sporangia on either side; spores 23-32 microns across; 2n=250-260. Native; very short turf by sea; local in Guernsey and Scillies. **RRR**

2. BOTRYCHIUM Sw. - *Moonwort*

Sterile blade pinnate; fertile blade a panicle of axes with sessile but not sunken sporangia.

1. B. lunaria (L.) Sw. - *Moonwort*. Leaves to 30cm; sterile blade 2-12cm, 1-pinnate, with asymmetric fan-shaped pinnules; fertile blade 1-5cm; 2n=90. Native; dry grassland, mostly in uplands; throughout Br and Ir, especially N & W Br.

6. OSMUNDACEAE - *Royal Fern family*

Rhizomes rather short, thick, not scaly, suberect to ascending, eventually forming solid, raised clumps; leaves in tufts at apices of rhizome-branches, spirally coiled when young, with stout ± erect petiole and large 2-pinnate blade, the lower parts sterile, the upper parts often fertile, not green, covered in masses of sporangia; homosporous; gametophytes green, surface-living.

The *Royal Fern*, with its large leaves sterile below and fertile above, is unmistakable.

1. OSMUNDA L. - *Royal Fern*

1. O. regalis L. - *Royal Fern*. Leaves (incl. petiole) (0.3)0.6-2(4)m, with 5-15 pairs of pinnae; pinnules oblong-lanceolate, 2-8 x 1-1.5cm, crenate to subentire; 2n=44. Native; fens, bogs, wet woods and heaths on peaty soil, also grown in gardens and natd in ditches, woods and hedgerows; throughout BI, common in parts of W Ir and W Br, absent from large areas of E Br.

7. ADIANTACEAE - *Maidenhair Fern family*
(Cryptogrammaceae, Gymnogrammaceae)

Rhizomes short to long, scaly; leaves tufted at apices of rhizome-branches or close together along rhizomes, spirally coiled when young, of 1 sort (bearing sporangia on normal pinnules) or of 2 sorts (fertile and sterile), 1-4-pinnate; sori linear along veins with no indusium and flat leaf-margin, or oblong to suborbicular near vein-endings and protected by reflexed leaf-margin; homosporous; gametophytes green, surface-living.

Not recognisable as a family on superficial characters, but each of the 3 genera is very distinctive.

1. CRYPTOGRAMMA R. Br. - *Parsley Fern*

Rhizome short; leaves tufted at apices of rhizome-branches, of 2 sorts (fertile and sterile), 2-4-pinnate; sori oblong, becoming contiguous when mature, protected by continuous reflexed leaf-margin.

1. C. crispa (L.) R. Br. ex Hook. - *Parsley Fern*. Leaves (incl. petiole) to 30cm,

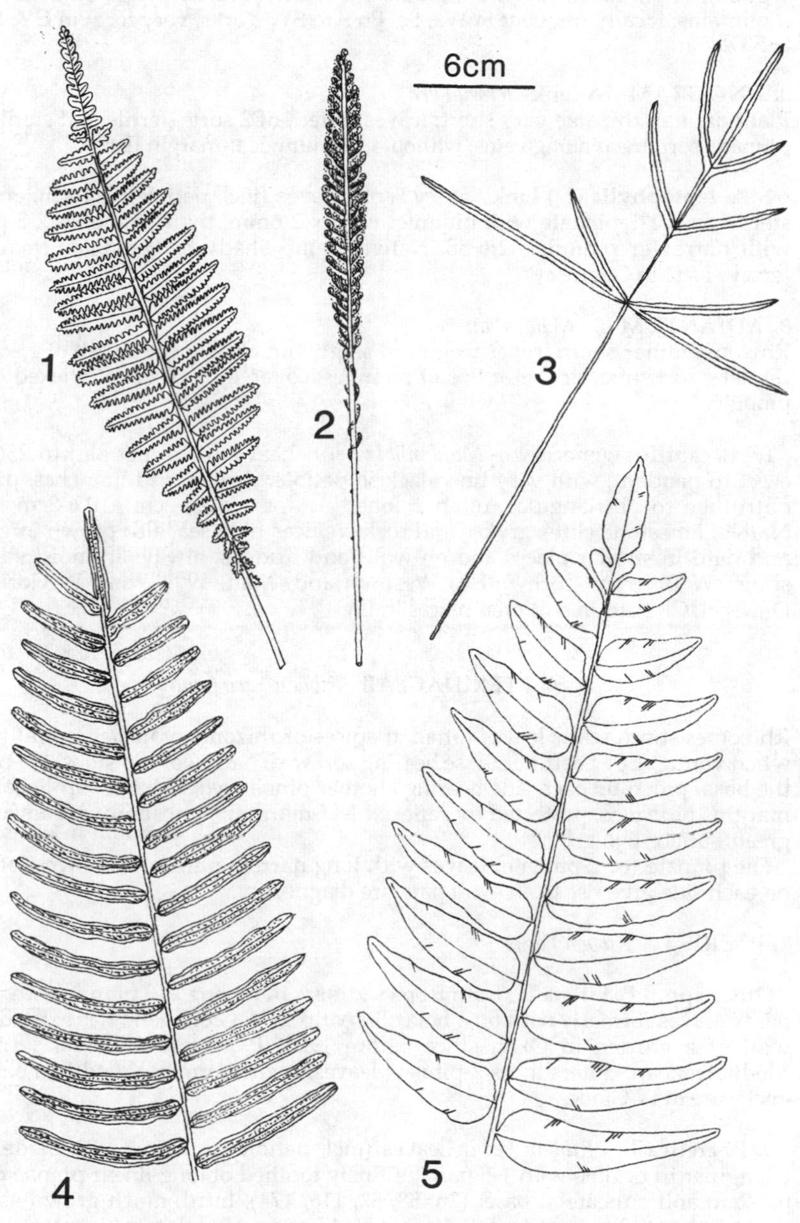

FIG 15 - Pteropsida. 1-2, ***Matteuccia struthiopteris***. 1, sterile. 2, fertile. 3, ***Pteris cretica***, fertile. 4-5, ***Blechnum cordatum***. 4, fertile. 5, sterile.

erect, the sterile ones 2-3-pinnate with pinnules c.5-10 x 3-7mm, the fertile ones 3-4-pinnate with narrower pinnules; 2n=120. Native; rocky places on acid soils on mountains; locally frequent in Wa, Sc, En S to SW Yorks, very local in C & N Ir, N & S Devon.

2. ANOGRAMMA Link - *Jersey Fern*
Plant annual; rhizome very short; leaves tufted, of 2 sorts (fertile and sterile), 1-3-pinnate; sori linear along veins, without indusium; leaf-margin flat.

1. A. leptophylla (L.) Link - *Jersey Fern*. Leaves (incl. petiole) to 10cm, erect, the **RR** sterile ones 1-2-pinnate with pinnules c.3-8 x 2-6mm, the fertile ones 2-3-pinnate with narrower pinnules; 2n=58. Native; damp shady hedgebanks; frequent in Jersey, 1 site in Guernsey.

3. ADIANTUM L. - *Maidenhair Fern*
Rhizome rather short; leaves produced singly but close together, all of 1 sort, 1-3-pinnate; sori suborbicular at tips of pinnules, covered by discrete reflexed flaps of pinnule.

1. A. capillus-veneris L. - *Maidenhair Fern*. Leaves (incl. petiole) to 25(45)cm, **R** erect to pendent, with very fine blackish petiole, rhachis and branches; pinnules obtrullate to obtriangular, often ± lobed at apex, c.0.6-3cm x 0.6-3cm; 2n=60. Native; limestone cliffs, grykes and rock crevices near sea, also grown in gardens and natd in similar places and on walls and bridges, always in moist sheltered spots; W Br from Cornwall to Westmorland, Man, W Ir from W Cork to W Donegal, CI, natd in scattered places in En.

8. PTERIDACEAE - *Ribbon Fern family*

Rhizomes short, scaly; leaves tufted at apices of rhizome-branches, spirally coiled when young, ± of 1 sort but those bearing sori with narrower divisions, 1-2-pinnate, the basal pinna on each side bearing another pinna near its base; sori linear along margins of leaves, protected by reflexed leaf-margin; homosporous; gametophyte green, surface-living.

The pinnate (or 2-pinnate) leaves with long narrow pinnae, the lowest of which on each side give rise to an extra pair, are diagnostic.

1. PTERIS L. - *Ribbon Ferns*

Other spp. - P. vittata L., from tropics, differs in having ≥10 pinnae with cordate bases; it was formerly natd on a hot colliery tip in W Gloucs and has existed on the wall of a garden in Oxon since before 1924. **P. serrulata** Forssk., from W Mediterranean, differs in its 2-pinnate leaves; it was formerly natd in a basement enclosure in W Gloucs.

1. P. cretica L. - *Ribbon Fern*. Leaves (incl. petiole) to 75cm, with blade ovate- **15** triangular in outline with 1-5 pairs of finely toothed oblong-linear pinnae c.7-16 x 0.7-2cm and cuneate at base; (2n=58, 87, 116, 174). Intrd; much grown as a pot-plant or in gardens, natd on walls, old buildings and rock-faces in very sheltered places; scattered in S & C En; S Europe.

9. MARSILEACEAE - *Pillwort family*

Rhizomes long, thin, pubescent, not scaly; leaves borne singly but often close together, spirally coiled when young, of 2 sorts (sterile and fertile), the sterile ones filiform (?petioles only represented), the fertile ones small, globose, shortly stalked, completely enclosing the sporangia; heterosporous.

Pilularia could be mistaken only for an angiosperm, but the spirally coiled young leaves and globose spore-bearing leaves distinguish it.

1. PILULARIA L. - *Pillwort*

1. P. globulifera L. - *Pillwort*. Leaves erect, 3-8(15)cm x c.0.5mm; fertile leaves c.3mm across, on stalks c.1mm; 2n=26. Native; on silty mud by lakes, ponds and reservoirs, submerged for at least part of year; scattered throughout most of BI but much less common than formerly, now frequent only in CW Ir, CS En and parts of Wa. **R**

10. HYMENOPHYLLACEAE - *Filmy-fern family*

Rhizomes thin, glabrous or pubescent but not scaly, surface-running; leaves borne singly along rhizome, spirally coiled when young, of 1 sort, 1-3-pinnate, very thin, membranous and translucent; sori in tubular or valve-like indusia at vein-ends, protruding from edge of leaf; homosporous; gametophyte green, surface-living.

The very thin leaves with protruding marginal sori in tubular or valve-like indusia are unique.

1. HYMENOPHYLLUM Sm. - *Filmy-ferns*

Rhizomes filiform, glabrous; petiole not winged; leaves 1-2-pinnate; indusium of 2 valves, without protruding bristle.

1. H. tunbrigense (L.) Sm. - *Tunbridge Filmy-fern*. Leaves (incl. petiole) 2-5(10)cm, the blade usually elliptic-oblong in outline, the veins ending just short of leaf-margin (x10 lens), with cells mostly c.1.5x as long as wide; indusium valves conspicuously dentate; spores 40-50 microns across; 2n=26. Native; shaded, damp rock-faces and tree-trunks; local in Ir and W Br from Cornwall to N Ebudes, E Sussex. **12**

2. H. wilsonii Hook. - *Wilson's Filmy-fern*. Leaves (incl. petiole) 3-10(20)cm, the blade usually narrowly elliptic-lanceolate in outline, the veins mostly reaching leaf-margin (x10 lens), with cells mostly ≥2x as long as wide; indusium valves entire; spores 60-75 microns across; 2n=36. Native; similar places to *H. tunbrigense*, often with it; similar distribution in Ir and W Br but commoner and N to Shetland, not E Sussex. **12**

2. TRICHOMANES L. - *Killarney Fern*

Rhizomes thin, pubescent; petiole winged at least distally; leaves 2-3-pinnate; indusium tubular, with protruding bristle when mature.

1. T. speciosum Willd. - *Killarney Fern*. Leaves (incl. petiole) 7-45cm, the blade usually triangular-ovate in outline, the veins ending just short of leaf-margin (x10 lens); 2n=144. Native; very sheltered, damp rock-faces, often near waterfalls or at cave entrances; extremely local in SW, NW & CN En, Wa, SW Sc and Ir (mainly SW), formerly commoner and still surviving as gametophytes in parts of N & SW En and N Wa where sporophytes no longer develop. **RRR**

11. POLYPODIACEAE - *Polypody family*

Rhizomes extended, scaly; leaves borne singly along rhizome, spirally coiled when young, of 1 sort, 1-pinnate to simple but very deeply pinnately lobed, the lobes/pinnae linear, entire to shallowly serrate; sori orbicular to elliptic, on leaf lowerside; indusium 0; homosporous; gametophyte green, surface-living.

The only ferns with pinnate or deeply pinnately lobed leaves with ± parallel-sided pinnae/lobes and orbicular to elliptic sori without indusia.

1. POLYPODIUM L. - *Polypodies*

Leaves pinnate to very deeply pinnately lobed right to base of blade; sori in row on either side of midribs of pinnae, not sunken.

Microscopic examination is necessary for certain identification except with extreme or very typical examples. In the hybrids all or most of the spores are empty and shrivelled; plants with all or most spores full and turgid can be separated as in the key. Ten sporangia per plant should be measured to obtain mean figures.

1 Leaf-blades mostly ≤2x as long as wide; pinnae usually narrowly acute, often markedly serrate; sporangia mixed with hairs (paraphyses) which are ≥0.5mm **3. P. cambricum**

1 Leaf-blades >2x (up to c.6x) as long as wide; pinnae rounded to acute, usually subentire; sporangia without paraphyses 2

 2 Sori usually orbicular; mature leaves parallel-sided in proximal 1/3-1/2; annulus at yellow sporangium stage dark orange-brown, with mean of 10-14 thick-walled cells **1. P. vulgare**

 2 Sori usually broadly elliptic; mature leaves scarcely parallel-sided; annulus at yellow sporangium stage pale buff to golden-brown, with mean of 7-9 thick-walled cells **2. P. interjectum**

1. P. vulgare L. - *Polypody.* Leaves (incl. petiole) to 25cm, usually narrowly 37
oblong, mostly 3-6x as long as wide with 12-30 pinnae on each side; rhizome-scales mostly <6mm, triangular to narrowly so; thickened cells of annulus 7-17 (mean per plant 10-14); 2n=148 (tetraploid). Native; on rocks, walls, tree-trunks and banks, often on acid soils; frequent to common ± throughout Br and Ir, the most exposure- and acid-tolerant sp.

1 x 2. P. vulgare x P. interjectum = P. x mantoniae Rothm. & U. Schneid. occurs in scattered localities throughout Br and Ir (the commonest hybrid); it has a mean of 9-10 thickened annulus cells, is intermediate in other characters, has no paraphyses, and is sterile; 2n=185 (pentaploid).

1 x 3. P. vulgare x P. cambricum = P. x font-queri Rothm. occurs in a few scattered places in Br N to S Sc; it has a mean of 11-14 thickened annulus cells, is intermediate in other characters, has no paraphyses, and is sterile; (2n=111) (triploid).

2. P. interjectum Shivas (*P. vulgare* ssp. *prionodes* (Asch.) Rothm.) - *Intermediate* 37
Polypody. Leaves (incl. petiole) to 40cm, usually narrowly oblong-lanceolate, mostly 2-4x as long as wide with 12-30 pinnae on each side; rhizome-scales mostly 3.5-11mm, triangular-acuminate to narrowly so; thickened cells of annulus 4-13 (mean per plant 7-9); 2n=222 (hexaploid). Native; similar places to *P. vulgare* but more calcicole and usually in more shaded places; similar distribution to *P. vulgare*, but commoner in S (and in CI) and rare in N (not in Orkney or Shetland).

2 x 3. P. interjectum x P. cambricum = P. x shivasiae Rothm. occurs in very scattered places in Ir, W Br, E Sussex and Guernsey; it has a mean of 7-8 thickened annulus cells, is intermediate in other characters, has paraphyses in some but not all plants, and is sterile; (2n=148) (tetraploid).

3. P. cambricum L. (*P. australe* Fée, *P. vulgare* ssp. *serrulatum* F.W. Schultz ex 37
Arcang.) - *Southern Polypody.* Leaves (incl. petiole) to 40cm, usually triangular-ovate to oblong-ovate, mostly 1.25-2x as long as wide with 9-22 pinnae on each side; rhizome-scales mostly 5-16mm, lanceolate; thickened cells of annulus 4-19 (mean per plant 5-10); 2n=74 (diploid). Native; mostly on base-rich rocks,

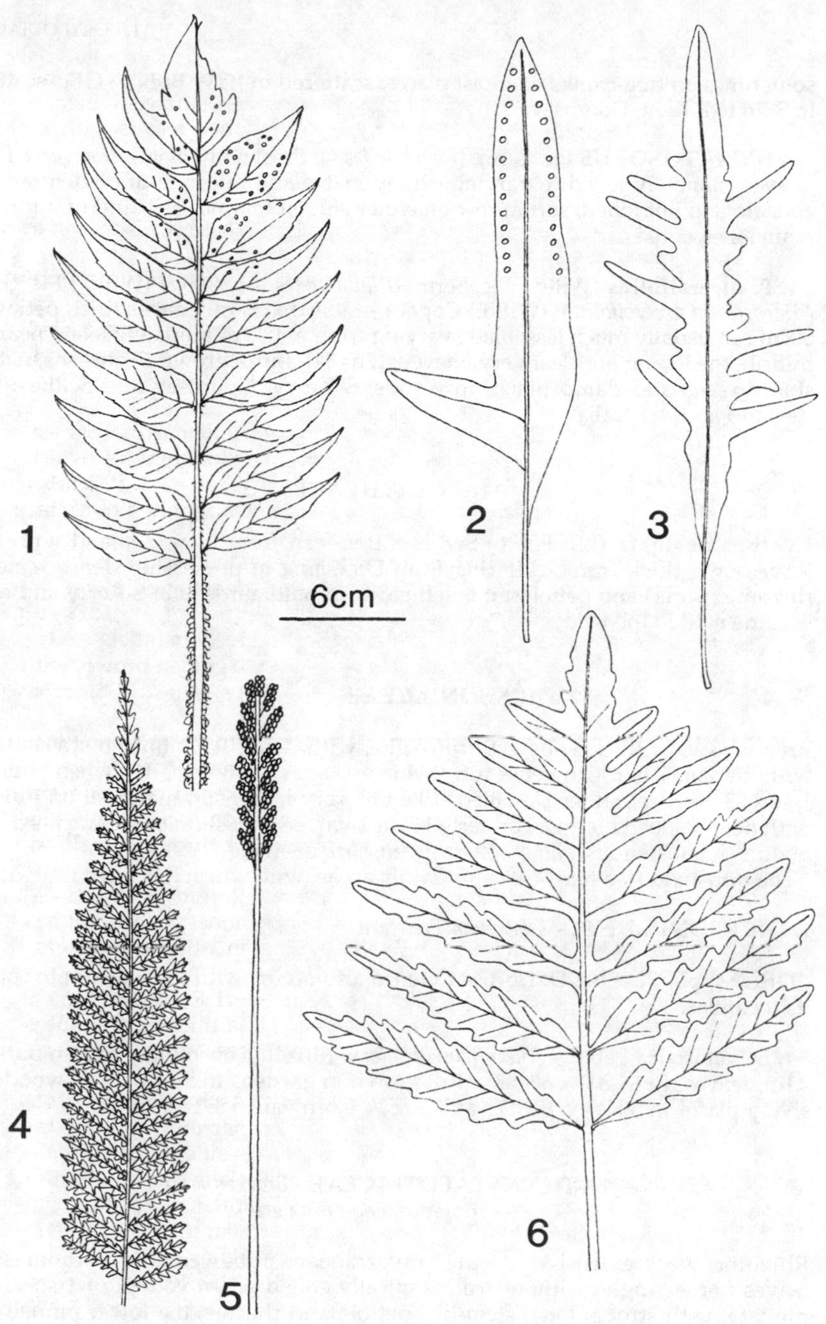

FIG 19 - Pteropsida. 1, *Cyrtomium falcatum*. 2-3, *Phymatosorus diversifolius*. 4, *Dicksonia antarctica* (small piece of 1 pinna). 5-6, *Onoclea sensibilis*. 5, fertile. 6, sterile.

sometimes on tree-trunks, in moist places; scattered in Ir, W Br N to C Sc, scattered in S En to E Kent, Guernsey.

2. PHYMATOSORUS Pic. Serm. (*Phymatodes* C. Presl nom. illeg.) - *Kangaroo Fern*
Leaves pinnately lobed to varying extent on 1 plant, but base of blade gradually cuneate and not lobed; sori in row on either side of main midrib and of midribs of main lobes, sunken.

1. P. diversifolius (Willd.) Pic. Serm. (*Phymatodes diversifolia* (Willd.) Pic. Serm., **19**
Microsorum diversifolium (Willd.) Copel.) - *Kangaroo Fern*. Leaves (incl. petiole) to 60cm but usually much less; blade varying from entire to pinnately lobed nearly to midrib, the lobing on 1 leaf very uneven; (2n=74). Intrd; grown in gardens, natd on shady walls and damp places in woods; S Kerry, Guernsey and Scillies; New Zealand and Australia.

11A. CYATHEACEAE

Cyathea dealbata (J.R. Forst.) Sw. is a tree-fern from New Zealand with huge leaves on a thick trunk, differing from *Dicksonia* in possessing dense scales on rhizome (trunk) and petioles; it is self-sown in wild gardens in S Kerry and might become natd in the wild.

12. DICKSONIACEAE - *Tree-fern family*

Rhizomes very thick, suberect, trunk-like, with dense hairs but no scales; leaves without scales, borne in dense tuft at rhizome apex, spirally coiled when young, of 1 sort, 3-pinnate, rather *Dryopteris*-like but very large; sori marginal on pinnules with deeply lobed margins, protected by a 2-valved indusium becoming hard when mature; homosporous; gametophyte green, surface-living.

The only natd tree-ferns (see also Cyatheaceae, with which it is often united).

1. DICKSONIA L'Hér. - *Australian Tree-fern*

Other spp. - Possibly **D. fibrosa** Colenso also occurs with *D. antarctica* in some of its localities.

1. D. antarctica Labill. - *Australian Tree-fern*. Rhizome eventually to 15m but often **19**
<1m; leaves up to 200 x 60cm. Intrd; grown in gardens in SW, natd in woods and shady places in S Kerry, Scillies and E & W Cornwall; S & E Australia.

13. DENNSTAEDTIACEAE - *Bracken family*
(Hypolepidaceae *pro parte*)

Rhizomes very extensive, deeply subterranean, pubescent but without scales; leaves borne singly, without scales, spirally coiled when young, of 1 sort, (2-)3-pinnate, with strong, long, stem-like petiole and rhachis, the lower pinnae often remote; sori marginal, ± continuous, enfolded by reflexed margin of pinnule; homosporous; gametophyte green, surface-living.

Bracken is unmistakable in appearance and in the scent of its crushed fresh leaves.

1. PTERIDIUM Gled. ex Scop. - *Bracken*
1. P. aquilinum (L.) Kuhn - *Bracken*. Leaves (incl. petiole) stiffly erect, to 3(5)m,

usually pubescent below and on lowerside, the blade usually rather leathery and shiny on upperside. Native.

Very variable, often variously divided into spp., sspp. or vars., but a world-wide study is essential before a useful classification can be adopted. 2 taxonomic groupings are recognisable in the N Hemisphere: the *P. aquilinum* and **P. latiusculum** (Desv.) Hieron. ex Fr. complexes. The latter is characteristic of coniferous forests in high latitudes. According to C.N. Page both complexes contain several spp., incl. 1 from each in BI: *P. aquilinum* and **P. pinetorum** respectively. Page further recognises 3 sspp. within *P. aquilinum* (sspp. **aquilinum, atlanticum** C.N. Page and **fulvum** C.N. Page) and 2 within *P. pinetorum* (sspp. **pinetorum** and **osmundaceum** (H. Christ) C.N. Page). *P. aquilinum* ssp. *aquilinum* is the common widespread taxon; ssp. *atlanticum* occurs mostly on limestone in W Br from SW En to CW Sc (S to W Africa); the other 3 taxa are all restricted in Br to small parts of the highlands of Sc (ssp. *fulvum* is endemic). So far as is known all these taxa are interfertile and their hybrids are fertile. The compromise solution adopted here is to recognise the two complexes as sspp. and to consider the above segregates as local ecotypes.

a. Ssp. aquilinum. Leaves often robust, sometimes small and wiry but with >1 pair of pinnae, the pinnae unfolding sequentially acropetally in spring; pinna midribs horizontal to downward-curved at least when young; petioles of emerging leaves with abundant white hairs and some reddish hairs, all gradually lost during maturity. 2n=104, 156. Woods, heaths and moors, often dominant over large areas, usually on acid dry soils, rarely on calcareous ones; abundant throughout BI.

b. Ssp. latiusculum (Desv.) C.N. Page (*P. pinetorum* C.N. Page & R.R. Mill). **R**
Leaves relatively small and wiry, ≤1m, often with only 1 main pair of pinnae each ± as large as rest of leaf, the 3 portions unfolding ± simultaneously in spring; pinna midribs upturned and ± straight; petioles of emerging leaves with 0 or few white hairs but abundant reddish hairs, all soon lost. Presumed relic of native *Pinus* woodland in open woodland and on moors; very scattered in C & N Sc S to M Perths.

14. THELYPTERIDACEAE - *Marsh Fern family*

Rhizomes long and slender or short and thick, with (sometimes scarce) scales; leaves borne singly or in dense terminal tufts, with sparse or very sparse scales, spirally coiled when young, of 1 or 2 sorts, almost 2-pinnate, with entire to crenate or sinuate pinnules; sori in a row on pinnule lowerside usually near to margin, with indusium 0 or small, thin and soon withering; homosporous; gametophyte green, surface-living.

Recognizable by the submarginal ± round sori with 0 or flimsy and quickly withering indusia on somewhat *Dryopteris*-like leaves.

1. THELYPTERIS Schmidel - *Marsh Fern*

Rhizomes long and slender, bearing leaves singly at intervals or sometimes in sparse tufts; leaves ± of 2 sorts, straight, the fertile usually longer and with narrower pinnules recurved at margins; lowest pair of pinnae usually ≥1/2 as long as longest pair and parallel with it; petiole c. as long as blade or longer.

1. T. palustris Schott (*T. thelypteroides* Michx. ssp. *glabra* Holub) - *Marsh Fern*. **R**
Leaves to 80cm (sterile) or 1.5m (fertile), the blade lanceolate to narrowly elliptic in outline, subglabrous, pale green; 2n=70. Native; marshes and fens, usually shaded among taller herbs or shrubs; scattered in Br and Ir N to C Sc, frequent only in E Anglia, decreasing.

2. PHEGOPTERIS (C. Presl) Fée - *Beech Fern*
Rhizomes long and slender, bearing leaves singly at intervals; leaves of 1 sort, with blade bent sharply away from plane of petiole; lowest pair of pinnae ± as long as longest pair, reflexed; petiole as long as blade or longer.

1. P. connectilis (Michx.) Watt (*Thelypteris phegopteris* (L.) Sloss.) - *Beech Fern.* Leaves to 50cm, the blade triangular to ovate-triangular in outline with strongly acuminate pinnae, minutely pubescent, pale green; 2n=90. Native; damp woods, shady rocky places and banks, on acid soils; rather common in W & N Br, absent from most of C, S & E En, scattered in Ir.

3. OREOPTERIS Holub - *Lemon-scented Fern*
Rhizomes short and stout, bearing leaves in tuft at apex; leaves of 1 sort, straight; lowest pair of pinnae extremely short, much <1/2 as long as longest; petiole much shorter than blade.

1. O. limbosperma (Bellardi ex All.) Holub (*Thelypteris oreopteris* (Ehrh.) Sloss., *T. limbosperma* (Bellardi ex All.) H.P. Fuchs) - *Lemon-scented Fern.* Leaves to 1.2m, the blade oblanceolate in outline, with pinnae gradually reducing in length basally to very short basal ones, with numerous minute glands on lowerside giving lemon scent when fresh leaf is bruised, pale green; 2n=68. Native; damp shady places and woods on acid soil; similar distribution to *Phegopteris* but more common, and also frequent in SE & SW En.

15. ASPLENIACEAE - *Spleenwort family*

Rhizomes short or very short, with scales; leaves borne in tufts at end of rhizomes, with scales (often sparse), spirally coiled when young, of 1 sort, simple and entire to 3-pinnate; sori oblong to long and linear, on leaf lowerside, with indusium the same shape as sorus and attached to side of it, or without indusium but whole leaf lowerside covered with scales; homosporous; gametophyte green, surface-living.

Distinguished by the elongated (often very long) sori either with a straight-sided indusium or with indusium 0 but whole leaf lowerside covered with scales; most spp. are small and/or have relatively little-divided leaf-blades.

1. PHYLLITIS Hill - *Hart's-tongue*
Leaves simple, entire, with few or 0 scales on lowerside; sori linear, long, strictly parallel with each other and with lateral veins, each apparent sorus actually 2 lying adjacent, the pair with a linear indusium on each side with facing openings.

1. P. scolopendrium (L.) Newman (*Asplenium scolopendrium* L.) - *Hart's-tongue.* Leaves (incl. petiole) to 60cm but often much less; blade linear-oblong, cordate at base, up to 7cm wide; petiole ≤1/2 as long as blade; 2n=72. Native; shady moist rocky places, banks, walls and woods; common throughout BI except very scattered in N Sc.

1 x 2. PHYLLITIS x ASPLENIUM = X ASPLENOPHYLLITIS Alston
The 3 combinations that have been found in BI are all very rare and highly sterile, and all show intermediacy between *Phyllitis* and the *Asplenium* parent. The blades are pinnate below and simple above, the latter character separating them from the somewhat similar *Asplenium marinum*. In most plants there is a mixture of single sori (as in *Asplenium*) and paired sori (as in *Phyllitis*).

1 Pinnae in lower part of blade with conspicuous gaps between, not or scarcely longer than wide **3. X A. confluens**

1 Pinnae in lower part of blade without or with very small gaps between, often overlapping, usually c.2x as long as wide 2
2 Longest pinna (2)3-4(8) from bottom; sori 2-3(4)mm, usually confined to outer 1/2 of blade and near to pinna-margin **2. X A. microdon**
2 Longest pinna the lowest or next to lowest; sori 4-6mm, usually confined to inner 1/2 of blade and near to pinna-midrib **1. X A. jacksonii**

1. X A. jacksonii Alston (*Asplenium x jacksonii* (Alston) Lawalrée; *P. scolopendrium* x *A. adiantum-nigrum*). Leaves (incl. petiole) up to 16(20)cm; (2n=108). Native; found in W Cornwall, N Devon and CI in 19th Century; endemic. RR

2. X A. microdon (T. Moore) Alston (*Asplenium x microdon* (T. Moore) Lovis & Vida; *P. scolopendrium* x *A. obovatum*). Leaves (incl. petiole) up to 20(30)cm; 2n=108. Native; found in W Cornwall and Guernsey in 19th Century, rediscovered on hedgebanks in Guernsey in 1965; endemic. RR

3. X A. confluens (T. Moore ex E.J. Lowe) Alston (*Asplenium x confluens* (T. Moore ex E.J. Lowe) Lawalrée; *P. scolopendrium* x *A. trichomanes*). Leaves (incl. petiole) up to 10(20)cm. Native; found in NE Yorks, Westmorland and N Kerry in 19th Century, refound in S Kerry in 1982. RR

2. ASPLENIUM L. - *Spleenworts*

Leaves 1-3-pinnate or sparsely and irregularly divided into linear segments, with few or 0 scales on lowerside; sori oblong to linear, usually parallel with lateral veins of pinnae or pinnules, single and each with 1 indusium.

Hybrids are quite numerous but all are very rare and scarcely contribute to identification problems. They are all sterile and ± obviously intermediate.

1 Leaves irregularly and sparsely divided into linear segments **8. A. septentrionale**
1 Leaves 1-3-pinnate; pinnules or pinnae not linear 2
2 Leaves 1-pinnate 3
2 Leaves 2-3-pinnate at least in part 5
3 Distal part of rhachis conspicuously green-winged; larger pinnae >12mm, ≥2x as long as wide **4. A. marinum**
3 Rhachis not or scarcely winged; pinnae ≤12mm, ≤2x as long as wide 4
4 Rhachis blackish, with very narrow scarious wings **5. A. trichomanes**
4 Rhachis green, not winged or with extremely narrow green wings **6. A. viride**
5 Longest pinnae at c. middle of blade, the basal pair slightly to considerably shorter; petioles usually much shorter than blade **3. A. obovatum**
5 Longest pinnae clearly the basal ones; petioles nearly as long as to longer than blade 6
6 Petioles green; blade 1-2-pinnate; indusia with fringed margins **7. A. ruta-muraria**
6 Petioles reddish-brown to blackish; blade (2-)3-pinnate; indusia subentire 7
7 Blade and pinnae acute to very shortly acuminate; spores (32)39-45(52) microns across; widespread **1. A. adiantum-nigrum**
7 Blade and pinnae very long-acuminate; spores (25)30-35(39) microns across; S Ir only **2. A. onopteris**

Other spp. - There is no good evidence for the existence of either **A. fontanum** (L.) Bernh. or **A. cuneifolium** Viv. in the wild in BI as sometimes claimed. Specimens of the former were probably in gardens or temporarily escaped just outside, and records of the latter from serpentine rocks in Sc were errors for *A.*

adiantum-nigrum with unusually wide, blunt ultimate leaf-segments.

1. A. adiantum-nigrum L. (*A. cuneifolium* auct. non Viv.) - *Black Spleenwort.* 25
Leaves 2-3-pinnate, triangular-ovate in outline, to 50cm incl. petiole c. as long as blade; pinnules ovate to lanceolate, with obtusely cuneate base, with small dark scales (x10 lens) on lowerside; sori linear to linear-oblong; 2n=144. Native; rocky places in woods, banks and open sites, walls; common throughout most of BI.

1 x 2. A. adiantum-nigrum x A. onopteris = A. x ticinense D.E. Mey. occurs in W & M Cork and Co. Kilkenny.

1 x 3. A. adiantum-nigrum x A. obovatum = A. x sarniense Sleep occurs in Guernsey.

1 x 8. A. adiantum-nigrum x A. septentrionale = A. x contrei Callé, Lovis & Reichst. was found in Caerns in 19th Century; (2n=144).

2. A. onopteris L. (*A. adiantum-nigrum* ssp. *onopteris* (L.) Heufl.) - *Irish* RR
Spleenwort. Leaves 3-pinnate, triangular in outline, to 50cm incl. petiole usually 25
longer than blade; pinnules lanceolate to linear-lanceolate, with acutely cuneate base, without scales; sori linear to linear-oblong; 2n=72. Native; dry banks and rock-faces mostly on limestone and near sea; very local in N & S Kerry, W & M Cork and Co Kilkenny, formerly elsewhere in S Ir.

3. A. obovatum Viv. (*A. billotii* F.W. Schultz) - *Lanceolate Spleenwort.* Leaves 2- R
pinnate at least at base, oblong-lanceolate in outline, to 30cm incl. petiole usually 25
much shorter than blade; pinnules ovate to suborbicular; sori oblong; 2n=144. Native; rocks, hedgebanks and walls, especially near sea; common in CI and SW En, frequent in W Wa, very scattered in Ir and W Br N to W Sutherland, formerly W Kent and E Sussex. Our plant is ssp. **lanceolatum** (Fiori) P. Silva.

4. A. marinum L. - *Sea Spleenwort.* Leaves 1-pinnate, oblong-lanceolate in outline, to 40cm (but often much less) incl. petiole much shorter than blade; pinnae oblong to ovate-oblong or narrowly so; sori linear; 2n=72. Native; walls, cliffs and rock-crevices close to sea (often sea-sprayed); frequent on coasts of BI except E & S coast of En from S Hants to SE Yorks, formerly E Sussex.

5. A. trichomanes L. - *Maidenhair Spleenwort.* Leaves 1-pinnate, linear in outline, to 20(40)cm incl. petiole much <1/2 as long as blade; sori linear to oblong-linear. Native. The 3 sspp. are often distinguishable only with difficulty; spore size is most reliable for separating the 2 common sspp.

1 Pinnae oblong-triangular, usually attached to rhachis in centre of base, usually conspicuously serrate, minutely pubescent on lowerside **c. ssp. pachyrachis**

1 Pinnae suborbicular to oblong, often attached to petiole at proximal corner, usually subentire, ± glabrous 2

2 Pinnae suborbicular to rhombic, often asymmetric due to uneven basal bulges, up to 8mm, flat to concave on upperside; sori ≤2mm, ≤6(9) per pinna; rhizome-scales ≤3.5mm; calcifuge **a. ssp. trichomanes**

2 Pinnae oblong, usually ± symmetric, up to 12mm, convex to flat on upperside; sori ≤3mm, ≤9(12) per pinna; rhizome-scales ≤5mm; calcicole **b ssp. quadrivalens**

a. Ssp. trichomanes. Smaller-leaved and more delicate than ssp. *quadrivalens*, 25
with thinner petiole and rhachis; pinnae distinctly stalked, often asymmetric in overall shape but often attached to rhachis ± in centre of base; spores (23)29-36(42) microns; 2n=72. Non-calcareous rocks and walls; scattered in Sc, local in Wa and Lake District, Co Down.

b. Ssp. quadrivalens D.E. Mey. Less delicate and longer-leaved than ssp. 25
trichomanes, with thicker petiole and rhachis; pinnae subsessile, usually symmetric in overall shape but attached to rhachis near proximal corner; spores (27)34-43(50) microns; 2n=144. Calcareous or neutral rocks or walls (incl. mortar in walls in acidic regions); common over most of BI. Hybridises with ssp. *trichomanes* (=nothossp. **x lusaticum** (D.E. Mey.) Lawalrée) and ssp. *pachyrachis* (=nothossp. **x**

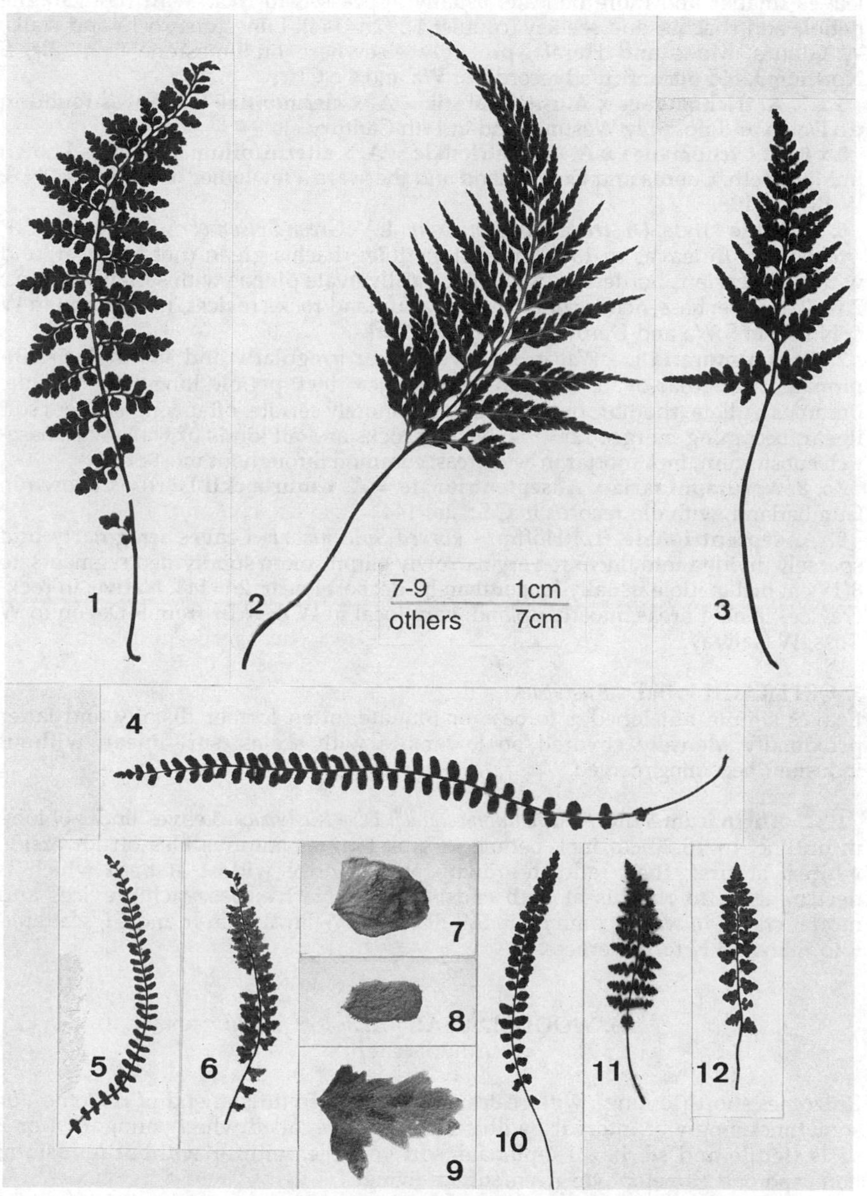

FIG 25 - Pteropsida. 1-10, ***Asplenium***. 1, ***A. obovatum***. 2, ***A. onopteris***. 3, ***A. adiantum-nigrum***. 4-6, ***A. trichomanes***. 4, ssp. ***quadrivalens***. 5, ssp. ***trichomanes***. 6, ssp. ***pachyrachis***. 7-9, pinnae of ***A. trichomanes***. 7, ssp. ***trichomanes***. 8, ssp. ***quadrivalens***. 9, ssp. ***pachyrachis***. 10, ***A. viride***. 11-12, ***Woodsia***. 11, ***W. ilvensis***. 12, ***W. alpina***.

staufferi Lovis & Reichst.).

c. Ssp. pachyrachis (H. Christ) Lovis & Reichst. Differs from ssp. *quadrivalens* in RR
leaves smaller and more delicate, usually appressed to rock, with rigid, fragile 25
petiole and rhachis; and see key (couplet 1); (2n=144). Limestone rocks and walls; W Gloucs, Mons and Herefs, probably elsewhere on limestone in W Br, S Northumb, old unconfirmed records for Wa and Co Clare.

5 x 7. A. trichomanes x A. ruta-muraria = A. x clermontiae Syme was found in Co Down and possibly Westmorland in 19th Century.

5 x 8. A. trichomanes x A. septentrionale = A. x alternifolium Wulfen is known in Merioneth, Caerns and Cumberland and there are a few other records from N & W Br; 2n=108.

6. A. viride Huds. (*A. trichomanes-ramosum* L.) - *Green Spleenwort.* Differs from *A.* 25
trichomanes in leaves to 15(20)cm incl. petiole; rhachis green (not blackish) and without 2 scarious borders; and asymmetrically ovate pinnae with serrate margins; 2n=72. Native; base-rich (mainly limestone) upland rock-crevices; rather local in W & N Br S to S Wa and Derbys, formerly to Warks.

7. A. ruta-muraria L. - *Wall-rue.* Leaves rather irregularly and sparingly 1-2(3)-pinnate, triangular-ovate in outline, to 8(15)cm incl. petiole longer than blade; pinnules trullate, rhombic or obtriangular, minutely serrate, often long-stalked; sori linear, becoming merged; 2n=144. Native; rocks and all kinds of walls with base-rich substratum, incl. mortar in acid areas; common throughout most of BI.

7 x 8. A. ruta-muraria x A. septentrionale = A. x murbeckii Dörfl. is known in Cumberland, with old records in C Sc; 2n=144.

8. A. septentrionale (L.) Hoffm. - *Forked Spleenwort.* Leaves irregularly and R
sparsely divided into linear to very narrowly elliptic often subdivided segments, to 8(15)cm incl. petiole usually longer than blade; sori linear; 2n=144. Native; in rock-crevices in acid areas, mostly upland; very local in W & N Br from S Devon to W Ross, W Galway.

3. CETERACH Willd. - *Rustyback*

Leaves simple but lobed ± to base or pinnate, often former distally and latter proximally, densely covered on lowerside with scales; sori linear, without indusium, becoming merged.

1. C. officinarum Willd. (*Asplenium ceterach* L.) - *Rustyback.* Leaves linear-oblong in outline, to 15(20)cm incl. petiole <1/2 as long as blade; scales on lowerside whitish at first, then reddish-brown; lobes/pinnae widest at base which is decurrent on to rhachis at both ends; 2n=144. Native; base-rich crevices and mortar cracks in walls; common in SW Br N to NW En, and in Ir and CI, scattered E to E En and N to Easterness.

16. WOODSIACEAE - *Lady-fern family*
(Athyriaceae)

Rhizomes short (to long), with scales; leaves borne in tufts at end of rhizomes, or sometimes singly at intervals, with scales, spirally coiled when young, of 1 or 2 sorts (fertile and sterile), 1-4-pinnate; sori variable, with or without indusium; homosporous; gametophyte green, surface-living.

The family in our flora most defying definition except by reference to vascular architecture; the 6 genera differ widely in superficial characters.

1. MATTEUCCIA Tod. - *Ostrich Fern*

Rhizomes short; leaves borne in apical tufts, of 2 sorts; sterile leaves 2-pinnate or nearly so; fertile leaves shorter, with longer petiole, 1-pinnate, not green; sori in 1-2 contiguous rows on each pinna, protected by tightly inrolled leaf-margin.

1. M. struthiopteris (L.) Tod. - *Ostrich Fern.* Sterile leaves up to 1.5m incl. petiole 15
much <1/2 as long as blade, 1-pinnate with deeply lobed pinnae or just 2-pinnate, with pinnae gradually reducing in length basally to very short basal ones (superficially closely resembling *Oreopteris limbosperma*); fertile leaves up to 60cm, with pinnae very narrow due to margin inrolling; (2n=78, 80). Intrd; grown in gardens, natd in shady places; scattered in S & C Sc, Westmorland, Salop and N Ir; Europe.

2. ONOCLEA L. - *Sensitive Fern*
Rhizomes long; leaves borne singly, of 2 sorts; sterile leaves 1-pinnate or nearly so with winged rhachis and entire to lobed pinnae; fertile leaves shorter, 2-pinnate or nearly so, with non-green pinnae; sori in compact groups protected by tightly inrolled leaf-margin.

1. O. sensibilis L. - *Sensitive Fern.* Sterile leaves up to 1m incl. petiole much longer 19
than blade, 1-pinnate at base, simple and deeply lobed above, the lower pinnae lobed, the lowest pinna the longest; fertile leaves up to 80cm, with pinnae very narrow due to inrolled margins; (2n=74). Intrd; grown in gardens, natd in shady places; scattered in CW & SW Sc, W Wa and NW & S En, Jersey; E N America and E Asia.

3. ATHYRIUM Roth - *Lady-ferns*
Rhizomes short; leaves borne in apical tufts, of 1 sort, 2-pinnate with deeply lobed to crenate pinnules; sori on lowerside of pinnules, not protected by inrolled leaf-margin, with 0, inconspicuous or well-developed indusium.

1 Sori oblong to curved with well-developed J- or C-shaped indusium **1. A. filix-femina**
1 Sori orbicular, with 0 or vestigial indusium 2
2 Leaves oblanceolate to narrowly elliptic, up to 30(70)cm incl. petiole c.1/8-1/4 as long as blade; sori commoner near base of blade **2. A. distentifolium**
2 Leaves irregularly narrowly oblong-elliptic, up to 20(30)cm incl. petiole ≤1/8 as long as blade; sori commoner near apex of blade **3. A. flexile**

1. A. filix-femina (L.) Roth - *Lady-fern.* Leaves to 1(1.5)m, suberect, straight, with 29
petiole usually 1/4-1/3 as long as blade; blade sometimes ± 3-pinnate; indusium well-developd, J- or C-shaped; 2n=80. Native; damp woods, shady hedgebanks and rocky places, mountain screes, marshes; common throughout most of BI.

2. A. distentifolium Tausch ex Opiz (*A. alpestre* (Hoppe) Rylands ex T. Moore R
non Clairv.) - *Alpine Lady-fern.* Differs from small plants of *A. filix-femina* only as in 29
key (couplet 1); 2n=80. Native; acid gullies, boulder-slopes and scree, sometimes with *A. filix-femina*, rarely below 600m; local in C & N Sc.

3. A. flexile (Newman) Druce (*A. alpestre* var. *flexile* (Newman) Milde, *A.* RR
distentifolium var. *flexile* (Newman) Jermy) - *Newman's Lady-fern.* Differs from *A.* 29
distentifolium in leaves bent at or just above base of blade to produce spreading, often ± flexuous (not suberect, straight) leaves; and see key (couplet 2); 2n=80. Native; damp acid rocky places at 1040-1140m, often with *A. distentifolium;* very local in C Sc; endemic. Plants given this name possibly represent a chance combination of characters rather than a taxon, or perhaps are only a var. of *A. distentifolium*.

4. GYMNOCARPIUM Newman - *Oak Ferns*
Rhizomes long; leaves borne singly, of 1 sort, 2-3-pinnate, with petiole much longer than blade, with ± triangular blade; sori on underside of pinnules rather near

margin, orbicular to elliptic, without indusium.

1. G. dryopteris (L.) Newman (*Thelypteris dryopteris* (L.) Sloss.) - *Oak Fern.* 29
Leaves to 40cm (incl. blackish petiole); blade yellowish- to mid-green, non-glandular; basal pinnae each nearly as large as rest of blade, these 3 units rolled separately as 3 'balls' in the young leaf; 2n=160. Native; damp woods and shady rocks, banks and ravines, often in ± acid humus-rich soil and often with *Phegopteris connectilis;* frequent in N & W Br S to Severn-Humber estuaries, rare and scattered SE of that line and in N Ir.

2. G. robertianum (Hoffm.) Newman (*Thelypteris robertiana* (Hoffm.) Sloss.) - R
Limestone Fern. Leaves to 50cm (incl. greenish-brown petiole); blade dull green, with 29
small glands that are also on rhachis and top of petiole; basal pinnae each c.1/2 as large as rest of blade, each pinnule and pinna rolled separately and these in turn rolled into 1 'ball' in the young leaf; 2n=160-168. Native; open or partly shaded scree-slopes or rocky places on limestone; local in En and Wa, rare and very scattered in Sc and W Ir, rather frequently natd on walls elsewhere, especially E En.

5. CYSTOPTERIS Bernh. - *Bladder-ferns*
Rhizomes long with leaves borne singly, or short with leaves in a terminal tuft; leaves of 1 sort, 2-3(4)-pinnate; sori orbicular, on lowerside of pinnules, not protected by inrolled leaf-margin, with flap-like indusium that becomes reflexed to expose sporangia.

1 Rhizomes elongated, bearing leaves singly; leaves triangular-ovate, the lowest pair of pinnae the longest **3. C. montana**
1 Rhizome short, bearing terminal tuft of leaves; leaves narrowly oblong to lanceolate, the longest pinnae near the middle of leaf 2
  2 Spores rugose; adjacent pinnules and pinnae strongly overlapping **2. C. dickieana**
  2 Spores spinose; adjacent pinnules and pinnae scarcely overlapping **1. C. fragilis**

1. C. fragilis (L.) Bernh. (*C. regia* (L.) Desv.) - *Brittle Bladder-fern.* Leaves tufted, 29
to 25(45)cm (incl. petiole shorter than blade), lanceolate to oblanceolate or narrowly oblong, 2-pinnate with toothed to deeply lobed pinnules or 3(-4)-pinnate; 2n=168, 252. Native; shady rocks and walls or rocky woods on basic soils, incl. mortar in acid areas; common in N & W Br, very scattered in S & E, frequent but scattered in Ir.

2. C. dickieana R. Sim - *Dickie's Bladder-fern.* Leaves to 20(25)cm (incl. petiole RRR
shorter than blade), 2(-3)-pinnate; differs from *C. fragilis* as in key; 2n=168. Native; 29
basic rocks in sea-caves and stream-gorges; Kincardines, Easterness and M Perth. Plants given this name possibly represent a chance combination of characters rather than a taxon.

3. C. montana (Lam.) Desv. - *Mountain Bladder-fern.* Leaves borne singly, to 45cm RR
(incl. petiole longer than blade), triangular-ovate, 3(-4)-pinnate; (2n=84, 168). 29
Native; shady wet basic rock-ledges, gullies and scree above 700m; local in C Sc, formerly Westmorland.

6. WOODSIA R.Br. - *Woodsias*
Rhizomes short, with leaves borne in a terminal tuft; leaves of 1 sort, 1-pinnate with deeply lobed pinnae or 2-pinnate on proximal pinnae; petiole with a visible joint c.1/3 way up (an eventual abscission point); sori orbicular, on lowerside of pinnules, not protected by inrolled leaf-margin, with an indusium consisting of basal ring of hairs or narrow scales.

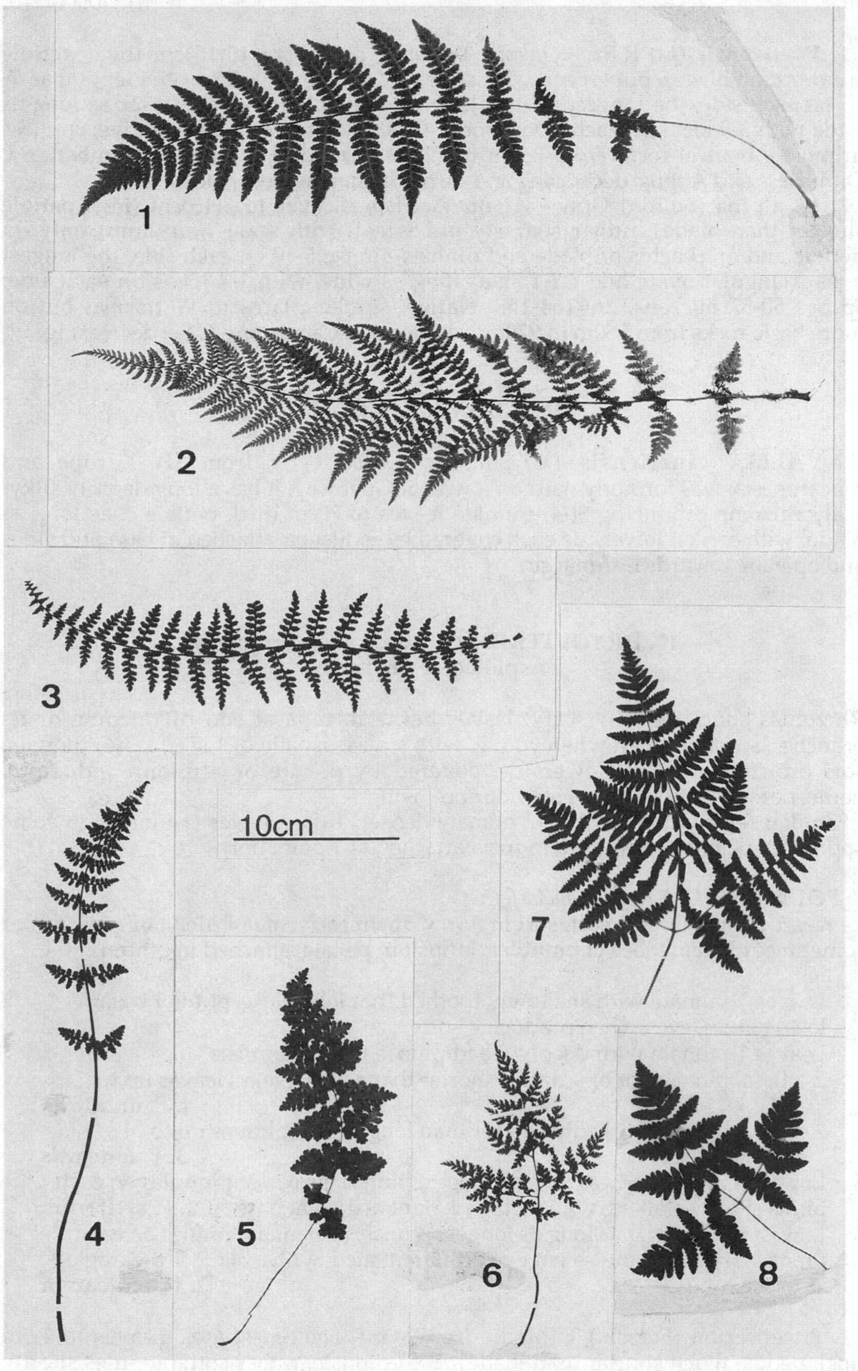

FIG 29 - Pteropsida. 1-3, *Athyrium.* 1, *A. filix-femina.* 2, *A. distentifolium.* 3, *A. flexile.* 4-6, *Cystopteris.* 4, *C. fragilis.* 5, *C. dickieana.* 6, *C. montana.* 7-8, *Gymnocarpium.* 7, *G. robertianum.* 8, *G. dryopteris.*

1. W. ilvensis (L.) R.Br. - *Oblong Woodsia*. Leaves to 10(15)cm (incl. petiole **RRR**
shorter than blade), pubescent, with scales (c.2-3mm) on leaf lowerside; pinnae 7- **25**
15 on each side, the longest ones oblong or ovate-oblong and c.1.5-2x as long as wide with 3-8 lobes on each side; spores 42-50 microns; 2n=82-84. Native; crevices in mostly neutral rocks from 360m to 720m; very local in Caerns, Cumberland, Dumfriess and Angus, decreasing and formerly more widespread.

2. W. alpina (Bolton) Gray - *Alpine Woodsia*. Leaves to 8(15)cm (incl. petiole **RRR**
shorter than blade), rather sparsely pubescent, with scales (c.1-2mm) only on **25**
petiole and on rhachis of blade and pinnae; pinnae 5-10 on each side, the longest ones triangular-ovate and c.1-1.5x as long as wide, with 1-4 lobes on each side; spores 50-57 microns; 2n=164-168. Native; similar places to *W. ilvensis* but on more basic rocks from 580m to 920m; very local in Caerns and C Sc, decreasing.

16A. DAVALLIACEAE

DAVALLIA canariensis (L.) Sm., *Hare's-foot Fern*, from SW Europe and Macaronesia, was formerly natd on a wall in Guernsey; it has a long densely silky-scaly rhizome producing 3(-4)-pinnate leaves to 70cm (incl. petiole c. as long as blade) with sori on lowerside each covered by indusium attached at base and sides and opening towards leaf-margin.

17. DRYOPTERIDACEAE - *Buckler-fern family*
(Aspidiaceae *pro parte*)

Rhizomes short, densely scaly; leaves borne in tufts at end of rhizome or its branches, spirally coiled when young, with scales, usually of 1 sort, 1-3(4)-pinnate; sori orbicular, on leaf lowerside, covered by peltate or reniform indusium; homosporous; gametophyte green, surface-living.

Familiar ferns with usually 2-3-pinnate densely tufted leaves (1-pinnate in some spp.) with peltate or reniform indusia attached at 1 point only.

1. POLYSTICHUM Roth - *Shield-ferns*

Leaves 1-2-pinnate, of 1 sort; sori in a row down each side of pinna or pinnules or sometimes of main lobes of pinnules; indusium peltate, attached in centre.

1 Leaves 1-pinnate with shallowly toothed (not lobed, except for 1 basal lobe) pinnae, even when producing sori 2
1 Leaves 1-pinnate with deeply lobed pinnae to 2-(3)-pinnate 3
  2 Basal pinnae not or scarcely shorter than longest ones; leaves up to 120 x 15cm **4. P. munitum**
  2 Basal pinnae markedly shorter than longest ones; leaves up to 30(60) x 6cm **3. P. lonchitis**
3 Lowest pinnae nearly as long as longest pinnae; proximal pinnules on each pinna shortly stalked, with a blade ± right-angled at base **1. P. setiferum**
3 Lowest pinnae c.1/2 as long as longest pinnae; proximal pinnules on each pinna sessile or pinnules not even differentiated, with a blade acute-angled at base **2. P. aculeatum**

1. P. setiferum (Forssk.) T. Moore ex Woyn. - *Soft Shield-fern*. Leaves to 1.5m **34**
(incl. petiole much shorter than blade), 2(-3)-pinnate, rather soft; pinnules shortly stalked, their blades right-angled or nearly so at base with the edge nearer the leaf rhachis ± at right-angles to the pinna rhachis; 2n=82. Native; woods and hedgebanks in moist places; frequent in CI, S & W Br and Ir, becoming rarer in NE and absent from most of N & E Sc.

1 x 2. P. setiferum x P. aculeatum = P. x bicknellii (H. Christ) Hahne is scattered throughout BI N to C Sc and probably frequent though overlooked; it is intermediate and sterile; (2n=123).

1 x 3. P. setiferum x P. lonchitis = P. x lonchitiforme (Halácsy) Bech. occurs at Glende, Co Leitrim, with the parents and *P. x illyricum;* it is intermediate and sterile; 2n=82.

1 x 4. P. setiferum x P. munitum occurs as 1 plant with the parents in the Surrey locality for *P. munitum;* ?endemic.

2. P. aculeatum (L.) Roth - *Hard Shield-fern.* Leaves to 1m (incl. petiole much 34
shorter than blade), 1-2-pinnate, rather hard in texture; pinnules sessile, their blades acute-angled at base with the edge nearer the leaf rhachis at an acute angle to the pinna rhachis; 2n=164. Native; similar places to *P. setiferum* but more upland and northern; frequent in Br and Ir, common in N & W Br but less common than *P. setiferum* in SW En and S Wa. Juvenile plants are often very similar to mature ones of *P. lonchitis*, but lack sori.

2 x 3. P. aculeatum x P. lonchitis = P. x illyricum (Borbás) Hahne occurs with the parents in Co Leitrim and scattered in N & C Sc; it is intermediate and sterile; (2n=123).

3. P. lonchitis (L.) Roth - *Holly-fern.* Leaves to 30(60)cm (incl. petiole much 34
shorter than blade), 1-pinnate, hard in texture; lowest pinnae much shorter than longest ones; pinnae usually with 1 basal lobe but otherwise only shortly toothed, asymmetrically and narrowly triangular- to oblong-ovate; 2n=82. Native; basic rock-crevices, scree and ravines mostly above 600m (but down to 150m); local in N En, C & N Sc and W Ir, Caerns, Dumfriess, natd on bridge in Northants.

4. P. munitum (Kaulf.) C. Presl - *Western Sword-fern.* Leaves to 1.2m (incl. petiole 34
much shorter than blade), 1-pinnate, hard in texture; lowest pinnae scarcely shorter than longest ones; pinnae asymmetrically ± linear, much narrower than in *P. lonchitis*, usually with 1 basal lobe but otherwise only shortly toothed; (2n=82). Intrd; garden escape on shady laneside bank since 1980; Surrey; W N America.

2. CYRTOMIUM C. Presl (*Phanerophlebia* C. Presl) - *House Holly-fern*
Leaves 1-pinnate, of 1 sort; sori scattered all over lowerside of pinnae; indusium peltate, attached in centre.

1. C. falcatum (L.f.) C. Presl (*Phanerophlebia falcata* (L.f.) Copel., *Polystichum* 19
falcatum (L.f.) Diels) - *House Holly-fern.* Leaves to 1.2m (incl. petiole shorter than 34
blade), very hard in texture; pinnae usually with 1 basal lobe but otherwise entire to very shortly obtusely toothed, asymmetrically triangular-ovate; (2n=82, 123). Intrd; grown in conservatories etc., natd on and by walls, among maritime rocks and in other shady places; scattered in W Br from Scillies to WC Sc, also in Middlesex, E & W Kent, W Cork and CI; E Asia.

3. DRYOPTERIS Adans. - *Buckler-ferns*
Leaves 2-3(4)-pinnate, usually of 1 sort; sori in a row down each side of pinnules or sometimes of main lobes of pinnules; indusium reniform, attached at the notch.

A difficult genus, especially due to the extent of hybridization. Most of the hybrids are sterile, but some produce good spores with an unreduced chromosome number that develop into gametophytes that reproduce apogamously (grow directly into new sporophytes without fertilization). Such taxa, e.g. *D. affinis* and *D. remota*, are best treated as spp.; *D. x complexa* is theoretically in the same category, but its 2 parents are already so close that recognition of a third (intermediate) species of very sporadic occurrence is unrealistic. The apogamous taxa can hybridize with sexual taxa by producing male gametes that fertilize the female gametes of the sexual taxa, but they do not themselves produce female gametes. Such hybrids are usually also apogamous (*D. affinis* and its hybrids demonstrate this). Hybrids between sexual spp. are sterile.

1 Leaves 1-pinnate with deeply lobed pinnae, to 2-pinnate with pinnules lobed to c.1/2 way to midrib 2

1 Leaves 2-pinnate with pinnules lobed nearly to midrib, to 3(-4)-pinnate 7

2 Leaves ± of 2 sorts, the fertile longer and more erect, lanceolate-oblong, parallel-sided for most of length; pinnae with <15 pinnules or lobes each side; pinnules with mucronate teeth **7. D. cristata**

2 Leaves of 1 sort, lanceolate-elliptic, scarcely parallel-sided; pinnae with >15 pinnules each side; pinnules with acute, obtuse or 0 teeth 3

3 Leaf dull green, ± mealy due to many minute glands on both surfaces; lowest pinna with proximal 3-4 pinnules on both sides ± same size **6. D. submontana**

3 Leaf clear green (of various shades), not or slightly glandular; lowest pinna with pinnules successively smaller from base distally, or just the 2 proximal ± same size 4

4 Pinnules evenly lobed nearly 1/2 way to midrib; petiole c.1/3-1/2 as long as blade; extremely rare **4. D. remota**

4 Pinnules entire to lobed distinctly <1/2 way to midrib; petiole usually ≤1/3 as long as blade; common 5

5 Pinnules parallel-sided for most of length, broadly rounded to ± truncate (but often toothed) at apex; pinnae with dark blotch where they join rhachis; petioles with dense golden scales **3. D. affinis**

5 Pinnules distinctly tapering, rounded to ± obtuse at apex; pinnae without dark blotches at base; petioles with sparse to ± dense greyish- or pale-brown scales 6

6 Rhizome not or little branched, hence leaf-tufts scattered; pinnules with erect or convergent teeth at apex; sori c.1.5mm across, along ± whole length of pinnules, with non-glandular indusia with edges not tucked under; widespread **2. D. filix-mas**

6 Rhizome well branched, hence leaf-tufts in groups; pinnules with teeth spreading fan-wise at apex; sori c.1mm across, ± confined to proximal 1/2 of pinnules with ± glandular indusia with edge well tucked under before sporangial dehiscence; mountains **1. D. oreades**

7 Leaf dull green, ± mealy due to many minute glands on both surfaces; lowest pinna with proximal 3-4 pinnules on both sides ± same size **6. D. submontana**

7 Leaf clear green (of various shades), not or slightly glandular on upperside; lowest pinna with pinnules successively smaller from base distally 8

8 Petioles with dense golden scales at least in proximal 1/2; pinnules evenly lobed ≤5/6 way to midrib; extremely rare **4. D. remota**

8 Petioles variously clothed but not with dense golden scales; leaves 2-3(4)-pinnate, if only 2-pinnate then at least lowest pinnules on lowest pinnae lobed ± to base; common 9

9 Pinnules distinctly concave on upperside, with numerous minute glands on lowerside and sometimes upperside **5. D. aemula**

9 Pinnules flat or convex on upperside, glands 0 or rare on lowerside (except sometimes on indusia) and 0 on upperside 10

10 Scales on petiole usually with very distinct dark centre (sometimes uniformly pale in small upland plants); pinnules usually convex on upperside; indusia glandular **9. D. dilatata**

10 Scales on petiole uniformly pale or dark brown, or somewhat darker in centre (the dark suffusing outwards, not in a distinct zone); pinnules usually flat; indusia glandular or not (if glandular then scales not pale) 11

11 Leaf-blades ovate-triangular, with all pinnae ± in 1 plane; petioles with scales mid- to dark-brown or with darker centres; lowest pinnae with

lowest pinnule on basal side usually ≥1/2 as long as its pinna
10. D. expansa

11 Leaf-blades narrowly triangular-lanceolate, with lower pinnae twisted into ± horizontal plane; petioles with pale brown scales; lowest pinnae with lowest pinnule on basal side <1/2 as long as its pinna
8. D. carthusiana

1. D. oreades Fomin (*D. abbreviata* auct. non DC.) - *Mountain Male-fern.* Leaves to 34
0.5(-1.2)m (incl. petiole 1/8-1/4 as long as blade), 1-pinnate with deeply divided pinnae to 2-pinnate, with crenate lobes/pinnules; petiole with rather numerous dull pale brown scales; blade dull mid-green, without dark blotch at base of pinna, with lowest pinnae <1/2 as long as longest; 2n=82. Native; rocky places on mountains, in open or slight shade, scree-slopes; frequent above 240m in Wa, N En S to MW Yorks, Sc, S Kerry, Co Down.

1 x 2. D. oreades x D. filix-mas = D. x mantoniae Fraser-Jenk. & Corley is scattered in N Wa, Lake District and Sc.

1 x 5. D. oreades x D. aemula = D. x pseudoabbreviata Jermy was found on Mull (M Ebudes) in 1967; endemic.

2. D. filix-mas (L.) Schott - *Male-fern.* Leaves to 1.2m (incl. petiole 1/4-1/3 as long 34
as blade), 1-pinnate with deeply divided pinnae to 2-pinnate, with usually acutely toothed lobes/pinnules; petiole with rather numerous pale brown to straw-coloured scales; blade pale mid-green, without dark blotch at base of pinna, with lowest pinnae c.4/5 as long as longest; 2n=164. Native; woods, hedgebanks, ditches, mountains in open or shade; common throughout BI.

2 x 3. D. filix-mas x D. affinis = D. x complexa Fraser-Jenk. (*D. x tavelii* auct. non Rothm.) probably occurs throughout BI where the parents co-exist. Hybrids occur with all 3 sspp. of *D. affinis*: nothossp. **x complexa** (with ssp. *affinis*); nothossp. **x contorta** Fraser-Jenk. (with ssp. *cambrensis*); and nothossp. **x critica** Fraser-Jenk. (with ssp. *borreri*). All 7 taxa occur in 1 small wood in Leics. 2n=164, 205.

2 x 8. D. filix-mas x D. carthusiana = D. x brathaica Fraser-Jenk. & Reichst. (*D. x remota* auct. non (A. Braun ex Döll) Druce) was found in Westmorland once (c.1854) and is still in cultivation. Elsewhere found only in France (1989).

3. D. affinis (Lowe) Fraser-Jenk. - *Scaly Male-fern.* Leaves to 1.5m (incl. petiole usually <1/4 as long as blade), 1-pinnate with deeply divided pinnae to 2-pinnate, with subentire to acutely or obtusely toothed lobes/pinnules; petiole with very dense golden-brown scales; blade yellowish-green, with dark blotch at base of pinna. Native; similar places to *D. filix-mas* and often with it; frequent to common throughout BI (but less so than *D. filix-mas*). *D. affinis* consists of apogamous diploids or triploids derived from hybridization between *D. oreades*, the non-British *D. caucasica* (A. Braun) Fraser-Jenk. & Corley, and at least 1 other ancestral diploid. 3-6 sspp. have been recognized, but their characters and delimitation are still under study. The 3 following are best understood at present, but recognition is possible only after considerable experience; precise distributions are unknown, but sspp. *affinis* and *cambrensis* are commoner in W and ssp. *borreri* commoner in E.

a. Ssp. affinis (*D. pseudomas* (Woll.) Holub & Pouzar). The most extreme (i.e. 34
least like *D. filix-mas*) ssp., with shiny leaves with very densely golden-scaly petioles, with lowest pinnae c.1/2 as long as longest; pinnae parallel- and straight-sided for proximal 1/2; pinnules with subtruncate apex with short obtuse teeth and subentire sides, and the lowest with a slight, rounded basal lobe; 2n=82.

b. Ssp. cambrensis Fraser-Jenk. (ssp. *stilluppensis* auct. non (Sabr.) Fraser-Jenk.). 34
Intermediate between ssp. *affinis* and *D. oreades*, with rather shiny leaves with densely reddish-golden-scaly petioles, with lowest pinnae usually <1/2 as long as longest; pinnae tapering from base to apex; pinnules with very broadly rounded apex with obtuse teeth and obtusely toothed often revolute sides, and the lowest with a substantial rounded basal lobe often overlapping leaf rhachis; 2n=123. Probably mostly in N & W.

FIG 34 - Pteropsida. 1-6, basal pinna of ***Dryopteris***. 1, ***D. remota***. 2, ***D. filix-mas***. 3-5, ***D. affinis***. 3, ssp. ***affinis***. 4, ssp. ***borreri***. 5, ssp. ***cambrensis***. 6, ***D. oreades***. 7-10, basal pinna of ***Polystichum***. 7, ***P. setiferum***. 8, ***P. aculeatum***. 9, ***P. lonchitis***. 10, ***P. munitum***. 11, basal pinna of ***Cyrtomium falcatum***.

c. Ssp. borreri (Newman) Fraser-Jenk. (ssp. *stilluppensis* (Sabr.) Fraser-Jenk., *D.* 34
borreri (Newman) Newman ex Oberh. & Tavel, *D. tavelii* Rothm., *D. woynarii* auct. non Rothm.). Closest to *D. filix-mas*, with scarcely shiny leaves with moderately densely light-golden-scaly petioles, with lowest pinnae >1/2 as long as longest; pinnae parallel-sided for proximal 1/2 but uneven due to various-lengthed pinnules; pinnules with rounded apex often with a large tooth on each 'shoulder' and well-toothed sides, and the lowest with a large pointed basal lobe; 2n=123. Probably the most widespread ssp.

4. D. remota (A. Braun ex Döll) Druce (*D. woynarii* Rothm.) - *Scaly Buckler-fern*. RR
Leaves to 75cm (incl. petiole c.1/3-1/2 as long as blade), 2-pinnate with pinnules 34
lobed c.1/3-5/6 way to midrib; petiole with dense golden-brown scales often with darker centre; blade narrowly ovate, with lowest pinnae c.4/5 as long as longest; 2n=123. Native; formerly in woods in N Kerry and SE Galway, possibly Dunbarton; extinct. Apogamous hybrid derivative of *D. affinis* x *D. expansa*, although *D. expansa* is not known in Ir. Plants from Ir and Sc are still in cultivation.

5. D. aemula (Aiton) Kuntze - *Hay-scented Buckler-fern*. Leaves to 75cm (incl. petiole >1/2 as long as blade), 3(-4)-pinnate; petiole with few pale scales; blade triangular-ovate, with lowest pinnae the longest; 2n=82. Native; moist shady places in woods, ravines and hedgebanks; local in Ir and W Br from W Cornwall to Outer Hebrides and Orkney, very scattered in E Br except frequent in acid parts of Kent and Sussex.

6. D. submontana (Fraser-Jenk. & Jermy) Fraser-Jenk. (*D. villarii* (Bellardi) Woyn. R
ex Schinz & Thell. ssp. *submontana* Fraser-Jenk. & Jermy) - *Rigid Buckler-fern*. Leaves to 75cm (incl. petiole >1/2 as long as blade), 2-pinnate with pinnules lobed 1/3-3/4 way to midrib; petiole with rather sparse pale brown scales; blade narrowly ovate- to lanceolate-triangular, with lowest pinnae the longest; 2n=164. Native; in limestone crevices, grykes and scree; very locally frequent in NW En, very rare in Derbys, N Wa and S Wa, formerly Arran (Clyde Is).

7. D. cristata (L.) A. Gray - *Crested Buckler-fern*. Leaves to 60(100)cm (incl. petiole RR
1/3-1/2 as long as blade), 1-pinnate with very deeply divided pinnae to just 2-pinnate, with mucronate-toothed lobes/pinnules; petiole with sparse, pale brown scales; blade lanceolate-oblong, with lowest pinnae nearly as long as longest; 2n=164. Native; wet heaths, dune-slacks, marshes and fens, often with *Thelypteris palustris*; very local and decreasing in Surrey, Berks, E Suffolk, E & W Norfolk and Renfrews, formerly scattered elsewhere in En.

7 x 8. D. cristata x D. carthusiana = D. x uliginosa (A. Braun ex Döll) Kuntze ex Druce occurs in E & W Norfolk, formerly in other sites of *D. cristata*.

8. D. carthusiana (Vill.) H.P. Fuchs (*D. lanceolatocristata* (Hoffm.) Alston, *D. spinulosa* Kuntze) - *Narrow Buckler-fern*. Leaves to 80(100)cm (incl. petiole c. as long as blade), 2-3-pinnate; petiole with sparse, pale brown scales; blade narrowly ovate-oblong, with lowest pinnae ± as long as longest; 2n=164. Native; damp or wet woods, marshes, fens and wet heaths; frequent throughout most of Br and Ir.

8 x 9. D. carthusiana x D. dilatata = D. x deweveri (J.T. Jansen) Wacht. occurs frequently with the parents scattered over Br and Ir; 2n=164.

8 x 10. D. carthusiana x D. expansa = D. x sarvelae Fraser-Jenk. & Jermy occurs in Westerness and Kintyre, discovered in 1978; 2n=123.

9. D. dilatata (Hoffm.) A. Gray (*D. austriaca* Woyn. ex Schinz & Thell. non Jacq.) - *Broad Buckler-fern*. Leaves to 1(1.5)m (incl. petiole 1/4-2/3 as long as blade), 3(-4)-pinnate; petiole with numerous scales with dark centres and paler edges; blade ovate to triangular-ovate, with lowest pinnae the longest; 2n=164. Native; woods, hedgebanks, ditches, shady places on heaths and mountains; common throughout BI.

9 x 10. D. dilatata x D. expansa = D. x ambroseae Fraser-Jenk. & Jermy occurs in N & W Br S to Caerns; 2n=123.

10. D. expansa (C. Presl) Fraser-Jenk. & Jermy (*D. assimilis* S. Walker) - *Northern Buckler-fern*. Leaves to 80(100)cm (incl. petiole c. as long as blade), 3-4-pinnate;

petiole with fairly numerous pale- to reddish-brown scales often with darker centres; blade triangular-ovate, with lowest pinna usually the longest; 2n=82. Native; cool, often damp places in woods and mountain crevices and scree; locally frequent in Sc, Wa and En S to Westmorland.

18. BLECHNACEAE - *Hard-fern family*

Rhizomes long or short, scaly; leaves coriaceous, borne in tufts at ends of rhizome branches, spirally coiled when young, sparsely scaly, of 2 sorts (of 1 sort in *Woodwardia*), 1-pinnate with entire pinnae (with deeply acutely lobed pinnae in *Woodwardia*); sori linear, continuous either side of pinna midrib (stout, discrete, in row either side of pinna-lobe midribs in *Woodwardia*), covered by indusium same shape as sorus and opening towards midrib; homosporous; gametophyte green, surface-living.

The sori having openings towards the midrib and the leaves being coriaceous and 1-pinnate (or nearly 2-pinnate in *Woodwardia*) are diagnostic, as are the 1-pinnate separate sterile and fertile leaves of *Blechnum*.

Other genera - WOODWARDIA radicans (L.) Sm. (*Chain Fern*), from SW Europe, regenerates in gardens in S Kerry and W Cornwall and might escape into the wild; it differs from *Blechnum* as above, with leaves to 2m that root at tips from a scaly bud.

1. BLECHNUM L. - *Hard-ferns*

Other spp. - B. penna-marina (Poir.) Kuhn, from extreme S S America and Australasia, has been natd for a short time in a few places and might well be again; it resembles a very small *B. spicant*, with leaves $\leq$20 x 1.5cm and a slender creeping rhizome.

1. B. spicant (L.) Roth - *Hard-fern*. Leaves to 50(75) x 4cm, the fertile more erect and slightly longer than the sterile; pinnae entire, the sterile up to 2 x 0.5cm, the fertile as long but much narrower, attached to rhachis over whole width of their base; 2n=68. Native; woods, heaths, moors, grassy and rocky slopes on acid soils, often in rather dry places; common throughout most of BI but absent from much of C & CE En.

2. B. cordatum (Desv.) Hieron. (*B. chilense* (Kaulf.) Mett.) - *Chilean Hard-fern*. 15
Leaves to 100 x 20cm, the fertile scattered among the sterile; pinnae subentire, the sterile up to 10 x 2.5cm, the fertile as long but much narrower, greatly narrowed at base and attached to rhachis over short width. Intrd; grown in gardens, natd in shady places and by streams in SW En, W Sc and SW Ir; S S America. The identity of our plant is not certain; possibly >1 sp. is present.

19. AZOLLACEAE - *Water Fern family*

Stems very slender, branching, floating on water, with hanging, simple roots, without scales; leaves on 2 opposite sides of stem, small, 2-lobed, not spirally coiled when young, of 1 sort; sori borne on lower lobe of 1st leaf of each branch; heterosporous.

The only floating fern.

1. AZOLLA Lam. - *Water Fern*

1. A. filiculoides Lam. - *Water Fern*. Stems 1-5(10)cm; leaves up to 2.5 x 1.5mm in surface view; plant green in early part of season but becoming bright red in autumn;

(2n=44, 66). Intrd; natd on ponds and in canals and dykes, often covering water surface like a *Lemna* (and often with it); fairly frequent but often sporadic in C & S Br N to MW Yorks, Man, E S Ir, CI; tropical America.

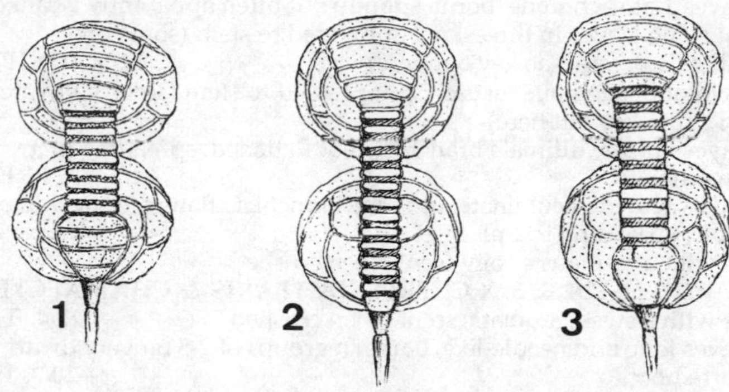

FIG 37 - Dehisced sporangia of ***Polypodium***, to show indusium.
1, ***P. cambricum***. 2, ***P. vulgare***. 3, ***P. interjectum***. Drawings by R.H. Roberts.

P I N O P S I D A - CONIFERS, GYMNOSPERMS

(Coniferopsida, Coniferae, Gymnospermae, Taxopsida)

Trees or shrubs with simple, usually evergreen leaves. Male sporangia borne on sporophylls arranged in male cones. Female sporangia borne in naked ovules either borne singly and terminally or borne on cone-scales in female cones. Female gametophyte greatly reduced and retained in ovule. Fertilized ovule (seed) retained on sporophyte until ripe.

VEGETATIVE KEY TO GENERA OF PINOPSIDA

1 Some leaves at least 1cm wide **23/1. ARAUCARIA**

1 All leaves <1cm wide 2

  2 Leaves in opposite pairs or in threes at each node 3

  2 Leaves 1 at each node, borne spirally but often apparently 2-ranked 6

3 At least some leaves in threes, not appressed to stem (some other cultivated Cupressaceae key out here) **22/5. JUNIPERUS**

3 Leaves in opposite pairs, usually ± appressed to stem (some cultivated *Juniperus* spp. key out here) 4

  4 Leaves obtuse; ultimate branchlets not flattened, spreading in 3 dimensions **22/1. CUPRESSUS**

  4 Leaves acute to acuminate; ultimate branchlets flattened, mainly or wholly spreading in 1 plane 5

5 Foliage with a resinous or oily scent when crushed **22/2 & 3. X CUPRESSOCYPARIS & CHAMAECYPARIS**

5 Foliage with a sweet aromatic scent when crushed **22/4. THUJA**

  6 Leaves long and needle-like, borne in groups of 2-5 on very dwarf short-shoots **20/7. PINUS**

  6 Leaves all borne singly on long-shoots or most borne in clusters of >10 on short-shoots 7

7 Most leaves borne in dense clusters on short-shoots 8

7 All leaves borne singly on long-shoots 9

  8 Leaves deciduous, dorsiventrally flattened **20/5. LARIX**

  8 Leaves evergreen, 3-5-angled in section **20/6. CEDRUS**

9 Leaves tapered from broad base to narrowly acute apex, not dorsiventrally flattened 10

9 Leaves narrowed at base and apex, flattened or not 11

  10 Trunk with thick spongy outer bark; leaves wider than thick **21/2. SEQUOIADENDRON**

  10 Trunk with thin stringy outer bark; leaves thicker than wide **21/3. CRYPTOMERIA**

11 Trunk with thick spongy outer bark; leading shoots with reduced ± scale-like leaves **21/1. SEQUOIA**

11 Trunk with thin scaly outer bark; leading shoots with ± full-sized leaves 12

  12 Winter vegetative buds green; leaves and wood without resin-ducts **24/1. TAXUS**

  12 Winter vegetative buds brown; leaves and wood normally with resin-ducts 13

13 Leaves with distinct slender, short, green petiole ± appressed to twig;

resin-duct 1 per leaf **20/3. TSUGA**
13 Leaves sessile, though often much narrowed at base and sometimes borne on brown, petiole-like peg; resin-ducts usually >1 per leaf, sometimes 0-1 14
14 Leaves borne on distinct brown, petiole-like pegs remaining on twig when leaf falls **20/4. PICEA**
14 Leaves sessile on twigs or on slightly raised cushions 15
15 Winter buds conical, shiny, sharply pointed **20/2. PSEUDOTSUGA**
15 Winter buds rounded at apex **20/1. ABIES**

KEY TO FAMILIES OF PINOPSIDA

1 Some leaves at least 1cm wide **23. ARAUCARIACEAE**
1 All leaves <1cm wide 2
2 Leaves in opposite pairs or in 3s at each node **22. CUPRESSACEAE**
2 Leaves 1 at each node and borne spirally (but often apparently 2-ranked), or tightly clustered in groups of 2-many on short-shoots 3
3 Vegetative buds with green bud-scales; usually dioecious; ovules solitary, surrounded by red succulent upgrowth at seed maturity **24. TAXACEAE**
3 Vegetative buds with brown bud-scales or proper bud-scales 0; monoecious; ovules borne in cones that become woody at seed maturity 4
4 Vegetative buds with brown bud-scales; bracts and cone-scales distinct **20. PINACEAE**
4 Vegetative buds without proper bud-scales; bracts and cone-scales completely fused, their distinction difficult **21. TAXODIACEAE**

20. PINACEAE - *Pine family*

Evergreen or deciduous resiniferous trees; vegetative buds with brown bud-scales; leaves borne spirally, ± entire, linear to needle-like, borne singly on long-shoots or in clusters of 2-many on short-shoots. Monoecious; male sporangia 2 per sporophyll; female cones with spirally arranged cone-scales bearing 2 ovules; seeds winged and becoming detached from cone-scale at maturity; cone-scales becoming woody, each with a distinct bract below it.

Distinguished from other coniferous families by the combination of spirally borne, very narrow leaves, scaly buds, and ovules borne in cones in which each cone-scale has a distinct bract below it.

Some genera bear short lateral stems (short-shoots) of very limited length, in addition to the normal extension shoots (long-shoots). 'Twigs' refers to stems of the previous 1-2 years' growth.

1 Leaves borne in groups of 2-5 on very dwarf short-shoots borne in axils of small scale-leaves **7. PINUS**
1 Leaves all borne singly on long-shoots, or most borne in dense clusters on short-shoots 2
2 Most leaves borne in dense clusters on short-shoots 3
2 All leaves borne singly on long-shoots 4
3 Leaves deciduous, flattened; female cones <5cm, eventually falling whole **5. LARIX**
3 Leaves evergreen, 3-5-angled in section; female cones >5cm, eventually disintegrating on tree **6. CEDRUS**
4 Leaves with a distinct slender, short, green petiole appressed to twig; female cones <3cm **3. TSUGA**
4 Leaves sessile though usually narrowed at base and sometimes

borne on a brown petiole-like peg; female cones >3cm 5
5 Leaves borne on distinct brown, petiole-like pegs; female cones pendent, falling whole; bracts not protruding beyond cone-scales **4. PICEA**
5 Leaves sessile or borne on slightly raised cushions 6
6 Winter buds conical, sharply pointed; female cones pendent, falling whole; bracts 3-lobed, protruding beyond cone-scales but not reflexed **2. PSEUDOTSUGA**
6 Winter buds rounded to obtuse; female cones erect, disintegrating on tree; bracts not protruding beyond cone-scales, or protruding, 1-lobed and reflexed **1. ABIES**

1. ABIES Mill. - *Firs*
Evergreen; leaves borne only on long-shoots, single, sessile, falling to leave disc-like scars, with 2 resin-ducts, dorsiventrally flattened; female cones erect, the scales deciduous at maturity.

1 Leaves on upperside of twigs all strongly divergent, leaving centre parting exposing the twig axis 2
1 Leaves on upperside of twigs not divergent so as to expose the twig axis 3
2 Twigs conspicuously pubescent (hairs >0.25mm); buds not resinous; female cones with exserted bracts **1. A. alba**
2 Twigs minutely pubescent (hairs <0.25mm); buds resinous; female cones with included bracts **3. A. grandis**
3 Leaves grey-green on upperside, bent or curved, appressed to twig near base and then widely divergent; female cones mostly >15cm **4. A. procera**
3 Leaves dark to bright green on upperside, ± straight, those along midline not divergent from twig axis and overlying it; female cones mostly <15cm **2. P. nordmanniana**

Other spp. - **A. cephalonica** Loudon (*Greek Fir*), from Greece, differs from all other spp. in its very rigid leaves projecting all around the stem; it is planted on a very small scale and self-sown seedlings appear rarely in S En.

1. A. alba Mill. - *European Silver-fir*. Tree to 48m; twigs conspicuously pubescent; leaves 15-30mm, those on upperside of twigs laterally spreading and leaving a distinct parting, dark green on upperside, with 2 whitish stripes on lowerside; female cones 10-15(20)cm, with exserted bracts; (2n=24). Intrd; formerly much planted for timber, especially in N and W, and also in parks, often self-sown; mountains of C Europe.

2. A. nordmanniana (Steven) Spach - *Caucasian Fir*. Tree to 44m; twigs very sparsely pubescent to ± glabrous; leaves 15-40mm, those on upperside of twig forward-pointing and not leaving a parting, dark green on upperside, with 2 silvery bands on lowerside; female cones 10-15cm, with exserted bracts; (2n=24). Intrd; now planted for amenity more often than *A. alba* and being used in some forestry trials as well as for use as 'Christmas trees' (for the last use it might replace *Picea abies* since its leaves are not readily deciduous indoors); mainly in W Br, self-sown in Surrey and Cards; Caucasus.

3. A. grandis (Douglas ex D. Don) Lindl. - *Giant Fir*. Tree to 60m; twigs minutely **47**
pubescent; leaves 20-60mm, those on upperside of twigs laterally spreading and leaving a distinct parting, bright green on upperside, with 2 whitish stripes on lowerside; female cones 5-10cm, with included bracts; (2n=24). Intrd; increasingly used in forestry, especially in N and W, occasionally self-sown; W N America.

4. A. procera Rehder - *Noble Fir*. Tree to 47m; twigs densely reddish pubescent; **42**
leaves 10-35mm, those on upperside of twigs short, forward and upward pointing, not leaving a parting, grey-green with paler bands near apex on upperside, rather similar on lowerside; female cones 12-25cm, with exserted bracts; (2n=24). Intrd;

used on a small scale in forestry mostly in W and in parks in N and W, occasionally self-sown; W N America.

2. PSEUDOTSUGA Carrière - *Douglas Fir*
Evergreen; vegetative buds sharply pointed; leaves borne only on long-shoots, sessile, single, falling to leave slightly raised cushions, with 2 resin-ducts, dorsiventrally flattened; female cones pendent, falling whole, with exserted, forward-pointing, 3-toothed bracts.

1. P. menziesii (Mirb.) Franco (*P. taxifolia* Britton nom. illeg.) - *Douglas Fir*. Tree **42**
to 65m (tallest tree of any sp. in BI); twigs sparsely pubescent; leaves 20-35mm, those on upperside of twigs laterally spreading and leaving a distinct parting, light to dark green on upperside, with 2 whitish to pale green stripes on lowerside; female cones 5-10cm; (2n=26). Intrd; very widely planted for timber, and also in parks, occasionally self-sown; W N America.

3. TSUGA (Antoine) Carrière - *Hemlock-spruces*
Evergreen; vegetative buds subacute to rounded; leaves borne only on long-shoots, single, distinctly petiolate, falling to leave short brown pegs, with 1 resin-duct, dorsiventrally flattened; female cones pendent, falling whole, with included bracts.

Other spp. - T. canadensis (L.) Carrière (*Eastern Hemlock-spruce*), from E & C N America, is frequent in parks and has been used in small-scale forest plots. It differs from *T. heterophylla* in its leaves tapering from near base to apex and with narrower white stripes on lowerside.

1. T. heterophylla (Raf.) Sarg. - *Western Hemlock-spruce*. Tree to 46m; twigs **42**
densely pubescent; leaves 6-20mm, ± parallel-sided, dark green on upperside, with 2 very broad whitish stripes on lowerside, those laterally spreading distinctly longer than those on upperside of twigs; female cones 1.5-2.5cm; (2n=24). Intrd; frequent in plantations, often mixed with hardwoods, often self-sown; W N America.

4. PICEA A. Dietr. - *Spruces*
Evergreen; vegetative buds acute to rounded; leaves borne only on long-shoots, single, sessile, falling singly to leave distinct brown pegs, with 0-2 resin-ducts, dorsiventrally flattened or 4-angled in section; female cones pendent, falling whole, with minute bracts. Spp. with flattened leaves have the leaf twisted at the base so that the morphologically true lower surface is uppermost.

1 Leaves dorsiventrally flattened 2
1 Leaves about as thick as wide or thicker than wide, 4-angled in section 3
2 Twigs glabrous; leaves mostly 15-25mm, sharply pointed at apex; female cones 6-10cm **1. P. sitchensis**
2 Twigs pubescent; leaves 8-18mm, obtuse to rounded but shortly mucronate; female cones 3-6cm **2. P. omorika**
3 Leaves bright to dark green with faint stripes, with a resinous smell when crushed; female cones 10-20cm **3. P. abies**
3 Leaves bluish-green with rather conspicuous stripes, with an unpleasant smell when crushed; female cones ≤7.5cm 4
4 Twigs glabrous; leaves 8-18mm **4. P. glauca**
4 Twigs pubescent; leaves 15-25mm **5. P. engelmannii**

1. P. sitchensis (Bong.) Carrière - *Sitka Spruce*. Tree to 55m; twigs glabrous; leaves **47**
(8)15-25mm, flattened, dark green on true upperside, with 2 broad whitish stripes on true lowerside; female cones 6-10cm; (2n=24). Intrd; abundant forest tree in W

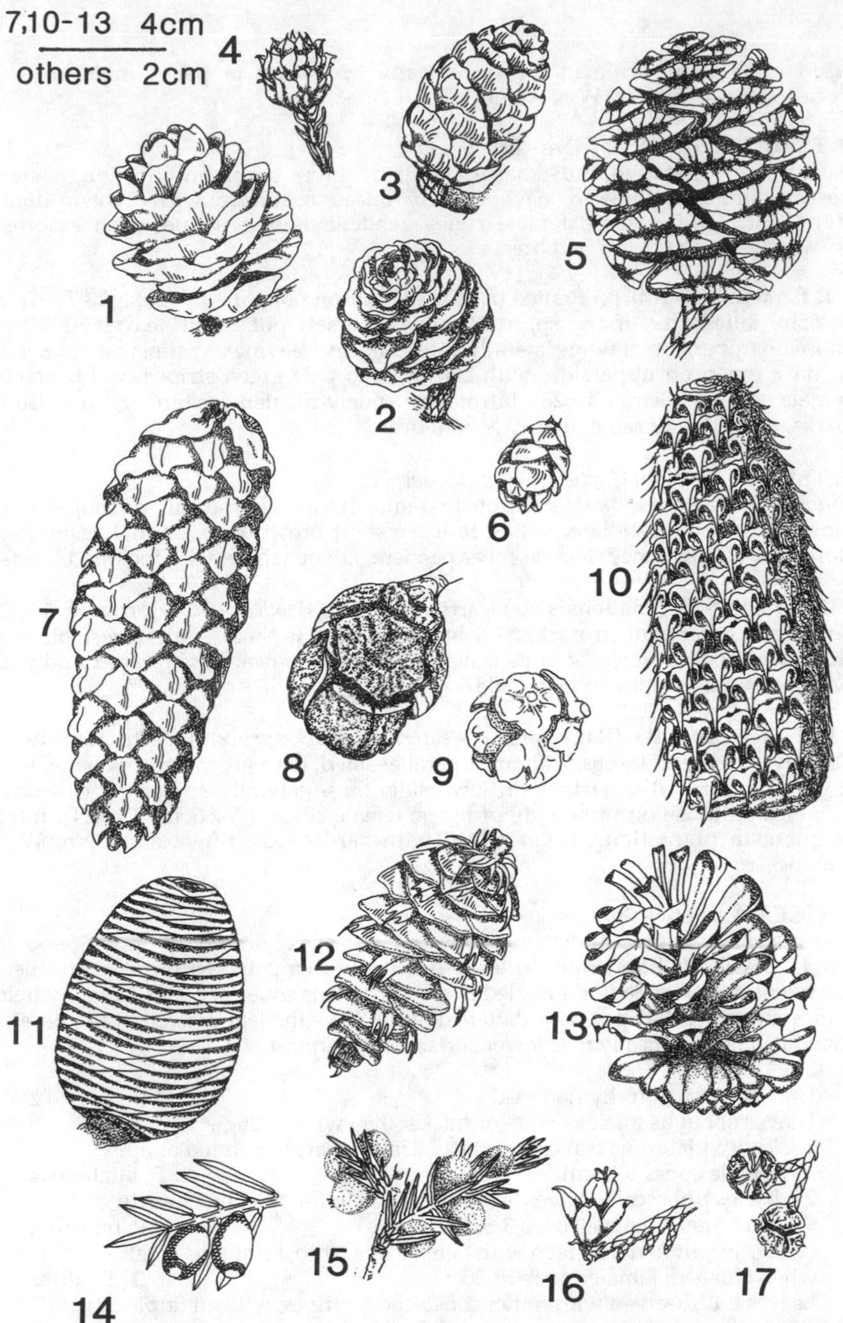

FIG 42 - Female cones of **Pinopsida**. 1, *Larix x marschlinsii*. 2, *L. kaempferi*. 3, *L. decidua*. 4, *Cryptomeria japonica*. 5, *Sequoiadendron giganteum*. 6, *Tsuga heterophylla*. 7, *Picea abies*. 8, *Cupressus macrocarpa*. 9, *X Cupressocyparis leylandii*. 10, *Abies procera*. 11, *Cedrus libani*. 12, *Pseudotsuga menziesii*. 13, *Pinus contorta*. 14, *Taxus baccata*. 15, *Juniperus communis*. 16, *Thuja plicata*. 17, *Chamaecyparis lawsoniana*.

and also in parks, often self-sown; W N America.

2. P. omorika (Pancic) Purk. - *Serbian Spruce*. Tree to 31m; twigs pubescent; leaves 8-18mm, flattened, yellowish- or bluish-green on true upperside, with 2 broad whitish stripes on true lowerside; female cones 3-6cm; (2n=24). Intrd; planted on a small scale in plantations in W; Jugoslavia.

3. P. abies (L.) H. Karst. - *Norway Spruce*. Tree to 46m; twigs usually glabrous, **42**
sometimes pubescent; leaves 10-25mm, 4-angled, bright to dark green; female cones **47**
10-20cm; (2n=24). Intrd; abundant forest tree and in shelter-belts and parks, occasionally self-sown; Europe and W Asia.

4. P. glauca (Moench) Voss - *White Spruce*. Tree to 31m; twigs glabrous; leaves 8-18mm, 4-angled, bluish-green; female cones 2.5-6cm; (2n=24). Intrd; used in shelter-belts and very small-scale forestry in W; N N America.

5. P. engelmannii W. Parry ex Engelm. - *Engelmann Spruce*. Tree to 27m; twigs pubescent; leaves 15-25mm, 4-angled, bluish-green; female cones 3.5-7.5cm; (2n=24). Intrd; used in shelter-belts and very small-scale forestry in W; W N America.

5. LARIX Mill. - *Larches*

Deciduous; leaves borne singly on leading long-shoots and in dense clusters on lateral short-shoots, sessile, falling singly, with 2 resin-ducts, dorsiventrally flattened; male cones borne singly, <2cm; female cones erect, eventually falling whole, with bracts exserted at anthesis but ± concealed at maturity.

1 Female cone-scales erect, not with recurved tips; leaves with inconspicuous greenish stripes on lowerside **1. L. decidua**
1 Female cone-scales with patent or somewhat recurved tips; leaves with rather conspicuous greyish or whitish stripes on lowerside 2
  2 Female cones ovoid, usually c.1.25-1.5x as long as wide **2. L. x marschlinsii**
  2 Female cones broadly ovoid to subglobose, usually 1-1.25x as long as wide **3. L. kaempferi**

1. L. decidua Mill. - *European Larch*. Tree to 46m; young shoots not glaucous; **42**
leaves with inconspicuous greenish stripes on lowerside; female cones (1.5)2-3.5(4.5)cm, ovoid, usually c.1.25-1.5x as long as wide, the cone-scales often slightly wavy but not recurved at tip; (2n=24). Intrd; much planted for forestry and in parks, commonly self-sown; mountains of C Europe.

2. L. x marschlinsii Coaz (*L. x henryana* Rehder, *L. x eurolepis* A. Henry nom. illeg.; **42**
L. decidua x *L. kaempferi*) - *Hybrid Larch*. Tree to 30m, fertile, intermediate between parents in most characters, but female cones more similar to those of *L. decidua* in shape and cone-scales more similar to those of *L. kaempferi* in tip recurving; (2n=24). First noticed at Dunkeld, E Perth, in 1904 among progeny from seed collected from *L. kaempferi* growing in mixed stands; now planted for forestry more than either parent, occasionally backcrossing and often originating anew.

3. L. kaempferi (Lindl.) Carrière (*L. leptolepis* (Siebold & Zucc.) Endl.) - *Japanese* **42**
Larch. Tree to 37m; young shoots glaucous; leaves with conspicuous whitish stripes on lowerside; female cones 1.5-3.5cm, broadly ovoid to subglobose, usually c.1-1.25x as long as wide, the cone-scales distinctly recurved at tip; (2n=24). Intrd; much planted for forestry and for land reclamation schemes, sometimes self-sown; Japan.

6. CEDRUS Trew - *Cedars*

Evergreen; leaves borne singly on leading long-shoots and in dense clusters on lateral short-shoots, sessile, falling singly, with 2 resin-ducts, 3-5-angled in section; male cones borne singly, >2cm; female cones erect, disintegrating on tree, with minute bracts.

1 Tip of tree pendent; leaves on short-shoots mostly 3-3.5cm **1. C. deodara**
1 Tip of tree stiffly curved to 1 side or erect; leaves on short-shoots mostly 1.5-2.5cm 2
2 Leaves mostly 1.5-2.5cm, abruptly tapered to c.0.2mm ± translucent tip; twigs usually glabrous; female cones mostly 7-10cm **2. C. libani**
2 Leaves mostly 1.5-2cm, rather gradually tapered to c.0.5mm translucent tip; twigs very shortly pubescent; female cones mostly 5-8cm **3. C. atlantica**

1. C. deodara (Roxb. ex D. Don) G. Don - *Deodar*. Tree to 37m with pendent tip; 47
leaves bright green, mostly 3-3.5cm on short-shoots, rather gradually tapered to translucent tip ≤0.4mm; female cones rather rare, 8-12cm; (2n=24). Intrd; formerly used in shelter-belts and small-scale forestry, persisting and occasionally self-sown; Afghanistan to W Himalayas.

2. C. libani A. Rich. - *Cedar-of-Lebanon*. Tree to 40m with tip stiffly curved to 1 42
side or erect; leaves dark green to glaucous, mostly 1.5-2.5cm on short-shoots, 47
abruptly tapered to scarcely translucent tip c.0.2mm; female cones mostly 7-10cm; (2n=24). Intrd; commonly planted in parks, etc., for ornament; Lebanon to Turkey.

3. C. atlantica (Endl.) Carrière (*C. libani* ssp. *atlantica* (Endl.) Batt. & Trab.) - 47
Atlas Cedar. Tree to 39m with tip stiffly curved to 1 side or erect; leaves dark green to glaucous (more often so than in *C. libani*); differing from *C. libani* in less well developed low horizontal branches and in characters in key, of which the longer (c.0.5mm) translucent leaf-tip is probably most reliable. Intrd; widely planted in parks, etc., mainly as cv. '**Glauca**', but less common than *C. libani*, self-sown in Surrey, Middlesex and Herts; Morocco.

7. PINUS L. - *Pines*

Evergreen; leaves borne in groups of 2, 3 or 5 on very dwarf short-shoots, sessile, with 2-many resin-ducts, semicircular to variously angled in section; male cones borne in clusters; female cones eventually falling whole, with minute bracts.

On very young plants long leaves are borne singly on long-shoots, but otherwise long-shoots bear only scale-leaves.

1 Leaves in pairs 2
1 Leaves in threes or fives 7
2 Leaves mostly >10cm 3
2 Leaves mostly <10cm 4
3 Bud-scales recurved at apex; female cones 8-22cm **5. P. pinaster**
3 Bud-scales appressed at apex; female cones 3-9cm **2. P. nigra**
4 Upper part of trunk pale orange-red; cone-scales ± without prickle on outer face; leaves glaucous, twisted **1. P. sylvestris**
4 Upper part of trunk grey-brown to red-brown; cone-scales usually with distinct prickle on outer face; leaves green, twisted or not 5
5 Leaves distinctly twisted; cone-scales with rather long slender prickle **4. P. contorta**
5 Leaves scarcely twisted; cone-scales with very short stout prickle 6
6 Buds obtuse to subacute; leaves ≤8cm; female cones ≤5cm; resin-ducts just below leaf surface **3. P. mugo**
6 Buds acuminate; leaves often >8cm; female cones often >5cm; resin-ducts deep-seated in leaf **2. P. nigra**
7 Leaves in threes 8
7 Leaves in fives 9
8 Leaves 10-15cm, bright green, rarely >1mm wide **6. P. radiata**
8 Leaves (10)15-25cm, dull green, ≤2mm wide **7. P. ponderosa**
9 Twigs with minute hairs at base of short-shoots; tip of female cone-scales scarcely thickened **8. P. strobus**

9 Twigs glabrous; tip of female cone-scales distinctly thickened 10
10 Leaves 7-12cm, directed towards tip of shoot; cones 8-15cm **9. P. peuce**
10 Leaves 8-20cm, pendent; cones mostly 15-25cm **10. P. wallichiana**

Other spp. - Many other spp. are grown for ornament. **P. muricata** D. Don (*Bishop Pine*), from SW USA, is a 2-leaved sp. with leaves mostly >10cm, bud-scales appressed and cones c.8cm, very persistent (as in *P. radiata*) and with long sharp prickles on ends of cone-scales; it has been reported as producing seedlings in CI, but the record needs checking.

1. P. sylvestris L. - *Scots Pine*. Tree to 36m with trunk pale orange-red above; R
leaves in pairs, 2-8cm, glaucous, twisted; female cones 2.5-7.5cm; 2n=24. Native; 47
forming pure or mixed woodland in Highlands of Sc S to M Perth; very widely planted in BI for timber and ornament and commonly natd, especially on sandy soils. Native plants retain their pyramidal shape until late in life and have short leaves and cones; they are often distinguished as ssp. **scotica** (P.K. Schott) E.F. Warb.

2. P. nigra J.F. Arnold - see sspp. for English names. Tree to 42m with trunk dark grey above; leaves in pairs, 8-18cm, not glaucous, scarcely twisted; female cones 3-9cm. Intrd.

a. Ssp. nigra - *Austrian Pine*. Crown wide with long side branches; leaves dark green, 8-12cm, rather stiff; (2n=24). Commonly planted in shelter-belts and for ornament, often self-sown; C & SE Europe.

b. Ssp. laricio Maire (*P. nigra* var. *corsicana* (Loudon) Hyl., var. *maritima* auct. non (Aiton) Melville) - *Corsican Pine*. Crown columnar with short side branches; leaves bright green, 10-18cm, rather flexible; (2n=24). Commonly planted in shelter-belts and for ornament, especially near sea in E, and also for forestry on sandy soils in S, often self-sown; Corsica, Sicily and S Italy. Perhaps better united under ssp. **salzmannii** (Dunal) Franco

3. P. mugo Turra - *Dwarf Mountain-pine*. Shrub to 4m with irregularly spreading branches; leaves in pairs, 3-8cm, bright green, ± straight; female cones 2-5cm; (2n=24). Intrd; very hardy and planted in upland areas as windbreak, especially for young plantations; C & SE Europe.

4. P. contorta Douglas ex Loudon - *Lodgepole Pine*. Tree to 25m with trunk dark 42
red-brown above; leaves in pairs, 3-8.5cm, bright green, twisted; female cones 2-6cm, with characteristic long prickles on ends of cone-scales; (2n=24). Intrd; now one of the most frequently planted forestry conifers but rarely used for other purposes, self-sown in Cards and Scillies; W N America.

5. P. pinaster Aiton - *Maritime Pine*. Tree to 35m with trunk dark red-brown above; leaves in pairs, (10)15-20(25)cm, dark to greyish green, sometimes twisted; female cones 8-22cm; (2n=24). Intrd; used in small-scale forestry and shelter-belts, especially on sandy soil or by sea in S, often self-sown; Mediterranean and SW Europe. Most trees grown in BI are referable to ssp. **pinaster** (ssp. *atlantica* Villar), but distinctness of sspp. among cultivated plants is dubious.

6. P. radiata D. Don - *Monterey Pine*. Tree to 41m; leaves in threes, 10-25cm, bright green, very slender; female cones 7-14cm, very persistent on branches, very asymmetrical, with small non-persistent point at apex of cone-scales; (2n=24). Intrd; commonly planted in W Wa, SW En and CI in parks, field borders and small plantations, especially near sea, self-sown in Cornwall; California.

7. P. ponderosa Douglas ex P. & C. Lawson - *Western Yellow-pine*. Tree to 40m; leaves in threes, (10)15-25cm, deep yellow-green, stout; female cones 7-15cm, ± symmetrical, with strong persistent point at apex of cone-scales; (2n=24). Intrd; frequently planted in parks and occasionally used in experimental forestry plantations in W; W N America.

8. P. strobus L. - *Weymouth Pine*. Tree to 40m; twigs with tuft of hairs at base of short-shoots; leaves in fives, 5-14cm; female cones 8-20cm, with tips of cone-scales scarcely thickened; (2n=24). Intrd; occasionally grown in trial plantations in W, and in parks, occasionally self-sown; C & E N America.

9. P. peuce Griseb. - *Macedonian Pine.* Tree to 30m; twigs glabrous; leaves in fives, 7-12cm; female cones 8-15cm, with tips of cone-scales distinctly thickened; (2n=24). Intrd; occasionally grown in trial plantations in uplands in W, and in parks; Balkans.

10. P. wallichiana A. B. Jack. (*P. chylla* Lodd. nom. nud.) - *Bhutan Pine.* Tree to 35m; twigs glabrous; leaves in fives, pendent, 8-20cm; female cones 15-25(30)cm, with tips of cone-scales distinctly thickened; (2n=24). Intrd; planted for ornament in parks, etc., self-sown in Surrey; Himalayas.

21. TAXODIACEAE - *Redwood family*

Evergreen or deciduous resiniferous trees; vegetative buds without proper bud-scales; leaves borne spirally, entire, linear, needle-like or scale-like, borne singly on long-shoots. Monoecious; male sporangia 2-8 per sporophyll; female cones with woody, spirally arranged cone-scales bearing 2-12 ovules; seeds winged or not, becoming detached from cone-scale at maturity; cone-scales and bracts wholly or partially fused.

Distinguished from Pinaceae in the indistinct bracts fused to the cone-scales; the usually >2 male and female sporangia; and the buds lacking proper bud-scales.

1 Leaves dorsiventrally flattened, not decurrent at base, spreading in 1 plane on opposite sides of twigs **1. SEQUOIA**

1 Leaves tapering from decurrent base to apex, not flattened, disposed ± equally around stem 2

  2 Trunk with thick, spongy outer bark; leaves wider than thick; female cones >4cm, without recurved points at apex of scale **2. SEQUOIADENDRON**

  2 Trunk with thin, stringy outer bark; leaves thicker than wide; female cones <3cm, with recurved points at apex of scales **3. CRYPTOMERIA**

Other genera - TAXODIUM Rich. and **METASEQUOIA** Miki resemble *Sequoia* in their dorsiventrally flattened leaves spreading in 1 plane on opposite sides of the twigs, but the leaves and ultimate lateral shoots are deciduous and they have a stringy bark. **T. distichum** (L.) Rich. (*Swamp Cypress*), from SE United States, has apparently alternate leaves, and **M. glyptostroboides** Hu & W.C. Cheng (*Dawn Redwood*), from China, has apparently opposite leaves.

1. SEQUOIA Endl. - *Coastal Redwood*

Evergreen; trunk with thick, spongy, red-brown outer bark; leaves dorsiventrally flattened, not decurrent; male cones solitary at apex of twigs; female cones without recurved point at apex of scale, with 15-20 cone-scales.

1. S. sempervirens (D. Don) Endl. - *Coastal Redwood.* Tree to 43m; leaves of 51
lateral twigs 6-25mm, widely spreading, those of leading and cone-bearing twigs much shorter; female cones 18-25mm; (2n=66). Intrd; grown as for *Sequoiadendron* but rarely in large groups or rows; self-sown (or suckering?) in S Somerset, Surrey and Cards; Californian coast (where examples exceed 110m).

2. SEQUOIADENDRON Buchholz - *Wellingtonia*

Evergreen; trunk with thick, spongy, red-brown outer bark; leaves tapering from decurrent base to pointed apex; male cones solitary at apex of twigs; female cones without recurved point at apex of scale, with 25-40 cone-scales.

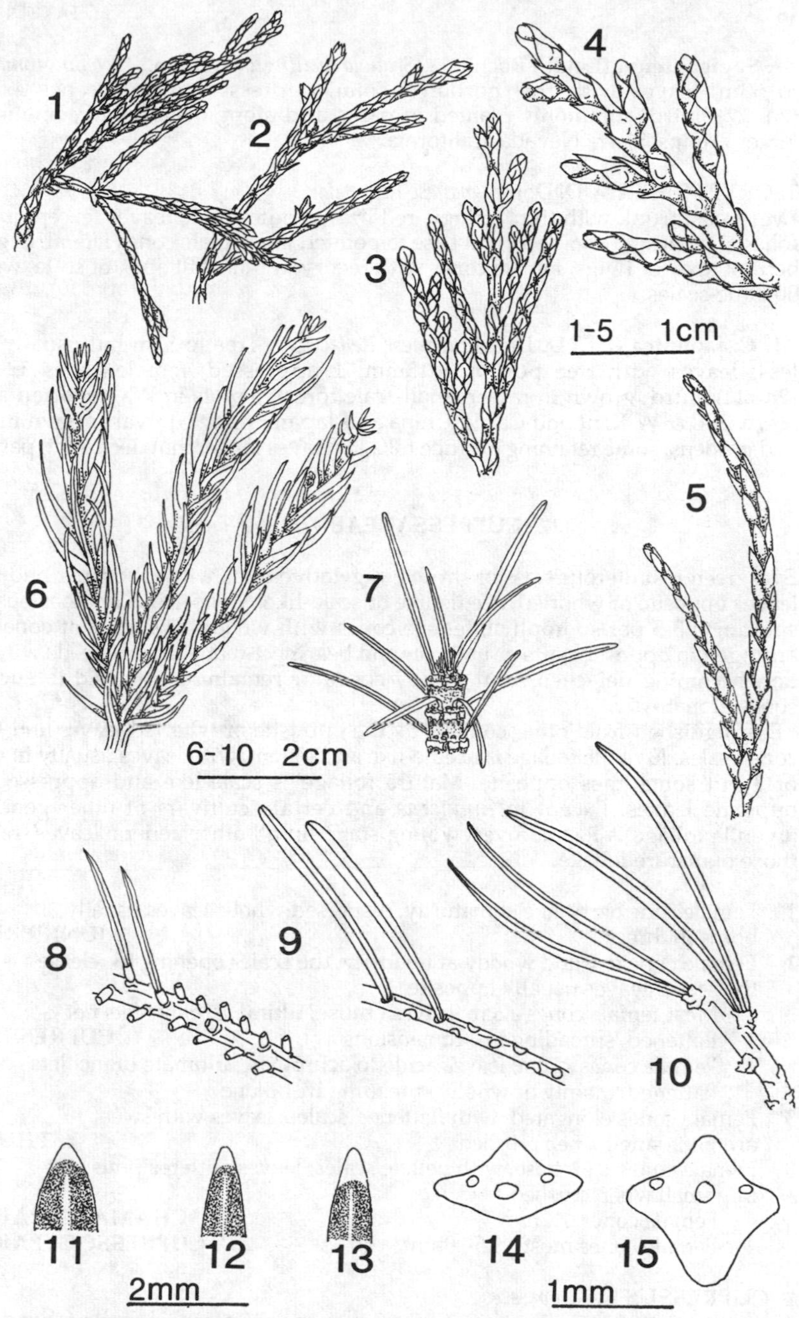

FIG 47 - Pinopsida. 1-10, leafy shoots. 1, *Cupressus macrocarpa*. 2, *Chamaecyparis lawsoniana*. 3, *X Cupressocyparis leylandii*. 4, *Thuja plicata*. 5, *Sequoiadendron giganteum*. 6, *Cryptomeria japonica*. 7, *Cedrus atlantica*. 8, *Picea sitchensis*. 9, *Abies grandis*. 10, *Pinus sylvestris*. 11-13, leaf-apices of *Cedrus*. 11, *C. libani*. 12, *C. deodara*. 13, *C. atlantica*. 14-15, leaf-sections of *Picea*. 14, *P. sitchensis*. 15, *P. abies*.

1. S. giganteum (Lindl.) Buchholz (*Sequoia wellingtonia* Seem.) - *Wellingtonia*. Tree 42
to 53m; leaves with free portion 3-7mm, appressed; female cones 4.5-8cm; 47
(2n=22). Intrd; commonly planted in parks and along drives, and sometimes in larger groups; Sierra Nevada, California.

3. CRYPTOMERIA D. Don - *Japanese Red-cedar*
Evergreen; trunk with thin, stringy, red-brown outer bark; leaves (except juvenile foliage) tapering from decurrent base to pointed apex; male cones lateral, in groups behind apex of twigs; female cones with recurved points at apex of scale, with 20-30 cone-scales.

1. C. japonica (L.f.) D. Don - *Japanese Red-cedar*. Tree to 37m but usually much 42
less; leaves with free portion 6-15mm, ± appressed; female cones 12-30cm; 47
(2n=22). Intrd; grown in rather small-scale forestry plots in Wa, reported as self-sown in E & W Kent and Cards; China and Japan. Many cultivars grown in parks and gardens, some retaining juvenile foliage (leaves longer, not decurrent, patent).

22. CUPRESSACEAE - *Juniper family*

Evergreen resiniferous trees or shrubs; vegetative buds without proper bud-scales; leaves opposite or whorled, needle-like or scale-like. Monoecious or dioecious; male sporangia 3-5 per sporophyll; female cones with woody or succulent cone-scales arranged in opposite pairs or in threes and bearing 1-many ovules; seeds winged or not, becoming detached from woody cones or remaining attached to succulent cones at maturity.

Distinguished from other conifers by the opposite or whorled leaves and female cone-scales. Juvenile foliage is needle-like and patent, with leaves usually in whorls of 3 but sometimes opposite. Mature foliage is scale-like and appressed with opposite leaves. Except in *Juniperus* and certain cultivars of other genera the juvenile foliage is lost at a very young stage. In all other genera 'leaves' refers to those of mature foliage.

1 Female cone berry-like at maturity, dispersed whole; leaves usually borne in threes **5. JUNIPERUS**
1 Female cone dry and woody at maturity, the scales opening to release the seeds; leaves usually opposite 2
2 Most female cones ≥2cm; leaves obtuse; ultimate branchlets not flattened, spreading in 3 dimensions **1. CUPRESSUS**
2 Female cones <2cm; leaves acute to acuminate; ultimate branchlets flattened, mainly or wholly spreading in 1 plane 3
3 Female cones elongated, with flattened scales; leaves with sweet aromatic smell when crushed **4. THUJA**
3 Female cones ± globose, with peltate scales; leaves with resinous or oily smell when crushed 4
4 Female cones ≤12mm **3. CHAMAECYPARIS**
4 Female cones mostly 15-20mm **2. X CUPRESSOCYPARIS**

1. CUPRESSUS L. - *Cypresses*
Twigs spreading in 3 dimensions, not flattened in 1 plane; juvenile foliage lost at very young stage; mature foliage with leaves opposite, scale-like, obtuse, appressed to twig; monoecious; female cones ripening in second year, ± globose, with 4-7 decussate pairs of peltate, woody cone-scales each with 8-20 narrowly winged seeds.

1. C. macrocarpa Hartw. ex Gordon - *Monterey Cypress.* Tree to 37m; leaves dark 42

green, 1-2mm; female cones 20-35mm, with stout conical protuberance on each 47
cone-scale; (2n=22). Intrd; commonly planted in parks, village greens, roadsides, etc., especially near coast in SW, self-sown in Scillies and Jersey; California.

2. X CUPRESSOCYPARIS Dallim. (*Cupressus* x *Chamaecyparis*) - *Leyland Cypress* (See *Chamaecyparis* key)
Variously intermediate between the parental genera; more similar to *Chamaecyparis* in vegetative characters but near half-way in cone characters; ultimate branchlets flattened, mainly spreading in 1 plane but some leaving it; female cones usually with 4 pairs of cone-scales and c.5-6 seeds per scale.

1. X C. leylandii (A. B. Jack. & Dallim.) Dallim. (*Cu. macrocarpa* x *Ch. nootkatensis*) 42
- *Leyland Cypress*. Tree to 34m; foliage like that of *Ch. nootkatensis* but ultimate 47
branchlets patent to erect, not pendent, and not spreading entirely in 1 plane; female cones 15-20mm, with a strong conical protuberance on each cone-scale. Arose in 1888 in a female cone of *Ch. nootkatensis* growing near *Cu. macrocarpa* at Welshpool, Monts; now very commonly planted as a fast-growing screen in parks, around playing fields, factories, etc., and beside roads.

3. CHAMAECYPARIS Spach - *Cypresses*
Twigs spreading in 1 plane, flattened; juvenile foliage usually lost at very young stage; mature foliage with leaves opposite, ± scale-like, acute to acuminate, ± appressed to twig; monoecious; female cones usually ripening in first year, ± globose, with 3-6 decussate pairs of peltate, woody cone-scales each with 2-5 narrowly winged seeds.

Key to Chamaecyparis and X Cupressocyparis

1 Female cones with small prickle on each cone-scale; ultimate branchlets with white markings beneath 2
1 Female cones with prominent central conical spine on each cone-scale; ultimate branchlets without white markings beneath 3
 2 Leaves usually acuminate, with inconspicuous glands on back; female cones c.5-6mm **2. Ch. pisifera**
 2 Leaves usually acute or narrowly acute, the dorsal row with a conspicuous translucent gland on back; female cones c.7-9mm **1. Ch. lawsoniana**
3 Ultimate branchlets pendent; female cones 8-12mm **3. Ch. nootkatensis**
3 Ultimate branchlets patent to erect; female cones mostly 15-20mm **X Cu. leylandii**

1. C. lawsoniana (A. Murray bis) Parl. - *Lawson's Cypress*. Tree to 41m, extremely 42
variable in habit and colour with numerous cultivars; branchlets marked with white 47
beneath; leaves ≤c.2mm, acute to narrowly acute; male cones pinkish-red; (2n=22). Intrd; very widely planted in groves and as windbreaks and screens, and mixed or alone in forestry plantations, commonly self-sown; W United States.

2. C. pisifera (Siebold & Zucc.) Siebold & Zucc. - *Sawara Cypress*. Tree to 26m, with many variable cultivars, several retaining juvenile foliage; differing from *C. lawsoniana* in its acuminate leaves with less conspicuous dorsal gland and smaller female cones with smaller apical spine on cone-scales; (2n=22). Intrd; commmonly planted in parks and occasionally self-sown; Japan.

3. C. nootkatensis (Lamb.) Spach - *Nootka Cypress*. Tree to 30m; differing from other 2 spp. in lack of white markings on lowerside of branchlets, leaves ≤c.3mm, and larger cones which do not open until the second year and have a prominent conical spine on end of cone-scales. Intrd; frequent in parks and sometimes found as experimental forestry plantings; W N America.

4. THUJA L. - *Red-cedars*
Differing from *Chamaecyparis* in the elongated female cones with usually 10-12 flattened cone-scales each with a recurved apical spike and 2-3 seeds.

1. T. plicata Donn ex D. Don - *Western Red-cedar*. Tree to 42m; vegetatively similar 42
to *Chamaecyparis lawsoniana* but leaves longer (≤6mm) and with a sweet aromatic 47
smell when crushed; male cones blackish, becoming yellowish when ripe; female cones 10-12mm; (2n=22). Intrd; very commonly planted in parks and as windbreaks, etc., and usually mixed in forestry plantations, frequently self-sown; W N America.

5. JUNIPERUS L. - *Junipers*
Twigs spreading in 3 dimensions, not flattened in 1 plane; juvenile foliage with erect to patent leaves usually borne in whorls of 3; adult foliage with appressed scale-like opposite leaves; usually dioecious; female cones berry-like, with succulent ± fused cone-scales; seeds not winged, 1 per cone-scale, retained within cone at dispersal.

Other spp. - Several spp. and a great many cultivars are grown in parks and on roadsides as ground-cover or bushes in shrubberies and rockeries, etc. Most of these spp. (notably **J. chinensis** L. (*Chinese Juniper*), from China and Japan) bear both juvenile (3-ranked) and adult (opposite) foliage, but often the former is eventually lost or, in many cultivars, the latter is never developed. In some cultivated spp. the juvenile leaves are borne in opposite pairs.

1. J. communis L. - *Common Juniper*. Tree or shrub to 7(16)m, never attaining 42
adult foliage; leaves 4-20mm, linear to linear-oblong, with a single broad white band on upperside; female cones 5-10mm, ± globose. Native; little planted.

1 Leaves up to 15mm, erecto-patent, acute to obtuse with a mucronate tip **c. ssp. nana**
1 Leaves up to 20mm, ± patent, acuminate with a sharp point 2
  2 Leaves 1-1.5mm wide, loosely and irregularly spaced, with a greyish-white stomatal band **a. ssp. communis**
  2 Leaves 1.3-2mm wide, closely and regularly spaced, with a pure white stomatal band **b. ssp. hemisphaerica**

a. Ssp. communis. Spreading shrub to erect tree; leaves mostly 8-20 x 1-1.5mm, ± patent, linear, gradually tapered to sharp point so that branchlets are prickly to touch; stomatal band greyish-white; 2n=22. Very local but often common throughout Br and Ir, on both limestone and acid soils.

b. Ssp. hemisphaerica (J. & C. Presl) Nyman. Low compact shrub; leaves mostly RR
8-20 x 1.3-2mm, patent, linear-oblong, gradually tapered to sharp point so that branchlets are prickly to touch; stomatal band pure white. Confined to maritime low cliffs in W Cornwall and Pembs. Identity of our plants needs confirmation.

c. Ssp. nana (Hook.) Syme (ssp. *alpina* Celak. nom. illeg.). Procumbent matted shrub, leaves mostly 4-12 x 1-2mm, erecto-patent, linear-oblong to very narrowly triangular, abruptly tapered to short sometimes blunt point so that branchlets are scarcely prickly to touch; stomatal band pure white; 2n=22. NW Wa, NW En, W Ir and C & NW Sc, on rocks and moorland mostly in upland areas.

There are frequent intermediates between ssp. *communis* and the other 2 sspp.

23. ARAUCARIACEAE - *Monkey-puzzle family*

Evergreen resiniferous trees; vegetative buds without proper buds-scales; leaves borne spirally, broad and flat. Normally dioecious; male sporangia 5-15 per sporophyll; female cones with woody, spirally arranged cone-scales bearing 1

ovule; seeds not winged, remaining attached to cone-scale which falls at maturity.
Differs from other conifers in the broad many-veined leaves.

1. ARAUCARIA Juss. - *Monkey-puzzle*
1. A. araucana (Molina) K. Koch - *Monkey-puzzle*. Tree to 30m; younger branches densely clothed in erecto-patent leaves; leaves 2.5-5cm, leathery, lanceolate-triangular, at least some >1cm wide at base, narrowed to sharp apical spine; male cones c.10-12cm; female cones c.15-20 x 12-15cm; (2n=26). Intrd; commonly planted in rows or groups by drives, wood-borders, etc., very rarely self-sown; Chile and W Argentina. **51**

24. TAXACEAE - *Yew family*

Evergreen non-resiniferous trees or shrubs; vegetative buds with green bud-scales; leaves borne spirally, linear. Normally dioecious; male sporangia 4-9 per sporophyll; ovules borne singly (not in cones), terminal on very short lateral branches; seed surrounded except at apex by succulent upgrowth (aril), falling with it at maturity.
Distinguished from other conifers by the lack of resin and the single ovules with a succulent aril after fertilization.

1. TAXUS L. - *Yew*
1. T. baccata L. - *Yew*. Large bush or spreading tree to 28m, often with multiple trunks; leaves 10-30(45) x 2-3mm, dark green with two pale stripes on lowerside; aril c.1cm, red; 2n=24. Native; very local but often common throughout En, Wa and Ir, but very rare in Sc; on well-drained limestone and also locally on acid sandstone, sometimes dominant; also very widely grown and often self-sown. Numerous cultivars, with varied habit and leaf and aril colour, are grown. **42**

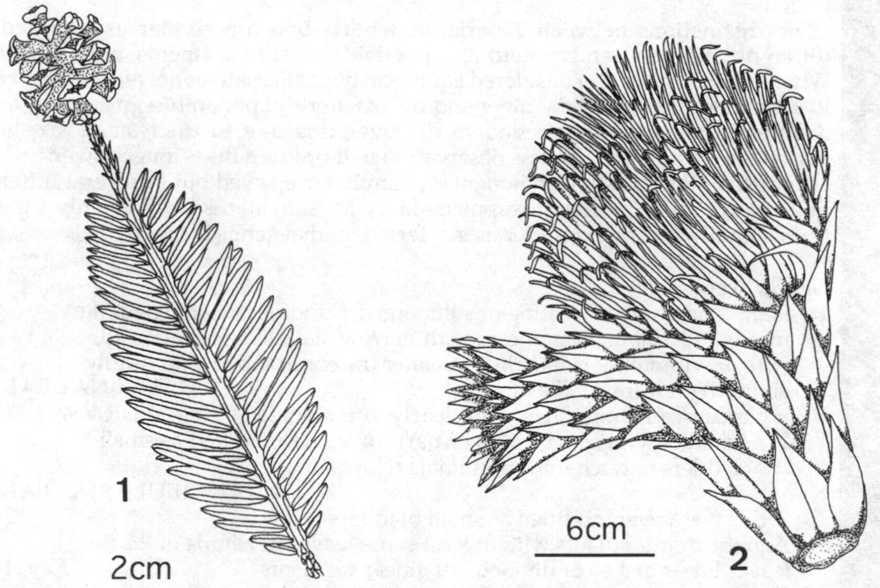

FIG 51 - Shoot and female cone. 1, *Sequoia sempervirens*. 2, *Araucaria araucana*.

MAGNOLIOPSIDA — ANGIOSPERMS, FLOWERING PLANTS

(Angiospermopsida, Angiospermae, Anthophyta)

Trees, shrubs, climbers, herbs or variously reduced plants of extremely varied growth-form. Male sporangia borne in specialized organs (stamens) grouped 1-many in male or bisexual flowers. Female sporangia borne in ovules enclosed 1-many together in carpels; carpels grouped 1-many in female or bisexual flowers. Female gametophyte greatly reduced and retained in ovule. Fertilized ovule (seed) retained within carpel on sporophyte until ripe, then dispersed separately or within carpel (fruit).

KEYS TO FAMILIES OF MAGNOLIOPSIDA

(Dicotyledons and Monocotyledons)

Before using these keys the section on Identification Keys in the Introduction should be read carefully. In particular, it is essential to work out fully the structure of the flowers *before* starting on the keys. The keys have been made as user-friendly as possible, but it must be admitted that pitfalls still abound. Only after considerable experience should the supplementary keys (A, B, etc.) be used directly; it is advisable to start always with the General key. In either case it is of course of paramount importance to arrive at the correct supplementary key; hence the General key should be used very carefully and a good working knowledge of it be built up.

The distinctions between 2 perianth whorls that are similar as opposed to different, and between sepaloid and petaloid perianth segments, are subjective. Wherever the answer is considered equivocal both alternatives are allowed for here. In some cases, however, the inner and outer whorls of perianth segments are quite different, but a lens is needed to discover this due to their small size (e.g. *Empetrum, Ruscus*) and careless observation will produce the wrong answer.

Because of the variation shown, many families are keyed out in several different positions in up to 5 different supplementary keys. In all cases where only 1 genus from a family containing >1 genus is referred to, that genus is stated.

General key

1 Plants consisting of floating or submerged ± undifferentiated pad-like fronds ≤10(15)mm, sometimes with narrow stalk-like part at 1 end, with or without roots dangling in water (rarely stranded temporarily on mud) (beware *Azolla*) **152. LEMNACEAE**

1 If plants free-floating then with clearly differentiated stems and leaves 2

2 Aquatics with some leaves or parts of leaves modified as small bladders to catch minute animals (*Utricularia*) **132. LENTIBULARIACEAE**

2 Leaves never modified as small bladders 3

3 Aquatic or mud plants with at least some leaves in whorls of ≥3, the leaves linear or ± so or divided into linear segments *Key A*

3 If aquatic or on mud then leaves not whorled and/or not linear or with linear segments 4

4 Woody plant parasitic on aerial parts of trees, the roots buried in

living host branches **89. VISCACEAE**

4 If growing on aerial parts of trees then merely epiphytic, with roots not buried in living host branches 5

5 Trees with unbranched stem and terminal rosettes of huge pinnate or palmate leaves; or seedlings with leaves ribbed alternately on each surface **150A. ARECACEAE**

5 If trees then not with single terminal rosette of compound leaves; if seedlings then not with leaves ribbed alternately on each surface 6

6 Plant consisting of 1-few rosettes of many linear simple leaves usually ≥1m, either borne on ground or at tips of woody branches **164. AGAVACEAE**

6 If leaves all in 1 or few rosettes then much <1m and often not linear 7

7 Plant consisting of dense hemispherical mass often ≥1m across, with narrow pineapple-like leaves with strongly spiny margins **160. BROMELIACEAE**

7 If plant in a dense hemispherical spiny mass then without pineapple-like leaves 8

8 All inflorescences entirely replaced by vegetative propagules (bulbils or plantlets) *Key B*

8 At least some inflorescences bearing flowers (or fruits) 9

9 Perianth 0, or of 1 whorl, or of ≥2 whorls or a spiral of similar segments 10

9 Perianth of 2 (rarely more) distinct whorls or rarely a spiral, the inner and outer differing markedly in shape/size/colour 12

10 Perianth ± corolla-like, usually white or distinctly (often brightly) coloured *Key C*

10 Perianth ± calyx-like or bract-like (greenish to brownish, or scarious or much reduced) or 0 11

11 Trees or shrubs, sometimes very short or procumbent but then with woody stems producing growth in subsequent years *Key D*

11 Herbs, with non-woody (or only basally woody) stems dying to ground level after 1(-few) year(s) *Key E*

12 Flowers all male or all female, the male or female parts 0 or extremely vestigial *Key F*

12 At least some flowers bisexual or both male and female flowers present 13

13 Petals fused at base for varying distances to apex 14

13 Petals free, or rarely ± fused just at basal point, or rarely fused above base but free at base 15

14 Ovary superior (hypogynous or perigynous) *Key G*

14 Ovary inferior (epigynous) or partly so *Key H*

15 Ovary inferior (epigynous) or partly so *Key I*

15 Ovary superior (hypogynous or perigynous) 16

16 Carpels and styles free, or carpels fused just at base *Key J*

16 Carpels and/or styles fused wholly or for greater part, or carpel 1 *Key K*

Key A - Aquatic or mud plants with whorled leaves, the leaves either linear or divided into linear segments (beware *Equisetum*, with whorled lateral branches)

1 Leaves simple, sometimes prominently toothed 2

1 Leaves divided (forked or pinnate) 6

2 Tip of stems erect and emergent at flowering; leaves in whorls of 4-12 3

2 Stems always submerged (unless stranded by drought); leaves opposite or in whorls of 3-6(8) 4

3 Leaves entire, 6-12 per whorl **123. HIPPURIDACEAE**

3 Leaves with few to many pricklets on margins, 4-8 per whorl (*Galium*) **134. RUBIACEAE**
4 Leaves with stipules free from leaf-base **149. ZANNICHELLIACEAE**
4 Leaves without stipules, often with sheathing leaf-base or with 2 minute scales at base in axil 5
5 Leaves with distinctly widened base shortly sheathing stem **148. NAJADACEAE**
5 Leaves wider near base than near apex but narrowed at extreme base, sometimes slightly clasping stem but not sheathing it **142. HYDROCHARITACEAE**
6 Leaves forked 1-4 times **29. CERATOPHYLLACEAE**
6 Leaves 1-2-pinnate 7
7 Leaves with flat segments; flowers conspicuous, >15mm across (*Hottonia*) **71. PRIMULACEAE**
7 Leaves with filiform segments; flowers inconspicuous, <6mm across (*Myriophyllum*) **81. HALORAGACEAE**

Key B - Plants with all inflorescences entirely replaced by vegetative propagules

1 Leaves reniform, palmately lobed (*Saxifraga*) **76. SAXIFRAGACEAE**
1 Leaves linear to lanceolate, entire 2
2 Inflorescences proliferating by producing small plantlets in place of flowers 3
2 Inflorescences replaced by axillary solitary or terminal clusters of small solid structures ('bulbils') 4
3 Stems with a central pith; leaves unifacial, flattened-cylindrical (*Juncus*) **155. JUNCACEAE**
3 Stems hollow; leaves bifacial, flat or inrolled or infolded **157. POACEAE**
4 Bulbils sessile, forming ± globose compact terminal head (*Allium*) or ± globose mass at ground level (*Gagea*) **162. LILIACEAE**
4 Bulbils on slender stalks, solitary in leaf-axils (*Lysimachia*) **71. PRIMULACEAE**

Key C - Perianth of 1 whorl, or of ≥2 whorls or a spiral of ± similar segments, ± corolla-like (usually white or distinctly coloured) (beware taxa with distinct inner and outer perianth whorls but 1 or other soon falling, e.g. Papaveraceae)

1 Flowers bisexual or monoecious 2
1 Flowers dioecious 32
2 Ovary inferior (flowers epigynous) 3
2 Ovary superior (flowers hypogynous to perigynous) 14
3 Stamens ≥6 4
3 Stamens ≤5 5
4 Leaves ovate or broadly so, cordate at base; ovary 6-celled; styles/stigmas 6 **26. ARISTOLOCHIACEAE**
4 Leaves narrow, gradually tapered to base; ovary 3-celled; styles/stigmas 1 or 3 **162. LILIACEAE**
5 Stamens 1-3 6
5 Stamens 4-5 (some may drop very early) 8
6 Perianth with 5 segments; ovule 1 (or 0) per cell **137. VALERIANACEAE**
6 Perianth with 6 segments; ovules numerous per cell 7
7 Style obvious, with 3 obvious branches or 3 separate stigmas; stamens 3 **163. IRIDACEAE**
7 Styles 0; stamens 1 or 2 **166. ORCHIDACEAE**
8 Leaves simple, in whorls of ≥4 **134. RUBIACEAE**
8 Leaves not in whorls of ≥4 9

9 Petals free, inserted on top of ovary 10

9 Petals fused into tube at least proximally 12

10 Stamens 4; carpels and styles 1; herbaceous perennial (*Sanguisorba*) **77. ROSACEAE**

10 Stamens 5; carpels 2-5; styles 2 or 5, or if 1 then woody evergreen 11

11 Fruit a 2-celled dry schizocarp; styles 2; mostly herbaceous, if woody then leaves simple, entire and with a clear midrib **111. APIACEAE**

11 Fruit a berry; styles 1 or 5; mostly woody **110. ARALIACEAE**

12 Flowers borne in open raceme-like cymes **88. SANTALACEAE**

12 Flowers borne in dense capitula 13

13 Stamens 4, free, ± exserted; ovary and fruit surrounded by epicalyx **138. DIPSACACEAE**

13 Stamens 5, their anthers fused into tube round style; ovary and fruit not surrounded by epicalyx **139. ASTERACEAE**

14 Tree or shrub or woody climber 15

14 Herb 19

15 Tepals 10; stamens 5 **98. STAPHYLEACEAE**

15 Tepals 3-8; stamens 6-many or flowers female 16

16 Climber or scrambler 17

16 Strongly self-supporting shrub 18

17 Leaves simple; stipules scarious, fused round stem; stamens 6-9 **49. POLYGONACEAE**

17 Leaves usually ternate or pinnate; stipules 0; stamens >10 (*Clematis*) **30. RANUNCULACEAE**

18 Tepals free, creamy- to greenish-yellow; flowers dioecious; leaves evergreen, sweetly scented when crushed **25. LAURACEAE**

18 Tepals fused into tube (or arising from tubular hypanthium), purple; flowers bisexual; leaves deciduous, not sweetly scented **84. THYMELAEACEAE**

19 Carpels ≥2, free or ± so 20

19 Carpels 1, or >1 and fused 25

20 Aquatic plant with finely dissected submerged and entire floating leaves **28. CABOMBACEAE**

20 Not an aquatic plant with dissected submerged and entire floating leaves 21

21 Carpels each with 1-2 ovules 22

21 Carpels each with several to many ovules 23

22 Flowers in racemes; fruit succulent **43. PHYTOLACCACEAE**

22 Flowers solitary or in cymes; fruit not succulent **30. RANUNCULACEAE**

23 Tepals 1(-2), white; inflorescence a forked spike just above water-surface **143. APONOGETONACEAE**

23 Tepals ≥5, usually coloured; inflorescence not a forked spike 24

24 Leaves simple, linear, without petiole; inflorescence an umbel **140. BUTOMACEAE**

24 Leaves compound and/or with well developed petiole; inflorescence not an umbel **30. RANUNCULACEAE**

25 Stamens >10 26

25 Stamens 3-9 27

26 Leaves compound **30. RANUNCULACEAE**

26 Leaves simple, entire **43. PHYTOLACCACEAE**

27 Ovary 1-3-celled, each cell with 2-many ovules 28

27 Ovary 1-celled (or 3-celled with 2 cells ± aborted and empty), with 1 ovule 29

28 Ovary 1-celled; tepals and stamens 5 (*Glaux*) **71. PRIMULACEAE**

28 Ovary 3-celled; tepals and stamens 4 or 6 **162. LILIACEAE**

29 Leaves pinnate (*Sanguisorba*) **77. ROSACEAE**
29 Leaves simple, usually entire 30
30 Perianth >15mm across, with tube >2cm **43A. NYCTAGINACEAE**
30 Perianth <10mm across, with tube <1cm (often 0) 31
31 Emergent aquatic with blue flowers; tepals and stamens always 6 **161. PONTEDERIACEAE**
31 Flowers not blue; tepals mostly 5; stamens (4)8(-9) **49. POLYGONACEAE**
32 Flowers in dense capitula closely surrounded by ≥1 row of bracts **139. ASTERACEAE**
32 Flowers not in dense capitula 33
33 Tepals 4 or 6 34
33 Tepals 5 35
34 Tree or shrub; tepals 4; leaves elliptic **25. LAURACEAE**
34 Herb; tepals 6; leaves reduced to scales, replaced by cylindrical cladodes (*Asparagus*) **162. LILIACEAE**
35 Leaves simple, entire or ± so **49. POLYGONACEAE**
35 At least leaves on stem pinnate or deeply pinnately lobed 36
36 Tepals free; basal leaves pinnate (*Trinia*) **111. APIACEAE**
36 Tepals fused into tube proximally; basal leaves simple and ± entire (*Valeriana*) **137. VALERIANACEAE**

Key D - Perianth 0 or of 1 or more whorls or a spiral of ± similar segments, ± calyx-like or bract-like (usually greenish or brownish or scarious); plants with woody stems

1 Leaves pinnate or ternate 2
1 Leaves simple, often deeply lobed 7
2 Stems spiny (*Aralia*) **110. ARALIACEAE**
2 Stems not spiny 3
3 Shrubs <2m; flowers in dense capitula 4
3 Trees >2m; flowers various 5
4 Stems ± erect; flowers in clusters of ± bell-shaped capitula (*Artemisia*) **139. ASTERACEAE**
4 Woody stems procumbent; flowers in spherical capitula solitary at apex of erect herbaceous stem (*Acaena*) **77. ROSACEAE**
5 Leaves alternate or spiral; fruit a nut, sometimes with wing ± surrounding it; monoecious **39. JUGLANDACEAE**
5 Leaves opposite or ± so; fruit 1 or 2 achenes each with elongated wing on 1 side; often dioecious or ± so 6
6 Stamens c.8; fruit of 2 achenes each with elongated wing **101. ACERACEAE**
6 Stamens 2; fruit of 1 achene with elongated wing **127. OLEACEAE**
7 Leaves opposite or ± so 8
7 Leaves alternate or spiral 12
8 Plants dioecious; male and female flowers in elongated dense catkins (*Salix*) **63. SALICACEAE**
8 Plants dioecious or not; flowers sometimes in pendent panicles (if so not dioecious) but not in compact catkins 9
9 Leaves entire 10
9 Leaves palmately lobed or serrate 11
10 Most leaves >5cm; tepals fused into tube proximally (*Coprosma*) **134. RUBIACEAE**
10 Leaves all <3cm; tepals 0 or ± free (*Buxus*) **92. BUXACEAE**
11 Leaves palmately lobed, the lobes entire to serrate; fruit of 2 winged achenes **101. ACERACEAE**
11 Leaves not lobed, serrate; fruit a black berry (*Rhamnus*) **94. RHAMNACEAE**

12 Stems scrambling or climbing 13
12 Plant a self-supporting shrub or tree, but sometimes dwarf 14
13 Leaves evergreen, palmately veined and often palmately lobed; fruit a black or rarely yellow berry (*Hedera*) **110. ARALIACEAE**
13 Leaves deciduous, pinnately veined, not lobed; fruit and achenes surrounded by white succulent tepals (*Muehlenbeckia*) **49. POLYGONACEAE**
14 Flowers borne on inside of hollow receptacles that become succulent in fruit (*Ficus*) **37. MORACEAE**
14 Flowers not borne on inside of hollow receptacles 15
15 At least male flowers (if dioecious then female flowers also) in pendent or rigid catkins or in pendent tassels or globular heads 16
15 Flowers not in catkins, if in tight groups then not pendent 22
16 Male and female flowers in separate, spherical, pendent capitula; leaves with petiole hollow at base and forming cap over axillary bud **34. PLATANACEAE**
16 Flowers not in spherical pendent capitula; base of petiole not concealing axillary bud 17
17 Leaves densely mealy; male and female flowers in same catkin (*Atriplex*) **45. CHENOPODIACEAE**
17 Leaves glabrous to pubescent, not mealy; male and female flowers in separate catkins or heads (or dioecious) 18
18 Leaves dotted with translucent glands, with strong aromatic scent when crushed **40. MYRICACEAE**
18 Leaves not gland-dotted 19
19 Fresh stems and leaves with latex; fruits red to black, succulent (*Morus*) **37. MORACEAE**
19 Latex absent; fruits not succulent 20
20 Ovary 1-celled, with many ovules; fruit a capsule with many plumed seeds; or flowers all male **63. SALICACEAE**
20 Ovary 2-6-celled, each cell with 1-2 ovules; fruit a nut, sometimes borne in husk formed from enlarged scales 21
21 Ovary 3- or 6-celled, with 3-9 styles **41. FAGACEAE**
21 Ovary 2-celled, with 2 styles **42. BETULACEAE**
22 Leaves palmately lobed 23
22 Leaves not palmately lobed 24
23 Leaves peltate; fruit a capsule; ovary superior (*Ricinus*) **93. EUPHORBIACEAE**
23 Leaves not peltate; fruit a berry; ovary inferior (*Fatsia*) **110. ARALIACEAE**
24 Leaves with dense ± sessile scales at least on lowerside, appearing mealy **80. ELAEAGNACEAE**
24 Leaves without scales, glabrous to pubescent 25
25 Leaves <20 x 2mm, succulent (*Suaeda*) **45. CHENOPODIACEAE**
25 At least most leaves >20mm and >2mm wide, not succulent 26
26 Flowers in compound umbels, epigynous (*Bupleurum*) **111. APIACEAE**
26 Flowers not in compound umbels, hypogynous or perigynous 27
27 Stamens 8-12, or flowers all female with 1-celled ovary 28
27 Stamens 4-5, or flowers all female with 2-4-celled ovary 29
28 Dioecious; tepals longer than perianth-tube and falling from it after flowering **25. LAURACEAE**
28 Flowers bisexual; tepals shorter than perianth-tube and not falling separately from it **84. THYMELAEACEAE**
29 Flowers all bisexual; ovary 1-celled; fruit a winged achene; leaves deciduous **35. ULMACEAE**
29 Flowers monoecious or bisexual to dioecious; ovary 2-4-celled or 0; fruit succulent or ± so; leaves evergreen 30

30 Shrub to 5m; fruit black, without persistent styles (*Rhamnus*) **94. RHAMNACEAE**
30 Dwarf shrub to 25cm; fruit whitish, with 2 persistent horn-like styles(*Pachysandra*) **92. BUXACEAE**

Key E - Perianth 0 or of 1 or more whorls or a spiral of ± similar segments, ± calyx-like or bract-like (usually greenish or brownish or scarious); herbs

1 Flowers numerous in dense capitula closely surrounded by sepal-like bracts **139. ASTERACEAE**
1 Flowers not in dense capitula, or if so then not closely surrounded by sepal-like bracts 2
2 Leaves in whorls of ≥4 (*Rubia*) **134. RUBIACEAE**
2 Leaves not whorled or in whorls of 3 3
3 Leaves at least partly opposite or whorled; aquatic or marsh plants with floating, procumbent or very weakly ascending stems 4
3 Leaves all alternate or all basal, or if some or all opposite then plant not aquatic, or if so then stems self-supporting 12
4 Leaves fused in opposite pairs, forming succulent sheath round stem **45. CHENOPODIACEAE**
4 Leaves not fused in succulent sheath round stem 5
5 Tepals 0; stamens 1-4 (or flowers female) 6
5 Tepals 4-6 on at least some flowers; stamens 4-12 (or flowers female) 8
6 Flowers bisexual; stamens 4; fruits on stalks >10mm; only upper leaves opposite **147. RUPPIACEAE**
6 Monoecious; male flowers with 1-2 stamens; fruits on stalks ≤10mm; ± all leaves opposite or in 3s 7
7 Stigmas linear, 2 per ovary; ovary developing into 4 nutlets **124. CALLITRICHACEAE**
7 Stigmas peltate, 1 per carpel; 1-4(more) carpels per flower developing into nutlet **149. ZANNICHELLIACEAE**
8 Tepals 5-6 (or 0 in female flowers) 9
8 Tepals 4 10
9 Flowers monoecious, hypogynous; tepals 5 in male flowers, 0 in female flowers; female flowers and fruit with 2 prominent basal bracteoles **45. CHENOPODIACEAE**
9 Flowers bisexual, perigynous; tepals 6; flowers without bracteoles **83. LYTHRACEAE**
10 Flowers in terminal flat-topped cymes; stamens 8 (*Chrysosplenium*) **76. SAXIFRAGACEAE**
10 Flowers solitary or in spikes in leaf-axils; stamens 4 11
11 Flowers in long-stalked axillary spikes **146. POTAMOGETONACEAE**
11 Flowers solitary and sessile in leaf-axils (*Ludwigia*) **86. ONAGRACEAE**
12 Flowers greatly reduced, arranged in units largely composed of leafy or membranous scaly bracts, with perianth 0 or represented by bristles or minute scales, aerial; leaves linear, grass-like, sheathing the stem proximally 13
12 Flowers with obvious structure, mostly with perianth, if greatly reduced with 0 or obscure perianth then not arranged in units as above and often subaquatic or on water surface; leaves various 14
13 Flowers with bract above as well as below (if not then stems hollow); stems usually with hollow internodes, circular or rarely compressed or ± quadrangular in section; leaf-sheaths usually with free overlapping margins **157. POACEAE**
13 Flowers never with bract above; stems usually with solid internodes, often ± triangular in section; leaf-sheaths usually cylindrical, with fused margins **156. CYPERACEAE**

14 Aquatic or marsh plants with linear leaves 15
14 If leaves linear then plants not in water or marshes; if aquatic then leaves not linear 26
15 Leaves all basal; inflorescence a tight capitate mass on long leafless stem **154. ERIOCAULACEAE**
15 If leaves all basal, then inflorescence not a single terminal tight capitate mass 16
16 Flowers very small, many tightly packed in dense spherical or elongated conspicuous clusters 17
16 Flowers not many together in dense clusters 19
17 Flowers bisexual; fresh leaves with strong spicy scent when crushed (*Acorus*) **151. ARACEAE**
17 Flowers unisexual, the male and female in clearly separated parts of inflorescence; leaves without spicy scent 18
18 Flowers in globose heads **158. SPARGANIACEAE**
18 Flowers in cylindrical spikes **159. TYPHACEAE**
19 Leaves very thin, ribbon- or thread-like, mostly subaquatic 20
19 Leaves thicker, not ribbon- or thread-like 22
20 Flowers bisexual, borne in stalked spikes **146. POTAMOGETONACEAE**
20 Flowers dioecious or monoecious, borne in stalked or sessile spathes 21
21 Flowers dioecious, in short- or long-stalked spathes; tepals 3; freshwater (*Vallisneria*) **142. HYDROCHARITACEAE**
21 Flowers monoecious, in sessile spathes; tepals 0; marine **150. ZOSTERACEAE**
22 Tepals 4-5, or 0 23
22 Tepals 6 24
23 Tepals 5 (or 0 in female flowers); leaves alternate **45. CHENOPODIACEAE**
23 Tepals 4; leaves all basal (*Subularia*) **64. BRASSICACEAE**
24 Flowers in branched cymes, sometimes compact **155. JUNCACEAE**
24 Flowers in a simple terminal raceme 25
25 Leaves on stems 0 or few and near base, without pore at apex; flowers numerous, without bracts **145. JUNCAGINACEAE**
25 Leaves several on stems, with prominent pore at apex; flowers <12, with bracts **144. SCHEUCHZERIACEAE**
26 Flowers small, in dense spike on axis, the axis sometimes extended distally as sterile projection, with large spathe at base often partly or wholly obscuring flowers **151. ARACEAE**
26 If flowers in single dense spike then without large spathe at base 27
27 Stems entirely rhizomatous, producing large (usually >1m across) leaves and huge (usually >50cm) elongated dense panicles **82. GUNNERACEAE**
27 If stems entirely rhizomatous then leaves <10cm and flowers solitary or in whorls or umbels 28
28 Inflorescence consisting of units arranged in umbels, each unit consisting of several male flowers (each of 1 stamen) and 1 female flower (of 1 stalked ovary) all surrounded by 4-5 conspicuous glands; plants with copious white latex (*Euphorbia*) **93. EUPHORBIACEAE**
28 Inflorescence not consisting of units as above; plants without copious white latex 29
29 Leaves opposite; stems procumbent to weakly ascending 30
29 If leaves opposite then stems self-supporting and ± erect 32
30 Tepals 5, free **48. CARYOPHYLLACEAE**
30 Tepals 3-4, fused 31
31 Leaves ≤15mm; perianth 4-lobed, greenish (*Nertera*) **134. RUBIACEAE**
31 Leaves >20mm; perianth 3-lobed, brownish-purple (*Asarum*) **26. ARISTOLOCHIACEAE**

32 Plants aquatic or in wet bogs; leaves simple, entire or ± so **146. POTAMOGETONACEAE**
32 Plants usually on dry ground, if in marshes or bogs then leaves not simple and entire 33
33 Perianth of 6 lobes or segments, 3 in outer and 3 in inner whorl 34
33 Perianth lobes or segments 1, or 2-5, or 4-5 in each of 2 whorls, rarely 6 and then not in 2 whorls of 3 37
34 Leaves or leaf-like organs linear, without basal lobes 35
34 Leaves broader than linear, or if not then with basal lobes 36
35 Dioecious; leaves reduced to scales, their normal function replaced by clusters of 4-10(more) cladodes (*Asparagus*) **162. LILIACEAE**
35 Flowers bisexual; leaves reduced to scales or not, but not replaced by clusters of cladodes **155. JUNCACEAE**
36 Twining climber; ovary inferior; fruit a red berry **165. DIOSCOREACEAE**
36 Not climbing; ovary superior, fruit an achene (*Rumex*) **49. POLYGONACEAE**
37 Flowers epigynous, semi-epigynous, or perigynous with deeply concave hypanthium 38
37 Flowers hypogynous, or perigynous with flat to saucer-shaped hypanthium, or flowers all male (dioecious) 44
38 Ovary with 3-8 cells; stigmas 3-8 39
38 Ovary with 1-2 cells; styles/stigmas 1-2 40
39 Leaves simple, entire; inflorescence small, axillary (*Tetragonia*) **44. AIZOACEAE**
39 Leaves 1-2-pinnate, inflorescence large, conspicuous, terminal (*Aralia*) **110. ARALIACEAE**
40 Tepals 4, or 4 plus 4 epicalyx segments beneath; stamens 1-4, or ≥8 41
40 Tepals 5; stamens 5 42
41 Stamens 1-4, or ≥10; fruit 1-many achenes **77. ROSACEAE**
41 Stamens 8; fruit a capsule (*Chrysosplenium*) **76. SAXIFRAGACEAE**
42 Ovary 2-celled; fruit a 2-celled schizocarp **111. APIACEAE**
42 Ovary 1-celled; fruit a 1-celled achene 43
43 Leaves linear or ± so **88. SANTALACEAE**
43 Leaves ovate to lanceolate or deltate (*Beta*) **45. CHENOPODIACEAE**
44 Tepals 2; stamens 12-18, conspicuous (*Macleaya*) **32. PAPAVERACEAE**
44 Tepals 1, or 3-5, or 5 with 5 epicalyx segments beneath, or sometimes 0 in female flowers, rarely 2 and then stamens also 2 45
45 Leaves opposite; flowers often dioecious 46
45 Leaves alternate; flowers rarely dioecious 50
46 Leaves deeply palmately lobed to ± palmate **36. CANNABACEAE**
46 Leaves simple, at most toothed 47
47 Leaves ≥1.5cm; stigmas branched or conspicuously papillose; at least male flowers in axillary spikes (catkins) 48
47 Leaves <1.5cm; stigmas ± smooth or minutely papillose; flowers not in spikes 49
48 Leaves usually with strong stinging hairs; tepals 4; fruit an achene (*Urtica*) **38. URTICACEAE**
48 Leaves without stinging hairs; tepals 3; fruit a 2-celled capsule (*Mercurialis*) **93. EUPHORBIACEAE**
49 Tepals 3; leaves with conspicuous stipules (*Koenigia*) **49. POLYGONACEAE**
49 Tepals 4-5; leaves without stipules **48. CARYOPHYLLACEAE**
50 Gynoecium composed of 1 carpel with 1 ovule, producing a 1-seeded fruit, or all flowers male 51

50 Gynoecium composed of 2-many (often fused) carpels, with many ovules in total, producing a many-seeded fruit or many 1-seeeded fruits 55

51 Leaves palmately lobed to base or ± so (*Cannabis*) **36. CANNABACEAE**

51 Leaves simple, entire or pinnately lobed but not ± to base 52

52 Leaves with well-developed stipules fused into short tube round stem **49. POLYGONACEAE**

52 Leaves without stipules 53

53 Tepals (2)3-5, scarious, with 3-5 often similar bracteoles just below; fruit often dehiscent **46. AMARANTHACEAE**

53 Tepals 0-5, if >1 then herbaceous; flowers often without bracteoles; fruit always indehiscent 54

54 Tepals 1 or 3-5 or 0; styles 2-3, with smooth to papillate stigmas **45. CHENOPODIACEAE**

54 Tepals 4; style 1; stigma much branched **38. URTICACEAE**

55 Ovary of 2 carpels completely fused; tepals 4 **64. BRASSICACEAE**

55 Ovary of 2-many carpels, free or <1/2 fused, if >1/2 fused then tepals 5 56

56 Each carpel with several to many ovules, producing several to many seeds 57

56 Each carpel with 1-2 ovules, producing 1 seed 58

57 Leaves palmate or ± so; flowers >1cm across (*Helleborus*) **30. RANUNCULACEAE**

57 Leaves 2-ternate to 2-pinnate; flowers <1cm across (*Astilbe*) **76. SAXIFRAGACEAE**

58 Carpels fused, elongated distally into sterile column ending in 1 style with 5 stigmas **106. GERANIACEAE**

58 Carpels free, each with 1 stigma 59

59 Tepals 5, with 5 epicalyx segments just beneath (*Sibbaldia*) **77. ROSACEAE**

59 Tepals 4-5, without epicalyx **30. RANUNCULACEAE**

Key F - Perianth of 2 (rarely more) distinct whorls or rarely a spiral, the inner and outer differing markedly in shape/size/colour; flowers all male or all female

1 Tree or shrub, sometimes procumbent but with distinctly woody stems 2

1 Herb 11

2 Leaves pinnate; tree with white flowers (*Fraxinus*) **127. OLEACEAE**

2 Leaves simple, or if pinnate then shrub with yellow flowers 3

3 Leaves all scale-like, functionally replaced by leaf-like stem outgrowths bearing flowers on their faces (*Ruscus*) **162. LILIACEAE**

3 Leaves photosynthetic, not bearing flowers on their faces 4

4 Stems procumbent; leaves *Erica*-like, with revolute margins **66. EMPETRACEAE**

4 Stems ascending, spreading or erect; leaves not *Erica*-like 5

5 Leaves stipulate, either pinnate or simple and entire **77. ROSACEAE**

5 Leaves without stipules, or if stipulate then simple and serrate 6

6 Leaves palmately lobed (*Ribes*) **74. GROSSULARIACEAE**

6 Leaves entire to serrate or shallowly spinose-pinnately lobed 7

7 Leaves gland-dotted, <2cm **70A. MYRSINACEAE**

7 Leaves not gland-dotted, most or all >2cm 8

8 Corolla white; at least some leaves usually with spine-tipped teeth **91. AQUIFOLIACEAE**

8 Corolla purple or greenish-yellow; leaves not spiny 9

9 Flowers ± hypogynous, with large nectar-secreting disc at base **90. CELASTRACEAE**

9 Flowers clearly epigynous or perigynous, the inferior ovary or hypanthium distinct even in male flowers 10

10 Leaves closely serrate, usually stipulate (*Rhamnus*) **94. RHAMNACEAE**
10 Leaves entire or remotely serrate, without stipules **87. CORNACEAE**
11 Climbing plant bearing tendrils (*Bryonia*) **62. CUCURBITACEAE**
11 Plant not climbing, without tendrils 12
12 Aquatic or marsh plants with all leaves in basal rosette(s) 13
12 If aquatic or marsh plants then all leaves not in basal rosettes 14
13 Petals 3, free, conspicuous, white **142. HYDROCHARITACEAE**
13 Petals 4, fused, inconspicuous, ± scarious (female flowers present low down in leaf-rosette, easily missed) (*Littorella*) **125. PLANTAGINACEAE**
14 Sepals 3; petals 3 15
14 Sepals and petals not both 3 16
15 Leaves reduced to scales, replaced by bunches of ± cylindrical cladodes (*Asparagus*) **162. LILIACEAE**
15 Leaves flat, with 2 basal lobes (*Rumex*) **49. POLYGONACEAE**
16 Flowers in dense capitula closely surrounded by ≥1 row of bracts **139. ASTERACEAE**
16 Flowers not in dense capitula though sometimes crowded 17
17 At least some leaves ternate to pinnate 18
17 All leaves simple (sometimes lobed) 21
18 Basal leaves simple; stem-leaves 1-pinnate (*Valeriana*) **137. VALERIANACEAE**
18 All leaves compound, at least some 2-3-pinnate or 1-3-ternate 19
19 All leaves 1-ternate (*Fragaria*) **77. ROSACEAE**
19 At least some leaves 2-3-pinnate or 2-3-ternate 20
20 Stamens ≤10; carpels usually 2; plant <1m (*Astilbe*) **76. SAXIFRAGACEAE**
20 Stamens >10; carpels usually 3; plant usually ≥1m (*Aruncus*) **77. ROSACEAE**
21 Leaves petiolate, truncate to cordate at base 22
21 Leaves sessile or ± so, rounded to cuneate at base 23
22 Leaves not lobed, entire; tepals 5, all ± white; plant >(25)100cm (*Fallopia*) **49. POLYGONACEAE**
22 Leaves lobed, serrate; sepals 5, green; petals 5, white; plant <25cm (*Rubus*) **77. ROSACEAE**
23 Leaves alternate; petals 4; stamens 8 or ovaries 4 (*Sedum*) **75. CRASSULACEAE**
23 Leaves opposite; petals 5; stamens usually 10 or ovary 1 **48. CARYOPHYLLACEAE**

Key G - Perianth of 2 (rarely more) distinct whorls or rarely a spiral, the inner and outer differing markedly in shape/size/colour; petals fused at base for varying distances to apex; ovary present in at least some flowers, superior

1 Stems twining, not green; leaves reduced to small scales, not green **116. CUSCUTACEAE**
1 If stems twining then leaves expanded and green 2
2 Plant wholly lacking green colour, yellow to brown, sometimes red- or purple-tinged **129. OROBANCHACEAE**
2 Plant with obvious green colouring 3
3 Leaves linear to subulate, all in basal rosette under water or on mud; flowers inconspicuous, solitary on pedicels or densely clustered on leafless scapes 4
3 Leaves not both linear to subulate and all in basal rosette 6
4 Flowers bisexual, solitary; stamens included in corolla (*Limosella*) **128. SCROPHULARIACEAE**

4 Flowers in dense heads or spikes on scapes, or if solitary then unisexual and stamens exserted from corolla 5

5 Tepals 4, densely fringed at apex; flowers in capitate clusters, unisexual **154. ERIOCAULACEAE**

5 Tepals >4, not fringed at apex; flowers either bisexual and in spikes, or unisexual and solitary **125. PLANTAGINACEAE**

6 Basal leaves peltate, succulent, glabrous; inflorescence a terminal raceme (*Umbilicus*) **75. CRASSULACEAE**

6 Leaves not peltate and succulent 7

7 Flowers pea-like, zygomorphic; petals 5, 1 upper, 2 lateral, and 2 lower fused to form keel; stamens 10 **79. FABACEAE**

7 Flowers not pea-like with 5 petals and 10 stamens 8

8 Sepals 5 (3 outer small, 2 inner large); petals 3; stamens 8 **97. POLYGALACEAE**

8 Not with the combination sepals 5, petals 3, stamens 8 9

9 Stamens >10 10

9 Stamens ≤10 12

10 Carpels 5-many, usually ± fused; leaves stipulate **55. MALVACEAE**

10 Carpel 1; leaves without stipules 11

11 Tree or shrub; flowers numerous in dense clusters, each with perianth <1cm **78. MIMOSACEAE**

11 Herb; flowers in elongated raceme or sometimes few, each with perianth >1cm (*Consolida*) **30. RANUNCULACEAE**

12 Stamens 2 13

12 Stamens >2 16

13 Leaves all in a basal rosette; flowers solitary on erect pedicels (*Pinguicula*) **132. LENTIBULARIACEAE**

13 Plant not with all leaves in a basal rosette and flowers solitary on erect pedicels 14

14 Ovary 4-celled, each cell with 1 ovule **122. LAMIACEAE**

14 Ovary 2-celled, each cell with 2-many ovules 15

15 Perianth actinomorphic; fruit a berry or capsule with ≤4 seeds **127. OLEACEAE**

15 Perianth slightly to strongly zygomorphic; fruit a capsule with usually >4 seeds **128. SCROPHULARIACEAE**

16 Ovary 4-celled with 1 ovule per cell; fruit a cluster of 1-seeded nutlets; plant bisexual 17

16 Ovary not 4-celled with 1 ovule per cell, or if so then fruit a berry and plant dioecious 19

17 Leaves alternate; flowers usually in cymes spirally coiled when young; stems not square in section **120. BORAGINACEAE**

17 Leaves opposite; flowers not in spirally coiled cymes; stems usually ± square in section 18

18 Ovary scarcely lobed at flowering, with terminal style and capitate stigma **121. VERBENACEAE**

18 Ovary deeply lobed at flowering, with usually basal style and (1-)2 linear stigmas **122. LAMIACEAE**

19 Sepals 2; petals fused only at base **47. PORTULACACEAE**

19 Sepals usually >2, if 2 then petals fused for >1/2 of length 20

20 Tree or shrub (sometimes very dwarf) 21

20 Herb; if stems woody then climbing or trailing 31

21 Leaves opposite or whorled 22

21 Leaves alternate or spiral 26

22 Stamens 5-10 23

22 Stamens 4 24

23 Stamens 5; leaves mostly >1cm wide; flowers >15mm across **113. APOCYNACEAE**

23 Stamens 8 or 10, or if 5 then leaves all <5mm wide and flowers <10mm across **67. ERICACEAE**

24 Flowers dull, brownish (*Plantago*) **125. PLANTAGINACEAE**

24 Flowers white or brightly coloured 25

25 Flowers ≥2.5cm, in loose inflorescences **128. SCROPHULARIACEAE**

25 Flowers <2.5cm, in dense inflorescences **126. BUDDLEJACEAE**

26 Dioecious, either stamens or ovary rudimentary 27

26 Flowers bisexual, with functional stamens and ovary 28

27 Leaves gland-dotted, <2cm, serrate; fruit purple **70A. MYRSINACEAE**

27 Leaves not gland-dotted, most >2cm, entire or with spiny teeth; fruit red to yellow **91. AQUIFOLIACEAE**

28 Stamens 4 **126. BUDDLEJACEAE**

28 Stamens 5-10 29

29 Plant a dense cushion <10cm with solitary flowers on erect stalks <5cm; stamens 5, with anthers opening by slits **70. DIAPENSIACEAE**

29 Plant rarely a dense cushion, if so then stamens 8 or 10 with anthers opening by apical pores 30

30 Stamens 5; flowers actinomorphic **114. SOLANACEAE**

30 Stamens 8 or 10, or if 5 then flowers slightly zygomorphic **67. ERICACEAE**

31 Inner perianth-segments (corolla-lobes) and stamens on same radius 32

31 All or most inner perianth-segments (corolla-lobes) and stamens on alternating radii 33

32 Styles 5; stigmas linear **50. PLUMBAGINACEAE**

32 Styles 1; stigma capitate **71. PRIMULACEAE**

33 Aquatic or bog plant with showy corollas densely fringed distally with hairs or narrow serrations **117. MENYANTHACEAE**

33 If aquatic or marsh plant then corollas not fringed or fringed at base of lobes 34

34 Leaves all in basal rosette; corolla violet with yellow centre and yellow anthers **130. GESNERIACEAE**

34 If leaves all in basal rosette then corolla not violet with yellow centre 35

35 Flowers in compact spikes; perianth scarious and brownish; stamens long-exserted (*Plantago*) **125. PLANTAGINACEAE**

35 If flowers in compact spikes then not with both scarious perianth and long-exserted stamens 36

36 Flowers in axils of spiny bracts; sepals 4, 2 large and 2 small; corolla with 3-lobed lower lip and 0 upper lip **131. ACANTHACEAE**

36 If bracts spiny then sepals not 2 large and 2 small and corolla not with large lower and 0 upper lip 37

37 Leaves opposite 38

37 Leaves alternate 41

38 Ovary 1-celled 39

38 Ovary 2-celled, or ovaries 2 and each 1-celled 40

39 Ovule 1 **43A. NYCTAGINACEAE**

39 Ovules many **112. GENTIANACEAE**

40 Ovaries 2, with common style expanded into ring below stigma **113. APOCYNACEAE**

40 Ovary 1, with styles not expanded into ring below stigma **128. SCROPHULARIACEAE**

41 Stamens normally 6-8; inner or outer tepals with keel on abaxial side, otherwise similar to others; ovary 1-celled, with 1 ovule (*Fallopia*) **49. POLYGONACEAE**

41 Stamens 3-5; outer tepals sepaloid, inner petaloid; ovary 1-5-celled, with

total of ≥4 ovules 42
42 Ovary 3-celled; stigmas 3 on 1 style **118. POLEMONIACEAE**
42 Ovary 1-5-celled, if 3-celled then stigmas 1-2 per style 43
43 Stamens 3 or 4, at least in most flowers **128. SCROPHULARIACEAE**
43 Stamens 5, at least in most flowers 44
44 Styles 2, or style 1 but divided distally into 2 branches/long stigmas 45
44 Style 1, with 1-2 short stigmas at apex 46
45 Ovary 2-lobed, each lobe with 1 style; corolla 1.5-2.5mm (*Dichondra*) **115. CONVOLVULACEAE**
45 Ovary not lobed, with 1 style divided distally; corolla 6-10mm **119. HYDROPHYLLACEAE**
46 Ovary 1-2(3)-celled, with total of 4(-6) ovules; stems trailing or climbing **115. CONVOLVULACEAE**
46 Ovary 2(-5)-celled, each cell with many ovules; stems usually not trailing or climbing 47
47 Flowers distinctly zygomorphic; corolla divided >1/2 way to base; anthers not forming close cone round style (*Verbascum*) **128. SCROPHULARIACEAE**
47 Flowers actinomorphic or slightly zygomorphic; corolla divided <1/2 way to base or if >1/2 way to base then flowers actinomorphic and anthers forming close cone round style **114. SOLANACEAE**

Key H - Perianth of 2 (rarely more) distinct whorls or rarely a spiral, the inner and outer differing markedly in shape/size/colour; petals fused at base for varying distances to apex; ovary present in at least some flowers, inferior or partly so

1 Stamens 8 to numerous (or 4-5 but appearing 8-10 because filaments split to base with each 1/2 bearing one 1/2-anther) 2
1 Stamens ≤5 (sometimes alternating with staminodes) 4
2 Tree (*Eucalyptus*) **85. MYRTACEAE**
2 Herb or shrub <2m 3
3 Flowers 5 in 1 terminal cluster, greenish; leaves ternate; herb **136. ADOXACEAE**
3 Flowers 1-several, axillary, white to red; leaves simple; shrub (*Vaccinium*) **67. ERICACEAE**
4 Flowers in dense heads surrounded by row(s) of sepal-like bracts 5
4 Flowers not in dense heads, or if so then with only 2 bracts at base 8
5 Ovary 2-celled; sepals 4-5, fused proximally 6
5 Ovary 1-celled; calyx represented by a cup, or by often >5 narrow teeth, bristles, scales or hairs 7
6 Leaves alternate; stamens borne on receptacle; ovules numerous **133. CAMPANULACEAE**
6 Leaves whorled; stamens borne on corolla-tube; ovules 1 per cell (*Sherardia*) **134. RUBIACEAE**
7 Anthers usually 5, fused into tube round style; ovary and fruit not enclosed in epicalyx **139. ASTERACEAE**
7 Anthers 4, not fused; ovary and fruit enclosed in tubular epicalyx **138. DIPSACACEAE**
8 Stamens 1-3 (vestigial if flowers female) 9
8 Stamens 4-5 (0 or vestigial if flowers female) 11
9 Flowers numerous, ± closely packed in flat or domed inflorescences, <1cm across **137. VALERIANACEAE**
9 Flowers rather few in spikes or racemes, or solitary, >2cm across 10
10 Stamens 3, with obvious filaments and anthers **163. IRIDACEAE**
10 Stamens 2, with sessile anthers (*Cypripedium*) **166. ORCHIDACEAE**
11 Leaves opposite or in whorls 12

11 Leaves alternate or spiral 13
12 Leaves ≥4 per node, or 2 per node and each stipulate **134. RUBIACEAE**
12 Leaves 2 per node, without stipules **135. CAPRIFOLIACEAE**
13 Flowers unisexual; stamens 3 (0 in female flowers), 2 with 2 pollen-sacs and 1 with 1 **62. CUCURBITACEAE**
13 Flowers bisexual; stamens 5, all with 4 pollen-sacs 14
14 Stamens opposite corolla-lobes; stigma 1 (*Samolus*) **71. PRIMULACEAE**
14 Stamens alternating with corolla-lobes; stigmas 2-5 **133. CAMPANULACEAE**

Key I - Perianth of 2 (rarely more) distinct whorls or rarely a spiral, the inner and outer differing markedly in shape/size/colour; petals free, or rarely fused just at basal point, or fused near apex but free at base; ovary present in at least some flowers, inferior or partly so.

1 Petals >8 2
1 Petals ≤6 (unless *flore pleno*) 3
2 Aquatic plant with floating flowers and leaves (*Nymphaea*) **27. NYMPHAEACEAE**
2 Terrestrial plant; leaves very succulent **44. AIZOACEAE**
3 Sepals and petals each 3 4
3 Sepals and petals each 2 or 4-6 7
4 Flowers zygomorphic; stamens 1-2 **166. ORCHIDACEAE**
4 Flowers actinomorphic; stamens 3-12 or 0 in female flowers 5
5 Outer whorl of tepals sepaloid; stamens 9-12 or 0 in female flowers; plant aquatic **142. HYDROCHARITACEAE**
5 Both whorls of tepals petaloid; stamens 3-6; plant terrestrial 6
6 Stamens 6 (*Galanthus*) **162. LILIACEAE**
6 Stamens 3 (*Libertia*) **163. IRIDACEAE**
7 Tree or shrub 8
7 Herb, rarely woody at base 15
8 Stamens 4-5 or 0 in female flowers 9
8 Stamens 8-many 11
9 Flowers in true umbels, arising at 1 point, sometimes the umbels further aggregated **110. ARALIACEAE**
9 Flowers solitary or in racemes or panicles, sometimes in corymbose umbel-like clusters but not arising at 1 point 10
10 Origin of sepals separated from ovary wall by saucer- to cup-shaped hypanthium **74. GROSSULARIACEAE**
10 Flowers without hypanthium (sepals arising direct from ovary wall) **87. CORNACEAE**
11 Leaves alternate 12
11 Leaves opposite or whorled 13
12 Fruit a woody capsule; style 1 (*Leptospermum*) **85. MYRTACEAE**
12 Fruit surrounded by succulent or pithy hypanthium; styles (1)2-many **77. ROSACEAE**
13 Fruit a capsule; styles 2-4, free or united proximally **73. HYDRANGEACEAE**
13 Fruit a berry; style 1 14
14 Hypanthium extending 5-16mm beyond ovary apex; flowers pendent on long pedicels (*Fuchsia*) **86. ONAGRACEAE**
14 Hypanthium not or scarcely extending beyond ovary apex; flowers not pendent nor on long pedicels (*Luma*) **85. MYRTACEAE**
15 Leaves simple, all or most >1m across; inflorescence an elongated very dense panicle >50cm **82. GUNNERACEAE**

15 If leaves >50cm then compound; if inflorescence >30cm then not dense and elongated 16
16 Flowers in umbels 17
16 Flowers not in umbels, but sometimes corymbose 19
17 Styles 2; fruit a dry 2-celled schizocarp **111. APIACEAE**
17 Styles 1 or 5; fruit a succulent drupe or berry 18
18 Plant with a single terminal umbel; fruit red; leaves simple (*Cornus*) **87. CORNACEAE**
18 Plant with umbels in large panicles; fruit dark purple to black; leaves 1-2-pinnate (*Aralia*) **110. ARALIACEAE**
19 Sepals 5 20
19 Sepals 2 or 4 21
20 Petals bright yellow; leaves pinnate **77. ROSACEAE**
20 Petals not bright yellow, or if so then leaves simple **76. SAXIFRAGACEAE**
21 Sepals 2; petals yellow, <1cm; fruit a capsule opening transversely (*Portulaca*) **47. PORTULACACEAE**
21 If sepals 2 then petals white; if petals yellow then ≥1cm; if fruit a capsule then not opening transversely 22
22 Fruit an achene without bristles, with persistent sepals; leaves subsessile (*Haloragis*) **81. HALORAGACEAE**
22 Fruit a many-seeded capsule, or if a 1-2-celled achene then with many hooked bristles and without persistent sepals, and leaves long-petiolate **86. ONAGRACEAE**

Key J - Perianth of 2 (rarely more) distinct whorls or rarely a spiral, the inner and outer differing markedly in shape/size/colour; petals free or rarely fused just at basal point, or fused near apex but free at base; >1 ovary present in at least some flowers, superior, free or fused just at extreme base

1 Tree with flowers borne in pendent unisexual globose clusters **34. PLATANACEAE**
1 Herb or shrub; flowers not in pendent unisexual globose clusters 2
2 Sepals and petals each 3 3
2 Sepals and petals each >3 5
3 Carpels 3; stamens 3; flowers <3mm across (*Crassula*) **75. CRASSULACEAE**
3 Carpels ≥6; stamens ≥6 or 0 in female flowers; flowers >5mm across 4
4 Carpels with numerous ovules, forming follicles in fruit; sepals purple-tinged green; leaves linear **140. BUTOMACEAE**
4 Carpels with 1 ovule forming achenes in fruit, or if with >1 ovules forming follicles in fruit then sepals green and leaves ovate with cordate base **141. ALISMATACEAE**
5 Leaves opposite and/or very succulent **75. CRASSULACEAE**
5 Leaves all basal, alternate or spiral, not or only slightly succulent 6
6 Shrub **77. ROSACEAE**
6 Herb, sometimes with ± woody surface rhizome 7
7 Flowers perigynous; origin of sepals (often of stamens and petals too) separated from base of ovary by obvious flat or saucer- to bowl-shaped hypanthium 8
7 Flowers hypogynous or ± so, the stamens, petals and sepals free and arising directly at base of carpels 9
8 Carpels 2, each with many ovules; fruit a pair of follicles; leaves simple, unlobed; petals bright pink (*Bergenia*) **76. SAXIFRAGACEAE**
8 Carpels 3-many, or if 2 then each with 1 ovule; fruit 1-many achenes; plant not with both simple unlobed leaves and pink petals **77. ROSACEAE**

9 Carpels with 1 ovule; fruit an achene **30. RANUNCULACEAE**

9 Carpels with few to many ovules; fruit a follicle 10

10 Flowers ≥7cm across; seeds black (fertile) and red (sterile) usually mixed in each follicle **51. PAEONIACEAE**

10 Flowers <7cm across; seeds not red and black mixed together 11

11 Carpels 5-many **30. RANUNCULACEAE**

11 Carpels 2-4 12

12 Stamens 5-10 (or 0 in female flowers); carpels rarely >2 **76. SAXIFRAGACEAE**

12 Stamens >10 (or 0 in female flowers); carpels usually >2 13

13 Leaves palmate or deeply palmately lobed **30. RANUNCULACEAE**

13 Leaves 2-3-pinnate (*Aruncus*) **77. ROSACEAE**

Key K - Perianth of 2 (rarely more) distinct whorls or rarely a spiral, the inner and outer differing markedly in shape/size/colour; petals free, or rarely fused just at basal point, or fused near apex but free at base; ovary present in at least some flowers, superior, of 1 carpel or of >1 wholly or mostly fused carpels

1 Plant yellowish-brown, lacking green pigment in all parts **69. MONOTROPACEAE**

1 Plant with at least leaves or stems at least partly green 2

2 Leaves modified to form tubular 'pitchers' with small blade ('hood') at entrance, all in basal rosette **56. SARRACENIACEAE**

2 Leaves not modified as 'pitchers' 3

3 Leaves covered on upperside with very sticky glandular hairs, reddish, all basal **57. DROSERACEAE**

3 Leaves not all basal and covered with very sticky glandular hairs 4

4 Flowers zygomorphic 5

4 Flowers actinomorphic 19

5 At least 1 sepal or petal with conspicuous basal spur or pouch 6

5 Flowers without basal spur(s) 10

6 Spur(s) formed from petal(s); ovary 1-celled 7

6 Spur(s) formed from sepal(s); ovary 3-5-celled 8

7 Leaves pinnate or ternate, without stipules; sepals 2; petals 4 **33. FUMARIACEAE**

7 Leaves simple, stipulate (the stipules sometimes deeply divided or pinnate); sepals 5; petals 5 **59. VIOLACEAE**

8 Sepals 3; petals 5 but apparently 3 due to fusion of 2 pairs of laterals **109. BALSAMINACEAE**

8 Sepals 5; petals 5 9

9 Upper sepal with spur fused to pedicel; ovary 5-celled, with distal sterile beak (*Pelargonium*) **106. GERANIACEAE**

9 Upper 1-3 sepals with free spurs; ovary 3-celled, without distal beak **108. TROPAEOLACEAE**

10 Stamens 8 or 10, all or all but 1 with filaments fused into tube 11

10 Stamens 3-many, free 12

11 Flowers with sepal uppermost (on top-line); stamens 8; anthers opening by pores **97. POLYGALACEAE**

11 Flowers with petal uppermost (on top-line); stamens 10; anthers opening by slits **79. FABACEAE**

12 Tree, or shrub >1m high; ovary 3-celled 13

12 Herb, or shrub <1m high; ovary 1-2-celled 14

13 Leaves palmate, opposite; flowers white to red **100. HIPPOCASTANACEAE**

13 Leaves 1-2-pinnate, alternate; flowers yellow **99. SAPINDACEAE**

14 Stamens 3; petals 3, filiform, brown (*Tolmiea*) **76. SAXIFRAGACEAE**

14 Stamens 5-many; petals >3, not filiform, white or pink 15

15 Ovary 2-celled ; petals 4, the lower 2 and upper 2 forming 2 different pairs 16

15 Ovary 1-celled; petals 4-6(8), if only 4 then upper one, lateral two and lower one of 3 different forms 17

16 Leaves palmate; capsule >2cm **63A. CAPPARACEAE**

16 Leaves simple to pinnate; capsule <2cm **64. BRASSICACEAE**

17 Petals 4-6(8), at least some deeply lobed ; stamens 7-many (*Reseda*) **65. RESEDACEAE**

17 Petals 5, or apparently 4 due to fusion of lower 2, entire to very shallowly lobed or toothed 18

18 Leaves paripinnate **78A. CAESALPINIACEAE**

18 Leaves ternate (*Thermopsis*) **79. FABACEAE**

19 Stamens >12, >2x as many as petals 20

19 Stamens 1-12, ≤2x as many as petals 28

20 Petals ≥9 21

20 Petals ≤6 22

21 Aquatic plant with leaves and flowers at or just above water surface (*Nuphar*) **27. NYMPHAEACEAE**

21 Terrestrial plant with succulent leaves **44. AIZOACEAE**

22 Stamens with filaments fused into tube round styles **55. MALVACEAE**

22 Stamens free or ± united into bundles, but not forming tube round styles 23

23 Ovary with 1-2(5) ovules per cell 24

23 Ovary with several to many ovules per cell 25

24 Ovary 5-celled; inflorescence stalks fused to narrowly oblong papery bract **54. TILIACEAE**

24 Ovary 1-celled; inflorescence stalks not fused to long papery bract **77. ROSACEAE**

25 Leaves simple, entire 26

25 Leaves simple and toothed or lobed, or compound 27

26 Leaves with translucent and/or coloured sessile glands; styles 3 or 5 **53. CLUSIACEAE**

26 Leaves without sessile glands, style 0 or 1 **58. CISTACEAE**

27 Sepals 2(-3), sepaloid, usually falling early **32. PAPAVERACEAE**

27 Sepals (3)4-5, petaloid, not falling early **30. RANUNCULACEAE**

28 Perianth in 3-6 whorls each of 2-4 segments **31. BERBERIDACEAE**

28 Perianth in 2 whorls each of ≥2 segments 29

29 Stems herbaceous, sometimes woody just at base 30

29 Stems wholly or mostly woody 54

30 Sepals 2; petals 5 31

30 Sepals ≥3; petals as many as sepals or fewer 32

31 Stems twining or sprawling, >1m; underground tubers present; capsule indehiscent **47A. BASELLACEAE**

31 Stems decumbent to erect, <50cm; underground tubers 0; capsule dehiscent **47. PORTULACACEAE**

32 Sepals 3; petals 2-3 33

32 Sepals >3; petals >3 35

33 Ovary 1-celled, with 1 ovule; sepals and petals both sepaloid or petaloid (*Rumex*) **49. POLYGONACEAE**

33 Ovary 3-celled, each cell with 2-many ovules; petals petaloid; sepals sepaloid 34

34 Leaves opposite, stipulate, <1cm; flowers <5mm across **52. ELATINACEAE**

34 Leaves alternate, with sheathing base but without stipules, >1cm; flowers ≥10mm across **153. COMMELINACEAE**

35 Ovary 1-celled, at least apically, with 1-many ovules 36

35 Ovary 2-10-celled throughout with 1-many ovules per cell 42

36 Stamens 5, alternating with 5 conspicuous deeply divided staminodes (*Parnassia*) **76. SAXIFRAGACEAE**

36 Deeply divided staminodes 0 37

37 Ovary with 1 ovule 38

37 Ovary with few to many ovules 40

38 Leaves without stipules; flowers showy, usually pink or blue, >4mm across **50. PLUMBAGINACEAE**

38 Leaves stipulate; flowers inconspicuous, green, yellowish-green or white 39

39 Leaves simple, entire **48. CARYOPHYLLACEAE**

39 Leaves compound or conspicuously lobed **77. ROSACEAE**

40 Style 1, ≥1mm, with 1 or 5 stigmas at apex **68. PYROLACEAE**

40 Styles ± 0 or 2-5, free, if 1 then either <1mm or divided into 3 distally 41

41 Ovary with ovules on parietal placentas; styles fused proximally; leaves without stipules **61. FRANKENIACEAE**

41 Ovary with ovules on free-central placenta; styles free, or 0, or rarely fused proximally and then leaves stipulate **48. CARYOPHYLLACEAE**

42 Flowers perigynous, with tubular or cup- or bowl-shaped hypanthium bearing sepals and petals at apex 43

42 Flowers hypogynous, with sepals and petals borne at base of ovary 45

43 Hypanthium becoming hard and protective at fruiting; carpels 2, each with 1 ovule **77. ROSACEAE**

43 Hypanthium not becoming hard and protective; carpels 2(-4), each with many ovules 44

44 Sepals and petals each usually 6, with epicalyx segments outside; leaves ± all on stem, at least the lower opposite or in whorls of 3 **83. LYTHRACEAE**

44 Sepals and petals each 5, without epicalyx segments; leaves all or nearly all basal (*Rodgersia*) **76. SAXIFRAGACEAE**

45 Stem-leaves all in single whorl of 3-8, with 1 flower above (*Paris*) **162. LILIACEAE**

45 Stem-leaves (if present) not all in 1 whorl 46

46 Petals and sepals each 4, at least in most flowers 47

46 Petals and sepals each 5 in all flowers 50

47 Stamens 4 or 6 48

47 Stamens 8 49

48 Stamens 4; capsule dehiscing by 8 valves (*Radiola*) **96. LINACEAE**

48 Stamens (4-)6; capsule dehiscing by 2 valves or breaking transversely or indehiscent **64. BRASSICACEAE**

49 Leaves simple, entire, stipulate **52. ELATINACEAE**

49 Leaves 2-3-pinnately lobed, without stipules (*Ruta*) **104. RUTACEAE**

50 Fruit a 5-celled schizocarp with 1 seed per cell, elongated distally into sterile column 51

50 Fruit a 5- to many-seeded capsule, without a sterile column 52

51 Leaves stipulate; petals white to red, blue or purple **106. GERANIACEAE**

51 Leaves without stipules; petals yellow **107. LIMNANTHACEAE**

52 Ovary 2-celled; styles 0 or 2 (*Saxifraga*) **76. SAXIFRAGACEAE**

52 Ovary 5-celled; styles 5 53

53 Capsule opening by 5 valves; leaves ternate or palmate **105. OXALIDACEAE**

53 Capsule opening by 10 valves; leaves simple, entire **96. LINACEAE**

54 Sepals 3; petals 2-3 55

54 Sepals and petals each >3 57

55 Woody climber (*Fallopia*) **49. POLYGONACEAE**

55 Procumbent to erect shrub 56

56 Leaves <3mm wide; flowers in small axillary clusters **66. EMPETRACEAE**

56 Leaves replaced by leaf-like flat cladodes >3mm wide, bearing 1-2 flowers on their surface (*Ruscus*) **162. LILIACEAE**

57 Stamens 2x as many as petals 58

57 Stamens <2x as many as petals at least on most flowers 62

58 Woody stems ± procumbent; leaves >6cm, simple, serrate (*Bergenia*) **76. SAXIFRAGACEAE**

58 Woody stems erect or ± so; leaves usually compound or lobed 59

59 Leaves simple, serrate; ovules many in each cell **65A. CLETHRACEAE**

59 Leaves compound, or simple and entire, or lobed with the lobes entire; ovules 1-2 per cell 60

60 Ovary 2-celled; leaves opposite **101. ACERACEAE**

60 Ovary 4-6-celled; leaves alternate or opposite 61

61 Fruit a 4(-5)-celled capsule, not winged; leaves simple, or ternate, or pinnate or pinnately lobed with segments <1cm wide **104. RUTACEAE**

61 Fruit a group of 1-6 winged achenes; leaves pinnate with leaflets mostly >1cm wide **103. SIMAROUBACEAE**

62 Leaves opposite 63

62 Leaves alternate 67

63 Stems procumbent; ovary 1-celled **61. FRANKENIACEAE**

63 Stems erect; ovary >1-celled 64

64 Leaves simple and not lobed 65

64 Leaves pinnate 66

65 Fruit a dehiscent capsule with orange seeds; leaves without stipules (*Euonymus*) **90. CELASTRACEAE**

65 Fruit a black non-dehiscent berry; leaves with herbaceous stipules at least when young **94. RHAMNACEAE**

66 Sepals 5, free; stamens 5; ovules many in each cell; fruit an inflated capsule **98. STAPHYLEACEAE**

66 Sepals 4, fused into tube proximally; stamens 2; ovules 2 in each cell; fruit a winged achene (*Fraxinus*) **127. OLEACEAE**

67 Petals and sepals each 4; stamens 6; ovules many in each cell **64. BRASSICACEAE**

67 Petals and sepals each 5, or if 4 then stamens 4 and ovules 1-2 per cell 68

68 Leaves <5mm, ± scale-like **60. TAMARICACEAE**

68 Leaves >5mm, not scale-like 69

69 Leaves with strongly revolute margins, densely rusty-tomentose on lowerside; stamens mostly 6-8 (*Ledum*) **67. ERICACEAE**

69 Leaves with margins not or scarcely revolute, not rusty-tomentose (sometimes white-tomentose) on lowerside; stamens 4-5 70

70 Leaves palmate or palmately lobed **95. VITACEAE**

70 Leaves pinnate, pinnately lobed or unlobed 71

71 Petals purple; ovary 1-celled with many ovules **72. PITTOSPORACEAE**

71 Petals not purple; ovary 1-5-celled, each cell with 1-2 ovules 72

72 Ovary 1-celled with 1 ovule **102. ANACARDIACEAE**

72 Ovary 2-5-celled, each cell with 1-2 ovules 73

73 Leaves evergreen, coriaceous and shiny, at least some with strong marginal spines; petals white **91. AQUIFOLIACEAE**

73 Leaves deciduous, not coriaceous or shiny, without spines on margin; petals yellowish-green 74

74 Fruit a black non-dehiscent berry; leaves with herbaceous stipules at least when young **94. RHAMNACEAE**

74 Fruit a dehiscent capsule with red seeds; leaves with spinose stipules at least when young (*Celastrus*) **90. CELASTRACEAE**

MAGNOLIIDAE — DICOTYLEDONS
(Dicotyledonidae)

Often trees or shrubs; commonly with secondary thickening from a permanent vascular cambium; vascular bundles usually in a ring in the stem; primary root commonly persisting; leaves usually with pinnate or palmate major venation and reticulate minor venation; flower parts mostly in fours or fives; pollen grains mostly radially symmetrical, commonly with 3 pores and/or furrows; cotyledons normally 2; endosperm typically nuclear or cellular. Numerous exceptions to all the above occur.

25. LAURACEAE - *Bay family*

Trees or shrubs; leaves aromatic, simple, evergreen, entire, usually petiolate, alternate, without stipules. Flowers solitary or in few-flowered clusters in leaf-axils, dioecious, perigynous, actinomorphic; perianth of 1 whorl of 4 ± free lobes; male flowers with 8-12 stamens; female flowers with 2-4 staminodes and a 1-celled ovary with 1 ovule; style 1; stigma capitate, fruit a 1-seeded berry.

Easily recognized by the aromatic leaves and dioecious flowers with 4-lobed perianth.

1. LAURUS L. - *Bay*

1. L. nobilis L. - *Bay*. Shrub or tree exceptionally to 18m; leaves 5-10cm, laurel-like, glabrous, usually acute, with distinctive aromatic smell; flowers c.1cm across, cream; (2n=42, 48). Intrd; widely planted in small numbers for culinary use, sometimes persisting in wild places in S, natd (probably bird-sown) in scrub and on cliffs near sea in SW Ir, S Wa, SW En and Jersey, occasionally elsewhere in S En, 1 shrub in S Lancs since 1984; Mediterranean.

26. ARISTOLOCHIACEAE - *Birthwort family*

Perennial herbs; leaves simple, entire, strongly cordate, without stipules. Flowers solitary and terminal or 1-8 in leaf-axils, bisexual, epigynous, actinomorphic or zygomorphic; perianth of 1 whorl, tubular with 1 or 3 terminal lobes; stamens 6-12 in 1-2 whorls, arising from base or side of stylar column; ovary 6-celled, each cell with many ovules; styles 6, united into a column with 6-lobed stigma; fruit a capsule.

2 distinctive genera easily recognized by their cordate leaves and weird flowers.

1 Aerial stem ≤10cm, densely pubescent; petiole much longer than lamina; flower solitary, terminal; perianth 3-lobed **1. ASARUM**
1 Aerial stem >10cm, ± glabrous; petiole shorter than lamina or ± absent; flowers 1-8, axillary; perianth 1-lobed **2. ARISTOLOCHIA**

1. ASARUM L. - *Asarabacca*
Rhizomatous herb with aerial stems bearing scales, usually 2 apparently opposite foliage leaves, and 1 terminal flower; flowers actinomorphic; perianth bell-shaped, 3-lobed; stamens 12, in 2 whorls, arising from base of stylar column.

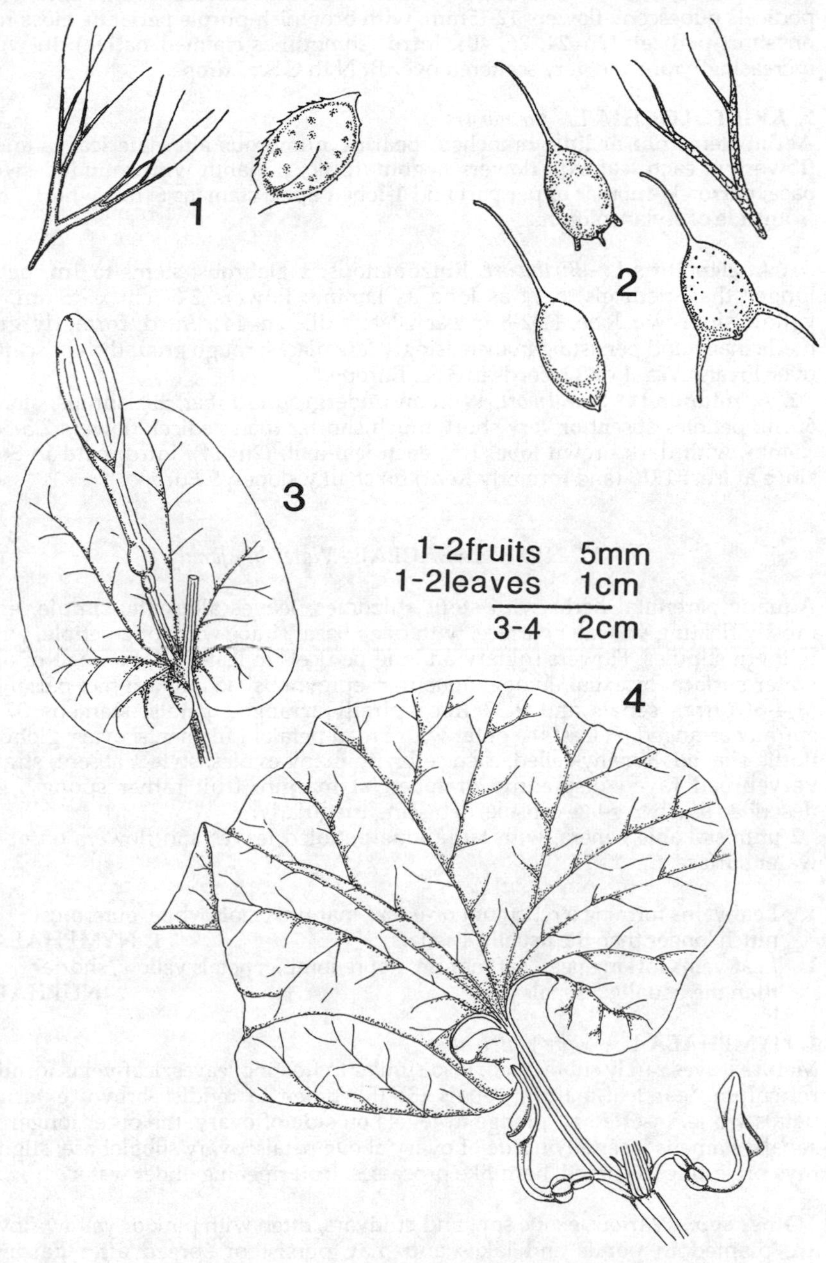

FIG 73 - ***Ceratophyllum, Aristolochia.*** 1-2, leaves and fruits of ***Ceratophyllum.***
1, *C.* ***submersum.*** 2, *C.* ***demersum*** (3 fruit types).
3-4, flowering nodes of ***Aristolochia.*** 3, *A.* ***rotunda.*** 4, *A.* ***clematitis.***

1. A. europaeum L. - *Asarabacca*. Rhizome on soil surface; stems, petioles and pedicels pubescent: flowers 12-15mm, with brownish-purple perianth, close to soil on short pedicel; (2n=24, 26, 40). Intrd (sometimes claimed native); in woods, increasingly rare and very scattered over Br N to C Sc; Europe.

2. ARISTOLOCHIA L. - *Birthworts*
Aerial stems not or little branched, bearing numerous alternate leaves and 1-8 flowers in each leaf-axil; flowers zygomorphic; perianth with rounded swollen base, narrowly tubular upper part and 1-lobed apex; stamens 6, in 1 whorl, arising from side of stylar column.

1. A. clematitis L. - *Birthwort*. Rhizomatous; ± glabrous; stems to 1m; petioles longer than pedicels, c.1/2 as long as lamina; flowers 2-3.5cm x <5mm, with yellowish-brown lobe, (1)2-8 in each leaf-axil; (2n=14). Intrd; formerly grown medicinally and persisting in increasingly few places; rough ground, very scattered over En and Wa, 1 or 2 records in S Sc; Europe. 73

2. A. rotunda L. - *Smearwort*. With an underground tuber; ± glabrous; stems to 60cm; petioles absent or very short, much shorter than pedicel; flowers 2.5-5cm x <5mm, with dark brown lobe, 1 in each leaf-axil; (2n=14). Intrd; natd in Surrey since at least 1918 (and formerly Kent) on chalky slopes; S Europe. 73

27. NYMPHAEACEAE - *Water-lily family*

Aquatic perennial herbs with stout rhizomes; leaves alternate, simple, entire, mostly floating on water surface, with deep basal sinus, with long petiole, with or without stipules. Flowers solitary on long pedicels in leaf-axils, borne at or above water surface, bisexual, hypogynous to ± epigynous, actinomorphic; perianth of (3)4-6(7) free sepals and 9-33 free spirally-arranged petals; stamens 37-200, spirally arranged, at least the outer with broad petaloid filaments; ovary globose to bottle-shaped, 8-many-celled, each cell with many ovules; style ± absent; stigma a very broad rayed disc entire or lobed at margin; fruit rather spongy, often described as a berry-like capsule, dehiscing irregularly.

2 unmistakable genera, with large long-petioled leaves and flowers on or near water surface.

1 Leaf-veins forming a reticulum near leaf-margin; petals white, outermost much longer than the usually 4 sepals **1. NYMPHAEA**
1 Leaf-veins forking near leaf-margin, not re-joining; petals yellow, shorter than the usually 5 sepals **2. NUPHAR**

1. NYMPHAEA L. - *White Water-lilies*
Mature leaves rarely submerged, if so similar to floating leaves; leaf-veins forming a reticulum near leaf-margin; sepals (3)4(5), green to reddish-brown externally; petals white, inserted at a range of levels on side of ovary, the outer longer than sepals; stamens inserted on side of ovary above petals; ovary subglobose; stigmatic rays projecting as curved, horn-like processes; fruit ripening under water.

Other spp. - Various exotic spp. and cultivars, often with pink or yellow flowers, are planted in ponds and lakes and may persist or spread after gardening activities cease. Their identity needs investigating; most are referable to **N. marliacea** Lat.-Marl., covering various hybrids of *N. alba*.

1. N. alba L. - *White Water-lily*. Petals 12-33, the outer 2-8.5cm; stamens 46-125; stigmatic rays (and carpels) 9-25. Native; in lakes, ponds, dykes and slow-flowing rivers.

a. Ssp. alba. Leaves 9-30cm; flowers 9-20cm across, opening wide; pollen-grains usually with projections of varying lengths; stamens borne almost to top of ovary and leaving scars to top of fruit; stigmatic rays usually >14; fruit usually obovoid; 2n=84. Throughout BI except CI, but absent from several areas and replaced by ssp. *occidentalis* in parts of N and W.

b. Ssp. occidentalis (Ostenf.) Hyl. Leaves 9-13cm; flowers 5-12cm across, usually never opening wide; pollen-grains usually with projections of ± uniform length; stamens not borne on upper part of ovary; stigmatic rays usually <16; fruit usually subglobose, without stamen-scars in upper part. W Ir and N & W Sc, especially Hebrides and Shetland. Intermediates between the sspp. occur and are not confined to areas where their ranges meet.

2. NUPHAR Sm. - *Yellow Water-lilies*

Mature leaves usually floating and submerged, the latter thinner and with undulate margins; leaf-veins forking and not rejoining near leaf-margin; sepals 5(-7), yellowish-green; petals 9-26, yellow, shorter than sepals; petals and stamens inserted at base of ovary; stamens 37-200; ovary bottle-shaped; stigmatic rays 7-24, not projecting or projecting as flat, obtuse bulges; fruit ripening above water.

1 Leaf-blades erect, held above water surface; petioles ± terete **4. N. advena**
1 Leaf-blades horizontal, floating on water surface; petioles dorsiventrally compressed or triangular in section with rounded angles 2
  2 Leaves with 23-28 lateral veins on each side; stigmatic disc 10-15mm across, circular or slightly crenate at margin, with 9-24 rays; flowers 3-6cm across **1. N. lutea**
  2 Leaves with 11-22 lateral veins on each side; stigmatic disc 6-11mm across, crenate to lobed at margin, with 7-14 rays; flowers 1.5-4cm across 3
3 Leaves with 15-22 lateral veins on each side; stigmatic disc 7.5-11mm across, crenate at margin, with 9-14 rays; stamens 60-100; pollen <25% fertile **2. N. x spenneriana**
3 Leaves with 11-18 lateral veins on each side; stigmatic disc 6-8.5mm across, distinctly lobed at margin, with 7-12 rays; stamens 37-65; pollen >90% fertile **3. N. pumila**

1. N. lutea (L.) Sm. - *Yellow Water-lily.* Floating leaves to 40 x 30cm, with 23-28 lateral veins on each side, with a very narrow basal sinus; petiole triangular in section with rounded angles; stigmatic disc 10-15mm across, circular or slightly crenate at margin; 2n=34. Native; in lakes, ponds, dykes and rivers; frequent throughout BI except CI and extreme N Sc.

2. N. x spenneriana Gaudin (*N. lutea* x *N. pumila*) - *Hybrid Water-lily.* **RR**
Intermediate between the parents; floating leaves to 18 x 14cm, with 15-22 lateral veins on each side; stigmatic disc 7.5-11mm across, crenate at margin; pollen <25% fertile, seed fertility c.20%; 2n=34. Native; scattered in extreme N En and S & C Sc largely outside area of *N. pumila*, and 1 record in Merioneth, natd in Surrey; introgression of *N. lutea* into *N. pumila* occurs sometimes.

3. N. pumila (Timm) DC. - *Least Water-lily.* Floating leaves to 17 x 12.5cm, with **R**
11-18 lateral veins on each side, with a broader sinus than in *N. lutea*; petiole dorsiventrally compressed, ± keeled beneath; stigmatic disc 6-8.5mm across, distinctly lobed at margin; 2n=34. Native; in ponds and lochs in highland areas but often at low altitude; local in N, C & SW Sc, 1 locality in Salop, planted and natd in Surrey.

4. N. advena (Aiton) W.T. Aiton - *Spatter-dock.* Leaves erect, held above water, rarely floating, to 40 x 30cm, with 20-38 lateral veins on each side, with narrowly to broadly triangular basal sinus; petiole ± terete; stigmatic disc (6-)12-20mm across, ± circular at margin; flowers sometimes red-tinged; (2n=34). Intrd; planted as ornament and ± natd in scattered places from Surrey to Easterness, perhaps overlooked; E N America.

28. CABOMBACEAE - *Water-shield family*

Aquatic perennial herbs; leaves mostly submerged, opposite, deeply and finely palmately dissected, petiolate; floating leaves associated with flowers, peltate, entire, petiolate, usually alternate; stipules 0. Flowers bisexual, slightly emergent, solitary in leaf-axils, on long pedicels, actinomorphic, hypogynous; perianth of 6 similar, free, white tepals in 2 whorls; stamens 6; carpels usually 3, free, each with usually 3 dorsal ovules, a short style and capitate stigma; fruit an indehiscent usually 3-seeded follicle.

The opposite, palmately dissected, submerged leaves, peltate floating leaves and flowers with 6 white tepals are unique.

1. CABOMBA Aubl. - *Water-shields*

1. C. caroliniana A. Gray - *Carolina Water-shield*. Stems to 2m; submerged leaves 3-5cm, superficially resembling those of some aquatic *Ranunculus* but opposite and with petioles 1-3cm; floating leaves 0.6-2cm, linear to narrowly elliptic, with petiole joining in centre; flowers 6-10mm; (2n=24, c.96, 104). Intrd; aquarium throwout natd and spreading in canal in N Hants since 1990, formerly Dunbarton; E N America.

29. CERATOPHYLLACEAE - *Hornwort family*

Submerged aquatic perennial herbs; leaves in whorls of (3)6-8(12), regularly forked into linear segments, without stipules. Flowers monoecious, subaquatic, sessile, solitary in leaf-axils, male and female at different nodes, actinomorphic, hypogynous; perianth of numerous free green, narrow segments in 1 whorl; male flowers wth 10-25 stamens; female flowers with 1-celled ovary with 1 ovule; style 1; stigma minutely bifid; fruit an achene.

Distinguished from other subaquatic plants by the whorled, bifid leaves.

1. CERATOPHYLLUM L. - *Hornworts*

1. C. demersum L. - *Rigid Hornwort*. Stems branched, up to 1m; leaves dark 73
green, rigid, forked 1-2x, with conspicuously denticulate segments; fruit 4-5mm, rather rarely produced, smooth or slightly warty, apiculate to long-spined at apex, often also with 2 long spines at base or these missing (var. **inerme** Gay ex Radcl.-Sm.) or reduced to tubercles (var. **apiculatum** (Cham.) Asch.); (2n=24, 28, 34, 38, 48, 72). Native; in ponds, ditches and slow rivers; scattered over En and Wa, rare and very local in Sc, Ir and CI.

2. C. submersum L. - *Soft Hornwort*. Leaves lighter green, less rigid, forked 3-4x, 73
with rather sparsely denticulate segments; fruit 4-5mm, rather commonly produced (Aug-Oct), conspicuously warty, apiculate to obtuse at apex, without basal spines; (2n=24, 40, 72). Native; in ponds and ditches, mostly near sea; very local in C & SE Br N to Westmorland, Co Down, CI

30. RANUNCULACEAE - *Buttercup family*

Herbaceous annuals or perennials, sometimes woody climbers; leaves borne spirally or sometimes opposite or whorled, simple or variously compound, usually petiolate, usually without stipules. Flowers variously arranged, bisexual, hypogynous, usually actinomorphic, sometimes zygomorphic; perianth of 1-2 whorls of free segments, the outer (sepals) often petaloid, of various colours, the inner (petals or honey-leaves) bearing nectaries and often reduced or absent, sometimes (when petals 0) a whorl of sepal-like bracts outside the petaloid sepals; stamens usually numerous, rarely as few as sepals; carpels 1-many, sometimes partially or rarely fully fused, if many usually spirally arranged, with 1-many ovules; fruit usually an achene or follicle, rarely a berry or capsule.

Very variable in floral morphology, but most genera have spirally arranged leaves without stipules and produce a head of achenes or follicles from each flower, which is always hypogynous and often with the sepals more conspicuous than the petals.

1 Woody climber; leaves opposite; perianth of 4 segments in 1 whorl **12. CLEMATIS**
1 Herbaceous, not climbing; leaves alternate or whorled; perianth rarely of 4 segments, often in 2 whorls 2
 2 Ovary with many ovules; fruit a follicle, capsule or berry 3
 2 Ovary with 1 ovule; fruit an achene 11
3 Flowers with 1 carpel 4
3 Flowers with (2)3 or more carpels 5
 4 Flowers actinomorphic, whitish; sepals not spurred; fruit a berry **8. ACTAEA**
 4 Flowers zygomorphic, usually blue, sometimes white or pink; the upper sepal with a conspicuous spur; fruit a follicle **7. CONSOLIDA**
5 At least 1 of the petals or sepals conspicuously hooded or spurred 6
5 Petals and sepals not spurred or hooded 7
 6 Flowers zygomorphic, the upper sepal hooded, only the 2 upper petals spurred **6. ACONITUM**
 6 Flowers actinomorphic, the sepals not hooded, each of the 5 petals spurred **16. AQUILEGIA**
7 Flowers white to reddish, bluish or greenish, sometimes tinged purple 8
7 Flowers yellow 9
 8 Annuals; sepals bluish; follicles fused up to apex; leaves divided into fine linear entire segments **5. NIGELLA**
 8 Perennials; sepals green, white, violet or purple; follicles free or fused at base only; leaves divided into wider, toothed segments **3. HELLEBORUS**
9 Stem-leaves 3, in a whorl just below flower **4. ERANTHIS**
9 Stem-leaves 1-many, not whorled and usually not just below flower 10
 10 Leaves deeply lobed; perianth of 2 whorls, the inner consisting of small nectaries **2.TROLLIUS**
 10 Leaves simple, finely toothed; perianth of 1 whorl **1. CALTHA**
11 Perianth in a single whorl 12
11 Perianth apparently in two whorls 14
 12 Stamens longer than perianth; flowering stems with alternate leaves and many flowers **17. THALICTRUM**
 12 Stamens shorter than perianth; flowering stems with 1-few flowers and whorled leaves or bracts 13
13 Styles scarcely elongating in fruit; whole plant glabrous to shortly pubescent **9. ANEMONE**
13 Styles greatly elongating in fruit and becoming feathery; whole plant with long hairs **11. PULSATILLA**
 14 Outer whorl of apparent perianth of 3 segments; inner whorl of perianth segments blue or white **10. HEPATICA**
 14 Outer whorl of apparent perianth usually of 5 segments, if of 3 then inner whorl yellow 15
15 Flowers solitary on leafless stem; leaves all linear, in a basal rosette;

sepals spurred **15. MYOSURUS**
15 Flowers on usually branched, leafy stems; lowest leaves not linear; sepals not spurred 16
16 Petals red with a blackish basal blotch **14. ADONIS**
16 Petals yellow or white, without dark basal blotch **13. RANUNCULUS**

1. CALTHA L. - *Marsh-marigold*
Herbaceous perennials; leaves spirally arranged, simple, without stipules; flowers solitary or in few-flowered cymes, without a whorl of bracts below, actinomorphic; perianth of 1 whorl of 5-8(10) petaloid sepals; stamens numerous; carpels 5-15, free, spirally arranged; fruit a follicle.

1. C. palustris L. (*C. radicans* T. F. Forst.) - *Marsh-marigold.* Rhizomatous, 79
glabrous; basal leaves long-petioled, cordate, denticulate; flowers 1-5cm across, saucer-shaped; sepals golden yellow; 2n=48, 52, 54, 56, 64, 72, c.80. Native; in marshes and ditches and beside ponds and streams. Plants with procumbent stems rooting at nodes and little-branched, eventually turning up to produce usually 1 small flower, are found in N Wa, N En, Sc and Ir in upland areas, and are best recognised as var. **radicans** (T. F. Forst.) Hook. (ssp. *minor* auct. non Mill.). Many cultivated ornamentals have very large, small or pale (even white) flowers, or may be *flore pleno*, and may escape.

2. TROLLIUS L. - *Globeflower*
Herbaceous perennials; leaves borne spirally, palmate or deeply palmately lobed, without stipules; flowers solitary or in few-flowered cymes, without a whorl of bracts below, actinomorphic; perianth of 2 whorls; sepals 5-15, petaloid; petals 5-15, in the form of narrow, small nectaries; stamens numerous; carpels numerous, free, spirally arranged; fruit a follicle.

1. T. europaeus L. - *Globeflower*. Aerial stems to 70cm, glabrous, erect, not or little 79
branched, arising from stout stock; basal leaves long-petioled, palmate or nearly so, with 3-5 lobed leaflets; flowers 2.5-5cm across, ± globose; sepals pale yellow, hiding petals; 2n=16. Native; damp places in grassland or woods, often upland; local in Wa, N En, Sc and NW Ir.

3. HELLEBORUS L. - *Hellebores*
Herbaceous perennials; leaves spirally arranged or all basal, ternate to palmate with long, toothed leaflets, the lateral ones joined at base, without stipules; flowers in few- to many-flowered cymes, or solitary, without whorl of bracts below, actinomorphic, appearing very early in season; perianth of 2 whorls; sepals 5, usually not brightly coloured; petals 5-12, in the form of small tubular nectaries; stamens numerous; carpels 2-5, usually slightly fused at base; fruit a follicle.

1 Stems lasting from 1 spring to next, many-flowered; bracts ± entire; leaves all on stem 2
1 Stems lasting from winter to late spring, 2-4-flowered; bracts deeply divided; leaves all basal 3
2 Leaves palmate, the leaflets with short acute to obtuse teeth; flowers bell-shaped; fresh plant stinking when crushed **1. H. foetidus**
2 Leaves ternate, the leaflets with subulate ± sharp teeth; flowers cup-shaped; fresh plant not stinking when crushed **2. H. argutifolius**
3 Follicles fused for c.1/4 their length, sessile; flowers mostly 3-5cm across, pale green **3. H. viridis**
3 Follicles free to base, shortly stalked; flowers mostly 5-7cm across, yellowish-green to purplish **4. H. orientalis**

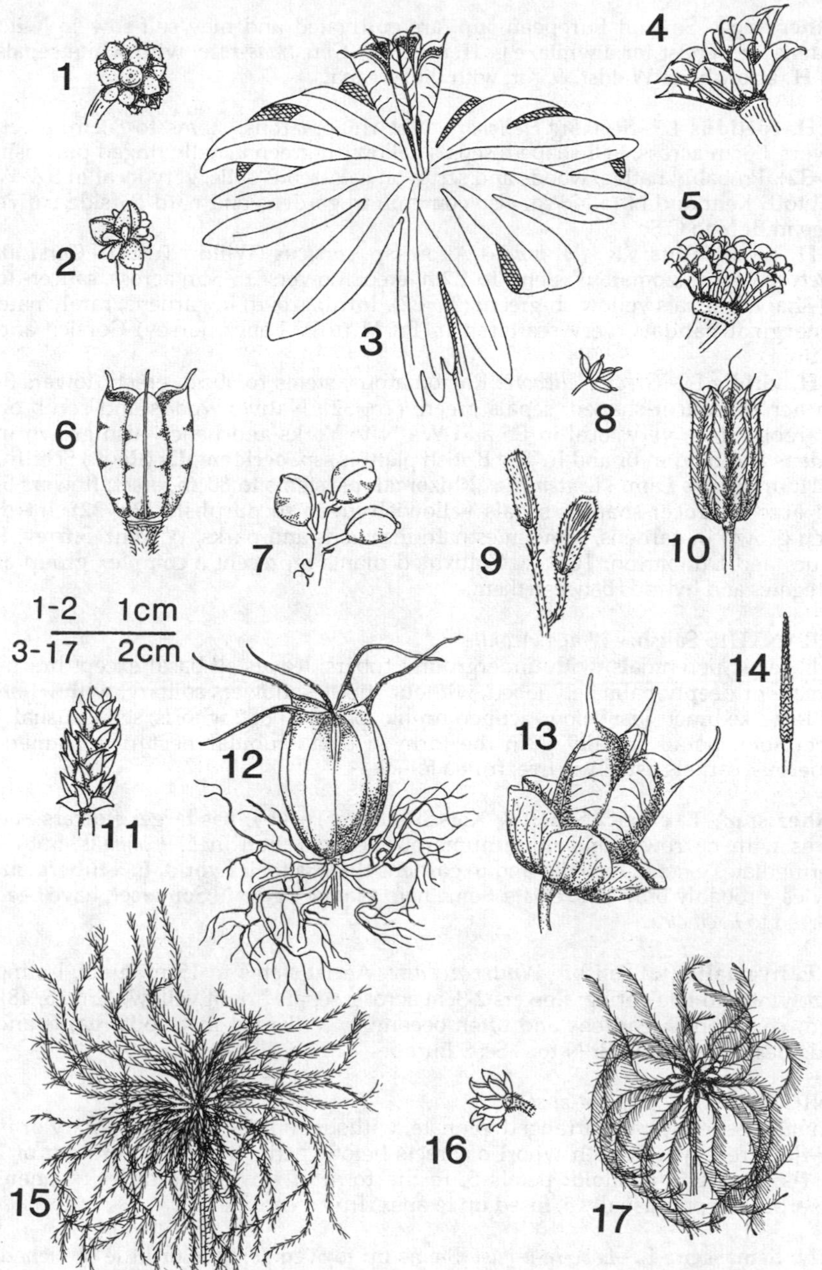

FIG 79 - Fruits of **Ranunculaceae**. 1-2, *Ranunculus ficaria*. 1, ssp. *ficaria*. 2, ssp. *bulbilifer*. 3, *Eranthis hyemalis*. 4, *Caltha palustris*. 5, *Trollius europaeus*. 6, *Aconitum napellus*. 7, *Actaea spicata*. 8, *Thalictrum minus*. 9, *Consolida ajacis*. 10, *Aquilegia vulgaris*. 11, *Adonis annua*. 12, *Nigella damascena*. 13, *Helleborus foetidus*. 14, *Myosurus minimus*. 15, *Pulsatilla vulgaris*. 16, *Anemone nemorosa*. 17, *Clematis vitalba*.

Other spp. - Several European spp. are cultivated and may self-sow in 'wild gardens' or persist for a while, e.g. **H. niger** L. (*Christmas-rose*) with white sepals and **H. atrorubens** Waldst. & Kit. with violet sepals.

1. H. foetidus L. - *Stinking Hellebore*. Not rhizomatous; stems to 80cm, erect; R
flowers 1-3cm across, bell-shaped; sepals yellowish-green usually tinged purplish; 79
(2n=32). Probably native; woods and scrub on calcareous soils; very local in S & W Br E to E Kent and N to N Wa, also common in gardens and natd outside native range in Br N to C Sc.

2. H. argutifolius Viv. (*H. lividus* Aiton ssp. *corsicus* (Willd.) Tutin) - *Corsican Hellebore*. Not rhizomatous; stems to 1.2m, erect; flowers 2.5-5cm across, saucer- to cup-shaped; sepals yellowish-green; (2n=32). Intrd; grown in gardens, rarely natd in marginal habitats; very scattered in En N to W Lancs, Jersey; Corsica and Sardinia.

3. H. viridis L. - *Green Hellebore*. Rhizomatous; stems to 40cm, erect; flowers 3-5cm across, saucer-shaped; sepals green; (2n=32). Native; woods and scrub on calcareous soils; very local in En and Wa N to Yorks and Lancs, also grown in gardens and natd in Br and Ir. The British plant is ssp. **occidentalis** (Reut.) Schiffn.

4. H. orientalis Lam. - *Lenten-rose*. Rhizomatous; stems to 60cm, erect; flowers 5-7cm across, saucer-shaped; sepals yellowish-green to purplish; (2n=32). Intrd; much grown in gardens, sometimes natd in woods and parks; W Kent, Surrey, E Gloucs and Dunbarton; Turkey. Cultivated plants represent a complex group of segregates and hybrids between them.

4. ERANTHIS Salisb. - *Winter Aconite*
Herbaceous perennials, with underground tubers; leaves all basal except bracts, palmate or deeply palmately lobed, without stipules; flowers solitary, with whorl of 3 leaf-like bracts just below, actinomorphic; perianth of 2 whorls; sepals usually 6, petaloid; petals usually 6, in the form of small tubular nectaries; stamens numerous; carpels usually 6, free; fruit a follicle.

Other spp. - **E. cilicica** Schott & Kotschy, from Turkey, has larger flowers and leaves with narrower and more numerous leaflets than in *E. hyemalis*, but all intermediates exist in the wild and in gardens there is their hybrid, **E. x tubergenii** Bowles; probably only 1 sp. exists. Some natd plants, e.g. in N Somerset, have been referred to *E. cilicica*.

1. E. hyemalis (L.) Salisb. - *Winter Aconite*. Aerial stems to 15cm, erect, lasting 79
from winter to late spring; flowers 2-3cm across; sepals bright yellow; (2n=16, 48). Intrd; common in gardens and often becoming well natd in woods, parks and roadsides; scattered in Br N to C Sc; S Europe.

5. NIGELLA L. - *Love-in-a-mist*
Annuals; leaves spirally arranged, pinnate, without stipules; flowers solitary or in few-flowered cymes, with whorl of bracts below, actinomorphic; perianth of 2 whorls; sepals 5, petaloid; petals 5, in the form of clawed nectaries; stamens numerous; carpels usually 5, fused up to apex; fruit a capsule.

1. N. damascena L. - *Love-in-a-mist*. Stems up to 50cm, simple or little branched, 79
glabrous; leaves finely divided with linear segments, at least 3 in a cluster just below each terminal flower; sepals blue; capsule strongly inflated, ± globose; (2n=12). Intrd; commonly grown in gardens and often persisting on waste ground and rubbish tips; En and Wa; S Europe.

6. ACONITUM L. - *Monk's-hoods*
Herbaceous perennials; leaves spirally arranged, palmate or deeply palmately

lobed, without stipules; flowers in terminal racemes, each in axil of small bract, zygomorphic; perianth of 2 whorls; sepals 5, petaloid, the upper one forming an elongated hood; petals 2-10, in the form of nectaries, the 2 upper large and enclosed in the sepal-hood, the others very small or absent; stamens numerous; carpels 3(-5), ± fused at base; fruit a follicle.

1 Flowers yellow or cream; upper sepal >2x as high as wide; leaves not divided to base **3. A. lycoctonum**
1 Flowers blue or blue and white; upper sepal <2x as high as wide; at least lower leaves divided to base 2
  2 Pedicels pubescent to densely pubescent; flowers usually blue; upper sepal ± as high as wide, gradually tapered into forward-projecting spur **1. A. napellus**
  2 Pedicels glabrous to sparsely pubescent; flowers blue and white or blue; upper sepal distinctly higher than wide, abruptly narrowed into forward-projecting spur **2. A. x cammarum**

Other spp. - All records of **A. variegatum** L. and many of **A. compactum** (Rchb.) Gáyer are errors for *A. x cammarum.*

1. A. napellus L. (*A. anglicum* Stapf) - *Monk's-hood.* Stems to 1.5m, erect; leaves **R**
divided to base, with narrow deeply divided segments; pedicels conspicuously **79**
appressed-pubescent; flowers blue to violet; upper sepals ± as wide as high; pollen **84**
full; seeds fertile; 2n=32. Native; very local in shady places by streams; probably only in SW En and S Wa, but commonly cultivated and natd sparsely over much of Br in native-type habitats as well as waste places. Cultivated plants are very variable in habit, leaf-lobing, flower colour and flowering time; they may be referred to ≥4 sspp. but distinctions between them break down in garden material. *A. anglicum* comes under ssp. **napellus** and true *A. compactum* under ssp. **vulgare** (DC.) Rouy & Foucaud; the latter has less tapering apices to leaf-lobes, a more compact and less or not branched inflorescence, pedicels with bracteoles mostly >3mm (not <3mm), and flowers later (late summer and autumn, not early to mid summer).

2. A. x cammarum L. (*A. napellus* x *A. variegatum* L.) - *Hybrid Monk's-hood.* Stems **84**
to 1.5m, erect; leaves as in *A. napellus;* pedicels glabrous to sparsely appressed-pubescent; flowers blue to violet or variegated with white; upper sepal distinctly higher than wide; pollen empty; seeds sterile. Intrd; grown in gardens and natd in damp shady places; frequent in Sc, rare in Wa and En. Most natd plants are cv. 'Bicolor', with variegated flowers, of garden origin.

3. A. lycoctonum L. (*A. vulparia* Rchb.) - *Wolf's-bane.* Stems to 1m, erect; leaves **84**
not divided to base and with broader and less divided segments than in *A. napellus;* pedicels conspicuously appressed-pubescent; flowers yellow; upper sepals c.3x as high as wide; (2n=16). Intrd; grown in gardens and occasionally natd by streams and in woods; C En to N Sc; C Europe. Our plant is ssp. **vulparia** (Rchb.) Nyman.

7. CONSOLIDA (DC.) Gray (*Delphinium* L. pro parte) - *Larkspurs*

Annuals; leaves spirally arranged, palmate with finely divided segments, without stipules; flowers in terminal racemes, each in axil of bract and with bracteoles on pedicel, zygomorphic; perianth of 2 whorls; sepals 5, petaloid, the upper one long-spurred; petals 4, elaborately shaped, the upper two fused and with a nectariferous spur enclosed in the sepal-spur; stamens numerous; carpel 1; fruit a follicle.

Other spp. - **C. orientalis** (J. Gay) Schroedinger (*Delphinium orientale* J. Gay) (*Eastern Larkspur*) differs from *C. ajacis* in its upper bracteoles overlapping the base of the flower (not so in *C. ajacis*) and its <12mm sepal-spur. **C. regalis** Gray (*Delphinium consolida* L.) (*Forking Larkspur*) differs from the other 2 spp. in having glabrous follicles, and branches arising at wide angles. Both come from Europe and formerly occurred as casuals.

1. C. ajacis (L.) Schur (*C. ambigua* auct. non (L.) P.W. Ball & Heywood, **79**
Delphinium ambiguum auct. non L.) - *Larkspur*. Stems to 1m, simple or with **84**
branches arising at rather narrow angles; at least lower bracts deeply divided; flowers blue, pink or white; sepal-spur 12-18mm; follicle pubescent; (2n=16). Intrd; much grown in gardens and a common escape in BI, mainly S; formerly a corn-field weed natd in E Anglia; Mediterranean.

8. ACTAEA L. - *Baneberry*

Herbaceous perennials; leaves borne spirally, ternate or pinnate, without stipules; flowers in terminal racemes, each in axil of a small bract, actinomorphic; perianth usually of 2 whorls; sepals (3)4(5), petaloid; petals 4-6 or 0, petaloid, without nectar; stamens numerous; carpel 1; fruit a berry.

1. A. spicata L. - *Baneberry*. Stems to 60cm, simple or little branched; leaves with **R**
broad leaflets; flowers small and crowded, white; berry 10-13mm, black; (2n=16). **79**
Native, limestone pavements and sparse woodland on limestone; local in Yorks, Westmorland and Lancs., intrd in scattered sites over much of Br.

9. ANEMONE L. - *Anemones*

Herbaceous perennials; leaves all basal, palmate, palmately lobed or ternate, without stipules; flowers solitary or few, with a whorl of 3 leaf-like bracts some way below, or in terminal few-flowered cymes, actinomorphic; perianth of 1 whorl of 5-20 petaloid sepals; stamens numerous; carpels numerous, free, spirally arranged; fruit an achene, the style remaining shorter than fertile portion.

1 Stem >30cm, usually branched with several flowers; sepals densely silky-pubescent on lowerside; autumn flowering **5. A. x hybrida**
1 Stem <30cm, simple, with 1(-3) flowers; sepals glabrous to sparsely pubescent on lowerside; spring flowering 2
  2 Sepals yellow, mostly 5 **4. A. ranunculoides**
  2 Sepals white to pink or blue, mostly 6 or more 3
3 Sepals (5)6-7(9), mostly white or pinkish **1. A. nemorosa**
3 Sepals (8)10-15(20), mostly blue 4
  4 Sepals and basal leaves sparsely pubescent on lowerside; head of ripe achenes erect **2. A. apennina**
  4 Sepals and basal leaves glabrous on lowerside; head of ripe achenes pendent **3. A. blanda**

1. A. nemorosa L. - *Wood Anemone*. Rhizomes elongated; stems 5-30cm, erect, **79**
with whorl of 3 leaf-like bracts 1-6cm below the solitary flower; sepals (5)6-7(9), glabrous or sparsely pubescent near base on lowerside, usually white, often variously tinged with pink or purple, sometimes pale blue; head of achenes pendent; 2n=30 (16-46). Native; woodland, hedgerows, and open grassland in wetter districts; throughout BI but local in S Ir.

2. A. apennina L. - *Blue Anemone*. Differs from *A. nemorosa* in its sepals (8)10-15(18), narrower, usually blue (sometimes pink or white), sparsely pubescent near base on lowerside; anthers paler yellow; head of achenes erect; (2n=16). Intrd; commonly cultivated and frequently persisting as a throwout or escape in woodland, hedgerows and rough ground; scattered in Br N to C Sc; C Mediterranean.

3. A. blanda Schott & Kotschy - *Balkan Anemone*. Differs from *A. nemorosa* in its short, tuberous rhizome; sepals 10-20, narrower, usually blue (sometimes pink or white), glabrous on lowerside; anthers paler yellow; (2n=16, 32). Intrd; grown and natd as for *A. apennina* but more rarely natd; very scattered in En, Man; Balkans.

4. A. ranunculoides L. - *Yellow Anemone.* Differs from *A. nemorosa* in its fewer (sometimes 0) basal leaves with ± sessile or incomplete (not stalked) main leaf-segments, and 1(-3) flowers with usually 5 yellow sepals; (2n=32). Intrd; grown in gardens and sometimes persisting as a throwout or escape in shady places; very scattered in En and Sc; Europe.

5. A. x hybrida Paxton (*A. x japonica* auct.; *A. hupehensis* (Lemoine) Lemoine x *A. vitifolia* Buch.-Ham. ex DC.) - *Japanese Anemone.* Stems 40-150cm, erect, rather sparsely branched above with whorls of ± leaf-like bracts at each node and 1 flower terminating each branch; sepals up to c.30, the inner white to purple, densely silky-pubescent on lowerside; (2n=16). Intrd; much cultivated and persisting on old garden sites or as an escape or throwout; rare throughout BI; garden origin.

10. HEPATICA Mill. - *Liverleaf*

Herbaceous perennials; leaves all basal, distinctively 3-lobed, without stipules; flowers solitary, actinomorphic; perianth apparently of 2 whorls, but inner whorl of 6-10 coloured segments is calyx and outer whorl of 3 sepal-like segments is bracts; stamens numerous; carpels numerous, free, spirally arranged; fruit an achene, the style remaining shorter than fertile portion.

1. H. nobilis Schreb. - *Liverleaf.* Leaves on petioles to 15cm, cordate at base, with entire lobes; flowers on peduncles to 15cm arising from leaf-rosette; flowers 15-25mm across, blue, rarely white; (2n=14). Intrd; grown in shady places in wild gardens, natd in E Gloucs and SE Yorks; Europe.

11. PULSATILLA Mill. (*Anemone* subg. *Pulsatilla* (Mill.) Thomé) - *Pasqueflower*

Herbaceous perennials; leaves all basal, 2-pinnate, without stipules; flowers solitary, with a whorl of 3 almost leaf-like bracts just below, actinomorphic; perianth of 1 whorl of 6 petaloid sepals; stamens numerous, the outer sterile and nectariferous; carpels numerous, free, spirally arranged; fruit an achene, the style becoming greatly elongated and feathery.

1. P. vulgaris Mill. (*Anemone pulsatilla* L.) - *Pasqueflower.* Stems to 30cm, simple, **R**
with ± spreading hairs; sepals 6, 2-5cm, deep violet-purple, densely silky- **79**
pubescent on lowerside; style becoming 3-5cm in fruit; 2n=32. Native; dry calcareous grassland; very local in C & E En, from W Gloucs and S Wilts to Cambs and N Lincs, formerly further N.

12. CLEMATIS L. - *Traveller's-joys*

Woody climbers; leaves opposite, pinnate or ternate or rarely simple, without stipules, the petioles and rhachis twining round supports; flowers in axillary cymes, often with 2 small opposite bracteoles below, actinomorphic; perianth of 1 whorl of usually 4 petaloid sepals; stamens numerous; carpels numerous, free, spirally arranged; fruit an achene, the style becoming greatly elongated and feathery or not so.

1 Leaves ternate, the primary divisions serrate **4. C. montana**
1 Leaves pinnate, the primary divisions divided further or not 2
  2 Flowers yellow **3. C. tangutica**
  2 Flowers white to blue or purple 3
3 Flowers blue to purple; styles glabrous, not elongating in fruit **5. C. viticella**
3 Flowers white or cream; styles pubescent, elongating in fruit 4
  4 Leaves 1-pinnate; sepals pubescent on both surfaces **1. C. vitalba**

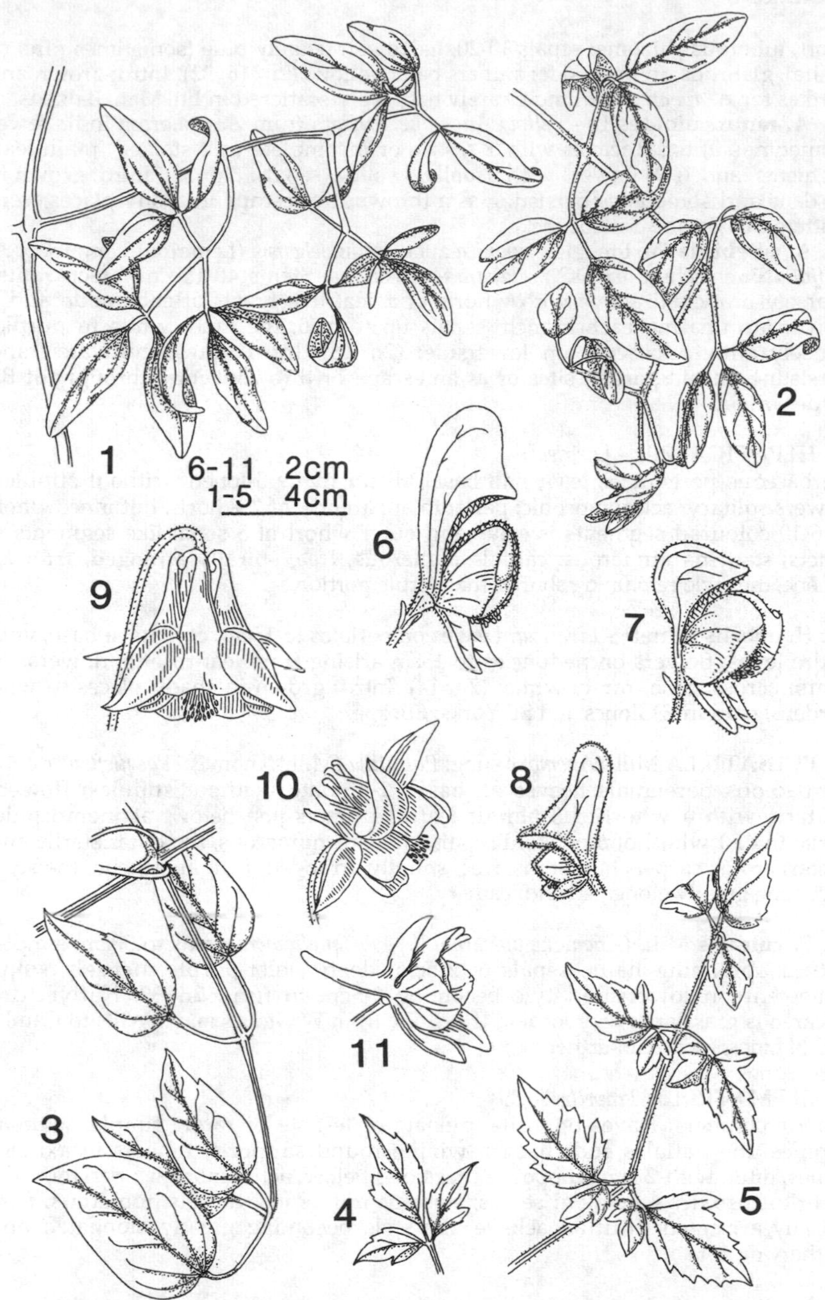

FIG 84 - Ranunculaceae. 1-5, leaves of *Clematis*. 1, *C. flammula*. 2, *C. viticella*. 3, *C. vitalba*. 4, *C. montana*. 5, *C. tangutica*. 6-8, flowers of *Aconitum*. 6, *A. x cammarum*. 7, *A. napellus*. 8, *A. lycoctonum*. 9-10, flowers of *Aquilegia*. 9, *A. vulgaris*. 10, *A. pyrenaica*. 11, flower of *Consolida ajacis*.

4 Leaves 2-pinnate; sepals pubescent only on lowerside **2. C. flammula**

Other spp. - **C. cirrhosa** L., from S. Europe, grows outside gardens on walls in Guernsey; it is an evergreen with simple, lobed or ternate leaves, nodding whitish flowers, and fused bracteoles.

1. C. vitalba L. - *Traveller's-joy*. Rampant deciduous climber to 30m; leaves 1- **79**
pinnate; leaflets rounded to cordate at base; flowers in dense clusters; sepals **84**
c.1cm, creamish- or greenish-white; style becoming long and feathery in fruit; 2n=16. Native; hedgerows, scrub and woodland on base-rich soils; locally abundant in En and Wa N to SW Yorks, also natd in scattered localities N to C Sc, Man and over most of Ir and CI.

2. C. flammula L. - *Virgin's-bower*. Sprawling deciduous climber to 6m; leaves 2- **84**
pinnate; leaflets rounded to cuneate at base; flowers in dense clusters; sepals c.1cm, white; style becoming long and feathery in fruit; (2n=16). Intrd; natd on cliffs and dunes near sea; E Kent since 1927, S Hants, Dorset, W Cornwall and Caerns; Mediterranean.

3. C. tangutica (Maxim.) Korsh. - *Orange-peel Clematis*. Deciduous climber to 4m; **84**
leaves pinnate with primary divisions mostly ternate; leaflets cuneate to cordate at base; flowers usually 1 in each leaf-axil; sepals 3-5cm, yellow; style becoming long and feathery in fruit; (2n=16). Intrd; natd in several localities and habitats in Br for short periods, but since 1990 perhaps only in Man; China.

4. C. montana Buch.-Ham. ex DC. - *Himalayan Clematis*. Vigorous deciduous **84**
climber to 6m; leaves ternate; leaflets cuneate to rounded at base; flowers in groups of 1-6; sepals 1.5-2.5cm, white to pink or bluish; style becoming long and feathery in fruit; (2n=16). Intrd; much grown in gardens and rarely escaping or persisting over hedges and walls; very scattered in Br N to Dunbarton, Man; Afganistan to Taiwan.

5. C. viticella L. - *Purple Clematis*. Deciduous climber to 4m; leaves pinnate with **84**
primary divisions ternate; leaflets cordate to broadly cuneate at base; flowers 1-3 in each leaf-axil; sepals 1.5-3cm, blue to purple; style glabrous, elongating little in fruit; (2n=16). Intrd; natd in hedges; very scattered, Dorset and Surrey to Salop; S Europe.

13. RANUNCULUS L. - *Buttercups*

Herbaceous perennials or annuals, some aquatic; leaves spirally arranged, simple to much divided, with or without stipules; flowers usually in cymes, sometimes solitary and leaf-opposed, without a whorl of bracts below, actinomorphic; perianth of 2 whorls; sepals 3 or 5, sepaloid; petals 5 or fewer by reduction (to 0) or 7-12, petaloid, usually with a small nectar-secreting pit (nectar-pit) on inner face; stamens usually numerous, sometimes 5-10; carpels numerous, free, spirally arranged; fruit an achene, the style remaining shorter than fertile portion.

General key

1 Sepals 3; petals 7-12; many roots modified as whitish, swollen tubers with rounded apices **18. R. ficaria**

1 Sepals 5; petals usually 5, sometimes <5 or many; root tubers rarely present, if so then with finely tapering apices 2

  2 Petals yellow ***Key A***

  2 Petals white, often yellow at base 3

3 Plant erect, terrestrial; largest leaves basal; achenes not transversely ridged, often *flore pleno* **17. R. aconitifolius**

3 Plant rarely erect, often aquatic; basal leaves 0; achenes transversely ridged, never *flore pleno* ***Key B***

Key A - Sepals 5; petals yellow, usually 5

1 Leaves entire or toothed at margin, unlobed 2
1 At least some leaves divided at least 1/4 way to base 7
2 Flowers 2-5cm across; achenes c.2.5mm **12. R. lingua**
2 Flowers <2(2.5)cm across; achenes 1-2(2.3)mm 3
3 Lower leaves cordate at base 4
3 All leaves cuneate at base 5
4 Stems erect; achenes tuberculate **16. R. ophioglossifolius**
4 Stems procumbent to decumbent; achenes smooth **13. R. flammula**
5 Stems erect to ± procumbent, usually rooting only at lower nodes; beak c.1/8-1/10 as long as rest of achene; widest petals usually >4mm across **13. R. flammula**
5 Stems procumbent, rooting at all or most nodes; beak c.1/3-1/6 as long as rest of achene; widest petals usually <4mm across 6
6 Beak c.1/5 -1/6 as long as rest of achene; basal leaves usually >1.5mm wide **14. R. x levenensis**
6 Beak c.1/3-1/4 as long as rest of achene; basal leaves usually <1.2mm wide **15. R. reptans**
7 Sepals strongly reflexed at anthesis 8
7 Sepals not reflexed at anthesis 13
8 Plant perennial with swollen stem-base **3. R. bulbosus**
8 Plant annual, without swollen stem-base 9
9 Achenes c.1mm, on elongated receptacle, ± smooth on sides **11. R. sceleratus**
9 Achenes >2mm, on ± spherical receptacle, usually with tubercles or spines on sides 10
10 Achenes 5-8mm including beak of 2-3mm; longest spines on sides of achene at least 1mm **6. R. muricatus**
10 Achenes ≤5mm including beak of <1.5mm; longest tubercles on sides of achene <1mm or tubercles 0 11
11 Flowers <8mm across; receptacle glabrous **7. R. parviflorus**
11 Flowers >9mm across; receptacle pubescent 12
12 Achenes with tubercles (if present) confined to edge of faces, close to border; beak 0.2-0.5(0.6)mm **4. R. sardous**
12 Achenes with tubercles (if present) covering faces; beak (0.5)0.6-1mm **5. R. marginatus**
13 Plant with procumbent stems rooting at nodes **2. R. repens**
13 Stems usually not procumbent, not rooting at nodes 14
14 Plant annual; receptacle pubescent; achenes with conspicuous spines **8. R. arvensis**
14 Plant perennial; receptacle glabrous; achenes glabrous or pubescent but without spines 15
15 Some roots swollen into spindle-shaped tubers; achenes borne in an elongated head **9. R. paludosus**
15 All roots thin; achenes borne in a ± spherical head 16
16 Basal leaves with a reniform outline, glabrous to sparsely appressed-pubescent; achenes pubescent **10. R. auricomus**
16 Basal leaves with a polygonal or polygonal-rounded outline, conspicuously pubescent; achenes glabrous **1. R. acris**

Key B - Sepals 5; petals white, 5; basal leaves 0

1 Laminar (floating or aerial) leaves the only ones present 2
1 Capillary (normally submerged) leaves present 6
2 Receptacle pubescent; leaves divided usually >1/2 way into 3(-5) main lobes 3
2 Receptacle glabrous; leaves divided <1/2 way into 3-5(7) main lobes 5

3 Petals >5.5mm, contiguous at anthesis; achenes narrowly winged and borne on ovoid receptacle when completely mature **23. R. baudotii**

3 Petals <6mm, not contiguous at anthesis; achenes not winged, borne on ± spherical receptacle 4

4 Leaves usually 3-lobed, rarely 5-lobed; petals ≤4.5mm, c.1.5x as long as sepals; pedicels strongly reflexed when fruit ripe **22. R. tripartitus**

4 Leaves often 5-lobed; petals ≤6mm, c.2x as long as sepals; pedicels remaining erect when fruit ripe **21. R. x novae-forestae**

5 Leaf-lobes broadest at their base; leaf-sinuses very open, obtuse to subacute; petals <4.5mm, little longer than sepals **19. R. hederaceus**

5 Leaf-lobes broadest above their base; leaf-sinuses narrowly acute; petals >4.5mm, 2-3x as long as sepals **20. R. omiophyllus**

6 Plant with both laminar and capillary leaves 7

6 Plant with only capillary leaves 13

7 Petals ≤6mm, not contiguous at anthesis 8

7 Petals 5-20mm, contiguous at anthesis 9

8 Laminar leaves usually 3-lobed, rarely 5-lobed; petals ≤4.5mm, c.1.5x as long as sepals; pedicels strongly reflexed when fruit ripe **22. R. tripartitus**

8 Laminar leaves often 5-lobed; petals ≤6mm, c.2x as long as sepals; pedicels remaining erect when fruit ripe **21. R. x novae-forestae**

9 Achenes glabrous when immature, narrowly winged and borne on ovoid receptacle when completely mature **23. R. baudotii**

9 Achenes usually pubescent when immature, not winged at maturity, borne on ± spherical receptacle 10

10 Pedicel in fruit rarely >50mm, shorter than petiole of opposed laminar leaf; petals <10mm, with circular nectar-pit **25. R. aquatilis**

10 Pedicel in fruit usually >50mm, usually longer than petiole of opposed laminar leaf; petals usually >10mm, with pear-shaped nectar-pit 11

11 Leaves intermediate between laminar and capillary usually common, mostly as capillary leaves with some slightly flattened segments; mature achenes not formed **27. R. x kelchoensis**

11 Leaves intermediate between laminar and capillary rarely present; fertile achenes regularly produced 12

12 Capillary leaves rigid or flaccid, shorter than adjacent stem internode, with markedly divergent segments **26. R. peltatus**

12 Capillary leaves flaccid, usually longer than adjacent stem internode, with ± parallel segments **28. R. penicillatus**

13 Leaves rigid, circular in outline, the segments all lying in 1 plane **31. R. circinatus**

13 Leaves with segments not lying in 1 plane 14

14 Achenes narrowly winged and borne on ovoid receptacle when completely mature **23. R. baudotii**

14 Achenes not winged at maturity, borne on ± spherical receptacle 15

15 Petals usually <6mm, with lunate nectar-pit, not contiguous at anthesis **24. R. trichophyllus**

15 Petals >5mm, with circular or pear-shaped (rarely lunate) nectar-pit, contiguous at anthesis 16

16 Receptacle pilose to glabrous; mature leaves longer than adjacent internode, rarely >4x forked, with ± parallel segments **30. R. fluitans**

16 Receptacle densely pubescent; mature leaves shorter to longer than adjacent internode, usually some >4x forked, with divergent to ± parallel segments 17

17 Pedicel in fruit rarely >50mm; petals <10mm, with circular nectar-pit **25. R. aquatilis**

17 Pedicel in fruit usually >50mm; petals usually >10mm, with pear-shaped

(rarely lunate) nectar-pit 18
18 Mature achenes not formed; pedicels not elongating **29. R. x bachii**
18 Fertile achenes regularly produced from each flower; pedicels elongating in fruit 19
19 Ultimate segments of well-developed leaves <100 **26. R. peltatus**
19 Ultimate segments of well-developed leaves >100, often >200 **28. R. penicillatus**

Subgenus* 1 - *RANUNCULUS (spp. 1-17). Petals yellow except in *R. aconitifolius* (white), normally 5; sepals 5; achenes not transversely ridged; roots rarely modified as tubers and then not with rounded apices.

1. R. acris L. - *Meadow Buttercup*. Erect perennial to 1m; basal leaves deeply palmately lobed, pubescent; flowers 15-25mm across; sepals not reflexed; achenes 2-3.5mm, glabrous, smooth, with short hooked beak; 2n=14. Native; grassland, especially damp and calcareous. Very variable; 3 vars may be usefully recognised. Var. **villosus** (Drabble) S.M. Coles is common in undisturbed areas of N Sc (including the islands) and W & C Ir; it differs from var. **acris** in having little-branched and few-flowered stems, leaves with relatively broad lobes, and early-formed basal leaves with hairs mostly >1.2mm (not mostly <1.2mm), but intermediates occur in N Sc. Var. **pumilus** Wahlenb. (ssp. *pumilus* (Wahlenb.) Á. & D. Löve, ssp. *borealis* auct.) has similar stems but early-formed basal leaves are glabrous and the leaves are relatively shallowly divided; it occurs in Cairngorms, C Sc. **89**

2. R. repens L. - *Creeping Buttercup*. Perennial with strong creeping stems rooting at nodes and ± erect flowering stems to 60cm; basal leaves triangular-ovate in outline, with 3 main segments, the middle one long-stalked and borne above the 2 laterals, usually pubescent; flowers 20-30mm across; sepals not reflexed; achenes 2.5-3.8mm, glabrous, smooth, with short curved beak; 2n=32. Native; wet grassland, woods, streamsides, marshes and duneslacks, and as a weed; abundant throughout BI. **89**

3. R. bulbosus L. - *Bulbous Buttercup*. Erect perennial to 40cm with swollen corm-like stem-base; basal leaves ovate in outline, with 3 main lobes, the middle one sessile or stalked, pubescent; flowers 15-30mm across; sepals strongly reflexed at anthesis; achenes 2-4mm, glabrous, finely pitted, with short hooked beak; 2n=16. Native; dry grassland and fixed dunes; common in most of BI but absent from parts of Sc and Ir. Ssp. **bulbifer** (Jord.) P. Fourn. is continuously connected to ssp. **bulbosus** and probably not worthy of recognition. A variant described from maritime dunes in CI and W BI as var. **dunensis** Druce has large flowers (25-30mm across), rather short stout stems, and stems and upper leaves with dense, long, white, patent hairs, but similar plants predominate all over Ir and parts of BI and do not merit ssp. status. **89**

4. R. sardous Crantz - *Hairy Buttercup*. Erect annual to 40cm; leaves similar to those of *R. bulbosus*; flowers 12-25mm across, paler yellow than in *R. bulbosus*; sepals strongly reflexed at anthesis; achenes 2-3(4)mm, glabrous, smooth apart from few tubercles just inside border, with curved beak ≤0.6mm; (2n=16). Probably native; grassland and cultivated land; frequent near S & E coasts of En and in CI, very local (formerly commoner) elsewhere in Br N to C Sc. **89**

5. R. marginatus d'Urv. - *St Martin's Buttercup*. Erect annual to 40cm; lower leaves with 3 broad incomplete lobes, sparsely pubescent; flowers 12-25mm across; sepals strongly reflexed at anthesis; achenes 3.5-5mm, glabrous, densely tuberculate on sides or without tubercles, with curved beak 0.5-1mm; (2n=32). Intrd; ± natd as weed of cultivated ground in extreme SW En, especially Scillies, rare casual elsewhere; E Mediterranean. Scillonian and some casual examples are referable to var. **trachycarpus** (Fisch. & C.A. Mey.) Azn.; var. **marginatus** has smooth achenes and has been found since 1991 in scattered sites in En as a **89**

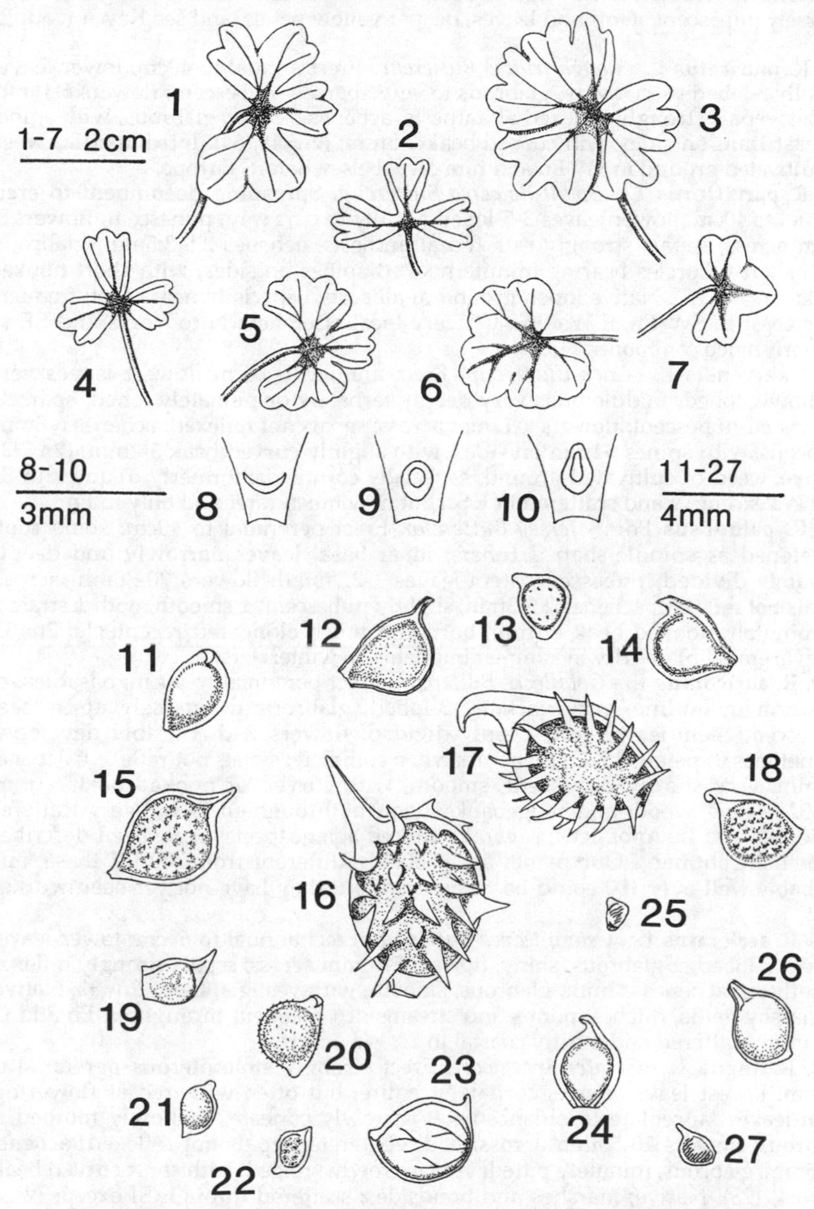

FIG 89 - ***Ranunculus***. 1-7, leaves. 1, ***R. peltatus***. 2, ***R. baudotii***. 3, ***R. aquatilis***. 4, ***R. tripartitus***. 5, ***R. penicillatus***. 6, ***R. omiophyllus***. 7, ***R. hederaceus***. 8-10, nectar-pits at base of petals. 8, lunate. 9, circular. 10, pear-shaped. 11-27, achenes. 11, ***R. acris***. 12, ***R. repens***. 13, ***R. sardous***. 14, ***R. bulbosus***. 15, ***R. marginatus***. 16, ***R. arvensis***. 17, ***R. muricatus***. 18, ***R. parviflorus***. 19, ***R. paludosus***. 20, ***R. auricomus***. 21, ***R. flammula***. 22, ***R. ophioglossifolius***. 23, ***R. aconitifolius***. 24, ***R. lingua***. 25, ***R. sceleratus***. 26, ***R. ficaria***. 27, ***R. aquatilis***.

constituent of sown 'wild' flower seed. Differs from *R. sardous* in much more sparsely pubescent stems and leaves, deeper yellow petals, and see Key A (couplet 12).

6. R. muricatus L. - *Rough-fruited Buttercup*. Erect annual to 40cm; lower leaves **89**
usually 3-lobed to <1/2 way, glabrous to very sparsely pubescent; flowers 6-16mm across; sepals strongly reflexed at anthesis; achenes 5-8mm, glabrous, with spines at least 1mm on sides, wih curved beak 2-3mm; (2n=48, 64). Intrd; natd as weed of cultivated ground in SW En and rare casual elsewhere; S Europe.

7. R. parviflorus L. - *Small-flowered Buttercup*. Spreading decumbent to erect **89**
annual to 40cm;; lower leaves 3-5-lobed usually to c.1/2 way, pubescent; flowers 3-6mm across; sepals strongly reflexed at anthesis; achenes 2.5-3.5mm, glabrous, with short tubercles bearing minute hooked spines on sides, with short hooked beak; (2n=14, 28). Native; open ground of all sorts, especially near coast; frequent near coast in SW En, S Wa and CI, very local elsewhere N to Yorks and SE Ir, formerly much commoner.

8. R. arvensis L. - *Corn Buttercup*. Erect annual to 60cm; lowest leaves very **89**
shallowly lobed, middle ones very deeply ternately or pinnately lobed, sparsely appressed-pubescent; flowers 4-12mm across; sepals not reflexed; achenes 6-8mm, glabrous, with spines >1mm on sides, with slightly curved beak 3-4mm; (2n=32). Native; weed of cultivated ground, especially cornfields; formerly frequent in En and Wa except W and scattered in E Sc, but now much rarer and only in En.

9. R. paludosus Poir. - *Jersey Buttercup*. Erect perennial to 40cm, some roots **RR**
developed as spindle-shaped tubers; inner basal leaves narrowly and deeply **89**
ternately divided, pubescent; stem-leaves 1-2, small; flowers 20-30mm across; sepals not reflexed; achenes 2.5-3mm, slightly pubescent, ± smooth, with ± straight or minutely hooked beak c.1mm, borne on much elongated receptacle; 2n=32. Native; grassy places dry in summer but damp in winter; Jersey.

10. R. auricomus L. - *Goldilocks Buttercup*. Erect perennial to 40cm; basal leaves **89**
reniform in outline, very variably 3-lobed, glabrous to sparsely appressed-pubescent; stem-leaves few, deeply divided; flowers with variably developed (sometimes 0) petals, 15-25mm across when complete; sepals not reflexed; achenes 3-4mm, very shortly pubescent, smooth, with curved or hooked beak >1mm; 2n=32. Native; woods and hedgebanks; frequent throughout En but very scattered in Sc, Wa and Ir. Apomictic; several hundred agamospecies have been described from the Continent. Our plants are probably different from any of these, and probably well over 100 could be recognized, but they have not yet been worked out.

11. R. sceleratus L. - *Celery-leaved Buttercup*. Erect annual to 60cm; lower leaves **89**
deeply 3-lobed, ± glabrous, shiny; flowers 5-10mm across; sepals strongly reflexed at anthesis; achenes c.1mm, glabrous, smooth, with vestigial beak; 2n=32. Native; in marshy fields, ditches, ponds and streamsides; frequent throughout En and CI but more scattered and mostly coastal in Sc, Wa and Ir.

12. R. lingua L. - *Greater Spearwort*. Erect strongly stoloniferous perennial to **89**
120cm; lowest leaves ovate, cordate, ± entire, but often withered at flowering; stem-leaves lanceolate to oblanceolate, narrowly cuneate, shallowly toothed, ± glabrous; flowers 20-50mm across; pedicels terete; sepals not reflexed; achenes c.2.5mm, glabrous, minutely pitted, very narrowly winged, with short curved beak; (2n=64, 128). Native; marshes and pondsides; scattered through BI except N Sc; commonly grown and often ± natd.

13. R. flammula L. - *Lesser Spearwort*. Erect to procumbent perennial to 50cm; **89**
stem-leaves lanceolate to oblanceolate or linear, narrowly cuneate, ± glabrous; flowers 7-20(25)mm across; pedicels furrowed; sepals not reflexed, achenes 1-2(2.3)mm, glabrous, minutely pitted, not winged, plus beak c.1/8-1/10 as long. Native; all kinds of wet places; throughout BI.

1 Lowest leaves with lamina cordate at base **b. ssp. minimus**
1 Lowest leaves with lamina cuneate at base or 0 2

2 Lowest leaves persistent, with well developed lamina **a. ssp. flammula**
2 Lowest leaves falling early, with lamina 0 or much reduced **c. ssp. scoticus**

a. Ssp. flammula. Erect to procumbent, often rooting at lower nodes; lowest leaves oblong-lanceolate, cuneate, ± entire; achenes 1.1-1.6x as long as wide; 2n=32. Range of sp.

b. Ssp. minimus (A. Benn.) Padmore. Decumbent to procumbent, rooting only at RR
lower nodes; lowest leaves broadly ovate, cordate, ± entire; flowers >15mm across; achenes 1-1.4x as long as wide. N Sc (mainland and Outer Isles) and Clare.

c. Ssp. scoticus (E.S. Marshall) A.R. Clapham. Erect; lowest leaves consisting of RR
long petiole with reduced or no lamina, falling early; flowers usually 1, mostly 10-15mm across; achenes as in ssp. *flammula*. NW Sc and NW Ir.

14. R. x levenensis Druce ex Gornall (*R. flammula* x *R. reptans*) - *Loch Leven* RR
Spearwort. Differs from *R. flammula* ssp. *flammula* in its procumbent stems rooting at most nodes; flowers c.5-12mm across, solitary on upturned flowering stem; achenes 1-1.5(1.9)mm plus beak c.1/5-1/6 as long; and see key; 2n=32. Native; barish lake shores on pebbly or silty substrata; local in Br from Lake District to N Aberdeen, mostly in absence of *R. reptans* but usually with *R. flammula*, with which it backcrosses.

15. R. reptans L. - *Creeping Spearwort*. Differs from *R. flammula* and *R. x levenensis* RR
in its procumbent stems rooting at ± all nodes; solitary flowers c.5(-10)mm across; achenes 1-1.5mm plus beak c.1/3-1/4 as long; and see key; (2n=32). Native; same habitats and localities as *R.* x *levenensis* and giving rise to it, now extinct except for occasional non-persistent reintroductions (probably by geese).

16. R. ophioglossifolius Vill. - *Adder's-tongue Spearwort*. Erect annual to 40cm; RRR
lowest leaves ovate, cordate, glabrous, ± entire; stem-leaves broadly to narrowly 89
elliptic, shallowly and distantly toothed; flowers 5-9mm across; sepals not reflexed; achenes c.1.5mm, glabrous, with small tubercles on sides, with very short beak; 2n=16. Native; marshy ground; Gloucs, formerly Dorset, S Hants and Jersey.

17. R. aconitifolius L. - *Aconite-leaved Buttercup*. Erect perennial to 60cm; lowest 89
leaves deeply palmately 3-7-lobed; flowers 10-25mm across; sepals not reflexed; petals white; achenes 2-4mm, glabrous, smooth apart from raised veins, with short hooked beak; (2n=14, 16). Intrd; grown in gardens, often as *flore pleno*, and natd in a few damp places and by streams northwards from Yorks; Europe.

Subgenus 2 - FICARIA (Schaeff.) L.D. Benson (sp. 18). Petals yellow, 7-12; sepals 3; achenes not transversely ridged; many roots modified as tubers with rounded apices.

18. R. ficaria L. - *Lesser Celandine*. Glabrous ascending perennial to 25cm; basal leaves ovate, cordate, with long petiole; flowers 10-30(50)mm across; sepals not reflexed; achenes c.2-5mm, shortly pubescent, smooth, with very short beak. In damp meadows, lawns, woods and hedgebanks, and beside streams; common throughout BI. Many cultivated ornamentals have very large, orange or pale (even white) flowers, or may be *flore pleno*, and may escape.

1 Tubers absent from leaf-axils after flowering period 2
1 Cream-coloured tubers present in leaf-axils after flowering period 3
2 Petals 10-20 x 4-9mm **a. ssp. ficaria**
2 Petals 18-25 x 9-15mm **d. ssp. chrysocephalus**
3 Well-developed achenes 0-6 per head; petals mostly <15mm; pollen-grains mostly empty **b. ssp. bulbilifer**
3 All or ± all achenes well developed; petals mostly >15mm; pollen-grains mostly full **c. ssp. ficariiformis**

a. Ssp. ficaria (var. *fertilis* A.R. Clapham nom. nud.). Tubers not formed in leaf- 79
axils; full head of ripe achenes usually produced; achenes mostly 2.5-3.5 x 1.7- 89
2.2mm; petals mostly 10-20 x 4-9mm; 2n=16 (diploid); mean no. chloroplasts per

stomatal guard-cell c.13-17. Native; throughout BI.

b. Ssp. bulbilifer Lambinon (ssp. *bulbifer* Lawalrée nom. illeg., var. *bulbifera* 79
Marsden-Jones nom. illeg.). Tubers formed in leaf-axils after anthesis; few (rarely >6) ripe achenes produced in each head; petals mostly 6-11 x 2-5mm; 2n=32 (tetraploid); mean no. chloroplasts per stomatal guard-cell c.24-28. Native; almost throughout BI, but apparently absent from Shetland and CI.

c. Ssp. ficariiformis (F.W. Schultz) Rouy & Foucaud. Tubers formed in leaf-axils after anthesis; full head of ripe achenes usually produced; achenes mostly 4-5 x 2.5-3.5mm; petals mostly 17-26 x 4-12mm; 2n=32 (tetraploid). Intrd; garden escape fairly frequent in CI, SW En and SW Wa, very scattered elsewhere in S En. Possibly native in extreme SW En and SW Wa and in CI.

d. Ssp. chrysocephalus P.D. Sell. Tubers not formed in leaf-axils; full head of ripe achenes usually produced; achenes mostly 3-4 x 2-2.5mm; petals mostly 18-25 x 9-15mm; 2n=32 (tetraploid). Intrd; garden escape so far found natd only in Herts but probably overlooked; E Mediterranean.

Other differences between the sspp. are less reliable. Ssp. *ficaria* tends to have larger flowers with wider petals and larger, entire (not shallowly lobed) leaves than ssp. *bulbilifer*. Both sspp. occur commonly almost throughout BI, but locally 1 may be absent or much less common than the other. Habitats of the 2 do not differ overall, but locally their occurrence is often mutually exclusive. Ssp. *bulbilifer* is less common in Ir and W Br, but in E Br is less tolerant of open conditions. Triploids (2n=24), recognized by extremely small flowers, total sterility, usually 0 (rarely few) axillary tubers, and c.20-22 chloroplasts per stomatal guard-cell, occur widely, usually close to diploids or tetraploids or both, and are probably hybrids. Ssp. *ficariiformis* is usually recognizable by its large size, but poorly grown or late-developing plants fall well within the size range of ssp. *ficaria*. The same is probably true of ssp. *chrysocephalus*.

Subgenus 3 - BATRACHIUM (DC.) A. Gray (spp. 19-31). Petals white, normally 5; sepals 5; achenes transversely ridged; roots not modified as tubers; stems with broadly lobed (*laminar*) normally floating leaves, very finely divided (*capillary*) normally submerged leaves, or both (*heterophyllous*); transitional leaves rare except in hybrids.

19. R. hederaceus L. - *Ivy-leaved Crowfoot*. Procumbent annual or perennial; 89
leaves all laminar, divided <1/2 way into 3-5(7) lobes widest at base; petals (1.2)2.5-3.5(4.3)mm, not contiguous at anthesis, with lunate nectar-pit; sepals not reflexed; receptacle glabrous; immature achenes glabrous; 2n=16. Native; on mud and in shallow water; frequent throughout BI.

20. R. omiophyllus Ten. (*R. lenormandii* F.W. Schultz) - *Round-leaved Crowfoot*. 89
Procumbent annual or perennial; leaves all laminar, divided <1/2 way into 3-5(7) lobes widest above base; petals (4)5-6(7)mm, with lunate nectar-pit; sepals reflexed; receptacle glabrous; immature achenes glabrous; 2n=32. Native; locally frequent on wet mud and in shallow ponds and streams; Wa, W & S En, SW Sc, S Ir and Sark (CI).

20 x 26. R. omiophyllus x R. peltatus = R. x hiltonii H. & J. Groves was known in E Sussex between 1896 and 1926 in a stream with both parents. It was ± intermediate, with the lower leaves finely divided but only subcapillary, but fertile with large pollen grains. It might have been an amphidiploid. Recently a sterile hybrid has been found in S Hants and there is an old record for E Cornwall. Endemic.

21. R. x novae-forestae S.D. Webster (*R. lutarius* auct. non (Revel) Bouvet; RR
R.omiophyllus x *R. tripartitus*) - *New Forest Crowfoot*. Like *R. tripartitus* but laminar leaves often 5-lobed with shallower sinuses; petals ≤6mm; pedicels often erect when fruit ripe; achenes ≤c.60% fertile. In and by pools in New Forest, S Hants, formerly also Glam.

22. R. tripartitus DC. - *Three-lobed Crowfoot*. Procumbent or subaquatic annual or R perennial; heterophyllous or leaves all laminar; laminar leaves divided >1/2 way 89 into 3(-5) lobes widest well above base; petals 1.25-4.5mm, not contiguous at anthesis, with lunate nectar-pit; sepals reflexed, blue-tipped; receptacle pubescent; immature achenes glabrous; 2n=48. Native; wet mud, ditches and ponds; decreasing in S & W En and Wa from E Sussex to Pembs (formerly to Cheshire), formerly W Cork.

22 x 25. R. tripartitus x R. aquatilis is found in W Cornwall usually near both parents. It is sterile, weak and non-persistent, with many transitional leaves, and ± intermediate. Endemic.

23. R. baudotii Godr. (*R. peltatus* ssp. *baudotii* (Godr.) Meikle ex C.D.K. Cook) - 89 *Brackish Water-crowfoot*. Procumbent or subaquatic annual or perennial; heterophyllous, or with capillary leaves only, or rarely with laminar leaves only; laminar leaves divided >1/2 way into 3(-5) lobes; capillary leaves with rigid divergent segments; petals 5.5-10mm, contiguous at anthesis, with lunate nectar-pit; sepals reflexed, usually blue-tipped (sometimes so in other spp.); receptacle pubescent, becoming ovoid in fruit; immature achenes glabrous, becoming very narrowly winged at full maturity; 2n=32. Native; ditches and ponds near sea, often brackish; scattered round coasts of BI.

23 x 24. R. baudotii x R. trichophyllus = R. x segretii A. Félix is found in coastal regions near the parents N to Flints and S Lincs. It is sterile and intermediate, but closer to *R. trichophyllus*, and develops transitional leaves.

23 x 25. R. baudotii x R. aquatilis = R. x lambertii A. Félix is found with both parents in W Cornwall. It is sterile and intermediate, with transitional leaves.

23 x 26. R. baudotii x R. peltatus has probably been found recently in S Devon. It resembles *R. baudotii* more closely but is sterile and has elongated nectar-pits; 2n=32. Endemic.

24. R. trichophyllus Chaix - *Thread-leaved Water-crowfoot*. Tufted or subaquatic annual or perennial; leaves all capillary with flaccid or rigid divergent segments; petals 3.5-6(6.5)mm, not contiguous at anthesis, with lunate nectar-pit; sepals not reflexed; receptacle pubescent; immature achenes pubescent; 2n=32. Native; ponds, ditches, canals and slow rivers; scattered but often common throughout BI.

24 x 25. R. trichophyllus x R. aquatilis = R. x lutzii A. Félix is scattered through Br N to Derbys and Denbs with both parents. It is intermediate and sterile, with transitional leaves.

24 x 26. R. trichophyllus x R. peltatus is known from Warks and perhaps elsewhere in En with both parents. It is intermediate and sterile.

24 x 31. R. trichophyllus x R. circinatus (= *R. x glueckii* A. Félix nom. nud.) occurs with the parents in W Suffolk. It is sterile and ± intermediate.

25. R. aquatilis L. - *Common Water-crowfoot*. Tufted or subaquatic annual or 89 perennial; heterophyllous or with capillary leaves only; laminar leaves divided slightly >1/2 way into (3-)5(-7) lobes, with acute basal sinus; capillary leaves with flaccid or rigid divergent segments; petals 5-10mm, contiguous at anthesis, with circular nectar-pit; sepals not reflexed; receptacle pubescent; immature achenes pubescent; 2n=48. Native; ponds, ditches, canals and slow rivers; frequent throughout most of BI, commonest sp. of subgenus.

25 x 26. R. aquatilis x R. peltatus = R. x virzionensis A. Félix has been found in Warks and there are other unconfirmed records from En.

26. R. peltatus Schrank - *Pond Water-crowfoot*. Tufted or subaquatic annual or 89 perennial; heterophyllous or with capillary leaves only; laminar leaves divided slightly >1/2 way into (3-)5(-7) lobes, with obtuse basal sinus; capillary leaves with flaccid or rigid divergent segments; petals (9)12-15(20)mm, contiguous at anthesis, with pear-shaped nectar-pit; sepals not reflexed; receptacle pubescent; immature achenes pubescent; 2n=48. Native; ponds, ditches, canals and slow rivers; frequent throughout most of BI.

27. R. x kelchoensis S.D. Webster (*R. peltatus* x *R. fluitans*) - *Kelso Water-crowfoot*. R

Intermediate between the parents but receptacle pubescent; laminar leaves like those of *R. peltatus*; transitional leaves usually produced; very robust; sterile; 2n=40. Native; slow rivers; scattered in Br N to Berwicks, especially in EC En and SE Sc, Co Antrim, often in absence of both parents. Might have given rise to *R. penicillatus* ssp. *penicillatus*.

28. R. penicillatus (Dumort.) Bab. - *Stream Water-crowfoot.* Subaquatic perennial, heterophyllous or with flaccid capillary leaves only; petals (5)10-15(20)mm, contiguous at anthesis, with pear-shaped nectar-pit; sepals not reflexed; receptacle pubescent; immature achenes pubescent. Native; rivers, usually swift-flowing.

a. Ssp. penicillatus. Heterophyllous; laminar leaves like those of *R. peltatus*; **89**
capillary leaves longer than adjacent stem internode, with flaccid ± parallel segments; 2n=48. Ir, Wa and W En.

b. Ssp. pseudofluitans (Syme) S.D. Webster (var. *calcareus* (Butcher) C.D.K. Cook, var. *vertumnus* C.D.K. Cook, *R. pseudofluitans* (Syme) Newbould ex Baker & Foggitt, *R. peltatus* ssp. *pseudofluitans* (Syme) C.D.K. Cook nom. inval.). Leaves all capillary, shorter to longer than adjacent stem internode, usually c.6-8x forked with flaccid or rigid, ± parallel or divergent segments; 2n=32, 48. En, Wa, N Ir and S Sc. Vegetatively variable. *R. penicillatus* is often very difficult to distinguish from *R. x bachii*, *R. peltatus* x *R. fluitans*, *R. peltatus* or *R. fluitans*.

29. R. x bachii Wirtg. (*R. fluitans* x *R. trichophyllus* & R. *fluitans* x R. *aquatilis; R. x bachii* probably strictly refers to former) - *Wirtgen's Water-crowfoot.* Very robust perennials closely resembling *R. penicillatus* ssp. *pseudofluitans* (i.e. with capillary leaves only), but highly sterile. The 2 combinations are indistinguishable morphologically, but former has 2n=24, latter 2n=40. Native; rivers, usually swift-flowing; very scattered in En, Wa and S Sc, often replacing *R. fluitans*. Might have given rise to *R. penicillatus* ssp. *pseudofluitans*.

30. R. fluitans Lam. - *River Water-crowfoot.* Subaquatic perennial; leaves all capillary, rarely <8cm, rarely >4x forked, usually longer than adjacent stem internode, with flaccid ± parallel segments; petals 7-13mm, contiguous at anthesis, with pear-shaped nectar-pit; sepals not reflexed; receptacle pilose to glabrous; immature achenes sparsely pubescent to glabrous; 2n=16, 32. Native; mostly in larger rivers of moderate flow-rate; scattered in En, especially in N, Wa and S Sc, Co Antrim; decreasing.

30 x 31. R. fluitans x R. circinatus occurs in River Blackadder, Berwicks, with *R. circinatus*. It is ± intermediate but has a pubescent receptacle.

31. R. circinatus Sibth. - *Fan-leaved Water-crowfoot.* Subaquatic annual to perennial; leaves all capillary, with short, rigid, divergent segments lying in 1 plane; petals 4-10(12)mm, barely contiguous at anthesis, with lunate nectar-pit; sepals not reflexed; receptacle pubescent; immature achenes pubescent; 2n=16. Native; in ponds, canals and slow-flowing rivers, usually base-rich; sparsely scattered in En, Wa, Ir and S Sc.

14. ADONIS L. - *Pheasant's-eye*

Annuals; leaves spirally arranged, 3-pinnate, without stipules; flowers ± solitary, without a whorl of bracts below, actinomorphic; perianth of 2 whorls; sepals 5, sepaloid; petals 5-8, petaloid, not nectariferous; stamens numerous; carpels numerous, free, spirally arranged; fruit an achene, the style remaining shorter than fertile portion.

1. A. annua L. - *Pheasant's-eye.* Stems simple to branched, to 40cm; leaves much **RR**
divided, with narrow segments; flowers 15-25mm across; petals bright scarlet with **79**
dark basal spot; receptacle elongating as fruits ripen; (2n=32). Intrd; weed of cultivated and waste ground, formerly locally natd in cornfields in S En but now almost extinct; a rare casual in S En, Wa and CI, formerly S Ir; S Europe.

15. MYOSURUS L. - *Mousetail*
Annuals; leaves all basal, simple, without stipules; flowers solitary, without a whorl of bracts below, actinomorphic; perianth of 2 whorls; sepals 5(-7), ± sepaloid, each with a small basal spur; petals 5(-7), in the form of tubular nectaries; stamens 5-10; carpels numerous, free, spirally arranged; fruit an achene, the style remaining shorter than fertile portion.

1. M. minimus L. - *Mousetail.* Leaves 1-8cm, linear, in a basal rosette; flowers 79
solitary on bare scapes to 10cm; petals 3-4mm, greenish, inconspicuous; receptacle becoming much elongated, finally c.2-7cm; (2n=16). Probably native; damp arable ground; En N to Durham, rare and declining, formerly Guernsey.

16. AQUILEGIA L. - *Columbines*
Herbaceous perennials; leaves spirally arranged, 2-ternate, without stipules; flowers in cymes, without a whorl of bracts below, actinomorphic; perianth of 2 whorls; sepals 5, petaloid; petals 5, petaloid, each with a long, backwards-directed nectariferous spur; stamens numerous, the inner c.10 flat and sterile; carpels 5(-10), free; fruit a follicle.

Other spp. - Some garden escapes passing as *A. vulgaris* might actually represent related spp. or hybrids; study of these is needed. 1 such is **A. olympica** Boiss., from SW Asia, which has slightly larger flowers with blue sepals and petal-spurs and whitish petal-blades, and follicles 20-30mm; it was natd in 1980s in W Kent.

1. A. vulgaris L. - *Columbine.* Stems to 1m, branched, usually with several well 79
developed leaves and flowers; leaves pubescent on lowerside; flowers usually blue, 84
sometimes white or pink to purple; sepals 15-30mm; petal-spur 15-22mm, strongly hooked at end; follicles 15-20mm; 2n=14. Native; woods, fens and damp calcareous grassland and scree; local in BI N to C Sc; also much grown and a frequent escape, becoming natd in some areas; native populations are usually all violet-blue-flowered.

2. A. pyrenaica DC. - *Pyrenean Columbine.* Stems to 25cm, often simple, usually 84
with 0-1 well developed leaves and 1-3 flowers; leaves glabrous to sparsely pubescent on lowerside; flowers blue; sepals 20-35mm; petal-spur 10-16mm, gently curved; follicles 12-17mm; (2n=14). Intrd; planted on rock-ledges at c.900m, Caenlochan Glen, Angus, known since 1895; Pyrenees.

17. THALICTRUM L. - *Meadow-rues*
Herbaceous perennials; leaves spirally arranged, pinnate to ternate, stipulate; flowers in racemes of compound inflorescences without a whorl of bracts below, actinomorphic; perianth of 1 whorl of 4 small but ± petaloid sepals; stamens numerous, more conspicuous than sepals; carpels 2-15, free; fruit an achene, the style remaining shorter than fertile portion.

1 Inflorescence a simple raceme; plant rarely >15cm **5. T. alpinum**
1 Inflorescence branched with >1 flower per branch; plant usually >15cm 2
2 Filaments thickened, wider than anthers, white to lilac or pink **1. T. aquilegiifolium**
2 Filaments thin, narrower than anthers, yellowish 3
3 Sepals pink to lilac, c. as long as stamens **2. T. delavayi**
3 Sepals yellow, much shorter than stamens 4
4 Inflorescence diffuse; stamens ± pendent; leaflets not or little longer than wide; achenes with 8-10 ribs **4. T. minus**
4 Inflorescence ± dense; stamens held stiffly erect to patent; leaflets much longer than wide; achenes with 6 ribs **3. T. flavum**

Other spp. - 1 plant of **T. lucidum** L., from E Europe, was reported natd on a river-bank in Berwicks; it differs from *T. flavum* in its achenes with 8-10 ribs and lanceolate to linear (not obovate to oblong) leaflets of upper leaves.

1. T. aquilegiifolium L. - *French Meadow-rue.* Scarcely rhizomatous; stems to 1m, erect, usually simple; leaves 2-4-ternate; infloresence compound; flowers in dense clusters, whitish to lilac or pink, with erect to patent stamens with filaments coloured and wider than anthers; (2n=14). Intrd; grown in gardens and ± natd in grassy places as a throwout in a few places in En, Man and Sc; Europe.

2. T. delavayi Franch. - *Chinese Meadow-rue.* Scarcely rhizomatous; stems to 1m, erect, branched; leaves 3-5-pinnate; inflorescence compound; flowers in ± dense clusters; sepals 6-15mm, pink to lilac; filaments yellowish, narrow; (2n=28, 42). Intrd; natd in grassy area in Cambs; China.

3. T. flavum L. - *Common Meadow-rue.* Strongly rhizomatous; stems to 1.2m, erect, simple or little-branched; leaves 2-3-pinnate; inflorescence compound; flowers in dense clusters, bright yellow, with erect to patent stamens with narrow yellowish filaments; 2n=84. Native; fens, streamsides and wet meadows; scattered and declining through Br N to S Sc, mostly in E En, very scattered in Ir.

4. T. minus L. - *Lesser Meadow-rue.* Scarcely to moderately rhizomatous; stems to 79
1.2m, erect or spreading, often zigzag, simple or branched; leaves 3-4-ternate to -pinnate; inflorescence compound; flowers in diffuse panicles, pale yellow, with ± pendent stamens with narrow yellowish filaments; 2n=42. Native; in varied, usually calcareous habitats such as dunes, limestone cliffs and pavement, grassy banks and hedgerows, scrubland, and lakesides; scattered in Br and Ir, locally common but absent from large areas including C & SE En; grown in gardens and a frequent persister or throwout outside native range. A very variable and little understood sp.; up to 8 spp. or sspp. have been segregated in Br, based mainly on characters of fruit, habit and indumentum, but until properly investigated they are not worth recognizing. The most distinctive are plants with glaucous leaves with dense stalked glands on lowerside that occur on coastal dunes in W & N Br; they have been called ssp. **arenarium** (Butcher) A.R. Clapham. Garden plants increase the range of variation found.

5. T. alpinum L. - *Alpine Meadow-rue.* Rhizomatous; stems rarely >15cm, erect, very thin, simple; leaves 2-pinnate to -ternate; inflorescence simple; flowers well spaced, pale yellow, with dangling stamens with very thin filaments; 2n=14. Native; grassy and rocky places on mountains; W & C Sc, very local in N Wa, NW Ir and extreme N En; formerly MW Yorks.

31. BERBERIDACEAE - *Barberry family*

Shrubs with yellow wood or rarely herbaceous perennials; leaves alternate, simple to pinnate or ternate, entire to toothed, without stipules; petioles present or 0. Flowers axillary or terminal, actinomorphic, bisexual, hypogynous, solitary or in usually racemose inflorescences; perianth usually of 4-6 whorls of 2-3 free segments each, 1-2 outer whorls sepaloid, the rest petaloid, usually yellow; stamens 4 or 6 in 2 whorls; ovary 1-celled, with 1-many ovules; style short or 0; stigma capitate; fruit a capsule or berry.

Usually distinguishable by its shrubby often spiny habit, yellow wood, perianth of several whorls, and 1-celled ovary.

1 Stems with spines; leaves simple **1. BERBERIS**
1 Stems without spines; leaves pinnate **2. MAHONIA**

Other genera - EPIMEDIUM L. differs in being herbaceous with 2-ternate leaves, perianth of 6 whorls of 2 (apparently 3 whorls of 4) segments, and the fruit a capsule. **E. alpinum** L. (*Barren-wort*) from S Europe is grown in gardens; it formerly escaped in woodland etc. and still occasionally persists in neglected gardens and estates. It has 2-ternate leaf on the flowering stem and compound glandular-pubescent inflorescences. Most modern records of garden throwouts or relics probably refer to other taxa; notably **E. pinnatum** Fisch. ssp. **colchicum** Boiss. from the Caucasus, with leafless flowering stems and simple glabrous or sparsely glandular-pubescent inflorescences; and **E. x versicolor** C. Morren (*E. grandiflorum* C. Morren x *E. pinnatum*), with leafless to biternate-leaved flowering stems, and mostly simple (often branched at base), ± glandular-pubescent inflorescences. The last taxon appears to be the commonest garden plant, often with red flowers.

1. BERBERIS L. - *Barberries*

Shrubs; stems with (1)3(-7)-partite spines bearing short-shoots in their axils; leaves deciduous or evergreen, simple; flowers in axillary racemes, fascicles or panicles, or solitary; perianth of 4-5 whorls each of 3 segments, various shades of yellow; stamens 6; fruit a few-seeded, red to purple-black, often bloomed berry.

1	Leaves deciduous to semi-evergreen, entire to serrate, the apex and teeth without spines or with weak spines much <1mm; fruit red	2
1	Leaves evergreen, entire to spinose-toothed, but apex on most or all leaves with pungent spine ≥1mm; fruit bluish-black	5
2	Leaves 2.5-6cm, with >10 teeth on each side; flowers in pendent racemes	**1. B. vulgaris**
2	Leaves 1-3cm, entire or with <10 teeth on each side; flowers in 1-many-flowered fascicles or panicles	3
3	Leaves with several teeth on each side; flowers numerous in dense panicles	**5. B. aggregata**
3	Leaves usually entire; flowers in loose fascicles of 1-6	4
4	Twigs minutely pubescent; spines mostly 3-partite; leaves oblanceolate	**4. B. wilsoniae**
4	Twigs glabrous; spines mostly simple; leaves obovate	**2. B. thunbergii**
5	Flowers in fascicles without or with very short common peduncle	6
5	Flowers in racemes with common peduncle	9
6	Leaves >3cm, with >3 spinose teeth on each margin	7
6	Leaves <3cm, with <3 spinose teeth on each margin	8
7	Leaves narrowly elliptic; twigs ± terete; fruit with nearly sessile stigma	**6. B. gagnepainii**
7	Leaves linear to linear-lanceolate; twigs with raised ridges; fruit with distinct style	**7. B. julianae**
8	Leaves elliptic to obovate, with flat margins	**8. B. buxifolia**
8	Leaves narrowly elliptic, with revolute margins	**10. B. x stenophylla**
9	Spines <1cm, 3-7-partite; leaves 1-3cm; flowers orange	**9. B. darwinii**
9	Spines 1-3cm, 1-3-partite; leaves 2.5-8cm; flowers yellow	**3. B. glaucocarpa**

Other spp. - Many spp., hybrids and cultivars are grown in gardens and several are mass planted in parks, whence occasional bird-sown bushes (often not coming true from seed) on walls and in hedges etc. can originate. **B. manipurana** Ahrendt, from NE India, has been reported as isolated bushes in S Hants and W Kent; it would key to couplet 7 but has terete twigs, and thinner, flatter, narrowly elliptic leaves rather abruptly narrowed to a short petiole <5mm (not very gradually tapered to a longer petiole).

1. B. vulgaris L. - *Barberry*. Deciduous shrub to 3m; spines mostly 3-partite; **98**
leaves 2.5-6cm, elliptic to obovate, with numerous small subspinose teeth; flowers **99**
in pendent racemes 30-50mm, yellow; fruit red; (2n=28). Probably intrd; long natd
in hedges and rough ground throughout most of BI, but very scattered; Europe.

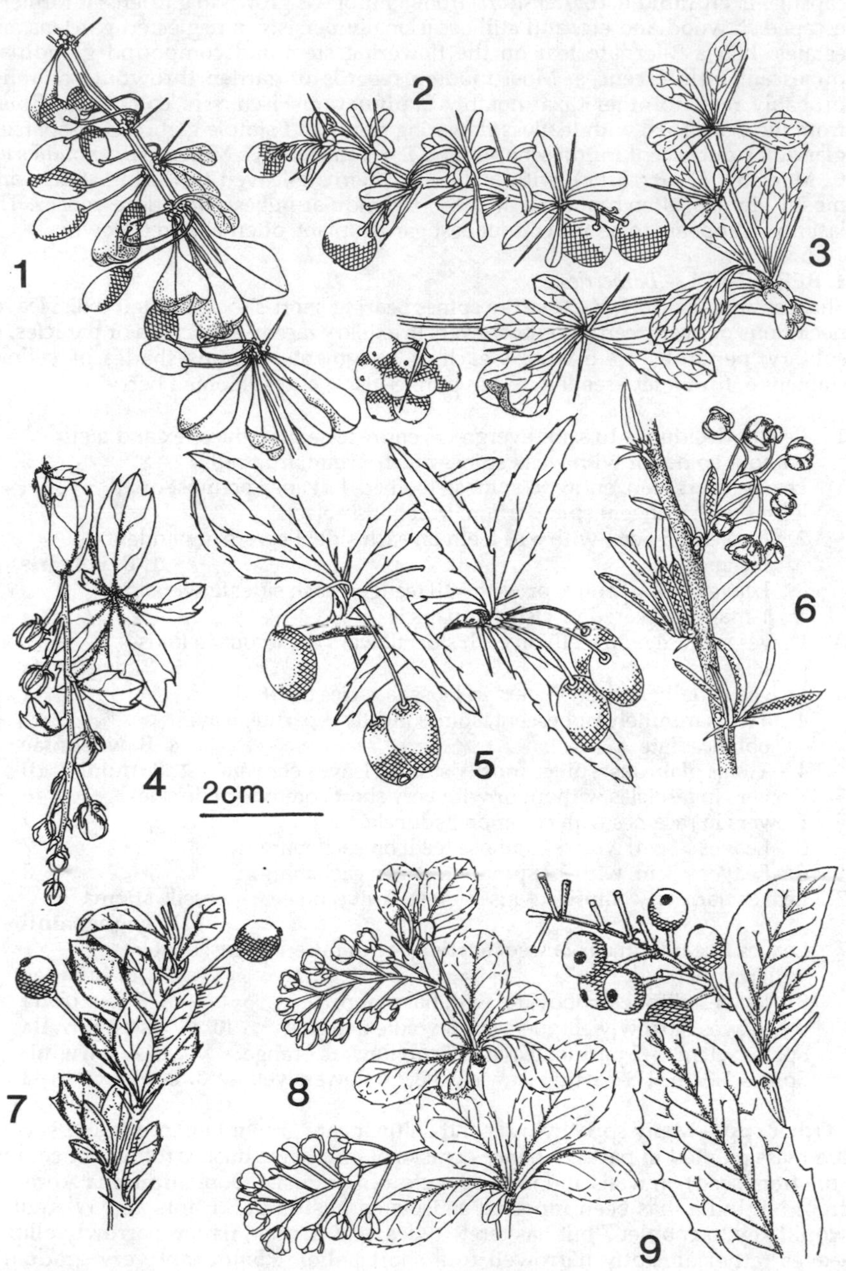

FIG 98 - Shoots of *Berberis*. 1, *B. thunbergii*. 2, *B. wilsoniae*. 3, *B. aggregata*. 4, *B. darwinii*. 5, *B. gagnepainii*. 6, *B. x stenophylla*. 7, *B. buxifolia*. 8, *B. vulgaris*. 9, *B. glaucocarpa*.

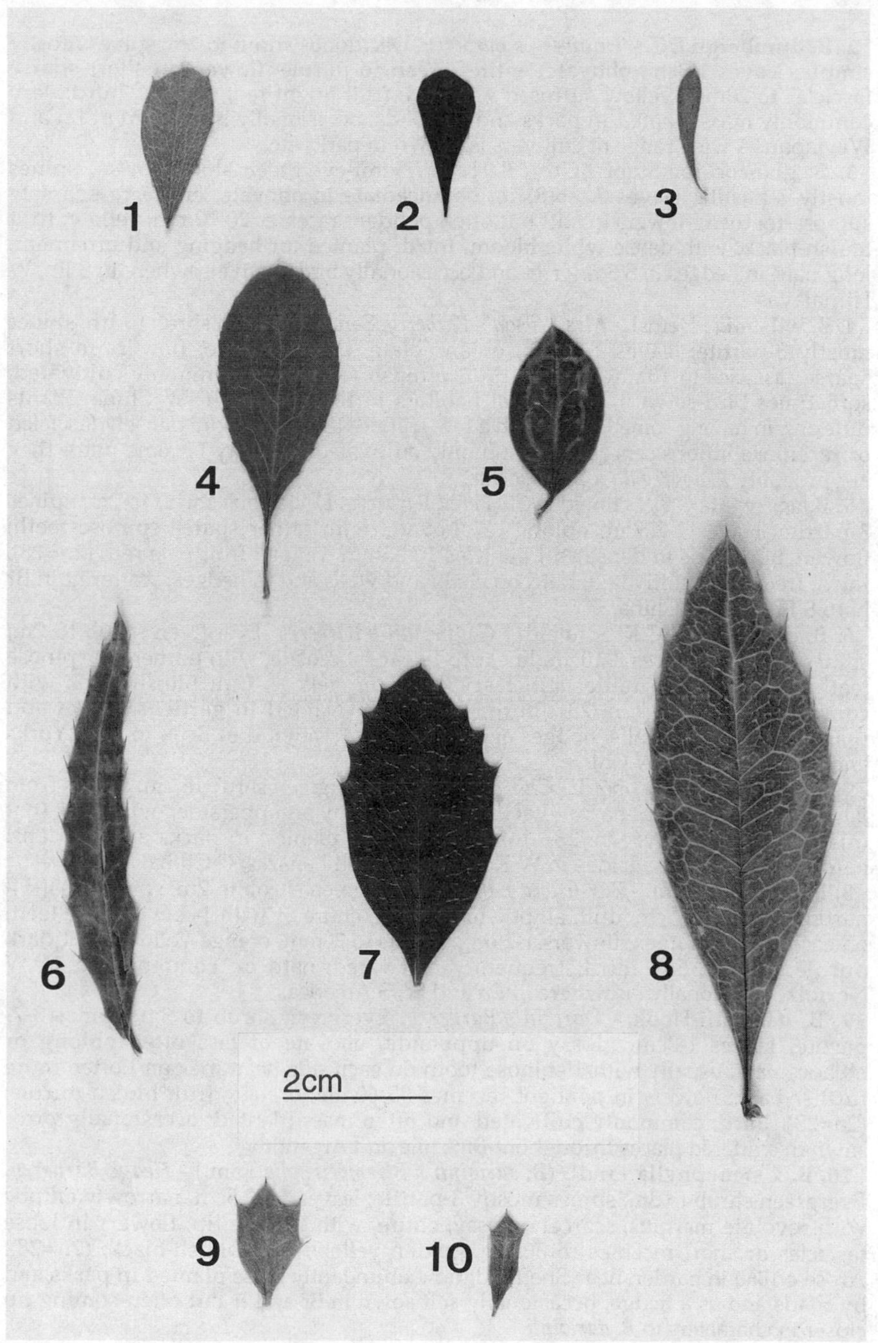

FIG 99 - Leaves of *Berberis*. 1, *B. aggregata*. 2, *B. thunbergii*. 3, *B. wilsoniae*. 4, *B. vulgaris*. 5, *B. buxifolia*. 6, *B. gagnepainii*. 7, *B. glaucocarpa*. 8, *B. julianae*. 9, *B. darwinii*. 10, *B. x stenophylla*.

2. B. thunbergii DC. - *Thunberg's Barberry*. Deciduous shrub to 2m; spines mostly **98**
simple; leaves 1-3cm, obovate, entire, green to purple; flowers in short sparse **99**
fascicles to 20mm, yellow suffused with red; fruit bright red; (2n=28). Intrd; very commonly mass planted in parks and by roads; occasionally bird-sown in En and Wa; Japan. A wide range of cultivars is grown in parks etc.

3. B. glaucocarpa Stapf - *Great Barberry*. Semi-evergreen shrub to 3m; spines **98**
mostly 3-partite; leaves 2.5-6(8)cm, oblanceolate to obovate, entire to sparsely **99**
spinose-toothed; flowers in stiff but often pendent racemes 20-70mm, yellow; fruit bluish-black, with dense white bloom. Intrd; planted for hedging and ornament; long natd in hedges in S Somerset and occasionally bird-sown elsewhere in S En; W Himalayas.

4. B. wilsoniae Hemsl. - *Mrs Wilson's Barberry*. Semi-evergreen shrub to 1m; spines **98**
mostly 3-partite; leaves 1-2.5cm, oblanceolate, usually entire; flowers in short **99**
sparse fascicles to 10mm, yellow; fruit pinkish red. Intrd; commonly cultivated; sometimes bird-sown in a range of habitats in En, Wa and Ir; W China. Plants differing in having some leaves with a few spinose teeth and more densely fascicled or racemose inflorescences are commonly cultivated and may become natd; they are probably *B. aggregata* x *B. wilsoniae*.

5. B. aggregata C.K. Schneid. - *Clustered Barberry*. Deciduous shrub to 2m; spines **98**
3-partite, leaves 1-2.5cm, oblong to obovate, with rather sparse spinose teeth; **99**
flowers numerous in dense stiff panicles 10-35mm, yellow; fruit pale red; (2n=28). Intrd; frequently cultivated; natd on chalk and walls and in hedges; scattered in Br N to S Ebudes; W China.

6. B. gagnepainii C.K. Schneid. - *Gagnepain's Barberry*. Evergreen shrub to 2m; **98**
spines 3-partite; leaves 3-10cm, linear to linear-lanceolate, with numerous spinose **99**
teeth; flowers 3-9 in rather stiff fascicles <3cm, yellow; fruit bluish-black with conspicuous bloom; (2n=28). Intrd; frequently planted in gardens, parks and roadsides; natd on walls, hedges and riverbanks; scattered in Br N to MW Yorks and Kirkcudbrights; W China.

7. B. julianae C.K. Schneid. - *Chinese Barberry*. Evergreen shrub to 2m; differs from **99**
B. gagnepainii in leaves narrowly elliptic, more shiny on upperside; ovules 1-2 (not (3)4-15); and see key; (2n=28). Intrd; frequently planted in parks and gardens; natd on walls and roadsides in W Kent, Surrey and Lanarks; C China.

8. B. buxifolia Lam. - *Box-leaved Barberry*. Evergreen shrub to 2m; spines (1)3(-5)- **98**
partite; leaves 1-2.5cm, dull, elliptic to obovate, entire or with 1-few spinose teeth **99**
on each side near apex; flowers 1-2 on pedicels to 20mm, orange-yellow; fruit dark purple; (2n=28, 56). Intrd; frequently cultivated; natd on commonland in W Norfolk, occasionally elsewhere in En and Sc; S America.

9. B. darwinii Hook. - *Darwin's Barberry*. Evergreen shrub to 3m; spines 3-7- **98**
partite; leaves 1-3cm, glossy on upperside, obovate or less often oblong or **99**
oblanceolate, usually with 1 spinose tooth on each side near apex and often more further back; flowers in pendent racemes 35-60mm, orange; fruit bluish-purple; (2n=28). Intrd; commonly cultivated and often mass planted; occasionally bird-sown in scattered places throughout BI; Chile and Argentina.

10. B. x stenophylla Lindl. (*B. darwinii* x *B. empetrifolia* Lam.) - *Hedge Barberry*. **98**
Evergreen shrub to 3m, spines mostly 3-partite; leaves 1.5-2.5cm, narrowly elliptic **99**
with revolute margins, scarcely glossy, entire, with spinose tip; flowers in loose fascicles or short racemes to 30mm, golden yellow; fruit bluish-black; (2n=28). Arose c.1860 in garden near Sheffield; now abundantly mass planted in parks and by roads and as a hedge; occasionally self-sown in Br and Ir but often coming up closer in characters to *B. darwinii*.

2. MAHONIA Nutt. - *Oregon-grapes*

Shrubs; stems without spines; leaves evergreen, pinnate; flowers in axillary fasciculate racemes, yellow, structure as in *Berberis*; fruit a blue-black, bloomed berry with few seeds.

Other spp. - Some natd *Mahonia* determined as *M. aquifolium* might be **M. x wagneri** (Jouin) Rehder (*M. aquifolium* x *M. pinnata* (Lag.) Fedde), with 7-13 leaflets >2x as long as wide.

1. M. aquifolium (Pursh) Nutt. - *Oregon-grape*. Somewhat stoloniferous shrub with ascending stems to 1.5m; leaflets 5-9, 3-8cm, c.2x as long as wide, glossy on upperside, not papillose on lowerside, with c.5-15 spinulose teeth on each side; (2n=28). Intrd; commonly cultivated and often mass planted for ground and game cover; natd in scrub, woodland and hedges etc. throughout Br N to C Sc; W N America.

2. M. x decumbens Stace (*M. aquifolium* x *M. repens* (Lindl.) G. Don - *Newmarket Oregon-grape*. Sprawling, strongly stoloniferous shrub to 50cm, with decumbent stems; leaflets 5-7, <2x as long as wide, dull on upperside, minutely papillose on lowerside, with c.8-22 spinulose teeth on each side. Intrd; frequently grown for ground cover and natd in woodland in Wilts, Cambs and Man. *M. repens* and hybrid occur within the range of *M. aquifolium* in N America, but crossing also occurs in cultivation. *M. repens* might also occur wild in Br.

32. PAPAVERACEAE - *Poppy family*

Herbaceous annuals or perennials usually with white or yellow latex; leaves spiral, shallowly pinnately lobed to 2-pinnate, without stipules; petioles present or 0. Flowers terminal or axillary, actinomorphic, bisexual, hypogynous or rarely perigynous, solitary or in umbellate or paniculate inflorescences; sepals 2(-3), free or fused, normally very falling early; petals 4(-6, or 0), free, often large and brightly coloured and crumpled at first, normally falling early; stamens numerous; ovary of 2-many fused carpels, 1(-2)-celled, with (2-)many ovules; style 1, short or 0; stigma usually large and ± capitate or peltate, lobed or rayed or ± divided; fruit a capsule, dehiscent or not.

The 2 sepals, 4(-6) showy petals, and distinctive latex usually distinguish this family, but *Macleaya* lacks petals.

1 Petals 0; flowers numerous in large panicles **7. MACLEAYA**

1 Petals conspicuous; flowers solitary or in <10-flowered inflorescences 2

2 Sap watery; sepals fused, shed as a hood as flower opens; receptacle raised above base of ovary (flowers perigynous) **6. ESCHSCHOLZIA**

2 Sap a white to orange latex; sepals often adherent but not fused; flowers hypogynous 3

3 Capsule >10x as long as wide; stigma with 2 lobes 4

3 Capsule <6x as long as wide; stigma with >3 lobes or rays 5

4 Flowers >3cm across, 1-2 per leaf-axil; capsule >10cm, 2-celled **4. GLAUCIUM**

4 Flowers <3cm across, mostly in umbels of >3; capsule <6cm, 1-celled **5. CHELIDONIUM**

5 Lobes and teeth of leaves each ending in a long weak spine **3. ARGEMONE**

5 Leaves not spiny 6

6 Petals white to red or mauve; style absent; stigma a 4-20-rayed disk **1. PAPAVER**

6 Petals yellow; style present; stigma 4-6-lobed **2. MECONOPSIS**

Other genera - A sp. of **ROEMERIA** Medik., **R. hybrida** (L.) DC. (*Violet Horned-poppy*), from Europe, with violet petals and a long linear capsule, was formerly natd in E Anglia but is now apparently extinct even as a casual.

1. PAPAVER L. - *Poppies*

Annuals or perennials with white latex unless otherwise stated; leaves glabrous to pubescent; flowers solitary; petals red or mauve to white; capsule 1-celled, with 4-20 incomplete placentae projecting inwards, opening by pores just below the persistent stigma; stigma sessile, a 4-20-rayed flat disk.

1 Tufted perennials with basal rosette of leaves at flowering 2
1 Annuals usually without basal leaves at flowering 3
 2 Petals <45mm; capsule narrowly obovoid to clavate, widest immediately below stigma **2. P. atlanticum**
 2 Petals >45mm; capsule obovoid, widest c.3/4 way from base to apex **1. P. pseudoorientale**
3 Stem-leaves strongly glaucous, clasping stem at base, toothed or lobed but not to near midrib **3. P. somniferum**
3 Stem-leaves green to slightly glaucous, not clasping stem, mostly divided as far as midrib 4
 4 Capsule with long stiff hairs; filaments dilated distally 5
 4 Capsule glabrous; filaments not dilated 6
5 Capsule <1.5cm, c. as long as wide **6. P. hybridum**
5 Capsule >1.5cm, at least some >2x as long as wide **7. P. argemone**
 6 Capsule <2x as long as wide **4. P. rhoeas**
 6 Capsule >2x as long as wide **5. P. dubium**

Other spp. - Records of **P. lateritium** K. Koch (*Armenian Poppy*), from Armenia, were errors for *P. atlanticum*.

1. P. pseudoorientale (Fedde) Medw. (*P. orientale* auct. non L.) - *Oriental Poppy*. **104**
Tufted perennial to 1m; pedicel with conspicuous appressed hairs; flowers with or without large green bracts immediately under the flower; sepals usually 3, with many stiff hairs; petals (4-)6, 45-80mm, pale pink to orange-red, usually with dark basal blotch; anthers violet; capsule ≤40mm, subglobose to obovoid, glabrous; stigma 7-16(20)-rayed, c. as wide as capsule. Intrd; common in gardens; natd in various habitats on well-drained soils in En and Sc; SW Asia. Some natd plants probably belong to **P. bracteatum** Lindl., which has leafy bracts, unblotched petals and sepals with broad-based stiff hairs, and to **P. orientale** L., which has no bracts and unblotched petals. The 3 spp. are ± sympatric and intermediates occur; they might be conspecific.

2. P. atlanticum (Ball) Coss. - *Atlas Poppy*. Tufted perennial to 60cm; pedicel **104**
with appressed to erecto-patent hairs; petals 20-40mm, orange-red; anthers yellow; capsule ≤25mm, narrowly obovoid to clavate, glabrous; stigma (5-)6(-8)-rayed, c. as wide as capsule; (2n=14). Intrd; grown in gardens and natd on walls, roadsides and rough ground sparsely throughout Br; Morocco. Capsule usually characteristically transversely wrinkled.

3. P. somniferum L. - *Opium Poppy*. Erect strongly glaucous annual to 50(100)cm; **104**
pedicel glabrous to sparsely bristly; petals 25-50mm, white to deep mauve, sometimes red or variegated; anthers yellowish or dark purplish; capsule ≤90mm, globose to obovoid, glabrous; stigma 5-12(18)-rayed, c. as wide as capsule; (2n=22).

a. Ssp. somniferum (ssp. *hortense* (Hussenot) Syme). Glabrous or very sparsely bristly; leaves rather shallowly lobed; stigma-rays never overlapping; very variable, especially in petal colour, size and shape and capsule size and shape. Intrd; cultivated in many parts of world for latex (opium), seeds and ornament; common casual over most of BI, grown for ornament and on small scale for seed (*Blue Poppy*)

b. Ssp. setigerum (DC.) Arcang. (*P. setigerum* DC.). Stems, leaves and sepals conspicuously bristly; leaves rather deeply lobed; capsule relatively narrow, with overlapping stigma-rays. Intrd; rare casual on tips etc. in Br; S Europe. Recent work suggests this might be a distinct sp.

4. P. rhoeas L. (*P. strigosum* (Boenn.) Schur, *P. commutatum* Fisch. & C.A. Mey.) - 104
Common Poppy. Erect annual to 60(80)cm; latex usually white, sometimes yellow; pedicels usually with patent hairs but sometimes with appressed hairs distally; petals 30-45mm, usually bright scarlet, often with black blotch at base, sometimes white, pink, mauvish or variegated; anthers bluish-black; capsule ≤20mm, obovoid to subglobose, glabrous; stigma (5)8-12(18)-rayed, at least as wide as capsule; 2n=14. Native; arable ground, roadsides and waste places; throughout BI and often common, but rare in most of N and W; grown in gardens (*Shirley Poppy*) and often escaping.

4 x 5. P. rhoeas x P. dubium = P. x hungaricum Borbás (*P. x expectatum* Fedde) is often recorded but usually in error for abnormal plants of *P. rhoeas*. There are a few scattered records of the hybrid from S En; it is intermediate in capsule shape and, like some plants of *P. rhoeas*, sterile.

5. P. dubium L. - *Long-headed Poppy.* Erect annual to 60cm; pedicels with 104
appressed hairs distally; petals 15-35mm, pink to red (usually paler than in *P. rhoeas*), sometimes with dark blotch at base; anthers bluish-black; capsule ≤25mm, narrowly obovoid to clavate, glabrous; stigma (4)7-9(12)-rayed, slightly less wide than capsule; seeds bluish-black. Probably native; similar places to *P. rhoeas*.

a. Ssp. dubium. Latex white or cream, becoming brown to black when dry; upper leaves with ultimate lobes often >1.5mm wide; anthers brownish- to bluish-black; 2n=28, 42. Most of BI, rarer than *P. rhoeas* in S & E but more widespread in W & N.

b. Ssp. lecoqii (Lamotte) Syme (*P. lecoqii* Lamotte). Latex yellow or quickly turning yellow on exposure to air, becoming reddish when dry; upper leaves with ultimate lobes rarely >1.5mm wide; anthers often yellow; 2n=28. Very sparsely scattered in En, Wa and Ir, absent from most of Sc, frequent only on chalk in S En.

6. P. hybridum L. - *Rough Poppy.* Erect annual to 50cm; pedicels with appressed 104
hairs distally; petals 10-25mm, crimson, with dark blotch at base; anthers blue; capsule ≤15mm, subglobose to broadly obovoid or ellipsoidal, densely patent- to appressed-bristly; stigma 4-8-rayed, much less wide than capsule; 2n=14. Probably native; arable fields and waste places; formerly widespread in BI except Sc, now much less common and ± confined to E & S En on calcareous soils.

7. P. argemone L. - *Prickly Poppy.* Erect or ascending annual to 45cm; pedicels 104
with appressed hairs; petals 15-25mm, pale scarlet, sometimes with dark blotch at base; anthers blue; capsule ≤25mm, narrowly obovoid, sparsely ± appressed-bristly; stigma 4-6-rayed, less wide than capsule; 2n=42. Probably native; arable fields and waste places on light soils; formerly widespread in BI, now much less common and largely confined to C & S En.

2. MECONOPSIS Vig. - *Welsh Poppy*

Perennials with yellow latex; leaves nearly glabrous; flowers solitary; petals yellow; capsule 1-celled, with 4-6 incomplete placentae projecting inwards, opening by elongated pores at apex; style short but distinct; stigma ± capitate, 4-6-lobed.

1. M. cambrica (L.) Vig. - *Welsh Poppy.* Stems to 60cm, erect, very sparsely R
pubescent; leaves pinnate with pinnately lobed leaflets; flowers 50-80mm across; 104
anthers yellow; capsule 20-40mm, narrowly obovoid; 2n=22. Native; shady places among rocks or under trees especially in hilly country; Wa, SW En and scattered parts of Ir, but also grown in gardens and extensively natd mainly in N En and Sc.

3. ARGEMONE L. - *Mexican Poppy*

Annuals with yellow latex; leaves with a weak spine at tip of lobes and teeth; flowers solitary, with usually 2 leaf-like bracts just below; petals 4-6, yellow; capsule 1-celled, with 4-6 incomplete placentae projecting inwards, opening by 4-6 elongated pores at apex; style short but distinct; stigma ± capitate, 4-6-lobed.

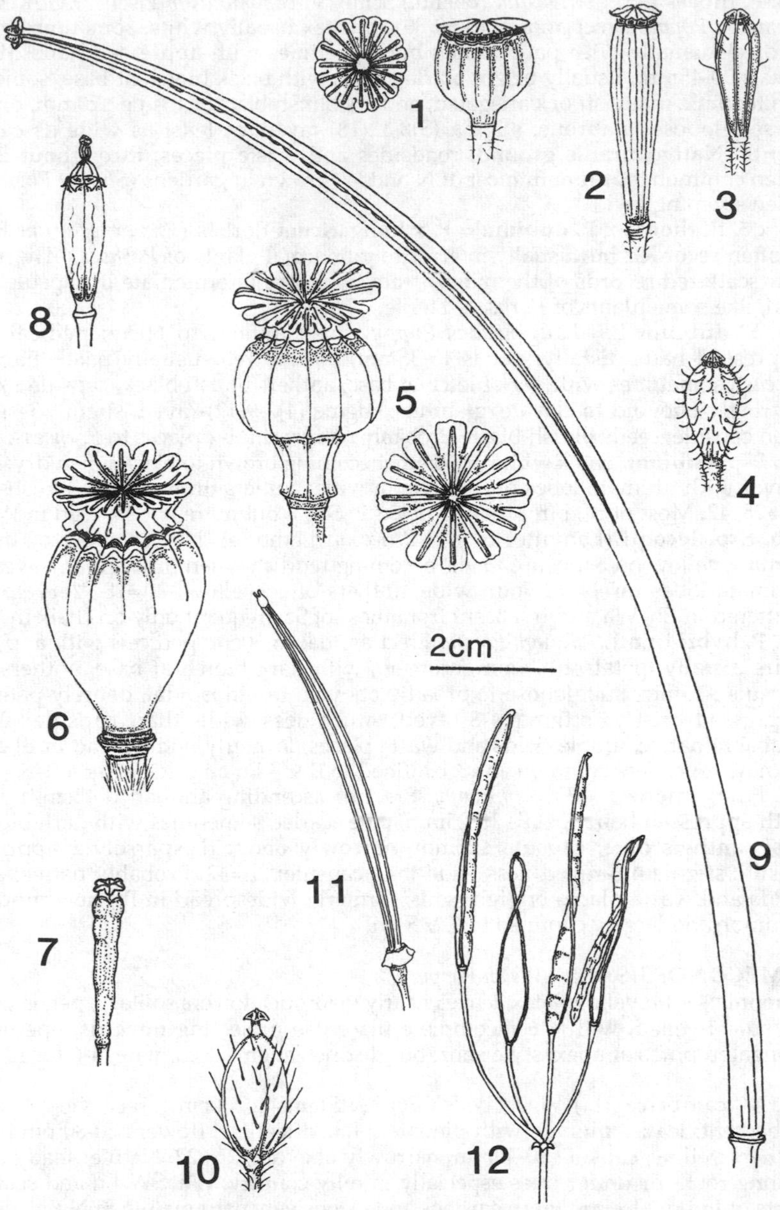

FIG 104 - Fruits of **Papaveraceae**. 1, *Papaver rhoeas*. 2, *P. dubium*. 3, *P. argemone*. 4, *P. hybridum*. 5, *P. somniferum*. 6, *P. pseudoorientale*. 7, *P. atlanticum*. 8, *Meconopsis cambrica*. 9, *Glaucium flavum*. 10, *Argemone mexicana*. 11, *Eschscholzia californica*. 12, *Chelidonium majus*.

1. A. mexicana L. - *Mexican Poppy*. Stems much branched, weakly spiny, 104 spreading, to 90cm but usually much less; leaves deeply pinnately lobed; flowers 40-60mm across; capsule 25-45mm, spinose, ellipsoidal; (2n=28, 112). Intrd; wool-alien, grain-alien and cultivated plant escaping as a casual; very scattered in S & C En; C America.

4. GLAUCIUM Mill. - *Horned-poppies*
Annuals to perennials with yellow latex; leaves glaucous, the lower pubescent; flowers solitary; petals yellow or red; capsule 2-celled, opening from above ± along its length by 2 valves and leaving the seeds embedded in septum; style ± 0; stigma ± capitate, 2-lobed.

Other spp. - G. corniculatum (L.) Rudolph (*Red Horned-poppy*), an annual from S Europe, has red petals and pubescent stems; it formerly occurred as a casual and was occasionally natd.

1. G. flavum Crantz - *Yellow Horned-poppy*. Biennial to perennial; stems glabrous, 104 much branched, spreading, to 90cm; lower leaves deeply pinnately lobed, the upper shallowly lobed, much less hairy and clasping stem at base; flowers 6-9cm across; petals yellow; capsule 15-30cm, linear, glabrous; 2n=12. Native; on maritime shingle and less often other substrata; common on coasts of En, Wa and CI, local in Ir and very local in S Sc.

5. CHELIDONIUM L. - *Greater Celandine*
Perennials with orange latex; leaves nearly glabrous; flowers in umbels of (2-)3-6; petals yellow; capsule 1-celled, opening from below along whole length by 2 valves; style short but distinct; stigma ± capitate, 2-lobed.

1. C. majus L. - *Greater Celandine*. Stems sparsely pubescent, ± spreading, to 104 90cm; leaves pinnate with broad, lobed leaflets; flowers 15-25mm across; capsule 3-5cm, linear, glabrous; 2n=12. Possibly native; hedgerows, walls and other marginal habitats, often near habitation (formerly cultivated); throughout BI but rare in N Sc.

6. ESCHSCHOLZIA Cham. - *Californian Poppy*
Annuals to perennials with watery sap; leaves glabrous, glaucous; flowers solitary; petals yellow to orange; capsule 1-celled, opening from below along its whole length by 2 valves; style very short; stigma deeply 4-6-lobed.

1. E. californica Cham.- *Californian Poppy*. Rarely perennating in BI; stems 104 glabrous, little branched, to 60cm, erect to spreading; leaves compoundly pinnate with linear leaflets; flowers 2-12cm across; capsule 7-10cm, linear, glabrous; (2n=12). Intrd; commonly grown as summer bedding plant and often found as casual on tips and roadsides; natd and perennating on dunes, walls and cliff-tops in Guernsey, and in quarries and by railways in Kent; SW N America.

7. MACLEAYA R. Br. - *Plume-poppies*
Perennials with orange latex; leaves glabrous or sparsely pubescent, white-glaucous on lowerside; flowers small, in crowded, large, terminal panicles; petals 0; capsule 1-celled, opening from above by 2 valves; style very short; stigma deeply 2-lobed. Perhaps not distinct from **Bocconia** L., which has priority.

1. M. x kewensis Turrill (*M. cordata* (Willd.) R. Br. x M. *microcarpa* (Maxim.) Fedde) - *Hybrid Plume-poppy*. Rhizomatous; stems sparsely pubescent to glabrous, glaucous, erect, to 2.5m; leaves petiolate, cordate at base, pinnately lobed; stamens 12-18, pinkish-buff; ovules 2-4; capsule not developed. Intrd; grown in gardens and found as escape and throwout, long persisting if undisturbed, in En, Ir and S Sc; garden origin from E Asian parents.

33. FUMARIACEAE - *Fumitory family*

Herbaceous annuals or perennials with watery sap; leaves spiral or all basal, pinnate or ternate, without stipules; petioles present. Flowers in simple or compound racemes, zygomorphic, bisexual, hypogynous; sepals 2, free, small, falling early; petals 4, free or adherent, white to pink or yellow, upper or upper and lower with basal spur; stamens 2, tripartite; ovary 1-celled, with 1-many ovules; style 1; stigma ± 2-lobed; fruit an achene or capsule.

The distinctive flowers are unique, but the family is linked to Papaveraceae by intermediates and is probably best united with it.

1 Flowers with 2 spurred petals (the upper and lower) **1. DICENTRA**
1 Flowers with 1 spurred petal (the upper) 2
2 Fruit a 1-seeded achene **5. FUMARIA**
2 Fruit a dehiscent capsule, usually with >1 seed 3
3 Annual; leaves with tendrils; flowers <8mm **4. CERATOCAPNOS**
3 Perennial,; leaves without tendrils; flowers >8mm 4
4 Stems branched, arising from ± cylindrical stock; flowers cream to yellow **3. PSEUDOFUMARIA**
4 Stems simple, arising from ± globose tuber; flowers white to purple **2. CORYDALIS**

1. DICENTRA Bernh. - *Bleeding-hearts*
Perennials, with short branched rhizome; aerial stems with terminal inflorescence; leaves all or mostly basal; flowers pink; upper and lower petals spurred at base; fruit a capsule; seeds several, with an aril.

Other spp. - D. spectabilis (L.) Lem., from E Asia, has been recorded either in error for *D. formosa* or as a very rare and impermanent casual; it has simple racemes and leaves on all the stems.

1. D. formosa (Haw.) Walp. - *Bleeding-heart.* Leaves all basal, 2-4-ternate; flowering stems to 30cm, leafless; flowers 15-20mm, pink, in rather dense compound racemes; (2n=16, 24). Intrd; commonly grown in gardens; natd in shady places especially by streams mainly in W & N Br; W N America. Some natd *Dicentra* might be **D. eximia** (Ker Gawl.) Torr. (*Turkey-corn*), from E N America, or *D. formosa* x *D. eximia*. The latter sp. differs in its corolla with a more narrowed upper part, with more prominent inner petals, and separating to below the middle. The hybrid can be triploid and sterile, or diploid and fertile.

2. CORYDALIS DC. - *Corydalises*
Perennials; stems usually simple; inflorescence 1, terminal; leaves present on aerial stem; flowers white to purple or yellow; upper petal spurred at base; fruit a capsule; seeds several, with aril.

1 Flowers yellow; leaves 2-4-pinnate, fern-like; subterranean tuber 0 **3. C. cheilanthifolia**
1 Flowers white to purple; leaves 2-3-ternate, not fern-like; stems arising from subterranean tuber 2
2 Bracts narrowly lobed; stem with large scale just below lowest leaf **1. C. solida**

2 Bracts ± entire; stem without large scale **2. C. cava**

1. C. solida (L.) Clairv. (*C. bulbosa* (L.) DC. nom. illeg.) - *Bird-in-a-bush*. Stems ± erect, to 20cm, arising from ± globose solid tuber; flowers 15-25(30)mm incl. basal spur c.1/2 total length; (2n=16, 24, 32). Intrd; grown in gardens and natd in woods and hedges; scattered over En and Wa; Europe. A variant with entire bracts occurs but is not recorded from BI.

2. C. cava (L.) Schweigg. & Körte (*C. bulbosa* auct. non (L.) DC. nom. illeg.) - *Hollowroot*. Differs from *C. solida* in hollow tuber; and see key; (2n=16, 32). Intrd; in similar places to *C. solida*; En and Wa but much less common; Europe.

3. C. cheilanthifolia Hemsl. - *Fern-leaved Corydalis*. Stems erect, to 25cm, usually shorter than basal leaves; flowers 10-16mm incl. basal spur c.1/3 total length; (2n=12, 16). Intrd; grown in gardens, natd on walls; Surrey and S Hants; China.

3. PSEUDOFUMARIA Medik. - *Corydalises*

Perennials, arising from ± cylindrical stock; stems branched; inflorescences several, terminal and leaf-opposed; flowers cream to yellow; upper petal spurred at base; fruit a capsule; seeds several, with aril.

1. P. lutea (L.) Borkh. (*Corydalis lutea* (L.) DC.) - *Yellow Corydalis*. Stems erect to spreading or hanging, to 30cm; leaves 2-3-pinnate or -ternate; petioles ridged but not winged; flowers 12-18mm including basal spur 2-4mm, yellow; seeds shiny; (2n=28, 56). Intrd; commonly cultivated and natd on walls and in stony places over most of BI; S Alps.

2. P. alba (Mill.) Lidén (*Corydalis ochroleuca* Koch nom. illeg., *P. ochroleuca* (Koch) Holub nom. illeg.) - *Pale Corydalis*. Differs from *P. lutea* in its very narrowly winged petioles; flowers 10-18mm including basal spur 1-2mm, cream with yellow tip; seeds matt; (2n=28, 32). Intrd; similar places to *P. lutea* but much rarer and only in S Br; CS Europe.

4. CERATOCAPNOS Durieu - *Climbing Corydalis*

Annuals; stems branched, climbing by means of leaf-tendrils; inflorescences several, leaf-opposed; flowers pale cream; upper petal spurred at base; fruit a capsule; seeds (1)2-3, with aril.

1. C. claviculata (L.) Lidén (*Corydalis claviculata* (L.) DC.) - *Climbing Corydalis*. Stems to 75cm, scrambling; leaves 2-pinnate to -ternate, ending in branched tendril; flowers 4-6mm including basal spur to 1mm; (2n=32). Native; woods and other shady places often on rocks; scattered over most of BI.

5. FUMARIA L. - *Fumitories*

Annuals to 1m but often much less; stems much-branched, scrambling, thin; leaves all on stem, 2-4-pinnate; inflorescences leaf-opposed racemes; flowers white to purple; upper petal spurred at base; upper and lateral petals darker coloured at tip; fruit a 1-seeded achene; seeds without aril.

All spp. are similar in appearance and are distinguished by inflorescence, flower and fruit characters. The upper petal has a dorsal ridge and lateral margins which may be bent upwards to ± conceal the ridge or spread laterally to reveal it. The lower petal appears parallel-sided to strongly spathulate in top or bottom view according to the relative expansion of the margins distally. Flower colours given ignore the very dark petal tips. Flower length is measured from end of basal spur to tip of longest petal on fresh material; dried material has smaller floral parts (especially sepals). Fruit shape and measurements refer to those seen in widest profile on mature, dried fruits. It is essential to base determinations on well-grown, non-shaded material in early or mid flowering season. Late or shade-grown plants may be very atypical, with short (often cleistogamous) paler petals, longer narrower sepals, relatively long bracts and less or more recurved fruiting pedicels.

1 Flowers ≥9mm; sepals (2)3-6.5mm; lower petal usually ± parallel-sided, rarely subspathulate 2
1 Flowers ≤9mm; sepals <4mm; lower petal distinctly spathulate 9
 2 Lower petal with broad margins; fruit 2.75-3 x 2.75-3mm; flowers 12-14mm **2. F. occidentalis**
 2 Lower petal with narrow margins; fruit ≤2.75 x 2.5mm; flowers usually <12mm 3
3 Fruiting pedicels rigidly recurved to patent; sepals 4-6(6.5)mm 4
3 Fruiting pedicels erecto-patent, often not rigid; sepals (2)3-5mm 5
 4 Petals usually creamy-white, sometimes pink to red; upper petal with spreading margins not concealing dorsal ridge **1. F. capreolata**
 4 Petals purple; upper petal with erect margins ± concealing tip of dorsal ridge **6. F. purpurea**
5 Lower petal subspathulate 6
5 Lower petal ± parallel-sided 7
 6 Flowers 9-10mm **5. F. muralis**
 6 Flowers 10-11mm **F. x painteri**
7 Sepals 2-3 x 1-2mm, with sharp forward-pointing teeth all round margin; fruit distinctly rugose when dry **3. F. bastardii**
7 Sepals 3-5 x 1.5-3mm, with outward-pointing teeth in basal part of margin only or subentire; fruit smooth to finely rugose when dry 8
 8 Flowers 9-11(12)mm; sepals usually dentate near base; racemes c. as long as peduncles **5. F. muralis**
 8 Flowers (10)11-13mm; sepals subentire to denticulate near base; racemes distinctly longer than peduncles **4. F. reuteri**
9 Flowers ≥6mm; sepals ≥1.5mm 10
9 Flowers 5-6mm; sepals 0.5-1.5mm 13
 10 Bracts at least as long as pedicels, often longer **8. F. densiflora**
 10 Bracts shorter than pedicels 11
11 Fruits truncate to retuse at apex, distinctly wider than long **7. F. officinalis**
11 Fruits rounded to subacute at apex, narrower than to as wide as long 12
 12 Sepals 2-3(3.5) x 1-2mm, with sharp forward-pointing teeth all round margin; fruit distinctly rugose when dry **3. F. bastardii**
 12 Sepals (2.7)3-5 x 1.5-3mm, with outward-pointing teeth in basal part of margin only; fruit smooth to finely rugose when dry **5. F. muralis**
13 Corolla white to pale pink; bracts at least as long as fruiting pedicels **9. F. parviflora**
13 Corolla pink; bracts shorter than fruiting pedicels **10. F. vaillantii**

1. F. capreolata L. - *White Ramping-fumitory*. Flowers 10-13(14)mm, usually **109**
creamy-white suffused with pink, rarely reddish; upper petal with narrow spreading margins; sepals 4-6 x (2)2.5-3(4)mm, dentate mostly near base; fruit 2-2.5 x 2.5mm, orbicular, ± smooth when dry. Native; arable and waste ground and hedges. The 2 sspp. are rather doubtfully distinct.

a. Ssp. capreolata. Bracts usually shorter than fruiting pedicels; fruit c.2 x 2mm, **RR**
with rounded to very obtuse apex; (2n=56). CI; the Continental ssp.

b. Ssp. babingtonii (Pugsley) P.D. Sell. Bracts ± equalling fruiting pedicels; fruit c.2.5 x 2.5mm, with truncate apex; 2n=64. Scattered over much of Br and Ir, mainly near coasts, but absent from most of Sc and E En; endemic.

2. F. occidentalis Pugsley - *Western Ramping-fumitory*. Flowers 12-14mm, whitish **RR**
at first, later pink; upper and lower petals with broad margins; sepals 4-5.5 x 2- **109**
3(3.5)mm, dentate mostly near base; fruit 2.75-3 x 2.75-3mm, orbicular, minutely retuse at apex, rugose when dry; 2n=c.112. Native; arable land and waste places; Cornwall and Scillies; endemic.

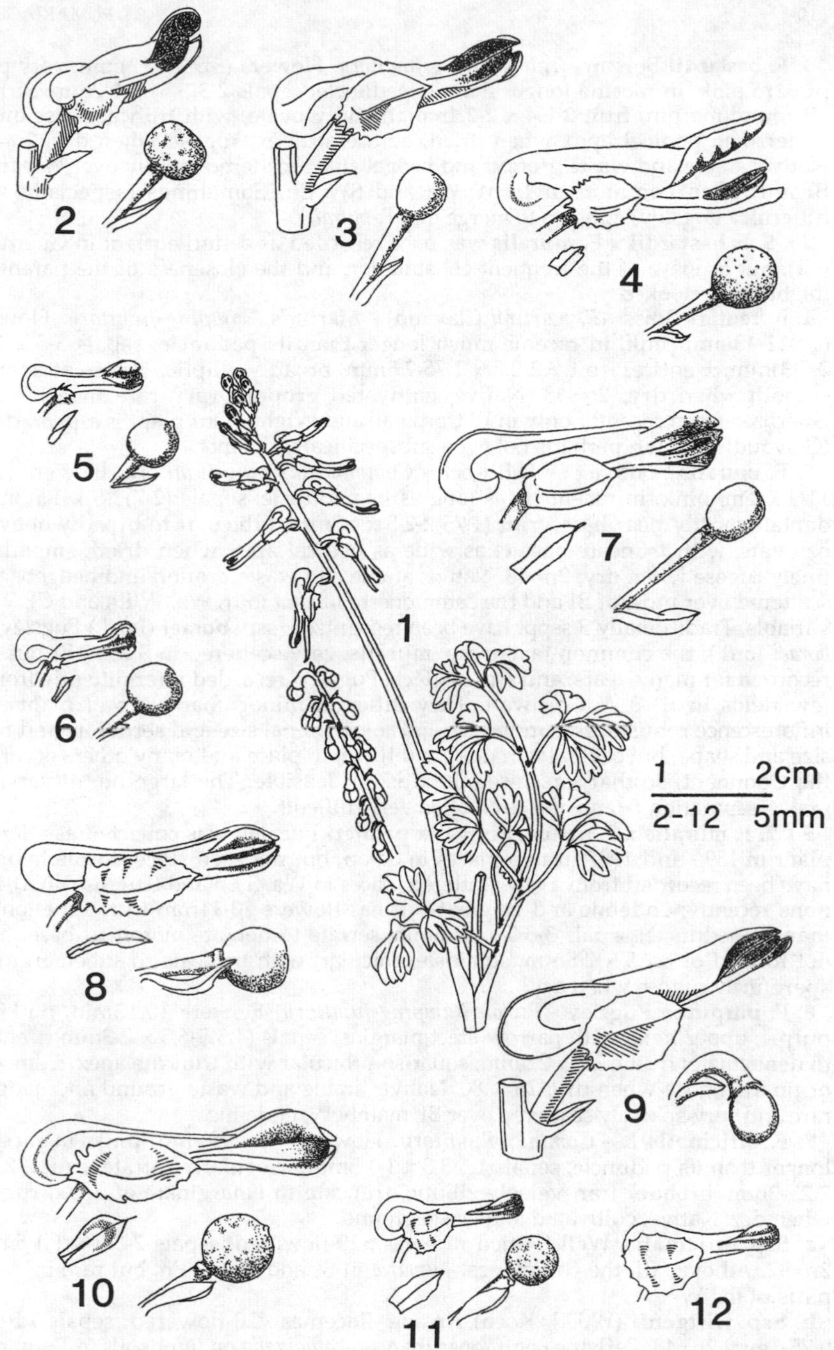

FIG 109 - *Fumaria.* 1, flowering node and fruiting raceme of *F. capreolata.* 2-12, flowers, bracts and fruits. 2, *F. officinalis* ssp. *officinalis.* 3, *F. muralis* ssp. *boraei.* 4, *F. bastardii.* 5, *F. parviflora.* 6, *F. vaillantii.* 7, *F. reuteri.* 8, *F. purpurea.* 9, *F. capreolata* ssp. *capreolata.* 10, *F. occidentalis.* 11-12, *F. densiflora.*

3. F. bastardii Boreau - *Tall Ramping-fumitory.* Flowers (8)9-11(12)mm, very pale **109**
pink to pink, in raceme longer than its peduncle; sepals 2-3(3.5) x 1-2mm, serrate all round margin; fruit 2-2.4 x 2-2.4mm, broadly ovate, with truncate base much wider than pedicel apex when dried, obtuse at apex, rugose when dry; 2n=48. Native; arable and waste ground and hedgebanks; scattered mainly over W part of BI and often common in Ir, W Wa and SW En. Sometimes (especially var. **hibernica** Pugsley) appears to merge into *F. muralis.*

3 x 5. F. bastardii x F. muralis was once recorded as a sterile plant in Guernsey (CI), but in view of the frequent cohabitation and the closeness of the parents it might be overlooked.

4. F. reuteri Boiss. (*F. martinii* Clavaud) - *Martin's Ramping-fumitory.* Flowers **RRR**
(10)11-13mm, pink, in raceme much longer than its peduncle; sepals 3-5 x 1.5- **109**
2.5(3)mm, ± entire; fruit 2.2-2.7 x 1.75-2.5mm, broadly elliptic, obtuse at apex, ± smooth when dry; 2n=48. Native; cultivated ground; very rare in S En and Guernsey (seen recently only in W Cornwall and Wight). Our plant is ssp. **martinii** (Clavaud) A. Soler, perhaps not even subspecifically distinct.

5. F. muralis Sond. ex W.D.J. Koch - *Common Ramping-fumitory.* Flowers (8)9- **109**
11(12)mm, pink, in raceme c. as long as its peduncle; sepals (2.7)3-5 x 1.5-3mm, dentate mostly near base; fruit (1.75)2-2.5 x c.2mm, orbicular to broadly obovate or ovate, with truncate base c. as wide as pedicel apex when dried, smooth to finely rugose when dry; 2n=48. Native; arable and waste ground and hedgebanks; scattered over most of BI and the commonest *Fumaria* in Ir, Wa, W En and CI. Very variable. Traditionally 3 sspp. have been recognized: ssp. **boraei** (Jord.) Pugsley (*F. boraei* Jord.), the common taxon; ssp. **muralis**, very scattered in S & W Br but not recorded for many years; and ssp. **neglecta** Pugsley, recorded intermittently from a few fields in E & W Cornwall. They differ in minor characters of plant and inflorescence robustness, corolla size and colour, sepal size and serration, and fruit size and shape, but several variants are difficult to place and many others occur on the Continent, so that ssp. recognition is not feasible. The large no. of variants makes separation from *F. bastardii* often very difficult.

5 x 7. F. muralis x F. officinalis = F. x painteri Pugsley was collected as a fertile plant in 1896 and 1907 in two places in Salop, but not seen since; sterile hybrids have been recorded from a few scattered places in Wa, S En and Guernsey (CI), but none recently; endemic and enigmatic. It has flowers 10-11mm in raceme longer than its peduncle; sepals 3-3.5 x c.1.5mm, serrate to dentate mainly at base; fruit not formed or c.2.5 x 2.5mm, squarish-orbicular, with truncate to subemarginate apex, finely rugose when dry.

6. F. purpurea Pugsley - *Purple Ramping-fumitory.* Flowers 10-13mm, pinkish- **R**
purple; upper petal with narrow erect margins; sepals (4.5)5-6.5 x 2-3mm, ± entire **109**
to denticulate; fruit c.2.5 x 2.5mm, squarish-orbicular with truncate apex, ± smooth or finely rugose when dry; 2n=c.80. Native; arable and waste ground and hedges; rare and very sparsely scattered over BI, mainly W; endemic.

7. F. officinalis L. - *Common Fumitory.* Flowers (6)7-8(9)mm, pink, in raceme longer than its peduncle; sepals 1.5-3.5 x 1-1.5mm, irregularly serrate; fruit 2-2.5 x 2.25-3mm, broadly transversely elliptic, truncate to emarginate at apex, rugose when dry. Native; cultivated and waste ground.

a. Ssp. officinalis. Well-formed racemes >20-flowered; sepals 2-3.5 x 1-1.5mm; **109**
2n=32. All over BI, the commonest *Fumaria* in Sc and E & C En, but rare in many parts of the W.

b. Ssp. wirtgenii (W.D.J. Koch) Arcang. Racemes <20-flowered; sepals 1.5-2 x 0.75-1mm; 2n=48. Perhaps commoner than ssp. *officinalis* on light soils in E En, rare elsewhere in BI.

7 x 8. F. officinalis x F. densiflora has been found as a sterile plant in 4 localities in SE En, only once recently; endemic.

7 x 9. F. officinalis x F. parviflora was recorded in 1910 in Surrey; endemic.

8. F. densiflora DC. (*F. micrantha* Lag.) - *Dense-flowered Fumitory.* Flowers 6-7mm, **R**

pink, densely packed in raceme much longer than its peduncle; sepals (2)2.5-3.5 x **109**
(1)2-3mm, subentire to irregularly toothed; fruit 2-2.5 x 2-2.5mm, orbicular, ± rounded at apex, rugose when dry; (2n=28). Native; arable land, frequent on well-drained soils in SE En, very sparsely scattered and rare over rest of Br and Ir.

9. F. parviflora Lam. - *Fine-leaved Fumitory*. Flowers 5-6mm, white or very pale **R**
pink, in almost sessile raceme; sepals 0.5-0.75(1) x 0.5-0.8mm, irregularly serrate; **109**
fruit 1.7-2.3 x 1.75-2.5mm, broadly ovate to orbicular, rounded to subacute at apex, verrucose-rugose when dry; 2n=32. Native; arable land usually on chalk; E, SE & SC En, formerly rare in E Sc.

10. F. vaillantii Loisel. - *Few-flowered Fumitory*. Flowers 5-6mm, pale pink, in lax **R**
raceme much longer than its peduncle; sepals 0.5-0.75-1(1.5) x 0.25-0.5mm, **109**
irregularly serrate; fruit 2-2.25 x 2-2.25mm, ± orbicular, rounded to truncate or slightly emarginate at apex, verrucose-rugose when dry; 2n=32. Native; arable land usually on chalk; SE & SC En, casual N to Angus.

34. PLATANACEAE - *Plane family*

Trees; leaves simple, deciduous, palmately lobed, petiolate, alternate, stipulate when young. Flowers monoecious, in stalked, dense spherical clusters 2-several together on pendent unisexual stalks, hypogynous, actinomorphic; perianth small, fused or not, of 1-2 whorls of usually 3-4 segments; male flowers with usually 3-4 stamens; female flowers usually with 3-4 staminodes and 5-8 free carpels each with 1(-2) ovules; style 1; stigma linear; fruit an achene with long hairs at base.

An unmistakable tree.

1. PLATANUS L. - *Planes*

1. P. x hispanica Mill. ex Münchh. (*P.* x *hybrida* Brot., *P. x acerifolia* (Aiton) Willd.; *P. occidentalis* L. x *P. orientalis* L.) - *London Plane*. Tree to 44m; bark with conspicuous large peeling plates; leaves with hollow petiole-base concealing axillary bud, with sharply pointed leaf-lobes; fruits in spherical pendent clusters 2-3.5cm across, breaking up in spring; (2n=16, 42). Intrd; abundantly planted as street and park tree, especially in S En, and often producing seedlings. Of uncertain horticultural origin; either of above parentage or a variant of *P. orientalis*.

35. ULMACEAE - *Elm family*

Trees; leaves simple, deciduous, serrate, usually asymmetrical at base, petiolate, alternate, stipulate when young. Flowers in small axillary clusters produced before leaves, hypogynous, actinomorphic, bisexual; perianth inconspicuous, bell-shaped, 4-5-lobed; stamens 4-5; ovary 1-celled, with 1 ovule; styles 2; stigma linear; fruit an achene with 2 wide wings extending beyond both base and apex.

Unmistakable flowers, fruits and asymmetric leaf-bases.

1. ULMUS L. - *Elms*

An extremely difficult genus, having been interpreted in widely different ways, with 2-7 spp. and 1-c.12 interspecific hybrid combinations recognised. The complex hybrid origin of many taxa postulated by R. Melville may be correct in many instances, but does not form the basis of a practical classification. The 2-species concept of R.H. Richens is not sufficiently discriminating to be of taxonomic value. This account treats the 4 most distinctive native taxa as spp. and is a reasonable compromise between the 2 extreme opinions. Only binary hybrids are mentioned; the existence of ternary and quaternary hybrids is highly contentious, as is the history of the various taxa in Br, but hybridization is (or has been) undoubtedly very widespread and frequent. Dutch Elm Disease has killed most trees in much of S and C Br, including the Midlands where the greatest variation occurs. Over large areas the elm population exists solely or largely as hedgerow suckers, and is now largely unidentifiable.

Only leaves from the middle of short-shoots in high summer should be used for identification; leaves from long-shoots, suckers, epicormic shoots or Lammas shoots must be avoided. Leaf lengths below are measured from the base of the longer side of the lamina to its apex. Tree outlines refer to mature, solitary specimens. All taxa except *U. glabra* and some *U.* x *vegeta* produce abundant suckers.

1 Pedicels ≥3x as long as flowers and most ≥1x as long as fruits; leaves glabrous or very softly pubescent **7. U. laevis**
1 Pedicels shorter than flowers and very much shorter than fruits; leaves glabrous or harshly pubescent 2
 2 Rust-coloured hairs abundant on buds; leaves >7cm, very rough on upperside, with >12 pairs of lateral veins; petiole <3mm, most of it overlapped by base of long-side of lamina **1. U. glabra**
 2 Rust-coloured hairs 0 or present on buds; leaves usually <7cm, if >7cm smooth on upperside, with c.5-18 pairs of lateral veins; petiole usually >(3)5mm, not or partly overlapped by base of lamina 3
3 Leaves usually >7cm, with length x width >28(cm); rust-coloured hairs often present on buds 4
3 Leaves usually <7cm, with length x width <28(cm); rust-coloured hairs 0 on buds (except in *U. glabra* hybrids) 5
 4 Leaves almost 2x as long as wide, acuminate at apex, with 12-18 pairs of lateral veins; tree outline broadly obovate to orbicular, with long branches from low down **2. U. x vegeta**
 4 Leaves distinctly <2x as long as wide, acute to shortly acuminate at apex, with 10-14 pairs of lateral veins; tree outline obovate, narrow below but with spreading branches above **3. U. x hollandica**
5 Leaf width/length ratio >0.75; leaves usually rough on upperside; tree outline obovate to oblong, with strong branches at all levels **4. U. procera**
5 Leaf width/length ratio <0.75; leaves usually smooth on upperside, if rough then tree outline very narrow with no strong branches 6
 6 Tree outline narrow but irregular, with leading shoot arching or pendent; strong branches 0, all partly pendent; short-shoots mostly continuing growth as long-shoots; leaves usually rough on upperside **6. U. plotii**
 6 Tree outline various but never as last, with erect leading shoot; some strong branches usually present, only the lower or 0 pendent; short-shoots rarely continuing growth; leaves smooth on upperside **5. U. minor**

Other spp. - Since the decimation of *Ulmus* by Dutch Elm Disease in the 1970s several exotic spp. thought to be resistant to the fungus have been planted. These include **U. thomasii** Sarg. (*Rock Elm*), from NE N America; they might be expected in the wild in the future.

1. U. glabra Huds. - *Wych Elm.* Tree to 37m; outline ± orbicular; trunk dividing low down into many long spreading branches; leaves 8-16cm, very rough on upperside, asymmetric at base, with 12-18 pairs of lateral veins; petiole ≤3mm, most of it overlapped by base of long-side of lamina; 2n=28. Native; woods and hedgerows, sometimes dominant, especially on limestone; throughout BI, but much commoner in N & W and not native in much of SE. 2 ill-marked sspp. are sometimes recognized: ssp. **glabra**, with broadly obovate leaves, more southern in distribution; and ssp. **montana** Hyl., with narrowly obovate leaves and typical of 113

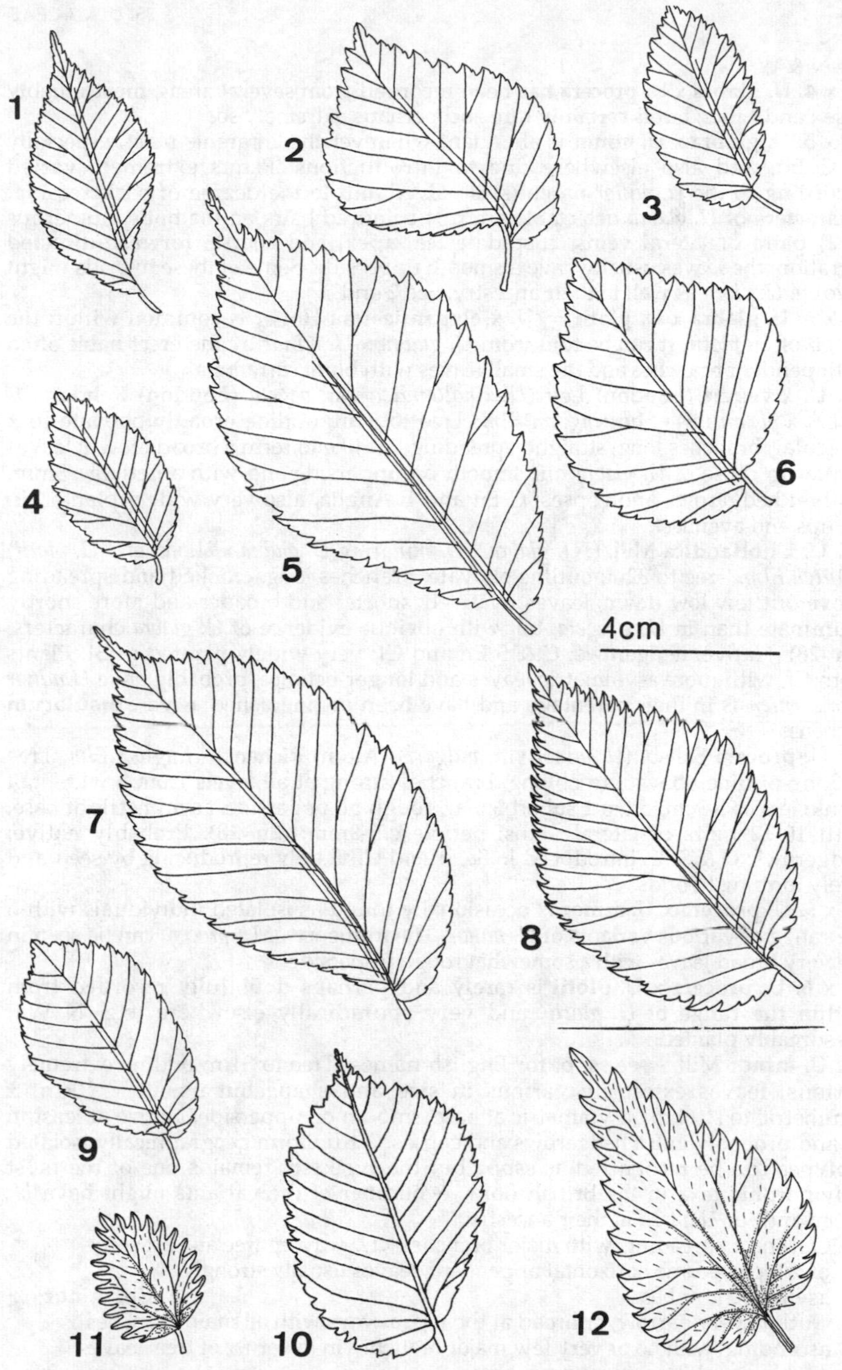

FIG 113 - Leaves of *Ulmus*, *Urtica*. 1, ***Ulmus carpinifolia***. 2, ***U. coritana***. 3, ***U. minor*** ssp. ***sarniensis***. 4, ***U. plotii***. 5, ***U. x vegeta***. 6, ***U. procera***. 7, ***U. glabra***. 8, ***U. x hollandica***. 9, ***U. minor*** ssp. ***angustifolia***. 10, ***U. laevis***. 11, ***Urtica urens***. 12, ***Urtica dioica***.

the N & W.

1 x 4. U. glabra x U. procera has been recorded from several areas, most reliably Essex and Lincs, but is certainly rare and perhaps extremely so.

1 x 5. U. glabra x U. minor is abundant wherever the 2 parents meet, especially in C En, and also elsewhere due to introductions. Plants extremely varied according to the *U. minor* parent(s) involved and to the degree of backcrossing. Characters of *U. glabra* detectable are rust-coloured hairs on the buds, numerous (>12) pairs of lateral veins, cuspidate leaf-apex, and coarse, forward-directed serration; the leaves usually have a smooth upperside. Some of these hybrids might involve *U. plotii* as well in their ancestry. See 2 and 3.

1 x 6. U. glabra x U. plotii = ?U. x elegantissima Horw. is common within the range of *U. plotii*. It can be told from *U. glabra* x *U. minor* by the erect habit often with pendent branches and the small leaves with blunt serrations.

2. U. x vegeta (Loudon) Ley (*U. x hollandica* var. *vegeta* (Loudon) Rehder; *U.* **113**
glabra x *U. minor*) - *Huntingdon Elm*. Tree to 32m; outline broadly obovate to ± orbicular; branches long, straight, spreading fan-like to form a broad crown; leaves similar to those of *U. glabra* but smooth on upperside and with a petiole >5mm. Native; hedgerows and copses; C En and E Anglia; also very widely planted in groups and avenues.

3. U. x hollandica Mill. (*?U. glabra* x *U. minor*, or *U. glabra* x *U. minor* x *U. plotii*) **113**
- *Dutch Elm*. Tree to 32m; outline obovate; branches long, crooked and spreading above but few low down; leaves relatively shorter and broader and more shortly acuminate than in *U. x vegeta*, but with obvious evidence of *U. glabra* characters; (2n=28). Native; hedgerows; C & S En and CI; very widely planted in BI. Plants from CI, with more asymmetric leaves and longer petioles, probably have *U.minor* ssp. *sarniensis* in their parentage and have been distinguished as var. **insularum** Richens.

4. U. procera Salisb. (*U. minor* var. *vulgaris* (Aiton) Richens) - *English Elm*. Tree **113**
to 33m; outline obovate to oblong; branches strong at all levels from a wide, tall trunk; leaves 5-9cm, often ± suborbicular, rough on upperside, asymmetric at base, with 10-12 pairs of lateral veins; petiole c.5-8mm; (2n=28). Probably native; hedgerows; C & S Br, introduced in Sc, Ir and CI. Rarely reproducing by seed and rarely forming hybrids.

4 x 5. U. procera x U. minor is occasionally found as isolated individuals within the range of various variants of *U. minor*. The influence of *U. procera* can be seen in the very broad leaves with a somewhat rough upperside.

4 x 6. U. procera x U. plotii is rarely and perhaps doubtfully recorded from within the range of *U. plotii*, and very sporadically elsewhere (e.g. N Wa) presumably planted.

5. U. minor Mill. - see sspp. for English names. Tree to 31m; outline extremely various; leaves extremely various in size and shape but usually <10cm, ± symmetric to strongly asymmetric at base, smooth on upperside. Native at least in CI and probably in Br; hedgerows and copses. 2 ± uniform geographically isolated biotypes can be recognized as sspp., but the type ssp. remains one of the most polymorphic taxa in the British flora. A number of the variants might have *U. glabra* and/or *U. plotii* in their ancestry.

1 Outline ± spreading, with major branches <1/2 way up tree at least some of which become horizontal or pendent; leaves usually strongly asymmetric at base **a. ssp. minor**
1 Outline very narrow, or broad at the top, always with all main branches ascending, with no or very few major branches in lower 1/2 of tree; leaves usually weakly asymmetric at base 2
  2 Trunk persisting to tree apex; branches numerous, slightly ascending **c. ssp. sarniensis**
  2 Trunk ending short of tree apex; branches few, the lowest steeply ascending **b. ssp. angustifolia**

a. Ssp. minor (*U. carpinifolia* Gled., *U. coritana* Melville, *U. diversifolia* Melville) - *Small-leaved Elm*. Outline variously spreading; trunk usually dividing <1/2 way up tree. Kent to S Wa and C En. The 3 synonyms given represent widely divergent biotypes, but there are many others as well and complete intergradation occurs. The following are distinctive in their extreme form: **U. carpinifolia** has narrow leaves with the margin of the long-side ± straight for the lower 1/3-1/2; **U. coritana** has broad leaves often similar to those of *U. procera* in shape but with a very unequal base and very large patches of hairs around the main vein axils on lowerside. **113**

b. Ssp. angustifolia (Weston) Stace (*U. angustifolia* (Weston) Weston, *U. stricta* (Aiton) Lindl., *U. minor* var. *cornubiensis* (Weston) Richens) - *Cornish Elm*. Outline narrow; lower branches few and steeply ascending; leaves 4-8cm, narrowly to broadly ovate or obovate, weakly asymmetric to subequal at base, with 8-12 pairs of lateral veins; petiole 4-7mm. Cornwall to S Hants. The isolated biotype in S Hants and eastern Dorset (*U. stricta* var. *goodyeri* Melville) may merit separate subspecific recognition; it differs in its much broader crown (but still with steeply ascending branches) and leaves with a broader base and petiole 6-12mm. If so *Goodyer's Elm* would become ssp. *angustifolia* and *Cornish Elm* would need a new ssp. name. **113**

c. Ssp. sarniensis (C.K. Schneid.) Stace (*U. sarniensis* (C.K. Schneid.) H.H. Bancr., *U. minor* var. *sarniensis* (C.K. Schneid.) Richens) - *Jersey Elm*. Outline narrow; trunk extending to apex of tree; branches slightly ascending, progressively shorter towards tree apex; leaves similar to those of ssp. *angustifolia*. Native; Guernsey, probably intrd into rest of CI and Br, especially as a common roadside tree. **113**

5 x 6. U. minor x U. plotii = ?U. x viminalis Lodd. ex Loudon occurs in E Anglia and C En with the parents and is planted elsewhere. The influence of *U. plotii* is most often shown by the elongating short-shoots.

6. U. plotii Druce (*U. minor* var. *lockii* (Druce) Richens) - *Plot's Elm*. Outline narrow; leading shoot arching or pendent; branches all weak, pendent at ends; short-shoots mostly continuing growth as long-shoots; leaves 3-7cm, elliptic, ± rough on upperside, subequal at base, with 7-10 pairs of lateral veins; petiole 3-6mm, slender. Native; hedgerows; C En, possibly E Wa; endemic. **R** **113**

7. U. laevis Pall. - *European White-elm*. Tree to 21m; outline ± orbicular; trunk often with conspicuous burs, dividing low down into long spreading branches; leaves 4-12cm, obovate to broadly so, glabrous to densely softly pubescent on lowerside, variably and often strongly asymmetric at base, with 10-18 pairs of lateral veins; petiole c.3-9mm. Intrd; formerly planted but now rare, vigorously suckering from old trees in wood in 1 site in Cards; C & E Europe. The long pedicels and yellowish-green rather than reddish flowers are diagnostic; winter buds are also more acute than in the native spp. **113**

36. CANNABACEAE - *Hop family*

Annual or perennial herbs; leaves palmately lobed to ± palmate, petiolate, alternate or opposite, stipulate. Flowers small and inconspicuous, in varied inflorescences, usually dioecious, hypogynous, actinomorphic; perianth small to ± absent, of 1 whorl of 5 completely or partly fused segments; male flowers with 5 stamens; female flowers with 1-celled ovary with 1 ovule; styles 2; fruit an achene subtended by persistent bract.

Both representatives are very well known and distinctive.

1 Erect annual; at least upper leaves alternate **1. CANNABIS**
1 Rampant perennial climber; leaves all opposite **2. HUMULUS**

1. CANNABIS L. - *Hemp*
Upper leaves alternate, lower sometimes opposite; inflorescences axillary clusters towards stem apex, the male looser, the female ripening to form irregular groups of achenes and bracts.

1. C. sativa L. - *Hemp*. Erect unbranched or little-branched annual to 2.5m; leaves palmately divided to base or almost so into 3-9 lanceolate lobes; (2n=20). Intrd; formerly grown for fibre and still illicitly on small scale for drug, now again being grown in trials as a fibre crop; imported in mixed birdseed and a common casual on tips, and in parks and farms where birds are fed; much of BI, especially urban areas in S; S & W Asia.

2. HUMULUS L. - *Hop*
Leaves all opposite; male inflorescences loose spreading axillary panicles; female inflorescences small dense capitate clusters ripening to cone-like fruiting heads, in rather loose panicles.

1. H. lupulus L. - *Hop*. Scrambling perennial climber to 8m; leaves palmately divided c.3/4 to base into 3-5 ovate lobes; fruiting heads commonly 3-5cm, up to c.10cm in cultivars; 2n=20. Native; hedgerows, scrub and fen-carr; S Br and CI; cultivated especially in SE En and SW Midlands for brewing industry and very widely natd almost throughout BI.

37. MORACEAE - *Mulberry family*

Trees or shrubs with latex, leaves simple, deciduous, often palmately lobed, petiolate, alternate, stipulate when young. Flowers small and inconspicuous, crowded into dense heads or into hollow receptacles (*figs*), monoecious, hypogynous, actinomorphic; perianth reduced, of 1 whorl of 4-5 free segments; male flowers with 4-5 stamens; female flowers with 1-celled ovary with 1 ovule; styles 1-2; stigma linear or capitate; fruiting head a mass of drupes surrounded by succulent perianth or succulent receptacle.
Both representatives are very well known and distinctive.

1 Fruiting head raspberry-like in appearance; stipule-scar on 1 side of stem; latex watery **1. MORUS**
1 Fruiting head ± pear-shaped; stipule-scar completely encircling stem; latex milky **2. FICUS**

1. MORUS L. - *Mulberries*
Leaves mostly unlobed; stipules separate, their scars not encircling stem; fruiting head raspberry-like.

1. M. nigra L. - *Black Mulberry*. Upright tree to 14m; leaves 6-20cm, usually simple, sometimes palmately lobed, cordate at base, pubescent; fruiting heads raspberry-like, sessile, red when mature; (2n=308). Intrd; waste ground and walls, bird-sown from cultivated trees; natd by Thames in Middlesex, frequently grown in parks and seedlings rarely found in scattered localities; C Asia.

2. FICUS L. - *Fig*
Latex milky; leaves mosty deeply palmately lobed; stipules fused, their scars encircling stem; fruiting head pear-shaped.

1. F. carica L. - *Fig*. Spreading shrub or small tree to 10m; leaves 10-20cm, cordate at base, with 3-5 obtuse to rounded lobes, sparsely and roughly hairy;

fruiting head the characteristic 'fig', green to blackish at maturity; (2n=26). Intrd; waste ground and walls, especially by rivers; natd in S Br, S Ir and CI, and rarely as far N as C Sc, bird-sown or from imported fruit; SW Asia.

38. URTICACEAE - *Nettle family*

Annual or perennial herbs; leaves simple, opposite or alternate, with or without stipules. Flowers axillary, small and inconspicuous, solitary or in crowded inflorescences, monoecious or dioecious, hypogynous, actinomorphic; perianth of 1 whorl of 4 often partly fused greenish to brownish segments; male flowers with 4 stamens; female flowers with 1-celled ovary with 1 ovule; style 1; stigma much branched; fruit an achene.

The 3 genera appear very different vegetatively, but are characterized by their inconspicuous, unisexual flowers with 4 perianth segments, 4 stamens, 1-celled superior ovary with 1 ovule, 1 style and densely branched stigma.

1 Leaves opposite, usually toothed and with stinging hairs, stipulate; stems erect **1. URTICA**
1 Leaves alternate, entire, without stinging hairs, without stipules; stems procumbent to decumbent 2
 2 Stems decumbent but not rooting at nodes; leaves mostly >10mm; flowers crowded **2. PARIETARIA**
 2 Stems procumbent and rooting at nodes; leaves rarely >6mm; flowers solitary **3. SOLEIROLIA**

1. URTICA L. - *Nettles*
Annual or perennial with erect stems; leaves opposite, stipulate, toothed and normally with stinging hairs; flowers in dense axillary usually elongate inflorescences, monoecious or dioecious; perianth of free segments, 2 inner longer than 2 outer and enclosing fruit.

Other spp. - U. incisa Poir. is a rare wool-alien from Australia and New Zealand. It is a monoecious perennial, glabrous apart from rather sparse stinging hairs, vegetatively rather similar to some subglabrous variants of *U. dioica* but lacking rhizomes; usually lower inflorescences are female and upper male. **U. pilulifera** L. (*Roman Nettle*), from S Europe, used to occur as a casual.

1. U. dioica L. (*U. galeopsifolia* Wierzb. ex Opiz) - *Common Nettle*. Strongly 113
rhizomatous and/or stoloniferous usually dioecious perennial to 1.5m; leaves and stems usually with abundant stinging hairs and more numerous smaller non-stinging hairs; terminal leaf-tooth longer than adjacent laterals; 2n=48, 52. Native; in many habitats, especially woodland, fens, cultivated ground and where animals defecate; abundant throughout BI. Very variable, especially in leaf-shape and pubescence; stingless, subglabrous and monoecious variants are known. Plants from Wicken Fen, Cambs, have stingless leaves densely pubescent on lowerside and have been referred to **U. galeopsifolia**, but this taxon is of very doubtful status.

2. U. urens L. - *Small Nettle*. Monoecious annual to 60cm; inflorescences each 113
with many female and few male flowers; leaves and stems with usually abundant stinging hairs but otherwise glabrous to sparsely hairy; terminal leaf-tooth about as long as adjacent laterals; 2n=24. Probably native; cultivated and waste ground; frequent throughout BI but commoner in E.

2. PARIETARIA L. - *Pellitories-of-the-wall*
Perennial with decumbent stems; leaves alternate, entire, softly pubescent, without

stipules; flowers in dense axillary short inflorescences, mostly unisexual, usually some bisexual; perianth of equal segments fused at least at base and enclosing fruit.

1. P. judaica L. (*P. diffusa* Mert. & W.D.J. Koch) - *Pellitory-of-the-wall.* Stems to 40(110)cm, much-branched, usually procumbent to ascending; leaves ovate to elliptic, acuminate, ≤4(6)cm; bracts fused at base; perianth of female flowers 2-2.3mm in fruit; perianth of bisexual flowers becoming tubular and ≥3mm in fruit; achenes 1-1.2mm; 2n=26. Native; on walls, rocks, cliffs and steep hedgebanks; frequent in BI, except rare in Sc and not in N.

2. P. officinalis L. - *Eastern Pellitory-of-the-wall.* Differs from *P. judaica* in more robust, to 70(160)cm, usually erect; leaves up to 12(20)cm, longer-acuminate and with longer petioles; bracts free; perianth of female flowers 2.7-3mm in fruit; perianth of bisexual flowers remaining bell-shaped and ≤3mm in fruit; achenes 1.5-1.8mm; (2n=14). Intrd; natd in old neglected woods and hedgerows; 1 locality each in Middlesex (since 1992) and S Essex (since 1966); C & SE Europe.

3. SOLEIROLIA Gaudich. (*Helxine* Req. non (L.) L.) - *Mind-your-own-business*
Perennial with procumbent stems rooting at nodes; leaves alternate, entire, sparsely pubescent, without stipules; flowers solitary in leaf-axils, monoecious; perianth of equal segments fused at least at base and enclosing fruit.

1. S. soleirolii (Req.) Dandy (*Helxine soleirolii* Req.) - *Mind-your-own-business.* Stems to 20cm, very slender; leaves suborbicular, <6mm; (2n=20). Intrd; natd on damp shady walls and banks; frequent in S En and CI, very scattered elsewhere and very rare in Sc; W Mediterranean islands.

39. JUGLANDACEAE - *Walnut family*

Trees; leaves pinnate, deciduous, alternate, without stipules. Flowers monoecious, the male in pendent catkins, the female in pendent catkins or 1-few in stiff clusters, epigynous, ± actinomorphic, in axil of a bract and with 2 bracteoles; perianth small and inconspicuous, 1-whorled, 1-5-lobed; male flowers with 3-many stamens; female flowers with 1-celled ovary with 1 ovule, styles 2; stigmas branched; fruit a drupe or winged nut.

Easily recognized trees with pinnate leaves and distinctive flowers and fruits.

1 Leaflets entire, aromatic when crushed; fruits 1-few in rigid clusters, with green husk and hard-shelled 'nut' inside **1. JUGLANS**
1 Leaflets serrate, not aromatic; fruits several-many in pendent clusters, with broad suborbicular wing around nut **2. PTEROCARYA**

1. JUGLANS L. - *Walnut*
Leaves with 3-9 entire, aromatic, ovate leaflets; female flowers 1-few in stiff clusters; fruits drupes, with green outer husk and hard-shelled edible 'nut' inside.

1. J. regia L. - *Walnut.* Tree to 24m, not suckering at base; winter buds sessile, glabrous, with bud-scales; fruits ellipsoid, mostly >3cm; (2n=32). Intrd; commonly planted and often surviving in wild places, but self-sown only in warmer parts; most of Br; SE Europe and Asia.

2. PTEROCARYA Kunth - *Wingnuts*
Leaves with (7)15-27(41) serrate, non-aromatic, lanceolate leaflets; female flowers in pendent catkins; fruits nuts with broad suborbicular wings derived from bracteoles.

1. P. fraxinifolia (Poir.) Spach - *Caucasian Wingnut*. Tree to 35m, usually forming thickets due to suckers if left undisturbed; winter buds stalked, pubescent, without bud-scales; fruits suborbicular, c.2cm across, broadly winged; (2n=32). Intrd; planted on field borders and embankments and occasionally running wild; SE En and SW Sc; Caucasus.

40. MYRICACEAE - *Bog-myrtle family*

Shrubs; leaves simple, deciduous, alternate, petiolate, without stipules, strongly aromatic. Flowers normally dioecious, sometimes monoecious, in stiff catkins, hypogynous, actinomorphic, in axil of a bract, the female with 2 bracteoles; perianth 0; male flowers with usually 4 stamens; female flowers with 1-celled ovary with 1 ovule; styles 2; stigmas linear; fruit a drupe or narrowly winged nut.

Easily recognized by the aromatic foliage and distinctive catkins.

1. MYRICA L. - *Bog-myrtles*

1. M. gale L. - *Bog-myrtle*. Erect shrub to 1.5(2)m; current season's twigs very sparsely pubescent; leaves 2-6cm, oblanceolate to linear-oblanceolate, serrate towards apex, very sparsely pubescent; catkins appearing before leaves, on twigs of previous season that do not continue growth; fruit a very narrowly 2-winged nut; 2n=48. Native; wet moorland and heathland, bogs and fens; throughout most of Br and Ir but mostly in NW, there often abundant.

2. M. pensylvanica Loisel. ex Duhamel (*M. cerifera* auct. non L., *M. caroliniensis* auct. non Mill.) - *Bayberry*. Shrub to 2m; current season's twigs densely pubescent; leaves 3-8cm, oblanceolate to obovate or elliptic, usually serrate or dentate towards apex, sparsely pubescent; catkins appearing before leaves, on twigs of previous season that continue forward growth and leaf production; fruit a globose drupe c.3-5mm, with waxy bloom; (2n=16). Intrd; planted for cover and winter effect and natd on wet heathland in N & S Hants; E N America.

41. FAGACEAE - *Beech family*

Trees; leaves simple, deciduous or evergreen, alternate, petiolate, stipulate at least when young. Flowers monoecious, inconspicuous, the male in pendent or stiff catkins or 1-numerous in heads, the female in small groups of 1-few surrounded by numerous small scales, epigynous, actinomorphic; perianth 1-whorled, 4-7-lobed; male flowers with 4-40 stamens; female flowers with 3- or 6-celled ovary with 2 ovules per cell; styles 3-9; stigmas linear; fruit a nut, 1-3(6) surrounded by a *cupule* formed from the fused scales.

Catkin-bearing trees distinguished from other such families by the 3- or 6-celled ovary with 3-9 styles.

1 Male flowers in elongated catkins; nuts terete or with rounded corners; winter buds obtuse to rounded at apex — 2
1 Male flowers 1-many in short heads; nuts sharply 3-angled; winter buds acute at apex — 3
2 Male flowers in stiff catkins; cupule strongly spiny, completely enclosing 1-3(6) nuts during development — **3. CASTANEA**
2 Male flowers in pendent catkins; cupule not spiny, enclosing only lower part of 1 nut — **4. QUERCUS**
3 Male flowers 1-3 in stiff clusters; nuts usually 3 per cupule — **2. NOTHOFAGUS**
3 Male flowers numerous in pendent tassels; nuts usually 2 per cupule — **1. FAGUS**

1. FAGUS L. - *Beech*
Leaves deciduous; male flowers numerous in pendent heads, with 8-16 stamens; female flowers usually 2 together in erect, pedunculate clusters; nuts sharply 3-angled, 1-2 per cupule.

1. F. sylvatica L. - *Beech*. Tree to 46m; winter buds long and slender, finely 121
pointed; leaves 4-9cm, ovate to elliptic, entire, long-pubescent on margin; fruiting cupule with stiff pointed subulate scales; 2n=24. Native; well drained soils, often forming pure woods on chalk and soft limestone and sometimes acid sandstone; SE Wa and S En; also very widely planted and natd over all BI.

2. NOTHOFAGUS Blume - *Southern Beeches*
Leaves deciduous or evergreen, often slightly asymmetric at base; male flowers 1-3 in stiff clusters, with 8-40 stamens; female flowers 1-3 in ± sessile clusters; nuts sharply 3-angled, usually 3 per cupule.

Other spp. - Of the several spp. now being tried for forestry, **N. dombeyi** (Mirb.) Blume (*Coigue*), from Chile and W Argentina, with evergreen leaves up to 4cm, is now found in a few plantations in SW En and Wa.

1. N. obliqua (Mirb.) Blume - *Roble*. Deciduous tree to 30m; winter buds c.4mm; 121
leaves coarsely and doubly serrate, ± glabrous, with 7-11 pairs of lateral veins; cupules with short stiff scales; (2n=26). Intrd; now widely planted for forestry in W En, W Sc and Wa, often producing self-sown offspring; Chile and W Argentina.

1 x 2. N. obliqua x N. nervosa occurs spontaneously in mixed plantations mainly in SW En; endemic.

2. N. nervosa (Phil.) Krasser (*N. procera* Oerst.) - *Rauli*. Deciduous tree to 26m; 121
winter buds c.10mm; leaves 4-8cm, ovate-oblong, acute to rounded at apex, finely serrate to subentire or crenate, pubescent on veins on lowerside, with 14-24 pairs of lateral veins; cupules with deeply laciniate scales. Intrd; now widely planted and self-sowing as for *N. obliqua*; Chile and W Argentina.

3. CASTANEA Mill. - *Sweet Chestnut*
Leaves deciduous; flowers in long rather stiff insect-pollinated catkins, mostly male but female at base, the male with 10-20 stamens, the female usually in groups of 3; nuts with 2-4 rounded angles, 1-3(more) per cupule.

1. C. sativa Mill. - *Sweet Chestnut*. Tree to 35m; leaves 10-30cm, oblong- 121
lanceolate, regularly coarsely serrate with finely pointed teeth, ± glabrous at maturity; fruiting cupule densely covered with sharp spines; (2n=22, 24). Intrd; planted throughout BI, especially as coppiced woodland in SE En; setting seed and natd ± only in S En and CI; S Europe.

4. QUERCUS L. - *Oaks*
Leaves deciduous or evergreen; male flowers in pendent catkins, with 4-12 stamens; female flowers 1-few in stiff pedunculate or sessile clusters; nuts (acorns) terete, 1 per cupule.

1 Fruit cupule with erecto-patent to reflexed scales; at least terminal buds surrounded by persistent stipules 2
1 Fruit cupule with short appressed scales; buds not surrounded by persistent stipules 3
  2 Tree without green leaves in winter; leaf-lobes obtuse to acute or shortly apiculate **1. Q. cerris**
  2 Tree usually with some green leaves in winter; leaf-lobes mucronate to aristate **2. Q. x crenata**

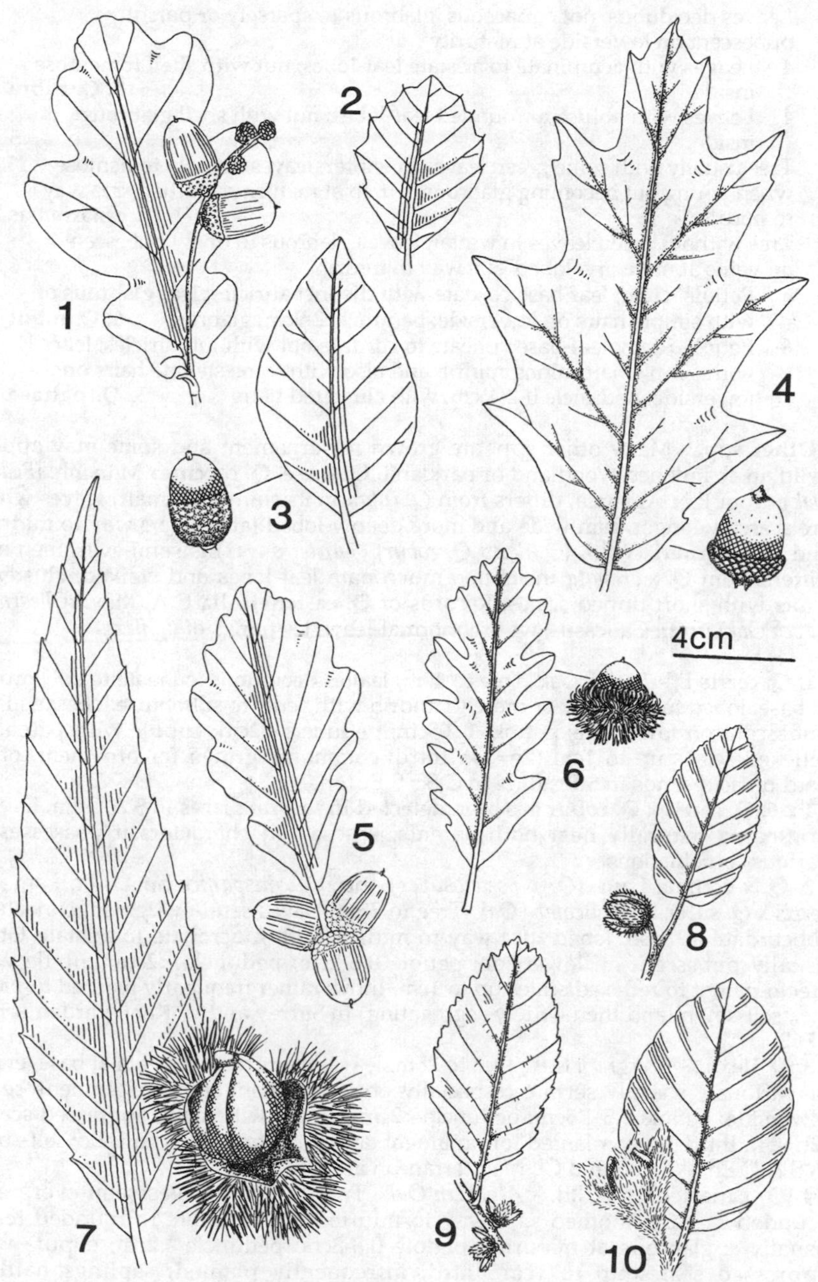

FIG 121 - Leaves and fruits of **Fagaceae**. 1, *Quercus robur*. 2-3, *Q. ilex*. 4, *Q. rubra*. 5, *Q. petraea*. 6, *Q. cerris*. 7, *Castanea sativa*. 8, *Nothofagus nervosa*. 9, *N. obliqua*. 10, *Fagus sylvatica*.

3 Leaves evergreen, coriaceous, grey-tomentose on lowerside even at maturity **3. Q. ilex**

3 Leaves deciduous, not coriaceous, glabrous to sparsely or patchily pubescent on lowerside at maturity 4

 4 Leaves with acuminate to aristate leaf-lobes; nut with shell tomentose inside **7. Q. rubra**

 4 Leaves with obtuse to rounded leaf-lobes; nut with shell glabrous inside 5

5 Tree usually with some green leaves in winter; leaves densely tomentose when young but becoming glabrous or ± so at maturity, toothed <1/3 way to midrib **4. Q. canariensis**

5 Tree without green leaves in winter; leaves glabrous to finely pubescent on veins at maturity, lobed ≥1/3 way to midrib 6

 6 Petiole <1cm; leaf-base cordate with distinct auricles; leaf glabrous or with simple hairs on lowerside; peduncle 2-9cm, glabrous **6. Q. robur**

 6 Petiole >1cm; leaf-base cuneate to cordate but without auricles; leaf with simple hairs along midrib and also with some stellate hairs on lowerside; peduncle 0-2(4)cm, with clustered hairs **5. Q. petraea**

Other spp. - Many other spp. are grown for ornament and some may appear 'wild' in abandoned woodland or parkland. Of these, **Q. coccinea** Münchh. (*Scarlet Oak*), from E N America, differs from *Q. rubra* in its shiny (not matt) leaves which are scarcely longer than wide and more deeply lobed (at least 2/3 way to midrib); and **Q. x turneri** Willd. (*Q. ilex* x *Q. robur*) (*Turner's Oak*) is semi-evergreen and differs from *Q.* x *crenata* in its non-mucronate leaf-lobes and stalked, clustered fruits with short-tipped scales. Records of **Q. castaneifolia** C.A. Mey. (*Chestnut-leaved Oak*), from Caucasus, were abnormal-leaved saplings of *Q. ilex*.

1. Q. cerris L. - *Turkey Oak*. Tree to 40m; leaves deciduous, cuneate to subcordate 121
at base, lobed to c.1/2 way or more to midrib with acute to subobtuse lobes, usually pubescent on lowerside; petiole 1-2.5cm; peduncle <2cm; cupule with patent to reflexed scales up to 1cm (2n=24). Intrd; commonly grown for ornament, often natd on acid sands in S Br; BI N to C Sc; S Europe.

1 x 6. Q. cerris x Q. robur has been detected in several places in S & C En, having originated naturally near both parents, whose leaf characters it possesses in various combinations.

2. Q. x crenata Lam. (*Q. x pseudosuber* Santi, *Q. x hispanica* auct. non Lam.; *Q. cerris* x *Q. suber* L.) - *Lucombe Oak*. Tree to 35m; leaves semi-evergreen, cuneate to subcordate at base, lobed <1/2 way to midrib, with mucronate to aristate lobes, usually pubescent on lowerside; petiole 0.5-2cm; peduncle <2cm; cupule with erecto-patent to reflexed scales up to 1cm. Intrd; rather frequently planted in parks etc., self-sown (and then widely segregating) in Surrey and W Kent; garden origin 1762.

3. Q. ilex L. - *Evergreen Oak*. Tree to 25m; leaves evergreen, cuneate at base, entire 121
or variously sharply serrate (especially on juvenile shoots), grey-tomentose on lowerside; petiole 0.5-1.5cm; peduncle <2cm; cupule with small appressed scales; (2n=24). Intrd; much planted for ornament, and often for shelter in E En; self-sown in S & C En, Wa, S Ir and CI; Mediterranean and SW Europe.

4. Q. canariensis Willd. - *Algerian Oak*. Tree to 31m; leaves semi-evergreen, rounded at base, toothed <1/3 way to midrib with subacute to rounded teeth, usually ± glabrous at maturity; petiole 0.8-3cm; peduncle <2cm; cupule with appressed scales up to 1cm. Intrd; infrequently planted, saplings natd in woodland in Surrey, perhaps now gone; W Mediterranean.

5. Q. petraea (Matt.) Liebl. - *Sessile Oak*. Tree to 42m, of less spreading habit and 121
with a taller, straighter trunk than *Q. robur*; leaves deciduous, usually ± elliptic, cuneate to cordate with no or weak auricles at base, lobed ± as in *Q. robur* but

usually with 5-8 pairs of less deep lobes, with frequent simple and some stellate hairs on lowerside; petiole 13-25mm; peduncle 0-2(4)cm; cupule with small appressed scales; 2n=24. Native; almost throughout Br and Ir and often abundant and forming dense woodland, epecially on shallow, sandy, acid soils and in N and W, often over 300m altitude, but absent or only intrd in some areas.

5 x 6. Q. petraea x Q. robur = Q. x rosacea Bechst. occurs throughout BI in areas where 1 or both parents occur, occasionally being commoner than either. It combines in various ways the leaf and fruit characters of the 2 spp., and is fertile.

6. Q. robur L. - *Pedunculate Oak.* Tree to 40m; leaves deciduous, usually obovate, **121**
cordate and with well-marked auricles at base, lobed <1/2 way to midrib with 3-6 pairs of rounded lobes, glabrous or with few simple hairs on lowerside; petiole (0)2-3(7)mm; peduncle 2-9cm; cupule with small appressed scales; 2n=24. Native; almost throughout BI and often abundant and forming dense woodland, on a wide range of soils but especially deep rich ones, rarely over 300m altitude, not native in some areas in N and W and on acid shallow soils.

7. Q. rubra L. (*Q. borealis* F. Michx.) - *Red Oak.* Tree to 34m; leaves deciduous, **121**
often turning red in autumn, cuneate at base, lobed c.1/2 way to midrib with acuminate to aristate lobes, ± glabrous; petiole 2-5cm; peduncle <2cm; cupule with small appressed scales; (2n=24). Intrd; much planted for ornament and sometimes for forestry or screening, especially on shallow sandy soils, often self-sowing; En, Wa and CI; E N America.

42. BETULACEAE - *Birch family*
(Corylaceae)

Trees or shrubs; leaves simple, deciduous, alternate, petiolate, stipulate when young. Flowers monoecious, inconspicuous, epigynous, actinomorphic; perianth 0 or 1-whorled and very small; male flowers numerous in mostly pendent catkins, 1 or 3 per bract, with 0, 2 or 4 bracteoles per group, with 2-14 stamens; female flowers in small groups on erect or pendent catkins, 2-3 per bract, with 2 or 4 bracteoles per group, with 2-celled ovary with 1 ovule per cell; styles 2; stigmas linear; fruit a nut, winged or not.

Distinguished from other monoecious catkin-bearing trees by the simple leaves and 2-celled ovary with 2 styles. Corylaceae (*Corylus* and *Carpinus*) are often separated off.

1 Fruits winged, in compact cone-like structure formed from dried-out bracts; male flowers 3 per bract — 2
1 Fruits not winged but each with enlarged lobed or laciniate bracts at base; male flowers 1 per bract — 3
  2 Bracts of fruiting cones falling from axis with the fruits, distinctly 3-lobed; male catkins opening with the leaves; stamens 2, the lobes of each well separated — **1. BETULA**
  2 Bracts of fruiting cones persistent, falling with the whole cone, obscurely 5-lobed; male catkins opening before the leaves; stamens 4, the lobes of each slightly separated — **2. ALNUS**
3 Winter buds rounded at apex; fruits 1-several in short clusters, each surrounded by cupule of laciniate bracts — **4. CORYLUS**
3 Winter buds acute; fruits several in pendent catkins; each with large 3-lobed bract — **3. CARPINUS**

1. BETULA L. - *Birches*
Trees or shrubs; male flowers 3 per bract, with 2 bracteoles per group, with minute perianth and 2 stamens; female flowers in stiff erect catkin, 3 per bract, with 2 bracteoles per group, without perianth; fruits winged, in compact cone-like

structure which disintegrates from axis at maturity to release fruits and strongly 3-lobed bracts.

1 Leaves <2cm, ± orbicular, obtuse to truncate at apex; petiole <5mm; male catkins not exposed in winter, erect at least until anthesis, <1cm **3. B. nana**
1 Leaves >2cm, ± ovate, subacute to acuminate at apex; petiole >5mm; male catkins exposed in winter, pendent, >2cm 2
 2 Leaves acuminate at apex, distinctly doubly serrate with prominent primary teeth, truncate to broadly cuneate at base, those on mature shoots glabrous **1. B. pendula**
 2 Leaves acute to subacute at apex, rather evenly or irregularly serrate without obvious primary teeth, rounded to cuneate at base, glabrous to pubescent **2. B. pubescens**

Other spp. - Several spp. are grown in parks and on roadside verges, especially the N N American **B. papyrifera** Marshall (*Paper Birch*) and the Himalayan **B. jacquemontii** Spach (*Jacquemont's Birch*). Both have rather large doubly serrate leaves and whiter barks than the native spp.; the former has glabrous and the latter pubescent shoots.

1. B. pendula Roth - *Silver Birch.* Tree to 30m; bark silver-white above, erupting 125
into irregular blackish fissures below; twigs often pendent, glabrous; leaves ovate, doubly serrate, acuminate, glabrous; each wing of fruit ≥2x as wide as body, extending well beyond stigmas at apex of body; 2n=28. Native; forming woods on light, mostly acid soils, especially heathland, usually in drier places than *B. pubescens*; throughout almost all Br and Ir, commoner in S. Young saplings and regrowth can have pubescent twigs and leaves.

1 x 2. B. pendula x B. pubescens = B. x aurata Borkh. occurs in many areas, often in absence of 1 or both parents. It varies in characters between the 2 parents, in chromosome number (2n=30-48, 56), and in fertility (fully fertile to highly sterile). Its frequency is hard to determine due to uncertainty of parental limits. Many 'intermediates' have 2n=56 and could be hybrids or *B. pubescens. B. pendula*
and *B. pubescens* are most reliably separated by the Atkinson Discriminant 125
Function: Positive values = *B. pendula*; negative values = *B. pubescens*; 93% certainty of correct answer, most errors being *B. pubescens* with slightly positive values. Experimental hybrids are usually closer to *B. pubescens* in diagnostic characters and many would be determined as that in the field.

2. B. pubescens Ehrh. - *Downy Birch.* Tree to 24m; bark variously brown, grey or 125
white but rarely white with strongly contrasting black fissures below; twigs usually not pendent, pubescent to glabrous; leaves ovate, singly or irregularly to somewhat doubly serrate, acute or subacute to slightly acuminate, pubescent to glabrous; each wing of fruit c.1-1.5x as wide as body, not extending beyond stigmas at apex of body; 2n=56. Native; in similar places to *B. pendula* but favouring wetter and more peaty soils, especially upland; throughout almost all Br and Ir, commoner in N.

a. Ssp. pubescens. Usually a tree; young twigs and petioles usually pubescent, sometimes glabrous; leaves mostly >3cm; each wing of fruit wider than body. Throughout BI but ± confined to lowland in N.

b. Ssp. tortuosa (Ledeb.) Nyman (ssp. *carpatica* (Willd.) Asch. & Graebn., ssp. *odorata* sensu E.F. Warb.). Usually shrubby; young twigs and petioles pubescent or ± glabrous; leaves mostly <3cm; each wing of fruit not or little wider than body. Upland areas of N Br, replacing ssp. *pubescens*. A rather ill-defined taxon, but worth recognising due to its distinctive distribution. Small-leaved variants of it might have arisen by introgression from *B. nana*.

2 x 3. B. pubescens x B. nana = B. x intermedia Thomas ex Gaudin occurs within the range of *B. nana* in Sc. It varies considerably between the parents, especially in leaf characters, but seems only slightly fertile, though introgression

Atkinson Discriminant Function = 12LTF + 2DFT – 2LTW – 23
(means of 5 short-shoot leaves)

LTF = Leaf Tooth Factor - number of teeth projecting beyond line connecting tips of main teeth at ends of 3rd and 4th lateral veins from apex, subtracted from total number of teeth between these 2 main teeth.
LTW = Leaf Tip Width - width in mm of leaf 1/4 distance from apex to base.
DFT = Distance to First Tooth - distance in mm from apex of petiole to first tooth.

Leaf 1 (***Betula pubescens***) = –35 Leaf 3 (***Betula pendula***) = +21

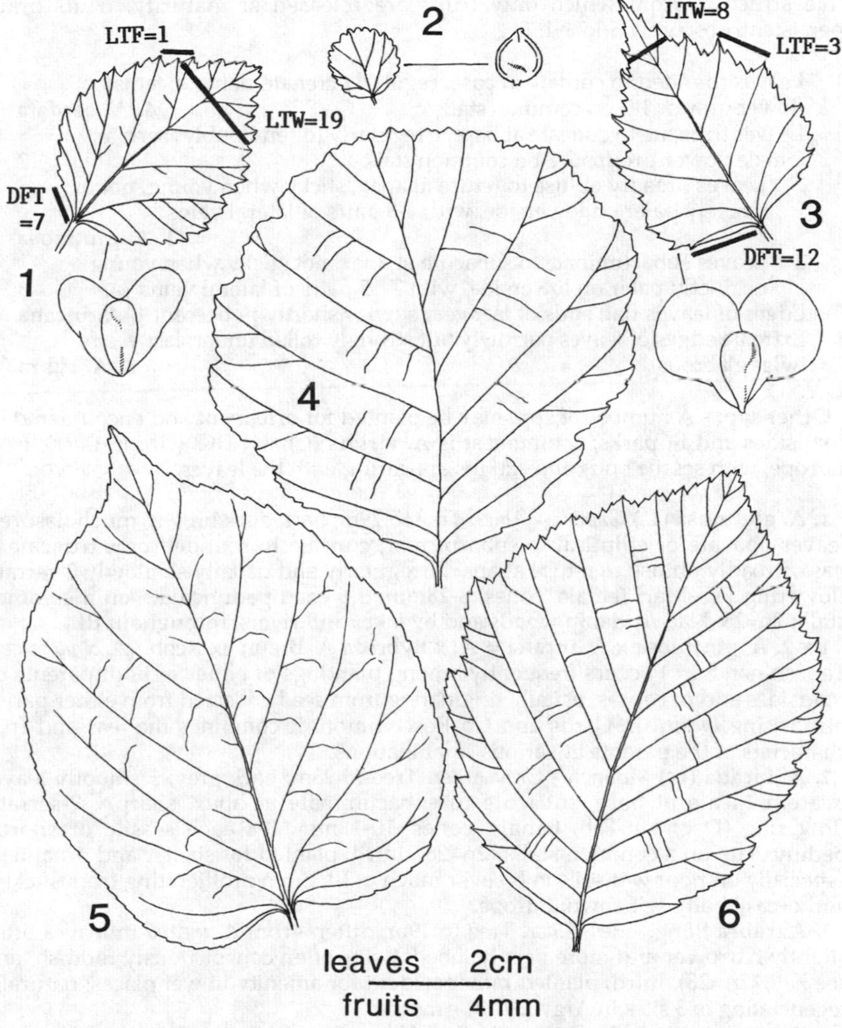

FIG 125 - ***Betula, Alnus.*** 1-3, leaves and fruits of ***Betula,*** showing Atkinson Discriminant Function. 1, ***B. pubescens.*** 2, ***B. nana.*** 3, ***B. pendula.*** 4-6, leaves of ***Alnus.*** 4, ***A. glutinosa.*** 5, ***A. cordata.*** 6, ***A. incana.***

might occur.

3. B. nana L. - *Dwarf Birch*. Shrub to 1m, with procumbent to ascending stems; R
twigs stiff, pubescent; leaves ± orbicular, ± rounded at both ends, deeply and 125
regularly crenate, glabrous at maturity; each wing of fruit much narrower than body and not extending beyond it; 2n=28. Native; upland moors and bogs on peat; NW Yorks, S Northumb and C & N mainland Sc, formerly Lanarks.

2. ALNUS Mill. - *Alders*

Trees; male flowers 3 per bract, with 4 bracteoles per group, with minute perianth and 4 stamens; female flowers in small stalked groups, 2 per bract, with 4 bracteoles per group, without perianth; fruits narrowly winged, in compact cone-like structure from which only fruits are released at maturity; fruits bracts persistent, obscurely 5-lobed.

1 Leaves rounded to cordate at base, regularly crenate-dentate; female flower-groups 1-3 on common stalk **4. A. cordata**

1 Leaves truncate to cuneate at base, irregularly (often doubly) serrate; female flower-groups 3-8 on common stalk 2

  2 Leaves broadly obtuse to retuse at apex, sticky when young, not or scarcely paler on lowerside, with 4-8 pairs of lateral veins **1. A. glutinosa**

  2 Leaves subacuminate to subacute at apex, not sticky when young, distinctly paler on lowerside, with 7-15 pairs of lateral veins 3

3 Edges of leaves flat; ends of last year's twigs shortly pubescent **2. A. incana**

3 Extreme edges of leaves narrowly but strongly rolled under; last year's twigs glabrous **3. A. rubra**

Other spp. - A number of spp. may be planted for ornament and encountered on roadsides and in parks; commonest is **A. viridis** (Chaix) DC. (*Green Alder*), from Europe, with sessile buds and catkins appearing with the leaves.

1. A. glutinosa (L.) Gaertn. - *Alder*. Tree to 29m; bark dark brown, much fissured; 125
leaves obovate or elliptical to suborbicular, cuneate to rounded or ± truncate at base, broadly obtuse to retuse at apex, irregularly and usually shallowly 2-serrate; flowering Feb-Mar; female 'cones' 8-28mm, 3-6 each pedunculate on a common stalk; 2n=28. Native; damp woods and by lakes and rivers; throughout BI.

1 x 2. A. glutinosa x A. incana = A. x hybrida A. Braun ex Rchb. (*A. x pubescens* Tausch non Sart.) occurs frequently among plantings of either or both parents on roadsides and in copses, usually originating from seed collected from either parent but arising *in situ* in Cards and Co Roscommon. It combines the leaf and fruit characters of the parents in various combinations.

2. A. incana (L.) Moench - *Grey Alder*. Tree to 23m; bark grey, ± smooth; leaves 125
ovate, cuneate at base, subacute to subacuminate at apex, sharply 2-serrate; flowering (Dec) Jan-Feb; female 'cones' 10-18mm, 3-8 each sessile or shortly pedunculate on a common stalk; (2n=28). Intrd; planted for shelter and ornament, especially on poor wet soils in N, over much of BI; often proliferating from suckers and occasionally self-sown; Europe.

3. A. rubra Bong. - *Red Alder*. Tree to 19m; differs from *A. incana* in leaves often 236
slightly narrower and more deeply lobed; twigs often conspicuously reddish; and see key; (2n=28). Intrd; planted rather seldom for amenity in wet places; naturally regenerating in 3 sites in Argyll; W N America.

4. A. cordata (Loisel.) Duby - *Italian Alder*. Tree to 28m; bark greyish, rather 125
smooth; leaves ovate, cordate to rounded at base, obtuse to shortly acuminate at apex, regularly crenate-dentate; flowering Feb-Mar; female 'cones' 15-30mm, 1-3 each pedunculate on a common stalk; (2n=28, 42). Intrd; rather frequently roadside or mass planted for ornament over much of BI, especially in S, rarely self-sown but

natd in N Somerset; Italy and Corsica.

3. CARPINUS L. - *Hornbeam*
Trees; male flowers 1 per bract, without bracteoles, without perianth, with c.10 stamens; female flowers in pendent catkin, 2 per bract, with 2 bracteoles per group, with minute irregularly toothed perianth; fruits not winged, in conspicuous pendent catkin, each nut subtended by much enlarged 3-lobed bract.

1. C. betulus L. - *Hornbeam*. Tree to 32m; leaves ovate, rounded to cordate at base, acute to acuminate at apex, sharply 2-serrate, with 7-15 pairs of lateral veins, glabrous except on veins on lowerside; male catkins opening just before leaves (Apr); nuts subtended by 3-lobed bract to 5cm; (2n=16, 32, 64). Native; forming woods and copses on clay soils in SE En extending to Mons and Cambs; much planted on roadsides and as hedging over rest of BI, often as a fastigiate cultivar.

4. CORYLUS L. - *Hazels*
Small trees or large shrubs; male flowers 1 per bract, with 2 bracteoles (fused to bract) per group, without perianth, with c.4 stamens; female flowers in small sessile, bud-like group, 2 per bract, with 2 bracteoles per group, with minute irregularly lobed perianth; fruits not winged, large and edible, each nut surrounded by much enlarged, laciniate, fused bracts.

1 Bracts around ripe nut forming ± cylindrical structure slightly narrowed beyond nut apex **2. C. maxima**
1 Bracts around ripe nut forming ± bell-shaped structure not reaching or not narrowed beyond nut apex 2
 2 Bracts slightly shorter to slightly longer than ripe nut, laciniate <1/2 way to base, with shaggy hairs on outside only near base **1. C. avellana**
 2 Bracts c.2x as long as ripe nut, laciniate >1/2 way to base, with long shaggy hairs on outside nearly to apex **3. C. colurna**

1. C. avellana L. - *Hazel*. Several-stemmed shrub to 6(12)m; leaves suborbicular, usually cordate at base, obtuse to cuspidate at apex, sharply 2-serrate, softly pubescent; male catkins up to 8cm, opening well before leaves (Jan-Mar); nuts 1-few in cluster, up to 2cm, globose to broadly ovoid, each surrounded by girdle of fused, laciniate bracts slightly shorter than to slightly longer than nut and not becoming reflexed when dry; 2n=22 (22, 28). Native; hedgerows, scrub and woodland; whole of BI.

1 x 2. C. avellana x C. maxima, with intermediate fruits, has been reported from Suffolk; many (perhaps most) *Kentish Cobs* are of this parentage.

2. C. maxima Mill. - *Filbert*. Several-stemmed shrub or sometimes small tree to 10m; differs from *C. avellana* in the ovoid nuts usually >2cm, with the fused surrounding bracts much longer than nut and contracted beyond its apex; (2n=22, 28). Intrd; grown for its nuts (*Kentish Cobs*) in orchards in Kent and in gardens elsewhere, sometimes found as a relic and rarely self-sown; SE Europe and SW Asia. Fertile fruits develop equally well following pollination by *C. maxima* or *C. avellana*.

3. C. colurna L. - *Turkish Hazel*. Tree to 26m with single trunk; leaves narrower, more acuminate and more lobed, lighter or brighter green, and with longer petiole than in *C. avellana*; male catkins up to 12cm; nuts up to 2cm, usually distinctly flattened, with thicker shell than in *C. avellana*, with girdle of deeply laciniate bracts much longer than nut and becoming strongly reflexed when dry; (2n=28). Intrd; grown in plantation on dunes in S Lancs for many years, now being planted as street tree in En; SE Europe and SW Asia.

43. PHYTOLACCACEAE - *Pokeweed family*

Herbaceous perennials, ± woody at base; leaves alternate, simple, entire, petiolate, without stipules. Flowers many in leaf-opposed racemes, bisexual, hypogynous, actinomorphic; perianth of 1 whorl of 5 free segments, usually petaloid; stamens 7-16; carpels 6-10 in 1 whorl, ± free to fused, each with 1 ovule and 1 short style; fruit succulent, berry-like, 6-10-lobed from close appression or fusion of the carpels.

Distinctive in habit and in flower and fruit structure.

1. PHYTOLACCA L. - *Pokeweeds*

Other spp. - 2 other Asian spp., **P. latbenia** (Buch.-Ham.) H. Walter and **P. esculenta** Van Houtte, differing from *P. acinosa* and *P. polyandra* in minor features of flower colour, stamen number, and pubescence, are grown in gardens and may also appear bird-sown outside. The distinctness of all 4 remains to be confirmed. **P. americana** L. (*American Pokeweed*), from N America, with narrower leaves, longer racemes ± pendent in fruit and carpels fused except at tips, is the sp. usually recorded, but usually and perhaps always in error for an Asian sp.

1. P. acinosa Roxb. - *Indian Pokeweed*. Stems branching, erect, to 1.5(2)m; leaves ovate to ovate-elliptic, ≤30cm; racemes erect even in fruit, many-flowered, c. as long as leaves; perianth 3-4mm, whitish-green to red; stamens 7-15, in 1-2 whorls; carpels 6-9, free or united just at base; fruit c.5-8mm, blackish, with red juice; (2n=18, 36, 72). Intrd; grown in gardens and bird-sown usually as individual plants outside; occasional in S En, S Wa and CI; Himalayas to Japan.

2. P. polyandra Batalin (*P. clavigera* W.W. Sm.) - *Chinese Pokeweed*. Differs from *P. acinosa* mainly in carpels ± completely united (best observed in flower); stamens 12-16, in 2 whorls. Intrd; grown and bird-sown as for *P. acinosa*; natd in Jethou (CI) since 1985, formerly in E Suffolk; China. Distinction from *P. acinosa* not certain.

43A. NYCTAGINACEAE

MIRABILIS jalapa L. (*Marvel-of-Peru*), from Tropical America, is grown in gardens in S En and CI and is occasionally found on rubbish-tips. It is an erect, branched perennial to 1m, with opposite, simple, entire, leaves without stipules and a trumpet-shaped, variously coloured (usually red) perianth consisting of a narrow tube c.3cm and a limb of 5 spreading lobes c.25mm across.

44. AIZOACEAE - *Dewplant family*
(Tetragoniaceae)

Annuals or slightly to moderately woody perennials, usually glabrous; leaves opposite or alternate, simple, usually entire, usually thick and succulent, without stipules; petioles present or 0. Flowers solitary or in cymes, actinomorphic, usually bisexual, epigynous to perigynous; sepals 4-5(6), free or fused at base to form short tube above ovary, usually succulent; 'petals' (actually petaloid staminodes) 0 or numerous and ± free and in 1-several rows, usually linear, often brightly coloured; stamens 3-numerous, usually in several rows; ovary with c.3-20 cells, with 1-many variously arranged ovules in each cell; styles as many as carpels, free or fused at base; stigmas minute or linear; fruit a hard, dehiscent or indehiscent capsule with 3-many seeds, or succulent with many seeds.

Most spp. are easily recognized by their succulent leaves ('ice-plants') and colourful, many-petalled daisy-like flowers. There remain many taxonomic problems at both generic and specific levels. Identification is much easier in the fresh state.

1 Leaves >15mm wide, strongly flattened, abruptly petiolate 2
1 Leaves <15mm wide, usually <2x as wide as thick, sessile or gradually petiolate 3
2 Leaves alternate; petals 0 **9. TETRAGONIA**
2 Leaves opposite; petals present **1. APTENIA**
3 Fruit succulent; seeds embedded in mucilage; stigmas c.8-20 **8. CARPOBROTUS**
3 Fruit woody, without copious mucilage; stigmas <8 4
4 Leaves c. as long as wide, triangular in section, sparsely but conspicuously toothed on all 3 angles **4. OSCULARIA**
4 Leaves >2x as long as wide, triangular in section to terete, entire or minutely toothed 5
5 Leaves widest and thickest near apex, with 3 acute angles with a conspicuous (c.0.5mm wide), translucent, shallowly and irregularly toothed border on each **7. EREPSIA**
5 Leaves not or scarcely wider or thicker near apex, terete or with 3 rounded or rarely acute angles each with a very narrow, ± entire, translucent border 6
6 Main stems long-procumbent, often rooting at nodes, mat-forming 7
6 Stems upright to ascending or rarely ± procumbent, stiff and strongly woody below, forming upright or spreading shrub 8
7 Leaves covered with rounded whitish papillae; young stems with white, patent or reflexed hairs **6. DROSANTHEMUM**
7 Leaves not papillose; stems not hairy **5. DISPHYMA**
8 All leaves <3cm long and/or <5mm thick; styles stout (c.0.4mm wide at mid length); stigmas tuft-like; flowers (20)30-50mm across **3. LAMPRANTHUS**
8 Some leaves ≥3cm long and ≥5mm thick; styles filiform (c.0.1mm wide at mid length); stigmas tapered; flowers 18-30mm across **2. RUSCHIA**

Other genera - **MESEMBRYANTHEMUM** L. differs from all the above in that the whole plant is densely covered with crystal-like vesicles. **M. crystallinum** L. (*Common Ice-plant*), from Mediterranean, has flat ovate petiolate leaves and is a rare wool-alien.

1. APTENIA N.E. Br. - *Heart-leaf Ice-plant*

Perennial; leaves flattened, succulent, entire, petiolate; flowers axillary and terminal, solitary; sepals 4; petals numerous; stigmas 4; capsule woody, opening by 4 valves, with axile placentation.

1. A. cordifolia (L.f.) Schwantes - *Heart-leaf Ice-plant*. Stems woody below, freely **131**
branched, spreading, to 60cm; leaves ovate, cordate to cuneate at base, acute to obtuse at apex, papillose; flowers (5)10-18mm across, purplish-red; (2n=18). Intrd; grown as pot-plant or summer bedding plant and escaping on walls and dry ground in Scillies, W Cornwall and CI; S Africa.

2. RUSCHIA Schwantes - *Shrubby Dewplant*

Dwarf shrubs, strongly woody below; leaves triangular in section with rounded angles to ± terete, entire, sessile, opposite pairs united into common sheath; flowers terminal, usually 3-4 together; sepals 5; petals numerous; stigmas 4-5; capsule woody, opening by 4-5 wingless valves, with parietal placentation.

1. R. caroli (L. Bolus) Schwantes - *Shrubby Dewplant*. Stems ± erect to ± 131
procumbent, to 80cm; leaves 1.5-7cm, 2-8mm wide and thick, acute at apex, glaucous with green dots; flowers 18-30mm across, purplish-red. Intrd; grown for ornament and escaping on sea-cliffs in Scillies; S Africa.

3. LAMPRANTHUS N.E. Br. - *Dewplants*
Dwarf shrubs, strongly woody below; leaves triangular in section with acute angles to ± terete, entire or nearly so, sessile, opposite pairs shortly united; flowers terminal, 1-many together; sepals 5; petals numerous; stigmas 5; capsule woody, opening by 5 winged valves, with parietal placentation. A large critical S African genus.

Other spp. - L. conspicuus (Haw.) N.E. Br., with leaves 5-7cm and purplish flowers c.5cm across, has been reported from Scillies, and **L. scaber** (L.) N.E. Br., with leaves 2-3cm and with roughly papillose edges and purplish flowers c.3cm across, has been reported from Jersey.

1. L. falciformis (Haw.) N.E. Br. - *Sickle-leaved Dewplant*. Erect, bushy, to 30cm; 131
leaves (6)10-15(20) x 1.5-5mm, falcate, triangular in section with acute to obtuse angles, markedly mucronate at apex, conspicuously dotted; flowers 3.5-4.5cm across, pale pink; (2n=36). Intrd; cultivated near sea and natd on walls, hedgebanks, cliffs and quarries; Pembs, Scillies, ?CI; S Africa.

2. L. roseus (Willd.) Schwantes (*?L. multiradiatus* (Jacq.) N.E. Br.) - *Rosy* 131
Dewplant. Erect bushy shrub to 60cm, similar to *L. falciformis;* differing in leaves (10)20-40mm, not or slightly falcate, acute or acuminate to obtuse or obscurely mucronate at apex, less conspicuously dotted; flowers (2)3-5cm across; (2n=18). Intrd; in similar places to *L. falciformis*; W Cornwall, E & W Cork, CI; S Africa.

4. OSCULARIA Schwantes - *Deltoid-leaved Dewplant*
Perennial; leaves sharply triangular in section, with conspicuous distant teeth on all 3 angles, opposite pairs shortly united; flowers terminal, 1-3 together; sepals 5; petals numerous; stigmas 5; capsule woody, opening by 5 narrowly winged valves, with parietal placentation.

1. O. deltoides (L.) Schwantes (*Lampranthus deltoides* (L.) Glen) - *Deltoid-leaved* 131
Dewplant. Stems woody below, with spreading branches to 50cm; leaves 6-18mm, scarcely less wide and thick, narrowed towards base, acute to truncate at apex, glaucous; flowers 10-20mm across, pink; (2n=18). Intrd; grown on walls and banks and escaping or persisting; Scillies, very rare in Guernsey; S Africa.

5. DISPHYMA N.E. Br. - *Dewplants*
Perennial, with procumbent, succulent, somewhat woody main stems; leaves triangular in section with rounded angles to ± terete, entire, smooth, sessile, opposite pairs very shortly united; flowers terminal on short, ± erect, lateral branches, mostly solitary; sepals 5; petals numerous; stigmas 5; capsule rather spongy, opening by 5 winged valves, with parietal placentation.

Other spp. - D. australe (Sol. ex G. Forst.) J.M. Black, from Australia and New Zealand, has been reported from Scillies, but is only doubtfully distinct from *D. crassifolium*.

1. D. crassifolium (L.) L. Bolus - *Purple Dewplant*. Main stems to 1m, rooting at 131
nodes, well branched; leaves 12-40 x c.5mm, dark green, with translucent dots, obtuse to mucronate at apex; flowers 2.5-4(5)cm across, reddish-purple. Intrd; grown for ornament near sea and natd on walls, cliffs and sandy places; W Cornwall, Scillies, E Sussex, W Suffolk, Anglesey, CI; S Africa.

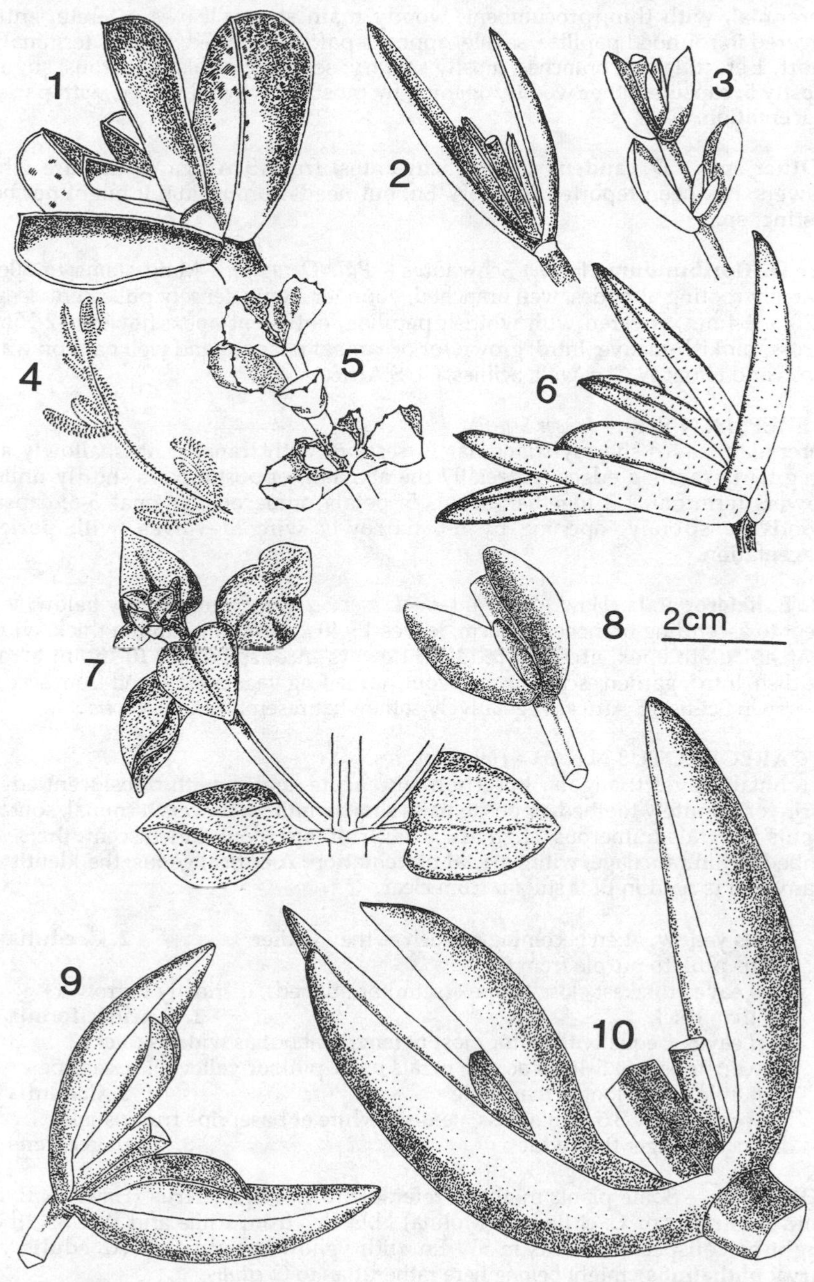

FIG 131 - Aizoaceae. 1, *Erepsia heteropetala.* 2, *Lampranthus roseus.* 3, *L. falciformis.* 4, *Drosanthemum floribundum.* 5, *Oscularia deltoides.* 6, *Ruschia caroli.* 7, *Aptenia cordifolia.* 8, *Disphyma crassifolium.* 9, *Carpobrotus glaucescens.* 10, *C. acinaciformis.*

6. DROSANTHEMUM Schwantes - *Pale Dewplant*

Perennial, with thin, procumbent, woody main stems; leaves ± terete, entire, covered in rounded papillae, sessile, opposite pairs not united; flowers terminal on short, ± erect, lateral branches, mostly solitary; sepals 5; petals numerous; stigmas mostly 5; capsule rather woody, opening by mostly 5 winged valves, with parietal placentation.

Other spp. - **D. candens** (Haw.) Schwantes, from S Africa, with pure white flowers, has been reported from SW En, but needs confirming; it might not be a distinct sp.

1. D. floribundum (Haw.) Schwantes - *Pale Dewplant*. Main stems to 80cm, 131
scarcely rooting at nodes, well branched; younger stems densely pubescent; leaves 6-20 x 2-4mm, covered with whitish papillae, obtuse at apex; flowers 12-25mm across, pinkish-mauve. Intrd; grown for ornament near sea and well natd on walls, rocks and cliffs; W Cornwall, Scillies, CI; S Africa.

7. EREPSIA N.E. Br. - *Lesser Sea-fig*

Perennial; leaves sharply triangular in section, with translucent, shallowly and irregularly toothed edges (especially the abaxial), opposite pairs shortly united; flowers terminal, 1-3 together; sepals 5; petals numerous; stigmas 5-6; capsule woody to spongy, opening by 5-6 narrowly winged valves, with parietal placentation.

1. E. heteropetala (Haw.) Schwantes - *Lesser Sea-fig*. Stems woody below, with 131
erect to ascending branches to 30cm; leaves 15-40 x 5-10mm, 6-15mm thick, widest near apiculate apex, green to reddish; flowers inconspicuous, 10-15mm across, reddish. Intrd; garden escape or throwout, spreading vegetatively and from seed; in quarry in Scillies; S Africa. Vegetatively somewhat resembles *Carpobrotus*.

8. CARPOBROTUS N.E. Br. - *Hottentot-figs*

Perennial; leaves triangular in section with acute angles, with translucent edges, entire or minutely toothed, opposite pairs shortly united; flowers terminal, solitary; sepals 5; petals numerous; stigmas c.8-20; fruit succulent, indehiscent, the seeds embedded in mucilage, with parietal placentation. A critical genus; the identity of many plants natd in BI is still far from clear.

1 Petals yellow, often becoming pinkish as they wither **2. C. edulis**
1 Petals pink to purple from first 2
  2 Leaves thickest close to apex (scimitar-shaped), distinctly narrower than thick **1. C. acinaciformis**
  2 Leaves ± equally thick for most of length, about as wide as thick 3
3 Flowers mostly 4.5-10cm across; petals paler pink or yellow at base; ripe fruit little or not longer than wide **2. C. edulis**
3 Flowers mostly 3.5-6cm across; petals ± white at base; ripe fruit usually distinctly longer than wide **3. C. glaucescens**

Other spp. - Some plants might be referable to **C. aequilaterus** (Haw.) N.E. Br., from Australia, or **C. chilensis** (Molina) N.E. Br., from Chile and W USA; these might be conspecific. Plants in SW En with yellow-based petals (**C. edulis** var. **chrysophthalmus**), might belong here rather than to *C. edulis*.

1. C. acinaciformis (L.) L. Bolus - *Sally-my-handsome*. Stems procumbent, woody, 131
angled, to 2m; leaves 4-10cm, thickest near apex with strongly curved abaxial angle, much thicker than wide; flowers mostly 7-10(12)cm across; petals pinkish-purple; ovary flat to depressed on top; receptacle abruptly narrowed into pedicel.

Intrd; on rocks, cliffs and sand near sea; rare in Devon, Cornwall and Scillies; S Africa. Much confused with purple-flowered plants of *C. edulis*.

2. C. edulis (L.) N.E. Br. - *Hottentot-fig.* Stems procumbent, woody, angled, to 3m; leaves 5-12cm, equally thick for most of length, c. as thick as wide; flowers mostly 4.5-10cm across; petals yellow (often but not always fading pinkish) (var. **edulis**), pinkish-purple (var. **rubescens** Druce) or pink with yellow bases (var. **chrysophthalmus** C.D. Preston & P.D. Sell); ovary flat to raised on top; receptacle gradually tapered into pedicel; (2n=18). Intrd; on rocks, cliffs and sand near sea; CI, Scillies to N Wa and E Suffolk, very local in W Lancs, Man and S & E Ir; S Africa.

3. C. glaucescens (Haw.) Schwantes - *Angular Sea-fig.* Similar to *C. edulis* but **131**
smaller; leaves 2-7cm; flowers mostly 3.5-6cm across; petals pinkish-purple, white or very pale pink at base; and see key. Intrd; rocks and cliffs by sea; CI, E Suffolk, Wigtowns; E Australia.

9. TETRAGONIA L. - *New Zealand Spinach*

Annual; leaves alternate, flat, ± succulent, entire, abruptly narrowed to long petiole; flowers axillary, mostly solitary; sepals 4-5; petals 0; stigmas 3-8; fruit ± woody, indehiscent, ridged, with 1 seed in each of 3-8 cells.

1. T. tetragonioides (Pall.) Kuntze - *New Zealand Spinach.* Stems much branched, procumbent to ascending, to 1m; leaves to 10cm, ovate to rhombic, cuneate to hastate at base, acute to obtuse at apex, papillose; flowers very inconspicuous, yellow-green; (2n=16, 32). Intrd; cultivated as a leaf-vegetable and found on rubbish-tips and waste ground in S En; Australia and Japan to S America.

45. CHENOPODIACEAE - *Goosefoot family*

Herbaceous annuals or perennials or shrubs; leaves usually alternate, rarely opposite, sometimes succulent, simple, without stipules, petiolate or sessile. Flowers small and greenish, sometimes with bracteoles, usually borne in cymes, the cymes axillary or in panicles, rarely solitary, bisexual or unisexual, hypogynous or semi-inferior, actinomorphic; perianth herbaceous, of 1 whorl of 1-5 free or partly fused tepals, sometimes 0; stamens 1-5; ovary 1-celled, with 1 ovule; styles 2-3; stigma linear or feathery; fruit an achene.

A family of mainly dull-coloured weedy plants usually recognizable by the 1-whorled herbaceous perianth, 1-celled superior or semi-inferior ovary with 1 ovule, and 2-3 styles.

1 Leaves fused in opposite pairs, forming succulent sheath round stem, with 0 or very short free part ... 2
1 Leaves not fused to form succulent sheath round stem, usually alternate, with distinct free lamina ... 3
  2 Annual, easily uprooted in entirety ... **8. SALICORNIA**
  2 Perennial, with procumbent rhizomes at or just below soil surface giving rise to aerial stems ... **7. SARCOCORNIA**
3 Leaves <5mm wide, thick and succulent, entire ... 4
3 Leaves not both <5mm wide and succulent, mostly neither, flat, often lobed or toothed at margin ... 5
  4 Leaves ending in a spine, plant bristly ... **10. SALSOLA**
  4 Leaves acute to obtuse but without a spine; plant glabrous ... **9. SUAEDA**
5 Flowers unisexual; fruits surrounded by 2 enlarged bracteoles ... 6
5 Flowers bisexual, or bisexual and female; fruits without bracteoles, usually surrounded by persistent tepals ... 7
  6 Stigmas 2; bracteoles almost free to ± completely fused, if fused

>1/2 way then leaves mealy-white and cuneate at base **5. ATRIPLEX**
6 Stigmas 4-5; bracteoles ± completely fused; leaves green, at least some ± truncate at base **4. SPINACIA**
7 Tepal 1; achene laterally compressed, narrowly winged **3. CORISPERMUM**
7 Tepals (2)4-5; achene not compressed, not winged 8
8 Tepals at fruiting with very short transverse wing or tubercle abaxially **2. BASSIA**
8 Tepals at fruiting without transverse tubercle or wing abaxially, but often with longitudinal keel 9
9 Ovary semi-inferior; receptacle becoming swollen at fruiting **6. BETA**
9 Ovary superior; receptacle not becoming swollen at fruiting **1. CHENOPODIUM**

Other genera - AXYRIS L. would key out as *Atriplex* but female flowers have 3 tepals, the achenes have a terminal wing and the leaves are densely stellate-pubescent. **A. amaranthoides** L. (*Russian Pigweed*), from Russia, is an annual to 80cm and occasionally occurs in waste places.

1. CHENOPODIUM L. (*Blitum* L.) - *Goosefoots*
Annual or perennial herbs; leaves flattened, often mealy, entire, toothed or lobed; bracteoles 0; flowers bisexual or some female ; tepals mostly 4-5, persistent and surrounding fruit, with or without abaxial longitudinal keel.

Vegetatively extremely plastic, especially in habit and leaf shape. Testa sculpturing is important and sometimes essential for identification. It can be examined under >x20 magnification after removal of the pericarp, which may be effected either by rubbing in the hand or sometimes only after boiling and dissection. The orientation of the seed is also important: either 'vertical', with long axis parallel to the length of the flower; or 'horizontal', with it at right angles to the length of the flower. Several spp. were formerly commoner than now and are included only for comparison, since they are often represented in herbaria. *C. album* is extremely variable and may very closely resemble typical variants of spp. 21, 22 and 24-27, which are distinguished only with difficulty and by a combination of characters. They often occur mixed with *C. album* and are often frosted before seeding. Spp. 21, 24, 26 and 27, and perhaps 22 and 25, are doubtfully distinct from *C. album*.

1 Stems glandular-pubescent, at least towards apex; plant aromatic 2
1 Stems glabrous or mealy, not glandular-pubescent; plant not aromatic, sometimes stinking 6
2 Tepals fused >1/2 way, net-veined on outside **2. C. multifidum**
2 Tepals not fused or fused <1/2 way, not net-veined on outside 3
3 Flower clusters in ± sessile axillary racemes or panicles; at least some seeds horizontal; larger leaves rarely <3cm **1. C. ambrosioides**
3 Flower clusters sessile and solitary in leaf axils; seeds vertical; larger leaves rarely >3cm 4
4 Tepals rounded abaxially, not keeled, not meeting at margins and only partially concealing fruit **3. C. pumilio**
4 Tepals prominently keeled abaxially, ± concealing the fruit 5
5 Tepals ± truncate and variably toothed at apex in side view; keel entire **4. C. carinatum**
5 Tepals with long beak at apex; keel deeply laciniate along most of length **5. C. cristatum**
6 Stems woody at least below; inflorescence branchlets ending in bare weakly spinose points **7. C. nitrariaceum**
6 Stems herbaceous; branchlets not bare and spinose at tips 7
7 Flowers in a spike of globose sessile axillary heads >5mm across;

perianth turning red and succulent at fruiting **6. C. capitatum**

7 Flowers in racemes or panicles of heads usually <5mm across; perianth not turning red and succulent at fruiting 8

8 Rhizomatous perennial; stigmas 0.8-1.5mm **8. C. bonus-henricus**

8 Annual; stigmas <0.8mm 9

9 Fruiting perianths mostly longer than wide, with vertical or oblique seeds; inflorescence glabrous 10

9 Fruiting perianths wider than long, with horizontal seeds; inflorescence mealy or glabrous 12

10 Leaves green on upperside, mealy-grey on lowerside **9. C. glaucum**

10 Leaves green (to reddish) on both surfaces 11

11 Tepals of lateral fruits in each cluster fused <1/2 way; leaves usually strongly toothed or lobed **10. C. rubrum**

11 Tepals of lateral fruits fused to near apex; leaves usually entire to sparsely lobed or toothed **11. C. chenopodioides**

12 Leaf-blades weakly cordate at junction with petiole, at least on some leaves **14. C. hybridum**

12 Leaf-blades cuneate at junction with petiole 13

13 Seeds distinctly acutely keeled at edges; tepals minutely denticulate **16. C. murale**

13 Seeds with subacute to rounded unkeeled edges; tepals entire 14

14 Leaves entire or at most with 1 obscure tooth or lobe on each side 15

14 At least lower leaves distinctly toothed and/or lobed 18

15 Leaves green (or reddish) on both surfaces, not mealy; stems square in section **12. C. polyspermum**

15 Leaves mealy-grey at least on lowerside; stems ± terete to ridged 16

16 Leaves ovate-trullate, <2.5cm; tepals rounded abaxially; plant stinking like rotten fish **13. C. vulvaria**

16 Leaves linear to triangular-ovate, rarely ovate-trullate, the largest usually >2.5cm; tepals keeled abaxially; plant not stinking 17

17 Leaves linear to linear-oblong, densely mealy-grey on lowerside, distinctly mucronate at apex, mostly with only 1(-2) pairs of lateral veins visible; petiole <1cm **17. C. desiccatum**

17 Leaves usually oblong to ovate, trullate or triangular, less densely mealy-grey on lowerside, rarely mucronate at apex, mostly with >2 pairs of lateral veins visible; longest petioles usually >1cm 24

18 Stems, leaves and flowers not or very sparsely mealy; leaves trullate to triangular **15. C. urbicum**

18 At least flowers and small branchlets conspicuously mealy; leaves various 19

19 Testa with conspicuous regular pits delimited by pronounced reticulum of ridges 20

19 Testa irregularly pitted to almost featureless, often with radial and/or tangential furrows, sometimes with regular reticulum of slightly raised ridges but with flat (not concave) areas within 22

20 Lower leaves rarely distinctly 3-lobed; tepals with strong wing-like keel along whole length; testa with almost isodiametric honeycomb-like pitting **20. C. berlandieri**

20 Lower leaves usually distinctly 3-lobed; tepals weakly keeled or strongly so in distal part only; testa with radially-elongated pitting 21

21 Plant not stinking like rotten fish; lower leaves with elongated central lobe 2-3x as long as side lobes; seeds <1.4mm across **18. C. ficifolium**

21 Plant stinking like rotten fish; lower leaves with short central lobe little longer than side lobes; seeds >1.4mm across **19. C. hircinum**

22 Plant to 2m; young shoots usually extensively coloured reddish-purple; larger leaves ≤14cm, ± always some >6cm, ovate-trullate

to ovate-triangular **27. C. giganteum**

22 Plant usually less robust, if >1.5m leaves <6cm; plant green or variously red-tinged or -striped, but not with young shoots extensively reddened 23

23 Seeds (1.3)1.5-2mm in longest diameter, rarely <1.5mm in shortest diameter; perianth often blackish-green, rather sparsely mealy; leaves trullate **21. C. bushianum**

23 Seeds ≤1.5mm in longest diameter, <1.5mm in shortest diameter; perianth usually green to grey, variously mealy; leaves various 24

24 Leaves not or little longer than wide, ovate-trullate, not lobed or with 2 basal shallow lobes; tepals often fused to c.1/2 way; infloresence densely mealy **22. C. opulifolium**

24 Leaves distinctly longer than wide; tepals usually fused to <1/2 way; inflorescence variously mealy 25

25 Leaves narrowly oblong, with ± parallel sides, often obtuse, with shallow teeth but not lobed; seeds with ratio of longest to shortest diameters usually >1.15 **24. C. strictum**

25 Leaves various but not with parallel sides, usually acute; seeds with ratio of longest to shortest diameters usually <1.15 26

26 Seeds with subacute edges as seen in narrowest profile **23. C. album**

26 Seeds with obtuse to rounded edges as seen in narrowest profile 27

27 Plant often >1m, frequently tinged with red; seeds usually <1.25mm in longest diameter but flowering very late, usually frosted before seeding; tepals slightly keeled **26. C. probstii**

27 Plant <1m, not tinged with red (except rarely in leaf-axils); seeds often >1.25mm in longest diameter, usually produced well before frosts; tepals usually well keeled **25. C. suecicum**

Other spp. - About 40 additional spp. have been reliably recorded from Br, especially as grain- and wool-aliens. **C. quinoa** Willd. (*Quinoa*), from S America, is becoming cultivated in gardens as a grain-crop and is likely to occur on tips; it has large seeds as in C. *bushianum*, but very dense inflorescences showing the conspicuously pale straw-coloured pericarps and perianths. **C. botrys** L. (*Sticky Goosefoot*), from warm parts of Eurasia, resembles C. *multifidum* but the sepals are free above 1/2 way and the axillary flower clusters are elongated to equal or exceed the bracts; it is a rare wool-alien.

Section **1** - ***AMBRINA*** (Spach) Hook. f. (sp. 1). Annuals with glandular hairs, not mealy, flowers in dense axillary clusters arranged in terminal leafy panicles; tepals fused <1/2 way, rounded abaxially; stamens (3-)5; seeds <1mm across, vertical and horizontal.

1. C. ambrosioides L. - *Mexican-tea*. Strongly aromatic; stems upright, to 1m; leaves usually lanceolate, entire to deeply dentate; (2n=32). Intrd; casual or rarely natd in waste places and tips, from wool, soyabean waste, birdseed and other sources; occasional in S Br; warm parts of America.

Section **2** - ***ROUBIEVA*** (Moq.) Rouy (sp. 2). Annuals with glandular hairs, not mealy; flowers in sessile axillary clusters; tepals fused >1/2 way, rounded abaxially; stamens 5; seeds mostly <1mm, vertical.

2. C. multifidum L. - *Scented Goosefoot*. Aromatic; stems procumbent to erect, to 50cm; leaves narrowly oblong, with narrow deep regular lobes; (2n=32). Intrd; casual on tips and in fields as wool-alien; rather rare in S Br; S America.

Section **3** - ***ORTHOSPORUM*** R. Br. (spp. 3-5). Annuals with glandular hairs, not

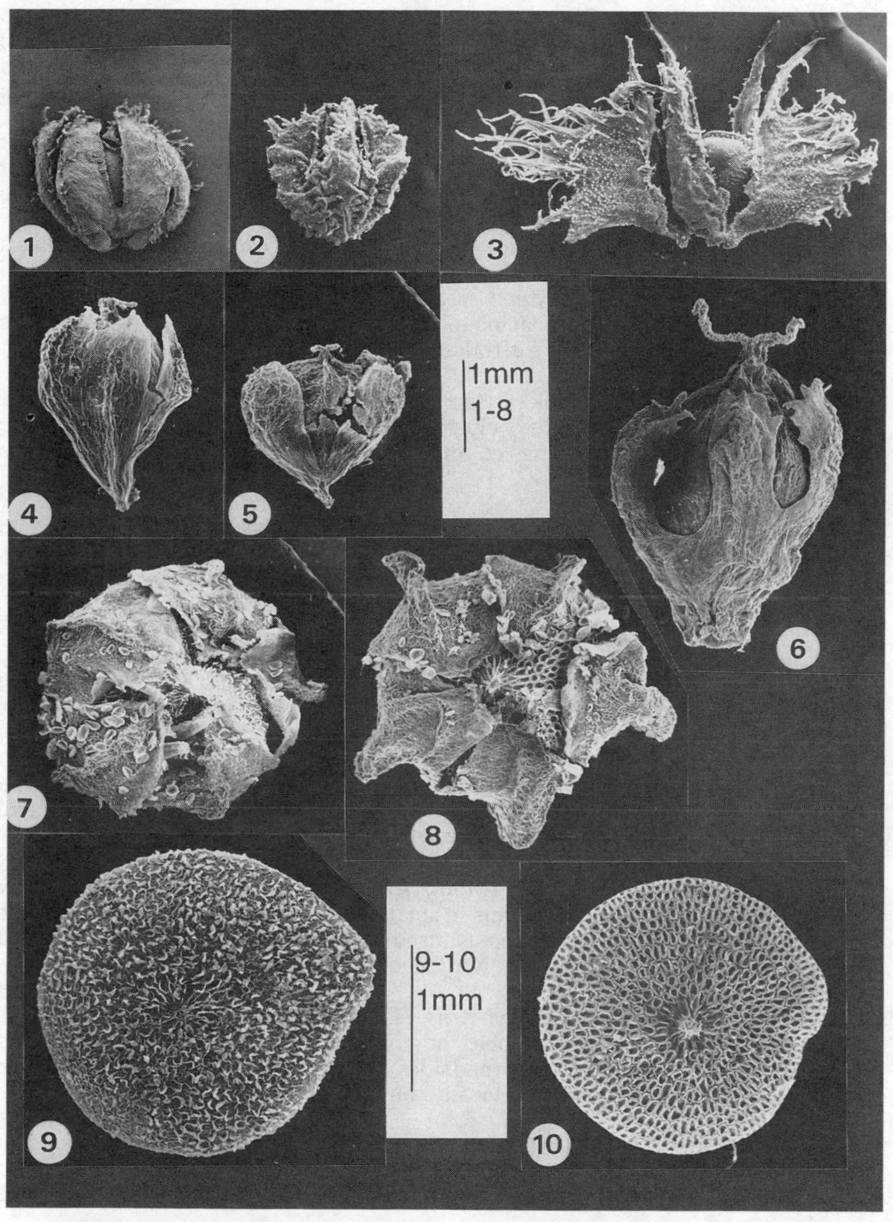

FIG 137 - ***Chenopodium***. 1-6, fruiting perianths from side. 1, *C.* ***pumilio***. 2, *C.* ***carinatum***. 3, *C.* ***cristatum***. 4, *C.* ***chenopodioides***. 5, *C.* ***rubrum***. 6, *C.* ***bonus-henricus***. 7-8, fruiting perianths from apex. 7, *C.* ***album***. 8, *C.* ***berlandieri***. 9-10, fruits from apex. 9, *C.* ***album***. 10, *C.* ***berlandieri***.

mealy; flowers in sessile axillary clusters; tepals fused <1/2 way, rounded or keeled abaxially; stamen 1; seeds <1mm, vertical.

3. C. pumilio R. Br. - *Clammy Goosefoot*. Aromatic; stems procumbent to 137
ascending, to 50cm; leaves elliptic-oblong, with shallow to medium rounded lobes; tepals narrow, not fully concealing fruit, rounded and without keel abaxially; (2n=18). Intrd; casual or rarely persisting on tips and in fields as wool-alien; occasional in S Br, Midlothian; Australia.

4. C. carinatum R.Br. - *Keeled Goosefoot*. Similar to *C. pumilio* but tepals widest 137
near middle, shallowly to conspicuously toothed distally, with ± entire prominent keel; (2n=16). Intrd; casual on tips and in fields as wool-alien; rather rare in S En; Australia.

4 x 5. C. carinatum x C. cristatum = C. x bontei Aellen occurs rarely in S En as a wool-alien; it has intermediate tepals.

5. C. cristatum (F. Muell.) F. Muell. - *Crested Goosefoot*. Similar to *C. carinatum* but 137
tepals with prominent, deeply laciniate keel. Intrd; casual on tips and in fields as wool-alien; rather rare in S En; Australia.

Section 4 - ***BLITUM*** (L.) Hook. f. (sect. *Morocarpus* Graebn.) (sp.6). Annuals, ± glabrous; flowers in sessile axillary clusters, the uppermost not subtended by bract; tepals fused <1/2 way, becoming red and succulent in fruit; stamen often 1; seeds c.1mm, mostly vertical.

6. C. capitatum (L.) Asch. (*Blitum capitatum* L.) - *Strawberry-blite*. Stems erect, to 50cm; leaves triangular-ovate, usually with prominent lobe on each side near base, sometimes with more distal lobes; (2n=18). Intrd; casual or sometimes persisting in fields and waste places; sporadic in Br and Ir; widespread in warm areas.

Section 5 - ***RHAGODIOIDES*** Benth. (sp. 7). Divaricately branching, ± mealy shrub; ultimate branches bare and ± spinose; flowers in small clusters in terminal spicate to subpaniculate inflorescences; tepals fused <1/2 way, rounded abaxially; stamens 5; seeds c.1mm, vertical.

7. C. nitrariaceum (F. Muell.) F. Muell. ex Benth. - *Nitre Goosefoot*. Straggly shrub to 1m; leaves entire, linear to oblanceolate or narrowly oblong; (2n=36). Intrd; casual in fields and on tips as wool-alien; rather rare in En; Australia.

Section 6. - ***AGATHOPHYTON*** (Moq.) Asch. (sp. 8). Rhizomatous, sparsely mealy herbaceous perennials; flowers in small clusters arranged in terminal spicate or subpaniculate inflorescences; tepals fused <1/2 way, rounded or slightly keeled abaxially; stamens 4-5; stigmas >0.8mm; seeds >1mm, mostly vertical.

8. C. bonus-henricus L. - *Good-King-Henry*. Stems erect, to 50cm; leaves 137
triangular with prominent basal lobes, otherwise entire or sinuate or sparsely and shallowly lobed; 2n=36. Native; roadsides, pastures and by farm buildings in nitrogen-rich places; scattered and locally common over most of BI.

Section 7 - ***PSEUDOBLITUM*** (Gren.) Asch. (spp. 9-10). Annuals, glabrous or mealy; flowers in clusters arranged in terminal leafy panicles; tepals fused <1/2 way, slightly or not keeled abaxially; stamens 2-5; seeds mostly <1mm, vertical and horizontal.

9. C. glaucum L. - *Oak-leaved Goosefoot*. Stems much branched, procumbent to **RR**
erect, to 50cm, ± glabrous; leaves narrowly elliptic to elliptic, shallowly but rather **141**
regularly lobed, green on upperside, mealy-grey on lowerside; plant rarely red-tinged; 2n=18. Possibly native; waste places on rich soils, often near sea, and

casual on tips and in dockland; very local in Br, mostly in S & E En, very rare in Ir.

10. C. rubrum L. - *Red Goosefoot*. Stems much to little branched, procumbent to **137**
erect, to 80cm; leaves ovate to triangular or elliptic, variably but often strongly **141**
lobed; plant often red-tinged, especially at fruiting; 2n=36. Native; cultivated and waste ground, often near sea, and on tips; frequent or common in much of En, local or rare elsewhere.

Section 8 - *DEGENIA* Aellen (sp. 11). Annuals, ± glabrous; flowers in clusters arranged in terminal bracteate spikes or panicles; tepals of lateral flowers fused >1/2 way at fruiting, slightly keeled abaxially; stamens 1-5; seeds <1mm, vertical and horizontal.

11. C. chenopodioides (L.) Aellen (*C. botryodes* Sm.) - *Saltmarsh Goosefoot*. **RR**
Resembles a small (to 30cm), usually much branched and procumbent to ascending **137**
C. rubrum, but leaves usually triangular and with only the basal lobes well **141**
developed; diagnosed by perianth (see key); (2n=18). Native; by dykes and in barish pastures near sea; local in SE En and CI.

Section 9 - *CHENOPODIUM* (sect. *Leprophyllum* Dumort.) (spp. 12-27). Annuals, glabrous to densely mealy; flowers in clusters arranged in leafy or leafless spikes or panicles; tepals fused <1/2 way, rounded to keeled abaxially; stamens 5; seeds mostly >1mm, horizontal.

12. C. polyspermum L. - *Many-seeded Goosefoot*. Plant ± glabrous; stems usually **141**
much branched, decumbent to ascending, to 1m; leaves ± glabrous, ovate to elliptic, ± entire; tepals rounded abaxially; testa with raised radial, sinuate striations; 2n=18. Native; waste and cultivated ground; common in CI and C & S En, local or rare and not native in Wa, S Sc and Ir.

13. C. vulvaria L. - *Stinking Goosefoot*. Plant stinking of rotten fish, mealy-grey; **RRR**
stems much branched from base, mostly ascending, to 40cm; leaves ovate-trullate, **141**
± entire; tepals rounded abaxially; testa with faint radial furrows; 2n=18. Probably native; barish places near sea; rare in S En and CI, formerly frequent but very much reduced; now most often as casual in waste places in C & S BI.

14. C. hybridum L. - *Maple-leaved Goosefoot*. Plant glabrous to very slightly mealy; **141**
stems erect, to 1m; leaves ovate-triangular, cordate at base, with few acute lobes; tepals rounded or slightly keeled abaxially; testa covered with small, regular, deep pits; 2n=18. Probably intrd; waste and arable ground; rare in Br, mainly S and usually casual; Europe.

15. C. urbicum L. - *Upright Goosefoot*. Plant ± glabrous; stems erect, to 1m; leaves **141**
ovate-trullate to -triangular, with variable mostly acute lobes; tepals ± rounded abaxially; testa with rather faint reticulate furrows; 2n=36. Probably intrd; waste and cultivated ground, usually from grain; now much rarer than formerly; rare and very scattered in En and Wa and usually only casual; Europe. Distinguished from *C. rubrum* by the horizontal seeds with black (not brown) testa.

16. C. murale L. - *Nettle-leaved Goosefoot*. Similar to *C. urbicum* but inflorescence **R**
much looser, more branched and somewhat mealy; seeds with acute keel around **141**
margin; tepals minutely denticulate; testa covered with minute rounded pits; 2n=18. Probably native near sea in S En; waste and cultivated ground; rarely natd very scattered casual over much of BI, locally frequent in SE En and CI.

17. C. desiccatum A. Nelson (*C. pratericola* Rydb.) - *Slimleaf Goosefoot*. Plant **141**
mealy; stems well branched, ascending to erect, to 1m; leaves linear to linear-oblong, ± entire; tepals strongly keeled; testa with reticulum of furrows; (2n=36). Intrd; tips and waste ground; rather rare casual, mainly from grain, S Br; N America. *C. pratericola* (our plant) is possibly a distinct sp.

18. C. ficifolium Sm. - *Fig-leaved Goosefoot*. Plant mealy; stems ascending to erect, **141**
to 1m; leaves distinctly 3-lobed, the central lobe narrow and 2-3x as long as

laterals; tepals rather weakly keeled abaxially; testa with deep conspicuous pits ± arranged in radial rows and slightly radially elongated; 2n=18. Native; waste and arable ground; CI, S & E En and S Wa, casual in rest of Br and Ir.

19. C. hircinum Schrad. - *Foetid Goosefoot*. Plant stinking of rotten fish, mealy; 141
stems erect to ascending, to 1m; leaves distinctly 3-lobed, the central lobe the widest but little longer than laterals; tepals keeled abaxially, usually strongly so towards apex; testa ± as in *C. ficifolium* but pits strongly radially elongated; (2n=36). Intrd; casual on tips and waste ground from wool and birdseed, formerly fairly frequent but now rare; En and Wa; S America.

20. C. berlandieri Moq. (*C. zschackei* Murr) - *Pitseed Goosefoot*. Plant mealy; stems 137
erect, to 1.5m; leaves trullate-ovate, variously toothed and lobed, sometimes the 141
lower ± 3-lobed; tepals with strong wing-like keel abaxially; testa with deep, conspicuous ± honeycomb-like pits; 2n=36. Intrd; tips and waste ground, formerly fairly frequent as grain-alien but now very rare; En and S Wa; N America. Often very closely resembles *C. album* but has quite different testa markings.

21. C. bushianum Aellen - *Soyabean Goosefoot*. Like *C. album* but only moderately 141
mealy; leaves triangular to trullate, usually very shallowly toothed; tepals with moderate keel abaxially; (2n=36). Intrd; tips and waste ground, mainly from soyabean waste; S En; N America.

22. C. opulifolium Schrad. ex W.D.J. Koch & Ziz - *Grey Goosefoot*. Like *C. album* 141
but densely mealy; stems never reddish-purple; leaves ovate-trullate, entire to shallowly lobed or toothed; tepals slightly keeled abaxially; 2n=54. Intrd; tips and waste ground from many sources, often grain-alien; scattered in Br, mainly in S; Europe.

23. C. album L. (*C. reticulatum* Aellen, *C. album* ssp. *reticulatum* (Aellen) Beauge 137
ex Greuter & Burdet) - *Fat-hen*. Plant variably mealy; stems erect to ascending, 141
often reddish-purple suffused or in axils, to 1.5m; leaves lanceolate to ovate, trullate or triangular, ± entire to shallowly lobed or toothed; tepals with moderate to indistinct keel abaxially; testa ± smooth to weakly radially furrowed or rarely (var. **reticulatum** (Aellen) Uotila) with prominent reticulate ridges; 2n=54. Native; waste and cultivated ground; throughout BI. Extremely variable, especially in height, branching, leaf-shape, mealiness and even testa markings.

C. album has been recorded as hybridising with a number of other spp., notably *C. ficifolium* (=**C. x zahnii** Murr), *C. berlandieri* (=**C. x variabile** Aellen), *C. opulifolium* (=**C. x preissmannii** Murr) and *C. suecicum* (=**C. x fursajewii** Aellen & Iljin). These 4 hybrids have been reported from several parts of Br, but many if not all plants were probably 1 or other putative parent, the variation of which makes the determination of hybrids extremely hazardous.

24. C. strictum Roth (*C. striatum* (Krasan) Murr - *Striped Goosefoot*. Like *C. album* 141
but stems usually red-striped; leaves narrowly oblong with ± parallel sides, shallowly toothed but not lobed; tepals scarcely keeled; (2n=36). Intrd; tips and waste places mainly from grain and wool; rather rare in Br, mainly S, 1 plant in Ir in 1988; Europe.

25. C. suecicum Murr - *Swedish Goosefoot*. Like *C. album* but stems rarely with 141
reddish-purple; leaves ovate-trullate, usually sharply toothed; tepals moderately to prominently keeled abaxially; seeds with rather more strongly furrowed testa and blunt edge; 2n=18. Intrd; tips, cultivated ground and waste places; scattered in Br; N Europe. Much confused with *C. album* and true distribution unknown; it is much rarer than the latter but has been confirmed post-1990 as natd in widely separated parts of En.

26. C. probstii Aellen - *Probst's Goosefoot*. Like *C. album* but often to 2m; stems 141
and leaves usually tinged reddish-purple; lower leaves usually with distinct basal lobe on each side, sharply toothed; tepals slightly keeled abaxially; (2n=54). Intrd; tips and waste places from wool, birdseed, soyabean and other sources; scattered in Br, rarely ± natd in W Kent; N America.

27. C. giganteum D. Don (*C. album* ssp. *amaranticolor* Coste & A. Reyn.) - *Tree* 141

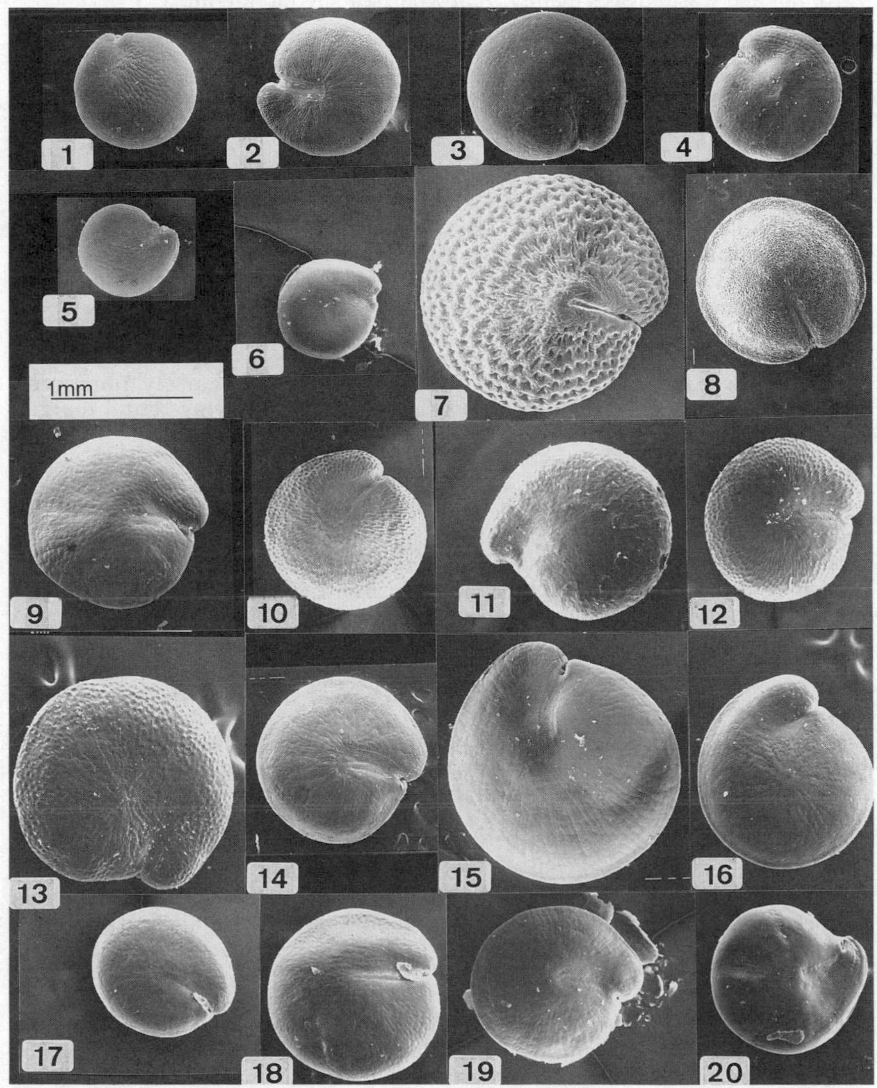

FIG 141 - Seeds of *Chenopodium*. 1, *C. glaucum*. 2, *C. polyspermum*. 3, *C. vulvaria*. 4, *C. urbicum*. 5, *C. rubrum*. 6, *C. chenopodioides*. 7, *C. hybridum*. 8, *C. murale*. 9, *C. desiccatum*. 10, *C. ficifolium*. 11, *C. hircinum*. 12, *C. berlandieri*. 13, *C. bushianum*. 14, *C. opulifolium*. 15, *C. album* var. *album*. 16, *C. album* var. *reticulatum*. 17, *C. strictum*. 18, *C. suecicum*. 19, *C. probstii*. 20, *C. giganteum*.

Spinach. Like *C. album* but usually much larger and extensively reddish-purple on stems and leaves; stems to 2m; leaves ovate-trullate to ovate-triangular, irregularly toothed but scarcely lobed, ≤14cm; tepals with indistinct keel abaxially; (2n=54). Intrd; tips and waste places mainly from wool; scattered in En; India.

2. BASSIA All. (*Kochia* Roth) - *Summer-cypress*
Annual herb; leaves flat, not mealy, entire, at least the lower pubescent; bracteoles 0; flowers bisexual or some female; tepals 5, developing small transverse wing or tubercle abaxially at fruiting.

1. B. scoparia (L.) Voss (*Kochia scoparia* (L.) Schrad.) - *Summer-cypress*. Stem erect, much branched, forming dense ovoid bushy plant to 1m; leaves linear to lanceolate, sessile, the larger ones 3-veined, ≤5cm but usually <3cm; whole plant often becoming purplish-red in autumn; (2n=18). Intrd; much grown in gardens and also contaminant in seed and wool; tips and waste places; throughout much of BI but mostly S; temperate Asia.

3. CORISPERMUM L. - *Bugseed*
Annual herb; leaves flat, not mealy, entire, glabrous or sparsely pubescent; bracteoles 0; flowers bisexual; tepal 1; achene with narrow wing round margin, laterally compressed.

1. C. leptopterum (Asch.) Iljin - *Bugseed*. Stems erect to ascending, to 60cm; leaves linear to very narrowly elliptic or linear-lanceolate, 1-veined, ± sessile, <5cm; whole plant often becoming purplish-red in autumn; (2n=18). Intrd; casual or rarely persistent in sandy, mostly coastal places; scattered in S Br, recurrent in Warks since 1962; S & C Europe. Resembles a small *Bassia scoparia* or a non-succulent *Suaeda maritima*, but the leaves are 1-veined and flat and the 1 tepal and winged achenes are diagnostic. *C. leptopterum* has often been placed under *C. hyssopifolium* L. or *C. intermedium* Schweigg.

4. SPINACIA L. - *Spinach*
Usually annual herb; leaves flat, not mealy, glabrous; bracteoles present with female flowers, developing abaxial spine at fruiting; flowers unisexual, male in dense spikes or panicles, female axillary; tepals 4-5 in male flowers, 0 in female flowers.

1. S. oleracea L. - *Spinach*. Stems erect, well branched, to 1m; leaves rather variable, ovate to triangular, often with 1 basal acute lobe each side; petiole distinct; (2n=12). Intrd; grown as vegetable (summer spinach) but less so now than formerly; rather rare on tips and in waste places in En, over-recorded for *Beta vulgaris*; Asia.

5. ATRIPLEX L. (*Halimione* Aellen) - *Oraches*
Annual herbs or perennial shrubs; leaves flattened, often mealy, entire, toothed or lobed; flowers mostly unisexual, the male with 5 tepals, the female usually with 0 tepals but 2 bracteoles enlarging and partly concealing fruit.

Vegetatively extremely plastic, especially in habit, mealiness and leaf-shape. Young plants often closely resemble *Chenopodium* spp., but in fruit the genus is very distinct. A problem in identification of annual spp. is that both lower leaves, which are often lost before fruit ripens, and also ripe fruit and surrounding bracteoles, are important diagnostic characters. However, populations usually contain a range of individuals of different ages.

1 Shrubs 2
1 Annual herbs 3

2 Lower leaves opposite; bracteoles fused to >1/2 way **13. A. portulacoides**

2 All leaves alternate; bracteoles fused only at base **12. A. halimus**

3 Bracteoles with small apical lobe much exceeded by 2 adjacent laterals, fused ± to apex **14. A. pedunculata**

3 Bracteoles without 3 such apical lobes, rarely fused to >1/2 way 4

4 Bracteoles orbicular to broadly elliptic, entire, papery, present with only some female flowers **1. A. hortensis**

4 Bracteoles not orbicular to broadly elliptic, angled or toothed, herbaceous to ± woody, present with all female flowers 5

5 Bracteoles hardened (cartilaginous) in basal part at fruiting; ultimate venation of leaves thick, dark green against lighter background (fresh material) 6

5 Bracteoles not hardened at fruiting, remaining herbaceous or becoming spongy; ultimate venation of leaves very thin, not green 7

6 Bracteoles 2-5mm, with 3-9 acute teeth in distal 1/2, not tuberculate abaxially; wool-alien **10. A. suberecta**

6 Bracteoles 6-7mm, with irregular mostly obtuse teeth around middle, usually tuberculate abaxially; coastal beaches **11. A. laciniata**

7 Lower leaves linear to linear-lanceolate, entire to toothed but without distinct basal lobes; coastal except as casual **8. A. littoralis**

7 Lower leaves lanceolate to triangular or trullate, with distinct basal lobes; coastal or inland 8

8 Bracteoles fused for >1/3 their length, often to c.1/2 way 9

8 Bracteoles fused at base only, for <1/4 way 11

9 Lower leaves lanceolate to trullate, acutely cuneate at base, with forwardly directed basal lobes; bracteoles herbaceous at base; coastal and inland **9. A. patula**

9 Lower leaves triangular to trullate, truncate to obtusely cuneate at base, with laterally or forwardly directed basal lobes; bracteoles thickened and spongy at base; coastal only 10

10 Bracteoles sessile, 4-10mm, not foliaceous distally **4. A. glabriuscula**

10 Some bracteoles stalked, up to 20mm and foliaceous distally **5. A. x taschereaui**

11 Some bracteoles >10mm and foliaceous distally, with stalks ≥5mm 12

11 Bracteoles all <10mm, rarely foliaceous distally, with stalks ≤5mm 13

12 Bracteoles to 25mm, strongly foliaceous distally, united only at base, with stalks to 25(30)mm **6. A. longipes**

12 Bracteoles to 20mm, sometimes foliaceous distally, the smaller ones often united nearly to 1/2 way, with stalks to 10mm **5. A. x taschereaui**

13 Lower leaves trullate, cuneate at base; lower littoral zone of coasts only **7. A. praecox**

13 Lower leaves triangular, ± truncate at base; coastal and inland 14

14 Bracteoles 2-6(8)mm, sessile **2. A. prostrata**

14 Bracteoles 3.5-9mm, always some with stalks to 1mm and often to 5mm **3. A. x gustafssoniana**

Other spp. - About 20 additional spp. have been reliably recorded from Br, especially as wool-aliens, but all are rarer than the above. **A. muelleri** Benth., from Australia, has been much confused with *A. suberecta*, but differs in having leaves truncate to emarginate and mucronate at apex and bracteoles obovate rather than rhombic. **A. sagittata** Borkh. (*A. nitens* Schkuhr nom. illeg.), from E Europe, differing from *A. hortensis* in its leaves white on lowerside and oblong-cordate bracteoles, is now only a rare casual.

1. A. hortensis L. - *Garden Orache.* Erect annual to 2m; leaves often >10cm, 145 triangular, ± truncate at base, basal lobes present or 0, entire or toothed, green or more often purplish-red; bracteoles orbicular to broadly elliptic, 5-15mm, entire, fused only at base, papery, ± sessile; (2n=18). Intrd; sometimes grown for ornament (rarely as leaf-vegetable) and escaping, also a birdseed-alien, occasional on tips and in waste places in much of BI, especially S; ?Asia.

2. A. prostrata Boucher ex DC. (*A. hastata* auct. non L.) - *Spear-leaved Orache.* **145** Erect to procumbent annual to 1m; lower leaves triangular, ± truncate with laterally directed lobes at base, usually shallowly toothed, green to strongly mealy; bracteoles triangular, 2-6(8)mm, herbaceous to spongy at base, fused only at base, entire to dentate, sessile; seeds with radicle positioned basally or sub-basally and directed laterally or obliquely; 2n=18. Native; waste places and cultivated ground, often in saline habitats inland and by sea; throughout BI.

2 x 4. A. prostrata x A. glabriuscula occurs rarely on beaches in En and Sc with both parents, but has been much over-recorded. It is fertile and has bracteoles spongy and fused like those of *A. glabriuscula*, but smaller and in denser inflorescences.

2 x 8. A. prostrata x A. littoralis = A. x hulmeana Tascher. occurs locally on coasts in E Anglia and N En with both parents. It is fertile and has narrowly trullate, succulent lower leaves and spongy bracteoles; 2n=18.

3. A. x gustafssoniana Tascher. (*A. longipes* ssp. *kattegatensis* Turesson; *A. prostrata* x *A. longipes*) - *Kattegat Orache.* **145** Variously intermediate and fertile; var. **kattegattensis** (Turesson) Tascher. has bracteoles 3.5-5mm with variously fused margins and stalks to only 1mm; other nothomorphs have bracteoles to 9mm, some with stalks to 5mm. Native; coastal estuarine and sometimes inland saline areas, often without 1 or both parents, commoner than *A. longipes* and as common as *A. glabriuscula;* scattered round coasts of Br, but var. *kattegatensis* only in N Sc.

4. A. glabriuscula Edmondston - *Babington's Orache.* **145** Procumbent to rarely erect annual; differs from *A. prostrata* in more mealy stems and leaves, in 4-10mm bracteoles more spongy at base and with margins fused to c.1/2 way, and in seeds with radicle positioned laterally and directed apically; 2n=18. Native; sandy or shingly beaches all round coasts of BI, but rare in many places.

4 x 7. A. glabriuscula x A. praecox occurs in N Sc with both parents, also recorded without either in Cheviot; most plants resemble *A. praecox* in leaf-shape but *A. glabriuscula* in bracteole characters; fertile.

5. A. x taschereaui Stace (*A. glabriuscula* x *A. longipes*) - *Taschereau's Orache.* **145** Vegetatively close to *A. glabriuscula* but some bracteoles to 20mm and foliaceous at apex and with stalks to 10mm; fertile. Native; exposed coastal beaches in Sc, N En and Man, usually with *A. glabriuscula* but often not with *A. longipes* and commoner than it.

6. A. longipes Drejer - *Long-stalked Orache.* Erect to procumbent annual to 90cm; **R** lower leaves narrowly triangular or trullate, cuneate with laterally or forwardly **145** directed lobes at base, entire to shallowly toothed, not mealy; smaller bracteoles triangular to trullate, ± entire, fused only at base, herbaceous, 5-10mm, sessile or with stalks to 1mm; larger bracteoles to 25mm with stalks to 25(30)mm, foliaceous at apex; seeds with basal, laterally directed radicle; 2n=18. Native; in taller saltmarsh vegetation; very local on coasts of Br N to C Sc.

7. A. praecox Hülph. - *Early Orache.* Procumbent to erect annual to 10(15)cm; **R** lower leaves ovate or lanceolate to trullate, cuneate with laterally or forwardly **145** directed lobes at base, ± entire, often red-tinged, not mealy; bracteoles triangular to ovate, ± entire, 3-5mm, herbaceous, united only at base, sessile or with stalks to 1.5mm; seeds with sub-basal laterally or obliquely directed radicle; 2n=18. Native; margins of sea inlets just above *Fucus* zone; N & W Sc S to Kirkcudbrights, Cheviot.

8. A. littoralis L. - *Grass-leaved Orache.* Usually erect annual to 1.5m; lower leaves **145** linear to linear-lanceolate, entire to toothed, without basal lobes, not mealy; bracteoles triangular to trullate with slender acute apex, entire to toothed, 3-6mm, spongy at base, fused only at base, sessile; seeds with sub-basal laterally to

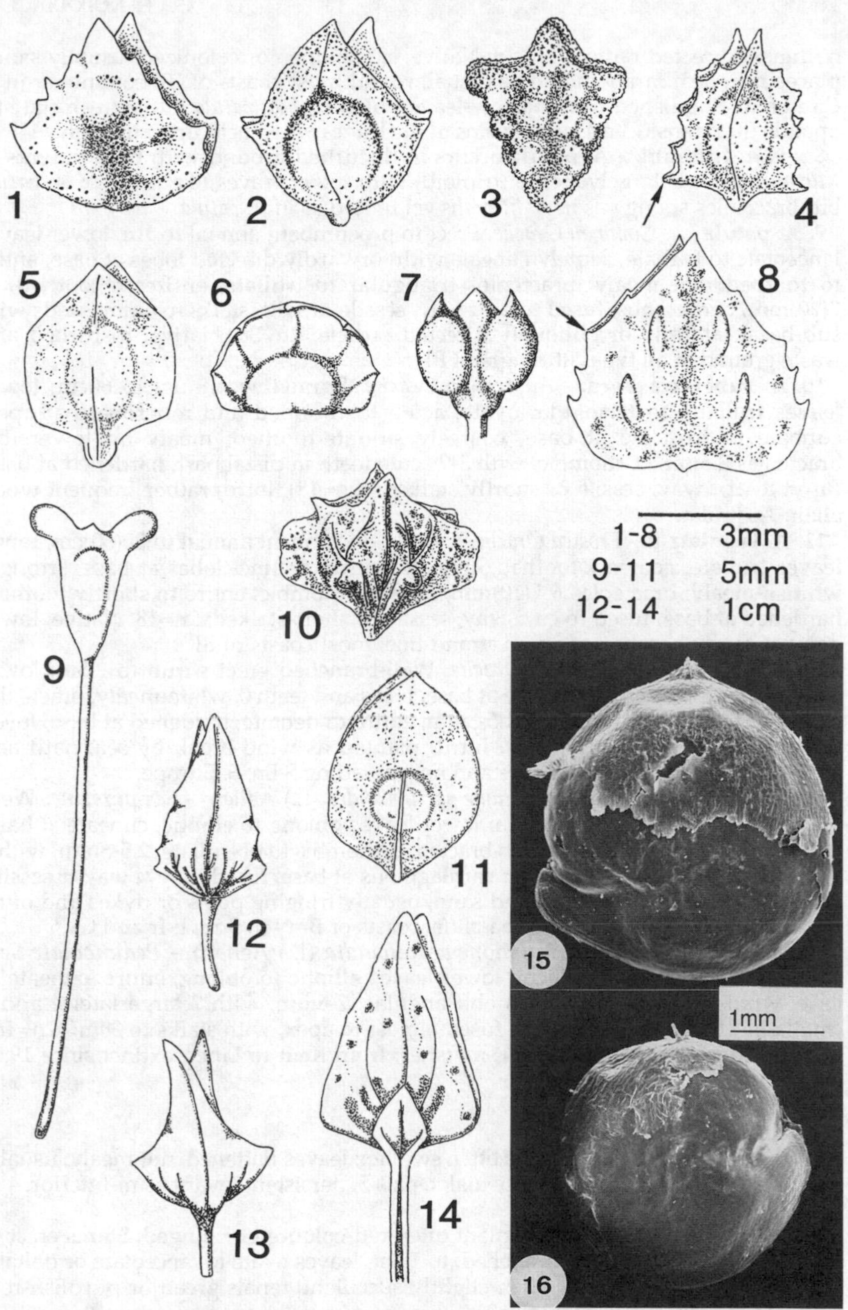

FIG 145 - *Atriplex*. 1-14, bracteoles (arrows indicate limit of fusion). 1, *A. littoralis*. 2, *A. patula*. 3, *A. portulacoides*. 4, *A. prostrata*. 5, *A. praecox*. 6, *A. halimus*. 7, *A. suberecta*. 8, *A. glabriuscula*. 9, *A. pedunculata*. 10, *A. laciniata*. 11, *A. hortensis*. 12, *A. x taschereaui*. 13, *A. x gustafssoniana*. 14, *A. longipes*. 15-16, fruits with lower part of pericarp removed. 15, *A. prostrata*. 16, *A. glabriuscula*.

obliquely directed radicle; 2n=18. Native; saline open or colonized, usually sandy places near sea, rarely inland as casual; round most coasts of BI, commoner in E. Can be distinguished from narrow-leaved plants of *A. patula* in the fresh state by opaque (not translucent) lateral veins of the leaves when held up to the light.

8 x 9. A. littoralis x A. patula occurs in disturbed ground with both parents in Midlothian. It is largely sterile (triploid) with lower leaves like those of *A. patula* but bracteoles spongy as in *A. littoralis* yet united as in *A. patula*.

9. A. patula L. - *Common Orache*. Erect to procumbent annual to 1m; lower leaves **145**
lanceolate to trullate, acutely cuneate with forwardly directed lobes at base, entire to toothed, not mealy; bracteoles triangular to trullate, entire to toothed, 3-7(20)mm, herbaceous, fused to c.1/2 way, sessile or with stalks to 4mm; seeds with sub-basal laterally or obliquely directed radicle; 2n=36. Native; disturbed and waste ground of all types; throughout BI.

10. A. suberecta Verd. - *Australian Orache*. Sprawling annual to 60cm; lower **145**
leaves trullate- to triangular-ovate, acute to rounded and mucronate at apex, cuneate to truncate at base, coarsely sinuate-toothed, mealy on lowerside; bracteoles 2-5mm, ± rhombic, with 3-9 acute teeth in distal part, hardened at base, fused to ≥1/2 way, sessile or shortly stalked; (2n=18). Intrd; rather frequent wool-alien; Australia.

11. A. laciniata L. - *Frosted Orache*. Usually decumbent annual to 30(50)cm; lower **145**
leaves trullate, coarsely toothed, cuneate with ± distinct lobes at base, strongly whitish-mealy; bracteoles 6-7(10)mm, broadly rhombic, entire to shortly toothed, hardened at base, fused to c.1/2 way, sessile or short-stalked; 2n=18. Native; lower parts of sandy beaches, often on strand-line; most coasts of BI.

12. A. halimus L. - *Shrubby Orache*. Well-branched erect shrub to 2.5m; lower **145**
leaves oblong to elliptic, cuneate at base, lobes and teeth 0, white-mealy; bracteoles orbicular to ovate or reniform, 1.5-3mm, entire to dentate, hardened at base, fused only at base, ± sessile; (2n=18). Intrd; planted as wind-break by sea, natd and spreading vegetatively in CI, rare and not spreading S En; S. Europe.

13. A. portulacoides L. (*Halimione portulacoides* (L.) Aellen) - *Sea-purslane*. Well- **145**
branched sprawling shrub to 1m; lower leaves oblong to elliptic, cuneate at base, lobes and teeth 0, whitish-mealy; bracteoles rhombic to obtrullate, 2.5-5mm, with 3 large lobes near apex, somewhat cartilaginous at base, fused to >1/2 way, ± sessile; 2n=36. Native; in saline mud and sand, usually fringing pools or dykes and often flooded at high tide, rarely on sea-cliffs; coasts of Br N to S Sc, E Ir and CI.

14. A. pedunculata L. (*Halimione pedunculata* (L.) Aellen). - *Pedunculate Sea-* **RRR**
purslane. Erect annual to 30cm; lower leaves elliptic to oblong, entire, cuneate at **145**
base, whitish-mealy; bracteoles obtriangular, 2-6mm, with 2 large lateral and 1 small apical lobe, ± herbaceous, fused almost to apex, with stalks to 3cm; (2n=18). Native; drier, barish parts of salt-marshes; from Kent to Lincs, extinct since 1938, refound S Essex 1987.

6. BETA L. - *Beets*

Annual to perennial herbs; roots often swollen; leaves flattened, not mealy, usually ± entire; bracteoles 0; flowers bisexual; tepals 5, persistent; ovary semi-inferior.

1. B. vulgaris L. - *Beet*. Whole plant often red-coloured or -tinged. Stems erect to decumbent, little- to much-branched, to 1.5m; leaves ovate to lanceolate or deltate, cordate to cuneate at base, often slightly succulent; tepals green or purplish-red, incurved in fruit; stigmas usually 2.

1 Usually sprawling perennials; lower leaves mostly <10cm; lower bracts mostly 10-35mm; maritime **a. ssp. maritima**
1 Usually erect annuals or biennials; lower leaves mostly >10cm; lower bracts mostly 2-20mm; cultivated 2
    2 Grown for large foliage; roots not to moderately swollen **b. ssp. cicla**
    2 Grown for greatly swollen roots **c. ssp. vulgaris**

a. Ssp. maritima (L.) Arcang. - *Sea Beet*. Usually much-branched sprawling perennial; root not strongly swollen; lower leaves usually <10cm; 2n=18. Native; shores and waste ground near sea; round coasts of BI except most of N & C Sc.

b. Ssp. cicla (L.) Arcang. - *Foliage Beet*. Usually little-branched, erect annual to biennial; root mostly slightly swollen; lower leaves usually >20cm; (2n=18). Intrd; cultivated for its foliage and a common casual or relic. Includes var. **cicla** L. (*Spinach Beet*) and var. **flavescens** (Lam.) Lam. (*Swiss Chard*).

c. Ssp. vulgaris - *Root Beet*. Usually little-branched, erect annual to biennial; root strongly swollen; lower leaves usually >10cm; (2n=18). Intrd; cultivated for its roots and a common casual or relic. Includes *Beetroot*, *Sugar Beet*, *Fodder Beet* and *Mangel-wurzel*.

2. B. trigyna Waldst. & Kit. - *Caucasian Beet*. Erect, usually little-branched perennial to 1m; leaves ovate, usually cordate at base; tepals whitish-yellow, erect in fruit; stigmas 3; (2n=36, 54). Intrd; grown for ornament and persisting or relict in waste places and on tips, also seed contaminant; S En, Co Dublin; SE Europe.

7. SARCOCORNIA A. J. Scott - *Perennial Glasswort*

Dwarf subshrubs; leaves fused in opposite pairs, forming succulent sheath round stem which appears composed of short segments; flower ± immersed in row of segments at ends of main stem ('terminal spike') and branches in 2 opposite groups in each segment, each group with 3 flowers in a ± straight transverse row, the centre one completely separating and c. as large as the laterals.

Variously included in *Salicornia* or *Arthrocnemon* Moq.

1. S. perennis (Mill.) A. J. Scott (*Salicornia perennis* Mill., *Arthrocnemon perenne* **R**
(Mill.) Moss) - *Perennial Glasswort*. Aerial stems to 30cm, some fertile, some not, **149**
erect to decumbent, usually little-branched, becoming yellowish to reddish, arising from thin extensive rhizomes; terminal spike 10-40mm; fertile segments 3-4mm, 3-4.5mm wide at narrowest point; anthers c.1.5mm; 2n=18. Native; mostly middle and upper parts of salt-marshes; scattered in En and Wa N to N Lincs and Merioneth, formerly to E Lothian, frequent only in SE En, Wexford.

8. SALICORNIA L. - *Glassworts*

Like *Sarcocornia* but annuals and with 1-3 flowers in each group, the central 1 of each group of 3 extending much further apically than the laterals, so that the group is triangular in shape.

An extremely difficult genus, the problems arising mainly from great phenotypic plasticity and the inbreeding nature of the plants, which tend to form numerous distinctive local populations. At least 20-30 'sorts' can be distinguished in SE En; possibly only 3 spp. (*S. pusilla*; *S. europaea* agg.; *S. procumbens* agg.) should be recognized. Identification of segregates within the latter 2 aggregates should be attempted only on several fresh well-grown plants from unshaded populations developing ripe fruit (Sep-Oct).

1 Flowers 1 per group; fertile segments disarticulating when fruit ripe, <2 x 2mm **1. S. pusilla**
1 Flowers mostly 3 per group; fertile segments not disarticulating, >2 x 2mm 2
  2 Anthers 0.2-0.5(0.6)mm; stamens 1(-2); central flower distinctly larger (c.2x) than 2 laterals; fertile segments with distinctly convex sides; seeds 1-1.7mm 3
  2 Anthers (0.5)0.6-0.9mm; stamens (1-)2; all 3 flowers about same size; fertile segments with straight or sometimes slightly convex or concave sides; seeds (1.3)1.5-2.3mm 5
3 Apical edge of fertile segments with scarious border 0.1-0.2mm wide; plants deep shiny-green becoming reddish-purple **2. S. ramosissima**
3 Apical edge of fertile segments with scarious border ≤0.1mm wide; plants

lighter duller green becoming yellowish-green sometimes suffused with pinkish-purple 4

4 Plant glaucous, matt, not reddening or only slightly so around flowers; branches simple, the lowest usually <1/2 as long as main stem, curving upwards distally **4. S. obscura**

4 Plant clear green, not matt, usually reddening; branches usually branched, the lowest usually >1/2 as long as main stem, ± straight **3. S. europaea**

5 Lower fertile segments ≤3(3.5)mm, ≤3.5(4)mm wide at narrowest point, plant becoming brownish-purple or -orange **5. S. nitens**

5 Lower fertile segments 3-6mm, 3-6mm wide at narrowest point; plant becoming pale green to yellowish, sometimes tinged purple 6

6 Terminal spike ± cylindrical, of 6-15(22) fertile segments; plant becoming yellowish-green to bright yellow **6. S. fragilis**

6 Terminal spike usually tapering, of 12-30 fertile segments; plant becoming dull green, dull yellow or yellowish-brown **7. S. dolichostachya**

1. S. pusilla Woods - *One-flowered Glasswort*. Erect to procumbent, simple to much-branched, to 25cm, becoming orangy- or purplish-pink; branches ± straight; terminal spike <10mm; lower fertile segments 1-1.5mm, 1-1.5mm wide at narrowest point; 2n=18. Native; drier parts of salt-marshes; S Br and S Ir N to N Lincs, Pembs and Co Dublin. R 149

1 x 2. S. pusilla x S. ramosissima has been found in S En and N Lincs close to both parents; it has 1-3 flowers per group but resembles *S. pusilla* more closely in segment-shape and colour and is fertile; endemic.

2-4. S. europaea L. agg. Anthers 0.2-0.5(0.6)mm; stamens 1(-2); central flower distinctly larger than 2 laterals; fertile segments with distinctly convex sides; seeds 1-1.7mm; 2n=18 (diploid).

2. S. ramosissima Woods - *Purple Glasswort*. Erect to procumbent, simple to much-branched, to 40cm, usually becoming dark purple; branches ± straight; terminal spike (5)10-30(40)mm; lower fertile segments 1.9-3.5mm, 2-4mm wide at narrowest point; 2n=18. Native; mostly middle and upper parts of salt-marshes; round coasts of BI. Particularly variable in habit and colour; recent work suggests that this sp. is not distinct from *S. europaea*. 149

3. S. europaea L. - *Common Glasswort*. Usually erect, much-branched, to 35cm, usually becoming yellowish-green suffused with pink or red; branches ± straight; terminal spike 10-50(60)mm; lower fertile segments 2.5-4mm, 3-4.5mm wide at narrowest point; 2n=18. Native; at all levels in salt-marshes; round coasts of BI. 149

4. S. obscura P.W. Ball & Tutin - *Glaucous Glasswort*. Usually erect and little-branched, to 40cm, glaucous, becoming dull yellowish-green; branches curving upwards distally; terminal spike 10-40(45)mm; lower fertile segments 2.5-4.5mm, 2.8-4(5)mm wide at narrowest point; 2n=18. Native; on bare mud in salt-pans and beside channels; S & E coasts of En from S Lincs to N Somerset. 149

5-7. S. procumbens Sm. agg. Anthers (0.5)0.6-0.9mm; stamens (1-)2; all 3 flowers c. same size; fertile segments with straight or slightly convex or concave sides; seeds (1.3)1.5-2.3mm; 2n=36 (tetraploid).

5. S. nitens P.W. Ball & Tutin - *Shiny Glasswort*. Usually erect and little-branched, to 25cm, becoming light brownish-purple to brownish-orange with diffuse red tinge; terminal spike 12-40mm; lower fertile segments (1.8)2-3(3.5)mm, 1.8-3.5mm wide at narrowest point; 2n=36. Native; in middle and upper parts of salt-marshes; scattered in S Br and S Ir, Cheviot. The French **S. emerici** Duval- 149

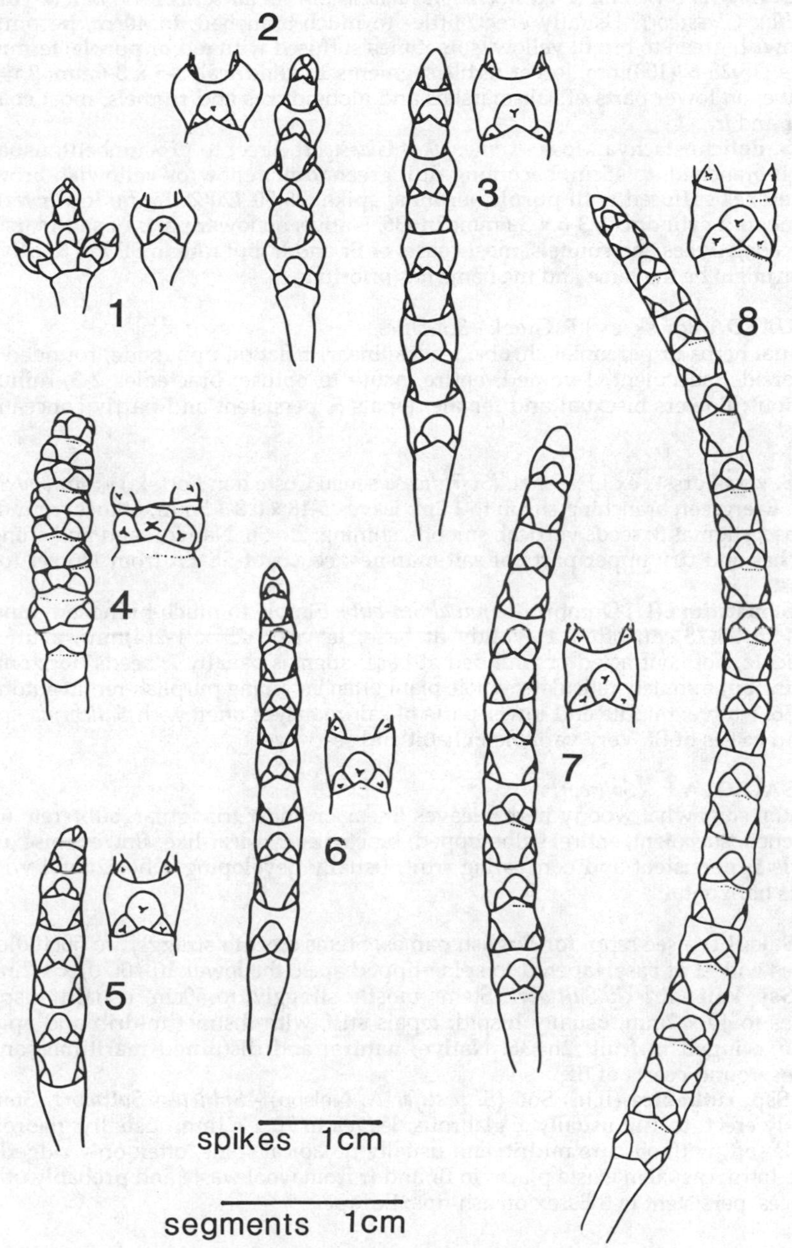

FIG 149 - Fruiting terminal spikes and fruiting segments of *Salicornia, Sarcocornia*. 1, ***Salicornia pusilla***. 2, ***S. ramosissima***. 3, ***S. europaea***. 4, ***Sarcocornia perennis***. 5, ***Salicornia obscura***. 6, ***S. nitens***. 7, ***S. fragilis***. 8, ***S. dolichostachya***. Drawings by D.H. Dalby.

Jouve might be the same and the name has priority.

6. S. fragilis P.W. Ball & Tutin (?*S. procumbens* Sm., *S. lutescens* P.W. Ball & Tutin) - *Yellow Glasswort*. Usually erect, little- to much-branched, to 40cm, becoming yellowish-green to bright yellow, sometimes suffused with red or purple; terminal spike (15)25-80(100)mm; lower fertile segments ± cylindrical, 3-5 x 3-6mm; 2n=36. Native; on lower parts of salt-marshes and along dykes and runnels; most coasts of Br and Ir. **149**

7. S. dolichostachya Moss - *Long-spiked Glasswort*. Erect to procumbent, usually much-branched, to 45cm, becoming dull green, dull yellow or yellowish-brown, sometimes suffused with purple, terminal spike (25)50-120(200)mm; lower fertile segments ± cylindrical, 3-6 x 3-6mm; 2n=36. Native; on lower parts of salt-marshes and along dykes and runnels; most coasts of Br and Ir, but rare in SE En. **S. oliveri** Moss might be the same and the name has priority. **149**

9. SUAEDA Forssk. ex J.F. Gmel. - *Sea-blites*

Annual herbs or perennial shrubs; leaves linear, ± flat on upperside, rounded on lowerside, succulent, 1-veined, entire, acute to obtuse; bracteoles 2-3, minute, scarious; flowers bisexual and female; tepals 5, persistent and partly concealing fruit.

1. S. vera Forssk. ex J.F. Gmel. (*S. fruticosa* sensu Coste non Forssk.) - *Shrubby Sea-blite*. Evergreen branching shrub to 1.2m; leaves 5-18 x 0.8-1.5mm, obtuse, rounded at base; stigmas 3; seeds vertical, smooth, shining; 2n=36. Native; sand and shingle beaches and dry upper parts of salt-marshes; coasts of SE En from Dorset to N Lincs. **R**

2. S. maritima (L.) Dumort. - *Annual Sea-blite*. Simple to much-branched annual herb to 30(75)cm, often ± woody at base; leaves 3-25 x 1-2(4)mm, acute to subacute, not contracted or rounded at base; stigmas usually 2; seeds horizontal, shining but minutely reticulate; whole plant often becoming purplish-red in autumn; 2n=36. Native; middle and lower parts of salt-marshes, often with *Salicornia* spp.; round coasts of BI. Very variable in habit and seed-size.

10. SALSOLA L. - *Saltworts*

Annual somewhat woody herbs; leaves linear or linear-triangular, subterete to ± flattened, succulent, entire, spine-tipped; bracteoles 2, ± leaf-like; flowers bisexual; tepals 5, persistent and concealing fruit, usually developing a horizontal wing; seeds horizontal.

1. S. kali L. - see sspp. for English names. Stems erect to straggly, to 50(100)cm; leaves widest at base, tapered to spine-tipped apex, the lower 10-40(70) x 1-2mm.

a. Ssp. kali - *Prickly Saltwort*. Stems mostly straggly, to 50cm, usually hispid; leaves to 40 x 2mm, usually hispid; tepals stiff, with distinct midrib and apical spine, winged in fruit; 2n=36. Native; natural and disturbed maritime sandy places; round coasts of BI.

b. Ssp. ruthenica (Iljin) Soó (*S. pestifer* A. Nelson) - *Spineless Saltwort*. Stems mostly erect, to 1m, usually ± glabrous; leaves to 70 x c.1mm, usually glabrous; tepals soft, with obscure midrib and usually no apical spine, often only ridged in fruit. Intrd; casual in waste places in Br and Ir from wool waste and probably other sources, persistent in S Essex on ash-tips; Europe.

46. AMARANTHACEAE - *Pigweed family*

Herbaceous annuals or perennials; leaves alternate, simple, entire, without stipules, the lower petiolate. Flowers small and usually brownish, with 3-5 bracteoles, usually borne in cymes, the cymes axillary or densely clustered at stem apex, mostly monoecious, sometimes dioecious, hypogynous, actinomorphic; perianth scarious, of 1 whorl of usually (2)3-5 free tepals, sometimes 0; stamens as many as tepals; ovary 1-celled, with 1 ovule; styles 2-3; stigma linear; fruit an achene or usually a 1-seeded capsule.

Distinguished from Chenopodiaceae by the brownish, scarious perianth and often dehiscent fruit.

Other genera - CELOSIA L. differs in having several-seeded capsules. **C. argentea** L. (*Cockscomb*), from the Tropics, with gaudy wax-like inflorescences, is probably the world's ugliest plant; it is much grown as a park bedding-plant and occasionally occurs on tips etc.

1. AMARANTHUS L. - *Pigweeds*

A difficult, wholly alien genus which should be collected late in the year or grown on to obtain fruit. Even so it is often very difficult to distinguish between spp. 2-6 & 8, which are a wild complex and their domesticated derivatives in which the growth-form of fully mature plants is a valuable character. It is important to distinguish between bracteoles and tepals (which can be similar), and male and female flowers. In key and descriptions 'tepals' refers to those of female flowers only.

1 Flowering stems leafy to apex, the flowers borne in axillary clusters 2
1 Flowering stems leafless towards apex, the flowers borne in dense spike-like terminal panicles (often also in axillary clusters further back) 7
 2 Bracteoles c.2x as long as perianth, with ± spiny tip **12. A. albus**
 2 Bracteoles shorter than perianth, not with spiny tip 3
3 Tepals 4 5 4
3 Tepals 3 5
 4 Tepals subequal, obovate to spathulate; fruit indehiscent **17. A. standleyanus**
 4 Tepals unequal, ovate to elliptic; fruit transversely dehiscent **13. A. blitoides**
5 Tepals shorter than fruit, with apical point <0.5mm and ± straight **16. A. graecizans**
5 Tepals longer than fruit, with apical point >0.5mm and usually bent outwards or hooked 6
 6 Tepals with fine, straight to outwardly curved (rarely >90°) apical point **14. A. thunbergii**
 6 Tepals with stiffly hooked (mostly 180-360°) apical point **15. A. capensis**
7 Plant wholly male **9. A. palmeri**
7 Plant with female (and usually male) flowers 8
 8 Tepals (2-)3, much shorter than fruit; fruit indehiscent 9
 8 Tepals (3-)5, nearly as long as to longer than fruit; fruit dehiscent or not 10
9 Leaves obtuse; stems very shortly pubescent at least near apex; fruit inflated (seed much smaller than cavity) **10. A. deflexus**
9 Leaves usually retuse; stems ± glabrous; fruit not inflated **11. A. blitum**
 10 Fruits indehiscent or irregularly dehiscent 11
 10 Fruits transversely dehiscent 12
11 Tepals tapered to very acute apex; bracteoles longer than perianth **7. A. bouchonii**
11 Tepals rounded to retuse at apex, obovate to spathulate; bracteoles shorter than perianth **17. A. standleyanus**
 12 Tepals all tapered to acute apex 13
 12 At least the inner tepals obovate to spathulate, rounded to retuse

at apex though often with mucro 16

13 Terminal inflorescence often >30cm, heavy, usually brightly (green, red or yellow) coloured; seeds pale or dark brown; bracts not exceeding styles; longest bracteoles of female flowers 1-1.5x as long as perianth 14

13 Terminal inflorescence <30cm, not massive, usually green, rarely red; seeds dark brown; bracts usually exceeding styles; longest bracteoles of female flowers c.2x as long as perianth 15

14 Inflorescence stiff; bracts with stout midrib; styles thickened at base **8. A. hypochondriacus**

14 Inflorescence lax; bracts with slender midrib; styles not thickened at base **3. A. cruentus**

15 Inflorescence stiff; bracts 5-5.5mm, with stout midrib; styles thickened at base; tepals 3-5 **6. A. powellii**

15 Inflorescence lax; bracts 3-5mm, with slender midrib; styles not thickened at base; tepals 5 **2. A. hybridus**

16 Inflorescence long, pendent or trailing, often red; tepals obovate or broadly spathulate to elliptic, strongly overlapping at margins **5. A. caudatus**

16 Inflorescence erect or weakly pendent, rarely red; tepals narrowly oblong to spathulate, not or weakly overlapping at margins 17

17 Tepals with midrib ending below apex (though apex often mucronate); stems at base of flowering region pubescent to densely so **1. A. retroflexus**

17 Tepals with midrib extending beyond apex into mucro; stems at base of flowering region glabrous to sparsely pubescent 18

18 All flowers female; tepals dissimilar, the outer longer and tapering to acute apex **9. A. palmeri**

18 Male and female flowers mixed; tepals all similar **4. A. quitensis**

Other spp. - About 16 other spp. have been reliably recorded from Br as rare casuals. Least rare are **A. spinosus** L., **A. crispus** (Lesp. & Thévenau) N. Terracc. and **A. viridis** L. from tropical and S America. *A. spinosus* (*Spiny Amaranth*) has paired spines in the leaf axils and dehiscent or indehiscent fruits. *A. crispus* has leafy inflorescences, undulate leaf-margins, 5 tepals and indehiscent fruits. *A. viridis* has terminal spikes and indehiscent or irregularly dehiscent fruits not exceeding the 3 tepals. Of additional spp. recently found with soyabean waste a dioecious American taxon with long narrow, terminal, interrupted, leafless inflorescences is most common; it might be **A. arenicola** I. M. Johnst.

1. A. retroflexus L. - *Common Amaranth.* Annual to 1m, usually ± erect; main 153
inflorescence terminal, thick, with few side branches, very rarely red; tepals 5, narrowly oblong to spathulate, sometimes with mucro but midrib usually ending short of apex; fruit dehiscent; (2n=32, 34). Intrd; frequent casual in waste places and on tips from many sources, including wool, birdseed and soyabean waste, sometimes natd; scattered in BI, mainly S; N America.

1 x 2. A. retroflexus x A. hybridus = A. x ozanonii (Thell.) C. Schust. & M. 153
Goldschm. (*A. x adulterinus* Thell.) has been found as a casual in a few places in S En from 1959 onwards; it has intermediate tepals and is partially fertile.

2. A. hybridus L. (*A. patulus* Bertol.) - *Green Amaranth.* Annual to 1m, usually ± 153
erect; main inflorescence as in *A. retroflexus* but usually less dense and more branched; bracteoles c.1.5-2x as long as perianth; tepals (3-)5, lanceolate, tapering to acute apex; fruit dehiscent; (2n=24, 32, 34). Intrd; frequent casual from many sources, especially wool, birdseed and soyabean waste, very rarely becoming natd; Br, CI and rarely Ir; America. *A. hybridus* and its 2 close relatives *A. quitensis* and *A. powellii* are thought to be the progenitors of the domesticated spp. *A. caudatus*, *A. cruentus* and *A. hypochondriacus*. These 6 taxa have been variously amalgamated by different authorities but in absence of a consensus are best kept separate.

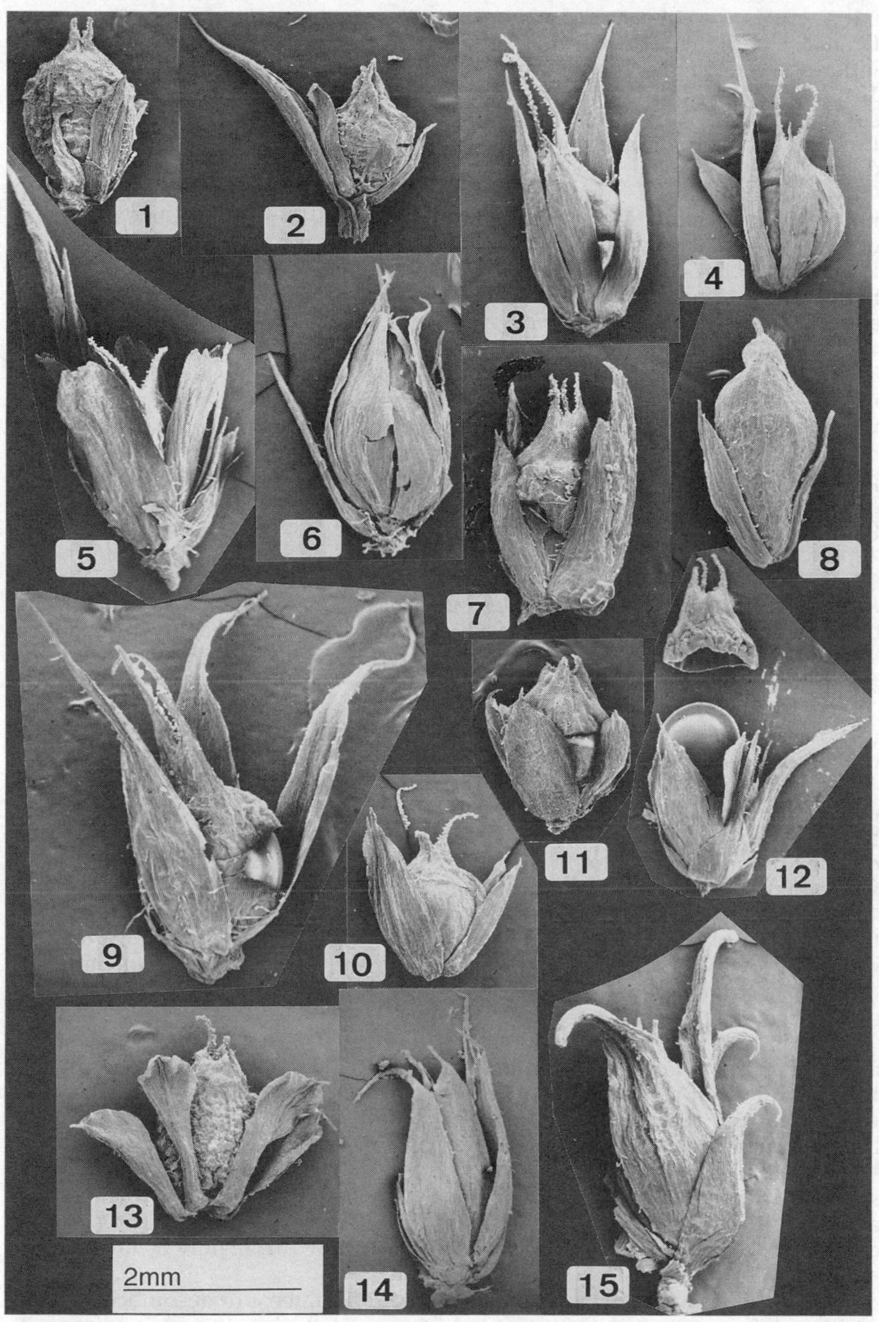

FIG 153 - Fruiting perianths of *Amaranthus*. 1, *A. graecizans*. 2, *A. albus*. 3, *A. hybridus*. 4, *A. x ozanonii*. 5, *A. retroflexus*. 6, *A. quitensis*. 7, *A. blitoides*. 8, *A. deflexus*. 9, *A. thunbergii*. 10, *A. bouchonii*. 11, *A. caudatus*. 12, *A. cruentus*. 13, *A. standleyanus*. 14, *A. palmeri*. 15, *A. capensis*.

3. A. cruentus L. (*A. hybridus* ssp. *cruentus* (L.) Thell., ssp. *incurvatus* (Timeroy ex 153
Gren. & Godr.) Brenan, ? *A. paniculatus* L.) - *Purple Amaranth.* Like *A. hybridus* but inflorescence often red, bracteoles c.1-1.5x as long as perianth and see key (couplet 13); (2n=32, 34). Intrd; fairly frequent casual from several sources, including soyabean waste, very rarely becoming natd; S En; originated in C America. Probably the domesticated derivative of *A. hybridus*, and perhaps not specifically distinct.

4. A. quitensis Kunth - *Mucronate Amaranth.* Annual to 1m, usually ± erect; main 153
inflorescence terminal, large, with many side-branches, green; tepals 5, narrowly oblong to spathulate, with midrib continued into mucro; fruit dehiscent; (2n=32). Intrd; infrequent casual from wool, soyabean, birdseed and other sources; Br, mainly S; S America. Flowers very late and never sets seed. Perhaps not specifically distinct from *A. hybridus.*

5. A. caudatus L. - *Love-lies-bleeding.* Arching annual to 80cm; main inflorescence 153
terminal, very long, thick, pendent, usually red, sometimes green, yellow or white; tepals 5, obovate or broadly spathulate to elliptic, strongly overlapping at margins; fruit dehiscent; (2n=32, 34, 64). Intrd; commonly grown as ornamental and frequent on tips and in waste places in Br and CI, mainly S. A domesticated sp. probably originating from *A. quitensis* in S America.

6. A. powellii S. Watson - *Powell's Amaranth.* Differs from *A. hybridus* as in key (couplet 15); (2n=32, 34). Intrd; infrequent casual from soyabean waste; S En; America.

7. A. bouchonii Thell. (*A. hybridus* ssp. *bouchonii* (Thell.) O. Bolòs & Vigo) - 153
Indehiscent Amaranth. Like *A. hybridus* but fruit indehiscent; (2n=32). Intrd; casual or sometimes natd in E Anglia, Scillies, very rare but probably overlooked elsewhere in Br. Probably a sporadic mutant of *A. powellii* originating in Europe and possibly not specifically distinct.

8. A. hypochondriacus L. - *Prince's-feather.* Differs from *A. cruentus* as in key (couplet 14); (2n=32). Intrd; rather infrequent casual from several sources; S En. A domesticated sp. probably originating from *A. powellii* in N America.

9. A. palmeri S. Watson - *Dioecious Amaranth.* Dioecious annual to 1m, usually ± 153
erect; inflorescence a short terminal spike-like panicle with many axillary clusters below; tepals 5, dissimilar, the outer tapering-acute, the inner ± spathulate and obtuse to rounded or retuse; fruit dehiscent; (2n=32, 34). Intrd; infrequent casual from grain, soyabean waste and other sources; S En; N America.

10. A. deflexus L. - *Perennial Pigweed.* Procumbent to ascending perennial herb 153
(only annual where casual) to 60cm; inflorescence mostly terminal, green to reddish-brown; tepals (2-)3, c.1/2 as long as fruit; fruit indehiscent, inflated; (2n=34). Intrd; infrequent casual from several sources in S & C Br and CI, established as a perennial weed in CI; S America.

11. A. blitum L. (*A. lividus* L.) - *Guernsey Pigweed.* Procumbent to ascending annual to 60cm; inflorescence terminal and axillary, greenish; tepals (2-)3, c.2/3 as long as fruit; fruit indehiscent, scarcely inflated; (2n=16, 34). Intrd; casual on tips, waste and cultivated ground; frequent in 19th century, later very rare, now becoming more frequent in S Br, annually in Guernsey; S Europe.

12. A. albus L. - *White Pigweed.* Erect to procumbent annual with stiff whitish 153
stems to 60cm; inflorescences axillary; tepals 3, shorter than fruit, linear-lanceolate, acute; fruit dehiscent; (2n=32, 34). Intrd; frequent casual from many sources, especially wool and birdseed, occasionally persisting; Br, mainly S, rarely Ir; N America.

13. A. blitoides S. Watson - *Prostrate Pigweed.* Procumbent to decumbent annual 153
to 50cm; inflorescences axillary; tepals 4-5, unequal, the longer c. as long as fruit, oblong-ovate, acute; fruit dehiscent; (2n=32). Intrd; infrequent casual from several sources, mainly birdseed and wool; Br, mainly S, rarely Ir; N America.

14. A. thunbergii Moq. - *Thunberg's Pigweed.* Procumbent to ascending annual to 153
50cm; inflorescences axillary; tepals 3, longer than fruit, with long fine usually

outwardly curved apex; fruit dehiscent. Intrd; frequent wool-alien in En and Sc; S Africa. The leaves often have a dark blotch on upperside.

15. A. capensis Thell. (*A. dinteri* auct. non Schinz) - *Cape Pigweed.* Procumbent to 153
decumbent annual to 40cm; inflorescences axillary; tepals 3, longer than fruit, with fine, rigidly hooked apices; fruit dehiscent. Intrd; infrequent casual, mostly as wool-alien; En; S Africa. English material is referable to ssp. **uncinatus** (Thell.) Brenan (*A. dinteri* var. *uncinatus* Thell.).

16. A. graecizans L. - *Short-tepalled Pigweed.* Procumbent to decumbent annual to 153
70cm; inflorescences axillary; tepals 3, shorter than fruit, narrowly ovate to elliptic, mucronate; fruit dehiscent; (2n=32, 34). Intrd; infrequent casual from several sources including birdseed and wool; S Br; Mediterranean and Africa. Ssp. **sylvestris** (Vill.) Brenan, with broader leaves than the type, also occurs in Br, but is not worth ssp. rank.

17. A. standleyanus Parodi ex Covas - *Indehiscent Pigweed.* Erect to decumbent 153
annual to 70cm; inflorescences all axillary or some condensed into leafless apical spike-like panicle; tepals 5, c. as long as fruit, obovate to spathulate, mucronate; fruit indehiscent; (2n=34). Intrd; infrequent casual from several sources; S Br; S America.

47. PORTULACACEAE - *Blinks family*

Herbaceous annuals to perennials; leaves mostly basal or alternate to opposite, simple, somewhat succulent, entire, with or without stipules, petiolate or not. Flowers solitary or in terminal or axillary cymes, bisexual, hypogynous or semi-inferior, actinomorphic or nearly so; sepals 2, free or united below; petals 4-6, free or fused proximally; stamens 3-c.12; ovary 1-celled, with 1-many ovules; styles 1-3(6); stigmas 3-6, linear; fruit a 1-many seeded capsule.

Distinguished from Caryophyllaceae by the 2 (not 5) sepals.

1 Petals yellow; stamens 6-c.12; capsule opening transversely; seeds >5 **1. PORTULACA**
1 Petals white to pink; stamens 3-5; capsule opening vertically; seeds 1-3 2
  2 Stem-leaves 1 pair, opposite **2. CLAYTONIA**
  2 Stem-leaves several pairs, opposite or alternate **3. MONTIA**

Other genera - CALANDRINIA Kunth has alternate leaves, flowers with 5 red petals, and many-seeded capsules opening vertically. **C. ciliata** (Ruiz & Pav.) DC. (*Red-maids*), from W N America, occurs occasionally on sandy soils mainly in S En and CI, probably as a grain- and seed-alien.

1. PORTULACA L. - *Common Purslane*

Stems with several pairs of opposite to subopposite (or some alternate) leaves with usually bristle-like stipules; flowers sessile, in groups of 1-3 with group of leaves just below; sepals and petals falling before fruiting; stamens 6-c.12; ovary semi-inferior; fruit a many-seeded capsule, opening transversely.

1. P. oleracea L. - *Common Purslane.* Procumbent to erect, ± succulent annual to 50cm; petals (4-)5(6), c.4-8mm, yellow, free or joined at extreme base; (2n=18, 54). Intrd; natd weed of arable ground in CI and Scillies, rare casual in En perhaps from birdseed; Mediterranean. There seem to be no records in BI of the pot-herb ssp. **sativa** (Haw.) Celak.

2. CLAYTONIA L. - *Purslanes*

Stems with 1 pair of opposite leaves; stipules 0; flowers stalked, several in terminal cyme; sepals persistent to after seed dispersal; petals 5, free; stamens 5;

ovary superior; fruit a 1-seeded capsule, opening vertically.

1. C. perfoliata Donn ex Willd. (*Montia perfoliata* (Donn ex Willd.) Howell) - *Springbeauty*. Annual to 30cm; stem-leaves fused to form cup-like structure at base of inflorescence; petals <5mm, white, entire or slightly notched; 2n=36. Intrd; weed of cultivated and waste ground; scattered throughout BI and CI, locally common but rare in W Br and in Ir; W N America.

2. C. sibirica L. (*Montia sibirica* (L.) Howell) - *Pink Purslane*. Annual to 40cm; stem-leaves sessile but not fused; petals >5mm, pink or sometimes white, deeply notched; (2n=24, 36, 48). Intrd; barish damp shady places; scattered through Br and CI, commoner in N & W, very rare in Ir; W N America.

3. MONTIA L. - *Blinks*

Stems with several alternate or opposite leaves; stipules 0; flowers stalked, in terminal or axillary groups of 1-3 or several in terminal cyme; sepals persistent to after seed dispersal; petals 5, fused at base or free; stamens 3-5; ovary superior; fruit a usually 3-seeded capsule, opening vertically.

1. M. fontana L. - *Blinks*. Annual to perennial; stems branched, erect to procumbent or floating, 1-20(50)cm; stem-leaves opposite; basal leaves 0; petals <2mm, white; seeds black to dark brown, smooth or with projecting cells ('tubercles'); 2n=20. Native; many kinds of damp places, from streams to seasonally damp hollows. The following 4 taxa might be worth only varietal status; intermediates occur but are rare.

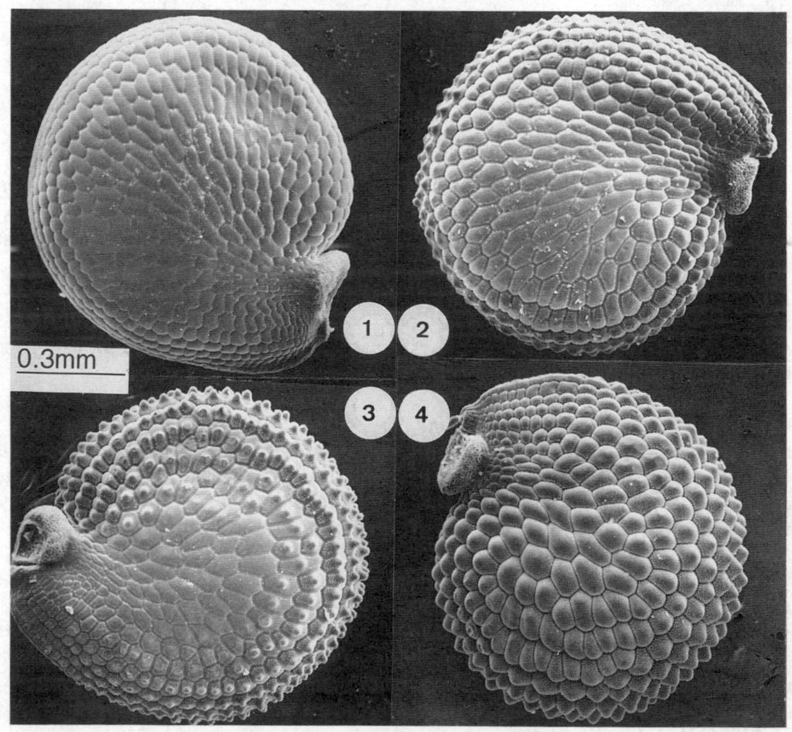

FIG 156 - Seeds of *Montia fontana*.
1, ssp. *fontana*. 2, ssp. *variabilis*. 3, ssp. *amporitana*. 4, ssp. *chondrosperma*.

1 Seeds smooth on faces and margin, very shiny **a. ssp. fontana**
1 Seeds tuberculate at least on margin, dull or somewhat shiny 2
  2 Seeds with broad rounded tubercles on faces and margin, dull **d. ssp. chondrosperma**
  2 Seeds smooth near centre of faces, somewhat shiny 3
3 Seeds with ≥3 rows of narrow long-pointed tubercles along margin **c. ssp. amporitana**
3 Seeds with usually only 1-4 rows of broad short-pointed tubercles along margin **b. ssp. variabilis**

a. Ssp. fontana. Locally common in N En, W Wa, Sc and Ir; the commonest ssp. in NW. **156**

b. Ssp. variabilis Walters. Scattered over most of Br and Ir, but rare in S & E En. **156**

c. Ssp. amporitana Sennen (ssp. *intermedia* (Beeby) Walters). Scattered over most of BI, locally common in SW, rare in Sc. **156**

d. Ssp. chondrosperma (Fenzl) Walters (ssp. *minor* W.R. Hayw. nom. illeg.). Scattered over most of BI, rare in N, locally common in S & C Br. **156**

2. M. parvifolia (Moç.) Greene - *Small-leaved Blinks*. Stoloniferous perennial; flowering stems erect to ascending, unbranched, to 20cm; stem-leaves alternate; basal leaves long-petiolate; petals >6mm, pink; seeds black, minutely rough; (2n=22, 36). Intrd; wet banks and rocks by river; natd by R. Cart, Lanarks; W N America.

47A. BASELLACEAE

ANREDERA cordifolia (Ten.) Steenis (*Boussingaultia cordifolia* Ten. non (Moq.) Volkens, *B. baselloides* Hook. non Kunth) (*Madeira-vine*), from S America, is a twining or sprawling herbaceous perennial with tuber-bearing rhizomes, ovate-cordate leaves, and flowers in branching racemes, 1 plant of which was established on waste ground in Guernsey for a few years. The flowers are small (c.2mm) and whitish, with 2 sepals and 5 petals; it differs from Portulacaceae in the solitary ovule and seed and indehiscent fruit.

48. CARYOPHYLLACEAE - *Pink family*
(Illecebraceae)

Herbaceous annuals to perennials, sometimes woody at base; leaves opposite or sometimes whorled or alternate, simple, usually entire, without stipules or with scarious stipules. Flowers variously arranged but usually in cymes, bisexual or sometimes dioecious or variable, hypogynous or sometimes perigynous, actinomorphic or nearly so; perianth of 2 whorls or petals absent; sepals 4-5, free, green; petals usually 4-5, free, white or coloured, sometimes greenish or 0; stamens 2x as many as sepals, sometimes fewer; carpels fused, with 1 cell with many or sometimes 1-few ovules, with usually free-central placentation; styles 2-5, sometimes very short; stigmas linear to capitate; fruit a many-seeded capsule, sometimes a berry or 1-seeded achene.

Mostly recognisable by the opposite leaves without stipules, 4-5 sepals and petals, 8 or 10 stamens, and many-seeded capsule with free-central placentation, but exceptions to all these occur. The leaves are usually narrow and the petals white and often bifid.

General key

1 Stipules present, at least partly scarious *Key A*
1 Stipules 0 2
  2 Sepals free or joined only at extreme base *Key B*

2 Sepals joined to form distinct calyx-tube *Key C*

Key A - Stipules present, at least partly scarious

1 Leaves alternate **12. CORRIGIOLA**

1 At least lower leaves opposite or whorled 2

2 Petals conspicuous, about as wide as sepals or wider 3

2 Petals inconspicuous and much narrower than sepals 4

3 Styles 5; capsule opening by 5 valves; petals white **16. SPERGULA**

3 Styles 3; capsule opening by 3 valves; petals usually pink, sometimes white **17. SPERGULARIA**

4 Stigmas 3; fruit with >1 seed **15. POLYCARPON**

4 Stigmas 2; fruit with 1 seed 5

5 Sepals conspicuous, white, spongy; fruit a dehiscent capsule **14. ILLECEBRUM**

5 Sepals inconspicuous, greenish to brownish, thin; fruit an indehiscent achene **13. HERNIARIA**

Key B - Stipules 0; sepals free or joined only at extreme base

1 Flowers perigynous; styles 2; fruit an indehiscent achene **11. SCLERANTHUS**

1 Flowers hypogynous; styles >2; fruit with >1 seed, usually dehiscent 2

2 Capsule-teeth or -valves as many as styles (but sometimes slightly bifid), or flowers male 3

2 Capsule-teeth or -valves 2x as many as styles 6

3 Leaves and capsules succulent; seeds at least 3mm; maritime **3. HONCKENYA**

3 Plant not succulent; seeds <3mm 4

4 Petals bifid >3/4 way to base **8. MYOSOTON**

4 Petals entire or slightly emarginate, or 0 5

5 Most flowers with 3 styles and most capsules with 3 teeth, or flowers male **4. MINUARTIA**

5 Styles 4-5; capsule with 4-5 valves **10. SAGINA**

6 Petals 0 7

6 Petals present 8

7 Styles 3 **5. STELLARIA**

7 Styles 5 **7. CERASTIUM**

8 Petals irregularly toothed or jagged **6. HOLOSTEUM**

8 Petals entire to deeply bifid 9

9 Petals bifid ≥1/4 way to base 10

9 Petals entire to slightly emarginate 12

10 Styles 4-5 **7. CERASTIUM**

10 Styles 3 11

11 Petals divided ≤1/2 way; alpine **7. CERASTIUM**

11 Petals divided ≥1/2 way **5. STELLARIA**

12 Styles 4 **9. MOENCHIA**

12 At least most flowers with 3 styles 13

13 Seeds smooth, shiny, with small, lobed oil-body near hilum **2. MOEHRINGIA**

13 Seeds tuberculate or papillose, without oil-body **1. ARENARIA**

Key C - Stipules 0; sepals joined to form distinct calyx-tube

1 Styles 2 (rarely 3 on few flowers) 2

1 Styles 3-5 (rarely 2 on few flowers), or flowers male 6

2 Calyx-tube scarious at joints between lobes 3

2 Calyx-tube herbaceous all round 4

3 Calyx cylindrical or bell-shaped; seeds flattened, with convex and

concave sides, the hilum on the latter; bracteoles placed close to base of calyx and forming epicalyx **24. PETRORHAGIA**

3 Calyx bell-shaped; seeds reniform or comma-shaped, the hilum in the notch; bracteoles not close to base of calyx and hence not forming epicalyx **25. GYPSOPHILA**

4 Calyx-tube with 5 longitudinal wings **23. VACCARIA**

4 Calyx-tube without wings 5

5 Calyx-tube with 2-6 sepal-like bracteoles appressed around base; scales absent from base of petal-limb **26. DIANTHUS**

5 Bracteoles absent; 2 small scales present on inner face of base of petal-limb **22. SAPONARIA**

6 Calyx-teeth longer than -tube, extending beyond petals **19. AGROSTEMMA**

6 Calyx-teeth shorter than -tube, falling short of petal tips 7

7 Fruit a black berry **21. CUCUBALUS**

7 Fruit a capsule 8

8 Styles (2)3(-5), or flowers male; capsule with 2x as many teeth as styles **20. SILENE**

8 Styles 5; capsule with 5 teeth **18. LYCHNIS**

Other genera - **TELEPHIUM** L. has alternate leaves and is related to *Corrigiola*, from which it differs in its woody stems and many-seeded dehiscent fruit; both may be better placed in Molluginaceae. **T. imperati** L., from the Mediterranean, is a rare casual, formerly more common. **PARONYCHIA** Mill. differs from *Herniaria* in its terminal (not axillary) inflorescence and large silvery stipules. **P. polygonifolia** (Vill.) DC., from S Europe, has a similar status to *T. imperati*.

SUBFAMILY 1 - ALSINOIDEAE (genera 1-11). Leaves opposite, without stipules; sepals free or joined only at extreme base; petals mostly 4-5, small to medium; carpophore 0; fruit a capsule or achene.

1. ARENARIA L. - *Sandworts*

Annuals to perennials; sepals 5; petals 5, white, entire or ± so; stamens 10; styles 3(-5); fruit a capsule opening by 6(10) teeth; seeds without oil-body.

1 Petals shorter than sepals **1. A. serpyllifolia**

1 Petals longer than sepals 2

2 Leaves all petiolate, almost all borne on procumbent stems rooting at nodes **4. A. balearica**

2 Upper leaves sessile, borne on ± erect stems; stems not rooting 3

3 Petals >10mm; leaves 10-20mm; sepals pubescent over whole surface on lowerside **5. A. montana**

3 Petals <10mm; leaves 3-6mm; sepals glabrous or pubescent at base 4

4 Leaves slightly succulent, with obscure midrib, the margins glabrous or pubescent only in lower 1/3; sepals glabrous or with a few basal hairs **2. A. norvegica**

4 Leaves not succulent, with distinct midrib, the margins pubescent for >1/3 from base; sepals with hairs on lower part of margin and back **3. A. ciliata**

1. A. serpyllifolia L. - *Thyme-leaved Sandwort*. Decumbent to erect or stiffly spreading annual or biennial to 30cm; leaves broadly ovate to narrowly ovate; flowers numerous; petals shorter than sepals; styles 3. Native; open ground on well-drained soils, especially sand and limestone. Recent work has shown the 2 sspp. to differ constantly in chromosome number and capsule anatomy, and suggests that they should be treated as separate spp.

a. Ssp. serpyllifolia (ssp. *lloydii* (Jord.) Bonnier, ssp. *macrocarpa* F.H. Perring & P.D. Sell). Relatively robust with dense to diffuse inflorescence at fruiting; sepals 3-4mm; petals 1.6-2.7mm; capsule flask-shaped, with 'neck', with brittle walls after seed dispersal, 3-3.5 x 1.5-2.5mm; seeds c.0.5-0.6.5mm; 2n=40. Throughout BI. A dune ecotype scattered round coasts of BI, mostly in SW, is often separated as ssp. **lloydii**, but the differences (slightly larger seeds and wider capsules, shorter pedicels and denser inflorescences) are not constant and the chromosome number is the same. **167**

b. Ssp. leptoclados (Rchb.) Nyman (*A. leptoclados* (Rchb.) Guss.). Relatively slender with diffuse inflorescence at fruiting; sepals 2.5-3.1(3.5)mm; petals 1.1-1.6mm; capsule conical, straight-sided, with flexible walls after seed dispersal, 2.5-3 x 1.2-1.5mm; seeds c.0.4-0.5mm; 2n=20. Throughout Bl, probably less common than ssp. *serpyllifolia*, especially in N. **167**

2. A. norvegica Gunnerus - see sspp. for English names. Annual to perennial well branched below with ascending stems to 6cm; flowers 1-2(4) per stem; petals 4-5.5mm; styles 3-5.

a. Ssp. norvegica - *Arctic Sandwort*. Perennial with many non-flowering shoots; leaves obovate; flowers c.9-10mm across; styles 3-5; 2n=80. Native; base-rich scree and river shingle; extremely local in NW Sc and W Ir. **RRR**

b. Ssp. anglica G. Halliday - *English Sandwort*. Annual or biennial with few non-flowering shoots; leaves narrowly elliptic to narrowly ovate; flowers c.11-23mm across; styles 3; 2n=80. Native; bare limestone in MW Yorks; endemic. **RRR**

3. A. ciliata L. - *Fringed Sandwort*. Similar to *A. norvegica* ssp. *norvegica* but differs in key characters, and in usually oblanceolate leaves; styles 3; petals 5-7.5mm; and flowers 12-16mm across; 2n=40. Native; limestone cliffs in Sligo. The Irish plant is ssp. **hibernica** Ostenf. & O.C. Dahl (var. *hibernica* (Ostenf. & O.C. Dahl) Druce); endemic. **RRR**

4. A. balearica L. - *Mossy Sandwort*. Procumbent much-branched perennial rooting at nodes, with solitary flowers on long up-turned pedicels, ascending to 5cm; leaves ovate-elliptic to suborbicular; petals c.2x as long as sepals; styles 3; (2n=18). Intrd; well natd garden plant in damp rocky places and on paths and walls, scattered over most of BI; W Mediterranean islands.

5. A. montana L. - *Mountain Sandwort*. Perennial with procumbent sterile and erect to ascending flowering stems to 20(30)cm; leaves linear- to oblong-lanceolate; flowers 1-few per stem; petals 10-20mm; styles 3; (2n=20, 28). Intrd; garden plant sometimes natd; among *Pteridium* on Guernsey since 1990; SW Europe.

2. MOEHRINGIA L. - *Three-nerved Sandwort*
Mostly annuals; sepals 5; petals 5, white, entire; stamens 10; styles 3(-4); fruit a capsule opening by 6(-8) teeth; seeds with distinctive oil-body.

1. M. trinervia (L.) Clairv. - *Three-nerved Sandwort*. Stems decumbent to erect, diffusely branched, to 40cm; leaves ovate, 3-5-veined; petals shorter than sepals; 2n=24. Native; shady places in woods and hedgebanks; throughout most of BI.

3. HONCKENYA Ehrh. - *Sea Sandwort*
Subdioecious ± succulent perennials; sepals 5; petals 5, greenish-white, entire; stamens 10 in male flowers; styles 3(-5) in female flowers; fruit a globose capsule opening by 3 valves.

1. H. peploides (L.) Ehrh. - *Sea Sandwort*. Decumbent to erect flowering stems to 25cm arising from extensive stolons or rhizomes; leaves ovate, succulent; petals c. as long as sepals in male flowers, much shorter in female flowers; 2n=48, c.64, 68. Native; bare maritime sand and sandy shingle; common round coasts of BI.

4. MINUARTIA L. (*Cherleria* L.) - *Sandworts*

Annuals or perennials; sepals usually 5; petals usually 5 or 0, white, entire; stamens 10 or fewer; styles 3(-5); fruit a capsule opening by 3 teeth.

1 Slender annual without non-flowering shoots **5. M. hybrida**
1 Mat- or cushion-forming perennials with non-flowering shoots 2
2 Petals usually 0 or minute, rarely ≥1/2 as long as sepals; nectaries 10, conspicuous at base of sepals **6. M. sedoides**
2 Petals ≥1/2 as long as sepals; nectaries vestigial 3
3 Leaves indistinctly 1-veined; pedicels glabrous **4. M. stricta**
3 Leaves distinctly 3-veined; pedicels glandular-hairy 4
4 Petals shorter than sepals **3. M. rubella**
4 Petals as long as or longer than sepals 5
5 Sepals 3-veined; usually laxly tufted **2. M. verna**
5 Sepals 5(-7)-veined; usually densely tufted; SW Ireland **1. M. recurva**

Other spp. - M. rubra (Scop.) McNeill is one of several European alpine plants said to have been collected in Sc over 100 years ago but never refound; most if not all are spurious records.

1. M. recurva (All.) Schinz & Thell. - *Recurved Sandwort*. Densely tufted perennial, woody below, to 5cm; leaves 4-8mm, linear, recurved; petals longer than sepals; styles 3; (2n=30). Native; non-calcareous rocks above 500m in W Cork and S Kerry, discovered 1964. RRR

2. M. verna (L.) Hiern - *Spring Sandwort*. Laxly tufted to loosely-carpeting perennial, scarcely woody below, to 15cm; leaves 6-20mm, linear, ± straight; petals usually longer than sepals; styles 3; 2n=24. Native; base-rich rocky places and sparse grassland, often on lead-mine spoil; locally abundant in N En, N Wa and W Ir, extremely scattered in Sc, N Ir, C & S Wa, Man and SW En. R

3. M. rubella (Wahlenb.) Hiern - *Mountain Sandwort*. Tufted perennial, slightly woody below, to 6cm; leaves 4-8mm, linear, mostly ± straight; petals c.2/3 as long as sepals; styles 3-4(5); 2n=24. Native; base-rich rocks on mountains; very local in N & C Sc. RR

4. M. stricta (Sw.) Hiern) - *Teesdale Sandwort*. Laxly tufted perennial, slightly woody below, to 10cm; leaves 6-12mm, linear, mostly ± straight; petals usually slightly shorter than sepals; styles 3; 2n=22, 30. Native; calcareous flushes in upper Teesdale, Durham, above 450m. RRR

5. M. hybrida (Vill.) Schischk. - *Fine-leaved Sandwort*. Slender erect annual to 20cm; leaves 5-15mm, linear, straight or recurved; petals distinctly shorter than sepals; styles 3; 2n=46. Native; dry bare stony places, walls and arable land; scattered in En, Ir, E Wa and CI, frequent only in E Anglia and CS En but formerly much commoner. Our plant is ssp. **tenuifolia** (L.) Kerguélen. R

6. M. sedoides (L.) Hiern (*Cherleria sedoides* L.) - *Cyphel*. Dense yellow-green moss-like cushion, woody below, to 8cm; leaves 3.5-6mm, linear-lanceolate, crowded; petals 0 or ± so, rarely equalling sepals; styles 3; 2n=51-52. Native; rocky damp ledges and slopes, mostly on mountains but down to sea-level; N & C Sc. R

5. STELLARIA L. - *Stitchworts*

Annuals or perennials; sepals 5; petals 5, sometimes fewer or 0, bifid, white; stamens 10 or fewer; styles 3; fruit a capsule opening by 6 teeth.

1 At least lower leaves petiolate; stems often terete to only weakly and smoothly ridged, sometimes square in section 2
1 All leaves sessile; stems strongly ridged to square in section 6
2 Petals >1.5x as long as sepals; stems ± equally pubescent all round **1. S. nemorum**

2 Petals <1.5x as long as sepals; stems glabrous or with 1(-2) lines of hairs down each internode 3

3 Stems glabrous, square in section; sepals tapered-acute **8. S. uliginosa**

3 Stems with 1(-2) lines of hairs down each internode, terete or weakly ridged; sepals abruptly acute to obtuse 4

4 Sepals mostly 5-6.5mm; stamens ≥8; seeds mostly >1.3mm **4. S. neglecta**

4 Sepals mostly <5mm; stamens mostly <8; seeds mostly <1.3mm 5

5 Stamens 3-8; sepals mostly >3mm; seeds mostly >0.8mm; petals usually present **2. S. media**

5 Stamens 1-3; sepals mostly <3mm; seeds mostly <0.8mm; petals usually absent **3. S. pallida**

6 Bracts entirely herbaceous; petals bifid c.1/2 way to base **5. S. holostea**

6 Bracts entirely scarious or with wide scarious margins; petals bifid much >1/2 way to base 7

7 Petals distinctly shorter than sepals; leaves mostly <15mm **8. S. uliginosa**

7 Petals as long as to longer than sepals; leaves mostly >15mm 8

8 Bracts and outer sepals pubescent at margins; flowers c.5-12mm across **7. S. graminea**

8 Bracts and sepals glabrous; flowers c.12-18mm across **6. S. palustris**

1. S. nemorum L. - *Wood Stitchwort*. Stoloniferous perennial with decumbent to ascending stems to 60cm; stems pubescent; lower leaves long-petiolate, ovate, cordate; petals mostly c.2x as long as sepals; 2n=26. Native; damp woods and shady streamsides.

a. Ssp. nemorum. Bracts gradually decreasing in size at each higher node of inflorescence, those at 2nd node ≥1/3 as long as those at lst; seeds with rounded hemispherical marginal papillae. Scattered in N & W Br, locally common in N En. **172**

b. Ssp. montana (Pierrat) Berher (ssp. *glochidisperma* Murb.). Bracts abruptly shorter above 1st node of inflorescence, those at 2nd node ≤1/3 as long as those at 1st; seeds with cylindrical marginal papillae. Scattered in Wa. **172**

2. S. media (L.) Vill. - *Common Chickweed*. Sprawling much-branched annual (often over-wintering); stems procumbent, to 50cm, with 1 line of hairs; lower leaves petiolate, ovate to elliptic, sometimes ± cordate at base; sepals 2.7-5.2mm, pubescent or glandular-pubescent; petals 1.1-3.1mm, rarely 0 or minute; stamens 3-5(8); seeds 0.8-1.4mm; 2n=44. Native; ubiquitous weed of cultivated and open ground. Stamen number and seed size are the best characters to separate *S. media*, *S. pallida* and *S. neglecta*.

3. S. pallida (Dumort.) Crép. - *Lesser Chickweed*. Sprawling, often yellowish-green, much-branched, short-lived annual; like *S. media* but sepals 2.1-3.6mm, glabrous to pubescent; petals 0 or <1mm; stamens 1-2(3); seeds 0.6-0.9mm; 2n=22. Native; coastal dunes and shingle, inland on bare sandy soil; locally frequent in En, Wa and CI, rare and scattered in Sc and Ir.

4. S. neglecta Weihe - *Greater Chickweed*. Usually ascending annual to perennial to 80cm; like *S. media* but sepals 5-6.5mm, glandular-pubescent; petals 2.5-4mm; stamens 10; seeds 1.1-1.7mm; 2n=22. Native; shady, usually damp places; scattered through most of Br, mainly in S, very rare in W Ir.

5. S. holostea L. - *Greater Stitchwort*. Scrambling to ascending perennial to 60cm; stems with rough angles, glabrous or pubescent above; leaves lanceolate, sessile; bracts herbaceous; petals much longer than sepals; 2n=26. Native; woods and shady hedgerows; common throughout most of BI.

6. S. palustris Retz. - *Marsh Stitchwort*. Ascending to ± erect perennial to 60cm; stems with smooth angles, glabrous; leaves linear-lanceolate, sessile; bracts scarious except for green midline; petals slightly to much longer than sepals; 2n=c.130. Native; usually base-rich marshes and fens; scattered in Br S from C Sc, C Ir.

7. S. graminea L. - *Lesser Stitchwort*. Decumbent to ascending perennial to 80cm; stems with smooth angles, glabrous; leaves lanceolate to linear-lanceolate, sessile; bracts scarious; petals c. as long as to longer than sepals; 2n=26. Native; grassy, often dry places; throughout BI.

8. S. uliginosa Murray (*S. alsine* Grimm nom. inval.) - *Bog Stitchwort*. Decumbent to ascending perennial to 40cm; stems with smooth angles, glabrous; leaves narrowly ovate to elliptic, mostly sessile but those on sterile shoots petiolate; bracts scarious except for green midline; petals shorter than sepals; 2n=24. Native; streamsides, ditches, wet tracks, depressions, often on acid soils; throughout BI.

6. HOLOSTEUM L. - *Jagged Chickweed*
Annuals; sepals 5; petals 5, irregularly toothed, white; stamens usually 3-5; styles 3; fruit a capsule opening by 6 teeth.

1. H. umbellatum L. - *Jagged Chickweed*. Stems ± erect, little-branched, to 20cm, glandular-pubescent; leaves oblanceolate to elliptic, the lowest shortly petiolate; flowers in umbel on different-lengthed pedicels; petals longer than sepals; (2n=20). Possibly native; arable fields, banks and walls on well-drained soils; E Anglia and SE En, but extinct since 1930. **RR**

7. CERASTIUM L. - *Mouse-ears*
Annuals or perennials; sepals 4-5; petals 4-5, sometimes 0, retuse or emarginate to bifid to 1/2 way, white; stamens 4, 5 or 10; styles 3-5(6); fruit a capsule opening by 2x as many teeth as styles.

1 Perennial, with usually ± procumbent non-flowering shoots as well as more erect flowering shoots 2
1 Annual, with all shoots producing flowers 8
  2 Styles mostly 3; capsule-teeth mostly 6 **1. C. cerastoides**
  2 Styles 5; capsule-teeth 10 3
3 Petals ≤1.7(2)x as long as sepals **7. C. fontanum**
3 Petals >(1.7)2x as long as sepals 4
  4 Leaves and upper parts of stems with predominantly densely matted or long (>1mm) shaggy hairs 5
  4 Leaves and stems with predominantly short (<1mm) ± straight hairs; long shaggy hairs less frequent or 0 6
5 Leaves linear to narrowly elliptic, densely tomentose with matted hairs; most stems with >4 flowers; lowland **3. C. tomentosum**
5 At least lower leaves elliptic to ovate or obovate, with many long shaggy hairs; stems with 1-4 flowers; arctic-alpine **4. C. alpinum**
  6 Leaves linear to narrowly oblong; flowering stems with short leafy shoots in some leaf axils; lowland **2. C. arvense**
  6 Leaves narrowly to broadly elliptic; flowering stems without conspicuous axillary shoots; arctic-alpine 7
7 Leaves narrowly elliptic to elliptic, either sparsely pubescent or with both glandular and non-glandular hairs; not Shetland **5. C. arcticum**
7 Leaves elliptic to broadly elliptic, densely glandular-pubescent; Shetland **6. C. nigrescens**
  8 Sepals with long eglandular hairs projecting beyond sepal apex 9
  8 Sepals without eglandular hairs projecting beyond sepal apex 10
9 Inflorescence compact at fruiting; fruiting pedicels mostly shorter than sepals; glandular hairs abundant on sepals **8. C. glomeratum**
9 Inflorescence lax at fruiting; fruiting pedicels mostly longer than sepals; glandular hairs 0 **9. C. brachypetalum**
  10 Bracts completely herbaceous; petals and stamens usually 4; capsule-teeth usually 8 **10. C. diffusum**

10 Bracts with scarious tips; petals and stamens usually 5; capsule-teeth usually 10 11

11 Uppermost bracts scarious for at least apical 1/3; petals distinctly shorter (c.2/3 as long) than sepals **12. C. semidecandrum**

11 Uppermost bracts scarious for at most apical 1/4; petals c. as long as sepals **11. C. pumilum**

1. C. cerastoides (L.) Britton - *Starwort Mouse-ear*. Mat-forming perennial with R ascending to decumbent vegetative and flowering shoots to 15cm, almost glabrous apart from short glandular hairs on pedicels and sepals; petals nearly 2x as long as sepals; styles 3(-6); 2n=38. Native; moist rocky places on mountains; mainland N Sc S to E Perth and Westerness.

2. C. arvense L. - *Field Mouse-ear*. Loose mat-forming perennial with ascending to decumbent vegetative and flowering shoots to 30cm, shortly pubescent (often sparsely so) with mostly eglandular but some glandular hairs; petals c.2x as long as sepals; styles 5; 2n=72. Native; dry grassland; most of Br but rare in W, local in Ir.

2 x 3. C. arvense x C. tomentosum (= *C. x maureri* M. Schulze nom. nud., *C. decalvans* auct. non Schloss. & Vuk.) occurs in E & SE En and C Sc near both parents; it is intermediate in pubescence and produces some good seed; 2n=72. Probably overlooked; likely to spread away from its parents.

2 x 7. C. arvense x C. fontanum = C. x pseudoalpinum Murr has occurred in 3 places in Co Durham and S Lincs in rough grassland with the parents; it resembles *C. arvense* but with smaller flowers, wider leaves and denser pubescence, and is sterile.

3. C. tomentosum L. - *Snow-in-summer*. Rampant mat-forming perennial with strong ascending to decumbent vegetative and flowering shoots to 40cm, densely whitish-tomentose with matted elgandular hairs; petals c.2x as long as sepals; styles 5; 2n=72. Intrd; commonly cultivated in rock gardens; natd widely in BI in dry places; Italy and Sicily. Part of a difficult complex of taxa, others of which, especially **C. biebersteinii** DC., from Crimea, and **C. decalvans** Schloss. & Vuk., from Balkans, have been recorded in error. *C. tomentosum* is extremely variable in Italy and Sicily and several of the variants occur in BI.

4. C. alpinum L. - *Alpine Mouse-ear*. Tufted or matted perennial with decumbent R to ascending vegetative and flowering shoots to 20cm, pubescent with many long and short, soft, whitish, eglandular hairs and few or 0 glandular hairs; petals c.2x as long as sepals; styles 5; 2n=72, 108. Native; rock outcrops and ledges on mountains; mainland C & N Sc, also extremely local in S Sc, NW En and N Wa. The British plant is ssp. **lanatum** (Lam.) Cesati.

4 x 5. C. alpinum x C. arcticum (*C. x blyttii* auct. non Baen.) occurs in N Wa and Sc near its parents; it is intermediate in pubescence and bract-form and is largely sterile.

4 x 7. C. alpinum x C. fontanum = C. x symei Druce occurs on mountains in Sc with both parents; it is intermediate in pubescence and flower-size and -number and is sterile.

5. C. arcticum Lange (*C. nigrescens* ssp. *arcticum* (Lange) P.S. Lusby) - *Arctic* R *Mouse-ear*. Tufted or matted perennial with ascending to decumbent vegetative and flowering shoots to 15cm, pubescent with ± 0 to frequent short hairs, 0 to frequent (var. **alpinopilosum** Hultén) long wavy hairs, and few to many glandular hairs; leaves elliptic to narrowly elliptic; sepals acute to subacute; petals c.2x as long as sepals; styles 5; 2n=108. Native; rocky ledges and outcrops on mountains; N Wa, C & N mainland Sc. Var. *alpinopilosum*, which might represent introgression from *C. alpinum*, is much the commoner var. Plants from 1 area of the Cairngorms, S Aberdeen, almost entirely lack non-glandular hairs and are closer to var. **arcticum**, but differ from it and might represent a new taxon.

5 x 7. C. arcticum x C. fontanum = C. x richardsonii Druce has been found with the parents in N Wa and N Sc; it is intermediate but with petals as long as those of *C. arcticum* and is largely sterile.

6. C. nigrescens (H.C. Watson) Edmondston ex H.C. Watson (*C. arcticum* ssp. *edmondstonii* (Edmondston) Á. & D. Löve) - *Shetland Mouse-ear*. Differs from *C. arcticum* in leaves elliptic to broadly elliptic, usually dark green and purplish tinged, with abundant glandular and sparse to frequent short non-glandular hairs; sepals mostly obtuse; 2n=72. Native; serpentine rocks on Unst, Shetland; endemic. **RR**

7. C. fontanum Baumg. - *Common Mouse-ear*. Tufted or matted perennial with decumbent vegetative and ascending flowering shoots to 50cm, sparsely to densely pubescent with few or 0 glandular hairs; upper bracts usually with narrow scarious margins; petals mostly slightly shorter to slightly longer than sepals, rarely nearly 2x as long; styles 5; 2n=72, 108, 144. Native; grassland, and open waste and cultivated ground. Very variable; other taxa than the 3 following occur, but all might be better placed in 1 ssp.

1 Seeds 0.8-1mm, with tubercles c.0.1mm across at base; petals c.1.3-1.7x as long as sepals **c. ssp. scoticum**
1 Seeds 0.4-0.9mm, with tubercles c.0.05mm across at base; petals c.0.8-1.2(2)x as long as sepals 2
  2 Lower stem internodes with hairs all round; leaves pubescent on both sides **a. ssp. vulgare**
  2 Lower stem internodes glabrous or with 1-2 lines of hairs; leaves ± glabrous or very sparsely pubescent with most hairs on margin or lowerside midrib **b. ssp. holosteoides**

a. Ssp. vulgare (Hartm.) Greuter & Burdet (ssp. *triviale* (Spenn.) Jalas). See key for characters. Common through BI.

b. Ssp. holosteoides (Fr.) Salman, Ommering & de Voogd (ssp. *glabrescens* (G. Mey.) Salman, Ommering & de Voogd, *C. holosteoides* Fr., incl. var. *glabrescens* (G. Mey.) Hyl.). Differs from ssp. *vulgare* in less pubescent stems and leaves, mostly narrower upper leaves (length/width ratio mostly >4) and mostly slightly longer calyx (5-8mm, not 4-7mm) and capsules (8-16mm, not 7-11mm). Mostly wet places; probably throughout BI but much commoner in N.

c. Ssp. scoticum Jalas & P.D. Sell. Differs from other sspp. in larger seeds, usually longer petals, and more prominently keeled sepals; pubescence usually ± intermediate between that of other 2 sspp. Rocky mountain sites above 900m; very local in Angus; endemic. Records from the W Sutherland coast refer to large-flowered ssp. *vulgare*.

8. C. glomeratum Thuill. - *Sticky Mouse-ear*. Erect or ascending annual to 45cm, with glandular and abundant eglandular hairs; bracts entirely herbaceous; inflorescence compact at fruiting; flowers 5-merous; petals c. as long as sepals; 2n=72. Native; open places in natural and artificial habitats; common throughout BI.

9. C. brachypetalum Pers. - *Grey Mouse-ear*. Erect annual to 30cm, with 0 glandular and abundant patent eglandular hairs; bracts entirely herbaceous; inflorescence diffuse at fruiting; flowers 5-merous; petals c.1/2 as long as sepals; 2n=90. Possibly native; chalk grassland and barish places on banks; Beds (discovered 1947), Northants (1973) and W Kent (1978). **RR**

10. C. diffusum Pers. (*C. atrovirens* Bab.) - *Sea Mouse-ear*. Decumbent to erect annual to 30cm, with abundant glandular and some eglandular hairs; bracts entirely herbaceous; inflorescence diffuse at fruiting; flowers 4-(5-)merous; petals c.3/4 as long as sepals; 2n=72. Native; dry, open sandy places, common round coasts of BI and very local inland.

11. C. pumilum Curtis - *Dwarf Mouse-ear*. Erect to ascending annual to 12cm, with abundant glandular and non-glandular hairs; bracts with scarious tips ≤1/4 total length; inflorescence diffuse at fruiting; flowers 5-merous; petals c. as long as sepals; 2n=72, 90. Native; calcareous bare ground or open grassland; very scattered, En N to Leics and Wa N to Caerns and Denbs. **R**

12. C. semidecandrum L. - *Little Mouse-ear.* Erect to ascending annual to 20cm, with abundant glandular and non-glandular hairs; bracts with scarious tips ≥1/3 total length; inflorescence diffuse at fruiting; flowers 5-merous; petals c.2/3 as long as sepals; 2n=36. Native; dry open places on sandy or calcareous soils, especially dunes; frequent throughout most of BI but rare in N Sc and C Ir.

8. MYOSOTON Moench - *Water Chickweed*
Perennials; sepals 5; petals 5, white, bifid almost to base; stamens 10; styles 5; fruit a capsule opening by 5 slightly bifid teeth.

1. M. aquaticum (L.) Moench (*Stellaria aquatica* (L.) Scop.) - *Water Chickweed.* Plant straggling, decumbent to ascending, to 1m, glabrous below, with abundant glandular hairs above; leaves ovate, cordate at base; petals longer than sepals; 2n=28. Native; marshes, ditches and banks of water-courses; fairly common in En and Wa N to Durham, formerly S Sc.

9. MOENCHIA Ehrh. - *Upright Chickweed*
Annuals; sepals 4; petals 4, entire, white; stamens 4; styles 4; fruit a capsule opening by 8 teeth.

1. M. erecta (L.) P. Gaertn., B. Mey. & Scherb. - *Upright Chickweed.* Erect to procumbent, glabrous, ± glaucous annual to 12cm; leaves linear; petals shorter than sepals; 2n=36. Native; barish places on sandy or gravelly ground; local in En and Wa N to Cheviot and Caerns, formerly more frequent, CI.

10. SAGINA L. - *Pearlworts*
Annuals or perennials; leaves linear; sepals 4-5; petals 4-5, sometimes 0, entire, white or whitish; stamens 4, 5, 8 or 10; styles 4-5; fruit a capsule opening by 4-5 valves.

1	Annual, with all shoots producing flowers	2
1	Perennial, tufted or mat-forming, with short non-flowering shoots	3
2	Leaves obtuse, sometimes minutely mucronate with point <0.1mm	**9. S. maritima**
2	Leaves abruptly protracted into fine point >0.1mm	**8. S. apetala**
3	Petals 0 or <1/2 as long as sepals	4
3	Petals 3/4-2x as long as sepals	5
4	Plant a densely compact small cushion with rigidly recurved leaves hiding the stems; seed rarely produced	**7. S. boydii**
4	Stems usually partly visible; leaves not or weakly recurved; seed abundantly produced	**6. S. procumbens**
5	Leaves at uppermost nodes usually >3x shorter than those at lower nodes; petals c.2x as long as sepals	**1. S. nodosa**
5	Leaves at uppermost nodes usually <2x shorter than those at lower nodes; petals <1.25x as long as sepals	6
6	Sepals with glandular hairs	**3. S. subulata**

FIG 167 - Caryophyllaceae. 1-4, fruiting perianths of *Scleranthus*. 1, *S.* ***perennis*** ssp. ***prostratus***. 2, ssp. ***perennis***. 3, *S.* ***annuus*** ssp. ***polycarpos***. 4, ssp. ***annuus***. 5-6, flowers of ***Silene***. 5, *S.* ***vulgaris***. 6, *S.* ***uniflora***. 7-10, capsules of ***Silene***. 7, *S.* ***italica***. 8, *S.* ***nutans***. 9, *S.* ***latifolia***. 10, *S.* ***dioica***. 11-13, capsules of ***Arenaria serpyllifolia***. 11, ssp. ***serpyllifolia***. 12, ssp. ***leptoclados***. 13, ssp. ***serpyllifolia*** (ssp. *lloydii*). 14-16, shoots of ***Herniaria***. 14, *H.* ***hirsuta***. 15, *H.* ***ciliolata***. 16, *H.* ***glabra***. 17-20, leaves and capsules of ***Sagina***. 17, *S.* ***subulata***. 18, *S.* ***procumbens***. 19, *S.* ***apetala*** ssp. ***erecta***. 20, ssp. ***apetala***.

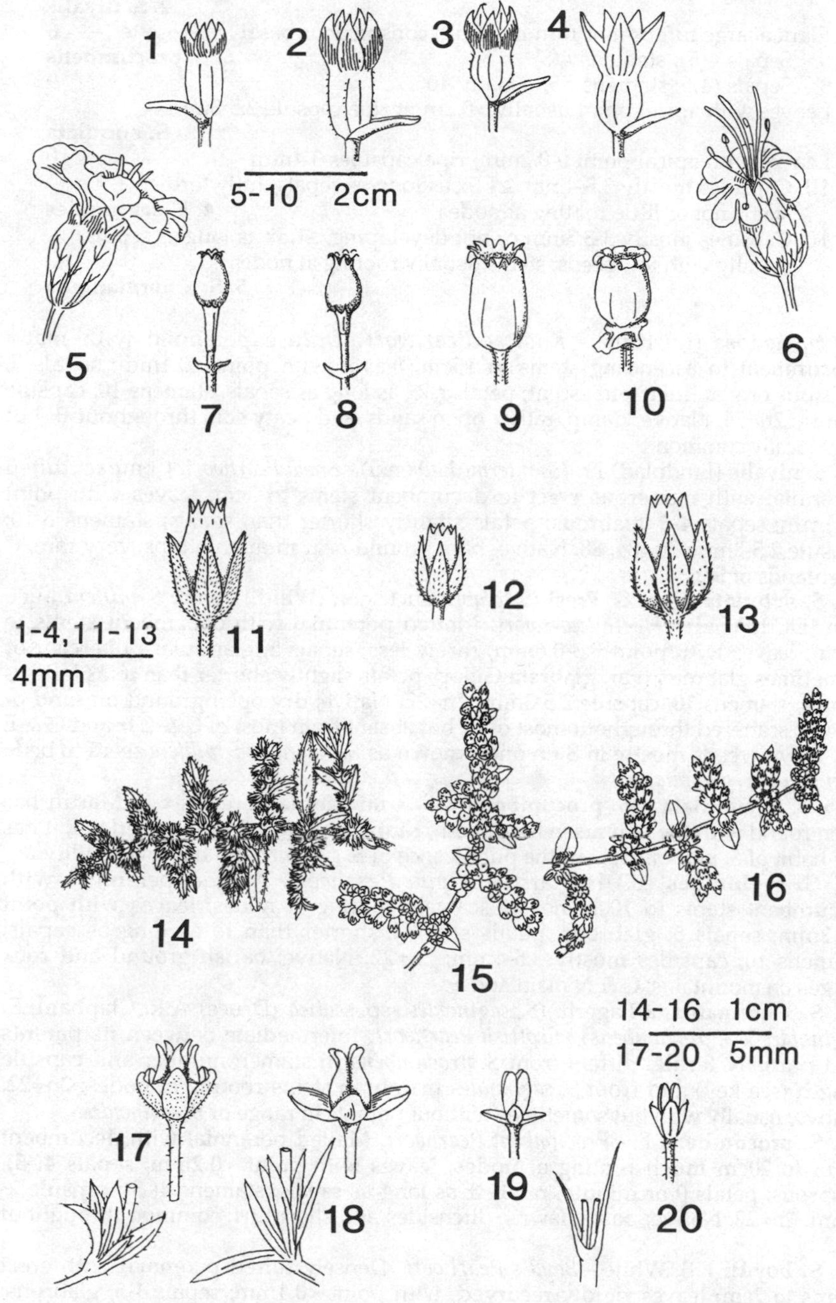

FIG 167 - see caption opposite

6 Sepals glabrous 7

7 Plant compactly tufted, with basal leaf-rosette only in first year **2. S. nivalis**

7 Plant a large tuft or mat-forming, with conspicuous basal leaf-rosette 8

8 Sepals 4(5); stamens 4(5) **6. S. procumbens**

8 Sepals (4)5; stamens >5, usually 10 9

9 Leaves with apical point usually >0.2mm; ripe capsule 2.5-3mm **3. S. subulata**

9 Leaves with apical point 0-0.2mm; ripe capsules 3-4mm 10

10 Capsules mostly 3.5-4mm, ≥1.5x as long as sepals, fully fertile; stems not or little rooting at nodes **4. S. saginoides**

10 Capsules mostly 3-3.5mm or not developing, ≤1.5x as long as sepals, usually with few seeds; stems usually rooting at nodes **5. S. x normaniana**

1. S. nodosa (L.) Fenzl - *Knotted Pearlwort*. Diffuse perennial with many procumbent to ascending stems to 15cm; leaves with point ≤0.1mm; sepals 5, glabrous or glandular-pubescent; petals c.2x as long as sepals; stamens 10; capsule c.4mm; 2n=56. Native; damp, rather open sandy and peaty soil; throughout BI but only locally common.

2. S. nivalis (Lindblad) Fr. (*S. intermedia* Fenzl) - *Snow Pearlwort*. Compact tufted **RR** perennial with numerous erect to decumbent stems to 3cm; leaves with point <0.1mm; sepals 4-5, glabrous; petals slightly shorter than sepals; stamens 8-10; capsule 2.5-3mm; 2n=84, 88. Native; bare ground near mountain tops; very rare, C Highlands of Sc.

3. S. subulata (Sw.) C. Presl (*S. glabra* auct. non (Willd.) Fenzl, *S. pilifera* auct. **167** non (DC.) Fenzl) - *Heath Pearlwort*. Matted perennial with decumbent stems to 10cm; leaves with point 0.3-0.6mm, rarely less; sepals 5, glandular-pubescent or sometimes glabrous (var. **glabrata** Gillot); petals slightly shorter than to as long as sepals; stamens 10; capsule 2.5-3mm; 2n=22. Native; dry open ground on sand or gravel; scattered throughout most of BI but absent from most of C & E Ir and C & E En. Lawn-weeds mostly in S En once known as *S. glabra* or *S. pilifera* seem to be *S. subulata* var. *glabrata*.

3 x 6. S. subulata x S. procumbens = S. x micrantha Boreau ex E. Martin has been found with the parents in Merioneth, Kintyre and (doubtfully) Shetland; it has the habit of *S. procumbens* but the pubescence of *S. subulata* and has low fertility.

4. S. saginoides (L.) H. Karst. - *Alpine Pearlwort*. Matted perennial with **R** decumbent stems to 10cm not or scarcely rooting at nodes; leaves with point <0.2mm; sepals 5, glabrous; petals slightly shorter than to as long as sepals; stamens 10; capsules mostly 3.5-4mm; 2n=22. Native; barish ground and rock ledges on mountains; C & N mainland Sc.

5. S. x normaniana Lagerh. (*S. saginoides* ssp. *scotica* (Druce) A.R. Clapham; *S.* **RR** *saginoides* x *S. procumbens*) - *Scottish Pearlwort*. Intermediate between its parents and partially fertile; differs from *S. procumbens* in stamen number and capsule length (see key) and from *S. saginoides* mainly in stems rooting at nodes; 2n=22. Native; usually with but sometimes without parents in range of *S. saginoides*.

6. S. procumbens L. - *Procumbent Pearlwort*. Matted perennial with decumbent **167** stem to 20cm much rooting at nodes; leaves with point <0.2mm; sepals 4(-5), glabrous; petals 0 or minute, rarely ± as long as sepals; stamens 4(-5); capsule 2-3mm; 2n=22. Native; paths, lawns, ditchsides and short turf; common throughout BI.

7. S. boydii F.B. White - *Boyd's Pearlwort*. Densely tufted perennial with erect **RR** stems to 2cm; leaves rigidly recurved, with point <0.1mm; sepals 4-5, glabrous; petals 0 or minute; stamens 5-10; capsule extremely rarely developing; 2n=22. Enigmatic; presumed to have been collected in S Aberdeen in 1878; not seen wild since but still in cultivation; endemic. Comes true from seed.

8. S. apetala Ard. - *Annual Pearlwort*. Diffusely branched ± erect annual to 15cm; leaves with point 0.1-0.4mm; sepals usually 4, glabrous or glandular-pubescent; petals 0 or minute; stamens usually 4; capsule 1.6-2.5mm. Native.

a. Ssp. apetala (*S. ciliata* Fr., *S. filicaulis* Jord.). Sepals erect to erecto-patent in fruit, at least the outer subacute; most seeds >1/3mm; 2n=12. Dry bare ground on heaths and paths; scattered through most of BI. **S. filicaulis**, more slender, with smaller parts and glandular rather than glabrous pedicels and sepals, may merit equal rank. **167**

b. Ssp. erecta F. Herm. (*S. apetala* auct. non Ard.). Sepals patent in fruit, subobtuse; most seeds <1/3mm; 2n=12. Paths, walls and bare cultivated ground; common through most of BI but rare in N Sc. Possibly these 2 sspp. should be recognized as 2 or 3 vars. **167**

9. S. maritima Don - *Sea Pearlwort*. Diffusely branched ± erect annual to 15cm; leaves with point <0.1mm; sepals 4, obtuse, glabrous, erecto-patent in fruit; petals 0 or minute; stamens 4; capsules 2-2.8mm; 2n=28. Native; damp sand-dunes, rocky places and cliff-ledges on barish soil; round coasts of BI and rarely slightly inland on mountains in Sc.

11. SCLERANTHUS L. - *Knawels*

Annuals to perennials; leaves linear, fused at base in opposite pairs; flowers greenish, inconspicuous; sepals usually 5; petals 0; fertile stamens (2)5-10; styles 2; fruit an achene.

1. S. perennis L. - *Perennial Knawel*. Biennial to perennial to 20cm with woody basal parts and some sterile shoots at flowering time; sepals subacute to obtuse, with white border c.0.3-0.5mm wide (nearly as wide as green central part), the tips parallel or incurved over ripe achene; fertile stamens 10. Native. **RRR**

a. Ssp. perennis. Stems ascending to erect; achene (incl. sepals) (3)3.5-4.5mm. Doloritic rocks in Rads. **167**

b. Ssp. prostratus P.D. Sell. Stems procumbent to ± ascending; achene (incl. sepals) 2-3(3.5)mm; 2n=22. Very local and decreasing on sandy heaths in Norfolk and Suffolk; endemic. **167**

2. S. annuus L. - *Annual Knawel*. Decumbent to erect annual or biennial to 20cm, without sterile shoots at flowering time; sepals acute to subacute, with whitish border c.0.1mm wide (1/2 as wide as green central part); fertile stamens 2-10; 2n=44. Native; dry open sandy ground.

a. Ssp. annuus. Achene (incl. sepals) 3.2-4.5(5.5)mm, with divergent sepals when ripe. Scattered over most of BI but rare in Ir and N Sc. **167**

b. Ssp. polycarpos (L.) Bonnier & Layens (*S. polycarpos* L.). Achene (incl. sepals) **R**
2.2-3.0(3.8)mm, with parallel to convergent sepals when ripe. Frequent on sandy **167**
ground in Suffolk.

SUBFAMILY 2 - PARONYCHIOIDEAE (genera 12-17).

Leaves opposite, alternate or whorled, stipulate; sepals 5, free; petals 5, often very small; carpophore 0; fruit a capsule or achene.

12. CORRIGIOLA L. - *Strapwort*

Leaves alternate; flowers slightly perigynous; petals ± equalling sepals, white, sometimes red-tipped, entire; stamens 5; stigmas 3, sessile, capitate; fruit an achene. This and *Telephium* might be better removed to Molluginaceae.

Other spp. - C. telephiifolia Pourr. (*C. litoralis* ssp. *telephiifolia* (Pourr.) Briq.) was natd for some years recently at Gloucester Docks, E Gloucs, but has now gone; it is a perennial with ± 0 bracts in the inflorescence but often difficult to distinguish.

1. C. litoralis L. - *Strapwort*. Glabrous ± glaucous annual with 1-many decumbent **RRR**

stems to 25cm; leaves linear to oblanceolate, obtuse; inflorescence branches with leaf-like bracts; 2n=16, 18. Native; on sand and gravel by ponds; S Devon, formerly W Cornwall and Dorset, infrequent casual elsewhere mainly by railways in En.

13. HERNIARIA L. - *Ruptureworts*

Leaves opposite; petals shorter than sepals, filiform; stamens 5; stigmas 2, capitate, sessile or on very short styles or common style; fruit an achene.

1 Calyx and leaves with many hairs on surfaces **3. H. hirsuta**
1 Calyx and leaves glabrous or with hairs at margins only 2
 2 Fruit acute to subacute, distinctly exceeding sepals; seeds c.0.5-0.6mm **1. H. glabra**
 2 Fruit obtuse to subobtuse, scarcely or not exceeding sepals; seeds c.0.7-0.8mm **2. H. ciliolata**

1. H. glabra L. - *Smooth Rupturewort*. Annual to perennial with numerous procumbent branched stems to 30cm; stems usually hairy all round, not or slightly woody at base; 2n=18. Native; dry sandy ground; E En, extinct or rare casual elsewhere in Br and Ir. **RR** **167**

2. H. ciliolata Melderis - *Fringed Rupturewort*. Evergreen perennial; differs from *H. glabra* in stems strongly woody below; and see key. Inflorescence shape, leaf pubescence and sepal length and pubescence, often used to separate *H. ciliolata* and *H. glabra*, seem unreliable. Native; maritime sandy and rocky places. **RR** **167**

a. Ssp. ciliolata. Stems hairy on 1 (usually upper) side; leaves ovate to obovate; 2n=72. W Cornwall, Guernsey and Alderney; endemic.

b. Ssp. subciliata (Bab.) Chaudhri (*H. ciliolata* var. *angustifolia* (Pugsley) Melderis). Stems hairy all round; leaves narrowly elliptic; 2n=108. Jersey; endemic.

3. H. hirsuta L. (*H. cinerea* DC.) - *Hairy Rupturewort*. Annual similar to *H. glabra* but whole plant surface with patent hairs; 2n=36. Intrd; open waste ground; occasional casual from several sources, sometimes natd; En; Europe. Extreme **H. cinerea** (*H. hirsuta* ssp. *cinerea* (DC.) Cout.) is distinct but all intermediates occur. **167**

14. ILLECEBRUM L. - *Coral-necklace*

Leaves opposite; flowers hypogynous; petals much shorter than sepals, filiform; stamens 5; stigmas 2, sessile, capitate; fruit a 1-sided capsule opening by 5 valves adherent at apex.

1. I. verticillatum L. - *Coral-necklace*. Glabrous annual with procumbent to decumbent stems to 20cm, often reddish; leaves obovate, obtuse; flowers densely clustered in leaf-axils; sepals white, thick, succulent, hooded, with finely pointed apex; 2n=10. Native; damp sandy open ground; rare in extreme SC & SW En, rarely natd or casual elsewhere. **R**

15. POLYCARPON L. - *Four-leaved Allseed*

Leaves opposite or 4-whorled; flowers hypogynous; petals much shorter than sepals, often emarginate; stamens (1)3-5; stigmas 3, on separate styles; fruit a capsule opening by 3 valves.

1. P. tetraphyllum (L.) L. (*P. diphyllum* Cav.) - *Four-leaved Allseed*. Almost glabrous annual with erect to ascending much-branched stems to 25cm; leaves obovate, obtuse to rounded; flowers numerous, small and greenish; 2n=54. Native; open sandy and waste ground near sea; CI, Dorset, S Devon, Cornwall and Scillies, casual elsewhere. Plants with leaves only paired, contracted inflorescences, and stamens 1-3 (not 3-5) are distinguishable as var. **diphyllum** (Cav.) DC. (ssp. *diphyllum* (Cav.) O. Bolòs & Font Quer). **RR**

16. SPERGULA L. - *Spurreys*
Leaves opposite but often appearing whorled due to axillary leaf clusters; flowers hypogynous; petals c. as long as sepals or slightly longer, white, entire; stamens 5-10; stigmas 5, on separate styles; fruit a capsule opening by 5 valves.

1. S. arvensis L. - *Corn Spurrey*. Glandular-pubescent annual with erect to 172
ascending branched stems to 40(60)cm; leaves linear, furrowed on lowerside; seeds little flattened, without wing or with circum-equatorial wing <1/10 as wide as actual seed; 2n=18. Native; calcifuge on usually sandy cultivated ground or rarely in short maritime turf; throughout BI. Var. **arvensis** has seeds with 0 or extremely narrow wing and the surface covered by clavate papillae; it is usually sparsely glandular and commoner in SE. Var. **sativa** (Boenn.) Mert. & W.D.J. Koch has seeds with 0 papillae and a wider wing; it is usually densely glandular and commoner in NW. Other alien vars might occur but have not been studied. Var. **nana** E.F. Linton (ssp. *nana* (E.F. Linton) D.C. McClint.), a dwarf variant of var. *arvensis* hardly worthy of separation, occurs in CI.

2. S. morisonii Boreau - *Pearlwort Spurrey*. Annual with slenderer but more rigid 172
stems to 30cm, ± glabrous to sparsely glandular-pubescent; leaves linear, not furrowed on lowerside; seeds strongly flattened, with circum-equatorial wing >1/4 as wide as actual seed; 2n=18. Intrd; natd on sandy or peaty arable land; E Sussex and Kildare; C Europe.

17. SPERGULARIA (Pers.) J. & C. Presl - *Sea-spurreys*
Leaves opposite; flowers hypogynous; petals shorter to longer than sepals, (white to) pink, entire; stamens (0)5-10; stigmas 3, on separate styles; fruit a capsule opening by 3 valves.

1 At least some seeds with distinct circum-equatorial wing 2
1 All seeds unwinged 3
  2 Flowers (7)10-12(13)mm across; stamens (0-)10, if fewer than 10 the remainder represented as staminodes **2. S. media**
  2 Flowers (4)5-8mm across; stamens (0)2-7(10); staminodes 0(-3) **3. S. marina**
3 Flowers mostly >8mm across; sepals mostly >4mm; capsule mostly >5mm 4
3 Flowers mostly <8mm across; sepals mostly <4mm; capsule mostly <5mm 5
  4 Plant glabrous to glandular-pubescent only in inflorescence, not woody below **2. S. media**
  4 Plant glandular-pubescent over most of stem, woody below **1. S. rupicola**
5 Seeds ≥0.6mm; stipules on young shoots fused for ≥1/3 their length **3. S. marina**
5 Seeds <0.6mm; stipules on young shoots fused for <1/4 their length 6
  6 Upper bracts not much different from stem-leaves; most pedicels much longer than capsules **4. S. rubra**
  6 Upper bracts much shorter than stem-leaves or represented ± only by stipules; most pedicels shorter than capsules **5. S. bocconei**

1. S. rupicola Lebel ex Le Jol. - *Rock Sea-spurrey*. Perennial with ± woody base; 172
stems to 35cm, decumbent, glandular-pubescent to well below inflorescence; sepals 4-4.5mm, about as long as petals; capsule 4.5-7mm; seeds not winged; 2n=36. Native; walls and rocky maritime places; locally common on coasts of CI, Ir, Wa, W En and SW Sc.

1 x 3. S. rupicola x S. marina has been found in mixed populations of the parents in Co Dublin, E Cornwall, Dorset and Monts; it is a completely sterile vigorous floriferous perennial intermediate in morphological details.

2. S. media (L.) C. Presl (*S. marginata* Kitt. nom. illeg., *S. maritima* (All.) Chiov.) - 172

FIG 172 - Seeds of **Caryophyllaceae**. 1, ***Spergularia media*** (winged). 2, ***S. marina*** (winged). 3, ***S. rupicola***. 4, ***S. media*** (unwinged). 5, ***S. marina*** (unwinged). 6, ***S. bocconei***. 7, ***S. rubra***. 8, ***Stellaria nemorum*** ssp. ***montana***. 9, ***S. nemorum*** ssp. ***nemorum***. 10, ***Spergula morisonii***. 11, ***S. arvensis*** var. ***sativa***. 12, ***S. arvensis*** var. ***arvensis***.

Greater Sea-spurrey. Perennial with non-woody base; stems to 40cm, decumbent to ascending, glabrous or glandular-pubescent only in inflorescence; sepals (3.5)4-6mm, usually slightly shorter than petals; capsule (4)7-9mm; seeds 0.7-1mm (excl. wing), all or many (very rarely 0) winged; 2n=18. Native; sandy and muddy maritime places; common round coasts of Br and Ir, very rarely inland.

3. S. marina (L.) Griseb. (*S. salina* J. & C. Presl) - *Lesser Sea-spurrey*. Annual or sometimes perennial; stems decumbent to procumbent, to 35cm, usually glandular-pubescent in inflorescence and often some way below; sepals 2.5-4mm, usually longer than petals; capsule 3-6mm; seeds 0.6-0.8mm (excl. wing), wingless, winged, or mixed; 2n=36. Native; sandy and muddy maritime places and inland saline areas; common round coasts of BI and locally frequent inland. **172**

4. S. rubra (L.) J. & C. Presl - *Sand Spurrey*. Annual or biennial; stems to 25cm, decumbent, glandular-pubescent in inflorescence and usually some way below; sepals 3-4(5)mm, usually longer than petals; capsule 3.5-5mm; seeds not winged; 2n=36. Native; calcifuge of sandy or gravelly ground; throughout most of BI and often common. **172**

5. S. bocconei (Scheele) Graebn. - *Greek Sea-spurrey*. Annual; stems to 20cm, decumbent; densely glandular-pubescent in inflorescence and some way below; sepals 2-4mm, longer than petals; capsule 2-4mm; seeds not winged; 2n=36. Possibly native; dry sandy and waste places by sea; frequent in CI, rare and probably intrd in Cornwall, rare casual elsewhere in S En and S Wa. **RR** **172**

SUBFAMILY 3 - CARYOPHYLLOIDEAE (*Silenoideae*) (genera 18-25). Leaves opposite, without stipules; sepals fused to form calyx-tube and -lobes; petals 5, mostly medium to large, usually differentiated into a narrow proximal claw within and widened distal limb above calyx-tube; stamens 10; petals, stamens and ovary usually elevated above receptacle on sterile axis (*carpophore*); fruit a capsule or berry.

18. LYCHNIS L. (*Viscaria* Röhl., *Steris* Adans.) - *Catchflies*
Bracteoles not forming epicalyx; petal-limb entire or 2-4-lobed, with 2 scales at base, scarlet, purple or rarely white; styles 5; fruit a capsule opening by 5 teeth. Sometimes subdivided, but perhaps best amalgamated with *Silene*.

1 Leaves and stems with dense white woolly hairs **1. L. coronaria**
1 Leaves and stems glabrous to pubescent but not covered with white hairs 2
  2 Petal-limb divided much >1/2 way into 4 narrow lobes **2. L. flos-cuculi**
  2 Petal-limb divided up to c.1/2 way into 2 lobes 3
3 Stem-leaves ovate, sparsely pubescent, with cordate base **3. L. chalcedonica**
3 Stem-leaves linear to elliptic-lanceolate or -oblanceolate, glabrous or pubescent only on margins, not cordate at base 4
  4 Petal-limb divided much <1/2 way, with conspicuous scales at base; carpophore >3mm at fruiting **4. L. viscaria**
  4 Petal-limb divided c.1/2 way, with minute knob-like scales at base; carpophore <2mm at fruiting **5. L. alpina**

1. L. coronaria (L.) Murray (*Silene coronaria* (L.) Clairv.) - *Rose Campion*. Erect branched perennial to 1m, covered with dense white woolly hairs; flowers on long individual stalks; petal-limb purple, entire to emarginate; (2n=24). Intrd; commonly grown in gardens, frequently escaping and sometimes natd in waste places; En and Wa; rare casual elsewhere; SE Europe.

2. L. flos-cuculi L. (*Silene flos-cuculi* (L.) Clairv.) - *Ragged-Robin*. Erect simple or branched perennial to 75cm, glabrous to sparsely pubescent; flowers loosely clustered in open inflorescence; petal-limb pale purple, deeply 4-lobed; 2n=24. Native; marshy fields and other damp places; throughout BI.

3. L. chalcedonica L. - *Maltese-Cross*. Erect ± pubescent perennial to 1m; flowers ± densely clustered in compact head; petal-limb scarlet, 2-lobed c.1/3 way to base; (2n=24, 48). Intrd; much grown in gardens and escaping to rough and marginal ground; casual and occasionally natd in S & C Br; Russia.

4. L. viscaria L. (*Viscaria vulgaris* Bernh., *Steris viscaria* (L.) Raf.) - *Sticky Catchfly*. **RR** Erect simple perennial to 60cm, glabrous to sparsely pubescent; stems very sticky below each node; flowers loosely clustered in open inflorescence; petal-limb purple, emarginate to shortly 2-lobed; 2n=24. Native; cliffs and rocky places; extremely local in Wa and Sc; grown in gardens and rarely escaping.

5. L. alpina L. (*Viscaria alpina* (L.) Don, *Steris alpina* (L.) Sourk. nom. inval.) - **RRR** *Alpine Catchfly*. Erect simple perennial to 20cm, glabrous or very sparsely pubescent; stems not sticky; flowers ± densely clustered in compact head; petal-limb pale purple, 2-lobed to c.1/2 way; 2n=24. Native; mountain serpentine or heavy-metalliferous rocks; extremely rare in Cumberland and Angus.

19. AGROSTEMMA L. - *Corncockle*

Bracteoles not forming epicalyx; petal limb ± rounded to retuse, without scales, purple or rarely white; carpophore 0; styles 5; fruit a capsule opening by 5 teeth.

Other spp. - A. gracile Boiss., from E Mediterranean, has since 1989 been found as a casual originating from wild flower seed in a few localities. It might be under-recorded. It differs from *A. githago* mainly in its calyx-lobes shorter than the petals.

1. A. githago L. - *Corncockle*. Erect simple to slightly branched annual to 1m, with abundant appressed hairs; flowers on long individual stalks; petal-limb purple with dark streaks, exceeded by calyx-lobe; 2n=48. Intrd; cultivated and waste ground, formerly common in cornfields; formerly scattered over most of BI but now very rare and only casual.

20. SILENE L. (*Melandrium* Röhl.) - *Campions*

Bracteoles not forming epicalyx; petal-limb entire to 2-lobed, with or without scales at base, white to red or purple or yellow; styles 3 or 5; fruit a capsule opening by 2x as many teeth as styles. Often subdivided or amalgamated with related genera.

1 Calyx 3-6mm; flowers arranged in whorl-like groups at each inflorescence node **3. S. otites**

1 Calyx 5-30mm, flowers not in whorl-like groups 2

2 Plant with only male flowers 3

2 Plant with at least some female or bisexual flowers 6

3 Calyx 20-veined, strongly inflated 11

3 Calyx 10-veined, not or scarcely inflated 4

4 Dwarf cushion-plant with 1 flower per stem **6. S. acaulis**

4 Stems branched, with >1 flower 5

5 Upper part of inflorescence sticky; calyx-teeth with broad scarious margins 14

5 Upper part of inflorescence not or scarcely sticky; calyx-teeth with 0 or narrow scarious margins 8

6 Styles 5; capsule-teeth 10 7

6 Styles 3; capsule-teeth 6 9

7 Flowers all bisexual; carpophore >5mm **12. S. coeli-rosa**

7 Flowers dioecious (sometimes smut-infected stamens present in female flowers); carpophore <2mm 8

8 Corolla white; capsule-teeth erect **10. S. latifolia**

8 Corolla red or pink (rarely white, see text); capsule-teeth revolute **11. S. dioica**

9 Calyx 20-30-veined, strongly inflated at fruiting 10

9 Calyx 10-veined, usually not or scarcely inflated at fruiting 12
10 Calyx 30-veined, cross-connecting veins not apparent; annual **15. S. conica**
10 Calyx 20-veined, wih conspicuous cross-connecting veins; perennial 11
11 Bracts largely herbaceous; capsule with patent to revolute teeth **5. S. uniflora**
11 Bracts scarious; capsule with erect to erecto-patent teeth **4. S. vulgaris**
12 Perennial, with non-flowering shoots at flowering time 13
12 Annual, without non-flowering shoots 16
13 Dwarf cushion-plant with 1 flower per stem **6. S. acaulis**
13 Stems branched, with >1 flower 14
14 Calyx 4-7mm; petal-limb shortly 4-lobed or -toothed **7. S. quadrifida**
14 Calyx 9-24mm; petal-limb deeply bifid 15
15 Carpophore about as long as capsule; petal-limb with small knob-like scales at base **1. S. italica**
15 Carpophore <1/2 as long as capsule; petal-limb with conspicuous acute scales at base **2. S. nutans**
16 Stems, leaves and calyx glabrous **8. S. armeria**
16 Stems, leaves and calyx pubescent 17
17 Calyx >20mm **9. S. noctiflora**
17 Calyx <18mm 18
18 Calyx >12mm, with short sparse hairs **13. S. pendula**
18 Calyx <12mm, with long ± patent hairs **14. S. gallica**

Other spp. - Many spp. are grown as ornamentals or occur as impurities in birdseed and appear occasionally on tips and waste ground. The European annuals **S. dichotoma** Ehrh. (*Forked Catchfly*), **S. csereii** Baumg., **S. behen** L., **S. cretica** L., and **S. muscipula** L. formerly occurred frequently as casuals but are now rare. *S. dichotoma* resembles *S. gallica* but has horizontal flowers with whitish petals and calyx 8-15mm. *S. csereii* resembles *S. vulgaris* but the calyx has 10 long and 10 short veins (not 20 long veins). *S. muscipula* and *S. cretica* resemble *S. armeria* but have acute calyx-teeth and all the leaves tapered to a narrow base; the carpophore is pubescent and glabrous respectively. *S. behen* is even closer to *S. armeria* but the carpophore is only 1-2mm (not >4mm). **S. fimbriata** (Adams ex F. Weber & D. Mohr) Sims (*S. multifida* (F. Weber & D. Mohr) Rohrb.) and **S. schafta** S.G. Gmel. ex Hohen. are garden perennials from Caucasus. *S. fimbriata* was formerly natd in W Sc; it has erect stems, a strongly inflated 10-veined calyx and much-divided white petals. *S. schafta* is a rare escape formerly natd in S Hants and NE Yorks; it has several ± simple decumbent stems to 15cm, each with few flowers, a narrow calyx >20mm, and broadly obovate dentate red petals. **S. conoidea** L., from S Europe, differs from *S. conica* in being less pubescent and larger in most parts (calyx ≥15mm, capsule ≥12mm, seeds >1mm); it is a rare casual and plants reported to be natd in CI were actually variants of *S. conica*.

1. S. italica (L.) Pers. - *Italian Catchfly*. Pubescent perennial with ascending to erect stems to 70cm; stem-leaves linear to oblanceolate; flowers in lax dichasia, sometimes unisexual; calyx 14-21mm; petals whitish, yellowish or pinkish, deeply bifid; (2n=24). Intrd; roadside banks and quarries; natd since 1863 near Greenhithe, W Kent, a rare casual elsewhere in Br; Europe. **167**

2. S. nutans L. - *Nottingham Catchfly*. Pubescent perennial with erect stems to 80cm; similar to *S. italica* but calyx 9-12mm, and see key; 2n=24. Native; dry grassy or bare places; very scattered but very locally common in Br N to N Wa and Peak District (formerly to Yorks and Lancs), E Sc, CI. British plants are all ssp. **smithiana** (Moss) Jeanm. & Bocq., but this might not be distinct. **R** **167**

3. S. otites (L.) Wibel - *Spanish Catchfly*. Shortly pubescent perennial with erect stems to 80cm; stem-leaves linear to oblanceolate; flowers numerous in whorl-like **RR**

clusters at each node, ± dioecious; calyx 4-6mm; petals pale yellowish-green, entire; 2n=24. Native; dry grassy heathland; E Anglia, rare casual elsewhere.

4. S. vulgaris Garcke - *Bladder Campion*. Glabrous to pubescent perennial with decumbent to erect stems to 80cm; flowers in lax dichasia, often unisexual; calyx 10-18mm, strongly inflated, narrowed at mouth; petals deeply bifid. Grassy places, open and rough ground. **167**

a. Ssp. vulgaris. Underground stolons absent; lower stem-leaves elliptic to narrowly ovate; calyx broadly ellipsoid to subglobose; petals white, capsule 6-9(11)mm; seeds 1-1.6mm; 2n=24. Native; throughout most of BI but rare in N & W Sc and N Ir.

b. Ssp. macrocarpa Turrill (*S. linearis* auct. non Sweet). Underground stolons present; lower stem-leaves linear-lanceolate to lanceolate; calyx oblong to ellipsoid-petals pinkish to greenish; capsule 11-13mm; seeds 1.6-2.1mm. Intrd; natd at Plymouth, S Devon, since 1921; Mediterranean.

4 x 5. S. vulgaris x S. uniflora occurs when the parents meet, which is rather rarely, in scattered places in Br. Hybrids are completely fertile and have intermediate characters, but precise identification is difficult except in mixed populations.

5. S. uniflora Roth (*S. maritima* With., *S. vulgaris* ssp. *maritima* (With.) Á. & D. Löve) - *Sea Campion*. Glabrous or sparsely pubescent perennial with procumbent to ascending stems to 30cm; similar to *S. vulgaris* but flowers slightly larger and rarely >4 per stem; stem-leaves linear to linear-lanceolate; calyx often not narrowed at apex; and see key; 2n=24. Native; rocky and shingly coasts, and cliffs, lake-shores and streamsides in mountains; round most coasts of BI and rather rare in mountains of N En, Wa and Sc. **167**

6. S. acaulis (L.) Jacq. - *Moss Campion*. Sparsely pubescent, densely tufted perennial with erect 1-flowered stems to 10cm; leaves linear; flowers often ± dioecious; calyx 5-9mm; petals pink, emarginate; 2n=24. Native; mountain rock-ledges and scree; N Wa, Lake District, C & NW Sc, rare in N Ir.

7. S. quadrifida (L.) L. (*S. alpestris* Jacq.) - *Alpine Campion*. Sparsely pubescent perennial with ± woody base and erect or ascending stems to 30cm; stem-leaves narrowly elliptic-oblanceolate, narrowed to base; flowers few in lax cymes with ± sticky axes; calyx 4-7mm; petals white, with 4 short lobes or teeth; (2n=24). Intrd; old plantings now natd on rocks in uplands; M Perth since 1974, MW Yorks; E Alps.

8. S. armeria L. - *Sweet-William Catchfly*. Erect glabrous annual to 40cm; stems sticky towards apex; stem-leaves lanceolate to narrowly ovate, cordate, clasping the stem; flowers in lax to dense corymbose cymes; calyx 12-15mm, with obtuse teeth; petals pink, emarginate; (2n=24). Intrd; infrequent garden escape on tips and in waste places; scattered in En and Jersey; Europe.

9. S. noctiflora L. - *Night-flowering Catchfly*. Erect glandular-pubescent annual to 50cm; stem-leaves ovate to lanceolate, narrowed to stalk-like base clasping the stem; flowers in sparse cymes; calyx 20-30mm, with long narrow teeth; petals whitish-yellow, deeply bifid; 2n=24. Native; sandy arable soils; scattered through En, Wa, E Sc and E Ir, now much reduced and rare except in E En; also an infrequent casual.

10. S. latifolia Poir. (*S. alba* (Mill.) E.H.L. Krause nom. illeg., *S. pratensis* (Rafn) Godr. & Gren.) - *White Campion*. Erect, pubescent, ± glandular, rarely glabrous (annual to) perennial to 1m; stem-leaves similar to those of *S. noctiflora*; flowers dioecious, in open dichasia, more numerous on male plants; calyx 15-22mm and 10-veined (male), 20-30mm and 20-veined (female), with narrow teeth; petals white, deeply bifid; 2n=24. Native; banks, roadsides, waste and cultivated ground, mostly on light soils in the open; throughout most of BI but rare or absent from much of W. British plants are ssp. **alba** (Mill.) Greuter & Burdet, but ssp. **latifolia** (*S. alba* ssp. *divaricata* (Rchb.) Walters, *S. pratensis* ssp. *divaricata* (Rchb.) McNeill & H.C. Prent., *S. macrocarpa* (Boiss. & Reut.) E.H.L. Krause) from S Europe, differing **167**

in its more finely tapered calyx-teeth and divergent to recurved capsule-teeth, occurs as a rare casual.

10 x 11. S. latifolia x S. dioica = S. x hampeana Meusel & K. Werner (*S. x intermedia* (Schur) Philp nom. illeg. non (Lange) Bocq.) occurs commonly whenever the parents meet over most of BI, especially in lowland areas; it is highly fertile and intermediate in characters, notably the pale pink petals.

11. S. dioica (L.) Clairv. - *Red Campion*. Erect pubescent, sometimes glandular, **167**
rarely glabrous perennial to 1m; very like *S. latifolia* but capsule-teeth revolute and petals red or pink (rarely white); 2n=24. Native; woods and hedgerows, usually in shady places in lowlands but on open cliffs and scree-slopes in mountains; throughout BI but very scattered in S Ir. White-flowered plants can usually be told from *S. latifolia* by the lack of anthocyanin in stem and leaves as well, but pale pink-flowered plants of *S. dioica* seem sometimes identical with certain hybrids. Some plants in Shetland have large deep red flowers and strong densely pubescent stems and have been recognized as ssp. **zetlandica** (Compton) A.R.Clapham nom. inval. (*Melandrium dioicum* var. *zetlandicum* Compton). However they represent only extreme variants with one combination of characters not worthy of ssp. rank.

12. S. coeli-rosa (L.) Godr. - *Rose-of-heaven*. Erect glabrous annual to 50cm; leaves linear; flowers few per stem; calyx 15-28mm, clavate; petals bright pink, bifid, conspicuous; (2n=24). Intrd; grown in gardens and an occasional escape on tips and in waste places; very scattered in S Br; SW Europe.

13. S. pendula L. - *Nodding Catchfly*. Pubescent ascending annual to 45cm; leaves obovate to elliptic; flowers very few per stem; calyx 13-18mm, ± inflated; petals bright pink, emarginate to shallowly bifid; (2n=24). Intrd; grown in gardens, readily self-sowing and an occasional escape; very scattered in S En; S Europe.

14. S. gallica L. - *Small-flowered Catchfly*. Pubescent erect annual to 45cm; leaves R
oblanceolate (lower) to linear (upper); flowers ± erect in monochasia; calyx 7-10mm with long teeth; petals yellowish-white to pink or red-blotched, emarginate to retuse; 2n=24. Native; waste places, cultivated land and open sandy ground; scattered in S Br and CI N to Yorks and in Ir, formerly much commoner and N to C Sc.

15. S. conica L. - *Sand Catchfly*. Densely glandular-pubescent decumbent to erect R
annual to 35cm; leaves linear to linear-lanceolate; flowers rather few per stem; calyx 10-15mm, becoming strongly ovoid, with long fine teeth; petals pink, emarginate to shallowly bifid; 2n=20, 24. Native; sandy places, especially maritime dunes; CI and E Anglia (perhaps native only there), also scattered from S En to C Sc.

21. CUCUBALUS L. - *Berry Catchfly*

Bracteoles not forming an epicalyx; petal-limb bifid, with 2 scales at base, white; styles 3; fruit a berry.

1. C. baccifer L. - *Berry Catchfly*. Shortly pubescent perennial with much-branched sprawling stems to 1m; leaves ovate, acuminate, with short petiole; flowers in open dichasia; calyx 8-15mm, the petals longer; berry 6-8mm across, black, not concealed by calyx; 2n=24. Intrd; rough grassy places; rare casual, natd in W Norfolk; Europe.

22. SAPONARIA L. - *Soapworts*

Bracteoles not forming an epicalyx; petal-limb ± entire, with 2 scales at base, pink (to white); styles 2; fruit a capsule opening by 4 teeth.

1. S. officinalis L. - *Soapwort*. Glabrous perennial with erect or ascending stems to 90cm; leaves ovate to elliptic, the largest >5cm; flowers in rather compact corymbose cymes, c.2.5cm across; calyx glabrous, 15-20mm; 2n=28. Probably intrd; grown in gardens and a common escape and throwout on waste ground, roadsides and grassy places, often *flore pleno*, often well natd and perhaps native in SW En; throughout most of BI but rare in N Sc; Europe.

2. S. ocymoides L. - *Rock Soapwort*. Pubescent perennial with procumbent to decumbent stems to 50cm; leaves narrowly elliptic to oblanceolate, <3cm; flowers in rather lax dichasia, c.1cm across; calyx pubescent, 7-12mm; (2n=28). Intrd; grown in rock-gardens and sometimes escaping on walls and stony banks, natd in a few places; En and Wa; S Europe.

23. VACCARIA Wolf - *Cowherb*
Bracteoles not forming an epicalyx; petal-limb entire or shortly bifid, without scales, pink; styles 2; fruit a capsule opening by 4 teeth.

1. V. hispanica (Mill.) Rauschert (*V. pyramidata* Medik.) - *Cowherb*. Glabrous glaucous erect annual to 60cm; leaves lanceolate to ovate, clasping the stem; flowers in lax dichasia; calyx becoming inflated in fruit, with 5 acute angles or wings; (2n=30). Intrd; casual mainly from birdseed on tips and waysides and in parks, occasionally natd; scattered in BI, mainly S; S & C Europe.

24. PETRORHAGIA (Ser. ex DC.) Link (*Kohlrauschia* Kunth) - *Pinks*
Bracteoles close to base of calyx, forming epicalyx; calyx-tube scarious at joints between lobes; petal-limb emarginate, without scales, pink; styles 2; fruit a capsule opening by 4 teeth.

1 Mat-forming perennials; flowers not in compact heads; epicalyx of (2-)4(5) whitish bracts; calyx <8mm, ± bell-shaped **3. P. saxifraga**
1 Erect annuals; flowers in compact heads of (1)3-11; epicalyx of several pairs of brown bracts; calyx >8mm, cylindrical 2
  2 Seeds tuberculate on surface **1. P. nanteuilii**
  2 Seeds reticulate on surface **2. P. prolifera**

1. P. nanteuilii (Burnat) P.W. Ball & Heywood (*Kohlrauschia nanteuilii* (Burnat) RRR P.W. Ball & Heywood, *K. prolifera* auct. non (L.) Kunth) - *Childing Pink*. Stems to 50cm, glabrous or shortly pubescent; flowers in compact ovoid heads, opening 1 at a time; leaf-sheaths c. as long as wide; calyx 10-13mm; seeds tuberculate; 2n=60. Native; dry grassy places; very rare in SC En (perhaps now confined to W Sussex), S Wa and CI, casual rarely natd elsewhere.

2. P. prolifera (L.) P.W. Ball & Heywood (*Kohlrauschia prolifera* (L.) Kunth) - RR *Proliferous Pink*. Differs from *P. nanteuilii* in its reticulate seed-surface; stem usually more densely pubescent and leaf-sheaths up to 2x as long as wide; 2n=30. Probably native; dry banks; Beds and W Norfolk, possibly elsewhere in En.

3. P. saxifraga (L.) Link (*Kohlrauschia saxifraga* (L.) Dandy) - *Tunicflower*. Mat-forming perennial with decumbent to ascending, glabrous to pubescent stems to 35cm; calyx 3-6mm; seeds ± tuberculate; (2n=60). Intrd; waste places; rare casual in Br, natd near Tenby, Pembs; Europe.

25. GYPSOPHILA L. - *Baby's-breath*
Bracteoles not close to base of calyx and forming epicalyx; calyx-tube scarious at joints between lobes; petal-limb retuse or entire, without scales, usually white, sometimes pink; styles 2; fruit a capsule opening by 4 teeth.

Other spp. - **G. elegans** M. Bieb., an annual with calyx 3-5mm and petals c.2x as long, is the least rare of several occasional casuals.

1. G. paniculata L. - *Baby's-breath*. Erect perennial to 1m with deep, strong tap-root; stems much-branched above forming wide, diffuse inflorescence; calyx 1.5-2mm; petals c.2x as long as calyx; (2n=34). Intrd; much grown in gardens, escape in rough grassy places on sandy soils; natd near Sandwich (E Kent) for many years, also Man and M Ebudes, formerly in E Suffolk, casual elsewhere; E & C Europe.

26. DIANTHUS L. - *Pinks*
Bracteoles close to base of calyx, forming epicalyx; calyx-tube not scarious at joints between lobes; petal-limb variously divided, without scales, usually pink to red; styles 2; fruit a capsule opening by 4 teeth.

1 Flowers mostly in compact cymose clusters of usually >3 surrounded by involucre of leaf-like bracts; epicalyx nearly as long as to longer than calyx 2
1 Flowers solitary or in lax cymes, sometimes 2-3 close together but not surrounded by involucre of leaf-like bracts; epicalyx <3/4 as long as calyx 3
  2 Annual to biennial, without sterile shoots at flowering time; inflorescence pubescent **7. D. armeria**
  2 Perennial, with sterile shoots at flowering time; inflorescence glabrous **6. D. barbatus**
3 Petal-limb divided >1/4 way to base into narrow lobes 4
3 Petal-limb divided <1/4 way to base into narrow or triangular teeth 5
  4 Leaves on sterile shoots mostly >2cm, narrowly acute; inner bracteoles of epicalyx broadly obovate, shortly apiculate at apex **3. D. plumarius**
  4 Leaves on sterile shoots mostly <1.5cm, obtuse to subacute; inner bracteoles of epicalyx oblong to obovate, apiculate to cuspidate at apex **4. D. gallicus**
5 Stems shortly pubescent below; flowers scentless; inner bracteoles of epicalyx cuspidate **5. D. deltoides**
5 Stems glabrous; flowers scented; inner bracteoles of epicalyx apiculate 6
  6 Flowers <30mm across; calyx <20mm; stems rarely >20cm **1. D. gratianopolitanus**
  6 Flowers >30mm across; calyx >20mm; stems rarely <20cm **2. D. caryophyllus**

Other spp. - **D. carthusianorum** L., from Europe, is occasionally found as a garden escape; it is a perennial with the flowers packed into tight clusters, epicalyx c.1/2 as long as calyx, and usually dark red petals. **D. superbus** L. (*Large Pink*), from Europe, is grown in gardens and was found in W Norfolk in 1993 and suggested as native, but it has not appeared since and was probably planted or sown; it is very fragrant and has deeply divided petal-limbs but differs from *D. gallicus* and *D. plumarius* in its ovate (not obovate or oblong) inner bracteoles of the epicalyx and longer and narrower petals.

1. D. gratianopolitanus Vill. - *Cheddar Pink*. Mat-forming, glabrous, glaucous perennial with ± erect flowering-stems to 20cm; flowers <30mm across, usually solitary, fragrant, pale to deep pink; calyx <20mm; 2n=90. Native; limestone rock crevices and cliff-ledges; Cheddar Gorge, N Somerset, rare escape sometimes planted or natd elsewhere. RRR

1 x 2. D. gratianopolitanus x D. caryophyllus,
1 x 2 x 3. D. gratianopolitanus x D. caryophyllus x D. plumarius, and
1 x 3. D. gratianopolitanus x D. plumarius
are the parentages of a number of escaped garden Pinks, some of which are occasionally natd on walls etc. in S & C Br. Characters of the 3 spp. occur in varying combinations.

2. D. caryophyllus L. - *Clove Pink*. Tufted, glabrous, glaucous perennial with ± erect flowering stems to 60cm; flowers >30mm across, 1-5 in lax cymes, fragrant, usually pink; calyx >20mm; (2n=30). Intrd; grown in gardens and occasionally natd on old walls; S & C Br, ephemeral outcast elsewhere; S Europe.

2 x 3. D. caryophyllus x D. plumarius occurs in the same manner as the above 3 hybrids.

3. D. plumarius L. - *Pink*. Tufted or mat-forming, glabrous, glaucous perennial with ± erect flowering stems to 30cm; flowers 20-40mm across, 1-5 in lax cymes, fragrant, white to deep pink; calyx >20cm; 2n=90. Intrd; commonly grown in gardens and sometimes natd on old walls or banks; similar distribution to *D. caryophyllus* but commoner; SE Europe.

4. D. gallicus Pers. - *Jersey Pink*. Mat-forming glaucous perennial, shortly pubescent below, with ± erect flowering stems to 25cm; flowers 15-25mm across, 1-3 together, fragrant, pink; calyx 20-25mm; 2n=60. Intrd; grassy coastal dunes; natd in Jersey since 1892; W coast of Europe.

5. D. deltoides L. - *Maiden Pink*. Loosely tufted, shortly pubescent, green to glaucous perennial with ± erect flowering stems to 45cm; flowers 12-20mm across, 1(-3) together, scentless, white to pink, usually spotted; calyx 12-18mm; 2n=30. Native; dry grassland; scattered and decreasing in Br N to Banffs, formerly CI. R

6. D. barbatus L. - *Sweet-William*. Mat-forming, ± glabrous, green perennial with erect flowering stems to 50cm; flowers <20mm across, in dense clusters at stem apex, fragrant, pink to red, often spotted; calyx 12-18mm; (2n=30). Intrd; commonly grown in gardens, often escaping or as a throwout on tips and waste places; Br, mainly S, rarely persisting; S Europe.

7. D. armeria L. - *Deptford Pink*. Erect, pubescent, green annual or biennial to 60cm; flowers <15mm across, in ± dense clusters at stem apex, scentless, bright pink, often spotted; calyx 15-20mm; 2n=30. Native; dry grassy places; very scattered and decreasing in S Br N to N Wa and N Lincs, intrd further N, W Cork, CI. R

49. POLYGONACEAE - *Knotweed family*

Herbaceous annuals to perennials or woody climbers; leaves alternate (rarely subopposite), simple, usually entire, with fused often scarious stipules sheathing stem. Flowers in simple or branched racemes, sometimes few or solitary, bisexual to dioecious or variable, hypogynous, actinomorphic; perianth of 1-2 usually ± similar whorls, of 3-6 tepals (2-3 per whorl) free or fused below, greenish, brownish, white or pink, persistent in fruit; stamens (3)6-9; ovary 1-celled, with 1 basal ovule; stigmas 2-3, sessile or on styles, capitate to finely divided; fruit an achene.

Usually easily recognized by the distinctive stipules and tepals and by the single basal ovule.

1 Tepals 3; tiny annual with mostly subopposite leaves **2. KOENIGIA**
1 Tepals mostly ≥4; annual to perennial with alternate leaves 2
2 Tepals 4; leaves reniform **9. OXYRIA**
2 Tepals mostly 5-6; leaves very rarely reniform 3
3 Tepals 6, in 2 whorls of 3 4
3 Tepals mostly 5 5
4 Stamens 6; inner 3 tepals enlarging and enclosing achene; achene not winged; leaves pinnately veined **8. RUMEX**
4 Stamens 9; tepals remaining small in fruit; achene 3-winged; leaves palmately veined **7. RHEUM**
5 Stems woody, twining 6
5 Stems mostly herbaceous though often shrub-like 7
6 Leaves <2cm, rounded at base **6. MUEHLENBECKIA**
6 Leaves >2cm, truncate to cordate at base **5. FALLOPIA**

7 Outer 3 tepals with longitudinal wing or strong keel **5. FALLOPIA**
7 Tepals not winged, rarely weakly keeled 8
8 Inflorescences ≤6-flowered, all axillary **4. POLYGONUM**
8 Inflorescences mostly >6-flowered, some or all terminal 9
9 Achene >2x as long as perianth, not winged; leaves sagittate; filaments flattened or winged **3. FAGOPYRUM**
9 Achene usually <2x as long as perianth, if >2x as long then strongly winged; leaves very rarely sagittate; filaments not winged or flat **1. PERSICARIA**

1. PERSICARIA Mill. (*Aconogonum* Rchb., *Bistorta* (L.) Adans., *Polygonum* sects. *Aconogonon* Meissner, *Persicaria* (L.) Meissner, *Echinocaulon* Meissner, *Cephalophilon* Meissner, *Bistorta* (L.) D. Don) - *Knotweeds*

Annuals to perennials, rhizomatous, stoloniferous or neither; inflorescence many-flowered, terminal and axillary, spike-like to subcapitate or paniculate; tepals mostly 5, not winged, petaloid, not enlarging in fruit; stamens 8 (or 4-7 by reduction); style 1 and divided into 2 or 3, or 3; stigmas small, capitate to clavate; achene lenticular (biconvex to biconcave), 3-angled or 3-winged.

1 Stigmas 3 in all flowers; stamens 8; perennials 2
1 Stigmas 2 in all or most flowers; stamens often <8 in at least some flowers; annuals (perennial in *P. amphibia*) 9
2 Inflorescence unbranched; stamens exserted 3
2 Inflorescence a branched panicle; stamens included 5
3 Lower part of inflorescence with bulbils instead of flowers; lower leaves cuneate at base **8. P. vivipara**
3 Bulbils 0; lower leaves truncate to cordate at base 4
4 Petioles of basal and lower stem-leaves winged above; flowers usually pale pink; stems unbranched **6. P. bistorta**
4 Petioles unwinged; flowers usually red; stems usually branched **7. P. amplexicaulis**
5 Tepals <2.5mm at flowering 6
5 Tepals >2.5mm at flowering 7
6 Leaves tomentose on lowerside, with densely matted, twisted hairs; achene with 3 wings; perianth shrivelling in fruit **4. P. weyrichii**
6 Leaves glabrous to densely pubescent on lowerside, with ± straight hairs; achene not winged; perianth becoming succulent in fruit and ≤3mm **5. P. mollis**
7 Tepals fused to c.1/4-1/2 way, usually pinkish **2. P. campanulata**
7 Tepals free ± to base, white 8
8 Styles (+ stigmas) <0.5mm; lower leaves cuneate at base, <3.5cm wide **1. P. alpina**
8 Styles (+ stigmas) >0.5mm; lower leaves truncate to cordate at base, >3.5cm wide **3. P. wallichii**
9 Stems with recurved prickles **17. P. sagittata**
9 Stems without prickles, usually glabrous 10
10 Flowers in ± globose heads; styles 3 **10. P. nepalensis**
10 Flowers in cylindrical or tapering inflorescences; styles 2(-3) 11
11 Perennial with strong rhizomes or stolons, often aquatic; stamens often (not always) exserted at anthesis **9. P. amphibia**
11 Annual, often rooting at lower nodes; stamens always included 12
12 Sessile or stalked glands present on perianth and/or peduncle 13
12 Glands 0 on perianth and peduncle 16
13 Glands on peduncle with stalks much longer than heads **13. P. pensylvanica**
13 Glands on peduncle ± 0 or sessile or with stalks shorter than heads 14
14 Inflorescence lax, the flowers mostly separated; leaves with sharp

peppery taste (usually strong, sometimes faint) when fresh **14. P. hydropiper**

14 Inflorescence dense, the flowers crowded; leaves without peppery taste 15

15 Achenes biconcave to planoconcave or rarely (<1%) 3-angled **12. P. lapathifolia**

15 Achenes biconvex to planoconvex or (c.10-60%) 3-angled **11. P. maculosa**

16 Inflorescence dense, the flowers crowded; leaves often with dark blotch **11. P. maculosa**

16 Inflorescence lax, the flowers mostly separated; leaves never with dark blotch 17

17 Lower leaves usually <5x as long as wide, 12-30mm wide; achene 2.5-3.5mm **15. P. mitis**

17 Leaves usually >5x as long as wide, 2-15mm wide; achene 2-2.5mm **6. P. minor**

Other spp. - P. capitata (Buch.-Ham. ex D. Don) H. Gross (*Polygonum capitatum* Buch.-Ham. ex D. Don), from Himalayas, is a perennial similar to *P. nepalensis* but differs in its conspicuously glandular-pubescent stems and stipules, abruptly acute leaves with short petiole with curious lobe at base, and no tuft of white hairs at base of stipules (present in *P. nepalensis*); it is an infrequent garden or pot-plant escape not hardy in Br. Records of **Polygonum lichiangense** W. Sm. are errors.

1. P. alpina (All.) H. Gross (*Polygonum alpinum* All.) - *Alpine Knotweed*. Stems erect, to 1m; leaves ≤15 x 3.5cm, lanceolate, sparsely pubescent; perianth 2.5-3.5mm; styles <0.5mm; achene 4-5mm, slightly exceeding perianth, 3-angled; (2n=20, 22). Intrd; garden escape natd on river shingle; by R. Dee near Ballater, S Aberdeen, less permanent elsewhere in Sc; Europe. A plant natd by a road near Huddersfield, SW Yorks, might be a variant of this sp., a related sp. or a hybrid; it differs in wider leaves and achene with 3 thick wings c.2x as long as perianth.

2. P. campanulata (Hook. f.) Ronse Decr. (*Polygonum campanulatum* Hook. f.) - *Lesser Knotweed*. Stems ascending, to 1m; leaves ≤15 x 7cm, lanceolate to ovate, tomentose on lowerside; perianth 3.5-5mm; heterostylous, with styles c.0.5mm or c.1mm; achene 3-4mm, included in perianth; (2n=22, c.36, c.64, c.70). Intrd; garden escape natd in damp shady places; sparsely scattered over BI; Himalayas.

3. P. wallichii Greuter & Burdet (*P. polystachya* (Wall. ex Meisn.) H. Gross non Opiz, *Polygonum polystachyum* Wall. ex Meisn.) - *Himalayan Knotweed*. Stems erect, to 1.5m; leaves ≤20 x 8cm, lanceolate to narrowly ovate, almost glabrous to densely pubescent on lowerside; perianth (2.5)3-4mm; heterostylous, with styles c.0.6-0.8 or 0.8-1.2mm; achene 1.5-3mm, 3-angled, included in perianth; (2n=22). Intrd; garden escape natd in grassy places and roadsides; scattered over Br and Ir; Himalayas. Densely pubescent plants are best separated from *P. mollis* by the unequal tepals.

4. P. weyrichii (F. Schmidt ex Maxim.) Ronse Decr. (*Polygonum weyrichii* F. Schmidt ex Maxim.) - *Chinese Knotweed*. Stems erect, to 1.5m; leaves ≤20 x 6cm, lanceolate to narrowly ovate, densely tomentose on lowerside; perianth 1.5-2mm; styles <0.5mm; achene 4-6mm, broadly 3-winged, >2x as long as perianth; (2n=20). Intrd; garden escape natd in rough ground; Wastwater, Cumberland; E Asia.

5. P. mollis (D. Don) H. Gross (*Polygonum molle* D. Don, *P. rude* Meisn., *P. paniculatum* Blume) - *Soft Knotweed*. Stems erect, to 1.5m; leaves ≤25 x 8cm, lanceolate to narrowly ovate, almost glabrous to densely pubescent on lowerside; perianth 1.5-2.2mm; styles <0.5mm; achene 2-2.5mm, 3-angled, included in perianth which becomes succulent and blackish; (2n=20). Intrd; garden escape natd in rough ground; near Coylet, Argyll and Tunbridge Wells, W Kent; Himalayas.

6. P. bistorta (L.) Samp. (*Polygonum bistorta* L.) - *Common Bistort*. Stems erect, simple, to 80(100)cm; leaves truncate to cordate at base, the lower with winged petioles; flowers pink; inflorescences dense, 1 per stem; 2n=48. Native; grassy places; throughout most of Br and Ir but very common only in NW En and intrd in Ir and much of S Br.

7. P. amplexicaulis (D. Don) Ronse Decr. (*Polygonum amplexicaule* D. Don) - *Red Bistort*. Stems erect, usually branched, to 1m; leaves cordate at base, the upper clasping the stem; petioles not winged; flowers red; inflorescences dense, usually >1 per stem; 2n=40. Intrd; cultivated in gardens, natd in grassy and rough places; C & W Ir, CI and very scattered in Br; Himalayas.

8. P. vivipara (L.) Ronse Decr. (*Polygonum viviparum* L.) - *Alpine Bistort*. Stems erect to ascending, simple, to 30cm; leaves cuneate at base; petioles not winged; flowers pink; inflorescences dense, 1 per stem, lower part (rarely ± all) occupied by dark purple bulbils; 2n=66-110. Native; grassland and rock-ledges on mountains; Sc and N En, very rare in N Wa and SW Ir.

9. P. amphibia (L.) Gray (*Polygonum amphibium* L.) - *Amphibious Bistort*. Rhizomatous perennial of very varied habit, in water with glabrous floating leaves, on dry land with ± erect sparsely pubescent stems to 60cm, intermediate habits common; lower leaves cordate to truncate at base; inflorescence cylindrical, dense, obtuse; peduncle glandular or not; perianth with or without glands; achene 2-3mm, shiny, lenticular to subglobose; 2n=88. Native; in water, wet places, river banks and a weed on rough ground; throughout BI.

10. P. nepalensis (Meisn.) H. Gross (*Polygonum nepalense* Meisn.) - *Nepal Persicaria*. Decumbent, glabrous to sparsely pubescent annual to 40cm; leaves tapering-acute, the lower narrowed to winged petiole which is often expanded and cordate at base; inflorescence ± globose, with leafy bract at base; peduncles with stalked glands; perianth eglandular, pinkish; achene 1.5-2mm, lenticular, dull; (2n=48). Intrd; cultivated and waste ground, rare casual in Br, natd in rough ground in S Somerset, Dorset and S Hants; Himalayas.

11. P. maculosa Gray (*Polygonum persicaria* L.) - *Redshank*. Decumbent to erect ± glabrous annual to 80cm; leaves lanceolate; inflorescence cylindrical, dense, obtuse; peduncle and perianth usually eglandular, rarely glandular; achene 2-3.2mm, shiny, lenticular (biconvex to planoconvex) or (c.10-60%) 3-angled; 2n=22, 44. Native; waste, cultivated and open ground; throughout BI.

11 x 12. P. maculosa x P. lapathifolia = P. x pseudolapathum (Schur) D.H. Kent (*P. x lenticularis* (Hy) Soják, *Polygonum x lenticulare* Hy);

11 x 14. P. maculosa x P. hydropiper = P. x intercedens (Beck) Soják (*Polygonum x intercedens* Beck);

11 x 15. P. maculosa x P. mitis = P. x condensata (F.W. Schultz) Soják (*Polygonum x condensatum* (F.W. Schultz) F.W. Schultz; and

11 x 16. P. maculosa x P. minor = P. x brauniana (F.W. Schultz) Soják (*Polygonum x braunianum* F.W. Schultz) have been recorded occasionally in S & C Br with apparently intermediate characters and varying sterility. They are probably genuine but all rare, probably over-recorded, and need checking.

12. P. lapathifolia (L.) Gray (*Polygonum lapathifolium* L., *P. nodosum* Pers.) - *Pale Persicaria*. Decumbent to erect ± glabrous to tomentose-leaved annual to 1m; leaves lanceolate; inflorescence cylindrical, dense, obtuse; peduncle and perianth with glands with stalks no longer than heads; achene 2-3.3mm, shiny, lenticular (biconcave to planoconcave) or (<1%) 3-angled; 2n=22. Native; waste, cultivated and open, especially damp ground; throughout BI, but very scattered in N Sc. Very variable in habit, flower colour (greenish to deep pink) and pubescence.

12 x 14. P. lapathifolia x P. hydropiper = P. x figertii (Beck) Soják (*Polygonum x metschii* Beck) has been recorded from Surrey, Cambs and Hunts, but needs checking.

13. P. pensylvanica (L.) M. Gómez (*Polygonum pensylvanicum* L.) - *Pinkweed*. Usually erect annual to 1m, glabrous below but glandular-pubescent above; leaves lanceolate; inflorescence cylindrical, dense, obtuse; peduncle with glands with stalks much longer than heads; perianth without glands; achene 2.5-3.5mm, shiny, lenticular (biconcave to planoconcave) or 3-angled; (2n=22). Intrd; on tips and waste ground, frequent with soyabean waste; S En; E N America.

14. P. hydropiper (L.) Spach (*Polygonum hydropiper* L.) - *Water-pepper*. Decumbent to ± erect, ± glabrous annual to 75cm; leaves lanceolate to linear-lanceolate; inflorescence tapering, rather lax, often nodding; peduncle glandular or not; perianth with many (rarely few) sessile glands; achene 2.5-3.8mm, dull, lenticular or 3-angled; (2n=20, 22). Native; damp places and shallow water, often shaded; throughout BI but very scattered in N Sc.

14 x 15. P. hydropiper x P. mitis = P. x hybrida (Chaub. ex St.-Amans) Soják (*Polygonum x oleraceum* Schur) has been reported from Surrey, W Gloucs and Co Cavan, but needs confirmation.

14 x 16. P. hydropiper x P. minor = P. x subglandulosa (Borbás) Soják (*Polygonum x subglandulosum* Borbás) has been reported from En N to MW Yorks, but needs confirmation.

15. P. mitis (Schrank) Opiz ex Assenov (*Polygonum mite* Schrank, *Persicaria laxiflora* (Weihe) Opiz) - *Tasteless Water-pepper*. Annual resembling *P. hydropiper* but without the sharp taste; glands 0 or very sparse on perianth and peduncle; bristles at tip of stipules >3mm (not <3mm); achene 2.5-3.5mm, ± shiny; (2n=40). Native; similar places to *P. hydropiper*; rare and very scattered over En, Wa and NE Ir, and over-recorded. **R**

15 x 16. P. mitis x P. minor = P. x wilmsii (Beck) Soják (*Polygonum x wilmsii* Beck) has been recorded from Tyrone, Co Antrim, Berks and Oxon, but needs confirming.

16. P. minor (Huds.) Opiz (*Polygonum minus* Huds.) - *Small Water-pepper*. Decumbent to ascending, ± glabrous annual to 40cm; leaves linear to narrowly elliptic; inflorescence tapering, rather lax; peduncle and perianth eglandular; achene 2-2.5mm, shiny, lenticular; 2n=40. Native; damp fields, ditches and pondsides; very scattered and rather rare over most of Br and Ir except N Sc.

17. P. sagittata (L.) H. Gross ex Nakai (*Polygonum sagittatum* L.) - *American Tear-thumb*. Sprawling annual to 1m, glabrous but with recurved prickles on stems, petioles and midribs; leaves sagittate; inflorescence rather sparse, ± capitate, peduncle and perianth eglandular; achene 2-4mm, shiny, 3-angled; (2n=40). Intrd; natd by streams near Castle Cove, S Kerry, since 1989, now nearly extinct; E N America.

2. KOENIGIA L. - *Iceland-purslane*

Annuals; inflorescence of 1-few flowers, terminal and axillary; tepals 3, not winged, not enlarging in fruit; stamens 3; styles 2; stigmas capitate; achene with 3 rounded angles.

Superficially like *Montia* or *Peplis*, but quite different in all details.

1. K. islandica L. - *Iceland-purslane*. Stems to 6cm, erect, branched, usually reddish; leaves subopposite, <5mm; flowers very inconspicuous; (2n=28). Native; damp stony and gravelly ground >500m; Skye and Mull (N & M Ebudes), discovered 1934. **RR**

3. FAGOPYRUM Mill. - *Buckwheats*

Annuals or herbaceous perennials; flowers in terminal and axillary panicles; tepals 5, petaloid, not winged or keeled, not enlarging in fruit; stamens 8; styles 3, long; stigmas capitate, small; achene with 3 acute angles, far exserted.

Other spp. - F. tataricum (L.) Gaertn. (*Green Buckwheat*) is a casual from Asia occasionally occurring as an impurity in *F. esculentum* seed and rarely sown for gamebirds; it differs from the latter in its short (c.2mm) greenish-white tepals and achenes with undulate margins.

1. F. esculentum Moench - *Buckwheat*. Erect, very sparsely pubescent, little-branched annual to 60cm; leaves sagittate; perianth 2.5-4mm; achene 5-7mm, with straight margins; (2n=16). Intrd; casual on tips and waste ground, formerly widely cultivated but now only rarely so; formerly common, now infrequent, over most of BI; Asia.

2. F. dibotrys (D. Don) H. Hara - *Tall Buckwheat*. Perennial with herbaceous, erect little-branched stems to 1(2)m; very like *F. esculentum* but perianth 2.5-3.5mm; achene 6-8mm. Intrd; rare garden plant natd by road; Pembs; Asia.

4. POLYGONUM L. (*Polygonum* sects. *Polygonum, Avicularia* Meisn.) - *Knotgrasses*
Annuals or perennials, with strong tap-root; leaves small, narrowed at base; inflorescences ≤6-flowered, axillary; tepals 5, ± petaloid, not or slightly keeled, not enlarging in fruit; stamens 8; stigmas 3, capitate, small, almost sessile; achene with 3 rounded angles.

For correct identification plants must possess ripe achenes but not be so old that all lower leaves are gone.

1 Uppermost bracts not exceeding flowers, usually partly scarious — 2
1 All bracts exceeding flowers, leaf-like — 3
  2 Achene c.3mm, shiny; tepals erect, green to apex in midline; stems often erect — **7. P. patulum**
  2 Achene c. 2mm, ± dull; tepals divergent and wholly pink at apex; stems procumbent to ascending — **8. P. arenarium**
3 Stems strongly woody below; stipules at upper nodes at least as long as internodes, with 6-12 branched veins — **1. P. maritimum**
3 Stems not or slightly woody below; stipules at upper nodes much shorter than internodes, with wholly or mostly unbranched veins — 4
  4 Achene shiny, slightly to much longer than perianth — **2. P. oxyspermum**
  4 Achene dull, shorter than to slightly longer than perianth — 5
5 Leaves of main and lateral stems similar in size; tepals fused for ≥1/3; achene 1.5-2.5mm, with 2 convex and 1 concave sides — **3. P. arenastrum**
5 Leaves of lateral stems much smaller than those of main stem; tepals fused for ≤1/4; achene 2.5-4.5mm, with 3 concave sides — 6
  6 Leaves <4mm wide; tepals narrowly oblong, gaping near apex to reveal achene — **6. P.rurivagum**
  6 Larger leaves >5mm wide; tepals oblong-obovate, overlapping almost to apex — 7
7 Leaves narrowly ovate to narrowly elliptic, with petioles ± included within stipules; achene ≤3.5mm — **4. P. aviculare**
7 Larger leaves obovate to narrowly so, with petioles well exserted from stipules; achene >3mm — **5. P. boreale**

Other spp. - P. cognatum Meisn. (*Indian Knotgrass*), a perennial from Asia differing from *P. maritimum* in its shorter achene (c.3mm) and perianth fused for c.1/2 (not ≤1/4), is now a rare grain-casual but was formerly natd near a few docks and breweries in S En.

1. P. maritimum L. - *Sea Knotgrass*. Procumbent glaucous perennial; stems woody **RRR** below, to 50cm; stipules conspicuous, silvery, longer than upper internodes; achene 4-4.5mm, as long as or slightly longer than perianth, shiny; 2n=20. Native; low down on sandy beaches; very rare and sporadic in CI, E & W Cornwall, E Sussex and Waterford, extinct in other S En sites.

2. P. oxyspermum C.A.Mey. & Bunge ex Ledeb. (*P. raii* Bab.) - *Ray's Knotgrass*. Procumbent annual or sometimes perennial; stems sometimes ± woody below, to 1m; stipules shorter than internodes, brownish with silvery tips; achene 3.5-5.5mm, slightly to much longer than perianth, shiny; 2n=40. Native; low down on sandy beaches; scattered and decreasing round coasts of BI, absent from most of E Br and N Sc. Our plant is ssp. **raii** (Bab.) D.A. Webb & Chater. Ssp. **oxyspermum** has been reported on beaches in CE Sc as a casual, perhaps having arrived naturally from the Baltic; it differs in its narrower leaves and much paler brown achenes >5mm.

3. P. arenastrum Boreau - *Equal-leaved Knotgrass.* Usually procumbent annual to 30(50)cm; leaves of main and lateral stems similar in size; achene 1.5-2.5mm, shorter than to very slightly longer than perianth, dull; 2n=40. Native; all sorts of open ground; common throughout BI but less so than *P. aviculare* except in N Sc.

4. P. aviculare L. - *Knotgrass.* Procumbent to scrambling heterophyllous annual to 2m; achene 2.5-3.5mm, shorter than to very slightly longer than perianth, dull; 2n=60. Native; all sorts of open ground; commonest sp. of genus throughout BI except in N Sc.

5. P. boreale (Lange) Small - *Northern Knotgrass.* Procumbent to scrambling heterophyllous annual to 1m; achene 3-4.5mm, shorter to very slightly longer than perianth, dull; 2n=40. Native; similar places to *P. aviculare*; Shetland and Orkney (commonest sp. of genus there), very scattered S to SW Sc. **R**

6. P. rurivagum Jord. ex Boreau - *Cornfield Knotgrass.* Usually ± erect slender heterophyllous annual to 30cm; achene 2.5-3.5mm, usually very slightly longer than perianth, dull; 2n=60. Native; cornfields and other arable land; rare and decreasing in S & SE En, extremely rare and scattered elsewhere in Br N to C Sc. Possibly best amalgamated with *P. aviculare.*

7. P. patulum M. Bieb. - *Red-knotgrass.* Usually ± erect annual to 1m; bracts not leaf-like towards stem apex; achene c.3mm, shorter than perianth, shiny; (2n=20). Intrd; rather frequent grain- and wool-alien on tips, waste places and shoddy-fields; scattered in Br; S & C Europe. Much confused with several spp. incl. *P. arenarium,* but apparently separable from latter on characters in key; relative distributions in Br unknown.

8. P. arenarium Waldst. & Kit. - *Lesser Red-knotgrass.* Procumbent or scrambling annual to 50cm; similar to *P. patulum* but see key; (2n=20). Intrd; status in Br similar to that of *P. patulum;* Mediterranean region. Plants in Br belong to ssp. **pulchellum** (Lois.) Thell.

5. FALLOPIA Adans. (*Bilderdykia* Dumort., *Reynoutria* Houtt., *Polygonum* sects *Tiniaria* Meisn., *Pleuropterus* (Turcz.) Benth.) - *Knotweeds*

Annuals to robust herbaceous or woody perennials; inflorescences terminal and axillary, simple to paniculate; tepals 5, ± petaloid, the outer 3 keeled or winged and enlarging to protect fruit; stamens 8; styles 3; stigmas ± capitate or much divided; achene with 3 rounded angles.

1 Rhizomatous herbaceous perennial; stigmas finely divided, flowers functionally dioecious 2
1 Twining; annual or woody perennial; rhizomes 0; stigmas capitate; flowers all bisexual 3
  2 Leaves rarely >12cm, truncate at base, cuspidate **1. F. japonica**
  2 Leaves often >12cm, cordate to cordate-truncate at base, acute to ± acuminate at apex **2. F. sachalinensis**
3 Woody perennial; the larger inflorescences well branched **3. F. baldschuanica**
3 Annual; inflorescences with 1 main axis 4
  4 Fruiting pedicels 1-3mm; achene 4-5mm, dull **4. F. convolvulus**
  4 Fruiting pedicels 3-8mm; achene 2.5-3mm, shiny **5. F. dumetorum**

1. F. japonica (Houtt.) Ronse Decr. (*Reynoutria japonica* Houtt., *Polygonum cuspidatum* Siebold & Zucc.) - *Japanese Knotweed.* Stems erect to arching, to 2m, often forming dense thickets; leaves broadly ovate, ≤12(18) x 10(13)cm;

inflorescences ≤15cm. Intrd; waste places, tips and by roads, railways and rivers; frequent to common over BI, originally a garden escape 1st found in wild in 1886; Japan. Almost all plants in BI are ± female octoploids; 2n=88. Almost all seed set is hybrid. Var. **compacta** (Hook. f.) J.P. Bailey (*R. japonica* var. *compacta* (Hook. f.) Buchheim) is a dwarf (<1m) tetraploid, 2n=44, with thick leaves ± as wide as long with ± undulate margins and usually red-tinged inflorescences; both sexes are still cultivated and natd plants are scattered throughout Br.

1 x 2. F. japonica x F. sachalinensis = F. x bohemica (Chrtek & Chrtková) J.P. Bailey (*Reynoutria* x *bohemica* Chrtek & Chrtková) occurs usually with 1 or both parents, but sometimes without either due to independent vegetative dispersal, in scattered sites in BI, mostly C & S En; it has intermediate leaf-shape and -size. Most plants are hexaploid, 2n=66, but some with 2n=44 are probably derived from *F. japonica* var. *compacta*.

1 x 3. F. japonica x F. baldschuanica is the parentage of much seed produced by *F. japonica* near plants of *F. baldschuanica*. The seed is viable but only 1 wild hybrid has been found (waste ground by railway, Middlesex, 1987); it has woody stems with rhizomes and scarcely climbing stems, and intermediate leaf-shape, perianth and stigmas; 2n=54.

2. F. sachalinensis (F. Schmidt ex Maxim.) Ronse Decr. (*Reynoutria sachalinensis* (F. Schmidt ex Maxim.) Nakai, *Polygonum sachalinense* F. Schmidt ex Maxim.) - *Giant Knotweed*. Similar to *F. japonica* but often taller (to 3m); leaves ovate-oblong, ≤38 x 28cm; inflorescences often shorter (<10cm) and denser; 2n=44. Intrd; in similar places to *F. japonica* but rarer; scattered over Br and Ir, locally common, 1st recorded in wild in 1896. Most plants in BI are ± female but ± male plants are not rare.

3. F. baldschuanica (Regel) Holub (*F. aubertii* (L. Henry) Holub, *Polygonum aubertii* L. Henry, *P. baldschuanicum* Regel, *Bilderdykia aubertii* (L. Henry) Moldenke, *B. baldschuanica* (Regel) D.A. Webb) - *Russian-vine*. Stems woody below, twining and scrambling for many m; leaves ovate-triangular, cordate, obtuse to acuminate; fruiting pedicels ≤8mm; outer tepals broadly winged in fruit; achenes 4-5mm, shiny; 2n=20. Intrd; commonly cultivated and a persistent throwout or relic in waste scrubby places or hedges; scattered over most of BI but rarely well natd; C Asia. *F. aubertii* perhaps differs in its smaller achenes and flowers, more papillose inflorescence-branches and more undulate leaf-margins; it appears to be rare in cultivation but is very doubtfully specifically distinct, and other differences claimed are not constant.

4. F. convolvulus (L.) Á. Löve (*Polygonum convolvulus* L.) - *Black-bindweed*. Annual with trailing or climbing stems to 1(1.5)m; leaves ovate-triangular, cordate to sagittate, obtuse to acuminate; fruiting pedicels 1-3mm; outer tepals keeled to narrowly winged (var. **subalatum** (Lej. & Courtois) D.H. Kent) in fruit; achenes 4-5mm, dull; 2n=40. Native; waste and arable ground; common in most of BI.

5. F. dumetorum (L.) Holub (*Polygonum dumetorum* L.) - *Copse-bindweed*. Very R
like *F. convolvulus* but stems climbing to 2m; leaves more narrowly acuminate; fruiting pedicels 3-8mm; outer tepals broadly winged in fruit; achenes 2.5-3mm, shiny; 2n=20. Native; in hedges and thickets; rare and very local in S En, formerly N to Caerns. Probably not distinct from the American **F. scandens** (L.) Holub, which has priority.

6. MUEHLENBECKIA Meisn. - *Wireplant*

Woody sprawling or climbing perennials; inflorescences short axillary or terminal racemes, dioecious; tepals 5, fused for ≥1/4 from base, enlarging and becoming white and succulent in fruit, not keeled or winged; stamens 8, styles 3, stigmas much divided; achene with 3 rounded angles.

1. M. complexa (A. Cunn.) Meisn. - *Wireplant*. Stems to several m; leaves <2cm, oblong to suborbicular, with distinct petiole, deciduous; flowers in autumn;

(2n=20). Intrd; garden escape natd on cliffs, walls and rough ground and in hedges; CI, Scillies, extreme SW En, S Hants and formerly E Suffolk, relic in Man; New Zealand.

7. RHEUM L. - *Rhubarbs*

Tall rhizomatous herbaceous perennials; leaves large, mostly basal, palmately veined; flowers in large terminal and axillary panicles; tepals 6, ± petaloid, not winged or keeled, not enlarging in fruit; stamens usually 9; anthers versatile; stigmas 3, subsessile, capitate, papillate; achene with 3 acute angles, with 3 broad membranous wings.

Other spp. - R. officinale Baill., from China, is an ornamental differing from *R. palmatum* in its 5-lobed leaves and channelled (not cylindric) petioles; it has been found in similar situations but apparently not recently.

1. R. x hybridum Murray (*R.* x *cultorum* Thorsrud & Reis. nom. nud., *R. rhaponticum* auct. non L., *R. rhabarbarum* auct. non L.) - *Rhubarb*. Basal leaves often 1m, glabrous, cordate, very shallowly lobed with entire, obtuse to rounded lobes and thick, edible petioles; flowering stems to 1.5m, glabrous, leafless; flowers cream; achenes 6-12mm with wing 2-3mm wide; (2n=44). Intrd; commonly grown vegetable, on field-scale especially in N En, often persisting as relic or throwout; scattered throughout BI; garden origin, probably from Siberian parents.

2. R. palmatum L. - *Ornamental Rhubarb*. Habit similar to that of *R.* x *hybridum* but leaves sparsely pubescent, distinctly lobed with >5 acute, dentate lobes; flowering stems pubescent; flowers reddish; (2n=22). Intrd; grown in gardens as ornament and formerly medicinally, occasionally found as a relic or outcast in grassy places; very scattered in En; NE Asia.

8. RUMEX L. - *Docks*

Usually herbaceous perennials (rarely annual or biennial), sometimes ± rhizomatous; inflorescences terminal and axillary racemes or panicles with whorled flowers; tepals 6, ± sepaloid, not keeled or winged, the inner usually enlarging in fruit and often with swollen tubercle on face; stamens 6; anthers basifixed; styles 3; stigmas deeply divided; achene with 3 acute angles.

'Tepals' refer to the inner 3 at fruiting. The keys deal only with species, not hybrids. All the hybrids are to some degree sterile, most highly so, with undeveloped achenes; they are not rare, but mostly occur as single or few plants and almost always with 1 or both parents.

General Key

1	Lower leaves sagittate to hastate, acid-tasting; flowers mostly unisexual	2
1	Leaves not sagittate or hastate, not acid-tasting; flowers bisexual	4
2	All leaves with distinct petiole, most c. as wide as long	**2. R. scutatus**
2	Upper leaves sessile, most distinctly longer than wide	3
3	Upper leaves clasping stem; basal lobes pointed ± basally (sagittate); tepals becoming much longer than achene	**3. R. acetosa**
3	Upper leaves not clasping stem; basal lobes mostly pointed laterally or ± forward (hastate); tepals not or scarcely longer than achene	**1. R. acetosella**
4	All tepals lacking swollen tubercle	*Key A*
4	At least 1 tepal with distinct swollen tubercle on outer face	5
5	Tepals with 1-several teeth (each >0.5mm) on each side	*Key B*
5	Tepals entire to crenate	*Key C*

Key A - All tepals lacking swollen tubercle

1	Tepals with long hooked teeth	**17. R. brownii**
1	Tepals entire or nearly so	2

2 Tepals distinctly longer than wide **7. R. aquaticus**

2 Tepals c. as long as wide 3

3 Plant rhizomatous; lower leaves <1.5x as long as wide **6. R. pseudoalpinus**

3 Plant not rhizomatous; leaves mostly >2x as long as wide **8. R. longifolius**

Key B - At least one tepal with distinct swollen tubercle; all tepals with 1-several teeth

1 Tepals mostly <4mm 2

1 Tepals mostly >4mm 3

2 Tepals mostly <3mm, with teeth c. as long (>2mm); tubercle acute distally; anthers 0.4-0.6mm **23. R. maritimus**

2 Tepals mostly >3mm, with teeth much shorter (<2mm); tubercle obtuse distally; anthers 0.9-1.3mm **22. R. palustris**

3 Tepals broadly ovate to suborbicular, >5mm wide (excl. teeth) **11. R. cristatus**

3 Tepals ovate-triangular, <4mm wide (excl. teeth) 4

4 Tubercles on tepals coarsely warty 5

4 Tubercles on tepals ± smooth 6

5 Perennial; leaves usually constricted just below middle (violin-shaped), rounded to cordate at base; at least some branches arising at >60° **18. R.pulcher**

5 Annual to biennial; leaves rarely violin-shaped, rounded to cuneate at base; branches arising at <60° **21. R. obovatus**

6 Annual with basal leaves to 12cm; usually all 3 tepals with well-developed tubercle **20. R. dentatus**

6 Perennial with basal leaves to 40cm; usually only 1 tepal with well-developed tubercle **19. R. obtusifolius**

Key C - At least 1 tepal with distinct swollen tubercle; all tepals entire to crenate

1 Part of pedicel above joint shorter than tepals 2

1 Part of pedicel above joint c. as long as or longer than tepals 3

2 Tepals entire, with ± smooth tubercles; branches few, arising at <45 degrees from main stem **5. R. frutescens**

2 Tepals with some short teeth, with warty tubercles; branches numerous, arising at ≥45° from main stem **18. R. pulcher**

3 Tepals mostly <5mm 4

3 Tepals mostly >5mm 9

4 Tepals <3mm 5

4 Tepals mostly >3mm 6

5 All 3 tepals with well-developed oblong tubercle **14. R. conglomeratus**

5 1 tepal with well-developed ± globose tubercle, other 2 with 0 or rudimentary tubercle **15. R. sanguineus**

6 Lower leaves ovate-oblong, strongly cordate at base; tepals often with some short teeth **19. R. obtusifolius**

6 Lower leaves narrowly oblong, narrowly elliptic or lanceolate, cuneate to subcordate at base; tepals entire 7

7 Tepals not or scarcely wider than tubercles **16. R. rupestris**

7 Tepals much wider than tubercles 8

8 Lower leaves tightly undulate; stems with short ± erect branches, erect **13. R. crispus**

8 Leaves not undulate; stems with long branches flowering later than main stem, often procumbent at base **4. R. salicifolius**

9 Tepals nearly as wide as long to wider, ± rounded at apex, only 1 with well-developed tubercle 10

9 Tepals distinctly longer than wide, tapered to rounded to obtuse apex, usually all 3 with ± well-developed tubercle 12

10 Basal leaves <1.5x as long as wide, deeply cordate at base; petiole longer than leaf **9. R. confertus**
10 Basal leaves >2x as long as wide, cuneate to cordate at base; petiole shorter than leaf 11
11 Lower leaves ± cordate at base; lateral veins arising at >60° from midrib **11. R. cristatus**
11 Lower leaves cuneate to truncate at base; lateral veins arising at <60 degrees from midrib **12. R. patientia**
12 Tubercles >3mm; robust waterside plant with basal leaves >60cm **10. R. hydrolapathum**
12 Tubercles <3mm; basal leaves <50cm 13
13 Basal leaves ovate-oblong, strongly cordate at base **19. R. obtusifolius**
13 Basal leaves narrowly oblong, narrowly elliptic or lanceolate, cuneate to subcordate at base **13. R. crispus**

Other spp. - c.24 other spp. have been found as rare casuals in BI. Of these **R. stenophyllus** Ledeb., from C & E Europe, was formerly natd at Avonmouth Docks, W Gloucs; it would key out as *R. obtusifolius* (tepals toothed and with tubercles) but in habit and leaf-shape more closely resembles *R. crispus.*

Subgenus* 1 - *ACETOSELLA Raf. (sp. 1). Dioecious; some roots horizontal and producing aerial shoots; leaves sagittate, acid-tasting; tepals remaining shorter than achene, without tubercles.

1. R. acetosella L. - *Sheep's Sorrel.* Stems procumbent to erect; leaves linear to oblong-lanceolate, with narrow, laterally or forward-directed basal lobes; achene 1-1.5mm; 2n=28, 42. Native; heathy open ground, short grassland and cultivated land, mostly on acid sandy soils; throughout BI. Distribution of the 2 sspp. in BI has not been worked out, but ssp. *pyrenaicus* is probably mainly in S and absent from N. Extent of overlap is unknown; both occur in C En and probably intermediates exist, as in areas of overlap in C Europe. **193**

a. Ssp. acetosella (*R. tenuifolius* (Wallr.) Á. Löve). Tepals forming a loose cover round ripe achene (easily rubbed off by rolling achenes between finger and thumb). Small plants with narrowly linear leaves and occurring on very dry sands are worth no more than varietal rank as var. **tenuifolius** Wallr.

b. Ssp. pyrenaicus (Pourr.) Akeroyd (ssp. *angiocarpus* auct. non (Murb.) Murb., *R. angiocarpus* auct. non Murb.). Tepals tightly adherent to ripe achene (not able to be rubbed off).

Subgenus* 2 - *ACETOSA Raf. (spp. 2-3). Dioecious; aerial stems arising from short rhizomes; leaves sagittate to hastate, acid-tasting; tepals enlarging to much longer than achene, each with or without tubercle.

2. R. scutatus L. - *French Sorrel.* Stems much-branched, erect to spreading, ± woody below, to 50cm; leaves broadly ovate, ± hastate, with very wide out-turned basal lobes; tepals 5-8mm, orbicular, cordate, without tubercles; (2n=20). Intrd; surviving on banks, old walls and rough ground in very scattered places in Br, mainly in N; C & S Europe. **193**

3. R. acetosa L. - *Common Sorrel.* Stems not or little-branched, usually erect, herbaceous, to 1m; leaves ovate or obovate to narrowly triangular, sagittate, with very acute backward-directed basal lobes; tepals 2.5-4mm, suborbicular, ± cordate, each with small tubercle near base; 2n=14 (female), 15 (male). **193**

1 Plant papillose to shortly pubescent (hairs <0.3mm) on all vegetative parts **b. ssp. hibernicus**
1 Plant usually ± glabrous, with papillae ± confined to basal margin of leaves 2

2 Inflorescence with well-branched branches **d. ssp. ambiguus**
2 Inflorescence with simple branches 3
3 Leaves thick, succulent; basal leaves c.2x as long as wide; stem-leaves (1)2-4; coastal **c. ssp. biformis**
3 Leaves thin, not succulent; basal leaves usually 2-4x as long as wide; stem-leaves often >4; widespread **a. ssp. acetosa**

a. Ssp. acetosa. Plant to 60(100)cm; leaves rarely succulent; inflorescence usually with several simple branches. Native; in wide range of grassy places; common throughout BI.

b. Ssp. hibernicus (Rech. f.) Akeroyd (*R. hibernicus* Rech. f.). Plant to 30(50)cm; R
leaves succulent; inflorescence with few simple branches. Native; coastal dunes; NW, W & S Ir, N Sc, SW En, SW Wa; endemic.

c. Ssp. biformis (Lange) Valdés Berm. & Castrov. Plant to 20(30)cm; leaves RR
succulent; inflorescence with few simple branches. Native; sea-cliffs; W Cornwall, Cards and Co Clare, but perhaps overlooked.

d. Ssp. ambiguus (Gren.) Á. Löve (*R. rugosus* Campd.). Plant to 1.2m; leaves large, thin; inflorescence with well branched branches. Intrd; grown as vegetable, natd in Herts and E Suffolk; origin unknown.

Subgenus* 3 - *RUMEX (spp. 4-23). Bisexual; aerial stems usually erect, arising from tap-root or sometimes rhizomes; leaves cordate to cuneate, larger than in other subgenera; inner tepals enlarging to much longer than achene, 0, 1 or 3 with tubercle. 'Leaves' refers to lower stem-leaves and basal leaves; 'tepals' refers to inner tepals at fruiting. Hybrids are frequent in mixed populations; they mostly possess predictably intermediate characters.

4. R. salicifolius Weinm. (*R. triangulivalvis* (Danser) Rech. f.) - *Willow-leaved Dock*. 193
Decumbent to erect perennial to 50(100)cm, often with long branches from near base; leaves lanceolate to linear-lanceolate; inflorescence branched, open; tepals 3-4mm, ovate-triangular, ± entire, all with narrow warty tubercle; (2n=20). Intrd; natd on waste land by docks, railways, canals and tips, originating with grain; very scattered over Br and Ir, especially SE En, often only casual; N America. Our plant is ssp. **triangulivalvis** Danser.

5. R. frutescens Thouars - *Argentine Dock*. Rhizomatous perennial with ascending 193
to erect unbranched stems to 30cm; leaves obovate to oblanceolate, leathery; inflorescence dense, not or shortly branched; tepals 4-5mm, narrowly ovate-triangular, entire, all with large ± smooth tubercle; (2n=40). Intrd; natd on coastal dunes in SW En and S Wa, casual near docks in Sc, Wa and W En; S America.

5 x 14. R. frutescens x R. conglomeratus = R. x wrightii Lousley was found in 1952 at Braunton Burrows, N Devon, and in 1994 at Phillack Towans, W Cornwall, with *R. frutescens;* endemic.

5 x 19. R. frutescens x R. obtusifolius = R. x cornubiensis Holyoak was found at Phillack Towans, W Cornwall, with both parents in 1994; endemic.

6. R. pseudoalpinus Höfft (*R. alpinus* L. 1759 non 1753) - *Monk's-rhubarb*. 193
Rhizomatous perennial with erect stems to 70(100)cm; leaves broadly ovate, cordate; inflorescence dense, with erect branches; tepals 5-6mm, broadly ovate-triangular, entire, without tubercles; (2n=20). Intrd; natd relic of old cultivation in grassy places by roads, streams and old buildings; scattered in Sc and En S to Staffs; Europe.

7. R. aquaticus L. - *Scottish Dock*. Erect perennial to 2m; leaves triangular-ovate, RR
cordate; inflorescence rather open, with erecto-patent branches; tepals 5-8mm, 193
ovate-triangular, entire, without tubercles; (2n=c.200). Native; seasonally flooded ground by Loch Lomond, Stirlings and Dunbarton.

7 x 13. R. aquaticus x R. crispus = R. x conspersus Hartm. is found with both parents in Dunbarton and Stirlings.

7 x 15. R. aquaticus x R. sanguineus = R. x dumulosus Hausskn. was found near both parents in Stirlings in 1989.

7 x 19. R. aquaticus x R. obtusifolius = R. x platyphyllos Aresch. (*R. x schmidtii* Hausskn.) occurs with both parents in Stirlings and Dunbarton.

8. R. longifolius DC. - *Northern Dock.* Erect perennial to 1.2m; leaves lanceolate **193**
to narrowly ovate; inflorescence dense, with short erect branches; tepals 4-5.5mm, suborbicular, cordate, entire, without tubercles; 2n=60. Native; damp open and grassy ground often by water; Sc and En S to S Lancs, Staffs.

8 x 13. R. longifolius x R. crispus = R. x propinquus Aresch. is frequent in Sc wherever the parents meet.

8 x 19. R. longifolius x R. obtusifolius = R. x hybridus Kindb. (*R. x arnottii* Druce) is frequent in Sc wherever the parents meet (often near *R.* x *propinquus* too), and sparse in N En.

9. R. confertus Willd. - *Russian Dock.* Erect perennial to 1.2m; leaves broadly **193**
ovate, cordate; inflorescence fairly dense, with erect branches; tepals 6-9mm, suborbicular, cordate, ± entire, 1 with small tubercle; (2n=38, 40, 60). Intrd; natd on roadside in E Kent, formerly Surrey and Oxon; E Europe.

9 x 13. R. confertus x R. crispus = R. x skofitzii Blocki occurred in Surrey in 1954-5 with *R. confertus.*

9 x 19. R. confertus x R. obtusifolius = R. x borbasii Blocki occurred in Surrey in 1954 with *R. confertus.*

10. R. hydrolapathum Huds. - *Water Dock.* Erect perennial to 2m; leaves **193**
lanceolate to narrowly ovate, cuneate; inflorescence large, rather open, with many erecto-patent branches; tepals 5-8mm, ovate-triangular, ± entire, all with elongated smooth tubercle; 2n=130. Native; by lakes, rivers, canals, ditches and marshes; scattered through BI N to Banffs.

10 x 13. R. hydrolapathum x R. crispus = R. x schreberi Hausskn. has been found in E Suffolk, Notts, Caerns and Co Down near both parents.

10 x 14. R. hydrolapathum x R. conglomeratus = R. x digeneus Beck has been found with both parents in S En.

10 x 19. R. hydrolapathum x R. obtusifolius = R. x lingulatus Jungner (*R. x weberi* Fisch.-Benz.) has been found several times in S En and S Wa, also E Cork.

11. R. cristatus DC. - *Greek Dock.* Erect perennial to 2m; leaves narrowly ovate, **193**
cordate; inflorescence dense but with long erecto-patent branches; tepals 5-8mm, broadly ovate to suborbicular, cordate, denticulate to dentate with teeth to 1mm, 1 with large rounded smooth tubercle; (2n=80). Intrd; natd on waste ground in SE En and S Wa, casual elsewhere in S En; CS Europe. Perhaps only a ssp. of *R. patientia;* has been confused with *R. obtusifolius* x *R. patientia.*

11 x 13. R. cristatus x R. crispus = R. x dimidiatus Hausskn. has occurred with the parents near R. Thames in W Kent, S Essex and Middlesex.

11 x 14. R. cristatus x R. conglomeratus was found in S Essex in 1985; endemic.

11 x 19. R. cristatus x R. obtusifolius = R. x lousleyi D.H. Kent occurs with the parents in SE En and SE Wa; endemic.

11 x 22. R. cristatus x R. palustris has been found in S Essex; endemic.

12. R. patientia L. - *Patience Dock.* Erect perennial to 2m; leaves narrowly ovate, **193**
cuneate to truncate at base; inflorescence ± as in *R. cristatus;* tepals 5-8mm, broadly ovate to suborbicular, cordate, ± entire, 1 with small rounded smooth tubercle; (2n=60). Intrd; natd in a few waste places by docks and breweries, scattered casual elsewhere in En; Europe. Several sspp. based on tepal shape are probably

---

FIG 193 - Fruiting tepals of ***Rumex***. 1, ***R. obovatus***. 2, ***R. longifolius***. 3, ***R. pseudoalpinus***. 4, ***R. scutatus***. 5, ***R. aquaticus***. 6, ***R. confertus***. 7, ***R. cristatus***. 8, ***R. frutescens***. 9, ***R. patientia***. 10, ***R. pulcher***. 11, ***R. dentatus***. 12, ***R. hydrolapathum***. 13, ***R. crispus***. 14, ***R. x pratensis***. 15, ***R. obtusifolius*** var. ***obtusifolius***. 16, var. ***microcarpus***. 17, ***R. salicifolius***. 18, ***R. brownii***. 19, ***R. acetosella***. 20, ***R. acetosa***. 21, ***R. maritimus***. 22, ***R. rupestris***. 23, ***R. conglomeratus***. 24, ***R. sanguineus***. 25, ***R. palustris***.

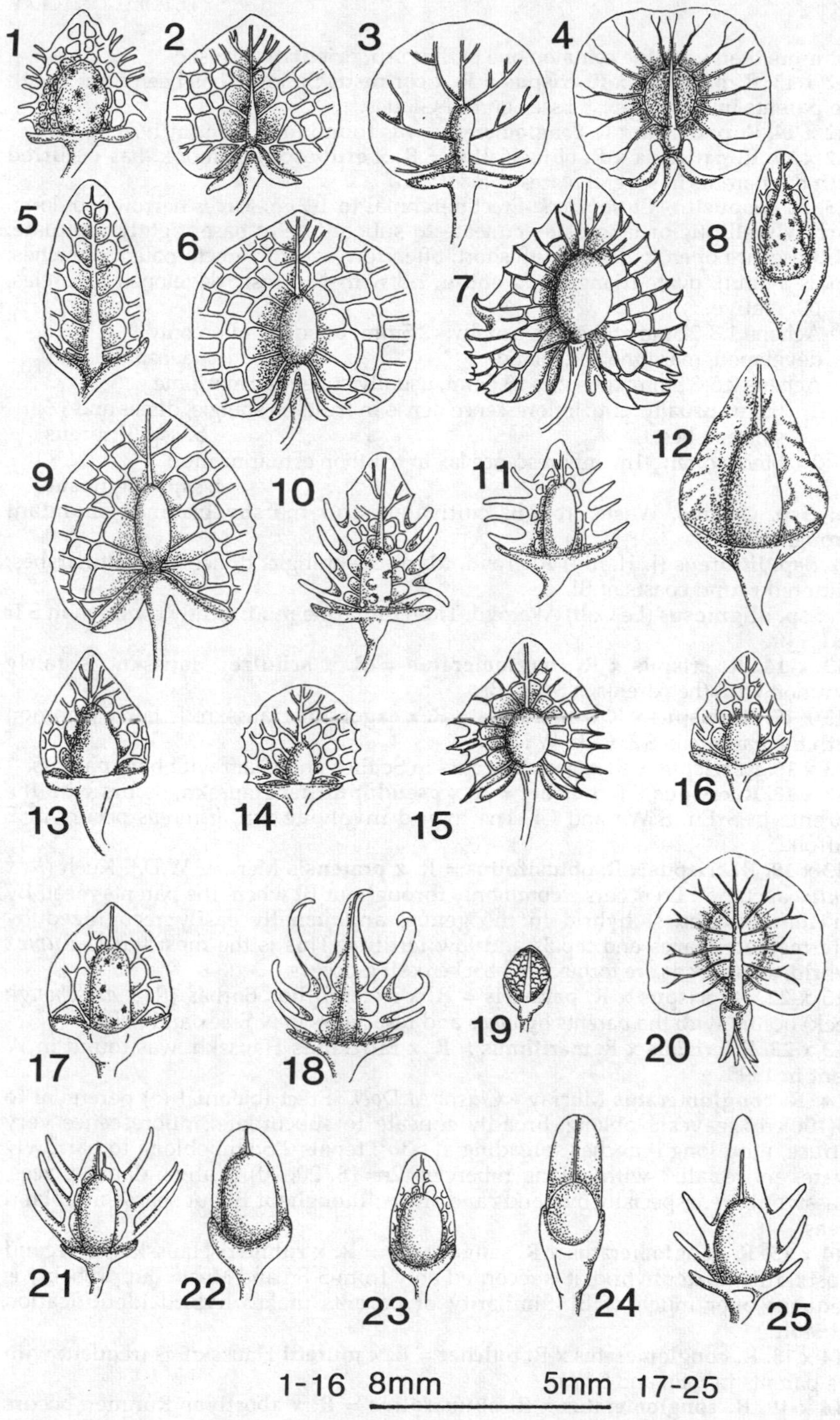

FIG 193 - see caption opposite

not more than vars; the common one in BI is ssp. **orientalis** Danser.

12 x 13. R. patientia x R. crispus = R. x confusus Simonk. has been found with the parents in Middlesex, S Essex and W Gloucs.

12 x 14. R. patientia x R. conglomeratus was found in W Kent in 1978; endemic.

12 x 19. R. patientia x R. obtusifolius = R. x erubescens Simonk. has occurred with the parents in several places in S & C En.

13. R. crispus L. - *Curled Dock*. Erect perennial to 1(2)m; leaves narrowly oblong, **193**
narrowly elliptic or lanceolate, cuneate to subcordate at base, tightly undulate; inflorescence open to dense, with short, often few, erect to erecto-patent branches; tepals 3-6mm, ovate-triangular, ± entire, 1-3 with variously developed tubercles; 2n=60. Native.

1 Achene 1.3-2.5mm; tubercles usually <2.5mm, unequal, often only 1 developed; maritime and inland **a. ssp. crispus**
1 Achene 2.5-3.5mm; tubercles ≤3.5mm, usually subequal; maritime 2
 2 Stems usually <1m; inflorescence dense in fruit; on shingle, dunes and saltmarshes **b. ssp. littoreus**
 2 Stems often >1m; inflorescence lax in fruit; on estuarine mud **c. ssp. uliginosus**

a. Ssp. crispus. Waste, rough, cultivated and marshy ground; abundant throughout BI.

b. Ssp. littoreus (J. Hardy) Akeroyd. Maritime shingle, dunes and saltmarshes; scattered round coasts of BI.

c. Ssp. uliginosus (Le Gall) Akeroyd. Tidal estuarine mud; locally common in S Ir and S Br.

13 x 14. R. crispus x R. conglomeratus = R. x schulzei Hausskn. is fairly common with the parents in S & C Br.

13 x 15. R. crispus x R. sanguineus = R. x sagorskii Hausskn. is fairly common with the parents in S & C Br, W Mayo.

13 x 16. R. crispus x R. rupestris occurs in Scillies and Glam with both parents.

13 x 18. R. crispus x R. pulcher = R. x pseudopulcher Hausskn. occurs with the parents in S En, S Wa and CI. The hybrid involving ssp. *littoreus* occurs in E Suffolk.

13 x 19. R. crispus x R. obtusifolius = R. x pratensis Mert. & W.D.J. Koch (*R. x* **193**
acutus auct. non L.) occurs ± commonly throughout BI where the parents meet. By far the commonest hybrid in the genus, and usually easily recognized by intermediate leaves and tepals and low fertility. This is the most fertile *Rumex* hybrid; some seeds are formed and backcrossing occurs.

13 x 22. R. crispus x R. palustris = R. x heteranthos Borbás (*R. x areschougii* Beck) occurs with the parents by lakes and reservoirs in N Essex and Leics.

13 x 23. R. crispus x R. maritimus = R. x fallacinus Hausskn. was found in W Kent in 1995.

14. R. conglomeratus Murray - *Clustered Dock*. Erect (biennial to) perennial to **193**
60(100)cm; leaves ± oblong, broadly cuneate to subcordate; inflorescence very diffuse, with long branches spreading at >30°; tepals 2-3mm, oblong to narrowly ovate, entire, all 3 with oblong tubercle; (2n=18, 20, 40). Native; damp places, grassy or bare, especially by ponds and rivers; thoughout BI but sparse in upland areas.

14 x 15. R. conglomeratus x R. sanguineus = R. x ruhmeri Hausskn. is second most fertile *Rumex* hybrid; it is recorded only from S En and Meath but probably is frequent over much of BI. Similarity of parents makes hybrid identification difficult.

14 x 18. R. conglomeratus x R. pulcher = R. x muretii Hausskn. is frequent with the parents in S En and S Wa.

14 x 19. R. conglomeratus x R. obtusifolius = R. x abortivus Ruhmer occurs frequently with the parents scattered throughout Br.

14 x 22. R. conglomeratus x R. palustris = R. x wirtgenii Beck occurs with the parents in C & SE En.

14 x 23. R. conglomeratus x R. maritimus = R. x knafii Celak. occurs with the parents scattered in En.

15. R. sanguineus L. - *Wood Dock*. Erect perennial to 60(100)cm; leaves and **193**
inflorescence as in *R. conglomeratus* but branches arising at <30(45)° and tubercle usually 1, ± globose; 2n=20. Native; damp shady places, mostly in woods or hedgerows or by water; common throughout most of BI but rare in N Sc. The common plant is var. **viridis** (Sibth.) W.D.J. Koch; var. **sanguineus** has blood-red leaf-veins and is a rare garden escape or casual.

15 x 18. R. sanguineus x R. pulcher = R. x mixtus Lamb. occurs with the parents in S En.

15 x 19. R. sanguineus x R. obtusifolius = R. x dufftii Hausskn. occurs with the parents scattered over BI.

16. R. rupestris Le Gall - *Shore Dock*. Erect perennial to 50(70)cm; leaves thick, **RRR**
narrowly oblong to lanceolate, undulate, cuneate to subcordate; inflorescence **193**
diffuse, with long branches spreading at 25-50°; tepals 3-4mm, oblong to narrowly ovate, entire, all with large tubercle; (2n=20). Native; damp places on sand or rocks by sea; Anglesey, S Wa, SW En and CI.

16 x 18. R. rupestris x R. pulcher = R. x trimenii E.G. Camus has occurred with *R. rupestris* in Cornwall.

17. R. brownii Campd. - *Hooked Dock*. Rhizomatous perennial with erect stems to **193**
60cm; leaves lanceolate to ovate, tightly undulate, cordate to cuneate; inflorescence very diffuse, with few long branches; tepals 2-3.5mm, ovate, with long hooked teeth, without tubercles; (2n=40). Intrd; wool-alien scattered in En and Sc and sometimes persisting; Australia.

18. R. pulcher L. - *Fiddle Dock*. Erect to spreading perennial to 40(50)cm; leaves **193**
oblong-obovate, usually strongly constricted just above cordate to rounded base; inflorescence very diffuse, with long branches spreading at 45-90°; tepals 4-5.5mm, narrowly to broadly ovate, usually with well-developed teeth but sometimes with few short ones and rarely entire, (1-)3 with elongate warty tubercle; (2n=20). Native; dry grassy places; locally frequent in BI N to N Lincs, Anglesey, Co Carlow and W Cork, rare casual further N. Several sspp. based on tepal-shape are probably not more than vars; only ssp. **pulcher** is native.

18 x 19. R. pulcher x R. obtusifolius = R. x ogulinensis Borbás occurs with *R. pulcher* in S En.

19. R. obtusifolius L. - *Broad-leaved Dock*. Erect perennial to 1(1.2)m; leaves **193**
ovate-oblong, cordate; inflorescence rather open, with erecto-patent branches; tepals 3-6mm, triangular- to oblong-ovate, with variable teeth as in *R. pulcher*, 1(-3) with smooth tubercle; 2n=40. Native; grassland, by roads and rivers, waste and cultivated ground; abundant throughout BI. 3 taxa based on tepal-shape are best treated as vars: var. **obtusifolius** has large tepals with strong teeth and 1 tubercle; var. **microcarpus** Dierb. (ssp. *sylvestris* (Wallr.) Celak.) has small tepals with 0 or few short teeth and 3 tubercles; var. **transiens** (Simonk.) Kubát (ssp. *transiens* (Simonk.) Rech.f.) is intermediate. Only the 1st is native.

19 x 22. R. obtusifolius x R. palustris = R. x steinii Becker occurs with *R. palustris* in SE En.

19 x 23. R. obtusifolius x R. maritimus = R. x callianthemus Danser occurs with *R. maritimus* in SE En.

20. R. dentatus L. - *Aegean Dock*. Erect annual to 70cm; leaves oblong-lanceolate, **193**
rounded to cordate; inflorescence very diffuse, with few long erecto-patent branches; tepals 4-6mm, triangular-ovate, with mostly long teeth, (1-)3 with smooth tubercle; (2n=40). Intrd; rather rare wool-alien; very scattered in Br; SE Europe, Asia. Plants in Br are ssp. **halacsyi** (Rech.) Rech. f.

21. R. obovatus Danser - *Obovate-leaved Dock*. Erect annual or biennial to **193**
40(70)cm; leaves mostly obovate, rounded to broadly cuneate at base; inflorescence rather diffuse but very leafy, with few erecto-patent branches; tepals

4-5mm, ± as in *R. dentatus* but all with warty tubercle; (2n=60). Intrd; grain-alien near mills, wharves and warehouses; very scattered in Br; Argentina and Paraguay.

22. R. palustris Sm. - *Marsh Dock.* Erect biennial to perennial to 60(100)cm; leaves lanceolate to narrowly elliptic, cuneate; inflorescence diffuse, pale brown in fruit, with many long, widely spreading then incurved branches; tepals 3-4mm, narrowly ovate-triangular, with few long rather rigid teeth, all with large tubercle; 2n=60. Native; edges of ponds, ditches, gravel-pits and in marshy fields, usually flooded at times; very local in S & E En N to MW Yorks, Mons, rare casual elsewhere. Possibly hybridises with *R. maritimus*. **193**

23. R. maritimus L. - *Golden Dock.* Erect annual to perennial to 40(100)cm, turning golden at fruiting; similar to *R. palustris* but branches shorter and straighter and tepals 2.5-3mm, with longer more flexible teeth and narrower tubercle; 2n=40. Native; similar places to *R. palustris* but rarely with it, more widespread but equally scarce; scattered in Br N to S Sc, very scattered in Ir. **193**

9. OXYRIA Hill - *Mountain Sorrel*

Herbaceous perennials; inflorescence a terminal panicle; tepals 4, sepaloid, not keeled or winged, the inner enlarging in fruit but without tubercles; stamens 6; anthers versatile; styles 2; stigmas deeply divided; achene biconvex.

1. O. digyna (L.) Hill - *Mountain Sorrel.* Stems to 30cm, erect, little branched, with 0(-2) leaves; basal leaves long-stalked, reniform; achene 3-4mm, broadly winged; 2n=14. Native; damp rocky places on mountains; N Wa, NW En, Sc, rare in Ir.

50. PLUMBAGINACEAE - *Thrift family*

Perennial herbs; leaves all basal, narrowed to base, simple, entire, without stipules. Flowers in branched cymes or hemispherical heads, bisexual, hypogynous, actinomorphic, 5-merous; calyx tubular below, with free lobes above, scarious at least above; corolla of 5 pink to blue petals fused at base; stamens 5, borne on base of corolla; ovary 1-celled, with 1 basal ovule; styles 5; stigmas linear; fruit a 1-seeded capsule.

Two unmistakable, usually coastal genera.

1 Flowers in dense hemispherical heads with tubular sheath of fused scarious bracts beneath **2. ARMERIA**
1 Flowers in branching cymes, the ultimate units of 1-5 flowers with 3 scale-like bracts **1. LIMONIUM**

1. LIMONIUM Mill. - *Sea-lavenders*

Aerial stems branched, the ultimate branches consisting of small clusters (spikelets) of 1-5 flowers with 3 bracts (outer, middle, inner) below, the spikelets aggregated into spikes occupying ends of branches; flowers blue to purple or lilac; styles glabrous.

Each plant produces pollen of 1 of 2 sorts (A, coarsely reticulate; B; finely reticulate) and possesses stigmas of 1 of 2 sorts (Cob, with rounded papillae; Papillate, with prominent papillae). Species may be dimorphic (A/Cob and B/Papillate plants) and self-incompatible; monomorphic (all plants A/Papillate) and self-compatible; or monomorphic (various combinations) and apomictic. **197**

1 Leaves distinctly pinnately veined 2
1 All obvious veins (1-9) arising separately from petiole-like base; vein-branches from midrib 0 or indistinct 5
 2 Upper part of stem and leaf-midribs pubescent **3. L. latifolium**
 2 Plant glabrous 3

3 Leaves rounded to emarginate at apex, dying off before autumn; cliffs **7. L. hyblaeum**

3 Leaves acute to obtuse, and mucronate, at apex, dying in autumn or winter; salt-marshes 4

4 Spikes mostly 1-2cm, with >4 spikelets/cm; outer bract 1.7-3mm; anthers yellow **1. L. vulgare**

4 Longest spikes 2-5cm, with ≤3 spikelets/cm; outer bract 3-4mm; anthers reddish-brown **2. L. humile**

5 Stems with numerous well-branched non-flowering lateral branches below; outer bract scarious except on midline **4. L. bellidifolium**

5 Stems with 0 or few or little-branched non-flowering lateral branches below; outer bract herbaceous for most of width 6

6 Leaves linear-oblong to oblanceolate-spathulate, rarely obovate-spathulate, 5-15(25)mm wide, with 1-3(5) obvious veins; inflorescence widest below top and tapered above; widespread **8-16. L. binervosum agg.**

6 Leaves obovate-spathulate, 11-25mm wide, with 5-7(9) obvious veins; inflorescence widest and ± flat at top; Channel Islands only 7

7 Outer bract (1.8)1.9-2.4(2.9)mm; calyx (3.8)4-4.6(5.5)mm **5. L. auriculae-ursifolium**

7 Outer bract (2.6)3-4(4.2)mm; calyx (4.1)4.8-5.5(6.3)mm **6. L. normannicum**

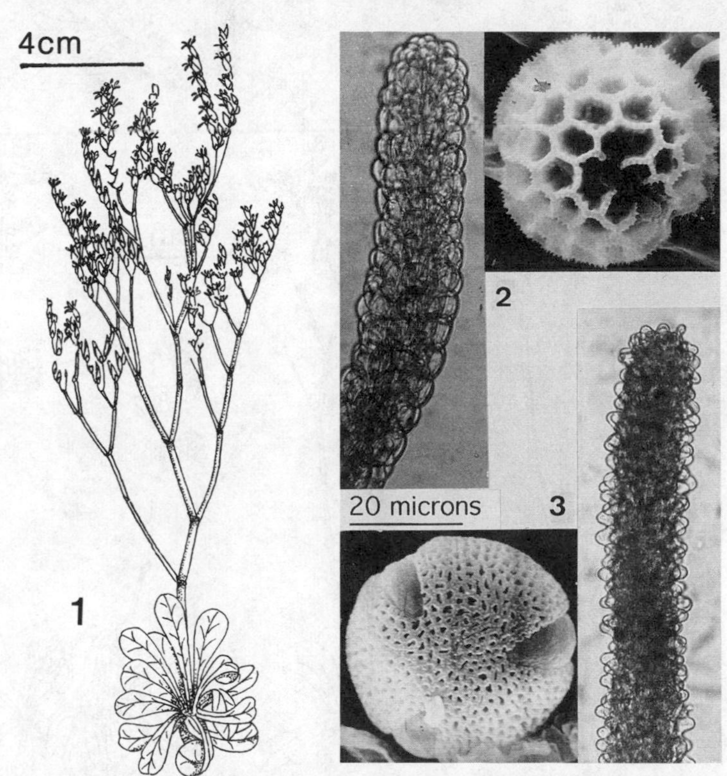

FIG 197 - *Limonium*. 1, *L. hyblaeum*. 2-3, pollen- and stigma-types. 2, cob stigma and 'A' pollen. 3, papillate stigma and 'B' pollen. Photographs courtesy of M.J. Ingrouille.

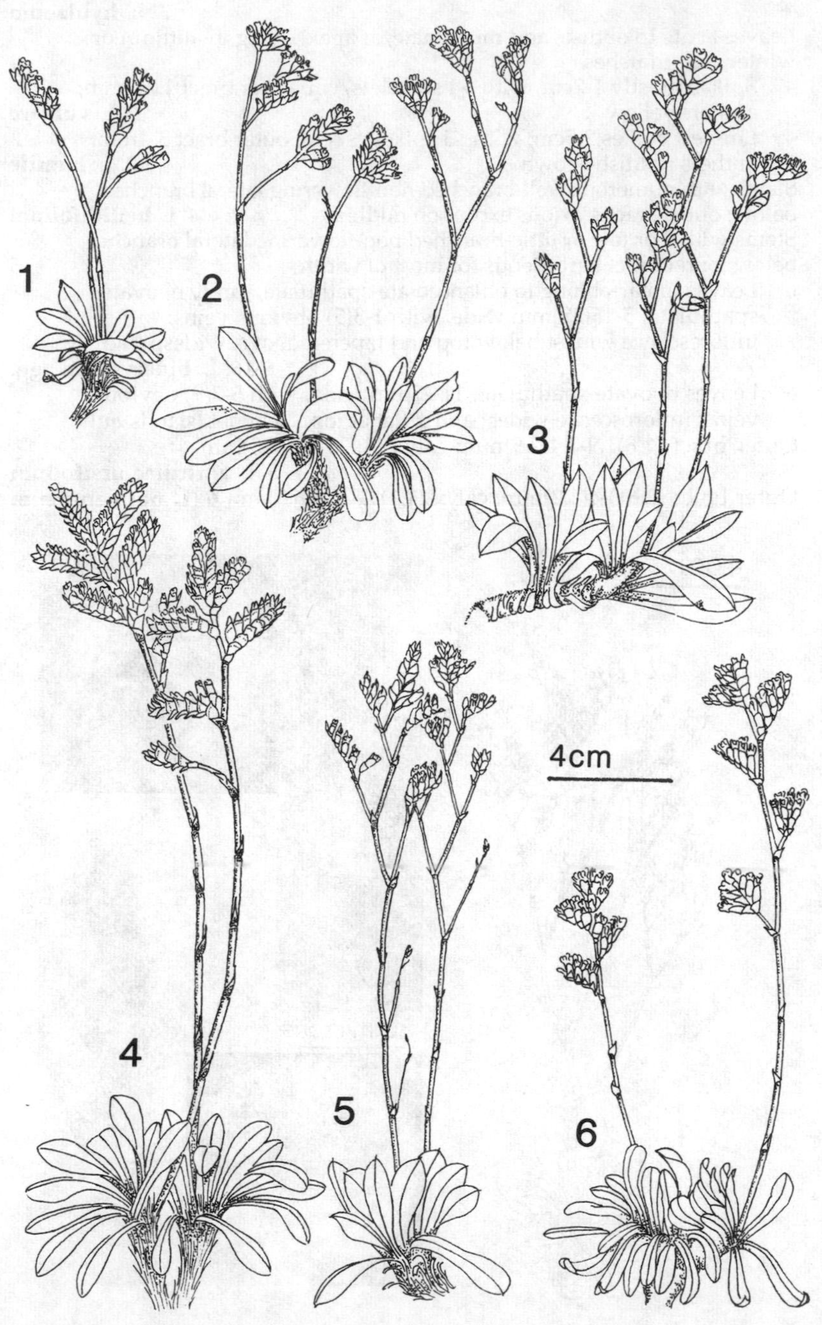

FIG 198 - *Limonium binervosum* agg. 1, *L. parvum*. 2, *L. britannicum*. 3, *L. paradoxum*. 4, *L. recurvum*. 5, *L. loganicum*. 6, *L. transwallianum*.

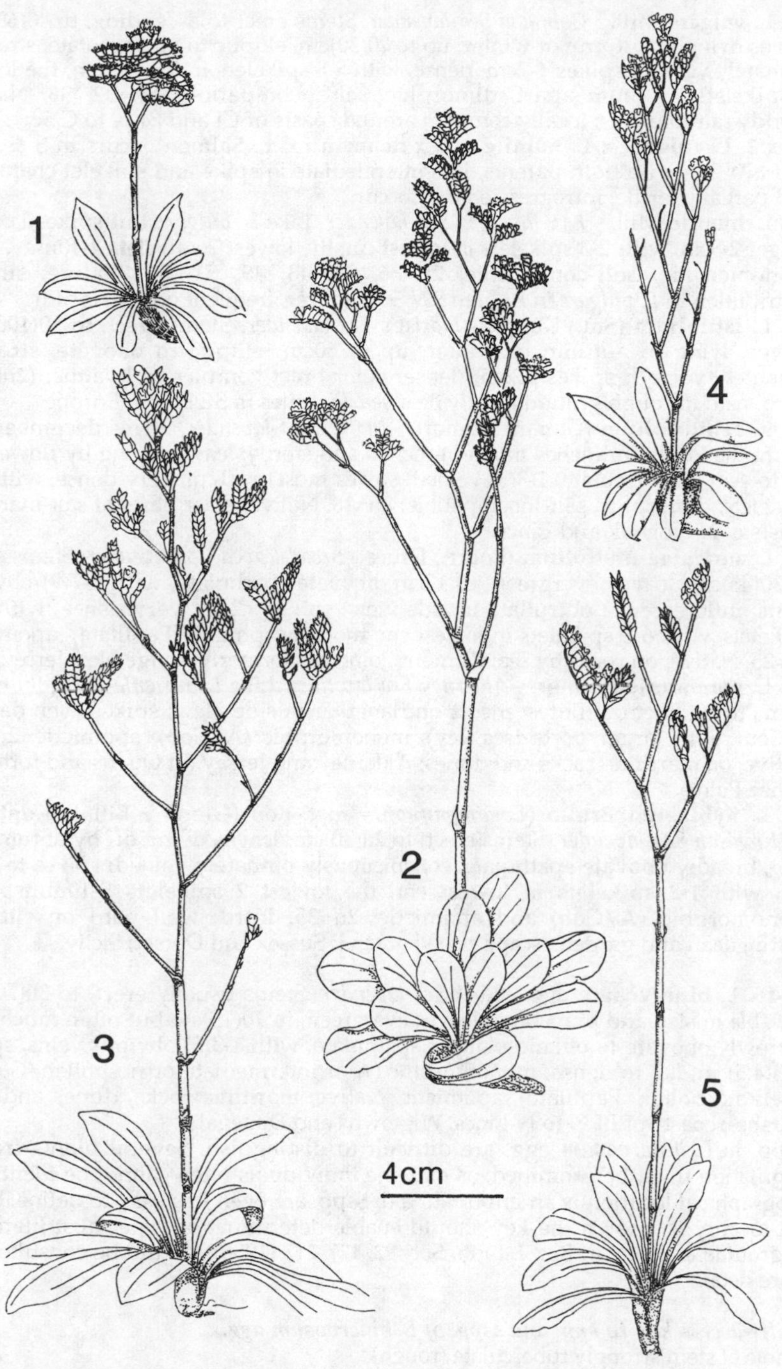

FIG 199 - *Limonium*. 1, *L. normannicum*. 2, *L. auriculae-ursifolium*. 3-5, *L. binervosum* agg. 3, *L. procerum*. 4, *L. binervosum*. 5, *L. dodartiforme*.

1. L. vulgare Mill. - *Common Sea-lavender*. Stems erect to ascending, to 40(60)cm; leaves dying in autumn or winter, up to 20(30)cm, elliptic to oblanceolate, strongly pinnately veined; spikes 1-2cm, dense, with 5-8 spikelets in lowest cm, the lowest 2 spikelets 1.5-3mm apart; dimorphic, self-incompatible; 2n=32, 36. Native; muddy salt-marshes; locally common around coasts of CI and Br N to C Sc.

1 x 2. L. vulgare x L. humile = L. x neumanii C.E. Salmon occurs in S & E En and NW Wa with both parents. It is intermediate in spike and spikelet characters and partially fertile; introgression may occur.

2. L. humile Mill. - *Lax-flowered Sea-lavender*. Like *L. vulgare* but spikes lax, the **R**
longer 2-5cm, with 2-3 spikelets in lowest cm, the lowest 2 spikelets 4-10mm apart; monomorphic, self-compatible; 2n=36, 38, 48, 49, 51, 54. Native; similar distribution to *L. vulgare* in Br N to SW Sc but rarer, frequent on coasts of Ir.

3. L. latifolium (Sm.) Kuntze - *Florist's Sea-lavender*. Stems erect, to 80(100)cm; leaves dying in autumn or winter, up to 60cm, elliptic to obovate, strongly pinnately veined; spikes ≤2cm, dense; outer bract completely hyaline; (2n=18). Intrd; natd in rough ground mostly near sea; few sites in SE En; SE Europe.

4. L. bellidifolium (Gouan) Dumort. - *Matted Sea-lavender*. Stems decumbent, to **RR**
30cm, the lowest branches much-branched and sterile; leaves dying by flowering, up to 4cm, oblanceolate, 1-3(5)-veined; spikes mostly <1cm, very dense, with <10 spikelets; dimorphic, self-incompatible; 2n=18. Native; drier parts of salt-marshes: coasts of N Norfolk and Lincs.

5. L. auriculae-ursifolium (Pourr.) Druce - *Broad-leaved Sea-lavender*. Stems erect, **RR**
to 30(45)cm; leaves evergreen, to 12cm, obovate-spathulate, with 5-7(9) obvious **199**
veins; inflorescence obtrullate in side view; spikes <2cm, very dense, with <13 spikelets, with 6-8 spikelets in lowest cm; monomorphic (B/Papillate), apomictic; 2n=25. Native; on rocks by sea; Plemont Point and formerly Rouge Nez, Jersey.

6. L. normannicum Ingr. - *Alderney Sea-lavender*. Like *L. auriculae-ursifolium* but **RR**
stems to 20(25)cm; inflorescence ± obtriangular in side view; spikes even denser, **199**
<1.5cm, with larger parts (see key); monomorphic (A/Cob), apomictic; 2n=25. Native; on maritime rocks and dunes; Alderney and Jersey (St Ouens and formerly Ronez Point).

7. L. hyblaeum Brullo (*L. companyonis* auct. non (Gren. & Billot) Kuntze) - **197**
Rottingdean Sea-lavender. Stems erect, to 20(30)cm; leaves dying off by autumn, to 5cm, broadly obovate-spathulate, conspicuously pinnately veined; spikes to 6cm, lax, with 1-2 spikelets in lowest cm, the lowest 2 spikelets 5-10mm apart; monomorphic (A/Cob) and apomictic; 2n=35. Intrd; well natd on cliffs at Rottingdean and garden escape elsewhere in E Sussex and Dorset; Sicily.

8-16. L. binervosum agg. - *Rock Sea-lavender*. Stems usually erect, to 30(70)cm, variable in size and branching; leaves evergreen, to 10(15)cm but often much less, narrowly obovate to oblanceolate or spathulate, with 1-3(5) obvious veins; spikes to 3(4.5)cm, lax to dense; monomorphic (A or intermediate or no pollen/Cob, or rarely no pollen/Papillate), apomictic. Native; maritime rocks, dunes and salt-marshes; coasts of BI N to N Lincs, Wigtowns and Donegal.

Spp. in *L. binervosum* agg. are difficult to distinguish. Several plants from a population must be examined, as extreme individuals often cannot be identified; geographical location is an important aid. Sspp. are even less easy to define, but if the locality is known the key should enable determination. Specialist literature (Ingrouille & Stace, in Bot. J. Linn. Soc. 92: 177-217 (1986)) should be consulted for more details.

Multi-access key to spp. and sspp. of L. binervosum agg.

Whole of stem strongly tuberculate (rough) A
Stem not tuberculate (smooth) or tuberculate only above B
    ± all stems branched from low down C
    Some stems unbranched in lower half D

Flowers borne on upper 2/3 of stem	E
Flowers borne on upper 1/3(-1/2) of stem	F
Mean spike length <15mm	G
Mean spike length >15mm or spikes with <5 spikelets	H
Spikes dense, with 6-10 spikelets in lowest cm	I
Spikes lax, with 3-5 spikelets in lowest cm, or with <5 spikelets	J
Mean length outer bract >3mm	K
Mean length outer bract <3mm	L
Mean length inner bract <5mm	M
Mean length inner bract >5mm	N
Mean width petal <1.5mm	O
Mean width petal >1.5mm	P

ACF(GH)ILMP		**16. L. recurvum**
(GH)	Portland (tip), Dorset	**a. ssp. recurvum**
(GH)	Portland (other than tip), Dorset; Kerry	**b. ssp. portlandicum**
(GH)	Donegal; Cumberland; Wigtowns	**c. ssp. humile**
G	Clare	**d. ssp. pseudotranswallianum**
ACFGJLMP	Cornwall	**13. L. loganicum**
BC(EF)H(IJ)KNP	Dorset	**15. L. dodartiforme**
BCEGILMO	Pembs	**14. L. transwallianum**
BCF(GH)(IJ)KMP	Pembs	**9. L. paradoxum**
BCF(GH)(IJ)LMP		**11. L. britannicum**
G I	Cornwall(N)	**a. ssp. britannicum**
G I	Cornwall(S); Devon(S)	**b. ssp. coombense**
G I	Devon(N); Pembs	**c. ssp. transcanalis**
H J	Anglesey to Lancs	**d. ssp. celticum**
BCFHJL(MN)P		**8. L. binervosum**
M	Norfolk; Lincs	**c. ssp. anglicum**
M	Essex	**d. ssp. saxonicum**
N	Kent (S E); Sussex	**a. ssp. binervosum**
N	Kent (N E)	**b. ssp. cantianum**
N	Devon	**e. ssp. mutatum**
N	Channel Islands	**f. ssp. sarniense**
BDFGILMP	Pembs	**12. L. parvum**
BDFHI(KL)NP		**10. L. procerum**
	Leaves mostly >8mm wide; branches and stems ± straight; E Sussex and widespread in W	**a. ssp. procerum**
	Leaves mostly >8mm wide; branches and stems wavy; S Devon	**b. ssp. devoniense**
	Leaves mostly <8mm wide; branches and stems ± straight; Pembs	**c. ssp. cambrense**

8. L. binervosum (G. E. Sm.) C.E. Salmon. Stems relatively tall, to 50(70)cm, R
well-branched from low down, often with many sterile branches below, smooth at 199
least below; leaves 7-22mm wide; spikes (8)13-25(45)mm, rather lax, with 2-5(6) spikelets in lowest cm; outer bract 2.1-3mm; inner bract 4-5.3mm. The only sp. in E En, and the only sp. in NW France.

a. Ssp. binervosum. Stems to 50(70)cm, mostly <2.5mm thick, with only sterile or sterile and fertile branches in lower half; leaves usually <10mm wide, acute at apex; 2n=35. Chalk cliffs and salt-marshes; E Sussex and S E Kent.

b. Ssp. cantianum Ingr. Stems to 50(70)cm, mostly 2.5-3.5mm thick, with only sterile branches in lower half; leaves usually >10mm wide, acute to rounded at apex; 2n=35. Chalk cliffs and salt-marshes; N E Kent; endemic.

c. Ssp. anglicum Ingr. Stems to 40(50)cm, mostly 2.5-3.5mm thick, with mostly fertile branches in lower half; leaves usually >10mm wide, acute to obtuse at apex; 2n=35. Salt-marshes; E Norfolk to S N Lincs; endemic.

d. Ssp. saxonicum Ingr. Stems to 50(60)cm, mostly >3mm thick, with mostly fertile branches in lower half; leaves usually >10mm wide, acute to obtuse at apex; 2n=35. Salt-marshes; N Essex; endemic.

e. Ssp. mutatum Ingr. Stems to 30(40)cm, mostly <2.5mm thick, with mostly fertile branches in lower half; leaves usually >10mm wide, acute at apex. Seaside rocks; Lannacombe, S Devon; endemic.

f. Ssp. sarniense Ingr. Stems to 40(50)cm, mostly <2.5mm thick, with only sterile branches in lower half; leaves usually <10mm wide, acute at apex; 2n=35. Seaside cliffs and rocks; frequent in CI (all main islands).

9. L. paradoxum Pugsley. Stems relatively short, to 20(32)cm, rarely >2mm thick, **RR**
with short branches, with few or no sterile branches, smooth below but often rough **198**
above; leaves 3-13mm wide, obtuse at apex; spikes (5)10-25(33)mm, rather lax or dense, with 1-few spikelets in lowest cm; outer bract 2.4-3.7mm; inner bract 3.6-5.4mm.; the outer bract concealing inner bract near end of spike is diagnostic; 2n=33, 36. Rocks; St David's Head, Pembs; endemic.

10. L. procerum (C.E. Salmon) Ingr. Stems relatively tall, to 50(70)cm, mostly 3- **R**
3.5mm thick, usually branched only in upper half and with few sterile branches, **199**
smooth; leaves 5-25mm wide, acute to obtuse at apex; spikes (5)10-25(40)mm, lax or dense, with (3)5-9 spikelets in lowest cm; outer bract 2.3-3.6mm; inner bract 4.7-5.9mm. Endemic.

a. Ssp. procerum. Often >26cm; stems and branches straight; leaves ≥8mm wide; spikes lax or dense, with (3)5-9 spikelets in lowest cm; 2n=35, 36. Cliffs and salt-marshes; E Ir, Wa, SW En, and Rottingdean, E Sussex (possibly intrd in last).

b. Ssp. devoniense Ingr. Often >26cm; stems and branches wavy; leaves ≥8mm wide; spikes lax, with 3-5 spikelets in lowest cm; 2n=35. Cliffs; near Torquay, S Devon.

c. Ssp. cambrense Ingr. Never >26cm; stems and branches straight; leaves <8mm wide; spikes dense, with 5-9 spikelets in lowest cm; 2n=35. Limestone cliff near Pembroke, Pembs.

11. L. britannicum Ingr. Stems relatively short, to 30cm, up to 1.6mm thick, **R**
branched in lower half or not, with 0 to several sterile branches, smooth at least **198**
below; leaves up to 17(21)mm wide, obtuse at apex; spikes (4)7-15(29)mm, dense to somewhat lax, with (4)6-8(10) spikelets in lowest cm; outer bract 2.1-3.3mm; inner bract 4.1-5.3mm; leaves shorter and broader than in *L. procerum*. Endemic.

a. Ssp. britannicum. Stems branched in lower half, with several sterile branches; spikes up to 14mm, dense; 2n=36. Cliffs and promontories; N coast of E & W Cornwall.

b. Ssp. coombense Ingr. Stems unbranched in lower half, with very few sterile branches; spikes up to 20mm, dense; 2n=35. Cliffs; S Devon and S coast of E Cornwall.

c. Ssp. transcanalis Ingr. Stems unbranched in lower half, with 0 or very few sterile branches; spikes up to 19mm, dense; 2n=35. Cliffs, pebble beach and salt-marsh; N Devon and Pembs.

d. Ssp. celticum Ingr. Stems branched in lower half, with several sterile branches; spikes up to 29mm, usually lax; 2n=35, 36. Cliffs and salt-marshes; Anglesey, Caerns, Cheshire, W Lancs and Westmorland.

12. L. parvum Ingr. Stems very short and thin, to 7(13.5)cm, up to only 0.8mm **RR**
thick, with few branches and 0 to very few sterile ones, smooth; leaves up to **198**
9.5mm wide, acute at apex; spikes (5)8-19mm, dense, with 4-10 spikelets in lowest cm; outer bract 1.9-2.6mm; inner bract 3.5-4.9mm; a very small and delicate plant; 2n=35. Limestone cliff; Saddle Point, Pembs; endemic.

13. L. loganicum Ingr. Stems of medium size, to 35cm, up to 1(1.5)mm thick, **RR**
well-branched, with sterile and fertile branches in lower half, rough; leaves up to **198**

10mm wide, rounded at apex; spikes 9-24mm, rather lax, with 3-7 spikelets in lowest cm; outer bract 1.9-2.7mm; inner bract 4-4.5mm; 2n=36. On rocks and scree; S of Land's End, W Cornwall; endemic.

14. L. transwallianum (Pugsley) Pugsley. Stems of medium size, up to 39cm, up **RR**
to 1.2mm thick, often well-branched with fertile branches in lower half but sterile **198**
branches usually 0, smooth; leaves up to 12mm wide but often relatively long, obtuse at apex; spikes up to 19mm, dense, with often >10 spikelets in lowest cm; outer bract 1.8-2.2mm; inner bract 3.1-4.2mm; the petals only 1.2-1.5mm wide (≥1.9mm wide in all other taxa) are diagnostic; 2n=35. Limestone cliffs; Giltar Point, Pembs; endemic.

15. L. dodartiforme Ingr. Stems relatively tall, to 40cm, up to 2.8mm thick, well- **RR**
branched from lower half, with 0 to few sterile branches, rough at least above; **199**
leaves up to 22mm wide, obtuse at apex; spikes 10-35mm, rather lax, with 3-6(7) spikelets in lowest cm; outer bract 2.4-3.2mm; inner bract 4.2-6.1mm; the obovate-spathulate leaves to 22mm wide are diagnostic; 2n=35. Chalk cliffs and shingle; Dorset; endemic.

16. L. recurvum C.E.Salmon. Stems relatively short to medium, to 36cm, 1.2- **RR**
1.7mm thick, branched in lower half or not, with or without sterile branches, rough; **198**
leaves up to 13.5mm wide, obtuse to rounded at apex; spikes 7-24(29)mm, dense, with 3-11 spikelets in lowest cm; outer bract 2.3-3.7mm; inner bract 4.2-5.7mm. The most westerly and most northerly sp.; endemic.

a. Ssp. recurvum. Stems to 17(26)cm, with or without sterile branches in lower half; spikes 10-24mm; petals 2.5-3mm wide; 2n=27. Low limestone cliffs; tip of Portland, Dorset.

b. Ssp. portlandicum Ingr. Stems to 36cm, with sterile branches in lower half; spikes 8-24mm; petals 2.7-3mm wide; 2n=27. Limestone and salt-marshes; elsewhere on Portland, Dorset; Barrow Harbour and Banna Strand, N Kerry.

c. Ssp. pseudotranswallianum Ingr. Stems to 13cm, with or without sterile branches in lower half; spikes 8-16mm; petals 2.2-2.6mm wide; 2n=27. Limestone cliffs and rocks; Co Clare.

d. Ssp. humile (Girard) Ingr. Stems to 35cm, with sterile branches in lower half; spikes 7-23mm; petals 2-2.3(2.7)mm wide; 2n=27. Cliffs and scree; E & W Donegal, St Bees Head (Cumberland), Galloway (Wigtowns).

2. ARMERIA Willd. - *Thrifts*

Leaves numerous, very narrow; aerial stems unbranched, terminating in dense hemispherical inflorescence with tubular sheath of fused bracts beneath; flowers pink (rarely white); styles pubescent below.

Both species are dimorphic and self-incompatible, with pollen- and stigma-types as in *Limonium vulgare.*

Other spp. - A. pseudarmeria (Murray) Mansf. (*Estoril Thrift*), from Portugal, is grown in gardens; it was formerly natd on seaside cliffs in S Hants, but is now extinct. It differs from *A. alliacea* in being even more robust with flower-heads 3-5cm across and calyx-teeth with terminal points >1mm.

1. A. maritima Willd. - *Thrift.* Leaves linear, <2mm wide, 1(-3)-veined, usually with hairs at least on margins; flower-heads 15-25mm wide; calyx teeth ≤1mm incl. very short mucro.

a. Ssp. maritima. Stems usually pubescent, to 30cm; bract-sheath ≤15mm; outermost bracts (excl. sheath) shorter than inner; 2n=18. Native; salt-marshes, saline turf, rocks and cliffs by sea, and inland on mountain rocks; common round coasts of BI and on mountains in N Wa, N En and Sc.

b. Ssp. elongata (Hoffm.) Bonnier. Stems glabrous, the longer 20-55cm; bract- **RR**
sheath 12-25mm; outermost bracts (excl. sheath) usually as long as or slightly longer than inner; 2n=18. Native; lowland rough pasture; near Ancaster, S Lincs,

formerly elsewhere in S Lincs and Leics.

Other variants, notably those inland on mountains or on serpentine or heavy-metal spoil-heaps, are not sufficiently distinct for taxonomic recognition.

1 x 2. A. maritima x A. arenaria occurs in W Jersey where habitats of the parents meet. It is fertile and intermediate in leaf width, pubescence, and length of outer bracts and calyx-teeth; endemic.

2. A. arenaria (Pers.) Schult. (*A. alliacea* auct. non (Cav.) Hoffmanns. & Link) - *Jersey Thrift*. Leaves linear-oblanceolate, some >3mm wide, 3-5(7)-veined, glabrous; stems glabrous, to 60cm; flower-heads 20-30mm wide; bract-sheath 20-40mm; outermost bracts longer than inner, often ± leaf-like and longer than flower-head; calyx-teeth 1.5-2mm, incl. terminal point c.1/2 that; (2n=18). Native; fixed dunes; W & S Jersey. **RR**

51. PAEONIACEAE - *Peony family*

Perennial herbs; some roots strongly tuberous; leaves basal and spiral, (bi)pinnate to (bi)ternate, the segments entire, petiolate, without stipules. Flowers large, solitary or few per stem, terminal, bisexual, hypogynous, actinomorphic; sepals 5, free; petals 5-8, free, red, less often white or pink; stamens very numerous; carpels 2-5, free, each with several ovules; style 0; stigmas red or pink, hooked or coiled; fruit a follicle with black fertile and red sterile seeds.

Unmistakable flowers and compound leaves.

1. PAEONIA L. - *Peonies*

Other spp. - P. lactiflora Pall., from E Asia, has been found in similar situations to *P. officinalis* but rarely recently; it differs from *P. mascula* and *P. officinalis* in its yellow filaments, white or pink petals and flowers ≥2 per stem, and has glabrous to pubescent follicles.

1. P. mascula (L.) Mill. - *Peony*. Stems to 60cm, simple, erect, usually several with several basal leaves forming a clump; basal leaves mostly 2-ternate, with 9(-20) elliptic to ovate segments; flowers 8-14cm across, 1 per stem; petals and filaments red; follicles (2)3-5, 2-5cm, tomentose, ± recurved; (2n=20). Intrd; natd on limestone on Steep Holm Island, N Somerset since at least 1803, on Flat Holm Island, Glam since 1980s; S Europe. Other records are errors, especially for *P. officinalis.*

2. P. officinalis L. - *Garden Peony*. Differs from *P. mascula* in leaves more subdivided, with c.17-30 narrowly elliptic to lanceolate segments; follicles 2-3, 2-3.5cm; (2n=20). Intrd; much grown in gardens, persistent relic or throwout in rough and marginal ground; scattered in En and N to C Sc; S Europe.

52. ELATINACEAE - *Waterwort family*

Small ± aquatic annuals; leaves opposite, simple, entire, ± petiolate, stipulate. Flowers minute, solitary in leaf-axils, bisexual, hypogynous, actinomorphic; sepals 3-4, membranous, free or united at base; petals 3-4, free, pinkish-white; stamens 2x as many as petals; ovary 3-4-celled, each with numerous ovules on axile placentation; styles 3-4, very short; stigmas capitate; fruit a 3-4-celled capsule.

Superficially like Portulacaceae, but differs in many floral characters; distinguished from Caryophyllaceae by the completely compartmented ovary and herbaceous stipules (absent or scarious in Caryophyllaceae).

1. ELATINE L. - *Waterworts*

1. E. hexandra (Lapierre) DC. - *Six-stamened Waterwort*. Stems procumbent, R
rooting at nodes, to 10(20)cm; leaves c.6mm, elliptic to spathulate; pedicels at least as long as flowers; sepals and petals 3(-4); stamens 6(8); capsule with 3(-4) valves; (2n=72). Native; in ponds and on wet mud; very local and widely scattered in BI, mostly in W & SE. Can live for several years submerged without flowering.

2. E. hydropiper L. - *Eight-stamened Waterwort*. Like *E. hexandra* but pedicels 0 or R
very short; sepals and petals 4; stamens 8; capsule with 4 valves; (2n=40). Native; in ponds and small lakes; rare and very local in C Sc, NW Wa, WC En and NE Ir, formerly S En.

53. CLUSIACEAE - *St John's-wort family*
(Guttiferae, Hypericaceae)

Perennial or rarely annual herbs or small shrubs; leaves opposite, simple, ± entire, with translucent and/or coloured glands, without stipules; petioles absent or very short. Flowers solitary or in terminal cymes, actinomorphic, bisexual, hypogynous; sepals 5, free, often glandular; petals 5, free, yellow, often glandular; stamens numerous, usually partially fused into 3 or 5 bundles; ovary 1-, 3- or 5-celled, each cell with many ovules; styles 3 or 5; stigmas capitate; fruit a capsule or becoming succulent and berry-like.

Easily recognized by entire, opposite leaves without stipules, 5 yellow petals, and numerous stamens grouped in bundles (except in *H. canadense*).

1. HYPERICUM L. - *St John's-worts*

1 Stems and leaves conspicuously pubescent 2
1 Stems glabrous; leaves glabrous to slightly pubescent 3
 2 Stems procumbent and rooting at nodes below, soft; glands on sepals reddish **19. H. elodes**
 2 Stems erect, rooting only near base, stiff; glands on sepals black **17. H. hirsutum**
3 Glands on leaves, sepals and petals all translucent and inconspicuous, or 0 4
3 Some glands on leaves, sepals and/or petals black 11
 4 Delicate herb; leaves rarely >1.5 x 0.5cm **20. H. canadense**
 4 Stems woody; leaves rarely <2 x 1cm 5
5 Styles 5; petals >(16)20mm; stamens c.1/2-3/4 as long as petals 6
5 Styles 3; petals <20mm; stamens c. as long as or longer than petals 8
 6 Rhizomatous; aerial stems to 60cm, not or little branched **1. H. calycinum**
 6 Not rhizomatous; aerial stems to 1.7m, much branched 7
7 Sepals acute to obtuse; stems persistently 4-ridged **2. H. pseudohenryi**
7 Sepals rounded at apex, sometimes apiculate; stems soon becoming terete **3. H. forrestii**
 8 Petals >15mm; styles >3x as long as ovary at flowering; leaves strongly smelling of goats when bruised **6. H. hircinum**
 8 Petals <15mm; styles <3x as long as ovary at flowering; leaves often scented but without strong goat-like smell when bruised 9
9 Petals shorter than to c. as long as sepals; styles <5mm, c.1/2 as long as ovary; fruit succulent when ripe **4. H. androsaemum**
9 Petals longer than sepals; styles >5mm, at least as long as ovary; fruit dry when ripe 10
 10 Petals <4.5mm wide; sepals <2.5mm wide, not overlapping laterally **7. H. xylosteifolium**

10 Petals >4.5mm wide; sepals >2mm wide, overlapping laterally **5. H. x inodorum**
11 Stems with 4 ridges, sometimes 2 strong and 2 weak; ridges often winged 12
11 Stems with 0-2 ridges; ridges not winged 15
12 Stems with 2 weak and 2 strong ridges **9. H. x desetangsii**
12 Stems square in section, with 4 strong ridges 13
13 Petals pale yellow, not red-tinged, <7.5mm, c. as long as sepals; stems broadly winged (0.25-0.5mm) **12. H. tetrapterum**
13 Petals bright yellow, sometimes red-tinged, >7.5mm, ≥2x as long as sepals; stems not or narrowly winged (≤0.25mm) 14
14 Stems not winged; sepals obtuse **10. H. maculatum**
14 Stems narrowly winged; sepals acute **11. H. undulatum**
15 Leaves at least sparsely pubescent on lowerside **18. H. montanum**
15 Leaves glabrous 16
16 Sepals unequal, 3 longer and wider than the other 2; petals <2x as long as sepals **13. H. humifusum**
16 Sepals ± equal; petals ≥2x as long as sepals 17
17 Stems not ridged; leaves without black glands 18
17 Stems with 2 ridges; leaves with black glands near margin 19
18 Leaves on main stem triangular-ovate, widest near base, ± sessile, ± clasping stem **15. H. pulchrum**
18 Leaves broadly elliptic to orbicular, widest near middle, distinctly petiolate, not clasping stem **16. H. nummularium**
19 Margins of sepals fringed with stalked black glands; leaves ± without translucent glands **14. H. linariifolium**
19 Margins of sepals without stalked black glands; leaves with several to numerous translucent glands 20
20 Margins of sepals entire, acute at apex **8. H. perforatum**
20 Margins of sepals denticulate towards apex, with an apical apiculus **9. H. x desetangsii**

Other spp. - A few other spp. grown in gardens as shrubs or rock-garden plants may be found as relics or throwouts. **H. 'Hidcote'**, a sterile cultivar of unknown, probably hybrid origin, is commonly mass-planted in public places. It differs from *H. pseudohenryi* in its stamens being <1/2 (not c.3/4) as long as petals and in its rounded (not acuminate to obtuse) sepals, and from *H. forrestii* in its styles longer (not shorter) than ovary and orange (not yellow) anthers. **H. olympicum** L. (sect. *Olympia* (Spach) Nyman), from the Balkans, is a densely bushy plant with stems to 50cm and woody near base, glaucous elliptic-oblong leaves without black glands, and flowers 25-60mm across; it has been found self-sown in a few places in En and Sc.

Section* 1. - *ASCYREIA Choisy (sect. *Eremanthe* (Spach) Boiss.) (spp. 1-3). Glabrous shrubs; black glands 0; sepals entire; petals and stamens deciduous after anthesis; stamens in 5 bundles; styles 5.

1. H. calycinum L. - *Rose-of-Sharon*. Strongly rhizomatous glabrous evergreen shrub with erect little-branched 4-lined stems to 60cm; flowers 5-8cm across. Spreads vegetatively but rarely sets seed; (2n=20). Intrd; cultd in gardens and mass planted in parks and on roadsides; natd on hedgebanks and in shrubberies usually near gardens; scattered through most of Br and Ir, especially S; Turkey and Bulgaria.

2. H. pseudohenryi N. Robson - *Irish Tutsan*. Erect semi-evergreen glabrous shrub with much-branched 4-ridged stems to 1.7m; flowers 3-5.5cm across. Intrd; natd in riverside woodland in W Cork; China.

3. H. forrestii (Chitt.) N. Robson - *Forrest's Tutsan*. Erect semi-evergreen glabrous shrub with much-branched ± terete stems to 1.5m; flowers 3.5-6cm across; (2n=36, 38). Intrd; persistent garden escape in rocky or shady places; scattered in Br N to Kirkcudbrights; China.

Section* 2 - *ANDROSAEMUM (Duhamel) Godr. (spp. 4-6). Glabrous shrubs; black glands 0; sepals entire; petals and stamens deciduous after anthesis; stamens in 5 bundles; styles 3.

4. H. androsaemum L. - *Tutsan*. Erect ± deciduous glabrous shrub with branched 2-ridged stems to 0.8m; flowers 1.5-2.5cm across; stamens c. as long as petals; 2n=40. Native; damp woods and shady hedgebanks; locally frequent throughout most of BI, especially in W, but almost absent from NE Sc; cultivated and sometimes natd.

5. H. x inodorum Mill. (*H. x elatum* Aiton; *H. androsaemum* x *H. hircinum*) - *Tall Tutsan*. Erect ± deciduous glabrous shrub with much-branched 2-ridged stems to 2m; flowers 1.5-3cm across; stamens slightly longer than petals; partially fertile; (2n=40). Intrd; grown in gardens and mass-planted in parks and on roadsides (especially cv. 'Elstead' for its red fruits); rarely natd in shady places; very sparsely scattered throughout much of BI, especially in SW; probably spontaneous in SW Europe but not in BI.

6. H. hircinum L. - *Stinking Tutsan*. Erect ± deciduous glabrous shrub with much-branched 4-ridged stems to 1.5m; flowers 2.5-4cm across; stamens slightly longer than petals; (2n=40). Intrd; grown in gardens and rarely natd in shady places; scattered through much of Bl, especially SW; Mediterranean. Our plant is ssp. **majus** (Aiton) N. Robson.

Section* 3 - *INODORA Stef. (sp. 7). Glabrous shrubs; black glands 0; sepals entire; petals and stamens persistent after anthesis; stamens in 5 bundles; styles 3.

7. H. xylosteifolium (Spach) N. Robson - *Turkish Tutsan*. Rhizomatous evergreen glabrous shrub; stems to 1.5m, erect, much-branched, 4-lined; flowers 1.5-3cm across; stamens c. as long as petals; (2n=40). Intrd; sometimes grown in gardens and natd in 3 shady places in W Lancs and MW Yorks; Caucasus.

Section* 4 - *HYPERICUM (spp. 8-12). Glabrous herbs with 2-4-lined or 4-winged stems; black glands present on stems, leaves and flowers; petals with surface black glands; sepals entire to glandular-denticulate with round-topped glands; petals and stamens persistent after anthesis; stamens in 3 bundles; styles 3; seeds reticulate-pitted.

8. H. perforatum L. - *Perforate St John's-wort*. Erect glabrous rhizomatous perennial with 2-ridged stems to 80cm; leaves with abundant translucent glands; flowers 15-25mm across; sepals entire, acute at apex; petals bright yellow; black glands few on sepals and petals, sessile; 2n=32. Native; dryish grassland, banks and open woodland; common throughout most of BI and the commonest *Hypericum* in En, but rare in N Sc. Very variable, especially in leaf-shape.

9. H. x desetangsii Lamotte (*H. perforatum* x *H. maculatum*) - *Des Etangs' St John's-wort*. Partially fertile and backcrossing to give a range of intermediates; variously intermediate between its parents (especially in its 2-4 stem-ridges, rather few translucent leaf-glands, and denticulate sepal margins with an apiculate apex); often found in absence of 1 or both parents. Native; grassland of varying dampness; sparsely scattered through En, Wa and S Sc but easily overlooked. All but 2 records refer to nothossp. **desetangsii**, 2n=32, with *H. maculatum* ssp. *obtusiusculum* as 1 parent; nothossp. **carinthiacum** (A. Fröhl.) N. Robson (2n=24, 40), with *H. maculatum* ssp. *maculatum* as parent, was found in Kintyre in 1899 and Lanarks in 1984. The latter nothossp. has less widely branched stems and entire sepals.

10. H. maculatum Crantz - *Imperforate St John's-wort.* Erect rhizomatous glabrous perennial; stems to 60cm, square in section, wingless; leaves with 0 or very few translucent glands; flowers 15-25mm across; petals bright yellow; black glands on leaves, petals and sometimes sepals, sessile. Native; grassy, usually moist places.

a. Ssp. maculatum. Inflorescence-branches arising at c.30°; black glands on petals mainly as superficial dots; sepals (1.7)2-3mm wide, entire; 2n=16. Scattered in C Sc; rare and very scattered in En and perhaps intrd there. **R**

b. Ssp. obtusiusculum (Tourlet) Hayek (*H. dubium* Leers). Inflorescence-branches arising at c.50°; black glands on petals mainly as superficial lines; sepals 1.2-2mm wide, denticulate at edges; 2n=32. Scattered over most of Br and Ir and locally frequent; much commoner than ssp. *maculatum* and often treated as a separate sp.

11. H. undulatum Schousb. ex Willd. - *Wavy St John's-wort.* Erect rhizomatous glabrous perennial; stems to 60cm, square in section, with narrow wings ≤0.25mm wide; leaves undulate, with abundant translucent glands; flowers 12-20mm across; petals bright yellow, red-tinged on lowerside; black glands on leaves, sepals and sometimes petals, sessile; 2n=16. Native; marshy fields and streamsides; very local in SW En and W Wa. **R**

12. H. tetrapterum Fr. (*H. quadrangulum* L.) - *Square-stalked St John's-wort.* Erect rhizomatous glabrous perennial; stems to 60cm, square in section, with distinct wings 0.25-0.5mm wide; leaves with many small translucent glands; flowers 9-13mm across; petals pale yellow; black glands on leaves and sometimes petals or sepals, sessile; 2n=16. Native; marshes, riverbanks and damp meadows; frequent throughout BI except N Sc.

Section* 5 - *OLIGOSTEMMA (Boiss.) Stef. (spp. 13-14). Glabrous herbs with 2-lined stems; black glands present on leaves and flowers; sepals glandular-denticulate with round-topped glands; petals with marginal black glands; petals and stamens persistent after anthesis; stamens in 3 bundles; styles 3; seeds reticulate-pitted.

13. H. humifusum L. - *Trailing St John's-wort.* Procumbent to ascending glabrous perennial with thin 2-lined stems to 20cm; leaves ≤15mm, obovate or ovate to oblanceolate or lanceolate; flowers 8-12mm across; petals <2x as long as sepals, bright yellow; black glands on leaves and rather sparse on sepals and petals, sessile and/or stalked; 2n=16. Native; open woods, hedgebanks and dry heathland mostly on acid soils; frequent throughout most of BI but rare in N Sc.

13 x 14. H. humifusum x H. linariifolium is the probable parentage of intermediates (notably in petal length) that occur in CI with both parents. Possibly endemic.

14. H. linariifolium Vahl - *Toadflax-leaved St John's-wort.* Erect to ascending glabrous perennial with thin faintly 2-lined stems to 40cm; leaves ≤30mm, linear-lanceolate to linear; flowers 15-20mm across, petals ≥2x as long as sepals, bright yellow; black glands sessile on leaves, stalked on sepals and petals; 2n=16. Native; rocky acid slopes; rare in SW En, Wa and CI. **RR**

Section* 6 - *TAENIOCARPIUM Jaub. & Spach (spp. 15-17). Glabrous or pubescent herbs with terete stems; black glands present only on flowers except 2 at leaf-apex in *H. nummularium*; sepals glandular-denticulate with round-topped glands; petals with marginal black glands; petals and stamens persistent after anthesis; stamens in 3 bundles; styles 3; seeds rugose to papillose.

15. H. pulchrum L. - *Slender St John's-wort.* Erect glabrous perennial with thin stems to 60cm; main stem-leaves triangular-ovate; flowers 12-18mm across; petals bright yellow, red-tinged on lowerside; black glands on sepals and petals stalked; 2n=18. Native; dry open woodland, hedgebanks and heathland, usually on acid soils; locally common throughout BI.

16. H. nummularium L. - *Round-leaved St John's-wort.* Decumbent to erect glabrous perennial with thin stems to 30cm; leaves broadly elliptic to orbicular; flowers 18-30mm across; petals bright yellow, red tinged; black glands on sepals and petals ± stalked, on leaves sessile. Intrd; natd on limestone quarry rock-ledge in NW Yorks probably since c.1920; mountains of SW Europe.

17. H. hirsutum L. - *Hairy St John's-wort.* Erect pubescent perennial with stems to 1m; flowers 15-22mm across; petals pale yellow; black glands on sepals and petals stalked; 2n=18. Native; open woodland, river-banks and damp grassland; frequent throughout most of Br but rare in W, very local in Ir.

Section 7 - ADENOSEPALUM Spach (sp. 18). ± glabrous herbs with terete stems; black glands present on leaves and sepals; sepals glandular-dentate with flat-topped glands; petals and stamens persistent after anthesis; stamens in 3 bundles; styles 3; seeds reticulate-pitted.

18. H. montanum L. - *Pale St John's-wort.* Erect perennial to 1m; petals pale yellow; black glands sessile on leaves, stalked on sepals; (2n=16). Native; open woodland, hedgebanks and rocky slopes, usually on calcareous soils; local but widespread in En and Wa.

Section 8 - ELODES (Adans.) W.D.J. Koch (sp. 19). Pubescent herbs with terete stems; black glands 0; red glands present on sepals; sepals glandular-denticulate with flat-topped glands; petals with 3-fid nectary at base; petals and stamens persistent after anthesis; stamens in 3 bundles, alternating with 3 short processes; styles 3; seeds longitudinally ribbed.

19. H. elodes L. - *Marsh St John's-wort.* Densely pubescent stoloniferous perennial with ascending to erect flowering stems to 40cm; flowers 12-20mm across; red glands on sepals stalked; 2n=16, 32. Native; bogs, pondsides and streamsides on acid soil; locally frequent in suitable places throughout BI in W & S, but rare or absent in C & E.

Section 9 - SPACHIUM (R. Keller) N. Robson (sp. 20). Glabrous herbs with 4-lined stems; black and red glands 0; sepals entire; petals and stamens persistent after anthesis; stamens in a continuous narrow ring; styles 3; seeds longitudinally ribbed.

20. H. canadense L. - *Irish St John's-wort.* Glabrous annual or perennial with basal buds; stems to 20cm, erect, very slender, square in section; flowers <1cm across; petals bright yellow; 2n=16. Probably intrd; very rare in 2 areas in W and SW Ir; N America. **RR**

54. TILIACEAE - *Lime family*

Trees; leaves deciduous, alternate, simple, broadly ovate, cordate, serrate, petiolate, with stipules falling early. Flowers fragrant, 2-25 in cymes whose stalk is fused to large, narrowly oblong papery persistent bracteole dispersed with fruit, actinomorphic, bisexual, hypogynous; sepals and petals 5 each, free, creamish-yellow; stamens numerous, ± coherent in 5 bundles; ovary 5-celled, each cell with 2 ovules; style 1; stigma capitate, ± lobed; fruit a nut with 1-3 seeds.

The persistent bracteole is diagnostic.

1. TILIA L. - *Limes*

1 Leaves pubescent on lowerside; flowers 2-4(6) per cyme; fruit strongly

ribbed **1. T. platyphyllos**
1 Leaves glabrous except for dense hair-tufts in vein axils on lowerside; flowers 4-15 per cyme; fruit not or slightly ribbed 2
  2 Cymes held obliquely erect above foliage; leaves mostly 3-6cm, with scarcely prominent tertiary veins on upperside **3. T. cordata**
  2 Cymes pendent among foliage; leaves mostly 6-9cm, with prominent tertiary veins on upperside **2. T. x europaea**

Other spp. - Several spp. may be found planted in parks, etc. **T. 'Petiolaris'** (*Pendent Silver-lime*), easily the commonest, has densely pubescent young twigs, whitish-tomentose lowerside to leaves and petioles >1/2 as long as leaves; **T. tomentosa** Moench (*Silver-lime*), from SE Europe, is similar to *T.* 'Petiolaris' but has short petioles <1/2 as long as leaves and much less pendent branches; and **T. x euchlora** K. Koch (*Caucasian Lime*), from Caucasus, differs from *T. x europaea* in its glossy leaf upperside. *T. tomentosa* has been reported as producing seedlings.

1. T. platyphyllos Scop. - *Large-leaved Lime*. Tree to 34m; young twigs pubescent; leaves mostly 6-12cm, thinly pubescent on lowerside, especially on veins; cymes pendent among leaves, with 2-4(6) flowers; (2n=82). Native; in woods and copses on base-rich soils; very local and scattered in En and Wa, more widespread as intrd plants but status often doubtful. Our plant is said to be ssp. **cordifolia** (Besser) C.K. Schneid., but perhaps not all intrd plants are this. R

2. T. x europaea L. (*T. x vulgaris* Hayne; *T. platyphyllos* x *T. cordata*) - *Lime*. Tree to 46m; young twigs soon glabrous; leaves mostly 6-9cm, glabrous or ± so on lowerside except for vein-axils; cymes pendent among leaves, with 4-10 flowers; (2n=82). Native; rare, in a few woods with both parents from Herefs to NE Yorks, widely planted and sometimes natd throughout BI; 1 of the commonest planted trees. Partially fertile. R

3. T. cordata Mill. - *Small-leaved Lime*. Tree to 38m; young twigs soon glabrous; leaves mostly 3-6cm, glabrous on lowerside except for vein-axils; cymes obliquely erect, held above foliage, with 4-10(15) flowers; 2n=82, 86. Native; woods on rich soils; En (mostly C) and Wa, locally common as a native and also planted and ± natd more widely.

55. MALVACEAE - *Mallow family*

Annual to perennial herbs or sometimes shrubs; leaves alternate, usually palmately veined, often palmately lobed and sometimes ± palmate, petiolate, stipulate. Flowers in racemes or small panicles or solitary and axillary, actinomorphic, bisexual or sometimes gynomonoecious, hypogynous, usually with an epicalyx of 3-c.13 sepal-like segments below calyx; sepals 5, free or fused below; petals 5, ± free but often fused at extreme base, mostly pink to purple; stamens numerous, the filaments united below into tube, divided above with each branch bearing a 1-celled anther-lobe; ovary 5-many-celled, each cell with 1-many ovules; styles free or united below, 5-many; stigmas linear or capitate; fruit a capsule or breaking into 1-several-seeded nutlets.

Easily recognized by the stamens united into a tube round the carpels.

1 Epicalyx absent 2
1 Calyx-like epicalyx present below true calyx 5
  2 Flowers pink to purple; stigmas linear **7. SIDALCEA**
  2 Flowers yellow or white; stigmas capitate 3
3 Leaves palmately veined; nutlets with several seeds **8. ABUTILON**
3 Leaves pinnately veined; nutlets 1-seeded 4
  4 Petals white; calyx not curved over fruit; carpels winged **2. HOHERIA**

4 Petals yellow; calyx curved over fruit; carpels not winged **1. SIDA**
5 Carpels 5; fruit a capsule **9. HIBISCUS**
5 Carpels ≥6; fruit breaking into nutlets 6
6 Epicalyx-segments 6-10 7
6 Epicalyx-segments 3 8
7 Staminal-tube terete, pubescent **5. ALTHAEA**
7 Staminal-tube 5-angled, glabrous **6. ALCEA**
8 Epicalyx-segments free to base **3. MALVA**
8 Epicalyx-segments fused below **4. LAVATERA**

Other genera - **MALOPE** L. would key to *Malva* but the mericarps form a globose head and the epicalyx-segments are wider than the sepals; **M. trifida** Cav. (*Mallow-wort*), from W Mediterranean, an annual to 1.5m with mallow-like flowers, is a rare wool- and birdseed-alien. **ANODA** Cav. would key to couplet 2 but has blue to mauve flowers, capitate stigmas and 1-seeded nutlets; **A. cristata** (L.) Schltdl., from SE N America, is an erect annual to 80cm found as a rare birdseed-, wool- and oilseed-alien.

1. SIDA L. - *Queensland-hemps*
Annuals to perennials, woody or not; epicalyx 0; petals yellowish; carpels 5-14; stigmas capitate; fruit breaking into numerous 1-seeded nutlets.

Other spp. - **S. cordifolia** L., from tropics, is only slightly woody and has pedicels shorter than petioles and leaves broadly rounded-cordate at base as in *S. spinosa*, but has 8-12 carpels and flowers clustered in leaf-axils; it is a rare wool-alien.

1. S. spinosa L. - *Prickly Mallow*. Annual to perennial; stems to 70cm, often woody at base, erect; leaves widest near base, rounded and cordate at base; petioles ≤3cm, those on larger leaves with small spine; carpels 5; (2n=14, 28). Intrd; very local casual on tips, but characteristic introduction with soyabeans and sometimes with birdseed; C & S En; tropics. **213**

2. S. rhombifolia L. - *Queensland-hemp*. Usually woody perennial to 2m, but smaller annual here; leaves widest near middle, narrowed to base but often minutely cordate at extreme base; petiole <1cm; carpels 8-10(14); (2n=14, 28). Intrd; frequent on tips with soyabean waste, but less common than *S. spinosa*; S En; tropics. **213**

2. HOHERIA A. Cunn. - *New Zealand Mallow*
Woody shrubs; epicalyx 0; petals white; carpels 5(-6); stigmas capitate; fruit breaking into numerous 1-seeded nutlets.

1. H. populnea A. Cunn. - *New Zealand Mallow*. Evergreen shrub potentially to 10m; leaves 7-14cm, broadly to narrowly ovate or elliptic, acuminate at apex, rounded to cordate at base, 2-serrate, with petiole 1-2cm; petals c.10mm; (2n=42). Intrd; rare garden plant found as relic or self-sown in old gardens or on rough and marginal ground; Man; New Zealand.

3. MALVA L. - *Mallows*
Annual to perennial herbs; epicalyx of 3 segments free to base; petals pink to purple (or white); carpels numerous; stigmas linear; fruit breaking into numerous 1-seeded nutlets.

Fully mature fruit is essential for correct determination, especially of spp. 5-8.

1 Nutlets smooth or faintly reticulate with low rounded ridges 2
1 Nutlets strongly reticulate with sharp ridges 5
2 Petals bright pink (or white), >16mm, ≥3x as long as sepals; upper

leaves usually very deeply divided 3
2 Petals pinkish- or whitish-mauve, <16mm, <3x as long as sepals; leaves shallowly lobed 4
3 Epicalyx-segments >3x as long as wide; calyx, epicalyx and pedicels with only simple hairs **1. M. moschata**
3 Epicalyx-segments <3x as long as wide; calyx, epicalyx and pedicels with many stellate hairs **2. M. alcea**
4 Nutlets smooth, shortly pubescent; petals usually 2-3x as long as sepals **7. M. neglecta**
4 Nutlets obscurely ridged, glabrous or nearly so; petals usually <2x as long as sepals **8. M.verticillata**
5 Perennial; petals (12)20-30mm, >2x as long as sepals, usually bright pinkish-purple **3. M. sylvestris**
5 Annual or biennial; petals 4-12mm, <2x as long as sepals, usually pale pinkish- or whitish-mauve 6
6 Epicalyx-segments narrowly ovate, ≤3x as long as wide; staminal-tube pubescent **4. M. nicaeensis**
6 Epicalyx-segments linear-lanceolate, >3x as long as wide; staminal-tube glabrous or nearly so 7
7 Calyx with marginal hairs c.1mm; some pedicels >1cm at fruiting; angle of nutlets between dorsal and lateral surfaces sharp but not winged **6. M. pusilla**
7 Calyx glabrous or with marginal hairs <0.5mm; pedicels all <1cm at fruiting; angle of nutlets between dorsal and lateral surfaces forming narrow wavy wing **5. M. parviflora**

1. M. moschata L. - *Musk-mallow*. Perennial with erect stems to 80cm; upper leaves usually very deeply divided, but sometimes not so; petals >16mm, bright pink or white; nutlets smooth, with long hairs, rounded between lateral and dorsal surfaces; 2n=42. Native; grassy banks and fields, especially on rich soils; throughout most of BI but rare in N & W Sc and parts of Ir, probably not native in most of Ir and N Br. **213**

2. M. alcea L. - *Greater Musk-mallow*. Perennial with erect or spreading stems to 1.2m; similar to *M. moschata* but with stellate, not simple, hairs on all vegetative parts; nutlets glabrous or pubescent; and see key; (2n=84). Intrd; grassy and waste places, sometimes natd; SE En and Glam; Europe. **213**

3. M. sylvestris L. - *Common Mallow*. Spreading perennial with erect to decumbent stems to 1m; leaves with shallow rounded lobes; petals >(12)20mm, bright pinkish-purple with dark stripes; nutlets strongly reticulate, usually glabrous, with sharp angle between dorsal and lateral surfaces; 2n=42. Native; waste and rough ground, by roads and railways; common throughout lowland En, Wa and CI, scattered elsewhere. **213**

4. M. nicaeensis All. - *French Mallow*. Annual or biennial with ascending to decumbent stems to 50cm; similar to *M. sylvestris* but petals 10-12mm, pale mauve; and see key; (2n=42, c.64). Intrd; waste places, rather rare casual from wool and other sources; mainly S En, very rare elsewhere; S Europe. **213**

5. M. parviflora L. - *Least Mallow*. Annual with erect to decumbent stems to 50cm; leaves with shallow rounded lobes; petals ≤5mm, pale mauve; nutlets strongly reticulate, glabrous or pubescent, the angle between dorsal and lateral surfaces raised to form narrow wavy wing; (2n=42). Intrd; waste places, casual from several sources, especially wool; very scattered over Br; S Europe. **213**

6. M. pusilla Sm. (*M. rotundifolia* L.) - *Small Mallow*. Similar to *M. parviflora* but see key; nutlets strongly reticulate, glabrous or pubescent, the angle between dorsal and lateral surface sharp and jagged but not winged; (2n=42). Intrd; waste places, casual from several sources, especially wool, sometimes persistent; very scattered over Br, rare in Ir; Europe. **213**

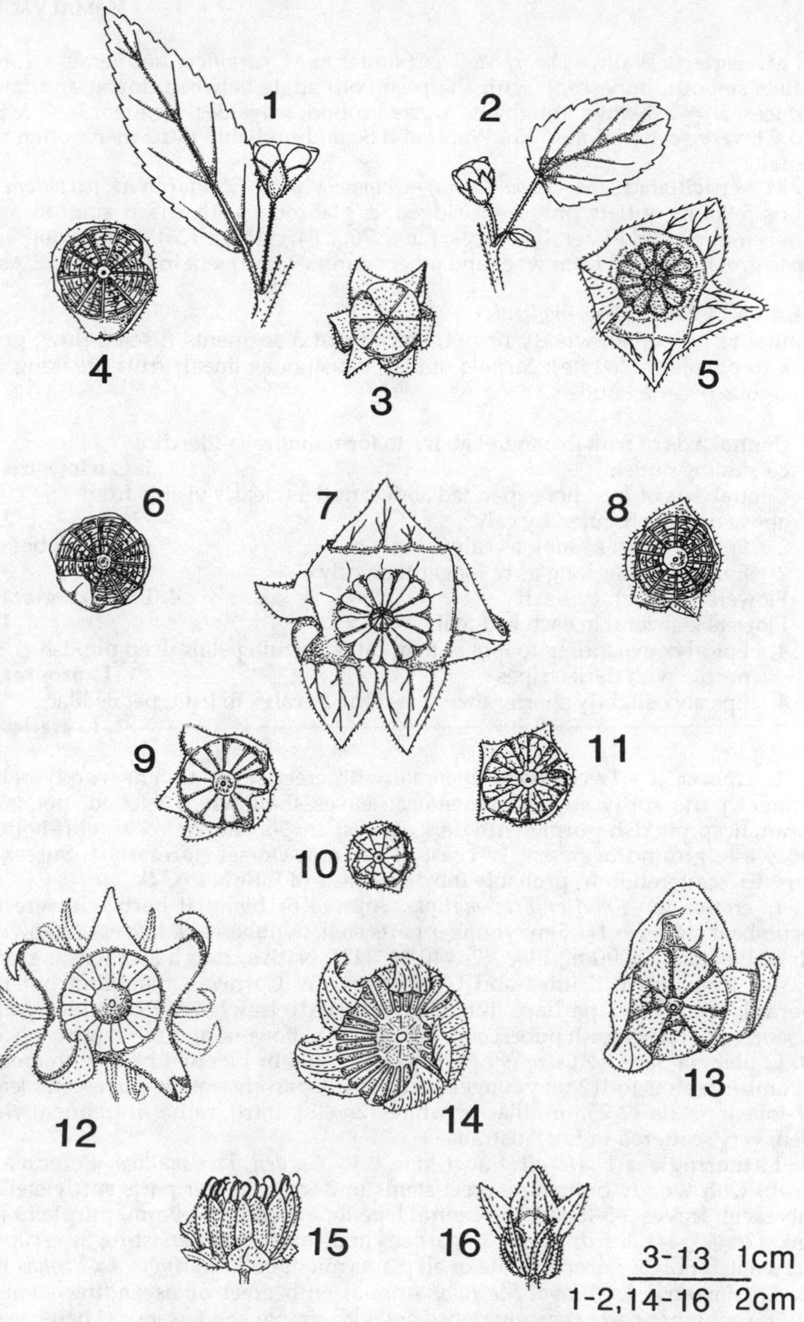

FIG 213 - Malvaceae. 1-2, flowering nodes of *Sida*. 1, ***S. rhombifolia***. 2, ***S. spinosa***. 3-16, fruits. 3, ***Sidalcea malviflora***. 4, ***Malva nicaeensis***. 5, ***M. moschata***. 6, ***M. parviflora***. 7, ***M. alcea***. 8, ***M. sylvestris***. 9, ***M. neglecta***. 10, ***M. verticillata***. 11, ***M. pusilla***. 12, ***Althaea officinalis***. 13, ***Lavatera arborea***. 14, ***Alcea rosea***. 15, ***Abutilon theophrasti***. 16, ***Hibiscus trionum***.

7. M. neglecta Wallr. - *Dwarf Mallow*. Similar to *M. parviflora* but petals 9-13mm; 213
nutlets smooth, pubescent, with sharp smooth angle between dorsal and lateral surfaces; 2n=42. Native; rough and waste ground, waysides; frequent in C & S Br and CI, very scattered in N En, Wa, Ir and Sc and probably intrd there, often only casual.

8. M. verticillata L. (*M. crispa* (L.) L.) - *Chinese Mallow*. Similar to *M. parviflora* but 213
petals 5-9mm; nutlets obscurely ridged, ± glabrous, with sharp smooth angle between dorsal and lateral surfaces; (2n=c.76, c.84, c.112, c.126). Intrd; rough and waste ground, casual from wool and other sources; infrequent in C & S En; E Asia.

4. LAVATERA L. - *Tree-mallows*

Annual to perennial, woody or not; epicalyx of 3 segments fused below; petals pink to purple (or white); carpels numerous; stigmas linear; fruit breaking into numerous 1-seeded nutlets.

1 Central axis of fruit expanded above to form umbrella-like disc concealing nutlets **5. L. trimestris**
1 Central axis of fruit not expanded above; nutlets clearly visible from above unless obscured by calyx 2
  2 Epicalyx <3/4 as long as calyx **3. L. plebeia**
  2 Epicalyx 3/4 as long as to longer than calyx 3
3 Flowers 1 in each leaf-axil **4. L. thuringiaca**
3 Flowers 2-several in each leaf-axil 4
  4 Epicalyx expanding to longer than calyx in fruit; petals deep pinkish-purple with dark stripes **1. L. arborea**
  4 Epicalyx slightly shorter than to as long as calyx in fruit; petals lilac **2. L. cretica**

1. L. arborea L. - *Tree-mallow*. Biennial with erect stems to 3m, woody below; 213
younger parts softly stellate-tomentose; leaves shallowly 5-7-lobed; petals 14-20mm, deep pinkish-purple with dark stripes; 2n=36. Native; rocks, cliff-bottoms and waste ground near sea; W coast of Br from Dorset (formerly E Sussex) to Ayrs, CI, scattered in Ir, probably intrd on coasts of E Br N to C Sc.

2. L. cretica L. - *Smaller Tree-mallow*. Annual or biennial herb with erect to RR
decumbent stems to 1(1.5)m; younger parts stellate-pubescent; leaves shallowly 5-7-lobed; petals 10-20mm, lilac; (2n=40-44, 112). Native; rough and waste ground by sea; frequent in Scillies and CI, formerly W Cornwall mainland but now sporadic there and perhaps intrd, rare casual elsewhere. Resembles *Malva sylvestris*, but has greyish pubescence and dull lilac flowers, and see generic key.

3. L. plebeia Sims - *Australian Hollyhock*. Annual or biennial herb with erect to decumbent stems to 1(2)m; younger parts rather sparsely stellate-pubescent; leaves 3-7-lobed; petals 12-25mm, lilac or white; (2n=43). Intrd; rather infrequent wool-alien; very scattered in En; Australia.

4. L. thuringiaca L. (*L. olbia* auct. non L.) - *Garden Tree-mallow*. Perennial (± shrub) with woody branching erect stems to 2.5m; younger parts softly stellate-pubescent; leaves 3-5-lobed, the central lobe long; petals 15-30mm, purple to pale pink; (2n=40, 44). Intrd; grown in gardens and sometimes persisting as escape or throwout; S En; SE Europe. Some or all plants might be *L. thuringiaca* x *L. olbia* L.

5. L. trimestris L. - *Royal Mallow*. Annual with erect or ascending stems to 1(1.2)m; younger parts sparsely pubescent with simple and few-rayed hairs; leaves shallowly 3-7-lobed; petals 20-45mm, pink or white; (2n=14). Intrd; grown in gardens and sometimes found as casual, escape or throwout, also grain- and wool-alien; S En; Mediterranean.

5. ALTHAEA L. - *Marsh-mallows*

Annual to perennial herbs; epicalyx of 6-10 segments fused below; petals pink to purple; staminal-tube terete, pubescent; carpels numerous; stigmas linear; fruit breaking into numerous 1-seeded nutlets.

1. A. officinalis L. - *Marsh-mallow*. Perennial with erect stems to 1.5m; younger parts softly stellate-tomentose; leaves shallowly 3-5-lobed; petals 15-20mm, pale pink; 2n=42. Native; brackish ditches, banks and grassland near sea; locally common round coasts of Br N to Lincs and S Wa, formerly CI, scattered elsewhere in Br and Ir but intrd. **R** **213**

2. A. hirsuta L. - *Rough Marsh-mallow*. Annual with erect to decumbent stems to 60cm, coarsely pubescent with long simple and short stellate hairs; lower leaves shallowly, upper deeply 3-5-lobed or palmate; petals 12-16mm, lilac; (2n=36). Probably intrd; field and wood borders, N Somerset and W Kent (since 1792), casual or persistent in scattered places over most of En and Wa; Europe. **RRR**

6. ALCEA L. - *Hollyhock*

Biennial to perennial herbs; epicalyx of 6-7 segments fused below; petals various shades of red, yellow or white; staminal-tube 5-angled, glabrous; carpels numerous; stigmas linear; fruit breaking into numerous 1-seeded nutlets.

1. A. rosea L. (*A. ficifolia* L., *Althaea rosea* (L.) Cav., *A. ficifolia* (L.) Cav., *A. cultorum* Bergmans) - *Hollyhock*. Stems erect, to 3m; younger parts rather softly stellate-tomentose; leaves shallowly to rather deeply 3-9-lobed; petals 25-50mm; (2n=26, 40, 42, 56). Intrd; much grown in gardens and frequent escape or throwout on tips and waste ground, usually casual; scattered over much of BI; garden origin from W Asian parents. *A. ficifolia* has rather deeply divided leaves and usually yellow flowers, but is not specifically distinct. **213**

7. SIDALCEA A. Gray ex Benth. - *Greek Mallows*

Perennial herbs; epicalyx 0; petals pinkish-purple or white; carpels numerous; stigmas linear; fruit breaking into numerous 1-seeded nutlets.

Other spp. - A plant natd by a stream in Shetland has been tentatively named **S. hendersonii** S. Watson, but needs checking.

1. S. malviflora (DC.) A. Gray ex Benth. - *Greek Mallow*. Stems erect, to 1.5m, somewhat pubescent above; lower leaves shallowly, upper deeply 3-7-lobed or palmate, pubescent on upperside; flowers in simple racemes; sepals 8-12mm; petals 10-25mm, pinkish-purple, rarely white; (2n=20, 40, 60). Intrd; grown for ornament and sometimes persisting as throwout or escape; very scattered over Br; SW N America. Plants in Br are cultivars of unknown origin and parentage; few or none are pure *S. malviflora*, but probably include it in their parentage. **213**

2. S. candida A. Gray - *Prairie Mallow*. Stems erect, to 80cm; differs from *S. malviflora* in stems and leaf upperside ± glabrous; sepals 4-6mm; petals 8-15mm, white to cream; (2n=20). Intrd; garden plant grown and natd as for *S. malviflora* but much more rarely; 1 large patch in Dunbarton since 1992; W N America.

8. ABUTILON Mill. - *Velvetleaf*

Annual herb; epicalyx 0; petals yellow; carpels numerous; stigmas capitate; fruit composed of 5 almost separate, several-seeded nutlets.

1. A. theophrasti Medik. - *Velvetleaf*. Erect annual to 1m; younger parts softly tomentose with stellate and simple hairs; leaves suborbicular, acuminate, cordate, not lobed; petals 7-13mm; nutlets each with slender beak 2-3mm; (2n=42). Intrd; casual in waste places and tips, frequent birdseed-, oilseed- and wool-alien; scattered over En and Wa; SE Europe. **213**

9. HIBISCUS L. - *Bladder Ketmia*
Annual herb; epicalyx of 10-13 segments free ± to base; petals pale yellow with violet patch at base; carpels 5, fused; stigmas capitate; fruit a dehiscent capsule with 5 many-seeded cells.

1. H. trionum L. - *Bladder Ketmia*. Erect to decumbent annual to 50cm; rather sparsely stellate-pubescent; leaves mostly palmate with 3(-5) lobes; petals 15-25mm; calyx enlarging and inflated round fruit; (2n=28, 56). Intrd; casual in waste places and tips; frequent birdseed-, oilseed- and wool-alien; scattered over En, Wa and CI; SE Europe, Asia, Africa. **213**

56. SARRACENIACEAE - *Pitcherplant family*

Herbaceous perennials; leaves all in basal rosette, consisting of tubular insectivorous 'pitchers' with small leaf-blade ('hood') at entrance. Flowers solitary, terminal, on leafless pedicels from ground level, actinomorphic, bisexual, hypogynous, with 3 small bracteoles resembling epicalyx; sepals 5, ± petaloid, free; petals 5, showy, free; stamens numerous, ovary 5-celled, with many ovules in each cell; style 1, short; stigma greatly expanded, peltate; fruit a capsule.
Unmistakable pitcherplants.

1. SARRACENIA L. - *Pitcherplant*

Other spp. - S. flava L. (*Trumpets*), from SE N America, has been planted in bogs in Surrey and S Hants; it differs from *S. purpurea* in its pitchers being straight and ± erect, with an arching hood, and in its yellow petals.

1. S. purpurea L. - *Pitcherplant*. Pitchers ≤30cm, green, marbled with red, curved, decumbent; hood erect; pedicels ≤60cm; flowers nodding, c.5cm across, with purplish-red petals and sepals; stigma c.3cm across, greenish; (2n=26). Intrd; planted in wet peat-bogs and well natd; C Ir (planted in Roscommon in 1906, later from there to other places), a few less permanent sites scattered in En and Sc; NE N America.

57. DROSERACEAE - *Sundew family*

Herbaceous perennials, leaves all in basal rosette, reddish and covered with sticky hairs (insectivorous), stipulate, petiolate. Flowers in simple cyme, on leafless peduncle from ground level, actinomorphic, bisexual, hypogynous, usually remaining closed; sepals 5-8, ± fused at base; petals 5-8, 4-6mm, white, free; stamens 5-8; ovary 1-celled, with many ovules; styles 2-6, often deeply divided above or fused below; stigmas linear; fruit a capsule.
Unmistakable insectivorous plants.

1. DROSERA L. - *Sundews*

1 Leaf-blade ± orbicular, abruptly narrowed into pubescent petiole **1. D. rotundifolia**
1 Leaf-blade obovate to linear-oblong, gradually narrowed into ± glabrous petiole 2
  2 Peduncles straight, rising from near centre of leaf-rosette, often much longer than leaves **2. D. anglica**
  2 Peduncles curved at base, arising from sides of leaf-rosette, often little longer than leaves **3. D. intermedia**

Other spp. - 2 S Hemisphere spp. have been planted on bogs in Surrey in 1970s and 1980s and are surviving. **D. capensis** L., from S Africa, has linear-oblong leaves mostly 10-15cm, often a short ± erect aerial stem, and purple flowers; **D. binata** Labill., from Australia and New Zealand, has even larger leaves 1-few times forked into linear lobes ≤15cm, and white petals ≤1cm.

1. D. rotundifolia L. - *Round-leaved Sundew*. Inflorescences to 10(25)cm, the peduncles straight and arising centrally as in *D. anglica* (see key); leaves (incl. petiole) ≤5cm, the blades ± orbicular, ≤10 x 10mm; 2n=20. Native; wet acid peaty places with little shade; in suitable places over most of BI, often common in N & W but absent from much of C, E & S En.

1 x 2. D. rotundifolia x D. anglica = D. x obovata Mert. & W.D.J. Koch is similar to *D. intermedia* in leaf characters but has a straight, centrally arising inflorescence and is sterile; it is rare but has scattered records over much of W Br and Ir, especially NW Sc.

1 x 3. D. rotundifolia x D. intermedia = D. x belezeana E.G. Camus has a leaf-shape closer to that of *D. rotundifolia* but curved, lateral inflorescences and is sterile; it has been recorded in S Hants, W Norfolk and Cards.

2. D. anglica Huds. (*D. longifolia* L.) - *Great Sundew*. Inflorescences to 18(30)cm; leaves (incl. petiole) ≤13cm, the blades narrowly obovate to linear-oblong, ≤30 x 10mm; 2n=40. Native; similar places (or wetter) and distribution to *D. rotundifolia*, but much more local and absent from CI and most of En, Wa and E Sc.

3. D. intermedia Hayne - *Oblong-leaved Sundew*. Inflorescences to 5(10)cm; leaves (incl. petiole) ≤5cm, the blades obovate to narrowly so, ≤10 x 5mm; 2n=20. Native; similar places to *D. rotundifolia*; very locally common in scattered areas of Br and Ir, mostly in W.

58. CISTACEAE - *Rock-rose family*

Annuals or woody evergreen dwarf perennials; leaves opposite (at least below), simple, entire, sessile to shortly petiolate, stipulate or not. Flowers in terminal mostly simple cymes, actinomorphic, bisexual, hypogynous; sepals 5, free, 2 outer smaller than 3 inner (except in *Cistus*); petals 5, free, predominantly yellow or white; stamens numerous; ovary 1-celled, with many ovules on 3 placentae (5-celled in *Cistus*); style 0 or 1; stigma 1 (and then 3-lobed, or 5-lobed in *Cistus*) or 3; fruit a capsule with 3 valves (5 in *Cistus*).

Recognizable by the distinctive sepals and gynoecium, and stellate hairs on vegetative parts.

1 Annuals; style very short or 0 **1. TUBERARIA**
1 Woody perennials; style at least as long as ovary **2. HELIANTHEMUM**

Other genera - CISTUS L. differs from other 2 genera in contrasts given in family description and is a shrub. **C. laurifolius** L. and **C. incanus** L., from S Europe, are grown for ornament and occasionally self-sown individuals persist for a short while outside gardens. In both spp. the flowers are 4-6cm across, white or pink respectively.

1. TUBERARIA (Dunal) Spach - *Spotted Rock-rose*
Annual with basal leaf-rosette; leaves with 3 obvious veins; stipules 0; style 0 or very short.

1. T. guttata (L.) Fourr. - *Spotted Rock-rose*. Stems ± procumbent to erect, to 30cm but often <10cm; leaves pubescent or sparsely so on lowerside; flowers 8-15mm across; petals yellow, usually with brownish-red blotch at base; 2n=36. Native; dry **RR**

barish ground near sea; W & SW Ir, NW Wa, Jersey and Alderney. Plants from Wa and Ir differ from those in CI and France in having wider leaves, shorter internodes and flowers with bracts. They have been separated as ssp. **breweri** (Planch.) E.F. Warb., but are probably not worth ssp. status.

2. HELIANTHEMUM Mill. - *Rock-roses*

Dwarf straggly or bushy woody perennials; leaves with 1 obvious vein; stipules present or 0; style at least as long as ovary.

1 Stipules 0; style strongly S-shaped, shorter than stamens; flowers <15(20)mm across **3. H. oelandicum**
1 Stipules present; style slightly kinked near base, longer than stamens; flowers mostly >(15)20mm across 2
  2 Petals white; leaves grey-tomentose on upperside **2. H. apenninum**
  2 Petals yellow; leaves green-pubescent on upperside **1. H. nummularium**

1. H. nummularium (L.) Mill. (*H. chamaecistus* Mill.) - *Common Rock-rose.* Stems procumbent or decumbent, wiry, to 50cm; leaves elliptic-oblong to narrowly so, green and sparsely pubescent on upperside, whitish-tomentose on lowerside; flowers 1-12 per cyme, mostly 20-25mm across; petals yellow; 2n=20. Native; base-rich grassland; common in suitable places over most of Br, but absent from many areas, 1 place in E Donegal. Garden plants natd in S Br, some with orange petals, might represent other sspp.

1 x 2. H. nummularium x H. apenninum = H. x sulphureum Willd. ex Schltdl. occurs in the N Somerset and formerly in the S Devon site of *H. apenninum* with both parents; it has pale yellow petals and intermediate leaf characters and is usually highly fertile.

2. H. apenninum (L.) Mill. - *White Rock-rose.* Similar to *H. nummularium* but leaves usually narrower (narrowly oblong to linear) and grey-tomentose on upperside; petals white; 2n=20. Native; in dry limestone grassland; 1 area in each of N Somerset and S Devon. **RR**

3. H. oelandicum (L.) Dum. Cours. - *Hoary Rock-rose.* Stems procumbent to ascending, often rather twiggy below, to 25cm; leaves green on upperside, whitish-tomentose on lowerside; flowers 1-6(10) per cyme, mostly 10-15mm across; petals yellow. Native; rocky limestone pastures.

1 Leaves glabrous to subglabrous on upperside; flowers 1-3(5) per cyme **c. ssp. levigatum**
1 Leaves pubescent to sparsely so on upperside; flowers (2)3-6 per cyme 2
  2 Leaves ± persistent on lower parts of non-flowering shoots, usually sparsely pubescent on upperside, mostly >10mm **b. ssp. piloselloides**
  2 Leaves not persistent on lower parts of non-flowering shoots, usually pubescent on upperside, often <10mm **a. ssp. incanum**

a. Ssp. incanum (Willk.) G. López (ssp. *canum* (L.) Bonnier & Layens, *H. canum* (L.) Hornem.). Leaves more pubescent than in other sspp.; 2n=22. Very local in N Wa, S Wa and NW En. **R**

b. Ssp. piloselloides (Lapeyr.) Greuter & Burdet (*H. canum* ssp. *piloselloides* (Lapeyr.) M. Proctor). Leaves larger and less pubescent than in ssp. *incanum;* 2n=22. W Ir. Possibly distinct from Pyrenean ssp. *piloselloides.* **RR**

c. Ssp. levigatum (M. Proctor) D.H. Kent (*H. canum* ssp. *levigatum* M. Proctor). Stems shorter, leaves smaller and less pubescent, and inflorescence fewer flowered than in ssp. *incanum;* 2n=22. Cronkley Fell, NW Yorks; endemic. **RR**

59. VIOLACEAE - *Violet family*

Annual or perennial herbs; leaves alternate or ± all basal, simple, toothed, petiolate, stipulate. Flowers solitary, axillary or from basal rosette, zygomorphic, bisexual, hypogynous; sepals 5, free, with appendages below their insertion; petals 5, free, the lower one with a backwards-directed spur, blue or yellow to white; stamens 5, the 2 lower with spur inserted into petal spur; ovary 1-celled with many ovules on 3 placentas; style 1, thickened above, usually bent; stigma 1, capitate; fruit a capsule with 3 valves.

The well-known pansies and violets.

1. VIOLA L. - *Violets*

1 Stipules ovate to linear-lanceolate, finely toothed; lateral 2 petals spreading horizontally; style not or gradually thickened distally, often hooked (subg. *Viola* - *Violets*) 2

1 Stipules ± leaf-like, at least some deeply lobed; lateral 2 petals directed upwards; style with ± globose apical swelling with hollow in one side (subg. *Melanium* (DC.) Hegi - *Pansies*) 10

2 Style straight above, with oblique apex; leaves orbicular, obtuse to rounded at apex **9. V. palustris**

2 Style hooked or with lateral beak at apex; leaves acute to obtuse at apex 3

3 Leaves (petioles and blades) and usually capsules pubescent 4

3 Leaves and capsules glabrous 6

4 Sepals acute **3. V. rupestris**

4 Sepals obtuse to ± rounded 5

5 Creeping stolons present; flowers sweet-scented; hairs on petioles mostly <0.3mm, reflexed or appressed **1. V. odorata**

5 Stolons 0; flowers not scented; hairs on petioles mostly 0.3-1mm, patent **2. V. hirta**

6 Plant with basal leaf-rosette; leaves ≤1.3x as long as wide 7

6 Basal leaf-rosette 0; leaves ≥1.4x as long as wide 8

7 Sepal-appendages >1.5mm, corolla-spur paler than petals **4. V. riviniana**

7 Sepal-appendages <1.5mm; corolla-spur darker than petals **5. V. reichenbachiana**

8 Corolla-spur <2x as long as sepal-appendages; roots creeping underground, sending up stems at intervals **8. V. persicifolia**

8 Corolla-spur >2x as long as sepal-appendages; all stems arising from central tuft 9

9 Petals clear blue; leaves ovate to narrowly ovate, truncate to cordate at base **6. V. canina**

9 Petals cream to greyish-violet; leaves lanceolate to narrowly ovate, rounded to cuneate at base **7. V. lactea**

10 Spur 10-15mm **10. V. cornuta**

10 Spur <7mm 11

11 Flowers usually >3.5cm vertically across, with strongly overlapping petals **13. V. x wittrockiana**

11 Flowers <2.5(3.5)cm vertically across, with not or slightly overlapping petals 12

12 Plant perennial, with stems creeping underground 13

12 Plant annual to perennial; creeping underground stems 0 14

13 Flowers >(1.5)2cm vertically across; terminal segment of stipules scarcely wider than others, entire **11. V. lutea**

13 Flowers often <2cm vertically across; terminal segment of stipules distinctly wider than others, often slightly crenate **12. V. tricolor**

14 Corolla 4-8mm vertically across, concave; Scillies and CI only **15. V. kitaibeliana**

14 Corolla ≥8mm vertically across, ± flat when fresh; widespread 15

15 Corolla 8-20mm vertically across, usually yellow or cream (rarely with blue), usually shorter than calyx; terminal segment of stipules usually strongly crenate, ± leaf-like; projection below stigma-hollow 0 or indistinct **14. V. arvensis**

15 Corolla 10-25(35)mm vertically across, usually with some blue or violet, usually longer than calyx; terminal segment of stipules entire to obscurely crenate, scarcely leaf-like; projection below stigma-hollow distinct **12. V. tricolor**

1. V. odorata L. - *Sweet Violet*. Perennial with leaves and flowers from central tuft; creeping stolons present; leaves pubescent, broadly ovate to ovate-orbicular, deeply cordate; flowers violet, or white with violet or purple spur, rarely pinkish; 2n=20. Native; woodland, scrub and hedgerows, mostly on base-rich soils; most of BI, but local or very local in Sc, Ir, Wa and CI and often not native.

1 x 2. V. odorata x V. hirta = V. x scabra F. Braun (*V. x permixta* Jord.) is found in Br N to Durham and in Roxburghs with the parents; it is intermediate in most characters, incl. stolon development and flower-scent, and partially fertile.

2. V. hirta L. - *Hairy Violet*. Perennial with leaves and flowers from central tuft; stolons 0; leaves pubescent, ovate, deeply cordate; flowers blue-violet; 2n=20. Native; calcareous pastures and open scrub; suitable places in Br N to C Sc, rare and scattered in C Ir. Ssp. **calcarea** (Bab.) E.F. Warb. differs in being smaller in most parts, especially the shorter petal-spur and narrower petals, but is probably only a var.

3. V. rupestris F.W. Schmidt - *Teesdale Violet*. Perennial usually with leafy stems (short or 0 in dwarfed plants) to 5(10)cm and basal leaves from central tuft, pubescent ± all over; leaves broadly ovate to ovate-orbicular, cordate; flowers pale blue-violet; 2n=20. Native; short turf or barish places on limestone; Durham, Westmorland and MW Yorks. **RR**

3 x 4. V. rupestris x V. riviniana = V. x burnatii Gremli occurs with the parents in Upper Teesdale (Durham); it is intermediate in most characters, including a very fine pubescence, and highly sterile.

4. V. riviniana Rchb. - *Common Dog-violet*. Perennial with leafy stems (short or 0 in dwarfed plants) to 20cm and basal leaves from central tuft, glabrous to slightly pubescent on most parts; leaves broadly ovate to ovate-orbicular, obtuse, deeply cordate; flowers pale blue-violet; 2n=40, 48. Native; wide range of woods and grassland; common all over BI. Ssp. **minor** (Murb. ex Greg.) Valentine differs in being smaller in all parts; it is an ecotype of more exposed places, best treated as var. *minor* (Murb. ex Greg.) Valentine.

4 x 5. V. riviniana x V. reichenbachiana = V. x bavarica Schrank is intermediate in length of sepal-appendages but has a dark spur and is highly (but not fully) sterile; it is only sparsely scattered in En, Wa and Ir despite the frequent cohabitation of the parents.

4 x 6. V. riviniana x V. canina = V. x intersita Beck (*V. x weinhartii* W. Becker) is scattered in Br and Ir where the parents meet on heaths and dunes; it is intermediate in habit and leaf characters and highly sterile.

4 x 7. V. riviniana x V. lactea (= *V. x lambertii* H. Lév. nom. nud.) is frequent in En, Wa and Ir in the range of *V. lactea*; it is intermediate in habit and leaf and flower characters, and quite highly sterile; 2n=50. It often forms large clonal patches.

5. V. reichenbachiana Jord. ex Boreau - *Early Dog-violet*. ± glabrous perennial with leafy stems to 20cm and basal leaves from central tuft; leaves ovate to broadly ovate, subacute, deeply cordate; flowers bluish-mauve; 2n=20. Native; woods and hedgebanks; common in En, scattered in Wa and Ir, rare in Sc.

5 x 6. V. reichenbachiana x V. canina = V. x mixta A. Kern. (*V. x borussica* (Borbás) W. Becker) has been recorded from S En but is very rare as the parents rarely meet; it is intermediate and fairly highly sterile, but would be extremely difficult to tell from *V. riviniana* x *V. canina*

6. V. canina L. - *Heath Dog-violet.* Glabrous or sparsely pubescent perennial with leafy stems to 30(40)cm; basal leaf-rosette 0; leaves ovate to narrowly ovate; flowers blue with whitish-yellow spur. Native.

a. Ssp. canina. Stems procumbent to ascending; leaves <2x as long as wide, cordate; stipules of middle leaves <1/3 as long as petiole; 2n=40. Dry or wet heaths and dunes, fens; local throughout BI.

b. Ssp. montana (L.) Hartm. Stems erect; leaves c.2x as long as wide, truncate to subcordate; stipules of middle leaves ≤1/2 as long as petiole; flowers relatively large; 2n=40. Fens; Hunts and Cambs.

6 x 7. V. canina x V. lactea (= *V. x militaris* Savouré nom.nud.) occurs quite frequently within the range of *V. lactea* in En, Wa and Ir; it is intermediate in leaf-shape and flower-colour, shows c.10 per cent fertility and is vigorous.

6 x 8. V. canina x V. persicifolia = V. x ritschliana W. Becker occurs in C En and C Ir with the parents; it is intermediate in leaf chracters but close to *V. canina* in flowers and to *V. persicifolia* in habit. It is sterile and often very vigorous.

7. V. lactea Sm. - *Pale Dog-violet.* Subglabrous perennial with leafy stems to 20cm; basal leaf-rosette 0; leaves lanceolate to narrowly ovate, rounded to cuneate at base; flowers cream to greyish-violet; 2n=58. Native; dry heaths; very local in S En, S & W Wa and C & S Ir, decreasing, formerly Man. **R**

8. V. persicifolia Schreb. (*V. stagnina* Kit.) - Fen Violet. Subglabrous perennial with leafy stems to 25cm from central tuft and from spreading roots; basal leaf-rosette 0; leaves narrowly ovate, truncate to subcordate at base; flowers bluish-white to white, with greenish spur; (2n=20). Native; fens; Cambs and Hunts, formerly N to SW Yorks, C & W Ir. **RRR**

9. V. palustris L. - *Marsh Violet.* Perennial with thin rhizomes or sometimes stolons producing leaves and flowers; aerial stems 0; leaves orbicular, deeply cordate; flowers bluish-lilac with darker veins. Native; bogs, fens, marshes, wet heaths and woods.

a. Ssp. palustris. Leaves rounded at apex, glabrous; bracteoles below middle of pedicel; 2n=48. Throughout most of Br and Ir, but absent from much of C & E En.

b. Ssp. juressi (Link ex Wein) P. Fourn. Leaves obtuse to subacute, usually with pubescent petioles; bracteoles near middle of pedicel. Local in Ir, S & W Wa and W En.

10. V. cornuta L. - *Horned Pansy.* Perennial with slender rhizome; aerial stems to 30cm; stipules serrate or lobed to c.1/2 way to midrib, with ± triangular apical lobe; flowers 2-4cm across, violet or lilac, fragrant; spur 10-15mm; (2n=20, 22, 42). Intrd; grown in gardens and natd in grassy places; frequent in Sc, rare and mostly ± casual elsewhere in Br; Pyrenees. Hybrids between this and *V. x wittrockiana* are grown (*Bedding Viola, Violetta*) and might also escape.

11. V. lutea Huds. - *Mountain Pansy.* Perennial with slender rhizome; aerial stems to 20cm; stipules lobed nearly to midrib, with linear-elliptic entire apical lobe; flowers (1.5)2-3.5cm across, usually yellow, less often blue or purple or blotched, spur 3-6mm; most pollen-grains with 4 pores; 2n=48, 50. Native; upland pastures and rocky places, often on base-rich or heavy-metalliferous soils; upland areas of Wa, Sc and N En, very local in S Somerset and C & S Ir.

11 x 12. V. lutea x V. tricolor is known for certain only from Derbys and S Northumb but is probably overlooked; it is difficult to determine without detailed study. Hybrids are intermediate or closer to *V. lutea* in appearance and partially fertile, and probably backcross.

12. V. tricolor L. - *Wild Pansy.* Annual to perennial; stipules lobed ± to midrib, with narrowly elliptic to oblanceolate or narrowly obovate entire to crenate apical lobe; flowers 1-2.5(3.5)cm across, usually purple to violet or blotched with yellow, less often all yellow; spur 3-6.5mm; most pollen-grains with 4 pores. Native.

a. Ssp. tricolor. Annual, or perennial with rhizomes 0 or ill-developed; 2n=26. Common on waste, marginal and cultivated ground throughout most of BI.

Perennial plants occurring in hilly areas of N have been referred to ssp. **saxatilis** (F.W. Schmidt) E. Warb. (*V. lepida* Jord.), but probably are variants of ssp. *tricolor* or its hybrids with *V. arvensis* or *V. lutea*.

b. Ssp. curtisii (E. Forst.) Syme. Perennial with well-developed rhizomes; 2n=26. Frequent on maritime dunes in W Br and most of Ir, and inland by lakes in N Ir and on heaths of Norfolk and Suffolk.

12 x 14. V. tricolor x V. arvensis = V. x contempta Jord. occurs scattered throughout BI, but its frequency is uncertain. It is intermediate in most characters, especially stylar flap and corolla-size and -colour, and partially fertile; 2n=26, 28, 30. Variants of both parents, but mainly *V. arvensis*, resemble it.

13. V. x wittrockiana Gams ex Kappert - *Garden Pansy*. Probably derived from various *V. tricolor* x *V. arvensis* crosses and recognisable by flowers 3.5-10cm across with strongly overlapping petals; (2n=48-50). Intrd; much grown in gardens and parks and often escaping on rough or cultivated land or on tips; scattered throughout lowland Br, mainly in S. Wild populations often show segregation and/or backcrossing to *V. tricolor* or less often to *V. arvensis*.

14. V. arvensis Murray - *Field Pansy*. Annual with erect to decumbent stems to 40cm; stipules lobed nearly to base, with leaf-like elliptic to obovate (to narrowly so) crenate to serrate apical lobe; flowers 8-20mm across, usually yellow or cream, less often with violet blotches or suffusion; spur 2-4mm; most pollen-grains with 5 spores; 2n=34. Native; weed of cultivated and waste ground; common throughout most of BI.

15. V. kitaibeliana Schult. - *Dwarf Pansy*. Annual with erect to decumbent stems **RR**
to 10cm; stipules similar to those of *V. arvensis* but much smaller; flowers 4-8mm across, usually cream suffused with violet; spur 1-2mm, often violet; 2n=48. Native; short turf on sandy soil by sea; Scillies and CI.

60. TAMARICACEAE - *Tamarisk family*

Deciduous shrubs or small trees; leaves alternate, simple, entire, small and ± scale-like, without stipules, sessile. Flowers small, in long catkin-like racemes, actinomorphic, bisexual, hypogynous; sepals 5, free; petals 5, free, white to pink, ovate to trullate; stamens 5, joined at base to lobed or stellate nectariferous disc; ovary 1-celled, with many ovules; styles 3-4; stigmas capitate; fruit a capsule; seeds with apical tuft of hairs.

Easily recognizable by the distinctive habit and flower structure.

1. TAMARIX L. - *Tamarisks*

1. T. gallica L. (*T. anglica* Webb) - *Tamarisk*. Shrub or small tree to 3m; leaves 1.5-3mm, narrowly triangular-ovate, acute to acuminate, green to glaucous; racemes 2-5cm x 4-5mm; petals 1.5-2mm; (2n=24). Intrd; planted as windbreak and stabilizer in sandy places by sea, persistent and appearing natd but very rarely if ever self-sown; coasts of CI and Br N to N Wa and Suffolk, rarely elsewhere in Br and Ir; SW Europe.

2. T. africana Poir. - *African Tamarisk*. Differs mainly in larger racemes (3-7cm x 5-9mm) and petals (2.5-3mm); (2n=24). Intrd; same habitats as *T. gallica*, but much rarer and never self-sown; S coast of En; Mediterranean.

61. FRANKENIACEAE - *Sea-heath family*

Evergreen perennial with stems woody at base; leaves opposite, simple, entire, small and heather-like, without stipules, sessile. Flowers small, solitary, terminal or in branchlet-forks, sessile, actinomorphic, bisexual, hypogynous; sepals usually 5, fused to >1/2 way; petals usually 5, free, pink, with small scale at junction of claw and expanded limb; stamens usually 6 in 2 whorls of 3; ovary 1-celled, with many ovules; style 1, divided into 3 near apex; stigmas 3, elongate; fruit a capsule.

Easily recognisable by the distinctive leaves and flower structure.

1. FRANKENIA L. - *Sea-heath*

R **1. F. laevis** L. - *Sea-heath*. Stems procumbent, mat-forming, to 35cm; leaves 3-7mm, with strongly revolute margins, appearing linear to narrowly oblong with widened basal part pubescent at margins; flowers 5-6mm across; 2n=30. Native; on sandy or silty barish ground on drier parts of salt-marshes; coasts of CI and SE Br from S Hants to N Lincs, sporadic and probably intrd elsewhere in En and Wa.

62. CUCURBITACEAE - *White Bryony family*

Annuals or herbaceous perennials; stems scrambling or climbing, usually with tendrils; leaves alternate, simple, usually palmately veined, usually palmately lobed, without stipules, petiolate. Flowers 1-several in axillary racemes or irregular groups, actinomorphic, epigynous, unisexual (dioecious or monoecious); hypanthium present, bearing sepals, petals and stamens; sepals 5, free; petals 5, united at least at base; stamens apparently 3 (rarely 5), 2 with 2 pollen-sacs and 1 with 1; ovary 1-celled but often appearing (2)3(-5)-celled due to deeply intrusive parietal placentae, with few to many ovules; style 1; stigmas 2-3, ± capitate; fruit succulent and indehiscent, berry-like or cucumber-like.

Recognizable by the scrambling or climbing habit, unisexual epigynous flowers with unusual stamens, and distinctive fruit.

1 Dioecious perennials; fruit <1cm, red **1. BRYONIA**
1 Monoecious annuals or perennials; fruit >2cm, green to yellow 2
2 Tendrils absent; fruit densely hispid when ripe **2. ECBALLIUM**
2 Tendrils present; fruit glabrous to sparsely pubescent or hispid when ripe 3
3 Male flowers several together; central axis of anther extending beyond pollen-sacs **3. CUCUMIS**
3 Male flowers solitary; central axis of anther not extending beyond pollen-sacs 4
4 Leaves pinnately lobed; corolla divided much >1/2 way **4. CITRULLUS**
4 Leaves palmately lobed; corolla divided <1/2 way **5. CUCURBITA**

Other genera - **SICYOS** L. resembles *Cucumis* in its clustered male flowers and deeply divided corolla, but has branched tendrils, bristly fruits and 5 ± adherent stamens. **S. angulatus** L. (*Bur Cucumber*), from E N America, occurs as an infrequent casual mainly with soyabean waste.

1. BRYONIA L. - *White Bryony*

Dioecious perennials with tuberous roots; stems climbing, with simple tendrils; flowers several together; corolla divided >1/2 way, greenish-white; fruit a red globose, ± glabrous berry.

1. B. dioica Jacq. (*B. cretica* L. ssp. *dioica* (Jacq.) Tutin) - *White Bryony*. Stems to c.4m; leaves deeply palmately lobed, ± hispid; corolla 5-12mm; fruit 5-9mm; 2n=20. Native; in scrub and hedgerows, mostly on well-drained base-rich soils; En but absent from most of N and SW, Wa but rare except in E, intrd elsewhere.

2. ECBALLIUM A. Rich. - *Squirting Cucumber*

Monoecious perennials with tuberous roots; stems trailing; tendrils 0; male flowers several together; female flowers solitary; corolla divided <1/2 way, yellowish; fruit ellipsoid to narrowly so, green, densely hispid, explosive when ripe.

1. E. elaterium (L.) A. Rich. - *Squirting Cucumber*. Stems to 50cm, hispid; leaves not or shallowly and irregularly palmately lobed, densely and coarsely pubescent; corolla 10-20mm; fruit 3-5cm, usually pendent on ± erect pedicel; (2n=18, 24). Intrd; rare casual in waste and cultivated land in CI and S Br, sometimes persisting (and perennial) in Jersey and extreme S En; S Europe.

3. CUCUMIS L. - *Cucumbers*
Monoecious annuals with non-tuberous roots; stems climbing with simple tendrils; male flowers several together; female flowers solitary; corolla divided <1/2 way, yellow; fruit ellipsoid to cylindrical, very variable in surface texture, green to yellow, glabrous to pubescent.

Other spp. - C. myriocarpus Naudin (*Gooseberry Cucumber*), from S Africa, differs from C. *melo* in its corolla only 4-5mm and pubescent globose fruits 2-2.5cm; it is a rare wool-alien.

1. C. melo L. - *Melon*. Stems to 1m, hispid; leaves not or shallowly palmately lobed, hispid to softly pubescent; corolla-tube 8-28mm; fruit 3->30cm, green to yellow, usually ellipsoid, pubescent at first; (2n=24). Intrd; common casual on tips and at sewerage works; mainly S En; Africa.

2. C. sativus L. - *Cucumber*. Stems to 1(2)m, hispid; leaves shallowly to rather deeply palmately lobed, pubescent and scabrid; corolla-tube 34-65mm, yellow; fruit 10->30cm, green, usually glabrous, smooth or warty, much longer than wide; (2n=14). Intrd; rare casual on tips and at sewerage works, sometimes grown as crop on field-scale; S En; India.

4. CITRULLUS Schrad. - *Water Melon*
Monoecious annuals with non-tuberous roots; stems climbing with simple or branched tendrils; male and female flowers solitary; corolla divided >1/2 way, yellow; fruit globose to slightly elongated, green to yellow, smooth, ± glabrous.

1. C. lanatus (Thunb.) Matsum. & Nakai - *Water Melon*. Stems to 1m, woolly when young; leaves with 3 main lobes, the central one forming most of leaf, all 3 pinnately lobed, pubescent when young, scabrid later; corolla 5-17mm; fruit 2->30cm; (2n=22). Intrd; rare casual on tips, frequent at sewerage works; S Br; S Africa.

5. CUCURBITA L. - *Marrows*
Monoecious annuals with non-tuberous roots; stems climbing with usually branched tendrils; male and female flowers solitary; corolla divided <1/2 way, yellow; fruit globose to cylindrical, green to yellow, smooth to warty, finally glabrous.

1. C. pepo L. - *Marrow*. Stems to 2m, hispid; leaves shallowly to deeply palmately lobed, hispid; corolla 30-110mm; fruit extremely variable, often very large, with thickened and fluted pedicel; (2n=10, 22, 24, 28, 40, 44). Intrd; common casual on tips and at sewerage works, sometimes grown on field scale; mostly S Br; N C America. Most casual plants in Br result from use as vegetables (*marrow*, *courgette*), but some perhaps from use as ornamentals (*gourds*).

2. C. maxima Duchesne ex Lam. - *Pumpkin*. Differs from C. *pepo* in its narrow terete female pedicels and usually huge depressed-globose fruits; (2n=40). Intrd; casual on tips and sewerage farms from use as vegetable; rather rare in S En but recurrent at 1 site in Herts; C America. The thin pedicel is diagnostic; some cultivars of C. *pepo* have depressed-globose fruits.

63. SALICACEAE - *Willow family*

Deciduous trees or shrubs; leaves alternate, rarely ± opposite, simple, usually serrate, stipulate, petiolate. Flowers much reduced, in racemose catkins, normally dioecious, each in axil of bract, each with either cup-like perianth or 1-2 small nectaries at base; male flowers with 1-many stamens; female flowers with 1-celled ovary with numerous ovules; styles 1(-2), short; stigmas 2, often bifid, often large; fruit a 2-valved capsule; seeds with long plume of hairs arising from base.

Dioecious, catkin-bearing trees or shrubs with simple flowers and plumed seeds borne in a capsule.

1 Flowers with 1-2 nectaries; cup-like perianth 0; bracts entire; winter buds with 1 outer scale **2. SALIX**
1 Flowers with cup-like perianth; nectaries 0; bracts toothed; winter buds with several outer scales **1. POPULUS**

1. POPULUS L. - *Poplars*

Trees; winter buds with several outer scales; flowers appearing before leaves, with toothed or deeply divided bracts, with cup-like perianth; nectaries 0; stamens 5-c.60.

The most important characters are the general habit of the tree and the shape of leaves on the dwarf lateral shoots (short-shoots) or extension shoots (long-shoots); leaves on suckers or epicormic shoots are often very different. The sex of the tree can be important in identification. Apart from the need for mature leaves, most characters are best observed in late spring, when shoot pubescence and colour, sex and tree-shape are all obvious.

1 Leaves of long-shoots densely tomentose on lowerside, sometimes becoming glabrous late in season 2
1 Leaves of long-shoots glabrous to slightly pubescent on lowerside 3
2 Leaves of long-shoots distinctly palmately 3-5-lobed; bracts subentire to dentate; trees usually female **1. P. alba**
2 Leaves of long-shoots not or scarcely palmately lobed, coarsely and bluntly dentate; bracts laciniate; trees mostly male **2. P. x canescens**
3 Leaves of long- and short-shoots coarsely and bluntly sinuate-dentate; bracts with long silky hairs **3. P. tremula**
3 Leaves shortly and regularly serrate; bracts glabrous or shortly pubescent 4
4 Not balsam-scented; leaves only slightly paler on lowerside; petioles strongly laterally compressed (*Black-poplars*) 5
4 Buds and young leaves distinctly balsam-scented; leaves much paler on lowerside; petioles nearly terete (*Balsam-* and *Black-/Balsam-poplars*) 6
5 Leaves without sessile glands on lamina near top of petiole, broadly cuneate (to truncate) at base, not minutely pubescent at margin (often slightly pubescent on face); tree either fastigiate with erect branches or with spreading branches and usually large burrs on trunk **4. P. nigra**
5 Many leaves with 1-2 small sessile glands on lamina near top of petiole, truncate or subcordate to broadly cuneate at base, minutely pubescent at margin when young; tree narrow or broad but not fastigiate and not with burrs on trunk **5. P. x canadensis**
6 Leaves serrate, the tip of each serration reaching nearer leaf apex than the base of the notch distal to it (*Black-/Balsam-poplar* hybrids) 7
6 Leaves crenate, the tip of each crenation not reaching nearer leaf apex than the base of the depression distal to it (*Balsam-poplars*) 9
7 Leaves ovate-elliptic to -rhombic, rounded to cuneate at base **6. P. x berolinensis**
7 Leaves broadly to triangular-ovate, truncate to cordate at base 8

8 Young branchlets conspicuously angled or ridged; young branchlets and petioles not or sparsely pubescent; suckers not produced; trees male or female **7. P. x generosa**

8 Young branchlets ± rounded; young branchlets and petioles conspicuously pubescent; mature trees producing abundant suckers; trees all female **8. P. x jackii**

9 Tree with subfastigiate outline; lower branches arising at c.45°, with few or no pendent twigs; trees all female **10. P. 'Balsam Spire'**

9 Tree rather narrowly pyramidal; some lower branches spreading and with pendent twigs; trees male or female 10

10 Young branchlets conspicuously angled or ridged; leaves usually widest not much above petiole apex and abruptly rounded there, with mostly truncate to subcordate base; suckers scarce **9. P. trichocarpa**

10 Young branchlets scarcely angled or ridged; leaves usually widest well above petiole apex and gradually rounded there, with mostly rounded base; suckers often abundant **11. P. balsamifera**

Other spp. - Several other spp. are grown in parks and some are now being used in groups or rows. Most common are 2 balsam-poplars: **P. laurifolia** Ledeb. (*Laurel-leaved Poplar*), from C Asia, differing from the others in its spreading habit, narrowly ovate leaves with rounded bases and acuminate apices, and very angled branchlets; and **P. maximowiczii** A. Henry (*Japanese Balsam-poplar*), from E Asia, with large ovate-elliptic leaves with very narrow cordate bases and apiculate twisted apices, and rounded branchlets. Cultivars of *P. maximowiczii* x *P. trichocarpa* (especially 'Androscoggin') have been grown for timber. **P. deltoides** Marshall, from N America and much used in hybridizations, is rarely planted in BI.

1. P. alba L. - *White Poplar*. Tree to 24m with broad spreading crown, many suckers and smooth grey bark; leaves palmately lobed, densely white-tomentose on lowerside; males very rare; (2n=38, 57). Intrd; much planted and often well natd from dense sucker growth, especially on coastal dunes; scattered in BI, especially in C & S Br and BI; Europe. **228**

2. P. x canescens (Aiton) Sm. (*P.* x *hybrida* M. Bieb.; *P. alba* x *P. tremula*) - *Grey Poplar*. Tree to 46m, similar to *P. alba* but more vigorous; leaves coarsely and bluntly sinuate-dentate, densely tomentose on lowerside at first, often much less so or subglabrous later; female rather rare; partially fertile; (2n=38, 57). Possibly native; much planted but rarely natd, probably never arising anew in BI but often alone or in groups among native taxa, especially in damp woods and by streams, perhaps reaching Br naturally as a hybrid; scattered throughout BI but absent from much of Sc; Europe. **228**

3. P. tremula L. - *Aspen*. Tree to 24m, rather similar in habit to *P. alba*; leaves suborbicular, coarsely and bluntly sinuate-dentate, glabrous or rarely sparsely pubescent at maturity; sucker leaves ovate-cordate, pubescent; 2n=38. Native; woods and hedgerows on all sorts of soils, often forming suckering thickets; throughout BI. **228**

4. P. nigra L. - *Black-poplar*. Trees of various habit; leaves rhombic-ovate to trullate, broadly cuneate (to truncate) at base, without glands, serrate with many small teeth.

1 Tree with spreading crown; trunk with large burrs **a. ssp. betulifolia**

1 Tree fastigiate, with ± erect branches; trunk without burrs 2

2 Young leaves and stems sparsely pubescent; male **d. 'Plantierensis'**

2 Young leaves and stems glabrous 3

3 Tree very narrowly fastigiate; male **b. 'Italica'**

3 Tree often more widely fastigiate; female **c. 'Gigantea'**

a. Ssp. betulifolia (Pursh) Dippel. Spreading tree to 33m with burred trunk; **228**

young leaves and stems sparsely pubescent; 2n=38. Native; in fields by streams and ponds, typically in river flood-plains; scattered throughout most of En and Wa, frequent in E Wa, C En and E Anglia, also in C Ir, planted elsewhere, mostly male.

b. 'Italica' (*P. nigra* var. *italica* Münchh.) - *Lombardy-poplar*. Narrowly fastigiate tree to 36m; trunk not burred; shoots glabrous; male only. Intrd; much planted in parks and as screening or windbreaks, not natd; throughout much of BI; garden origin.

c. 'Gigantea' - *Giant Lombardy-poplar*. Like 'Italica' but female and usually broader as in 'Plantierensis'. Intrd; grown in same way as 'Italica' but less common.

d. 'Plantierensis' (*P. nigra* ssp. *betulifolia* x *P.* 'Italica'). Similar to 'Italica' but often less narrowly fastigiate; young shoots pubescent; male only. Intrd; used similarly to 'Italica' and now as common; garden origin.

4 x 7. P. nigra x P. x jackii. In Leeds, MW Yorks, *P. x jackii* has been pollinated by *P. nigra* to give strongly regenerating hybrids on waste ground, first found 1983. The hybrids show variable characters of the parents on 1 plant: c.1/2 of the petioles are glabrous, 1/2 pubescent; c.1/2 the leaf-margins are serrate, 1/2 crenate; leaf-base is cuneate (as in *P. nigra*) in most cases but c.20% are rounded (shape not in either parent); leaf lowerside colour varies between those of the parents. This is effectively *P. balsamifera* x *P. deltoides* x *P. nigra*.

5. P. x canadensis Moench (*P. x euramericana* (Dode) Guinier; *P. nigra* x *P. deltoides*) - *Hybrid Black-poplar*. Trees of various habit, to >40m, those with spreading crowns differing from *P. nigra* in absence of burrs on trunks; usually largely up-swept, not down-curved, lower branches (not twigs); and leaves usually with sessile glands near junction with petiole; (2n=38). Intrd; many cultivars are grown for ornament, screening and forestry, only the more frequent are treated here; garden origin.

Multi-access key to cultivars of P. x canadensis

Crown of tree with rounded outline — A
Crown of tree narrow (but not fastigiate) — B
  Twigs of lowest branches stiff, many up-pointed — C
  Twigs of lowest branches mostly pendent, often out- or up-turned at tip — D
Tree male — E
Tree female — F
  Young twigs glabrous — G
  Young twigs pubescent — H
Young leaves green almost from start — I
Young leaves bronze-coloured at first — J
  Mature leaves mostly truncate to very broadly cuneate at base — K
  Mature leaves mostly distinctly cuneate at base — L
Early leafing — M
Leafing in mid-season — N
Late leafing — O
Very late leafing — P

ACEGJKO — **b. 'Gelrica'**
ACEGJKP — **a. 'Serotina'**
ACFGJKN — **c. 'I78'**
ADFGJKO — **d. 'Regenerata'**
ADFHILN — **h. 'Marilandica'**
BCEGJKM — **f. 'Heidemij'**
BCEGJLN — **g. 'Eugenei'**
BCEHJKM — **e. 'Robusta'**

a. 'Serotina' (*P. nigra* ssp. *nigra* x *P. deltoides*) - *Black-Italian Poplar*. Grown for **228**
ornament and screening in all sorts of situations; the commonest *Populus* in such places throughout BI; arose France early 1700s.

b. 'Gelrica'. Similar to 'Serotina' but bark persistently whitish-grey until becoming

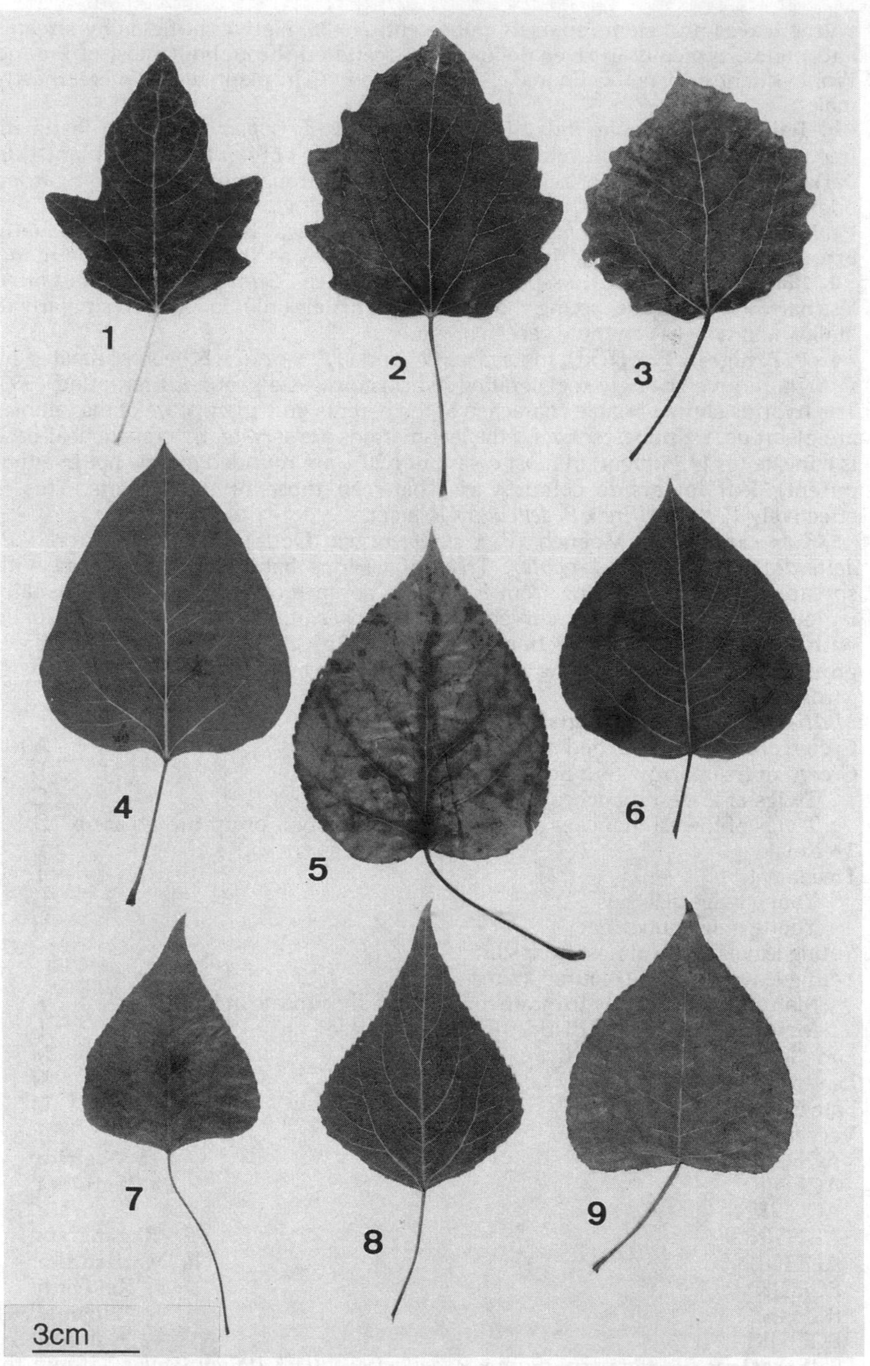

FIG 228 - Leaves of ***Populus***. 1, ***P. alba***. 2, ***P. x canescens***. 3, ***P. tremula***. 4, ***P. trichocarpa***. 5, ***P. x jackii***. 6, ***P.* 'Balsam Spire'**. 7, ***P. nigra***. 8, ***P. x canadensis* 'Marilandica'**. 9, ***P. x canadensis* 'Serotina'**.

fissured; grown mostly for timber in small plantations; arose Holland late 1800s.

c. 'I78'. Similar to 'Gelrica' and 'Serotina' but female and earlier in leaf, now much more grown than either for timber, but only in S Br; spontaneous in Italy in 1920s.

d. 'Regenerata' (*P*. 'Marilandica' x *P*. 'Serotina') - *Railway Poplar*. Crown with characteristically asymmetrically ascending branches and without obvious main axis; grown for ornament and screening, especially by railways in SE En; arose France 1814.

e. 'Robusta' (*P. deltoides* 'Cordata' x *P. nigra* 'Plantierensis'). Grown as timber in small plantations and as screening by roads, factories and canals; much of Br, especially S; arose France 1895.

f. 'Heidemij'. Apparently told from 'Robusta' only by the glabrous young shoots, and similarly grown; arose Netherlands 1891. Although both are 'early leafers', 'Heidemij is slightly later than 'Robusta'.

g. 'Eugenei' (*P*. 'Regenerata' x *P. nigra* 'Italica') - *Carolina Poplar*. Grown in same situations as 'Robusta'; similar distribution but less common; arose France 1832.

h. 'Marilandica' (*P. nigra* ssp. *nigra* x *P*. 'Serotina'). Crown narrower and more 228
symmetrical than in 'Regenerata' and with obvious main axis; planted as ornamental and for screening over much of Br, especially S; arose Europe c.1800.

Hybrid Black Poplars also form an extensive young natural population on damp rough ground at Hackney, Middlesex. The parentage is probably *P*. x *canadensis* 'Marilandica' x *P*. x *canadensis* 'Serotina'.

5 x 7. P. x canadensis 'Serotina' x P. x jackii occurs in the Hackney hybrid population and also nearby in S Essex; the plants are obviously Black-Balsam poplar hybrids and the suggested parentage is the only likely one in the area. This has the same basic parentage as *P. nigra* x *P. x jackii*, but in different proportions and from different genotypes.

6. P. x berolinensis (K. Koch) Dippel (*P. nigra* 'Italica' x *P. laurifolia*) - *Berlin Poplar*. Narrow, subfastigiate tree to 33m, not suckering; leaves ovate-elliptic to -rhombic, rounded to cuneate at base, acuminate at apex; petioles sparsely pubescent; male and female; (2n=38). Intrd; formerly much planted for amenity and still found surviving; Br, mainly S; arose in Berlin in 1860s.

7. P. x generosa A. Henry (*P. deltoides* x *P. trichocarpa*) - *Generous Poplar*. Spreading tree to 40m, not suckering; leaves triangular-ovate, cordate to truncate at base, abruptly acuminate at apex; petioles subglabrous; male and female; (2n=38). Intrd; formerly much planted for amenity and still found surviving; Br, mainly S; arose in England in 1912. Several more modern and superior cultivars are now being grown for timber, e.g. 'Beaupré'.

8. P. x jackii Sarg. (*P. x candicans* auct. non Aiton, *P. x gileadensis* Rouleau; *P.* 228
deltoides x *P. balsamifera*) - *Balm-of-Gilead*. Spreading tree to 30m with many suckers and very viscid buds; leaves broadly to triangular-ovate, truncate to cordate at base, abruptly acuminate to cuspidate at apex; petioles pubescent; female only; (2n=38). Intrd; planted for ornament in damp woods and by rivers and ponds, often natd by suckering; Br, mainly S; artificially produced, also native in N America. Commonest cultivar is 'Gileadensis', but 'Aurora', with large variegated leaves, is now much grown in parks and by roads. Fruiting catkins (c.10-18cm) are ≥2x as long as those of other balsam-poplars except *P. maximowiczii*.

9. P. trichocarpa Torr. & A. Gray ex Hook. - *Western Balsam-poplar*. Tree to 35m 228
with narrow outline but spreading lower branches, with few or no suckers; leaves ovate to narrowly so, mostly truncate to subcordate at base, acuminate at apex; petioles inconspicuously and very shortly (but often densely) pubescent; mostly male; (2n=38). Intrd; frequently planted for ornament and for timber in small plots; scattered in Br; W N America. Several new cultivars are now being grown, e.g. 'Trichobel', with very fast growth-rates.

10. P. 'Balsam Spire' ('Tacatricho 32'; *P. trichocarpa* x *P. balsamifera*) - *Hybrid* 228
Balsam-poplar. Tree to >30m with very narrow outline, without suckers; leaves similar to those of *P. balsamifera*; female only. Used for screening, as windbreak and

in plantations for timber, now 1 of the most planted poplars in Br; arose N America 1930s.

11. P. balsamifera L. (*P. tacamahacca* Mill.) - *Eastern Balsam-poplar*. Tree to 25m with narrow outline, often with many suckers; leaves ovate to broadly so, mostly rounded at base, abruptly acuminate at apex; petioles as in *P. trichocarpa*; male and female; (2n=38, 76). Intrd; planted as for *P. trichocarpa* but less common; E N America.

2. SALIX L. - *Willows*

Trees or dwarf to tall shrubs; winter buds with 1 outer scale; flowers appearing before, with or after leaves, with entire bracts, with 1-2 nectaries; cup-like perianth 0; stamens 1-5(12).

Identification is often made difficult by the extensive degree of hybridization (64 combinations known in BI, of which 10 are hybrids between 3 spp. and 1 is a hybrid between 4 spp.). The 14 hybrids treated fully here are not necessarily the most common, but those likely to be found without parents nearby. Most of these are crosses between the taller lowland species that have been planted for ornament and basket-work.

In many cases catkins and both young and mature leaves are desirable for identification; keys based on catkins alone are not satisfactory. Mature leaves should be collected in Jul-Sep, before senescence; abnormally vigorous shoots (e.g. suckers) should be avoided. In this account the so-called catkin-scales (scales associated with each flower) are termed bracts, which they truly are. 'Twigs' refer to those that have completed 1-2 years' growth, not current growth. Except locally, where plants have been clonally propagated, the spp. are generally represented by both sexes roughly equally, but many of the hybrids are much more commonly, or only, female. Bisexual catkins are not rare, especially in hybrids, including some of those noted as 'female only'.

General key

1 Shrubs <80(100)cm high (stems may be longer but not erect) ***Key A***
1 Tree or shrub >1m high 2
  2 Shrub <1.5(2)m high ***Key B***
  2 Tree or shrub >2m high 3
3 Leaves closely and finely serrate (>(3)5 serrations/cm at leaf midpoint); bracts yellowish (except *S. daphnoides* and *S. acutifolia*) ***Key C***
3 Leaves entire, obscurely crenate-serrate, or coarsely and irregularly serrate (<4(7) serrations/cm at leaf midpoint); bracts dark or dark-tipped (except *S. x mollissima*, *S. elaeagnos* and *S. udensis*) 4
  4 Leaves <3x as long as wide, or mostly so ***Key D***
  4 Leaves >3x as long as wide, or mostly so ***Key E***

Key A - Shrubs <80(100)cm high

1 Stems usually <10cm; leaves not or scarcely longer than wide, rounded to emarginate at apex; catkins appearing when leaves ± mature, all or most from terminal buds 2
1 Stems usually >10cm; leaves longer than wide, usually obtuse to acute at apex; catkins appearing before leaves mature, mostly from lateral buds 3
  2 Leaves entire, with distinctly impressed veins on upperside; petioles mostly >10mm **37. S. reticulata**
  2 Leaves crenate-serrate; veins not impressed on leaf upperside; petioles mostly <4mm **36. S. herbacea**
3 Leaves densely pubescent to tomentose at least until maturity at least on lowerside 4
3 Leaves glabrous to sparsely pubescent at maturity on both sides 7

4 Stipules mostly large and ± persistent at maturity 5
4 Stipules small and mostly soon dropping off 6
5 Leaves ± entire; catkins densely silky with yellow hairs; Scottish Highlands only **33. S. lanata**
5 Leaves crenate-serrate; catkins with greyish to whitish hairs; widespread **27. S. aurita**
6 Leaves silky-pubescent on lowerside; widespread **31. S. repens**
6 Leaves tomentose on lowerside; mountains of N Br only **32. S. lapponum**
7 Leaves cordate at base **28. S. eriocephala**
7 Leaves cuneate at base 8
8 Stipules mostly large and persistent to maturity 9
8 Stipules mostly small and soon dropping off 10
9 Leaves shining green on both sides, not blackening when dried; ovary usually pubescent **35. S. myrsinites**
9 Leaves green on upperside, ± glaucous on lowerside, blackening when dried; ovary mostly ± glabrous **29. S. myrsinifolia**
10 Leaves mostly <2.5 x 1.5cm **34. S. arbuscula**
10 Leaves mostly >2.5 x 1.5cm **30. S. phylicifolia**

Key B - Shrubs 1-1.5(2)m high
1 Extension growth malformed, with contorted, flattened and laterally fused (fasciated) branches **18. S. udensis**
1 Extension growth not malformed and fasciated 2
2 Leaves glabrous to sparsely pubescent at maturity on both sides 3
2 Leaves densely pubescent to tomentose at least until maturity at least on lowerside 5
3 Leaves cordate at base **28. S. eriocephala**
3 Leaves cuneate at base 4
4 Stipules mostly large and persistent at maturity; leaves blackening when dried; ovary mostly ± glabrous, with pedicel c.1mm; twigs usually pubescent **29. S. myrsinifolia**
4 Stipules mostly small and soon dropping off; leaves not blackening when dried; ovary mostly pubescent, with 0 to very short pedicel; twigs usually ± glabrous **30. S. phylicifolia**
5 Leaves mostly >4x as long as wide, tomentose on lowerside; female only **23. S. x fruticosa**
5 Leaves <4x as long as wide, or if more then appressed-pubescent on lowerside 6
6 Stipules mostly large and ± persistent to maturity **27. S. aurita**
6 Stipules small and mostly soon dropping off 7
7 Leaves silky-pubescent on lowerside; widespread **31. S. repens**
7 Leaves tomentose on lowerside; mountains of N Br only **32. S. lapponum**

Key C - Trees or shrubs >2m high; leaves closely and finely serrate; bracts yellowish (except dark-tipped in *S. daphnoides* and *S. acutifolia*)
1 Twigs dark purple, with conspicuous whitish bloom easily rubbed off; bracts dark-tipped 2
1 Twigs variously coloured, rarely dark purple, without bloom; bracts yellowish 3
2 Leaves >5x as long as wide; ultimate branchlets ± pendent **15. S. acutifolia**
2 Leaves usually <5x as long as wide; branchlets not pendent **14. S. daphnoides**
3 Branchlets pendent (weeping willows) 4
3 Branchlets erect or spreading 5

4 Leaves pubescent to silky-pubescent ± until maturity; ovary little longer than subtending bract **8. S. x sepulcralis**
4 Leaves glabrous to sparsely pubescent even when very young; ovary much longer than subtending bract **6. S. x pendulina**
5 Leaves elliptic to ovate, 2-4x as long as wide; stamens ≥4 **1. S. pentandra**
5 Leaves narrowly elliptic to lanceolate, >(3)4x as long as wide; stamens ≤4 6
6 Stipules mostly large and ± persistent to maturity; bark smooth, flaking off in large patches or peels 7
6 Stipules mostly 0 or small and soon dropping off; bark furrowed, not flaking off 8
7 Young leaves and twigs glabrous; twigs with strong ridges or angles **9. S. triandra**
7 Young leaves and twigs pubescent; twigs terete or obscurely ridged or angled **10. S. x mollissima**
8 Mature leaves silky-pubescent, or ± glabrous and dull on upperside 9
8 Mature leaves glabrous and rather glossy on upperside 10
9 Leaves subglabrous on both sides at maturity **5. S. x rubens**
9 Leaves ± silky-pubescent at least on lowerside at maturity **7. S. alba**
10 Twigs and leaves glabrous from the first 11
10 Twigs and leaves sparsely pubescent or leaves pubescent at margin at first 12
11 Leaves mostly with 3-6 serrations/cm; twigs very fragile at branches; buds shiny; male flowers with 2-3 stamens **4. S. fragilis**
11 Leaves mostly with 6-10 serrations/cm; twigs scarcely fragile at branches; buds scarcely shiny; male flowers mostly with 3-4 stamens **2. S. x meyeriana**
12 Twigs yellowish-orange to reddish **5. S. x rubens**
12 Twigs brownish 13
13 Leaves mostly with 6-10 serrations/cm; male flowers mostly with 3-4 stamens; male only **3. S. x ehrhartiana**
13 Leaves mostly with 3-6 serrations/cm; male flowers with 2-3 stamens **4. S. fragilis**

Key D - Trees or shrubs >2m high; leaves entire, obscurely crenate-serrate, or coarsely and irregularly serrate, <3x as long as wide; bracts dark or dark-tipped

1 Catkins developing with leaves, usually with small leaf-like bracts on peduncle 2
1 Catkins developing before leaves on bare twigs, with reduced or 0 bracts on peduncle 3
2 Stipules mostly large and persistent at maturity; leaves blackening when dried; ovary mostly ± glabrous, with pedicel c.1mm; twigs usually pubescent **29. S. myrsinifolia**
2 Stipules mostly small and soon dropping off; leaves not blackening when dried; ovary mostly pubescent, with 0 or very short pedicel; twigs usually ± glabrous **30. S. phylicifolia**
3 Wood of twigs without ridges under bark 4
3 Wood of twigs with longitudinal ridges under bark 5
4 Leaves glabrous, >2x as long as wide **11. S. purpurea**
4 Leaves pubescent on lowerside, ≤2x as long as wide **24. S. caprea**
5 Leaves very sparsely pubescent on lowerside (mostly on veins) at maturity; stipules small, soon falling off **26. S. x laurina**
5 Leaves distinctly pubescent on lowerside at maturity; stipules large, persistent to maturity 6

6 Leaves mostly >8cm; catkins mostly >3cm; female only **20. S. x calodendron**
6 Leaves mostly <8cm; catkins mostly <3cm 7
7 Leaves distinctly rugose, undulate at margin, without rust-coloured hairs on lowerside **27. S. aurita**
7 Leaves scarcely rugose or undulate, usually with some rust-coloured hairs on lowerside **25. S. cinerea**

Key E - Trees or shrubs >2m high; leaves entire, obscurely crenate-serrate, or coarsely and irregularly serrate, >3x as long as wide; bracts dark or dark-tipped (except yellowish in *S. x mollissima* and *S. elaeagnos*)
1 Extension growth malformed, with contorted, flattened and laterally fused (fasciated) branches **18. S. udensis**
1 Extension growth not malformed and fasciated 2
2 Leaves glabrous to sparsely pubescent on lowerside at maturity 3
2 Mature leaves pubescent to tomentose on lowerside 6
3 Leaves glabrous even when young, commonly opposite or subopposite; male flowers apparently with 1 stamen **11. S. purpurea**
3 Leaves pubescent when young, alternate; male flowers with 2 free or only partially fused stamens 4
4 Leaves glabrous at maturity **13. S. x forbyana**
4 Leaves sparsely pubescent on lowerside at maturity 5
5 Catkins appearing with leaves; bracts yellowish; male flowers with 2-3 free stamens **10. S. x mollissima**
5 Catkins appearing before leaves; bracts dark-tipped; male flowers with 2 free or partly united stamens **12. S. x rubra**
6 Leaves linear or nearly so, mostly >(6)10x as long as wide 7
6 Leaves lanceolate to narrowly elliptic, mostly <6x as long as wide 8
7 Leaves densely silky-pubescent (hairs appressed) on lowerside; catkin-scales reddish-brown, darker at apex; filaments free; ovary densely pubescent **16. S. viminalis**
7 Leaves tomentose (hairs matted) on lowerside; catkin-scales yellowish, not darker at apex; filaments fused at base, often ± to 1/2 way; ovary glabrous **17. S. elaeagnos**
8 Wood of twigs without ridges under bark **19. S. x sericans**
8 Wood of twigs with longitudinal ridges under bark 9
9 Leaves pubescent on lowerside at maturity, but not densely so or soft to touch **22. S. x smithiana**
9 Leaves densely pubescent, velvety or silky to touch 10
10 Most leaves <4x as long as wide, not undulate at margin **20. S. x calodendron**
10 Most leaves >4x as long as wide, strongly to slightly (or not) undulate at margin 11
11 Leaves usually slightly undulate at margin, velvety-tomentose on lowerside **21. S. x stipularis**
11 Leaves usually strongly undulate at margin, pubescent with appressed or crisped hairs on lowerside **23. S. x fruticosa**

Other spp. - Several spp. are cultivated for ornament and may rarely appear as isolated plants in the wild. **S. babylonica** L., from China, the original *Weeping Willow,* is rarely grown in BI but much mis-recorded for its hybrids with *S. alba* and *S. fragilis*. A number of exotic spp. are being planted for biomass production (along with several native or natd taxa, q.v.) and might soon become part of our wild flora; commonest are **S. bebbiana** Sarg. from N America, and **S. disperma** Roxb. ex D. Don and **S. wilhelmsiana** M. Bieb. (*S. spaethii* Koopmann) from Asia.

Section* 1 - *SALIX (spp. 1-10). Trees or tall shrubs; leaves finely and closely serrate, mostly lanceolate to narrowly elliptic, acuminate; catkins long and slender, borne on leafy peduncles arising from lateral buds; bracts yellowish; stamens 2-8(12), free; male flowers with 2 nectaries, female with 1-2; ovary glabrous except in some hybrids.

1. S. pentandra L. - *Bay Willow*. Tree to 10(18)m; twigs reddish-brown, glossy, **238**
glabrous; leaves 5-12 x 2-5cm, glabrous; stamens (4)5-8(12); 2n=76. Native; wet ground and by ponds and streams; N & C Br and Ir, planted and sometimes ± wild further S.

2. S. x meyeriana Rostk. ex Willd. (*S. pentandra* x *S. fragilis*) - *Shiny-leaved Willow*. **238**
Tree to 15m; twigs brown, glossy, glabrous; leaves 5-12 x 1.5-4cm, glabrous; stamens (2)3-4(5). Native; damp areas and by rivers; frequent in N En, N Wa and N Ir but there and also outside range of *S. pentandra* often of cultivated origin.

3. S. x ehrhartiana Sm. (*S. pentandra* x *S. alba*) - *Ehrhart's Willow*. Tree to **238**
15(25)m; twigs brownish, slightly glossy, glabrous; leaves 6-10 x 1.3-2.5cm, appressed-pubescent at first, becoming glabrous; stamens (2)3-4(5); male only. Probably intrd; planted and often ± wild; scattered in En, especially SE; garden origin but native in Europe.

4. S. fragilis L. (*S. decipiens* Hoffm.) - *Crack-willow*. Tree to 25m; twigs very brittle **238**
at branches, often sparsely pubescent at first but glabrous later, pale brown; leaves 9-15 x 1.5-3(4)cm, at first sparsely pubescent, soon glabrous, ± glossy on upperside, glaucous on lowerside, more coarsely serrate than other taxa in subg. *Salix*; stamens 2(-3); 2n=76. Native; common in damp places over most of lowland BI, but often planted. Variable; the commonest variant (var. **russelliana** (Sm.) W.D.J. Koch - *Bedford Willow*) is a female tree with long narrow leaves with rather uneven teeth; var. **furcata** Ser. ex Gaudin is a male tree with forked catkins and rather wide leaves; var. **decipiens** (Hoffm.) W.D.J. Koch is also male but has unbranched catkins and shiny very pale twigs; all 3 are presumably of garden origin. Native plants are var. **fragilis**, of both sexes, with leaves at first sparsely pubescent and with even serration.

4 x 9. S. fragilis x S. triandra = S. x alopecuroides Tausch (*S. x speciosa* Host) occurs very sparsely in Br and Ir; it is intermediate, with 2-3 stamens; female unknown.

5. S. x rubens Schrank (*S. x basfordiana* Scaling ex Salter; *S. alba* x *S. fragilis*) - **238**
Hybrid Crack-willow. Tree to 30m; represented by whole range of variants and cultivars linking *S. fragilis* with *S. alba* var. *caerulea* (nothovar. **rubens**) and with *S. alba* var. *vitellina* (nothovar. **basfordiana** (Scaling ex S.J.A. Salter) Meikle); the latter have yellowish-orange twigs; 2n=76. Native; frequent over most of lowland BI but perhaps most often of cultivated origin.

6. S. x pendulina Wender. (*S. x blanda* Andersson; *S. fragilis* x *S. babylonica*) - **238**
Weeping Crack-willow. Tree to 15m with distinctive weeping habit; twigs pale brown, glabrous, somewhat brittle at branches, glossy; leaves 10-12 x 1.5-2cm, glabrous almost from first; catkins female or bisexual. Intrd; frequently planted and found in wild in BI, but less common than *S. x sepulcralis*.

7. S. alba L. - *White Willow*. Tree to 33m; twigs pubescent when young, becoming **238**
glabrous and ± glossy, brown, or yellowish to reddish (var. **vitellina** (L.) Stokes (ssp. *vitellina* (L.) Arcang.) - *Golden Willow*); leaves 5-10(12) x 0.5-1.5cm, densely whitish-pubescent at first, becoming sparsely so at maturity, or 6-12 x 1.5-2.5cm and becoming ± glabrous on upperside at maturity (var. **caerulea** (Sm.) Dumort. (ssp. *caerulea* (Sm.) Rech. f.) - *Cricket-bat Willow*); stamens 2; 2n=76. Native; marshes, wet hollows and by streams and ponds; common over most of lowland BI, rarer in W and N Sc, but often planted. *S. alba* 'Britzensis' is a much planted cultivar of var. *vitellina* with bright orange-red twigs.

8. S. x sepulcralis Simonk. (*S. x chrysocoma* Dode; *S. alba* x *S. babylonica*) - *Weeping* **238**
Willow. Tree to 22m with distinctive weeping habit; twigs sparsely pubescent at

first, soon glabrous and ± glossy, brownish, or yellowish (nothovar. **chrysocoma** (Dode) Meikle); leaves 7-12 x 0.7-1.8cm, sparsely pubescent at first, soon ± glabrous; stamens 2; catkins often bisexual; (2n=76). Intrd; much planted, often in masses and often persisting in wild places; lowland BI; garden origin.

9. S. triandra L. - *Almond Willow*. Small tree or shrub to 10m; twigs glabrous, pale 238
brown, glossy, ridged; leaves (2)4-11(15) x 1-3(4)cm, glabrous; stamens 3; 2n=38. Native; damp places; frequent in C & S En, much less so in Wa, N En, Sc and Ir and probably always intrd; much planted for basketry.

9 x 11. S. triandra x S. purpurea (= *S.* x *leiophylla* auct. non E.G. & A. Camus) is known only as a male from an osier-bed in Notts.

10. S. x mollissima Hoffm. ex Elwert (*S. triandra* x *S. viminalis*) - *Sharp-stipuled* 238
Willow. Shrub to 5m; twigs glabrous or becoming so, brownish, dull; leaves 8-13 x 1-1.5cm; stamens 2-3; 2n=38. Possibly native; damp places; scattered over Br and Ir, frequent in C & S En but very local elsewhere. Nothovar. **hippophaifolia** (Thuill.) Wimm. is closer to *S. viminalis*, with subentire or obscurely serrate leaves sparsely pubescent on lowerside, and with catkins <3.5cm (both sexes); nothovar. **undulata** (Ehrh.) Wimm. is closer to *S. triandra*, with closely serrate ± glabrous leaves and longer narrower catkins (female only). The latter is now being planted for biomass production.

Section* 2 - *VETRIX Dumort. (spp. 11-34). Shrubs or small trees; leaves mostly entire to coarsely and irregularly serrate, orbicular to linear; catkins short and wide, sessile or borne on leafy peduncles, arising from lateral buds; bracts dark or with dark tips; stamens 2, free or variously fused, sometimes appearing 1; nectary 1; ovary usually pubescent.

11. S. purpurea L. - *Purple Willow*. Shrub 1.5-5m; twigs glabrous, yellowish to 238
purplish-brown, ± glossy; leaves often opposite to subopposite, (2)3-8(10) x 0.5-3cm, glabrous, or pubescent at first, entire or obscurely serrate; stamen apparently 1; 2n=38. Native; damp places; scattered throughout BI, locally common, often planted. The type has linear-oblong to linear-oblanceolate ± entire leaves. Plants with broader (oblanceolate to narrowly obovate) serrulate leaves are sometimes recognized as ssp. **lambertiana** (Sm.) A. Neumann ex Rech. f.; on the Continent this is commoner in the lowlands, but ssp. **purpurea** is the commoner in BI and many intermediates occur.

11 x 16 x 24 x 25. S. purpurea x S. viminalis x S. caprea x S. cinerea = S. x taylorii Rech. f. has been reported, with some doubt, from Angus; endemic.

11 x 25. S. purpurea x S. cinerea = S. x pontederiana Willd. (*S. x sordida* A. Kern.) is very scattered in Br.

11 x 25 x 27. S. purpurea x S. cinerea x S. aurita = S. x confinis E.G. & A. Camus has been recorded from M Perth.

11 x 27. S. purpurea x S. aurita = S. x dichroa Döll is known from Cheviot.

11 x 27 x 30. S. purpurea x S. aurita x S. phylicifolia = S. x sesquitertia F.B. White has been found in Dumfriess; endemic.

11 x 29. S. purpurea x S. myrsinifolia = S. x beckiana L.C. Beck occurs in MW Yorks and S Northumb.

11 x 30. S. purpurea x S. phylicifolia = S. x secernata F.B. White is known from S Northumb, Dumfriess and Dunbarton; endemic.

11 x 31. S. purpurea x S. repens = S. x doniana G. Anderson ex Sm. has been found in S Lancs and M Perth.

12. S. x rubra Huds. (*S. purpurea* x *S. viminalis*) - *Green-leaved Willow*. Shrub or 238
small tree to 7m; twigs glabrous when mature, yellowish-brown, ± glossy; leaves 4-12(15) x 0.8-1(1.5)cm, at first densely pubescent, becoming glabrous or remaining pubescent, ± entire to remotely serrate; stamens free or partly united; 2n=38. Native; wet places, often with parents, but also relic of cultivation; throughout most of Br and Ir.

13. S. x forbyana Sm. (*S. purpurea* x *S. viminalis* x *S. cinerea*) - *Fine Osier*. Erect shrub to 5m; twigs glabrous, yellowish, glossy; leaves 3-12 x 0.8-2.5cm, at first pubescent, becoming glabrous, obscurely serrate; stamens free or partly united (male very rare). Probably intrd; wet places, often as relic of osier cultivation; scattered in En, Wa, S Sc and N Ir; garden origin. 238

14. S. daphnoides Vill. - *European Violet-willow*. Tall shrub or slender tree to 10(18)m; twigs glabrous, violet-brown, glossy but with dense whitish bloom, not pendent; leaves (4)7-12(14) x (1)2-3(4)cm, soon glabrous, closely and finely serrate; stamens free; 2n=38. Intrd; widely planted (both sexes) and sometimes found in wild places; very sparsely scattered in Br; Scandinavia. 238

15. S. acutifolia Willd. (*S. daphnoides* ssp. *acutifolia* (Willd.) Zahn) - *Siberian Violet-willow*. Very similar to *S. daphnoides* but leaves 6-16 x 1-2cm; twigs very slender, ± pendent at ends; 2n=38. Intrd; widely planted (male only) and found in similar places to *S. daphnoides;* U.S.S.R. 238

16. S. viminalis L. - *Osier*. Erect shrub or small tree to 6(-10)m; twigs pubescent at first, becoming glabrous, rather dull, yellowish-brown to brown; leaves 10-15(18) x 0.5-1.5(2.5)cm, densely silky-pubescent on lowerside, dull and sparsely pubescent on upperside, with ± entire revolute margins; stamens free; 2n=38. Native; mostly damp places; common in lowland BI, but perhaps not native in N & W. Several clones of *S. viminalis* and of some of its hybrids are now being planted as short-rotation coppice for biomass production, and will become commoner in the wild. They are known mainly under cultivar names. 239

16 x 25 x 31. S. viminalis x S. cinerea x S. repens = S. x angusensis Rech. f. occurs on sand-dunes in Angus; endemic.

16 x 25 x 27. S. viminalis x S. cinerea x S. aurita = S. x hirtei Strähler has not been recorded from the wild in BI, but might occur; several cultivars are being planted for biomass production. *Salix x hirtei*, *S. x calodendron* and *S. x stipularis* form a perplexing group of hybrids.

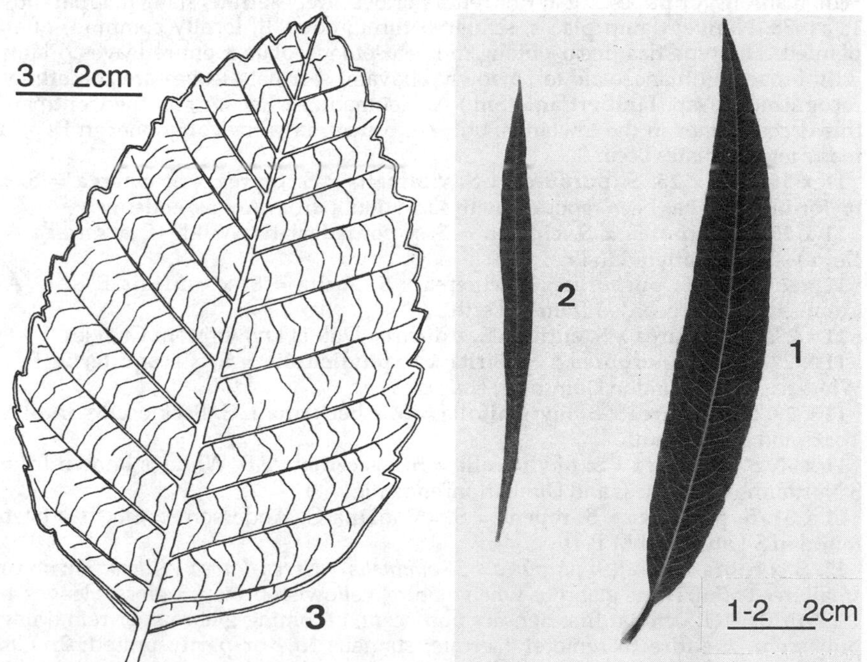

FIG 236 - Leaves of ***Salix, Alnus***. 1, ***Salix udensis***. 2, ***S. elaeagnos***. 3, ***Alnus rubra***.

16 x 29. S. viminalis x S. myrsinifolia = S. x seminigricans E.G. & A. Camus was found in M Perth in 1985.

16 x 31. S. viminalis x S. repens = S. x friesiana Andersson occurs in S Lancs, Westmorland and Sutherland.

17. S. elaeagnos Scop. - *Olive Willow*. Erect shrub to 6m; twigs densely pubescent 236
in first year, glabrous later, yellowish- or reddish-brown; leaves 5-15 x 0.3-0.8cm, tomentose on lowerside, dull and pubescent on upperside, with ± entire strongly revolute margins; stamen filaments fused at base up to 1/2 way; (2n=38). Intrd; grown in gardens and becoming ± natd where neglected or thrown out; scattered in En N to MW Yorks; S Europe. Vegetatively close to *S. viminalis*, but differs in pubescence of leaf lowerside, and also in 3 catkin characters (see Key E, couplet 7).

18. S. udensis Trautv. & C.A. Mey. (*S. sachalinensis* F. Schmidt) - *Sachalin Willow*. 236
Shrub or small tree (to 30m in Asia); twigs glabrous, shiny, reddish- to yellowish-brown; leaves 6-15 x 0.8-3cm, sparsely pubescent on lowerside at maturity, crenate-serrate; stamens free or fused for <1/2 way; (2n=38). Intrd; garden throwout often thriving where dumped or planted in wet ground; several sites NW En to C Sc; E Asia. Only the distinctive contorted and fasciated male clone 'Sekka' is grown and natd here.

19. S. x sericans Tausch ex A. Kern. (*S. x laurina* auct. non Sm.; *S. viminalis* x *S.* 239
caprea) - *Broad-leaved Osier*. Shrub or small tree to 9m; twigs at first pubescent, soon ± glabrous, ± glossy, yellowish to reddish; leaves 6-12 x 1.3-3cm, densely grey-pubescent on lowerside, becoming glabrous on upperside, subentire to remotely serrate; stamens free. Native; often with parents but also relic of cultivation; common in most of BI. Now being planted for biomass production.

20. S. x calodendron Wimm. (*S. x dasyclados* auct. non Wimm.; *S. viminalis* x *S.* 239
caprea x *S. cinerea*) - *Holme Willow*. Erect shrub or small tree to 12m; twigs persistently densely pubescent; leaves 7-18 x 2.5-5cm, densely grey-pubescent on lowerside, sparsely pubescent on upperside, subentire to remotely serrate; female only; (2n=76). Native; in damp places not always near any of its parents, probably often intrd; frequent over most of BI. Now being planted for biomass production.

21. S. x stipularis Sm. (*S. viminalis* x *S. caprea* x *S. aurita*) - *Eared Osier*. Erect 239
shrub or small tree to 10m; twigs densely greyish-pubescent; leaves 6-11 x 1-2.5cm, densely grey-pubescent on lowerside, sparsely pubescent on upperside, entire, often ± undulate; female only; (2n=114). Native; damp places, often planted, scattered in N Br and N Ir.

22. S. x smithiana Willd. (*S. x geminata* J. Forbes; *S. viminalis* x *S. cinerea*) - *Silky-* 239
leaved Osier. Erect shrub or small tree to 9m; twigs densely and persistently pubescent; leaves 6-11 x 0.8-2.5cm, pubescent on lowerside, becoming glabrous on upperside; stamens free; (2n=38). Native; often with parents but sometimes not; common throughout most of BI.

23. S. x fruticosa Döll (*S. viminalis* x *S. aurita*) - *Shrubby Osier*. Erect shrub to 5m, 239
twigs sparsely to densely grey-pubescent; leaves 4-10 x 0.7-2cm, grey-pubescent on lowerside, subglabrous to sparsely pubescent on upperside, subentire or irregularly crenate, often undulate; female only. Native; often with parents but sometimes not; scattered over Br and Ir, especially in N & W. Now being planted for biomass production.

24. S. caprea L. - *Goat Willow*. Shrub or small tree to 10(19)m; twigs pubescent to 239
sparsely so at first, becoming glabrous, glossy or dull; stamens free; 2n=38. Native.

a. Ssp. caprea. Twigs quickly becoming glabrous; leaves 5-12 x 2.5-8cm, densely pubescent on lowerside, subglabrous to sparsely pubescent on upperside, irregularly undulate-serrate. Damp and rough ground, hedges and open woodland; locally common to abundant throughout BI.

b. Ssp. sphacelata (Sm.) Macreight (ssp. *sericea* (Andersson) Flod., *S. coaetanea* (Hartm.) Flod.). Twigs and leaves remaining pubescent for longer; leaves 3-7 x 1.5-4.5cm, densely appressed-pubescent on lowerside, entire to obscurely serrate. Damp ground on mountains; Sc.

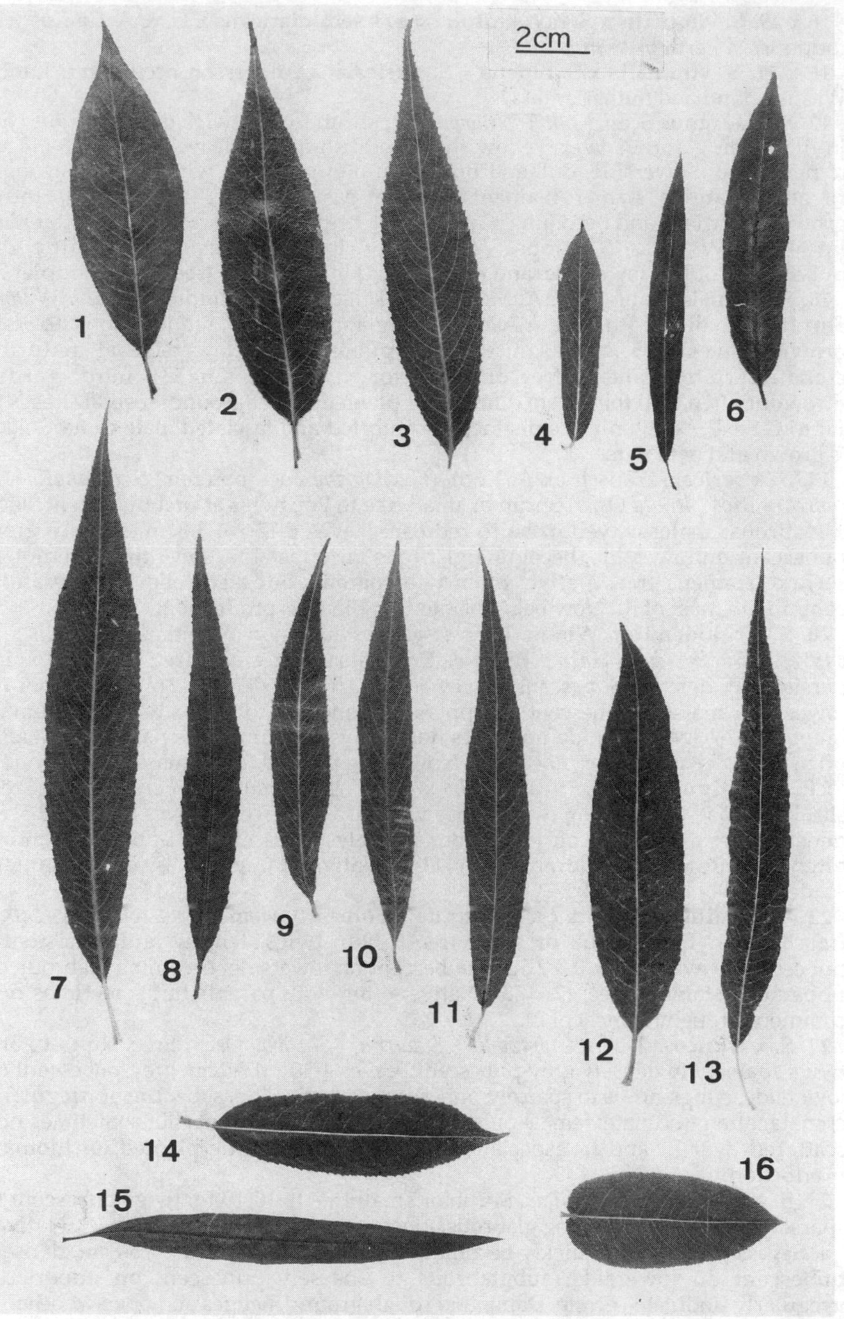

FIG 238 - Leaves of *Salix*. 1, *S. **pentandra***. 2, *S. **x meyeriana***. 3, *S. **x ehrhartiana***. 4, *S. **purpurea***. 5, *S. **x rubra***. 6, *S. **x forbyana***. 7, *S. **fragilis***. 8, *S. **x rubens***. 9, *S. **x pendulina***. 10, *S. **alba***. 11, *S. **x sepulcralis***. 12, *S. **daphnoides***. 13, *S. **acutifolia***. 14, *S. **triandra***. 15, *S. **x mollissima***. 16, *S. **eriocephala***.

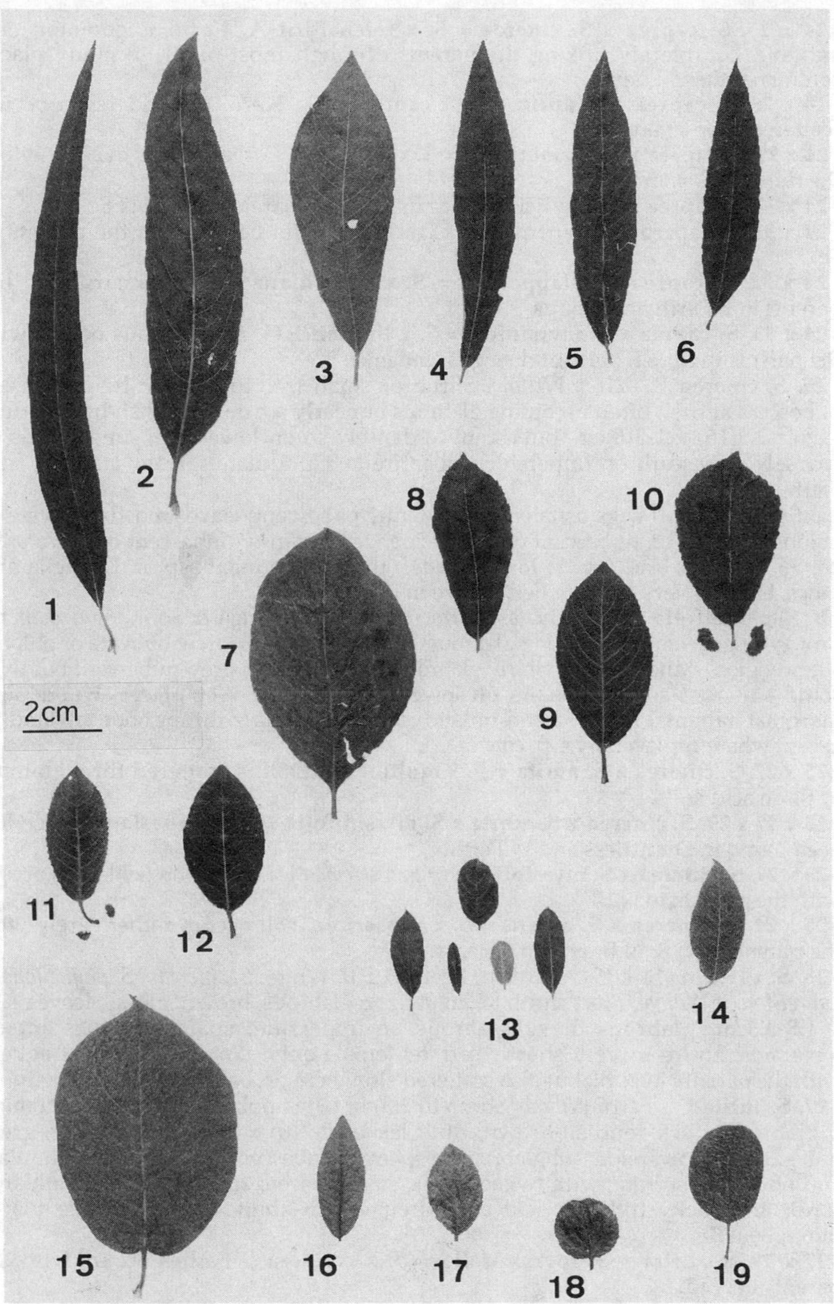

FIG 239 - Leaves of *Salix*. 1, ***S. viminalis***. 2, ***S. x sericans***. 3, ***S. x calodendron***. 4, ***S. x stipularis***. 5, ***S. x smithiana***. 6, ***S. x fruticosa***. 7, ***S. caprea***. 8, ***S. cinerea***. 9, ***S. x laurina***. 10, ***S. aurita***. 11, ***S. myrsinifolia***. 12, ***S. phylicifolia***. 13, ***S. repens***. 14, ***S. lapponum***. 15, ***S. lanata***. 16, ***S. arbuscula***. 17, ***S. myrsinites***. 18, ***S. herbacea***. 19, ***S. reticulata***.

24 x 25. S. caprea x S. cinerea = S. x reichardtii A. Kern. is common and variable, completely linking the parents, through most of BI, in many places commoner than *S. caprea*.

24 x 26. S. caprea x S. aurita = S. x capreola Jos. Kern. ex Andersson occurs scattered over Br and Ir.

24 x 29. S. caprea x S. myrsinifolia = S. x latifolia J. Forbes occurs in N Br within the range of *S. myrsinifolia*.

24 x 30. S. caprea x S. phylicifolia occurs with the parents in C & N Sc.

24 x 31. S. caprea x S. repens = S. x laschiana Zahn occurs with the parents in Sc.

24 x 32. S. caprea x S. lapponum = S. x laestadiana Hartm. occurs with the parents in E Perth and Angus.

24 x 35. S. caprea x S. myrsinites = S. x lintonii E.G. & A. Camus occurs with the parents in M & E Perth and Angus; endemic.

25. S. cinerea L. - *Grey Willow*. Shrub or small tree to 6(15)m; twigs densely pubescent at first, often becoming glabrous or nearly so, dark reddish-brown, dull; leaves 2-9(16) x 1-3(5)cm, pubescent to densely so on lowerside, subglabrous or sparsely pubescent on upperside, subentire to ± undulate-serrate; stamens free. Native. **239**

a. Ssp. cinerea. Twigs usually ± persistently pubescent; leaves mostly obovate or oblong, dull and ± pubescent on upperside, densely grey-pubescent on lowerside; 2n=76. Marshes and fens at low altitudes; the predominant ssp. in E Anglia and Lincs, but also very scattered elsewhere in Br and Ir.

b. Ssp. oleifolia Macreight (ssp. *atrocinerea* (Brot.) Pirajá & Sobr., *S. atrocinerea* Brot.). Twigs usually becoming glabrous; leaves mostly narrowly obovate or oblong, slightly glossy and usually nearly glabrous on upperside, grey-pubescent but with some stiff rust-coloured hairs on lowerside; (2n=76). Wet places, woods and marginal habitats, lowland and upland; commonest *Salix* throughout lowland BI except where replaced by ssp. *cinerea*.

25 x 27. S. cinerea x S. aurita = S. x multinervis Döll is scattered through most of BI on acid soils.

25 x 27 x 29. S. cinerea x S. aurita x S. myrsinifolia = S. x forbesiana Druce has been found in Dumfriess and W Perth.

25 x 29. S. cinerea x S. myrsinifolia (= *S. x strepida* J. Forbes non Schleich.) occurs with the parents in N BI.

25 x 31. S. cinerea x S. repens = S. x subsericea Döll occurs rather rarely with the parents in C & N Br and in S Hants.

26. S. x laurina Sm. (*S. x wardiana* Leefe ex F.B. White; *S. cinerea* x *S. phylicifolia*) - *Laurel-leaved Willow*. Erect shrub to 6m; twigs ± glabrous, brown, glossy; leaves 4-10 x 1.8-3.5cm, glabrous or subglabrous on upperside, sparsely pubescent on lowerside, entire to very shortly serrate; female only; (2n=76). Native; frequent with the parents in N BI, but also scattered elsewhere probably of garden origin. **239**

27. S. aurita L. - *Eared Willow*. Shrub to 2(3)m; twigs pubescent at first, becoming ± glabrous, dark reddish-brown, dull; leaves 2-4(6) x 1-3(4)cm, rugose, grey-pubescent on lowerside, subglabrous to sparsely pubescent on upperside, undulate and obscurely serrate, with twisted apex; stamens free; 2n=76. Native; heathland, scrub and rocky hills on acid soils; frequent to abundant in suitable places throughout BI. **239**

27 x 29. S. aurita x S. myrsinifolia = S. x coriacea J. Forbes occurs with the parents in N BI.

27 x 29 x 30. S. aurita x S. myrsinifolia x S. phylicifolia = S. x saxetana F.B. White occurs with parents in S & C Sc.

27 x 30. S. aurita x S. phylicifolia = S. x ludificans F.B. White occurs with the parents in N Br; 2n=42-64.

27 x 31. S. aurita x S. repens = S. x ambigua Ehrh. occurs commonly throughout BI wherever the parents meet.

27 x 31 x 36. S. aurita x S. repens x S. herbacea = S. x grahamii Borrer ex Baker has been found in W Sutherland and W Donegal.

27 x 32. S. aurita x S. lapponum = S. x obtusifolia Willd. occurs with the parents in C & N Sc.

27 x 36. S. aurita x S. herbacea = S. x margarita F.B. White occurs with the parents in C & N Sc.

28. S. eriocephala Michx. (*S. rigida* Muhl., *S. cordata* Muhl. non Michx.) - *Heart-leaved Willow*. Rhizomatous shrub to 2m; twigs very brittle at branches, glabrous, greenish-brown; leaves 5-12 x 1.5-4cm, glabrous, finely serrate, strongly cordate at base; (2n=44). Intrd; bogs; natd in E Sussex, Cards, W Lancs and Warks; N America. **238**

29. S. myrsinifolia Salisb. (*S. nigricans* Sm.) - *Dark-leaved Willow*. Shrub to 4m, often much less; twigs densely pubescent at first, slowly becoming subglabrous, dull, brown or greenish; leaves 2-6.5 x 1.5-3.5cm, sparsely pubescent at first, becoming subglabrous, irregularly crenate-serrate; stamens free; 2n=114. Native; by ponds and streams and in damp rocky places; frequent in N En and Sc, formerly N Ir, scattered intrd plants in C & S Br, but perhaps native in CE En. **239**

29 x 30. S. myrsinifolia x S. phylicifolia = S. x tetrapla Walker (*S. x tenuifolia* Sm.) occurs commonly in N Br and often completely links the parents wherever they meet; it occurs with *S. myrsinifolia* over wide areas of central highlands of Sc where *S. phylicifolia* is not present.

29 x 30 x 31. S. myrsinifolia x S. phylicifolia x S. repens has been found in S Aberdeen, E Perths and Angus and doubtfully elsewhere in Sc; endemic. Original records were of *S. phylicifolia* x *S. repens*, but *S. x tetrapla* rather than *S. phylicifolia* occurs in the area.

29 x 30 x 32. S. myrsinifolia x S. phylicifolia x S. lapponum has been recorded from S Aberdeen and Angus, and doubtfully from M Perths; endemic. Original records were of *S. phylicifolia* x *S. lapponum*, but *S. x tetrapla* rather than *S. phylicifolia* occurs in the area.

29 x 30 x 35. S. myrsinifolia x S. phylicifolia x S. myrsinites occurs with the parents in CE Sc and probably includes plants from there previously named *S. phylicifolia* x *S. myrsinites*, since *S. x tetrapla* rather than *S. phylicifolia* occurs in the area.

29 x 31. S. myrsinifolia x S. repens = S. x felina Buser ex E.G. & A. Camus occurs with the parents in E Perth.

29 x 32. S. myrsinifolia x S. lapponum (= *S. x dalecarlica* auct. non Rouy) was recorded from Angus in 1892 but not confirmed, and was confirmed in S Aberdeen in 1959; endemic.

29 x 35. S. myrsinifolia x S. myrsinites = S. x punctata Wahlenb. occurs with the parents in C Sc.

29 x 36. S. myrsinifolia x S. herbacea = S. x semireticulata F.B. White occurs with the parents in M Perth; endemic.

30. S. phylicifolia L. (*S. hibernica* Rech. f.) - *Tea-leaved Willow*. Shrub to 4(5)m; twigs subglabrous to sparsely pubescent at first, soon ± glabrous, reddish-brown, ± glossy; leaves 2-6 x 1-5cm, often sparsely pubescent at first, glabrous at maturity, entire to irregularly serrate, of thicker texture than in *S. myrsinifolia*; stamens free; 2n=88. Native; similar places to *S. myrsinifolia* but more confined to base-rich soils; frequent in N En and Sc, and in Leitrim and Sligo. **239**

30 x 31. S. phylicifolia x S. repens (= *S. x schraderiana* auct. non Willd.) has been found in Orkney; endemic. *S. myrsinifolia* does not occur there.

30 x 32. S. phylicifolia x S. lapponum = S. x gillotii E.G. & A. Camus has been recorded from S Aberdeen, M Perth and Angus, but the plants were probably *S. myrsinifolia* x *S. phylicifolia* x *S. lapponum*, since *S. x tetrapla* rather than *S. phylicifolia* occurs in the region.

30 x 35. S. phylicifolia x S. myrsinites = S. x notha Andersson has been recorded in M Perth and Angus, but the plants were probably *S. myrsinifolia* x *S.*

phylicifolia x *S. myrsinites*, since *S. x tetrapla* rather than *S. phylicifolia* occurs in the region.

31. S. repens L. - *Creeping Willow*. Procumbent to erect shrub to 1.5(2)m, but usually <1m, twigs glabrous to densely silky-pubescent and variously coloured; leaves 1-3.5 x 0.4-2.5cm, glabrous to pubescent or silvery-silky-pubescent on both sides, but usually ± pubescent on lowerside, entire to obscurely crenate-serrate; stamens free; 2n=38. Native; acid heaths and moors both dry and wet, fens, dunes and dune-slacks; in suitable places throughout BI. Extremely variable, especially in leaf-shape. The commonest variant (var. **repens**) is rhizomatous, with procumbent to decumbent soon subglabrous stems and small leaves soon subglabrous on upperside; it occurs mostly on heaths and moors. Var. **argentea** (Sm.) Wimm. & Grab. (ssp. *argentea* (Sm.) E.G. & A. Camus, ssp. *arenaria* (L.) Hiitonen, *S. arenaria* L.) occurs mostly in dune-slacks and differs in its taller, strong habit with silky-pubescent stems and larger leaves persistently densely silky-pubescent on both sides. Var. **fusca** Wimm. & Grab. is shortly rhizomatous with ± erect stems but leaves as in var. *repens*, and occurs in fens in E Anglia. Numerous intermediates occur. **239**

31 x 32. S. repens x S. lapponum = S. x pithoensis Rouy occurs with the parents in E & M Perth.

31 x 34. S. repens x S. arbuscula was found in M Perth in 1994; endemic.

31 x 36. S. repens x S. herbacea = S. x cernua E.F. Linton occurs with the parents in C & N Sc.

32. S. lapponum L. - *Downy Willow*. Low shrub to 1(1.5)m; twigs pubescent at first, glabrous and rather glossy dark reddish-brown later; leaves 1.5-7 x 1-2.5cm, subglabrous to pubescent on upperside, usually densely pubescent on lowerside, entire or subentire; stamens free; 2n=38. Native; rocky mountain slopes and cliffs; N & C mainland Sc, isolated localities in Westmorland and S Sc. **R** **239**

32 x 33. S. lapponum x S. lanata (=*S. x stuartii* hort.) has been recorded from S Aberdeen in 19th century, Angus in 1928 and E Perth in 1994, but there doubts about the identification in all cases.

32 x 34. S. lapponum x S. arbuscula = S. x pseudospuria Rouy occurs with the parents in C & N Sc.

32 x 36. S. lapponum x S. herbacea = S. x sobrina F.B. White occurs with the parents in CE Sc.

32 x 37. S. lapponum x S. reticulata = S. x boydii E.F. Linton occurred with the parents in Angus; endemic. There is, however, considerable doubt that this was the correct parentage for *S. x boydii*.

33. S. lanata L. - *Woolly Willow*. Low shrub to 1m; twigs pubescent at first, soon glabrous, ± glossy, brown; leaves 3.5-7 x 3-6.5cm, densely white-tomentose at first, becoming subglabrous with age, entire or subentire; stamens free; 2n=38. Native; damp mountain rock-ledges; local, highlands of C Sc. **RR** **239**

33 x 36. S. lanata x S. herbacea = S. x sadleri Syme occurs with the parents in CE Sc.

34. S. arbuscula L. - *Mountain Willow*. Low shrub to 80cm; twigs sparsely pubescent at first, soon glabrous, reddish-brown, ± glossy; leaves 1.5-3(5) x 1-1.5(3)cm, glabrous on upperside, densely appressed-pubescent on lowerside at first, becoming ± glabrous, crenate-serrate; stamens free; (2n=38). Native; damp rocky slopes and mountain ledges; locally abundant on mountains in C (formerly S) Sc. **R** **239**

34 x 36. S. arbuscula x S. herbacea = S. x simulatrix F.B. White occurs with the parents in M Perth and Argyll.

34 x 37. S. arbuscula x S. reticulata = S. x ganderi Huter ex Zahn occurs with the parents in M Perth.

Section 3* - *CHAMAETIA Dumort. (spp. 35-37). Dwarf shrubs; leaves entire to coarsely serrate, ovate to orbicular; catkins short and wide, on leafless peduncles, mostly arising from terminal buds; bracts pale or dark; stamens 2, free; nectaries 1-2, often lobed; ovary glabrous or pubescent.

35. S. myrsinites L. - *Whortle-leaved Willow.* Low spreading shrub <50cm; twigs R
sparsely pubescent at first, soon becoming glabrous, ± glossy, reddish-brown; 239
leaves 1.5-7 x 0.5-2.5(3)cm, sparsely pubescent at first, soon glabrous, glossy on upperside, crenate-serrate; bracts dark brown; nectary 1; ovary densely pubescent with iridescent hairs; 2n=38. Native; rocky ledges and slopes on mountains; local in N & C Sc, rare in S Sc. Possibly better placed in subg. *Vetrix*.

35 x 36. S. myrsinites x S. herbacea was found in S Aberdeen in 1995; ?endemic. 239

36. S. herbacea L. - *Dwarf Willow.* Dwarf shrub <10cm; twigs very sparsely pubescent at first, soon glabrous, glossy, reddish-brown; leaves 0.3-2 x 0.3-2cm, sparsely pubescent at first, soon glabrous, ± glossy, rounded to emarginate at apex, crenate-serrate; bracts yellowish or red-tinged; nectaries 1-2, usually lobed; ovary glabrous; 2n=38. Native; rock-ledges and rocky mountain-tops; locally common in Sc, N En and parts of Wa and Ir.

37. S. reticulata L. - *Net-leaved Willow.* Dwarf shrub <20cm; twigs sparsely R
pubescent at first, soon glabrous, dark reddish-brown; leaves 1.2-4(5) x 1- 239
2.5(3.5)cm, densely pubescent at first, becoming glabrous at least on upperside, rounded at apex, entire; bracts purplish-brown, suborbicular; nectaries usually in form of lobed cup; ovary densely pubescent; 2n=38. Native; wet rock-ledges and mountain-slopes; very local in C & N mainland Sc.

63A. CAPPARACEAE

CLEOME sesquiorygalis Naudin ex C. Huber (*C. hassleriana* Chodat) (*Spiderflower*), from S America, is an erect annual to 1m with palmate leaves often prickly on petioles and with 5-7 leaflets, showy scented flowers with 4 white or pink petals arranged in a bracteate raceme, and a narrow capsule ≤12cm x 5mm. It is grown as a bedding plant and sometimes occurs as a relic on tips in S Br.

64. BRASSICACEAE - *Cabbage family*
(Cruciferae)

Herbaceous annuals to perennials, rarely dwarf shrubs; leaves alternate, simple to pinnate, petiolate or sessile, without stipules. Flowers usually in racemes, usually much elongating after flowering, sometimes in panicles, normally bisexual, hypogynous, usually actinomorphic; sepals 4, free; petals 4, free, rarely 0; stamens usually 6 (2 shorter outer, 4 longer inner), sometimes 4 by loss of 2 outer, rarely only 2 in *Coronopus* and *Lepidium*; ovary usually 2-celled, each with 1-many ovules; style 1; stigma 2, often bilobed, usually capitate; fruit dry, usually opening from below by 2 valves, sometimes breaking into segments transversely, sometimes indehiscent.

Easily recognized by the distinctive flowers with 4 sepals and 4 petals in decussate pairs, the (4-)6 stamens, and the characteristic fruits.

It is difficult to construct keys without both ripe fruits and a knowledge of petal colour. Caution is needed when assessing flower colour of dried specimens, as sometimes yellow can fade to white and white discolour to yellow; the keys allow for this in most but perhaps not all cases. Hair-type is also very important, and a strong lens is often needed; both leaves and stems should be examined, as hair-type can differ on these 2 organs (branched hairs are commoner on leaves). Plants that are sterile for some reason (hybridity, lack of suitable pollen, genetic mutation) are often impossible to key out, or will key out wrongly due to mis-shapen fruits or fewer than normal seeds; these possibilities must always be borne in mind. In this work 16 groups of distinctive genera are first defined; some of these rarely produce ripe fruits and would not be identifiable with the main keys which, however, do cover all the genera. Fruit lengths include beak, style and stigma unless otherwise stated.

Distinctive genera

Stems woody for most of length **16. AUBRIETA, 18. ALYSSUM, 31. IBERIS**
Fresh plants smelling of garlic when crushed **3. ALLIARIA, 30. THLASPI**
Flowers distinctly zygomorphic, 2 petals on 1 side much shorter than 2 on other **29. TEESDALIA, 31. IBERIS**
Erect plants with white scaly rhizome, pinnate basal leaves, pink flowers, and purple bulbils in leaf-axils **14. CARDAMINE**
Robust plants with strong tap-root, large dock-like leaves sometimes partly pinnately divided, and small white flowers in large panicles **13. ARMORACIA**
Small aquatic, often submerged, plants with leaves all subulate and in basal rosette; inflorescence with few small flowers or short swollen fruits **34. SUBULARIA**
Low straggly annuals with pinnate leaves, small inflorescences borne opposite the leaves, and bilobed 2-seeded fruits **33. CORONOPUS**
Erect plants with small yellow flowers in panicles, and pendent, winged, indehiscent, 1-seeded fruits **5. ISATIS**
Leaves succulent **23. COCHLEARIA, 37. BRASSICA, 43. CAKILE, 45. CRAMBE**
Stem-leaves 0 **14. CARDAMINE, 21. DRABA, 22. EROPHILA, 29. TEESDALIA, 34. SUBULARIA, 41. COINCYA**
Petals 0 **14. CARDAMINE, 26. CAPSELLA, 32. LEPIDIUM, 33. CORONOPUS, 34. SUBULARIA, 35. CONRINGIA**
Petals bifid >1/3 way to base **19. BERTEROA, 22. EROPHILA**
Stamens 2-4 **4. ARABIDOPSIS, 14. CARDAMINE, 21. DRABA, 32. LEPIDIUM, 33. CORONOPUS**
Fruits mostly 1-seeded **5. ISATIS, 6. BUNIAS, 25. NESLIA, 43. CAKILE, 44. RAPISTRUM, 45. CRAMBE**
Fruits with stalk-like base >1mm above sepal-scars **1. SISYMBRIUM, 17. LUNARIA, 36. DIPLOTAXIS, 37. BRASSICA**
Fruits with seed-containing beak beyond dehiscent lower segment **37. BRASSICA, 38. SINAPIS, 41. COINCYA, 42. HIRSCHFELDIA**

General key

1 Fruit splitting longitudinally by 2 valves to release 1-many seeds on each side 2
1 Fruit indehiscent, or splitting transversely, or splitting longitudinally but not releasing seeds separately *Key A*
2 Fruit >3x as long as wide 3
2 Fruit ≤3x as long as wide 4
3 Petals pale yellow to golden *Key B*
3 Petals white, pink, purple, mauve or reddish, or rarely 0 *Key C*
4 Fruit distinctly compressed, the septum at right angles to plane of compression *Key D*
4 Fruit distinctly compressed, the septum parallel with the plane of compression, or fruit scarcely compressed *Key E*

Key A - Fruit indehiscent, or splitting transversely, or splitting longitudinally but not releasing seeds separately

1 Fruit pendent, flattened, winged, 1-seeded **5. ISATIS**
1 Fruit not pendent, not winged 2

2 Petals yellow 3
2 Petals white to pink or purple 6
3 Fruit >12mm; fresh root smelling of radish **46. RAPHANUS**
3 Fruit <12mm; fresh root not smelling of radish 4
4 Fruit composed of 2 segments separated by transverse constriction **44. RAPISTRUM**
4 Fruit composed of 1 segment or with slight longitudinal constriction 5
5 Fruit irregularly warty, 5-8mm; at least some leaves pinnately lobed **6. BUNIAS**
5 Fruit minutely tuberculate, <5mm; leaves entire to dentate **25. NESLIA**
6 Fruit clearly divided longitudinally into 2 1-seeded halves 7
6 Fruit clearly divided transversely or not obviously divided 8
7 Leaves simple; inflorescence a terminal panicle **32. LEPIDIUM**
7 Leaves pinnate; inflorescences mostly simple leaf-opposed racemes **33. CORONOPUS**
8 Plant with stiff hairs below; leaves not succulent; fresh root smelling of radish **46. RAPHANUS**
8 Plant glabrous; leaves succulent; fresh root not smelling of radish 9
9 Perennial with *Brassica*-like leaves; lower fruit-segment sterile, stalk-like; filaments of inner stamens toothed **45. CRAMBE**
9 Annual with narrow, pinnately-lobed leaves; lower fruit-segment usually 1-seeded; filaments of inner stamens not toothed **43. CAKILE**

Key B - Fruits >3x as long as wide, splitting longitudinally from base into 2 valves; flowers pale yellow to golden

1 Stem-leaves distinctly clasping stem at their base 2
1 Stem-leaves not clasping stem, petiolate or narrowed to base 7
2 Hairs present at least on lowest leaves, branched **15. ARABIS**
2 Hairs 0 or simple on all parts 3
3 Fruit with distinct stalk 0.5-1mm between valves and sepal-scars; at least lower flowers with lobed bracts **40. ERUCASTRUM**
3 Fruit with base of valves immediately above sepal-scars; flowers with entire or toothed bracts or without bracts 4
4 Fruit with broad-based beak ≥4mm **37. BRASSICA**
4 Fruit with narrowly cylindrical beak ≤4mm 5
5 Leaves all simple and entire **35. CONRINGIA**
5 At least lower leaves pinnate or pinnately lobed 6
6 Valves of fruit with prominent midrib; seeds in 1 row under each valve **11. BARBAREA**
6 Valves of fruit with midrib not or scarcely discernible; seeds ± in 2 rows under each valve **12. RORIPPA**
7 Leaves linear, all in tight basal rosette **21. DRABA**
7 Leaves rarely all linear, not in tight basal rosette 8
8 Fruit terminated by distinct beak >3mm; beak either wide at base and tapered distally or wide for most of length, sometimes with seeds 9
8 Fruit terminated by sessile stigmas or by short narrow cylindrical style ≤4mm 13
9 Seeds in 2 rows under each valve; petals pale yellow with conspicuous violet veins **39. ERUCA**
9 Seeds in 1 row under each valve; petals without conspicuous violet veins 10
10 Fruit <2cm, closely appressed to stem **42. HIRSCHFELDIA**
10 Fruit >2cm, not closely appressed to stem 11
11 Each valve of fruit with 1 prominent vein; beak long-conical **37. BRASSICA**
11 Each valve of fruit with 3(-7) prominent parallel veins; beak long-conical or flat and sword-like 12

12 Sepals erect; beak of fruit flat and sword-like **41. COINCYA**
12 Sepals patent (to erecto-patent); beak of fruit flat and sword-like or long-conical **38. SINAPIS**
13 Hairs present at least below, at least some stellate or 2-3-armed (arms often closely appressed to plant) 14
13 Hairs 0, or present and all simple 16
14 Leaves 2-3x finely divided almost to midrib; petals c. as long as sepals **2. DESCURAINEA**
14 Leaves entire to simply toothed or lobed; petals c.2x as long as sepals 15
15 Compact basal leaf-rosette present; basal leaves pinnately lobed **15. ARABIS**
15 Basal leaf-rosette 0 or very ill-defined; lower leaves entire to toothed **7. ERYSIMUM**
16 Valves of fruit with midrib not or scarcely discernible **12. RORIPPA**
16 Valves of fruit with conspicuous midrib 17
17 Seeds in 2 rows under each valve in well developed fruits **36. DIPLOTAXIS**
17 Seeds in 1 row under each valve 18
18 Each valve of fruit with 3 prominent parallel veins **1. SISYMBRIUM**
18 Each valve of fruit with 1 prominent vein 19
19 Seeds ± globose; flowers all without bracts; valves of fruit rounded **37. BRASSICA**
19 Seeds distinctly longer than wide; lower flowers usually with small bracts; valves of fruit keeled at midrib **40. ERUCASTRUM**

Key C - Fruits >3x as long as wide, splitting longitudinally from base into 2 valves; flowers white, mauve, pink, purple or red

1 Fruit with flat, sword-like beak at apex; petals whitish with violet veins **39. ERUCA**
1 Fruit with beak 0 or cylindrical; petals rarely whitish with violet veins 2
2 Basal and lower stem-leaves pinnate or ternate 3
2 Basal and lower stem-leaves entire to lobed, but lobes not reaching midrib 4
3 Valves of fruit convex (fruit little wider than thick), opening by separating laterally from base **12. RORIPPA**
3 Valves of fruit ± flat (fruit much wider than thick), opening by suddenly spiralling upwards from base **14. CARDAMINE**
4 Petals bifid c.1/2 way to base; small annual with leaves confined to basal rosette (or ± so) **2. EROPHILA**
4 Petals entire or notched to <1/4 way to base; stems usually bearing leaves 5
5 Plant glabrous or with simple hairs only 6
5 Plant pubescent; all or some hairs branched or stellate, sometimes sparse 11
6 Stem-leaves strongly clasping stem at base **35. CONRINGIA**
6 Stem-leaves not or scarcely clasping stem 7
7 Basal leaves cordate at base; plant smelling of garlic when crushed **3. ALLIARIA**
7 Basal leaves cuneate at base; plant not smelling of garlic 8
8 Petals >15mm; stigma deeply 2-lobed **8. HESPERIS**
8 Petals <15mm; stigma capitate or shortly 2-lobed 9
9 Perennial of mountains of N & W; valves of fruit ± flat **15. ARABIS**
9 Annual lowland weed; valves of fruit convex 10
10 Seeds in 1 row under each valve; petals shorter than stamens, white only after drying **1. SISYMBRIUM**
10 Seeds in 2 rows under each valve in well developed fruits; petals longer than stamens, always white **36. DIPLOTAXIS**

11 Stigma deeply 2-lobed, the lobes visible at apex of fruit (though often closed up together) 12

11 Stigma capitate or slightly notched 15

  12 Seeds narrowly to broadly winged 13

  12 Seeds not winged 14

13 Hairs branching from basal stalk in tree-like form; seeds broadly winged all round **10. MATTHIOLA**

13 Hairs with 2 arms, both appressed to plant body, without common stalk; seeds narrowly winged, often not all round **7. ERYSIMUM**

  14 Spreading annual; most hairs with stalk 0 and 2-4 arms appressed close to plant surface **9. MALCOLMIA**

  14 Erect biennial or perennial; most hairs simple or with stalk and 2 arms, patent **8. HESPERIS**

15 Fruit <8x as long as wide (excl. style), all or most >2mm wide 16

15 Fruit >8x as long as wide (excl. style), all or most <2mm wide 17

  16 Flowers white; fruit flat, usually twisted; stems erect **21. DRABA**

  16 Flowers rarely white; fruit scarcely compressed, not twisted; stems procumbent to decumbent **16. AUBRIETA**

17 Basal leaves pinnately lobed **15. ARABIS**

17 Basal leaves entire to dentate 18

  18 Annual; fruit circular to square in section **4. ARABIDOPSIS**

  18 Biennial to perennial; fruit strongly flattened **15. ARABIS**

Key D - Fruits <3x as long as wide, obviously compressed, with septum at right angles to plane of compression, splitting longitudinally from base into 2 valves to release 1 or more seeds on each side

1 Corolla distinctly zygomorphic, with 2 petals on 1 side much shorter than 2 on other 2

1 Corolla actinomorphic, with 4 equal petals 3

  2 Seeds 2 under each valve; style ± 0; shorter petals c. as long as sepals **29. TEESDALIA**

  2 Seed 1 under each valve; style distinct; shorter petals ≥2x as long as sepals **31. IBERIS**

3 Fruit distinctly winged at edges (i.e. valve midribs) 4

3 Fruit sometimes keeled but not winged at edges (i.e. valve midribs) 6

  4 Basal leaves strongly cordate at base **30. THLASPI**

  4 Basal leaves 0 or cuneate to rounded at base 5

5 Seeds 3-8 under each valve **30. THLASPI**

5 Seeds 1(-2) under each valve **32. LEPIDIUM**

  6 Fruit obtriangular **26. CAPSELLA**

  6 Fruit orbicular to elliptic, oblong or obovate 7

7 Flowers appearing solitary, on long pedicels arising from axils of basal leaves **28. JONOPSIDIUM**

7 Flowers in racemes, all or mostly without bracts 8

  8 Fruits distinctly keeled at edges (i.e. valve midribs) **32. LEPIDIUM**

  8 Fruits obtuse to rounded at edges (i.e. valve midribs) 9

9 Basal leaves deeply pinnately lobed; flowers <2mm across **27. HORNUNGIA**

9 Basal leaves entire to palmately angled or shallowly lobed; flowers >2mm across **23. COCHLEARIA**

Key E - Fruits <3x as long as wide, either not compressed or obviously compressed and with septum parallel with plane of compression, splitting longitudinally to release 1 or more seeds on each side

1 Leaves all confined to basal rosette 2

1 At least some leaves borne on stem 5

2 Petals yellow **21. DRABA**
2 Petals white 3
3 Aquatic, often submerged; leaves subulate **34. SUBULARIA**
3 Not aquatic; leaves flat 4
4 Annual; petals bifid c.1/2 way to base; fruit glabrous **22. EROPHILA**
4 Perennial; petals notched <1/4 way to base; fruit pubescent **21. DRABA**
5 Petals yellow 6
5 Petals white to pink or purple 10
6 Plant glabrous or with simple hairs 7
6 Plant pubescent; all or many hairs branched 8
7 Fruit not or scarcely keeled at junction of valves **12. RORIPPA**
7 Fruit strongly keeled (± winged) at junction of valves **24. CAMELINA**
8 Petals bifid nearly 1/2 way to base **19. BERTEROA**
8 Petals entire or notched <1/4 way to base 9
9 Fruit not or scarcely flattened; seeds >2 under each valve **24. CAMELINA**
9 Fruit strongly flattened; seeds 2 under each valve **18. ALYSSUM**
10 Robust perennial with strong tap-root and large dock-like basal leaves (sometimes pinnately lobed) **13. ARMORACIA**
10 Plant without dock-like basal leaves 11
11 Fruits >2 x 1cm, very flat **17. LUNARIA**
11 Fruits <2cm, <1cm wide, flat or not 12
12 Plant glabrous or with simple hairs 13
12 Plant with at least some branched or stellate hairs 14
13 Basal leaves abruptly contracted into long petiole **23. COCHLEARIA**
13 Basal leaves ± sessile, gradually narrowed to base **24. CAMELINA**
14 Fruit scarcely compressed 15
14 Fruit strongly compressed 16
15 Flowers rarely white; procumbent to decumbent much-branched perennial **16. AUBRIETA**
15 Flowers white; erect little-branched annual **24. CAMELINA**
16 Petals bifid nearly 1/2 way to base **19. BERTEROA**
16 Petals entire to notched <1/4 way to base 17
17 Fruit c. as long as wide, notched at apex **18. ALYSSUM**
17 Fruit longer than wide, rounded to obtuse at apex 18
18 Seeds >2 under each valve; basal leaf-rosette present at flowering **21. DRABA**
18 Seeds 1 under each valve; basal leaf-rosette 0 at flowering **20. LOBULARIA**

Other genera - Many other genera have been recorded as casuals in BI. The following 13 genera, stated in recent British and European Floras to occur in BI, either do not nowadays occur or are far too rare to merit treatments here: **AETHIONEMA** R. Br., **BOREAVA** Jaub. & Spach, **CALEPINA** Adans., **CHORISPORA** R. Br. ex DC., **ENARTHROCARPUS** Labill., **EUCLIDIUM** R. Br., **GOLDBACHIA** DC., **IONDRABA** Rchb. (cf. *Biscutella* L.), **LYCOCARPUS** O.E. Schulz, **MICROSISYMBRIUM** O.E. Schulz, **MORICANDIA** DC., **SUCCOWIA** Medik., **TETRACME** Bunge. Nearly 30 spp. mentioned under 'Other spp.' in various genera as too rare to be treated here come into the same category; in many cases the generic descriptions do not cover these, and often even listing them has been considered too wasteful of space. 3 Mediterranean genera are slightly less rare casuals. **CARRICHTERA** DC. has short fruits of 2 ± equal halves, a basal dehiscent segment with strong curved bristles and 3-4 seeds per valve, and a distal sterile flattened bristleless segment with a rounded apex; **C. annua** (L.) DC. (*Cress Rocket*) is a rare annual wool- and grain-alien with pale yellow petals, deeply dissected leaves and fruits 6-8mm. **MYAGRUM** L. has sessile ± entire stem-leaves with pronounced auricles clasping the stem and short indehiscent obovate to broadly club-shaped fruits with 3 segments, 2 apical side-by side and sterile, 1 basal and fertile; **M. perfoliatum** L. (*Mitre Cress*) is a glabrous, glaucous annual with yellow petals and fruits 5-8mm. **ERUCARIA** Gaertn. has mauve petals, deeply pinnately lobed leaves and narrowly club-shaped fruits with 2 segments, the lower cylindrical, 2-4-seeded and dehiscent, the distal flattened, 1-3-seeded and indehiscent; **E. hispanica** (L.) Druce (*E. myagroides* (L.) Halácsy) is a sparsely pubescent annual with fruits 1-2cm and petals 10-15mm.

TRIBE 1 - SISYMBRIEAE (genera 1-6). Hairs branched or unbranched, sometimes glandular; petals yellow or white; fruit usually beakless, >3x as long as wide and with small stigma, sometimes 1-2-seeded and indehiscent.

1. SISYMBRIUM L. - *Rockets*

Annuals or perennials; basal leaves simple, entire to very deeply lobed; hairs unbranched; petals yellow, sometimes very pale; fruit beakless, >3x as long as wide, with convex valves; seeds in 1 row under each valve.

1 Leaves all entire to dentate — 2
1 Lowest leaves lobed >1/4 way to midrib — 4
2 Annual; petals ≤3.5mm, not longer than sepals or stamens — **7. S. erysimoides**
2 Perennial; petals >5mm, longer than sepals and stamens — 3
3 Leaves pubescent on lowerside, narrowly acute to acuminate at apex — **1. S. strictissimum**
3 Leaves ± glabrous on lowerside, subacute to apiculate at apex — **4. S. volgense**
4 Fruits ≤2cm, strongly appressed to stem — **8. S. officinale**
4 At least most fruits >2cm, patent to ± erect but not appressed to stem — 5
5 Upper stem-leaves pinnate with linear divisions — **5. S. altissimum**
5 Upper stem-leaves variously divided but usually not to midrib and never with linear segments — 6
6 Pedicels c.1mm wide, >2/3 as wide as ripe fruit — 7
6 Pedicels c.0.3-0.6mm wide, <2/3 as wide as ripe fruit — 8
7 Petals ≤3.5mm, not longer than sepals or stamens; fruit 2-5cm — **7. S. erysimoides**
7 Petals >6mm, much longer than sepals and stamens; fruit (2.5)5-12cm — **6. S. orientale**
8 Lower part of stem with ± dense patent to reflexed stiff hairs >1mm — **3. S. loeselii**
8 Lower part of stem glabrous to sparsely pubescent, or ± densely pubescent with hairs <0.5mm — 9
9 Petals pale yellow, <5mm; anthers <1mm — **2. S. irio**
9 Petals bright yellow, mostly >5mm; anthers >1mm — 10
10 Rhizomatous perennial; most fruits >3cm — **4. S. volgense**
10 Annual; most fruits <3cm — **3. S. loeselii**

Other spp. - **S. austriacum** Jacq., **S. runcinatum** Lag. ex DC. and **S. polyceratium** L., from S Europe, have occurred as casuals but now either do not occur or are extremely rare.

1. S. strictissimum L. - *Perennial Rocket*. Perennial with strong rootstock; stems little branched, to 1.2m, erect; leaves entire to dentate; fruits 3-9(11)cm, erect to erecto-patent; (2n=28, 32). Intrd; walls, rough and waste ground; natd in a very few places in En from Surrey to Durham, occasionally a casual; Europe. **251**

2. S. irio L. - *London-rocket*. Annual; stems well branched, to 60cm, erect to **251**

ascending; lower leaves variously lobed but some deeply so; fruits 2-6cm, patent to ± erect; (2n=14, 28, 42, 56). Intrd; walls, roadsides and waste places; ± natd in a few places in Br and Ir, a more frequent casual, especially wool-alien; Europe.

3. S. loeselii L. - *False London-rocket*. Annual; stems rather little branched, to 251
1(1.5)m, erect; lower leaves deeply and regularly lobed; fruits 1-3(4.5)cm, erect or erecto-patent; (2n=14). Intrd; waste places and tips; fairly frequent casual in S Br and natd in London area, 1 plant in Ir in 1988; Europe.

4. S. volgense M. Bieb. ex E. Fourn. - *Russian Mustard*. Rhizomatous perennial; 251
stems to 75cm, erect to ascending; lower leaves variously lobed, often deeply so, rarely not; fruits 1.5-6cm, erect to erecto-patent; (2n=14). Intrd; waste ground; natd in a few places widely scattered in En; Russia.

5. S. altissimum L. - *Tall Rocket*. Annual; stems to 1m, ± erect; leaves deeply 251
divided, the upper pinnate with long linear lobes; fruit 3-11cm, erecto-patent; (2n=14). Intrd; waste places; frequently natd or persistent casual scattered over much of BI; Europe.

6. S. orientale L. - *Eastern Rocket*. Annual; stems to 1m, erect to ascending; leaves 251
deeply divided, the upper with only 0-2 basal lobes each side; fruit 2.5-12cm, erecto-patent; (2n=14). Intrd; waste places; frequently natd or persistent casual scattered over much of BI; Europe.

7. S. erysimoides Desf. - *French Rocket*. Annual; stems to 60cm, erect; leaves 251
variously divided, often deeply so, rarely unlobed; fruits 2-5cm, patent to erecto-patent; (2n=14). Intrd; a fairly regular wool-alien in En; W Mediterranean. Petals often fade to white in dried material.

8. S. officinale (L.) Scop. - *Hedge Mustard*. Annual or biennial; stems to 1m, erect 251
or spreading; leaves deeply lobed; fruits 1-2cm, erect and closely appressed to stem; (2n=14). Native; waste places, rough and cultivated ground, hedges and roadsides; very common throughout BI except N Sc.

2. DESCURAINIA Webb & Berthel. - *Flixweed*
Annuals or biennials; basal leaves 2-3-pinnate or nearly so; hairs both unbranched and branched, ± patent; petals pale yellow; fruit beakless, >3x as long as wide, with convex valves; seeds in 1 row under each valve.

1. D. sophia (L.) Webb ex Prantl - *Flixweed*. Stems to 1m, erect, branched above; 251
leaves 2-3x divided ± to midrib, with linear segments; fruits 1-3.5cm, ± erect to erecto-patent on long pedicels; (2n=28, 56). Possibly native; roadsides, rough and waste ground; throughout much of Br, locally frequent in E, absent from much of N & W, very local in Ir.

3. ALLIARIA Heist. ex Fabr. - *Garlic Mustard*
Biennials; basal leaves simple, toothed; hairs unbranched; petals white; fruit beakless, >3x as long as wide, with angled valves; seeds in 1 row under each valve.

1. A. petiolata (M. Bieb.) Cavara & Grande - *Garlic Mustard*. Fresh plant smelling 251
of garlic; stems to 1.2m, erect, little-branched; leaves petiolate, ovate, the lower cordate, coarsely dentate; fruits (2)3-7.5cm, erect to erecto-patent on short thick pedicels; 2n=42. Native; rough ground, hedgerows and shady places; throughout most of BI but local in Sc and Ir.

4. ARABIDOPSIS (DC.) Heynh. - *Thale Cress*
Annuals; basal leaves simple; hairs both unbranched and branched; petals white; fruit beakless, >3x as long as wide, with convex to angled valves; seeds in 1 row under each valve.

1. A. thaliana (L.) Heynh. - *Thale Cress*. Stems erect, to 30(50)cm, simple or 251
branched, with few leaves; basal leaves entire to dentate; stem-leaves ± entire,

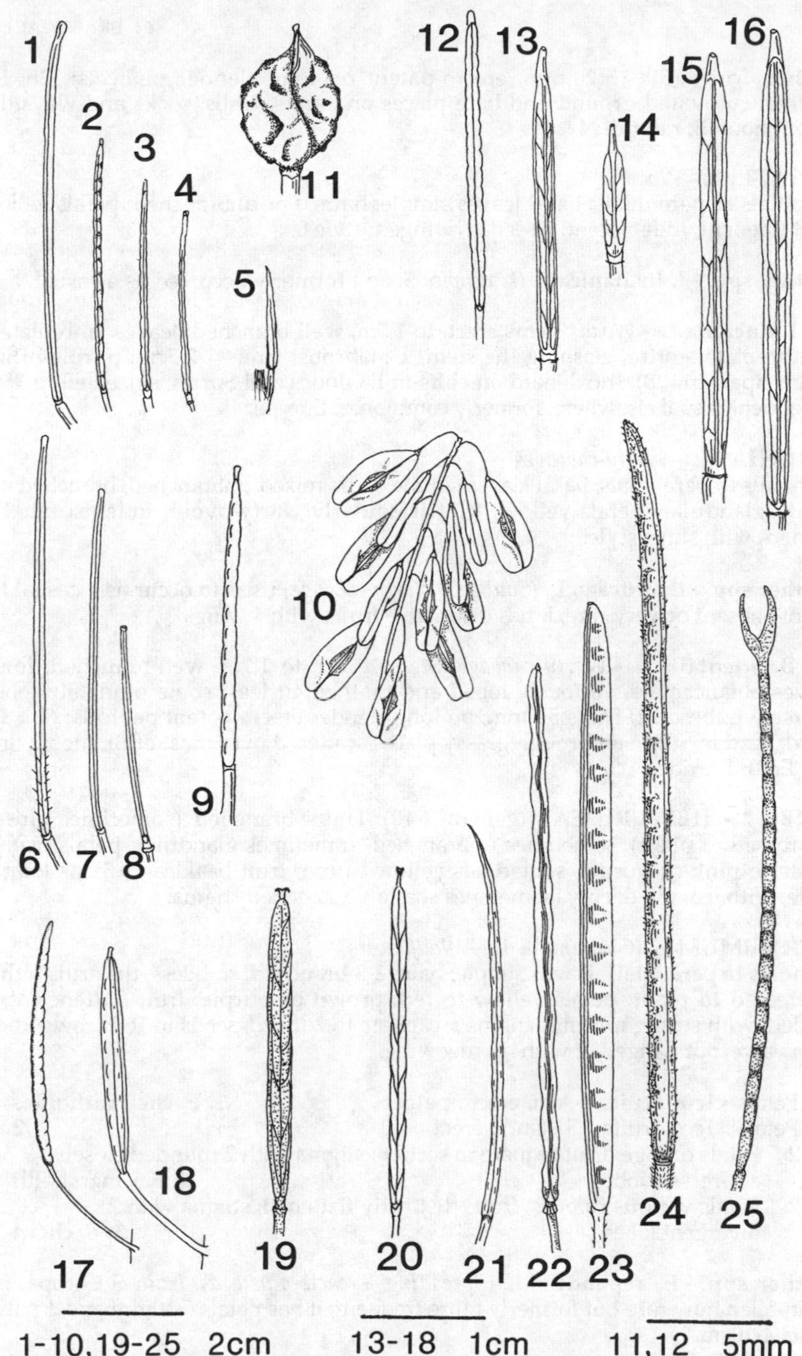

FIG 251 - Fruits of **Brassicaceae**. 1, ***Sisymbrium altissimum***. 2, ***S. irio***. 3, ***S. erysimoides***. 4, ***S. loeselii***. 5, ***S. officinale***. 6, ***S. orientale***. 7, ***S. volgense***. 8, ***S. strictissimum***. 9, ***Alliaria petiolata***. 10, ***Isatis tinctoria***. 11, ***Bunias orientalis***. 12, ***Arabidopsis thaliana***. 13, ***Barbarea stricta***. 14, ***B. vulgaris***. 15, ***B. intermedia***. 16, ***B. verna***. 17, ***Descurainia sophia***. 18, ***Erysimum cheiranthoides***. 19, ***E. cheiri***. 20, ***E. x marshallii***. 21, ***Malcolmia maritima***. 22, ***Hesperis matronalis***. 23, ***Matthiola incana***. 24, ***M. sinuata***. 25, ***M. longipetala***.

sessile; fruits (5)8-15(20)mm, erecto-patent on long slender pedicels; (2n=10). Native; cultivated ground and bare places on banks, walls, rocks and waysides; throughout BI, rarer in N & W.

5. ISATIS L. - *Woad*
Biennials to perennials; basal leaves simple; hairs 0 or unbranched; petals yellow; fruit pendent, indehiscent, 1-seeded, winged; style 0.

Other spp. - I. lusitanica L. (*I. aleppica* Scop.) formerly occurred as a casual.

1. I. tinctoria L. - *Woad*. Stems erect, to 1.5m, well-branched; leaves lanceolate to **RR**
oblanceolate, entire, clasping the stem, ± glabrous; fruits 1-2.5cm, purple-brown **251**
when ripe; (2n=28). Intrd; natd on cliffs in E Gloucs and Surrey since before 1800, infrequent casual elsewhere, formerly commoner; Europe.

6. BUNIAS L. - *Warty-cabbages*
Biennials to perennials; basal leaves simple; hairs mixed unbranched, branched and warty-glandular; petals yellow; fruit irregularly warty-ovoid, indehiscent, 1-2-seeded, with short style.

Other spp. - B. erucago L. (*Southern Warty-cabbage*) used to occur as a casual but is now absent or very rare; it has distinctive fruits with 4 wings.

1. B. orientalis L. - *Warty-cabbage*. Stems erect, to 1.2m, well-branched; lower **251**
leaves oblanceolate, variously lobed and toothed, at least some pinnately lobed, sparsely pubescent; fruits 5-8mm, on long slender erecto-patent pedicels; (2n=14). Intrd; natd in waste and rough grassy places scattered over most of BI, mostly in C & S En; E Europe.

TRIBE 2 - HESPERIDEAE (genera 7-10). Hairs branched (sometimes closely appressed to plant), sometimes unbranched, sometimes glandular; petals usually white to pink or purple, sometimes yellow to red; fruit beakless, >3x as long as wide, with usually deeply (sometimes shallowly) 2-lobed stigma.

7. ERYSIMUM L. (*Cheiranthus* L.) - *Wallflowers*
Annuals to perennials; leaves simple; hairs 2-3-branched, stalkless, the arms tightly appressed to plant; petals yellow to red, brown or purple; fruit flattened to 4-angled, with strong midribs; stigma ± capitate to 2-lobed; seeds in 1(-2) rows under each valve, not winged or with narrow wing.

1 Petals <1cm; fruits 1-3cm, erecto-patent **1. E. cheiranthoides**
1 Petals >1cm; fruits 2.5-9cm, ± erect 2
  2 Petals orange; fruits square in section; stigma with 2 rounded closely appressed lobes **2. E. x marshallii**
  2 Petals various colours; fruits distinctly flattened; stigma with 2 divergent lobes **3. E. cheiri**

Other spp. - E. repandum L. (*Spreading Treacle-mustard*), from S Europe, is a grain-alien now rare but formerly more frequent; it has petals 6-10mm and ± patent fruits 4-10cm.

1. E. cheiranthoides L. - *Treacle-mustard*. Usually annual; stems erect, usually **251**
branched, to 60(100)cm; lower leaves elliptic, entire to shallowly toothed; petals 3-6mm, yellow; fruits 1-3cm, sparsely pubescent, erecto-patent; (2n=16). Intrd; cultivated and waste ground; scattered over much of BI but rare or local except in C & S En; Europe.

2. E. x marshallii (Henfr.) Bois (*E. allionii* hort. nom. illeg., *Cheiranthus allionii* 251
hort.; *E. decumbens* (Schleich. ex Willd.) Dennst. x *E. perofskianum* Fisch. & C.A. 254
Mey.) - *Siberian Wallflower.* Usually biennial; stems often branched below, erect to ascending, to 50cm; leaves narrowly elliptic, remotely dentate; petals 15-30mm, orange; fruits 3-9cm, appressed-pubescent, nearly erect; (2n=42). Intrd; much grown in gardens, not natd but fairly frequent on tips etc.; S En; probably garden origin. Distinguished from *E. cheiri* with orange flowers by its bilobed stigma and fruits square in section.

3. E. cheiri (L.) Crantz (*Cheiranthus cheiri* L.) - *Wallflower.* Usually perennial; 251
stems well-branched and woody below, erect to decumbent, to 50(80)cm; leaves 254
narrowly elliptic, entire; flowers characteristically scented; petals 15-35mm, yellow to red, brown or purple; fruits 2.5-8cm, densely appressed-pubescent, nearly erect; (2n=14). Intrd; much grown and often natd on walls, banks and other dry places; scattered throughout BI except N Sc; E Mediterranean.

8. HESPERIS L. - *Dame's-violet*

Biennial to perennial; hairs mixed simple and branched, or 0; petals white to pink or purple; fruit ± cylindrical, slightly constricted, with ± strong midribs; stigma 2-lobed, seeds in 1 row under each valve, not winged.

Other spp. - **H. laciniata** All., from S Europe, formerly occurred as a casual.

1. H. matronalis L. - *Dame's-violet.* Stems erect, to 1.5m, branched above, 251
glabrous to sparsely pubescent; stem-leaves ovate, petiolate, sharply serrate; petals 15-30mm; fruits 2-11.5cm, ± glabrous, erecto-patent to patent; (2n=14, 16, 24, 26, 28, 32). Intrd; much grown and a common escape natd in waste places, banks, grassland, hedges and verges; frequent throughout BI; Europe.

9. MALCOLMIA W.T. Aiton - *Virginia Stock*

Annuals; basal leaves simple; hairs branched, with 2-4 appressed arms and stalk 0; petals pink to purple; fruit constricted, with ± rounded valves with 3 veins; stigma 2-lobed; seeds in 1 row under each valve, not winged.

1. M. maritima (L.) W.T. Aiton - *Virginia Stock.* Stems decumbent to erect, much- 251
branched from base, to 50cm; leaves obovate to oblanceolate, entire or toothed; petals 10-21mm; fruits 1.5-8cm, pubescent, ± erect; (2n=14-16). Intrd; much grown in gardens and commonly escaping on tips and in waste places; S & C Br and CI; Balkans.

10. MATTHIOLA W.T. Aiton - *Stocks*

Annuals to perennials; basal leaves simple; hairs branched, stalked; petals white to pink or purple; fruit not or sometimes ± constricted, with 1-veined flattened to ± rounded valves; stigma strongly 2-lobed, each lobe with dorsal horn-like or hump-like process; seeds in 1 row under each valve, broadly winged.

1 Fruit ± cylindrical, ± constricted between seeds, with apical horn-like processes (2)5-7(10)mm **3. M. longipetala**

1 Fruit ± compressed, not constricted between seeds; with apical horn-like processes <3mm 2

  2 Leaves all entire; fruits without glands **1. M. incana**

  2 Lower leaves sinuate to lobed; fruits with large yellow to black sessile glands **2. M. sinuata**

Other spp. - **M. tricuspidata** (L.) W.T. Aiton and **M. fruticulosa** L.) Maire (*M. tristis* R. Br.) have occurred as casuals, but very rarely or not nowadays.

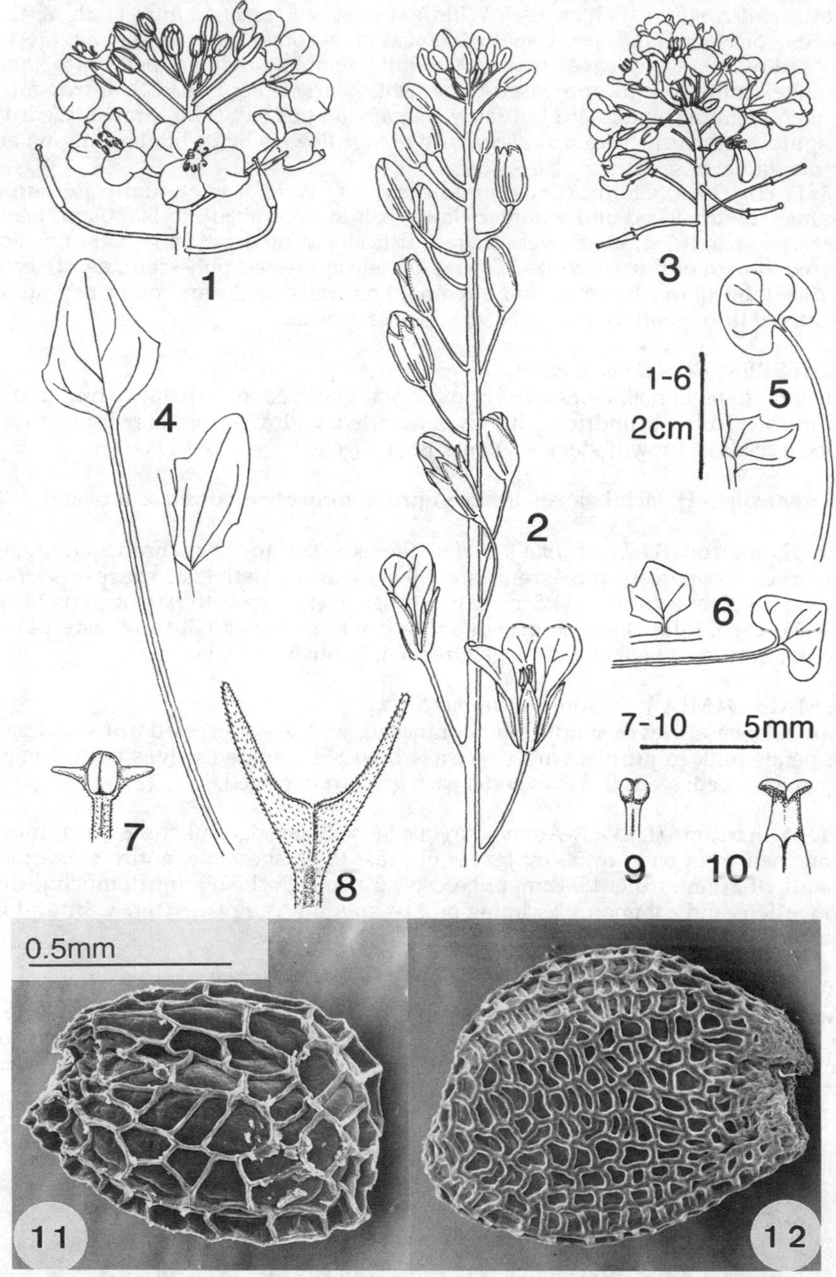

FIG 254 - Brassicaceae. 1-3, inflorescences of *Brassica*. 1, *B. napus*. 2, *B. oleracea*. 3, *B. rapa*. 4-6, basal and stem-leaves of *Cochlearia*. 4, *C. anglica*. 5, *C. officinalis*. 6, *C. danica*. 7-10, stigmas. 7, *Matthiola incana*. 8, *M. longipetala*. 9, *Erysimum x marshallii*. 10, *E. cheiri*. 11-12, seeds of *Rorippa*. 11, *R. nasturtium-aquaticum*. 12, *R. microphylla*.

1. M. incana (L.) W.T. Aiton - *Hoary Stock.* Annual or perennial; stems woody RR
below, erect to ascending, branched from base, to 80cm; leaves lanceolate to 251
oblanceolate, entire; petals 20-30mm; fruits (3)4.5-15cm, erecto-patent; 2n=14. 254
Possibly native on sea-cliffs in S En; elsewhere a garden escape, often natd on walls and banks etc. in En, Wa, Man and CI.

2. M. sinuata (L.) W.T. Aiton - *Sea Stock.* Biennial; stems woody below, erect, RR
branched or not, to 1m; leaves lanceolate to oblanceolate, the lower sinuate-lobed; 251
petals 15-25mm; fruits 5-15cm, erecto-patent; 2n=14. Native; sand-dunes and sea-cliffs; S Ir, S Wa, SW En and CI, very local and decreasing.

3. M. longipetala (Vent.) DC. (*M. bicornis* (Sibth. & Sm.) DC., *M. oxyceras* DC.) - 251
Night-scented Stock. Annual; stems much-branched, diffuse, to 50cm; leaves linear 254
to narrowly elliptic, entire to distantly toothed; petals 15-25mm; fruits 3.5-15cm, ± patent, with upcurved horn-like stigmatic processes (2)5-7(10)mm; (2n=14). Intrd; grown in gardens and a frequent casual on tips and in waste places in En; Balkans. Our plants are mostly ssp. **bicornis** (Sibth. & Sm.) P.W. Ball.

TRIBE 3 - ARABIDEAE (genera 11-16). Hairs branched or unbranched, sometimes stellate, not glandular; petals yellow or white; fruit beakless, with small stigma, >3x or sometimes <3x as long as wide.

11. BARBAREA W.T. Aiton - *Winter-cresses*

Biennials or perennials; basal leaves pinnate; hairs 0 or few, unbranched; petals yellow; fruit >3x as long as wide, with angled valves; seeds in 1 row under each valve.

1 Uppermost stem-leaves simple, toothed or lobed to <1/2 way to midrib (ignore basal lobes); seeds 1-1.8mm 2
1 Uppermost stem-leaves ± pinnate to pinnately lobed to >1/2 way to midrib; seeds 1.6-2.4mm 3
 2 Fruit with style (1.7)2-3.5(4)mm; flowers buds glabrous ..... **1. B. vulgaris**
 2 Fruit with style 0.5-1.8(2.3)mm; flower buds with hairs at apex of sepals ..... **2. B. stricta**
3 At least some fruits >4cm; fresh petals >5.6mm ..... **4. B. verna**
3 Fruits <4cm; fresh petals ≤5.6(6.3)mm ..... **3. B. intermedia**

1. B. vulgaris W.T. Aiton (*B. arcuata* (Opiz ex J. & C. Presl) Rchb.) - *Winter-cress.* 251
Stems erect, to 1m; basal leaves usually at least as wide at most distal lateral leaflets as at terminal leaflet; fruits (0.7)1.5-3.2cm with style (1.7)2-3.5(4)mm; 2n=16. Native; hedges, streamsides, roadsides and waste places, often on damp soil; throughout most of BI.

2. B. stricta Andrz. - *Small-flowered Winter-cress.* Stems erect, to 1m; basal leaves 251
usually much wider at terminal leaflet than at any lateral leaflets; fruits 1.3-2.8(3.5)cm with style 0.5-1.8(2.3)mm; (2n=14, 16, 18). Probably intrd; similar places to *B. vulgaris*; very scattered and mostly casual in En, Wa and S Sc, natd in parts of S En and S Wa; Europe.

3. B. intermedia Boreau - *Medium-flowered Winter-cress.* Stems erect, to 60cm; 251
basal leaves with 2-6 pairs of lateral leaflets; fruit 1.5-3.5(4)cm with style 0.6-1.7mm; (2n=16). Intrd; natd in waste, open and cultivated ground and by roads and streams; sparsely scattered in most of BI; Europe.

4. B. verna (Mill.) Asch. - *American Winter-cress.* Stems erect, to 1m; basal leaves 251
with(3)4-10 pairs of lateral leaflets; fruit (2.8)3.5-7.1cm with style 0.6-2(2.3)mm; (2n=16). Intrd; natd in waste, cultivated and open ground and by roads; sparsely scattered through most of BI but rare in Sc; Europe.

12. RORIPPA Scop. (*Nasturtium* W.T. Aiton) - *Water-cresses*

Annuals to perennials; basal leaves simple and unlobed to pinnate; hairs 0 or

unbranched; petals white or yellow; fruit >3x or <3x as long as wide, with convex valves and indistinct midribs; seeds in (1 or) 2 rows under each valve.
Hybrids are not common but are effectively spread vegetatively and may occur in absence of both parents. *R. sylvestris*, *R. amphibia* and *R. austriaca* are highly self-sterile and therefore often do not set seed; hence sterility is not diagnostic for hybridity in the yellow-flowered taxa.

1 Petals white 2
1 Petals yellow 4
  2 Seeds per fruit <5; pollen <40% with full contents **2. R. x sterilis**
  2 Seeds per fruit >10; pollen >80% with full contents 3
3 Seeds in 2 rows under each valve, with c.6-12 cells showing across broadest width **1. R. nasturtium-aquaticum**
3 Seeds mostly in 1 row under each valve, with c.12-20 cells showing across broadest width **3. R. microphylla**
  4 Petals c. as long as sepals; stems not rooting 5
  4 Petals ≥1.5x as long as sepals; stems often rooting when contacting ground or water 6
5 Fruit 2-3x as long as pedicels; sepals <1.6mm **4. R. islandica**
5 Fruit 0.8-2x as long as pedicels; sepals ≥1.6mm **5. R. palustris**
  6 Stem-leaves with auricles at base, ± clasping stem 7
  6 Stem-leaves with no or very small auricles, not clasping stem 9
7 Leaves toothed but not lobed; fruit (excl. style) <1.5x as long as wide **11. R. austriaca**
7 Leaves strongly lobed; fruit (excl. style) >1.5x as long as wide 8
  8 Petals mostly <3.5mm when fresh; stem-leaves c.4x as long as wide **6. R. x erythrocaulis**
  8 Petals mostly >3.5mm when fresh; stem-leaves 2-3x as long as wide **9. R. x armoracioides**
9 Upper stem-leaves toothed but not lobed; fruit (excl. style) 2-2.5(3)x as long as wide **10. R. amphibia**
9 Upper stem-leaves lobed; fruit (excl. style) (2)2.5-7.5x as long as wide 10
  10 Stem-leaves with terminal segment <1/4 of total length; fruit 7-23 x 0.9-1.8mm, plus style ≤1.2mm **7. R. sylvestris**
  10 Stem-leaves with terminal segment >1/4 of total length; fruit 3-10 x 1.2-2.5mm, plus style usually ≥1.2mm 11
11 Fruiting pedicels mostly reflexed **8. R. x anceps**
11 Fruiting pedicels patent to erecto-patent **9. R. x armoracioides**

1. R. nasturtium-aquaticum (L.) Hayek (*Nasturtium officinale* W.T. Aiton) - 254
Water-cress. Perennial with procumbent to ascending stems to 1m; leaves pinnate, 261
entire to sinuate; petals white, 3.5-6.6mm when fresh; stamens dehiscing inwardly; fruit 11-19mm (mean per plant) x 1.9-2.7mm, with pedicels 7-12mm (mean per plant); seeds in 2 rows under each valve, with c.25-50(70) cells showing in lateral view, c.6-12 across broadest width; 2n=32. Native; in and by streams, ditches and marshes; frequent to common over most of BI.

2. R. x sterilis Airy Shaw (*R. nasturtium-aquaticum* x *R. microphylla*) - *Hybrid* 261
Water-cress. Differs from *R. nasturtium-aquaticum* in stamens dehiscing mostly outwardly; fruit c.5-11mm, irregularly shaped, distorted due to only 0-3 full seeds variously arranged, with pedicels 7-10mm (mean per plant); full seeds with c.50-100(120) cells showing in lateral view, c.10-14 across broadest width; 2n=48. Native; in similar places to parents but often without either near; scattered over most of BI.

3. R. microphylla (Boenn.) Hyl. ex Á. & D. Löve (*Nasturtium microphyllum* 254
(Boenn.) Rchb.) - *Narrow-fruited Water-cress*. Differs from *R. nasturtium-aquaticum* in 261
stamens dehiscing outwardly; fruit 16-23mm (mean per plant) x 1.3-1.8mm, with

pedicels 14-19mm (mean per plant); seeds in 1 row under each valve, with c.90-120(190) cells showing in lateral view, c.12-20 across broadest width; 2n=64. Native; similar habitat and distribution to *R. nasturtium-aquaticum* but slightly less common in S and more common in N.

4. R. islandica (Oeder ex Gunnerus) Borbás - *Northern Yellow-cress*. Annual or short-lived perennial with procumbent to ascending stems to 15(30)cm; stem-leaves deeply pinnately lobed or ± pinnate; petals yellow, 1-1.7mm when fresh; fruit 5-12 x 1.7-3mm, c.2-3x as long as patent to reflexed pedicels; 2n=16. Native; open pondsides and other damp places; very scattered in W Ir, Wa, Man and Sc, mostly near sea. **RR**

5. R. palustris (L.) Besser (*R. islandica* auct. non (Oeder ex Gunnerus) Borbás) - *Marsh Yellow-cress*. Differs from *R. islandica* in often ascending to erect stems to 60cm; petals 1.4-2.8mm when fresh; fruit 4-12 x 1.3-3mm, c.0.8-2x as long as pedicels; 2n=32. Native; open damp and waste ground; frequent to common in most of BI but rare in N Sc. **261**

5 x 7. R. palustris x R. sylvestris has been recorded from BI on several occasions but needs confirmation.

6. R. x erythrocaulis Borbás (*R. palustris* x *R. amphibia*) - *Thames Yellow-cress*. Vigorous perennial to 1m; stem-leaves pinnately lobed; petals yellow, 2.5-3.6mm when fresh; fruit 3.5-5.5 x 1.5-2mm, much shorter than patent to reflexed pedicels; triploid and sterile, or tetraploid and fertile; 2n=24, 32. Native; by R. Thames in Surrey and Middlesex and by R. Avon in Warks with parents, by R. Avon in W Gloucs without parents. **RR**

7. R. sylvestris (L.) Besser - *Creeping Yellow-cress*. Spreading perennial with shoots arising from creeping roots; stems to 60cm, decumbent to ± erect; stem-leaves pinnate to deeply pinnately lobed; petals yellow, 2.2-5.5mm when fresh; fruit 7-23 x 0.9-1.8mm, c.2x as long as patent to erecto-patent pedicels; 2n=32, 48. Native; in damp places and disturbed ground; frequent in most of BI but rare in N Sc. **261**

8. R. x anceps (Wahlenb.) Rchb. (*R. sylvestris* x R. *amphibia*) - *Hybrid Yellow-cress*. Perennial with habit of *R. sylvestris*; stem-leaves intermediate in lobing; petals yellow, 3.5-5.5mm when fresh; fruit 3-10 x 1.2-2.5mm, c. as long as mostly reflexed pedicels; 2n=32, 40. Native; the commonest hybrid yellow-cress, often fertile and backcrossing but often not with parents; scattered in En, Wa and Ir.

9. R. x armoracioides (Tausch) Fuss (*R. sylvestris* x *R. austriaca*) - *Walthamstow Yellow-cress*. Perennial with habit of *R. sylvestris*; stem-leaves intermediate in lobing; petals yellow, 3-4.5mm when fresh; fruit 3-9 x 1.5-2mm, c. as long as patent to erecto-patent pedicels, with some full seeds; (2n=32). Native; damp waste ground; S Essex, Middlesex, Cheshire and Argyll. **RR**

10. R. amphibia (L.) Besser - *Great Yellow-cress*. Perennial with shoots arising from axils of creeping stems; stems to 1.2m, ascending to erect; lower stem-leaves pinnately lobed, upper only toothed; petals yellow, 3.3-6.2mm when fresh; fruit 2.5-7.5 x 1.5-3mm, much shorter than mostly reflexed pedicels; 2n=16, 32. Native; in and by rivers, ponds and ditches; frequent in En, local in Wa and Ir, rare and probably intrd in S Sc. **261**

10 x 11. R. amphibia x R. austriaca = R. x hungarica Borbás was found in 1987 as 1 plant on a river-bank in S Essex; it is intermediate (petals 4.8-5.7mm; sepals 3.2-3.6mm) and sterile.

11. R. austriaca (Crantz) Besser - *Austrian Yellow-cress*. Perennial with shoots arising from creeping roots; stems to 1m, ascending to erect; leaves toothed, not lobed; petals yellow, 3-5mm when fresh; fruit 2.5-3.5 x 1.5-2.5mm, much shorter than patent to erecto-patent pedicels; 2n=16. Intrd; waste ground, and by roads and rivers; natd in scattered localities in En and S Wa; Europe. **261**

13. ARMORACIA P. Gaertn., B. Mey. & Scherb. - *Horse-radish*
Perennials with deep strong roots; basal leaves simple; hairs ± 0; petals white; fruit not or very rarely ripening, the best developed being ± terete, <3x as long as wide,

with few seeds in 2 rows under each valve.

1. A. rusticana P. Gaertn., B. Mey. & Scherb. - *Horse-radish.* Forming extensive 261
patches; stems to 1.5m, with narrow unlobed leaves; basal leaves to 50(100)cm, *Rumex*-like, deeply lobed on young plants, later ± entire; inflorescences densely branched panicles; (2n=32). Intrd; grassy places and waste ground as relic of cultivation or where dumped; frequent in En, Wa and CI, scattered in Ir and Sc; ? W Asia.

14. CARDAMINE L. (*Dentaria* L.) - *Bitter-cresses*

Annuals to perennials; basal leaves pinnate or ternate; hairs 0 or unbranched; petals white to pink or purple, sometimes 0; fruit compressed, >3x as long as wide, with valves that spring open suddenly to release seeds; seeds in 1 row under each valve.

1 Upper stem-leaves with purple bulbils in axils **1. C. bulbifera**
1 Stem-leaves without axillary bulbils 2
  2 All leaves with (1-)3 leaflets **3. C. trifolia**
  2 At least some lower leaves with ≥5 leaflets 3
3 Petals ≤5.5 x 2.8mm, white, sometimes 0 4
3 Petals ≥5.5 x 2.8mm, white to pink or purple 7
  4 Stem-leaves with small basal auricles clasping stem **7. C. impatiens**
  4 Stem-leaves without auricles 5
5 Most flowers with 4(5) stamens **9. C. hirsuta**
5 Most flowers with 6 stamens 6
  6 Stems with 0-3 leaves, glabrous to sparsely pubescent; leaflets 3-7; petals 2-2.8mm wide **10. C. corymbosa**
  6 Stems with (3)4-7(10) leaves, conspicuously pubescent; leaflets 5-17; petals 1-2(2.2)mm wide **8. C. flexuosa**
7 Plant with whitish succulent rhizome with succulent scale-leaves **2. C. heptaphylla**
7 Rhizome 0 or present, not whitish or with succulent scale-leaves 8
  8 Anthers blackish-violet; petals usually white; stigma tapered from style **4. C. amara**
  8 Anthers yellow; petals usually pale to deep pink; stigma minutely capitate 9
9 All leaves large, with apical leaflet much wider than lateral ones **5. C. raphanifolia**
9 Upper leaves with much narrower leaflets than lower leaves, the apical leaflet not or scarcely wider than lateral ones **6. C. pratensis**

1. C. bulbifera (L.) Crantz (*Dentaria bulbifera* L.) - *Coralroot*. Perennial with white R
succulent rhizome; stems erect, unbranched, to 75cm; leaves with 1-7 entire to serrate leaflets; petals 10-17mm, pinkish-purple; (2n=32, 48, 96). Native; deciduous woodland; very local in C & SE En N to Staffs, rarely natd elsewhere in En, Sc and Ir.

2. C. heptaphylla (Vill.) O.E. Schulz (*Dentaria pinnata* (Lam.) R.Br.) - *Pinnate Coralroot*. Perennial with white succulent rhizome; stems erect, unbranched, to 60cm; leaves with 5-11 serrate leaflets; petals 11-20mm, white or pinkish-purple; (2n=48). Intrd; persistent relic of cultivation in woodland; W Sussex and W Gloucs to Durham; Europe.

3. C. trifolia L. - *Trefoil Cress*. Perennial with thin rhizome; stems erect to ascending, unbranched, to 30cm; leaves mostly ternate, those on stem few and reduced; petals 4.5-11mm, white; (2n=16). Intrd; natd in shady places as relic of cultivation; scattered over En; Europe.

4. C. amara L. - *Large Bitter-cress*. Perennial with short rhizome; stems little or not branched, erect to ascending, to 60cm; leaves with 5-11 sinuate-crenate leaflets; petals 5.5-12mm, white or rarely pinkish-purple; 2n=16. Native; streamsides, marshes and flushes; locally common over En and Sc, very local in Wa and N Ir.

5. C. raphanifolia Pourr. (*C. latifolia* Vahl non Lej.) - *Greater Cuckooflower*. Perennial with long strong rhizome; stems little or not branched, erect, to 70cm; leaves with 3-13 sinuate-crenate leaflets; petals 8-14mm, purplish-red, rarely white; (2n=46). Intrd; natd in damp often shady places; very scattered over Br, mostly in N & W; S Europe.

6. C. pratensis L. (*C. nymanii* Gand., *C. matthioli* Moretti ex Comolli, *C. hayneana* (Rchb.) Fritsch, *C. palustris* (Wimm. & Grab.) Peterm., *C. rivularis* Schur, *C. crassifolia* Pourr.) - *Cuckooflower*. Perennial with 0 or short rhizome; stems little or not branched, ascending to erect, to 60cm; leaves very variable, with 3-c.21 entire to crenate leaflets; petals 6-18mm, pale to deep pink, rarely white; 2n=30, 56-58, c.64, 72 (16-96). Native; wet grassy places; common throughout BI. Many clones are sterile, but reproduce by rooting from leaves. Extremely variable but impossible to subdivide satisfactorily. On the Continent plants with middle and upper stem-leaves pinnately lobed and with thin leaves with raised veins have lower ploidy levels (diploid to heptaploid) and are *C. pratensis sensu stricto* (ssp. **pratensis**). Plants with higher ploidy levels (heptaploid to dodecaploid) may resemble *C. pratensis* but have all leaves pinnate and usually larger flowers (ssp. **dentata** (Schult.) Celak., *C. palustris*), or may have thick leaves with impressed veins and smaller flowers as in ssp. *pratensis* (ssp. **polemonioides** Rouy, *C. nymanii*). The latter 2 are found in more northern and upland areas. These 3 taxa are those most likely to occur in BI, but identity of British variants with taxa named in Europe is very dubious.

6 x 8. C. pratensis x C. flexuosa = C. x fringsii F. Wirtg. (*C. x haussknechtiana* O.E. Schulz) occurs in scattered localities in Br with the parents; it is intermediate and sterile, but reproduces vegetatively as in *C. pratensis* 2n=32. At least 2 variants exist: 1 with white petals nearer in size to those of *C. flexuosa*; and 1 with pink petals nearer in size to those of *C. pratensis*.

7. C. impatiens L. - *Narrow-leaved Bitter-cress*. Annual or biennial; stems erect, to R
80cm, little or not branched, with many leaves with 7-23 entire to serrate leaflets; petals 1.5-3.5mm, white, sometimes 0; stamens 6; (2n=16). Native; damp woods and river-banks; locally frequent in Br N to S Sc but absent from SW, CE & NE En, Westmeath.

8. C. flexuosa With. - *Wavy Bitter-cress*. Annual to short-lived perennial; stems 261
flexuous, ascending to erect, often well branched, to 50cm, with (3)4-7(10) leaves with 5-15 entire to dentate leaflets; petals (2.2)2.5-4(5) x 0.8-2(2.2)mm, white; stamens mostly 6; 2n=32. Native; marshes, streamsides and sometimes cultivated ground; common throughout BI.

8 x 9. C. flexuosa x C. hirsuta = C. x zahlbruckneriana O.E. Schulz has been recorded near the parents in Monts; it resembles *C. flexuosa* but is sterile; (2n=24).

9. C. hirsuta L. - *Hairy Bitter-cress*. Annual to biennial; stems erect to ascending, often well branched at base, to 30cm, with (0)1-4(5) leaves with 3-c.11 entire to dentate leaflets; petals 2.5-4(5) x 0.8-2(2.2)mm, white, sometimes 0; stamens mostly 4; 2n=16. Native; open and cultivated ground, rocks and walls; common throughout BI.

10. C. corymbosa Hook. f. (*C. uniflora* (Hook. f.) Allan non Michx.) - *New Zealand Bitter-cress*. Annual but often propagating by rooting at leaf-tips; stems decumbent to ascending, simple, to 10cm, with 0-3 leaves with 3-7 leaflets; petals 3-5.5 x 2-2.8mm, white; stamens 6; (2n=48). Intrd; spread as horticultural contaminant to paths, rockeries and pavement cracks; scattered in C & N Br and NE Ir, first recorded 1985; New Zealand. Stems frequently have 0 leaves and only 1 flower.

15. ARABIS L. (*Turritis* L., *Cardaminopsis* (C.A. Mey.) Hayek) - *Rock-cresses*

Annuals to perennials; basal leaves simple, deeply lobed to ± entire; hairs usually branched and unbranched, rarely 0 or all simple; petals white, pale yellow or pinkish-purple; fruit compressed or 4-angled, >3x as long as wide; seeds in 1 or 2 rows under each valve.

1 Stem-leaves strongly clasping stem at base; auricles distinctly longer than stem-width 2
1 Stem-leaves not or scarcely clasping stem at base; auricles 0 or much shorter than stem-width 6
2 Non-flowering shoots 0 or forming ± sessile rosettes in compact clump 3
2 Non-flowering shoots elongated, some becoming stolons and mat-forming 5
3 Fruits >7cm, patent and curved downwards when ripe **4. A. turrita**
3 Fruits ≤7cm, erect 4
4 Petals pale yellow; seeds in 2 rows under each valve **3. A. glabra**
4 Petals white; seeds in 1 row under each valve **7. A. hirsuta**
5 Lowest leaves with petiole c. as long as blade; stem-leaves with pointed to narrowly rounded auricles longer than wide; widespread **6. A. caucasica**
5 Lowest leaves with petiole much shorter than blade; stem-leaves with broadly rounded auricles c. as long as wide; Skye **5. A. alpina**
6 Ripe fruits patent to erecto-patent, arising at ≤45°; basal leaves mostly deeply sinuate, with long petioles 7
6 Ripe fruits erect or nearly so, arising at much <45°; basal leaves mostly shallowly sinuate with short petioles 8
7 Plant with branching rhizome; mountains of N & W BI **1. A. petraea**
7 Plant with only root underground; alien in C & S En **2. A. arenosa**
8 Basal leaves entire to slightly toothed; petals white; pedicels mostly <8mm **7. A. hirsuta**
8 Basal leaves sinuate-lobed; petals white, pink or yellow, if pure white then most pedicels >8mm 9
9 Petals pale yellow; pedicels mostly <7mm **9. A. scabra**
9 Petals white to pink; pedicels mostly >7mm **8. A. collina**

1. A. petraea (L.) Lam. (*Cardaminopsis petraea* (L.) Hiitonen) - *Northern Rock-cress.* R
Loosely mat-forming perennial with branching rhizome; stems ascending to erect, to 261
25cm; basal leaves long-stalked, deeply lobed; flowers few; petals white to purplish; 2n=16. Native; mountain rock-ledges and crevices; NW Wa, N & C Sc, extremely local in Ir.

2. A. arenosa (L.) Scop. (*Cardaminopsis arenosa* (L.) Hayek) - *Sand Rock-cress.* Tufted annual to perennial; stems erect, to 40cm; basal leaves long-stalked, deeply lobed; flowers usually numerous; petals white or pinkish-purple; (2n=16, 28, 32). Intrd; waste and other open ground; C & S En; Europe.

3. A. glabra (L.) Bernh. (*Turritis glabra* L.) - *Tower Mustard.* Tufted biennial; stems R
erect, to 1m, pubescent below, glabrous above; basal leaves entire to sinuate-lobed; 261
flowers numerous; petals pale yellow; 2n=12. Native; dry grassy, rocky and waste places; very local and decreasing in En, casual in Sc.

4. A. turrita L. - *Tower Cress.* Tufted biennial or perennial; stems erect, to 70cm, pubescent; basal leaves sinuate-toothed; flowers numerous; petals pale yellow; (2n=16). Intrd; on old walls; Cambridge, formerly elsewhere; Europe.

5. A. alpina L. - *Alpine Rock-cress.* Mat-forming perennial with stolons producing RRR
rosettes; stems ± erect, to 40cm; basal leaves short-stalked, sinuate-toothed; flowers numerous; petals white; (2n=16). Native; rock-ledges at 820-850m; Skye (N Ebudes), relic of old planting in MW Yorks.

6. A. caucasica Willd. ex Schltdl. (*A. alpina* ssp. *caucasica* (Willd. ex Schltdl.) Briq.). - *Garden Arabis.* Differs from *A. alpina* in its long-stalked basal leaves with fewer teeth; larger petals (mostly >10mm, not <10mm) white to purplish; and leaves more densely whitish-grey (not grey-green) pubescent; (2n=16). Intrd; garden

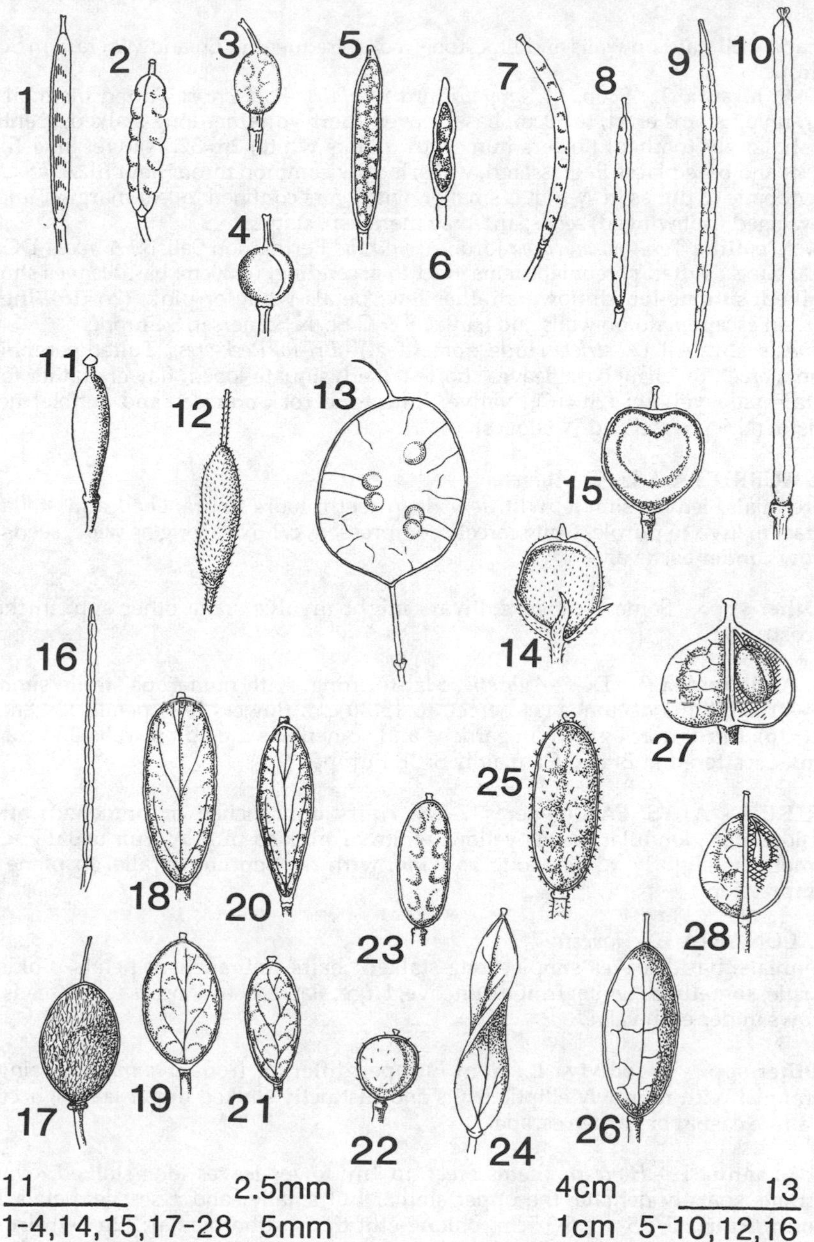

FIG 261 - Fruits of **Brassicaceae**. 1, *Rorippa sylvestris*. 2, *R. palustris*. 3, *R. amphibia*. 4, *R. austriaca*. 5, *R. nasturtium-aquaticum*. 6, *R. x sterilis*. 7, *R. microphylla*. 8, *Arabis petraea*. 9, *A. hirsuta*. 10, *A. glabra*. 11, *Armoracia rusticana*. 12, *Aubrieta deltoidea*. 13, *Lunaria annua*. 14, *Alyssum alyssoides*. 15, *A. saxatile*. 16, *Cardamine flexuosa*. 17, *Berteroa incana*. 18, *Erophila verna* var. *verna*. 19, var. *praecox*. 20, *E. majuscula*. 21, *E. glabrescens*. 22, *Lobularia maritima*. 23, *Draba muralis*. 24, *D. incana*. 25, *D. norvegica*. 26, *D. aizoides*. 27, *Lepidium draba* ssp. *draba*. 28, ssp. *chalepense*.

escape well natd on walls and limestone rocks; frequent in En and Wa, rare in Sc; S Europe.

7. A. hirsuta (L.) Scop. (*A. brownii* Jord.) - *Hairy Rock-cress.* Tufted biennial to **261**
perennial; stems erect, to 60cm; basal leaves short- to rather long-stalked, ± entire to shallowly toothed; flowers numerous; petals white; 2n=32. Native; limestone rocks and bare places in grassland, walls; locally common throughout BI. *A. brownii* is endemic to dunes in W Ir; it is smaller, with hairs confined to leaf-margins, and ± unwinged (not winged) seeds, and may merit ssp. status.

8. A. collina Ten. (*A. muricola* Jord., *A. muralis* Bertol. non Salisb., *A. rosea* DC.) - *Rosy Cress.* Tufted perennial; stems erect to ascending, to 30cm; basal leaves short-stalked, sinuate-lobed; flowers rather few; petals white or pink; (2n=16). Intrd; garden escape natd on walls and banks; S & C Sc, N Somerset; S Europe.

9. A. scabra All. (*A. stricta* Huds. nom. illeg.) - *Bristol Rock-cress.* Tufted perennial; **RRR**
stems erect, to 25cm; basal leaves short-stalked, sinuate-lobed; flowers rather few; petals pale yellow; (2n=16). Native; limestone rock-crevices and rubble; near Bristol (N Somerset and W Gloucs).

16. AUBRIETA Adans. - *Aubretia*

Perennials; leaves simple, with few deep teeth; hairs unbranched and stellate; petals mauve to purple, fruit scarcely compressed, c.2-5x as long as wide; seeds in 2 rows under each valve.

Other spp. - Some escaped cultivars might involve some other spp. in their ancestry.

1. A. deltoidea (L.) DC. - *Aubretia*. Mat-forming, with numerous sterile shoots; **261**
flowering stems decumbent to erect; to 15(30)cm; flowers few; petals 12-28mm; (2n=16). Intrd; much grown in gardens and sometimes ± natd on walls and rocky banks; scattered in Br and CI, mainly S; SE Europe.

TRIBE 4 - ALYSSEAE (genera 17-22). Hairs unbranched or branched, often stellate, not glandular; petals yellow, white or pinkish-purple; fruit usually <3x, sometimes slightly >3x, as long as wide, with the septum parallel to plane of compression.

17. LUNARIA L. - *Honesty*

Biennials; basal leaves simple, long-stalked; hairs unbranched; petals pinkish-purple, sometimes white; fruit distinctive, large, flat, c.2x as long as wide; seeds in 2 rows under each valve.

Other spp. - **L. rediviva** L., from Europe, differing from *L. annua* in being a perennial with narrowly elliptic fruits and distinctly stalked upper leaves, occurs as a rare casual or garden escape.

1. L. annua L. - *Honesty*. Stems erect, to 1m; lower leaves long-stalked, ovate- **261**
cordate, sharply dentate, the upper similar but smaller and ± sessile; petals 15-25mm; fruit 2.5-7.5 x 1.5-3.5cm, oblong-elliptic to suborbicular; (2n=30). Intrd; very commonly grown and often escaping on tips, roadsides, waste ground etc, occasionally persistent; scattered in Br and CI; SE Europe.

18. ALYSSUM L. (*Aurinia* Desv.) - *Alisons*

Annuals to perennials; hairs branched and unbranched, often ± stellate; leaves simple, ± entire; petals yellow; fruit <3x as long as wide, not or slightly inflated; seeds 2 under each valve.

1. A. alyssoides (L.) L. - *Small Alison*. Densely pubescent annual or biennial with **261**

erect to ascending often well-branched stems to 30cm; basal leaves withered by flowering time; petals pale yellow, fading to white when dried; fruit (2)3-4(5)mm, ± orbicular, pubescent, slightly inflated; (2n=32). Intrd; grassy and arable fields; formerly widespread in S & E Br, now ± confined to Suffolk; Europe.

2. A. saxatile L. (*Aurinia saxatilis* (L.) Desv.) - *Golden Alison.* Densely pubescent **261**
perennial to 45cm; stems branched and ± woody at base, apically producing biennial flowering stems with basal leaves persistent at flowering; petals usually bright yellow; fruit 3-8mm, ± orbicular, glabrous, ± flat; (2n=16). Intrd; commonly grown in gardens and seeding abundantly, sometimes escaping on to walls and dry banks; scattered in Br; C & SE Europe.

19. BERTEROA DC. - *Hoary Alison*
Annuals to perennials; hairs stellate; leaves simple, ± entire; petals white, often discolouring yellow when dried, bifid nearly 1/2 way to base; fruit <3x as long as wide, slightly inflated; seeds 2-6 under each valve.

1. B. incana (L.) DC. - *Hoary Alison*. Grey-pubescent; stems erect, to 60cm; fruit **261**
4-10mm, with style (1)1.5-3.5mm; (2n=16). Intrd; rough grassy waste places, waysides; scattered as casual in C & S Br, natd in few areas in S; Europe.

20. LOBULARIA Desv. - *Sweet Alison*
Annuals to perennials; hairs 2-armed with 0 stalk; leaves simple, entire; petals white, sometimes purplish; fruit <3x as long as wide, slightly inflated; seeds 1 under each valve.

1. L. maritima (L.) Desv. - *Sweet Alison*. Plant grey-pubescent; stems much **261**
branched, decumbent to ascending, to 30cm; fruit obovate, broadly elliptic or suborbicular, 2-4.2mm; (2n=24). Intrd; much grown in gardens and commonly escaping on walls and other dry places, well natd on coastal sands in S Br and CI; scattered over most of BI; S Europe.

21. DRABA L. - *Whitlowgrasses*
Annuals to perennials; hairs unbranched or branched, often stellate; leaves simple; petals white or yellow; fruit <3x or sometimes slightly >3x as long as wide, ± not inflated; seeds in 2 rows under each valve.

1 Petals yellow; style ≥1mm in fruit; hairs all unbranched **1. D. aizoides**
1 Petals white; style <1mm in fruit; at least some hairs branched and stellate 2
  2 Annual; stem-leaves <2x as long as wide, cordate and clasping stem at base **4. D. muralis**
  2 Perennial; at least some stem-leaves >2x as long as wide or stem-leaves 0, tapered to rounded at base and ± not clasping stem 3
3 Stems with 0-2(3) leaves, <8cm; fruit 3-8mm, not twisted **2. D. norvegica**
3 Stems normally with >3 leaves, normally >10cm; fruit 5.5-12mm, usually twisted **3. D. incana**

1. D. aizoides L. - *Yellow Whitlowgrass*. Tufted perennial with all leaves in basal **RR**
rosettes; leaves linear, entire; stems to 10(15)cm, erect, few-flowered; petals **261**
yellow; fruit 5-10mm, with style 1-3mm; 2n=16. Probably native; limestone rocks and walls near sea; Gower peninsula, Glam (known since 1795); European mountains.

2. D. norvegica Gunnerus (*D. rupestris* W.T. Aiton) - *Rock Whitlowgrass*. Tufted **R**
perennial with all leaves in basal rosettes or 1-3 on flowering stems; leaves linear to **261**
narrowly elliptic, entire or nearly so; stems to 6cm, erect, few-flowered; petals white; fruit 3-8mm, with style ≤0.5mm; (2n=c.48). Native; bare places near

mountain-tops, usually calcareous; very local in C & N Sc. *D. rupestris*, from Sc and Faroes, might be distinct.

3. D. incana L. - *Hoary Whitlowgrass*. Tufted perennial with leafy many-flowered 261
erect stems to 40cm and sterile basal rosettes; leaves narrowly elliptic to narrowly ovate, usually coarsely toothed; petals white; fruit 5.5-12mm, with style <1mm; 2n=32. Native; rock-ledges and soil-pockets mainly on limestone, and sand-dunes in N & W Sc; very local in N Wa, N En, N & W Ir and C & N Sc. Dwarf plants are easily confused with *D. norvegica*, which usually has few stellate and more simple hairs.

4. D. muralis L. - *Wall Whitlowgrass*. Annual with basal leaf rosette and sparsely R
leafy many-flowered erect stem to 40cm; basal leaves obovate; stem-leaves ovate- 261
cordate, coarsely toothed; petals white; fruit 3-6mm, with style <0.5mm; (2n=32). Native; soil-pockets on limestone rocks and cliffs, also natd on walls and in gardens; scattered over much of BI except N Sc and CI, but perhaps native only in SW & N En.

22. EROPHILA DC. - *Whitlowgrasses*

Early flowering ephemerals; hairs unbranched or branched, often stellate; leaves confined to basal rosette, simple, entire or toothed; petals white, bifid ±1/2 way to base; fruit <3x or sometimes slightly >3x as long as wide, inflated or not; seeds in 2 rows under each valve; flowering in very early spring.

A much misunderstood genus. Many traditionally used characters, such as fruit size and shape and ovule number, do not correlate well with cytological characters or breeding behaviour and separate only pure-breeding lines. This account is based upon the work of S.A. Filfilan and T.T. Elkington, which largely confirmed the earlier conclusions of O. Winge.

1 Leaves and lower parts of stems densely grey-pubescent; petioles 1/5-1/2 as long as laminas; seeds 0.3-0.5mm; petals bifid ≤1/2 way to base **1. E. majuscula**

1 Leaves and lower parts of stems subglabrous to moderately pubescent, green; petioles ≥1/2 as long as laminas; seeds 0.5-0.8mm 2

  2 Petioles 0.5-1x as long as laminas; petals bifid 1/2-3/4 way to base; usually pubescent **2. E. verna**

  2 Petioles 1.5-2.5x as long as laminas; petals bifid ≤1/2 way to base; usually subglabrous **3. E. glabrescens**

1. E. majuscula Jord. - *Hairy Whitlowgrass*. Plant densely pubescent, with 2-many 261
stems to 9cm; fruit oblong to elliptic, ± flat, 2.5-6mm, 1.5-4x as long as wide, with 15-60(70) seeds; 2n=14. Native; all sorts of open, dry ground, especially on calcareous soils, but rarely or not on dunes; scattered sparsely through BI except N & W Sc.

2. E. verna (L.) DC. (*E. spathulata* Láng, *E. praecox* (Steven) DC.) - *Common* 261
Whitlowgrass. Plant sparsely to moderately pubescent, with 1-many stems to 10(25)cm; fruit oblanceolate to oblong or elliptic (broadly elliptic to ± orbicular in var. **praecox** (Steven) Diklic (ssp. *praecox* (Steven) Walters)), ± flat or ± inflated, 1.5-9mm, 1.5-3x as long as wide, with 15-50 seeds; 2n=30-46. Native; all sorts of open, dry ground, especially calcareous, rocks, walls, open grassland and dunes; locally common throughout BI.

3. E. glabrescens Jord. - *Glabrous Whitlowgrass*. Plant subglabrous to sparsely 261
pubescent with 1-many stems to 9cm; fruit oblanceolate to elliptic, ± flat, 3-6mm, 1.5-4x as long as wide, with 20-60 seeds; 2n=48-56. Native; habitat and distribution as in *E. verna*, but less common.

TRIBE 5 - LEPIDEAE (genera 23-34). Hairs 0, unbranched or (less often) branched; petals white, sometimes pink, yellow or 0; fruit ≤3x as long as wide, with septum at right angles to or sometimes parallel with plane of compression, sometimes indehiscent.

23. COCHLEARIA L. - *Scurvygrasses.*
Annuals to perennials (mostly biennials); basal leaves simple, long-stalked; hairs 0 or unbranched; petals white or mauve; fruit ≤3x as long as wide, inflated or ± compressed and then with septum at right angles to plane of compression; seeds in 2 rows under each valve.

1 Stems >50cm tall; pedicels >3x as long as ripe fruits **1. C. megalosperma**
1 Stems <50cm tall; pedicels <3x as long as ripe fruits 2
  2 Basal leaves cuneate at base; fruit compressed, with septum >3x as long as wide and at right angles to plane of compression **2. C. anglica**
  2 Basal leaves cordate to very broadly cuneate at base; fruit scarcely compressed, the septum <2(3)x as long as wide 3
3 All or all except extreme uppermost stem-leaves petiolate; flowers ≤5(6)mm across; fruits with ≤12(16) seeds **6. C. danica**
3 Upper stem-leaves sessile, often clasping stem; flowers mostly >5mm across; fruits with ≤8 seeds 4
  4 Perennial with ± woody base; at least some fruits acute at both ends and widest at or just below middle, with veins not or scarcely forming reticulation when dry **5. C. micacea**
  4 Biennial to perennial, rarely woody at base; fruits rarely acute at both ends, if so widest above middle, with veins forming distinct reticulation when dry 5
5 Upper stem-leaves slightly or not clasping stem at base; flowers 5-8mm across; inland sp. in upland areas **3. C. pyrenaica**
5 Upper stem-leaves distinctly clasping stem at base; flowers (5)8-15mm across; maritime habitats and inland by some roads **4. C. officinalis**

1. C. megalosperma (Maire) Vogt (*C. glastifolia* L. var. *megalosperma* Maire) - *Tall Scurvygrass.* Annual to short-lived perennial; stems erect, to 1(1.5)m; basal leaves oblong to elliptic, cuneate at base; stem-leaves sessile, strongly clasping stem; fruit 2-3.5mm, ± globose, not compressed. Intrd; natd in cultivated and waste ground in Nottingham (Notts) since c.1974; SW Europe.

2. C. anglica L. - *English Scurvygrass.* Biennial to perennial with erect to ascending **254**
stems to 40cm; basal leaves cuneate at base; stem-leaves stalked or not, the latter **269**
clasping stem; fruit 7-14mm, compressed; 2n=37, 48-50. Native; muddy shores and estuaries, often in very wet places; coasts of most of BI but local in Ir.

2 x 4. C. anglica x C. officinalis = C. x hollandica Henrard occurs probably wherever the parents meet but in Ir is much commoner than *C. anglica* and often occurs without it. It is intermediate in fruit compression and shape of basal leaves, is vigorous and fertile, and backcrosses to both parents; (2n=36).

3-5. C. officinalis L. agg. Biennial to perennial; basal leaves rounded to cordate at base; upper stem-leaves sessile, clasping stem or not; fruit (2.5)3-7(9)mm, not compressed. A difficult group due to cryptic speciation and considerable plasticity, with several different treatments recognizing 1-6 British spp.; this account is based upon the work of G.M. Fearn and J.J.B. Gill modified by the views of D.H. Dalby.

3. C. pyrenaica DC. - *Pyrenean Scurvygrass.* Stems to 30cm, procumbent to erect; basal leaves cordate at base; upper stem-leaves slightly or not clasping stem; flowers 5-8mm across; fruits 3-5mm, with veins forming distinct reticulation when dry. Native; barish damp mostly base-rich soils, rocks and spoil-heaps in upland areas.

a. Ssp. pyrenaica (*C. officinalis* ssp. *pyrenaica* (DC.) Bonnier & Layens). Leaves not succulent; fruits cuneate at base; 2n=12. N En (Derbys to Cumberland), N Ebudes.

b. Ssp. alpina (Bab.) Dalby (*C. alpina* (Bab.) H.C. Watson, *C. officinalis* ssp. *alpina* (Bab.) Hook. f.). Leaves usually succulent; fruits rounded at base; 2n=24. Often on less basic substrata; N Somerset, N Wa, W Ir and mountains in Sc. This taxon has often been associated with *C. officinalis* rather than with *C. pyrenaica*.

4. C. officinalis L. - *Common Scurvygrass*. Stems to 30(50)cm, decumbent to erect; basal leaves cordate to rounded at base; upper stem-leaves usually clasping stem; flowers (5)8-15mm across; fruit (2.5)3-7(9)mm, with veins forming distinct reticulation when dry. Native.

a. Ssp. officinalis (*C. atlantica* Pobed., *C. islandica* auct. non Pobed.). Stems to 254
30(50)cm; basal leaves cordate at base, 1.5-3cm wide; petals mostly 4-8mm, 269
usually white; fruits mostly 3-7mm; 2n=24. Salt-marshes, cliffs and other habitats by or near sea and by salt-treated roads inland; round the coasts of BI and by some roads in En and Wa. Very variable. Small plants in W Sc ± similar to ssp. *scotica* but with white flowers c.10mm across and darker green leaves have been separated as **C. atlantica**; they might simply be intermediates or merit separate ssp. rank.

b. Ssp. scotica (Druce) P.S. Wyse Jacks. (*C. scotica* Druce, *C. groenlandica* auct. non L.). Stems to 10cm; basal leaves rounded to shallowly cordate at base, 0.6-1.6cm wide; petals mostly 2-4mm, often lilac; fruits mostly 2.5-3.5mm; 2n=24. Rocky and sandy coasts of Sc, N & W Ir, Man, SW En and perhaps Wa.

4 x 6. C. officinalis x C. danica occurs probably frequently in Br where the parents meet; it is intermediate in leaf and flower characters and fertile; 2n=24-34.

5. C. micacea E.S. Marshall - *Mountain Scurvygrass*. Perennial with woody base RR
producing short stolons and procumbent to erect stems to 20cm; basal leaves cordate at base; stem-leaves stalked below, sessile above, rarely clasping stem; flowers 5-8mm across; fruit 3-6mm, with veins not or scarcely forming reticulation when dry; 2n=26. Native; on micaceous schists above 800m; M & E Perth, Angus and W Ross (records from Shetland were errors); ?endemic.

6. C. danica L. - *Danish Scurvygrass*. Annual to biennial with ascending to erect 254
stems to 25cm; basal leaves cordate at base; stem-leaves ivy-shaped, stalked or extreme uppermost sessile but not clasping stem; fruit 3-5(7)mm, not compressed; 2n=42. Native; sandy and pebbly shores, banks and walls near sea, and by railways and salt-treated roads inland; coasts of most of BI and widespread inland in En, Wa and NE Ir. Some plants from Shetland are much more robust with very succulent leaves, and might deserve taxonomic recognition.

24. CAMELINA Crantz - *Gold-of-pleasures*

Annuals or sometimes biennials; leaves simple, lanceolate to narrowly elliptic, entire or nearly so; hairs branched and unbranched; petals yellow; fruit <3x as long as wide, ± inflated, ± smooth, with septum parallel with plane of compression, with keeled to ± winged margin and long persistent style, on patent to erecto-patent pedicels; seeds in 2 rows under each valve.

Other spp. - C. alyssum (Mill.) Thell. (*C. sativa* ssp. *alyssum* (Mill.) Hegi & Em. Schmid, *C. macrocarpa* Wierzb. ex Rchb.) and **C. rumelica** Velen. have been recorded as casuals, but very rarely if ever occur nowadays.

1. C. sativa (L.) Crantz - *Gold-of-pleasure*. Plant subglabrous or sometimes 278
pubescent; stems to 70(100)cm, erect, petals 3.4-6mm; fruit 5.5-10.3 x (3.5)4-8mm (excl. style); style <1/3x as long as rest of fruit; seeds 1.2-2(2.5)mm; (2n=40). Intrd; frequent casual from many sources, especially birdseed, formerly common in arable fields, now mostly on tips; scattered over Br and Ir; S & E Europe.

2. C. microcarpa Andrz. ex DC. (*C. sativa* ssp. *microcarpa* (Andrz. ex DC.) Thell.) 278
- *Lesser Gold-of-pleasure*. Differs from *C. sativa* in being always pubescent; petals 2.2-4.6mm; fruit 4-7(7.5) x (2)2.5-4.5(4.8)mm (excl. style); style >1/3x as long as rest of fruit; seeds 0.9-1.4mm; (2n=40). Intrd; similar places to *C. sativa* but much less common; sporadic in Br; S & E Europe.

25. NESLIA Desv. - *Ball Mustard*
Annuals; leaves simple, narrowly ovate to elliptic, entire or nearly so, clasping stem at base; hairs ± stellate; petals yellow; fruit <3x as long as wide, indehiscent, reticulately ridged, unkeeled, with long persistent style, with septum parallel with plane of compression; seed usually 1.

Other spp. - N. apiculata Fisch., C.A. Mey. & Avé-Lall. (*N. paniculata* ssp. *thracica* (Velen.) Bornm.) has fruits with a short stalk-like portion apically and basally and 4 longitudinal ridges; it is rare and perhaps not worth separating.

1. N. paniculata (L.) Desv. - *Ball Mustard.* Plant pubescent; stems to 80cm, erect; 278
petals 2.2-3.2mm; fruit 1.5-3.5 x 1.5-3.5mm on patent to erecto-patent pedicels to 13mm; (2n=14). Intrd; casual from a variety of sources, once common now rare, on tips and waste places; sporadic in Br and Ir; S & E Europe.

26. CAPSELLA Medik. - *Shepherd's-purses*
Annuals to biennials; basal leaves simple, entire to deeply pinnately lobed; stem-leaves clasping stem at base; hairs unbranched; petals white or sometimes red-tinged or 0; fruit <3x as long as wide, ± obtriangular, compressed, with septum at right angles to plane of compression; seeds in 2 rows under each valve.

Genetically similar to *Erophila* in producing many pure-breeding lines at more than 1 ploidy level, and as with the latter it is possible with difficulty to recognize the different ploidy levels as spp., at least in BI.

1. C. bursa-pastoris (L.) Medik. - *Shepherd's-purse.* Plant glabrous to sparsely 269
pubescent; stems to 50cm, usually erect; petals white, 1.5-3.5mm, sometimes 0; fruit 4-10 x 3.4-7.5mm, ± obtriangular, with straight to slightly convex sides, variably emarginate at apex, with style ≤0.8mm, the pedicels patent to erecto-patent; 2n=32. Native; cultivated and other open ground; common throughout BI. Extremely variable, especially in leaf and fruit shape; c.25 segregates have been recognised in BI.

1 x 2. C. bursa-pastoris x C. rubella = C. x gracilis Gren. has been found sporadically in En with the parents; it is intermediate in fruit characters and produces very long inflorescences with very little seed, but note that *C. bursa-pastoris* often forms sterile capsules in colder weather.

2. C. rubella Reut. - *Pink Shepherd's-purse.* Differs from *C. bursa-pastoris* in its 269
petals 1.5-2mm (scarcely longer than sepals) and usually (like the sepals) red-tinged; fruit 5-7 x 4-6mm, with concave sides forming more gradually tapered base; (2n=16). Intrd; cultivated and waste ground; sporadic in Br and CI, mainly S & C En; S Europe.

27. HORNUNGIA Rchb. - *Hutchinsia*
Annuals; leaves deeply pinnately lobed or ± pinnate; hairs stellate; petals white; fruit <3x as long as wide, ± compressed, with septum at right angles to plane of compression; seeds (1-)2 under each valve.

1. H. petraea (L.) Rchb. - *Hutchinsia.* Early-flowering ephemeral; stems erect or R
ascending, to 10(15)cm; petals 0.5-1mm, scarcely longer than sepals; fruit (1.8)2-3 269
x 1-1.7mm, on patent to erecto-patent pedicels 2-6mm; 2n=12. Native; bare places becoming desiccated in summer, on carboniferous limestone and calcareous dunes;

very local in N & SW En, Wa and CI.

28. JONOPSIDIUM Rchb. - *Violet Cress*
Annuals; leaves simple, entire to 3-lobed; hairs 0; petals pink to purple, sometimes white; fruit <3x as long as wide, slightly compressed, with septum at right angles to plane of compression; seeds 2-5 under each valve.

1. J. acaule (Desf.) Rchb. - *Violet Cress*. Stems ± absent to short and congested; **278**
leaves appearing as if in a rosette, ovate to orbicular, entire, c.0.5cm, on long thin petioles; flowers solitary in leaf-axils on long thin pedicels; petals 5-8mm, ± not notched; fruit 4-6mm, with style exceeding apical notch; (2n=24). Intrd; grown as garden annual and sometimes found self-sown on tips, rough ground and roadsides; S En and Man; Portugal.

29. TEESDALIA W.T. Aiton - *Shepherd's Cress*
Annuals; leaves deeply pinnately lobed or ± pinnate; hairs unbranched or 0; petals white, the 2 abaxial c.2x as long as the 2 adaxial ones; fruit <3x as long as wide, compressed, with septum at right angles to plane of compression, ± keeled to very narrowly winged round edges; seeds 2 under each valve.

Other spp. - T. coronopifolia (J.P. Bergeret) Thell. was recorded in error from Eigg, N Ebudes.

1. T. nudicaulis (L.) W.T. Aiton - *Shepherd's Cress*. Most leaves in basal rosette; **269**
stem-leaves few and reduced; stems erect to ascending, to 25cm, shortly pubescent; fruit 2.5-5 x 2-4mm, rounded at base, emarginate at apex, with minute style, on patent pedicels 2-6mm; (2n=36). Native; open sand, gravel or shingle; scattered very locally through Br and CI, NE Ir.

30. THLASPI L. (*Pachyphragma* (DC.) Rchb.) - *Penny-cresses*
Annuals to perennials, sometimes with rhizome; leaves simple, entire to dentate; hairs unbranched or 0; petals white; fruit <3x as long as wide, compressed, with septum at right angles to plane of compression, with narrow to broad wing round edges; seeds (1)3-8 under each valve.

1 Plant with strong rhizome; stem-leaves all petiolate **5. T. macrophyllum**
1 Plant without rhizome; stem-leaves sessile, clasping stem 2
  2 Fruit ≥9mm, with wing >1mm wide at midpoint **1. T. arvense**
  2 Fruit ≤10mm, with wing <1mm wide at midpoint 3
3 Biennial to perennial, usually with non-flowering leaf-rosettes; style equalling or exceeding apical notch of fruit **4. T. caerulescens**
3 Annual; style c.1/2 as long as apical notch of fruit 4
  4 Plant glabrous, not smelling of garlic; stem-leaves with rounded auricles **3. T. perfoliatum**
  4 Plant smelling of garlic when crushed; stem sparsely pubescent at base; stem-leaves with acute auricles **2. T. alliaceum**

1. T. arvense L. - *Field Penny-cress*. Glabrous annual, stinking (but scarcely of **269**
garlic) when crushed; stems erect, to 60cm; stem-leaves with pointed auricles; fruit 9-20mm, broadly winged all round, with deep apical notch much exceeding style length; 2n=14. Possibly native; weed of waste and arable land; scattered over most of BI but much commoner in E and absent from much of Sc, Wa and Ir.

2. T. alliaceum L. - *Garlic Penny-cress*. Annual smelling of garlic when crushed; **269**
stems erect, to 60cm, pubescent at base of stem; stem-leaves sessile, clasping stem with pointed auricles; fruit 4-10mm, narrowly winged mainly apically, with shallow apical notch but even shorter style; (2n=14). Intrd; natd weed of arable

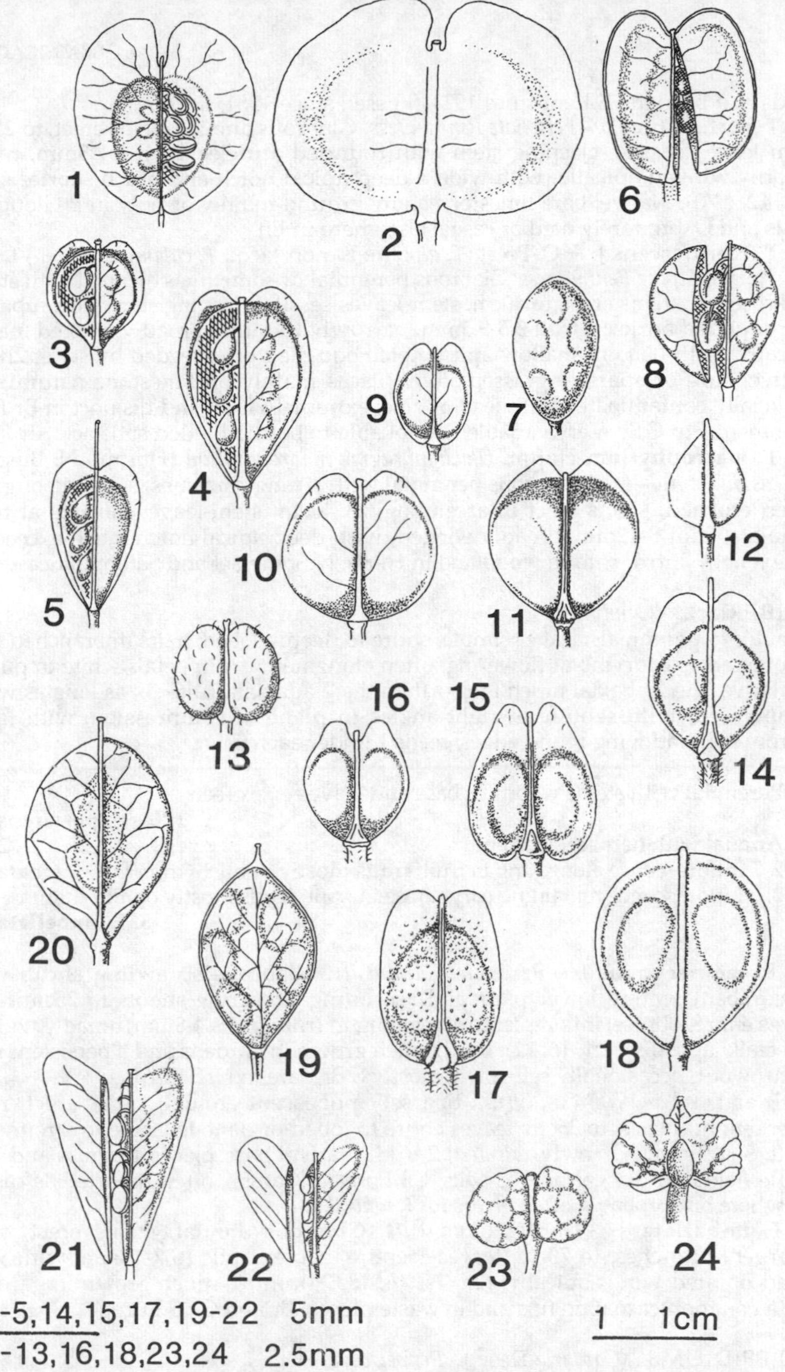

FIG 269 - Fruits of **Brassicaceae.** 1, ***Thlaspi arvense.*** 2, ***T. macrophyllum.*** 3, ***T. perfoliatum.*** 4, ***T. alliaceum.*** 5, ***T. caerulescens.*** 6, ***Teesdalia nudicaulis.*** 7. ***Subularia aquatica.*** 8, ***Hornungia petraea.*** 9, ***Lepidium hyssopifolium.*** 10, ***L. bonariense.*** 11, ***L. virginicum.*** 12, ***L. graminifolium.*** 13, ***L. latifolium.*** 14, ***L. heterophyllum.*** 15, ***L. sativum.*** 16, ***L. ruderale.*** 17, ***L. campestre.*** 18, ***L. perfoliatum.*** 19, ***Cochlearia officinalis.*** 20, ***C. anglica.*** 21, ***Capsella bursa-pastoris.*** 22, ***C. rubella.*** 23, ***Coronopus didymus.*** 24, ***C. squamatus.***

fields and borders; E Kent since 1923, S Essex since 1951; Europe.

3. T. perfoliatum L. - *Perfoliate Penny-cress.* Glabrous annual; stems erect, to 25cm; **RRR**
stem-leaves sessile, clasping stem with rounded auricles; fruit 3-7.5mm, rather **269**
broadly winged apically, with wide ± deep apical notch and much shorter style; (2n=42, c.70). Native; bare limestone stony ground mainly or only in E Gloucs, N Wilts and Oxon, rarely natd or casual elsewhere in En.

4. T. caerulescens J. & C. Presl (*T. alpestre* L. non Jacq., *T. calaminare* (Lej.) Lej. & **R**
Courtois) - *Alpine Penny-cress.* Glabrous perennial or sometimes biennial with sterile **269**
leaf-rosettes; stems erect, to 40cm; stem-leaves sessile, clasping stem with subacute to ± rounded auricles; fruit 3.5-9.5mm, narrowly to rather broadly winged mainly apically, with usually shallow apical notch equalled or exceeded by style; 2n=14. Native; bare or sparsely grassed stony places mainly on limestone naturally or artificially contaminated with lead or zinc; extremely local and disjunct in Br from N Somerset to C Sc. Very variable but not able to be subdivided satisfactorily.

5. T. macrophyllum Hoffm. (*Pachyphragma macrophyllum* (Hoffm.) N. Busch) - **269**
Caucasian Penny-cress. Glabrous perennial with strong rhizomes, smelling of garlic when crushed; stems erect to ascending, to 40cm; stem-leaves cordate at base, petiolate; fruit 8-12mm, broadly winged, with deep apical notch much exceeding style length. Intrd; natd in woodland in Herts, N Somerset and Salop; Caucasus.

31. IBERIS L. - *Candytufts*

Annuals to perennials; leaves simple, entire to deeply lobed; hairs unbranched or 0; inflorescence a corymb at flowering, often elongate in fruit; petals white to purple or mauve, the 2 abaxial much longer than the 2 adaxial; fruit <3x as long as wide, compressed, with septum at right angles to plane of compression, with fairly narrow to broad wing round edges; seeds 1 under each valve.

1 Perennial with stems woody at base and leaves evergreen **1. I. sempervirens**
1 Annual with herbaceous stem 2
 2 Inflorescence elongating in fruit; fruits mostly or all 3-6mm **2. I. amara**
 2 Inflorescence remaining corymbose in fruit; fruits mostly or all 7-10mm **3. I. umbellata**

1. I. sempervirens L. - *Perennial Candytuft.* Glabrous sprawling shrub with **278**
procumbent woody lower parts and ascending flowering shoots to 25cm high; leaves entire, obtuse; inflorescence elongating in fruit; fruits 4-8mm, broadly winged especially apically; (2n=16, 22). Intrd; much grown in gardens and a persistent relic or throwout, occasionally self-sown; mostly S Br, rare in N; S Europe.

2. I. amara L. - *Wild Candytuft.* Sparsely pubescent annual; stems erect, with **R**
divergent branches, to 35cm; leaves entire to lobed or dentate, acute to ± rounded; **278**
fruits 3-6(7)mm, narrowly winged; 2n=14. Native; bare places in grassland and arable fields on dry calcareous soils; CS En N to Cambs, often reported as casual elsewhere but probably often errors for *I. umbellata.*

3. I. umbellata L. - *Garden Candytuft.* Glabrous annual; stems erect, with **278**
divergent branches, to 70cm; leaves ± entire, acute; fruits (6)7-10mm, with very broad pointed wings apically; (2n=14, 16, 18, 34). Intrd; much grown in gardens and a common casual on tips and in waste places; Br and CI; S Europe.

32. LEPIDIUM L. (*Cardaria* Desv.) - *Pepperworts*

Annuals to perennials; leaves simple to 2-3-pinnate; hairs unbranched or 0; inflorescence a raceme or panicle; petals usually white, sometimes reddish, yellowish or 0; fruit <3x as long as wide, compressed, with septum at right angles to plane of compression, sometimes ± indehiscent, strongly keeled to winged round edges; seeds 1(2) under each valve but sometimes not developing.

1 Fruit not or scarcely dehiscent; inflorescence a ± corymbose panicle **11. L. draba**
1 Fruit readily dehicent; inflorescence a raceme or a racemose to pyramidal panicle 2
  2 Lower and upper stem-leaves strikingly different, the lower finely 2-3-pinnate, the upper ovate, entire, ± encircling stem; petals yellow **8. L. perfoliatum**
  2 Lower and upper stem-leaves often different but with ± gradual transition; petals white, reddish or 0 3
3 Fruit ≥(4)4.5mm, usually at least as long as pedicel; basal 1/2 of style usually fused with wings of fruit 4
3 Fruit ≤4.2mm, usually shorter than pedicel; style free from wings of fruit or wings 0 6
  4 Lower and middle stem-leaves very deeply pinnately lobed to 2-pinnate, not clasping stem **1. L. sativum**
  4 Lower and middle stem-leaves simple, toothed to very shallowly lobed, clasping stem at base 5
5 Fruit covered with scale-like vesicles; style not exceeding apical notch of fruit **2. L. campestre**
5 Fruit with few or 0 vesicles; style exceeding apical notch of fruit **3. L. heterophyllum**
  6 Perennial; fruit unwinged, rounded to subacute at apex, with style projecting beyond 7
  6 Normally annual or biennial; fruit usually winged at least apically, notched at apex, with style not or scarcely projecting beyond notch 8
7 Upper stem-leaves linear; fruit 2-4.2mm, in lax branched racemes **10. L. graminifolium**
7 Upper stem-leaves elliptic to narrowly so; fruit 1.5-2.7mm, in congested panicles **9. L. latifolium**
  8 Middle and upper stem-leaves deeply pinnately lobed to 2-pinnate **5. L. bonariense**
  8 Middle and upper stem-leaves entire to dentate 9
9 Fruit 2.5-4mm wide; apical notch >2mm deep, c.1/10 of fruit length **4. L. virginicum**
9 Fruit 1.4-2.3mm wide; apical notch <2mm deep, <1/10 of fruit length 10
  10 Upper stem-leaves entire, rounded to subacute at apex; basal leaves (gone before fruiting) pinnate to 2-pinnate **6. L. ruderale**
  10 Upper stem-leaves entire to dentate, acuminate to subacute at apex; basal leaves (gone before fruiting) dentate to deeply pinnately lobed **7. L. hyssopifolium**

Other spp. - >30 other spp. have been recorded as casuals in BI, most as wool-aliens or seed-aliens, all rare or no longer occurring. Least rare is **L. fasciculatum** Thell., from Australia, which has very short or 0 petals but differs from other annual spp. with this feature in its short ± corymbose inflorescences.

1. L. sativum L. - *Garden Cress*. Erect annual to 50(100)cm; stem-leaves not clasping stem, lower ones 1-2-pinnate, upper ones entire to lobed; fruit 4.5-7mm, broadly winged apically with deep notch and shorter style; (2n=16, 24). Intrd; commonly grown as seedling salad-plant and a frequent contaminant of grain and birdseed, found on tips and waysides; frequent casual in much of BI; W Asia originally. **269**

2. L. campestre (L.) W.T. Aiton - *Field Pepperwort*. Erect to decumbent annual to biennial to 60cm; stem-leaves simple, shortly dentate, with acute auricles clasping stem; fruit (4)4.5-6.8mm, broadly winged apically with ± deep notch and shorter style; 2n=16. Native; open grassland, banks, walls, waysides and arable fields; **269**

scattered but locally common in much of Br and Ir, mostly in S.
3. L. heterophyllum Benth. (*L. smithii* Hook., *L. pratense* (J. Serres ex Gren. & **269**
Godr.) Rouy & Foucaud, *L. villarsii* Gren. & Godr., *L. hirtum* (L.) Sm.) - *Smith's Pepperwort*. Erect to decumbent perennial to 50cm; stem-leaves similar to those of *L. campestre*; fruit 4.5-8.6mm, broadly winged apically with rather shallow notch and longer style; 2n=16. Native; similar places to *L. campestre*; scattered but locally common in much of BI, mostly in W.

4. L. virginicum L. (*L. neglectum* Thell., *L. ramosissimum* A. Nelson, *L. densiflorum* **269**
Schrad.) - *Least Pepperwort*. Erect annual to biennial to 50cm; stem-leaves simple, not clasping stem, lower ones ± lobed, middle and upper ones dentate to ± entire; fruit 2.3-4mm, narrowly winged apically with shallow notch and shorter style; (2n=32). Intrd; rather infrequent casual probably mainly from wool and birdseed; scattered and sporadic in Br and Ir; N America. **L. densiflorum** (incl. *L. neglectum*) is here considered to be a short- or non-petalled variant of *L. virginicum*, but many authors keep them separate; there seem to be no other important differences.

5. L. bonariense L. - *Argentine Pepperwort*. Erect annual or biennial to 60cm; **269**
stem-leaves not clasping stem, 2-pinnate to deeply pinnately lobed; fruit 2-3.5mm, narrowly winged apically with shallow notch and shorter style; (2n=64). Intrd; infrequent casual on tips and in fields, mainly from wool and birdseed; very scattered in Br, mostly S En; S America.

6. L. ruderale L. - *Narrow-leaved Pepperwort*. Erect often much-branched annual or **269**
biennial to 45cm; stem-leaves not clasping stem, lower ones pinnate, upper ones simple, linear; fruit 1.5-2.7mm, narrowly winged apically with shallow notch and shorter style; (2n=16, 32). Probably intrd; common casual of waste places, waysides and tips, natd and possibly native in open ground especially near sea; locally common in E & SE En, scattered elsewhere in BI, mainly En.

7. L. hyssopifolium Desv. (*L. africanum* (Burm. f.) DC., *L. divaricatum* Sol. ssp. **269**
linoides (Thunb.) Thell.) - *African Pepperwort*. Erect annual or biennial to 45cm; stem-leaves not clasping stem, lower ones deeply lobed, upper ones entire to dentate, linear; fruit 1.8-3(3.7)mm, narrowly winged apically, with shallow notch and shorter style; (2n=16). Intrd; infrequent casual on tips and in fields, mainly from wool; very scattered in Br; S Africa. A difficult complex, from which **L. africanum** and **L. divaricatum** might be separable.

8. L. perfoliatum L. - *Perfoliate Pepperwort*. Erect annual or biennial to 45cm; basal **269**
and lower stem-leaves 2-3-pinnate, not clasping stem, upper stem-leaves entire, ovate, ± encircling stem; fruit 3-4.5mm, narrowly winged apically, with shallow notch and style usually ± equalling it; (2n=16). Intrd; infrequent casual on tips and by docks, mostly from grain and and grass-seed; very scattered in Br; Europe.

9. L. latifolium L. - *Dittander*. Erect rhizomatous perennial to 1.5m; basal leaves **R**
ovate, cordate to broadly cuneate, serrate, with long petiole, merging into narrowly **269**
elliptic, entire, sessile upper stem-leaves; fruit 1.5-2.7mm, not winged or notched, with slightly protruding stigma; 2n=24. Native; damp barish ground near sea; on or near coasts of Br N to S Wa and E Norfolk (formerly CE Sc), Guernsey, casual or natd on waste land elsewhere on coast and inland.

10. L. graminifolium L. - *Tall Pepperwort*. Erect perennial (but often not persistent **269**
in BI) to 60cm; stem-leaves elliptic to linear, lobed to entire; fruit 2-4.2mm, not winged or notched, with slightly protruding stigma; (2n=16, 48). Intrd; a casual mostly near docks, sometimes persistent but decreasing; very sporadic in S Br; S Europe.

11. L. draba L. - *Hoary Cress*. Erect rhizomatous perennial to 60cm; basal leaves cuneate, entire to dentate, petiolate; stem-leaves sessile, elliptic, clasping stem at base; fruit 3-4.8mm, obtuse at apex with projecting style 0.7-1.8mm. Intrd; waste ground, by roads, paths and railways, arable land, sandy ground near sea.

a. Ssp. draba (*Cardaria draba* (L.) Desv.). Leaves greyish-green; fruit at least as **261**
wide as long, truncate to cordate at base, usually reticulately ridged; (2n=64). Throughout most of BI, but absent from large areas of N Br and Ir, first record

1829; S Europe.

b. Ssp. chalepense (L.) Thell. (*Cardaria chalepensis* (L.) Hand.-Mazz., *C. draba* 261
ssp. *chalepensis* (L.) O.E. Schulz). Leaves brighter green; fruit usually longer than wide, rounded to broadly cuneate at base, usually smooth; (2n=80). Rare casual formerly natd in Staffs, E Cornwall and Cumberland, now only at Sharpness (W Gloucs) and elsewhere in S En; SW Asia.

33. CORONOPUS Zinn - *Swine-cresses*

Annuals or biennials with inflorescences mostly opposite leaves; leaves all deeply pinnately lobed; hairs unbranched or 0; petals white or 0; fruit <3x as long as wide, only slightly compressed, with septum at right angles to plane of compression, indehiscent or breaking into 2 halves; seed 1 under each valve.

1. C. squamatus (Forssk.) Asch. - *Swine-cress*. Stems glabrous or nearly so, 269
procumbent, to 30cm; petals 1-2mm, longer than sepals; fertile stamens 6; pedicels shorter than fruit; fruit 2-3.5mm, strongly ridged or warty, rounded to slightly retuse at apex with style protruding, indehiscent(2n=32). Probably native; waste ground, paths and round trodden gateways; throughout much of BI, but rare and absent from large areas NW of Yorks to S Wa.

2. C. didymus (L.) Sm. - *Lesser Swine-cress.* Strong smelling when crushed; stems 269
usually ± pubescent, procumbent to ascending, to 40cm; petals 0 or c.0.5mm, shorter than sepals; fertile stamens 2(-4); pedicels longer than fruit; fruit 1.2-1.7mm, reticulate, emarginate at apex with very short included style; breaking into 2 halves; 2n=32. Intrd; cultivated and waste ground; frequent in S BI, scattered N to C Sc; S America.

34. SUBULARIA L. - *Awlwort*

Aquatic annuals or biennials; leaves confined to basal rosette, subulate, entire; hairs 0; petals white, rarely 0; fruit <3x as long as wide, ± inflated, with septum parallel with plane of compression; 2-7 seeds in 2 rows under each valve.

1. S. aquatica L. - *Awlwort*. Leaves ± erect, to 4(7)cm; flowering stems erect, to 269
8(12)cm, with 2-8(12) flowers; fruit (1)2-5mm; (2n=28, c.36). Native; in stony or gravelly base-poor lakes, usually totally submerged, sometimes exposed in droughts; very local in Wa, NW En, N & W Ir, Sc. Often grows with and confused with *Isoetes, Littorella, Lobelia dortmanna, Eleocharis acicularis* and *Juncus bulbosus.*

TRIBE 6 - BRASSICEAE (genera 35-46). Hairs 0 or unbranched; petals yellow, white or pink to purple or mauve; fruit usually >3x as long as wide and many-seeded, beaked or unbeaked, usually dehiscing longitudinally to release seeds, sometimes dehiscing transversely but not releasing seeds separately, sometimes indehiscent, sometimes only 1-4-seeded and transversely dehiscent.

35. CONRINGIA Heist. ex Fabr. - *Hare's-ear Mustard*

Annuals; leaves simple, entire, basal ones petiolate, upper ones sessile, with large auricles clasping stem; petals greenish- to yellowish-white, rarely 0; fruit >3x as long as wide, unbeaked, longitudinally dehiscent; seeds in 1 row under each valve.

Other spp. - C. austriaca (Jacq.) Sweet, from SE Europe, differing from *C. orientalis* in its 8-angled fruits (3 prominent veins per valve) and paler, smaller petals, formerly occurred as a casual.

1. C. orientalis (L.) Dumort. - *Hare's-ear Mustard*. Glabrous, glaucous; stems erect, 278
to 60cm; leaves ovate to obovate; fruit 4.5-15cm, ± erect, 4-angled, each valve with 1 prominent vein; (2n=14). Intrd; casual in arable land and waste places, often near sea, once frequent, now very sporadic; scattered in Br and CI; E

Mediterranean. Often overlooked as *Brassica rapa*, which has beaked fruits.

36. DIPLOTAXIS DC. - *Wall-rockets*
Annuals to perennials; leaves deeply pinnately lobed, sometimes not so in *D. erucoides*, strongly smelling when crushed; petals yellow, rarely white; fruit >3x as long as wide, unbeaked, longitudinally dehiscent; seeds in 2 rows under each valve in well-developed fruits; valves with 1 strong vein.

1 Petals white; fruit with beak 2-4(6)mm **3. D. erucoides**
1 Petals yellow; fruit with beak 1-3(3.5)mm 2
  2 Fruit with distinct stalk (0.5-6.5mm) between sepal-scars and base of valves; petals 8-15mm **1. D. tenuifolia**
  2 Fruit with base of valves immediately above sepal-scars; petals 4-8(8.5)mm **2. D muralis**

Other spp. - D. viminea (L.) DC. and **D. catholica** (L.) DC. from S Europe, and **D. tenuisiliqua** Del. from N Africa, formerly occurred as casuals.

1. D. tenuifolia (L.) DC. - *Perennial Wall-rocket*. Glabrous perennial with branching leafy stems to 80cm; leaves with lobes >3x as long as wide; petals 8-15mm; fruit 1.5-6cm, erect on erecto-patent pedicel nearly as long as to longer than fruit; (2n=22). Possibly native; dry waste places, bare ground, banks and walls; scattered through Br and CI N to C Sc, locally common in S En and CI. **278**

2. D. muralis (L.) DC. - *Annual Wall-rocket*. Glabrous or sparsely pubescent annual or sometimes short-lived perennial with branched stems to 60cm leafy only near base; leaves with lobes <3x as long as wide; petals 4-8(8.5)mm; fruit (1.2)1.5-4.2cm, erecto-patent, with pedicel usually <1/2 as long as fruit; (2n=42). Intrd; dry waste places, rocks, walls and arable land; similar distribution to *D. tenuifolia* but scattered in Ir; Europe. **278**

3. D. erucoides (L.) DC. - *White Wall-rocket*. Rather sparsely pubescent annual with branching leafy stems to 50cm; leaves variably (sometimes scarcely) lobed; petals 7-13mm; fruit 1-5cm, patent to erecto-patent on pedicel c. 2/5-1/2 as long, with stalk between sepal-scars and base of valves 0.3-1mm; (2n=14). Intrd; casual of waste ground and by paths; formerly frequent, then rare, now occasional in En; S Europe. **278**

37. BRASSICA L. - *Cabbages*
Annuals to perennials; leaves crenate to deeply pinnately lobed; petals yellow; fruit >3x as long as wide, beaked or unbeaked, longitudinally dehiscent; seeds in 1 row under each valve; valves with 1 strong vein.

1 Stem-leaves distinctly clasping stem at base 2
1 Stem-leaves not clasping stem, petiolate or narrowed to base 4
  2 Sepals erect in flower; flowering part of inflorescence elongated, the buds greatly overtopping open flowers; plant glabrous **1. B. oleracea**
  2 Sepals erecto-patent to patent in flower; flowering part of inflorescence scarcely elongated, the buds slightly overtopping or overtopped by uppermost open flowers; lowest leaves usually with some hairs 3
3 Buds slightly overtopping open flowers, forming convex 'dome'; petals mostly >11mm, bright pale yellow **2. B. napus**
3 Buds overtopped by open flowers, forming concave 'bowl'; petals mostly <12mm, bright deep yellow **3. B. rapa**
  4 Fruit terminated by distinct ± conical beak ≥(4)5mm and sometimes with 1(-3) seeds 5
  4 Fruit terminated by slender beak ≤4mm not or scarcely wider at base

and seedless 7
5 Lowest leaves with >3 pairs of lateral lobes; fruit with beak ≥10mm, some with 1(-3) seeds **4. B. tournefortii**
5 Lowest leaves with ≤3 pairs of lateral lobes; fruit with seedless beak ≤10mm 6
6 Lowest leaves with 1-3 pairs of lateral lobes; fruit with beak (4)5-9(12)mm; sepals 4.5-7mm **5. B. juncea**
6 Lowest leaves with 0-1 pairs of lateral lobes; fruit with beak 2.5-6(7)mm; sepals 7-10mm **6. B. carinata**
7 Fruit closely appressed to stem; pedicels 3-8mm **8. B. nigra**
7 Fruit erecto-patent; pedicels (6)8-18mm 8
8 Fruit (1)1.5-2.5mm wide, with distinct stalk (1)1.5-5mm between sepal-scars and base of valves, with style ≤2(3)mm **7. B. elongata**
8 Fruit 3-9mm wide, with base of valves within 1mm of sepal-scars, with style ≥2.5mm **6. B. carinata**

Other spp. - B. fruticulosa Cirillo, from S Europe, used to be found as a casual but is now absent or extremely rare.

1. B. oleracea L. - *Cabbage*. Glabrous biennial to perennial to 2m; stems often R
decumbent and woody below, with numerous leaf-scars, erect above; roots never 254
tuberous (unless diseased); basal leaves crenate to deeply lobed; stem-leaves clasping stem at base; sepals erect; petals 12-30mm; fruit (2.5)5-10cm, with a conical beak 3-10mm with 0-2 seeds; 2n=18. Possibly native on sea-cliffs scattered round Br, mostly in S; common casual on tips, neglected gardens and roadsides throughout BI. Wild plants are var. **oleracea** (*Wild Cabbage*); the commonest crop-plants are placed in var. **capitata** L. (*Cabbuge*), var. **sabauda** L. (*Savoy Cabbage*), var. **viridis** L. (Kale), var. **botrytis** L. (Cauliflower, Broccoli), var. **gemmifera** DC. (*Brussels-sprout*), and var. **gongylodes** L. (*Kohl-rabi*).

2. B. napus L. - *Rape*. Annual to biennial with glaucous, ± glabrous, erect stems to 254
1.5m; roots sometimes tuberous; basal leaves crenate to deeply lobed, usually 278
sparsely pubescent; stem-leaves clasping stem at base; sepals erecto-patent; petals 11-18mm; fruit (2)5-10(11)cm, with a conical beak 4-16mm with 0-1 seed; 2n=38 (derived from *B. oleracea* x *B. rapa*). Intrd; frequent relic of cultivation and from seed importation; throughout BI and now greatly increasing.

a. Ssp. oleifera (DC.) Metzg. - *Oil-seed Rape*. Root slender; 2n=38. Grown for its oil-bearing seeds and as a seedling salad-plant (substitute for *Sinapis alba*), and rarely for fodder and green manure; now the commonest casual *Brassica* by roads, field-margins and on tips; also a birdseed-alien and found near oil-processing factories.

b. Ssp. rapifera Metzg. (var. *napobrassica* (L.) Rchb.) - *Swede*. Root swollen into a yellow-fleshed tuber; (2n=38). Grown for its root-tubers and rarely for fodder; frequent relic of cultivation. Other sspp., including the weedy ssp. *napus*, may occur but are much rarer.

2 x 3. B. napus x B. rapa = B. x harmsiana O.E. Schulz occurs sporadically in crops of *B. napus* when exposed to pollen from *B. rapa;* it closely resembles *B. napus* and can be told only by its sterility and chromosome number (2n=29).

3. B. rapa L. - *Turnip*. Differs from *B. napus* in its green usually more pubescent 254
basal leaves; petals 6-12(14)mm; patent sepals; 2n=20; and see key. Probably intrd; occasional by streams and rivers, and as a relic of cultivation; throughout BI, probably less common nowadays.

a. Ssp. campestris (L.) A.R. Clapham (ssp. *sylvestris* (Lam.) Janch., *B. campestris* L.) - *Wild Turnip*. Root slender; seeds <1.6mm, grey to blackish; 2n=20. Occasional by streams and rivers, often with *B. nigra*.

b. Ssp. oleifera (DC.) Metzg. - *Turnip-rape*. Root slender; seeds mostly >1.6mm, red-brown; (2n=20). Grown as fodder or oilseed crop; mostly a birdseed or oil-processing alien.

c. Ssp. rapa - *Turnip*. Root swollen into a white-fleshed tuber; (2n=20). Grown for its root-tubers and sometimes as fodder or green manure; frequent relic of cultivation but less common than formerly.

4. B. tournefortii Gouan - *Pale Cabbage*. Usually annual; stems erect, to 50cm; 278
basal leaves deeply lobed with 5-10 pairs of lobes, hispid; stem-leaves much reduced, not clasping stem; sepals ± erect; petals 4-7mm; fruit 3-7cm, with tapering-conical beak 10-16(20)mm with (0)1(-3) seeds; (2n=20). Intrd; a rather frequent wool-alien, very rarely from other sources; sporadic in Br; Mediterranean region.

5. B. juncea (L.) Czern. (*B. integrifolia* (H. West) Rupr.) - *Chinese Mustard*. Annual; stems erect, to 1m; basal leaves deeply lobed with 1-3 pairs of lobes, sparsely hispid; stem-leaves deeply dentate to ± entire, petiolate; sepals erecto-patent; petals 9-14mm; fruit 2-6cm, with conical seedless beak (4)5-9(12)mm 1/5-1/4 as long as valves; (2n=36) (derived from *B. nigra* x *B. rapa*). Intrd; a frequent birdseed-alien, sometimes from wool and other sources; scattered in BI; S & E Asia. Often superficially closely resembling *Sinapis arvensis*, but see sepal and fruit characters.

6. B. carinata A. Braun (*B. integrifolia* auct. non (West) Rupr.) - *Ethiopian Rape*. Differs from *B. juncea* in petals 13-17mm; fruit 2.5-6cm, with seedless beak 2.5-6(7)mm; (2n=34) (derived from *B. oleracea* x *B. nigra*); and see key. Intrd; occasional birdseed-alien, sporadic in En, formerly natd in Middlesex; Abyssinia.

7. B. elongata Ehrh. - *Long-stalked Rape*. Biennial to perennial; stems erect, to 1m; basal leaves pinnately lobed to shallowly toothed, hispid; stem-leaves much reduced, not clasping stem; sepals ± erect; petals 6-8.5(10)mm; fruit 1-4cm, with stalk-like base above sepal-scars 0.8-5mm, with seedless narrow beak 0.5-3mm; (2n=22). Intrd; an occasional casual of waste places; sporadic in Br; SE Europe.

8. B. nigra (L.) W.D.J. Koch - *Black Mustard*. Annual; stems erect, to 2m; basal 278
leaves pinnately lobed, hispid; stem-leaves lobed below, entire above, petiolate; sepals erecto-patent; petals (7)9.5-13mm; fruit 8-25(33)mm, with seedless narrow beak 1.5-5(6)mm; 2n=16. Probably native; sea-cliffs, river banks, rough ground and waste places; frequent in Br and CI N to S Sc, very scattered in Ir.

38. SINAPIS L. - *Mustards*

Annuals; leaves crenate to deeply pinnately lobed; sepals patent; petals yellow; fruit >3x as long as wide, longitudinally dehiscent, with a distinct beak usually >1/3 as long as valves; seeds in 1 row under each valve; valves with 3(-7) strong veins.

Other spp. - **S. flexuosa** Poir. (*S. hispida* Schousb.), from Spain, formerly occurred as a casual; differs from *S. alba* in having <10 seeds per fruit and more hispid leaves and stems.

1. S. arvensis L. - *Charlock*. Plant to 1(1.5)m, simple to well-branched, ± glabrous 278
to hispid; leaves lobed or not, if so the terminal lobe much the largest; sepals patent; petals 7.5-17mm; fruit 2.2-5.7cm, with 4-24 seeds, with 0-1-seeded conical beak 7-16mm and 1/3-3/4 as long as valves; 2n=18. Probably native; arable and waste land, tips and roadsides; throughout BI. See *Brassica juncea* for differences.

2. S. alba L. - *White Mustard*. Plant to 70(100)cm, differing from *S. arvensis* in its 278
deeply pinnately lobed leaves; petals 7.5-14mm; fruit 2-4.2cm, with 2-8 seeds, with 0-1-seeded strongly flattened beak 10-24(30)mm and 1-1.5x as long as valves. Intrd; S Europe.

a. Ssp. alba. Leaves deeply pinnately lobed, the terminal lobe much the largest; fruit usually hispid; (2n=24). Grown as fodder or green manure or for mustard-seed or seedling salad-plant, natd or casual in arable or waste land and on waysides and tips, especially on calcareous soils; scattered over most of BI but absent from much of N.

b. Ssp. dissecta (Lag.) Bonnier (*S. dissecta* Lag.). Leaves 2-pinnately lobed, the terminal lobe little larger than largest laterals; fruit glabrous to slightly pubescent; (2n=24). Infrequent casual on waste land; sporadic in Br.

39. ERUCA Mill. - *Garden Rocket*
Annuals; leaves deeply pinnately lobed; sepals erect; petals white to pale yellow with conspicuous violet veins; fruit >3x as long as wide, longitudinally dehiscent, with a distinct beak usually >1/3 as long as valves; seeds in 2 rows under each valve; valves with 1 strong vein.

1. E. vesicaria (L.) Cav. (*E. sativa* Mill.) - *Garden Rocket*. Stems usually much branched, decumbent to erect, to 1m, glabrous to hispid; fruit 1.2-3.5(4)cm, with seedless strongly flattened beak 4-11mm; (2n=22). Intrd; infrequent casual, rarely persisting, on waste land; scattered over Br and CI; S Europe. Most or all plants are ssp. **sativa** (Mill.) Thell., with sepals falling early and of which only the outer 2 are saccate at base; ssp. **vesicaria**, with persistent sepals all saccate at base might sometimes occur. **278**

40. ERUCASTRUM C. Presl - *Hairy Rocket*
Annuals to perennials; leaves deeply pinnately lobed; sepals erect to ± patent; petals yellow; fruit >3x as long as wide, beakless or shortly beaked, longitudinally dehiscent, slightly constricted between seeds; seeds in 1 row under each valve; valve with 1 strong vein.

Other spp. - **E. nasturtiifolium** (Poir.) O.E. Schulz, from SW Europe, formerly occurred as a casual; it differs from *E. gallicum* in its stem-leaves having their basal lobes clasping the stem; flowers all without bracts; sepals ± patent; and fruit with c.0.3-1mm stalk above sepal-scars and with 3-6.5mm beak with (0)1(-2) seeds.

1. E. gallicum (Willd.) O.E. Schulz - *Hairy Rocket*. Pubescent annual or biennial; stems erect, to 60cm; stem-leaves ± pinnate; sepals ± erect; fruit (1.6)2-4.5cm, with seedless beak (1.5)2-4mm; (2n=30). Intrd; infrequent, formerly frequent, casual of arable and waste land, rarely persisting; scattered in BI, mainly S; Europe. **278**

41. COINCYA Rouy (*Hutera* Porta, *Rhynchosinapis* Hayek) - *Cabbages*
Annuals to perennials; leaves pinnately lobed to pinnate; sepals erect; petals yellow; fruit >3x as long as wide, longitudinally dehiscent, with distinct beak usually 1/5-1/3 as long as valves and with (0)1-4(5) seeds; seeds in 1 row under each valve; valves with 3 strong veins.

1. C. monensis (L.) Greuter & Burdet - see sspp. for English names. Annuals to perennials, subglabrous to pubescent below; basal leaves with 3-9 pairs of lateral lobes, the terminal lobe not much larger than laterals; fruit (2.5)3.5-8(8.5)cm incl. beak 5-24(34)mm, 1.2-3mm wide.

a. Ssp. monensis (*Hutera monensis* (L.) Gomez-Campo, *Rhynchosinapis monensis* (L.) Dandy ex A.R. Clapham) - *Isle of Man Cabbage*. Plant to 60cm; stems procumbent to ascending, glabrous to sparsely hispid below; seeds 1.3-2mm; 2n=24. Native; sandy ground near sea; Man and W Br from S Lancs to Clyde Is (formerly M Ebudes), rare casual elsewhere; endemic. **R** **278**

b. Ssp. cheiranthos (Vill.) Aedo, Leadlay & Muñoz Garm. (ssp. *recurvata* (All.) Leadlay, *C. cheiranthos* (Vill.) Greuter & Burdet, *Hutera cheiranthos* (Vill.) Gomez-Campo, *Rhynchosinapis cheiranthos* (Vill.) Dandy, *R. erucastrum* Dandy ex A.R. Clapham pro parte excl. typ.) - *Wallflower Cabbage*. Plant to 1m; stems usually erect, hispid to rather sparsely so below; seeds 0.8-1.6mm; 2n=48. Intrd; casual on sandy ground, waste places and roadsides in SW Br, natd in Mons since 1975,

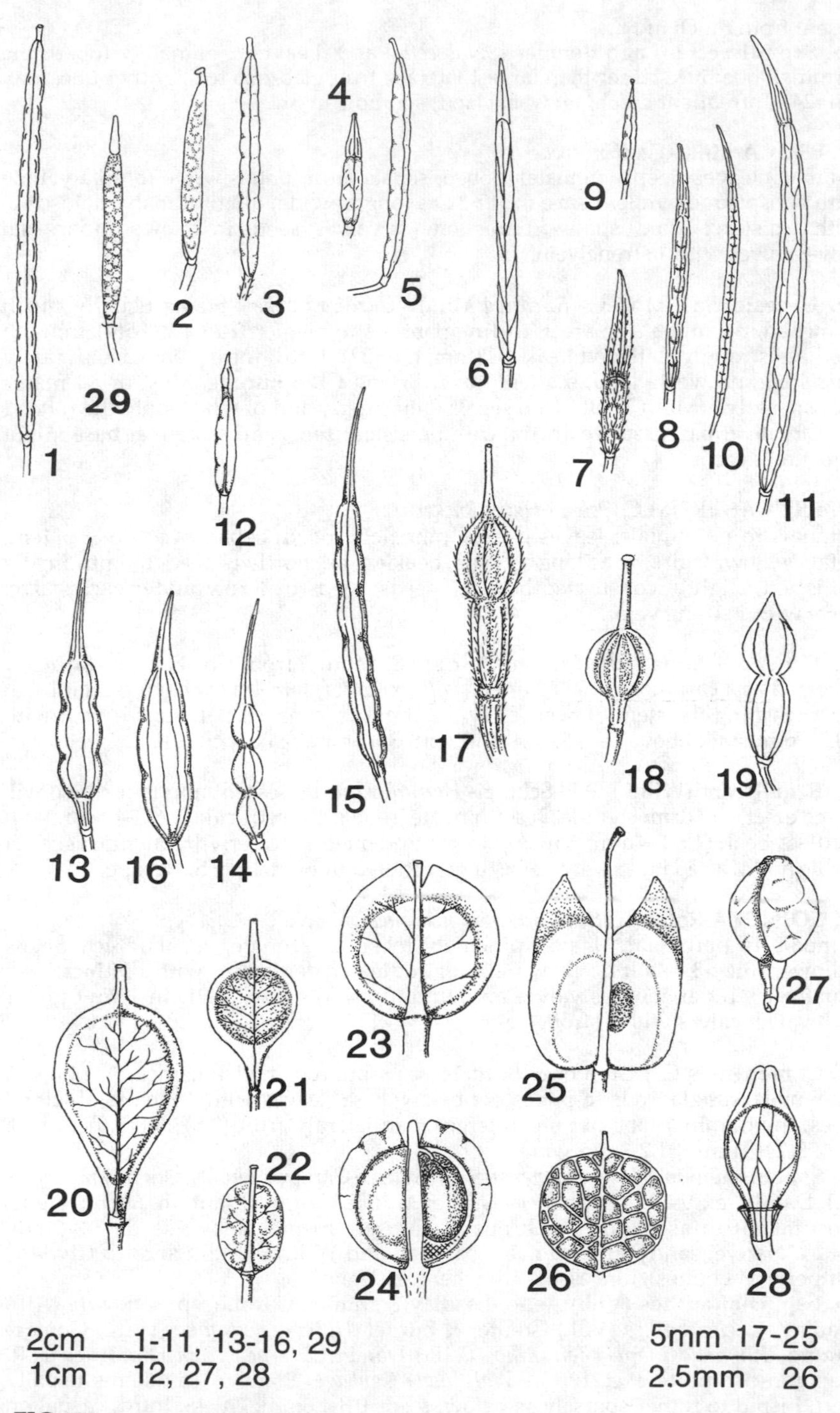

FIG 278 - see caption opposite

Jersey since 1832; W Europe.

2. C. wrightii (O.E. Schulz) Stace (*Hutera wrightii* (O.E. Schulz) Gomez-Campo, *Rhynchosinapis wrightii* (O.E. Schulz) Dandy ex A.R. Clapham) - *Lundy Cabbage*. Biennials to perennials to 1m, often woody near base, pubescent over all or most of stem incl. inflorescence; basal leaves with 2-5(6) pairs of lateral lobes, the terminal lobe much larger than laterals; fruit 2-8cm incl. beak 7-16mm, 1.7-4mm wide; seeds 1.2-1.9mm; 2n=24. Native; cliffs and slopes on SE part of Lundy Island, N Devon; endemic. **RRR**

42. HIRSCHFELDIA Moench - *Hoary Mustard*

Annuals to short-lived perennials; lower leaves pinnate to deeply pinnately lobed; sepals ± erect; petals yellow; fruit >3x as long as wide, longitudinally dehiscent, with distinct beak usually c.1/2 as long as valves with (0)1(-2) seeds; seeds in 1 row under each valve; valves with 1-3 ± strong veins.

1. H. incana (L.) Lagr.-Foss. - *Hoary Mustard*. Stems erect, to 1.3m, whitish pubescent below with short stiff hairs; fruit 6-17mm, appressed to stem, with beak 3-6.5mm, swollen round seeds and abruptly narrowed distally; 2n=14. Intrd; waste places and waysides; BI, especially S, often casual but increasingly natd, also frequent wool-alien; S Europe. **278**

43. CAKILE Mill. - *Sea Rocket*

Glabrous annuals; leaves entire to pinnately lobed; sepals erect; petals mauve to pink or white; fruit breaking transversely into 2 portions, the proximal (0-)1-seeded, the distal 1(-2)-seeded, longer and wider, both keeled laterally and with prominent veins and margin.

1. C. maritima Scop. (*C. edentula* auct. non (Bigelow) Hook.) - *Sea Rocket*. Stems procumbent to ± erect, to 50cm; leaves ± glaucous, ± succulent; fruit 12-25mm; proximal segment 4-9mm; distal segment 8-20mm, ovoid; 2n=18. Native; near sea drift-line on sand and sometimes shingle; around coasts of BI. Variable in leaf- and fruit-shape, even in our plants, which are all referable to ssp. **integrifolia** (Hornem.) Hyl. ex Greuter & Burdet. Records of *C. edentula* from BI were errors. **278**

44. RAPISTRUM Crantz - *Cabbages*

Annuals to perennials; leaves dentate to deeply pinnately lobed; sepals erecto-patent; petals yellow; fruit breaking transversely into 2 ± equal-lengthed portions, the proximal 0-1(3)-seeded, the distal 1-seeded, narrowed at apex into persistent style and variously ribbed or wrinkled.

1. R. rugosum (L.) J.P. Bergeret (*R. hispanicum* (L.) Crantz, *R. orientale* (L.) Crantz, *R. rugosum* ssp. *orientale* (L.) Arcang., spp. *linnaeanum* (Coss.) Rouy & Foucaud) - *Bastard Cabbage*. Erect annual to 1m, hispid at least below; fruit 3-12mm; distal segment abruptly narrowed into (0.8)1-3.5(5)mm style, ribbed and rugose, glabrous to densely hispid; proximal segment usually much narrower than **278**

FIG 278 - Fruits of **Brassicaceae**. 1, ***Conringia orientalis***. 2, ***Diplotaxis tenuifolia***. 3, ***D. muralis***. 4, ***Eruca vesicaria***. 5, ***Erucastrum gallicum***. 6, ***Coincya monensis***. 7, ***Sinapis alba***. 8, ***S. arvensis***. 9, ***Brassica nigra***. 10, ***B. tournefortii***. 11, ***B. napus***. 12, ***Hirschfeldia incana***. 13, ***Raphanus raphanistrum*** ssp. ***landra***. 14, ssp. ***maritimus***. 15, ssp. ***raphanistrum***. 16, ***R. sativus***. 17, ***Rapistrum rugosum*** ssp. ***rugosum***. 18, ssp. ***linnaeanum***. 19, ***R. perenne***. 20, ***Camelina sativa***. 21, ***C. microcarpa***. 22, ***Jonopsidium acaule***. 23, ***Iberis sempervirens***. 24, ***I. amara***. 25, ***I. umbellata***. 26, ***Neslia paniculata***. 27, ***Crambe maritima***. 28, ***Cakile maritima***. 29, ***Diplotaxis erucoides***.

distal, mostly 0(1)-seeded; (2n=16). Intrd; casual in waste and arable land, on tips, waysides and open grassland (where it can become invasive); frequent and increasing in S & C Br and CI, natd in parts of S En; S Europe. Fruits very variable in relative sizes of pedicel and 2 segments, pubescence and degree of sculpturing, but the variants are probably not more than vars. All or most of our plants are referable to ssp. **linnaeanum**, with long thin appressed pedicels (1.5-5 x 0.3-0.7mm) and seedless lower fruit segments.

2. R. perenne (L.) All. - *Steppe Cabbage*. Erect or spreading biennial or perennial 278
to 80cm, hispid at least below; fruit 5-10mm; distal segment gradually narrowed into 0.5-1.2mm style, longitudinally ribbed, glabrous; proximal segment usually similar in size to distal but less or not ribbed, mostly 1-seeded; (2n=16, 18). Intrd; similar places to *R. rugosum*; scattered in S & C Br, much rarer than *R. rugosum* but well natd in some places; C & E Europe.

45. CRAMBE L. - *Sea-kale*

Large perennials with thick long roots; leaves large and irregularly lobed or toothed; sepals ± patent; petals white; fruit breaking transversely into 2 portions, the proximal small, sterile and stalk-like, the distal 1-seeded.

1. C. maritima L. - *Sea-kale*. Glabrous, glaucous, densely branched cabbage-like 278
plant to 75cm; basal leaves undulate at margins; stem-leaves many, similar; distal portion of fruit 6-14 x 6-11mm, broadly ellipsoid to globose; 2n=60. Native; on sand, rocks and cliffs but mostly shingle, by sea; coasts of BI N to E Ross and Ayrs, but absent from many areas.

2. C. cordifolia Steven - *Greater Sea-kale*. Sparsely pubescent erect perennial to 2m; basal leaves plane at margins; stem-leaves very few, much smaller; distal portion of fruit 4-6 x 2.6-6.8mm, ovoid; (2n=c.120). Intrd; garden throwout with persistent roots conspicuous on tips and waste ground; scattered in En; Caucasus.

46. RAPHANUS L. - *Radishes*

Annuals to perennials with distinctive radish-like smell when crushed; leaves shallowly pinnately lobed to pinnate; sepals erect; petals white, mauve or yellow, often with darker veins; main part (upper segment) of fruit indehiscent or transversely dehiscent into 1-10 1-seeded portions (mericarps), with short, inconspicuous, 0-1-seeded lower segment and with long persistent narrow beak.

1. R. raphanistrum L. - see sspp. for English names. Hispid annual to perennial, with slender root; petals white, mauve or yellow; fruit strongly constricted between seeds, at least partly transversely dehiscent.

1 Fruit with cylindrical or oblong mericarps usually longer than wide, with beak (2.5)3-6x as long as most apical mericarp **a. ssp. raphanistrum**

1 Fruit with ± globose mericarps c. as long as wide, with beak 1-3(4)x as long as most apical mericarp 2

2 Leaves with crowded lateral lobes; petals (14)15-22(25)mm when fresh **b. ssp. maritimus**

2 Leaves with ± distant lateral lobes; petal 8-15mm when fresh **c. ssp. landra**

a. Ssp. raphanistrum - *Wild Radish*. Annual; stems ascending to erect, to 75cm, 278
usually little branched; petals white, mauve or yellow; fruit 2.5-9cm, 2-5.5(6)mm wide, with (1)3-8(10) mericarps and beak 6-30mm; 2n=18. Probably intrd; cultivated and rough ground, waste places and tips; frequent throughout BI, yellow-petalled plants commonest in N & W; Europe.

b. Ssp. maritimus (Sm.) Thell. (*R. maritimus* Sm.) - *Sea Radish*. Biennial to 278
perennial; stems often much branched and very leafy, to 80cm; petals usually yellow, often white in CI; fruit (1)1.5-5.5cm, 4.5-10mm wide, with 1-5(6) mericarps and beak 5-25mm; 2n=18. Native; sea-shores and maritime cliffs and waste

places; coasts of BI N to NW Sc, absent from most of E Br. Intermediates (probably hybrids) with ssp. *raphanistrum* occur in SW En and CI.

c. Ssp. landra (Moretti ex DC.) Bonnier & Layens (*R. landra* Moretti ex DC.) - 278
Mediterranean Radish. Like ssp. *maritimus* but annual to perennial, less robust; petals yellow or white; fruit 2.5-6cm, with beak 15-40mm; and see key. Intrd; rather rare casual of waste places, mostly from grain; sporadic in En and Wa; Mediterranean.

1 x 2. R. raphanistrum x R. sativus = R. x micranthus (Uechtr.) O.E. Schulz occurs sporadically where *R. raphanistrum* has occurred as a weed near *R. sativus.* It is partially fertile, usually white-flowered, and variously intermediate, usually with partly dehiscent fruits and thin roots.

2. R. sativus L. - *Garden Radish.* Hispid annual or biennial to 1m, usually with 278
swollen often reddish root; petals white or mauve; fruit not or scarcely constricted between seeds, indehiscent, 2-9cm x 5-15mm, with (1)5-12 seeds, with lower segment 0-1-seeded; (2n=18). Intrd; grown as salad plant or as animal fodder (*Fodder Radish,* with thin root and much-branched leafy stems) and often escaping or persisting; fields, gardens and tips; sporadic throughout BI; origin probably Mediterranean.

65. RESEDACEAE - *Mignonette family*

Herbaceous annuals to perennials; leaves alternate, simple to pinnate, sessile or petiolate, ± without stipules (stipules often present as minute glandular teeth). Flowers in racemes, bisexual, hypogynous, zygomorphic; sepals 4-6(8), free; petals as many as sepals, free, the upper 1-2, lateral 2 and lower 1-2 different in form, the upper largest, at least some with entire proximal and deeply lobed distal regions; stamens 7-c.25, inserted on nectar-secreting disc; ovary 1-celled, open at top, composed of 3-4 carpels, with many ovules; styles 0; stigmas borne 1 on apical lobe of each carpel; fruit a capsule, open at top from start.

Easily recognized by the zygomorphic flowers with open-topped ovary and 4-6 white or yellowish distinctive petals.

1. RESEDA L. - *Mignonettes*

1 At least upper and middle leaves deeply pinnately lobed 2
1 All leaves entire to minutely toothed, or a few with 1-2 lateral lobes 3
  2 Carpels 3; petals yellowish; filaments falling after flowering; seeds smooth **3. R. lutea**
  2 Carpels 4; petals white; filaments persistent until fruit ripe; seeds tuberculate **2. R. alba**
3 Sepals and petals 4; fruits crowded, stiffly erecto-patent, <7mm, on pedicels <4mm; seeds smooth **1. R.luteola**
3 Sepals and petals 6; fruits well spaced, pendent, ≥7mm, on pedicels >5mm; seeds rugose 4
  4 Capsules 7-11mm; sepals ≤5mm at fruiting **5. R. odorata**
  4 Most mature capsules 11-15mm; many sepals >5mm at fruiting **4. R. phyteuma**

1. R. luteola L. - *Weld.* Glabrous biennial; main stem stiffly erect, to 1.5m; leaves entire, linear to lanceolate or oblanceolate; sepals and (yellow) petals 4; capsule 3-6mm, of 3 carpels; 2n=26. Native; open grassland, disturbed, waste and arable land mostly on base-rich soils; throughout most of BI except much of N & W Sc.

2. R. alba L. - *White Mignonette.* Glabrous annual to perennial; stem well branched, erect to ascending, to 75cm; leaves pinnately lobed; sepals and (white) petals 5-6; capsule 6-15mm, of 4 carpels; (2n=20). Intrd; casual on waste ground, often near sea, occasionally persisting (natd in Mons since 1968) but less common than formerly; S Br and CI; S Europe.

3. R. lutea L. (*R. stricta* auct. non Pers.) - *Wild Mignonette.* Shortly pubescent biennial to perennial; stem well branched, decumbent to erect, to 75cm; leaves pinnately lobed; sepals and (yellowish) petals 6; capsule 7-20mm, of 3 carpels; (2n=48). Native; disturbed, waste and arable land, especially on calcareous soils; through much of BI but much commoner in E and absent from large areas of Sc, Wa and Ir.

4. R. phyteuma L. - *Corn Mignonette.* Shortly pubescent annual to biennial; stems much branched below, procumbent to ascending, to 30cm; leaves mostly simple, sometimes some with 1-2 lateral lobes; sepals and (white) petals 6; capsule mostly 11-25mm, of 3 carpels; (2n=12). Intrd; rare and decreasing casual of waste ground, ± natd in cornfields and field margins in a few places in S En; S Europe.

5. R. odorata L. - *Garden Mignonette.* Similar to *R. phyteuma* but see key; (2n=12). Intrd; grown in gardens for its scented flowers and occasionally found as casual on tips or waste places in S Br; often overlooked for *R. phyteuma* or *R. lutea*; SE Mediterranean.

65A. CLETHRACEAE

CLETHRA arborea Aiton (*Lily-of-the-valley-tree*), from Madeira, is an evergreen tree that produces self-sown seedlings on warm walls in CI and occurs ± wild in woodland in Scillies. It has terminal panicles of white flowers similar to those of *Arbutus*, but differing in the ± free petals and pollen grains dispersed singly.

66. EMPETRACEAE - *Crowberry family*

Dwarf, heather-like, evergreen shrubs; leaves whorled to spiral, simple, entire, shortly petiolate, without stipules. Flowers in axillary clusters of 1-3 with 3 small bracts, small and inconspicuous, bisexual to dioecious, hypogynous, actinomorphic; sepals 3, free; petals 3, free, pinkish; stamens 3; ovary 6-9-celled, with 1 ovule in each cell; style 1, short; stigmas in form of 6-9 broad toothed rays; fruit a drupe with ≤9 seeds.

Distinctive in its heather-like habit, 3-merous flowers and black fruits.

1. EMPETRUM L. - *Crowberry*

1. E. nigrum L. - *Crowberry.* Stems and leaf-margins glandular when young, glabrous later; leaves strongly revolute obscuring abaxial surface, 3-7 x 1-2mm; fruit black, 4-8mm across, subglobose. Native; peaty and rocky moors, bogs and mountain-tops.

a. Ssp. nigrum. Stems to 1.2m, procumbent, slender, rooting along length; leaves ± parallel-sided, mostly 3-5x as long as wide; flowers dioecious, rarely bisexual; 2n=26. Frequent and often abundant in suitable places in Br and Ir NW of line from Devon to NE Yorks, lowland to c.800m.

b. Ssp. hermaphroditum (Hagerup) Böcher (*E. hermaphroditum* Hagerup). Stems to 50cm, less procumbent, stiff, not rooting; leaves with curved sides, mostly 2-4x as long as wide; flowers bisexual (remains of stamens usually visible at base of some fruits); (2n=52). Usually at higher altitudes (>650m) than *E. nigrum* but overlapping, often in drier places; Caerns, Lake District, highlands of Sc.

67. ERICACEAE - *Heather family*

Deciduous or evergreen trees or dwarf shrubs; leaves whorled, opposite or alternate, simple, petiolate or not, without stipules. Flowers variously arranged, bisexual or functionally dioecious, hypogynous to epigynous, actinomorphic to slightly zygomorphic; sepals free or fused, 4-5; petals as many as sepals, partly or completely fused or rarely free; stamens from lx to 2x as many as petals, usually borne on receptacle or disc, usually with anthers opening by pores; pollen-grains released in tetrads. Ovary 4-5(10)-celled with axile placentation and 1-many ovules per cell; style 1; stigma capitate; fruit a capsule, berry or drupe.

Very variable in superficial flower characters, but usually recognizable by the woody often evergreen habit, usually fused petals, stamens borne on receptacle (not on corolla), and anthers usually opening by pores. All our spp., except *Arbutus unedo*, *Erica terminalis* and *E. x darleyensis*, are calcifuges.

1 Ovary inferior; fruit a berry with persistent calyx-lobes at apex **13. VACCINIUM**
1 Ovary superior; fruit various, if succulent then calyx deciduous or persistent at base of fruit 2
  2 Petals free, white; leaves tomentose with rust-coloured hairs on lowerside **1. LEDUM**
  2 Petals fused at least at base; leaves tomentose or not, if so then hairs not rust-coloured 3
3 Most leaves opposite or whorled; anthers with 0 or basal appendages 4
3 Leaves alternate or spiral; anthers with 0 or terminal appendages 7
  4 Sepals and petals 4; stamens 8; corolla persistent around ripe fruit 5
  4 Sepals and petals 5; stamens 5 or 10; corolla falling before fruiting 6
5 Leaves opposite; corolla shorter than calyx, divided >1/2 way to base **11. CALLUNA**
5 Leaves mostly in whorls of 3-4(5); corolla longer than calyx, normally divided <1/2 way to base **12. ERICA**
  6 Leaves <15mm, with strongly revolute margins; stamens 5; corolla divided c.1/2 way to base, without pouches **3. LOISELEURIA**
  6 Leaves >15mm, with flat margins; stamens 10; corolla divided much <1/2 way to base, with 10 small pouches on inside near base **4. KALMIA**
7 Corolla widened distally, >15mm, slightly zygomorphic **2. RHODODENDRON**
7 Corolla narrowed distally, <15mm, actinomorphic 8
  8 Fruit succulent or surrounded by a succulent calyx 9
  8 Fruit a dry capsule with a dry calyx 12
9 Leaves spine-tipped **8. GAULTHERIA**
9 Leaves without spines 10
  10 Calyx becoming succulent and surrounding fruit when ripe **8. GAULTHERIA**
  10 Calyx remaining dry and small at base of succulent fruit 11
11 Erect tree or shrub flowering in late autumn; fruit orange to red, very warty **9. ARBUTUS**
11 Procumbent shrub flowering in summer; fruit red or black, smooth, ± glossy **10. ARCTOSTAPHYLOS**
  12 Petals and sepals 4; leaves white-tomentose on lowerside **6. DABOECIA**
  12 Petals and sepals 5; leaves glabrous but often white on lowerside 13
13 Calyx and pedicels with glandular hairs; corolla purple; anthers without appendages **5. PHYLLODOCE**
13 Calyx and pedicels glabrous; corolla pink; anthers with horn-like appendages at apex **7. ANDROMEDA**

Other genera - **PIERIS** D. Don would key out in couplet 13 but differs from

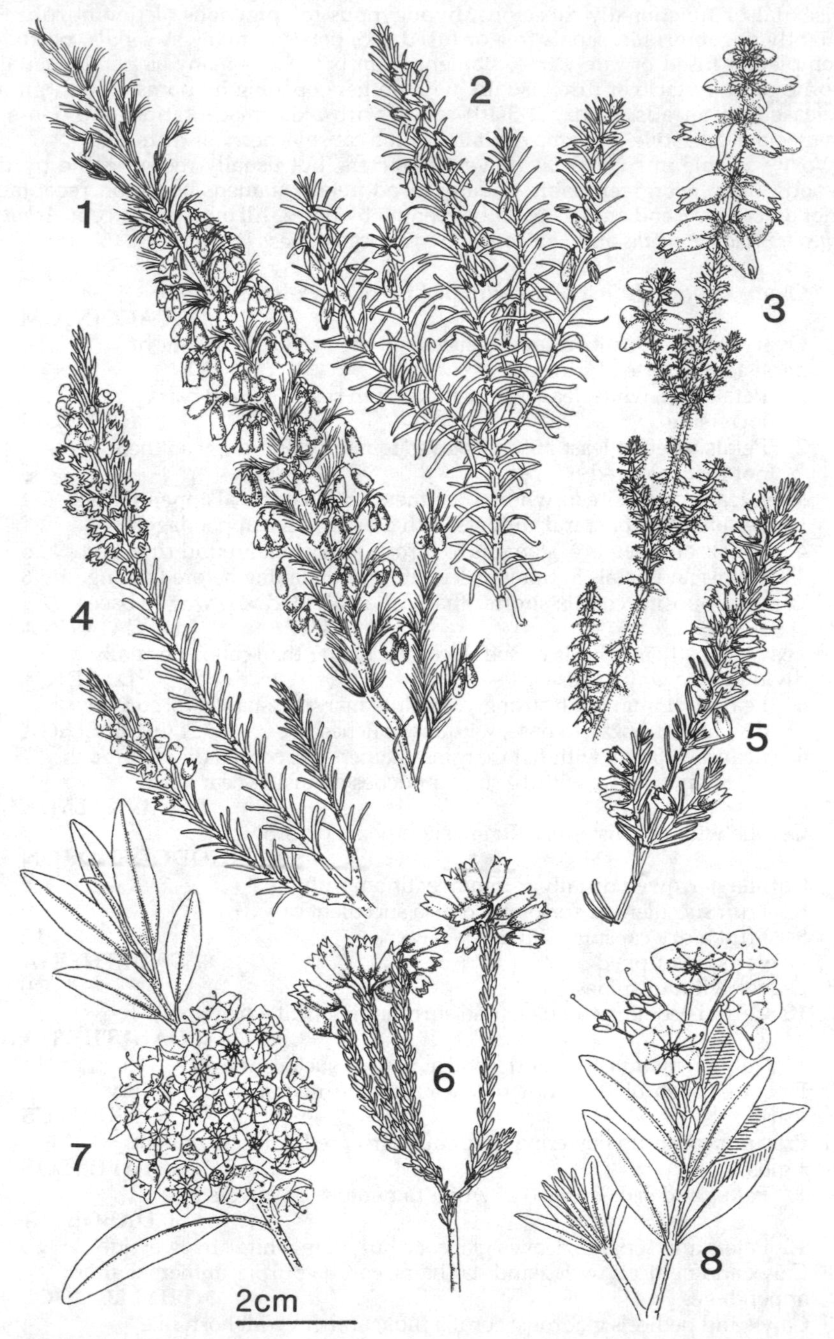

FIG 284 - Ericaceae. 1, *Erica lusitanica*. 2, *E. x darleyensis*. 3, *E. ciliaris*. 4, *E. vagans*. 5, *E. erigena*. 6, *E. terminalis*. 7, *Kalmia angustifolia*. 8, *K. polifolia*.

Phyllodoce and *Andromeda* in its flat, crenate-serrate leaves. **P. japonica** (Thunb.) D. Don, from Japan, along with 2 similar spp., is grown for ornament, and seedlings occur in a cemetery in Surrey; it is an evergreen shrub to 2m with leaves 3-8cm and white flowers in ± pendent panicles.

1. LEDUM L. - *Labrador-tea*
Leaves alternate, evergreen; flowers in dense terminal racemes, petals 5, free, white; stamens (5)6-8(10); anthers without appendages; ovary superior; fruit a capsule.

1. L. palustre L. (*L. groenlandicum* Oeder) - *Labrador-tea*. Well-branched ± upright 290
shrub to 1.2m; leaves 1.5-5cm, entire, with strongly revolute margins, densely rusty-tomentose on lowerside, elliptic to narrowly so; flowers 8-16mm across. Intrd; well natd in bogs and other wet peaty ground; scattered from S En to C Sc; N Europe and N America. Despite claims that ssp. **palustre** (2n=52), from N Europe, occurs in Br, all our plants are ssp. **groenlandicum** (Oeder) Hultén (2n=26) or perhaps intermediate; ssp. *palustre* has narrower, less densely tomentose leaves and has stamens (7-)10(11).

2. RHODODENDRON L. - *Rhododendrons*
Leaves alternate, deciduous or evergreen; flowers in dense terminal racemes; petals 5, fused to form bell-shaped lobed corolla; stamens 5 or 10; anthers without appendages; ovary superior; fruit a capsule.

Other spp. - A very large number of spp., hybrids and cultivars are grown and may persist in neglected parks and gardens; some records of the following 2 might belong to some of these.

1. R. ponticum L. - *Rhododendron*. Densely branched, spreading to upright shrub to 5m; leaves 6-20cm, evergreen, ± flat, elliptic to oblong or narrowly so, entire, glabrous; flowers mauvish-purple, c.4-6cm across; stamens 10; ovary glabrous but glandular; 2n=26. Intrd; extensively natd by seeding and suckering on sandy and peaty soils and on rocks both in woods and in open; suitable places throughout BI. Our plants are closest to ssp. **ponticum**, from SE Europe and SW Asia, but some claim that most of them are in fact **R. ponticum** x **R. maximum** L. *R. maximum* has pubescent ovaries and leaf lowersides and orange spots on the corolla.

2. R. luteum Sweet - *Yellow Azalea*. Upright shrub to 2m; leaves 6-12cm, deciduous, flat, oblong-lanceolate, shallowly serrate, slightly pubescent; flowers yellow, c.5cm across; stamens 5; ovary glabrous but glandular; (2n=26). Intrd; natd in woods by suckering and seeding; scattered in Br; E Europe and W Asia.

3. LOISELEURIA Desv. - *Trailing Azalea*
Leaves opposite, evergreen; flowers 1-several in terminal apparent umbels; petals 5, fused to form bell-shaped lobed corolla; stamens 5; anthers without appendages; ovary superior; fruit a capsule.

1. L. procumbens (L.) Desv. - *Trailing Azalea*. Densely branched domed or trailing shrub to 25cm high; leaves <1cm, entire, elliptic to narrowly so, glabrous, with strongly revolute margins; flowers pink, 3-6mm aross; (2n=24). Native; rocky and peaty moors and mountains above 400m; locally frequent in highlands of Sc, rare in Orkney and Shetland.

4. KALMIA L. - *Sheep-laurels*
Leaves alternate, whorled or opposite, evergreen; flowers in small corymbose racemes; petals 5, fused to form saucer-shaped slightly lobed corolla with 10 small pouches on inside near base; stamens 10, enclosed in corolla-pouches before dehiscence; anthers without appendages; ovary superior; fruit a capsule.

The frequently used character 'inflorescence terminal or lateral' is misleading; in all cases the inflorescence arises from leaf-axils near the apex of the previous year's growth, which might or might not become overtopped by a current season's shoot. The former is normal in only *K. angustifolia*, but even there is not constant.

1 Leaves alternate or irregularly arranged, finely acute; flowers 20-25mm across **3. K. latifolia**
1 Leaves opposite or in whorls of 3, obtuse to subacute; flowers 6-16mm across 2
  2 Leaves sessile or with petioles <4mm; pedicels glabrous; flowers 10-16mm across; inflorescence apparently terminal **1. K. polifolia**
  2 Leaves with petioles c. 4-8mm; pedicels very shortly pubescent; flowers 6-12mm across; inflorescence usually becoming overtopped by current season's shoot and appearing lateral **2. K. angustifolia**

1. K. polifolia Wangenh. - *Bog-laurel*. Straggling to erect shrub to 70cm; leaves 284
mostly opposite, 1-3(4)cm, narrowly to very narrowly elliptic to oblong, glabrous, entire, with revolute margins, whitish on lowerside; inflorescence apparently terminal; flowers pink, 10-16mm across; (2n=44, 48). Intrd; natd in wet peaty bogs and moors; few places in SE & N En and C Sc; N America.

2. K. angustifolia L. - *Sheep-laurel*. Erect to ascending shrub to 1m; leaves 284
opposite or in whorls of 3, 2-5(7)cm, elliptic to oblong to narrowly so, entire, glabrous, with flat or slightly revolute margins, dull pale green on lowerside; inflorescence usually apparently lateral; flowers pink, 6-12mm across; (2n=24). Intrd; in similar places to *K. polifolia*; scattered from S to N En; E N America.

3. K. latifolia L. - *Mountain-laurel*. Erect shrub (or tree) to 3(12)m; leaves mostly alternate, 5-12cm, narrowly elliptic to ovate-elliptic, entire, with flat margins, pale yellowish-green on lowerside; inflorescence apparently terminal; flowers pink, 20-25mm across; (2n=24). Intrd; natd in a few wet acid places; SE En, Man; E N America.

5. PHYLLODOCE Salisb. - *Blue Heath*
Leaves alternate, evergreen; flowers few on long pedicels in subterminal clusters; petals 5, fused to form tubular corolla narrowed and shortly lobed distally; stamens 10; anthers without appendages; ovary superior; fruit a capsule.

1. P. caerulea (L.) Bab. - *Blue Heath*. Domed shrub to 20cm high; leaves <15mm, RRR
strongly revolute at margins, minutely serrate on apparent margins, linear to linear-oblong; young leaves, pedicels and calyx glandular; flowers 7-12mm, mauvish-purple; (2n=24). Native; rocky moorland at 680-840m; very local in Westerness and M Perth. Irregular flowering behaviour; when sterile easily overlooked for *Loiseleuria* or *Empetrum*, with which it often grows, but details of leaves are quite different.

6. DABOECIA D. Don - *St Dabeoc's Heath*
Leaves alternate, evergreen; flowers in lax terminal racemes; petals 4, fused to form tubular corolla narrowed and shortly lobed distally; stamens 8; anthers without appendages; ovary superior; fruit a capsule.

1. D. cantabrica (Huds.) K. Koch - *St Dabeoc's Heath*. Straggly or loosely domed R
shrub to 50(70)cm high; leaves <15mm, entire, revolute at margins, sublinear to elliptic, white-tomentose on lowerside; flowers 8-14mm, pinkish-purple; 2n=24. Native; peaty and rocky moorland; locally common in Connemara (E & W Mayo and W Galway).

7. ANDROMEDA L. - *Bog-rosemary*
Leaves alternate, evergreen; flowers in small terminal umbel-like clusters; petals 5, fused to form tubular corolla narrowed and shortly lobed distally; stamens 10; anthers each with 2 long terminal appendages; ovary superior; fruit a capsule.

1. A. polifolia L. - *Bog-rosemary*. Straggly glabrous shrub; stems to 35cm; leaves 1-4cm, entire, revolute at margins, sublinear to elliptic, white-glaucous on lowerside; flowers 5-8mm, pale pink; 2n=48. Native; wet peaty places; locally common, C En and S Wa to C Sc, Ir, formerly more widespread in En.

8. GAULTHERIA L. (*Pernettya* Gaudich.) - *Aromatic Wintergreens*
Leaves alternate, evergreen; flowers solitary and axillary or in terminal and subterminal racemes; petals 5, fused to form tubular corolla narrowed and shortly lobed distally; stamens 10; anthers with 4 short terminal appendages; ovary superior; fruit a capsule surrounded by succulent, berry-like swollen calyx, or a succulent berry with calyx remaining small and dry at its base.

1 Leaves ≤2cm, spine-tipped; fruit a succulent berry with calyx remaining small and dry at its base; functionally dioecious **3. G. mucronata**
1 Leaves ≥2cm, not spine-tipped; fruit a capsule surrounded by succulent berry-like swollen calyx; plants bisexual 2
2 Leaves cuneate at base; flowers solitary in leaf-axils; fruit bright red **2. G. procumbens**
2 Leaves rounded to cordate at base; flowers in terminal and subterminal racemes; fruit purplish-black **1. G. shallon**

1. G. shallon Pursh - *Shallon*. Thicket-forming shrub to 1.5m; leaves 5-10cm, **290**
rounded to cordate at base, closely minutely serrate; flowers in terminal and subterminal racemes 5-12cm, with dense stalked glands on pedicels and rhachis, 7-10mm, white to pink; fruit purplish-black; (2n=22, 48, 88). Intrd; planted as cover and food for birds, natd in woodland and shrubberies especially on sand and peat; scattered through En, Ir, Sc and CI; W N America.

1 x 3. G. shallon x G. mucronata = G. x wisleyensis D.J. Middleton (= X *Gaulnettya wisleyensis* Marchant nom. nud.) is a well-known garden plant but has arisen on heathy ground in S Hants (found in 1981 with both parents). It is intermediate in leaf characters, but the wild plant is closer to *G. shallon* than is the commonest garden cultivar.

2. G. procumbens L. - *Checkerberry*. Dwarf ground-covering shrub to 15cm; leaves 2-5cm, cuneate at base, rather remotely minutely serrate; flowers solitary in leaf-axils, with pubescent but not glandular pedicels, 5-7mm, white to pink; fruit bright red; (2n=24, c.44). Intrd; planted as ground cover and natd in woodland; few places in Sc and S En; E N America.

3. G. mucronata (L.f.) Hook. & Arn. (*Pernettya mucronata* (L.f.) Gaudich. ex **290**
Spreng.) - *Prickly Heath*. Erect or spreading suckering shrub to 1.5m; leaves 8-20mm, cuneate to subcordate at base, remotely and minutely serrate; flowers 1-few in subterminal axillary clusters, with subglabrous to shortly pubescent pedicels, 2.5-5mm, white; berry white to purple; (2n=66). Intrd; natd in open woodland and shrubberies on sandy soil; scattered in Br and Ir; Chile.

9. ARBUTUS L. - *Strawberry-tree*
Leaves alternate, evergreen; flowers in terminal panicles; petals 5, fused to form tubular corolla narrowed and shortly lobed distally; stamens 10; anthers with 2 long terminal appendages; ovary superior; fruit a warty, globose berry.

1. A. unedo L. - *Strawberry-tree*. Shrub or tree to 5(11)m; leaves 4-11cm, ovate to **RR**
obovate or narrowly so, serrate; flowers 6-11mm, white or pink-tinged; fruit red,

1.5-2cm across; flowering in autumn together with fruits from previous year; 2n=26. Native; rocky ground in scrub and young woodland; S & N Kerry, W Cork and Sligo; rarely natd (bird-sown) on mostly chalk or limestone slopes in En and Wa.

10. ARCTOSTAPHYLOS Adans. (*Arctous* (A. Gray) Nied.) - *Bearberries*
Leaves alternate, deciduous or evergreen; flowers 1-few in terminal clusters; petals 5, fused to form tubular corolla narrowed and shortly lobed distally; stamens 10; anthers with 2 terminal appendages; ovary superior; fruit a smooth, globose, berry-like drupe.

1. A. uva-ursi (L.) Spreng. - *Bearberry.* Procumbent shrub with stems to 1.5m; leaves 1-3cm, entire, evergreen, obovate to oblanceolate, without gland-dots but with conspicuous fine vein reticulation on lowerside; flowers 4-6mm, white, pink-tinged; stamens with reflexed appendages c. equalling anthers; fruit c.8-10mm across, bright red; 2n=52. Native; peaty and rocky moorland in lowlands and mountains; locally common in Sc, N En and N & W Ir. See *Vaccinium vitis-idaea* for differences.

2. A. alpinus (L.) Spreng. (*Arctous alpinus* (L.) Nied.) - *Arctic Bearberry.* R
Procumbent shrub with stems to 60cm; leaves 1-2.5cm, serrate, dying in autumn but persistent until next spring, obovate; flowers 3-5mm, white or green-tinged; stamens with erect appendages much shorter than anthers; fruit c.6-10mm across, black; 2n=26. Native; mountain moorland; local in N & NW Sc from Shetland to M Perth.

11. CALLUNA Salisb. - *Heather*
Leaves opposite, evergreen, sessile; flowers in long usually terminal racemes or panicles; petals 4, fused for basal 1/4 or less; stamens 8; anthers with 2 basal appendages; ovary superior; fruit a capsule dehiscing along line of fusion of carpels.

1. C. vulgaris (L.) Hull - *Heather*. Decumbent to erect shrub to 60(150)cm; leaves 2-3.5mm, sessile, entire, very strongly revolute at margins making the cross-section triangular, with 2 pointed auricles at base; flowers 3-4.5mm, pink (or white); 2n=16. Native; heaths, moors, rocky places, bogs and open woodland, mainly on sandy or peaty soil; abundant in suitable places throughout BI.

12. ERICA L. - *Heaths*
Leaves in whorls of 3-4(5), evergreen, shortly petiolate; flowers in various terminal and/or axillary clusters; petals 4, fused for basal 1/2 or more; stamens 8, anthers with 0 or 2 basal appendages; ovary superior; fruit a capsule dehiscing between lines of fusion of carpels.

1 Anthers at least partly exserted from corolla 2
1 Anthers included in corolla 4
  2 Pedicels longer than calyx; corolla-lobes divergent distally; summer-flowering **10. E. vagans**
  2 Pedicels shorter than calyx; corolla-lobes ± parallel distally; winter- to spring-flowering 3
3 Stems to 1.2(-2)m, with well-developed main stems; young twigs with flanges of tissue running <1/2 way from leaf-bases to next lower node; flowering Mar-Jun **8. E. erigena**
3 Stems to 60cm, without well-developed main stems; young twigs with flanges of tissue running >1/2 way from leaf-bases to next lower node; flowering Nov-Jun **9. E. x darleyensis**
  4 Revolute leaf-margins meeting closely under leaf, entirely obscuring

lowerside; flowers usually in panicles 5
4 Revolute leaf-margins not meeting or meeting only distally under leaf, revealing at least proximal part of lowerside; flowers in racemes or apparent umbels 7
5 Some bracteoles borne near apex of pedicel, overlapping calyx; shrub <80cm, summer-flowering **5. E. cinerea**
5 Bracteoles borne only on proximal part of pedicel, not overlapping calyx; shrub often >80cm, spring-flowering 6
6 Corolla 4-5mm; stigma red; all hairs smooth **6. E. lusitanica**
6 Corolla 2.5-4mm; stigma white; some hairs on young twigs with rough surface **7. E. arborea**
7 Flowers in terminal ± elongated racemes; anthers with 0 basal appendages **1. E. ciliaris**
7 Flowers in terminal umbel-like clusters; anthers with basal appendages 8
8 Lowerside of leaves green; sepals glabrous or with only short hairs; anthers with triangular appendages **4. E. terminalis**
8 Lowerside of leaves whitish; sepals with long hairs; anthers with linear appendages 9
9 Sepals and uppersides of leaves usually glabrous except for long hairs; most of leaf lowerside exposed; ovary and fruit glabrous **2. E. mackaiana**
9 Sepals and uppersides of leaves usually with dense short hairs as well as long hairs; most of leaf lowerside obscured; ovary and fruit pubescent **3. E. tetralix**

1. E. ciliaris L. - *Dorset Heath*. Straggly shrub to 60cm; leaves in whorls of 3(-4), 2-4mm, with long usually glandular hairs; flowers in elongated racemes; corolla 8-12mm, bright reddish-pink (or white), oblique-ended; (2n=24). Native; heaths, often damp; very locally frequent in S Hants, Dorset, S Devon, W Cornwall and W Galway. Flowers Jun-Sep. **RR** **284**

1 x 3. E. ciliaris x E. tetralix = E. x watsonii Benth. occurs throughout the range of *E. ciliaris* in BI; it has leaves more like those of *E. tetralix* and flowers nearer those of *E. ciliaris* and is highly sterile, though backcrossing occurs in Dorset.

2. E. mackaiana Bab. - *Mackay's Heath*. Straggly to compact shrub to 60cm; leaves in whorls of 4, 2-4.5mm, with long glandular hairs, with most of lowerside exposed; flowers in terminal umbel-like clusters; corolla 5-7mm, purplish-pink; (2n=24). Native; peaty bogs; very local in W Galway, W Mayo and W Donegal. Flowers Aug-Sep. **RR**

2 x 3. E. mackaiana x E. tetralix = E. x stuartii (Macfarl.) Mast. (*E. x praegeri* Ostenf.) occurs near *E. mackaiana* in all its sites; it is intermediate in leaf-shape and pubescence and is completely sterile.

3. E. tetralix L. - *Cross-leaved Heath*. Straggly shrub to 70cm; leaves in whorls of 4, 2-5mm, with dense short and usually few long usually glandular hairs, with most (but not all) of lowerside obscured; flowers in terminal umbel-like clusters; corolla 5-9mm, pink (or white); 2n=24. Native; bogs and usually wet heaths and moors; suitable places throughout BI. Flowers Jun-Sep.

3 x 9. E. tetralix x E. vagans = E. x williamsii Druce has been found as small patches on several occasions on Lizard Peninsula, W Cornwall, with the parents; it resembles *E. vagans* in most characters but is sterile and has pubescent young leaves. It is common in cultivation.

4. E. terminalis Salisb. - *Corsican Heath*. Bushy or erect shrub to 1m (rarely more); leaves in whorls of 4, 3-6mm, shortly pubescent when young, without long hairs; ± glabrous later; flowers in terminal umbel-like clusters; corolla 5-7mm, bright pink. Intrd; natd on sand-dunes; Magilligan, Londonderry, known since c.1900; W Mediterranean. Flowers Jun-Sep. **284**

5. E. cinerea L. - *Bell Heather*. Straggly shrub to 60cm; leaves in whorls of 3, 4-7mm, glabrous; flowers in terminal racemes or panicles, or sometimes in ± umbel-

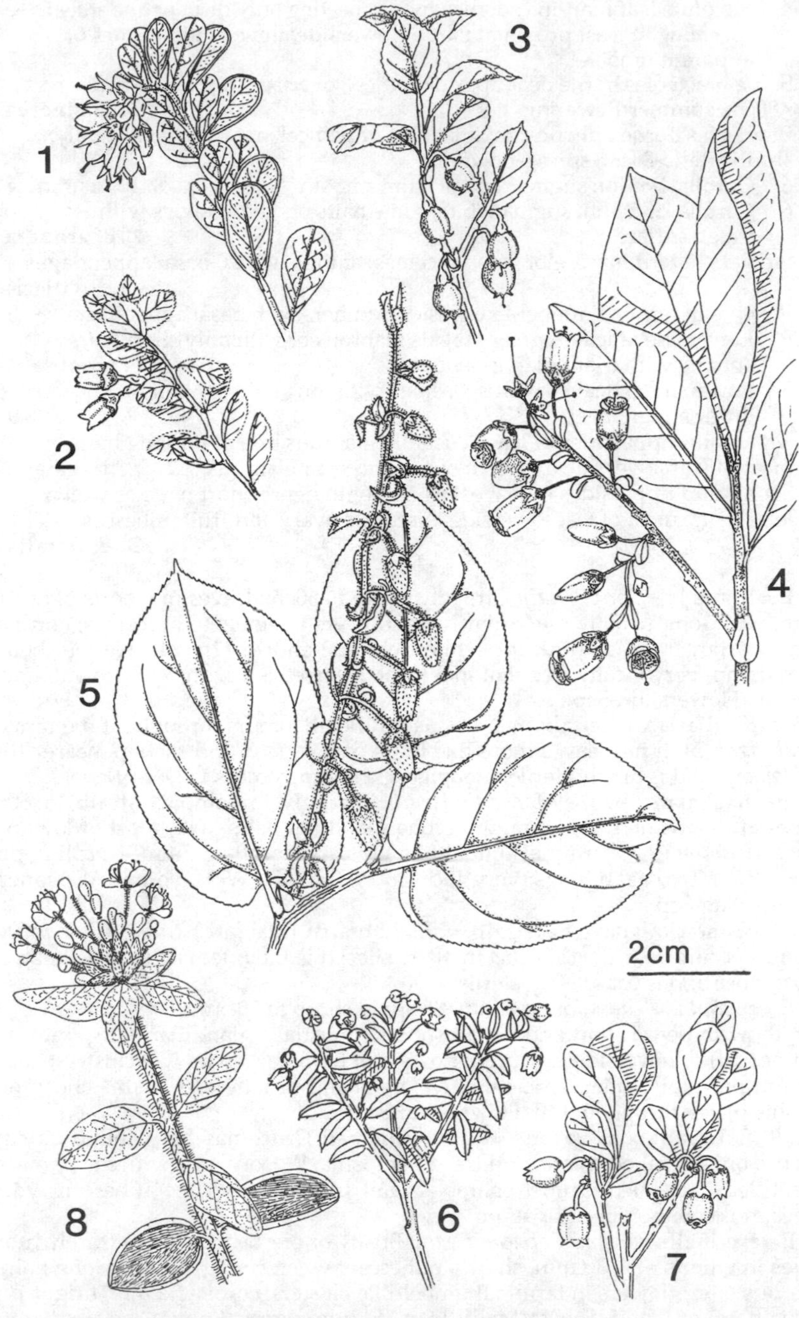

FIG 290 - Ericaceae. 1, *Vaccinium vitis-idaea.* 2, *V. x intermedium.* 3, *V. myrtillus.* 4, *V. corymbosum.* 5, *Gaultheria shallon.* 6, *G. mucronata.* 7, *Vaccinium uliginosum.* 8, *Ledum palustre* ssp. *groenlandicum.*

like clusters; corolla 4-6mm, bright reddish-purple (or white); 2n=24. Native; usually dry heaths and moors; suitable places throughout BI. Flowers Jul-Sep.

6. E. lusitanica Rudolphi - *Portuguese Heath.* Erect shrub to 2(-3)m; leaves in whorls of 3-4, 5-7mm, glabrous; flowers in terminal panicles; corolla 4-5mm, pinkish-white. Intrd; garden escape natd on heaths and railway-banks; E & W Cornwall (known since 1920) and Dorset; SW Europe. Flowers Jan-Apr.

7. E. arborea L. - *Tree Heath.* Differs from *E. lusitanica* as in key; (2n=24). Intrd; garden escape natd in hedgerows and open woodland; less common than *E. lusitanica*, Surrey, Scillies and Man; Mediterranean. Flowers Jan-Apr.

8. E. erigena R. Ross (*E. hibernica* (Hook & Arn.) Syme non Utinet, *E. mediterranea* **R**
auct. non L., *E. herbacea* L. ssp. *occidentalis* (Benth.) Laínz) - *Irish Heath.* Erect shrub **284**
to 1.2(-2)m; leaves in whorls of 4, 5-8mm, glabrous; flowers in terminal racemes or panicles; corolla 5-7mm, purplish-pink (or white); (2n=24). Native; in usually well-drained parts of bogs; W Galway and W Mayo. Flowers Mar-Jun.

9. E. x darleyensis Bean (*E. erigena* x *E. carnea* L.) - *Darley Dale Heath.* Differs **284**
from *E. erigena* in key characters. Intrd; planted in 1930s on bank in E Suffolk and well established until 1980s, now in Surrey and W Cornwall; garden origin. Flowers Nov-Jun.

10. E. vagans L. - *Cornish Heath.* Straggly shrub to 80cm; leaves in whorls of 4-5, **RR**
5-10mm, glabrous; flowers in subterminal racemes, on long pedicels; corolla 2.5- **284**
3.5mm, pale pink to white; 2n=24. Native; dry heaths, often relatively base-rich; W Cornwall, mostly on Lizard Peninsula; commonly cultivated and sometimes persisting elsewhere, well natd or perhaps native in Fermanagh. Flowers Jul-Aug.

13. VACCINIUM L. (*Oxycoccus* Hill) - *Bilberries*

Leaves alternate, deciduous or evergreen; flowers solitary or clustered, terminal or axillary; petals 4-5, fused at base or for most part to form variously shaped corolla; stamens 8 or 10; anthers with or without terminal appendages; ovary inferior; fruit a berry.

1 Corolla divided >3/4 way to base; pedicels erect, filiform; leafy stems procumbent for most part 2
1 Corolla divided <2/3 way to base; pedicels not erect or filiform; leafy stems erect to decumbent 4
  2 Leaves narrowly oblong, at least some >1cm; bracteoles above middle of pedicel, mostly >1mm wide; leafy shoot continuing growth beyond flower cluster in same year **3. V. macrocarpon**
  2 Leaves ovate-elliptic or narrowly so, rarely >1cm; bracteoles at or below middle of pedicel, <0.5mm wide; flowers in terminal groups of 1-c.5 3
3 Pedicels minutely pubescent **1. V. oxycoccos**
3 Pedicels glabrous or almost so **2. V. microcarpum**
  4 Some leaves >3cm (to 8cm) **7. V. corymbosum**
  4 Leaves <3cm 5
5 Leaves serrate to serrulate; stems acutely angled **6. V. myrtillus**
5 Leaves entire to obscurely crenulate; stems terete 6
  6 Leaves evergreen; flowers in terminal racemes; corolla widest at mouth, divided c.1/2 way to base; fruit red **4. V. vitis-idaea**
  6 Leaves deciduous; flowers in axillary clusters of 1-4; corolla narrowed at mouth, divided ≤1/4 way to base; fruit bluish-black **5. V. uliginosum**

1. V. oxycoccos L. (*Oxycoccus palustris* Pers.) - *Cranberry.* Procumbent shrub with stems to 30(80)cm; leaves (2)5-10(12) x 2-6mm, entire, evergreen, often widest near or not much below middle; flowers in terminal groups of 1-4; pedicels minutely pubescent; corolla bright pink, with strongly reflexed lobes; fruit red, (6)8-10(15)mm across, globose to pear-shaped; (2n=48, 72). Native; bogs and very wet heaths; locally frequent in much of Br and Ir, but absent from most of S En, S Ir and N Sc.

2. V. microcarpum (Turcz. ex Rupr.) Schmalh. (*Oxycoccus microcarpus* Turcz. ex R
Rupr.) - *Small Cranberry*. Differs from *V. oxycoccos* in its leaves 2-6(8) x 1-2.5mm, often widest near base; flowers in groups of 1-2; pedicels glabrous or almost so; fruit 5-8(10)mm across, pear-shaped or ellipsoid; (2n=24). Native; bogs; C & N mainland Sc, S Northumb and Cheviot. Intermediates with *V. oxycoccos* occur and the 2 taxa might be better considered as vars or sspp. of the latter.

3. V. macrocarpon Aiton (*Oxycoccus macrocarpus* (Aiton) Pursh) - *American Cranberry*. Differs from *V. oxycoccos* in being more robust; leaves 6-18 x 2-5mm, oblong; flowers in groups of 1-5(10); fruit 9-14(20)mm across, globose; and see key; (2n=24). Intrd; grown for its fruit and natd from bird-sown seed, mostly in peaty places; scattered in Br from S En to W Sc; E N America.

4. V. vitis-idaea L. - *Cowberry*. Erect to decumbent shrub to 30cm; young stems 290
terete, minutely pubescent; leaves 10-30mm, obovate to elliptic, entire to obscurely crenulate, evergreen; flowers in short terminal racemes, pinkish white; fruit red, 6-10mm across, globose; 2n=24. Native; moors and open peaty woods; locally abundant in Br from S Wa and C En northwards, scattered in Ir. Superficially resembles *Arctostaphylos uva-ursi* vegetatively, but leaf lowerside has many minute gland-dots and very inconspicuous vein reticulation.

4 x 6. V. vitis-idaea x V. myrtillus = V. x intermedium Ruthe occurs very locally 290
with the parents in Staffs, Derbys and Yorks, with old or unconfirmed records elsewhere, but is absent from most areas where the parents cohabit; it is convincingly intermediate in all leaf, stem, flower and fruit characters and sets some good seed; 2n=24.

5. V. uliginosum L. - *Bog Bilberry*. Erect to ascending shrub to 50(80)cm; young 290
stems terete, glabrous to minutely pubescent; leaves 8-25mm, obovate, entire, deciduous; flowers in groups of 1-4 on short side-branches, pale pink; fruit bluish-black with whitish bloom, 6-10mm across, globose; 2n=39, 48. Native; moors; locally common in C & N Sc, very local in S Sc and N En. Some plants from Shetland resemble ssp. **microphyllum** Lange, from N Europe, which differs in its smaller parts, less revolute corolla-lobes and diploid (2n=24) chromosome number; they need further study.

6. V. myrtillus L. - *Bilberry*. Erect to ascending shrub to 50(100)cm; stems acutely 290
angled, glabrous; leaves 10-30mm, ovate to elliptic, serrate to serrulate, deciduous; flowers 1(-2) in leaf-axils, pinkish-red; fruit bluish-black with whitish bloom, 6-10mm across, ± globose; 2n=24. Native; heaths, moors and woods; common in suitable places throughout Br and Ir, but absent from much of C & E En.

7. V. corymbosum L. - *Blueberry*. Erect shrub to 1(-2.5)m; young stems terete, 290
minutely pubescent; leaves 10-80mm, elliptic, entire to serrate, deciduous; flowers in terminal or lateral stalked racemes, white or pink-tinged; fruit bluish-black with whitish bloom, 5-12mm across, ± globose; (2n=48). Intrd; natd on heathland from bird-sown seeds; S Hants and Dorset; E N America.

68. PYROLACEAE - *Wintergreen family*

Herbaceous rhizomatous perennials; leaves opposite, whorled, alternate or ± all basal, simple, petiolate, without stipules. Flowers terminal, solitary or in a raceme, bisexual, hypogynous, actinomorphic; sepals 5, ± free, petals 5, free, ± white; stamens 10, with anthers opening by 2 apparently apical pores; ovary incompletely 5-celled with axile placentation and many ovules per cell; style 1; stigma 5-lobed to ± capitate; fruit a capsule dehiscing between lines of fusion of carpels.

Herbaceous plants with ± white petals and with flowers similar in structure to those of Ericaceae.

1 Flowers solitary **3. MONESES**
1 Flowers in terminal raceme 2
 2 Flowers all turned to 1 side of axis; anther pores borne on main body of anther; petioles <2cm **2. ORTHILIA**
 2 Flowers facing all directions; anther pores borne at end of short tubular anther outgrowths; longest petioles >2cm **1. PYROLA**

1. PYROLA L. - *Wintergreens*
Flowers in terminal raceme, facing all directions; anthers with pores borne on very short tubes; pollen-grains released in tetrads. Leaf characters do not reliably separate any of the taxa in this genus.

1 Style strongly curved; flowers saucer- to cup-shaped **3. P. rotundifolia**
1 Style ± straight; flowers ± globose 2
 2 Style 1-2mm, included in flower, not widened below stigma **1. P. minor**
 2 Style 4-6mm, just exserted from flower, widened immediately below stigma-lobes **2. P. media**

1. P. minor L. - *Common Wintergreen*. Flowering stem erect, to 20(30)cm, with all leaves near base; leaves 2.5-6cm, usually ovate-elliptic, obtuse to rounded at apex, crenate-serrate, and with petioles slightly shorter; flowers 4-7mm across; petals pinkish-white; style 1-2mm, straight, not widened below stigma; (2n=46). Native; mostly on leaf-mould in woods in S, also on damp rock-ledges and peaty moors in N, rarely sand-dunes; very scattered over most of Br and Ir.

2. P. media Sw. - *Intermediate Wintergreen*. Differs from *P. minor* in leaves usually broadly elliptic to orbicular, with petioles c. as long or longer; flowers 7-11mm across; style 4-6mm, widened immediately below stigma; 2n=46. Native; humus-rich moors and woods; frequent in C & N Sc (not outer islands), extremely local in N Ir, S Sc and N En, Worcs, formerly Sussex. **R**

3. P. rotundifolia L. - *Round-leaved Wintergreen*. Differs from *P. minor* in leaves usually ovate-elliptic to orbicular, with petioles longer; flowers 8-12mm across; petals white; style 4-10mm, strongly curved, widened immediately below stigma. Native. The 2 sspp. are of doubtful value; all characters, except possibly those of the sepals, break down.

a. Ssp. rotundifolia. Leaves rarely orbicular; scale-leaves on stems above true leaves 1-2; pedicels 4-8mm; sepals triangular-lanceolate, acute; anthers 2.2-2.8mm; style 6-10mm; (2n=46). Damp rock-ledges, woods, bogs and fens; very local in En, Sc and C Ir. **R**

b. Ssp. maritima (Kenyon) E.F. Warb. Leaves usually orbicular; scale-leaves 2-5; pedicels 2-5mm; sepals oblong-lanceolate, obtuse; anthers 1.9-2.4mm; style 4-6mm; 2n=46. Damp hollows in sand-dunes; W coast of Br from Cumberland to N Devon, Co Wexford, doubtfully reported from E Br except probably correctly so in W Kent. **R**

2. ORTHILIA Raf. (*Ramischia* Opiz ex Garcke) - *Serrated Wintergreen*
Stem-leaves alternate; flowers in terminal raceme, all turned to 1 side; anthers with pores borne on main body of anther; pollen-grains released singly.

1. O. secunda (L.) House (*Ramischia secunda* (L.) Garcke, *Pyrola secunda* L.) - *Serrated Wintergreen*. Flowering stem erect, often curved at top, to 10(20)cm, with all leaves near base; leaves 2-4cm, ovate to elliptic-ovate, acute to obtuse at apex, finely serrate, with shorter petiole; flowers 4.5-6mm across; petals greenish-white; style 4.5-6mm, straight; (2n=38). Native; woods and damp rock-ledges; rather local from N En to N Sc, very local in Wa and N & C Ir.

3. MONESES Salisb. ex Gray - *One-flowered Wintergreen*
Stem-leaves opposite or in whorls of 3; flowers single, terminal, pendent or turned to 1 side; anthers with pores borne on short tubes; pollen-grains released in tetrads.

1. M. uniflora (L.) A. Gray - *One-flowered Wintergreen.* Flowering stem erect, to 10(15)cm, with all leaves near base; leaves 6-25mm, orbicular, crenate-serrate, rounded at apex, with shorter petiole; flowers 12-20mm across; petals white; style 5-7mm, straight; (2n=26). Native; on leaf-litter in pinewoods; very local in NE Sc. **RR**

69. MONOTROPACEAE - *Bird's-nest family*

Saprophytic ± chlorophyll-less herbaceous perennials; leaves alternate, sessile, scale-like. Flowers in terminal raceme, bisexual, hypogynous, actinomorphic; sepals 4-5, free; petals 4-5, free, pale brownish-yellow; stamens 8 or 10, opening by longitudinal slits; ovary and fruit as in Pyrolaceae.

Easily recognized by its brownish-yellow aerial parts (utterly different in flower structure from the similarly chlorophyll-less *Neottia, Orobanche* and *Lathraea*), but very close to and perhaps best in Pyrolaceae.

1. MONOTROPA L. - *Yellow Bird's-nest*

1. M. hypopitys L. - *Yellow Bird's-nest.* Stems to 30cm, pendent at apex in flower, erect in fruit, arising from dense ± globose underground organ; scale-leaves numerous, brownish-yellow, to 13mm; flowers pale dull yellow, usually glabrous on outside. Native; on leaf-litter in woods (especially *Pinus* and *Fagus*) and on sand-dunes.

a. Ssp. hypopitys. Flowers ≤11; petals 9-13mm; stamens, carpels and inside of petals pubescent; style equalling or longer than ovary; 2n=48. Very scattered in En, ? elsewhere.

b. Ssp. hypophegea (Wallr.) Holmboe (*M. hypophegea* Wallr., *M. hypopitys* var. *glabra* Roth). Flowers ≤8; petals 8-10mm; ovary glabrous; stamens, style and inside of petals glabrous or pubescent; style equalling or shorter than ovary; 2n=16. Local in Br and Ir and absent from much of N & W.

So-called intermediates from Fife resemble ssp. *hypophegea* in all characters except chromosome number and pubescence and suggest that only those 2 characters are reliable.

70. DIAPENSIACEAE - *Diapensia family*

Cushion-like, evergreen dwarf shrub; leaves alternate, simple, entire, tapered to base, without stipules. Flowers solitary and terminal; bisexual, hypogynous, actinomorphic; sepals 5, ± free; petals 5, fused to about 1/2 way, white; stamens 5, alternating with petals; ovary 3-celled with axile placentation and many ovules per cell; style 1; stigma capitate or ± 3-lobed; fruit a capsule.

Easily recognisable by its habit, 5 stamens and 3-locular ovary.

1. DIAPENSIA L. - *Diapensia*

1. D. lapponica L. - *Diapensia.* Plant forming dome-shaped cushions to 6cm high; leaves 5-10mm, obovate, obtuse; flowers 1-2cm across, on stems to 3cm; (2n=12). Native; exposed mountain at c.760m; on hill NW of Fort William, Westerness, first found in 1951, 2nd record nearby was an error. **RRR**

70A. MYRSINACEAE

MYRSINE africana L., from Africa to Nepal, is an evergreen shrub to 1.5(2)m that is used as a hedging plant in Guernsey and Scillies, where it can survive after neglect. It has small obovate leaves toothed distally and gland-dotted, clusters of 3-6 subsessile very small dioecious flowers with 4-5 petals fused into a very short tube, and a 1-seeded pale blue succulent fruit.

71. PRIMULACEAE - *Primrose family*

Herbaceous annuals or perennials; leaves variously arranged, simple to pinnate, petiolate or not, without stipules. Flowers variously arranged, bisexual, hypogynous (semi-epigynous in *Samolus*), actinomorphic; sepals free to fused, 5(-9); petals fused at least at base, as many as sepals (0 in *Glaux*); stamens as many as sepals, borne on corolla if present; ovary 1-celled with free-central placentation and many ovules; style 1; stigma capitate; fruit a capsule.

Very variable in superficial flower characters, but usually recognizable by the herbaceous habit, fused petals, and 1-celled ovary with free-central placentation.

1 Leaves pinnate; plant a submerged aquatic **2. HOTTONIA**
1 Leaves simple; plant not submerged 2
  2 All leaves basal 3
  2 Some or all leaves on stems 4
3 Corolla-lobes patent to erecto-patent; underground tuber 0 **1. PRIMULA**
3 Corolla-lobes strongly reflexed; underground corm present **3. CYCLAMEN**
  4 Corolla yellow **4. LYSIMACHIA**
  4 Corolla white to red, blue or purple, or 0 5
5 All or most leaves in single apparent whorl at top of stem; corolla-lobes mostly 6-7 **5. TRIENTALIS**
5 Leaves opposite or alternate along stem; corolla-lobes 5 6
  6 Ovary 1/2-inferior; corolla white, longer than calyx **8. SAMOLUS**
  6 Ovary superior; corolla usually coloured or 0 or shorter than calyx 7
7 Corolla present; calyx-lobes ± free; capsule dehiscing transversely **6. ANAGALLIS**
7 Corolla 0; calyx-lobes fused for >1/4; capsule dehiscing longitudinally **7. GLAUX**

1. PRIMULA L. - *Primroses*

Perennials; leaves all basal, simple; calyx-tube longer than lobes; corolla-lobes 5, patent to erecto-patent; corolla-tube c. as long as lobes or longer; capsule dehiscing by 5 teeth or valves. Most spp. are usually heterostylous, some plants having stigmas higher than anthers (pin-eyed) and others vice versa (thrum-eyed).

1 Pale mealy coating present on various parts of leaves, scape, pedicels and flowers 2
1 Mealy coating 0 6
  2 Corolla ≥(12)15mm across, yellow 3
  2 Corolla ≤15mm across, lilac to purple, rarely white 5
3 Leaves with narrow whitish border with short-stalked glands, often mealy or minutely pubescent on upperside **6. P. auricula**
3 Leaves green to margin, glabrous, not mealy 4
  4 Leaves cuneate at base **8. P. sikkimensis**
  4 Leaves truncate to cordate at base **7. P. florindae**
5 Flowers heterostylous; corolla-lobes usually lilac, with gaps between at least near base (N En) **4. P. farinosa**

5 Flowers homostylous (anthers and stigma ± at same level); corolla-lobes usually purple, overlapping or contiguous (N Sc) **5. P. scotica**

 6 Flowers borne in 2 or more whorled tiers up scape; corolla purplish-red or white **9. P. japonica**

 6 Flowers borne singly from base of plant or in a single umbel on scape; corolla usually yellow, rarely white or purplish-red 7

7 Pedicels with long shaggy hairs; ripe capsules lying near or on ground, with viscid seeds; flowers usually borne singly from base of plant **1. P. vulgaris**

7 Pedicels with short fine hairs; ripe capsules held ± erect, with dry seeds; flowers borne in umbel on scape 8

 8 Corolla usually <15mm across, with folds in throat; calyx uniformly pale green, with acute or obtuse and apiculate teeth; capsule enclosed in calyx **3. P. veris**

 8 Corolla usually >15mm across, without folds in throat; calyx pale green with dark green midribs, with ± acuminate teeth; capsule c. as long as or longer than calyx **2. P. elatior**

Other spp. - Several spp. are grown in gardens or are planted outside and may persist for a while without attention, sometimes self-sowing. **P. pulverulenta** Duthie, from W China, resembles *P. japonica* but is mealy; there are recent records from Surrey, Berks and W Perth. **P. 'Wanda'** (= *P. juliae* Krecz. x *P. vulgaris*), of garden origin, resembles *P. vulgaris* but has wine-red flowers and is ± glabrous. **P. denticulata** Sm. (*Drumstick Primrose*), from Himalayas, is a mealy plant with a close spherical head of mauve or white flowers.

1. P. vulgaris Huds. (*P. acaulis* (L.) Hill) - *Primrose*. Plant with long shaggy hairs; leaves gradually tapered to base; flowers borne singly on pedicels to 12cm arising from leaf-rosette, or rarely in an umbel on a scape, c.2-4cm across, usually pale yellow, rarely white or reddish-pink; calyx ± uniformly pale green; 2n=22. Native; woods, hedgebanks and in damper areas in grassland, often on heavy soils; throughout BI but rare or 0 locally.

1 x 2. P. vulgaris x P. elatior = P. x digenea A. Kern. occurs frequently around the area of *P. elatior* in E Anglia. It is intermediate in leaf, flower and pubescence characters; the flowers are normally borne in an umbel on a scape. It is fertile and hybrid swarms arise.

1 x 2 x 3. P. vulgaris x P. elatior x P. veris = P. x murbeckii Lindq. occurs very rarely near *P. elatior* in W Suffolk, both from wild parents and by pollination of wild *P. elatior* by garden (sometimes purple-flowered) *P. x polyantha*.

1 x 3. P. vulgaris x P. veris = P. x polyantha Mill. (*P. x tommasinii* Gren. & Godr., *P.* x *variabilis* Goupil non Bastard) occurs sporadically where the parents meet over most of BI, often in scrubby areas. It is intermediate in leaf, flower and pubescence characters; the flowers are normally borne in an umbel on a scape. It is partially fertile but backcrossing and introgression are rare. The garden *Polyanthus* is probably of this origin, and occasionally this escapes from cultivation or persists where planted; such plants may vary considerably in flower colour, from white to orange, purple or mauve. This hybrid, *P. elatior*, and plants of *P. vulgaris* with a scape are often confused; careful attention to leaves, flowers and pubescence is needed to avoid this.

2. P. elatior (L.) Hill - *Oxlip*. Plant with rather short crisped hairs; leaves less **R**
gradually narrowed at base than in *P. vulgaris* and often ± abruptly so; flowers in an umbel on a scape to 30cm, c.15-20mm across, pale yellow; calyx with darker green midribs; 2n=22. Native; very locally abundant in woods on clay in E Anglia, in area ± lacking *P. vulgaris*, and 2 small outlying areas in Bucks, rare escape elsewhere.

2 x 3. P. elatior x P. veris = P. x media Peterm. occurs rarely in the area of *P.*

elatior in E Anglia; it is intermediate in diagnostic characters and partially fertile, but backcrossing has not been detected.

3. P. veris L. - *Cowslip.* Plant with short straight hairs; leaves abruptly contracted to petiole at base; flowers in an umbel on a scape to 30cm, c.8-15mm across, deep or brownish-yellow, rarely reddish; calyx uniformly pale green; 2n=22. Native; locally common in grassy places usually on light base-rich soils; throughout most of BI, but rare in much of N.

4. P. farinosa L. - *Bird's-eye Primrose.* Much of plant with mealy covering, but R
hairs 0; leaves gradually tapered at base, obovate; flowers in an umbel on a scape to 15cm, c.7-15mm across, usually pinkish-lilac; heterostylous; 2n=18. Native; damp grassy, stony or peaty ground on limestone; locally frequent in N En from Yorks and Lancs to Cumberland, formerly S Sc.

5. P. scotica Hook. - *Scottish Primrose.* Sometimes biennial; differs from *P. farinosa* R
in its usually wider leaves, longer narrower corolla-tube, and more overlapping and purple corolla-lobes; homostylous; 2n=54. Native; damp grassy places near sea on cliffs, dunes and pastures; W Sutherland, Caithness and Orkney; endemic.

6. P. auricula L. - *Auricula.* Inflorescence with mealy covering; leaves gradually or abruptly contracted to short petiole, mealy or not, with short-stalked glandular hairs; flowers in an umbel on a scape to 15cm, c.(12)15-25mm across, dull yellow (other colours in garden plants); (2n=54-72). Intrd; planted on rock-ledge in Caenlochan Glen, Angus, known since 1880, also persistent on old estate in MW Yorks; mountains of CS Europe.

7. P. florindae Kingdon-Ward - *Tibetan Cowslip.* Inflorescence with mealy 298
covering; leaves abruptly contracted to long petiole, with cordate or truncate base, not mealy, with 0 hairs; flowers in an umbel on a scape to 60cm, c.(12)15-30mm across, yellow; (2n=22). Intrd; planted by ponds and streams and in marshes; natd in several places in N En and Sc; Tibet.

8. P. sikkimensis Hook. f. - *Sikkim Cowslip.* Differs from *P. florindae* in its leaves gradually tapered to cuneate base; (2n=22). Intrd; planted on mountains in Caerns in 1921 and still there, often not flowering; Himalayas to W China.

9. P. japonica A. Gray - *Japanese Cowslip.* Plant not mealy, glabrous; leaves 298
gradually tapered to base; flowers borne in 2-several tiered whorls on a scape to 50cm, c.15-25mm across, purplish-red or white; (2n=44). Intrd; planted in shady moist places; natd in a few places in En, W Sc and W Ir; Japan.

2. HOTTONIA L. - *Water-violet*

Perennials with submerged vegetative parts and emergent inflorescences; leaves ± whorled, pinnate with linear lobes; calyx divided nearly to base; corolla lobes 5, patent, longer than tube; capsule dehiscing by 5 valves.

1. H. palustris L. - *Water-violet.* Stems to 1m or more, floating under water, rooting at nodes; flowers on tiered whorls on erect peduncle, 15-25mm across, lilac with yellow throat; (2n=20). Native; shallow ponds and ditches; scattered in En and Wa, locally common E En, very locally natd in Ir.

3. CYCLAMEN L. - *Sowbreads*

Perennials with large underground corm; leaves all basal, simple, with long petiole; flowers borne singly on long pedicels from corm; calyx-tube shorter than lobes; corolla-lobes 5, strongly reflexed, longer than tube; capsule dehiscing by 5 valves which become reflexed and on pedicel which becomes tightly spiralled at maturity.

1 Flowers appearing in late summer and autumn, usually before the leaves; corolla with conspicuous flaps or bulges at point of reflexion of lobes from tube; corm rooting mainly from upperside **1. C. hederifolium**

1 Flowers appearing in spring with the leaves; corolla without lobes or bulges at point of reflexion; corm rooting only from lowerside 2

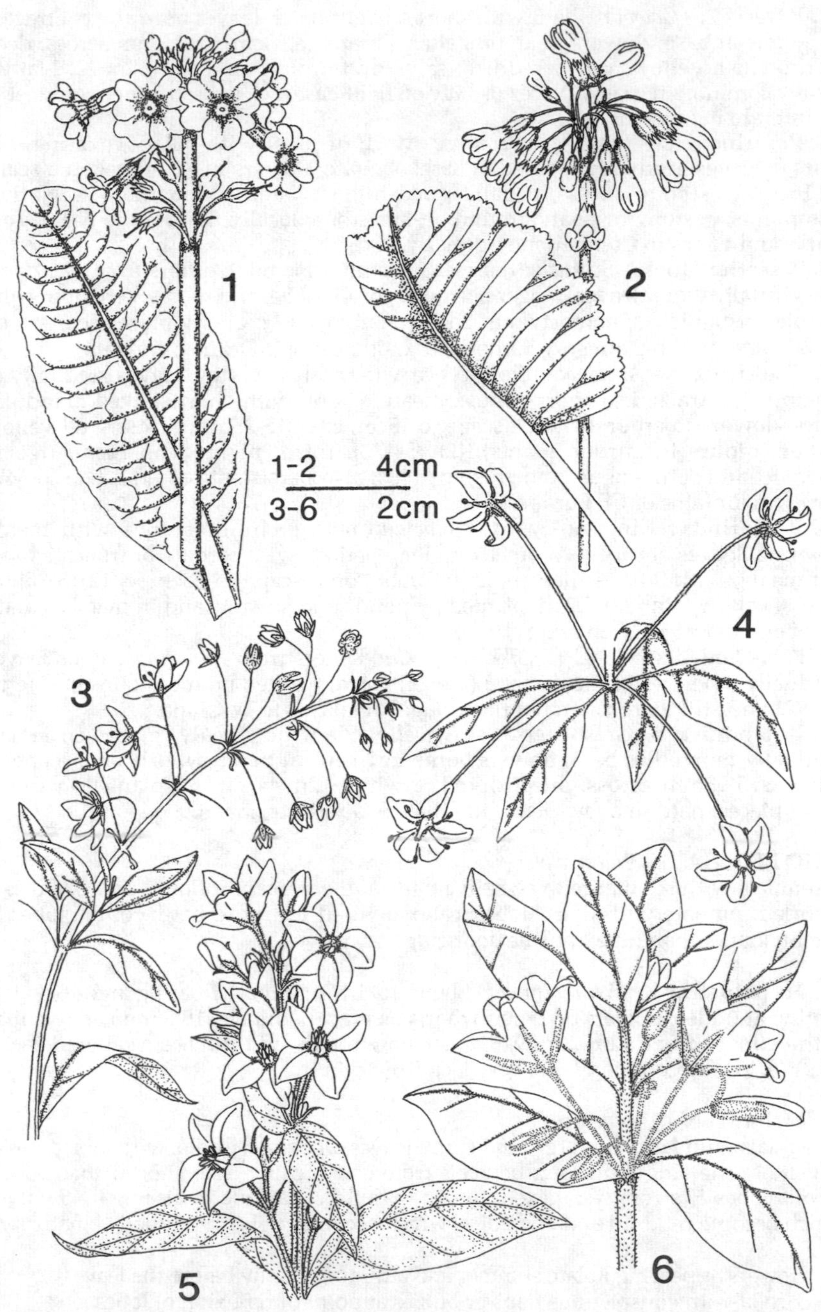

FIG 298 - Primulaceae. 1, *Primula japonica*. 2, *P. florindae*. 3, *Lysimachia terrestris*. 4, *L. ciliata*. 5, *L. vulgaris*. 6, *L. punctata*.

2 Corolla-lobes 7-15mm, with prominent violet blotch at base; leaves usually only undulate or toothed **2. C. coum**
2 Corolla-lobes 15-30mm, without violet blotch at base; leaves usually angled or shallowly lobed **3. C. repandum**

Other spp. - Several spp. are grown in 'wild' gardens and may locally persist after neglect. **C. graecum** Link, from Greece, is an autumn-flowering sp. differing from *C. hederifolium* in the tuber rooting only from lowerside and the leaves not or scarcely angled.

1. C. hederifolium Aiton (*C. neapolitanum* Ten.) - *Sowbread*. Leaves 2.5-10cm, broadly ovate, cordate, ± angled; corolla-lobes pale pink or rarely white each with dark basal blotch, 12-25mm; (2n=24, 32, 34, 36). Intrd; natd in woods and hedgerows; very scattered in CI, Man and Br N to S Sc, known in E Kent since 1778; S Europe.

2. C. coum Mill. - *Eastern Sowbread*. Leaves 3-7cm, usually ± reniform, not angled; corolla-lobes bright pink each with dark basal blotch, 7-15mm; (2n=28, 30). Intrd; natd on roadside verges and in grassy places; scattered in S En; E Mediterranean.

3. C. repandum Sibth. & Sm. - *Spring Sowbread*. Leaves 4-13cm, broadly ovate, cordate, angled or lobed; corolla-lobes purplish- to pale pink, often darker at base but not with distinct blotch, 15-30mm; (2n=20, 28). Intrd; natd on roadside verges and in woods and grassy places; scattered in SW & SE En; C & E Mediterranean.

4. LYSIMACHIA L. (*Naumburgia* Moench) - *Loosestrifes*
Perennials; leaves opposite or whorled, simple; calyx divided nearly to base; corolla-lobes 5-7, patent to ± erect, longer than tube, yellow; capsule opening by 5 valves.

1 Stems procumbent or decumbent; flowers borne singly in axils of normal leaves 2
1 Stems erect; at least some main leaf-axils with >1 flower, or flowers borne in terminal or axillary racemes 3
2 Leaves obtuse to rounded at apex, dotted with usually black glands; calyx-lobes ovate **2. L. nummularia**
2 Leaves acute to subacute at apex, not glandular; calyx-lobes subulate to linear **1. L. nemorum**
3 Flowers all borne in axils of much reduced bracts in terminal raceme; bulbils up to 2cm x 2mm usually produced in leaf-axils late in season **6. L. terrestris**
3 At least lower flower clusters or racemes borne in axils of ± normal leaves; bulbils 0 4
4 Flowers <10mm across, borne in dense axillary racemes; petals linear **7. L. thyrsiflora**
4 Flowers >15mm across, borne in few-flowered axillary clusters or in terminal panicles 5
5 Pedicels >2cm; calyx and leaves glabrous **4. L. ciliata**
5 Pedicels ≤2cm; calyx glandular-pubescent; leaves pubescent 6
6 Corolla-lobes glandular-pubescent at margins; calyx-teeth uniformly green **5. L. punctata**
6 Corolla-lobes glabrous at margins; calyx-teeth with conspicuous orange margin **3. L. vulgaris**

1. L. nemorum L. - *Yellow Pimpernel*. Stems to 40cm, procumbent to decumbent; leaves 1-3cm, opposite, ovate to broadly so, glabrous; flowers solitary in leaf-axils; corolla 5-8mm, glabrous, with 5 lobes; 2n=16, 18. Native; woods and copses; throughout most of BI.

2. L. nummularia L. - *Creeping-Jenny*. Stems to 60cm, procumbent; leaves 1.5-3cm, opposite, broadly ovate-elliptic to orbicular, dotted with black glands, otherwide glabrous; flowers solitary in leaf-axils; corolla 8-18mm, dotted with black glands, with 5 lobes; 2n=43 (30, 32, 36, 43). Native; damp places, often in shade; throughout most of BI N to C Sc, but a natd garden escape in many localities, especially in N.

3. L. vulgaris L. - *Yellow Loosestrife*. Stems to 1.5m, erect; rhizomes and/or stolons **298** extensive, giving rise to scattered stems; leaves 3-10cm, lanceolate to narrowly elliptic, opposite or in whorls of 3-4, pubescent and with minute glands; flowers in terminal panicles composed of axillary flower-clusters; corolla 8-15mm, dotted with pale glands, with 5 lobes; 2n=84 (42, 84). Native; ditches, marshes and by lakes and rivers; scattered through most of BI except N Sc.

4. L. ciliata L. - *Fringed Loosestrife*. Stems to 1.2m, erect; rhizomes extensive; **298** leaves 4-12cm, lanceolate to ovate, opposite or in whorls of 3-4, ± glabrous; flowers in loose axillary groups; corolla 9-15mm, densely covered with sessile glands towards base, with 5 lobes; (2n=34, 72, 92, 98, 100, 108, 112). Intrd; natd in rough ground and damp or shady places; rare garden escape scattered in Br mainly in N En and Sc; N America. Confused with *L. punctata*.

5. L. punctata L. - *Dotted Loosestrife*. Stems to 1.2m, erect; rhizomes short, **298** forming dense patches; leaves 4-12cm, lanceolate to ovate, opposite or in whorls of 3-4, pubescent and gland-dotted; flowers in dense axillary groups; corolla 10-16mm, with dense short-stalked glands especially at margin, with 5 lobes; (2n=30). Intrd; natd in rough ground and damp places; common and increasing garden escape scattered over most of BI; SE Europe. Rarely or never produces seed.

6. L. terrestris (L.) Britton, Sterns & Poggenb. - *Lake Loosestrife*. Stems to 80cm, **298** erect; leaves 2-8cm, narrowly elliptic to lanceolate, opposite, dotted with black glands, otherwise glabrous; flowers (often not produced) in axils of much reduced bracts in terminal raceme; corolla 4-7mm, with black glandular streaks, with 5 lobes; (2n=84). Intrd; damp places on shore of Lake Windermere, Westmorland since 1885; N America.

7. L. thyrsiflora L. (*Naumburgia thyrsiflora* (L.) Rchb.) - *Tufted Loosestrife*. Stems to **R** 70cm, erect; leaves 7-10cm, linear-lanceolate to narrowly elliptic, opposite, dotted with black glands but otherwise glabrous; flowers in dense, stalked, axillary

FIG 300 - Apex of corolla-lobe and single glandular hair of ***Anagallis arvensis***. 1, ssp. ***arvensis***. 2, ssp. ***foemina***. Drawings by C.A. Stace.

racemes with much reduced bracts; corolla 4-6mm, with black glandular dots, with 5-7 lobes; (2n=42). Native; wet places in marshes and by ditches and canals; scattered in N En and C & S Sc, E Donegal, rarely natd in C & S En.

5. TRIENTALIS L. - *Chickweed-wintergreens*
Glabrous perennials; leaves mostly in an apparent whorl at apex of stem, simple; calyx divided almost to base; corolla-lobes (5)6-7(9), erecto-patent, longer than tube, white; capsule opening by 5 valves.

Other spp. - T. borealis Raf., from N America, is sold in nurseries and has been found to persist in and near them in W Kent and S Hants; easily confused with *T. europaea* but has the larger leaves up to 8(10)cm, ovate to elliptic and tapering acute to acuminate, with only scale-leaves up stem.

1. T. europaea L. - *Chickweed-wintergreen.* Stems to 20(25)cm, simple, erect; leaves (3)5-8(10), 1-5(8)cm, the larger obovate, acute to subobtuse, a few much reduced but not scale-like up the stem; flowers 1-2, each on erect pedicel 2-7cm; corolla 6-10mm; 2n=c.160. Native; on humus in open pine-woods and heather-moors; Lancs and Yorks to N Sc, E Suffolk, locally common in Sc.

6. ANAGALLIS L. (*Centunculus* L.) - *Pimpernels*
Glabrous annuals or perennials; leaves opposite or alternate, simple; calyx divided ± to base; corolla-lobes 5, erect to patent, scarcely longer to much longer than tube; capsule opening by transverse line of dehiscence.

1 Upper leaves alternate; corolla divided <3/4 way to base, <2mm, much shorter than calyx-lobes **3. A. minima**
1 Leaves all opposite; corolla divided almost to base, the lobes >3mm, c. as long as or longer than calyx-lobes 2
  2 Stems procumbent, rooting at nodes; corolla-lobes >2x as long as calyx-lobes; leaves suborbicular **1. A. tenella**
  2 Stems decumbent to ascending, not rooting at nodes; corolla-lobes <2x as long as calyx-lobes; leaves ovate **2. A. arvensis**

1. A. tenella (L.) L. - *Bog Pimpernel.* Perennial; stems to 20cm, procumbent; leaves <1cm, opposite; flowers solitary in leaf-axils on erect pedicels to 3.5cm; corolla 6-10mm, pale pink, the lobes erecto-patent, entire, glabrous; 2n=22. Native; bogs and damp peaty ground; scattered over BI, common in parts of W, absent from much of E.

2. A. arvensis L. - see sspp. for English names. Usually annual; stems to 40cm, not rooting at nodes; leaves often >1cm, opposite; flowers solitary in leaf-axils on pedicels to 3.5cm; corolla 4-7(10)mm, the lobes patent, denticulate or with glands at margin. Native; arable and waste land and open ground.

a. Ssp. arvensis - *Scarlet Pimpernel.* Corolla-lobes usually red, sometimes **300**
variously pink or white or blue, entire to crenulate, with numerous minute hairs with 3 cells (incl. basal one), the most distal globose and glandular; 2n=40. Most of BI, common in S, rare in N Sc.

b. Ssp. foemina (Mill.) Schinz & Thell. (*A. foemina* Mill., *A. arvensis* ssp. *caerulea* **R**
Hartm. nom. illeg.) - *Blue Pimpernel.* Corolla-lobes blue, crenulate to denticulate, **300**
with sparse minute hairs with 4 cells (incl. basal one), the most distal ellipsoid and glandular; (2n=40). Much rarer than ssp. *arvensis*; usually in arable land and mostly in C & S En. Sterile red-flowered hybrids (*A. x doerfleri* Ronniger) occur between the 2 sspp. but are very rare.

3. A. minima (L.) E.H.L. Krause (*Centunculus minimus* L.) - *Chaffweed.* Annual; stems erect to decumbent, to 5(8)cm; leaves 3-5mm, the upper ones alternate; flowers solitary in leaf-axils, subsessile; corolla <2mm, white to pink, the lobes ±

erect, entire, glabrous; (2n=22). Native; bare damp sandy ground on heaths and in woodland rides; scattered over most of BI, mainly coastal in W.

7. GLAUX L. - *Sea-milkwort*
Glabrous slightly succulent rhizomatous perennials; leaves opposite, simple; calyx divided c.1/2 way to base into 5 lobes; corolla 0; capsule opening by 5 valves.

1. G. maritima L. - *Sea-milkwort.* Stems procumbent to suberect, to 30cm; leaves 4-12mm, often reduced below; flowers solitary in leaf-axils, sessile; calyx 3-5mm, white to pink; 2n=30. Native; saline sandy, muddy, rocky or grassy places; round coasts of BI and in a few inland salt-marshes.

8. SAMOLUS L. - *Brookweed*
Glabrous perennials; leaves in basal rosette and alternate up stem, simple; calyx-lobes ± free but fused to ovary; corolla-lobes 5, erect, longer than tube, white; capsule opening by 5 teeth.

1. S. valerandi L. - *Brookweed.* Stems to 45cm, erect; leaves 1-8cm, obovate; flowers 2-4mm, in terminal and axillary racemes, each with 1 small bracteole on pedicel; 2n=26. Native; wet places, especially by streams and flushes near the sea; coasts of BI, except NE Sc, frequent in E En but otherwise rare inland.

72. PITTOSPORACEAE - *Pittosporum family*

Shrubs; leaves evergreen, alternate, simple, entire, without stipules, petiolate. Flowers solitary in leaf-axils or few in terminal clusters, bisexual or monoecious, hypogynous, actinomorphic; sepals 5, fused at base; petals 5, free, purplish; stamens 5; ovary 1-celled with numerous ovules on 2-4 parietal placentas; style 1; stigma capitate; fruit a 2-4-celled capsule.

Easily recognized by the evergreen entire leaves, purple petals and 1-celled, 2-4-carpellary ovary.

1. PITTOSPORUM Banks ex Sol.- *Pittosporums*

1. P. crassifolium Banks & Sol. ex A. Cunn. - *Karo.* Dense shrub or tree to 5(8)m; **304**
leaves 5-8cm, obovate-oblong, upperside dark green, lowerside white-tomentose, with revolute margin; flowers male and female mixed in terminal clusters, 10-15mm; capsule with 3(-4) valves; (2n=24). Intrd; planted as screen or windbreak by sea; persisting and sometimes self-sown in Scillies, rarely in W Cornwall and Jersey; New Zealand. Resembles *Olearia traversii* (Asteraceae) when sterile but leaves are alternate and 1st-year twigs ± terete.

2. P. tenuifolium Gaertn. - *Kohuhu.* Shrub or tree to 5(10)m; leaves 1-7cm, **304**
elliptic-oblong, mid-green on both surfaces, ± glabrous when mature, with undulate margin; flowers bisexual, solitary, axillary, 10-15mm; capsule with 2 valves; (2n=24). Intrd; planted for ornament or screen by sea; self-sown in W Cornwall; New Zealand. Resembles *Olearia paniculata* (Asteraceae) when sterile but latter has thicker leaves with white lowerside.

73. HYDRANGEACEAE - *Mock-orange family*

Shrubs; leaves deciduous, opposite, simple, without stipules, petiolate. Flowers in raceme-like or corymbose cymes terminal on lateral or main shoots, bisexual, 1/2-epigynous to epigynous, actinomorphic; sepals 4-5, free; petals 4-5, free; stamens 9-numerous; ovary 2-4-celled, with numerous ovules on axile placenta; styles 2-4, united below or free; stigmas clavate; fruit a 2-4-valved or -toothed capsule.

The only opposite-leaved shrubs with free petals, 9-numerous stamens and 1/2-inferior or inferior ovary.

1 Flowers in corymbs, at least the outer ones with much-enlarged sepals and sterile **3. HYDRANGEA**
1 Flowers in racemose or paniculate cymes, all similar and fertile 2
 2 Petals and sepals 4; styles 4, united >1/2 way; stamens ≥20 **1. PHILADELPHUS**
 2 Petals and sepals 5; styles usually 3, ± free; stamens 10 **2. DEUTZIA**

1. PHILADELPHUS L. - *Mock-oranges*
Flowers in raceme-like cymes terminal on lateral shoots, all similar and fertile; sepals 4; petals 4 (or *flore pleno*), white or creamy-white; stamens c.20-30; ovary 4-celled with 4 styles united ≥1/2 way to apex; fruit a 4-valved capsule.

Other spp. - Several other spp. from N America and China and hybrids often of complex parentage are grown in gardens, and some wild plants might be referable to them. Small-flowered, very fragrant, glabrous plants with almost entire leaves are probably hybrids involving **P. microphyllus** A. Gray.

1. P. coronarius L. - *Mock-orange.* Shrub to 3m; leaves 5-10cm, ovate to elliptic-oblong, serrulate, glabrous to sparsely pubescent on lowerside; flowers 2.5-5cm across, fragrant, glabrous, with ± patent petals; (2n=26). Intrd; commonly grown and frequently found as a relic in hedges and copses but rarely self-sown; scattered in En, Man; Europe.
2. P. x virginalis Rehder (?*P. coronarius* x *P. microphyllus* x *P. pubescens* Loisel.). - **304**
Hairy Mock-orange. Differs from *P. coronarius* in leaves pubescent on lowerside; calyx pubescent; flowers cup-shaped, often *flore pleno*; (2n=26). Intrd; now commoner than *P. coronarius* in gardens and most recent relics in the wild are probably this hybrid; scattered in Br; garden origin.

2. DEUTZIA Thunb. - *Deutzia*
Flowers in raceme-like cymes terminal on lateral shoots, all similar and fertile; sepals 5; petals 5 (or *flore pleno*), white or tinged with pink; stamens 10; ovary 3-celled with 3 free styles; fruit a 3-valved capsule.

Other spp. - Many garden plants belong to related spp. or complex hybrids and the natd ones may well include some of these.

1. D. scabra Thunb. - *Deutzia.* Shrub to 3m; leaves 4-8cm, ovate, scabrid with minute pubescence; flowers 1-2cm across, not fragrant, with erect to erecto-patent petals; (2n=130). Intrd; grown and natd as for *Philadelphus*, but less commonly; scattered in Br N to C Sc; China and Japan.

3. HYDRANGEA L. - *Hydrangeas*
Flowers in terminal corymbs, at least the outer ones with much-enlarged sepals and sterile; sepals 4-5; petals 4-5, usually varying shades of red to blue or white; stamens 9-20; ovary 2-4-celled with 2-4 short but free styles; fruit a capsule with 2-4 teeth.

Other spp. - Several other spp. are grown in gardens and some persist or spread on to walls. **H. petiolaris** Siebold & Zucc. and **H. anomala** D. Don both climb by aerial rootlets like *Hedera*; *H. petiolaris*, from Japan, has 15-20 stamens, and *H. anomala*, from China, has 9-15 stamens and more deeply toothed leaves. **H. sargentiana** Rehder, from China, is an erect shrub to 3m with stems densely covered in hairs and bristles and with pubescent leaves; it was formerly ± natd in S

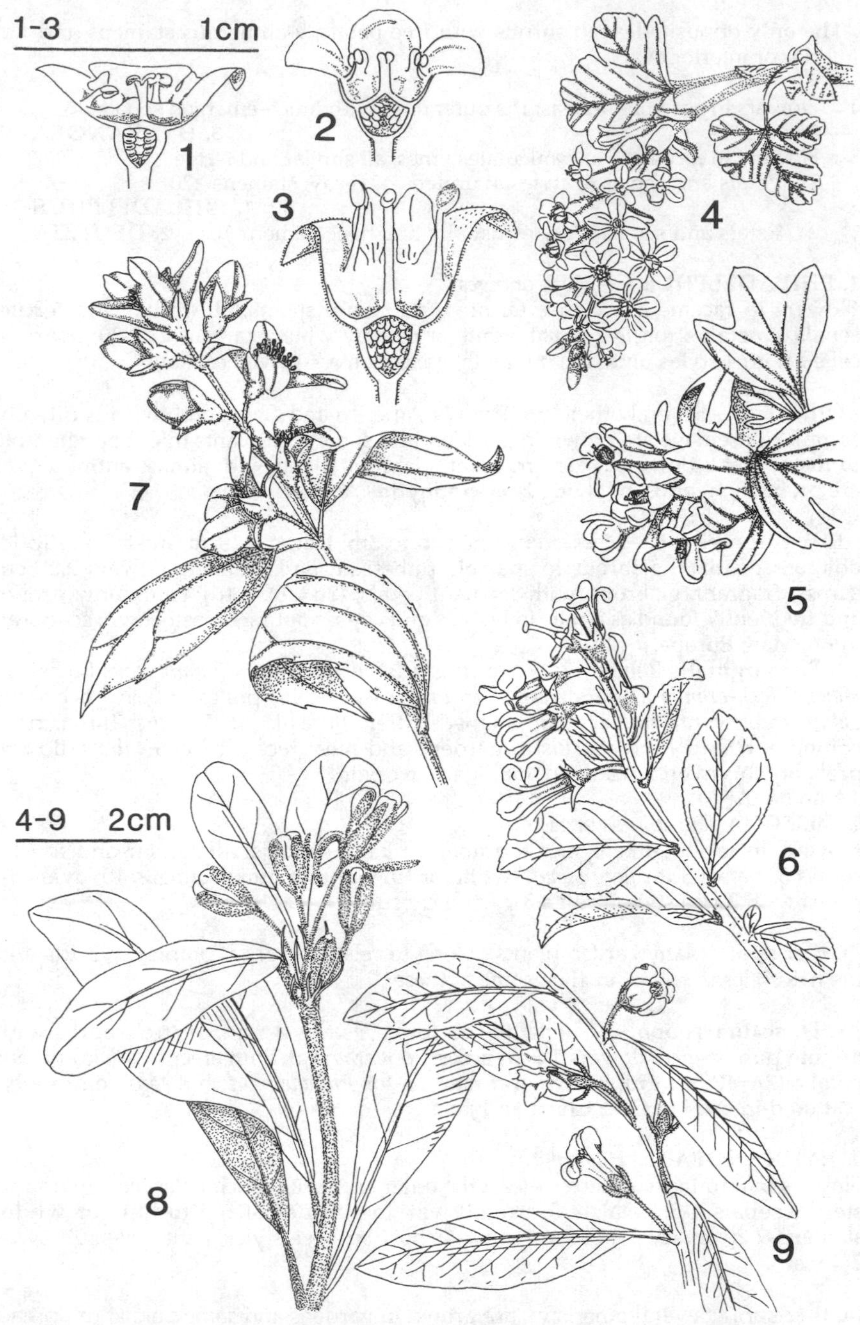

FIG 304 - Grossulariaceae, Hydrangeaceae, Pittosporaceae. 1-3, flowers of *Ribes*. 1, *R. rubrum*. 2, *R. spicatum*. 3, *R. nigrum*. 4, *R. sanguineum*. 5, *R. odoratum*. 6, *Escallonia macrantha*. 7, *Philadelphus x virginalis*. 8, *Pittosporum crassifolium*. 9, *P. tenuifolium*.

Kerry.

1. H. macrophylla (Thunb.) Ser. - *Hydrangea*. Soft-wooded shrub to 1m; leaves 10-20cm, ovate to obovate, ± glabrous; fertile flowers <1cm across; stamens 10, longer than petals; (2n=36). Intrd; much grown in gardens, persistent where thrown out or neglected; scattered in S & W Br, CI; Japan. Often killed to ground level by frost.

74. GROSSULARIACEAE - *Gooseberry family*
(Escalloniaceae)

Shrubs; leaves alternate, simple, often lobed, without stipules, petiolate. Flowers solitary or in racemes or panicles, bisexual to dioecious, epigynous, with hypanthium, actinomorphic; sepals 5, arising from hypanthium; petals 5, free; stamens 5; ovary 1-locular, with numerous ovules on 2 parietal placentas; styles 2 with capitate stigmas or style 1 with bilobed or capitate stigma; fruit a berry or capsule.

Distinguishable by shrubby habit, 5 sepals, petals and stamens arising from hypanthium, and inferior 1-celled ovary with 2 parietal placentas.

1 Leaves evergreen, not lobed; fruit a capsule; petals longer than sepals **1. ESCALLONIA**

1 Leaves deciduous, palmately lobed; fruit a berry; petals shorter than sepals **2. RIBES**

1. ESCALLONIA Mutis ex L.f. - *Escallonia*

Leaves evergreen, simple, serrate; flowers in terminal racemes or panicles; petals much longer than calyx, with distinct claw; fruit a capsule.

Other spp. - Many spp., hybrids and cultivars are grown, occasionally producing seedlings; some wild plants might not be *E. macrantha*, but this needs checking. The most likely extra taxon is **E. x langleyensis** Veitch (*E. macrantha* x *E. virgata* (Ruiz & Pav.) Pers.), of garden origin, with petals with claws shorter than limbs and not forming a tube and usually pale pink.

1. E. macrantha Hook. & Arn. (*E. rubra* (Ruiz & Pav.) Pers. var. *macrantha* (Hook. **304**
& Arn.) Reiche) - *Escallonia*. Shrub to 3m (rarely more); young growth glandular-viscid; leaves 1-8cm, elliptic to obovate, gland-dotted, scented when crushed; flowers 12-20mm; petals pink to deep red, with claws longer than limbs and arranged into false corolla-tube; (2n=24). Intrd; planted for hedging and ornament near sea; common persistent relic in SW En, Wa (mainly S), Man, W Ir and CI, also Dunbarton, rarely self-sown; Chile.

2. RIBES L. - *Gooseberries*

Leaves deciduous, palmately lobed, variously toothed; flowers solitary or in racemes, on short lateral branches; petals shorter than sepals, not forming a tube; fruit a berry.

1 Spines present on branches; flowers solitary or in short racemes of 2(-3) **7. R. uva-crispa**

1 Spines 0; flowers in racemes of >4 2

2 Flowers bright pink to red, bright yellow, or (rarely) white; hypanthium tubular, longer than wide 3

2 Flowers green to yellowish-green, sometimes tinged purplish; hypanthium disk- to cup-shaped, wider than long 4

3 Flowers bright pink to red, rarely white; leaves pubescent, scented when crushed **4. R. sanguineum**
3 Flowers bright yellow; leaves glabrous, not scented **5. R. odoratum**
  4 Leaves with sessile orange glands on lowerside, scented when crushed; fruit black **3. R. nigrum**
  4 Leaves with mostly stalked reddish glands, not scented; fruit red or rarely whitish 5
5 Dioecious; bracts >4mm **6. R. alpinum**
5 Flowers bisexual; bracts <2mm 6
  6 Hypanthium cup-shaped; anther-lobes contiguous **2. R. spicatum**
  6 Hypanthium saucer-shaped; anther-lobes distinctly separated by connective **1. R. rubrum**

1. R. rubrum L. (*R. sylvestre* (Lam.) Mert. & W.D.J. Koch) - *Red Currant*. Shrub to 2m; leaves 3-10cm, glabrous to sparsely pubescent, not scented; flowers in pendent racemes, 4-6mm across, greenish-yellow, ± glabrous; berry 6-10mm across, red or rarely whitish; 2n=16. Probably intrd; woods, hedges and scrub, much grown and often obviously relict or escaped; throughout most of BI but rare in many places; Europe. **304**

2. R. spicatum E. Robson - *Downy Currant*. Differs from *R. rubrum* in key characters; usually more pubescent but other differences are not constant; (2n=16). Native; woods on limestone, mostly in uplands; very local from Lancs and Yorks to Caithness, rarely natd further S. **R** **304**

3. R. nigrum L. - *Black Currant*. Shrub to 2m; leaves as in *R. rubrum* but with more pointed main lobes, scented when crushed; flowers in pendent racemes, 6-10mm across, greenish-yellow or tinged with purple; hypanthium deeply cup-shaped, pubescent; berry mostly 10-15mm across, black, without bloom; 2n=16. Probably intrd; woods, hedges and shady streamsides, much grown and usually obviously relict or escaped; throughout most of BI; Europe. **304**

4. R. sanguineum Pursh - *Flowering Currant*. Shrub to 2.5m; leaves 3-10cm, pubescent, scented when crushed; flowers in pendent racemes, 6-10mm across, bright pink to red, rarely white; berry 6-10mm across, purplish-black with whitish bloom; (2n=16). Intrd; much grown in gardens and a frequent relic, sometimes self-sown, especially in W; scattered throughout BI; W N America. **304**

5. R. odoratum H.L. Wendl. (*R. aureum* auct. non Pursh) - *Buffalo Currant*. Shrub to 2.5m; leaves 2-5cm, glabrous, not scented; flowers in pendent racemes, 6-10mm across, bright yellow, fragrant, glabrous; berry 6-10mm across, dark red to purplish-black, without bloom; (2n=16). Intrd; commonly grown and sometimes found relict or self-sown in hedgerows, roadsides and scrub; scattered in S En, Man, Lanarks; C USA. **304**

6. R. alpinum L. - *Mountain Currant*. Shrub to 2m, sometimes pendent on rock-faces; leaves 2-5cm, sparsely pubescent, not scented; flowers in ± erect racemes, dioecious, the female fewer per raceme, 1.5-3mm across, greenish-yellow, glabrous to sparsely pubescent; berry 6-10mm across, red; (2n=16). Native; limestone woods, often on rocks or cliffs, also an escape in other shady places; native for certain only N Wa and N En, common only in Peak District, widespread as escape in Br except extreme N & S. **R**

7. R. uva-crispa L. - *Gooseberry*. Spiny shrub to 1(-1.5)m; leaves 2-5cm, usually pubescent, not scented; flowers 1-3 in stiff groups, 6-12mm across, greenish-yellow or red-tinged, pubescent; berry 10-20mm across, greenish-yellow, sometimes reddish, much larger in cultivars; 2n=16. Probably native; hedges, scrub and open woods, often obviously relict or escaped; over most of BI but not native in Ir.

75. CRASSULACEAE - *Stonecrop family*

Annual to perennial herbs or rarely woody; leaves often succulent, spiral or alternate, less often opposite, simple, sessile or petiolate, without stipules. Flowers usually in terminal cymes, less often in terminal racemes or solitary and axillary, bisexual or rarely dioecious, hypogynous, actinomorphic; sepals free to fused, 3-c.18 (mostly 5); petals free to fused, as many as sepals; stamens as many or 2x as many as petals; carpels as many as petals, free or slightly fused at base, with 2-many ovules, tapering to small stigma; fruit a group of follicles.

Easily recognized by the free (or ± free) carpels as many as sepals and petals, stamens as many or 2x as many as petals, and usually succulent leaves.

1 Petals fused to form tube for >1/2 their length; basal leaves peltate **2. UMBILICUS**
1 Petals free or fused only at base; leaves not peltate 2
  2 Stamens as many as petals; leaves opposite **1. CRASSULA**
  2 Stamens 2x as many as petals; leaves usually alternate or spiral 3
3 Flowers with 4-5 petals and sepals **5. SEDUM**
3 Flowers with 6 or more petals and sepals 4
  4 Leaves about as thick as wide **5. SEDUM**
  4 Leaves distinctly wider than thick, distinctly flat on upperside 5
5 Petals yellow **4. AEONIUM**
5 Petals dull pink to purplish **3. SEMPERVIVUM**

1. CRASSULA L. (*Tillaea* L.) - *Pigmyweeds*

Aquatic or terrestrial annuals to perennials, glabrous or nearly so; leaves opposite, often fused in pairs at base, succulent or ± so, entire; flowers <5mm, 3-5-merous; petals free or ± so, white to pink; stamens as many as petals.

1 Leaves strongly succulent, obovate; flowers in heads on common peduncle; petals 5, with apical dorsal appendage **5. C. pubescens**
1 Leaves weakly succulent, linear to ovate or elliptic; flowers borne singly in leaf-axils; petals 3-4, without apical appendage 2
  2 Flowers sessile or ± so (pedicels <1mm) 3
  2 Flowers on ≥2mm pedicels 4
3 Leaves 1-2mm; petals mostly 3, shorter than sepals **1. C. tillaea**
3 Leaves 3-5mm; petals 4, longer than sepals **2. C. aquatica**
  4 Terrestrial; petals shorter than to c. as long as sepals; stems to 12cm **4. C. decumbens**
  4 Aquatic or on mud; petals longer than sepals; stems often >12cm **3. C. helmsii**

1. C. tillaea Lest.-Garl. (*Tillaea muscosa* L.) - *Mossy Stonecrop*. Annual; stems R
procumbent to ascending, to 5cm; leaves closely set on stem, usually red, 1-2mm, ovate or elliptic; flowers 1-2mm, solitary in leaf-axils, sessile or ± so, 3(-4)-merous; (2n=16, 32, 64). Native; sandy or gravelly ground in open places; S En, E Anglia, Notts, CI.

2. C. aquatica (L.) Schönland (*Tillaea aquatica* L.) - *Pigmyweed*. Annual; stems RRR
procumbent to decumbent, lax, to 5cm; leaves ± distant on stem, 3-5mm, linear; 311
flowers 1-2mm, solitary in leaf-axils, sessile or ± so, 4-merous; (2n=42). Probably native; muddy pool-margin, MW Yorks, found 1921, gone by 1945; on mud by water, Westerness, found 1969.

3. C. helmsii (Kirk) Cockayne (*C. recurva* (Hook. f.) Ostenf. non N.E. Br., *Tillaea* 311
recurva (Hook. f.) Hook. f.) - *New Zealand Pigmyweed*. Perennial; stems trailing in water or ascending from it or decumbent in mud, lax, to 30cm; leaves ± distant on stem, 4-15(20)mm, linear to lanceolate; flowers 1-2mm, solitary in leaf-axils, on

pedicels 2-8mm, 4-merous. Intrd; grown by aquarists and discarded or planted in ponds; well natd in many places in S En and CI, scattered N to C Sc, Co Down, rapidly spreading; Australia and New Zealand.

4. C. decumbens Thunb. (*C. macrantha* (Hook. f.) Diels & Pritz.) - *Scilly Pigmyweed*. Annual; stems decumbent to ascending, to 12cm; leaves ± distant on stem, 4-7mm, linear-lanceolate; flowers 2-4mm, solitary in leaf-axils, on pedicels 4-10(15)mm, 4-merous; (2n=64). Intrd; natd weed in damp sandy bulbfields and tracksides in Scillies (first found 1959), occasional wool-alien in S En; S Africa, Australasia. **311**

5. C. pubescens Thunb. (*C. radicans* (Haw.) D. Dietr.) - *Jersey Pigmyweed*. Perennial; stems erect to procumbent and stoloniferous, succulent, to 20cm; leaves crowded in 1.5-2.5cm across rosettes near stem tips, 3-10mm, obovate, very succulent; flowers 3-5mm, 6-10 subsessile in crowded heads on 3-4cm axillary peduncles, 5-merous; (2n=28). Intrd; natd on sandy ground in 1 place in Jersey, found 1970; S Africa. Our plant is probably ssp. **radicans** (Haw.) Toelken, but its identity needs checking. **311**

2. UMBILICUS DC. - *Navelwort*

Glabrous perennials; leaves alternate on stem and in basal rosette, succulent, the lower and basal ones peltate, crenate; flowers >5mm, 5-merous; petals fused >1/2 way from base; stamens 2x as many as petals, fused to corolla-tube.

1. U. rupestris (Salisb.) Dandy - *Navelwort*. Stem usually erect, to 30(50)cm; basal leaves orbicular, 1-7cm across, with long petiole; flowers in simple raceme occupying most of stem, 7-10mm, greenish-white to pinkish-brown; 2n=48. Native; rocks, walls and stony hedgebanks; frequent in Ir, CI and W Br N to C Sc, rare or absent in E & C Br.

3. SEMPERVIVUM L. - *House-leeks*

Glandular-pubescent perennials; leaves narrow, in dense basal rosette and alternate up stem, succulent, entire; flowering stems erect, arising from centre of mature rosette which then dies, with cymes of flowers at apex; flowers >5mm, 8-18(mostly 13)-merous; petals ± free, dull pink to purplish, narrow; stamens 2x as many as petals.

1. S. tectorum L. - *House-leek*. Rosettes mostly >3cm across; stems to 50cm; leaves 2-4cm, glabrous, often reddish; flowers 15-30mm across; petals 8-12mm; 2n=72. Intrd; grown on wall-tops and roofs, rarely sand-dunes, very persistent but hardly natd; scattered over Br; mountains of C & S Europe.

2. S. arachnoideum L. - *Cobweb House-leek*. Rosettes mostly <2cm across; stems to 12cm; leaves 0.7-1.2cm, pubescent and with apical tuft of long web-like hairs matted over rosette, often reddish; flowers 12-20mm across; petals 7-10mm; (2n=32, 64). Intrd; on barn roof and walls; 1 site in W Norfolk since 1993; mountains of SW Europe.

4. AEONIUM Webb & Berthel. - *Aeonium*

Almost glabrous perennials, sometimes woody below; leaves in large dense rosette and alternate up stem, succulent, ± entire; flowers >5mm, 8-11-merous; petals ± free, yellow; stamens 2x as many as petals.

Other spp. - A few other spp. are grown in SW En and are very noticeable, but not truly natd. **A. arboreum** (L.) Webb & Berthel., from Morocco, differs from *A. cuneatum* in having the smaller leaf-rosettes (10-18cm across) borne at the apex of thick branching woody stems.

1. A. cuneatum Webb & Berthel. - *Aeonium*. Plant mostly herbaceous, woody very near base, leaf-rosettes near ground, saucer-shaped, to 50cm across; leaves oblong-oblanceolate, bright green; flowering stems to 80(120)cm, ± erect; flowers 1-2cm across, in dense mass of cymes at stem-apex; (2n=36). Intrd; grown on walls in SW En and conspicuous and ± natd in Scillies; Canary Isles.

5. SEDUM L. (*Rhodiola* L.) - *Stonecrops*

Annuals or (usually) perennials; leaves alternate, sometimes crowded and ± in a rosette, succulent, entire or toothed; flowers >3mm, 4-9-merous; petals free, various in colour; stamens 2x (or c.2x) as many as petals.

1 Leaves ± flat and distinctly dorsiventral 2

1 Leaves ± terete, or rounded on lowerside and flattened on upperside 11

2 Rhizome thick, succulent, scaly; flowers dioecious, usually 4-merous **1. S. rosea**

2 Rhizome 0, or not scaly and non-succulent; flowers bisexual, 5(-6)-merous 3

3 Petals yellow 4

3 Petals pink to purplish-red, rarely white 7

4 Leaves conspicuously white-bloomed, forming flat dense rosette at stem apex **8. S. spathulifolium**

4 Leaves not white-bloomed, not forming flat dense rosette at stem apex 5

5 Stems thin, annual **7. S. kamtschaticum**

5 Stems stout, woody at least at base 6

6 Leaves mostly >4.5cm, 3-4x as long as wide; petals c.4x as long as wide **2. S. praealtum**

6 Leaves mostly <4.5cm, c.2x as long as wide; petals c.3x as long as wide **3. S. confusum**

7 Stamens distinctly longer than petals and sepals **4. S. spectabile**

7 Stamens shorter than to c. as long as petals 8

8 Petals 3-5mm; stems rooting only near base 9

8 Petals 5-12mm; stems rooting along length 10

9 Leaves entire; non-flowering shoots procumbent; roots not tuberous **6. S. anacampseros**

9 Leaves toothed, non-flowering shoots ± erect; roots tuberous **5. S. telephium**

10 Petals mostly >8mm; leaves narrowed to base but scarcely petiolate; flowers mostly pedicellate **9. S. spurium**

10 Petals mostly <8mm; leaves with distinct petiole; flowers ± sessile **10. S. stoloniferum**

11 Petals yellow 12

11 Petals white to pink or red 16

12 Most leaves >7mm, acute or apiculate; ripe follicles erect; flowers 5-9-merous 13

12 Leaves <7mm, obtuse; ripe follicles ± patent; flowers 5-merous 15

13 Sterile shoots with terminal tassel-like cluster of living leaves and persistent dead ones below; leaves flattened, abruptly apiculate; filaments and follicles smooth; sepals subacute to obtuse **13. S. forsterianum**

13 Sterile shoots with long terminal region of living leaves; dead leaves not persistent; leaves subterete, acute to acuminate; base of filaments and inner side of follicles minutely papillose; sepals acute 14

14 Inflorescence erect in bud; leaves mostly >3mm wide **11. S. nicaeense**

14 Inflorescence pendent in bud; leaves mostly <3mm wide **12. S. rupestre**

15 Leaves ovoid, broadest near base, with acrid taste when fresh **14. S. acre**

15 Leaves ± cylindrical, ± parallel-sided, not acrid **15. S. sexangulare**

16 Leaves, pedicels and sepals with small glandular hairs 17

16 Plant glabrous 19
17 Leaves mostly opposite **19. S. dasyphyllum**
17 Leaves mostly alternate 18
18 Petals 5, pink; ripe follicles erect **20. S. villosum**
18 Petals mostly 6-7, white, with pink midrib; ripe follicles patent **21. S. hispanicum**
19 Inflorescence with 2(-3) main branches each with 3-6 flowers **18. S. anglicum**
19 Inflorescence dense, corymbose, with several branches 20
20 Inflorescence <20-flowered **17. S. lydium**
20 Inflorescence >20-flowered **16. S. album**

Other spp. - Several other spp. are grown in gardens and may be persistent relics or throwouts. **S. hybridum** L. (sect. *Aizoon*), from E Asia, resembles *S. kamtschaticum* but has procumbent stems and over-wintering sterile shoots. **S. cepaea** L. and **S. stellatum** L. (sect. *Epeteium*) are annuals from S Europe; *S. cepaea* is white-flowered with erect stems to 25cm ± entirely occupied by the pyramidal diffuse inflorescence and was formerly natd in Bucks; *S. stellatum* is pink-flowered with a flat-topped inflorescence and has recently been reported from N Somerset. **S. mexicanum** Britton, from Mexico, is a yellow-flowered sp. of sect. *Sedum* with linear lower leaves 6-15mm and in whorls of 3-5; it was formerly a garden escape in Midlothian.

*Section 1 - **RHODIOLA*** (L.) Scop. (*Rhodiola* L.) (sp. 1). Dioecious; petals 4(-5), yellow; rhizome tuberous, ± succulent; leaves dorsiventral.

1. S. rosea (L.) Scop. (*Rhodiola rosea* L.) - *Roseroot*. Glabrous rhizomatous perennial; stems erect, to 35cm; leaves alternate, 1-4cm, dorsiventral, serrate; flowers in dense subcorymbose panicle; petals 4(-5), 2-4mm, greenish-yellow; 2n=22. Native; mountain rocks and sea cliffs; Wa, Ir, Sc, N En, rare garden escape elsewhere.

*Section 2 - **DENDROSEDUM*** Berger (spp. 2-3). Bisexual; petals 5, yellow; evergreen dwarf shrubs with ± succulent stems; leaves dorsiventral.

2. S. praealtum A. DC. (*S. dendroideum* auct. non Sessé & Moç. ex DC.) - *Greater Mexican-stonecrop*. Bushy, glabrous, evergreen shrub to 75cm; leaves near ends of branches, alternate, 5-7cm, dorsiventral but very thick, entire; flowers in large panicle; petals 5, 6-9mm, bright yellow; (2n=68). Intrd; natd on cliffs in Jersey since 1920, occasional relic on banks in Guernsey and S Devon; Mexico.

3. S. confusum Hemsl. - *Lesser Mexican-stonecrop*. Differs from *S. praealtum* in smaller size (to 40cm); shorter leaves (see key) with rounded (not obtuse) apex; smaller denser inflorescence (c.5cm, not c.10cm); and shorter petals (see key); (2n=68). Intrd; persistent spreading relic of cultivation on banks in Guernsey and W Cornwall; Mexico.

*Section 3 - **TELEPHIUM*** Gray (spp. 4-6). Bisexual; petals (4-)5, pink; rhizome tuberous, ± succulent; stems slightly succulent; leaves dorsiventral.

4. S. spectabile Boreau - *Butterfly Stonecrop*. Glabrous perennial; stems erect, to **311**
50cm; leaves opposite or in whorls of 3, 4-10cm, dorsiventral, crenate-serrate; flowers in large corymbose panicle; petals 5, 5-8.5mm, pale pink to purplish-red; (2n=50, 51). Intrd; common in gardens and a persistent relic or throwout; S & C En, natd in woodland in N Wilts since 1930; China and Japan. Also grown in gardens is **S. 'Herbstfreude'** ('Autumn Joy'), said to be *S. spectabile* x *S. telephium*; it has narrower, more deeply serrate leaves and might be overlooked as an escape.

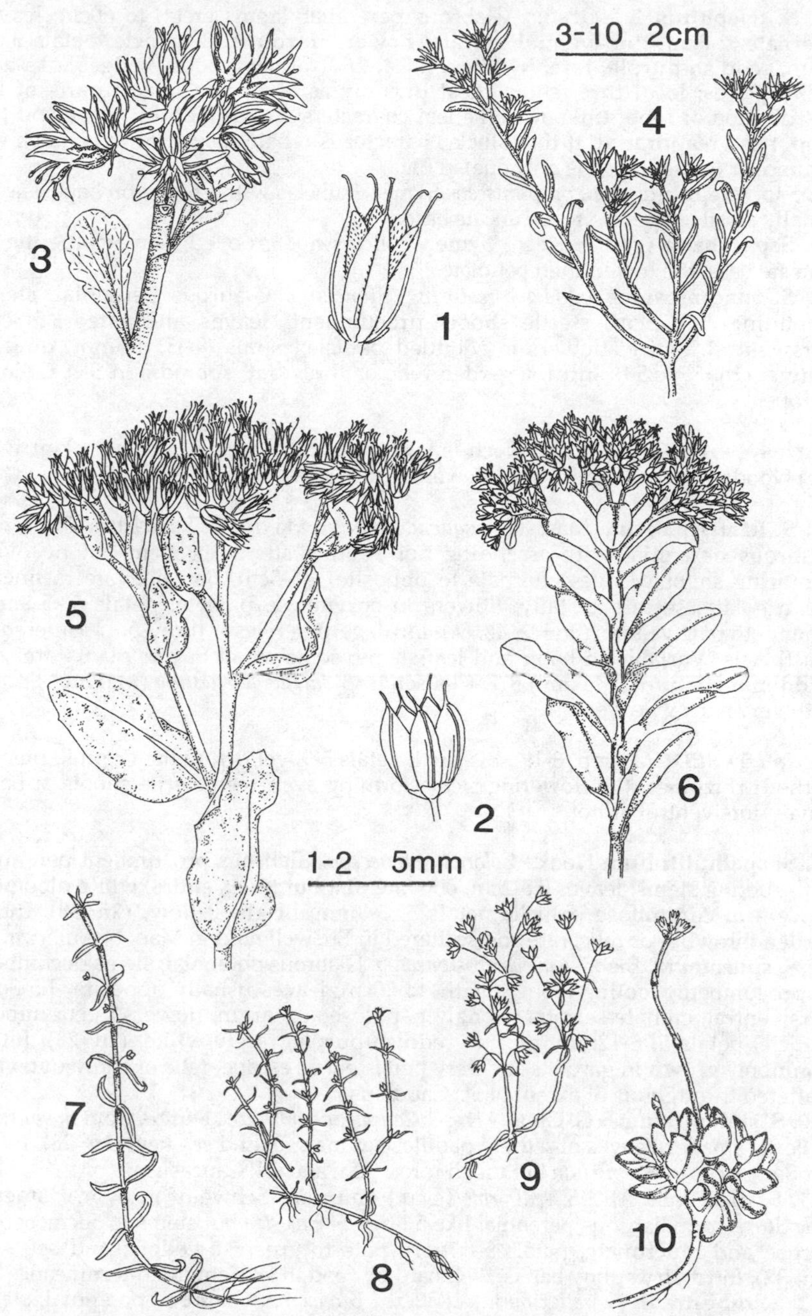

FIG 311 - Crassulaceae. 1, fruits of *Sedum rupestre*. 2, fruits of *S. forsterianum*. 3, ***Sedum spurium***. 4, ***S. hispanicum***. 5, *S. spectabile*. 6, *S. anacampseros*. 7. ***Crassula helmsii***. 8, *C. aquatica*. 9, *C. decumbens*. 10, *C. pubescens*.

5. S. telephium L. - *Orpine*. Glabrous perennial; stems erect, to 60cm; leaves alternate, 2-8cm, dorsiventral, serrate; flowers in rounded panicle; petals 5, 3-5mm, reddish-purple, rarely white; 2n=24, 26. Native; woods, hedgebanks and rocky places; local throughout most Br, only as escape in Ir and parts of Br. Distribution of sspp. unknown; the leaf characters are of doubtful value and the sspp. need reappraisal. If the follicle character is reliable ssp. *fabaria* is much the commoner (or perhaps the only one) in BI.

a. Ssp. telephium (ssp. *purpurascens* Syme). Follicles with groove on back; leaves usually sessile, tapering to ± truncate base.

b. Ssp. fabaria (W.D.J. Koch) Syme. Follicles not grooved; leaves tapering to cuneate base, the lower often petiolate.

6. S. anacampseros L. - *Love-restoring Stonecrop*. Glabrous perennial; stems **311**
ascending, to 25cm; sterile shoots procumbent; leaves alternate, 1.5-3cm, dorsiventral, entire; flowers in rounded panicle; petals (4-)5, 4-5mm, pink to mauve; (2n=36, c.50). Intrd; a garden relic or throwout; sporadic in S & C En; S Europe.

Section 4 - AIZOON W.D.J. Koch (sp. 7). Bisexual; petals 5, yellow; plant with thin woody rhizomes and annual flowering stems; leaves dorsiventral.

7. S. kamtschaticum Mast. (*S. middendorfianum* Maxim.) - *Kamchatka Stonecrop*. Glabrous perennial with ascending non-rooting stems to 30cm but no over-wintering shoots; leaves alternate to opposite, 1.5-5cm, flat, obovate to linear-oblanceolate, serrate distally; flowers in corymbose panicle; petals 5, 5-8mm, orange- to pale yellow; (2n=32, 48, 64). Intrd; garden relic or throwout; scattered in En; E Asia. Variable in habit and leaf-shape; some or all of our plants are var. **middendorfianum** (Maxim.) R.T. Clausen (ssp. *middendorfianum* (Maxim.) Fröd.), with very narrow leaves.

Section 5 - SEDUM (spp. 8-19). Bisexual; petals 5-9, yellow, white or pink; plant ± without rhizomes, the flowering stems forming evergreen sterile shoots at base; leaves dorsiventral or not.

8. S. spathulifolium Hook.- *Colorado Stonecrop*. Glabrous procumbent perennial with rooting stems; leaves 0.8-2cm, obovate, flat but thick, entire, white-bloomed; flowers in corymbose panicle; petals 5, 5-8mm, bright yellow; (2n=30). Intrd; garden throwout or relic; rare and scattered in En, well natd in Man; N America.

9. S. spurium M. Bieb. - *Caucasian-stonecrop*. Glabrous perennial; stems decumbent **311**
or procumbent, rooting along length, to 20cm; leaves usually opposite, 1.5-3cm, dorsiventral, crenate-serrate, strongly papillose at margin; flowers in corymbose panicle; petals 5, 8-12mm, pink to reddish-purple, rarely white; (2n=28). Intrd; commonly grown in gardens and very persistent as escape, relic or throwout; very scattered throughout BI except N Sc; Caucasus.

10. S. stoloniferum S.G. Gmel. - *Lesser Caucasian-stonecrop*. Differs from *S. spurium* in being smaller; leaves obscurely papillose at margin; and see key; (2n=28). Intrd; similar places to *S. spurium* but much rarer; S En and CI; Caucasus.

11. S. nicaeense All. (*S. sediforme* (Jacq.) Pau non (Schweinf.) Raym.-Hamet) - *Pale Stonecrop*. Glabrous perennial like a large *S. rupestre* but stems to 50cm; sepals obtuse and mucronate; petals 5-8, very pale to greenish-yellow; and see key; (2n=32). Intrd; dry sunny banks; well natd by road in W Kent; Mediterranean.

12. S. rupestre L. (*S. reflexum* L.) - *Reflexed Stonecrop*. Glabrous perennial; stems **311**
erect to ascending, to 35cm; sterile shoots decumbent, rooting; leaves spiral, 12-20mm, ± terete, linear; flowers in corymbose to rounded panicle; petals (5)6-7(9), 6-7mm, bright yellow; (2n=34, c.56, 68, c.112). Intrd; natd on walls, rocks and stony banks; locally common over BI except N Sc; Europe.

13. S. forsterianum Sm. (*S. elegans* Lej., *S. rupestre* auct. non L.) - *Rock Stonecrop*. **R**

Usually less robust than *S. rupestre* and differs constantly in all key characters; 2n=c.90. Native; rocks and screes, either dry in open or wet in woods; local in Wa and SW En, also grown in gardens and natd as for *S. rupestre.* Populations differ in various characters but do not fit into 2 sspp. (ssp. **forsterianum** and ssp. **elegans** (Lej.) E.F. Warb.) as sometimes claimed. 311

14. S. acre L. - *Biting Stonecrop*. Glabrous, acrid-tasting perennial; stems procumbent, rooting, sending up ascending to erect flowering stems to 10cm and shorter sterile shoots; leaves alternate, 3-5mm, ± terete, ovoid; flowers in small cymes; petals 5, 6-8mm, bright yellow; 2n=80. Native; walls, rocks, open grassland and maritime sand and shingle; throughout most of BI.

15. S. sexangulare L. - *Tasteless Stonecrop*. Differs from *S. acre* in being often taller (to 25cm); leaves 3-6mm; petals 4-6mm; and see key; (2n=72, 74, 108). Intrd; natd on walls and rocks; scattered in En and Wa; Europe.

16. S. album L. - *White Stonecrop*. Glabrous perennial; stems procumbent, rooting, sending up ascending to erect flowering stems to 20cm and shorter sterile shoots; leaves alternate, 4-12mm, ± terete, ovoid to cylindrical; flowers in dense subcorymbose panicle; petals 5, 2-4mm, white, sometimes tinged pink; (2n=32, 64, 68, 102, 136). Probably intrd; walls, rocks and stony ground; scattered through most of BI, possibly native in SW and WC En; Europe. Ssp. **micranthum** (Bastard) Syme, with smaller flowers and shorter ovoid leaves than ssp. **album**, is only an extreme variant probably not worth ssp. rank.

17. S. lydium Boiss. - *Least Stonecrop*. Glabrous perennial; similar vegetatively to *S. sexangulare* but leaves not spurred at base as in latter; flowers in small dense corymbose panicle 1-2cm across; petals 5, 2-4mm, white with red midrib; (2n=12). Intrd; garden outcast or escape natd in a few places in En, Sc and Ir; Turkey.

18. S. anglicum Huds. - *English Stonecrop*. Glabrous perennial; similar to *S. acre* but leaves grey-green often tinged red (not bright green) and not acrid; petals 2.5-4.5m, white tinged pink; follicles ± erect (not ± patent); 2n=120-130, 160. Native; rocks, sand and shingle; common in much of CI, Ir and W Br, very local and mainly coastal in C & E Br.

19. S. dasyphyllum L. - *Thick-leaved Stonecrop*. Glandular-pubescent perennial; stems ascending, rooting at base, to 10cm; leaves mostly opposite, 3-6mm, ± terete, ovoid to ellipsoid; flowers in rather few-flowered cymes; petals 5-6, 2.5-4mm, white tinged pink; (2n=28, 42, 56). Intrd; natd on walls and rocks; very scattered in En, Wa and Ir; Europe.

Section 6 - EPETEIUM Boiss. (spp. 20-21). Bisexual; petals 5-9, pink; plant without rhizomes, ± monocarpic (or annual), forming 0-few sterile shoots; leaves (dorsiventral or) not.

20. S. villosum L. - *Hairy Stonecrop*. Glandular-pubescent biennial to perennial; R
stems erect to ascending, rooting near base, to 10(15)cm; leaves alternate, 3-8mm, semi-terete (flat on upperside), often reddish, linear-oblong; flowers in rather few-flowered cymes; petals 5, 3.5-5mm, pink; (2n=30). Native; streamsides and stony flushes in hilly areas; N En, C & S Sc.

21. S. hispanicum L. - *Spanish Stonecrop*. Glandular-pubescent, usually perennial; 311
stems decumbent to ± erect, to 10cm; leaves alternate, 4-15mm, semi-terete; linear to narrowly ellipsoid; flowers in diffuse groups of cymes; petals (5)6-7(9), 3.5-6mm, white with red midrib; (2n=14, 28, 30, 40, 42). Intrd; natd on walls and stony ground; rare in S & C En; SE Europe.

76. SAXIFRAGACEAE - *Saxifrage family*
(Parnassiaceae)

Annual to perennial herbs, rarely woody at base; leaves alternate or all basal, rarely opposite, simple to compound, sessile or petiolate, stipulate or not. Flowers solitary or in various cymes or racemes, bisexual or sometimes unisexual in various arrangements, variously hypogynous, perigynous or epigynous, actinomorphic or less often zygomorphic; sepals 4-5, free to fused at base but often borne on hypanthium; petals 4-5, free, less often 0; stamens 3, 5, 8 or 10; carpels normally 2, sometimes 4, fused only at base or for varying distances ± to top, with many ovules on axile or parietal placentas, tapering to small stigma; fruit 2 follicles variously fused to form a capsule.

Very variable in vegetative and floral characters, but distinguishable by the 2 carpels fused only at base or for varying distances to apex; *Parnassia* is distinct in its staminodes.

1 Leaves compound, or simple and divided >1/2 way to base 2
1 Leaves simple, divided <1/2 way to base 4
  2 Leaves (incl. petioles) <5cm; inflorescence few-flowered, <5cm **5. SAXIFRAGA**
  2 Leaves >10cm; inflorescence many-flowered, >5cm 3
3 Leaves palmate, with 5-9 leaflets **2. RODGERSIA**
3 Leaves ternate to pinnate, the main divisions ternate to pinnate **1. ASTILBE**
  4 Leaves peltate **4. DARMERA**
  4 Leaves not peltate (petiole joining lamina at edge) 5
5 Petals 0; sepals 4 **9. CHRYSOSPLENIUM**
5 Petals present; sepals 5 6
  6 Five large divided staminodes present, alternating with 5 stamens; flowering stems with 1 flower and 1 leaf **10. PARNASSIA**
  6 Staminodes 0; flowering stems with >1 flower and/or >1 leaf 7
7 Stamens as many as sepals or fewer 8
7 Stamens 2x as many as sepals 9
  8 Petals 4, brown; stamens 3 **7. TOLMIEA**
  8 Petals 5, pink to red; stamens 5 **6. HEUCHERA**
9 Petals fringed with long narrow lobes **8. TELLIMA**
9 Petals entire to minutely toothed 10
  10 Thick surface rhizome present; at least some leaves >10cm; petals pink to red **3. BERGENIA**
  10 Rhizome 0 or thin; all leaves <10cm (if petals pink or red then leaves <1cm) **5. SAXIFRAGA**

1. ASTILBE Buch.-Ham. ex D. Don - *False-buck's-beards*
Perennials; leaves ternate to pinnate, the primary divisions ternate to pinnate; inflorescence a many-flowered terminal panicle; flowers bisexual or unisexual (dioecious to variously arranged), ± hypogynous; sepals 5; petals 0 or 5; stamens (5-)10; carpels 2(-3), ± free to united at base to form 2-celled ovary with axile placentation.

Often confused with *Spiraea* (always shrubs) or *Aruncus* (carpels 3, stamens >10) (both Rosaceae), but differs in floral details. Female plants have short sterile stamens.

1 Petals 0; plant usually >1m **3. A. rivularis**
1 Petals 5, longer than sepals; plant <1m 2
  2 Stems with short whitish to brown hairs; petals white to pale pink **1. A. japonica**
  2 Stems with dense long shaggy brown hairs; petals pink to red **2. A. x arendsii**

Other spp. - Several spp. and hybrids are grown in gardens and some wild plants might be other than the following.

1. A. japonica (C. Morren & Decne.) A. Gray - *False-buck's-beard.* Stems to 80cm, erect, rather sparsely pubescent with very short white to brown hairs; leaves 2-3-ternate; inflorescence up to 30cm; petals 1.5-2x as long as sepals, linear, white or pale pink; (2n=14). Intrd; grown in gardens and natd in wild usually in damp places; very scattered in Br, mostly in N En and Sc; Japan.

2. A. x arendsii Arends (*A. x rosea* Hort.; ?*A. chinensis* (Maxim.) Franch. x *A.* **316**
japonica) - *Red False-buck's-beard.* Differs from *A. japonica* in stems with dense long shaggy brown hairs; primary leaf divisions sometimes pinnate; inflorescence denser; petals 2-3x as long as sepals, pink to red. Intrd; similar situations to *A. japonica*; very scattered in En and Sc; garden origin.

3. A. rivularis Buch.-Ham. ex D. Don - *Tall False-buck's-beard.* Stems to 1.6m, erect, glabrous to sparsely pubescent with very short brownish hairs; leaves 2-3-ternate or -pinnate, with sparse but long, wispy brown hairs at branch-points; inflorescence up to 60cm; (2n=14, 28). Intrd; grown rather rarely in gardens, natd by streams and tracks in forest plantations in Kintyre and Argyll; C Asia.

2. RODGERSIA A. Gray - *Rodgersia*

Perennials with short stout rhizome; leaves palmate with 5-9 simple leaflets; inflorescence a many-flowered terminal panicle; flowers bisexual, shallowly perigynous; sepals 5; petals 5; stamens 10; carpels 2, united for most part to form 2-celled ovary with axile placentation.

1. R. podophylla A. Gray - *Rodgersia.* Stems to 1.3m, erect, pubescent; leaves with long petiole and 5-9 distally acutely lobed leaflets each up to 30cm; inflorescence up to 25cm; flowers c.5mm across, yellowish-white; (2n=30). Intrd; planted in damp places by ponds and rivers, sometimes persistent and spreading vegetatively (rarely sets seed); natd in Surrey, Monts and Dunbarton, perhaps elsewhere; Korea and Japan.

3. BERGENIA Moench - *Elephant-ears*

Glabrous perennials with stout scaly rhizome usually on soil surface; leaves large, simple, thick, serrate-crenate; inflorescence a many-flowered panicle on erect leafless stem; flowers perigynous with cup-shaped hypanthium; sepals 5; petals 5, pink; stamens 10; carpels 2, united only at base to form 2-celled ovary with axile placentation.

Other spp. - The following is the most commonly cultivated sp., but some escapes or relics might be the related **B. cordifolia** (Haw.) Sternb., from Siberia, which differs in its ± orbicular cordate leaves, less pendent flowers and petals with ± orbicular limb; or the hybrid between the 2; or the hybrid *B. crassifolia* x *B. ciliata* (Haw.) Sternb. = **B. x schmidtii** (Regel) Silva Tar., which has the ciliate leaf-margins of the latter parent.

1. B. crassifolia (L.) Fritsch - *Elephant-ears.* Leaves ovate to obovate, 6-20cm with petiole c.1/2 as long, cuneate to subcordate at base; flowering stem 10-40cm; flowers 15-25mm across, ± pendent; petals with ovate to obovate limb; (2n=34). Intrd; very persistent garden relic or throwout; scattered in Br and CI; Siberia. Rarely sets seed.

4. DARMERA Voss ex Post & Kuntze (*Peltiphyllum* (Engl.) Engl. non *Peltophyllum* Gardner) - *Indian-rhubarb*

Pubescent perennials with stout rhizome; leaves simple, peltate, palmately lobed and sharply serrate; inflorescence a ± corymbose panicle on erect leafless stem; flowers almost hypogynous; sepals 5; petals 5, pink to whitish; stamens 10; carpels 2, ± free.

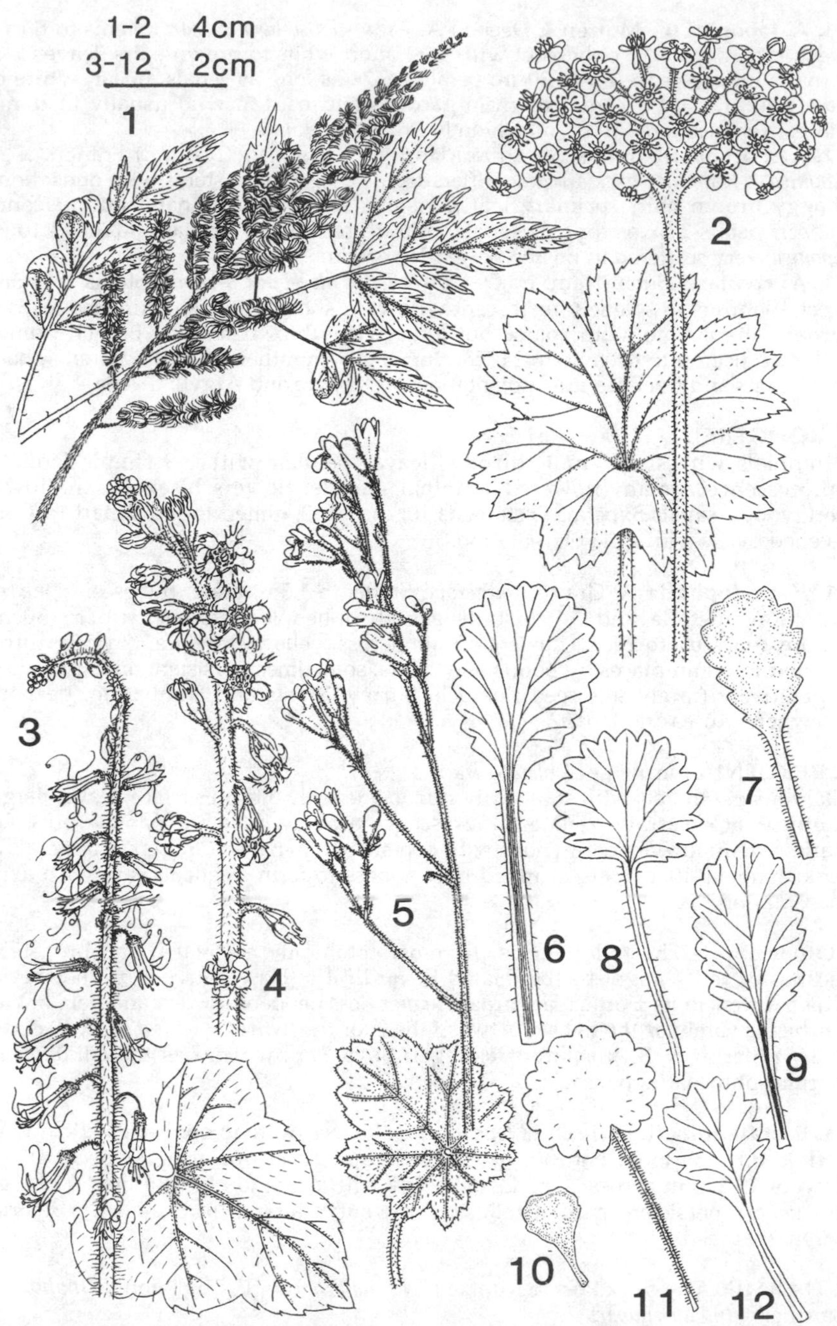

FIG 316 - Saxifragaceae. 1, *Astilbe x arendsii*. 2, *Darmera peltata*. 3, *Tolmiea menziesii*. 4, *Tellima grandiflora*. 5, *Heuchera sanguinea*. 6-12, leaves of *Saxifraga*. 6, *S. x geum*. 7, *S. umbrosa*. 8, *S. x polita*. 9, *S. x urbium*. 10, *S. cuneifolia*. 11, *S. hirsuta*. 12, *S. spathularis*.

1. D. peltata (Torr. ex Benth.) Voss ex Post & Kuntze (*Peltiphyllum peltatum* **316**
(Torr. ex Benth.) Engl.) - *Indian-rhubarb*. Leaves 5-40cm across, on petioles up to 1m; flowering stem to 1(1.5)m, branched only near top; flowers 10-15mm across; (2n=34). Intrd; grown in damp places and natd where planted or outcast; scattered over Br and Ir; W USA. Rarely sets seed.

5. SAXIFRAGA L. - *Saxifrages*

Pubescent annuals or perennials; leaves simple to almost compound; flowers in simple or compound cymes, sometimes solitary; ovary superior to ± completely inferior; hypanthium ± 0; sepals 5; petals 5; stamens 10; carpels 2, fused at least at base to form 2-celled ovary with axile placentation.

Our 20 spp. and 3 fully treated hybrids fall into 11 different sections, 7 of which contain only 1 sp. and 2 others only 2 each, so they are not utilized here; spp. 7-13 belong to sect. *Gymnopera* D. Don, and spp. 19-23 belong to sect. *Saxifraga*.

1 Leaves opposite; petals purple **14. S. oppositifolia**

1 Leaves alternate (spiral); petals yellow or white 2

2 Some or all flowers replaced by reddish bulbils **18. S. cernua**

2 Bulbils in inflorescence 0 3

3 Petals bright yellow 4

3 Petals white to cream 6

4 Annuals; leaves ± orbicular, abruptly contracted to petiole **2. S. cymbalaria**

4 Perennials; leaves linear to oblanceolate, gradually contracted to petiole or ± sessile 5

5 Ovary superior; flowers 1(-3) per stem **1. S. hirculus**

5 Ovary semi-inferior; flowers usually >3 per stem **16. S. aizoides**

6 Flowers strongly zygomorphic, the lower petals >2x as long as 3 upper ones **5. S. stolonifera**

6 Flowers actinomorphic or ± so 7

7 Ovary superior, the sepals arising from underneath it 8

7 Ovary partly inferior, the sepals arising from its side or top 16

8 Leaves ± sessile or with petiole <1/2 as long as blade **4. S. stellaris**

8 Basal leaves with petiole almost as long as to longer than blade 9

9 Stems leafy; sepals erecto-patent **6. S. rotundifolia**

9 Stems leafless; sepals reflexed 10

10 Petals without red spots; leaf-laminas entire for ≥ basal 1/3 of margin **7. S. cuneifolia**

10 Petals usually with small red spots; leaf-laminas toothed for ≥ apical 3/4 of margin 11

11 Petioles subterete; laminas sparsely pubescent over both surfaces, cordate at base **13. S. hirsuta**

11 Petioles distinctly flattened; laminas glabrous or nearly so at least on lowerside 12

12 Petioles densely pubescent on lateral margins; glabrous on upperside **8. S. umbrosa**

12 Petioles ± glabrous to rather sparsely pubescent 13

13 Laminas with acute teeth; petioles subglabrous **11. S. spathularis**

13 Laminas with subacute to rounded teeth, and/or petioles distinctly pubescent on margins 14

14 Petioles usually much longer than lamina; laminas sparsely pubescent to glabrous 15

14 Petioles usually c. as long as to slightly longer than laminas; laminas glabrous **9. S. x urbium**

15 Leaves mostly c. as long as wide, with scarcely visible translucent border (<0.2mm) and rather sharp teeth **12. S. x polita**

15 Many leaves distinctly longer than wide, with conspicuous translucent border (≥0.2mm) and low, blunt teeth **10. S. x geum**

16 Leaves all in basal rosette or just reduced ones on stem 17

16 Normal leaves present on stem 18

17 Leaves sessile, finely serrate, encrusted with lime on margins **15. S. paniculata**

17 Leaves petiolate, crenate-serrate, not lime-encrusted **3. S. nivalis**

18 Annual, without perennating organs **23. S. tridactylites**

18 Perennial, with sterile rosettes, stolons, rhizomes or basal bulbils 19

19 Basal bulbils usually present; stolons terminating in leafy rosettes 0 20

19 Bulbils 0; procumbent stolons terminating in dense leafy rosettes present 21

20 Flowers mostly >3 per stem; petals >(6)9mm; basal leaves mostly ≥7-lobed **19. S. granulata**

20 Flowers 1-3 per stem; petals <6mm; basal leaves 3-7-lobed **17. S. rivularis**

21 Leaf-lobes acuminate to narrowly acute, apiculate to aristate; flower-buds ± pendent **20. S. hypnoides**

21 Leaf-lobes rounded, obtuse or acute, shortly mucronate or not; flower-buds erect 22

22 Leaf-lobes rounded, obtuse or subacute; petals dull creamy- or greenish-white; Sc and Wa **22. S. cespitosa**

22 Leaf-lobes subacute to acute; petals pure white; Ir and Wa **21. S. rosacea**

1. S. hirculus L. - *Marsh Saxifrage.* Stoloniferous perennial; stems ascending to erect, leafy, to 20cm; leaves lanceolate to oblanceolate, the lowest tapered to long petiole, entire; flowers 1(-3), terminal; petals yellow, 10-15mm; ovary superior; (2n=16, 28, 32). Native; wet places on moors; very local in N En, N Ir and S & C Sc, decreasing and formerly more widespread. **RRR**

2. S. cymbalaria L. - *Celandine Saxifrage.* Annual; stems decumbent to suberect, to 20cm; leaves orbicular to transversely elliptic, cordate to rounded at base, with 3-7 broad shallow lobes, with long petiole; flowers in sparse diffuse cymes; petals yellow, 3-5mm; ovary superior; (2n=18). Intrd; weed escaping locally from nurseries and gardens in shady places; scattered in Br, Co Antrim; E Mediterranean. Our plant is var. **huetiana** (Boiss.) Engl. & Irmsch.

3. S. nivalis L. - *Alpine Saxifrage.* Perennial with basal leaf-rosette and leafless erect stem to 15cm; leaves obovate to suborbicular, on long petioles, closely crenate-serrate; flowers in dense panicle; petals white, 2-3mm; ovary semi-inferior; (2n=60). Native; mountain rocks and cliffs; very local in N Wa, NW En, NW Ir and Sc. **R**

4. S. stellaris L. - *Starry Saxifrage.* Stoloniferous perennial with basal leaf-rosette and ± leafless erect stem to 20cm; leaves obovate or obtrullate to narrowly so, scarcely petiolate, remotely serrate or dentate; flowers in loose panicle or cyme, petals white, 4-6mm; ovary superior; 2n=28. Native; wet rocks and stony places, in flushes and by streams in mountains; frequent in N Wa, N En and Sc, local in Ir.

5. S. stolonifera Curtis (*S. sarmentosa* L. f.) - *Strawberry Saxifrage.* Perennial with long thin stolons producing new plants at apex; leaves orbicular, cordate, crenate-dentate, with long petioles; flowers in large diffuse panicles on leafless stems to 60cm; petals white, the 2 lower petals much longer than 3 upper; ovary <1/2-superior; (2n=36, 54). Intrd; natd on shady walls in Cornwall; Japan and China.

6. S. rotundifolia L. - *Round-leaved Saxifrage.* Perennial with basal leaf-rosette and erect leafy stem to 40cm; leaves orbicular, thin, cordate, coarsely dentate, basal ones with long petiole; flowers in loose panicle; petals white with red spots, 6-11mm; ovary superior; (2n=22). Intrd; grown in gardens and natd by shady streams in N En and Sc; Europe.

7. S. cuneifolia L. - *Lesser Londonpride.* Stoloniferous perennial with basal leaf- **316**

rosette and leafless erect stem to 30cm; leaves suborbicular-spathulate, thick, on long petioles, dentate in distal 1/2; flowers in loose panicle; petals white, 2.5-4mm; ovary superior; (2n=28). Intrd; natd on rocks and old walls; N Somerset, Man, MW Yorks and Westerness, rare short-lived escape elsewhere; S Europe.

8. S. umbrosa L. - *Pyrenean Saxifrage.* Differs from *S. spathularis* in larger obovate-oblong crenate-dentate leaves with 9-21 obtuse to rounded teeth and translucent border 0.25-0.4mm; petioles densely pubescent on margins; petals 3-4mm; 2n=28. Intrd; formerly much grown in gardens, natd on shady limestone rocks; MW Yorks since 1792, perhaps Derbys and Dunbarton; Pyrenees. Many records refer to *S.* x *urbium.* **316**

9. S. x urbium D.A. Webb (*S. umbrosa* x *S. spathularis*) - *Londonpride.* Differs from *S. spathularis* in larger less orbicular leaves with usually 19-25 subacute to obtuse teeth and translucent border 0.2-0.35mm; petioles pubescent on margins; petals 4-5mm. Intrd; much grown in gardens, natd in waste places, in woods, by streams and on walls and rocks; throughout BI; garden origin. Usually sterile but fertile plants exist. **316**

10. S. x geum L. (*S. umbrosa* x *S. hirsuta*) - *Scarce Londonpride.* Fertile and forms spectrum connecting parents in all leaf characters. Intrd; rather rare in gardens and rarely natd in shady and damp, often rocky places; (2n=28). W Cornwall, SE & MW Yorks, Man, Dunbarton and Fife; Pyrenees. Many records refer to *S. hirsuta* or *S.* x *urbium.* **316**

11. S. spathularis Brot. - *St Patrick's-cabbage.* Stoloniferous perennial with basal leaf-rosette and leafless erect stem to 40cm; leaves ± orbicular to obovate, thick, with translucent border ≤0.2mm, glabrous, acutely dentate with usually 9-15 teeth; petiole long, flattened, pubescent only at base; flowers in loose panicle; petals white with red spots, 3-5mm; ovary superior; 2n=28. Native; damp rocks in mountains; locally common in W & SW Ir, rare elsewhere in Ir. **R 316**

12. S. x polita (Haw.) Link (*S. spathularis* x *S. hirsuta*) - *False Londonpride.* Fertile and forms spectrum connecting the parents; intermediate in leaf shape, toothing and pubescence and in petiole shape and pubescence. Native; mostly with parents and commoner than *S. hirsuta* in SW Ir, frequent in W Mayo and W Galway without *S. hirsuta* and sometimes without *S. spathularis,* also natd garden escape in S Wa, N En and C Sc. **R 316**

13. S. hirsuta L. (*S. geum* L. 1762 non L. 1753) - *Kidney Saxifrage.* Stoloniferous perennial with basal leaf-rosette and leafless erect stem to 40cm; leaves orbicular, thin, cordate, pubescent on both sides, crenate-dentate, with (11)15-25 ± rounded teeth; petiole long, scarcely flattened, pubescent all round; flowers in loose panicle; petals white with red spots, 3-5mm; ovary superior; 2n=28. Native; damp rocks in mountains; locally common in N & S Kerry and W Cork, grown in gardens and natd in S & C Sc, N En and Man. **R 316**

14. S. oppositifolia L. - *Purple Saxifrage.* Procumbent mat-forming perennial; leaves opposite, ovate to obovate, entire; flowers solitary, terminal on shortly upturned stems; petals purple, 5-10mm; ovary semi-inferior; 2n=26. Native; damp mountain rocks and scree; locally common in N & S Wa, N & W Ir, NW En and Sc.

15. S. paniculata Mill. - *Livelong Saxifrage.* Shortly stoloniferous perennial with ± hemispherical basal leaf-rosettes and erect stems to 40cm; leaves obovate to oblong-oblanceolate, finely serrate, lime-encrusted at edges, sessile; inflorescence a small terminal panicle with glandular-pubescent branches; petals white to cream, sometimes red-spotted, 3-6mm; ovary 3/4-inferior; (2n=28). Intrd; natd in crack in limestone scar in MW Yorks since at least 1988, formerly Lanarks; mountains of Europe.

16. S. aizoides L. - *Yellow Saxifrage.* Perennial, with sterile and fertile decumbent to ascending stems to 25cm; leaves linear to narrowly oblong-elliptic, scarcely petiolate, with small distant teeth; flowers in diffuse cyme; petals yellow, 3-6mm; ovary semi-inferior; 2n=26. Native; wet rocks and streamsides in mountains, down to sea-level on dunes in N Sc; locally common in N En and N & C Sc, rare in N Ir, extinct in Wa.

17. S. rivularis L. - *Highland Saxifrage*. Stoloniferous perennial usually with basal RR
bulbils; stems slender, ascending, to 12cm, with few leaves; leaves orbicular to transversely elliptic, cordate, with 3-7 deep lobes; flowers 1-3 per stem; petals white, 3-5mm; ovary semi-inferior; (2n=26, 52). Native; wet mountain rocks above 900m; rare in C & N Sc.

18. S. cernua L. - *Drooping Saxifrage*. Perennial with basal leaf-rosette bearing RRR
axillary bulbils; stem erect, to 15cm, with axillary bulbils replacing all flowers or sometimes all but the terminal one; leaves orbicular to transversely elliptic, cordate, with 3-5 rather deep lobes; petals white, up to 12mm; ovary nearly superior; (2n=36, 54, 60, 64, 70). Native; basic mountain rocks above 900m; very rare in C Sc.

19. S. granulata L. - *Meadow Saxifrage*. Perennial with basal leaf-rosette bearing axillary bulbils; stems erect, to 50cm, leafy; leaves ± orbicular, cordate, mostly ≥7-lobed, the basal with long petioles; flowers in loose cyme, gynodioecious; petals white, (6)9-16(20)mm; ovary semi-inferior; 2n=c.32, 48. Native; moist but well-drained base-rich grassland; locally common throughout most of Br, very rare in E Ir, also natd garden escape (often as *flore pleno*) elsewhere in BI.

20. S. hypnoides L. (*S. platypetala* Sm.) - *Mossy Saxifrage*. Stoloniferous laxly mat-forming perennial; differs from *S. rosacea* ssp. *rosacea* in linear leaf-lobes acuminate and ± aristate at apex; flower-buds pendent; 2n=26. Native; damp rock-ledges, boulders and dunes and by mountain streams; locally common in Br from Derbys, N Somerset and Wa northwards, very local in Ir, rare garden escape elsewhere. Garden Mossy Saxifrages vary greatly in habit, leaf dissection and width of leaf-segments, and petal shape and colour (red or yellow to white); they are known under the invalid name *S. x arendsii* hort. and occasionally persist as relics or throwouts. They are complex hybrids of *S. hypnoides*, *S. rosacea* and other spp.

20 x 21. S. hypnoides x S. rosacea is intermediate in habit and leaf characters and occurs with the parents in S Tipperary and Clare; some persistent garden escapes in Br might be of the same parentage. Certain examples of *S. hypnoides* in N Wa also suggest past hybridization with *S. rosacea*. Endemic.

20 x 23. S. hypnoides x S. tridactylites was found with the parents in 1906 on limestone in MW Yorks; endemic.

21. S. rosacea Moench (*S. decipiens* Ehrh. nom. nud.) - *Irish Saxifrage*. Stoloniferous cushion- or mat-forming perennial with leaf-rosettes and erect nearly leafless stems to 20cm; leaves with 3-5(7) lobes acute to subacute and often mucronate at apex; flowers 2-5(8) in lax cyme; petals white, 6-8mm; ovary semi-inferior. Native; damp cliffs, rocks and streamsides on mountains. Very variable; 1 population deserves ssp. rank.

a. Ssp. rosacea. Leaves with glandular and non-glandular hairs, many of which RR
exceed 0.5mm; 2n=52. Locally common in S & W Ir, 1 place in Caerns, possibly formerly in Angus; natd in N Somerset.

b. Ssp. hartii (D.A. Webb) D.A. Webb (*S. hartii* D.A. Webb). Leaves with hairs RRR
all glandular and <0.5mm; plant more robust; 2n=52. Arranmore Island, W Donegal; endemic.

22. S. cespitosa L. - *Tufted Saxifrage*. Stoloniferous cushion-forming perennial RRR
with basal leaf-rosettes and erect nearly leafless stems to 10cm; leaves (entire to) 3(-5)-lobed, with many short glandular hairs, the lobes rounded to subacute at apex; flowers 1-3(5) in lax cyme; petals off-white, 4-5mm; ovary semi-inferior; 2n=78. Native; mountain rocks above 600m; rare and very local in C Sc and N Wa.

23. S. tridactylites L. - *Rue-leaved Saxifrage*. Annual; stems erect, to 10(16)cm; leaves oblanceolate, the lower ones mostly with 3-5 deep lobes, otherwise entire; flowers in diffuse cymes; petals white, 2-3mm; ovary semi-inferior; (2n=22). Native; bare dry ground on walls, rocks and sand, mostly calcareous; locally common throughout most of BI.

6. HEUCHERA L. - *Coralbells*
Pubescent perennials; leaves all basal, palmately lobed, serrate; inflorescence a panicle on erect leafless stem; flowers 1/2- or more-inferior with bell-shaped hypanthium above; sepals 5; petals 5; stamens 5; carpels 2, fused to form 1-celled ovary with parietal placentation.

Other spp. - Some wild plants might be hybrids of the following with other spp. of this or related genera.

1. H. sanguinea Engelm. - *Coralbells*. Leaves 2-6cm, broadly ovate to orbicular, cordate, on petioles 7-15cm; flowering stem to 50cm; flowers bright pinkish-red, 6-13mm; petals smaller than sepals; (2n=14). Intrd; grown in gardens and persistent on waste ground and tips; scattered in SE En; S N America. Rarely sets seed. **316**

7. TOLMIEA Torr. & A. Gray - *Pick-a-back-plant*
Pubescent perennials; leaves mostly basal, palmately lobed, serrate, producing plantlets at junction with petiole in moist conditions; inflorescence a simple raceme on ± leafy stem; flowers zygomorphic, perigynous, with tubular hypanthium; sepals 5; petals 4(-5), filiform, the lowest usually missing; stamens 3, opposite the 3 upper sepals; carpels 2, fused for most part to form 1-celled ovary with parietal placentation.

1. T. menziesii (Pursh) Torr. & A. Gray - *Pick-a-back-plant*. Leaves 4-10cm, broadly ovate to ± orbicular, cordate, on petioles 10-30cm; flowering stems to 70cm; flowers brown, 6-15mm; petals much narrower than sepals; (2n=28). Intrd; grown as pot-plant and in gardens, natd in damp shady places and persistent on tips and waste ground; scattered through most of Br; W N America. Rarely sets seed. **316**

8. TELLIMA R. Br. - *Fringecups*
Pubescent perennials; leaves mostly basal, palmately lobed, serrate; inflorescence a simple raceme on ± leafy stem; flowers 1/4-1/2-inferior, with bell-shaped hypanthium above; sepals 5; petals 5, broad and fringed with long narrow lobes; stamens 10; carpels 2, fused for most part to form 1-celled ovary with parietal placentation.

1. T. grandiflora (Pursh) Douglas ex Lindl. - *Fringecups*. Leaves 4-10cm, broadly ovate to orbicular, cordate, on petioles 5-20cm; flowering stems to 70cm; flowers green, usually pink- or red-tinged, 6-15mm; (2n=14). Intrd; grown in gardens and well natd in woods and damp hedgerows; scattered through most of Br, Co Longford; W N America. Sets abundant seed. **316**

9. CHRYSOSPLENIUM L. - *Golden-saxifrages*
Sparsely pubescent perennials with procumbent sterile and erect leafy flowering stems; leaves orbicular, crenate, petiolate; inflorescence of dichotomous subcorymbose cymes; flowers golden-yellow, epigynous; hypanthium 0; sepals 4; petals 0; stamens 8; carpels 2, fused to form 1-celled ovary with parietal placentation.

1. C. oppositifolium L. - *Opposite-leaved Golden-saxifrage*. Sterile shoots leafy; flowering stems to 15cm; leaves opposite, up to 2cm, cuneate to rounded at base, with petiole up to as long as blade; 2n=42. Native; wet places by streams, in flushes and boggy woods, on mountain ledges; throughout BI but rare in parts of E En.

2. C. alternifolium L. - *Alternate-leaved Golden-saxifrage*. Sterile shoots with only scale-leaves; flowering stems to 20cm; leaves alternate, up to 2.5cm, cordate at base, the lowest with petiole much longer than blade; 2n=48. Native; similar places to C. *oppositifolium* and often with it; local over most of Br.

10. PARNASSIA L. - *Grass-of-Parnassus*
Glabrous perennials; leaves mostly basal, simple, entire, petiolate; flower solitary, terminal, hypogynous; sepals 5; petals 5, white; stamens 5, alternating with 5 large much divided staminodes; carpels 4, fused to form 1-celled ovary with parietal placentation at least below.

1. P. palustris L. - *Grass-of-Parnassus.* Flowering stems to 30cm, often much less, with 1 sessile leaf; basal leaves 1-5cm, ovate, cordate, petiolate; flowers 15-30mm across; 2n=18, 36. Native; marshes, damp grassland and dune-slacks; local in much of Br and Ir, but absent or very rare in most of S Wa and S En.

77. ROSACEAE - *Rose family*

Trees, shrubs or annual to perennial herbs; leaves alternate, simple to compound, usually petiolate, stipulate (stipules often falling early) or rarely without stipules. Flowers sometimes solitary, usually in various, often dense and compound, cymes or racemes, bisexual or sometimes unisexual in various arrangements, usually perigynous or epigynous, the hypanthium variously developed but sometimes ± 0 and flowers ± hypogynous, actinomorphic; sepals (4)5(-10), free but usually borne on hypanthium; epicalyx often present outside calyx; petals (4)5(-16), free, sometimes 0; stamens usually 2-4x as many as sepals, sometimes more or fewer; carpels 1 to many, free or fused, each usually with 2, sometimes 1 or >2, ovules; style 1 per carpel, usually free, sometimes ± fused into column; stigmas capitate; fruit of 1 or more achenes, drupes, follicles or rarely a capsule, in epigynous flowers surrounded by succulent or pithy hypanthium.

Extremely variable in most characters, but usually recognizable by the alternate stipulate leaves (stipules often present only in young state or on leader shoots in woody spp.); the perigynous or epigynous flowers with usually 5 free petals, numerous stamens, and 1-many free or fused 2-ovuled carpels each with a separate style; and the 1-2-seeded fruits often aggregated into false-fruits. Exceptions to virtually all the above occur.

General key
1 Herbs; stems annual, sometimes woody at base ***Key C***
1 Trees or shrubs; stems woody, biennial to perennial 2
  2 Flowers hypogynous to perigynous; hypanthium variously developed and often enclosing carpels, but not fused with them; fruit dry (but surrounded by succulent hypanthium in *Rosa*) or 1-several drupes ***Key A***
  2 Flowers epigynous; hypanthium completely enclosing carpels, becoming succulent or ± so and fused with them at fruiting ***Key B***

Key A - Woody plants with hypogynous to perigynous flowers
1 Fruits a cluster of small drupes on strongly convex receptacle with ± flat hypanthium around **8. RUBUS**
1 Fruits dry (but surrounded by succulent hypanthium in *Rosa*), or if drupes then hypanthium cup-shaped (and drupe usually 1) 2
  2 Leaves simple 3
  2 Leaves pinnate 9
3 Petals yellow **7. KERRIA**
3 Petals white to bright pink 4
  4 Petals >6; fruit terminated by long feathery appendage (style) **14. DRYAS**

4 Petals normally 5; fruit without feathery appendage 5

5 Fruit 1-5 succulent drupes 6

5 Fruit dry achenes or follicles 7

6 Carpel, style and drupe 1; leaves serrate to crenate **22. PRUNUS**

6 Carpels and styles 5; drupes 1-5; leaves entire **23. OEMLERIA**

7 Carpels with 1-2 ovules; fruits indehiscent, 1-seeded (achenes) **5. HOLODISCUS**

7 Carpels with >2 ovules; fruits dehiscent, >1-seeded (follicles) 8

8 Stipules 0; carpels free; follicles not inflated, dehiscing along 1 side **3. SPIRAEA**

8 Stipules present; carpels fused at base; follicles inflated, dehiscing along 2 sides **2. PHYSOCARPUS**

9 Stems spiny; fruits enclosed in succulent hypanthium **21. ROSA**

9 Stems not spiny; fruits dry, not enclosed in hypanthium 10

10 Flowers in dense globose heads; petals 0; hypanthium usually with spines at maturity **18. ACAENA**

10 Flowers not in dense globose heads; petals present; hypanthium never spiny 11

11 Flowers <15mm across, many in each group; petals white; leaflets >9; fruit follicles **1. SORBARIA**

11 Flowers >15mm across, 1-few in each group; petals yellow; leaflets <9; fruit achenes **9. POTENTILLA**

Key B - Woody plants with epigynous flowers

1 Walls of carpels (within hypanthium) becoming stony at fruiting 2

1 Walls of carpels (within hypanthium) becoming cartilaginous at fruiting 5

2 Flowers ≥2cm across; sepals ≥1cm; fruit ≥2cm, brown **34. MESPILUS**

2 Flowers <2cm across; sepals <1cm; fruit <2cm, orange to red or purple to black 3

3 Stems not spiny; leaves entire **32. COTONEASTER**

3 Stems usually spiny; leaves toothed or lobed 4

4 Leaves evergreen; stipules minute, falling early **33. PYRACANTHA**

4 Leaves deciduous; stipules persistent at least on leading shoots **35. CRATAEGUS**

5 Carpels with ≥4 ovules, later usually with >2 seeds 6

5 Carpels with 1-2 ovules and seeds 7

6 Leaves entire, dull green on upperside, pubescent on lowerside; styles free **24. CYDONIA**

6 Leaves serrate, shiny green on upperside, glabrous; styles fused below **25. CHAENOMELES**

7 Flowers in racemes **30. AMELANCHIER**

7 Flowers in umbels or corymbs 8

8 Inflorescence a simple umbel or corymb, rarely producing >3 fruits 9

8 Inflorescence a branched corymb, usually producing >3 fruits 10

9 Flesh (hypanthium) of fruit with gritty groups of stone-cells; styles free; fruit usually pear-shaped **26. PYRUS**

9 Flesh of fruit without stone-cells; styles fused at base; fruit usually apple-shaped **27. MALUS**

10 Leaves entire, evergreen; hypanthium not enclosing apex of carpels and easily separated from carpels at maturity **31. PHOTINIA**

10 Leaves serrate to pinnate, deciduous; hypanthium wholly enclosing and strongly adherent to carpels at maturity 11

11 Styles 2-4(5); leaves simple or pinnate, the main veins running straight to leaf margin **30. SORBUS**

11 Styles 5; leaves simple, the main veins curved near apex and running parallel to (not reaching) leaf margin **29. ARONIA**

Key C - Herbaceous plants

1 Petals 0 2
1 Petals present 6
2 Leaves simple or palmate 3
2 Leaves pinnate 5
3 Annuals; flowers in leaf-opposed clusters; stamens 1(-2) **20. APHANES**
3 Perennials; flowers in terminal clusters; stamens 4-5(10) 4
4 Carpel 1; leaves palmate or palmately lobed **19. ALCHEMILLA**
4 Carpels 5-12; leaves ternate **10. SIBBALDIA**
5 Hypanthium usually with 4 spines at apex; stamens 2; plant with at least some procumbent stems **18. ACAENA**
5 Hypanthium not spiny; stamens 4-many; stems erect **17. SANGUISORBA**
6 Petals >6 **14. DRYAS**
6 Petals 4-6 7
7 Epicalyx 0 8
7 Calyx-like epicalyx present behind true sepals 11
8 Petals yellow; fruit with hooked bristles **15. AGRIMONIA**
8 Petals white to red or purple; fruit without bristles 9
9 Leaves 2-3-pinnate; stipules 0 ; carpels 3, several-seeded (or plant male) **4. ARUNCUS**
9 Leaves simple, 1-pinnate or 1-ternate; stipules present; carpels >5, 1-seeded (or plant male) 10
10 Flowers numerous in inflorescence; fruit a head of achenes **6. FILIPENDULA**
10 Flowers 1-6(10) in inflorescence; fruit a head of drupes **8. RUBUS**
11 Leaves pinnate 12
11 Leaves palmate or ternate 14
12 Carpels and achenes enclosed in hypanthium; stamens 5-10 **16. AREMONIA**
12 Hypanthium not covering carpels and achenes; stamens 10-numerous 13
13 Styles strongly hooked, persistent in fruit **13. GEUM**
13 Styles not hooked, deciduous before fruiting **9. POTENTILLA**
14 Receptacle becoming red and succulent in fruit (or not forming fruits) 15
14 Receptacle remaining dry in fruit 16
15 Petals yellow; epicalyx segments serrate at apex **12. DUCHESNEA**
15 Petals white to pinkish; epicalyx segments entire **12. FRAGARIA**
16 Petals ≤2mm; leaflets with (usually) 3 teeth at extreme apex **10. SIBBALDIA**
16 Petals ≥(1.5)2mm; if <2.5mm then leaflets serrated round apical ≥1/2 of margin **9. POTENTILLA**

Other genera - 2 woody genera of Spiraeoideae are often grown in gardens and mass-planted in public places, and may occasionally persist. **STEPHANANDRA** Siebold & Zucc. (represented mainly by **S. incisa** (Thunb.) Zabel) and **NEILLIA** D. Don (mainly **N. sinensis** Oliv.) are both Asian and have simple, lobed, stipulate leaves, and differ from *Physocarpus* in having non-inflated follicles dehiscent down 1 side only. *Stephanandra* has flowers in corymbs and 1 1-2-seeded follicle; *Neillia* has racemes and 1-2 usually 5-seeded follicles.

SUBFAMILY 1 - SPIRAEOIDEAE (genera 1-5). Stipules present or 0; shrubs, sometimes herbs; hypanthium cup- or saucer-shaped, not enclosing carpels and fused to them only at base; epicalyx 0; stamens >10; flowers 5-merous; carpels 3-5, often >2-seeded; fruit dry, a group of follicles or sometimes of achenes or a capsule; chromosome base-number 8 or 9.

1. SORBARIA (Ser. ex DC.) A. Braun - *Sorbarias*
Deciduous shrubs; leaves pinnate with ≥11 leaflets, stipulate; flowers crowded in large terminal panicles; petals white; carpels 5; fruit several-seeded follicles dehiscing along 1 margin. The spp. are often mis-identified.

1 Panicle rather dense; follicles pubescent, on erect stalks **1. S. sorbifolia**
1 Panicle rather diffuse; follicles glabrous, on recurved stalks 2
  2 Styles arising from apex of carpel; longest stamens c. as long as petals **2. S. tomentosa**
  2 Style arising from well below apex of carpel; longest stamens nearly 2x as long as petals **3. S. kirilowii**

Other spp. - S. grandiflora (Sweet) Maxim., from E Siberia, has recently been found ± wild in W Kent; it differs from *S. sorbifolia* in its stem to only 1m, smaller leaves, and longest stamens c. as long as (not 2x as long as) petals.

1. S. sorbifolia (L.) A. Braun - *Sorbaria*. Stems to 2.5m, ± erect; leaflets 2-9cm, **326**
lanceolate, acuminate, 2-serrate; panicle 7-34cm, with dense flowers and branches arising at narrow angle; (2n=36). Intrd; commonly grown and extensively suckering, natd on walls and waste ground; SE En; N Asia.

2. S. tomentosa (Lindl.) Rehder (*S. aitchisonii* (Hemsl.) Hemsl. ex Rehder) - **326**
Himalayan Sorbaria. Differs from *S. kirilowii* in leaves glabrous to pubescent along veins on lowerside, without stellate hairs; and see key. Intrd; grown in gardens and sometimes natd on walls and in scrub; C & S Sc, C & S En; W Himalayas.

3. S. kirilowii (Regel) Maxim. (*S. arborea* C.K. Schneid., *S. assurgens* M. Vilm. & Bois) - *Chinese Sorbaria*. Stems to 6m, spreading; leaflets 3-13cm, lanceolate, acuminate, 2-serrate, ± glabrous to sparsely pubescent on lowerside with simple hairs in vein axils and sometimes more extensive stellate hairs; panicle 11-42cm, rather diffuse with widely spreading branches; (2n=36). Intrd; grown in gardens and sometimes natd; SE En, Jersey; China.

2. PHYSOCARPUS (Cambess.) Maxim. - *Ninebark*
Deciduous shrubs; leaves simple, stipulate when young; flowers crowded in ± corymbose racemes; petals white to pinkish; carpels (3)4-5; fruit 2- to several-seeded inflated follicles dehiscing along both margins.

1. P. opulifolius (L.) Maxim. - *Ninebark*. Stems to 2(3)m, erect to arching; leaves **326**
broadly ovate, with 3-5 serrate lobes, ± glabrous; flowers c.1m across; fruit on long pedicels, erect, glabrous; (2n=18). Intrd; natd as relic of cultivation in shrubberies, rough ground and by streams; C & N En and Sc, Tyrone; E N America.

3. SPIRAEA L. - *Brideworts*
Deciduous shrubs; leaves simple, without stipules; flowers in dense or fairly dense panicles or corymbs; petals white to pinkish-purple; carpels 5; fruit several-seeded follicles dehiscing along 1 margin.

1 Inflorescence terminating each leafy shoot corymbose (hemispherical to flat-topped, at least as wide as long) 2
1 Inflorescence terminating each leafy shoot cylindrical to conical, longer than wide, sometimes comprised of a number of lateral corymbs 5
  2 Inflorescences simple corymbs 3
  2 At least the larger inflorescences compound corymbs 4
3 Leaves cuneate at base, obovate, at least some 3(-5)-lobed distally; petals longer than stamens **13. S. x vanhouttei**
3 Leaves rounded at base, ovate, not distinctly lobed; petals shorter than stamens **11. S. chamaedryfolia**

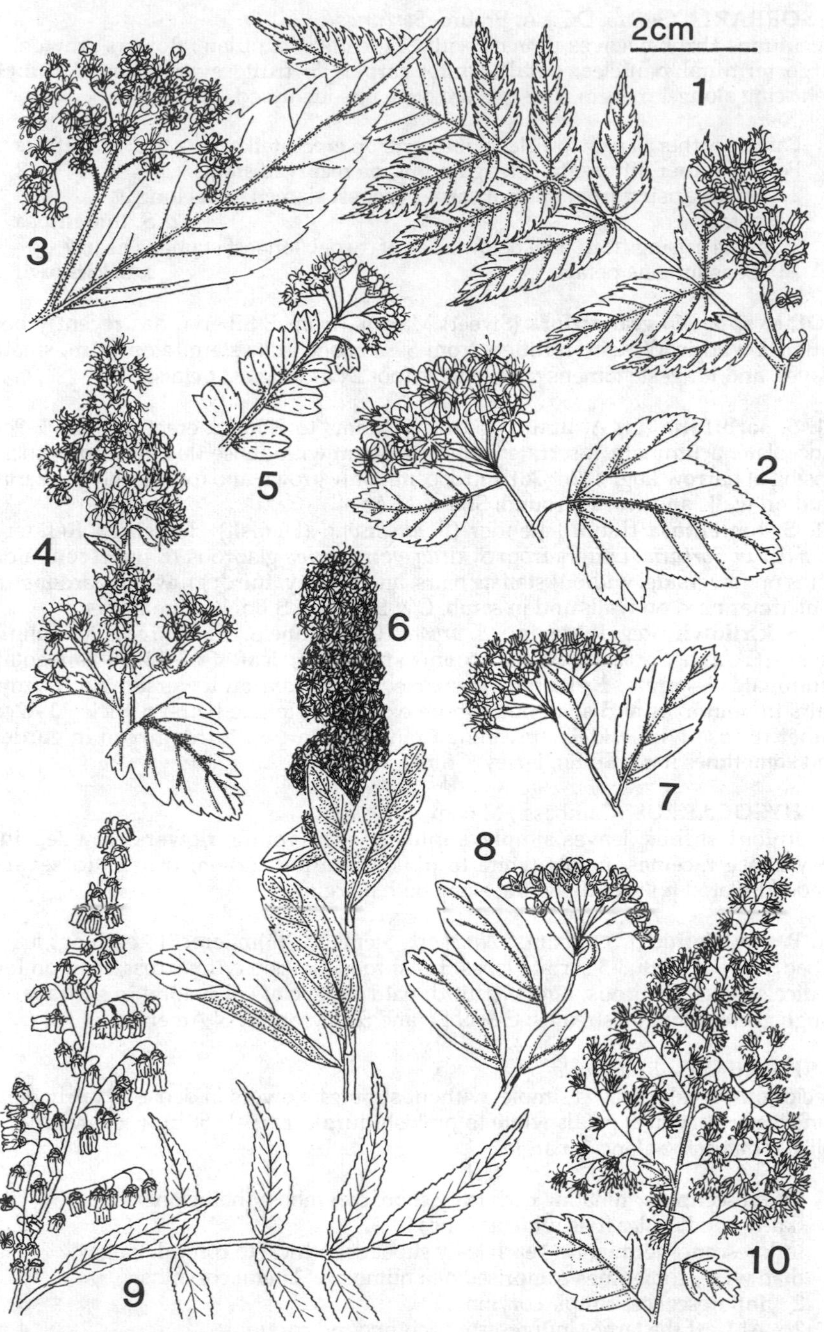

FIG 326 - Rosaceae: Spiraeoideae. 1, *Sorbaria sorbifolia*. 2, *Physocarpus opulifolius*. 3, *Spiraea japonica*. 4, *S. alba*. 5, *S. canescens*. 6, *S. douglasii*. 7, *S. chamaedryfolia*. 8, *S. x vanhouttei*. 9, *Sorbaria tomentosa*. 10, *Holodiscus discolor*.

4 Petals white; branchlets strongly angled; leaves mostly <2cm; stamens <1.2x as long as petals; follicles pubescent **10. S. canescens**

4 Petals pink; branchlets ± terete; leaves mostly >2cm; stamens >1.2x as long as petals; follicles glabrous **9. S. japonica**

5 Petals much longer than stamens, white **14. S. x arguta**

5 Petals shorter than stamens or ± as long, pink or white 6

6 Leaves glabrous or nearly so when mature 7

6 Leaves tomentose to slightly pubescent on lowerside even when mature 10

7 Inflorescence axis glabrous; leaves serrate only in apical 1/2; sepals reflexed in fruit **12. S. media**

7 Inflorescence axis pubescent; leaves serrate for most of length; sepals erect in fruit 8

8 Panicles broadly conical, with long branches near base; leaves mostly widest above mid-way; petals usually white; stamens c. as long as petals **4. S. alba**

8 Panicles ± cylindrical, with short branches; leaves mostly widest below mid-way; petals usually pink; stamens usually distinctly longer than petals 9

9 Panicle-branches usually pubescent; leaves usually lanceolate; petals usually bright pink; pollen >90 per cent fertile **1. S. salicifolia**

9 Panicle-branches usually sparsely pubescent; leaves usually narrowly ovate; petals usually pale pink; pollen <20% fertile **2. S. x rosalba**

10 Panicles usually both terminal and lateral, the terminal ones broadly conical, c. as long as wide; leaves of flowering shoots mostly <4cm **8. S. x brachybotrys**

10 Panicles terminal only, much longer than wide; leaves of flowering shoots mostly >4cm 11

11 Leaves tomentose on lowerside 12

11 Leaves pubescent on lowerside, the leaf surface showing through 13

12 Leaves oblong, white- to pale grey-tomentose on lowerside, ± entire in proximal c.1/2 of margin; follicles glabrous **6. S. douglasii**

12 Leaves ovate, grey- to buff-tomentose on lowerside, serrate for distal ≥2/3 of margin; follicles pubescent **7. S. tomentosa**

13 Leaves oblong to narrowly so, ± entire in proximal c.1/2 of margin; pollen >90 per cent fertile **6. S. douglasii**

13 Leaves ovate to elliptic to narrowly so, serrate for distal ≥2/3 of margin; pollen <20 per cent fertile 14

14 Panicles narrowly conical; petals often very pale pink **5. S. x billardii**

14 Panicles subcylindrical; petals pink **3. S. x pseudosalicifolia**

1. S. salicifolia L. - *Bridewort*. Strongly suckering; stems ± erect, to 2m; leaves 4-8cm, elliptic-oblong to lanceolate, sharply serrate except at extreme proximal end, glabrous; panicle ± cylindrical; petals pink; (2n=36). Intrd; formerly much grown, natd in hedges and rough ground; throughout BI but over-recorded and now very rare; C Europe.

2. S. x rosalba Dippel (*S. x rubella* Dippel; *S. salicifolia* x *S. alba*) - *Intermediate Bridewort*. Differs from *S. salicifolia* in narrowly conical panicle; usually very pale pink petals; and see key. Intrd; garden hybrid natd in hedges, etc.; probably throughout Br and commoner than *S. salicifolia* in Sc.

3. S. x pseudosalicifolia Silverside (*S. x billardii* auct. non Hérincq; *S. salicifolia* x *S. douglasii*) - *Confused Bridewort*. Intermediate in leaf shape, serration and pubescence; inflorescence ± cylindrical; petals pink. Intrd; garden hybrid natd in hedges, etc.; throughout Br.

4. S. alba Du Roi (*S. latifolia* (Aiton) Borkh.) - *Pale Bridewort*. Differs from *S.* 326
salicifolia in leaf- and panicle-shape (see key); petals usually white, sometimes pink;

(2n=36). Intrd; natd from gardens in hedges, etc.; frequent in Sc, Man, N Ir and probably N En; E N America.

5. S. x billardii Hérincq (*S. alba* x *S. douglasii*) - *Billard's Bridewort*. Intermediate in leaf shape, pubescence and serration, panicle shape and flower colour (see key). Intrd; garden hybrid natd in hedges, etc.; rather rare in Br and Man.

6. S. douglasii Hook. - *Steeple-bush*. Strongly suckering; stems ± erect, to 2m; **326**
leaves 4-8cm, oblong, sharply serrate in distal 1/2, ± entire in proximal 1/2; panicle ± cylindrical; petals pink; (2n=36). Intrd; commonly grown in gardens, natd in hedges, etc.; commoner than *S. salicifolia* in Br and under-recorded, Man, ?Ir; W N America.

a. Ssp. douglasii. Leaves whitish- or pale greyish-tomentose on lowerside. Common.

b. Ssp. menziesii (Hook.) Calder & Roy L. Taylor. Leaves subglabrous to pubescent on veins on lowerside. Very rarely natd.

7. S. tomentosa L. - *Hardhack*. Differs from *S. douglasii* in leaf shape, pubescence and serration, and follicle pubescence (see key); (2n=36). Intrd; grown in gardens, rather rarely natd in En; E N America.

8. S. x brachybotrys Lange (*S. douglasii* x *S. canescens*) - *Lange's Spiraea*. Stems ± erect, to 2m; leaves 1.5-4cm, elliptic-oblong, toothed only near apex, greyish-tomentose to -pubescent on lowerside; panicle broadly conical; petals pale pink. Intrd; garden hybrid rarely natd in Sc.

9. S. japonica L. f. - *Japanese Spiraea*. Stems ± erect, to 1.5m; leaves 5-12cm, **326**
elliptic to narrowly so, coarsely serrate, glabrous to slightly pubescent on veins; panicle ± flat-topped; petals pink; (2n=14, 18, 34). Intrd; grown in gardens, natd in hedges, etc.; occasional in Br, Man and CI; Japan. **S. 'Anthony Waterer'** is the commonest clone in gardens and might be so in the wild too.

10. S. canescens D. Don - *Himalayan Spiraea*. Stems arching, to 2m; leaves 1- **326**
2.5cm, obovate, crenate-dentate in distal 1/2, pubescent on lowerside; panicle round-topped; petals white; (2n=36). Intrd; grown in gardens, natd in scrub; occasional in S Br; Himalayas.

11. S. chamaedryfolia L. - *Elm-leaved Spiraea*. Stems arching, to 2m; leaves 2-7cm, **326**
ovate, coarsely serrate except near base, glabrous or nearly so; corymb simple, round-topped; petals white; (2n=18, 32, 36). Intrd; grown in gardens, natd in river gorge in Angus since before 1966; SE Europe. Our plant is ssp. **ulmifolia** (Scop.) J. Duvign.

12. S. media Schmidt - *Russian Spiraea*. Stems arching, to 2m; leaves 2-5cm, elliptic-oblong, serrate in apical 1/2, glabrous or nearly so; inflorescence an elongated panicle of lateral corymbs; petals white; (2n=10, 18, 20, 36). Intrd; garden plant natd by road and in open woodland in Westmorland since 1989; Russia and E Europe.

13. S. x vanhouttei (Briot) Carrière (*S. cantoniensis* Lour. x *S. trilobata* L.) - *Van* **326**
Houtte's Spiraea. Stems arching, to 2m; leaves 1.5-4cm, obovate, coarsely serrate and at least some 3(-5)-lobed in distal part, glabrous; corymb simple, nearly flat-topped; petals white. Intrd; grown in gardens; natd in hedges, etc.; scattered in En and Man; garden origin.

14. S. x arguta Zabel (*S. thunbergii* Sieb. ex Blume x *S. x multiflora* Zabel) - *Bridal-spray*. Stems arching, to 2m; leaves 15-40mm, oblanceolate to oblong-obovate, serrate, glabrous at maturity; small lateral corymbs closely set along branches forming terminal arching sprays; petals white. Intrd; garden plant natd in shrubberies, rough ground and by paths and roads; scattered in En and Sc, Man; garden origin.

4. ARUNCUS L. - *Buck's-beard*

Herbaceous perennials; leaves 2-3-pinnate, without stipules; flowers ± dioecious, in very dense terminal panicles; petals white; carpels 3; fruit 2- to several-seeded follicles dehiscing along 1 margin.

See *Astilbe* (Saxifragaceae) for differences.

1. A. dioicus (Walter) Fernald (*A. sylvester* Kostel., *A. vulgaris* Raf.) - *Buck's-beard*. Stems to 2m, erect; basal leaves to 1m, the leaflets ovate, sharply 2-serrate, glabrous or nearly so; inflorescence up to 50cm, much-branched; pedicels strongly recurved in fruit; follicles glabrous; (2n=18). Intrd; much planted in gardens, very persistent in woodland and by water, but ± never seeding; scattered throughout En and Sc, especially N; Europe.

5. HOLODISCUS (K. Koch) Maxim. - *Oceanspray*
Deciduous shrubs; leaves simple, without stipules; flowers in dense terminal panicles; petals creamy-white; carpels 5; fruit 1-seeded indehiscent achenes.

1. H. discolor (Pursh) Maxim. - *Oceanspray*. Stems to 4m, erect to arching; leaves **326**
ovate to broadly so, coarsely toothed and ± lobed, greyish-pubescent to -tomentose on lowerside; panicles arching to pendent, up to 30cm; achenes pubescent; (2n=18). Intrd; grown in gardens, natd in hedges and scrub and on walls; scattered in En and Sc; W N America.

SUBFAMILY 2 - ROSOIDEAE (genera 6-21). Stipulate herbs, sometimes shrubs; hypanthium variably concave, sometimes flat, not or sometimes enclosing carpels but fused to them only at base; epicalyx present or 0; stamens 1-numerous; flowers mostly 4-6-merous; carpels 1-numerous, 1-2-seeded; fruit a head of achenes or drupes, sometimes borne on succulent receptacle or surrounded by succulent or dry hypanthium; chromosome base-number 7, 8 or 9.

6. FILIPENDULA Mill. - *Meadowsweets*
Herbaceous perennials; leaves pinnate or reduced to terminal lobe only; flowers in terminal ± flat-topped panicles, 5-6-merous; epicalyx 0; hypanthium ± flat to saucer-shaped; stamens numerous; carpels 4-12; fruit a head of achenes each with 1-2 seeds.

1 Basal leaves with terminal and 8-30 pairs of main leaflets all 0.5-2cm; petals usually 6 **1. F. vulgaris**
1 Basal leaves with terminal and 0-5 pairs of main leaflets all >2cm; petals usually 5 2
2 Carpels 6-10, spirally twisted, glabrous; leaves with 2-5 pairs of large lateral leaflets **2. F. ulmaria**
2 Carpels 4-6, straight, pubescent on edges; leaves without large lateral leaflets **3. F. kamtschatica**

1. F. vulgaris Moench - *Dropwort*. Stems erect, to 50(100)cm; leaves with 8-30 pairs of main leaflets and smaller ones between, glabrous; petals 5-9mm, creamy-white; achenes 6-12, erect, pubescent; 2n=14, 15. Native; calcareous grassland; very locally frequent in Br N to C Sc, grown in gardens and sometimes natd elsewhere.

2. F. ulmaria (L.) Maxim. - *Meadowsweet*. Stems erect, to 1.2m; leaves with 2-5 pairs of main leaflets and smaller ones between, tomentose to glabrous on lowerside; flowers >5mm across, white; 2n=14, 16. Native; all sorts of wet and damp places; common throughout BI.

3. F. kamtschatica (Pall.) Maxim. - *Giant Meadowsweet*. Stems erect, to 3m; leaves densely pubescent on lowerside, with 1 large, lobed terminal leaflet and 1-2 pairs of very small laterals; flowers <5mm across, red, pink or white; (2n=14, 18, 28, 42, 44). Intrd; planted in damp places and sometimes natd; very scattered in Br, especially Sc and N En, W Ir; E Asia. Plants with pink to red flowers are often referred uncritically to **F. purpurea** Maxim., from Japan or to **F. rubra** (Hill)

Robinson, from N America, but these are ± glabrous and our plants are probably cultivars of *F. kamtschatica* or hybrids of it. *F. purpurea* and *F. rubra* are grown in gardens.

7. KERRIA DC. - *Kerria*

Deciduous shrubs; leaves simple; flowers solitary, terminal on lateral branches, 5-merous; epicalyx 0; hypanthium ± flat; stamens numerous; carpels 5-8; fruit a head of achenes.

1. K. japonica (L.) DC. - *Kerria*. Stems to 2.5m, erect, bright green above; leaves ovate to lanceolate, coarsely serrate, sparsely pubescent; flowers 2.5-5cm across, yellow; (2n=18). Intrd; much grown in gardens, persistent in neglected shrubberies and old garden sites, usually as *flore pleno*, but self-sown in Middlesex; very scattered in Br, Man; China.

8. RUBUS L. - *Brambles*

Deciduous or semi-evergreen shrubs, often spiny, or herbaceous perennials; leaves simple, ternate, pinnate or palmate; flowers solitary or in racemes or panicles, usually 5-merous; epicalyx 0; hypanthium flat, with receptacle usually extended upwards from centre; stamens numerous; carpels usually numerous; fruit a head of (1)2-many 1-seeded drupes.

Most of the taxa of subg. *Rubus* (taxa 12-25) form an extremely complex, largely apomictic group (sect. *Glandulosus*), often known collectively as **R. fruticosus** L. agg. 2 other sections (*Rubus* and *Corylifolii*) are often included within this aggregate, but they are probably derived from ancient and some recent hybrids between it and *R. idaeus* and *R. caesius* respectively, and are here treated separately. *R. caesius* forms a 4th section. Over 400 microspp. (320 currently) have been recognized in BI in these 4 sections together. In this work the microspp. of sects. *Rubus*, *Corylifolii* and *Glandulosus* are not treated in full but, following the view of A. Newton and E.S. Edees, 11 rather ill-defined series, representing the main nodes in the spectrum of variation, are recognized in sect. *Glandulosus*. These are keyed out in couplets 15-24, but most of the characters used are relative and a high level of success will be achieved only after much experience.

1 Leaves simple — 2
1 Leaves pinnate, palmate or ternate — 5
  2 Stipules arising direct from stem; receptacle strongly convex — 3
  2 Stipules fused to petiole proximally; receptacle flat — 4
3 Stems herbaceous, erect; flowers solitary, dioecious — **1. R. chamaemorus**
3 Stems arching or procumbent, rooting at tips; flowers in clusters, bisexual — **2. R. tricolor**
  4 Petals white — **6. R. parviflorus**
  4 Petals reddish-purple — **5. R. odoratus**
5 Stipules arising direct from stem; stems all annual, producing flowers in 1st year; receptacle flat — 6
5 Stipules fused to petiole proximally; stems at least biennial, producing flowers in 2nd year; receptacle conical — 7
  6 Stoloniferous, without rhizomes; petals <6mm, white — **3. R. saxatilis**
  6 Rhizomatous, without stolons; petals >6mm, pink — **4. R. arcticus**
7 Stems densely covered with white bloom — **10. R. cockburnianus**
7 Stems green to red, sometimes glaucous but not with dense white bloom — 8
  8 Stems usually not rooting at tips; leaves ternate or pinnate; fruit usually red to orange, usually separating from receptacle at maturity — 9
  8 Stems usually rooting at tips; leaves ternate or palmate; fruit usually black, sometimes red, coming away with extension of receptacle at maturity (subg. *Rubus*) — 12

9 At least lower leaves pinnate with 5(-7) leaflets; petals white 10
9 Leaves normally ternate; petals pink to purple 11
10 Stems ± erect; fruit separating from receptacle at maturity; stem with weak prickles **7. R. idaeus**
10 Stems arching and eventually rooting at tips; fruit coming away with extension of receptacle at maturity; stem with moderate prickles **11. R. loganobaccus**
11 Leaves ± glabrous; flowers >2cm across, 1-few per group, appearing in spring; glandular hairs 0 **9. R. spectabilis**
11 Leaves white-tomentose on lowerside; flowers <1.5cm across, many per group, appearing in summer; glandular bristles dense on stems and flower-stalks **8. R. phoenicolasius**
12 Fruit with glaucous bloom, composed of rather few, large, only loosely coherent drupes; leaflets 3, the lateral ones ± sessile **25. R. caesius**
12 Fruit without glaucous bloom, usually composed of many tightly coherent drupes; leaflets often >3 13
13 Leaflets ± overlapping, the basal pair ± sessile; stipules lanceolate; inflorescence usually a ± simple corymb **24. R. sect. Corylifolii**
13 Leaflets mostly ± not overlapping, the basal pair usually stalked; stipules linear; inflorescence compound or ± racemose 14
14 Stems suberect, usually not rooting at tips; suckers often produced from roots; fruits often red to purple, sometimes black **12. R. sect. Rubus**
14 Stems procumbent to arching, usually rooting at tips; suckers not produced; fruits black (sect. *Glandulosus*) 15
15 Stalked glands 0 or rare and inconspicuous on lst-year stems 16
15 Stalked glands on lst-year stems obvious 18
16 Leaflets variously pubescent on lowerside but not, or only the upper ones, tomentose **13. R. series Sylvatici**
16 Leaflets all tomentose on lowerside 17
17 Leaflets greyish-white-tomentose on lowerside; few stalked glands sometimes present in inflorescence **14. R. series Rhamnifolii**
17 Leaflets chalky-white-tomentose on lowerside; stalked glands 0 **16. R. series Discolores**
18 Stamens shorter than styles; stalked glands usually few **15. R. series Sprengeliani**
18 Stamens as long as or longer than styles; stalked glands usually many 19
19 Main prickles ± confined to angles of stem, distinct from smaller pricklets and acicles 20
19 Main prickles occurring all round stem, grading into pricklets and acicles 23
20 Stems conspicuously pubescent, the stalked glands less conspicuous **17. R. series Vestiti**
20 Stems glabrous to pubescent, the stalked glands more conspicuous than hairs 21
21 Terminal leaflet obovate, often broadly so, with short cuspidate apex and serrulate margin **18. R. series Mucronati**
21 Terminal leaflet usually ovate to elliptic, usually less abruptly narrowed at apex, usually more serrate on margin 22
22 Pricklets, acicles and stalked glands on stems of similar lengths **19. R. series Radulae**
22 Pricklets, acicles and stalked glands on stems of different lengths **20. R. series Micantes**
23 Prickles and stalked glands in very variable quantities on same plant **21. R. series Anisacanthi**
23 Prickles and stalked glands not markedly variable in quantity on same plant 24

24 Prickles strong; pricklets often more numerous than stalked glands **22. R. series Hystrices**
24 Prickles weak; stalked glands and acicles often very numerous **23. R. series Glandulosi**

Other spp. - R. coreanus Miq. (*Korean Bramble*)(subg. *Idaeobatus*), from E Asia, has white stems like *R. cockburnianus* but leaves with only 5(-7) leaflets the terminal one of which is rhombic and subacute (not ovate and acuminate); it was formerly natd in Yorks. **R. deliciosus** Torr. (*Rocky Mountain Raspberry*)(subg. *Anoplobatus*), from W N America, has simple leaves with 3-5 obtuse lobes and solitary white flowers 5-6cm across; it was formerly natd in E Suffolk.

Subgenus 1 - CHAMAEMORUS (Hill) Focke (sp. 1). Stems herbaceous, without prickles or acicles; rhizomes present; stolons 0; leaves simple; fruits orange.

1. R. chamaemorus L. - *Cloudberry.* Stems to 20cm, erect, annual, clothed only with hairs; leaves simple, ± orbicular, palmately 5-7-lobed; flowers dioecious, solitary, terminal, white, 20-30mm across; fruit orange when ripe, of 4-20 large drupes; 2n=56. Native; peaty moors and bogs on mountains; Br N from N Wa and Derbys, Tyrone.

Subgenus 2 - DALIBARDASTRUM Focke (sp. 2). Stems ± woody, stoloniferous, without prickles but with stiff hairs and some acicles; rhizomes 0; leaves simple; fruits red.

2. R. tricolor Focke - *Chinese Bramble.* Stems to several m, trailing, rooting at tips, 333
with dense brownish bristles; leaves simple, ovate, pinnately 3-5-lobed; flowers in racemes, 20-25mm across, white; fruit red, but plant often not flowering; (2n=28). Intrd; grown as ground-cover but spreading and natd in hedges and shrubberies; very scattered in En; China.

Subgenus 3 - CYLACTIS (Raf.) Focke (spp. 3-4). Stems herbaceous, without or with few weak prickles and acicles; rhizomes or stolons present (not both); leaves (simple to) ternate; fruits red.

3. R. saxatilis L. - *Stone Bramble.* Flowering stems to 40cm, erect, with hairs and very weak prickles; stolons longer, often rooting at apex but the rest dying in winter; leaves ternate, with ovate to elliptic leaflets; flowers few, in terminal corymbs, white, 8-15mm across; fruit red, of 1-6 drupes; 2n=28. Native; woods, screes and mountain slopes on basic soils; rather scattered in Ir, Sc, Wa and C, N & W En.

4. R. arcticus L. - *Arctic Bramble.* Stems to 30cm, erect, annual, ± glabrous; RR
rhizomes present; leaves mostly ternate, with ovate to obovate leaflets; flowers 1-3, terminal, pink, 15-25mm across; fruit dark red, of numerous drupes; (2n=14). Probably native; several records from highlands of Sc, the last in 1841; extinct.

Subgenus 4 - ANOPLOBATUS Focke (spp. 5-6). Stems woody, biennial, without prickles or acicles; rhizomes present; stolons 0; leaves simple; fruits red.

5. R. odoratus L. - *Purple-flowered Raspberry.* Stems to 3m, erect, with glandular and non-glandular hairs; leaves simple, ± orbicular, with 3-5 acute lobes; flowers in panicles of usually >6, purple, 3-5cm across; fruit red; (2n=14). Intrd; grown for ornament, occasionally natd in rough places very scattered in S En; E N America.

5 x 6. R. odoratus x R. parviflorus = R. x fraseri Rehder has been natd in roadside scrub in W Sussex since 1985; garden origin.

6. R. parviflorus Nutt. - *Thimbleberry.* Stems to 2m, erect, with glandular and non- 333

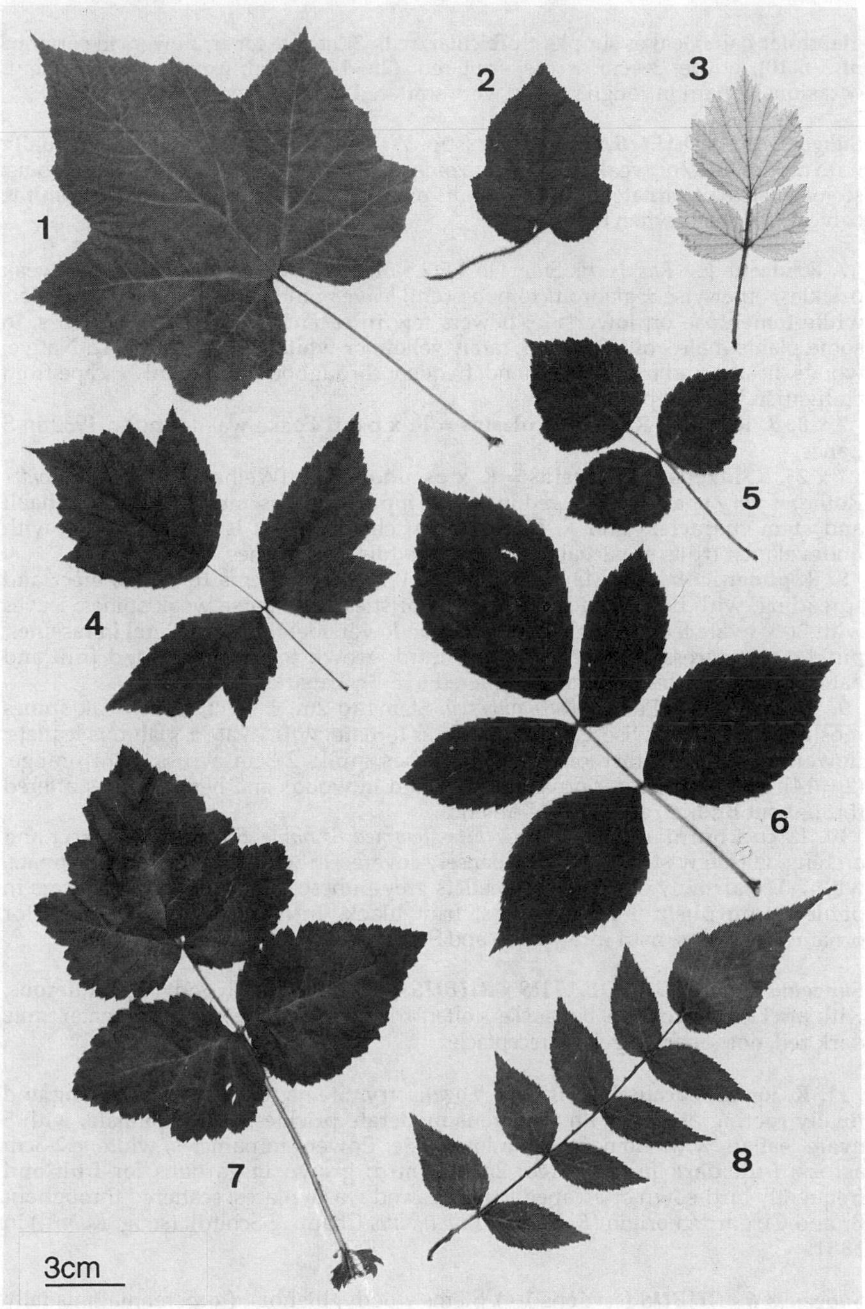

FIG 333 - Leaves of ***Rubus***. 1, ***R. parviflorus***. 2, ***R. tricolor***. 3, ***R. phoenicolasius***. 4, ***R. spectabilis***. 5, ***R. caesius***. 6, ***R. idaeus***. 7, ***R. loganobaccus***. 8, ***R. cockburnianus***.

glandular hairs; leaves simple, ± orbicular, with 3-5 acute lobes; flowers in corymbs of 3-6(10), white, 3-6cm across; fruit red; (2n=14). Intrd; grown for ornament, occasionally natd in rough ground; very scattered in Br, N Kerry; W N America.

Subgenus 5 - IDAEOBATUS Focke (spp. 7-10). Stems woody, biennial, usually with acicles and/or weak prickles; rhizomes 0 but suckers often arising from roots; stolons 0; leaves ternate to pinnate; fruits orange, red or black, separating from the conical receptacle when fully ripe.

7. R. idaeus L. - *Raspberry*. Stems to 1.5(2.5)m, erect, with few to numerous weak **333**
prickles, otherwise ± glabrous to pubescent; leaves pinnate, with 3-7 ovate leaflets **336**
white-tomentose on lowerside; flowers few in racemes, white, c.1cm across, in some plants male only; fruit red, rarely yellow or white; 2n=14, 21, 28, 42. Native; woods, heaths and marginal ground; frequent throughout BI, but only escape from cultivation in some places.

7 x 8. R. idaeus x R. phoenicolasius = R. x paxii Focke was found in 1930 in S Lancs.

7 x 25. R. idaeus x R. caesius = R. x pseudoidaeus (Weihe) Lej. (*R. x idaeoides* Ruthe) is very sparsely scattered in En, N Tipperary; it resembles *R. caesius* in habit and stem characters and *R. idaeus* in leaf characters. It is largely sterile, with undeveloped fruits or partially developed reddish-black ones.

8. R. phoenicolasius Maxim. - *Japanese Wineberry*. Stems to 2(3)m, erect and **333**
spreading, with dense reddish glandular bristles and sparse weak spines; leaves with 3(-5) ovate leaflets white-tomentose on lowerside; flowers several in racemes, pink, c.1cm across; fruit red; (2n=14). Intrd; grown for ornament and fruit and natd in rough places and scrub; scattered in S Br, Lanarks; E Asia.

9. R. spectabilis Pursh - *Salmonberry*. Stems to 2m, ± erect, with weak spines **333**
mostly below, otherwise ± glabrous; leaves ternate, with ovate ± glabrous leaflets; flowers usually solitary on lateral branches, pink, 2-3cm across; fruit orange; (2n=14). Intrd; grown for ornament and natd in woods and hedgerows; scattered throughout Br, Man and Ir; W N America.

10. R. cockburnianus Hemsl. - *White-stemmed Bramble*. Stems to 5m, erect and **333**
arching, with few strong prickles, densely covered in white bloom; leaves pinnate, with 7-11 narrowly elliptic-ovate leaflets grey-pubescent on lowerside; flowers in panicles, purplish, 1-1.8cm across; fruit black. Intrd; grown in gardens for ornamental stems, natd rarely in En and Sc; China.

Subgenus 5 x 6 - IDAEOBATUS x RUBUS (sp. 11). Stems woody, stoloniferous, with prickles; rhizomes 0 but suckers often arising from roots; leaves pinnate; fruits dark red, not separating from receptacle.

11. R. loganobaccus L.H. Bailey - *Loganberry*. Stems to several m, arching and **333**
finally rooting at tips, with numerous moderate prickles; leaves pinnate, with 5 ovate leaflets white-tomentose on lowerside; flowers in panicles, white, c.2-3cm across; fruit dark purplish-red; 2n=49. Intrd; grown in gardens for fruit and frequently bird-sown or escaped in hedges and waste places; scattered throughout Br and CI; garden origin (*R. idaeus* x *R. vitifolius* Cham. & Schltdl. (subg. *Rubus*)) in 1881.

Subgenus 6 - RUBUS (sections 1-4). Stems woody, biennial (to perennial), usually stoloniferous, usually with strong prickles and acicles; rhizomes 0; leaves palmate; fruits usually black, sometimes red, not readily separating from receptacle.

R. ulmifolius is a sexual diploid with 2n=14, but all other microspp. are facultative apomicts (mostly tetraploids but also triploids, pentaploids, hexaploids and heptaploids) that require pollination in order to produce even apomictic seed (pseudogamy). Sexual seed can also be produced, and hybridization is common. 320 microspp. are currently recognized. Although abundant and characteristic of much of C & S BI, this subgenus is much less common in Sc; only c.16 microspp. reach Outer Hebrides and no native ones reach Orkney or Shetland.

12. R. sect. 1 Rubus (sect. *Suberecti* Lindl.). Variously intermediate between *R. idaeus* and *R. fruticosus* agg.; stems usually suberect, often not rooting at tips, with weakly to well developed prickles and varying quantities (mostly few) of pricklets, acicles, stalked glands and hairs; suckers often produced from roots; leaflets 5(3-7), usually not overlapping, the basal ones usually stalked; stipules linear-filiform; flowers mostly in racemose inflorescences, white to pink; fruit red to black, without glaucous bloom; 2n=21, 28, 42. Native; sunny and partly shaded places; throughout BI, mostly En and Wa, only 7 microspp. in Ir, 4 in Sc, 3 in Man, 1 in CI; 7 endemic. 4 of the microspp. (**R. fissus** Lindl., **R. nessensis** Hall, **R. plicatus** Weihe & Nees and **R. scissus** W.C.R. Watson) are probably among the 9 most widely distributed of the subgenus in BI. 20 microspp. currently placed here, incl. the natd N American spp. **R. canadensis** L. and **R. allegheniensis** Porter. A natural recent hybrid between *R. idaeus* and *R. fruticosus* agg. was recorded from Berks in 1922, it had stems rooting at the tips and red fruit tasting of raspberries and separating from the receptacle (not so in subgenus *Rubus*). *R. loganobaccus* (q.v.) and more recent crop species (e.g. *Tayberry, Boysenberry*) are also of similar parentage. **336**

13-23. R. sect. 2 Glandulosus Wimm. & Grab. (subsect. *Hiemales* E.H.L. Krause; *R. fruticosus* L. agg.) - *Bramble*. Stems procumbent to arching, (potentially) rooting at tips, with well-developed prickles and varying quantities of pricklets, acicles, stalked glands and hairs; roots not producing suckers; leaflets (3-)5(-7), usually not overlapping, the basal ones usually stalked; stipules linear-filiform; flowers in compound or racemose inflorescences, white to pink or red; fruit black, without glaucous bloom. Native; all sorts of habitats both natural and man-made, but much less common on calcareous soils; throughout BI, but very local in N Sc. Very variable; almost all characters of inflorescence and vegetative parts are of value in distinguishing the microspp. and series, but probably the indumentum and armature are the most important. 276 microspp. are currently placed here.

13. R. series Sylvatici (P.J. Müll.) Focke (sect. *Sylvatici* P.J. Müll., subsect. *Virescentes* Genev.). Stems arching or procumbent, glabrous to pubescent; usually without stalked glands or acicles; prickles confined to angles of stem; stalked glands (but 0 acicles) sometimes present in inflorescence; leaflets ± green and variously pubescent but not tomentose on lowerside; 2n=21, 28, 42. Throughout BI, 8 microspp. reaching Sc, 3 reaching CI; c.39 endemic. 58 microspp. currently placed here. The alien **R. laciniatus** Willd. (origin unknown), with distinctive dissected leaves, is grown for its fruits and is frequently natd over most of BI. **R. lindleianus** Lees is 1 of the most widespread microspp. in the subgenus in BI. **336**

14. R. series Rhamnifolii (Bab.) Focke (subsect. *Discoloroides* Genev. ex Sudre). Stems arching or procumbent, glabrous to pubescent; usually without stalked glands or acicles; prickles confined to angles of stem; stalked glands (but 0 acicles) sometimes present in inflorescence; leaflets greyish-white-tomentose on lowerside; 2n=28. Throughout BI, 13 microspp. reaching Sc, 4 reaching CI; c.27 endemic. 41 microspp. currently placed here, incl. the alien **R. elegantispinosus** (Schumach.) H.E. Weber, from Europe. **R. nemoralis** P.J. Müll. and **R. polyanthemus** Lindeb. are 2 of the most widespread microspp. in the subgenus in BI. **336**

15. R. series Sprengeliani Focke (sect. *Sprengeliani* (Focke) W.C.R. Watson nom. inval.). Stems low-arching to procumbent, sparsely pubescent, with usually few stalked glands but 0 acicles; prickles confined to angles of stem; stalked glands and acicles present in inflorescence; leaflets ± green and sparsely pubescent on **336**

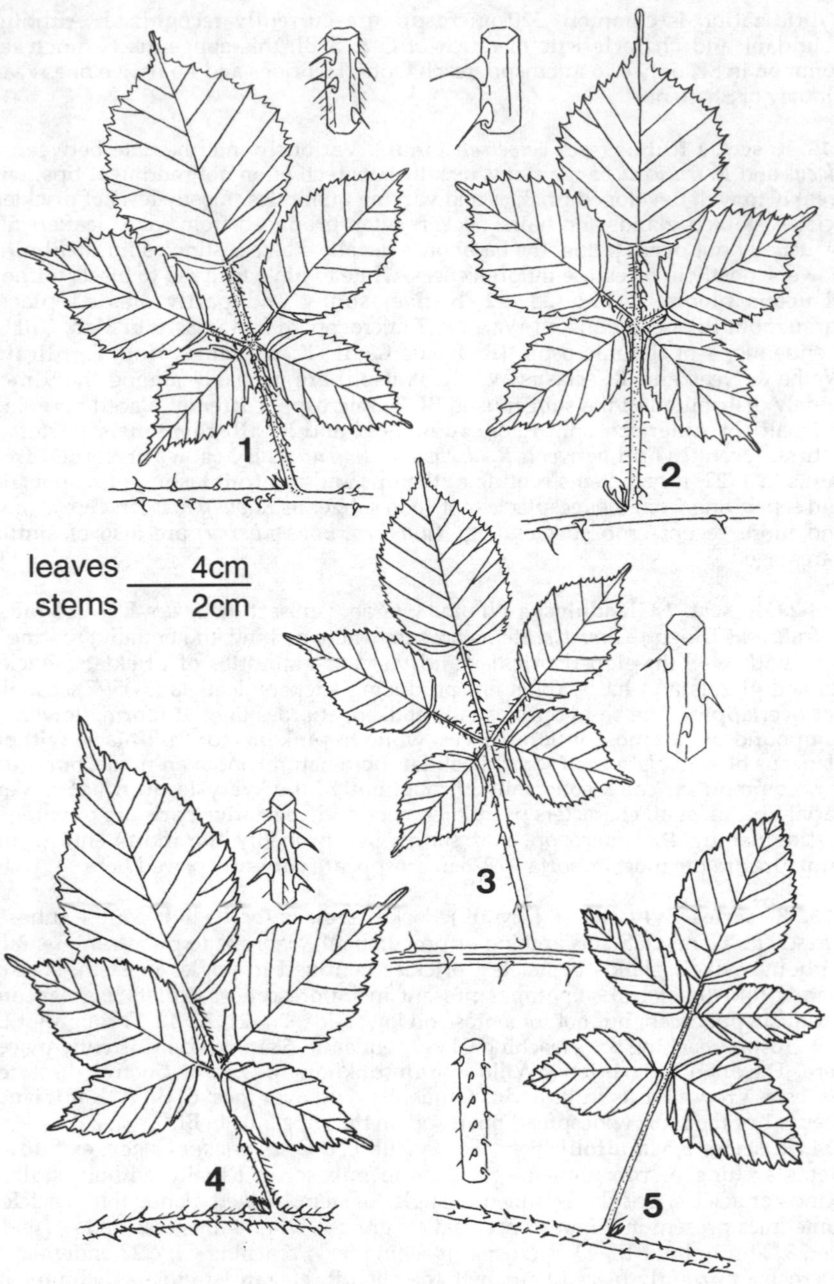

Fig 336 - Leaves and first-year stems of *Rubus*. 1, ***R. calvatus*** (series ***Sylvatici***). 2, ***R. nobilissimus*** (sect. ***Rubus***). 3, ***R. cardiophyllus*** (series ***Rhamnifolii***). 4, ***R. sprengelii*** (series ***Sprengeliani***). 5, ***R. idaeus*** (subgenus ***Idaeobatus***).

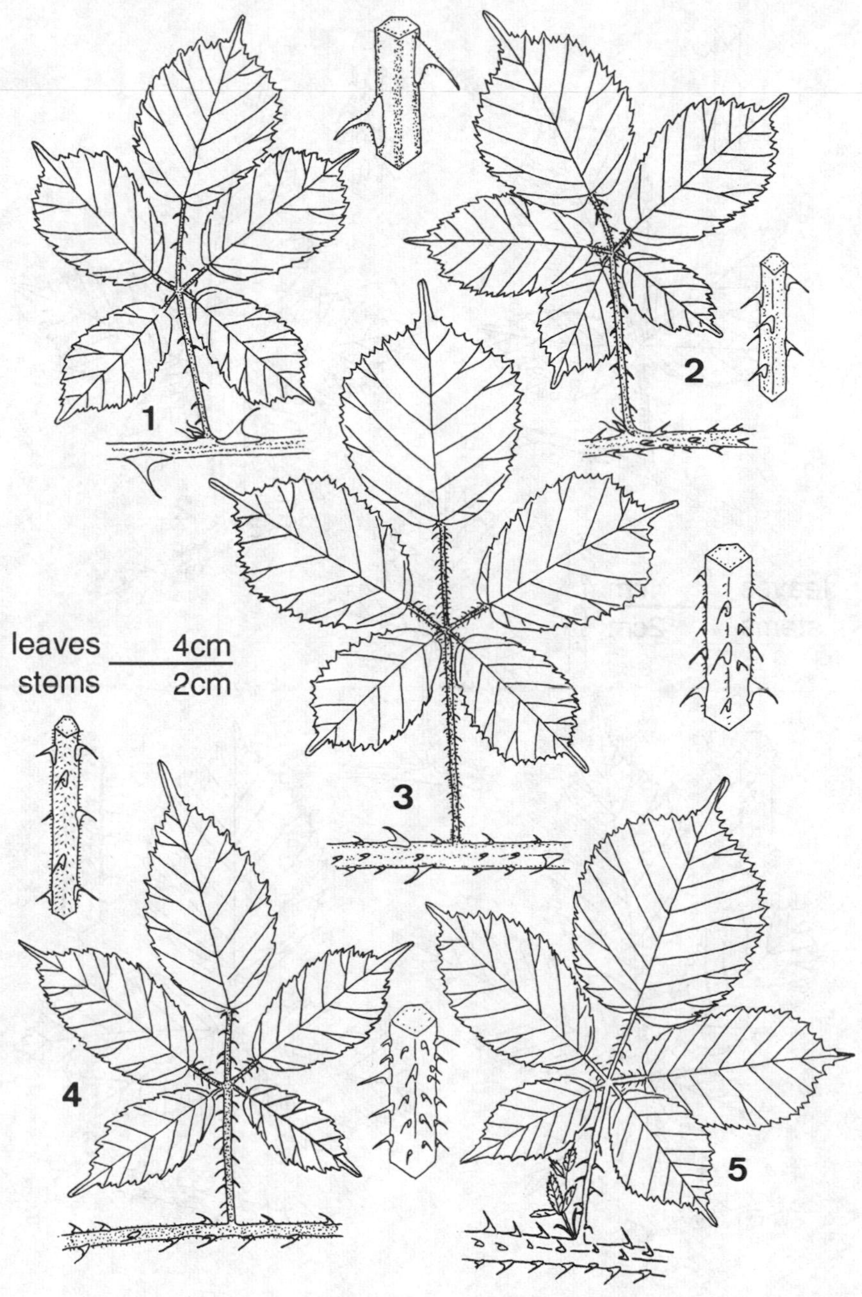

FIG 337 - Leaves and first-year stems of *Rubus*. 1, ***R. ulmifolius*** (series *Discolores*). 2, ***R. mucronulatus*** (series *Mucronati*). 3, ***R. boraeanus*** (series *Vestiti*). 4, ***R. wedgwoodiae*** (series *Micantes*). 5, ***R. leyanus*** (series *Anisacanthi*).

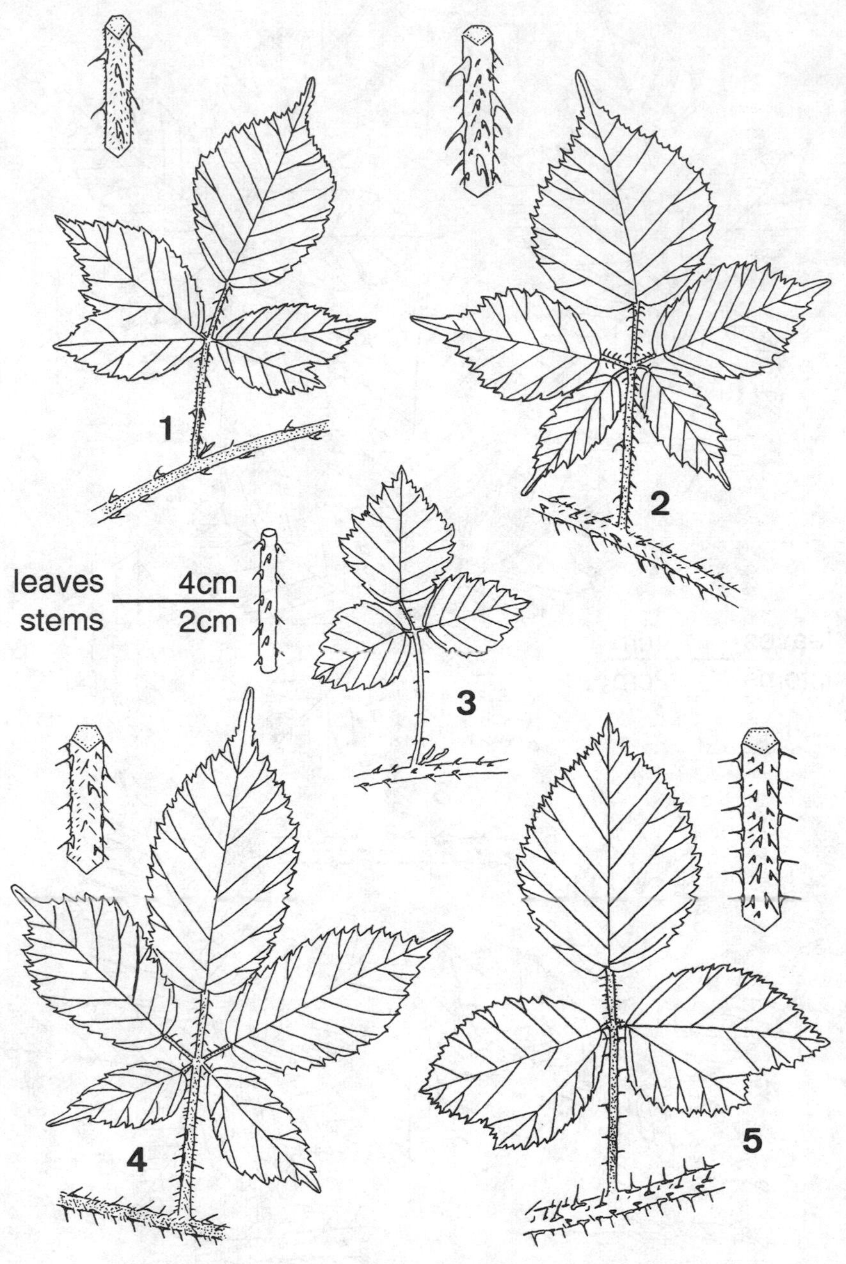

FIG 338 - Leaves and first-year stems of *Rubus*. 1, ***R. fuscicaulis*** (series *Radulae*). 2, ***R. atrebatum*** (series *Hystrices*). 3, ***R. caesius*** (sect. *Caesii*). 4, ***R. hylonomus*** (series *Glandulosi*). 5, ***R. tuberculatus*** (sect. *Corylifolii*).

lowerside; stamens usually shorter than styles (usually longer in all other series); 2n=21, 28. Br N to SW Sc, S Ir, CI; 2 microspp. endemic. 4 microspp. currently placed here; **R. sprengelii** Weihe with the distribution of the series, the other 3 only in S En.

16. R. series Discolores (P.J. Müll.) Focke (Sect. *Discolores* P.J. Müll.). Stems often 337
scrambling, glabrous to pubescent, without stalked glands or acicles; prickles confined to angles of stem; stalked glands and acicles 0 in inflorescence; leaflets chalky-white-tomentose on lowerside; 2n=14, 21, 28, 35, 49. Throughout most of BI, 2 microspp. reaching Sc, 3 reaching Ir, 2 reaching CI; 5 endemic. 11 microspp. currently placed here, incl. the alien **R. armeniacus** Focke (*R. procerus* auct.), from Europe, much grown for its fruit as 'Himalayan Giant'. **R. ulmifolius** Schott (*R. inermis* auct. non Pourr.), is the only diploid and only wholly sexual sp. of subg. *Rubus* in BI. It is widespread in BI, especially on chalk and clay where few other spp. occur, and hence is the commonest sp. in several areas incl. CI. It hybridizes with many other spp. as pollen parents; hybrids may be fertile or sterile. It is easily recognized by its stems with a whitish bloom and very short pubescence but otherwise only large subequal prickles, rather small elliptic to obovate leaflets white-tomentose on lowerside, flowers produced relatively late (from Aug) in long narrow inflorescences and with rounded pink petals, and small fruit.

17. R. series Vestiti (Focke) Focke (group. *Vestiti* Focke). Stems usually arching, 337
conspicuously pubescent, with stalked glands and acicles; prickles confined to angles of stem, subequal and distinct from pricklets and acicles; stalked glands and usually acicles present in inflorescence; leaflets pubescent to greyish-white-tomentose on lowerside; 2n=28, 35, 42. Throughout most of BI except Sc, where only the most widespread sp. (**R. vestitus** Weihe) reaches; c.17 endemic microspp. 22 microspp. currently placed here.

18. R. series Mucronati (Focke) H.E. Weber (sect. *Rotundifolii* W.C.R. Watson 337
nom. inval., sect. *Appendiculati* (Genev.) Sudre). Stems arching, glabrous or less often pubescent, with stalked glands and acicles; prickles confined to angles of stem, subequal and distinct from pricklets and acicles; stalked glands and acicles present in inflorescence; leaflets ± green and sparsely pubescent to pubescent on lowerside, the terminal one obovate, cuspidate at apex and with serrulate margin; 2n=28. Scattered throughout most of BI except CI; 8 endemic microspp. 11 microspp. currently placed here.

19. R. series Radulae (Focke) Focke (subsect. *Rudes* Sudre pro parte, series *Pallidi* 338
W.C.R. Watson). Stems usually arching, glabrous or pubescent, with stalked glands and acicles; prickles confined to angles of stem, subequal and distinct from pricklets and acicles; pricklets, acicles and stalked glands on stems of similar lengths; stalked glands and acicles present in inflorescence; leaflets green and pubescent to grey-tomentose on lowerside, the terminal one usually elliptic to ovate, usually acuminate at apex and with serrate margin; 2n=28, 35. Throughout most of BI, 6 microspp. reaching Sc, 1 reaching CI; c.27 endemic. 40 microspp. currently placed here. **R. radula** Weihe ex Boenn. is 1 of the most widespread microspp. in the subgenus in BI.

20. R. series Micantes Sudre ex Bouvet (series *Apiculati* Focke pro parte, series 337
Grandifolii Focke). Stems usually arching, glabrous or pubescent, with stalked glands and acicles; prickles confined to angles of stem, subequal and distinct from pricklets and acicles; pricklets, acicles and stalked glands on stems of different lengths; stalked glands and acicles present in inflorescence; leaflets usually ± green and sparsely pubescent to pubescent (sometimes ± grey-tomentose) on lowerside, the terminal one usually elliptic to ovate, usually acuminate at apex and with serrate margin; 2n=28, 35. Local in En and Wa, 1 microsp. in Sc, 5 very local in Ir, 2 in CI; c.24 endemic. 29 microspp. currently placed here.

21. R. series Anisacanthi H.E. Weber (series *Apiculati* Focke pro parte, series 337
Dispares W.C.R. Watson pro parte nom. inval.). Stems arching, glabrous to pubescent, with varying numbers of stalked glands and acicles; prickles borne all

round stem, unequal in length and grading into pricklets and acicles; stalked glands and acicles present in inflorescence; prickles, pricklets, acicles and stalked glands varying in abundance on different parts of same plant; leaflets green and sparsely pubescent to pubescent to grey-tomentose on lowerside, the terminal one obovate to ovate, cuspidate to acuminate at apex; 2n=28, 35. Throughout most of BI, 4 microspp. reaching Sc, 1 reaching CI; c. 11 endemic. 19 microspp. currently placed here.

22. R. series Hystrices Focke. Stems arching to procumbent, glabrous to 338
pubescent, with stalked glands and acicles but pricklets usually more conspicuous than either; prickles borne all round stem, unequal in length and grading into pricklets and acicles; stalked glands and acicles present in inflorescence; prickles, pricklets, acicles and stalked glands not varying in abundance on different parts of same plant; leaflets green and sparsely pubescent to pubescent to grey-tomentose on lowerside, the terminal one obovate to ovate, cuspidate to acuminate at apex; 2n=28, 35, 42. Throughout most of BI except CI, 3 microspp. reaching Sc, 2 reaching Ir and Man; c.29 endemic. 34 microspp. currently placed here. **R. dasyphyllus** (W.M. Rogers) E.S. Marshall is 1 of the most widespread microspp. in the subgenus in BI.

23. R. series Glandulosi (Wimm. & Grab.) Focke (sect. *Glandulosus* Wimm. & 338
Grab., series *Euglandulosi* W.C.R. Watson nom. inval.). Stems usually low-arching to procumbent, glabrous to pubescent, with stalked glands and acicles usually more conspicuous than pricklets; prickles borne all round stem, unequal in length and grading into pricklets and acicles; stalked glands and acicles present in inflorescence; prickles, pricklets, acicles and stalked glands not varying in abundance on different parts of same plant; leaflets usually green and sparsely pubescent to pubescent on lowerside, the terminal one usually ovate to elliptic, acuminate at apex; 2n=28, 35. Scattered in C & S Br, 1 microsp. reaching N En, 1 in Ir; 2 endemic. 7 microspp. currently placed here.

24. R. sect. 3 Corylifolii Lindl. (sect. *Triviales* P.J. Müll.). Variously intermediate 338
between *R. caesius* and *R. fruticosus* agg.; stems low-arching to procumbent, (potentially) rooting at tips, often with glaucous bloom, with variable indumentum and armature but usually only moderate prickles; leaflets 3-5, usually overlapping, the basal pair sessile or nearly so; stipules lanceolate; flowers usually large, often in ± corymbose inflorescences, white to pink or red; fruits black, without glaucous bloom, often with relatively large, few drupelets; 2n=28, 35, 42. Native; open places; throughout most of Br but only 6 microspp. in Sc, 6 scattered in Ir, 3 in Man, 1 in CI; c.18 endemic. 23 microspp. are currently placed here. Some of these, and other very local, un-named plants, have highly imperfect fruits and are probably recent hybrids. The hybrid *R. ulmifolius* x *R. caesius* has been recorded from several places.

25. R. sect. 4 Caesii Lej. & Courtois (sect. *Glaucobatus* Dumort.) - *Dewberry*. Stems 333
low-arching to procumbent, (potentially) rooting at tips, with glaucous bloom, with 338
moderate prickles but 0 or few hairs or glands; leaflets 3, overlapping, the 2 basal sessile or nearly so; stipules lanceolate; flowers few, in corymbs, white, large (2-3cm across); fruits with large, few drupelets, black, with glaucous bloom; 2n=28, 35. Native; disturbed ground, grassland, scrub and sand-dunes, often on clayey or basic soils; throughout C & S BI but scattered and local in much of W and in Sc. The only sp. is **R. caesius** L., which is easily distinguished by its ternate leaves; thin, procumbent, glaucous stems with short (1-2mm), slender prickles; large flowers with broad, white petals; and glaucous fruits with few, large drupelets.

9. POTENTILLA L. (*Comarum* L.) - *Cinquefoils*

Herbaceous (annuals to) perennials or rarely deciduous shrubs; leaves pinnate, ternate or palmate, or upper ones simple; flowers solitary or few in cymes, (4-)5-

merous; epicalyx present; hypanthium flat to saucer-shaped, with receptacle slightly to strongly convex; stamens ≥(5)10; carpels 4-numerous; fruit a head of achenes.

1 Lower leaves pinnate 2

1 Lower leaves ternate or palmate 5

2 Shrub; leaflets entire; achenes pubescent **1. P. fruticosa**

2 Herb; leaflets toothed; achenes glabrous 3

3 Petals purple; plant with long woody rhizome **2. P. palustris**

3 Petals yellow or white; plant without long rhizome 4

4 Petals white; flowers in terminal cyme **4. P. rupestris**

4 Petals yellow; flowers solitary, axillary **3. P. anserina**

5 Petals white; achenes pubescent on 1 side **17. P. sterilis**

5 Petals yellow; achenes glabrous (but receptacle often pubescent) 6

6 Leaves grey- to white-tomentose on lowerside 7

6 Leaf-surface visible through pubescence on lowerside 8

7 Leaves grey-tomentose on lowerside, with flat margin; upper part of stem with mixed straight and woolly hairs; petals 5-7mm **6. P. inclinata**

7 Leaves white-tomentose on lowerside, with revolute margin visible from lowerside as narrow green edge; upper part of stem with dense woolly hairs; petals 4-5mm **5. P. argentea**

8 At least some flowers with 4 petals and sepals 9

8 All flowers with 5 petals and sepals 11

9 Plant highly sterile (with 0-few achenes per flower); petioles of stem-leaves >1cm, all ± same length **15. P. x mixta**

9 Plant fertile (many achenes per flower); stem-leaves ± sessile to petiolate, if latter petioles decreasing markedly in size towards stem apex 10

10 Carpels <20; stem-leaves sessile or with stalks <5mm; all or nearly all leaves ternate (ignore stipules); stems not rooting at nodes; flowers ± all 4-merous **13. P. erecta**

10 Carpels >20, lower stem-leaves with stalks >10mm; some leaves with 4-5 leaflets; stems rooting at nodes late in season; some flowers 5-merous **14. P. anglica**

11 Flowers solitary in leaf-axils; main stems procumbent and rooting at nodes **16. P. reptans**

11 Flowers (often few) in terminal cymes; main stems not rooting at nodes 12

12 Flowering stem arising laterally from side of terminal leaf-rosette, usually <2mm wide, with only reduced (usually simple) leaves 13

12 Flowering stem arising from centre of leaf-rosette (the latter often withered by flowering-time), usually >2mm wide, with several well-developed leaves 14

13 Vegetative stems long, procumbent, often rooting, mat-forming; free part of stipules of basal leaves linear-triangular; flowers mostly <15mm across **12. P. neumanniana**

13 Vegetative stems short, not rooting or mat-forming; free part of stipules of basal leaves narrowly ovate; flowers mostly >15mm across **11. P. crantzii**

14 Petals >6mm, longer than sepals **7. P. recta**

14 Petals <5mm, shorter than to as long as sepals 15

15 Petals c.1/2 as long as sepals; stamens 5-10; achenes <0.8mm, smooth **10. P. rivalis**

15 Petals c.3/4-1x as long as sepals; stamens c.20; achenes >0.8mm, minutely rugose 16

16 Leaves ± all ternate; epicalyx-segments longer than sepals in fruit **9. P. norvegica**

16 Most lower leaves with 5 leaflets; epicalyx-segments shorter than

sepals in fruit **8. P. intermedia**

Other spp. - P. thuringiaca Bernh. ex Link, from C & S Europe, was formerly ± natd in a few places; it is similar to *P. crantzii* but more robust with narrower leaflets with the terminal tooth shorter than the 2 on either side. **P. eriocarpa** Wall. ex Lehm., from Himalayas, was found in a quarry in Kirkcudbrights in 1993 and might persist; it is a low-growing perennial with 5 large yellow petals, densely pubescent receptacle and achenes, and ternate leaves with the leaflets toothed only near apex as in *Sibbaldia*.

1. P. fruticosa L. - *Shrubby Cinquefoil.* Deciduous erect or spreading shrub to 1m; **RR** leaves pinnate, with (3)5-7(9) leaflets; flowers 1-many in cymes, functionally dioecious; petals yellow, 6-16mm, much longer than sepals; 2n=28. Native; rock-ledges, river- and lake-margins in full sun; extremely local in 2 areas of N En and Co Clare, NE Galway and E Mayo, W Ir. Numerous garden cultivars, with white to orange petals, are widely planted in public places and may occur as relics; some might involve the white-petalled **P. davurica** Nestl. (*P. glabrata* Willd.), from E Asia.

2. P. palustris (L.) Scop. (*Comarum palustre* L.) - *Marsh Cinquefoil.* Herbaceous perennial with long woody rhizome and ascending stems to 50cm; leaves pinnate, with (3)5-7 leaflets; flowers 1-few in terminal cymes; petals purple, much shorter than sepals; 2n=35, 40, 64. Native; fens, marshes and bogs; common over most of BI but very local in S & C Br and CI.

3. P. anserina L. - *Silverweed.* Perennial with long procumbent stolons and terminal leaf-rosettes; leaves pinnate, with (3)7-12 pairs of narrowly elliptic-oblong main leaflets alternating with small ones; flowers solitary in leaf-axils on stolons, on erect pedicels to 25cm; petals yellow, 7-10mm, c.2x as long as sepals; 2n=28, 42. Native; waste places, waysides, pastures and sand-dunes; common throughout BI.

4. P. rupestris L. - *Rock Cinquefoil.* Perennial with erect flowering stems to 60m **RRR** arising from leaf-rosette; leaves pinnate, with 2-4 pairs of broadly elliptic leaflets; flowers in terminal cymes; petals white, longer than sepals; (2n=14). Native; on basic rocks; 1 site each in Monts and Rads, 2 in E Sutherland, occasional escape from cultivation elsewhere.

5. P. argentea L. - *Hoary Cinquefoil.* Perennial with decumbent to ascending stems **343** to 30cm arising from leaf-rosette; leaves palmate with 5 leaflets white-tomentose on lowerside; flowers in terminal cymes; petals yellow, 4-5mm, c. as long as sepals; 2n=14. Native; sandy grassland and waste ground; local and decreasing in CI and Br N to C Sc, common only in E En.

6. P. inclinata Vill. (*P. canescens* Besser) - *Grey Cinquefoil.* Perennial with erect to **343** ascending stems to 50cm arising from leaf-rosette; differs from *P. argentea* in larger size and see key; (2n=42). Intrd; waste places; occasional casual in Br, mainly SE En; C & S Europe.

7. P. recta L. - *Sulphur Cinquefoil.* Perennial with erect stems to 10cm arising from **343** leaf-rosette; leaves palmate with 5-7 leaflets; flowers in terminal cymes; petals yellow, 6-12mm, longer than sepals; (2n=28, 42). Intrd; grown in gardens and natd in waste ground, roadside banks and grassy places; scattered over C & S En and CI, rare casual elsewhere; S Europe.

8. P. intermedia L. - *Russian Cinquefoil.* Biennial or perennial with erect to **343** ascending stems to 50cm arising from leaf-rosette; leaves palmate, the lower with 5 leaflets; flowers in terminal cymes; petals yellow, 4-5mm, c. as long as sepals; (2n=28, 42, 56). Intrd; casual in waste and grassy places, sometimes ± natd; scattered in En; Russia.

9. P. norvegica L. - *Ternate-leaved Cinquefoil.* Annual to short-lived perennial with **343** erect to ascending stems to 50cm; leaves ternate; flowers in terminal cymes; petals yellow, 4-5mm, c. as long as sepals or shorter; (2n=42, 56, 70). Intrd; casual or

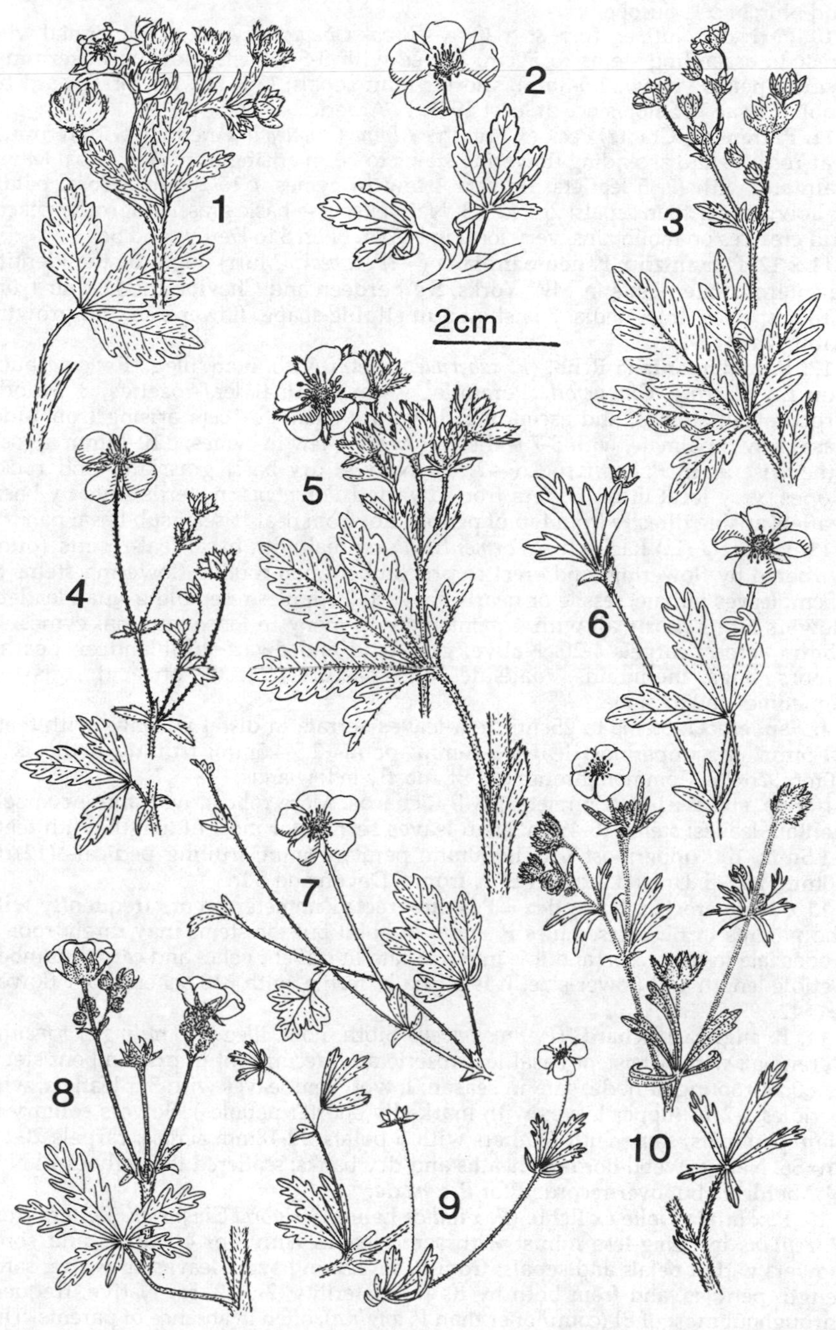

FIG 343 - *Potentilla*. 1, *P. norvegica*. 2, *P. reptans*. 3, *P. intermedia*. 4, *P. x suberecta*. 5, *P. recta*. 6, *P. anglica*. 7, *P. x mixta*. 8, *P. inclinata*. 9, *P. erecta*. 10, *P. argentea*.

sometimes natd in waste places; very scattered in En and Wa, rare casual in Sc and N Ir; N & C Europe.

10. P. rivalis Nutt. ex Torr. & A. Gray - *Brook Cinquefoil.* Annual to biennial with erect to ascending stems to 50cm; leaves with 3-5 leaflets; flowers in terminal cymes; petals yellow, 1.5-3mm, shorter than sepals; (2n=14, 70). Intrd; natd by pool at 1 site in Salop since at least 1976; N America.

11. P. crantzii (Crantz) Beck ex Fritsch - *Alpine Cinquefoil.* Perennial with terminal R
leaf-rosettes and ascending flowering stems to 20cm arising from side; basal leaves palmate, with (3-)5 leaflets; flowers 1-few in cymes, c.15-20mm across; petals yellow, longer than sepals; 2n=42, 63. Native; sparse basic grassland, rocky places and crevices on mountains; very local in N Wa, N En S to Derbys and Sc.

11 x 12. P. crantzii x P. neumanniana (= ?*P. x beckii* Murr) might be the identity of intermediate plants in MW Yorks, S Aberdeen and Cheviot, the last far from either species. Intermediacy is shown in stipule-shape, flower-size and growth-habit; 2n=61-64.

12. P. neumanniana Rchb. (*P. tabernaemontani* Asch. nom. illeg., *P. verna* auct. R
non L.) - *Spring Cinquefoil.* Perennial with terminal leaf-rosettes, ± woody procumbent stolons, and ascending flowering stems to 10cm arising from side; basal leaves palmate, with 5-7 leaflets; flowers 1-few in cymes, c.10-15mm across, otherwise as in *P. crantzii*; 2n=42, 49. Native; dry basic grassland and rocky slopes; very local in Br. Differs from dwarfed *P. reptans* in sterile state by basal leaflets arising directly from top of petiole, not from near base of sub-basal pair.

13. P. erecta (L.) Raeusch. - *Tormentil.* Perennial with basal leaf-rosette (often 343
withered by flowering) and erect to procumbent non-rooting flowering stems to 45cm; leaves ternate, sessile or nearly so, with 2 stipules resembling small leaflets; flowers all or nearly all with 4 petals, few to many in loose terminal cymes, 7-15mm across; carpels 4-20. Native; grassland and dwarf-shrubland on heaths, moors, bogs, mountains, roadsides and pastures, mostly on acid soils but sometimes on limestone.

a. Ssp. erecta. Stems to 25cm; stem-leaves serrate in distal 1/2 only, with teeth <1.5mm, the uppermost leaf c.6-16mm; petals 2.5-4.5mm; fruiting pedicels 6-30mm; 2n=28. Common throughout Bl, mostly in lowlands.

b. Ssp. strictissima (Zimmeter) A.J. Richards. More robust, with more coarsely dentate leaves; stems 15-45cm; stem-leaves serrate for most of length, with teeth >1.5mm, the uppermost leaf 12-30mm; petals 4-6mm; fruiting pedicels (12)20-50mm; 2n=28. Upland areas of BI N from S Devon and S Ir.

13 x 14. P. erecta x P. anglica = P. x suberecta Zimmeter occurs frequently with 343
the parents in BI. It resembles *P. erecta* in habit but the stems may rarely root at nodes late in the season and it is intermediate in leaflet-, petal- and carpel-number, petiole-length and flower-size. It is partially fertile, with <10 achenes per flower; 2n=42.

14. P. anglica Laichard. (*P. procumbens* Sibth. nom. illeg.) - *Trailing Tormentil.* 343
Perennial, with persistent basal leaf-rosette and decumbent to procumbent stems to 80cm rooting at nodes late in season; lower stem-leaves with 3-5 leaflets, with petioles 1-2cm (upper leaves with markedly shorter petioles); flowers solitary in stem-leaf axils, some with 4 others with 5 petals, 12-18mm across; carpels 20-50; 2n=56. Native; wood-borders, heaths and dry banks; scattered throughout Bl N to N Aberdeen, but over-recorded for *P. x mixta.*

15. P. x mixta Nolte ex Rchb. (*P. x italica* Lehm.) - *Hybrid Cinquefoil.* Differs from 343
P. reptans in being less robust with some leaves with 3 or 4 leaflets and some flowers with 4 petals and sepals; from *P. anglica* in having leaves all with ± same length petioles, and from both by its high sterility; 2n=42, 56. Native; frequent throughout most of BI (commoner than *P. anglica*), often in absence of parents. This taxon is derived from both *P. erecta* x *P. reptans* and *P. anglica* x *P. reptans.*

16. P. reptans L. - *Creeping Cinquefoil.* Perennial with persistent basal leaf-rosette 343
and procumbent rooting flowering stems to 1m; stem-leaves all palmate with 5

leaflets and petioles all >1cm and ± equal-lengthed; flowers solitary in stem-leaf axils, all with 5 petals and sepals, 15-25mm across; carpels 60-120; 2n=28. Native; rough ground, hedgebanks, sand-dunes and open grassland; common in BI N to S Sc, very local further N.

17. P. sterilis (L.) Garcke - *Barren Strawberry*. Perennial with procumbent stolons and terminal leaf-rosettes; leaves ternate, with broadly obovate leaflets; flowers 1-3 in cymes on decumbent axillary flowering stems to 15cm; petals white, c. as long as sepals; 2n=28. Native; wood-margins and clearings, scrub and hedgebanks; common throughout BI except extreme N. Distinguished from *Fragaria vesca* by the grey-green leaflets with terminal tooth shorter than adjacent 2, and petals with wide gaps between.

10. SIBBALDIA L. - *Sibbaldia*

Herbaceous perennials; differ from *Potentilla* in leaves ternate; flowers 5-merous, in compact heads; petals ≤2mm or 0; stamens (4)5(-10); carpels 5-12.

1. S. procumbens L. (*Potentilla sibbaldii* Haller f.) - *Sibbaldia*. Leaves in basal rosette with obovate-obtriangular leaflets mostly with 3 apical teeth; flowering stems 1-5cm, procumbent to ascending; flowers densely clustered, c.5mm across; petals 0 or 5, 1.5-2mm, pale yellow; (2n=14). Native; grassy and rocky slopes and rock-crevices above 470m; frequent in C & N Sc, Westmorland record probably an error. **R**

11. FRAGARIA L. - *Strawberries*

Herbaceous perennials, usually stoloniferous; leaves ternate; flowers in cymes on ± leafless stems arising from axils of leaf-rosette, 5-merous; epicalyx present, with entire segments; hypanthium ± flat, with strongly convex receptacle; petals white or flushed pink; stamens and carpels numerous (sometimes flowers ± dioecious); fruit a head of achenes borne on outside of enlarged, red, succulent receptacle.

1 Leaflets glabrous or nearly so on upperside; most fruiting heads >15mm wide, usually with sepals appressed; achenes sunk into surface of ripe receptacle **3. F. x ananassa**
1 Leaflets pubescent to sparsely so on upperside; fruiting heads <15mm wide, with sepals not appressed; achenes prominent from surface of ripe receptacle 2
 2 Flowers bisexual; uppermost pedicel in each cyme with apically directed hairs at fruiting; leaves rather glossy on upperside when fresh **1. F. vesca**
 2 Flowers functionally dioecious; uppermost pedicel in each cyme with many patent hairs at fruiting; leaves dull on upperside when fresh **2. F. moschata**

Other spp. - F. chiloensis (L.) Duchesne (*Beach Strawberry*), from W N America, was natd on a bank in Sark but now exists only in gardens from that source; it has large flowers and appressed fruiting sepals as in *F. x ananassa* but dark green coriaceous leaves with impressed veins on upperside, and is dioecious (only male plants in Sark).

1. F. vesca L. - *Wild Strawberry*. Stolons 0 to abundant; leaflets 1-6cm, elliptic-obovate, acute at base; flowering stems about as long as rosette-leaves, to 30cm; flowers 10-20mm across; fruit c.1cm across, with achenes raised above surface of ripe receptacle; 2n=14. Native; woods, scrub and hedgerows; common throughout BI. A robust variant without stolons and with flower and fruit produced continuously until the frosts is known as *Alpine Strawberry*, and may escape. See *Potentilla sterilis* for leaf differences.

2. F. moschata (Duchesne) Duchesne (*F. muricata* sensu D.H. Kent non L.) - *Hautbois Strawberry*. Differs from *F. vesca* in larger size (stems to 40cm, longer than leaves); terminal leaflet usually subacute to obtuse at base; flowers 15-30mm across; and see key; (2n=42). Intrd; formerly cultivated, natd in scrub and hedgerows; scattered throughout Br but now rare and over-recorded for large *F. vesca* or small *F. x ananassa*; C Europe.

3. F. x ananassa (Duchesne) Duchesne - *Garden Strawberry*. Differs from *F. vesca* in larger size of parts; many stolons; terminal leaflet usually obtuse to rounded at base; flowers 20-35mm across; and see key; (2n=56). Intrd; much cultivated, frequently natd in waste places and field-borders; scattered throughout BI; garden origin.

12. DUCHESNEA Sm. - *Yellow-flowered Strawberry*
Stoloniferous perennials; differ from *Fragaria* in solitary flowers; epicalyx-segments 3-toothed at apex, much larger than sepals; petals yellow. Possibly should be united wih *Potentilla*.

1. D. indica (Jacks.) Focke - *Yellow-flowered Strawberry*. Stolons to 50cm, slender, bearing ternate leaves and solitary axillary flowers on erect pedicels 3-10cm; flowers c.10-18mm across; fruit ± globose, 8-16mm across, dry, tasteless; (2n=42, 84). Intrd; cultivated as curiosity and occasionally escapes; very scattered in Br; S & E Asia.

13. GEUM L. - *Avens*
Herbaceous perennials; leaves pinnate; flowers in terminal cymes on stems arising from leaf-rosette, rarely solitary, 5-(rarely more)merous; epicalyx present; hypanthium ± flat to saucer-shaped, with strongly convex receptacle; petals yellow to purple; stamens and carpels numerous; fruit a head of achenes with long styles terminating in hook (after apical segment has fallen off).

1 Petals creamy-pink to purplish, with long claw at base; ripe achenes carried up from flower centre on stalk >5mm **1. G. rivale**
1 Petals yellow, scarcely clawed at base; ripe achenes ± sessile in flower centre 2
  2 Achenes <150, in globose head, the receptacle with dense hairs >1mm **2. G. urbanum**
  2 Achenes >150, in ovoid head, the receptacle with sparse hairs <1mm **3. G. macrophyllum**

Other spp. - Garden plants resembling *G. rivale* but with bright red petals are cultivars of **G. quellyon** Sweet (*G. chiloense* Balb. ex DC. nom. inval., *G. coccineum* auct. non Sibth. & Sm.), from Chile; they occasionally persist on tips for a while.

1. G. rivale L. - *Water Avens*. Stems to 50cm, erect; stipules mostly <1cm; flowers pendent, with erect, creamy-pink to pinkish-purple petals 8-15mm; achenes in globose head stalked above sepals, with style with long hairs and short glands near base; receptacle with dense long hairs; 2n=42. Native; marshes, streamsides, mountain rock-ledges and open woodland; throughout Br and Ir but very local in S and absent from large areas, also garden escape in non-native areas. Dwarf plants with large flowers and more deeply and roughly incised leaves from N Sc have been called ssp. **islandicum** Á. & D. Löve, but all intermediates occur.

1 x 2. G. rivale x G. urbanum = G. x intermedium Ehrh. is common wherever the parents meet. It is intermediate in all respects and highly fertile, forming a complete spectrum between the parents.

2. G. urbanum L. - *Wood Avens*. Stems to 70cm, erect or spreading; stipules 1-3cm; flowers erect, with patent, yellow petals 4-7mm; achenes in sessile globose head, with style glabrous or sometimes with glands near base; receptacle with dense long hairs; 2n=42. Native; woods and hedgerows; common throughout BI except far N.

3. G. macrophyllum Willd. - *Large-leaved Avens.* Stems to 1m, erect; stipules 6-15mm; flowers erect, with patent, yellow petals 3.5-7mm; achenes in sessile oblong head, with style glabrous or sometimes with glands near base; receptacle with sparse short hairs; (2n=42). Intrd; natd in woods and by paths; Cards and several parts of Sc; N America and NE Asia.

14. DRYAS L. - *Mountain Avens*

Perennials, woody at base, herbaceous above; leaves ± evergreen, simple, bluntly serrate; flowers solitary, axillary, 7-10-merous; epicalyx 0; hypanthium saucer-shaped, with convex receptacle; petals white, mostly 8; stamens and carpels numerous; fruit a head of achenes with long feathery styles.

1. D. octopetala L. - *Mountain Avens.* Stems to 50cm, procumbent, rooting; leaves **R**
6-25mm, dark glossy green on upperside, white-tomentose on lowerside; petals 7-17mm; styles in fruit 2-3cm; 2n=18. Native; base-rich rock-crevices and -ledges on mountains; very local in NW Wa, N Ir and N En, local in MW Ir and N & W Sc.

15. AGRIMONIA L. - *Agrimonies*

Herbaceous perennials; leaves pinnate; flowers in long terminal racemes, 5-merous; epicalyx 0; hypanthium deeply concave, narrow at mouth, surrounding carpels; petals yellow; stamens 5-20; carpels 2; fruit of 1-2 achenes enclosed in woody hypanthium which has ring of hooked bristles distally.

Vegetative differences between the 2 spp. are often exaggerated; *A. eupatoria* is fragrant when crushed and is often glandular on leaf lowersides, though glands may be concealed by denser pubescence.

1. A. eupatoria L. - *Agrimony.* Stems ± erect, to 1m, diffusely branched, **349**
pubescent with long and short non-glandular hairs; main leaflets 3-6 pairs, rather bluntly serrate, with 0 to rather few sessile shining glands on lowerside; fruiting hypanthium obconical, deeply grooved ± to apex, with outermost bristles patent to erecto-patent; 2n=28. Native; grassy places in fields and hedgerows; throughout BI except most of N Sc.

1 x 2. A. eupatoria x A. procera = A. x wirtgenii Asch. & Graebn. is intermediate in indumentum (glands and hairs) and leaflet toothing; it does not form fruit; 2n=42. It was found in the late 1940s in S Northumb, but other records have not been confirmed cytologically.

2. A. procera Wallr. (*A. odorata* auct. non (L.) Mill., *A. repens* auct. non L.) - **349**
Fragrant Agrimony. Differs from *A. eupatoria* in more leafy stems with ± all hairs ≥2mm; more deeply and acutely serrate, less pubescent leaflets with abundant sessile glands on lowerside; fruiting hypanthium bell-shaped, with grooves extending <3/4 way to apex and outermost bristles reflexed; 2n=56. Native; same habitats and distribution as *A. eupatoria* but much more scattered.

16. AREMONIA Neck. ex Nestl. - *Bastard Agrimony*

Herbaceous perennials; leaves pinnate; flowers in small terminal cymes, 5-merous; epicalyx of 5 lobes, surrounded by 8-12 fused bracts; hypanthium deeply concave, surrounding carpels; petals yellow; stamens 5-10; carpels 2; fruit of 1-2 achenes enclosed in woody pubescent hypanthium without hooked bristles but concealed by bracts.

1. A. agrimonioides (L.) DC. (*Agrimonia agrimonioides* L.) - *Bastard Agrimony.* Stems to 30cm, decumbent, scarcely or not longer than basal leaves; leaves 2-3 pairs of main leaflets, the apical 3 much the largest; fruiting hypanthium 5-6mm, ±

globose; (2n=42). Intrd; natd in woods and shady roadsides; C Sc; S & C Europe.

17. SANGUISORBA L. (*Poterium* L.) - *Burnets*
Herbaceous perennials; leaves pinnate; flowers in very dense terminal spikes, all bisexual or bisexual and unisexual mixed; sepals 4; epicalyx 0; hypanthium deeply concave, surrounding carpels; petals 0; stamens 4 or numerous; carpels 1-2; stigmas papillate or tasselled; fruit 1-2 achenes enclosed in hard hypanthium.

1 Upper flowers female, others bisexual or male; stamens numerous; stigmas 2, tasselled **3. S. minor**
1 All flowers bisexual; stamens 4; stigma 1, papillate 2
 2 Flower-heads 1-3cm; sepals and stamens dull purplish; stamens c.4mm **1. S. officinalis**
 2 Flower-heads 3-16cm; sepals green and white; stamens white, c.10mm **2. S. canadensis**

1. S. officinalis L. (*Poterium officinale* (L.) A. Gray) - *Great Burnet*. Stems to 1.2(1.7)m, erect; basal leaves with 3-7 pairs of leaflets; flower-heads subglobose to oblong-ellipsoid, purplish; 2n=28, 56. Native; damp unimproved grassland; locally frequent in Br N to C Sc, N Ir.

2. S. canadensis L. - *White Burnet*. Stems to 2m, erect; basal leaves with 5-10 pairs of leaflets; flower-heads cylindrical, white; (2n=56). Intrd; grown in gardens and sometimes natd; frequent in C Sc, very scattered elsewhere in Sc and En; N America.

3. S. minor Scop. (*Poterium sanguisorba* L.) - see sspp. for English names. Stems erect, to 50(80)cm; leaves with 3 12 pairs of leaflets; flower-heads 6-15mm, globose or slightly elongated, sepals green and purple; stamens yellow, c.6mm.

a. Ssp. minor - *Salad Burnet*. Fruiting hypanthium 3-4.5mm, with thickened but **349**
scarcely winged ridges on angles, the faces distinctly reticulate with finer ridges; 2n=28. Native; calcareous or sometimes neutral grassland and rocky places; locally common throughout BI N to C Sc.

b. Ssp. muricata (Gremli) Briq. (*Poterium sanguisorba* ssp. *muricatum* (Gremli) **349**
Rouy & Camus, *P. polygamum* Waldst. & Kit.) - *Fodder Burnet*. Usually more robust and leafy with more deeply and sharply toothed leaflets; fruiting hypanthium 3-5mm, with often undulate wings on angles, the faces smooth to irregularly rugose. Intrd; formerly grown for fodder and still natd in grassy places; very scattered in BI, frequent only in C & S En; S Europe.

18. ACAENA Mutis ex L. - *Pirri-pirri-burs*
Perennials, with stems woody at base but herbaceous distally; leaves pinnate; flowers in globose heads, bisexual; sepals 4; epicalyx 0; hypanthium deeply concave, surrounding carpels; petals 0; stamens 2; carpels 1-2; stigmas long-papillate; fruit 1-2 achenes enclosed in dry hypanthium which usually develops barbed spines at apex.

1 Apical pair of leaflets c. as long as wide; spines on hypanthium 0, or imperfect, or not barbed **4. A. inermis**
1 Apical pair of leaflets 1.2-2.5x as long as wide; spines on hypanthium 2-4, barbed at apex 2
 2 Apical pair of leaflets each with (11)17-23 teeth; hypanthium with 2 spines **3. A. ovalifolia**
 2 Apical pair of leaflets each with 5-12(15) teeth; hypanthium with up to 4 spines 3
3 Leaflets glossy green on upperside; apical pair of leaflets 1.8-2.5x as long as wide **1. A. novae-zelandiae**
3 Leaflets matt green on upperside, often edged and veined with brown;

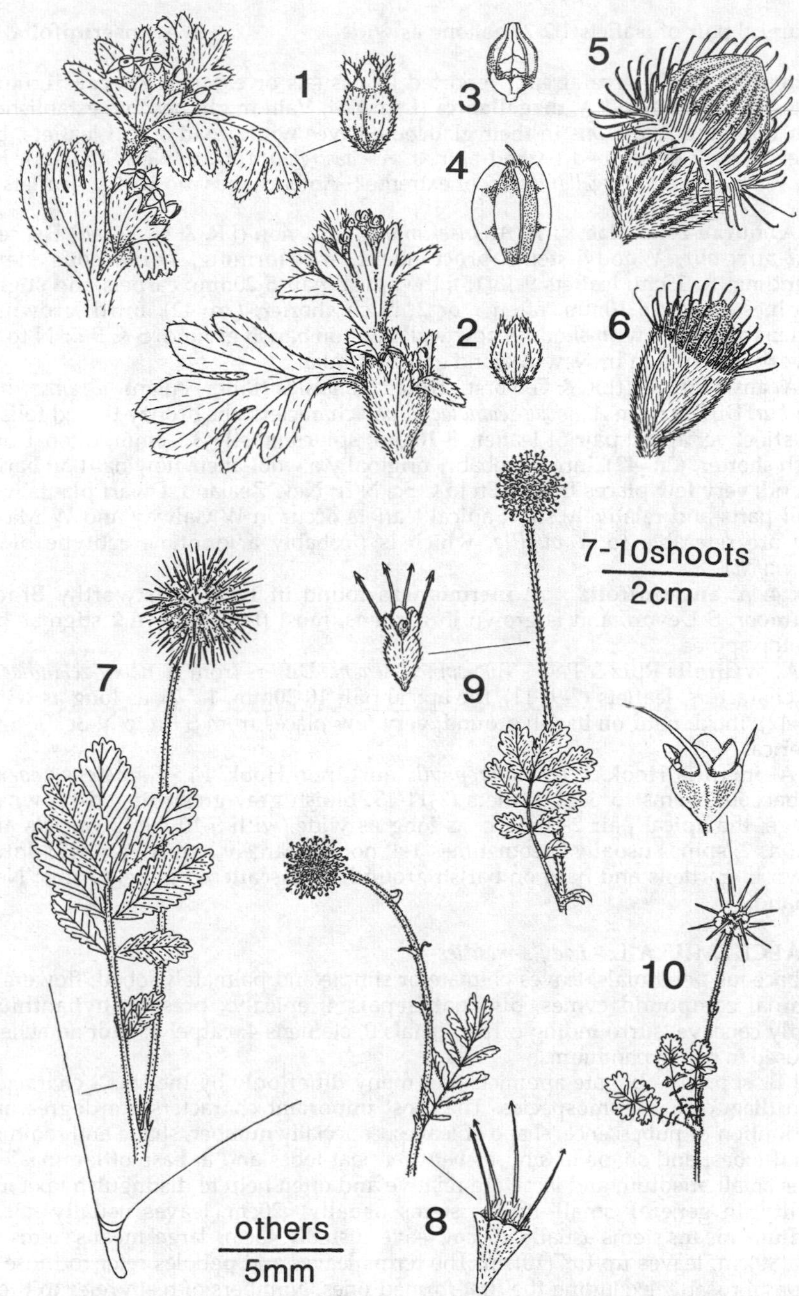

FIG 349 - Rosaceae. 1-2, shoots and fruiting hypanthia of *Aphanes*. 1, *A. **arvensis***. 2, *A. **australis***. 3-4, fruiting hypanthia of ***Sanguisorba minor***. 3, ssp. ***minor***. 4, ssp. ***muricata***. 5-6, fruiting hypanthia of ***Agrimonia***. 5, *A. **procera***. 6, *A. **eupatoria***. 7-10, shoots and fruiting hypanthia of ***Acaena***. 7, *A. **ovalifolia***. 8, *A. **novae-zelandiae***. 9, *A. **anserinifolia***. 10. *A. **inermis***.

apical pair of leaflets 1.2-2x as long as wide **2. A. anserinifolia**

Other spp. - Of several spp. reported as casuals or escapes, **A. caesiiglauca** (Bitter) Bergmans and **A. magellanica** (Lam.) M. Vahl might become established. Both resemble *A. inermis* in their glaucous leaves with broad apical leaflets, but have 2-4 well-developed barbed spines. *A. caesiiglauca*, from New Zealand, has only 7-9 leaflets; *A. magellanica*, from extreme S America, has mostly ≥11 leaflets.

1. A. novae-zelandiae Kirk (*A. anserinifolia* auct. non (J.R. & G. Forst.) Druce) - *Pirri-pirri-bur*. Woody stems procumbent, mat-forming; herbaceous stems ascending, to 15cm; leaflets 9-13(15), the apical pair 5-20mm; carpels and stigma 1; spines mostly 6-10mm, often 1 or 2 much shorter; (2n=42). Intrd; grown in gardens and intrd with shoddy, now well natd on barish ground; S & E Br N to SE Sc, very scattered in Ir; New Zealand and Australia. **349**

2. A. anserinifolia (J.R. & G. Forst.) Druce (*A. pusilla* (Bitter) Allan) - *Bronze Pirri-pirri-bur*. Differs from *A. novae-zelandiae* in key characters; the bronzy-tinged foliage is distinctive; apical pair of leaflets 3-10mm; spines mostly 3.5-6mm, often 1 or 2 much shorter; (2n=42). Intrd; probably originally a wool-alien, now natd on barish ground; very few places from S En to C Sc, N Ir; New Zealand. Dwarf plants with small parts and relatively short apical leaflets occur in W Galway and W Mayo; they are referable to *A. pusilla*, which is probably a montane ecotype of *A. anserinifolia*. **349**

2 x 4. A. anserinifolia x A. inermis was found in 1967 at Batworthy Brook, Dartmoor, S Devon, and is grown in gardens; most flowers have 2 stigmas but develop spines.

3. A. ovalifolia Ruiz & Pav. - *Two-spined Acaena*. Differs from *A. novae-zelandiae* in key characters; leaflets (7)9(-11), the apical pair 10-30mm, 1.7-2x as long as wide; (2n=42). Intrd; natd on barish ground; very few places from S En to N Sc, SE Ir; S America. **349**

4. A. inermis Hook. f. (*A. microphylla* auct. non Hook. f.) - *Spineless Acaena*. Herbaceous stems to 6cm; leaflets (7)11-13, bluish grey-green tinged brown or orange, the apical pair 2-8mm, c. as long as wide, with 5-10 teeth; carpels and stigmas 2; spines usually 0, sometimes 1-4, not or scarcely barbed; (2n=42). Intrd; grown in gardens and natd on barish ground; very scattered in Sc, Carms; New Zealand. **349**

19. ALCHEMILLA L. - *Lady's-mantles*

Herbaceous perennials; leaves palmate or simple and palmately lobed; flowers in terminal compound cymes, bisexual; sepals 4; epicalyx present; hypanthium deeply concave, surrounding carpel; petals 0; stamens 4; carpel 1; fruit an achene enclosed in dry hypanthium.

All Br spp. are obligate apomicts and many differ only by the small characters often diagnosing agamospecies. The most important characters are degree and distribution of pubescence; shape of leaves, especially number, shape and toothing of leaf-lobes; and shape of sinuses between leaf-lobes and at base of lamina. The terms small, medium and large are relative and often help to distinguish taxa in 1 locality; in general 'small' means stems usually <20cm, leaves usually <3cm; 'medium' means stems usually <50cm, leaves usually <5cm; 'large' means stems up to 60(80)cm, leaves up to 7(10)cm. The terms leaves and petioles refer to those of the basal rosette, excluding the first-formed ones. Numbers of teeth refer to those on the mid-lobe of each leaf.

1 Leaves palmately divided >1/2 way to base, densely silver-silky-pubescent on lowerside 2
1 Leaves palmately divided <1/2 way to base, variously pubescent but never silver-silky-pubescent 3

2 Leaves divided ± to base, the leaflets mostly <6mm wide **1. A. alpina**

2 Leaves divided 3/5-4/5 way to base, the lobes mostly >6mm wide **2. A. conjuncta**

3 Petioles and stem glabrous, subglabrous or with appressed or subappressed hairs 4

3 Petioles and lower part of stem with erecto-patent, patent or reflexed hairs 7

4 Stems up to and including 1st inflorescence-branches pubescent; leaves pubescent on upperside 5

4 Stems pubescent only on lowest 2 or 3 internodes; leaves glabrous on upperside 6

5 Hypanthium tapered to cuneate base **8. A. micans**

5 Hypanthium rounded at base **12. A. glomerulans**

6 Sinuses between leaf-lobes toothed to base; teeth in middle of each side of leaf-lobes larger than those above or below **14. A. glabra**

6 Sinuses between leaf-lobes with toothless region at base c.2x as long as a tooth; teeth on leaf-lobes subequal **13. A. wichurae**

7 Epicalyx-segments c. as long as sepals; hypanthium much shorter than mature achene **15. A. mollis**

7 Epicalyx-segments distinctly shorter than sepals; hypanthium as long as mature achene 8

8 Pedicels and hypanthia both pubescent, sometimes sparsely 9

8 Pedicels glabrous; hypanthia usually glabrous 11

9 Leaf-lobes usually 5, with sinuses between them with toothless region at base ≥2x as long as a tooth; stems rarely >5cm **11. A. minima**

9 Leaf-lobes (5)7-9, with sinuses between them toothed to base; stems usually >5cm 10

10 Base of stems and petioles tinged wine-red; leaf basal sinus open (≥45°) **10b. A. filicaulis ssp. vestita**

10 Base of stems and petioles brownish; leaf basal sinus very narrow (<30°) to closed **3. A. glaucescens**

11 Leaves glabrous on upperside (or with very sparse hairs in folds) 12

11 Leaves pubescent on upperside, sometimes only in folds 13

12 Leaf-lobes rounded, with subequal teeth **9. A. xanthochlora**

12 Leaf-lobes ± straight-sided then rounded to subtruncate at apex, with teeth in middle of each side larger than those above or below **7. A. acutiloba**

13 Leaves ± densely pubescent on both surfaces 14

13 Leaves rather sparsely pubescent on upperside, with hairs usually ± confined to folds (or frequent only there) 16

14 Hypanthium tapered to cuneate base; stem-hairs mostly erecto-patent **8. A. micans**

14 Hypanthium rounded at base; stem-hairs patent to reflexed 15

15 Flowers mostly >2.5mm across; petioles and stems with patent hairs **4. A. monticola**

15 Flowers mostly <2.5mm across; petioles and stems with many reflexed hairs **5. A. tytthantha**

16 Leaf-lobes ± straight-sided then rounded to subtruncate at apex, with teeth in middle of each side larger than those above or below **7. A. acutiloba**

16 Leaf-lobes ± rounded, with teeth ± equal to unequal but not with large ones in middle of each side 17

17 Leaf-lobes usually 5, with sinuses between them with toothless region at base ≥2x as long as a tooth; stems rarely >5cm **11. A. minima**

17 Leaf-lobes (5)7-9, with sinuses between them toothed to base; stems usually >5cm 18

18 Hypanthium often with some hairs; stems and petioles with patent hairs, tinged wine-red at base **10a. A. filicaulis ssp. filicaulis**
18 Hypanthium glabrous; stems and petioles usually with some reflexed hairs, brownish at base **6. A. subcrenata**

Other spp. - A. venosa Juz., from Caucasus, resembles *A. mollis* but has lower parts appressed-pubescent, upper parts subglabrous and more deeply lobed leaves; it was formerly natd in 2 places but is grown in gardens and might be overlooked for *A. mollis* elsewhere.

1. A. alpina L. - *Alpine Lady's-mantle.* Stems to 20cm, ascending; leaves palmate **353**
or almost so, with 5-7 narrowly oblong-elliptic leaflets, green and glabrous on upperside, densely silver-silky-pubescent on lowerside; (2n=c.120, c.128, c.140, c.152). Native; grassland, scree and rock-crevices on mountains; Lake District, C & N Sc, Kerry and Wicklow, intrd in Derbys.

2. A. conjuncta Bab. - *Silver Lady's-mantle.* Stems to 30cm, ascending; leaves **353**
simple, palmately lobed 3/5-4/5 way to base, with 7-9 elliptic lobes, green and glabrous on upperside, densely silver-silky-pubescent on lowerside; (2n=120-128). Intrd; much grown in gardens and sometimes escaping; scattered in N & C Br, with some very old records in N Wa and Sc once considered native but probably planted; Alps.

3-15. A. vulgaris L. agg. - *Lady's-mantle.* Stems to 60(80)cm, often much less, decumbent to ascending; leaves up to 7(10)cm, often much less, simple, palmately lobed <1/2 way to base, glabrous to densely pubescent. Damp rich grassland, woodland margins and rides, rock-ledges; throughout most of Br and Ir but rare or absent in most of SE En.

3. A. glaucescens Wallr. (*A. minor* auct. non Huds.). Plant small; leaf-lobes (5)7- **RR**
9, rounded, with 9-11 straight subobtuse teeth; basal sinus ± closed; whole plant **353**
with ± dense patent to erecto-patent hairs; 2n=c.110. Native; very local in N En, NW Sc, NW Ir, rare escape elsewhere in En.

4. A. monticola Opiz (*A. gracilis* Opiz). Plant medium; leaf-lobes 9-11, rounded, **RR**
with 15-19 slightly incurved acute teeth; basal sinus ± closed; petioles, leaves, **353**
stems and lowest inflorescence-branches with ± dense patent hairs; ultimate inflorescence-branches and pedicels glabrous; hypanthium glabrous to sparsely pubescent; (c.101-109). Native; very local in NW Yorks and Co Durham, rare escape elsewhere in En.

5. A. tytthantha Juz. Plant medium; leaf-lobes usually 9, short, rounded to **353**
obtuse, with 13-17 straight acute teeth; basal sinus narrow; stems and petioles with patent and reflexed hairs; leaves with dense patent hairs; upper part of inflorescence, pedicels and hypanthia glabrous. Intrd; natd in C & S Sc, probably from botanic gardens; Crimea.

6. A. subcrenata Buser. Plant medium; leaf-lobes 7-9, long, rounded at apex, **RR**
with 13-17 coarse subobtuse teeth; basal sinus narrow; petioles and lower parts of **353**
stems with patent and usually reflexed hairs; upper part of stem and whole inflorescence glabrous; leaves pubescent, rather sparsely so on upperside; (2n=c.90, c.96, 104-110). Native; very local in NW Yorks and Co Durham, 1st found 1951.

7. A. acutiloba Opiz. Plant large; leaf-lobes 9-11(13), straight-sided with narrow **RR**
apex, with (13)15-19(21) straight acute teeth; basal sinus wide; petioles and lower **353**
parts of stems with dense patent hairs; upper part of stem and whole inflorescence glabrous; leaves densely pubescent on lowerside, glabrous to sparsely pubescent on upperside; (2n=c.100, c.105-109). Native; quite widespread in Co Durham (discovered 1946), natd in Lanarks since 1992.

8. A. micans Buser (*A. gracilis* auct. non Opiz). Plant medium; leaf-lobes 9(-11), **RR**
rounded but with rather narrow apex, with 11-15 acute unequal teeth; basal sinus **353**

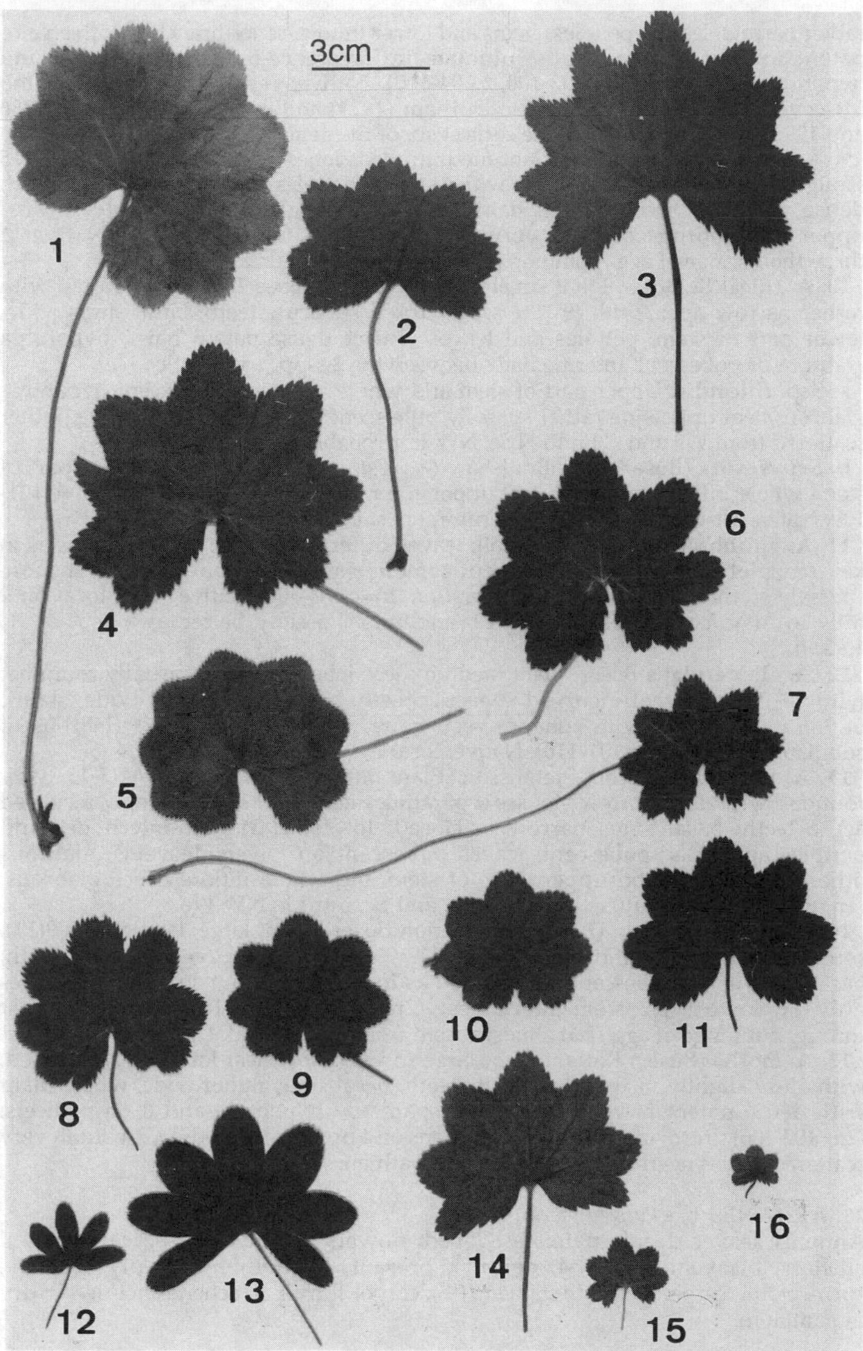

FIG 353 - Leaves of *Alchemilla*. 1, *A. mollis*. 2, *A. xanthochlora*. 3, *A. acutiloba*. 4, *A. glomerulans*. 5, *A. tytthantha*. 6, *A. subcrenata*. 7, *A. micans*. 8, *A. wichurae*. 9, *A. monticola*. 10, *A. glaucescens*. 11, *A. glabra*. 12, *A. alpina*. 13, *A. conjuncta*. 14-15, *A. filicaulis*. 16, *A. minima*.

rather narrow; leaves, petioles, stems and lower inflorescence-branches with erecto-patent to subappressed hairs; ultimate inflorescence-branches, pedicels and hypanthia glabrous; (2n=c.93-100, c.104-110). Native; very local in S Northumb (discovered 1976), also once in Co Durham (1924) and casual in Lanarks in 1986 and 1992. In S Northumb it is the earliest sp. of the genus to flower.

9. A. xanthochlora Rothm. Plant medium; leaf-lobes 7-11, rounded, with 11-15 **353**
straight acute teeth; basal sinus wide; stems, petioles and leaf lowerside with dense patent or erecto-patent hairs; inflorescence glabrous or nearly so; leaf upperside glabrous or with very sparse hairs in folds; 2n=c.107. Native; ± throughout range of agg., common in N, very local in SE En and S Ir.

10. A. filicaulis Buser. Plant small to medium; leaf-lobes 7(-9), rounded but with **353**
rather narrow apex, with (9-)11(-13) incurved subacute teeth; basal sinus wide; lower part of stem, petioles and leaves with ± dense patent hairs; hypanthia glabrous or pubescent. Intermediates between the 2 sspp. occur in Sc.

a. Ssp. filicaulis. Upper part of stem and whole inflorescence except hypanthia glabrous; leaf upperside rather sparsely pubescent; 2n=c.103, c.150. Native; rather scattered from Wa and C En to N Sc, NW Ir, probably under-recorded.

b. Ssp. vestita (Buser) M.E. Bradshaw (*A. vestita* (Buser) Raunk.). Upper part of stem, whole inflorescence and leaf upperside rather densely pubescent; 2n=c.101-109. Native; distribution of agg., the commonest taxon in genus.

11. A. minima Walters. Plant small; leaves differ from those of *A. filicaulis* as in **RR**
key (couplet 17); whole plant except sometimes pedicels pubescent, but more **353**
sparsely so than in *A. filicaulis* ssp. *vestita*; 2n=c.103-108. Native; very local on 2 hills in MW Yorks, described 1949; endemic. Possibly better as a ssp. of *A. filicaulis*.

12. A. glomerulans Buser. Plant medium; leaf-lobes usually 9, usually rounded, **R**
with 13-19 somewhat incurved subacute teeth; basal sinus rather wide; stems, **353**
petioles and leaves with subappressed hairs; whole inflorescence glabrous to subglabrous; (2n=64, c.101-110). Native; local in N En and Sc.

13. A. wichurae (Buser) Stefánsson. Plant rather small; leaf-lobes 7-11, very **R**
rounded with deep narrow sinuses separating them, with 15-19 strongly incurved **353**
acute teeth; basal sinus narrow to closed; lowest 2(-3) stem-internodes and petioles appressed-pubescent; leaves pubescent on lowerside veins, glabrous otherwise; middle and upper parts of stem and whole inflorescence glabrous; (2n=64, c.103-107). Native; local in N En and Sc, intrd in NW Wa.

14. A. glabra Neygenf. (*A. obtusa* auct. non Buser). Plant large; leaf-lobes 7-9(11), **353**
rounded or with straightish sides, with 11-19 somewhat incurved subacute teeth; basal sinus wide; pubescence distributed as in *A. wichurae* but sparser, sometimes only on lowest 1-2 stem-internodes; (2n=96-100, 102-110). Native; almost throughout range of agg., but almost absent from En S of Peak District.

15. A. mollis (Buser) Rothm. Plant large to very large; leaf-lobes 9-11, rounded, **353**
with 15-19 slightly incurved subacute teeth; basal sinus rather wide; whole plant with dense patent hairs except rather sparse on hypanthia and 0 on pedicels; (2n=102-106). Intrd; much grown in gardens and prolifically seeding; ± natd in very scattered places nearly throughout Br; Carpathians.

20. APHANES L. - *Parsley-pierts*

Annuals; leaves deeply palmately lobed; flowers in small dense, leaf-opposed clusters, bisexual; sepals 4; epicalyx present; hypanthium deeply concave, surrounding carpel; petals 0; stamen 1(-2); carpel 1; fruit an achene enclosed in dry hypanthium.

1. A. arvensis L. (*Alchemilla arvensis* (L.) Scop.) - *Parsley-piert*. Stems decumbent **349**
to nearly erect, to 10(20)cm; leaves up to 1cm; stipules at fruiting nodes fused into a leaf-like cup with ovate-triangular teeth at apex c.1/2 as long as entire portion; fruiting hypanthium 2-2.6mm incl. erect sepals c.0.6-0.8mm, with a slight

constriction where hypanthium and sepals meet, reaching ± to apex of stipules at maturity; 2n=48. Native; cultivated and other bare ground on well-drained soils; frequent ± throughout BI.

2. A. australis Rydb. (*A. inexspectata* W. Lippert, *A. microcarpa* auct. non (Boiss. & Reut.) Rothm., *Alchemilla microcarpa* auct. non Boiss. & Reut.) - *Slender Parsley-piert*. **349**
Differs from *A. arvensis* in stipule-teeth at fruiting nodes ovate-oblong, c. as long as entire portion; fruiting hypanthium 1.4-1.9mm incl. convergent sepals c.0.3-0.5mm, the sepals continuing curved outline of hypanthium, falling well short of apex of stipules at maturity; 2n=16. Native; similar places to *A. arvensis* but rarely on base-rich soils and less often in arable ground; frequent ± throughout BI.

21. ROSA L. - *Roses*

Deciduous or rarely evergreen shrubs with spiny stems and petioles; leaves pinnate; flowers solitary or in few-(many-)flowered corymbs, usually 5-merous; epicalyx 0; hypanthium deeply concave, surrounding carpels, narrowed at apex; stamens and carpels numerous; fruit (actually a false fruit) a head of achenes enclosed by succulent hypanthium.

An extremely complex genus, much hybridized and selected in cultivation. Half of our spp. (sect. *Caninae* DC., nos. 11-21) contribute unbalanced gametes, the male one having 7 chromosomes and the female ones 21, 28, 35 or 42 (from plants with 2n=28, 35, 42, 56 respectively). Hybrids are very common and reciprocal crosses can produce very different offspring. The ripe fruits and leaves provide the most important characters, but a collection from the same plant at flowering is often also desirable. 'Disc' refers to the thickened rim at top of the hypanthium at fruiting, in centre of which is the 'orifice' through which the styles project. Only the spp. are keyed here, but even for these access to accurately named material is often a prerequisite for successful determination; in some areas hybrids are commoner than either parent, but can often be determined if the parents present in the area are known. The classification of sect. *Caninae* adopted here reflects the views of G.G. Graham and A.L. Primavesi.

The application of many binomials to hybrid combinations is uncertain; further literature research and typification will undoubtedly uncover names earlier than several of those used here.

1 Styles exserted and fused into a column, sometimes becoming free at fruiting — 2
1 Styles exserted or not, free (may appear fused in dried material) — 6
 2 Leaflets 3(-5) — **2. R. setigera**
 2 Leaflets 5-9 — 3
3 Styles pubescent; semi-evergreen; stems ± procumbent — **3. R. luciae**
3 Styles glabrous; deciduous; stems trailing to strongly arching — 4
 4 Flowers 2-3cm across, in groups of >(6)10; stipules lobed >1/2 way to petiole — **1. R. multiflora**
 4 Flowers mostly 3-5cm across, in groups of 1-6(10); stipules not lobed or lobed <1/2 way to petiole — 5
5 Styles as long as stamens; top of hypanthium flat; inner sepals entire, outer with very few lobes — **4. R. arvensis**
5 Styles shorter than stamens; top of hypanthium conical; sepals pinnately lobed — **12. R. stylosa**
 6 Sepals entire, ± tapering to apex, erect or suberect and persistent until after fruit ripe — 7
 6 At least some sepals lobed or with strongly expanded tips, if ± entire then patent to reflexed and/or falling before fruit ripe — 10
7 Fruit blackish when ripe; flowers all solitary, without bract — **5. R. pimpinellifolia**
7 Fruit red when ripe; flowers 1-several, with 1 or more much reduced

leaves (bracts) at base of pedicels 8
8 Stems and prickles ± glabrous; flowers 3-5cm across; pedicels glandular-pubescent **18. R. mollis**
8 Stems and prickles tomentose; flowers 6-9cm across; pedicels tomentose 9
9 Fruits 1.5-2.5cm, usually wider than long; leaflets bullate, rather shiny **6. R. rugosa**
9 Fruits 0.8-1.5cm, usually longer than wide; leaflets scarcely bullate, matt **7. R.'Hollandica'**
10 Stems with many mixed prickles, pricklets and acicles; flowers often solitary **9. R. gallica**
10 Stems without acicles; flowers usually >1 per branch 11
11 Flowers 6-8cm across, *flore pleno* to some degree, usually white **10. R. x alba**
11 Flowers (2)3-5(6)cm across, very rarely *flore pleno* 12
12 Sepals entire or some with few very narrow lateral lobes, some or all falling before fruit ripe 13
12 Outer sepals on ± all flowers with lateral lobes; sepals falling early or persistent 14
13 Leaves strongly red-tinged; petals usually shorter than sepals; pedicels, fruits and sepals glabrous to very sparsely glandular-pubescent **11. R. ferruginea**
13 Leaves green; petals longer than sepals; pedicels, fruits and sepals densely glandular-pubescent **8. R. virginiana**
14 Leaflets glabrous, sometimes with few stalked glands on midrib but without eglandular hairs 15
14 Leaflets tomentose to pubescent with eglandular hairs on lowerside, at least on midrib 16
15 Orifice of disc c.1/5 its total width; styles glabrous, pubescent or woolly, forming ± loose group; sepals mostly patent to reflexed after flowering, falling before fruit ripe **13. R. canina**
15 Orifice of disc c.1/3 its total width; styles woolly, forming dense mass ± obscuring disc; sepals mostly erect to erecto-patent after flowering, usually persistent until fruit ripe **14b. R. caesia ssp. glauca**
16 Leaflets with prominent, ± sticky, sessile and short-stalked glands on lowerside, giving fresh leaf fruity smell when rubbed, pubescent to sparsely so with eglandular hairs on lowerside 17
16 Leaflets with 0 or ± inconspicuous glands on lowerside, the glands ± confined to veins or if over whole surface then with no or with resinous smell and leaves usually ± tomentose on lowerside 19
17 Pedicels glabrous; leaflets cuneate at base **21. R. agrestis**
17 Pedicels glandular-pubescent; leaflets rounded at base 18
18 Stems erect; prickles unequal; styles pubescent; sepals mostly erect to patent, persistent until fruit reddens **19. R. rubiginosa**
18 Stems arching; prickles ± equal; styles glabrous or nearly so; sepals mostly reflexed, falling before fruit reddens **20. R. micrantha**
19 Leaflets without glands or with few on midrib on lowerside, 1-2-serrate with teeth not or variably gland-tipped 20
19 Leaflets glandular on lowerside, at least on midrib and lateral veins, 2-serrate with gland-tipped teeth 22
20 Orifice of disc c.1/3 its total width; styles woolly, forming dense mass ± obscuring disc; sepals mostly erect to erecto-patent after flowering, usually persistent until fruit ripe **14a. R. caesia ssp. caesia**
20 Orifice of disc c.1/5 its total width; styles glabrous, pubescent or woolly, forming ± loose group; sepals mostly patent to reflexed after flowering, falling before fruit ripe 21
21 Lobes on outer sepals narrow, usually entire; prickles moderately

hooked, longer than width of base; leaves usually eglandular on lowerside **13. R. canina**

21 Lobes on outer sepals broad, usually lobed or toothed; prickles strongly hooked, c. as long as width of base; leaves usually glandular on lowerside of midrib **15. R. obtusifolia**

22 Orifice of disc c.1/5 its total width; styles glabrous to pubescent 23

22 Orifice of disc c.1/3-1/2 its total width; styles woolly 24

23 Prickles strongly hooked, c. as long as width of base; pedicels 5-15mm; pedicels, fruits and sepals glabrous or sparsely glandular-pubescent **15. R. obtusifolia**

23 Prickles ± straight to arched, longer than width of base; pedicels (10)15-25mm; pedicels, fruits and sepals glandular-pubescent to densely so **16. R. tomentosa**

24 Sepals erect or suberect, entire or with few lateral lobes, persistent until fruit decays; orifice of disc c.2/5-1/2 its total width; prickles ± all straight **18. R. mollis**

24 Sepals erect to erecto-patent, with lateral lobes, falling from ripe fruit; orifice of disc c.1/3 its total width; at least some prickles curved **17. R. sherardii**

Other spp. - Innumerable spp., cultivars and (often complex) hybrids are grown in gardens and may be very persistent in hedges, waste ground, parks and estates. The cultivars and hybrids are best named as cultivars without reference to their parentage, e.g. *Rosa* **'Queen Elizabeth'**; in many cases the parentage is in fact unknown. These cultivars are not included in this work; they include nearly all *flore pleno* plants (but most of the non-native spp. also occur in this form). **R. sempervirens** L. (sect. *Synstylae*) (*Evergreen Rose*) was formerly natd in Worcs; it is evergreen with glabrous styles, red fruits, leaves with 5-7 leaflets and unlobed stipules, and flowers 2.5-6cm across. **R. majalis** Herrm. (*R. cinnamomea* L. 1759 non 1753) (sect. *Cassiorhodon*) and **R. pendulina** L. (*R. cinnamomea* L. 1753 non 1759) (sect. *Synstylae*) are sometimes found in semi-wild places; the former has depressed-globose fruits with glabrous pedicels; the latter ovoid to obovoid fruits with glandular-pubescent pedicels recurved in fruit. **R. sericea** Lindl. (*R. omeiensis* Rolfe) (sect. *Pimpinellifoliae*) and **R. gymnocarpa** Nutt. ex Torr. & A. Gray var. **willmottiae** (Hemsl.) P.V. Heath (*R. willmottiae* Hemsl.) (sect. *Gymnocarpae* Thory) have recently been recorded from hedges in MW Yorks; the former has very broad-based prickles, 9-13 leaflets and solitary white flowers with 4 petals; the latter has straight, paired prickles, 7-9 glabrous leaflets <15mm, and solitary pinkish-purple flowers. **R. nitida** Willd. (sect. *Carolinae*) differs from *R. virginiana* in its narrower leaflets and abundant hairs and stalked glands; it has been doubtfully recorded in similar situations.

Section **1** - *SYNSTYLAE* DC. (*spp.* 1-4). Leaflets 3-9; flowers 1-many, with bracts; sepals entire or nearly so, falling before fruit ripe; styles fused into column ± as long as stamens.

1. R. multiflora Thunb. ex Murray - *Many-flowered Rose*. Scrambler, to 5m; leaves **360**
with (5)7(-9) leaflets, pubescent to sparsely so but eglandular on lowerside, 1-serrate with glandular teeth; flowers usually >10, usually white, 2-3cm across; fruits ovoid to globose, <1cm; (2n=14). Intrd; grown in gardens for ornament and as stock for rambler-roses; natd in hedges and copses; scattered in CI and Br; E Asia.

1 x 19. R. multiflora x R. rubiginosa was recorded on the Common, Herm, CI, but apparently in error for *R. micrantha*.

2. R. setigera Michx. - *Prairie Rose*. Scrambler, to 5m; leaves with 3(-5) leaflets, **360**
glabrous to ± pubescent but eglandular or ± so on lowerside, 1-serrate with

eglandular teeth; flowers few-several, white to pink, 4-6cm across; fruits ± globose, <1cm; (2n=14). Intrd; well natd in scrub; Jersey and Guernsey; E & C N America.

3. R. luciae Franch. & Rochebr. ex Crép. (*R. wichurana* Crép.) - *Memorial Rose.* 360
Semi-evergreen, ± procumbent, to 4m; leaves with (5)7-9 leaflets, ± glabrous, 1-serrate, eglandular; flowers few-several, white, 3-5cm across; fruits ovoid to globose, ≤1cm; (2n=14). Intrd; grown as several cultivars (and hybrids) and natd in open places, low scrub and beaches mostly near sea; Dunbarton, E Kent, Guernsey and Jersey; E Asia.

4. R. arvensis Huds. - *Field-rose.* Weakly trailing, to 1(2)m; leaves with 5-7 360
leaflets, glabrous or almost so on lowerside, usually 1-serrate, usually glandless; flowers 1-few, white, 3-5cm across; fruit globose to ellipsoid, 0.8-1.5cm; 2n=14. Native; low scrub, hedgerows, woods and open places; frequent in En, Wa and much of Ir, very rare in Man and Sc. The following hybrids occur; those with *R. arvensis* as female parent usually resemble that sp. in habit but have pink petals and other characters (often pubescent leaves) indicating the male parent; those with a sp. of *Caninae* as female parent usually resemble the latter in habit (and pubescence) but reveal various features of *R. arvensis* on inspection.

4 x 6. R. arvensis x R. rugosa = R. x paulii Rehder is a garden plant (white-flowered like its *R. rugosa* parent) found ± natd in London (Surrey).

4 x 12. R. arvensis x R. stylosa = R. x pseudorusticana Crép. ex Preston (*R. x bibracteoides* Wolley-Dod) (4 x 12 and 12 x 4); S Br N to Worcs, Co Dublin.

4 x 13. R. arvensis x R. canina = R. x verticillacantha Mérat (*R. x kosinskiana* Besser, *R. x wheldonii* Wolley-Dod, *R. x deseglisei* Boreau) (4 x 13 and 13 x 4); BI N to S Sc.

4 x 14. R. arvensis x R. caesia ssp. **glauca** (14 x 4 only); C En.

4 x 15. R. arvensis x R. obtusifolia = R. x rouyana Duffort ex Rouy (15 x 4 only); C Wa and C En.

4 x 16. R. arvensis x R. tomentosa (5 x 16 only); Mons.

4 x 17. R. arvensis x R. sherardii (4 x 17 and 17 x 4); Leics.

4 x 19. R. arvensis x R. rubiginosa = R. x consanguinea Gren. (4 x 19 and 19 x 4); En and Wa N to Derbys and Carms.

4 x 20. R. arvensis x R. micrantha = R. x vituperabilis Duffort ex Rouy (*R. x inelegans* Wolley-Dod) (20 x 4 only); C En, C Wa and S Ir.

Section* 2 - *PIMPINELLIFOLIAE DC. (sp. 5). Leaflets (5)9-11(13); flower 1, without bracts; sepals entire, persistent until fruit decay; styles free.

5. R. pimpinellifolia L. - *Burnet Rose.* Strongly suckering; stems erect, to 360
50(100)cm, with numerous slender prickles and acicles; leaves glabrous or sparsely pubescent on lowerside, 1(-2)-serrate, with 0 or sparse glands; flowers white, rarely pale pink, 2-4cm across; fruit ± globose, blackish-purple, rather dry, with erect sepals, 0.5-1.5(2)cm; 2n=28. Native; dry sandy places near sea, on inland heaths and limestone; round most coasts of BI, very local inland.

The following hybrids occur; they can be recognized in the same way as *R. arvensis* hybrids with spp. of Caninae. Several resemble *R. pimpinellifolia* but have red fruits.

5 x 13. R. pimpinellifolia x R. canina = R. x hibernica Templeton (5 x 13 and 13 x 5); very scattered in BI N to S Sc.

5 x 14. R. pimpinellifolia x R. caesia = R. x margerisonii (Wolley-Dod) Wolley-Dod (*R. x setonensis* Wolley-Dod) (5 x 14 and 14 x 5; both sspp. of *R. caesia*); N En, Sc and N Ir.

5 x 16. R. pimpinellifolia x R. tomentosa = R. x coronata Crép. ex Reut. (5 x 16 and 16 x 5); N Wa, N Ir and S En N to Warks.

5 x 17. R. pimpinellifolia x R. sherardii = R. x involuta Sm. (5 x 17 and 17 x 5); most of BI but rare in S En.

5 x 18. R. pimpinellifolia x R. mollis = R. x sabinii Woods (5 x 18 and 18 x 5); scattered over BI but rare in S & C En.

5 x 19. R. pimpinellifolia x R. rubiginosa = R. x biturigensis Boreau (*R. x cantiana* (Wolley-Dod) Wolley-Dod, *R. x moorei* (Baker) Wolley-Dod) (5 x 19 and 19 x 5); very scattered in En, Sc and Ir.

5 x 21. R. pimpinellifolia x R. agrestis = R. x caviniacensis Ozanon was found in Fermanagh in 1900.

Section* 3 - *CASSIORHODON Dumort. (sect. *Cinnamomeae* auct. non Ser.) (spp. 6-7). Leaflets 5-9; flowers 1-several, with bracts; sepals entire or nearly so, persistent until after fruit ripe; styles free.

6. R. rugosa Thunb. ex Murray - *Japanese Rose*. Strongly suckering; stems erect, to **360**
1.5(2)m, tomentose, with numerous tomentose prickles and acicles; leaflets bullate, green, rather shiny, tomentose to pubescent and glandular on lowerside, 1-serrate; flowers 1(-3), white to red, 6-9cm across; fruit globose to depressed-globose, 1.5-2.5cm; (2n=14, 28). Intrd; much grown for ornament and as stock for cultivars, natd on dunes, rough ground and banks, often mass-planted; scattered through much of BI; E Asia.

6 x 13. R. rugosa x R. canina = R. x praegeri Wolley-Dod (6 x 13 only); described from Co Antrim in 1927 and found more recently in W Norfolk and W Galway.

R. rugosa occasionally forms spontaneous hybrids with other native spp., but the other sp. is often very difficult to determine. Such hybrids sometimes resemble *R.* 'Hollandica'; some possibly involving *R. mollis* and *R. caesia* have been recorded in En and Sc. See also *R. arvensis* x *R. rugosa*.

7. R. 'Hollandica' - *Dutch Rose*. Differs from *R. rugosa* in lighter green, matt, **360**
scarcely bullate, pubescent to sparsely pubescent leaves; flowers 2-5 together, with dark or mauvish-red petals; and globose to ovoid fruits 0.8-1.5cm and often ± pendent. Intrd; much used as rootstock, ± natd outcast or relic in hedges and waste ground; scattered in most of Br, mainly Sc. Garden origin; hybrid between *R. rugosa* and some other sp.

Section* 4 - *CAROLINAE Crép. (sp. 8). Leaflets (5)7-9: flowers 1-few, with bracts; sepals entire or nearly so, falling before fruit ripe; styles free.

8. R. virginiana Herrm. (*R. lucida* Ehrh.) - *Virginian Rose*. Variably suckering; **360**
stems ± erect, to 2m, with few curved prickles and 0 or few hairs and stalked glands; leaflets glabrous or sparsely pubescent on lowerside, 1-serrate, eglandular; flowers pink or white, 4-6cm across; fruit subglobose, 1-1.5cm; (2n=28). Intrd; formerly much grown and still natd in scrub and hedgerows; scattered in En and Wa; E N America.

Section* 5 - *ROSA (sect. *Gallicanae* DC.) (spp. 9-10). Leaflets 3-5(7); flowers 1-3, without bracts; sepals pinnately lobed, falling before fruit ripe; styles free.

9. R. gallica L. - *Red Rose* (*of Lancaster*). Strongly suckering; stems erect, to 1.5m, **360**
with slender prickles, acicles and glandular hairs; leaflets pubescent and glandular on lowerside, mostly 1-serrate; flowers pink to red, 5-7(9)cm across; sepals reflexed after flowering; fruit ellipsoid to globose, 1-2cm; pedicels and fruit glandular-pubescent; (2n=28). Intrd; formerly much grown, still natd in scrub and hedges; very scattered in En, Man and Guernsey; Europe. Its hybrid derivatives **R. centifolia** L. and **R. damascena** Mill. are sometimes found in semi-wild places.

10. R. x alba L. (*R. x collina* Jacq. non Woods; *R. gallica* x *R. arvensis* and/or *R.* **360**
canina) - *White Rose* (*of York*). Stems erect, to 2m, with slightly curved prickles; leaflets pubescent at least on veins on lowerside, sharply 1-serrate; flowers white (to pink), fully or partially *flore pleno*, 6-8(9)cm across; sepals as in *R. gallica*; fruit ovoid, 1.5-2cm; pedicels and fruit glandular-pubescent; (2n=28, 42). Intrd; natd in

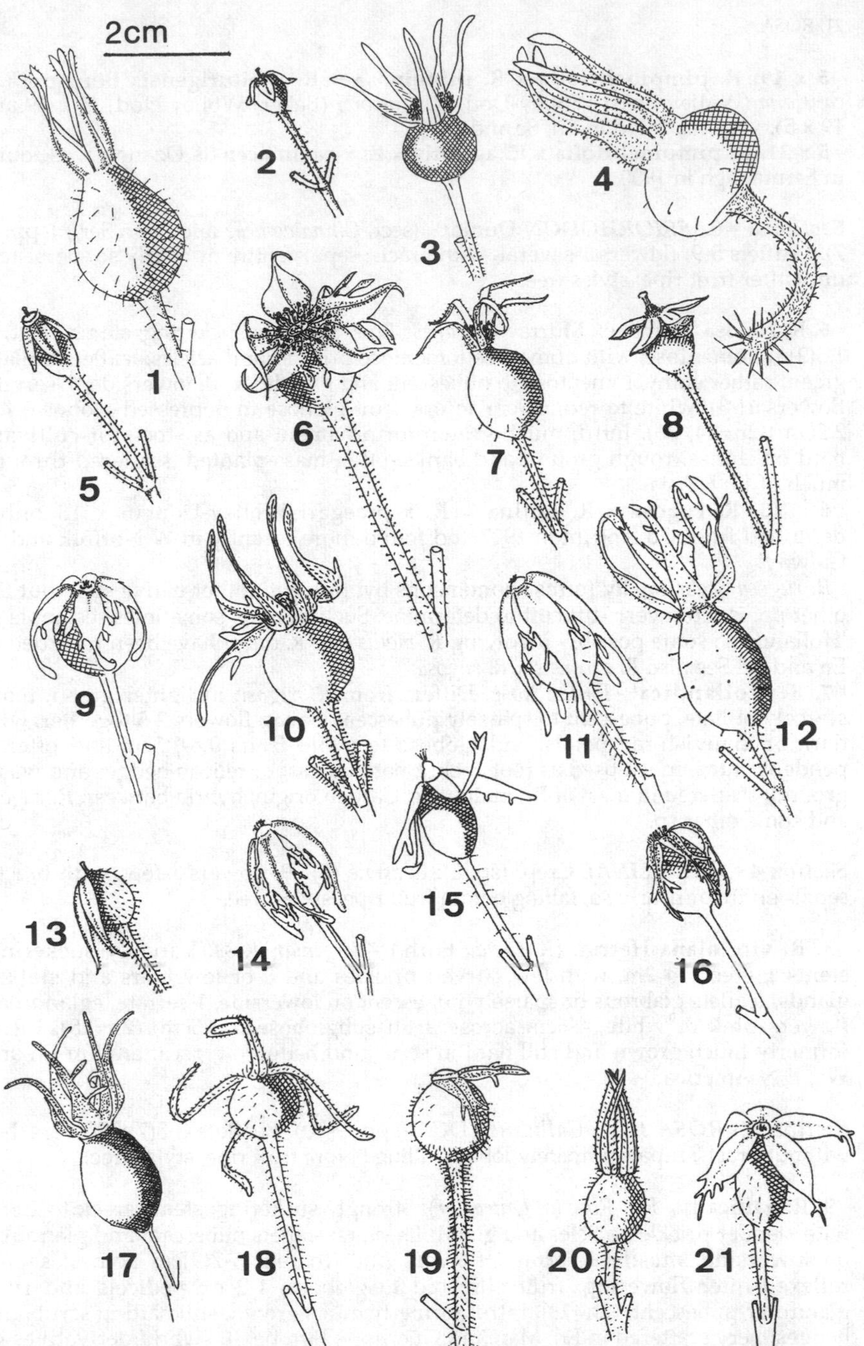

FIG 360 - Fruits of *Rosa*. 1, *R. mollis*. 2, *R. multiflora*. 3, *R. pimpinellifolia*. 4, *R. rugosa*. 5, *R. luciae*. 6, *R. gallica*. 7, *R. ferruginea*. 8, *R. arvensis*. 9, *R. canina*. 10, *R. rubiginosa*. 11, *R. micrantha*. 12, *R. sherardii*. 13, *R. setigera*. 14, *R. obtusifolia*. 15, *R. stylosa*. 16, *R. agrestis*. 17, *R. caesia*. 18, *R. virginiana*. 19, *R. tomentosa*. 20, *R.* **'Hollandica'**. 21, *R. x alba*.

hedges and other marginal sites; SE Yorks, Kirkcudbrights and Man; garden origin.

Section 6 - CANINAE DC. (spp. 11-21). Leaflets (3)5-7(9); flowers 1-several, with bracts; sepals usually pinnately lobed, falling before or persistent until after fruit ripe; styles free or (*R. stylosa* only) fused into column shorter than stamens.

11. R. ferruginea Vill. (*R. glauca* Pourr. non Vill. ex Loisel., *R. rubrifolia* Vill. nom. illeg.) - *Red-leaved Rose*. Stems erect, to 3m, glabrous, with few prickles, with numerous acicles on suckers; leaflets ± flat, red-tinged, glabrous, 1-serrate, glandless; flowers (1)few-several, deep pink, 3-4.5cm across; fruit ellipsoid to subglobose, 1-1.8cm; (2n=28). Intrd; much grown and natd via birds; very scattered in Sc, Man and En, but probably under-recorded; C Europe. **360**

12. R. stylosa Desv. - *Short-styled Field-rose*. Stems arching, to 3(4)m, with hooked prickles; leaflets pubescent on lowerside, 1-serrate, eglandular; flowers white to pale pink, 3-5cm across; fruit ovoid, 1-2cm, ± glabrous, pedicels 2-4cm, usually glandular-pubescent; sepals reflexed after flowering, falling before fruit ripe; 2n=35, 42. Native; hedges, scrub, wood-borders; Br and Ir S of a line from Co Dublin to E Suffolk, but now rare. **360**

Hybrids with *R. arvensis* (q.v.) and spp. of *Caninae* occur:

12 x 13. R. stylosa x R. canina = R. x andegavensis Bastard (*R. x rufescens* Wolley-Dod) (12 x 13 and 13 x 12); most of En, Wa and Ir, much commoner in S.

12 x 14. R. stylosa x R. caesia ssp. **glauca** (12 x 14 only); SW En N to Hunts and Worcs.

12 x 15. R. stylosa x R. obtusifolia (15 x 12 only); Br N to Mons and Cambs.

12 x 17. R. stylosa x R. sherardii (12 x 17 only); Wa and Co Carlow.

12 x 19. R. stylosa x R. rubiginosa = R. x bengyana Rouy (19 x 12 only); E Gloucs.

12 x 21. R. stylosa x R. agrestis (=*R. x belnensis* auct. non Ozanon) (12 x 21 and 21 x 12); N & S Somerset and Surrey.

13. R. canina L. (*R. dumetorum* auct. non Thuill., *R. squarrosa* auct., ?(A. Rau) Boreau, *R. corymbifera* Borkh.) - *Dog-rose*. Stems arching, to 3(4)m, with usually strongly curved to hooked prickles; leaflets glabrous to pubescent on lowerside, 1-2-serrate, eglandular or with glandular teeth or sometimes some glands on lowerside veins; flowers white to pink, 3-6cm across; fruit (globose to) ovoid, 1-2.5cm, usually glabrous; pedicels 1.5-2.5cm, glabrous to sparsely glandular-pubescent; sepals reflexed after flowering, falling before fruit ripe; 2n=35. Native; hedges, scrub, wood-borders; common throughout most of BI except sparse in Sc and absent from much of N. Extremely variable. **360**

Pubescent plants (*R. dumetorum* auct., *R. corymbifera* Borkh.) are often separated as a sp., but every intermediate exists; they are best treated as an informal group '**Pubescentes**'. 3 other groups have been recognized: '**Lutetianae**' (*R. canina* sensu stricto), with 1-serrate leaves and with glands confined to bracts; '**Dumales**' (*R. dumalis* auct. non Bechst.), with 2-serrate leaves and with glands on leaf-teeth, bracts, stipules and leaf-rhachis; and '**Transitoriae**', intermediate. Pubescent plants are usually 1-serrate, but often 2-serrate though rarely with glandular teeth. The numerous intermediates and poor character-correlation suggest only 1 sp. should be recognized; there are no distributional differences.

R. canina hybridises with all other native spp. of *Rosa*:

13 x 14. R. canina x R. caesia = R. x dumalis Bechst. (*R. x subcanina* (H. Christ) Dalla Torre & Sarnth., *R. x subcollina* (H. Christ) Dalla Torre & Sarnth.) (13 x 14 and 14 x 13; both sspp. of *R. caesia*); most of BI, often common.

13 x 15. R. canina x R. obtusifolia = R. x dumetorum Thuill. (*R. x subobtusifolia* Wolley-Dod, *R. x concinnoides* Wolley-Dod) (13 x 15 and 15 x 13); BI N to MW Yorks and Co Londonderry.

13 x 16. R. canina x R. tomentosa = R. x scabriuscula Sm. (*R. x curvispina* Wolley-Dod, *R. x aberrans* Wolley-Dod) (13 x 16 and 16 x 13); most of En, Wa and

Ir.

13 x 17. R. canina x R. sherardii = R. x rothschildii Druce (13 x 17 and 17 x 13); most of BI but rare in SE En.

13 x 18. R. canina x R. mollis = R. x molletorum Hesl.-Harr. (13 x 18 and 18 x 13); Sc and En S to Staffs.

13 x 19. R. canina x R. rubiginosa = R. x nitidula Besser (*R. x latebrosa* Déségl., *R. x latens* Wolley-Dod) (13 x 19 and 19 x 13); Br N to Durham, Co Antrim.

13 x 20. R. canina x R. micrantha = R. x toddiae Wolley-Dod (13 x 20 and 20 x 13); scattered in Br N to E Suffolk and N Wa, W Cork.

13 x 21. R. canina x R. agrestis = R. x belnensis Ozanon (21 x 13 only); E Norfolk.

14. R. caesia Sm. (*R. dumalis* auct. non Bechst.) - see sspp. for English names. **360**
Stems arching, to 2(3)m, with strongly curved to hooked prickles; leaflets mostly 1-serrate, with few or 0 glands; flowers and fruit as in *R. canina* except flowers 3-5cm across, fruit 2-3cm, pedicels 0.5-1.5cm, glabrous and see key (couplets 15 or 20). Native; hedges, scrub and wood-borders; throughout most of N 3/4 of BI, rare and very scattered in S. Distribution of 2 sspp. similar, but ssp. *glauca* is much commoner in N & NW Sc.

a. Ssp. caesia (*R. coriifolia* Fr.) - *Hairy Dog-rose*. Stems green or somewhat red; leaflets rugose, scarcely glaucous, pubescent on lowerside; 2n=35.

b. Ssp. glauca (Nyman) G.G. Graham & Primavesi (*R. afzeliana* Fr. nom. illeg., *R. vosagiaca* N.H.F. Desp., *R. glauca* Vill. ex Loisel. non Pourr.) - *Glaucous Dog-rose*. Stems often strongly red-coloured and glaucous; leaflets scarcely rugose, glabrous, glaucous; 2n=35. Isolated bushes have a characteristic open, not dense, appearance, usually enhanced by the leaflets being folded down at the midrib.

The following additional hybrids have been recorded:

14 x 15. R. caesia x R. obtusifolia (15 x 14 only; ssp. *glauca* only); C En.

14 x 16. R. caesia x R. tomentosa = R. x cottetii (H. Christ) Lagger & Puget ex Cottet (*R. x rogersii* Wolley-Dod) (14 x 16 and 16 x 14; ssp. *glauca* only); C & N En.

14 x 17. R. caesia x R. sherardii (?= *R. x alpestris* S. Rapin ex Reut.) (14 x 17 and 17 x 14; both sspp. of *R. caesia*); BI S to Leics, Brecs and Meath.

14 x 18. R. caesia x R. mollis = R. x glaucoides Wolley-Dod (14 x 18 and 18 x 14; both sspp. of *R. caesia*); Sc and En S to Derbys.

14 x 19. R. caesia x R. rubiginosa (=*R. x obovata* (Baker) Ley non Raf.) (14 x 19 and 19 x 14; both sspp. of *R. caesia*); Sc and En S to Derbys.

14 x 20 R. caesia x R. micrantha = R. x longicolla Ravaud ex Rouy (14 x 20 only; ssp. *glauca* only); Brecs.

15. R. obtusifolia Desv. - *Round-leaved Dog-rose*. Stems arching, to 2(3)m, with **360**
strongly and abruptly hooked prickles; leaflets pubescent on lowerside (often also on upperside), 2-serrate with glandular teeth and often glandular on lowerside, more rounded overall and at base than in *R. canina*; flowers white (to pale pink), 3-4cm across; fruit globose to ovoid, 1-1.5(2)cm, usually glabrous; pedicels 0.5-1.5cm, usually glabrous; sepals strongly lobed, reflexed after flowering, falling before fruit ripe; 2n=35. Native; hedges and scrub; En, Wa and Ir, but rare in many areas, frequent only in C En.

The following additional hybrids have been recorded:

15 x 16. R. obtusifolia x R. tomentosa (15 x 16 and 16 x 15); Leics and Laois.

15 x 19. R. obtusifolia x R. rubiginosa = R. x timbalii Crép. (*R. x tomentelliformis* Wolley-Dod) (15 x 19 only); Northants and Cheshire.

15 x 20. R. obtusifolia x R. micrantha (20 x 15 only); N Somerset.

16. R. tomentosa Sm. (*R. scabriuscula* auct. non Sm.) - *Harsh Downy-rose*. Stems **360**
arching, to 3m, with ± straight to arching prickles; leaflets pubescent to tomentose on lowerside (often also on upperside), 2-serrate with glandular teeth, with glands on lowerside; flowers white to pink, 3-5cm across; fruit ovoid, 1-2cm, (glabrous or) glandular-pubescent; orifice of disc c.1/5 its total width; pedicels 1.5-3.5cm, glandular-pubescent; sepals pinnately lobed, suberect to patent, falling before fruit

ripe; 2n=35. Native; hedges, scrub and open woods; frequent in BI N to S Sc, very rare in C & N Sc.

The following additional hybrids have been recorded:

16 x 17. R. tomentosa x R. sherardii = R. x suberectiformis Wolley-Dod (16 x 17 and 17 x 16); BI S to S Lancs, Cards and Co Kildare.

16 x 18. R. tomentosa x R. mollis (16 x 18 only); S Lancs and Staffs.

16 x 19. R. tomentosa x R. rubiginosa = R. x avrayensis Rouy (16 x 19 and 19 x 16); S Wa and CS En.

16 x 20. R. tomentosa x R. micrantha (16 x 20 only); S En and Cards.

16 x 21. R. tomentosa x R. agrestis (16 x 21 only); N Tipperary.

17. R. sherardii Davies - *Sherard's Downy-rose.* Differs from *R. tomentosa* in more 360
compact habit, to 1.5(2)m; prickles more slender; flowers 2.5-4cm across, usually deep pink, sometimes white; fruit 1.5-2.5cm, usually globose to obovoid; orifice of disc c.1/3-1/2 its total width; pedicels 0.5-1.5cm; sepals erect to suberect, persistent until fruit ripe but falling before its decay; and see key (couplet 22); 2n=28, 35, 42. Native; scrub, hedges and open woods, throughout most of Br and Ir but common in Sc and rare in SE En.

The following additional hybrids have been recorded:

17 x 18. R. sherardii x R. mollis = ?R. x shoolbredii Wolley-Dod (17 x 18 and 18 x 17); BI S to NW Yorks, S Wa and Co Louth.

17 x 19. R. sherardii x R. rubiginosa = R. x suberecta (Woods) Ley (*R. x burdonii* Wolley-Dod) (17 x 19 and 19 x 17); BI S to Leics, Flints and M Cork.

17 x 20. R. sherardii x R. micrantha (17 x 20 only); M Wa and W Cork.

17 x 21. R. sherardii x R. agrestis (17 x 21 and 21 x 17); C Ir.

18. R. mollis Sm. (*R. villosa* auct. non L.) - *Soft Downy-rose.* Differs from *R.* 360
tomentosa and *R. sherardii* in more compact habit with erect stems to 1.5(2)m; fresh leaves often more strongly resinous-scented; prickles more slender, straight; flowers 3-4.5cm across, usually very deep pink, rarely white; fruit 1.5-3cm, usually globose; orifice of disc c.1/3-1/2 its total width; pedicels 0.5-1.5cm; sepals erect, simple to slightly lobed, persisting until fruit decays; and see key (couplets 22 & 24); 2n=28, 56. Native; similar habitats and range to *R. sherardii*, but S to only Glam and Derbys.

The following additional hybrid has been recorded:

18 x 19. R. mollis x R. rubiginosa = R. x molliformis Wolley-Dod (18 x 19 and 19 x 18); scattered in Sc.

19. R. rubiginosa L. - *Sweet-briar.* Stems erect, ± straight, to 2m, with hooked, 360
unequal prickles; leaflets pubescent at least on veins and glandular on lowerside, ± 2-serrate with glandular teeth; flowers pink, 2.5-4cm across; fruit subglobose to ovoid, 1-2.5cm, glabrous or sparsely glandular-pubescent; pedicels 0.8-1.5cm, glandular-pubescent; sepals erect to patent, persistent until fruit reddens; 2n=35. Native; mostly in scrub on calcareous soils; scattered throughout BI, ± common on chalk in SE En, rare in C En and W Sc.

The following additional hybrid has been recorded:

19 x 20. R. rubiginosa x R. micrantha = R. x bigeneris Duffort ex Rouy (20 x 19 only); SE En and E Cork.

20. R. micrantha Borrer ex Sm. - *Small-flowered Sweet-briar.* Differs from *R.* 360
rubiginosa in stems arching, often scrambling, to 3m; flowers 2-3.5cm across; fruit 1-2cm; pedicels 1-2cm, rarely glabrous; and see key (couplet 18); 2n=35, 42. Native; similar places to *R. rubiginosa* but often not on calcareous soils; scattered throughout En, Wa and S Ir, local in CI, very rare in Sc.

The following additional hybrid has been recorded:

20 x 21. R. micrantha x R. agrestis = R. x bishopii Wolley-Dod (20 x 21 and 21 x 20); S En.

21. R. agrestis Savi (*R. elliptica* auct. non Tausch) - *Small-leaved Sweet-briar.* RR
Differs from *R. rubiginosa* in stems erect but somewhat flexuous, to1.5(2)m; prickles 360
± equal; flowers 2-4cm across, white to pale pink; fruit 1-2cm, usually glabrous;

pedicels 1-1.5cm, glabrous; sepals usually reflexed, falling before fruit reddens; 2n=35, 42. Native; scrub, mostly on calcareous soils; very scattered and mostly rare in Br and Ir N to S Sc, frequent mainly in parts of C Ir. Plants identified as **R. elliptica** Tausch were probably *R. agrestis* and/or hybrids of the latter or perhaps of *R. micrantha*.

SUBFAMILY 3 - AMYGDALOIDEAE (Prunoideae) (genera 22-23). Stipulate trees or shrubs; hypanthium concave, not enclosing carpels and fixed to them only at base; epicalyx 0; stamens >10; flowers 5-merous; carpels 1(or 5), 1-2-seeded; fruit 1(-5) drupes; chromosome base-number 8.

22. PRUNUS L. - *Cherries*
Trees or shrubs; leaves simple, serrate to crenate; flowers solitary or in racemes, corymbs or umbels; carpel 1; fruit a drupe.

1 Flowers usually >10 in elongated racemes 2
1 Flowers solitary or 2-c.10 in umbels or corymbs 5
  2 Leaves coriaceous, evergreen; racemes without leaves at base; fruit conical-acute at apex 3
  2 Leaves herbaceous, deciduous; racemes usually with 1-few leaves (often reduced) near base; fruit rounded to apiculate at apex 4
3 Petioles and lst-year stems deep red; leaves serrate; racemes mostly longer than leaves **14. P. lusitanica**
3 Petioles and lst-year stems green; leaves crenate to obscurely serrate; racemes mostly shorter than leaves **15. P. laurocerasus**
  4 Petals >5mm; sepals falling before fruit ripe **12. P. padus**
  4 Petals <5mm; sepals persistent until fruit ripe **13. P. serotina**
5 Ovary and fruit pubescent; leaves mostly lanceolate to oblanceolate 6
5 Ovary and fruit glabrous; leaves mostly ovate to obovate 7
  6 Drupe becoming dry and splitting at maturity, much wider than thick, with pitted stone **2. P. dulcis**
  6 Drupe becoming very succulent and not splitting at maturity, subglobose, with deeply grooved stone **1. P. persica**
7 Flowers in short subcorymbose racemes with green bracts on proximal part of axis **8. P. mahaleb**
7 Flowers solitary or in umbels, or rarely very short corymbose racemes without green bracts on axis 8
  8 Flowers (1)2-6(10) together, with group of large (>5mm) often green or reddish bud-scales at base of cluster (not in *P. pensylvanica*); ripe fruit not pruinose, with longer pedicel (*cherries*) 9
  8 Flowers 1-3 together, with 0 or small (<3mm) brown bud-scales at base of cluster; ripe fruit pruinose, with usually shorter pedicel (*plums*) 13
9 Leaves with acuminate to aristate teeth 10
9 Leaves with acute to obtuse teeth 11
  10 Leaves <8cm, with acuminate teeth; flowers not pendent **11. P. incisa**
  10 Some leaves >8cm, with aristate teeth; flowers usually pendent **10. P. serrulata**
11 Bud-scales at base of inflorescence falling before flowers fully open; flowers 1-2cm across **9. P. pensylvanica**
11 Bud-scales at base of inflorescence persistent; flowers 2-3.5cm across 12
  12 Hypanthium cup-shaped, not constricted at opening; some bud-scales at base of flowers usually green and leaf-like; never a large tree **7. P. cerasus**
  12 Hypanthium bowl-shaped, constricted at opening; bud-scales not green and leaf-like; often a large tree **6. P. avium**

13 lst-year twigs green, shiny, glabrous **3. P. cerasifera**
13 lst-year twigs brown to grey, dull, often pubescent 14
14 Fruit <2cm, blue-black; flowers appearing before leaves; petals 5-8mm; twigs very spiny **4. P. spinosa**
14 Fruit usually >2cm, blue-black, red or yellow-green; flowers appearing with leaves; petals 7-12mm; twigs not or sparsely spiny **5. P. domestica**

Other spp. - Many spp. and cultivars are grown for ornament and occasionally persist in semi-wild conditions; most cultivars are grafted on to stock of wild British spp. and often do not produce fruit, so do not regenerate sexually or vegetatively.

1. P. persica (L.) Batsch - *Peach*. Deciduous tree to 6m; leaves lanceolate to oblanceolate, 5-15cm; petioles mostly 1-1.5cm; flowers 1(-2); petals 10-20mm, deep pink; fruit the familiar peach; (2n=16). Intrd; frequent on tips and waste ground in towns, from discarded stones, but rarely reaching maturity; S & C En; China.

2. P. dulcis (Mill.) D.A. Webb (*P. amygdalus* Batsch) - *Almond*. Vegetatively similar to *P. persica* but tree to 8m; leaves 4-12cm; petioles mostly 1.5-2.5cm; flowers (1-)2; petals white to deep pink; fruit the familiar almond; (2n=16). Intrd; extremely common street- and park-tree, also frequent on tips and waste places in S Br; SW Asia.

3. P. cerasifera Ehrh. - *Cherry Plum*. Deciduous, sometimes spiny shrub or tree to 8(12)m; leaves 3-7cm, ovate to obovate, glabrous, green or purplish; flowers appearing with or before leaves, the earliest in the genus, 1-2(3); petals white (or pink in cultivars); fruit ± globose, 2-3cm, dark red or yellow, scarcely bloomed, with little-flattened stone; (2n=16). Intrd; common in hedges and as street-tree, planted for hedging and ornament (mostly purplish-leaved: white- to very pale pink-flowered var. **pissardii** (Carrière) L.H. Bailey (*P.* 'Atropurpurea'), or pink-flowered **P. 'Nigra'**), spreading by suckers and often used as stock for other *Prunus* cultivars, not commonly fruiting: CI and Br N to C Sc; SE Europe and SW Asia.

4. P. spinosa L. - *Blackthorn*. Deciduous, dense, spiny shrub to 4m; leaves obovate to oblanceolate, 1-3(4)cm, ± pubescent; flowers appearing before leaves, 1(-2); petals white; fruit nearly globose, 8-15mm, bluish-black with dense bloom, with not or scarcely flattened stone; 2n=32. Native: hedges, scrub and woods; common almost throughout BI.

4 x 5. P. spinosa x P. domestica = P. x fruticans Weihe occurs in hedges sporadically throughout En; it is intermediate, fertile and variable. *P. spinosa* var. *macrocarpa* Wallr. probably belongs here.

5. P. domestica L. (*P. insititia* L.) - *Wild Plum*. Deciduous large shrub or tree to 8(12)m; leaves 3-8cm, obovate to elliptic; flowers appearing with leaves, mostly 2-3; petals white; fruit varied in size, shape and colour, with weak to strong bloom; (2n=48). Intrd; hedges, copses, scrub and waste ground; throughout most of BI; SW Asia. 2 or 3 sspp. are often recognized: ssp. **domestica** (*Plum*) with sparsely pubescent spineless twigs and usually large fruits with very flattened stone; ssp. **insititia** (L.) Bonnier & Layens (*Bullace, Damson*), with densely pubescent often spiny twigs and small fruits with less flattened stone; and ssp. **italica** (Borkh.) Gams ex Hegi (*P. x italica* Borkh.) (*Greengage*), ± intermediate. However, these have been so much hybridized that character-correlation has partly broken down and the sspp. are often scarcely discernible.

6. P. avium (L.) L. - *Wild Cherry*. Deciduous tree to 31m; leaves 6-15cm, obovate to elliptic; flowers 2-6 in umbels, on pedicels 15-45mm, usually shallowly cup-shaped; fruit ± globose, 9-12mm, black, red or yellow; 2n=16. Native; hedgerows, wood-borders and copses; throughout BI.

7. P. cerasus L. - *Dwarf Cherry* Deciduous shrub or small tree to 8m; differs from *P. avium* in saucer-shaped flowers fewer per umbel (mostly 2-4) with shorter

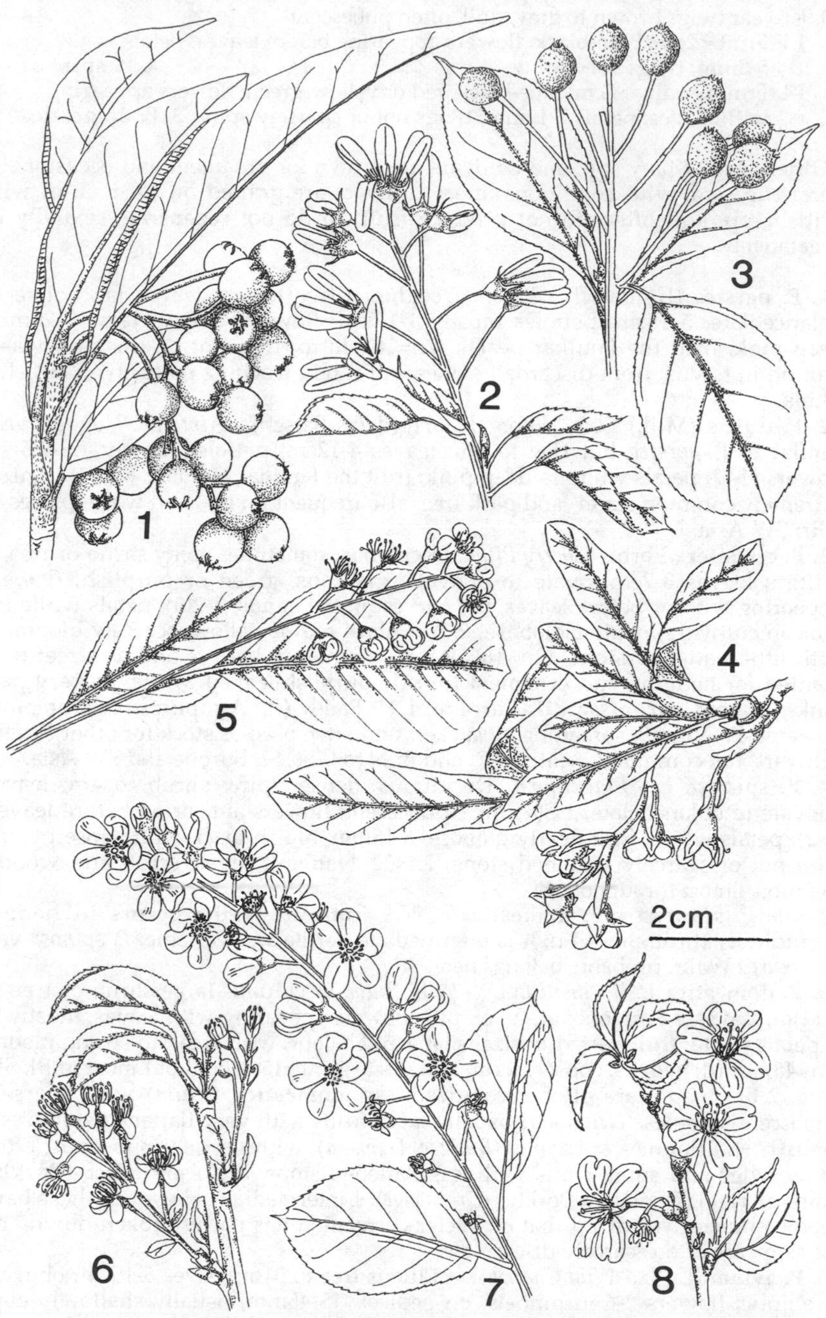

FIG 366 - **Rosaceae**. 1, *Photinia davidiana*. 2, *Amelanchier lamarckii*. 3, *Aronia melanocarpa*. 4, *Oemleria cerasiformis*. 5, *Prunus serotina*. 6, *P. mahaleb*. 7, *P. padus*. 8, *P. incisa*.

pedicels (1-4cm) hence mixed more among the leaves than projecting beyond; fruit bright red, never sweet; and see key; (2n=32). Intrd; hedges and copses; throughout most of BI N to C Sc, but much confused with *P. avium* and distribution uncertain; SW Asia.

8. P. mahaleb L. - *St Lucie Cherry*. Deciduous shrub or small tree to 6(10)m; 366
leaves 3-7cm, ovate to broadly so, rounded to subcordate at base; flowers up to 10 in subcorymbose racemes on pedicels up to 15mm; petals white; fruit ≤1cm, broadly ovoid, black; (2n=16). Intrd; well natd by railways, in grassland and in woods; several places in S En; S Europe.

9. P. pensylvanica L.f. - *Pin Cherry*. Deciduous tree to 12m; leaves 6-12cm, ovate to obovate or narrowly so, cuneate to rounded at base; flowers up to 6(10) in umbels or very short corymbose racemes on pedicels up to 15mm; petals white; fruit c.6mm, subglobose, red; (2n=16, 32). Intrd; natd in and by woodland (self-sown trees to 9m) in Surrey; N America.

10. P. serrulata Lindl. - *Japanese Cherry*. Deciduous tree to 12m; leaves 5-12cm, ovate to obovate or narrowly so, long-acuminate, with aristate teeth; flowers 2-6 in umbels; petals white or pink; fruit ± globose, black, but rarely formed; (2n=16). Intrd; much planted by roads and in parks, and often found as relic in wild places in much of BI; Japan and China. Many cultivars, often *flore pleno* (commonest, with pink petals, is 'Kanzan') and usually grafted on to *P. avium* as stock; not natd.

11. P. incisa Thunb. ex Murray - *Fuji Cherry*. Deciduous shrub or small tree to 366
6(10)m; leaves 4-8cm, ovate to narrowly so, long-acuminate, 2-serrate, with acuminate teeth; flowers 1-4, mostly 3; petals pink; fruit 6-8mm, ovoid, purplish-black; (2n=16). Intrd; natd in oakwoods on clay-with-flints, Chinnor Hill, Oxon; Japan.

12. P. padus L. - *Bird Cherry*. Deciduous shrub or tree to 19m; leaves 5-10cm, 366
obovate to elliptic, glabrous or with white hairs in tufts along lowerside midrib; flowers in elongate ± erect to pendent racemes; pedicels 8-15mm in flower; petals 6-9mm, white; fruit 6-8mm, ± globose,shiny-black; 2n=32. Native; woods and scrub; Br from C En and S Wa to N Sc, very scattered in Ir, also much planted and natd in S & C En.

13. P. serotina Ehrh. - *Rum Cherry*. Differs from *P. padus* in leaves glabrous or 366
with rows of brown hairs along lowerside midrib; shorter stiffer racemes with closer-packed flowers; pedicels 3-7mm in flower (-10mm in fruit); petals 3-4.5mm; fruit 8-10mm, purplish-black; (2n=32). Intrd; natd in woods and commons; S En and Wa; E N America.

14. P. lusitanica L. - *Portugal Laurel*. Evergreen shrub or tree to 12m; leaves narrowly ovate to oblong-ovate, 6-13cm; flowers in 10-25cm racemes; petals white; fruit conical-ovoid, mostly 8-10mm, purplish-black; (2n=64). Intrd; commonly planted, sometimes natd in woods, shrubberies and waste land; scattered in Br N to C Sc; SW Europe.

15. P. laurocerasus L. - *Cherry Laurel*. Evergreen shrub or tree to 10m; leaves usually oblong-ovate, 5-15cm; flowers in 7-13cm racemes; petals white; fruit conical-ovoid, mostly10-12mm, purplish-black to ± black; (2n=144, 170-180). Intrd; abundantly planted, sometimes natd in woods and shrubberies, throughout most of BI; SE Europe.

23. OEMLERIA Rchb. (*Osmaronia* Greene) - *Osoberry*

Deciduous shrubs; leaves simple, entire; flowers dioecious or partially so, in racemes; carpels 5; fruit a cluster of 1-5 drupes.

1. O. cerasiformis (Torr. & A. Gray ex Hook. & Arn.) J.W. Landon (*Osmaronia* 366
cerasiformis (Torr. & A. Gray ex Hook. & Arn.) Greene) - *Osoberry*. Suckering shrub with erect stems to 2(3)m; leaves lanceolate to narrowly elliptic, 7-11cm; flowers c.5-10 in short pendent racemes, scented; petals greenish-white; fruit 10-15mm, bluish-black, bloomed, rarely produced; (2n=12). Intrd; natd on rough ground in

Mons, Middlesex and W Kent; W N America.

SUBFAMILY 4 - MALOIDEAE (Pomoideae) (genera 24-35). Trees or shrubs with stipules; hypanthium concave, enclosing carpels and fused to them all round; epicalyx 0; stamens >10; flowers 5-merous; carpels 1-5, 1-2(several)-seeded; fruit consisting of fused carpels surrounded by succulent hypanthium; chromosome base-number 17.

24. CYDONIA Mill. - *Quince*
Leaves simple, entire, deciduous; flowers solitary; stamens 15-25; carpels 5, with free styles and numerous ovules, the walls cartilaginous in fruit.

1. C. oblonga Mill. - *Quince*. Spineless shrub or tree to 3(6)m; leaves 5-10cm, ovate, grey-tomentose on lowerside; flowers 4-5cm across, white or pink; fruit up to 12cm, yellow; (2n=34). Intrd; not or rarely regenerating but very persistent in hedges and woods, sometimes used as stock for *Pyrus*; scattered in S Br, Man; Asia.

25. CHAENOMELES Lindl. - *Japanese Quinces*
Leaves simple, serrate, deciduous, flowers in clusters of 1-4; stamens 40-60; carpels 5, with styles fused at base and numerous ovules, the walls cartilaginous in fruit.

Other spp. - *C. speciosa* x *C. japonica* = **C. x superba** (Frahm) Rehder is grown in gardens and may persist as a relic or throwout; difficult to distinguish from *C. speciosa*. The large garden shrub is mostly *C. speciosa* but usually known as *Japonica*.

1. C. speciosa (Sweet) Nakai - *Chinese Quince*. Open ± spiny shrub to 3m; leaves 4-10cm, ovate to obovate, glabrous; flowers 3.5-4.5cm across, usually red; fruit up to 7.5cm, yellowish-green; (2n=34). Intrd; not or rarely regenerating but very persistent in hedges and woods; scattered in CI and S En; China.

2. C. japonica (Thunb.) Spach - *Japanese Quince*. Open ± spiny shrub to 1m; twigs becoming glabrous and distinctly warty; leaves smaller (2.5-6cm) and more coarsely toothed than in *C. speciosa*; flowers 3.5-4cm across, usually pink; fruit up to 5cm; (2n=34). Intrd; persistent as in *C. speciosa* but apparently less so; rare in S En; Japan.

26. PYRUS L. - *Pears*
Leaves simple, serrate to crenate, deciduous; flowers in simple corymbs; petals white; stamens 20-30; carpels 2-5, with free styles and 2 ovules, the walls cartilaginous in fruit; fruit with groups of gritty stone-cells in hypanthium.

1 Fruit >(5)6cm, soft and sweet when mature, usually pear-shaped **3. P. communis**
1 Fruit <5cm, hard and sour even when mature, not pear-shaped 2
  2 Fruit with calyx falling early, <1.5(2)cm; petals 6-10mm; inflorescence rhachis >1cm **1. P. cordata**
  2 Fruit with persistent calyx, 1.5-4cm; petals 10-17mm; inflorescence rhachis <1cm **2. P. pyraster**

Other spp. - **P. salicifolia** Pall. (*Willow-leaved Pear*), from the Caucasus, is popular in gardens and sometimes persists as a relic; it has narrowly elliptic to linear-lanceolate, densely silvery-pubescent leaves, pendent twigs and small brown hard pear-shaped fruits.

1. P. cordata Desv. - *Plymouth Pear*. Spiny shrub to 8m; leaves 1-4.5cm, ovate or **RRR**

broadly so; fruits 8-20mm, globose or obovoid, brownish-red to blackish. Probably native; 2 hedges near Plymouth, S Devon (known since 1870), 3 sites near Truro, W Cornwall (found 1989).

2. P. pyraster (L.) Burgsd. - *Wild Pear*. Usually ± spiny shrub or tree to 15m; leaves 2.5-7cm, ovate to broadly so; fruits 1.5-4cm, globose to obovoid or obconical, yellow to reddish- or dark-brown; (2n=34). Intrd; hedges and wood-margins; scattered through S & C Br, CI and Ir; Europe. 2 sspp. recently recognized (ssp. **pyraster**, with ± glabrous leaves and globose to obconical fruits; ssp. **achras** (Wallr.) Terpó, with pubescent leaves and obconical to obovoid fruits) seem scarcely distinct in practice. Many wild pears are probably stocks of formerly cultivated trees.

3. P. communis L. - *Pear*. Differs from *P. pyraster* in usually non-spiny tree to 20m; fruit >(5)6cm, edible, usually pear-shaped; (2n=34). Intrd; very commonly grown and found in the wild in hedges and waste ground from old trees or discarded seeds; garden origin. Often scarcely distinguishable from *P. pyraster* except in fruit, and perhaps not specifically distinct.

27. MALUS Mill. - *Apples*

Leaves simple, serrate, deciduous; flowers in simple corymbs; stamens 15-50; carpels 3-5, with styles fused below and with 2 ovules, the walls cartilaginous in fruit; fruit without groups of gritty stone-cells.

1 Leaves purplish-green to purple; fruit dark reddish-purple, c. as long as pedicel **3. M. x purpurea**
1 Leaves green; fruits green to yellow or red, longer than pedicel 2
2 Leaves glabrous when mature; pedicels and outside of calyx glabrous **1. M. sylvestris**
2 Leaves pubescent on lowerside; pedicels and outside of calyx pubescent **2. M. domestica**

Other spp. - Many spp. and cultivars are grown for their ornamental flowers or fruit, and sometimes persist in wild places or produce odd seedlings. The following are street trees with pedicels longer than fruit: **M. baccata** (L.) Borkh. (*Siberian Crab*) with white petals, mostly 5 styles and red fruits ≤1cm; and **M. floribunda** Sieb. ex Van Houtte (*Japanese Crab*), with pink and white petals, mostly 4 styles, and red or yellow fruits >1cm. Many taxa are best known by their cultivar names only, e.g. *M.* **'John Downie'**.

1. M. sylvestris (L.) Mill. - *Crab Apple*. Tree to 10m, often spiny; twigs glabrous; leaves 3-5cm, ovate to elliptic, glabrous when mature; petiole 1.5-3cm; pedicels and outside of calyx glabrous; petals pinkish-white; fruit apple-shaped, yellowish-green, c.2-3cm; (2n=34). Native; woods, hedges and scrub; probably throughout BI N to Shetland but very rare in N Sc. Much over-recorded for *M. domestica* and often very difficult to separate from it; intermediates are frequent and may be hybrids. The 2 are perhaps not specifically distinct.

2. M. domestica Borkh. (*M. sylvestris* ssp. *mitis* (Wallr.) Mansf.) - *Apple*. Tree to 10(20)m, not spiny; similar to *M. sylvestris* but larger in most parts; leaves up to 15cm, pubescent on lowerside, with relatively shorter petiole; fruit up to 12cm, variously coloured; (2n=34). Intrd; much grown and often natd in hedges, scrub and waste ground; throughout BI and much commoner than *M. sylvestris*. Self-sown plants usually have small, yellowish, sour fruits.

3. M. x purpurea (E. Barbier) Rehder (*M. niedzwetzkvana* Dieck x *M. atrosanguinea* (Späth) C.K. Schneid.) - *Purple Crab*. Tree to 11m, not spiny; leaves up to 10cm, ovate to elliptic, pubescent; petiole 2-3.5cm; petals deep pink; fruit ± globose, dull reddish-purple, 2.5-3.5cm. Intrd; grown in parks and by roads, rarely self-sown; natd in W Kent, perhaps elsewhere in S En; garden origin c.l900.

28. SORBUS L. - *Whitebeams*

Leaves pinnate, or simple and serrate to pinnately lobed, deciduous; flowers in compound corymbs; petals white; stamens 15-25; carpels 2-4(5), with styles free or fused below and 2 ovules, the walls cartilaginous in fruit.

A difficult genus, consisting of several well-defined but variable sexual species and a number of apomictic ones, many of which are of hybrid origin. The apomicts mostly fall into 3 groups: those similar to *S. aria* (*S. aria* agg.); those intermediate between either *S. torminalis* or *S. aria* agg. and *S. aucuparia* (*S. intermedia* agg.); and those intermediate between *S. aria* agg. and *S. torminalis* (*S. latifolia* agg.). Most diagnostic characters concern leaves and fruits, both of which are required for identification by beginners. 'Leaves' refers to the broader leaves of the short-shoots (not those on leading shoots). It is important that plants are surveyed as fully as possible and that the means of each character-state are used for identification.

Recent suggestions that *Sorbus* should be split into 5 genera (4 British) are best not followed until unequivocal evidence is produced.

1 Leaves pinnate, at least proximally, with ≥1 pair of completely free leaflets 2
1 Leaves not to deeply lobed, but without free leaflets 5
  2 Leaves completely pinnate, with ≥4 pairs of leaflets and a single terminal leaflet 3
  2 Leaves pinnate only proximally, with 1-2(3) pairs of leaflets and a distal lobed part of leaf (if ≤5 pairs of leaflets, see **2x8** and **2x10**) 4
3 Buds glabrous, viscid, greenish to pale brown; stipules 6-14mm, forked at or below 1/2 way, soon falling; fruits mostly >2cm across, greenish-brown sometimes red-tinged **1. S. domestica**
3 Buds tomentose, not viscid, brown; stipules 4-8mm, simple or fan-shaped, persistent on long-shoots; fruits <15mm across, scarlet (to yellow) **2. S. aucuparia**
  4 Leaves 5.5-8.5cm, rather thinly grey-tomentose on lowerside, with mostly 1 pair of free leaflets; fruit longer than wide; petals c.4mm **3. S. pseudofennica**
  4 Leaves 7.5-10.5cm, densely whitish-grey-tomentose on lowerside, with mostly 2 pairs of free leaflets; fruit globose; petals c.6mm **4. S. hybrida**
5 Mature leaves sparsely pubescent on lowerside, at least some lobed ≥1/3 way to midrib proximally; fruit brown **25. S. torminalis**
5 Mature leaves tomentose on lowerside (wearing off with age), if lobed ≥1/3 way to midrib then fruit red 6
  6 Leaves white-tomentose on lowerside; fruit red **10-18. S. aria** agg.
  6 Leaves grey- to yellowish-tomentose on lowerside; fruit red, orange or brown 7
7 Fruit red **5-9. S. intermedia** agg.
7 Fruit orange to brown **19-24. S. latifolia** agg.

Other spp. - Many spp. are grown for ornament and may persist in wild places, especially: **S. hupehensis** C.K. Schneid. (*Hupeh Rowan*), from China, with pinnate leaves and ± white fruit; and several taxa in *S. aria* agg.

1. S. domestica L. - *Service-tree*. Tree to 23m; trunk deeply and closely fissured; leaves pinnate, with 6-8 pairs of leaflets; styles 5(-6); fruit 20-25mm across, obovoid or pear-shaped, greenish-brown, sometimes red-tinged, with numerous large lenticels, with groups of gritty stone-cells; (2n=34). Probably native; in 3 sites on limestone sea-cliffs in Glam, first found 1983, as stunted (to 3(5)m) non-fruiting plants; planted and sometimes self-sown in a few places in S En, possibly formerly native in Worcs. **RR**

2. S. aucuparia L. - *Rowan*. Tree to 18m; trunk smooth or shallowly and sparsely fissured; leaves pinnate, with (4)5-7(9) pairs of leaflets; styles 3-4; fruit 6-9(14)mm, ± globose, scarlet (to yellow in cultivars), with few inconspicuous lenticels, without groups of gritty stone-cells; 2n=34. Native; woods, moors, rocky places except on heavy soils; throughout Br and Ir, intrd in CI.

2 x 8. S. aucuparia x S. intermedia occurs as occasional spontaneous trees near the parents in Br N to Staffs and Merioneth. It has partially pinnate ovate-oblong leaves with usually (1)2-3(5) pairs of free, well-separated leaflets and a total of 10-12 pairs of lateral veins, sterile pollen, and scarlet fruits with no seed; 2n=51.

2 x 10. S. aucuparia x S. aria = S. x thuringiaca (Ilse) Fritsch (*S. x semipinnata* **372**
(Roth) Hedl. non Borbás) occurs as occasional spontaneous trees near the parents scattered over Br mostly in N, and as planted trees. It closely resembles *S. aucuparia* x *S. intermedia* but has brownish-red fruits with some viable seed, some fertile pollen, and usually wider, more oblong leaves; 2n=34.

3. S. pseudofennica E.F. Warb. - *Arran Service-tree.* Tree to 7m; leaves partially **RR**
pinnate, differing from those of above 2 hybrids in having total of 7-9(10) pairs of **372**
lateral veins and 1(-2) pairs of broader ± overlapping free leaflets; styles 2-3; fruits scarlet, with few inconspicuous lenticels; fully fertile; 2n=51. Native; steep granite stream-bank; Glen Catacol, Arran (Clyde Is); endemic.

4. S. hybrida L. - *Swedish Service-tree.* Tree to 10m; leaves partially pinnate, **372**
differing from those of above 2 hybrids in having total of 8-10 pairs of lateral veins and (1-)2(-3) pairs of free well-separated leaflets; styles 2-3; fruits scarlet, with small sparse lenticels; fully fertile. Intrd; frequently grown in gardens and parks and sometimes self-sown; natd in W Kent and N Aberdeen; Scandinavia.

5-9. S. intermedia agg. - *Swedish Whitebeam.* Leaves variable but often with lowest 3-6 pairs of lateral veins ending in distinct leaf-lobe divided ≥1/6 way to midrib, grey- to yellowish-tomentose on lowerside; fruit scarlet to crimson; styles 2.

Multi-access key to spp. of S. intermedia agg.

Leaves mostly with 7-8 pairs of lateral veins	A
Leaves mostly with 8-10 pairs of lateral veins	B
Leaves mostly lobed >1/3 way to midrib	C
Leaves mostly lobed <1/3 way to midrib	D
Petals mostly c.4mm	E
Petals mostly c.5mm	F
Petals mostly c.6mm	G
Fruit mostly 6-8mm	H
Fruit mostly 8-11mm	I
Fruit mostly 11-15mm	J
Fruit ± globose or wider than long	K
Fruit longer than wide	L
ACEIL	**5. S. arranensis**
BCFIK	**6. S. leyana**
BDEHK	**7. S. minima**
BDG(IJ)K	**9. S. anglica**
B(CD)GJL	**8. S. intermedia**

5. S. arranensis Hedl. Tree to 7.5m; leaves narrowly elliptic, 1.5-2.1(2.6)x as long **RR**
as wide, acute at base and apex, lobed (1/3)1/2-3/4 way to midrib, with 7-8(9) pairs **373**
of veins; fruit 8-10mm, with few inconspicuous lenticels; 2n=51. Native; steep granite stream-banks; Arran (Clyde Is); endemic.

6. S. leyana Wilmott. Shrub to 3m, rarely tree to 12m; leaves elliptic-oblong, 1.3- **RR**
1.7(1.9)x as long as wide, acute to subacute at base and apex, lobed 1/3-1/2 way to **373**
midrib, with (8)9-10 pairs of veins; fruit c.10mm, with few small lenticels. Native; carboniferous limestone crags; near Merthyr Tydfil (Brecs); endemic.

7. S. minima (Ley) Hedl. Shrub to 3m; leaves narrowly elliptic or narrowly **RR**
oblong-elliptic, (1.4)1.8-2.2x as long as wide, acute at apex, acute to subacute at **373**

FIG 372 - Leaves of *Sorbus*. 1, *S. aria*. 2, *S. leptophylla*. 3, *S. eminens*. 4, *S. hibernica*. 5, *S. wilmottiana*. 6, *S. lancastriensis*. 7, *S. rupicola*. 8, *S. vexans*. 9, *S. porrigentiformis*. 10, *S. pseudofennica*. 11, *S. hybrida*. 12, *S. x thuringiaca*. 13, *S. x vagensis*.

FIG 373 - Leaves of *Sorbus*. 1, ***S. latifolia***. 2, ***S. decipiens***. 3, ***S. bristoliensis***. 4, ***S. devoniensis***. 5, ***S. croceocarpa***. 6, ***S. leyana***. 7, ***S. intermedia***. 8, ***S. minima***. 9, ***S. subcuneata***. 10, ***S. anglica*** (Llangollen). 11, ***S. arranensis***. 12, ***S. anglica*** (typical).

base, lobed 1/5-1/3(1/2) way to midrib, with (7)8-9(10) pairs of veins; fruit 6-8mm, with few small lenticels; 2n=51. Native; carboniferous limestone crags; near Crickhowell (Brecs); endemic.

8. S. intermedia (Ehrh.) Pers. Tree to 10(18)m; leaves ovate-oblong to elliptic-oblong, (1.5)1.6-1.8(1.9)x as long as wide, subacute to obtuse at apex, rounded at base, lobed 1/4-1/3(1/2) way to midrib, with 7-9 pairs of veins; fruit 12-15mm, with few small lenticels; 2n=68. Intrd; much planted and frequently self-sown in copses and on rough ground; scattered over most of Br and Ir; Baltic region. **373**

9. S. anglica Hedl. Shrub to 3m; leaves obovate to ovate-oblong, 1.3-1.7(1.8)x as long as wide, subacute to obtuse at base and apex, lobed 1/6-1/4 way to midrib, with (7)8-10(11) pairs of veins; fruit 7-12mm, ± crimson (scarlet in rest of agg.), with few to many usually small lenticels; 2n=68. Native; woods and rocky places mostly on carboniferous limestone; very local in Wa, SW En and N Kerry; endemic. A variant from near Llangollen (Denbs) with more narrowly cuneate leaf-bases might be a separate sp. **RR** **373**

10-18. S. aria agg. - *Common Whitebeam.* Leaves toothed, lobed ≤1/5 way to midrib, white- or whitish-tomentose on lowerside; fruit scarlet to crimson; styles 2.

Multi-access key to spp. of S. aria agg.

Fruit longer than wide	A
Fruit ± globose or wider than long	B
Fruit mostly 8-12mm	C
Fruit mostly 12-15mm	D
Fruit mostly 15-22mm	E
Fruit scarlet	F
Fruit crimson or pinkish-scarlet	G
Leaves mostly ≤1.5x as long as wide	H
Leaves mostly 1.5-2x as long as wide	I
Leaves mostly ≥2x as long as wide	J
Leaves mostly with ≥10 pairs of lateral veins	K
Leaves mostly with ≤10 pairs of lateral veins	L

(AB)(CD)F(HIJ)K		**10. S. aria**
A(CD)GIL		**12. S. wilmottiana**
ADFIL		**18. S. vexans**
AEFIK		**11. S. leptophylla**
AEGHK		**13. S. eminens**
BCG(HI)L		**15. S. porrigentiformis**
B(DE)GH(KL)	Ireland only	**14. S. hibernica**
BDG(HI)L	NW England only	
	Leaves <1.75(1.9)x as long as wide	**16. S.lancastriensis**
BDG(IJ)L	Leaves >(1.7)1.8x as long as wide	**17. S. rupicola**

10. S. aria (L.) Crantz. Tree to 15(23)m; leaves elliptic to ovate or oblong (rarely obovate), 1.1-1.6(2.4)x as long as wide, the sides ± curved throughout and entire for basal ≤1/5, densely white-tomentose on lowerside, with (9)10-14(15) pairs of veins; leaf-lobes with apical tooth scarcely projecting beyond its neighbours; fruit 8-15mm, scarlet, with few to numerous small evenly spread lenticels; 2n=34. Native; woods, scrub, rocky places, mostly on calcareous soils; probably native only in En N to Beds and E Gloucs, and in Galway, but commonly planted and ± natd throughout most of Br and Ir. Sexual and very variable, incl. several cultivars, overlapping to some extent with most other spp. in the agg. **372**

10 x 17. S. aria x S. rupicola has been reported from N Somerset and W Gloucs.

10 x 25. S. aria x S. torminalis = S. x vagensis Wilmott (?*S. x rotundifolia* (Bechst.) Hedl.) occurs as occasional trees near the parents in N Somerset, W Gloucs, Mons and Herefs. It is rather variable and both fertile and sterile trees are known. It has brownish-orange fruits and differs from *S. devoniensis* and *S.* **372**

subcuneata in the longer than wide (not subglobose) fruit with few small (not many large) lenticels.

11. S. leptophylla E.F. Warb. Shrub to 3m; leaves obovate, (1.3)1.5-1.7(1.9)x as long as wide (longer on long-shoots), usually ± straight-sided for apical and basal 1/3, entire for basal ≤1/5, greenish-white-tomentose on lowerside, with (9-)11(-13) pairs of veins; leaf-lobes with sharper and deeper teeth than in *S. aria* and apical tooth projecting c.3mm beyond its neighbours; fruit c.20mm, scarlet, with few lenticels, the larger ones towards base; 2n=51, 68. Native; carboniferous limestone crags; 2 areas in Brecs; endemic. **RR** **372**

12. S. wilmottiana E.F. Warb. Shrub or small tree to 6m; leaves elliptic to ovate-elliptic, 1.6-2(2.7)x as long as wide, the sides ± curved throughout and entire for basal ≤1/5, greenish-white-tomentose on lowerside, with 8-9(10)) pairs of veins; leaf-lobes with apical tooth scarcely projecting beyond its neighbours; fruit 10-13mm, crimson, with few rather large lenticels mostly towards base; 2n=51. Native; rocky carboniferous limestone woodland and scrub in Avon Gorge (N Somerset and W Gloucs); endemic. **RR** **372**

13. S. eminens E.F. Warb. Shrub or small tree to 6m; leaves orbicular-ovate to -obovate, 1.1-1.4(1.7)x as long as wide, the sides ± curved throughout and entire for basal ≤1/5, greenish-white-tomentose on lowerside, with (9)10-11(12) pairs of veins; leaf-lobes with apical tooth conspicuously projecting beyond its neighbours; fruit c.20mm, crimson, with ± numerous small and large lenticels mostly towards base; 2n=34, 51. Native; rocky carboniferous limestone woodland; Avon Gorge and Wye Valley (N Somerset, W Gloucs, Mons and Herefs); endemic. **RR** **372**

14. S. hibernica E.F. Warb. Tree to 6m; leaves elliptic-ovate to -obovate, (1.1)1.2-1.5(1.8)x as long as wide, the sides ± curved throughout and entire for basal ≤1/5, greenish-white-tomentose on lowerside, with (8)9-11(12) pairs of veins; leaf-lobes with apical tooth conspicuously projecting beyond its neighbours; fruit c.15mm, pinkish-scarlet, with rather few lenticels, the larger ones towards base. Native; rocky carboniferous limestone scrub; Ir, mostly C; endemic. **R** **372**

15. S. porrigentiformis E.F. Warb. Shrub or small tree to 5m; leaves obovate, 1.3-1.7x as long as wide, ± straight-sided in lower 1/3(-1/2), entire for basal ≤1/5, white-tomentose on lowerside, with (7)8-10(11) pairs of veins; leaf-lobes with apical tooth conspicuously projecting beyond its neighbours; fruit 8-12mm, crimson, with rather few large lenticels mostly towards base; 2n=34, 68. Native; rocky limestone woods; SW Br from S Devon to Brecs; endemic. **R** **372**

15 x 25. S. porrigentiformis x S. torminalis has been reported from W Gloucs; endemic.

16. S. lancastriensis E.F. Warb. Shrub or small tree to 5m; leaves obovate to elliptic-obovate, (1.4)1.5-1.8(2)x as long as wide, ± straight-sided in lower 1/3, ± entire for basal 1/4, greyish-white-tomentose on lowerside, with (7)8-10 pairs of veins; leaf-lobes with apical tooth scarcely projecting beyond its neighbours; fruit 12-15mm, crimson, with ± numerous lenticels, the larger ones towards base; 2n=68. Native; rocky scrub and woodland, usually on carboniferous limestone; W Lancs and Westmorland; endemic. **RR** **372**

17. S. rupicola (Syme) Hedl. Shrub or small tree to 6(10)m; leaves narrowly obovate, (1.4)1.6-2.1(2.4)x as long as wide, ± straight-sided in lower 1/2, ± entire for basal 1/3-1/2, densely white-tomentose on lowerside, with (6)7-9(10) pairs of veins; leaf-lobes with apical tooth scarcely projecting beyond its neighbours; fruit 12-15mm, deep red, with numerous small to medium evenly spread lenticels; 2n=34, 51, 68. Native, rocky woodland, scrub and cliffs, usually on limestone; scattered over Br and Ir but not in E, SE or SC En. **R** **372**

17 x 25. S. rupicola x S. torminalis has been recorded from Glam and Brecs.

18. S. vexans E.F. Warb. Small tree to 6m; leaves obovate, (1.4)1.5-1.9(2)x as long as wide, ± straight-sided in lower 1/2, ± entire for basal 1/4-1/3, white-tomentose on lowerside, with 8-9(10) pairs of veins; leaf-lobes with apical tooth scarcely projecting beyond its neighbours; fruit 12-15mm, scarlet, with few **RR** **372**

lenticels, the larger ones towards the base; 2n=68. Native; rocky woods near coast of Bristol Channel (N Devon and S Somerset), not on limestone; endemic.

19-24. S. latifolia agg. - *Broad-leaved Whitebeam*. Leaves with ± triangular, acute or acuminate lobes usually divided >1/6 way to midrib, grey-tomentose on lowerside; fruit orange to orange-brown, often ± brown when fully ripe; styles 2, often fused at base.

Multi-access key to spp. of S. latifolia agg.

Fruit orange	A
Fruit orange-brown, sometimes turning brown	B
Fruits ≤13mm	C
Some fruits >13mm	D
Fruits mostly longer than wide	E
Fruits mostly subglobose	F
Anthers cream	G
Anthers pink	H
Leaves mostly with ≥10 pairs of lateral veins	I
Leaves mostly with ≤10 pairs of lateral veins	J
Leaves mostly ≥1.3x as long as wide	K
Leaves mostly ≤1.3x as long as wide	L
Largest leaf-lobes divided ≤1/6 way to midrib	M
Largest leaf-lobes divided ≥1/6 way to midrib	N

A(CD)EGIKN		**19. S. decipiens**
ACEHJKM		**23. S. bristoliensis**
AD(EF)GJLN	Leaf l/w ratio 1.1-1.3	**24. S. latifolia**
ADFG(IJ)(KL)M	Leaf l/w ratio 1.2-1.6	**22. S. croceocarpa**
B(CD)FGJKM	Leaf l/w ratio 1.3-1.6 Fruit with many large lenticels	**21. S. devoniensis**
BCEGJ(KL)N	Leaf l/w ratio 1.2-1.7 Fruit with few small lenticels	**10 x 25. S. aria x S. torminalis**
BCFGJKN	Leaf l/w ratio 1.6-1.9 Fruit with many large lenticels	**20. S. subcuneata**

19. S. decipiens (Bechst.) Irmisch. Tree to 10m; leaves ovate to elliptic, 1.3-1.8(2.5)x as long as wide, subacute to rounded at base, with 10-13 pairs of veins; leaf-lobes acute to acuminate, the lower ones ± patent, extending ≤1/5 way to midrib; fruit 8-17mm, orange, with scattered large and medium lenticels. Intrd; natd in Avon Gorge (W Gloucs) and at Achnashellach (W Ross), and planted elsewhere; C Europe. **373**

20. S. subcuneata Wilmott. Tree to 10m; leaves elliptic to ovate-elliptic, (1.5)1.6-1.9(2.5)x as long as wide, rounded at base, with (6)8-9(10) pairs of veins; leaf-lobes obtuse, forward-pointing, extending ≤1/4(1/3) way to midrib; fruit 10-13mm, brownish-orange becoming brown, with numerous lenticels, large towards base, smaller above. Native; open rocky *Quercus* woods near coast; N Devon and S Somerset; endemic. **RR 373**

21. S. devoniensis E.F. Warb. Tree to 15m; leaves broadly oblong-elliptic, 1.3-1.6(1.8)x as long as wide, rounded at base, with 7-9 pairs of veins; leaf-lobes acute, the lower ones erecto-patent, extending 1/8(1/6) way to midrib; fruit 10-15mm, brownish-orange becoming brown, with numerous lenticels, large towards base, smaller above. Native; woods and hedges on well-drained soils; widespread in Devon, very local in E Cornwall, S Somerset, Man and SE & NE Ir; endemic. **R 373**

22. S. croceocarpa P.D. Sell. Tree to 21m; leaves ovate or broadly so to ovate-elliptic, 1.2-1.6(1.8)x as long as wide, rounded at base, with (8)9-11 pairs of veins; leaf-lobes usually vestigial, sometimes extending ≤1/10 way to midrib, subacute, forward-pointing; fruit 11-22mm, bright orange, with numerous small to large **373**

lenticels, the larger ones towards the base. Intrd; frequently planted, natd in Br from S En to C Sc; origin unknown. Leaves least deeply lobed in the agg.

23. S. bristoliensis Wilmott. Tree to 10m; leaves elliptic-oblong to -obovate, 1.4- **RR**
1.7(2)x as long as wide, subacute to obtuse or slightly rounded at base, with (7)8- **373**
9(10) pairs of veins; leaf-lobes obtuse, forward-pointing, extending ≤1/6 way to midrib; fruit 9-11mm, bright orange, with ± numerous small and medium lenticels mostly towards base; 2n=51. Native; rocky woods and scrub on carboniferous limestone; Avon Gorge (N Somerset and W Gloucs); endemic.

24. S. latifolia (Lam.) Pers. Tree to 20m; leaves broadly ovate, 1.1-1.3x as long as **373**
wide, rounded at base, with 7-9 pairs of veins; leaf-lobes acute to abruptly acuminate, the lower ones ± patent, extending ≤1/4 of way to midrib; fruit 14-17mm, yellowish- to deep-orange, with few to numerous rather large lenticels; 2n=51, 68. Intrd; frequently planted, natd in Br from S En to C Sc; SW Europe.

25. S. torminalis (L.) Crantz - *Wild Service-tree.* Tree to 27m; leaves broadly ovate, 0.9-1.3x as long as wide, with triangular acute lobes divided ≤1/2 way to midrib, slightly pubescent on lowerside; fruit 12-16mm, brown, longer than wide, with numerous large lenticels; styles 2, fused to c.1/2 way; (2n=34). Native; woods, scrub and hedgerows mostly on clay or limestone; thoughout most of En and Wa, but local.

29. ARONIA Medik. - *Chokeberries*

Leaves simple, serrate, deciduous; flowers in compound corymbs; petals white; stamens c.20; carpels 5, with 5 styles fused near base, the walls cartilaginous at fruiting. Very few records, but likely to increase in future.

Possibly better united under *Photinia.*

Other spp. - *A. arbutifolia* x *A. melanocarpa* = **A. x prunifolia** (Marshall) Rehder, with purple fruits, is grown and should be sought in the wild.

1. A. arbutifolia (L.) Pers. - *Red Chokeberry*. Suckering shrub to 3m; leaves narrowly obovate to elliptic, densely pubescent on lowerside; fruit red; (2n=34). Intrd; natd in woodland on sandy soil; Surrey; E N America.

2. A. melanocarpa (Michx.) Elliott - *Black Chokeberry.* Suckering shrub to 1.5m; **366**
leaves obovate, subglabrous; fruit black; (2n=34). Intrd, natd in boggy areas; Dorset and Caerns; E N America.

30. AMELANCHIER Medik. - *Juneberry*

Leaves simple, serrate, deciduous; flowers in racemes; petals white; stamens 10-20; carpels 5, with 5 styles fused near base, the walls cartilaginous at fruiting.

1. A. lamarckii F. G. Schroed. (*A. laevis* auct. non Wiegand, *A. confusa* auct. non **366**
Hyl., *A. grandiflora* auct. non Rehder, *A. intermedia* auct. non Spach, *A. canadensis* auct. non (L.) Medik.) - *Juneberry.* Tree to 10m; leaves mostly oblong, subglabrous; fruit purplish-black, often not formed. Intrd; natd on mainly sandy soils in woodland and scrub; frequent in SC & SE En, very sparsely scattered elsewhere in En and in Jersey; N America.

31. PHOTINIA Lindl. (*Stranvaesia* Lindl.) - *Stranvaesia*

Leaves simple, entire, evergreen; flowers in compound corymbs; petals white; stamens c.20; carpels 5, with 5 styles fused to ≥1/3 way, the walls cartilaginous at fruiting; hypanthium not quite reaching apex of carpels at fruiting.

1. P. davidiana (Decne.) Cardot (*Stranvaesia davidiana* Decne.) - *Stranvaesia.* **366**
Shrub or tree to 3(8)m; leaves lanceolate to oblanceolate, subglabrous; fruit scarlet; (2n=34). Intrd; grown in gardens, bird-sown in rough ground in S, planted in

forestry plantations in N; extremely scattered in En, Man and Sc; China.

32. COTONEASTER Medik. - *Cotoneasters*

Leaves simple, entire; flowers in compound corymbs, small clusters, or solitary; stamens 10-20; carpels 1-5, with 1-5 ± free styles, the walls stony at fruiting.

The genus parallels *Sorbus* in that it contains both very variable sexual spp. (all apparently diploid, 2n=34) and much less variable (in this case ± invariable) apomictic spp. (triploid to hexaploid, 2n=51, 68, 85 or 102); the precise extent of apomixis is unknown but perhaps c.95% of spp. exhibit it. Much misidentification has occurred of both garden and wild plants. The sp. concept adopted here is that of B. Hylmö and J. Fryer.

A large genus becoming increasingly natd via bird-sown seed from garden or roadside ornamentals. Until familiar with the genus, flowers, ripe fruit and summer leaves (after flowering but before fruit ripe) are necessary for determination, and a knowledge of the degree of leaf retention in winter is also desirable. Fruit colours are often diagnostic, but are difficult to describe, so only 6 colours are defined here: dark purple to black; yellow (to yellowish-orange); orange-red; bright red; crimson (a deep red mildly tinted with blue); and maroon (often becoming brownish). The number of carpels ('stones') per fruit should be counted in at least 5 fruits; closely adherent stones are counted as separate. Leaf-sizes and pubescence refer to those of fully grown summer leaves; at flowering they may be much smaller and more densely pubescent in deciduous spp.

The following keys first use 8 characters to separate 12 groups in a multi-access key. 10 of these 12 groups are then dealt with by dichotomous keys; the other 2 groups each contain only *C. cooperi*.

Multi-access general key

Petals patent, usually white A
Petals erect to erecto-patent, usually pink B
  Leaves deciduous or mostly so, most dropped by Jan C
  Leaves evergreen D
Fruits dark purple to black E
Fruits yellow to bright red, crimson or maroon F
  Veins deeply impressed on leaf upperside G
  Veins not or slightly impressed on leaf upperside H
Summer leaves densely pubescent to tomentose on lowerside, largely or wholly obscuring surface I
Summer leaves glabrous to pubescent on lowerside, leaving most of surface exposed J
  Hypanthium and calyx glabrous to sparsely pubescent K
  Hypanthium and calyx pubescent L
  Hypanthium and calyx densely pubescent to tomentose M
Flowers mostly 1-2(3) together N
Flowers mostly 3-10(12) together O
Flowers mostly (10)12-many together P
  Anthers white (to pale yellow) Q
  Anthers pigmented (pink, mauve, violet, purple or blackish) R

ACEH(IJ)(KLM)(OP)(QR) ***Key A***
ACFH(IJ)KPR ***Key B***
ACF(GH)(IJ)(LM)PR ***Key C***
ADEH(IJ)KPR **10. C. cooperi**
ADF(GH)(IJ)M(OP)R ***Key D***
ADFH(IJ)(KL)(NO)R ***Key E***
ADFH(IJ)KPR **10. C. cooperi**
BCE(GH)(IJ)(KLM)(NOP)Q ***Key F***

BCF(GH)I(KLM)(OP)Q *Key G*
BCFGJ(KL)PQ *Key H*
BCFHJ(KL)(NO)Q *Key I*
BDF(GH)I(KLM)O(QR) *Key J*

Key A - Petals patent; leaves deciduous; fruits dark purple to black

1 Leaves <5(6)cm; flowers mostly <20 per inflorescence; anthers white or purplish-black 2
1 Some leaves >6cm; flowers mostly >(15)20 per inflorescence; anthers pink or red to violet or mauve 4
 2 Anthers purplish-black; fruits 8-11mm, oblong-ellipsoid, maroon; leaves c.1.25-1.5x as long as wide **1. C. monopyrenus**
 2 Anthers white; fruits 6-9mm, subglobose, bluish-black; leaves c.1-1.25x as long as wide 3
3 Shrub to 2m; leaves 1.5-4cm; flowers in loose inflorescence; hypanthium and calyx pubescent **4. C. hissaricus**
3 Shrub or tree to 6m; leaves 2.5-5(6)cm; flowers in compact inflorescence; hypanthium and calyx tomentose **3. C. ellipticus**
 4 Hypanthium and calyx densely pubescent to tomentose; leaves sparsely to densely pubescent on lowerside in summer; midrib lowersides, peduncles and pedicels pubescent to densely so at fruiting 5
 4 Hypanthium and calyx sparsely pubescent to pubescent (sometimes hypanthium densely so); leaves subglabrous on lowerside in summer; midrib lowersides, peduncles and pedicels very sparsely pubescent at fruiting 6
5 Flowers mostly <15 per inflorescence; fruiting inflorescences mostly <3 x 3cm **6. C. affinis**
5 Flowers mostly >15 per inflorescence; fruiting inflorescences mostly >3 x 3cm **5. C. ignotus**
 6 Larger leaves mostly <2x as long as wide, very obtuse to rounded and apiculate at apex **7. C. obtusus**
 6 Larger leaves mostly ≥2x as long as wide, acuminate to obtuse and apiculate at apex 7
7 Fruits red, turning to dark purple when fully ripe, not exposing stones at apex; inflorescences 5-20-flowered **10. C. cooperi**
7 Fruits soon becoming purplish- to brownish-black when ripe, usually exposing stones at apex; inflorescences 15-30-flowered 8
 8 Fruits with dense whitish bloom, with very open apex exposing stones **8. C. bacillaris**
 8 Fruits without or with sparse whitish bloom, with scarcely open apex ± not exposing stones **9. C. transens**

Key B - Petals patent; leaves deciduous to ± so; fruits yellow to bright red; hypanthium and calyx glabrous to sparsely pubescent

1 Leaves pubescent to densely so on lowerside, <2x as long as wide; anthers purplish-black **1. C. monopyrenus**
1 Leaves sparsely pubescent to subglabrous on lowerside, ≥2x as long as wide; anthers mauve 2
 2 Fruit eventually brownish-black, with stones slightly showing at apex; inflorescences 5-20-flowered **9. C. transens**
 2 Fruit eventually purplish-black, with stones not showing at apex; inflorescences 15-30-flowered **10. C. cooperi**

Key C - Petals patent; leaves deciduous to ± so; fruits yellow to bright red or crimson; hypanthium and calyx pubescent to tomentose

1 Leaves flat on upperside, very dull, the veins not impressed 2

1 Leaves with slightly to strongly impressed veins on upperside, usually somewhat shiny 3

2 Leaves 2-5cm, obovate to suborbicular; fruits 8-13mm, mostly with 1 stone; inflorescences c.6-20-flowered **2. C. tomentellus**

2 Leaves 6-15cm, elliptic-oblong; fruits 4-6mm, with 2 stones; inflorescences mostly >20-flowered **11. C. frigidus**

3 Leaves mostly 2.5-3x as long as wide; flowers ≥20 per inflorescence; fruits with 2(-5) stones, 5-8mm **12. C. x watereri & 15. C. henryanus**

3 Leaves mostly 2-2.5x as long as wide; flowers <15(20) per inflorescence; fruits with 1-2 stones, 6-10mm **6. C. affinis**

Key D - Petals patent; leaves evergreen or ± so; fruits yellow to bright red or crimson; hypanthium and calyx densely pubescent to tomentose

1 Leaves most or all <3cm; flowers <15 per inflorescence **19. C. pannosus**

1 Leaves most or all >3cm; flowers >15 per inflorescence 2

2 Leaves flat on upperside, the veins not impressed, usually semi-deciduous **11. C. frigidus**

2 Leaves with somewhat to strongly impressed veins on upperside, usually ± evergreen 3

3 Leaves obtuse to rounded at apex; fruits usually with 2 stones **20. C. lacteus**

3 Leaves acute to subacute at apex; fruits with 2-5 stones 4

4 Leaves sparsely pubescent on lowerside **14. C. 'Hybridus Pendulus'**

4 Leaves densely pubescent to tomentose on lowerside 5

5 Leaves oblanceolate to narrowly elliptic; fruits mostly 4-5mm, often slightly wider than long, with 3-5 stones **13. C. salicifolius**

5 Leaves narrowly obovate to elliptic; fruits mostly 5-8mm, often slightly longer than wide, with 2-3(5) stones 6

6 Petals pink; leaves with 5-7 pairs of lateral veins **16. C. hylmoei**

6 Petals white; larger leaves with >7 pairs of lateral veins **12. C. x watereri & 15. C. henryanus**

Key E - Petals patent; leaves evergreen; fruits bright red to crimson or orange; hypanthium and calyx glabrous to sparsely pubescent or pubescent

1 Inflorescences with >10 flowers 2

1 Inflorescences with <10 flowers 3

2 Erect shrub to 8m; fruits with 2 stones **10. C. cooperi**

2 Procumbent shrub; fruits with 3-5 stones **14. C. 'Hybridus Pendulus'**

3 Some leaves >1.5cm; fruits with 2-5 stones 4

3 Leaves all <1(1.5)cm, or if >1.5mm then fruits with 2(-3) stones 6

4 Leaves all <2.5cm; at least some stems erect or arching; fruits with 2-4 stones **18. C. x suecicus**

4 Some leaves >2.5cm; stems all procumbent; fruits often with 5 stones 5

5 Leaf apex acute to acuminate **14. C. 'Hybridus Pendulus'**

5 Leaf apex obtuse to rounded **17. C. dammeri**

6 Fruits orange, with 1(-2) stones **24. C. sherriffii**

6 Fruits bright red to crimson, with 2(-3) stones 7

7 Flowers mostly 2-5(7) per cyme 8

7 Flowers mostly 1(-2) per cyme except at ends of branches 9

8 Leaves 10-25mm, ± matt on upperside, densely pubescent on lowerside **29. C. marginatus**

8 Leaves 5-12mm, shiny on upperside, rather sparsely pubescent on lowerside **21. C. microphyllus**

9 Leaves mostly ≥2x as long as wide 10

9 Leaves <2x as long as wide 12

10 Fruits bright red, shiny; leaves mid-green and slightly shiny on

upperside **22. C. conspicuus**

10 Fruits crimson, dull; leaves dark green and very shiny on upperside 11

11 Leaves 7-15 x 3-6mm with petioles 2-4mm; fruits 8-10mm; flowers c.11mm across **31. C. integrifolius**

11 Leaves 4-7 x 1-3mm with petioles 0.5-1.5mm; fruits 4-5mm; flowers c.5mm across **32. C. linearifolius**

12 Leaves mid-green and matt on upperside **30. C. congestus**

12 Leaves medium- to dark-green on upperside 13

13 Larger leaves >15mm; shrub to 2m, often ± erect **25. C. rotundifolius**

13 All leaves <15mm; procumbent to arching shrub rarely >0.5m 14

14 Leaves densely pubescent to tomentose on lowerside **23. C. astrophoros**

14 Leaves sparsely to very sparsely pubescent on lowerside 15

15 Fruits 4-6mm **27. C. cashmiriensis**

15 Fruits 6-10mm 16

16 Some branches arching above ground to 50cm high; fruits bright red **28. C. prostratus**

16 Branches all flat to ground, rarely >20cm high; fruits crimson-red **26. C. cochleatus**

Key F - Petals erect to erecto-patent; leaves deciduous; fruits dark purple to black

1 Leaves <2cm, obtuse at apex; flowers 1-3(4) together; stamens 10 **42. C. nitens**

1 Leaves all or nearly all >2cm, acute to acuminate at apex; flowers 3-30 together; stamens 15-20 2

2 Leaves strongly bullate with deeply impressed veins on upperside; inflorescences mostly 12-30-flowered; fruits with 3-5 stones 3

2 Leaves not bullate but veins slightly to strongly impressed on upperside; inflorescences mostly 3-15-flowered; fruits with 2-3 stones 4

3 Leaves 3-5cm, acute to shortly acuminate at apex; fruits maroon, with 3-5 stones **56. C. obscurus**

3 Leaves 4-12cm, long-acuminate at apex; fruits purplish-black, with 5 stones **57. C. moupinensis**

4 Leaves 2-5.5(7)cm, most or all <5cm 5

4 Leaves 3-11cm, most or many >5cm 6

5 Calyx and hypanthium glabrous or nearly so on outside at flowering, glabrous at fruiting; leaves shiny on upperside **43. C. lucidus**

5 Calyx and hypanthium pubescent at flowering, sparsely pubescent at fruiting; leaves matt on upperside **46. C. pseudoambiguus**

6 Leaves tapering-acuminate 7

6 Leaves acute to shortly and ± abruptly acuminate 8

7 Leaves with not or scarcely impressed lateral veins on upperside, pubescent on lowerside; inflorescences with 3-9(15) flowers; fruits mostly with 2 stones **45. C. laetevirens**

7 Leaves with distinctly impressed lateral veins on upperside, very sparsely pubescent on lowerside; inflorescences with 7-15 flowers; fruits mostly with 3 stones **47. C. hummelii**

8 Leaves with apex abruptly acuminate to fine point; calyx-lobes pubescent only at base; fruits 10-11mm, globose, with 2-3 stones **48. C. hsingshangensis**

8 Leaves acute to shortly and gradually acuminate apex; calyx-lobes tomentose; fruits 8-10mm, broadly obovoid, with 2(-3) stones **44. C. villosulus**

Key G -Petals erect to erecto-patent; leaves deciduous; fruits bright red to orange-red; leaves densely pubescent to tomentose on lowerside

1 Hypanthium and calyx glabrous or nearly so **49. C. cambricus**

1 Hypanthium and calyx pubescent to tomentose, the pubescence persisting on calyx until fruit ripe 2

2 Fruits pendent, with 2 stones; some leaves usually >2.5cm 3

2 Fruits held stiffly, with 3-4 stones; usually ± all leaves <2.5cm 4

3 Fruits strongly obovoid to almost pear-shaped; often some leaves >3cm **67. C. zabelii**

3 Fruits globose to ellipsoid or slightly obovoid; leaves ± all <3cm **68. C. fangianus**

4 Fruits bright red, 6-8mm, ± globose; branches arching-erect, long and whip-like; plant usually >1.5m **65. C. dielsianus**

4 Fruits orange-red, 8-10mm, usually slightly longer than wide; branches stiffly spreading; plant usually <1.5m **66. C. splendens**

Key H - Petals erect; leaves deciduous; fruit bright red to orange-red or maroon; leaves pubescent to sparsely so on lowerside, bullate or at least with veins deeply impressed on upperside

1 Leaves 1.5-3cm; inflorescences with 3-7(15) flowers **60. C. mairei**

1 Most or all leaves >3cm; inflorescences with 9-30 flowers 2

2 Fruits maroon; leaves 3-5cm **56. C. obscurus**

2 Fruits bright red to orange-red; leaves 3-15cm 3

3 Fruits orange-red, with 3-4(5) stones; leaves 3-6cm **55. C. boisianus**

3 Fruits bright red, with (4-)5 stones; leaves 3.5-15cm 4

4 Leaves 3.5-7cm, ± bullate, pubescent on lowerside at flowering; fruits mostly <8mm **53. C. bullatus**

4 Leaves 5-12cm, strongly bullate, sparsely pubescent on lowerside at flowering; fruits mostly >8mm **54. C. rehderi**

Key I - Petals erect to erecto-patent; leaves deciduous; fruit bright red to orange-red; leaves pubescent to sparsely so on lowerside, not bullate and with veins not or scarcely impressed on upperside

1 Leaves most or all >1.3x as long as wide, acute or acuminate to obtuse at apex 2

1 Leaves most or all ≤1.3x as long as wide, mostly rounded to broadly obtuse (sometimes subacute or apiculate) at apex 5

2 Fruits parallel-sided, oblong in side view (sausage-shaped); stamens 10(-15) **41. C. divaricatus**

2 Fruits subglobose to broadly ellipsoid or obovoid, with curved sides; stamens (15-)20 3

3 Fruits orange-red; leaves 1.5-3cm **51. C. simonsii**

3 Fruits bright red; leaves 2-5.5cm, most >3cm 4

4 Leaves ± tomentose on lowerside; flowers in groups of 1-4; fruits with 2-3 stones **50. C. mucronatus**

4 Leaves pubescent on lowerside; flowers in groups of (3)5-7(9); fruits with 3-4(5) stones **52. C. tengyuehensis**

5 Shrub to 3m; flowers pendent, with pedicels c.5mm; stamens (15-)20 **33. C. nitidus**

5 Shrub to 1(-2)m; flowers usually erect, with pedicels to c.2mm; stamens 10-13 6

6 Leaves matt or ± so on upperside, ± undulate; fruits bright red 7

6 Leaves shiny on upperside, flat or less often undulate; fruits orange-red to bright red 8

7 Leaves mostly 1-2.5cm; flowers 2-4 together; fruits mostly 10-12mm; plant to 1m **36. C. nanshan**

7 Leaves mostly 0.5-1.5cm; flowers 1(-2) together; fruits mostly 6-7mm; plant to 25cm **34. C. adpressus**

8 Petals red with purplish-black base and narrow white fringe

38. C. atropurpureus

8 Petals pink and red and/or white 9

9 Sepals glabrous on outer surface; leaves 1-2cm 10

9 Sepals pubescent on outer surface; leaves 0.6-1.5cm 11

10 Leaves apiculate to very shortly acuminate at apex; fruits 10-12mm, bright red **35. C. apiculatus**

10 Leaves obtuse to rounded or mucronate at apex; fruits 6-8mm, orange-red **39. C. hjelmqvistii**

11 Branches forming regular herring-bone pattern; fruits orange-red **37. C. horizontalis**

11 Branches irregular, not forming herring-bone pattern; fruits bright red **40. C. ascendens**

Key J - Petals erect to erecto-patent; leaves evergreen; fruits orange-red to bright red

1 Most leaves >2cm 2

1 Most leaves ≤2cm 7

2 Leaves pubescent on lowerside; anthers white **52. C. tengyuehensis**

2 Leaves tomentose on lowerside; anthers white or pigmented 3

3 Most leaves >3cm 4

3 Most leaves <3cm 5

4 Anthers pale mauve; fruits with 2 stones **59. C. wardii**

4 Anthers white; fruits with (2-)3 stones **61. C. sternianus**

5 Anthers white; inflorescences mostly with 3-7 flowers; leaves semi-evergreen **60. C. mairei**

5 Anthers pink to mauve or pale purple; inflorescences with 5-15 flowers; leaves fully evergreen 6

6 Fruits orange-red, with (2-)3 stones; leaves with veins deeply impressed on upperside, silvery- to yellowish-tomentose on lowerside **58. C. franchetii**

6 Fruits bright red, with 2-3 stones; leaves with veins slightly impressed on upperside, greyish-white-tomentose on lowerside **62. C. vilmorinianus**

7 Leaves 0.5-1.3cm, shiny on upperside, subglabrous on lowerside **33. C. nitidus**

7 Leaves 1-2(2.5)cm, scarcely shiny on upperside, tomentose on lowerside 8

8 Veins deeply impressed on leaf upperside; flowers 1-4 together; fruit with 3-4(5)stones; anthers white **63. C. insculptus**

8 Veins not or scarcely impressed on leaf upperside; flowers 6-10 together; fruit with 2-3 stones; anthers pinkish-purple **64. C. amoenus**

Other spp. - Over 100 spp. are widely cultivated and any of them can be expected to become natd as the fruits are so attractive to birds, but many spp. have been recorded in error. Natd plants occur wherever birds defecate, notably on waste and rough ground, on banks and walls, in open woodland and scrub, and on grassland becoming colonized by shrubs, especially on chalk and limestone.

Section **1** - *CHAENOPETALUM* Koehne (spp. 1-32). Petals patent, white (except pink in *C. hylmoei*); each inflorescence with all flowers opening simultaneously or over a very short time; flowers often in groups of 15-many (often not); leaves often evergreen (often not); fruits with 1-2(3) stones (less often 3-5).

1. C. monopyrenus (W.W. Sm.) Flinck & B. Hylmö (*C. veitchii* auct. non (Rehder **385**
& E.H. Wilson) G. Klotz, *C. multiflorus* auct. non Bunge) - *One-stoned Cotoneaster*. **387**
Deciduous shrub to 3m with arching branches; leaves 2.5-5cm, flat, pubescent to densely so on lowerside; inflorescences mostly 6-20-flowered; anthers purplish-

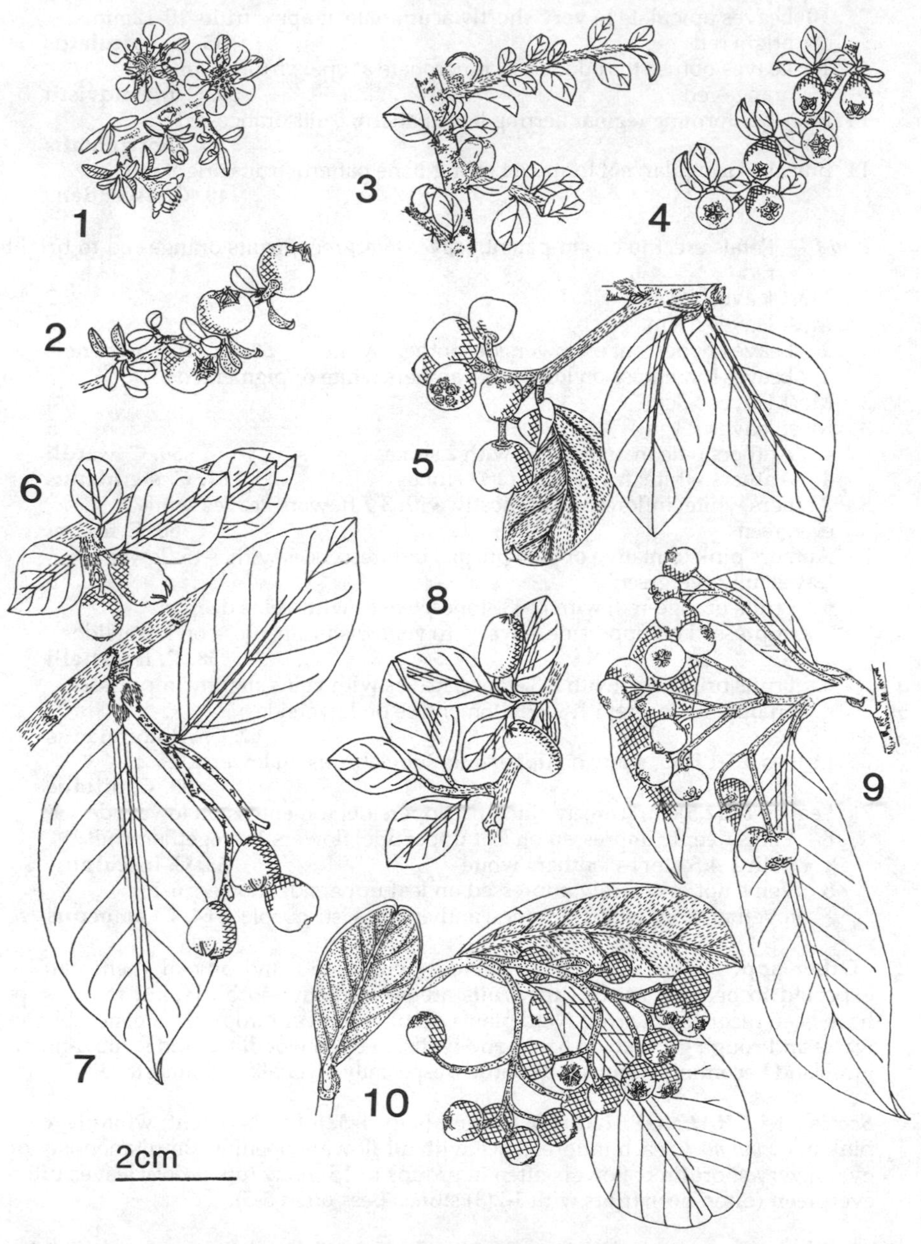

FIG 384 - *Cotoneaster*. 1-2, *C. integrifolius* (1, flowering. 2, fruiting). 3-4, *C. horizontalis* (3, flowering. 4, fruiting). 5, *C. sternianus*. 6, *C. simonsii*. 7, *C. rehderi*. 8, *C. divaricatus*. 9, *C. x watereri*. 10, *C. lacteus*.

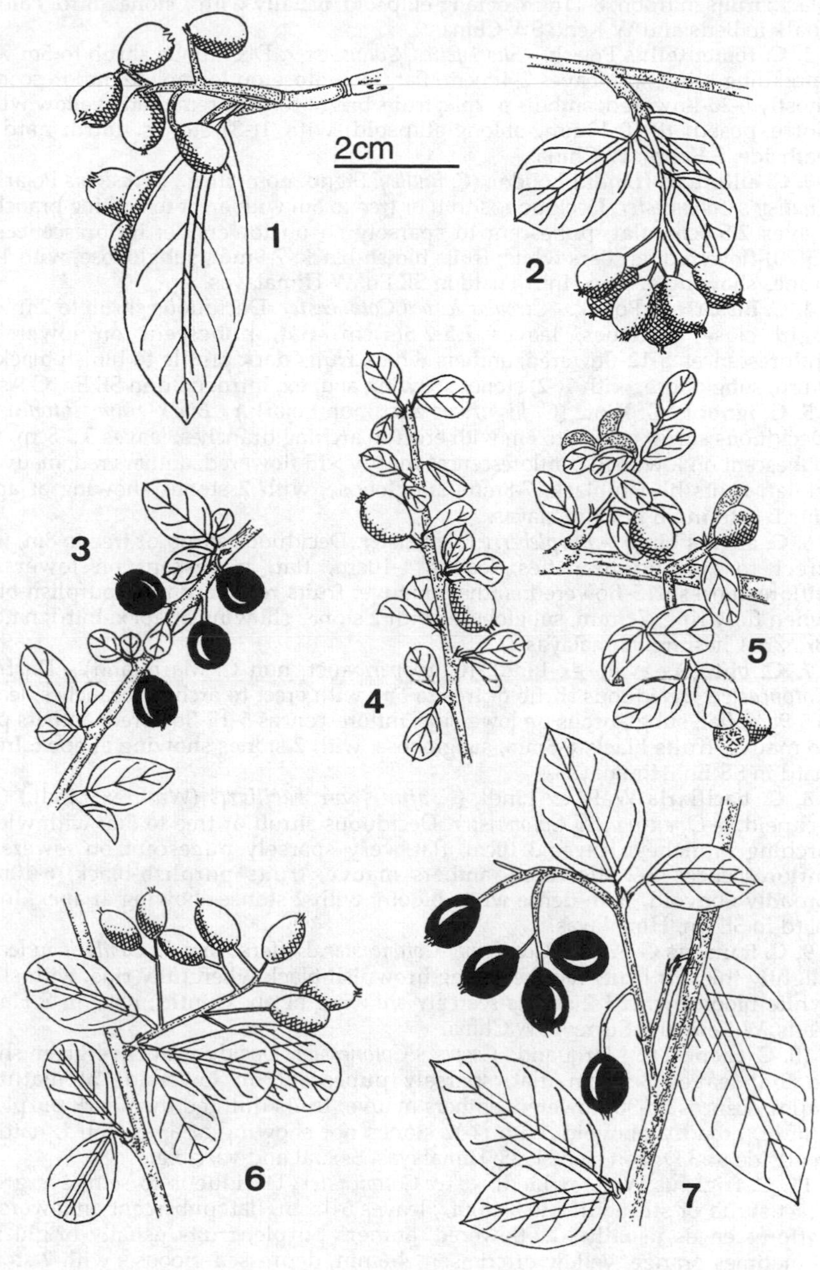

Fig 385 - *Cotoneaster*. 1, *C. bacillaris*. 2, *C. zabelii*. 3, *C. nitens*. 4, *C. nitidus*. 5, *C. x suecicus* 'Coral Beauty'. 6, *C. monopyrenus*. 7, *C. lucidus*.

black; fruits maroon, 8-11mm, oblong-ellipsoid, usually with 1 stone. Intrd; natd on
chalk in Beds and W Kent; SW China.

2. C. tomentellus Pojark. - *Short-felted Cotoneaster*. Deciduous shrub to 5m with 387
spreading branches; leaves 2-4(5)cm, flat, tomentose on lowerside; inflorescences
mostly 6-20-flowered; anthers purple; fruits bright red but remaining yellow where
not exposed, (8)10-13mm, oblong-ellipsoid, with 1(-3) stones. Intrd; natd on
pathside, S Lancs; W China.

3. C. ellipticus (Lindl.) Loudon (*C. lindleyi* Steud. nom. illeg., *C. insignis* Pojark.) - 387
Lindley's Cotoneaster. Deciduous shrub or tree to 6m with erect to arching branches,
leaves 2.5-6cm, flat, pubescent to sparsely so on lowerside; inflorescences 5-
15(20)-flowered; anthers white; fruits bluish-black, 7-9mm, subglobose, with 1(-2)
stones showing at apex. Intrd; natd in SE En; W Himalayas.

4. C. hissaricus Pojark. - *Circular-leaved Cotoneaster*. Deciduous shrub to 2m with 387
rigid close branches; leaves 1.5-2.5(4)cm, flat, pubescent on lowerside;
inflorescences 5-12-flowered; anthers white; fruits dark purple to bluish-black, 6-
9mm, subglobose, with 1(-2) stones showing at apex. Intrd; natd in SE En; C Asia.

5. C. ignotus G. Klotz (*C. hissaricus* auct. non Pojark.) - *Black-grape Cotoneaster*. 387
Deciduous shrub or tree to 6m with erect to arching branches; leaves 3.5-8cm, flat,
pubescent on lowerside; inflorescences mostly >15-flowered; anthers red, mauve or
violet; fruits bluish-black, 7-8mm, subglobose, with 2 stones showing at apex.
Intrd; natd in SE En; Himalayas.

6. C. affinis Lindl. - *Purpleberry Cotoneaster*. Deciduous shrub or tree to 8m, with 387
erect to arching branches; leaves 4-10cm, flat, pubescent on lowerside;
inflorescences ≤15-flowered; anthers mauve; fruits red, becoming purplish-black
when fully ripe, 7-9mm, subglobose, with 2 stones showing at apex. Intrd; natd in
Br N to Cheshire; Himalayas.

7. C. obtusus Wall. ex Lindl. (*C. cooperi* auct. non C. Marquand) - *Dartford* 387
Cotoneaster. Deciduous shrub or tree to 5m, with erect to arching branches, leaves
3.5-9cm, flat, subglabrous on lowerside; inflorescences 5-15-flowered; anthers pink
to mauve; fruits black, 6-8mm, subglobose, with 2 stones showing at apex. Intrd;
natd in SE En; Himalayas.

8. C. bacillaris Wall. ex Lindl. (*C. affinis* var. *bacillaris* (Wall. ex Lindl.) C.K. 385
Schneid.) - *Open-fruited Cotoneaster*. Deciduous shrub or tree to 5m, with widely 387
arching branches; leaves 3-10cm, flat, very sparsely pubescent on lowerside;
inflorescences 5-20-flowered; anthers mauve; fruits purplish-black, 6-10mm,
broadly obovoid, with dense white bloom, with 2 stones showing at apex. Intrd;
natd in SE En; Himalayas.

9. C. transens G. Klotz - *Godalming Cotoneaster*. Differs from *C. bacillaris* in leaves 387
slightly thicker; fruits red, becoming brownish-black when fully ripe, with slight
white bloom, with 1-2 stones scarcely showing at apex. Intrd; natd in S Hants,
Beds, W Kent and Surrey; SW China.

10. C. cooperi C. Marquand - *Cooper's Cotoneaster*. Deciduous to evergreen shrub 387
to 8m; leaves 3.5-8cm, flat, sparsely pubescent on lowerside at maturity;
inflorescences 15-30-flowered; anthers mauve; fruits dull red then dark purple, 6-
10mm, broadly obovoid, with (1-)2 stones not showing at apex. Intrd; natd on
roadside in S Devon since 1985; Himalayas. Sexual and variable.

11. C. frigidus Wall. ex Lindl. - *Tree Cotoneaster*. Deciduous to semi-evergreen, 387
erect shrub or strong tree to 8(18)m, ; leaves 6-15cm, flat, pubescent on lowerside;
inflorescences usually >20-flowered; anthers purple; fruits usually bright red,
sometimes orange, yellow or crimson, 4-6mm, depressed-globose, with 2 stones.
Intrd; natd in many places in BI; Himalayas.

12. C. x watereri Exell (*C. frigidus* x *C. salicifolius*) - *Waterer's Cotoneaster*. Usually 384
semi-evergreen erect shrub to 8m; many cultivars and their seedlings variously 387
intermediate between the parents especially in leaves, with range in fruit colour as
for *C. frigidus*, but fruits often larger than in either parent (5-8mm) and with 2-3(5)
stones. Intrd; natd in many parts of BI; garden origin. See note under *C. henryanus*.

FIG 387 - Leaves of *Cotoneaster*. 1, *C. **monopyrenus***. 2, *C. **tomentellus***. 3, *C. **ellipticus***. 4, *C. **hissaricus***. 5, *C. **ignotus***. 6, *C. **affinis***. 7, *C. **obtusus***. 8, *C. **bacillaris***. 9, *C. **transens***. 10, *C. **cooperi***. 11, *C. **frigidus***. 12, *C. x **watereri***. 13, *C. **salicifolius***. 14, *C.* **'Hybridus Pendulus'**. 15, *C. **henryanus***. 16, *C. **hylmoei***. 17, *C. **dammeri***. 18, *C. x **suecicus***. 19, *C. **pannosus***.

13. C. salicifolius Franch. - *Willow-leaved Cotoneaster*. Erect, arching or ± 387
procumbent evergreen shrub to 5m; leaves 3-10cm, shiny and ± bullate on upperside, tomentose on lowerside; inflorescences >(20)30-flowered; anthers purple to black; fruits bright red, 4-5mm, subglobose to depressed-globose, with 3-5 stones. Intrd; natd in many parts of BI; W China. Sexual and variable.

14. C. 'Hybridus Pendulus' (*C. salicifolius* x *C. dammeri*) - *Weeping Cotoneaster*. 387
Procumbent (in the wild) evergreen shrub with branches to 1m; leaves 2-5cm, with veins slightly impressed on upperside, very sparsely pubescent on lowerside; inflorescences few- to many-flowered; anthers purple; fruits bright red, 4-7mm, subglobose, with 3-5 stones. Intrd; natd Glam, probably overlooked elsewhere; garden origin. Garden plants are usually grown pendent on standards.

15. C. henryanus (C.K. Schneid.) Rehder & E.H. Wilson - *Henry's Cotoneaster*. 387
Semi-evergreen shrub to 5m; leaves 4-12cm, slightly bullate on upperside, tomentose on lowerside; inflorescences usually >20-flowered; anthers purple; fruits bright red, 6-7mm, subglobose, with 2-3 stones. Intrd; natd in a few places in S En, C Sc and N Ir; C China. This comes very close to some variants of *C. x watereri* that are closer to *C. salicifolius* than to *C. frigidus*, and might be involved in its parentage. Considerable experience is necessary for certain determination and experimental work with wild-collected material is needed to clarify relationships in this group.

16. C. hylmoei Flinck & J. Fryer - *Hylmö's Cotoneaster*. Erect, arching, evergreen 387
shrub to 3m; leaves 3-7cm, with impressed veins on upperside, tomentose on lowerside; inflorescences 15-50-flowered; anthers purplish-black; fruits bright red, 5-6mm, subglobose, with 2-3(4) stones. Intrd; natd on rough ground in Lanarks and Offaly; C China. Our only sp. of subg. *Chaenopetalum* with pink petals.

17. C. dammeri C.K. Schneid. - *Bearberry Cotoneaster*. Procumbent evergreen shrub 387
with branches to 3m; leaves 1.5-3cm, shiny and ± flat on upperside, glabrous to sparsely pubescent on lowerside; inflorescences 1-2(4)-flowered; anthers purple; fruits bright red, 6-8mm, globose to very broadly obovoid, usually with 5 stones. Intrd; natd in Br N to C Sc; C China.

18. C. x suecicus G. Klotz (?*C. dammeri* x *C. conspicuus*) - *Swedish Cotoneaster*. 385
Stems arching to 60cm high, trailing to 2m; leaves evergreen, 1-2.5cm, otherwise 387
similar to those of *C. dammeri;* inflorescences 1-4(6)-flowered; anthers purple; fruits as in *C. dammeri* but less bright red and with 3-4 stones. Intrd; natd in Br N to C Sc; garden origin. Cultivars **'Skogholm'** and **'Coral Beauty'**, often mass-planted, belong here; seedlings are often very like their parent.

19. C. pannosus Franch. - *Silverleaf Cotoneaster*. Stems erect, long, slender, slightly 387
arching, to 3(5)m; leaves evergreen, 1-2.5(3.5)cm, dull and nearly flat on upperside, tomentose on lowerside; inflorescences 6-12-flowered; anthers purplish-black; fruits dull red, 5-8mm, subglobose to ellipsoid, often with conspicuous erect sepals as in *C. amoenus*, with 2 stones. Intrd; natd in W Kent; SW China.

20. C. lacteus W.W. Sm. - *Late Cotoneaster*. Evergreen spreading shrub to 5m; 384
leaves 3.5-9cm, slightly shiny and with deeply impressed veins on upperside, 393
tomentose on lowerside; inflorescences >30-flowered; anthers purplish-black; fruits bright red to crimson-red, 5-6mm, ± globose, with 2 stones. Intrd; natd in Br N to C Sc, planted as field-hedges in E Anglia; SW China. 1 of the latest flowering and fruiting spp.; fruits rarely ripen before Nov.

21. C. microphyllus Wall. ex Lindl. - *Small-leaved Cotoneaster*. Evergreen, 392
procumbent to ascending shrub to 1m; leaves 5-8(12)mm, shiny and flat on upperside, appressed-pubescent on lowerside; inflorescences 1-5-flowered; flowers c.6mm across; sepals obtuse; anthers purple to black; fruits crimson, 5-8mm, depressed-globose to globose, with 2 stones. Intrd; natd in several places in C Ir, surely overlooked in Br; Himalayas. Material so determined from Br has so far all proved to be *C. integrifolius*.

22. C. conspicuus C. Marquand - *Tibetan Cotoneaster*. Evergreen stiffly erect to 392
spreading shrub to 1.5m; differs from *C. microphyllus* in leaves 5-20mm, much less shiny on upperside; inflorescences mostly 1-flowered; flowers c.10mm across;

sepals acute to acuminate; fruits bright red, shiny, 6-9mm; and from other relatives
as in Key E. Intrd; natd in Br N to C Sc; Tibet. Sexual and variable.

23. C. astrophoros J. Fryer & E.C. Nelson - *Starry Cotoneaster*. Evergreen arching **392**
and spreading shrub to 0.3m; differs from C. *microphyllus* in leaves 6-8mm,
tomentose on lowerside; inflorescences 1-flowered; flowers c.10mm across; sepals
acute; fruits broadly oblong-ellipsoid, dull red but ± shiny; and from other relatives
as in Key E. Intrd; natd by railway in Lanarks; W & SW China. Cultivated as **C.
'Donard Gem'**.

24. C. sherriffii G. Klotz - *Sherriff's Cotoneaster*. Evergreen stiffly erect to **392**
spreading shrub to 2m; differs from C. *microphyllus* in leaves 5-15mm, ± matt on
upperside; inflorescences 1-flowered; flowers c.10mm across; sepals acute; fruits
broadly oblong-ellipsoid, 7-10mm, orange, with 1(-2) stones. Intrd; natd in
scattered sites in Br; SE Tibet. Cultivated as **C. 'Highlight'**.

25. C. rotundifolius Wall. ex Lindl. - *Round-leaved Cotoneaster*. Evergreen stiffly **392**
erect to spreading shrub to 2m; differs from C. *microphyllus* in leaves 7-20mm,
sparsely pubescent on lowerside; inflorescences 1-flowered; flowers 10-13mm
across; sepals acute; fruits 7-10mm, globose to depressed-globose; and from other
relatives as in Key E. Intrd; natd in scattered sites in Br N to Westmorland;
Himalayas.

26. C. cochleatus (Franch.) G. Klotz - *Yunnan Cotoneaster*. Evergreen procumbent **392**
shrub to 0.2m; differs from C. *microphyllus* in leaves 5-14mm, sparsely pubescent
on lowerside; inflorescences 1-flowered; flowers 8-10mm across; fruits 6-10mm,
subglobose, red to crimson; and from other relatives as in Key E. Intrd; natd in
Midlothian; SW China.

27. C. cashmiriensis G. Klotz (C. *cochleatus* auct. non (Franch.) G. Klotz) - **392**
Kashmir Cotoneaster. Evergreen procumbent shrub to 0.3m; differs from C.
microphyllus in leaves 4-11mm; inflorescences 1-flowered; flowers c. 7-9mm across;
fruits 4-6mm, bright red, subglobose; and from other relatives as in Key E. Intrd;
natd in E Kent and formerly Lanarks; Kashmir. Commonly grown, usually as C.
cochleatus

28. C. prostratus Baker (C. *buxifolius* auct. non Wall. ex Lindl.) - *Procumbent* **392**
Cotoneaster. Evergreen procumbent shrub to 0.5m; differs from C. *microphyllus* in
leaves 6-13mm; inflorescences 1-2-flowered; flowers c.10mm across; fruits bright
red, 9-10mm, subglobose; and from other relatives as in Key E. Intrd; natd in W
Kent and N Somerset; Himalayas.

29. C. marginatus (Loud.) Schltdl. - *Fringed Cotoneaster*. Evergreen erect to **392**
procumbent shrub to 3m; differs from C. *microphyllus* in leaves 10-25mm, ± matt on
upperside, rather densely pubescent on lowerside, ciliate; inflorescences 2-7-
flowered; flowers 8-10mm across; fruits 8-10mm, with 2(-3) stones. Intrd; natd in
numerous sites in Br N to Westmorland; Himalayas. **392**

30. C. congestus Baker - *Congested Cotoneaster*. Tightly branched evergreen shrub
to 70cm; differs from C. *microphyllus* in leaves 5-14mm, matt and slightly rugose on
upperside, very sparsely pubescent on lowerside; inflorescences 1-flowered;
flowers 7-9mm across; fruits with 2(-3) stones; and from other relatives as in Key
E. Intrd; natd in W Kent and Man; Himalayas. Sexual and variable.

31. C. integrifolius (Roxb.) G. Klotz (C. *thymifolius* Wall. ex Lindl., C. **384**
microphyllus auct. non Wall. ex Lindl.) - *Entire-leaved Cotoneaster*. Procumbent to **392**
arching evergreen shrub to 1m; differs from C. *microphyllus* in leaves 7-15mm;
inflorescences usually 1-flowered; flowers c.11mm across; fruits 8-10mm; and from
other relatives as in Key E. Intrd; frequently natd over much of BI; Himalayas to W
China. This and C. *microphyllus* are commonly grown and much confused. Sexual
and variable.

32. C. linearifolius (G. Klotz) G. Klotz (C. *thymifolius* auct. non Wall. ex Lindl.) - **392**
Thyme-leaved Cotoneaster. Evergreen procumbent or arching shrub to 60cm; differs
from C. *microphyllus* in leaves 4-7mm; inflorescences 1-flowered; flowers c. 5mm
across; fruits 4-5mm; and from other relatives as in Key E. Intrd; natd in S En, W Ir

and M Ebudes; Himalayas. Sexual and variable; perhaps better included in *C. integrifolius*.

Section 2 - COTONEASTER (sect. *Orthopetalum* Koehne) (spp. 33-68). Petals erect to suberect, pink or pink-tinged; each inflorescence with flowers opening in sequence over a long period; flowers usually in groups of 1-10(12) (sometimes more); leaves deciduous (less often evergreen); fruits with 2-5 stones.

33. C. nitidus Jacques (*C. distichus* Lange) - *Distichous Cotoneaster*. Erect, evergreen **385**
or semi-evergreen shrub to 3m, with wide-spreading branches; leaves 0.5-1.3cm, **392**
shiny and flat on upperside, subglabrous on lowerside; flowers solitary; anthers white; fruits bright red to orange-red, 7-11mm, broadly obovate to subglobose, with 3 stones. Intrd; natd in S Hants; Himalayas.

34. C. adpressus Bois - *Creeping Cotoneaster*. Usually procumbent, deciduous **392**
shrub to 0.3m, with irregular branching; leaves 0.5-1.5cm, matt on upperside, ± undulate, very sparsely pubescent on lowerside; flowers 1(-2) together; anthers white; fruits bright red, 6-7mm, subglobose, with 2 stones. Intrd; natd in W Kent and Lanarks; ?W China.

35. C apiculatus Rehder & E.H. Wilson - *Apiculate Cotoneaster*. Procumbent to **392**
ascending deciduous shrub to 1m; leaves 1-2cm, ± flat, shiny on upperside, very sparsely pubescent on lowerside; flowers 1(-2) together; anthers white; fruits bright red, 10-12mm, obovoid, with usually 3 stones. Intrd; natd on wall and rough ground in Renfrews and Lanarks; W China.

36. C. nanshan M. Vilm. ex Mottet (*C. adpressus* var. *praecox* Bois & Berthault) - **392**
Dwarf Cotoneaster. Spreading deciduous shrub to 0.5(1)m; leaves 1-2.5cm, undulate, sparsely pubescent on lowerside; flowers 2-4 together; anthers white; fruits bright red, 9-12mm, very broadly ellipsoid, with usually 2 stones. Intrd; natd in W Kent; W China.

37. C. horizontalis Decne. - *Wall Cotoneaster*. Arching to horizontal, deciduous **384**
shrub to 1(3)m, often vertical on walls, with very regular herring-bone branching; **392**
leaves 0.6-1.2cm, shiny and flat on upperside, subglabrous on lowerside; flowers (1)2-3 together; anthers white; fruits orange-red, 4-6mm, ± globose, with 3 stones. Intrd; commonly natd in much of BI; W China.

38. C. atropurpureus Flinck & B. Hylmö (*C. horizontalis* 'Prostratus') - *Purple-* **392**
flowered Cotoneaster. Differs from *C. horizontalis* in branching less regular to irregular; leaves 0.9-1.4cm, thin, slightly undulate at margin; flowers (1-)3 together; fruits mostly 6-9mm, with 2-3 stones; and see Key I. Intrd; natd in S Br and C Ir; C China.

39. C. hjelmqvistii Flinck & B. Hylmö (*C. horizontalis* 'Robustus') - *Hjelmqvist's* **392**
Cotoneaster. Differs from *C. horizontalis* in stronger growth to 1.5(4.5)m with less regular herring-bone branches; leaves to 1.5(2)cm, ± orbicular; fruits 6-8mm, with 2 stones; and see Key I. Intrd; natd in Br N to C Sc; W China.

40. C. ascendens Flinck & B. Hylmö (*C. horizontalis* var. *wilsonii* Havemeyer ex **392**
E.H. Wilson) - *Ascending Cotoneaster*. Differs from *C. horizontalis* in branches ascending to 1(2)m, with less regular pattern; leaves 0.6-1.5cm; flowers 1-3 together; fruits bright red, with 2-3 stones. Intrd; natd in open woodland, W Lancs; C China.

41. C. divaricatus Rehder & E.H. Wilson - *Spreading Cotoneaster*. Deciduous shrub **384**
to 2m with wide spreading or arching branches; leaves 0.8-2.5cm, flat and shiny on **392**
upperside, very sparsely pubescent on lowerside; flowers (1)2-3(4) together; anthers white; fruits (7)9-12mm, bright red, oblong-ellipsoid to cylindric, with 2 stones. Intrd; natd scattered in En, Lanarks; C China.

42. C. nitens Rehder & E.H. Wilson - *Few-flowered Cotoneaster*. Erect, densely **385**
branched, deciduous shrub to 3.5m; leaves 0.8-2.2cm, flat and shiny on upperside, **392**
sparsely pubescent on lowerside; flowers 1-3(4) together; anthers white; fruits 7-9mm, black, subglobose, with 2(-3) stones. Intrd; natd in S & SE En; W China.

43. C. lucidus Schltdl. - *Shiny Cotoneaster*. Spreading to erect, deciduous shrub to 385
3m; leaves 2-5(7)cm, shiny and with slightly impressed veins on upperside, 393
pubescent on lowerside, with brilliant autumn colours; flowers 3-15 together; anthers white; fruits 8-10mm, shiny-black, subglobose, with (2-)3 stones. Intrd; natd in scattered places in En; Siberia and Mongolia.

44. C. villosulus (Rehder & E.H. Wilson) Flinck & B. Hylmö (*C. acutifolius* auct. 393
non Turcz.) - *Lleyn Cotoneaster*. Erect, deciduous shrub to 5m; leaves 3-8(10)cm, acute to shortly acuminate, slightly shiny on upperside and with impressed veins, rather shaggy-pubescent on lowerside; flowers 3-9(15) together; anthers white; fruits 8-10mm, shiny-black, broadly obovoid, with 2(-3) stones. Intrd; natd in scattered places in Br; C China.

45. C. laetevirens (Rehder & E.H. Wilson) G. Klotz (*C. ambiguus* auct. non Rehder 393
& E.H. Wilson) - *Ampfield Cotoneaster*. Differs from *C. villosulus* in shrub to 3m; leaves tapering-acuminate, ± matt on upperside and with scarcely impressed veins. Intrd; natd in S Hants; W China.

46. C. pseudoambiguus J. Fryer & B. Hylmö - *Kangting Cotoneaster*. Differs from 393
C. villosulus in leaves 3.5-5.5cm, matt on upperside; flowers 4-7 together; fruits with 2-3 stones. Intrd; natd in Westmorland; NW China.

47. C. hummelii J. Fryer & B. Hylmö - *Hummel's Cotoneaster*. Differs from *C.* 393
villosulus in leaves 5-11cm, subglabrous on lowerside, finely acuminate; flowers 7-15 together; fruits 9-13mm, with 2-3 stones. Intrd; natd in S Hants; NW China.

48. C. hsingshangensis J. Fryer & B. Hylmö - *Hsing-Shan Cotoneaster*. Differs 393
from *C. villosulus* in shrub to 3m; leaves abruptly acuminate; flowers 3-9 together; fruits 10-11mm, globose, with 2-3 stones. Intrd; natd in wood and by road in S Hants and Westmorland; C China.

49. C. cambricus J. Fryer & B. Hylmö (*C. integerrimus* auct. non Medik.) - *Wild* **RRR**
Cotoneaster. Irregularly branched, spreading, deciduous shrub to 1.5m; leaves 1- 393
4cm, flat and matt on upperside, tomentose on lowerside; flowers 1-4(7) together; anthers white; fruits 7-11mm, bright red, globose, with 2-3 stones; 2n=68. Native; very few plants on limestone of Great Orme's Head, Caerns, known since 1783; endemic. The 6 plants remaining in 1983 have since been increased by re-introduction of native material.

50. C. mucronatus Franch. - *Mucronate Cotoneaster*. Erect, deciduous shrub to 4m; 393
leaves 2-5cm, flat and matt on upperside but often with undulate edges, densely pubescent on lowerside; flowers 1-4 together; anthers white; fruits 8-12mm, bright red, broadly ellipsoid or obovoid, with 2-3 stones. Intrd; natd in S Br; W China.

51. C. simonsii Baker - *Himalayan Cotoneaster*. Erect, deciduous shrub to 3(4)m; 384
leaves 1.5-2.5(3)cm, shiny and flat on upperside, rather sparsely pubescent on 393
lowerside; flowers 1-4 together; anthers white; fruits (6)8-11mm, orange-red, globose to broadly obovoid, with 3-4 stones. Intrd; natd rather commonly thoughout most of BI; Himalayas.

52. C. tengyuehensis J. Fryer & B. Hylmö - *Tengyueh Cotoneaster*. Erect evergreen 393
to semi-evergreen shrub to 2.5m; leaves 3.5-5.5cm, shiny and with slightly impressed veins on upperside, pubescent on lowerside; flowers 3-9 together; anthers white; fruits 7-10mm, bright red, broadly obovoid to oblong-ellipsoid, with 3-4(5) stones. Intrd; natd in W Kent; SW China.

53. C. bullatus Bois - *Hollyberry Cotoneaster*. Arching, deciduous shrub to 4m; 393
leaves 3.5-7cm, shiny and ± bullate on upperside, rather densely pubescent on lowerside; flowers 12-30 together; anthers white; fruits mostly 6-8mm, bright shiny red, subglobose to obovoid, with (4-)5 stones. Intrd; natd frequently in Br and Ir, Man; W China. Fruits ripen very early (Aug). Over-recorded for *C. rehderi*.

54. C. rehderi Pojark. (*C. bullatus* var. *macrophyllus* Rehder & E.H. Wilson) - 384
Bullate Cotoneaster. Arching, deciduous shrub to 5m; differs from *C. bullatus* in 393
leaves 5-15cm, extremely bullate on upperside, pubescent to rather sparsely so on lowerside; calyx less pubescent; fruits mostly 8-11mm. Intrd; natd frequently in Br and Ir; W China. Fruits also ripen in Aug.

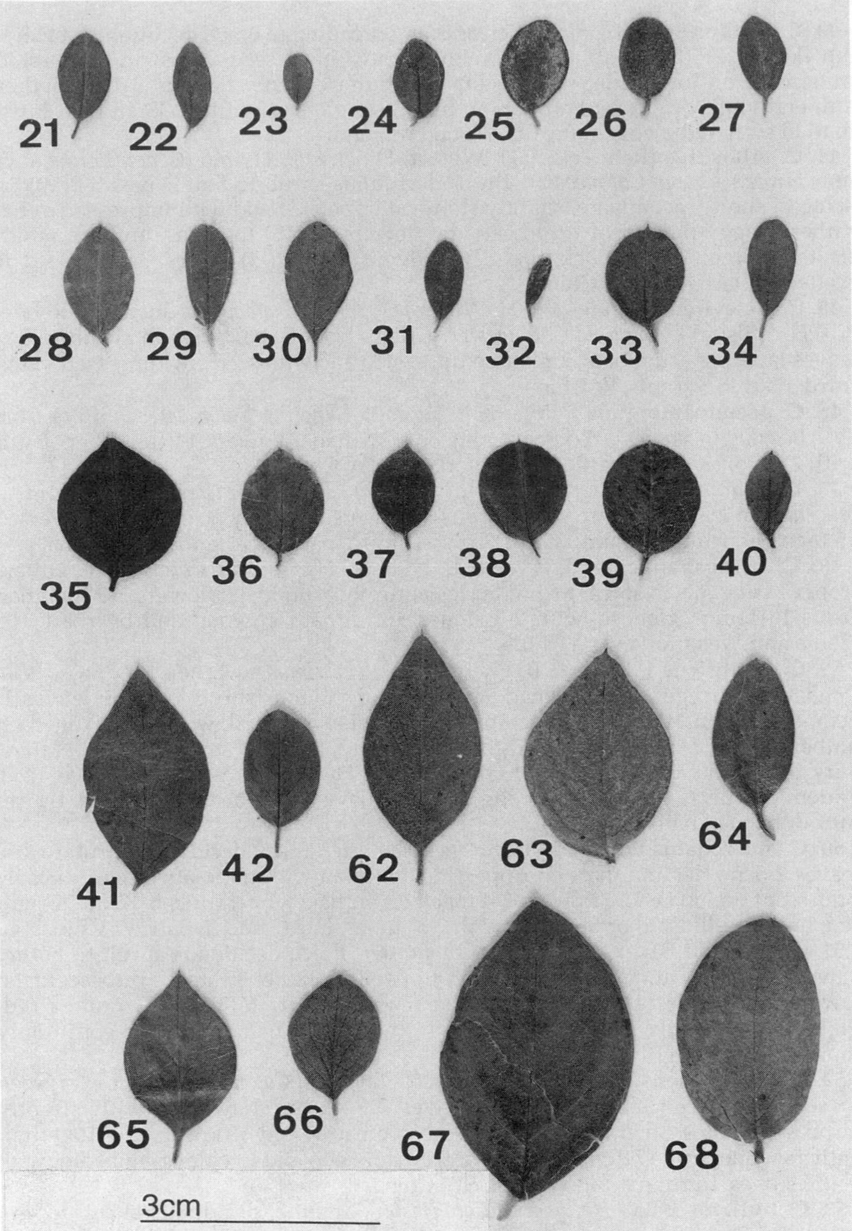

FIG 392 - Leaves of *Cotoneaster*. 21, *C. microphyllus*. 22, *C. conspicuus*. 23, *C. astrophoros*. 24, *C. sherriffii*. 25, *C. rotundifolius*. 26, *C. cochleatus*. 27, *C. cashmiriensis*. 28, *C. prostratus*. 29, *C. marginatus*. 30, *C. congestus*. 31, *C. integrifolius*. 32, *C. linearifolius*. 33, *C. nitidus*. 34, *C. adpressus*. 35, *C. apiculatus*. 36, *C. nanshan*. 37, *C. horizontalis*. 38, *C. atropurpureus*. 39, *C. hjelmqvistii*. 40, *C. ascendens*. 41, *C. divaricatus*. 42, *C. nitens*. 62, *C. vilmorinianus*. 63, *C. insculptus*. 64, *C. amoenus*. 65, *C. dielsianus*. 66, *C. splendens*. 67, *C. zabelii*. 68, *C. fangianus*.

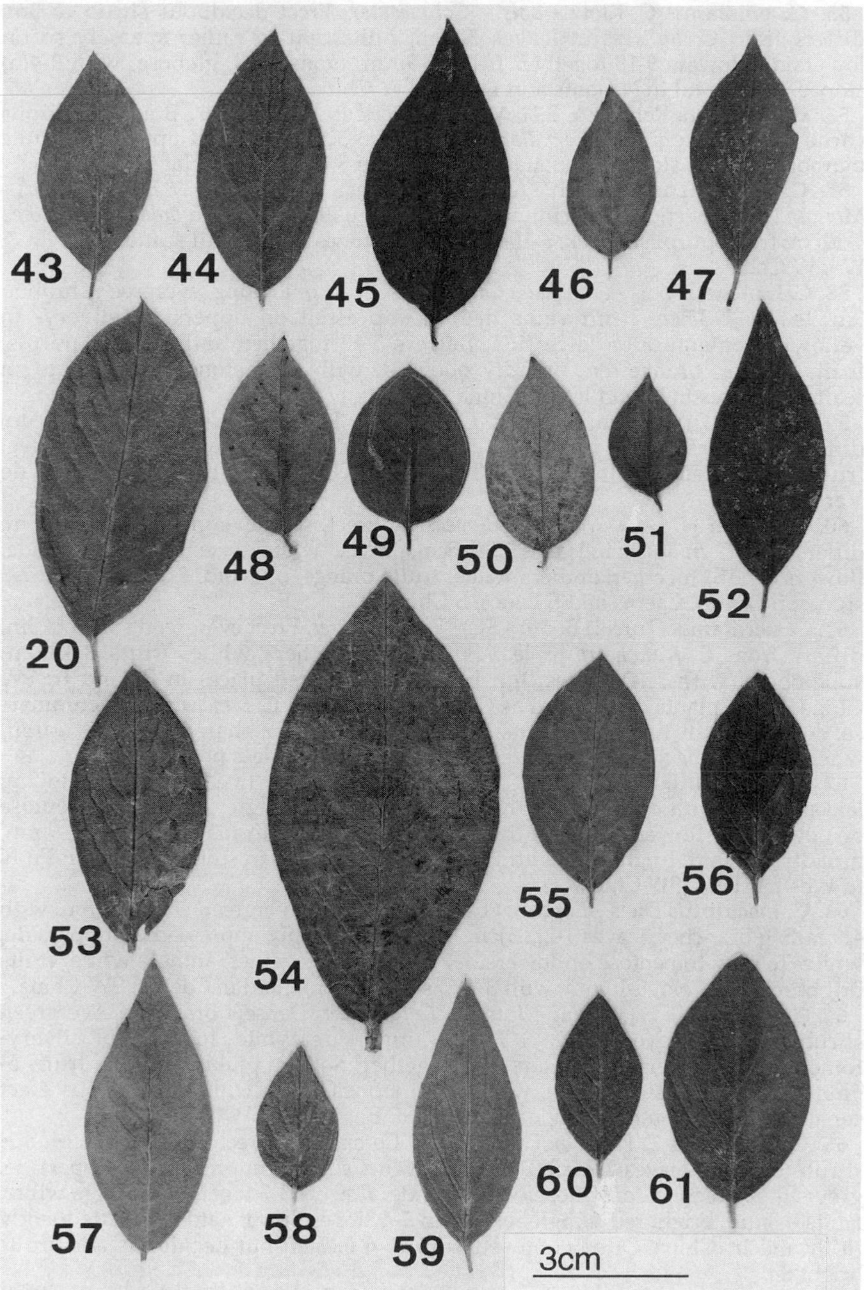

FIG 393 - Leaves of *Cotoneaster*. 20, *C. lacteus*. 43, *C. lucidus*. 44, *C. villosulus*. 45, *C. laetevirens*. 46, *C. pseudoambiguus*. 47, *C. hummelii*. 48, *C. hsingshangensis*. 49, *C. cambricus*. 50, *C. mucronatus*. 51, *C. simonsii*. 52, *C. tengyuehensis*. 53, *C. bullatus*. 54, *C. rehderi*. 55, *C. boisianus*. 56, *C. obscurus*. 57, *C. moupinensis*. 58, *C. franchetii*. 59, *C. wardii*. 60, *C. mairei*. 61, *C. sternianus*.

55. C. boisianus G. Klotz - *Bois's Cotoneaster*. Erect deciduous shrub to 3m; 393
differs from *C. bullatus* in leaves 3-6cm, pubescent to rather sparsely so on
lowerside; flowers 9-18 together; fruits 7-9mm, orange-red, globose, with 3-4(5)
stones. Intrd; natd in N Hants and Lanarks; W China.

56. C. obscurus Rehder & E.H. Wilson - *Obscure Cotoneaster*. Erect deciduous 393
shrub to 2m; differs from *C. bullatus* in leaves 3-5cm, with matt upperside; fruits
maroon, with 3-5 stones. Intrd; natd in hedges in S Lancs and S Hants; W China.

57. C. moupinensis Franch. (*C. foveolatus* auct. non Rehder & E.H. Wilson) - 393
Moupin Cotoneaster. Erect deciduous shrub to 3m; differs from *C. bullatus* in leaves
4-12cm; fruits purplish-black, 6-10mm, with 5 stones. Intrd; natd scattered in En &
Wa; W China.

58. C. franchetii Bois - *Franchet's Cotoneaster*. Erect or arching, evergreen shrub to 393
3m; leaves 2-3.5cm, with veins deeply impressed on upperside, silvery- to
yellowish-tomentose on lowerside; flowers 5-15 together; anthers pale purple;
fruits 6-9mm, orange-red, broadly obovoid, with (2-)3 stones. Intrd; natd in
scattered places in Br and Ir; SW China.

59. C. wardii W.W. Sm. - *Ward's Cotoneaster*. Erect evergreen shrub to 2.5m; 393
differs from *C. franchetii* in leaves 3-5cm, thinner in texture; anthers pale mauve;
fruits with 2 stones. Intrd; natd in Offaly; Tibet. Grown as **C.** '**Gloire de
Versailles**'.

60. C. mairei H. Lév. - *Maire's Cotoneaster*. Erect, semi-evergreen shrub to 3m; 393
differs from *C. franchetii* in leaves 1.5-3.5cm, greyish-white tomentose on lowerside;
flowers 3-7(15) together; anthers white; fruits orange, obovoid, 8-10mm, with 2-3
stones; S Devon, Caerns and S Lancs; S China.

61. C. sternianus (Turrill) Boom - *Stern's Cotoneaster*. Erect evergreen shrub to 3m; 384
differs from *C. franchetii* in leaves 2.5-5cm; anthers white; fruits 8-10mm, 393
subglobose, with 3(-4) stones. Intrd; natd in scattered places in Br and Ir; SW
China. Many plants determined as *C. franchetii* are this. *C. franchetii* has acuminate
calyx-lobes with hair-less points ≥0.5mm, a character shared with *C. wardii*,
whereas *C. sternianus* has apiculate calyx-lobes with hair-less points c.0.2mm

62. C. vilmorinianus G. Klotz - *Vilmorin's Cotoneaster*. Evergreen shrub to 2m; 392
leaves 2-3cm, with veins slightly impressed on upperside, greyish-white tomentose
on lowerside; flowers 7-15(20) together; anthers pink to mauve; fruits 8-9mm,
broadly obovoid, bright red, with 2-3 stones. Intrd; natd in scattered sites in En N
to Westmorland; SW China.

63. C. insculptus Diels - *Engraved Cotoneaster*. Erect evergreen shrub to 3m, with 392
spreading branches; leaves 1-2(2.5)cm, with veins deeply impressed on upperside,
white- to grey-tomentose on lowerside; flowers 1-4 together; anthers white; fruits
bright red, obovoid, 6-10mm, with 3-4(5) stones. Intrd; natd in Caerns; SW China.

64. C. amoenus E.H. Wilson - *Beautiful Cotoneaster*. Densely branched, evergreen 392
shrub to 1.5m; leaves 1-2cm, ± flat on upperside, white- to grey- or silvery-
tomentose on lowerside; flowers 6-10 together; anthers pinkish-purple; fruits 5-
6mm, bright red, subglobose, with 2-3(4) stones, often with conspicuous erect
sepals as in *C. pannosus*. Intrd; natd in S & SE En, Offaly; SW China.

65. C. dielsianus E. Pritz. ex Diels - *Diels' Cotoneaster*. Erect, arching, deciduous 392
shrub to 2(3)m; leaves 1.2-2.5(3)cm, with veins slightly impressed on upperside,
greyish- or greenish-tomentose on lowerside; flowers 3-7 together; anthers white;
fruits 6-8mm, bright red, subglobose, with 3-4 stones. Intrd; natd rather frequently
in Br and Ir; China. Often confused with *C. franchetii* but deciduous and fruits
bright red. 392

66. C. splendens Flinck & B. Hylmö - *Showy Cotoneaster*. Widely spreading,
deciduous shrub to 1(2)m; leaves 1-2(2.5)cm, ± flat on upperside, greyish- or
yellowish-tomentose on lowerside; flowers 2-3(8) together; anthers white; fruits 8-
11mm, orange to orange-red, broadly obovoid to subglobose, with usually 4 stones.
Intrd; natd in SW & SE En; W China.

67. C. zabelii C.K. Schneid. - *Cherryred Cotoneaster*. Erect, deciduous shrub to 3m; 385

leaves 1.5-4cm, with veins slightly impressed on upperside, densely pubescent to ± 392
tomentose on lowerside; flowers 4-12 together; anthers white; fruits 7-9mm, bright red, strongly obovoid to ± pear-shaped, with 2 stones. Intrd; natd in W Kent, Surrey and Man; China.

68. C. fangianus T.T. Yu - *Fang's Cotoneaster*. Erect deciduous shrub to 3m; leaves 392
1-3cm, with veins slightly impressed on upperside, white- to grey-tomentose on lowerside; flowers 5-15 together; anthers white; fruits 7-9mm, bright red, broadly obovoid to globose, with 2 stones. Intrd; natd in S Lancs; C China.

33. PYRACANTHA M. Roem. - *Firethorns*

Leaves simple, serrate, evergreen; flowers in compound corymbs; petals white; stamens 20; carpels 5, with 5 free styles, the walls stony at fruiting.

Other spp. - *P. coccinea* x *P. rogersiana* is also grown in gardens under cultivar names and some of these might be natd too, possibly overlooked.

1. P. coccinea M. Roem. - *Firethorn*. Spiny shrub to 2(6)m; leaves 2-7cm, narrowly obovate to oblanceolate, with pubescent petioles; inflorescence-stalks pubescent; fruits globose to depressed-globose, yellow, orange or scarlet; (2n=34). Intrd; very commonly grown and natd as bird-sown plants on banks and walls and in rough ground; frequent in S & C Br; S Europe.

2. P. rogersiana (A.B. Jacks.) Coltm.-Rog. - *Asian Firethorn*. Differs from *P. coccinea* in ± glabrous petioles and inflorescence branches and smaller leaves up to 3(5)cm. Intrd; commonly grown but known natd only in E & W Kent, probably overlooked; China.

34. MESPILUS L. - *Medlar*

Leaves simple, ± entire, deciduous; flowers solitary; petals white; stamens 30-40; carpels 5, with 5 free styles, the walls stony at fruiting.

1. M. germanica L. - *Medlar*. Shrub or tree to 9m, sometimes spiny; leaves 5-12cm, elliptic-oblong, pubescent; flowers 3-5cm across excl. sepals which project beyond petals; fruit subglobose, 2-3cm (to 6cm in cultivars), solitary, with persistent erect to erecto-patent sepals; (2n=34). Intrd; natd in hedges for at least 4 centuries; local in CI and S Br, sporadic in C & N En; SE Europe.

34 x 35. MESPILUS X CRATAEGUS = X CRATAEMESPILUS E.G. Camus - *Haw-medlar*

34/1 x 35/1. x C. grandiflora (Sm.) E.G. Camus (*M. germanica* x *C. laevigata*) occurs sporadically as isolated trees of uncertain origin very scattered through Br; leaves elliptic to obovate, serrate, usually slightly lobed; fruits brown, obovoid, <2cm, with 2-3 styles and carpels, some usually 2-3 together; sepals persistent but <Icm. This binomial might apply to *M. germanica* x *C. monogyna* (X *C. gillotii* Beck) which possibly also occurs. Not to be confused with the graft-hybrid + **Crataegomespilus dardarii** Simon-Louis ex Bellair, which has very variable leaves on one plant and usually branches of the pure spp. here and there, and clustered fruits with sepals like those of *Mespilus*.

35. CRATAEGUS L. - *Hawthorns*

Leaves simple, serrate, lobed or not, deciduous; flowers in corymbs; petals white to pink or red; stamens (5)10-20; carpels 1-5, with as many free styles, the walls stony at fruiting.

1 At least some leaves usually lobed >1/3 way to midrib; both apices of lobes and sinuses between lobes on at least some leaves reached by major vein from midrib 2

1 Leaves not lobed or lobed usually <1/3 way to midrib; only apices of lobes or main teeth (not sinuses) reached by major lateral vein 5
2 Leaves densely pubescent; styles and nutlets (3)4-5 **10. C. orientalis**
2 Leaves glabrous to sparsely pubescent; styles and nutlets 1-3 3
3 Leaves varying from unlobed to deeply lobed on 1 tree, some narrowly oblong to oblanceolate and entire in basal 1/2 **9. C. heterophylla**
3 Leaves not strongly heterophyllous, none narrowly oblong, unlobed and entire in basal 1/2 4
4 Styles and nutlets 1(-2); deepest sinus between leaf-lobes reaching >2/3 way to midrib; leaf-lobes (3)5-7, the lowest pair acute **7. C. monogyna**
4 Styles and nutlets (1)2-3; deepest sinus between leaf-lobes reaching <2/3 way to midrib; leaf-lobes usually 3, the lateral pair obtuse **8. C. laevigata**
5 Styles and nutlets mostly 2-3 6
5 Styles and nutlets mostly 4-5 8
6 Leaves and inflorescences glabrous; nutlets without hollows on inner surfaces **4. C. crus-galli**
6 Leaves pubescent on lowerside veins; inflorescence-stalks pubescent; nutlets with hollows on inner surfaces 7
7 Stamens 10-15; leaves cuneate (<90°) at base **5. C. persimilis**
7 Stamens 15-20; leaves broadly cuneate (>90°) at base **6. C. succulenta**
8 Leaves pubescent on lowerside at first, only on veins later; inflorescence-stalks tomentose; stamens 10 **1. C. submollis**
8 Leaves glabrous to pubescent only on veins at first, glabrous later; inflorescence-stalks glabrous or with sparse shaggy hairs 9
9 Stamens 20; fruits subglobose; flowers >2cm across **2. C. coccinioides**
9 Stamens 10; fruits ellipsoid to pear-shaped; flowers ≤2cm across **3. C. pedicellata**

Other spp. - Increasing numbers of spp. are grown for ornament and more might become natd, especially N American spp. with scarcely lobed leaves and large fruits; some are hybrids and do not breed true. Odd individuals in woods and hedges may be bird-sown but are often planted, frequently as grafts on common spp. There are odd records for at least 10 spp. **C. punctata** Jacq., from N America, related to *C. crus-galli*, was formerly natd.

1. C. submollis Sarg. (*C. coccinea* auct. non L.) - *Hairy Cockspurthorn*. Tree to 8m; spines 5-7cm, thin; leaves ovate, doubly serrate and shallowly lobed; fruits pear-shaped, 15-20mm; (2n=68). Intrd; frequently planted, self-sown in hedges and rough ground; W Kent and S Devon; E N America.

2. C. coccinioides Ashe - *Large-flowered Cockspurthorn*. Tree to 7m; spines 3-5cm, strong; leaves triangular-ovate, doubly serrate and shallowly lobed, with acuminate teeth; fruits subglobose, 15-20mm. Intrd; planted and natd as for *C. submollis* but much less common; Surrey and Yorks; C USA.

3. C. pedicellata Sarg. (*C. coccinea* auct. non L.) - *Pear-fruited Cockspurthorn*. Tree to 7m; spines 3-5cm, medium; leaves broadly ovate, doubly serrate and shallowly lobed; fruits pear-shaped to ellipsoid, 15-20mm; (2n=68). Intrd; planted and natd as for *C. submollis;* SE En and CI; E N America.

4. C. crus-galli L. - *Cockspurthorn*. Tree to 6m; spines 3-8cm, medium; leaves elliptic to obovate, doubly serrate, not lobed; fruits subglobose, c.10mm; (2n=51, 68). Intrd; formerly much planted and still natd in hedges in S & C Br, over-recorded for *C. persimilis;* E N America.

5. C. persimilis Sarg. (*C. prunifolia* Pers. non (Marshall) Baumg.) - *Broad-leaved Cockspurthorn*. Differs from *C. crus-galli* in pubescence on leaves and inflorescence, wider leaves, and fruit falling in autumn (not spring); (2n=68). Intrd; abundantly planted, often grafted on *C. monogyna* stock, sometimes self-sown in S En, often misnamed *C. crus-galli;* our plant is cultivar **'Prunifolia'**.

6. C. succulenta Schrad. - *Round-fruited Cockspurthorn.* Tree to 6m; spines 3-5cm, strong; leaves broadly obovate to elliptic, doubly serrate, slightly lobed near apex; fruits globose, 12-15mm; (2n=34, 51). Intrd; occasionally planted, natd in woods and hedges; S En; E N America.

7. C. monogyna Jacq. (*C. laciniata* Ucria)- *Hawthorn.* Shrub or tree to 10(15)m; spines 1-2.5cm, strong to medium; leaves ovate to broadly so, (3)5-7-lobed ≥2/3 way to midrib, the lobes acutely serrate near apex; fruits subglobose to broadly ellipsoid, (6)8-10(13)mm; 2n=34. Native; wood-borders, scrub and hedges; abundant throughout BI. Very varied in growth-form, from procumbent shrub to erect tree. Some cultivated examples have pink flowers. Our plants have been referred to ssp. **nordica** Franco; plants of ssp. **azarella** (Griseb.) Franco, from S Europe, with pubescent (not glabrous or nearly so) twigs and leaves, are commonly grown in parks and gardens and occur as bird-sown escapes, but the value of these sspp. is uncertain.

7 x 8. C. monogyna x C. laevigata = C. x media Bechst. (*C. x macrocarpa* auct. non Hegetschw., *C. x ovalis* Kitt.) is common throughout the range of *C. laevigata* and even beyond (where the latter presumably once occurred); natd in Co Antrim. It is fertile and covers the whole spectrum of intermediacy.

7 x 9. C. monogyna x C. heterophylla has arisen naturally in a wooded cemetery in Surrey with both parents and apparently backcrosses; endemic.

8. C. laevigata (Poir.) DC. (*C. oxyacanthoides* Thuill.) - *Midland Hawthorn.* Tree to 10m; differs from *C. monogyna* in less stiff, less spiny twigs; and see key; 2n=34. Native; woods, often well shaded, and hedges; common in C & SE En, scattered W to Wa and N to N En, but extent of native area uncertain, natd in SW & NE Ir. Most cultivars with pink or red flowers and/or *flore pleno* belong to this sp. or to *C. x media* rather than to *C. monogyna.*

9. C. heterophylla Flüggé - *Various-leaved Hawthorn.* Tree to 6m; spines very sparse; leaves strongly heterophyllous, narrowly oblong to elliptic, some with 3-5 forward-directed lobes, some unlobed; fruits oblong, 10-18mm, bright red, with 1 nutlet and style. Intrd; planted in parks, rarely self-sown; natd in urban woodland in Middlesex and Surrey; Caucasus.

10. C. orientalis Pall. ex M. Bieb. (*C. laciniata* auct. non Ucria) - *Oriental Hawthorn.* Tree to 6m, spines very sparse; leaves obovate to oblong-obovate, 3-7-lobed >3/4 (often nearly whole) way to midrib, the lobes narrow with few apical teeth; fruits globose to pear-shaped, 15-20mm, brick-red to orange; (2n=68). Intrd; frequently grown, natd in hedges and on banks in S En N to Derbys; SE Europe and SW Asia.

78. MIMOSACEAE - *Australian Blackwood family*
(Leguminosae subfam. Mimosoideae)

Suckering trees; leaves alternate, of 2 sorts, the juvenile bipinnate with numerous leaflets, the adult simple; stipules ± 0. Flowers in dense racemes of spherical heads, bisexual, hypogynous, actinomorphic; sepals mostly 5, fused into tube; petals mostly 5, fused into tube proximally; stamens numerous, longer than petals; carpel 1, with several ovules in row; style 1; stigma capitate; fruit a legume.

At once recognizable by the small spherical pom-poms of flowers of which the stamens are the most conspicuous part.

Other genera - ALBIZIA Durazz. differs from *Acacia* in the stamens fused below and in the tuft-like, not spherical, flower clusters. **A. lophantha** Benth., from Australia, with large bipinnate leaves, is planted for ornament in Scillies and gives rise to seedlings.

1. ACACIA Mill. - *Australian Blackwood*

Other spp. - A few spp. are grown for ornament in SW En and may sometimes spread very locally by suckers or produce young seedlings, but are not truly natd.

1. A. melanoxylon R. Br. - *Australian Blackwood*. Tree to 15m; adult leaves 6-13(20)cm, lanceolate to oblanceolate, slightly curved, with 3-5 longitudinal veins; flower-heads c.10mm across, few per raceme, cream to yellow; (2n=26). Intrd; grown for ornament; locally ± natd in S Devon and Scillies; Australia.

78A. CAESALPINIACEAE

(Leguminosae subfam. Caesalpinioideae)

SENNA Mill., from the tropics, is a fairly frequent genus of casuals from soyabean waste that rarely reach flowering. They are erect herbaceous perennials with paripinnate leaves up to 20cm and yellow flowers up to 3cm across in racemes; sepals 5; petals 5, slightly zygomorphic; stamens 10, ranging from large to very reduced; carpel 1, with many ovules; fruit a ± indehiscent legume 5-13cm. **S. occidentalis** (L.) Link (*Cassia occidentalis* L.) (*Coffee Senna*) has 4-5 pairs of ovate leaflets and a large gland near the petiole base; **S. obtusifolia** (L.) H.S. Irwin & Barneby (*Cassia obtusifolia* L.) (*American Sicklepod*) has 3 pairs of obovate leaflets and 1 or 2 glands near the leaf rhachis base.

79. FABACEAE - *Pea family*

(Papilionaceae; Leguminosae subfam. Papilionoideae, subfam. Lotoideae)

Annual to perennial herbs, shrubs or trees, sometimes spiny; leaves alternate, simple to palmate or pinnate, often with tendrils, usually stipulate. Flowers solitary or variously grouped, bisexual, hypogynous, zygomorphic, always like that of the pea in organization; sepals usually 5, usually fused into tube, often differentiated into upper lip of 2 sepals and lower lip of 3 sepals; petals 5, the upper (*standard*), 2 free laterals (*wings*) and 2 fused lower (*keel*), the last ± concealing the stamens and carpel; stamens 10, usually all fused into tube below or the uppermost free and the 9 lower fused, rarely all 10 free; carpel 1, with 1-many ovules in row; style 1, stigma capitate; fruit basically a legume, but very variably modified, usually dehiscent along 2 sides but often a schizocarp (breaking transversely into 1-seeded units).

The flowers, like those of a pea, are diagnostic; fruits and leaves are very variable.

General key

1 Leaves simple, sometimes reduced to a tendril, spine or scale, sometimes 0 — ***Key A***
1 At least some leaves with at least 2 leaflets — 2
  2 Leaves with 1-many pairs of leaflets, with or without an odd terminal leaflet, if with then pairs of leaflets >1, if without then often with tendrils — ***Key C***
  2 Leaves ternate or palmate, without tendrils — ***Key B***

Key A - Leaves simple, sometimes reduced to tendril, spine or scale, or 0

1 Herbaceous annuals or perennials — 2
1 Woody shrubs — 5
  2 Fruit opening along 2 sides like a pea-pod — **21. LATHYRUS**

2 Fruit indehiscent, or breaking transversely between seeds 3

3 Fruit 1(-2)-seeded, enclosed in calyx **11. ANTHYLLIS**

3 Fruit >2-seeded, exserted from calyx 4

4 Plant glabrous; fruit curved, smooth **15. CORONILLA**

4 Plant slightly pubescent; fruit spiralled, with tubercles or weak spines **18. SCORPIURUS**

5 Plant spiny, at least some spines branched 6

5 Spines 0 or simple 7

6 Upper calyx-lip with 2 short teeth; small bracteole present on either side of flower **35. ULEX**

6 Upper calyx-lip divided >1/3 way to base; bracteoles 0 **34. GENISTA**

7 Flowers white **32. CYTISUS**

7 Flowers yellow to reddish 8

8 Twigs strongly angled or grooved; upper calyx-lip with 2 short teeth (divided <1/5 way to base) **32. CYTISUS**

8 Twigs finely grooved; upper calyx-lip divided ≥1/4 way to base 9

9 Spines 0; upper calyx-lip divided nearly to base, the 2 halves inclined downwards near lower lip **33. SPARTIUM**

9 Spines 0 or present; calyx with distinct upper and lower lips, the upper divided 1/4-3/4 way to base **34. GENISTA**

Key B - Leaves ternate or palmate, without tendrils

1 Leaves palmate, with >4 leaflets **30. LUPINUS**

1 Leaves ternate 2

2 Woody trees or shrubs 3

2 Herbaceous annuals or perennials 8

3 Stems spiny; corolla yellow **35. ULEX**

3 Stems not spiny, or if spiny corolla not yellow 4

4 Fruit spiral **27. MEDICAGO**

4 Fruit ± straight 5

5 Leaflets toothed; flowers usually pinkish-purple, rarely white **24. ONONIS**

5 Leaflets entire; flowers usually yellow, rarely white 6

6 Leaflets mostly >3cm; racemes usually pendent **31. LABURNUM**

6 Leaflets mostly <3cm; flowers 1-few or in ± erect racemes 7

7 Upper lip of calyx deeply bifid **34. GENISTA**

7 Upper lip of calyx with 2 short teeth **32. CYTISUS**

8 Main lateral veins of leaflets running whole way to margin; leaflets often toothed 9

8 Main lateral veins of leaflets not reaching margin; leaflets rarely toothed 17

9 Calyx with glandular (and often non-glandular) hairs; all 10 stamens fused into tube **24. ONONIS**

9 Calyx without glandular hairs; 9 stamens fused into tube, the 10th free 10

10 Flowers in elongated racemes; fruits ≤7mm, all with 1-2 seeds 11

10 Flowers few, or in short dense heads, or if in elongated racemes then fruits >7mm and at least some with >2 seeds 12

11 Flowers yellow or white; fruits exserted from calyx-tube **25. MELILOTUS**

11 Flowers cream or pink to purple; fruits included in calyx-tube **28. TRIFOLIUM**

12 Fruits spiralled into >1/2 complete coil, often spiny **27. MEDICAGO**

12 Fruits straight to curved (≤1/2 complete coil), never spiny 13

13 Fruits >3cm, plus beak >1cm **26. TRIGONELLA**

13 Fruits <3cm, incl. beak <1cm 14

14 Flowers yellow; fruits ± curved, >7mm, at least some >2-seeded 15

14 Flowers not yellow, or if yellow then fruits straight, <7mm, 1-2-seeded 16

15 Ripe fruits pendent; plant annual **26. TRIGONELLA**
15 Ripe fruits ± erect; plant perennial **27. MEDICAGO**
16 Fruits inflated, exserted from calyx and forming a compact naked head **26. TRIGONELLA**
16 Fruits not or scarcely inflated, usually at least partly covered by calyx or persistent corolla, not forming a compact naked head **28. TRIFOLIUM**
17 Leaflets with small stipule-like outgrowths at base 18
17 Leaflets without stipule-like outgrowths at base 20
18 Fruit erect to patent; stipules broadly ovate **3. VIGNA**
18 Fruit pendent; stipules triangular to narrowly so 19
19 Corolla c. as long as calyx; common peduncle 0 to very short; plant with brown patent hairs **4. GLYCINE**
19 Corolla much longer than calyx; common peduncle long; plant glabrous to rather sparsely pubescent with whitish hairs **2. PHASEOLUS**
20 Petals blue to white; leaflets dentate **5. PSORALEA**
20 Petals yellow; leaflets entire 21
21 All 10 stamens free **29. THERMOPSIS**
21 9(-10) stamens fused into a tube 22
22 Fruit 1(-2)-seeded, enclosed in calyx **11. ANTHYLLIS**
22 Fruit >2-seeded, exserted 23
23 Fruit curved, breaking transversely between seeds at maturity **15. CORONILLA**
23 Fruit ± straight, dehiscing longitudinally along 2 sides, sometimes tardily 24
24 Fruit with 4 longitudinal wings **13. TETRAGONOLOBUS**
24 Fruit not winged **12. LOTUS**

Key C - Leaves with 1-many pairs of leaflets, if with 1 then without odd terminal leaflet, often with tendrils
1 Leaves with even no. of leaflets, terminated by point or tendril 2
1 Leaves with odd no. of leaflets, terminated by single leaflet 6
2 Stem winged, and/or leaflets parallel-veined **21. LATHYRUS**
2 Stem not or scarcely winged; leaflets pinnately veined 3
3 At least some stipules >2cm, larger than leaflets **22. PISUM**
3 Stipules <2cm, smaller than leaflets 4
4 Calyx-teeth equal, >2x as long as tube **20. LENS**
4 Calyx-teeth usually unequal, 2-5 of them <2x as long as tube 5
5 Style glabrous, or pubescent all round, or pubescent only on lowerside **19. VICIA**
5 Style pubescent only on upperside **21. LATHYRUS**
6 Woody shrubs or trees 7
6 Herbaceous, sometimes ± woody at base 10
7 Tree; corolla white **1. ROBINIA**
7 Shrub; corolla pale yellow to orange 8
8 Fruit strongly inflated, indehiscent or dehiscing longitudinally; flowers in racemes **7. COLUTEA**
8 Fruit ± not inflated, breaking transversely between seeds; flowers in umbels 9
9 Claws of wings and standard 2-3x as long as calyx; fruits 5-11cm; stems ridged or furrowed **16. HIPPOCREPIS**
9 Claws of wings and standard 1-1.3x as long as calyx; fruits 1-5cm; stems terete **15. CORONILLA**
10 Flowers in racemes 11
10 Flowers solitary or in umbels 14
11 Corolla pink to purple; fruits with 1 seed **10. ONOBRYCHIS**
11 Corolla white, yellow or blue; fruits with ≥2 seeds 12

12 Keel beaked at apex **9. OXYTROPIS**
12 Keel subacute to rounded at apex 13
13 Fruit not partitioned internally, terete, glabrous; uppermost stamen fused to stamen-tube for part of its length **6. GALEGA**
13 Fruit longitudinally partitioned by internal membrane variously shaped but not terete and glabrous; uppermost stamen free **8. ASTRAGALUS**
14 Fruit with 2 longitudinal sutures, usually dehiscing along them, without transverse sutures 15
14 Fruit breaking along transverse sutures between seeds 17
15 Fruits >3x as long as wide, with >3 seeds **12. LOTUS**
15 Fruits <3x as long as wide, with 1-2 seeds 16
16 Fruits longer than calyx; leaflets toothed **23. CICER**
16 Fruits enclosed within calyx; leaflets entire **11. ANTHYLLIS**
17 Fruit-segments and seeds horseshoe-shaped **16. HIPPOCREPIS**
17 Fruit-segments and seeds oblong-ellipsoid 18
18 Strong perennials; flowers usually ≥10 per inflorescence, 8-15mm **17. SECURIGERA**
18 Annuals; flowers ≤8(12) per inflorescence, 3-9mm **14. ORNITHOPUS**

Other genera - DORYCNIUM Mill. (Loteae), from S Europe, differs from *Lotus* in the tall semi-shrubby habit and densely crowded white to pink small flowers with a dark red to black keel; **D. rectum** (L.) Ser. was formerly natd in Surrey and **D. pentaphyllum** Scop. (*D. gracile* Jord.) in E Kent. **SESBANIA** Adans. (Sesbanieae) is a tall erect glabrous annual with pinnate leaves with very numerous leaflets and few yellow flowers in axillary racemes; **S. exaltata** (Raf.) Cory (*Colorado River-hemp*), from S N America, occurs occasionally as a casual from various food-plant sources. **ARACHIS** L., (Aeschynomeneae), originally from Brazil, is a small annual with pinnate leaves with 2 pairs of large leaflets and solitary yellow flowers whose stalks elongate and bury under the ground where the fruit develops; **A. hypogaea** L. (*Ground-nut*) is occasional on rubbish-tips but usually does not reach flowering. **CARAGANA** Fabr. is a shrub or small tree with paripinnate leaves with persistent spinose stipules, large yellow flowers in small clusters and fruits 3-6cm; **C. arborescens** Lam. (*Siberian Pea-tree*), from N Asia, was formerly a natd garden escape in W Kent and Middlesex.

TRIBE 1 - ROBINEAE (genus 1). Trees; leaves imparipinnate, with entire leaflets; flowers in pendent racemes; 9 stamens forming tube, 10th free; fruit longitudinally dehiscent, ≥3-seeded.

1. ROBINIA L. - *False-acacia*

1. R. pseudoacacia L. - *False-acacia*. Deciduous tree to 29m; twigs with stipular spines; leaves ± glabrous, with elliptic leaflets; flowers numerous, white; fruit 5-10cm, with 3-10 seeds; (2n=20). Intrd; banks, scrub and woodland, natd by suckering and (less often) seeding; scattered over most of BI but natd ± only in S; N America.

TRIBE 2 - PHASEOLEAE (genera 2-4). Herbs; leaves ternate, with entire leaflets, with stipule-like outgrowths below each leaflet; flowers in axillary racemes or clusters; 9 stamens forming tube, 10th free or also part of tube; fruit longitudinally dehiscent, often tardily so, ≥2-seeded.

2. PHASEOLUS L. - *Beans*

Herbs, often climbing, subglabrous to shortly or appressed-pubescent; flowers in racemes with long common peduncle; corolla much longer than calyx; fruit pendent, many-seeded.

1. P. vulgaris L. - *French Bean*. Annual, with spirally climbing stem to 3m or (usually) not climbing; flowers ≤6, white, sometimes purplish; cotyledons borne above ground; (2n=22). Intrd; much grown as vegetable and frequent on tips and in waste places; scattered throughout lowland Br; S America.

2. P. coccineus L. - *Runner Bean*. Perennial with tuberous roots but rarely surviving winter, with climbing stem to 5m; flowers ≤20, bright red, rarely white; cotyledons subterranean; (2n=22). Intrd; much grown as vegetable and casual as for *P. vulgaris*; tropical America.

3. VIGNA Savi - *Mung-bean*
Herbs, not climbing, with patent hairs; flowers with long common peduncle; corolla much longer than calyx; fruit erect to patent, (3)8-14-seeded.

1. V. radiata (L.) Wilczek (*V. mungo* auct. non (L.) Hepper, *Phaseolus aureus* Roxb.) - *Mung-bean*. Differs from *Glycine max* in generic characters; (2n=22). Intrd; occurrence as for *G. max;* Asia.

4. GLYCINE Willd. - *Soyabean*
Herbs, not climbing, with long patent brown hairs; flowers with 0 or very short common peduncle; corolla c. as long as calyx; fruit pendent, 2-4-seeded.

1. G. max (L.) Merr. (*G. soja* auct. non Siebold & Zucc.) - *Soyabean*. Erect annual to 60(100)cm; flowers ≤8, often many fewer, white to purple, inconspicuous; (2n=40). Intrd; where seed is spilled on tips, waste land and near docks and factories; very scattered in S Br; origin uncertain.

TRIBE 3 - PSORALEEAE (genus 5). Herbs; leaves ternate, with dentate leaflets; flowers in erect axillary racemes; 10 stamens forming tube; fruit 1-seeded, indehiscent.

5. PSORALEA L. - *Scurfy Pea*
1. P. americana L. - *Scurfy Pea*. Erect perennial (but not surviving winter) to 50cm; **403**
leaflets ovate-orbicular, 8-50mm; flowers fairly dense, c.8mm, white tinged violet; **410**
(2n=22). Intrd; birdseed-alien on tips; sporadic in S Br; W Mediterranean.

TRIBE 4 - GALEGEAE (genera 6-9). Herbs or shrubs; leaves imparipinnate, with entire leaflets; flowers in axillary racemes; 9 stamens forming tube, 10th free or fused to others for c.1/2 way; fruit longitudinally dehiscent or ± indehiscent, few- to many-seeded.

6. GALEGA L. - *Goat's-rue*
Flowers numerous in erect racemes; 10th stamen partly fused to other 9; fruit ± terete, erect, not inflated, dehiscent.

1. G. officinalis L. - *Goat's-rue*. Erect, glabrous or sparsely pubescent perennial to **403**
1.5m; leaves with 9-17 oblong-elliptic to ovate or narrowly so leaflets; flowers white to bluish-mauve; (2n=16). Intrd; much grown for ornament and frequent on tips and in waste and grassy places; frequent in S & C Br; Europe.

7. COLUTEA L. - *Bladder-sennas*
Flowers c.2-8 in ± erect racemes; 10th stamen free; fruit greatly inflated, ± pendent.

1. C. arborescens L. - *Bladder-senna*. Deciduous shrub to 4m; leaves with 7-13 **403**
elliptic to broadly elliptic leaflets, sparsely pubescent; flowers pale to deep yellow, with beakless keel; fruit 5-7cm, indehiscent; (2n=16). Intrd; much grown in gardens and natd in waste and grassy places, on roadsides and railway banks; frequent in

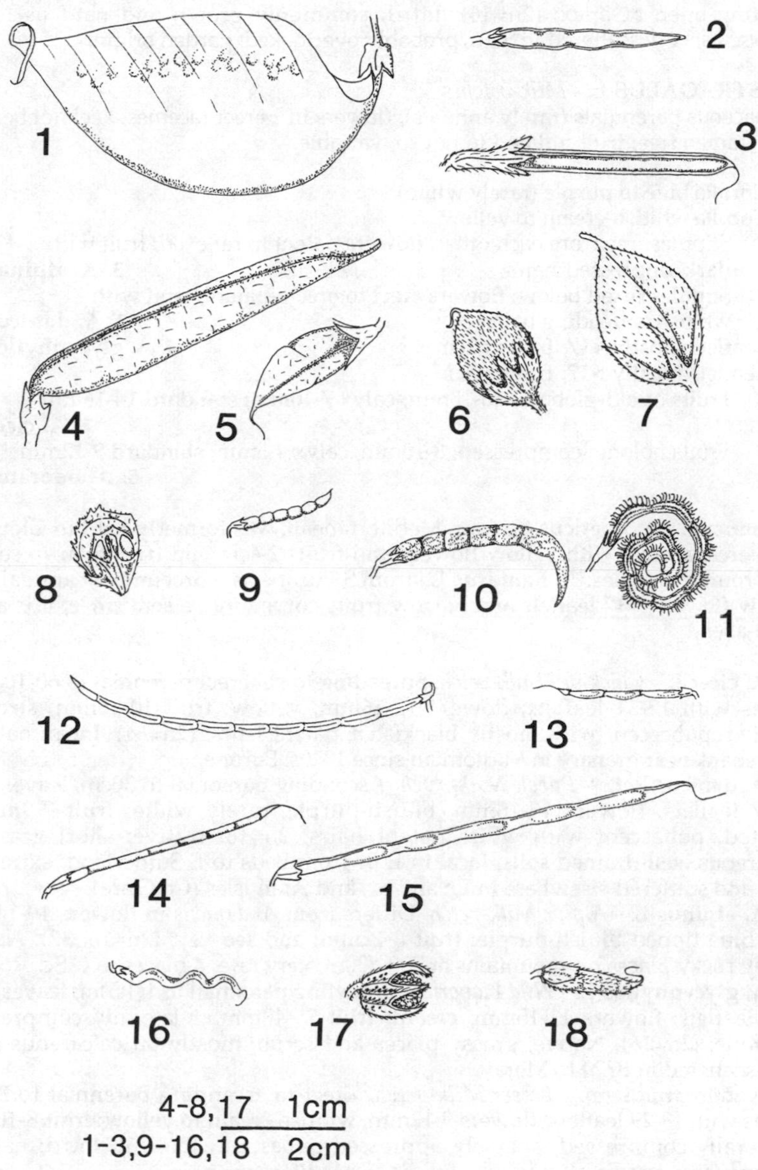

FIG 403 - Fruits of **Fabaceae**. 1, *Colutea arborescens*. 2, *Galega officinalis*. 3, *Tetragonolobus maritimus*. 4, *Astragalus glycyphyllos*. 5, *A. odoratus*. 6, *A. danicus*. 7, *A. cicer*. 8, *Onobrychis viciifolia*. 9, *Ornithopus perpusillus*. 10, *O. compressus*. 11, *Scorpiurus muricatus*. 12, *Coronilla scorpioides*. 13, *C. valentina*. 14, *Securigera varia*. 15, *Hippocrepis emerus*. 16, *H. comosa*. 17, *Psoralea americana*. 18, *Oxytropis campestris*.

S Br, especially SE En, Man; Europe.

2. C. x media Willd. (*C. arborescens* x *C. orientalis* Mill.) - *Orange Bladder-senna*. Differs from *C. arborescens* in orange-bronze flowers with beaked keel; fruits splitting open at apex; (2n=16). Intrd; commonly grown and natd as for *C. arborescens*; very scattered in S En, probably overlooked; garden origin.

8. ASTRAGALUS L. - *Milk-vetches*

Herbaceous perennials (rarely annuals); flowers in ± erect racemes; keel not beaked; 10th stamen free; fruit inflated to not so, variable.

1 Corolla blue to purple (rarely white) 2
1 Corolla whitish-cream to yellow 3
  2 Stipules free from each other; flowers patent to reflexed; fruit with dark, appressed hairs **3. A. alpinus**
  2 Stipules fused below; flowers erect to erecto-patent; fruit with whitish, spreading hairs **2. A. danicus**
3 Leaflets mostly <17; fruits >2cm **4. A. glycyphyllos**
3 Leaflets mostly >17; fruits <2cm 4
  4 Fruits ovoid-globose, 10-15mm; calyx 7-10mm; standard 14-16mm **1. A. cicer**
  4 Fruits oblong, compressed, 8-10mm; calyx 4-5mm; standard 9-12mm **5. A. odoratus**

Other spp. - A. boeticus L., from Mediterranean, was formerly natd in Gloucs; it is an erect annual with yellow flowers and fruits 2-4cm and triangular in section with rounded angles. **A. hamosus** L., from S Europe, is a procumbent annual with yellow flowers, >17 leaflets and narrow fruits curved into a semicircle; it is a rare grain-alien.

1. A. cicer L. - *Chick-pea Milk-vetch*. Spreading to suberect perennial to 60(100)cm; 403
leaves with 17-31 leaflets; flowers 12-16mm, yellow; fruit 10-15mm, strongly 410
inflated, pubescent with mostly blackish ± patent hairs; (2n=64). Intrd; natd on hedgebank near granary in Midlothian since 1920s; Europe.

2. A. danicus Retz. - *Purple Milk-vetch*. Ascending perennial to 30cm; leaves with 403
13-27 leaflets, flowers 15-18mm, bluish-purple, rarely white; fruit 7-9mm, ± inflated, pubescent with white patent hairs; 2n=16. Native; short grass on calcareous well-drained soils; local in E Br from Beds to E Sutherland, extremely local and scattered elsewhere in En and Sc, and Aran Isles (Co Clare).

3. A. alpinus L. - *Alpine Milk-vetch*. Differs from *A. danicus* in flowers 10-14mm, RR
pale blue tipped bluish-purple; fruit 8-12mm; and see key; (2n=16, 32). Native; grassy rocky places on mountains at 700-800m; very rare, 4 places in C Sc.

4. A. glycyphyllos L. - *Wild Liquorice*. Sprawling perennial to 1(1.5)m; leaves with 403
7-15 leaflets; flowers 11-15mm, cream; fruit 25-40mm, ± laterally compressed, glabrous; (2n=16). Native; grassy places and scrub, mostly on calcareous soils; very scattered in Br N to Moray.

5. A. odoratus Lam. - *Lesser Milk-vetch*. Erect to ascending perennial to 30cm; 403
leaves with 19-29 leaflets; flowers 9-12mm, whitish-cream to yellow; fruit 8-10mm, ± laterally compressed, sparsely appressed-pubescent; (2n=16). Intrd; natd in grassy places; very scattered in C & S En; E Mediterranean.

9. OXYTROPIS DC. - *Oxytropises*

Herbaceous perennials; flowers in erect racemes; keel beaked at apex; 10th stamen free; fruit elongated, grooved abaxially, slightly inflated, erect.

1. O. halleri Bunge ex W.D.J. Koch - *Purple Oxytropis*. Pubescent perennial with RR
leaves and leafless peduncle to 30cm arising from dense tuft; leaves with 21-31

leaflets; flowers 15-20mm, usually pale purple, rarely white; fruit 15-20(25)mm, pubescent, divided internally by septa from both adaxial and abaxial sutures and ± bilocular; 2n=32. Native; grassy rocky places; very local in SW, C & N mainland Sc.

2. O. campestris (L.) DC. - *Yellow Oxytropis*. Differs from *O. halleri* in pale yellow RR
flowers often strongly tinged with purple; fruit 14-18mm, divided internally by **403**
septum from abaxial suture only and semi-bilocular; (2n=32, 48). Native; cliffs and rock-ledges; rare and extremely local in C and SW Sc. Other characters used to separate the spp. are unreliable; even flower colour is fallible.

TRIBE 5 - HEDYSAREAE (genus 10). Herbaceous perennials; leaves imparipinnate, with entire leaflets; flowers in axillary racemes; 9 stamens forming tube, 10th free; fruit indehiscent, 1-seeded.

10. ONOBRYCHIS Mill. - *Sainfoin*

1. O. viciifolia Scop. - *Sainfoin*. Stems suberect, to 60(80)cm; leaves with 13-29 **403**
leaflets; flowers numerous in erect racemes, pinkish-red, with wings <1/2 as long as other petals; fruit 5-8mm, pubescent, compressed, reticulately ridged, toothed; (2n=28). Possibly native; grassland and bare places mostly on chalk and limestone; locally frequent in Br N to Yorks, scattered casual or natd alien elsewhere in Br and Ir.

TRIBE 6 - LOTEAE (genera 11-13). Annuals or herbaceous perennials; leaves ternate to imparipinnate, with entire leaflets; flowers solitary or in umbels; 9 stamens forming tube, 10th free or variably and loosely fused with others; fruit dehiscent or not, 1-many-seeded.

11. ANTHYLLIS L. - *Kidney Vetch*

Lower leaves with large terminal and 0-3 pairs of small lateral leaflets, grading variably to upper leaves with ≤15 ± equal leaflets; stipules small, falling early; 10th stamen variably and loosely fused with other 9; calyx inflated, enclosing indehiscent 1(-2)-seeded fruit.

1. A. vulneraria L. - *Kidney Vetch*. Erect to procumbent perennial to 60cm; inflorescences usually paired, with leaf-like bracts at base; calyx densely white-pubescent, dispersed with fruit; petals yellow to red. The following sspp. are not well differentiated and a population should be sampled to obtain a set of characters. Perhaps only 2 sspp. (a-c and d-e) should be recognised.

1 Calyx (4.5)5-7mm wide, the lateral teeth not appressed to upper ones; upper leaves with large terminal and (0)1-4 pairs of smaller lateral leaflets 2
1 Calyx 2-4(5)mm wide, the lateral teeth appressed to upper ones and ± obscured; upper leaves with 4-7 pairs of lateral leaflets scarcely smaller than terminal one 3
  2 Calyx-hairs appressed, sparse **d. ssp. carpatica**
  2 Calyx-hairs ± patent **e. ssp. lapponica**
3 Stem-hairs all patent **c. ssp. corbierei**
3 At least upper part of stem with appressed hairs 4
  4 Calyx usually with red tip; stems usually with appressed hairs throughout or semi-appressed ones at base **a. ssp. vulneraria**
  4 Calyx usually without red tip; lower part of stems with patent hairs **b. ssp. polyphylla**

a. Ssp. vulneraria. Stems appressed-pubescent throughout or with semi-appressed or rarely patent hairs at base; upper leaves with 4-7 pairs of leaflets scarcely or not smaller than terminal one; calyx usually red-tipped; corolla yellow to orange, rarely red; 2n=12. Native; grassland, dunes, cliff-tops, waste ground, usually calcareous; locally common throughout BI. Red-flowered plants on sea-cliffs in W Cornwall and Pembs are var. **coccinea** L. Pink-flowered plants occur in Caerns. Plants with stem-branches exceeding their subtending leaves (not so in var. **vulneraria**) on coasts of En, Wa, Ir and CI are var. **langei** Jalas (*A. maritima* auct. non Schweigg.); they are intermediate between sspp. *vulneraria* and **iberica** (W. Becker) Jalas, from W and SW Europe.

b. Ssp. polyphylla (Ser.) Nyman. Stems with some patent hairs near base, with only appressed hairs above; leaves as in ssp. *vulneraria* but generally larger, hence plant more leafy; calyx usually not red-tipped; corolla pale yellow; (2n=12). Intrd; natd in grassy places in E Sc and S En; C & E Europe.

c. Ssp. corbierei (C.E. Salmon & Travis) Cullen (var. *sericea* Bréb.). Stems with all hairs patent; leaves ± succulent, the upper with 4-5 pairs of leaflets scarcely or not smaller than terminal one; calyx not red-tipped; corolla yellow. Native; sea-cliffs in Anglesey, W Cornwall and CI (Sark and Guernsey); endemic. **RR**

d. Ssp. carpatica (Pant.) Nyman (ssp. *vulgaris* (W.D.J. Koch) Corb., ssp. *pseudovulneraria* (Sagorski) J. Duvign.). Stems sparsely appressed-pubescent; upper leaves with large terminal leaflet and (0)1-4 pairs of smaller lateral leaflets; calyx red-tipped or not; corolla pale yellow. Intrd; natd in marginal and disturbed ground; scattered over Br, Ir and Man; C Europe. Our plant is var. **pseudovulneraria** (Sagorski) Cullen.

e. Ssp. lapponica (Hyl.) Jalas. Stems appressed-pubescent or with patent hairs below; upper leaves as in ssp. *carpatica*; calyx usually extensively red-tipped; corolla yellow; (2n=12). Native; banks, cliffs and rock-ledges; mostly mountainous areas of N En, W & NW Ir and Sc. **R**

12. LOTUS L. - *Bird's-foot-trefoils*

Leaves with 5 leaflets, the lowest 2 at base of rhachis and resembling stipules; stipules minute, falling or withering early; 10th stamen free; calyx not enclosing fruit; fruit several-seeded, longitudinally dehiscent, not ridged or angled.

1 Annuals, with dense patent hairs; fruit mostly <2mm wide 2
1 Perennials, glabrous to variously pubescent; fruit mostly >2mm wide 3
  2 Fruit 12-30mm, ≥3x as long as calyx, with >12 seeds; keel with ± right-angled bend c.1/2 way along lower edge of limb **5. L. angustissimus**
  2 Fruit 6-15mm, ≤3x as long as calyx, with ≤12 seeds; keel with obtuse-angled bend near base of lower edge of limb **4. L. subbiflorus**
3 Stem hollow; calyx-teeth recurved in bud, the upper 2 with acute sinus between them **3. L. pedunculatus**
3 Stem solid or rarely hollow; calyx-teeth erect in bud, the upper 2 with obtuse sinus between them 4
  4 Leaflets of upper leaves mostly >4x as long as wide, acute to acuminate **1. L. glaber**
  4 Leaflets of upper leaves mostly <3(4)x as long as wide, subacute to obtuse **2. L. corniculatus**

1. L. glaber Mill. (*L. tenuis* Waldst. & Kit. ex Willd., L. *corniculatus* ssp. *tenuis* (Waldst. & Kit. ex Willd.) Syme, ssp. *tenuifolius* (L.) Hartm.) - *Narrow-leaved Bird's-foot-trefoil.* Glabrous to sparsely pubescent perennial with sprawling to suberect stems to 90cm; leaflets linear to linear-lanceolate; flowers (1)2-4(6), 6-12mm, yellow; fruit 15-30mm; 2n=12. Native; dry grassy places; scattered in Br N to C Sc, N & E Ir, CI, frequent only in S & E En and probably intrd in Ir, Sc and much of N & W En.

2. L. corniculatus L. - *Common Bird's-foot-trefoil.* Glabrous to sparsely pubescent perennial with procumbent to ascending stems to 50cm; leaflets lanceolate or oblanceolate to suborbicular; flowers (1)2-7, 10-16mm, yellow to orange, often streaked with red; fruit 15-30mm; 2n=24. Native; grassy and barish places, mostly on well-drained soils; common throughout BI. Native plants have solid stems; robust erect plants with large leaflets, light-coloured keels and sometimes hollow stems, mostly on roadsides, are probably intrd and referable to var. **sativus** Chrtková.

3. L. pedunculatus Cav. (*L. uliginosus* Schkuhr) - *Greater Bird's-foot-trefoil.* Glabrous to pubescent perennial with erect to ascending stems to 1m; leaflets ovate to obovate; flowers 5-12, 10-18mm, yellow; fruits 15-35cm; 2n=12. Native; damp grassy places, marshes and pondsides; frequent throughout most of BI.

4. L. subbiflorus Lag. (*L. suaveolens* Pers., *L. parviflorus* auct. non Desf., *L. hispidus* auct. non Desf. ex DC.) - *Hairy Bird's-foot-trefoil.* Pubescent annual with procumbent to decumbent stems to 30(80)cm; leaflets narrowly ovate to obovate; flowers (1)2-4, 5-10mm, yellow; 2n=24. Native; dry grassy places near sea; very local in SW En E to S Hants, extreme SW Wa, SW & SE Ir, CI. **R**

5. L. angustissimus L. - *Slender Bird's-foot-trefoil.* Pubescent annual; differs from *L. subbiflorus* in flowers 1-3, 5-12mm; and see key; 2n=12. Native; dry grassy places near sea; very local in S En and CI. **RR**

13. TETRAGONOLOBUS Scop. - *Dragon's-teeth*

Leaves with 3 leaflets; stipules herbaceous, persistent; 10th stamen free; calyx not enclosing fruit; fruits several-seeded, tardily longitudinally dehiscent, square in section with wing c.1mm wide on each angle.

1. T. maritimus (L.) Roth - *Dragon's-teeth.* Sparsely pubescent perennial with decumbent stems to 30cm; leaflets obovate to oblanceolate; flowers solitary, with leaf-like bract at base, on peduncle longer than subtending leaf, 25-30mm, yellow with brownish streaks; fruits 25-60mm; (2n=14). Intrd; very locally but well natd in rough calcareous grassland; S En, rare casual elsewhere in S & C Br, known since 1875; C & S Europe. **403**

TRIBE 7 - CORONILLEAE (genera 14-18). Herbaceous annuals or perennials or shrubs; leaves simple, ternate or imparipinnate, with entire leaflets; flowers solitary or in umbels; 9 stamens forming tube, 10th free; fruit dehiscing transversely between seeds or ± indehiscent, few- to many-seeded.

14. ORNITHOPUS L. - *Bird's-foots*

Annuals; leaves imparipinnate with ≥3 pairs of lateral leaflets; fruits curved to ± straight, beaked, not or slightly constricted between the 3-12 cylindrical to oblong-ellipsoid segments.

1 Flower-heads without bract at base, or with minute scarious bracts **4. O. pinnatus**
1 Flower-heads with leaf-like bract at base 2
 2 Corolla yellow; fruits not or scarcely constricted between segments **1. O. compressus**
 2 Corolla white to pink; fruits distinctly constricted between segments 3
3 Corolla >5.5mm; bract c.1/2 as long as flowers **2. O. sativus**
3 Corolla <5.5mm; bract at least as long as flowers **3. O. perpusillus**

1. O. compressus L. - *Yellow Serradella.* Pubescent; stems to 50cm, procumbent to decumbent; leaves with 13-37 leaflets; flowers 3-5, with leafy bracts at least as long, yellow, 5-6mm; fruits 2-5cm, curved, ± compressed, not or scarcely constricted between the 5-8 segments; (2n=14). Intrd; barish sandy banks; 1 locality in W Kent sporadically since 1957, rare casual in S Br and CI; S Europe. **403**

2. O. sativus Brot. (*O. roseus* Dufour) - *Serradella.* Pubescent; stems to 70cm, procumbent to ascending; leaves with 13-37 leaflets; flowers 2-5, with leafy bracts c.1/2 as long, white to pink, 6-9mm; fruits 1.2-2.5cm, ± straight, compressed, slightly constricted between the 3-7 segments; (2n=14). Intrd; planted and natd on china-clay waste in E Cornwall since 1978, rare casual in S Br; SW Europe.

3. O. perpusillus L. - *Bird's-foot*. Pubescent; stems to 30cm, procumbent to decumbent; leaves with 9-27 leaflets; flowers 3-8, with leafy bracts at least as long, white to pink, 3-5mm; fruits 1-2cm, curved, compressed, slightly constricted between the 4-9 segments; 2n=14. Native; dry barish sandy and gravelly ground; locally common in much of BI, especially S & E, but absent from much of Ir and Sc. **403**

4. O. pinnatus (Mill.) Druce - *Orange Bird's-foot*. Glabrous to sparsely pubescent; stems to 50cm, decumbent to ascending; leaves with 5-15 leaflets; flowers 1-2(5), with 0 or minute bracts, yellow veined with red, 6-8mm; fruits 2-3.5cm, curved, terete, very slightly constricted between the 5-8(12) segments; (2n=14). Native; short turf or open ground on sandy soil; CI (all islands) and Scillies. **RR**

15. CORONILLA L. - *Scorpion-vetches*

Annuals or shrubs, with terete branches; leaves simple, ternate or imparipinnate; fruits curved to ± straight, beaked, not or scarcely constricted between the (1)2-11 ± cylindrical segments.

1. C. valentina L. (*C. glauca* L.) - *Shrubby Scorpion-vetch*. Shrub to 1m; leaves with 5-13 leaflets, each 10-20mm, obovate; flowers 4-8(12), 7-12(14)mm, yellow; fruits 1-5cm, with (1)2-4(10) segments; (2n=12, 24). Intrd; natd on cliffs near Torquay, S Devon, and on promenade, Eastbourne, E Sussex; Mediterranean. Our plant is ssp. **glauca** (L.) Battand. **403**

2. C. scorpioides (L.) W.D.J. Koch - *Annual Scorpion-vetch*. Glaucous, glabrous, erect to ascending annual to 40cm; leaves with 1-3 leaflets, rarely all with 1, the terminal leaflet much the largest (up to 4cm), broadly elliptic; flowers 2-5, 3-8mm, yellow; fruits 2-6cm, with 2-11 segments; (2n=12). Intrd; fairly frequent casual, mostly from birdseed; S Br; S Europe. **403** **410**

16. HIPPOCREPIS L. - *Horseshoe Vetches*

Herbaceous perennials or shrubs, with furrowed or ridged stems; leaves imparipinnate; fruits ± straight to curved, beaked, with horseshoe-shaped segments, strongly compressed.

1. H. emerus (L.) Lassen (*Coronilla emerus* L.) - *Scorpion Senna*. Shrub to 1.5(2)m; leaves with 5-9 leaflets, each 10-20mm, obovate; flowers 2-5(7), (12)14-20mm, yellow; fruits 5-11cm, with 3-12 segments; (2n=14). Intrd; natd on roadsides and banks in several places in En N to S Lincs; Europe. **403** **410**

2. H. comosa L. - *Horseshoe Vetch*. Herbaceous perennial; stems procumbent to suberect, to 30(50)cm; leaves with 7-25 leaflets, each (2)4-8(16)mm, oblong to obovate; flowers (2)4-8(12), 5-10(14)mm, yellow; fruits 10-30mm, with 3-6 segments; 2n=14, 28. Native; dry calcareous grassland and cliff-tops; local in Br N to Westmorland, frequent in S & E En, NW Jersey. **403**

17. SECURIGERA DC. - *Crown Vetch*

Herbaceous perennials with furrowed or ridged stems; leaves imparipinnate; fruits ± straight to slightly curved, beaked, not or scarcely constricted between the 3-8(12) cylindrical segments.

1. S. varia (L.) Lassen (*Coronilla varia* L.) - *Crown Vetch*. Stems sprawling, to 1.2m; leaves with 11-25 leaflets, each 6-20mm, elliptic to oblong; flowers (5)10-20, 8-15mm, white to pink or purple; fruits 2-6(8)cm, with 3-8(12) segments; (2n=24). Intrd; natd in grassy places and rough ground; scattered through Br N to C Sc, E Ir, Guernsey; Europe. **403**

18. SCORPIURUS L. - *Caterpillar-plant*
Annuals with furrowed or ridged stems; leaves simple; fruits much curved, spiralled or variously contorted, longitudinally ridged, variously ornamented with spines and tubercles, beaked, indehiscent or tardily dehiscent.

1. S. muricatus L. (*S. sulcatus* L., *S. subvillosus* L.) - *Caterpillar-plant*. Stems procumbent to suberect, to 80cm; leaves obovate to oblanceolate; flowers (1)2-5, 5-12mm, yellow, often red-tinged; fruits with 3-10 segments; (2n=28). Intrd; birdseed- or wool-alien on tips, rough ground and in gardens and parks; C & S Br, mainly S En; S Europe. **403**

TRIBE 8 - FABEAE (Vicieae) (genera 19-22). Herbaceous annuals or perennials; leaves usually paripinnate, often with tendril(s) at apex, rarely simple or reduced to a tendril, with usually entire (rarely dentate) leaflets; flowers solitary or in axillary racemes; 9 stamens forming tube, 10th free; fruit dehiscing longitudinally, (1)2-many seeded.

19. VICIA L. - *Vetches*
Stem ± not winged (often ridged); leaves paripinnate with (1)2-many pairs of pinnately-veined leaflets, usually with terminal tendril(s); stipules smaller than leaflets; at least 2 calyx-teeth <2x as long as tube; style glabrous, or pubescent all round, or pubescent only on lowerside.

1 All leaves without tendrils, terminated by small point 2
1 At least upper leaves terminated by tendril(s) 4
2 Perennials; flowers ≥6; peduncles >3cm; leaflets >5 pairs **1. V. orobus**
2 Annuals; flowers ≤6; peduncles 0 or <2cm; leaflets <5 pairs 3
3 Flowers <1cm; leaflets <2cm; fruits <3cm **13. V. lathyroides**
3 Flowers >1cm; leaflets >2cm; fruit >5cm **17. V. faba**
4 Peduncle 0, or shorter than each flower 5
4 Peduncle longer than each flower 11
5 Standard pubescent on back **11. V. pannonica**
5 Standard glabrous on back 6
6 Flowers 6-9mm, solitary; seeds tuberculate **13. V. lathyroides**
6 Flowers 9-25(30)mm, 1-several; flowers rarely cleistogamous and <9mm but then seeds smooth 7
7 Leaflets 1-3 pairs; stipules c.1cm 8
7 At least some leaves with >3 pairs leaflets; stipules <8mm 9
8 Fruit pubescent all over, ≤1cm wide; leaflets >2x as long as wide **15. V. bithynica**
8 Fruit pubescent only along 2 sutures, ≥1cm wide; leaflets <2x as long as wide **16. V. narbonensis**
9 Perennial; seeds with hilum >1/2 total circumference; lower calyx-teeth longer than upper but shorter than tube **10. V. sepium**
9 Annual; seeds with hilum <1/2 total circumference; all calyx-teeth equal, or unequal and lower longer than tube 10
10 Calyx-teeth unequal (the lower longer); seeds with hilum 1/3-1/2 total circumference; corolla usually yellow **14. V. lutea**
10 Calyx-teeth ± equal; seeds with hilum 1/4-1/3 total circumference; corolla usually pink to purple **12. V. sativa**
11 Flowers 2-8(9)mm, white to purple (not blue), 1-8 per raceme 12
11 Flowers 8-20mm, if <10mm then blue and >8 per raceme 14
12 Calyx-teeth equal, all at least as long as tube; fruits usually 2-seeded **7. V. hirsuta**
12 Calyx-teeth unequal, at least the upper shorter than tube; fruits 3-6-seeded 13

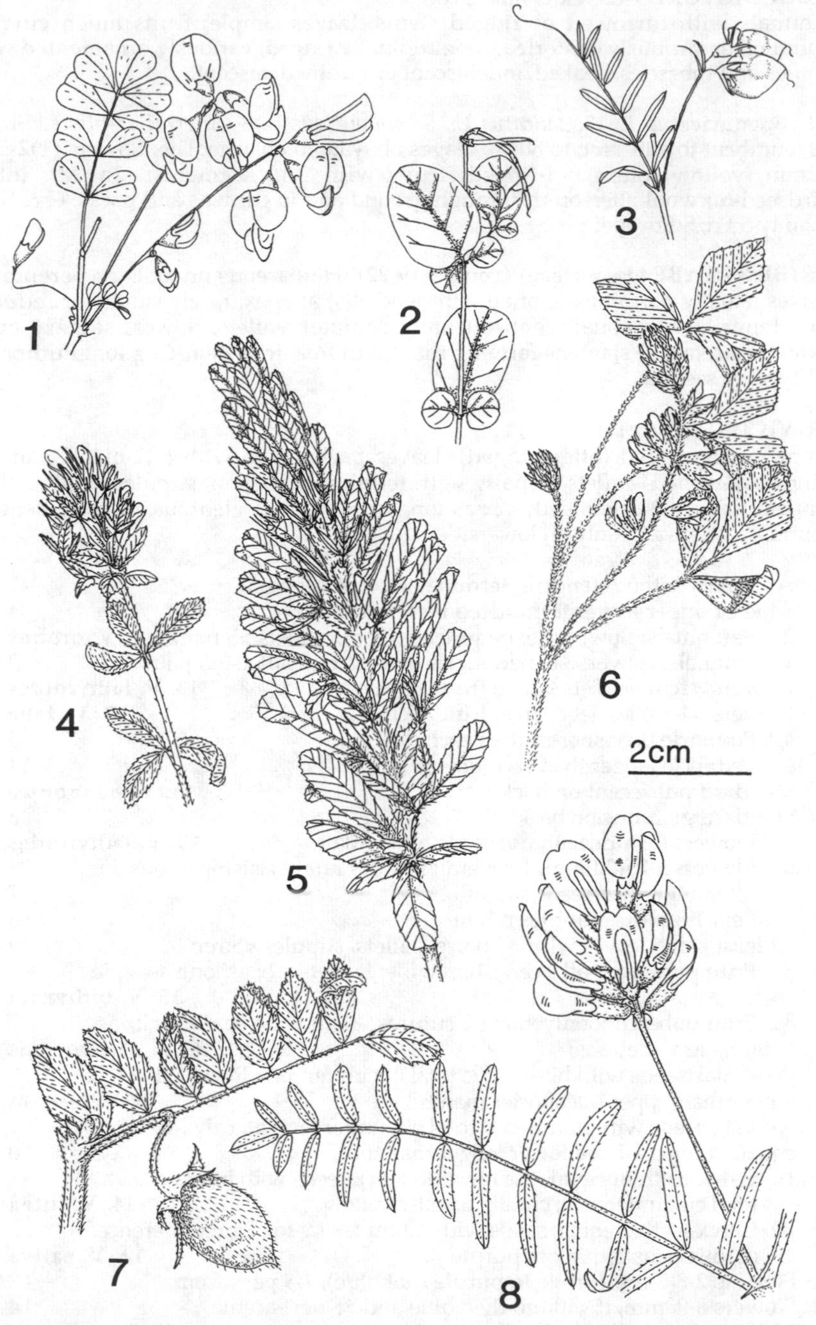

FIG 410 - Fabaceae. 1, *Hippocrepis emerus*. 2, *Coronilla scorpioides*. 3, *Lens culinaris*. 4, *Ononis mitissima*. 5, *O. alopecuroides*. 6, *Psoralea americana*. 7, *Cicer arietinum*. 8, *Astragalus cicer*.

13 Flowers 1-2; seeds (3)4(-5), with hilum >2x as long as wide and c.1/5 seed circumference **9. V. tetrasperma**
13 Flowers 1-4(5); seeds 4-6(8), with hilum little longer than wide and <1/3 seed circumference **8. V. parviflora**
14 Leaves with 2-3 pairs leaflets; flowers 1-3 **15. V. bithynica**
14 Leaves with ≥4 pairs leaflets; flowers usually >4 15
15 Standard with limb c.1/2 as long as claw; calyx very asymmetrical at base, with large bulge on upperside 16
15 Standard with limb c. as long as or longer than claw; calyx only slightly asymmetrical at base 17
16 Corolla reddish-purple with blackish tip **6. V. benghalensis**
16 Corolla blue to violet or purple, sometimes with white or yellow wings **5. V. villosa**
17 Lower lobe of stipules strongly toothed; corolla white with blue or purple veins; seeds with hilum >1/2 total circumference **4. V. sylvatica**
17 Lower lobe of stipules entire; corolla blue to purple or violet; seeds with hilum <1/2 total circumference 18
18 Corolla 8-12(13)mm; limb of standard c. as long as claw; seeds with hilum 1/4-1/3 total circumference **2. V. cracca**
18 Corolla (10)12-18mm; limb of standard longer than claw; seeds with hilum 1/5-1/4 total circumference **3. V. tenuifolia**

Other spp. - **V. cassubica** L. (*Danzig Vetch*), from Europe, resembles *V. cracca* but has fruits 6-8mm wide (not 4-6mm) and with 1-3 seeds; it was natd in W Kent from 1931 to 1964. **V. hybrida** L. (*Hairy Yellow-vetch*), from S Europe, resembles *V. pannonica* but has unequal (not subequal) calyx-teeth, hilum c.1mm (not c.2mm), and yellow corolla 18-30mm; it was formerly natd in N Somerset. Both are still rare casuals, as is **V. monantha** Retz., from Mediterranean; this resembles *V. villosa* with only 1-2 purple flowers but has deeply 2-lobed stipules with entire lobes and yellow (not brown) fruits.

1. V. orobus DC. - *Wood Bitter-vetch*. Perennial with erect stems to 60cm; leaflets 6-15 pairs; tendrils 0; flowers 6-20, 12-15(20)mm, white with purple veins; fruits 20-30mm, with 4-5 seeds; 2n=12. Native; grassy and rocky places and scrub; scattered through W En, Sc, Wa and C & N Ir.

2. V. cracca L. - *Tufted Vetch*. Scrambling or climbing perennial to 2m; leaflets 5-15 pairs; tendrils branched; flowers 10-30(40), 8-12(13)mm, bluish-violet; fruits 10-25mm, 2-6(8)-seeded; 2n=28. Native; grassy and bushy places and hedgerows; common throughout BI.

3. V. tenuifolia Roth - *Fine-leaved Vetch*. Differs from *V. cracca* in flowers (10)12-18mm, bluish-lilac to purple; and see key; (2n=24). Intrd; natd in grassy places and rough ground; scattered throughout most of En and Sc, often only casual; Europe. Perhaps only a ssp. of *V. cracca*.

4. V. sylvatica L. - *Wood Vetch*. Scrambling or climbing perennial to 2m; leaflets 4-12 pairs; tendrils branched; flowers 4-15(20), 12-20mm, white with purple veins; fruits 25-30mm, 4-5-seeded; 2n=14. Native; open woods and wood-borders, scree, scrub, maritime cliffs and shingle; scattered throughout much of Br and Ir but local.

5. V. villosa Roth (*V. varia* Host, *V. dasycarpa* auct., ?Ten.) - *Fodder Vetch*. Scrambling or climbing annual to 2m; leaflets 4-12 pairs; tendrils branched; flowers 10-30, 10-20mm, blue to purple or violet, wings sometimes white or yellow; fruits 20-40mm, 2-8-seeded; (2n=14). Intrd; natd in grassy places, tips, rough and waste ground, more often casual; throughout much of Br; Europe. Several sspp., based on pubescence, flower number, size and colour, and calyx-teeth size, are often recognized; 5 have been recorded from Br but are of doubtful taxonomic value.

6. V. benghalensis L. - *Purple Vetch*. Differs from *V. villosa* in stems to 80cm; leaflets 5-9 pairs; flowers 2-20, reddish-purple with blackish tip; fruits 3-5-seeded;

(2n=12, 14). Intrd; occasional casual and sometimes briefly persisting on tips and rough and waste ground; sporadic since c.1910 in C & S Br; Mediterranean.

7. V. hirsuta (L.) Gray - *Hairy Tare*. Scrambling annual to 80cm; leaflets 4-10 pairs; tendrils branched; flowers (1)2-7(9), (2)3-5mm, whitish tinged purple; fruits 6-11mm, almost all 2-seeded; 2n=14. Native; rough ground and grassy places; throughout lowland BI but rare in N Sc.

8. V. parviflora Cav. (*V. laxiflora* Brot. nom. illeg., V. *tenuissima* auct. non (M. Bieb.) Schinz & Thell., *V. tetrasperma* ssp. *gracilis* Hook. f.) - *Slender Tare*. Scrambling annual to 60cm; leaflets 2-4(5) pairs; tendrils usually simple; flowers 1-4(5), (5)6-9mm, pale bluish-purple; fruits 12-17mm, 4-6(8)-seeded; (2n=14). Native; grassy places; local in S En N to Hunts, decreasing, 1 record (?intrd) in N Lincs. **R**

9. V. tetrasperma (L.) Schreb. - *Smooth Tare*. Differs from *V. parviflora* in leaflets 3-6(8) pairs; flowers 1-2(4), 4-8mm; fruits 9-16mm, (3)4(-5)-seeded; peduncles shorter than to as long as (not longer than) leaves; 2n=14. Native; grassy places, CI, En and Wa, very scattered and mostly intrd in Sc and Ir.

10. V. sepium L. - *Bush Vetch*. Climbing or sprawling perennial to 60(100)cm; leaflets (3)5-9 pairs; at least lower tendrils branched; flowers 2-6, 12-15mm, dull purple; fruits 20-35mm, 3-10-seeded; 2n=14. Native; grassy places, hedges, scrub and wood-borders, rarely sand-dunes in NW Sc and NW Ir; throughout BI.

11. V. pannonica Crantz - *Hungarian Vetch*. Climbing annual to 60cm; leaflets 4-19 pairs; tendrils branched; flowers 1-4, 14-22mm, dirty brownish-yellow; fruits 20-35mm, 2-8-seeded, appressed-pubescent; (2n=12). Intrd; frequent casual in waste places and roadside banks in En and Wa, natd in W Kent since 1971; Europe. Natd and most casual plants are ssp. **pannonica**. Ssp. **striata** (M. Bieb.) Nyman, with purplish flowers, is a rare casual.

12. V. sativa L. - *Common Vetch*. Climbing, sprawling or procumbent annual to 1.5m; leaflets 3-8 pairs; tendrils usually branched; flowers 1-2(4), variously pink to purple, rarely white or yellow; fruits 4-12-seeded, glabrous to sparsely pubescent.

1 Plant heterophyllous, the leaflets of upper leaves much (and abruptly) narrower than those of lower leaves; flowers ± concolorous, usually bright pinkish-purple **a. ssp. nigra**
1 Plant ± isophyllous, the leaflets of upper leaves little (and gradually) narrower than those of lower leaves; flowers usually bicolorous, the standard much paler than wings 2
2 Fruits smooth, usually glabrous, brown to black **b. ssp. segetalis**
2 Fruits slightly constricted between seeds, often pubescent, yellowish to brown **c. ssp. sativa**

a. Ssp. nigra (L.) Ehrh. (ssp. *angustifolia* (L.) Gaudin, *V. angustifolia* L. ssp. *angustifolia*). Plant slender, often procumbent, to 75cm, strongly heterophyllous; flowers 14-19mm, ± concolorous, usually bright pinkish-purple; fruits 23-38mm, brown to black, smooth, glabrous; 2n=12. Native; sandy banks, heathland, maritime sand and shingle; throughout most of BI, but often intrd inland.

b. Ssp. segetalis (Thuill.) Gaudin (*V. angustifolia* ssp. *segetalis* (Thuill.) Arcang.). Plant more robust, to 1m, ± isophyllous; flowers 9-26mm, usually bicolorous; fruits 28-70mm, as in ssp. *nigra*; 2n=12. Probably intrd; in grassy and rough places and field-borders (sometimes cultivated for fodder); commonest ssp. throughout most of BI; Europe.

c. Ssp. sativa. Very robust, to 1.5m, ± isophyllous; flowers 11-26(30)mm, bicolorous; fruits 36-70(80)mm, yellowish to brown, constricted between seeds, often pubescent; 2n=12. Intrd; uncommon casual, rarely persisting, in waste places, field-borders (formerly cultivated for fodder); very scattered in Br, formerly CI and Ir; Europe.

13. V. lathyroides L. - *Spring Vetch*. Procumbent to weakly climbing annual to 20cm; leaflets 2-4 pairs; tendrils simple or scarcely developed; flowers solitary, (5)6-9mm, dull purple; fruits 15-30mm, 6-12-seeded; 2n=10, 12. Native; maritime sand and inland sandy heaths; scattered over most of BI except W & S Ir and NW Sc. Possibly over-recorded for *V. sativa* ssp. *nigra*.

14. V. lutea L. (*V. laevigata* Sm.) - *Yellow-vetch*. Procumbent to sprawling annual R
to 60cm; leaflets 3-8 pairs; tendrils branched; flowers 1-2, 15-25(30)mm, pale dull yellow; fruits 20-40mm, 4-8(10)-seeded; (2n=14). Native; maritime shingle and cliffs; very scattered round coasts of CI and Br N to S (formerly C) Sc, mainly S En and CI, rather frequent casual inland in Br.

15. V. bithynica (L.) L. - *Bithynian Vetch*. Climbing or scrambling annual to 60cm; R
leaflets (1)2-3 pairs; tendrils branched; flowers 1-2-(3), on very short to long peduncles, 16-20mm, standard purple, wings and keel white or very pale; fruits 25-50mm, 4-8-seeded; 2n=14. Probably native; scrub, rough grassland and hedges; rare and decreasing by or near coast in CI and Br N to Wigtowns, mainly S En, also fairly frequently intrd (casual or natd).

16. V. narbonensis L. - *Narbonne Vetch*. Erect climbing annual to 60cm; leaflets 1-3 pairs, often dentate; tendrils branched, usually only on upper leaves; flowers 1-3(6), 10-30mm, purple; fruits 30-50(70)mm, 4-7-seeded; (2n=14). Intrd; rather frequent casual in waste land from grain; scattered in S Br; S Europe.

17. V. faba L. - *Broad Bean*. Erect annual to 1m; leaflets (1)2-3 pairs; tendrils 0, replaced by weak point; flowers 1-6, usually white with blackish wings; fruits up to 30cm, 4-8-seeded; (2n=12). Intrd; widely cultivated as crop and common casual in BI, mainly in S Br; origin uncertain.

20. LENS Mill. - *Lentil*

Stems scarcely winged, markedly ridged; leaves paripinnate, with 3-8 pairs of pinnately-veined leaflets; simple tendril present on upper leaves; stipules smaller than leaflets; calyx-teeth all >2x as long as tube; style pubescent on upperside.

1. L. culinaris Medik. - *Lentil*. Weakly climbing annual to 40cm; leaflets (3)5-8 410
pairs; flowers 1-3, white to pale mauve; fruits (10)12-14(16)mm, nearly as deep, strongly compressed, 1-2(3)-seeded; (2n=14). Intrd; fairly frequent grain-alien on tips and rough ground; scattered in S Br; SW Asia.

21. LATHYRUS L. - *Peas*

Stems angled or winged; leaves usually paripinnate with 1-many pairs of pinnately- or parallel-veined leaflets, sometimes reduced to a simple blade or a simple tendril; terminal tendril present or 0; stipules variable; calyx-teeth variable; style pubescent only on upperside.

1 Leaves reduced to single blade or tendril 2
1 Leaves with 1-many pairs of leaflets; terminal tendril present or 0 3
  2 Leaf reduced to single blade; stipules <3mm; flowers reddish **15. L. nissolia**
  2 Leaf reduced to simple tendril; stipules >3mm, leaf-like; flowers yellow **16. L. aphaca**
3 Leaves (±) all with >1 pair of leaflets 4
3 Leaves (±) all with 1 pair of leaflets 7
  4 Stem angled, not winged 5
  4 Stem, at least above, with wings ≥1/2 as wide as stem 6
5 Stipules ovate-triangular; procumbent plant of maritime shingle; leaves usually with tendrils **2. L. japonicus**
5 Stipules lanceolate; erect garden escape; leaves without tendrils **1. L. niger**
  6 Tendrils 0; fruit scarcely compressed **3. L.linifolius**
  6 Tendrils well developed; fruit strongly compressed **5. L. palustris**
7 Stem angled, not winged 8
7 Stem, at least above, with wings ≥1/2 as wide as stem 10
  8 Flowers yellow; leaflets acute; fruits strongly compressed **4. L. pratensis**

8 Flowers pink to purple; leaflets obtuse to rounded or retuse at apex; fruits scarcely compressed 9

9 Plant minutely pubescent; flowers >25mm **7. L. grandiflorus**

9 Plant glabrous to very sparsely pubescent; flowers <25mm **6. L. tuberosus**

10 Perennials; flowers 3-many 11

10 Easily uprooted annuals; flowers 1-3(4) 13

11 Stipules <1/2 as wide as stem; all calyx-teeth shorter than tube **8. L. sylvestris**

11 Stipules >1/2 as wide as stem (often as wide); lowest calyx-tooth as long as or longer than tube 12

12 Flowers 12-22mm; leaflets >4x as long as wide; ovules ≤15 per ovary **10. L. heterophyllus**

12 Flowers 15-30mm; leaflets <4x as long as wide; ovules ≥16 per ovary **9. L. latifolius**

13 Plant glabrous (or with glands on young fruits) 14

13 Plant pubescent at least on pedicels, calyx and fruits 15

14 Corolla white, pink or blue; fruit with 2 narrow wings along dorsal suture **12. L. sativus**

14 Corolla yellow; fruit not winged **13. L. annuus**

15 Flowers >20mm; leaflets ovate-oblong **11. L. odoratus**

15 Flowers <20mm; leaflets linear-oblong **14. L. hirsutus**

Other spp. - **L. vernus** (L.) Bernh. (*Spring Pea*), from Europe, is a garden perennial sometimes found as a relic; it would key out as *L. niger* but has stems ≤40cm, leaflets with acuminate (not obtuse) apex, and brown (not black) fruits. 3 annuals from Mediterranean are sometimes found as birdseed-aliens. **L. ochrus** (L.) DC. has 0-2 pairs of leaflets, very broadly winged petioles and pale yellow flowers. **L. cicera** L. and **L. inconspicuus** L. have 1 pair of narrow leaflets, narrowly winged petioles and purple flowers; in *L. cicera* the stem is winged and the corolla is bright reddish-purple, while in *L. inconspicuus* the stem is unwinged and the corolla is pale to dull purple.

1. L. niger (L.) Bernh. - *Black Pea.* Erect perennial to 80cm; stems not winged; leaflets 3-6(10) pairs, lanceolate to elliptic; tendrils 0; flowers 2-10, 10-15mm, purple becoming bluish; (2n=14). Intrd; garden escape natd in grassy, rocky and scrubby places; rare and very scattered in Sc and En; Europe.

2. L. japonicus Willd. (*L. maritimus* (L.) Bigelow) - *Sea Pea.* Procumbent perennial R
to 90cm; stems not winged; leaflets 2-5(6) pairs, elliptic; tendrils simple or branched, sometimes 0; flowers 2-10(15), (12)15-25mm, purple becoming blue; 2n=14. Native; maritime shingle or rarely sand, very local on coasts of Br from Scillies and E Kent to Shetland, frequent in SE En, Guernsey. Formerly decreasing, perhaps no longer so. Our plant is ssp. **maritimus** (L.) P.W. Ball.

3. L. linifolius (Reichard) Bässler (*L. montanus* Bernh.) - *Bitter-vetch.* Erect perennial to 40cm; stems winged; leaflets 2-4 pairs, linear to elliptic; tendrils 0; flowers 2-6, 10-16mm, reddish-purple turning blue; 2n=14. Native; wood-borders, hedgerows, scrub; throughout Br and Ir but absent from much of CE En and C Ir. Our plant is var. **montanus** (Bernh.) Baessler.

4. L. pratensis L. - *Meadow Vetchling.* Climbing perennial to 1.2m; stems not winged; leaflets 1 pair, linear-lanceolate to elliptic; tendrils simple or branched; flowers (2)5-12,10-18mm, yellow; 2n=14, 28. Native; grassy places and rough ground; common throughout BI.

5. L. palustris L. - *Marsh Pea.* Climbing perennial to 1.2m; stems winged; leaflets R
2-3(5) pairs, narrowly elliptic-oblong; tendrils branched; flowers 2-6(8), 12-20mm, purplish-blue; 2n=42. Native; fens and tall damp grassland; very locally scattered in En, Wa, SW Sc and Ir, mainly E Anglia, decreasing.

6. L. tuberosus L. - *Tuberous Pea.* Climbing perennial to 1.2m; stems not winged;

leaflets 1 pair, elliptic to narrowly obovate; tendrils branched; flowers 2-7, 12-20mm, reddish-purple; 2n=14. Intrd; cornfields, hedgerows and roadsides; natd in N Essex since 1859, scattered through Br N to C Sc but casual or shortly persisting only in most places; Europe.

7. L. grandiflorus Sm. (*L. tingitanus auct.* non L.) - *Two-flowered Everlasting-pea.* Climbing perennial to 2m; stems not winged; leaflets 1 pair, obovate; tendrils branched; flowers 1-3(4), 25-35mm, pinkish-purple; (2n=14). Intrd; persistent garden escape in hedges, waste ground near old gardens; scattered in Br N to C Sc, Alderney; C Mediterranean.

8. L. sylvestris L. - *Narrow-leaved Everlasting-pea.* Climbing or scrambling perennial to 2m; stems winged; leaflets 1 pair, linear to narrowly elliptic (>4x as long as wide); tendrils branched; flowers 3-8(12), 12-20mm, dull pinkish-purple; ovules 9-15 per ovary; (2n=14). Native; scrub, wood-borders, hedgerows, rough ground; scattered through Br N to C Sc, but intrd in many places, especially in N.

9. L. latifolius L. - *Broad-leaved Everlasting-pea.* Differs from *L. sylvestris* in stems to 3m; leaflets narrowly to broadly elliptic (<4x as long as wide); flowers 3-12, 15-30mm, brighter coloured; ovules 16-23 per ovary; and see key; 2n=14. Intrd; very persistent garden escape in hedges, roadsides and railway banks, and rough ground; scattered through Br N to C Sc; Europe.

10. L. heterophyllus L. - *Norfolk Everlasting-pea.* Differs from *L. sylvestris* in flowers 12-22mm, brighter coloured; and see key; (2n=14). Intrd; natd since 1949 in damp hollows of sand-dunes in W Norfolk, since 1980s in N Hants; Europe. On the Continent plants mostly have upper leaves with 2-3 pairs of leaflets, but our plants (var. **unijugus** W.D.J. Koch) have only 1 pair.

11. L. odoratus L. - *Sweet Pea.* Climbing annual to 2.5m; stems winged; leaflets 1 pair, ovate-oblong; tendrils branched; flowers 1-3(4), 20-35mm, white, pink or mauve to purple; (2n=14). Intrd; frequent casual on tips and in waste places; En and Wa, mostly S; S Italy.

12. L. sativus L. - *Indian Pea.* Scrambling or climbing annual to 1m; stems winged; leaflets 1 pair, linear; tendrils branched; flowers solitary, 12-24mm, white, pink or bluish; (2n=14). Intrd; fairly frequent on tips, mostly from birdseed; sporadic in S En; Mediterranean.

13. L. annuus L. - *Fodder Pea.* Scrambling or climbing annual to 1m; stems winged; leaflets 1 pair, linear; tendrils branched; flowers 1-3, 12-18mm, yellow to orange-yellow; (2n=14). Intrd; tips and waste places from birdseed and other sources; sporadic in S En; Mediterranean.

14. L. hirsutus L. - *Hairy Vetchling.* Scrambling or climbing annual to 1m; stems winged; leaflets 1 pair, linear to narrowly elliptic-oblong; tendrils branched; flowers 1-3, 8-15mm, purple with bluish wings; 2n=14. Intrd; grassy and rough ground and waste places, formerly natd but now perhaps only casual; very scattered in En and S Sc; Europe.

15. L. nissolia L. - *Grass Vetchling.* Erect or ascending annual to 90cm; stems not winged; leaves reduced to a simple grass-like blade; stipules to 2mm; tendrils 0; flowers 1-2, 8-18mm, crimson; 2n=14. Native; grassy places; local in En and S Wa N to N Lincs, rare casual elsewhere, frequent only in SE En.

16. L. aphaca L. - *Yellow Vetchling.* Scrambling annual to 40(100)cm; stems not R
winged; leaves on mature plants reduced to simple tendril with leaf-like ovate-hastate stipules 6-50mm; flowers solitary, 10-13mm, yellow; 2n=14. Doubtfully native; dry banks, grassy places and rough ground; locally natd in S & SE En and CI, casual in Br N to C Sc, Ir; Europe.

22. PISUM L. - *Garden Pea*

Stems not winged, ± terete; leaves paripinnate with 1-3 pairs of pinnately-veined leaflets, with terminal branched tendril; stipules larger than leaflets; calyx-teeth broad, ± leafy; style pubescent only on upperside, proximal part with reflexed margins.

1. P. sativum L. (*P. arvense* L.) - *Garden Pea*. Climbing or sprawling annual to 2m (usually much less); flowers 1-3, 15-35mm, white to purple; fruits 3-12cm, not compressed when ripe; (2n=14). Intrd; grown on field-scale and a common casual by roads and fields, in waste places and on tips; scattered throughout much of BI; S Europe. Purple-flowered plants (var. **arvense** (L.) Poiret) usually have smaller parts and are grown for fodder (*Field Pea*).

TRIBE 9 - CICEREAE (genus 23). Annuals; leaves imparipinnate, with sharply serrate leaflets; tendrils 0; flowers solitary, axillary; 9 stamens forming tube, 10th free; fruit dehiscing longitudinally, 1-2-seeded.

23. CICER L. - *Chick Pea*

1. C. arietinum L. - *Chick Pea*. Erect, glandular-pubescent annual to 35(60)cm; 410
leaflets 3-8 pairs, ovate to obovate; calyx-teeth subequal, c.2x as long as tube; corolla 10-12mm, white to pale purple; fruits 17-30mm, inflated; (2n=14, 16). Intrd; tips and waste places from seed imported as food; scattered in Br, mainly En; Mediterranean.

TRIBE 10 - TRIFOLIEAE (genera 24-28). Annual or perennial herbs or rarely shrubs; leaves ternate, main lateral veins of leaflets running whole way to often toothed margin; flowers solitary or in racemes (often greatly condensed); 9 stamens forming tube, 10th free, or all 10 forming tube; fruits indehiscent to longitudinally dehiscent, with 1-many seeds.

24. ONONIS L. - *Restharrows*

Flowers solitary or in terminal racemes; calyx with glandular hairs; all 10 stamens forming tube; fruits straight, 1-many seeded, dehiscing longitudinally.

1 Perennials with stems woody at least below, sometimes spiny 2
1 Herbaceous annuals, not spiny 4
2 Corolla yellow, often streaked red; fruits 12-25mm, with 4-10 seeds **1. O. natrix**
2 Corolla pink, rarely white; fruits 5-10mm, with 1-2(4) seeds 3
3 Stems equally hairy all round, procumbent to ascending; leaflets <3x as long as wide, obtuse to emarginate **4. O. repens**
3 Stems mainly hairy along 1 side or 2 opposite sides, ascending to erect; leaflets >3x as long as wide, acute to subacute **3. O. spinosa**
4 Fruits >8-seeded; flowers 5-10mm; pedicels >5mm **2. O. reclinata**
4 Fruits 2-3(6)-seeded; flowers 9-17mm; pedicels <2mm 5
5 Racemes not leafy, their lower bracts with 0-1 leaflet, borne on bare peduncles **7. O. baetica**
5 Racemes leafy, their lower bracts with 3 leaflets, borne on leafy peduncles 6
6 Calyx 6.5-9mm, with ovate-acuminate teeth; seeds 1.5-2mm, tuberculate **5. O. mitissima**
6 Calyx 10-12mm, with linear-lanceolate teeth; seeds 2-2.3mm, smooth **6. O. alopecuroides**

1. O. natrix L. - *Yellow Restharrow*. Dwarf shrub to 60cm, without spines; leaves mostly with 3 leaflets; flowers in lax leafy panicles, 6-15(20)mm, yellow, often with red streaks; fruits 12-25mm, 4-10-seeded; (2n=28, 32, 64). Intrd; natd since 1947 in rough ground in Berks; S & W Europe.

2. O. reclinata L. - *Small Restharrow*. Erect to procumbent annual to 15cm; leaves RRR
all ternate; flowers in loose leafy terminal racemes, 5-10mm, pink; fruits 8-14mm, pendent, 10-20-seeded; 2n=60. Native; barish sand or limestone; rare and very local in S Devon, Glam, Pembs, Wigtowns, Guernsey (extinct) and Alderney.

3. O. spinosa L. (*O. campestris* W.D.J. Koch, *O. repens* ssp. *spinosa* Greuter) - *Spiny*

Restharrow. Erect or ascending usually spiny shrub to 70cm; differs from *O. repens* in flowers 10-20mm; fruits 6-10mm; seeds 2(-4); and see key; 2n=30. Native; grassy places and rough ground on mostly well-drained soils; locally frequent in Br N to S Sc, mostly S & C En.

3 x 4. O. spinosa x O. repens = O. x pseudohircina Schur occurs with both parents in a few places from Cambs to Durham; it is intermediate in leaflet-shape, stem-pubescence and fruit-size, and fertile; 2n=c.48.

4. O. repens L. (*O. spinosa* ssp. *maritima* (Dumort.) P. Fourn.) - *Common Restharrow.* Rhizomatous perennial; stems procumbent to ascending, woody at base, to 60cm, with or without spines; leaves usually with 1 and 3 leaflets on same plant; flowers in loose leafy terminal racemes, (7)12-20mm, pink; fruits 5-8mm, 1-2-seeded; 2n=60. Native; rough grassy places on well-drained soils, especially coastal; locally common in BI except N & NW Sc. Small, densely pubescent to tomentose plants with flowers 7-12mm (not 12-20mm) growing on maritime sands in N & S Devon are referable to ssp. **maritima** (Gren. & Godr.) Asch. & Graebn.; their taxonomic status needs investigating.

5. O. mitissima L. - *Mediterranean Restharrow.* Erect to procumbent annual to **410**
60cm; leaves ternate; flowers in ± dense leafy terminal racemes; corolla 9-12mm, pink; lower bracts ternate, the upper ones with 1 leaflet; fruit 5-7mm, 2-3-seeded; (2n=30). Intrd; occasional birdseed-alien on tips; sporadic in S En; Mediterranean.

6. O. alopecuroides L. (*O. salzmanniana* Boiss. & Reuter, *O. baetica* auct. non **410**
Clemente) - *Salzmann's Restharrow.* Erect to procumbent annual to 75cm; leaves with 1 leaflet; flowers in ± dense leafy terminal racemes; corolla 13-15mm, pink; bracts with 3 leaflets; fruits 6-8mm, 2-3-seeded; (2n=30, 32). Intrd; frequent birdseed-alien on tips and in waste places; sporadic in S En; W Mediterranean.

7. O. baetica Clemente - *Andalucian Restharrow.* Erect to ascending annual to 50cm; leaves ternate; flowers in short ± leafless terminal racemes; calyx 4-8mm; corolla (7)10-17mm, pink; bracts with 0(-1) leaflet; fruits 5-7mm, with c.5 seeds. Intrd; occasional birdseed-alien on tips; sporadic in S En; W Mediterranean.

25. MELILOTUS Mill. - *Melilots*

Flowers in elongated racemes, calyx without glandular hairs; 9 stamens forming tube, 10th free; fruits straight, 1-2-seeded, indehiscent or very tardily dehiscent longitudinally.

1 Corolla white **2. M. albus**
1 Corolla yellow 2
  2 Fruits with strong concentric ridges; wings shorter than keel **5. M. sulcatus**
  2 Fruits with weak to strong transverse or reticulate ridges; wings as long as or longer than keel 3
3 Flowers 2-3.5mm; fruits <3mm **4. M. indicus**
3 Flowers >4mm; fruits >3mm 4
  4 Fruits >5mm, mostly 2-seeded, black when ripe, pubescent; keel ± equalling wings **1. M. altissimus**
  4 Fruits <5mm, mostly 1-seeded, brown when ripe, glabrous; keel shorter than wings **3. M. officinalis**

1. M. altissimus Thuill. *Tall Melilot.* Erect biennial or perennial to 1.5m; flowers **418**
5-7mm, yellow; standard, wings and keel ± equal; fruits 5-7mm, reticulately or transversely ridged, black when ripe, pubescent; (2n=16). Intrd; natd in open grassland and rough ground, casual in waste places; frequent in S & C En, scattered W to E & S Ir and N to C Sc; Europe.

2. M. albus Medik. - *White Melilot.* Erect annual or biennial to 1.5m; flowers 4- **418**
5mm, white; wings and keel ± equal, standard longer; fruits 3-5mm, reticulately or transversely ridged, brown when ripe, glabrous; (2n=16). Intrd; similar distribution

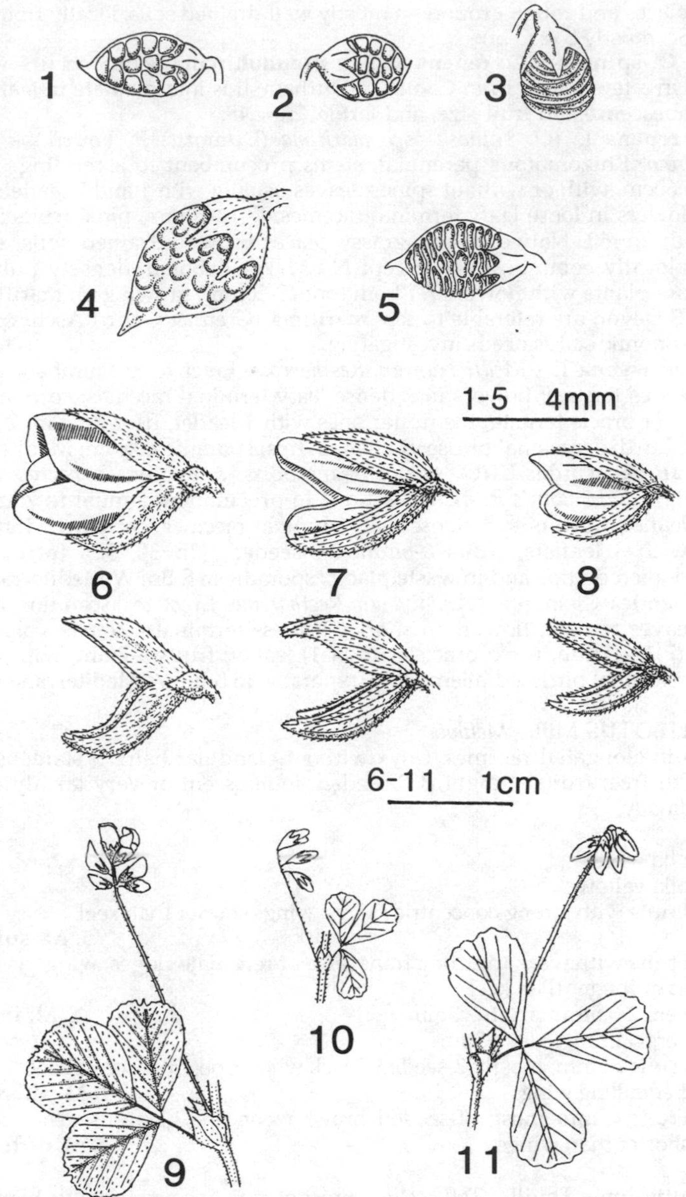

FIG 418 - Fabaceae. 1-5, fruits of *Melilotus*. 1, ***M. albus***. 2, ***M. indicus***. 3, ***M. sulcatus***. 4, ***M. altissimus***. 5, ***M. officinalis***. 6-8, flowers and calyces of ***Ulex***. 6, ***U. europaeus***. 7, ***U. gallii***. 8, ***U. minor***. 9-11, flowering nodes of ***Medicago*** and ***Trifolium***. 9, ***M. lupulina***. 10, ***T. micranthum***. 11, ***T. dubium***.

to *M. altissimus*, also CI; Europe.

3. M. officinalis (L.) Pall. - *Ribbed Melilot*. Erect to decumbent biennial to 1.5m; flowers 4-7mm, yellow; standard and wings ± equal, keel shorter; fruits 3-5mm, transversely ridged, brown when ripe, glabrous; (2n=16). Intrd; similar distribution to *M. altissimus*, also CI; Europe. **418**

4. M. indicus (L.) All. - *Small Melilot*. Erect to ascending annual to 40cm; flowers 2-3.5mm, yellow; wings and keel ± equal, standard longer; fruits 1.5-3mm, reticulately or transversely ridged, olive-green when ripe, glabrous; (2n=16). Intrd; rough ground and waste places, usually casual (often from birdseed or wool) but sometimes natd; scattered through BI N to C Sc, locally common in CI; S Europe. **418**

5. M. sulcatus Desf. - *Furrowed Melilot*. Usually erect annual to 40cm; flowers 3-4.5mm, yellow, standard and keel ± equal, wings much shorter; fruits 2.5-4mm, concentrically ridged, yellowish- or orangy-brown when ripe, glabrous; (2n=16). Intrd; fairly frequent birdseed-alien on tips and waste land; sporadic in S En; Mediterranean. **418**

26. TRIGONELLA L. - *Fenugreeks*

Annuals; flowers solitary in leaf-axils or in axillary racemes; calyx without glandular hairs; 9 stamens forming tube, 10th free; fruits straight to curved, (1)2-many-seeded, dehiscent longitudinally, often tardily.

1 Flowers >10mm, 1(-2) in leaf-axils; fruits >6cm, >9-seeded **3. T. foenum-graecum**

1 Flowers <8mm, in axillary racemes; fruits <2cm, <9-seeded 2

  2 Racemes elongated; corolla yellow; fruits >8mm, 4-8-seeded **1. T. corniculata**

  2 Racemes subcapitate; corolla white to blue; fruits <8mm, 1-3-seeded **2. T. caerulea**

Other spp. - T. procumbens (Besser) Rchb., from E Europe, is a rare grain-alien much confused with *T. caerulea*; it differs in its solid stems, narrower leaflets, more elongated racemes and fruit gradually (not abruptly) narrowed into beak.

1. T. corniculata (L.) L. - *Sickle-fruited Fenugreek*. Stems erect to procumbent, to 50cm; flowers yellow, in elongated *Melilotus*-like racemes, 5-7mm; fruits 10-18mm (excl. beak), curved, reflexed, linear, transversely ridged; (2n=16). Intrd; frequent birdseed-alien on tips and waste ground; S Br; Mediterranean. **422**

2. T. caerulea (L.) Ser. - *Blue Fenugreek*. Stems hollow, erect to decumbent, to 50cm; flowers white to blue, in short subcapitate racemes, 5-7mm; fruits 3.5-5mm (excl. beak), straight to slightly curved, patent, obovoid, mainly longitudinally ridged; (2n=16). Intrd; rather sparse wool-, grain- and birdseed-alien on tips and waste land, formerly commoner; very scattered in Br; E Mediterranean. **422**

3. T. foenum-graecum L. - *Fenugreek*. Stems erect to spreading, to 50cm; flowers yellowish-white, 1(-2) in leaf-axils, 12-18mm; fruits 50-110mm plus beak 10-35mm, straight or curved, patent, linear, mainly longitudinally ridged; (2n=16). Intrd; frequent birdseed- and spice-alien on tips and waste land, now grown on small scale for seedlings and seed; sporadic in C & S En; E Mediterranean. **422**

27. MEDICAGO L. - *Medicks*

Flowers 1-many in axillary racemes; calyx without glandular hairs; 9 stamens forming tube, 10th free; fruits slightly curved to spiral with several complete turns, 1-many-seeded, indehiscent, often spiny.

Spp. 3-8 have quite distinct fruits but the differences are difficult to describe; the illustrations should be consulted. Spine length is of virtually no value; in most spp. they can be very short to longer than the coil diameter, and rare spineless or tuberculate variants exist. The margin (outer edge) of each coil is occupied by a variously thickened vein or 'border'. Just inside this, on each face of the coil, is a variously thickened 'submarginal vein', and between these 2 thickened veins, in each face, is the (usually channelled) 'submarginal border'. The base of each spine straddles this border and originates from both the thickened veins. The 'face' of each coil between the coil centre and the submarginal border is often characteristically veined.

1 Fruits curved to spiralled, without spines or tubercles; if flowers yellow then fruits curved to spiralled in ≤1.5 turns 2
1 Fruits spiralled in >(1.5)2.5 turns, usually spiny, rarely tuberculate or smooth; flowers yellow 3
  2 Fruits <3mm in longest plane; flowers <4mm, yellow; seed 1 **1. M. lupulina**
  2 Fruits >4mm in longest plane; flowers >6mm, yellow, white, green or purplish; seeds >1 **2. M. sativa**
3 Leaflets usually each with dark blotch; fruit border grooved along centre, hence (with the 2 submarginal borders) forming 3 grooves at edge of each coil **8. M. arabica**
3 Leaflets not dark-blotched; fruit border not grooved, hence edge of each coil with 0 or 2 grooves 4
  4 Fruit border thinner than submarginal veins, scarcely contributing to origin of spines; spines ± not grooved at base; fruits with sparse long hairs **3. M. truncatula**
  4 Fruit border as thick or thicker than submarginal veins, contributing strongly to origin of spines; spines hence deeply grooved at least at base; fruits usually ± glabrous 5
5 Fruit border 0.5-0.8mm thick in lateral view, at least as thick as rest of each coil and ± obscuring it in lateral view **6. M. praecox**
5 Fruit border <0.4mm thick in lateral view, thinner than rest of each coil and not completely obscuring it in lateral view 6
  6 Stipules entire to denticulate, with teeth much shorter than entire part; leaves and stems ± densely pubescent **5. M. minima**
  6 Stipules deeply incised, with teeth much longer than entire part; leaves and stems glabrous to sparsely pubescent 7
7 Submarginal border broadly grooved, wider than border; fruit face with curved veins anastomosing to form reticulum adjacent to submarginal vein; wings longer than keel **7. M. polymorpha**
7 Submarginal border narrowly grooved, narrower than border; fruit face with sigmoid veins not or scarcely anastomosing before joining submarginal vein; keel longer than wings **4. M. laciniata**

Other spp. - 25-30 other Mediterranean spp., mainly spiny-fruited, have been found as aliens in wool and other sources, but all less commonly than the 6 treated here. The least rare of the rest are **M. intertexta** (L.) Mill., with large strongly spiny barrel-shaped fruits in 6-10 coils, and **M. orbicularis** (L.) Bartal., with distinctive smooth spineless discoid ('flying-saucer') fruits in 4-6 coils. 1 plant of **M. arborea** L. (*Tree Medick*), from Mediterranean, has survived on a cliff in N Somerset since 1973; it is a silvery-appressed-pubescent shrub to 1.5m with spineless fruits coiled in c.1 complete turn.

1. M. lupulina L. - *Black Medick*. Procumbent to scrambling annual or short-lived **418**
perennial to 80cm; leaves pubescent; stipules shallowly serrate; flowers numerous **422**
in compact short racemes, yellow, 2-3mm; fruits in c.1 coil, spineless, pubescent or glabrous, 1.5-3mm across, black when ripe, 1-seeded; 2n=16. Native; grassy places and rough ground; common throughout most of BI except parts of Sc and N Ir. See *Trifolium dubium* for differences.

2. M. sativa L. - see sspp. for English names. Erect to decumbent perennial to 90cm; leaves pubescent; stipules entire to shallowly serrate; flowers numerous in ± compact racemes, 5-12mm; fruits slightly curved (8-11mm long) to spiralled (5-9mm across) in 2-3(4) turns, spineless, pubescent or glabrous, 2-20-seeded.

1 Fruits spiralled in 2-3(4) complete turns; coils ± closed in centre; flowers mauve to violet **c. ssp. sativa**

1 Fruits curved or spiralled in ≤1.5 complete turns; coils ± open in centre; flowers yellow or white to purple or green or blackish 2

2 Fruits nearly straight to curved in arc ≤1/2 circle; flowers yellow **a. ssp. falcata**

2 Fruits curved or spiralled in 0.5-1.5 complete turns; flowers yellow or other colours **b. ssp. varia**

a. Ssp. falcata (L.) Arcang. (*M. falcata* L.) - *Sickle Medick*. Flowers yellow, 6-9mm; R
fruits nearly straight to curved in arc ≤1/2 circle; seeds 2-5; (2n=16, 32). Native; 422
grassy places and rough or waste ground; native locally in E Anglia, casual or sometimes natd elsewhere in Br scattered N to C Sc, mostly in SE En.

b. Ssp. varia (Martyn) Arcang. (*M. x varia* Martyn) - *Sand Lucerne*. Flowers variously yellow, pale mauve to purple, green or blackish, 7-10mm; fruits curved or spiralled in 0.5-1.5 complete turns; seeds 3-8, or abortive. Native; established or casual on sandy or rough ground, arising *in situ* or intrd as hybrid seed; scattered in Br N to C Sc, especially E Anglia, Dublin. Of hybrid origin from other 2 sspp., but deserves ssp. status due to its independent existence in many localities and use as a crop in Europe; partly fertile and backcrosses.

c. Ssp. sativa - *Lucerne*. Flowers pale mauve to violet, 8-11mm; fruits spiralled in 422
2-3(4) complete turns; seeds usually 10-20; 2n=32. Intrd; formerly much planted as crop, now less so, common relic of cultivation, on field-margins, roadsides, rough grassland and waste places; common in CI and S & C En, scattered throughout rest of BI except W and N Sc; Mediterranean.

3. M. truncatula Gaertn. (*M. tribuloides* Desr.) - *Strong-spined Medick*. Procumbent 422
to scrambling annual to 50cm, with long hairs; stipules deeply dentate to laciniate; flowers 1-3, 5-6mm, yellow; fruits of 3-6 coils, with sparse long hairs, with very stout ungrooved spines; (2n=16). Intrd; casual in waste places near origin of seed; rather scarce wool-alien, common tan-bark alien; scattered and sporadic in Br; Mediterranean.

4. M. laciniata (L.) Mill. (*M. aschersoniana* Urb.) - *Tattered Medick*. Procumbent 422
annual to 40cm, sparsely pubescent; stipules laciniate; flowers 1-2, 3.5-5mm, yellow; fruits of 3-7 coils, usually ± glabrous, usually with moderate spines grooved near base; (2n=16). Intrd; tips and rough ground, common wool-alien and from other sources, rarely persisting; sporadic throughout most of Br; N Africa.

5. M. minima (L.) Bartal. - *Bur Medick*. Procumbent annual to 20(40)cm, densely R
pubescent; stipules entire to denticulate; flowers 1-6, 2.5-4.5mm, yellow; fruits of 422
3-5 coils, often sparsely pubescent, usually with slender spines grooved near base; (2n=16). Native; sandy heaths and dunes and shingle by sea; very local in E En from Kent to Norfolk, Jersey; common casual throughout most of Br, often as wool-alien.

6. M. praecox DC. - *Early Medick*. Procumbent annual to 20cm, sparsely 422
pubescent; stipules laciniate; flowers 1-2, 2-5mm, yellow; fruits of 2.5-5 coils, usually ± glabrous, with moderate spines grooved near base; (2n=16). Intrd; casual on tips and rough ground, mostly from wool; sporadic throughout most of Br; Mediterranean.

7. M. polymorpha L. (*M. hispida* Gaertn. nom. illeg., *M. nigra* (L.) Krock.) - R
Toothed Medick. Procumbent to scrambling annual to 60cm, ± glabrous when 422
mature; stipules laciniate; flowers 1-8, 3-4.5mm, yellow; fruits of 1.5-5(6) coils, glabrous, usually with rather slender spines grooved near base; (2n=14, 16). Native; very local in sandy ground near sea; CI, S En N to S Somerset and W Norfolk; common casual throughout BI in waste places, often in wool. The sp. most

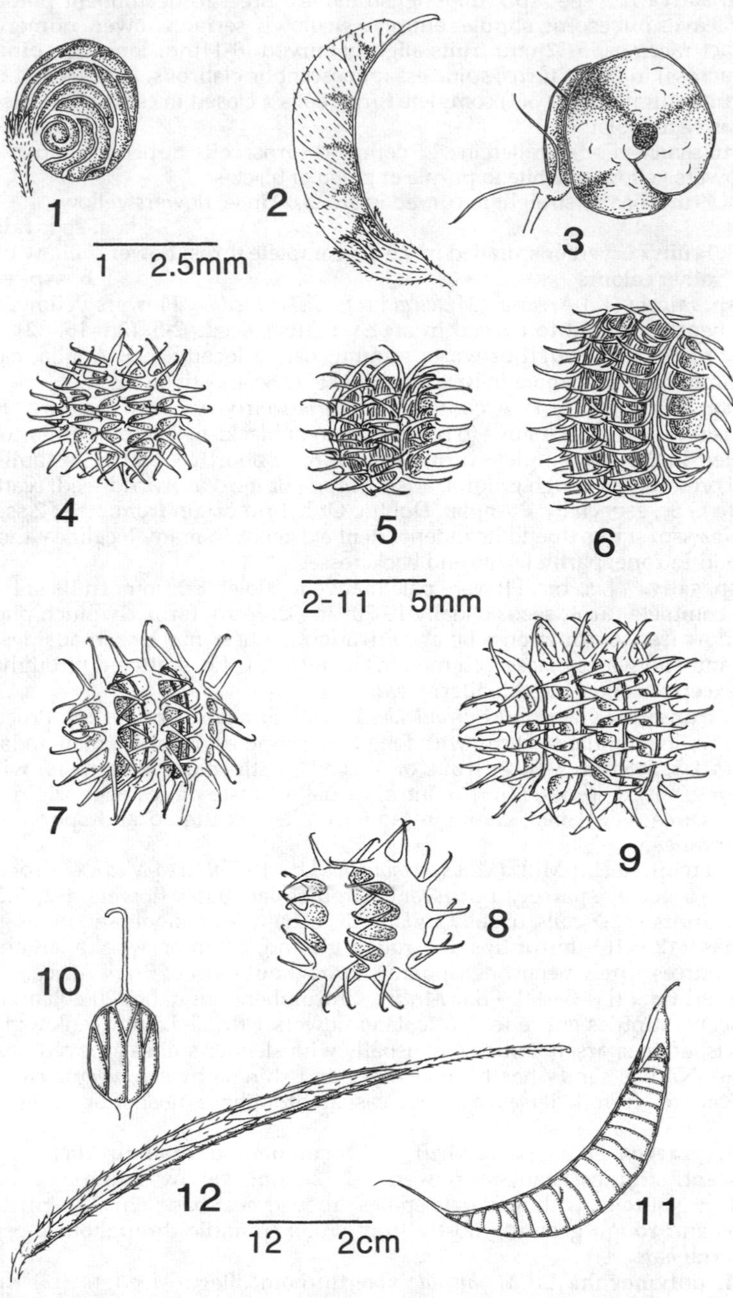

FIG 422 - Fruits of **Fabaceae**. 1, *Medicago lupulina*. 2-3, *M. sativa*. 2, ssp. *falcata*. 3, ssp. *sativa*. 4, *M. laciniata*. 5, *M. minima*. 6, *M. arabica*. 7, *M. polymorpha*. 8, *M. praecox*. 9, *M. truncatula*. 10, *Trigonella caerulea*. 11, *T. corniculata*. 12, *T. foenum-graecum*.

often lacking fruit spines.

8. M. arabica (L.) Huds. - *Spotted Medick.* Procumbent to scrambling annual to 60cm, ± glabrous when mature; stipules dentate usually <1/2 way to midrib; flowers 1-5, 4-6mm, yellow; fruits of 3-5(6) coils, glabrous, usually with rather slender spines deeply grooved >1/2 way; 2n=16. Native; grassy and barish places, especially near sea; CI, S & C Br; throughout BI in waste places as common casual, often in wool. **422**

28. TRIFOLIUM L. (see sections for synonyms) - *Clovers*

Flowers 1-many in axillary congested racemes; calyx without glandular hairs; 9 stamens forming tube, 10th free; fruits indehiscent, longitudinally dehiscent, or dehiscent by apex falling off, ± straight, 1-9-seeded, often partly or wholly enclosed in calyx or persistent corolla.

The keys require flowering and fruiting racemes, usually both present for a long period in the season. Calyx-tube vein-number is important; in spp. with densely pubescent calyx-tubes it is best observed by splitting the tube up one side and observing the veins from the inside.

General key

1 Racemes with sterile corolla-less flowers mixed with normal ones, becoming turned down and thrust into ground as fruit ripens **32. T. subterraneum**
1 Racemes wholly of sexual flowers with corollas, not becoming subterranean 2
  2 Calyx becoming greatly inflated in fruit, the inflation confined to upper lip with 2 teeth *Key A*
  2 Calyx not becoming inflated in fruit, or only moderately and symmetrically so 3
3 Calyx-tube with 5 veins; corolla yellow, ≤8mm; fruit stalked, 1-seeded *Key B*
3 Calyx-tube with (5)10-20 veins; corolla mostly white or pink to purple, if yellow then >8mm; fruit sessile, 1-9-seeded 4
  4 Throat of calyx (remove corolla) at least partly closed by a thickening or a ring of hairs; flowers without bracts ± sessile *Key C*
  4 Throat of calyx glabrous and open; each flower with small bract at base, or the bracts variably fused; flowers sessile or distinctly pedicellate *Key D*

Key A - Calyx asymmetrically strongly inflated in fruit (sect. *Vesicaria*)

1 Perennials; stems rooting at nodes; flowers with standard uppermost **9. T. fragiferum**
1 Annuals; stems not rooting at nodes; corolla upside-down so that standard is lowermost 2
  2 Fruiting calyx densely woolly, with calyx-teeth ± completely obscured; leaflets 4-12mm **11. T. tomentosum**
  2 Fruiting calyx sparsely to densely pubescent, with 2 prominent divergent calyx-teeth at apex; leaflets 10-20(25)mm **10. T. resupinatum**

Key B - Calyx not inflated or moderately and symmetrically so, with 5 veins; corolla yellow, <1cm (sect. *Chronosemium*)

1 Apical leaflet with stalk >0.5mm, distinctly longer than that of lateral leaflets 2
1 All leaflets with stalks <0.5mm 3
  2 Corolla 3-4mm, with standard folded longitudinally over fruit; racemes mostly <25-flowered **14. T. dubium**
  2 Corolla 4-7mm, with standard ± flat over fruit; racemes mostly

>25-flowered **13. T. campestre**
3 Corolla 5-8mm; pedicels c.1mm; racemes mostly >20-flowered **12. T. aureum**
3 Corolla 1.5-3mm; pedicels c.1.5mm; racemes <10-flowered **15. T. micranthum**

Key C - Calyx not inflated or moderately and symmetrically so, with 10-20 veins, its throat ± closed by thickening or hairs; bracts 0 (sect. *Trifolium*)

1 Firmly rooted perennials with habit of *T. pratense* or rhizomatous; flowers ≥12mm 2
1 Easily uprooted annuals; flowers often <12mm 5
2 Free part of stipules of stem-leaves triangular-ovate, abruptly narrowed to brown bristle-like point **16. T. pratense**
2 Free part of stipules of stem-leaves linear to lanceolate, green ± to apex 3
3 Corolla reddish-purple (rare albinos); lowest calyx-lobe c.1.5x as long as upper lobes and as calyx-tube **17. T. medium**
3 Corolla whitish-yellow, (?sometimes pale pink); lowest calyx-lobe ≥2x as long as upper lobes and as calyx-tube 4
4 Flowers 15-18mm; leaflets elliptic to oblong or obovate; racemes ± subsessile **18. T. ochroleucon**
4 Flowers 20-25mm; leaflets elliptic to linear-oblong; some racemes pedunculate **19. T. pannonicum**
5 Calyx-tube with c.20 veins 6
5 Calyx-tube with 10 veins 7
6 Calyx-tube with many long hairs; corolla >10mm **25. T. hirtum**
6 Calyx-tube without long hairs, usually glabrous; corolla <10mm **26. T. lappaceum**
7 Free part of stipules of stem-leaves ovate-triangular, abruptly narrowed to obtuse, acute or shortly apiculate apex 8
7 Free part of stipules of stem-leaves gradually tapered to long narrow apex 9
8 Racemes ± globose; stipules sharply serrate **20. T. stellatum**
8 Racemes cylindrical; stipules obscurely denticulate **21. T. incarnatum**
9 Racemes ± cylindrical in fruit, >2x as long as wide 10
9 Racemes globose to ovoid in fruit <1.5(2)x as long as wide 11
10 Flowers 10-13mm; calyx-lobes unequal, the lowest much the longest **28. T. angustifolium**
10 Flowers 3-6mm; calyx-lobes subequal **27. T. arvense**
11 At least the 2 uppermost leaves on each stem-branch exactly opposite; flowers (5)7-12mm 12
11 All leaves alternate (sometimes 2 uppermost close together, but not exactly opposite); flowers 4-7mm 14
12 Calyx with ring of hairs in throat; fruit exserted from calyx-tube **30. T. alexandrinum**
12 Calyx with bilateral swellings in throat, completely closed; fruit enclosed in calyx-tube 13
13 Calyx-lobes with 1 vein, or with 3 veins just at base **31. T. echinatum**
13 Calyx-lobes with 3 veins for >1/2 length **29. T. squamosum**
14 Leaflets with lateral veins thickened and arched-recurved at leaf-margin **24. T. scabrum**
14 Leaflets with lateral veins thin and straight or slightly forward-curved at leaf-margin 15
15 Leaflets ± glabrous on upperside; fruiting calyx-tube not inflated **23. T. bocconei**
15 Leaflets pubescent on upperside; fruiting calyx-tube inflated **22. T. striatum**

Key D - Calyx not inflated or slightly and symmetrically so, with (5)10(-12) veins, its throat open; bracts present (sects. *Lotoidea* & *Paramesus*)

1 Flowers 1-4(5), with pedicels ≥1mm; seeds 5-9 **1. T. ornithopodioides**

1 Flowers (3)10-many, if <10 then ± sessile; seeds 1-4 2

2 Flowers ± sessile, not reflexed after pollination; fruit shorter than calyx (incl. lobes) 3

2 Flowers with pedicels ≥1mm, strongly reflexed after pollination; fruit longer than calyx (incl. lobes) 5

3 Stipules serrate; teeth of leaflets and stipules gland-tipped **8. T. strictum**

3 Stipules usually entire, sometimes serrate; teeth of stipules and leaflets not gland-tipped 4

4 Corolla 3-4mm, shorter than calyx (incl. lobes), whitish; many (often ± all) racemes congested at base of plant **7. T. suffocatum**

4 Corolla 4-7mm, longer than calyx (incl. lobes), purplish; racemes dispersed along stems **6. T. glomeratum**

5 Flowers 4-5mm; upper racemes with peduncles <1cm, often subsessile; annual **5. T. cernuum**

5 Flowers (5)7-13mm; all racemes with peduncles >1cm; perennial 6

6 Stems not rooting at nodes, usually erect to ascending **4. T. hybridum**

6 Stems procumbent, rooting at nodes 7

7 Leaflets usually >10mm, obovate, often with light or dark markings, with veins translucent when fresh; petioles glabrous; calyx-lobes triangular-lanceolate; standard rounded at apex **2. T. repens**

7 Leaflets usually <10mm, suborbicular, without light or dark markings, with veins not translucent when fresh; petioles sparsely pubescent; calyx-lobes triangular-ovate; standard emarginate at apex **3. T. occidentale**

Other spp. - T. spumosum L., from Mediterranean, is a small annual with pink flowers and a much inflated calyx, but the latter is glabrous; it is a rare casual from grain and grass-seed.

Section **1** - *LOTOIDEA* Crantz (sect. *Ornithopoda* (Mall.) Tutin nom. inval., sect. *Trifoliastrum* Gray, *Amoria* C. Presl, *Falcatula* Brot.) (spp. 1-7). Flowers all fertile, not becoming subterranean; calyx not becoming inflated in fruit, with (5)10(-12) veins, glabrous and open at throat; fruits sessile, 1-9-seeded; bracts present; leaves without glands.

1. T. ornithopodioides L. (*Falcatula ornithopodioides* (L.) Brot. ex Bab.) - *Bird's-foot Clover*. Procumbent ± glabrous annual to 20cm; racemes 1-4(5)-flowered, stalked, axillary; flowers 6-8mm, pink to white or both; fruits exserted, 5-9-seeded; 2n=18. Native; sandy semi-open ground, mainly near sea; scattered round coasts of BI N to Norfolk, S Lancs and Co Dublin, also Man and Midlothian. **428**

2. T. repens L. (*Amoria repens* (L.) C. Presl) - *White Clover*. Subglabrous perennial with procumbent stems to 50cm rooting at nodes; racemes ± globose, on erect axillary peduncles up to 20cm; flowers 7-12mm, usually white, sometimes pale pink, rarely red or mauve; fruits exserted, 3-4-seeded; flowers scented when fresh; 2n=32. Native; grassy and rough ground; common throughout BI. **428**

3. T. occidentale Coombe (?*T. prostratum* Biasol.) - *Western Clover*. Differs from *T. repens* in less robust; leaflets thicker; flowers not scented; and see key; 2n=16. Native; short turf by sea; CI, SW En, S Wa, E Ir. **R** **428**

4. T. hybridum L. (*Amoria hybrida* (L.) C. Presl) - *Alsike Clover*. Erect to decumbent subglabrous perennial to 40(70)cm; racemes ± globose, mostly axillary, stalked; flowers 7-10mm, usually white and pink, sometimes either one; fruits slightly exserted, 2-4-seeded; (2n=16). Intrd; commonly natd in grassy and rough ground; throughout most of BI; Europe. Ssp. **hybridum** is the cultivated plant. Ssp. **elegans** (Savi) Asch. & Graebn., with solid (not hollow), less erect stems and **428**

smaller racemes (<2cm across) is probably the wild progenitor, but doubtfully a ssp.

5. T. cernuum Brot. - *Nodding Clover*. Procumbent to ascending subglabrous 428
annual to 25cm; racemes ± globose, mostly axillary, the lower stalked, the upper subsessile; flowers 3-5mm, pink; fruit exserted, 1-4 seeded; (2n=16). Intrd; casual from wool and other sources, sometimes persisting for few years; sporadic in C & S En; SW Europe.

6. T. glomeratum L. (*Amoria glomerata* (L.) Soják) - *Clustered Clover*. Procumbent R
to ascending subglabrous annual to 25cm; racemes ± globose, terminal and axillary, 428
sessile; flowers 4-5mm, purple; fruit enclosed in calyx, (1-)2-seeded; 2n=16. Native; grassy places on sandy soil mostly near sea; CI, S & E coast of En N to Norfolk, formerly commoner, rare casual elsewhere in En.

7. T. suffocatum L. (*Amoria suffocata* (L.) Soják) - *Suffocated Clover*. Procumbent R
subglabrous annual to 3(8)cm; racemes ± globose, terminal and axillary, sessile, 428
most or all densely crowded near root; flowers 3-4mm, whitish; fruit ± enclosed in calyx, 2-seeded; 2n=16. Native; similar places and distribution to *T. glomeratum*, but N to SE Yorks and Man.

Section 2 - PARAMESUS (C. Presl) Endl. (sect. *Involucrarium* auct. non Hook., *Paramesus* C. Presl) (sp. 8). Flowers all fertile, not becoming subterranean; calyx slightly and symmetrically swollen in fruit, with 10 veins, glabrous and open at throat; fruits sessile, 2-seeded; bracts present; leaf-teeth gland-tipped.

8. T. strictum L. (*Paramesus strictus* (L.) C. Presl) - *Upright Clover*. Erect to RR
ascending glabrous annual to 15(25)cm; racemes ± globose, terminal and axillary, 428
stalked; flowers 5-7mm, pinkish-purple; fruit ± enclosed in calyx, 2-seeded; (2n=16). Native; rocky grassy places, extremely local; Jersey, W Cornwall, Rads, formerly Guernsey, casual in 1994 in Man.

Section 3 - VESICARIA Crantz (sect. *Vesicastrum* Ser., sect. *Fragifera* W.D.J. Koch, *Galearia* C. Presl) (spp. 9-11). Flowers all fertile, not becoming subterranean; calyx becoming asymmetrically much inflated in fruit, with reticulate network of veins at fruiting, glabrous and open at throat at flowering but becoming narrowed in fruit; fruits sessile, 1-2-seeded; bracts present; leaves without glands.

9. T. fragiferum L. (*Galearia fragifera* (L.) C. Presl) - *Strawberry Clover*. Subglabrous 428
perennial with procumbent stems to 30cm rooting at nodes; racemes ± globose, on erect axillary peduncles up to 10(20)cm; flowers 5-7mm, pink; fruits enclosed in pubescent calyx, (1-)2-seeded; 2n=16. Native; grassy places, often on heavy or brackish soils; rather scattered but locally common in BI N to S Sc. Vegetatively resembles *T. repens* but lateral veins of leaflets are thickened and recurved distally almost as in *T. scabrum* (not so in *T. repens*), and stipules are gradually narrowed, not abruptly narrowed, to apex. Ssp. **fragiferum** is the common plant. Ssp. **bonannii** (C. Presl) Soják occurs in S En; it is said to have smaller calyces (4-6mm, not 8-10mm, in fruit) with exserted corollas and slightly elongated larger racemes, but its status is very doubtful.

10. T. resupinatum L. (*Galearia resupinata* (L.) C. Presl) - *Reversed Clover*. 428
Glabrous, procumbent to suberect annual to 30cm; racemes ± globose, terminal and axillary, stalked; flowers 5-8mm, pink to purple; fruits enclosed in pubescent calyx, 1-seeded; (2n=14, 16, 32). Intrd; rather frequent casual from wool and other sources, sometimes persisting; mainly S Br; S Europe.

11. T. tomentosum L. - *Woolly Clover*. Differs from *T. resupinatum* in stems 428
procumbent, to 15cm; racemes subsessile; flowers 3-6mm; fruits 1-2-seeded; and see key; (2n=16). Intrd; casual from wool and other sources; sporadic mainly in S Br; Mediterranean.

Section* 4 - *CHRONOSEMIUM Ser. (*Chrysaspis* Desv.) (spp. 12-15). Flowers all fertile, not becoming subterranean; calyx not becoming inflated in fruit, with 5 veins, with glabrous and open throat; fruits stalked, exserted from calyx, protected by standard, 1-2-seeded; bracts 0 or minute; leaves without glands.

12. T. aureum Pollich (*Chrysaspis aurea* (Pollich) B.D. Greene) - *Large Trefoil.* Erect **428**
to ascending, sparsely pubescent annual to 30cm; racemes 12-20mm, globose to obovoid, terminal and axillary, stalked; flowers >20, 5-8mm, golden-yellow; fruits 1-seeded; (2n=14, 16). Intrd; natd in grassy and rough ground, formerly frequent, now very local; scattered throughout much of Br, but now well natd ± only in C & S Sc; Europe.

13. T. campestre Schreb. (*Chrysaspis campestris* (Schreb.) Desv.) - *Hop Trefoil.* **428**
Procumbent to suberect sparsely pubescent annual to 30cm; racemes 8-15mm, ± globose, terminal and axillary, stalked; flowers >20, 4-7mm, pale yellow; fruits 1-seeded; 2n=14. Native; grassy and barish places; frequent throughout most of BI.

14. T. dubium Sibth. (*Chrysaspis dubia* (Sibth.) Desv.) - *Lesser Trefoil.* Procumbent **418**
to suberect very sparsely pubescent annual to 25cm; racemes 5-9mm, ± globose, **428**
terminal and axillary, stalked; flowers (3)5-20, 3-4mm, yellow; fruits 1-seeded; pedicels c.1mm; 2n=32. Native; grassy and open ground; common ± throughout BI. Often confused with *Medicago lupulina* before fruiting, but has glabrous (not pubescent) calyx, subglabrous (not pubescent) leaflets without (not with) an apical apiculus, and usually deeper yellow flowers with persistent corolla.

15. T. micranthum Viv. (*Chrysaspis micrantha* (Viv.) Hendrych) - *Slender Trefoil.* **418**
Procumbent to ascending subglabrous annual to 15cm; differs from *T. dubium* in **428**
racemes c.4mm; flowers (1)2-6(10), 1.5-3mm; fruits 1-2-seeded; pedicels c.1.5mm; (2n=16). Native; short turf, especially close-cut lawns; common in CI and SE En, scattered to W Ir and S Sc, casual in C & N Sc.

Section* 5 - *TRIFOLIUM (spp. 16-31). Flowers all fertile, not becoming subterranean; calyx not becoming inflated in fruit or becoming only moderately and symmetrically so, with 10-20 veins, with throat wholly or partly closed by ring of hairs or by thickenings; fruit sessile, enclosed in calyx, 1-seeded; bracts 0; leaves without glands.

16. T. pratense L. - *Red Clover.* Decumbent to erect, pubescent, tufted perennial **429**
to 60cm; leaflets elliptic to obovate; racemes globose to ellipsoid, terminal, ± sessile; flowers 12-18mm, pinkish-purple, sometimes pale pink, white or cream; 2n=14. Native; grassy places, waste and rough ground; common throughout BI. Agricultural variants are more robust and have hollow stems and less denticulate leaflets; they have been separated as var. **sativum** Schreb.

17. T. medium L. - *Zigzag Clover.* Ascending, sparsely pubescent, rhizomatous **429**
perennial to 50cm; leaflets narrowly elliptic; racemes globose to obovoid, terminal, stalked at maturity; flowers 12-20mm, reddish-purple; 2n=80. Native; grassy places, hedgerows and wood-borders; frequent ± throughout BI. Easily told from *T. pratense* by the stipules (see key), rhizomes and narrower, less pubescent leaflets.

18. T. ochroleucon Huds. - *Sulphur Clover.* Ascending to erect, pubescent, tufted **R**
but shortly rhizomatous perennial to 50cm; leaflets elliptic to obovate; racemes **429**
globose to ellipsoid, terminal, shortly stalked or subsessile; flowers 15-20mm, whitish-yellow; 2n=16. Native; grassy places on heavy soils; very local and decreasing in E Anglia W to Northants, casual elsewhere in En.

19. T. pannonicum Jacq. - *Hungarian Clover.* Differs from *T. ochroleucon* in stems ± **429**
erect, to 60(80)cm; and see key; (2n=c.96, c.130, c.180). Intrd; rather scarce casual mainly as grass-seed contaminant, rarely ± natd, as in Bucks; sporadic in S En; SE Europe.

20. T. stellatum L. - *Starry Clover.* Erect pubescent annual to 20(30)cm; leaflets **429**
obovate to obcordate; racemes ± globose, terminal, stalked; flowers 12-18mm,

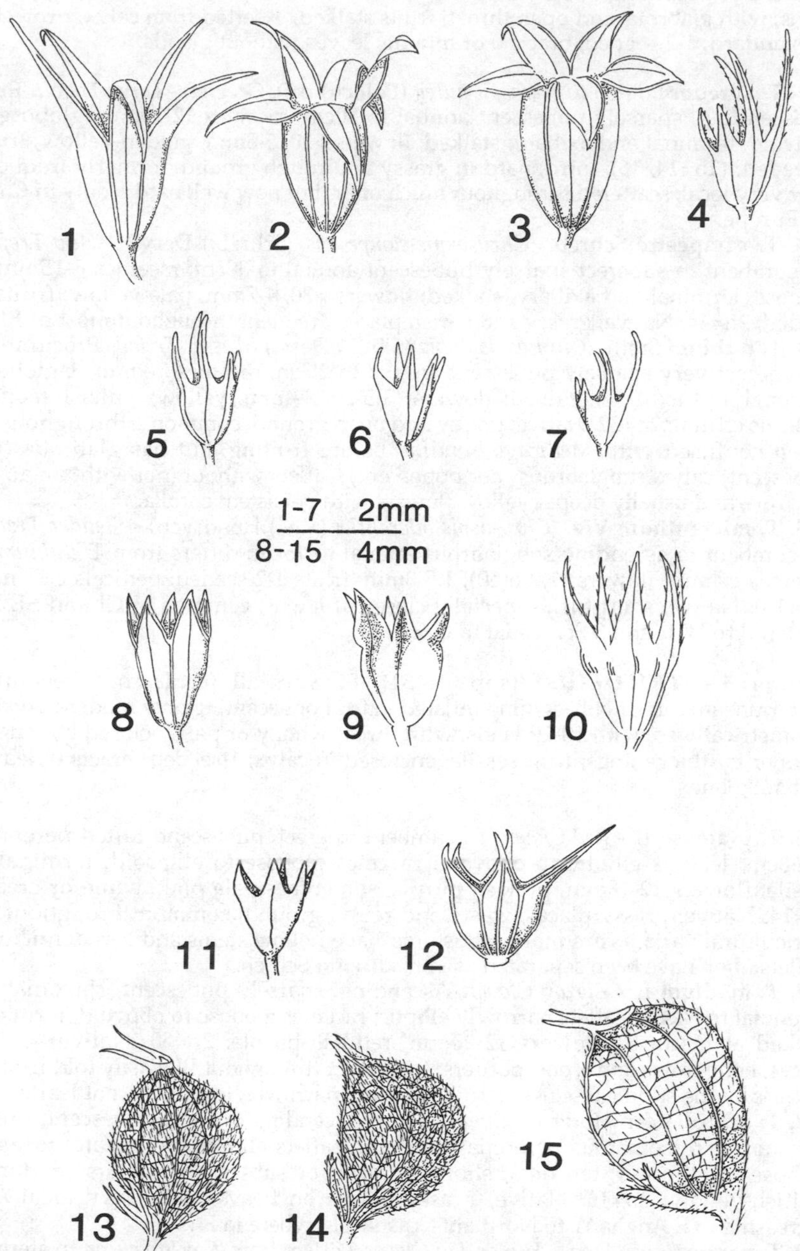

FIG 428 - Fruiting calyces of *Trifolium*. 1, *T. cernuum*. 2, *T. suffocatum*. 3, *T. glomeratum*. 4, *T.aureum*. 5, *T. dubium*. 6, *T. micranthum*. 7, *T. campestre*. 8, *T. repens*. 9, *T. occidentale*. 10, *T. ornithopodioides*. 11, *T. hybridum*. 12, *T. strictum*. 13, *T. resupinatum*. 14, *T. tomentosum*. 15, *T. fragiferum*.

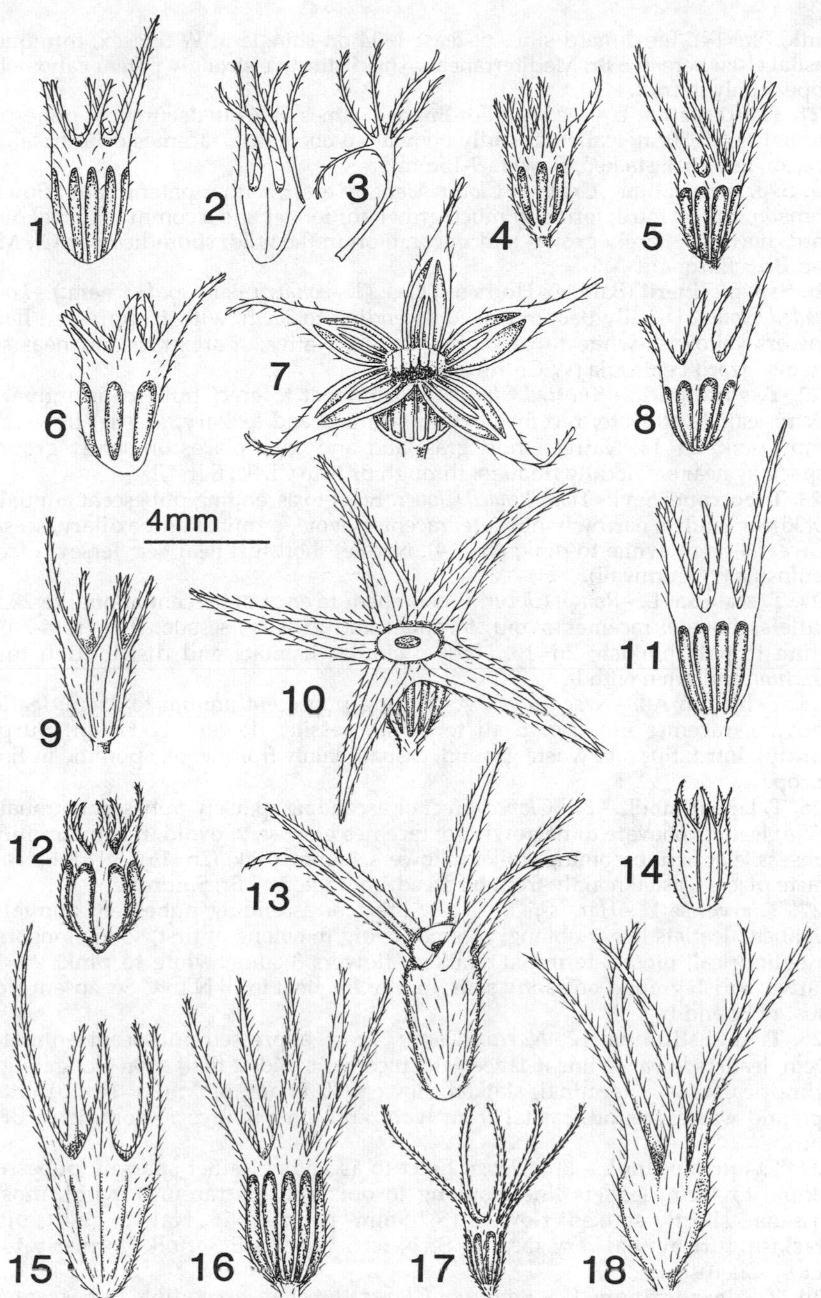

FIG 429 - Fruiting calyces of *Trifolium*. 1, *T. ochroleucon*. 2-3, *T. subterraneum* (fertile, sterile). 4, *T. scabrum*. 5, *T. alexandrinum*. 6, *T. squamosum*. 7, *T. lappaceum*. 8, *T. echinatum*. 9, *T. pratense*. 10, *T. stellatum*. 11, *T. medium*. 12, *T. striatum*. 13, *T. angustifolium*. 14, *T. bocconei*. 15, *T. pannonicum*. 16, *T. incarnatum*. 17, *T. arvense*. 18, *T. hirtum*.

pink; (2n=14). Intrd; natd since at least 1804 on shingle in W Sussex, infrequent casual elsewhere in S Br; Mediterranean. The distinctive strongly patent calyx-lobes appear only in fruit.

21. T. incarnatum L. - see sspp. for English names. Erect to decumbent pubescent 429
annual to (30)50cm; leaflets broadly obovate to obcordate; racemes cylindrical, up to 6cm, terminal, stalked; flowers 9-15mm.

a. Ssp. incarnatum - *Crimson Clover*. Usually erect, with ± patent hairs; flowers crimson; 2n=14. Intrd; formerly much grown for fodder and a common casual often natd, now very rarely grown and uncommon in the wild; sporadic over CI, Man and Br; S Europe.

b. Ssp. molinerii (Balb. ex Hornem.) Ces. (*T. molinerii* Balb. ex Hornem.) - *Long-* RR
headed Clover. Usually decumbent to ascending to 20cm, with ± appressed hairs; flowers yellowish-white to pale pink; 2n=14. Native; short grassland near sea; Jersey, Lizard Peninsula (W Cornwall).

22. T. striatum L. - *Knotted Clover*. Procumbent to erect pubescent annual to 429
30cm; leaflets obovate; racemes ovoid, terminal and axillary, sessile; flowers 4-7mm, pink; 2n=14. Native; short grassland and open places on sandy ground, especially near sea; locally frequent through Br N to CE Sc, E Ir, CI.

23. T. bocconei Savi - *Twin-headed Clover*. Erect to ascending pubescent annual to RR
20(30)cm; leaflets narrowly obovate; racemes ovoid, terminal and axillary, sessile; 429
flowers 4-6mm, white to pink; (2n=14). Native; short turf near sea; Jersey, Lizard Peninsula (W Cornwall).

24. T. scabrum L. - *Rough Clover*. Procumbent to erect pubescent annual to 20cm; 429
leaflets obovate; racemes ovoid, terminal and axillary, sessile; flowers 4-7mm, white to pale pinkish; 2n=10. Native; similar habitats and distribution to *T. striatum*, and often with it.

25. T. hirtum All. - *Rose Clover*. Ascending pubescent annual to 35cm; leaflets 429
obovate; racemes globose, ± all terminal, sessile; flowers 12-17mm, purple; (2n=10). Intrd; tips and waste ground, casual mainly from wool; sporadic in En; S Europe.

26. T. lappaceum L. - *Bur Clover*. Erect or ascending sparsely pubescent annual to 429
35cm; leaflets obovate to narrowly so; racemes globose to ovoid, mostly terminal, subsessile at first, becoming stalked; flowers 4-8mm, pink; (2n=16). Intrd; tips and waste places, casual mainly from birdseed; sporadic in S Br; S Europe.

27. T. arvense L. - *Hare's-foot Clover*. Erect or ascending pubescent annual to 429
20(40)cm; leaflets linear-oblong; racemes ovoid to oblong at first, soon elongating to cylindrical, mostly terminal, stalked; flowers 3-6mm, white to pink; 2n=14. Native; barish ground on sandy soils; locally frequent in BI N to C Sc, absent from most of Sc and Ir.

28. T. angustifolium L. - *Narrow Clover*. Erect, appressed-pubescent annual to 429
50cm; leaflets linear to linear-lanceolate; racemes ovoid at first, soon elongating to cylindrical, mostly terminal, stalked; flowers 10-13mm, pink; (2n=14, 16). Intrd; tips and waste ground, casual from wool and other sources; sporadic in Br; S Europe.

29. T. squamosum L. - *Sea Clover*. Erect to ascending rather sparsely pubescent R
annual to 40cm; leaflets linear-oblong to oblanceolate; racemes ovoid, mostly 429
terminal, shortly stalked; flowers (5)7-9mm, pink; 2n=16. Native; short, often brackish, turf by sea; very local in Br N to S Wa and E Suffolk (formerly to N Lincs), Guernsey.

30. T. alexandrinum L. - *Egyptian Clover*. Erect to ascending pubescent (to 429
sparsely so) annual to 60cm; leaflets elliptic or oblong to narrowly so; racemes ovoid, terminal and axillary, stalked; flowers 8-13mm, cream; (2n=16). Intrd; casual mainly from grass-seed mixtures, in newly sown grass by roads and in parks; Guernsey, rare in S En; SE Mediterranean. The much rarer **T. constantinopolitanum** Ser., from SW Asia, is often confused with *T. alexandrinum*, but would key out as *T. echinatum*, from which it differs in the calyx-tube at

fruiting being constricted, not gradually widened, distally.

31. T. echinatum M. Bieb. - *Hedgehog Clover*. Procumbent to erect, sparsely 429
pubescent annual to 50cm; leaflets obovate or elliptic to narrowly so; racemes ovoid, terminal and axillary, stalked; flowers 8-15mm, cream, often purple-tinged; (2n=16). Intrd; casual of tips and rough ground, mostly from birdseed; sporadic in En; SE Europe and SW Asia.

Section* 6 - *TRICHOCEPHALUM W.D.J. Koch (*Calycomorphum* C. Presl) (sp. 32). Racemes with sterile corolla-less flowers mixed with normal ones, becoming turned down and thrust into ground at fruiting; calyx becoming symmetrically swollen in fruit, with >20 very fine veins, with glabrous and open throat; fruit sessile, 1-seeded; bracts 0; leaves without glands.

32. T. subterraneum L. (*Calycomorphum subterraneum* (L.) C. Presl). - 429
Subterranean Clover. Decumbent to procumbent pubescent annual to 20(80)cm; racemes axillary, with 2-5(7) fertile flowers and numerous sterile ones consisting only of palmately-divided calyx, long-stalked; fertile flowers 8-14mm, whitish; 2n=16. Native; short turf and barish places on sandy soils, especially by sea; scattered in Br N to N Lincs and Man, CI, Wicklow; also frequent wool-alien in Br, and then often very robust (var. **oxaloides** (Bunge) Rouy).

TRIBE 11 - THERMOPSIDEAE (genus 29). Rhizomatous perennial herbs; leaves ternate, with entire leaflets; flowers in erect raceme with small ovate bracts; all 10 stamens free; fruits longitudinally dehiscent, with 2-7 seeds.

29. THERMOPSIS R. Br. - *False Lupin*

1. T. montana Nutt. ex Torr. & A. Gray - *False Lupin*. Stems erect, to 70(100)cm, pubescent; leaflets up to 10cm, elliptic to narrowly so; stipules ovate to lanceolate; flowers 5-60 per raceme, yellow, 20-25mm; fruits erect, pubescent, 4-7cm; (2n=18). Intrd; grown in gardens and natd on old sites or rough grassy places; Fetlar (Shetland) from c.1978, formerly Northants; W N America.

TRIBE 12 - GENISTEAE (genera 30-35). Perennial or annual herbs or woody shrubs or trees; leaves simple, entire, ternate or palmate; flowers in axillary or terminal racemes, sometimes reduced to 1 or 2; all 10 stamens forming tube; fruits longitudinally dehiscent, with 2-many seeds.

30. LUPINUS L. - *Lupins*

Herbaceous annuals or perennials, sometimes woody at base; leaves palmate, with long petiole; flowers in terminal erect racemes, variously coloured; fruits several-seeded, erecto-patent.

1 Stems woody towards base, not dying down to ground in winter **1. L. arboreus**
1 Stems herbaceous, dying down to ground in winter 2
 2 Annuals, easily uprooted 3
 2 Tuft-forming perennials 4
3 Leaflets oblanceolate; upper lip of calyx very shallowly 2-lobed or ± entire; seeds mostly >7mm **5. L. albus**
3 Leaflets linear; upper lip of calyx deeply bifid; seeds <7mm **6. L. angustifolius**
 4 Basal leaves absent at flowering time; upper part of stem and petioles usually shaggy-hairy; lower lip of calyx 7-13mm **4. L. nootkatensis**
 4 Basal leaves present at flowering time; stem and petioles with rather sparse short hairs; lower lip of calyx 3-8mm 5
5 Stems unbranched, with 1 inflorescence; flowers blue; leaflets obtuse to

acute; lower lip of calyx 3-6mm **3. L. polyphyllus**
5 Stems mostly branched, with >1 inflorescence; flowers various shades of blue, pink, purple, yellow or white; leaflets acute to acuminate; lower lip of calyx 5-8mm **2. L. x regalis**

Other spp. - **L. luteus** L., from Mediterranean, is an annual with yellow flowers and calyx with deeply 2-lobed upper lip; it is a rare escape from cultivation.

1. L. arboreus Sims - *Tree Lupin*. Stems erect, much branched, to 2m, with rather sparse short hairs; leaflets 5-10(12), mostly <6cm, oblanceolate to linear-oblanceolate; lower lip of calyx 7-11mm; corolla usually yellow, often blue-tinged, sometimes blue or whitish; (2n=40, 48). Intrd; garden escape natd in waste places and on maritime shingle or sand; very scattered mostly on coasts of Br N to C Sc, S Ir, CI; California. Killed or cut back in heavy frosts, so populations fluctuate.

2. L. x regalis Bergmans (*L. arboreus* x *L. polyphyllus*) - *Russell Lupin*. Much closer to *L. polyphyllus*; stems erect, usually at least sparingly branched, to 1.5m, with rather sparse short hairs; leaflets (7)9-15, mostly >6cm, oblanceolate to narrowly elliptic; lower lip of calyx 5-8mm; corolla various colours. Native; commonly grown and frequently well natd on rough ground, banks of roads and railways; throughout lowland Br; garden origin, but spontaneous hybrids occur in Moray, Wigtowns and Midlothian with the parents, and perhaps also backcrosses from natd *L.* x *regalis* to *L. arboreus*. Garden escapes are usually recorded as *L. polyphyllus*, but this is now very rarely grown and probably all modern records outside Sc are *L.* x *regalis*.

2 x 4. L. x regalis x L. nootkatensis probably occurs on river-shingle with the parents in Moray and M Perth, but is difficult to identify; it is intermediate and fertile.

3. L. polyphyllus Lindl. - *Garden Lupin*. Differs from *L. x regalis* in key characters; (2n=48). Intrd; formerly grown in gardens, natd by rivers and railways, sometimes in waste places; scattered in Sc; W N America. Flowers ?sometimes pink.

3 x 4. L. polyphyllus x L. nootkatensis = L. x pseudopolyphyllus C.P. Sm. occurs with the parents on river-shingle in Moray and M Perth; it is intermediate and fertile.

4. L. nootkatensis Donn ex Sims - *Nootka Lupin*. Stems erect, to 1m, usually with long shaggy hairs; leaflets 6-9(12), mostly <6cm, oblanceolate; lower lip of calyx 7-13mm; corolla bluish-purple, sometimes whitish tinged purplish; 2n=48. Intrd; natd on riverside shingle and moorland since at least 1862; C & N Sc from M Perth to Orkney, especially by R. Tay and R. Dee, NW Ir; NW N America and NE Asia.

5. L. albus L. - *White Lupin*. Stems erect to ascending, sparingly branched, to 60(100)cm, moderately pubescent; leaflets 5-9, <6cm, oblanceolate to narrowly obovate; lower lip of calyx 8-9mm; corolla white, usually variably tinged bluish-violet; (2n=40, 48, 50). Intrd; becoming grown as a seed-crop and appearing on tips, at docks and in waste places; sporadic in S En; Mediterranean. Will become much commoner.

6. L. angustifolius L. - *Narrow-leaved Lupin*. Stems erect, well branched, to 60(100)cm, with rather sparse short hairs; leaflets 5-9, <4cm, linear; lower lip of calyx 6-7mm; corolla blue; (2n=40, 48). Intrd; imported for grain and being used in trials as seed-crop, scarce casual at docks and waste places; sporadic in Br, mainly S; Mediterranean. Formerly grown for fodder.

31. LABURNUM Fabr. - *Laburnums*

Deciduous non-spiny trees; leaves ternate, with long petiole; flowers yellow, in pendent racemes on short-shoots; fruits several-seeded, pendent.

1. L. anagyroides Medik. - *Laburnum*. Tree to 8m; young twigs, petioles, leaf lowerside, peduncle, pedicels and young fruits appressed-pubescent, densely silvery so when very young; racemes mostly 15-30cm; flowers 17-23mm; fruits with narrow dorsal ridge thick and ± truncate in section; (2n=48). Intrd; much planted and frequently self-sown in rough ground and on banks by roads and railways; scattered through BI, formerly planted for hedging and still persistent in parts of the W; mountains of SC Europe.

1 x 2. L. anagyroides x L. alpinum = L. x watereri (Wettst.) Dippel (*L. x vossii* hort.) is now more commonly planted than either sp. but much less or never self-sown due to sterility; it is intermediate in pubescence to almost glabrous, intermediate in fruit shape, has the longer racemes with more flowers of *L. alpinum*, the larger (15-21mm) more densely arranged flowers of *L. anagyroides*, and usually forms few good seeds. Has apparently been used for hedging in the W.

2. L. alpinum (Mill.) J. Presl - *Scottish Laburnum*. Tree to 13m; differs from *L. anagyroides* in glabrous to sparsely pubescent parts; racemes mostly 25-35cm with more but less densely crowded flowers 13-18mm; fruits with distinctly winged dorsal ridge narrow and acute in section; (2n=48). Intrd; similar occurrence to *L. anagyroides* but much confused and exact relative abundance unknown (certainly not frequent in the wild or in cultivation in S & C Br), perhaps commoner in N Br and Ir, reported as having been used for hedging in Cards, Tyrone and Co Londonderry but perhaps over-recorded for *L. x watereri*; mountains of SC Europe.

32. CYTISUS Desf. (*Sarothamnus* Wimm., *Lembotropis* Griseb.) - *Brooms*
Non-spiny shrubs; leaves simple or ternate, sessile or petiolate; flowers yellow, in terminal or axillary racemes or 1-few in leaf-axils; calyx with upper lip 2-toothed or deeply bifid; fruits several- to many-seeded, erect to patent.

1 Flowers in terminal leafless racemes; all leaves ternate, most with petioles >1cm **1. C. nigricans**
1 Flowers 1-few in lateral groups with (often reduced) leaves; upper leaves usually simple, lower with petioles <1cm 2
  2 Flowers white, 7-12mm; calyx ≤3mm **2. C. multiflorus**
  2 Flowers yellow to red, 10-25mm; calyx >3mm 3
3 Fruits covered with dense white shaggy hairs; twigs c.10-angled, fragile **3. C. striatus**
3 Fruits with long hairs on sutures, ± glabrous on faces; twigs c.5-angled, pliable **4. C. scoparius**

1. C. nigricans L. (*Lembotropis nigricans* (L.) Griseb.) - *Black Broom*. Stems erect, to 434
1.5m, terete to obscurely ridged, pubescent; leaves all ternate, most with petioles >1cm, flowers in terminal leafless racemes, 7-12mm, yellow; fruits 1.5-3.5cm, appressed-pubescent; (2n=48). Intrd; natd on waste ground by railway and in gravel-pits since 1970; Middlesex, E Kent and W Ross; C & SE Europe.

2. C. multiflorus (L'Hér. ex Aiton) Sweet - *White Broom*. Stems erect or arching, to 434
2m, deeply c.8-ridged, glabrous after 1st year; leaves ternate below, simple above, sessile or nearly so; flowers 1-3 in lateral groups, 7-12mm, white; fruits 1-3.2cm, appressed-pubescent; (2n=48, 96). Intrd; natd on banks of roads and railways; very scattered in N Wa and S En, Berwicks; Iberian Peninsula.

3. C. striatus (Hill) Rothm. - *Hairy-fruited Broom*. Stems erect or arching, to 3m, 434
deeply c.10-ridged, glabrous after 1st year; leaves ternate and petiolate below, simple above; flowers 1-2 in lateral groups, 10-20mm, yellow (usually paler than in *C. scoparius*); calyx usually pubescent; fruits 1.5-3.5cm, slightly compressed when mature, long-patent-pubescent all over; (2n=46, 48). Intrd; natd on roadside banks; very scattered in En, Sc and Wa; Iberian Peninsula.

4. C. scoparius (L.) Link (*Sarothamnus scoparius* (L.) W.D.J. Koch) - *Broom*. Stems to 2.5m, deeply c.5-ridged, glabrous after 1st year; leaves as in *C. striatus*; flowers 1-2 in lateral groups, 15-20mm, deep yellow or with dark red to mauve areas; calyx usually glabrous; fruits 2.5-5cm, strongly compressed, long-patent-pubescent

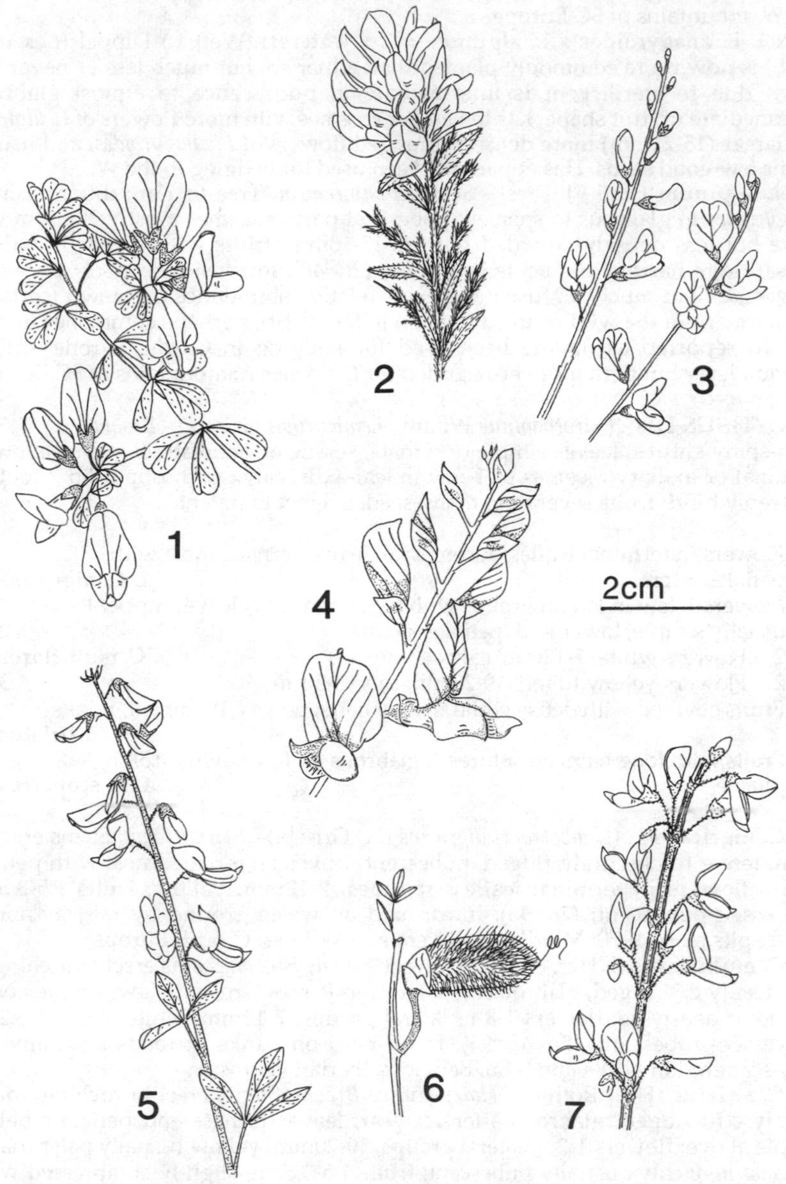

FIG 434 - Fabaceae: Genisteae. 1, *Genista monspessulana.* 2, *G. hispanica.* 3, *G. aetnensis.* 4, *Spartium junceum.* 5, *Cytisus nigricans.* 6, *C. striatus.* 7, *C. multiflorus.*

on sutures, glabrous on faces. Native.

a. Ssp. scoparius. Plant erect or arching, to 2.5m high; 2n=46, 48. Calcifuge of heathland, sandy banks, open woodland, rough ground; throughout most of BI.

b. Ssp. maritimus (Rouy) Heywood (*Sarothamnus scoparius* ssp. *maritimus* (Rouy) RR Ulbr.). Plant ± procumbent, to 50cm high; young branches more densely silky-pubescent than in ssp. *scoparius*; 2n=24, 46, 48. Maritime cliffs and perhaps shingle; W Wa, SW Ir, SW En, CI; plants on shingle at Dungeness (E Kent) do not come quite true from seed.

33. SPARTIUM L. - *Spanish Broom*

Non-spiny shrubs; leaves simple, shortly petiolate to sessile; flowers yellow, in leafless terminal racemes; calyx with upper lip divided nearly to base; fruits many-seeded, erecto-patent.

1. S. junceum L. - *Spanish Broom*. Stems erect, to 3m, terete with wide soft pith, 434 glabrous; leaves linear to oblanceolate; flowers 20-28mm, yellow; fruits 4-10cm, strongly compressed, pubescent; (2n=48, 52, 54). Intrd; natd on sandy roadside banks and rough ground; scattered in C & S Br; Mediterranean.

34. GENISTA L. (*Teline* Medik.) - *Greenweeds*

Spiny or non-spiny shrubs; leaves simple or ternate, sessile or shortly petiolate; flowers yellow, in terminal racemes or lateral clusters; calyx with upper lip divided 1/4-3/4 way to base; fruits 1-several seeded, erect to patent.

1	Leaves all ternate	**1. G. monspessulana**
1	Leaves all simple, or 0 due to early fall	2
2	Plant with branched spines	**5. G. hispanica**
2	Plant with 0 or simple spines	3
3	Seeds 1-2; leaves most or all fallen by flowering time	**6. G. aetnensis**
3	Seeds 3-12; leaves present at flowering time	4
4	Calyx, corolla and leaves pubescent	**3. G. pilosa**
4	Calyx, corolla and leaves ± glabrous	5
5	Flowers 7-10mm; fruits 12-20mm; plant usually spiny	**4. G. anglica**
5	Flowers 10-15mm; fruits (15)20-30mm; plant never spiny	**2. G. tinctoria**

1. G. monspessulana (L.) L.A.S. Johnson (*Cytisus monspessulanus* L., *Teline* 434 *monspessulana* (L.) K. Koch) - *Montpellier Broom*. Stems not spiny, erect, to 2(3)m; leaves ternate, pubescent, obovate, petiolate; flowers 9-12mm, in lateral clusters, sparsely silky-pubescent on keel; fruits 15-25mm, densely pubescent, 3-6-seeded; (2n=48). Intrd; natd on banks, roadsides and rough ground; scattered in CI and S En; Mediterranean.

2. G. tinctoria L. - *Dyer's Greenweed*. Stems not spiny, erect to procumbent, to 60(100)cm; leaves simple, glabrous, sessile; flowers 10-15mm, in terminal and lateral clusters, ± glabrous; fruits 15-30mm, (3)4-10-seeded; 2n=96. Native.

a. Ssp. tinctoria. Plant usually erect to ascending; leaves narrowly elliptic, mostly >4x as long as wide; fruits glabrous. Grassy places, banks and rough ground; locally common in Br N to S Sc, Jersey (not Cornwall).

b. Ssp. littoralis (Corb.) Rothm. Plant procumbent; leaves elliptic-oblong, mostly RR <4x as long as wide; fruits pubescent or glabrous. Grassy cliff-tops; N Devon, Cornwall, SW Pembs. On cliffs in Cornwall and N Devon procumbent plants with broad leaves but glabrous fruits occur; these are often referred to ssp. *tinctoria* but are probably better placed in ssp. *littoralis*.

3. G. pilosa L. - *Hairy Greenweed*. Stems not spiny, procumbent or scrambling, to RR 50cm; leaves simple, pubescent, ovate to obovate, sessile; flowers 7-11mm, in terminal and lateral clusters, pubescent on standard and keel; fruits 14-22mm, pubescent, 3-8-seeded; (2n=24). Native; cliff-tops and heathland; extremely local and decreasing in W Cornwall, Pembs, Merioneth, Brecs and E Sussex, formerly W Suffolk and W Kent.

4. G. anglica L. - *Petty Whin.* Stems usually with simple spines, sometimes without (var. **subinermis** (D. Legrand) Rouy), erect to spreading, to 1m; leaves simple, glabrous, elliptic (those on spines ± linear), sessile; flowers 7-10mm, in terminal and lateral clusters, glabrous; fruits 12-20mm, glabrous, (3)4-12-seeded; 2n=42. Native; sandy and peaty heaths and moors; in suitable places throughout most of Br.

5. G. hispanica L. - *Spanish Gorse.* Stems with branched spines, erect, to 70cm; **434**
leaves simple, pubescent, lanceolate to oblanceolate, sessile; flowers 8-14mm, in short dense terminal racemes, pubescent on keel; fruits 8-11mm, glabrous to pubescent, 1-2-seeded; (2n=36). Intrd; natd on sandy and rocky hills and roadsides; Cards (since 1927), S En and Man; SW Europe. Our plants are nearly all ssp. **occidentalis** Rouy, with appressed hairs on leaves and stems and standard 6-8mm, but ssp. **hispanica**, with patent hairs and standard 8-11mm, has been reported from W Kent.

6. G. aetnensis (Raf. ex Biv.) DC. - *Mount Etna Broom.* Stems not spiny, erect, to **434**
3(5)m; leaves simple, pubescent, <5mm, on young growth only; flowers 8-13mm, in lax terminal racemes, pubescent on keel; fruits 8-14mm, pubescent, 1-2-seeded; (2n=54). Intrd, natd on waste ground; W Kent and Surrey; Sicily and Sardinia. Flowers later than other spp., mostly Jul-Aug.

35. ULEX L. - *Gorses*

Shrubs with branched spines; leaves ternate on young plants, simple and reduced to scales or weak spines on mature plants; flowers yellow, 1-few in lateral clusters, with small bracteole on each side between calyx-lips; calyx with upper lip shortly 2-toothed; fruits (1)2-6(8)-seeded, >1/2 enclosed in calyx.

1 Bracteoles 1.8-4.5 x 1.5-4mm, ≥2x as wide as pedicels **1. U. europaeus**
1 Bracteoles <1.5 x 1mm, ≤2x as wide as pedicels 2
  2 Teeth of lower calyx-lip parallel to convergent; calyx 9-13(15)mm; standard (12)13-18(22)mm **2. U. gallii**
  2 Teeth of lower calyx-lip divergent; calyx 5-9.5(10)mm; standard 7-12(13)mm **3. U. minor**

1. U. europaeus L. - *Gorse.* Densely spiny spreading shrub to 2(2.5)m; spines **418**
very strong, deeply grooved; flowering mainly winter-spring (autumn flowers accompanied by many buds); calyx 10-16(20)mm, with convergent teeth; standard 12-18mm; ovules (8)9-14(16); fruits (12)14-17(19)mm, dehiscing in summer; 2n=96. Native; grassy places, heathland, open woods, mostly on sandy or peaty soil; throughout BI.

1 x 2. U. europaeus x U. gallii occurs in W Br and Ir where the 2 spp. overlap, but precise range unknown; it is highly fertile and variously intermediate, especially in calyx, corolla and bracteole size, and ovule number (7-10), and flowers in autumn and winter.

2. U. gallii Planch. - *Western Gorse.* Densely spiny spreading shrub to 1.5(2)m; **418**
spines moderately grooved; flowering mainly summer (autumn flowers accompanied by many withered ones); bracteoles 0.5-0.8 x 0.6-0.8mm; ovules (3)4-6(7); fruits (8)9-13(14)mm, dehiscing in spring; 2n=64, 96. Native; habitat as for *U. europaeus*; mainly in CI, W 1/2 of En, Wa, Man, SW Sc, C & S Ir, but scattered in E En and E Sc.

3. U. minor Roth - *Dwarf Gorse.* Densely spiny spreading shrub to 1(1.5)m; spines **418**
weakly grooved; flowering as in *U. gallii*; bracteoles 0.6-0.8 x 0.4-0.6mm; ovules (3)4-6(7); fruits 6-8.5mm, dehiscing in spring; 2n=32. Native; heaths; mainly in S En from E Kent to Dorset and Wilts, very scattered N to Flints and Notts, Cumberland, Argyll, formerly CI, sometimes grown for ornament and natd outside

that area.

80. ELAEAGNACEAE - *Sea-buckthorn family*

Deciduous or evergreen, usually spiny shrubs; leaves alternate, simple, entire or ± so, ± sessile to petiolate, without stipules, with dense silvery to reddish-brown scales. Flowers in small axillary clusters, dioecious or bisexual, deeply perigynous, actinomorphic; sepals 2 or 4, fused below; petals 0; stamens as many as sepals, inserted on base of calyx-tube; ovary 1-celled, with 1 basal ovule; style 1, long; stigma linear to capitate; fruit a drupe-like achene surrounded by succulent hypanthium.

Easily recognized by the entire, silver-scaly leaves, petal-less, 2- or 4-sepalled flowers, and succulent 1-seeded fruits.

1 Leaves linear-lanceolate; petiole <2mm; flowers dioecious, appearing before leaves **1. HIPPOPHAE**
1 Leaves elliptic to broadly ovate; petiole >5mm; flowers bisexual, appearing with leaves **2. ELAEAGNUS**

1. HIPPOPHAE L. - *Sea-buckthorn*
Flowers dioecious, wind-pollinated; sepals and stamens 2.

1. H. rhamnoides L. - *Sea-buckthorn*. Spiny, spreading and suckering shrub to 3(9)m; leaves deciduous, linear-lanceolate, 2-8 x 0.2-1.3cm, entire; flowers 3-4mm, early spring; fruits translucent-orange with sparse scales when ripe, broadly ellipsoid to globose, 6-10mm; 2n=24. Native; dunes and other sandy places by sea; round coasts of Br and Ir but perhaps native only in E from E Sussex to C Sc; widely planted by sea and along roads inland and often self-sown. R

2. ELAEAGNUS L. - *Oleasters*
Flowers bisexual, insect-pollinated; sepals and stamens 4.

Other spp. - E. pungens Thunb., from Japan, is the common evergreen garden sp. with oblong-elliptic, crenulate-undulate, often variegated leaves; it can persist in neglected plantings, especially as relics of hedges.

1. E. umbellata Thunb. - *Spreading Oleaster*. Spreading, often spiny shrub to 3(6)m; leaves deciduous or ± so, elliptic to ovate or narrowly so, 3-10 x 1.5-4cm, entire; flowers May-Jun; calyx 6-8mm, funnel-shaped, the lobes <1/2 as long as tube; fruits very scaly at first, then red, ellipsoid-oblong to globose, 6-15mm; (2n=28). Intrd; frequently grown in gardens and by roads; bird-sown plants natd in scattered places in S Br, used as a stock for *E. macrophylla* at least in Guernsey and Sark and often surviving when the latter dies; E Asia.

2. E. macrophylla Thunb. - *Broad-leaved Oleaster*. Spreading, sometimes spiny shrub to 3m; leaves evergreen, elliptic to ovate or broadly so, 4-13 x 2-7cm, entire; flowers Oct-Nov; calyx 6-10mm, bell-shaped, the lobes >1/2 as long as tube; fruits white-scaly, oblong, 15-20mm. Intrd; frequently grown as hedging (often on *E. umbellata* as stock) in Guernsey and Sark and long persistent, rarely self-sowing in S En; Japan and Korea. In Guernsey sometimes difficult to distinguish from *E. pungens*, and some or even most plants might be **E. x ebbingei** Door., a hybrid of *E. macrophylla* and either *E. pungens* or 1 of its hybrids.

81. HALORAGACEAE - *Water-milfoil family*

Perennial, mainly subaquatic herbs with weak trailing stems; leaves simple and opposite, or in whorls of 3-6 and finely pinnate, ± sessile, without stipules. Flowers small and inconspicuous, whorled, opposite or alternate in terminal spikes, variously dioecious to bisexual, epigynous, with 2 minute bracteoles and 1 bract at base; sepals 4, minute, ± free; petals 4, free, small in male or bisexual flowers, minute or 0 in female flowers; stamens (4-)8 (0 in female flowers); ovary 1- or 4-celled, with 1 apical ovule in each cell (0 in male flowers); styles 4 or 0; stigmas 4, clavate to feathery; fruits a nut or a group of ≤4 small nuts.

The inconspicuous flowers with 4 sepals, 4 petals and 8 stamens, and fruits of 1 or 4 nutlets are diagnostic.

1 Leaves opposite, simple, not or slightly toothed **1. HALORAGIS**
1 Leaves in whorls of 3-6, finely pinnate **2. MYRIOPHYLLUM**

1. HALORAGIS J.R. & G. Forst. - *Creeping Raspwort*
Plant growing on peat surface; leaves opposite, simple, entire or weakly serrate; ovary 1-celled; fruit a single nutlet.

1. H. micrantha (Thunb.) R. Br. ex Siebold & Zucc. - *Creeping Raspwort*. Stems decumbent, to 20cm; leaves broadly ovate to suborbicular, 3-10mm, subsessile; fruits c.0.5-1mm incl. persistent sepals; (2n=12). Intrd; natd on bare peat in bog in W Galway, discovered 1988; SE Asia to Australasia.

2. MYRIOPHYLLUM L. - *Water-milfoils*
Plants normally subaquatic, sometimes on mud; leaves in whorls of 3-6, finely pinnate; ovary 4-celled; fruit a group of ≤4 nutlets.

Vegetative characters, especially numbers of leaf-segments, are not sufficient for certain identification; in *M. verticillatum*, for example, there are usually 24-35 segments, but in some mud-growing plants there can be only 4.

1 Leaves many or mostly 5 in a whorl; uppermost bracts deeply serrate to pinnately dissected 2
1 Leaves (3)4(-5) in a whorl; uppermost bracts simple, entire or minutely serrate 3
 2 Emergent leaves with dense sessile glands; only female flowers present, whitish **2. M. aquaticum**
 2 Emergent leaves with sparse sessile glands; each plant with upper flowers male, lower female, and usually some bisexual between; flowers reddish **1. M. verticillatum**
3 All flowers in whorls, reddish; leaves usually with 13-38 segments; usually in base-rich water **3. M. spicatum**
3 Upper flowers opposite or alternate, all yellowish; leaves usually with 6-18 segments; usually in base-poor water **4. M. alterniflorum**

Other spp. - 2 other spp. formerly occurred as natd aliens. **M. heterophyllum** Michx., from E N America, occurred until recently in a canal in SW Yorks; it differs from all other spp. in its 4 (not 8) stamens and simple, entire to serrate emergent leaves. **M. verrucosum** Lindl., from Australia, occurred in gravel-pits in Beds as a wool-alien; it most closely resembles *M. spicatum* but has leaves mostly in threes and rarely >1cm and flowers all bisexual.

1. M. verticillatum L. - *Whorled Water-milfoil*. Stems to 3m; leaves (4)5(-6) in a whorl, usually with 24-35 segments; male, female and usually bisexual flowers present, mostly 5 in a whorl; 2n=28. Native mostly base-rich ponds, lakes and **R**

slow rivers in lowlands; very scattered in En, Wa and Ir, mostly E En, CI. Perennation is by turions produced late in the year. These are short (often c.5cm) lateral shoots with the leaves larger towards the apex and appressed to the stem, forming a clavate structure; they are diagnostic for *M. verticillatum*.

2. M. aquaticum (Vell.) Verdc. (*M. brasiliense* Cambess.) - *Parrot's-feather*. Stems to 2m; leaves 4-6 in a whorl, usually with 8-30 segments; dioecious, female only in Br; flowers mostly 4-6 in a whorl. Intrd; commonly grown by aquarists and becoming natd where thrown out; scattered in Br N to S Sc, Jersey; S America.

3. M. spicatum L. - *Spiked Water-milfoil*. Stems to 2.5m; leaves (3)4(-5) in a whorl, usually with 13-38 segments; spikes erect from first; male, female and bisexual flowers present, 4 in a whorl; 2n=42. Native; mostly base-rich ponds, lakes, slow rivers and ditches, mostly lowland; locally common over BI.

4. M. alterniflorum DC. - *Alternate Water-milfoil*. Stems to 1.2m; leaves (3-)4 in a whorl, usually with 6-18 segments; spikes pendent at first, then erect; male, female and bisexual flowers present, mostly 2-4-whorled below, the upper ones (male) opposite to alternate; 2n=14. Native; mostly base-poor lakes, ponds, slow streams and ditches, often upland; locally frequent over BI, mostly in N & W.

82. GUNNERACEAE - *Giant-rhubarb family*

Huge herbaceous perennials; stems wholly rhizomatous; leaves alternate but clustered, rhubarb-like, simple, palmately 5-9-lobed with jagged-serrate lobes, with long stout petioles, without stipules. Flowers small, in huge compound erect catkin-like panicles, usually male, female and bisexual mixed, epigynous; sepals 2, minute, free; petals 2, free, small; stamens (1-)2; ovary 1-celled, with 1 apical ovule; style ± 0; stigmas 2, linear; fruit a small drupe.

Unique in its huge rhubarb-like leaves and erect, compact, many-flowered inflorescences.

1. GUNNERA L. - *Giant-rhubarbs*

1. G. tinctoria (Molina) Mirb. - *Giant-rhubarb*. Leaves ≤2m across, cordate at base; petioles ≤1.5m, with pale bristles and weak spines; inflorescences ≤1m, with stout branches <8cm; (2n=34). Intrd; planted by lakes etc. and often self-sown where long-established; natd in scattered places throughout much of lowland BI; W S America.

2. G. manicata Linden ex André - *Brazilian Giant-rhubarb*. Leaves often >2m across, peltate; petioles ≤2m, with reddish bristles and spines; inflorescences ≤1.2m, with slender branches >10cm; (2n=34). Intrd; planted as for *G. tinctoria* and fertile but not recorded as self-sown; scarcely natd but persistent throughout much of lowland BI; S Brazil.

83. LYTHRACEAE - *Purple-loosestrife family*

Annuals or herbaceous perennials; leaves opposite or in whorls of 3, or upper ones alternate, simple, entire, sessile or petiolate, without stipules. Flowers solitary or clustered, in leaf-axils towards stem-apex, bisexual, perigynous, actinomorphic, monomorphic or trimorphic; hypanthium tubular to funnel- or cup-shaped, bearing (4-)6 sepals and (4-)6 epicalyx-segments at apex; petals usually 6, sometimes 0-5, free, borne near apex of hypanthium; stamens 6 or 12 (sometimes fewer); ovary 2-celled, each cell with many ovules on axile placentas; style 1; stigma 1, capitate; fruit a capsule, opening by 2 valves.

Distinguished by the perigynous flowers bearing 6 petals, sepals and epicalyx-segments near the hypanthium mouth.

1. LYTHRUM L. (*Peplis* L.) - *Purple-loosestrifes*

1 Petals >4mm; stigma or some stamens exceeding sepals; flowers trimorphic on separate plants 2
1 Petals <4mm; stigma and stamens not reaching apex of sepals; flowers monomorphic 3
 2 Petals >7mm; flowers clustered in whorls **1. L. salicaria**
 2 Petals <7mm; flowers 1-2 in each axil **2. L. junceum**
3 Leaves obovate-spathulate; hypanthium funnel- to cup-shaped; capsule subglobose **4. L. portula**
3 Leaves linear-oblong; hypanthium tubular; capsule cylindrical **3. L. hyssopifolium**

1. L. salicaria L. - *Purple-loosestrife*. Erect perennial to 1.5m; leaves sessile, lanceolate, all opposite or whorled; flowers clustered in whorls, trimorphic (styles long, medium or short; six stamens at each of other 2 levels); petals 8-10mm, purple; 2n=60. Native; by water and in marshes and fens; common throughout most of BI except N Sc.

2. L. junceum Banks & Sol. - *False Grass-poly*. Erect to decumbent perennial (but annual in Br) to 70cm; leaves sessile, linear- to elliptic-oblong, opposite below, alternate above; flowers 1(-2) in leaf-axils, trimorphic; petals 5-6mm, purple; (2n=10). Intrd; birdseed-alien in parks and waste places, sometimes from other sources; sporadic but frequent in S & C Br; Mediterranean. Often misdetermined as *L. hyssopifolium*.

3. L. hyssopifolium L. - *Grass-poly*. Erect to decumbent usually annual to 25cm; leaves sessile, linear to linear-oblong, opposite below, alternate above; flowers 1(or 2) in leaf-axils, monomorphic; petals 2-3mm, pink; stamens usually 4-6; (2n=20). Native; seasonally wet bare ground; extremely local in S En (5 sites) and Jersey (2 sites), rare casual elsewhere as for *L. junceum*. **RRR**

4. L. portula (L.) D.A. Webb (*Peplis portula* L.) - *Water-purslane*. Procumbent annual to 25cm, rooting at nodes; leaves obovate, petiolate, all opposite; flowers 1 per leaf-axil, monomorphic; petals 0-6, c.1mm, purplish; stamens usually 6; 2n=10. Native; open or bare ground by or in water, or in damp trackways; scattered throughout most of BI. Epicalyx-segments vary from c.0.3 to 2mm; plants at the latter end of the range have been called ssp. **longidentatum** (J. Gay) P.D. Sell.

84. THYMELAEACEAE - *Mezereon family*

Early-flowering, glabrous, poisonous shrubs; leaves alternate, simple, entire, sessile or with short petiole, without stipules. Flowers clustered in leaf-axils, bisexual, perigynous, actinomorphic or ± so; hypanthium tubular, with 4 concolorous sepals at apex; petals 0; stamens 8; ovary 1-celled, with 1 apical ovule; style 0 or short; stigma 1, capitate; fruit a 1-seeded drupe.

Distinctive in the perigynous flowers with 4 sepals, 0 petals and ovary with 1 ovule.

1. DAPHNE L. - *Mezereons*

1. D. mezereum L. - *Mezereon*. Erect, deciduous shrub to 2m; leaves light green; flowers bright pink, in groups of 2-4 in axils of last year's fallen leaves, pubescent on outside, very fragrant; fruit ellipsoid, 8-12mm, bright red; 2n=18. Probably native in some places but often bird-sown; calcareous woods; very local in Br N to Yorks and Lancs, sporadic alien elsewhere. **R**

1 x 2. D. mezereum x D. laureola = D. x houtteana Lindl. & Paxton has been found with the parents in N Somerset, W Sussex and MW Yorks, but not since 1954; it is sterile and intermediate in leaf-retention and flower-colour; endemic, but a well-known garden plant.

2. D. laureola L. - *Spurge-laurel*. Erect to decumbent evergreen shrub to 1.5m; leaves dark green; flowers yellowish-green, in racemes of 2-10 in axils of leaves, glabrous, not or slightly scented; fruit ellipsoid, 10-13mm, black; 2n=18. Native; woods mostly on calcareous or clayey soils; locally frequent in En, Wa and CI, sporadic alien elsewhere.

85. MYRTACEAE - *Myrtle family*

Evergreen trees or shrubs; leaves opposite or alternate, simple, entire or nearly so, sessile or petiolate, without stipules, with aromatic glands. Flowers solitary or clustered, axillary or terminal, bisexual, epigynous to semi-epigynous, actinomorphic; hypanthium not or slightly extended above ovary, bearing 4-5 sepals and petals; sepals minute or ± free; petals free or united into hood covering unopened flower; stamens numerous; ovary 2-5-celled with numerous ovules on axile placentas; style 1; stigma 1, capitate; fruit a many-seeded capsule or berry.

Recognized by the combination of evergreen, aromatic, entire leaves, inferior ovary, numerous stamens and 4-5 petals.

1 All leaves opposite; ovary (2-)3-celled; fruit a berry **3. LUMA**
1 Adult leaves alternate; ovary 4-5-celled; fruit a capsule 2
2 Leaves <2cm; petals free; shrubs **1. LEPTOSPERMUM**
2 Leaves >3cm; petals united into a hood covering flower in bud; trees **2. EUCALYPTUS**

1. LEPTOSPERMUM J.R. & G. Forst. - *Tea-trees*
All leaves alternate, <2cm; flowers solitary, 0.8-1.8cm across; petals 5, white or sometimes pink, free; ovary 5-celled; fruit a woody capsule, 6-10mm wide.

1. L. scoparium J.R. & G. Forst. - *Broom Tea-tree*. Shrub to 5m, glabrous or ± so; leaves 4-20 x 2-6mm, tapered to sharp point, borne rather densely on twigs; sepals falling as soon as flowers fade; (2n=22). Intrd; confined by frost to extreme SW BI but there freely self-sowing; natd on Tresco (Scillies); Australia and New Zealand.

2. L. lanigerum (Aiton) Sm. - *Woolly Tea-tree*. Differs from *L. scoparium* in leaves 4-15 x 2-4mm, abruptly pointed and scarcely sharply so; sepals, hypanthium and lowerside of leaf white-pubescent; sepals persistent on fruit; (2n=22). Intrd; natd on Tresco but less so than *L. scoparium*; Australia.

2. EUCALYPTUS L'Hér. - *Gums*
Leaves on adult shoots alternate, those on juvenile shoots opposite and of very different shape, >3cm; flowers solitary or in small umbel-like clusters; petals united into a hood covering flower in bud and breaking off transversely; ovary mostly 4-celled; fruit a woody capsule.

1 Flowers and fruits solitary **3. E. globulus**
1 Flowers and fruits in groups of 3 or more 2
2 Flowers in groups of 5-12 **6. E. pulchella**
2 Flowers in groups of 3 3
3 Juvenile leaves lanceolate to narrowly elliptic; fruit 5-6mm **5. E. viminalis**
3 Juvenile leaves orbicular to elliptic or ovate; fruit ≥7mm 4
4 Flower buds and fruit with 2-4 longitudinal ribs, sessile **4. E. johnstonii**
4 Flower buds and fruit ± terete, usually stalked 5
5 Mature leaves 4-7cm; flower buds 6-8mm; fruit 7-10mm **1. E. gunnii**
5 Mature leaves 8-18cm; flower buds c.12mm; fruit c.17mm **2. E. urnigera**

Other spp. - c.35 spp. have been tried for forestry purposes, mainly in Ir, of which experimental plots of c.10 still exist. The commonest of these are 1, 2, 4 and 5 below, but they rarely produce self-sown offspring. Of the others, **E. dalrympleana** Maiden (*Broad-leaved Kindlingbark*), from Tasmania and SE Australia, has been planted more recently and might become as frequent; it differs from *E. gunnii* in its sessile flowers and fruits and lanceolate mature leaves 10-22cm.

1. E. gunnii Hook. f. - *Cider Gum.* Tree to 30m; juvenile leaves orbicular to elliptic or ovate; mature leaves 4-7cm, lanceolate to narrowly ovate; flower buds 6-8mm, shortly pedicellate; fruit 7-10 x 8-9mm, hemispherical to bell-shaped, ± terete; (2n=22). Intrd; planted for small-scale forestry and for ornament widely in Ir and S & W Br, persistent also in E N Essex since 1887; the most hardy sp., sometimes self-sowing; Tasmania.

2. E. urnigera Hook. f. - *Urn-fruited Gum.* Tree to 30m; juvenile leaves orbicular to ovate; mature leaves 8-18cm, linear-lanceolate, dark green; flower buds c.12mm, pedicellate; fruit c.17 x 10mm, bowl-shaped and narrowed distally, ± terete. Intrd; planted for small-scale forestry in Ir; Tasmania.

3. E. globulus Labill. - *Southern Blue-gum.* Tree to 45m; juvenile leaves ovate to narrowly ovate; mature leaves 10-30cm, lanceolate to linear-lanceolate, often falcate, dark green; flower buds up to 30mm, sessile; fruit 10-15 x 15-30mm, obconical, 4-ribbed; (2n=22). Intrd; planted for ornament and rarely forestry, self-sown in W Ir, Scillies and Man; Tasmania.

4. E. johnstonii Maiden (*E. muelleri* T. Moore non Miq. nec Naudin) - *Johnston's Gum.* Tree to 50m; juvenile leaves orbicular; mature leaves 5-10cm, oblong-ovate to lanceolate; flower buds c.14 x 10mm, sessile; fruit c.10 x 12mm, hemispherical to obconical, 2-4-ribbed; (2n=22). Intrd; planted for small-scale forestry in Ir; Tasmania.

5. E. viminalis Labill. - *Ribbon Gum.* Tree to 50m; juvenile leaves ovate to narrowly so; mature leaves 11-18cm, linear to linear-lanceolate, pale green; flower buds 5-8mm, sessile to shortly pedicellate; fruit 5-8 x 5-8mm, spherical to obconical, ± terete; (2n=22). Intrd; planted for small-scale forestry in Ir; Tasmania and SE Australia.

6. E. pulchella Desf. (*E. linearis* Dehnh.) - *White Peppermint-gum.* Tree to 15m; juvenile leaves linear to linear-lanceolate; mature leaves 5-12cm, linear (to linear-lanceolate); flower-buds 4-5mm, shortly pedicellate; fruit 4-5.5 x 5-6.5mm, obovoid, ± terete; (2n=22). Intrd; planted for ornament in extreme SW En, self-sown in Scillies; Tasmania.

3. LUMA A. Gray (*Amomyrtus* auct. non (Burret) D. Legrand & Kausel) - *Myrtles*
Leaves all opposite; flowers 1(-3) in leaf-axils; petals 4(-5), white, free; ovary with (2-)3 cells; fruit a berry.

1. L. apiculata (DC.) Burret (*Myrtus luma* auct. non Molina, *Amomyrtus luma* auct. non (Molina) D. Legrand & Kausel) - *Chilean Myrtle.* Shrub or tree rarely to 18m; leaves 1.5-3cm; flowers 2-3cm across, with strongly concave petals; fruit dark purple, globose, 6-10mm. Intrd; thriving only in very mild areas but self-sown in semi-natural woodland; SW En, Guernsey, SW Ir, Man; S America.

86. ONAGRACEAE - *Willowherb family*

Herbaceous annuals, biennials or perennials or rarely shrubs; leaves opposite or alternate, simple, sessile or petiolate, without or with small early-falling stipules. Flowers solitary and axillary or in terminal racemes, bisexual, epigynous, actinomorphic to weakly zygomorphic; hypanthium 0 or a short to very long tube, with 2 or 4 free sepals at apex; petals 2 or 4 or rarely 0, free; stamens 2, 4 or 8; ovary 1-, 2- or 4-celled, each cell with 1-many ovules on axile placentas; style 1; stigma capitate to clavate or 4-lobed; fruit a 4-celled capsule or berry, or 1-2-seeded nut.

Distinguished by the combination of epigynous flowers with 2 or 4 sepals, 2, 4 or 8 stamens, and (1)2-4-celled ovary.

1 Sepals 2; petals 2; stamens 2; ovary 1-2-celled; fruit with hooked bristles **7. CIRCAEA**
1 Sepals 4; petals 4 (rarely 0); stamens 4 or 8; ovary 4-celled; fruit without hooked bristles 2
 2 Shrub; fruit a berry **6. FUCHSIA**
 2 Herb; fruit a capsule 3
3 Petals 0; stamens 4 **3. LUDWIGIA**
3 Petals 4; stamens 8 4
 4 Petals yellow (sometimes streaked or tinged reddish); hypanthium ≥20mm **4. OENOTHERA**
 4 Petals pink, red or purple, sometimes white, never yellow; hypanthium ≤11mm 5
5 Seeds without hairy plume; hypanthium 2-11mm; all leaves alternate **5. CLARKIA**
5 Seeds with hairy plume; hypanthium 0-c.3mm; lower leaves often opposite 6
 6 At least lowest leaves opposite; flowers ± erect when open, actinomorphic **1. EPILOBIUM**
 6 All leaves alternate; flowers held horizontally, slightly zygomorphic **2. CHAMERION**

1. EPILOBIUM L. - *Willowherbs*
Perennial herbs; at least lower leaves opposite or rarely whorled; flowers in loose terminal racemes, actinomorphic; hypanthium very short; sepals 4; petals 4, pink to purple, rarely white; stamens 8; ovary 4-celled; fruit a linear capsule; seeds with a hairy plume.

Plants vary greatly in stature, leaf-size, and degree of branching and of pubescence, but the *type* of hairs and certain aspects of leaf-shape are relatively constant. Seed-coat ornamentation is highly diagnostic, as is the presence of a terminal appendage, but a high magnification (x≥20) is required.

Hybrids occur commonly where 2 or more spp. occur together, especially in quantity for several years in disturbed ground. Hybrids are often recognizable by their larger and more-branched stature, longer flowering season, unusually large or small flowers markedly more darkly coloured at petal-tips, and partially or entirely abortive fruits. Most seeds are abortive but some are fertile; backcrossing and even triple-hybrids rarely occur. Most hybrids are variously intermediate in diagnostic characters, notably stigma-form and type of pubescence.

Multi-access Key to spp. of Epilobium
Stigma 4-lobed A
Stigma clavate B
 Stem hairs all ± appressed C
 Some stem-hairs patent or otherwise spreading, at least near stem apex D
Spreading hairs 0 or all glandular E
Some spreading hairs eglandular F
 Seeds truncate to gradually rounded at hairy end G
 Seeds with extra appendage 0.05-0.2mm long at hairy end H
Seeds minutely uniformly papillose I
Seeds with longitudinal papillose ridges J

Seeds obscurely reticulate, not papillose K
Stems decumbent to erect, at least stem apex and inflorescence axis upturned L
Stems procumbent, only individual flowers erect M

ADFGIL	Petals 10-16mm, purplish-pink; leaves slightly clasping stem at base, white subterranean rhizomes produced **1. E. hirsutum**
	Petals 5-9mm, paler; leaves not clasping stem at base; green surface short stolons produced **2. E. parviflorum**
ADEGIL	Lower leaves ovate, rounded at base, abruptly delimited from petiole 2-6mm **3. E. montanum**
	Lower leaves narrowly elliptic, cuneate at base, gradually narrowed to petiole 3-10mm **4. E. lanceolatum**
BCEGIL	Patent glandular hairs 0; perennating by ± sessile lax leaf-rosettes; capsules (5.5)6.5-8(10)cm **5. E. tetragonum**
	Patent glandular hairs present on hypanthium and sometimes capsule; perennating by elongated leafy stolons; capsules (3)4-6(6.5)cm **6. E. obscurum**
BDEGIL	Petioles 4-15mm; plant perennating by sessile leafy rosettes **7. E. roseum**
BDEHIL	Petioles ≤4mm; plant perennating by long slender stolons ending in tight bud **9. E. palustre**
BDEHJL	Petioles ≤4mm; plant perennating by sessile leafy rosettes **8. E. ciliatum**
B(CD)EHKL	Leaves narrowly elliptic-oblong, entire to denticulate; stolons on soil surface, with green leaves **10. E. anagallidifolium**
	Leaves ovate to narrowly ovate, distinctly dentate; stolons below soil surface, with yellowish scale-leaves **11. E. alsinifolium**
B(CD)EGIM	Leaves entire or nearly so, usually green and with obscure veins on upperside **12. E. brunnescens**
	Leaves distinctly dentate, green, with obscure veins on upperside **13. E. pedunculare**
B(CD)EGKM	Leaves entire, bronzy, with ± prominent veins on upperside **14. E. komarovianum**

1. E. hirsutum L. - *Great Willowherb.* Rhizomatous; stems erect, to 1.8m, densely shaggy-pubescent with spreading non-glandular and glandular hairs; leaves lanceolate to narrowly oblong, ± clasping stem and slightly decurrent at base; petals usually bright pinkish-purple, (6)10-16(18)mm; 2n=36. Native; in all sorts of wet or damp places; common throughout lowland BI except N & C Sc.

1 x 2. E. hirsutum x E. parviflorum = E. x subhirsutum Gennari has been found scattered in En, Wa, N Ir and Guernsey.

1 x 3. E. hirsutum x E. montanum = E. x erroneum Hausskn. has been found scattered in En, Wa and C Sc.

1 x 5. E. hirsutum x E. tetragonum = E. x brevipilum Hausskn. has been found scattered in En.

1 x 7. E. hirsutum x E. roseum = E. x goerzii Rubner has been found scattered in En.

1 x 8. E. hirsutum x E. ciliatum = E. x novae-civitatis Smejkal has been found scattered in En, C Sc and Co Dublin.

1 x 9. E. hirsutum x E. palustre = E. x waterfallii E.S. Marshall has been found scattered in En.

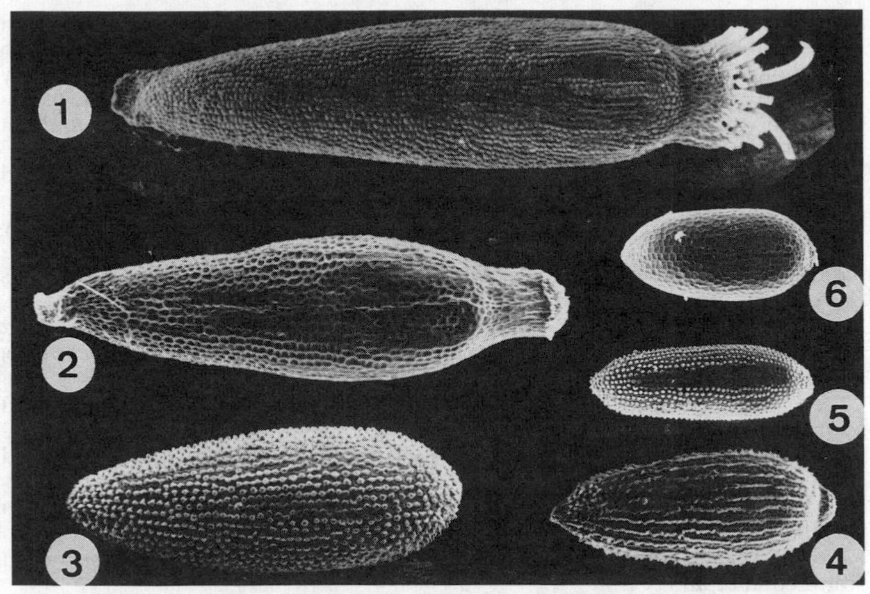

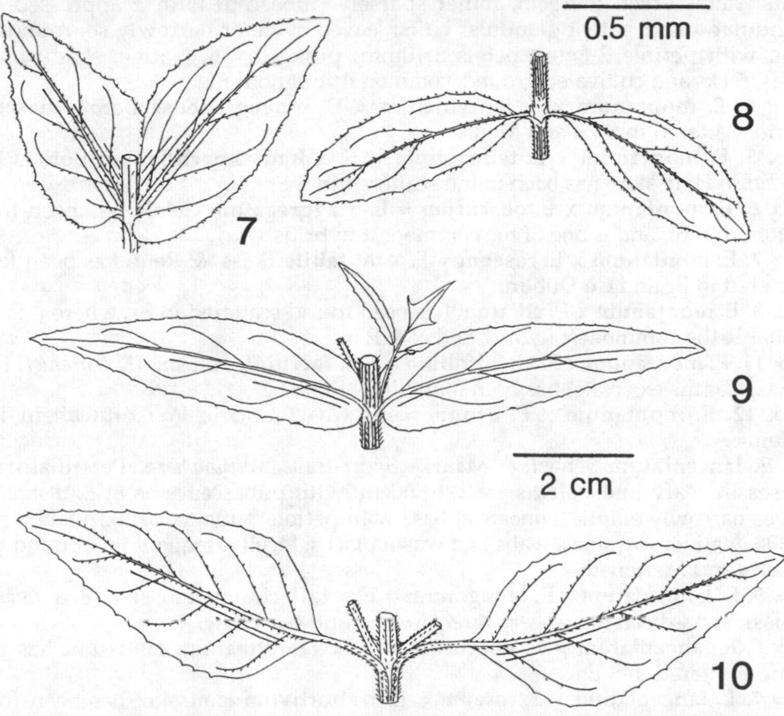

FIG 445 - *Epilobium*. 1-6, seeds. 1, *E. palustre*. 2, *E. alsinifolium*. 3, *E. montanum*. 4, *E. ciliatum*. 5, *E. brunnescens*. 6, *E. komarovianum*. 7-10, leaves. 7, *E. montanum*. 8, *E. lanceolatum*. 9, *E. ciliatum*. 10, *E. roseum*.

2. E. parviflorum Schreb. - *Hoary Willowherb.* Perennating by leaf rosettes or short leafy stolons; stems erect, to 75cm, densely rather matted-pubescent with spreading non-glandular and glandular hairs; leaves oblong-lanceolate to narrowly so, sessile, rounded at base and not clasping stem; petals pale pinkish-purple, 5-9mm; 2n=36. Native; in all sorts of wet or damp places; frequent throughout lowland BI.

2 x 3. E. parviflorum x E. montanum = E. x limosum Schur has been found scattered in Br and N Ir, and is one of the commonest hybrids. *E. parviflorum* x *E. montanum* x *E. obscurum* and *E. parviflorum* x *E. montanum* x *E. roseum* probably also occur.

2 x 4. E. parviflorum x E. lanceolatum = E. x aschersonianum Hausskn. was found last century in S Devon.

2 x 5. E. parviflorum x E. tetragonum = E. x palatinum F.W. Schultz (*E. x weissenbergense* F.W. Schultz) has been found scattered in S & C En. *E. parviflorum* x *E. tetragonum* x *E. obscurum* possibly also occurs.

2 x 6. E. parviflorum x E. obscurum = E. x dacicum Borbás has been found scattered in much of BI.

2 x 7. E. parviflorum x E. roseum = E. x persicinum Rchb. has been found scattered in En and Wa.

2 x 8. E. parviflorum x E. ciliatum has been found scattered in En and Sc and in Co Sligo and Guernsey.

2 x 9. E. parviflorum x E. palustre = E. x rivulare Wahlenb. has been found scattered in Br and N Ir.

3. E. montanum L. - *Broad-leaved Willowherb.* Perennating by subsessile leafy 445
buds; stems erect, to 75cm, rather sparsely pubescent with ± appressed non-glandular and ± patent glandular hairs; leaves ovate or narrowly so, rounded at base with petiole 2-6mm; petals 8-10mm, pink; 2n=36. Native; shady places, walls, rocks and cultivated ground; common throughout BI.

3 x 4. E. montanum x E. lanceolatum = E. x neogradense Borbás has been found scattered in S En and Mons.

3 x 5. E. montanum x E. tetragonum = E. x haussknechtianum Borbás (*E. x beckhausii* Hausskn.) has been found scattered in S En.

3 x 6. E. montanum x E. obscurum = E. x aggregatum Celak. has been found scattered in BI, and is one of the commonest hybrids.

3 x 7. E. montanum x E. roseum = E. x mutabile Boiss. & Reut. has been found scattered in Br and Co Dublin.

3 x 8. E. montanum x E. ciliatum has been found scattered in Br, where it is now probably the commonest hybrid, and N & E Ir.

3 x 11. E. montanum x E. alsinifolium = E. x facchinii Hausm. (*E. x grenieri* Rouy & E.G. Camus ex Focke) has been found in N En & Sc.

3 x 12. E. montanum x E. brunnescens was found in W Cornwall in 1995; endemic.

4. E. lanceolatum Sebast. & Mauri - *Spear-leaved Willowherb.* Perennating by 445
subsessile leafy buds; stems erect, to 60cm; with pubescence as in *E. montanum*; leaves narrowly elliptic, cuneate at base with petiole 3-10mm; petals 6-8mm, pink; 2n=36. Native; waysides, walls and waste places; locally frequent in S Br and CI N to Leics and Merioneth.

4 x 5. E. lanceolatum x E. tetragonum = E. x fallacinum Hausskn. (*E. x ambigens* Hausskn.) has been found very rarely in S En and Guernsey.

4 x 6. E. lanceolatum x E. obscurum = E. x lamotteanum Hausskn. has been found scattered in S En.

4 x 7. E. lanceolatum x E. roseum = E. x abortivum Hausskn. has been found rarely in S & C En.

4 x 8. E. lanceolatum x E. ciliatum has been found rarely in S En; endemic?

5. E. tetragonum L. (*E. adnatum* Griseb., *E. lamyi* F.W. Schultz, *E. semiadnatum* Borbás) - *Square-stalked Willowherb.* Perennating by subsessile loose leaf-rosettes;

stems erect, to 75cm; stems and inflorescences with usually dense, white, appressed hairs alone; leaves narrowly oblong to oblong-lanceolate; petals 5-7mm, pale purplish-pink; 2n=36. Native; hedgerows, open woods, by water, cultivated and waste ground; locally common in CI, S & C Br, very scattered in Ir and N Br. Plants sometimes separated as ssp. **lamyi** (F.W. Schultz) Nyman have shortly petiolate, non-decurrent leaves, rather than ± sessile leaves ± decurrent to stem-ridges, but all intermediates occur.

5 x 6. E. tetragonum x E. obscurum = E. x semiobscurum Borbás (*E. x thuringiacum* Hausskn.) has been found scattered in En, Wa and Jersey.

5 x 7. E. tetragonum x E. roseum = E. x borbasianum Hausskn. (*E. x dufftii* Hausskn.) has been found rarely in S En.

5 x 8. E. tetragonum x E. ciliatum (=*E. x iglaviense* Smejkal nom. nud.) has been recorded from Carms, Surrey and Bucks.

5 x 9. E. tetragonum x E. palustre = E. x laschianum Hausskn. (*E. x probstii* H. Lév.) has been recorded from E Kent and Surrey.

6. E. obscurum Schreb. - *Short-fruited Willowherb.* Perennating by ± elongated leafy stolons; stems erect, to 75cm; pubescence as in *E. tetragonum* but patent glandular hairs present on hypanthium and sometimes a few on fruit; leaves narrowly elliptic-ovate to lanceolate, ± sessile and ± decurrent on to stem; petals 4-7mm, pinkish-purple; 2n=36. Native; same habitats as *E. tetragonum*; frequent ± throughout BI.

6 x 7. E. obscurum x E. roseum = E. x brachiatum Celak. has been found scattered in En.

6 x 8. E. obscurum x E. ciliatum has been found scattered in Br, where it is now one of the commonest hybrids, and S & NW Ir.

6 x 9. E. obscurum x E. palustre = E. x schmidtianum Rostk. has been found scattered in Br and Ir.

6 x 10. E. obscurum x E. anagallidifolium = E. x marshallianum Hausskn. has been found in Stirlings and W Sutherland.

6 x 11. E. obscurum x E. alsinifolium = E. x rivulicola Hausskn. has been found in W Perth and Banffs.

6 x 12. E. obscurum x E. brunnescens was found in 1981 on a spoil-heap in Co Antrim, and in 1992 in Tyrone; endemic. The erect parent was previously misdetermined as *E. montanum* or *E. ciliatum.*

7. E. roseum Schreb. - *Pale Willowherb.* Perennating by subsessile leaf-rosettes; **445**
stems erect, to 75cm, with often abundant but sometimes very sparse patent glandular and appressed non-glandular hairs; leaves ovate-elliptic to narrowly so, gradually narrowed to petiole 4-15mm; petals 4-7mm, usually pale pink; 2n=36. Native; shady places, damp ground, cultivated and waste land; scattered throughout most of Br and Ir and locally frequent, apparently decreasing.

7 x 8. E. roseum x E. ciliatum (=*E. x nutantiflorum* Smejkal nom. nud.) has been found scattered in En and Sc.

7 x 9. E. roseum x E. palustre = E. x purpureum Fr. was collected once in Surrey.

8. E. ciliatum Raf. (*E. adenocaulon* Hausskn.) - *American Willowherb.* Perennating **445**
by ± sessile leaf-rosettes; stems erect, to 75(100)cm, with pubescence as in *E. roseum;* leaves oblong-lanceolate, rounded to subcordate at base with petiole 1.5-4mm; petals 3-6mm, pinkish-purple; 2n=36. Intrd; waste and cultivated ground, by roads, rivers and railways, on walls; 1st found in 1891, still spreading, now over most of BI and the commonest sp. in much of S Br; N America. Some claim **E. adenocaulon** (incl. all our plants) is distinct, but this name seems predated by **E. watsonii** Barbey.

8 x 9. E. ciliatum x E. palustre (=*E. x fossicola* Smejkal nom. nud.) has been recorded from N Hants, Lanarks, Dunbarton and Co Down.

8 x 12. E. ciliatum x E. brunnescens was found in W & E Cornwall and Caerns in 1995; endemic.

9. E. palustre L. - *Marsh Willowherb.* Perennating by small tight buds on ends of **445**

long filiform rhizomes; stems erect, to 60cm, with often sparse appressed (or ± so) non-glandular hairs and fewer patent glandular hairs near top, with scarcely any raised lines; leaves lanceolate to linear-lanceolate, narrowed at base, sessile or with petiole up to 4mm; petals pale pink, 4-7mm; 2n=36. Native; marshes, fens and ditches, often with *E. parviflorum*; frequent throughout BI.

9 x 10. E. palustre x E. anagallidifolium = E. x dasycarpum Fr. has been found in Easterness and W Sutherland.

9 x 11. E. palustre x E. alsinifolium = E. x haynaldianum Hausskn. has been found scattered in N En and Sc.

9 x 12. E. palustre x E. brunnescens was found in Cards in 1995; endemic.

10. E. anagallidifolium Lam. - *Alpine Willowherb*. Perennating by leafy stolons; stems ascending to decumbent, to 20cm, with sparse appressed non-glandular hairs; patent glandular hairs very sparse, mostly on hypanthium and capsules; leaves narrowly elliptic-oblong, gradually narrowed to short petiole-like base; petals 3-4.5mm, pinkish-purple; 2n=36. Native; mountain flushes and streamsides; locally frequent in N En and Sc.

10 x 11. E. anagallidifolium x E. alsinifolium = E. x boissieri Hausskn. has been found on mountains in C & N Sc.

11. E. alsinifolium Vill. - *Chickweed Willowherb*. Perennating by slender ± **445** subterranean stolons; stems (erect or) ascending to decumbent, to 25(30)cm; pubescence as in *E. anagallidifolium* or with slightly more glandular hairs; leaves narrowly ovate to lanceolate, rounded to short petiole at base; petals 7-10mm, reddish-purple; 2n=36. Native; similar habitats to *E. anagallidifolium* and similar distribution, but also Caerns.

12. E. brunnescens (Cockayne) P.H. Raven & Engelhorn (*E. pedunculare* auct. non **445** A. Cunn., *E. nerteroides* auct. non A. Cunn.) - *New Zealand Willowherb*. Stems procumbent, to 20cm, with sparse minute glandular and non-glandular hairs; leaves ± orbicular, often purplish on lowerside, 3-7(10)mm, with petioles 0.5-3mm; flowers solitary and erect in leaf-axils; petals 2.5-4mm, white to pale pink; 2n=36. Intrd; all sorts of damp barish ground, especially gravelly hillsides, railway sidings, waste-tips; first collected in 1908, still spreading, now over most of BI; New Zealand.

13. E. pedunculare A. Cunn. (*E. linnaeoides* Hook. f.) - *Rockery Willowherb*. Similar to a robust *E. brunnescens*, but with leaves 3-10(14)mm, acutely dentate, not purplish on lowerside; petals 3-5mm; (2n=36). Intrd; weed of barish damp ground natd by roads in W Perth, W Galway and W Mayo, 1st found 1953, rare garden weed in En and Sc; New Zealand.

14. E. komarovianum H. Lév. (*E. inornatum* Melville) - *Bronzy Willowherb*. Similar **445** to a small *E. brunnescens*, but with entire, broadly ovate-elliptic leaves 2-6(10)mm, usually green on lowerside, bronzy on upperside; (2n=36). Intrd; barely natd garden weed in few places in En, Sc and N Ir; New Zealand.

2. CHAMERION (Raf.) Raf. (*Chamaenerion* Ség. nom. illeg.) - *Rosebay Willowherb*
Perennial rhizomatous herbs; all leaves alternate; flowers in dense terminal racemes, slightly zygomorphic; hypanthium ± 0; sepals 4; petals 4, pinkish-purple, rarely white; stamens 8; ovary 4-celled; fruit a linear capsule; seeds with a hairy plume.

Other spp. - C. dodonaei (Vill.) Holub (*Chamaenerion dodonaei* (Vill.) Schur, *Epilobium dodonaei* Vill.), from C & S Europe, differs in its linear leaves but is only a very rare casual or escape.

1. C. angustifolium (L.) Holub (*Epilobium angustifolium* L., *Chamaenerion angustifolium* (L.) Scop. nom. illeg.) - *Rosebay Willowherb*. Stems erect, to 1.5m; leaves narrowly elliptic to oblong-lanceolate, ± sessile; flowers held horizontally; 2 upper petals wider than 2 lower ones; style deflexed; stigma 4-lobed; 2n=36.

Native; waste ground, woodland-clearings, embankments, rocky places and screes on mountains; throughout BI and often abundant.

3. LUDWIGIA L. - *Hampshire-purslane*
Annual to perennial ± aquatic herbs; all leaves opposite; flowers solitary in leaf-axils, actinomorphic; hypanthium 0; sepals 4; petals 0; stamens 4; ovary 4-celled; fruit a short, cylindrical, scarcely dehiscent capsule retaining sepals; seeds not plumed.

1. L. palustris (L.) Elliott - *Hampshire-purslane.* Stems procumbent to decumbent or ascending, sometimes floating at ends, to 30(60)cm; leaves ovate-elliptic, petiolate; flowers 2-5mm, inconspicuous; stigma capitate; 2n=16. Native; acid pools; extremely local in New Forest (S Hants) and Epping Forest (S Essex), formerly E & W Sussex and Jersey, found in 1991 in dew-pond on chalk in E Sussex but probably intrd, found intrd in Surrey in 1995. **RR**

4. OENOTHERA L. - *Evening-primroses*
Annual to biennial (rarely perennial) herbs; all leaves alternate; flowers in terminal racemes, ± actinomorphic; hypanthium 20-45mm, a narrow tube; sepals 4; petals 4, yellow, sometimes streaked or tinged red; stamens 8; ovary 4-celled; fruit a long ± cylindrical capsule; seeds not plumed.

A critical genus where species limits are a matter of opinion; that of K. Rostanski is followed here. Flower measurements refer to the lower (first opened) ones of the inflorescence (later ones may be much smaller); fruit pubescence should also be noted on the lower part of the inflorescence. Hybrids often occur wherever 2 or more spp. occur together; reciprocal crosses often have different characters. The hybrids are fertile and may backcross and form triple-hybrids (e.g. *O. biennis* x *O. glazioviana* x *O. cambrica*), but these are very difficult to identify.

1 Capsule widest near apex, c.2-4mm wide near base; petals tinged reddish when withering; seeds not angled — **5. O. stricta**
1 Capsule widest near base, there c.6-8mm wide; petals always yellow; seeds sharply angled — 2
 2 Petals 3-5cm; style longer than stamens, the stigmas held above anthers — **1. O. glazioviana**
 2 Petals 1.5-3cm; style shorter than stamens, the stigmas held ± at same level as anthers — 3
3 Green parts of stems and fruits without red bulbous-based hairs — 4
3 Green parts of stems and fruits with red bulbous-based hairs — 5
 4 Petals 1.5-3cm, wider than long; capsules 2-3(3.5)cm, all with glandular hairs — **3. O. biennis**
 4 Petals 2-3cm, c. as wide as long; capsules 3-4(4.5)cm, only upper with glandular hairs — **4. O. cambrica**
5 Sepals red-striped; rhachis reddish towards apex; capsules 2-3cm, all with glandular hairs — **2. O. x fallax**
5 Sepals and rhachis green; capsules 3-4(4.5)cm, only upper with glandular hairs — **4. O. cambrica**

Other spp. - 10 other spp. have been recorded as casuals; the 3 following are the least rare. **O. perangusta** R.R. Gates and **O. rubricaulis** Kleb. both have petals 10-20mm but differ from *O. biennis* in having red bulbous-based hairs on green parts of stems and capsules. *O. perangusta*, from N America, has the hypanthium c.30-32mm and glabrous patches on capsules; *O. rubricaulis*, probably originating in Europe, has the hypanthium 15-25mm and capsules pubescent all over. **O. rosea** L'Hér. ex Aiton, from C America, has a capsule with c.1mm wide wings, pink to violet petals only 5-10mm and a hypanthium <10mm. Several spp. have been

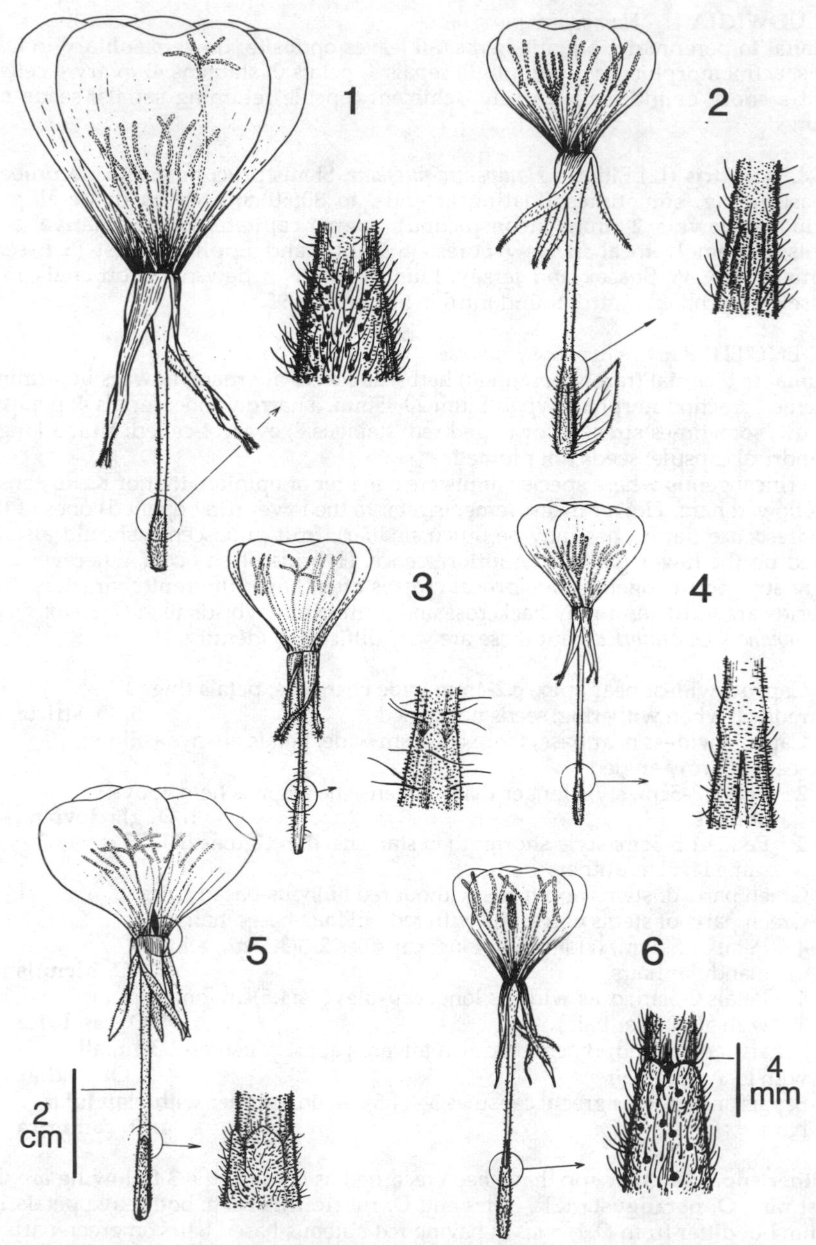

FIG 450 - Flowers of *Oenothera*. 1, *O. glazioviana*. 2, *O. x fallax*. 3, *O. stricta*. 4, *O. biennis*. 5, *O. x britannica*. 6, *O. cambrica*. Drawings by J. Zygmunt.

recorded erroneously (e.g. **O. grandiflora** Aiton and **O. ammophila** Focke), and several much over-recorded as errors for the 5 below (e.g. **O. parviflora** L. for *O. cambrica*). The genus is now being grown as an oilseed crop, and other alien spp. may occur in future; *O. glazioviana*, *O. biennis*, *O. parviflora* and **O. renneri** H. Scholz seem to be the spp. most utilized.

1. O. glazioviana P. Micheli ex Mart. (*O. erythrosepala* Borbás) - *Large-flowered* 450
Evening-primrose. Stems erect, to 1.8m; green parts of stems and fruits with many red bulbous-based hairs; rhachis red towards apex; sepals red-striped; petals 3-5cm, wider than long; (2n=14). Intrd; natd on sand-dunes, waste ground, waysides; common in suitable places in BI except N & C Sc; N America.

1 x 4. O. glazioviana x O. cambrica = O. x britannica Rostanski occurs as 1 x 4 450
and 4 x 1 with the parents in C & S En and S Wa; ?endemic.

2. O. x fallax Renner (*O. glazioviana* x *O. biennis*) - *Intermediate Evening-primrose*. 450
Stems erect, to 1.5m; usually differs from *O. glazioviana* in shorter style with stigmas at level of anthers and petals 2-3cm; (2n=14). Intrd or originating in situ; same habitats as *O. glazioviana*; scattered in Br N to C Sc, CI. A stable derivative of female *O. glazioviana* x male *O. biennis* is sometimes found in absence of both parents; it is ± constant in appearance, unlike the reciprocal cross, which has petals often >3cm and style often exceeding stamens, and is treated by Rostanski as a sp. Backcrosses to both parents occur.

2 x 4. O. x fallax x O. cambrica, effectively a triple-hybrid, has been found as both 2 x 4 and 4 x 2 in N Hants, Warks and Guernsey; ?endemic.

3. O. biennis L. - *Common Evening-primrose*. Stems erect, to 1.5m; green parts of 450
stems and fruits without red bulbous-based hairs (these present on red blotches on stems); rhachis green at tip, sepals green; petals 1.5-3cm, wider than long; (2n=14). Intrd; same habitats as *O. glazioviana*; frequent in suitable places in Br N to C Sc, perhaps CI; originated in Europe.

3 x 4. O. biennis x O. cambrica occurs as both 3 x 4 and 4 x 3 with the parents in C & S En and S Wa; ?endemic.

4. O. cambrica Rostanski (*O. novae-scotiae* auct. non R.R. Gates) - *Small-flowered* 450
Evening-primrose. Stems erect, to 1.2m; green parts of stems and fruits usually with, sometimes without (but probably because of hybridization with *O. biennis*), red bulbous-based hairs; rhachis green at tip; sepals green; petals 2-3cm, c. as wide as long; 2n=14. Intrd; natd in maritime sandy places and inland where intrd from coast; frequent by coast in Br N to Yorks and Cumberland, CI; probably N America.

5. O. stricta Ledeb. ex Link - *Fragrant Evening-primrose*. Stems erect to ascending, 450
to 1m; green parts of stems and fruits without red bulbous-based hairs; rhachis usually reddened towards tip; sepals red-striped and -suffused; petals 1.5-3.5cm, c. as wide as long; style c. as long as stamens; (2n=14). Intrd; natd in sandy places mostly on coasts; locally frequent in CI and Br N to Norfolk and S Wa, rare casual further N; Chile.

5. CLARKIA Pursh (*Godetia* Spach) - *Clarkias*
Annuals; all leaves alternate; flowers in loose terminal racemes, actinomorphic; hypanthium 2-11mm, narrowly tubular; sepals 4; petals 4, pink to purple, rarely white; stamens 8; ovary 4-celled; fruit a linear capsule; seeds not plumed.

1. C. unguiculata Lindl. - *Clarkia*. Stems erect to ascending, to 50(80)cm; leaves lanceolate to ovate, glabrous; flower-buds pendent; hypanthium 2-5mm; sepals 10-16mm; petals 1-2cm, with claw c. as long as limb; often *flore pleno*; (2n=18). Intrd; commonly grown in gardens and frequent casual on tips, in parks and waste places; scattered in Br; California.

2. C. amoena (Lehm.) A. Nelson & J.F. Macbr. - *Godetia*. Stems erect to ascending, to 50(80)cm; leaves lanceolate, minutely pubescent; flower-buds erect; hypanthium

5-30mm; sepals 12-30mm; petals 1-4(6)cm, with claw much shorter than limb; often *flore pleno*; (2n=14). Intrd; found as for *C. unguiculata;* W N America.

6. FUCHSIA L. - *Fuchsias*
Deciduous shrubs; all leaves opposite or sometimes in whorls of 3-4; flower solitary in leaf-axils, actinomorphic, pendent on long pedicels; hypanthium 5-16mm, broadly tubular; sepals 4; petals 4, pink to purple or violet, rarly white; stamens 8; ovary 4-celled; fruit a ± cylindrical black berry.

1. F. magellanica Lam. - *Fuchsia*. Spreading shrub to 1.5(3)m; leaves ovate to elliptic, 2.5-5.5cm; sepals 12-24 x 4-10mm, bright red; petals 6-12mm, violet; berry black, 15-22mm; (2n=22, 44). Intrd; planted as hedging in Ir, Man and W Br; natd in Ir (mainly S & W) and W Br N to Orkney, planted or non-persistent outcast elsewhere, scarcely natd in CI; S Chile and Argentina. More than 1 taxon occurs, but they have not been worked out. Possibly the commonest taxon natd in W BI is **F. 'Riccartonii'**, also known as var. **macrostema** (Ruiz & Pav.) Munz, with fatter buds and wider sepals than *F. magellanica*, and not seeding; it is of garden origin. The narrow-sepalled plant is sometimes self-sown in W Ir and Man.

2. F. 'Corallina' (?*F*. 'Exoniensis') - *Large-flowered Fuchsia*. Differs from *F. magellanica* in larger parts: leaves 3-8cm; sepals 25-40mm; petals 15-25mm. Intrd; natd as for *F. magellanica* in Lleyn (Caerns) and Lundy Island (N Devon); garden origin.

7. CIRCAEA L. - *Enchanter's-nightshades*
Perennial rhizomatous and/or stoloniferous herbs; all leaves opposite; flowers in loose terminal racemes, ± zygomorphic, ± horizontally held; hypanthium very short; sepals 2; petals 2, deeply 2-lobed, white or pinkish; stamens 2; ovary 1-2-celled, each cell with 1 seed; fruit a 1-2-seeded achene.

1 Open flowers crowded at inflorescence apex; pedicels, hypanthia and sepals glabrous; ovary 1-celled **3. C. alpina**
1 Open flowers on elongated raceme; pedicels, hypanthia and sepals with glandular hairs; ovary 2-celled 2
 2 Stolons produced from lower leaf-axils; petioles pubescent on upperside, subglabrous on lowerside; ovary with 1 large and 1 small cell; fruit not ripening **2. C. x intermedia**
 2 Stolons 0; petioles pubescent all round; ovary with 2 equal cells; fruit ripening **1. C. lutetiana**

1. C. lutetiana L. - *Enchanter's-nightshade*. Stems erect, to 60cm; leaves 4-10cm, ovate, truncate to weakly cordate at base, acuminate at apex, remotely denticulate; hypanthium 1-1.2mm; nectariferous disc conspicuous round style-base, 0.2-0.4mm high; petals 2-4mm; filaments 2.5-5.5mm; fruits 3-4 x 2-2.5mm; 2n=22. Native; woods, hedgerows and other shady places; common throughout BI but rare in N Sc.

2. C. x intermedia Ehrh. (*C. lutetiana* x *C. alpina*) - *Upland Enchanter's-nightshade*. Stems erect, to 45cm; differs from *C. lutetiana* in leaves cordate at base, abruptly acuminate, dentate; hypanthium 0.5-1.2mm; nectariferous disc inconspicuous around style-base, ≤0.2mm high; petals 1.8-4mm; filaments 2-5mm; fruits ≤2 x 1.2mm; and see key; 2n=22. Native; woods and shady rocky places, often on mountains; locally frequent in N & W Br S to S Wa and Derbys, very scattered in N & C Ir, often in absence of *C. alpina* or both parents.

3. C. alpina L. - *Alpine Enchanter's-nightshade*. Stems erect, to 30cm; stolons R
produced from lower leaf-axils; leaves cordate at base, acute to abruptly acuminate, strongly dentate; hypanthium 0.1-0.2mm; disc 0; petals 0.6-1.4mm; filaments 1-1.5mm; fruits c.2 x 1mm; 2n=22. Native; same habitats as *C. x intermedia*, much over-recorded for it, but more upland; very scattered in Wa, Lake

District and Sc.

87. CORNACEAE - *Dogwood family*

Deciduous or evergreen shrubs or perennial herbs; leaves opposite or alternate, simple, sessile or petiolate, without stipules. Flowers in terminal panicles, often ± umbel-like, bisexual or dioecious, epigynous, actinomorphic; hypanthium 0; sepals and petals 4 or 5, free; stamens 4 or 5, alternate with petals; ovary 1-2-celled, each cell with 1 apical ovule, or ovary 3-celled with 2 cells empty and 1 with apical ovule; styles 1 or 3; stigma capitate; fruit a drupe with 1 1-2-celled stone.

Distinguished by the usually opposite leaves, inferior ovary with 1 ovule per cell, 4 or 5 free petals, and drupe.

1 Leaves deciduous, not thick and glossy; flowers bisexual **1. CORNUS**
1 Leaves evergreen, thick and glossy; dioecious 2
2 Leaves opposite; petals dark purple, 4 **2. AUCUBA**
2 Leaves alternate; petals yellowish-green, 5 **3. GRISELINIA**

1. CORNUS L. (*Swida* Opiz, *Thelycrania* (Dumort.) Fourr., *Chamaepericlymenum* Hill) - *Dogwoods*

Perennial herbs or deciduous shrubs; leaves opposite, ± entire; flowers in corymbs or umbels, bisexual; sepals and petals 4; style 1; ovary 2-celled; fruit a drupe with 1 2-celled stone.

1 Rhizomatous herbs; petals (not the petal-like bracts) purple **5. C. suecica**
1 Shrubs; petals white to yellow 2
2 Inflorescences appearing before leaves, with 4 yellow petal-like bracts at base **4. C. mas**
2 Inflorescences appearing after leaves, without petal-like bracts at base; petals whitish 3
3 Fruit purplish-black; leaves rarely with >5 pairs of lateral veins; petals 4-7mm **1. C. sanguinea**
3 Fruit white to cream; many larger leaves with 6(-7) pairs of lateral veins; petals 2-4mm 4
4 Stone ellipsoid, tapered to flat base; leaves shortly and abruptly acuminate to acute **3. C. alba**
4 Stone subglobose, rounded at base; leaves tapering acuminate **2. C. sericea**

1. C. sanguinea L. (*Thelycrania sanguinea* (L.) Fourr., *Swida sanguinea* (L.) Opiz) - *Dogwood*. Shrub to 4m; bark of 1st year twigs dark red (at least on 1 side) in winter; leaves ovate to elliptic, 4-8cm, abruptly acuminate; fruit purplish-black, 5-8mm; 2n=22. Native; woods and scrub on limestone or base-rich clays; common in most of S & C lowland Br, very local in S Ir, escape elsewhere.

2. C. sericea L. (*C. stolonifera* Michx., *Thelycrania sericea* (L.) Dandy, *Swida sericea* (L.) Holub) - *Red-osier Dogwood*. Shrub to 3m; bark of 1st year twigs dark red or greenish-yellow in winter; leaves ovate to elliptic, 4-10cm, acuminate; fruit white to cream, 4-7mm, often not ripening; (2n=22). Intrd; much grown in parks and on roadsides, frequently natd by suckers; scattered in most of lowland BI; N America.

3. C. alba L. (*Thelycrania alba* (L.) Pojark.) - *White Dogwood.* Shrub to 3m; bark of 1st year twigs usually bright red in winter; differs from *C. sericea* in key characters, and less extensively suckering; (2n=22). Intrd; grown as for *C. sericea*, but less well natd; very scattered in lowland Br; E Asia. Possibly not a separate sp. from *C. sericea*.

4. C. mas L. - *Cornelian-cherry.* Shrub or small tree to 4(8)m; bark of 1st year twigs dull greenish-grey in winter; leaves 4-10cm, ovate to elliptic, acute to abruptly acuminate; fruit 12-15mm, red, usually not ripening; (2n=18, 54). Intrd; grown in hedges and roadside verges and often long persistent, but scarcely natd; scattered in S Br, but rarely N to C Sc, Man; Europe.

5. C. suecica L. (*Chamaepericlymenum suecicum* (L.) Asch. & Graebn.) - *Dwarf Cornel.* Stems erect, to 20cm, with terminal inflorescence subtended by 4 white bracts longer than flower-cluster; leaves ± sessile, ovate to elliptic, 1-3cm, with 3-5 veins all from base; fruits 5-10mm, bright red; (2n=22). Native; upland moors among low shrubs; extremely local in N En, locally frequent in W & C mainland Sc.

2. AUCUBA Thunb. - *Spotted-laurel*
Evergreen shrubs; leaves opposite, entire to remotely serrate; flowers dioecious, the male in erect terminal panicles, the female in small terminal clusters; sepals and petals 4; ovary 1-celled; style 1; fruit a 1-seeded drupe.

1. A. japonica Thunb. - *Spotted-laurel.* Shrub to 5m; leaves 8-20cm, lanceolate to narrowly ovate, tapering-acute, dark green but often with yellow blotches; drupes ellipsoid, 10-15mm, bright scarlet; (2n=32). Intrd; very commonly planted in shrubberies but rarely self-sown; very scattered in Br, mainly W, N to MW Sc, Man; Japan.

3. GRISELINIA G. Forst. - *New Zealand Broadleaf*
Evergreen shrubs; leaves alternate, entire; flowers dioecious, in axillary racemes or panicles; sepals and petals 5; ovary 1-2-celled; styles 3; fruit a 1-seeded drupe.

1. G. littoralis (Raoul) Raoul - *New Zealand Broadleaf.* Shrub to 3m or rarely tree to 15m; leaves 3-10cm, broadly ovate to broadly elliptic, rather yellowish-green; drupes 6-7mm, dark purple; (2n=36). Intrd; commonly planted in S & W, especially near sea, persistent and sometimes self-sown; very scattered in S & W Br N to C Sc, Man, Co Down; New Zealand.

88. SANTALACEAE - *Bastard-toadflax family*

Semi-parasitic herbaceous perennials; leaves alternate, simple, entire, sessile, without stipules. Flowers each with 3 bracteoles, in simple or branched terminal raceme-like cymes, bisexual, epigynous, actinomorphic; hypanthium short, funnel-shaped, with 5 free tepals at apex; stamens 5; ovary 1-celled, with 3 ovules; style 1, stigma capitate; fruit a 1-seeded nut.

Easily recognized by the habit, and the epigynous flowers with 5 tepals which are retained on the nut.

1. THESIUM L. - *Bastard-toadflax*
1. T. humifusum DC. - *Bastard-toadflax.* Stems procumbent to weakly ascending, R
to 20cm; leaves yellowish-green, linear, 1-veined, 5-25mm; flowers yellowish-green, 2-3mm, elongating to c.4mm in fruit; 2n=c.26. Native; chalk and limestone grassland, a root parasite on various herbs; very local in En N to S Lincs and E Gloucs, Jersey and Alderney.

89. VISCACEAE - *Mistletoe family*
(Loranthaceae *pro parte*)

Semi-parasitic evergreen shrubs growing on tree-branches; leaves opposite, simple, ± sessile, without stipules. Flowers 3-5 in tiny apical cymes, ± sessile, dioecious, epigynous, actinomorphic; hypanthium ± 0; tepals (sepaloid petals) 4, nearly free (female flowers also with 4 rudimentary sepals); male flowers with 4 anthers sessile on tepals; female flowers with 1-celled ovary with 2 undifferentiated ovules fused to massive placenta and 1 capitate sessile stigma; fruit a 1-seeded berry.

Unmistakable in habit.

1. VISCUM L. - *Mistletoe*

1. V. album L. - *Mistletoe.* Stems green, divergently branching to form ± spherical loose mass to 2m across; leaves yellowish-green, 2-8cm, oblanceolate to narrowly obovate, rounded at apex; flowers very inconspicuous, Feb-Apr; fruit white, globose, 6-10mm, Nov-Dec; 2n=20. Native; on many spp. of tree, especially *Malus, Tilia, Crataegus* and *Populus*; En and Wa N to Yorks, mostly local but common in S part of En/Wa borders, rare introduction elsewhere.

90. CELASTRACEAE - *Spindle family*

Evergreen or deciduous shrubs or woody climbers; leaves opposite or alternate, simple, petiolate, with small or without stipules. Flowers greenish-yellow, small, in axillary cymes, variously dioecious to bisexual, hypogynous to slightly epigynous, actinomorphic; sepals 4-5, free or fused at base; petals 4-5, free; stamens 4-5; ovary 3-5-celled, each cell with (1-)2 ovules on axile placenta; style 1; stigma capitate or 3-lobed; fruit an often ± succulent 3-5-angled dehiscent capsule with 1(-2) seeds in each cell; seeds covered in bright orange to red aril.

The distinctive brightly coloured seeds are diagnostic; otherwise distinguished by the woody habit and ± hypogynous flowers with 4-5 sepals, petals and stamens, with 3-5 fused carpels, and with a large nectar-secreting disc at base.

1 Erect shrub with opposite leaves; capsule 4-5-celled — **1. EUONYMUS**
1 Woody climber with alternate leaves; capsule 3-celled — **2. CELASTRUS**

1. EUONYMUS L. - *Spindles*

Erect non-spiny shrubs with opposite leaves; ovary 4-5-celled; stigma capitate; capsule 4-5-celled, with 4-5 distinct rounded to winged lobes, pinkish-red or creamish-yellow on outside; aril orange.

1 Leaves leathery, evergreen; fruit with rounded lobes — **3. E. japonicus**
1 Leaves thin, deciduous; fruit with obtuse to winged lobes — 2
2 Terminal buds (Jul-Mar) <5mm; flowers mostly with 4 sepals and stamens; fruits with mostly 4 obtuse angles — **1. E. europaeus**
2 Terminal buds (Jul-Mar) >5mm; flowers mostly with 5 sepals and stamens; fruits with mostly 5 winged angles — **2. E. latifolius**

Other spp. - A single bush of **E. hamiltonianus** Wall., from Himalayas, has persisted on a common in Surrey; it differs from *E. europaeus* in being semi-evergreen and having purple (not yellow) anthers. **E. fortunei** (Turcz.) Hand.-Mazz., from China, has evergreen, non-shiny, sometimes variegated leaves and trailing stems rooting on contact with soil; it is much grown for ground-cover and sometimes occurs as a relic.

1. E. europaeus L. - *Spindle.* Much-branched shrub or small tree to 5(8)m; leaves deciduous, 3-8(12)cm, elliptic to narrowly so or narrowly obovate, acuminate, entire to serrulate; fruits with mostly 4 obtuse lobes, 8-15mm across, pinkish-red; 2n=32. Native; hedges, scrub and open woods on calcareous or base-rich soils; frequent in Br and Ir N to C Sc, planted elsewhere.

2. E. latifolius (L.) Mill. - *Large-leaved Spindle.* Differs from *E. europaeus* in leaves 7-16cm, elliptic to obovate, rather abruptly acuminate, serrulate; fruits 15-25mm across; and see key; (2n=64). Intrd; planted in hedges and gardens and sometimes natd by bird-sown seeds; scattered in C & S En; Europe.

3. E. japonicus L. f. - *Evergreen Spindle*. Bushy shrub or small tree to 5(8)m; leaves evergreen, shiny on upperside, 2-7cm, usually obovate, obtuse to very abruptly acuminate, crenate-serrate, often variegated; fruits with 4 rounded lobes, 6-10mm across, pink; (2n=32). Intrd; much planted for hedging, especially near sea in S, very persistent but self-sown only in extreme S and Man; frequent relic by sea in S & W Br N to S Wa, CI, rare elsewhere; Japan. Variegated plants often have creamish-yellow capsules.

2. CELASTRUS L. - *Staff-vine*

Woody climbers with alternate leaves each with 2 spinose stipules in young state; ovary 3-celled; stigma 3-lobed; capsule 3-celled, subglobose, brownish- to greenish-yellow on outside, golden-yellow inside; seeds with red aril.

1. C. orbiculatus Thunb. ex Murray - *Staff-vine*. Stems twining, climbing potentially to 12m; leaves deciduous, 5-12cm, obovate to suborbicular, obtuse to abruptly acuminate, crenate-serrate; fruits with 3 valves, 6-10mm across; (2n=46). Intrd; garden plant spreading into woodland; 1 site in Surrey since 1985; E Asia.

91. AQUIFOLIACEAE - *Holly family*

Evergreen trees or shrubs; leaves alternate, simple, usually at least some with very spiny margins, petiolate, without stipules. Flowers in small axillary cymes, usually dioecious, hypogynous, actinomorphic; sepals 4, free; petals 4, fused at base to ± free, white; male flowers with 4 stamens; female flowers with 4 usually abortive stamens and a 4-celled ovary with 1(-2) apical ovules per cell; stigma 4-lobed, sessile; fruit a (2-)4-seeded drupe.

Unmistakable in habit, even in entire-leaved variants.

1. ILEX L. - *Hollies*

1. I. aquifolium L. - *Holly*. Shrub or tree to 23m; leaves glossy, 5-12cm, ovate to elliptic, at least the lower undulate and strongly spinose at margins; fruits 6-10mm, ± globose, scarlet, sometimes yellow or orange; 2n=40. Native; woods, hedges and scrub; common almost throughout BI. Many cultivars exist, some with variegated and/or ± spineless leaves.

2. I. x altaclerensis (Loudon) Dallim. (*I. aquifolium* x *I. perado* Aiton) - *Highclere Holly*. Differs from *I. aquifolium* in its usually slightly larger flowers, fruit and leaves; leaves mostly <2x as long as wide, ± flat, without lateral spines or with few ± forwardly-pointed ones. Intrd; planted as many cultivars, often variegated, and occurring in hedges and woodland as relics or bird-sown plants, the latter often not coming true from seed; scattered in Br and Man, probably under-recorded; garden origin. Merging into *I. aquifolium*.

92. BUXACEAE - *Box family*

Evergreen shrubs or small trees; leaves opposite or alternate, simple, entire or dentate, petiolate, without stipules. Flowers in early spring, monoecious, male and female together in small yellowish to white clusters, hypogynous, actinomorphic; petals 0; male flowers with 4 sepals and 4 stamens; female flowers with ≥4 sepal-like bracteoles and 2-3-celled ovary with 2 apical ovules per cell; styles 2-3, short; stigmas linear or bilobed; fruit a 2-3-celled capsule with 2 seeds per cell.

Recognized by the evergreen leaves, distinctively arranged unisexual flowers with 4 stamens and 2-3-celled ovary, and 2- or 3-horned fruits.

1 Erect shrub or tree with opposite, entire leaves; fruit 3-horned **1. BUXUS**
1 Stoloniferous dwarf shrub with alternate, dentate leaves; fruit 2-horned **2. PACHYSANDRA**

1. BUXUS L. - *Box*
Erect shrub or tree with opposite, entire, glabrous to sparsely pubescent leaves; inflorescences axillary, usually with 1 apical female and several lower male flowers; ovary 3-celled, with 3 styles; stigma bilobed; fruit a dry capsule.

1. B. sempervirens L. - *Box*. Shrub or small tree to 5(11)m; leaves 1-2.5cm, elliptic to oblong, rounded to retuse at apex; flowers pale yellow; fruits 7-11mm, green, with persistent styles as 3 horns; 2n=28. Native; woods and scrub on chalk and limestone; extremely local in W Kent, Surrey, Berks, Bucks and W Gloucs, rarely natd in hedges and woods elsewhere in S En. **RR**

2. PACHYSANDRA Michx. - *Carpet Box*
Dwarf stoloniferous shrub with alternate glabrous leaves dentate distally; inflorescence terminal on previous year's growth, with terminal male and lower female flowers; ovary 2-celled, with 2 styles; stigma linear; fruit a ± succulent drupe.

1. P. terminalis Siebold & Zucc. - *Carpet Box*. Stems ascending, to 25cm; leaves 5-10cm, obovate to obtrullate, subacute; flowers white; fruits c.10mm, whitish, with persistent styles as 2 horns; (2n=48). Intrd; much grown in public places as ground-cover, sometimes 'running wild'; few sites in W Kent since 1968; Japan.

93. EUPHORBIACEAE - *Spurge family*

Annual to perennial herbs or rarely woody annuals, often with white latex; leaves opposite or alternate, sometimes whorled, simple, entire or serrate, rarely palmately lobed, petiolate or sessile, with or without stipules. Flowers variously arranged, monoecious or dioecious, hypogynous, actinomorphic; perianth 0 or of 3(-5) sepal-like free lobes; male flowers with 1-many stamens, the filaments simple, jointed or branched; female flowers with 2-3-celled ovary with 1 ovule per cell; styles 2-3; stigmas strongly papillose or branched; fruit a 2-3-celled capsule.

3 extremely distinct genera with superficially little in common, but all with monoecious or dioecious flowers with 2-3-celled ovary with 1 ovule per cell and 2-3 strongly papillose or branched stigmas.

1 Leaves palmately lobed **2. RICINUS**
1 Leaves not lobed, entire to serrate 2
2 Plant with copious white latex, monoecious; ovary and fruit 3-celled; stamen 1 **3. EUPHORBIA**
2 Plant with watery sap, usually dioecious; ovary and fruit 2-celled; stamens numerous **1. MERCURIALIS**

1. MERCURIALIS L. - *Mercuries*
Herbs with watery sap; leaves opposite, unlobed, serrate; flowers usually dioecious, the male in catkin-like ± erect axillary spikes, the female in smaller axillary clusters; tepals 3, green; stamens numerous, with free, simple filaments; ovaries 2-celled.

1. M. perennis L. - *Dog's Mercury*. Rhizomatous perennial; stems erect, simple, to 40cm, pubescent; leaves pubescent, ovate to elliptic or narrowly so, 3-8cm; male spikes up to c.12cm; female flower-clusters on stalks usually >1cm; 2n=64. Native;

woods, hedgerows and shady places among rocks; common over much of Br but absent from Man, Hebrides, Orkney and Shetland, very local in Jersey and Ir (mainly intrd).

2. M. annua L. - *Annual Mercury*. Differs from *M. perennis* in annual with fibrous root system; stems often branched; stems and leaves glabrous or nearly so, usually paler green; female flowers fewer and subsessile; (2n=16). Possibly native; cultivated ground and waste places; frequent in S En and CI, scattered in Br N to C Sc and in Ir.

2. RICINUS L. - *Castor-oil-plant*
Annual herb or shrub with watery sap; leaves alternate, lobed, serrate; flowers monoecious, in branched axillary groups, the male below the female; tepals 3-5, membranous; stamens numerous, on branched filaments; ovary 3-celled.

1. R. communis L. - *Castor-oil-plant*. Stems simple or branched, to 2(4)m; leaves long-petiolate, peltate, palmately lobed, up to 60cm; (2n=20). Intrd; casual on tips and in waste places, often not reaching fruiting or even flowering, as garden throwout or oilseed-alien; scattered in S Br and CI; tropics.

3. EUPHORBIA L. - *Spurges*
Annual, biennial or perennial herbs with white latex; leaves opposite, alternate or whorled, unlobed, entire or serrate; flowers monoecious, in distinctive small units composed of 1 female and few male flowers together in a cup-shaped *cyathium*, which has 4-5 conspicuous glands at top; the cyathia solitary in leaf-axils, or (usually) in terminal compound cymes with paired or whorled branches each subtended by a bract which is leafy but often different in shape from the leaves; perianth 0; stamen 1, with jointed filament; ovary 3-celled.

1 Plants usually procumbent; stipules present; bracts and leaves similar, markedly unequal at base 2
1 Plants usually erect; stipules 0; bracts and leaves often different, ± equal at base 3
 2 Stems and capsules glabrous **1. E. peplis**
 2 Stems and capsules pubescent **2. E. maculata**
3 Leaves on main stems opposite **10. E. lathyris**
3 Leaves on main stems alternate 4
 4 Glands on cyathia rounded on outer edge 5
 4 Glands on cyathia concave on outer edge, prolonged into 2 points 11
5 Ovary and capsule smooth to granulose, but sometimes pubescent 6
5 Ovary and capsule conspicuously warty or papillose 7
 6 Ovary and capsule pubescent; leaves pubescent at least on lowerside, oblong to oblong-lanceolate **3. E. corallioides**
 6 Ovary and capsule glabrous; leaves glabrous, obovate **9. E. helioscopia**
7 Annuals with simple root system 8
7 Rhizomatous perennials 9
 8 Capsules with hemispherical papillae; umbel with 5 main branches, the bracts at that node similar to leaves below but markedly different from bracts at next higher node **7. E. platyphyllos**
 8 Capsules with cylindrical papillae; umbel with 2-5 main branches, the bracts at that node intermediate between leaves below and bracts at next higher node **8. E. serrulata**
9 Capsule with most papillae c. as long as wide (± hemispherical) **6. E. oblongata**
9 Capsule with many papillae c.2x as long as wide (± cylindrical) 10
 10 Stems without scales near base; capsule 5-6mm; bracts yellowish **4. E. hyberna**

10 Stems with scales near base; capsule (2)3-4mm; bracts green **5. E. dulcis**
11 Opposite pairs of bracts fused at base; stems pubescent 12
11 Opposite pairs of bracts not fused at base; stems glabrous 13
12 Capsule glabrous; primary branches of topmost whorl of inflorescence 4-12 **21. E. amygdaloides**
12 Capsule densely pubescent; primary branches of topmost whorl of inflorescence 10-20 **22. E. characias**
13 Annuals to perennials; rhizomes 0 (stems sometimes buried in sand) 14
13 Rhizomatous perennials 17
14 Annuals with thin leaves, rarely on maritime sands; bracts and leaves similar 15
14 Biennials to perennials with ± succulent leaves, on maritime sands; bracts and leaves markedly different 16
15 Leaves linear to narrowly oblong, sessile **11. E. exigua**
15 Leaves ovate to obovate, petiolate **12. E. peplus**
16 Midrib prominent on leaf lowerside; seeds pitted **13. E. portlandica**
16 Midrib obscure on leaf lowerside; seeds smooth **14. E. paralias**
17 Leaves ≤2(3)mm wide, often linear, those of lateral shoots crowded and conifer-like **20. E. cyparissias**
17 Some or all leaves ≥(2)3mm wide, not linear, those on lateral shoots not conifer-like 18
18 Leaves oblanceolate to oblong-oblanceolate, widest above middle, attenuate at base 19
18 Leaves linear- to oblong-lanceolate, widest at or below middle, not attenuate at base 20
19 Leaves ≤4mm wide **19. E. x pseudoesula**
19 At least some leaves >4mm wide **18. E. esula**
20 Leaves rounded to broadly cuneate-rounded at base **15. E. waldsteinii**
20 Leaves abruptly narrowed to cuneate base 21
21 Leaves mostly c.2-3mm wide **17. E. x gayeri**
21 Leaves mostly c.4-5mm wide **16. E. x pseudovirgata**

Other spp. - **E. villosa** Waldst. & Kit. ex Willd. (*E. pilosa* auct. non L.) (*Hairy Spurge*), from Europe, was natd near Bath, N Somerset, from 1576 to c.1924; it would key to couplet 6 but is a rhizomatous perennial with stems scaly near ground. **E. ceratocarpa** Ten., from Italy, was once natd in Glam; it would key to couplet 9 but is glabrous and has flat-conical papillae on capsule. For *E. boissieriana* see *E. esula* agg. **E. mellifera** Aiton is a tree-spurge from the Canaries with a grey bark and leaves clustered at branch-ends; it has produced seedlings on walls, etc. in W Cornwall (incl. Scillies).

1. E. peplis L. - *Purple Spurge*. Glabrous, often purplish, annual with procumbent stems to 10cm; cyathia solitary in stem-forks and leaf-axils; capsules smooth, glabrous; seeds smooth. Native; on sandy or shingly beaches, probably extinct; last record 1976 in Alderney; formerly in S Br from E Kent to Cards, Waterford, CI. **RR**

2. E. maculata L. - *Spotted Spurge*. Pubescent annual, often with dark blotches on leaves, with procumbent stems to 50cm; cyathia 1-few in leaf-axils and stem-forks; capsules smooth, pubescent; seeds with shallow transverse furrows; (2n=28). Intrd; ± natd weed of nurseries and in a quarry; very scattered in En, S Wa, Jersey; N. America, natd in Europe.

3. E. corallioides L. - *Coral Spurge*. Pubescent perennial with 0 or weak rhizomes; stems erect, to 60cm; leaves oblong to oblong-oblanceolate, ± sessile, serrulate; capsules finely granulose, pubescent; seeds smooth; 2n=26. Intrd; natd in woods and hedgerows in W Sussex since c.1808, along laneside in N Somerset, formerly in E Sussex and Oxon; Italy. **460**

4. E. hyberna L. - *Irish Spurge*. Sparsely pubescent or rarely glabrous perennial **R**

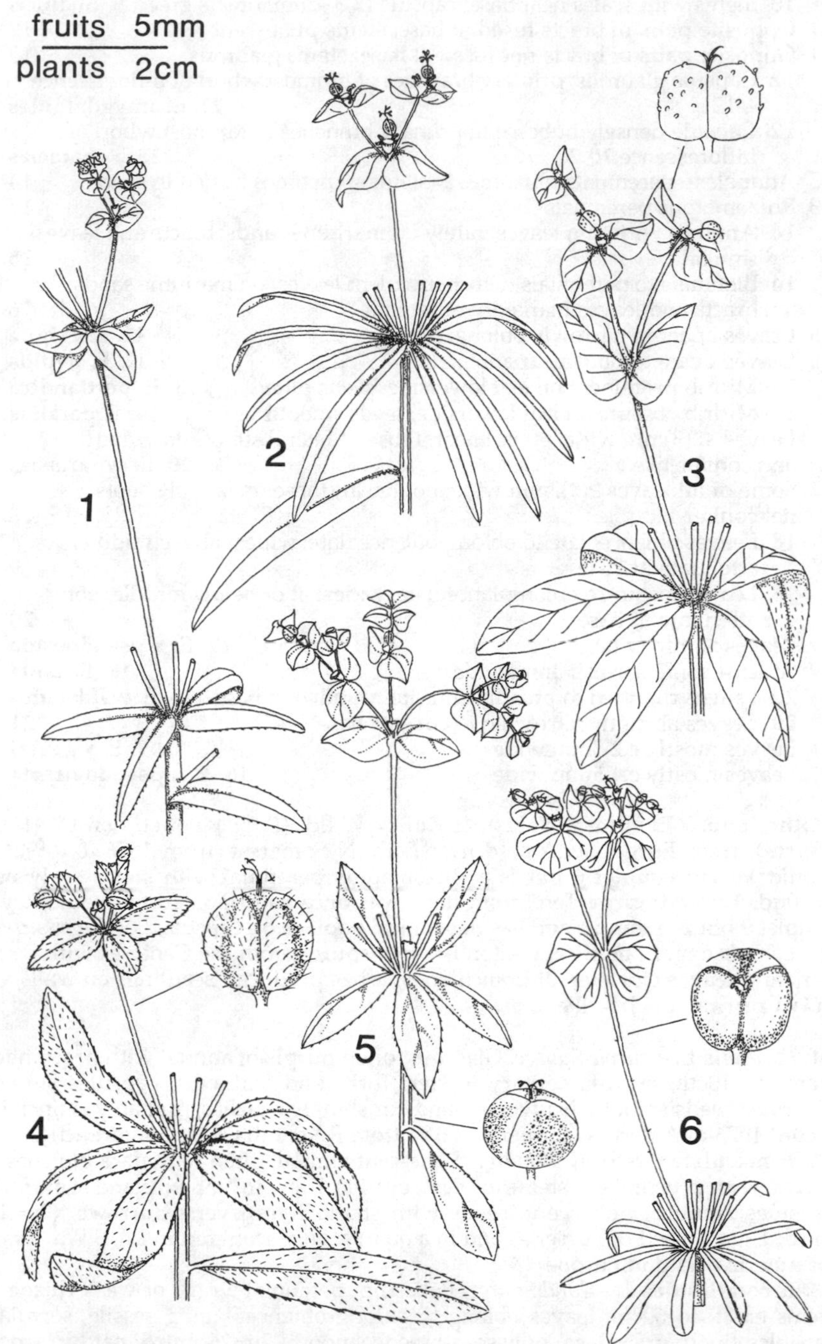

FIG 460 - Shoots and fruits of ***Euphorbia***. 1, *E. serrulata*. 2, *E.* x *pseudovirgata*. 3, *E. dulcis*. 4, *E. corallioides*. 5, *E. platyphyllos*. 6, *E. esula*.

with strong rhizomes; stems erect, to 60cm; leaves narrowly elliptic-oblong, sessile, entire; capsules with prominent cylindrical papillae, glabrous; seeds smooth; (2n=36). Native; woods, hedgerows, grassy places and stream-banks; extremely local in W Cornwall, N Devon and S Somerset, frequent in most of SW Ir, natd in Cards.

5. E. dulcis L. - *Sweet Spurge*. Sparsely pubescent to subglabrous perennial with **460**
strong rhizomes; stems erect, to 50cm; leaves oblong- or elliptic-oblanceolate, tapered to ± sessile base, entire to serrulate; capsules with prominent cylindrical papillae, glabrous; seeds smooth; (2n=12, 18, 24, 28). Intrd; natd in shady places; very scattered in Br, in E Ross since 1894; Europe.

6. E. oblongata Griseb. - *Balkan Spurge*. Densely pubescent perennial with strong rhizomes; stems erect, to 90cm; leaves narrowly oblong-ovate or -obovate, sessile, serrulate; capsules with conspicuous mostly ± hemispherical papillae, glabrous; seeds smooth. Intrd; well natd on grassy bank in S Hants since at least 1993, also Wight and N Essex; Balkans and Aegean.

7. E. platyphyllos L. - *Broad-leaved Spurge*. Glabrous or pubescent erect annual to **R**
80cm; leaves obovate to elliptic, sessile, cordate at extreme base, serrulate; **460**
capsules with small hemispherical papillae, glabrous; seeds smooth; 2n=30. Native; cultivated and rough ground; formerly locally frequent in S & E Br, now very local in S En N to Cambs and Worcs.

8. E. serrulata Thuill. (*E. stricta* L. nom. illeg.) - *Upright Spurge*. Glabrous erect **RR**
annual to 80cm; leaves as in *E. platyphyllos* but narrower though often cordate at **460**
base; capsules with prominent cylindrical papillae, glabrous; seeds smooth; (2n=20, 28). Native; limestone woods in c.10 places in W Gloucs and Mons, natd rarely in S En from S Somerset to W Kent.

9. E. helioscopia L. - *Sun Spurge*. Glabrous (or ± so) erect annual to 50cm; leaves obovate, tapered to base, serrulate; capsules smooth, glabrous; seeds with reticulate ridges; 2n=42. Native; cultivated ground and waste places; common over lowland BI.

10. E. lathyris L. - *Caper Spurge*. Glabrous biennial; stems to 1m in 1st year, producing inflorescence from top and to 2m in 2nd year; leaves linear to narrowly oblong-lanceolate, sessile, entire, ± glaucous; capsules smooth, glabrous; seeds rugose; 2n=20. Possibly native in shady places in S En, frequent casual or natd alien in waste places and gardens over much of Br and CI.

11. E. exigua L. - *Dwarf Spurge*. Glabrous erect annual to 20(30)cm; leaves linear to narrowly oblong, sessile, entire, ± glaucous; capsules smooth, with ridge on midline of each valve, glabrous; seeds closely rugose; 2n=24. Probably native; arable land, rarely elsewhere; common in S & E En, scattered elsewhere in most of BI except N Sc.

12. E. peplus L. - *Petty Spurge*. Glabrous erect annual to 30(40)cm; leaves ovate to obovate, ± petiolate, entire; capsules smooth, with 2 narrow wings near midline of each valve, glabrous; seeds pitted; 2n=16. Native; cultivated and waste ground; common throughout most of BI.

13. E. portlandica L. - *Portland Spurge*. Glabrous, erect biennial to perennial to **R**
40(50)cm; rhizomes 0; leaves slightly succulent, obovate to oblanceolate, sessile, entire; inflorescence arising from top of last year's stems; capsules granulose near midline of each valve, glabrous; seeds pitted; 2n=16. Native; maritime sand-dunes; rather local on coasts of Ir, CI and S & W Br from S Hants to Kintyre (formerly S Ebudes).

13 x 14. E. portlandica x E. paralias has been found with both parents in Wa and Wexford; it is partially sterile and intermediate in leaf characters; endemic.

14. E. paralias L. - *Sea Spurge*. Differs from *E. portlandica* in more robust stems to 60cm; leaves more succulent, ovate-oblong; capsules rugose-granulose; seeds smooth; and see key; 2n=16. Native; maritime sand-dunes, often with *E. portlandica*; rather local on coasts of Ir, CI and Br N to W Norfolk and Wigtowns.

15-20. E. esula L. agg. Glabrous, rhizomatous perennials; stems erect; leaves linear to lanceolate or oblanceolate, sessile, entire; capsules granulose near midline of each valve, glabrous; seeds smooth. A very difficult group of (in Br) 3 spp. and their 3 hybrids, all of which are fertile. Virtually the only diagnostic characters are the leaf-shapes: oblanceolate in *E. esula*, lanceolate in *E. waldsteinii* and linear in *E. cyparissias*. All are found natd in hedgerows and grassy and waste places; probably all but *E. cyparissias* and *E. x pseudovirgata* are under-recorded. Over 60 spp. have been recognized in the agg., and some others than the 3 treated here might occur in BI. **E. boissieriana** (Woronow) Prokh. and its hybrid with *E. esula* have been recorded; they differ from *E. waldsteinii* and its hybrid with *E. esula* ± only in their wider and less acuminate leaves

15. E. waldsteinii (Soják) Czerep. (*E. virgata* Waldst. & Kit. non Desf.) - *Waldstein's Spurge*. Stems to 1m, relatively robust; leaves linear- to oblong-lanceolate, widest below middle, mostly c.4-5mm wide, rounded to broadly cuneate-rounded at base, usually cuspidate to acuminate at apex. Intrd; natd in scattered places in En and N to Lanarks; C & SE Europe to SW Asia.

16. E. x pseudovirgata (Schur) Soó (*E. uralensis* auct. non Fisch. ex Link, *E. esula* **460**
ssp. *tommasiniana* auct. non (Bertol.) Kusmanov, *E. virgata* auct. non Waldst. & Kit. nec Desf.; *E. waldsteinii* x *E. esula*) - *Twiggy Spurge*. Stems to 1m; leaves linear to lanceolate or sometimes ± oblanceolate, mostly widest at or below middle, mostly c.4-5mm wide, not or abruptly narrowed to cuneate base, acuminate to acute at apex; 2n=60. Intrd; frequently natd in Br N to C Sc, Jersey; Europe. Easily the commonest member of the agg. in Br.

17. E. x gayeri Boros & Soó (*E. waldsteinii* x *E. cyparissias*). - *Gáyer's Spurge*. Stems to 60cm; leaves linear- to oblong-lanceolate, widest at or below middle, mostly 2-3mm wide, not or abruptly narrowed to cuneate base, acuminate to acute at apex. Intrd; natd in Cumberland and Brecs; Europe.

18. E. esula L. - *Leafy Spurge*. Stems to 60cm, with more delicate habit than *E. x* **460**
pseudovirgata or *E. waldsteinii*; leaves oblanceolate, widest above middle, (3)5-10mm wide, tapered to narrowly cuneate base, rounded at apex; 2n=60. Intrd; natd in scattered places in Br but more frequent in parts of C Sc; Europe.

19. E. x pseudoesula Schur (*E. esula* x *E. cyparissias*) - *Figert's Spurge*. Stems to 60cm; leaves oblanceolate to oblong-oblanceolate, widest above middle, mostly (2)3-4mm wide, tapered to cuneate base, rounded at apex. Intrd; natd in W Suffolk, Surrey and S Wa; Europe.

20. E. cyparissias L. - *Cypress Spurge*. Stems to 50cm; leaves linear, crowded (especially on lateral branches), ≤2(3)mm wide, scarcely narrowed at base or apex; 2n=20, 40. Possibly native in chalk grassland in E Kent and perhaps elsewhere in SE En, natd in rough grassland and waste places scattered throughout BI.

21. E. amygdaloides L. - *Wood Spurge*. Pubescent, tufted or rhizomatous perennial; stems biennial, with inflorescences arising from tops in 2nd year, to 90cm; leaves of lst-year stems obovate to oblanceolate or narrowly elliptic, tapered to base, entire; capsule smooth or minutely punctate, glabrous; seeds smooth.

a. Ssp. amygdaloides. Rhizomes short or 0; leaves of 1st-year stems herbaceous, dull, pale- to mid-green, pubescent on lowerside and margins; 2n=20. Native; woods and shady hedgerows; common in CI and much of S Br N to Flints and E Norfolk, rare alien further N and in Ir.

b. Ssp. robbiae (Turrill) Stace (*E. robbiae* Turrill, *E. amygdaloides* var. *robbiae* (Turrill) Radcl.-Sm.). Rhizomes long; leaves of lst-year stems ± coriaceous, ± shiny, dark green, ± glabrous. Intrd; natd in woods and other shady places; scattered in SE & C En; NW Turkey.

22. E. characias L. - *Mediterranean Spurge*. Densely pubescent, tufted perennial; stems biennial, with inflorescences arising from tops in 2nd year, to 1.5m; leaves of lst-year stems oblanceolate, entire, pubescent; capsule smooth, pubescent; seeds smooth; (2n=20). Intrd; grown in gardens, natd on old garden sites and waste ground.

a. Ssp. characias. Glands on cyathia dark reddish-brown, with short points. Surrey; W Mediterranean.

b. Ssp. wulfenii (Hoppe ex W.D.J. Koch) Radcl.-Sm. Glands on cyathia yellowish, with long points. N Somerset since 1953; E Mediterranean.

94. RHAMNACEAE - *Buckthorn family*

Evergreen or deciduous shrubs or small trees; leaves alternate or subopposite, simple, petiolate, stipulate. Flowers small, yellowish-green, in axillary cymes or solitary in leaf-axils, variously dioecious to bisexual, perigynous, actinomorphic; hypanthium short, ± bell-shaped; sepals 4-5, free; petals 0 or 4-5, free; stamens 4-5, abortive in female flowers; ovary 2-4-celled, each cell with 1 basal ovule; style 1 with 2-3-lobed capitate stigma or divided into 2-4 distally, each arm with capitate stigma; fruit a berry with 2-4 seeds, eventually black.

Recognizable by the shrubby habit, simple stipulate leaves, small 4-5-merous perigynous flowers, 2-4-celled ovary with 1 basal ovule per cell, and black berry.

1 Leaves serrate; winter buds with scales **1. RHAMNUS**
1 Leaves entire; winter buds without scales **2. FRANGULA**

1. RHAMNUS L. - *Buckthorns*
Evergreen or deciduous shrubs with alternate or subopposite serrate leaves; flowers 4-5-merous; style divided into 3 or 4 distally.

1. R. cathartica L. - *Buckthorn*. Deciduous, usually spiny shrub to 8m; leaves 4- **469**
9cm, mostly with 2-4(5) pairs of major lateral veins; petiole 6-25mm; sepals and petals mostly 4; fruit c.6-10mm, ± globose, with 3-4 seeds; 2n=24. Native; hedgerows, scrub and open woods on peat and base-rich soils; locally common in En, scattered in Wa and Ir, rare escape elsewhere.

2. R. alaternus L. - *Mediterranean Buckthorn*. Evergreen, non-spiny shrub to 5m; **469**
leaves 1-6cm, mostly with 3-6 pairs of major lateral veins; petiole 3-10mm; sepals 5; petals 0; fruit 4-6mm, obovoid, with 2-3 seeds; (2n=24). Intrd; well natd in scrub near sea in Caerns and Denbs; Mediterranean.

2. FRANGULA Mill. - *Alder Buckthorn*
Deciduous shrubs with alternate entire leaves; flowers 5-merous; style not divided, with 2-3-lobed stigma.

1. F. alnus Mill. - *Alder Buckthorn.* Non-spiny shrub to 5m; leaves 2-7cm, mostly **469**
with 6-10 pairs of major lateral veins; petiole 8-14mm; fruit 6-10mm, obovoid, with 2-3 seeds; 2n=20. Native; scrub, bogs and open woods usually on damp peaty soils, often base-poor but not always; locally common in En and Wa, commoner in W than *Rhamnus cathartica*, very scattered in Ir and Sc.

95. VITACEAE - *Grape-vine family*

Deciduous woody climbers with leaf-opposed tendrils; leaves alternate, simple and palmately lobed or palmate, petiolate, stipulate. Flowers small, reddish to greenish, in leaf-opposed cymes, bisexual or mostly so, hypogynous, actinomorphic; sepals 5, very short, fused into ± lobed rim; petals 5, free or fused distally; stamens 5; ovary 2-celled, each cell with 2 nearly basal ovules; style 1; stigma capitate; fruit a berry with up to 4 seeds.

The woody climbing habit with leaf-opposed tendrils and palmate or palmately-lobed leaves is diagnostic; differs from Rhamnaceae also in hypogynous flowers with fused sepals in leaf-opposed cymes.

1 Leaves simple; tendrils not ending in discs; petals fused distally, falling as flowers open **1. VITIS**
1 Leaves palmate or simple, if simple then tendrils ending in discs; petals free **2. PARTHENOCISSUS**

1. VITIS L. - *Grape-vine*
Leaves simple, palmately lobed; petals fused distally, forming cap in bud which drops as flowers open.

Other spp. - V. coignetiae Pulliat, from Japan, is grown in gardens and has been found as a relic in SE En; it has scarcely lobed leaves and reddish-brown hairs on leaves and stems.

1. V. vinifera L. - *Grape-vine.* Woody vine potentially >10m; tendrils branched, lacking discs; leaves orbicular, cordate, with 5-7 palmate lobes; fruit green to red or black, up to 2cm, broadly ellipsoid; (2n=38, 76). Intrd; increasingly grown on field-scale in S En, natd in hedges and scrub and by tips; scattered in CI, S En and S & WC Wa; Europe.

2. PARTHENOCISSUS Planch. - *Virginia-creepers*
Leaves simple and palmately lobed or palmate; petals free, remaining for while after flowers open.

1 At least some leaves simple, 3-lobed **3. P. tricuspidata**
1 All leaves palmate, most or all with 5 leaflets 2
  2 Tendrils with 5-8(12) branches each ending in adhesive disc **1. P. quinquefolia**
  2 Tendrils with 3-5 branches not ending in adhesive disc **2. P. inserta**

1. P. quinquefolia (L.) Planch. - *Virginia-creeper.* Woody vine potentially >20m; leaves palmate, the (3-)5(7) stalked leaflets dull green on lowerside; fruit bluish-black, <1cm, globose; (2n=40). Intrd; much planted and natd on old walls and tips and in hedges and scrub; scattered in Br N to SW Sc, CI; N America.

2. P. inserta (A. Kern.) Fritsch - *False Virginia-creeper.* Differs from *P. quinquefolia* in leaves more acutely serrate, shiny green on lowerside; and see key; (2n=40). Intrd; similar places to *P. quinquefolia* but rarer; scattered in S Br; N America.

3. P. tricuspidata (Siebold & Zucc.) Planch. - *Boston-ivy.* Differs from *P. quinquefolia* in most leaves simple and 3-lobed (some simple and unlobed, often some palmate with 3 leaflets); (2n=40). Intrd; similar places to *P. quinquefolia* but rarer; scattered in SE En, W Cornwall, Jersey; E Asia.

96. LINACEAE - *Flax family*

Herbaceous annuals or perennials; leaves opposite or alternate, simple, entire, sessile, without stipules. Flowers in terminal cymes, bisexual, hypogynous, actinomorphic; sepals 4-5, free; petals 4-5, free; stamens 4 without staminodes or 5 usually alternating with filiform staminodes; ovary 4-5-celled with 2 ovules per cell on axile placenta, or ± 8- or 10-celled with 1 ovule per cell; styles 4 or 5; stigmas capitate; fruit a capsule opening by 8 or 10 valves.

Distinguished by the 4-5 free sepals, petals and stamens, entire leaves without stipules, and 8- or 10-valved capsule with 8 or 10 seeds.

1 Sepals, petals and stamens 5; sepals entire to minutely serrate at apex; capsule with 10 valves **1. LINUM**
1 Sepals, petals and stamens 4; sepals deeply 2-4-toothed at apex; capsule with 8 valves **2. RADIOLA**

1. LINUM L. - *Flaxes*
Glabrous annuals to perennials; leaves opposite or alternate; sepals, petals and stamens 5; sepals entire to minutely serrate at apex; petals white or blue, much longer than sepals; capsule with 10 valves.

1 Leaves opposite; petals white, <7mm **4. L. catharticum**
1 Leaves alternate; petals usually blue, >7mm 2
  2 Sepals c.1/2 as long as ripe capsule, at least the 2 inner rounded and apiculate at apex; stigmas capitate, either higher or lower than anthers **3. L. perenne**
  2 Sepals c. as long as ripe capsule, all abruptly acuminate at apex; stigmas elongate-clavate, c. as high as anthers 3
3 Stems usually >1; sepals and capsules 4-6mm **1. L. bienne**
3 Stem usually 1; sepals and capsules 6-9mm **2. L. usitatissimum**

1. L. bienne Mill. - *Pale Flax*. (Annual,) biennial or perennial; stems several, ascending to erect, to 60cm; leaves linear to narrowly elliptic-oblong, 0.5-1.5mm wide, 1-3-veined; sepals 4-6mm; petals usually blue, 8-12mm; capsule 4-6mm; 2n=30. Native; dry grassy places; local in BI and mostly coastal in W, S from Notts, Man and Meath.

2. L. usitatissimum L. - *Flax*. Annual; stem usually 1, erect, to 85cm; differs from *L. bienne* in leaves 1.5-3(4)mm wide, 3-veined; sepals 6-9mm; petals 12-20mm; capsule 6-9mm; (2n=30). Intrd; formerly much grown for linen (tall unbranched cultivars) or linseed-oil (shorter branched cultivars) and a frequent casual by fields, now a frequent casual from birdseed on tips and in fields left for game, becoming more commonly grown again for oil; throughout most of BI; cultivated origin.

3. L. perenne L. (*L. anglicum* Mill.) - *Perennial Flax*. Perennial; stems >1, R
decumbent to suberect, to 60cm; differs from *L. bienne* in leaves 1-3.5mm wide; sepals 3.5-6.5mm; petals 13-20mm; capsule 5.5-7.5mm; 2n=36. Native; calcareous grassland; very local in mainly E En from N Essex to Durham and Kirkcudbrights. Our plant is the endemic ssp. **anglicum** (Mill.) Ockendon.

4. L. catharticum L. - *Fairy Flax*. Annual; stems erect, to 25cm; leaves elliptic-oblong, 1-veined; sepals 2-3mm, acute to acuminate; petals white, 4-6mm; capsule 2-3mm; 2n=16. Native; dry calcareous or sandy soils, also moorland and mountains; frequent throughout BI.

2. RADIOLA Hill - *Allseed*
Annuals; leaves opposite; sepals, petals and stamens 8; sepals deeply 2-4-toothed at apex; petals white, c. as long as sepals; capsule with 8 valves.

1. R. linoides Roth - *Allseed*. Stems much branched, extremely slender, ± erect, to 6(10)cm; leaves elliptic, 1-veined; sepals and petals c.1mm; capsule 0.7-1mm; 2n=18. Native; seasonally damp, bare, peaty or sandy, acid ground in open places or in woodland rides; scattered over most BI but mostly near coast.

97. POLYGALACEAE - *Milkwort family*

Small herbaceous perennials often woody at base; leaves opposite or alternate, simple, entire, sessile or shortly petiolate, without stipules. Flowers in usually terminal racemes, bisexual, hypogynous, zygomorphic; sepals 5, free, the 2 inner much larger than the 3 outer; petals 3, fused, 2 upper entire, 1 lower dissected distally; stamens 8, their filaments fused into a cleft tube that is also fused to the petals; ovary 2-celled with 1 apical ovule per cell; style 1, with stigma and a sterile lobe at apex; fruit a 2-seeded capsule.

The strange flowers, with 3 petals and 8 stamens fused together, are unique.

1. POLYGALA L. - *Milkworts*

1 Leaves near base of stems smaller than those above, ± acute, not congested into a rosette; inner sepals with veins anastomosing around edges **2**
1 Leaves near base of stems larger than those above, ± obtuse, congested into a rosette; inner sepals with veins not anastomosing or sparingly so and not around edges **3**
  2 Lower stem-leaves (sometimes lost by fruiting - see scars left) opposite **2. P. serpyllifolia**
  2 All leaves alternate **1. P. vulgaris**
3 Flowers 6-7mm; stems with ± leafless portion below leaf-rosette **3. P. calcarea**
3 Flowers 2-5mm; stems with leaf-rosette at or very near base **4. P. amarella**

1. P. vulgaris L. - *Common Milkwort.* Stems woody at base, procumbent or scrambling to erect, to 30cm; flowers various shades of blue, pink or white, 4-7mm, mostly >10 per main raceme; inner sepals slightly shorter than corolla, with anastomosing well-branched veins, c.3/4 as wide to slightly wider than capsule, acute to rounded-apiculate; 2n=c.56, 68. Native; calcareous or acid grassland, heathland and dunes.

a. Ssp. vulgaris. Inner sepals 6-8.5 x 3.5-5mm, c. as wide as capsule, with 6-20 inter-veinlet areolae; style c. as long as fruit apical notch. Frequent throughout BI.

b. Ssp. collina (Rchb.) Borbás (*P. oxyptera* auct. non Rchb.). Inner sepals 4-6 x 2-3.5mm, distinctly narrower than capsule, with 8-16(22) inter-veinlet areolae; style longer than fruit apical notch. Scattered throughout much of Br but distribution very uncertain.

1 x 3. P. vulgaris x P. calcarea has been recorded from scattered localities in S En; it is intermediate and sterile; endemic.

1 x 4. P. vulgaris x P. amarella (= *P. x skrivanekii* auct. non Podp.) has been found in E Kent with both parents; it has large lower leaves with a bitter taste as in *P. amarella*, but is much more vigorous, has intermediate corolla-size, and is partially fertile.

2. P. serpyllifolia Hosé - *Heath Milkwort.* Stems not or scarcely woody at base, procumbent to scrambling, to 25cm; flowers various shades of blue, pink or white, 4.5-6mm, mostly <10 per main raceme; inner sepals like those of *P. vulgaris* but usually ± acute; 2n=34. Native; acid grassland and heathland; frequent throughout BI.

3. P. calcarea F.W. Schultz - *Chalk Milkwort.* Stems woody at base, procumbent below leaf-rosette then erect to ascending, to 20cm; flowers usually blue, rarely pink or white, 3-6mm, 6-20 per main raceme; inner sepals slightly shorter than corolla, with rather sparsely branched, not or sparingly anastomosing veins, c.3/4 as wide as capsule, obtuse; 2n=34. Native; chalk and limestone grassland; local in S En N to S Lincs.

4. P. amarella Crantz (*P. amara* auct. non L., *P. austriaca* Crantz) - *Dwarf* **RR**
Milkwort. Plants bitter-tasting; stems ± woody at base, erect to ascending, to 10(16)cm; flowers blue or pink in N, blue or greyish-white in S, 2-5.5mm, 7-30 per main raceme; inner sepals longer than corolla, with sparsely-branched, non-anastomosing veins, c.1/2 as wide as capsule, acute to subacute; 2n=34. Native; chalk and limestone grassland; very local in E & W Kent, MW & NW Yorks,

Durham and Westmorland.

98. STAPHYLEACEAE - *Bladdernut family*

Deciduous shrubs; leaves opposite, pinnate, petiolate, stipulate when young. Flowers in small terminal panicles, bisexual, hypogynous, actinomorphic; sepals 5, free, petal-like; petals 5, free; stamens 5; ovary 2-3-celled with carpels free distally, ovules numerous on axile placentas; styles 2-3; stigmas capitate; fruit a much-inflated 2-3-celled capsule with many seeds.

Vegetatively very like *Sambucus nigra*, but without the characteristic smell to the crushed leaves and with diagnostic bladder-like 2-3-lobed capsules.

1. STAPHYLEA L. - *Bladdernut*

1. S. pinnata L. - *Bladdernut*. Shrub to 5m; leaflets (3)5(-7), 5-10cm, ovate, acuminate, glabrous; flowers in pendent panicles 5-10cm, 6-12mm, whitish; fruit 2.5-4cm, subglobose; (2n=24, 26). Intrd; natd in hedges and in banks; rare in S En, formerly commoner, now supplanted in gardens by other spp.; C Europe.

99. SAPINDACEAE - *Pride-of-India family*

Deciduous trees or shrubs; leaves alternate, pinnate to 2-pinnate, petiolate, without stipules. Flowers in large terminal panicles, functionally monoecious, hypogynous, zygomorphic; sepals 5, unequal, fused proximally; petals 4, free, all upturned, with basal appendages; stamens 8, with hairy filaments; ovary 3-celled, each cell with 1 ovule on axile placenta; style simple; stigmas 3, minute; fruit a much-inflated 3-celled capsule with 3 seeds.

Resembles Staphyleaceae in its pinnate leaves and inflated capsule, but leaves are alternate and flowers differ in many features (see also *Colutea*, Fabaceae).

1. KOELREUTERIA Laxm. - *Pride-of-India*

1. K. paniculata Laxm. - *Pride-of-India*. Tree to 16m; leaves mostly pinnate, 15-50cm, with 9-15 ovate, serrate leaflets; flowers numerous, bright yellow, 10-15mm across; fruits 3-5cm, ovoid-conical; (2n=22, 30). Intrd; frequently grown in S & SE En, natd saplings on waste land in W Kent, Surrey and Middlesex; China.

100. HIPPOCASTANACEAE - *Horse-chestnut family*

Deciduous trees; leaves opposite, palmate, petiolate, without stipules, with long petiole. Flowers in large terminal panicles, bisexual and male in each panicle, hypogynous, zygomorphic; sepals 5, fused for most part; petals (4-)5, unequal, free; stamens 5-9; ovary 3-celled, each cell with 2 ovules on axile placenta; style simple; stigma minute; fruit a large capsule with 3 valves and 1(-3) large seeds.

The only trees with opposite, palmate leaves; the fruits and flowers are also unique. Resemblance of the fruits to those of *Castanea* (Fagaceae) is purely superficial; in the latter the prickly husk is a cupule containing fruits (nuts).

1. AESCULUS L. - *Horse-chestnuts*

1 Petals 4; stamens exceeding petals by c.2cm; leaflets with stalks c.1cm; fruits obovoid to pear-shaped **3. A. indica**
1 Petals 4-5, always some flowers with 5; stamens exceeding petals by ≤1cm; leaflets sessile or with stalks <1cm; fruits globose 2
2 Petals pink to red; fruits with 0 or few blunt protuberances **2. A. carnea**

2 Petals predominantly white; fruits with many conical pointed protuberances **1. A. hippocastanum**

1. A. hippocastanum L. - *Horse-chestnut*. Wide-spreading tree to 39m; winter-buds large, very sticky; leaflets 5-7, obovate, sessile, 10-25cm, abruptly acuminate; flowers white with yellow to pink blotch at base of petals, in stiffly erect conical to cylindrical panicle 15-30cm; fruits 5-8cm, with numerous conical, pointed, protuberances; 2n=40. Intrd; abundantly planted for ornament and often self-sown in grassy places, copses and rough ground; throughout lowland BI; Balkans.

2. A. carnea J. Zeyh. - *Red Horse-chestnut*. Tree to 28m; differs from *A. hippocastanum* in smaller parts; winter buds scarcely or not sticky; flowers bright pink to red; leaflets sometimes shortly stalked; fruits without or with few blunt protuberances; (2n=80). Intrd; much planted for ornament, often grafted on to *A. hippocastanum*, in parks and by roads, self-sown in W Kent, Surrey (reached flowering) and N Hants; garden origin from *A. hippocastanum* x *A. pavia* L.

3. A. indica (Cambess.) Hook. - *Indian Horse-chestnut*. Tree to 19m; winter buds not sticky; leaflets 5-9, rather narrowly elliptic-obovate, stalked, 12-30cm, gradually acuminate; flowers white, to varying degrees tinged and marked with red, pink and yellow, in panicles as in *A. hippocastanum*; fruit 5-8cm, without protuberances; (2n=40). Intrd; planted in parks and by roads; rarely self-sown in W Sussex and Middlesex; Himalayas. Flowers 1 month later than other 2 spp.

101. ACERACEAE - *Maple family*

Deciduous trees; leaves opposite, palmately lobed or ternate to pinnate (rarely simple and unlobed), petiolate, without stipules. Flowers in terminal corymbs or raceme-like panicles, mostly functionally monoecious or dioecious, hypogynous or the male slightly perigynous, actinomorphic; sepals (4-)5, free; petals 0 or (4-)5, free, ± sepaloid; stamens usually 8, ovary 2-celled (rarely more-celled), ± flattened, with 2 basal ovules per cell; styles 2; stigmas long, linear, 1-sided; fruits of 2 parts, each 1-seeded with a long wing developed from the style.

The fruit is diagnostic; the opposite, palmately lobed leaves of all spp. except *A. negundo* (and *A. tataricum*) occur elsewhere in only *Viburnum opulus* (Caprifoliaceae).

1. ACER L. - *Maples*

1 Leaves ternate or pinnate; trees dioecious **6. A. negundo**
1 Leaves simple, palmately lobed; trees usually bisexual 2
 2 Leaves white on lowerside; flowers in small stiff compact clusters **5. A. saccharinum**
 2 Leaves green on lowerside; flowers in erect or pendent panicles 3
3 Leaf-lobes serrate; panicles pendent **4. A. pseudoplatanus**
3 Leaf-lobes entire to irregularly dentate; flowers in stiff ± corymbose panicles 4
 4 Leaf-lobes obtuse; body of fruit convex **3. A. campestre**
 4 Leaf-lobes acuminate; body of fruit flat 5
5 Leaf-lobes entire **2. A. cappadocicum**
5 Leaf-lobes with few acuminate teeth or sub-lobes **1. A. platanoides**

Other spp. - Many other spp. are grown in parks and by roads, and some occasionally produce seedlings in shrubberies, etc. In the latter category are **A. mono** Maxim., from E Asia, like *A. cappadocicum*, but with twigs rough as in *A. platanoides* (not remaining smooth for some years); **A. tataricum** L. (*Tartar Maple*), from SE Europe and SW Asia, with usually unlobed, ovate-oblong, serrate leaves;

FIG 469 - Leaves of *Acer*, **Rhamnaceae**. 1, *Acer saccharinum*. 2, *A. cappadocicum*. 3, *A. platanoides*. 4, *A. pseudoplatanus*. 5, *A. negundo*. 6, *A. campestre*. 7, *Rhamnus cathartica*. 8, *R. alaternus*. 9, *Frangula alnus*.

A. opalus Mill. (*Italian Maple*), from SW Europe, with leaves rather as in *A. campestre* but very shallowly lobed; **A. rufinerve** Siebold & Zucc. (*Grey-budded Maple*), from Japan, with serrate leaves with just 1 short lobe on each side and a decorative 'snakebark' trunk; **A. rubrum** L. (*Red Maple*), from E N America, with leaves silver on lowerside and with small stiff clusters of flowers before the leaves as in *A. saccharinum* but leaves similar in shape to those of *A. pseudoplatanus*; and **A. saccharum** Marshall (*Sugar Maple*), from E N America, with leaves like those of *A. platanoides* but with the teeth on the sides of the main lobes with rounded (not acuminate) tips.

1. A. platanoides L. - *Norway Maple.* Tree to 30m; leaves simple, with 5-7 **469**
acuminate lobes each with few large acuminate teeth or sub-lobes; flowers in ± erect, yellowish-green corymbs appearing ± before leaves; fruits with widely divergent to ± horizontal wings; (2n=26). Intrd; abundantly planted and often self-sown in rough grassland, scrub, hedges and woodland; throughout lowland BI; Europe.

2. A. cappadocicum Gled. (*A. pictum* auct. non Thunb.) - *Cappadocian Maple.* Tree **469**
to 26m; leaves simple, with 5-7 acuminate, entire lobes; flowers in ± erect, yellowish-green, subcorymbose panicles appearing ± before leaves; fruits as in *A. platanoides*; (2n=26). Intrd; frequently planted in parks and by roads; self-sown and extensively suckering in SE En, rarely N to MW Yorks; SW Asia.

3. A. campestre L. - *Field Maple.* Tree to 25m; leaves simple, with 3-5 obtuse to **469**
rounded lobes each entire or with few obtuse to rounded teeth or sub-lobes; flowers in ± erect, yellowish-green, subcorymbose panicles appearing with leaves; fruits with ± horizontal wings; 2n=26. Native; woods, scrub and hedgerows on calcareous or clay soils; common in En and Wa N to Co Durham, planted elsewhere.

4. A. pseudoplatanus L. - *Sycamore.* Tree to 35m; leaves simple, with usually 5 ± **469**
acute coarsely serrate lobes widest at base; flowers in ± cylindrical, pendent, yellowish-green panicles appearing with leaves; fruits with wings diverging at c.90°; (2n=52). Intrd; fully natd and 1 of the most abundant trees in wide range of habitats throughout BI; Europe.

5. A. saccharinum L. - *Silver Maple.* Tree to 31m; leaves with 5 acute irregularly **469**
toothed lobes narrowed at base; flowers in small, stiff, compact yellowish-green clusters appearing well before leaves, males and females in separate clusters on same or different trees; petals 0; fruits with widely divergent to ± horizontal wings; (2n=52). Intrd; much (and increasingly) planted for ornament in parks and by roads; rarely setting seed but self-sown in London area; N America.

6. A. negundo L. - *Ashleaf Maple.* Tree to 17m; leaves ternate to pinnate with 3- **469**
5(7) ovate, acute, slightly toothed leaflets; male flowers in corymbs with pendent stamens, the female in small pendent racemes, both appearing well before leaves; petals 0; fruits with wings diverging at <90°; (2n=26). Intrd; commonly planted for ornament in parks and by roads and railways; sometimes self-sown where both sexes occur in SE En; N America.

102. ANACARDIACEAE - *Sumach family*

Deciduous shrubs; leaves alternate, pinnate or simple, petiolate, without stipules. Flowers in large terminal panicles, dioecious or variously mixed, hypogynous, actinomorphic, small; sepals 5, fused at base; petals 5, ± free; stamens 5; ovary 1-celled, with 1 basal ovule; styles 3; stigmas capitate; fruit a small 1-seeded drupe.

Rhus is distinct in its thick, pithy, pubescent, little-branched stems with large pinnate leaves and large reddish inflorescences.

Other genera - COTINUS Mill. differs from *Rhus* in its simple, entire, rounded leaves (often purple); thin, glabrous twigs; and long hairy pedicels. **C. coggygria** Scop. (*Rhus cotinus* L.) (*Smoke-tree*), from S Europe, is commonly grown in gardens and in parks and on road and railway banks, and very occasionally seedlings are found in the London area.

1. RHUS L. - *Stag's-horn Sumach*

1. R. typhina L. (*R. hirta* (L.) Sudw.) - *Stag's-horn Sumach*. Shrub to 5(10)m; twigs thick, densely pubescent; leaflets (7)11-15(21), oblong-lanceolate, acute to acuminate, serrate, 5-12cm; inflorescence stiffly erect, 10-20cm, greenish in flower, then deep red; (2n=30). Intrd; much planted on verges and banks by roads and railways; extensively suckering but very rarely or never self-sown in S Br, scattered N to Cumberland; N America.

103. SIMAROUBACEAE - *Tree-of-heaven family*

Deciduous trees; leaves alternate, pinnate, petiolate, without stipules. Flowers in terminal panicles, small, functionally monoecious and bisexual mixed, hypogynous, actinomorphic; sepals 5, fused proximally; petals 5, free; stamens 10; ovary of 5(-6) carpels loosely fused, each carpel with 1 axile ovule; styles 5(-6); stigmas peltate; fruit a group of 1-5(6) long, winged achenes.

Leaves and fruits resemble those of *Fraxinus*, but leaves are alternate and fruits are usually >1 per flower and with seeds in middle (not at base) of wing.

1. AILANTHUS Desf. - *Tree-of-heaven*

1. A. altissima (Mill.) Swingle - *Tree-of-heaven*. Tree to 26m; leaves up to 90cm with up to 41 narrowly ovate, acuminate, serrate leaflets; panicles 10-20cm, greenish-white; achenes pendent, 3-4cm, reddish then whitish; (2n=64, 80). Intrd; much planted in SE En, especially in Greater London and there frequently extensively suckering and self-sown; China.

104. RUTACEAE - *Rue family*

Deciduous or evergreen shrubs; leaves opposite or alternate, simple, ternate, pinnate or deeply pinnately lobed, petiolate, without stipules. Flowers variously arranged, bisexual, hypogynous, actinomorphic, with a well-developed nectariferous disc; sepals 4-5, free or fused, petals as many as sepals, free, or fused above base but free at base and apex; stamens 2x as many as sepals; ovary of fused carpels (as many as sepals, visible as obvious lobes), each carpel with 2-many ovules; style 1; stigma 1 or as many as carpels, capitate; fruit a 4-, 5- or 8-lobed capsule.

Very variable vegetatively but all shrubs; the 4-5 sepals, free attractive petals and carpels, the lobed disc, the 8 or 10 stamens, and the 4-5-lobed capsule are diagnostic.

1 Leaves ternate, glabrous; flowers erect; petals white **1. CHOISYA**
1 Leaves simple, tomentose on lowerside; flowers pendent; petals greenish-yellow **2. CORREA**

Other genera - **CITRUS** spp. (*Orange, Lemon*, etc.) are often found as unidentifiable seedlings on rubbish tips; they can be told by their glossy simple leaves articulated upon the winged petioles. **RUTA graveolens** L. (*Rue*), from E Mediterranean, is a much-grown small deciduous shrub with bluish-grey-green deeply 2-3-pinnately lobed leaves with strong distinctive smell and numerous yellow 4-5-petalled flowers 15-20mm across and with 1 stigma; it sometimes produces seedlings in shrubberies, waste places and wall-cracks in S En.

1. CHOISYA Kunth - *Mexican Orange*
Evergreen shrub; leaves ternate; flowers several to many in erect corymbose cymes; petals 5, free, white; stamens 10; carpels 5, each 2-lobed and with 2 ovules; stigma 5-lobed.

1. C. ternata Kunth - *Mexican Orange*. Shrub to 2(3)m; leaflets glabrous, glossy, 3-7.5cm, elliptic, entire; flowers sweetly scented, 2-3cm across, with patent, white petals. Intrd; grown in shrubberies and on estates, sometimes self-sowing and natd in SE En and Man; Mexico.

2. CORREA J. Kenn. - *Tasmanian-fuchsia*
Evergreen shrub; leaves simple; flowers 1-3 in pendent groups; petals 4, free at base and apex (4-lobed corolla), but fused into cylindrical tube in middle part, greenish-yellow; stamens 8; carpels 4, each 1-lobed and with 2 ovules; stigma 4-lobed.

1. C. backhousiana Hook. - *Tasmanian-fuchsia*. Shrub to 2(3)m; leaves ± glabrous and glossy on upperside, densely tomentose on lowerside, 1.5-3cm, elliptic, entire; flowers c.2.5cm long; (2n=32). Intrd; grown as hedging and well natd in woods on Tresco (Scillies); Tasmania.

105. OXALIDACEAE - *Wood-sorrel family*

Perennial, rarely annual, often slightly succulent herbs, often with bulbs and/or rhizomes; leaves all basal or alternate, usually ternate, sometimes palmate, petiolate, with or without stipules. Flowers 1-several in axillary, often umbellate cymes, bisexual, hypogynous, actinomorphic, often trimorphic; sepals 5, free; petals 5, free or ± so; stamens 10, sometimes not all with anthers; ovary 5-celled, each cell with many ovules on axile placentas; styles 5; stigmas minute; fruit a 5-celled capsule.

The ternate or less often palmate leaves and conspicuous actinomorphic flowers are diagnostic.

1. OXALIS L. - *Wood-sorrels*
Most bulbous spp. are trimorphic and self-incompatible, and reproduce mainly or wholly vegetatively. The different clones may show morphological differences, but those present in BI represent only a small part of their whole range and a relatively broad view of sp. limits is taken here. Many spp. are only marginally natd, occurring ± wholly in cultivated ground, but can be very persistent. Corolla colours given are those in the fresh state; after drying the red/pink colours often fade or become more bluish.

1 Petals yellow 2
1 Petals red, pink, mauve or white 8
  2 3 sepals cordate; leaves succulent **7. O. megalorrhiza**
  2 No sepals cordate; leaves thin, not succulent 3
3 Aerial stem 0; bulbils present at or below soil level **14. O. pes-caprae**
3 Aerial stems present; bulbils 0 4
  4 Stems procumbent, rooting freely at nodes 5
  4 Stems decumbent to erect, not or very sparsely rooting 6
5 Inflorescences always 1-flowered; capsules 3-4.5mm, with 3-4 seeds per cell; usually 5 stamens with and 5 without anthers **4. O. exilis**
5 At least most inflorescences 2-8(12)-flowered; capsules (4)8-20mm,

with >4 seeds per cell; usually all 10 stamens with anthers **3. O. corniculata**

6 Capsules <2x as long as wide; petals 10-15mm,with purple veins **1. O. valdiviensis**

6 Capsules >3x as long as wide; petals 5-11mm, not purple-veined 7

7 Pedicels (but not capsules) patent or reflexed in fruit; inflorescence an umbel; vegetative parts with only white simple hairs **5. O. dillenii**

7 Pedicels erect in fruit; inflorescence cymose; vegetative parts with translucent septate hairs as well as white simple ones **6. O. stricta**

8 Leaves with 5-10 leaflets **13. O. decaphylla**

8 Leaves with 3 leaflets 9

9 Stem aerial, ± erect 10

9 Stem 0 or a rhizome at or below soil level 11

10 Bulbs 0; inflorescences >1-flowered **2. O. rosea**

10 Stem arising from bulb and producing axillary aerial bulbs; inflorescences 1-flowered **15. O. incarnata**

11 Bulbs 0; stem a rhizome 12

11 Leaves arising from bulb at or below soil level; bulb often producing thin rhizomes 13

12 Flowers solitary; rhizome slender, with distant succulent scales **9. O. acetosella**

12 Flowers in umbels; rhizome thick, with dense papery scales **8. O. articulata**

13 Leaves with 4 leaflets **12. O. tetraphylla**

13 Leaves with 3 leaflets 14

14 Leaflets widest about the middle, with submarginal orange or dark dots on lowerside **10. O. debilis**

14 Leaflets widest at or near apex, without submarginal dots **11. O. latifolia**

Other spp. - A pink-flowered bulbous plant found in gardens in Wigtowns and Co Antrim has 3 dark-blotched leaflets and has been variously determined as **O. bulbifera** R. Knuth, **O. drummondii** A. Gray or a variant of *O. tetraphylla;* it might be an unusual clone of *O. latifolia.* **O. magellanica** G. Forst., from S America, Australia and New Zealand, is like a small delicate *O. acetosella* with leaflets only c.5 x 5mm and suffused bronze and flowers only 4-10mm and pure white; it occurs as a relic of cultivation in flower-beds in Man.

1. O. valdiviensis Barnéoud - *Chilean Yellow-sorrel.* Bulbs 0; stems erect to decumbent, to 30cm, not or little branched, rather succulent; leaflets 3, obcordate, with rounded margins and acute sinus, glabrous; flowers yellow with purple veins, 10-15mm, in forked cymes; (2n=18). Intrd; garden weed, reproducing by seed; very scattered in C & S En; Chile.

2. O. rosea Jacq. - *Annual Pink-sorrel.* Bulbs 0; stems erect to ascending, to 25cm, **474**
branched or not; leaflets 3, obcordate, with rounded margins and obtuse sinus, subglabrous; flowers pale mauve-pink, 11-17mm, in sparsely branched cymes; (2n=12, 14). Intrd; garden weed, reproducing by seed; Jersey and W Cornwall; Chile.

3. O. corniculata L. - *Procumbent Yellow-sorrel.* Bulbs 0; stems mostly procumbent, to 50cm, rooting at nodes, much-branched; leaflets 3, obcordate, with rounded margins and acute sinus, sparsely pubescent especially on margins, often purple; flowers yellow, 4-7.5mm, (1)2-8(12) in umbels; pedicels patent or reflexed in fruit; capsule with dense appressed hairs; seeds >4 per cell; 2n=24. Intrd; pernicious weed of gardens, paths, walls and waste ground, reproducing mainly by seed; common in most of En, Wa, Man and CI, very scattered in Sc and Ir; warmer parts of world.

4. O. exilis A. Cunn. (*O. corniculata* var. *microphylla* Hook. f.) - *Least Yellow-sorrel.*

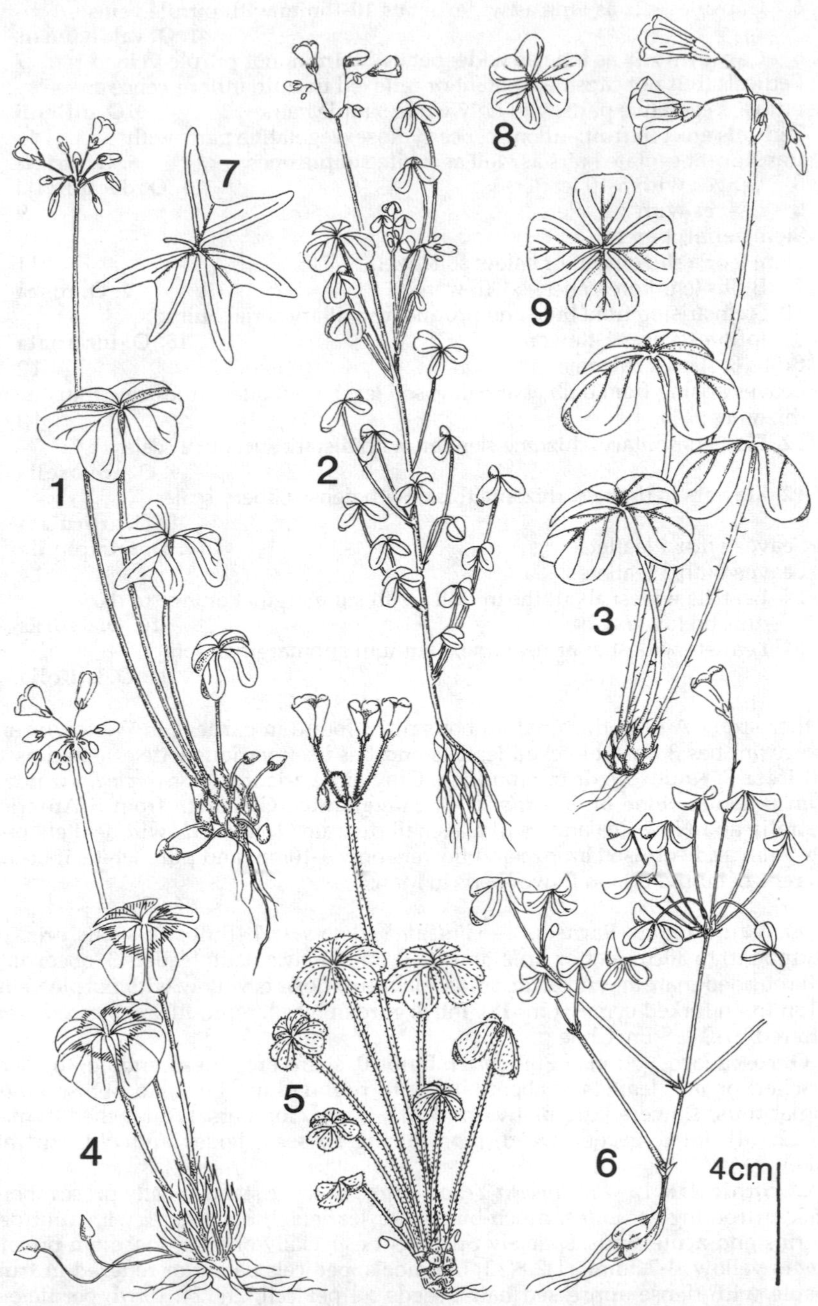

FIG 474 - *Oxalis*. 1, *O. latifolia* (ex Jersey). 2, *O. rosea*. 3, *O. debilis*. 4, *O. tetraphylla*. 5, *O. articulata*. 6, *O. incarnata*. 7-9, leaves of *Oxalis*. 7, *O. vespertilionis*. 8. *O. latifolia* (ex Cornwall). 9, *O. deppei*.

Differs from *O. corniculata* in filiform stems and all parts smaller; leaves always green; seeds 2-4 per cell; and see key. Intrd; similar places and distribution to *O. corniculata*, usually but not everywhere less common; New Zealand and Tasmania.

5. O. dillenii Jacq. (*O. stricta* auct. non L.) - *Sussex Yellow-sorrel.* Stems erect to decumbent, to 20cm, little-branched; leaves as in *O. corniculata* but always green; flowers yellow, 6-11mm, (1)2-3(4) in umbels; pedicels and capsule as in *O. corniculata*; seeds with white patches; (2n=16-24). Intrd; weed in sandy arable fields, reproducing by seed; known 1950-1984 in W Sussex, also Herm (CI); N America.

6. O. stricta L. (*O. fontana* Bunge, *O. europaea* Jord.) - *Upright Yellow-sorrel.* Differs from *O. dillenii* in stems to 40cm; thin subterranean rhizomes sometimes produced; leaves sometimes purple; flowers 5-9(15)mm, 2-5 in umbels; seeds without white patches; and see key; (2n=18-28). Intrd; weed of gardens and arable fields, reproducing mainly by seed; scattered over most of BI, common in parts of S En; N America.

7. O. megalorrhiza Jacq. (*O. carnosa* auct. non Molina) - *Fleshy Yellow-sorrel.* Stems erect to ascending, subterranean and aerial, to 20cm, thick and succulent, ± unbranched; leaflets 3, succulent, obovate, rounded to very shallowly notched at apex, glabrous, covered with translucent cells on lowerside; flowers bright yellow, 12-15mm, 1-3(5) in umbels; (2n=14). Intrd; natd on walls and banks in Scillies, known since c.1936, 1 place in W Cornwall; very frost sensitive; Chile.

8. O. articulata Savigny (*O. floribunda* auct. non Lehm.) - *Pink-sorrel.* Stem a thick **474**
brown-scaly horizontal to oblique rhizome; leaflets 3, obcordate, with rounded margins and acute sinus, appressed-pubescent, with orange warts especially near margin; petioles up to 25cm; peduncles up to 35cm; flowers deep pink, rarely white or pale pink, 10-15mm, c.3-25 in (sometimes partly compound) umbels; (2n=14). Intrd; much grown in gardens and established escape in waste and stony and sandy ground on roadsides, banks and seashores, often in closed vegetation, reproducing by seed and rhizomes; frequent in SW En and CI, scattered in C & S Br and Ir; E S America. Ssp. **rubra** (A. St.-Hil.) Lourteig differs in its shorter hairs, elliptic (not linear-lanceolate) sepals, and more pubescent petals; some of our plants might belong to it.

9. O. acetosella L. - *Wood-sorrel.* Stems horizontal thin rhizomes with distant succulent scale leaves; leaflets 3, obcordate, with rounded margins and acute to obtuse sinus, sparsely appressed-pubescent; petioles and peduncles up to 10cm; flowers white with mauve veins, rarely pink or mauve, 8-15mm, solitary; 2n=22. Native; woods, hedgebanks, shady rocks, often on humus; common ± throughout BI.

10. O. debilis Kunth (*O. corymbosa* DC.) - *Large-flowered Pink-sorrel.* Leaves and **474**
peduncles arising from scaly bulb which produces many sessile bulblets and 0-2 swollen succulent roots; leaflets 3, obcordate, with acute to closed sinus and rounded margins, very sparsely appressed-pubescent on lowerside, with orange or dark warts only or especially near margin; petioles up to 15(20)cm; peduncles up to 20(30)cm; flowers pinkish-mauve, 15-20mm, c.3-25 in contracted cymes; (2n=14, 28). Intrd; weed of gardens and other open ground, reproducing by bulblets only; frequent in S En and CI, scattered elsewhere in C & S Br and E Ir; S America. The above description applies to var. **corymbosa** (DC.) Lourteig; var. **debilis** differs in its salmon-pink to brick-red flowers, smaller bulblets and glandular hairs present on styles; it occurs in a few gardens in S En.

11. O. latifolia Kunth (*O. vespertilionis* Zucc.) - *Garden Pink-sorrel.* Differs from *O.* **474**
debilis in bulblets often formed at end of rhizomes to 3cm; leaflets with variously rounded to ± pointed lobes, with obtuse to rounded sinus, without warts; flowers pink, sometimes white, 8-13mm; (2n=14, 24). Intrd; habitat and distribution of *O. debilis*; C & S America. Variable in leaf-shape: the commonest plants have 'fishtail-shaped' (obsagittate) leaflets with elongated lobes rounded at ends and rounded to obtuse sinus; some plants from Devon, Cornwall and Guernsey have much more

rounded leaflet-lobes with obtuse sinus: plants known as **O. vespertilionis** (very rare in Br) have 'V-shaped leaflets' with very elongated lobes often subacute at extreme ends and rounded sinus.

12. O. tetraphylla Cav. (*O. deppei* Lodd. ex Sweet) - *Four-leaved Pink-sorrel.* 474
Differs from *O. debilis* in bulblets often formed at end of rhizomes to 6cm; leaflets 4, obsagittate, with ends of lobes rounded and sinus very obtuse, without warts; flowers pinkish-purple; (2n=56). Intrd; weed of gardens and arable land, reproducing by bulblets only; Jersey and Scillies; Mexico. **O. deppei** possibly occurs also; it differs in its scarcely or not lobed leaflets with ± truncate apex, brick-red flowers, and sessile bulblets, and might be a separate sp.

13. O. decaphylla Kunth (*O. lasiandra* auct. non Zucc.) - *Ten-leaved Pink-sorrel.* Leaves and peduncles arising from scaly bulb which produces many bulblets at the ends of thin rhizomes to 15cm; leaflets 5-10, narrowly obovate to oblanceolate, with apex rounded or notched with acute to subacute sinus, glabrous; flowers pale mauvish-pink, 10-14mm, (1)2-6 in umbels; (2n=28). Intrd; weed in public flower-beds in Man; Mexico.

14. O. pes-caprae L. - *Bermuda-buttercup.* Leaves and flowers arising at soil level among group of bulblets at top of short underground stem arising from deep-seated main bulb; leaflets 3, obcordate, with rounded margins and acute to subacute sinus, appressed-pubescent on lowerside; petioles and peduncles up to 30cm; flowers yellow, 20-25mm, c.3-25 in umbels; (2n=28, 34). Intrd; common weed of arable land (often bulbfields) in Scillies, rare in S Devon and CI, reproducing by bulblets and swollen roots; S Africa.

15. O. incarnata L. - *Pale Pink-sorrel.* Bulb producing annual, erect, branched stem 474
to 20cm with axillary sessile bulblets; leaflets 3, obcordate, with rounded margins and obtuse to subacute sinus, ± glabrous; flowers pale mauvish-pink with darker veins, 12-20mm, solitary; (2n=14). Intrd; weed of cultivated ground, walls and banks; frequent in SW En and CI, scattered elsewhere in BI; S Africa.

106. GERANIACEAE - *Crane's-bill family*

Herbaceous annuals to perennials, sometimes woody below; leaves alternate, simple and variously (often very deeply) palmately or pinnately lobed, or palmate to pinnate, at least the lower petiolate, stipulate. Flowers 1-several in terminal or axillary, often umbellate cymes, bisexual, hypogynous, actinomorphic to (*Pelargonium*) zygomorphic; sepals 5, free but in *Pelargonium* the upper one with a spur fused to pedicel; petals 5, free, rarely 0; stamens 5, or 10 but sometimes 3 or 5 without anthers, or rarely 15; ovary 5-celled, each cell with 2 ovules, elongated distally into a sterile column; style 1; stigmas 5, linear; fruit a dry 5-celled schizocarp, with 1 seed per cell, each mericarp and its apical sterile beak separating from the column and splitting open at maturity in a variety of ways (see below).

The distinctive 5-seeded fruits and usually conspicuous actinomorphic (or nearly so) flowers are diagnostic.

1 Corolla zygomorphic, the 2 upper petals much wider than others; uppermost sepal with spur tightly fused to pedicel (but spur inconspicuous) **4. PELARGONIUM**
1 Corolla actinomorphic to weakly zygomorphic, the petals scarcely different in width; sepals not spurred 2
  2 Stamens 15, all with anthers, fused to c.1/2 way into 5 groups of 3 each **2. MONSONIA**
  2 Stamens 10, or 5 alternating with anther-less staminodes, not fused 3
3 Beaks of mericarps curved or loosely spirally twisted for 1-2 turns at maturity; leaves palmate or palmately lobed **1. GERANIUM**

3 Beaks of mericarps becoming tightly spiralled along their own axis at maturity; leaves pinnate or pinnately or rarely ternately lobed **2. ERODIUM**

1. GERANIUM L. - *Crane's-bills*

Annuals to perennials; leaves simple and palmately lobed, or palmate; flowers actinomorphic or ± so; stamens 10, free, sometimes the outer 5 anther-less; fruits dispersing variously, but only in *G. phaeum* with the mericarp separating whole from the column and its beak twisting spirally, and then the spirals only 1-2 and in large loops.

'Leaves' refers to lower leaves with long petioles. Most taxa occur as rare white-petalled mutants; this aspect of variation is not mentioned for each sp.

1 Petals narrowed at base to distinct claw 1/2 as long as limb to longer than limb 2

1 Petals without claw or with claw <1/2 as long as limb 8

2 Most petals >14mm 3

2 Most petals <14mm 4

3 Perennial with thick rhizome **20. G. macrorrhizum**

3 Biennial with thin fibrous roots **24. G. rubescens**

4 Leaves divided ≤3/4 to base 5

4 Leaves divided ± to base (ternate to palmate) 7

5 Leaves glossy green to purple, very sparsely pubescent; sepals strongly keeled on back **21. G. lucidum**

5 Leaves grey-green, pubescent; sepals rounded on back 6

6 Mericarps appressed-pubescent, smooth; outer 5 stamens lacking anthers **18. G. pusillum**

6 Mericarps (excluding beaks) glabrous, usually ridged; all stamens with anthers **19. G. molle**

7 Anthers orange or purple (pale yellow in albinos); petals 8-14mm; mericarps with sparse fine ridges and 0-1(2) deep collar-like ridges at apex **22. G. robertianum**

7 Anthers yellow; petals 5-9mm; mericarps with dense wrinkle-like ridges and 2-3(4) deep collar-like ridges at apex **23. G. purpureum**

8 Annuals to biennials, or sometimes perennials ± without rhizome; petals mostly ≤10mm, rarely to 22mm 9

8 Perennials with distinct thick and/or elongated rhizome; petals >10mm, rarely less 17

9 Petals ≥15mm 10

9 Petals ≤10mm 11

10 Plant usually <60cm; flowers ≤3cm across; petal-claw 6-7mm **24. G. rubescens**

10 Plant usually >60cm; flowers >3cm across; petal-claw 2-2.5mm **25. G. maderense**

11 Seeds smooth 12

11 Seeds pitted or reticulately ridged 14

12 Mericarps (excl. beaks) glabrous, usually ridged **19. G. molle**

12 Mericarps pubescent, smooth 13

13 Petals <5mm; sepals <3mm; outer 5 stamens lacking anthers **18. G. pusillum**

13 Petals >6mm; sepals >3mm; all 10 stamens with anthers **17. G. pyrenaicum**

14 Petals ± rounded at apex; leaves divided <3/4 way to base, sepals with apiculus <0.5mm **5. G. rotundifolium**

14 Petals distinctly notched at apex; leaves divided ≥3/4 way to base; sepals with apiculus >0.5mm 15

15 Mericarps glabrous to sparsely pubescent; most pedicels >2.5cm **11. G. columbinum**

15 Mericarps pubescent; pedicels ≤2.5cm 16

16 Leaves divided almost to base; stalked glands frequent on upper parts of plant **12. G. dissectum**

16 Leaves divided c.3/4-7/8 way to base; stalked glands 0 **13. G. submolle**

17 Flowers all solitary on pedicel + peduncle **10. G. sanguineum**

17 At least most flowers in pairs, with 2 pedicels on a common peduncle 18

18 Mericarps pointed at base, with 2-4 collar-like ridges at apex; petals not notched, usually apiculate or with small triangular point at apex, sometimes ruffled or subentire 19

18 Mericarps rounded at base, with 0-1 collar-like ridge at apex; petals rounded to notched at apex, sometimes with small point in notch 20

19 Petals c. as long as wide, patent to slightly reflexed, usually purplish-black **26. G. phaeum**

19 Petals c.1.5x as long as wide, strongly reflexed, never purplish-black **27. G. x monacense**

20 Stalked glands 0 or <0.3mm long 21

20 Sepals and pedicels, and usually peduncles and upperparts of stems, with stalked glands >0.3mm long 26

21 Base of mericarps without tuft of bristles on inside; petals violet-blue **14. G. ibericum**

21 Base of mericarps with tuft of apically pointed bristles on inside directed on to seed or into cavity; petals whitish to bright- or purplish-pink 22

22 Hairs on pedicels, peduncles and upper parts of stems <0.2mm, appressed; main leaf-lobes toothed, with teeth up to c.5mm long **4. G. nodosum**

22 Pedicels, peduncles and upper parts of stem with patent hairs >0.5mm; main leaf-lobes with sub-lobes or deep teeth >(5)10mm long 23

23 Petals with veins darker than ground-colour; beaks of mericarps with hairs c.0.1mm 24

23 Petals with veins paler than or same colour as ground-colour; beaks of mericarps with some hairs >(0.2)0.5mm 25

24 Petals curved outwards at apex (flowers trumpet-shaped), with white to very pale pink ground-colour **3. G. versicolor**

24 Petals not curved outwards at apex (flowers funnel-shaped), ground-colour pink **2. G. x oxonianum**

25 Fruit with style (between tip of column and base of stigmas) 2.5-3(4)mm; petals deep bright pink **1. G. endressii**

25 Fruit with style (3)4-6mm; petals usually mid-pink, often variable on 1 plant **2. G. x oxonianum**

26 Stalked glands sparse, confined to floral parts and pedicels Return to 25

26 Stalked glands abundant on pedicels, peduncles and upper parts of stems 27

27 Petals conspicuously notched; base of mericarps (if ripening) without tuft of bristles on inside 28

27 Petals rounded at apex; base of mericarps with tuft of apically pointed bristles on inside directed on to seed or into cavity 29

28 Basal leaves divided c.1/2 way to base, their main lobes broadest near apex; flowers pointing horizontally; seeds ripening **16. G. platypetalum**

28 Basal leaves divided c.2/3-7/8 way to base, their main lobes broadest well below apex; flowers pointing upwards; seeds not ripening **15. G. x magnificum**

29 Petals blue to violet-blue, with white base; flowers and immature fruits

pointing sideways or ± downwards; fruits with styles >4mm 30
29 Petals pinkish-purple to magenta, with black or white base; flowers and immature fruits pointing obliquely or vertically upwards; fruits with styles <4mm 31
30 Sepals with point >1/5 as long as main part; leaves divided >5/6 way to base **8. G. pratense**
30 Sepals with point <1/5 as long as main part; leaves divided <5/6 way to base **9. G. himalayense**
31 Petals magenta with black base, >16mm, with few hairs on either side at base **7. G. psilostemon**
31 Petals pinkish-purple with white base, <16mm, with abundant hairs at base on upperside **6. G. sylvaticum**

Other spp. - G. canariense Reut., from Canaries, was misrecorded as *G. maderense* in Guernsey, but is there only self-sown in and near flower-beds. It differs in plant usually ≤60cm tall, flowers <3cm across and petal-claw 4.5-5.5mm; and from both *G. maderense* and *G. rubescens* in mid-lobe of leaves sessile (not stalked), styles white (not pink or purple) and throat of corolla much paler (not darker) in colour than petal-limb. **G. brutium** Gasp., from C & E Mediterranean, is a rare garden escape resembling *G. molle* but with leaves greener and those in the inflorescence much shorter, and petals longer (6-11mm).

Subgenus* 1 - *GERANIUM (spp. 1-16). Mature fruit dehiscing by mericarps springing from column and ejecting seeds explosively; beak becoming curved, remaining attached to pericarp, and remaining attached or not to column; mericarps rounded at base; petals ± without claw.

1. G. endressii J. Gay - *French Crane's-bill.* Extensively rhizomatous ± erect perennial to 70cm; leaves 5-lobed c.4/5-5/6 way to base; stalked glands usually present on sepals and tops of pedicels; petals 16-22mm, deep bright pink, slightly retuse at apex; (2n=26, 28). Intrd; much grown in gardens, frequently natd in grassy places and waste ground; scattered over most of Br, rare in Ir; Pyrenees.

2. G. x oxonianum Yeo (*G. endressii* x *G. versicolor*) - *Druce's Crane's-bill.* Fertile hybrid segregating and forming spectrum between 2 parents; petals 20-26mm, pale to deep pink, with or without dark veins, retuse at apex; and see key (couplets 23-25). Intrd; often grown in gardens, natd independently of parents in grassy places; distributed as for *G. endressii* but rarer, also Guernsey; garden origin.

3. G. versicolor L. - *Pencilled Crane's-bill.* Shortly rhizomatous ± erect perennial to 60cm; leaves 5-lobed c.2/3-4/5 way to base, the lobes less dissected than in *G. endressii*; stalked glands 0; petals 13-18mm, white or nearly so with magenta veins, retuse at apex; (2n=28). Intrd; grown and natd as for *G. endressii*, but rarer in both situations; scattered in C & S Br, Ir and CI; C Mediterranean.

4. G. nodosum L. - *Knotted Crane's-bill.* Shortly rhizomatous ± erect perennial to 50cm; leaves 5-lobed c.2/3 way to base, the lobes less dissected than in *G. versicolor*; stalked glands 0; petals 13-18mm, purplish-pink with darker veins, retuse at apex; (2n=28). Intrd; grown in gardens, natd in hedgerows and woodland; rare and very scattered in C & S Br; S Europe.

5. G. rotundifolium L. - *Round-leaved Crane's-bill.* Erect to decumbent annual to **481**
40cm; leaves 5-9-lobed ≤1/2 way to base; glandular hairs abundant on most of plant; petals 5-7mm, pink, rounded or slightly retuse at apex; (2n=26). Native; banks, walls and stony ground; local in C & S En, S Wa, S Ir and CI.

6. G. sylvaticum L. - *Wood Crane's-bill.* Compact ± erect perennial to 70cm; leaves (5-)7(9)-lobed 3/4-4/5 way to base; glandular hairs abundant on most of plant; petals 12-16mm, purplish-pink to mauvish with white base, rounded or slightly retuse at apex. Native; woods and hedges in lowlands, rock-ledges and meadows in uplands; locally common in N Br S to Yorks, very local in C En, Wa and N Ir, rarely natd elsewhere.

7. G. psilostemon Ledeb. - *Armenian Crane's-bill.* Compact ± erect perennial to 120cm; leaves 7-lobed c.4/5 way to base; glandular hairs abundant on most of plant; petals 16-20mm, magenta with black base, rounded or slightly retuse at apex. Intrd; commonly grown in gardens, natd in grassy places; extremely scattered in En and Sc; SW Caucasus and NE Turkey.

8. G. pratense L. - *Meadow Crane's-bill.* Compact ± erect perennial to 1m; leaves 7-9-lobed ≥6/7 way to base; glandular hairs abundant on most of plant; petals 16-24mm, blue to violet-blue, rounded at apex; 2n=28. Native; meadows, roadsides, open woodland, often in damp places; abundant in much of Br but absent from N Sc and parts of Wa and S En, very local in N Ir, occasionally natd elsewhere.

9. G. himalayense Klotzsch - *Himalayan Crane's-bill.* Extensively rhizomatous ± erect perennial to 60cm; leaves 7-lobed c.4/5 way to base, with much more obtuse teeth than in *G. pratense;* stalked glands abundant on most of plant; petals 20-30mm, blue to violet-blue, rounded at apex. Intrd; grown in gardens, natd in grassy places; extremely scattered in En and Sc, perhaps overlooked for *G. pratense* or *G.* x *magnificum;* Himalayas. **G. 'Johnson's Blue'**, also grown in gardens, is the sterile hybrid *G. pratense* x *G. himalayense*, and some of our plants might be this.

10. G. sanguineum L. - *Bloody Crane's-bill.* Shortly rhizomatous, erect to procumbent perennial to 40cm; leaves 5-7-lobed ≥3/4 way to base, the lobes deeply sublobed; stalked glands 0; petals 14-22mm, bright purplish-red, sometimes (var. **striatum** Weston (var. *lancastriense* (Mill.) Gray)) pink, retuse at apex; 2n=84. Native; grassland, rocky places, sand-dunes, open woods on calcareous soils; local in N & W Br and C Ir, mainly coastal, natd from gardens elsewhere.

11. G. columbinum L. - *Long-stalked Crane's-bill.* Erect to ascending or scrambling **481**
annual to 60cm; leaves 5-7-lobed almost to base, the lobes deeply sublobed; stalked glands 0; petals 7-10mm, reddish-pink, rounded, truncate or shallowly retuse at apex; 2n=18. Native; grassy places, banks and scrub, mostly on calcareous soils; locally frequent in much of BI, but rare in N.

12. G. dissectum L. - *Cut-leaved Crane's-bill.* Erect to procumbent annual to 60cm; **481**
leaves (5-)7-lobed almost to base, the lobes deeply sublobed; stalked glands frequent on upper parts; petals 4.5-6mm, pink, deeply retuse at apex; 2n=22. Native; grassy and stony ground, waste places and cultivated ground; common throughout most of BI.

13. G. submolle Steud. (*G. core-core* auct. non Steud.) - *Alderney Crane's-bill.* **481**
Perennial without rhizome, decumbent to ascending, to 60cm; leaves 5-7-lobed c.3/4-7/8 way to base, the lobes deeply sublobed; stalked glands 0; petals 4.5-6.5mm, pink, retuse at apex. Intrd; hedgerows and grassy and waste places; natd in Guernsey since 1926, Alderney since 1938, rare casual in Jersey; S America. The identity of our plant is far from certain; other names than the 2 above have been suggested.

14. G. ibericum Cav. - *Caucasian Crane's-bill.* Shortly rhizomatous to compact ± erect perennial to 50cm; leaves 9-11-lobed 2/3-7/8 way to base; stalked glands 0; petals 24-26mm, violet-blue, retuse at apex; (2n=56). Intrd; rather rarely grown in gardens, natd for many years in old churchyard in Cards, also E Lothian; Caucasus.

15. G. x magnificum Hyl. (*G. ibericum* x *G. platypetalum*) - *Purple Crane's-bill.* Differs from *G. ibericum* in stems to 75cm; leaves lobed rarely >4/5 way to base; stalked glands and non-glandular hairs ± equally abundant on most parts; petals 22-24mm, purplish-violet. Intrd; much grown in gardens, frequently natd in grassy places, by roads and on waste land, perhaps under-recorded for *G. pratense;* scattered throughout much of Br, Guernsey; garden origin. Sterile.

16. G. platypetalum Fisch. & C.A. Mey. - *Glandular Crane's-bill.* Differs from *G. ibericum* and *G. x magnificum* in stems to 40cm; leaves lobed c.1/2 way to base; stalked glands abundant on most parts, commmoner than non-glandular hairs on upper parts of stem; petals 16-22mm, violet-blue; flowers pointing sideways (not

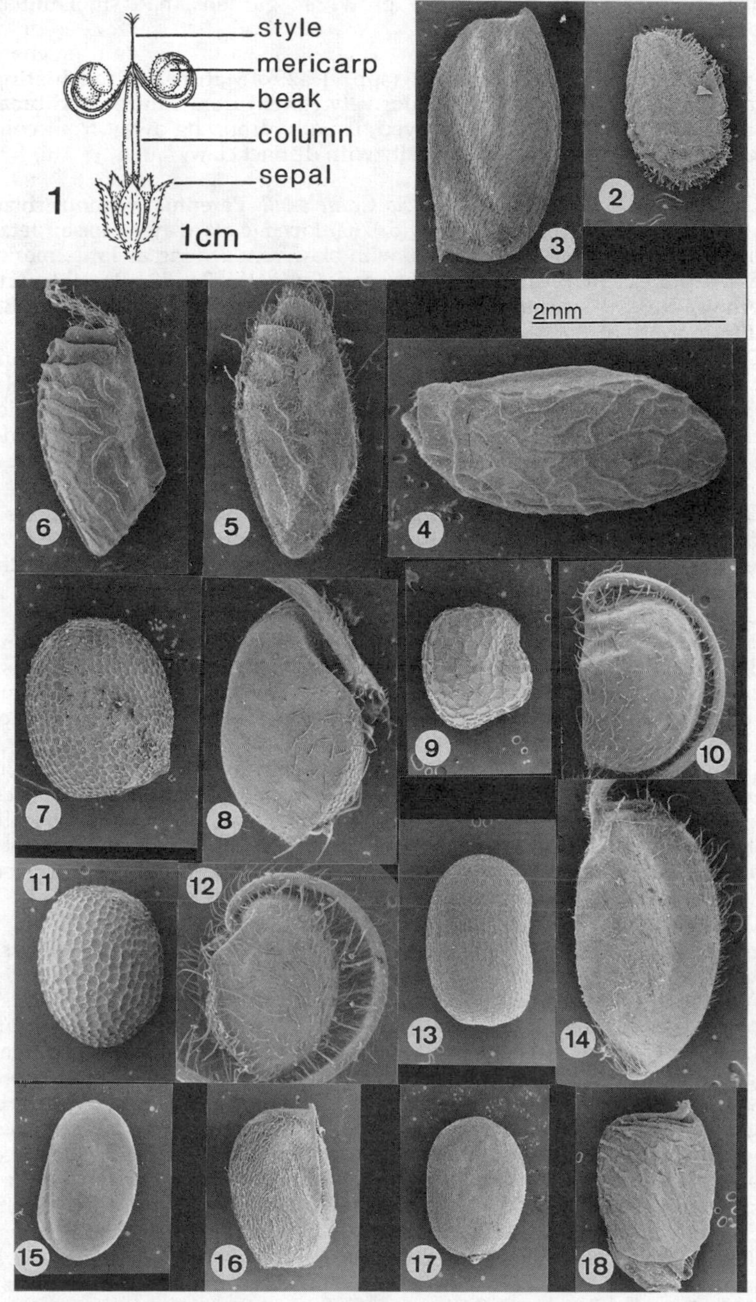

FIG 481 - *Geranium*. 1, labelled fruit of *G. rotundifolium*. 2-6, mericarps. 2, *G. lucidum*. 3, *G. pyrenaicum*. 4, *G. rubescens*. 5, *G. robertianum*. 6, *G. purpureum*. 7-18, seeds and mericarps. 7-8, *G. columbinum*. 9-10, *G. rotundifolium*. 11-12, *G. dissectum*. 13-14, *G. submolle*. 15-16, *G. pusillum*. 17-18, *G. molle*.

upwards); (2n=28, 42). Intrd; rarely grown in gardens, natd in Dunbarton; Caucasus.

Subgenus* 2 - *ROBERTIUM Picard (spp. 17-25). Mature fruit dehiscing by mericarps containing seeds being explosively ejected from column and breaking away from beak; beak becoming curved, usually dropping away from column; mericarps rounded at base; petals usually with distinct claw.

17. G. pyrenaicum Burm. f. - *Hedgerow Crane's-bill.* Perennial without rhizome, **481**
erect to ascending, to 60cm; leaves (5-)7(9)-lobed c.2/3 way to base; petals 7-10mm, pinkish-purple, retuse at apex, with claw <1/2 as long as limb; mericarps appressed-pubescent, nearly smooth; seeds smooth; 2n=28. Possibly native; hedgerows, grassy places and rough ground; locally frequent in much of BI but absent from much of N & W.

18. G. pusillum L. - *Small-flowered Crane's-bill.* Decumbent to ascending annual to **481**
40cm; leaves 7-9-lobed 1/2-2/3 way to base; pedicels with hairs all short; petals 2.5-4mm, mauvish-pink, retuse at apex, with claw ≤1/2 as long as limb; mericarps appressed-pubescent, ± smooth; seeds smooth; (2n=26). Native; cultivated and waste land and barish places among grass; frequent to scattered throughout most of BI.

19. G. molle L. - *Dove's-foot Crane's-bill.* Differs from *G. pusillum* in pedicels with **481**
some long as well as many short hairs; petals 4-6mm, more deeply retuse at apex; and see key (couplet 6), but mericarps rarely smooth (var. **aequale** Bab.); 2n=26. Native; similar places to *G. pusillum;* common all over BI.

20. G. macrorrhizum L. - *Rock Crane's-bill.* Shortly rhizomatous perennial, ± erect, to 50cm; leaves 7-lobed 2/3-5/6 way to base; petals 12-18mm, pinkish-purple, rounded at apex, with claw c. as long as limb; mericarps glabrous, with sharp horizontal ridges near apex; seeds smooth; (2n=46, c.92). Intrd; commonly grown in gardens, natd on walls and banks and in grassy places; extremely few places in En and Sc, natd since 1890 in S Devon; S Europe.

21. G. lucidum L. - *Shining Crane's-bill.* Very sparsely pubescent, shining, often **481**
red, erect to ascending annual to 40cm; leaves 5-lobed c.2/3 way to base; petals 8-10mm, deep pink, rounded at apex, with claw longer than limb; mericarps glabrous except minutely pubescent at apex, ridged; seeds smooth; 2n=20, 40. Native; bare ground, rocks, walls and stony banks, mostly on calcareous ground; locally common in most of BI except N Sc, also garden escape in S En.

22. G. robertianum L. - *Herb-Robert.* Procumbent to erect, strong-smelling, often **481**
red annual to biennial to 50cm; leaves ternate to palmate with 3-5 leaflets, the leaflets much divided ± to midribs; petals 8-14mm, usually deep pink, rounded at apex, with claw c. as long as limb; mericarps glabrous to sparsely pubescent, with rather sparse branching ridges and 0-1(2) collar-like ridges at apex; seeds smooth; 2n=32, 64. Native; woods, hedgerows, banks, scree and maritime shingle; common throughout BI. Very variable and often divided into sspp., but characters used for these cut across others and probably do not define meaningful taxa. Most often segregated is ssp. **maritimum** (Bab.) H.G. Baker, around coasts of BI (especially S & W) on shingle, with procumbent habit, often glabrous stems and leaves, usually glabrous fruits, and slightly smaller flowers, but only the prostrateness is constant. Ssp. **celticum** Ostenf. occurs in crevices of limestone rocks by or near coast in S Wa and CW Ir; it is distinct in the restriction of anthocyanin to the stem-nodes and petiole bases, and usually has large pale flowers, red or purple anthers, and large pubescent fruits.

22 x 23. G. robertianum x G. purpureum might occur with the parents on maritime shingle and it was reported from a hedgebank in E Cork in 1988. It is possible that *G. purpureum* ssp. *forsteri* arose from this cross.

23. G. purpureum Vill. (*G. robertianum* ssp. *purpureum* (Vill.) Nyman) - *Little-* **RR**
Robin. Differs from *G. robertianum* in key characters (couplet 7); 2n=32. Native; **481**

rocky and stony places, and on shingle and cliffs, usually near sea; very local in SW Br from W Sussex to Carms, M Cork, Co Waterford, CI; over-recorded for *G. robertianum* ssp. *maritimum*. Procumbent to decumbent plants of maritime shingle have been separated as ssp. **forsteri** (Wilmott) H.G. Baker; they approach *G. robertianum* in fruit characters.

24. G. rubescens Yeo - *Greater Herb-Robert*. Erect, reddish, faintly scented **481**
biennial to 60cm due to long petioles and large leaves, but with stem usually much shorter; differs from *G. robertianum* in larger leaves and flowers and fruits; petals 18-22mm, with claw c.1/2 as long as limb; fruits glabrous, with dense reticulate ridges. Intrd; natd on rough ground in Guernsey since 1968, Man since 1930s; Madeira.

25. G. maderense Yeo - *Giant Herb-Robert*. Erect perennial to 1(2)m, like a giant *G. rubescens* with stem up to 5cm wide; petioles >30cm; leaves >25cm; inflorescence >50cm across; flowers >3cm across; and petals 15-22 x 10-18mm with claw only 2-2.5mm. Intrd; well natd and increasing on cliffs in dense low vegetation in Scillies; Madeira. Records from Guernsey were errors for *G. canariense*.

Subgenus* 3 - *ERODIOIDEA Yeo (spp. 26-27). Mature fruit dehiscing by mericarps containing seeds and attached to beaks being explosively ejected from column; beak attaining 1-2 loose spirals; mericarps pointed at base and with transverse ridges near apex; petals without claw.

26. G. phaeum L. - *Dusky Crane's-bill*. Shortly rhizomatous to compact, ± erect perennial to 80cm; leaves 7(-9)-lobed c.2/3 way to base; stalked glands abundant on most parts of plant but only c.0.1mm long, with much longer simple hairs; petals 8-12(14)mm, often with small point at apex, usually dark purplish-black, rarely pinkish-mauve (var. **lividum** (L'Hér.) Pers.) or white; (2n=14, 28). Intrd; commonly grown, natd in shady places in hedges and wood-borders; scattered throughout most of Br, very local in Ir; C Europe.

27. G. x monacense Harz (*G. reflexum* auct. non L.; *G. phaeum* x *G. reflexum* L.) - *Munich Crane's-bill*. Differs from *G. phaeum* in petals much narrower, more reflexed and with more hairs at base and resembling those of *G. phaeum* var. *lividum* (presumably 1 parent) in colour. Intrd; natd on roadside verge in E Sussex since 1975, Midlothian, perhaps overlooked for *G. phaeum*; garden origin.

2. MONSONIA L. - *Dysentery-herbs*

Annuals; leaves simple, pinnately toothed; flowers actinomorphic, stamens 15, all with anthers, with filaments fused for c.1/2 their length into 5 groups of 3; seeds dispersed inside mericarps with beaks attached, the beaks tightly spiralled (twisted) on their own axes.

1. M. brevirostrata R. Knuth - *Short-fruited Dysentery-herb*. Sparsely pubescent, branched, ascending annual to 40cm; leaves 16-30mm, lanceolate to narrowly ovate; flowers 1-3 on common peduncle; petals 5-8mm, slightly longer than sepals, rounded at apex, bluish; fruits pubescent, 22-30mm. Intrd; fairly frequent wool-alien in fields and waste places; scattered in En; S Africa.

3. ERODIUM L'Hér. - *Stork's-bills*

Annuals to perennials, leaves simple and pinnately to ternately lobed or pinnate; flowers actinomorphic or slightly zygomorphic; stamens 5, free, alternating with staminodes; fruits dispersing as in *Monsonia*, with 2 pits at apex of mericarp, 1 either side of base of beak.

1 Leaves pinnate 2
1 Leaves simple, shallowly to deeply lobed, or if ± compound then ternate 5
2 Primary leaflets divided ≤3/4 way to midrib; apical pits of mericarp

with sessile glands **8. E. moschatum**

2 Primary leaflets divided nearly to base; apical pits of mericarp glandless 3

3 Petals 15-20mm; bracts green, fused into cupule; fruits with column 4-7cm **11. E. manescavii**

3 Petals 4-12mm; bracts brown, several, free; fruits with column 1-4cm 4

4 Apical pits of mericarp separated from main part of mericarp by sharp ridge and groove, not overarched by hairs; flowers 3-7 per peduncle, mostly >10mm across **9. E. cicutarium**

4 Apical pits of mericarp not delimited by sharp ridge and groove, overarched by hairs from main part of mericarp; flowers 2-4(5) per peduncle, mostly <10mm across **10. E. lebelii**

5 Beak of fruit 0.8-1cm; petals not exceeding sepals or 0 **3. E. maritimum**

5 Beak of fruit >1.5cm; petals exceeding sepals 6

6 Petals pinkish-purple; fruits with column ≤4cm; apical pits of mericarps delimited by very distinct groove and with conspicuous sessile glands, or with no groove and no glands 7

6 Petals blue to violet-purple; fruits with column >4cm; apical pits of mericarps delimited by distinct groove, without glands 8

7 Hairs on sepals and pedicels eglandular, appressed; apical pits of mericarp without glands, not delimited by groove **1. E. chium**

7 Most hairs on sepals and pedicels glandular, patent; apical pits of mericarp with sessile glands, delimited by very distinct groove **2. E. malachoides**

8 Lower leaves pinnately lobed, with ≥2 basal pairs of lobes not very different in size 9

8 Lower leaves ternately lobed, with 1 pair of basal lobes greatly exceeding all others 10

9 Apical pits of mericarp with sparse bristles, bounded below by 1 blunt-rimmed groove **5. E. brachycarpum**

9 Apical pits of mericarps completely glabrous, bounded below by (1)2-3 sharp-rimmed grooves **4. E. botrys**

10 Sepals and pedicels with many patent glandular hairs **7b. E. cygnorum ssp. glandulosum**

10 Sepals and pedicels without glandular hairs 11

11 Pedicels glabrous or with hairs near apex only; sepals with only ± appressed hairs <0.3mm **7a. E. cygnorum ssp. cygnorum**

11 Pedicels sparsely pubescent along length; sepals with short appressed and some longer ± patent hairs >0.5mm **6. E. crinitum**

Other spp. - Over 29 alien spp. have been recorded, many as wool-aliens, of which the genus *Erodium* is one of the most characteristic components. The 6 spp. treated here are by far the most common. Besides these the least rare is **E. stephanianum** Willd., from C & E Asia; this would key as *E. chium* but the leaves are deeply and finely pinnately divided and apical pit of mericarp is overarched by hairs.

1. E. chium (L.) Willd. - *Three-lobed Stork's-bill.* Suberect to ascending annual to 40cm; leaves simple, the lowest deeply 3-lobed; sepals and pedicels with ± appressed eglandular hairs; petals pinkish-purple; mericarps with beak 2-4cm, with eglandular pits not delimited by groove and not overarched by hairs; (2n=20, 40). Intrd; infrequent wool-alien; scattered in En and Wa; Mediterranean. **486**

2. E. malachoides (L.) L'Hér. - *Soft Stork's-bill.* Suberect to ascending annual to 40cm; leaves simple, the lowest toothed to shallowly 3-lobed; sepals and pedicels with patent glandular and some eglandular hairs; petals pinkish-purple; mericarps with beak 2-3.5cm, with conspicuously glandular pits delimited by deep groove; **486**

(2n=20, 40). Intrd; infrequent wool-alien; scattered in En and Wa; S Europe.

3. E. maritimum (L.) L'Hér. - *Sea Stork's-bill.* Procumbent to decumbent annual to **486**
10(20)cm; leaves simple, the lowest toothed to shallowly pinnately lobed; sepals and pedicels with erecto-patent eglandular hairs; petals usually 0, or pink and not exceeding calyx; mericarps with beak 0.8-1cm, with eglandular pits overarched by hairs from below and delimited by distinct groove; 2n=20. Native; fixed maritime dunes and barish places in short grassland, rarely inland; coasts of CI, E & S Ir, W Br from W Cornwall and E Sussex to Wigtowns. Formerly inland in Worcs.

4. E. botrys (Cav.) Bertol. - *Mediterranean Stork's-bill.* Suberect to ascending **486**
annual to 50cm; leaves simple, deeply pinnately lobed; sepals and pedicels with numerous patent glandular hairs and some ± patent eglandular hairs; petals bluish; mericarps with beak 6-9(11)cm, with glabrous eglandular pits delimited by usually 2-3 sharp-rimmed grooves; (2n=40). Intrd; common wool-alien; scattered in Br; Mediterranean.

5. E. brachycarpum (Godr.) Thell. (*E. obtusiplicatum* (Maire, Weiller & Wilczek) **486**
Howell) - *Hairy-pitted Stork's-bill.* Differs from *E. botrys* in usually slightly shorter mericarps and beaks (but with much overlap); apical pits with some bristles and usually 1 blunt-rimmed groove; (2n=40). Intrd; frequent wool-alien; scattered in En, perhaps overlooked for *E. botrys*; W Mediterranean. A doubtfully distinct sp.; only the key character is reliable.

6. E. crinitum Carolin - *Eastern Stork's-bill.* Decumbent to ascending annual with **486**
thick ± succulent root (perennial in native area) to 50cm; leaves simple with 3 very deep lobes or ± ternate; sepals and pedicels with erecto-patent eglandular hairs; petals bluish, with yellow or white veins; mericarps with beak 4-7cm, with eglandular pits delimited by groove; (2n=40). Intrd; frequent wool-alien; scattered in En; C & E Australia.

7. E. cygnorum Nees - *Western Stork's-bill.* Habit as in *E. crinitum*; leaves simple **486**
with 3 deep lobes; mericarps with beak 5-10cm, with eglandular pits delimited by groove; (2n=60). Intrd; frequent wool-alien; very scattered in En; W Australia. Both sspp. (possibly best as spp.) are frequent.

a. Ssp. cygnorum. Petals bluish, with yellow or white veins; mericarps rather sparsely pubescent; and see key (couplet 10).

b. Ssp. glandulosum Carolin. Petals bluish, with red veins; mericarps densely pubescent; and see key (couplet 10).

8. E. moschatum (L.) L'Hér. - *Musk Stork's-bill.* Suberect to procumbent annual to **R**
60cm; leaves pinnate, with toothed or shallowly lobed leaflets, smelling musky **486**
when bruised; sepals and pedicels with patent glandular hairs; petals pinkish-purple; mericarps with beak 2-4.5cm, with glandular pits delimited by deep groove; 2n=20. Native; rough ground and barish places in short grassland, mainly near sea, also frequent wool-alien; coasts of CI, W Br from W Cornwall and W Sussex to Man, Ir, widespread in Br as casual.

9. E. cicutarium (L.) L'Hér. - *Common Stork's-bill.* Suberect to procumbent annual **486**
to 60cm; leaves pinnate, with deeply pinnately lobed to ± pinnate leaflets; sepals and pedicels variably with glandular and/or eglandular hairs; petals pinkish-purple, rarely white, the upper 2 often with black basal spot; mericarps with beak 1.5-4cm, with eglandular pits not overarched by hairs from below and delimited by groove; 2n=40. Native; barish places in grassland, waste and rough ground, on sandy or chalky soils, also common wool-alien; locally scattered over most of BI. Very variable, especially as wool-alien and on coastal dunes; ssp. **dunense** Andreas is a dwarf coastal ecotype (sometimes wrongly determined as *E. lebelii* or as a hybrid), but intermediates are too common for its recognition.

9 x 10. E. cicutarium x E. lebelii = E. x anaristatum Andreas is known from coastal dunes in Wa and S Lancs; it is intermediate and sterile; (2n=30).

10. E. lebelii Jord. (*E. glutinosum* Dumort., *E. cicutarium* ssp. *bipinnatum* auct. non **R**
(Cav.) Tourlet) - *Sticky Stork's-bill.* Differs from *E. cicutarium* in stems to 15(25)cm; **486**
sepals and pedicels always with dense patent glandular hairs; petals pale pink to

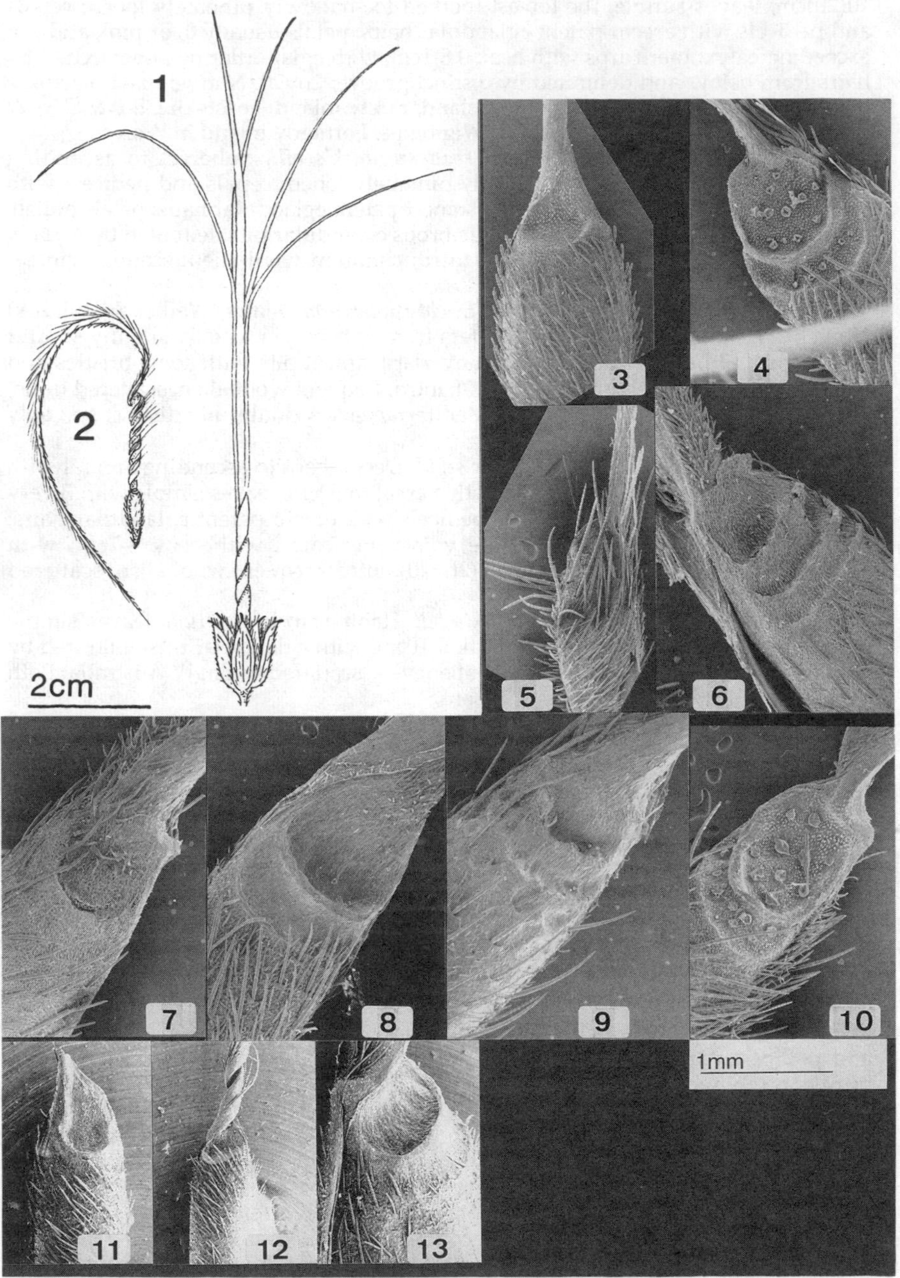

FIG 486 - ***Erodium***. 1-2, fruit and single mericarp of ***E. botrys***. 3-12, apical pit region of mericarps. 3, ***E. chium***. 4, ***E. malachoides***. 5, ***E. maritimum***. 6, ***E. botrys***. 7, ***E. brachycarpum***. 8, ***E. crinitum***. 9, ***E. cygnorum***. 10, ***E. moschatum***. 11, ***E. cicutarium***. 12, ***E. lebelii***. 13, ***E. manescavii***.

white; mericarps with beak up to 2.2cm; and see key (couplet 4); 2n=20. Native; barish places on fixed dunes; coasts of Br from E Kent to Wigtowns, Ir from W Cork to Co Antrim, Jersey and Sark.

11. E. manescavii Coss. - *Garden Stork's-bill.* Perennial with peduncle and leaves 486
arising from base; leaves pinnate, with deeply pinnately lobed leaflets; sepals and pedicels with patent glandular hairs; petals purple, each often with basal dark spot; mericarps with beak 4-7cm, with eglandular pits delimited by sharp-rimmed but shallow groove; (2n=40). Intrd; garden escape ± natd in hedgerows and open ground; W Kent, formerly Surrey and N Somerset; Pyrenees.

4. PELARGONIUM L'Hér. ex Aiton - *Geraniums*

Perennials, woody or ± so at base; leaves simple, tomentose, very shallowly to deeply palmately lobed; flowers zygomorphic, with the upper 2 petals wider than the others and the upper sepal with a backward-directed spur tightly fused to pedicel; stamens 10, free, but only 7 with anthers; fruits dispersing as in *Monsonia.*

Other spp. - The familar *Scarlet Geranium,* with characteristically scented leaves with zonal dark markings, is **P. x hybridum** Aiton (*P. inquinans* (L.) Aiton x *P. zonale* (L.) Aiton), of garden origin; it sometimes persists until the frosts on rubbish-tips and waste ground where discarded.

1. P. tomentosum Jacq. - *Peppermint-scented Geranium.* Scrambling to 1m; leaves up to 12cm, peppermint-scented, deeply palmately lobed; flowers up to 20 in umbel; petals whitish-pink, red at base, 6-9mm, 1.25-2x as long as sepals; fruit column 15-25mm; (2n=44). Intrd; formerly natd among *Pteridium,* now gone from there but natd in woodland clearing, Tresco (Scillies); S Africa.

107. LIMNANTHACEAE - *Meadow-foam family*

Annual, slightly succulent, ± glabrous herbs; leaves alternate, pinnate or deeply pinnately lobed, petiolate, without stipules. Flowers solitary in leaf-axils, bisexual, hypogynous, actinomorphic; sepals 5, free; petals 5, free; stamens 10; gynoecium similar to that of Geraniaceae but ovary with 1 ovule per cell, cells of fruit remaining closed and dispersed as separate whole mericarps after column has withered.

The buttercup-like flowers, but with quite different stamens and carpels, are distinctive; differs from Geraniaceae in lacking stipules and in above gynoecium characters.

1. LIMNANTHES R.Br. - *Meadow-foam*

1. L. douglasii R. Br. - *Meadow-foam.* Stems erect to ascending, to 35cm; flowers 16-25mm across, fragrant; petals white with conspicuous yellow basal part; beak of fruit 5-9mm; (2n=10). Intrd; much grown for ornament and frequent on tips and by roads, seashores and lakes, rarely briefly persisting; scattered in Br; California.

108. TROPAEOLACEAE - *Nasturtium family*

Annual or perennial, sometimes climbing, somewhat succulent herbs; leaves alternate, peltate, simple to very deeply palmately lobed (± palmate), petiolate, without stipules. Flowers solitary in leaf-axils, bisexual, slightly perigynous, zygomorphic, yellow to red, showy; sepals 5, ± free, the upper 1 or 3 with single backward-directed free spur; petals 5, free, the upper 2 different from lower 3; stamens 8; ovary 3-celled, each cell with 1 apical ovule; style 1; stigmas 3, linear; fruit a 3-celled schizocarp breaking into 3 indehiscent ± succulent segments.

The zygomorphic flowers with 5 free petals and sepals, 8 stamens and 3-celled ovary are unmistakable.

1. TROPAEOLUM L. - *Nasturtiums*

Other spp. - **T. peregrinum** L. (*T. canariense* hort.) (*Canary-creeper*), from S America, differs from *T. speciosum* in its yellow flowers and 5-lobed leaves; it has occasionally been reported to stray outside gardens.

1. T. speciosum Poepp. & Endl. - *Flame Nasturtium*. Rhizomatous perennial to several m, climbing by long petioles; leaves deeply 5-6-lobed, almost palmate; flowers scarlet, c.18-25mm across, with spur c.20-30mm. Intrd; natd clambering through bushes and in hedges; E Sussex, Man, Ayrs, Colonsay (M Ebudes), Co Antrim and Co Londonderry; Chile.

2. T. majus L. - *Nasturtium*. Annual of many cultivars, remaining dwarf or ± climbing to 2m; leaves slightly angled, otherwise entire; flowers various shades of yellow or red, usually orange, c.25-60mm across, with spur c.25-40mm; (2n=28). Intrd, commonly cultivated, frequent casual on tips and waste ground; scattered in Br and CI; Peru. *T. majus* probably arose as the spontaneous hybrid *T. ferreyae* Sparre x *T. minus* L. Some of our garden plants might also have some **T. peltophorum** Benth. in their parentage. *T. minus* has also been recorded.

109. BALSAMINACEAE - *Balsam family*

Glabrous, somewhat succulent, annual herbs; leaves alternate, opposite or in whorls of 3, simple, petiolate, without stipules or with stipules represented by basal glands. Flowers 1-several in axillary or terminal cymes, bisexual, sometimes some cleistogamous, hypogynous, zygomorphic; sepals 3, ± petaloid, free, the lowest sac-like with a conspicuous backward-directed or variously bent spur; petals 5, but apparently 3, the uppermost free, each of the 2 lower ones fused to each of the 2 lateral ones to form 2 compound flanges; stamens 5, with filaments fused distally and anthers fused round ovary; ovary 5-celled, each cell with numerous ovules on axile placentas; style 1; stigmas 1 or 5, minute to linear; fruit an explosive 5-celled capsule with many seeds.

The flowers, with spurred lower sepal and 2 others, fused pairs of lateral petals, and distally fused stamens, are unique.

1. IMPATIENS L. - *Balsams*

1 Flowers deep pinkish-purple to white; leaves opposite or in whorls **4. I. glandulifera**
1 Flowers pale yellow to orange 2
 2 Flowers (incl. spur) <1.5(2)cm; larger leaves with >20 teeth on each side **3. I. parviflora**
 2 Flowers (incl. spur) all or many >2cm; larger leaves with <20 teeth on each side 3
3 Flowers yellow; sepal-spur held at c.90° to rest of sepal in fresh state **1. I. noli-tangere**
3 Flowers orange; sepal-spur held ± parallel to rest of sepal in fresh state **2. I. capensis**

Other spp. - **I. balfourii** Hook. f., from Himalayas, has large pink flowers but differs from *I. glandulifera* in alternate leaves and sepal-spur >10mm and much less bent; it is grown in gardens and has been reported as an escape.

1. I. noli-tangere L. - *Touch-me-not Balsam.* Stems erect, to 1m; flowers mostly 2-3.5cm, yellow with small brownish spots; lowest sepal 10-20mm plus spur 6-12mm held at c.90°; (2n=20, 40). Native; damp places in woods; very local in Br N to C Sc, frequent in Lake District, probably native only there and in C Wa. **R**

2. I. capensis Meerb. - *Orange Balsam.* Stems erect, to 1.5m; flowers mostly 2-3.5cm, orange with large brownish blotches; lowest sepal 10-20mm plus spur 5-10mm held parallel to rest of sepal; (2n=14). Intrd; banks of rivers and canals; locally frequent in Br N to SW Yorks; N America.

3. I. parviflora DC. - *Small Balsam.* Stems erect, to 60(100)cm; flowers (incl. spur) mostly 0.6-1.8cm, pale yellow; lowest sepal gradually tapered into ± straight spur, in total 4-12mm; 2n=24, 26. Intrd; damp shady places in woods and hedges and disturbed or cultivated ground; scattered through most of Br, locally common; C Asia.

4. I. glandulifera Royle - *Indian Balsam.* Stems erect, to 2m; flowers 2.5-4cm, deep to pale pinkish-purple or white; lowest sepal 12-20mm plus spur 2-7mm bent at ≥90°; 2n=18. Intrd; banks of rivers and canals, damp places and waste ground; locally common throughout most of BI; Himalayas.

110. ARALIACEAE - *Ivy family*

Evergreen, woody climbers, herbaceous perennials or evergreen or deciduous shrubs; leaves alternate, simple and usually palmately lobed or 1-2-pinnate, petiolate, without stipules. Flowers in terminal umbels often grouped into panicles, white to greenish, bisexual or andromonoecious, epigynous, actinomorphic; hypanthium 0; sepals represented by 5 small teeth near top of ovary; petals 5, free; stamens 5; ovary 2-5-celled, with 1 apical ovule per cell; styles 1 or 5; stigma(s) minute; fruit a usually black berry with 2-5 seeds.

The umbellate inflorescence and 2-5-celled berries are diagnostic; the growth-forms of each of the genera are unmistakable.

1 Deciduous shrub or herbaceous; leaves 1-2-pinnate **3. ARALIA**
1 Woody evergreen; leaves simple, usually palmately lobed 2
  2 Upright shrub; bisexual flowers with 5 free styles; most or all leaves >25cm across **2. FATSIA**
  2 Climbing or scrambling; flowers with 1 style; leaves ≤25cm across **1. HEDERA**

1. HEDERA L. - *Ivies*

Evergreen non-spiny, woody climbers with numerous short roots borne along climbing stems; leaves ≤25cm, simple, usually palmately lobed; inflorescence a simple umbel or group of umbels; flowers autumn; fruits spring; style 1; fruits black unless otherwise stated.

In all spp. leaves on creeping or climbing stems differ from those on non-rooting, flowering stems; the former are more lobed, and are those referred to in key and descriptions as they and their hairs are the diagnostic ones.

1 Hairs on leaves and young stems semi-peltate, orange-brown, with rays mostly 8-25 and fused for 1/4-1/2 their length; leaves often >10cm wide, scarcely or not lobed or with 3 rather shallow lobes 2
1 Hairs on leaves and young stems stellate, white to yellowish-brown, with rays mostly 4-8(10) and fused only at extreme base; at least some leaves well lobed, rarely >10cm wide 3
  2 Leaves with matt surface, with ± cordate base, usually not or scarcely lobed, mostly >15cm across; hairs with mostly 15-25 rays **1. H. colchica**
  2 Leaves with glossy surface, with truncate to rounded base, usually

some 3-lobed, mostly <15cm across; hairs with mostly 8-18 rays **2. H. algeriensis**

3 Hairs whitish, with rays lying parallel to leaf surface and also projecting away from it; leaves usually <8cm across, often lobed >1/2 way to base **3a. H. helix ssp. helix**

3 Hairs often pale yellowish-brown, with rays ± all lying parallel to leaf surface; leaves often >8cm across, usually lobed <1/2 way to base **3b. H. helix ssp. hibernica**

Other spp. - Many cultivars of all 4 above taxa are grown, often obscuring their morphological and distributional limits.

1. H. colchica (K. Koch) K. Koch - *Persian Ivy.* Stems creeping, scrambling and weakly climbing to c.10m; leaves mostly 15-25cm wide, unlobed or slightly 3-5-lobed, entire or denticulate, with orange-brown semi-peltate hairs with mostly 15-25 rays held parallel to surface; (2n=192). Intrd; much grown in shrubberies and on walls and often long-persistent and spreading; scattered over Br, Man; Caucasus.

2. H. algeriensis Hibberd (*H. canariensis* auct. non Willd.) - *Algerian Ivy.* Stems variously creeping and climbing to c.5m; leaves mostly 6-15cm wide, usually some 3-lobed, entire, with orange-brown semi-peltate hairs with mostly 8-18 rays held parallel to surface; (2n=96). Intrd; grown in conservatories and on patios and walls, sometimes persisting and spreading; scattered in S & W Br N to CW Sc, Man; N Africa. The commonest indoor variegated ivy, usually mis-named *H. canariensis*; usually with distinctive long red petioles.

3. H. helix L. - *Ivy.* Stems variously creeping and climbing, often to great height; leaves rarely >15cm wide, distinctly 3-5-lobed, entire, with white to pale yellowish-brown stellate hairs (avoid old or rubbed leaves). Native; on trees, banks, rocks and sprawling over the ground.

a. Ssp. helix - *Common Ivy.* Leaves usually dark green, often with light marbling, often lobed >1/2 way to base; and see key; 2n=48. Common over E, C & N Br, uncertain further W.

b. Ssp. hibernica (G. Kirchn.) D.C. McClint. (*H. hibernica* G. Kirchn.) - *Atlantic Ivy.* Leaves usually mid-green, rarely marbled, usually lobed <1/2 way to base; and see key; 2n=96. The commoner taxon in CI, Ir, and W & SW Br N to SW Sc, but with much overlap. **H. 'Hibernica'** (*Irish Ivy*) is a well-known cultivar with large leaves and usually not or weakly climbing.

H. helix ssp. **poetarum** Nyman (f. *poetarum* (Nyman) McAllister & A. Rutherf.), from C & E Mediterranean, has brighter green, usually less deeply lobed leaves than ssp. *helix* and slightly larger yellowish-orange berries; it is grown and may persist in wild places.

2. FATSIA Decne. & Planch. - *Fatsia*

Evergreen, non-spiny shrubs, without roots along stems; leaves simple, mostly >25cm, deeply palmately lobed; inflorescence a large (20-40cm) panicle of umbels, the lateral ones male; flowers autumn; fruits spring; styles 5; fruits black.

1. F. japonica (Thunb.) Decne. & Planch. - *Fatsia.* Shrub to 3(-5)m; leaves 15-40cm, palmately divided >1/2 way into 7-9 lobes, with petiole c.10-30cm; (2n=24, 48). Intrd; much used in formal plantings; surviving as relics in SE En, relics and self-sown plants in Man; Japan.

3. ARALIA L. - *Angelica-trees*

Herbaceous perennials or deciduous, spiny shrubs, without roots along stems; leaves 1-2-pinnate, mostly >50cm; inflorescence a large (20-45cm) panicle of umbels; flowers summer; fruits autumn; styles 5; fruits dark purplish to black.

1 Spineless herbaceous perennial to 2m **3. A. racemosa**
1 Spiny deciduous shrubs often >2m 2
2 Inflorescence a ± conical panicle; leaves glabrous on lowerside except on veins **1. A. chinensis**
2 Inflorescence a ± umbrella-shaped panicle; leaves pubescent on lowerside **2. A. elata**

1. A. chinensis L. - *Chinese Angelica-tree.* Suckering shrub to 3(6)m, with thick spiny branches; leaves 2-pinnate, with ovate, serrate leaflets; petioles spiny; inflorescence a panicle of umbels. Intrd; planted on banks and in shrubberies and spreading by suckers; natd in W Kent, Man and Midlothian; China. Our plant is var. **nuda** Nakai.

2. A. elata (Miq.) Seem. - *Japanese Angelica-tree.* Shrub or tree to 5(10)m; differs from *A. chinensis* in leaflets narrower and less closely serrate; stems more spiny; and see key; (2n=24). Intrd; natd as for *A. chinensis*; Cheshire, Man and Jersey; far E Asia.

3. A. racemosa L. - *American-spikenard.* Herbaceous, spineless perennial to 2m; leaves 1-2-pinnate, with ovate, serrate leaflets; inflorescence an elongated panicle of umbels; (2n=24). Intrd; planted in shrubberies and open woodland; natd in Salop; E N America.

111. APIACEAE - *Carrot family*
(Umbelliferae, Hydrocotylaceae)

Herbaceous annuals to perennials, rarely shrubs; leaves alternate, simple to palmate, pinnate or ternate (often several times so), sessile or petiolate; stipules present in *Hydrocotyle*, 0 in others but petiole often much widened and sheathing stem at base. Flowers in terminal and lateral umbels, the umbels sometimes simple but usually compound, in *Hydrocotyle* often with whorls of flowers below main umbel, bisexual or frequently functionally andromonoecious, rarely dioecious, epigynous, actinomorphic or with zygomorphic petals (those away from centre of umbel larger); hypanthium 0; sepals 5, usually represented by small teeth near top of ovary, or 0; petals 5, usually white, pink or yellow; stamens 5; ovary 2-celled, with 1 apical ovule per cell; styles 2, often arising from swelling (*stylopodium*) on top of ovary; stigma minute to capitate; fruit a dry 2-celled schizocarp, the 2 mericarps usually separating from the sterile *carpophore* (or carpophore lacking) but each remaining indehiscent at maturity.

Most spp. are unmistakable 'umbellifers', with compound (less often simple) umbels and distinctive fruits; genera not so conforming (*Eryngium, Hydrocotyle*) are individually highly distinctive. Too many genera are recognized. Good candidates for amalgamation are *Chaerophyllum, Anthriscus* and *Myrrhis*; and *Peucedanum, Pastinaca, Ligusticum, Levisticum, Selinum* and *Angelica*, among others.

Fully ripe fruits are extremely important for identification. Important points are the degree and pattern of longitudinal ridging and the shape and position of sub-surface oil-bodies on the mericarps. The face of the mericarps where they join is the *commissure*, and the outer face is dorsal. Fruits may be compressed to varying degrees: either dorsally (with commissure as wide as fruit) or laterally (with commissure through short axis). Bracteoles (when present) subtend the flowers at the base of the pedicels; bracts (when present) subtend the main branches (*rays*) of the compound umbel. Fruit length excludes the stylopodium; width is that in widest view.

In this work 11 distinctive groups are first defined, followed by the keys which cover all taxa except for male plants of *Trinia*. The list of distinctive taxa and then the keys should be tackled successively.

Distinctive genera

Fresh plants smelling of aniseed when crushed
8. MYRRHIS, 21. FOENICULUM, 22. ANETHUM

Plant with subterranean swollen tubers (not swollen tap-roots)
11. BUNIUM, 12. CONOPODIUM, 19. OENANTHE

Base of stem sheathed by mass of fibres (remains of old petioles)
3. ASTRANTIA, 13. PIMPINELLA, 18. SESELI, 23. SILAUM, 24. MEUM, 28. TRINIA, 38. CARUM, 44. PEUCEDANUM

Plant entirely male or female **28. TRINIA**

Stem procumbent; leaves ± orbicular, shallowly lobed, stipulate
1. HYDROCOTYLE

Plant with subaquatic leaves much more finely divided than aerial ones
15. SIUM, 19. OENANTHE, 30. APIUM

Leaves all simple, not divided or divided <1/2 way to base
1. HYDROCOTYLE, 4. ERYNGIUM, 10. SMYRNIUM, 27. BUPLEURUM

Leaves divided >1/2 way to base but at most ternate to palmate or ternately to palmately lobed, the leaflets/lobes toothed to shallowly lobed
2. SANICULA, 3. ASTRANTIA, 4. ERYNGIUM, 14. AEGOPODIUM, 37. FALCARIA, 40. LIGUSTICUM, 44. PEUCEDANUM

Inflorescence with lobed or divided bracts
2. SANICULA, 4. ERYNGIUM, 16. BERULA, 28. TRINIA, 29. CUMINUM, 31. TRACHYSPERMUM, 32. PETROSELINUM, 36. AMMI, 38. CARUM, 44. PEUCEDANUM, 49. DAUCUS

Fruits with hooked or barbed bristles **2. SANICULA, 4. ERYNGIUM, 6. ANTHRISCUS, 48. TORILIS, 49. DAUCUS**

Fruits tuberculate (not smooth, scaly, spiny or pubescent)
27. BUPLEURUM, 31. TRACHYSPERMUM, 48. TORILIS

General key

1 Leaves all simple and entire **27. BUPLEURUM**
1 Leaves simple to compound; if simple at least toothed 2
  2 Stem-leaves spiny **4. ERYNGIUM**
  2 Leaves not spiny 3
3 Leaves stipulate, all simple, ± orbicular, shallowly lobed, with long petiole **1. HYDROCOTYLE**
3 Leaves without stipules, if all simple then either lobed >1/2 way to base or sessile 4
  4 Leaves all simple, with small teeth only **10. SMYRNIUM**
  4 Leaves simple to compound, if all simple then divided >1/2 way to base 5
5 Basal leaves all ternately or palmately lobed almost to base 6
5 Basal leaves ternate, palmate or pinnate, often compoundly so 7
  6 Flowers subsessile, with inconspicuous bracteoles; fruits with hooked bristles **2. SANICULA**
  6 Flowers distinctly pedicellate, with bracteoles at least as long; fruits covered with bifid scales **3. ASTRANTIA**
7 Fruit with apical beak >2x as long as seed-bearing part **7. SCANDIX**
7 Fruit beakless or with beak shorter than seed-bearing part 8
  8 Stem and often basal leaves with white, flexuous subterranean part arising from brown tuber 9
  8 Stem and basal leaves (if present) arising from roots at ground level, or from rhizome, or plant aquatic 10
9 Stem hollow at fruiting; fruit with ± erect styles gradually narrowed from stylopodium **12. CONOPODIUM**
9 Stem solid at fruiting; fruit with reflexed styles suddenly contracted from stylopodium **11. BUNIUM**

10 Petals yellow *Key A*

10 Petals white to pink or purplish, or greenish-white 11

11 Fruits with spikes, bristles, hairs or conspicuous tubercles *Key B*

11 Fruits glabrous, ± smooth 12

12 Fruits strongly compressed dorsally, distinctly wider in dorsal view than in lateral view *Key C*

12 Fruits not compressed dorsally or scarcely so 13

13 Base of stem sheathed by mass of fibres (remains of old petioles) 14

13 Basal mass of petiole-fibres 0 17

14 Basal leaves 1-pinnate, each lobe palmately divided ± to base into filiform segments appearing as if whorled **38. CARUM**

14 Basal leaves 1-4-pinnate, if 1-pinnate then lobes not divided ± to base into filiform segments 15

15 Fruit 4-10mm; basal leaves 3-4-pinnate with filiform ultimate lobes **24. MEUM**

15 Fruit 2-4mm; basal leaves 1-3-pinnate with ovate to linear ultimate lobes 16

16 Plant usually dioecious; all leaf-segments linear; umbels with <10 rays **28. TRINIA**

16 Plant bisexual; leaf-segments variable, some often ovate, very rarely all linear; umbels usually with ≥10 rays **13. PIMPINELLA**

17 Fruits >(2)2.5x as long as widest width *Key D*

17 Fruits ≤2x as long as widest width 18

18 Sepals 0, or minute teeth, or vestigial rim, not or scarcely visible at top of fruit *Key F*

18 Sepals ≥0.2mm, distinctly visible at top of fruit *Key E*

Key A - Petals yellow

1 Fruits strongly compressed dorsally, i.e. distinctly wider in dorsal view than in lateral view 2

1 Fruits not compressed dorsally, or scarcely so 5

2 Bracts >3; bracteoles fused at base **42. LEVISTICUM**

2 Bracts 0-3; bracteoles 0 or free 3

3 Leaves 1-pinnate, with ovate lobes **45. PASTINACA**

3 Leaves 2-several times pinnate or ternate, with linear lobes 4

4 Stem surrounded by fibrous remains of petioles at base; stem-leaves with narrow petiole; fruits 5-8mm **44. PEUCEDANUM**

4 Stem not surrounded by fibrous remains of petioles; stem-leaves with wide sheathing petiole; fruits 10-16mm **43. FERULA**

5 Ultimate leaf-lobes filiform, <0.5mm wide 6

5 Ultimate leaf-lobes flat, linear to ovate, >0.5mm wide 8

6 Firmly rooted perennials **21. FOENICULUM**

6 Easily uprooted annuals 7

7 Fruits dorsally compressed, with lateral wings; fresh plant smelling of aniseed **22. ANETHUM**

7 Fruits laterally compressed, not winged; fresh plant not smelling of aniseed **33. RIDOLFIA**

8 Fruit laterally compressed, c.2x as wide in lateral view as in dorsal view 9

8 Fruit not or scarcely compressed, c. as wide in lateral view as in dorsal view 10

9 Bracts and bracteoles 0-2, very short; mericarps with 3 acute dorsal ridges **19. SMYRNIUM**

9 Bracts 1-3, often lobed; bracteoles >3; mericarps with 3 rounded dorsal ridges **32. PETROSELINUM**

10 Leaves succulent; bracts >4; mericarps not winged **17. CRITHMUM**

10 Leaves not succulent; bracts 0-3; mericarps with narrow lateral wings **23. SILAUM**

Key B - Petals not yellow; fruits with spines, bristles, hairs or conspicuous tubercles

1 Fruits strongly compressed dorsally, i.e. distinctly wider in dorsal view than in lateral view (incl. projections) 2

1 Fruits not compressed dorsally, or scarcely so 3

2 Stems 1.5-5.5m; umbels with >30 rays; fruits >8mm **46. HERACLEUM**

2 Stems <1.5m; umbels with <20 rays; fruits <7mm **47. TORDYLIUM**

3 Fruits with conspicuous tubercles **31. TRACHYSPERMUM**

3 Fruit with spines, bristles or hairs 4

4 Fruits with hairs or weak or minute bristles 5

4 Fruits with usually stout, terminally hooked or barbed, spines 7

5 Fruits >3x as long as wide; fresh plant smelling of aniseed **8. MYRRHIS**

5 Fruits <3x as long as wide; fresh plant not smelling of aniseed 6

6 Slender annual without basal fibres; bracts and bracteoles each <5; fruits >3.5mm **29. CUMINUM**

6 Perennial with base of stem sheathed by mass of fibres; bracts and bracteoles each >5; fruits <3.5mm **18. SESELI**

7 Bracts deeply pinnately or ternately divided **49. DAUCUS**

7 Bracts 0 or simple 8

8 Fruits without beak, with spines up to base of stylopodium, with persistent sepals **48. TORILIS**

8 Fruits with spine-less but ridged beak below stylopodium; sepals 0 **6. ANTHRISCUS**

Key C - Petals not yellow; fruits without spines, bristles, hairs or conspicuous warts, strongly dorsally compressed (see also *Levisticum* and *Pastinaca* petals yellow, Key A)

1 Leaves and/or stems pubescent to hispid **46. HERACLEUM**

1 Leaves and main parts of stems glabrous or nearly so, sometimes coarsely papillose 2

2 Easily uprooted annuals 3

2 Firmly rooted perennials 4

3 Fruits ovoid, with strong ridges; outer petals <1.5mm; sepals 0 **20. AETHUSA**

3 Fruits ± globose, scarcely ridged; outer petals >2mm; sepals persistent **9. CORIANDRUM**

4 Leaf-lobes <2 x 1cm 5

4 Most leaf-lobes >2 x 1cm 6

5 Stems solid; bracts 0 or few and soon falling; bracteoles not or weakly reflexed **39. SELINUM**

5 Stems hollow; bracts >3, reflexed; bracteoles reflexed **43. PEUCEDANUM**

6 Larger leaves 2-3-pinnate (smaller ones sometimes 2-3-ternate or pinnate-ternate) **41. ANGELICA**

6 All leaves 1-2-ternate 7

7 Umbels with <20 rays; bracts 1-5; fruits with 3 prominent acute dorsal ridges **40. LIGUSTICUM**

7 Umbels with >20 rays; bracts 0(-2); fruits with low, obtuse dorsal ridges **44. PEUCEDANUM**

Key D - Petals not yellow; fruits without spines, bristles, hairs or conspicuous warts, not or scarcely dorsally compressed, >(2)2.5x as long as wide

1 Sepals 0 or minute or a vestigial rim, not or scarcely visible at top of fruit 2

1 Sepals ≥0.2mm, distinctly visible at top of fruit 4

2 Fruit not ridged in mid or basal regions; stems hollow **6. ANTHRISCUS**

2 Fruit ridged along length; stems solid or hollow 3

3 Stems solid; fruits with low, rounded ridges; fresh plant not aniseed-scented **5. CHAEROPHYLLUM**

3 Stems hollow; fruits with sharp, prominent ridges; fresh plant aniseed-scented **8. MYRRHIS**

4 Umbels with 1-5 rays; at least some bracts >1/2 as long as rays **29. CUMINUM**

4 Umbels with ≥6 rays, if fewer then bracts 0 or much <1/2 as long as rays 5

5 Fruits laterally compressed; lobes of lower leaves linear-lanceolate, >5cm, with regularly and sharply serrate margins; not aquatic **37. FALCARIA**

5 Fruits not compressed or slightly dorsally so; leaf-lobes various, but if linear or lanceolate and >5cm then entire to distantly and irregularly toothed; often aquatic **19. OENANTHE**

Key E - Petals not yellow; fruits without spines, bristles, hairs or conspicuous warts, not or scarcely dorsally compressed, ≤2x as long as wide; sepals ≥0.2mm, distinctly visible at top of fruit

1 Fruits subglobose, in lateral view c. as wide as long 2

1 Fruits in lateral view distinctly longer than wide 4

2 Easily uprooted annuals; mericarps remaining fused even at maturity **9. CORIANDRUM**

2 Firmly rooted perennials; mericarps splitting apart at maturity 3

3 Stems solid; fruits >2.5mm; petioles not widened at base, nor sheathing lateral stems **25. PHYSOSPERMUM**

3 Stems hollow; fruits <2.5mm; petioles widened at base and sheathing lateral stems **35. CICUTA**

4 Not aquatic; lobes of lower leaves linear-lanceolate, >5cm, with regularly and sharply serrate margins **37. FALCARIA**

4 Often aquatic; leaf-lobes various, but if linear or lanceolate and >5cm then entire to distantly or irregularly toothed 5

5 Fruits laterally compressed; lobes of lower leaves mostly >4 x 2cm **15. SIUM**

5 Fruits not compressed or slightly dorsally so; lobes of lower leaves <4 x 2cm **19. OENANTHE**

Key F - Petals not yellow; fruits without spines, bristles, hairs or conspicuous warts, not or scarcely dorsally compressed, ≤2x as long as wide, with sepals 0 or scarcely visible

1 Lower leaves simply pinnate, the lobes not divided as far as midrib 2

1 Lower leaves 2-4-pinnate or 1-2-ternate 8

2 Bracts 0(-2), if constantly present then stem mostly procumbent with ± only leaves and peduncles erect 3

2 Bracts 2-c.8; at least apical part of stem erect to ascending 5

3 Plant often aquatic, at least lower part of stem procumbent and rooting **30. APIUM**

3 Plant ± never aquatic; stem erect, not rooting 4

4 Fruit <2mm; plant glabrous **30. APIUM**

4 Fruit ≥2mm, or if <2mm then at least rays hispid **13. PIMPINELLA**

5 Styles at fruiting at least as long as stylopodium 6

5 Styles at fruiting much shorter than stylopodium 7

6 Lowest leaves with 2-5 pairs of leaflets each 3-6cm; all bracts <1/2 as long as all rays **34. SISON**

6 Lowest leaves with 4-12 pairs of leaflets each 0.5-3.5cm; longest bracts >1/2 as long as shortest rays **32. PETROSELINUM**

7 Stems solid; bracts divided to base into linear to filiform lobes, the longest >1/2 as long as rays; upper leaves >1-pinnate **36. AMMI**
7 Stems hollow; bracts lobed but usually not to base, <1/2 as long as rays; all leaves 1-pinnate **16. BERULA**
8 Bracts ≥4 9
8 Bracts 0-2(3) 10
9 Bracts <1/4 as long as rays, undivided; stems hollow; fruits with prominent ± wavy-edged ridges **26. CONIUM**
9 Bracts >1/2 as long as rays, deeply divided; stems solid; fruits with low smooth ridges **36. AMMI**
10 Plant in water or on mud; stems procumbent and rooting at least near base; styles at fruiting much shorter than stylopodium **30. APIUM**
10 Plant not aquatic; stems erect but sometimes rhizomes produced; styles at fruiting at least as long as stylopodium 11
11 Plant rhizomatous; leaves 1-2-ternate **14. AEGOPODIUM**
11 Plant not rhizomatous; leaves 2-3-pinnate 12
12 Easily uprooted annuals; bracteoles long, strongly reflexed **20. AETHUSA**
12 Firmly rooted biennials or perennials; bracteoles 0 or scarcely reflexed 13
13 Usually some umbels with 1-2(3) bracts; all ultimate leaf-lobes linear to linear-lanceolate; styles ± appressed to stylopodium **38. CARUM**
13 Bracts 0; usually at least some leaf-lobes ovate, if all linear to linear-lanceolate then styles not appressed to stylopodium **13. PIMPINELLA**

Other genera - CAUCALIS L. and **TURGENIA** Hoffm. are both annuals from S Europe with beakless, spiny fruits, allied to *Torilis*. They differ in having fruits with non-persistent sepals and with broad-based spines arranged in rows on the ridges. **C. platycarpos** L. (*Small Bur-parsley*) has 0(-2) bracts and ± actinomorphic corollas; **T. latifolia** (L.) Hoffm. (*Caucalis latifolia* L.) (*Greater Bur-parsley*) has (2)3-5 bracts and strongly zygomorphic corollas. Both are now extremely rare casuals, formerly more common. **LASER** Borkh. resembles *Tordylium* in its distinctive thick-winged, strongly dorsally-compressed fruits, but differs in its 2-3-ternate leaves and glabrous fruits. **L. trilobum** (L.) Borkh., from S Europe, was formerly natd in Cambs. **BIFORA** Hoffm. has distinctive fruits with subglobose mericarps joined at a narrow commissure and strongly zygomorphic marginal flowers (as in *Tordylium*). **B. radians** M. Bieb., from Europe, is a slender, white-flowered annual with narrowly divided leaves found rarely as a grain-alien.

SUBFAMILY 1 - HYDROCOTYLOIDEAE (genus 1). Leaves simple, ± orbicular, not or shallowly lobed, stipulate; flowers in simple axillary umbels, often with whorls below; fruit with a woody inner wall, but no carpophore and no oil-bodies; chromosome base-number 8.

1. HYDROCOTYLE L. - *Pennyworts*
Perennials with thin procumbent rooting stems; leaves on thin, usually erect petioles as long as to much longer than axillary peduncles; flowers very small, dull; fruits strongly laterally compressed, ± orbicular in lateral view.

Other spp. - H. sibthorpioides Lam., from the tropics, resembles a small *H. moschata* (or *Sibthorpia europaea* vegetatively) but with glabrous stems, patent-pubescent petioles 3-10(30)mm, and glabrous leaves 3-12(20)mm lobed 1/3-1/2 way to base; it has survived between paving slabs in a school garden in Staffs since c.1970.

1 Plant in water or bogs or on mud, glabrous at least on leaf uppersides

and on stems; fruits orbicular or slightly wider than long 2
1 Plant in turf or on banks, with pubescent leaves and stems; fruits longer than wide 3
 2 Leaves peltate, scarcely lobed **1. H. vulgaris**
 2 Leaves not peltate, with deep basal sinus, lobed ±1/2 way to base **2. H. ranunculoides**
3 Fruits ≥2 x 1.8mm; leaf-lobes crenate **4. H. novae-zeelandiae**
3 Fruits ≤1.6 x 1.5mm; leaf-lobes serrate **3. H. moschata**

1. H. vulgaris L. - *Marsh Pennywort*. Very variable in size; stems to 30cm, but usually much less; petioles 1-25cm, glabrous or with long hairs especially near apex; leaves peltate, glabrous or sparsely pubescent only on lowerside, 8-35mm across; fruit 1.5-1.8 x 1.8-2.3mm, glabrous, with thin low ridges; 2n=96. Native; in bogs, fens and marshes and at sides of lakes; locally common throughout BI. **499**

2. H. ranunculoides L. f. - *Floating Pennywort*. Similar to *H. vulgaris* in size and habit, glabrous; stems often floating; leaves not peltate, up to 70mm across; fruit suborbicular, 2-3mm; (2n=24). Intrd; natd in rivers, canals and ponds, first recorded 1990; several places in SE En and apparently rapidly spreading; N America. **499**

3. H. moschata G. Forst. - *Hairy Pennywort*. Similar to small-sized *H. vulgaris*; differs in pubescent stems; more deeply lobed (≤2/5 way to base) leaves 5-20mm across; fruits 1.3-1.6 x 1-1.5mm; and see key; (2n=48). Intrd; in lawns and on grassy banks; well natd on Valencia Island, S Kerry, less so in E Cornwall, E Sussex and Ayrs; New Zealand. **499**

4. H. novae-zeelandiae DC. (*H. microphylla* auct. non A. Cunn.) - *New Zealand Pennywort*. Differs from *H. moschata* in fruit 2-3.5 x 1.8-2.5mm; and leaves crenate; (2n=48). Intrd; on lawns and golf-courses, natd in turf; W Cornwall and Angus; New Zealand. **499**

SUBFAMILY 2 - SANICULOIDEAE (genera 2-4). Leaves mostly simple, palmately or ternately lobed, rarely pinnately lobed and then spiny, without stipules; flowers in simple umbels or capitula; fruit with soft inner wall and with oil-bodies, but without carpophore; chromosome base-number 7 or 8.

2. SANICULA L. - *Sanicle*

± glabrous perennials; most leaves in basal rosette; leaves palmately lobed almost to base; petioles long; umbels with pedicellate male and sessile bisexual flowers; bracteoles several, shorter than flowers; fruits laterally compressed, covered with long, hooked bristles; sepals persistent.

1. S. europaea L. - *Sanicle*. Stems erect, to 40(60)cm; leaf-lobes usually 5, shallowly lobed and toothed; flowers white or pinkish-white, in small, loosely-aggregated umbels; fruits 2.2-3mm; 2n=16. Native; in deciduous woods on leaf-mould; locally common throughout Br and Ir. **498**

3. ASTRANTIA L. - *Astrantia*

± glabrous perennials; most leaves in basal rosette; leaves ternately or palmately lobed almost to base; petioles long; umbels with pedicellate male and bisexual flowers; bracteoles numerous, lanceolate to oblanceolate, at least as long as flowers; fruits scarcely compressed, covered with swollen scales; sepals persistent.

Other spp. - Some other spp. now supplant *A. major* in gardens and may occur as escapes in future; the most common is **A. maxima** Pall., from Caucasus, with larger umbels and bracts, less deeply divided leaves, and fruits 5-6mm. Some garden plants of *A. major* represent ssp. **carinthiaca** (Hoppe ex W.D.J. Koch) Arcang., from SW Europe, and ssp. **elatior** (Friv.) K. Maly, from E Europe, and

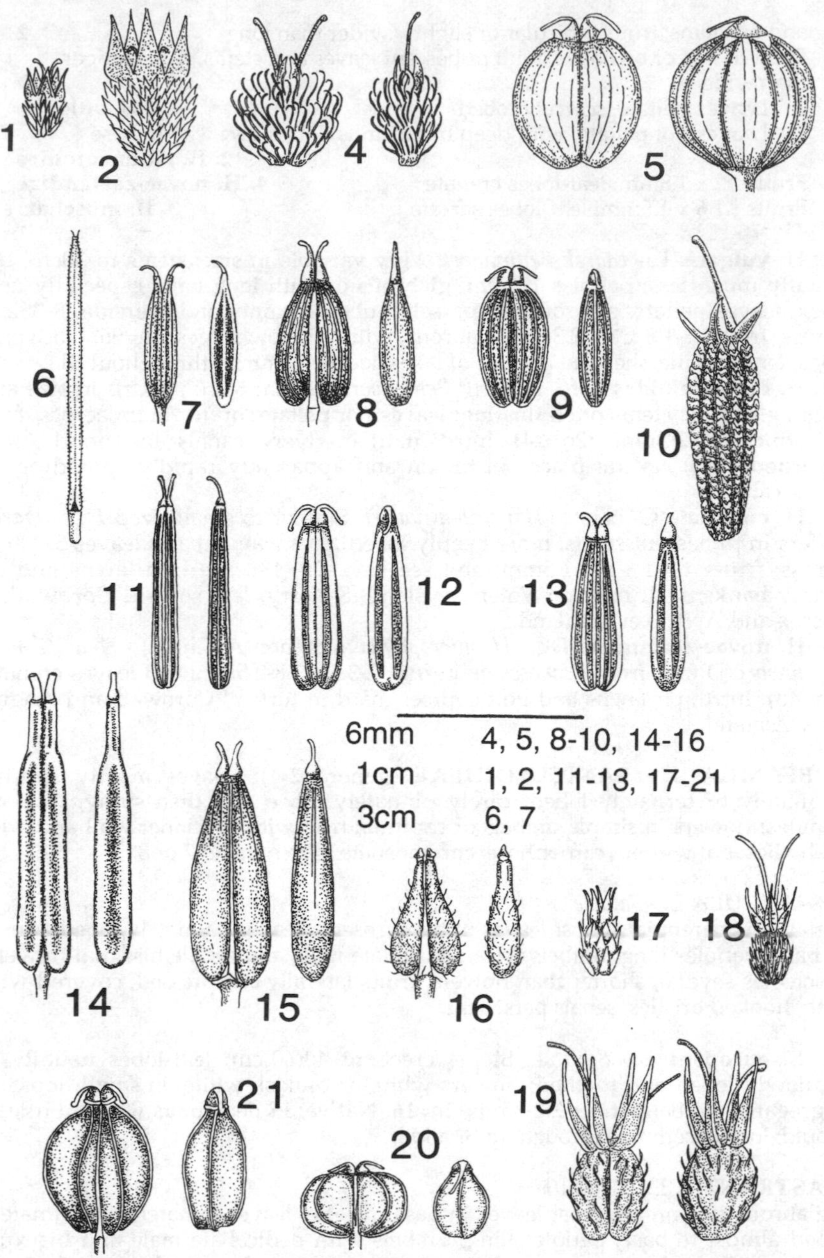

FIG 498 - Fruits (mostly with lateral and dorsal views) of **Apiaceae**. 1, ***Eryngium amethystinum***. 2, ***E. giganteum***. 4, ***Sanicula europaea***. 5, ***Coriandrum sativum***. 6, ***Scandix pecten-veneris***. 7, ***Myrrhis odorata***. 8, ***Conopodium majus***. 9, ***Bunium bulbocastanum***. 10, ***Astrantia major***. 11, ***Chaerophyllum hirsutum***. 12, ***C. aureum***. 13, ***C. temulum***. 14, ***Anthriscus cerefolium***. 15, ***A. sylvestris***. 16, ***A. caucalis***. 17, ***Eryngium planum***. 18, ***E. campestre***. 19, ***E. maritimum***. 20, ***Smyrnium perfoliatum***. 21, ***S. olusatrum***.

might occur as escapes.

1. A. major L. - *Astrantia*. Stems erect, to 80(100)cm; leaf-lobes 3-5, scarcely lobed, toothed; flowers and bracteoles whitish tinged with green and/or pink, in umbels up to 3.8cm across; fruits much longer than wide, 5-7mm; (2n=14, 28). Intrd; garden escape natd in grassy and shady places; scattered in Br, mostly in N & W; Europe. **498**

4. ERYNGIUM L. - *Sea-hollies*

± glabrous perennials; basal leaves petiolate, simple to ± pinnate; stem-leaves various, at least the upper spiny; flowers sessile, in ± globose to ovoid capitula with ± leaf-like spiny bracts at base; bracteoles 3-lobed or entire, with 1-3 spines, longer than flowers; fruits scarcely compressed, densely scaly or bristly; sepals persistent.

1 Basal leaves and lower stem-leaves pinnately or ternately divided or lobed almost to midrib 2
1 Basal leaves and lower stem-leaves not lobed or lobed ≤1/2 way to midrib 3
  2 Basal leaves with petiole unwinged, c. as long as its leaf; plant not blue **4. E. campestre**
  2 Basal leaves with petiole broadly winged, much shorter than its leaf; plant strongly blue-tinged **5. E. amethystinum**
3 Upper stem-leaves and bracts palmate or palmately lobed almost to base **2. E. planum**
3 Upper stem-leaves toothed, or lobed <1/2 way to base 4
  4 Basal leaves and at least lower stem-leaves shallowly toothed, not spiny **1. E. giganteum**
  4 Basal leaves and stem-leaves with deep, strongly spiny teeth **3. E. maritimum**

Other spp. - Several other spp. are grown for ornament and may persist for a while, but rarely or never become established. **E. x tripartitum** Desf., a garden hybrid of uncertain parentage, resembles *E. amethystinum* in habit and colour but has 3-lobed, less narrowly divided basal leaves.

1. E. giganteum M. Bieb. - *Tall Eryngo*. Stems erect, to 1.2m; basal leaves triangular-ovate, cordate at base, serrate; upper stem-leaves and bracts sharply **498**

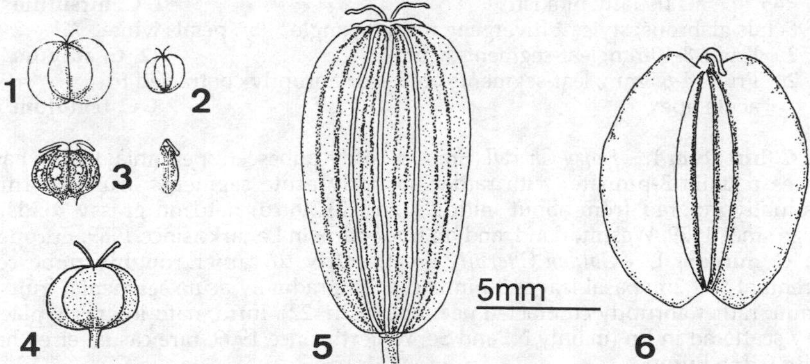

FIG 499 - Fruits of **Apiaceae**. 1, ***Hydrocotyle vulgaris***. 2, ***H. novae-zeelandiae***. 3, ***H. moschata***. 4, ***H. ranunculoides***. 5, ***Ferula communis***. 6, ***Angelica pachycarpa***. 1, lateral & dorsal views. 2-4, lateral view. 5-6, dorsal view.

toothed, scarcely spiny; capitula grey to bluish, oblong-ovate, 1-2.5cm across; fruits densely scaly; (2n=16). Intrd; grown for ornament, well natd on waste ground near Otley, MW Yorks since 1986; Caucasus.

2. E. planum L. - *Blue Eryngo*. Stems erect, to 50(100)cm; basal leaves oblong- **498**
ovate, cordate at base, serrate; upper stem-leaves deeply lobed, spiny; bracts linear-lanceolate; capitula blue, ovoid, 1-1.5cm across; fruits densely scaly; (2n=16). Intrd; grown for ornament and occasionally natd in waste places; scattered in S Br, notably on sandy ground at Littlestone, E Kent since before 1965, Lanarks; C & SE Europe.

3. E. maritimum L. - *Sea-holly*. Stems erect, to 60cm; basal leaves ovate, variously **498**
lobed, truncate to cordate at base, strongly spiny; stem-leaves similar but sessile; bracts ovate; capitula pale blue, ovoid, 1.5-3cm across; fruits with hooked bristles; 2n=16. Native; on maritime sand and shingle; formerly around all coasts of BI, now gone from most of Sc and NE En.

4. E. campestre L. - *Field Eryngo*. Stems erect, to 75cm; basal leaves pinnate or **RRR**
almost so, strongly spiny; upper stem-leaves deeply lobed, spiny; bracts linear- **498**
lanceolate; capitula green to brownish-green, ovoid to globose, 1-1.5cm across; fruits densely scaly; (2n=14, 28). Probably native in Kent, doubtfully elsewhere; grassland or open places, especially calcareous, mostly near sea; very local in extreme SE, S & SW Br, scattered and mostly casual elsewhere in S & C Br and CI.

5. E. amethystinum L. - *Italian Eryngo*. Stems erect, to 45cm; basal leaves pinnate **498**
or almost so, strongly spiny; upper stem-leaves deeply lobed, spiny; bracts linear-lanceolate; capitula blue, globose to ovoid, 1-2cm across; fruits sparsely scaly; (2n=14, 16). Intrd; grown for ornament and occasionally natd; dunes in Caerns since 1963, rare relic elsewhere; Italy and Balkans.

SUBFAMILY 3 - APIOIDEAE. (genera 5-48). Leaves various, often much divided, not spiny, without stipules; flowers usually in compound umbels; fruit with soft inner wall, carpophore and usually oil-bodies; chromosome base-number various, often 11.

5. CHAEROPHYLLUM L. - *Chervils*

Biennials or perennials; stems solid; leaves 2-3-pinnate; bracts 0 or present, entire; bracteoles present; sepals ± 0; petals white or pink, actinomorphic; fruits slightly compressed laterally, >3x as long as wide, glabrous, with low, wide, rounded ridges.

1 Petals minutely pubescent at margin; styles suberect, forming angle <45°; petals usually pinkish **1. C. hirsutum**
1 Petals glabrous; styles ± divergent, forming angle >45°; petals white 2
  2 Fruits 8-10mm; leaf-segments acute **2. C. aureum**
  2 Fruits 4-6.5mm; leaf-segments obtuse or abruptly contracted to acute apex **3. C. temulum**

1. C. hirsutum L. - *Hairy Chervil*. Erect, ± softly pubescent perennial to 1m; basal **498**
leaves usually 3-pinnate, with rather abruptly acute segments; fruits 8-11mm, gradually tapered from about middle; (2n=22). Intrd; natd on grassy roadside verge since 1979, Westmorland, and on river-bank in Lanarks since 1989; Europe.

2. C. aureum L. - *Golden Chervil*. Erect, softly to rather roughly pubescent **498**
perennial to 1.2m; basal leaves 3-pinnate, with gradually acute segments; fruits 7-10mm, rather abruptly contracted near apex; (2n=22). Intrd; natd in grassy places; very scattered in En (mainly N) and Sc, W Perth since 1909, rare casual elsewhere in Br; C & S Europe.

3. C. temulum L. (*C. temulentum* L.) - *Rough Chervil*. Erect, roughly pubescent **498**
biennial to 1m; basal leaves 2-3-pinnate, with obtuse to abruptly acute segments; fruits 4-6.5mm, gradually tapered from about middle; (2n=14, 22). Native; grassy

places, hedgerows and wood-borders; common over much of BI but sparse in CI and Ir and absent from most of W & N Sc.

6. ANTHRISCUS Pers. - *Chervils*
Annuals to perennials; stems hollow; leaves 2-3-pinnate; bracts 0(-1); bracteoles present; sepals ± 0; petals white, actinomorphic; fruits slightly compressed laterally, >3x as long as wide, glabrous, ridged only near apex.

1 Fruits <5mm, with abundant hooked bristles **3. A. caucalis**
1 Fruits >5mm, without bristles 2
  2 Annual; rays pubescent; fruits with well differentiated beak 1-4mm **2. A. cerefolium**
  2 Perennial; rays ± glabrous; fruits with scarcely differentiated beak ≤1mm **1. A. sylvestris**

1. A. sylvestris (L.) Hoffm. - *Cow Parsley*. Erect, pubescent perennial to 1.5m; basal leaves 3-pinnate, with acute segments; rays glabrous; fruits 6-10mm, glabrous, smooth, with scarcely differentiated beak ≤1mm; 2n=16. Native; grassy places, hedgerows and wood-margins; abundant throughout most of BI. **498**

2. A. cerefolium (L.) Hoffm. - *Garden Chervil*. Erect to spreading, sparsely pubescent annual to 70cm; leaves 2-3-pinnate, with subacute segments; rays pubescent; fruits 6-10mm, glabrous, smooth, with strongly differentiated beak 1-4mm; (2n=18). Intrd; formerly cultivated as herb, now rarely so, natd on rock-face in Herefs since at least 1867, decreasing casual in waste and marginal places; very scattered in S & C Br, Guernsey; SE Europe. **498**

3. A. caucalis M. Bieb. - *Bur Chervil*. Erect to decumbent, sparsely pubescent annual to 70cm; leaves 2-3-pinnate, with subacute segments; rays glabrous; fruits 2.9-3.2mm, with very strongly differentiated beak <1mm; 2n=14. Native; waste places, open ground and open hedgerows on sandy or shingly soils, especially near sea; common in parts of E En, scattered over most of rest of BI except N & NW Sc. **498**

7. SCANDIX L. - *Shepherd's-needle*
Annuals; stems hollow at fruiting; leaves 2-4-pinnate; bracts 0 or umbels simple; bracteoles present; sepals very small (<0.5mm) but persistent; petals white, zygomorphic; fruit ± not compressed, many times longer than wide, with beak 3-5x as long as seed-bearing part, glabrous, with wide low rounded ridges.

1. S. pecten-veneris L. - *Shepherd's-needle*. Very sparsely pubescent; stems **R**
usually erect, to 50cm; leaf-segments linear; fruits 3-7cm, scabrid with tiny apically **498**
directed bristles, with seed-bearing part c. as wide as beak, c.1cm; (2n=16). Possibly native; weed of arable land and waste places; formerly common in En, scattered over rest of BI, now rare and decreasing and only in C & S En.

8. MYRRHIS Mill. - *Sweet Cicely*
Perennials smelling of aniseed when crushed; stems hollow; leaves 2-4-pinnate; bracts 0; bracteoles present; sepals ± 0; petals white, slightly zygomorphic; fruit slightly laterally compressed, >3x as long as wide, glabrous or with minute bristles, with acute well developed ridges.

1. M. odorata (L.) Scop. - *Sweet Cicely*. Softly pubescent; stems erect, to 1.8m; leaves with whitish patches, with acute segments; fruits 15-25mm, contracted to short beak; 2n=22. Intrd; banks, pathsides, waste and grassy places; common in much of Br N from Derbys and Monts, rare in S Wa, S En and N & C Ir. **498**

9. CORIANDRUM L. - *Coriander*
Annuals; stems solid; leaves simple to 3-pinnate; bracts 0-1(2), entire; bracteoles

present; sepals conspicuous, persistent; petals white to purplish, zygomorphic; fruits ± globose, glabrous, the mericarps not separating at maturity, with low ± rounded ridges.

1. C. sativum L. - *Coriander*. Glabrous; stems erect, to 50cm, ± unridged; basal leaves simple with 3 or more wide lobes, or variously ternate to pinnate; stem-leaves 2-3-pinnate, with linear lobes; fruits 2-6mm; (2n=22). Intrd; casual of tips and waste places, mostly as birdseed-alien, now also a small-scale crop for fruits or as green salad; scattered in Br; E Mediterranean. **498**

10. SMYRNIUM L. - *Alexanders*

Glabrous biennials (to perennials?); stems solid; leaves simple to 3-pinnate or -ternate; bracts and bracteoles 0-few, entire; sepals minute; petals yellow, actinomorphic; fruits laterally compressed, not or scarcely longer than wide, glabrous, with prominent sharp ridges.

1. S. olusatrum L. - *Alexanders*. Stems erect, to 1.5m; leaves 2-3-pinnate or -ternate, dark glossy green, with wide, shallowly toothed lobes, all petiolate; stem-leaves with expanded sheathing petioles; fruit 6.5-8mm, c. as long as wide or slightly longer, blackish when ripe; 2n=22. Intrd; fully natd on cliffs and banks, by roads and ditches and in waste places, mostly near sea; common on coasts of BI N to C Sc, very scattered inland; Europe. **498**

2. S. perfoliatum L. - *Perfoliate Alexanders*. Stems erect, to 60(100)cm; leaves yellowish-green, the basal ones 2-3-pinnate or -ternate; stem-leaves simple, deeply cordate, sessile, strongly clasping stem, shallowly toothed; fruits 2-3.5mm, wider than long, dark brown when ripe; (2n=22). Intrd; natd in grassy places and flower-borders; few places in SE En, 1 place in Co Durham; S Europe. **498**

11. BUNIUM L. - *Great Pignut*

Glabrous perennials; stems solid, arising from subterranean ± globose tuber; leaves 2-3-pinnate, with linear lobes; bracts and bracteoles several, entire; sepals ± 0; petals white, actinomorphic; fruits slightly laterally compressed, <2x as long as wide, glabrous, with low, rounded ridges.

1. B. bulbocastanum L. - *Great Pignut*. Stems erect, to 50(80)cm; leaves mostly **RR**
withered by flowering time; fruits 3-4.5mm; (2n=20). Native; chalk grassland and **498**
banks; very local in Herts, Bucks, Beds and Cambs.

12. CONOPODIUM W.D.J. Koch - *Pignut*

Glabrous perennials; stems hollow after flowering, arising from subterranean ± globose tuber; leaves 2-3-pinnate, with linear lobes; bracts 0(-2); bracteoles several; sepals 0; petals white, actinomorphic; fruits slightly laterally compressed, <2x as long as wide, glabrous, with low, rounded ridges.

1. C. majus (Gouan) Loret - *Pignut*. Stems erect, to 40(75)cm; basal leaves mostly withered by fruiting time; fruits 3-4.5mm; 2n=22. Native; grassland, hedgerows, woods; common ± throughout BI. **498**

13. PIMPINELLA L. - *Burnet-saxifrages*

Perennials; stems hollow or solid; leaves 1-2(3)-pinnate; bracts and bracteoles 0; sepals 0; petals white or pinkish-white, actinomorphic; fruits slightly laterally compressed, slightly longer than wide, glabrous, with narrow low ridges.

Other spp. - P. affinis Ledeb., from SW Asia, has been natd at Southampton, S Hants, since 1995, but appears to be dying out; it differs from *P. saxifraga* in its hispid rays and fruit only c.1.5mm with an elongated-conical stylopodium.

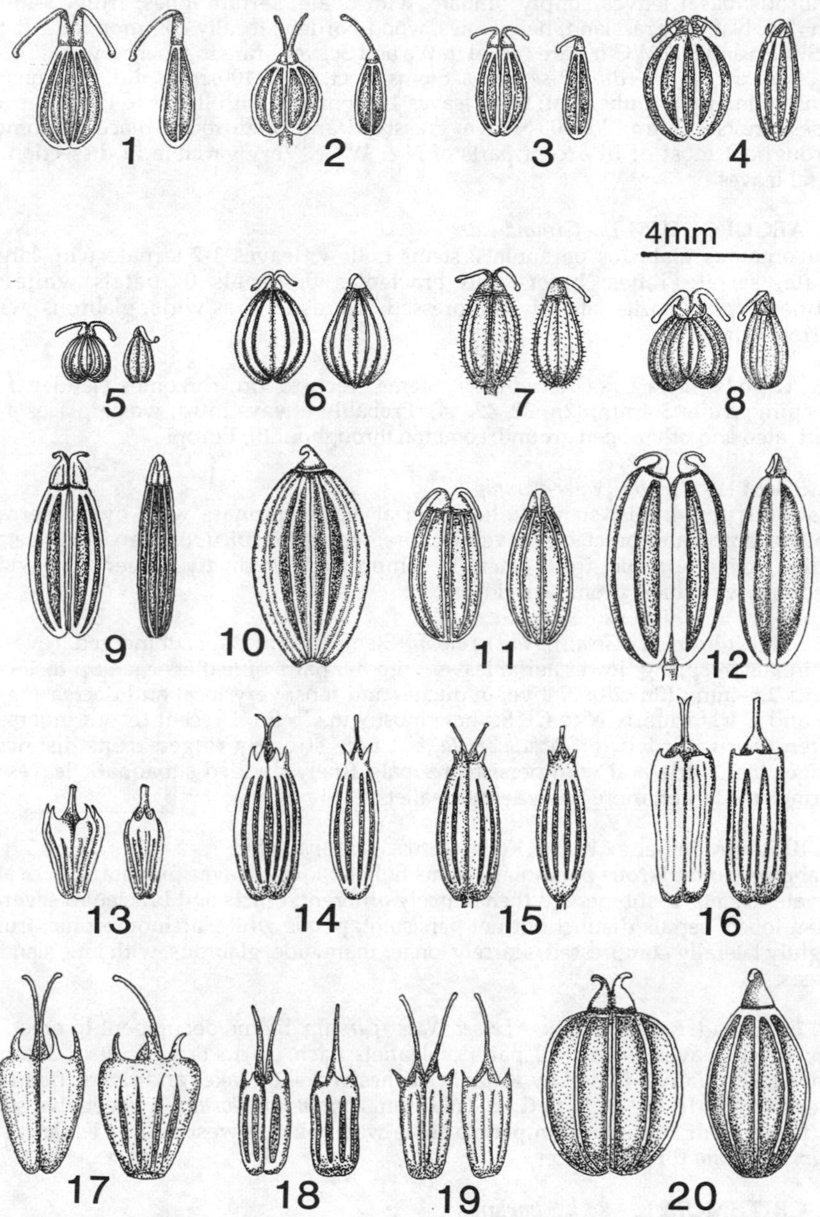

FIG 503 - Fruits (lateral and dorsal views) of **Apiaceae**. 1, ***Aegopodium podagraria***. 2, ***Pimpinella saxifraga***. 3, ***P. major***. 4, ***Sium latifolium***. 5, ***Berula erecta***. 6, ***Aethusa cynapium***. 7, ***Seseli libanotis***. 8, ***Physospermum cornubiense***. 9, ***Foeniculum vulgare***. 10, ***Anethum graveolens***. 11, ***Silaum silaus***. 12, ***Meum athamanticum***. 13, ***Oenanthe lachenalii***. 14, ***O. fluviatilis***. 15, ***O. aquatica***. 16, ***O. crocata***. 17, ***O. fistulosa***. 18, ***O. pimpinelloides***. 19, ***O. silaifolia***. 20, ***Crithmum maritimum***.

1. P. major (L.) Huds. - *Greater Burnet-saxifrage*. Stems erect, to 1(2)m, hollow, ± 503
glabrous; basal leaves simply pinnate, with ovate, serrate lobes; fruits 3-4mm; (2n=20). Native; grassland, hedgerows, wood-borders; locally common in C, E, SE & SW En and S & WC Ir, rare casual in Wa and Sc, very rare in Guernsey.

2. P. saxifraga L. - *Burnet-saxifrage*. Stems erect, to 70(100)cm, solid, glabrous to 503
densely minutely pubescent; basal leaves 1-2-pinnate, with linear to ovate-serrate lobes; fruits 2-3mm; 2n=40. Native; grassland and open rocky places; common throughout most of BI except parts of N & W Sc. Very variable in dissection of basal leaves.

14. AEGOPODIUM L. - *Ground-elder*

Rhizomatous glabrous perennials; stems hollow; leaves 1-2-ternate with large, ovate, serrate lobes; bracts and bracteoles 0; sepals 0; petals white, ± actinomorphic; fruits laterally compressed, <2x as long as wide, glabrous, with narrow ridges.

1. A. podagraria L. - *Ground-elder*. Stems erect, to 1m; rhizomes slender, far- 503
creeping; fruits 3-4mm; (2n=22, 42, 44). Probably always intrd; waste places and cultivated and other open ground; common throughout BI; Europe.

15. SIUM L. - *Greater Water-parsnip*

Glabrous perennials; stems hollow; aerial leaves pinnate with ovate, serrate leaflets; bracts and bracteoles several, entire; sepals conspicuous, persistent; petals white, actinomorphic; fruits laterally compressed, distinctly longer than wide, glabrous, with thick, prominent ridges.

1. S. latifolium L. - *Greater Water-parsnip*. Stems erect, to 2m; submerged leaves 2- **R**
3-pinnate in spring; lower aerial leaves with 3-8 pairs of leaflets each up to 15cm; 503
fruits 2.5-4mm; (2n=20). Native; in ditches and fens; very local and decreasing in En and C Ir, formerly N to CE Sc, now mostly in CE En, 1 recent record in Jersey. Often over-recorded for *Berula erecta*, but note strongly ridged fruits distinctly longer than wide and with persistent sepals; finely divided subaquatic leaves in spring; and larger, more finely serrate leaflets.

16. BERULA Besser ex W.D.J. Koch - *Lesser Water-parsnip*

Glabrous stoloniferous perennials; stems hollow; lower leaves pinnate with ovate, serrate leaflets, if submerged then scarcely different; bracts and bracteoles several, often lobed; sepals distinct but not persistent; petals white, actinomorphic; fruits slightly laterally compressed, scarcely longer than wide, glabrous, with low, slender ridges.

1. B. erecta (Huds.) Coville - *Lesser Water-parsnip*. Stems decumbent to erect, to 503
1m; lower leaves with 5-10 pairs of leaflets each up to 6cm; fruits 1.3-2mm; (2n=12, 18). Native; in and by water in ditches, marshes, lakes and rivers; frequent over much of Ir and Br N to C Sc. Told from *Apium nodiflorum* in vegetative state by presence of ring-mark on petiole some way below lowest pair of leaflets; see *Sium latifolium* for differences.

17. CRITHMUM L. - *Rock Samphire*

Glabrous perennials; stems solid, ± woody near base; leaves 2-3-pinnate, with linear, succulent lobes; bracts and bracteoles several, entire; sepals minute; petals yellowish-green, actinomorphic; fruits not compressed, each mericarp ± triangular in section, slightly longer than wide, spongy when fresh, glabrous, with thick, prominent ridges.

1. C. maritimum L. - *Rock Samphire*. Stems erect to decumbent, to 45cm; fresh 503 leaves smelling of furniture-polish when crushed; fruits 3.5-5mm; 2n=20. Native; cliffs, rocks and less often sand and shingle by sea; coasts of BI but N only to E Suffolk on E coast.

18. SESELI L. - *Moon Carrot*
Minutely pubescent biennial to monocarpic perennial; stems solid; leaves 2-3-pinnate, with rather narrow lobes; bracts and bracteoles numerous, entire; sepals small, sometimes ± persistent; petals white, actinomorphic; fruits not compressed, slightly longer than wide, minutely pubescent, with thick, prominent ridges.

1. S. libanotis (L.) W.D.J. Koch - *Moon Carrot*. Stems erect, to 60cm, with dense **RR**
sheath of fibres at base; fruits 2.5-3.5mm; (2n=18, 22). Native; grassland or rough **503**
ground on chalk; very local in E Sussex, Herts, Cambs and Beds, and decreasing.

19. OENANTHE L. - *Water-dropworts*
Glabrous annuals to perennials, often with tuberous roots; stems hollow or solid; leaves 1-4-pinnate; bracts 0-several, entire; bracteoles usually numerous; sepals conspicuous, persistent; petals white, ± actinomorphic to slightly zygomorphic; fruits not or very slightly dorsally or laterally compressed, up to c.2.5x as long as wide, glabrous, with obscure to rather prominent ridges.

1 Some umbels leaf-opposed, with peduncles shorter than rays; styles ≤1/4 as long as mature fruit 2
1 All umbels terminal, with peduncles longer than rays; styles >1/4 as long as mature fruit 3
2 Fruit ≤4.5mm; stems ascending to erect, often terrestrial **7. O. aquatica**
2 Fruit ≥5mm; stems usually floating at least at base **6. O. fluviatilis**
3 Ultimate clusters of ripe fruits globose; all fruits sessile; all leaves usually with petioles longer than divided part **1. O. fistulosa**
3 Ultimate clusters of ripe fruits not globose; some fruits stalked; all leaves usually with petioles shorter than divided part 4
4 Fruits ≥4mm; segments of mid stem-leaves ovate to ± orbicular, <2x as long as wide **5. O. crocata**
4 Fruits ≤3.5mm; segments of mid stem-leaves ± linear, >3x as long as wide 5
5 Bracts 0 at least on most umbels; rays ≥1mm thick at fruiting; stems at maturity hollow, straw-like, with walls c.0.5mm thick **2. O. silaifolia**
5 Bracts (0)1-c.5; rays <1mm thick at fruiting; stems at maturity solid to hollow with walls >0.5mm thick 6
6 Rays and pedicels thickened at fruiting, the pedicels >0.5mm thick; root-tubers ellipsoid,the proximal part of the root not thickened **3. O. pimpinelloides**
6 Rays and pedicels scarcely thickened at fruiting, the pedicels <0.5mm thick; root-tubers cylindrical to spindle-shaped, gradually widening ± from base of root **4. O. lachenalii**

1. O. fistulosa L. - *Tubular Water-dropwort*. Erect, stoloniferous perennial to 80cm; **503**
root-tubers spindle-shaped; stems hollow, with thin walls; leaves 1-3-pinnate, with ± linear segments; rays and pedicels thickened in fruit; bracts 0; fruits 3-3.5mm, with styles c. as long; 2n=22. Native; marshes, ditches and other wet places; locally frequent throughout BI N to C Sc, common in much of E En.

2. O. silaifolia M. Bieb. - *Narrow-leaved Water-dropwort*. Erect perennial to 1m; **R**
root-tubers spindle-shaped; stems hollow, with thin walls; leaves 1-4-pinnate, with **503**
linear to narrowly elliptic segments; rays and pedicels thickened in fruit; bracts usually 0; fruits 2.5-3.5mm, with styles nearly as long; 2n=22. Native; marshes, dykes and ditches; scattered and decreasing in C & S En N to SE Yorks.

3. O. pimpinelloides L. - *Corky-fruited Water-dropwort.* Erect perennial to 1m; root-tubers ellipsoid; stems solid (with pith) to ± hollow, with thick walls; leaves 1-3-pinnate, with linear to ovate segments; rays and pedicels thickened in fruit; bracts 1-5; fruits 3-3.5mm, with styles c. as long; 2n=22. Native; in dry or damp grassland, often by ditches or ponds; local in En S from Worcs and E Suffolk, Mons, Co Clare, formerly E Cork. **503**

4. O. lachenalii C.C. Gmel. - *Parsley Water-dropwort.* Erect perennial to 1m; root-tubers cylindrical to spindle-shaped; stems solid (with pith) to ± hollow, with thick walls; leaves 1-3-pinnate, with linear (rarely to ovate) segments; rays and pedicels not thickened in fruit; bracts (0)1-c.5; fruits 2.5-3mm, with shorter styles; 2n=22. Native; ditches, marshes and dykes, mostly near sea and often brackish; coasts of BI except N & E Sc, scattered inland in En. **503**

5. O. crocata L. - *Hemlock Water-dropwort.* Erect perennial to 1.5m; root-tubers spindle-shaped; stems hollow; leaves 1-3(4)-pinnate, with mostly ovate segments; bracts 3-6; fruits 4-5.5mm, with styles c.1/2 as long; 2n=22. Native; ditches, pondsides and other wet places; locally common in BI, but absent from much of C Ir and E Br. **503**

6. O. fluviatilis (Bab.) Coleman - *River Water-dropwort.* Erect to ascending or floating perennial to 1m; root-tubers 0 at maturity; stems hollow; leaves 1-3-pinnate, with linear to ovate segments; bracts usually 0; fruits 5-6.5mm, with styles <1/4 as long; 2n=22. Native; in slow rivers; scattered in S & E En N to MW Yorks, C & E Ir, Denbs. **503**

7. O. aquatica (L.) Poir. - *Fine-leaved Water-dropwort.* Erect to ascending annual to biennial to 1.5m; root-tubers 0 at maturity; stems hollow; leaves 2-4-pinnate, with linear to ovate segments; bracts usually 0; fruits 3-4.5mm, with styles ≤1/4 as long; 2n=22. Native; ditches and ponds, often drying up in summer; scattered in Br and Ir N to SE Sc. **503**

20. AETHUSA L. - *Fool's Parsley*

Glabrous annuals; stems hollow; leaves 2-3-pinnate; bracts usually 0; bracteoles usually 3-4; sepals 0; petals white, ± actinomorphic; fruits slightly dorsally compressed, slightly longer than wide, glabrous, with prominent, wide, keeled ridges.

1. A. cynapium L. - *Fool's Parsley.* Stems erect, to 1(1.5)m; bracteoles all on outer side of each flower-cluster, strongly reflexed; fruits (2.5)3-4mm. Native; cultivated and waste ground; throughout BI except much of C & N Sc. **503**

a. Ssp. cynapium. Stems to 1m; longest pedicels mostly <1/2 as long as bracteoles, mostly c.2x as long as fruits; 2n=20. The common plant of gardens and waste places, also arable land.

b. Ssp. agrestis (Wallr.) Dostál. Stems to 20cm; longest pedicels c. as long as bracteoles, mostly shorter than fruits. Arable land; distribution uncertain but mainly or only in S Br and CI; probably intrd.

Ssp. **cynapioides** (M. Bieb.) Nyman, from C Europe, a tall plant (to 1.5m) with shorter bracteoles than ssp. *cynapium*, and terete (not grooved) stems and narrower leaf-segments, has been recorded from Jersey and W Kent, and might become established.

21. FOENICULUM Mill. - *Fennel*

Glabrous perennials smelling strongly of aniseed; stems solid at first, becoming ± hollow; leaves 3-4-pinnate, with long filiform segments; bracts 0; bracteoles 0; sepals 0; petals yellow, actinomorphic; fruits scarcely compressed, c.2-4x as long as wide, glabrous, with prominent thick ribs.

1. F. vulgare Mill. - *Fennel.* Stems erect, ± glaucous, to 2.5m; fruits often not developing or developing very late in season, 4-5mm; (2n=22). Probably intrd; **503**

open ground and waste places, especially near the coast; well natd in BI N to N Ir, Man and Yorks, scarce and mainly casual further N; Europe. Nowadays rarely grown on field-scale.

22. ANETHUM L. - *Dill*
Glabrous annuals smelling strongly of aniseed; stems hollow; leaves as in *Foeniculum*; bracts 0; bracteoles 0; sepals 0; petals yellow, actinomorphic; fruits strongly dorsally compressed, c.2x as long as wide, glabrous, with prominent slender dorsal ridges and conspicuously winged lateral ones.

1. A. graveolens L. - *Dill*. Stems erect, to 60cm; differs from *Foeniculum* in annual habit; completely hollow stems; and totally different fruits 3-5mm; (2n=22). Intrd; rather frequent casual from birdseed or grain in waste and cultivated ground; scattered in Br; W & C Asia. **503**

23. SILAUM Mill. - *Pepper-saxifrage*
Glabrous perennials; stems solid; leaves 1-4-pinnate; bracts 0-3, entire; bracteoles numerous; sepals 0; petals yellowish, actinomorphic; fruits scarcely compressed, c.2x as long as wide, glabrous, with prominent slender ridges.

1. S. silaus (L.) Schinz & Thell. - *Pepper-saxifrage*. Stems erect, to 1m; leaf- **503**
segments linear to lanceolate; fruits 4-5mm; 2n=22. Native; grassy places; locally frequent in Br N to C Sc, mainly in E.

24. MEUM Mill. - *Spignel*
Glabrous perennials; stems hollow; leaves 3-4-pinnate; bracts 0-few, entire; bracteoles several; sepals 0; petals white to pinkish, actinomorphic; fruits scarcely compressed, <2x as long as wide, glabrous, with prominent, rather narrow ridges.

1. M. athamanticum Jacq. - *Spignel*. Plant sweetly aromatic; stems erect, to 60cm, **R**
with dense sheath of fibres at base; most leaves basal, with filiform segments; **503**
fruits (4)5-7(10)mm; 2n=22. Native; mountain grassland; local in N En, N Wa, C & S Sc.

25. PHYSOSPERMUM Cusson ex Juss. - *Bladderseed*
Almost glabrous perennials; stems solid; basal leaves 2-ternate; stem-leaves simple to 1-ternate; bracts and bracteoles several, entire; sepals conspicuous, persistent; petals white, actinomorphic; fruits ± inflated, wider than long, glabrous, with narrow low ridges.

1. P. cornubiense (L.) DC. - *Bladderseed*. Stems erect, to 1.2m; leaves mostly **RR**
basal, with ± ovate, deeply serrate segments; fruits 2.5-4mm; (2n=22). Native; **503**
arable fields, hedgebanks, scrub and woods; very local in E Cornwall and S Devon, natd in Bucks.

26. CONIUM L. - *Hemlock*
Glabrous biennials; stems hollow; leaves 2-4-pinnate; bracts and bracteoles several, entire; sepals 0; petals white, actinomorphic; fruits scarcely compressed, c. as wide as long, glabrous, with very prominent, narrow, ± undulate ridges.

1. C. maculatum L. - *Hemlock*. Stems erect, to 2.5m, usually purple-spotted; **509**
leaves with ovate, deeply serrate segments; fruits 2-3.5mm; 2n=22. Native; damp ground, roadsides, ditches, waste ground; common over most of BI except W & C Sc.

27. BUPLEURUM L. - *Hare's-ears*
Annual to perennial herbs, rarely shrubs; stems hollow or solid; leaves simple, entire; bracts 0 to several, entire; bracteoles several; sepals 0; petals yellow, actinomorphic; fruits slightly laterally compressed, 1-1.5x as long as wide, glabrous, sometimes papillose, strongly to scarcely ridged.

1 Upper leaves fused right around stem; bracts 0 — 2
1 Leaves not fused around stem; bracts present — 3
  2 Umbels with 4-8 rays; fruit smooth between ridges; leaves mostly <2x as long as wide — **5. B. rotundifolium**
  2 Umbels with 2-3 rays; fruit strongly papillose between ridges; leaves mostly >2x as long as wide — **6. B. subovatum**
3 Firmly rooted perennial; at least some leaves often >1cm wide — 4
3 Easily uprooted annual; leaves all <1cm wide — 5
  4 Evergreen shrub; leaves with strong midrib and many lateral veins — **1. B. fruticosum**
  4 Herb, often woody at extreme base; leaves with ≥3 equally strong ± parallel main veins — **2. B. falcatum**
5 Bracteoles linear, not concealing flowers or fruits; fruit tuberculate — **3. B. tenuissimum**
5 Bracteoles lanceolate, concealing flowers and fruits; fruits smooth — **4. B. baldense**

Other spp. - B. odontites L. (*B. fontanesii* Guss.), from Mediterranean region, is an annual somewhat resembling *B. baldense* but with much larger umbels, 5-7 rays, and bracteoles with conspicuous, abruptly recurved cross-veins; it is a rare birdseed-alien.

1. B. fruticosum L. - *Shrubby Hare's-ear*. Evergreen shrub to 2.5m, but often much 509
less due to annual frost-damage; leaves oblong to narrowly obovate, with 1 main vein, most >4 x 1.5cm; bracts several but often fallen before fruiting, ovate to obovate, shorter than rays; rays 5-many; fruits 4.5-6mm, with prominent, slender ridges, smooth; (2n=14). Intrd; grown in gardens, natd on roadside and railway banks since before 1909; E Suffolk, W Kent, S Devon and Worcs; S Europe.

2. B. falcatum L. - *Sickle-leaved Hare's-ear*. Herbaceous ± erect perennial to 1m; RRR
leaves oblanceolate to linear-falcate, with 5 main veins, the lower usually >5 x 1cm; 509
bracts several, linear to lanceolate, shorter than rays; rays 5-c.11; fruits 2.5-3.5mm, with slender, prominent ridges, smooth; (2n=16, 26, 28, 32). Possibly native; damp roadsides, ditches, hedgebanks and field-borders; 1 locality in S Essex, discovered 1831, eradicated 1956, then reintroduced from different source, rarely natd in C & N En.

3. B. tenuissimum L. - *Slender Hare's-ear.* Erect to procumbent slender annual to R
50cm; leaves linear to very narrowly oblanceolate, with 3 main veins, the lower 509
mostly >2cm x <5mm; bracts 3-5, often longer than shortest rays, ± linear; rays 1-3, unequal; fruit 1.5-2mm, with slender ridges, tuberculate; (2n=16). Native; grassy or barish brackish ground; coasts of S Br from Mons (formerly Glam) to SE Yorks (formerly Durham), Flints, formerly rare inland.

4. B. baldense Turra - *Small Hare's-ear*. Erect annual to 25cm, usually much RRR
<10cm; leaves linear to very narrowly oblanceolate, with 3-5 main veins, the lower 509
mostly >2cm x <0.3mm; bracts c.4, longer than shortest rays, lanceolate; rays 1-4, unequal; fruits 1.5-2.5mm, scarcely ridged, smooth; (2n=16). Native; barish ground on fixed dunes and cliffs by sea; very local in CI, S Devon and E Sussex, rare casual elsewhere. Often very dwarfed and superficially resembling *Euphorbia exigua*, with which it sometimes grows.

5. B. rotundifolium L. - *Thorow-wax*. Erect annual to 30cm; leaves elliptic to 509
suborbicular, 2-6 x 1.5-4cm, with >5 main veins; bracts 0; bracteoles ovate, longer

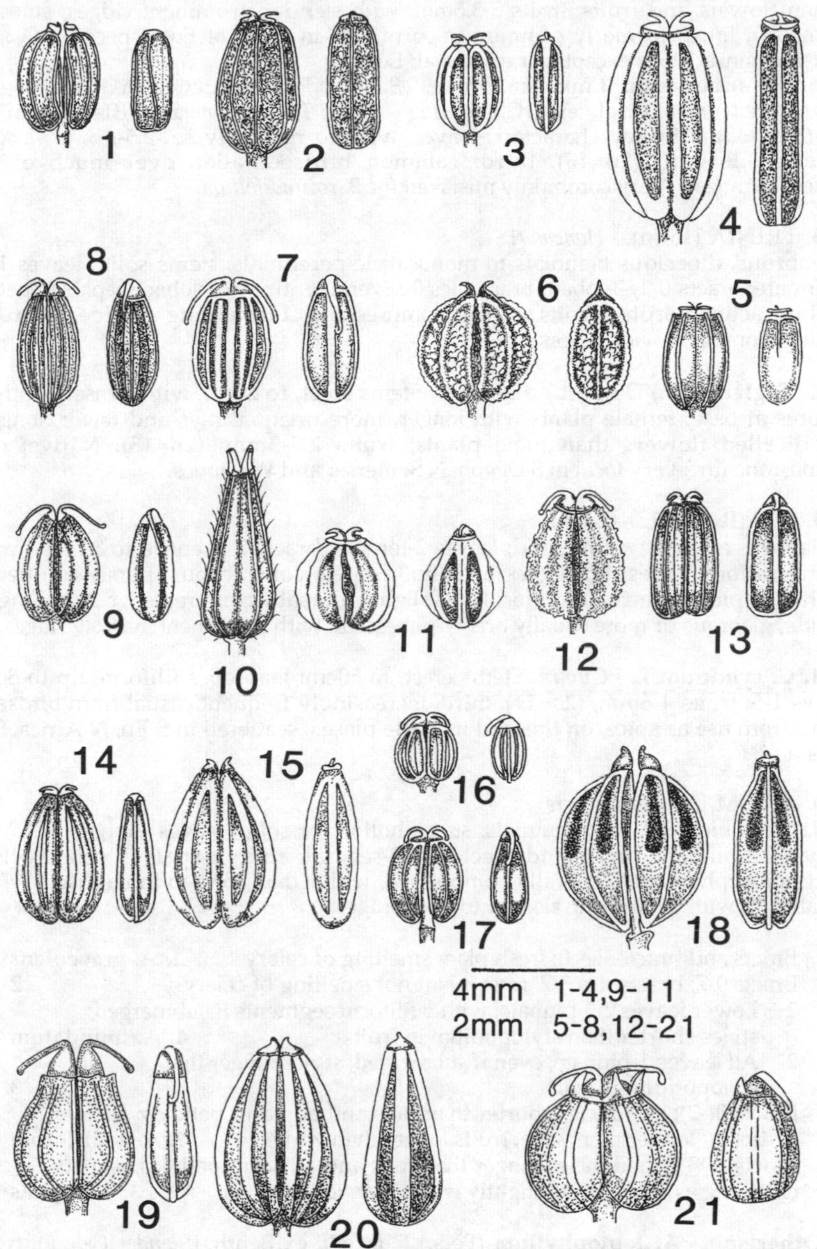

FIG 509 - Fruits (mostly with lateral and dorsal views) of **Apiaceae**. 1, ***Bupleurum rotundifolium***. 2, ***B. subovatum***. 3, ***B. falcatum***. 4, ***B. fruticosum***. 5, ***B. baldense***. 6, ***B. tenuissimum***. 7, ***Ammi visnaga***. 8, ***A. majus***. 9, ***Trinia glauca***. 10, ***Cuminum cyminum***. 11, ***Conium maculatum***. 12, ***Trachyspermum ammi***. 13, ***Ridolfia segetum***. 14, ***Apium nodiflorum***. 15, ***A. inundatum***. 16, ***A. repens***. 17, ***A. graveolens***. 18, ***Sison amomum***. 19, ***Petroselinum crispum***. 20, ***P. segetum***. 21, ***Cicuta virosa***.

than flowers and fruits; fruits 3-3.5mm, with slender prominent ridges, smooth; (2n=16). Intrd; formerly common in cornfields in most of En, especially C & S, extinct since 1960s except as rare casual; Europe.

6. B. subovatum Link ex Spreng. (*B. lancifolium* auct. non Hornem., *B. intermedium* (Loisel. ex DC.) Steud. - *False Thorow-wax*. Differs from *B. rotundifolium* in key characters; leaves ovate to narrowly so, 2.5-6 x 1.5-2.8cm; fruits 3.5-5mm; (2n=16). Intrd; common birdseed-alien over much of BI; Mediterranean. Very commonly mistaken for *B. rotundifolium*. **509**

28. TRINIA Hoffm. - *Honewort*

Glabrous, dioecious biennials to monocarpic perennials; stems solid; leaves 1-3-pinnate; bracts 0-1, 3-lobed; bracteoles 0-several, entire to 3-lobed; sepals 0; petals white, actinomorphic; fruits scarcely compressed, <1.5x as long as wide, glabrous, with prominent, wide ridges.

1. T. glauca (L.) Dumort. - *Honewort*. Stems erect, to 20cm, with dense sheath of fibres at base; female plants with longer, more unequal rays and fewer, longer-pedicelled flowers than male plants; fruits 2.3-3mm; (2n=18). Native; dry limestone turf; very local in S Devon, N Somerset and W Gloucs. **RR** **509**

29. CUMINUM L. - *Cumin*

Glabrous annuals; stems solid; leaves 2-ternate; bracts 2-4, entire to 3-lobed with long filiform lobes; bracteoles usually 3; sepals conspicuous, persistent; petals white to pinkish, actinomorphic; fruits slightly dorsally compressed, c.2x as long as wide, glabrous or more usually bristly-pubescent, with prominent, narrow ridges.

1. C. cyminum L. - *Cumin*. Stems erect, to 50cm; leaf-lobes filiform, up to 5cm; rays 1-5; fruits 4-6mm; (2n=14). Intrd; increasingly frequent casual from birdseed and from use as spice, on tips and in waste places; scattered in S En; N Africa, SW Asia. **509**

30. APIUM L. - *Marshworts*

Glabrous biennials to perennials; stems hollow or solid; leaves pinnate or lower ones 2-3-pinnate; bracts and bracteoles 0-several, entire; sepals 0; petals white, actinomorphic; fruits laterally compressed, wider than long to longer than wide, glabrous, with prominent, slender to thick ridges.

1 Bracts and bracteoles 0; fresh plant smelling of celery **1. A. graveolens**
1 Bracts 0-7; bracteoles 3-7; fresh plant not smelling of celery 2
  2 Lower leaves 2-3-pinnate, with ± filiform segments if submerged; styles shorter than stylopodium in fruit **4. A. inundatum**
  2 All leaves 1-pinnate, even if submerged; styles longer than stylopodium in fruit 3
3 Bracts 0(-2); peduncles shorter than rays and adjacent petioles; leaflets distinctly longer than wide; fruits longer than wide **2. A. nodiflorum**
3 Bracts (1)3-7; peduncles longer than rays and adjacent petioles; leaflets c. as long as wide; fruits slightly wider than long **3. A. repens**

Other spp. - A. leptophyllum (Pers.) F. Muell. ex Benth. (*Slender Celery*), from American tropics, has umbels and fruits similar to those of *A. nodiflorum* but 0 bracts and bracteoles and all leaves finely divided into filiform segments; it is a rare wool- and grain-alien. Sometimes segregated as *Ciclospermum leptophyllum* (Pers.) Britton & E.H. Wilson.

1. A. graveolens L. - *Wild Celery*. Usually erect biennial to 1m; stems solid; leaves 1-pinnate, rarely 1-ternate or 2-pinnate, the basal ones with stalked leaflets; fruits **509**

1-1.5mm, with slender ridges; 2n=22. Native; damp barish usually brackish places usually near sea; coasts of BI N to S Sc, very scattered inland. Var. **dulce** (Mill.) DC. (*Celery*), with fat, ± succulent petioles, and var. **rapaceum** (Mill.) DC. (*Celeriac*), with greatly swollen stem-bases, are both cultivated as vegetables but rarely or never escape.

2. A. nodiflorum (L.) Lag. - *Fool's-water-cress*. Stems to 1m, suberect to 509
procumbent, rooting at lower nodes, hollow; leaves 1-pinnate, with sessile lobes distinctly longer than wide; bracts 0(2); peduncles shorter than rays; rays 3-15; fruits 1.5-2.5mm, longer than wide, with thick ridges; 2n=22. Native; ditches, marshes and by lakes and rivers; common in BI N to S Sc, very local in C & N Sc. See *Berula erecta* for differences. Some growth-forms mimic *A. repens* very closely.

2 x 3. A. nodiflorum x A. repens (?= *A. x longipedunculatum* (F.W. Schultz) Rothm.) has been found with *A. nodiflorum* in some of the Oxon localities of *A. repens*, and in Cambs, SE Yorks, Fife and perhaps elsewhere where *A. repens* has become extinct; it usually has a more procumbent and extensively rooting stem than *A. nodiflorum*, leaflets longer than wide, peduncles shorter than to c. as long as rays, and 1-4 bracts; sterile; 2n=20; ?endemic.

2 x 4. A. nodiflorum x A. inundatum = A. x moorei (Syme) Druce occurs with the parents scattered over most of Ir and locally in CE En; all leaves 1-pinnate with obovate leaflets; rays usually 2-3; sterile; endemic.

3. A. repens (Jacq.) Lag. - *Creeping Marshwort*. Differs from dwarfest plants of *A.* RRR
nodiflorum in stems procumbent, rooting at ± all nodes; fruits 0.7-1mm, with 509
narrow ridges; and see key; 2n=18. Native; open wet places; perhaps in pure state now only in 1 site in Oxon, formerly scattered in E En and C & S Sc.

4. A. inundatum (L.) Rchb.f. - *Lesser Marshwort*. Stems decumbent to procumbent, 509
to 50cm, hollow, often largely submerged, rooting at lower nodes; lower leaves 2-3-pinnate, with narrow segments; bracts 0; bracteoles 3-6; rays 2(-4); fruits 2.5-3mm, with thick ridges; 2n=22. Native; in still, usually shallow water and on bare mud nearby; scattered over most BI.

31. TRACHYSPERMUM Link - *Ajowan*

Almost glabrous annuals; stems hollow; leaves mostly 2-pinnate; bracts several, entire or lobed; bracteoles several; sepals small, not persistent; petals white, actinomorphic; fruits laterally compressed, slightly longer than wide, densely tuberculate, with rather obscure, wide ridges.

1. T. ammi (L.) Sprague - *Ajowan*. Stems erect, to 30cm; leaves with filiform 509
segments to 2cm; fruits 1.5-2mm, very characteristically tuberculate; (2n=18). Intrd; casual on tips and waste ground from birdseed and use as spice, perhaps increasing; scattered in En, mainly S; E Mediterranean.

32. PETROSELINUM Hill - *Parsleys*

Glabrous annuals or biennials; stems solid; leaves 1-3-pinnate, with wide lobes; bracts 1-several, entire or lobed; bracteoles several; sepals 0 or very small; petals white or yellow, actinomorphic; fruits laterally compressed, somewhat longer than wide, glabrous, with prominent thick to narrow ridges.

1. P. crispum (Mill.) Nyman ex A.W. Hill - *Garden Parsley*. Erect biennials to 509
75cm, with characteristic parsley smell; lower leaves 3-pinnate, shiny; bracts 1-3, often lobed or leaf-like; petals yellow; fruits 2-2.5mm, with styles ± as long as stylopodium; 2n=22. Intrd; commonly grown on small scale and frequent escape on tips and in waste places; scattered in BI except most of C & N Sc; E Mediterranean.

2. P. segetum (L.) W.D.J. Koch - *Corn Parsley*. Erect annuals or biennials to 1m; 509
lower leaves 1-pinnate, with 4-12 pairs of leaflets, matt; bracts 2-5, sometimes lobed; petals white; fruits 2.3-3mm, with styles much shorter than stylopodium;

(2n=16). Native; barish or grassy places in arable fields, pastures and hedgerows and on banks; local in S Br N to SE Yorks and Cards, Jersey. Easily mistaken for *Sison*, but with extremely unequal-lengthed rays, more leaflets, thicker ridges on fruits, and without the characteristic smell.

33. RIDOLFIA Moris - *False Fennel*
Glabrous annuals; stems solid; leaves 3-4-pinnate, with filiform segments; bracts 0; bracteoles 0; sepals 0; petals yellow, actinomorphic; fruits laterally compressed, longer than wide, glabrous, with slender, rather low ridges.

1. R. segetum (Guss.) Moris - *False Fennel*. Stems erect, to 1m; leaves with filiform **509**
segments, resembling those of *Anethum* but lacking aniseed smell; fruits 1.5-2.5mm; (2n=22). Intrd; rather infrequent birdseed-alien on tips and waste ground; very scattered in S En; Mediterranean.

34. SISON L. - *Stone Parsley*
Glabrous biennials, stems solid; leaves 1-2-pinnate, the lower with wide lobes; bracts 2-4 and bracteoles 2-4, entire; sepals 0; petals white, actinomorphic; fruits laterally compressed, somewhat longer than wide, glabrous, with narrow prominent ridges.

1. S. amomum L. - *Stone Parsley*. Fresh plant smelling rather like petrol when **509**
crushed; stems erect, to 1m; lower leaves 1-pinnate, with 2-5 pairs of leaflets; most rays of ± same length, often 1 much shorter; fruits 1.5-3mm; 2n=14. Native; hedgebanks, grassland, roadsides; locally frequent in Br N to Cheshire and SE Yorks, rare casual further N. See *Petroselinum segetum* for differences.

35. CICUTA L. - *Cowbane*
Glabrous perennials; stems hollow; leaves 2-3-pinnate; bracts 0; bracteoles numerous; sepals conspicuous, persistent; petals white, actinomorphic; fruits laterally compressed but each mericarp ± globose, wider than long, glabrous, with wide inconspicuous ridges.

1. C. virosa L. - *Cowbane*. Stems erect, to 1.5m; leaves with narrowly elliptic to **R**
linear-lanceolate segments, up to 5(9)cm; fruits 1.2-2mm; 2n=22. Native; ditches, **509**
marshy fields, pondsides; very local in N & C Ir, E Anglia, WC En, very rare (formerly less so) elsewhere in BI.

36. AMMI L. - *Bullworts*
Glabrous annuals to biennials; stems solid; leaves 1-3(4)-pinnate; bracts several, mostly pinnately divided with linear to filiform lobes; bracteoles numerous; sepals 0; petals white, very slightly zygomorphic; fruits slightly laterally compressed; somewhat longer than wide, glabrous, with slender prominent ridges.

1. A. majus L. - *Bullwort*. Stems erect, to 1m; lower leaves 1-2-pinnate, with **509**
elliptic to narrowly elliptic, serrate lobes; upper leaves 2(-3)-pinnate; rays remaining slender and bracts not becoming strongly reflexed in fruit; fruits 1.5-2mm; (2n=22). Intrd; rather infrequent casual mainly from birdseed and wool on tips and waste ground and in fields; very scattered in CI, C & S (rarely N) Br; S Europe.

2. A. visnaga (L.) Lam. - *Toothpick-plant*. Stems erect, to 1m; lower leaves 1-2- **509**
pinnate; upper leaves 2-3(4)-pinnate; all leaves with linear to filiform segments; rays becoming rigid, thick and erect, and bracts becoming strongly reflexed, in fruit; fruits 2-2.8mm; (2n=20, 22). Intrd; casual in S En N to SW Yorks from same sources as *A. majus*, but rarer; Mediterranean.

37. FALCARIA Fabr. - *Longleaf*
Glabrous, glaucous, perennial; stems solid; leaves 1-2-ternate; bracts and bracteoles numerous, entire; sepals conspicuous, persistent; petals white, actinomorphic; fruits laterally compressed, 2-4x as long as wide, glabrous, with wide, low ridges.

1. F. vulgaris Bernh. - *Longleaf*. Stems erect, to 60cm; leaf-segments c.10-30cm, **516**
linear-lanceolate, serrate; fruits 2.5-4(5)mm; (2n=22). Intrd; natd in grassy and waste places and scrub; very local in CI and S & C Br, especially E Anglia; Europe.

38. CARUM L. - *Caraways*
Glabrous biennials or perennials; stems hollow; leaves 1-3-pinnate; bracts and bracteoles 0-numerous; sepals 0 or minute, not persistent; petals white, actinomorphic or nearly so; fruits laterally compressed, 1.2-1.7x as long as wide, glabrous, with narrow, ± prominent ridges.

1. C. carvi L. - *Caraway*. Erect biennial to 60cm, without fibres at base; leaves 2-3- **516**
pinnate, the lower with linear to lanceolate segments; bracts and bracteoles 0-few, the former sometimes ± leaf-like; fruits 3-4mm, with distinctive smell when crushed; (2n=20). Intrd; natd in fields, roadsides and waste places, sparsely scattered throughout BI, cultivated as flavouring less than formerly; Europe.

2. C. verticillatum (L.) W.D.J. Koch - *Whorled Caraway*. Erect perennial to 60cm, **516**
with dense sheath of fibres at base; leaves 1-pinnate, each leaflet deeply palmately divided to base into filiform segments; bracts and bracteoles numerous, entire; fruits 2-3mm; 2n=?18, (20, 22). Native; marshes, damp meadows and streamsides; locally frequent in SW & N Ir, W parts of Br from Cornwall to Westerness, Jersey, very rare elsewhere.

39. SELINUM L. - *Cambridge Milk-parsley*
Glabrous perennials; stems solid; leaves 2-3-pinnate; bracts 0-few, entire, soon falling; bracteoles numerous; sepals ± 0; petals white, actinomorphic; fruits dorsally compressed, c. as long as wide, glabrous, with conspicuous winged ridges.

1. S. carvifolia (L.) L. - *Cambridge Milk-parsley*. Stems erect, to 1m; lower leaves **RRR**
with lanceolate to ovate, deeply lobed segments <10mm; fruits 3-4mm; (2n=22). **516**
Native; fens and damp meadows; 3 localities in Cambs, formerly Notts and N Lincs. See *Peucedanum palustre* for differences.

40. LIGUSTICUM L. - *Scots Lovage*
Glabrous perennials; stems hollow; leaves 1-2-ternate, with stalked, wide leaflets; bracts 1-several, entire; bracteoles several; sepals conspicuous, persistent; petals greenish-white, actinomorphic; fruits strongly dorsally compressed, c.2x as long as wide, glabrous, with prominent, ± winged ridges.

1. L. scoticum L. - *Scots Lovage*. Stems erect, to 60(90)cm; leaves bright green, with **516**
ovate-trullate, serrate segments >2 x 1.5cm; fruits 4-7mm; 2n=22. Native; cliffs and rocky places near sea; frequent around whole coast of Sc, local in N & W Ir, formerly Cheviot.

41. ANGELICA L. - *Angelicas*
Minutely pubescent to ± glabrous perennials; stems hollow; leaves 2-3-pinnate, the upper with very strongly inflated petioles; bracts 0-few, entire, soon falling; bracteoles numerous; sepals very small; petals white, greenish-white or pinkish-white, actinomorphic; fruits strongly dorsally compressed, slightly longer than wide, glabrous, with low dorsal and conspicuously winged lateral ridges.

1 Fruits with thin membranous wings, 4-5mm **1. A. sylvestris**
1 Fruits with thick corky wings, 5-10mm 2
  2 Fruits 5-6mm; leaflets acute to acuminate **2. A. archangelica**
  2 Fruits 8-10mm; leaflets acute to subacute **3. A. pachycarpa**

1. A. sylvestris L. - *Wild Angelica.* Stems erect, to 2.5m, often much less, usually **516**
somewhat purplish; leaflets ovate, abruptly acute, closely and finely serrate; peduncles and rays minutely pubescent; petals white to pinkish-white; fruits 4-5mm; 2n=22. Native; damp grassy places, fens, marshes, by streams, ditches and ponds, in damp open woods; common throughout BI.

2. A. archangelica L. - *Garden Angelica.* Stems erect, to 2m, usually green; leaflets **516**
narrowly ovate, acute to acuminate, coarsely but sharply serrate; peduncles glabrous; rays minutely pubescent; petals greenish-white; fruits 5-6mm; (2n=22). Intrd; natd on river-banks and waste places, now cultivated less than formerly; scattered in Br, ± frequent in London area; N & E Europe.

3. A. pachycarpa Lange - *Portuguese Angelica.* Stems erect, to 1m; leaflets ovate, **499**
acute to subacute, coarsely and bluntly serrate; rays minutely pubescent; petals greenish-white; fruits 8-10mm; (2n=22). Intrd; natd in hedgebank, Guernsey, since 1993; NW Spain and Portugal.

42. LEVISTICUM Hill - *Lovage*

Almost glabrous perennials smelling of celery when crushed; stems hollow; leaves 2-3-pinnate; bracts and bracteoles numerous, entire, the latter fused at base; sepals ± 0; petals yellow, actinomorphic; fruits strongly dorsally compressed, <2x as long as wide, glabrous; with very prominent dorsal and winged lateral ridges.

1. L. officinale W.D.J. Koch - *Lovage.* Stems erect, to 2.5m; leaves with trullate to **516**
rhombic, sparsely but deeply serrate leaflets; fruits 4-7mm; (2n=22). Intrd; natd, usually not permanently and sometimes only a relic, in rough ground, by walls and paths; very scattered in Br, mostly in Sc and N En but S to Surrey and W Kent; Iran.

43. FERULA L. - *Giant Fennel*

Erect, glabrous perennials; stems solid; leaves 4-6-pinnate; bracts 0; bracteoles few, soon falling; petals yellow, actinomorphic; fruits strongly dorsally compressed, longer than wide, glabrous, with low dorsal and winged lateral ridges.

1. F. communis L. - *Giant Fennel.* Stems erect, to 3m; leaves with ultimate **499**
segments 10-50mm, linear, with very conspicuous, wide, sheathing petioles; fruits 10-16mm; (2n=22). Intrd; natd on roadside verge in W Suffolk since 1988, formerly Northants 1956-1988, occasionally elsewhere in En; Mediterranean.

44. PEUCEDANUM L. - *Hog's Fennels*

Erect, glabrous to nearly glabrous biennials to perennials; stems hollow or solid; leaves 1-6-ternate or 2-4-pinnate; bracts 0-several, entire; bracteoles few to several; sepals ± 0 to conspicuous, persistent or not; petals white or yellow, actinomorphic; fruits strongly dorsally compressed, somewhat longer than wide, glabrous, with low dorsal and winged lateral ridges.

The 3 spp. in this genus seem at least as different as several other genera; probably it should be united with a number of related genera.

1 Petals yellow; stems solid, with dense sheath of fibres at base; ultimate leaf-segments >3cm, linear **1. P. officinale**
1 Petals white; stems hollow, without basal fibres; ultimate leaf-segments not linear, or if so <2cm 2
  2 Leaves 2-4-pinnate, with ultimate segments <1.5 x 0.5cm; bracts

≥3, reflexed **2. P. palustre**
2 Leaves 1-2-ternate, with ultimate segments >2 x 1cm; bracts 0(-2) **3. P. ostruthium**

1. P. officinale L. - *Hog's Fennel*. Stems to 2m; leaves 3-6-ternate, with ultimate RR
segments c.4-10cm, linear; petals yellow; fruits 5-8mm; (2n=66). Native; rough 516
brackish grassland, banks of creeks and pathsides near sea; extremely local in E Kent and N Essex.

2. P. palustre (L.) Moench - *Milk-parsley*. Stems to 1.5m; leaves 2-4-pinnate, with R
ultimate segments linear to narrowly oblong-lanceolate, c.0.5-1.5cm; petals white; 516
fruits 3-5mm; (2n=22). Native; fens and marshes; local in En N to SE Yorks, very rare outside fens of E Anglia. Often with *Selinum* which differs in solid stems, patent bracts, and dorsally winged fruits.

3. P. ostruthium (L.) W.D.J. Koch - *Masterwort*. Stems to 1m; leaves 1-2-ternate, 516
with ultimate segments ovate, serrate, c.5-12cm; petals white; fruits 3-5mm; (2n=22). Intrd; natd in grassy places, marshy fields and riversides, nowadays rarely cultivated; scattered and decreasing in N Ir and Br N from S Lincs and Staffs, formerly S Ir and Wa; C & S Europe.

45. PASTINACA L. - *Parsnips*

Somewhat pubescent biennials with strong characteristic smell; stems hollow or solid; leaves 1-pinnate, with large ovate leaflets; bracts and bracteoles 0-2, entire, soon falling; sepals 0; petals yellow, actinomorphic; fruits strongly dorsally compressed, somewhat longer than wide, glabrous, with low dorsal and winged lateral ridges.

1. P. sativa L. - see vars. for English names. Stems erect, to 1.8m; fruits 4-7mm; 516
2n=22. Both our vars. belong to ssp. *sativa*.

a. Var. hortensis Gaudin - *Parsnip*. Rather sparsely pubescent; hairs on stems and petioles straight; root often swollen-conical. Intrd; frequent escape from cultivation, when established often losing its swollen root feature; scattered over BI; cultivated origin.

b. Var. sativa (var. *sylvestris* (Mill.) DC., ssp. *sylvestris* (Mill.) Rouy & E.G. Camus) - *Wild Parsnip*. Pubescent; hairs on stems and petioles longer and wavy; root not swollen. Native; grassland, roadsides, rough ground, especially on chalk and limestone; common in En SE of line from Humber to Severn, very local elsewhere in En and Wa.

46. HERACLEUM L. - *Hogweeds*

Erect, pubescent biennials to often monocarpic perennials; stems hollow; leaves simple and pinnately or ternately divided, or 1(-2)-pinnate, or ternate; bracts 0-several, entire; bracteoles several; sepals minute or conspicuous and persistent; petals white to purplish or greenish-white, zygomorphic to scarcely so; fruits strongly dorsally compressed, slightly to somewhat longer than wide, glabrous or pubescent, with very low dorsal and winged lateral ridges.

Other spp. - Plants named *H. mantegazzianum* are variable and might represent >1 sp. **H. persicum** Desf. ex Fisch., from Turkey to Iran, is 1 related taxon; it is smaller, to 2.5m, with more divided leaves with less sharp serrations.

1. H. sphondylium L. - *Hogweed*. Stems to 2(3)m, hispid; leaves 1(-2)-pinnate, 516
with large, hispid, usually lobed leaflets; bracts 0-few; fruits glabrous, without persistent sepals, with linear oil-bodies scarcely widened (<0.4mm wide) at proximal end. Native; grassy places, rough ground, roadsides and banks.

a. Ssp. sphondylium. Petals white or pinkish-white to purplish, the outer ones on outermost flowers of umbel bilobed and c.2x inner unlobed ones; fruits 6-10mm;

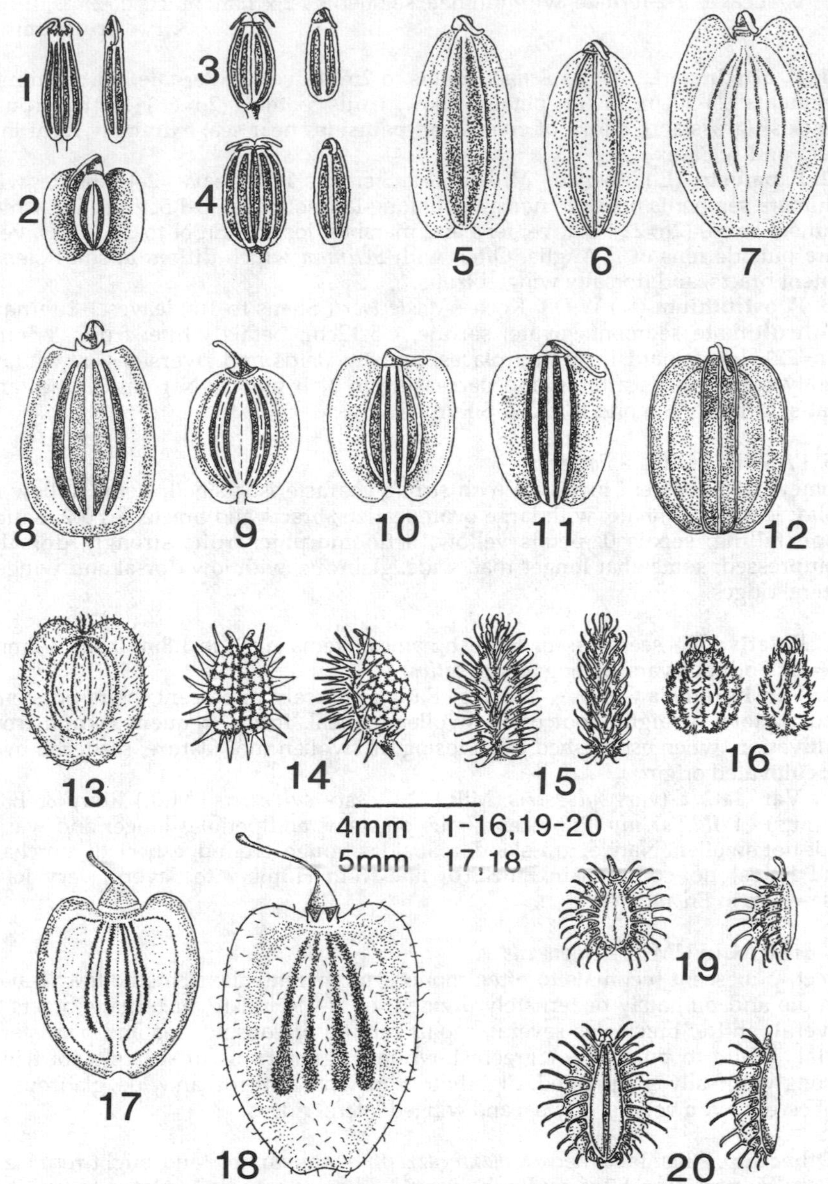

FIG 516 - Fruits (some with lateral and dorsal views) of **Apiaceae**. 1, ***Falcaria vulgaris***. 2, ***Selinum carvifolia***. 3, ***Carum verticillatum***. 4, ***C. carvi***. 5, ***Ligusticum scoticum***. 6, ***Levisticum officinale***. 7, ***Pastinaca sativa***. 8, ***Peucedanum officinale***. 9, ***P. palustre***. 10, ***P. ostruthium***. 11, ***Angelica sylvestris***. 12, ***A. archangelica***. 13, ***Tordylium maximum***. 14, ***Torilis nodosa***. 15, ***T. arvensis***. 16, ***T. japonica***. 17, ***Heracleum sphondylium***. 18, ***H. mantegazzianum***. 19-20, ***Daucus***, lateral view shows 1 mericarp only. 19, ***D. carota***. 20, ***D. glochidiatus***.

2n=22. Common throughout BI.

b. Ssp. sibiricum (L.) Simonk. Petals greenish-white, scarcely zygomorphic; fruits 4-6mm (?always). NE parts of E Norfolk (?intrd). **RR**

1 x 2. H. sphondylium x H. mantegazzianum is scattered in En, Ir and Sc, especially SE Sc and the London area; it is intermediate in size, pubescence, leaf-shape and fruit characters, and has very low fertility; 2n=22; endemic.

2. H. mantegazzianum Sommier & Levier - *Giant Hogweed*. Stems to 5.5m, rather softly pubescent; leaves pinnate to ternate, or simple and ternately to pinnately lobed, the lowest up to 2.5m; bracts several; petals white, zygomorphic; fruits 9-14mm, glabrous or pubescent, with persistent sepals, with conspicuous oil-bodies swollen to 0.6-1mm wide at proximal end; 2n=22. Intrd; natd on waste ground, roadside and riverside banks, rough grassland; scattered throughout BI and locally frequent; SW Asia. **516**

47. TORDYLIUM L. - *Hartwort*

Hispid annuals or biennials; stems hollow or ± solid; leaves 1-pinnate (basal ± simple but gone by flowering); bracts and bracteoles several, entire; sepals conspicuous, persistent; petals white, zygomorphic; fruits strongly dorsally compressed, scarcely longer than wide, hispid, with very low dorsal and broadly whitish-winged lateral ridges.

1. T. maximum L. - *Hartwort*. Stems erect, to 1m; lower leaves with 2-5 pairs of lanceolate, coarsely serrate leaflets; fruits 4.5-6(8)mm; (2n=22). Possibly native; rough scrubby grassland; near Thames in S Essex since 1875, formerly elsewhere as casual in SE En; S Europe. **516**

48. TORILIS Adans. - *Hedge-parsleys*

Hispid annuals (rarely biennials); stems solid; leaves 1-3-pinnate; bracts 0-numerous, entire; bracteoles several; sepals persistent but inconspicuous at fruiting due to spines; petals white to purplish-white, not or slightly zygomorphic; fruits ± not compressed, somewhat longer than wide, variously furnished with curved or hooked spines.

1 Fruits with dimorphic mericarps, 1 with spines, 1 tuberculate; peduncles ≤1cm; rays <5mm, ± hidden by flowers or fruits **3. T. nodosa**
1 Both mericarps with spines; peduncles >1cm; rays >5mm, conspicuous 2
 2 Bracts 0-1; fruits 3-4mm (excl. spines), with ± straight spines minutely hooked at end **2. T. arvensis**
 2 Bracts >2; fruits 2-2.5mm (excl. spines), with curved spines not hooked at end **1. T. japonica**

Other spp. - T. leptophylla (L.) Rchb.f., from S Europe, is a slender annual to 40cm similar to *T. arvensis* but with shorter peduncles, 2-3(4) rays (not 3-6(12)) and 0 bracts; it is a rare birdseed-, grain- and wool-alien.

1. T. japonica (Houtt.) DC. - *Upright Hedge-parsley*. Stems erect, to 1.2m; leaves 1-3-pinnate; bracts 4-6(12); petals scarcely zygomorphic; fruits 2-2.5mm; both mericarps with stout, tapering, curved spines; 2n=16. Native; grassy places, hedgerows, wood-borders and -clearings; frequent throughout BI except N & NW Sc. **516**

2. T. arvensis (Huds.) Link - *Spreading Hedge-parsley*. Stems erect, with wide-spreading branches, to 50cm; leaves 1-2-pinnate; bracts 0-1; petals somewhat zygomorphic; fruits 3-4mm; both mericarps with slender, ± straight spines minutely hooked at apex; 2n=12. Probably intrd; weed of arable land; formerly frequent in S & C Br, now rare and mainly casual in S En; Europe. **R** **516**

3. T. nodosa (L.) Gaertn. - *Knotted Hedge-parsley*. Stems procumbent to ascending, **516**

to 50cm; leaves 1-2-pinnate; bracts 0; petals actinomorphic; fruits 2.5-3.5mm, the outer mericarp with minutely hooked, straight, stout spines, the inner tuberculate; 2n=24. Native; arable and barish ground, especially near sea; rather scattered in BI N to SE Sc, mostly in E En.

49. DAUCUS L. - *Carrots*

Glabrous annuals to biennials with strong characteristic smell (especially in crushed root); stems solid; leaves (1)2-3-pinnate; bracts numerous, usually longer than rays, pinnately divided into filiform lobes; bracteoles numerous; sepals small, scarcely visible in fruit; umbel often with 1 dark purple central flower; other petals white, slightly zygomorphic; fruits strongly dorsally compressed, somewhat longer than wide; mericarps each with 2 lateral and 2 secondary dorsal ridges each with row of terminally barbed spines, the 3 primary dorsal ridges alternating with the secondaries and bearing only short weak bristles.

1. D. carota L. - see sspp. for English names. Usually biennial; stems erect to procumbent, scarcely branched to strongly branched with widely spreading branches, to 1m; rays >10; fruits 2-3mm, with stout, ± straight spines. **516**

1 Umbels convex to slightly concave in fruit **c. ssp. gummifer**
1 Umbels very contracted in fruit, very concave 2
 2 Root swollen in 1st year, usually orange; leaves usually bright green **b. ssp. sativus**
 2 Root not swollen, whitish; leaves usually grey-green **a. ssp. carota**

a. Ssp. carota - *Wild Carrot*. Root not swollen, whitish; stems erect, usually narrowly branching; leaves usually dark to rather grey-green, usually thin, hispid-pubescent; umbels very contracted and concave in fruit, with sparsely hispid to subglabrous rays; 2n=18. Native; grassy and rough ground, mostly on chalky soils and near sea (there often very stunted); throughout most of BI but mainly coastal in N & W Br.

b. Ssp. sativus (Hoffm.) Arcang. - *Carrot*. Root swollen in 1st year, usually orange; leaves usually bright green, thin, usually rather sparsely pubescent; stems and umbels as in ssp. *carota*; (2n=18). Intrd; casual in waste places and tips and a relic where planted; scattered over BI; garden origin.

c. Ssp. gummifer (Syme) Hook. f. - *Sea Carrot*. Root not swollen, whitish; stems erect, rarely >25cm, usually widely branching; leaves dark green, usually thick and ± succulent, hispid-pubescent; umbels not contracted in fruit, hence convex to slightly concave, with hispid rays; (2n=18). Native; cliffs, dunes and rocky places by sea; coasts of CI, S & SE Ir, S & W Br from Anglesey to E Kent.

2. D. glochidiatus (Labill.) Fisch., C.A. Mey. & Avé-Lall. - *Australian Carrot*. **516**
Slender annual; stems erect, ± glabrous, to 40cm, little branched; leaves sparsely pubescent; rays <8, very uneven lengthed, slender, each with ≤6 flowers, or sometimes umbels simple; fruits 3-5mm, with dense, rather slender spines; (2n=44). Intrd; rather frequent wool-alien; very scattered in En: Australia.

112. GENTIANACEAE - *Gentian family*

Glabrous annuals to herbaceous perennials; leaves opposite, simple, entire, without stipules, sessile or ± so. Flowers in terminal dichasial cymes, sometimes monochasial cymes or solitary, actinomorphic, bisexual, hypogynous; sepals 4-5(8), fused at least at base; petals 4-5(8), variously coloured, fused into corolla-tube proximally, persistent in fruit; stamens as many as petals, borne on corolla-tube; ovary 1-celled, each cell with many ovules on 2 parietal placentas; styles 1-2, if 1 usually divided into 2 distally, sometimes ± 0; stigmas (1-)2, capitate to variously expanded; fruit a capsule dehiscing into 2 valves.

Easily recognized by the opposite, entire, glabrous leaves, 4-5(8) fused petals with as many stamens borne on corolla-tube, and 1-celled ovary with many ovules forming 2-celled capsule.

1 Petals purple to blue, rarely white, very rarely pinkish; style and stigmas persistent in fruit 2
1 Petals pink, yellow or white; stigmas and sometimes style falling before fruiting 3
  2 Corolla with small inner lobes alternating with main ones; distal part of calyx-tube membranous between the calyx-lobe origins **6. GENTIANA**
  2 Corolla without small inner lobes but with long fringes; calyx-tube without membranous part **5. GENTIANELLA**
3 Pairs of stem-leaves fused at base; flowers 6-8-merous **4. BLACKSTONIA**
3 Stem-leaves not fused in pairs; flowers 4-5-merous 4
  4 Calyx-lobes shorter than calyx-tube; corolla yellow **1. CICENDIA**
  4 Calyx-lobes longer than calyx-tube; corolla pink or white 5
5 Calyx-lobes (4-)5, keeled; corolla-lobes (4-)5, ≥2mm; anthers twisting after flowering **3. CENTAURIUM**
5 Calyx-lobes 4, not keeled; corolla-lobes 4, <2mm; anthers not becoming twisted **2. EXACULUM**

1. CICENDIA Adans. - *Yellow Centaury*
Annuals; flowers 4-merous; calyx-lobes triangular, shorter than calyx-tube; corolla yellow; anthers not becoming twisted; style simple; stigma 1, peltate.

1. C. filiformis (L.) Delarbre - *Yellow Centaury*. Stems erect, very slender, to 10(18)cm; leaves few, linear, ≤6mm; corolla 3-7mm, opening only in the sun; (2n=26). Native; damp sandy and peaty barish ground mostly near coast; very local in W Ir, S & SW Br N to Caerns and E to E Sussex, formerly (to 1928) W Norfolk and Lincs. R

2. EXACULUM Caruel - *Guernsey Centaury*
Annuals; flowers 4-merous; calyx-lobes linear, flat, longer than calyx-tube; corolla pale pink; anthers not becoming twisted; style 1, divided near apex; stigmas 2.

1. E. pusillum (Lam.) Caruel - *Guernsey Centaury*. Stems procumbent to ascending, very slender, to 4(10)cm; leaves linear, ≤7mm; corolla 3-6mm, with lobes c.1.5mm; (2n=20). Native; with *Cicendia* in short ± open turf in dune-slacks; few places in Guernsey. RR

3. CENTAURIUM Hill - *Centauries*
Annuals, biennials or perennials; flowers (4-)5-merous; calyx-lobes linear, keeled, longer than calyx-tube; corolla pink or white; anthers becoming twisted at fruiting; style 1, divided near apex; stigmas 2.

1 Perennials with procumbent to decumbent non-flowering stems; corolla-lobes >7mm **1. C. scilloides**
1 Annuals to biennials without procumbent to decumbent non-flowering stems; corolla-lobes ≤7mm 2
  2 Usually biennials, normally with basal leaf-rosette at flowering; flowers with 1-2 bracts at base of calyx, the stalk between 0-1mm 3
  2 Usually annuals, normally without basal leaf-rosette at flowering; flowers with stalks 1-4mm between base of calyx and bracts 4
3 Stem-leaves narrowly oblong-elliptic, almost parallel-sided, obtuse to rounded at apex; calyx usually >3/4 as long as corolla-tube; stigmas broadly rounded to nearly flat at apex **3. C. littorale**

3 Stem-leaves ovate to elliptic, acute to subacute at apex; calyx usually <3/4 as long as corolla-tube; stigmas narrowly rounded to nearly conical at apex **2. C. erythraea**

4 Main stem with 2-4 internodes; all branches arising at c.30-45 degrees and forming rather open inflorescence; corolla usually pink **4. C. pulchellum**

4 Main stem with 5-9 internodes; upper branches arising at c.20-30 degrees and forming rather dense inflorescence; corolla usually white **5. C. tenuiflorum**

1. C. scilloides (L.f.) Samp. (*C. portense* (Brot.) Butcher) - *Perennial Centaury.* RR
Perennial with procumbent to decumbent sterile stems and ascending flowering stems to 30cm; leaves on sterile stems shortly petiolate, broadly elliptic; flowers in groups of 1-6; corolla-lobes 6-9mm; 2n=20. Native; grassy cliff-tops and dunes by sea in Pembs and W Cornwall, formerly E Cornwall, natd as lawn-weed from garden plants in W Kent and E Sussex.

2. C. erythraea Rafn (C. *latifolium* (Sm.) Druce, C. *capitatum* (Willd. ex Cham.) Borbás, *C. minus* auct. non Moench) - *Common Centaury.* Erect biennial (rarely annual) to 50cm; flowers numerous in rather dense to tightly congested inflorescences; corolla-lobes 4.5-6mm; 2n=40. Native; grassy and rather open ground on well-drained soils; frequent throughout BI except local and mostly coastal in Sc. Very variable, often dwarfed in exposed places. Very dwarf plants with very dense inflorescences and stamens inserted at base (not apex) of corolla-tube (var. **capitatum** (Willd. ex Cham.) Melderis) occur on the coast locally in En and Wa. **C. latifolium** was known from the coast of S Lancs from 1804 to 1872 (endemic). It differed in its corolla-lobes only 3-4mm and its broadly elliptic to suborbicular basal leaves; like var. *capitatum*, it was probably a mutant of *C. erythraea*.

2 x 3. C. erythraea x C. littorale = C. x intermedium (Wheldon) Druce occurs with the parents on coastal dunes in S & W Lancs, Anglesey and Merioneth. F1 plants are intermediate in all 3 key characters and like both parents have 2n=40; they have low fertility. In Lancs backcrossing to both parents has occurred, producing fertile plants with 2n=40 appearing as introgressed C. *erythraea* and, in other populations, fertile plants with 2n=60 closer to C. *littorale*. The latter could be treated as a distinct new sp., but the parents themselves are so close that this is not feasible. **C. intermedium** strictly refers to this taxon.

2 x 4. C. erythraea x C. pulchellum occurs with the parents on the coast in N Somerset, S Essex and W Lancs; it is intermediate in all characters, notably pedicel and corolla-lobe lengths, and highly fertile

3. C. littorale (Turner ex Sm.) Gilmour (*C. minus* Moench) - *Seaside Centaury.* Erect R
biennial to 26cm; flowers in rather dense inflorescences; corolla-lobes 5-6.5mm; 2n=40. Native; coastal dunes and sandy turf; local in Br N from S Wa and NE Yorks, Co Londonderry. Leaf-shape is diagnostic and differs from that of even the most narrow-leaved variants of *C. erythraea*, to which records from S Hants refer.

4. C. pulchellum (Sw.) Druce - *Lesser Centaury.* Erect annual to 20cm but often <6cm; smallest plants unbranched with 1-few flowers, larger ones with wide-spreading branches and diffuse inflorescence; corolla-lobes 2-4mm; 2n=c.34, 36. Native; woodland rides, damp grassy or ± open ground, especially near sea; local in Br N to Cumberland and NE Yorks, CI, S & E Ir.

5. C. tenuiflorum (Hoffmanns. & Link) Fritsch - *Slender Centaury.* Erect annual to RRR
35cm; branches usually many, the upper usually rather narrowly divergent, forming rather dense inflorescence; corolla-lobes 2-4mm; (2n=40). Native; damp grassy places near sea; very rare in Dorset, formerly Wight, CI pre-1840. Very similar to C. *pulchellum* and specific distinctness needs confirming, but our plants normally have white corollas.

4. BLACKSTONIA Huds. - *Yellow-wort*
Annuals; flowers 6-8-merous; calyx divided almost to base into linear, flat lobes; corolla yellow; anthers not becoming twisted; style 1, divided near apex; stigmas 2.

1. B. perfoliata (L.) Huds. - *Yellow-wort*. Stems erect, to 50cm, glaucous; stem-leaves triangular-ovate, pairs fused round stem at base, glaucous; corolla-lobes longer than tube, 5-10mm; 2n=40. Native; calcareous grassland, bare chalk and dunes; locally frequent in BI N to Co Sligo, Westmorland and S Northumb.

5. GENTIANELLA Moench - *Gentians*
Annuals to biennials; flowers 4-5-merous; calyx-lobes shorter or longer than tube; corolla blue or dark- to whitish-purple, rarely pink, the lobes with long fringes at margins or at base on inner face; anthers not becoming twisted; style scarcely distinct; stigmas 2.

Spp. 3-6 are variable in size and habit. The existence of hybrids and of diminutive annuals (with smaller than normal floral parts) in normally biennial taxa can make determination difficult.

1 Corolla-lobe with long narrow fringes along sides, not at base on inner side **1. G. ciliata**
1 Corolla-lobes with long narrow fringes at base on inner side, not along sides 2
  2 Flowers 4-merous; calyx with 2 lobes several times wider than other 2 lobes **2. G. campestris**
  2 Flowers 4-5-merous (often both on same plant); calyx with 4-5 lobes, the widest ≤2x as wide as others 3
3 Corolla (15)25-35mm, ≥2x as long as calyx; plant with 9-15 internodes **3. G. germanica**
3 Corolla 12-22mm, ≤2x as long as calyx; plant with 2-11 internodes 4
  4 Internodes (2)4-9(11); apical pedicel <1/4 total height to pedicel apex **4. G. amarella**
  4 Internodes 0-3(5); apical pedicel usually ≥1/2 total height to pedicel apex (except in Cornwall and N Devon) 5
5 Stem-leaves lanceolate to oblong- or linear-lanceolate; calyx-lobes ± equal, appressed to corolla **5. G. anglica**
5 Stem-leaves ovate to narrowly ovate; some calyx-lobes usually distinctly longer and wider than others, somewhat divergent from corolla **6. G. uliginosa**

1. G. ciliata (L.) Borkh. (*Gentiana ciliata* L.) - *Fringed Gentian*. Erect biennial to **RRR**
30cm; corolla 25-40mm, blue, with ovate- to rhombic-oblong lobes 10-18mm; **522**
corolla-tube widest at apex; (2n=44). Possibly native; chalk grassland; 1 place in Bucks, reported in 1875, disregarded, refound 1982, 1 record (1892) in S Wilts; Europe.

2. G. campestris (L.) Börner (*Gentiana campestris* L.) - *Field Gentian*. Erect annual **522**
or biennial to 30cm; corolla 15-25(30)mm, bluish-purple to ± white, with narrowly oblong-ovate lobes 6-11mm; corolla-tube ± cylindrical; (2n=36). Native; grassland and dunes; scattered over Br and Ir, locally common in N, absent from most of S Ir and S & C Br.

3. G. germanica (Willd.) Börner (*Gentiana germanica* Willd.) - *Chiltern Gentian*. **R**
Erect (annual or) biennial to 40cm; corolla (15)25-35mm, bright bluish-purple, with **522**
narrowly triangular-ovate lobes 6-11mm; corolla-tube widest at apex; (2n=36). Native; chalk grassland, mostly sheltered or among scrub; very local in SC En from N Hants to Herts and Beds, formerly more widespread.

3 x 4. G. germanica x G. amarella = G. x pamplinii (Druce) E.F. Warb. is intermediate in corolla-shape, -size and -colour and is ≥50% fertile; it occurs near

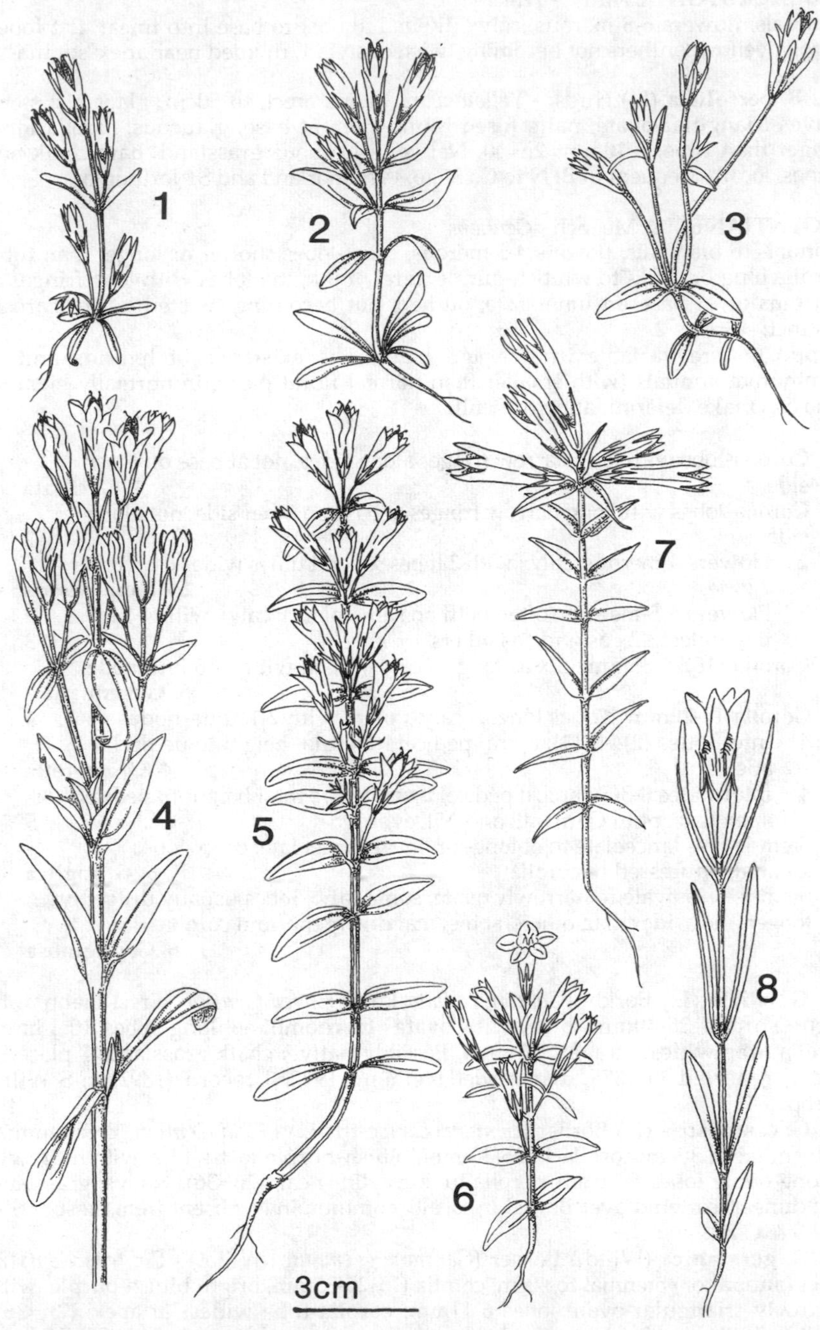

FIG 522 - *Gentianella*. 1-2, *G. anglica*. 1, ssp. *anglica*. 2, ssp. *cornubiensis*. 3, *G. uliginosa*. 4, *G. campestris*. 5-6, *G. amarella*. 5, ssp. *amarella*. 6, ssp. *septentrionalis*. 7, *G. germanica*. 8, *G. ciliata*.

most populations of *G. germanica*. Backcrossing to *G. amarella* occurs, and introgressed *G. amarella* colonies exist from E Kent to Oxon, partly outside the present area of *G. germanica*.

4. G. amarella (L.) Börner (*Gentiana amarella* L.) - *Autumn Gentian*. Erect (annual or) biennial to 30cm; corolla 12-22mm, with narrowly oblong-ovate lobes 4-7mm; corolla-tube ± cylindrical; (2n=36). Native; basic pastures and dunes. Flowering Jul-Oct.

1 Corolla creamy-white, suffused purplish-red on outside, (12)14-17mm **c. ssp. septentrionalis**
1 Corolla usually dull purple, rarely pale blue, pink or whitish, (14)16-22mm 2
  2 Corolla (14)16-18(20)mm; Britain **a. ssp. amarella**
  2 Corolla (17)19-22mm; Ireland **b. ssp. hibernica**

a. Ssp. amarella. Stems with 4-9(11) internodes. Locally frequent in Br S from 522
Angus and S Ebudes.

b. Ssp. hibernica N.M. Pritch. Stems with 7-11 internodes; leaves usually R
narrower and less tapering than in ssp. *amarella*. Frequent over most of Ir; endemic.

c. Ssp. septentrionalis (Druce) N.M. Pritch. (ssp. *druceana* N.M. Pritch., *G.* R
septentrionalis (Druce) E.F. Warb.). Stem with 2-7 internodes. Locally frequent in Sc 522
N from N Aberdeen and M Perth, outlying sites in Cheviot and MW Yorks; probably endemic. Ssp. **druceana** usually has 2-5(6) (not 6-7) internodes and narrower leaves; it is more eastern in distribution but is doubtfully worth ssp. recognition.

4 x 5. G. amarella x G. anglica occurs with the parents in W Cornwall, N Devon, Wight, E Sussex and W Kent and is fertile, showing all grades of intermediacy; endemic.

4 x 6. G. amarella x G. uliginosa occurs with most colonies of *G. uliginosa* in S Wa and is fertile, showing all grades of intermediacy, sometimes pure *G. uliginosa* becoming rare or absent.

5. G. anglica (Pugsley) E.F. Warb. - *Early Gentian*. Erect (annual or) biennial to RRR
20cm, usually much less; corolla 13-20mm, dull purple to whitish, with narrowly oblong-ovate lobes 5-8mm; corolla-tube ± cylindrical. Native; endemic. Flowering Mar-Jul(Aug).

a. Ssp. anglica. Stem with 2-3(4) internodes, the uppermost usually longer than 522
others; terminal pedicel usually ≥1/2 total height to pedicel apex; corolla 13-16mm. Chalk and limestone grassland and fixed dunes; local in C & S En from S Devon to S Lincs and E Kent.

b. Ssp. cornubiensis N.M. Pritch. Stem with 3-5 ± equal internodes; terminal 522
pedicel <1/3 total height to pedicel apex; corolla (15)17-20mm. Cliff-top turf and fixed dunes; N W Cornwall and N Devon. There is good circumstantial evidence that this taxon is in fact referable to *G. anglica* x *G. amarella*.

6. G. uliginosa (Willd.) Börner (*Gentiana uliginosa* Willd.) - *Dune Gentian*. Erect RRR
annual or biennial to 15cm, with 0-2(3) internodes; terminal pedicel >1/2 total 522
height to pedicel apex; corolla 9-22mm, dull purple, with narrowly oblong-ovate lobes 3-7mm; corolla-tube ± cylindrical; (2n=c.54). Native; coastal dunes and dune-slacks; Glam, Carms and Pembs, perhaps formerly N Devon.

6. GENTIANA L. - *Gentians*

Annuals to perennials; flowers 5-merous; calyx-lobes shorter than to c. as long as tube, with small membranous connexion at base of sinuses; corolla blue, rarely white or pink, the lobes not fringed, with 5 small lobes alternating with the 5 large ones; anthers, styles and stigmas as in *Gentianella*.

1 Corolla-tube >1cm wide, widening distally; leaves all or most >15(20)mm 2
1 Corolla-tube <8mm wide, ± cylindrical; leaves all <15mm 4
  2 Leaves crowded in basal rosette, few reduced ones up stem **3. G. clusii**

2 Leaves spread ± evenly up stem 3
3 Leaves linear, <1cm wide, with 1 vein **2. G. pneumonanthe**
3 Leaves lanceolate (to ovate), >1cm wide, with 3-5 veins **1. G. asclepiadea**
4 Rhizomatous perennial with several rosettes of leaves; corolla lobes >8mm **4. G. verna**
4 Annual, with or without 1 basal leaf-rosette; corolla-lobes <6mm **5. G. nivalis**

1. G. asclepiadea L. - *Willow Gentian*. Perennial; stems erect, to 60cm, with 1-30 flowers; leaves lanceolate to ovate, 2-12cm; corolla blue, usually with purple spots inside, 35-50mm; (2n=32, 36, 44). Intrd; grown in gardens and self-sowing freely, natd by streams and in shady places; E & W Sussex, MW Yorks and C Sc; C Europe.

2. G. pneumonanthe L. - *Marsh Gentian*. Perennial; stems erect, to 40cm, with 1-10(28) flowers; leaves linear, 15-40mm; corolla blue with 5 green lines outside, 25-50mm; (2n=26). Native; wet heathland; very local in Br from Dorset and E Sussex (formerly W Kent) to NE Yorks and Westmorland (formerly Cumberland), decreasing. R

3. G. clusii E.P. Perrier & Songeon - *Trumpet Gentian*. Perennial; stems erect, to 6cm, each with 1 flower; basal leaves lanceolate to oblanceolate, 15-40mm; corolla blue, 40-70mm; (2n=36). Intrd; planted and natd since 1960 in chalk grassland in 3 nearby places in Surrey; Alps.

4. G. verna L. - *Spring Gentian*. Perennial; stems erect, 0.5-7cm, each with 1 flower; basal leaves ovate to elliptic, 8-15(20)mm; corolla-tube 15-25mm; corolla-lobes patent at anthesis, forming brilliant blue limb 17-31mm across; 2n=28. Native; grassland on limestone, calcareous glacial drift and fixed dunes; extremely local in N En and W Ir. RRR

5. G. nivalis L. - *Alpine Gentian*. Annual; stems erect, to 15cm, with 1-10 flowers; leaves 2-10mm, the basal ones ovate to obovate, the upper ones narrower; corolla-tube 10-15mm; corolla-lobes patent at anthesis, forming brilliant blue limb 7-10mm across; (2n=14). Native; rock-ledges above 730m; very local and rare in M Perth and Angus. RRR

113. APOCYNACEAE - *Periwinkle family*

Slightly woody perennials; leaves evergreen, opposite, simple, entire, shortly petiolate, without stipules. Flowers solitary in leaf-axils, actinomorphic, bisexual, hypogynous; sepals 5, fused at base; petals 5, blue, rarely white, fused into corolla-tube proximally; stamens 5, inserted on corolla-tube, not exserted; ovaries 2, free, each with many ovules, united by common single style; stigma 1, capitate-peltate, complexly ornamented; fruit rarely produced, of 2 follicles.

The opposite, evergreen entire leaves and blue flowers of characteristic shape are diagnostic.

1. VINCA L. - *Periwinkles*

1 Margin of leaves and calyx-lobes minutely pubescent **3. V. major**
1 Margin of leaves and calyx-lobes glabrous 2
2 Corolla-tube 9-11mm; corolla-limb 25-30mm across; calyx-lobes 3-4(5)mm **1. V. minor**
2 Corolla-tube 12-18mm; corolla-limb 30-45mm across; calyx-lobes 5-14mm **2. V. difformis**

Other spp. - Records of the entirely herbaceous **V. herbacea** Waldst. & Kit., from E Europe, are errors for variants of *V. major*, which often is cut to the ground by frosts, or of *V. minor*.

1. V. minor L. - *Lesser Periwinkle.* Vegetative stems procumbent to arching, rooting at tips, to 1m; leaves ovate to elliptic or narrowly so, 15-45mm; flowering stems erect to ascending, to 20cm; flower dimensions as in key; corolla-lobes usually sky-blue, obliquely truncate; 2n=46. Probably intrd; well natd in woods, hedgebanks and other shady places; scattered throughout most of BI; Europe.

2. V. difformis Pourr. - *Intermediate Periwinkle.* Similar to *V. minor* in habit but more robust, with vegetative stems to 2m and flowering stems to 30cm; leaves lanceolate to ovate, 25-70mm; flower dimensions as in key; corolla-lobes usually pale blue, very obliquely truncate to acute; (2n=46). Intrd; natd on bank in W Kent and (white-flowered) W Cornwall; SW Europe.

3. V. major L. - *Greater Periwinkle.* Vegetative stems ascending-arching then often procumbent, to 1.5m; leaves ovate, 25-90mm; flowering stems as in *V. difformis*; calyx-lobes 7-17mm; corolla-tube 12-15mm; corolla-limb 30-50mm across; corolla-lobes usually purplish-blue, obliquely truncate (violet-blue, narrower and acute in var. **oxyloba** Stearn (ssp. *hirsuta* auct. non (Boiss.) Stearn)); 2n=92. Intrd; natd in hedgebanks, shrubberies and rough ground; most of BI, frequent in S, rare in N; Mediterranean. Var. *oxyloba* rarely natd in S Br.

114. SOLANACEAE - *Nightshade family*

Annuals to herbaceous perennials or shrubs; leaves alternate, rarely opposite, simple (entire to deeply lobed) or pinnate, without stipules, usually petiolate. Flowers solitary or in axillary or terminal cymes or racemes, actinomorphic or slightly zygomorphic, bisexual, hypogynous; sepals 5, fused into tube proximally, sometimes 2-lipped; petals 5, fused into tube proximally, variously coloured; stamens 5, borne on corolla-tube; ovary 2(-5)-celled, each cell with many ovules on axile placentas; style 1; stigmas 1-2, ± capitate; fruit a berry (sometimes ± dry) or 2(-4)-valved capsule.

Distinguished from all but the very distinctive *Verbascum* in its ± actinomorphic 5-merous flowers with fused calyx and corolla and usually 2-celled ovary with many ovules.

1 Open flowers with anthers touching laterally, forming cone-shaped group around style — 2
1 Open flowers with anthers separated laterally, not forming cone-shaped group around style — 4
  2 Stamens opening by apical pores; fruit succulent; if corolla yellow then plant spiny — **10. SOLANUM**
  2 Stamens opening by longitudinal slits; if fruit very succulent then corolla yellow and plant not spiny — 3
3 Leaves simple; corolla white to purple; fruit rather dry — **8. CAPSICUM**
3 Leaves pinnate; corolla yellow; fruit very succulent — **9. LYCOPERSICON**
  4 Woody shrub; corolla blue to purple, rarely white; fruits red or yellowish-green — 5
  4 Stems herbaceous, or if woody towards base then flowers and fruits both whitish — 6
5 Stems usually spiny; corolla <2cm, divided c.1/2 way; stamens exserted — **2. LYCIUM**
5 Stems not spiny; corolla >2cm, divided much <1/2 way; stamens not exserted — **3. IOCHROMA**
  6 Calyx toothed, the teeth <1/4 total length at flowering — 7
  6 Calyx lobed, ≥1 lobe ≥1/3 total length at flowering (often not so at fruiting) — 8

7 Calyx in fruit funnel-shaped, persistent, c.2x as long as fruit; capsule opening by lid; corolla often with network of dark veins **5. HYOSCYAMUS**

7 Calyx in fruit tubular to bell-shaped, withering, <2x as long as fruit; capsule opening by longitudinal valves; corolla without network of dark veins **11. DATURA**

  8 Fruit a capsule; corolla tubular to trumpet-shaped 9

  8 Fruit a berry; corolla cup-, bowl-, bell- or star-shaped 10

9 Calyx-teeth ≥3/4 total length of calyx; flowers solitary in leaf-axils **13. PETUNIA**

9 Most or all calyx teeth ≤2/3 total length of calyx; flowers opposed to or in axils of much reduced bracts **12. NICOTIANA**

  10 Fruit ± completely enclosed in enlarged calyx 11

  10 Fruit well exposed from calyx 12

11 Fruiting calyx with lobes much longer than tube; ovary and fruit 3-5-celled **1. NICANDRA**

11 Fruiting calyx with tube much longer than lobes; ovary and fruit 2-celled **7. PHYSALIS**

  12 Corolla brownish- to greenish-purple, >20mm; largest leaves >5cm **4. ATROPA**

  12 Corolla whitish, <12mm; largest leaves <5cm **6. SALPICHROA**

Other genera - SCOPOLIA Jacq. resembles a dwarf (to 60cm) *Atropa* with reddish-brown or sometimes yellow pendent flowers, but has a shallowly lobed calyx and a capsule with a lid. **S. carniolica** Jacq., from C Europe, is grown in gardens and can exist after neglect as does 1 small clump in Surrey; it has erect, ± unbranched stems to only 60cm. The genus is often placed in Scrophulariaceae. **VESTIA** Willd. is an evergreen shrub with pendent, tubular, pale yellow flowers. **V. foetida** (Ruiz & Pav.) Hoffmanns. (*V. lycioides* Willd.) (*Huevil*), from Chile, is grown in Cornwall and may persist in hedges.

1. NICANDRA Adans. - *Apple-of-Peru*
Glabrous annuals; leaves simple, toothed to ± lobed; flowers solitary, axillary; calyx deeply 5-lobed, later enlarging and enclosing fruit; corolla bell-shaped, shallowly lobed; ovary 3-5-celled; fruit a rather dry berry.

1. N. physalodes (L.) Gaertn. - *Apple-of-Peru*. Stems erect, to 80cm; corolla 25- 529
40mm long and across, blue to mauve; fruit globose, brownish, 12-20mm; fruiting calyx 25-35mm; (2n=20). Intrd; frequent casual in waste and cultivated ground and on tips, also alien from wool and birdseed, sometimes persistent; scattered in CI and Br, mainly S; Peru.

2. LYCIUM L. - *Teaplants*
Almost glabrous, usually spiny, deciduous shrubs with arching branches; leaves simple, entire; flowers axillary, 1-few together; calyx irregularly 2-lipped, not enclosing fruit; corolla funnel-shaped, rather deeply lobed; ovary 2-celled; fruit a berry.

1. L. barbarum L. (*L. halimifolium* Mill.) - *Duke of Argyll's Teaplant*. Scrambling shrub to 2.5m; leaves narrowly elliptic, widest near middle, 2-10cm; calyx c.4mm; corolla purplish, 10-15mm, divided to c.1/2 way; berry bright red, ellipsoid, 10-20mm; (2n=24). Intrd; grown as hedging and natd in rough ground, hedges and on walls; frequent in CI, En, Wa and Man, very scattered in Ir and Sc; China.

2. L. chinense Mill. - *Chinese Teaplant*. Differs from *L. barbarum* in leaves lanceolate to narrowly ovate, widest below middle; calyx c.3mm; corolla 10-15mm, divided to >1/2 way; (2n=24). Intrd; similar habitats to *L. barbarum*; China. Much confused with *L. barbarum*, distinction from it and and relative abundance extremely uncertain.

3. IOCHROMA Benth. - *Argentine-pear*
Pubescent, non-spiny, deciduous shrubs; leaves simple, entire; flowers axillary, (1)2-5 together; calyx bell-shaped, with 5 lobes, enlarging but not enclosing fruit; corolla narrowly bell-shaped, with 5 shallow lobes; ovary 2-celled; fruit a berry.

1. I. australe Griseb. (*Acnistus australis* (Griseb.) Griseb.) - *Argentine-pear*. Shrub to 5m; leaves narrowly elliptic, 4-8cm; corolla 25-45cm, deep blue, sometimes white; fruit globose, c.12mm across, yellowish-green; (2n=24). Intrd; grown for pharmaceutical purposes since at least 1975, now natd in waste places; Nottingham (Notts); Argentina.

4. ATROPA L. - *Deadly Nightshade*
Nearly glabrous to glandular-pubescent perennials; leaves simple, entire; flowers solitary, axillary; calyx rather deeply 5-lobed, slightly enlarging later but not enclosing fruit; corolla bell-shaped, shallowly lobed; ovary 2-celled; fruit a berry.

1. A. belladonna L. - *Deadly Nightshade*. Stems erect, to 2m; leaves ovate to elliptic, 8-20cm; corolla 24-30mm, greenish- or brownish-purple; fruit globose to depressed-globose, 15-20mm across, shiny black; 2n=72. Native; woods, scrub, rough and cultivated ground; locally frequent in C & S En, scattered in Ir and elsewhere in Br, probably native only in C & S Br on chalk and limestone.

5. HYOSCYAMUS L. - *Henbane*
Glandular-pubescent stinking annuals to biennials; leaves simple, toothed to ± lobed; flowers solitary, axillary, forming 2 rows on 1 side of stem; calyx funnel-shaped, enlarging later and becoming swollen at base to accommodate fruit, broadly 5-toothed; corolla broadly funnel-shaped, rather deeply lobed; ovary 2-celled; fruit a capsule dehiscing by lid.

Other spp. - H. albus L., from S Europe, differs from *H. niger* in its petiolate stem-leaves and corolla lacking purple veins; it is a rare casual.

1. H. niger L. - *Henbane*. Stems erect, to 80cm; leaves ovate-oblong, those on stem sessile, 6-20cm; corolla 2-3cm, yellowish with strong purple reticulate venation, 2-3cm across; capsule enclosed by calyx, c.1cm, broadly ovoid; 2n=34. Native; maritime sand and shingle, inland rough and waste ground, especially manured by rabbits or cattle; scattered in Br and Ir, mainly C & S.

6. SALPICHROA Miers - *Cock's-eggs*
Pubescent perennials, somewhat woody below; leaves simple, entire, with petiole c. as long; flowers solitary, axillary; calyx cup-shaped, divided nearly to base, not enlarging; corolla bowl-shaped, with rather short lobes; ovary 2-celled; fruit an ovoid berry.

1. S. origanifolia (Lam.) Thell. - *Cock's-eggs*. Stems much-branched, sprawling, to **529**
1.5m; leaves suborbicular to ovate-rhombic or -trullate, 15-25mm; corolla 6-10mm, whitish, with reflexed lobes; berry 10-15mm, whitish; (2n=24). Intrd; grown for ornament, natd in rough ground and open places; few places on S & SE coasts of En since 1927, Guernsey since 1946, rare and impermanent elsewhere; S America.

7. PHYSALIS L. - *Japanese-lanterns*
Subglabrous to pubescent annuals to perennials; leaves simple, entire to coarsely dentate; flowers solitary, axillary; calyx bell-shaped, 5-lobed, the calyx-tube later enlarging to enclose fruit; corolla broadly bell-shaped to funnel-shaped, shallowly

to rather deeply lobed; ovary 2-celled; fruit a globose berry.

1 Fruiting calyx and berry red to orange; corolla whitish **1. P. alkekengi**
1 Fruiting calyx green to yellowish-green; berry yellow, green or purple; corolla yellowish 2
  2 Sparsely pubescent; leaves cuneate to ± rounded at base; annual **3. P. ixocarpa**
  2 Densely pubescent; leaves cordate at base; perennial **2. P. peruviana**

1. P. alkekengi L. (*P. franchetii* Mast.) - *Japanese-lantern*. Rather sparsely 536
pubescent, erect, rhizomatous perennial to 60cm; leaves broadly cuneate to subcordate at base, entire to coarsely dentate; fruiting calyx 2.5-5cm; fruit 12-17mm, not filling calyx; (2n=24). Intrd; grown for ornament, natd on waste land, roadsides and in shrubberies; scattered in En and Wa, mainly S; Europe.

2. P. peruviana L. - *Cape-gooseberry*. Densely pubescent, erect, rhizomatous 536
perennial to 1m; leaves cordate at base, entire to obscurely dentate; fruiting calyx 3-5cm; fruit 12-20mm, yellow, not filling calyx; (2n=24, 48). Intrd; imported as minor fruit and casual on tips, natd in Herts; occasional in S Br and E Ir; S America.

3. P. ixocarpa Brot. ex Hornem. - *Tomatillo*. Sparsely pubescent, erect annual to 536
60cm; leaves cuneate to rounded at base, entire to coarsely dentate; fruiting calyx 3-5cm, often dark-veined; fruit 13-40mm, green to purple, completely filling calyx; (2n=24). Intrd; casual mostly as wool-alien, perhaps sometimes from use as minor fruit; scattered in En and Wa; N & S America. The closely related **P. philadelphica** Lam., with larger flowers and fruit and curved (not straight) anthers, is the sp. more usually used for fruit, but is only a rare casual.

8. CAPSICUM L. - *Sweet Pepper*

Glabrous annuals; leaves simple, entire; flowers solitary, axillary; calyx bell-shaped, shallowly toothed, slightly enlarging; corolla star-shaped, deeply lobed; ovary 2-3(5)-celled; fruit a rather dry ovoid berry, with large cavities when mature.

1. C. annuum L. - *Sweet Pepper*. Stems erect, to 60cm; leaves ovate to narrowly so, 8-15cm; corolla white to purplish, 2-3cm across; fruit green, yellow or red, (1)3-15(25)cm; (2n=12, 24). Intrd; imported as birdseed, as fruit for cooking and, increasingly, as green salad, now occasional casual on tips and sewerage works; scattered in S En; tropical America.

9. LYCOPERSICON Mill. - *Tomato*

Glandular-pubescent annuals; leaves pinnate with mixed large and small leaflets; flowers in leaf-opposed cymes; calyx star-shaped, lobed nearly to base, slightly enlarged and lobes reflexed at fruiting; corolla star-shaped, deeply lobed; ovary 2-3(5)-celled; fruit a succulent, depressed-globose to globose berry.

1. L. esculentum Mill. (*L. lycopersicum* (L.) Karsten) - *Tomato*. Stems erect to decumbent or scrambling, to 2m; leaves 20-40cm, with toothed to lobed leaflets; corolla yellow, 18-25mm across; fruit usually red, rarely yellow to orange, 2-10cm across; (2n=24). Intrd; much used as salad-fruit and vegetable and common on tips and in sewerage works and waste places, grown on field scale in S En and CI; frequent throughout BI; C & S America.

10. SOLANUM L. - *Nightshades*

Annual to perennial herbs or shrubs; leaves simple and entire to pinnate; flowers in axillary or leaf-opposed cymes or solitary; calyx star- to cup-shaped, usually deeply lobed; corolla star-shaped, deeply to scarcely lobed; ovary 2(-4)-celled; fruit a succulent to dry berry.

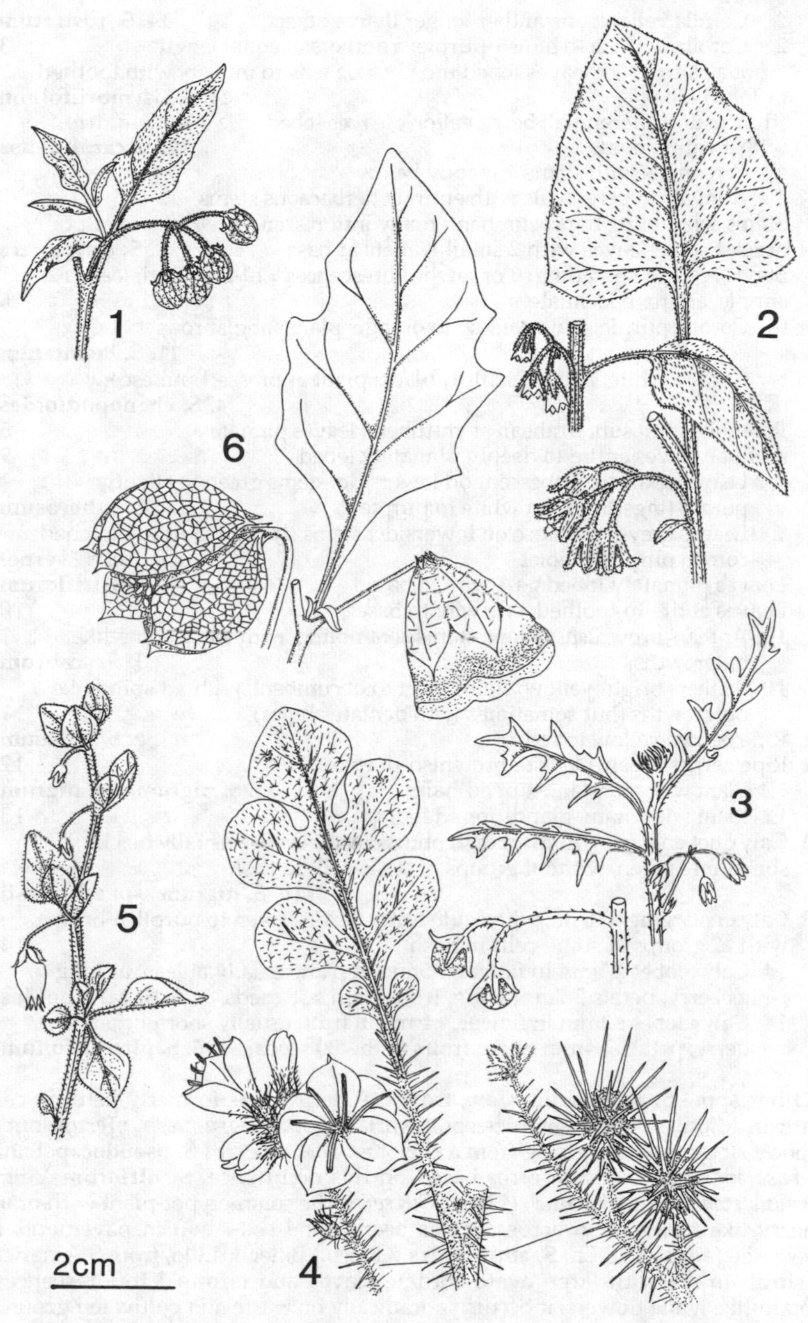

FIG 529 - Solanaceae. 1, *Solanum chenopodioides*. 2, *S. physalifolium*. 3, *S. triflorum*. 4, *S. rostratum*. 5, *Salpichroa origanifolia*. 6, *Nicandra physalodes*.

1 Stems and leaves with strong spines 2

1 Spines 0 4

2 Corolla yellow; one anther longer than 4 others **14. S. rostratum**

2 Corolla whitish to bluish-purple; 5 anthers of equal length 3

3 Annual; berry red; leaves lobed mostly >1/2 way to midrib, with toothed or lobed lobes **13. S. sisymbriifolium**

3 Rhizomatous perennial; berry yellow; leaves lobed <1/2 way to midrib, with ± entire lobes **12. S. carolinense**

4 Perennials with stems ± woody below 5

4 Annuals to perennials with entirely herbaceous stems 7

5 Stems scrambling to procumbent; many inflorescences >10-flowered; at least some leaves with 2 small leaflets at base **8. S. dulcamara**

5 Stems erect to spreading; 0 or few inflorescences >10-flowered; leaves simple, entire to laciniate 6

6 Corolla purple; fruits yellow to orange; plant subglabrous **11. S. laciniatum**

6 Corolla white; fruits purplish-black; plant appressed-pubescent **4. S. chenopodioides**

7 Perennial with subterranean stem-tubers; leaves pinnate 8

7 Annual; leaves entire to deeply pinnately lobed 9

8 Leaves sparsely pubescent on lowerside; stems green to slightly purple-tinged; corolla white to purple **9. S. tuberosum**

8 Leaves grey-tomentose on lowerside; stems strongly purple-blotched; corolla purplish-violet **10. S. vernei**

9 Leaves pinnately lobed >3/4 way to base **7. S. triflorum**

9 Leaves entire to toothed <1/2 way to base 10

10 Anthers brownish-yellow; stems scrambling, with weak spine-like outgrowths **3. S. scabrum**

10 Anthers bright yellow; stems erect to decumbent, without spine-like outgrowths (but sometimes with dentate angles) 11

11 Ripe berries yellow to red **2. S. villosum**

11 Ripe berries green, black or brownish-purple 12

12 Plant without gland-tipped hairs **1a. S. nigrum ssp. nigrum**

12 Plant with many gland-tipped hairs 13

13 Calyx not enlarging in fruit, with obtuse teeth; berries usually black, sometimes green, without groups of stone-cells in flesh **1b. S. nigrum ssp. schultesii**

13 Calyx enlarging in fruit, with acute teeth; berries green to purplish-brown, with ≥2 groups of stone-cells in flesh 14

14 Calyx-lobes ≥3mm in flower, >5mm in fruit, usually at least as long as berry; petals 5-7mm wide; fruits with >50 seeds **6. S. sarachoides**

14 Calyx-lobes ≤2mm in flower, <4mm in fruit, usually shorter than berry; petals 2-4mm wide; fruits with <30 seeds **5. S. physalifolium**

Other spp. - **S. pygmaeum** Cav., from S America, was formerly natd on canal-path in Middlesex; it resembles some variants of *S. nigrum* ssp. *nigrum* but has woody stems to 20cm arising from a rhizome. The record of **S. pseudocapsicum** L. (*Jerusalem-cherry*) was an error based on this occurrence. **S. diflorum** Vell. (*S. capsicastrum* Link ex Schauer) (*Winter-cherry*), is the common pot-plant with orange, cherry-like fruits c.2cm across; it has been found self-sown in pavements and sewerage farms in SE En. **S. abutiloides** (Griseb.) Bitter & Lillo, from S America, is a shrub to 2m with large ovate cordate leaves and terminal inflorescences of potato-like white flowers; it occurs sporadically on waste and cultivated ground in Nottingham, Notts. **S. americanum** Mill. (*Small-flowered Nightshade*), from tropics, resembles *S. nigrum* but has ± umbellate cymes, smaller corolla (5-9mm across) and seeds (1-1.5mm) and erect fruiting pedicels; it is a rare wool- and oilseed-alien.

1. S. nigrum L. - *Black Nightshade*. Sparsely to densely pubescent, erect to decumbent annual to 70cm; leaves entire to coarsely dentate; flowers (3)5-10; corolla white; fruit black, rarely green, slightly depressed-globose, 6-10mm across.

a. Ssp. nigrum. Hairs eglandular, mostly appressed, often sparse; 2n=72. Native; waste and cultivated ground; common in CI, most of En and S Wa, local in rest of En and Wa and Man, very scattered casual in Ir and Sc.

b. Ssp. schultesii (Opiz) Wessely. Hairs frequent to abundant, mostly glandular, mostly patent. Intrd; waste and cultivated ground, locally natd; sporadic in S & E En N to SW Yorks, mainly SE; S Europe.

1 x 3. S. nigrum x S. physalifolium = S. x procurrens A.C. Leslie occurs with the parents in cultivated ground in W Kent, W Suffolk, Cambs and Beds; it is intermediate in calyx characters and has black berries with 0-few seeds; 2n=48; endemic.

2. S. villosum Mill. (*S. luteum* Mill.) - *Red Nightshade*. Differs from *S. nigrum* in flowers 3-5 per cluster; fruit yellow to red, subglobose; (2n=48). Intrd; casual from wool, birdseed and oilseed, very rarely natd; S Europe.

a. Ssp. villosum. Many glandular, patent hairs present; stems with rounded, entire ridges. Infrequent casual in S Br, natd in Nottingham (Notts) from pharmaceutical use.

b. Ssp. miniatum (Bernh. ex Willd.) Edmonds (ssp. *alatum* (Moench) Edmonds, ssp. *puniceum* (Kirschl.) Edmonds, *S. luteum* ssp. *alatum* (Moench) Dostál, *S. miniatum* Bernh. ex Willd.). Hairs eglandular, mostly appressed, often sparse; stems with angled, slightly dentate ridges. Infrequent casual in S Br.

3. S. scabrum Mill. (*S. melanocerasum* All., *S. nigrum* var. *guineense* L.) - *Garden Huckleberry*. Differs from *S. nigrum* in stems usually scrambling, sparsely pubescent, to 1m, with strongly dentated to weakly spinous angles; fruit shiny-black, 12-17mm across; (2n=48). Intrd; rare casual from fruit or cultivated plants, on tips, sewerage farms and fields spread with sludge; S En, natd in Herts; ?Africa. The brownish-yellow rather than bright yellow anthers are diagnostic.

4. S. chenopodioides Lam. (*S. sublobatum* Willd. ex Roem. & Schult.) - *Tall* 529
Nightshade. Erect perennial to 1.6m, woody at least below, with ± appressed eglandular hairs; leaves entire to sparsely and shallowly dentate; flowers (1)3-8; corolla white; fruit as in *S. nigrum* but blackish-purple; (2n=24). Intrd; natd on rough ground; Guernsey since 1958, Jersey, London area since 1989, very rare casual elsewhere; S America.

5. S. physalifolium Rusby (*S. nitidibaccatum* Bitter, *S. sarachoides* auct. non 529
Sendtn.) - *Green Nightshade*. Annual with decumbent stems to 1(2)m, to 40cm tall, with many patent glandular hairs; leaves as in *S. nigrum*; flowers (3)4-8(10); corolla white; fruit as in *S. nigrum* but green to purplish-brown, partly concealed by sepals; 2n=16. Intrd; natd in cultivated and waste ground in C & S En and CI, casual from wool and other sources elsewhere in Br; S America. Our plant is var. **nitidibaccatum** (Bitter) Edmonds.

6. S. sarachoides Sendtn. - *Leafy-fruited Nightshade*. Annual with erect to decumbent stems to 2m, to 60cm tall; differs from *S. physalifolium* in fruit green, ± fully enveloped by sepals; and see key; 2n=24. Intrd; casual on tips and waste ground in S En, natd on tip at Dagenham, S Essex; S America.

7. S. triflorum Nutt. - *Small Nightshade*. Diffusely branched, sparsely pubescent 529
annual to 60cm; leaves deeply pinnately lobed; flowers (1)2-3; corolla white; fruit marbled green and white, globose, 10-15mm; (2n=24). Intrd; casual in cultivated and rough ground scattered in En and Wa, natd in W Norfolk and Cheviot; W N America.

8. S. dulcamara L. - *Bittersweet*. Scrambling (procumbent in var. **marinum** Bab.) perennial to 3(7)m, woody below, glabrous to pubescent; leaves ovate to lanceolate, often simple and entire, ± succulent in var. *marinum*, at least some with 2 small lobes or leaflets near base; flowers usually ≥10; corolla usually purple; fruit bright red, ovoid-ellipsoid, 8-12mm; 2n=24. Native; walls, hedges, woods, ditches, fens, pondsides, rough ground and shingle beaches; common throughout lowland BI except N Sc.

9. S. tuberosum L. - *Potato.* Sparsely pubescent herbaceous perennial with erect to decumbent stems to 1m; subterranean stems bearing large terminal tubers (the only overwintering part); leaves pinnate, with mixed large and small leaflets; flowers (20)25-35(40)mm across, few to many; corolla white to purple or mauve; fruit greenish to purplish, depressed-globose, 2-4cm across; (2n=48). Intrd; much cultivated vegetable, casual and often persistent on tips, waste ground and in fields as a relic, rarely reproducing by seed; scattered throughout BI; S America.

10. S. vernei Bitter & Wittm. - *Purple Potato.* Differs from *S. tuberosum* in stems to 1(2)m, strongly purple-blotched; leaflets grey-pubescent on lowerside; flowers (25)35-45(50)mm across, corolla purplish-violet; (2n=24). Intrd; weed in flower-beds and by paths in Berks, reproducing by seed and tubers; S America.

11. S. laciniatum Aiton (*S. aviculare* auct. non G. Forst.) - *Kangaroo-apple.* Glabrous (or ± so) shrub to 1(2)m; leaves entire and lanceolate, and deeply lobed and ovate (the latter commoner) on same plant; flowers 3-12; corolla purple; fruits yellow to orange, ellipsoid, 14-25mm; (2n=48, 92). Intrd; rough ground, tips and maritime sand, mostly casual but sometimes natd; W Cornwall and CI; Australia.

12. S. carolinense L. - *Horse-nettle.* Erect perennial to 80(100)cm, with branched hairs; rather sparse strong spines present on stems and leaves; leaves ovate, shallowly lobed; flowers 3-8; corolla white to purple; fruit yellow, globose, 10-15mm; (2n=24). Intrd; tips and waste places, or frequent soyabean-alien; S En; N America. Some plants so-recorded might be **S. dimidiatum** Raf., from Australia.

13. S. sisymbriifolium Lam. - *Red Buffalo-bur.* Erect annual to 1m, with branched and glandular simple hairs; many spines present on stems, leaves and calyx; leaves deeply pinnately lobed; flowers (1)2-10; corolla white or purplish; fruit red, globose, 12-20mm, enclosed but not completely obscured by calyx; (2n=24). Intrd; waste places and cultivated ground, casual from wool and other sources; scattered in S Br; S America.

14. S. rostratum Dunal (*S. cornutum* auct. non Lam) - *Buffalo-bur.* Similar in **529**
appearance to *S. sisymbriifolium* but stems to 60cm; glandular hairs 0; corolla yellow; fruit entirely concealed by calyx; and see key; (2n=24). Intrd; rather frequent casual from wool, birdseed and other sources in arable fields, tips and waste places; scattered in En, Wa and CI; N America.

11. DATURA L. - *Thorn-apples*

Glabrous to sparsely pubescent annuals; leaves simple, coarsely toothed to lobed; flowers solitary, axillary; calyx tubular, with 5 teeth; corolla trumpet-shaped, very shallowly lobed; ovary 2-celled towards apex, 4-celled towards base; fruit a usually spiny capsule dehiscing by 4 valves.

1. D. stramonium L. (*D. tatula* L., *D. inermis* Juss. ex Jacq.) - *Thorn-apple.* Stems erect, to 1(1.5)m; leaves 5-18cm, elliptic-ovate; calyx 3-5cm, with teeth (3)5-10mm; corolla 5-10cm, white (purple in var. **chalybaea** W.D.J. Koch (var. *tatula* (L.) Torr.)); capsule (2.5)3.5-7cm incl. spines, with slender spines 2-15mm (spineless in var. **inermis** (Juss. ex Jacq.) Schinz & Thell.); (2n=24). Intrd; casual on tips and waste and cultivated ground, especially manured places, from several sources incl. birdseed, wool and soyabean; sporadic ± throughout BI; America.

2. D. ferox L. - *Angels'-trumpets.* Differs from *D. stramonium* in calyx 2.5-4cm, with teeth 3-5mm; corolla 4-6cm, white; capsule 5-8cm incl. spines, with stout spines 1-3cm, some broad-based and >2cm; (2n=24). Intrd; mainly wool-alien on tips and in fields; occasional casual in En and Wa (mainly S); E Asia.

12. NICOTIANA L. - *Tobaccos*

Glandular pubescent, ± sticky (except *N. goodspeedii*) annuals; leaves simple, ± entire; flowers in terminal panicle- or raceme-like cymes; calyx tubular, with 5 unequal lobes c.1/4-2/3 total calyx length; corolla tubular with expanded limb to trumpet-shaped; ovary 2-celled; fruit a capsule with 2 short valves, each 2-lobed at apex.

1 Petioles not winged **1. N. rustica**
1 Petioles broadly winged, the wing clasping stem 2
2 Inflorescence a cymose panicle; corolla-tube with distal wider part c.1/4 total length **2. N. tabacum**
2 Inflorescence a simple raceme-like cyme; corolla-tube with distal wider part ≤1/8 total length 3
3 Corolla-limb red to purple on upperside; filaments inserted in basal 1/2 of corolla-tube **4. N. forgetiana**
3 Corolla-limb white on upperside; filaments inserted in apical 1/2 of corolla-tube **3. N. alata**

Other spp. - **N. sylvestris** Speg. & Comes, from Argentina, is also grown in gardens but less commonly and is a rare casual; it has large white flowers in short dense panicles clustered at stem apex. **N. goodspeedii** H.-M. Wheeler, from Australia, differs from all others in its ± glabrous leaves and stems below inflorescence and narrowly winged non-clasping petioles; it has corolla-tube 8-20mm and -limb 5-12mm across and is a rare wool-alien.

1. N. rustica L. - *Wild Tobacco*. Erect annual to 1.5m; leaves with unwinged petioles; inflorescence a cymose panicle; 4 calyx-lobes shorter than -tube, 1 c. as long; corolla-tube 12-17mm; corolla-limb 9-16mm across, scarcely lobed, greenish-yellow; (2n=48). Intrd; once grown for tobacco, now occasional casual on tips; scattered in S En; N America.

2. N. tabacum L. - *Tobacco*. Erect annual to 2(3)m; leaves with winged petioles, the wings clasping stem; inflorescence a cymose panicle; calyx-lobes c. as long as -tube or slightly shorter; corolla-tube 30-55mm; corolla-limb 20-30mm across, lobed c.1/2 way, whitish to dingy red; (2n=48). Intrd; rarely grown for tobacco, now rare casual on tips and as relic, once much commoner; scattered in S En; S & C America.

3. N. alata Link & Otto - *Sweet Tobacco*. Erect annual to 1.5m; leaves with winged petioles, the wings clasping stem; inflorescence a simple raceme-like cyme; calyx-lobes c. as long as -tube or somewhat longer; corolla-tube 50-100mm; corolla-limb 35-60mm across, lobed ≥1/2 way, white; (2n=18). Intrd; much grown for ornament and frequent casual on tips and rough ground; scattered in S Br; S America.

3 x 4. N. alata x N. forgetiana = N. x sanderae W. Watson is grown in gardens like its parents and similarly occurs as a casual; it is intermediate in all characters, with a range of flower colours.

4. N. forgetiana Hemsl. - *Red Tobacco*. Differs from *N. alata* in stems to 1m; some calyx-lobes shorter than -tube; corolla-tube 20-33mm; corolla-limb 25-40mm across, red; (2n=18). Intrd; grown and found as casual as for *N. alata*; S America.

13. PETUNIA Juss. - *Petunia*

Glandular-pubescent, ± sticky annuals (at least in BI); leaves simple, ± entire; flowers solitary, axillary; calyx divided ≥3/4 way to base into 5 narrow lobes; corolla trumpet-shaped; ovary 2-celled; fruit a capsule with 2 valves, each slightly notched.

1. P. x hybrida (Hook.) Vilm. (*P. axillaris* (Lam.) Britton, Sterns & Poggenb. x *P. integrifolia* (Hook.) Schinz & Thell.) - *Petunia*. Stems procumbent to erect, to 60cm; leaves ovate-elliptic, sessile to shortly petiolate; corolla 5-12cm, often equally wide distally, white to red, mauve or purple; (2n=14, 28). Intrd; much grown in gardens and frequent on tips and rough ground, sometimes self-sown; S Br; garden origin.

115. CONVOLVULACEAE - *Bindweed family*

Annuals to herbaceous perennials, with twining or procumbent stems; leaves alternate, simple, entire, petiolate, without stipules. Flowers 1-few in leaf axils, actinomorphic, bisexual, hypogynous, usually with 2 bracteoles near base; sepals 5, free; petals 5, fused almost whole way to form funnel- to trumpet-shaped corolla, sometimes fused only to c.1/2 way, variously coloured; stamens 5, borne on corolla-tube; ovary 1-celled with 4 basal ovules, or 2-3-celled with 2 basal ovules per cell, or (*Dichondra*) the 2 cells almost as separate ovaries; style 1, or (*Dichondra*) 2; stigmas 1-2 per style, globose to linear or 2-3-lobed; fruit a capsule, usually without proper dehiscence mechanism.

All but the distinctive *Dichondra* are distinguishable by the large funnel- to trumpet-shaped corolla, usually 2-carpellary ovary with usually 2 stigmas and 2 ovules per cell, and often twining stems.

1 Corolla <5mm, divided to c.1/2 way; capsule deeply bilobed **1. DICHONDRA**
1 Corolla ≥10mm, normally not or scarcely lobed; capsule ± not lobed 2
  2 Bracteoles ovate, often pouched, partly or wholly obscuring sepals **3. CALYSTEGIA**
  2 Bracteoles linear, distant from sepals 3
3 Stigmas linear; sepals obtuse to retuse, rarely acuminate **2. CONVOLVULUS**
3 Stigmas ± globose; sepals acute to acuminate **4. IPOMOEA**

1. DICHONDRA J.R. & G. Forst. - *Kidneyweed*
Perennials with procumbent stems rooting at nodes; leaves orbicular to reniform; flowers 1(-few); corolla divided to c.1/2 way, c. as long as calyx; styles 2; stigmas 1 per style, capitate; ovary and fruit deeply bilobed.

1. D. micrantha Urb. (*D. repens* auct. non J.R. & G. Forst.) - *Kidneyweed*. Stems very thin, to 50cm; leaves 4-20(30)mm, appressed-pubescent; petioles 5-50mm; corolla 1.5-2.5mm, yellowish- or greenish-white, on stalk 3-20mm. Intrd; dry fixed sand-dunes; natd near Hayle, W Cornwall, 1955-1979, but now gone; E Asia.

2. CONVOLVULUS L. - *Field Bindweed*
Rhizomatous perennials with trailing or climbing stems, or weak annuals; leaves triangular or ovate-oblong to linear, hastate to sagittate at base (not in *C. tricolor*); flowers 1-few, with 2 linear bracteoles some way below; corolla usually scarcely divided, much longer than calyx; style 1; stigmas 2, linear; ovary and fruit unlobed.

Other spp. - C. tricolor L., from Mediterranean, is a weak annual with sessile obovate leaves and tricoloured (blue, white, yellow basipetally-zoned) corolla; it is a rare birdseed-alien and garden escape.

1. C. arvensis L. - *Field Bindweed*. Stems to 1(2)m, often much less; leaves 2-6cm, glabrous or pubescent; corolla 10-25mm, white, pink or striped pink-and-white, rarely deeply 5-lobed (var. **stonestreetii** Druce); 2n=48. Native; waste and cultivated ground, waysides, banks and rough or short grassland; common throughout most of lowland BI except N Sc.

3. CALYSTEGIA R. Br. - *Bindweeds*
Rhizomatous perennials with trailing or climbing stems; leaves triangular and sagittate, or reniform; flowers usually 1, with 2 ovate, often ± pouched bracteoles partly or wholly concealing calyx; corolla usually scarcely divided, much longer than calyx; style 1; stigmas 2, ellipsoid; ovary and fruit not lobed.

1 Leaves reniform; stems not or weakly climbing **1. C. soldanella**
1 Leaves triangular and sagittate; stems usually strongly climbing 2
  2 Bracteoles 10-18mm wide when flattened, not or little overlapping at edges, not or little obscuring sepals in lateral view; ratio of midrib-to-midrib to edge-to-edge distances in natural condition 0.4-1.1 **2. C. sepium**
  2 Bracteoles 18-45mm wide when flattened, strongly overlapping at edges, completely or nearly obscuring sepals in lateral view; ratio of midrib-to-midrib to edge-to-edge distance in natural condition 1.1-2.2 3
3 Corolla pink or pink-and-white striped; pedicels shortly pubescent (often only sparsely or partly so), usually with narrow wavy-edged wing near apex **3. C. pulchra**
3 Corolla white (sometimes narrowly pink-striped on outside only); pedicels glabrous, without wing **4. C. silvatica**

1. C. soldanella (L.) R. Br. - *Sea Bindweed*. Stems trailing or weakly climbing, to 1m; bracteoles pouched, slightly overlapping laterally; corolla 3-5.5cm, trumpet-shaped, pink with 5 white stripes, yellowish in centre; 2n=22. Native; on sand-dunes and sometimes shingle, by sea; coasts of BI N to C Sc.

2. C. sepium (L.) R. Br. - *Hedge Bindweed*. Stems strongly climbing, to 2(3)m; **536**
bracteoles flat to strongly keeled and slightly pouched, not or slightly overlapping laterally; corolla 3-6cm, funnel- to trumpet-shaped, rarely deeply 5-lobed (f. **schizoflora** (Druce) Stace), white, or pink with 5 white stripes. Native; hedges, ditches, fens, marshes, by water and on rough and waste ground.

a. Ssp. sepium. Glabrous; corolla 3-5(5.5)cm, white, or pink with 5 white stripes (f. **colorata** (Lange) Dörfl.); stamens 15-25mm; 2n=22. Throughout BI, but local and perhaps intrd in N Sc.

b. Ssp. roseata Brummitt. Stems, petioles and pedicels sparsely short-pubescent; **R**
corolla 4-5.5cm, pink with 5 white stripes; stamens 17-25mm; 2n=22. Local near W coast of Br and Ir, CI, occasionally elsewhere but perhaps intrd.

Ssp. **spectabilis** Brummitt, from N Eurasia, was formerly natd at 1 place in Merioneth; it has pink corollas and differs from the above 2 sspp. in its rounded (not acute) basal leaf-sinus, and slightly larger parts (corolla 5-6cm; stamens 20-30mm). It may be overlooked elsewhere.

2 x 3. C. sepium x C. pulchra = C. x scanica Brummitt is intermediate between the parents and partially fertile; corolla 4.5-6.5cm, white to pale pink; stamens 23-28mm. Scattered in C & S Br, CI.

2 x 4. C. sepium x C. silvatica = C. x lucana (Ten.) G. Don is intermediate between the parents and highly fertile; corolla 4-6.5cm, white or very pale pink (according to *C. sepium* parent); stamens 20-25mm. Scattered over C & S Br and CI, frequent in some parts of SE En, especially Greater London, Kirkcudbrights.

3. C. pulchra Brummitt & Heywood (*C. sepium* ssp. *pulchra* (Brummitt & **536**
Heywood) Tutin nom. inval., *C. dahurica* auct. non (Herb.) G. Don) - *Hairy Bindweed*. Stems strongly climbing to 3(5)m; bracteoles strongly pouched, strongly overlapping laterally; stems, petioles and pedicels sparsely short-pubescent; pedicels with narrow wavy wing; corolla 5-7.5cm, pink with white stripes; stamens 25-35mm; 2n=22. Intrd; natd in hedges and on rough and waste ground; scattered ± throughout BI; origin uncertain. This sp. is rarely highly fertile and, since it shares characters (pubescence, colour) with some sspp. of *C. sepium*, might be derived by hybridization between the latter and some other sp.

3 x 4. C. pulchra x C. silvatica = C. x howittiorum Brummitt is intermediate between the parents and partially fertile; corolla 5.3-6.5cm, pink; stamens 25-30mm. Very scattered in En; endemic.

4. C. silvatica (Kit.) Griseb. (*C. sepium* ssp. *silvatica* (Kit.) Batt.) - *Large Bindweed*. **536**
Glabrous; stems strongly climbing, to 3(5)m; bracteoles strongly pouched, strongly overlapping laterally; corolla(5)6-9cm, trumpet-shaped, white, rarely deeply 5-

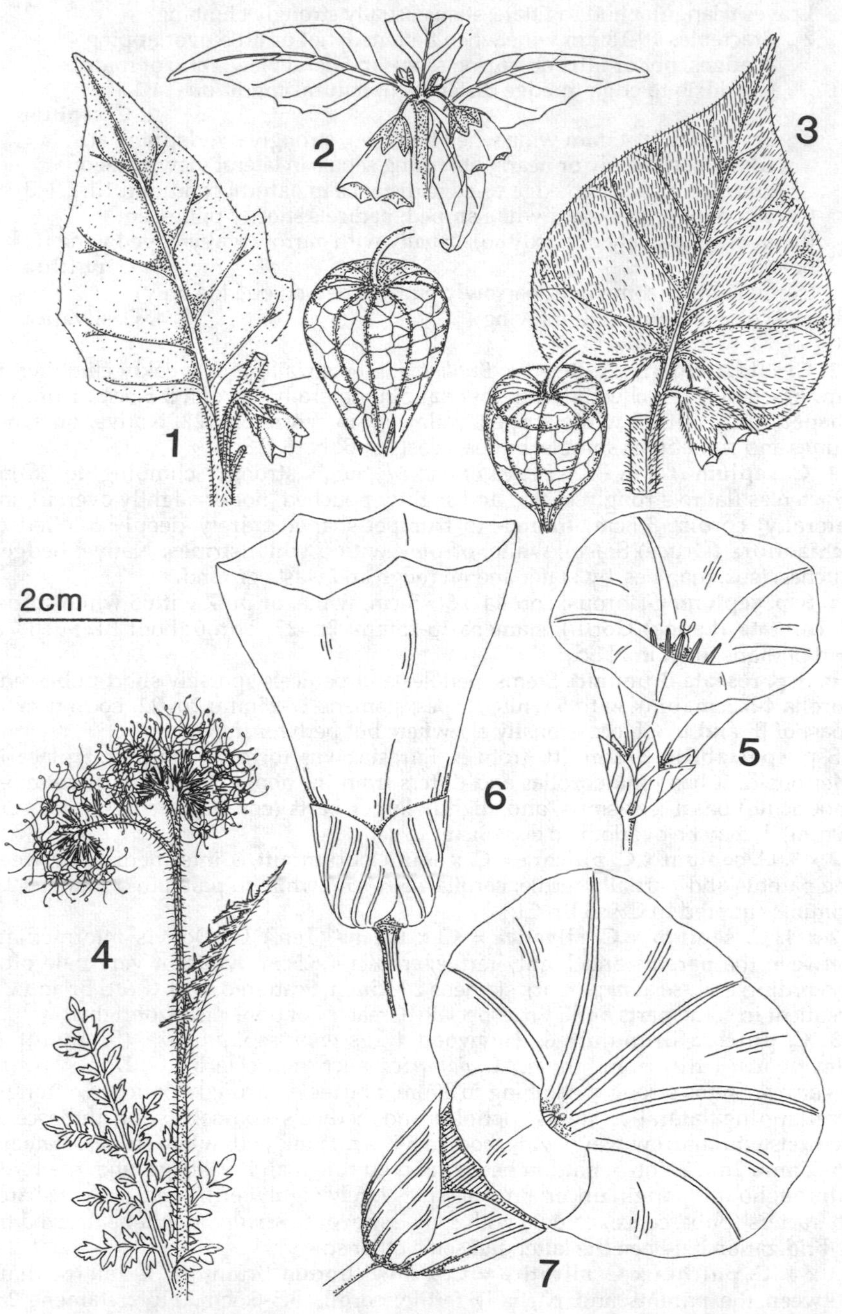

FIG 536 - Solanaceae, Hydrophyllaceae, Convolvulaceae.
1, *Physalis alkekengi*. 2, *P. ixocarpa*. 3, *P. peruviana*. 4, *Phacelia tanacetifolia*. 5, *Calystegia sepium*. 6, *C. pulchra*. 7, *C. silvatica*.

lobed (var.**quinquepartita** N. Terracc.); stamens 23-35mm; 2n=22. Intrd; natd as for *C. pulchra*, but much commoner; S Europe.

4. IPOMOEA L. - *Morning-glories*
Annuals or tuber-bearing perennials with usually strongly climbing stems; leaves ovate, cordate, entire to deeply 3-lobed; flowers 1-few, with 2 linear bracteoles some way below; corolla scarcely to shallowly lobed, much longer than calyx, funnel- to trumpet-shaped; style 1; stigma 1, 2-3-lobed; ovary and fruit not lobed.

1 Flower-stalks shorter than petioles, with 0-few patent to forwardly directed hairs; corollas usually white (rarely purple) **3. I. lacunosa**
1 Flower-stalks longer than petioles, with reflexed hairs; corollas usually blue, fading or drying to pinkish-purple (rarely white) 2
2 Corolla <5cm; sepals abruptly long-acuminate; most leaves deeply 3-lobed **2. I. hederacea**
2 Corolla ≥5cm; sepals acute; most leaves entire **1. I. purpurea**

Other spp. - I. batatas (L.) Lam. (*Sweet-potato*), from tropical America, with weakly to non-climbing stems, edible underground tubers, sepals 7-12mm and white or dull violet to purple corolla 3-5cm, rarely persists on tips from discarded vegetables.

1. I. purpurea Roth - *Common Morning-glory*. Stems to 3m; leaves mostly entire, sometimes a few 3-lobed; sepals (10)12-16(20)mm; corolla 4-7cm, usually blue when fresh; ovary 3-celled; stigma 3-lobed; (2n=30). Intrd; tips and waste places, casual from soyabeans, birdseed and other sources; sporadic in S En; N America.

2. I. hederacea Jacq. - *Ivy-leaved Morning-glory*. Stems to 2m; leaves mostly deeply 3-lobed, sometimes a few entire; sepals 15-30mm, corolla 2-4cm, usually blue when fresh; ovary 3-celled; stigma 3-lobed; (2n=30). Intrd; occurrence as for *I.purpurea* but less common; N America. The similar **I. nil** (L.) Roth, with corollas 5-7.5cm, has been much confused with *I. hederacea* and might occur in Br.

3. I. lacunosa L. - *White Morning-glory*. Stems to 2m; leaves entire or rather shallowly 3-lobed; sepals 6-13mm, shortly acuminate; corolla 1.5-2.5cm, usually white; ovary 2-celled; stigma 2-lobed; (2n=30). Intrd; a rather constant casual from soyabean waste; very local and sporadic in S En; SE N America.

116. CUSCUTACEAE - *Dodder family*

Herbaceous, annual to perennial, rootless parasites without visible chlorophyll; stems twining and adherent to host plants by haustoria, very thin; leaves alternate, reduced to minute scales, without stipules, sessile. Flowers in dense, sessile, ± globose heads, very small, actinomorphic, bisexual, hypogynous; sepals 4-5, fused at base, petals 4-5, fused into tube proximally; stamens 4-5, borne on corolla-tube, with small corolla-scale just below each; ovary 2-celled with 2 basal ovules per cell; styles 2; stigmas linear or capitate; fruit a capsule dehiscing transversely.

The very thin, chlorophyll-less, rootless stems with haustoria and small globose inflorescences are unique.

1. CUSCUTA L. - *Dodders*

1 Stigmas capitate (style ending in distinct knob); stems yellowish **1. C. campestris**
1 Stigmas linear (style scarcely thickened distally); stems reddish 2
2 Styles + stigmas shorter than ovary; corolla-scales bifid, minute, or 0 **2. C. europaea**

2 Styles + stigmas longer than ovary; corolla-scales not bifid, reaching to base of filaments **3. C. epithymum**

Other spp. - C. epilinum Weihe (*Flax Dodder*), from Europe, is parasitic on *Linum usitatissimum* and used to occur sporadically in Br and Ir as a casual in flax fields; it has been rare since 1900 and is now extinct. It differs from *C. europaea* in its nearly unbranched stems, 5-merous flowers and acute sepals.

1. C. campestris Yunck. - *Yellow Dodder*. Stems yellow; flowers pentamerous; sepals obtuse; corolla-scales not bilobed, laciniate, reaching above insertion of stamens; stamens exserted; (2n=56). Intrd; on a range of cultivated plants, especially carrot; scattered in En and Wa; N America.

2. C. europaea L. - *Greater Dodder*. Stems reddish; flowers 4-5-merous; sepals obtuse; corolla-scales deeply bilobed, each lobe entire or sparsely laciniate, ending well short of stamen insertion; stamens included; (2n=14). Native; on a range of hosts but usually primarily on *Urtica dioica*, often near water; scattered and rare in En N to Northants, formerly commoner. **R**

3. C. epithymum (L.) L. - *Dodder*. Stems reddish; flowers 5-merous; sepals acute; corolla-scales not bifid, laciniate, reaching ± to stamen insertion; stamens ± exserted; 2n=14. Native; on a wide range of hosts, most commonly on *Ulex* spp. and *Calluna* on heathland; frequent in S Br and CI, scattered elsewhere in BI N to C Sc but mostly casual.

117. MENYANTHACEAE - *Bogbean family*

Glabrous, stoloniferous, aquatic or semi-aquatic perennials; leaves alternate, simple or ternate, entire or with ± entire leaflets, without stipules but with flat, sheathing petiole-bases. Flowers showy, in axillary small clusters or elongated racemes, actinomorphic, bisexual, hypogynous; sepals 5, fused only at base; petals 5, fused into short tube with 5 lobes; stamens 5, borne on corolla-tube, sometimes alternating with short fringed scales; ovary 1-celled, with many ovules on 2 parietal placentas; style 1; stigma 2-lobed; fruit a capsule.

Easily recognized as aquatics with showy white to pink or yellow flowers with 5 fringed petals.

1 Leaves ternate; corolla white to pink **1. MENYANTHES**
1 Leaves simple; corolla yellow **2. NYMPHOIDES**

1. MENYANTHES L. - *Bogbean*

Leaves ternate, all alternate, held above water level; flowers monomorphic, in erect racemes; corolla white to pink, with many long fringes on inner side of lobes; capsule dehiscing by 2 valves.

1. M. trifoliata L. - *Bogbean*. Stems procumbent or floating, to 1.5m; racemes ≤30cm; flowers 1.5-2cm across; 2n=54. Native; in shallow water, bogs and fens; throughout Br and Ir, local in many parts of En, formerly CI.

2. NYMPHOIDES Ség. - *Fringed Water-lily*

Leaves simple, alternate on vegetative stems, opposite on flowering stems, cordate, floating on water; flowers dimorphic, in small axillary groups, on long pedicels; corolla yellow, with fringes on margins of lobes; capsule dehiscing irregularly. The flowers are heterostylous, as in *Primula*, with pin and thrum morphs.

1. N. peltata Kuntze - *Fringed Water-lily*. Stems floating, to 1.5m; leaves up to c.12 x 10cm; pedicels ≤8cm; flowers 3-4cm across; 2n=54. Possibly native **R**

(presumably only where dimorphic); in ponds and slow rivers; fens of E Anglia and Thames basin, natd in scattered places elsewhere N to C Sc.

118. POLEMONIACEAE - *Jacob's-ladder family*

Herbaceous perennials; leaves opposite and simple or alternate and pinnate, without stipules, petiolate to ± sessile. Flowers showy, in terminal corymbose racemes, actinomorphic, mostly bisexual, hypogynous; sepals 5, fused into tube proximally; petals 5, fused into short or long tube, white to purple or blue; stamens 5, borne on corolla-tube; ovary 3-celled, with several ovules per cell on axile placentas; style 1; stigmas 3; fruit a capsule opening by 3 valves.

Told from other families with 5 fused sepals and petals and actinomorphic, hypogynous flowers (except the distinctive Diapensiaceae) by the 3-celled ovary and 3 stigmas.

1 Leaves pinnate; corolla-tube much shorter than -lobes, with long-exserted stamens **1. POLEMONIUM**
1 Leaves simple; corolla-tube longer than -lobes, with anthers at apex of corolla-tube **2. PHLOX**

1. POLEMONIUM L. - *Jacob's-ladder*
Leaves pinnate, petiolate, with 6-15 pairs of entire leaflets; corolla-lobes much longer than -tube; stamens ± equal-lengthed, well exserted, with hairy base to filaments.

1. P. caeruleum L. - *Jacob's-ladder*. Stems erect, to 1m; flowers blue (or white), 2-3cm across; 2n=18. Native; limestone grassland, scree, rock-ledges, wood-borders; locally frequent in Peak District, Yorkshire Dales, 1 place in S Northumb, sporadic garden escape elsewhere in Br. **RR**

2. PHLOX L. - *Phlox*
Leaves simple, ± sessile, entire; corolla-lobes shorter than tube, patent; stamens with anthers at different heights, the longest at mouth of corolla-tube, glabrous.

1. P. paniculata L. - *Phlox*. Stems erect, woody at base, to 1.5m; flowers white to pink, purple or mauve, 2-3cm across; (2n=14). Intrd; much grown in gardens and natd on rough and waste ground; sporadic in En; N America.

119. HYDROPHYLLACEAE - *Phacelia family*

Annuals; leaves alternate, deeply pinnately lobed to ± pinnate, without stipules, petiolate. Flowers in terminal clustered spiralled cymes, actinomorphic, bisexual, hypogynous; sepals 5, ± free; petals 5, fused into tube longer than lobes, bluish; stamens 5, borne on corolla-tube, well exserted; ovary 1-celled, with ovules borne on 2 intruded parietal placentas; style 1, divided into 2 distally; stigmas minute; fruit a 2-valved capsule.

Similar to Boraginaceae in its flowers in scorpioid cymes, but with 2-valved capsule and deeply divided style.

1. PHACELIA Juss. - *Phacelia*

Other spp. - P. ciliata Benth., from W N America, is a rare grain-casual differing from *P. tanacetifolia* in its oblong to ovate calyx-lobes with conspicuous venation.

1. P. tanacetifolia Benth. - *Phacelia*. Pubescent, erect to ascending annual to 70(100)cm; flowers numerous; calyx with linear lobes; corolla blue or pale mauve, 6-10mm; (2n=22). Intrd; grown in gardens for ornament and small-scale in fields for bees, also contaminant of crop- and grass-seed, casual on tips, waste ground and among crops and new grass; very scattered but increasing in En and Wa; California. 536

120. BORAGINACEAE - *Borage family*

Annual to perennial herbs, often hispid or scabrid; leaves alternate, simple, entire or ± so, without stipules, sessile or petiolate. Flowers in often spiralled cymes, actinomorphic to weakly zygomorphic, bisexual, hypogynous; sepals 5, united into tube with 5 lobes or teeth; petals 5, fused into tube with distal limb, the latter with 5 lobes, mostly blue to pink, often with a knob, scale or hair-tuft at throat of tube; stamens 5, borne on corolla-tube; ovary 4-celled, deeply 4-lobed, with 1 ovule per cell; style 1 (bifid at apex in *Echium*), arising from base of ovary where the 4 cells meet (terminal in *Heliotropium*); stigma 1 (2 in *Echium*), capitate or bilobed; fruit a cluster of 4 1-seeded nutlets (*schizocarp*).

Like Verbenaceae and Lamiaceae in its 4-celled ovary with 1 ovule per cell and a fruit of 4 nutlets, but differing from both in usually alternate leaves, and spiralled cymose inflorescence.

Much value is placed in many keys on the presence or absence of folds, scales or bands of hairs at the throat of the corolla-tube; these are often difficult to make out and very little use is made of them here. When they are well developed they may meet in the centre or around the style and the corolla-tube appears closed.

1 Style bifid at apex; flowers distinctly zygomorphic, with unequal stamens and corolla-lobes **2. ECHIUM**
1 Style simple; flowers actinomorphic to weakly zygomorphic, with equal stamens and ± equal corolla-lobes 2
  2 All anthers completely exserted 3
  2 All anthers completely included or only tips exserted 4
3 Annual; filaments glabrous; anthers longer than filaments; calyx divided nearly to base **9. BORAGO**
3 Rhizomatous perennial; filaments pubescent; anthers shorter than filaments; calyx divided c.1/2 way **10. TRACHYSTEMON**
  4 Calyx-lobes with some small teeth between 5 main ones, enlarging greatly in fruit and forming 2-lipped covering **14. ASPERUGO**
  4 Calyx-lobes 5, entire, not or slightly enlarging in fruit 5
5 Nutlets with hooked or barbed bristles 6
5 Nutlets smooth to warty, ridged or pubescent 7
  6 Flowers and fruits all or mostly without bract; nutlets >4.5mm **18. CYNOGLOSSUM**
  6 Flowers and fruits all or mostly with bract; nutlets <4.5mm **16. LAPPULA**
7 Plant glabrous, often very glaucous **11. MERTENSIA**
7 Plant bristly to (sometimes appressed-)pubescent, not or scarcely glaucous 8
  8 At least lower leaves opposite **13. PLAGIOBOTHRYS**
  8 All leaves alternate (rarely uppermost pair opposite in *Myosotis*) 9
9 Open flowers pendent, with exserted stigma **4. SYMPHYTUM**
9 Open flowers erect, with stigma included or at throat of corolla-tube 10
  10 Ripe nutlets smooth (sometimes pubescent or with keel round edge) 11
  10 Ripe nutlets tuberculate to strongly warty and/or with variously branched ridges 15

11 Basal and all or most stem-leaves petiolate **17. OMPHALODES**
11 All or most stem-leaves sessile 12
12 Corolla-tube plus -lobes >10mm 13
12 Corolla-tube plus -lobes <10mm 14
13 Calyx-lobes divided nearly to base **1. LITHOSPERMUM**
13 Calyx-lobes fused ≥1/2 way **3. PULMONARIA**
14 Corolla-tube longer than -lobes; calyx-hairs straight; throat of corolla partially closed by hairy folds **1. LITHOSPERMUM**
14 Corolla-tube usually shorter than -lobes, if longer then calyx-hairs hooked; throat of corolla closed by glabrous or papillate scales **15. MYOSOTIS**
15 Basal leaves strongly cordate at base **5. BRUNNERA**
15 Basal leaves gradually to abruptly cuneate at base 16
16 Leaves ovate to obovate, at least most basal ones >5cm wide 17
16 Leaves lanceolate to oblanceolate or linear-oblong, <5cm wide 18
17 Corolla-lobes rounded; corolla-scales closing throat of corolla-tube; nutlets stalked **8. PENTAGLOTTIS**
17 Corolla-lobes acute; corolla-scales not closing throat of corolla-tube; nutlets sessile **9. BORAGO**
18 Nutlets tuberculate to strongly warty, not ridged apart from marginal keel, without collar-like base 19
18 Nutlets tuberculate and with strong branching ridges, with distinct collar-like base at point of attachment 20
19 Corolla yellow to orange; nutlets coarsely warty **12. AMSINCKIA**
19 Corolla white to bluish-purple; nutlets minutely tuberculate **1. LITHOSPERMUM**
20 Corolla-tube longer than -limb **6. ANCHUSA**
20 Corolla-tube shorter than -limb **7. CYNOGLOTTIS**

Other genera - NONEA Medik. resembles *Anchusa* but has very small corolla-scales and a ± tubular corolla. **N. rosea** (M. Bieb.) Link, from Caucasus, with a pinkish-purple corolla 15-18mm, is a rare casual. **N. lutea** (Desr.) DC., from Russia, with a yellow corolla 7-12mm, was formerly established in Caerns. **HELIOTROPIUM** L. somewhat resembles *Myosotis*, with small white flowers in strongly coiled cymes, but there are small lobes between the 5 main corolla-lobes and unlike all our other genera the style is terminal, not basal. **H. europaeum** L. (*European Turnsole*), from Europe, is a pubescent annual to 40cm and is a rare wool- and oilseed-alien.

1. LITHOSPERMUM L. (*Buglossoides* Moench) - *Gromwells*
Pubescent to hispid annuals to perennials; leaves lanceolate to narrowly elliptic, narrowed to base; flowers solitary in leaf-axils, congested at flowering, distant at fruiting; calyx divided ± to base; corolla actinomorphic, with narrow tube at least as long as expanded limb, purplish-blue or white to yellowish; stamens equal, included; style simple, included; nutlets smooth to warty, without collar-like base.

1 Corolla purplish-blue, 11-16mm **1. L. purpureocaeruleum**
1 Corolla usually white to yellowish, <10mm 2
2 Leaves with lateral veins apparent on lowerside; nutlets white, smooth **2. L. officinale**
2 Leaves without lateral veins apparent; nutlets brown, tuberculate **3. L. arvense**

1. L. purpureocaeruleum L. (*Buglossoides purpureocaerulea* (L.) I.M. Johnst.) - **RR**
Purple Gromwell. Rhizomatous perennial with procumbent sterile and erect flowering stems to 60cm; corolla 11-16mm, purplish-blue; nutlets white, smooth,

shining; (2n=16). Native; scrub and wood-margins on chalk and limestone; very local in SW En, S & N Wa, formerly W Kent, rare casual elsewhere.

2. L. officinale L. - *Common Gromwell.* Shortly rhizomatous perennial with erect stems to (80)100cm; corolla 3-6mm, yellowish- or greenish-white; nutlets white, smooth, shining; 2n=28. Native; grassy and bushy places, hedgerows and wood-borders mostly on basic soils; locally frequent in En, very local in Wa and Ir, rare casual elsewhere.

3. L. arvense L. (*Buglossoides arvensis* (L.) I.M. Johnst.) - *Field Gromwell.* Erect annual to 50(80)cm; corolla 5-9mm, whitish (rarely bluish); nutlets pale brown, tuberculate, shining on tubercles; 2n=40. Native; arable fields, rough ground and open grassy places; locally frequent in En, very scattered and often casual in Wa, Sc, Ir and CI.

2. ECHIUM L. - *Viper's-buglosses*

Plants hispid, monocarpic (usually biennials); leaves lanceolate to oblanceolate or the lower ovate, tapered to base; cymes terminal and lateral, forming compound narrow panicle; calyx divided nearly to base; corolla zygomorphic, pink, purple or blue with tube shorter than limb, the latter with unequal lobes; stamens unequal, at least some exserted; apex of style bifid and exserted; nutlets without collar-like base, warty to ridged.

1 Shrubs with unbranched woody stem to 75 x 3-5cm and with terminal panicle up to 3.5m; leaves up to 50cm, crowded below panicle **4. E. pininana**

1 Stems herbaceous or ± so, ≤1m (incl. panicle) x 1cm; leaves scattered up stem, the panicle not sharply delimited from rest of stem 2

2 Corolla pubescent on veins and margins only; usually 2 stamens exserted **2. E. plantagineum**

2 Corolla ± uniformly pubescent on outside; usually 3-5 stamens exserted 3

3 Corolla blue when fully open; stems stiffly erect, with narrow rather dense inflorescence with bracts scarcely exceeding cymes **1. E. vulgare**

3 Corolla pinkish-violet when fully open; stems ascending, with rather loose inflorescence with conspicuous leafy bracts **3. E. rosulatum**

Other spp. - Further giant Canarian spp. and hybrids related to *E. pininana* (e.g. *E. pininana* x *E. webbii* Coincy = **E. x scilloniensis** hort.) have been found self-sown in waste places in Scillies; they seldom persist.

1. E. vulgare L. - *Viper's-bugloss.* Stems usually erect, to 1m; corolla 10-19mm, blue; usually 4-5 stamens exserted; nutlets irregularly and densely ridged; 2n=32. Native; open grassy places, cliffs, dunes, shingle, rough ground, usually on light, often calcareous soils; locally frequent to common in BI, especially S & E En.

2. E. plantagineum L. (*E. lycopsis* auct. non L.) - *Purple Viper's-bugloss.* Stems erect to ascending, to 75cm; corolla 18-30mm, purple; usually 2 stamens exserted; nutlets warty; (2n=16). Native; disturbed or open grassy, sandy ground near sea; frequent in Jersey, very local in Scillies and W Cornwall, rare escape or casual elsewhere in S En. **RR**

3. E. rosulatum Lange (*E. humile* auct. non Desf.) - *Lax Viper's-bugloss.* Stems ascending, to 75cm; corolla 11-20mm, pinkish-violet; usually 3-4 stamens exserted; nutlets irregularly and densely ridged; (2n=32). Intrd; natd on waste ground at Barry Docks, Glam, since 1927 but perhaps now gone; Portugal and NW Spain.

4. E. pininana Webb & Berthel. - *Giant Viper's-bugloss.* Stems erect, to 4m; corolla c.13mm, blue; 5 stamens well exserted; nutlets sharply warty; (2n=16). Intrd; self-sown garden escapes on rough ground in CI, Man, Caerns, W Cornwall and Scillies; Canaries. Plants usually exist for several years before flowering, followed by seeding and death.

3. PULMONARIA L. - *Lungworts*
Pubescent to slightly hispid tufted perennials; leaves lanceolate to ovate, abruptly to gradually contracted at base; flowers in rather dense terminal clusters of cymes; calyx divided <1/2 way to base; corolla actinomorphic, blue, red or purple, with tube slightly shorter to slightly longer than limb; stamens equal, included; style simple, included; nutlets smooth, sparsely pubescent, with collar-like base.

Leaf-characters must be observed on basal leaves that develop during the flowering season and reach maturity during the summer. There are 5 conspicuous hair-tufts at the throat of the corolla; pubescence of inside of corolla-tube refers to region below these tufts. The flowers are heterostylous, as in *Primula*, with pin and thrum morphs.

1 Basal leaves developing at flowering cordate to broadly cuneate at base, abruptly contracted into petiole — 2
1 Basal leaves developing at flowering gradually cuneate at base, tapered into petiole — 4
  2 Corolla bright red when open; inside of corolla-tube pubescent below hair-tufts; basal leaves abruptly cuneate to rounded at base — **3. P. rubra**
  2 Corolla reddish- to bluish-violet when open; inside of corolla-tube glabrous below hair-tufts; basal leaves cordate at base — 3
3 Basal leaves with large white spots — **1. P. officinalis**
3 Basal leaves unspotted or with faint pale green spots — **2. P. obscura**
  4 Stalked glands 0 to very sparse in inflorescence; leaves hispid on upperside, the basal ones white-spotted; flowers pin or thrum — **5. P. longifolia**
  4 Stalked glands very frequent in inflorescence; leaves softly pubescent on upperside, unspotted; flowers always thrum — **4. P. 'Mawson's Blue'**

Other spp. - **P. angustifolia** L., from Europe, is grown in gardens and might escape; it has been much confused with *P. longifolia*, but has unspotted basal leaves mostly >5cm wide, a less dense inflorescence, and leaf-hairs of uniform (not variable) length.

1. P. officinalis L. - *Lungwort*. Stems erect, to 30cm; basal leaves ovate, cordate to rounded at base, white-spotted; stalked glands fairly frequent in inflorescence; calyx-lobes subacute; corolla reddish- to bluish-violet when open; (2n=16). Intrd; much grown in gardens, natd on banks and in scrub, woods and rough ground; scattered throughout Br; Europe.

2. P. obscura Dumort. (*P. officinalis* ssp. *obscura* (Dumort.) Murb.) - *Suffolk* **RR**
Lungwort. Differs from *P. officinalis* in darker green, unspotted leaves, and sparser **544**
stalked glands in inflorescence; 2n=14. Native; woods and woodland rides and clearings; 3 sites in E Suffolk within 4 sq km, formerly W Suffolk, perhaps natd elsewhere but confused with *P. officinalis*. Leaf/petiole ratio seems not to distinguish the 2 spp.

3. P. rubra Schott - *Red Lungwort*. Stems erect, to 50cm; basal leaves ovate or narrowly so, rounded to very broadly cuneate at base, usually unspotted; stalked glands very frequent in inflorescence; calyx-lobes acute; corolla red when open; (2n=14). Intrd; grown in gardens, natd in grassy places, hedges and scrub; scattered in N En, C & S Sc; SE Europe.

4. P. 'Mawson's Blue' - *Mawson's Lungwort*. Stems ascending to erect, to 30cm; **544**
basal leaves elliptic to narrowly so; stalked glands very frequent in inflorescence; calyx-lobes obtuse; corolla blue when open, glabrous on inside of tube below hair-tufts. Intrd; grown in gardens, natd in shady places; very scattered in S & C En, S & C Sc; uncertain garden origin.

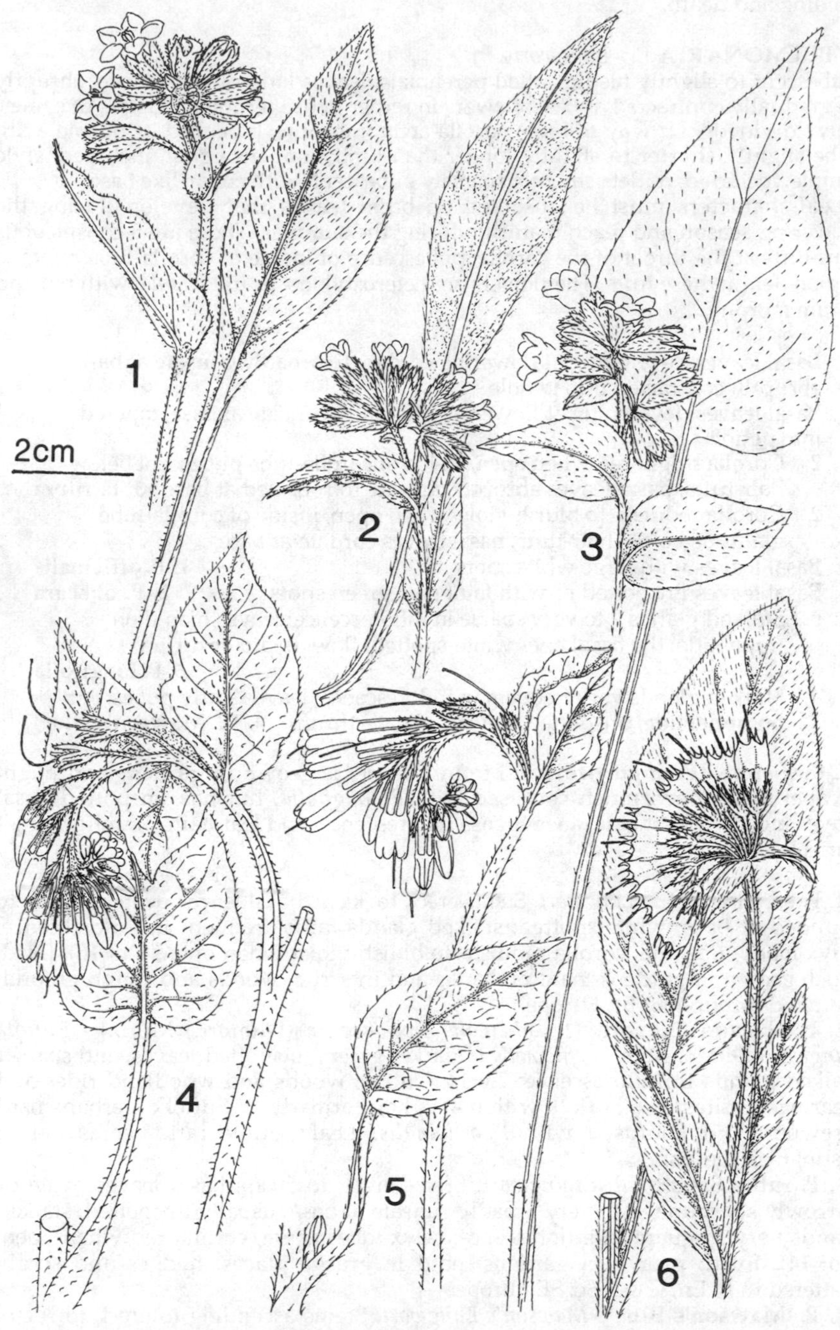

FIG 544 - Boraginaceae. 1, ***Pulmonaria*** **'Mawson's Blue'.** 2, ***P. longifolia.*** 3, ***P. obscura.*** 4, ***Symphytum*** **'Hidcote Blue'.** 5, ***S. grandiflorum.*** 6, ***S. bulbosum.***

5. P. longifolia (Bastard) Boreau (*P. angustifolia* auct. non L.) - *Narrow-leaved Lungwort*. Stems erect, to 40cm; basal leaves narrowly elliptic, mostly <5cm wide; stalked glands 0 to very sparse in inflorescence; calyx-lobes acute; corolla blue to violet when open, glabrous on inside of tube below hair-tufts; (2n=14). Native; woods and scrub; extremely local in Dorset, S Hants and Wight, perhaps rare escape elsewhere. R 544

4. SYMPHYTUM L. - *Comfreys*

Hispid perennials; leaves ovate-elliptic, subcordate to broadly cuneate at base, the basal long-petiolate; flowers in rather dense cymes forming terminal panicle; calyx lobed c.1/5-9/10 way to base; corolla actinomorphic, various colours, with limb little wider than tube and c. as long, the limb with short lobes; stamens equal, included, alternating with 5 long corolla-scales; style simple, exserted; nutlets smooth to granulate, sometimes also ridged, with collar-like base.

1 Plant with decumbent to procumbent leafy stolons 2
1 Plant without stolons 3
  2 Corolla pale yellow when open, often flushed reddish on outside; flowering stems unbranched **6. S. grandiflorum**
  2 Corolla predominantly blue or pink when open; larger flowering stems branched **5. S. 'Hidcote Blue'**
3 Nutlets ± smooth, shining; stem-leaves strongly decurrent, forming wings on stem extending down for >1 internode **1. S. officinale**
3 Nutlets minutely tuberculate, dull; stem-leaves not to moderately decurrent, the wings rarely extending for >1 internode 4
  4 Corolla pink, purple or blue 5
  4 Corolla pale yellow to white 7
5 Calyx divided <1/2 way to base **9. S. caucasicum**
5 Calyx divided ≥1/2 way to base 6
  6 Calyx-hairs almost all broad-based whitish bristles, with some much finer and smaller hairs; upper stem-leaves shortly petiolate, not decurrent or clasping stem **3. S. asperum**
  6 Calyx-hairs a mixture of broad-based bristles and finer and smaller hairs and all intermediates; upper stem-leaves sessile, shortly decurrent or clasping stem **2. S. x uplandicum**
7 Corolla-scales exserted for >1mm **10. S. bulbosum**
7 Corolla-scales included 8
  8 Calyx divided <1/2 way to base; corolla pure white **8. S. orientale**
  8 Calyx divided >1/2 way to base; corolla yellow to pale yellow 9
9 Stems not or little branched; middle and upper stem-leaves sessile, the upper shortly decurrent; rhizomes with swollen tubers present **4. S. tuberosum**
9 Stems well-branched; middle stem-leaves petiolate, uppermost ones sessile but none decurrent; rhizomes 0 **7. S. tauricum**

1. S. officinale L. - *Common Comfrey*. Stems erect, well-branched, to 1.5m, from thick, vertical root; stem-leaves sessile, long-decurrent; calyx divided c.2/3-4/5 way to base; corolla purplish or pale creamy-yellow; 2n=24, 44, 48. Native; by streams and rivers, in fens and marshy places, also roadsides and rough ground; locally frequent in BI, but less common than and over-recorded for *S. x uplandicum*. The flowers are often wrongly described as white; except for very rare albinos they are pale creamy-yellow (or purplish).

2. S. x uplandicum Nyman (*S. officinale* x *S. asperum*) - *Russian Comfrey*. Differs from *S. officinale* in more bristly stems, leaves and calyx; corolla blue to violet or purplish when open; and see key (couplet 3); 2n=36, 40. Intrd originally as fodder, or possibly arisen anew in a few places; roadsides, rough and damp ground, wood-borders; frequent over most of BI. Fertile and backcrosses to *S. officinale*, forming a spectrum of intermediates.

2 x 4. S. x uplandicum x S. tuberosum occurs near the parents in very scattered localities in En and Sc; it is intermediate in all characters, with yellow corollas tinged with blue or purple and tuberous rhizomes, and is at least partially sterile; endemic. Some plants might be *S. officinale* x *S. tuberosum*

3. S. asperum Lepech. - *Rough Comfrey*. Stems erect, well-branched, to 1.5m, from thick, vertical root; stem-leaves shortly petiolate, not decurrent; calyx divided c.2/3-4/5 way to base, enlarging somewhat in fruit, with less sharply acute lobes than in *S. officinale*; corolla sky blue when open; 2n=32. Intrd; natd in rough and waste ground; formerly occasional, now very rare, scattered over Br, Co Sligo, much over-recorded for *S. x uplandicum*; SW Asia. The shortly petiolate upper stem-leaves and white, sub-spiny bristles, are diagnostic.

4. S. tuberosum L. - *Tuberous Comfrey*. Stems erect, little or not branched, to 60cm, from rhizomes with thick swollen regions; stem-leaves sessile, shortly decurrent; calyx divided c.3/4-9/10 way to base; corolla pale yellow; 2n=96. Native; damp woods, ditches and river banks; frequent in lowland Sc, scattered in En, Wa and Ir and perhaps only intrd.

5. S. 'Hidcote Blue' (*S. grandiflorum* x ?*S. x uplandicum*) - *Hidcote Comfrey*. **544**
Ascending to erect flowering stems to 50(100)cm and procumbent to decumbent stolons arising from rhizomes; stem-leaves mostly petiolate, not or scarcely decurrent; calyx divided c.3/5-2/3 way to base; corolla blue (often flushed red on outside) when open, pink earlier. Intrd; grown in gardens, natd in hedges and woodland; W Kent, Surrey, Salop, E Lothian and Guernsey; garden origin. A similar plant with presumably the same parentage but with pink, not blue, flowers (**S. 'Hidcote Pink'**) has been found natd in Surrey.

6. S. grandiflorum DC. (*S. ibericum* Steven) - *Creeping Comfrey*. Differs from *S.* **544**
'Hidcote Blue' in unbranched stems to 40cm; corolla pale yellow when open, pinkish-red earlier; (2n=24, 60). Intrd; common in gardens and well natd in woods and hedges; scattered in C & S Br, rarely N to C Sc; Caucasus.

7. S. tauricum Willd. - *Crimean Comfrey*. Stems erect, well-branched, to 60cm, from thick, vertical root; stem-leaves petiolate to sessile but not decurrent; calyx divided c.3/5-5/6 way to base; corolla pale yellow; (2n=14, 18, 80). Intrd; natd on hedgebank; 1 place in Cambs since 1973; SE Europe.

8. S. orientale L. - *White Comfrey*. Stems erect, little-branched, to 70cm, from thick branched roots; stem-leaves petiolate to sessile but not decurrent; calyx divided c.1/4-2/5 to base; corolla white; (2n=32). Intrd; natd in hedgerows and other shady places, often self-sown; frequent in E & S En and C Sc, very scattered elsewhere; W Russia and Turkey.

9. S. caucasicum M. Bieb. - *Caucasian Comfrey*. Stems erect, to 60cm, from thick, branched roots; stem-leaves sessile, shortly decurrent; calyx divided c.1/4-1/2 way to base; corolla blue; (2n=24, 36, 48). Intrd; natd in hedgerows and other shady places; very scattered, and often not persistent, in SE En, Flints; Caucasus.

10. S. bulbosum K.F. Schimper - *Bulbous Comfrey*. Stems simple, erect, to 50cm, **544**
from rhizome with subglobose tubers; stem-leaves petiolate to sessile, somewhat decurrent; calyx divided c.1/2-3/4 way to base; corolla pale yellow; (2n=c.72, 84, 120). Intrd; natd in woods and by streams; very scattered in C & S Br; SC & SE Europe.

5. BRUNNERA Steven - *Great Forget-me-not*

Appressed-pubescent densely tufted perennials; basal and lower stem-leaves ovate-cordate, petiolate; flowers in bractless dense cymes in terminal subcorymbose panicles; calyx divided nearly to base; corolla actinomorphic, blue, with tube shorter than limb, the latter with patent lobes; stamens equal, included; style simple, included; nutlets ridged, with collar-like base.

1. B. macrophylla (Adams) I.M. Johnst. - *Great Forget-me-not*. Stems erect, to 50cm, the basal leaves plus petioles often not much shorter; calyx c.1mm in flower, corolla c.3-4mm across, resembling a *Myosotis*; (2n=12). Intrd; much grown in gardens, very persistent throwout, sometimes self-sown, in woods and on rough ground and tips; scattered in Br; Caucasus. See *Omphalodes verna* for differences. 552

6. ANCHUSA L. (*Lycopsis* L.) - *Alkanets*
Hispid annuals to perennials; leaves lanceolate to oblanceolate, sessile or narrowed to short petiole; flowers in terminal, branched, spiralled cymes; calyx divided c.1/3 way to nearly wholly to base; corolla actinomorphic to slightly zygomorphic, blue to purple, or yellow, with tube longer than limb; stamens equal, included; style simple, included; nutlets ridged and tuberculate, with collar-like base.

1 Corolla with curved tube and 5 slightly unequal lobes **4. A. arvensis**
1 Corolla with straight tube and 5 equal lobes 2
  2 Corolla yellow; calyx-lobes obtuse to rounded **1. A. ochroleuca**
  2 Corolla blue to purple; calyx-lobes acute 3
3 Calyx divided nearly to base; nutlets >5mm in longest plane **3. A. azurea**
3 Calyx divided c.1/2 way to nearly to base; nutlets <5mm in longest plane **2. A. officinalis**

1. A. ochroleuca M. Bieb. - *Yellow Alkanet*. Rather softly pubescent, erect perennial to 50cm; calyx divided c.1/3-1/2 way to base, with obtuse to rounded lobes; corolla actinomorphic, yellow, with tube 5-10mm, with limb 7-15mm across; (2n=24). Intrd; natd on rough ground on Phillack Towans (W Cornwall) since at least 1922, probably from animal feed, rare impermanent escape elsewhere; E Europe. 552

1 x 2. A. ochroleuca x A. officinalis = A. x baumgartenii (Nyman) Gusul. occurs in W Cornwall with both parents; it is intermediate in pubescence and calyx-lobe tips, has pale blue and/or greyish-yellow corollas, has calyx divided <1/2 way to base, and is partially fertile. It first appeared in 1950; since 1990 it has formed backcrosses with segregating characters but mostly yellow corollas.

2. A. officinalis L. - *Alkanet*. Hispid, erect perennial to 1.5m; calyx divided c.1/2 way to nearly to base; corolla actinomorphic, purplish-violet, with tube 5-7mm, with limb 7-15mm across; (2n=16). Intrd; grown in gardens, rather impermanent escape on rough and waste ground and tips, sometimes natd; scattered over lowland Br; Europe. White- or yellow-flowered plants occur on the Continent, and perhaps in gardens, but have not been reported from the wild in BI.

3. A. azurea Mill. - *Garden Anchusa*. Differs from *A. officinalis* in corolla bright blue, with tube 6-10mm, with limb 8-20mm across; and see key; (2n=32). Intrd; the commonest *Anchusa* grown in gardens, rather infrequent escape on waste and rough ground and tips, also birdseed-alien, very rarely natd; scattered over lowland Br, natd in W Cornwall c.1922-1972; S Europe. 552

4. A. arvensis (L.) M. Bieb. (*Lycopsis arvensis* L.) - *Bugloss*. Hispid, procumbent to erect annual to 50cm; leaves undulate, crenate; calyx divided nearly to base; corolla zygomorphic (see key), blue, with tube 4-7mm, with limb 4-6mm across; (2n=48). Native; weed of arable and rough ground on light, acid or calcareous soils; locally common throughout lowland Br, CI, E & N Ir.

7. CYNOGLOTTIS (Gusul.) Vural & Kit Tan - *False Alkanet*
Hispid perennials; differ from *Anchusa* in calyx divided >1/2 way to base; corolla actinomorphic, blue, with tube much shorter than limb.

1. C. barrelieri (All.) Vural & Kit Tan (*Anchusa barrelieri* (All.) Vitman) - *False Alkanet*. Plant appressed-hispid, erect, to 70cm; calyx with dense, white, appressed, stiff hairs, with lobes truncate to rounded or very broadly obtuse;

corolla-tube 1-2mm; corolla-limb 7-10mm across; (2n=16, 18, 24). Intrd; grown in gardens, infrequent but persistent escape; very scattered in S En; SE Europe.

8. PENTAGLOTTIS Tausch - *Green Alkanet*
Hispid perennials with deep, thick roots; leaves ovate, abruptly contracted at base; flowers in terminal and lateral dense cymes; calyx divided >3/4 way to base; corolla actinomorphic, blue, with tube shorter than limb; stamens equal, included; style simple, included; nutlets ridged, with knob-like, stalked base.

1. P. sempervirens (L.) Tausch ex L.H. Bailey - *Green Alkanet*. Stems erect, to 1m; basal leaves up to 40cm, long-petiolate; corolla 8-10mm across; 2n=22. Intrd; natd in hedges and wood-borders and on rough ground: frequent over much of BI; SW Europe.

9. BORAGO L. - *Borages*
Hispid annuals or perennials; leaves lanceolate to ovate or obovate, the lower abruptly tapered to petiole; cymes terminal, rather lax; calyx divided nearly to base; corolla actinomorphic, blue (rarely white), with lobes longer than rest of limb plus tube; stamens equal, exserted or included; style simple, included in corolla or between anthers; nutlets with collar-like base, ridged.

1. B. officinalis L. - *Borage*. Erect annual to 60cm; calyx 7-15mm at flowering, up to 20mm in fruit; corolla with patent to reflexed lobes 7-15mm; stamens completely exposed; nutlets 7-10mm; (2n=16). Intrd; still grown as herb and persistent on tips, rough ground and waysides; scattered over much of BI; S Europe.

2. B. pygmaea (DC.) Chater & Greuter (*B. laxiflora* (DC.) Fisch. non Poir.) - *Slender Borage*. Decumbent to ascending perennial to 60cm; calyx 4-6mm at flowering, up to 8mm in fruit; corolla with ± erect lobes 5-8mm; stamens included; nutlets 3-4mm; (2n=32). Intrd; natd on heathy ground, Jethou (CI) since 1932, less permanent on rough ground and by paths in very scattered places in Wa and S En; Corsica and Sardinia.

10. TRACHYSTEMON D. Don - *Abraham-Isaac-Jacob*
Hispid, rhizomatous perennials; leaves ovate, the basal long-petiolate and cordate to rounded at base; cymes dense, several in terminal panicle; calyx divided c.1/2 way to base; corolla actinomorphic, blue, white near base, with lobes longer than rest of limb plus tube; stamens equal, completely exposed; style simple, included between closely appressed stamens; nutlets with collar-like base, ridged.

1. T. orientalis (L.) G. Don - *Abraham-Isaac-Jacob*. Stems erect, to 40cm; calyx 4-7mm at flowering, up to 10mm in fruit; corolla with patent and revolute lobes 9-12mm; (2n=56). Intrd; natd on shady banks and in dampish woods; scattered in Br N to C Sc; Caucasus and Turkey.

11. MERTENSIA Roth - *Oysterplants*
Glabrous, usually glaucous perennials; leaves papillose, obovate or elliptic to oblanceolate, the middle and lower ones with winged petioles, the upper sessile; flowers in terminal, rather dense cymes; calyx divided ± to base; corolla actinomorphic, blue or blue and pink, pink in bud, with tube shorter than limb, the latter bell-shaped and lobed c.1/2 way to base; stamens equal, included or slightly exserted; style simple, included; nutlets slightly flattened, smooth, succulent then papery on outside.

Other spp. - M. virginica (L.) Pers. (*Virginia-bluebells*), from E N America, is not glaucous and has much larger blue corollas 18-25mm; it is a rare garden escape. **M. ciliata** Don, from W N America, is confused with this, is also grown and perhaps may also escape; it has a more diffuse inflorescence, corollas 10-17mm and calyx divided nearly to base (not to c.1/2 way).

1. M. maritima (L.) Gray - *Oysterplant*. Stems usually decumbent, to 60cm; R
pedicels elongating in fruit up to 2.5cm; corolla 4-6mm, c.6mm across; 2n=24. Native; on bare shingle or shingly sand by sea; local on coasts of N Ir and N Br S to Denbs, formerly S to E Norfolk, Cards and N Kerry, decreasing but sporadic.

12. AMSINCKIA Lehm. - *Fiddlenecks*
Hispid annuals; leaves linear to oblong or lanceolate, sessile; flowers in terminal, spiralled, bracteate or bractless cymes; calyx divided nearly to base; corolla actinomorphic, yellow to orange, with tube longer than limb; stamens equal, included; style simple, included; nutlets keeled, warty, without collar-like base.

1. A. lycopsoides (Lehm.) Lehm. - *Scarce Fiddleneck*. Plant to 70cm, erect, little- to much-branched, with abundant patent white bristles on most parts; fruiting calyx (6)8-11(15)mm; corolla 5-8mm, with pubescent scales at apex of tube visible from above; stamens inserted c.1/2 way up corolla-tube or just below; (2n=30). Intrd; natd on rough ground in Farne Islands (Cheviot) since 1922, becoming increasingly frequent over much of En, especially E, on sandy soils; W N America.

2. A. micrantha Suksd. (*A. menziesii* auct., ?(Lehm.) A. Nelson & J.F. Macbr., *A.* 552
intermedia auct., ?Fisch. & C.A. Mey., *A. calycina* auct., ?(Moris) Chater) - *Common Fiddleneck*. Differs from *A. lycopsoides* in fruiting calyx 5-6(9)mm; corolla paler yellow, 3-5mm, without hairs or scales inside, with stamens borne on upper 1/2 of tube; (2n=32). Intrd; recently increased casual in arable land and sandy rough ground, often persistent; frequent in E En and E Sc N to E Ross, especially E Anglia, rare in C & W Br; W N America.

13. PLAGIOBOTHRYS Fisch. & C.A. Mey. - *White Forget-me-not*
Appressed-pubescent annuals; leaves linear-oblong, the lower ones opposite, sessile; flowers in terminal, semi-bractless, spiralled cymes; calyx divided >1/2 way to base; corolla actinomorphic, white, with tube shorter than limb, the latter divided c.1/2 way into 5 rounded lobes; stamens equal, included; style simple, included; nutlets keeled, ridged, minutely tuberculate, without collar-like base.

1. P. scouleri (Hook. & Arn.) I.M. Johnst. - *White Forget-me-not*. Stems erect to decumbent, to 20cm, often well-branched; leaves up to 60 x 5mm; corolla 2-4mm across; flowers resembling a white *Myosotis*. Intrd; casual, sometimes persistent, with grass-seed; Caithness, Shetland, S Wilts and S Hants; W N America.

14. ASPERUGO L. - *Madwort*
Hispid annuals; leaves lanceolate to oblanceolate, tapered to base; flowers 1-2 in leaf-axils; calyx with small teeth between 5 main ones, at fruiting much enlarged and forming 2-lipped structure around nutlets; corolla actinomorphic, purplish-blue, with tube shorter than limb; stamens equal, included; style simple, included; nutlets strongly compressed, without collar-like base, with dense low tubercles.

1. A. procumbens L. - *Madwort*. Stems procumbent to scrambling, to 60cm; calyx growing to several times original length at fruiting, with very distinctive 2-lipped shape; corolla 1.5-3mm, inconspicuous. Intrd; arable fields, waste and rough ground, occasionally natd; very scattered over Br especially Sc, Angus since 1884, formerly much commoner; Europe.

15. MYOSOTIS L. - *Forget-me-nots*
Pubescent annuals to perennials; leaves mostly narrowly oblong to oblanceolate, the basal ones tapered to petiole-like base, the upper ones sessile; flowers in

terminal spiralled cymes; calyx divided c.1/4 way to nearly to base; corolla actinomorphic, blue or sometimes white or yellow, usually pink in bud, with tube shorter to longer than limb, with limb divided ± to base; stamens equal, included; style simple, usually included; nutlets slightly compressed, with distinct keel, smooth.

Maximum corolla-size is of diagnostic value, but larger-flowered spp. often produce flowers with unusually small corollas.

1 Calyx with all hairs ± straight and closely appressed 2
1 Calyx with some hairs patent and distally hooked or at least strongly curved 7
 2 Nutlets ≤1mm, shining olive-brown; annual without sterile shoots; calyx divided <1/2 way to base at flowering; Jersey only **5. M. sicula**
 2 Nutlets ≥1.2mm, mid-brown to black, shining or not, or if <1.2mm then plants with axillary stolons; calyx often divided ≥1/2 way to base at flowering; widespread 3
3 Style longer than calyx-tube and often exceeding calyx-lobes at flowering; calyx divided <1/2 way to base at flowering, with broad teeth forming equilateral triangle **1. M. scorpioides**
3 Style shorter than calyx-tube at flowering; calyx often divided ≥1/2 way to base at flowering, with narrow teeth forming isosceles triangle with base shorter than sides 4
 4 Lower part of stem with ± patent hairs 5
 4 Stem with only appressed hairs 6
5 Pedicels eventually 2.5-5x as long as fruiting calyx; nutlets <2mm **2. M. secunda**
5 Pedicels <2x as long as fruiting calyx; nutlets usually >2mm **6. M. alpestris**
 6 Stolons produced from lower nodes; leaves rarely >3x as long as wide **3. M. stolonifera**
 6 Stolons 0; larger leaves >(3)4x as long as wide **4. M. laxa**
7 Perennial, with sterile basal shoots at fruiting 8
7 Annual, without sterile shoots at fruiting 10
 8 Fruiting calyx narrowed and acute to subacute at base; nutlets obtuse to rounded at apex; only on hills >700m **6. M. alpestris**
 8 Fruiting calyx rounded to broadly obtuse at base; nutlets acute to subacute at apex; lowland or upland 9
9 Corolla ≤8mm across, with ± flat limb; calyx-teeth erecto-patent, exposing ripe nutlets **7. M. sylvatica**
9 Corolla ≤5mm across, with saucer-shaped limb; calyx-teeth erect, ± appressed and concealing ripe nutlets (often squashed open when pressed) **8. M. arvensis**
 10 Pedicels 1.2-2x as long as fruiting calyx **8. M. arvensis**
 10 Pedicels shorter than to c. as long as fruiting calyx 11
11 Corolla cream to yellow at first (rarely white), with tube eventually longer than calyx; calyx-lobes oblong-lanceolate **10. M. discolor**
11 Corolla blue (rarely white) from start, with tube shorter than calyx; calyx-lobes narrowly to broadly triangular **9. M. ramosissima**

Other spp. - M. decumbens Host has been doubtfully recorded from mountains in M Perth and needs confirming; it differs from *M. sylvatica* in its corolla-tube longer (not shorter) than calyx and nutlets c.2mm (not ≤1.8mm).

1. M. scorpioides L. - *Water Forget-me-not*. Erect to ascending, appressed or sometimes partly patent-pubescent perennial to 70cm, with rhizomes and/or stolons; calyx with rather sparse appressed hairs, lobed <1/2 way to base at flowering; corolla up to 8mm across; 2n=64. Native; by or in edges of ponds and rivers, in damp fields; common throughout most of BI except very rare in CI (Jersey only). Distinguished by its long style (see key).

1 x 4. M. scorpioides x M. laxa = M. x suzae Domin has been found with the parents very scattered in En and Wa but perhaps overlooked; it is partially fertile and intermediate, with corolla 5-7mm across and style c. as long as calyx-tube.

2. M. secunda Al. Murray - *Creeping Forget-me-not.* Erect to ascending annual to perennial to 50cm, with stolons, with hairs appressed above but patent below; calyx with appressed hairs, lobed c.1/2 way or slightly more to base at flowering; corolla up to 6mm across; 2n=48. Native; wet, often acidic places by streams and in pools and bogs; common in many parts of BI, especially upland and N & W, rare or absent in most of C & E En. Distinguished by its long pedicels (see key) and patent hairs.

3. M. stolonifera (DC.) J. Gay ex Leresche & Levier (*M. brevifolia* Salmon) - *Pale Forget-me-not.* Erect, appressed-pubescent perennial to 20(30)cm, with numerous leafy stolons; calyx with appressed hairs, lobed >1/2 way to base at flowering; corolla up to 5mm across; 2n=24. Native; wet flushes and streamsides on hills; local in N En and S Sc. Distinguished by its short leaves (see key) and abundant stolons. **R**

4. M. laxa Lehm. (*M. caespitosa* Schultz) - *Tufted Forget-me-not.* Erect to ascending, appressed-pubescent annual to biennial; calyx with rather sparse appressed hairs, lobed more or less than 1/2 way to base at flowering; corolla up to 5mm across; (2n=88). Native; same places as *M. scorpioides* and often with it; fairly common ± throughout BI. Our plant is ssp. **caespitosa** (Schultz) Hyl. ex Nordh. Best told from *M. scorpioides* by its short style (see key).

5. M. sicula Guss. - *Jersey Forget-me-not.* Erect to decumbent, appressed-pubescent annual to 20cm; calyx with appressed hairs, lobed <1/2 way to base at flowering; corolla up to 3mm across; (2n=36, 46). Native; damp grassland and by pond; very local, 2 places in Jersey, discovered 1922. **RR**

6. M. alpestris F.W. Schmidt - *Alpine Forget-me-not.* Erect, ± rhizomatous, patent-pubescent perennial to 25cm; calyx with dense appressed to erecto-patent curved hairs, with or without some patent hooked ones on tube; corolla up to 10mm across; 2n=48. Native; mountain slopes and ledges, 700-1200m; very local in c.11 sites in NW Yorks, Westmorland and M Perth, ?intrd in Angus. Often mis-recorded elsewhere for upland variants of *M. sylvatica*; see key (couplet 8) for differences. **RR**

7. M. sylvatica Hoffm. - *Wood Forget-me-not.* Erect to ascending, tufted, patent-pubescent perennial to 50cm; calyx with dense appressed to erecto-patent curved hairs and many patent hooked ones; corolla up to 8mm across; 2n=18. Native; woods, scree and rock-ledges; locally common in C & N Br, very local in S Br, frequent garden escape elsewhere. The common garden sp.

8. M. arvensis (L.) Hill - *Field Forget-me-not.* Erect to ascending, tufted, patent-pubescent annual to perennial to 40cm, more greyish-green than *M. sylvatica*; calyx with many patent-hooked and some suberect hairs; corolla up to 3(5)mm across; (2n=36, 48, 52). Native; open, well-drained ground in many places, including gardens; common throughout most of BI. Var. **sylvestris** Schltdl. (ssp. *umbrata* (Mert. & W.D.J. Koch) O. Schwarz, var. *umbrosa* Bab.) is said to have calyx ≤7mm (not ≤5mm) in fruit; hooked hairs ≤0.6mm (not ≤0.4mm); nutlets ≤2.5mm (not ≤2mm); and a larger corolla 3-5mm across; it is often mistaken for *M. sylvatica* (see key, couplet 9).

9. M. ramosissima Rochel - *Early Forget-me-not.* Erect to decumbent, patent-pubescent annual to 25cm; calyx with many patent-hooked and some suberect hairs; corolla up to 3mm across, blue (rarely white), with tube shorter than calyx; (2n=48). Native; dry open places on sandy or limestone soils; locally common over most of lowland BI, especially En and CI. Ssp. **globularis** (Samp.) Grau (var. *mittenii* (Baker) ined.) has calyx ≤2.5mm in fruit (not ≤4mm) with broadly (not narrowly) triangular segments, but scarcely merits var. status. Ssp. **lebelii** (Godr.)

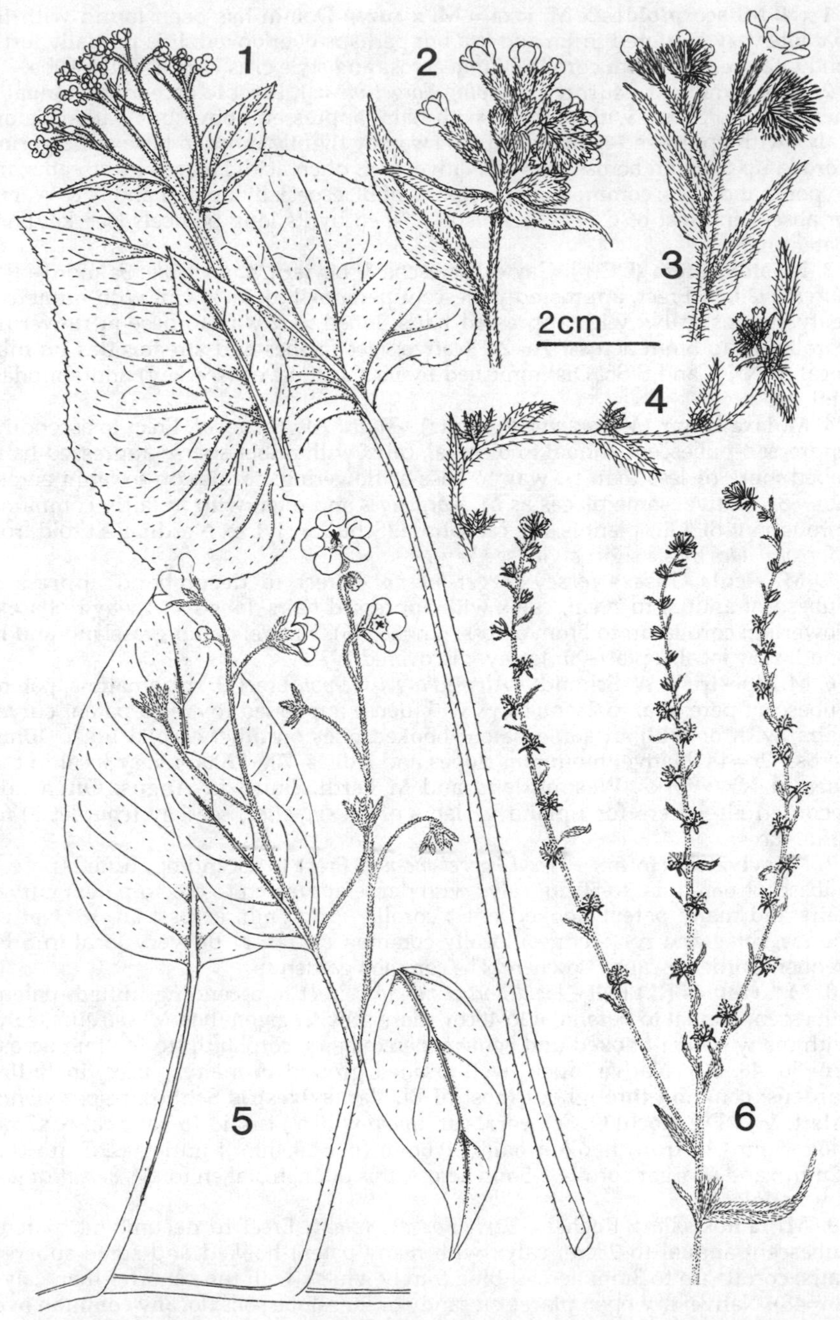

FIG 552 - Boraginaceae. 1, *Brunnera macrophylla*. 2, *Anchusa ochroleuca*. 3, *A. azurea*. 4, *Amsinckia micrantha*. 5, *Omphalodes verna*. 6, *Lappula squarrosa*.

Blaise is said to be intermediate between the 2.

10. M. discolor Pers. - *Changing Forget-me-not*. Erect, patent-pubescent annual to 25cm; differs from *M. ramosissima* in corolla up to 2mm across, yellow or cream at first, then pink to blue, with tube usually lengthening to longer than calyx; and see key; (2n=24, 72). Native; similar places to *M. ramosissima* but common also in Sc, Wa and Ir, and in N Br also in damp places, marshes and dune-slacks. Ssp. **dubia** (Arrond.) Blaise differs in its small corolla (≤2mm across) cream (not yellow) at first and the 2 uppermost leaves on main stem not opposite; both sorts often occur together and deserve at most varietal rank.

16. LAPPULA Gilib. - *Bur Forget-me-not*

Pubescent annuals or biennials; leaves linear-lanceolate to oblong-lanceolate, the basal ones tapered to base, the upper ones sessile; flowers in terminal, spiralled cymes; calyx divided nearly to base; corolla actinomorphic, blue, with tube shorter than limb, the latter 5-lobed, nearly flat; stamens equal, included; style simple, included; nutlets covered with hook-tipped spines.

1. L. squarrosa (Retz.) Dumort. (*L. myosotis* Moench) - *Bur Forget-me-not*. Stems erect, to 50cm, with dense appressed and patent white hairs; corolla 2-4mm across; nutlets 2.5-4mm; (2n=48). Intrd; casual from wool, birdseed, grass-seed and grain, in waste places, rough ground and tips; very scattered in C & S Br and E Ir; Europe. **552**

17. OMPHALODES Mill. - *Blue-eyed-Mary*

Rhizomatous and stoloniferous, rather sparsely pubescent perennials; leaves ovate, rounded to cordate at base, all or nearly all long-petiolate; flowers in few-flowered terminal cymes; calyx divided nearly to base; corolla actinomorphic, blue, with tube shorter than limb, the latter 5-lobed, nearly flat; stamens equal, included; style simple, included; nutlets smooth, pubescent.

1. O. verna Moench - *Blue-eyed-Mary*. Stems erect to ascending, to 25cm; calyx c.4mm; corolla 10-15mm across; (2n=42, 48). Intrd; grown in gardens, persistent relic or throwout mostly in woods; very scattered over Br, especially in N and Wa, Man, much rarer than formerly; SE Europe. Over-recorded for *Brunnera macrophylla*, which has much smaller, more densely congested flowers, coarsely-pubescent leaves, no stolons and ridged, glabrous nutlets. **552**

18. CYNOGLOSSUM L. - *Hound's-tongues*

Pubescent biennials; leaves ovate to linear-lanceolate, the lower petiolate, the upper sessile; flowers in spiralled terminal and lateral cymes; calyx divided nearly to base; corolla actinomorphic, reddish-purple, with tube about as long as funnel-shaped limb; stamens equal, included; style simple, included; nutlets large (5-9mm in longest plane), covered with hooked spines.

Other spp. - C. amabile Stapf & J.R. Drumm. (*Chinese Hound's-tongue*), from China, is a rare garden escape; it has blue, white or pink corollas only c.5mm across.

1. C. officinale L. - *Hound's-tongue*. Stems erect, to 60(90)cm; leaves grey-green, densely pubescent; pedicels <5mm; corolla 6-10mm across; nutlets with distinct thickened rim, uniformly spiny; (2n=24). Native; rather open ground mostly on sand, shingle or limestone, and waste ground; locally frequent in S, C & E Br, CI and E Ir, rare casual elsewhere.

2. C. germanicum Jacq. - *Green Hound's-tongue*. Differs from *C. officinale* in stems more slender, with longer, more diffuse branches; leaves green, sparsely pubescent; pedicels mostly >5mm; nutlets without thickened rim, but spines longer and denser **RRR**

at edges; (2n=24). Native; woods and hedgerows; rare and decreasing in E Gloucs, Oxon, Bucks and Surrey, formerly widespread but very local in S & C En, very rare casual elsewhere.

121. VERBENACEAE - *Vervain family*

Herbaceous perennials or rarely annuals; stems square in section; leaves opposite, simple, serrate to deeply pinnately lobed, without stipules, sessile to petiolate. Flowers in terminal elongated to corymbose spikes, zygomorphic, bisexual, hypogynous; sepals 5, fused into tube with 5 teeth; corolla a tube with 5 slightly unequal lobes arranged in upper (2-lobed) and lower (3-lobed) lips, lilac to purple or blue; stamens 4, included, 2 borne at each of 2 levels in corolla-tube; ovary 4-celled, scarcely lobed, with 1 ovule per cell; style 1, terminal; stigma 1, capitate; fruit a cluster of 4 1-seeded nutlets.

Told from Boraginaceae by square-sectioned stems, opposite leaves and 4 stamens, and from Lamiaceae by scarcely lobed ovary, terminal style and capitate stigma.

1. VERBENA L. - *Vervains*

1 At least lower stem-leaves petiolate, deeply pinnately lobed; spikes long, slender **1. V. officinalis**
1 Stem-leaves sessile, at most sharply serrate; spikes short, very dense, forming subcorymbose panicle 2
2 Bracts shorter than to as long as calyx; corolla mostly <2x as long as calyx **2. V. bonariensis**
2 Bracts longer than calyx; corolla mostly >2x as long as calyx **3. V. rigida**

Other spp. - The dwarf annual bedding plant known as *Verbena* is **V. x hybrida** Voss, of uncertain parentage, sometimes found on tips, as are a few other ornamental spp. **V. litoralis** Kunth, from S America, is closely related to *V. bonariensis* but differs in its corolla 3-3.5mm x 1.5-2mm with stamens borne well above middle; it is a rare wool-alien.

1. V. officinalis L. - *Vervain*. Stems erect, to 75cm; lower and mid stem-leaves petiolate, deeply pinnately lobed; spikes elongated (up to 25cm) in fruit; corolla pinkish-lilac, 3.5-5mm, 3-5mm across; 2n=14. Native; barish ground and rough grassy places, on well-drained often calcareous soils; locally common in S Br, more scattered elsewhere in BI N to N En and C Ir.

2. V. bonariensis L. - *Argentinian Vervain*. Stems erect, to 1(1.5)m; stem-leaves oblong-lanceolate, sharply serrate, sessile; spikes rarely >2cm in fruit; calyx mostly <3mm; corolla blue to purple, 5-6.5mm, 3-3.5mm across; (2n=28). Intrd; wool-alien and garden escape on tips and waste ground; very scattered in En; S America.

3. V. rigida Spreng. (*V. venosa* Gillies & Hook.) - *Slender Vervain*. Stems erect, to 50cm; differs from *V. bonariensis* in leaves mostly obovate-elliptic; spikes usually >2cm in fruit, usually shorter-stalked and more crowded; calyx mostly >3mm; and see key; (2n=42). Intrd; garden escape on tips; occasional in En and Sc; S America.

122. LAMIACEAE - *Dead-nettle family*
(Labiatae)

Herbaceous annuals to perennials or dwarf shrubs, often aromatic; stems usually square in section; leaves opposite, simple, entire to serrate or rarely deeply lobed, without stipules, sessile to petiolate. Flowers 1-many in contracted cymes in axils of leaf-like to much reduced bracts, the opposite pairs of cymes often forming false whorl of flowers, sometimes the whole forming terminal spike-like inflorescence, zygomorphic, bisexual or male-sterile, hypogynous; sepals 5, fused into tube with usually 5 teeth, often separated distally into upper (usually 3-toothed) and lower (usually 2-toothed) lips; petals 5, fused into tube with 3-5 lobes scarcely or strongly grouped into upper (usually 2-lobed) and lower (usually 3-lobed) lips, sometimes upper lip scarcely lobed, or much reduced, or its 2 lobes incorporated into lower lip, variously coloured; stamens 4, 2 long and 2 short, sometimes 2, borne on corolla-tube; ovary 4-celled, each cell with 1 ovule, deeply 4-lobed; style 1, usually arising from base of ovary where 4 cells meet, sometimes ± apical; stigmas usually 2, sometimes 1, linear; fruit a cluster of 4 1-seeded nutlets.

Like Verbenaceae and Boraginaceae in its 4-celled ovary with 1 ovule per cell and fruit of 4 nutlets, but differing as under those 2 families. Some Scrophulariaceae resemble Lamiaceae vegetatively, but have totally different ovaries.

Several genera in both subfamilies may produce male-sterile flowers on same plant as or on different plants from functionally bisexual flowers; such male-sterile flowers usually have smaller corollas and much reduced, pollen-less stamens, often included in the corolla-tube. They are not covered in the following keys, but usually occur with bisexual flowers or plants. Corolla-lengths refer to length from base of calyx to tip of longest corolla-lobe on fresh flowers; considerable shortening occurs on drying.

8 groups of distinctive genera are first defined, but all are included in the main keys which follow.

Distinctive genera

Plants annual **1. STACHYS, 5. LAMIUM, 6. GALEOPSIS, 11.TEUCRIUM, 12. AJUGA, 18. CLINOPODIUM, 26.SALVIA**

Shrubs >40cm high **7. PHLOMIS, 17. SATUREJA, 19. HYSSOPUS, 21. THYMUS, 24. LAVANDULA, 25. ROSMARINUS**

Flowers with only 2 fertile stamens **22. LYCOPUS, 25. ROSMARINUS, 26. SALVIA**

Corolla yellow or yellowish **1. STACHYS, 4. LAMIASTRUM, 6. GALEOPSIS, 7. PHLOMIS, 11. TEUCRIUM, 12. AJUGA, 15. PRUNELLA, 16. MELISSA**

Calyx-teeth spiny tipped **1. STACHYS, 3. LEONURUS, 6. GALEOPSIS, 7. PHLOMIS, 22. LYCOPUS**

Calyx-teeth 10, hooked at apex **9. MARRUBIUM**

Leaves lobed >1/2 way to midrib **3. LEONURUS, 11. TEUCRIUM, 12. AJUGA, 15. PRUNELLA, 22. LYCOPUS**

Stamens longer than corolla (incl. lips) **19. HYSSOPUS, 20. ORIGANUM, 21. THYMUS, 22. LYCOPUS, 23. MENTHA, 25. ROSMARINUS**

General key

1 Corolla with well-developed lower lip; upper lip 0 or represented by 1-2 short lobes 2
1 Corolla with upper and lower lips well developed, or ± actinomorphic (4-5-lobed) 3
 2 Corolla with ring of hairs inside tube; lower lip 3-lobed (central lobe often ± bifid); upper lip of 1-2 short lobes **12. AJUGA**
 2 Corolla without ring of hairs inside tube; lower lip 5-lobed (central lobe sometimes slightly bifid); upper lip 0 **11. TEUCRIUM**
3 Stamens 2 4
3 Stamens 4 (often very reduced in female flowers) 6
 4 Shrub; leaves entire **25. ROSMARINUS**
 4 Herbaceous annual or perennial; leaves crenate to pinnately lobed,

rarely some entire 5
5 Calyx and corolla both distinctly 2-lipped **26. SALVIA**
5 Calyx with 5 equal lobes; corolla with 4 subequal lobes (the uppermost slightly wider and emarginate) **22. LYCOPUS**
6 Calyx with 2 entire lips, the upper with a dorsal outgrowth **10. SCUTELLARIA**
6 Calyx with 5-10 lobes or teeth, often 3 forming upper and 2 lower lip, without dorsal outgrowth 7
7 Calyx-teeth 10, hooked at apex **9. MARRUBIUM**
7 Calyx-teeth (4-)5, not hooked at apex 8
8 Corolla ± actinomorphic, indistinctly 2-lipped, or distinctly 2-lipped with upper lip ± flat; stamens (except in female flowers) usually fully exposed from front view of flower, sometimes longer than corolla ***Key A***
8 Corolla distinctly 2-lipped, with upper lip distinctly hooded and usually at least partially concealing stamens from front view ***Key B***

Key A **-** Calyx with 5 teeth or lobes; stamens 4, not included within corolla-tube (but often very reduced in female flowers); corolla ± actinomorphic to zygomorphic and 2-lipped with upper lip ± flat and not concealing stamens from front view.

1 Corolla ± actinomorphic, with 4 ± equal lobes or 1 lobe shortly bifid **23. MENTHA**
1 Corolla strongly 2-lipped, or weakly zygomorphic with upper lip of 2 and lower lip of 3 lobes 2
2 Upper calyx-tooth with widened apical appendage; shrub with flowers crowded in long-stalked spikes **24. LAVANDULA**
2 Upper calyx-tooth without appendage; if a shrub, flowers not crowded in long-stalked spikes 3
3 Calyx-teeth ± equal, not forming 2 lips 4
3 Calyx-teeth unequal, the upper 3 differing markedly in length and/or breadth from lower 2 9
4 Evergreen shrubs up to c.60cm; leaves entire 5
4 Herbaceous perennials; leaves often crenate to serrate 6
5 Stamens divergent, the longer 2 longer than corolla; corolla bluish-violet (very rarely white) **19. HYSSOPUS**
5 Stamens convergent, shorter than corolla; corolla pinkish-purple (very rarely white) **17. SATUREJA**
6 Flowers few in axillary clusters; plant with long procumbent stolons **14. GLECHOMA**
6 Flowers numerous in terminal inflorescences; plant without stolons 7
7 Leaves entire or nearly so **20. ORIGANUM**
7 Leaves conspicuously crenate to serrate 8
8 Calyx 15-veined; stem-leaves many pairs **13. NEPETA**
8 Calyx 5-10-veined; stem-leaves ≤4(5) pairs **1. STACHYS**
9 Corolla ≥25mm; calyx with upper lip 2-3-toothed, with lower lip with 2 much deeper but scarcely narrower lobes **8. MELITTIS**
9 Corolla ≤22mm; calyx with upper lip 3-lobed, with lower lip with 2 much narrower lobes 10
10 Stems woody, either erect to ascending or filiform and procumbent to decumbent **21. THYMUS**
10 Stems herbaceous, usually ± erect 11
11 Corolla-tube curved; stigmas ± equal; fresh plant lemon-scented **16. MELISSA**
11 Corolla-tube straight; stigmas distinctly unequal; fresh plant variously scented but not of lemon **18. CLINOPODIUM**

Key B - Calyx with 5 teeth or lobes; stamens 4, not included within corolla-tube (but often very reduced in female flowers); corolla distinctly 2-lipped, with upper lip distinctly hooded and often at least partially concealing stamens from front view.

1 Plant with stolons >10cm 2
1 Plant without stolons 5
  2 Corolla yellow **4. LAMIASTRUM**
  2 Corolla variously bluish, purplish or white 3
3 Flowers few in axillary clusters; bracts all leaf-like **14. GLECHOMA**
3 Flowers in dense whorls forming terminal inflorescence; at least upper bracts much reduced 4
  4 Lateral lobes of lower lip of corolla obscure or pointed, if rounded much <1/2 as large as terminal lobe; terminal lobe bifid for ≥1/3 length; carpels and mericarps truncate at apex **5. LAMIUM**
  4 Lateral lobes of lower lip of corolla conspicuous, rounded, usually c.1/2 as large as terminal lobe; terminal lobe not bifid or bifid for <1/3 length; carpels and mericarps rounded at apex **1. STACHYS**
5 Calyx-teeth unequal, the upper 3 differing markedly in length and/or breadth from lower 2 6
5 Calyx-teeth ± equal, not forming 2 lips 8
  6 Corolla >2cm; calyx ≥12mm **8. MELITTIS**
  6 Corolla <2cm; calyx <12mm 7
7 Flowers forming dense terminal head; upper lip of calyx nearly entire, with 3 small teeth; fresh plant not strongly scented **15. PRUNELLA**
7 Flowers in axillary whorls, only the uppermost of which merge; upper lip of calyx with 3 distinct lobes; fresh plant strongly lemon-scented **16. MELISSA**
  8 Lower leaves lobed >1/2 way to midrib **3. LEONURUS**
  8 Leaves entire to serrate 9
9 Calyx-teeth and associated bracteoles spine-tipped; lower lip of corolla with conical projection at base of each of 2 lateral lobes **6. GALEOPSIS**
9 Calyx-teeth and bracteoles not spine-tipped, or former sometimes so; lower lip of corolla without 2 conical projections 10
  10 Stigmas distinctly unequal; leaves whitish-tomentose on lowerside **7. PHLOMIS**
  10 Stigmas ± equal; leaves rarely white-tomentose on lowerside 11
11 Lateral lobes of lower lip of corolla obscure or pointed, if rounded much <1/2 as large as terminal lobe; terminal lobe bifid for ≥1/3 length; carpels and mericarps truncate at apex **5. LAMIUM**
11 Lateral lobes of lower lip of corolla conspicuous, rounded, usually c.1/2 as large as terminal lobe; terminal lobe not bifid or bifid for <1/3 length; carpels and mericarps rounded at apex 12
  12 Upper lip of corolla scarcely hooded; calyx 15-veined **13. NEPETA**
  12 Upper lip of corolla strongly hooded; calyx 5-10-veined 13
13 Calyx-teeth c. as long as wide; calyx-tube trumpet-shaped (cylindrical or ± so for most part but conspicuously expanded in distal 1/2) **2. BALLOTA**
13 Calyx-teeth usually distinctly longer than wide; calyx-tube cylindrical to obconical, not conspicuously expanded in distal 1/2 **1. STACHYS**

Other genera - **DRACOCEPHALUM** L. resembles *Prunella* in h calyx has 15 (not 10) veins and its upper lip has long (not very sho middle 1 of which is much wider than the other 2. **D. parviflorum** Nutt. *Dragon-head*), from N America, is a rare birdseed- and wool-alien occas. shortly persisting. It is an erect perennial with serrate leaves and a blue weakly lipped corolla scarcely exceeding calyx.

SUBFAMILY 1 - LAMIOIDEAE (Stachyoideae, Ajugoideae, Scutellarioideae) (genera 1-12). Plants mostly not pleasantly scented; male-sterile plants rather rare; upper lip of corolla usually conspicuously hooded (not in *Marrubium*) or 0; stamens 4, shorter than corolla; cotyledons usually at least as long as wide; seeds with endosperm; pollen with 3 furrows, 2-celled at dispersal.

1. STACHYS L. (*Betonica* L.) - *Woundworts*
Herbaceous annuals or perennials; leaves crenate to serrate; calyx with 5 ± equal acute (sometimes ± spine-tipped) lobes; corolla yellow, pink to purple, or white, with hooded upper lip, with 3-lobed lower lip; stamens 4, shorter than upper lip of corolla; whorls distant in leaf-axils, or congested in axils of reduced bracts.

1 Corolla cream to yellow 2

1 Corolla purplish, very rarely white 3

2 Annual; flowers ≤6 per node; corolla 10-16mm **9. S. annua**

2 Biennial to perennial; flowers usually ≥6 per node; corolla 15-20mm **8. S. recta**

3 Annual; corolla <10mm **10. S. arvensis**

3 Perennial; corolla >10mm 4

4 Bracteoles 0 or minute 5

4 Bracteoles mixed in with flowers, at least as long as calyx 7

5 Middle and upper stem-leaves sessile **7. S. palustris**

5 All leaves up to 1st inflorescence node petiolate 6

6 Petioles of middle and upper stem-leaves 1/10-1/5 total leaf + petiole length; few or 0 fruits ripening **6. S. x ambigua**

6 Petioles of middle and upper stem-leaves 1/4-2/5(1/2) total leaf + petiole length; all or most fruits ripening **5. S. sylvatica**

7 Corolla-tube longer than calyx; calyx very sparsely pubescent to glabrous on outside; sterile leaf-rosettes present at flowering **1. S. officinalis**

7 Corolla-tube shorter than calyx; calyx densely pubescent on outside; sterile leaf-rosettes 0 at flowering 8

8 Upper part of stem and calyx with stalked glands as well as eglandular hairs **4. S. alpina**

8 Stem and calyx with only eglandular hairs 9

9 Leaves cuneate at base; stem and both leaf-surfaces densely white-tomentose **2. S. byzantina**

9 At least lower leaves cordate at base; green upper leaf-surface showing through pubescence **3. S. germanica**

1. S. officinalis (L.) Trevis. (*Betonica officinalis* L.) - *Betony*. Erect, sparsely pubescent perennial to 75cm, with sterile leaf-rosettes and only 2-4 pairs of stem-leaves; corolla reddish-purple, 12-18mm; 2n=16. Native; hedgebanks, grassland, heaths, avoiding heavy soils; common in En and Wa, local in Jersey, extremely so in Sc and Ir. Often very dwarf on grassy cliff-tops.

2. S. byzantina K. Koch (*S. lanata* Jacq. non Crantz) - *Lamb's-ear*. Erect to ascending, densely white-woolly perennial to 75cm, with thick surface rhizomes and many sterile leaf-rosettes; corolla pinkish-purple, 15-25mm; (2n=30). Intrd; much grown in gardens, persistent throwout and sometimes self-sown on tips and waste ground; scattered in Br N to C Sc; SW Asia.

3. S. germanica L. - *Downy Woundwort*. Erect, densely white-pubescent to RRR
-subtomentose biennial to 80(100)cm, without rhizomes; corolla pinkish-purple, 12-20mm; (2n=30). Native; rough open grassland, hedgebanks and wood-borders on calcareous soils; extremely local over c.20 x 7km in Oxon, formerly elsewhere in S & C Br.

4. S. alpina L. - *Limestone Woundwort*. Erect, softly pubescent perennial to 1m, RRR
with stalked glands in upper parts, without rhizomes; corolla as in *S. germanica*;

2n=30. Native; open woods; 1 site each in Denbs and W Gloucs.

5. S. sylvatica L. - *Hedge Woundwort.* Erect, strong-smelling, harshly pubescent 561
perennial to 1m, with stalked glands in upper parts, with strong ± surface rhizomes; leaves all petiolate, ovate, cordate at base; corolla reddish-purple, 13-18mm; 2n=64. Native; woods, hedgerows, rough ground; common over most of BI.

6. S. x ambigua Sm. (*S. sylvatica* x *S. palustris*) - *Hybrid Woundwort.* Intermediate 561
between parents in most characters; leaves all petiolate, middle and upper stem-leaves with petioles 1/10-1/5 total length; few or no fruits ripening; 2n=83. Native; in habitats of either parent, but often in absence of 1 or both, distributed by fragmented rhizomes; scattered over whole of BI.

7. S. palustris L. - *Marsh Woundwort.* Erect, slightly-smelling, pubescent perennial 561
to 1m, with stalked glands in upper parts, with strong ± surface rhizomes developing slight swellings at tips late in year; upper and middle stem-leaves sessile, narrowly oblong-ovate, cordate to rounded at base; corolla pinkish-purple, 12-15mm; 2n=102. Native; damp places, by rivers and ponds, and on rough ground; common over most of BI.

8. S. recta L. - *Perennial Yellow-woundwort.* Erect to ascending, shortly pubescent perennial to 70cm, without rhizomes, without glandular hairs; corolla pale yellow, 15-20mm; (2n=32, 34, 48). Intrd; waste ground; natd at Barry Docks, Glam, since 1923; Europe.

9. S. annua (L.) L. - *Annual Yellow-woundwort.* Erect, shortly pubescent annual to 30cm, with or without glandular hairs; corolla pale yellow, 10-16mm; (2n=34). Intrd; waste ground; rather rare casual in C & S Br; Europe.

10. S. arvensis (L.) L. - *Field Woundwort.* Erect to ascending, pubescent annual to 25cm, without glandular hairs; corolla dull pinkish-purple, 6-8mm; 2n–10. Native; arable ground on non-calcareous soils; scattered throughout much of BI.

2. BALLOTA L. - *Black Horehound*
Herbaceous perennials; leaves serrate; calyx with 5 ± equal acuminate lobes; corolla reddish-mauve, with hooded upper lip, with 3-lobed lower lip; stamens 4, shorter than upper lip of corolla; whorls distant, in leaf-axils.

1. B. nigra L. - *Black Horehound.* Erect perennial to 1m, with unpleasant smell when bruised; leaves petiolate, ovate, subcordate to broadly cuneate at base; corolla 9-15mm; 2n=22. Native; hedgerows, waysides, rough ground; common in most of En, Wa and CI, very local in Sc and Ir and mostly intrd. Our plant is ssp. **meridionalis** (Bég.) Bég. (ssp. *foetida* (Vis.) Hayek); reports of ssp. **nigra**, with narrower and longer calyx-teeth, are all or nearly all errors.

3. LEONURUS L. - *Motherwort*
Herbaceous perennials; lower and middle stem-leaves and basal leaves deeply palmately to ternately lobed; calyx with 5 ± equal spine-tipped patent lobes; corolla pinkish-purple, with hooded upper lip, with 3-lobed lower lip; stamens 4, shorter than upper lip of corolla; whorls distant in leaf-axils.

1. L. cardiaca L. - *Motherwort.* Rhizomatous; stems erect, to 1.2m, shortly pubescent; corolla 8-12mm, densely pubescent outside on upper lip; 2n=18. Intrd; formerly grown medicinally, natd in waste places and waysides; thinly scattered over much of BI, becoming scarcer; Europe.

4. LAMIASTRUM Heist. ex Fabr. (*Galeobdolon* Adans.) - *Yellow Archangel*
Stoloniferous herbaceous perennials; leaves serrate; calyx with 5 ± equal triangular-acuminate lobes; corolla yellow, with hooded upper lip, with 3-lobed lower lip; stamens 4, shorter than upper lip of corolla; whorls distant in leaf-axils. Differs from *Lamium* in glabrous (not pubescent) anthers.

1. L. galeobdolon (L.) Ehrend. & Polatschek (*Lamium galeobdolon* (L.) L., *Galeobdolon luteum* Huds.) - *Yellow Archangel.* Flowering stems erect, to 60cm; stolons leafy, rooting at nodes, often >1m.

1 Most leaves with large conspicuous whitish blotches for whole year; fruiting calyx ≥(11.5)12mm; upper lip of corolla ≥(7.5)8mm wide **c. ssp. argentatum**

1 Leaves without whitish blotches, or some with blotches for some of year; fruiting calyx ≤12(12.5)mm; upper lip of corolla ≤8(8.5)mm 2

2 Bracts 1-2(2.2)x as long as wide, obtusely serrate; flowers ≤8(9) per node, at ≤4(5) nodes; flowering stems with hairs ± confined to 4 angles **a. ssp. galeobdolon**

2 Bracts (1.5)1.7-3.6x as long as wide, acutely serrate; flowers ≥(7)10 per node, at ≥(3)4 nodes; flowering stems with hairs on faces as well as angles **b. ssp. montanum**

a. Ssp. galeobdolon (*Lamium galeobdolon* ssp. *galeobdolon*, *Galeobdolon luteum* ssp. **RR**
luteum). Leaves rarely whitish-blotched; differs from ssp. *montanum* in less robust habit with less extensive stolons; and see key; 2n=18. Native; woods, wood-borders and hedgerows; very local in small area of N Lincs, in absence of ssp. *montanum.*

b. Ssp. montanum (Pers.) Ehrend. & Polatschek (*Lamium galeobdolon* ssp. *montanum* (Pers.) Hayek, *Galeobdolon luteum* ssp. *montanum* (Pers.) Dvoráková). Differs from ssp. *galeobdolon* in more robust habit with very long stolons; and see key; 2n=36. Native; woods, wood-borders and hedgerows; common over most of En and Wa, very local in SW Sc, E Ir and CI.

c. Ssp. argentatum (Smejkal) Stace (*Galeobdolon argentatum* Smejkal). Differs from other 2 sspp. in large conspicuous whitish blotches on all leaves at all seasons; and see key; (2n=36). Intrd; much grown in gardens, natd in shrubberies and waysides; scattered over BI; origin uncertain. Perhaps only a cultivar of ssp. *montanum.*

5. LAMIUM L. - *Dead-nettles*

Annuals or herbaceous perennials; leaves serrate to deeply so, rarely ± entire, calyx with 5 ± equal narrowly triangular-acuminate lobes; corolla white, or pink to purple or mauve, with hooded upper lip, with ± 1-lobed lower lip of which lateral lobes are much reduced and pointed or rounded; stamens 4, shorter than upper lip of corolla; whorls distant in leaf-axils, or ± congested in axils of modified leaves.

1 Perennials with rhizomes and/or stolons; corolla-tube curved 2

1 Annuals; corolla-tube straight 3

2 Corolla white; leaves never blotched whitish; lower lip of corolla with 2-3 teeth each side **1. L. album**

2 Corolla usually pinkish-purple; leaves usually blotched whitish; lower lip of corolla with 1 tooth on each side **2. L. maculatum**

3 ± all leaves petiolate 4

3 Middle and upper leaves subtending whorls sessile 5

4 Leaves subtending whorls serrate to crenate-serrate, with teeth <2mm long **3. L. purpureum**

4 Leaves subtending whorls deeply serrate, with many teeth >2mm long **4. L. hybridum**

5 Calyx 5-7mm at flowering, densely white- ± patent-pubescent, the teeth erect to convergent at fruiting; lower lip of corolla <3mm **6. L. amplexicaule**

5 Calyx 8-12mm at flowering, ± appressed-pubescent, the teeth divergent at fruiting; lower lip of corolla >3mm **5. L. confertum**

1. L. album L. - *White Dead-nettle.* Rhizomatous or sometimes stoloniferous perennial; calyx 9-15mm; corolla white, 18-25mm, with tube 9-14mm; 2n=18.

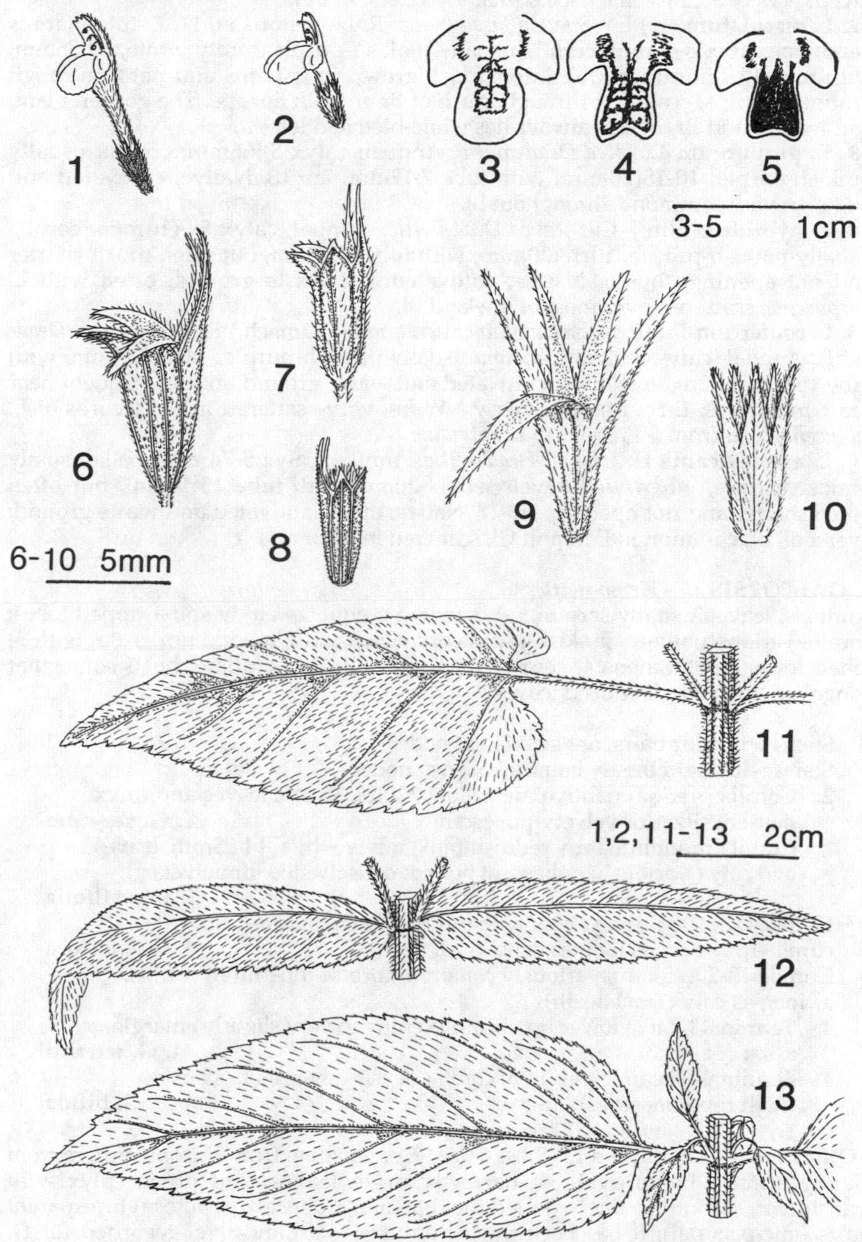

FIG 561 - Lamiaceae. 1-2, *Galeopsis* flowers. 1, *G. segetum*. 2, *G. angustifolia*. 3-5, *Galeopsis* lower lips of corolla. 3, *G. tetrahit*. 4-5, *G. bifida*. 6-8, *Clinopodium* calyces. 6, *C. menthifolium*. 7, *C. ascendens*. 8, *C. calamintha*. 9-10, *Lamium* calyces. 9, *L. confertum*. 10, *L. amplexicaule*. 11-13, *Stachys* leaves. 11, *S. sylvatica*. 12, *S. palustris*. 13, *S. x ambigua*. 3-5 drawn by C.A. Stace.

Native; hedgebanks, waysides, rough ground; common in most of lowland Br except N & W Sc, rare in CI, local (mainly NE & CE) in Ir.

2. L. maculatum (L.) L. - *Spotted Dead-nettle.* Rhizomatous and/or stoloniferous perennial; calyx 8-15mm; corolla usually pinkish-purple, rarely white, 20-35mm, with tube 10-18mm; (2n=18). Intrd; much grown in gardens and natd on rough ground and tips; scattered through much of Br and CI; Europe. The garden plant, and that natd in Br, nearly always has white-blotched leaves.

3. L. purpureum L. - *Red Dead-nettle.* Annual; calyx 5-7.5mm; corolla usually pinkish-purple, 10-18(20)mm, with tube 7-12mm; 2n=18. Native; cultivated and waste ground; common ± throughout BI.

4. L. hybridum Vill. - *Cut-leaved Dead-nettle.* Annual; calyx 5-7(10)mm; corolla usually pinkish-purple, 10-18(20)mm, with tube 7-12mm, but often much shorter and not opening; 2n=36. Native; cultivated and waste ground, often with *L. purpureum*; scattered over most of lowland BI.

5. L. confertum Fr. (*L. molucellifolium* auct. non (Schumach.) Fr.) - *Northern Dead-* **561**
nettle. Annual; calyx 8-12mm; corolla usually pinkish-purple, 14-20(25)mm, with tube 10-15mm; 2n=36. Native; cultivated and waste ground; locally frequent near coast in N, W & E Sc, Man, formerly NW En, very scattered in Ir. Records of *L. molucellifolium* from S En refer to *L. hybridum*.

6. L. amplexicaule L. - *Henbit Dead-nettle.* Annual; calyx 5-7mm; corolla usually **561**
pinkish-purple, when well-developed 14-20mm with tube 10-14mm, but often much shorter and not opening; 2n=18. Native, open, cultivated and waste ground; over ± all BI, common in E Br and CI, scattered in W Br and Ir.

6. GALEOPSIS L. - *Hemp-nettles*

Annuals; leaves usually serrate; calyx with 5 ± equal, weakly spine-tipped lobes; corolla variously white, pinkish-purple or yellow, with hooded upper lip, with 3-lobed lower lip; stamens 4, shorter than upper lip of corolla; whorls somewhat congested, in axils of reduced leaves.

1 Stems with soft hairs, not swollen at nodes 2
1 Stems with rigid bristly hairs, swollen at nodes 3
  2 Corolla predominantly pale yellow, (20)25-30mm; leaves and calyx densely silky- or velvety-pubescent **1. G. segetum**
  2 Corolla predominantly reddish-pink, rarely white, 14-25mm; leaves and calyx variously pubescent but not densely silky or velvety **2. G. angustifolia**
3 Corolla (22)27-35mm, yellow with purple blotch on lower lip; corolla-tube c.2x as long as calyx (incl. teeth) **3. G. speciosa**
3 Corolla 13-20(25)mm, variously coloured; corolla-tube rarely >1.5x as long as calyx (incl. teeth) 4
  4 Terminal lobe of lower lip of corolla entire to very slightly emarginate, ± flat **4. G. tetrahit**
  4 Terminal lobe of lower lip of corolla clearly emarginate, convex (with revolute sides) **5. G. bifida**

Other spp. - G. ladanum L. (*Broad-leaved Hemp-nettle*), from Europe, differs from *G. angustifolia* in its ovate to narrowly ovate leaves, and green calyx with translucent, smooth to finely dotted (not opaque, ± densely papillose) transparent hairs (microscope!); it has been much confused with and over-recorded for *G. angustifolia*, but has never been more than an infrequent casual and is now very sporadic.

1. G. segetum Neck. - *Downy Hemp-nettle.* Stems erect, to 50cm; leaves lanceolate **RR**
to ovate; calyx 7-10mm; corolla (20)25-30(35)mm, pale yellow; (2n=16). Native in **561**
arable land in Caerns, once constant, now sporadic; casual in arable and waste

ground, scattered and sporadic in En and Wa.

2. G. angustifolia Ehrh. ex Hoffm. - *Red Hemp-nettle.* Stems erect, to 50cm; leaves linear-lanceolate to narrowly ovate, rarely ovate; calyx 8-13mm; corolla 14-25mm, reddish-pink, rarely white; (2n=16). Native; arable land, open ground mostly on calcareous soils or maritime sand or shingle; very scattered in SC En N to SE Yorks, extinct in Sc, SW En and most of N En and Wa, decreasing. R 561

3. G. speciosa Mill. - *Large-flowered Hemp-nettle.* Stems erect to ascending, to 1m; leaves ovate; calyx 12-17mm; corolla (22)27-35mm, yellow with purple blotch on lower lip (sometimes whole lower lip purplish); 2n=16. Native; arable land, often on peaty soil with root-crops, and waste places; locally common in C & N Br and N Ir, rare and scattered in S Br and C & S Ir.

4. G. tetrahit L. - *Common Hemp-nettle.* Stems erect, to 1m; leaves narrowly ovate to ovate; calyx 12-14mm; corolla 13-20(25)mm, pinkish-red with darker lines and blotches on lower lip, the darker markings falling well short of tip of terminal lobe, often white but perhaps never yellow; 2n=32. Native; arable land, rough ground, woodland clearings, damp places; common over most of Br and Ir, rare in CI. 561

4 x 5. G. tetrahit x G. bifida = G. x ludwigii Hausskn. is intermediate in corolla shape and has 20-70 per cent pollen fertility and low seed-set; it has been recorded from Wa and C En, but is probably frequent.

5. G. bifida Boenn. - *Bifid Hemp-nettle.* Differs from *G. tetrahit* in usually shorter corolla (13-16mm) coloured like *G. tetrahit* or like *G. speciosa* or white; darker markings on lower lip very extensive to near margin (sometimes whole terminal lobe dark); and see key; 2n=32. Native; in similar places to *G. tetrahit* and probably at least as common and widespread. 561

7. PHLOMIS L. - *Sages*

Perennial herbs or shrubs; leaves entire to crenate, white-tomentose; calyx with 5 ± equal, spine-tipped teeth; corolla yellow, rarely purplish, with strongly hooded upper lip, with 3-lobed lower lip; stamens 4, shorter than upper lip of corolla; whorls somewhat congested, in axils of reduced leaves.

Other spp. - P. samia L., from Greece, is grown in gardens and has been recorded as a persistent outcast in the past; it differs from *P. russeliana* in its purplish-mauve corolla and glandular hairs on stem and bracteoles (0 in *P. russeliana*).

1. P. russeliana (Sims) Benth. - *Turkish Sage.* Herb to 1m; basal leaves ovate, cordate at base, 6-20cm, with longer petiole; calyx 20-25mm; corolla 30-35mm, yellow. Intrd; banks by roads and railways, rough ground; persistent or ± natd in few places in En and Sc; Turkey. 577

2. P. fruticosa L. - *Jerusalem Sage.* Evergreen shrub to 1.3m; basal leaves 0; lower leaves elliptic-ovate, cuneate to truncate at base, 3-9cm, with shorter petiole; calyx 10-20mm; corolla 23-35mm, yellow; (2n=20). Intrd; grown in gardens and ± natd on sea-cliffs, banks and rough ground in S En, Man and Ir, rare throwout elsewhere in Br; Mediterranean. 577

8. MELITTIS L. - *Bastard Balm*

Perennial herbs; leaves serrate; calyx with 2 lips, the upper with 2-3 short lobes, the lower with 2 deeper lobes; corolla white, or pink to mauve or purple, with flat or slightly hooded upper lip, with 3-lobed lower lip; stamens 4, shorter than upper lip of corolla; whorls distant, in leaf-axils.

1. M. melissophyllum L. - *Bastard Balm.* Stems erect, to 70cm; lower leaves ovate, cordate to rounded at base, petiolate; corolla 25-40mm, with lower lip longer than upper; 2n=30. Native; woods and hedgerows; very local in SW Wa, SW & S En E to W (formerly E) Sussex, rare escape from gardens elsewhere in Wa and S & C En. R

9. MARRUBIUM L. - *White Horehound*
Perennial herbs; leaves serrate; calyx with 10 equal teeth, each hooked at end and patent at fruiting; corolla white, with erect, flat, bilobed upper lip, with 3-lobed lower lip; stamens 4, all included in corolla-tube; whorls distant, in leaf-axils.

1. M. vulgare L. - *White Horehound*. Plant whitish-tomentose or densely pubescent; stems erect or ascending, to 60cm; leaves ovate to suborbicular, petiolate; flowers numerous and dense in each whorl; 2n=34. Native; short grassland, open or rough ground and waste places; sparsely scattered in BI to Moray, becoming much rarer, probably native only near sea in S & W Br from E Sussex to Denbs. **R**

10. SCUTELLARIA L. - *Skullcaps*
Perennial herbs; leaves entire to serrate; calyx 2-lipped, both lips entire, the upper with a dorsal outgrowth; corolla pinkish to blue or purple, with hooded upper lip, with rather obscurely 3-4-lobed lower lip; stamens 4, shorter than to c. as long as upper lip of corolla; flowers 2 at each node, in leaf-axils.

1 Flowers in axils of bracts markedly smaller than foliage leaves; lower leaves 5-15cm, with petioles >1cm **1. S. altissima**
1 Flowers in axils of foliage leaves; leaves very rarely >5cm, with petioles <1cm 2
2 Corolla 6-10mm, pale pinkish-purple, with nearly straight tube **4. S. minor**
2 Corolla 10-20mm, blue, with strongly bent tube 3
3 Calyx with glandular hairs; leaves entire at least in distal 1/2, hastate at base **3. S. hastifolia**
3 Calyx with eglandular or 0 hairs; leaves crenate along length, cordate to rounded at base **2. S. galericulata**

1. S. altissima L. - *Somerset Skullcap*. Stems erect, to 80cm; leaves ovate-cordate, serrate, petiolate; calyx with sparse usually glandular and eglandular hairs; corolla 12-18mm, blue with white lower lip; (2n=34). Intrd; natd in hedgerows and wood borders since 1929 in N Somerset and since 1972 in Surrey; S & E Europe.

2. S. galericulata L. - *Skullcap*. Stems erect to decumbent, to 50cm; leaves narrowly ovate-elliptic to lanceolate, cordate to rounded at base, crenate, sessile to shortly petiolate; calyx with rather sparse eglandular hairs; corolla 10-18mm, blue; 2n=30, 32. Native; fens, wet meadows, by ponds and rivers; locally common throughout most of BI.

2 x 4. S. galericulata x S. minor = S. x hybrida Strail occurs locally near the parents in S En and S Ir; it is intermediate in corolla and leaf characters and highly sterile, but vegetatively vigorous.

3. S. hastifolia L. - *Norfolk Skullcap*. Stems erect to decumbent, to 40cm; at least middle stem-leaves ovate-triangular, hastate at base, with 0-2 extra serrations near base on each side, shortly petiolate; calyx rather densely glandular-pubescent; corolla 12-20mm, blue; (2n=32). Intrd; natd in woodland at 1 site in W Norfolk 1948-1980, but now gone; Europe.

4. S. minor Huds. - *Lesser Skullcap*. Stems erect to decumbent, to 25cm; leaves narrowly ovate, entire or with 1-2 serrations on each side near base, rounded to cordate or subhastate at base, shortly petiolate; calyx with eglandular hairs or glabrous; corolla 6-10mm, pale pinkish-purple; 2n=(28), c.32. Native; wet heaths and open woodland on acid soils; locally frequent in S & W Br and S Ir, rare and very scattered elsewhere in En and Jersey.

11. TEUCRIUM L. - *Germanders*
Annual or perennial herbs or very low shrubs; leaves serrate to deeply lobed; calyx

rather unequally 5-lobed, not 2-lipped; corolla pinkish-purple or greenish-cream, with 5-lobed lower lip and 0 upper lip; stamens 4, shorter than lower lip; whorls distant in leaf-axils to congested in axils of reduced bracts.

1 Leaves divided much >1/2 way to midrib; annual **4. T. botrys**
1 Leaves serrate much <1/2 way to midrib; perennial 2
 2 Corolla greenish-cream; upper calyx-tooth much wider than other 4 **1. T. scorodonia**
 2 Corolla pinkish-purple (rarely white); upper calyx-tooth scarcely different from other 4 3
3 Very dwarf evergreen shrub; whorls forming a terminal inflorescence **2. T. chamaedrys**
3 Entirely herbaceous; whorls spaced down stem **3. T. scordium**

1. T. scorodonia L. - *Wood Sage*. Herbaceous perennial; stems erect, to 50cm; leaves serrate, cordate at base, petiolate; corolla greenish-cream; 2n=32, 34. Native; woods, hedgerows, hilly areas, fixed shingle and dunes, on acidic or alkaline usually well-drained soils; common in suitable places throughout most of BI, but rare in much of C Ir and CE En.

2. T. chamaedrys L. - *Wall Germander*. Dwarf evergreen shrub; stems suberect to decumbent, to 40cm; leaves serrate, cuneate at base, shortly petiolate; corolla pinkish-purple; (2n=32, 58, 60, 62, 64, 96). Native; chalk grassland at 1 site in E Sussex, discovered 1945; also grown in gardens and natd on old walls and dry banks, very scattered in BI N to C Sc, mainly S En, decreasing; Europe. Garden plants are often more robust and less hairy and are probably the hybrid *T. chamaedrys* x *T. lucidum* L.; some or most natd plants might be this also. **RR**

3. T. scordium L. - *Water Germander*. Herbaceous perennial; stems ascending to decumbent, to 50cm; leaves crenate-serrate, broadly crenate to rounded at base, ± sessile; corolla pinkish-mauve; 2n=32. Native; fens, dune-slacks and riverbanks on calcareous soils; very local, 1 site each in N Devon and Cambs, locally frequent in WC Ir, formerly scattered in En N to NW Yorks and in Guernsey, decreasing. **RRR**

4. T. botrys L. - *Cut-leaved Germander*. Erect annual to 30cm; leaves dissected nearly to midrib, petiolate; corolla pinkish-purple; 2n=32. Native; bare chalk and chalky fallow fields on downs; c.5 places in N Hants, W Kent, Surrey and E Gloucs, formerly elsewhere in S En, decreasing. **RRR**

12. AJUGA L. - *Bugles*

Annual or perennial herbs; leaves subentire to deeply divided; calyx ± equally 5-lobed; corolla pink, blue, white or yellow, with 3-lobed lower lip (terminal lobe often bifid), with very short 1-2-lobed upper lip; stamens 4, shorter than lower lip; whorls distant to slightly congested, in leaf-axils.

1 Leaves divided much >1/2 way to midrib; annual; corolla yellow **3. A. chamaepitys**
1 Leaves subentire or serrate much <1/2 way to midrib; perennial with stolons and/or rhizomes; corolla blue, pink or white 2
 2 Plant with stolons; upper bracts shorter than flowers; upper part of stem hairy only on 2 opposite sides **1. A. reptans**
 2 Plant without stolons, with rhizomes; all bracts longer than flowers; stem hairy all round **2. A. pyramidalis**

Other spp. - A. genevensis L., from Europe, was formerly natd in chalk grassland in Berks and on dunes in W Cornwall; it differs from *A. reptans* in absence of stolons and in stems hairy all round, and from *A. pyramidalis* in upper bracts shorter than flowers and without the purple or violet tint.

1. A. reptans L. - *Bugle*. Stolons present; flowering stems erect, to 30cm; bracts green or blue-tinged (a copper-leaved cultivar exists); corolla blue, rarely pink or white; 2n=32. Native; woods, shady places, damp grassland; common throughout most of BI.

1 x 2. A. reptans x A. pyramidalis = A. x pseudopyramidalis Schur (*A. x hampeana* A. Braun & Vatke) has occurred near the parents in Co Clare, W Sutherland, E Ross and Orkney; it is sterile and intermediate in most characters, forming stolons late in the season. Possibly now extinct.

2. A. pyramidalis L. - *Pyramidal Bugle*. Rhizomes present; flowering stems erect, to 30cm; bracts rather pale green strongly tinged with mauve or purple; corolla blue (rarely pink or white abroad); 2n=32. Native; rock crevices in hilly areas; very local in N, NW & S Sc, WC Ir, Westmorland and Co Antrim. **R**

3. A. chamaepitys (L.) Schreb. - *Ground-pine*. Stems erect to decumbent, usually branched low down, to 20cm; leaves divided nearly to midrib in often linear segments; corolla yellow; 2n=28. Native; bare chalk and chalky arable fields on downs; local in S & SE En N to W Suffolk, formerly Northants, decreasing. **RRR**

SUBFAMILY 2 - NEPETOIDEAE (Lavanduloideae, Rosmarinoideae) (genera 13-26). Plants mostly pleasantly scented; male-sterile plants very common; upper lip of corolla usually ± flat (hooded in *Prunella*, *Rosmarinus* and *Salvia*) or corolla ± actinomorphic; stamens 2 or 4, sometimes exceeding corolla; cotyledons usually wider than long; seeds without endosperm; pollen with 6 furrows, 3-celled at dispersal.

13. NEPETA L. - *Cat-mints*
Perennial herbs; leaves serrate; calyx with 5 subequal teeth; corolla white to blue, with flat to slightly hooded upper lip, with 3-lobed lower lip; stamens 4, shorter than upper lip of corolla.

1. N. cataria L. - *Cat-mint*. Plant softly densely grey-pubescent; stems erect, to 1m; leaves ovate, cordate at base, the lower usually ≥4cm; whorls densely crowded into terminal inflorescences; corolla white with small purple spots, the tube shorter than calyx; 2n=34, 36. Probably native; open grassland, waysides, rough ground on calcareous soils; rather scattered in En and Wa, once more common, rarely natd or casual in Ir.

2. N. x faassenii Bergmans ex Stearn (*N. mussinii* auct. non Spreng. ex Henckel; *N. racemosa* Lam. (*N. mussinii* Spreng. ex Henckel) x *N. nepetella* L.) - *Garden Cat-mint*. Plant softly grey-tomentose; stems ascending, to 1.2m; leaves narrowly ovate-oblong, mostly truncate at base, rarely >3cm; whorls in lax elongated inflorescences; corolla blue, the tube longer than calyx. Intrd; much grown in gardens, frequent throwout and sometimes natd on tips and rough ground; scattered in En, Wa and CI; garden origin. Said to be sterile, but fertile plants exist; perhaps some plants are *N. racemosa*.

14. GLECHOMA L. - *Ground-ivy*
Perennial herbs with long trailing stolons; leaves crenate-serrate; calyx with 5 subequal teeth; corolla blue, rarely pink or white, as in *Nepeta*; stamens as in *Nepeta*; flowers only 2-4 per node, the whorls ± distant in leaf-axils.

1. G. hederacea L. - *Ground-ivy*. Stolons often >1m; flowering stems suberect, to 30cm; leaves broadly ovate to orbicular, cordate at base, those on stolons with long petioles; corolla 15-20mm in bisexual flowers, smaller in female ones; 2n=36. Native; woods, hedgerows, rough ground, often on heavy soils; common throughout BI except N Sc.

15. PRUNELLA L. - *Selfheals*
Perennial herbs; leaves entire to divided ± to midrib; calyx 2-lobed, the upper lip ± truncate with 3 very short teeth, the lower lip with 2 long teeth; corolla yellow, blue, pink or white, with 3-lobed lower lip, with ± entire, strongly hooded upper lip; stamens 4, shorter than upper lip of corolla; whorls congested, in axils of strongly modified bracts.

Other spp. - P. grandiflora (L.) Scholler, from Europe, is grown in gardens and rarely occurs as a short-term relic or escape; it differs from *P. vulgaris* mainly in the corolla 18-30mm and less leafy inflorescence.

1. P. vulgaris L. - *Selfheal.* Stems erect to decumbent, to 30cm; leaves entire or shallowly toothed; corolla 10-15mm, bluish-violet, rarely pink or white; 2n=28. Native; grassland, lawns, wood-clearings, rough ground; common throughout BI.

1 x 2. P. vulgaris x P. laciniata = P. x intermedia Link (*P. x hybrida* Knaf) occurs scattered in S & C En wherever *P. laciniata* occurs or has occurred; it is variably intermediate in leaf-shape and corolla-colour; 2n=28. It is said to be sterile, but intermediates of all degrees occur, suggesting backcrossing and/or segregation.

2. P. laciniata (L.) L. - *Cut-leaved Selfheal.* Differs from *P. vulgaris* in leaves **R**
variably shaped, often some entire but at least the upper deeply divided ± to midrib; corolla usually 15-17mm, creamy-yellow or -white, rarely pale blue; (2n=28). Possibly native; calcareous grassland; scattered in S & C En N to N Lincs.

16. MELISSA L. - *Balm*
Perennial herbs; leaves serrate; calyx 2-lipped, the upper lip ± truncate with 3 short teeth, the lower lip with 2 long teeth; corolla pale yellow, becoming whitish or pinkish, with 3-lobed lower lip, with 2-lobed flat or slightly hooded upper lip; stamens 4, shorter than upper lip of corolla; whorls distant, in leaf-axils.

1. M. officinalis L. - *Balm.* Plant lemon-scented when fresh; stems erect, to 1m; leaves ovate, the lower with cordate to truncate base and petiole ≥1/2 as long as lamina; corolla 8-15mm; (2n=32, 34, 64). Intrd; much grown in gardens (often variegated) and a frequent throwout or self-sown and natd; scattered in BI N to Co Cavan, Man and Cumberland; S Europe.

17. SATUREJA L. - *Winter Savory*
Evergreen low shrubs; leaves entire; calyx with 5 subequal teeth; corolla pinkish-purple, rarely white, with 3-lobed lower lip, with shallowly 2-lobed ± flat upper lip; stamens 4, shorter than upper lip; stigmas ± equal; flowers in contracted cymes in axils of reduced leaves, the whorls slightly congested.

1. S. montana L. - *Winter Savory.* Stems erect to ascending, to 50cm; leaves linear- **570**
lanceolate, finely acute; corolla 6-12mm; (2n=30). Intrd; grown in gardens and ± natd on old walls in N Somerset and S Hants, less persistent elsewhere in S Br; S Europe.

18. CLINOPODIUM L. (*Calamintha* Mill., *Acinos* Mill.) - *Calamints*
Herbaceous annuals or perennials; leaves subentire to serrate, calyx 2-lipped, the lower lip with 2 longer and narrower teeth than the 3 teeth of upper lip; corolla pinkish-purple or violet to pale lilac or almost white, with short 3-lobed lower lip, with shallowly 2-lobed flat upper lip; stamens 4, shorter than corolla; stigmas markedly unequal; whorls distant to congested, in axils of reduced leaves.

1 Axillary flower-clusters very dense, without common stalk; calyx-tube asymmetrically curved (upper side convex, lower side concave near apex); flowers pinkish-purple to violet (rarely white) 2

1 Axillary flower-clusters in contracted cymes with common stalk; calyx-tube straight or slightly curved symmetrically on upper and lower sides; corolla very pale lilac to mauvish-pink 3
  2 Most or all whorls with >8 flowers; calyx-tube not or scarcely swollen **4. C. vulgare**
  2 Whorls with ≤6(8) flowers; calyx-tube strongly swollen near base on lower side, especially in fruit **5. C. acinos**
3 Teeth of lower calyx-lobe 1-2mm, with hairs all 0.1mm or very few longer; hairs in throat of calyx protruding beyond tube **3. C. calamintha**
3 Teeth of lower calyx-lobe 2-4mm, with many hairs ≥0.2mm; hairs in throat of calyx usually entirely included 4
  4 Teeth of lower calyx-lobe 3-4mm; corolla 15-22mm; leaves often >4cm, with 6-10 teeth on each side **1. C. menthifolium**
  4 Teeth of lower calyx-lobe 2-3(3.5)mm; corolla 10-16mm; leaves rarely >4cm, with 3-8 teeth on each side **2. C. ascendens**

Other spp. - C. grandiflorum (L.) Stace (*Calamintha grandiflora* (L.) Moench, *Satureja grandiflora* (L.) Scheele) (*Greater Calamint*), from S Europe, differs from *C. sylvaticum* in its larger flowers (calyx 10-16mm, corolla 25-40mm); it is grown in gardens and persists rarely as a throwout.

1. C. menthifolium (Host) Stace (*Calamintha sylvatica* Bromf., *Satureja menthifolia* RRR
(Host) Fritsch) - *Wood Calamint*. Stems erect, to 60cm; stem-leaves 3-7cm; calyx 6- 561
10mm, with lower teeth 3-4mm; corolla 15-22mm in bisexual flowers; 2n=24. Native; scrubby laneside bank on chalk; 1 site in Wight.

2. C. ascendens (Jord.) Samp. (*Calamintha ascendens* Jord., *C. sylvatica* ssp. 561
ascendens (Jord.) P.W. Ball, *Satureja ascendens* (Jord.) K. Maly) - *Common Calamint*. Stems erect, to 60cm; stem-leaves 1.5-4(5)cm; calyx 5-8mm, with lower teeth 2-3(3.5)mm; corolla 10-16mm in bisexual flowers; 2n=48. Native; dry banks and rough grassland, usually calcareous; local in BI N to E Donegal, Man and Durham, decreasing.

3. C. calamintha (L.) Stace (*Calamintha nepeta* (L.) Savi incl. ssp. *glandulosa* (Req.) R
P.W. Ball, *Satureja calamintha* (L.) Scheele) - *Lesser Calamint*. Stems erect, to 60cm; 561
stem-leaves 1-2(2.5)cm; calyx 3-6mm, with lower teeth 1-2mm; corolla 10-15mm in bisexual flowers; 2n=48, 72. Native; dry banks and rough grassland, usually calcareous; local in CE and SE En, extending W to Pembs, CI, formerly Yorks, decreasing. Also has more-branched flower-clusters and more strongly toothed leaves than *C. ascendens*.

4. C. vulgare L. (*Satureja vulgaris* (L.) Fritsch) - *Wild Basil*. Stems erect, to 75cm; stem-leaves 1.5-5cm; calyx 7-9.5mm, with lower teeth 2.5-4mm; corolla 12-22mm in bisexual flowers, pinkish-purple; 2n=20. Native; hedgerows, wood-borders and scrubby grassland on light soils; frequent in Br N to C Sc, very scattered in Ir, Alderney.

5. C. acinos (L.) Kuntze (*Acinos arvensis* (Lam.) Dandy, *Satureja acinos* (L.) Scheele) - *Basil Thyme*. Usually annual; stems erect to decumbent, to 25cm; stem-leaves 5-15mm; calyx 4.5-7mm, with lower teeth 1.5-2.8mm; corolla 7-10mm, violet; 2n=18. Native; bare or rocky ground, arable fields on dry, usually calcareous soils; rather local in Br N to C Sc, sparse in N and decreasing there, very local (?intrd) in C & SE Ir.

19. HYSSOPUS L. - *Hyssop*

Evergreen low shrubs; leaves entire; calyx with 5 equal teeth; corolla blue, rarely white, with 3-lobed lower lip and shallowly 2-lobed ± flat upper lip; stamens 4, longer than corolla; styles ± equal; whorls distant below, fairly congested above, in axils of reduced leaves.

1. H. officinalis L. - *Hyssop.* Stems erect to ascending, to 60cm; leaves narrowly oblong-elliptic to ± linear, rounded to obtuse or acuminate; corolla 7-12mm, the stamens ≤4mm longer; (2n=12). Intrd; grown in gardens and ± natd on old walls in Dorset, E Gloucs and Berks (no longer S Hants), rare casual elsewhere; S Europe. **570**

20. ORIGANUM L. - *Wild Marjoram*

Herbaceous perennials; leaves entire or remotely crenate-denticulate; calyx with 5 subequal teeth; corolla reddish-purple, rarely white, with short 3-lobed lower lip, with shallowly 2-lobed flat upper lip; stamens 4, longer than corolla in bisexual flowers; inflorescence a mass of dense cymes forming corymbose panicle, with large purple bracteoles.

1. O. vulgare L. - *Wild Marjoram.* Stems erect, to 50(80)cm; leaves ovate, 1-4cm; corolla 4-7mm; 2n=30. Native; dry grassland, hedgebanks and scrub, usually on calcareous soils; locally common in BI N to C Sc. Grown as the herb oregano; the herb marjoram is **O. majorana** L. (*Pot Marjoram*), a rare casual from N Africa and SW Asia.

21. THYMUS L. - *Thymes*

Dwarf evergreen shrubs; leaves entire; calyx 2-lipped, the upper lip with 3 short teeth, the lower lip with 2 long teeth; corolla pinkish-purple or mauve to white, with 2 ill-defined lips, the upper of 1 emarginate lobe, the lower of 3 lobes; stamens 4, longer than corolla in bisexual flowers; whorls crowded into dense terminal heads or the lower ones more distant, in axils of reduced leaves.

1 Leaf margin revolute so that leaves are linear to narrowly oblong-elliptic in outline; plant without procumbent stems rooting at nodes **1. T. vulgaris**
1 Leaf margins not or scarcely revolute; leaves elliptic or elliptic-oblong to narrowly so; plant usually with procumbent stems rooting at nodes 2
2 Lower internodes of flowering stems with hairs all or nearly all on the 4 angles **2. T. pulegioides**
2 Lower internodes of flowering stems with hairs mainly on 2 or 4 faces 3
3 Lower internodes of flowering stems with hairs on all faces ± evenly distributed **4. T. serpyllum**
3 Lower internodes of flowering stems with hairs on 2 opposite faces, the 2 other faces glabrous or nearly so **3. T. polytrichus**

Other spp. - T. x citriodorus Pers. (*T. vulgaris* x T. *pulegioides*) (*Lemon Thyme*) is grown in gardens and might occur as a throwout or relic; it has the upright habit of *T. vulgaris* but wider, scarcely revolute leaves and a distinct lemon scent when fresh.

1. T. vulgaris L. - *Garden Thyme.* Stems erect to decumbent, not or scarcely rooting along length, to 40cm; leaves 3-8 x 0.5-2.5mm, grey-green with very short hairs; corolla pale purple to very pale mauve; (2n=30). Intrd; grown in gardens as herb and natd on old walls and stony banks; very scattered in CI and S En; W Mediterranean. **570**

2. T. pulegioides L. - *Large Thyme.* Vegetative stems procumbent to ascending, rooting but not forming dense mats; flowering stems suberect to decumbent, to 25cm, with hairs ± only on the 4 angles; leaves ≤12 x 6mm, usually glabrous, green; corolla pinkish-purple; 2n=28. Native; short fine turf or barish places in coarser turf on well-drained chalky or sandy soils; locally frequent in S & C En, scattered N to SE Yorks, very rare and scattered in Ir and Sc. More robust than next 2 spp and flowering c.1 month later in S En. The only sp. on heaths and dunes in SE En. **570**

3. T. polytrichus A. Kern. ex Borbás (*T. drucei* Ronniger, *T. praecox* auct. non Opiz, *T. serpyllum* auct. non L.) - *Wild Thyme.* Vegetative stems procumbent, **570**

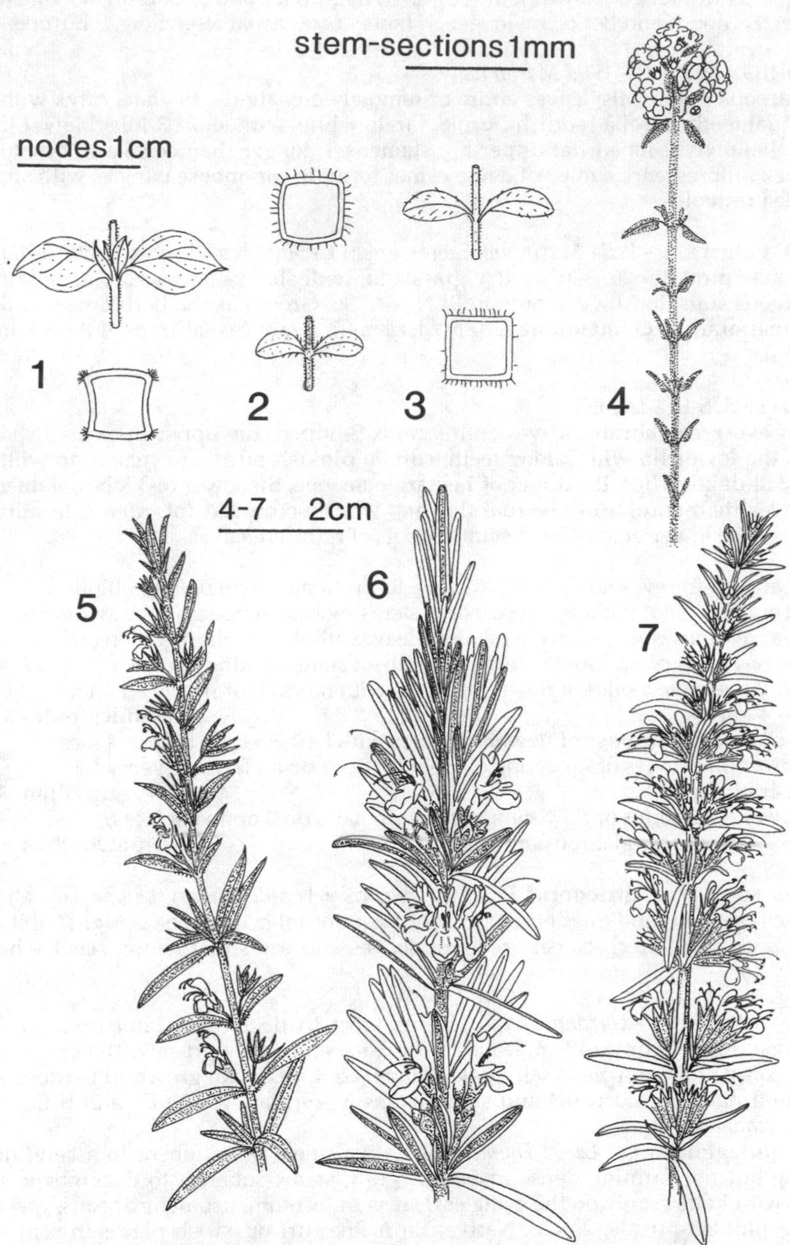

FIG 570 - Lamiaceae. 1-3, nodes and stem-sections of *Thymus*. 1, *T. pulegioides*. 2, *T. serpyllum*. 3, *T. polytrichus*. 4, *Thymus vulgaris*. 5, *Satureja montana*. 6, *Rosmarinus officinalis*. 7, *Hyssopus officinalis*.

abundantly rooting, usually forming dense mats; flowering stems decumbent to ascending, to 10cm, with hairs ± only on 2 opposite faces; leaves ≤8 x 4mm, glabrous or pubescent, green; corolla pink to pinkish-purple; 2n=50-56. Native; short fine turf or open sandy or rocky places; common over BI in suitable places; more confined than *T. pulegioides* to chalk in SE En, but much more catholic in N En, Sc, Ir and CI. Our plant is ssp. **britannicus** (Ronniger) Kerguélen (*T. praecox* ssp. *britannicus* (Ronniger) Holub, ssp. *arcticus* (Durand) Jalas).

4. T. serpyllum L. - *Breckland Thyme.* Closely resembles small *T. polytrichus* but RR
differs in flowering stems with hairs ± equally on all faces; leaves ≤5 x 1.5mm, with 570
lateral veins mostly ending short of margin (in *T. polytrichus* lateral veins usually curve round parallel to margin and join at leaf apex); 2n=24. Native; sandy heaths in c.22 sites over c.30km in W Suffolk and W Norfolk, formerly Cambs.

22. LYCOPUS L. - *Gypsywort*

Herbaceous perennials; leaves sharply serrate to deeply and acutely lobed; calyx with 5 equal teeth; corolla white with small purple dots, nearly actinomorphic, with 4 subequal lobes, the uppermost usually wider and shallowly bifid; stamens 2, longer than corolla; whorls remote, in leaf-axils.

1. L. europaeus L. - *Gypsywort.* Stems erect, to 1m; lower leaves partly divided >1/2 way to midrib into very acute lobes, tapered to short petiole; flowers densely clustered; 3-5mm; 2n=22. Native; fens, wet fields, by lakes and rivers; common over most of En and Wa, much more scattered in CI, Ir and Sc.

23. MENTHA L. - *Mints*

Herbaceous rhizomatous and/or stoloniferous perennials with characteristic scents when fresh; leaves entire to serrate; calyx with 5 equal to rather unequal teeth; corolla pinkish to bluish-mauve or white, nearly actinomorphic, with 4 subequal lobes, the uppermost usually wider and shallowly bifid; stamens 4, longer than corolla in bisexual flowers of most fertile taxa; whorls all distant in leaf-axils, or the upper congested in axils of reduced leaves.

Taxonomically difficult due to well marked plasticity, widespread hybridization, and the clonal propagation of mutants and nothomorphs by the strongly developed rhizomes. The stamens are typically exserted, but are included in *M. requienii*, some plants of *M. pulegium*, female flowers of all other spp., and most (but not all) hybrids. With practice the scent of fresh plants is very helpful, but difficult to describe.

Taxa 7-11 (*M. spicata* group), involving *M. spicata* (tetraploid), *M. suaveolens* (diploid) and their hybrids with each other and with the non-British *M. longifolia* (diploid), are particularly difficult, especially owing to the great variation of *M. spicata*, which is itself derived from *M. longifolia* x *M. suaveolens*. The 2 triploid hybrids are sterile, but the diploid hybrid and the species are fertile in bisexual plants or in female plants open to a pollen source. Pubescent plants that are not *M. suaveolens* are often impossible to name for certain. Characters of *M. suaveolens* often seen in its hybrids are the broad, obtuse, very rugose leaves with teeth partly folded under the margin and with patchy or clumped indumentum on lowerside; of *M. longifolia* are the lanceolate-oblong, acute, flat leaves with sharp patent teeth and with felted grey indumentum; and of *M. spicata* are the lanceolate to ovate, acute, not to slightly rugose leaves with usually forward-directed teeth and relatively coarse pubescence. *M. spicata* is the most variable; very broad-leaved plants, and strongly rugose-leaved plants occur, but this is the only sp. of the 3 that can be glabrous to sparsely pubescent and the only sp. that can smell of spearmint. Hybrids (especially *M. x villosa*) can be very variable, showing many combinations of characters not always connected by intermediates, but they do not exactly duplicate the combinations shown by any of the spp.

1 Stems filiform, procumbent, rooting, mat-forming; flowers <6 per node **13. M. requienii**

1 Stems sometimes procumbent but not rooting along length and mat-forming; flowers usually >6 per node 2

2 Whorls usually all axillary, the axis terminated by leaves, or by a reduced whorl; bracts like the leaves but reduced 3

2 Upper whorls contracted into terminal long or rounded head; upper bracts much reduced, unlike leaves 7

3 Calyx with hairs in throat; lower 2 calyx-teeth narrower and slightly longer than 3 upper **12. M. pulegium**

3 Calyx without hairs in throat; calyx-teeth ± equal 4

4 Calyx 1.5-2.5mm, incl. triangular teeth ≤0.5mm, pubescent all over; usually fertile **1. M. arvensis**

4 Calyx 2-4mm, incl. narrowly triangular to subulate teeth 0.5-1.5mm, glabrous or pubescent; usually sterile 5

5 Calyx 2-3.5mm, incl. teeth usually ≤1mm, the tube <2x (usually c.1.5x) as long as wide **4. M. x gracilis**

5 Calyx 2.5-4mm, incl. teeth usually ≥1mm, the tube c.2x as long as wide 6

6 Plant subglabrous; calyx mostly >3.5mm; stamens usually exserted **3. M. x smithiana**

6 Plant pubescent; calyx mostly <3.5mm; stamens usually included **2. M. x verticillata**

7 Leaves distinctly petiolate; flower-head 12-25mm across 8

7 Leaves sessile or ± so; flower-head 5-15mm across 10

8 Leaves and calyx-tube glabrous to very sparsely pubescent **6. M. x piperita**

8 Leaves and calyx-tube pubescent 9

9 Leaves ovate to narrowly ovate; flowers in rounded heads; plant normally fertile **5. M. aquatica**

9 Leaves lanceolate to narrowly ovate; flowers usually in elongate ± pyramidal heads; plant normally sterile **6. M. x piperita**

10 Leaves suborbicular to ovate, strongly rugose, pubescent, obtuse to ± rounded at apex, with teeth bent under and hence appearing as crenations from above; corolla whitish; fresh plant with sickly scent **11. M. suaveolens**

10 Leaves lanceolate to ovate-oblong, rugose or not, glabrous to densely pubescent, acute to subobtuse at apex, with teeth not bent over and hence appearing acute from above; if leaves close to those of *M. suaveolens* then corolla pinkish and fresh plant with sweet scent 11

11 Plant subglabrous to densely pubescent; corolla white to pinkish; leaves with acute, usually forwardly-directed teeth unless leaves broadly ovate to suborbicular 12

11 Plant pubescent to densely so; corolla pinkish; leaves with often acuminate teeth that curve outwards and become patent, especially near leaf-base, never broadly ovate to suborbicular 13

12 Plant normally fertile; leaves lanceolate to broadly ovate, rarely both rugose and pubescent **7. M. spicata**

12 Plant sterile; leaves usually ovate to suborbicular, often both rugose and pubescent **9. M. x villosa**

13 Plant normally fertile; leaves oblong-lanceolate to -ovate, with nearly parallel sides and broad rounded base; mostly Scotland **10. M. x rotundifolia**

13 Plant sterile; leaves lanceolate-elliptic, broadest near middle, narrowed into rounded to subcuneate base; widespread **8. M. x villosonervata**

Other spp. - M. longifolia (L.) Huds. (*Horse Mint*), from Europe, has often been

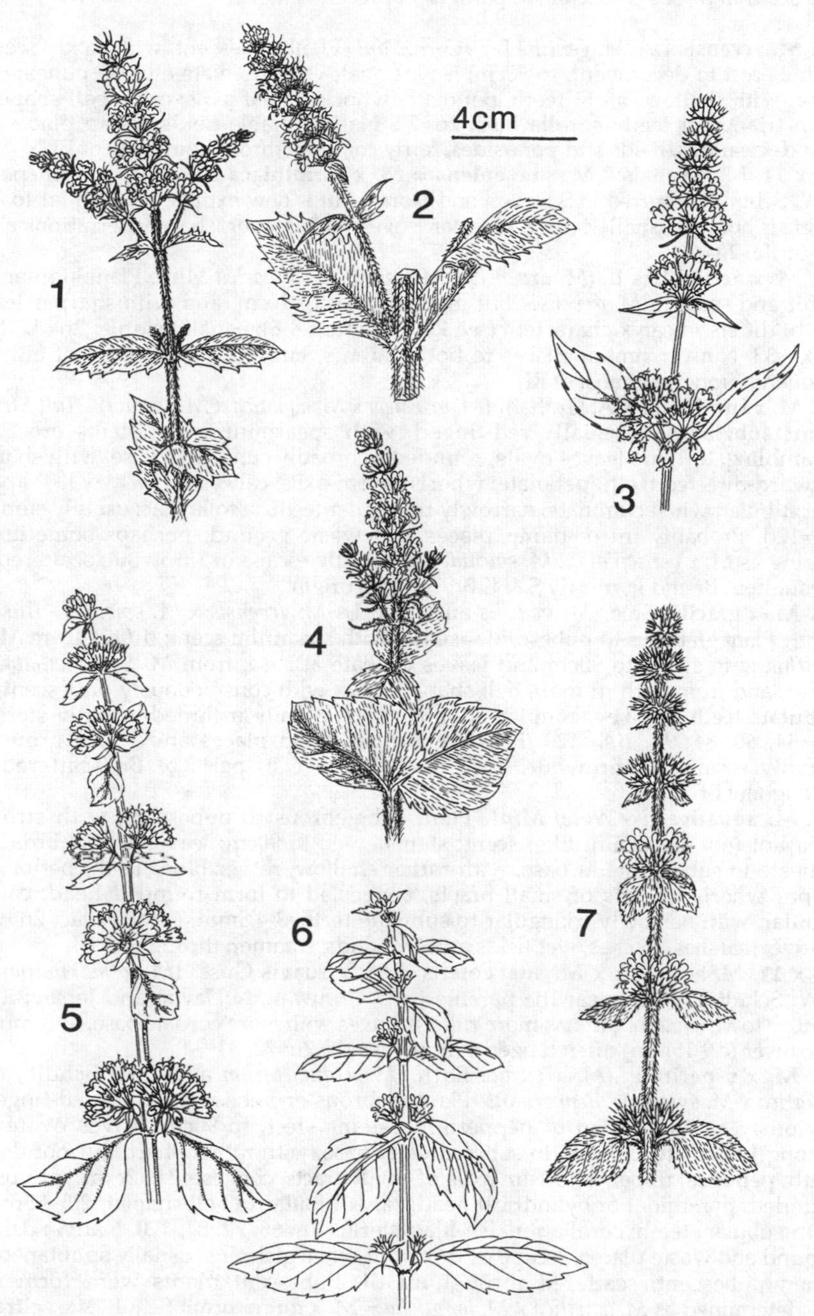

FIG 573 - *Mentha*. 1, *M. x rotundifolia*. 2, *M. x villosonervata*. 3, *M. x piperita*. 4, *M. x villosa*. 5, *M. x smithiana*. 6, *M. x gracilis*. 7, *M. x verticillata*.

misdetermined for pubescent plants of *M. spicata* or *M. x villosonervata*, but does not occur in BI; see characters in paragraph preceding key.

1. M. arvensis L. (?*M. gentilis* L.) - *Corn Mint*. Plant pubescent, with sickly scent; stems erect to decumbent, to 60cm; leaves ovate- to lanceolate-elliptic, cuneate at base, with shallow, blunt teeth, petiolate; whorls in leaf-axils; calyx bell-shaped, with triangular teeth; corolla lilac; 2n=72. Native; arable fields, damp places in wood- clearings, fields and pondsides; fairly common throughout most of BI.

1 x 11. M. arvensis x M. suaveolens = M. x carinthiaca Host (*M. x muelleriana* F.W. Schultz) occurred in S Devon and Dorset but is now extinct; it is similar to *M. arvensis* but has usually larger, broader leaves with deeper, sharper serrations and is sterile; 2n=60.

2. M. x verticillata L. (*M. arvensis* x *M. aquatica*) - *Whorled Mint*. Plant similar in **573**
habit and scent to *M. arvensis* but more robust, to 90cm, and with sharper leaf-teeth; differs in calyx characters (see key, couplets 4-6); usually sterile; 2n=42, 84, 120, 132. Native; similar places to both parents, but often in absence of either; frequent throughout most of BI.

3. M. x smithiana R.A. Graham (*M. arvensis* x *M. aquatica* x *M. spicata*) - *Tall Mint*. **573**
Plant subglabrous, usually red-tinged, with spearmint scent; stems erect to scrambling, to 1.5m; leaves ovate, rounded to broadly cuneate at base, with sharp, forward-directed teeth, petiolate; whorls in leaf-axils; calyx narrowly bell-shaped to ± tubular, with acuminate, narrowly triangular teeth; corolla lilac; usually sterile; 2n=120. Probably intrd; damp places and waste ground, perhaps sometimes arising as *M. x verticillata* x *M. spicata*, but usually escape or throwout; scattered ± throughout Br and Ir, mostly S & C Br; ?garden origin.

4. M. x gracilis Sole (*M. gentilis* auct. non L.; *M. arvensis* x *M. spicata*) - *Bushy* **573**
Mint. Plant glabrous to pubescent, usually with spearmint scent; differs from *M. x smithiana* in stems to 90cm and leaves cuneate at base, from *M. x verticillata* in scent, and from both in more bell-shaped calyx with conspicuously pubescent, ± subulate teeth (see key, couplets 4-6); stamens usually included; usually sterile; 2n=54, 60, 84, 96, 108, 120. Possibly native; damp places and waste ground, usually escape or throwout, but possibly native in parts of Br; scattered ± throughout BI.

5. M. aquatica L. - *Water Mint*. Plant subglabrous to pubescent, with strong pleasant (not spearmint-like) scent; stems erect, to 90cm; leaves ovate, broadly cuneate to subcordate at base, with rather shallow, rather blunt teeth, petiolate; upper whorls in axils of small bracts, congested to form rounded head; calyx tubular, with narrowly triangular to subulate teeth, 3-4.5mm; corolla lilac; 2n=96. Native; marshes, ditches, wet fields and by ponds; common throughout BI.

5 x 11. M. aquatica x M. suaveolens = M. x suavis Guss. (*M. x maximilianea* F.W. Schultz) occurs near the parents in W Cornwall, N Devon and Jersey; it is similar to *M. aquatica* but has more rugose leaves with more cordate base, and much narrower (c.9-15mm) often longer terminal heads; 2n=72, 84.

6. M. x piperita L. (*M. x citrata* Ehrh., *M. x dumetorum* auct. non Schult.; *M.* **573**
aquatica x *M. spicata*) - *Peppermint*. Plant glabrous or pubescent, often red-tinged, variously scented, often of peppermint; stems erect, to 90cm; leaves ovate to oblong-lanceolate, cuneate to subcordate at base, with rather sharp but not deep teeth, petiolate; upper whorls in axils of small bracts, congested to form variously rounded, pyramidal or cylindrical head; calyx tubular to bell-shaped, 2.5-4.5mm, with subulate teeth; corolla pinkish-lilac; sterile; 2n=66, 72, 84, 120. Native; damp ground and waste places, escape or throwout when glabrous, usually spontaneous when pubescent; scattered throughout BI. Pubescent plants were formerly misdetermined as *M. aquatica* x *M. longifolia* = **M. x dumetorum** Schult. Var. **citrata** (Ehrh.) Briq. (*Eau de Cologne Mint*) is a distinctive glabrous variant with scent of Eau de Cologne, ovate, subcordate leaves, and a rounded inflorescence; it is often grown and escapes.

7. M. spicata L. (*M. scotica* R.A. Graham, *M. longifolia* auct. non (L.) Huds.) - *Spear Mint*. Plant glabrous to tomentose, usually with characteristic spearmint scent; stems erect, to 90cm; leaves lanceolate to ovate; sometimes broadly ovate, usually not or slightly rugose, sometimes markedly so, usually with sharp, forwardly-directed teeth, sessile or ± so; upper whorls in axils of small bracts, congested to form narrow, long, not or little-branched, spike-like head; calyx bell-shaped, with narrowly triangular to subulate teeth, 1-3mm; corolla white to pink or lilac; 2n=36, 48. Intrd; much grown and natd in rough and waste ground; scattered throughout most of BI; ?garden origin. Fertile tetraploid derived from *M. longifolia* x *M. suaveolens*, variously approaching 1 or the other; often almost impossible to distinguish from its hybrids with these 2 spp., but the hybrids are sterile triploids. Pubescent variants have in the past been misidentified as **M. longifolia** (L.) Huds. **M. scotica** is a pubescent variant from E Sc with lanceolate, obtuse, shallowly serrate leaves with an unpleasant scent. A very distinctive glabrous variant with rugose, broadly ovate leaves and a strong spearmint scent occurs in SW Br; it was formerly referred to as glabrous *M. spicata* x *M. suaveolens*.

8. M. x villosonervata Opiz (*M. longifolia* var. *horridula* auct. non Briq.; *M. spicata* **573**
x *M. longifolia*) - *Sharp-toothed Mint*. Sterile triploids resembling pubescent, lanceolate-leaved *M. spicata* but with leaf-teeth acuminate and ± patent; corolla pink; stamens included or exserted; 2n=36, 48. Intrd; rough and waste ground; sparsely scattered in Br; ?garden origin.

9. M. x villosa Huds. (*M. x cordifolia* auct. ?non Opiz ex Fresen., *M. x niliaca* auct. **573**
non Juss. ex Jacq.; *M. spicata* x *M. suaveolens*) - *Apple-mint*. Very variable sterile triploids, varying in form from 1 parent to other; leaves glabrous (*M. x cordifolia* auct.) to pubescent (*M. x niliaca* auct.), lanceolate to suborbicular, with spearmint or various other scent; 2n=36. Intrd; much grown and natd in rough and waste ground; scattered ± throughout BI; ?garden origin. Var. **alopecuroides** (Hull) Briq. is the usual *Apple-mint* of gardens; it comes close to *M. suaveolens* but has spreading leaf-teeth, not folded under, and pink corolla. The typical var. **villosa** is more clearly intermediate. Var. **nicholsoniana** (Strail) Harley is natd in C to SE Wa and adjacent En; it has lanceolate-oblong, pubescent leaves with ± acuminate apex and sometimes ± patent teeth. None of the glabrous variants has received a varietal name.

10. M. x rotundifolia (L.) Huds. (*M. x niliaca* Juss. ex Jacq.; *M. longifolia* x *M.* **573**
suaveolens) - *False Apple-mint*. Close to some variants of *M. x villosa* but always pubescent; leaves oblong-lanceolate, broadly rounded at base, rugose, with subacute to obtuse apex and patent teeth; fertile; 2n=24. Intrd; damp places and rough ground; very scattered in Br, mainly Sc; Europe. Most of our plants are var. **webberi** (J. Fraser) Harley.

11. M. suaveolens Ehrh. (*M. rotundifolia* auct. non (L.) Huds.) - *Round-leaved Mint*. Plant pubescent, with sickly scent; stems erect, to 1m; leaves oblong-ovate to suborbicular, strongly rugose, with teeth bent under and hence appearing as blunt crenations, sessile or ± so; upper whorls in axils of bracts, congested to form very narrow often much-branched panicle of spike-like heads; calyx bell-shaped, 1-2mm; corolla usually whitish; 2n=24. Native; ditches and other damp places, waysides; locally frequent in W & S Wa, SW En and CI, natd sparsely elsewhere in BI.

12. M. pulegium L. - *Pennyroyal*. Plant (often sparsely) pubescent, with pungent **RRR**
scent, resembling a small *M. arvensis*; stems erect to procumbent, to 30cm; leaves elliptic, with few very small teeth to subentire, obtuse to rounded at apex, tapered to petiole; whorls in leaf-axils; calyx tubular to narrowly bell-shaped, 2-3mm; corolla lilac; 2n=20. Native; damp grassy or heathy places and by ponds; very local, often near sea, in S En, S Wa, Man, Jersey and Ir, formerly much more widespread N to C Sc, sometimes natd elsewhere and now increasing as a grass-seed contaminant.

13. M. requienii Benth. - *Corsican Mint*. Plant with pleasant pungent scent, glabrous to sparsely pubescent; stems filiform, procumbent, to 12cm; leaves ≤5mm, orbicular to broadly elliptic, entire, petiolate; whorls of 2-6 flowers, in leaf-axils; calyx 1-2mm, bell-shaped; corolla lilac; 2n=18. Intrd; natd from gardens on damp paths and rocky places; very scattered in S Br and 3 parts of Ir.

24. LAVANDULA L. - *Lavenders*
Evergreen shrubs; leaves entire, lanceolate to oblanceolate or linear; calyx with 5 subequal teeth, the uppermost with obcordate appendage at apex; corolla purple, weakly zygomorphic, with short 3-lobed lower lip, with short shallowly 2-lobed flat upper lip; stamens 4, included within corolla-tube; flowers on long peduncles, congested in spikes with much shorter bracts.

1. L. x intermedia Loisel. (*L. angustifolia* Mill. x *L. latifolia* Medik.) - *Garden Lavender*. Erect shrub to 1m, with characteristic scent; leaves 2-4cm, white-tomentose when young; peduncles up to 40cm, with 0-2 pairs of reduced leaves, simple or with 3 branches; spikes 2-8cm; corolla 9-12mm; (2n=48, 50). Intrd; much grown in gardens, rarely self-sown on walls and banks or persistent throwout; sporadic in S En, cultivated on field-scale in E Anglia; W Mediterranean. Most garden, crop and natd *Lavandula* plants, which exist in many cultivars, are probably this fertile hybrid, but certainly some garden plants and perhaps some of those natd might belong to 1 of the parents (particularly *L. angustifolia*).

25. ROSMARINUS L. - *Rosemary*
Evergreen shrubs: leaves linear, entire, with revolute margins; calyx 2 lipped, with large ± entire upper lip, with narrower 2-lobed lower lip; corolla pale to deep mauvish-blue, strongly zygomorphic, with 2-lobed upper lip, with 3-lobed lower lip; stamens 2, longer than corolla; whorls ± distant in leaf-axils.

1. R. officinalis L. - *Rosemary*. Usually erect shrub to 2m, with characteristic **570**
scent; leaves 1.5-4cm, tomentose on (mostly hidden) lowerside; corolla 10-15mm; (2n=24). Intrd; much grown in gardens, self-sown on walls and rough ground; sporadic in S En, Caerns, Cheshire and CI, not fully frost-hardy; Mediterranean.

26. SALVIA L. - *Claries*
Herbaceous annuals to perennials or rarely small shrubs; leaves simple, serrate to crenate; calyx 2-lipped, the lower with 2 longer teeth, the upper entire or with 3 shorter teeth; corolla blue to purple or pink, or yellow, rarely red or white, strongly zygomorphic, with shortly bilobed strongly hooded upper lip and 3-lobed lower lip; stamens 2, shorter than upper lip of corolla; whorls of 1-many flowers in axils of modified bracts in terminal interrupted spikes.

1 Corolla ≥18mm 2
1 Corolla ≤18mm 4
  2 Corolla yellow with reddish markings, >30mm; leaves often hastate at base **2. S. glutinosa**
  2 Corolla blue to violet, rarely pink or white, <30mm; leaves never hastate at base 3
3 Bracts green, often tinged with violet-blue, much shorter than flowers **3. S. pratensis**
3 Bracts pink or white, at least as long as flowers **1. S. sclarea**
  4 Flowers (8)15-30 in each whorl **7. S. verticillata**
  4 Flowers 1-6(8) in each whorl 5
5 Inflorescence with conspicuous tuft of green or coloured flower-less bracts at apex **6. S. viridis**
5 Inflorescence without terminal tuft of conspicuous flower-less bracts 6
  6 Eglandular; stem-leaves usually >3 pairs, linear to lanceolate-oblong,

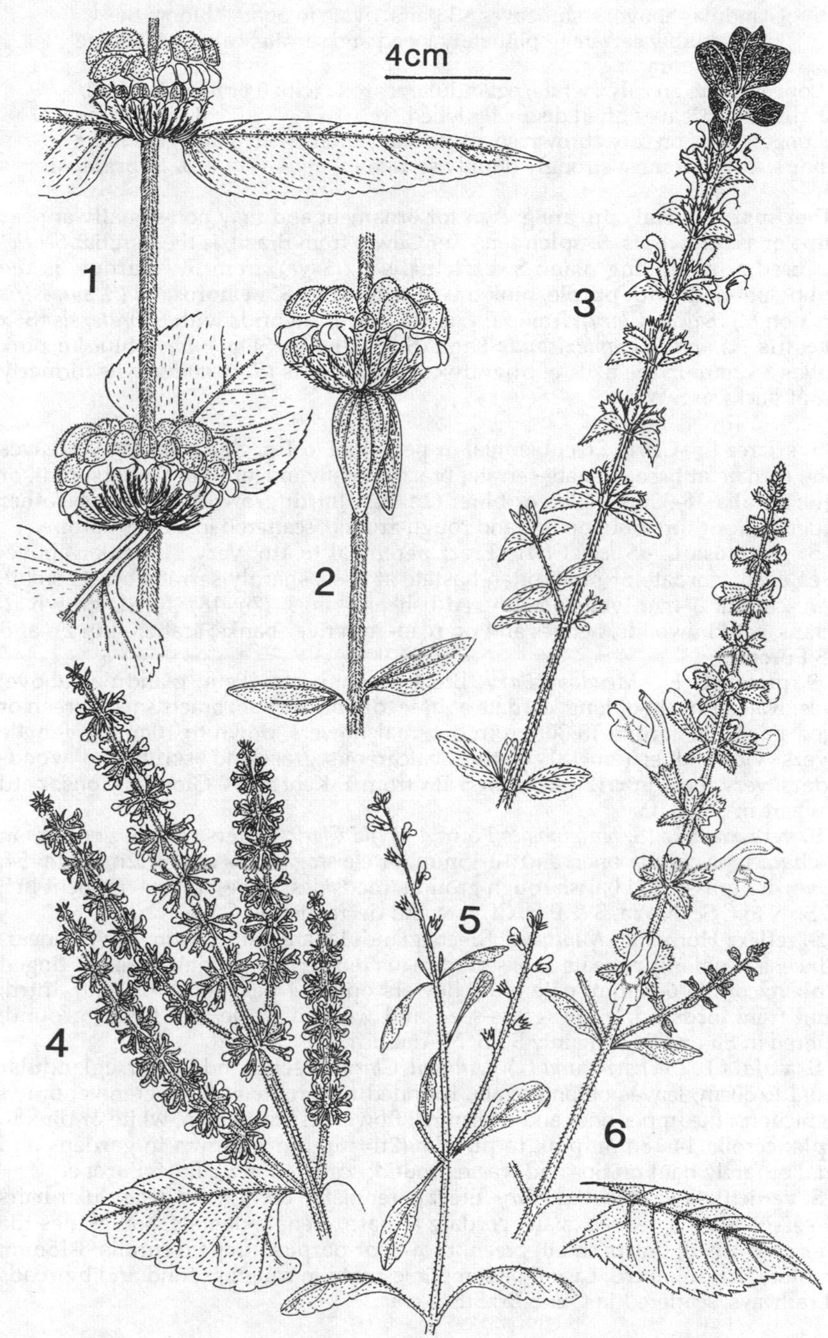

FIG 577 - Lamiaceae. 1-2, *Phlomis*. 1, *P. russeliana*. 2, *P. fruticosa*. 3-6, *Salvia*. 3, *S. viridis*. 4, *S. verticillata*. 5, *S. reflexa*. 6, *S. glutinosa*.

crenate; calyx with hairs <0.2mm **5. S. reflexa**
6 Glandular above; stem-leaves ≤3 pairs, ovate to ovate-oblong, the lower doubly serrate to pinnately lobed and serrate; calyx with some hairs >0.5mm 7
7 Longest hairs on calyx white, eglandular; corolla with 0 or few glandular hairs; lower leaves often distinctly lobed **4. S. verbenaca**
7 Longest hairs on calyx brownish, glandular; corolla with many glandular hairs; leaves at most strongly doubly serrate **3. S. pratensis**

Other spp. - Several spp. are grown for ornament and may occasionally appear on tips or as throwouts. **S. splendens** Ker Gawl., from Brazil, is the familiar *Scarlet Sage* used as a bedding plant; **S. officinalis** L. (*Sage*), from SW Europe, is the shrubby pot-herb with purple, pink or white flowers; **S. nemorosa** L. (*S. sylvestris* auct. non L.) (*Balkan Clary*), from SE Europe, and its hybrids with *S. pratensis* (**S. x sylvestris** L.) and *S. amplexicaulis* Lam. (**S. x superba** Stapf), have blue to pink corollas 8-12mm in the axils of brightly coloured bracts (*S. nemorosa* was formerly natd at docks in S Wa).

1. S. sclarea L. - *Clary*. Erect biennial or perennial to 1m, glandular above; leaves ovate, cordate at base, crenate-serrate; bracts usually longer than flowers, pink or white; corolla 18-30mm, lilac or blue; (2n=22). Intrd; grown in gardens, rather frequent relic or throwout on tips and rough ground; scattered in S Br; S Europe.

2. S. glutinosa L. - *Sticky Clary*. Erect perennial to 1m, very glandular above; **577** leaves ovate, cordate or more often hastate at base, sharply serrate; bracts small, green; corolla 3-4cm, yellow with reddish markings; (2n=16). Intrd; grown in gardens, natd in woods, hedges and on road- and river-banks; scattered in En and Sc; S Europe.

3. S. pratensis L. - *Meadow Clary*. Erect perennial to 80cm, glandular above; **RRR** leaves ovate or ovate-oblong, cordate at base, doubly serrate; bracts small, green or tinged purplish; corolla 15-30mm in bisexual flowers, down to 10mm in female flowers, violet-blue; (2n=18). Native; calcareous grassland, scrub and wood-borders; very local in c.12 places in S En from E Kent to W Gloucs, Mons, natd elsewhere in C & S Br.

4. S. verbenaca L. (*S. horminoides* Pourr.) - *Wild Clary*. Differs from *S. pratensis* in key characters; corolla open and 10-15mm, or cleistogamous and 6-12mm; 2n=54. Native; dry grassy and barish rough ground, roadsides, dunes; rather frequent in S & E Br N to C Sc, N Wa, S & E Ir, CI, rare and decreasing in C En.

5. S. reflexa Hornem. - *Mintweed*. Erect eglandular annual to 60cm; leaves linear- **577** to lanceolate-oblong, cuneate at base, entire to crenate; bracts small, green or tinged purplish; corolla 6-12mm, pale blue; flowers only 1-4 per whorl; (2n=20). Intrd; casual from birdseed, grain, grass-seed and wool in waste and rough ground; scattered in En and S Sc, mainly S En; N America.

6. S. viridis L. (*S. horminum* L.) - *Annual Clary*. Erect glandular or eglandular **577** annual to 50cm; leaves oblong-ovate, rounded to cordate at base, crenate; bracts conspicuous, the upper ones and a terminal flower-less tuft green, white or pink to purple; corolla 14-18mm, pink to purple; (2n=16). Intrd; grown in gardens and casual or rarely natd on tips and waste ground; scattered in SE En; S Europe.

7. S. verticillata L. - *Whorled Clary*. Erect perennial to 80cm, with eglandular hairs **577** and sessile glands; leaves ovate, cordate at base, crenate-dentate, sometimes the lower ones lobed; bracts small, green, brown or purplish-tinged; corolla 8-15mm, lilac-blue; (2n=16). Intrd; casual or sometimes natd in rough ground and by roads and railways; scattered in C & S Br; S Europe.

123. HIPPURIDACEAE - *Mare's-tail family*

Rhizomatous, perennial, aquatic or mud-dwelling herbs, with conspicuous air-cavities in stems; leaves in whorls of 6-12, simple, linear, sessile, without stipules. Flowers 1 per leaf-axil, very small, often male, female and bisexual on same plant with female ones more apical, epigynous; perianth reduced to minute rim at apex of ovary; stamen 1, borne on top of ovary if present; ovary 1-celled with 1 apical ovule; style/stigma 1, filiform; fruit a 1-seeded nut.

The only aquatic or mud-plant with linear simple entire leaves >4 in a whorl.

1. HIPPURIS L. - *Mare's-tail*

1. H. vulgaris L. - *Mare's-tail.* Usually aquatic, then with stems to 1(2)m, the apical part emergent and bearing flowers, and with leaves up to 8cm; sometimes on mud, then with stems c.4-20cm and leaves c.1-2cm; 2n=32. Native; in ponds and slow-flowing rivers, especially base-rich; locally frequent throughout BI.

124. CALLITRICHACEAE - *Water-starwort family*

Annual or perennial herbs, aquatic or on mud, with filiform stems; leaves opposite, simple, linear, entire or notched at apex, often forming terminal rosette, without stipules, ± sessile. Flowers very small, monoecious, 1 or 2 (and then 1 of each sex) per leaf-axil, each with 0 or 2 bracteoles; perianth 0; male flowers with 1 stamen; female flowers with 1 4-celled ovary, each cell with 1 apical ovule; styles/stigmas 2, filiform; fruit a group of 4 1-seeded nutlets, 2 pairs each more closely united.

Distinguished from other aquatics and mud-dwellers by the opposite, thin, narrow, often notched leaves, single stamen and distinctive fruits.

1. CALLITRICHE L. - *Water-starworts*

Vegetatively very variable and often shy-fruiting; difficult to identify certainly without a strong lens or microscope. The following 3 keys should be used only on fruiting material, preferably with some stamens available as well. Use of names for British plants was very confused before the work of H.D. Schotsman (1954 onwards); synonyms should be applied with great caution. Fruits should be mature. Plants normally reaching water surface but recently flooded can be told by their terminal rosettes of leaves different in shape from those further back. Fruit-wings are whitish or translucent when viewed in transmitted light. Leaf-shape is notoriously variable and misleading.

General key

1 Fruiting plants terrestrial or on wet mud — *Key C*
1 Fruiting plants mainly or completely submerged — 2
  2 Fruiting plants completely submerged — *Key A*
  2 Fruiting plants mainly submerged but with terminal leaf rosette at water surface — *Key B*

Key A - Fruiting plants completely submerged; terminal leaves not in distinct rosette; pollen grains with ± colourless unsculptured wall (beware re-submerged plants with terminal leaf-rosette).

1 Leaves transparent, all with 1 vein, without stomata; minute sessile hairs present only in leaf-axils, composed of a ± irregular cell-mass — 2
1 Leaves opaque, often some with >1 vein, often some with stomata; minute sessile hairs present on leaves, stems and in leaf-axils, composed of cells radiating from central point — 3
  2 Fruits ± orbicular in side view, c.1.4-2.2(3.3)mm, with conspicuous wing 0.1-0.5mm wide; leaves mostly ≥1cm, conspicuously notched at

apex, usually pale to mid green **1. C. hermaphroditica**

2 Fruits wider than long, c.1-1.2 x 1.4-1.6mm, not winged nor sharply keeled; leaves usually <1cm, truncate to shallowly notched at apex; usually dark green **2. C. truncata**

3 Leaves expanded suddenly just at apex with deep notch ± like a bicycle-spanner; fruits ± orbicular in side view, 1.2-1.5mm, with wing usually <0.1mm wide **7. C. hamulata**

3 Leaves not expanded suddenly just at apex, with variable, often shallow or asymmetric notch; fruits often longer than wide, 1-1.4 x 1-1.2mm, with wing usually ≥0.1mm wide **6. C. brutia**

Key B - Fruiting plants aquatic but with terminal rosette of leaves at water surface

1 Flowers submerged; pollen grains with ± colourless unsculptured wall; styles usually persistent, reflexed and appressed to sides of fruit 2

1 Flowers aerial; pollen-grains yellow, with sculptured wall; styles, if persistent, erect to patent 3

2 Submerged leaves expanded at apex with deep notch ± like a bicycle-spanner; fruits ± orbicular in side view, 1.2-1.5mm, with wing usually <0.1mm wide **7. C. hamulata**

2 Submerged leaves not expanded at apex, with variable, often shallow or asymmetric notch; fruits often longer than wide, 1-1.4 x 1-1.2mm, with wing usually ≥0.1mm wide **6. C. brutia**

3 Fruits without wing, rounded to obtuse at edges, distinctly longer than wide **5. C. obtusangula**

3 Fruits with wing 0.07-0.25mm wide, often nearly orbicular in side view 4

4 Fruits with wing 0.12-0.25mm wide; stamens c.2mm at anthesis, with anthers c.0.5mm wide; pollen grains ± globose **3. C. stagnalis**

4 Fruits with wing 0.07-0.1mm wide; stamens c.4mm at anthesis, with anthers c.1mm wide; many pollen grains mis-shapen or ellipsoid **4. C. platycarpa**

Key C - Fruiting plants terrestrial or on wet mud

1 Fruits without wing, rounded to obtuse at edges **5. C. obtusangula**

1 Fruits with wing 0.07-0.25mm wide 2

2 Fruits with a stalk 2-10mm **6. C. brutia**

2 Fruits sessile or with stalk <2mm 3

3 Styles usually persistent, reflexed and appressed to sides of fruit; pollen grains with ± colourless unsculptured wall **7. C. hamulata**

3 Styles, if persistent, erect to patent; pollen grains yellow, with sculptured wall 4

4 Fruits with wing 0.12-0.25mm wide; stamens c.2mm long, with anthers c.0.5mm wide; pollen grains ± globose **3. C. stagnalis**

4 Fruits with wing 0.07-0.1mm wide; stamens c.4mm at anthesis, with anthers c.1mm wide; many pollen grains mis-shapen or ellipsoid **4. C. platycarpa**

Other spp. - C. palustris L., a small fragile plant with obovate fruits 0.8-1 x 0.7-0.8mm usually narrowly winged just at apex, has been found for certain in BI once only (Surrey, 1877); other records are probable or certain errors. **C. cophocarpa** Sendtn. differs from *C. platycarpa* in the absence of fruit wings and from *C. obtusangula* in its nearly orbicular fruits; it has been said to occur in BI but such statements were based on supposition, not fact.

1. C. hermaphroditica L. - *Autumnal Water-starwort.* Submerged annual to 50cm; **581**
leaves 8-18mm, widest near base, tapering to emarginate apex; fruits common, ± orbicular in side view, c.1.4-2.2(3.3)mm, with wing 0.1-0.5mm wide; 2n=6. Native;

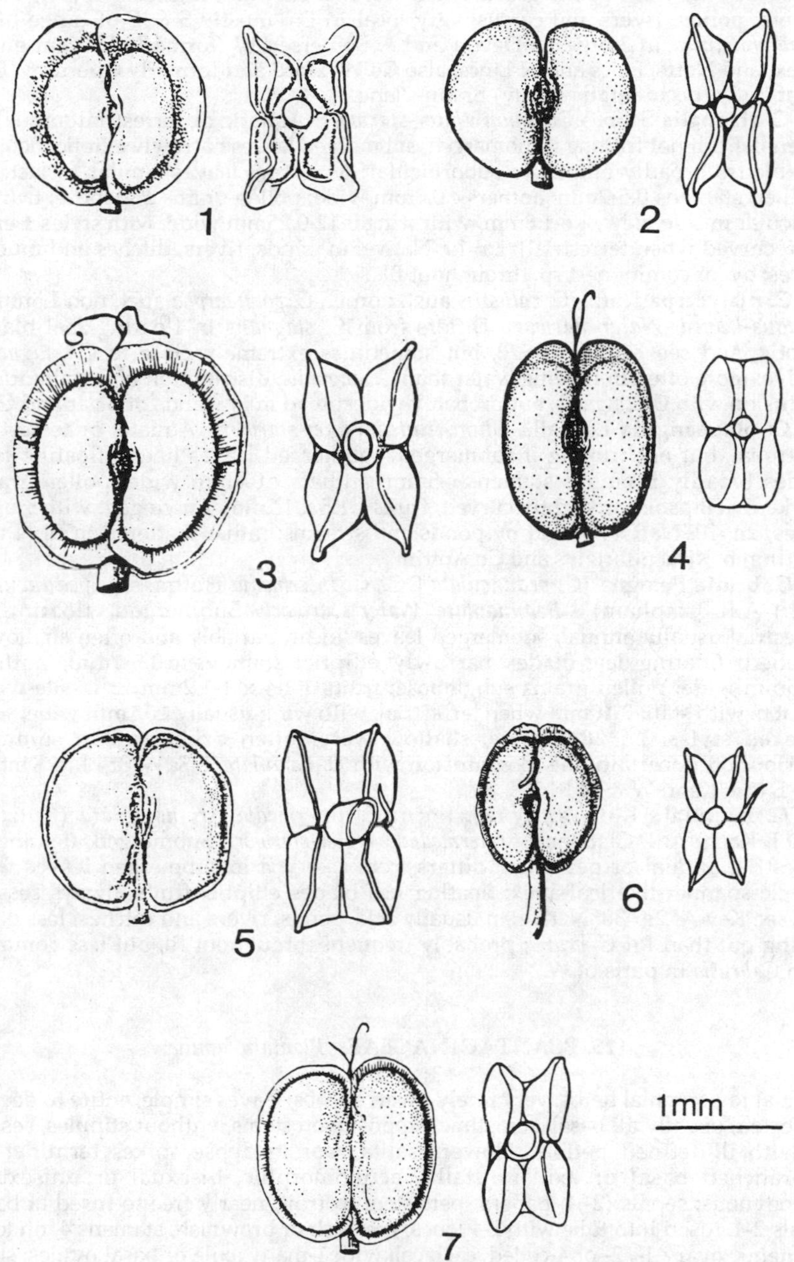

FIG 581 - Fruits of *Callitriche,* lateral and basal views. 1, *C. hermaphroditica.* 2, *C. truncata.* 3, *C. stagnalis.* 4, *C. obtusangula.* 5, *C. hamulata.* 6, *C. brutia.* 7, *C. platycarpa.* Courtesy of H.D. Schotsman.

lakes and rivers; scattered in BI N from S Lincs, Worcs, Brecs and W Cork, S Wilts.

2. C. truncata Guss. - *Short-leaved Water-starwort*. Differs from *C. hermaphroditica* R
in leaves 5-11mm, darker green when fresh; fruits rare; and see Key A; 2n=6. 581
Native; ponds, rivers and canals; very local in En, mostly S & E of range of *C. hermaphroditica* in 3 areas, S Devon and N Somerset, W (formerly E) Kent and S Essex, and Notts, Leics and N Lincs, also Co Wexford and formerly Guernsey. Our plant is ssp. **occidentalis** (Rouy) Braun-Blanq.

3. C. stagnalis Scop. - *Common Water-starwort*. Aquatic or terrestrial annual or 581
perennial, but not fruiting if submerged; submerged leaves narrowly elliptic; floating leaf-blades broadly elliptic to suborbicular; terrestrial leaves similar to last but smaller; stamens 0.5-2mm; anthers c.0.5mm wide; pollen grains ± globose; fruits ± orbicular in side view, 1.6-1.8mm, with wing 0.12-0.25mm wide, with styles ± erect (or recurved when terrestrial); 2n=10. Native; in ponds, rivers, ditches and muddy places; by far commonest sp. throughout BI.

4. C. platycarpa Kütz. (*C. palustris* auct. non L., *C. polymorpha* auct. non Lönnr.) - 581
Various-leaved Water-starwort. Differs from *C. stagnalis* in floating leaf-blades elliptic; and see Key B; 2n=20, but sometimes extremely close to *C. stagnalis*. Native; more often in flowing water than *C. stagnalis*; distribution uncertain due to confusion with *C. stagnalis*, but probably widespread in lowlands of Br, Ir and CI.

5. C. obtusangula Le Gall - *Blunt-fruited Water-starwort*. Aquatic or terrestrial 581
perennial, but not fruiting if submerged; submerged leaves linear; floating leaf-blades broadly rhombic; stamens c.5mm; anthers c.0.6mm wide; pollen grains markedly ellipsoid or slightly curved; fruits c.1.5 x 1.2mm, unwinged, with ± erect styles; 2n=10. Native; in and by ponds and streams; rather scattered in BI N to S Northumb, Kirkcudbrights and Co Antrim.

6. C. brutia Petagna (*C. pedunculata* DC., *C. intermedia* Hoffm. ssp. *pedunculata* 581
(DC.) A.R. Clapham) - *Pedunculate Water-starwort*. Submerged, floating or terrestrial usually annual; submerged leaves linear, variably and often shallowly notched; floating leaf-blades narrowly elliptic; stamens c.0.5-1mm; anthers c.0.5mm wide; pollen grains subglobose; fruits 1-1.4 x 1-1.2mm, ± sessile when aquatic, with stalk 2-10mm when terrestrial, with wing usually ≥0.1mm wide, with reflexed styles; 2n=28. Native; shallow water often drying up in summer; distribution uncertain due to confusion with *C. hamulata*; S & W Br N to Kintyre and E to E Kent, W & N Ir, ?CI.

7. C. hamulata Kütz. ex W.D.J. Koch (*C. intermedia* ssp. *hamulata* (Kütz. ex 581
W.D.J. Koch) A.R. Clapham) - *Intermediate Water-starwort*. Submerged, floating or terrestrial annual or perennial; differs from *C. brutia* in submerged leaves with bicycle-spanner-like leaf-apex; floating leaf-blades elliptic; fruits always sessile; and see Key A; 2n=38. Native; in usually acid ponds, rivers and ditches, less often drying out than for *C. brutia*; probably frequent throughout Bl, but less common than *C. brutia* in parts of W.

125. PLANTAGINACEAE - *Plantain family*

Annual to perennial herbs, very rarely dwarf shrubs; leaves simple, entire to deeply dissected, usually all basal, sometimes opposite on stems, without stipules, sessile or with ill-defined petiole. Flowers solitary or in dense spikes terminal on unbranched basal or axillary stalks, actinomorphic, bisexual or unisexual, hypogynous; sepals (2-)4, papery, persistent in fruit, nearly free to fused at base; petals 2-4, fused into tube with 2-4 lobes, greenish or brownish; stamens 4, on long filaments; ovary 1-, 2- or 4-celled, each cell with 1-many axile or basal ovules; style 1, terminal; stigma 1, linear; fruit a 2- or 4-celled transversely-dehiscing capsule or indehiscent 1-seeded nut.

Easily distinguished by the brownish spike-like inflorescence (or long-stalked solitary male flowers in *Littorella*) and leaves usually all in a basal rosette.

1 Flowers bisexual, in compact spikes; fruit a many-seeded dehiscent capsule **1. PLANTAGO**
1 Flowers unisexual, the male solitary on long stalks; fruit a 1-seeded nut **2. LITTORELLA**

1. PLANTAGO L. - *Plantains*
Stolons 0; flowers bisexual, in compact spikes, 4-merous; stamens borne on corolla-tube; ovary 2- or 4-celled, each cell with 2-many axile (and basal) ovules; capsule transversely dehiscent.

1 Spikes borne in axils of opposite stem-leaves 2
1 Spikes borne on leafless scapes 3
2 Inflorescence and peduncles with many glandular hairs; spikes with all bracts similar **7. P. afra**
2 Inflorescence and peduncles with 0 glandular hairs; spikes with lower bracts strongly differing from upper bracts **6. P. arenaria**
3 Scapes strongly furrowed **5. P. lanceolata**
3 Scapes not furrowed 4
4 Corolla-tube glabrous on outside; leaves narrowly elliptic to broadly ovate, at most bluntly and distantly toothed 5
4 Corolla-tube pubescent on outside; leaves linear to linear-elliptic, often sharply toothed or deeply and finely lobed 6
5 Capsule usually 4-seeded; seeds >2mm; stamens exserted >5mm; anthers >1.5mm **4. P. media**
5 Capsule usually >4-seeded; seeds <2mm; stamens exserted <5mm; anthers <1.5mm **3. P. major**
6 Capsule 2-celled, with 1-2 seeds per cell; seeds >1.5mm; corolla-lobes with conspicuous brown midrib; leaves usually entire, sometimes slightly toothed **2. P. maritima**
6 Capsule 3(-4)-celled, with 1-2 seeds per cell; seeds <1.5mm; corolla-lobes with 0 or inconspicuous midrib; leaves usually deeply and narrowly lobed, sometimes toothed, rarely all entire **1. P. coronopus**

Other spp. - P. sempervirens Crantz, from SW Europe, was formerly natd in Kent, but not since 1920; it superficially resembles *P. arenaria*, but is a woody perennial.

1. P. coronopus L. - *Buck's-horn Plantain*. Usually pubescent annual to perennial with 1-many rosettes; leaves linear and entire to toothed, or deeply and finely divided; scapes to 20cm; spikes up to 4(7)cm; 2n=10. Native; barish places or in very short turf on sandy or gravelly soils, sometimes on rocks, mostly near sea; common round coasts of BI, inland in scattered lowland places in En and SW Wa.

2. P. maritima L. - *Sea Plantain*. Usually glabrous perennial with 1-many rosettes; leaves linear to linear-elliptic, entire or sparsely toothed; scapes to 30cm; spikes up to 7(10)cm; 2n=12. Native; in salt-marshes, rock-crevices and short turf near sea, and wet rocky places on mountains; common round coasts of BI, inland on mountains in Sc, N En and N & W Ir, rare in inland salt-marshes or by salt-treated roads.

3. P. major L. - *Greater Plantain*. Glabrous to pubescent perennial with 1 rosette; leaves ovate to broadly so, abruptly narrowed to petiole usually ± as long, entire to weakly toothed; scapes to 40cm; spikes up to 20cm. Native.

a. Ssp. major. Leaves mostly with 5-9 veins, usually obtuse at apex, subcordate to rounded at base and subentire; capsules mostly with 4-15 seeds; seeds (1)1.2-1.8(2.1)mm; 2n=12. Open and rough ground, either cultivated or grassy, and on lawns; abundant throughout BI.

b. Ssp. intermedia (Gilib.) Lange (*P. intermedia* Gilib., *P. uliginosa* F.W. Schmidt).

Plant usually smaller with much shorter spikes; leaves mostly with 3-5 veins, usually subacute at apex, broadly cuneate at base and ± undulate-toothed near base; capsules mostly with (9)14-25(36) seeds; seeds (0.6)0.8-1.2(1.5)mm; 2n=12. Damp, usually slightly saline places near sea and less often inland; distribution uncertain but probably scattered through much of BI.

4. P. media L. - *Hoary Plantain.* Pubescent perennial with 1-few rosettes; leaves elliptic to broadly ovate, usually rather abruptly narrowed to usually short petiole, usually weakly toothed; scapes to 40cm; spikes up to 6(12)cm; 2n=24. Native; neutral and basic grassland; locally common in En, very local in Ir, Wa, CI and S & E Sc.

5. P. lanceolata L. - *Ribwort Plantain.* Glabrous to pubescent perennial with 1-several rosettes; leaves linear- to narrowly ovate-elliptic, very gradually narrowed to petiole, entire to sparsely and weakly toothed; scapes to 50cm; spikes up to 4(8)cm; 2n=12. Native; grassy places; abundant throughout BI.

6. P. arenaria Waldst. & Kit. (*P. scabra* Moench nom. illeg., *P. psyllium* L. nom. ambig., *P. indica* L. nom. illeg.) - *Branched Plantain.* Erect to decumbent pubescent annual to 30(50)cm, with opposite leaves and axillary peduncles; leaves linear, entire; peduncles 1-6cm; spikes 0.5-1.5cm; (2n=12). Intrd; open and rough ground on sandy soil, casual or sometimes natd; very scattered in S Br, especially E Anglia, less common than formerly, now often over-recorded for *P. afra;* S Europe.

7. P. afra L. (*P. psyllium* L. 1762 non 1753, *P. indica* auct. non L.) - *Glandular Plantain.* Differs from *P. arenaria* in key characters; especially note lower bracts lacking the long linear leaf-like apices of *P. arenaria;* (2n=12). Intrd; birdseed- and grain-alien of tips and waste places; scattered throughout much of Br, now commoner than *P. arenaria;* S Europe.

2. LITTORELLA P.J. Bergius - *Shoreweed*

Stoloniferous; flowers unisexual, the male solitary on long scapes and 3-4-merous, the female 1-few sessile at base of male scape and 2-4-merous; ovary 1-celled, with 1(-2) ovules; fruit a 1-seeded nut.

1. L. uniflora (L.) Asch. - *Shoreweed.* Leaves in rosette, semi-cylindrical, usually subulate, 1.5-10(25)cm; male scapes up to as long as leaves; stamens 1-2cm; 2n=24. Native; typically in shallow water at lake edges but often on exposed shore or down to 4m, on sandy or gravelly acid soils; in suitable places through much of BI, but very local and decreasing in lowlands. Not flowering unless exposed.

126. BUDDLEJACEAE - *Butterfly-bush family*

Deciduous or semi-evergreen shrubs; leaves simple, entire to serrate, without stipules, opposite or alternate, ± petiolate. Flowers in ± dense panicles, small but showy in mass, fragrant, actinomorphic, bisexual, hypogynous; sepals 4, fused into lobed tube; petals 4, fused into lobed tube, various colours; stamens 4, borne on corolla-tube; ovary 2-celled, each cell with many ovules on axile placentas; style 1; stigmas 2, linear; fruit a 2-valved capsule with many narrowly winged seeds.

Recognized by its shrubby habit and numerous small but brightly coloured flowers with 4 fused sepals and petals, 2-celled ovary and 4 stamens.

1. BUDDLEJA L. - *Butterfly-bushes*

1 Leaves alternate; flowers borne in small clusters along previous year's wood **1. B. alternifolia**
1 Leaves opposite; flowers borne in panicles on current year's growth 2
 2 Flowers in dense globose stalked clusters c.2cm across arranged in large open panicles **4. B. globosa**

2 Flowers in long dense narrowly pyramidal panicles, often interrupted mostly below and sometimes the segments ± globose 3
3 Flowers lilac to purple or white **2. B. davidii**
3 Flowers dull yellow to greyish- or purplish-yellow **3. B. x weyeriana**

1. B. alternifolia Maxim. - *Alternate-leaved Butterfly-bush.* Large shrub to 8m with long thin arching branches; leaves alternate, narrowly elliptic; flowers lilac, in ± sessile clusters borne along last year's wood; (2n=38). Intrd; grown in gardens, persistent and sometimes self-sown in woodland, hedges or banks; very scattered in En N to MW Yorks; China.

2. B. davidii Franch. - *Butterfly-bush.* Shrub to 5m with long ± arching branches; leaves opposite, lanceolate to narrowly ovate; flowers usually lilac, sometimes purple or white, in long pyramidal dense panicles borne on current year's wood; (2n=76). Intrd; natd in waste ground, walls, banks and scrub; common in S BI, decreasing northwards to C Sc; China.

3. B. x weyeriana Weyer (*B. davidii* x *B. globosa*) - *Weyer's Butterfly-bush.* Habit more of *B. davidii* but inflorescence interrupted, especially below, the segments forming rather loose globose sessile clusters; flowers usually dull yellow but varying through greyish to purple tinged with yellow. Intrd; grown and natd as for *B. globosa*; scattered records in S En; garden origin.

4. B. globosa Hope - *Orange-ball-tree.* Shrub to 5m with stiff erect to spreading branches; leaves opposite, lanceolate to narrowly ovate or elliptic; flowers orange-yellow, in dense globose stalked clusters arranged in stiff open panicles on current year's wood; (2n=38). Intrd; grown in gardens, natd rarely on roadsides and rough ground; very scattered in Br N to Anglesey and Yorks, Man; Chile and Peru.

127. OLEACEAE - *Ash family*

Trees, shrubs or woody trailers; leaves opposite, simple to pinnate, entire to serrate, without stipules, petiolate. Flowers variously arranged, often in dense clusters, small and dull to large and showy, various colours, actinomorphic, bisexual to dioecious or monoecious, hypogynous; sepals 0, or 4 and united into lobed tube; petals 0, or 4 (-6) and united into lobed tube (rarely free); stamens 2, borne on corolla-tube when that present; ovary 2-celled, each cell with 2 ovules on axile placentas; style 1; stigmas 1 and 2-lobed or 2, slightly elongated; fruit a 2-valved capsule, winged achene, or 2-4-seeded berry.

Distinguished by the woody habit and flowers with 4 fused sepals, 4-6 fused petals (0 or free in *Fraxinus*), 2-celled ovary and 2 stamens.

1 Corolla and calyx 0 (rarely both present and then petals free); flowers often unisexual; leaves usually pinnate; fruit a winged achene **3. FRAXINUS**
1 Corolla showy; calyx present; petals fused; flowers bisexual; leaves simple or with 1 main leaflet plus 1 or 2 small basal ones; fruit a capsule or berry 2
2 Flowers yellow, appearing before leaves 3
2 Flowers white, or lilac to mauve or red, appearing after leaves 4
3 Corolla-lobes mostly 4; flowers spring; leaves simple or with 1 main leaflet plus 1 or 2 small basal ones; fruit a capsule **1. FORSYTHIA**
3 Corolla-lobes (5-)6; flowers autumn to spring; leaves ternate; fruit a usually 2-lobed berry **2. JASMINUM**
4 Fruit a capsule; leaves truncate to cordate at base **4. SYRINGA**
4 Fruit a berry; leaves cuneate to rounded at base 5
5 Corolla <1cm across, white; leaves simple **5. LIGUSTRUM**
5 Corolla >1cm across; leaves ternate to pinnate, or if simple then corolla

red to pink **2. JASMINUM**

1. FORSYTHIA Vahl - *Forsythias*
Deciduous shrubs; leaves simple or with 2 small leaflets at base; petals 4(-6), united into 4(-6)-lobed tube, yellow, appearing before leaves; fruit a capsule.

Other spp. - **F. suspensa** (Thunb.) Vahl, an upright to rambling shrub more regularly with hollow branches and ternate leaves than *F. x intermedia*, and **F. giraldiana** Lingelsh., a large loose shrub with pale early-opening flowers, both from China, are common in gardens and some escapes may be these. The taxa are much confused, with many cultivars, and badly in need of revision.

1. F. x intermedia Zabel (*F. suspensa* x *F. viridissima* Lindl.) - *Forsythia*. Erect to arching shrub to 5m; leaves ovate to narrowly ovate, cuneate to rounded at base, serrate; flowers in small clusters on old wood, 2-3.5cm; (2n=28). Intrd; the most-grown garden taxon, occasional relic or throwout in rough ground or on road- or river-banks; few places in SE En, W Cornwall, Mons, Man; garden origin.

2. JASMINUM L. - *Jasmines*
Deciduous scrambling shrubs, rooting along stems; leaves simple, ternate or pinnate; petals 4-6, united into 4-6-lobed tube; fruit a usually 2-lobed black berry.

1 Flowers yellow, produced well before leaves, not fragrant **3. J. nudiflorum**
1 Flowers white to red, produced after leaves, fragrant 2
  2 Leaves simple; corolla red to pink, mostly 6-lobed **2. J. beesianum**
  2 Leaves pinnate; corolla white, often tinged pinkish-purple, mostly 4-5-lobed **1. J. officinale**

1. J. officinale L. - *Summer Jasmine*. Stems scrambling to 10(20)m; leaves pinnate; flowers produced in summer when leaves mature, in terminal ± corymbose cymes, fragrant; corolla mostly 4-5-lobed, white, often tinged pinkish-purple; (2n=26). Intrd; much grown in gardens, natd on walls and in marginal ground; S En; Caucasus to China.

2. J. beesianum Forrest & Diels - *Red Jasmine*. Erect or scrambling shrub to 2m; leaves simple; flowers produced in summer when leaves mature, 1-3 in terminal leaf-axils, fragrant; corolla mostly 6-lobed, bright pink to red; (2n=26). Intrd; not commonly grown (flowers not conspicuous) but natd as for *J. officinale*; En N to Cheshire; China.

3. J. nudiflorum Lindl. - *Winter Jasmine*. Stems scrambling to 5m; leaves ternate; flowers produced autumn to spring without leaves present, axillary and solitary, not fragrant; corolla mostly 6-lobed, yellow; (2n=26, 52). Intrd; very commonly grown, natd as for *J. officinale*; S En; China.

3. FRAXINUS L. - *Ashes*
Deciduous trees; leaves normally pinnate (rare form with 1 leaflet exists); petals 0 (rarely present, free, white); fruit an achene with a single long wing.

Other spp. - 2 spp. from S Europe are grown as street trees or on road-, railway- or canal-banks in many places and might appear wild. **F. angustifolia** Vahl (*Narrow-leaved Ash*) closely resembles *F. excelsior* but has narrower leaflets with fewer serrations and dark brown (not black) winter buds. **F. ornus** L. (*Manna Ash*) is a more delicate tree to 22m, with flowers with white petals appearing in large panicles with or after the leaves.

1. F. excelsior L. - *Ash*. Thick-twigged tree to 37m; leaflets serrate, lanceolate to ovate; flowers in dense axillary or terminal panicles appearing well before leaves, unisexual or bisexual, different sorts on same or different trees; 2n=46, 48. Native; woods, scrub and hedgerows, often the commonest tree, especially on damp or base-rich soils; common ± throughout BI.

4. SYRINGA L. - *Lilac*
Deciduous shrubs; leaves simple; petals usually 4, united into usually 4-lobed tube, white, or lilac to mauve or purple, appearing after leaves; fruit a capsule.

1. S. vulgaris L. - *Lilac*. Erect suckering shrub to 7m; leaves ovate, cordate to truncate at base, entire; flowers in large terminal pyramidal panicles, 1.5-2cm; (2n=44, 46, 48). Intrd; much grown in gardens, occasional relic or throwout in hedges and road- and railway-banks; scattered through most of BI, rarely self-sown; SE Europe.

5. LIGUSTRUM L. - *Privets*
Deciduous to evergreen shrubs; leaves simple; petals 4, united into 4-lobed tube, white, appearing after leaves; fruit a black berry.

Other spp. - **L. lucidum** W.T. Aiton, from E Asia, is a shrub or tree to 14m with fully evergreen coriaceous leaves and much larger, more diffuse panicles; it is grown in the S and sometimes produces seedlings.

1. L. vulgare L. - *Wild Privet*. Erect, semi-deciduous shrub to 3(5)m; 1-year-old stems and panicle branches densely but minutely pubescent; leaves lanceolate to oblanceolate, narrowly acute; corolla-tube c. as long as -limb; 2n=46. Native; hedgerows and scrub, especially on base-rich soils; throughout most of BI except parts of N Sc.

2. L. ovalifolium Hassk. - *Garden Privet*. Differs from *L. vulgare* in more (but not fully) evergreen leaves usually elliptic, acute to rounded at apex; young stems and panicle branches glabrous; corolla-tube distinctly longer than lobes; (2n=46). Intrd; abundantly planted for hedging, often yellow- or variegated-leaved, persistent in hedges and rough ground; scattered throughout much of BI, rarely self-sown; Japan.

128. SCROPHULARIACEAE - *Figwort family*

Herbaceous annuals to perennials or very rarely small shrubs or trees; stems usually round, sometimes square, in section; leaves alternate, opposite, whorled, or rarely all basal, simple, entire to deeply divided, without stipules, sessile to petiolate. Flowers single and axillary or in terminal or less often axillary racemes or sometimes in racemose or cymose panicles, zygomorphic (occasionally nearly actinomorphic), bisexual, hypogynous; calyx with (2)4-5-lobes fused just at base to most of length, never well differentiated into upper and lower lips; corolla strongly 4-5-lobed and sometimes nearly actinomorphic, to 2-lipped with lobes variously or not developed, variously coloured; stamens usually 4, sometimes with a staminode, sometimes 2 or 5 or rarely 3, borne on corolla-tube; ovary 2-celled, each cell with numerous ovules on axile placentas; style 1, terminal; stigmas 1-2, short, usually ± capitate; fruit a 2-celled capsule.

Distinguished from Lamiaceae, Verbenaceae and Boraginaceae by the totally different ovary and fruit, and from other families with zygomorphic flowers and 2-celled ovary by the herbaceous habit (woody in *Paulownia*, *Hebe* and *Phygelius*), 2 or 4 stamens (5 in *Verbascum*), and non-spiny bracts. See also Orobanchaceae.

1 Stamens 2 2
1 Stamens 4 or 5 4
  2 Corolla 2-lipped, the lower lip large and inflated, yellow to reddish-

brown; stamens included **7. CALCEOLARIA**

2 Corolla 4-lobed, without inflated portion, white to blue or pink; stamens exserted 3

3 Stems woody; leaves evergreen, entire **18. HEBE**

3 Herbaceous plants; stems not woody or woody only at base and then with serrate leaves **17. VERONICA**

4 Deciduous tree; flowers purplish-blue to violet **4. PAULOWNIA**

4 Herbaceous, or if a woody shrub then flowers red 5

5 Flowers 2.5-4cm, red, pendent in terminal panicles; stems woody **3. PHYGELIUS**

5 Flowers often <2cm, if >2cm and red then in racemes; stems herbaceous 6

6 Corolla with conspicuous basal spur or pouch on lowerside 7

6 Corolla not spurred or pouched at base 13

7 Leaves palmately veined and lobed 8

7 Leaves with single midrib, often also with lateral pinnate veins, entire to serrate 9

8 Plant glandular-pubescent; corolla ≥3cm, yellow with purple veins, pouched at base **11. ASARINA**

8 Plant glabrous to minutely pubescent, not glandular; corolla ≤3cm, mauve to purple, often with yellow centre, spurred at base **12. CYMBALARIA**

9 Corolla-tube with broad, rounded pouch (wider than long) at base 10

9 Corolla-tube with narrow, often pointed, spur (longer than wide) 11

10 Calyx-lobes ± equal, all shorter than corolla-tube; corolla >2.5cm **8. ANTIRRHINUM**

10 Calyx-lobes distinctly unequal, all longer than corolla-tube; corolla <2cm **10. MISOPATES**

11 Leaves ovate to obovate, rounded to cordate at base; capsule opening by detachment of 2 oblique lids leaving large pores **13. KICKXIA**

11 Leaves linear to lanceolate or oblanceolate, narrowed to base, rarely ovate to obovate and rounded at base and then capsule without detachable lids 12

12 Mouth of corolla completely closed by boss-like swelling on lower lip **14. LINARIA**

12 Mouth of corolla incompletely closed by small swelling **9. CHAENORHINUM**

13 Fertile stamens 5, at least in most flowers **1. VERBASCUM**

13 Fertile stamens 4, sometimes with a sterile staminode representing fifth 14

14 Leaves all in basal rosettes, linear to spathulate **6. LIMOSELLA**

14 At least some leaves borne on stems 15

15 Stems procumbent, rooting at nodes; leaves reniform, long-stalked; flowers solitary in leaf-axils **19. SIBTHORPIA**

15 Stems not procumbent and rooting at nodes; leaves not reniform; flowers in terminal inflorescences 16

16 Calyx 5-lobed or -toothed 17

16 Calyx 4-lobed or -toothed, rarely 2-5-lobed with toothed lobes and inflated tube 20

17 Leaves opposite 18

17 Leaves alternate 19

18 Calyx-tube shorter than -lobes; corolla purplish-brown to dull yellowish-green, with tube scarcely longer than wide **2. SCROPHULARIA**

18 Calyx-tube longer than -teeth; corolla bright yellow, often with red spots or blotches, with tube much longer than wide **5. MIMULUS**

19 Corolla distinctly zygomorphic, with tube >2x as long as calyx, lobes not strongly patent **15. DIGITALIS**

19 Corolla scarcely zygomorphic, with tube <2x as long as calyx, with strongly patent lobes **16. ERINUS**
20 Calyx irregularly 2-5-lobed, with toothed lobes; leaves divided almost to base, the lobes toothed **26. PEDICULARIS**
20 Calyx regularly 4-lobed, with entire lobes; leaves entire to simply toothed up to c.1/2 way to base 21
21 Calyx-tube inflated, especially at fruiting; seeds discoid, with marginal wing **25. RHINANTHUS**
21 Calyx-tube not inflated; seeds not discoid, without marginal wing 22
22 Lower lip of corolla with 3 distinctly emarginate lobes **21. EUPHRASIA**
22 Lower lip of corolla with 3 entire lobes, or sometimes middle lobe slightly emarginate 23
23 Mouth of corolla partially closed by boss-like swellings on lower lip; capsules with 1-4 seeds **20. MELAMPYRUM**
23 Mouth of corolla open; lower lip without swellings; capsules with >4 seeds 24
24 Corolla yellow 25
24 Corolla pink to dark purple, rarely white 26
25 Leaves serrate; seeds c.0.5mm, ± smooth **24. PARENTUCELLIA**
25 Leaves entire or ± so; seeds >1mm, ridged and grooved **22. ODONTITES**
26 Perennial; corolla dark purple, >12mm **23. BARTSIA**
26 Annual; corolla pink to reddish-purple, rarely white, <12mm **22. ODONTITES**

Other genera - NEMESIA Vent. (*Nemesias*) (tribe Hemimerideae), from S Africa, are commonly grown for ornament and rarely occur as casuals on tips, etc. They are branched annuals with variously coloured showy flowers with 4-lobed upper lip, large slightly bossed lower lip, and spurred or pouched tube. **N. strumosa** Benth., with pouched corolla-tube, is the most grown sp. For **SCOPOLIA** see Solanaceae.

TRIBE 1 - SCROPHULARIEAE (Verbasceae, Cheloneae pro parte) (genera 1-4). Herbaceous annuals or perennials or rarely shrubs; leaves alternate or opposite; flowers nearly actinomorphic to zygomorphic; corolla not to obscurely 2-lipped, not spurred or pouched at base; stamens 4-5.

1. VERBASCUM L. - *Mulleins*

Herbaceous biennials, less often annuals or perennials, with tall, erect, terete to ridged stems; leaves alternate; corolla yellow, purplish or white, with 5 ± equal lobes and short tube; stamens 5, rarely 4 in some flowers, all fertile, at least the upper 3 with very hairy filaments.

Hybrids arise frequently where 2 or more spp. occur together; they are usually highly but often not completely sterile. Characters may be as in 1 or other parent or intermediate. In hybrids between spp. with violet stamen-hairs and spp. with white stamen-hairs all the stamens may have all violet hairs, all have all pale violet hairs, all have mixed white and violet hairs, or the upper 3 have white and the lower 2 have violet hairs. Oblique-asymmetrical anthers of 1 parent may or may not appear in the hybrid. Hybrids between white-flowered *V. lychnitis* and yellow-flowered spp. are yellow-flowered. Combinations other than those listed often appear with the parents in gardens.

1 Anthers all reniform, symmetrical, placed transversely on filaments 2
1 Anthers of 3 upper stamens as above, of 2 lower stamens asymmetrical, placed obliquely or ± longitudinally on filaments and often ± decurrent on them 8
2 All hairs on filaments yellow or white 3

2 All or many hairs on filaments violet 5

3 Stems and leaves uniformly and persistently densely pubescent **11. V. speciosum**

3 Stems and leaves unevenly mealy- or powdery-pubescent at first, becoming less pubescent to glabrous later 4

4 Corolla usually white (rarely yellow); all or most pedicels >6mm; leaves sparsely pubescent to glabrous and green on upperside **13. V. lychnitis**

4 Corolla ± always yellow; all or most pedicels <6mm; leaves whitish-pubescent on upperside **12. V. pulverulentum**

5 Flowers 1 per node in axil of bract; bracteoles 0 6

5 Flowers 2-several per node in axil of bract; each pedicel with 2 small bracteoles 7

6 Corolla violet to purple; hairs all simple, mostly glandular; inflorescence usually simple **3. V. phoeniceum**

6 Corolla yellow; many hairs stellate; inflorescence much branched **4. V. pyramidatum**

7 Basal leaves truncate to rounded at base; pedicels all similar length, c. as long as calyx **9. V. chaixii**

7 Basal leaves cordate at base; pedicels variable in length, many c. as long as calyx, the longest ≥2x as long as calyx **10. V. nigrum**

8 Simple stalked glands present; branched hairs present or 0 9

8 Simple stalked glands 0; branched hairs always present 10

9 Flowers 1 per node; pedicels mostly longer than calyx; plant usually with stalked glands only in upper parts **1. V. blattaria**

9 Flowers usually >1 per node in lower parts of inflorescence; pedicels mostly shorter than calyx; plant usually with stalked glands ± throughout **2. V. virgatum**

10 Upper and middle stem-leaves distinctly decurrent 11

10 Stem-leaves not decurrent 12

11 Stigma capitate **8. V. thapsus**

11 Stigma elongated, spathulate **7. V. densiflorum**

12 Two lower stamens with filaments pubescent in lower 1/2, with anthers ≤4mm **5. V. bombyciferum**

12 Two lower stamens with glabrous to subglabrous filaments, with anthers ≥4mm **6. V. phlomoides**

Other spp. - V. sinuatum L., from S Europe, has strongly lobed basal leaves and stamens as in *V. nigrum;* it no longer occurs in BI.

1. V. blattaria L. - *Moth Mullein.* Annual to biennial to 1m, ± glabrous below and on leaves, with stalked glands above, inflorescence usually simple; flowers 1 per bract-axil; corolla usually yellow; 3 upper anthers reniform, 2 lower decurrent; filaments all with violet hairs (upper 3 or all 5 also with white hairs); (2n=30, 32). Intrd; waste and rough ground; scattered in C & S Br, formerly Ir and CI, rarely persistent; Europe.

1 x 10. V. blattaria x V. nigrum = V. x intermedium Rupr. ex Bercht. & Pfund has occurred in N Wilts and N Lincs.

2. V. virgatum Stokes - *Twiggy Mullein.* Erect biennial to 1m; differs from *V. blattaria* in key characters (couplet 9); 2n=c.64 (32, 64, 66). Native; fields, waste places and dry banks; locally frequent in Devon and Cornwall, fairly frequent casual in waste places elsewhere in Br, formerly Ir and CI. R

2 x 8. V. virgatum x V. thapsus = V. x lemaitrei Boreau has occurred in Warks as a casual.

3. V. phoeniceum L. - *Purple Mullein.* Perennial to 1m, with simple hairs below, with stalked glands above; inflorescence usually simple; flowers 1 per bract-axil;

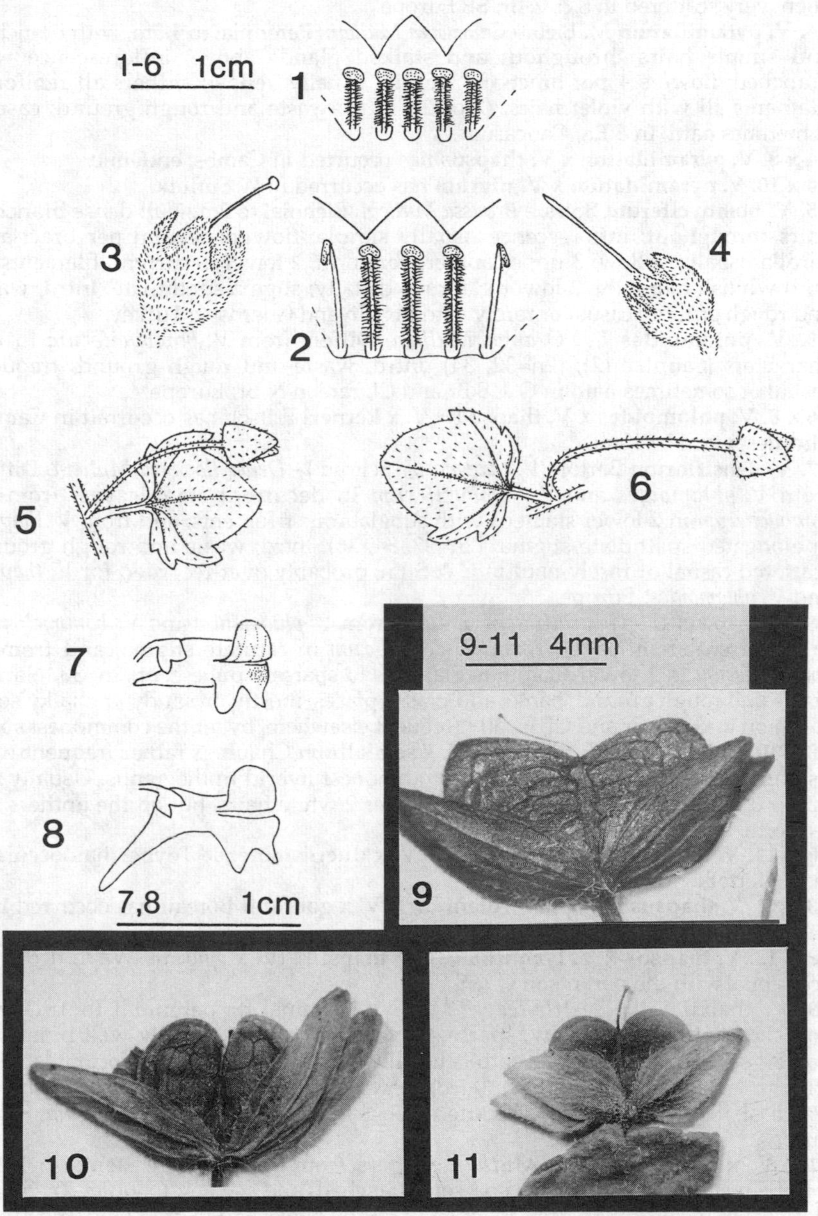

FIG 591 - Scrophulariaceae. 1-2, stamens of *Verbascum*. 1, *V. pulverulentum*. 2, *V. thapsus*. 3-4, calyx and style of *Verbascum*. 3, *V. thapsus*. 4, *V. phlomoides*. 5-6, fruiting nodes of *Veronica hederifolia*. 5, ssp. *hederifolia*. 6, ssp. *lucorum*. 7-8, corollas of *Linaria*. 7, *L. repens*. 8, *L. purpurea*. 9-11, fruits of *Veronica*. 9, *V. persica*. 10, *V. agrestis*. 11, *V. polita*.

corolla violet to purple; anthers all reniform; filaments all with violet hairs; (2n=32, 36). Intrd; grown in gardens, casual on tips and waste ground and as birdseed-alien; very scattered in S & C Br; SE Europe.

4. V. pyramidatum M. Bieb. - *Caucasian Mullein*. Perennial to 1.5m, with branched and simple hairs throughout and stalked glands above; inflorescence well branched; flowers 1 per bract-axil; corolla usually yellow; anthers all reniform; filaments all with violet hairs; (2n=32). Intrd; waste and rough ground; casual, sometimes natd, in S En; Caucasus.

4 x 8. V. pyramidatum x V. thapsus has occurred in Cambs; endemic.

4 x 10. V. pyramidatum x V. nigrum has occurred in W Suffolk.

5. V. bombyciferum Boiss. - *Broussa Mullein*. Biennial to 2m, with dense branched hairs throughout; inflorescence usually simple; flowers several per bract-axil; corolla usually yellow; 3 upper anthers reniform, 2 lower decurrent; filaments all with whitish hairs, the 2 lower glabrous distally; stigma spathulate. Intrd; waste and rough ground; casual or rarely natd in S En and Guernsey; Turkey.

6. V. phlomoides L. - *Orange Mullein*. Differs from *V. bombyciferum* in key **591**
characters (couplet 12); (2n=32, 34). Intrd; waste and rough ground; frequent casual or sometimes natd in C & S Br and CI, rare in N Br; Europe.

6 x 8. V. phlomoides x V. thapsus = V. x kerneri Fritsch has occurred in Cambs, Middlesex and W Kent.

7. V. densiflorum Bertol. (*V. thapsiforme* Schrad.) - *Dense-flowered Mullein*. Differs from *V. phlomoides* and *V. bombyciferum* in decurrent stem-leaves; from *V. bombyciferum* in 2 lower stamens with subglabrous filaments; and from *V. thapsus* in elongated spathulate stigma; (2n=32, 34, 36). Intrd; waste and rough ground; scattered casual or rarely natd in C & S En, probably over-recorded for *V. thapsus* and *V. phlomoides*; Europe.

8. V. thapsus L. - *Great Mullein*. Differs from *V. phlomoides* and *V. bombyciferum* **591**
in decurrent stem-leaves; from *V. densiflorum* in capitate stigma; and from *V. bombyciferum* in 2 lower filaments glabrous to sparsely pubescent; 2n=36. Native; waste and rough ground, banks and grassy places, mostly on sandy or chalky soils; common in C & S Br and CI, locally frequent elsewhere, by far the commonest sp.

8 x 10. V. thapsus x V. nigrum = V. x semialbum Chaub. is rather frequent with its parents in much of Br; by far the commonest hybrid in the genus. Usually the upper 3 filaments have violet and the lower 2 white hairs, but all the anthers are reniform.

8 x 11. V. thapsus x V. speciosum = V. x duernsteinense Teyber has occurred in W Norfolk.

8 x 12. V. thapsus x V. pulverulentum = V. x **godronii** Boreau has occurred in a few places in S En.

8 x 13. V. thapsus x V. lychnitis = V. x thapsi L. (*V. x spurium* W.D.J. Koch) is occasional with the parents in C & S Br.

9. V. chaixii Vill. - *Nettle-leaved Mullein*. Biennial or perennial to 1m, with branched hairs dense below, sparse above; inflorescence usually well branched; flowers several per bract-axil; corolla usually yellow; anthers all reniform; filaments all with violet hairs; (2n=30, 32). Intrd; grown in gardens, casual or natd escape on waste or rough ground; very scattered in S Br, overlooked as *V. nigrum*; S & C Europe.

10. V. nigrum L. - *Dark Mullein*. Differs from *V. chaixii* in stems to 1.2m; inflorescence simple to rather sparsely branched; and see key (couplet 7); 2n=30. Native; waste and rough ground, open places on banks and in grassland, mostly on soft limestone; locally common in CI and C & S Br, sparse casual elsewhere.

10 x 12. V. nigrum x V. pulverulentum = V. x mixtum Ramond ex DC. (*V. x wirtgenii* Franch.) occurs rarely with the parents in E Anglia and has occasionally been found elsewhere in En as a casual.

10 x 13. V. nigrum x V. lychnitis = V. x incanum Gaudin (*V. x schiedeanum* W.D.J. Koch) occurs sporadically in S En with the parents.

11. V. speciosum Schrad. - *Hungarian Mullein*. Biennial to 2m, with dense branched persistent hairs on all parts; inflorescence well branched; flowers several per bract-axil; corolla usually yellow; anthers all reniform; filaments all with white hairs; (2n=28). Intrd; grown in gardens (probably the finest sp.), casual or natd escape in W Kent, W Norfolk and (since 1930) E Suffolk; SE Europe.

12. V. pulverulentum Vill. - *Hoary Mullein*. Biennial to 1.5m; differs from *V.* R
speciosum in stems and leaves becoming glabrous later, the pubescence becoming 591
patchy and tufted as it wears off; 2n=32. Native; local on barish ground on chalky soils in E Anglia (mostly Norfolk) rare casual or natd escape elsewhere in Br.

12 x 13. V. pulverulentum x V. lychnitis = V. x regelianum Wirtg. has been recorded from N & S Devon as a casual; the parents are not sympatric as natives in BI.

13. V. lychnitis L. - *White Mullein*. Differs from *V. speciosum* and *V. pulverulentum* R
in key characters (couplets 3 & 4); stems to 1.5m; corolla usually white (var. **album** (Mill.) Druce), but yellow around Minehead, S Somerset; (2n=26, 32, 34). Native; similar places to *V. pulverulentum;* local in S En from Devon to E Kent, infrequent casual or natd escape elsewhere in Br, and then often yellow-flowered.

2. SCROPHULARIA L. - *Figworts*

Herbaceous perennials, less often biennials, with erect, usually 4-angled stems usually square in section; leaves opposite; corolla dull purplish, brownish or greenish-yellow, with 5 nearly equal lobes and short tube, the lobes usually obscurely organised into 2-lobed upper and 3-lobed lower lips; fertile stamens 4, with the 5th (uppermost) usually represented by a sterile staminode.

1 Staminode 0; corolla-lobes equal, not organized into upper and lower lips **5. S. vernalis**
1 Staminode present, larger than anthers; corolla-lobes unequal, distinctly forming 2 weak lips 2
2 Leaves and stems pubescent **4. S. scorodonia**
2 Leaves and stems ± glabrous 3
3 Stem-angles not or scarcely winged; sepals with narrow scarious border <0.3mm wide **1. S. nodosa**
3 Stem-angles distinctly winged; sepals with conspicuous scarious border >0.3mm wide 4
4 Staminode ± orbicular, entire; leaves broadly cuneate to subcordate at base **2. S. auriculata**
4 Staminode bifid or with 2 divergent lobes at apex; leaves cuneate to rounded at base **3. S. umbrosa**

Other spp. - S. canina L. (*French Figwort*), from S Europe, with leaves pinnately lobed ± to midrib and purplish corolla with 0 staminode, was once regular at docks in S Wa but no longer occurs.

1. S. nodosa L. - *Common Figwort*. Perennial to 1m, glabrous apart from stalked glands in inflorescence; stem-angles acute; leaves ovate, acute, serrate; corolla greenish to purplish-brown; staminode obovate, truncate to emarginate at apex; 2n=36. Native; damp open and shady places and in hedgerows; common throughout most of BI.

2. S. auriculata L. (*S. aquatica* auct. non L.) - *Water Figwort*. Perennial to 1.2m, usually glabrous (but sometimes slightly and minutely pubescent) apart from stalked glands in inflorescence; stem-angles distinctly winged; leaves ovate or elliptic-ovate, obtuse, crenate; corolla greenish to purplish-brown; staminode ± orbicular, rounded at apex; 2n=78, 80. Native; places similar to or wetter than those of *S. nodosa*; common throughout En, Wa and CI, frequent in Ir, rare in Sc and Man.

3. S. umbrosa Dumort. - *Green Figwort*. Differs from *S. auriculata* in stem-angles more broadly winged; leaves acute to obtuse, serrate; bracts often leaf-like; calyx-teeth more deeply serrate; and see key; (2n=26, 52). Native; damp shady places; very locally scattered in Br and Ir.

4. S. scorodonia L. - *Balm-leaved Figwort*. Perennial to 1m, pubescent on all parts and with stalked glands in inflorescence; stem-angles acute to obtuse; leaves ovate, acute to obtuse, doubly serrate; corolla dull purple; staminode as in *S. auriculata*; (2n=58). Native; grassy field-borders and hedgerows; very local, mostly near coast, in Devon and Cornwall, common (the commonest sp.) in CI, natd in S Wa. **RR**

5. S. vernalis L. - *Yellow Figwort*. Biennial or perennial to 50(80)cm, pubescent on all parts and with stalked glands in inflorescence; stem-angles obtuse; leaves broadly ovate, acute to obtuse, strongly serrate; corolla dull yellowish; staminode 0; (2n=28, 40). Intrd; waste and rough ground, hedges and woodland clearings; scattered throughout much of Br, locally natd; Europe.

3. PHYGELIUS E. Mey. ex Benth. - *Cape Figwort*

Evergreen shrubs with 4-angled stems; leaves opposite; corolla red, trumpet-shaped, with 5 nearly equal lobes not organized into 2 lips and with tube much longer than lobes; stamens 4, exserted.

1. P. capensis E. Mey. ex Benth. - *Cape Figwort*. Glabrous shrub to 1.5m; leaves ovate, crenate-serrate; flowers pendent, in large terminal erect panicles, 2.5-4cm, with short, patent lobes. Intrd; natd by rivers in Co Wicklow since at least 1970, ± natd by garden boundaries in Dunbarton, W Kent and W Galway; S Africa.

4. PAULOWNIA Siebold & Zucc. - *Foxglove-tree*

Deciduous tree with spreading crown; leaves opposite; corolla purplish-blue to violet, trumpet-shaped, with 5 nearly equal lobes scarcely organized into 2 lips, with tube much longer than lobes; stamens 4, included.

1. P. tomentosa (Thunb.) Steud. - *Foxglove-tree*. Tree to 26m; leaves broadly ovate, 12-25(50)cm, cordate at base, often shallowly 3-5-lobed, entire; flowers erect, in large terminal erect panicles, 3.5-6cm, with patent lobes; (2n=40). Intrd; ornamental planted in parks and by roads, natd in rough ground; Middlesex since at least 1990; China.

TRIBE 2 - GRATIOLEAE (genera 5-6). Herbaceous annuals to perennials; leaves opposite (at least below) or all in basal rosette; flowers nearly actinomorphic to strongly zygomorphic; corolla not to strongly 2-lipped, not spurred or pouched at base; stamens 4.

5. MIMULUS L. - *Monkeyflowers*

Perennials with leafy stolons; leaves opposite; flowers strongly zygomorphic, with well-defined upper and lower lips, showy, yellow to red.

M. guttatus and *M. luteus* agg. (incl. *M. variegatus* and *M. cupreus*) form a difficult, interfertile group; hybrids are frequent and at least 3 occur in the absence of either parent. Hybrids between *M. guttatus* and *M. luteus* agg. are sterile (pollen <40 per cent full; seeds 0 or few per capsule), but hybrids within *M. luteus* agg. are fertile (pollen >50 per cent full; seeds many per capsule). In this account *M. cupreus* is considered distinct from *M. luteus*, but *M. variegatus* is included in the latter, as in most American Floras.

1 Calyx-teeth ± equal; plant glandular-pubescent ± all over; corolla <2.5cm **1. M. moschatus**

1 Upper calyx-tooth distinctly longer than lower 4; plant glabrous below, often glandular pubescent above; corolla >2.5cm 2

2 Inflorescence with abundant stalked glands; small simple hairs present at least on keels of calyx, often also elsewhere in inflorescence; plant fertile or sterile 3
2 Inflorescence glabrous or with sparse stalked glands; small simple hairs 0 except inside calyx; plant fertile 5
3 Corolla copper-coloured, often also spotted or blotched red or purplish; calyx sometimes petaloid **4. M. x burnetii(*)**
3 Corolla yellow, often spotted or blotched with orange, red or purplish; calyx not petaloid 4
4 Throat of corolla ± closed by 2 boss-like swellings on lower lip; corolla wholly yellow or with red spots in throat, but with unmarked lobes; plant usually fertile **2. M.guttatus(**)**
4 Throat of corolla ± open, the boss-like swellings low or inconspicuous; corolla lobes with orange, red or purplish spots or blotches; plant sterile **3. M. x robertsii**
5 Leaves with even, triangular, flat teeth; corolla yellow, usually with coppery-orange spots or blotches, or mainly coppery **6. M. x maculosus**
5 Leaves with irregular, oblong, often twisted teeth; corolla yellow, with 1 or more dark red to purplish-brown blotches **5. M. luteus**

(*) *M. guttatus* x *M. luteus* x *M. cupreus* would also key out here; see text.
(**) *M. guttatus* is fertile; sterile plants keying out here are hybrids of *M. guttatus* otherwise ± indistinguishable from it.

Other spp. - M. cupreus Dombrain has been recorded from several places, but in error for *M. x burnetii.* All the spp. are natd in damp or wet places by streams and ponds, on river shingle, in flushes and damp patches in fields or open woods.

1. M. moschatus Douglas ex Lindl. - *Musk*. Whole plant glandular-pubescent; stems decumbent to ascending, to 40cm; corolla 1-2cm, rather pale yellow; 2n=32. Intrd; scattered over most of Br and Ir; W N America.

2. M. guttatus DC. - *Monkeyflower*. Plant glabrous below, densely glandular-pubescent above; stems erect to ascending, to 75cm; corolla 2.5-4.5cm, bright yellow, often with red spots on throat, its throat ± closed by 2 bosses on lower lip; fertile; 2n=28. Intrd; scattered and locally common over most of BI, commonest lowland taxon; W N America.

2 x 6. M. guttatus x (M. luteus x M. cupreus) occurs as a garden escape or sometimes spontaneously with its 2 parents; it differs from *M. x burnetii* in its corolla-lobes having dark blotches, and from *M. luteus* x *M. cupreus* in being glandular-pubescent above and sterile; 2n=45.

3. M. x robertsii Silverside (*M. luteus* auct. non L.; *M. guttatus* x *M. luteus*) - *Hybrid Monkeyflower*. Plant glabrous below, variably glandular-pubescent above; stems erect to ascending, to 50cm; corolla 2.5-4.5cm, bright yellow with orange to red or purplish-brown spots and blotches variably developed on throat and lobes, its throat ± open; sterile; 2n=44, 45, 46, 54. Intrd; scattered and locally common over much of Br and Ir, mostly in N & W and commonest upland taxon; garden origin. Hybrids also occur between *M. guttatus* and *M. smithii* (see under *M. luteus*).

4. M. x burnetii S. Arn. (*M. cupreus* auct. non Dombrain; *M. guttatus* x *M. cupreus* Dombrain) - *Coppery Monkeyflower*. Differs from *M. guttatus* x *M. luteus* in corolla copper-coloured with lighter throat marked with red spots but no dark blotches on corolla-lobes; 2n=45. Intrd; W & N Br N from S Somerset and Yorks; garden origin.

5. M. luteus L. (*M. variegatus* Lodd. nom. nud.) - *Blood-drop-emlets*. Plant glabrous or very sparsely glandular pubescent in inflorescence; stems decumbent to ascending, to 50cm; corolla 2.5-4.5cm, yellow, with red spots in throat and reddish blotches on lobes, its throat open; fertile; (2n=60, 62, 64). Intrd; rather uncommon in N Br S to Durham, much over-recorded for *M. x robertsii;* Chile. Our plant is var. **rivularis** Lindl. Hybrids between it and var. *luteus* come under **M. smithii** Paxton and are natd in N Br, but var. **luteus** itself (with no corolla blotches) does not occur.

6. M. x maculosus T. Moore (*M. luteus* x *M. cupreus*) - *Scottish Monkeyflower*. Differs from *M. luteus* in key characters; corolla variable, usually with coppery-orange spots or blotches but sometimes more coppery and sometimes as in unblotched forms of *M. luteus*; fertile. Intrd; Sc N from Peebles; garden origin. Hybrids between *M. cupreus* and *M. smithii* (see under *M. luteus*), much grown as bedding plants, are **M. x hybridus** Siebert & Voss.

6. LIMOSELLA L. - *Mudworts*
Glabrous annuals (usually) with thin leafless stolons; leaves ± erect in basal rosette; flowers ± actinomorphic, small and inconspicuous, solitary on pedicels from leaf-rosettes.

1. L. aquatica L. - *Mudwort*. Leaves (incl. petioles) to 6(10)cm, mostly with narrowly elliptic blades up to 2cm; flowers 2.5-3mm, scentless; calyx longer than corolla-tube; corolla white to pale mauve; 2n=40. Native; wet sandy mud by ponds, often dried out in summer; very scattered and decreasing in Br and Ir N to N Aberdeen. **R**

1 x 2. L. aquatica x L. australis occurs with the parents at Morfa Pools, Glam; it is sterile but more vigorous than either parent and often perennial, with a mixture of subulate and elliptic leaf-blades and calyx c. as long as corolla-tube; 2n=30; endemic.

2. L. australis R.Br. (*L. subulata* E. Ives) - *Welsh Mudwort*. Leaf-blades and petioles not differentiated, subulate, to 4cm; flowers 3.5-4mm, scented; calyx shorter than corolla-tube; corolla white; 2n=20. Native; similar places to *L. aquatica*; extremely local in Glam, Merioneth and Caerns (nowhere else in Europe). **RRR**

TRIBE 3 - CALCEOLARIEAE (genus 7). Herbaceous annuals; leaves opposite; flowers strongly zygomorphic; corolla strongly 2-lipped, not spurred or pouched at base, with greatly incurved 'slipper-like' lower lip; stamens 2.

7. CALCEOLARIA L. - *Slipperwort*

Other spp. - C. integrifolia L., from Chile, is a shrub with simple serrate leaves and yellow to reddish-brown corollas; it is grown in gardens and occasional seedlings appear on walls in CI.

1. C. chelidonioides Kunth - *Slipperwort*. Stems well-branched, to 40cm, glandular-pubescent; leaves pinnate, with serrate leaflets; inflorescence few-flowered; corolla yellow, 10-14mm; (2n=60). Intrd; tips, waste places, cultivated ground; casual, sometimes natd, scattered in Br N to S Sc, especially E Anglia; C & S America.

TRIBE 4 - ANTIRRHINEAE (genera 8-14). Herbaceous annuals to perennials, rarely ± woody at base; leaves usually opposite below, often mostly alternate; flowers strongly zygomorphic; corolla strongly 2-lipped, with boss-like swelling on lower lip and conspicuous spur or pouch at base of tube; stamens 4.

8. ANTIRRHINUM L. - *Snapdragon*
Tufted, perennating by means of basal shoots; leaves entire, with single midrib; calyx-lobes ± equal, shorter than corolla-tube; corolla with broad, rounded pouch at base, with mouth closed by boss-like swelling on lower lip; capsule opening by 3 apical pores.

1. A. majus L. - *Snapdragon*. Stems erect to ascending, glabrous or glandular-pubescent above, to 1m; corolla 3-4.5cm, usually pink to purple, sometimes white, yellow, orange or combinations; (2n=16). Intrd; rough ground, walls, rocks, buildings; frequently natd throughout most of BI, but often killed in winter; SW Europe.

9. CHAENORHINUM (DC. ex Duby) Rchb. - *Toadflaxes*
Annuals, or tufted perennials with basal new shoots; leaves entire, with single midrib; calyx with slightly unequal lobes >1/2 as long to c. as long as corolla-tube; corolla with narrow conical spur at base, with mouth not completely closed by low boss-like swelling on lower lip; capsule opening by irregular large apical pores or tears.

1. C. origanifolium (L.) Kostel. - *Malling Toadflax*. Glandular-pubescent biennials to perennials; stems decumbent to erect, to 30cm, often several; lower leaves elliptic; corolla (incl. spur) 8-15(20)mm, bluish-mauve with pale yellow boss; (2n=14). Intrd; well natd on old walls at West Malling, W Kent, since c. 1880, rare and impermanent in similar places elsewhere in En; SW Europe.

2. C. minus (L.) Lange - *Small Toadflax*. Glandular-pubescent or rarely glabrous annuals; stems erect, to 25cm, often 1; lower leaves oblong-oblanceolate; corolla (incl. spur) 6-9mm, pale purple with pale yellow boss; 2n=14. Native; arable land, waste places, railway tracks, in open ground; frequent over most of Br and Ir except N Sc.

10. MISOPATES Raf. - *Weasel's-snouts*
Annuals; leaves entire, with single midrib; calyx-lobes distinctly unequal, all longer than corolla-tube; corolla and capsule as in *Antirrhinum*.

1. M. orontium (L.) Raf. (*Antirrhinum orontium* L.) - *Weasel's-snout*. Stems erect, usually glandular-pubescent above, to 50cm; corolla 10-17mm, usually bright pink, rarely white; (2n=14, 16). Probably native; weed of cultivated ground; locally frequent in S Br and CI, very scattered and decreasing in C & N Wa and C & N En, casual in Sc and Ir.

2. M. calycinum Rothm. - *Pale Weasel's-snout*. Differs from *M. orontium* in plant usually glabrous; corolla pale pink to white, 18-22mm. Intrd; birdseed-alien on tips and waste ground; rather scarce in S En; W Mediterranean.

11. ASARINA Mill. - *Trailing Snapdragon*
Stoloniferous perennials; leaves crenate to shallowly lobed, with palmate main veins; calyx with slightly unequal lobes much shorter than corolla-tube; corolla as in *Antirrhinum*; capsule opening by 2 apical pores.

1. A. procumbens Mill. - *Trailing Snapdragon*. Stems procumbent, glandular-pubescent, to 60cm; corolla 3-3.5cm, pale yellow with pinkish-purple veins, the boss deep yellow; (2n=18). Intrd; natd on dry banks, walls and cliffs; scattered in Br N to Dumfriess; mountains of S France and NE Spain.

12. CYMBALARIA Hill - *Ivy-leaved Toadflaxes*
Stoloniferous perennials; leaves subentire to shallowly and obtusely lobed, with palmate main veins; calyx with unequal lobes shorter than corolla-tube; corolla with narrow cylindrical to conical spur at base, with mouth completely closed by boss-like swelling on lower lip; capsule opening by irregular apical longitudinal slits.

1 Corolla 9-15mm, incl. spur 1.5-3mm; stems long and trailing, often >20cm **1. C. muralis**
1 Corolla 15-25mm, incl. spur 4-9mm; stems usually not trailing, <20cm 2

2 Stems, leaves, petioles and calyx shortly and densely pubescent **2. C. pallida**
2 Plant glabrous or nearly so **3. C. hepaticifolia**

1. C. muralis P. Gaertn., B. Mey. & Scherb. - *Ivy-leaved Toadflax*. Stems long and trailing, to 60cm; corolla 9-15mm incl. spur 1.5-3mm, mauvish-violet on lower lip. Intrd.

a. Ssp. muralis. Plant glabrous, or sparsely pubescent on calyx and young parts; (2n=14). Well natd on walls, pavements, rocky or stony banks, first recorded 1640; frequent over most of BI; CS Europe.

b. Ssp. visianii (Kümmerle ex Jáv.) D.A. Webb. Plant pubescent on ± all parts; (2n=14). Natd on waste ground since 1970 in Surrey; Italy and Jugoslavia.

2. C. pallida (Ten.) Wettst. - *Italian Toadflax*. Stems <20cm, decumbent to suberect; plant shortly pubescent; corolla 15-25mm incl. spur 6-9mm, mauvish-violet with white boss on lower lip; (2n=14). Intrd; natd on walls, shingle and stony places; scattered in Br, mainly N En and Sc, less well natd in S; Italy.

3. C. hepaticifolia (Poir.) Wettst. - *Corsican Toadflax*. Differs from *C. pallida* in plant glabrous or nearly so; corolla 15-18mm incl. spur 4-5mm; (2n=98). Intrd; marginally natd in and near gardens and nurseries, perhaps overlooked for *C. pallida*; very scattered in En and Sc; Corsica.

13. KICKXIA Dumort. - *Fluellens*

Annuals; leaves ovate to elliptic, ± entire to remotely dentate, with pinnate venation; calyx with equal lobes c. as long as corolla-tube; corolla with narrowly conical spur at base, with mouth completely closed by boss-like swelling on lower lip; capsule opening by 2 large oblique lids.

1. K. elatine (L.) Dumort. - *Sharp-leaved Fluellen*. Stems procumbent to suberect, usually well-branched at base, to 50cm; whole plant except pedicels and corolla with patent hairs (pedicels pubescent immediately below flower); leaves hastate; corolla 7-12mm incl. spur c.1/2 total length, yellow with violet upper lip; 2n=36. Probably native; arable fields and field-borders on light, usually calcareous soils; locally common in Br N to N Wa and N Lincs but intrd or casual in much of N & W, CI, intrd in S & W Ir.

2. K. spuria (L.) Dumort. - *Round-leaved Fluellen*. Differs from *K. elatine* in usually more robust; pedicels pubescent; leaves rounded at base; corolla 8-15mm, with more curved spur and purple upper lip; 2n=18. Probably native; similar places to *K. elatine*, often with it; SE Br NW to N Lincs and S Wa.

14. LINARIA L. - *Toadflaxes*

Annuals or perennials, sometimes rhizomatous; leaves entire, with single midrib; calyx with usually unequal lobes shorter than corolla-tube; corolla with narrow conical spur at base, with mouth closed (or ± so) by boss-like swelling on lower lip; capsule opening by irregular apical longitudinal slits.

1 Spur longer than rest of corolla **8. L. maroccana**
1 Spur much shorter than to nearly as long as rest of corolla 2
2 Whole plant glandular-pubescent; corolla (incl. spur) 4-7mm **6. L. arenaria**
2 Plant glabrous below, glabrous to glandular-pubescent above; corolla (incl. spur) usually ≥8mm 3
3 Corolla predominantly yellow, sometimes very pale or with purplish tinge (if with violet veins, see *L. vulgaris* x *L. repens*) 4
3 Corolla predominantly mauve, violet, purple or pink, sometimes very pale but then with darker veins, sometimes with yellow to orange boss 6
4 Annual; stems decumbent, with conspicuous region below

inflorescence bare of leaves **5. L. supina**
4 Perennial; stems normally erect, with leaves ± up to inflorescence; rhizomes often present 5
5 Seeds disc-like, with broad wing round circumference; plant often glandular-pubescent above; leaves linear to narrowly elliptic-oblanceolate, cuneate at base **1. L. vulgaris**
5 Seeds angular, scarcely winged; plant always glabrous; at least some leaves lanceolate to ovate and subcordate at base **2. L. dalmatica**
6 Annual; capsule shorter than calyx; seeds disc-like, with broad wing round circumference **7. L. pelisseriana**
6 Perennial; capsule longer than calyx; seeds angular, not winged 7
7 Spur <1/2 as long as rest of corolla, straight, subacute to rounded at tip; corolla with orange patch on boss **4. L. repens**
7 Spur ≥1/2 as long as rest of corolla, usually curved, acute at tip; corolla without orange (sometimes with white) patch on boss **3. L. purpurea**

1. L. vulgaris Mill. - *Common Toadflax*. Erect to ascending perennial to 80cm, glabrous all over or glandular-pubescent above; corolla 18-35mm incl. spur 6-13mm, yellow (sometimes very pale) with orange boss; seeds discoid, with broad marginal wing; 2n=12. Native; rough and waste ground, stony places, banks, open grassland; common over most of BI, absent from parts of Ir and C & N Sc.

1 x 4. L. vulgaris x L. repens = L. x sepium G.J. Allman is frequent within the range of *L. repens* in BI N to Ayrs; it is intermediate in corolla colour and fertile, often forming hybrid swarms; (2n=12). The corolla is most often pale yellow with violet veins and intermediate in size and shape, but in hybrid swarms plants close to either parent may occur.

2. L. dalmatica (L.) Mill. (*L. genistifolia* (L.) Mill. ssp. *dalmatica* (L.) Maire & Petitm.) - *Balkan Toadflax*. Erect, glabrous perennial to 80cm; corolla 20-55mm, incl. spur 4-25mm, yellow; seeds angular, not or scarcely winged; (2n=12). Intrd; natd in waste places, waysides and by railways; scattered in S En, rare casual elsewhere; SE Europe.

3. L. purpurea (L.) Mill. - *Purple Toadflax*. Erect, glabrous perennial to 1m; corolla **591**
7-15mm incl. curved spur 3-6mm, mauve with heavy purplish-violet veins or wholly purplish-violet, rarely pink, with concolorous or whitish boss; seeds angular, not winged; (2n=12). Intrd; natd on rough ground, walls, banks; frequent to sparse throughout much of BI; Italy.

3 x 4. L. purpurea x L. repens = L. x dominii Druce occurs in scattered localities usually with 1 or both parents in Br N to Westmorland; it is intermediate in corolla characters and fertile, sometimes segregating to give hybrid swarms.

4. L. repens (L.) Mill. - *Pale Toadflax*. Glabrous, decumbent to erect perennial to **591**
80cm; corolla 8-15mm incl. spur 1-4mm, whitish to pale mauve with violet thin veins, with orange spot on boss; seeds angular, not winged; (2n=12). Native; stony places, rough ground, banks and walls; scattered over much of Br, CI and E Ir, but absent from many places and frequent only in parts of S & W Br.

4 x 5. L. repens x L. supina = L. x cornubiensis Druce was collected from Par, E Cornwall, with both parents in 1925 and 1930; it was said to be intermediate in corolla characters and sterile; endemic.

5. L. supina (L.) Chaz. - *Prostrate Toadflax*. Procumbent to decumbent annual to **RR**
20cm, glabrous below, glandular pubescent above; corolla 15-25mm incl. spur 7-11mm, pale yellow with yellowish-orange boss; seeds discoid, with broad marginal wing; (2n=12). Possibly native; sandy ground near Par, E Cornwall, natd in open waste ground mainly by railways in E & W Cornwall, S Devon and Carms, rare casual elsewhere; SW Europe.

6. L. arenaria DC. - *Sand Toadflax*. Erect, glandular-pubescent annual to 15cm; corolla 5-7mm incl. spur 1.5-3mm, yellowish with yellowish to violet spur; seeds discoid, with rather narrow marginal wing. Intrd; planted at Braunton Burrows, N

Devon c. 1893, now well natd on semi-fixed dunes; W France and NW Spain.

7. L. pelisseriana (L.) Mill. - *Jersey Toadflax*. Erect, glabrous annual to 30cm; corolla 11-20mm, incl. spur 5-9mm, purplish-violet with whitish boss; seeds discoid, with broad marginal wing; (2n=24). Native; rough ground, rocky places and hedgebanks; very rare and sporadic in Jersey, last seen 1955, perhaps not native, rare casual in C & S Br. **RR**

8. L. maroccana Hook. f. - *Annual Toadflax*. Erect annual to 50cm, glabrous below, glandular-pubescent above; corolla 17-30mm, incl. spur 9-17mm, usually purplish-violet with yellow boss but white, yellow, pink, red or variously variegated colour-forms occur; seeds angular, not winged; (2n=12). Intrd; grown in gardens, frequent casual on tips and in waste places; scattered in Br and CI, mainly S En; Morocco.

TRIBE 5 - DIGITALEAE (Veroniceae) (genera 15-19). Herbaceous annuals to perennials or sometimes shrubs; leaves opposite or alternate; flowers nearly actinomorphic to zygomorphic; corolla not to weakly 2-lipped, not spurred or pouched at base; stamens 2 or 4 (rarely 3 or 5).

15. DIGITALIS L. - *Foxgloves*

Herbaceous biennials to perennials; leaves alternate, with pinnate venation; flowers in terminal racemes; corolla showy, weakly 2-lipped, with 5 lobes shorter than tube; stamens 4.

Other spp. - D. grandiflora Mill. (*Yellow Foxglove*), with large yellow flowers, **D. lanata** Ehrh. (*Grecian Foxglove*), with whitish-yellow corolla with brownish veins, and **D. ferruginea** L. (*Rusty Foxglove*), with yellowish- or reddish-brown corolla with darker veins, all from S & C Europe, have all been reported as garden escapes but have not survived. *D. lutea* x *D. purpurea* = **D. x fucata** Ehrh. (*D. x purpurascens* Roth) is not rare as a spontaneous hybrid in gardens with the parents, but has not been recorded outside.

1. D. purpurea L. - *Foxglove*. Erect usually densely pubescent biennial to short-lived perennial to 2m; leaves ovate to lanceolate, ± rugose; corolla 40-55mm, pink to purple with dark spots inside, sometimes white; 2n=56. Native; many sorts of open places, especially woodland clearings, heaths and mountainsides, also waste ground, on acid soils; common throughout BI in suitable places, garden escape elsewhere.

2. D. lutea L. - *Straw Foxglove*. Erect glabrous to slightly pubescent usually perennial to 1m; leaves lanceolate- to oblanceolate-oblong, ± smooth; corolla 9-25mm, pale yellow; (2n=16, 56, 96, 112). Intrd; natd on waste ground, roadsides and walls; scattered in S En; W Europe.

16. ERINUS L. - *Fairy Foxglove*

Herbaceous perennials; leaves alternate, with pinnate venation; flowers in terminal raceme; corolla showy, not 2-lipped, with 5 ± equal, patent, emarginate lobes little shorter than tube; stamens 4.

1. E. alpinus L. - *Fairy Foxglove*. Stems to 20cm, ascending to suberect, pubescent; leaves oblanceolate to obovate, serrate; corolla-tube 3-7mm, corolla-limb 6-9mm across, purple, sometimes white; (2n=14). Intrd; natd on walls and stony places; scattered in BI, mainly N En to C Sc; mountains of SW Europe.

17. VERONICA L. - *Speedwells*

Herbaceous annuals to perennials, sometimes woody at base; leaves opposite, at least below, with pinnate or sometimes ± palmate venation; flowers solitary in leaf-axils or in terminal or axillary racemes; corolla showy to inconspicuous, not 2-

lipped, with 4 subequal lobes much longer than tube; stamens 2.

General key

1 Flowers (at least mostly) in axillary racemes (include plants with a single raceme in 1 of the most apical leaf-axils) *Key A*

1 Flowers in terminal racemes or solitary in leaf-axils 2

  2 Flowers in terminal racemes; bracts all or at least the upper very different from the foliage leaves, but the lower sometimes similar to them *Key B*

  2 Flowers solitary in axils of leaves closely resembling the foliage leaves, though the upper often smaller than them *Key C*

Key A - Flowers in axillary racemes

1 Leaves and stems glabrous, except stems sometimes glandular-pubescent in inflorescence 2

1 Leaves and stems pubescent 5

  2 Racemes 1 per node; capsule dehiscing into 2 valves **9. V. scutellata**

  2 Racemes mostly 2 per node; capsule dehiscing into 4 valves 3

3 Leaves all shortly petiolate; flowering stems procumbent or decumbent (to ascending) **10. V. beccabunga**

3 Upper leaves sessile; flowering stems usually erect (to ascending) 4

  4 Corolla pale blue; pedicels erecto-patent in fruit; capsule 2.5-4mm, ± orbicular **11. V. anagallis-aquatica**

  4 Corolla pinkish; pedicels patent in fruit; capsule 2-3mm, wider than long **12. V. catenata**

5 Stems pubescent along 2 opposite lines only; capsule shorter than calyx **7. V. chamaedrys**

5 Stems pubescent all round; capsule longer than calyx 6

  6 Petioles >6mm; capsules >6mm wide **8. V. montana**

  6 Petioles <6mm; capsules <6mm wide 7

7 Leaves linear-oblong to -lanceolate; pedicels >6mm in fruit, longer than bracts **9. V. scutellata**

7 Leaves lanceolate-oblong to ovate or elliptic; pedicels <6mm in fruit, up to as long as bracts 8

  8 Calyx-lobes usually 5, 1 much shorter than other 4; leaves sessile **5. V. austriaca**

  8 Calyx-lobes 4; at least lower leaves petiolate **6. V. officinalis**

Key B - Flowers in terminal racemes

1 Annuals with 1 root system, easily uprooted 2

1 Perennials, with non-flowering shoots and/or stems rooted more than just at base 8

  2 Plant glabrous **18. V. peregrina**

  2 Plant pubescent or minutely so, at least on capsules and inflorescence-axis 3

3 Bracts much longer than fruiting pedicels 4

3 Bracts shorter than to ± as long as fruiting pedicels 6

  4 At least some upper leaves lobed >1/2 way to midrib **17. V. verna**

  4 All leaves entire to crenate-serrate, toothed <1/2 way to midrib 5

5 All hairs glandular; leaves oblanceolate to narrowly oblong **18. V. peregrina**

5 Many hairs, at least below, non-glandular; leaves ovate **16. V. arvensis**

  6 Lower bracts and upper leaves lobed much >1/2 way to base **15. V. triphyllos**

  6 Bracts and leaves toothed <1/2 way to midrib 7

7 Capsule notched to c.1/2 way; pedicels >2x as long as calyx; seeds flat **13. V. acinifolia**
7 Capsule notched to ≤1/4 way; pedicels <2x as long as calyx; seeds cup-shaped **14. V. praecox**
8 Corolla-tube usually >2mm, longer than wide; racemes dense, long and many-flowered 9
8 Corolla-tube usually <2mm, wider than long; racemes lax, short and/or few-flowered 10
9 Leaves usually widest in middle 1/3, crenate to serrate with usually obtuse teeth, pubescent on both surfaces **26. V. spicata**
9 Leaves usually widest in basal 1/3, serrate to biserrate with acute to subacuminate teeth, glabrous to sparsely and often minutely pubescent on both surfaces **25. V. longifolia**
10 Corolla pink; style c.2x as long as capsule **2. V. repens**
10 Corolla white to blue; style shorter than capsule 11
11 Capsule wider than long **1. V. serpyllifolia**
11 Capsule longer than wide 12
12 Corolla >10mm across; stems woody at base; style >2mm; racemes with eglandular hairs **4. V. fruticans**
12 Corolla <10mm across; stems herbaceous; style <2mm; racemes with glandular hairs **3. V. alpina**

Key C - Flowers solitary in leaf-axils
1 Calyx-lobes apparently 2, each bilobed at apex **22. V. crista-galli**
1 Calyx-lobes 4, each acute to rounded at apex 2
2 Perennial; stems rooting at nodes along length; pedicels >2x as long as leaves + petioles **23. V. filiformis**
2 Annual; stems not rooting at nodes or doing so only near base; pedicels <2x as long as leaves + petioles 3
3 Calyx-lobes cordate at base; leaves with 3-7 shallow lobes or teeth **24. V. hederifolia**
3 Calyx-lobes cuneate to rounded at base; leaves crenate-serrate, most with >7 teeth 4
4 Lobes of capsule with apices diverging at c.90°; corolla mostly ≥8mm across **21. V. persica**
4 Lobes of capsule with apices ± parallel or diverging at narrow angle; corolla ≤8mm across 5
5 Capsule with patent glandular hairs only **19. V. agrestis**
5 Capsule with many short eglandular arched hairs and variable numbers of patent glandular hairs **20. V. polita**

Other spp. - Records of **V. paniculata** L. (*V. spuria* auct. non L.) as garden escapes seem to be all or mostly errors for *V. longifolia.* **V. cymbalaria** Bodard (*Pale Speedwell*), from S Europe, differs from *V. hederifolia* in its 5-9-lobed leaves, obtuse to rounded (not acute to subacute) calyx-lobes and pubescent (not glabrous) capsules; it occurred in W Cornwall in 1985 and might become more frequent.

Section* 1 - *VERONICASTRUM W.D.J. Koch (sect. *Berula* Dumort.) (spp. 1-4). Perennials; inflorescence a terminal raceme; corolla-tube wider than long; capsule dehiscing into 2 valves, usually only apically; seeds flat.

1. V. serpyllifolia L. - *Thyme-leaved Speedwell.* Stems to 30cm, herbaceous, rooting at nodes; leaves shortly petiolate, glabrous; corolla 5-10mm across; capsule wider than long, shorter than calyx, with style c. as long. Native.

a. Ssp. serpyllifolia. At least 1/2 of flowering stem upturned and erect; leaves ovate-elliptic; racemes ± glabrous or with eglandular hairs, usually with >12 flowers; pedicels c. as long as calyx; corolla 6-8mm across, whitish to pale blue with darker veins; 2n=14. Waste and cultivated ground, paths, lawns, open grassland, woodland rides and on mountains; common throughout BI.

b. Ssp. humifusa (Dicks.) Syme. Most of flowering stem procumbent; leaves ovate-orbicular; racemes with glandular hairs and often <12 flowers; pedicels longer than calyx; corolla 7-10mm across, bright blue; 2n=14. Rock-ledges, flushes and wet gravel in mountains; Scottish Highlands, less extreme plants in N En and N & S Wa. Mountain plants of ssp. *serpyllifolia* often approach ssp. *humifusa* in their bluer corollas, more procumbent stems, fewer flowers and presence of glandular hairs, and need investigating. R

2. V. repens Clarion ex DC. (*V. reptans* D.H. Kent nom. illeg.) - *Corsican Speedwell.* Differs from *V. serpyllifolia* ssp. *humifusa* in pedicels usually longer than bracts; corolla pink; style much longer than calyx; (2n=14). Intrd; natd weed in lawns in very few places in Sc and N En, less common than formerly; Corsica and S Spain.

3. V. alpina L. - *Alpine Speedwell.* Stems to 15cm, herbaceous, erect to ascending from short rooting portion; leaves ovate-elliptic, subsessile, ± glabrous; racemes glandular-pubescent; corolla 5-10mm across, dull blue; capsule longer than wide, longer than calyx, with much shorter style; 2n=18. Native; damp alpine rocks above 500m; local in mainland C Sc. R

4. V. fruticans Jacq. - *Rock Speedwell.* Stems woody below, to 20cm, erect to ascending; leaves obovate-elliptic, subsessile, glabrous; racemes minutely pubescent; corolla 10-15mm across, bright blue; capsule longer than wide, longer than calyx, with slightly shorter style; (2n=16). Native; alpine rocks above 500m; very local in C Sc. RR

Section* 2 - *VERONICA (spp. 5-9). Perennials: inflorescences opposite or alternate axillary racemes; corolla-tube wider than long; capsule dehiscing into 2 valves, usually only apically; seeds flat.

5. V. austriaca L. (*V. teucrium* L.) - *Large Speedwell.* Stems decumbent to erect, to 50cm, pubescent all round; leaves ovate-oblong, serrate, sessile, pubescent; corolla 10-15mm across, bright blue; capsule elliptic, longer than wide; (2n=64, 68). Intrd; grown in gardens and natd in open and rough ground, dunes; scattered in Br N to Angus; Europe. Our plant is ssp. **teucrium** (L.) D.A. Webb.

6. V. officinalis L. - *Heath Speedwell.* Stems procumbent to ascending, to 40cm, pubescent all round; leaves obovate-elliptic, serrate, petiolate, pubescent; corolla 5-9mm across, lilac; capsule obtriangular-obovate, c. as long as wide; 2n=36. Native; banks, open woods, grassland and heathland on well-drained soils; common throughout most of BI.

7. V. chamaedrys L. - *Germander Speedwell.* Stems erect to ascending, to 50cm, pubescent only along 2 opposite lines; leaves triangular-ovate, serrate, sessile or with petioles up to 5mm, pubescent; corolla 8-12mm across, bright blue; capsule obtriangular-obovate, wider than long; 2n=32. Native; woods, hedgerows, grassland in damper areas; common throughout BI.

8. V. montana L. - *Wood Speedwell.* Stems procumbent to ascending, to 40cm, pubescent all round; leaves ovate to broadly so, serrate, with petioles 5-15mm, pubescent; corolla 8-10mm across, pale lilac-blue; capsule broadly transversely elliptic; 2n=18. Native; dampish woods; scattered and locally frequent throughout Br and Ir except N Sc.

9. V. scutellata L. - *Marsh Speedwell.* Stems decumbent to scrambling-erect, to 60cm, glabrous or pubescent (var. **villosa** Schum.) all round; leaves linear-oblong to -lanceolate, entire to distantly serrate, sessile, glabrous or pubescent; corolla 5-8mm across, whitish to pale pinkish or lilac; capsule reniform; 2n=18. Native; bogs, marshes, wet meadows, by ponds and lakes, on bare ground or among tall vegetation; locally frequent throughout BI.

Section 3 - *BECCABUNGA* (Hill) Dumort. (spp. 10-12). Perennials (sometimes annual); inflorescences opposite, axillary racemes; corolla-tube wider than long; capsule dehiscing into 4 valves; seeds flat on 1 side, convex on other.

10. V. beccabunga L. - *Brooklime.* Stems procumbent to ascending, to 60cm; leaves ovate-oblong to elliptic or broadly so, obtuse to rounded, petiolate; pedicels ± patent in fruit, shorter to longer than bracts; corolla bright blue; capsule 2-4mm, c. as long as wide; 2n=18. Native; streams, ditches, marshes, pondsides and river-banks; common throughout BI except rare in N Sc.

11. V. anagallis-aquatica L. - *Blue Water-Speedwell.* Stems erect, to 50cm; leaves **623** often narrowly ovate and shortly petiolate below, always lanceolate and sessile above, acute or subacute; pedicels erecto-patent in fruit, at least as long as bracts at flowering; corolla pale blue with darker veins; 2n=36. Native; by ponds and streams, in marshes and wet meadows; scattered and locally common throughout most of BI.

11 x 12. V. anagallis-aquatica x V. catenata = V. x lackschewitzii J.B. Keller is frequent with the parents throughout their range but commonest in En; it is intermediate, but often more robust and with longer racemes than either parent, and partially fertile.

12. V. catenata Pennell - *Pink Water-Speedwell.* Differs from *V. anagallis-aquatica* **623** in leaves all narrow and sessile; pedicels usually shorter than bracts at flowering; and see Key A; (2n=36). Native; mostly in open muddy places with little or no flowing water; locally frequent in En, Wa and Ir, rare in Sc and CI.

Section 4 - *POCILLA* Dumort. (spp. 13-24). Annuals (except *V. filiformis*); flowers solitary in leaf-axils or in terminal racemes; corolla-tube wider than long; capsule dehiscing into 2 or 4 valves; seeds fairly flat, flat to convex on 1 side, flat to concave on other.

13. V. acinifolia L. - *French Speedwell.* Stems erect, to 15cm, glandular-pubescent throughout or above only and glabrous below; leaves ovate, obscurely serrate; flowers in racemes; corolla 2-3mm across, blue; capsule with patent glandular hairs; (2n=14, 16). Intrd; casual or persistent weed of gardens, nurseries and public flower-beds; scattered in S En; S Europe.

14. V. praecox All. - *Breckland Speedwell.* Stems erect to ascending, to 20cm, **RR** pubescent, glandular-pubescent above; leaves ovate, deeply serrate; flowers in racemes; corolla 2.5-4mm across, blue; capsule with patent glandular hairs; (2n=18). Perhaps native; sandy arable fields; very local in W Suffolk and W Norfolk, 1st recorded 1933, rare casual elsewhere, decreasing.

15. V. triphyllos L. - *Fingered Speedwell.* Stems erect to ascending, to 20cm, **RRR** glandular-pubescent; at least upper leaves and lower bracts lobed palmately almost to base with 3-7 lobes; flowers in racemes; corolla 3-4mm across, blue; capsule with patent glandular hairs; (2n=14). Native; sandy arable fields; very local in W Norfolk and E & W Suffolk, formerly scattered Surrey to MW Yorks, decreasing, perhaps extinct.

16. V. arvensis L. - *Wall Speedwell.* Stems erect to decumbent, to 30cm, pubescent or glandular-pubescent; leaves ovate, serrate to crenate-serrate; flowers in racemes; corolla 2-3mm across, blue; capsule glandular-pubescent; 2n=16. Native; walls, banks, open acid or calcareous ground and cultivated land; common throughout BI.

17. V. verna L. - *Spring Speedwell.* Stems erect, to 15cm; differs from *V. arvensis* **RR** in stems more glandular-pubescent above; capsule wider than long (not c. as long as wide); and see Key B; 2n=16. Native; open places in poor grassland on dry sandy soils, often with *V. arvensis*; very local in W Suffolk, formerly E Suffolk and E & W Norfolk.

18. V. peregrina L. - *American Speedwell.* Stems erect, to 25cm, usually glabrous; leaves oblanceolate to narrowly oblong, entire to distantly crenate; flowers in racemes; corolla 2-3mm across, blue; capsule glabrous; (2n=52). Intrd; casual or persistent weed of gardens, nurseries and public flower-beds; scattered in BI N to C Sc; N & S America. A glandular-pubescent variant has been recorded once as a casual in N Hants.

19. V. agrestis L. - *Green Field-speedwell.* Stems procumbent to ascending, to 30cm, pubescent; leaves ovate, serrate; flowers solitary in leaf-axils; corolla 3-8mm across, whitish to pale blue or pale lilac; capsules with patent glandular hairs only; 2n=28. Native; cultivated ground; frequent throughout most of BI. **591**

20. V. polita Fr. - *Grey Field-speedwell.* Differs from *V. agrestis* in lower leaves usually wider than long (not longer than wide), dull to greyish-green (not light to mid-green); calyx-lobes acute to subacute (not obtuse to subacute); corolla bright blue; and capsule with many short eglandular ± to strongly arched hairs and variable numbers of patent glandular hairs; 2n=14. Native; cultivated ground; frequent in most of BI but rare in C & N Sc. **591**

21. V. persica Poir. - *Common Field-speedwell.* Stems procumbent to decumbent, to 50cm, pubescent; leaves ovate, serrate; flowers solitary in leaf-axils; corolla 8-12mm across, bright blue; capsules with patent glandular and short eglandular hairs; (2n=28). Intrd; well natd in cultivated and waste ground throughout most of BI, 1st recorded 1825; SW Asia. **591**

22. V. crista-galli Steven - *Crested Field-speedwell.* Differs from *V. persica* in corolla 5-7mm across; calyx-segments fused in pairs forming conspicuous bilobed structures completely concealing much more shallowly notched capsules with non-divergent lobes; (2n=18). Intrd; occasional casual in cultivated and rough ground and waste places; very scattered in En and Wa, natd in N Somerset and formerly in W Sussex, well natd in SW Ir; Caucasus.

23. V. filiformis Sm. - *Slender Speedwell.* Perennial with procumbent, minutely pubescent stems to 50cm; leaves orbicular to reniform, crenate; flowers solitary in leaf-axils on erect long pedicels; corolla 8-15mm across, blue; capsule (seldom produced) with patent glandular hairs only, the lobes only slightly divergent; (2n=14). Intrd; well natd on streamsides, lawns, grassy paths, banks and roadsides throughout most of BI, 1st found 1838 (N Essex), then 1927 onwards; Turkey and Caucasus.

24. V. hederifolia L. - *Ivy-leaved Speedwell.* Stems procumbent to scrambling or ascending, to 60cm, pubescent, rather succulent; leaves orbicular to reniform, with 1-3 large teeth or lobes each side; flowers solitary in leaf-axils; corolla 4-9mm across; capsule glabrous. Native; cultivated and waste ground, open woods, hedgerows, walls and banks; common throughout BI except parts of Sc and Ir.

a. Ssp. hederifolia. Apical leaf-lobe usually wider than long; fruiting pedicels mostly 2-4x as long as calyx; calyx enlarging strongly after flowering, with marginal hairs mostly ≥0.9mm; corolla mostly ≥6mm across, whitish to blue; anthers blue, 0.7-1.2mm; 2n=54. Commoner in open places and cultivated ground. **591**

b. Ssp. lucorum (Klett & Richt.) Hartl (*V. sublobata* M.A. Fisch.). Apical leaf-lobe usually longer than wide; fruiting pedicels mostly 3.5-7x as long as calyx; calyx enlarging slightly after flowering, with marginal hairs mostly ≤0.9mm; corolla mostly ≤6mm across, whitish to pale lilac-blue; anthers whitish to pale blue, 0.4-0.8mm; (2n=36). Commoner in shady places. Determination of the ssp. is best using as many characters as possible; even so, up to 25% of plants may be difficult to name. **591**

Section* 5 - *PSEUDOLYSIMACHIUM W.D.J. Koch (spp. 25-26). Perennials; flowers in dense, long terminal racemes; corolla-tube longer than wide; capsule dehiscing into 2 valves usually only apically; seeds flat on 1 side, convex on other.

25. V. longifolia L. - *Garden Speedwell.* Stems erect, woody near base, to 1.2m, glabrous to minutely pubescent; leaves lanceolate, acute to acuminate, serrate to biserrate with acute to subacuminate teeth, glabrous or sparsely pubescent, 2-4 per node; calyx-lobes acute; 2n=68. Intrd; much grown in gardens and natd on waste and rough ground, banks, roadsides; scattered in Br, mainly S & C; N & C Europe.

25 x 26. V. longifolia x V. spicata is probably the parentage of many garden and some natd plants; the hybrid is intermediate and fertile.

26. V. spicata L. - *Spiked Speedwell.* Stems erect to ascending, woody near base, to **RRR** 60(80)cm, pubescent; leaves oblong-elliptic to narrowly so, acute to obtuse, serrate or crenate with obtuse teeth, pubescent, 2 per node; calyx-lobes obtuse; 2n=c.68 (34, 68). Native; rocks and short grassland on limestone and other basic soils; very local in W Br from N Somerset to Westmorland, 4 sites in Cambs, W Suffolk and W Norfolk. Plants in W Br are often separated as ssp. **hybrida** (L.) Gaudin; they usually differ in being taller and having larger, more extensively crenate-serrate leaves that are widest below (not at) the middle and are more abruptly narrowed into the petiole, but the differences are not constant and these races are only 2 of a large number in Europe.

18. HEBE Comm. ex Juss. - *Hedge Veronicas*

Evergreen shrubs; leaves opposite, with pinnate venation but usually with obscured lateral veins; flowers in axillary racemes; corolla showy, as in *Veronica;* stamens 2.

The terminal leaf-bud is composed of successively smaller developing leaves, without any modified bud-scales. The outermost 2 leaves enclose all the inner ones and their margins meet along 2 sides, but towards their base might (or might not) leave a gap (leaf-bud sinus) on either side due to the presence (or absence) of a distinct petiole.

1 Leaf-buds with distinct sinuses; leaves ± petiolate 2
1 Leaf-buds without sinuses; leaves ± sessile 4
 2 Leaves ≤2.5 x 0.8cm; racemes mostly <3cm **4. H. brachysiphon**
 2 Leaves ≥(2.5)3 x 0.7cm; racemes mostly >3cm 3
3 Leaves linear-lanceolate, c.8-12x as long as wide, narrowly acute to acuminate at apex; racemes mostly >10cm; corolla usually white **1. H. salicifolia**
3 Leaves oblanceolate to obovate or oblong-obovate, c.2-4x as long as wide, subacute to rounded at apex; racemes <10cm; corolla usually blue or pink to purple **3. H. x franciscana**
 4 Leaves obovate to oblong-obovate, often shortly acuminate at apex, 2-3x as long as wide **2. H. x lewisii**
 4 Leaves lanceolate to oblanceolate, subacute to obtuse at apex, 3-5x as long as wide 5
5 Leaves 4.5-8cm, mid-green; stem glabrous to finely and minutely pubescent, green **5. H. dieffenbachii**
5 Leaves 3-5cm, grey-green; stem glabrous, becoming purple **6. H. barkeri**

Other spp. - Apart from some dwarf montane spp. not natd in BI, no taxa are fully frost-hardy, and except in the extreme SW and in CI they are killed off in hard winters; however, viable seed can survive these winters in the ground. Many spp., hybrids and cultivars are grown, especially by the sea in S En, but none is as well natd as the following 6. Records of **H. elliptica** and **H. speciosa** apparently all refer to *H. x franciscana.*

1. H. salicifolia (G. Forst.) Pennell - *Koromiko*. Shrub to 2m; leaf-bud sinuses **607** present; leaves 7-12 x 0.7-1.5cm, linear-lanceolate, acuminate, subglabrous to pubescent at margin; racemes 10-20cm; corolla white or tinged pale lilac; (2n=40, 80). Intrd; natd by sea in Devon, Cornwall and Man, marginally so elsewhere in S En, N Wa, SW & WC Sc, W Ir and CI; New Zealand and Chile.

2. H. x lewisii (J.B. Armstr.) A. Wall (*H. salicifolia* x *H. elliptica* (G. Forst.) Pennell) **607** - *Lewis's Hebe*. Shrub to 1.5m; leaf-bud sinuses 0: leaves 4.5-7 x 1.8-2.5cm, obovate

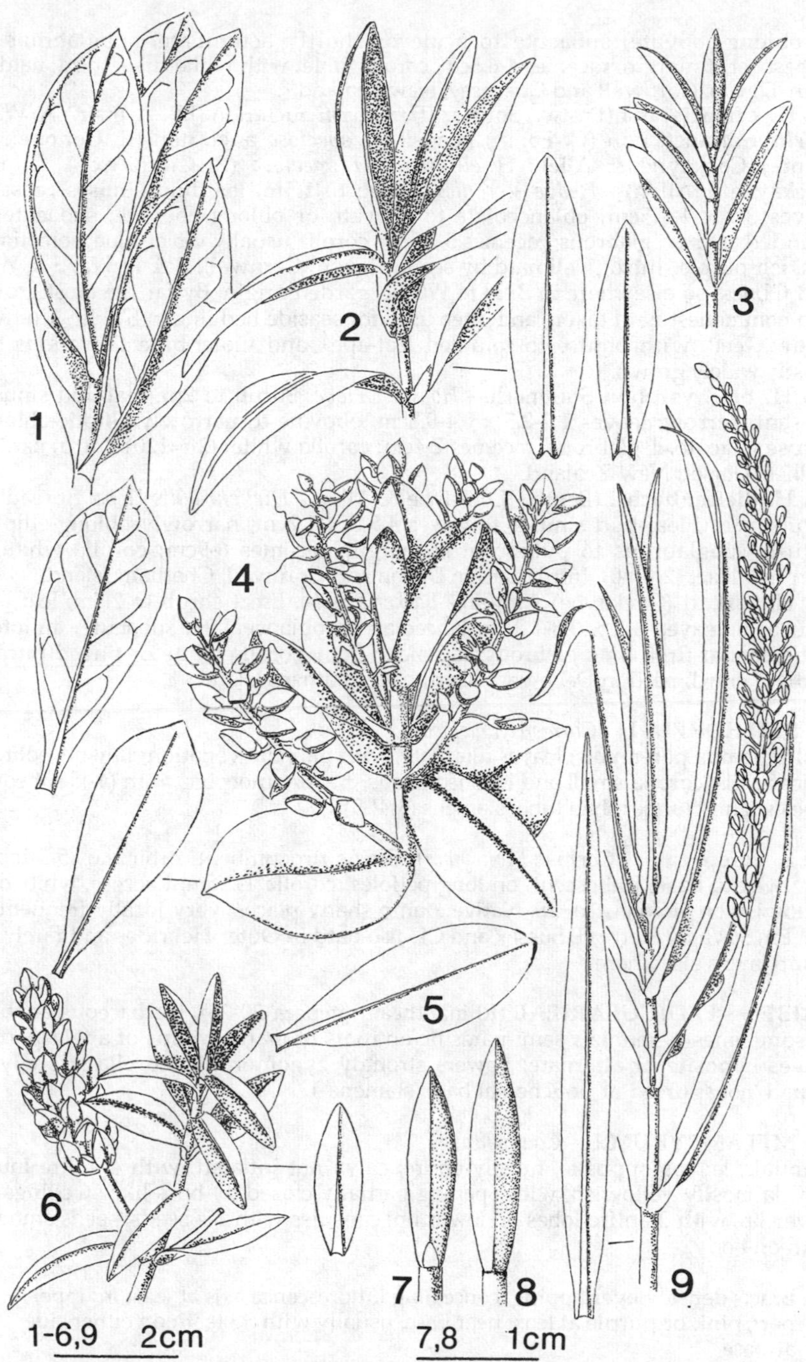

FIG 607 - *Hebe*. 1, *H. x lewisii*. 2, *H. dieffenbachii*. 3, *H. barkeri*. 4, *H. x franciscana*. 5, leaf of *H. x franciscana* **'Blue Gem'**. 6, *H. brachysiphon*. 7, shoot apex of *H. brachysiphon*. 8, shoot apex of *H. barkeri*. 9, *H. salicifolia*.

to oblong-obovate, subacute to acute or shortly acuminate, subglabrous to pubescent at margin; racemes 4-6.5cm; corolla violet with white tube. Intrd; natd in Man, Devon, Cornwall and Guernsey; New Zealand.

3. H. x franciscana (Eastw.) Souster (*H. x lewisii* auct. non (J.B. Armstr.) A. Wall, *H. elliptica* auct. non (G. Forst.) Pennell, *H. speciosa* auct. non (R. Cunn. ex A. Cunn.) Cockayne & Allen; *H. elliptica* x *H. speciosa* (R. Cunn. ex A. Cunn.) Cockayne & Allen) - *Hedge Veronica*. Shrub to 1.5m, leaf-bud sinuses present; leaves 3-9 x 1-2.5cm, oblanceolate to obovate or oblong-obovate, subacute to rounded at apex, glabrous; racemes 5-7.5cm; corolla usually violet-blue, sometimes pinkish-purple. Intrd; well natd by sea in Devon, Cornwall, S & N Wa, S & W Ir and CI, less so elsewhere in BI N to WC Sc; garden origin. By far the most grown and commonest natd taxon, and often used for seaside hedging in SW En. Cultivar **'Blue Gem'**, with obtuse to rounded leaf-apex and violet-blue flowers, is the mostly widely grown. **607**

4. H. brachysiphon Summerh. - *Hooker's Hebe*. Shrub to 2m; leaf-bud sinuses present, narrow; leaves 1.2-2.5 x 0.4-0.8cm, obovate to narrowly elliptic-oblong, obtuse to acute, ± glabrous; racemes 2-4cm; corolla white; (2n=120). Intrd; natd in Scillies, Dorset; New Zealand. **607**

5. H. dieffenbachii (Benth.) Cockayne & Allan - *Dieffenbach's Hebe*. Spreading shrub to 1m; leaf-bud sinuses 0; leaves 4.5-8 x 1.2cm, narrowly oblong-elliptic, obtuse, subglabrous to pubescent at margin; racemes 6-9cm; corolla white to purplish-lilac; (2n=40). Intrd; natd in Devon and Cornwall; Chatham Island. **607**

6. H. barkeri (Cockayne) A. Wall - *Barker's Hebe*. Erect shrub to 2.5m; leaf-bud sinuses 0; leaves 3-5 x 0.6-1.2cm, lanceolate to oblanceolate, subacute, minutely pubescent at first, then glabrous; racemes 5-7cm; corolla white or tinged purple; (2n=40). Intrd; natd in Devon and Cornwall; Chatham Island. **607**

19. SIBTHORPIA L. - *Cornish Moneywort*

Stoloniferous perennials; leaves alternate, with palmate venation; flowers solitary in leaf-axils; corolla small and inconspicuous, ± actinomorphic, with (4-)5 subequal lobes slightly longer than tube; stamens (3)4(-5).

1. S. europaea L. - *Cornish Moneywort*. Stems procumbent, orbicular, 5-9-lobed <1/2 way to base, pubescent, on long petioles; corolla 1-2.5mm across, whitish to yellowish or pinkish; 2n=18. Native; damp shady places; very locally frequent in SW En, S Wa, N Kerry, E Sussex and CI, also natd in Outer Hebrides and rarely on damp lawns elsewhere. **R**

TRIBE 6 - PEDICULARIEAE (Rhinantheae) (genera 20-26). Herbaceous annuals or sometimes perennials semi-parasitic on roots of many groups of angiosperms; leaves opposite or alternate; flowers strongly zygomorphic; corolla strongly 2-lipped, not spurred or pouched at base; stamens 4.

20. MELAMPYRUM L. - *Cow-wheats*

Annuals; leaves opposite, mostly entire; calyx not inflated, with 4 entire lobes; corolla mostly yellowish, with opening partially closed by boss-like swellings on lower lip, with 3 entire lobes on lower lip; capsules with 1-4 seeds; seeds smooth, with oil-body.

1 Bracts densely overlapping, concealing inflorescence axis at least in upper part, pink or purple at least near base, usually with >3 teeth on either side at base — 2
1 Bracts not or scarcely overlapping, with inflorescence axis well exposed, green, usually with <3 teeth on either side at base — 3
  2 Bracts cordate at base, strongly recurved, folded inwards along midrib proximally — **1. M. cristatum**

2 Bracts rounded to cuneate at base, not recurved, not folded inwards along midrib **2. M. arvense**
3 Lower lip of corolla not reflexed, its underside forming a straight line with lower edge of tube; lower 2 calyx-lobes appressed to corolla and upswept; fruit with four seeds **3. M. pratense**
3 Lower lip of corolla strongly reflexed (turned down); lower 2 calyx-lobes patent, not upswept; fruit with two seeds **4. M. sylvaticum**

1. M. cristatum L. - *Crested Cow-wheat.* Stems erect, to 50cm; inflorescence dense, R
4-sided; bracts strongly recurved, the lower part infolded along midrib and with many fine teeth on either side, bright purple towards base; corolla 12-16mm, pale yellow with purple and darker yellow areas; 2n=18. Native; wood-borders and scrub; very local in E Anglia and adjacent C En, formerly elsewhere, decreasing.

2. M. arvense L. - *Field Cow-wheat.* Stems erect, to 60cm; inflorescence rather RRR
dense, cylindrical; bracts not recurved or infolded, with >3 strong teeth on either side, pink at first; corolla 20-25mm, pink and yellow; (2n=18). Possibly native; cornfields and grassy field margins; very local in Wight, N Essex and Beds, formerly more widespread, decreasing.

3. M. pratense L. - *Common Cow-wheat.* Stems erect, to 60cm; inflorescence very 623
lax; bracts entire or with 1-2(3) pairs of teeth near base, green, not recurved or infolded; corolla 10-18mm, pale to golden yellow, often with purple marks near mouth; (2n=18). Native; woods, scrub, heathland.

a. Ssp. pratense. Uppermost leaves (below bracts) (1)2-8(11)cm x (1)2-10(20)mm, mostly 7-15x as long as wide. On acid soils in suitable places throughout most of Br and Ir.

b. Ssp. commutatum (Tausch ex A. Kern.) C.E. Britton. Uppermost leaves R
(below bracts) (3)4-7(10)cm x (4)8-20(27)mm, mostly 3-8x as long as wide. On calcareous soils in S En and SE Wa N to Herefs.

4. M. sylvaticum L. - *Small Cow-wheat.* Stems erect, to 35cm; inflorescence very R
lax; bracts entire or the upper ones with ≤2 pairs of small teeth, green, not recurved 623
or infolded; corolla 8-12mm, usually deep, often brownish-yellow; (2n=18). Native; upland woods and moorland; local in Sc N to E Ross, MW Yorks, Co Londonderry, Co Antrim, formerly Co Durham.

21. EUPHRASIA L. - *Eyebrights*

Annuals; leaves mostly opposite, conspicuously toothed; calyx not inflated, with 4 entire lobes; corolla white to purple, usually with darker veins and yellow blotch on lower lip, rarely yellow all over, with open mouth, with lower lip with 3 emarginate lobes; capsules with many seeds; seeds furrowed longitudinally, without oil-body.

A highly critical genus with >60 wild hybrids, for which the key does not allow. For a good chance of correct determination at least 5 or 6 well-grown (not stunted or spindly) and undamaged plants bearing some fruits as well as open flowers should be examined from a population. Ranges, rather than means, from these should be used. Single plants, or plants not agreeing with the above definition, are not allowed for in the key. The following key and accounts are based upon the classification of P.F. Yeo; other specialists, such as T. Karlsson, hold different views with a broader sp. limit that might eventually prove more valuable. Nodes are numbered from the base upwards, excluding the cotyledonary node. Corolla length is from base of tube to tip of upper lip in fresh state; dried specimens may have shrivelled or stretched (≤1mm) corollas. The name **E. officinalis** L. is often applied in an aggregate sense to the whole genus or to all spp. except *E. salisburgensis*. All spp. are known as *Eyebright*.

1 Middle and upper leaves with glandular hairs with stalk (6)10-12x as long as head 2
1 Middle and upper leaves without glandular hairs, or with glandular hairs

with stalk ≤6x as long as head 9
2 Capsule >2x as long as wide **5. E. arctica**
2 Capsule ≤2x as long as wide 3
3 Corolla ≤7mm 4
3 Corolla >7mm 5
4 Lowest flower at node 5-8; lower bracts 5-12mm, often longer than flowers; plant usually with 1-4 pairs of strong branches **3. E. anglica**
4 Lowest flower at node (2)3-5(6); lower bracts 3-6(7)mm, shorter than flowers; plant not branched or with 1-2 pairs of short branches **2. E. rivularis**
5 Lowest flower at node 2-5(6) 6
5 Lowest flower at node 5 or higher 7
6 Corolla 9-12.5mm; lower bracts 5-12(20)mm **1. E. rostkoviana**
6 Corolla ≤9mm; lower bracts 3-6(7)mm **2. E. rivularis**
7 Leaves dull greyish-green, often strongly suffused with violet or black; corolla usually lilac to purple **4. E. vigursii**
7 Leaves light or dark green, usually with little violet suffusion; corolla usually with at least lower lip white 8
8 Stem usually erect, with erect or divergent branches; lower internodes of inflorescence mostly 1.5-3x as long as bracts; corolla 8-12mm **1. E. rostkoviana**
8 Stem usually flexuous with flexuous or arched branches; lower internodes of inflorescence mostly <1.5x as long as bracts; corolla usually 6.5-8mm **3. E. anglica**
9 Capsule glabrous or with a few short hairs; at least 2 distal pairs of leaf-teeth (and sometimes all) not contiguous at base **21. E. salisburgensis**
9 Capsule with long ± numerous hairs in distal part; usually all leaf-teeth contiguous at base 10
10 Corolla >7.5mm 11
10 Corolla ≤7.5mm 16
11 Basal pair of teeth of lower bracts directed apically 12
11 Basal pair of teeth of lower bracts patent 13
12 Stem and branches slender, flexuous; capsule usually c. as long as calyx; lower bracts mostly alternate **9. E. confusa**
12 Stem stout, erect, with straight or evenly curved branches; capsule usually much shorter than calyx; lower bracts mostly opposite **10. E. stricta**
13 Lowest flower at node 8 or lower; capsule usually elliptic to obovate **5. E. arctica**
13 Lowest flower at node 9 or higher; capsule oblong to elliptic-oblong 14
14 Stem and branches flexuous; leaves near base of branches usually very small **9. E. confusa**
14 Stem and branches usually straight or gradually curved; leaves near base of branches not much smaller than others 15
15 Teeth of bracts acute to acuminate; capsule usually slightly shorter than calyx **7. E. nemorosa**
15 Teeth of bracts mostly aristate; capsule much shorter than calyx **8. E. pseudokerneri**
16 Calyx-tube whitish and membranous, with prominent green to blackish veins **17. E. campbelliae**
16 Calyx-tube green, not membranous 17
17 Lowest flower at node 6 or higher 18
17 Lowest flower at node 5(-6) or lower 38
18 Stem internodes mostly 2-6x as long as leaves 19
18 Stem internodes mostly ≤2x as long as leaves 32
19 Basal pair of teeth of lower bracts directed apically 20

19 Basal pair of teeth of lower bracts patent 23
20 Teeth of lower bracts obtuse to acute; corolla ≤6.5mm 21
20 Teeth of lower bracts acute to aristate; corolla ≥7mm 22
21 Leaves strongly purple-tinged, not darker on lowerside; corolla usually lilac to purple; capsule shorter than calyx **18. E. micrantha**
21 Leaves weakly or moderately purple-tinged, often darker on lowerside; corolla usually white; capsule at least as long as calyx **19. E. scottica**
22 Capsule ≤3x as long as wide **5. E. arctica**
22 Capsule ≥3x as long as wide **10. E. stricta**
23 Corolla ≥6.5mm 24
23 Corolla ≤6.5mm 25
24 Lowest flower at node 9 or higher; leaves usually without glandular hairs; lower bracts smaller than upper leaves **7. E. nemorosa**
24 Lowest flower at node 8 or lower; leaves usually with glandular hairs; lower bracts larger than upper leaves **5. E. arctica**
25 Leaves subglabrous to sparsely pubescent 26
25 Leaves densely pubescent 28
26 Stem and branches very slender, blackish; leaves strongly purple-tinged, not darker on lowerside; corolla usually lilac to purple **18. E. micrantha**
26 Stem and branches either stout or lightly pigmented; leaves weakly or moderately purple-tinged; corolla usually white 27
27 Lowest flower at node 8 or higher; stem stout; leaves not darker on lowerside; capsule usually shorter than calyx **7. E. nemorosa**
27 Lowest flower at node 7 or lower; stem slender; leaves usually light green on upperside and purplish on lowerside; capsule usually longer than calyx **19. E. scottica**
28 Lowest flower at node 9 or higher; stem ≤40cm; lower bracts often longer than wide **7. E. nemorosa**
28 Lowest flower at node 8 or lower; stem ≤15cm; lower bracts c. as long as wide 29
29 Leaves pubescent mainly near apex, obovate to narrowly ovate to elliptic **17. E. campbelliae**
29 Leaves ± uniformly pubescent, usually suborbicular, ovate or ovate-oblong 30
30 Teeth of lower bracts mostly wider than long; branches ≤3 pairs **16. E. rotundifolia**
30 Teeth of lower bracts mostly as long as wide; branches ≤5 pairs 31
31 Corolla 5.5-7mm; capsule usually >2x as long as wide **15. E. marshallii**
31 Corolla 4.5-6mm; capsule ≤2x as long as wide **14. E. ostenfeldii**
32 Basal pair of teeth of lower bracts directed apically 33
32 Basal pair of teeth of lower bracts patent 34
33 Teeth of lower bracts not much longer than wide Return to 29
33 Teeth of lower bracts much longer than wide Return to 12
34 Lowest flower at node 10 or higher 35
34 Lowest flower at node 9 or lower 36
35 Stem erect, stout, with stout ascending branches; lower bracts mostly opposite **7. E. nemorosa**
35 Stem and branches slender and flexuous; lower bracts mostly alternate **9. E. confusa**
36 Leaves with numerous eglandular hairs Return to 29
36 Leaves with few eglandular hairs 37
37 Capsule 5.5-7mm, often slightly curved, as long as or longer than calyx **20. E. heslop-harrisonii**
37 Capsule usually ≤5.5mm, straight, usually shorter than calyx **6. E. tetraquetra**

38 Stem internodes mostly ≥2.5x as long as leaves 39
38 Stem internodes mostly <2.5x as long as leaves 44
39 Capsule broadly elliptic to obovate-elliptic 40
39 Capsule oblong to narrowly elliptic 41
40 Teeth of lower bracts mostly subacute, not longer than wide; corolla 4.5-7mm; lowest flower at node 2-4(5) **11. E. frigida**
40 Teeth of lower bracts usually acute or acuminate, longer than wide; corolla ≥6.5mm; lowest flower usually at node 4 or higher **5. E. arctica**
41 Lower bracts deeply serrate, with basal pair of teeth directed apically **10. E. stricta**
41 Lower bracts crenate to shallowly serrate, with basal pair of teeth usually patent 42
42 Upper leaves elliptic-ovate to narrowly obovate **19. E. scottica**
42 Upper leaves suborbicular to broadly ovate or broadly obovate 43
43 Lowest flower at node 4 or lower; lower bracts often considerably larger than upper leaves **11. E. frigida**
43 Lowest flower at node 4 or higher; lower bracts scarcely larger than upper leaves Return to 29
44 Corolla ≥6mm 45
44 Corolla ≤6mm 47
45 Teeth of lower bracts usually very acute, all directed apically Return to 12
45 Teeth of lower bracts acute to subacute, the basal pair patent 46
46 Capsule at least as long as calyx, usually emarginate **11. E. frigida**
46 Capsule shorter than calyx, truncate to slightly emarginate **6. E. tetraquetra**
47 Lower bracts ovate to rhombic, with acute to aristate teeth, the basal pair directed forwards Return to 12
47 Lower bracts broadly ovate or rhombic to suborbicular, with obtuse to subacute teeth, the basal pair patent 48
48 Leaves with numerous hairs, all eglandular 49
48 Leaves with few eglandular hairs, sometimes with short glandular hairs 50
49 Lower bracts scarcely larger than upper leaves Return to 29
49 Lower bracts considerably larger than upper leaves **11. E. frigida**
50 Capsule elliptic to obovate, emarginate **13. E. cambrica**
50 Capsule oblong to elliptic-oblong, usually truncate 51
51 Capsule usually shorter than calyx; distal teeth of lower bracts not incurved **6. E. tetraquetra**
51 Capsule as long as or longer than calyx; distal teeth of lower bracts ± incurved 52
52 Capsule 4.5-5.5(7)mm, c.2x as long as wide, straight; upper leaves only obscurely petiolate, with margins of teeth not wavy **12. E. foulaensis**
52 Capsule (4.5)5.5-7mm, 2-3x as long as wide, often slightly curved; upper leaves ± distinctly petiolate, with margins of teeth wavy **20. E. heslop-harrisonii**

Other spp. - Records of **E. hirtella** Jord. ex Reuter refer to *E. anglica* or *E. rostkoviana* ssp. *rostkoviana*, and those of **E. brevipila** Burnat & Gremli ex Gremli to *E. arctica* ssp. *borealis*. **E. eurycarpa** Pugsley is either a variant of *E. ostenfeldii* or one of its hybrid segregates; **E. rhumica** Pugsley is either *E. micrantha* or a hybrid of it.

Hybrids - All spp. have been reported to form hybrids.
Crosses between spp. of Group 2 (tetraploids of subsection *Ciliatae*) are highly fertile and often common where 2 or more spp. occur together; 46 binary combinations have been reliably recorded in BI, as well as some triple hybrids. The

following 11 hybrids are frequent and may occur in the absence of 1 or both parents, locally replacing them: *E. arctica* x *E. nemorosa*, x *E. confusa* and x *E. micrantha*; *E. tetraquetra* x *E. confusa*; *E. nemorosa* x *E. pseudokerneri*, x *E. confusa* and x *E. micrantha*; *E. confusa* x *E. micrantha* and x *E. scottica*; *E. frigida* x *E. scottica*; and *E. micrantha* x *E. scottica*.

Hybrids within Group 1 (diploids of subsection *Ciliatae*) are much less common, mainly because the spp. tend to be ± allopatric. 3 combinations are known, resulting in intergradation between the parent spp., mostly in Wa and W En.

Hybrids between Groups 1 and 2 also occur; 10 combinations have been identified. First generation hybrids are probably highly but not totally sterile triploids, but most examined in the field are ± fertile diploid introgressants.

3 spp. of Group 2 (*E. arctica*, *E. nemorosa* and *E. micrantha*) have been found to hybridise with the tetraploid *E. salisburgensis* (Group 3; subsection *Angustifoliae*) in Ir; these hybrids are highly but not totally sterile.

Group* 1 - *Subsection CILIATAE Jörg.: diploids (2n=22) (spp. 1-4). Middle and upper leaves with long glandular hairs with stalk (6)10-12x as long as head; usually all leaf-teeth contiguous at base; capsule with long, ± numerous hairs in distal part.

1. E. rostkoviana Hayne. Stems stout, erect, to 35cm; branches 0-5(more) pairs, often again branched; corolla (6.5)8-12.5mm, with usually white lower and lilac upper lip. Native; grassland in hilly areas. Our largest flowered sp.

a. Ssp. rostkoviana (*E. officinalis* ssp. *rostkoviana* (Hayne) F. Towns.). Branches 1-5(more) pairs, ascending, divergent or erect; internodes shorter than to 3(4)x as long as leaves; lowest flower usually at node 6-10; corolla (6.5)8-12mm; 2n=22. Often in damper places and by rivers, sometimes lowland; locally frequent in Ir, Wa, N En and C & S Sc. **R** **614**

b. Ssp. montana (Jord.) Wettst. (*E. montana* Jord., *E. officinalis* ssp. *montana* (Jord.) Berher). Branches 0-3(4) pairs, erect; internodes 2-6(10)x as long as leaves; lowest flower usually at node 2-6; corolla (7)9-12.5mm. Usually in drier upland places; very local in N En and S Sc, possibly errors in Wa and Ir. Flowers earlier than and perhaps an aestival variant of the autumnal ssp. *rostkoviana*. **R** **614**

1 x 2. E. rostkoviana x E. rivularis; both sspp. of *E. rostkoviana* involved.

1 x 3. E. rostkoviana x E. anglica; only ssp. *rostkoviana* involved.

1 x 5. E. rostkoviana x E. arctica; only sspp. *rostkoviana* and *borealis* involved.

1 x 7. E. rostkoviana x E. nemorosa = E. x glanduligera Wettst.; only ssp. *rostkoviana* involved.

1 x 9. E. rostkoviana x E. confusa; only ssp. *rostkoviana* involved.

1 x 18. E. rostkoviana x E. micrantha; only ssp. *rostkoviana* involved.

1 x 19. E. rostkoviana x E. scottica; only ssp. *rostkoviana* involved.

2. E. rivularis Pugsley. Stems flexuous-erect, to 15cm; branches 0-2 pairs, short; lowest flower usually at node (2)3-5(6); corolla 6.5-9mm, with white or lilac lower and lilac upper lip; 2n=22. Native; damp mountain pastures and streamsides; very local in NW Wa, Lake District; endemic. Probably a stabilized hybrid segregate of *E. rostkoviana* x *E. micrantha*. **RR** **614**

3. E. anglica Pugsley (*E. officinalis* ssp. *anglica* (Pugsley) Silverside). Stems flexuous-erect, to 20(30)cm; branches (0)1-4(6) pairs, flexuous or arcuate, usually again branched; lowest flower at node 5-8; corolla (5)6.5-8(10)mm, with white or lilac lower and lilac upper lip; 2n=22. Native; in short turf on often damp soils, heathland; rather local in C & S Br N to SW Sc, Ir; ?endemic. Mostly replaces *E. rostkoviana* in S Br. **614**

3 x 4. E. anglica x E. vigursii.

3 x 5. E. anglica x E. arctica; only ssp. *borealis* involved.

3 x 7. E. anglica x E. nemorosa (= *E. x glanduligera* auct. non Wettst.).

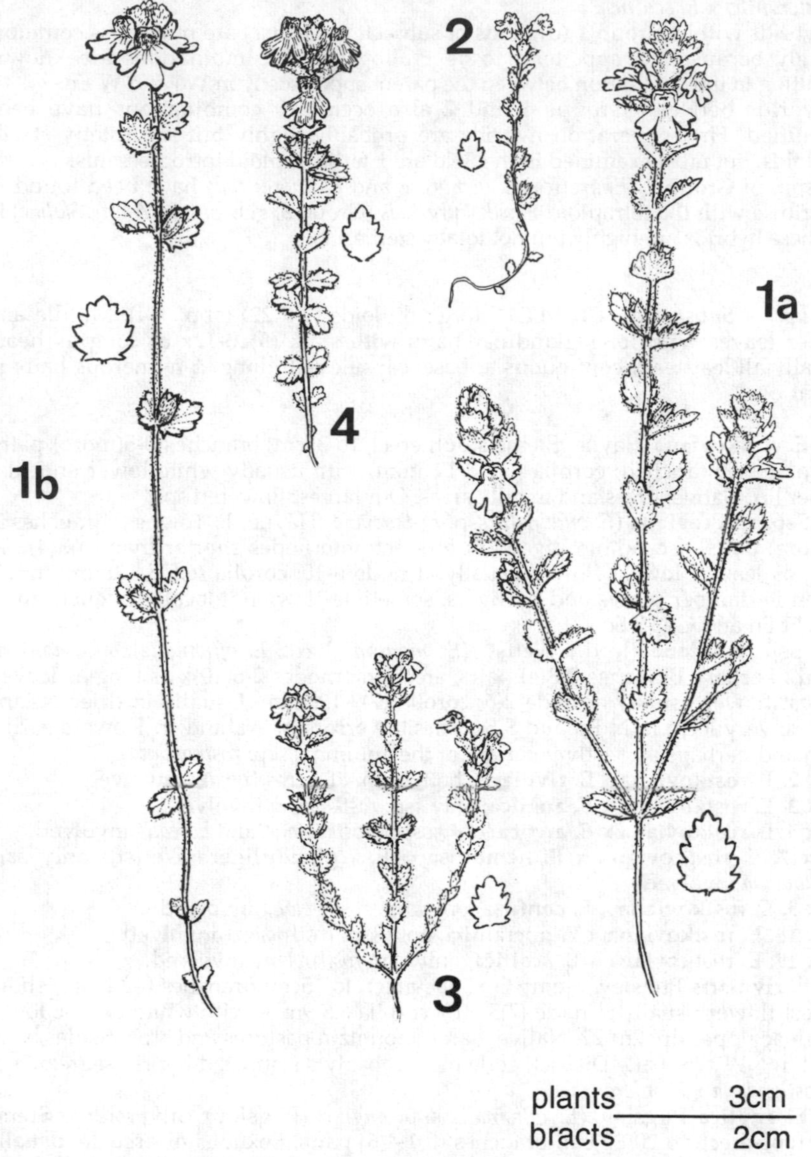

FIG 614 - Whole plant and lowest bract of ***Euphrasia***. 1a, ***E. rostkoviana*** ssp. ***rostkoviana***. 1b, ***E. rostkoviana*** ssp. ***montana***. 2, ***E. rivularis***. 3, ***E. anglica***. 4, ***E. vigursii***. Drawings by Olga Stewart.

3 x 9. E. anglica x E. confusa.

3 x 18. E. anglica x E. micrantha; 2n=22, c.33.

4. E. vigursii Davey. Stems erect, to 20(25)cm; branches 0-5(7) pairs, erect, often **RR**
again branched; lowest flower at node 7-10(12); corolla (6)7-8.5mm, with usually **614**
lilac to deep purple (occasionally white) lower and lilac to deep purple upper lip; 2n=22. Native; *Ulex gallii*/*Agrostis curtisii* heathland in Devon and Cornwall; endemic. Probably a stabilized segregate of *E. anglica* x *E. micrantha.*

4 x 6. E. vigursii x E. tetraquetra.

Group 2 - *Subsection CILIATAE* Jörg.: tetraploids (2n=44) (spp. 5-20). Middle and upper leaves without glandular hairs, or with glandular hairs with stalk <6x as long as head; usually all leaf-teeth contiguous at base; capsule with long, ± numerous hairs in distal part.

5. E. arctica Lange ex Rostrup. Stems erect at least above, to 30(35)cm, with 0-5(6) pairs of branches sometimes again branched; lowest flower at node (3)4-8(10); corolla 6-11(13)mm, with usually white (sometimes lilac) lower and lilac to purple (sometimes white) upper lip. Native; meadows and pastures, to some extent replacing *E. nemorosa* as commonest sp. in N & W. Usually with longer flowers and broader bracts than *E. nemorosa.*

a. Ssp. arctica (*E. borealis* auct. non (F. Towns.) Wettst.). Stem procumbent or **617**
flexuous at base, then erect; lower bracts suborbicular to broadly ovate; corolla 7-11(13)mm; capsule (5.5)6-7.5(8)mm. Orkney and Shetland.

b. Ssp. borealis (F. Towns.) Yeo (*E. borealis* (F. Towns.) Wettst., *E. brevipila* auct. **616**
non Burnat & Gremli ex Gremli). Stem erect from base; lower bracts narrowly to broadly ovate, rhombic or trullate; corolla 6-9(10)mm; capsule (4)4.5-6.5(7)mm; 2n=44. Ir, N & W Br except Shetland, very scattered in S & E Br. Flowers later than ssp. *arctica* in Orkney.

5 x 6. E. arctica x E. tetraquetra = E. x pratiuscula F. Towns.; only ssp. *borealis* involved.

5 x 7. E. arctica x E. nemorosa; both sspp. of *E. arctica* involved.

5 x 9. E. arctica x E. confusa; both sspp. of *E. arctica* involved.

5 x 12. E. arctica x E. foulaensis; both sspp. of *E. arctica* involved.

5 x 15. E. arctica x E. marshallii; both sspp. of *E. arctica* involved.

5 x 18. E. arctica x E. micrantha = E. x difformis F. Towns; both sspp. of *E. arctica* involved.

5 x 19. E. arctica x E. scottica = E. x venusta F. Towns.; only ssp. *borealis* involved.

5 x 20. E. arctica x E. heslop-harrisonii; only ssp. *borealis* involved.

5 x 21. E. arctica x E. salisburgensis; only ssp. *borealis* involved.

6. E. tetraquetra (Bréb.) Arrond. (*E. occidentalis* Wettst.). Stems erect, stout, to **616**
15(20)cm; branches 0-5(8) pairs, usually rather short and erect or ascending, but sometimes branched again, forming compact plant; lowest flower at node (3)5-7(9); corolla (4)5-7(8)mm, with usually white (sometimes lilac) lower and white or lilac upper lip; 2n=44. Native; short turf on cliffs and dunes by sea, limestone pasture inland; coasts of BI except most of E En and N Sc, inland in parts of SW En.

6 x 7. E. tetraquetra x E. nemorosa.

6 x 8. E. tetraquetra x E. pseudokerneri.

6 x 9. E. tetraquetra x E. confusa.

6 x 10. E. tetraquetra x E. stricta.

6 x 14. E. tetraquetra x E. ostenfeldii.

6 x 18. E. tetraquetra x E. micrantha.

7. E. nemorosa (Pers.) Wallr. (*E. curta* (Fr.) Wettst. pro parte). Stems erect, to **616**
35(40)cm; branches 1-9 pairs, ascending, often again branched; lowest flower at node (5)10-14; corolla 5-7.5(8.5)mm, coloured as in *E. tetraquetra*; 2n=44. Native;

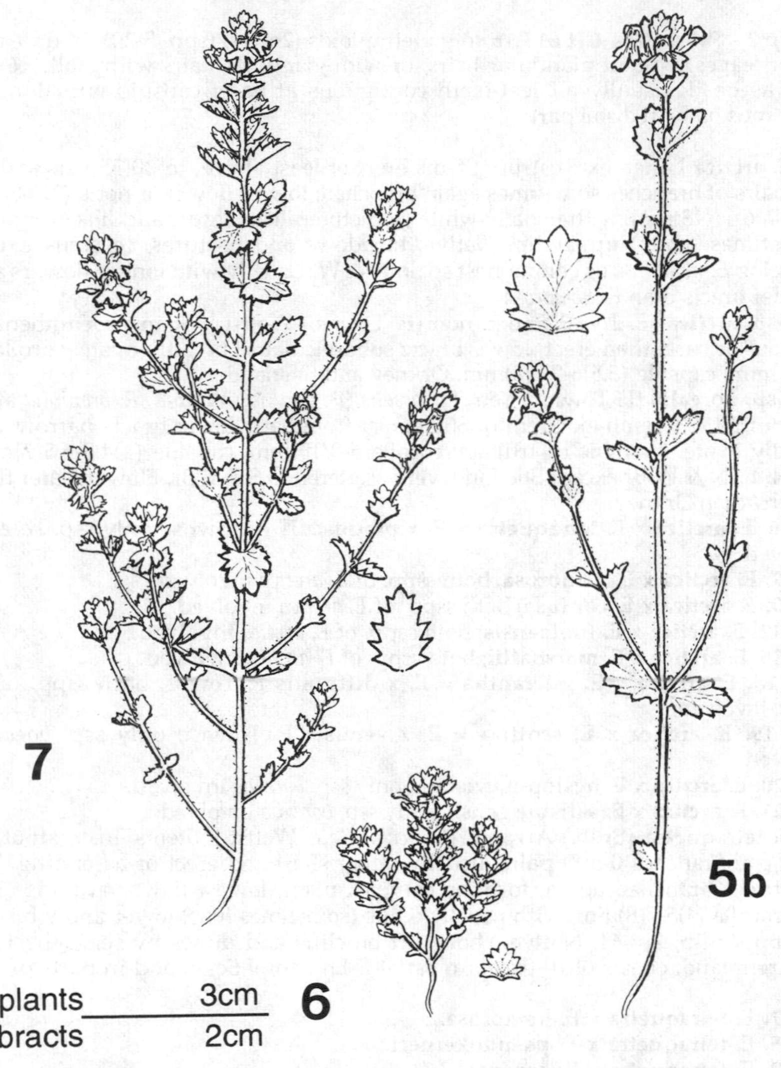

FIG 616 - Whole plant and lowest bract of ***Euphrasia***. 5b, ***E. arctica*** ssp. ***borealis***. 6, ***E. tetraquetra***. 7, ***E. nemorosa***. Drawings by Olga Stewart.

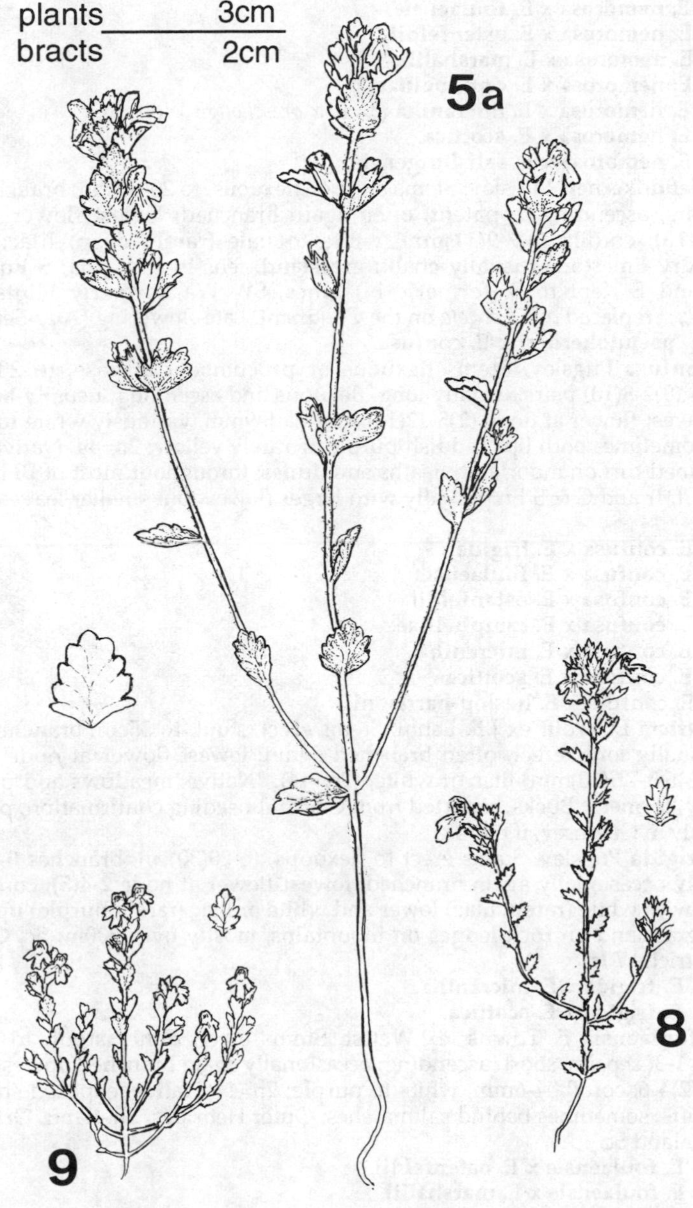

FIG 617 - Whole plant and lowest bract of ***Euphrasia***. 5a, ***E. arctica*** ssp. ***arctica***. 8, ***E. pseudokerneri***. 9, ***E. confusa***. Drawings by Olga Stewart.

pastures, scrub, woodland rides, marginal areas and heathland (dunes in Sc); throughout Bl, the commonest lowland sp.

7 x 8. E. nemorosa x E. pseudokerneri.

7 x 9. E. nemorosa x E. confusa.

7 x 10. E. nemorosa x E. stricta = E. x haussknechtii Wettst.

7 x 12. E. nemorosa x E. foulaensis.

7 x 14. E. nemorosa x E. ostenfeldii.

7 x 15. E. nemorosa x E. marshallii.

7 x 17. E. nemorosa x E. campbelliae.

7 x 18. E. nemorosa x E. micrantha (?= *E. x areschougii* Wettst.).

7 x 19. E. nemorosa x E. scottica.

7 x 21. E. nemorosa x E. salisburgensis.

8. E. pseudokerneri Pugsley. Stems erect or flexuous, to 20(30)cm; branches (0)3- **R**
8(10) pairs, ascending to patent, often again branched; lowest flower at node **617**
(5)10-16(18); corolla (6)7-9(11)mm, white to pale (rarely deep) lilac; 2n=44. Native; dry limestone (usually chalk) grassland, fens in E Anglia; S En from S Devon and E Kent to S (formerly N) Lincs, SW Wa, formerly Flints, W Ir; ?endemic, ± replaced by *E. stricta* on the Continent. Late-flowering (Aug-Sep).

8 x 9. E. pseudokerneri x E. confusa.

9. E. confusa Pugsley. Stems flexuous or procumbent at base, to 20(45)cm; **617**
branches (0)2-8(10) pairs, usually long, flexuous and ascending, usually branched again; lowest flower at node (2)5-12(14); corolla 5-9mm, variously white to lilac or mixed, sometimes both lips reddish-purple or rarely yellow; 2n=44. Native; short, well-drained turf on moorland, heaths and dunes; throughout most of BI but very local in CI, Ir and C & S En. Usually with larger flowers but smaller leaves than *E. nemorosa*.

9 x 11. E. confusa x E. frigida.

9 x 12. E. confusa x E. foulaensis.

9 x 14. E. confusa x E. ostenfeldii.

9 x 17. E. confusa x E. campbelliae.

9 x 18. E. confusa x E. micrantha.

9 x 19. E. confusa x E. scottica.

9 x 20. E. confusa x E. heslop-harrisonii.

10. E. stricta D. Wolff ex J.F. Lehm. Stems erect, stout, to 35cm; branches (0)2-6 **RR**
pairs, usually long, erect, often branched again; lowest flower at node (3)7-14; **619**
corolla (6)6.5-7.5(10)mm, lilac or white; (2n=44). ?Native; meadows and pastures; Guernsey, formerly Bucks, reported from C Sc but needing confirmation, probably native only in Guernsey, if there.

11. E. frigida Pugsley. Stems erect to flexuous, to 20(30)cm; branches 0-2 pairs, **R**
erect, only occasionally again branched; lowest flower at node 2-4(5); corolla 4.5- **619**
7(8)mm, with white (rarely lilac) lower and white or lilac (rarely purple) upper lip. Native; grassland on rock-ledges on mountains, mostly over 600m; Sc, Cheviot, Lake District, W Ir.

11 x 18. E. frigida x E. micrantha.

11 x 19. E. frigida x E. scottica.

12. E. foulaensis F. Towns. ex Wettst. Stems erect, rather stout, to 6(9)cm; **R**
branches 1-3(4) pairs, short, ascending, occasionally again branched; lowest flower **619**
at node (2)4-6; corolla 4-6mm, white to purple; 2n=44. Native; exposed short turf on sea-cliffs, sometimes behind saltmarshes; Outer Hebrides, Shetland, Orkney, N & W mainland Sc.

12 x 14. E. foulaensis x E. ostenfeldii.

12 x 15. E. foulaensis x E. marshallii.

12 x 16. E. foulaensis x E. rotundifolia.

12 x 18. E. foulaensis x E. micrantha.

12 x 19. E. foulaensis x E. scottica.

13. E. cambrica Pugsley. Stem flexuous, to 8cm; branches 0-2 pairs, flexuous; **RR**

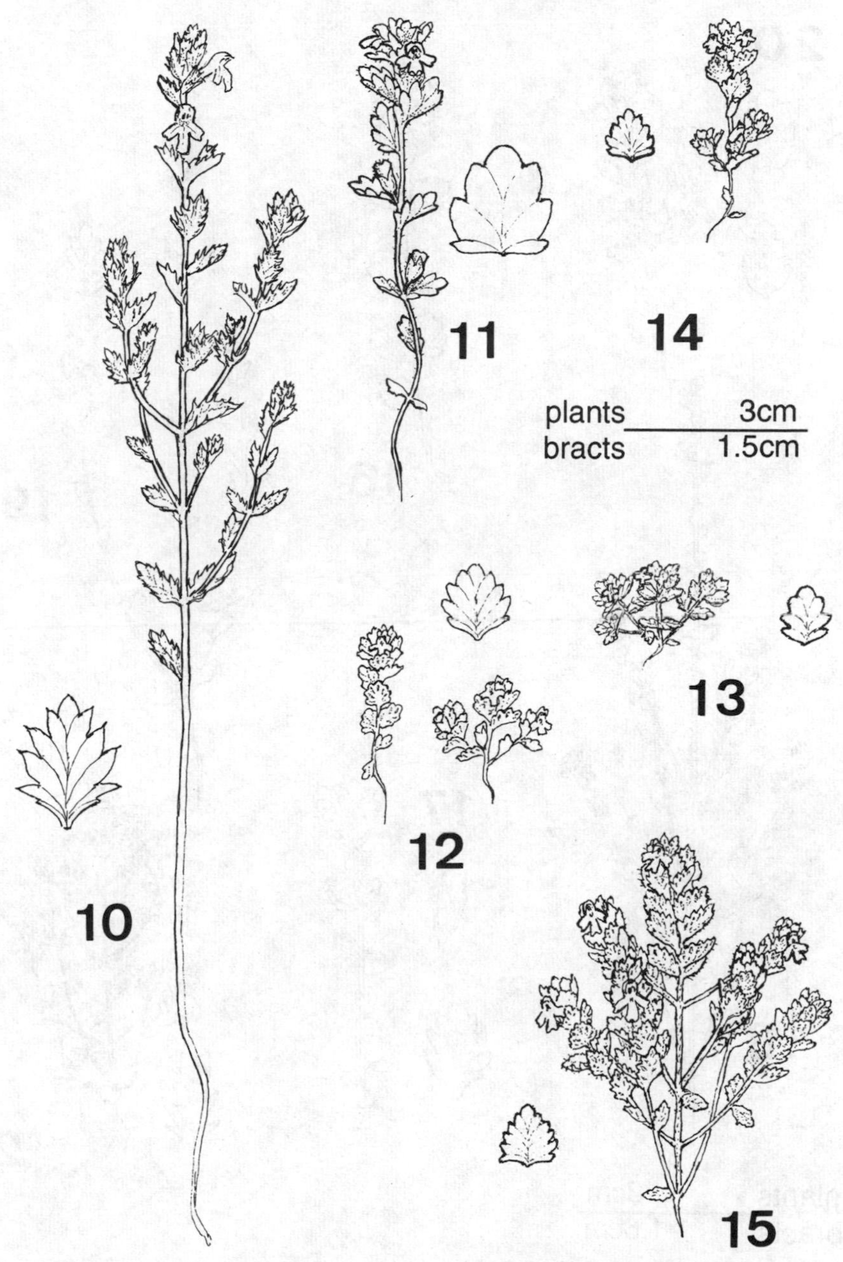

FIG 619 - Whole plant and lowest bract of ***Euphrasia***. 10, *E.* ***stricta***. 11, *E.* ***frigida***. 12, *E.* ***foulaensis***. 13, *E.* ***cambrica***. 14, *E.* ***ostenfeldii***. 15, *E.* ***marshallii***. Drawings by Olga Stewart.

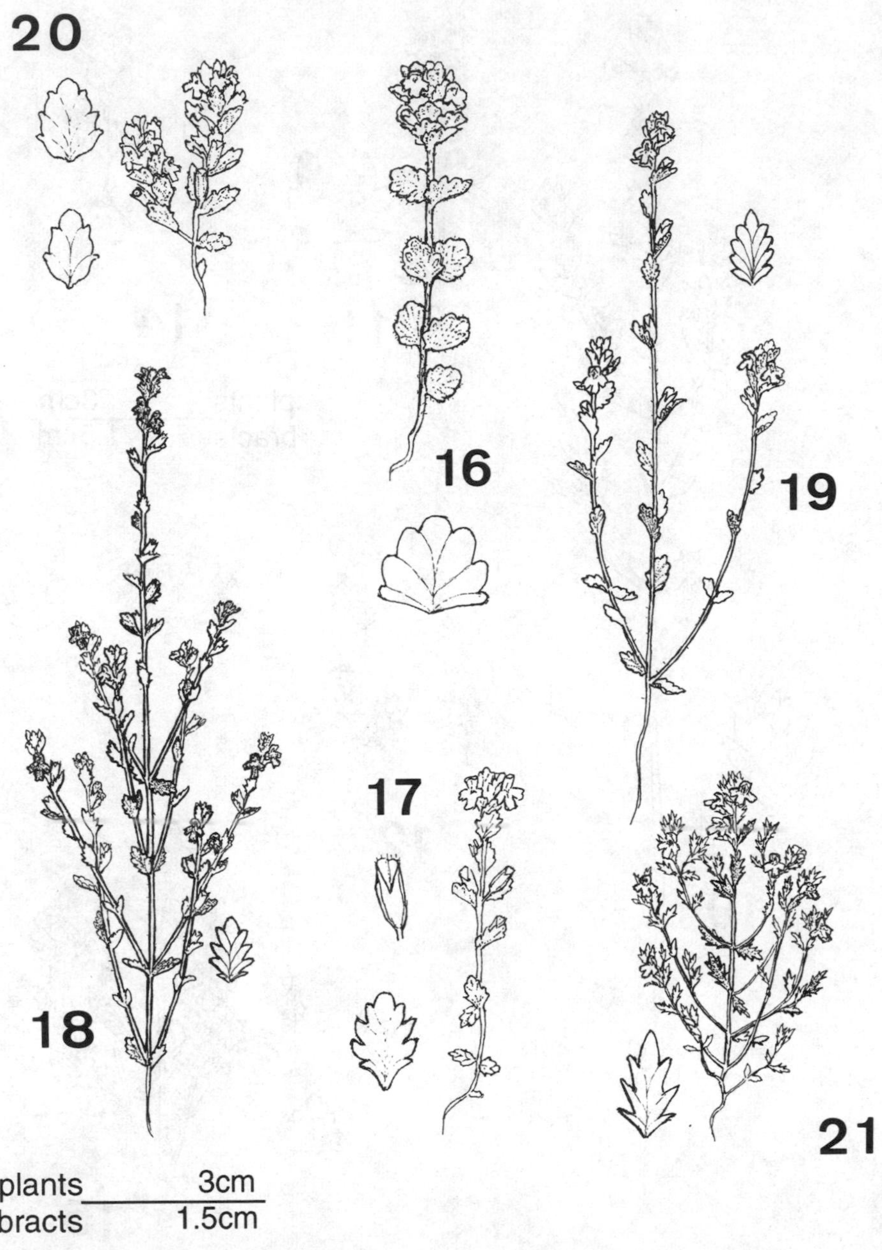

FIG 620 - Whole plant and lowest bract of ***Euphrasia***. 16, ***E. rotundifolia***. 17, ***E. campbelliae***. 18, ***E. micrantha***. 19, ***E. scottica***. 20, ***E. heslop-harrisonii***. 21, ***E. salisburgensis***. Drawings by Olga Stewart.

lowest flower at node 2-4; corolla 4-5.5mm, with white or yellowish-white lower 619
and white or lilac upper lip. Native; mountain grassland and rock-ledges; Caerns and Merioneth; endemic.

13 x 19. E. cambrica x E. scottica has been reported, but with some doubt.

14. E. ostenfeldii (Pugsley) Yeo (*E. curta* auct. non (Fr.) Wettst.). Stems erect or R
flexuous below, to 12(15)cm; branches 0-4(6) pairs, erect or ascending, sometimes 619
again branched; lowest flower at node (3)4-7(9); corolla (3.5)4.5-6mm, with white lower and white or lilac upper lip. Native; grassy, stony and sandy places, mostly near sea; N & NW Sc incl. Orkney and Shetland, Lake District, Caerns (?extinct). This and the next 2 spp. are usually greyish due to dense pubescence.

14 x 18. E. ostenfeldii x E. micrantha.

14 x 19. E. ostenfeldii x E. scottica.

15. E. marshallii Pugsley. Stems erect, to 12cm; branches (0)1-5 pairs, rather long, RR
erect, sometimes again branched; lowest flower at node (5)7-9; corolla 5.5-7mm, 619
white or lilac; 2n=44. Native; turf on sea-cliffs or dunes; extreme N Sc from N Ebudes and Outer Hebrides to Shetland; endemic.

15 x 16. E. marshallii x E. rotundifolia.

15 x 17. E. marshallii x E. campbelliae.

15 x 18. E. marshallii x E. micrantha.

15 x 19. E. marshallii x E. scottica.

16. E. rotundifolia Pugsley. Stems erect, to 10cm; branches 0-3 pairs, short, erect; RR
lowest flower at node 6-8(9); corolla 5-6mm, white or lilac or the upper lip 620
purplish, rarely pale yellow. Native; basic turf on sea-cliffs or dunes; similar distribution to *E. marshallii* and sometimes with it but flowering later; endemic. Possibly derived from *E. foulaensis* x *E. marshallii*.

17. E. campbelliae Pugsley. Stems erect, to 10cm; branches 0-2 pairs, short, erect; RR
lowest flower at node 5-7; corolla 5.5-7mm, usually with white lower and lilac 620
upper lip. Native; damp heathland near sea; Isle of Lewis (Outer Hebrides), ?Shetland; endemic. Probably of complex hybrid origin.

17 x 19. E. campbelliae x E. scottica.

18. E. micrantha Rchb. Stems erect, slender, to 25cm; branches (0)2-7(10) pairs, 620
slender, erect, usually again branched; lowest flower at node (4)6-14(16); corolla 4.5-6.5mm, lilac to purple or with white lower lip (rarely white all over); 2n=44. Native; heathland, usually with *Calluna*, sometimes in damp places; throughout most of BI but absent from most of C & E En.

18 x 19. E. micrantha x E. scottica = E. x electa F. Towns.

18 x 21. E. micrantha x E. salisburgensis.

19. E. scottica Wettst. Stems erect, to 25cm; branches 0-4 pairs, long, erect to 620
ascending; lowest flower at node (2)3-6(8); corolla (3.5)4.5-6.5mm, white or with lilac upper lip; 2n=44. Native; wet moorland; Wa, N W En, Sc and Ir. Close to *E. micrantha* and perhaps not distinct; usually flowers earlier and with less anthocyanin on corollas and vegetative parts except that leaves are often purple on lowerside.

20. E. heslop-harrisonii Pugsley. Stems erect from usually flexuous base, to 15cm; RR
branches 0-4(5) pairs, erect or patent, sometimes again branched; lowest flower at 620
node 4-7(8); corolla 4.5-6(6.5)mm, white or occasionally lilac. Native; turf in salt-marshes or drier grassy places; extreme N, NW & CW Sc; endemic. Close to the last 2 spp. and perhaps not distinct.

Group 3 - Subsection ANGUSTIFOLIAE (Wettst.) Jörg.: tetraploids (2n=44) (sp. 21). Middle and upper leaves without glandular hairs; at least 2 distal pairs of leaf-teeth (and sometimes all) not contiguous at base; capsule glabrous or with a few short hairs.

21. E. salisburgensis Funck. Stems erect or flexuous, to 12cm; branches (0)1-7 R
pairs, slender, erect or patent, often again branched; lowest flower at node 5-13; 620

corolla 4.5-6.5mm, white or upper lip sometimes lilac; 2n=44. Native; among limestone rocks and on dunes; W Ir from Co Limerick to E Donegal, probably mislocalized specimens from MW Yorks in 1885/6. Our plant is var. **hibernica** Pugsley. Glabrous capsules and distinctive leaf-toothing are diagnostic.

22. ODONTITES Ludw. - *Bartsias*
Annuals; leaves opposite, entire to toothed; calyx not inflated, with 4 entire lobes; corolla yellow or pinkish-purple, rarely white, with open mouth, with lower lip with 3 entire to slightly notched lobes; capsules with rather few seeds; seeds furrowed longitudinally, without oil-body. Intercalary leaves are those at nodes on the main stem between the topmost branches and the lowest bract.

1. O. jaubertianus (Boreau) D. Dietr. ex Walp. - *French Bartsia*. Stems erect, to 50cm; leaves linear, ≤4mm wide, entire to slightly toothed; calyx 4-5mm; corolla 7-9mm, yellow, often tinged pinkish; (2n=40). Intrd; natd on gravelly rough ground near Aldermaston, Berks, since 1965; France. Originally misdetermined as **O. luteus** (L.) Clairv. Our plant is ssp. **chrysanthus** (Boreau) P. Fourn.

2. O. vernus (Bellardi) Dumort. - *Red Bartsia*. Stems erect, to 50cm; leaves lanceolate or linear-lanceolate to oblong-lanceolate, serrate or crenate-serrate; calyx 5-8mm; corolla 8-10mm, pinkish-purple, rarely white. Native. Unbranched (starved) plants occur in most populations and are best ignored unless the prevalent sort; as for *Euphrasia*, a range of individuals should be measured for determination. Ssp. *pumilus* in the past has been used to cover ssp. *litoralis* and maritime variants of ssp. *serotinus*.

1 Branches ≤8 pairs, held at ≥50° to main stem; intercalary leaves 2-7 pairs; lowest flower at node 8-14 **b. ssp. serotinus**
1 Branches ≤4 pairs, held at ≤50° to main stem; intercalary leaves 0-1 pairs; lowest flower at node 4-9 2
 2 Calyx-teeth narrowly triangular, acute, as long as tube; style exserted from corolla at full anthesis **a. ssp. vernus**
 2 Calyx-teeth triangular, subacute to obtuse, shorter than tube; style included in upper lip of corolla of full anthesis **c. ssp. litoralis**

a. Ssp. vernus. Stems to 25cm, with (0)1-4 pairs of branches; lowest flower at node 6-9; (2n=40). Grassy places, arable and waste ground, waysides; recorded from most of BI but many errors in S, scattered in Ir, W & N Br, replacing ssp. *serotinus* in most of C & N Sc. Aestival. **623**

b. Ssp. serotinus (Syme) Corb. (*O. vulgaris* Moench, *O.vernus* ssp. *pumilus* (Nordst.) A. Pedersen). Stems to 50cm, with (0)2-8 pairs of branches; calyx-teeth usually as in ssp. *vernus* but as in ssp. *litoralis* in some maritime populations; style as in ssp. *vernus*; (2n=18, 20). In habitats of both sspp. *serotinus* and *litoralis*; frequent over most BI except C & N Sc, the only ssp. in most of C & S Br. Autumnal. **623**

c. Ssp. litoralis (Fr.) Nyman (*O. litoralis* Fr., *O. vernus* ssp. *pumilus* auct. non (Nordst.) A. Pedersen). Stems to 20cm, with 0-3 pairs of branches; lowest flower at node 4-8; (2n=18, 20). Gravelly and rocky sea-shores and in salt-marshes; coasts of N & W Sc from Arran to Shetland, ?Merioneth, the only ssp. in Shetland. Aestival. **R** **623**

23. BARTSIA L. - *Alpine Bartsia*
Perennials; leaves opposite, toothed; calyx not inflated, with 4 entire lobes; corolla dull purple, with open mouth, with lower lip with 3 entire lobes; capsules with few seeds; seeds with rather narrow longitudinal wings, without oil-body.

1. B. alpina L. - *Alpine Bartsia*. Stems erect, to 25cm; leaves and bracts glandular-pubescent, ovate; bracts all wholly or partly dark purple; corolla 15-20mm; 2n=24. Native; grassy places and rock-ledges on basic, often damp, soils in mountains; **RR**

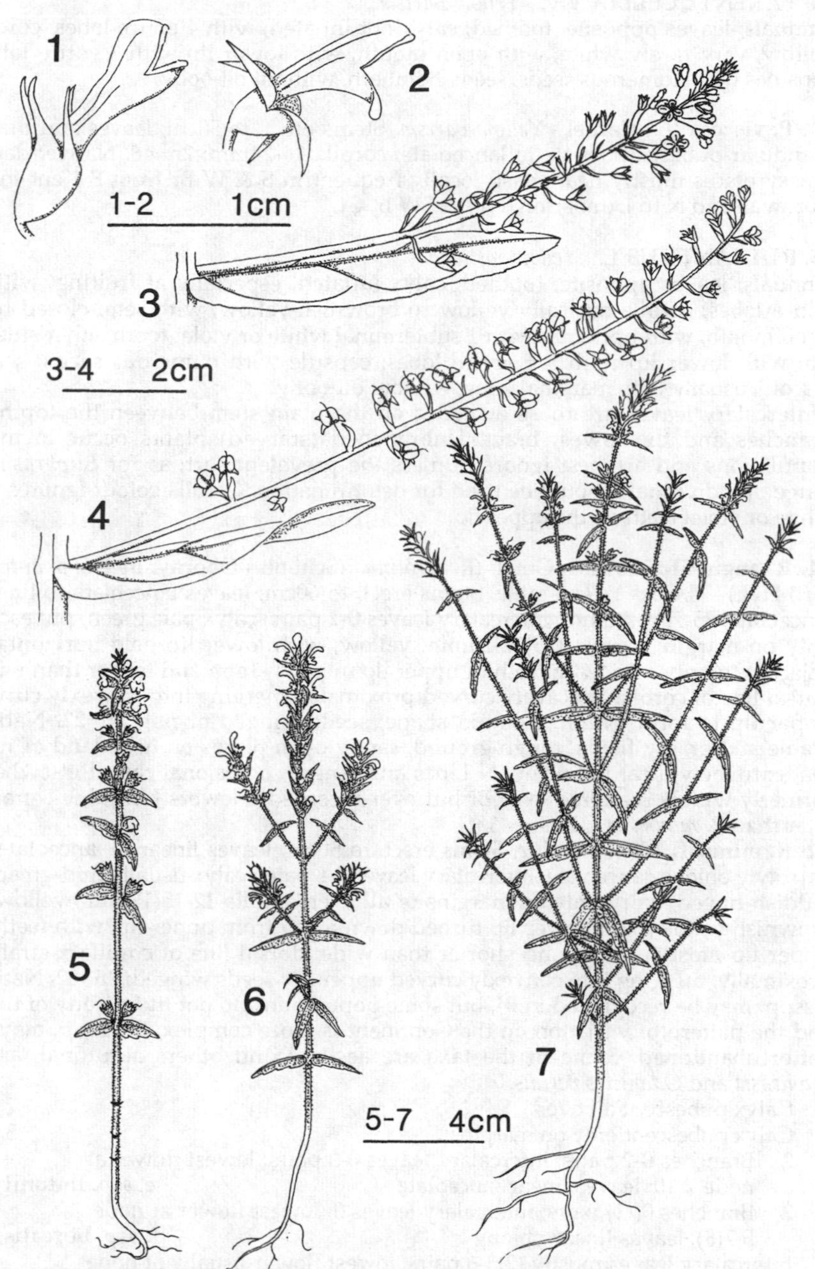

FIG 623 - Scrophulariaceae. 1-2, flowers of *Melampyrum*. 1, *M. pratense*. 2, *M. sylvaticum*. 3-4, racemes of *Veronica*. 3, *V. anagallis-aquatica*. 4, *V. catenata*. 5-7, *Odontites vernus*. 5, ssp. *litoralis*. 6, ssp. *vernus*. 7, ssp. *serotinus*.

very local in hills of N En and C Sc.

24. PARENTUCELLIA Viv. - *Yellow Bartsia*
Annuals; leaves opposite, toothed; calyx not inflated, with 4 entire lobes; corolla yellow, very rarely white, with open mouth, with lower lip with 3 entire lobes; capsules with numerous seeds; seeds ± smooth, without oil-body.

1. P. viscosa (L.) Caruel - *Yellow Bartsia.* Stems erect, to 50cm; leaves and bracts glandular-pubescent, ovate to lanceolate; corolla 16-24mm; 2n=48. Native; damp grassy places mostly near coast; locally frequent in S & W Br from E Kent to W Cornwall and N to Dunbarton, NW & SW Ir, CI.

25. RHINANTHUS L. - *Yellow-rattles*
Annuals; leaves opposite, toothed; calyx inflated, especially at fruiting, with 4 entire lobes; corolla basically yellow to brownish-yellow, with semi-closed to ± open mouth, with upper lip with 1 subterminal white or violet tooth either side of tip, with lower lip with 3 ± entire lobes; capsule with numerous seeds; seeds discoid, usually with marginal wing, without oil-body.

Intercalary leaves are those at nodes on the main stem between the topmost branches and the lowest bract. Unbranched (starved) plants occur in most populations and are best ignored unless the prevalent sort; as for *Euphrasia*, a range of individuals should be used for determination. Corolla colour ignores the white or violet teeth on the upper lip.

1. R. angustifolius C.C. Gmel. (*R. serotinus* (Schönh.) Oborny, incl. ssp. *apterus* **RRR**
(Fr.) Hyl.) - *Greater Yellow-rattle.* Stems erect, to 60cm; leaves lanceolate to linear- **625**
lanceolate, 25-70 x 3-8mm; intercalary leaves 0-2 pairs; calyx pale green, pubescent only on margin; corolla (15)17-20mm, yellow, with lower lip held horizontal ± adjacent to upper lip, with teeth of upper lip mostly >1mm and longer than wide; dorsal line of corolla concavely curved proximally, merging into convexly curved upper lip to form overall sigmoid shape; seeds winged or not; 2n=22. Native; arable and grassy fields, rough ground, sandy open places on heathland or near sea; extremely local in Surrey, N Lincs and Angus, occasional casual elsewhere, formerly widely scattered over Br but over-recorded. Flowers Jun-Jul, c.1 month later than *R. minor*, in C En.

2. R. minor L. - *Yellow-rattle.* Stems erect, to 50cm; leaves linear to lanceolate or **625**
narrowly oblong-lanceolate; intercalary leaves 0-6 pairs; calyx usually mid-green or reddish-tinged, hairy only on margins or all over; corolla 12-15(17)mm, yellow to brownish-yellow, with lower lip turned down away from upper lip, with teeth of upper lip mostly <1mm and shorter than wide; dorsal line of corolla ± straight proximally, merging into convexly curved upper lip; seeds winged; 2n=22. Native. 6 sspp. may be recognized in BI, but some populations do not fit into any of them and the pattern of variation on the Continent is more complex; the sspp. may be better abandoned. Some of the taxa are aestival and others autumnal, as in *Euphrasia* and *Odontites vernus*.

1 Calyx pubescent all over 2
1 Calyx pubescent only on margins 3
  2 Branches 0-2 pairs; intercalary leaves 0-3 pairs; lowest flower at node 7-10; leaves linear-lanceolate **e. ssp. lintonii**
  2 Branches 0(-1) pairs; intercalary leaves 0; lowest flower at node 5-7(8); leaves linear-oblong **f. ssp. borealis**
3 Intercalary leaves mostly (2)3-6 pairs; lowest flower usually at node 14-19; leaves mostly linear **d. ssp. calcareus**
3 Intercalary leaves mostly 0-2(4) pairs; lowest flower usually at node 6-13(15); leaves mostly linear-narrowly oblong to linear-lanceolate 4
  4 Intercalary leaves mostly 0(-1) pairs; lowest flower usually at node

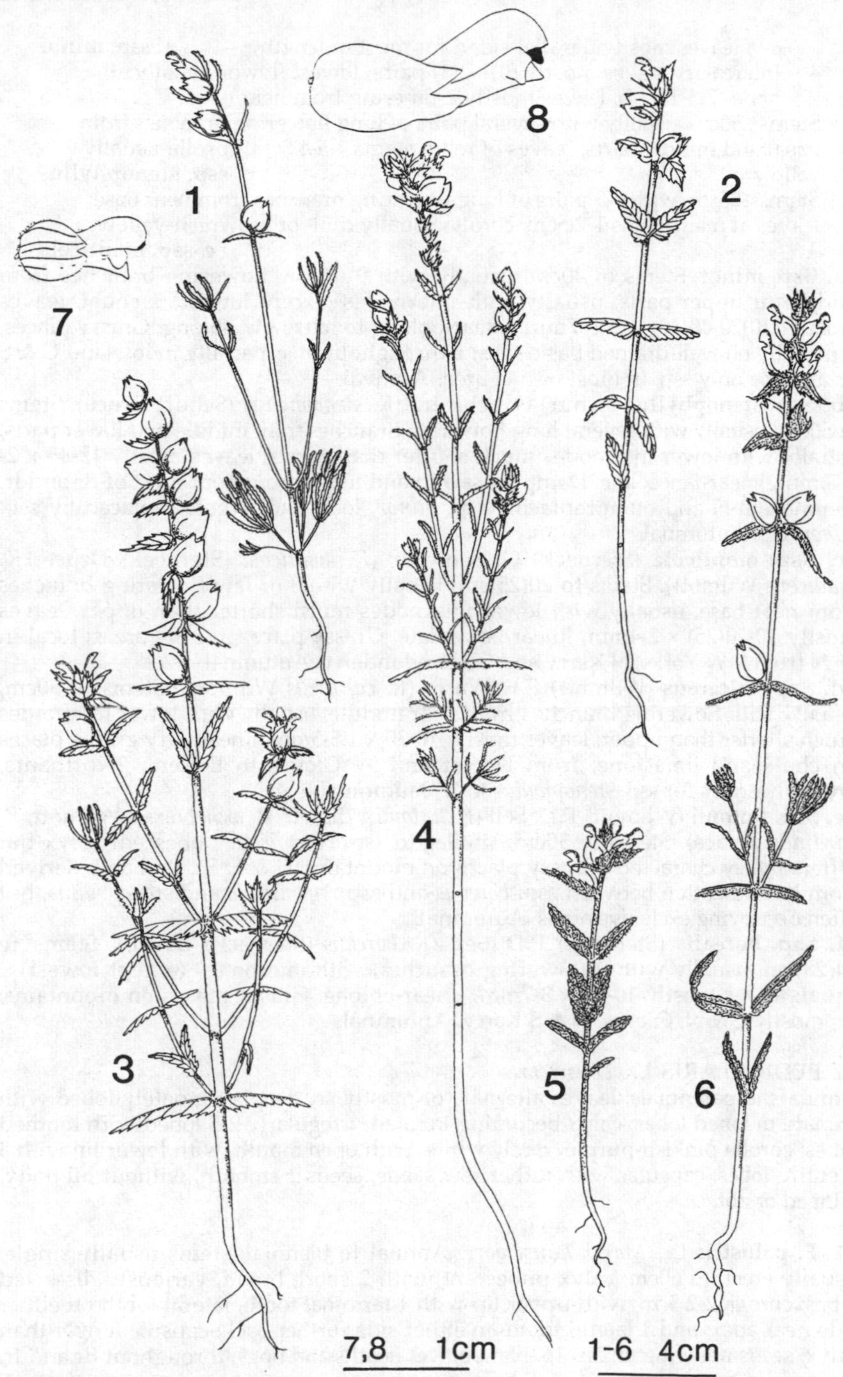

FIG 625 - *Rhinanthus*. 1-6, plants of *R. minor*. 1, ssp. ***monticola***. 2, ssp. ***lintonii***. 3, ssp. ***stenophyllus***. 4, ssp. ***calcareus***. 5, ssp. ***borealis***. 6, ssp. ***minor***. 7-8, flowers, courtesy of David Hambler. 7, *R. minor*. 8, *R. angustifolius*.

6-9; leaves mostly parallel-sided for most of length **a. ssp. minor**
4 Intercalary leaves mostly (0)1-2(4) pairs; lowest flower usually at node (7)8-13(15); leaves mostly ± tapering from near base 5
5 Stems ≤50cm, usually with several pairs of long flowering branches from basal and middle parts; leaves of main stems 1.5-4.5cm; corolla usually yellow **b. ssp. stenophyllus**
5 Stems ≤25cm, with 0-3 pairs of long flowering branches from near base; leaves of main stems 1-2.5cm; corolla usually dull- or brownish-yellow **c. ssp. monticola**

a. Ssp. minor. Stems to 40cm, usually with 0 or few flowering branches from **625**
middle or upper parts, usually with internodes (except lowest) ± equal; leaves mostly (10)20-40(50) x (3)5-7mm, linear-oblong to narrowly oblong. Grassy places, especially on well-drained basic soils; ± throughout BI, especially in lowland C & S Br and the only ssp. in most of that area. Aestival.

b. Ssp. stenophyllus (Schur) O. Schwarz (*R. stenophyllus* (Schur) Druce). Stems **625**
to 50cm, usually with several long flowering branches from middle and lower parts, usually with lower internodes much shorter than upper; leaves mostly 15-45 x 2-5(7)mm, linear-lanceolate. Damp grassland and fens; throughout most of Br and Ir, common in N and often replacing ssp. *minor*, local and largely replaced by ssp. *minor* in S. Autumnal.

c. Ssp. monticola (Sterneck) O. Schwarz (*R. monticola* (Sterneck) Druce, *R.* **R**
spadiceus Wilmott). Stems to 20(25)cm, usually with 0 or few flowering branches **625**
from near base, usually with lower internodes much shorter than upper; leaves mostly 10-20(25) x 2-4mm, linear-lanceolate. Grassy places in hilly areas; local in Br N from MW Yorks, N Kerry and Co Londonderry. Autumnal.

d. Ssp. calcareus (Wilmott) E.F. Warb. (*R. calcareus* Wilmott). Stems to 50cm, **R**
usually with flowering branches from near middle, usually with lower internodes **625**
much shorter than upper; leaves mostly 10-25 x 1.5-3mm, linear. Dry grassy places on chalk and limestone; from Dorset and W Gloucs to E Kent, ?Northants, probably errors for ssp. *stenophyllus* in Ir. ?autumnal.

e. Ssp. lintonii (Wilmott) P.D. Sell (*R. lintonii* Wilmott, *R. lochabrensis* Wilmott, *R.* **R**
gardineri Druce). Stems to 30cm; similar to ssp. *borealis* in pubescent calyx but **625**
differs in key characters. Grassy places on mountains; C & N Sc. Probably derived from hybridization between ssp. *borealis* and sspp. *monticola* and *stenophyllus*, but often occupying exclusive areas. Autumnal.

f. Ssp. borealis (Sterneck) P.D. Sell (*R. borealis* (Sterneck) Druce). Stems to **R**
20(28)cm, usually with 0 flowering branches, with internodes (except lowest) ± **625**
equal; leaves mostly 10-30 x 3-7mm, linear-oblong. Grassy places on mountains; Sc, mostly C & N, Caerns, N & S Kerry. Autumnal.

26. PEDICULARIS L. - *Louseworts*

Annuals to perennials; leaves alternate or mostly so, deeply pinnately lobed with crenate to lobed lobes; calyx becoming inflated, irregularly 2-5 lobed with toothed lobes; corolla pinkish-purple, rarely white, with open mouth, with lower lip with 3 ± entire lobes; capsules with rather few seeds; seeds ± smooth, without oil-body, winged or not.

1. P. palustris L. - *Marsh Lousewort*. Annual to biennial; stems usually single, usually erect, to 60cm; calyx pubescent, with 2 short, broad, variously dissected lobes; corolla 2-2.5cm, with upper lip with 1 terminal tooth, lateral tooth on either side near apex, and 1 lateral tooth on either side further back; capsule longer than calyx; seeds not winged; 2n=16. Native; wet heaths and bogs; throughout Br and Ir, common in N & W, rare in C & E En.

2. P. sylvatica L. - *Lousewort*. (Biennial to) perennial; stems several, procumbent to ascending or some suberect, to 25cm; calyx with 4 short dissected lobes; corolla 2-2.5cm, as in *P. palustris* but lacking the second pair of lateral teeth; capsule shorter than or equalling calyx; seeds usually partially winged; 2n=16. Native; similar places to *P. palustris* and sometimes with it, but often in drier habitats.

a. Ssp. sylvatica. Calyx and pedicels glabrous. Throughout BI except parts of C En and where replaced by ssp. *hibernica*.

b. Ssp. hibernica D.A. Webb. Calyx and pedicels pubescent. Prevalent in W Ir, very local in extreme W Sc, Lake District, E & S Ir and W Wa. Intermediates with ssp. *sylvatica* occur, especially in E Ir.

129. OROBANCHACEAE - *Broomrape family*

Brown to whitish, reddish or bluish, herbaceous, erect, perennial root-parasites; chlorophyll ± lacking; leaves ± scale-like, alternate, simple, entire, without stipules, sessile. Flowers in terminal racemes or spikes, in axils of bracts, zygomorphic, bisexual, hypogynous; calyx with 4 ± equal lobes fused to c.1/2 way, or with 2 lateral lips each not to ± deeply bifid; corolla tubular, with upper and lower lips, the upper lip ± entire to 2-lobed, the lower lip 3-lobed, variously coloured; stamens 4, borne on corolla-tube; ovary 1-celled, with numerous ovules borne on 4 (or 2 2-lobed) inwards-thrusted parietal placentas; style 1, terminal; stigmas 2, ± capitate; fruit a 2-celled capsule.

Distinguished from Scrophulariaceae by the chlorophyll-less aerial parts and 1-celled ovary. Either the whole family or just *Lathraea* is often placed in the Scrophulariaceae, the evidence for which is equivocal.

1 Plant rhizomatous; flowers pedicellate; calyx with 4 equal lobes **1. LATHRAEA**

1 Plant not rhizomatous; flowers sessile except rarely some near base of inflorescence; calyx with 2-4(5) teeth arranged in 2 lateral lips **2. OROBANCHE**

1. LATHRAEA L. - *Toothworts*

Plants with rhizome with ± succulent scales; flowers pedicellate; calyx with 4 equal lobes; 2 lips of corolla held nearly parallel to one another.

1. L. squamaria L. - *Toothwort*. Aerial stems whitish to cream or pale pink, to 30cm, pedicels shorter than calyx; calyx glandular-pubescent; corolla <2x as long as calyx, whitish-cream usually tinged with purple or pink, 14-20mm; capsule with numerous seeds; (2n=36). Native; in woods and hedgerows, usually on moist rich soils, on a range of woody plants especially *Ulmus* and *Corylus*; locally frequent in Br and Ir N to C Sc.

2. L. clandestina L. - *Purple Toothwort*. Aerial stems 0; flowers arising from axils of scale-leaves near apex of rhizome; pedicels c. as long as to longer than calyx; calyx glabrous; corolla ≥2x as long as calyx, purple-violet, 40-50mm; capsule with 4-5 seeds; 2n=42. Intrd; natd in damp places on *Salix* and *Populus*; scattered in En, N Wa, S Sc and E Ir, Guernsey; W & SW Europe.

2. OROBANCHE L. - *Broomrapes*

Rhizomes 0; flowers sessile, rarely lower ones pedicellate; calyx with 2-4(5) teeth arranged in 2 lateral lips, the lips usually open to the base on upper or both sides; 2 lips of corolla held apart, the lower turned down.

Pressed plants are difficult to determine because the corolla shape and colour and the stigma colour are lost or obscured. At collection some corollas should be opened out by slitting up 1 side and pressing, the shape of the corolla in side view and of the lower lip in front view should be recorded, and the colour of the stem, corolla and stigmas noted. The host sp. is a useful character, but often difficult to ascertain with certainty. Corolla-lengths are from base to tip of upper lip in a straight line.

1 Each flower with 2 bracteoles ± similar to the 4 calyx-teeth (1 bracteole and 2 calyx teeth each side of flower) in axil of each bract; stigmas white; capsule-valves free **1. O. purpurea**

1 Bracteoles 0; each flower with 2-4-toothed calyx (1-2 calyx-teeth on each side of flower); stigmas yellow, red or purplish, rarely white; capsule-valves coherent distally 2

2 Lower lip of corolla with minute glandular hairs at margins 3

2 Margins of lower lip of corolla glabrous or with very few glandular hairs, but latter often frequent elsewhere on corolla 5

3 Stigma-lobes yellow at flower opening; filaments glabrous in basal 1/3 **2. O. rapum-genistae**

3 Stigma-lobes red to purple at flower opening; filaments pubescent at least at base 4

4 Stigma-lobes separate; corolla not suffused dark red, mostly >20mm; each calyx-lip with 1-2 teeth, shorter than corolla-tube **3. O. caryophyllacea**

4 Stigma-lobes partly fused; corolla suffused dark red, mostly <20mm; each calyx-lip with 1 tooth, c. as long as corolla-tube **5. O. alba**

5 Calyx with 2 lateral lips partially fused on lowerside; stamens inserted (3)4-6mm above base of corolla-tube **4. O. elatior**

5 Calyx with 2 lateral lips free on both upperside and lowerside; stamens inserted 2-4(5)mm above base of corolla-tube 6

6 Corolla with sparse dark glands mostly distally, with very strongly curved back so that mouth is nearly at right angles to base **6. O. reticulata**

6 Corolla without dark glands (often with pale ones), with nearly straight to slightly curved back 7

7 Corolla 20-30mm, with 2 upper and 3 lower lobes with conspicuous and patent margins; filaments sparsely pubescent along length **7. O. crenata**

7 Corolla 10-22mm, with 0 or only 3 lower lobes with conspicuous and patent margins; filaments glabrous or pubescent only proximally 8

8 Corolla-tube constricted just behind mouth; lower lip of corolla with acute to subacute lobes; stigmas usually yellow, rarely purplish **8. O. hederae**

8 Corolla-tube not constricted; lower lip of corolla with obtuse to rounded lobes; stigmas usually purplish, rarely yellow 9

9 Filaments with long white hairs at base, usually inserted ≥3mm above base of corolla-tube; bract equal to or longer than corolla; all 4 calyx-teeth long and filiform **9. O. artemisiae-campestris**

9 Filaments glabrous to sparsely pubescent at base, usually inserted ≤3mm above base of corolla-tube; bract shorter than or equal to corolla; calyx-teeth various but not all 4 long and filiform **10. O. minor**

Other spp. - O. ramosa L. (*Hemp Broomrape*), from S Europe, was formerly natd, especially in CI, but not since 1928; it resembles a small, often branched, *O. purpurea* but the corolla is only 10-18(22)mm. Records of **O. cernua** Loefl. and **O. amethystea** Thuill. are errors.

1. O. purpurea Jacq. - *Yarrow Broomrape*. Stems to 45cm, tinged bluish; corolla 18-26mm, bluish-violet at least distally; filaments glabrous or ± so, inserted >5-8mm above base of corolla; stigma white or very pale blue; 2n=24. Native; on *Achillea millefolium* and perhaps other Asteraceae; very local from Pembs and Dorset to N Lincs, CI, formerly more widespread in S En and S Wa. **RR** **629**

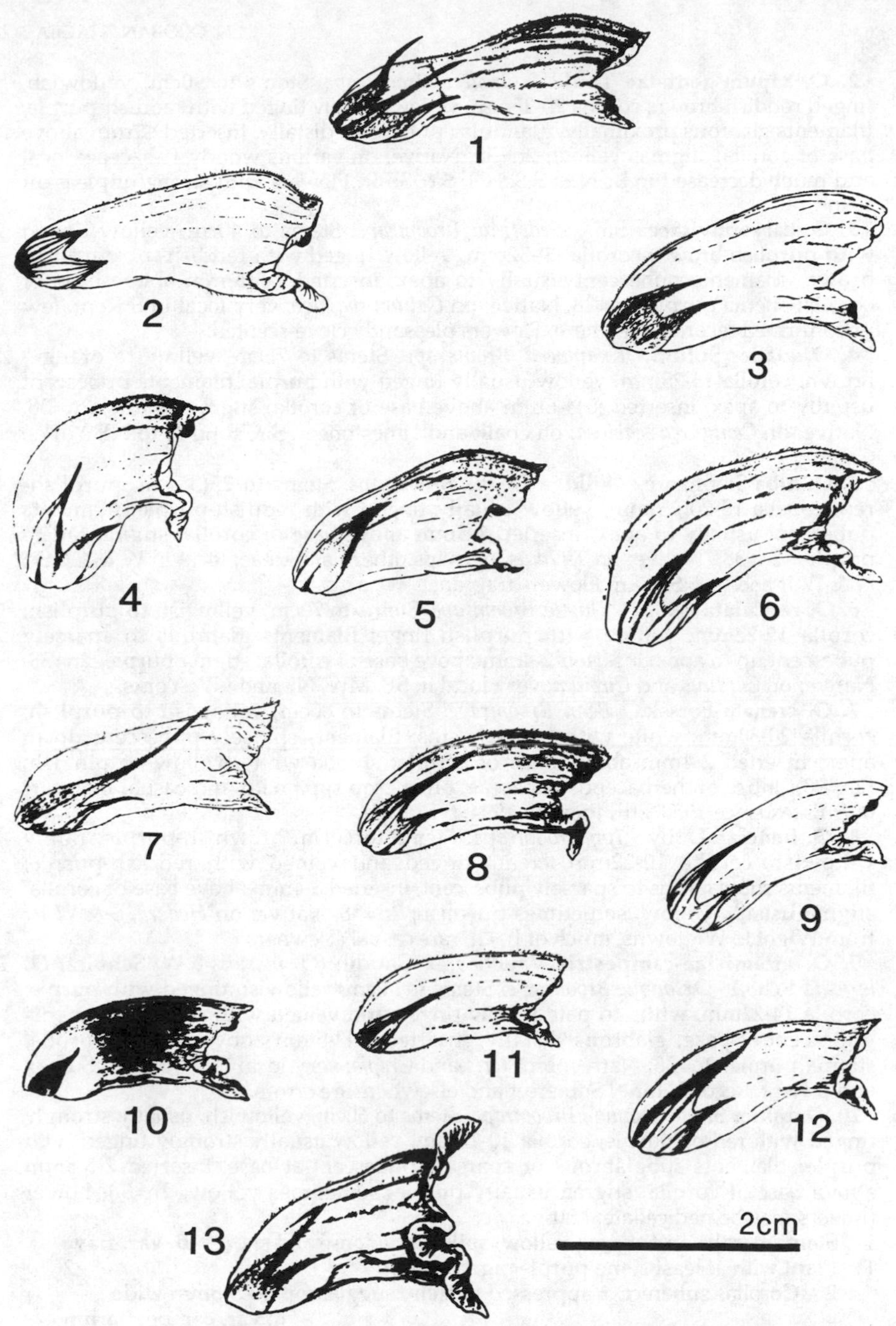

FIG 629 - Corollas of *Orobanche*. 1, *O. purpurea*. 2, *O. caryophyllacea*. 3, *O. rapum-genistae*. 4, *O. reticulata*. 5, *O. alba*. 6, *O. elatior*. 7, *O. artemisiae-campestris*. 8-11, *O. minor*. 8, var. *minor*. 9, var. *compositarum*. 10, var. *maritima*. 11, var. *flava*. 12, *O. hederae*. 13, *O. crenata*. Drawings by F.J. Rumsey.

2. O. rapum-genistae Thuill. - *Greater Broomrape*. Stems to 90cm, yellowish, R
tinged reddish-brown; corolla 20-25mm, yellow usually tinged with reddish-purple; 629
filaments glabrous proximally, glandular-pubescent distally, inserted ≤2mm above base of corolla; stigmas yellow; 2n=38. Native; on various woody Fabaceae; local and much decreased in Br N to S Sc, CI, S & SE Ir. Flowers with strong unpleasant scent.

3. O. caryophyllacea Sm. - *Bedstraw Broomrape*. Stems to 40cm, yellow tinged RRR
with purplish-brown; corolla 20-32mm, yellow tinged with reddish- or purplish- 629
brown; filaments pubescent usually to apex, inserted 1-3(5)mm above base of corolla; stigma purple; 2n=38. Native; on *Galium mollugo*; very local in E Kent, few unconfirmed records elsewhere. Flowers pleasantly clove-scented.

4. O. elatior Sutton - *Knapweed Broomrape*. Stems to 75cm, yellow to orange- 629
brown; corolla 18-25mm, yellow usually tinged with purple; filaments pubescent usually to apex, inserted (3)4-6mm above base of corolla; stigma yellow; 2n=38. Native; on *Centaurea scabiosa*; on chalk and 1imestone in S & E En N to NE Yorks, Glam.

5. O. alba Stephan ex Willd. - *Thyme Broomrape*. Stems to 25(35)cm, purplish- R
red; corolla 15-20(25)mm, yellow usually tinged with reddish-purple; filaments 629
pubescent usually to apex, inserted 1-3mm above base of corolla; stigma red to purple; (2n=38). Native; on *Thymus*, perhaps other Lamiaceae; local in W & CE Sc, N & W Ir and N & SW En. Flowers fragrant.

6. O. reticulata Wallr. - *Thistle Broomrape*. Stems to 70cm, yellowish to purplish; RRR
corolla 12-22mm, yellow with purplish tinge; filaments glabrous to sparsely 629
pubescent up to apex, inserted 2-4mm above base of corolla; stigma purple; 2n=38. Native; on *Carduus* and *Cirsium*; very local in SE, MW, NE and NW Yorks.

7. O. crenata Forssk. - *Bean Broomrape*. Stems to 80cm, yellowish to purplish; 629
corolla 20-30mm, white with mauve veins; filaments sparsely pubescent up to apex, inserted 2-4mm above base of corol]a; stigma white, yellow or pinkish; (2n=38). Intrd; on herbaceous Fabaceae, often crop spp.; natd and casual in 1 part of S Essex since 1950, still there in 1994; S Europe.

8. O. hederae Duby - *Ivy Broomrape*. Stems to 60cm, brownish-purple, rarely R
yellowish; corolla 10-22mm, cream tinged and veined with reddish-purple; 629
filaments subglabrous to sparsely pubescent, inserted 3-4mm above base of corolla; stigma usually yellow, sometimes purplish; 2n=38. Native; on *Hedera*; S & W Br from Wight to Wigtowns, much of Ir, CI, rare casual elsewhere.

9. O. artemisiae-campestris Vaucher ex Gaudin (*O. picridis* F.W. Schultz, *O.* RRR
loricata Rchb.) - *Oxtongue Broomrape*. Stems to 60cm, yellowish tinged with purple; 629
corolla 14-22mm, white to pale yellow tinged and veined with purple; filaments pubescent at base, glabrous distally, inserted (2)3-5mm above base of corolla; stigma purple; 2n=38. Native; on *Picris* and *Crepis*; very local in Wight, W Sussex and E Kent. Records for N Somerset and elsewhere are errors.

10. O. minor Sm. - *Common Broomrape*. Stems to 60cm, yellowish, usually strongly tinged with red or purple; corolla 10-18mm, yellow usually strongly tinged with purple; filaments subglabrous or sparsely pubescent at base, inserted 2-3.5mm above base of corolla; stigma usually purple, sometimes yellow; 2n=38. Lower flowers may be pedicellate. Native.

1 Stems, corolla and stigma yellow; spike very dense and short **b. var. flava**
1 Plant with at least some purple pigmentation 2
  2 Corollas suberect, ± appressed to stem, subglabrous, 3.5-5mm wide **d. var. compositarum**
  2 Corollas erecto-patent, not appressed to stem, subglabrous to glandular-pubescent, 5-8mm wide 3
3 Lower lip of corolla with large yellow bosses, with middle lobe the largest; lips of calyx each 1-lobed; stigma-lobes partly fused **c. var. maritima**
3 Lower lip of corolla without prominent bosses, with 3 ± equal lobes;

lips of calyx each usually 2-lobed; stigma-lobes separate **a. var. minor**

a. Var. minor. Stems and corollas usually strongly purple-tinged. On a very wide 629
range of dicotyledons, including most specific for other spp. and vars; throughout En, Wa, Ir and CI but absent from many areas.

b. Var. flava Regel. Differs from var. *minor* in key characters and 1-lobed calyx- 629
lips. On *Hypochaeris* and relatives; Newport Docks (Mons), formerly CI.

c. Var. maritima (Pugsley) Rumsey & Jury (*O. maritima* Pugsley, *O. amethystea* R
auct. non Thuill.). Differs from var. *minor* in key characters. On *Daucus, Plantago* 629
coronopus and *Ononis repens*; coasts of S En and CI, formerly S Wa.

d. Var. compositarum Pugsley. Differs from var. *minor* in key characters. Mostly 629
on Asteraceae: Lactuceae but sometimes other families; scattered throughout much of range of var. *minor*, but mostly in E Anglia.

130. GESNERIACEAE - *Pyrenean-violet family*

Herbaceous perennials; stem ± absent; leaves in basal rosette, simple, dentate, without stipules, petiolate. Flowers 1-few on long peduncles arising from rosette, nearly actinomorphic, bisexual, hypogynous; sepals 5, fused into tube proximally; petals 5, fused into tube proximally; stamens 5, borne on corolla-tube, exserted; ovary 1-celled, with many ovules on 2 parietal placentas; style 1; stigma 1, capitate; fruit a capsule dehiscing into 2 valves.

Easily recognized by the growth-form, the leaves with long brown hairs, and the attractive violet and yellow flowers distinguished from those of Scrophulariaceae by the 1-celled ovary.

1. RAMONDA Rich. - *Pyrenean-violet*

1. R. myconi (L.) Rchb. - *Pyrenean-violet.* Leaves elliptic to broadly so, with long brown hairs; flowers 3-4cm across, violet with yellow centre and yellow exserted anthers; (2n=48). Intrd; planted on rock-face in Cwm Glas, Caerns in 1921 and still there, often not flowering, also persistent on old estate in MW Yorks; Pyrenees.

131. ACANTHACEAE - *Bear's-breech family*

Herbaceous perennials; leaves in basal rosette, few opposite or alternate on stems, ± pinnate to simple and deeply pinnately lobed, without stipules, petiolate. Flowers large, in robust terminal spikes, zygomorphic, bisexual, hypogynous, each in axils of spiny-toothed bract and 2 long linear bracteoles; sepals 4, 2 small lateral and 2 large upper and lower, fused at base, persistent to fruiting; corolla with short tube and 3-lobed lower lip (upper lip 0); stamens 4, borne on corolla-tube, the filaments free but the 1-celled anthers fused in pairs; ovary 2-celled, each cell with many ovules on axile placentas; style 1; stigmas 2, slightly unequal, linear; fruit a 2-celled capsule with persistent style.

Easily distinguished by the robust growth-habit, large pinnately lobed leaves, spiny bracts and unique flower structure.

1. ACANTHUS L. - *Bear's-breeches*

1. A. mollis L. - *Bear's-breech.* Stems to 1m, erect, glabrous at least above; leaves deeply pinnately lobed with acutely toothed but not spiny lobes, glabrous; corolla 3.5-5cm, white with purple veins; bracts glabrous, often purplish; (2n=56, 80). Intrd; grown in gardens and long natd in waste places, roadside and railway banks, and scrub; scattered in S En, S Wa and CI, especially SW En; W & C Mediterranean.

2. A. spinosus L. - *Spiny Bear's-breech.* Differs from *A. mollis* in stems usually shorter (to 80cm), often pubescent; leaves usually pubescent, with spiny teeth on more dissected lobes; bracts usually pubescent; (2n=112). Intrd; similar places to *A. mollis* but much less common; very scattered in S En; C Mediterranean.

132. LENTIBULARIACEAE - *Bladderwort family*

Rootless aquatic, insectivorous, perennials, or rooted, stemless, rosette-forming insectivorous perennials; leaves simple and with many slime-oozing glands or alternate and divided into filiform to linear segments some of which are modified as small bladder-traps; flowers on erect stalks, in racemes or solitary, yellow, bluish or white, zygomorphic, bisexual, hypogynous; calyx of 5 subequal lobes or of 2 obscurely lobed lips, fused at base; corolla of 2 lips, the upper 2-lobed, the lower 3-lobed, spurred at base, yellow, violet or white; stamens 2, borne at base of corolla; ovary 1-celled, with many ovules on free-central placenta; style 0 or very short; stigma variously expanded; fruit a capsule.

Easily recognized by the very different insectivorous habit of the 2 genera, both of which have a 2-lipped, spurred corolla, 2 stamens and free-central placentation.

1 Plant rooted, with basal rosette of simple entire leaves; corolla white to violet **1. PINGUICULA**
1 Plant rootless, aquatic; leaves divided into linear to filiform segments; corolla yellow **2. UTRICULARIA**

1. PINGUICULA L. - *Butterworts*
Roots present; leaves simple, entire, in basal rosette, covered in slime; flowers solitary on erect pedicels; calyx of 5 subequal lobes; corolla white to violet, with open mouth; capsule opening by 2 valves.

1 Corolla white, with 1-2 yellow spots on lower lip; ?extinct **2. P. alpina**
1 Corolla pale lilac or pinkish to violet, very rarely white, sometimes with yellowish throat but never with yellow spots on lower lip 2
 2 Corolla 7-11mm incl. cylindrical spur 2-4mm, usually pale lilac **1. P. lusitanica**
 2 Corolla 14-35mm incl. tapering spur 4-14mm, usually violet with whitish throat 3
3 Corolla 14-22(25)mm incl. spur 4-7(10)mm; lobes of lower lip separated laterally **3. P. vulgaris**
3 Corolla 25-35mm incl. spur 10-14mm; lobes of lower lip overlapping laterally **4. P. grandiflora**

1. P. lusitanica L. - *Pale Butterwort*. Overwintering as a rosette; leaves oblong, 1-2.5cm; pedicels 3-15cm; corolla 7-11mm incl. spur 2-4mm, pale lilac with darker suffusion proximally, often with yellow in throat; 2n=12. Native; bogs and wet heaths; locally frequent in Ir, W & N Sc, Man, SW Wa and SW & SC En.

2. P. alpina L. - *Alpine Butterwort*. Overwintering as a bud; leaves elliptic-oblong, 2-5cm; pedicels 5-11cm; corolla 10-16mm incl. tapering spur 2-3mm, white with 1-2 yellow spots on lower lip; (2n=32). Formerly native; boggy places in E Ross from 1831 to c.1900. **RR**

3. P. vulgaris L. - *Common Butterwort*. Overwintering as a bud; leaves ovate-oblong, 2-8cm; pedicels 5-18cm; corolla 14-22(25)mm incl. spur 4-7(10)mm, violet with whitish throat; (2n=32, 64). Native; bogs, wet heathland and limestone flushes; locally common over much of Br and Ir, especially N & W, absent from most of C & S En.

3 x 4. P. vulgaris x P. grandiflora = P. x scullyi Druce occurs rarely with the parents in S Kerry and Clare, perhaps elsewhere; it is intermediate in flower morphology and largely, though probably not entirely, sterile.

4. P. grandiflora Lam. - *Large-flowered Butterwort.* Overwintering as a bud, but leaves often persisting or precociously developing; leaves ovate-oblong, 2-8cm; pedicels 5-18cm; corolla 25-35mm incl. spur 10-14mm, violet with whitish throat; (2n=32, 64). Native; bogs and damp moorland; locally common in SW Ir to Clare and E Cork, planted and persistent in scattered places in En, Merioneth and ?Co Carlow. R

2. UTRICULARIA L. - *Bladderworts*
Roots 0; plants free-floating or with lower stems in substratum; leaves divided into linear to filiform segments, some or all bearing tiny animal-catching bladders; flowers on erect racemes emerging from water; calyx of 2 obscurely-lobed lips; corolla yellow, with mouth ± closed by swollen upfolding of lower lip; capsule opening irregularly.

The leaf-segments usually have small marginal teeth (often extremely short) which bear 1 or more long bristles. The bladders are 1-4mm long and bear hairs on their inner and outer surfaces; those on the inner surface are 2- or 4-armed. Small circular or elliptic glands are present on the stems, leaves, outside of bladders and inside of corolla-spur. The presence of bristle-bearing teeth on the leaf-segment margins, the shape of the 4-armed bladder hairs ('quadrifids'), and the distribution of the glands on the inside of the corolla-spur are of diagnostic importance. The first can be seen with a strong lens, the second 2 need a microscope. Pressed, dried material is ideal for the first 2, but pressing distorts the morphology of the quadrifids. At least 5-10 quadrifids should be examined and the range noted. Since the basal corolla-spur is forward-directed, its abaxial side is the side nearer the lower lip.

1 Margins of leaf-segments without teeth with bristles; lower lip of corolla <8mm; spur 1-2mm **6. U. minor**
1 Margins of leaf-segments with teeth with bristles; lower lip of corolla ≥8mm; spur 3-10mm 2
  2 Stems of 1 sort, all bearing green leaves and bladders, free-floating; leaf-segments filiform; quadrifids with 2 long arms 1.2-2x as long as 2 short arms 3
  2 Stems of 2 sorts - free-floating ones bearing green leaves and 0 or few bladders, and ones often anchored in substratum and bearing very reduced non-green leaves and many bladders; leaf-segments linear; quadrifids with 2 long arms 1.8-2.8x as long as 2 short arms 4
3 Lower lip of corolla with reflexed margins (fresh material only); pedicel 8-15mm, recurved but not elongating after flowering; glands present on inside of only abaxial side of spur **1. U. vulgaris**
3 Lower lip of corolla with flat or slightly upturned margins; pedicel 8-15mm at flowering, becoming sinuous and 10-30mm after; glands present on inside of both abaxial and adaxial sides of spur **2. U. australis**
  4 Green leaves totally without traps; apex of leaf-segments usually obtuse; spur 8-10mm, c. as long as lower lip; quadrifids with 2 shorter arms ± parallel or diverging at ≤21(37)° **3. U. intermedia**
  4 Green leaves usually with some traps; apex of leaf-segments subulate; spur 3-5mm, c.1/2 as long as lower lip; quadrifids with 2 shorter arms diverging at >(30)52° 5
5 Margin of leaf-segments with 2-7 teeth with bristles; lower lip of corolla with flat or slightly upturned margins, 9-11 x 12-15mm; quadrifids with 2 shorter arms diverging at (30)52-97(140)° **4. U. stygia**
5 Margin of leaf-segments with 0-5 teeth with bristles; lower lip of corolla with flat margins at first, later with reflexed margins, c.8 x 9mm; quadrifids with 2 shorter arms diverging at (117)146-197(228)° **5. U. ochroleuca**

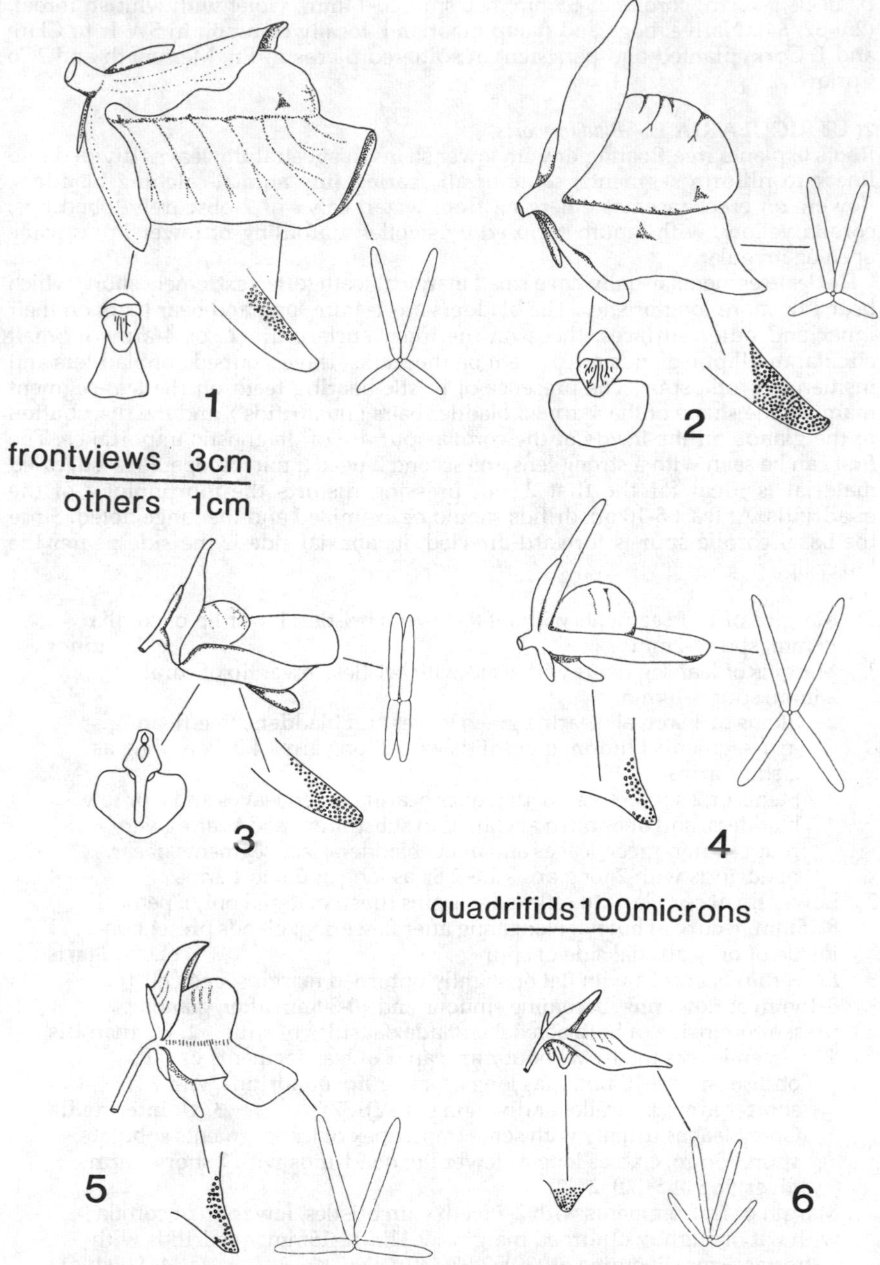

FIG 634 - Flowers (lateral and front views, spur showing gland distribution) and bladder quadrifids of ***Utricularia***. 1, ***U. vulgaris***. 2, ***U. australis***. 3, ***U. intermedia***. 4, ***U. stygia***. 5, ***U. ochroleuca***. 6, ***U. minor***. Courtesy of G. Thor.

1. U. vulgaris L. - *Greater Bladderwort*. Stems of 1 sort, free-floating, to 1m; quadrifids with long/short arm ratio 1.8-2.8, with 2 short arms diverging at mostly 85-130°; glands in corolla-spur on abaxial side only; corolla yellow to bright yellow; upper lip c.11 x 10mm; lower lip 12-15 x 14mm; spur 7-8mm; (2n=36, 44). Native; usually base-rich still or slow water; scattered in Br and Ir, commonest in E En, perhaps absent from N Sc. **634**

2. U. australis R. Br. (*U. neglecta* Lehm.) - *Bladderwort*. Differs from *U. vulgaris* in stems to 60cm; quadrifids with 2 short arms diverging at mostly 100-157°; corolla pale yellow to yellow; and see key; (2n=44). Native; usually acidic still or slow water; scattered throughout Br and Ir, commonest in W & N Br. **634**

Distributions of *U. vulgaris* and *U. australis* are uncertain as distinction is uncertain in absence of flowers, which are rarely produced in N Br.

3. U. intermedia Hayne - *Intermediate Bladderwort*. Stems of 2 sorts, the floating ones to 20(40)cm; quadrifids with long/short arm ratio 1.2-2, with 2 short arms diverging at (2)6-21(37)°; glands in corolla spur on both abaxial and adaxial sides; corolla yellow; upper lip 7-8 x c.7mm; lower lip 8-10 x 12-13mm; spur 8-10mm; (2n=44). Native; still shallow water in peaty bogs and marshes; very scattered in Ir, Sc, N & SC En, E Anglia and Caerns, much over-recorded for *U. ochroleuca* and perhaps for *U. stygia*. Flowers not found in BI. **634**

4. U. stygia G. Thor - *Nordic Bladderwort*. Stems of 2 sorts, the floating ones to 20cm; quadrifids with long/short arm ratio 1.2-2, with 2 short arms diverging at (30)52-97(140)°; glands in corolla-spur on both abaxial and adaxial sides; corolla yellow with reddish tinge; upper lip c.8 x 6mm; lower lip (9)10-11 x (12)13-15mm; spur 4-5mm. Native; similar places to *U. intermedia*; W Sc (Wigtowns to W Sutherland), probably under-recorded. **R** **634**

5. U. ochroleuca R.W. Hartm. - *Pale Bladderwort*. Differs from *U. stygia* in corolla pale yellow; upper lip c. 7 x 5mm; spur c.3mm; and see key; (2n=44). Native; similar places to *U. intermedia*; locally frequent in Sc, very scattered in Ir, N & SC En. Fruits not formed. **634**

U. intermedia, *U. stygia* and *U. ochroleuca* have been much confused in the past and their distributions are very unclear. *U. ochroleuca* is probably the commonest at least in Sc. All 3 flower very rarely but can be distinguished on microscopic vegetative characters.

6. U. minor L. - *Lesser Bladderwort*. Stems usually of 2 sorts, sometimes all floating, to 40cm; quadrifids with long/short arm ratio 1.2-2, with 2 short arms swept up to point same way as 2 long arms hence diverging at >180°; glands almost covering whole of inside of corolla-spur; corolla greenish- to light-yellow; upper lip c.4 x 3mm; lower lip c.7 x 6mm; spur 1-2mm; (2n=36). Native; in boggy pools and fen-ditches; scattered in suitable places over Br and Ir. **634**

133. CAMPANULACEAE - *Bellflower family*
(Lobeliaceae)

Herbaceous annuals to perennials, often with white latex; leaves alternate, sometimes mostly basal, simple, without stipules, petiolate or not. Flowers usually showy, solitary or in simple or branched racemes, corymbose racemes, congested spikes or heads, actinomorphic or zygomorphic, bisexual, epigynous; sepals 5, fused into tube proximally; petals 5, fused into tube proximally, the 5 lobes ± equal or organized into 2-lobed upper and 3-lobed lower lips, often blue but also other colours; stamens 5, borne around style-base (not on corolla), usually closely appressed round style and sometimes anthers or anthers and filaments fused laterally into a ring; ovary (1)2-5-celled, each cell with many ovules on axile placentas; style 1; stigmas usually as many as ovary-cells, capitate to filiform; fruit a 2-5-celled capsule opening variously or a berry.

Differs from other families with fused petals and inferior ovaries by the 5 stamens borne on the receptacle (not on corolla), and numerous ovules on axile placentae.

1 Flowers densely packed into flattish heads or globose to elongated spikes; corolla divided nearly to base, with linear lobes 2
1 Flowers not all packed into dense heads, or if so then corolla divided <2/3 way to base 3
2 Flowers in flattish heads with conspicuous region of flowerless bracts at base; each flower with 0 bract; flower buds straight; stigmas ± globose; stems pubescent **6. JASIONE**
2 Flowers in globose to elongated spikes, without flowerless bracts at base; each flower with 1 bract; flower buds curved; stigmas linear; stems ± glabrous **5. PHYTEUMA**
3 Corolla distinctly zygomorphic; filaments fused laterally at least distally to form tube round style 4
3 Corolla actinomorphic; filaments free though often close together round style (anthers sometimes fused laterally) 6
4 Stems procumbent to decumbent, rooting at nodes; leaves suborbicular; fruit a berry **8. PRATIA**
4 Stems erect to ascending, not rooting at nodes; leaves linear to obovate; fruit a capsule 5
5 Flowers and capsules pedicellate; ovary and capsules <1.5cm, widening distally, 2-celled **7. LOBELIA**
5 Flowers and capsules sessile; ovary and capsules >2cm, cylindrical, 1-celled **9. DOWNINGIA**
6 Ovary and fruits >3x as long as wide; corolla shorter than calyx; annual **2. LEGOUSIA**
6 Ovary and fruits <2(3)x as long as wide; corolla longer than (rarely ± as long as) calyx; biennial or perennial 7
7 Corolla-tube <2mm wide; style >1.5x as long as corolla (tube + lobes) **4. TRACHELIUM**
7 Corolla-tube >3mm wide; style not or scarcely longer than corolla 8
8 Stems filiform, procumbent, with solitary axillary flowers on erect stalks much longer than corolla; all leaves petiolate; capsule opening apically (i.e. within calyx) **3. WAHLENBERGIA**
8 Usually at least flowering stems erect to ascending, if all procumbent then not all pedicels longer than corolla; usually at least apical leaves sessile or ± so; capsule opening laterally or basally (i.e. outside calyx) **1. CAMPANULA**

1. CAMPANULA L. - *Bellflowers*
Biennials to perennials; flowers in racemes or panicles, sometimes in ± compact heads; corolla usually blue, actinomorphic, divided up to 1/2(2/3) way to base; filaments and anthers free; ovary 3-5-celled; style shorter than to slightly longer than corolla; stigmas 3-5, linear; capsule dehiscing by lateral or basal pores.

White-flowered variants of most spp. are not rare but have not been mentioned under each sp. Corolla-lengths given are those in the fresh state; considerable shrinking often occurs on drying.

1 Calyx with 5 sepal-like reflexed appendages alternating with 5 calyx-lobes 2
1 Calyx with 5 calyx-lobes but no extra appendages 3
2 Biennial; all leaves cuneate at base; ovary and fruit 5-celled; stigmas 5 **5. C. medium**
2 Perennial; basal and lower stem-leaves cordate at base; ovary and fruit 3-celled; stigmas 3 **6. C. alliariifolia**

3 Capsule with pores in apical 1/2 4
3 Capsule with pores at or near base 7
4 Calyx-lobes lanceolate to ovate, serrate; lower stem-leaves ovate to ovate-oblong **3. C. lactiflora**
4 Calyx-lobes linear to lanceolate, entire or with 1-2 basal small teeth; stem-leaves linear to obovate 5
5 Perennial, with non-flowering rosettes arising from rhizomes; corollas mostly >3cm; stigmas >1/2 as long as styles **4. C. persicifolia**
5 Usually biennial, without non-flowering rosettes; corollas mostly <3cm; stigmas <1/2 as long as style 6
6 Tap-root thickened; inflorescence narrowly pyramidal; basal leaves abruptly narrowed to distinct petiole **2. C. rapunculus**
6 Tap-root thin; inflorescence widely spreading; basal leaves gradually narrowed to indistinct petiole **1. C. patula**
7 Flowers sessile **7. C. glomerata**
7 Flowers with distinct pedicels 8
8 Calyx-teeth linear to filiform, <1mm wide at base 9
8 Calyx-teeth lanceolate or narrowly triangular to ovate-oblong or -triangular, >1mm wide at base 10
9 Middle stem-leaves ovate-oblong or narrowly so, rounded at base, serrate **15. C. rhomboidalis**
9 Middle stem-leaves linear to linear-elliptic, very gradually tapered to base, ± entire **16. C. rotundifolia**
10 Stems decumbent to ascending, ≤30(50)cm 11
10 Stems erect, usually >50cm 13
11 Corolla funnel-shaped (diameter at apex much < length), lobed 1/4-2/5 way to base **9. C. portenschlagiana**
11 Corolla broadly bell-shaped, ± star-shaped from above (diameter at apex c. equalling to wider than length), lobed 1/2-3/4 way to base 12
12 Basal leaves 2-serrate; corolla-lobes c.5mm wide at base **10. C. poscharskyana**
12 Basal leaves bluntly 1-serrate; corolla-lobes c.10mm wide at base **11. C. fragilis**
13 Capsules erect; plant glabrous; inflorescence dense, pyramidal or cylindrical **8. C. pyramidalis**
13 Capsules pendent; plant pubescent; inflorescence racemose 14
14 Plant patch-forming, with shoots arising from rhizomes and/or root-buds; calyx-teeth patent to reflexed **14. C. rapunculoides**
14 Plant tufted, without rhizomes or root-buds; calyx-teeth erect to erecto-patent 15
15 Middle and lower stem-leaves sessile, cuneate at base; stem bluntly ridged, softly pubescent to subglabrous **12. C. latifolia**
15 Middle and lower stem-leaves petiolate, cordate at base; stem sharply angled, sparsely hispid **13. C. trachelium**

Other spp. - **C. carpatica** Jacq., from the Carpathians, has subapical pores in the capsule, ovate-cordate glabrous leaves and large flowers on rather short weak stems; it is grown in gardens and may persist or self-sow on nearby walls and paths in S.

1. C. patula L. - *Spreading Bellflower.* Stems scabrid-pubescent, erect, to 60cm; calyx-lobes linear, erecto-patent; corolla 15-25mm, pale to purplish-blue, broadly funnel-shaped, lobed c.1/2 way to base; 2n=20. Native; open woods, wood-borders, hedgebanks; very local in S Br N to Salop and Leics, decreasing, also natd garden escape. **R**

2. C. rapunculus L. - *Rampion Bellflower.* Stems usually scabrid-pubescent, erect, **RR**

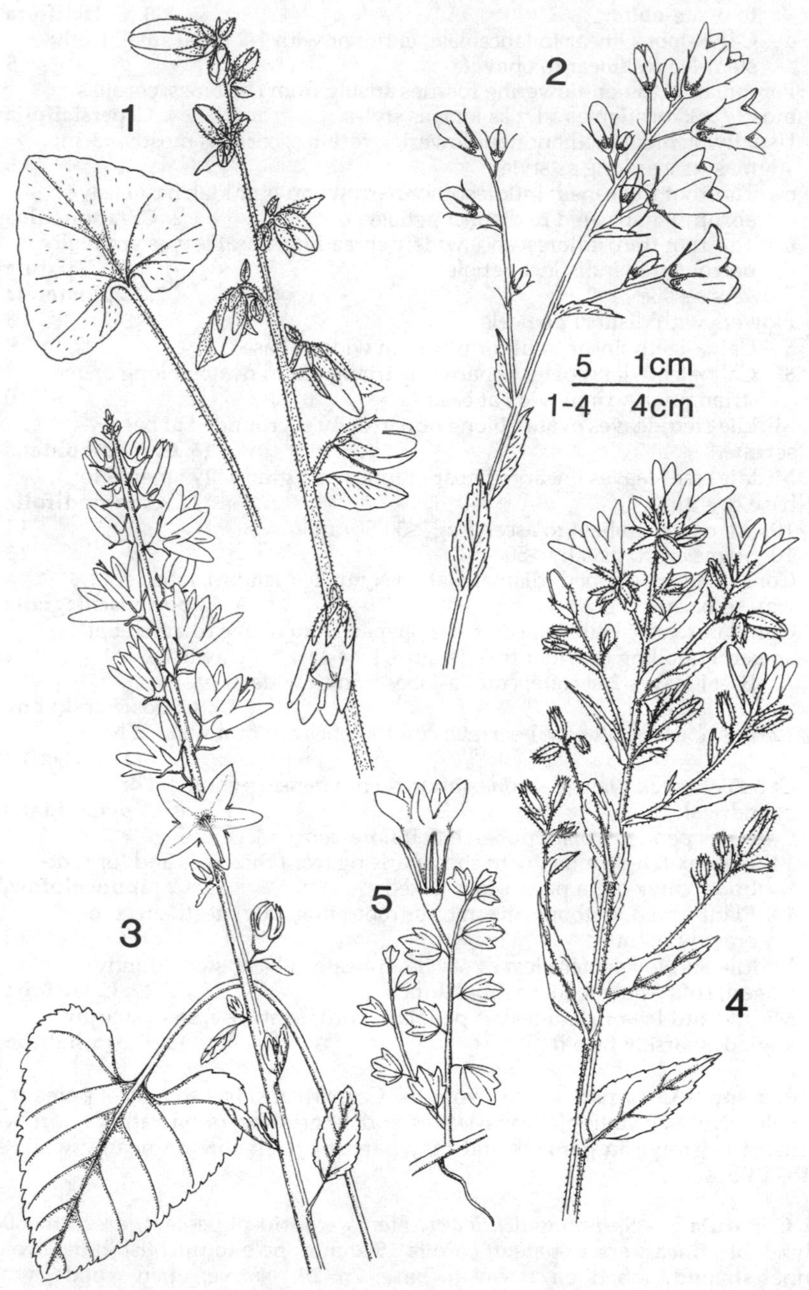

FIG 638 - Campanulaceae. 1, *Campanula alliariifolia*. 2, *C. rhomboidalis*. 3, *C. pyramidalis*. 4, *C. lactiflora*. 5, *Pratia angulata*.

to 80cm; calyx-lobes linear, erect to erecto-patent; corolla 10-22mm, pale blue, funnel-shaped; lobed c.1/3 way to base; (2n=20). Intrd; grown (now rarely) as ornament and salad-vegetable, natd in rough grassy fields and banks; SE En, now rare, formerly scattered over much of Br N to C Sc; Europe.

3. C. lactiflora M. Bieb. - *Milky Bellflower*. Stems sparsely scabrid-pubescent, **638**
erect, to 2m; calyx-lobes lanceolate to ovate, erecto-patent; corolla 15-25mm, very pale to bright blue, broadly bell-shaped, lobed 1/2-2/3 way to base; (2n=34, 36). Intrd; grown as ornament, natd in waste and rough ground, often in damp places; scattered in Br, especially N En and Sc; Turkey to Iran.

4. C. persicifolia L. - *Peach-leaved Bellflower*. Stems glabrous, erect, to 80cm; calyx-lobes lanceolate to linear-lanceolate, patent to erecto-patent; corolla 25-50mm, blue, broadly bell-shaped, divided <1/4 way to base; (2n=16). Intrd; waste and rough ground, grassy places and banks, often natd; scattered through most of Br; Europe.

5. C. medium L. - *Canterbury-bells*. Stems hispid-pubescent, erect, to 60cm; calyx-lobes lanceolate to narrowly ovate, alternating with reflexed appendages, erect; corolla 40-55mm, bright to violet-blue, bell-shaped, lobed <1/4 way to base; (2n=34). Intrd; much grown in gardens, casual and ± natd on waste and rough ground, grassy places and banks, scattered in Br, mainly C & S; Italy and SE France. Some cultivars are to varying degrees *flore pleno*.

6. C. alliariifolia Willd. - *Cornish Bellflower*. Stems pubescent, erect or ± so, to **638**
70cm; calyx-lobes lanceolate to narrowly ovate, erect, alternating with reflexed appendages; corolla 20-40mm, white, bell-shaped, lobed 1/4-1/3 way to base; (2n=34, 68). Intrd; grown in gardens, natd on banks and rough ground; S En, especially by railways in S & SW; Turkey and Caucasus.

7. C. glomerata L. - *Clustered Bellflower*. Stems pubescent, erect, to 80cm but often <20cm; calyx-lobes lanceolate, erect; corolla 12-25mm, violet- or purplish-blue, bell-shaped, lobed 1/4-1/2 way to base; 2n=30. Native; chalk and limestone grassland, scrub and open woodland, cliffs and dunes by sea, also casual or natd escape on rough ground; mainly S & E Br N to CE Sc, scattered escape elsewhere. Garden escapes are very variable (usually more robust); they might belong to other sspp. or to cultivars that resemble them or were derived from them.

8. C. pyramidalis L. - *Chimney Bellflower*. Stems glabrous, erect, to 1m; calyx-lobes **638**
lanceolate, patent; corolla 10-30mm, broadly bell-shaped, purplish-blue, lobed c.1/2 way to base; (2n=34). Intrd; grown in gardens, natd on walls; Guernsey and W Kent, rare casual elsewhere; Italy and Jugoslavia.

9. C. portenschlagiana Schult. - *Adria Bellflower*. Stems glabrous to sparsely pubescent, decumbent to ascending, to 30(50)cm; calyx-lobes lanceolate, erect to erecto-patent; corolla funnel-shaped, 15-25mm, violet-blue, lobed 1/4-2/5 way to base; (2n=34). Intrd; much grown on walls and rockeries in gardens; natd on walls and rocky banks; scattered in Br, mostly C & S, CI, probably over-recorded for *C. poscharskyana*; Jugoslavia.

10. C. poscharskyana Degen - *Trailing Bellflower*. Stems grey-pubescent, decumbent to ascending, to 30(50)cm; calyx-lobes lanceolate to narrowly ovate, erect to erecto-patent; corolla broadly bell-shaped with ± patent lobes, 15-25mm, slatey-blue, lobed 1/2-3/4 way to base; (2n=34). Intrd; grown and natd as for *C. portenschlagiana;* scattered in Br, mostly C & S; Jugoslavia.

11. C. fragilis Cirillo - *Italian Bellflower*. Stems glabrous to pubescent, decumbent to ascending, to 20(40)cm; calyx-lobes lanceolate, erect to erecto-patent; corolla broadly bell-shaped with ± patent lobes, 15-30mm, pale to purplish-blue, lobed c.1/2 way to base; (2n=32, 34). Intrd; grown on walls and rockeries in gardens; natd on wall; Guernsey since 1976; Italy.

12. C. latifolia L. - *Giant Bellflower*. Stems softly pubescent to subglabrous, erect, to 1.2m; calyx-lobes lanceolate to narrowly triangular, erect to erecto-patent; corolla bell-shaped, 35-55mm, pale to purplish-blue, lobed 1/3-1/2 way to base; (2n=34). Native; rich, often damp, mainly calcareous woods; most of Br but very rare to absent in S En and N Sc, NE Ir (?intrd).

13. C. trachelium L. - *Nettle-leaved Bellflower*. Differs from *C. latifolia* in stems to 80(100)cm, corolla 25-35mm, lobed 1/4-2/5 to base; and see key; 2n=34. Native; mainly base-rich woods and hedgebanks; frequent in Br N to N Lincs and N Wa, SE Ir, well natd from gardens elsewhere in Br and Ir.

14. C. rapunculoides L. - *Creeping Bellflower*. Stems sparsely pubescent, erect, to 80cm; calyx-lobes lanceolate to triangular-ovate, patent to reflexed; corolla bell-shaped to funnel-shaped, 20-30mm, violet-blue, lobed 1/3-1/2 way to base; (2n=68, 102). Intrd; grown in gardens, natd in fields, woods, banks and rough ground, very persistent; widely scattered in Br and Ir; Europe.

15. C. rhomboidalis L. - *Broad-leaved Harebell*. Stems sparsely pubescent, erect, to **638**
60cm; calyx-lobes linear, patent; corolla bell-shaped, 15-22cm, pale to bright blue, lobed 1/4-1/3 way to base; (2n=34). Intrd; grown in gardens, natd on shady river-bank in Dumfriess and roadside bank in Westmorland; W Alps.

16. C. rotundifolia L. (*C. giesekiana* auct., ?Vest) - *Harebell*. Stems glabrous to sparsely pubescent, erect to decumbent, to 50cm; calyx-lobes linear, patent; corolla bell-shaped, 12-20mm, pale to bright blue, lobed 1/4-1/3 way to base; 2n=68, 102. Native; grassy places, fixed dunes, rock-ledges, usually on acid often sandy soils; in suitable places throughout BI, but absent from CI and most S & E Ir. A difficult European complex; tetraploids and hexaploids are found in BI, the latter mainly in Ir, W Sc, Man and extreme SW En. The hexaploids often more closely resemble the Scandinavian **C. giesekiana** than tetraploid *C. rotundifolia* (often solitary, larger flowers, squat capsules and wider, blunter tipped upper stem-leaves), but *C. giesekiana* is diploid in Scandinavia and British hexaploids are not consistently separable from tetraploids. There are a few unconfirmed diploid counts for En that need following up.

2. LEGOUSIA Durande - *Venus's-looking-glasses*

Annuals; flowers in terminal cymes; corolla lilac to purple, actinomorphic, divided c.1/2 way to base; filaments and anthers free; ovary 3-celled; style shorter than corolla; stigmas 3, linear to ± globose; capsule dehiscing by subapical lateral pores.

1. L. hybrida (L.) Delarbre - *Venus's-looking-glass*. Stems hispid-pubescent, erect to decumbent, to 30cm; leaves sessile, narrowly oblong-obovate; flowers sessile, mostly forming terminal corymbs; calyx-lobes oblong-lanceolate, acute to obtuse, c.2x as long as corolla, c.1/2 as long as ovary at anthesis; corolla 4-10mm across, widely funnel-shaped; fruit 15-30mm, with persistent calyx; 2n=20. Native; arable fields; scattered in S, C & E En, mostly on calcareous soils, decreasing.

2. L. speculum-veneris (L.) Chaix - *Large Venus's-looking-glass*. Differs from *L. hybrida* in flowers forming terminal pyramidal inflorescence; calyx-lobes linear, acute to acuminate, slightly shorter than corolla, c. as long as ovary at anthesis; corolla 15-20mm across; fruit 10-15mm; (2n=20). Probably intrd; sporadic in cornfields in N Hants since 1916, also occasional grain-alien and garden escape in S En; Europe.

3. WAHLENBERGIA Schrad. ex Roth - *Ivy-leaved Bellflower*

Perennials; flowers solitary, axillary; corolla blue, actinomorphic, divided 1/3-1/2 way to base; filaments and anthers free; ovary 3-celled; style shorter than corolla; stigmas 3, linear; capsule dehiscing by apical pores.

1. W. hederacea (L.) Rchb. - *Ivy-leaved Bellflower*. Stems filiform, procumbent, to 30cm; leaves petiolate, broadly ovate to orbicular-reniform, angled or shortly lobed; flowers on long, filiform, erect stalks; corolla 6-10mm, bell-shaped, pendent; fruit c.3mm, with persistent calyx; 2n=36. Native; damp acid places on heaths and moors, in woods, by streams; W & S Br N to Argyll, S & SE Ir, formerly CI, common only in Wa and SW En, natd rarely elsewhere in wet lawns.

4. TRACHELIUM L. - *Throatwort*
Perennials; flowers numerous in terminal corymbose compound cymes; corolla blue, actinomorphic, with narrow tube and small limb divided ≥1/2 way to base; filaments and anthers free; ovary 2-3-celled; style longer than corolla; stigmas 2-3, capitate; capsule dehiscing by 2-3 sub-basal pores.

1. T. caeruleum L. - *Throatwort*. Stems glabrous, erect, to 1m; lower leaves petiolate, elliptic-ovate; flowers with tube 4-7 x c.0.5-1mm, with limb 2-3mm across; (2n=34). Intrd; natd on walls since 1892 in Guernsey, also Jersey, Middlesex and W Kent; W Mediterranean.

5. PHYTEUMA L. - *Rampions*
Perennials; flowers numerous in terminal congested heads; corolla blue or yellow, slightly zygomorphic, tubular and usually curved in bud but split nearly to base when open; filaments and anthers free but appressed around style; ovary 2-3-celled; style slightly shorter than corolla; stigmas 2-3, linear; capsule dehiscing by 2-3 lateral pores.

1 Corolla usually pale yellow, rarely blue; inflorescence oblong to cylindrical in flower **1. P. spicatum**
1 Corolla violet-blue; inflorescence globose to very shortly ovoid in flower 2
  2 Corolla nearly straight in bud; lowest bracts longer than inflorescence (though often reflexed); lowest leaves strongly cordate at base **3. P. scheuchzeri**
  2 Corolla strongly curved in bud; bracts shorter than inflorescence; lowest leaves rounded to rarely subcordate at base **2. P. orbiculare**

1. P. spicatum L. - *Spiked Rampion*. Stems glabrous, erect, to 80cm; lowest leaves ovate-cordate, with long petiole; inflorescence oblong to cylindrical, 3-8cm; corolla pale yellow in native plants, 7-10mm; stigmas usually 2; (2n=22). Native; woods, scrub and hedgerows on acid soils in area <20 x 10km in E Sussex, rare escape (usually blue-flowered) elsewhere in Br. **RR**

2. P. orbiculare L. (*P. tenerum* Rich. Schulz) - *Round-headed Rampion*. Stems glabrous to sparsely pubescent, erect, to 50cm (but often <15cm); lowest leaves ovate to narrowly so, rounded to rarely subcordate at base, with long petiole; inflorescence globose, 1-2cm; corolla violet-blue, 5-8mm; stigmas 2-3; 2n=22. Native; open chalk grassland; local in S En from N Wilts to E Sussex, formerly to E Kent. Our plant belongs to the segregate **P. tenerum**, sometimes considered a separate sp. **R**

3. P. scheuchzeri All. - *Oxford Rampion*. Stems glabrous, erect to decumbent or pendent, to 40cm; lowest leaves ovate-cordate, with long petiole; inflorescence globose to very shortly ovoid, 2-2.5cm; corolla blue, 8-12mm; stigmas 3; (2n=26). Intrd; natd in limestone cracks at Inchnadamph, W Sutherland, since at least 1992, formerly on walls and in pavements in Oxford, Oxon; S Alps.

6. JASIONE L. - *Sheep's-bit*
Annuals to perennials; flowers numerous in terminal, congested heads; corolla blue, actinomorphic, tubular and straight in bud but split nearly to base when open; filaments free, anthers slightly laterally fused; ovary 2-celled; style longer than corolla; stigmas 2, subcapitate; capsule dehiscing by 2 apical short valves.

Superficially closely resembles a *Scabious* (Dipsacaceae), but in the latter the stamens are well separated, the corolla is split ≤1/2 way to the base, and the fruit is indehiscent and 1-seeded.

1. J. montana L. - *Sheep's-bit*. Stems pubescent, suberect to decumbent, to 50cm; lower leaves narrowly oblong to oblanceolate, shortly petiolate; inflorescence depressed-globose, 0.5-3.5cm across; 2n=12. Native; grassy or sandy or rocky places on acid soils, walls, cliffs, banks; locally common in BI, mainly in W, absent from much of C & E Br and C Ir.

7. LOBELIA L. - *Lobelias*
Annuals or perennials; flowers in terminal racemes; corolla pale lilac to blue, zygomorphic, with tube (with deep dorsal split) and expanded limb, the latter with 5 lobes arranged 2 in upper and 3 in lower lip; filaments and anthers fused laterally around style; ovary 2-celled; style shorter than corolla; stigma capitate, 2-lobed; capsule dehiscing by 2 apical valves.

1 Leaves all basal, linear, entire; plant submerged (except inflorescence) or at lakeside **3. L. dortmanna**
1 Stem-leaves present, the lower ones obovate, serrate; plant terrestrial 2
  2 Corolla-lobes <2mm wide; pedicels <1cm **1. L. urens**
  2 Lower 3 corolla-lobes >2mm wide; at least lower pedicels >1cm **2. L. erinus**

Other spp. - **L. siphilitica** L., from N America, is grown in gardens and is a very rare escape; it is a strong perennial to 1m, with dense raceme of blue flowers with corollas 2-3.5cm.

1. L. urens L. - *Heath Lobelia.* Perennials; stems sparsely and shortly pubescent, **RR**
erect, to 80cm; flowers many, in long raceme with bracts much narrower than leaves, 10-15mm, purplish-blue; corolla-lobes 2-4 x <1mm, with recurved, acute apex; (2n=14). Native; acid heathy grassland, open woods, wood-borders; very local in S En from E Cornwall to W Kent, formerly E Kent and Herefs.

2. L. erinus L. - *Garden Lobelia.* Annuals (in BI); stems glabrous to sparsely pubescent, erect to decumbent, to 30cm; flowers rather few, in racemes with lower bracts leaf-like, 8-20mm, usually purplish-blue, sometimes white or pinkish; lower corolla-lobes c.3-8 x 2-6mm, with rounded, apiculate apex; upper corolla-lobes narrower; (2n=14, 28, 42). Intrd; much grown in gardens, frequent escape, sometimes self-sown, on tips and in pavement-cracks and rough ground; scattered in Br, CI; S Africa.

3. L. dortmanna L. - *Water Lobelia.* Perennials; stems glabrous, erect, to 70(120)cm; flowers (1)3-10 in very lax raceme with very reduced bracts, 12-20mm, pale lilac; lower corolla-lobes 5-10 x 1-3mm, acute, upper corolla-lobes slightly smaller; 2n=14. Native; in stony, acid mainly montane lakes, rarely on wet ground adjacent; locally common in N & W Br S to S Wa, W, N & E Ir.

8. PRATIA Gaudich. - *Lawn Lobelia*
Perennials; flowers solitary in leaf-axils; corolla white; differs from *Lobelia* in fruit a berry.

1. P. angulata (G. Forst.) Hook. f. - *Lawn Lobelia.* Stems procumbent, rooting at **638**
nodes, glabrous, to 15cm; leaves ≤12mm, suborbicular, shallowly lobed, shortly petiolate; flowers 7-20mm; (2n=70). Intrd; grown and becoming established on damp lawns; scattered in Sc, Surrey and W Kent; New Zealand.

9. DOWNINGIA Torr. - *Californian Lobelia*
Annuals; corolla blue with white centre; differs from *Lobelia* in corolla with lower lip only shallowly lobed, flowers sessile with long, pedicel-like ovary, ovary 1-celled, and capsule dehiscing by 3-5 longitudinal slits.

1. D. elegans (Douglas ex Lindl.) Torr. - *Californian Lobelia.* Stems erect, glabrous, to 25cm; leaves linear-lanceolate, sessile; flowers 8-18mm; (2n=20). Intrd; casual or perhaps natd in grassy places, probably intrd with grass-seed, first found 1978, apparently increasing; sporadic in SE En; W N America.

134. RUBIACEAE - *Bedstraw family*

Annual to perennial herbs, evergreen climbers or rarely shrubs; leaves opposite with 1-several stipules per leaf, the stipules usually leaf-like and as large as leaves, hence leaves apparently in whorls of 4 or more, simple, ± entire, usually sessile and narrow. Flowers small, in usually compound terminal and/or axillary cymes, often aggregated into terminal panicle, actinomorphic, bisexual or bisexual and male mixed (*Cruciata*) or ± dioecious (*Coprosma*), epigynous; sepals 0 or minute, 4-5, fused below; petals 4-5, fused into long or short tube below, various colours; stamens 4-5, borne at apex of corolla-tube (near base in *Coprosma*); ovary 2-celled, each cell with usually 1 ovule on axile placenta; styles 1-2, if 1 often branched into 2; stigmas 1 per style or style-branch, capitate; fruit mostly of 2 fused (later separating) 1-seeded nutlets, or succulent with 1-2 seeds.

Easily recognized by the small flowers with 4-5 petals fused into an (often very short) tube, 0 or minute calyx and inferior 2-celled ovary with 1 ovule per cell. Most spp. have apparently whorled leaves, 4 corolla-lobes and distinctive paired nutlets.

1 Leaves opposite, usually with stipules or smaller leaves also at same node — 2
1 Leaves in whorls of ≥4, ± all same size in 1 whorl — 4
  2 Evergreen shrub — **1. COPROSMA**
  2 Procumbent to ascending herb — 3
3 Leaves linear or nearly so; fruit of 2 nutlets — **5. ASPERULA**
3 Leaves ovate to suborbicular; fruit succulent — **2. NERTERA**
  4 Most or all flowers with 5 corolla-lobes — 5
  4 Most or all flowers with 4 corolla-lobes — 6
5 Procumbent to ascending annual; leaves ≥6 in a whorl; corolla pink; fruit dry — **4. PHUOPSIS**
5 Evergreen climber or scrambler; leaves 4-6 in a whorl; corolla yellowish-green; fruit succulent — **8. RUBIA**
  6 Calyx distinct, c.0.5-1mm at first, slightly enlarging in fruit — **3. SHERARDIA**
  6 Calyx absent or vestigial — 7
7 Corolla-tube >1mm — 8
7 Corolla-tube <1mm — 9
  8 Ovary and fruit smooth to papillose — **5. ASPERULA**
  8 Ovary and fruit covered with hooked bristles — **6. GALIUM**
9 At least some whorls with >4 leaves — **6. GALIUM**
9 All whorls with 4 leaves — 10
  10 Flowers in dense axillary whorls; ovary and fruit smooth — **7. CRUCIATA**
  10 Flowers in terminal panicles; ovary and fruit covered with hooked bristles — **6. GALIUM**

Other genera - CRUCIANELLA L. would key to couplet 8 but differs from all other genera in having flowers in compact spikes, with 1 flower per axil. **C. angustifolia** L., from S Europe, is an annual with leaves in whorls of 6-8 and a yellow corolla; it is a rare birdseed-alien.

1. COPROSMA J.R. & G. Forst. - *Tree Bedstraw*

Evergreen, usually dioecious shrubs; leaves opposite, with small stipules, petiolate;

flowers inconspicuous, <1cm, in axillary clusters with 2 partly fused bracts below; calyx minute; corolla greenish, 4-5-lobed, with long tube; fruit succulent, with 2 nuts.

1. C. repens A. Rich. (*C. baueri* auct. non Endl.) - *Tree Bedstraw*. Shrub to 3m; leaves broadly ovate-oblong, (2)5-8cm, very shiny on upperside, with recurved margins; fruit depressed-obovoid, c.10 x 8mm, orange-red; (2n=44). Intrd; planted as windbreak in Scillies and sometimes self-sown; New Zealand.

2. NERTERA Banks & Sol. ex Gaertn. - *Beadplant*
Herbaceous perennials; leaves opposite, with small stipules, shortly petiolate; flowers solitary, axillary and terminal, <5mm; calyx minute; corolla greenish, 4-lobed; fruit succulent, with 2 nuts.

1. N. granadensis (Mutis ex L.f.) Druce (*N. depressa* Banks & Sol. ex Gaertn.) - *Beadplant*. Stems procumbent, to 15cm, glabrous; leaves ovate to suborbicular, (3)5-8(15)mm, with recurved margins, glabrous; fruit globose, c.4mm, bright reddish-orange; (2n=44). Intrd; natd on damp lawns; few places in WC Sc; Australia, New Zealand, S America.

3. SHERARDIA L. - *Field Madder*
Annuals; leaves in whorls of 4-6, sessile; flowers 4-10 in dense terminal and axillary clusters with whorl of 8-10 leaf-like bracts at base, the clusters stalked; calyx 0.5-1mm at first, slightly enlarging in fruit, with 4-6 deeply toothed lobes; corolla pale to deep mauvish-pink, 4-lobed; fruit a pair of scabrid nutlets with persistent calyx on top.

1. S. arvensis L. - *Field Madder*. Stems procumbent to ascending, to 40cm, glabrous to pubescent; lower leaves obovate; upper leaves narrowly elliptic to oblanceolate, 5-18mm; corolla 4-5mm, with tube longer than lobes; 2n=22. Native; arable fields, waste places, thin grassland and lawns; frequent almost throughout BI but local in Sc.

4. PHUOPSIS (Griseb.) Hook. f. - *Caucasian Crosswort*
Annuals to perennials; leaves in whorls of 6-9, sessile; flowers many in dense terminal clusters with whorl of many leaf-like bracts at base; calyx minute; corolla deep pink, 5-lobed; fruit a pair of glabrous, papillose nutlets.

1. P. stylosa (Trin.) Benth. & Hook. f. ex B.D. Jacks. - *Caucasian Crosswort*. Stems procumbent to sprawling, to 30(70)cm, sparsely scabrid-pubescent; leaves linear to narrowly elliptic, 12-30mm; corolla 12-15mm, with narrow tube much longer than lobes; style long-exserted; (2n=20, 22). Intrd; grown in gardens and sometimes persistent outside on waste and rough ground and tips; scattered in C & S Br; Caucasus and Iran.

5. ASPERULA L. - *Woodruffs*
Annuals to herbaceous perennials; leaves in whorls of 4-8 and all equal, or in whorls of 4 with 2 long and 2 short, sessile; flowers in dense terminal clusters with whorl of leaf-like bracts at base or in loose terminal panicles; calyx minute; corolla various colours, 4-lobed; fruit a pair of smooth to papillose nutlets.

1 Leaves in whorls of 6-8; corolla blue **3. A. arvensis**
1 Leaves in whorls of 4, equal or 2 short and 2 long; corolla white to pink 2
  2 Leaves equal in all whorls, narrowly ovate to narrowly elliptic; inflorescence a compact cluster with whorl of leaf-like bracts below **2. A. taurina**

2 Leaves at upper whorls 2 long and 2 short, linear-oblanceolate; inflorescence a diffuse panicle **1. A. cynanchica**

1. A. cynanchica L. - *Squinancywort*. Subglabrous perennials; stems procumbent to ascending, to 50cm; at least upper whorls with 2 long and 2 short leaves; inflorescence a diffuse panicle; corolla white to pink, 2.5-5mm; 2n=40. Native.

a. Ssp. cynanchica. Rhizomes 0 or brown; leaves mostly linear, mostly 10-20 x 0.5-1mm; pedicels 0-1mm; corolla-lobes usually distinctly shorter than -tube. Limestone and chalk grassland, calcareous dunes; locally common in S Br and W Ir, scattered N to Westmorland and SE Yorks, formerly Jersey.

b. Ssp. occidentalis (Rouy) Stace (*A. occidentalis* Rouy). Rhizomes orange; leaves linear to oblanceolate, often <10mm and >1mm wide; pedicels usually 0; corolla-lobes usually c. as long as -tube. Calcareous dunes; S Wa, W Ir. Leaves of ssp. *cynanchica* in dune habitats resemble those of ssp. *occidentalis* rather closely, but the other characters are more constant. The sp. occurs in Alderney and needs checking for ssp. determination. **RR**

2. A. taurina L. - *Pink Woodruff*. Pubescent perennials; stems erect, to 50cm, all whorls with 4 equal leaves; inflorescence a dense cluster with whorl of leaf-like bracts at base; corolla white to yellowish-pink, 10-14mm, with tube much longer than lobes; (2n=22). Intrd; grown in gardens, natd in damp woods; locally natd in C Sc, rare and impermanent further S; S Europe.

3. A. arvensis L. - *Blue Woodruff*. Annuals, glabrous apart from bracts long-pubescent at margin; stems erect, to 50cm; all whorls with 6-8 equal linear leaves; inflorescence a dense cluster with whorl of leaf-like bracts at base; corolla blue, 5-6.5mm, with tube much longer than lobes; (2n=22). Intrd; casual on tips and in waste places mainly (?only) from birdseed; scattered in most of Br and CI, mainly S; Europe. Often confused with the garden plant **A. orientalis** Boiss. & Hohen. (*A. azurea* Jaub. & Spach), from S W Asia, which is a much rarer casual with a longer corolla (7-12(14)mm).

6. GALIUM L. - *Bedstraws*

Annuals to herbaceous perennials; leaves in whorls of 4-12 and all equal, sessile; flowers in terminal panicles or axillary cymes; calyx minute; corolla white to yellow, 4-lobed; fruit a pair of smooth to bristly nutlets, the bristles sometimes hooked.

1 Ovaries and fruits with hooked bristles 2
1 Ovaries and fruits smooth to rugose or papillose 5
 2 All whorls with 4 leaves **1. G. boreale**
 2 Most or all whorls with ≥5 leaves 3
3 Rhizomatous perennial; flowers in terminal panicles; corolla-tube >1mm **2. G. odoratum**
3 Annual; flowers in axillary cymes; corolla-tube <1mm 4
 4 Fruit >3mm (excl. bristles), its bristles with bulbous base; corolla >1.4mm across **11. G. aparine**
 4 Fruit <3mm (excl. bristles), its bristles wider towards base but not bulbous at base; corolla <1.4mm across **12. G. spurium**
5 Corolla bright yellow **6. G. verum**
5 Corolla white to pale cream (N.B. *G. x pomeranicum*), sometimes tinged pink 6
 6 Annuals of open ground, usually easily uprooted, with sparse rooting system 7
 6 Perennials, firmly rooted, often in grassland or wet ground 9
7 Plant much slenderer than *G. aparine*; leaves with forward-directed marginal prickles **14. G. parisiense**
7 Plant coarse, with habit of *G. aparine*; leaves with backward-directed

marginal prickles 8

8 Fruit smooth, <3mm; peduncles and pedicels divaricate at various angles at fruiting, but straight **12. G. spurium**

8 Fruit papillose, >3mm; peduncles and/or pedicels strongly recurved **13. G. tricornutum**

9 Leaves obtuse to acute, never apiculate or mucronate 10

9 Leaves apiculate to mucronate at apex 11

10 Leaves linear; pedicels scarcely divaricate at fruiting; inflorescence obconical, widest near top; rare **4. G. constrictum**

10 Leaves linear-oblong to oblanceolate or narrowly elliptic; pedicels strongly divaricate at fruiting; inflorescence usually conical to cylindrical, widest well away from apex; common **5. G. palustre**

11 Stems rough, with projecting minute papillae or pricklets **3. G. uliginosum**

11 Stems perfectly smooth 12

12 Corolla-lobes apiculate to strongly mucronate at apex, with point ≥0.2mm; fruit minutely wrinkled **7. G. mollugo**

12 Corolla-lobes acute to minutely apiculate at apex, with point <0.2mm; fruit minutely tuberculate 13

13 Leaf-margins with forward-directed prickles; leaves on flowering shoots oblanceolate **10. G. saxatile**

13 Leaf-margins with at least some backward-directed prickles; leaves on flowering shoots linear to linear-elliptic or -oblanceolate 14

14 Fruit with minute high-domed subacute tubercles; NW of line from Severn to Humber **9. G. sterneri**

14 Fruit with minute low-domed to rounded tubercles; almost entirely SE of line from Severn to Humber **8. G. pumilum**

Other spp. - G. verrucosum Huds. (*G. valantia* Weber, *G. saccharatum* All.), from S Europe, has been found as a casual but is very rare; it resembles *G. tricornutum* in its papillose fruits >3mm but has leaves with forward-directed marginal prickles.

1. G. boreale L. - *Northern Bedstraw*. Erect perennial to 45cm; leaves all 4 per whorl, with 3 main veins, lanceolate to narrowly ovate or narrowly elliptic; corolla white, with very short tube; fruit with hooked bristles; 2n=44. Native; damp grassy, rocky and gravelly places usually on hills and often by streams, also on sand-dunes; locally frequent in Br N from MW Yorks, scattered in W, C & N Ir, very local in N & S Wa.

2. G. odoratum (L.) Scop. - *Woodruff*. Erect perennial to 45cm; leaves 6-8(9) per whorl, with 1 main vein and distinct laterals, elliptic to oblanceolate; corolla white, with tube c. as long as lobes; fruit with hooked bristles; 2n=44. Native; damp, base-rich woods and hedgerows; frequent throughout most of BI except CI and Outer Isles, garden escape in Jersey.

3. G. uliginosum L. - *Fen Bedstraw*. Decumbent to ascending or scrambling, **648**
scabrid perennial to 60cm; leaves (4)5-8 per whorl, 1-veined, linear-oblanceolate to -elliptic, strongly mucronate at apex; corolla white, with very short tube; fruit with low-domed tubercles; 2n=22, 88 (22, 44). Native; fens and base-rich marshy places, scattered in BI except N Sc, N Ir and CI.

4. G. constrictum Chaub. (*G. debile* Desv. non Hoffmanns. & Link) - *Slender* **RR**
Marsh-bedstraw. Decumbent to ascending or scrambling, smooth or slightly scabrid **648**
perennial to 40cm; leaves 4-6 per whorl, 1-veined, linear, obtuse to subacute (but often revolute and appearing acute) at apex; corolla white, with very short tube; fruit with high-domed tubercles; 2n=24. Native; marshy places, ditches and pondsides; very local in SE Yorks, S Hants, S Wilts, S Devon, and CI.

5. G. palustre L. - *Common Marsh-bedstraw*. Decumbent to ascending or scrambling, smooth or more often scabrid perennial to 1m; leaves 4-6 per whorl, 1-veined, linear-oblong to oblanceolate or narrowly elliptic, rounded to subacute at apex; corolla white, with very short tube; fruit slightly wrinkled. Native; damp meadows, pondsides, ditches, marshes and fens; common throughout BI.

a. Ssp. palustre (ssp. *tetraploideum* A.R. Clapham). Most leaves <20mm; 648
inflorescence ± cylindrical; pedicels mostly <4mm at flowering; corolla mostly 2-3.5mm across; fruit c.1.6mm; 2n=24, 48.

b. Ssp. elongatum (C. Presl) Arcang. (*G. elongatum* C. Presl). Most leaves 648
>20mm; inflorescence ± conical; pedicels mostly >4mm at flowering; corolla mostly 3-4.5mm across; fruit c.1.9mm; 2n=96 (96, 144). More robust than ssp. *palustre* and commoner.

6. G. verum L. - *Lady's Bedstraw*. Procumbent to erect, smooth perennial to 1m; 648
leaves (6)8-12 per whorl, 1-veined, linear, apiculate to shortly mucronate at apex; corolla bright yellow, with very short tube, with acute to apiculate lobes; fruit smooth to minutely wrinkled; 2n=44. Native; dry grassy places especially on calcareous soils, often by sea; common throughout BI. Dwarf, procumbent plants with internodes shorter than leaves (var. **maritimum** DC.) are common on maritime dunes and cliff-tops.

6 x 7. G. verum x G. mollugo = G. x pomeranicum Retz. is frequent with the 648
parents in Br N to N Aberdeen and in CI, and rare in S & E Ir; it is intermediate in all characters, notably petal- and leaf-shape and flower-colour, and somewhat variable, suggesting backcrossing. Often *G. verum* var. *maritimum* is involved.

7. G. mollugo L. - *Hedge Bedstraw*. Decumbent to erect, smooth perennial to 1.5m; leaves 5-8 per whorl, 1-veined, oblong to oblanceolate, apiculate to strongly mucronate at apex; corolla white, with very short tube, with apiculate to strongly mucronate lobes; fruit smooth to wrinkled; 2n=44. Native; all sorts of grassy places, hedgerows, mainly on well-drained base-rich soils; throughout most of BI but rare in Sc, Wa and Ir, common in S En and CI.

a. Ssp. mollugo. Leaves mostly oblanceolate to narrowly obovate; inflorescence 648
broad, with branches mostly at ≥45°, rather lax; corolla 2-3mm across; pedicels strongly divaricate at fruiting.

b. Ssp. erectum Syme (*G. album* Mill.). Leaves mostly linear-oblanceolate to 648
oblanceolate; inflorescence rather narrow, with branches mostly at <45°, rather dense; corolla 2.5-5mm across; pedicels weakly divaricate at fruiting. Mostly on drier, more calcareous soils.

G. mollugo is very variable and the 2 sspp. are of doubtful value. On the Continent diploids (2n=22) and tetraploids (2n=44) occur and are usually referred to *G. mollugo* and *G. album* respectively. The correlation of these with the 2 sspp. is highly uncertain; in BI only the tetraploid has been recorded but no detailed survey has been carried out.

8. G. pumilum Murray (*G. fleurotii* auct., ?Jord.) - *Slender Bedstraw*. Decumbent to R
erect, smooth perennial to 40cm; leaves 5-9 per whorl, 1-veined, linear-oblanceolate 648
to oblanceolate, strongly mucronate at apex, with mainly backward-directed prickles on margins; corolla white, with very short tube; fruit with low-domed tubercles; 2n=88. Native; dry chalk and limestone grassland; scattered and local in En SE of line from Severn to Humber, extinct in several areas and decreasing. The plants at Cheddar, N Somerset, have been referred to **G. fleurotii**, but they have the same chromosome number as other populations and do not consistently differ by the characters said to distinguish Continental *G. fleurotii*.

9. G. sterneri Ehrend. - *Limestone Bedstraw*. Differs from *G. pumilum* in usually 648
many more non-flowering shoots, forming a mat; flowering stems to 30cm; fruit with high-domed subacute tubercles; 2n=22, 44. Native; limestone or other base-rich grassland or rocks; local in BI NW of line from Severn to Humber. Populations in NW Wa, N Sc and W Ir are diploid (2n=22); others counted are tetraploid, but scarcely any morphological differences exist.

9 x 10. G. sterneri x G. saxatile has been cytologically confirmed (2n=33 or 55 according to *G. sterneri* parent 2n=22 or 44) in Caerns, M Perth, Easterness and W Sutherland; it is intermediate in leaf-shape and marginal prickles and highly sterile.

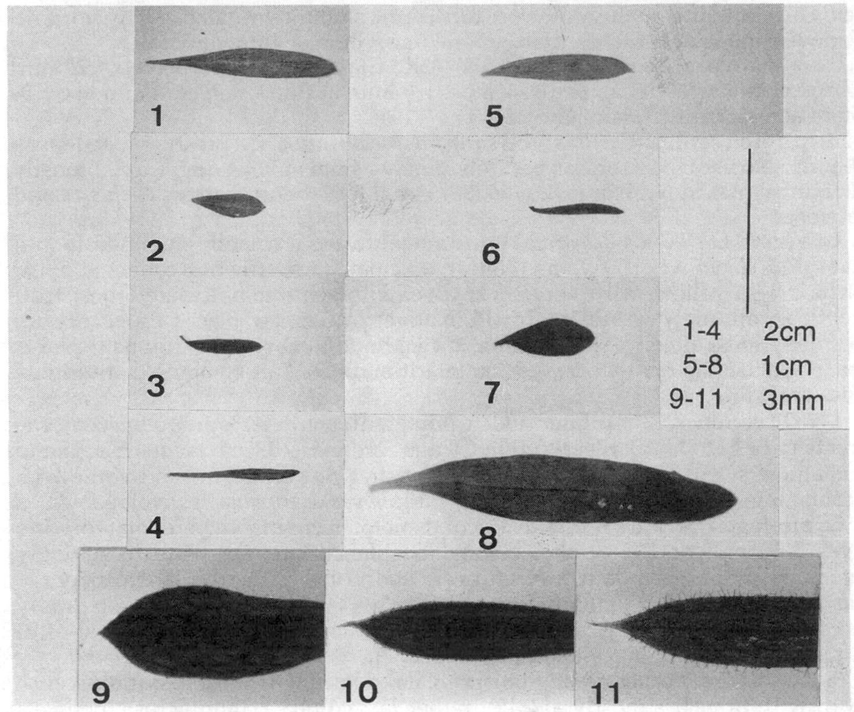

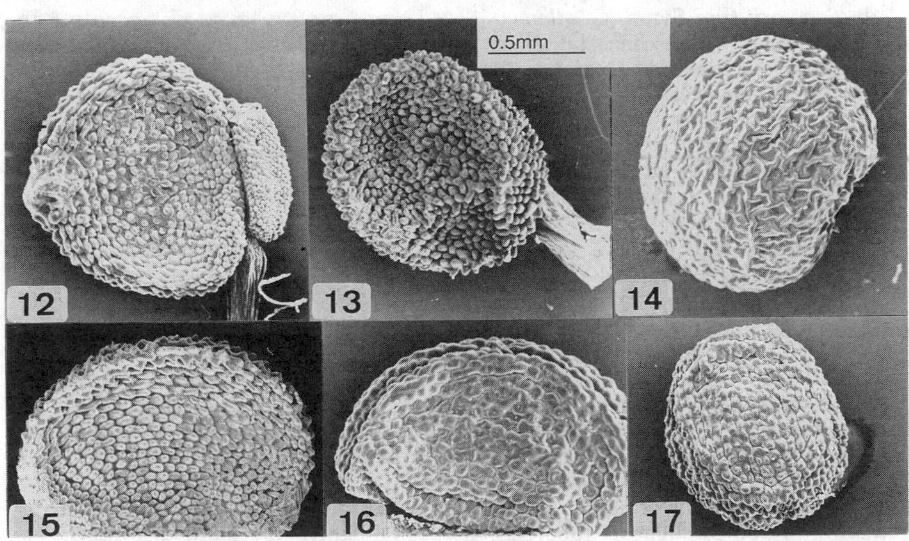

FIG 648 - *Galium*. 1-11, leaves. 1, *G. mollugo* ssp. *erectum*. 2, ssp. *mollugo*. 3, *G. x pomeranicum*. 4, *G. verum*. 5, *G. uliginosum*. 6, *G. constrictum*. 7, *G. palustre* ssp. *palustre*. 8, ssp. *elongatum*. 9, *G. saxatile*. 10, *G. sterneri*. 11, *G. pumilum*. 12-17, one nutlet. 12, *G. uliginosum*. 13, *G. constrictum*. 14, *G. palustre*. 15, *G. saxatile*. 16, *G. pumilum*. 17, *G. sterneri*.

10. G. saxatile L. - *Heath Bedstraw.* Decumbent to ascending, smooth perennial to 648
30cm; non-flowering shoots numerous, mat-forming; leaves 5-8 per whorl, 1-veined, oblanceolate to obovate, apiculate to shortly mucronate at apex, with forward-directed prickles on margins; corolla white, with very short tube; fruit with high-domed subacute tubercles; 2n=44. Native; dry grassland, rocky places and open woods on acid soils; common throughout most of BI.

11. G. aparine L. - *Cleavers.* Procumbent to scrambling-erect annual to 3m, with strongly recurved prickles on stems; leaves 6-8 per whorl, 1-veined, oblanceolate, strongly mucronate at apex; corolla white, with very short tube; fruit 3-5mm (excl. bristles), covered with hooked bristles; 2n=66. Native; cultivated and arable land, hedgerows and scrub, other open ground; common throughout BI.

12. G. spurium L. - *False Cleavers.* Differs from *G. aparine* in slightly more slender RR
habit; stems to 1m; leaves linear-lanceolate to lanceolate; corolla greenish-cream; fruit 1.5-3mm (excl. bristles), smooth or with hooked bristles; and see key (couplet 4); (2n=20, 44). Probably intrd; well natd as arable weed around Saffron Waldron, N Essex since 1844, elsewhere as casual or temporarily natd in Br N to S Sc; Europe. The N Essex plant and most others are var. **vaillantii** (DC.) Gren, with bristly fruits; the smooth-fruited var. **spurium** has occurred only as a very rare casual.

13. G. tricornutum Dandy - *Corn Cleavers.* Differs from *G. aparine* in stems to RR
60cm; flowers and fruits mainly in clusters of 3 (not 2-5); peduncles and/or pedicels strongly recurved in fruit; fruit acutely papillose; (2n=44). Probably intrd; once common in arable and waste places in S, C & E En, rare elsewhere in Br, now rare and sporadic in C & SE En; Europe.

14. G. parisiense L. - *Wall Bedstraw.* Slender, procumbent to ascending annual to R
30cm, with small recurved prickles on stems; leaves 5-7 per whorl, 1-veined, linear-oblanceolate to oblanceolate, mucronate at apex; corolla greenish-white tinged reddish, with very short tube; fruit 0.8-1.2mm, very finely papillose; (2n=22, 44, 66). Native; walls and sandy banks; scattered in E Anglia and SE En, rare casual elsewhere, decreasing.

7. CRUCIATA Mill. - *Crosswort*

Herbaceous perennials; leaves in whorls of 4, all equal, ± sessile; flowers in dense axillary whorls of cymes, the terminal flower bisexual and the laterals male in each cyme; calyx minute; corolla yellow, 4-lobed, with very short tube; fruit 1 or a pair of smooth nutlets.

1. C. laevipes Opiz (*C. chersonensis* auct. non (Willd.) Ehrend., *Galium cruciata* (L.) Scop.) - *Crosswort.* Stems erect, to 60cm, conspicuously pubescent; leaves elliptic to oblong- or ovate-elliptic, 10-20mm, yellowish-green; 2n=22. Native; grassy places, hedgerows, scrub and rough ground, mostly on calcareous soils; common in Br N to C Sc, intrd in Ir.

8. RUBIA L. - *Madders*

Evergreen scrambling perennials; leaves 4-6 per whorl, all equal, narrowed to ± sessile base; flowers in diffuse axillary and terminal panicles; calyx minute; corolla pale yellowish-green, 5-lobed, with very short tube; fruit succulent, with 1 seed.

Other spp. - R. tinctorum L. (*Madder*), from Asia, was formerly grown for its dye and used to occur as a casual and escape, but no longer; it differs in its light green leaves 2-10cm and its anthers 5-6x (not 1.3-2x) as long as wide.

1. R. peregrina L. - *Wild Madder.* Stems trailing to scrambling, to 1.5m, glabrous, with strong recurved prickles; leaves 1-5cm, elliptic to narrowly so, leathery, 1-veined, glabrous, with strong recurved prickles on margins; (2n=22, 44, 66, 88, c.110, c.132). Native; hedges, scrub, rocky places; locally common in S & W Br from E Kent to N Wa, S, E & C Ir, CI, mainly coastal.

135. CAPRIFOLIACEAE - *Honeysuckle family*

Deciduous or evergreen shrubs (small and procumbent in *Linnaea*), small trees or woody climbers, rarely herbaceous perennials; leaves opposite, simple (lobed or not) or pinnate, without stipules, petiolate. Flowers variously arranged, axillary or terminal, zygomorphic or actinomorphic, bisexual or some sterile, epigynous; sepals 5, fused into tube proximally; petals 5, fused into tube proximally, sometimes 2-lipped with 4-lobed upper and 1-lobed lower lip, usually white to yellow, sometimes pink to reddish; stamens 4-5, borne on corolla-tube; ovary 1-5-celled, sometimes including 2 sterile cells, each cell with 1 apical ovule or many ovules on axile placenta; style 0 or 1; stigmas 1-5, ± capitate; fruit succulent, 1-several seeded (an achene in *Linnaea*, a capsule in *Weigela*).

The only woody plants with fused petals, inferior ovary and stamens only 4 or 5.

1 Leaves pinnate **1. SAMBUCUS**
1 Leaves simple (sometimes deeply lobed) 2
 2 Flowers numerous in corymbose compound cymes; style ± 0 **2. VIBURNUM**
 2 Flowers 2-few, not corymbose; style conspicuous 3
3 Main stems procumbent, the flowers in pairs terminal on erect lateral stems **4. LINNAEA**
3 Main stems ± erect or climbing 4
 4 Bracts ≥15mm, leaf-like, purple, or green strongly tinged purple **5. LEYCESTERIA**
 4 Bracts <15mm, usually not purple 5
5 Deciduous shrub; ovary 4-celled, with 2 fertile and 2 sterile cells, the former each with 1 ovule; corolla actinomorphic, <10mm **3. SYMPHORICARPOS**
5 Ovary 2-3-celled, all cells fertile and with >1 ovule; corolla strongly zygomorphic to ± actinomorphic, if <10mm then plant an evergreen shrub or corolla strongly zygomorphic or both 6
 6 Fruit a capsule; corolla >20mm, weakly zygomorphic, scarcely 2-lipped **6. WEIGELA**
 6 Fruit a berry; if corolla >20mm then strongly zygomorphic and 2-lipped **7. LONICERA**

Other genera - Several other genera are grown as garden ornamentals. **KOLKWITZIA** Graebn. is similar to *Weigela* and shrubby *Lonicera* spp. but has fruit an achene; **K. amabilis** Graebn. (*Beauty-bush*), with pale pink flowers with a yellow throat, from China, occasionally produces seedlings in waste places in S En or persists as isolated bushes.

1. SAMBUCUS L. - *Elders*
Deciduous shrubs or herbaceous perennials; leaves pinnate; flowers numerous in corymbose or paniculate compound cymes, actinomorphic; stamens 5; ovary 3-5-celled, each cell with 1 ovule; style 0; stigmas as many as carpels; fruit a drupe with 3-5 seeds.

1 Inflorescence an ovoid to ± globose panicle; ripe fruits red; stipules represented by stalked glands **1. S. racemosa**
1 Inflorescence a flat or slightly convex corymb; ripe fruits black to purplish-black, rarely red or greenish-yellow or -white; stipules 0 or subulate to ovate 2

2 Rhizomatous herbaceous perennial; stipules conspicuous, ovate or narrowly so; anthers purple **4. S. ebulus**
2 Erect shrub; stipules small and subulate; anthers cream 3
3 Fruits black, rarely greenish-yellow or -white; 2nd year twigs with numerous lenticels; leaflets (3)5(-7); not rhizomatous **2. S. nigra**
3 Fruits purplish-black, rarely red; 2nd year twigs with few lenticels; leaflets (5)7(-11); rhizomatous **3. S. canadensis**

1. S. racemosa L. (*S. pubens* Michx., *S. sieboldiana* (Miq.) Graebn.) - *Red-berried Elder*. Shrub to 4m; leaves with (3)5-7 leaflets; stipules represented by stalked glands; flowers cream; fruits bright red; flowering Apr-May; (2n=36). Intrd; well natd in hedges, woods and shrubberies; frequent in Br N from Derbys and Cheshire, very scattered further S; N Temperate. Most natd plants are var. **racemosa**, from Europe. Plants with more diffuse panicles and a pubescent rhachis and sometimes leaf lowerside come from N America (var. **pubens** (Michx.) Koehne); those with more diffuse panicles, more finely serrate leaflets and smaller fruits (c. 3mm, not 4-5mm) come from E Asia (var. **sieboldiana** Miq.). Both the latter are natd in Sc, but all 3 vars hybridise and are difficult to delimit. Cultivars with greenish-yellow fruits, with dissected leaflets, or with variegated leaves exist.

2. S. nigra L. - *Elder*. Shrub or small tree to 10m; leaves with (3)5(-7) leaflets; stipules 0 or small and subulate; flowers creamish-white; fruits black, sometimes greenish-yellow; flowering Jun-Jul; 2n=36. Native; hedges, woods, shrubberies, waste and rough ground, especially on manured soils; common throughout BI but only intrd in Orkney and Shetland. Cultivars exist with dissected leaflets or with variegated leaves; both are natd.

3. S. canadensis L. - *American Elder*. Suckering shrub to 4m; leaves with (5)7(-11) leaflets; stipules 0 or small and subulate; flowers white; fruits purplish-black, rarely red; flowering Jul-Sep; (2n=36). Intrd; natd in scrub, rough ground and on railway banks; very scattered in Sc, N En and Surrey. Cultivars exist with greenish-yellow fruits, with dissected leaflets, or with variegated leaves; the last is natd and has red fruits. Perhaps better as a ssp. of *S. nigra*.

4. S. ebulus L. - *Dwarf Elder*. Rhizomatous perennial; stems herbaceous, erect, to 1.5m, leaves with (5)7-13 leaflets; stipules conspicuous, ovate to narrowly so; flowers white, sometimes pink-tinged; fruits black; flowering Jul-Aug; 2n=36. Possibly native; waysides, rough and waste ground; scattered over most of BI.

2. VIBURNUM L. - *Viburnums*

Deciduous or evergreen shrubs; leaves simple, sometimes lobed; flowers numerous in corymbose compound cymes, actinomorphic, sometimes some sterile; stamens 5; ovary 3-celled, but appearing 1-celled due to abortion of 2 cells, with 1 ovule; style 0; stigmas 3; fruit a drupe with 1 seed.

1 Leaves deciduous, lobed or serrate 2
1 Leaves evergreen, entire or obscurely denticulate 3
2 Leaves lobed; outer flowers sterile much larger than inner; fruits red, subglobose **1. V. opulus**
2 Leaves serrate; all flowers fertile, uniform in size; fruits red, then black, compressed **2. V. lantana**
3 Leaves smooth; first-year twigs glabrous or sparsely pubescent **3. V. tinus**
3 Leaves strongly wrinkled; first-year twigs tomentose **4. V. rhytidophyllum**

Other spp. - Several other spp., evergreen and deciduous, winter- and summer-flowering, are mass-planted on roadsides etc. and may produce occasional seedlings.

1. V. opulus L. - *Guelder-rose*. Deciduous shrub to 4m; leaves lobed, the lobes irregularly dentate; corymbs with large outer sterile flowers surrounding small fertile ones; corolla white; fruits globose, bright red; 2n=18. Native; woods, scrub and hedges; frequent throughout Br and Ir except N Sc.

2. V. lantana L. - *Wayfaring-tree.* Deciduous shrub to 6m; leaves regularly serrate; flowers all fertile, cream; fruits compressed, becoming red then black; 2n=18. Native; woods, scrub and hedges, especially on base-rich soils; common in Br SE of line from Glam to S Lincs, scattered elsewhere in BI probably always alien, much used as stock for cultivated spp. and often persisting when latter die or when suckering away from them.

2 x 4. V. lantana x V. rhytidophyllum = V. x rhytidophylloides J.V. Suringar is grown in gardens and plants have been found in the wild in W Kent and Surrey; it is intermediate in leaf-shape, -wrinkledness and -pubescence; garden origin; (2n=18).

3. V. tinus L. - *Laurustinus.* Evergreen shrub to 6m; leaves entire; flowers all fertile, white to pink; fruits subglobose, blue-black; (2n=36). Intrd; much grown in shrubberies and natd on cliffs, banks and rough ground; widespread in S En and S & N Wa, Man; S Europe.

4. V. rhytidophyllum Hemsl. - *Wrinkled Viburnum.* Evergreen shrub to 6m; leaves entire to nearly so; flowers all fertile, yellowish-white; fruits ± globose, becoming red then black; (2n=18). Intrd; much grown in shrubberies and natd or a relic in old woodland or parkland; very scattered in En N to Leics; China.

3. SYMPHORICARPOS Duhamel - *Snowberries*

Deciduous shrubs; leaves simple, entire, sometimes deeply lobed; flowers in dense terminal spikes, actinomorphic; stamens (4-)5; ovary 4-celled, with 2 fertile cells each with 1 ovule and 2 sterile cells; style present; stigma 1, capitate; fruit a drupe with 2 seeds.

1 Suckering from rhizomes; fruits pure white; style glabrous **1. S. albus**
1 Rooting from stoloniferous stem-tips; fruits pink or pink-flushed; style pubescent 2
2 Fruits ± uniformly pink, 4-6mm **3. S. orbiculatus**
2 Fruits white, flushed pink on exposed side, 6-10mm **2. S. x chenaultii**

Other spp. - Modern cultivars are mostly the so-called 'Doorenbos Hybrids', most of which are *S. albus* x *S.* x *chenaultii* and are variously intermediate or close to one or other parent; they usually have fruits coloured as in *S. x chenaultii* but show their *S. albus* parentage in leaf-shape and sparse pubescence. They might already escape and surely will in future. Records of *S. microphyllus* (see below) need checking.

1. S. albus (L.) S.F. Blake (*S. rivularis* Suksd., *S. racemosus* Michx.) - *Snowberry.* Strongly suckering from rhizomes, ± erect then arching shrub to 2m; leaves glabrous to sparsely pubescent, rounded to very broadly obtuse or apiculate at apex, those on strong sterile shoots often deeply lobed; corolla 5-8mm, pink, pubescent inside; fruit white, 8-15mm; (2n=36, 54, 72). Intrd; natd in woods, scrub, rough ground; frequent throughout BI; our plant is var. **laevigatus** (Fernald) S.F. Blake, from W N America.

2. S. x chenaultii Rehder (*S. microphyllus* Kunth x *S. orbiculatus* Moench) - *Hybrid Coralberry.* Arching shrub to 1.5m, with procumbent or arching then procumbent non-flowering stems (stolons) rooting at tips; leaves pubescent on lowerside, acute to obtuse at apex, not lobed; corolla 3-5mm, pinkish-white, sparsely pubescent inside; fruit white, flushed pink on 1 side, 6-10mm; (2n=18). Intrd; natd as for *S. albus* but much rarer; scattered in Br N to C Sc; garden origin.

3. S. orbiculatus Moench - *Coralberry.* Often differing little from *S. x chenaultii* except in key character; leaves more obtuse; corolla slightly shorter and less pubescent inside; (2n=18). Intrd; natd as for S. x *chenaultii* but even rarer and not commonly grown; open scrub in W Kent since at least 1984; E N America.

4. LINNAEA L. - *Twinflower*
Procumbent, evergreen, dwarf shrubs; leaves simple, crenate; flowers 2, each with 1 bract at base and 2 bracteoles at apex of pedicel, actinomorphic; stamens 4; ovary with 1 fertile cell with 1 ovule and 2 sterile cells; style present; stigma 1, bilobed; fruit an achene.

1. L. borealis L. - *Twinflower*. Stems procumbent, to 40cm; leaves broadly ovate to suborbicular, 4-16mm; flowers in pairs borne on erect leafless stems to 8cm; corolla 5-10mm, pink; fruit, calyx, bracteoles and pedicels glandular-pubescent; 2n=32. Native; on barish ground under shade of rocks or trees, mostly in woods, especially of *Pinus*; very local in E Sc N to Caithness, formerly S to NE Yorks, decreasing. Cultivated plants are often ssp. **americana** (Forbes) Hultén (var. *americana* (Forbes) Rehder), with corolla 10-16mm with distinct narrow tube at base, and might escape. R

5. LEYCESTERIA Wall. - *Himalayan Honeysuckle*
Deciduous shrub, often semi-herbaceous; leaves simple, entire to serrate; flowers in crowded terminal spikes with large purple or purple-green bracts, ± actinomorphic; stamens 5; ovary 5-celled, each cell with several ovules; style present; stigma 1, capitate; fruit a several-seeded berry.

1. L. formosa Wall. - *Himalayan Honeysuckle*. Stems erect, to 2m; leaves 5-18cm, ovate, acuminate at apex; bracts 12-35mm; corolla 10-20mm, pinkish-purple; fruit purple, subglobose, c.1cm; (2n=18). Intrd; natd in woods, shrubberies and rough ground, often appearing unexpectedly from bird-sown seed; scattered ± throughout BI; Himalayas.

6. WEIGELA Thunb. - *Weigelia*
Deciduous shrubs; leaves simple, serrate; flowers in small axillary cymes; corolla weakly zygomorphic, with 5 lobes scarcely arranged in 2 lips; stamens 5; ovary 2-celled, with many ovules per cell; style present; stigma 1, capitate or 2-lobed; fruit a capsule.

1. W. florida (Bunge) A. DC. - *Weigelia*. Erect arching shrub to 2(3)m; leaves 5-10cm, ovate to elliptic-oblong, acuminate at apex, sparsely pubescent; corolla 25-35mm, pink; (2n=36). Intrd; much grown in gardens, rarely persistent or self-sown in rough and marginal ground; very scattered in En and N to C Sc, Man; China. Most garden plants are hybrids involving this and c.4 other spp.; they often have red or more numerous flowers or less or more densely pubescent leaves, but the identity of wild plants has not been worked out.

7. LONICERA L. - *Honeysuckles*
Deciduous or evergreen shrubs or climbers; leaves simple, sometimes lobed, entire; flowers sessile, in pedunculate axillary pairs or in terminal heads; corolla zygomorphic with 4-lobed upper and 1-lobed lower lip, or ± actinomorphic with 5 lobes; stamens 5; ovary 2-3-celled with several ovules per cell; style long; stigma capitate or slightly lobed; fruit a several-seeded berry.

1 Flowers and fruit sessile in terminal and subterminal whorls; climbers 2
1 Flowers and fruit in pairs, sessile at apex of common axillary stalk, sometimes crowded near branch ends 4
  2 All leaves separate, not fused in pairs; berry red **8. L. periclymenum**
  2 At least most apical pair of leaves on each branch fused around

stem at base; berry orange 3
3 Bracteoles at base of each flower 0 or minute **9. L. caprifolium**
3 Bractoles c.1mm, obscuring base of ovary **10. L. x italica**
4 Stems twining 5
4 Stems not twining 6
5 Flowering nodes often clustered into terminal spikes; corolla glabrous on outside; 2 bracts at base of each flower-pair subulate **6. L. henryi**
5 Flowering nodes not clustered near branch apex; corolla pubescent on outside; 2 bracts at base of each flower-pair leaf-like **7. L. japonica**
6 Two bracts at base of each flower-pair (and bracteoles within) ovate, obscuring base of flower, purple and enlarging in fruit **3. L. involucrata**
6 Two bracts at base of each flower-pair subulate to linear-lanceolate, not obscuring ovaries, scarcely enlarging in fruit 7
7 Corolla distinctly 2-lipped; leaves deciduous, (20)30-80mm; berry red 8
7 Corolla ± actinomorphic; leaves evergreen, (4)6-32mm; berry violet 9
8 Corolla 8-15mm, pale yellow to cream, sometimes tinged pink; young stems and leaves pubescent **4. L. xylosteum**
8 Corolla 15-25mm, pink to red, sometimes white; young stems and leaves glabrous **5. L. tatarica**
9 Often >1m; leaves (4)6-16mm, mostly ovate, rounded to subcordate at base **2. L. nitida**
9 Rarely >1m; leaves (6)12-32mm, mostly oblong-elliptic to narrowly so, cuneate at base **1. L. pileata**

Other spp. - **L. trichosantha** Bureau & Franch., from China, differs from *L. xylosteum* in its leaves mostly widest below middle and corollas bright yellow; a few bushes occur as relics of planting in Clapham Woods, MW Yorks.

1. L. pileata Oliv. - *Box-leaved Honeysuckle*. Evergreen shrub to 1m, with spreading branches; leaves (6)12-32mm, oblong-elliptic to narrowly so; flowers in pairs in leaf-axils, ± actinomorphic, 6-8mm, cream; berry violet; (2n=18). Intrd; much grown in shrubberies and road-borders, sometimes self-sown; very scattered in Br N to C Sc, Man; China. **655**

2. L. nitida E.H. Wilson - *Wilson's Honeysuckle*. Evergreen shrub to 1.8m, with erect to arching branches; leaves (4)6-16mm, ovate; flowers in pairs in leaf-axils, ± actinomorphic, 5-7mm, cream; berry violet; (2n=18). Intrd; much grown for hedging and then very rarely flowering, but not rarely self-sown in scrub, hedges, woodland, banks and rough ground; scattered ± throughout BI; China. **655**

3. L. involucrata (Richardson) Banks ex Spreng. (*L. ledebourii* Eschsch.) - *Californian Honeysuckle*. Deciduous shrub to 2m, with spreading or arching branches; leaves 4-12cm, elliptic-oblong; flowers in pairs in leaf-axils, actinomorphic, 10-15mm, pale yellow often tinged with red; berry shining black, subtended by usually purple bracts; (2n=18). Intrd; frequent in gardens, sometimes bird-sown in rough and marginal ground; scattered in En (mostly N), Man, SW Sc and Ir; W N America. **655**

4. L. xylosteum L. - *Fly Honeysuckle*. Deciduous shrub to 2m; leaves 3-7cm, obovate to elliptic or broadly so; flowers in pairs in leaf-axils, zygomorphic, 8-15mm, pale yellow or cream, sometimes tinged pink; berry red; (2n=18). Possibly native; woods and scrub on chalk in W Sussex, also widely natd (bird-sown) in hedges, woods and scrub throughout much of Br and Ir. **RR**

5. L. tatarica L. - *Tartarian Honeysuckle*. Deciduous shrub to 4m; leaves 3-8cm, ovate to elliptic; flowers in pairs in leaf-axils, zygomorphic, 15-25mm, pink to red, sometimes white; berry red; (2n=18). Intrd; garden plant bird-sown in hedges and rough ground; very scattered in En N to Cheshire; W & C Asia.

6. L. henryi Hemsl. - *Henry's Honeysuckle*. Evergreen climber to 5(10)m; leaves 4-13cm, lanceolate to oblong-lanceolate; flowers in axillary pairs but clustered into **655**

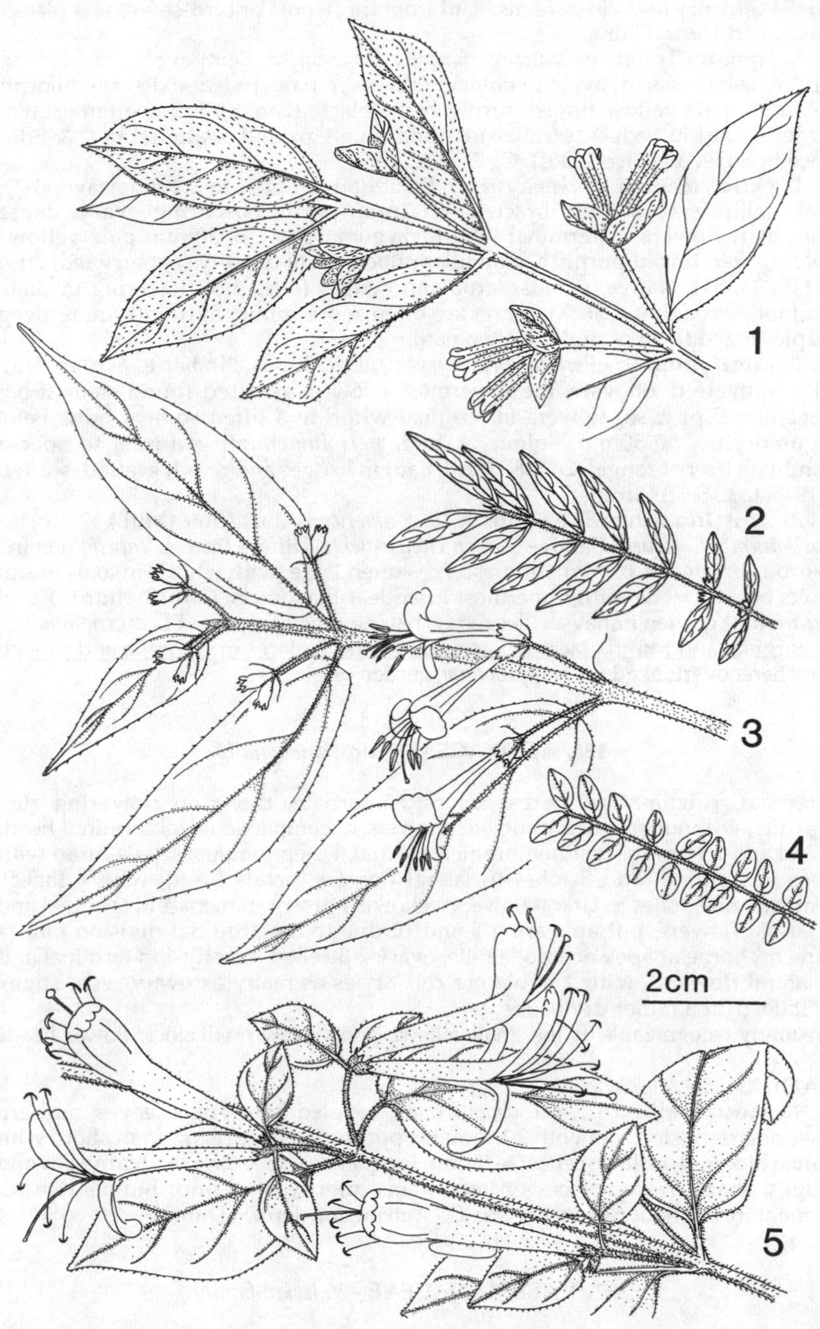

FIG 655 - *Lonicera*. 1, *L. involucrata*. 2, *L. pileata*. 3, *L. henryi*. 4, *L. nitida*. 5, *L. japonica*.

terminal groups, zygomorphic, 15-25mm, yellow tinged with red; berry black; (2n=54). Intrd; grown in gardens, natd from throwouts or bird-sown; few places in Surrey and Herts; China.

7. L. japonica Thunb. ex Murray - *Japanese Honeysuckle*. Semi-evergreen climber to 655
5(10)m; leaves 3-8cm, ovate to oblong; flowers in pairs in leaf-axils, zygomorphic, 30-50mm, pale yellow tinged purple; berry black; (2n=18). Intrd; often grown in gardens, natd in hedges, scrub, banks and rough ground; scattered in C & S Br (1 place in S Devon since 1930s), CI; E Asia.

8. L. periclymenum L. - *Honeysuckle*. Deciduous climber to 6(10)m; leaves 3-7cm, ovate, elliptic or oblong; bracteoles 1-2mm, partly obscuring ovary, densely glandular; flowers in terminal whorls, zygomorphic, 40-50mm, pale yellow to yellow, often tinged purplish, glandular pubescent to densely so; berry red; 2n=18, 54 (18, 36, 54). Native; woods, scrub and hedges (often not flowering in shade); common throughout BI. Cultivars are often more robust and with more deeply purple-tinged flowers, and are often natd.

9. L. caprifolium L. - *Perfoliate Honeysuckle*. Deciduous climber to 6(10)m; leaves 4-10cm, ovate to obovate, the uppermost 1-few pairs fused round stem at base; bracteoles 0 or ± so; flowers in terminal whorl and often some whorls below, zygomorphic, 40-50mm, colour as in *L. periclymenum*, glabrous to sparsely glandular; berry orange; (2n=18). Intrd; natd in hedges and rough ground; scattered in Br N to C Sc; S Europe.

10. L. x italica Schmidt ex Tausch (*L. x americana* auct. non (Mill.) K. Koch; *L. caprifolium* x *L. etrusca* Santi) - *Garden Honeysuckle*. Differs from *L. caprifolium* in its glabrous bracteoles c.1mm; inflorescences often large, with whorls in axils of small bracts beyond node with uppermost fused leaf-like bracts; (2n=18). Intrd; 1 of the commonest garden honeysuckles today, having mostly replaced *L. caprifolium*, natd in marginal and rough places in Surrey and several places in E Anglia and probably elsewhere, overlooked for *L. caprifolium*; garden origin.

136. ADOXACEAE - *Moschatel family*

Perennial, rhizomatous herbs; leaves 1-3-ternate, those on flowering stems opposite, without stipules, petiolate. Flowers in compact cubical terminal head, 4 lateral and 1 terminal, actinomorphic, bisexual, 1/2-epigynous; sepals fused with 2 lobes in terminal and 3 lobes in lateral flowers; petals fused, with 4 lobes in terminal and 5 lobes in lateral flowers, yellowish-green; stamens 4 in terminal and 5 in lateral flowers, but appearing 8 and 10 due to longitudinal division into 1/2-stamens, borne at apex of corolla tube; ovary 2-5-celled (mostly 4 in terminal and 5 in lateral flowers), with 1 ovule per cell; styles as many as ovary-cells; stigmas capitate; fruit a rather dry drupe.

Instantly recognizable by the small yellowish-green 'townhall clock' flower-head.

1. ADOXA L. - *Moschatel*

1. A. moschatellina L. - *Moschatel*. Long-petioled 2-3-ternate leaves, and erect flowering stems to 15cm with 2 1-ternate opposite leaves, arising from short, white, scaly rhizome; inflorescence 6-10mm long and wide; 2n=36. Native; woods, hedges, shady rocky places on mountains, mostly on damp, humus-rich soil; frequent throughout Br except N Sc, Co Antrim, intrd in Co Dublin.

137. VALERIANACEAE - *Valerian family*

Annual to perennial herbs; leaves simple or pinnate, opposite, without stipules, petiolate or sessile. Flowers numerous in terminal paniculate cymes, often ± corymbose, ± actinomorphic to zygomorphic, bisexual or dioecious, epigynous;

calyx represented by 0-many teeth, very small in flower, similar in fruit or developing long feathery appendages; petals 5, equal or slightly longer on abaxial side, fused into tube proximally, the tube straight or slightly pouched or with a long backward-directed spur at base; stamens 1 or 3, borne on corolla-tube; ovary 3-celled, 1 adaxial cell fertile with 1 ovule, 2 abaxial cells sterile and equally large to vestigial; style 1; stigma 1 and capitate or 3 and linear-oblong; fruit a 1-seeded nut.

Distinguished by the inferior ovary with 1 ovule but 2 other sterile cells (these often obscure), 0 or minute calyx at flowering, 5-lobed tubular corolla often pouched or spurred at base, and 1 or 3 stamens.

1 Stems forked into 2 at each node; calyx remaining minute at fruiting **1. VALERIANELLA**
1 Main stem simple or with lateral branches; calyx developing long feathery projections at fruiting 2
  2 Stamen 1 **3. CENTRANTHUS**
  2 Stamens 3 **2. VALERIANA**

1. VALERIANELLA Mill. - *Cornsalads*

Annuals with stems repeatedly forked; leaves simple, entire to serrate or sparsely lobed; flowers in rather lax to dense compound cymes, bisexual; calyx ± 0 or small, persistent but remaining small on top of fruit, usually unequal; corolla-tube not pouched or spurred; stamens 3; stigmas 3; sterile cells of ovary small to large.

Ripe fruits are essential for determination; the key does not require fruit sections to be cut, but their appearance in section is important and is mentioned in the diagnoses. Fruit lengths exclude the calyx.

1 Calyx in fruit absent or vestigial, <1/10 as long as rest of fruit 2
1 Calyx in fruit distinct, c.1/4 to nearly as long as rest of fruit 3
  2 Fruit c. as wide as thick, much longer than wide or thick, with a very deep groove on abaxial face **2. V. carinata**
  2 Fruit c.2x as thick as wide, scarcely longer than thick, shallowly grooved on abaxial face **1. V. locusta**
3 Calyx in fruit with short tube, with usually 6 teeth, >2/3 as long as rest of fruit, nearly as wide as fruit **5. V. eriocarpa**
3 Calyx in fruit with very short or 0 tube, with <6 (often 1) teeth, <1/2(2/3) as long as rest of fruit, <1/2 as wide as fruit 4
  4 Main tooth of calyx in fruit scarcely or not toothed; fruit ± smooth on all faces, with 2-6 fine grooves and/or longitudinal ridges, with easily broken walls **3. V. rimosa**
  4 Main tooth of calyx in fruit usually with 2 or more distinct teeth; fruit with 2 distinct ribs on abaxial face delimiting ovate ± flat area, with hard walls **4. V. dentata**

1. V. locusta (L.) Laterr. - *Common Cornsalad.* Stems erect, to 15(40)cm; **658**
inflorescences compact; calyx 0 or vestigial; fruits glabrous, 1.8-2.5mm, 1.8-2.5mm thick, 1-1.5mm wide; fertile cell with outer (adaxial) wall spongy, ± as thick as rest of cell; sterile cells each c. as large as fertile cell (excl. spongy layer) scarcely grooved between them; (2n=14, 16, 18). Native; arable and rough ground, bare places in grassland, on banks, walls, rocky outcrops and dunes; frequent throughout BI. Very dwarf, ± stemless plants on dunes in W Br are best recognized as var. **dunensis** D.E. Allen (ssp. *dunensis* (D.E. Allen) P.D. Sell).

2. V. carinata Loisel. - *Keeled-fruited Cornsalad.* Differs from *V. locusta* in fruits 2- **658**
2.7mm, 0.8-1.4mm thick and wide; fertile cell without spongy wall; sterile cells slightly smaller than fertile one, with deep groove between them; (2n=14, 16, 18). Native; similar places to *V. locusta* but (?) not dunes; scattered in Br N to Berwicks, CI and (intrd) S & E Ir.

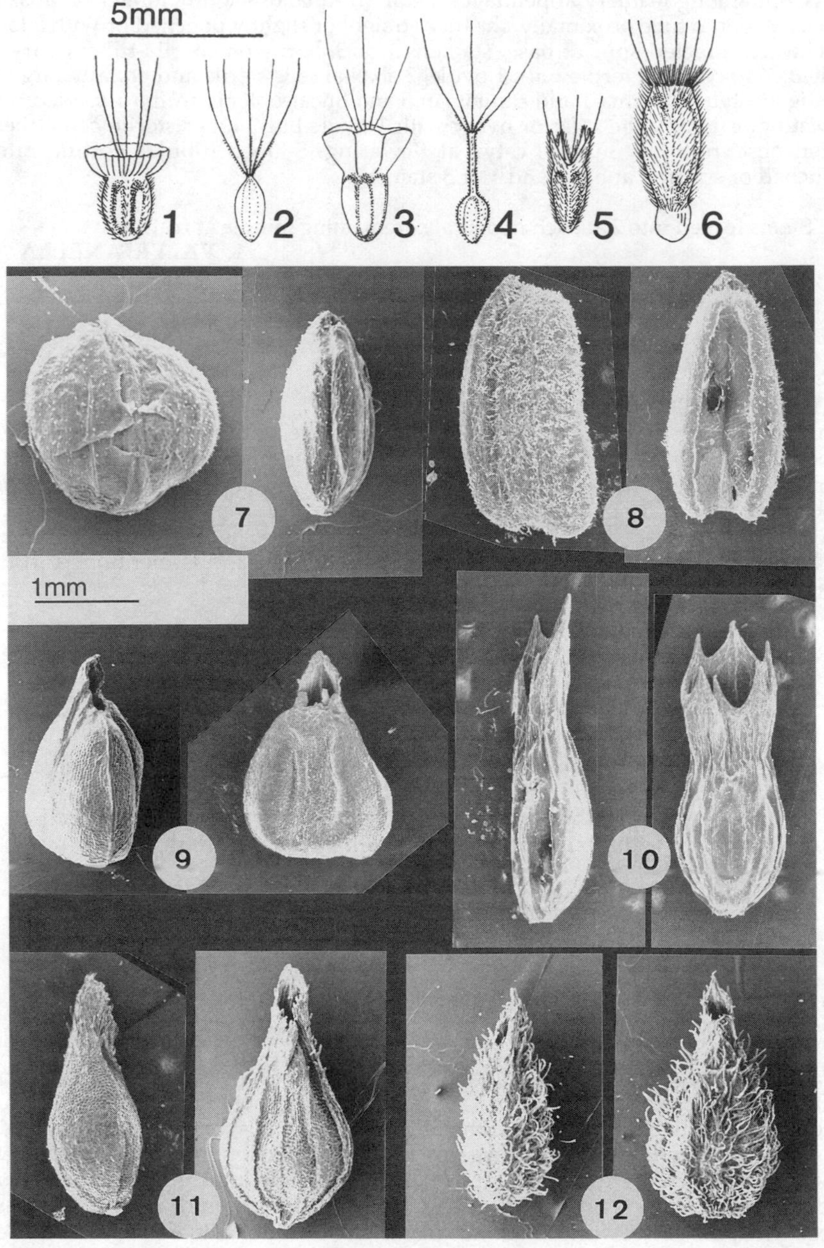

FIG 658 - Fruits of **Dipsacaceae**, **Valerianaceae**. 1-2, ***Scabiosa columbaria,*** with and without epicalyx. 3-4, ***S. atropurpurea,*** with and without epicalyx. 5, ***Succisa pratensis,*** with epicalyx. 6, ***Knautia arvensis,*** with epicalyx. 7-12, ***Valerianella,*** lateral and abaxial views. 7, ***V. locusta***. 8, ***V. carinata***. 9, ***V. rimosa***. 10, ***V. eriocarpa***. 11, ***V. dentata*** (glabrous). 2, ***V. dentata*** (pubescent).

3. V. rimosa Bastard - *Broad-fruited Cornsalad*. Stems erect, to 15(40)cm; inflorescences lax; calyx in fruit composed largely of 1 nearly entire tooth 0.5-1mm; fruits glabrous (?always), 1.5-2.5mm, 1-1.5mm thick, 1.5-2.5mm wide; fertile cell without spongy wall; sterile cells slightly larger than fertile cell, scarcely grooved between them; (2n=14, 16). Native; cornfields and rough ground; very local in S En and (?still) Ir, formerly widespread in Ir and Br N to C Sc. **RR** **658**

4. V. dentata (L.) Pollich - *Narrow-fruited Cornsalad*. Stems erect, to 15(40)cm; inflorescences rather lax to dense; calyx in fruit like that of *V. rimosa* but main tooth often toothed; fruits glabrous or pubescent, 1.5-2mm, 0.6-0.8mm thick, 0.7-1.2mm wide; fertile cell without spongy wall; sterile cells reduced to 2 abaxial ridges; (2n=14, 16). Native; cornfields and rough ground; scattered in Br and Ir N to Cheviot, formerly to C Sc and in Guernsey, decreasing. **658**

5. V. eriocarpa Desv. - *Hairy-fruited Cornsalad*. Stems erect, to 15(40)cm; inflorescences compact; calyx 0.7-1.3mm, with a very short tube and usually 6 unequal teeth; fruit similar to that of *V. dentata*; (2n=14, 16). Intrd; banks, walls and rough ground; very scattered in extreme S En and CI, probably decreasing; S Europe. **RR** **658**

2. VALERIANA L. - *Valerians*

Perennials with main and lateral stems; leaves pinnate and/or simple on each plant; flowers in rather dense compound cymes, bisexual or dioecious; calyx developing long feathery projections at fruiting; corolla-tube not or slightly pouched at base; stamens 3; stigmas 3; sterile cells of ovary scarcely discernible.

1 Stem-leaves and basal leaves pinnate, with several lateral leaflets as large as terminal one **1. V. officinalis**
1 At least basal leaves simple 2
 2 Basal leaves entire, cuneate to rounded at base; stem-leaves with several lateral leaflets as large as terminal one **3. V. dioica**
 2 Basal leaves dentate, cordate at base; stem-leaves simple or with 1 large terminal and 1-2 much smaller pairs of lateral leaflets **2. V. pyrenaica**

1. V. officinalis L. (*V. sambucifolia* J.C. Mikan ex Pohl) - *Common Valerian*. Stems erect, to 2m, sometimes with short stolons; all leaves (except very first-formed) pinnate, with linear-lanceolate to narrowly ovate, entire to serrate leaflets; flowers pink; 2n=28, 56, 70. Native; dry or damp grassy places and rough ground; frequent ± throughout Br and Ir. Very variable; short tetraploid (2n=28) plants with very narrow, ± entire leaflets (ssp. **collina** Nyman) occur on dry calcareous soils in S & C Br, while taller octoploid or rarely decaploid plants with broader leaflets (ssp. **sambucifolia** (J.C. Mikan ex Pohl) W.R. Hayw.) occur in damper places throughout Br and Ir, but every intermediate is found. The diploid (2n=14) ssp. **officinalis** has not been confirmed from BI.

2. V. pyrenaica L. - *Pyrenean Valerian*. Stems erect, to 1.2m; stolons 0; basal leaves simple, ovate to suborbicular, cordate, dentate; stem-leaves similar but often with 1-2 pairs of small lateral leaflets also; flowers pink; (2n=16). Intrd; damp woods and shady hedgebanks; natd in N & W Br from E Cornwall to N Aberdeen, frequent in S & C Sc, since 1782 in Fife, rare in W & NE Ir; Pyrenees.

3. V. dioica L. - *Marsh Valerian*. Stems erect, to 40cm; stolons well developed; basal leaves and those on stolons simple, entire; stem-leaves pinnate, with entire to slightly dentate leaflets; flowers dioecious, white to pink; 2n=16. Native; marshes, fens and bogs, rarer in S and there often in shade; frequent in Br N to S Sc.

3. CENTRANTHUS Neck. ex Lam. & DC. - *Red Valerian*

Annuals or perennials with main and usually lateral stems; leaves simple and entire to deeply pinnately lobed or ± pinnate; flowers in dense compound cymes, bisexual; calyx developing long feathery projections at fruiting; corolla-tube with backward-directed spur; stamen 1; stigma 1; sterile cells of ovary scarcely discernible.

1. C. ruber (L.) DC. - *Red Valerian*. Erect perennial to 80cm; leaves simple, somewhat glaucous, ovate to narrowly elliptic or lanceolate, entire or remotely dentate, those on stems mostly ± sessile; flowers red, pink or white; corolla-spur (2)5-12mm; (2n=32). Intrd; natd on walls, dry rocky or sandy places, cliffs and banks; common in CI and S & C Br, extending to N & E Ir and C Sc; Mediterranean region.

2. C. calcitrapae (L.) Dufr. - *Annual Valerian.* Erect annual to 30cm; leaves deeply pinnately lobed to ± pinnate; flowers pink; corolla-spur <2mm; (2n=32). Intrd; natd since 1982 on open ground in churchyard, Surrey; S Europe.

138. DIPSACACEAE - *Teasel family*

Biennial to perennial herbs; leaves simple or pinnate, opposite, without stipules, petiolate or sessile. Flowers numerous, borne on common receptacle in dense terminal heads (*capitula*), weakly zygomorphic, bisexual or gynodioecious, epigynous; calyx small, cup-shaped or divided into 4-8 teeth or bristles; petals 4-5, fused into tube proximally, the lobes usually larger on abaxial side, especially in outer flowers; stamens 4, free, ± exserted, borne on corolla-tube; ovary 1-celled with 1 ovule; style 1; stigma 2-lobed, or simple and capitate or oblique; fruit an achene.

The calyx, which arises from the top of the ovary outside the corolla, is often less conspicuous than the *epicalyx*, which arises at the base of the ovary but encloses the latter in a tubular structure and often expands into lobes around the calyx. The ripe fruit remains enclosed in the epicalyx. The receptacle bears bracts at the base of the capitulum and usually a bract associated with each flower within the capitulum.

The flowers with 4 free stamens and borne in a capitulum, and the ovary and fruit enclosed in a tubular epicalyx, are unique.

Before using the key a capitulum should be dissected carefully to distinguish between calyx and epicalyx.

1 Stems, and usually midribs on lowerside of leaves, prickly **1. DIPSACUS**
1 Stems and leaves glabrous to pubescent, not prickly 2
2 Corolla 5-lobed; epicalyx expanded at apex into membranous, veined funnel **5. SCABIOSA**
2 Corolla 4-lobed; epicalyx variously expanded at apex but not membranous 3
3 Corolla cream or blue; calyx without teeth or bristles **2. CEPHALARIA**
3 Corolla blue to purple or violet, rarely white or pinkish; calyx with 4-8 teeth or bristles 4
4 Flowers ± all equal-sized; calyx with 4-5 bristles; receptacle bearing bracts, 1 subtending each flower as well as some at base **4. SUCCISA**
4 Outer flowers much longer than inner ones; calyx with 8 bristles; receptacle bearing bracts at base but not subtending each flower **3. KNAUTIA**

1. DIPSACUS L. - *Teasels*

Biennials; stems prickly; leaves simple or pinnate; receptacle with spine-tipped bracts subtending each flower and very long spiny ones at base; flowers all ± 1 size; epicalyx 4-angled, scarcely toothed at apex; calyx cup-shaped, scarcely toothed; corolla 4-lobed.

1 Capitula ovoid-cylindrical; upper stem-leaves sessile, fused in pairs round stem at base 2
1 Capitula globose; upper stem-leaves petiolate, not fused in opposite pairs 4
2 Leaves pinnately dissected at least 1/2 way to midrib **3. D. laciniatus**
2 Leaves entire to dentate or serrate 3
3 Bracts on receptacle with stiff but flexible, straight apical spine **1. D. fullonum**
3 Bracts on receptacle with stiff, rigid, recurved apical spine **2. D. sativus**
4 Capitula 15-28mm across (incl. bracts 7-13mm) in full flower or fruit **4. D. pilosus**
4 Capitula 30-40mm across (incl. bracts 14-20mm) in full flower or fruit **5. D. strigosus**

1. D. fullonum L. (*D. sylvestris* Huds., *D. fullonum* ssp. *sylvestris* (Huds.) P. Fourn.) - *Wild Teasel.* Stems erect, to 2(3)m; leaves simple, entire to dentate or serrate, prickly on midrib on lowerside; capitula ovoid-cylindrical, 4-8cm, with basal bracts linear, curving upwards, often as long as capitulum; bracts on receptacle longer than flowers, with stiff but flexible, straight apical spine; corolla pinkish-purple to lilac; 2n=18. Native; marginal habitats and rough ground by roads, railways, streams, woods and fields; frequent in CI and Br N to C Sc, local in Ir.

1 x 2. D. fullonum x D. sativus occurs rarely on waste ground where the 2 parents occurred earlier; Beds pre-1953, N Somerset 1973, Co Dublin 1995.

1 x 3. D. fullonum x D. laciniatus = D. x pseudosilvester Schur (*D. x fallax* Simonk.) occurs with both parents in Oxon.

2. D. sativus (L.) Honck. (*D. fullonum* ssp. *sativus* (L.) Thell.) - *Fuller's Teasel.* Differs from *D. fullonum* in capitula with basal bracts ± patent; bracts on receptacle c. as long as flowers, with stiff, rigid, recurved, apical spine; (2n=18). Intrd; still grown for fulling in Somerset, formerly more widely so, now a frequent casual from birdseed on tips and waste ground; scattered in CI and Br, Co Dublin; uncertain origin.

3. D. laciniatus L. - *Cut-leaved Teasel.* Differs from *D. fullonum* in stems to 3(4)m, with less stout prickles; leaves lobed ≥1/2 way to midrib; basal bracts broader (linear-lanceolate); corolla pale pink; (2n=16, 18). Intrd; natd from birdseed and/or wild-flower mixtures in rough and marginal ground; Middlesex and Oxon, formerly Surrey; Europe.

4. D. pilosus L. - *Small Teasel.* Stems erect, to 1.5m; lower leaves simple, upper ones often pinnate with 1 pair of small lateral leaflets, serrate, not or slightly prickly on midrib on lowerside; capitula globose, with basal bracts c. as long as those on receptacle; corolla whitish; (2n=18). Native; damp places in open woods, hedgerows and by streams; rather scattered in En and Wa.

5. D. strigosus Willd. - *Yellow-flowered Teasel.* Differs from *D. pilosus* in larger capitula (see key); corolla pale yellow; (2n=18). Intrd; natd in rough ground and waste places; Cambs, rare casual elsewhere in En; Russia.

2. CEPHALARIA Schrad. - *Giant Scabious*

Bisexual perennials; stems sparsely pubescent; leaves simple and deeply pinnately lobed to pinnate; receptacle with rather leathery bracts subtending each flower and similar ones at base; outer flowers longer than inner; epicalyx 8-ridged, with 8 apical teeth; calyx cup-shaped, scarcely toothed; corolla 4-lobed.

Other spp. - C. syriaca (L.) Roem. & Schult., from SW Asia, differs from *C. gigantea* in being shorter (to 1m) and having a blue corolla and entire to weakly lobed leaves; it is a rare birdseed-alien or garden escape.

1. C. gigantea (Ledeb.) Bobrov - *Giant Scabious.* Stems erect, to 2m, with long branches; leaves all deeply pinnately lobed, the uppermost usually pinnate; capitula 4-10cm across; corolla pale yellow; (2n=36). Intrd; grown in gardens, natd or persistent in rough grassy places and waste ground; scattered in C & S En, rarely further N; Caucasus.

3. KNAUTIA L. - *Field Scabious*
Gynodioecious perennials; stems pubescent; leaves simple and crenate to pinnate; receptacle with herbaceous bracts at base but none subtending each flower; outer flowers longer than inner; epicalyx 4-ridged, scarcely toothed but with dense hairs at apex; calyx with 8 long bristles; corolla 4-lobed.

1. K. arvensis (L.) Coult. - *Field Scabious*. Stems erect to ascending, to 1m; lowest **658**
leaves simple and crenate, the upper ones pinnate, with all intermediates between; bisexual capitula 2.5-4cm and female ones 1.5-3cm across; corolla bluish-lilac; (2n=16, 20, 40). Native; dry grassy places on light soils; frequent over most of BI but very local in N Ir and N & W Sc.

4. SUCCISA Haller - *Devil's-bit Scabious*
Gynodioecious perennials; stems sparsely pubescent; leaves all simple, entire or distantly toothed; receptacle with herbaceous bracts at base and smaller ones subtending each flower; flowers all ± same size; epicalyx 4-angled, with 4 teeth; calyx with 4-5 bristle-tipped teeth; corolla 4-lobed.

1. S. pratensis Moench - *Devil's-bit Scabious*. Stems erect to ascending, to 1m; **658**
bisexual capitula 2-3cm and female ones 1.5-2.5cm across; corolla bluish-violet, sometimes pinkish-lilac; 2n=20. Native; many sorts of grassy places, wet or dry, acid or calcareous, in open or shade; common over most of BI.

5. SCABIOSA L. (*Sixalix* Raf.) - *Scabiouses*
Bisexual perennials; stems sparsely pubescent; leaves simple and serrate or lobed, to pinnate; receptacle with herbaceous bracts at base and narrower ones subtending each flower; outer flowers longer than inner; epicalyx 8-ridged, expanded at apex into membranous funnel; calyx with 5 long bristles at top of stalk (*hypanthium*); corolla 5-lobed.

1. S. columbaria L. - *Small Scabious*. Stems erect, to 70cm; capitula 1.5-3.5cm **658**
across, corolla bluish-lilac; fruiting capitula up to 1.5(2)cm long; epicalyx-tube 2-3mm, with 16-24-veined membranous funnel 0.8-1.5mm; hypanthium shorter than fruit and falling short of apex of epicalyx funnel; 2n=16. Native; dry calcareous grassland and rocky places; locally common in Br N to S Sc. Often very dwarf on exposed maritime cliff-tops.

2. S. atropurpurea L. (*Sixalix atropurpurea* (L.) Greuter & Burdet) - *Sweet Scabious*. **658**
Differs from *S. columbaria* in corolla dark purple to pale lilac; fruiting capitula up to 2.5(3)cm long; epicalyx-tube 1.5-2.5mm, with 8-ribbed funnel c. as long (shorter or longer) and inrolled at apex; hypanthium longer than fruit and reaching at least to apex of epicalyx funnel; (2n=16). Intrd; grown in gardens, natd on rough ground by sea in W Cornwall and E Kent (at Folkestone since 1862), rare casual elsewhere; S Europe.

139. ASTERACEAE - *Daisy family*
(Compositae)

Annual to perennial herbs, often woody near base, rarely shrubs; leaves simple and entire to pinnately or palmately lobed or variously pinnately compound, alternate or sometimes opposite, without stipules, petiolate or sessile. Flowers usually numerous, small, borne on common receptacle in dense terminal heads (*capitula*), zygomorphic or actinomorphic (often both in same capitulum), bisexual to variously monoecious or dioecious, epigynous; sepals 0 or represented by *pappus* of scales, teeth, bristles, hairs or a membranous ring, often enlarging in fruit; petals 5 (rarely 0 or 4), fused into tube with distal limb of either (a) 5 (rarely 4) actinomorphic (or nearly so) lobes or teeth (*tubular* flowers), or (b) unilateral strap-like *ligule* often with 3 or 5 apical teeth (*ligulate* flowers), the tube often extremely short in latter type; stamens 5, borne on corolla-tube, the anthers fused laterally into cylinder around style (not in *Xanthium*); ovary 1-celled with 1 ovule; style 1, usually branched, each branch with linear stigmatic surface, fruit an achene, often with persistent pappus.

The capitula bear around the outside of the flower-bearing area a series of often sepal-like bracts or *phyllaries*. They may also bear mixed in with the flowers (often 1 per flower) small *receptacular scales* or *bristles*. Each achene is often inserted on the receptacle into a minute *achene-pit*. Each capitulum may be *discoid*, with tubular flowers only, *ligulate*, with ligulate flowers (usually with 5-toothed ligules) only, or *radiate*, with a central region of tubular flowers (*disc flowers*) and an outer region of ligulate flowers (*ray flowers*) (usually with 3-toothed ligules). Sometimes each capitulum has very few flowers (very rarely only 1, e.g. *Olearia paniculata*). In *Echinops* the capitula have only 1 flower but are aggregated into large spherical compound heads. In *Ambrosia* and *Xanthium* the male and female capitula are separate on the same plant; the females contain only 1 or 2 flowers respectively. Otherwise, except in dioecious spp., the disc flowers are bisexual, but ray flowers may be bisexual, female or sterile.

The flowers borne in capitula, with 5 stamens with laterally fused anthers (except in *Xanthium*), and the fruit an achene, are a unique combination. The largest dicotyledon family.

Before using the keys a capitulum should be dissected to identify all its component parts. Rare variants without ray flowers, or with all the flowers ligulate (*flore pleno*), are not covered in the key, though more frequent variations in the presence or absence of ray flowers (e.g. in *Aster*, *Bidens*, *Senecio*) are.

General key

1 Shrubs, with new growth arising each year from older woody stems ***Key A***
1 Herbs; stems sometimes woody at base 2
  2 Flowers 1 per capitulum, the capitula aggregated into tight globose heads **1. ECHINOPS**
  2 Flowers >1 per capitulum at least in some capitula, the capitula not in tight globose heads 3
3 Male and female capitula separate on same plant, the male more apical and with several flowers, the female lower down and with 1-2 flowers 4
3 If male and female capitula separate then on different plants, both with several-many flowers 5
  4 Leaves alternate; fruiting heads with 2 prominent terminal processes, covered with stiff hooked bristles **84. XANTHIUM**
  4 Leaves mostly opposite; fruiting heads without terminal processes, with short straight spines or with tubercles **82. AMBROSIA**
5 Flowers all ligulate, the ligules usually 5-toothed at ends; milky latex usually present ***Key B***
5 Flowers all tubular or tubular and ligulate in same head, the ligules usually 3-toothed at ends; milky latex 0 6
  6 Style with ring of minute hairs, or with glabrous thickened ring, just below the branches; anthers with long 'tails' at base; leaves and/or phyllaries often spiny; flowers mostly pink or blue to purple, all tubular ***Key C***
  6 Style without ring of hairs or thickened zone below branches; anthers

mostly without basal 'tails'; plant rarely with spines; flowers various colours, often yellow, the marginal ones often ligulate 7
7 At least the lower leaves opposite *Key D*
7 All leaves basal or alternate 8
8 Capitula discoid, with ligules 0 or inconspicuous and not exceeding inner phyllaries *Key E*
8 Capitula radiate, with ≥1 obvious ligules exceeding inner phyllaries 9
9 Pappus of at least inner flowers of hairs (sometimes also of scales) *Key F*
9 Pappus 0, or of scales and/or few bristles 10
10 Receptacular scales or abundant bristles present *Key G*
10 Receptacular scales and bristles 0, but sometimes short fringes round achene-pits present *Key H*

Key A - Shrubs (all aliens)
1 Marginal flowers conspicuously ligulate 2
1 Flowers 1, or few to many and all tubular, the outer sometimes with very short ligule-like lobes not or scarcely exceeding phyllaries (but beware ligule-like inner phyllaries in *Helichrysum*) 5
2 Leaves pinnate or divided nearly to midrib, or if not then ligules purple **70. SENECIO**
2 Leaves entire to toothed <1/2 way to midrib; ligules white or yellow 3
3 Leaves dying in winter, very sticky; glandular hairs abundant **42. DITTRICHIA**
3 Leaves evergreen, not sticky; glandular hairs 0 4
4 Ligules white **53. OLEARIA**
4 Ligules yellow **74. BRACHYGLOTTIS**
5 Leaves divided >1/2 way to midrib, very fragrant when bruised 6
5 Leaves entire to shallowly lobed or toothed, usually not fragrant 7
6 Leaves all <5mm wide, with crowded incurved lobes; capitula >5mm wide, solitary on long stalks **59. SANTOLINA**
6 At least lower leaves >1cm wide, with spreading lobes; capitula <5mm wide, clustered **58. ARTEMISIA**
7 Stems procumbent, rooting, <60cm; inner phyllaries patent in flower, white, ligule-like **40. HELICHRYSUM**
7 Stems erect to scrambling, often >60cm; inner phyllaries not patent and ligule-like 8
8 Young stems and leaf lowersides densely white- to buff-tomentose 9
8 Stems and leaves glabrous or nearly so 11
9 At least larger leaves remotely sinuate-lobed **74. BRACHYGLOTTIS**
9 Leaves entire 10
10 Leaves green on upperside **53. OLEARIA**
10 Leaves white-tomentose on upperside **40. HELICHRYSUM**
11 Leaves linear, <5mm wide **49. CHRYSOCOMA**
11 Larger leaves lanceolate to orbicular, >1cm wide 12
12 Erect shrub; leaves rhombic to obovate, cuneate at base; flowers white, dioecious **54. BACCHARIS**
12 Scrambler; leaves ± ivy-shaped, cordate to hastate at base; flowers yellow, bisexual **72. DELAIREA**

Key B - Flowers all ligulate (tribe Lactuceae)
1 Pappus 0 or of scales 2
1 Pappus, at least in central flowers, of hairs 7
2 Leaves spiny; plant thistle-like **14. SCOLYMUS**
2 Leaves not spiny; plant not thistle-like 3
3 Pappus of scales 4
3 Pappus 0; achene often terminating in minute collar 6

4 Capitula <2cm across; ligules yellow **19. HEDYPNOIS**
4 Capitula >2cm across; ligules blue, rarely white 5
5 Receptacle without scales; capitula sessile or short-stalked, several forming terminal raceme; most or all leaves at flowering time on stems **15. CICHORIUM**
5 Receptacle with scales; capitula solitary on long stalks; most or all leaves at flowering time basal **16. CATANANCHE**
6 Leaves all basal; peduncles strongly dilated below capitula **17. ARNOSERIS**
6 Stems leafy; peduncles not or scarcely dilated below capitula **18. LAPSANA**
7 Pappus-hairs feathery, with slender lateral branches visible to naked eye, at least on some achenes 8
7 Pappus-hairs all simple, smooth or shortly toothed (lens) 12
8 Aerial stems with 0 or extremely reduced leaves 9
8 Aerial stems with well-developed leaves 10
9 Receptacular scales present among flowers **20. HYPOCHAERIS**
9 Receptacular scales 0 **21. LEONTODON**
10 Plant hispid, with harsh, hooked bristles **22. PICRIS**
10 Plant glabrous or with very soft hairs 11
11 Phyllaries c.8, all in 1 row and of 1 length **24. TRAGOPOGON**
11 Phyllaries >10, the outermost c.1/2 as long as innermost **23. SCORZONERA**
12 Leaves with strong spines; receptacular scales present, wrapped round achenes; pappus-hairs 2-4 **14. SCOLYMUS**
12 Leaves not or weakly spiny; receptacular scales usually 0; pappus-hairs numerous 13
13 Achenes distinctly flattened 14
13 Achenes not or scarcely flattened 17
14 Achenes with distinct narrow beak at apex, or at least markedly narrowed distally 15
14 Achenes without beak and scarcely narrowed distally 16
15 Pappus-hairs in 2 equal rows; phyllaries in several rows **27. LACTUCA**
15 Pappus-hairs in 2 unequal rows; phyllaries in 2 distinct unequal rows **29. MYCELIS**
16 Ligules yellow **26. SONCHUS**
16 Ligules blue to mauve **28. CICERBITA**
17 Capitulum 1 per stem; stems without leaves or scales; rhizomes and stolons 0 **30. TARAXACUM**
17 Capitula >1 per stem; *or* stems with leaves or scales; *or* rhizomes and/or stolons present 18
18 Pappus-hairs pure white 19
18 Pappus-hairs yellowish-white to pale brown 20
19 Capitulum 1 per stem; long thin rhizomes present **25. AETHEORHIZA**
19 Capitula normally >1 per stem; rhizomes 0 **31. CREPIS**
20 Stolons usually present; achenes ≤2(2.5)mm, each rib ending in a small point at apex of achene **32. PILOSELLA**
20 Stolons 0; achenes >(1.5)2.5mm, each rib ending in a smooth ring at apex of achene 21
21 Plant glabrous except for phyllaries; phyllaries in distinct inner and outer rows **31. CREPIS**
21 Plant with hairs (often dense) on some (often most) parts; phyllaries graduated between innermost and outermost **33. HIERACIUM**

Key C - Style with ring of minute hairs or with glabrous thickened ring just below branches (tribe Cardueae)

1 Phyllaries strongly hooked at tip, stiff, subulate **3. ARCTIUM**

1 Phyllaries not hooked at tip 2
 2 Corolla pale yellow to reddish-orange (beware inner phyllaries of *Carlina*) 3
 2 Corolla pink or blue to purple or mauve, rarely white 5
3 Pappus-hairs feathery, with slender lateral branches visible to naked eye or weak lens **6. CIRSIUM**
3 Pappus 0, or of narrow scales, or of simple to toothed hairs 4
 4 Outer phyllaries large and leaf-like **13. CARTHAMUS**
 4 Outer (and inner) phyllaries scale-like, with spiny apical portion **12. CENTAUREA**
5 Pappus-hairs feathery, with slender lateral branches visible to naked eye or weak lens 6
5 Pappus 0, of narrow scales, or of simple to toothed hairs 9
 6 Inner phyllaries with distinct apical portion (sometimes only a stout spine but always abruptly delimited from basal portion) **8. CYNARA**
 6 Phyllaries without distinct apical and basal portions, if with apical spine then gradually narrowed into it 7
7 Phyllaries all obtuse to rounded at apex; leaves ± entire to distantly toothed, the teeth not spinose or bristle-like **4. SAUSSUREA**
7 At least outer phyllaries spinose, mucronate or acuminate at apex; leaves spiny or at least with fine bristle-like teeth 8
 8 Outer phyllaries leaf-like; inner ones scarious, patent in dry weather, pale yellow on upperside, appearing like ligules **2. CARLINA**
 8 All phyllaries scale-like to ± subulate **6. CIRSIUM**
9 Leaves with sharp spines 10
9 Leaves without spines 12
 10 Receptacle glabrous, but achene-pits fringed with teeth **7. ONOPORDUM**
 10 Receptacle densely pubescent or bristly 11
11 Stem-leaves not decurrent down stem; stems not spiny; outer phyllaries with spine-tipped lateral lobes or teeth **9. SILYBUM**
11 Stem-leaves decurrent down stem in a spiny wing; phyllaries all entire, with terminal spine **5. CARDUUS**
 12 Phyllaries simple, entire **10. SERRATULA**
 12 At least inner phyllaries with distinct apical portion which is scarious, toothed, or spiny 13
13 All flowers bisexual and of same size; apical portion of phyllaries scarious, not separated from main part by constriction **11. ACROPTILON**
13 Marginal flowers functionally female though sometimes with sterile stamens, often longer than inner flowers; apical portion of phyllaries usually toothed or spiny, if merely scarious then separated from main part by constriction **12. CENTAUREA**

Key D - Plant herbaceous; at least lower leaves opposite
1 Capitula discoid, with only tubular flowers 2
1 Capitula radiate, with ≥1 marginal ligulate flower 6
 2 Terminal capitula male only, in elongated bractless racemes **82. AMBROSIA**
 2 All capitula bisexual, variously arranged 3
3 Pappus of hairs **98. EUPATORIUM**
3 Pappus 0 or of scales or stout bristles 4
 4 Leaves pinnate, or simple with narrowly cuneate base; pappus of barbed bristles **90. BIDENS**
 4 Leaves simple, broadly cuneate to cordate at base; pappus 0 or of scales 5
5 Flowers blue; receptacular scales 0; pappus of scales **99. AGERATUM**

5 Flowers greenish-white; receptacular scales present; pappus 0 **83. IVA**
6 Capitula with 1 ligulate flower **95. SCHKUHRIA**
6 Capitula with 3-numerous ligulate flowers 7
7 Plant with large underground tubers 8
7 Plant without underground tubers 9
8 Leaves (bi-)pinnate or (bi-)ternate; ligules often red to pink or purple, or white **93. DAHLIA**
8 Leaves simple; ligules yellow **88. HELIANTHUS**
9 Ligules pink or white 10
9 Ligules yellow to greenish- or brownish-yellow 12
10 Leaf-lobes linear to filiform; ligules usually pink, rarely white **92. COSMOS**
10 Leaf-lobes lanceolate to ovate; ligules white, rarely purplish 11
11 Capitula <7mm across excl. ligules; pappus of scales **89. GALINSOGA**
11 Capitula >7mm across excl. ligules; pappus of barbed strong bristles **90. BIDENS**
12 Pappus of barbed, strong, persistent bristles **90. BIDENS**
12 Pappus 0 or of weak deciduous bristles and/or of scales 13
13 Most leaves divided ± to base or to midrib 14
13 Leaves simple, subentire to toothed or shallowly lobed 15
14 Pappus of conspicuous scales; receptacular scales 0 **94. TAGETES**
14 Pappus 0, or minute, or of 2 small scales; receptacular scales present **91. COREOPSIS**
15 Ligules >1cm; outer phyllaries lanceolate to ovate, not or scarcely glandular 16
15 Ligules <1cm; outer phyllaries linear, with dense glandular hairs **86. SIGESBECKIA**
16 Annual or perennial; leaves petiolate, or sessile but not clasping stem; achenes flattened in radial plane **88. HELIANTHUS**
16 Annual; leaves sessile, clasping stem at base; achenes flattened in tangential plane **85. GUIZOTIA**

Key E - Plant herbaceous; leaves alternate or all basal; capitula discoid or with very inconspicuous ligules (excl. Cardueae)

1 Pappus 0, or of small scales or a few bristles 2
1 Pappus of hairs (somewhat expanded apically in male *Antennaria* and *Anaphalis*) 15
2 Leaves divided <1/2 way to midrib 3
2 Leaves divided >1/2 way to midrib 7
3 Receptacular scales present; plant densely white-woolly **60. OTANTHUS**
3 Receptacular scales 0; plant not densely woolly 4
4 Pappus of 1-8 stiff barbed bristles and sometimes also some small scales **46. CALOTIS**
4 Pappus 0 or a small membranous ring 5
5 Capitula very numerous, in racemes or panicles **58. ARTEMISIA**
5 Capitula 1-many, if many then in corymbs 6
6 Achenes ± compressed; stems procumbent to weakly erect, <30cm **69. COTULA**
6 Achenes scarcely compressed; stems strong, erect, usually >50cm **56. TANACETUM**
7 Capitula very numerous, in racemes or panicles; corolla brownish-, reddish- or greenish-yellow 8
7 Capitula 1-few, or if ± numerous then in corymbs and corolla bright yellow 9
8 Flowers all identical, with functional male and female parts **57. SERIPHIDIUM**

8 Outer flowers functionally female, with filiform corolla; inner ones bisexual and tubular **58. ARTEMISIA**
9 Corolla pubescent; stem and leaves densely pubescent **58. ARTEMISIA**
9 Corolla glabrous; stem and leaves glabrous to pubescent 10
10 Receptacular scales present 11
10 Receptacular scales 0 12
11 Corolla with small pouch at base, obscuring top of ovary in 1 plane **62. CHAMAEMELUM**
11 Corolla not pouched at base, not obscuring top of ovary **63. ANTHEMIS**
12 Achenes compressed; stems procumbent to weakly erect, <30cm **69. COTULA**
12 Achenes not or scarcely compressed; stems usually stiffly erect to ascending 13
13 Capitula flat, in corymbs; perennial **56. TANACETUM**
13 Capitula conical to convex, not in corymbs; annual 14
14 Achenes strongly 3-ribbed, with usually 2 resin-glands near apex **68. TRIPLEUROSPERMUM**
14 Achenes with 4-5 ribs, without resin-glands **67. MATRICARIA**
15 Leaves broadly ovate to orbicular, cordate at base, petiolate 16
15 Leaves linear to ovate, cuneate at base, petiolate or sessile 18
16 Plant with scrambling stems bearing ivy-shaped leaves **72. DELAIREA**
16 Plants with aerial stems bearing only reduced, bract-like leaves, the main leaves all basal 17
17 Capitula solitary **80. HOMOGYNE**
17 Capitula in inflorescences **79. PETASITES**
18 Plant densely glandular, very sticky **42. DITTRICHIA**
18 Plant glabrous to woolly, not obviously glandular, not sticky 19
19 Plant glabrous to pubescent, not woolly 20
19 Plant woolly at least in part, especially near tops of stems 25
20 Leaves conspicuously serrate to deeply lobed or ± pinnate; phyllaries in 1 main row with much shorter supplementary ones near base of capitulum **70. SENECIO**
20 Leaves entire to slightly or remotely serrate; phyllaries in several rows 21
21 Phyllaries totally scarious, coloured or white **40. HELICHRYSUM**
21 Phyllaries at least partly herbaceous and green 22
22 Stems glabrous, sometimes weakly scabrid 23
22 Stems pubescent 24
23 Pappus straw-coloured to pale reddish or brownish; at least lower leaves >2.5cm, either succulent or flat **48. ASTER**
23 Pappus pure white; all leaves <2.5cm, with margins rolled under **49. CHRYSOCOMA**
24 Flowers yellow; outer phyllaries with patent to recurved tips **41. INULA**
24 Flowers white to cream or pinkish; phyllaries appressed to flowers **51. CONYZA**
25 Annual with simple root system 26
25 Perennial with rhizomes or stolons (often short) 27
26 Marginal florets with receptacular scales; outer phyllaries herbaceous, woolly beyond 1/2 way to apex **36. FILAGO**
26 Receptacular scales 0; all phyllaries scarious, glabrous or woolly in lower 1/2 **39. GNAPHALIUM**
27 Capitula in elongate panicles; capitula bisexual **39. GNAPHALIUM**
27 Capitula in ± crowded terminal subcorymbose clusters; plants dioecious or subdioecious 28
28 Largest leaves forming basal rosettes, the stem-leaves much narrower; stems rarely >20cm; stoloniferous **37. ANTENNARIA**

28 Leaves not forming basal rosette, the largest ones being on the stems; stems rarely <25cm; rhizomatous **38. ANAPHALIS**

Key F - Plant herbaceous; leaves alternate or all basal; capitula radiate; pappus of hairs

1 Ligules white, or pink or blue to purple or mauve 2
1 Ligules yellow to orange 9
2 Main leaves all basal, broadly ovate to orbicular, cordate at base; flowering stems with only reduced bract-like leaves **79. PETASITES**
2 Leaves on flowering stems, or if ± all basal then not cordate at base 3
3 Outer phyllaries broad, green, ± leafy **52. CALLISTEPHUS**
3 Outer phyllaries similar to or smaller than inner ones 4
4 At least the larger leaves truncate to cordate at base 5
4 Leaves all cuneate at base 7
5 Phyllaries in a graded series of rows **48. ASTER**
5 Phyllaries in 1 main row, with smaller supplementary ones near base of capitulum 6
6 Leaves pinnately veined; phyllaries in 1 main row with smaller supplementary ones at base of capitulum; ligules white **70. SENECIO**
6 Leaves palmately veined; capitula without supplementary phyllaries; ligules white or coloured **71. PERICALLIS**
7 Ligules linear to narrowly elliptic, >1mm wide; phyllaries all green or more green in apical than basal 1/2 **48. ASTER**
7 Ligules filiform to linear, usually <0.6 but sometimes up to 1.5mm wide; phyllaries more green in basal than in apical 1/2 8
8 Ligules ≤1mm; central tubular flowers fewer than peripheral filiform ones **51. CONYZA**
8 Ligules >1mm; central tubular flowers more numerous than peripheral filiform ones (or the latter 0) **50. ERIGERON**
9 Main leaves all basal; flowering stems with 1 capitulum and only reduced bract-like leaves **78. TUSSILAGO**
9 Leaves on flowering stems, or if ± all basal then capitula >1 per stem 10
10 Phyllaries in 1 main row, often with smaller supplementary ones near base of capitulum 11
10 Phyllaries in 2 or more (often indistinct) rows 15
11 Capitula with only 2-4 disc and 2-4 ray flowers, numerous in cylindrical or pyramidal panicles 12
11 Capitula with >5 disc and ≥4 ray flowers, not in dense pyramidal panicles 13
12 Lower leaves deeply palmately lobed; petioles sheathing stem; ray flowers usually 2 **76. LIGULARIA**
12 Lower leaves deeply pinnately lobed; petioles not sheathing stem; ray flowers usually 3 **75. SINACALIA**
13 Stem-leaves and basal leaves with petioles with broad sheathing bases **76. LIGULARIA**
13 Stem-leaves and basal leaves without sheathing petioles 14
14 Phyllaries in 1 main row with supplementary smaller ones near base of capitulum **70. SENECIO**
14 Phyllaries in 1 row, with no supplementary small ones near base of capitulum **73. TEPHROSERIS**
15 Phyllaries in 2 distinct rows of ± equal length **77. DORONICUM**
15 Phyllaries in 2 or more indistinct rows, progressively longer towards the inside 16
16 Plant densely glandular, very sticky **42. DITTRICHIA**
16 Plant not or slightly glandular, not sticky 17
17 Pappus of inner row of hairs and outer row of small (often laterally

fused) scales **43. PULICARIA**
17 Pappus entirely of hairs 18
18 Capitula >(1.5)2cm across (incl. ligules); ligules >10mm; anthers with long filiform basal appendages **41. INULA**
18 Capitula <1.5(2)cm across (incl. ligules); ligules <10mm; anthers without basal appendages **47. SOLIDAGO**

Key G - Plant herbaceous; leaves alternate or all basal; capitula radiate; pappus 0, or of scales or bristles; receptacular scales present
1 Receptacle with abundant bristles only; pappus of 5-10 scales with long apical bristles **96. GAILLARDIA**
1 Receptacular scales present; pappus 0, or of small scales not bristle-tipped 2
2 Achenes strongly compressed, >2x as wide as thick, with lateral strong ribs or narrow wings 3
2 Achenes angular, or not or slightly compressed, <2x as wide as thick, without lateral ribs or wings 4
3 Capitula <2cm across (incl. ligules); ligules <1cm **61. ACHILLEA**
3 Capitula >2cm across (incl. ligules); ligules >1cm **91. COREOPSIS**
4 Ligules white 5
4 Ligules yellow to orange 6
5 Corolla of tubular flowers with small pouch at base, obscuring top of ovary in 1 plane **62. CHAMAEMELUM**
5 Corolla of tubular flowers not pouched, not obscuring top of ovary **63. ANTHEMIS**
6 Phyllaries with broad scarious margins and tips **63. ANTHEMIS**
6 Phyllaries entirely herbaceous 7
7 Ligules <2mm wide; receptacle slightly convex; lower leaves cordate at base **44. TELEKIA**
7 Ligules >2mm wide; receptacle conical; leaves all cuneate at base **87. RUDBECKIA**

Key H - Plant herbaceous; leaves alternate or all basal; capitula radiate; pappus 0, or of scales or bristles; receptacular scales 0
1 Capitula with 1 ligulate flower **95. SCHKUHRIA**
1 Capitula with outer row of ligulate flowers 2
2 Ligules yellow to orange or brownish-red, at least in part; pappus often of distinct scales or bristles 3
2 Ligules white to pink; pappus 0 or a minute rim 9
3 At least some achenes very strongly curved, very warty on outer face **81. CALENDULA**
3 Achenes not or slightly curved, if warty then not just on outer face 4
4 Pappus of 1-8 bristles, sometimes with minute scales as well 5
4 Pappus 0, or a minute rim, or of scales 6
5 Pappus bristles barbed, persistent; fruiting capitula <1cm across, not resinous **46. CALOTIS**
5 Pappus bristles smooth or forwardly serrated, deciduous; fruiting capitula >1cm across, very resinous **45. GRINDELIA**
6 Pappus 0 **64. CHRYSANTHEMUM**
6 Pappus of distinct scales 7
7 Leaves subglabrous to sparsely pubescent, decurrent on stems as wings; receptacle markedly convex **97. HELENIUM**
7 Leaves densely white-tomentose on lowerside, not decurrent; receptacle ± flat to slightly convex 8
8 Ligules pale yellow on upperside, purplish on lowerside; phyllaries free, glabrous to pubescent **34. ARCTOTHECA**

8 Ligules orange-yellow, with basal black blotch bearing central white spot, or rarely plain yellow; outer phyllaries fused into cup-like structure, white-tomentose **35. GAZANIA**
9 Rosette-plant; capitula solitary on leafless stems **55. BELLIS**
9 Flowering stems bearing leaves 10
10 Stem-leaves simple, shallowly to deeply toothed but not to midrib, the teeth simple 11
10 Stem-leaves pinnately lobed to midrib or nearly so, the lobes further lobed 13
11 Ligules <10mm **56. TANACETUM**
11 Ligules >10mm 12
12 Lowest (tubular) part of corolla of ray flowers with 2 narrow transparent wings **66. LEUCANTHEMUM**
12 Lowest (tubular) part of corolla of ray flowers not winged **65. LEUCANTHEMELLA**
13 Ultimate leaf-segments lanceolate to ovate, flat **56. TANACETUM**
13 Ultimate leaf-segments linear to filiform, not or scarcely flattened 14
14 Achenes strongly 3-ribbed, with usually 2 resin-glands near apex **68. TRIPLEUROSPERMUM**
14 Achenes with 4-5 ribs, without resin-glands **67. MATRICARIA**

Other genera - Many other genera occur as casuals or as non-persistent escapes or throwouts. The following are those most often claimed, mentioned under the key in which they would appear.

Key A (Calenduleae) - **OSTEOSPERMUM** L. (*Cape Daisies*) resembles a *Gazania* or *Arctotheca* but has stems woody below and ligules rarely yellow. It has recently become very popular in warm areas of Br and is sometimes found as a relic or throwout, especially near the sea. The most hardy spp. are **O. ecklonis** (DC.) Norl., with ligules white to pale lilac on upperside and purple or dark mauve on lowerside, and **O. jucundum** (E. Phillips) Norl., with ligules deep red on both sides, both from S Africa, but there now many garden hybrids with a wide range of colours.

Key B (Lactuceae) - **TOLPIS** Adans. is distinguished by its very long, very narrow, curved, semi-patent outer phyllaries; **RHAGADIOLUS** Scop. by its few, widely patent, long, narrow, pappus-less achenes. **T. barbata** (L.) Gaertn. and **R. stellatus** (L.) Gaertn. are rare birdseed-aliens from Europe.

Key C (Cardueae) - **CNICUS benedictus** L. (*Blessed Thistle*) is a subspinose annual with single yellow capitula partly concealed by leaves and large dimorphic phyllaries. **GALACTITES tomentosa** Moench resembles *Cirsium* (with feathery pappus-hairs) but has a densely pubescent receptacle and deeply dissected leaves densely white-woolly on lowerside. **MANTISALCA salmantica** (L.) Briq. & Cavill. is like a tall slender *Centaurea* with purple flowers but has a pappus of long scales surrounded by longer bristles and phyllaries with 1 apical spine. All 3 are rare birdseed-aliens from the Mediterranean, *Galactites* perhaps getting commoner.

Key D (Heliantheae) - **ZINNIA** L., **HELIOPSIS** Pers. and **SANVITALIA** Lam. are related ornamental American genera with ligules persistent on the fruiting capitula. **Z. elegans** Jacq. is an annual with red or orange ligules (often *flore pleno*) and sessile upper stem-leaves; **H. scabra** Dunal and **H. helianthoides** (L.) Sweet are *Helianthus*-like perennials; and **S. procumbens** Lam. is a ± procumbent annual with yellow ligules and purple disc flowers. **SPILANTHES** Jacq. resembles a small *Helianthus* but has a narrowly conical receptacle; **S. oleracea** L. is grown in warm countries as a salad plant (*Pará-cress*). **VERBESINA** L. resembles *Helianthus* but has strongly flattened and winged achenes and upper leaves with winged petioles and auricles. **V. encelioides** (Cav.) Benth. & Hook.f. ex A. Gray (*Golden Crownbeard*), from N America, is an annual to 1.3m with yellow solitary capitula 2.5-5cm across; it is a rather rare wool-alien.

Key E (Anthemideae) - **SOLIVA** Ruiz & Pav., from S America, would key out as *Cotula* but has winged achenes and sessile capitula; **S. anthemifolia** (Juss.) Sweet was formerly an occasional wool-alien. **ONCOSIPHON** Källersjö resembles a ligule-less *Matricaria* with tomentose phyllaries and large orange capitula; **O. grandiflorum** (Thunb.) Källersjö (*Pentzia grandiflora* (Thunb.) Hutch., *Matricaria grandiflora* (Thunb.) Fenzl ex Harv.), from Africa, is a scarce wool-alien.

Key G - **BUPHTHALMUM** L. (Inuleae) resembles *Telekia* but has dimorphic achenes and is a much smaller plant; **B. salicifolium** L., from Europe, is an impermanent garden escape with us but is increasing as such on the Continent. **ANACYCLUS** L. (Anthemideae), from Mediterranean, resembles *Anthemis* but the outer achenes are winged; 7 spp. have been recorded as casuals. **MADIA** Molina and **HEMIZONIA** DC. (*Spikeweed*) (Heliantheae), from N & C America, are coarse unattractive annuals with receptacular scales subtending only the outermost row of disc flowers. In *Madia* the achenes of ray flowers are completely enclosed by the inner phyllaries, while in *Hemizonia* they are only 1/2-enclosed; **M. capitata** Nutt., **M. glomerata** Hook. and **H. pungens** (Hook. & Arn.) Torr. & A. Gray are rare casuals.

Key H (Anthemideae) - **ISMELIA** Cass., from Africa, resembles *Chrysanthemum* but has flattened (and winged) disc-achenes and keeled phyllaries; **I. carinata** (Schousb.) Sch. Bip. (*Chrysanthemum carinatum* Schousb.) is similar to *C. coronarium*, with yellow ligules darker or lighter near base, but is a rare casual. **DENDRANTHEMA** (DC.) Des Moul., from the Orient, is the familiar *Florist's Chrysanthemum*, occasionally found on tips; several spp. and hybrids are involved.

SUBFAMILY 1 - LACTUCOIDEAE (Cichorioideae, Liguliflorae) (tribes 1-3; genera 1-35). Plants often producing white latex; stem-leaves usually spiral ('alternate'), sometimes 0; capitula ligulate, discoid or rarely (Arctotideae) radiate; tubular flowers (if present) usually with long narrow lobes, usually blue to red, rarely yellow; filaments joining anthers on back; 2 style branches usually each with 1 broad stigmatic surface on inner face; pollen grains ridged, spiny or both.

TRIBE 1 - CARDUEAE (Cynareae, Echinopeae, Carlineae) (genera 1-13). Plant not producing white latex, often spiny; capitula discoid, with the outermost florets often longer and with larger lobes and hence pseudo-radiate, usually red to blue (or white), rarely yellow.

1. ECHINOPS L. - *Globe-thistles*

Biennials to perennials; leaves deeply pinnately lobed, white- to grey-tomentose on lowerside, with teeth ending in rather weak spines; capitula each with 1 flower, many arranged in compact, globose heads ≥2.5cm; phyllaries in c.3 rows, herbaceous, long-pointed and with long fine teeth, with a zone of bristles outside them; corolla white to blue; pappus of partially fused scale-like bristles.

1 Phyllaries strongly recurved at tip; glandular hairs 0 **2. E. exaltatus**
1 Phyllaries erect or very slightly curved at tip; glandular hairs present at least on leaf upperside 2
2 Phyllaries without glandular hairs; corolla bluish; plant rarely >1.25m **3. E. bannaticus**
2 Phyllaries with abundant glandular hairs; corolla greyish; plant often >1.5m **1. E. sphaerocephalus**

Other spp. - Records of **E. ritro** L., from S Europe, are errors for *E. bannaticus*.

1. E. sphaerocephalus L. - *Glandular Globe-thistle*. Stems erect, to 2.5m, glandular-pubescent above; leaves glandular-pubescent on upperside; phyllaries with dense stalked glands; corolla white to greyish; (2n=30, 32). Intrd; grown in gardens, natd in waste places and rough ground, on road- and railway-banks; scattered throughout much of Br, probably over-recorded for other 2 spp.; S Europe. The hybrid *E. sphaerocephalus* x *E. exaltatus* = **E. x pallenzianus** Hügin & W. Lohmeyer is said to be commoner than either parent in Germany and might occur here.

2. E. exaltatus Schrad. (*E. commutatus* Jur.) - *Globe-thistle*. Stems erect, to 2.5m, with eglandular hairs; leaves with only eglandular hairs; phyllaries without stalked glands, strongly recurved at apex; corolla white to greyish; (2n=30). Intrd; grown and natd as for *E. sphaerocephalus*, but commoner; E & SE Europe.

3. E. bannaticus Rochel ex Schrad. - *Blue Globe-thistle*. Stems erect, to 1.25m, with eglandular hairs; leaves glandular-pubescent on upperside; phyllaries without stalked glands; corolla bluish; (2n=30). Intrd; grown and natd as for *E. sphaerocephalus*, the commonest sp. in Br; SE Europe.

2. CARLINA L. - *Carline Thistle*

Biennials; leaves pinnately lobed, very spiny; outer phyllaries ± leaf-like, innermost linear, entire, scarious, patent and ligule-like in dry weather, straw-yellow on upperside; corolla purple; pappus of feathery hairs often variously united proximally.

1. C. vulgaris L. - *Carline Thistle*. Stems erect, to 60cm, not spiny; leaves very spiny, ± glabrous to cottony-pubescent; capitula 1-6 in terminal corymb, c.1.5-3cm across when phyllaries patent; 2n=20. Native; open grassland, on usually calcareous but sometimes sandy soils, fixed dunes and cliff-tops; frequent in suitable places throughout most of BI, only coastal in N Ir and Sc and absent from Outer Isles.

3. ARCTIUM L. - *Burdocks*

Biennials; leaves simple, entire to remotely denticulate, the lower ovate-cordate and petiolate, not spiny; phyllaries very numerous, stiff, subulate, strongly hooked at apex, spreading to form subglobose head; corolla purple, rarely white; pappus of rough yellowish free hairs.

A difficult genus with varying taxonomic treatments; that of H. Duistermaat is followed here.

1 Petiole of basal leaves solid at least at base; capitula forming ± corymbose clusters on each main branch; capitula on distal part of each main branch with peduncles ≥2.5cm **1. A. lappa**

1 Petiole of basal leaves hollow at least near base; capitula on distal part of each main branch sessile or with peduncles usually ≤2cm, if >2cm then capitula forming clearly racemose clusters on each main branch 2

2 Capitula on distal part of each main branch sessile; middle phyllaries (1.6)1.7-2.5mm wide; phyllaries exceeding corolla by 1.2-6mm; corolla glabrous **3. A. nemorosum**

2 Capitula on distal part of each main branch sessile or pedunculate; middle phyllaries ≤1.6mm wide, or if ≤1.8mm wide then corolla equalling or exceeding phyllaries or corolla pubescent or both **2. A. minus**

Other spp. - A. tomentosum Mill., from Europe, is a rare casual; it differs from *A. lappa* in its basal petioles hollow or solid, inner phyllaries constricted above the middle, corolla usually glandular-pubescent and capitula often densely pubescent. The hybrid *A. tomentosum* x *A. lappa* = **A. x ambiguum** (Celak.) Nyman has also been reliably recorded from Beds.

1. A. lappa L. (*A. vulgare* (Hill) Druce) - *Greater Burdock*. Stems erect, well branched, to 2m; capitula of main branches in corymbose clusters, with peduncles 2.5-26cm, 24-47 x 20-30mm, glabrous to sparsely pubescent; phyllaries exceeding corolla by 1-5mm; middle phyllaries 0.9-1.7mm wide; corolla glabrous or glandular-pubescent distally; 2n=36. Native; waysides, field-borders, wood-clearings, waste places; rather scattered in S & C Br, very scattered in Ir.

1 x 2. A. lappa x A. minus = A. x nothum (Ruhmer) J. Weiss (*A. x debrayi* Senay) occurs infrequently in S Br with the parents but exact distribution and abundance is unknown; usually plants combine the corymbose inflorescence of *A. lappa* with the hollow petiole of *A. minus*, the corolla ± equals the phyllaries, and there is reduced achene fertility.

2. A. minus (Hill) Bernh. (*A. pubens* Bab., *A. minus* ssp. *pubens* (Bab.) P. Fourn., *A. vulgare* auct. non (Hill) Druce) - *Lesser Burdock*. Stems as in *A. lappa*; capitula of main branches in racemose or spikiform, sometimes subcorymbose clusters, with peduncles 0-9.5cm, 15-32 x 11-24mm, glabrous to densely pubescent; phyllaries exceeding corolla by ≤3.6mm to falling short of it by ≤2.5mm; middle phyllaries 0.6-1.8mm wide; corolla glabrous to glandular-pubescent; 2n=36. Native; similar places to *A. lappa*; apparently common throughout BI.

Much confused with *A. nemorosum*, but hybrids with it are unknown. Plants with pedunculate distal capitula, or with middle phyllaries <1.6mm wide, or with a glandular-pubescent corolla, or with corolla equalling or exceeding phyllaries, or with capitula ≤27 x 19mm are *A. minus*, but the reverse is not true.

3. A. nemorosum Lej. (*A. minus* ssp. *nemorosum* (Lej.) Syme, *A. vulgare* auct. non (Hill) Druce) - *Wood Burdock*. Stems as in *A. lappa*; capitula of main branches in racemose or spikiform clusters, the distal ones sessile, 27-40 x 19-29mm, glabrous to densely pubescent; phyllaries exceeding corolla by 1.2-6mm; middle phyllaries 1.6-2.5mm wide; corolla glabrous; 2n=36. Native; mostly in open woods and semi-shaded disturbed ground, especially on calcareous soils; distribution uncertain due to confusion with *A. minus*, but possibly rare or absent from SW Br, CI and Ir.

4. SAUSSUREA DC. - *Alpine Saw-wort*

Perennials; leaves simple, subentire to denticulate, not spiny; phyllaries in many rows, simple, entire, rounded to obtuse at apex; corolla purple; pappus of feathery hairs in 1 row, often ± united proximally, with an outer row of simple shorter hairs, the former often deciduous.

1. S. alpina (L.) DC. - *Alpine Saw-wort*. Stems erect, to 45cm; leaves ovate or **682**
elliptic to narrowly so, the basal and lower ones petiolate, densely white-pubescent on lowerside; capitula 15-20mm, 1-several in ± dense terminal cluster; 2n=52, 54. Native; mountain cliffs and scree, maritime in N Sc; local in N En, N Wa, N, W & S Sc, scattered in Ir.

5. CARDUUS L. - *Thistles*

Annuals or biennials; stems with spiny wings at least in part; leaves variously lobed, sharply and densely spiny; phyllaries simple, linear-subulate to narrowly ovate, spine-tipped, in many rows; corolla purple (or white); pappus of many rows of simple hairs united proximally.

1 Capitula subcylindrical, <14mm across (excl. flowers); corolla with 5 ± equal lobes 2

1 Capitula globose to bell-shaped, >14mm across (excl. flowers); corolla with 1 lobe distinctly more deeply delimited than other 4 3

  2 Capitula in clusters of 3-10; stems with spiny wings right up to base of capitula; phyllaries thin and transparent on margins, without strongly thickened midrib except sometimes near apex **1. C. tenuiflorus**

  2 Capitula in clusters of 1-3; stems with discontinuous spiny wings, with at least some peduncles unwinged distally; phyllaries with

strongly thickened margins and midrib for at least distal 1/2 **2. C. pycnocephalus**

3 Capitula 15-25(30)mm across (excl. flowers), in clusters of (1)2-4(5), ± erect; phyllaries linear-subulate, not narrowed just above base; corolla 12-15mm **3. C. crispus**

3 Capitula (20)30-60mm across (excl. flowers), usually solitary, pendent; phyllaries lanceolate, narrowed just above base; corolla 15-25mm **4. C. nutans**

Other spp. - **C. acanthoides** L., from C Europe, is the name that has often been given to *C. crispus* ssp. *multiflorus;* it differs in its longer-spined (c.5mm, not c.2-3mm) stem-wings, web-like pubescent (not subglabrous) stems, and usually larger (25-35mm across) solitary capitula. It might occur as a rare casual, but most plants that resemble it are extreme variants of *C. crispus.* **C. macrocephalus** Desf. (*Giant Thistle*), from C & E Mediterranean, is grown in gardens and was formerly natd in S En; it differs from *C. nutans* in its more densely pubescent stems to 2m, longer-stalked capitula, and shorter phyllaries with prominent (not obscure) midrib. **C. thoermeri** Weinm., from E Europe, also resembles *C. nutans* but has phyllaries widened below into an appendage 4-8mm wide; it is a rare grain- and birdseed-alien.

1. C. tenuiflorus Curtis - *Slender Thistle.* Stems erect, to 60(80)cm, continuously spiny-winged up to capitula; leaves grey- to white-cottony on lowerside; capitula ± sessile, in clusters of (1)3-10, 12-18 x 5-10mm (excl. flowers), ± erect; corollas mostly shorter than phyllaries; 2n=54. Native; waysides, rough and open ground; locally frequent by coasts in BI except most of N & W Sc, very scattered inland. 682

2. C. pycnocephalus L. - *Plymouth Thistle.* Differs from *C. tenuiflorus* in stems and leaves more densely white-cottony; capitula subsessile to stalked, in clusters of 1-3, 14-20 x 7-12mm (excl. flowers); corollas mostly longer than phyllaries; and see key; (2n=54). Intrd; natd on open limestone cliff at Plymouth (S Devon) since 1868, infrequent casual elsewhere from wool and birdseed; S Europe. 682

3. C. crispus L. (*C. acanthoides* auct. non L.) - *Welted Thistle.* Stems erect, to 1.5m, continuously spiny-winged up to or nearly up to capitula; leaves subglabrous on lowerside; capitula mostly stalked, in clusters of (1)2-4(5), 9-15 x 15-20(25)mm (excl. flowers), bell-shaped, ± erect; 2n=16. Native; hedgerows, ditchsides and streamsides, rough ground, especially on rich or basic soils; frequent in Br N to C Sc, scattered in Ir and W & N Sc. Our plant is ssp. **multiflorus** (Gaudin) Gremli.

3 x 4. C. crispus x C. nutans = C. x stangii H. Buek (?*C. x dubius* Balb., *C. x polyacanthus* Schleich. non Lam., *C. x orthocephalus* auct. non Wallr.) is found scattered in Br with the parents N to Yorks; it is partially fertile and intermediate in capitulum characters.

4. C. nutans L.- *Musk Thistle.* Stems erect, to 1m, discontinuously spiny-winged with peduncles unwinged distally; leaves pubescent on lowerside; capitula stalked, mostly solitary, 16-30 x 20-60mm (excl. flowers), depressed-globose, pendent, with strongly reflexed outer phyllaries; 2n=16. Native; grassy and bare places, waysides and rough ground mostly on calcareous soils; locally frequent in CI and Br N to S Sc, rare elsewhere in Sc and in Ir, sometimes casual.

6. CIRSIUM Mill. - *Thistles*

Biennials to perennials; stems with or without spiny wings; leaves denticulate to deeply lobed, spiny or at least with bristle-pointed teeth; phyllaries simple, ovate to linear-subulate, spine-tipped to mucronate or acuminate at apex, in many rows; corolla purple (or white), rarely yellow; pappus of many rows of feathery hairs united proximally.

1 Leaves with rigid bristles on upperside 2

1 Leaves glabrous or with soft hairs on upperside 3

2 Stem with discontinuous spiny wings **2. C. vulgare**

2 Stem not winged **1. C. eriophorum**

3 Flowers yellow 4

3 Flowers purple, rarely white 5

4 Upper part of stem with only very reduced, distant leaves; capitula pendent **5. C. erisithales**

4 Upper part of stem with large yellowish-green leaves exceeding the erect capitula **7. C. oleraceum**

5 Stems continuously spiny-winged; biennial with tap-root **9. C. palustre**

5 Stems not winged or with very short wings below each leaf; perennial with at least short rhizomes 6

6 Distal broad part of corolla c.1/2 as long as proximal narrow part, lobed ≥3/4 way to base; stem usually well branched **10. C. arvense**

6 Distal broad part of corolla c. as long as proximal narrow part, lobed c.1/2 way to base; stem usually unbranched 7

7 Stem 0-10cm, or if up to 30cm then with ≥1 well-developed leaf near top **8. C. acaule**

7 Stem >10cm, with only distant much reduced leaves in distal 1/4 8

8 Stem-leaves widened proximally to broad base clasping stem; capitula mostly >20mm to tip of uppermost phyllaries; mostly N En and Sc **6. C. heterophyllum**

8 Stem-leaves narrowed to base, not or scarcely clasping stem; capitula mostly <20mm to tip of uppermost phyllaries; mostly S En, Wa and Ir 9

9 Lower stem-leaves deeply lobed, the lobes deeply lobed, green on lowerside; some roots swollen into tubers **4. C. tuberosum**

9 Lower stem-leaves not or rather shallowly lobed, the lobes scarcely or not lobed, white on lowerside; roots not swollen **3. C. dissectum**

1. C. eriophorum (L.) Scop. - *Woolly Thistle.* Erect biennial to 1.5m; stems leafy to top, not winged; leaves deeply lobed, very strongly spiny; capitula 3-5 x 4-7cm (excl. flowers); flowers purple; 2n=34. Native; dry grassland, scrub and banks on calcareous soil; locally frequent in Br N to Co Durham.

1 x 2. C. eriophorum x C. vulgare = C. x grandiflorum Kitt. (*C. x gerhardtii* Sch. Bip.) has occurred rarely in En N to SE Yorks; it is intermediate in stem wingedness and capitulum characters and partially fertile.

2. C. vulgare (Savi) Ten. - *Spear Thistle.* Erect biennial to 1.5m; stems leafy to top, with spiny wings; leaves deeply lobed, very strongly spiny; capitula 2.5-4 x 2-5cm (excl. flowers); flowers purple; 2n=68. Native; grassland, waysides, cultivated, rough and waste ground; common throughout BI.

2 x 8. C. vulgare x C. acaule = C. x sabaudum M. Loehr has occurred rarely in En N to N Lincs; it has stems to c.30cm with short spiny wings below each leaf, a few bristles on the leaf uppersides, intermediate capitula and is slightly fertile.

2 x 9. C. vulgare x C. palustre = C. x subspinuligerum Peterm. was found in Merioneth in 1986.

3. C. dissectum (L.) Hill - *Meadow Thistle.* Erect perennial to 80cm; stems with only very reduced leaves in upper part, not winged; leaves scarcely to rather deeply lobed, softly spiny, white-pubescent on lowerside; capitula 1.3-2.5 x 1.2-3cm (excl. flowers); flowers purple; 2n=34. Native; fens, bogs, wet fields on peaty soil; local in En and Wa N to NE Yorks and Merioneth, throughout Ir, very local in SW Sc.

3 x 8. C. dissectum x C. acaule = C. x woodwardii (H.C. Watson) Nyman once occurred near Swindon, N Wilts.

3 x 9. C. dissectum x C. palustre = C. x forsteri (Sm.) Loudon is not uncommon with the parents throughout the range of *C. dissectum* and is the commonest hybrid

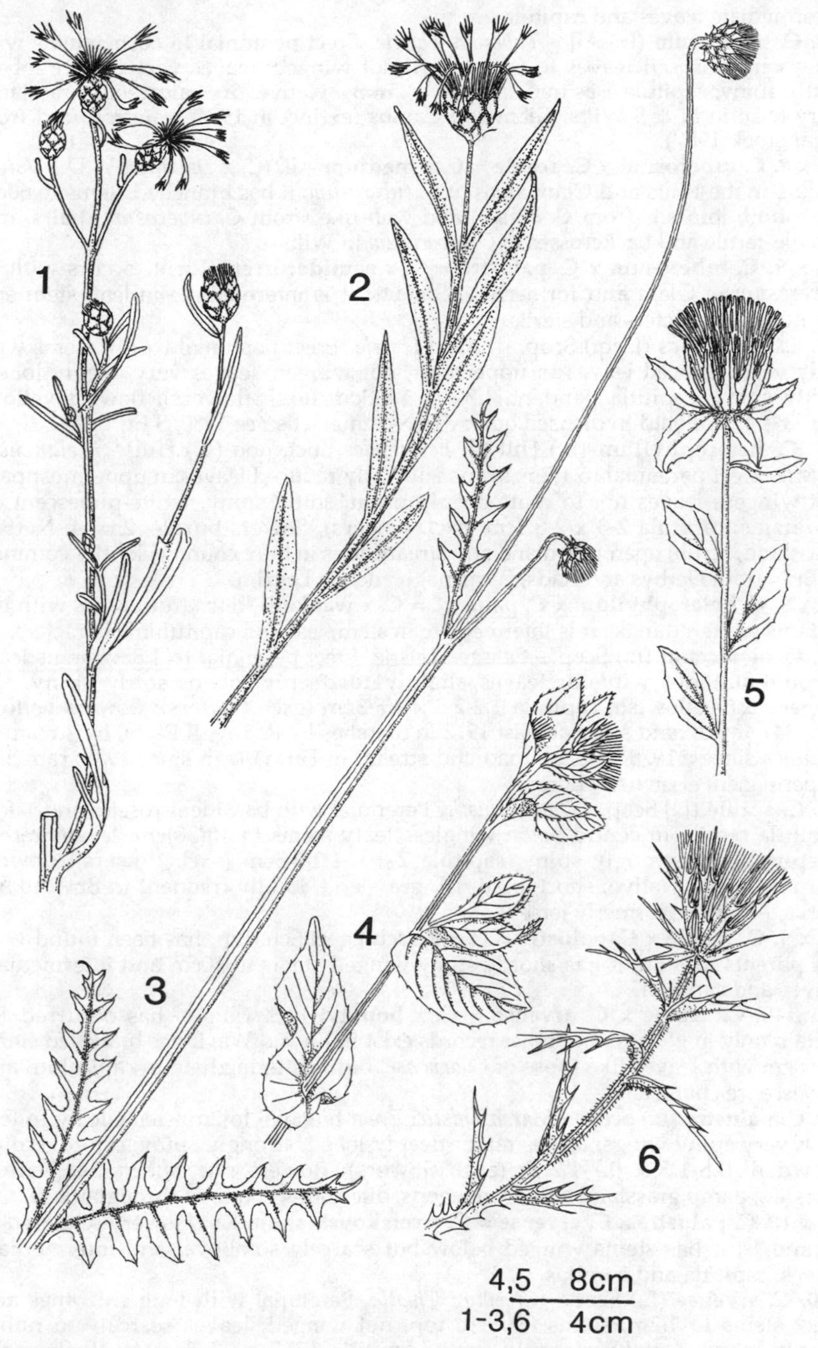

FIG 677 - Asteraceae: Cardueae. 1, *Centaurea diluta*. 2, *C. montana*. 3, *Cirsium erisithales*. 4, *C. oleraceum*. 5, *Carthamus tinctorius*. 6, *C. lanatus*.

thistle; it has discontinuous spiny-winged, cottony-pubescent stems, and intermediate leaves and capitula.

4. C. tuberosum (L.) All. - *Tuberous Thistle.* Erect perennial to 60cm; stems with only very reduced leaves in upper part, not winged; leaves very deeply lobed, softly spiny; capitula ± as in *C. dissectum*; 2n=34. Native; dry calcareous grassland; very local in N & S Wilts, Glam and Cambs (extinct in 1974, re-introduced from local stock 1987). **RR**

4 x 8. C. tuberosum x C. acaule = C. x medium All. (*C. x zizianum* W.D.J. Koch) occurs in the Wilts and Glam areas of *C. tuberosum*; it has branched stems to 60cm with both jointed (from *C. acaule*) and web-like (from *C. tuberosum*) hairs. It is female-fertile and backcrosses to *C. tuberosum* in Wilts.

4 x 9. C. tuberosum x C. palustre = C. x semidecurrens Richt. occurs with *C. tuberosum* in Glam and formerly in S Wilts; it is intermediate in leaf, stem and capitulum characters and sterile.

5. C. erisithales (Jacq.) Scop. - *Yellow Thistle.* Erect perennial to 1m; stems with only very reduced leaves in upper part, not winged; leaves very deeply lobed, softly spiny; capitula pendent, 1.3-2 x 1.5-3cm (excl. flowers); flowers yellow; (2n=34). Intrd; natd in disused quarry in N Somerset since 1980; S Europe. **677**

6. C. heterophyllum (L.) Hill (*C. helenioides* auct. non (L.) Hill) - *Melancholy Thistle.* Erect perennial to 1.2m; stems with only reduced leaves in uppermost part, not winged; leaves not to quite deeply lobed, softly spiny, white-pubescent on lowerside; capitula 2-3 x 2-3.5cm (excl. flowers); flowers purple; 2n=34. Native; grassland, scrub, open woodland and streamsides in hilly country; locally common in Br N from Derbys and Rads, Fermanagh and Co Leitrim.

6 x 9. C. heterophyllum x C. palustre = C. x wankelii Reichardt occurs with the parents scattered in Sc; it is intermediate in stem, leaf and capitulum characters.

7. C. oleraceum (L.) Scop. - *Cabbage Thistle.* Erect perennial to 1.5m; stems leafy throughout, not winged; leaves slightly to deeply lobed, softly spiny, the uppermost yellowish; capitula 1.2-2.5 x 1.5-3cm (excl. flowers); flowers yellow; (2n=34). Intrd; natd since at least 1912 in marshes by R. Tay, E Perth, by stream in S Lancs since c.1978, and by road and stream in Fermanagh since 1988, rare and impermanent elsewhere; Europe. **677**

8. C. acaule (L.) Scop. - *Dwarf Thistle.* Perennial with basal leaf-rosette and 1-few capitula sessile in centre or on wingless leafy stems to 10(30)cm; leaves rather deeply lobed, strongly spiny; capitula 2-3 x 1.2-2.5cm (excl. flowers); flowers purple; 2n=34. Native; short, base-rich grassland; locally frequent in Br N to NE Yorks, Alderney, formerly Jersey.

8 x 9. C. acaule x C. palustre = C. x kirschlegeri Sch. Bip. has been found with the parents in S En; it has shortly spiny-winged stems to 40cm and intermediate leaves and capitula.

8 x 10. C. acaule x C. arvense = C. x boulayi E.G. Camus has occurred for certain only in N Essex but other records exist in En and Wa; it has branched stems to 60cm with leaves like those of *C. arvense*, but is intermediate in capitulum and pubescence characters.

9. C. palustre (L.) Scop. - *Marsh Thistle.* Erect biennial to 2m; stems leafy to top, with very spiny wings; leaves rather deeply lobed, strongly spiny; capitula often crowded, 0.8-1.5 x 0.7-1.2cm (excl. flowers); flowers purple; 2n=34. Native; marshes, damp grassland and open woods, ditchsides; common throughout BI.

9 x 10. C. palustre x C. arvense = C. x celakovskianum Knaf is very scattered in Br and Ir; it has stems winged below but scarcely so above, and intermediate leaves, capitula and corollas.

10. C. arvense (L.) Scop. - *Creeping Thistle.* Perennial with long rhizomes and erect stems to 1.2m; stems leafy to top, not winged; leaves scarcely to rather deeply lobed, usually strongly spiny; capitula 1-2.2 x 0.8-2cm (excl. flowers); flowers purple; 2n=34. Native; grassland, hedgerows, arable, waste and rough ground; common throughout BI. Var. **incanum** (Fisch.) Ledeb., from S Europe, has leaves greyish-white pubescent on lowerside, and occurs as a casual.

7. ONOPORDUM L. - *Cotton Thistles*
Biennials; stems with spiny wings; leaves dentate to shallowly lobed, strongly spiny; phyllaries simple, linear-lanceolate, strongly spine-tipped, in many rows; corolla purple (rarely white); pappus of many rows of simple hairs united proximally.

1. O. acanthium L. - *Cotton Thistle*. Stems erect, to 2.5m; stems and leaves greyish-white with cottony hairs; stem-wings with rather weak scarcely reticulate veins; capitula 2-6cm wide; outer phyllaries 1.5-2.5mm wide at widest point; corolla 15-25mm; (2n=34). Probably intrd; natd or casual since at least 16th century in fields, marginal habitats, waste and rough ground; locally frequent in En (especially SE, possibly native in E Anglia), very scattered in Wa, Sc and CI; Europe.

2. O. nervosum Boiss. - *Reticulate Thistle*. Stems erect, to 2.5m; lower part of stem and lower leaves with pubescence as in *O. acanthium*; upper regions green, with sparse pubescence and strongly contrasting raised white veins on leaf lowerside; stem-wings with strong reticulate veins; capitula 3-5cm wide; outer phyllaries 4-6mm wide at widest point; corolla 30-35mm. Intrd; garden ornamental sometimes natd on waste and rough ground; few places in S En, Herts since at least 1986; Spain and Portugal.

8. CYNARA L. - *Globe Artichoke*
Perennials; stems not winged, spiny or not; leaves deeply lobed, spiny or not; phyllaries in many rows, the outer with or without apical spine, the inner with distinct apical portion with or without apical spine; corolla mauve or violet to blue; pappus of many rows of feathery hairs united proximally.

1. C. cardunculus L. (*C. scolymus* L.) - *Globe Artichoke*. Stems erect, to 1.8m; leaves spiny or not, sparsely pubescent on upperside, tomentose on lowerside; capitula 3.5-7.5 x 3.5-9.5cm (excl. flowers); (2n=34). Intrd; garden ornamental or vegetable long persisting after neglect or discard in derelict or rough ground; scattered in CI and En N to S Lancs; Mediterranean. Var. **cardunculus** (*Cardoon*) has strongly spiny leaves and phyllaries and the blanched young shoots are eaten; var. **scolymus** (L.) Fiori (*C. scolymus*) (*Globe Artichoke*) is spineless and the succulent receptacle and phyllary-bases are eaten; both are grown as ornamentals.

9. SILYBUM Adans. - *Milk Thistle*
Annuals to biennials; stems not spiny; leaves shallowly to rather deeply lobed, strongly spiny; phyllaries in many rows, the outer with spine-tipped lateral lobes or teeth and strong apical spine; corolla purple (rarely white); pappus of many rows of simple hairs united proximally.

1. S. marianum (L.) Gaertn. - *Milk Thistle*. Stems erect, to 1m; leaves bright green, usually veined or marbled with white; capitula 2.5-4 x 5-14cm (excl. flowers); (2n=34). Intrd; frequent casual, sometimes natd, in waste and rough ground since at least 17th century, now often a wool- or birdseed-alien; scattered throughout BI N to E Ross.

10. SERRATULA L. - *Saw-wort*
Perennials; stems and leaves not spiny; lower leaves usually lobed almost to midrib; phyllaries simple, acute to acuminate, not spiny, in many rows; corolla purple (rarely white); pappus of many rows of free, simple hairs, the outermost much shorter than the inner ones.

1. S. tinctoria L. - *Saw-wort*. Stems erect, to 70(100)cm, but often very short (<10cm) in very exposed places, usually well branched above; leaves dark green, subglabrous, with bristle-tipped teeth; 2n=22. Native; grassland, scrub, open woodland, cliff-tops and rocky streamsides on well-drained soils; local in Br N to SW Sc, Jersey, formerly Co Wexford. **682**

11. ACROPTILON Cass. - *Russian Knapweed*
Perennials; stems and leaves not spiny; leaves simple, entire to narrowly lobed; phyllaries with conspicuous broad apical scarious border, in many rows; corolla pink, the outermost no longer than the inner; pappus of hairs, soon falling.

1. A. repens (L.) DC. (*Centaurea repens* L.) - *Russian Knapweed*. Stems erect, well branched, to 70cm; leaves lanceolate to oblanceolate, densely grey-pubescent; phyllaries pale brown with paler, minutely pubescent, scarious border at apex, becoming jagged later; (2n=26). Intrd; natd on waste ground at Hereford, Herefs, since 1959; Russia. **682**

12. CENTAUREA L. - *Knapweeds*
Annuals to perennials; stems and leaves not spiny; leaves simple and entire to ± pinnate; phyllaries with distinct apical portion which is scarious, toothed or spiny, in many rows; corolla purple to pink or blue, white or yellow, that of outermost flowers often much longer than that of inner flowers (pseudo-radiate); pappus 0, or of many rows of simple to toothed free hairs, sometimes also with some scales.

1 Apical portion of phyllaries with ≥1 sharp spines 2
1 Apical portion of phyllaries merely scarious or variously toothed, not spiny 6
  2 Flowers yellow; stems strongly winged 3
  2 Flowers purple, rarely white; stems not winged 4
3 Apical phyllary spines <10mm, with lateral spines arranged pinnately along its proximal 1/2; corolla with minute sessile glands **7. C. melitensis**
3 At least some apical phyllary spines >10mm, with lateral spines arranged palmately at its base; corolla not glandular **6. C. solstitialis**
  4 Apical portion of phyllaries scarious, variously toothed, with 1(-few) terminal spines **8. C. diluta**
  4 Whole of apical portion of phyllaries modified into spines 5
5 Apical phyllary spine >10mm, >3x as long as longest laterals **4. C. calcitrapa**
5 Apical phyllary spine <5mm, <1.5x as long as longest laterals **5. C. aspera**
  6 Apical portion of phyllaries with 1(-few) distinct terminal subspinose teeth distinctly different from lateral teeth **8. C. diluta**
  6 Apical portion of phyllaries similarly toothed at apex and sides 7
7 Flowers yellow; apical portion of outer phyllaries >1cm wide **11. C. macrocephala**
7 Flowers purple to blue, rarely white; apical portion of outer phyllaries <1cm wide 8
  8 Apical portion of phyllaries strongly delimited from basal portion, with slight constriction between 9
  8 Apical portion of phyllaries ill-delimited from basal portion, the former decurrent down sides of latter 10
9 Apical portion of outer phyllaries dark brown to black, deeply and very regularly toothed; pappus present or absent **10. C. nigra**
9 Apical portion of outer phyllaries pale to dark brown, irregularly toothed or deeply jagged; pappus absent or ± so **9. C. x moncktonii**
  10 Easily uprooted annual; basal leaves 0 at flowering time **3. C. cyanus**
  10 Deeply rooted perennial; basal leaves or non-flowering shoots present

at flowering time 11
11 Leaves simple, entire, strongly decurrent and forming distinct wings on stem; strongly rhizomatous; flowers usually blue **2. C. montana**
11 Leaves usually deeply lobed, rarely simple and entire, not decurrent on stem; scarcely rhizomatous; flowers usually reddish-purple **1. C. scabiosa**

Other spp. - 2 European aliens formerly natd now seem to be extinct. **C. jacea** L. 682
(*Brown Knapweed*) differs from *C. nigra* in its pale brown irregularly jagged apical
portion of the phyllaries; it formerly occurred in grassy places in S En and CI, but
now only exists as its hybrid *C x moncktonii* (q.v.). **C. paniculata** L. (*Jersey* 682
Knapweed) was natd in Jersey from 1851 to 1981; it differs from *C. jacea* and *C. nigra* in its much smaller capitula (3-8mm across at level of phyllaries, not ≥10mm across), phyllaries with pale brown regularly toothed apical portion, and lower leaves very deeply divided into very narrow lobes. **C. hyalolepis** Boiss., from E Mediterranean, has spiny phyllaries but combines yellow flowers with non-winged stems; it is a rare grain-, wool- and birdseed-alien.

1. C. scabiosa L. - *Greater Knapweed*. Erect perennial to 1.2m; leaves mostly very 682
deeply lobed, rarely all or most simple (var. **succisifolia** E. Marshall); phyllaries with blackish-brown, decurrent, deeply toothed apical portion; capitula pseudo-radiate; corolla usually reddish-purple; 2n=20. Native; grassland, rough ground, cliffs and waysides mainly on calcareous soils; locally common in BI N to N En and C Ir, very local in mainland Sc.

2. C. montana L. - *Perennial Cornflower*. Erect rhizomatous perennial to 80cm; 677
leaves simple, entire; phyllaries with blackish-brown, decurrent, deeply toothed 682
apical portion; capitula pseudo-radiate; corolla usually blue; flowers with strong sweet scent; (2n=24, 40, 44). Intrd; much grown in gardens, natd in grassy places and rough ground throughout most of Br, especially C & N; mountains of C & S Europe.

3. C. cyanus L. - *Cornflower*. Erect annual to 80cm; leaves linear-lanceolate, the R
lower with few pinnate lobes, the upper entire; phyllaries with blackish-brown, 682
decurrent, deeply toothed apical portion; capitula pseudo-radiate; corolla usually blue, sometimes pink or white; (2n=24). Native; traditionally natd in cornfields, now mostly casual birdseed-alien or garden escape in waste places; scattered throughout BI, much scarcer than formerly.

4. C. calcitrapa L. - *Red Star-thistle*. Erect to spreading biennial to 60cm; leaves 682
shallowly to very deeply lobed; phyllaries with apical portion of 1 terminal strong spine >1cm and several much shorter laterals near its base; capitula ± discoid; corolla reddish-purple; (2n=20). Intrd; waste and rough ground, waysides, on well-drained soils; natd in very scattered places in S En, decreasing, casual from wool, birdseed and other sources elsewhere; Europe.

5. C. aspera L. - *Rough Star-thistle*. Erect to spreading perennial to 80cm; leaves 682
pinnately lobed to dentate; phyllaries with apical portion of 3-5 subequal spines <5mm; capitula ± discoid; corolla reddish-purple; 2n=22. Intrd; natd on maritime dunes in Jersey since at least 1839, Guernsey since 1788, casual or ± natd in few places in S Br; Europe.

6. C. solstitialis L. - *Yellow Star-thistle*. Erect to spreading annual to biennial to 682
60cm; leaves pinnately lobed to dentate; phyllaries with apical portion of 1 terminal strong spine >1cm and several much shorter laterals at its base; capitula ± discoid; corolla yellow; (2n=16). Intrd; formerly natd in arable fields (especially of lucerne), now more usually casual of waste ground mostly from wool and birdseed; scattered mainly in C & S Br; S Europe.

7. C. melitensis L. - *Maltese Star-thistle*. Differs from *C. solstitialis* in key 682
characters; (2n=22, 24, 36). Intrd; casual of tips and waste ground mainly from birdseed and wool; scattered mainly in S & C Br; S Europe.

8. C. diluta Aiton - *Lesser Star-thistle*. Erect perennial (often behaving as annual) 677

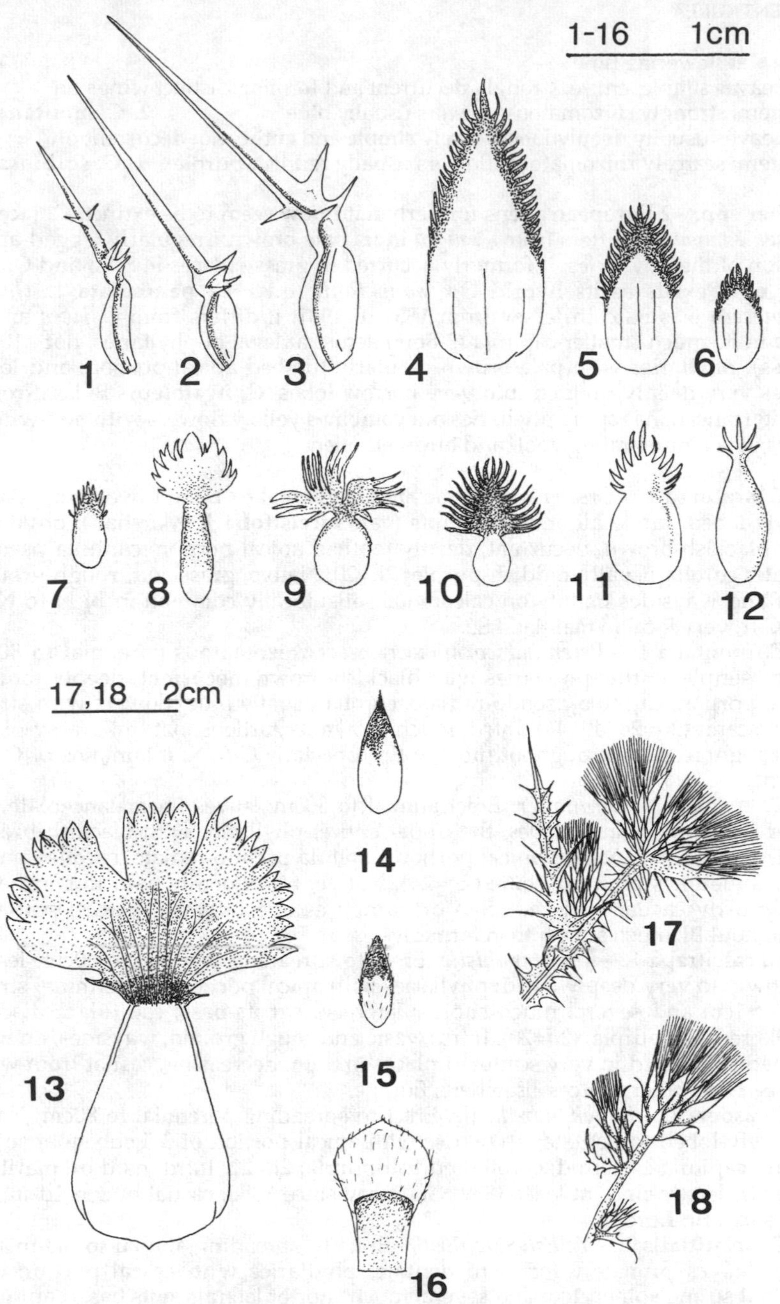

FIG 682 - Asteraceae: Cardueae. 1-16, phyllaries. 1, *Centaurea melitensis*. 2, *C. solstitialis*. 3, *C. calcitrapa*. 4, *C. montana*. 5, *C. scabiosa*. 6, *C. cyanus*. 7, *C. paniculata*. 8, *C. jacea*. 9, *C. x moncktonii*. 10, *C. nigra*. 11, *C. diluta*. 12, *C. aspera*. 13, *C. macrocephala*. 14, *Serratula tinctoria*. 15, *Saussurea alpina*. 16, *Acroptilon repens*. 17-18, terminal capitula of *Carduus*. 17, *C. tenuiflorus*. 18, *C. pycnocephalus*.

to 80cm; leaves shallowly lobed to dentate; phyllaries with slightly decurrent pale 682
brown apical portion with small lateral teeth and 1(-few) weak terminal spines; capitula pseudo-radiate; corolla pinkish-purple; (2n=20). Intrd; usually casual on waste ground and tips, mainly from birdseed; scattered mainly in S & C Br; W Mediterranean.

9. C. x moncktonii C.E. Britton (*C. x drucei* C.E. Britton; *C. jacea* x *C. nigra*) - 682
Hybrid Knapweed. Erect perennial to 60cm; leaves entire to medium-lobed; phyllaries with pale to dark brown irregularly toothed or deeply jagged apical portion; capitula usually pseudo-radiate; corolla reddish-purple; 2n=44. Native, from crosses between *C. nigra* and formerly natd *C. jacea*; grassy places; very scattered in S En and CI, formerly more widespread. Evidence of *C. jacea* genes in plants of *C. nigra* is possibly shown by apical portion of phyllaries pale brown, shiny, more irregularly toothed; outer flowers ray-like, pale in colour; pappus irregular or 0. However, these characters are also used to segregate *C. nemoralis* from *C. nigra sensu stricto*. *C. x moncktonii* exhibits the range of variation between its 2 parents; some plants, notably in Berks, approach *C. jacea* very closely.

10. C. nigra L. (*C. nemoralis* Jord., *C. nigra* ssp. *nemoralis* (Jord.) Gremli, *C.* 682
debeauxii Gren. & Godr. ssp. *nemoralis* (Jord.) Dostál) - *Common Knapweed*. Erect perennial to 1m; leaves entire to deeply lobed; phyllaries with dark brown to black finely and very regularly toothed apical portion; capitula usually discoid, sometimes pseudoradiate; corolla reddish-purple; 2n=22, 44. Native; grassy places, rough ground, waysides; common throughout BI. British botanists have been unable to maintain a distinction between *C. nigra* and **C. nemoralis**, which has narrower leaves, slenderer branches, smaller capitula, phyllaries with apical portions divided >3/4 (not <3/4) way to midrib, and shorter or no pappus. Such plants are largely confined to light soils in S Br, but occur with a wide range of intermediates.

11. C. macrocephala Muss. Puschk. ex Willd. - *Giant Knapweed*. Erect perennial to 682
1m; leaves ± entire, the upper decurrent on stem; phyllaries with large, pale brown, entire to dentate, papery apical portions that conceal whole of capitulum below flowers; capitula ± discoid; corolla yellow; (2n=18). Intrd; garden ornamental natd on waste and rough ground; abandoned orchard in E Suffolk since 1985; Caucasus.

13. CARTHAMUS L. - *Safflowers*

Annuals; leaves entire or with acute lobes and apex tipped with spines or bristles; phyllaries in many rows, the outer leaf-like, the inner usually spine-tipped; corolla yellow to orange; pappus 0 or of narrow pointed scales.

1. C. tinctorius L. - *Safflower*. Stems erect, glabrous, to 60cm; leaves and outer 677
phyllaries entire to shallowly lobed with bristly or softly spine-tipped lobes; corolla yellow to reddish-orange; pappus 0; (2n=24). Intrd; tips, waste and rough ground, casual from birdseed; rather frequent in S Br, very scattered further N and in Ir; SW Asia.

2. C. lanatus L. - *Downy Safflower*. Stems erect, densely pubescent, to 60cm; 677
leaves and outer phyllaries with apex and usually deep lobes with strong terminal spines; corolla yellow; pappus c. as long as achene; (2n=44). Intrd; similar habitats and distribution to *C. tinctorius* but mostly a wool-alien; S Europe.

TRIBE 2 - LACTUCEAE (Cichorieae) (genera 14-33). Plants producing white latex, rarely spiny; capitula ligulate, the flowers all bisexual and ligulate, usually with 5-toothed ligules, usually yellow.

14. SCOLYMUS L. - *Golden Thistle*

Thistle-like (annuals,) biennials or perennials; rhizomes and stolons 0; stems leafy, with several capitula; phyllaries in several rows; receptacular scales present, wrapped round achenes; pappus (0 or) of a few rigid hairs; ligules yellow; achenes flattened, not beaked.

Other spp. - **S. maculatus** L. is an annual with white-bordered leaves and pappus 0; it is a rare casual.

1. S. hispanicus L. - *Golden Thistle*. Stems erect, to 80cm, pubescent; leaves linear to ovate, lobed, with strong rigid spines; capitula 2.5-4cm across; (2n=20). Intrd; casual in waste places as grain-, birdseed- or shoddy-alien; very scattered in S En and Wa; S Europe. Resembles a thistle but flowers all ligulate and stems with latex. **685**

15. CICHORIUM L. - *Chicory*
Perennials; rhizomes and stolons 0; stems leafy, with many capitula; phyllaries in 2 rows; receptacular scales present, small; pappus of short scales; ligules blue, rarely white; achenes not flattened, somewhat angular, not beaked.

Other spp. - **C. endivia** L. (*Endive*), from S Europe, differs from *C. intybus* in its much more strongly thickened peduncles and pappus-scales 1/6-1/2 (not 1/10-1/8) as long as achene; it is eaten as salad (like a flat dissected-leaved lettuce) and occurs as a wool- and birdseed-alien and rarely as a relic of cultivation.

1. C. intybus L. - *Chicory*. Stems stiff, procumbent to erect, glabrous to pubescent, to 1m; lower leaves oblanceolate, deeply lobed to toothed; capitula 2.5-4cm across; 2n=18. Possibly native; roadsides, rough grassland, waste places; locally common, especially on calcareous soils, in C & S Br, scattered elsewhere in BI. Most if not all plants are of the cultivated variant, sometimes separated as ssp. **sativum** (Bisch.) Janch., whose young tight shoots are blanched for use as salad.

16. CATANANCHE L. - *Blue Cupidone*
Perennials; rhizomes and stolons 0; stems not or very sparsely leafy, branched sparingly, with 1 capitulum at end of each long branch; phyllaries in several rows, translucent-scarious except for dark midrib; receptacular scales present; pappus of scales; ligules blue, rarely white; achenes not flattened, 5-10-ribbed, not beaked.

1. C. caerulea L. - *Blue Cupidone*. Stems erect, to 80cm, pubescent; leaves mostly or all basal, to 30cm, linear to elliptic-linear, entire or with few narrow lateral lobes; capitulum 2.5-4cm across, with silvery papery phyllaries; (2n=18). Intrd; garden ornamental escaping or persistent on roadside; N Somerset since 1989; SW Europe.

17. ARNOSERIS Gaertn. - *Lamb's Succory*
Annuals; stems leafless, with small scale-like bracts, branched or not; phyllaries in 2 rows, the outer small and incomplete; receptacular scales 0; pappus 0; ligules yellow; achenes somewhat flattened, c.5-ribbed, not beaked.

1. A. minima (L.) Schweigg. & Körte - *Lamb's Succory*. Stems erect, to 30cm, ± glabrous, conspicuously dilated some way below capitula; leaves all basal, narrowly obovate to oblanceolate, lobed to toothed; capitula 7-12mm across; 2n=18. Probably native; sandy arable fields; formerly very local in E En from Dorset and E Kent to SE Yorks, and in E Sc, extinct since 1971. **RR**

18. LAPSANA L. - *Nipplewort*
Annuals to perennials; rhizomes and stolons 0; stems leafy, with many capitula; phyllaries in 2 rows, the outer small and incomplete; receptacular scales 0; pappus 0; ligules yellow; achenes slightly flattened, c.20-ribbed, not beaked.

1. L. communis L. - *Nipplewort*. Stems erect, to 1m, pubescent at least below;

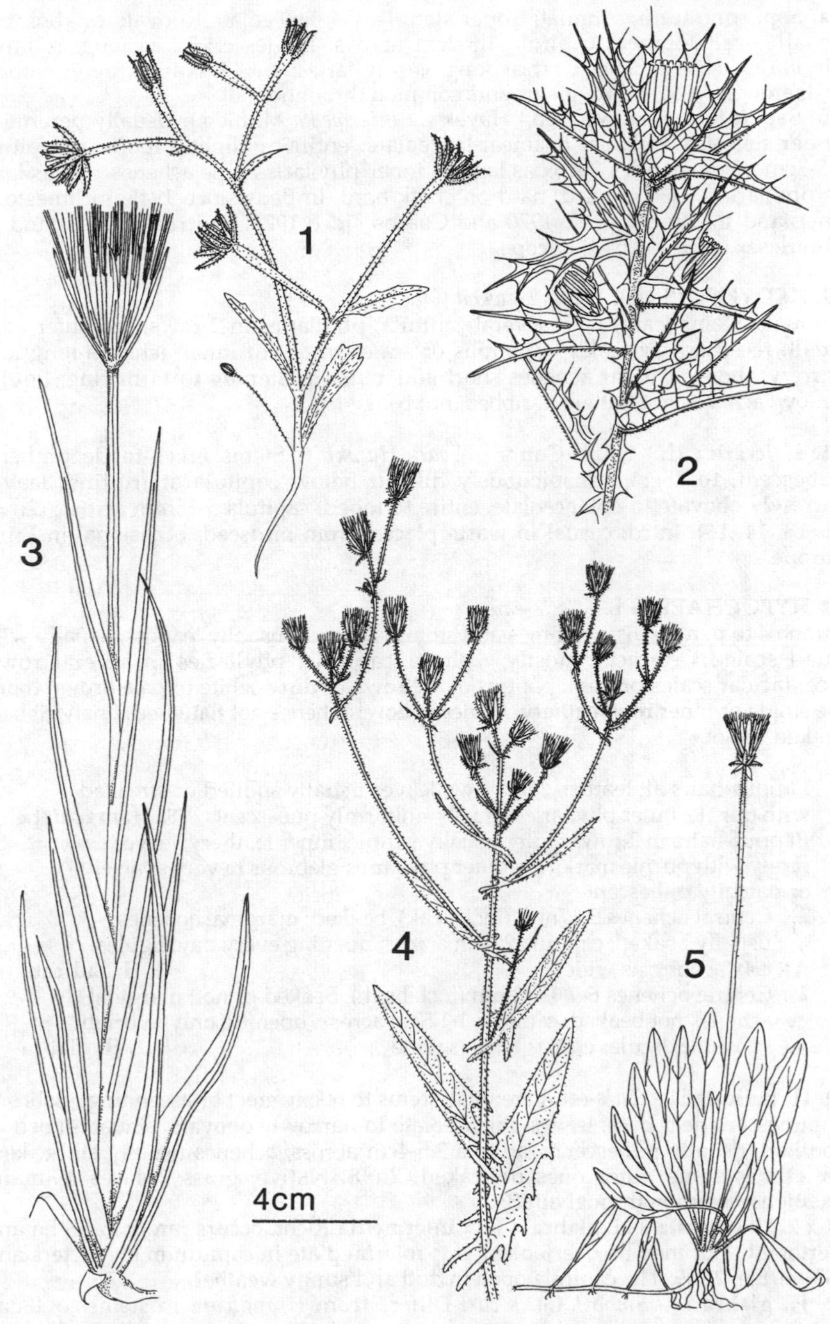

FIG 685 - Asteraceae: Lactuceae. 1, *Hedypnois cretica*. 2, *Scolymus hispanicus*. 3, *Tragopogon hybridus*. 4, *Crepis setosa*. 5, *Aetheorhiza bulbosa*.

lower leaves narrowly obovate to oblanceolate, ± pinnate with lateral leaflets much smaller than terminal one.

a. Ssp. communis. Annual; upper stem-leaves lanceolate to ovate or rhombic, usually well toothed; capitula 1.5-2cm across; ligules c.1.5x as long as inner phyllaries; ripe achenes c.1/2 as long as phyllaries; 2n=14. Native; open woods, hedgerows, waste and rough ground; common throughout BI.

b. Ssp. intermedia (M. Bieb.) Hayek (*L. intermedia* M. Bieb.). Usually perennial; upper stem-leaves linear to linear-lanceolate, entire or slightly toothed; capitula 2.5-3cm across; ligules 2-2.5x as long as inner phyllaries; ripe achenes <1/2 as long as phyllaries; 2n=14. Intrd; natd on chalk bank in Beds since 1945, in limestone grassland in Flints since c.1970 and Caerns since 1977, and rough grassland in Middlesex since 1982; SE Europe.

19. HEDYPNOIS Mill.- *Scaly Hawkbit*

Annuals; stems leafy, with several capitula; phyllaries in 2 rows, the outer very small; receptacular scales 0; pappus of scales, those of inner achenes long and narrow, those of outer achenes short and ± fused laterally to form ring; ligules yellow; achenes not flattened, ribbed, not beaked.

1. H. cretica (L.) Dum. Cours. - *Scaly Hawkbit*. Stems erect to decumbent, 685
pubescent, to 45cm, conspicuously dilated below capitula at fruiting; leaves narrowly obovate to oblanceolate, entire to lobed; capitula 5-15mm across; (2n=8, 12, 13, 14, 16). Intrd; casual in waste places, from birdseed; occasional in En; S Europe.

20. HYPOCHAERIS L. - *Cat's-ears*

Annuals to perennials; rhizomes and stolons 0; stems usually leafless, usually with small scale-like bracts, mostly with >1 capitula; phyllaries in several rows; receptacular scales present; pappus of 1-2 rows of dirty-white to pale brown hairs, the single or inner row feathery; ligules yellow; achenes not flattened, finely ribbed, beaked or not.

1 Pappus-hairs all feathery, in 1 row; leaves usually spotted or streaked with purple; outer phyllaries usually uniformly pubescent **3. H. maculata**

1 Pappus-hairs in 2 rows, outer usually simple, inner feathery; leaves rarely with purple markings; outer phyllaries glabrous to very sparsely or patchily pubescent 2

2 Central achenes 8-17mm (incl. beak), beaked; marginal achenes usually beaked; capitula 2-4cm across, opening every day; ligules c.4x as long as wide **1. H. radicata**

2 Central achenes 6-9(13.5)mm (incl. beak), beaked or not; marginal achenes not beaked; capitula 1-1.5cm across, opening only in bright sunshine; ligules c.2x as long as wide **2. H. glabra**

1. H. radicata L. - *Cat's-ear*. Perennial; stems to 60cm, erect or ascending, glabrous 701
or pubescent near base; leaves oblanceolate to narrowly obovate, sinuate-lobed or -toothed, usually pubescent; capitula 2.5-4cm across; achenes usually all beaked, sometimes some outer ones unbeaked; 2n=8. Native; grassy places in many situations; common throughout BI.

1 x 2. H. radicata x H. glabra = H. x intermedia Richt. occurs rarely in SE En and Merioneth, but may be overlooked; it is intermediate in capitulum characters and <5% fertile; 2n=9. The capitula open in dull and sunny weather.

2. H. glabra L. - *Smooth Cat's-ear*. Differs from *H. radicata* in stems to 40cm, R
glabrous; leaves usually glabrous; outer achenes beakless, inner usually beaked but 701
sometimes beakless; and see key; 2n=10. Native; grassy or open ground on sandy soils; frequent in E Anglia and CI, scattered elsewhere in Br N to NE Sc but over-

recorded for small *H. radicata*, decreasing.

3. H. maculata L. - *Spotted Cat's-ear*. Perennial; stems to 60cm, erect, often RR
simple, usually pubescent ± throughout; leaves as in *H. radicata* but usually more 701
shallowly lobed and purple-marked; capitula 2.5-4.5cm across, paler yellow than in *H. radicata*; achenes all beaked; 2n=10. Native; grassy or open ground mostly on calcareous or sandy soils and on maritime cliffs; very local in Br N to Westmorland, Jersey.

21. LEONTODON L. - *Hawkbits*

Perennials; rhizomes and stolons 0; stems leafless, usually with small scale-like bracts, simple or branched; phyllaries in several rows; receptacular scales 0; pappus of 1-2 rows of pale brown to dirty-white hairs, the single or inner row feathery, the outer row simple, or outer achenes with pappus a scaly ring; ligules yellow; achenes not flattened, finely ribbed, not or indistinctly beaked.

1 Leaves glabrous or with simple hairs; pappus of 1 row of feathery hairs; stems usually branched **1. L. autumnalis**
1 Leaves with at least some hairs forked; pappus of 2 rows of hairs, the inner feathery, the outer simple; stems simple 2
2 All achenes with pappus of hairs; phyllaries (9)11-13(15)mm; whole of stem and phyllaries usually conspicuously pubescent **2. L. hispidus**
2 Outer achenes with pappus a scaly ring; phyllaries 7-11mm; upper part of stem and phyllaries usually glabrous to sparsely pubescent **3. L. saxatilis**

1. L. autumnalis L. - *Autumn Hawkbit*. Stems to 60cm, usually at least some 701
branched, glabrous, or sparsely pubescent below; leaves oblanceolate, deeply lobed to (rarely) ± entire; phyllaries 7-12mm, glabrous to pubescent; outermost ligules usually reddish on lowerside; 2n=12. Native; grassy places in very many situations; very common throughout BI. Very variable; distinctive variants occur in saltmarshes (var. **salina** (Aspegren) Lange: stems mostly unbranched; phyllaries glabrous; leaves ± entire) and on mountains (ssp. **pratensis** (Hornem.) Gremli: stems mostly unbranched; phyllaries with dense long dark hairs), but their distinctness and status need study.

2. L. hispidus L. - *Rough Hawkbit*. Stems to 60cm, simple, usually conspicuously 701
pubescent; leaves oblanceolate, sinuate-dentate to deeply lobed; phyllaries usually conspicuously pubescent; outermost ligules usually reddish (rarely greyish-violet) on lowerside; 2n=14. Native; basic, often calcareous, grassland; common in Br N to C Sc, locally frequent in Ir; very scattered in Sc, not in CI or Man. Nearly glabrous plants (var. **glabratus** (W.D.J. Koch) Bisch.) occur but are rare.

2 x 3. L. hispidus x L. saxatilis occurs rarely in En but may be overlooked; it is intermediate in habit and pubescence, but resembles *L. saxatilis* in pappus-type and is <1% fertile; 2n=11; endemic.

3. L. saxatilis Lam. (*L. taraxacoides* (Vill.) Mérat nom. illeg.) - *Lesser Hawkbit*. 701
Differs from *L. hispidus* in stems to 40cm, glabrous or sparsely pubescent; outermost ligules usually greyish-violet on lowerside; and see key; 2n=8. Native; similar places and distribution to *L. hispidus* but frequent in CI and Man and less local in Ir.

22. PICRIS L. - *Oxtongues*

Annuals to perennials; rhizomes and stolons 0; stems leafy, with numerous stiff bristles hooked at apex, with several capitula; phyllaries in several rows; receptacular scales 0; pappus of 2 rows of white to off-white hairs, the inner or both rows feathery; ligules yellow; achenes somewhat flattened, weakly ribbed, transversely wrinkled, beaked or not.

1. P. echioides L. - *Bristly Oxtongue.* Annual or biennial; stems erect, to 80cm; leaves narrowly elliptic-oblanceolate, bristly, entire to dentate; inner phyllaries 12-20mm, the outer 3-5 ovate-cordate, >2x as wide as inner; achenes with beak ± as long as body; pappus white; 2n=10. Probably intrd; well natd in marginal, disturbed and rough ground and waste places; frequent in S & E En, more scattered elsewhere in Br N to N En and in CI, very local in S & E Ir; S Europe.

2. P. hieracioides L. - *Hawkweed Oxtongue.* Biennial to perennial; stems erect, to 1m; leaves narrowly elliptic to oblanceolate, bristly, sinuate-toothed to -lobed; inner phyllaries 11-13mm, the outer ones similar to inner but shorter and often patent; achenes with beak 0 or <1/2 as long as body; pappus off-white; 2n=10. Native; grassland and open or rough ground mostly on calcareous soils; similar distribution to *P. echioides*. Ssp. **spinulosa** (Bertol. ex Guss.) Arcang. (*P. spinulosa* Bertol. ex Guss.), from S Europe, was recorded from rough ground in N W Kent in 1935 and 1936, but the plants might have been variants of ssp. *hieracioides*; the true plant has subsessile crowded capitula with ± glabrous phyllaries.

23. SCORZONERA L. - *Viper's-grass*
Perennials; rhizomes and stolons 0; stems leafy, usually simple, sometimes with 1 or 2 branches; phyllaries in several rows; receptacular scales 0; pappus of several rows of dirty-white feathery hairs; ligules yellow, the outermost flushed crimson on lowerside; achenes slightly flattened, ribbed, not beaked.

Other spp. - S. hispanica L., from C & S Europe, provides the garden vegetable *Scorzonera* and sometimes persists briefly on tips and roadsides; it differs in its larger achenes 10-15mm with rugose (not smooth) ridges.

1. S. humilis L. - *Viper's-grass.* Stems erect, to 50cm, woolly when young, becoming glabrous; leaves entire, narrowly elliptic to linear-lanceolate; capitula 2-3cm across; achenes 7-11mm; 2n=14. Native; marshy fields, now confined to 1 site in Dorset, formerly 1 other site in Dorset and 1 site in Warks. **RRR**

24. TRAGOPOGON L. (*Geropogon* L.) - *Goat's-beards*
Annuals to perennials; rhizomes and stolons 0; stems leafy, simple or branched; phyllaries in 1 row; receptacular scales 0; pappus of 1 row of mainly feathery but some simple hairs, or the outer achenes with entirely simple hairs, dirty-white to pale brown; ligules yellow or purple; achenes not flattened, ribbed, beaked. The capitula open in the morning but close regularly about noon.

1 Ligules yellow; outer achenes often <3cm **1. T. pratensis**
1 Ligules pink to purple; outer achenes ≥3cm 2
  2 Marginal achenes with pappus of simple hairs; achenes very gradually narrowed into beak **3. T. hybridus**
  2 All achenes with pappus with some feathery hairs; achenes abruptly narrowed into beak **2. T. porrifolius**

1. T. pratensis L. - *Goat's-beard.* Annual to perennial; stems erect, to 75cm, glabrous or woolly when young, slightly dilated below capitula; leaves linear to linear-lanceolate, entire, glabrous; ligules yellow.

a. Ssp. pratensis. Ligules as long as or longer than phyllaries. Intrd; grassy places and open or rough ground; very scattered in S & C Br, perhaps over-recorded; Europe.

b. Ssp. minor (Mill.) Wahlenb. Ligules c.1/2-3/4 as long as phyllaries; 2n=12. Native; grassy places, roadsides, rough and cultivated ground; common in most of Br N to C Sc, scattered in N Sc and Ir, Alderney. **Ssp. orientalis** (L.) Celak., from E Europe, has been recorded but is only a rare casual; it has ligules as in ssp. *pratensis* but deeper yellow and achenes with a beak shorter than the body (about as long as body in other 2 sspp.).

1 x 2. T. pratensis x T. porrifolius = T. x mirabilis Rouy occurs rarely in C & S En near *T. porrifolius* but rarely persists; it has yellow ligules suffused purple distally, so that the capitulum appears purple with a yellow centre, and has a low level of fertility.

2. T. porrifolius L. - *Salsify*. Biennial; stems erect, to 1m, glabrous, strongly dilated below capitula; leaves as in *T. pratensis* but more widened at base; ligules purple, c.1/2-1x as long as phyllaries; 2n=12. Intrd; grown as vegetable, casual or rarely natd in waste and rough ground, waysides; very scattered in BI, mainly S En and CI; Mediterranean.

3. T. hybridus L. (*Geropogon glaber* L.) - *Slender Salsify*. Annual; stems erect, to 685
50cm, glabrous, strongly dilated below capitula; leaves linear, entire, glabrous; ligules pink to pinkish-purple, ≤1/2 as long as phyllaries; (2n=14). Intrd; occasional birdseed-alien in gardens, parks, tips and waste ground; scattered in S Br; S Europe.

25. AETHEORHIZA Cass. - *Tuberous Hawk's-beard*

Perennials with long thin rhizomes bearing large tubers; stems simple, with 1 capitulum and 0(-2) leaves; phyllaries in several rows, often weakly 2-rowed; receptacular scales 0; pappus of several rows of white, simple hairs; ligules yellow; achenes not flattened, with 4 deep grooves, not beaked.

1. A. bulbosa (L.) Cass. - *Tuberous Hawk's-beard*. Stems to 30cm, erect, glabrous; 685
leaves obovate to narrowly so, sinuate-dentate; phyllaries (13)14-15(16)mm; achenes 3-4.5mm; (2n=32). Intrd; grown in gardens and very persistent when neglected; very local in Co Wexford and Co Armagh; Mediterranean region.

26. SONCHUS L. - *Sow-thistles*

Annuals or perennials, sometimes with rhizomes; stems leafy, usually branched; phyllaries in several rows; receptacular scales 0; pappus of 2 or more rows of white, simple, hairs; ligules yellow; achenes distinctly flattened, distinctly ribbed, not beaked.

1 Plant annual or biennial, with main root and laterals 2
1 Plant perennial, rhizomatous or with thick ± erect underground portion 3
  2 Auricles of stem-leaves pointed; achenes transversely rugose **3. S. oleraceus**
  2 Auricles of stem-leaves rounded; achenes not rugose **4. S. asper**
3 Auricles of stem-leaves pointed; stems arising from thick underground root-like organ; achenes straw-coloured **1. S. palustris**
3 Auricles of stem-leaves rounded; plant strongly rhizomatous; achenes bright brown **2. S. arvensis**

1. S. palustris L. - *Marsh Sow-thistle*. Stems erect, to 2.5m, glabrous below, R
glandular-pubescent above with usually dark glands; lower leaves deeply pinnately lobed, upper ones unlobed except for auricles; achenes 3.5-4mm, weakly transversely rugose, usually with 6-10 ribs; 2n=18. Native; marshes, fens and riversides; SE En from W Kent to E Norfolk, S Hants, formerly N to N Lincs and W to Staffs and Oxon, intrd in parts of C En.

2. S. arvensis L. - *Perennial Sow-thistle*. Stems erect, to 1.5m; glabrous to densely glandular-pubescent above with usually yellowish glands; most leaves variably pinnately lobed, very rarely as deeply as lower ones or as little as upper ones of *S. palustris*; achenes 2.5-3.5mm, transversely rugose, usually with 12 ribs; 2n=54. Native; arable and waste land, waysides, dunes and shingle by sea, ditches and river-banks; common throughout lowland BI. Glabrous plants (ssp. **uliginosus** (M. Bieb.) Nyman) are said to have shorter phyllaries (mostly 12-16mm, not mostly 15-

18mm) and have (2n=36), but their status needs checking.

3. S. oleraceus L. - *Smooth Sow-thistle.* Stems erect, to 1.5m, usually glabrous except glandular-pubescent below capitula; leaves very variously divided, the middle ones usually deeply pinnately lobed; phyllaries usually glandular-pubescent, often also woolly at first; achenes 2.5-3.75mm, variably c.8-12-ribbed; 2n=32. Native; waste and cultivated ground, roadsides; abundant throughout lowland BI.

3 x 4. S. oleraceus x S. asper has occurred very rarely in C & S En, with more records doubtful or erroneous; it has leaf-auricles rounded and dentate as in *S. asper* but with one long pointed tooth, and is sterile; 2n=25.

4. S. asper (L.) Hill - *Prickly Sow-thistle.* Habit and pubescence as in *S. oleraceus*; leaves often less lobed than in *S. oleraceus* but with more sharply dentate margins, but sometimes very deeply lobed; achenes 2-3mm, usually with 3 thin ribs on each face and 2 wide wing-like marginal ones; 2n=18. Native; similar places and distribution to *S. oleraceus*, and often with it. Ssp. **glaucescens** (Jord.) Ball differs in being biennial (often with a basal leaf rosette) and in having better developed recurved spinules on the achene ribs; its status needs checking.

27. LACTUCA L. (*Mulgedium* Cass.) - *Lettuces*

Annuals or biennials, or perennials with rhizomes; stems leafy, branched at least above, with many capitula; phyllaries in several rows; receptacular scales 0; pappus of 2 rows of white, simple hairs; ligules yellow or blue; achenes distinctly flattened, ribbed, beaked.

1 Achenes with beak <1/2 as long as body and of same colour; ligules blue; perennial **5. L. tatarica**
1 Achenes with beak ≥1/2 as long as body and much lighter in colour; ligules yellow; annual to biennial 2
 2 Plant <1m; leaf midrib on lowerside glabrous to sparsely hispid; achenes without bristles 3
 2 Plant often 1-2m; leaf midrib on lowerside with strong prickles; achenes with minute bristles just below beak 4
3 Stem-leaves oblong-ovate, cordate and clasping stem at base; inflorescence subcorymbose **2. L. sativa**
3 Middle and upper stem-leaves linear-oblong, sagittate and clasping stem at base; inflorescence very narrow **4. L. saligna**
 4 Ripe achenes (excl. beak) (4)4.2-4.8(5.2)mm, maroon to blackish; stems and midribs strongly tinged maroon **3. L. virosa**
 4 Ripe achenes (excl. beak) (2.8)3-4(4.2)mm, olive-grey; stems and midribs greenish-white **1. L. serriola**

1. L. serriola L. - *Prickly Lettuce.* Stems erect, to 2m; leaves deeply pinnately lobed **691**
to (more often) unlobed, sharply dentate; achenes olive-grey, mostly 3-4mm, with very narrow lateral wings visible ± only near apex, with white beak c. as long; 2n=18. Probably native; waysides and waste and rough ground; frequent in En SE of line from Severn to Humber, especially in SE and E Anglia, very scattered elsewhere in En, Wa and CI.

2. L. sativa L. - *Garden Lettuce.* Stems erect, to 75cm; leaves unlobed or the basal **691**
pinnately lobed, entire to remotely denticulate; achenes brownish-grey, 3-4mm, with wings as in *L. serriola*, with white beak 1/2-1x as long; (2n=18). Intrd; casual on tips, waste ground and abandoned arable land, also birdseed-alien; scattered in Br, mainly S; probably arose from *L. serriola* in E Mediterranean.

3. L. virosa L. - *Great Lettuce.* Stems erect, to 2m; leaves as in *L. serriola* but more **691**
often well lobed, often undulate, and usually maroon-tinged; achenes maroon to blackish, mostly 4.2-4.8mm, with lateral wings clearly visible from base to apex, with white beak c. as long or shorter; 2n=18. Probably native; similar distribution

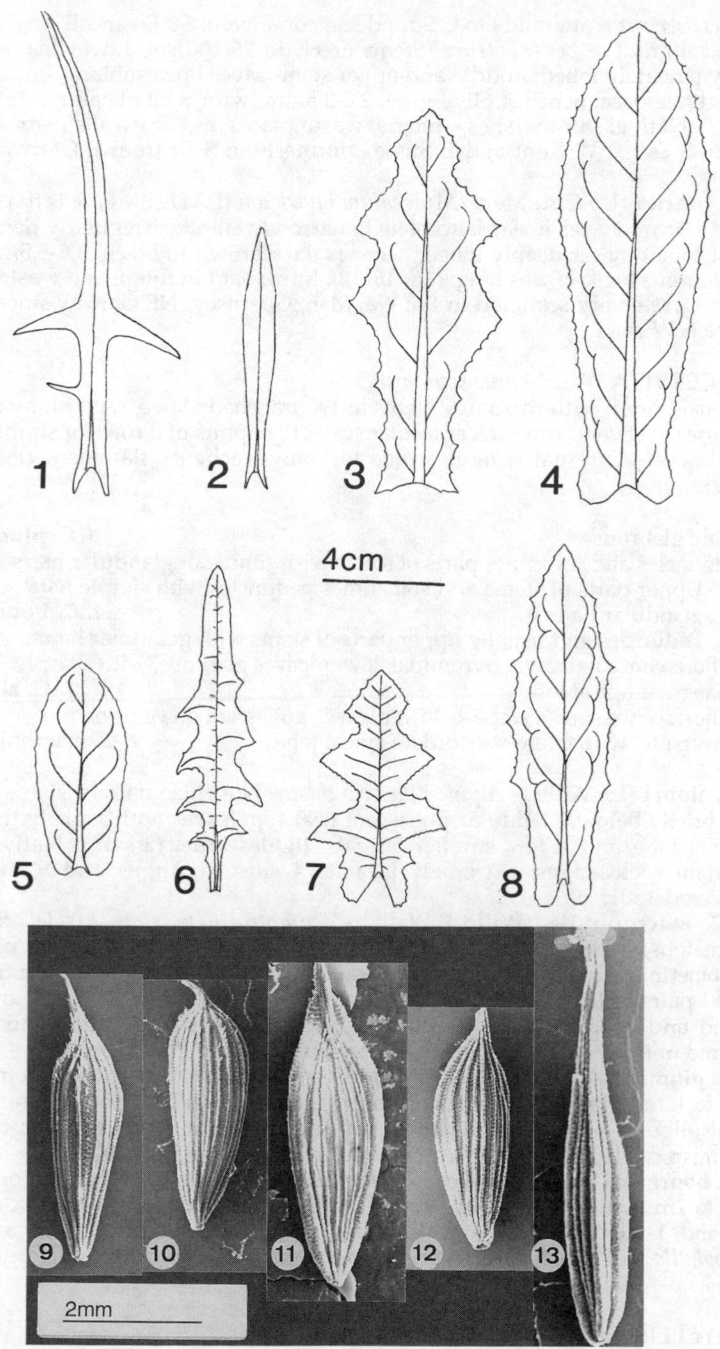

FIG 691 - *Lactuca*. 1-8, leaves. 1-2, *L. saligna*. 3-4, *L. virosa*. 5, *L. sativa*. 6, *L. tatarica*. 7-8, *L. serriola*. 9-13, achenes, pappus removed. 9, *L. serriola*. 10, *L. sativa*. 11, *L. virosa*. 12, *L. saligna*. 13, *L. tatarica*.

to *L. serriola* but scattered N to C Sc and less common in SE En and E Anglia.

4. L. saligna L. - *Least Lettuce.* Stems erect, to 75(100)cm; lower leaves rather RRR
deeply pinnately lobed; middle and upper stem-leaves linear-oblong, entire except 691
for sagittate base; achenes olive-grey, 2.8-3.5mm, with white beak 1/2-1x as long; (2n=18). Native; salt-marshes, shingle, waste places and sea-walls near sea; very local in S Essex, W Kent and E Sussex, formerly in S Br from E Cornwall to W Norfolk.

5. L. tatarica (L.) C.A. Mey. (*Mulgedium tataricum* (L.) DC.) - *Blue Lettuce.* Stems 691
erect, to 80cm; leaves linear-lanceolate to narrowly elliptic, irregularly dentate, the lowest sometimes ± deeply lobed; achenes dark brown to black, 3.5-5.5mm, with concolourous beak <1/2 as long; (2n=16, 18). Intrd; natd in rough and waste ground mostly by sea; very scattered in En, Wa, Man, Guernsey, NE Galway since 1923; E Europe & W Asia.

28. CICERBITA Wallr. - *Blue-sow-thistles*

Perennials, often with rhizomes; stems leafy, branched above, with many capitula; phyllaries in several rows; receptacular scales 0; pappus of 2 rows of simple, dirty- to yellowish-white hairs; ligules blue to mauve; achenes flattened, ribbed, not beaked.

1 Plant glabrous **3. C. plumieri**
1 Peduncles and/or upper parts of stems with simple or glandular hairs 2
  2 Upper parts of stems and sometimes peduncles with simple hairs; glandular hairs 0 **4. C. bourgaei**
  2 Peduncles and usually upper parts of stems with glandular hairs 3
3 Inflorescence narrowly pyramidal; lower leaves glabrous, with sharply triangular apical lobe **1. C. alpina**
3 Inflorescence subcorymbose; lower leaves pubescent on veins on lowerside, with ovate-subcordate apical lobe **2. C. macrophylla**

1. C. alpina (L.) Wallr. - *Alpine Blue-sow-thistle.* Not rhizomatous; stems erect, to RRR
1.3m, bristly below, glandular-pubescent near top; leaves with sharply triangular terminal lobe and a few smaller laterals; ligules blue; (2n=18). Native; moist mountain rock-ledges; extremely local in 4 sites in Angus and S Aberdeen, discovered 1801.

2. C. macrophylla (Willd.) Wallr. - *Common Blue-sow-thistle.* Strongly rhizomatous; stems erect, to 2m, glabrous below, glandular-pubescent on peduncles and sometimes upper parts of stems; leaves with ovate-subcordate terminal lobe and 0-1 pairs of laterals; ligules pale mauve. Intrd; well natd on rough and waste ground and roadsides, first recorded 1915; frequent throughout most of Br, scattered in Ir; Urals. Our plant is ssp. **uralensis** (Rouy) P.D. Sell.

3. C. plumieri (L.) Kirschl. - *Hairless Blue-sow-thistle.* Not rhizomatous; stems erect, to 1.3m, glabrous; leaves with triangular terminal lobe and several pairs of laterals; ligules blue; (2n=16). Intrd; natd as for *C. macrophylla*, much rarer but often misrecorded for it; very scattered in En and Sc; WC Europe.

4. C. bourgaei (Boiss.) Beauverd - *Pontic Blue-sow-thistle.* Not rhizomatous; stems erect, to 2m, with rather sparse simple hairs; leaves with ovate-rhombic terminal lobe and 1-few pairs of laterals; ligules pale mauve. Intrd; natd as for *C. macrophylla* but much rarer; very scattered in En, Sc and Man; Georgia & NE Turkey.

29. MYCELIS Cass. - *Wall Lettuce*

Perennials, often short-lived; stems leafy, branched above, with many capitula; phyllaries in 2 very unequal rows; receptacular scales 0; pappus of 2 unequal rows of white simple hairs; ligules yellow; achenes distinctly flattened, ribbed, very shortly beaked.

1. M. muralis (L.) Dumort. - *Wall Lettuce.* Usually suffused maroon; stems erect, to 1m, glabrous; leaves deeply pinnately lobed, sharply dentate; capitula 1-1.5cm across, with c.5 flowers; 2n=18. Native; shady places in woods, on walls and rocks and in hedgerows; locally common in En and Wa, scattered in Sc, Ir and Man.

30. TARAXACUM F.H. Wigg. - *Dandelions*

Perennials with tap-roots; stems usually leafless, with 1 capitulum; phyllaries in 2 often very different rows; receptacular scales 0; pappus of several rows of white, simple hairs; ligules yellow, usually with coloured stripe(s) on lowerside; achenes not flattened, finely ribbed, usually spinulose near apex, beaked.

A very critical genus in which apomixis is the rule; 229 microspp. are currently recognized in BI, of which >40 are probably endemic but <1/2 of the rest native. Triploids (2n=24), tetraploids (2n=32), pentaploids (2n=40) (rare) and hexaploids (2n=48) (rare) occur, of which the latter 3 are probably all obligately apomictic. In BI almost all triploids are also obligately apomictic, but in a few sexuality occurs very rarely, producing non-persistent sexual diploids that can hybridize with each other and with pollen from apomictic plants to produce diploid to triploid hybrids.

In this work the microspp. are not treated in full but are aggregated into 9 rather ill-defined sections, determination of which is often not easy even after much experience.

In most spp. the achene is spinulose near its apex, but between that region and the beak there is a short, usually pyramidal region known as the *cone*. Descriptions apply only to fully ripe achenes; *achene length* excludes cone and beak. Leaves produced in summer do not maintain all diagnostic characters, so determination should be attempted only with specimens collected during the first main flush of flowering (usually Apr to early May in the lowlands). Plants from shaded, heavily trodden or grazed, or mown areas should be avoided.

This account follows the sectional and microsp. delimitation recognized by A.J. Richards, C.C. Haworth and A.A. Dudman.

1 Plants delicate, usually with strongly dissected (often nearly pinnate) leaves; outer row of phyllaries mostly <8mm, with small outgrowth near apex on lowerside; capitula rarely >3cm across 2

1 Plants usually medium to robust, rarely with nearly pinnate leaves; outer row of phyllaries mostly >8mm, without subapical outgrowth; capitulum usually >3cm across 3

2 Achenes greyish-brown, with pyramidal cone c.0.4mm; leaves often with ≥6 pairs of lateral lobes **2. T. sect. Obliqua**

2 Achenes usually purplish-violet, reddish or yellowish-brown, with cylindrical cone mostly 0.6-1mm; leaves rarely with >6 pairs of lateral lobes **1. T. sect. Erythrosperma**

3 Outer row of phyllaries appressed, ovate, with broad scarious border; leaves very narrow, usually scarcely lobed **3. T. sect. Palustria**

3 Outer row of phyllaries appressed to recurved, linear to narrowly ovate, with no or with narrow to very narrow scarious border; leaves broader, usually distinctly lobed 4

4 Leaves and petioles green; rare plants of a few mountain cliffs in Sc **6. T. sect. Taraxacum**

4 Lowland plants, or if on mountain cliffs then leaves usually dark or blotched or spotted with purple and petiole usually purple 5

5 Achenes (excl. cone and beak) ≥4mm, nearly cylindrical; outer row of phyllaries erect to appressed; ligules usually with dark red stripes on lowerside; pollen usually 0 **4. T. sect. Spectabilia**

5 Achenes very rarely >4mm, narrowly top-shaped; outer row of phyllaries

rarely appressed; ligule stripes rarely dark red; pollen present or 0 6

6 Leaves with large dark spots covering >10% of surface **5. T. sect. Naevosa**

6 Leaves unspotted or with spots covering <10% of blade (beware leaves damaged or attacked by pathogens) 7

7 Petiole and midrib uppersides green or solid red or purple; outer row of phyllaries mostly 9-16mm, usually recurved, not dark on lowerside; leaves often complexly lobed and folded in 3 dimensions **9. T. sect. Ruderalia**

7 Petiole and midrib uppersides usually minutely (lens) striped red or purple; outer row of phyllaries mostly 7-12mm, often (often not) patent to erect and dark on lowerside; leaves ± flat, relatively simply lobed 8

8 Lateral leaf-lobes broad-based, with convex front and concave rear edge, commonly 4 pairs; outer row of phyllaries usually arched to varying degrees, often subobtuse **8. T. sect. Hamata**

8 Lateral leaf-lobes rarely as above, often 5-6 pairs; outer row of phyllaries erect to recurved all ± to same degree, often acute **7. T. sect. Celtica**

1. T. sect. Erythrosperma (H. Lindb.) Dahlst. (*T. laevigatum* (Willd.) DC. group, 695
T. simile Raunk. group, *T. fulvum* Raunk. group). Plants small (to medium), delicate; leaves very deeply dissected, very rarely spotted, usually with purplish petiole and midrib; ligules pale to deep yellow, usually striped red to purplish on lowerside; outer phyllaries mostly 5-9mm, appressed to recurved; achenes 2.5-3.5mm, variously coloured red or purple to brown or straw-coloured but seldom greyish-brown, with cone mostly 0.6-1mm. 30 microspp. currently placed here; triploids or tetraploids. Mostly native; dry exposed places, usually on well-drained soils, short grassland, heathland, dunes; throughout BI, commonest lowland section after *Hamata* and *Ruderalia*, mostly maritime in N; 8 endemics.

2. T. sect. Obliqua Dahlst. (*T. obliquum* (Fr.) Dahlst. group). Plants small, 696
delicate; leaves very deeply dissected, not spotted, with green petiole and midrib; ligules deep to orange-yellow, striped red on lowerside; outer phyllaries 6-7mm, appressed to erect; achenes c.3mm, greyish-brown, with cone c.0.4mm. 2 microspp. currently placed here; triploids. Native; open sandy turf by sea; local on coasts of BI, commonest in Sc, but S to CI. **T. platyglossum** Raunk. is more widespread than **T. obliquum**.

3. T. sect. Palustria (Lindeb.) Dahlst. (*T. palustre* (Lyons) Symons group). Plants 697
medium-sized; leaves narrow, scarcely or very shallowly lobed, not spotted, with purple petiole and midrib; ligules yellow to deep yellow, striped purplish or greyish on lowerside; outer phyllaries (3)6-7mm, appressed; achenes mostly 3.5-4.3mm, brown to straw-coloured, with cone 0.5-1mm. 4 microspp. currently placed here; tetraploids and pentaploids. Native; wet usually base-rich meadows and fen grassland; local, scattered in BI; 2 endemics. **T. palustre** is the best-known sp. with its distinctive ± unlobed linear leaves.

4. T. sect. Spectabilia (Dahlst.) Dahlst. (*T. spectabile* Dahlst. group). Plants 695
mostly medium-sized; leaves medium to scarcely lobed, usually spotted, with purplish petiole and midrib; ligules bright to deep yellow, striped purplish on lowerside; outer phyllaries 7-8mm, appressed (to erect); achenes 4-5mm, straw-coloured, with cone 0.3-0.4mm. 3 microspp. currently placed here; pentaploids. Native; damp or wet acidic grassy places, often in upland areas, also roadsides etc.; throughout BI; 2 Shetland endemics and **T. faeroense** (Dahlst.) Dahlst., which is possibly the commonest *Taraxacum* in native habitats in N Br.

5. T. sect. Naevosa M.P. Christ. (*T. naevosum* Dahlst. group, *T. praestans* H. 696
Lindb. group pro parte). Plants medium-sized to robust; leaves medium to deeply lobed, usually spotted, usually with purplish petiole and midrib; ligules mid to deep yellow, striped purplish or greyish on lowerside; outer phyllaries (6)9-14mm, patent (to erect); achenes 3-4mm, reddish-brown to straw-coloured, with cone 0.5-

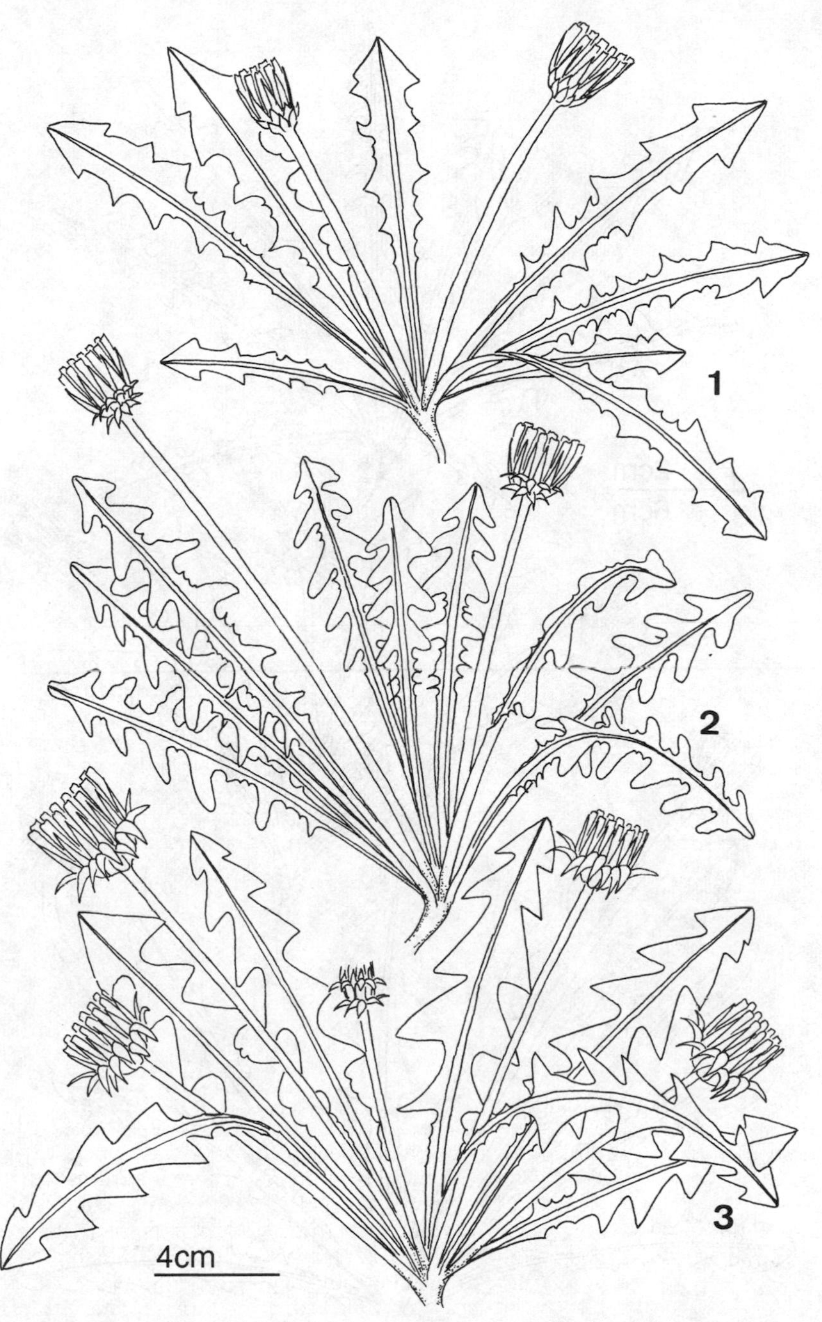

FIG 695 - *Taraxacum*. 1, *T. faeroense* (sect. *Spectabilia*). 2, *T. lacistophyllum* (sect. *Erythrosperma*). 3, *T. hamatum* (sect. *Hamata*).

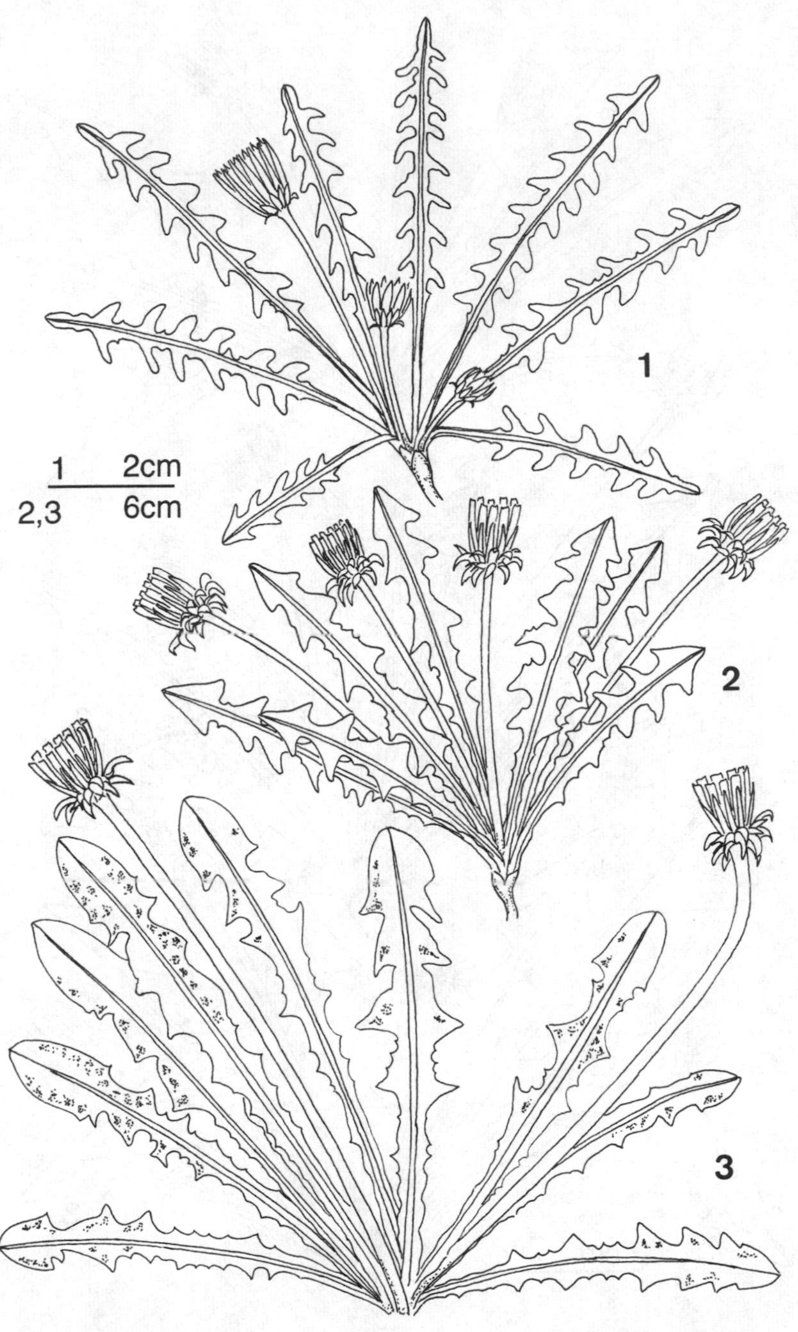

FIG 696 - *Taraxacum*. 1, ***T. obliquum*** (sect. ***Obliqua***).
2, ***T. croceiflorum*** (sect. ***Ruderalia***). 3, ***T. euryphyllum*** (sect. ***Naevosa***).

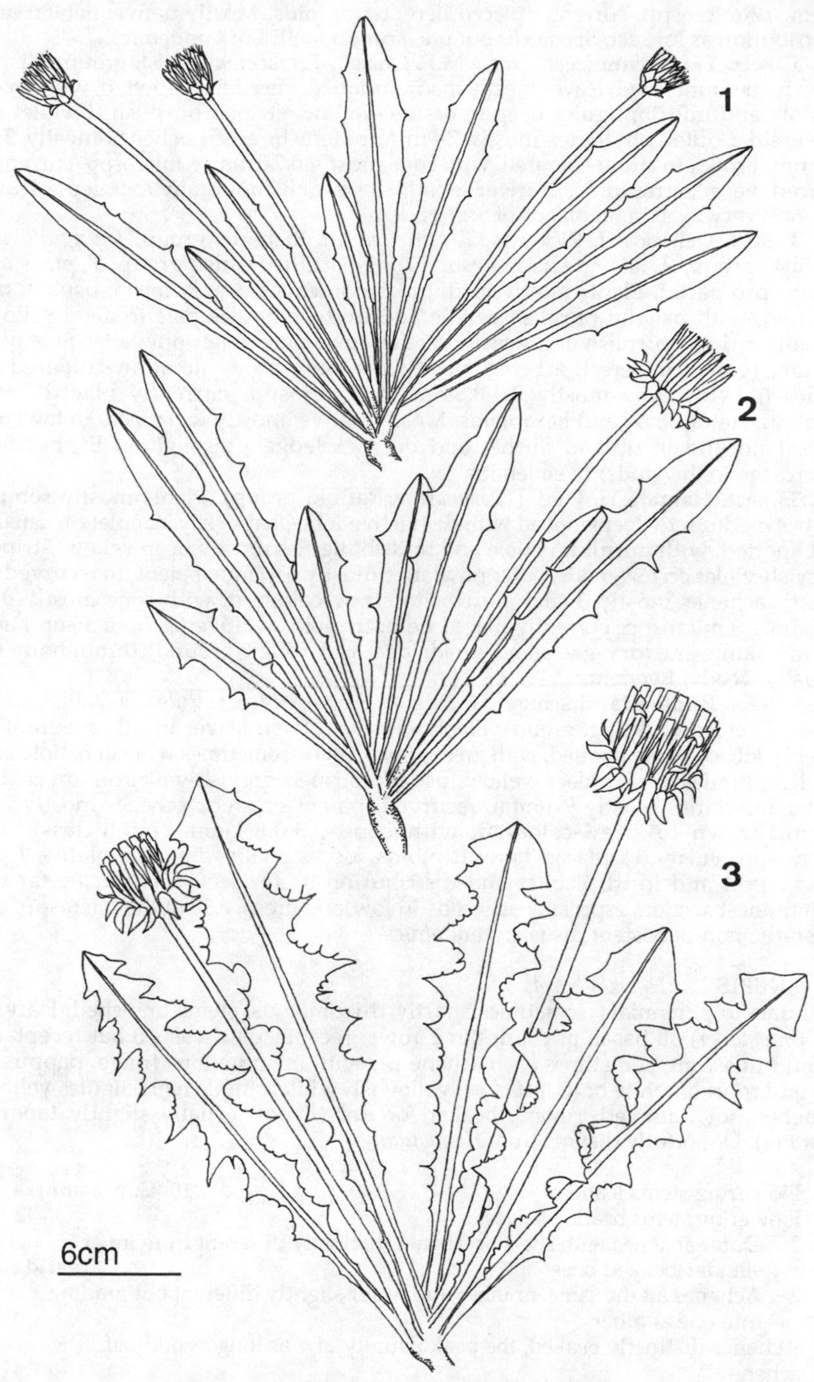

FIG 697 - *Taraxacum*. 1, *T. palustre* (sect. *Palustria*). 2, *T. ceratolobum* (sect. *Taraxacum*). 3, *T. duplidentifrons* (sect. *Celtica*).

1mm. 12 microspp. currently placed here; tetraploids. Mostly native; habitat and distribution as for sect. *Spectabilia* but uncommon in SE En; 4 endemics.

6. T. sect. Taraxacum (sect. *Crocea* M.P. Christ., *T. croceum* Dahlst. group). Plants **697**
mostly medium-sized; leaves mostly medium-lobed, very rarely spotted, with green petiole and midrib; ligules deep to orange-yellow, striped purplish to violet on lowerside; outer phyllaries mostly 7-9mm, patent to erect; achenes mostly 3.5-5.3mm, brown to straw-coloured, with cone mostly 0.7-1mm. 6 microspp. currently placed here; tetraploids. Native; mostly base-rich mountain rock-ledges and flushes; very local in highlands of Sc; 1 endemic.

7. T. sect. Celtica A.J. Richards (*T. celticum* A.J. Richards group, *T. unguilobum* **697**
Dahlst. group, *T. adamii* Claire group, *T. nordstedtii* Dahlst. group, *T. praestans* group pro parte). Plants mostly medium-sized; leaves deeply lobed, usually not spotted, with usually purplish petiole and midrib; ligules pale to deep yellow, usually striped purplish or greyish-violet on lowerside; outer phyllaries mostly 7-12mm, (patent to) erect; achenes mostly 2.8-4mm, brown to straw-coloured or reddish, with cone mostly 0.3-0.8mm. 34 microspp. currently placed here; triploids, tetraploids and hexaploids. Mostly native; mostly wet places in lowland grassland and in upland flushes and on rock-ledges; throughout BI, but few microspp. in lowlands; 18 endemics.

8. T. sect. Hamata H. Øllg. (*T. hamatum* Raunk. group). Plants mostly robust; **695**
leaves medium to deeply lobed with distinctive lobe-shape (key, couplet 8), usually not spotted, with purplish petiole and midrib; ligules mostly deep yellow, striped greyish-violet on lowerside; outer phyllaries mostly 8-13mm, patent (to recurved or erect); achenes mostly 3-4mm, brown to straw-coloured, with cone mostly 0.3-0.7mm. 18 microspp. currently placed here; triploids. Native (c. 7 microspp.) and intrd; damp and dry grassland, roadsides and rough ground; throughout BI, usually weedy; 1 endemic.

9. T. sect. Ruderalia Kirschner, H. Øllg. & Stepanek (sect. *Vulgaria* Dahlst. nom. **696**
illeg., *T. officinale* Wigg. group). Plants mostly robust; leaves mostly medium to deeply lobed, rarely spotted, with green, purplish or sometimes whitish petiole and midrib; ligules mid to deep yellow, usually striped greyish-violet on lowerside; outer phyllaries mostly 9-16mm, recurved, patent or erect; achenes mostly 2.5-4mm, brown to straw-coloured, with cone 0.3-0.8(1)mm (rarely less). 120 microspp. currently placed here; triploids, occasionally diploids. Native (c.26 microspp.) and intrd; habitat and distribution as for sect. *Hamata*. By far the commonest section, especially as weeds in lowland areas. c.50 of the microspp. are sporadic non-persistent casuals; 5 endemics.

31. CREPIS L. - *Hawk's-beards*

Annuals to perennials, sometimes shortly rhizomatous; stems branched, leafy or (*C. praemorsa*) all basal; phyllaries in 2 rows; receptacular scales 0 but receptacle often pubescent, sometimes each achene-pit with membranous fringe; pappus of several rows of white or (*C. paludosa*) yellowish-white, simple hairs; ligules yellow; achenes not flattened, ribbed, beaked or not (if not, usually slightly tapered distally). Doubtfully distinct from *Hieracium*.

1 Flowering stems leafless **10. C. praemorsa**
1 Flowering stems bearing leaves 2
  2 Outer achenes with short or 0 beak, distinctly different from inner slender-beaked ones **9. C. foetida**
  2 Achenes all the same, or inner and outer slightly different but grading into one another 3
3 Achenes distinctly beaked, the beak usually ≥1/2 as long as body of achene 4
3 Achenes not beaked, but often narrowed at apex 6
  4 Basal lobes of upper stem-leaves not clasping stem; achenes with

beak scarcely 1/2 as long as body **4. C. tectorum**
4 Basal lobes of upper stem-leaves clasping stem; achenes with beak c. as long as body 5
5 Upper parts of plant nearly always with many patent stiff bristles; achenes (incl. beak) 3-5.5mm **8. C. setosa**
5 Upper parts of plant without patent stiff bristles; achenes (incl. beak) (5)6-8(9)mm **7. C. vesicaria**
6 Phyllaries pubescent on their inner faces 7
6 Phyllaries glabrous on their inner faces 8
7 Achenes (2.5)3-4(4.5)mm, with 10 ribs distinctly rough towards apex **4. C. tectorum**
7 Achenes 4-7.5mm, with 13-20 ± smooth ribs **3. C. biennis**
8 Pappus-hairs yellowish-white, brittle **1. C. paludosa**
8 Pappus-hairs pure white, ± flexible 9
9 Achenes c.20-ribbed; perennial arising from short rhizome; rare in N En and Sc **2. C. mollis**
9 Achenes 10-ribbed; common annual or biennial with tap-root 10
10 Achenes 2.5-3.8mm; outer phyllaries patent to erecto-patent; receptacle with laciniate membranous fringes around achene-pits **5. C. nicaeensis**
10 Achenes 1.4-2.5mm; outer phyllaries appressed to inner; receptacle with a few hairs around achene-pits **6. C. capillaris**

1. C. paludosa (L.) Moench - *Marsh Hawk's-beard.* Erect subglabrous perennial to 80cm; stem-leaves toothed to shallowly lobed, clasping stem at base; phyllaries with woolly eglandular and straight glandular hairs; achenes 4-5.5mm, not beaked; 2n=12. Native; wet places in open woodland, grassland, fens; BI N from S Ir, S Wa and Leics, frequent in Sc and N En. 701

2. C. mollis (Jacq.) Asch. - *Northern Hawk's-beard.* Erect subglabrous to sparsely pubescent perennial to 60cm; stem-leaves entire to sinuate-toothed, tapering to broad base; phyllaries with glandular hairs; achenes 3-4.5mm, not beaked; 2n=12. Native; grassy, often damp slopes or hills; very local from MW Yorks to E Perth (formerly Banffs), formerly Denbs. R 701

3. C. biennis L. - *Rough Hawk's-beard.* Erect pubescent biennial to 1.2m; stem-leaves irregularly and sharply lobed, ± clasping stem at base; phyllaries with eglandular and often glandular hairs; achenes 4-7.5mm, not beaked; 2n=40. Probably native; rough grassy places, waysides; scattered in Br and Ir N to C Sc, frequent in SE En. 701

4. C. tectorum L. - *Narrow-leaved Hawk's-beard.* Erect to ascending subglabrous annual to 75cm; stem-leaves entire to sinuate-toothed, tapered to base; phyllaries with eglandular and glandular hairs, achenes 2.5-4(4.5)mm (incl. beak), strongly tapered at apex or shortly beaked; (2n=8). Intrd; occasional casual on roadsides, disturbed soil and re-seeded verges, a grain- and grass-seed-alien; scattered in En, Ir and Sc; Europe. 701

5. C. nicaeensis Balb. - *French Hawk's-beard.* Erect to ascending rather sparsely pubescent annual or biennial to 80cm; leaf-shape and capitula as in *C. capillaris* but latter usually larger; phyllaries with glandular and eglandular hairs; achenes 2.5-3.8mm, not beaked; (2n=8). Intrd; habitat, source and distribution as for *C. tectorum*, becoming rare except in S En; W & C Mediterranean. 701

6. C. capillaris (L.) Wallr. - *Smooth Hawk's-beard.* Erect to decumbent glabrous to sparsely pubescent annual or biennial to 75cm; stem-leaves subentire to deeply and sharply lobed, not clasping stem; phyllaries glabrous or pubescent, with (var. **glandulosa** Druce) or without glandular hairs; achenes 1.4-2.5mm, not beaked; 2n=6. Native; grassy places, rough and waste ground; common throughout BI, var. *glandulosa* commonest in the N, especially Sc. 701

7. C. vesicaria L. - *Beaked Hawk's-beard.* Erect, pubescent perennial to 80cm; stem- 701 leaves deeply sharply lobed, clasping stem at base; phyllaries usually with eglandular and often with glandular hairs; achenes 5-9mm incl. beak c. as long as body; 2n=8. Intrd; grassy places, waysides, walls and rough ground; common in S and CE En, scattered N to N En and in CI and Ir; Europe. Our plant is ssp. **taraxacifolia** (Thuill.) Thell. ex Schinz & R. Keller (ssp. *haenseleri* (Boiss. ex DC.) P.D. Sell).

8. C. setosa Haller f. - *Bristly Hawk's-beard*. Erect, usually hispid (very rarely **685**
subglabrous) annual or biennial to 75cm; stem-leaves usually toothed to sharply **701**
lobed, clasping stem at base; phyllaries with patent stiff eglandular hairs; achenes 3-5.5mm incl. beak c. as long as body; 2n=8. Intrd; occasional casual with crops or grass or in rough ground; scattered in Br, mostly C & S; S Europe.

9. C. foetida L. - *Stinking Hawk's-beard*. Erect, pubescent annual or biennial to **RRR**
60cm, stinking when fresh; stem-leaves few, usually deeply sharply lobed, ± **701**
clasping stem at base; phyllaries with eglandular and glandular hairs; inner achenes 10-15mm incl. beak c. as long as body; outer achenes 6-9mm incl. beak much shorter than body; (2n=8, 10). Native; waysides and rough ground; now probably only on shingle at Dungeness (E Kent), formerly scattered in SE En and intrd NW to Worcs.

10. C. praemorsa (L.) Walther - *Leafless Hawk's-beard*. Erect, sparsely pubescent **RR**
perennial to 60cm; stem-leaves 0; basal leaves narrowly obovate, not or shallowly **701**
toothed; phyllaries sparsely pubescent; achenes 3-4mm, not beaked; (2n=8). Possibly native; on natural calcareous grassy bank, Westmorland, discovered 1988.

32. PILOSELLA Hill (*Hieracium* subg. *Pilosella* (Hill) Gray) - *Mouse-ear-hawkweeds*
Perennials, usually stoloniferous; stems usually leafless, sometimes with few leaves, with 1-many capitula; basal leaves oblanceolate to narrowly obovate or narrowly elliptic, pubescent, subentire; phyllaries in several rows; receptacular scales 0 but achene-pits often variously fringed; pappus of 1 row of dirty-white to pale brown, simple hairs; ligules yellow to orange; achenes not flattened, 10-ribbed, not beaked, scarcely tapered towards apex. Doubtfully distinct from *Hieracium*.

Both apomictic and sexual plants occur, often within 1 sp; hybridization is frequent wherever 2 spp. occur together. Of the native spp., *P. peleteriana* is always diploid and sexual, *P. flagellaris* is hexaploid, whereas *P. officinarum* may be sexual (diploid to tetraploid) or apomictic (pentaploid or hexaploid); the sexual plants are commoner in the S and rare or 0 in Sc, and the diploids and triploids are known only in E Anglia.

1 All flowering stems with only 1 capitulum 2
1 At least some flowering stems with >1 capitulum 3
  2 Stolons elongated, slender, with spaced out small leaves, 0 or few ending in leaf-rosette **2. P. officinarum**
  2 Stolons few or 0, short (<5cm), stout, with full-sized ± crowded leaves, often ending in leaf-rosette **1. P. peleteriana**
3 Ligules orange-brown to brick-red, often turning purplish when dried **7. P. aurantiaca**
3 Ligules yellow, sometimes red-striped on lowerside 4
  4 Capitula (1)2-4(7) per stem, not crowded; phyllaries (8)9-12mm **3. P. flagellaris**
  4 Capitula (3)6-50 per stem, many closely crowded; phyllaries 5-9mm 5
5 Largest leaves ≤12(20)mm wide; phyllaries mostly <1mm wide, acute **4. P. praealta**
5 Largest leaves ≥(12)15mm wide; phyllaries mostly >1mm wide, obtuse 6
  6 Leaves scarcely glaucous; capitula >10 on well-developed inflorescences **5. P. caespitosa**
  6 Leaves distinctly glaucous; capitula <10 per inflorescence **6. P. x floribunda**

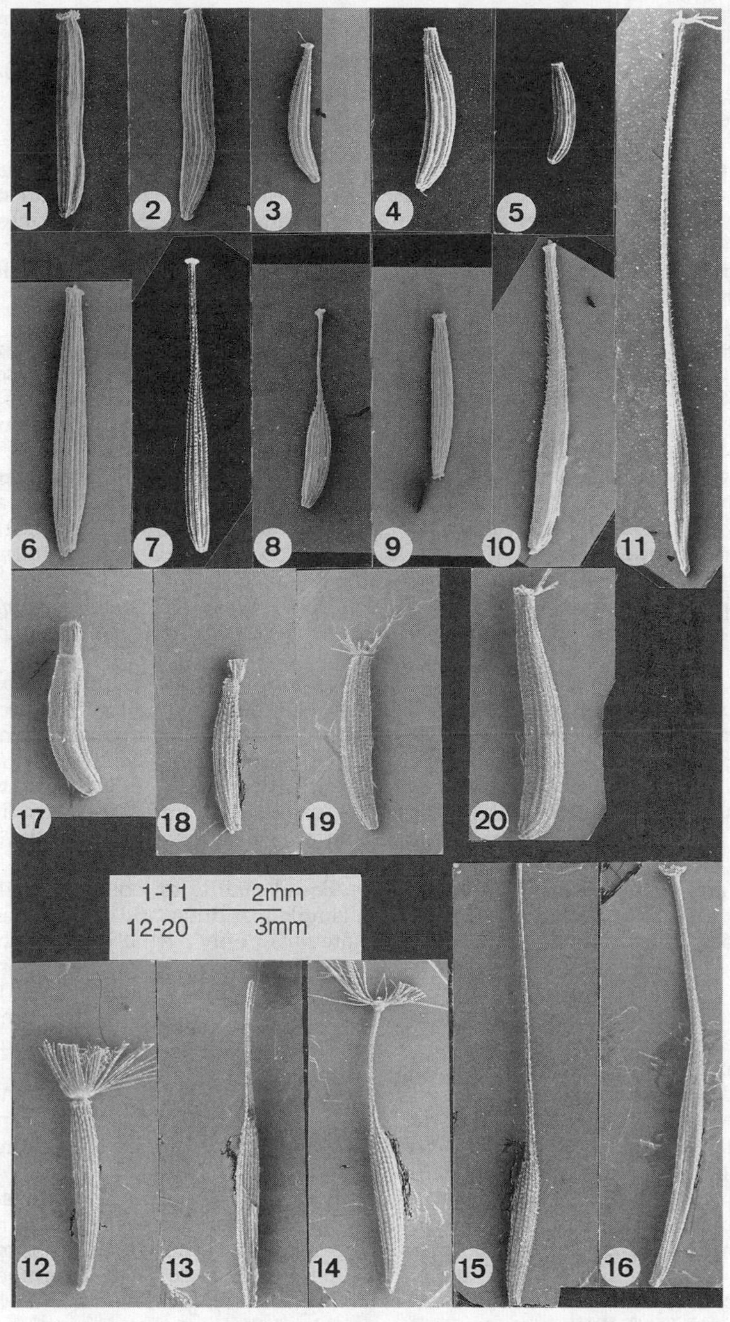

FIG 701 - Achenes of **Asteraceae: Lactuceae**, pappus-hairs removed. 1, ***Crepis paludosa***. 2, ***C. mollis***. 3, ***C. tectorum***. 4, ***C. nicaeensis***. 5, ***C. capillaris***. 6, ***C. biennis***. 7, ***C. vesicaria***. 8, ***C. setosa***. 9, ***C. praemorsa***. 10-11, ***C. foetida*** (outer, inner). 12-13, ***Hypochaeris glabra*** (outer, inner). 14-15, ***H. radicata*** (outer, inner). 16, ***H. maculata***.
17-20, ***Leontodon***. 17-18, ***L. saxatilis*** (outer, inner). 19, ***L. autumnalis***. 20, ***L. hispidus***.

Other spp. - **P. lactucella** (Wallr.) P.D. Sell & C. West (*H. lactucella* Wallr.), from Europe, has 1-7 capitula per stem and glaucous, glabrous to very sparsely pubescent leaves; it was formerly natd in S Wilts.

1. P. peleteriana (Mérat) F.W. Schultz & Sch. Bip. (*H. peleterianum* Mérat) - *Shaggy Mouse-ear-hawkweed.* Like a robust, large-headed *P. officinarum* with stolons 0 or short, thick and often ending in a leaf-rosette; phyllaries with dense long eglandular, few or 0 glandular, and usually few stellate hairs. Native; short grassland on well-drained soils, dunes; very local, but commoner in CI than *P. officinarum.* The sspp. are probably better as vars. RR

1 Scapes up to 12(18)cm; rosette-leaves 9-20mm wide, not or scarcely tapered at base **a. ssp. peleteriana**
1 Scapes (6)10-30cm; rosette-leaves 4-12(18)mm wide, long-tapered at base 2
 2 Phyllaries 11-15mm, lanceolate; capitula 12-17mm across excl. ligules **b. ssp. subpeleteriana**
 2 Phyllaries 10-12(13)mm, linear-lanceolate; capitula (9)10-12(14)mm across excl. ligules **c. ssp. tenuiscapa**

a. Ssp. peleteriana. Phyllaries 11-15mm, lanceolate; capitula 12-20mm across excl. ligules; 2n=18. CI (all islands), Dorset, Wight and E Kent (extinct in last).

b. Ssp. subpeleteriana (Nägeli & Peter) P.D. Sell (*H. peleterianum* ssp. *subpeleterianum* Nägeli & Peter). Monts (Craig Breidden).

c. Ssp. tenuiscapa (Pugsley) P.D. Sell & C. West (*H. peleterianum* ssp. *tenuiscapum* (Pugsley) P.D. Sell). 2n=18. Jersey, S Devon, Staffs, Derbys and MW Yorks.

1 x 2. P. peleteriana x P. officinarum = P. x longisquama (Peter) Holub (*H. x longisquamum* Peter, *P. x pachylodes* (Nägeli & Peter) Soják nom. illeg.) has occurred with the parents in Jersey, Guernsey, Staffs and E Kent; it is intermediate; (2n=27).

2. P. officinarum F.W. Schultz & Sch. Bip. (*H. pilosella* L.) - *Mouse-ear-hawkweed.* Stolons long, slender, with ± distant reduced leaves; scapes to 30(50)cm, with 1 capitulum, densely pubescent; ligules yellow; capitula 7-12mm across excl. ligules; 2n=18, 27, 36, 45, 54. Native; short grassland on well-drained soils, banks, rocky places; locally common throughout BI except Shetland. Very variable in scape height and robustness and in pubescence. Based mainly on colour, length and relative distribution of simple eglandular, simple glandular and stellate hairs on phyllaries, 7 taxa can be recognized. These are not or only partially geographically separated and are no more than vars.

2 x 6. P. officinarum x P. aurantiaca = P. x stoloniflora (Waldst. & Kit.) F.W. Schultz & Sch. Bip. (*H. x stoloniflorum* Waldst. & Kit.) occurs with the parents in scattered places from Guernsey to N Sc; it is intermediate in ligule colour and capitulum number per scape; 2n=c.63.

3. P. flagellaris (Willd.) P.D. Sell & C. West (*H. flagellare* Willd.) - *Shetland Mouse-ear-hawkweed.* Stolons long, stout, leafy; scapes to 40cm, with (1)2-4(7) capitula, with glandular and eglandular hairs; ligules yellow; phyllaries with numerous simple glandular and eglandular hairs and sparse stellate hairs.

a. Ssp. flagellaris. Scapes to 40cm; capitula 2-4(7); peduncles with simple eglandular hairs 2-3mm; phyllaries with few to numerous simple eglandular hairs ≤1.5mm; (2n=36, 42). Intrd; natd on grassy roadsides and railway banks as garden escape; scattered in C & CS En and CE SC, first recorded 1869; C & E Europe.

b. Ssp. bicapitata P.D. Sell & C. West (*H. flagellare* ssp. *bicapitatum* (P.D. Sell & C. West) P.D. Sell). Scapes to 18cm; capitula (1)2(-4); peduncles with simple eglandular hairs ≤7.5mm; phyllaries with dense simple eglandular hairs ≤2.5mm; 2n=54. Native; dry rocky pastures, rocky slopes and outcrops; 3 localities in Shetland; endemic, discovered 1962. RR

4. P. praealta (Vill. ex Gochnat) F.W. Schultz & Sch. Bip. (*H. praealtum* Vill. ex Gochnat) - *Tall Mouse-ear-hawkweed.* Stolons 0 to long; scapes to 65cm, with numerous capitula, pubescent above but glabrous to subglabrous below; ligules yellow; (2n=45). Intrd; natd garden escape on grassy roadsides, walls and railway banks; Europe.

a. Ssp. praealta. Stolons 0 or very short; phyllaries with numerous glandular and 0 or few eglandular hairs. Scattered localities in Br N to Ayrs, first recorded 1899.

b. Ssp. thaumasia (Peter) P.D. Sell (ssp. *arvorum* (Nägeli & Peter) P.D. Sell & C. West, ssp. *spraguei* (Pugsley) P.D. Sell & C. West, *H. pilosella* ssp. *thaumasium* (Peter) P.D. Sell). Stolons long and slender; phyllaries with numerous glandular and 0 to numerous eglandular hairs. Scattered localities in SC En and W Lothian, first recorded 1918.

5. P. caespitosa (Dumort.) P.D. Sell & C. West (*H. caespitosum* Dumort.) - *Yellow Fox-and-cubs*. Stolons strong, often above and below ground, with large leaves; scapes to 50(80)cm, with numerous crowded capitula, pubescent; ligules yellow; (2n=36). Intrd; natd garden escape on rough ground, walls and railway banks; scattered in Br and S Ir; N & E Europe. Our plant is ssp. **colliniformis** (Peter) P.D. Sell & C. West (*H. pilosella* ssp. *colliniforme* (Peter) P.D. Sell).

6. P. x floribunda (Wimm. & Grab.) Arv.-Touv. (*H. x floribundum* Wimm. & Grab., *P. lactucella* ssp. *helveola* (Dahlst.) P.D. Sell & C. West, *H. helveolum* (Dahlst.) Pugsley; *P. lactucella* x *P. caespitosa*) - *Irish Fox-and-cubs*. Differs from *P. caespitosa* in scapes to 35(45)cm; leaves less pubescent; fewer and less robust stolons; and see key. Intrd; natd in *Calluna*/*Erica* heathland; S Hants since 1991, formerly Co Antrim; N & C Europe.

7. P. aurantiaca (L.) F.W. Schultz & Sch. Bip. (*H. aurantiacum* L.) - *Fox-and-cubs*. Stolons strong, often above and below ground, with large leaves; scapes to 40(65)cm, with numerous crowded capitula, pubescent; ligules orange-brown to brick red. Intrd; natd garden escape (setting abundant seed) on rough ground, walls, roadsides and railway banks; N & C Europe. The 2 sspp. are of doubtful value.

a. Ssp. aurantiaca. Stolons mostly underground; basal leaves mostly 10-20 x 2-6cm; phyllaries 8-11mm; 2n=36. Very scattered in Br.

b. Ssp. carpathicola (Nägeli & Peter) Soják (ssp. *brunneocrocea* (Pugsley) P.D. Sell & C. West, *H. aurantiacum* ssp. *carpathicola* Nägeli & Peter, *H. brunneocroceum* Pugsley). Stolons mostly above ground; basal leaves mostly 6-10(16) x 1.2-2(3)cm; phyllaries 5-8mm. Frequent throughout BI.

33. HIERACIUM L. - *Hawkweeds*

Perennials, without stolons or rhizomes; stems leafy or sometimes not, with (1)few-several capitula, with or without basal rosette of leaves at flowering; phyllaries in several rows; receptacular scales 0 but achene-pits often variously fringed; pappus of 1 row of dirty-white to pale brown, simple hairs; ligules yellow; achenes not flattened, 10-ribbed, not beaked, scarcely tapered towards apex.

All the taxa are obligate apomicts and are triploids with 2n=27 or tetraploids with 2n=36 so far as is known, except for 1 un-named apomictic pentaploid with 2n=45 in sect. *Alpina* and the single sp. of section *Hieracioides* (*H. umbellatum* L.), which exists as diploid sexual plants with 2n=18 and triploid apomictic plants with 2n=27. 261 microspp. are currently recognized in BI, of which many are endemic and probably a considerable number are aliens. In this work they are not treated in full, but are aggregated into 15 sections that are recognizable after a little practice. The sectional classification adopted follows the views of P.D. Sell.

Plants often exhibit a second phase of flowering on new growth, either naturally or if the first growth is damaged. Only the first growth provides reliable diagnostic characters. As a rule of thumb, identification should not be attempted on plants with 0-1 stem-leaves after mid-Jun, on plants with 2-8 stem-leaves after mid-Jul, and on others after mid-Aug.

1 Stem-leaves 8-many except in dwarfed plants; rosette of leaves usually 0 at flowering 2

1 Stem-leaves 0-8(15); rosette of leaves usually present at flowering 7
2 Middle stem-leaves not or scarcely clasping stem at base 3
2 Middle stem-leaves distinctly clasping stem at base, though often very narrowly so 5
3 Leaves all sessile, often linear-lanceolate, ± all of similar shape, with recurved margins; phyllaries (except innermost) with recurved tips; styles yellow when fresh **1. H. sect. Hieracioides**
3 Lower leaves petiolate, usually broader, middle and upper ones sessile or nearly so, not with recurved margins; phyllaries very rarely with recurved tips; styles usually dark when fresh 4
4 Stem-leaves rarely <15, often crowded, upper ones with broad rounded bases **2. H. sect. Sabauda**
4 Stem-leaves usually <15, rarely crowded, upper ones narrowed to base **4. H. sect. Tridentata**
5 Middle stem-leaves slightly constricted just above the broad clasping base; peduncles with dense glandular hairs; achenes pale brown **5. H. sect. Prenanthoidea**
5 Middle stem-leaves not constricted, with narrow clasping base; peduncles with 0-few glandular hairs; achenes purplish- or blackish-brown 6
6 Stem-leaves c.10-30, the lower ones clasping stem at base to merely sessile; phyllaries sparsely pubescent and glandular; ligules glabrous at tip **3. H. sect. Foliosa**
6 Stem-leaves c.2-10(15), the lower ones subpetiolate; phyllaries moderately pubescent and glandular; ligules glabrous or pubescent at tip **6. H. sect. Alpestria**
7 Stem-leaves 1-7(12), clasping stem at base 8
7 Stem-leaves 0-8(12), not clasping stem at base 9
8 Stem-leaves yellowish-green; plant viscid-glandular **11. H. sect. Amplexicaulia**
8 Stem-leaves glaucous-green; plant not viscid-glandular **12. H. sect. Cerinthoidea**
9 Stems, leaves and phyllaries with dense, white, patent hairs; E Norfolk **13. H. sect. Andryaloidea**
9 White patent hairs not dense on stems, leaves and phyllaries 10
10 Leaves with small glandular hairs on margins and sometimes on surface; phyllaries usually with shaggy hairs; almost confined to Sc, N Wa and Lake District 11
10 Leaves without stalked glands; phyllaries without shaggy hairs; widespread 12
11 Stem-leaves 0-4, narrow and bract-like; capitula 1(-5); plants to 15(30)cm **15. H. sect. Alpina**
11 Stem-leaves (0)1-4, usually at least one leaf-like; capitula (1)2-5; plants to 45cm **14. H. sect. Subalpina**
12 Leaves usually bristly at least along margins; phyllaries erect in bud, without dense white stellate hairs 13
12 Leaves variously pubescent but not bristly; phyllaries incurved in bud, with dense white stellate hairs at least on margins 14
13 Stem-leaves 2-10(12); basal leaves few (mostly ≤4) **10. H. sect. Oreadea**
13 Stem-leaves 0-1(2); basal leaves numerous **9. H. sect. Stelligera**
14 Stem-leaves 0-2(3); basal leaves numerous **7. H. sect. Hieracium**
14 Stem-leaves 2-8(15); basal leaves few (commonly 2-4), often withering at flowering **8. H. sect. Vulgata**

1. H. sect. Hieracioides Dumort. (sect. *Umbellata* (Fr.) Gremli). Plants normally >30cm; basal leaves 0 at flowering; stem-leaves normally >15, ovate-lanceolate to **706**

linear, all sessile, not clasping stem at base; capitula few to many; phyllaries ± obtuse, at least outer ones patent to recurved; vegetative parts with rather sparse eglandular simple and stellate hairs; capitula and peduncles sparsely pubescent and with some small glandular hairs. 1 microsp. placed here (**H. umbellatum** L.). Native; sandy heathland, dunes and dry rocky places, often near the coast; scattered throughout BI but mostly in W and S. Ssp. **bichlorophyllum** (Druce & Zahn) P.D. Sell & C. West has broader leaves (the lower narrowly ovate to oblong, not linear to linear-lanceolate) and is confined to S & W Wa, SW En, W Ir and CI.

2. H. sect. Sabauda (Fr.) Gremli (*H. sabaudum* L. group). Plants normally >30cm; **706**
basal leaves 0 at flowering; stem-leaves normally >15, ovate to ovate-lanceolate, the lower ones petiolate, not clasping stem at base; capitula many; phyllaries ± obtuse, rarely patent or recurved; pubescence as in sect. *Hieracioides*. 5 microspp. currently placed here. Native; common in rough ground, grassy and marginal places and on roadside and railway banks; En and Wa, rather local in Sc, very local in E Ir. **H. sabaudum** (*H. perpropinquum* (Zahn) Druce) is the commonest many-leaved *Hieracium* of S & C En and S Wa; **H. vagum** Jord. is the commonest in N En and N Wa.

3. H. sect. Foliosa (Fr.) Dahlst. (*H. inuloides* Tausch group, *H. crocatum* Fr. group). **706**
Plants normally >30cm; basal leaves 0 at flowering; stem-leaves c.10-30, ovate-lanceolate to lanceolate, the lower ones petiolate, narrowly clasping stem at base; capitula few to many; phyllaries ± obtuse, rarely patent or recurved; vegetative parts with rather sparse eglandular simple and stellate hairs; capitula and peduncles with eglandular and glandular hairs. 10 microspp. currently placed here. Native; grassy and rocky places; locally common in N Br S to S Wa and Peak District, very local in E, N & W Ir. **H. subcrocatum** (E.F. Linton) Roffey is the commonest sp. of the section in N En and Wa.

4. H. sect. Tridentata (Fr.) Gremli (*H. laevigatum* Willd. group). Stems normally **707**
>30cm; basal leaves 0 or few at flowering; stem-leaves c.6-30, ovate-lanceolate to ± linear, the lower ones petiolate, not clasping stem at base; capitula few to many; phyllaries ± obtuse, appressed; pubescence as in sect. *Hieracioides*. 21 microspp. currently placed here. Native; grassy, rocky and marginal habitats; frequent in Br, rare in W & N Ir. Several spp. are locally but none widely common.

5. H. sect. Prenanthoidea W.D.J. Koch (*H. prenanthoides* Vill. group, *H. juranum* **707**
Fr. group). Stems normally >30cm; basal leaves 0 at flowering; stem-leaves c.6-30, the lower ones oblanceolate with winged petiole, the upper ones ovate to ovate-lanceolate and broadly clasping stem at base; capitula usually numerous; phyllaries ± obtuse, appressed; vegetative parts with fairly dense simple and stellate eglandular hairs and sometimes some glandular hairs above; capitula and peduncles with stellate and numerous glandular hairs but 0-few simple eglandular ones. 2 microspp. placed here, **H. prenanthoides** with the sectional distribution, **H. borreri** Syme formerly in Selkirks but now extinct. Native; grassy and rocky places, often on limestone; local in N Br S to S Wa and Peak District, Co Antrim.

6. H. sect. Alpestria (Fr.) Gremli (incl. 4 sp. groups). Plants normally >20cm; **707**
basal leaves 0 or few at flowering; stem-leaves c.2-15, mostly lanceolate-elliptic, the lower ones petiolate, narrowly or very narrowly clasping stem at base; capitula c.2-10; phyllaries ± obtuse, appressed; pubescence often fairly dense, usually with glandular and simple and stellate eglandular hairs on all parts. 19 microspp. currently placed here. Native; rocky places, cliffs, hillsides; 14 microspp. endemic to Shetland, others very local in Sc and N En.

7. H. sect. Hieracium (sects. *Bifida* (Arv.-Touv.) A.R. Clapham, *Glandulosa* **708**
(Pugsley) A.R. Clapham, *Sagittata* (Pugsley) A.R. Clapham; incl. 4 sp. groups). Plants mostly >(10)20cm; basal leaves normally numerous and present at flowering, variable in shape, long-petiolate, often violet on lowerside, sometimes glaucous, sometimes purple-spotted; stem-leaves 0-2(3), variable in shape, usually petiolate, not clasping stem at base; capitula mostly 2-20; phyllaries obtuse to acute, appressed; pubescence usually not dense, variously of glandular and simple

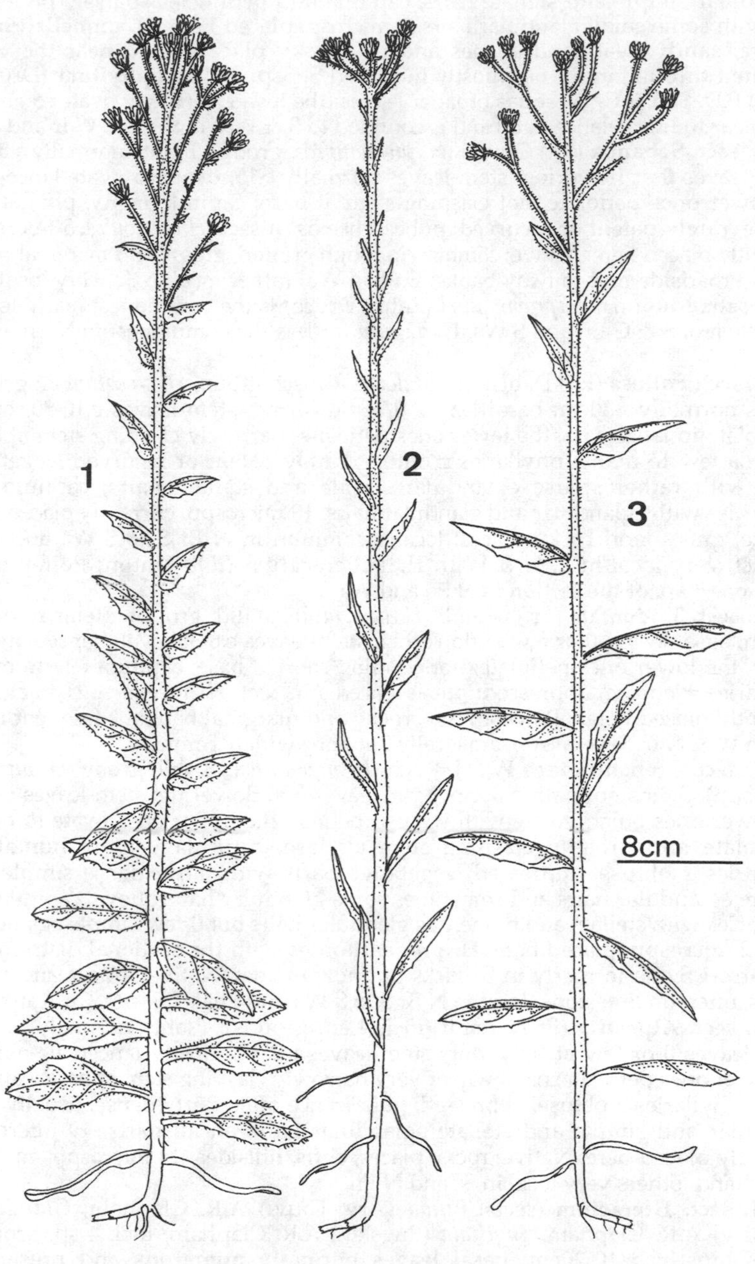

FIG 706 - *Hieracium*. 1, *H. vagum* (sect. *Sabauda*).
2, *H. umbellatum* (sect. *Hieracioides*). 3, *H. latobrigorum* (sect. *Foliosa*).

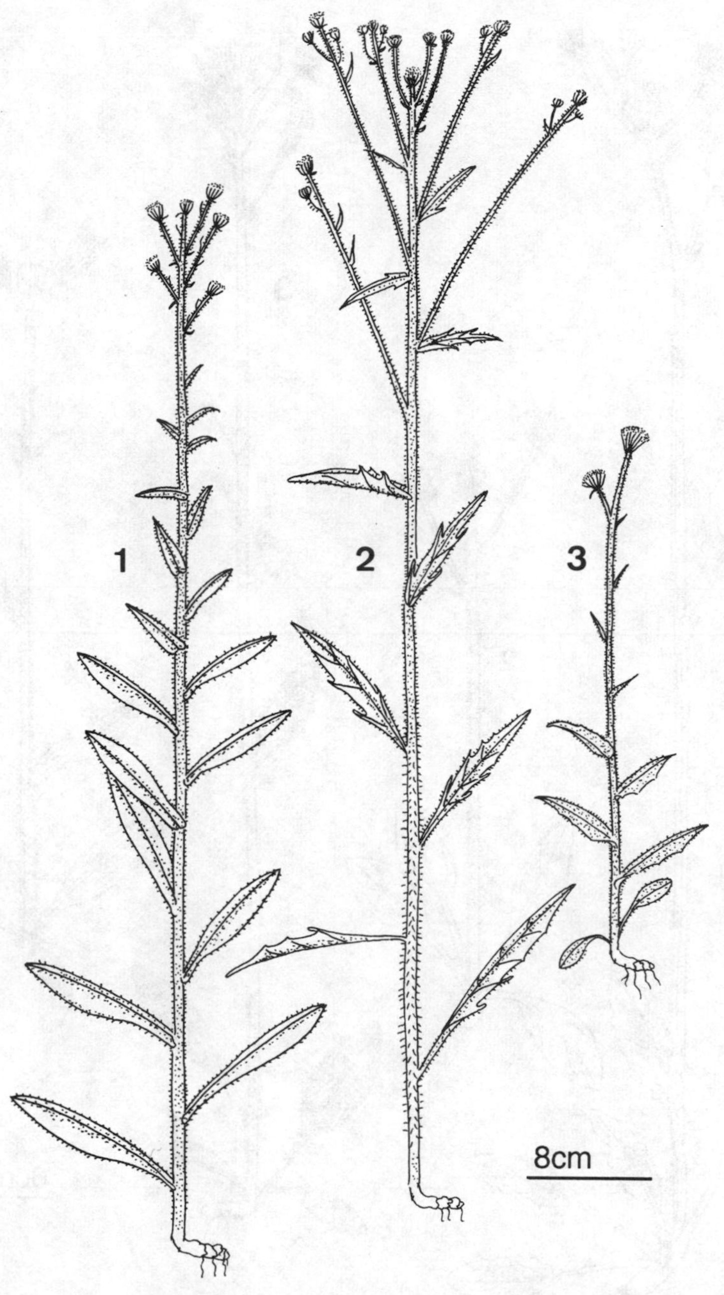

FIG 707 - *Hieracium*. 1, *H. prenanthoides* (sect. *Prenanthoidea*). 2, *H. trichocaulon* (sect. *Tridentata*). 3, *H. hethlandiae* (sect. *Alpestria*).

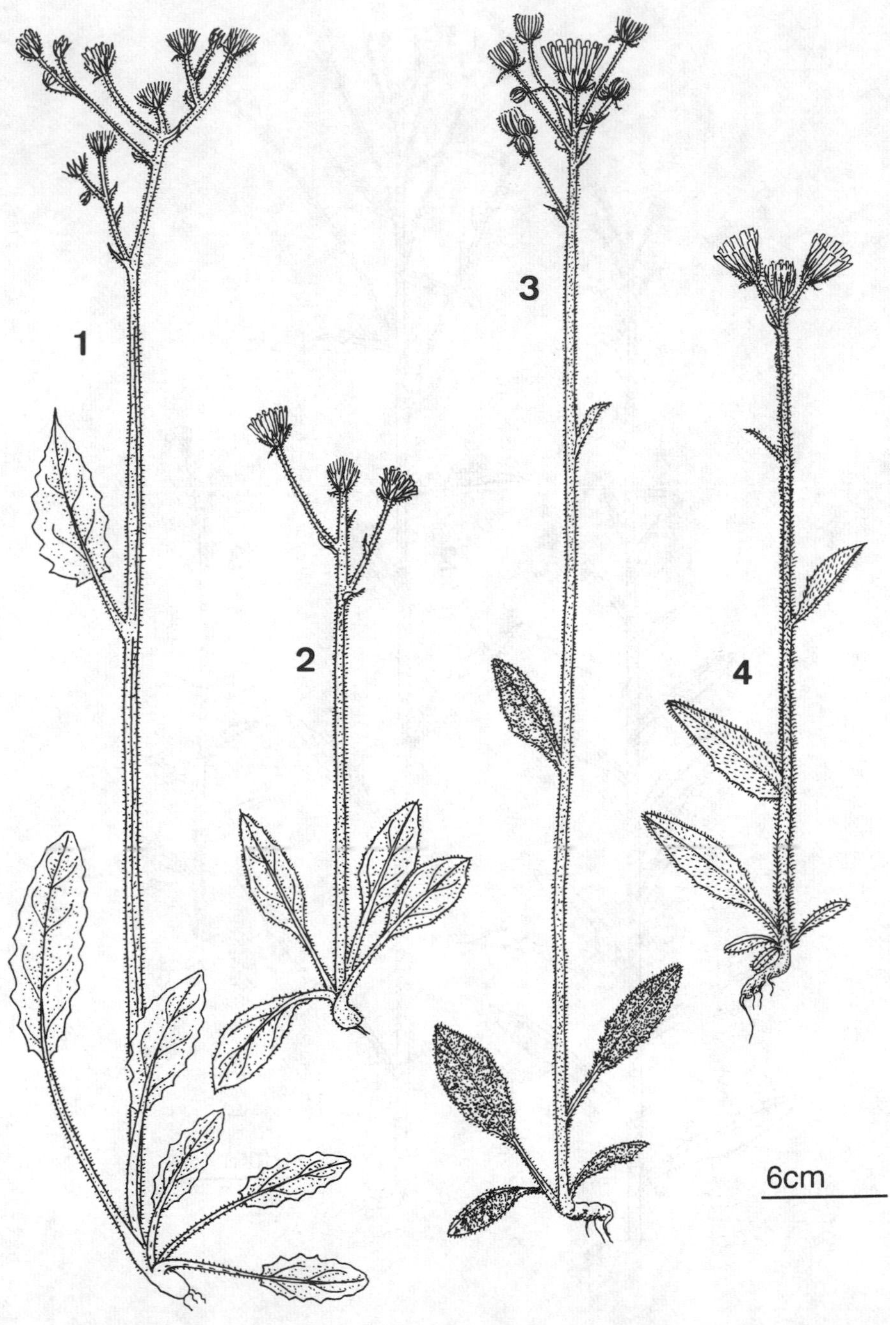

FIG 708 - ***Hieracium***. 1, ***H. exotericum*** (sect. ***Hieracium***). 2, ***H. leyi*** (sect. ***Stelligera***). 3, ***H. maculatum*** (sect. ***Vulgata***). 4, ***H. proximum*** (sect. ***Oreadea***).

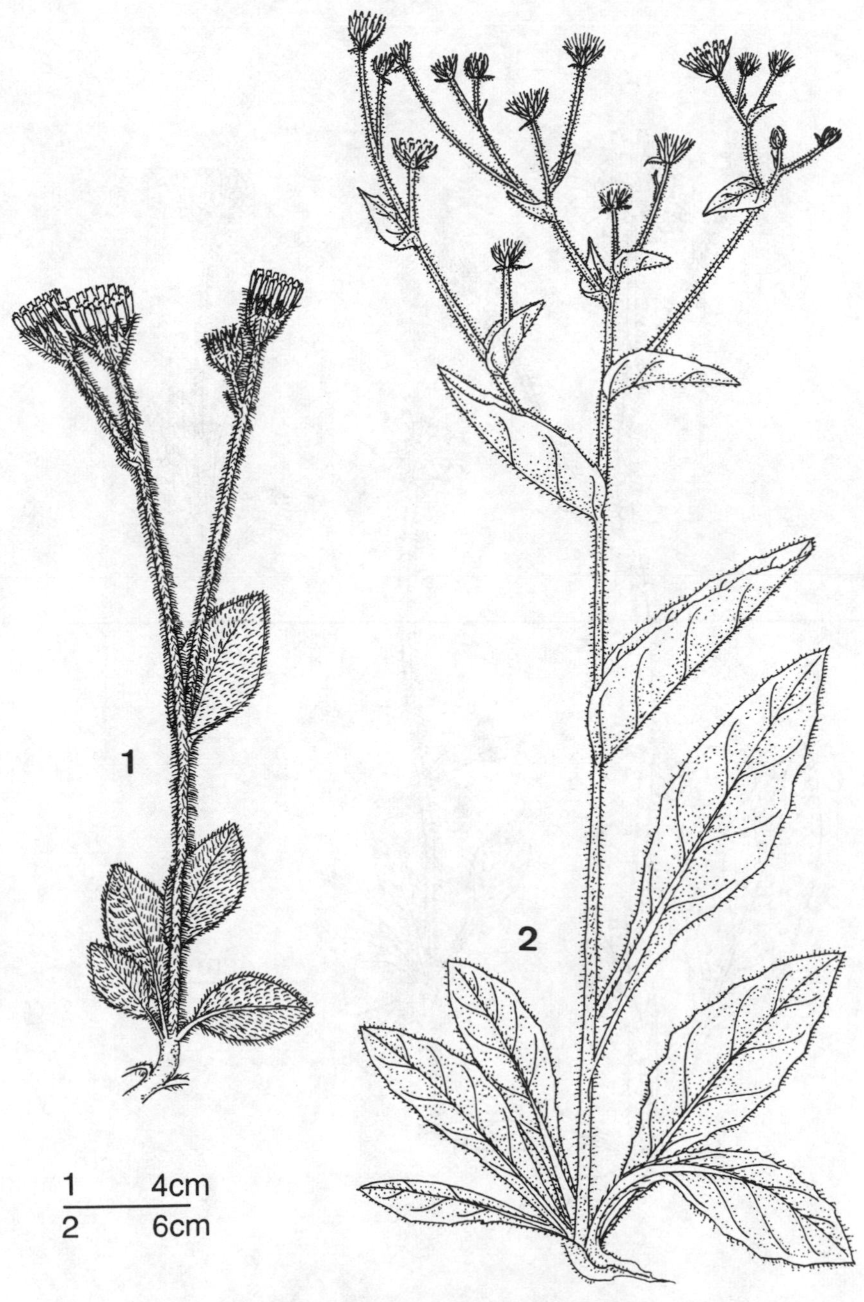

FIG 709 - ***Hieracium*****.**
1, ***H. lanatum*** (sect. ***Andryaloidea***). 2, ***H. speluncarum*** (sect. ***Amplexicaulia***).

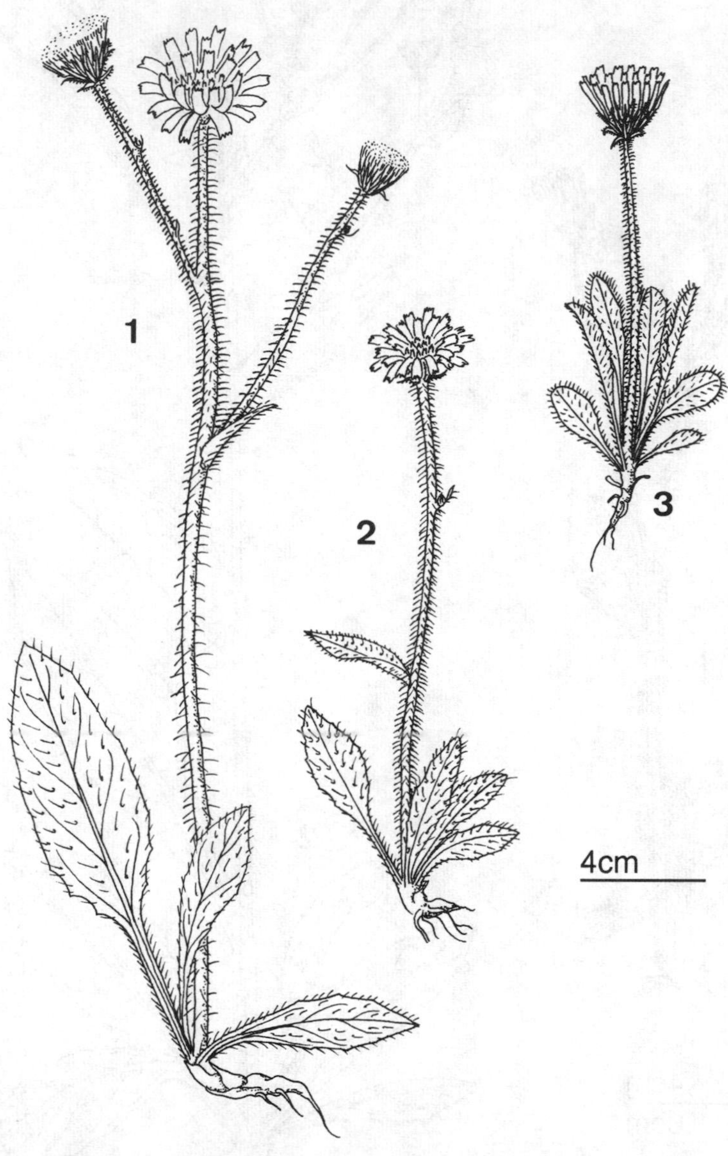

FIG 710 - *Hieracium*. 1, ***H. anglicum*** (sect. *Cerinthoidea*).
2, ***H. lingulatum*** (sect. *Subalpina*). 3, ***H. alpinum*** (sect. *Alpina*).

and stellate eglandular hairs but glandular hairs usually absent from leaves. 53 microspp. currently placed here. Native and intrd; rough ground, woodland, marginal habitats, cliffs and rocky places; throughout BI. The commonest microspp. of lowland Br belong here and in *Vulgata*, and are the earliest to flower. They are particularly characteristic of railway and roadside walls and banks.

8. H. sect. Vulgata (Griseb.) Willk. & Lange (sect. *Caesia* (Almq.) A.R. Clapham; **708**
incl. 5 sp. groups). Differs from sect. *Hieracium* in basal leaves fewer (commonly 2-4), sometimes withering at flowering; and stem-leaves 2-8(15). 30 microspp. currently placed here. Native and intrd; habitat and distribution as for sect. *Hieracium*. **H. maculatum** Sm. is the commonest sp. of the genus with heavily purple-blotched leaves.

9. H. sect. Stelligera Zahn (sect. *Suboreadea* Pugsley; incl. 4 sp. groups). Plants **708**
mostly 10-60cm; basal leaves normally numerous and present at flowering, ovate to lanceolate, long-petiolate, glaucous; stem-leaves mostly 0-1(2), often all or most much reduced, not clasping stem at base; capitula mostly 2-12; phyllaries acute, appressed; all parts variously with glandular and simple and stellate eglandular hairs. 30 microspp. currently placed here. Native; cliffs, rocky and grassy banks, often on limestone; scattered in Ir, Wa, Sc and W, C & N En. Despite the many spp. all are either scattered or local and none occurs in E or SE En.

10. H. sect. Oreadea (Fr.) Dahlst. (incl. 6 sp. groups). Differs from sect. *Stelligera* **708**
in basal leaves fewer, mostly ≤4; and stem-leaves 2-10(12). 15 microspp. currently placed here. Native; habitat and distribution as for sect. *Stelligera*. Differs from sect. *Stelligera* in same way that sect. *Vulgata* differs from sect. *Hieracium*, i.e. fewer basal leaves and more stem-leaves.

11. H. sect. Amplexicaulia (Griseb.) Scheele (*H. amplexicaule* L. group). Plants 10- **709**
60cm; basal leaves present at flowering, ovate to lanceolate, with winged petiole; stem-leaves 2-6(12), ovate-elliptic, broadly clasping stem at base; capitula 2-15(20); phyllaries acute to acuminate, appressed; whole plant with numerous viscid glandular as well as simple and stellate eglandular hairs. 3 microspp. placed here. Intrd; walls and rough ground; very scattered in En and Sc; C Europe to Pyrenees.

12. H. sect. Cerinthoidea Monnier (*H. alatum* Lapeyr. group). Plants 10-60cm; **710**
basal leaves normally present at flowering, ovate or obovate to narrowly elliptic, with ± winged petiole, glaucous; stem-leaves 1-7, same shape as basal, clasping stem at base; capitula (1)2-8(20), relatively large; phyllaries acute to acuminate, appressed; vegetative parts with simple eglandular hairs, and glandular and stellate hairs above; capitula and peduncles with glandular and simple and stellate eglandular hairs. 10 microspp. currently placed here. Native; cliffs and rocky streamsides; coastal and upland areas of Sc, N En, N & W Ir, 1 site in Wa. **H. anglicum** Fr. is the commonest sp. of the section in N En and Sc and the commonest *Hieracium* in Ir

13. H. sect. Andryaloidea Monnier. Plant 10-50cm; basal leaves present at **709**
flowering, ovate-elliptic, with winged petiole; stem-leaves 2-5(8), same shape as basal, sessile, not clasping stem at base; capitula (2)3-7(12); phyllaries acute to acuminate, appressed; vegetative parts and capitula with dense white simple or tufted eglandular hairs, usually without glandular or stellate hairs. 1 microsp. (**H. lanatum** Vill.) placed here. Intrd; natd on coastal dunes in E Norfolk since 1981; mountains of W Europe.

14. H. sect. Subalpina Pugsley (*H. senescens* Backh. f. group, *H. atratum* Fr. group, **710**
H. rohacsense Kit. ex Kanitz group). Plants 20-50cm; basal leaves present at flowering, narrowly to broadly elliptic, petiolate; stem-leaves (0)1-4, usually 1 leaf-like, the others much reduced, not clasping stem at base; capitula (1)2-5; phyllaries acute to obtuse, appressed; all parts variously with glandular and simple and stellate eglandular hairs, the simple hairs usually blackish-based. 31 microspp. currently placed here. Native; rock-ledges and rocky streamsides usually above 450m; local in mainland Sc, 2 microspp. extend to N En, 1 to Co Antrim, and 1 is

endemic to the Lake district.

15. H. sect. Alpina (Griseb.) Gremli (*H. alpinum* L. group, *H. nigrescens* Willd. 710
group). Plants 5-15(25)cm; basal leaves present at flowering, very narrowly elliptic to obovate, petiolate; stem-leaves 0-4, usually all much reduced, not clasping stem at base; capitula 1(-c.5); phyllaries acute to obtuse, appressed to erect; all parts variously with glandular and eglandular blackish-based simple hairs, stellate hairs only on vegetative parts. 30 microspp. currently placed here. Native; rock-ledges, barish slopes and scree, grassy banks, usually above 650m; 27 microspp. endemic to mainland Sc, 1 endemic to the Lake District, 1 (**H. holosericeum** Backh. f.) endemic to Sc (over the whole range of the section) and the Lake District and Snowdonia, and 1 (**H. alpinum** L.) in Sc and widespread in Europe.

TRIBE 3 - ARCTOTIDEAE (genera 34-35). Plants not producing white latex, not spiny; capitula radiate, with sterile ligulate flowers with yellow to orange, usually 3-toothed ligules.

34. ARCTOTHECA J.C. Wendl. (*Cryptostemma* R. Br.) - *Plain Treasureflower*
Annuals or perennials (annuals in BI); lower leaves deeply pinnately lobed, white-tomentose on lowerside; phyllaries in several rows, free, glabrous to sparsely pubescent, with conspicuous scarious, rounded to obtuse tips; receptacular scales 0; achenes densely pubescent; pappus of distinct scales.

1. A. calendula (L.) Levyns (*Cryptostemma calendulacea* (Hill) R. Br.) - *Plain Treasureflower*. Stems decumbent, leafy only near base, to 40cm, white-pubescent; all leaves usually deeply lobed; capitula on long peduncles; ligules pale yellow on upperside, purplish on lowerside; (2n=18). Intrd; rather frequent wool-alien in arable fields and waste places; scattered in Br; S Africa.

35. GAZANIA Gaertn. - *Treasureflower*
Differs from *Arctotheca* in outer phyllaries fused into cup-like structure, white-tomentose on lowerside, with acute to acuminate scarious tips; leaves sometimes all simple.

1. G. rigens (L.) Gaertn. (*G. uniflora* (L.f.) Sims, *G. splendens* Hend. & A.A. Hend.) - *Treasureflower*. Stems decumbent to ascending, to 50cm, often ± woody at base, white-tomentose below; lower leaves deeply pinnately lobed, upper ones ± entire; capitula on long peduncles; ligules orange-yellow with basal black blotch bearing central white spot, or rarely plain yellow (var. **uniflora** (L.f.) Roessler); (2n=10, 14). Intrd; grown in gardens, ± natd on walls, rocks and cliffs near sea; Scillies and CI; S Africa.

SUBFAMILY 2 - ASTEROIDEAE (Tubuliflorae) (tribes 4-10; genera 36-99). Plant not producing white latex; stem-leaves usually spiral ('alternate'), sometimes opposite or 0; capitula mostly radiate, sometimes discoid; tubular flowers most commonly yellow to orange, usually with short lobes or teeth; filaments joining anthers at base; 2 style branches each with 2 stigmatic zones, 1 near each margin of inner face; pollen grains spiny.

TRIBE 4 - INULEAE (genera 36-44). Annual to perennial herbs, rarely shrubs; leaves alternate, simple; capitula discoid or radiate; phyllaries in several rows, herbaceous to scarious; receptacular scales present or (usually) 0; pappus usually of hairs, sometimes of short scales or of scales and hairs; flowers usually yellow or brown to whitish.

36. FILAGO L. (*Logfia* Cass.) - *Cudweeds*
Annuals with stems and leaves ± covered with woolly hairs; capitula small,

brownish, borne in clusters of 2-c.40; phyllaries in few ill-defined rows, the outer herbaceous, the inner scarious; receptacle conical, with scales associated with outer (female) florets only; flowers all tubular, the inner bisexual, with wider corollas than the outer female; pappus of bisexual flowers of simple hairs, of female flowers of simple hairs or 0.

1 Capitula 2-7(14) in each cluster; outer phyllaries obtuse to subacute, patent in fruit 2
1 Capitula (5)10-c.40 in each cluster; outer phyllaries acuminate, erect in fruit 3
 2 Leaves 4-10mm, the most apical ones not overtopping clusters of capitula **4. F. minima**
 2 Leaves (8)12-20(25)mm, the most apical ones overtopping clusters of capitula **5. F. gallica**
3 Leaves widest in basal 1/2, the most apical ones not overtopping clusters of capitula; capitula in clusters of (15)20-40 **1. F. vulgaris**
3 Leaves widest in apical 1/2, the most apical ones usually overtopping clusters of capitula; capitula in clusters of (5)10-20(25) 4
 4 Clusters of capitula each overtopped by (0)1-2 leaves; outer phyllaries with erect red-tinged points; plant usually yellowish-woolly **2. F. lutescens**
 4 Clusters of capitula each overtopped by 2-4(5) leaves; outer phyllaries with recurved yellowish points; plant white-woolly **3. F. pyramidata**

Other spp. - **F. arvensis** L. (*Logfia arvensis* (L.) Holub), from Europe, once occurred regularly as a casual but is now rare; it differs from *F. minima* in its racemosely branching, not bifurcating stems and outer phyllaries woolly to the apex (not glabrous near apex).

1. F. vulgaris Lam. (*F. germanica* L. non Huds.) - *Common Cudweed.* Stems erect, to 40cm, branching below each cluster of capitula; clusters of capitula ± globose, c.10-12mm across; 2n=28. Native; barish places on sandy soils, e.g. heaths, waysides, sand-pits; throughout most of BI, but absent from most of N & W Sc.

2. F. lutescens Jord. (*F. apiculata* G.E. Sm. ex Bab.) - *Red-tipped Cudweed.* Differs from *F. vulgaris* in ± irregularly branched stems; yellowish (not whitish) hairs; apiculate (not acute) leaves; and see key (couplet 3); (2n=28). Native; similar places to *F. vulgaris*; very local in S & E En from S Hants to SE Yorks, formerly W to Worcs. **RRR**

3. F. pyramidata L. (*F. spathulata* auct. non C. Presl) - *Broad-leaved Cudweed.* Differs from *F. vulgaris* in leaves often apiculate (not acute); outer phyllaries with recurved (not erect) tips; and see key (couplet 3); (2n=28). Native; similar places to *F. vulgaris*; very local in S En, formerly N to N Lincs and in Jersey. **RRR**

4. F. minima (Sm.) Pers. (*Logfia minima* (Sm.) Dumort.) - *Small Cudweed.* Stems erect, to 25cm, irregularly branching; clusters of capitula ovoid, c.2-5mm across; (2n=28). Native; similar habitats and distribution to *F. vulgaris*.

5. F. gallica L. (*Logfia gallica* (L.) Coss. & Germ.) - *Narrow-leaved Cudweed.* Differs from *F. minima* in key characters; inflorescence appearing very leafy; (2n=28). Intrd; natd in sandy and gravelly ground; Sark since 1902, formerly S Hants to N Essex from 1696 to 1955; S & W Europe.

37. ANTENNARIA Gaertn. - *Mountain Everlasting*
Dioecious whitish-woolly perennials; capitula pale to deep pink, sometimes whitish, borne in terminal umbel-like clusters of 2-8, phyllaries in several rows, scarious, the outer ones of male capitula patent and perianth-like in flower, white to pink; receptacle flat, without scales; flowers all tubular, males with wider

corollas than females; pappus of female flowers of simple hairs, of male flowers of simple hairs widened distally.

1. A. dioica (L.) Gaertn. - *Mountain Everlasting*. Stems erect, to 20cm, with basal leaf-rosette; surface-creeping leafy stolons present; leaves green on upperside, white-woolly on lowerside; 2n=28. Native; heaths, moors, mountain slopes; common in much of N 1/2 of BI, scattered S to S Ir and W Cornwall, much reduced in S & E Br, no longer in CE or SE En except 1 place in Northants.

38. ANAPHALIS DC. - *Pearly Everlasting*
Dioecious to variably sexed white-woolly perennials; capitula white, with yellow flowers, borne in large terminal corymbose inflorescences; phyllaries in several rows, scarious, pearly white; receptacle convex, without scales; flowers all tubular, male and female variously arranged, the males with wider corollas; pappus of 1 row of hairs.

1. A. margaritacea (L.) Benth. - *Pearly Everlasting*. Stems erect to ascending, to 716
1m, without basal leaf-rosette; rhizomes present; leaves green on upperside, white-woolly on lowerside; 2n=28. Intrd; common in gardens and well natd as relic or throwout by rivers and in grassland, marginal and rough ground; scattered in BI, locally frequent in W: N America.

39. GNAPHALIUM L. (*Omalotheca* Cass., *Filaginella* Opiz, *Gamochaeta* Wedd., *Pseudognaphalium* Kirp.) - *Cudweeds*
Annuals or perennials ± covered with whitish woolly hairs; capitula small, yellowish to brown, variously arranged; phyllaries whitish to yellowish or brown, subherbaceous to scarious, in several rows; receptacle flat, without scales; flowers all tubular, the inner bisexual, with wider corollas than the outer female; pappus of simple hairs.

1 Capitula in terminal, subglobose to subcorymbose clusters; annual 2
1 Capitula in elongated, racemose clusters, sometimes few or rarely 1; annual to perennial 4
2 Phyllaries brown; clusters of capitula conspicuously leafy **5. G. uliginosum**
2 Phyllaries uniformly white to yellowish: clusters of capitula not leafy 3
3 Leaves white-woolly on both sides, not decurrent down stem **6. G. luteoalbum**
3 Leaves green on upperside, white-woolly on lowerside, decurrent down stem **7. G. undulatum**
4 Annual to biennial without rhizome; achenes <1mm, glabrous; phyllaries acute to acuminate **4. G. purpureum**
4 Perennial with short ± surface rhizome; achenes >1mm, pubescent; phyllaries obtuse to rounded or retuse 5
5 Capitula <10 per stem; pappus-hairs free, falling separately **3. G. supinum**
5 Capitula normally >10 per stem; pappus-hairs united at base, falling as a unit 6
6 Stem-leaves 1-(or indistinctly 3-)veined, steadily diminishing in size up the stem **2. G. sylvaticum**
6 Stem-leaves 3(-5)-veined, scarcely diminishing in size until above 1/2 way up stem **1. G. norvegicum**

1. G. norvegicum Gunnerus (*Omalotheca norvegica* (Gunnerus) Sch. Bip. & F.W. RR
Schultz) - *Highland Cudweed*. Stems erect, to 30cm; leaves grey-green on upperside; phyllaries grey-green in centre, brownish-scarious around edges; (2n=56). Native; mountain rocks and gravel; very local in C Sc.

2. G. sylvaticum L. (*Omalotheca sylvatica* (L.) Sch. Bip. & F.W. Schultz) - *Heath Cudweed.* Stems erect, to 60cm; leaves green on upperside; phyllaries green in centre, brownish-scarious around edges; 2n=56. Native; rather open ground on heaths, banks, woodland rides; locally frequent in much of Br and Ir, but much less common than formerly, especially in the W.

3. G. supinum L. (*Omalotheca supina* (L.) DC.) - *Dwarf Cudweed.* Stems erect, to 12(20)cm; leaves grey-woolly on upperside; phyllaries grey-green in centre, brownish-scarious around edges; 2n=28. Native; mountain rocks and gravel; local in C, N & W mainland Sc and Skye (N Ebudes).

4. G. purpureum L. (*G. pensylvanicum* Willd. nom. illeg.) - *American Cudweed.* **716**
Stems decumbent to erect, to 40cm; leaves woolly on upperside; phyllaries green, with brown tips; (2n=14, 18, 28). Intrd; natd in churchyard since 1940s; Surrey; USA.

5. G. uliginosum L. (*Filaginella uliginosa* (L.) Opiz) - *Marsh Cudweed.* Stems decumbent to erect, to 25cm; leaves woolly on upperside; phyllaries pale brown with dark brown tips; 2n=14. Native; damp places in fields and arable land and by ponds and paths; common throughout BI.

6. G. luteoalbum L. (*Pseudognaphalium luteoalbum* (L.) Hilliard & B.L. Burtt) - **RRR**
Jersey Cudweed. Stems erect, to 50cm; leaves woolly on upperside; phyllaries scarious, straw-coloured; 2n=14. Native; sandy fields, waste places and sand-dunes; very local in CI and W Norfolk, formerly E Norfolk, W Suffolk and Cambs.

7. G. undulatum L. (*Pseudognaphalium undulatum* (L.) Hilliard & B.L. Burtt) - **716**
Cape Cudweed. Stems erect, to 80cm; leaves green on upperside; phyllaries whitish-scarious. Intrd; rough ground, cliffs, marginal habitats; natd in CI (all main islands) since 1888, E Cornwall; S Africa. Resembles a small *Anaphalis.*

40. HELICHRYSUM Mill. - *Everlastingflowers*

Woody perennials or annuals; capitula conspicuous, terminal, usually solitary; phyllaries in several rows, scarious, white or coloured; receptacle slightly convex, without scales; flowers all tubular, the outer female, the inner bisexual and with wider corollas; pappus of 1 row of hairs.

Other spp. - H. bracteatum (Vent.) Andrews, from Australia, is the common annual *Everlastingflower* of florists; it has solitary capitula >20mm across with yellow, red, white or mixed-coloured phyllaries and is sometimes found on tips in S En. **H. petiolare** Hilliard & B.L. Burtt, from S Africa, is the silver-leaved small wiry shrub commonly used as summer ground-cover or in hanging baskets; it has many small, dull-coloured capitula on a long common stem with reduced leaves and has been found by a stone wall in the Scillies.

1. H. bellidioides (G. Forst.) Willd. - *New Zealand Everlastingflower.* Mat-forming **716**
dwarf evergreen shrub with decumbent stems to 60cm; leaves c.5-8mm, *Thymus*-like, silvery on lowerside; capitula on erect stems up to 10cm, c.2-3cm across; flowers yellow, surrounded by patent white inner phyllaries up to 1cm; (2n=28). Intrd; natd in rocky turf by stream; 1 place in Shetland, since 1975; New Zealand.

41. INULA L. - *Fleabanes*

Perennial herbs, sometimes woody at base; capitula 1-many, terminal, usually subcorymbose, usually showy, yellow, usually radiate, less often discoid; phyllaries in several rows, herbaceous; receptacle flat or slightly convex, without scales; pappus of 1 row of hairs.

1 Stem and leaves succulent, glabrous **5. I. crithmoides**
1 Stem and leaves not succulent, very sparsely to densely pubescent 2
  2 Ligules 0 or <1mm; capitula numerous on each stem **4. I. conyzae**
  2 Ligules conspicuous, >1cm; capitula 1-c.5 on each stem 3

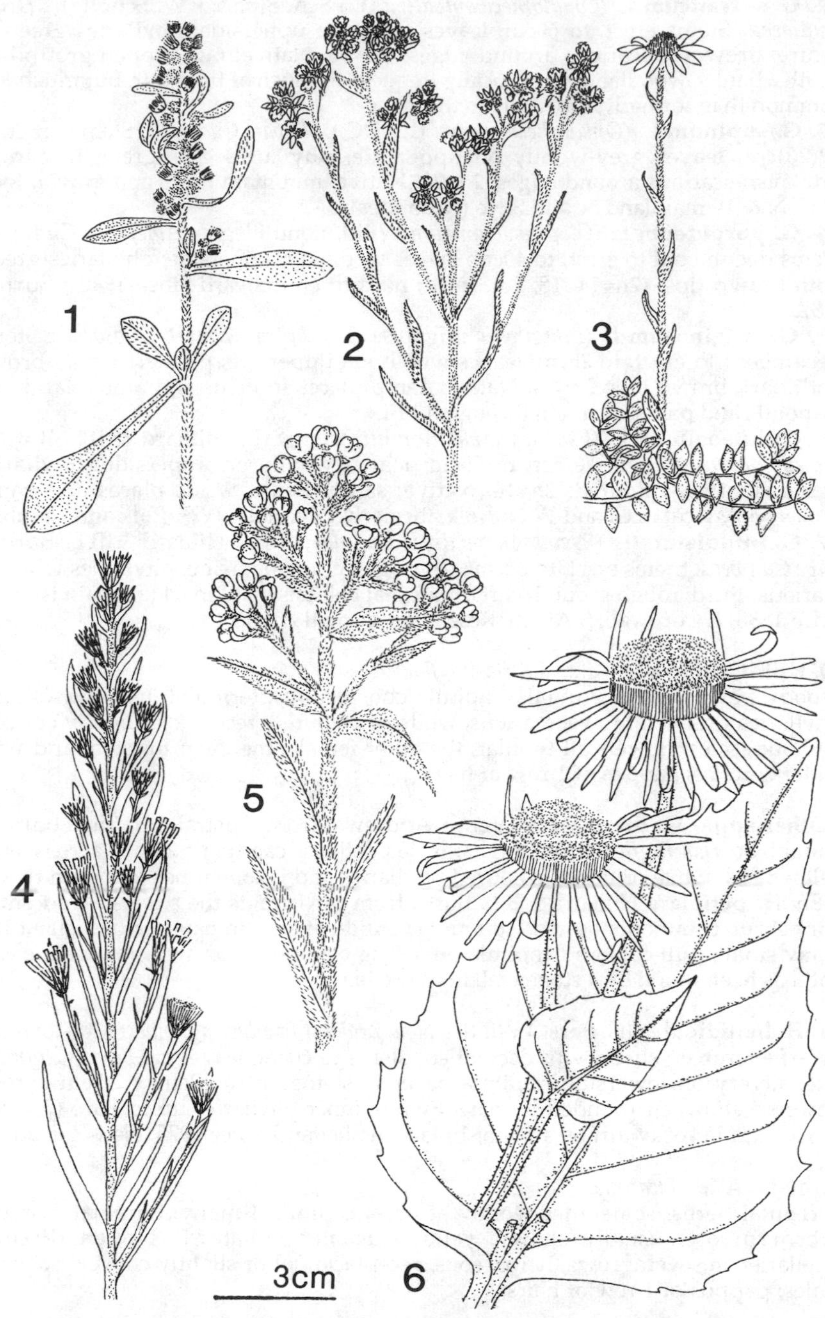

FIG 716 - Asteraceae: Inuleae. 1, ***Gnaphalium purpureum***. 2, ***G. undulatum***. 3, ***Helichrysum bellidioides***. 4, ***Dittrichia viscosa***. 5, ***Anaphalis margaritacea***. 6, ***Telekia speciosa***.

3 Outer phyllaries ovate; capitula >5cm across (incl. ligules); stems rarely <1m **1. I. helenium**
3 Outer phyllaries lanceolate; capitula <5cm across (incl. ligules); stems <1m 4
 4 Leaves subglabrous to sparsely pubescent on lowerside, prominently reticulate-veined on upperside; achenes glabrous **2. I. salicina**
 4 Leaves densely pubescent on lowerside, obscurely veined on upperside; achenes pubescent **3. I. oculus-christi**

Other spp. - I. britannica L., from Europe, was natd by a reservoir in Leics from 1894 to at least 1932; it resembles *I. oculus-christi* in stature and pubescence but the outer phyllaries are patent to deflexed, not erect.

1. I. helenium L. - *Elecampane*. Stems erect, to 2.5m, pubescent; stem-leaves ovate, cordate, pubescent; capitula few, c.6-9cm across (incl. ligules); 2n=20. Intrd; natd in fields, waysides, marginal habitats, rough ground; scattered throughout BI, less common than formerly since grown less; W & C Asia. See *Telekia speciosa* for differences.

2. I. salicina L. - *Irish Fleabane*. Stems erect, to 70cm, ± glabrous; stem-leaves RRR
narrowly elliptic to narrowly ovate, cordate, subglabrous to sparsely pubescent; capitula few, 2.5-4.5cm across (incl. ligules); 2n=16. Native; stony limestone shores of Lough Derg (N Tipperary), formerly SE Galway.

3. I. oculus-christi L. - *Hairy Fleabane*. Stems erect, to 60cm, densely pubescent; stem-leaves narrowly oblong- or elliptic-ovate, densely pubescent; capitula few, 2.5-3.5cm across (incl. ligules); (2n=30, 32). Intrd; garden ornamental natd in derelict gardens and rough ground; 2 sites in Man since 1960s; E Europe and W Asia.

4. I. conyzae (Griess.) Meikle (*I. conyza* DC.) - *Ploughman's-spikenard*. Stems erect, to 1.25m, pubescent; stem-leaves narrowly ovate to narrowly obovate, pubescent; capitula numerous, c.0.7-1.2cm across (incl. ligules, if present); 2n=32. Native; scrub, grassland and barish places on calcareous soils; locally common throughout En, Wa and CI.

5. I. crithmoides L. - *Golden-samphire*. Stems erect to decumbent, to 1m, glabrous; R
stem-leaves linear to oblanceolate, succulent, glabrous; capitula rather few, c.1.5-2.5cm across (incl. ligules); 2n=18. Native; salt-marshes, shingle, cliffs, rocks and ditchsides by sea; local on coasts of BI N to Co Louth, Wigtowns and N Lincs.

42. DITTRICHIA Greuter - *Fleabanes*

Low shrubs or annual herbs with glandular-sticky stems and leaves; capitula several in racemose inflorescences, rather showy, yellow, radiate but ligules often very short; phyllaries in several rows, herbaceous; receptacle flat or slightly convex, without scales; pappus of 1 row of hairs fused at base.

1. D. viscosa (L.) Greuter (*Inula viscosa* (L.) Aiton) - *Woody Fleabane*. Perennial 716
with resinous smell when crushed; stems ascending to erect, woody, to 1m; capitula 1-1.5(2)cm across; ligules ≤10mm, much longer than phyllaries; (2n=18). Intrd; ± natd in rough ground in E Suffolk and by harbour in E Sussex, casual elsewhere in S En; S Europe.

2. D. graveolens (L.) Greuter (*Inula graveolens* (L.) Desf.) - *Stinking Fleabane*. Annual with strong camphorous smell when crushed; stems erect, to 50cm; capitula 0.6-1.2cm across; ligules ≤3mm, not or scarcely exceeding phyllaries; (2n=18). Intrd; a rather frequent wool-alien in fields, etc.; scattered in En; S Europe.

43. PULICARIA Gaertn. - *Fleabanes*

Annuals or perennials; capitula several to many, terminal, usually subcorymbose, usually showy, yellow, radiate; phyllaries in several rows, herbaceous; receptacle flat, without scales; pappus of 1 row of hairs plus an outer row of free or fused scales.

1. P. dysenterica (L.) Bernh. - *Common Fleabane*. Densely pubescent perennial with extensive rhizomes; stems erect, to 1m; stem-leaves sessile, cordate at base; capitula 1.5-3cm across; ligules usually c.1.5x as long as phyllaries; 2n=18. Native; marshes, ditches, wet fields and hedgebanks; common in lowland BI N to N En, rare in S & C Sc and N Ir.

2. P. vulgaris Gaertn. - *Small Fleabane*. Pubescent annual; stems erect, to 45cm; **RRR**
stem-leaves sessile, rounded to cuneate at base; capitula 0.6-1.2cm across; ligules c. as long as phyllaries; 2n=18. Native; sandy places flooded in winter, often by ponds; formerly widespread in CI and En N to Leics, now very local in Surrey and S Hants.

44. TELEKIA Baumg. - *Yellow Oxeye*
Herbaceous perennials; capitula 1 to several, terminal, subcorymbose, showy, yellow, radiate; phyllaries in several rows, herbaceous; receptacle convex, with scales; pappus of fused scales. Probably best united with *Buphthalmum* L.

1. T. speciosa (Schreb.) Baumg. - *Yellow Oxeye*. Stems erect, pubescent, to 2m; **716**
lower stem-leaves petiolate and cordate at base, upper ones sessile and rounded to broadly cuneate at base; capitula 5-8cm across; ligules 1-2.5cm; (2n=20). Intrd: grown in gardens and natd in rough ground and by lakes and rivers; scattered throughout most of Br, mostly in N; C & SE Europe. Resembles *Inula helenium* in habit, but basal leaves deeply cordate (not narrowly cuneate), leaf-margins sharply dentate (not crenate-dentate), and note pappus.

TRIBE 5 - ASTEREAE (genera 45-55)
Annual to (usually) perennial herbs, rarely shrubs; leaves alternate or all basal, simple; capitula discoid or radiate; phyllaries in 2-several rows, usually herbaceous; receptacular scales 0; pappus usually of hairs, sometimes 0 or of strong bristles; flowers various colours.

45. GRINDELIA Willd. - *Gumplants*
Herbaceous perennials; stem-leaves sessile, clasping stem, serrate; capitula radiate, with yellow ray and disc flowers; phyllaries sticky, in several rows, herbaceous, with recurved tips; pappus of 2-8 stiff bristles.

Other spp. - Records of **G. squarrosa** (Pursh) Dunal and **G. rubricaulis** DC., from USA, are either old or errors.

1. G. stricta DC. - *Coastal Gumplant*. Stems erect, to 75cm, sparsely shaggy-pubescent; capitula 3-5cm across; ligules 8-15mm; (2n=24). Intrd; natd on sea-cliffs at Whitby, NE Yorks, since 1977; W coast of N America.

46. CALOTIS R. Br. - *Bur Daisy*
Annuals or perennials; stem-leaves various; capitula radiate or ± discoid, with yellow disc and yellow, white or mauve ray flowers; phyllaries in ± 2 rows, herbaceous; pappus of (1)3(-6) rigid barbed bristles and usually some extra shorter bristles or scales; fruiting capitula forming a globose bur 5-9mm across.

Other spp. - Several other spp. occur as wool-aliens, of which **C. lappulacea** Benth. and **C. hispidula** (F. Muell.) F. Muell., from Australia, both with short yellow ligules, are most common. The former is perennial with linear stem-leaves, glabrous bristles and ligules exceeding phyllaries; the latter is annual with oblanceolate stem-leaves, pubescent bristles and ligules shorter than phyllaries.

1. C. cuneifolia R. Br. - *Bur Daisy*. Perennial with branching, erect to procumbent 726
stems to 30(60)cm; leaves obtriangular, narrowed to petiole, dentate at distal end; capitula 1-2cm across in flower, with white or mauve ligules 3-9mm, much longer than phyllaries; (2n=16, 20, 32). Intrd; rather characteristic wool-alien, scattered in En; Australia.

47. SOLIDAGO L. (*Euthamia* Nutt.) - *Goldenrods*

Perennials; stem-leaves narrowly elliptic to oblanceolate or obovate, serrate, narrowed to base; capitula small, numerous, ± crowded, radiate, yellow; phyllaries in many rows, herbaceous; pappus of 1-2 rows of hairs.

The N American spp. are numerous and very difficult; they have possibly given rise in cultivation in Br to new taxa that add to the problems of identification.

1 Capitula sessile, in small clusters forming corymbose inflorescence; leaves gland-spotted, ligules 1-1.5mm **5. S. graminifolia**

1 At least most capitula stalked, forming pyramidal to ± cylindrical inflorescence; leaves not gland-spotted; ligules ≥1.5mm 2

2 Leaves with many pairs of short lateral veins (though often inconspicuous) 3

2 Leaves with 1(-2) pairs of main lateral veins from near base, running parallel with midrib for most of length 4

3 Most stem-leaves rounded to acute at apex; inner phyllaries >4.5mm; disc flowers ≥10; ligules 4-9mm **1. S. virgaurea**

3 Most stem-leaves acute to acuminate at apex; inner phyllaries ≤4.5mm; disc flowers ≤8; ligules 1.5-4mm **2. S. rugosa**

4 Leaves scabrid-pubescent on surfaces, stems pubescent at least in top 1/2 **3. S. canadensis**

4 Leaves glabrous on surfaces or pubescent only on lowerside veins; stems ± glabrous **4. S. gigantea**

1. S. virgaurea L. - *Goldenrod*. Stems erect, to 70(100)cm, but often much less, glabrous to pubescent; capitula in a raceme, or in a panicle with straight erect branches; disc flowers 10-30; ligules 6-12; 2n=18. Native; open woodland, grassland, hedgerows, rocky places, cliffs; frequent over most of BI except parts of C En, C Ir and CI. Very variable.

1 x 3. S. virgaurea x S. canadensis = S. x niederederi Khek was found on a railway bank in W Kent in 1979, in W Gloucs in 1979 and in Cheshire in 1995; it is closer to *S. canadensis* in inflorescence shape and capitulum size and to *S. virgaurea* in leaf venation, and is sterile.

2. S. rugosa Mill. (*S. altissima* Aiton non L.) - *Rough-stemmed Goldenrod*. Stems erect, to 1.5m, roughly pubescent; capitula on erecto-patent to patent curved branches forming terminal pyramidal inflorescence; disc flowers 3-8; ligules 6-11; (2n=18, 54). Intrd; natd as for *S. canadensis;* Renfrews and Dunbarton, formerly Argyll; N America.

3. S. canadensis L. (*S. altissima* L.) - *Canadian Goldenrod*. Stems erect, to 2.5m, pubescent; capitula on erecto-patent to patent curved branches forming terminal pyramidal inflorescence; disc flowers 2-8; ligules 6-15; (2n=18, 54). Intrd; much grown in gardens, fully natd on waste land, banks, waysides and rough grassland; frequent throughout C & S Br and CI, scattered in N Br and Ir; N America. **S. altissima** is sometimes recognized as a distinct sp., as ssp. **altissima** (L.) O. Bolòs & Vigo, or as var. **scabra** Torr. & A. Gray; it has stems pubescent throughout (not just in upper 1/2), phyllaries 2.5-4.5mm (not 2.1-3mm), ligules 2.4-4.1mm (not 1.5-2.8mm) and more sharply serrate leaves, and probably occurs with us, as might hybrids of *S. canadensis* with *S. rugosa* and *S. gigantea*. The precise identity of British material of *S. canadensis* in the sense used here is uncertain; much of it might be cultivars or hybrids that have arisen here.

4. S. gigantea Aiton (*S. serotina* Aiton non Retz.) - *Early Goldenrod*. Differs from *S. canadensis* in disc flowers 6-10(12); and see key; (2n=18, 36, 54). Intrd; natd as for *S. canadensis*, often with it but less common; N America. Our plant is ssp. **serotina** (O. Kuntze) McNeill (var. *serotina* (O. Kuntze) Cronquist, var. *leiophylla* Fernald).

5. S. graminifolia (L.) Salisb. (*Euthamia graminifolia* (L.) Elliott) - *Grass-leaved Goldenrod*. Stems erect, to 1.5m, glabrous to pubescent; leaves linear to linear-lanceolate, appearing 3(-5)-veined; inflorescence flat-topped; disc flowers (4)5-10(13); ligules 12-15; (2n=18). Intrd; natd as for *S. canadensis;* very scattered in En, older (?extant) records in C & S Sc; N America. **726**

48. ASTER L. (*Crinitaria* Cass.) - *Michaelmas-daisies*
Perennials; stem-leaves ovate to linear, entire, with various bases; capitula conspicuous, radiate or discoid, with yellow disc flowers and white to blue, pink or purple ligules; phyllaries in 2-several rows, herbaceous or partly membranous; pappus of 1-2 rows of hairs.

The cultivated *Michaelmas-daisies* that are found in the wild are difficult to determine due to hybridization between *A. novi-belgii* and 2 other spp. These 2 hybrids and *A. lanceolatus* appear to be the commonest taxa, and show every grade of variation from 1 parent to the other. The N American spp. are numerous and very difficult; they have possibly given rise in cultivation in Br to new taxa that add to the problems of identification.

1 Basal and lower stem-leaves with long petiole and cordate base **1. A. schreberi**
1 Basal leaves cuneate at base; stem-leaves cuneate at base and/or sessile 2
  2 Leaves all 1-veined, linear to very narrowly elliptic, gland-spotted, not succulent; maritime cliffs **9. A. linosyris**
  2 Leaves mostly with well-developed lateral veins, if all 1-veined then succulent, not gland-spotted but sometimes with stalked glands; widespread 3
3 Leaves succulent, with 0-few lateral veins mostly running parallel with midrib; mostly maritime **8. A. tripolium**
3 Leaves not succulent, usually with normally developed lateral veins, widespread 4
  4 Upper part of plant with abundant long patent hairs and shorter stalked glands **2. A. novae-angliae**
  4 Plant with 0 or rather sparse long patent hairs; stalked glands 0 5
5 Upper leaves tapering to base, not clasping stem; leaves rarely >1cm wide; phyllaries ≤5mm; ligules usually white (see also *A. x salignus*) **7. A. lanceolatus**
5 Upper leaves tapering to base or not, but distinctly (though often narrowly) clasping stem at base; some leaves >1cm wide; phyllaries >5mm; ligules usually coloured 6
  6 Phyllaries with wide white borders in basal 1/2 and narrow ones in apical 1/2, leaving elliptic to trullate green patch in centre near apex; outer phyllaries reaching ≤1/2 as high as inner ones 7
  6 Phyllaries wholly or mainly green in apical 1/2, hence appearing leafy near apex; outer phyllaries usually reaching ≥1/2 as high as inner ones 8
7 Leaves distinctly glaucous on upperside; outer phyllaries usually reaching distinctly <1/2 as high as inner ones; plant usually ≤1m **3. A. laevis**
7 Leaves not glaucous; outer phyllaries often reaching c.1/2 as high as inner ones; plant usually 1-2m **4. A. x versicolor**
  8 Middle stem-leaves mostly 2.5-5x as long as wide, conspicuously clasping stem; outer phyllaries usually c.1/2-3/4 as high as inner ones **4. A. x versicolor**

8 Middle stem-leaves mostly 4-10x as long as wide, usually very narrowly clasping stem; outer phyllaries nearly as long as inner ones 9

9 Outer phyllaries widest below middle, rather neatly appressed to capitulum **6. A. x salignus**

9 Outer phyllaries widest at or just above middle, with conspicuous leafy apical 1/2 loosely or unevenly appressed to capitulum **5. A. novi-belgii**

Other spp. - Many other spp. are grown in gardens and some may escape, but most records are suspect due to misidentification. Other spp. that might occur are **A. puniceus** L., with hispid stem and leaves and very leafy ± recurved tips to phyllaries; **A. dumosus** L., with a much-branched inflorescence, phyllaries ≤6mm and narrow non-clasping leaves (some of the dwarf *Michaelmas-daisies* may belong here or to hybrids of it); **A. concinnus** Willd., differing from last in wider (>1cm) leaves; and **A. foliaceus** Lindl., similar in leaf-shape to *A. laevis* but with leafy phyllaries, few, large capitula, and non-glaucous leaves. More distinct are the European spp. **A. amellus** L., with pubescent stems to 60cm, 2-6 capitula in a corymb, and pubescent ± petiolate lower stem-leaves, and **A. sedifolius** L., with 1-3-veined very narrow gland-spotted leaves, numerous capitula crowded into a corymb, and scabrid stems to 60cm; both have been found to persist for a few years.

1. A. schreberi Nees (*A. macrophyllus* auct. non L.) - *Nettle-leaved Michaelmas-daisy*. Stems erect, to 1m, minutely pubescent; phyllaries with narrow green midline and wide scarious margins, outer reaching <1/2 as high as inner; ligules whitish-grey tinged mauve; (2n=16, 18, 54). Intrd; natd on railway-bank in Renfrews 1931 to late 1980s, now gone; N America.

2. A. novae-angliae L. - *Hairy Michaelmas-daisy*. Stems erect, to 2m, with long patent hairs; phyllaries all green and/or purple, very glandular, ± equal; ligules usually bright pinkish-purple; (2n=10). Intrd; natd on waste and rough ground; very scattered in Br, mainly S & C; N America. The fresh inflorescence smells distinctly of *Calendula* when crushed.

3. A. laevis L. - *Glaucous Michaelmas-daisy*. Stems erect, to 1m, ± glabrous; **722**
phyllaries as in key (couplet 6); ligules usually bluish-purple; (2n=48, 54). Intrd; natd on waste and rough ground; rather rare and scattered in En, Tyrone; N America.

4. A. x versicolor Willd. (*A. novi-belgii* ssp. *laevigatus* (Lam.) Thell.; *A. laevis* x *A.* **722**
novi-belgii) - *Late Michaelmas-daisy*. Stems erect, to 2m, ± glabrous; phyllaries similar to those of *A. laevis* but more leafy at apex and less unequal; ligules usually bluish-purple. Intrd; natd on waste and rough ground; scattered in Br, probably under-recorded; garden origin. Most of the taller, larger flowered, late-flowering cultivars, often with dark red stems, belong here.

5. A. novi-belgii L. (*A. longifolius* auct. non Lam.) - *Confused Michaelmas-daisy*. **722**
Stems erect, to 1.5m, glabrous to sparsely pubescent; phyllaries as in key (couplets 8 & 9), with apical 1/2 ± entirely green; ligules usually mauve, but white to purple in cultivars; (2n=18, 48). Intrd; natd on waste and rough ground; scattered over Br, greatly over-recorded for *A. x salignus*; N America.

6. A. x salignus Willd. (*A. longifolius* auct. non Lam.; *A. novi-belgii* x *A.* **722**
lanceolatus) - *Common Michaelmas-daisy*. Differs from *A. novi-belgii* in leaves narrower and scarcely clasping stem; and see key (couplet 9). Intrd; natd on waste and rough, often damp ground; easily the commonest natd *Michaelmas-daisy* in BI, reproducing from seed and often weedy in appearance; garden origin.

7. A. lanceolatus Willd. - *Narrow-leaved Michaelmas-daisy*. Stems erect, to 1.2m, **722**
glabrous or sparsely pubescent; phyllaries with conspicuous membranous borders, not leafy at apex, the outer reaching c.1/2 as high as inner ones; ligules white or pale mauve; (2n=32, 48, 64). Intrd; natd on waste and rough ground; frequent throughout much of BI, under-recorded, less common than only *A.* x *salignus*; N

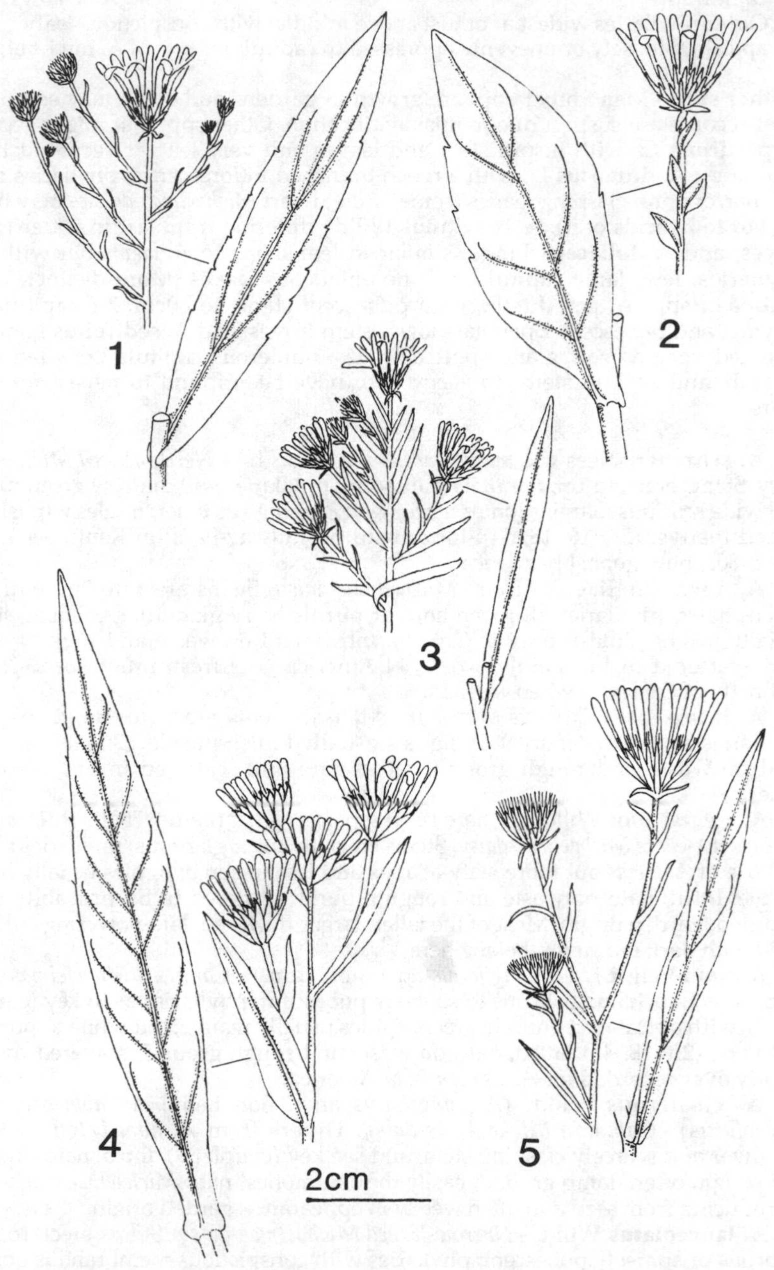

FIG 722 - Apex of inflorescence and mid stem-leaf of *Aster*. 1, *A. laevis*. 2, *A. x versicolor*. 3, *A. lanceolatus*. 4, *A. novi-belgii*. 5, *A. x salignus*.

America.

8. A. tripolium L. - *Sea Aster*. Biennial or sometimes annual; stems erect, to 1m, glabrous; phyllaries fewer and much blunter than in all other spp., unequal; ligules bluish-mauve or 0; 2n=18. Native; salt-marshes, less often cliffs and rocks on coasts around whole BI, rare in inland saline areas.

9. A. linosyris (L.) Bernh. (*Crinitaria linosyris* (L.) Less.) - *Goldilocks Aster*. Stems erect to decumbent, to 50cm, glabrous; phyllaries unequal, the outer ± wholly green and loosely appressed to capitulum; ligules 0; 2n=18. Native; limestone sea-cliffs; very local in W Br from S Devon to Westmorland. **RR**

49. CHRYSOCOMA L. - *Shrub Goldilocks*

Glabrous shrublets; stem-leaves linear to filiform, the edges rolled under, entire, sessile; capitula yellow, discoid, 1 on end of each branch; phyllaries in several rows, narrow, herbaceous with membranous margins; pappus of 1 row of hairs.

1. C. coma-aurea L. - *Shrub Goldilocks*. Stems thin, erect to ascending, to 60cm; leaves linear, absent from apical 1-3cm of stem below capitula; capitula c.10-15mm across; (2n=18). Intrd; grown in gardens, natd on walls, dunes and open ground; Scillies; S Africa.

2. C. tenuifolia P.J. Bergius - *Fine-leaved Goldilocks*. Differs from *C. coma-aurea* in leaves filiform, ascending to within <1cm of capitula; capitula c.5-10mm across. Intrd; fairly frequent wool-alien; scattered in En; S Africa.

50. ERIGERON L. - *Fleabanes*

Annuals to perennials; stem-leaves linear to obovate, entire or toothed, sessile or shortly petiolate; capitula 1-many per stem, radiate, with whitish to yellow disc flowers and whitish to pink or mauve ligules; central flowers tubular, bisexual, more numerous than peripheral female filiform flowers or the latter 0, the outermost female flowers with obvious ligules at least as long as tubular part; phyllaries in several rows, narrow, herbaceous with ± membranous margins; pappus of 1 row of hairs, sometimes with an outer row of very short hairs, the ray flowers sometimes with only very short hairs or narrow scales.

Merges into *Aster* at 1 extreme and *Conyza* at the other.

1 Ligules ≤4mm, not or scarcely exceeding pappus **6. E. acer**
1 Ligules ≥4mm, exceeding pappus by ≥2mm 2
  2 Capitula very showy, 3-5cm across, with bluish-mauve or pale mauve ligules 9-20mm; leaves succulent **1. E. glaucus**
  2 Capitula less showy, 1.5-3cm across, with white to pinkish (mauve in 1 alpine sp.) ligules 4-10mm; leaves not succulent 3
3 Stem procumbent to ascending; lower leaves with 1 pair of lateral lobes or teeth **4. E. karvinskianus**
3 Stem erect; lower leaves entire or toothed, but not regularly toothed as in last 4
  4 Stems rarely >20cm, with 1(-3) capitula; ligules mauve; rare alpine; leaves entire **3. E. borealis**
  4 Stems rarely <20cm, with numerous capitula; ligules white or pale pink or blue; lowland aliens; leaves often serrate 5
5 Leaves clasping stem; pappus of ray and of disc flowers of long hairs only **2. E. philadelphicus**
5 Leaves not clasping stem; pappus of disc flowers of long hairs and outer short scales, of ray flowers of short scales only **5. E. annuus**

Other spp. - The arctic **E. uniflorus** L., close to *E. borealis* but without filiform disc flowers, has been erroneously reported from Sc. **E. speciosus** (Lindl.) DC., from N America, is the familiar garden *Erigeron* with capitula 3-6cm across, mauve ligules 9-24mm, and glabrous, thin leaves; it is occasionally found as a relic or throwout.

1. E. glaucus Ker Gawl. - *Seaside Daisy*. Stems procumbent to ascending, pubescent, to 50cm, stem-leaves much smaller than basal ones, succulent, pubescent; ligules 9-15mm, mauve; (2n=18). Intrd; natd in rocky places and on cliffs in S En (especially Wight) and CI; N America.

2. E. philadelphicus L. - *Robin's-plantain*. Stems erect, pubescent, to 75cm; stem-leaves oblanceolate to narrowly obovate, irregularly dentate, pubescent; ligules 5-10mm, white to pink; (2n=18). Intrd; natd on walls and in rough ground; very scattered in Br; N America.

3. E. borealis (Vierh.) Simmons - *Alpine Fleabane*. Stems erect, pubescent, to 20cm; **RRR**
stem-leaves linear to lanceolate or oblanceolate, entire, pubescent; ligules 4-6mm, mauve; 2n=18. Native; mountain rock-ledges above 800m; very rare in M & E Perth, Angus and S Aberdeen.

4. E. karvinskianus DC. (*E. mucronatus* DC.) - *Mexican Fleabane*. Stems procumbent to ascending; sparsely pubescent, to 50cm; stem-leaves oblanceolate to obovate, entire or with 1 pair of lobes or large teeth, sparsely pubescent; ligules 5-8mm, white or pale mauve on upperside, pink to purple on lowerside; (2n=18, 32, 36, 54). Intrd; natd on walls, banks and stony ground; scattered in BI N to Co Armagh, Denbs and E Norfolk, especially SW En and CI; Mexico.

5. E. annuus (L.) Pers. (*E. strigosus* Muhl. ex Willd.) - *Tall Fleabane*. Stems erect, **726**
rather sparsely pubescent, to 70(100)cm; stem-leaves lanceolate to oblanceolate, entire to irregularly dentate, subglabrous; ligules 4-10mm, white to pale mauve; (2n=27). Intrd; natd in sandy places and rough ground; very scattered in SW En, rare casual elsewhere; N America. Ssp. **strigosus** (Muhl. ex Willd.) Wagenitz, with shorter stems, fewer narrower stem-leaves, and shorter ligules, occurred recently in N Hants; it is doubtfully distinct. The sp. is apomictic.

6. E. acer L. - *Blue Fleabane*. Stems erect, pubescent, to 60cm; stem-leaves lanceolate to oblanceolate, entire, pubescent; ligules 2-4mm, purplish-mauve; 2n=18. Native; barish sandy or calcareous soils, banks, walls and dunes; locally frequent in En, Wa and CI, rare in Ir, casual in Sc.

50 x 51. ERIGERON x CONYZA = X CONYZIGERON Rauschert

50/6 x 51/1. X C. huelsenii (Vatke) Rauschert (*Erigeron x huelsenii* Vatke; *E. acer* x *C. canadensis*) occurs sporadically with the parents in disturbed sandy places in S En; it is intermediate in pubescence and capitulum size (ligules pale mauve, 1-2mm), and sterile, somewhat resembling *C. bonariensis* (q.v.) in habit but with broader leaves with longer hairs and characteristically with a proportion of abortive capitula.

51. CONYZA Less. - *Fleabanes*

Annuals; stem-leaves linear to narrowly elliptic or oblanceolate, entire or toothed, sessile or shortly petiolate; capitula numerous, discoid or very inconspicuously radiate, with white to cream or pinkish flowers; central flowers tubular, bisexual; peripheral flowers female, more numerous, the outermost often with very short ligules (shorter than tubular part); phyllaries as in *Erigeron*; pappus of 1 row of hairs.

1 Phyllaries yellowish-green, glabrous to sparsely pubescent; disc flowers with 4-lobed corolla; inflorescence ± cylindrical (see also under *C. bilbaoana*) **1. C. canadensis**

1 Phyllaries greyish-green, pubescent to densely so; disc flowers with 5-lobed corolla; inflorescence pyramidal to corymbose 2

2 Inflorescence pyramidal; pappus yellowish-white or cream; phyllaries not or minutely red-tipped **2. C. sumatrensis**

2 Inflorescence with long lateral branches, usually subcorymbose; pappus greyish- or off-white; phyllaries usually conspicuously red-tipped **3. C. bonariensis**

1. C. canadensis (L.) Cronquist (*Erigeron canadensis* L.) - *Canadian Fleabane*. Stems erect, to 1m, green, rather sparsely pubescent; leaves green, linear to oblanceolate, the margins with well-spaced patent to erecto-patent hairs often >1mm; capitula 3-5mm wide; ligules c.0.5-1mm; pappus cream; 2n=18. Intrd; natd in waste and rough ground, walls, waysides and dunes on well-drained soils; common in SE En and CI, much sparser to N & W, rare in Sc, scattered in Ir; N America. Robust, more hispid plants with larger capitula, 5-lobed corolla in disc flowers and a subcorymbose inflorescence, 1st found in 1992 near Southampton, S Hants, have been called **C. bilbaoana** J. Rémy; their true identity, distinction from *C. canadensis* and relationship to the other spp. (with which they would key out above) needs further study.

1 x 3. C. canadensis x C. bonariensis was identified as a single sterile intermediate plant found in Middlesex in 1993.

2. C. sumatrensis (Retz.) E. Walker (*C. albida* Willd. ex Spreng., *C. floribunda* Kunth, *Erigeron sumatrensis* Retz.) - *Guernsey Fleabane*. Stems erect, to 1(2)m, dull green, pubescent; leaves dull green, linear-oblong to narrowly elliptic or narrowly obovate, the margins with many hooked hairs much <0.5mm; capitula mostly 5-8mm wide at fruiting; ligules ≤0.5mm; (2n=54). Intrd; natd in waste and rough ground in protected sunny spots; frequent in parts of London area and CI, scattered elsewhere in S En N to W Norfolk and in E Ir, increasing; S America.

3. C. bonariensis (L.) Cronquist (*Erigeron bonariensis* L.) - *Argentine Fleabane*. Stems erect, to 60cm, greyish-green, pubescent; leaves greyish-green, linear to oblanceolate, the margins with many hooked hairs much <0.5mm; capitula mostly 7-11mm wide at fruiting; ligules 0; (2n=54). Intrd; rather frequent casual in waste and cultivated ground, often as wool-alien, natd in Middlesex since 1993; scattered in CI, En and Sc; S America. The presence of glands or stickiness is of no value in separating this and the last sp.

52. CALLISTEPHUS Cass. - *China Aster*

Annuals; stem-leaves ovate, deeply toothed or lobed, the lower petiolate; capitula very conspicuous, radiate but often *flore pleno*, with yellow disc flowers when present, with white or blue to pink or purple ligules; phyllaries in several rows, the outer herbaceous and very leafy, the inner membranous-bordered; pappus of 2 rows of hairs.

1. C. chinensis (L.) Nees - *China Aster*. Stems erect, to 75cm, stiffly pubescent; capitula few, up to 12cm across, very showy; (2n=18, 36). Intrd; garden throwout on tips and waste ground; scattered in En; China.

53. OLEARIA Moench - *Daisy-bushes*

Strong shrubs to small trees, evergreen; leaves simple, alternate or opposite, white-tomentose on lowerside, entire to sharply toothed, petiolate; capitula numerous, ± crowded in lateral or terminal corymbose panicles, radiate or discoid, with yellow to reddish disc flowers and (if present) white ligules; phyllaries in several rows, rather scarious; pappus of 1 row of hairs.

1 Leaves with conspicuous acute teeth on margin **4. O. macrodonta**
1 Leaves entire 2
2 Leaves opposite **5. O. traversii**
2 Leaves alternate 3
3 Leaves undulate at margin; inflorescences axillary, flowering Oct-Nov; capitula with only 1 floret **1. O. paniculata**

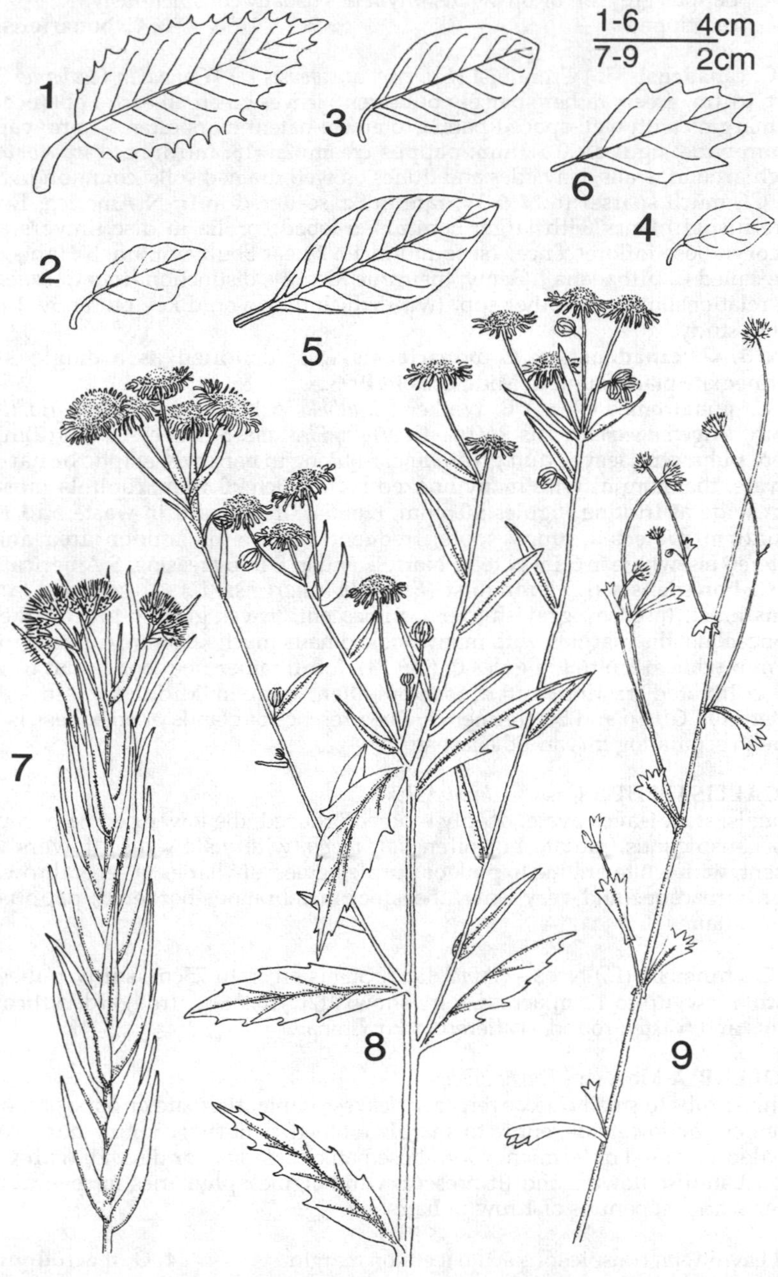

FIG 726 - Asteraceae: Astereae. 1-5, leaves of *Olearia*. 1, *O. macrodonta*. 2, *O. paniculata*. 3, *O. traversii*. 4, *O. x haastii*. 5, *O. avicenniifolia*. 6, leaf of *Baccharis halimiifolia*. 7, *Solidago graminifolia*. 8, *Erigeron annuus*. 9, *Calotis cuneifolia*.

3 Leaves flat; inflorescences terminal, flowering Jul-Aug; capitula with ≥2 florets 4
4 Leaves 4-10cm; capitula with 3-5 ray flowers **3. O. x haastii**
4 Leaves 1-3cm; capitula with 0-2 ray flowers **2. O. avicenniifolia**

Other spp. - **O. solandri** Hook. f. has opposite linear-obovate leaves up to 15mm and solitary capitula; **O. nummulariifolia** (Hook.f.) Hook.f. has broadly obovate to suborbicular leaves 5-10mm and solitary capitula; and **O. ilicifolia** (Hook.f.) Hook.f. has linear-oblong to -elliptic leaves >4x as long as wide with teeth as in *O. macrodonta*. All 3 are from New Zealand, are grown in gardens and have appeared semi-wild in W Cornwall, Co Louth and E Sussex respectively.

1. O. paniculata (J.R. & G. Forst.) Druce - *Akiraho*. Shrub to 3(6)m; leaves alternate, 2.5-10cm, entire, undulate-margined; inflorescences axillary, produced in autumn, pyramidal; capitula discoid, white, with only 1 flower; (2n=c.288). Intrd; grown in Guernsey for hedging, frequent relic, rarely self-sown, also ± natd in W Cornwall; New Zealand. See *Pittosporum tenuifolium* (Pittosporaceae) for differences. **726**

2. O. avicenniifolia (Raoul) Hook. f. - *Mangrove-leaved Daisy-bush*. Shrub to 6m; leaves alternate, 4-10cm, entire, flat; inflorescences terminal, produced in late summer, corymbose; capitula white, with 0-2 ray flowers and only 1-3 disc flowers; (2n=108). Intrd; grown in Scillies and natd on dunes; New Zealand. **726**

3. O. x haastii Hook. f. (?*O. avicenniifolia* (Raoul) Hook. f. x *O. moschata* Hook. f.) - *Daisy-bush*. Shrub to 2(3)m; leaves alternate, 1-3cm, entire, flat; inflorescences terminal, produced in summer, corymbose; capitula radiate, with white ligules and yellow disc flowers. Intrd; the hardiest sp. and much grown in gardens and shrubberies, often well established and rarely self-sown on walls and in open ground; very scattered in SW and WC Br; New Zealand. **726**

4. O. macrodonta Baker - *New Zealand Holly*. Shrub to 3(6)m; leaves alternate, 5-12cm, ovate, deeply and sharply dentate, undulate-margined; inflorescences terminal, produced in summer, corymbose; capitula radiate, with white ligules and reddish disc flowers. Intrd; grown by sea, sometimes self-sowing (the best natd sp.), in hedges, scrub, on banks and rough ground; scattered in Ir, Man and W Br N to Wigtowns, Guernsey; New Zealand. **726**

5. O. traversii (F. Muell.) Hook. f. - *Ake-ake*. Shrub or tree to 10m; leaves opposite, 2.5-7cm, entire, flat; inflorescences axillary, produced in summer, pyramidal; capitula discoid, greyish; (2n=108). Intrd; grown as hedging in Guernsey and Scillies, often well established but not self-sown, apparently self-sown and natd in W Cornwall and Man; Chatham Island. See *Pittosporum crassifolium* (Pittosporaceae) for differences. **726**

54. BACCHARIS L. - *Tree Groundsel*
Dioecious, deciduous shrubs; leaves simple, alternate, roughly toothed in distal 1/2, tapered to petiole; capitula ± numerous, in terminal, loose leafy panicles, small, discoid, whitish; phyllaries in several rows, herbaceous with scarious borders; pappus of 1 row of hairs, shorter in male plants.

1. B. halimiifolia L. - *Tree Groundsel*. Erect, ± sticky shrub to 4m; leaves obovate, glabrous; capitula in wide, terminal panicles, c.2mm across, white, produced in Oct; pappus conspicuous on female plants, white; (2n=18). Intrd; grown by sea in S due to salt-tolerance, natd in S Hants since 1942 but ?still; N America. **726**

55. BELLIS L. - *Daisy*
Herbaceous perennials; leaves all basal, in rosette, simple, toothed, petiolate; capitula single on leafless stalks, radiate, with white to pink or red ligules and yellow disc flowers, or *flore pleno*; phyllaries in 2 rows, herbaceous; pappus 0.

1. B. perennis L. - *Daisy*. Leaves obovate, irregularly serrate; stems procumbent to erect, to 12(20)cm, leafless, with 1 capitulum; capitula 12-25mm across, up to 80mm across and often *flore pleno* in cultivars; 2n=18. Native; abundant throughout BI mostly in short grassland.

TRIBE 6 - ANTHEMIDEAE (genera 56-69). Annual to perennial herbs, rarely shrubs; leaves alternate, simple to pinnate, often finely and deeply divided; capitula discoid or radiate; phyllaries in 2-several rows, herbaceous with scarious margins and apex; receptacular scales 0 or present; pappus usually 0, sometimes a short rim; usually with yellow disc flowers and white ligules but exceptions not rare.

56. TANACETUM L. (*Balsamita* Mill.) - *Tansies*
Strongly aromatic perennial herbs; leaves simple and toothed to deeply pinnately lobed or pinnate; capitula radiate or discoid, rarely *flore pleno*; disc flowers yellow; ligules white or 0; receptacular scales 0; pappus a very short rim.

1 Leaves toothed, divided much <1/2 way to midrib **4. T. balsamita**
1 All or most leaves pinnate, or pinnately lobed much >1/2 way to midrib 2
  2 Rhizomes 0; ultimate leaf-lobes obtuse to subacute, sometimes apiculate **1. T. parthenium**
  2 Rhizomatous; ultimate leaf-lobes acute to acuminate 3
3 Capitula discoid, >5mm across **3. T. vulgare**
3 Capitula radiate, ≤5mm across excl. ligules; ligules white **2. T. macrophyllum**

1. T. parthenium (L.) Sch. Bip. (*Chrysanthemum parthenium* (L.) Bernh.) - *Feverfew*. Stems erect, to 70cm; capitula usually radiate, rarely discoid or *flore pleno*, in lax corymbs, 15-23mm across (6-9mm excl. ligules); 2n=18. Intrd; natd on walls, waste ground and waysides; frequent throughout BI; Balkans.

2. T. macrophyllum (Waldst. & Kit.) Sch. Bip. - *Rayed Tansy*. Stems erect, to 730
1.2m; capitula radiate, in dense corymbs, 7-13mm across (3-5mm excl. ligules); (2n=14, 18). Intrd; grown for ornament, natd in grassy places and waysides since c.1912; very scattered in En and Sc; SE Europe. Has been misdetermined as *Achillea grandifolia*.

3. T. vulgare L. (*Chrysanthemum vulgare* (L.) Bernh.) - *Tansy*. Stems erect, to 1.2m; capitula discoid, in dense corymbs, 6-10mm across; 2n=18. Native; grassy places, waysides, rough ground; frequent throughout BI.

4. T. balsamita L. (*Chrysanthemum balsamita* (L.) Baill. non L., *Balsamita major* Desf.) - *Costmary*. Stems erect, to 1.2m; capitula discoid, in tight corymbs, 4-8mm across, rarely with few ligules 4-6mm; (2n=18, 54). Intrd; grown for ornament and cooking, rarely natd as outcast; very scattered in C & S Br; Caucasus.

57. SERIPHIDIUM (Besser ex Hook.) Fourr. - *Sea Wormwood*
Aromatic perennials; differ from *Artemisia* in flowers all similar and functionally bisexual.

1. S. maritimum (L.) Polj. (*Artemisia maritima* L.) - *Sea Wormwood*. Stems decumbent to erect, woody below; leaves white-woolly, 1-2-pinnate with linear ultimate segments; capitula yellowish- to reddish-brown, numerous in terminal panicle, 1.5-3.5mm across; 2n=54. Native; dry parts of saltmarshes, sea-walls and rough ground by sea; local on coasts of Br N to C Sc, especially E En, E & W Ir.

58. ARTEMISIA L. - *Mugworts*
Annual to perennial herbs or small shrubs, often aromatic; leaves entire to finely divided; capitula discoid, small, brownish overall; flowers usually yellowish, the

outer female, with filiform corolla, the inner bisexual, with tubular corolla; receptacular scales 0; pappus 0.

1 Leaves most or all entire **9. A. dracunculus**

1 Most or all leaves deeply divided 2

  2 Stems woody ± to top **6. A. abrotanum**

  2 Stems herbaceous, or woody only near base 3

3 Capitula 1-2(5); stems <10cm **5. A. norvegica**

3 Capitula normally >10; stems >10cm 4

  4 Annual or biennial with simple root system and 0 non-flowering shoots 5

  4 Perennial with strong underground portion and non-flowering shoots 6

5 Leaves in inflorescence projecting laterally well beyond capitula, with many primary divisions >(1.5)2cm x c.1-3mm **7. A. biennis**

5 Leaves in inflorescence extending laterally less far than capitula, with primary divisions <1(1.5)cm x c.0.5-1mm **8. A. annua**

  6 Mature leaves densely (often whitish-)pubescent on upperside 7

  6 Mature leaves glabrous or subglabrous on upperside (beware mildew) 8

7 Plant not aromatic, rhizomatous; receptacle glabrous; capitula 6-10 x 5-9mm excl. flowers **4. A. stelleriana**

7 Plant aromatic, at least when fresh, not rhizomatous; receptacle pubescent; capitula 1.5-3.5 x 3-5mm excl. flowers **3. A. absinthium**

  8 All leaf-lobes <2mm wide; plant not aromatic; achenes usually produced only by marginal flowers **10. A. campestris**

  8 Most or all leaf-lobes >2mm wide; plant aromatic, at least when fresh; achenes produced by all flowers 9

9 Plant not or scarcely rhizomatous; terminal untoothed portion of middle stem-leaves usually <3cm; stem with central (white) pith region occupying c.4/5 of total (white + green) pith diameter; flowers Jul-Sep **1. A. vulgaris**

9 Plant strongly rhizomatous; terminal untoothed portion of middle stem-leaves usually >3cm; stem with central (white) pith region occupying c.1/3 of total (white + green) pith diameter; flowers Oct-Dec **2. A. verlotiorum**

Other spp. - 11 other spp. recorded as casuals, incl. **A. afra** Jacq., **A. pontica** L. and **A. scoparia** Waldst. & Kit., are much rarer than the 10 spp. treated here.

1. A. vulgaris L. - *Mugwort*. Aromatic, tufted perennial to 1.5m; leaves glabrous on upperside, whitish-tomentose on lowerside, with lobes c.2.5-8mm wide; capitula numerous, 1.5-3.5mm across; 2n=16. Native; rough ground, waste places, waysides; common throughout lowland BI.

1 x 2. A. vulgaris x A. verlotiorum was discovered in Middlesex and S Essex in 1987 and Surrey in 1989; it is intermediate in all characters (white part of pith c.3/5 total pith width) and completely sterile (flowers appear Oct-Dec but have abortive stamens); 2n=34; endemic.

2. A. verlotiorum Lamotte - *Chinese Mugwort*. Rhizomatous perennial to 1.5m; differs from *A. vulgaris* in leaves darker on upperside with closer network of veins visible in fresh state; inflorescence more leafy; and see key; 2n=50, 52. Intrd; natd since 1908 in similar places to *A. vulgaris*; frequent in London area, especially near R Thames, very scattered elsewhere in S & C En, Caerns, Guernsey; China.

3. A. absinthium L. - *Wormwood*. Aromatic, tufted perennial to 1m; leaves greyish-pubescent on both surfaces, with lobes c.2-4mm wide; capitula numerous, 3-5mm across; 2n=18. Native; similar places to *A. vulgaris*; frequent in En, Wa and CI, very scattered in Sc and Ir.

4. A. stelleriana Besser - *Hoary Mugwort*. Non-aromatic, rhizomatous perennial

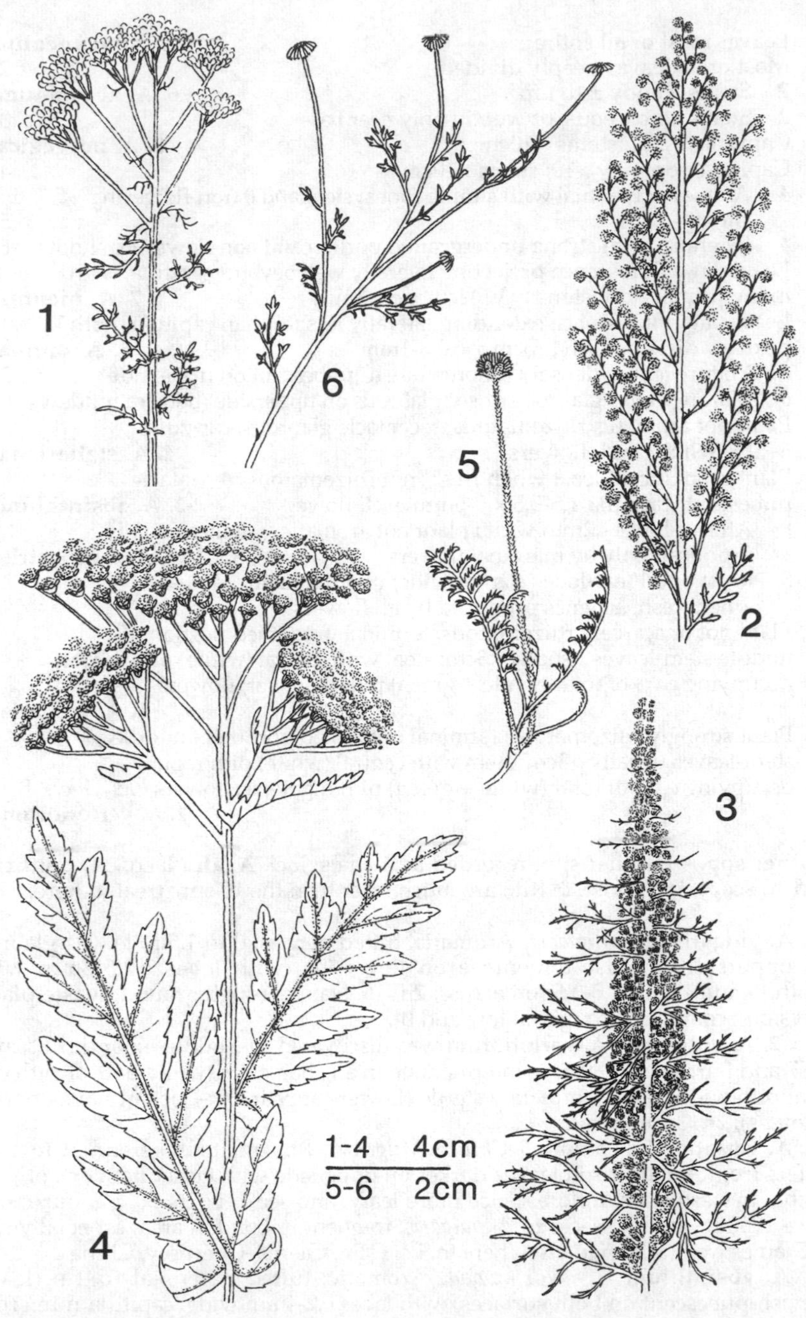

FIG 730 - Asteraceae: Anthemideae. 1, *Achillea ligustica*. 2, *Artemisia annua*. 3, *A. biennis*. 4, *Tanacetum macrophyllum*. 5, *Cotula squalida*. 6, *C. australis*.

to 60cm; leaves whitish-tomentose on both surfaces, with lobes c.3-8mm wide; capitula ± numerous, 5-9mm across; (2n=18). Intrd; natd on maritime dunes in Kirkcudbrights since 1979, Clyde Is since 1976, Ayrs since 1994, formerly W Cornwall and Co Dublin; NE Asia.

5. A. norvegica Fr. - *Norwegian Mugwort*. Aromatic, rosette-perennial to 8cm; leaves pubescent on both surfaces, few and reduced on stems, with lobes 0.5-1.5mm wide; capitula 1-2(5), 8-13mm across; 2n=18. Native; at 3 sites at c.800m on barish mountain-tops in E & W Ross, discovered 1950. Our plant has been named var. **scotica** Hultén. **RR**

6. A. abrotanum L. - *Southernwood*. Very aromatic shrub to 1.2m; leaves pubescent on both surfaces, with lobes c.0.6-1mm wide; capitula numerous, 3-4mm across; (2n=18). Intrd; much grown in gardens, rarely persistent on tips and waste ground; sporadic in S & C Br; origin unknown. Rarely or never flowers.

7. A. biennis Willd. - *Slender Mugwort*. Very aromatic erect annual (to biennial) to 1.5m; leaves glabrous, with primary lobes c.1-3mm wide excl. teeth; capitula numerous, 1.5-4mm across; (2n=18). Intrd; casual from grain and wool on waste ground and reservoir mud, ± natd in few sites in S Br; Asia and N America. **730**

8. A. annua L. - *Annual Mugwort*. Differs from *A. biennis* in key characters, but often confused with it; (2n=18). Intrd; casual from same sources as *A. biennis* but rarer and not natd; very scattered in S Br; SE Europe and Asia. **730**

9. A. dracunculus L. - *Tarragon*. Aromatic perennial to 1.2m; leaves glabrous, mostly linear to narrowly elliptic and entire, c.2-10mm wide; capitula numerous, 2-3mm across; (2n=18, 36, 54, 72). Intrd; grown for flavouring, rarely persistent on tips and waste ground; very scattered in S En; Russia.

10. A. campestris L. - *Field Wormwood*. Non-aromatic perennial to 75cm; leaves ± pubescent when young, ± glabrous when mature, with lobes c.0.3-1mm wide; capitula numerous, 2-4mm across; (2n=16, 18, 36). Native; grassy places by roads and on heathland; very local in W Suffolk and W Norfolk, formerly E Norfolk and Cambs, natd in Glam. **RRR**

59. SANTOLINA L. - *Lavender-cotton*

Evergreen shrubs; leaves neatly and closely pinnately lobed; capitula discoid, yellow; receptacular scales present; pappus 0.

1. S. chamaecyparissus L. - *Lavender-cotton*. Stems decumbent to suberect, to 60cm; whole plant white- to grey-tomentose; capitula 6-10mm across, solitary on erect stems; (2n=18, 36). Intrd; much grown in gardens, persistent on tips, rough ground, old gardens and rockeries, scattered in S & C Br, especially SW En, rarely elsewhere; Mediterranean.

60. OTANTHUS Hoffmanns. & Link - *Cottonweed*

Perennial, densely white-woolly herbs; leaves simple, crenate; capitula discoid with yellow flowers; phyllaries obscured by dense hairs; receptacular scales present; pappus 0.

1. O. maritimus (L.) Hoffmanns. & Link - *Cottonweed*. Stems erect to ascending, to 30cm; leaves oblong-obovate; capitula few, subcorymbose, 6-9mm across; 2n=18. Native; maritime fixed sand and shingle; now in 1 place in Co Wexford, formerly scattered in BI N to E Suffolk, Anglesey and Co Wicklow. **RR**

61. ACHILLEA L. - *Yarrows*

Perennial herbs; leaves simple and very shallowly toothed to deeply and finely dissected; capitula radiate, rarely *flore pleno*; disc flowers and ligules white to deep pink, rarely yellow; receptacular scales present; pappus 0.

1 Leaves simple, toothed much <1/2 way to midrib; capitula >1cm across **1. A. ptarmica**
1 Leaves compound, or simple and divided much >1/2 way to midrib; capitula ≤1cm across 2
 2 Middle stem-leaves <3x as long as wide; with <10 pairs of primary lateral lobes **2. A. ligustica**
 2 Middle stem-leaves >3x as long as wide, with >15 pairs of primary lateral lobes 3
3 Leaves ± flat in fresh state, the primary lateral lobes ± contiguous on the rhachis; inner phyllaries >3.5mm **4. A. distans**
3 Leaves with lobes spreading in 3 dimensions in fresh state, the primary lateral lobes separated by a length of winged rhachis; inner phyllaries ≤3.5mm **3. A. millefolium**

Other spp. - Several yellow-flowered spp. are grown in gardens and have been reported as rare escapes: **A. filipendulina** Lam., from W & C Asia, is the familiar border-plant >1m high with corymbs often >6cm across; **A. tomentosa** L., from SW Europe, is much smaller and is densely grey-pubescent. A similarly reported white- (or very pale yellow-)flowered sp. is the European **A. nobilis** L., differing from *A. ligustica* in its primary leaf-segments wider and more regularly subdivided; records of **A. grandifolia** Friv. are errors for *Tanacetum macrophyllum*.

1. A. ptarmica L. - *Sneezewort*. Stems erect, to 60cm; leaves linear to linear-lanceolate, finely and shallowly toothed; capitula <10(15), in lax corymbs, 12-20mm across, sometimes *flore pleno* in garden escapes; ligules white; 2n=18. Native; damp grassy places and marshy fields; frequent in Br and most of Ir, casual in CI.

2. A. ligustica All. - *Southern Yarrow*. Resembles *A. millefolium* but leaves shorter, **730**
wider and more finely divided (see key); capitula smaller (c.3mm across; inner phyllaries ≤3mm) and more numerous; ligules white; (2n=18). Intrd; natd in waste ground at Newport Docks, Mons, since 1953; Mediterranean.

3. A. millefolium L. - *Yarrow*. Stems erect, to 80cm; leaves very deeply divided into many, deeply divided lateral lobes; capitula >(25)50 in ± dense corymbs, c.4-6mm across; inner phyllaries c.3-3.5mm; ligules white (to deep pink); 2n=54. Native; grassland (usually short), banks and waysides; very common throughout BI.

4. A. distans Waldst. & Kit. ex Willd. - *Tall Yarrow*. Resembles *A. millefolium* but stems to 1.3m; leaves differ as in key; capitula larger (5-10mm across, inner phyllaries 3.5-5mm); (2n=54). Intrd; natd in grassy places; Derbys and MW Yorks; S & E Europe. Our plant is ssp. **tanacetifolia** Janch.

62. CHAMAEMELUM Mill. - *Chamomile*
Aromatic perennial herbs; leaves deeply and finely dissected; capitula radiate, rarely discoid; disc flowers yellow, with short pouch at base of tube; ligules white; receptacular scales present; pappus 0.

Other spp. - **C. mixtum** (L.) All., from Mediterranean, is non-aromatic, has densely (not sparsely) pubescent phyllaries, and has much less finely divided leaves; it is a scarce birdseed-, grain- and wool-alien.

1. C. nobile (L.) All. - *Chamomile*. Stems procumbent to ascending, to 30cm; **R**
receptacular scales oblong to narrowly obovate, acuminate; achenes weakly ridged on 1 face; 2n=18. Native; short grassy places on sandy soils; locally frequent in CI, S Br and SW Ir, scattered N to W Norfolk and Caerns but extinct in most of C Br, intrd N to C Sc.

63. ANTHEMIS L. - *Chamomiles*
Aromatic annual to perennial herbs; leaves deeply and finely dissected; capitula radiate, rarely discoid; disc flowers yellow; ligules white or yellow; receptacular scales present; pappus 0 or a short rim.

1 Ligules yellow, occasionally 0 **4. A. tinctoria**
1 Ligules white, very rarely 0 2
  2 Receptacular scales only on inner (upper) part of receptacle, linear-subulate; achenes tuberculate on ribs; fresh plant with unpleasant scent **3. A. cotula**
  2 Receptacular scales all over receptacle, at least the inner ones lanceolate to oblanceolate; achenes ribbed or scarcely so, but not tuberculate; fresh plant with sweet scent 3
3 Perennial, often woody near base and with non-flowering shoots; at least outer receptacular scales 3-toothed; achenes not or slightly ribbed **1. A. punctata**
3 Annual or biennial, not woody at base and usually without non-flowering shoots; receptacular scales with single slender apex; achenes strongly ribbed **2. A. arvensis**

Other spp. - A. austriaca Jacq., from E Europe, would key to couplet 3 and agrees with *A. arvensis* in habit and receptacular scale shape, but has scarcely ribbed achenes with a very short apical rim and the receptacular scales become rigid in fruit; it has appeared as a casual recently in S Br (probably from seed mixtures) and might be overlooked as *A. arvensis*. **A. ruthenica** M. Bieb., from E Europe, differs from *A. arvensis* in having broader receptacular scales with dentate-laciniate apex, narrower outer achenes and disc flowers with a swollen lower part; it is a scarce grain- and birdseed-alien.

1. A. punctata Vahl - *Sicilian Chamomile.* Perennial to 60cm; ligules white; at least outer receptacular scales 3-toothed at apex; achenes not or weakly ribbed, with apical rim 0.3-0.8mm; (2n=18). Intrd; grown in gardens, natd in rough and marginal ground and on cliffs, mostly near sea; very scattered in S Br; Sicily. Our plant is ssp. **cupaniana** (Tod. ex Nyman) R. Fern. **734**

2. A. arvensis L. - *Corn Chamomile.* Annual to 50cm; ligules white; receptacular scales oblong-lanceolate, cuspidate; achenes strongly ribbed, not tuberculate, with apical rim ≤0.5mm; 2n=18. Native; arable land, waste places and rough ground, usually on calcareous soils, also a grass-seed alien; locally frequent in S & C Br, rare and mainly casual in N Br and CI, extinct in Ir. **734**

3. A. cotula L. - *Stinking Chamomile.* Differs from *A. arvensis* in more stiffly erect stems; leaf segments linear and ± glabrous (not narrowly oblong and pubescent); ligules becoming reflexed (not remaining patent); achenes without apical rim; and see key; 2n=18. Native; similar habitats to *A. arvensis* but often on heavier soils; similar distribution to *A. arvensis* but commoner, still in S & E Ir. **734**

4. A. tinctoria L. - *Yellow Chamomile.* Biennial or perennial to 50cm; ligules yellow, occasionally 0; receptacular scales as in *A. arvensis*; achenes scarcely ribbed, not tuberculate, with apical rim <0.3mm; (2n=18). Intrd; natd or casual in waste places, rough and marginal land; rather frequent in S & C Br, rare in N; Europe. Ligule-less plants can be told from similar variants of other spp. by combination of receptacular scale and achene characters, larger capitula and very characteristic leaf-lobing. Sometimes mistaken for *Chrysanthemum segetum*, but this differs in many details, incl. leaf-lobing. **734**

63 x 68. ANTHEMIS x TRIPLEUROSPERMUM = X TRIPLEUROTHEMIS Stace
63/3 x 68/2. X T. maleolens (P. Fourn.) Stace (*X Anthemimatricaria celakovskyi* Geisenh. ex Domin nom. illeg.; *A. cotula* x *T. inodorum*) is intermediate in the

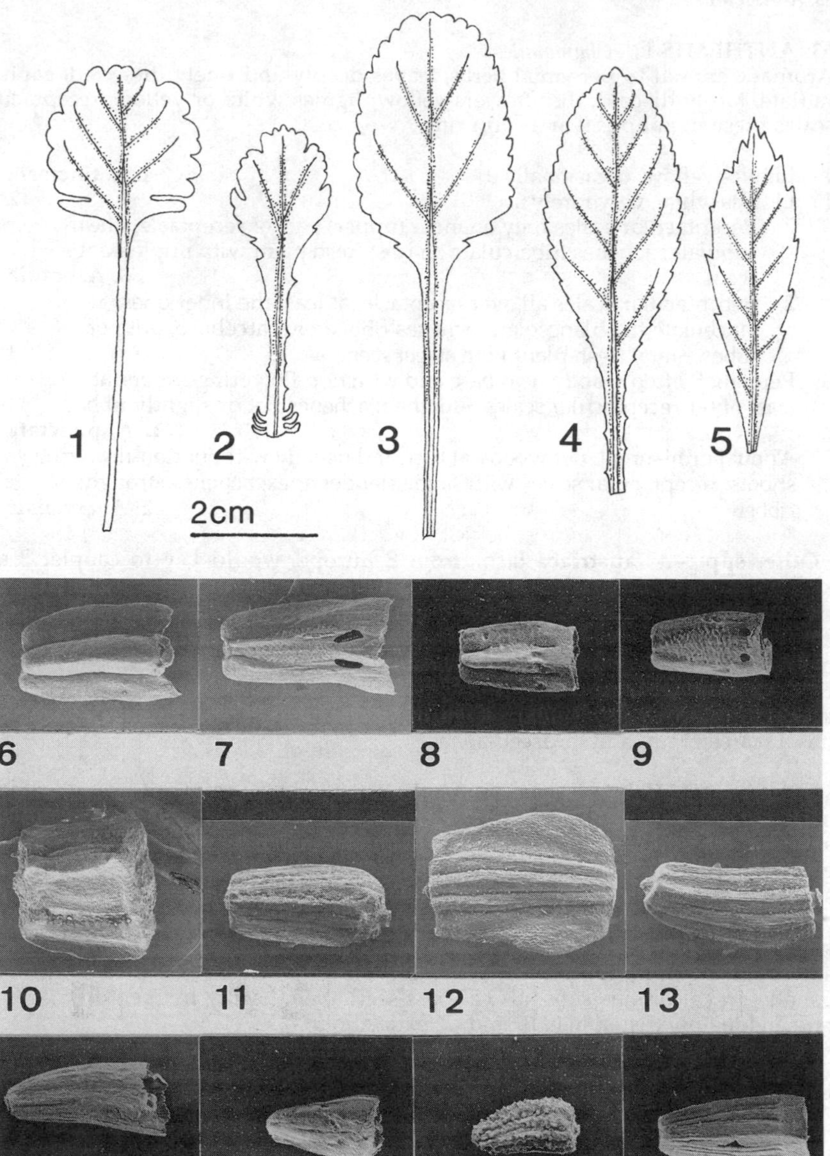

FIG 734 - Asteraceae: Anthemideae. 1-4, basal and stem-leaves of ***Leucanthemum***. 1-2, ***L. vulgare***. 3-4, ***L. x superbum***. 5, stem-leaf of ***Leucanthemella serotina***. 6-9, adaxial and abaxial faces of achenes of ***Tripleurospermum***. 6-7, ***T. maritimum***. 8-9, ***T. inodorum***. Oil-glands marked in black. 10-13, marginal and inner achenes of ***Chrysanthemum***. 10-11, ***C. coronarium***. 12-13, ***C. segetum***. 14-17, achenes of ***Anthemis***. 14, ***A. punctata***. 15, ***A. arvensis***. 16, ***A. cotula***. 17, ***A. tinctoria***.

irregular presence of receptacular scales that are intermediate between those of the *Anthemis* parent and the phyllaries, and in the sterile achenes with intermediate rib development and traces of subapical oil-glands. 1 plant in Berks (1966) and 2 in Salop (1969).

64. CHRYSANTHEMUM L. - *Crown Daisies*
Annual herbs; leaves simple, shallowly to deeply lobed; capitula radiate; disc flowers yellow; ligules yellow, cream or yellow and cream; receptacular scales 0; pappus 0.

1. C. segetum L. - *Corn Marigold*. Stems decumbent to erect, to 60cm; leaves **734**
glaucous, slightly toothed to deeply lobed, at least the upper usually lobed <1/2 way to midrib; capitula 3-7cm across; ligules yellow; achenes 2.5-3mm, deeply ridged, not winged; 2n=18. Intrd; natd or casual weed of arable fields, waste places and waysides; locally frequent throughout BI; Europe. See *Anthemis tinctoria* for differences.

2. C. coronarium L. - *Crown Daisy*. Stems ascending to erect, to 80cm; leaves **734**
green, lobed >1/2 way to midrib, often ± to midrib; capitula 4-8cm across; ligules cream, yellow, or cream and yellow; achenes 3-3.5mm, deeply ridged, the inner with adaxial wing, the marginal with 2 lateral and 1 adaxial wings; (2n=18). Intrd; similar places to *C. segetum* but much rarer, often as grain alien, not natd; very scattered in En and Wa; Europe.

65. LEUCANTHEMELLA Tzvelev - *Autumn Oxeye*
Perennial herbs; leaves simple, sharply serrate; capitula radiate; disc flowers yellow; ligules white; receptacular scales 0; pappus ± 0.

1. L. serotina (L.) Tzvelev (*Chrysanthemum uliginosum* (Waldst. & Kit. ex Willd.) **734**
Pers., *C. serotinum* L.) - *Autumn Oxeye*. Stems erect, to 1.5m; resembles *Leucanthemum x superbum* but leaves paler green and more sharply and deeply serrate (most serrations >3mm), flowers later (Sep-Oct), and tubular part of corolla of ray flowers unwinged. Intrd; garden escape or throwout natd on rough ground and by ditches and ponds; scattered in S En; SE Europe.

66. LEUCANTHEMUM Mill. - *Oxeye Daisies*
Differ from *Leucanthemella* in tubular part of ray flowers with 2 narrow translucent wings; and achenes with translucent secretory canals.

1. L. vulgare Lam. (*Chrysanthemum leucanthemum* L.) - *Oxeye Daisy*. Stems erect **734**
to ascending, to 75cm; basal and lower stem-leaves obovate-spathulate, abruptly contracted to broadly cuneate base; upper stem-leaves usually deeply serrate; capitula 2.5-6(7.5)cm across; 2n=18, 36. Native; grassy places, especially on rich soils; common throughout BI.

2. L. x superbum (Bergmans ex J.W. Ingram) D.H. Kent (*L. maximum* auct. non **734**
(Ramond) DC., *Chrysanthemum maximum* auct. non Ramond; *L. lacustre* (Brot.) Samp. x *L. maximum* (Ramond) DC.) - *Shasta Daisy*. Stems erect to ascending, to 1.2(1.5)m; basal and lower stem-leaves elliptic-oblong, gradually contracted to narrowly cuneate base; upper stem-leaves usually shallowly serrate to subentire; capitula (5)6-10cm across, often *flore pleno*. Intrd; abundant in gardens and fully fertile, well natd in waste and rough ground and grassy waysides; scattered throughout Br and CI; garden origin. Records of **L. lacustre** might, like all those of **L. maximum**, be errors for *L. x superbum*.

67. MATRICARIA L. (*Chamomilla* Gray) - *Mayweeds*
Annual herbs, differing from *Tripleurospermum* in much more conical, hollow (not solid) receptacle; ligules often 0; and achenes with 4-5 weak (not 3 strong) ribs and

without (not with) oil-glands.

1. M. recutita L. - *Scented Mayweed.* Superficially much like *Tripleurospermum inodorum* but usually (not always) more strongly and sweetly scented when fresh; phyllaries with very pale brown (not deep brown) scarious margins; ligules soon very strongly reflexed; and see generic characters above; 2n=18. Native; in similar places to and often with *T. inodorum* but less common and more restricted to arable ground on light soils; locally common in CI, En and Wa, very scattered in Sc, rare casual in Ir.

2. M. discoidea DC. (*M. matricarioides* (Less.) Porter nom. illeg., *M. suaveolens* (Pursh) Buchenau non L.) - *Pineappleweed.* Plant erect, to 35cm; ligules 0; differs from rare ligule-less plants of *M. recutita* in sweet pineapple-like scent; much wider, white scarious margins to phyllaries; and disc flowers with 4-lobed (not 5-lobed) corolla; 2n=18. Intrd; weed of barish places by paths and waste places; common throughout BI, 1st recorded 1871; widespread weed.

68. TRIPLEUROSPERMUM Sch. Bip. (*Matricaria* auct. non L.) - *Mayweeds*
Annual to perennial herbs; leaves deeply and finely dissected; capitula radiate, rarely discoid; disc flowers yellow; ligules white; receptacular scales 0; pappus a very short rim.

Other spp. - T. disciforme (C.A. Mey.) Sch. Bip. (*Matricaria disciformis* (C.A. Mey.) DC.) and **T. decipiens** (Fisch. & C.A. Mey.) Bornm. (*M. decipiens* (Fisch. & C.A. Mey.) K. Koch), from Turkey, are liguleless spp. that are rare casuals or have been briefly natd in the past; sometimes *Matricaria discoidea* is misrecorded for them.

1. T. maritimum (L.) W.D.J. Koch (*Matricaria maritima* L.) - *Sea Mayweed.* Erect to **734**
procumbent (biennial to) perennial to 60cm; leaf-segments succulent, (acute or) obtuse to rounded at apex; achenes 1.8-3.5mm, with 3 strong ribs ± touching laterally on 1 face, with 2 subapical distinctly elongated oil-glands on opposite face; 2n=18. Native; sand, shingle, rocks, walls, cliffs and waste ground near sea; locally common round most coasts of BI. Plants from N Sc (incl. Orkney and Shetland) have dark-bordered phyllaries and have been referred to ssp. **phaeocephalum** (Rupr.) Hämet-Ahti (*T. maritimum* var. *phaeocephalum* (Rupr.) Hyl., *M. maritima* ssp. *phaeocephala* (Rupr.) Rauschert, *M. maritima* var. *phaeocephala* (Rupr.) A.R. Clapham), but are not as extreme as the true Arctic taxon. Plants from S En with strong anthocyanin development and thinner leaf-segments are referable to var. **salinum** (Wallr.) Kay (*M. maritima* var. *salina* (Wallr.) A.R. Clapham), which has often been misplaced under *T. inodorum.*

1 x 2. T. maritimum x T. inodorum is intermediate in leaf and achene characters and is ≥80 per cent fertile (with backcrossing occurring); it is not infrequent in coastal areas and casts doubt on the distinction of the 2 parents at sp. level.

2. T. inodorum (L.) Sch. Bip. (*T. maritimum* ssp. *inodorum* (L.) Hyl. ex Vaar., **734**
Matricaria perforata Mérat) - *Scentless Mayweed.* Erect to ascending annual to 60cm; leaf-segments not succulent, acute and often bristle-tipped; achenes 1.3-2.2mm, with 3 strong ribs on 1 face separated by 2 distinct granular areas, with 2 subapical orbicular to angular oil-glands on opposite face; 2n=18. Native; waste, rough and cultivated land; common throughout lowland BI. See *Matricaria recutita* for differences.

69. COTULA L. (*Leptinella* Cass.) - *Buttonweeds*
Annual to perennial herbs: leaves entire to deeply pinnately divided; capitula discoid, bisexual or dioecious, yellow or white, with pedicellate flowers; in bisexual capitula outer flowers are female with 0 or minute corolla, inner ones bisexual with 4-lobed corolla; in dioecious capitula males and females both with minutely 4-

lobed corolla; receptacular scales 0; pappus 0.

The genus **Leptinella**, with female flowers possessing a corolla and functionally male flowers with an undivided (not 2-lobed) style, is sometimes segregated; *C. dioica* and *C. squalida* belong to it, but not all its spp. are dioecious.

1 Leaves entire to very irregularly pinnately lobed with usually <6 lobes, ± succulent; capitula 8-12mm across, bright yellow **1. C. coronopifolia**
1 Leaves regularly pinnately (to 2-pinnately) lobed with usually ≥6 lobes, not succulent; capitula 3-10mm across, white or dull yellow 2
 2 Annual; capitula bisexual, white, the female (outer) flowers with 0 corolla; phyllaries not purple-tinged **2. C. australis**
 2 Procumbent perennial with rooting stems; capitula dioecious, dull yellow, the female flowers with corolla; phyllaries strongly purple-tinged 3
3 Leaves with oblong-triangular abruptly apiculate teeth or shallow lobes **3. C. dioica**
3 Leaves lobed nearly to midrib, the lobes with lanceolate, acute to acuminate teeth **4. C. squalida**

1. C. coronopifolia L. - *Buttonweed*. Rather succulent glabrous annual to perennial with procumbent to ascending, often rooting stems to 30cm; capitula 8-12mm across, bright yellow; (2n=20, 40). Intrd; wet, usually saline places; natd in Cheshire since c.1880, MW & SW Yorks since 1959, S Hants since 1991, W Cork, rare casual elsewhere, perhaps increasing; S Africa and New Zealand.

2. C. australis (Sieber ex Spreng.) Hook. f. - *Annual Buttonweed*. Annual with 730
suberect to decumbent pubescent stems to 15cm; capitula 3-7mm across, white; (2n=18, 20, 36, 40). Intrd; rather frequent wool-alien, sometimes persisting in rough or arable ground, very scattered in En, natd in S Devon; Australia, New Zealand.

3. C. dioica (Hook. f.) Hook. f. (*Leptinella dioica* Hook. f.) - *Hairless Leptinella*. Dioecious procumbent perennial with rooting stems to 20cm; capitula 4-10mm across, the male dull yellow, the female ± enclosed by incurved phyllaries; (2n=260). Intrd; garden plant becoming natd in mown lawns, often not flowering; extremely scattered in BI; New Zealand.

4. C. squalida (Hook. f.) Hook. f. (*Leptinella squalida* Hook. f.) - *Leptinella*. Differs 730
from *C. dioica* as in key; (2n=260). Intrd; natd as for *C. dioica*, but commoner, scattered in Br and Ir, especially Sc; New Zealand.

TRIBE 7 - SENECIONEAE (genera 70-80). Annual to perennial herbs, sometimes shrubs, rarely weak climbers; leaves alternate or all basal; capitula discoid or radiate; phyllaries usually in 1 or 2 rows, often in 1 main row and 1 much shorter row, herbaceous; receptacular scales 0; pappus of 1-many rows of simple hairs; corolla most often yellow in both ray and disc flowers.

70. SENECIO L. - *Ragworts*

Annual to perennial herbs, rarely shrubby; leaves alternate, pinnately veined; capitula discoid or radiate; phyllaries in 1 main row with short supplementary ones at base of capitulum; corolla of disc flowers yellow, of ray flowers usually yellow (rarely white or purple).

1 Stems woody at least towards base 2
1 Stems entirely herbaceous 6
 2 Ligules purple **3. S. glastifolius**
 2 Ligules yellow 3
3 Stems and leaves ± glabrous 4
3 At least leaf lowersides grey- or white-pubescent 5
 4 Leaves linear, ± entire **5. S. inaequidens**

4 Leaves conspicuously toothed to deeply lobed **15. S. squalidus**

5 Leaves lobed ± to midrib, with obtuse to rounded lobes; phyllaries tomentose **1. S. cineraria**

5 Leaves serrate, acuminate at apex; phyllaries glabrous **2. S. pterophorus**

6 Ligules white or purple 7

6 Ligules yellow to orange, or 0 8

7 Leaves with linear lobes, divided >1/2 way to midrib: ligules usually purple **4. S. grandiflorus**

7 Leaves very shallowly and irregularly toothed: ligules white **11. S. smithii**

8 Leaves simple, entire to shallowly toothed 9

8 At least some leaves lobed ≥1/2 way to midrib 15

9 Ligules 4-8 10

9 Ligules >8 12

10 Middle and upper stem-leaves shortly but distinctly petiolate **7. S. ovatus**

10 Middle and upper stem-leaves sessile 11

11 Leaf-teeth divergent, often obtuse; phyllaries and peduncles usually glabrous **8. S. doria**

11 Leaf-teeth with acute, ± incurved apex; phyllaries and peduncles pubescent **6. S. fluviatilis**

12 Leaves linear, <5mm wide **5. S. inaequidens**

12 Leaves ovate or lanceolate to oblanceolate, most >5mm wide 13

13 Capitula 1-3(4); phyllaries 10-15mm **10. S. doronicum**

13 Capitula numerous; phyllaries 6-10mm 14

14 Phyllaries conspicuously black-tipped; leaves ± glabrous **15. S. squalidus**

14 Phyllaries not black-tipped; leaves pubescent on lowerside **9. S. paludosus**

15 Glandular hairs present at least on peduncles, often also on leaves and stems 16

15 Glandular hairs 0 17

16 Achenes glabrous; supplementary phyllaries at base of capitulum 1/3-1/2 as long as main ones; plant sticky **20. S. viscosus**

16 Achenes minutely pubescent; supplementary phyllaries ≤1/4(1/3) as long as main ones; plant not or scarcely sticky **19. S. sylvaticus**

17 Biennial to perennial, firmly rooted, usually with very short thick rhizome; phyllaries without black tips 18

17 Annual to perennial, usually easily uprooted, without rhizome; at least supplementary phyllaries with black tips 20

18 Supplementary phyllaries at base of capitulum c.1/2 as long as main ones; all achenes shortly pubescent; leaves grey-pubescent on lowerside **14. S. erucifolius**

18 Supplementary phyllaries c.1/4-2/5 as long as main ones; achenes of ray flowers glabrous; leaves glabrous to sparsely pubescent on lowerside 19

19 Achenes of disc flowers pubescent; stem-leaves with several pairs of lateral lobes and terminal lobe not much larger; corymbs dense **12. S. jacobaea**

19 Achenes of disc flowers glabrous to sparsely pubescent; stem-leaves with 1-few pairs of lateral lobes and terminal lobe much larger; corymbs lax **13. S. aquaticus**

20 Ligules <8mm or 0; capitula (excl. ligules) cylindrical in flower, c.2x as long as wide 21

20 Ligules usually ≥8mm, rarely shorter or 0; capitula (excl. ligules) bell-shaped in flower, <1.5x as long as wide 22

21 Ligules usually 0; achenes ≤2.5mm; pollen grains 20-25 microns across,

3-pored **17. S. vulgaris**

21 Ligules usually present; achenes >3mm; pollen grains 30-36 microns across, mostly 4-pored **16. S. cambrensis**

22 Leaves ± flat, usually with lateral lobes much longer than width of central undivided portion, usually glabrous or nearly so **15. S. squalidus**

22 Leaves usually undulate, with lateral lobes c. as long as width of central undivided portion, usually conspicuously pubescent **18. S. vernalis**

Other spp. - Records of **S. lautus** Sol. ex Willd., from Australia, are probably all errors for *S. inaequidens.*

1. S. cineraria DC. (*S. bicolor* (Willd.) Tod. ssp. *cineraria* (DC.) Chater) - *Silver Ragwort*. Densely silvery-pubescent spreading perennial with stems woody at least below; leaves deeply pinnately lobed, all but upper ones petiolate; ligules 10-13, 3-6mm, yellow; (2n=40). Intrd; natd on cliffs and rough ground mostly near sea; S & SW En, Wa, CI, Co Dublin; Mediterranean. Not to be confused with the pot-plant *Cineraria* (*Pericallis hybrida*).

1 x 12. S. cineraria x S. jacobaea = S. x albescens Burb. & Colgan occurs ± throughout the range of natd *S. cineraria* in BI and (from cultivated *S. cineraria*) scattered N to Dunbarton; it is intermediate in pubescence, leaf-shape and habit but has pubescent disc-achenes as in *S. jacobaea*. It is fertile and backcrosses, forming a range of intermediates.

1 x 14. S. cineraria x S. erucifolius = S. x thuretii Briq. & Cavill. (*S. x patersonianus* R.M. Burton) was found in E Kent in 1978; it differs from *S. x albescens* in having supplementary phyllaries c.1/3 (not ≤1/4) as long as the main ones and sometimes (?always) having short stolons.

2. S. pterophorus DC. - *Shoddy Ragwort*. Erect perennial to 1.5m (usually much less); stems woody at least below; leaves sharply and coarsely serrate, densely grey-pubescent on lowerside, subsessile; ligules c.8, 2-4mm, yellow; (2n=20). Intrd; occasional wool-alien in fields and waste land; very scattered in En; S Africa.

3. S. glastifolius L. f. - *Woad-leaved Ragwort*. Erect perennial to 1m; stems woody at least below; leaves coarsely dentate, ± glabrous, sessile, slightly clasping stem at base; ligules c.13, 12-18mm, purple. Intrd; garden escape natd in open woodland and on rubbish tip; Scillies; S Africa.

4. S. grandiflorus P.J. Bergius - *Purple Ragwort*. Erect perennial to 1.5m; leaves deeply pinnately lobed, sessile, clasping stem at base, sparsely pubescent: ligules c.8-13, 10-15mm, purple; (2n=40). Intrd; garden escape natd on rough ground in Guernsey; S Africa.

5. S. inaequidens DC. (*S. lautus* auct. non Sol. ex Willd.) - *Narrow-leaved Ragwort*. Subglabrous spreading perennial to 80cm, with stems often woody below; leaves linear, entire or nearly so, sessile; ligules 10-15, 5-8(10)mm, yellow; (2n=40). Intrd; rather frequent wool-alien in En and Sc, now natd on sandy beach in E Kent, perhaps soon to spread as in N France; S Africa.

6. S. fluviatilis Wallr. - *Broad-leaved Ragwort*. Erect perennial to 1.5m; leaves lanceolate to oblanceolate, shortly serrate, sessile, ± clasping stem at base; ligules mostly 6-8, 8-15mm, yellow; (2n=40). Intrd; natd by streams and in fens and swampy ground; scattered in Br and Ir N to C Sc, formerly more common; Europe.

7. S. ovatus (P. Gaertn., B. Mey. & Scherb.) Willd. (S. *fuchsii* C.C. Gmel., *S. nemorensis* L. ssp. *fuchsii* (C.C. Gmel.) Celak.) - *Wood Ragwort*. Erect perennial to 1.5m; leaves mostly narrowly elliptic, shortly serrate, distinctly though shortly petiolate; ligules 5-6, 12-15mm, yellow; (2n=40). Intrd; natd in damp shady places; MW Yorks and S & W Lancs; Europe. Our plant is probably ssp. **alpestris** (Gaudin) Herborg.

8. S. doria L. - *Golden Ragwort*. Erect perennial to 1.5m; leaves lanceolate to narrowly ovate, shortly serrate, sessile, clasping stem at base; ligules 4-6, 7-10mm, golden-yellow; (2n=40). Intrd; natd by streams and in wet meadows; Kirkcudbrights, S Ebudes, Offaly, Man, perhaps elsewhere; Europe.

9. S. paludosus L. - *Fen Ragwort.* Erect perennial to 1.5(2)m; leaves lanceolate to RRR
linear-lanceolate, shortly serrate, sessile, clasping stem at base; ligules 12-20, 10-14mm, yellow; (2n=40). Native; fenland ditches; formerly local in E En from N Lincs to Cambs, last seen 1857, refound in 1 place in Cambs in 1972.

10. S. doronicum (L.) L. - *Chamois Ragwort.* Erect perennial to 60cm; leaves elliptic-oblong (below) to linear-lanceolate, sessile and ± clasping stem (above), shortly to obscurely serrate; ligules 15-22, 12-20mm, golden-yellow; (2n=40, 80). Intrd; natd on river banks in M Perth since 1985; mountains of Europe.

11. S. smithii DC. - *Magellan Ragwort.* Erect perennial to 1m; leaves ovate-oblong to -triangular, shortly and irregularly dentate, the lower petiolate, the upper sessile; ligules 12-20, 15-25mm, white; (2n=40). Intrd; natd in grassy places and by streams; extreme N Caithness, Orkney, Shetlands; S Chile and S Argentina.

12. S. jacobaea L. - *Common Ragwort.* Erect perennial to 1.5m (often much shorter), very variably pubescent; leaves deeply pinnately lobed, the lowest lobes ± clasping stem; ligules 12-15, 5-9mm (rarely shorter), rarely 0 (var. **nudus** Weston), yellow; 2n=40. Native; grassland, waysides, waste ground, sand-dunes; common throughout BI. Very variable; dwarf coastal plants with dense web-like pubescence and 0 ligules have been called ssp. **dunensis** (Dumort.) Kadereit & P.D. Sell, but represent only 1 line of variation.

12 x 13. S. jacobaea x S. aquaticus = S. x ostenfeldii Druce occurs throughout much of Br and Ir, more commonly in W, with the parents; it is intermediate in leaf dissection, achene pubescence and inflorescence shape. It is <15% fertile but backcrosses and sometimes forms hybrid swarms.

13. S. aquaticus Hill (*S. erraticus* auct. non Bertol.) - *Marsh Ragwort.* Erect biennial to perennial to 80cm, glabrous to sparsely pubescent; differs from *S. jacobaea* in usually laxer habit, ray flowers always present, and see key; 2n=40. Native; marshes, damp meadows and streamsides; frequent throughout most of BI, common in much of W.

14. S. erucifolius L. - *Hoary Ragwort.* Erect perennial to 1.2m, with short rhizomes, densely pubescent when young; leaves deeply pinnately lobed, the lowest lobes ± clasping stem; ligules 12-15, 5-9mm, yellow; 2n=40. Native; grassy places, banks, waysides and field-borders, usually on base-rich soils; common in most of En and Wa, very local in E Ir, rare alien in Sc.

15. S. squalidus L. - *Oxford Ragwort.* Erect to ascending, ± glabrous annual to 741
perennial to 50cm, sometimes woody below; leaves usually deeply pinnately lobed, sometimes deeply serrate, the lower petiolate, the upper sessile: ligules 12-15, (6)8-10mm, yellow, rarely 0; 2n=20. Intrd; waste ground, walls and waysides; 1st recorded 1794, now common in En and Wa, local elsewhere in BI but still spreading; S Europe.

15 x 17. S. squalidus x S. vulgaris = S. x baxteri Druce is intermediate in leaf 741
and capitulum characters and highly sterile; 2n=30. It has been recorded from scattered places in Br and Ir but is probably over-recorded for radiate plants of *S. vulgaris*.

15 x 20. S. squalidus x S. viscosus = S. x subnebrodensis Simonk. (*S. x londinensis* Lousley) occurs sporadically on waste land in Br, mostly in S; it is intermediate in leaf-shape, pubescence and capitulum characters, and highly sterile; 2n=30.

16. S. cambrensis Rosser - *Welsh Groundsel.* Erect glabrous to sparsely pubescent RR
annual to 30(50)cm; leaves deeply pinnately lobed, the lower petiolate, the upper 741
sessile; ligules 8-15, 4-7mm, yellow; 2n=60. Native; waste ground and waysides; 1st found 1948 in Flints, now also in Denbs and Salop, found in Midlothian in 1982; originated as an amphidiploid of *S. squalidus* x *S. vulgaris*; endemic.

17. S. vulgaris L. - *Groundsel.* Usually erect, glabrous to webby-pubescent annual 741

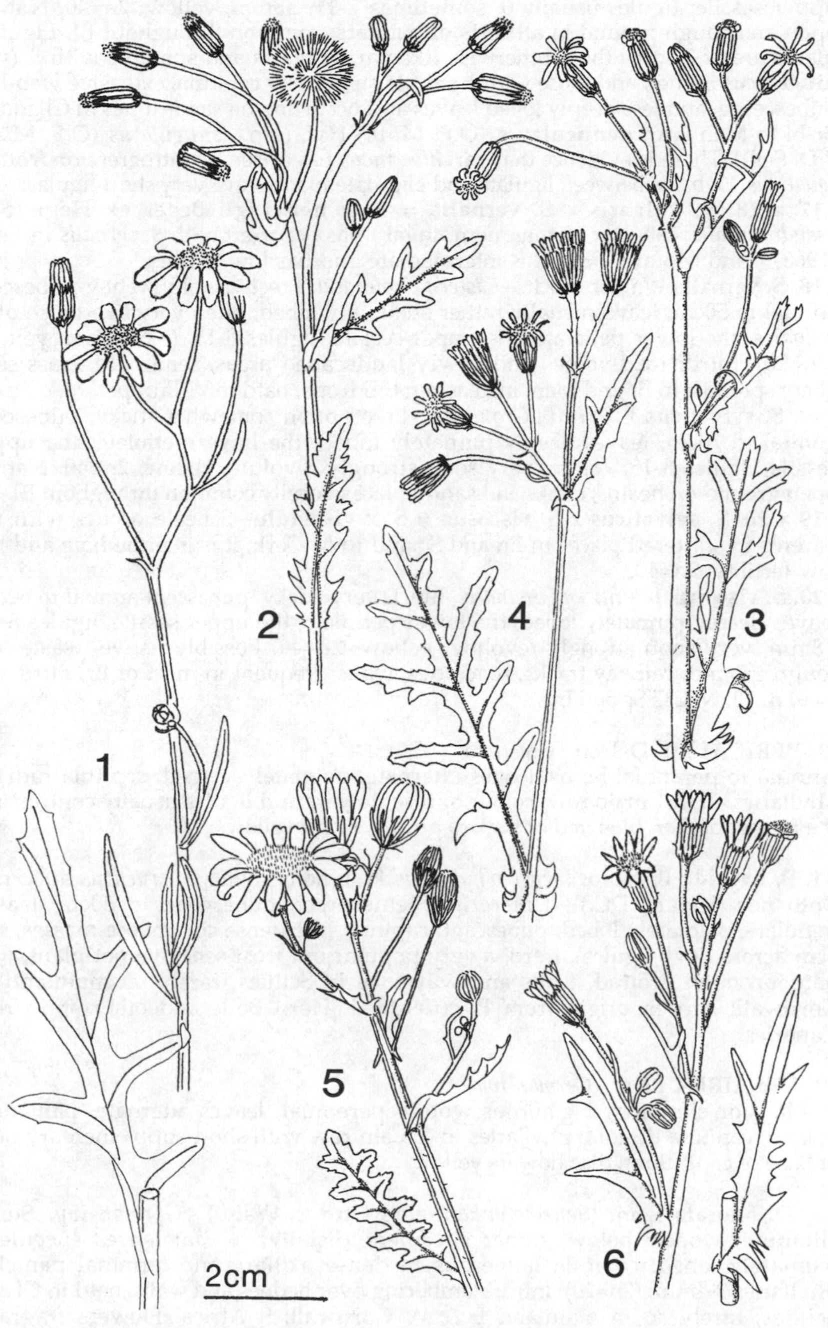

FIG 741 - *Senecio*. 1, *S. squalidus*. 2, *S. vulgaris* (rayless). 3, *S. vulgaris* (rayed). 4, *S. x baxteri*. 5, *S. vernalis*. 6, *S. cambrensis*.

to 30(45)cm; leaves shallowly to deeply pinnately lobed, the lower petiolate, the upper sessile; ligules usually 0, sometimes 7-11, ≤5mm, yellow; 2n=40. Native; open and rough ground in all sorts of habitats; common throughout BI. Ligulate plants are of 2 sorts: those otherwise like var. **vulgaris** and sporadic with it (var. **hibernicus** Syme); and those with ± simple stems, few capitula, extensive web-like pubescence, and less deeply lobed leaves and occurring on sand-dunes in CI and W Br N to Man (var. **denticulatus** (O.F. Müll.) Hyl. (ssp. *denticulatus* (O.F. Müll.) P.D. Sell)). There is evidence that var. *hibernicus* has arisen by introgression from *S. squalidus*. Hybrids between ligulate and eligulate plants have very short ligules.

17 x 18. S. vulgaris x S. vernalis = S. x helwingii Beger ex Hegi (*S. x pseudovernalis* Zabel ex Nyman nom. inval.) has occurred with *S. vernalis* in Leics (1968-9) and Moray (1983); it is intermediate and has low fertility.

18. S. vernalis Waldst. & Kit. - *Eastern Groundsel.* Erect, usually webby-pubescent **741**
annual to 50cm; leaves usually rather shallowly lobed, often very like those of *S. vulgaris*, the lower petiolate, the upper sessile; ligules 8-15, (6)8-10mm, yellow; (2n=20). Intrd; road-verges and newly landscaped areas, semi-natd grass-seed alien; sporadic in Br and Man, increasing; E Europe, natd in W Europe.

19. S. sylvaticus L. - *Heath Groundsel.* Erect, often somewhat sticky, pubescent annual to 70cm; leaves deeply pinnately lobed, the lower petiolate, the upper sessile; ligules 8-15, <6mm, very soon strongly revolute, yellow; 2n=40. Native; open ground on heaths, banks and sandy places; locally common throughout BI.

19 x 20. S. sylvaticus x S. viscosus = S. x viscidulus Scheele occurs with the parents in scattered places in En and Sc and in M Cork; it is intermediate and has low fertility; 2n=40.

20. S. viscosus L. - *Sticky Groundsel.* Erect, very sticky, pubescent annual to 60cm; leaves deeply pinnately lobed, the lower petiolate, the upper sessile; ligules 8-15, <8mm, very soon strongly revolute, yellow; 2n=40. Possibly native; waste and rough ground, railway tracks, roadsides, walls; frequent in most of Br, intrd and local in CI, N & C Sc and Ir.

71. PERICALLIS D. Don - *Cineraria*

Annual to perennial herbs; leaves alternate, palmately veined; capitula radiate; phyllaries all in 1 main row; colour of disc flowers and ligules usually contrasting, the former darker, blue, red or pink to purple, never yellow.

1. P. hybrida B. Nord. (*Senecio hybridus* Hyl. nom. inval., *S. cruentus* auct. non Roth nec (L'Hér.) DC.) - *Cineraria*. Stems erect, pubescent, to 80cm; leaves petiolate, palmately lobed, pubescent; capitula in ± dense corymbose masses, 1.5-4cm across (incl. ligules). Intrd; a very popular but frost-sensitive pot-plant, well natd on open ground, walls and waysides in Scillies, rarely on mainland W Cornwall; garden origin from **P. cruenta** (L.'Hér.) Bolle and other spp. from Canaries.

72. DELAIREA Lem. - *German-ivy*

Trailing or climbing, ± glabrous woody perennial; leaves alternate, palmately veined; capitula discoid; phyllaries in 1 main row with short supplementary ones at base of capitulum; disc flowers yellow.

1. D. odorata Lem. (*Senecio mikanioides* Otto ex Walp.) - *German-ivy*. Stems **744**
climbing, woody below, rather succulent distally, to 3m; leaves succulent, palmately lobed; capitula numerous in dense axillary and terminal panicles; phyllaries 3-4mm; (2n=20). Intrd; clambering over hedges and walls; natd in CI and Scillies, rarely so in mainland E & W Cornwall; S Africa. Flowers fragrant, appearing in Nov.

73. TEPHROSERIS (Rchb.) Rchb. - *Fleaworts*
(Biennial to) perennial herbs; leaves alternate, pinnately veined; capitula radiate; phyllaries all in 1 main row; disc flowers and ligules yellow.

1. T. integrifolia (L.) Holub (*Senecio integrifolius* (L.) Clairv.) - *Field Fleawort*. Erect, densely pubescent perennial to 60cm; basal leaves oblong-ovate, petiolate, entire to coarsely dentate; stem-leaves much smaller, lanceolate, sessile, rarely >10; capitula ≤12(15), 1.5-2.5cm across; ligules 12-15, 5-10mm. Native; short natural grassland.

a. Ssp. integrifolia. Stems to 30(40)cm; leaves entire to denticulate; stem-leaves usually ≤6; capitula rarely >6; phyllaries 6-8.5mm; 2n=48. On chalk and limestone; local in S En N to Cambs and E Gloucs, formerly to S Lincs. **R**

b. Ssp. maritima (Syme) B. Nord. (*Senecio spathulifolius* auct. non Griess., *S. integrifolius* ssp. *maritimus* (Syme) Chater). Stems to 60(90)cm; leaves usually dentate; stem-leaves often >6; capitula often >6; phyllaries 8-12mm; 2n=48. On glacial drift on sea-cliffs; extremely local on Holyhead Island (Anglesey); endemic. An extinct population in Westmorland probably represented an endemic, third ssp. **RR**

2. T. palustris (L.) Fourr. (*Senecio palustris* (L.) Hook. non Vell., *S. congestus* (R. Br.) DC.) - *Marsh Fleawort*. Erect, densely pubescent perennial to 1m; basal leaves usually withered before flowering; stem-leaves very numerous, lanceolate, sessile, dentate, ± clasping stem at base; capitula often >12, 2-3cm across; ligules c.21, 7-10mm; phyllaries 10-13mm; (2n=48). Native; fen ditches; formerly local in E En from W Sussex to MW Yorks; extinct (last recorded 1899). Our plant was ssp. **congesta** (R. Br.) Holub. **RR**

74. BRACHYGLOTTIS J.R. & G. Forst. - *Shrub Ragworts*
Evergreen shrubs; leaves alternate, pinnately veined, densely white-felted on lowerside; capitula discoid or radiate; phyllaries in 1 main row with short supplementary ones at base of capitulum; disc flowers cream or yellow; ligules yellow.

1 Many leaves >8cm, distantly sinuate-lobed; capitula <5mm across, cream; ligules 0; phyllaries with woolly hairs only at base **3. B. repanda**
1 Leaves <8cm, entire to denticulate or tightly undulate; capitula >1cm across, yellow; ligules conspicuous; phyllaries with woolly hairs along ± whole length 2
  2 Leaves <4cm, tightly crenate-undulate **2. B. monroi**
  2 Many leaves >4cm, entire to remotely denticulate **1. B. 'Sunshine'**

1. B. 'Sunshine' (*Senecio greyi* auct. non Hook. f.; ?*B. laxifolia* (Buchanan) B. Nord. x *B. compacta* (Kirk) B. Nord.) - *Shrub Ragwort*. Spreading shrub to 1(2)m; leaves up to 8cm, oblong-elliptic, entire to remotely denticulate; capitula 2-4.5cm across; ligules c.13, 10-15mm. Intrd; much grown in gardens and often mass planted by roads, etc., persistent on rough ground; scattered in Br and Man N to C Sc; garden origin. **744**

2. B. monroi (Hook. f.) B. Nord. (*Senecio monroi* Hook. f.) - *Monro's Ragwort*. Spreading shrub to 1m; leaves up to 4cm, elliptic-obovate, tightly crenate-undulate; capitula 1.5-4.5cm across; ligules c.13, 6-15mm; (2n=60). Intrd; persistent and ± natd in Man and on dunes near Llandudno, Caerns; New Zealand. **744**

3. B. repanda J.R. & G. Forst. - *Hedge Ragwort*. Shrub or small tree to 6m; leaves up to 25cm, remotely sinuate-lobed; capitula c.5mm across, in large dense panicles; ligules 0. Intrd; used as hedging in Scillies, often long persistent after neglect; New Zealand. **744**

75. SINACALIA H. Rob. & Brettell - *Chinese Ragwort*
Rhizomatous, ± glabrous herbaceous perennials; leaves alternate, pinnately veined;

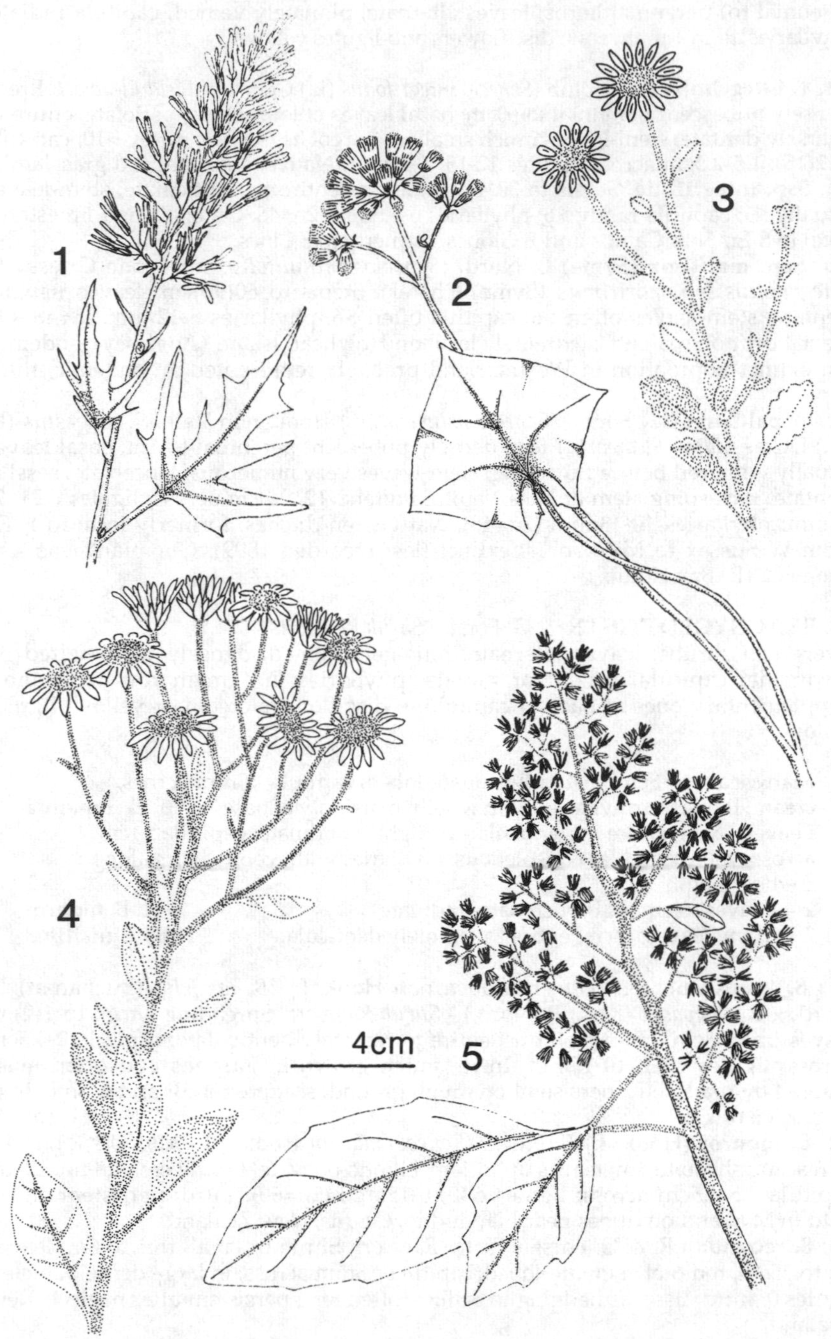

FIG 744 - Asteraceae: Senecioneae. 1, *Sinacalia tangutica*. 2, *Delairea odorata*. 3, *Brachyglottis monroi*. 4, *B.* **'Sunshine'**. 5, *B. repanda*.

capitula radiate, with 3 or 4 disc and 3 or 4 ray flowers; phyllaries all in 1 main row, but with small bracts some way below base of capitulum; disc flowers and ligules yellow.

1. S. tangutica (Maxim.) B. Nord. (*Senecio tanguticus* Maxim.) - *Chinese Ragwort*. 744
Stems erect, to 2m, unbranched except in inflorescence; leaves up to 20cm, ovate, deeply pinnately lobed; capitula numerous in large terminal panicles, <3mm wide excl. ligules; ligules 5-9mm. Intrd; grown in gardens, natd in damp shady places; scattered in N Wa, Man, N En and C Sc, Renfrews; China.

76. LIGULARIA Cass. - *Leopardplants*
Herbaceous perennials; leaves mostly basal, those on stems alternate, cordate at base, palmately veined, with sheathing petiole bases; capitula radiate; phyllaries all in 1 row; disc flowers brownish-yellow; ligules yellow to orange.

1. L. dentata (A. Gray) H. Hara (*L. clivorum* Maxim., *Senecio clivorum* (Maxim.) Maxim.) - *Leopardplant*. Stems erect, to 1.2m; basal leaves reniform, dentate, up to 50cm wide; capitula 4-10cm across, several in subcorymbose terminal cluster, with numerous disc flowers and 10-15 orange ligules 15-40mm; (2n=60). Intrd; grown in gardens, persistent in damp or shady places; very scattered in En and Sc, especially N; China and Japan.

2. L. przewalskii (Maxim.) Diels - *Przewalski's Leopardplant*. Stems erect, to 1.8m; basal leaves deeply palmately lobed, up to 50cm wide, the lobes lobed or toothed; capitula 1.5-3cm across, numerous in long narrow terminal raceme-like panicle, with usually only 3 disc flowers and 2 yellow ligules 6-15mm; (2n=60). Intrd; grown in gardens, persistent by R. Tyne in S Northumb; N China.

77. DORONICUM L. - *Leopard's-banes*
Rhizomatous herbaceous perennials; leaves alternate, ± palmately veined, the basal ones ± withered by flowering time; capitula radiate; phyllaries in 2 rows of equal length; disc flowers and ligules yellow.

1 Basal leaves all cuneate at base **3. D. plantagineum**
1 Most or all basal leaves cordate to rounded or truncate at base 2
  2 Petioles of basal leaves with many long (≥1mm) flexuous or patent hairs; capitula usually 3-8 per stem **1. D. pardalianches**
  2 Petioles of basal leaves with 0-very few long hairs; capitula 1-2(3) per stem 3
3 Basal leaves deeply cordate at base; all hairs on stems short (<1mm) and glandular **5. D. columnae**
3 Basal leaves shallowly cordate to truncate or rounded at base; stems usually with a few long (≥1mm) eglandular as well as short glandular hairs 4
  4 Basal leaves acute, mainly shallowly cordate at base, with prominent teeth >2mm **4. D. x excelsum**
  4 Basal leaves obtuse, mainly rounded to truncate at base, with less prominent teeth <2mm **2. D. x willdenowii**

1. D. pardalianches L. - *Leopard's-bane*. Stems erect, to 80cm, with long eglandular and short glandular hairs; basal leaves deeply cordate with narrow sinus, obscurely crenate-dentate; capitula c.3-8, mostly 3-4.5cm across; (2n=60). Intrd; well natd in woods and shady places; frequent throughout Br; W Europe.

2. D. x willdenowii (Rouy) A.W. Hill (*D. plantagineum* var. *willdenowii* (Rouy) A.B. Jacks.; ?*D. pardalianches* x *D. plantagineum*) - *Willdenow's Leopard's-bane*. Stems erect, to 1m, with short glandular and usually a few long eglandular hairs; basal leaves mostly rounded to truncate at base, obscurely crenate-dentate;

capitula 1-2(3) per stem, mostly 4.5-8cm across. Intrd; natd in woods and shady places; scattered in Br; W Europe or garden origin. Under-recorded for *D. plantagineum*.

3. D. plantagineum L. - *Plantain-leaved Leopard's-bane*. Stems erect, to 1m, with only short glandular hairs; basal leaves cuneate at base, obscurely crenate-dentate; capitula as in *D. x willdenowii*; (2n=120). Intrd; natd in woods and shady places; scattered in Br but over-recorded for last and next; W Europe.

4. D. x excelsum (N.E. Br.) Stace (*D. plantagineum* var. *excelsum* N.E. Br.; ?*D. pardalianches* x *D. plantagineum* x *D. columnae*) - *Harpur-Crewe's Leopard's-bane*. Stems erect, to 1m, with short glandular and usually a few long eglandular hairs; basal leaves mostly shallowly cordate at base, rather conspicuously dentate; capitula as in *D. x willdenowii*. Intrd; habitat and distribution as for *D. x willdenowii*; garden origin.

5. D. columnae Ten. (*D. cordatum* auct. non Lam.) - *Eastern Leopard's-bane*. Stems erect, to 60cm, with usually few short glandular hairs only; basal leaves deeply cordate with wide sinus, conspicuously dentate; capitulum usually 1, mostly 2.5-5cm across; (2n=60). Intrd; natd by path and on bank of reservoir in Surrey, perhaps overlooked; SE Europe.

78. TUSSILAGO L. - *Colt's-foot*

Rhizomatous, herbaceous perennials; leaves all basal, cordate at base, ± palmately veined, cottony-pubescent on lowerside; flowering stems bearing many bracts and 1 terminal capitulum, cottony-pubescent; capitula radiate; phyllaries all in 1 row; disc flowers and ligules yellow.

1. T. farfara L. - *Colt's-foot*. Stems erect, to 15cm, appearing before leaves; leaves broadly ovate, 20-30cm across, shallowly palmately lobed, the lobes dentate to denticulate; capitula 1.5-3.5cm across; 2n=60. Native: open or semi-open or disturbed ground in many habitats, including arable land and maritime sand and shingle; common throughout BI. Leaves differ from those of most spp. of *Petasites* in their distinct lobes and from *P. albus* in their black-tipped teeth.

79. PETASITES Mill. - *Butterburs*

Dioecious, rhizomatous, herbaceous perennials; leaves all basal, cordate at base, ± palmately veined, cottony-pubescent on lowerside; flowering stems bearing few to many bracts and a terminal raceme or panicle of capitula, cottony-pubescent; male capitula composed of male flowers with clavate sterile stigmas and sometimes some female-like sterile discoid or radiate flowers; female capitula composed of discoid female flowers and 1 or few central male-like sterile flowers; flowers white to purple or cream.

1 Marginal flowers ligulate (ligules <1cm); inflorescences appearing Nov-Feb, with basal leaves present, always male; leaves with small teeth all of 1 size **4. P. fragrans**

1 Marginal flowers tubular; inflorescences appearing Feb-Apr, before basal leaves, male or female; leaves unevenly dentate, with large teeth or short lobes dispersed among small teeth 2

2 Basal leaf sinus with parallel or divergent sides, bordered by 0-1 veins on each side; flowers pure white; leaves <30cm across **3. P. albus**

2 Basal leaf sinus with convergent sides, bordered by ≥2 veins on each side; flowers usually cream or with purplish tinge; leaves often >30cm across 3

3 Leaves distinctly but very shallowly lobed; mature inflorescences ± cylindrical; upper part of stem below inflorescence with bracts <1cm wide; phyllaries and/or florets usually with anthocyanin; corollas white to purple-tinged **1. P. hybridus**

3 Leaves scarcely or not lobed, unevenly dentate; mature inflorescences ± hemisperical; upper part of stem below inflorescence with bracts >1cm wide; anthocyanin absent; corollas cream in male capitula, whitish in female ones **2. P. japonicus**

1. P. hybridus (L.) P. Gaertn., B. Mey. & Scherb. - *Butterbur.* Leaves suborbicular, with convergent sides to basal sinus, up to 90cm across, obscurely lobed; petioles up to 1.5m; flowering stems erect, to 30cm (to 1m in fruit), with many narrow bracts; flowers white with purplish tinge; 2n=60. Native; by rivers and ditches, in damp fields and waysides, often in shade; male plant frequent throughout most of BI; female plant frequent in N & C En, very sporadic elsewhere.

2. P. japonicus (Siebold & Zucc.) Maxim. - *Giant Butterbur.* Leaves as in *P. hybridus* but see key; flowering stems erect, to 30cm, with many broad bracts; flowers cream to white; (2n=60). Intrd; natd by rivers and in damp places in open or shade; scattered throughout most of Br; Japan. The female plant is rarely or perhaps never natd here.

3. P. albus (L.) Gaertn. - *White Butterbur.* Leaves suborbicular, with divergent or parallel sides to basal sinus, up to 30cm across, with well-developed acute lobes; petioles up to 30cm; flowering stems erect, to 30cm (to 70cm in fruit), with few to rather many narrow to medium-broad bracts; flowers white; (2n=60). Intrd; natd in rough ground, waysides and woods; throughout Br, rare in S, common in N, N Ir; Europe. Female plant much less common than male here. See *Tussilago farfara* for leaf differences.

4. P. fragrans (Vill.) C. Presl (?*P. pyrenaicus* (L.) G. López) - *Winter Heliotrope.* Leaves suborbicular, with divergent sides to basal sinus, up to 20cm across, not lobed; petioles up to 30cm; flowering stems erect, to 30cm, with few medium-broad bracts; flowers white tinged purple, strongly vanilla-scented; 2n=52, 60. Intrd; natd on waste and rough ground and waysides; throughout BI, common in S, local in N; N Africa. Female plant unknown in BI.

80. HOMOGYNE Cass. - *Purple Colt's-foot*

Rhizomatous, herbaceous perennials; leaves all basal, cordate at base, palmately veined, rather sparsely pubescent on lowerside; flowering stems bearing few bracts and 1 terminal capitulum, cottony-pubescent; capitula discoid, the central flowers bisexual with 5-lobed corolla, the outermost row female with obliquely truncate tubular corolla; phyllaries ± in 1 row; flowers purple.

1. H. alpina (L.) Cass. - *Purple Colt's-foot.* Stems erect, to 35cm, appearing with leaves; leaves reniform-orbicular, up to 4cm across, shallowly crenate-dentate; capitula 10-15mm across; (2n=c.120, c.140). Probably intrd; 1 locality in Angus at c.600m, known c.1813, refound 1951, perhaps originally planted; C Europe. **RRR**

TRIBE 8 - CALENDULEAE (genus 81). Annual to perennial herbs; leaves alternate, simple, ± sessile; capitula radiate; phyllaries in 1-2 rows, herbaceous with scarious margin; receptacular scales 0; achenes varying in one capitulum in size, degree of curving and wartiness, all with pappus 0; flowers yellow to orange, sometimes *flore pleno.*

81. CALENDULA L. - *Marigolds*

1. C. officinalis L. - *Pot Marigold.* Perennial often behaving as annual, with distinctive scent; stems erect to procumbent, to 80cm; capitula 4-7cm across, pale yellow to deep orange; ligules 2x as long as phyllaries; (2n=28, 32). Intrd; much grown in gardens, frequent escape or throwout on tips and waste ground; scattered in BI, sometimes ± natd in S; ?garden origin.

2. C. arvensis L. - *Field Marigold.* Erect to procumbent annual to 30cm; differs from *C. officinalis* in capitula 1-2.5cm across; ligules <2x as long as phyllaries;

(2n=36, 44). Intrd; natd weed of cultivated ground in Guernsey and Scillies, rather uncommon casual elsewhere in S Br; S Europe.

TRIBE 9 - HELIANTHEAE (Helenieae, Tageteae) (genera 82-97). Annual to perennial herbs; leaves often opposite, sometimes alternate; capitula discoid or radiate; phyllaries in 1-several rows, all herbaceous or the inner scarious; receptacular scales usually present, sometimes 0; pappus 0 or minute, sometimes of barbed bristles; corolla yellow to brown in disc flowers, yellow to brown, orange, red, purple or white in ray flowers. *Ambrosia* and *Xanthium* have unusual aberrant floral structure (q.v.).

82. AMBROSIA L. - *Ragweeds*
Annuals or perennials; leaves all opposite or the upper ones alternate, variously deeply lobed, often nearly pinnate; phyllaries 5-12, ± herbaceous, fused proximally; capitula monoecious, <5mm across, discoid, the male in dense elongated terminal racemes without subtending bracts, the female just below in axils of leaf-like bracts and with only 1 flower, greenish; receptacle flat, with scales; pappus 0.

1 Leaves palmately 3-5-lobed or the upper ones not lobed, all opposite; larger leaf-lobes >1cm wide **3. A. trifida**
1 Leaves pinnately divided nearly to base, the upper ones alternate; leaf-lobes <1cm wide 2
  2 Annual; female phyllaries in fruit with small erect spines near or above the middle **1. A. artemisiifolia**
  2 Perennial, with stems arising from creeping roots; female phyllaries in fruit tuberculate, without spines **2. A. psilostachya**

1. A. artemisiifolia L. - *Ragweed*. Annual; stems erect, to 1m; leaves deeply 749
pinnately divided, the larger ones with the primary lobes divided again, green, appressed-pubescent but often sparsely so; (2n=36). Intrd; casual of waste ground and tips from several sources, incl. birdseed and oilseed; fairly frequent casual in En, Wa and CI, sometimes semi-natd; N America.

2. A. psilostachya DC. (*A. coronopifolia* Torr. & A. Gray) - *Perennial Ragweed*. 749
Differs from *A. artemisiifolia* in leaves usually grey with dense appressed hairs; and see key; (2n=72, 108, 144). Intrd; casual in waste places and on rough ground and dunes, natd locally (especially Ayrs and S & W Lancs), 1st recorded 1903; scattered in Br; N America.

3. A. trifida L. - *Giant Ragweed*. Annual; stems erect, to 2.5m; leaves palmately 749
divided c.3/4 way to base, very sparsely pubescent, green; (2n=24). Intrd; casual on tips and waste ground, mainly as oilseed- and soyabean-alien; scattered in Br; N America.

83. IVA L. - *Marsh-elder*
Annuals; leaves opposite below, alternate above, simple, ovate, petiolate; phyllaries 5, ± herbaceous, with 5 similar but membranous receptacular scales immediately within; capitula <5mm across, discoid, with 5 outer female and 8-20 inner male flowers, greenish; receptacle flat, with scales; pappus 0.

FIG 749 - Asteraceae: Heliantheae. 1-3, capitula of ***Tagetes***. 1, ***T. erecta***. 2, ***T. patula***. 3, ***T. minuta***. 4-6, leaves of ***Ambrosia***. 4, ***A. artemisiifolia***. 5, ***A. psilostachya***. 6, ***A. trifida***. 7-9, leaves of ***Xanthium***. 7, ***X. ambrosioides***. 8, ***X. spinosum***. 9, ***X. strumarium***. 10-13, achenes and receptacular scales of ***Galinsoga***. 10-11, ***G. parviflora***. 12-13, ***G. quadriradiata***. 14-19, achenes of ***Bidens***. 14, ***B. bipinnata***. 15, ***B. frondosa***. 16, ***B. pilosa***. 17, ***B. tripartita***. 18, ***B. cernua***. 19, ***B. connata***.

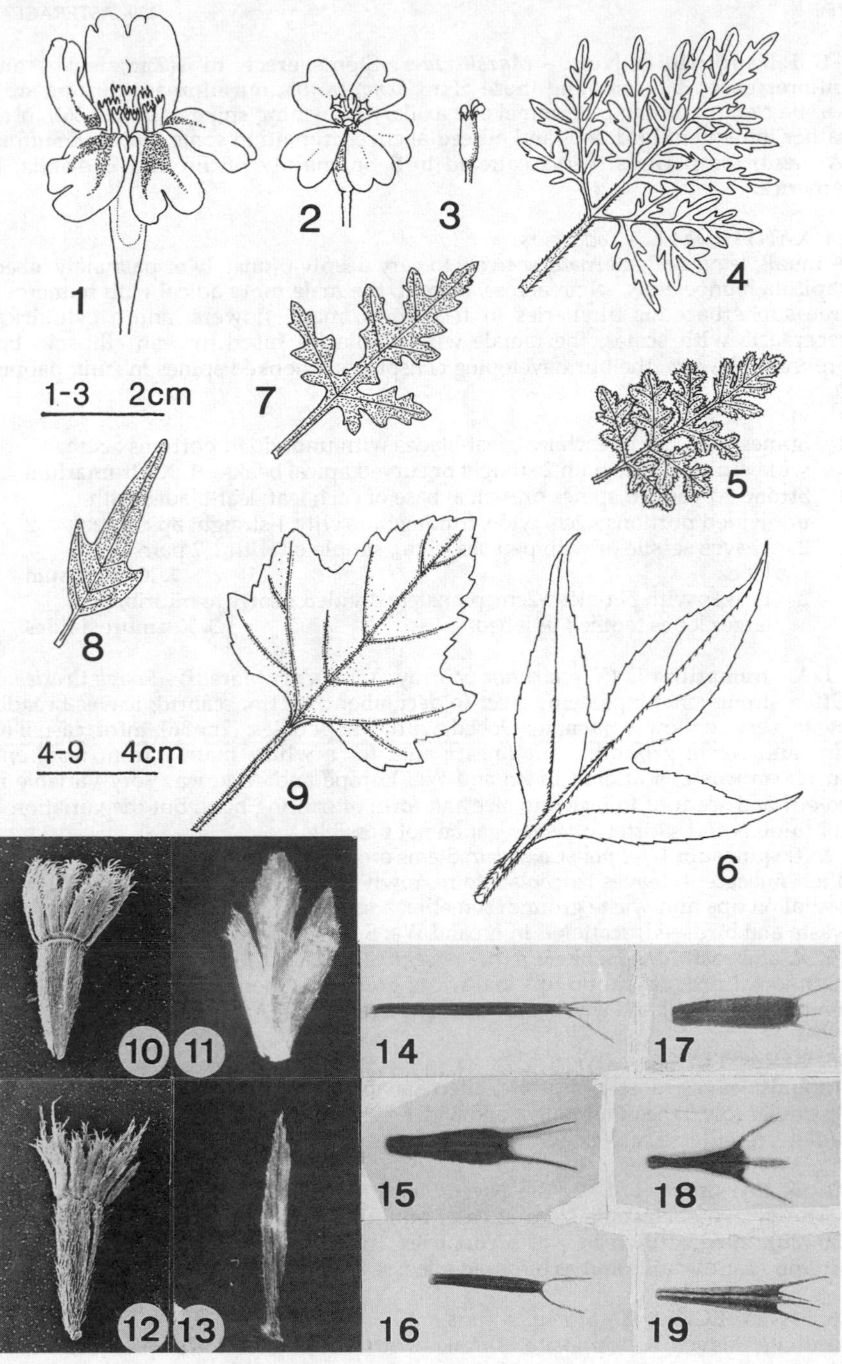

FIG 749 - see caption opposite

1. I. xanthiifolia Nutt. - *Marsh-elder*. Stems erect, to 1(2)m; leaves and inflorescences ± resembling those of a *Chenopodium*, but ultimate clusters are ± sessile capitula borne in terminal and axillary branching spikes; (2n=28, 36). Intrd; rather infrequent birdseed- and oilseed-alien, semi-natd in some places (E Suffolk, W Kent) for periods; very scattered in S Br, mainly SE En and E Anglia; N America. **752**

84. XANTHIUM L. - *Cockleburs*
Annuals; leaves all alternate, scarcely to very deeply pinnately or palmately lobed; capitula monoecious, <1cm across, discoid, the male more apical with numerous, free subherbaceous phyllaries in 1(-3) rows, many flowers, and a cylindrical receptacle with scales, the female with phyllaries fused into an ellipsoid bur around 2 flowers, the bur developing conspicuous hooked spines in fruit; pappus 0.

1 Spines 0 at base of each leaf; leaf-blades with undivided portions >3cm wide; fruiting burs with 2 straight or curved apical beaks **1. X. strumarium**
1 Strong 3-pronged spines present at base of each leaf; leaf-blades with undivided portions <2cm wide; fruiting burs with 1 straight apical beak 2
  2 Leaves sessile or with petioles <1cm, simple or with 1-2 pairs of lobes **2. X. spinosum**
  2 Leaves with petioles >2cm, pinnately divided nearly to midrib, the larger lobes toothed or lobed **3. X. ambrosioides**

1. X. strumarium L. (*X. echinatum* Murray, *X. italicum* Moretti) - *Rough Cocklebur*. Often strong-smelling; stems erect to decumbent, to 1m, scabrid; leaves broadly ovate, very shallowly palmately lobed, with long petioles; (2n=36). Intrd; casual on tips and waste ground, sometimes ± natd for a while, mainly from wool and soyabean waste; scattered in En and Wa; Europe and America. Very variable in colour and scent of foliage and size and form of fruiting burs, but the variation is continuous and satisfactory segregation not possible. **749**

2. X. spinosum L. - *Spiny Cocklebur*. Stems erect to decumbent, to 1m, smooth but often pubescent; leaves lanceolate to narrowly ovate, lobed or not; (2n=36). Intrd; casual on tips and waste ground, sometimes ± natd for a while, mainly from wool waste and birdseed; scattered in En and Wa; S America. **749**

3. X. ambrosioides Hook. & Arn. - *Argentine Cocklebur*. Differs from *X. spinosum* as in key. Intrd; casual on tips and waste ground, sometimes ± natd for a while, mainly from wool and soyabean waste; scattered in En; Argentina. **749**

85. GUIZOTIA Cass. - *Niger*
Annuals; leaves mostly opposite, alternate above, simple, sessile; phyllaries in 2 dissimilar rows, the outer longer and wider, herbaceous, the inner scarious; capitula radiate; receptacle convex, with scales; pappus 0; ligules c.8, yellow.

1. G. abyssinica (L. f.) Cass. - *Niger*. Stems erect, to 1(2)m; leaves lanceolate to narrowly ovate, clasping stem at base; capitula 2-4cm across, incl. ligules 1-2cm; (2n=30). Intrd; grown in warm countries as oilseed, casual on tips and waste ground near oil-mills and as birdseed-alien; scattered in Br and CI; E Africa. **752**

86. SIGESBECKIA L. - *St Paul's-worts*
Annuals; leaves all opposite, simple, with winged petioles; phyllaries in 2 dissimilar rows, the outer much longer and narrower, with many stalked glands, herbaceous; capitula ≤1cm across, radiate; receptacle flat or slightly convex, with scales; pappus 0; ligules numerous, yellow.

1. S. orientalis L. - *Eastern St Paul's-wort*. Stems to 1.2m, erect, often well branched; leaves triangular-hastate, irregularly dentate or lobed; petioles winged distally, the wings tapering proximally and ± absent at base; (2n=30). Intrd; occasional wool-alien, not natd; scattered in C & S En; tropical Asia. Over-recorded for *S. serrata*.

2. S. serrata DC. (*S. jorullensis* auct. non Kunth, *S. cordifolia* auct. non Kunth) - **752**
Western St Paul's-wort. Differs from *S. orientalis* in leaves ovate, cordate to broadly cuneate at base, shallowly crenate or serrate; petioles winged to base, the wings tapering proximally but widened at base and clasping stem; (2n=30). Intrd; locally natd in S Lancs as weed of waste and cultivated ground, occasional casual elsewhere; scattered in En; tropical America.

87. RUDBECKIA L. - *Coneflowers*

Perennials; leaves alternate, simple to deeply lobed; phyllaries in 2 or more rows, herbaceous; capitula radiate; receptacle conical, with scales partly enclosing achenes; pappus 0 or a short rim; ligules numerous, yellow to orange.

1. R. hirta L. (*R. serotina* Nutt.) - *Black-eyed-Susan*. Stems erect, to 80cm; leaves **752**
simple, roughly pubescent, entire or nearly so; capitula 5-10cm across; disc flowers brownish-purple; ligules yellow or orange; (2n=38). Intrd; grown in gardens, occasionally persistent in rough ground and waste places; scattered in S Br; N America.

2. R. laciniata L. - *Coneflower*. Stems erect, to 3m; leaves deeply divided, the lowest ± pinnate, glabrous or nearly so; capitula 7-14cm across; disc flowers greenish-yellow; ligules yellow; (2n=36, 38, 54, 72, 76, 102). Intrd; grown in gardens, natd on waste and rough ground, much commoner than *R. hirta*; scattered in Br, mainly S & C; N America.

88. HELIANTHUS L. - *Sunflowers*

Annuals to perennials; leaves opposite below, alternate above, simple; phyllaries in 2 or more rows, herbaceous; capitula radiate; receptacle flat or slightly convex, with scales partly enclosing achenes; pappus of 2 narrow scales soon falling off, sometimes with some shorter extra scales; ligules numerous, yellow (often *flore pleno*).

1 Plant annual, with simple tap-root 2
1 Plant perennial, clump-forming, with (often very short) rhizomes 3
  2 Phyllaries ovate, abruptly contracted to acuminate tip; central receptacular scales inconspicuously pubescent **1. H. annuus**
  2 Phyllaries lanceolate to narrowly ovate, gradually tapered to apex; central receptacular scales with conspicuous long white hairs at apex **2. H. petiolaris**
3 Stems ± glabrous in lower half; at least some phyllaries much exceeding edge of receptacle **3. H. x multiflorus**
3 Stems roughly pubescent almost to base; phyllaries not or scarcely exceeding edge of receptacle 4
  4 Rhizomes with swollen tubers; stems often >2m; phyllaries not or loosely appressed to receptacle **5. H. tuberosus**
  4 Rhizomes without swollen tubers; stems rarely >2m; phyllaries closely appressed to receptacle **4. H. x laetiflorus**

1. H. annuus L. - *Sunflower*. Annual; stems erect, to 3m, usually unbranched; capitula usually 1, with receptacle 2-30cm across; (2n=34). Intrd; grown as ornamental or on field scale as oilseed, common casual on tips and in waste places, also as birdseed-alien; throughout most of BI except most rural areas; N America.

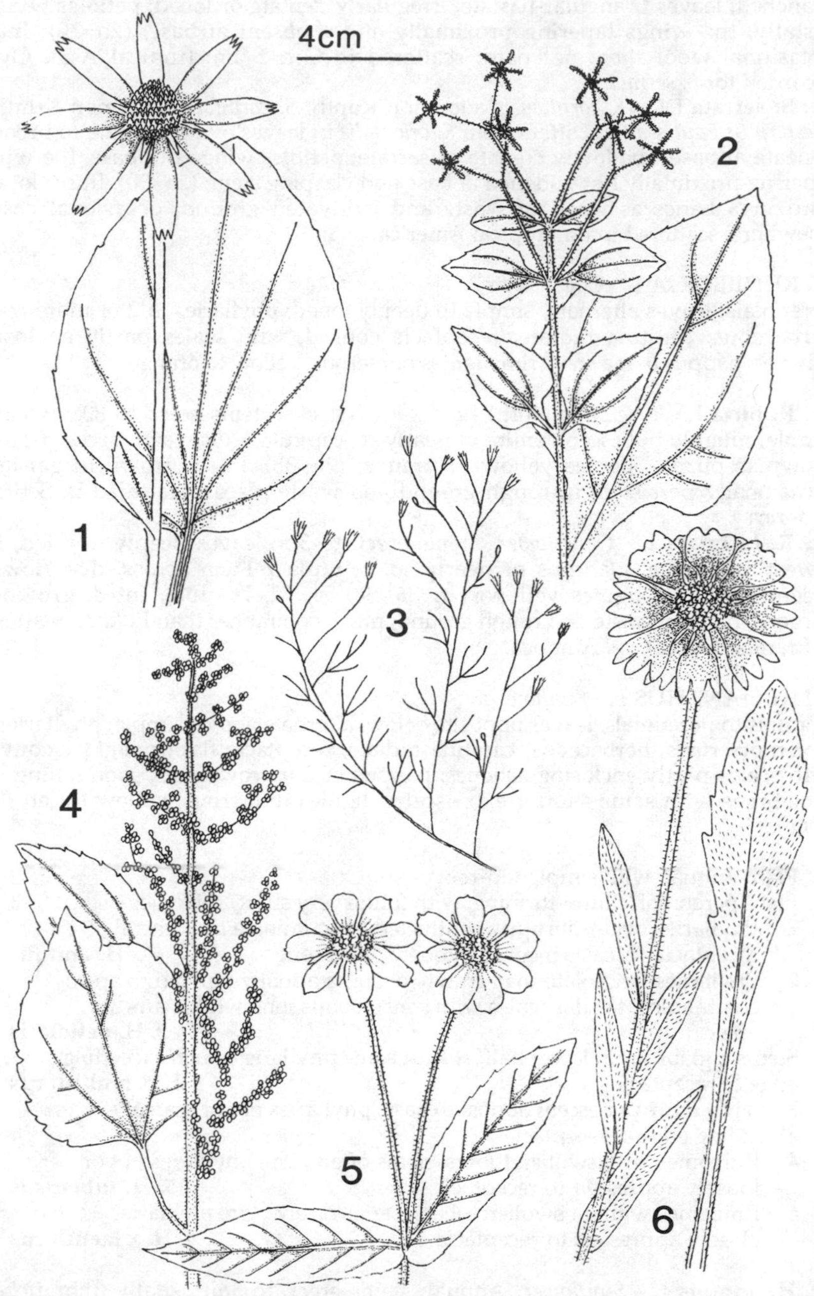

FIG 752 - Asteraceae: Heliantheae. 1, *Rudbeckia hirta*. 2, *Sigesbeckia serrata*. 3, *Schkuhria pinnata*. 4, *Iva xanthiifolia*. 5, *Guizotia abyssinica*. 6, *Gaillardia x grandiflora*.

2. H. petiolaris Nutt. - *Lesser Sunflower*. Resembles a small *H. annuus*; differs in stems usually $\leq$1m; receptacle 1-2.5cm across; and see key; (2n=34). Intrd; rather infrequent casual, especially from soyabean waste, perhaps overlooked; frequent in London area; N America.

3. H. x multiflorus L. (*H. annuus* x *H. decapetalus* L.) - *Thin-leaved Sunflower*. Perennial, stems erect, to 1.5(2)m, smooth and $\pm$ glabrous below; capitula usually $>$1, with receptacle 1-2cm across; often *flore pleno*. Intrd; grown in gardens, rather infrequent escape or throwout; scattered in Br, mainly S & C, but confused with *H. x laetiflorus*; garden origin. Possibly some records refer to *H. decapetalus*.

4. H. x laetiflorus Pers. (*H. pauciflorus* Nutt. (*H. rigidus* (Cass.) Desf.) x *H. tuberosus*) - *Perennial Sunflower*. Perennial; stems erect, to 1.5(2)m, scabrid-pubescent $\pm$ throughout; capitula usually $>$1, with receptacle 1-2.5cm across; often *flore pleno*; (2n=102). Intrd; much grown in gardens, frequent escape or throwout; throughout most of BI; probably garden origin. The commonest garden perennial sunflower. Some records may refer to *H. pauciflorus*, which has purplish-brown (not yellow) disc-flowers.

5. H. tuberosus L. - *Jerusalem Artichoke*. Perennial, with irregular tubers developing on rhizomes; stems erect, to 3m, scabrid-pubescent $\pm$ throughout; capitula usually several, with receptacle 1.5-2.5cm across, but stems usually frosted down before flowering; (2n=102). Intrd; grown as minor root-crop and very persistent in waste places; scattered throughout much of Br and CI; N America.

89. GALINSOGA Ruiz & Pav. - *Gallant-soldiers*
Annuals; leaves all opposite, simple, ovate, petiolate; phyllaries in 2 rows, largely herbaceous, the outer much shorter, the inner with membranous margins; capitula $<$1cm across, radiate; receptacle conical, with scales; pappus of scales; ligules few (usually 5), white or pinkish; disc flowers yellow.

1. G. parviflora Cav. - *Gallant-soldier*. Stems erect to ascending, to 80cm, glabrous **749**
or sparsely pubescent; peduncles rather sparsely pubescent with glandular and eglandular hairs c.0.2mm; receptacular scales mostly distinctly 3-lobed, the central lobe the largest; pappus-scales fringed with hairs, without a terminal projection; 2n=16. Intrd; well natd weed of cultivated and waste ground; locally frequent in CI and Br N to C Sc, especially in London and other large cities, first record 1860; S America.

2. G. quadriradiata Ruiz & Pav. (*G. ciliata* (Raf.) S.F. Blake) - *Shaggy-soldier*. **749**
Differs from *G. parviflora* in more densely pubescent stems; peduncles with many glandular and eglandular hairs c.0.5mm; receptacular scales mostly simple, some with 1 or 2 weak lateral lobes; pappus-scales fringed with hairs and with a fine terminal projection; 2n=32. Intrd; similar habitats and distribution to *G. parviflora*, often with it, also rare in Ir, first record 1909; S America. Often very distinct from *G. parviflora* in pubescence, but sometimes very close to it; pappus-scale apex is best character.

90. BIDENS L. - *Bur-marigolds*
Annuals; leaves all opposite, simple and toothed to pinnate; phyllaries in 2 dissimilar rows, the outer herbaceous, the inner $\pm$ membranous with a usually scarious border; capitula usually discoid, rarely radiate; receptacle flat or slightly convex, with scales; pappus of 2-5 barbed (forwardly or backwardly), strong bristles; ligules 0, rarely yellow, very rarely white.

The achenes provide important diagnostic characters, but some of those traditionally used, e.g. bristle number and direction of barbs on bristles, are sometimes unreliable. Throughout the account 'achenes' refers to the central ones in the capitulum; the outer ones may differ considerably. All spp. are usually eligulate, but *B. cernua* sometimes has conspicuous yellow rays and *B. frondosa* and *B. pilosa* white ones.

1 Leaves not lobed, or lobed <1/2 way to midrib 2
1 At least lower leaves pinnate, or lobed nearly to midrib 4
 2 Achenes scarcely flattened (<2x as wide as thick), the faces between the 4 ridges warty **3. B. connata**
 2 Achenes strongly flattened (>2x as wide as thick), the faces between the 2-4 ridges smooth 3
3 At least lower leaves distinctly petiolate, with 1(-2) pairs of distinct lobes **2. B. tripartita**
3 All leaves tapered to base but sessile, unlobed (but strongly serrate) **1. B. cernua**
 4 Leaflets lobed again to midrib or ± so **6. B. bipinnata**
 4 Leaflets unlobed 5
5 Petioles winged to base; apical (main) lobe of leaf scarcely stalked or with winged stalk; barbs on edge of achenes (?always) backward-directed **2. B. tripartita**
5 At least lower leaves with ± unwinged petioles and with apical (main) lobe with distinct ± wingless stalk; barbs on edge of achenes (but not on apical bristles) (?always) forward-directed 6
 6 Leaflet-teeth mostly wider than long; achenes ± parallel-sided, slightly tapered at each end **5. B. pilosa**
 6 Leaflet-teeth mostly longer than wide; achenes tapered ± from apex to base **4. B. frondosa**

Other ssp. - B. vulgata Greene, from N America, would key out as *B. frondosa* but has (10)13(-16) (not (5)8(-10)) outer phyllaries and rather pale yellow (not orange-yellow) disc flowers; it has been recorded as a rare casual and might be overlooked.

1. B. cernua L. - *Nodding Bur-marigold.* Stems erect, to 75cm; leaves not lobed, sessile; achenes tapered from apex to base, with 3-4 bristles, the achene and bristles with backward-directed barbs; 2n=24. Native; by ponds and streams and in ditches and marshy fields; locally common in C & S Br and Ir, rare and very scattered in N. Radiate plants (var. **radiata** DC.) are occasional in NW En, very rare elsewhere. 749

2. B. tripartita L. - *Trifid Bur-marigold.* Stems erect, to 75cm; many leaves usually with 1-2 pairs of deep lateral lobes, rarely all unlobed (var. **integra** W.D.J. Koch), with winged petiole; achenes tapered from apex to base, with 2-4 bristles, the achene and bristles with backward-directed barbs; 2n=48. Native; similar habitat and distribution to *B. cernua*, and often with it. 749

3. B. connata Muhl. ex Willd. - *London Bur-marigold.* Differs from *B. tripartita* in leaves not lobed; achenes (incl. bristles) with usually forward-directed (rarely backward-directed) barbs; and see key (couplet 2); (2n=72). Intrd; natd by canals; Middlesex, Herts, Bucks and W Kent; N America. 749

4. B. frondosa L. - *Beggarticks.* Stems erect, to 1m; many leaves pinnate, with 1-2 pairs of lateral leaflets; petioles and leaflet-stalks ± unwinged; achenes tapered from apex to base, with forward-directed barbs, the usually 2 bristles with backward-directed hairs; (2n=48). Intrd; natd by canals and rivers and on damp ground and waste places; scattered in En and Wa, frequent around Birmingham and London and in S Wa; N & S America. Sometimes with white ligules. 749

5. B. pilosa L. - *Black-jack.* Differs from *B. frondosa* in leaflets usually broader and less deeply serrate; bristles usually 2-3; and see key; (2n=24, 36, 48, 72). Intrd; a rather characteristic wool-alien; scattered in En; S America. 749

6. B. bipinnata L. - *Spanish-needles.* Similar to *B. pilosa* in capitulum size and achene morphology except bristles 2-4; leaves 1-2 pinnate, the primary leaflets divided again to midrib or nearly so; (2n=72). Intrd; a characteristic wool-alien; scattered in En and Sc; S America. 749

91. COREOPSIS L. - *Tickseeds*
Annuals to perennials; leaves (usually all) opposite, all or most pinnately or ternately divided to midrib or ± so, the primary divisions often divided again; phyllaries in 2 dissimilar rows, the outer narrower and shorter and ± herbaceous, the inner partially membranous; capitula radiate; receptacle flat or slightly convex, with scales; pappus 0 or of few very short teeth or bristles; ligules c.8, yellow, sometimes with dark basal blotch.

Other spp. - C. tinctoria Nutt., from USA, is an annual with disc flowers with 4-toothed (not 5-toothed), dark purple corollas and ligules with a brownish-purple blotch at base; it is grown in gardens and is a rare escape. **C. verticillata** L., from USA, has leaves finely divided from near base, so appearing whorled, and yellow capitula; it was established for a short while in N Hants.

1. C. grandiflora Hogg ex Sweet - *Large-flowered Tickseed*. Tufted perennial; stems erect, to 1m, well-branched above; capitula on long slender peduncles, uniformly yellow, 3-5cm across; (2n=26, 28). Intrd; grown in gardens, infrequently natd as escape or throwout; SE En; USA.

92. COSMOS Cav. - *Mexican Aster*
Annuals; leaves all opposite, 2-3-pinnate with linear to filiform segments; phyllaries in 2 dissimilar rows, the outer narrower, herbaceous with membranous border, the inner membranous; capitula radiate; receptacle flat, with scales; pappus of (0)2(-3) bristles with usually backward-directed barbs; ligules numerous, pinkish-purple, rarely white; disc flowers yellow.

1. C. bipinnatus Cav. - *Mexican Aster*. Stems erect, to 2m; capitula 4-9cm across, incl. ligules 1.5-4cm; (2n=24). Intrd; much grown in gardens, frequent persistent casual on tips and in waste places; scattered in Br, mostly C & S; Mexico, S USA.

93. DAHLIA Cav. - *Dahlia*
Perennials (but killed by first frosts); leaves all opposite, petiolate, (bi-)pinnate or (bi-)ternate, the uppermost simple, phyllaries in 2 dissimilar rows, the inner membranous, the outer ± herbaceous; capitula radiate (but most often *flore pleno*); receptacle flat or slightly convex, with scales; pappus 0 or of 2 obscure teeth; ligules numerous, yellow, white, pink or purple; disc flowers yellow.

1. D. pinnata Cav. (*D. variabilis* (Willd.) Desf.) - *Dahlia*. Stems erect, to 2m, rather succulent; capitula extremely variable (in different cultivars) in size (up to 30cm across but often <10cm), colour, shape of ligules, and degree to which they are *flore pleno*; (2n=32, 64). Intrd; much grown for ornament, frequent as throwout on tips and waste ground, not natd; scattered in En, mainly SE and SW; Mexico. The origin of cultivated *Dahlias* is uncertain; other spp., especially **D. coccinea** Cav., have probably shared in their parentage.

94. TAGETES L. - *Marigolds*
Aromatic annuals; leaves opposite below, alternate above, pinnate; phyllaries in 1 row, fused for most of length to form sheath round capitulum; capitula radiate; receptacle flat, without scales; pappus of unequal scales; ligules c.3-8, yellow to orange or brownish, often *flore pleno*.

1 Ligules <3mm; phyllary-sheath <4mm wide **3. T. minuta**
1 Ligules >(5)10mm; phyllary-sheath >5mm wide 2
  2 Peduncles conspicuously swollen below capitula; phyllaries mostly >1.5cm; ligules mostly 1-3cm **1. T. erecta**
  2 Peduncles not or scarcely swollen below capitula; phyllaries mostly

<1.5cm; ligules mostly (0.5)1-1.5cm **2. T. patula**

Other spp. - A few other spp. are grown in gardens and are rarely reported as escapes or throwouts.

1. T. erecta L. - *African Marigold*. Stems erect, to 50(100)cm; capitula very showy, 749
yellow to orange, 3-7cm across; (2n=24). Intrd; much grown in gardens, frequent casual as escape or throwout; scattered in En; Mexico.

2. T. patula L. - *French Marigold*. Stems erect, to 40cm; capitula showy, yellow to 749
orange or reddish-brown, often bicoloured, 2-4cm across; (2n=20, 24, 48). Intrd; occurrence as for *T. erecta*; Mexico.

3. T. minuta L. - *Southern Marigold*. Stems erect, to 1.2m; capitula inconspicuous, 749
<1cm across, with usually 3 yellowish-green ligules c.1-2mm; (2n=48). Intrd; rather characteristic wool-alien; scattered in En; S America.

95. SCHKUHRIA Roth - *Dwarf Marigold*
Annuals; leaves usually alternate, rarely some opposite, pinnate with linear to filiform leaflets with many minute sunken glands; phyllaries few in 1 overlapping row, herbaceous with membranous tips; capitula <1cm across, radiate with only 1 short yellow ligule; receptacle concave, without scales; pappus of scales.

1. S. pinnata (Lam.) Kuntze - *Dwarf Marigold*. Stems erect, to 60cm; capitula 752
numerous in subcorymbose panicle, obconical, 6-10 x 2-6mm (excl. ligule); (2n=20). Intrd; rather characteristic wool-alien; scattered in En; S & C America.

96. GAILLARDIA Foug. - *Blanketflower*
Annuals to perennials; leaves all alternate, simple, coarsely dentate, tapered to base but ± sessile; phyllaries in 2-3 rows, ± herbaceous, becoming reflexed before fruiting; capitula radiate; receptacle strongly convex to subglobose, with bristle-like scales; pappus of scales with apical bristle; ligules numerous, yellow or more often purple in proximal 1/4-3/4.

1. G. x grandiflora Van Houtte (*G. aristata* Pursh x *G. pulchella* Foug.) - 752
Blanketflower. Annual or short-lived perennial; stems erect, to 70cm; capitula 3.5-10cm across, very showy; (2n=72). Intrd; much grown in gardens, occasional casual on tips and rough ground, sometimes ± natd especially on sand and shingle by sea; local in S En; garden origin. Records for the 2 parents (both N American) are mostly (or all) errors for the hybrid.

97. HELENIUM L. - *Sneezeweed*
Tufted perennials; leaves all alternate, simple, subentire to shallowly dentate, tapered to base, decurrent on stem; phyllaries in 2-3 rows, herbaceous, becoming reflexed before fruiting; capitula radiate; receptacle convex, without scales; pappus of scales with apical bristle; ligules numerous, yellow to brownish-purple.

1. H. autumnale L. - *Sneezeweed*. Stems erect, to 1m; capitula 4-6.5cm across, the ligules soon becoming reflexed; disc and ray flowers yellow to brownish-purple in various combinations; (2n=34). Intrd; much grown in gardens, impermanent relic or throwout; sporadic in SE En; N America.

TRIBE 10 - EUPATORIEAE (genera 98-99). Annual to perennial herbs; leaves opposite; capitula discoid; phyllaries in several rows, herbaceous; receptacular scales 0; pappus of 1 row of hairs or pointed scales; corolla blue or pinkish-purple, sometimes white.

98. EUPATORIUM L. - *Hemp-agrimony*
Perennials; at least some leaves very deeply 3-(5-)lobed or palmate, cuneate at base; pappus of hairs; corolla pinkish-purple, rarely white.

1. E. cannabinum L. - *Hemp-agrimony*. Stems erect, to 1.5m; capitula 2-5mm across, very numerous in compound subcorymbose panicles; 2n=20. Native; all sorts of damp places and by water, in shade or open, sometimes in dry grassland or rough ground; common in most of En and Wa, frequent in Ir and CI, local and mostly coastal in Sc.

99. AGERATUM L. - *Flossflower*
Annuals; leaves simple, ovate, cordate; pappus of narrow scales; corolla usually blue, rarely pink or white.

1. A. houstonianum Mill. - *Flossflower*. Stems erect, to 60cm; capitula 6-10mm across, rather numerous in compound subcorymbose panicles; (2n=20, 40). Intrd; grown in gardens and escaping in waste ground or on tips; occasional in SE En, ± natd in hedgerow in Scillies; Mexico.

L I L I I D A E — MONOCOTYLEDONS
(Monocotyledonidae)

Very rarely trees, rarely shrubs; rarely with secondary thickening and never from a permanent vascular cambium; vascular bundles usually scattered through stem; primary root usually short-lived; leaves usually with parallel major venation, and minor venation scarcely or not reticulate; flower parts mostly in threes; pollen grains mostly bilaterally symmetrical, commonly with 1 pore and/or furrow; cotyledon normally 1; endosperm typically helobial. Numerous exceptions to all the above occur.

140. BUTOMACEAE - *Flowering-rush family*

Glabrous, aquatic perennials rooted in mud, emergent through water; leaves all basal, simple, linear, entire, sessile, without stipules. Flowers in terminal umbel with several bracts at base, bisexual, hypogynous, actinomorphic; sepals 3, free, purplish, tinged green; petals 3, free, pink; stamens 9; carpels 6, free except at extreme base, with numerous ovules all over inner side of wall; style 1; stigma slightly bilobed; fruit a group of 6 follicles.

Distinctive among petaloid monocots in its 6 ± free, dehiscent follicles, and similar sepals and petals.

1. BUTOMUS L. - *Flowering-rush*

1. B. umbellatus L. - *Flowering-rush.* Stems erect, to 1.5m; leaves usually slightly shorter; pedicels up to c.10cm; petals and sepals 1-1.5cm; (2n=16, 22, 24, 26, 30, 39, 40, 42). Native; in ponds, canals, ditches and river-edges; rather scattered in Br N to C Sc and in Ir, but often intrd.

141. ALISMATACEAE - *Water-plantain family*

Glabrous, aquatic annuals or perennials rooted in mud, often emergent through water; leaves usually all basal, simple, entire, sessile or petiolate, without stipules, sometimes produced in tufts on rooting stolons. Flowers in simple or compound umbels or in whorls, sometimes solitary, with bracts at base of umbel or whorl, bisexual or monoecious, hypogynous, actinomorphic; sepals 3, free, green; petals 3, free, white to pink; stamens 6 to numerous; carpels ≥6, free, with 1(-several) ovules; style 0 or very short; stigma not lobed, usually slightly elongated; fruit a group of achenes or few-seeded follicles.

Distinguished from Butomaceae in its very different petals and sepals, and 1-2(few)-seeded indehiscent fruits.

Submerged leaves are often ribbon-like, often very different from the diagnostically-shaped aerial leaves, and should be ignored.

1 Flowers monoecious, male and female in same inflorescence; stamens >6 **1. SAGITTARIA**
1 Flowers bisexual; stamens 6 2
 2 Stems procumbent or floating, rooting and producing tufts of leaves and inflorescences 3

2 Stems erect, leafless, all leaves basal 4
3 Floating and aerial leaves obtuse; carpels in an irregular whorl or flattish mass **3. LURONIUM**
3 Floating and aerial leaves acute; carpels spiral in a ± globose (*Ranunculus*-like) head **2. BALDELLIA**
4 Carpels spiral in ± globose (*Ranunculus*-like) head **2. BALDELLIA**
4 Carpels in a single (often irregular) whorl 5
5 Fruits curved inwards, ± unbeaked, 1-seeded **4. ALISMA**
5 Fruits divergent outwards, beaked, usually 2(-few)-seeded **5. DAMASONIUM**

1. SAGITTARIA L. - *Arrowheads*
Leaves all basal, linear and/or long-petiolate and sagittate; flowers conspicuous, usually in whorls, monoecious, the male flowers in upper and the female in lower whorls; petals white; stamens 7-numerous; carpels spiral in ± globose head, each with 1 ovule. In autumn stolons tipped by small bud-like propagules are formed.

Leaf-shape is notoriously variable in this genus and needs to be used with great caution.

1 Most or all emergent leaves strongly sagittate, with two long basal lobes 2
1 Most or all leaves linear to elliptic, rarely a few with short basal lobes 3
2 Achenes 4-6mm, with apical beak <1mm; petals usually with purple blotch at base; anthers purple **1. S. sagittifolia**
2 Achenes 2.5-4mm, with subapical beak >1mm; petals without basal purple blotch; anthers yellow **2. S. latifolia**
3 Flowers and leaves floating; many leaves linear, usually some or all floating ones elliptic; filaments glabrous **4. S. subulata**
3 Flowers and leaves emergent; emergent leaves elliptic or rarely some with short basal lobes; filaments with scale-like hairs **3. S. rigida**

1. S. sagittifolia L. - *Arrowhead.* Emergent leaves strongly sagittate; floating leaves often elliptic; submerged leaves linear; stems emergent, to 1m; flowers 2-3cm across; achenes 4-6mm, with beak <1mm; (2n=16, 20, 22). Native; in ponds, canals and slow rivers; frequent in En, very scattered in Wa and Ir.

2. S. latifolia Willd. - *Duck-potato.* Differs from *S. sagittifolia* in key characters; (2n=22). Intrd; natd in ponds and streamside in Surrey since 1941, Jersey since 1961; N America.

3. S. rigida Pursh - *Canadian Arrowhead.* Emergent and floating leaves elliptic, rarely with 2 short basal lobes; submerged leaves linear; stems emergent to 75cm; achenes 2.5-4mm, with beak >1mm; (2n=22). Intrd; 2 places in canal in S Devon since 1898; N America.

4. S. subulata (L.) Buchenau - *Narrow-leaved Arrowhead.* Most leaves linear and submerged, none emergent; floating leaves elliptic; stems to 30cm, producing flowers on water surface; achenes 1.5-2.5mm, with beak <1mm; (2n=22). Intrd; natd in acid pond in N Hants since 1962; N America.

2. BALDELLIA Parl. - *Lesser Water-plantain*
Varying in vegetative and inflorescence habit from *Alisma*-like to *Luronium*-like; petals pale mauve to ± white with yellow basal blotch; stamens 6; carpels spiral in ± globose head, each with 1 ovule.

Baldellia is very doubtfully distinct from the earlier **Echinodorus** Rich. ex Engelm., from N & S America and Africa, and should probably be placed under it.

1. B. ranunculoides (L.) Parl. - *Lesser Water-plantain.* Rosette-plant with erect stem to 20cm bearing inflorescence in simple whorl or with 1(-2) whorls below, or plant with trailing stems producing tufts of leaves and reduced (often 1-flowered) inflorescences; leaves linear or petiolate and narrowly elliptic; flowers 10-16mm across; achenes 2-3.5mm, tapered to acute apex; 2n=16. Native; wet places or shallow water in ditches, streamsides and pondsides; scattered over most of BI.

3. LURONIUM Raf. - *Floating Water-plantain*
Stems procumbent or floating, rooting at intervals and producing tufts of leaves and inflorescences; submerged leaves linear, floating ones petiolate, elliptic, obtuse; flowers solitary or 2-5 in simple umbels, bisexual; petals pale mauve to ± white with yellow basal blotch; stamens 6; carpels in irregular whorl or flattish mass, each with 1 ovule.

1. L. natans (L.) Raf. - *Floating Water-plantain.* Stems to 75cm, often much less; **RRR** floating leaves 1-2.5(4)cm; flowers 12-18mm across; achenes c.2.5mm, with apical, laterally pointed beak <1mm; (2n=42). Native; in acid ponds and canals; local in Wa and N & C En, decreasing.

4. ALISMA L. - *Water-plantains*
Leaves all basal, linear and/or long-petiolate and narrowly elliptic to elliptic-ovate; flowers in whorled panicles, or (in small plants) in simple whorls or umbels, bisexual; petals pale mauve to ± white with basal yellow blotch; stamens 6; carpels in single whorl with lateral or subterminal style, each with 1 ovule.

1 Leaves elliptic to ovate-elliptic, rounded to subcordate at base; style arising c.1/2 way up fruit **1. A. plantago-aquatica**
1 Leaves linear to narrowly elliptic or lanceolate-elliptic, cuneate at base; style arising in upper 1/2 of fruit 2
2 Fruits widest near middle, with ± straight, erect style **2. A. lanceolatum**
2 Fruits widest in upper 1/2, with strongly recurved style **3. A, gramineum**

1. A. plantago-aquatica L. - *Water-plantain.* Stems erect, to 1m; leaves elliptic to ovate-elliptic, rounded to subcordate at base; flowers 7-12mm across; achenes with straight, ± erect style; flowers open in afternoon; 2n=14. Native; in or by ponds, ditches, canals and slow rivers; common throughout BI except rare in N Sc.

1 x 2. A. plantago-aquatica x A. lanceolatum = A. x rhicnocarpum Schotsman is intermediate in all characters and sterile; there are records from Ir, Man, C Sc, London area and E Anglia, but all need confirming.

2. A. lanceolatum With. - *Narrow-leaved Water-plantain.* Differs from *A. plantago-aquatica* as in key (couplet 1); flowers open in morning (some overlap); 2n=28. Native; similar places to *A. plantago-aquatica*; scattered in Ir, in Br N to Yorks, and in Man.

3. A. gramineum Lej. - *Ribbon-leaved Water-plantain.* Differs from *A. lanceolatum* in **RRR** leaves linear or very narrowly elliptic; and see key; (2n=14). Possibly native, more likely sporadically intrd; in shallow ponds; 1 place in Worcs since 1920, for short periods in S Lincs, W Norfolk and Cambs. W Norfolk plants are said to belong to ssp. **wahlenbergii** Holm., with shorter achenes with thinner walls, and the others to ssp. **gramineum**, but the taxa are doubtfully distinct sspp.

5. DAMASONIUM Mill. - *Starfruit*
Leaves all basal, long-petiolate, submerged, floating or sometimes emergent, ovate-oblong with cordate base; flowers in whorls, bisexual; petals white, with basal yellow blotch; stamens 6; carpels 6-10 in 1 whorl, with terminal style, each with 2-several ovules.

1. D. alisma Mill. - *Starfruit.* Stems erect, to 30(60)cm; leaves 3-6(8)cm; flowers 5- **RRR** 9mm across; follicles 5-14mm, with long beak; (2n=42). Native; muddy margins of acid ponds; 1 place in Surrey, 2 in Bucks, formerly elsewhere in S & C En, apparently approaching extinction or at most sporadic occurrence.

142. HYDROCHARITACEAE - *Frogbit family*

Glabrous, aquatic perennials, free floating or rooted in mud; leaves submerged or floating, sometimes emergent, all basal or along stems, entire or serrate, sessile or petiolate, stipulate or not. Flowers 1-few in axillary inflorescences subtended by spathe, usually dioecious, epigynous, actinomorphic, conspicuous or inconspicuous; sepals 3, free; petals 3 or vestigial, white to purplish; stamens 2-12; ovary 1-celled, with 3-6 parietal placentas bearing many ovules; styles 3 or 6; stigmas usually linear, often bifid; fruit an irregularly opening capsule.

Distinctive among monocots in having an inferior ovary combined with 3 sepals, usually 3 petals and dioecious flowers.

1 Leaves with long petioles and suborbicular cordate blades **1. HYDROCHARIS**

1 Leaves sessile, tapering apically or at both ends 2

2 Leaves all in basal rosette 3

2 Leaves borne along stems 4

3 Leaves sharply serrate all along margins, rigid; petals conspicuous, white **2. STRATIOTES**

3 Leaves denticulate only near apex, flaccid; petals vestigial **7. VALLISNERIA**

4 Leaves variously whorled to spiral **6. LAGAROSIPHON**

4 Leaves all whorled or opposite 5

5 Middle and upper leaves in whorls of 3-6(8), with 2 minute (c.0.5mm) fringed scales at base **5. HYDRILLA**

5 Middle and upper leaves in whorls of 3-4(5), with or without 2 entire scales at base <0.5mm 6

6 Leaves in whorls of (3)4-5; petals >5mm, white; in industrially warmed water only **3. EGERIA**

6 Leaves in whorls of (2)3-4(5); petals <5mm, inconspicuous; widespread **4. ELODEA**

1. HYDROCHARIS L. - *Frogbit*

Plants usually floating, the roots hanging in water; leaves with long petioles and floating suborbicular cordate blades, all in basal rosette, entire, with large stipules; flowers mostly dioecious, c.5-10% monoecious but all of 1 sex from each rosette, arising on pedicels from a stalked spathe (1 in female, 1-3(4) in male), conspicuous, with petals much larger than sepals; stamens 9-12, some usually as sterile staminodes; female flowers with 6 staminodes only; styles 6, bifid.

1. H. morsus-ranae L. - *Frogbit*. Main stems are floating stolons to 50(100)cm forming over-wintering terminal buds in autumn; leaves 1.6-5cm across; flowers 2-3cm across; petals white with basal yellow blotch; fruits rarely forming; 2n=28. Native; in ponds, canals and ditches; locally frequent in En, very scattered in Wa and Ir, formerly CI, decreasing.

2. STRATIOTES L. - *Water-soldier*

Plant usually floating, mostly submerged, rising to surface at flowering; leaves sessile, linear-lanceolate, tapering from base, all in basal rosette, spinous-serrate, without stipules; flowers mostly dioecious, arising from a stalked spathe (1 ± sessile in female, several pedicelled in male), conspicuous, with petals much larger than sepals; stamens 12, with sterile staminodes surrounding them; female flowers with staminodes only; styles 6, bifid.

1. S. aloides L. - *Water-soldier*. Rosettes large, sturdy; leaves up to 50 x 2cm; flowers 3-4cm across; petals white; (2n=24). Native; ponds, dykes and canals, usually calcareous; now very local in E Anglia, N & S Lincs, SE Yorks and Denbs, formerly locally frequent in C Br, intrd in scattered places in Br N to C Sc. Only female plants occur in Br. R

3. EGERIA Planch. - *Large-flowered Waterweed*
Stems long, branched, rooted in mud, submerged; leaves sessile, in whorls of (3)4-5, narrowly oblong-linear, minutely serrate, without stipules; flowers dioecious, pedicellate, arising from sessile axillary spathe (1 in female, 2-4 in male), conspicuous; petals white, much larger than sepals; stamens 9; styles 3, bifid.

1. E. densa Planch. - *Large-flowered Waterweed*. Stems to 2m (?more); leaves 10-30 x 1.5-4mm, 0.5-1mm wide 0.5mm behind apex, with acute apex; flowers 1.2-2cm across; (2n=48). Intrd; in warmed water of canals and mill-lodges; very local in S Lancs, 1st recognized 1953, Glam (1988), Middlesex (1993); S America. Only male plants in Br, flowering rarely and only in well-warmed water. Closely resembles a robust *Elodea* when not in flower; leaves similar in width to those of *E. canadensis*, but longer and more acute.

4. ELODEA Michx. - *Waterweeds*
Stems long, branched, rooted in mud, submerged; leaves sessile, the lower opposite, the upper in whorls of 3-4(5), minutely serrate, with 2 minute entire basal scales; flowers dioecious, solitary from sessile axillary spathe, inconspicuous; petals whitish to reddish, c. as large as sepals; stamens 9; styles 3, bifid. Only female plants occur in BI.

1 Leaf apices obtuse to subacute; leaves (0.7)0.8-2.3mm wide 0.5mm behind apex **1. E. canadensis**
1 Leaf apices acute to acuminate; leaves 0.2-0.7(0.8)mm wide 0.5mm behind apex 2
2 Usually some leaves strongly recurved and/or twisted, with marginal teeth 0.05-0.1mm; root-tips white to greyish-green when fresh; sepals 1.6-2.5mm **2. E. nuttallii**
2 Usually no leaves strongly recurved or twisted, with marginal teeth usually 0.1-0.15mm; root-tips red when fresh; sepals 3-4.3mm **3. E. callitrichoides**

1. E. canadensis Michx. - *Canadian Waterweed*. Stems to 3m; leaves in whorls of (2)3(-4), 4.5-17 x 1.4-5.6mm, (0.7)0.8-2.3mm wide 0.5mm behind apex; 2n=c.24 (24, 48). Intrd; natd in ponds, lakes, canals, slow rivers; common throughout BI except extreme N & NW Sc; N America. 1st recorded 1836.

2. E. nuttallii (Planch.) H. St. John - *Nuttall's Waterweed*. Stems to 3m; leaves in whorls of (2)3-4(5), 5.5-35 x 0.8-3mm, 0.2-0.7(0.8)mm wide 0.5mm behind apex; 2n=c.48 (48). Intrd; habitat as for *E. canadensis*; locally common in En, very scattered in Wa, Sc, Ir and Jersey, rapidly spreading; N America. Leaves longer, narrower and more acute than in *E. canadensis*. 1st recorded 1966.

3. E. callitrichoides (Rich.) Casp. (*E. ernstiae* H. St. John) - *South American Waterweed*. Stems to 3m; leaves in whorls of 3, 9-25 x 0.7-2.2mm, 0.2-0.6mm wide 0.5mm behind apex; 2n=c.32 (16). Intrd; habitat as for *E. canadensis* but probably not fully natd; very local in S En and S Wa; S America. Differs from *E. nuttallii* as in key. 1st recorded 1948.

5. HYDRILLA Rich. - *Esthwaite Waterweed*
Stems long, branched, rooted in mud, submerged; leaves sessile, the lower opposite, the upper in whorls of 3-6(8), minutely serrate, with 2 minute fringed scales at base; flowers dioecious, solitary from sessile axillary spathe, inconspicuous; petals transparent with red streaks, c. as large as sepals; stamens 3; styles 3(-5), simple.

1. H. verticillata (L. f.) Royle (*Elodea nuttallii* auct. non (Planch.) H. St. John) - *Esthwaite Waterweed*. Stems to 1m (?more); leaves 5-20 x 0.7-2mm, 0.2-0.7mm wide 0.5mm behind apex, with narrowly acute to acuminate apex; 2n=16. Native; lakes; Esthwaite Water (Westmorland) 1914-c.1945, and Rusheenduff Lough (W Galway) from 1935 (no flowers found in former, female flowers once only in latter place). Closely resembles *Elodea nuttallii* but less robust and leaves more per node and with minute teeth ± to base (not just in distal 1/2). **RRR**

6. LAGAROSIPHON Harv. - *Curly Waterweed*
Stems long, branched, rooted in mud, submerged; leaves variously whorled to spiral, the lowest always spiral (not opposite), subentire to minutely denticulate, with 2 minute entire basal scales; flowers dioecious, inconspicuous, arising from sessile axillary spathe (1 in female, several in male); petals reddish, c. as large as sepals; stamens 3; styles 3, bifid.

1. L. major (Ridl.) Moss - *Curly Waterweed.* Stems to 3m; leaves 6-30 x 1-3mm, usually strongly recurved, 0.2-0.5mm wide 0.5mm behind apex, with narrowly acute to acuminate apex; (2n=22). Intrd; ponds, lakes, canals, slow rivers; locally frequent in CI and Br N to C Sc, M Cork, 1st recorded 1944; S Africa. Only female plants occur in BI.

7. VALLISNERIA L. - *Tapegrass*
Plant rooted in mud, submerged, with stems as stolons; leaves sessile, all in basal rosette, linear, denticulate near apex, without stipules; flowers dioecious (or monoecious?), solitary (female) or many (male) in stalked spathes, inconspicuous; petals 0 or vestigial; stamens (1)2(-3); styles 3, bifid.

1. V. spiralis L. - *Tapegrass*. Leaves 2-80cm x 1-10mm, ribbon-like; stalks of female spathes very long, spiralling after flowering; (2n=20, 30, 40). Intrd; natd (often not permanently) in slow rivers and canals, usually where water is heated; very local in S En, S Lancs, SW Yorks; warm regions of world.

143. APONOGETONACEAE - *Cape-pondweed family*

Glabrous, aquatic perennials with elongated stems rooted in mud at tuberous base; leaves floating, all basal, entire, long-petiolate with sheathing base. Flowers in a forked spike at water surface on long stalk, with deciduous spathe at base of spike, bisexual, hypogynous, actinomorphic except for perianth, conspicuous; tepals 1(-2), white; stamens 6-18; carpels 3, free, each with several ovules, short style, and linear stigma; fruit a group of follicles.

The forked, white inflorescence borne just above the water surface is unique.

1. APONOGETON L. f. - *Cape-pondweed.*

1. A. distachyos L. f. - *Cape-pondweed.* Leaves oblong-elliptic, up to 25 x 7cm; spikes up to 6cm, each with up to 10 flowers; tepals 10-20mm; (2n=16, 24, 32). Intrd; persistent or ± natd where planted in ponds; scattered in Br; S Africa.

144. SCHEUCHZERIACEAE - *Rannoch-rush family*

Glabrous, *Juncus*-like perennials with rhizomes clothed with old leaf-bases and erect leafy flowering stems; leaves alternate, linear, with sheathing base with ligule at top on adaxial side, entire, sessile. Flowers few in terminal raceme with bract at base of each, bisexual, hypogynous, actinomorphic, inconspicuous; tepals 6, sepaloid; stamens 6; carpels 3(-6), free, each with usually 2 ovules and sessile capitate stigma; fruit a group of follicles.

Juncus-like vegetatively but with free (usually 3) carpels each with usually only 2 seeds. The leaves are very distinctive in having a large apical pore.

1. SCHEUCHZERIA L. - *Rannoch-rush*

1. S. palustris L. - *Rannoch-rush*. Stems to 25(40)cm; upper stem-leaves usually overtopping inflorescence; tepals up to 3mm; follicles up to 7mm; 2n=22. Native; in pools or wet *Sphagnum* bogs; very rare in 2 places in M Perth, formerly scattered in Sc and N En and 1 place in Ir. **RR**

145. JUNCAGINACEAE- *Arrowgrass family*

Glabrous perennials with rhizomes and erect, ± leafless, flowering stems; leaves ± all basal, in rosette or a few alternate near stem base, linear, with sheathing base with ligule at top on adaxial side, entire, sessile. Flowers many in bractless simple terminal raceme, bisexual, hypogynous, inconspicuous; tepals 6, sepaloid; stamens 6; ovary 6-celled, all cells with 1 ovule or alternate cells small and sterile; styles ± 0; stigmas papillate to long-fringed; fruit 3 or 6 1-seeded units breaking apart at maturity.

Juncus-like or *Plantago*-like superficially, but the 3 or 6 1-seeded fruit segments are diagnostic.

1. TRIGLOCHIN L. - *Arrowgrasses*

1. T. palustre L. - *Marsh Arrowgrass*. Stems erect, to 60cm; leaves usually deeply furrowed on upperside near base; fruit with 3 fertile cells, 7-10mm, extremely narrowed at base; stigmas long-fringed; 2n=24. Native; marshy places and wet fields, sometimes at back of salt-marshes; throughout BI, commonest in N.

2. T. maritimum L. - *Sea Arrowgrass*. Stems erect, to 60cm; leaves usually flat on upperside near base; fruit with 6 fertile cells, 3-5mm, ± rounded at base; stigmas papillate; 2n=48. Native; in salt-marshes and salt-sprayed grassland; round coasts of whole BI, rare in inland salty areas. Ligule usually longer than wide (not wider than long) and more pointed than in *T. palustre*, but this and leaf character given above are not completely reliable.

146. POTAMOGETONACEAE - *Pondweed family*

Glabrous aquatic (or ± so) perennials, often with rhizomes, with leafy submerged flowering stems; leaves alternate or opposite, floating and/or submerged (or some ± aerial), linear to elliptic, entire or nearly so, petiolate or sessile, at least some with sheathing base which is either free and stipule-like or fused to leaf-base for most of length, forming sheath and free distal ligule. Flowers in short, axillary, stalked, bractless spikes, bisexual, hypogynous, actinomorphic, inconspicuous, submerged or aerial; tepals 4, sepaloid; stamens 4; carpels (1-)4, free, each with 1 ovule and ± sessile capitate stigma; fruits in a group of 1-4 achenes or drupes.

Distinguished from other pondweeds by the 4 tepals and 4 stamens.

1 All or most leaves alternate, all with membranous sheath or stipule **1. POTAMOGETON**
1 All leaves opposite (or some in whorls of 3), only the uppermost with membranous stipules **2. GROENLANDIA**

1. POTAMOGETON L. - *Pondweeds*

Leaves all alternate, or just those subtending inflorescences opposite, all with membranous sheath or stipules; fruits with thick pericarp, soft on outside but with bony inner layer.

For accurate identification it is essential first to examine thoroughly the leaf morphology, including range of leaf-shape, leaf-blade venation, and morphology of basal sheath/stipules. The key covers only mature plants that have reached flowering or are about to flower; beware recent flooding or drying out of habitat when distinguishing floating and submerged leaves. In taxa that can have floating leaves (whether actually present or not) the upper submerged leaves often approach the former in certain respects and may be quite different from the middle and lower submerged leaves, which are the diagnostic ones. In the keys and descriptions 'veins' refers to the midrib plus its laterals that run ± parallel to it for nearly the whole leaf length. Fruit lengths include the beak. Fresh material is best for determination; the stipule characters are diffcult to see in dried material, but are sometimes necessary for certain identification.

Although 25 hybrids have been found in BI, hybridization is not common and most hybrids are very rare. 8 that occur in the absence of both parents are treated fully. All hybrids except *P. x zizii* are sterile, with no fruit developed; the fruits of *P. x zizii* are of unproven viability. Hybrids are variously intermediate between the parents.

General key

1 Floating leaves present *Key A*
1 All leaves submerged 2
  2 Leaves slightly to strongly convex-sided, >6mm (usually >10mm) wide, often with ≥7 main veins *Key B*
  2 Leaves grass-like, parallel-sided for almost whole length, <5(7)mm wide, with 3-5(7) main veins *Key C*

Key A - Floating leaves present

1 Floating leaves with distinct hinge-like joint at junction with petiole; submerged leaves (if present) <3.5mm wide **1. P. natans**
1 Floating leaves merging gradually or abruptly into petiole, but without distinct joint; at least some submerged leaves (if present) >3.5mm wide 2
  2 Submerged leaves always present, sessile or with petiole <1cm 3
  2 Submerged leaves usually present, if so then distinctly petiolate with petiole >1cm 8
3 All submerged leaves strictly parallel-sided ± throughout length, mostly ≤8mm wide **16. P. epihydrus**
3 At least some submerged leaves convex-sided for at least part of length, often >8mm wide 4
  4 At least lower submerged leaves broadly rounded at base, ± clasping stem **10. P. x nitens**
  4 All submerged leaves narrowed to base, not clasping stem 5
5 Submerged leaves obtuse to rounded at apex 6
5 Submerged leaves acute, cuspidate or acuminate at apex 7
  6 Stem terete; at least some submerged leaves >1.5cm wide, with >7 veins; fertile **11. P. alpinus**
  6 Stem compressed; submerged leaves ≤1.5cm wide, all or most 5-7-veined; sterile **12. P. x olivaceus**

7 Submerged leaves ≤1.2cm wide, gradually acuminate to acute; stipules mostly 1-2.5cm on main stems **9. P. gramineus**
7 Submerged leaves ≥1cm wide, rather abruptly acuminate to cuspidate; stipules >2cm on main stems **7. P. x zizii**
 8 Submerged leaves always present, ≤1.2cm wide **2. P. x sparganiifolius**
 8 Submerged leaves usually present, if so ≥1cm wide 9
9 Floating leaves translucent, the vein network clearly visible (fresh or dried); fruits 1.5-1.9mm **4. P. coloratus**
9 Floating leaves opaque, the vein network difficult to see; fruits ≥1.9mm 10
 10 Most leaves floating (or ± aerial), at least some with rounded to cordate base; fruits 1.9-2.6mm **3. P. polygonifolius**
 10 Most leaves submerged; floating leaves with cuneate base; fruits 2.7-4.1mm 11
11 Submerged leaves with petiole c.1/2 as long as blade or longer; peduncles not or scarcely widened distally **5. P. nodosus**
11 Submerged leaves with petiole much <1/2 as long as blade; peduncles distinctly widened distally **7. P. x zizii**

Key B - All leaves submerged, convex-sided, >6mm wide, often with ≥7 veins
1 Leaves minutely serrate along ± whole margin to naked eye, usually regularly undulate; fruits with beak ≥1/2 as long as body **26. P. crispus**
1 Leaves entire even with x10 lens, or obscurely serrate just near apex, not regularly undulate; fruits with beak <1/2 as long as body 2
 2 Leaves with acute, acuminate or cuspidate apex 3
 2 Leaves with obtuse to rounded, sometimes hooded, apex 7
3 At least some leaves rounded at base and ± clasping stem; sterile hybrids 4
3 All leaves tapered to base, not clasping stem 5
 4 Stipules 2-7cm; leaves mostly >2cm wide **8. P. x salicifolius**
 4 Stipules 1-3cm; leaves mostly <2cm wide **10. x P. nitens**
5 Leaves ≤1.2cm wide, gradually acuminate to acute; stipules 1-3.5cm **9. P. gramineus**
5 Leaves ≥(1)1.5cm wide, rather abruptly acuminate to cuspidate; stipules >2cm 6
 6 Leaves all or mostly <12cm; only some leaves petiolate, the petiole usually narrowly winged; fruits 2.7-3.4mm **7. P. x zizii**
 6 Leaves all or mostly >12cm; all leaves petiolate, the petiole unwinged near base; fruits 3.2-4.5mm **6. P. lucens**
7 At least some leaves rounded at base, usually ± clasping stem 8
7 Leaves narrowed to base, not clasping stem 10
 8 Stipules conspicuous, >1cm (often much more so), persistent; leaves mostly >10cm, with ≥3 faint lateral strands between midrib and nearest strong lateral vein **13. P. praelongus**
 8 Stipules inconspicuous, at least some <1cm, soon disappearing; leaves all or most <10cm, with <3 faint lateral strands between midrib and nearest strong lateral vein 9
9 Stem compressed; leaves oblong-lanceolate, most or all <2cm wide; sterile **15. P. x cooperi**
9 Stem terete; leaves oblong-ovate, most or all >2cm wide; fertile **14. P. perfoliatus**
 10 Stem terete; leaves all with >7 veins; fertile **11. P. alpinus**
 10 Stem compressed; leaves all 3-7-veined; sterile 11
11 Leaves ≤5mm wide, with (3-)5 veins **18. P. x lintonii**
11 Leaves >5mm wide, with 5-7 veins **12. P. x olivaceus**

Key C - All leaves submerged, parallel-sided for almost whole length, <5(7)mm wide, with 3-5(7) veins

1 Stipules mostly >5cm; leaves actually blade-less petioles, opaque, without obvious midrib **1. P. natans**

1 Stipules <5cm; leaves translucent, with obvious midrib 2

2 Stipules fused to base of leaf, forming sheathing leaf-base, free distally to form ligule 3

2 Stipules free from leaf, forming stipule-like outgrowth from node 5

3 Plant either with leaf apex as in *P. pectinatus* and leaf-sheath as in *P. filiformis* or vice versa; fruit not formed **28. P. x suecicus**

3 Plant with both leaf apex and leaf-sheath as in *P. pectinatus* or *P. filiformis*; fruit well developed 4

4 Leaves acute to acuminate at apex; sheathing leaf-base not fused in tube round stem, but with margins overlapping; fruit 3.3-4.7mm **29. P. pectinatus**

4 Leaves obtuse to rounded at apex; sheathing leaf-base fused in tube round stem proximally when young; fruit 2.2-3.2mm **27. P. filiformis**

5 Most leaves <2mm wide, with 3(-5) veins 6

5 Most leaves >2mm wide, with 3-5(many) veins 9

6 Stipules fused in tube round stem proximally when young; fruits 1.8-2.3mm 7

6 Stipules not fused in tube round stem, but with margins overlapping; fruits 1.8-3.2mm 8

7 Leaves very gradually and finely acute to long-acuminate, rigid; C & N Sc only **19. P. rutilus**

7 Leaves acute to obtuse and abruptly mucronate, not rigid; widespread **20. P. pusillus**

8 Leaves acute to finely acute, mostly <1mm wide; carpels 1(-2) per flower; fruits usually warty or ± toothed near base **23. P. trichoides**

8 Leaves obtuse to subacute, often mucronate, mostly >1mm wide; carpels (3)4-5(7) per flower; fruits not warty **22. P. berchtoldii**

9 Leaves with 3 or 5 main veins and many finer strands between them 10

9 Leaves with 3 or 5(-7) main veins only 11

10 Leaves with 3 main veins, usually acuminate; fruit usually with basal wart or tooth, usually with erect symmetric beak **25. P. acutifolius**

10 Leaves with 5 main veins (2 submarginal), usually mucronate; fruit not warted, with asymmetric, curved beak **24. P. compressus**

11 Leaves usually minutely serrate along at least part of margin; fruits tapered to beak ≥1/2 as long as body **26. P. crispus**

11 Leaves entire or minutely serrate just near apex; fruits, if formed, with beak <1/2 as long as body 12

12 Stems with many lateral branches closely placed, forming fan-like sprays; stipules not fused in tube round stem **21. P. obtusifolius**

12 Stems without many closely placed lateral branches; stipules fused in tube round stem proximally when young 13

13 Leaves ≤5mm wide, usually serrulate distally, rounded to acute at apex; sterile **18. P. x lintonii**

13 Leaves ≤3.5mm wide, entire, mucronate at apex; fruits 2.4-3mm **17. P. friesii**

Subgenus 1 - *POTAMOGETON* (spp. 1-26). Leaves submerged and/or floating, sessile or petiolate, with basal sheaths free or nearly free from leaf-base, forming stipule-like outgrowth; flowers pollinated above water surface by wind.

1. P. natans L. - *Broad-leaved Pondweed*. Floating leaves opaque, elliptic to ovate- **770**
elliptic, up to 10(14) x 4.5(8)cm, very rarely 0; submerged leaves opaque, linear, **771**

≤3.5mm wide, often very long, rounded at apex, sessile; fruit 3.8-5mm; 2n=c.52 (42, 52). Native; lakes, ponds, rivers, ditches, usually over rich soils; frequent throughout BI.

1 x 3. P. natans x P. polygonifolius = P. x gessnacensis G. Fisch. has been found in Caerns and E Ross; other records are errors.

1 x 5. P. natans x P. nodosus = P. x schreberi G. Fisch. was discovered in 1992 in Dorset.

1 x 6. P. natans x P. lucens = P. x fluitans Roth has been found scattered in S En; (2n=52).

1 x 22. P. natans x P. berchtoldii = P. x variifolius Thore is known in W Mayo.

2. P. x sparganiifolius Laest. ex Fr. (*P. natans* x P. *gramineus*) - *Ribbon-leaved* R
Pondweed. Floating leaves opaque, elliptic to narrowly so, up to 11.5 x 3.5cm; 770
submerged leaves various, some linear to very narrowly elliptic and sessile, others 771
with narrowly elliptic blade and long petiole, acute at apex, up to 52 x ≤1.2cm. Native; lakes, ponds, canals, streams; scattered in Br and Ir.

3. P. polygonifolius Pourr. - *Bog Pondweed*. Floating (or emergent) leaves opaque, 770
similar in shape to those of *P. natans*, up to 10.5 x 7cm; submerged leaves narrowly 771
elliptic to oblanceolate, up to 16 x 2.5cm, petiolate, obtuse, sometimes 0; fruit 1.9-2.6mm; 2n=28. Native; shallow ponds, bogs, ditches, small streams, on acid soil; common throughout most of BI, but absent from parts of C & E En.

3 x 9. P. polygonifolius x P. gramineus = P. x lanceolatifolius (Tiselius) C.D. Preston has been found in Wigtowns and Easterness.

4. P. coloratus Hornem. - *Fen Pondweed*. Floating leaves ovate, up to 8.5 x 5.5cm, R
translucent, submerged leaves similar to those of *P. polygonifolius* but wider; fruit 770
1.5-1.9mm; 2n=c.26 (26). Native; ponds and pools, on base-rich peat; local 771
throughout BI except N & E Sc, mostly in C Ir and CE En.

4 x 9. P. coloratus x P. gramineus = P. x billupsii Fryer has been found in Cambs and Outer Hebrides.

4 x 22. P. coloratus x P. berchtoldii = P. x lanceolatus Sm. occurs in Anglesey, Co Clare and NE Galway, formerly in Cambs; endemic.

5. P. nodosus Poir. - *Loddon Pondweed*. Floating leaves elliptic to narrowly so, RR
translucent, up to 13 x 5cm; submerged leaves narrowly elliptic to oblanceolate, up 770
to 28 x 4mm, acute to obtuse at apex, petiolate; fruit 2.7-4.1mm; (2n=52). Native; 771
slow-flowing base-rich rivers; very local in S En, ?decreasing.

6. P. lucens L. - *Shining Pondweed*. Leaves all submerged, elliptic or obovate to 771
narrowly elliptic or oblanceolate, up to 20 x 6.5cm, acuminate to cuspidate at apex, petiolate; fruit 3.2-4.5mm; (2n=52). Native; lakes, ponds, canals, slow rivers on base-rich soil; locally common in En and Ir, rare in Wa, Sc and CI.

6 x 11. P. lucens x P. alpinus = P. x nerviger Wolfg. is known from Co Clare.

6 x 26. P. lucens x P. crispus = P. x cadburyae Dandy & G. Taylor was found in 1948 as a single plant in Warks; endemic. Possibly extinct.

7. P. x zizii W.D.J. Koch ex Roth (*P. lucens* x *P. gramineus*) - *Long-leaved Pondweed*. R
Floating leaves ± opaque, elliptic or narrowly so, up to 10.5 x 4cm, sometimes 0; 770
submerged leaves narrowly elliptic to oblanceolate, up to 13 x 3cm, sessile or 771
shortly petiolate (often both on 1 plant), cuspidate to mucronate at apex; fruit 2.7-3.4mm. Native; lakes, ponds, streams, canals; scattered throughout C & N Br and Ir, often without parents.

8. P. x salicifolius Wolfg. (*P. lucens* x *P. perfoliatus*) - *Willow-leaved Pondweed*. R
Leaves all submerged, elliptic-oblong to narrowly so, up to 21.5 x 4cm, sessile and 771
some clasping stem, acute to cuspidate at apex. Native; ponds, canals, rivers; scattered in C & S En, very local in E Sc, NE En and Ir.

9. P. gramineus L. - *Various-leaved Pondweed*. Floating leaves opaque, elliptic, up 770
to 9.5 x 3.4cm, sometimes 0; submerged leaves narrowly elliptic to oblanceolate, up 771
to 14 x 1.2cm, acute to acuminate or cuspidate at apex, sessile; fruit 2.4-3.1mm; (2n=52). Native; lakes, ponds, canals, streams, usually on acid soils; scattered throughout most of Br and Ir, but not in most of S En.

9 x 11. P. gramineus x P. alpinus = P. x nericius Hagstr. is known from S Aberdeen.

10. P. x nitens Weber (*P. gramineus* x *P. perfoliatus*) - *Bright-leaved Pondweed.* 770
Floating leaves ± opaque, elliptic or narrowly so, up to 6.5 x 2.3cm, often 0, 771
submerged leaves narrowly elliptic to lanceolate, up to 11.5 x 2.3cm, acute or acuminate at apex, at least lower ones sessile and clasping stem. Native; lakes, ponds, canals, streams; scattered through most of BI, but absent from CI and most of S Br.

11. P. alpinus Balb. - *Red Pondweed.* Plant often tinged reddish; floating leaves ± 770
translucent, elliptic to rather narrowly so, up to 9 x 2.5cm, sometimes 0; submerged 771
leaves lanceolate to narrowly elliptic, up to 22 x 3.3cm, obtuse at apex, all except the upper ones sessile; fruit 2.6-3.7mm; (2n=26, 52). Native; lakes, canals, streams, especially on peaty soil; fairly frequent ± throughout BI except SW En.

11 x 13. P. alpinus x P. praelongus = P. x griffithii A. Benn. has been found in Caerns, Westerness and W Donegal; endemic.

11 x 14. P. alpinus x P. perfoliatus = P. x prussicus Hagstr. has been found in Colonsay (S Ebudes), Benbecula (Outer Hebrides) and Co Kildare.

12. P. x olivaceus Baagøe ex G. Fisch. (*P. alpinus* x *P. crispus*) - *Graceful Pondweed.* R
Stem compressed; floating leaves reported sometimes, not seen by me; submerged 771
leaves narrowly oblong-linear, up to 12 x 1.5cm wide, rounded at apex, not or scarcely undulate, narrowed to sessile or shortly petiolate base. Native; rivers; very local in S Wa, E Sc and NE En.

13. P. praelongus Wulfen - *Long-stalked Pondweed.* Leaves all submerged, 771
narrowly oblong-ovate, up to 22 x 4cm, sessile and clasping stem, blunt and hooded at apex (bifid when pressed); fruit 4.5-5.5mm; (2n=52). Native; lakes, rivers, canals, streams; scattered throughout Br and Ir except S En.

13 x 14. P. praelongus x P. perfoliatus = P. x cognatus Asch. & Graebn. was found in N Lincs in 1943 and W Sutherland in 1948 (still in latter in 1993).

13 x 26. P. praelongus x P. crispus = P. x undulatus Wolfg. has been found in Rads (?extinct) and Co Antrim.

14. P. perfoliatus L. - *Perfoliate Pondweed.* Leaves all submerged, broadly ovate to 771
narrowly oblong-ovate, up to 11.5 x 4.2cm (often much shorter), obtuse to rounded and often ± hooded at apex, cordate and clasping stem; fruit 2.6-4mm; (2n=26, 40, 48, 52, 78). Native; lakes, ponds, canals, rivers, streams; frequent throughout Br and Ir.

15. P. x cooperi (Fryer) Fryer (*P. perfoliatus* x *P. crispus*) - *Cooper's Pondweed.* Stem R
compressed; leaves all submerged, narrowly oblong to narrowly oblong-ovate, up 771
to 8.5 x 2.5cm, rounded at apex, mostly rounded and clasping stem at base but the lower narrowed to sessile base. Native; lakes, ponds, canals, streams; very scattered in Br and Ir N to C Sc.

16. P. epihydrus Raf. - *American Pondweed.* Stem compressed; floating leaves RR
opaque, narrowly elliptic-oblong, up to 8 x 2.2cm; submerged leaves linear, up to 770
24 x 1.1cm, rounded to subacute at apex, tapered to sessile base; fruit 2.5-3.1mm; 771
(2n=26). Native; 3 lakes in S Uist (Outer Hebrides), discovered 1943, also natd in canals in S Lancs and SW Yorks since 1907.

17. P. friesii Rupr. - *Flat-stalked Pondweed.* Stem compressed; leaves all 775
submerged, linear, (1.5)3.5-4mm wide, mucronate at apex, (3)5(-7)-veined; fruit 2.4-3mm; (2n=26). Native; lakes, ponds, canals; frequent in En, scattered in Wa, Sc and Ir, locally frequent but absent from many areas.

17 x 25. P. friesii x P. acutifolius = P. x pseudofriesii Dandy & G. Taylor was found in 1952 in E Norfolk; endemic. Possibly extinct.

18. P. x lintonii Fryer (*P. friesii* x *P. crispus*) - *Linton's Pondweed.* Stem compressed; R
leaves all submerged, linear or very narrowly oblong-elliptic, 1.7-5mm wide, 771
rounded at apex, 3-5-veined. Native; rivers, canals, streams; very locally frequent in En, Monts, Kirkcudbrights, 4 sites in Ir.

19. P. rutilus Wolfg. - *Shetland Pondweed.* Stem compressed; leaves all submerged, RR

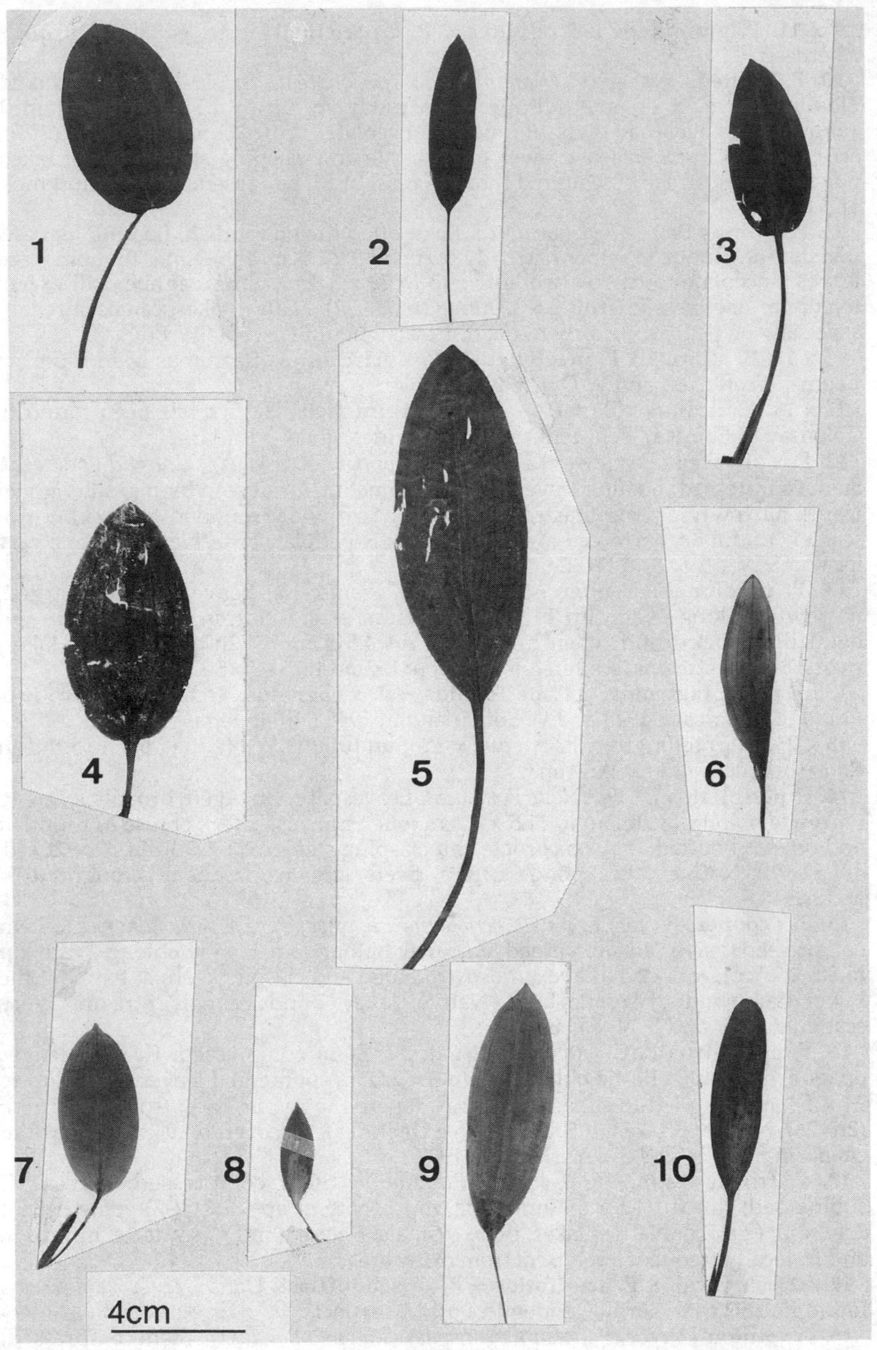

FIG 770 - Floating leaves of ***Potamogeton***. 1, ***P. natans***. 2, ***P. x sparganiifolius***. 3, ***P. polygonifolius***. 4, ***P. coloratus***. 5, ***P. nodosus***. 6, ***P. alpinus***. 7, ***P. gramineus***. 8, ***P. x nitens***. 9, ***P. x zizii***. 10, ***P. epihydrus***.

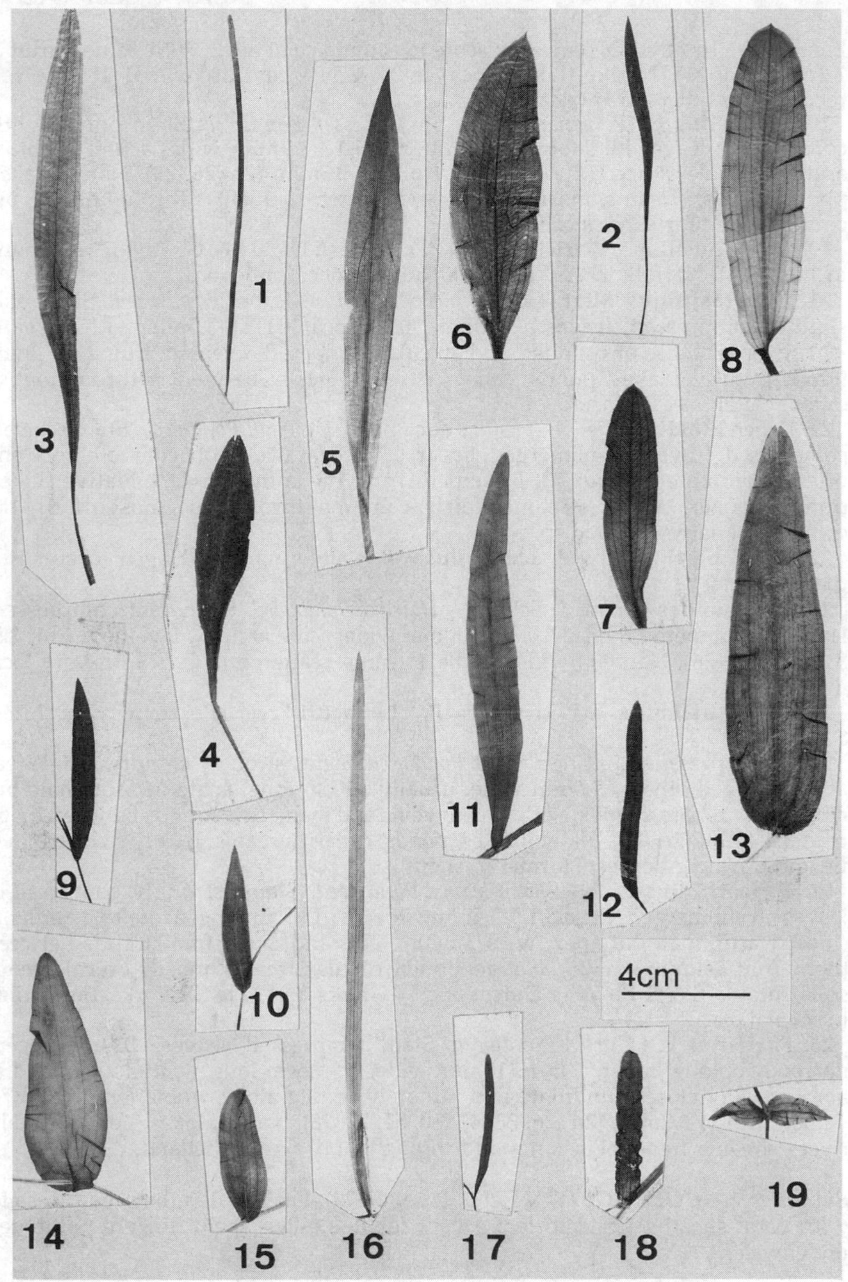

FIG 771 - Submerged leaves of **Potamogetonaceae**. 1, ***Potamogeton natans***. 2, ***P. x sparganiifolius***. 3, ***P. polygonifolius***. 4, ***P. coloratus***. 5, ***P. nodosus***. 6, ***P. lucens***. 7, ***P. x zizii***. 8, ***P. x salicifolius***. 9, ***P. gramineus***. 10, ***P. x nitens***. 11, ***P. alpinus***. 12, ***P. x olivaceus***. 13, ***P. praelongus***. 14, ***P. perfoliatus***. 15, ***P. x cooperi***. 16, ***P. epihydrus***. 17, ***P. x lintonii***. 18, ***P. crispus***. 19, ***Groenlandia densa***.

linear, 0.5-1.1mm wide, narrowly acute to acuminate at apex, 3(-5)-veined; fruit 2- **775**
2.1mm; (2n=26). Native; lakes; local in N & NW Sc, discovered 1890. Over-
recorded for other narrow-leaved spp.

20. P. pusillus L. (*P. panormitanus* Biv.) - *Lesser Pondweed*. Stem only slightly **775**
compressed; leaves all submerged, linear, 0.5-1.4(1.9)mm wide, acute or obtuse
and mucronate at apex, 3(-5)-veined; fruit 1.8-2.3mm; (2n=26, 28). Native; lakes,
ponds, canals, streams, mostly in base-rich water; frequent ± throughout BI, but
over-recorded for *P. berchtoldii*.

20 x 23. P. pusillus x P. trichoides = P. x grovesii Dandy & G. Taylor was found in 1 area in E Norfolk in 1897 and 1900, but not since; endemic.

21. P. obtusifolius Mert. & W.D.J. Koch - *Blunt-leaved Pondweed*. Stem only **775**
slightly compressed; leaves all submerged, linear, (1)2.5-3.5mm wide, usually
<10cm long, obtuse or rounded and apiculate at apex, 3-5-veined; fruit 2.6-3.2mm;
(2n=26). Native; lakes, ponds, canals, streams; locally frequent ± throughout BI
except Outer Isles, Man and CI.

22. P. berchtoldii Fieber (*P. pusillus* auct. non L.) - *Small Pondweed*. Stem scarcely **775**
compressed; leaves all submerged, linear, 0.5-2.3mm wide, subacute to obtuse and
often mucronate at apex, 3(-5)-veined; fruit 1.8-3mm; (2n=26). Native; lakes,
ponds, canals, rivers, streams; fairly common throughout most of BI, the
commonest narrow-leaved sp.

22 x 25. P. berchtoldii x P. acutifolius = P. x sudermanicus Hagstr. occurs in 1 place in Dorset.

23. P. trichoides Cham. & Schltdl. - *Hairlike Pondweed*. Stems not compressed; **R**
leaves all submerged, linear, 0.3-1(1.8)mm wide, acute at apex, 3-veined; fruit 2.5- **775**
3.2mm; (2n=26). Native; ponds, canals, streams; scattered in C & S Br, very local
in C Sc.

23 x 26. P. trichoides x P. crispus = P. x bennettii Fryer is extremely local in C Sc; endemic.

24. P. compressus L. - *Grass-wrack Pondweed*. Stems strongly compressed; leaves **R**
all submerged, linear, 3-6mm wide, usually >10cm long, acute to acuminate but **775**
mostly mucronate at apex, with 5 main veins and many fine strands between them;
fruit 3.4-4mm; (2n=26). Native; lakes, ponds, canals, streams; locally frequent in C
En and E Wa, S Aberdeen, formerly Angus.

25. P. acutifolius Link - *Sharp-leaved Pondweed*. Stems strongly compressed; **RR**
leaves all submerged, linear, 1.5-5.5mm wide, 3.5-13.5cm long, acute to acuminate **775**
(mostly acuminate) at apex, with 3 main veins and many fine strands between
them; fruit 3-4mm; (2n=26). Native; ponds, canals, streams, mostly on calcareous
soils; local in S & E En W to Dorset and W Gloucs and N to SE Yorks, but extinct
in several places.

26. P. crispus L. - *Curled Pondweed*. Stem compressed; leaves all submerged, **771**
narrowly oblong-linear, (3)5-12(18)mm wide, 2.5-9.5cm long, acute to rounded at
apex, usually closely undulate and minutely serrate along whole length, 3-5(7)-
veined; fruit 4-6.2mm; (2n=26, 36, 42, 50, 52, 72, 78). Native; lakes, ponds, canals,
rivers, streams; frequent to common ± throughout BI except Shetland.

Subgenus* 2 - *COLEOGETON Rchb. (spp. 27-29). Leaves all submerged, sessile, with basal sheaths fused to leaf except for ligule-like apex; flowers pollinated under water.

27. P. filiformis Pers. - *Slender-leaved Pondweed*. Stem not compressed; leaves all **R**
submerged, linear, 0.3-1.2mm wide, obtuse to rounded at apex, 3-veined but 2 **775**
laterals very faint and submarginal; fruit 2.2-3.2mm, with sessile stigma; (2n=66,
78). Native; lakes, rivers, streams, dykes, sometimes brackish; scattered in Sc and
N & W Ir, S Northumb, formerly Anglesey, mostly near coast.

28. P. x suecicus K. Richt. (*P. filiformis* x *P. pectinatus*) - *Swedish Pondweed*. Plants **R**
with leaf-apex of *P. pectinatus* and leaf-sheath and stigma of *P. filiformis*, or other

combinations. Native; local in MW & NW Yorks (outside range of *P. filiformis*), coasts of Sc and N & SW Ir.

29. P. pectinatus L. - *Fennel Pondweed*. Stem not compressed; leaves all 775
submerged, linear, 0.2-4mm wide, narrowly acute to subacute and mucronate at apex, 3-5-veined but laterals very faint; fruit 3.3-4.7mm, with short neck-like style; (2n=42, 78). Native; similar habitats to *P. filiformis*; frequent over most of BI.

29 x vag. P. pectinatus x P. vaginatus Turcz. = **P. x bottnicus** Hagstr. is the RR
probable parentage of plants found in 1994 in R. Tweed (Berwicks and Cheviot) close to *P. pectinatus*. *P. vaginatus*, from E Scandinavia, is the only European sp. of the genus not in BI; the hybrid might indicate former (or present) occurrence of *P. vaginatus* or might represent an independent introduction.

2. GROENLANDIA J. Gay - *Opposite-leaved Pondweed*

Leaves all opposite (or rarely some in whorls of 3), only the uppermost with 2 membranous stipules fused to edges of leaf-base; fruits with thin, papery pericarp.

1. G. densa (L.) Fourr. - *Opposite-leaved Pondweed*. Leaves all submerged, ovate to 771
lanceolate, up to 4.2 x 1.3cm, acute to obtuse at apex, clasping stem; fruit 3-4mm: (2n=30). Native; ponds, ditches, streams (often fast-flowing); locally frequent in En, scattered in S Wa and Ir, very rare in Sc.

147. RUPPIACEAE - *Tasselweed family*

Glabrous aquatic perennials rooted in substratum, with leafy submerged stems; leaves mostly alternate but the upper opposite, linear, sessile, with sheathing base, minutely denticulate near apex. Flowers (1)2-few in short subumbellate raceme on terminal peduncle, bisexual, hypogynous, inconspicuous, submerged; perianth 0; stamens 2 but each anther with 2 widely separate lobes, so appearing 4; carpels 4, free, each with 1 ovule and ± sessile peltate stigma; fruits in a group of up to 4 drupelets each becoming very long-stalked.

Distinguished from *Potamogeton* in its terminal inflorescence, lack of perianth, fruits becoming long-stalked and leaves with only a midrib.

1. RUPPIA L. - *Tasselweeds*

1. R. maritima L. - *Beaked Tasselweed*. Leaves 0.35-0.9mm wide, usually acute, 775
with slightly inflated sheath; peduncles ≤2.6cm, straight to curved or flexuous in fruit; anthers ≤1mm; drupelets 2-2.8mm, incl. beak 0.4-0.65mm, on stalks 0.3-3.5cm, asymmetrically pear-shaped; 2n=20. Native; in brackish ditches and pools; local round most coasts of Br and Ir, formerly Guernsey.

2. R. cirrhosa (Petagna) Grande (*R. spiralis* L. ex Dumort.) - *Spiral Tasselweed*. R
Leaves 0.2-1.4mm wide, usually obtuse to rounded, with strongly inflated sheath; 775
peduncles ≥4cm, usually spiral in fruit; anthers >1mm; drupelets 2.7-3.4mm, incl. beak 0.5-0.95mm, on stalks 0.4-3.2cm, pear-shaped or slightly asymmetrically so; (2n=20, 40). Native; similar places to *R. maritima*, sometimes with it but more local, frequent only in E & SE En and Shetland.

148. NAJADACEAE - *Naiad family*

Glabrous, aquatic annuals or perennials rooted in substratum, with leafy submerged stems; leaves opposite or whorled, linear, sessile, with sheathing base, minutely denticulate to conspicuously dentate or ± lobed. Flowers very inconspicuous, 1-3 sessile in each leaf-axil, monoecious or dioecious, hypogynous, submerged; male flowers each surrounded by 2 variously interpreted scales, with 1 sessile anther; female flowers without scales, with 1 carpel with 1 ovule and 2-4 elongated stigmas; fruit a sessile drupe.

Distinguished from *Potamogetonaceae* in its linear opposite (or whorled) leaves, with only a midrib, and unisexual sessile axillary flowers with 0 perianth and 1 carpel.

1. NAJAS L. - *Naiads*

Other spp. - N. graminea Delile, from the Tropics, occurred in a canal with warmed water in Manchester, S Lancs, between 1883 and 1947; it differs from *N. flexilis* in having long narrow auricles on its leaf-sheath, and most leaves densely packed on lateral branches.

1. N. flexilis (Willd.) Rostk. & W.L.E. Schmidt - *Slender Naiad*. Leaves 1-2.5(4)cm, RRR
<1mm wide incl. teeth, ± entire or minutely denticulate; fruit 2.5-3.5 x 1-1.5mm, plus long filiform style; 2n=24. Native; clean lakes; local in W Br from Westmorland to Outer Hebrides and in W Ir. Monoecious.

2. N. marina L. - *Holly-leaved Naiad*. Leaves 1-4.5cm, 1-6mm wide incl. teeth, RRR
conspicuously spinose-dentate; fruit (3)4-6(8) x 1.5-3mm, plus short thick style; (2n=12). Native; slightly brackish waterways; extremely local in NE E Norfolk. Apparently dioecious; fruits are common but male plants have not been recorded. Our plant is spp. **intermedia** (Wolfg. ex Gorski) Casper.

149. ZANNICHELLIACEAE - *Horned Pondweed family*

Glabrous, aquatic rhizomatous annuals or perennials rooted in substratum, with leafy submerged stems; leaves mostly opposite (sometimes alternate on sterile shoots), linear, sessile, with sheathing base ± free from leaf, entire. Flowers very inconspicuous, solitary in leaf-axils, monoecious, hypogynous, submerged; male flowers with 1-2 stamens, naked, long-stalked; female flowers with (2)4(-8) carpels with 1 ovule and peltate stigma on distinct style, in cup-shaped scale; fruits in a group of 1-several (often 4) achenes, each with persistent style.

Distinguished from *Potamogetonaceae* in its linear mostly opposite leaves and unisexual axillary flowers with 0 perianth; and from *Najas* in its several carpels and entire leaves.

1. ZANNICHELLIA L. - *Horned Pondweed*

1. Z. palustris L. - *Horned Pondweed*. Leaves 2-10cm x 0.4-1(2)mm, entire, acute 775
to obtuse at apex, 1(-3)-veined; fruits 3-6mm incl. style, variably stalked, variably winged and toothed on dorsal and ventral edges; 2n=36. Native; rivers, streams, ditches and ponds, fresh or brackish; frequent throughout most of BI.

150. ZOSTERACEAE - *Eelgrass family*

Glabrous, marine perennials rooted in substratum, with leafy submerged stems often exposed at low tide; leaves alternate, linear, sessile, with sheathing base with ligule at top, entire. Flowers inconspicuous, in congested compound cymes ± enclosed in leaf-sheath, monoecious, hypogynous; perianth 0; male flowers with 1 stamen; female flowers with 1 1-celled ovary with 1 ovule and 2 filiform stigmas on common style; fruit a sessile drupe.

Distinguished from other marine or brackish pondweeds in the complex, congested inflorescence ± enclosed in leaf-sheath.

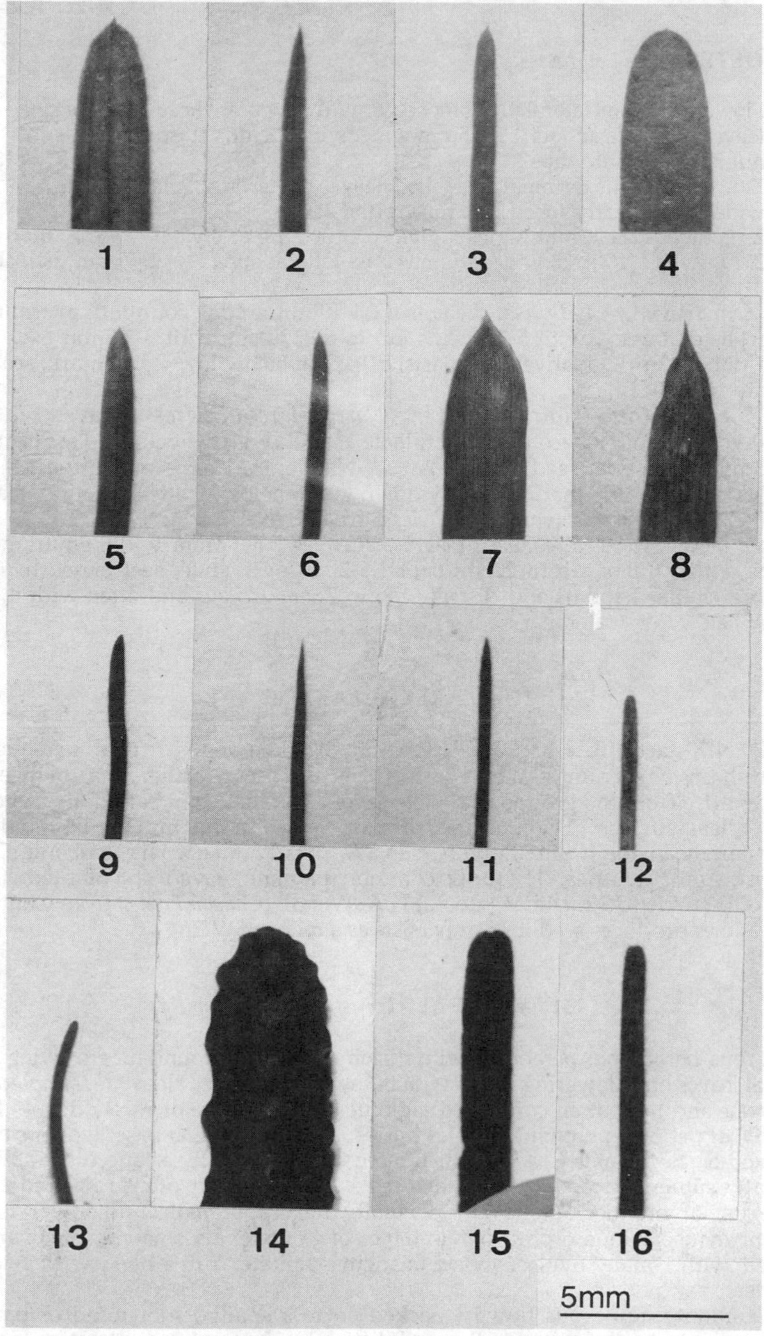

FIG 775 - Leaf-apices of *Potamogeton, Ruppia, Zannichellia, Zostera.* 1-10, ***Potamogeton***. 1, ***P. friesii***. 2, ***P. rutilus***. 3, ***P. pusillus***. 4, ***P. obtusifolius***. 5, ***P. berchtoldii***. 6, ***P. trichoides***. 7, ***P. compressus***. 8, ***P. acutifolius***. 9, ***P. filiformis***. 10, ***P. pectinatus***. 11-12, ***Ruppia***. 11, ***R. maritima***. 12, ***R. cirrhosa***. 13, ***Zannichellia palustris***. 14-16, ***Zostera***. 14, ***Z. marina***. 15, ***Z. angustifolia***. 16, ***Z. noltei***.

1. ZOSTERA L. - *Eelgrasses*

1 Flowering stems lateral, unbranched or with few branches near base; leaves of sterile shoots 0.5-1.5mm wide, with sheaths clasping stems but not fused into tube **3. Z. noltei**
1 Flowering stems terminal, branched; leaves of sterile shoots 1-10mm wide, with sheaths fused into tube round stem 2
 2 Leaves (2)4-10mm wide; stigmas c.2x as long as style **1. Z. marina**
 2 Leaves 1-2(3)mm wide; stigmas c. as long as style **2. Z. angustifolia**

1. Z. marina L. - *Eelgrass*. Leaves (2)4-10mm wide, rounded or rounded-mucronate at apex, with 5-11 veins, up to 50(120)cm; fruit 3-3.5mm excl. style; seed ribbed; 2n=12. Native; sea-coasts, c.0-4(9)m below low-water mark; scattered round coasts of BI. **R** **775**

2. Z. angustifolia (Hornem.) Rchb. - *Narrow-leaved Eelgrass*. Leaves 1-2(3)mm wide, rounded often becoming emarginate at apex, with 3-5 veins, up to 30cm; fruit 2.5-3mm excl. style; seed ribbed; 2n=12. Native; sea-coasts and estuaries, from half-tide to low-tide mark or rarely down to 4m below; scattered round coasts of BI. Perhaps only a var. (var. *angustifolia* Hornem.) of *Z. marina*. **R** **775**

3. Z. noltei Hornem. - *Dwarf Eelgrass*. Leaves 0.5-1.5mm wide, emarginate at apex, with 3 veins, up to 22cm; fruit 1.5-2mm excl. style; seed smooth; 2n=12. Native; similar habitats and distribution to *Z. angustifolia* and often with it, but ± never below low-water mark. **R** **775**

150A. ARECACEAE (Palmae)

PHOENIX dactylifera L. (*Date Palm*), from SW Asia and N Africa, is frequent on rubbish-tips as seedlings killed by first frosts; the 1-few, leathery, narrowly elliptic leaves arise from the ground and have several parallel veins raised alternately on either leaf surface. The familiar pinnate-leaved palm much planted in the Mediterranean and in small numbers in SW En is **P. canariensis** Chabaud (*Canary Palm*), from Canaries. The only common palmate-leaved sp. planted here is **TRACHYCARPUS fortunei** (Hook.) H.A. Wendl. (*Chusan Palm*), from China; this sometimes produces seedlings or persists as a relic in SW En.

151. ARACEAE - *Lords-and-Ladies family*

Glabrous, herbaceous perennials with rhizomes or underground tubers giving rise to aerial leaves and flowering stems with 0-few leaves; leaves alternate, simple, linear to ovate and then often cordate to sagittate at base, entire or rarely deeply lobed, sessile or petiolate, without stipules but usually sheathing at base. Flowers closely packed on sterile axis (*spadix*) which often extends distally as succulent *appendix*, usually subtended or partially enclosed by leaf-like but often coloured *spathe*, bisexual or monoecious, if latter usually the upper male and lower female, hypogynous, actinomorphic; perianth 0 or of 4 or 6 tepals; stamens 1-6; ovary 1-3-celled, with 1-many ovules; style ± 0; stigma capitate; fruit a berry with 1-several seeds.

The minute numerous flowers packed on to a spadix, subtended or partially enclosed by a spathe except in the very distinctive *Acorus*, are diagnostic.

1 Leaves linear, *Iris*-like; spadix apparently lateral and lacking spathe **1. ACORUS**
1 Leaves lanceolate to ovate, narrowed at base and usually petiolate; spadix terminal, with spathe 2

2 Flowers covering spadix to its apex 3
2 Spadix with succulent sterile region (appendix) distal to flowers 5
3 Spathe ± flat, not enclosing spadix even at extreme base **3. CALLA**
3 Spathe wrapped round basal part of spadix 4
4 Leaves truncate to cuneate at base, with petioles shorter than blade; tepals 4 **2. LYSICHITON**
4 Leaves cordate at base, with petiole longer than blade; tepals 0 **4. ZANTEDESCHIA**
5 Leaves palmately divided **6. DRACUNCULUS**
5 Leaves simple 6
6 Spathe fused into tube proximally, with distal filiform projection ≥5cm **7. ARISARUM**
6 Spathe overlapping at base, not fused into tube, no more than acuminate at apex **5. ARUM**

Other genera - COLOCASIA Schott, from tropical Asia, has large, entire, peltate long-petiolate leaves; **C. antiquorum** (L.) Schott (*Taro, Dasheen* or *Coco-yam*) is a rare ephemeral casual of rubbish-tips.

1. ACORUS L. - *Sweet-flags*
Rhizomatous; leaves linear, *Iris*-like, entire, sessile; spadix apparently lateral, without appendix; spathe apparently 0; flowers bisexual; tepals 6; stamens 6; ovary 2-3-celled: fruit not forming. Fresh leaves have strong spicy scent when bruised.

1. A. calamus L. - *Sweet-flag.* Leaves 50-125 x 0.7-2.5cm, with well-defined midrib, usually transversely wrinkled in places; spadix 5-9 x 0.6-1.2cm, yellowish-green; 2n=c.36 (18, 24, 36, 42, 44, 45). Intrd; in shallow water at edges of lakes, ponds, rivers and canals; scattered over most of BI, but frequent only in En; Asia and N America. Often not flowering.

2. A. gramineus Aiton - *Slender Sweet-flag.* Leaves 8-50 x 0.2-0.8cm, without obvious midrib; spadix 5-10 x 0.3-0.5cm; (2n=18, 22, 24). Intrd; natd by lake in Surrey since 1986; E Asia.

2. LYSICHITON Schott - *Skunk-cabbages*
Rhizomatous; leaves ovate-oblong, entire, truncate to cuneate at base, shortly petiolate; spadix terminal, without appendix; spathe wrapped round and concealing spadix at base, falling off after flowering; flowers bisexual; tepals 4; stamens 4; ovary (1-)2-celled; fruit a green berry with 2 seeds.

1. L. americanus Hultén & H. St. John - *American Skunk-cabbage.* Leaf-blades 30-150 x 25-70cm; flowers foul-smelling; spathe 10-35cm, yellow; spadix 3.5-12cm, greenish; tepals 3-4mm; anthers 0.9-2mm; (2n=28). Intrd; grown for ornament, persistent and spreading in swampy ground; scattered throughout Ir and S & W Br (but see next sp.); W N America.

2. L. camtschatcensis (L.) Schott - *Asian Skunk-cabbage.* Differs from *L. americanus* in ± scentless flowers; slightly smaller spathe and spadix; white spathe; tepals 2-3mm; and anthers 0.6-0.8mm; (2n=28). Intrd; similar places to *L. americanus*; distribution uncertain due to confusion with latter but probably rarer; extreme E Asia. Hybrids occur and might be natd; perhaps the 2 are better as sspp.

3. CALLA L. - *Bog Arum*
Rhizomatous; leaves ovate to broadly so, cordate, entire, with long petiole; spadix terminal, without appendix; spathe open, ± flat, not concealing spadix; flowers mostly bisexual but uppermost usually male; tepals 0; stamens 6; ovary 1-celled; fruit a red berry with several seeds.

1. C. palustris L. - *Bog Arum*. Leaf-blades 5-12 x 4-10cm; spathe 3-8 x 3-6cm, white or greenish-white; spadix 1-3 x 0.7-2cm; (2n=36, 72). Intrd; grown for ornament, persistent and spreading in marshy ground and shallow ponds, often in shade; very scattered in Br from SE En to C Sc; Europe.

4. ZANTEDESCHIA Spreng. - *Altar-lily*
Rhizomes short, tuberous; leaves ovate or broadly so, cordate, entire, with long petiole; spadix terminal, without appendix; spathe wrapped round and concealing spadix at base; flowers unisexual; tepals 0; stamens 2-3; ovary (1-)3-celled; fruit a yellow berry with several seeds; but very seldom (?never) produced.

1. Z. aethiopica (L.) Spreng. - *Altar-lily*. Leaf-blades 10-45 x 10-25cm; petiole up to 50(75)cm; spathe 10-25cm, pure white; spadix up to 15cm, bright yellow; (2n=24, 32). Intrd; grown for ornament, persistent and spreading in ditches, damp hedgerows and scrub, and neglected fields; CI, SW En, S Wa and W Ir; S Africa.

5. ARUM L. - *Lords-and-Ladies*
Rhizomes short, tuberous; leaves triangular-ovate, hastate to sagittate, entire, with long petiole; spadix terminal, with long appendix; spathe wrapped round and concealing spadix at base, pale greenish-yellow; flowers unisexual; tepals 0; stamens 3-4; ovary 1-celled; fruit a red berry with 1-several seeds.

1. A. maculatum L. - *Lords-and-Ladies*. Leaf-blades appearing in early spring, 7-20cm, often blackish-purple-spotted, with concolorous midrib; spadix appendage purple or yellow, usually reaching c.1/2 way up expanded part of spathe: spathe 10-25cm, sometimes with dark spots; fruiting spike 3-5cm; 2n=56, 84. Native; woods and hedgerows, usually on base-rich soils; frequent throughout BI N to C Sc, scattered and natd further N.

1 x 2. A. maculatum x A. italicum occurs rarely in S & SW En, S Wa and CI; it is intermediate and probably sterile, but has leaves appearing in early winter and often spotted; 2n=70.

2. A. italicum Mill. - *Italian Lords-and-Ladies*. Leaf-blades appearing in early winter, 15-35cm, with pale midrib; spadix appendage yellow, usually reaching c.1/3 way up expanded part of spathe; spathe 15-40cm, never spotted; fruiting spike 10-15cm.

a. Ssp. neglectum (F. Towns.) Prime. Leaves sometimes dark-spotted, with veins R
slightly paler than rest of leaf, with basal lobes somewhat convergent and sometimes overlapping; fruits with 1-2 seeds; 2n=84. Native; hedgerows, scrub and stony field-borders; extreme S & SW En and CI, formerly S Wa, very rarely natd elsewhere, common in CI and Scillies and the only sp. in latter place.

b. Ssp. italicum. Leaves never dark-spotted, with whitish veins, with basal lobes divergent; fruits with 2-4 seeds; (2n=84). Intrd; persistent garden throwout natd in similar places to ssp. *neglectum*; scattered in CI, Br N to Man, Dunbarton and S Lincs, E Ir, frequent and perhaps native in CI.

6. DRACUNCULUS Mill. - *Dragon Arum*
Rhizomes short, tuberous; leaves deeply ± palmately lobed, cordate at base, with entire lobes and long petiole; spadix terminal, with long appendix; spathe wrapped round and concealing spadix at base; flowers unisexual; tepals 0; stamens 2-4; ovary 1-celled; fruit a red berry with several seeds.

1. D. vulgaris Schott - *Dragon Arum*. Leaf-lobes up to 20cm; spadix appendage dark purple, nearly as long as spathe; spathe 25-40cm, dark purple; (2n=28, 32). Intrd; garden throwout natd in hedges, rough ground and old gardens; scattered in S & SE En and CI; E & C Mediterranean.

7. ARISARUM Mill. - *Mousetailplant*
Rhizomatous; leaves triangular-ovate, sagittate, entire, with long petiole; spadix terminal, with long appendix; spathe fused in tube round spadix and concealing most of it, extended apically into filiform projection; flowers unisexual; tepals 0; stamen 1; ovary 1-celled; fruit green, with several seeds.

1. A. proboscideum (L.) Savi - *Mousetailplant*. Leaf-blades 6-15cm; spadix appendage whitish, concealed within spathe; spathe 2-4cm excl. filiform projection 5-15cm, dark or greenish-brown; (2n=18, 28). Intrd; garden throwout natd in hedges, rough ground and old gardens; very scattered in S En; Spain and Italy.

152. LEMNACEAE - *Duckweed family*

Aquatic perennial plants reduced to ± undifferentiated pad-like *frond* to 15mm (but often much less) floating on or under water surface (sometimes stranded on mud), not or variously adhering together, with 0-21 roots per frond. Flowers rather rarely produced, very reduced, borne in (1-)2 hollows on frond, each hollow with 1-2 stamens and 1 ovary with 1-2 ovules and funnel-shaped stigma (variously interpreted as 1 flower or 1-2 male and 1 female flower) subtended or not by minute spathe; perianth 0; fruit of 1-2 seeds in thin sac.

The floating pad-like plants are unique.

Only well-grown spring or summer fronds should be used; poorly grown ones or those produced in autumn and over-wintering are often atypical, being smaller and often with fewer veins and fewer or 0 roots.

1 Fronds rootless and veinless, spherical to ellipsoid **3. WOLFFIA**
1 Each frond with (0)1-16(21) roots and 1-16(21) veins, ± flattened at least on upperside 2
 2 Each frond with (0-)1 root and 1-5(7) veins **2. LEMNA**
 2 Each frond with 7-16(21) roots and veins **1. SPIRODELA**

1. SPIRODELA Schleid. - *Greater Duckweed*
Fronds with 7-16(21) roots, with 7-16(21) veins, floating on water surface.

1. S. polyrhiza (L.) Schleid. (*Lemna polyrhiza* L.) - *Greater Duckweed*. Fronds 1.5-10 x 1.5-8mm, ± flattened on both surfaces; 2n=40. Native; canals, ditches and ponds; rather local in C & S Br, very scattered in N Br and Ir, formerly CI.

2. LEMNA L. - *Duckweeds*
Fronds with (0-)1 root, with 1-5(7) veins, floating on or below water surface.

1 Fronds narrowed to a stalk-like portion at 1 end, usually submerged, usually cohering in branched chains of 3-50 **3. L. trisulca**
1 Fronds orbicular to ellipsoid, without stalk-like portion, usually on water surface, cohering in small groups (not chains) 2
 2 Fronds usually strongly swollen on lowerside, ± hemispherical, with (3)4-5(7) veins originating from 1 point **1. L. gibba**
 2 Fronds ± flattened on both surfaces, with 1-3(5) veins, if 4 or 5 then 3 originating from 1 point and the outermost 1 or 2 extras branching from near base of inner laterals 3
3 Fronds 0.8-3(4)mm long, usually elliptic, with 1 vein **4. L. minuta**
3 Fronds (1)2-5(8)mm long, usually ovate, with 3(-5) veins **2. L. minor**

1. L. gibba L. - *Fat Duckweed*. Fronds 1-8 x 0.8-6mm, usually strongly swollen on lowerside (not in autumn-produced fronds, nor in starved plants), with (3)4-5(7) veins, with larger air-spaces (visible as reticulum on frond upperside) >0.3mm across; 2n=40. Native; ponds, ditches and canals, usually in rich, often brackish water; frequent in C & S Br, very scattered in CI and Ir.

2. L. minor L. - *Common Duckweed*. Fronds 1-8 x 0.6-5mm, ± flattened on both surfaces, with 3(-5) veins, with larger air-spaces ≤0.3mm across; 2n=40. Native; ponds, ditches, canals and slow parts of rivers and streams; common throughout BI except rare in N Sc.

3. L. trisulca L. - *Ivy-leaved Duckweed*. Fronds 3-15 (plus stalk 2-20) x 1-5mm, ± flattened on both surfaces, with (1-)3 veins, with 0-1 root; 2n=40, 44. Native; ponds, ditches and canals; frequent in most of BI, but very scattered in Sc.

4. L. minuta Kunth (*L. minuscula* Herter nom. illeg.) - *Least Duckweed*. Fronds 0.8-4 x 0.5-2.5mm, ± flattened on both surfaces, with 1 vein, with larger air-spaces ≤0.25mm across; 2n=40. Intrd; same habitats as *L. minor*; scattered in CI and Br N to E Norfolk and Flints, 1st recorded 1977, probably overlooked; N & S America.

3. WOLFFIA Horkel ex Schleid. - *Rootless Duckweed*
Fronds with 0 roots, with 0 veins, usually floating on water surface.

1. W. arrhiza (L.) Horkel ex Wimm. - *Rootless Duckweed*. Fronds 0.5-1.5 x 0.4-1.2mm, strongly swollen on both sides (thicker than wide); (2n=30, 40, 42, 44-46, 50, 60, 62, 63, 70, 80). Native; ponds and ditches; very local in S En and Mons, formerly Glam. **R**

153. COMMELINACEAE - *Spiderwort family*

Herbaceous perennials; leaves alternate, entire, with sheathing base, glabrous or pubescent just on sheath, sessile or nearly so, without stipules. Flowers few in terminal paired cymes with large leaf-like bract at base of each, bisexual, hypogynous, actinomorphic, showy; sepals 3, green, free; petals 3, white or coloured, free; stamens 6, with pubescent filaments; ovary 3-celled with 2 ovules per cell; style 1; stigma capitate; fruit a capsule (?never formed).

Easily recognized by the flowers with 3 green sepals, 3 white or coloured petals, and 6 stamens with conspicuous long hairs on filaments.

1. TRADESCANTIA L. (*Zebrina* Schnizl.) - *Spiderworts*

Other spp. - T. zebrina Bosse (*Zebrina pendula* Schnizl.), from Mexico, resembles *T. fluminensis* but has silver-striped leaves and pinkish-purple petals; it occasionally persists for a short while where thrown out.

1. T. virginiana L. - *Spiderwort*. Tufted, shortly rhizomatous perennial; stems erect, to 60cm; leaves linear, 15-35cm; flowers 2.5-3.5cm across; petals usually violet, sometimes white, pink or purple; (2n=12, 24). Intrd; grown in gardens, occasionally persisting on tips and waste ground where thrown out; rare in SE En; N America. The commonest garden plant is probably a hybrid of *T. virginiana* with 1 or more other spp.; it has no valid name (*T. x andersoniana* W. Ludwig & Rohw. nom. inval.).

2. T. fluminensis Vell. - *Wandering-jew*. Stems trailing, rooting at nodes, to 1m or more; leaves ovate-oblong, 1.5-5cm; flowers 1-1.5cm across, rarely produced; petals white; (2n=60). Intrd; much grown as pot-plant, sometimes persisting in shrubberies and frost-free rough ground where thrown out; rare in S En and CI; S America.

154. ERIOCAULACEAE - *Pipewort family*

Aquatic (or ± so) herbaceous perennials rooted in substratum; leaves appearing all basal, simple, subulate, entire, sessile, without stipules. Flowers in terminal, whitish, capitate mass on long leafless stem from basal rosette, unisexual with male flowers in centre of inflorescence and female around them, hypogynous, actinomorphic, with bracteoles below each; tepals 4, 2 outer fused (male) or ± free (female), 2 inner free, membranous and pubescent or fringed; stamens 4; ovary 2-celled, each cell with 1 ovule; style 1; stigmas 2, filiform; fruit a capsule with 2 seeds.

Unique in the subulate basal leaf-rosette and erect leafless stems bearing small capitate inflorescences of many unisexual flowers.

1. ERIOCAULON L. - *Pipewort*

1. E. aquaticum (Hill) Druce (*E. septangulare* With.) - *Pipewort*. Leaves clearly transversely septate, very finely pointed, up to 10cm; stems erect, usually emergent and varying in height according to water level, up to 20(150)cm; inflorescence 5-12(20)mm across; 2n=64. Native; in shallow lakes and pools or in bare wet peaty ground; extremely local in W Sc, local in W Ir. R

155. JUNCACEAE - *Rush family*

Annuals or herbaceous perennials, often ± aquatic; leaves alternate or all basal, grass-like (*bifacial*) to rush-like (cylindrical to flattened but *unifacial*), with sheathing base with membranous ligule at top of sheath, the blade simple, linear and entire or 0, without stipules. Flowers in various simple to complex, often congested cymes that are terminal but often appear lateral, bisexual, hypogynous, actinomorphic; tepals 6 (3 inner, 3 outer), greenish, brownish or membranous, free; stamens (3-)6; ovary 1-celled with 3 ovules, 1-celled with many ovules on parietal placentas, or 3-celled with many ovules per cell on axile placentas; style 0 or 1; stigmas 3, linear; fruit a capsule with 3 or numerous seeds.

Distinguished from other rush-, sedge- or grass-like plants by the flowers with 6 tepals, (3-)6 stamens and a single 1-3-celled ovary with 3-many ovules.

1 Leaves bifacial to unifacial, glabrous; ovary with many ovules; capsule with many seeds **1. JUNCUS**

1 Leaves bifacial, usually pubescent at least near base when young; ovary with 3 ovules; capsule with 3 seeds **2. LUZULA**

1. JUNCUS L. - *Rushes*

Annuals to perennials; leaves various, bifacial to unifacial, glabrous; ovary 1-3-celled, with many ovules; capsule with many seeds.

'Leaves' refers to stem-leaves and/or basal leaves, but excludes leaf-like bracts immediately below or within the inflorescence. In subg. *Genuini* the lowest inflorescence bract is cylindrical and stem-like, making the inflorescence appear lateral. A sharp scalpel or razor-blade is needed to cut longitudinal and transverse sections of stem and leaves to see the internal structure; dried material usually needs resuscitation by boiling in water. Seed length includes terminal appendages.

General key

1 Leaves distinctly bifacial, flat and ± grass-like with 2 opposite surfaces but sometimes inrolled, or subcylindrical and ± rush-like but with a distinct deep channel on upperside for most or all of length ***Key A***

1 Leaves unifacial or ± so, or apparently absent, cylindrical to flattened-cylindrical and rush-like, not deeply channelled or with deep channel

only near ligule and extending <1/2 way to leaf apex, sometimes with shallow grooves *Key B*

Key A - Leaves bifacial, flat, or subcylindrical but with a deep channel on upperside

1 Easily uprooted annual, with simple fibrous root-system 2

1 Rhizomatous perennial, usually firmly rooted (rhizomes often very short and plant densely tufted) 5

2 Stems unbranched, with basal leaves and leaf-like bracts at top but bare between **10. J. capitatus**

2 Stems branched, leafy (but leaves often short and very narrow) 3

3 Leaves usually >1.5mm wide; tepals usually with dark line either side of midrib; anthers 1.2-5x as long as filaments; seeds with longitudinal ridges (10-15 in side view) clearly visible with x20 lens **6. J. foliosus**

3 Leaves rarely >1.5mm wide; tepals rarely with dark line on either side of midrib; anthers usually 0.3-1.1x as long as filaments; seeds without longitudinal ridges visible with x20 lens (beware shrivelled seeds) 4

4 Inner tepals rounded to emarginate and mucronate at apex; capsule truncate at apex, at least as long as inner tepals **8. J. ambiguus**

4 Inner tepals acute to subacute at apex; capsule acute to obtuse at apex, rarely ± truncate, usually shorter than inner tepals **7. J. bufonius**

5 Outer 3 tepals obtuse to rounded at apex 6

5 Outer 3 tepals acute to acuminate at apex 7

6 Anthers 0.5-1mm, 1-2x as long as filaments; style 0.1-0.3mm, <1/2 as long as stigmas; seeds 0.35-0.5mm **3. J. compressus**

6 Anthers 1-2mm, 2-3x as long as filaments; style 0.5-0.8mm, c. as long as stigmas; seeds 0.5-0.7mm **4. J. gerardii**

7 Lowest 2 bracts of inflorescence leaf-like, usually far exceeding inflorescence; stem with (1)2-4(5) well-developed leaves (usually near base) 8

7 All bracts of inflorescence mostly scarious, usually much shorter than inflorescence; stem with 0(-1) well-developed leaves 9

8 Inflorescence of 1-3(4) flowers in tight cluster; anthers longer than filaments; seeds 0.9-1.6mm, with long appendage at each end **5. J. trifidus**

8 Inflorescence of 5-40 flowers, usually ± diffuse; anthers shorter than filaments; seeds 0.3-0.4mm, with short appendages **2. J. tenuis**

9 Leaves flat or ± inrolled; stamens 3 **9. J. planifolius**

9 Leaves rounded on lowerside, with deep channel on upperside; stamens 6 **1. J. squarrosus**

Key B - Leaves unifacial or ± so, cylindrical to flattened-cylindrical

1 Leaves on stems represented only by blade-less scarious sheaths near base 2

1 At least 1 stem-leaf with well-developed green blade 7

2 Stems strongly glaucous, with pith conspicuously and regularly interrupted at least in region just below inflorescence **25. J. inflexus**

2 Stems not glaucous, with pith well formed and conspicuous or ill-formed and irregular 3

3 Rhizomes extended, forming straight lines or diffuse patches of aerial stems; inflorescence usually <20-flowered; stems rarely >50cm 4

3 Rhizomes short, forming dense clumps of aerial stems (or large dense patches in very old plants); inflorescence usually >20-flowered; stems usually >50cm 5

4 Inflorescence in lower 2/3(-3/4) of apparent stem, ± globose; stem with subepidermal sclerenchyma girders, with fine longitudinal ridges

when dry **24. J. filiformis**

4 Inflorescence in upper 1/4 of apparent stem, usually elongated; stem without subepidermal sclerenchyma girders, not ridged when dry **23. J. balticus**

5 Fresh stems dull, ridged, with usually <35 ridges; main (stem-like) bract opened out and ± flat at base adjacent to inflorescence, causing it to hinge over backwards at end of season **27. J. conglomeratus**

5 Fresh stems glossy, smooth, becoming finely ridged with usually >35 ridges when dry; main bract scarcely opened out at base adjacent to inflorescence, not hingeing over backwards at end of season 6

6 Capsule as long as or longer than tepals, 2.5-3.5mm when fully fertile; stamens 6; larger stems usually >1.2m **28. J. pallidus**

6 Capsule shorter than tepals, 2-2.5mm when fully fertile; stamens 3(-6); stems usually <1.2m **26. J. effusus**

7 Leaves and main bract with very sharp apex, with subepidermal sclerenchyma girders, with vascular bundles scattered through pith 8

7 Leaves and main bract with soft apex, without subepidermal sclerenchyma girders, without vascular bundles in pith 9

8 Capsule 2.5-3.5mm, not or slightly longer than tepals; inner tepals obtuse to subacute, the extreme apex not exceeded by membranous margins **20. J. maritimus**

8 Capsule 4-6mm, much longer than tepals; inner tepals retuse, with membranous margins extended into lobes on each side of extreme apex and exceeding it **21. J. acutus**

9 Leaves cylindrical, with continuous pith within vascular cylinder; each flower with 2 small bracteoles immediately beneath tepals **22. J. subulatus**

9 Leaves cylindrical to flattened-cylindrical, with pith usually interrupted by transverse septa of stronger tissue and often with large cavities; flowers without bracteoles at base 10

10 Easily uprooted annual, with simple fibrous root system (W Cornwall) **16. J. pygmaeus**

10 Perennial, usually firmly rooted, either rhizomatous or with ± swollen stem-bases (widespread) 11

11 Anthers <1/3 as long as filaments; seeds with conspicuous whitish appendages at each end each c. as long as actual seed (alpine) 12

11 Anthers >1/3 (up to 2x) as long as filaments; seeds with at most minute points at each end (lowland or alpine) 14

12 Outer tepals acute; capsule ≥6mm; stems usually solitary from rhizome system; anthers >1mm **19. J. castaneus**

12 Outer tepals obtuse to rounded; capsule <6mm; stems usually in small tufts; anthers <1mm 13

13 Capsule 3.2-4.5mm (excl. style), retuse at apex; lowest bract usually exceeding inflorescence; flowers mostly 2 per inflorescence **17. J. biglumis**

13 Capsule 4-5.5mm (excl. style), obtuse at apex; lowest bract usually shorter than inflorescence; flowers mostly 3 per inflorescence **18. J. triglumis**

14 Leaves with >2 empty or loosely pith-filled longitudinal cavities separated by thin walls bearing a few vascular bundles 15

14 Leaves with 1 empty or loosely pith-filled longitudinal cavity 16

15 Rhizomatous; outer tepals obtuse, incurved at apex; flowers very rarely vegetatively proliferating **11. J. subnodulosus**

15 Not rhizomatous; stem-base usually swollen; outer tepals acute, not incurved; flowers commonly vegetatively proliferating **15. J. bulbosus**

16 Tepals acuminate, the outer with recurved apical points **14. J. acutiflorus**

16 Outer tepals obtuse to shortly acuminate with erect apical points; inner tepals acute to rounded 17

17 Outer tepals subacute to obtuse; inner tepals obtuse to rounded; capsule obtuse (ignore beak) **12. J. alpinoarticulatus**
17 Outer tepals acute; inner tepals acute to subacute; capsule acute (ignore beak) **13. J. articulatus**

Other spp. - Several other spp. of subg. *Genuini* occur as wool-aliens and have sometimes persisted in the past, but none so much as *J. pallidus*; their determination is often extremely difficult. **J. australis** Hook. f., from New Zealand, was persistent in a wool-manured field in W Kent in 1970s and appeared in a setaside field in Worcs in 1994; it resembles *J. inflexus* in its ± glaucous ridged stems with very sparse pith (with a few partial diaphragms only), but differs in its pale brown tepals and capsules and inflorescence consisting of dense sessile and stalked clusters of flowers. It might become more common.

Subgenus* 1 - *PSEUDOTENAGEIA V.I. Krecz. & Gontsch. (spp. 1-5). Rhizomatous (but sometimes densely tufted) perennials; leaves flat or strongly channelled, basal and along stem, not sharply pointed, with 1 subepidermal sclerenchyma girder at each margin (0 in *J. trifidus*), with cavities developing between vascular bundles but without pith; inflorescence terminal, very or rather compact, exceeded or not by 1-2 leaf-like or scale-like bracts; each flower with 2 small bracteoles; seeds with or without terminal appendages.

1. J. squarrosus L. - *Heath Rush.* Densely tufted; stems erect, to 50cm; leaves **786**
nearly all basal, rounded on lowerside, deeply channelled on upperside; lowest bract much shorter than inflorescence; 2n=40. Native; bogs, wet moors and heaths, on acid soil; common throughout Br and Ir where acid soils exist, absent from much of C Ir and C En.

2. J. tenuis Willd. (*J. dudleyi* Wiegand) - *Slender Rush.* Densely tufted; stems erect, **786**
to 80cm; leaves nearly all basal, flattened; lowest 1(-2) bracts usually much longer **792**
than inflorescence; (2n=32, 40, 80, 84). Intrd; damp barish ground on roadsides, tracks and paths; locally frequent throughout most BI, 1st recorded 1795/6; N & S America. Var. **dudleyi** (Wiegand) F.J. Herm. differs in its leaf-sheaths ending in brown auricles much wider than long (not scarious and much longer than wide); it is natd in M Perth.

3. J. compressus Jacq. - *Round-fruited Rush.* Loosely tufted to extensive patches; stems erect, to 50cm; leaves mostly basal, flattened, often ± inrolled; lowest bract usually exceeding inflorescence; (2n=40, 44). Native; marshes and wet meadows, often near sea; scattered in En and Wa, mostly C & E En. See *J. gerardii* for differences.

4. J. gerardii Loisel. - *Saltmarsh Rush.* Differs from *J. compressus* constantly only in key characters (see Key A, couplet 6); usually more extensively rhizomatous; lowest bract usually shorter than inflorescence; tepals usually dark (not light) brown; capsule usually subacute to obtuse (not obtuse to rounded) and scarcely (not greatly) exceeding tepals; 2n=84. Native; salt-marshes and inland saline areas; abundant round coasts of BI, very scattered inland. Stamens persist well after seed dispersal, flattened between tepals and capsule, so provide the best discriminator.

5. J. trifidus L. - *Three-leaved Rush.* Densely tufted; stems erect, to 30cm; leaves **786**
few, usually 0-1 at base and 1-2 plus leaf-like lowest bract near apex of stem, **792**
flattened and inrolled; lowest bract and 1-2 upper stem-leaves exceeding inflorescence; (2n=30). Native; barish places on mountains; locally frequent in C & W Sc, Shetland.

Subgenus* 2 - *POIOPHYLLI Buchenau (spp. 6-8). Annuals; leaves flat or inrolled, all on stem, those at base often withered by flowering time, not sharply pointed, with 1 subepidermal sclerenhyma girder at each margin, with cavities developing between vascular bundles but without pith; inflorescence terminal, usually very diffuse and occupying most of plant, interspersed with ± leaf-like bracts; each flower with 2 small bracteoles; seeds without appendages.

6. J. foliosus Desf. - *Leafy Rush*. Differs from *J. bufonius* in stems erect to 792
ascending; flowers well spaced; tepals usually with dark line on each side of midrib; and see Key A (couplet 3); 2n=26. Native; muddy margins of areas of fresh water, wet fields, marshes and ditches; scattered through W & S BI.

7. J. bufonius L. (*J. minutulus* V.I. Krecz. & Gontsch.) - *Toad Rush*. Stems erect to 786
procumbent, to 35(50)cm, variable in branching but often very diffuse; inflorescence 792
variable, with flowers well spaced to tightly bunched at branchlet ends; tepals rarely with dark lines; inner tepals acute to subacute, c. as long as to longer than acute to subacute or rarely truncate capsule; 2n=108. Native; all kinds of damp habitats, fresh-water and brackish, natural and artificial; common throughout BI.

8. J. ambiguus Guss. (*J. ranarius* Nees ex Songeon & E.P. Perrier) - *Frog Rush*. 792
Differs from *J. bufonius* in stems to 17cm; flowers 2-4(5) on each ultimate branch, with usually 2-3 bunched together at tip; inner tepals rounded to emarginate and mucronate at apex, shorter than to as long as truncate capsule; 2n=34. Native; damp brackish habitats near coast and inland, and on damp lime-waste; scattered in suitable habitats throughout BI.

Subgenus* 3 - *GRAMINIFOLII Buchenau (sp. 9). Shortly rhizomatous tufted perennials; leaves flat, basal, not sharply pointed, without subepidermal sclerenchyma girders, with cavities developing between vascular bundles but without pith; inflorescence terminal, rather compact, with short reduced main bract; flowers without bracteoles; seeds without appendages.

9. J. planifolius R. Br. - *Broad-leaved Rush*. Stems erect, to 30cm; leaves light green, 786
2-8mm wide; inflorescence with shape of that of *Luzula multiflora* ssp. *multiflora*. 792
Intrd; damp pathsides, lake shores and wet meadows; over c.40 square km in W Galway, discovered 1971; Australia, New Zealand, S America.

Subgenus* 4 - *JUNCINELLA V.I. Krecz. & Gontsch. (sp. 10). Annuals; leaves flat to strongly channelled, all basal, not sharply pointed, without subepidermal sclerenchyma girders, without cavities or pith region; inflorescence terminal, very compact, exceeded or not by 1-2 main bracts; flowers without bracteoles; seeds ± without appendages.

10. J. capitatus Weigel - *Dwarf Rush*. Stems erect, to 5cm and often much less; RR
leaves <1mm wide; (2n=18). Native; barish ground on heaths, usually where water 786
stands in winter; extremely local in W Cornwall and CI, formerly Anglesey. 792

Subgenus* 5 - *SEPTATI Buchenau (spp. 11-16). Annuals or rhizomatous or non-rhizomatous perennials; leaves terete or flattened-terete, all on stems or some basal, not sharply pointed, with 1-several central cavities, divided by transverse septa (often visible externally), without subepidermal sclerenchyma girders; inflorescence terminal, very compact to very diffuse, with ± leaf-like main bract usually shorter than it; flowers without bracteoles; seeds without appendages.

11. J. subnodulosus Schrank - *Blunt-flowered Rush*. Rhizomatous; stems erect, to 1.2m; leaves with very distinct transverse septa; inflorescence diffuse, with very widely spreading to ± reflexed branchlets; tepals 1.8-2.3mm, pale brown, with obtuse, incurved tips; 2n=40. Native; fens, marshes and dune-slacks on peaty base-rich soil; locally frequent in En, Wa and Ir, very local in Jersey, SW & CE Sc.

12. J. alpinoarticulatus Chaix (*J. alpinus* Vill. nom. illeg., *J. nodulosus* auct. non R
Wahlenb.) - *Alpine Rush*. Rhizomatous; stems erect to ascending, to 40cm; leaves with very distinct transverse septa; inflorescence diffuse but often rather sparse,

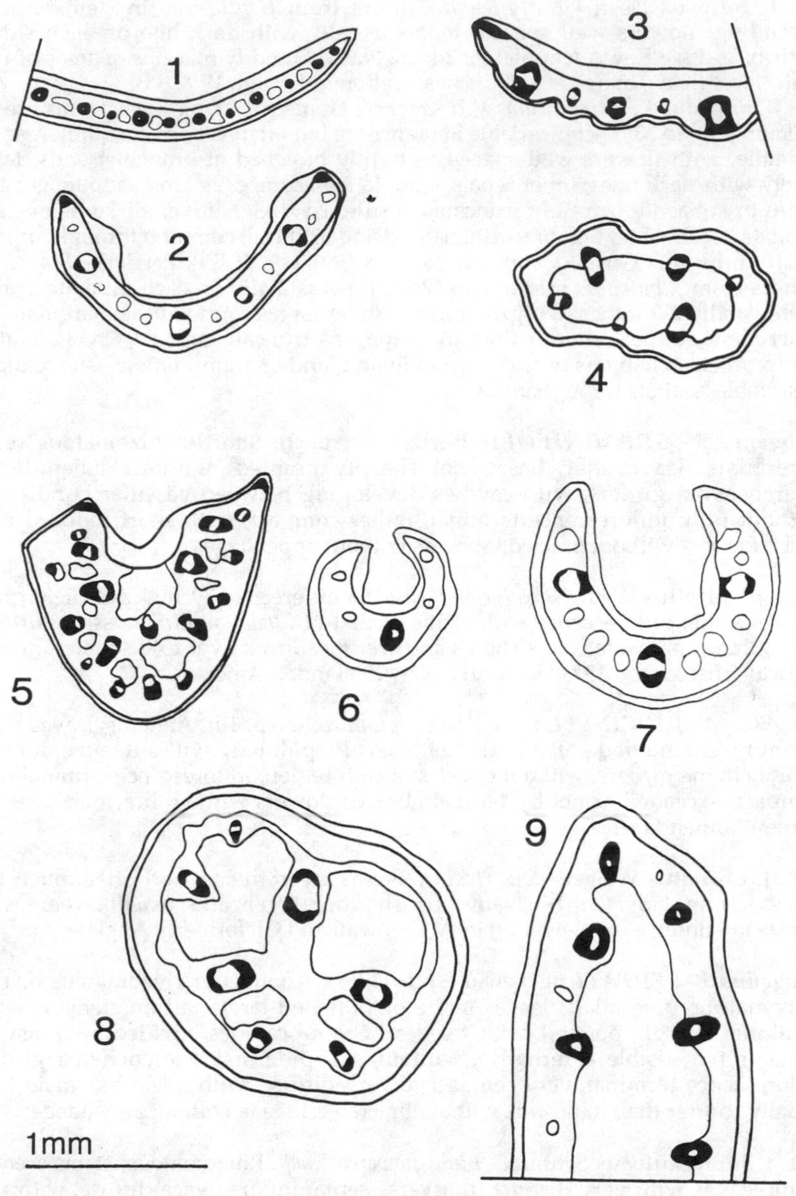

FIG 786 - Leaf-sections of ***Juncus***. 1, ***J. planifolius***. 2, ***J. bufonius***. 3, ***J. tenuis***. 4, ***J. biglumis***. 5, ***J. squarrosus***. 6, ***J. trifidus***. 7, ***J. capitatus***. 8, ***J. bulbosus***. 9, ***J. articulatus***. Sclerenchyma in black. Drawings by C.A. Stace.

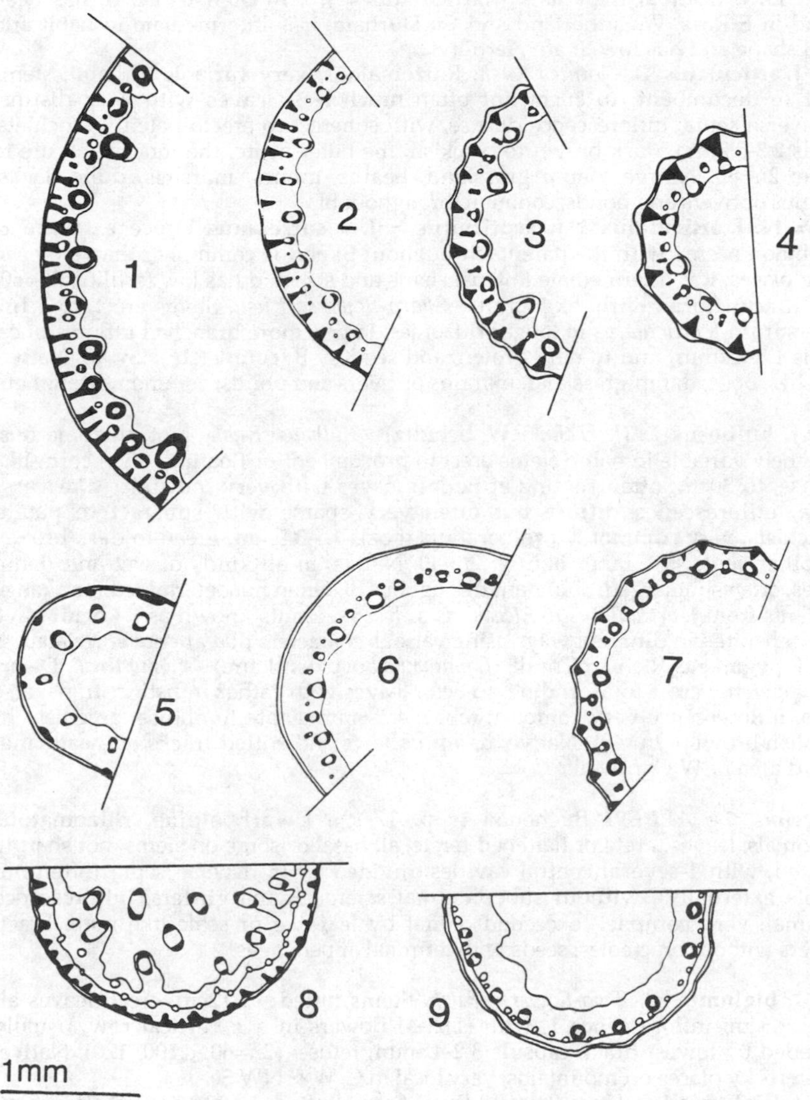

FIG 787 - Leaf-sections of *Juncus* (main bract in case of 1-7). 1, ***J. pallidus***. 2, ***J. effusus***. 3, ***J. x diffusus***. 4, ***J. inflexus***. 5, ***J. filiformis***. 6, ***J. balticus***. 7, ***J. conglomeratus***. 8, ***J. maritimus***. 9, ***J. subulatus***. Sclerenchyma in black. Drawings by C.A. Stace.

with suberect to erecto-patent branchlets; tepals 1.8-2.5mm, dark brown to blackish, the outer subacute to obtuse, the inner obtuse to rounded; 2n=40. Native; marshes, flushes and streamsides on mountains; local in mainland Br N from MW Yorks. Plants from E Ross and S Aberdeen have been referred to **J. nodulosus** Wahlenb. (probably better as a ssp. of *J. alpinoarticulatus*), but this is doubtful.

12 x 13. J. alpinoarticulatus x J. articulatus = J. x buchenaui Dörfl. has been found in E Ross, W Sutherland and Co Durham; it is intermediate in habit and tepal shape and has low, if any, fertility.

13. J. articulatus L. - *Jointed Rush.* Rhizomatous; very variable in habit; stems **786**
erect to decumbent, to 80cm but often much less; leaves with very distinct transverse septa; inflorescence diffuse, with suberect to erecto-patent branchlets; tepals 2.3-3.5mm, dark brown to blackish, the outer acute, the inner subacute to acute; 2n=80. Native; damp grassland, heaths, moors, marshes, dune-slacks, margins of rivers and ponds; common throughout BI.

13 x 14. J. articulatus x J. acutiflorus = J. x surrejanus Druce ex Stace & Lambinon occurs with the parents throughout BI and is commoner than either in some places; it is intermediate in tepal shape and size and has low fertility; 2n=60.

14. J. acutiflorus Ehrh. ex Hoffm. - *Sharp-flowered Rush.* Stems erect, to 1.1m; **792**
differs frorn *J. articulatus* in longer rhizomes; larger, more branched inflorescence; tepals 1.5-2.7mm, mid to dark brown; and see Key B (couplet 16); 2n=40. Native; marshes, bogs, damp grassland, margins of rivers and ponds; common throughout BI.

15. J. bulbosus L. (*J. kochii* F.W. Schultz) - *Bulbous Rush.* Not rhizomatous; **786**
extremely variable in habit; stems erect to procumbent or floating, often corm-like at base, to 30cm, often rooting at nodes; leaves with very indistinct transverse septa; inflorescence diffuse but often very sparse with suberect to patent branchlets, very commonly proliferating; tepals 1.5-3.5mm, green to dark brown, the outer acute, the inner obtuse; 2n=40. Native; in all kinds of wet and damp places, often submerged; abundant throughout BI. Often misdetermined for a range of plants from *J. articulatus* to *Eleocharis* or *Isoetes*. Plants known as **J. kochii** have been separated in different ways using various characters that are not correlated.

16. J. pygmaeus Rich. ex Thuill. (*J. mutabilis* auct. non Lam.) - *Pigmy Rush.* Dwarf **RR**
annual; stems erect to ascending, to 8cm; leaves with rather indistinct transverse septa; inflorescence very compact; tepals 4.5-6mm, acute to obtuse, greenish- or purplish-brown; (2n=40). Native; damp hollows and rutted tracks on heathland; Lizard area of W Cornwall.

Subgenus* 6 - *ALPINI Buchenau (spp. 17-19). Dwarf, alpine, rhizomatous perennials; leaves terete or flattened-terete, all basal or some on stems, not sharply pointed, with 1-several central cavities, divided by transverse septa (often not visible externally), without subepidermal sclerenchyma girders; inflorescence terminal, very compact, exceeded or not by leaf-like or scale-like main bract; flowers without bracteoles; seeds with terminal appendages.

17. J. biglumis L. - *Two-flowered Rush.* Stems tufted, to 12cm, erect; leaves all **R**
basal, ≤6cm; inflorescence 1, with (1)2(-4) flowers in a ± vertical row, usually **786**
exceeded by lowest bract; capsule 3.2-4.5mm, retuse; (2n=60, c.100, 120). Native; barish rocky places on mountains; very local in C, W & NW Sc.

18. J. triglumis L. - *Three-flowered Rush.* Stems tufted, to 20cm, erect; leaves all basal, ≤10cm; inflorescence 1, with (2)3(-5) flowers in ± horizontal row, usually not exceeded by lowest bract; capsule 4-5.5mm, obtuse to rounded; (2n=50, c.134). Native; boggy and rocky places on mountains; local in C, W & NW Sc, very local in N En and N Wa.

19. J. castaneus Sm. - *Chestnut Rush.* Stems not tufted, to 30cm; leaves basal and **R**
1-3 on stems, ≤20cm; inflorescences 1-3, each with 3-8(10) flowers in ± horizontal **792**
row, usually exceeded by lowest bract; capsule 6-7.5mm, obtuse to rounded;

(2n=40, 60). Native; boggy places and flushes on mountains; very local in C, W & NW Sc.

Subgenus* 7 - *JUNCUS (*Thalassii* Buchenau) (spp. 20-21). Robust maritime rhizomatous (but sometimes densely tufted) perennials; leaves cylindrical, basal and on stems, very sharply pointed, with central compact pith bearing scattered vascular bundles, not transversely septate, with very numerous subepidermal sclerenchyma girders; inflorescence terminal, with sharply pointed leaf-like main bract shorter or longer than it, very to somewhat compact; flowers without bracteoles; seeds with terminal appendages.

20. J. maritimus Lam. - *Sea Rush*. Stems densely to scarcely tufted, erect, very **787**
stiff, to 1m; inflorescence usually forming an interrupted panicle with erect to erecto-patent branches, usually exceeded by lowest bract; 2n=48. Native; saltmarshes; common round coasts of BI except extreme N Sc.

21. J. acutus L. - *Sharp Rush*. Stems very densely tufted, erect, extremely stiff, to **R**
1.5m; inflorescence usually a dense ± rounded head with erect to reflexed branches, **792**
usually exceeded by lowest bract; 2n=48. Native; sandy sea-shores and drier parts of saltmarshes; very local in BI N to W Norfolk, Caerns and Co Dublin, formerly NE Yorks.

Subgenus* 8 - *SUBULATI Buchenau (sp. 22). Rhizomatous maritime perennials; leaves cylindrical, basal and on stems, not sharply pointed, with central soft pith, not transversely septate, without subepidermal sclerenchyma girders; inflorescence terminal, with short much reduced main bract, diffuse; each flower with 2 small bracteoles; seeds with short terminal appendages.

22. J. subulatus Forssk. - *Somerset Rush*. Stems rather weak, ± erect, to 1m; **787**
inflorescence a rather diffuse panicle with suberect branches, much exceeding **792**
lowest bract; (2n=42). Intrd; saltmarsh in N Somerset, discovered 1957, and wet reclaimed land by docks in Stirlings, discovered 1983; Mediterranean.

Subgenus* 9 - *GENUINI Buchenau (spp. 23-28). Rhizomatous (but often densely tufted) perennials; leaves reduced to brown sheaths at base of stem and the lowest bract, which is stem-like and much exceeds inflorescence which appears lateral; main bract cylindrical, not or slightly sharply pointed, with central soft pith sometimes regularly interrupted, not transversely septate, with or sometimes without subepidermal sclerenchyma girders; inflorescence very compact to rather diffuse; each flower with 2 small bracteoles; seeds ± without appendages.

23. J. balticus Willd. - *Baltic Rush*. Strongly rhizomatous; stems erect, to 75cm, **R**
smooth and glossy when fresh, with continuous pith, without subepidermal **787**
sclerenchyma girders; inflorescence rather lax, with suberect branches; tepals dark brown; 2n=84. Native; maritime dune-slacks, rarely on upland river terraces, on bare or grassy ground; local in N Sc S to M Ebudes and Fife, S Lancs, formerly W Lancs.

23 x 25. J. balticus x J. inflexus occurs as 3 large patches of strongly rhizomatous completely sterile clones in S & W Lancs; 2 of them (S Lancs) are very tall (to 2m) and have an interrupted pith as in *J. inflexus*, the other (W Lancs) is close to *J. inflexus* in height but has a continuous pith; all 3 clones 2n=84; endemic.

23 x 26. J. balticus x J. effusus = J. x obotritorum Rothm. occurred as 3 patches, 1 large, of strongly rhizomatous completely sterile clones in S Lancs between 1933 and 1980 (transplanted portions still exist); they differ from *J. balticus* x *J. inflexus* in being much more slender and in minor anatomical characters; 2n=82.

24. J. filiformis L. - *Thread Rush*. Rather weakly rhizomatous; stems erect, to **R**
45cm, very slightly ridged when fresh, with continuous pith, with subepidermal **787**

sclerenchyma girders; inflorescence compact or very compact, with ≤10 flowers; tepals pale brown; (2n=40, 80). Native; on stony, silty edges of lakes and reservoirs; local in Br from Leics to Easterness, apparently spreading.

25. J. inflexus L. - *Hard Rush.* Densely tufted; stems erect, to 1.2m, glaucous, very **787**
strongly ridged when fresh, with interrupted pith, with very strong subepidermal sclerenchyma girders; inflorescence rather lax, with suberect branches; tepals dark brown; 2n=40, 42. Native; marshes, dune-slacks, wet meadows, ditches, by lakes and rivers, usually on neutral or base-rich soils; common throughout most of BI N to C Sc.

25 x 26. J. inflexus x J. effusus = J. x diffusus Hoppe occurs sporadically with **787**
the parents, usually as isolated plants, within the range of *J. inflexus*. The stems are not glaucous and have continuous pith, intermediate anatomy, and inflorescence shape ± as in *J. inflexus*; fertility low; 2n=42.

26. J. effusus L. - *Soft-rush.* Densely tufted or sometimes forming larger patches; **787**
stems erect, to 1.2m, smooth and glossy when fresh, with continuous pith, with **792**
subepidermal sclerenchyma girders; inflorescence lax, with suberect to widely divergent branches, or compact (var. **subglomeratus** DC. (var. *compactus* Lej. & Courtois)); tepals pale brown; 2n=40, 42. Native; marshes, ditches, bogs, wet meadows, by rivers and lakes, damp woods, mostly on acid soils; abundant throughout BI.

26 x 27. J. effusus x J. conglomeratus = J. x kern-reichgeltii Jansen & Wacht. ex Reichg. occurs sporadically with the parents in Br (mainly N & W) and E & W Cork, but many records are erroneous; it is intermediate in diagnostic characters, but due to its high fertility it is difficult to determine other than in the field with its parents.

27. J. conglomeratus L. (*J. subuliflorus* Drejer) - *Compact Rush.* Densely tufted; **787**
stems erect, to 1m, distinctly ridged and dull when fresh, with continuous pith, with subepidermal sclerenchyma girders; inflorescence usually very compact, sometimes of several stalked heads (var. **subuliflorus** (Drejer) Asch. & Graebn.); tepals pale brown; 2n=42. Native; similar places to *J. effusus*; common throughout BI.

28. J. pallidus R. Br. - *Great Soft-rush.* Densely tufted, like a very large *J. effusus*, **787**
with stems to 2(3)m; see Key B (couplet 6) for differences. Intrd; wool-alien formerly natd in Middlesex and Beds, now infrequently recurring in En; Australia and New Zealand.

J. pallidus x J. effusus and **J. pallidus x J. inflexus** occurred in Middlesex and Beds in the 1950s by hybridization between the native spp. and the wool-alien *J. pallidus*.

2. LUZULA DC. - *Wood-rushes*

Perennials, vegetatively grass-like; leaves bifacial, variously pubescent but rarely without hairs near base of leaf on margins; ovary 1-celled, with 3 ovules; capsules with 3 seeds.

1 Flowers all or most borne singly in inflorescence each on distinct pedicels >3mm, rarely some in pairs 2

1 Flowers mostly borne in groups of 2 or more, each one in a group sessile or with pedicels <2mm, often a few solitary 3

  2 Basal leaves rarely >4mm wide; inflorescence branches erect to widely erecto-patent in fruit; seeds with terminal appendage ≤1/2 as long as rest of seed, ± straight **1. L. forsteri**

  2 Some basal leaves usually >4mm wide; lower inflorescence branches reflexed in fruit; seeds with terminal appendage >1/2 as long as (often longer than) rest of seed, often curved or hooked **2. L. pilosa**

3 Tepals white to pale straw-coloured **4. L. luzuloides**

3 Tepals yellowish- to dark-brown 4

4 All or most basal leaves >8mm wide **3. L. sylvatica**

4 All leaves <8mm wide 5

5 Inflorescence drooping, spike-like, with the flower groups subsessile along main axis, or the lower themselves forming lateral spikes **9. L. spicata**

5 Inflorescence without single main axis; either all flower clusters congested in dense head or some or all with distinct stalks arising from short main axis near base of inflorescence 6

6 Leaves deeply channelled, glabrous or sparsely pubescent just near base; seeds with inconspicuous appendage ≤1/10 as long as rest of seed; Scottish mountains **8. L. arcuata**

6 Leaves ± flat, conspicuously pubescent; seeds with conspicuous whitish terminal appendage c.1/4-1/2 as long as rest of seed; widespread 7

7 Rhizomatous or stoloniferous; anthers (2.5)3-4x as long as filaments; style (excl. stigmas) (0.9)1.1-1.6mm **5. L. campestris**

7 Rhizomes and stolons 0; anthers 0.8-2.2(2.5)x as long as filaments; style (excl. stigmas) 0.2-0.8(0.9)mm 8

8 Outer tepals 2-2.6(2.8)mm; peduncles densely minutely papillose; seeds 0.5-0.6mm wide; style (excl. stigmas) 0.2-0.3mm; stigmas 0.5-0.6mm **7. L. pallidula**

8 Outer tepals 2.6-3.3(3.5)mm; peduncles smooth or distally sparsely papillose; seeds 0.7-0.9mm wide; style (excl. stigmas) 0.4-0.9mm; stigmas 1.2-2.4(3.1)mm **6. L. multiflora**

Other spp. - L. nivea (L.) DC. (*Snow-white Wood-rush*), from Europe, differs from *L. luzuloides* in its pure white tepals 4.5-5.5mm, c.2x as long as capsule (not 2.5-3.5mm, c. as long as capsule); it has been recorded in several places, but either in error or as non-persistent escapes.

1. L. forsteri (Sm.) DC. - *Southern Wood-rush*. Tufted, with very short rhizomes; 792
stems ± erect, to 35cm; inflorescence usually slightly 1-sided, with erect to widely erecto-patent branches bearing flowers singly (rarely in pairs); (2n=24). Native; woods and hedgerows; locally common in CI and Br N to Beds and Herefs.

1 x 2. L. forsteri x L. pilosa = L. x borreri Bromf. ex Bab. occurs frequently within the range of *L. forsteri* in Br, and formerly outside it in Co Wicklow; it is intermediate in inflorescence shape and leaf width, with very low fertility.

2. L. pilosa (L.) Willd. - *Hairy Wood-rush*. Differs from *L. forsteri* in leaves obtuse 792
to truncate at apex (with minute point in *L. forsteri*); and see key (couplet 2); 2n=62, 66. Native; woods, hedgerows, among heather on moors; common throughout most of Br, scattered in Ir.

3. L. sylvatica (Huds.) Gaudin - *Great Wood-rush*. Densely tufted and with long 792
rhizomes; stems erect, to 80cm; inflorescence with erect to reflexed subumbellate branches bearing flowers in groups of (2)3-5; 2n=12. Native; woods, moorland, shady streamsides; locally common throughout BI except parts of E En.

4. L. luzuloides (Lam.) Dandy & Wilmott - *White Wood-rush*. Tufted, with short rhizomes; stems ± erect, to 70cm; inflorescence of many erect to erecto-patent corymbose branches bearing flowers in groups of (1)2-5(8); (2n=12). Intrd; grown for ornament, natd in woods and by shady streams; scattered throughout most of Br; Europe.

5. L. campestris (L.) DC. - *Field Wood-rush*. Tufted, with short rhizomes; stems erect, to 15(25)cm; inflorescence of 1 sessile and 2-several stalked corymbose clusters of 3-12 flowers; 2n=12. Native; short grassland and similar places; very common throughout BI.

6. L. multiflora (Ehrh.) Lej. - *Heath Wood-rush*. Tufted, usually with 0 rhizomes; 792
stems erect, to 60cm; 2n=24, c.30, 36, 42. Native; grassland, heaths, moors, woods on acid soil.

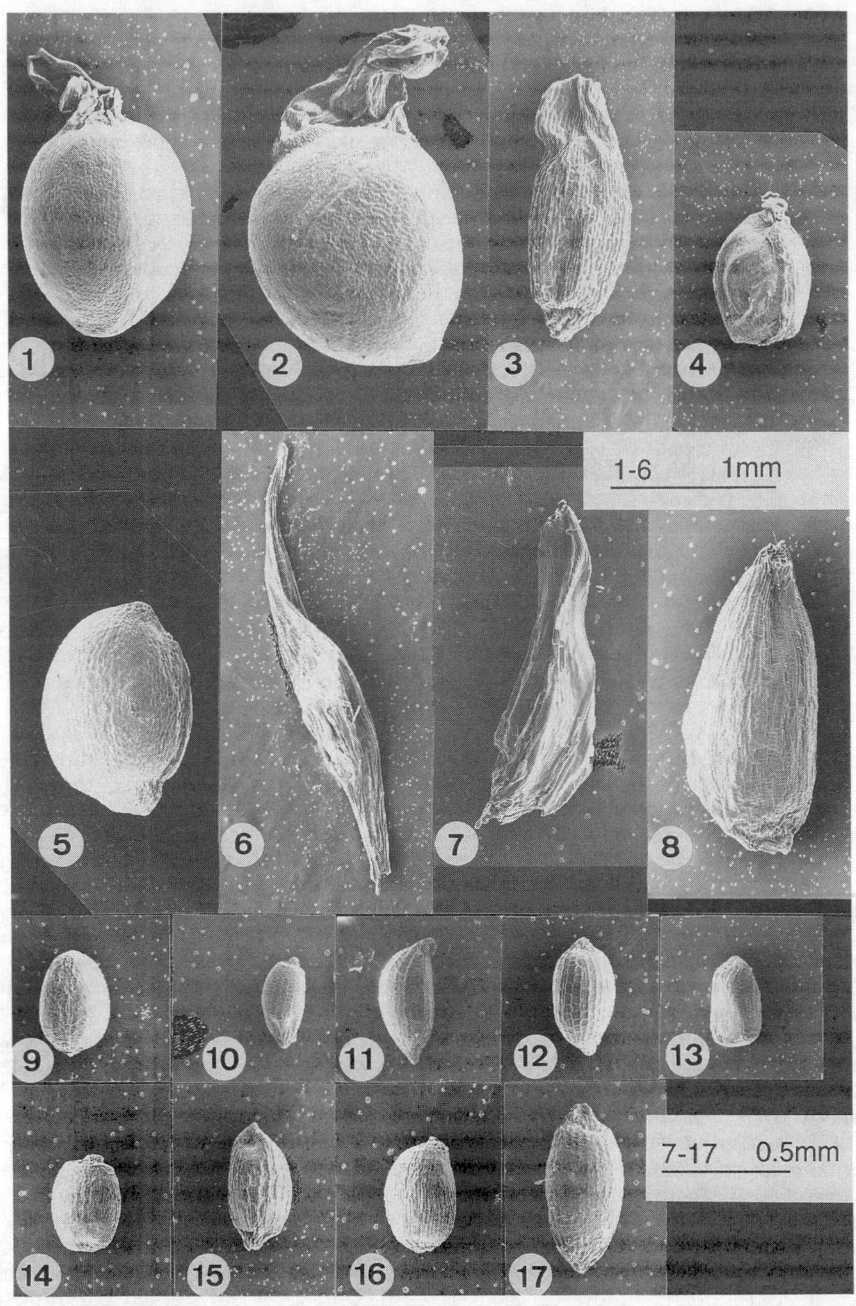

FIG 792 - Seeds of **Juncaceae**. 1-5, ***Luzula***. 1, ***L. forsteri***. 2, ***L. pilosa***. 3, ***L. multiflora***. 4, ***L. arcuata***. 5, ***L. sylvatica***. 6-17, ***Juncus***. 6, ***J. castaneus***. 7, ***J. acutus***. 8, ***J. trifidus***. 9, ***J. planifolius***. 10, ***J. capitatus***. 11, ***J. effusus***. 12, ***J. acutiflorus***. 13, ***J. tenuis***. 14, ***J. ambiguus***. 15, ***J. foliosus***. 16, ***J. bufonius***. 17, ***J. subulatus***.

1 Seeds 1.2-1.5 x 0.9-1mm (excl. appendage); all or most flower clusters subsessile in compact lobed head **c. ssp. congesta**
1 Seeds 0.8-1.2 x 0.7-0.9mm (excl. appendage); flowers in several mostly stalked corymbose clusters 2
2 Flower-clusters all on erect peduncles; seed appendages 0.4-0.5mm; most basal leaves >3mm wide; capsules 2-2.8mm **b. ssp. multiflora**
2 Some flower-clusters on recurved peduncles; seed appendages 0.2-0.3mm; most basal leaves <3mm wide; capsules 1.9-2.2mm **a. ssp. hibernica**

a. Ssp. hibernica Kirschner & T.C.G. Rich. Inflorescence of several stalked corymbose clusters of 5-10 straw- to mid-brown flowers; tepals c. as long as or slightly shorter than capsule; seeds 0.8-0.9 x 0.7mm plus appendage 0.2-0.3mm; 2n=24. Widespread in W & C Ir, largely replacing ssp. *multiflora*; endemic. Probably derived from *L. campestris* x *L. pallidula*.

b. Ssp. multiflora. Inflorescence of several stalked corymbose clusters of 5-18 reddish-brown to dark brown flowers; tepals c. as long as or slightly shorter than capsule; seeds 0.9-1.2 x 0.7-0.9mm plus appendage 0.4-0.5mm; 2n=36. Common throughout BI except parts of C & E En and C & W Ir.

c. Ssp. congesta (Thuill.) Arcang. (*L. congesta* (Thuill.) Lej.). Inflorescence with all or most flower clusters subsessile in compact lobed head; tepals longer than capsule, reddish-brown to dark brown; seeds 1.2-1.5 x 0.9-1mm plus appendage 0.4-0.6mm; 2n=48. Common throughout BI except parts of C & E En.

Sspp. *multiflora* and *congesta* often occur together; partially fertile intermediates with 2n=42 (= *L. x danica* H. Nordensk. & Kirschner) are found, especially in Sc.

7. L. pallidula Kirschner (*L. pallescens* auct. non Sw.) - *Fen Wood-rush.* Stems to 30cm; differs from *L. multiflora* sspp. *multiflora* and *hibernica* in pale yellowish-brown flowers; conspicuously unequal inner and outer tepals; and see key (couplet 8); 2n=12. Native; open grassy places in dry parts of fens; Hunts, Co Antrim and Offaly, but perhaps extinct in all. **RR**

8. L. arcuata Sw. - *Curved Wood-rush.* Tufted, with short rhizomes; stems erect, to 10cm; inflorescence of several variably-stalked clusters of 2-5 flowers, the longer stalks curved downwards; (2n=36, 42, 48). Native; open stony ground on high mountains; very local in C & N mainland Sc. **RR** **792**

9. L. spicata (L.) DC. - *Spiked Wood-rush.* Tufted, with short rhizomes; stems erect but pendent at apex, to 30cm; inflorescence with long main axis and many subsessile many-flowered clusters; 2n=24. Native; open stony ground on mountains; local in C & N Sc, formerly Lake District.

156. CYPERACEAE - *Sedge family*

Herbaceous, usually rhizomatous, perennials, rarely annuals, with usually solid, often 3-angled stems, mostly aquatic or in wet places; leaves alternate, grass-like to ± rush-like (flattened to subcylindrical), with sheathing base often with membranous ligule at top of sheath on adaxial side, the blade simple, linear and entire or 0, without stipules. Flowers much reduced, 1 each in axil of bract-like *glume*, 1-many in discrete units (*spikelets*), the spikelets terminal and solitary or in terminal spikes, racemes or panicles, often with extra sterile glumes, bisexual or unisexual (monoecious or sometimes dioecious), hypogynous, ± actinomorphic; perianth 0 or represented by bristles, sometimes elongating in fruit; stamens (1)2-3; ovary 1-celled with 1 ovule; style 0 or short; stigmas 2 or 3, elongated; fruit a nut.

Easily told from other grass- or rush-like plants except Poaceae by the (often unisexual) very reduced flowers with perianth 0 or in the form of bristles, and the 1-celled, 1-ovuled ovary, and from Poaceae by the absence of a bract (*palea*) *above* each flower and usually solid, often 3-angled stems. Some Poaceae lack a palea but all these have hollow stems; the only member of the Cyperaceae with this combination is *Cladium*.

1 Stems hollow; leaves usually with fierce saw-edged margins and lowerside of midrib, easily cutting the skin **14. CLADIUM**

1 Stems solid (centre often occupied by very soft pith); leaves not saw-edged or very mildly so 2

2 Perianth represented by bristles which elongate and greatly exceed glumes at fruiting, forming a whitish cottony head 3

2 Perianth 0 or represented by inconspicuous bristles shorter than glumes 4

3 Perianth-bristles 4-6 per flower; spikelet 1, terminal, <1cm excl. bristles (extinct) **2. TRICHOPHORUM**

3 Perianth-bristles numerous per flower; spikelets 1-several, >1cm excl. bristles **1. ERIOPHORUM**

4 Flowers all unisexual, the male and female in different spikes or different parts of the same spike, or rarely on different plants; ovary and fruit enclosed or closely enfolded in membranous innermost glume 5

4 Flowers all bisexual; ovary and fruit not enclosed or closely enfolded in innermost glume 6

5 Ovary and fruit entirely enclosed in fused membranous glume usually ending in a short or long beak; male and female flowers in same or different spikes or on different plants, the spikes variously crowded or distant, and stalked or sessile; stigmas and stamens 2 or 3 **16. CAREX**

5 Ovary and fruit closely enfolded in innermost glume which is not fused, leaving fruit exposed at top; male and female flowers in same spikes, the spikes crowded and ± sessile; stigmas and stamens 3 **15. KOBRESIA**

6 Inflorescence of 1 terminal spikelet; lowest bract not leaf-like or stem-like, shorter than spikelet 7

6 Inflorescence of ≥2 spikelets, or sometimes of 1 but then lowest bract leaf-like or stem-like and exceeding spikelet 9

7 Most or all leaf-sheaths on stems with leafy blades **9. ELEOGITON**

7 Most or all leaf-sheaths on stems without blades 8

8 Uppermost leaf-sheath on stem with short blade **2. TRICHOPHORUM**

8 Uppermost leaf-sheath on stem (and all or most below it) without a blade **3. ELEOCHARIS**

9 Inflorescence with ≥2 bifacial leaf-like bracts very close together at base 10

9 Inflorescence with basal bracts stem-like or leaf-like, if leaf-like then either 1 or ≥2 and well spaced out 12

10 Spikelets flattened, with glumes on 2 opposite sides of axis **11. CYPERUS**

10 Spikelets ± terete, with glumes spirally arranged 11

11 Inflorescence dense; spikelets >8mm **4. BOLBOSCHOENUS**

11 Inflorescence diffuse; spikelets <5mm **5. SCIRPUS**

12 Inflorescence a flattened compact terminal head, with spikelets only on 2 opposite sides of main axis **10. BLYSMUS**

12 Inflorescence various, if a compact terminal head then spikelets not only on 2 opposite sides of axis 13

13 Spikelets flattened, with glumes on 2 opposite sides of axis **12. SCHOENUS**

13 Spikelets terete, with glumes spirally arranged 14

14 Inflorescence obviously terminal with leaf-like main bract; stems with several well-developed leaf-blades **13. RHYNCHOSPORA**

14 Inflorescence usually apparently lateral, with main bract ± stem-like and continuing stem apically; stems with 0-1(2) reduced leaf-blades 15

15 Stems very slender, <1mm wide, rarely >20cm **8. ISOLEPIS**

15 Stems stouter, >1.5mm wide, rarely <30cm 16
16 Inflorescence composed of (1-)several sessile to stalked globose apparent spikelets **6. SCIRPOIDES**
16 Inflorescence composed of (1-)several sessile to stalked ovoid spikelets **7. SCHOENOPLECTUS**

1. ERIOPHORUM L. - *Cottongrasses*
Perennials with long or short rhizomes; stems terete to triangular in section, leafy; leaves variously shaped in section; inflorescence of 1-several large spikelets in a terminal umbel; lowest bract leaf-like or glume-like; flowers bisexual; perianth of numerous (>6) bristles elongating to form conspicuous white, cottony head in fruit; stamens 3; ovary not enfolded or enclosed by glume; stigmas 3.

1 Spikelet 1, erect, without leaf-like bract at base; leaf-blades ± triangular in section, 0 or very reduced on uppermost stem leaf-sheath **4. E. vaginatum**
1 Spikelets ≥(1)2, ± pendent in fruit, with 1-3 ± leaf-like bracts at base; leaf-blades flat to V-shaped in section, well developed on uppermost stem leaf-sheath 2
2 Stalks of spikelets smooth; stems ± terete to very bluntly 3-angled; anthers >2mm **1. E. angustifolium**
2 Stalks of spikelets with numerous minute forward-pointed bristles; stems distinctly 3-angled; anthers ≤2mm 3
3 Leaf-blades 0.5-2mm wide; glumes with midrib plus several shorter parallel veins on either side; plant with long rhizomes, with solitary stems **3. E. gracile**
3 Leaf-blades 3-8mm wide; glumes with only midrib; plant loosely tufted **2. E. latifolium**

1. E. angustifolium Honck. - *Common Cottongrass*. Rhizomes long; stems scattered, erect, to 60cm; leaves 2-6mm wide, V-shaped in section; inflorescence of (1)3-7 pendent spikelets; perianth-bristles 2.5-5cm; 2n=58. Native; wet usually acid bogs; common in suitable places throughout BI.

2. E. latifolium Hoppe - *Broad-leaved Cottongrass*. Differs from *E. angustifolium* in stems loosely tufted; leaves 3-8mm wide, ± flat; perianth-bristles 1.5-3cm, minutely toothed (not entire) at apex (microscope); and see key; (2n=54, 58, 72). Native; wet base-rich marshes and flushes; scattered throughout Br and Ir in suitable places, much less common than *E. angustifolium* and extinct in much of C En.

3. E. gracile W.D.J. Koch ex Roth - *Slender Cottongrass*. Differs from *E. angustifolium* in leaves 0.5-2mm wide; perianth-bristles 1-2.5cm; and see key; (2n=60, 76). Native; similar places to *E. angustifolium*; very local in S Br from Surrey to Caerns, C & W Ir, formerly NW Yorks, Northants and E Norfolk. **RRR**

4. E. vaginatum L. - *Hare's-tail Cottongrass*. Rhizomes very short; stems densely tufted, often tussock-forming, erect, to 50cm; leaves 0.5-1mm wide, triangular in section; inflorescence of 1 erect spikelet; perianth-bristles 2-3cm; 2n=58. Native; wet peaty places, especially on moorland bogs; common in Ir and W, C & N Br, very local in C, E & S En.

2. TRICHOPHORUM Pers. (*Scirpus* sect. *Baeothryon* auct. non (Ehrh. ex A. Dietr.) Benth. & Hook. f.) - *Deergrasses*
Tufted perennials; stems with 3 rounded angles to terete, with only the uppermost leaf-sheath with a blade; leaves thick-crescent-shaped in section, very narrow; inflorescence of 1 terminal spikelet; lowest bract glume-like; flowers bisexual; perianth of 4-6 bristles, elongating or not in fruit; stamens 3; ovary not enfolded or enclosed by glume; stigmas 3.

1. T. alpinum (L.) Pers., (*Scirpus hudsonianus* (Michx.) Fernald) - *Cotton Deergrass*. **RR**

Rather diffusely tufted; stems erect, to 20(30)cm, very slender, with 3 rounded angles, slightly scabrid near apex; perianth-bristles elongating to 10-25mm, forming white cottony head, at fruiting; (2n=58). Native; bog in Angus from 1791 to c.1813. **799**

2. T. cespitosum (L.) Hartm. (*Scirpus cespitosus* L.) - *Deergrass*. Densely tufted; stems erect, to 35cm, very slender, ± terete, smooth; perianth-bristles remaining shorter than glumes, pale brown; (2n=104). Native; bogs, wet moors and heaths. **799**

a. Ssp. cespitosum. Forming smaller tufts with weaker stems and smaller spikelets than ssp. *germanicum*; spikelets with 3-10 flowers; uppermost leaf-sheath (pull out terminal stem!) with oblique, elliptic opening c.2-3 x 1mm, with leaf-blade c. 2x as long as opening; stems without aerenchyma. Usually in wetter places than ssp. *germanicum*; S Northumb and Cheviot, probably overlooked, first recognized 1988. **RR**

b. Ssp. germanicum (Palla) Hegi (*Scirpus cespitosus* ssp. *germanicus* (Palla) Brodd.). Spikelets with 8-20 flowers; uppermost leaf-sheath with transverse ± orbicular opening c.1 x 1mm, with leaf-blade c. 5-10x as long as opening; stem sections with aerenchyma gaps clearly visible in photosynthetic tissue. Common throughout Br and Ir in suitables places, mostly on solid peat, but absent from most of C & E En.

Other characters of sheaths and bristles often quoted seem unreliable. The aerenchyma in the green photosynthetic tissue is visible in stem sections at x20, but is better observed at x50.

3. ELEOCHARIS R. Br. - *Spike-rushes*

Perennials with long or short stout and/or slender rhizomes; stems terete to ridged, with blade-less leaf-sheaths; leaf-blades 0 (some spikelet-less stems may resemble basal leaves); inflorescence of 1 terminal spikelet; lowest bract glume-like; flowers bisexual; perianth of 0-6 bristles, not elongating in fruit; stamens 3; ovary not enfolded or enclosed by glume; stigmas 2 or 3.

1 Lowest glume >(2/5-)1/2 as long as spikelet; spikelets 3-12-flowered 2
1 Lowest glume <2/5(-1/2) as long as spikelet; spikelets 10-many-flowered 4
  2 Glumes greenish; uppermost stem leaf-sheath delicate, inconspicuous; very slender whitish rhizomes ending in small whitish tubers (c.2-5mm) present **7. E. parvula**
  2 Glumes brown (often with green midrib); uppermost stem leaf-sheath conspicuous, brownish; rhizomes brownish, not bearing tubers 3
3 Stems <0.5mm wide, usually 4-ridged; spikelets 2-5mm; lowest glume 1.5-2.5mm **6. E. acicularis**
3 Stems ≥0.5mm wide, ± terete; spikelets 4-10mm; lowest glume 2.5-5mm **5. E. quinqueflora**
  4 Lowest glume ± completely encircling spikelet at base 5
  4 Lowest glume ≤1/2(-3/4)-encircling spikelet at base 6
5 Uppermost stem leaf-sheath oblique (c.45°) at apex; stigmas 3; nuts 3-angled **4. E. multicaulis**
5 Uppermost stem leaf-sheath ± truncate at apex; stigmas 2; nuts biconvex **3. E. uniglumis**
  6 Stems with 10-16 vascular bundles, showing up as fine ridges in dried state; bristles (4)5(-6); spikelets conical; swollen style-base on top of fruit c.1-1.5x as long as wide, slightly constricted at base **2. E. austriaca**
  6 Stems with ≥20 vascular bundles, showing up as fine ridges in dried state; bristles (0-)4; spikelets cylindrical- to ellipsoid-conical; swollen style-base on top of fruit wider than long, strongly constricted at base **1. E. palustris**

1. E. palustris (L.) Roem. & Schult. - *Common Spike-rush*. Stems loosely to rather densely tufted, to 75cm, often much less; uppermost stem leaf-sheath ± truncate at **799**

apex; spikelets 5-30mm, the lowest glume not fully encircling base; stigmas 2. Native; in or by ponds, marshes, ditches, riversides.

a. Ssp. vulgaris Walters. Spikelets usually with <40 flowers; glumes from middle of spikelet 3.5-4.5mm; nut (1.3)1.5-2mm; 2n=37, 38. Frequent throughout BI.

b. Ssp. palustris (ssp. *microcarpa* Walters). Spikelets usually with >40 flowers; glumes from middle of spikelet 2.7-3.5mm; nut 1.2-1.4(1.5)mm; 2n=16. S & C En from E Kent to S Hants, Worcs and Notts, but even there much rarer than ssp. *vulgaris*.

Partially fertile hybrids between the 2 sspp. have been found in Oxon; 2n=27.

1 x 3. E. palustris x E. uniglumis occurs within populations of *E. uniglumis* near to those of *E. palustris* on the W coasts of Br; it is intermediate and fertile; 2n=38-52.

2. E. austriaca Hayek - *Northern Spike-rush*. Differs from *E. palustris* in stems **RR**
rather fragile (not very flexible); spikelets 8-20mm; and see key (couplet 6); **799**
(2n=16). Native; wet, marshy and flushed areas in or by rivers; local in N En and S Sc from MW Yorks and S Lancs to Selkirks, first found 1947. The broadly conical spikelets with glumes becoming ± deciduous as fruits ripen are usually diagnostic.

3. E. uniglumis (Link) Schult. - *Slender Spike-rush*. Differs from *E. palustris* in **799**
shorter, slenderer, shinier stems; spikelets 5-12mm; and see key (couplet 4); 2n=40, 46-48, 50, 88-92. Native; marshes and dune-slacks; scattered throughout Br and Ir, mostly coastal.

4. E. multicaulis (Sm.) Desv. - *Many-stalked Spike-rush*. Stems densely tufted, to **799**
40cm; uppermost stem leaf-sheath acutely oblique at apex; spikelets 5-15mm, the lowest glume not fully encircling base; stigmas 3; 2n=20. Native; bogs and wet peaty places, usually on acid soils; throughout BI, common in W, sparse in E.

5. E. quinqueflora (Hartmann) O. Schwarz - *Few-flowered Spike-rush*. Stems **799**
loosely to rather densely tufted, to 30cm; uppermost stem leaf-sheath obtusely oblique at apex; spikelets 4-10mm, the lowest glume ± fully encircling base; stigmas 3; 2n=20. Native; wet places in fens, dune-slacks and moorland; throughout BI, commonest in NW.

6. E. acicularis (L.) Roem. & Schult. - *Needle Spike-rush*. Stems sparsely tufted, to **799**
10cm (or more if submerged and sterile); uppermost stem leaf-sheath acutely to obtusely oblique at apex; spikelets 2-5mm, the lowest glume ± fully encircling base; stigmas 3; (2n=20, 56). Native; in and by pond and lake margins; scattered throughout Br and Ir.

7. E. parvula (Roem. & Schult.) Link ex Bluff, Nees & Schauer - *Dwarf Spike-rush*. **RR**
Stems sparsely tufted, to 8cm; uppermost stem leaf-sheath obtusely oblique at **799**
apex; spikelets 2-4mm, the lowest glume ± fully encircling base; stigmas 3; (2n=8, 10). Native; wet muddy places by sea and in estuaries; very local in SW Br from S Hants to Caerns, formerly (?still) Ir.

4. BOLBOSCHOENUS (Asch.) Palla (*Scirpus* sect. *Bolboschoenus* (Asch.) Beetle) - *Sea Club-rush*

Strongly rhizomatous perennials; stems with 3 acute angles, leafy; leaves flattened, widely V-shaped in section; inflorescence of (1)3-many spikelets either sessile or variously clustered on 1-several stalks; lowest bract leaf-like; flowers bisexual; perianth-bristles 1-6, not elongating in fruit; stamens 3; ovary not enfolded or enclosed by glume; stigmas 2-3.

1. B. maritimus (L.) Palla (*Scirpus maritimus* L.) - *Sea Club-rush*. Stems strong, erect, to 1m; leaves long, 2-10mm wide; spikelets 10-30mm, dark brown; (2n=40, 55-60, 80, 86, 96, 104, 110). Native; wet muddy places in estuaries or by sea; common round coasts of BI except extreme N Sc, rarely inland.

5. SCIRPUS L. - *Wood Club-rush*

Strongly rhizomatous perennials; stems with 3 rounded angles, leafy; leaves flat;

inflorescence of very numerous spikelets 1-several on ends of diffusely branching panicle; lowest bract leaf-like; flowers bisexual; perianth-bristles 6, not elongating in fruit; stamens 3; ovary not enfolded or enclosed by glume; stigmas 3.

1. S. sylvaticus L. - *Wood Club-rush*. Stems strong, erect, to 1.2m; leaves long, 5-20mm wide; spikelets 3-4mm, greenish-brown; 2n=62. Native; by streams and in marshes and damp spots in woods or shady places; locally frequent over Br and Ir N to C Sc.

6. SCIRPOIDES Ség. (*Holoschoenus* Link, *Scirpus* sect. *Holoschoenus* (Link) W.D.J. Koch) - *Round-headed Club-rush*
Strongly rhizomatous perennials; stems terete; leaf-sheaths mostly blade-less but uppermost 1(-2) with well-developed blade; leaves semi-circular in section; inflorescence of (1)5-many variously stalked or sessile globular heads (apparent spikelets) each actually of numerous tightly packed spikelets; lowest bract ± stem-like, making inflorescence appear lateral; flowers bisexual; perianth-bristles 0 stamens 3; ovary not enfolded or enclosed by glume; stigmas 3.

1. S. holoschoenus (L.) Soják (*Scirpus holoschoenus* L., *Holoschoenus vulgaris* Link) **RR**
- *Round-headed Club-rush*. Stems erect, to 1.5m; leaves terete but ± flat on upperside, rather stem-like, 0.5-2mm wide; heads 3-10mm across, with spikelets 2.5-4mm, brown; (2n=c.164, 168). Native; damp sandy places near sea; very rare in N Devon and N Somerset, occasional elsewhere in S Br as introduction.

7. SCHOENOPLECTUS (Rchb.) Palla (*Scirpus* sects. *Schoenoplectus* (Rchb.) Benth. & Hook. f. & *Pterolepis* (Schrad.) Asch. & Graebn.) - *Club-rushes*
Strongly rhizomatous perennials; stems terete to triangular in section with acute angles; leaf-sheaths mostly blade-less but uppermost 1(-3) with rather short blade; leaves crescent-shaped in section; inflorescence of (1-)few to numerous variously stalked or sessile ovoid spikelets 5-8mm; lowest bract ± stem-like, making inflorescence appear lateral; flowers bisexual; perianth-bristles 0-6, not elongating in fruit; stamens 3; ovary not enfolded or enclosed by glume; stigmas 2-3.

1 Stems terete — 2
1 Stems triangular in section with acute angles — 3
 2 Glumes (except apical projection) smooth; stigmas mostly 3; nut 2.5-3mm, mostly 3 angled — **1. S. lacustris**
 2 Glumes minutely (x20 lens) but densely papillose at least near midrib and apex; stigmas 2; nut 2-2.5mm, biconvex or planoconvex — **2. S. tabernaemontani**
3 Glumes with rounded to obtuse lobe on either side of apical projection; stems with uppermost 1(-2) leaf-sheaths with blades; perianth-bristles 6, >1/2 as long as nut — **3. S. triqueter**
3 Glumes with acute to subacute lobe on either side of apical projection; stems with uppermost 2-3 leaf-sheaths with well-developed blades; perianth-bristles 0-6, <1/2 as long as nut — **4. S. pungens**

1. S. lacustris (L.) Palla (*Scirpus lacustris* L.) - *Common Club-rush*. Stems erect, to **799**

FIG 799 - Cyperaceae. 1-12, inflorescences. 1, ***Eleocharis palustris***. 2, ***E. austriaca***. 3, ***E. uniglumis***. 4, ***E. multicaulis***. 5, ***E. quinqueflora***. 6, ***E. acicularis***. 7, ***E. parvula***. 8, ***Trichophorum cespitosum***. 9, ***T. alpinum***. 10, ***Eleogiton fluitans***. 11, ***Isolepis setacea***. 12, ***I. cernua***. 13-16, glumes of **Schoenoplectus**. 13, ***S. lacustris***. 14, ***S. tabernaemontani***. 15, ***S. triqueter***. 16, ***S. pungens***. 17-22, inflorescences. 17, ***Rhynchospora alba***. 18, ***R. fusca***. 19, ***Blysmus compressus***. 20, ***B. rufus***. 21, ***Schoenus nigricans***. 22, ***S. ferrugineus***.

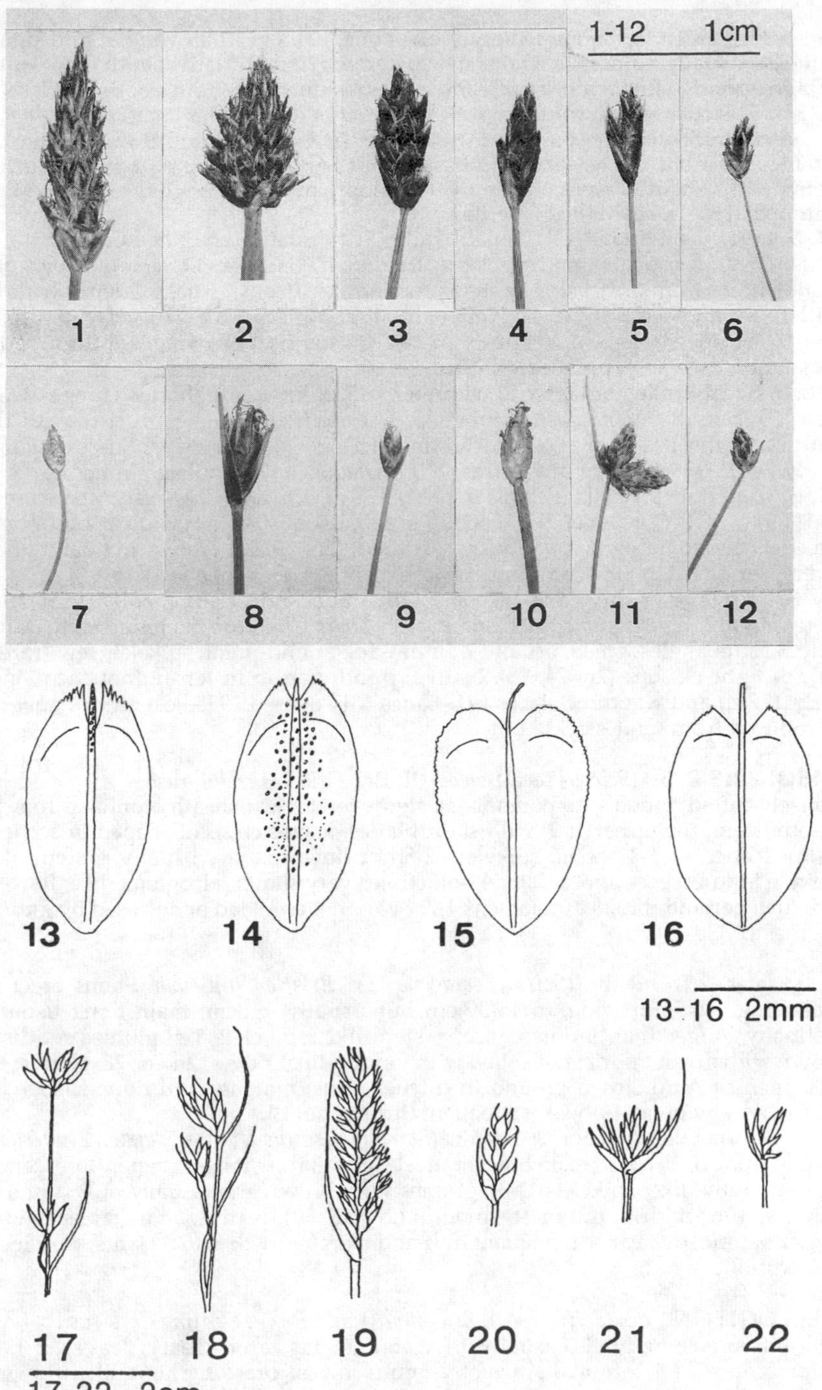

FIG 799 - see caption opposite

3m, terete, green, sometimes slightly glaucous, usually ≥1cm wide at mid-point; spikelets usually numerous, variously clustered; (2n=42). Native; in shallow water of lakes, ponds, slow rivers, canals and dykes; frequent over most of Br and Ir.

1 x 3. S. lacustris x S. triqueter = S. x carinatus (Sm.) Palla (*Scirpus x carinatus* Sm.) occurred with the parents in R Tamar (E Cornwall and S Devon) and R Thames (W Kent, Surrey and Middlesex), but is now very rare or extinct; it has stems with 3 acute angles above, terete below, and smooth glumes with rather rounded apical lobes, and is ± sterile.

2. S. tabernaemontani (C.C. Gmel.) Palla (*Scirpus tabernaemontani* C.C. Gmel., *S. lacustris* ssp. *glaucus* Hartm., ssp. *tabernaemontani* (C.C. Gmel.) Syme) - *Grey Club-rush.* Differs from *S. lacustris* in stems to 1.5m, glaucous, usually 3-8mm wide at mid-point; and see key; 2n=42. Native; in similar places to *S. lacustris* but also in marshes, dune-slacks and wet peaty places, mostly near sea; frequent throughout most of BI, except very scattered inland. **799**

2 x 3. S. tabernaemontani x S. triqueter = S. x kuekenthalianus (Junge) D.H. Kent (*Scirpus x kuekenthalianus* Junge, *S. x scheuchzeri* Brügger) occurs with the parents in the R Tamar (E Cornwall and S Devon), R Arun (W Sussex) and R Medway (E & W Kent); it differs from *S. x carinatus* in its papillose glumes.

3. S. triqueter (L.) Palla (*Scirpus triqueter* L.) - *Triangular Club-rush.* Stems erect, to 1(1.5)m, with 3 acute angles, usually green, usually 3-8mm wide at mid-point; spikelets few to numerous, variously clustered; (2n=40, 42). Native; in tidal mud of rivers; very local in S En (possibly extinct except in S Devon) and W Ir. **RRR** **799**

4. S. pungens (Vahl) Palla (*S. americanus* auct. non (Pers.) Volkart, *Scirpus pungens* Vahl, *S. americanus* auct. non Pers.) - *Sharp Club-rush.* Stems erect, to 60cm, with 3 acute angles, green, usually 2-5mm wide at mid-point; spikelets few (rarely >6), in tight cluster; (2n=74, 78). Native; pond-margin in Jersey (not seen since early 1970s) and wet dune-slacks in S Lancs (discovered 1928, lost c.1980, now re-introduced from same stock). **RR** **799**

8. ISOLEPIS R. Br. (*Scirpus* sect. *Isolepis* (R. Br.) Griseb.) - *Club-rushes*
Densely tufted annuals (to perennials); stems terete; leaf-sheaths confined to near base of stem, the upper 1(-2) with short blades; leaves crescent-shaped in section; inflorescence of 1-4 sessile spikelets 2-5mm; lowest bract usually ± stem-like, making inflorescence appear lateral, sometimes very short and ± glume-like; flowers bisexual; perianth-bristles 0; stamens 1-2; ovary not enfolded or enclosed by glume; stigmas (2-)3.

1. I. setacea (L.) R. Br. (*Scirpus setaceus* L.) - *Bristle Club-rush.* Stems erect to ascending, <0.5mm wide, to 15(30)cm but usually <10cm; main bract usually distinctly longer than inflorescence, ± stem-like; spikelets 1-4; glumes reddish-brown with green midrib; nut shiny, with longitudinal ridges; 2n=26, 28. Native; on wet open or semi-closed ground in ditches, fens, marshes and dune-slacks, on heaths, and by ponds and lakes; frequent throughout BI. **799**

2. I. cernua (Vahl) Roem. & Schult. (*Scirpus cernuus* Vahl) - *Slender Club-rush.* Differs from *I. setacea* in main bract at most only slightly longer than inflorescence, often ± glume-like; spikelets 1(-3); glumes with brown area usually only a blotch either side of midrib; nut matt, smooth; (2n=48, 60). Native; similar places to *I. setacea* but mostly near sea; frequent in Ir and extreme W Br E to S Hants, very local in E Norfolk. **799**

9. ELEOGITON Link (*Scirpus* sect. *Eleogiton* (Link) Pax) - *Floating Club-rush*
Stoloniferous perennials, usually in water; stems terete, leafy; leaves ± flat; inflorescence of 1 terminal spikelet 2-5mm; lowest bract ± glume-like; flowers bisexual; perianth-bristles 0; stamens 3; ovary not enfolded or enclosed by glume; stigmas 2-3.

1. E. fluitans (L.) Link (*Scirpus fluitans* L.) - *Floating Club-rush.* Stems usually 799
floating, rooting at nodes, to 50cm, sometimes on mud or gravel and much shorter, very leafy; spikelets green to pale brown; 2n=60. Native; in or by peaty ponds, lakes and ditches; fairly frequent throughout BI, commoner in W.

10. BLYSMUS Panz. ex Schult. - *Flat-sedges*
Rhizomatous perennials; stems subterete, leafy; leaves flat to strongly inrolled; inflorescence a flattened ± compact terminal head with spikelets on 2 opposite sides of axis, the spikelets 4-10mm; lowest bract leaf-like to ± glume-like; flowers bisexual; perianth-bristles 0-6; stamens 3; ovary not enfolded or enclosed by glume; stigmas 2.

1. B. compressus (L.) Panz. ex Link - *Flat-sedge.* Stems erect, to 40cm; leaves flat 799
to slightly keeled, rough, grass-like; spikelets usually 10-20, reddish-brown; perianth-bristles 3-6, longer than nut; nut 1.5-2mm; (2n=44, 78). Native; marshy, rather open ground; locally frequent in En but extinct in many places and in Wa, very local in Sc.

2. B. rufus (Huds.) Link - *Saltmarsh Flat-sedge.* Differs from *B. compressus* in 799
leaves strongly inrolled, smooth, rush-like; spikelets usually 3-8, dark brown; perianth-bristles 3-6, shorter than nut, or 0; nut 3-4mm; 2n=80. Native; saltmarshes and dune-slacks in turf; locally frequent on coasts of Br and Ir S to N Lincs and Pembs, common in W Sc.

11. CYPERUS L. - *Galingales*
Rhizomatous perennials or tufted annuals or perennials; stems triangular in section with acute to rounded angles, leafy at base; leaves flat to keeled, grass-like; inflorescence a simple or more often compound umbel or umbel-like raceme, with grass-like many-flowered spikelets usually clustered on ultimate branches or all clustered in ± dense head; lowest 2-10 bracts leaf-like, often much exceeding inflorescence; flowers bisexual; perianth-bristles 0; stamens 1-3; ovary not enfolded or enclosed by glume; stigmas (2-)3.

1 Tufted annual; leaves <3(-5)mm wide; glumes <1.5mm **3. C. fuscus**
1 Tufted or rhizomatous perennial; widest leaves ≥(2-)4mm wide; glumes >1.5mm 2
 2 Spikelets reddish-brown, ≤2mm wide; inflorescence ± diffuse; rhizomes long; stamens 3 **1. C. longus**
 2 Spikelets greenish- to yellowish-brown, ≥2mm wide; inflorescence ± compact; rhizomes very short; stamen 1 **2. C. eragrostis**

Other spp. - Several spp. occur as wool-aliens, but all are much rarer than *C. eragrostis.*

1. C. longus L. - *Galingale.* Shortly rhizomatous perennial; stems erect, to 1m; R
inflorescence diffuse, with very numerous linear-oblong reddish-brown spikelets 4-25mm; (2n=c.120). Native; marshes, pondsides and ditches; very local near coast in CI and SW Br E to W Sussex (formerly E Kent and E Suffolk) and N to Caerns, intrd in scattered places elsewhere in S & C En N to S Lancs.

2. C. eragrostis Lam. (*C. vegetus* Willd.) - *Pale Galingale.* Very shortly rhizomatous perennial; stems erect, to 60cm; inflorescence rather compact, with numerous elliptic-oblong greenish- to yellowish-brown spikelets 8-13mm; (2n=42). Intrd; grown for ornament and escaping to roadsides, rough ground and by water, also a frequent wool-alien and grass-seed-alien; scattered in CI and S Br, Cheshire, well natd in Guernsey; tropical America.

3. C. fuscus L. - *Brown Galingale.* Tufted annual; stems erect, to 20cm; RRR
inflorescence usually very compact, with numerous narrowly oblong dark brown

spikelets 3-6mm; 2n=36. Native; damp barish ground by ponds and in ditches; very rare in S Hants, N Somerset and Middlesex, formerly elsewhere in S En and CI, refound Jersey 1989.

12. SCHOENUS L. - *Bog-rushes*
Densely tufted perennials; stems terete, with leaf-sheaths only at or near base and bearing short or long blades; leaves very thickly crescent-shaped in section to subterete; inflorescence a compact head of 1-4-flowered flattened spikelets; lowest bract leaf-like to ± glume-like; flowers bisexual; perianth-bristles 0 or up to 6; stamens 3; ovary not enfolded or enclosed by glume; stigmas 3.

1. S. nigricans L. - *Black Bog-rush.* Stems erect, to 75cm; leaves shorter than to c. **799**
as long as stems; inflorescence of (2)5-10 spikelets, with lowest bract usually conspicuously exceeding it; glumes minutely rough (x20 lens) on keel; 2n=44. Native; damp peaty places, serpentine heathland, bogs, saltmarshes, fens, flushes; locally frequent in BI, especially near W coasts and in E Anglia, absent from most of En, Wa and E Sc.
2. S. ferrugineus L. - *Brown Bog-rush.* Stems erect, to 40cm; leaves shorter than **RR**
(often <1/2 as long as) stems; inflorescence of 1-3 spikelets, with lowest bract **799**
shorter than to c. as long as it; glumes smooth on keel; (2n=76). Native; semi-open ground in base-rich flushes, formerly also by lake; 4 places (2 transplants) in M & E Perth, first found 1884.

13. RHYNCHOSPORA Vahl - *Beak-sedges*
Tufted to creeping rhizomatous perennials; stems with 3 rounded angles or terete, leafy; leaves channelled; inflorescence of 1-few rather compact heads each of several 1-3-flowered spikelets; lowest bract leaf-like to ± glume-like; flowers bisexual; perianth-bristles 5-13; stamens 2-3; ovary not enfolded or enclosed by glume; stigmas 2, the common style-base persistent and forming beak to fruit.

1. R. alba (L.) Vahl - *White Beak-sedge.* Stems ± tufted, erect, to 40cm, often **799**
forming bulbil-like buds towards base; inflorescence whitish at flowering; lowest bract of terminal head not or sometimes slightly longer than head; (2n=26, 42). Native; wet acid peaty places, locally common in Br and Ir, but absent from most of En, E & S Wa and E Sc.
2. R. fusca (L.) W.T. Aiton - *Brown Beak-sedge.* Stems ± scattered, erect, to 30cm, **R**
without bulbils; inflorescence brown at flowering; lowest bract of terminal head **799**
>1cm longer than head; (2n=26, 32). Native; similar places to *R. alba* and usually with it; very local in W & C Ir and S & W Br from S Hants to W Sutherland.

14. CLADIUM P. Browne - *Great Fen-sedge*
Rhizomatous vigorous perennials; stems with 3 rounded angles or terete, leafy; leaves channelled, usually with fiercely serrate edges and keel; inflorescence much-branched, with many rather compact heads each of several 1-3-flowered spikelets; lowest bract of each head leaf-like to glume-like, but inflorescence branches with leaf-like long bracts at base; flowers bisexual; perianth-bristles 0; stamens 2(-3); ovary not enfolded or enclosed by glume; stigmas (2-)3.

1. C. mariscus (L.) Pohl - *Great Fen-sedge.* Stems erect, to 3m: leaves up to 2m x 2cm, grey-green: inflorescence up to 70 x 10cm; (2n=36, c.60). Native; wet, base-rich areas in fens and by streams and ponds; locally common but very scattered in BI. A rare variant in parts of Ir has apple-green less stiff leaves with only mildly serrate edges.

15. KOBRESIA Willd. - *False Sedge*
Rather densely tufted perennials: stems with 3 rounded angles, leafy at extreme

base; leaves channelled; inflorescence of 1-flowered spikelets arranged in terminal cluster of 3-10 spikes; lowest bract a sheath with short usually brown blade; flowers unisexual, the upper spikelets male and lower female in each spike; perianth-bristles 0; stamens 3; female flowers with an extra inner glume folded (but not fused) around ovary; stigmas 3.

1. K. simpliciuscula (Wahlenb.) Mack. - *False Sedge*. Stems erect, to 20cm; leaves 0.5-1.5mm wide; inflorescence 1-2.5cm x 2-6mm; (2n=72, 76). Native; flushed grassy or barish areas on mountains: very local in Upper Teesdale (NW Yorks and Durham) and M Perth, formerly in adjacent vice-counties. **RR**

16. CAREX L. - *Sedges*
Extensively rhizomatous to densely tufted perennials; stems erect, usually leafy but often only at extreme base, triangular in section with acute to rounded angles or terete; leaves flat to channelled or inrolled; inflorescence of 1-flowered spikelets grouped in variously arranged spikes, all except the terminal subtended by a bract; lowest bract leaf-like to glume-like; flowers unisexual, each in axil of 1 glume, the sexes variously arranged from mixed in 1 spike to dioecious, but commonly the upper spikes entirely male and the lower entirely female; perianth-bristles 0; stamens 2 or 3; female flowers with an extra inner glume (*utricle*) completely fused around ovary, forming a false fruit enclosing nut and usually with long or short distal beak; stigmas 2 or 3.

Ripe fruits are essential for keying down *Carex* spp.; the length of the utricle includes any beak. Two important diagnostic characters might present some difficulties. Genuinely 1-spiked inflorescences should not be confused with those with several congested spikes forming a single ± lobed head. In the former case there is 1 simple axis bearing flowers or fruits directly upon it; in the latter case the axis has lateral (often very short) branches, usually with a bract at the base of each. Depauperate stems of several spp. may rarely possess only 1 spike, but more normal stems should also be available. The number of stigmas (2 or 3) is important. In material with ripe fruits the stigmas might have disappeared, but the shape of the nut and often that of the utricle then provides the clue (see General Key, couplet 4). The beak of the utricle is often bifid, and these projections must not be mistaken for the stigmas. The glumes subtending the male and female flowers are called 'male glumes' and 'female glumes' respectively. The length of the sheath of the lowest bract refers to only the portion fused round the stem.

Many hybrids have been recorded but none except *C. hostiana* x *C. viridula* is common; all are variously intermediate and highly sterile, with empty though often well-developed utricles, except for partial fertility exhibited in hybrids between spp. 64-70.

General key
1 Spike 1, terminal ***Key A***
1 Spikes >1 (sometimes very close together) 2
2 Spikes all ± similar in appearance, often very close together and forming single lobed head, the terminal spike usually at least partly female ***Key B***
2 Spikes dissimilar in appearance, the upper ≥1 all or mostly male, the lower ≥1 all or mostly female, usually clearly separate and sometimes remote 3
3 Utricles pubescent on part or whole of main body (excl. beak and edges) (x10 lens) ***Key C***
3 Utricles glabrous on main body (sometimes pubescent on beak or along edges, sometimes papillose on main body) 4
4 Stigmas 2; utricles usually biconvex or plano-convex; nuts biconvex ***Key D***

4 Stigmas 3; utricles usually 3-angled to terete; nuts 3-angled 5
5 At least lowest spike pendent, stalked ***Key E***
5 All spikes erect to patent, often sessile ***Key F***

Key A - Spike 1, terminal
1 Spike all male 2
1 Spike female at least at base 3
2 Plant densely tufted; stems usually scabrid and with 3 rounded angles above (extinct) **19. C. davalliana**
2 Plant rhizomatous; stems usually smooth and terete **18. C. dioica**
3 Stigmas 2; utricles usually biconvex or plano-convex; nuts biconvex 4
3 Stigmas 3; utricles usually 3-angled to terete; nuts 3-angled 6
4 Utricles 4-6mm, not ribbed, strongly reflexed at maturity; plants monoecious, with spikes male at apex, female at base **75. C. pulicaris**
4 Utricles 2.5-4.5mm, distinctly ribbed, not or weakly reflexed at maturity; plants usually dioecious, sometimes monoecious 5
5 Utricles 2.5-3.5mm, abruptly contracted to scabrid beak; see also couplet 2 **18. C. dioica**
5 Utricles 3.5-4.5mm, gradually contracted to smooth beak; see also couplet 2 (extinct) **19. C. davalliana**
6 Utricles erecto-patent to erect when ripe, obovoid, usually ≤3.5mm; leaves curved or curly **74. C. rupestris**
6 Utricles patent to reflexed when ripe, narrowly ovoid to narrowly ellipsoid, 3.5-7.5mm; leaves ± straight 7
7 Utricles 3.5-5(6)mm, with fine bristle (as well as style base) protruding 1-2mm from beak; female glumes c.2mm **72. C. microglochin**
7 Utricles 5-7.5mm, with only style-base protruding from beak; female glumes c.4mm **73. C. pauciflora**

Key B - Spikes >1, all ± similar in appearance
1 Stigmas 3; utricles usually 3-angled to terete; nuts 3-angled 2
1 Stigmas 2; utricles usually biconvex or plano-convex; nuts biconvex 5
2 Lowest spike erect on short rigid stalk 3
2 Lowest spike pendent to patent on distinct flexible stalk 4
3 All spikes clustered and greatly overlapping; utricles 1.8-2.5mm, greenish-brown, minutely papillose, longer than acute female glumes **63. C. norvegica**
3 Spikes not or scarcely overlapping, the lowest arising ≥1cm below next; utricles 3-4.5mm, pale green, smooth, shorter than acuminate female glumes **62. C. buxbaumii**
4 Lowest bract with sheath 0-3mm; terminal spike female at top, male below **61. C. atrata**
4 Lowest bract with sheath ≥5mm; terminal spike male at top, female below **57. C. atrofusca**
5 Rhizomes long; stems very loosely tufted or scattered 6
5 Rhizomes short; stems densely tufted 13
6 Terminal spike (not necessarily that extending highest) female at least at apex 7
6 Terminal spike male at least at apex 8
7 Utricles 4-5.5(7)mm, reddish-brown, narrowly winged **11. C. disticha**
7 Utricles 2-3mm, yellowish-green, not winged **22. C. curta**
8 Utricles narrowly winged on body 9
8 Utricles not winged on body 10
9 Terminal spike all male; glumes (male and female) ≥5mm; leaf-sheaths hyaline on side opposite blade **10. C. arenaria**
9 Terminal spike male only at apex; glumes (male and female) ≤5mm;

leaf-sheaths herbaceous on side opposite blade, except for apical hyaline rim **11. C. disticha**

10 Stems ± terete, smooth 11

10 Stems triangular in section with acute to rounded angles, rough on angles near top 12

11 Stems rarely >15cm; leaves usually curved, reaching or nearly reaching inflorescence, crescent-shaped in section when fresh; coastal **14. C. maritima**

11 Stems rarely <15cm; leaves usually straight, falling well short of inflorescence, flat or V-shaped in section when fresh; not coastal **12. C. chordorrhiza**

12 Lowest bract leaf-like, mostly at least as long as whole inflorescence; beak of utricle <1/2 as long as body; utricles pale brown **13. C. divisa**

12 Lowest bract not leaf-like, much shorter than inflorescence; beak of utricle >1/2 as long as body; utricles blackish-brown **3. C. diandra**

13 All spikes with female flowers in apical part 14

13 At least some spikes (sometimes only uppermost or lowermost) with male flowers in apical part 19

14 Lowest bract easily exceeding inflorescence, leaf-like **15. C. remota**

14 Lowest bract shorter than inflorescence, usually not leaf-like 15

15 Spikes not longer than wide, each with <10 utricles; utricles patent **17. C. echinata**

15 Spikes longer than wide, each usually with >10 utricles; utricles erect to erecto-patent 16

16 Utricles winged in upper 1/2 **16. C. ovalis**

16 Utricles not winged 17

17 Utricles pale green to yellowish- or pale brownish-green **22. C. curta**

17 Utricles reddish- to dark-brown 18

18 Spikes (5)8-12(18); utricles divaricate, without slit in beak; lowland wet places **20. C. elongata**

18 Spikes (2)3-4(6); utricles appressed, with slit down back of beak; high mountains **21. C. lachenalii**

19 Utricles 2-2.6mm **6. C. vulpinoidea**

19 Utricles ≥2.7mm 20

20 Utricles biconvex (weakly to strongly convex on adaxial and strongly convex on abaxial side) 21

20 Utricles plano-convex (flat on adaxial and weakly convex on abaxial side) 23

21 Utricles conspicuously winged in upper 1/2 **1. C. paniculata**

21 Utricles not or scarcely winged 22

22 Lowest leaf-sheaths remaining whole; all spikes usually sessile **3. C. diandra**

22 Lowest leaf-sheaths decaying with fibres; lowest spikes usually stalked **2. C. appropinquata**

23 Stems >2mm wide; leaves mostly ≥4mm wide; utricles with distinct, often prominent veins 24

23 Stems <2mm wide; leaves mostly ≤4mm wide; utricles with obscure veins 25

24 Ligule truncate; leaf-sheaths transversely wrinkled on side opposite blade; utricles matt, papillose, with slit down back of beak **4. C. vulpina**

24 Ligule acute; leaf-sheaths not wrinkled on side opposite blade; utricles shiny, smooth, without slit in beak **5. C. otrubae**

25 Roots and often base of plant purple-tinged; ligule acute to obtuse, distinctly longer than wide; utricles thickened and corky at base **7. C. spicata**

25 Plant not purple-tinged; ligule rounded at apex, c. as long as wide; utricles not thickened at base 26

26 Lowest 2-4 spikes or clusters of spikes separated by a gap >2x their own length; ripe utricles appressed to axis **9a. C. divulsa ssp. divulsa**

26 Lowest spikes separated by a gap ≤2x their own length; ripe utricles divaricate from axis 27

27 Utricles 4.5-5mm, cuneate at base; inflorescence 3-5(8)cm **9b. C. divulsa ssp. leersii**

27 Utricles 2.6-4.5mm, truncate to rounded at base; inflorescence (1)2-3(4)cm 28

28 Utricles (3.5)4-4.5mm; female glumes shorter and darker than the utricles **8a. C. muricata ssp. muricata**

28 Utricles 2.6-3.5(4)mm; female glumes nearly as long as and similar in colour to or paler than the utricles **8b. C. muricata ssp. lamprocarpa**

Key C - Spikes >1, the upper (male) different in appearance from the lower (female); utricles pubescent

1 Utricles with conspicuously bifid beak ≥0.5mm 2

1 Utricles with truncate to notched beak 0-0.5mm 3

2 Lower leaf-sheaths pubescent; utricles 4.5-7mm, with beak 1.5-2.5mm **23. C. hirta**

2 Leaf-sheaths glabrous; utricles 3.5-5mm, with beak 0.5-1mm **24. C. lasiocarpa**

3 Rhizomes extended; stems not or loosely tufted, often borne singly 4

3 Rhizomes very short; stems densely tufted 7

4 Male spikes (1)2-3; lowest female spike clearly stalked, pendent to erecto-patent **36. C. flacca**

4 Male spike 1; lowest female spike sessile or with concealed stalk, erect 5

5 Leaves ± glaucous, erect; lowest living leaf-sheaths reddish-brown; stems usually >20cm **53. C. filiformis**

5 Leaves not glaucous, usually ± recurved; lowest living leaf-sheaths mid- to dark-brown; stems usually <20cm 6

6 Lowest bract with sheath 3-5mm; female glumes acute, green to brown, with 0 or narrow scarious border **52. C. caryophyllea**

6 Lowest bract with sheath 0-2mm; female glumes obtuse to rounded, purplish-black, with wide scarious border **54. C. ericetorum**

7 Inflorescence occupying >1/2 of stem length; stems much shorter than leaves; female spikes with 2-4 flowers **51. C. humilis**

7 Inflorescence occupying <1/4 of stem length; stems usually longer than leaves; female spikes usually with >4 flowers 8

8 Flowering stems arising laterally, from leaf axils, leafless; female spikes ≤3mm wide, overtopping male 9

8 Flowering stems terminal, leafy at base; female spikes ≥4mm wide, falling short of top of male 10

9 Utricles 3-4mm; female glumes purplish-brown, c. as long as utricles; female spikes arising 1 above the other; basal leaf-sheaths crimson **49. C. digitata**

9 Utricles 2-3mm; female glumes pale brown, much shorter than utricles; female spikes all arising at ± same point; basal leaf-sheaths brown **50. C. ornithopoda**

10 Lowest bract with sheath 3-5mm **52. C. caryophyllea**

10 Lowest bract with sheath 0-2mm 11

11 Lowest bract usually green, leaf-like; female glumes brown or reddish-brown; beak of utricle 0.3-0.5mm **56. C. pilulifera**

11 Lowest bract brown, glume-like or bristle-like; female glumes purplish-black; beak of utricle ≤0.3mm 12

12 Female glumes obtuse, with scarious minutely pubescent margin; utricles 2-3mm; leaves mostly >2mm wide, rigid, recurved **54. C. ericetorum**

12 Female glumes subacute (to obtuse) and mucronate, with hyaline (but not scarious) glabrous margin; utricles 3-4.5mm; leaves mostly <2mm wide, soft, ± erect **55. C. montana**

Key D - Spikes >1, the upper (male) different in appearance from the lower (female); utricles glabrous; stigmas 2

1 Utricles with distinct forked or notched beak >0.3mm 2

1 Utricles with 0 or indistinct truncate or minutely notched beak ≤0.3mm 5

2 Female glumes almost entirely hyaline so that female spikes are silvery-white **71. C. buchananii**

2 Female glumes hyaline only at edges; female spikes not silvery-white 3

3 Utricles not inflated; female glumes 3-4mm, acute to acuminate; female spikes up to 5cm **25. C. acutiformis**

3 Utricles inflated; female glumes 2-3mm, subacute; female spikes up to 3cm 4

4 Utricles 3-3.5mm, containing nut, ± not ribbed **31. C. saxatilis**

4 Utricles 4-5mm, empty, distinctly ribbed **30. C. x grahamii**

5 At least some female glumes with apical points >1/2 as long as rest of glume; glumes ≥2x as long as utricles **64. C. recta**

5 Female glumes rounded or obtuse to acuminate with 0 or short apical point (except sometimes the lowest 1(-3)); glumes ≤1.5x as long as utricles 6

6 Leaf-sheaths breaking into conspicuous ladder-like fibres on side opposite blade; stems densely tufted, often forming large tussocks **69. C. elata**

6 Leaf-sheaths not breaking into fibres; stems usually scattered, sometimes tufted but not tussock-forming 7

7 Utricles 3.5-5mm, prominently veined; female glumes 3-veined (extinct) **67. C. trinervis**

7 Utricles 2-3.5mm; female glumes 1-veined, or 3-veined and then utricles veinless 8

8 Utricles without visible veins 9

8 Utricles with distinct faint to ± prominent veins 10

9 Lowest bract exceeding inflorescence; stems usually >25cm, with 3 rounded angles, brittle; female glumes often 3-veined **65. C. aquatilis**

9 Lowest bract shorter than inflorescence; stems usually <25cm, with 3 acute angles, not brittle; female glumes 1-veined **70. C. bigelowii**

10 Leaves 1-3(5)mm wide, the margins rolling inwards on drying; lowest bract rarely as long as inflorescence; male spike usually 1 **68. C. nigra**

10 Leaves 3-10mm wide, the margins rolling outwards on drying; lowest bract usually exceeding inflorescence; male spikes usually 2-4 **66. C. acuta**

Key E - Spikes >1, the upper (male) different in appearance from the lower (female); utricles glabrous; stigmas 3; lowest spike pendent

1 Lower leaf-sheaths and lowerside of blades pubescent **48. C. pallescens**

1 Leaf-sheaths and blades glabrous 2

2 Utricles with distinct forked or notched beak usually ≥1mm (<1mm in *C. atrofusca* and *C. acutiformis*) 3

2 Utricles with beak 0 or ≤1mm and with truncate, oblique or very slightly notched apex 12

3 Male spikes ≥2 4

3 Male spike 1 8

4 Rhizomes extended; stems not or loosely tufted, often borne singly 5

4 Rhizomes very short; stems densely tufted 7

5 Female glumes 6-10mm, exceeding utricles **26. C. riparia**

5 Female glumes 4-6mm, mostly shorter than utricles 6

6 Utricles 3.5-5mm, with beak <1mm **25. C. acutiformis**

6 Utricles (4)5-8mm, with beak 1.5-2.5mm **29. C. vesicaria**

7 Female spikes 6-8mm wide, on peduncles with exposed portion usually shorter than spike **40. C. laevigata**

7 Female spikes 3-5mm wide, on peduncles with exposed portion usually much longer than spike **33. C. sylvatica**

8 Female glumes (except midrib) and utricles both purplish-black **57. C. atrofusca**

8 Female glumes and/or utricles brownish or greenish 9

9 Utricles with smooth beaks; lower female spikes with peduncles >1/2 exposed 10

9 Utricles with scabrid beaks; lower female spikes with peduncles >1/2 ensheathed 11

10 Female spikes 3-5mm wide; female glumes 3-5mm; ligules <5mm **33. C. sylvatica**

10 Female spikes 6-10mm wide; female glumes 5-10mm; ligules >5mm **27. C. pseudocyperus**

11 Leaves 5-12mm wide; female glumes acuminate; ligules 7-15mm **40. C. laevigata**

11 Leaves 2-5(7)mm wide; female glumes obtuse and mucronate; ligules 1-2mm **41. C. binervis**

12 Male spikes (1)2-3; utricles papillose **36. C. flacca**

12 Male spike usually 1; utricles not papillose 13

13 Rhizomes very short; stems densely tufted 14

13 Rhizomes extended; stems not or loosely tufted, often borne singly 16

14 Female spikes ≤2.5cm, arising very close together; all leaves <3mm wide; plant rarely >30cm **34. C. capillaris**

14 Female spikes ≥2.5cm, well spaced out along stem; largest leaves >4mm wide; plant rarely <30cm 15

15 Female spikes <3mm wide, with peduncle c.1/2 exposed **35. C. strigosa**

15 Female spikes >3mm wide, with peduncle ± entirely ensheathed **32. C. pendula**

16 Female glumes distinctly narrower than utricles, acuminate at apex, 5-6.5mm, >1.5x as long as utricles; lowest spike with 1-2 male flowers at base **60. C. magellanica**

16 Female glumes at least as wide as utricles, acute to obtuse (sometimes mucronate) at apex, 3-4.5mm, <1.5x as long as utricles; lowest spike entirely female 17

17 Female spikes 3-4mm wide, with 5-8 flowers; utricles ± beakless; stems usually smooth **59. C. rariflora**

17 Female spikes 5-7mm wide, with 7-20 flowers; utricles with distinct beak 0.1-0.5mm; stems usually rough distally **58. C. limosa**

Key F - Spikes >1, the upper (male) different in appearance from the lower (female); utricles glabrous; stigmas 3; lowest spike erect to patent

1 Lower leaf-sheaths and lowerside of blades pubescent **48. C. pallescens**

1 Leaf-sheaths and blades glabrous 2

2 Utricles papillose **36. C. flacca**

2 Utricles not papillose 3

3 Lowest bract not sheathing at base, or with sheath ≤2mm 4

3 Lowest bract with distinct sheathing base >3mm 9

4 Male glumes 7-9mm, acuminate; female glumes longer than utricles **26. C. riparia**

4 Male glumes 3-6mm, obtuse to acute; female glumes shorter than utricles 5

5 Beak >1mm 6

5 Beak <1mm 7

6 Utricles 3.5-6.5mm, usually patent, abruptly contracted into beak 1-1.5mm; female glumes acute **28. C. rostrata**

6 Utricles (4)5-8mm, usually erecto-patent, gradually contracted into beak 1.5-2.5mm; female glumes acuminate **29. C. vesicaria**

7 Stems solid, triangular in section with concave faces and acute angles; all or most leaves >5mm wide; female spikes 2-5cm **25. C. acutiformis**

7 Stems hollow, triangular in section with flat to convex faces and rounded angles; all or most leaves <5mm wide; female spikes 1-3cm 8

8 Utricles 3-3.5mm, containing nut, ± not ribbed **31. C. saxatilis**

8 Utricles 4-5mm, empty, distinctly ribbed **30. C. x grahamii**

9 Rhizomes extended; stems not or loosely tufted 10

9 Rhizomes very short; stems densely tufted 14

10 Utricles ≥5mm, with beak ≥1.5mm 11

10 Utricles ≤5mm, with beak <1.5mm 12

11 Female spikes <10-flowered; utricles with beak ≥2.5mm; male spike 1 **39. C. depauperata**

11 Female spikes >10-flowered; utricles with beak <2.5mm; male spikes 2-4 **29. C. vesicaria**

12 Utricles with scabrid, clearly bifid beak **45. C. hostiana**

12 Utricles with smooth, truncate to shallowly notched, often very short beak 13

13 Sheaths of lowest bract inflated, loose from stem; utricles with beak 0.5-1mm; leaves green or yellowish-green **38. C. vaginata**

13 Sheaths of lowest bract not inflated, close to stem; utricles with beak <0.5mm; leaves glaucous **37. C. panicea**

14 At least 1/2 of female spikes close-set to terminal male spike; 1-several bracts far exceeding inflorescence 15

14 At most 1 female spike close-set to terminal male spike; bracts usually shorter than inflorescence, sometimes just exceeding it 19

15 Utricles erecto-patent, greyish-green often purple-blotched; leaves dark- or greyish-green, deeply channelled and/or with inrolled margins **44. C. extensa**

15 At least lower utricles in each spike patent to reflexed, bright- or yellowish-green; leaves bright- or yellowish-green, flat or V-shaped 16

16 Beaks of utricles curved or bent, usually ≥1/2 as long as the usually ± curved body 17

16 Beaks of utricles straight, usually <1/2 as long as and continuing line of ± straight body at least when fresh 18

17 Utricles 5.5-6.5mm; male spike usually subsessile; leaves ≤7mm wide, ≥2/3 as long as stems (rare) **46. C. flava**

17 Utricles 3.5-5mm; male spike usually clearly stalked; leaves ≤4mm wide, <2/3 as long as stems (common) **47a. C. viridula ssp. brachyrrhyncha**

18 Utricles 3-4mm, with beak 0.8-1.3mm; male spike usually clearly stalked; lowest female spike usually distant from others **47b. C. viridula ssp. oedocarpa**

18 Utricles 1.75-3.5mm, with beak 0.25-1mm; male spike usually sessile; lowest female spike usually bunched with others **47c. C. viridula ssp. viridula**

19 Female spikes 2-3mm wide, lax-flowered; apex of sheath of uppermost stem-leaves (not bracts) truncate on side opposite blade **35. C. strigosa**

19 Female spikes ≥4mm wide, dense-flowered; apex of sheath of uppermost stem-leaves (not bracts) either concave on side opposite blade or with an apical projection 20
20 Utricles patent 21
20 Utricles erecto-patent to erect 22
21 Utricles yellowish-green, not shiny, contrasting with female glumes (dark brown with broad scarious margins) **45. C. hostiana**
21 Utricles pale green, often minutely dark-dotted, shiny, scarcely contrasting with female glumes (pale brown with narrow scarious margins) **43. C. punctata**
22 Leaves 5-12mm wide; female glumes acuminate; ligules 7-15mm **40. C. laevigata**
22 Leaves 2-5(7)mm wide; female glumes acute to obtuse, often mucronate or apiculate; ligules 1-3mm 23
23 Female spikes 1.5-4.5cm; female glumes dark reddish- or blackish-brown; utricles with 2 conspicuous green lateral ribs distinct from others **41. C. binervis**
23 Female spikes 1-2cm; female glumes either pale reddish-brown or dark brown with broad scarious margins; utricles with several ± equally prominent ribs 24
24 Leaf-blades and lowest bract rather abruptly contracted to narrow parallel-sided point; utricles with beak 0.8-1.2mm; female glumes dark brown with broad scarious margins **45. C. hostiana**
24 Leaf-blades and lowest bract gradually contracted to apex; utricles with beak 0.7-1mm; female glumes pale- to mid-brown with narrow scarious margins **42. C. distans**

Other spp. - **C. crawfordii** Fernald, from N America, is a rare wool-alien formerly ± natd in SE En; it resembles *C. ovalis* but differs in more numerous (7-15) spikes, shorter female glumes (2.5-3mm), and longer leaves (± equalling stems). 3 arctic spp. were recorded from S Uist (Outer Hebrides) and Rhum (N Ebudes) in 1940s but have not been seen recently and were probably planted: **C. capitata** L., **C. bicolor** All. and **C. glacialis** Mack.

Subgenus* 1 - *VIGNEA (P. Beauv. ex F. Lestib.) Kük. (spp. 1-22). Spikes ≥2, all similar in appearance though distribution of male and female flowers varies, and some spikes may be all male or all female; lowest bract glume-like or bristle-like and shorter than inflorescence unless otherwise stated, without sheath; spikes sessile or nearly so, often forming compact, lobed inflorescence, the lateral ones without scale between bract and lowest glume; stigmas 2; nut biconvex. In *C. dioica* and *C. davalliana* there is a single dioecious spike; these 2 are often placed in subg. *Primocarex*.

1. C. paniculata L. - *Greater Tussock-sedge.* Densely tufted, often forming large 812
tussocks; stems to 1.5m, with acute angles, rough; utricles 3-4mm, greenish- to blackish-brown, winged towards apex, with beak 1-1.5mm; (2n=60, 62, 64). Native; by lakes and streams, in marshes and fens and wet woods, on usually base-rich soils; frequent throughout most of BI.

1 x 2. C. paniculata x C. appropinquata = C. x rotae De Not. (*C. x solstitialis* Figert) occurs in fens with the parents in W Suffolk, E Norfolk, Westmeath and formerly Offaly.

1 x 3. C. paniculata x C. diandra = C. x beckmannii Keck ex F.W. Schultz occurs in SE Yorks, Ayrs, Argyll and M Cork.

1 x 15. C. paniculata x C. remota = C. x boenninghausiana Weihe is scattered throughout Br and Ir.

1 x 22. C. paniculata x C. curta = C. x ludibunda J. Gay has been found with the parents in E Sussex, Pembs, Caerns and Brecs.

2. C. appropinquata Schumach. - *Fibrous Tussock-sedge.* Differs from *C. paniculata* R
in stems to 1m; leaves 1-3mm (not 4-7mm) wide; old leaf-sheaths becoming very 812
fibrous; and utricles 2.7-3.7mm, greyish-brown, unwinged, with beak 0.7-1.2mm; (2n=64). Native; similar places to *C. paniculata*; very local in E Anglia, N (formerly S) En, S Sc and C Ir.

3. C. diandra Schrank - *Lesser Tussock-sedge.* Usually with rhizomes extended, 812
sometimes tufted but rarely forming tussocks; differs from *C. appropinquata* in stems to 60cm, leaves 1-2mm wide; old leaf-sheaths not becoming fibrous; utricles 2.7-3.5mm, blackish-brown, with beak 1-1.5mm; (2n=60). Native; wet peaty and acid places in ditches, meadows, marshes and scrub; scattered throughout most of Br and Ir.

4. C. vulpina L. - *True Fox-sedge.* Densely tufted; stems to 1m, stout, with acute ± RR
winged rough angles; utricles 4-5mm, with beak 1-1.5mm, with epidermal cells ± 812
isodiametric (microscope); (2n=65, 68). Native; wet places on heavy soils, in ditches and marshes and by streams; local in En N to SW (formerly NE) Yorks, mainly in SE.

5. C. otrubae Podp. (*C. cuprina* (I. Sándor ex Heuff.) Nendtv. ex A. Kern. nom. 812
inval.) - *False Fox-sedge.* Differs from *C. vulpina* in utricles 4.5-6mm with beak 1-2mm and ± winged (making utricle more gradually tapered), with epidermal cells markedly elongated (microscope); stems with unwinged angles; and see Key B (couplet 24); 2n=58. Native; wet places on heavy soils in a range of habitats; frequent throughout most of BI but ± entirely coastal in Sc and very rare in N Sc.

5 x 9. C. otrubae x C. divulsa was found in W Sussex and Worcs last century and in E Kent and Oxon in 1980s.

5 x 15. C. otrubae x C. remota = C. x pseudoaxillaris K. Richt. occurs with the parents scattered in BI N to Midlothian.

6. C. vulpinoidea Michx. - *American Fox-sedge.* Densely tufted; stems to 1m, with 812
acute angles, slightly rough; utricles 2-2.6mm, with beak 0.7-1.2mm; (2n=52, 54). Intrd; wool-alien sometimes becoming natd in rough ground; scattered in Br N to C Sc; N America.

7. C. spicata Huds. (*C. contigua* Hoppe) - *Spiked Sedge.* Densely tufted; stems to 812
80cm, with acute angles, rough; utricles 4-5mm, with beak 1-2mm; (2n=58). Native; damp grassy places in fields, on banks and waysides and by rivers and ponds; frequent throughout most of En, very scattered in Wa, C & S Sc, Ir and Guernsey.

8. C. muricata L. - *Prickly Sedge.* Differs from *C. spicata* in utricles 2.6-4.5mm, with beak 0.7-1.3mm; and see Key B (couplet 25). Native.

a. Ssp. muricata. Utricles (3.5)4-4.5mm, greyish-green then dark reddish-brown; RR
female glumes 2.5-3.5mm, markedly shorter than utricles, dark- or reddish-brown; 812
(2n=56, 58). Steep, dry, limestone slopes; very rare, 5 places in Br from W Gloucs to Berwicks.

b. Ssp. lamprocarpa Celak. (*C. pairii* F.W. Schultz). Utricles 2.6-3.5(4)mm, 812
yellowish-green then dark brown; female glumes 3-4.5mm, somewhat shorter than utricles, pale brown later fading; (2n=56). Open grassy places usually on dry acid soils; frequent throughout Br N to C Sc, CI, S & E Ir.

8 x 9. C. muricata x C. divulsa was found in E Cork in 1987, later in M Cork; ?endemic.

9. C. divulsa Stokes - *Grey Sedge.* Densely tufted; stems with rounded angles, rough; 2n=58. Native.

a. Ssp. divulsa. Leaves usually greyish- to dark-green; inflorescence 5-18cm, with 812
lower spikes well spaced out; utricles 3-4(4.5)mm, usually greenish-brown when ripe; (2n=58). Hedgerows, wood borders, grassy rough ground; frequent in S Br, scattered N to N En, scattered in Ir and CI.

b. Ssp. leersii (Kneuck.) W Koch (*C. polyphylla* Kar. & Kir.). Leaves usually 812
yellowish-green; inflorescence 4-8cm, with lower spikes <2cm apart; utricles 4-4.5(4.8)mm, usually reddish-brown when ripe; (2n=52, 58). Similar places to ssp.

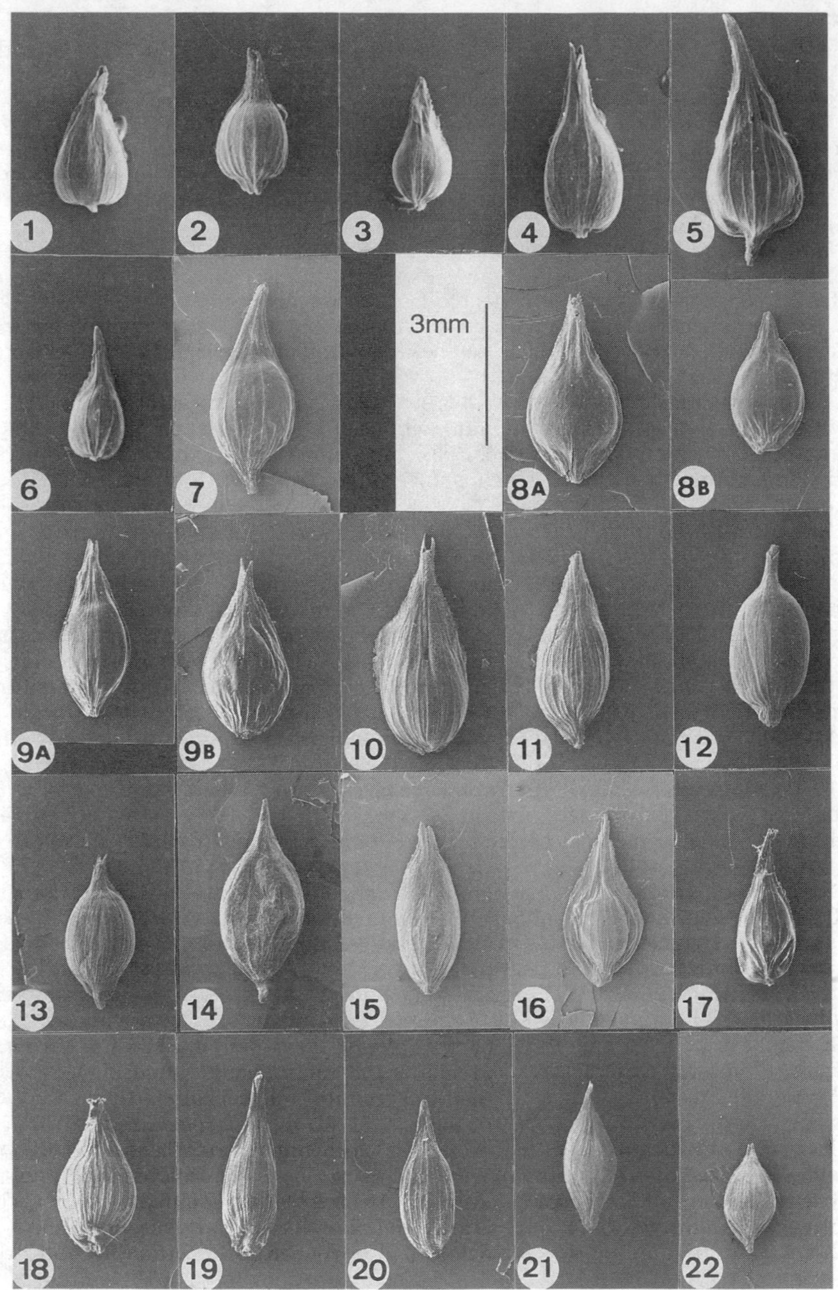

FIG 812 - Utricles of *Carex* subgenus *Vignea*. 1, *C. paniculata*. 2, *C. approprinquata*. 3, *C. diandra*. 4, *C. vulpina*. 5, *C. otrubae*. 6, *C. vulpinoidea*. 7. *C. spicata*. 8A, *C. muricata* ssp. *muricata*. 8B, *C. muricata* ssp. *lamprocarpa*. 9A, *C. divulsa* ssp. *divulsa*. 9B, *C. divulsa* ssp. *leersii*. 10, *C. arenaria*. 11, *C. disticha*. 12, *C. chordorrhiza*. 13, *C. divisa*. 14, *C. maritima*. 15, *C. remota*. 16, *C. ovalis*. 17, *C. echinata*. 18, *C. dioica*. 19, *C. davalliana*. 20, *C. elongata*. 21, *C. lachenalii*. 22, *C. curta*.

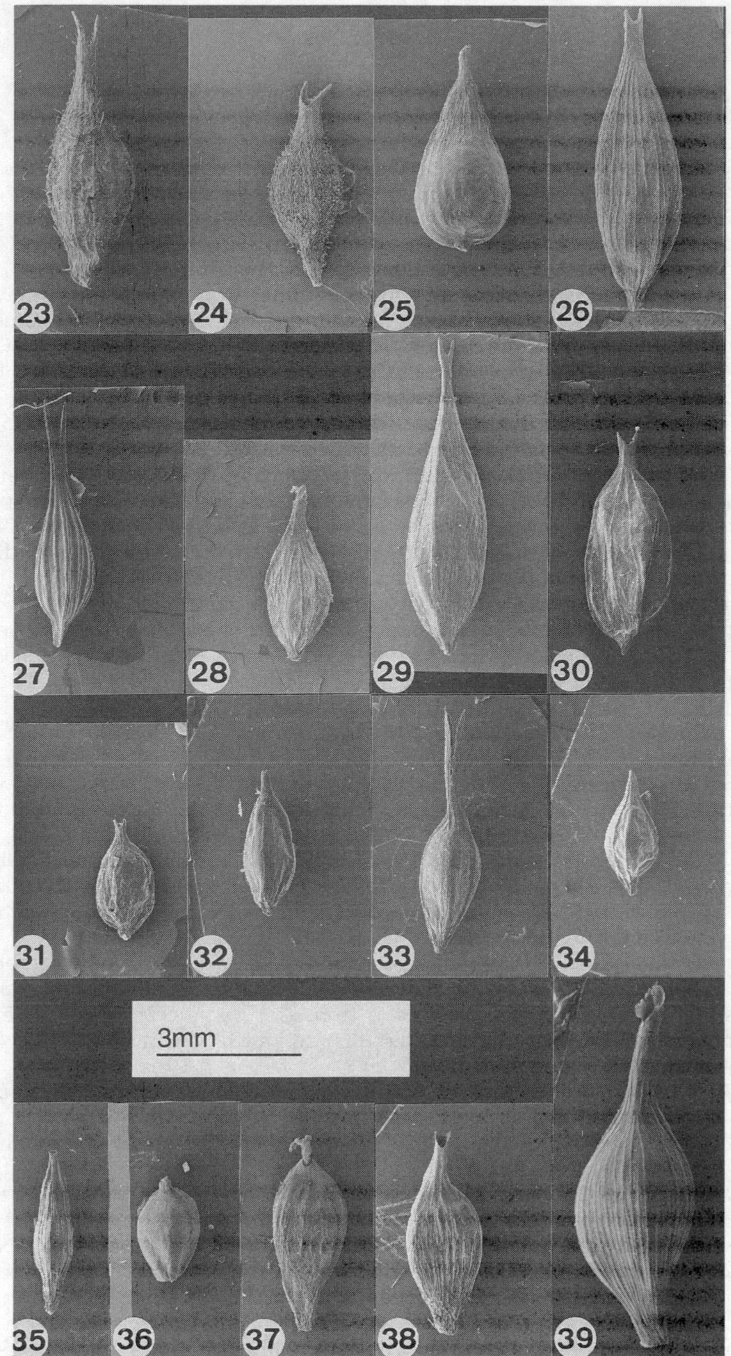

FIG 813 - Utricles of *Carex* subgenus *Carex*. 23, *C. hirta*. 24, *C. lasiocarpa*. 25, *C. acutiformis*. 26, *C. riparia*. 27, *C. pseudocyperus*. 28, *C. rostrata*. 29, *C. vesicaria*. 30, *C. x grahamii*. 31, *C. saxatilis*. 32, *C. pendula*. 33, *C. sylvatica*. 34, *C. capillaris*. 35, *C. strigosa*. 36, *C. flacca*. 37, *C. panicea*. 38, *C. vaginata*. 39, *C. depauperata*.

divulsa but only on chalk and limestone; local in Br N to Midlothian.

9 x 15. C. divulsa x C. remota = C. x emmae L. Gross has been found in E Sussex and M Cork.

10. C. arenaria L. - *Sand Sedge.* Very extensively rhizomatous, with stems borne 812
singly, to 90cm (often much less), with acute angles, slightly rough, utricles 4-5.5mm, winged, with beak 1-2mm; (2n=28, 58, 60-64). Native; bare or grassy maritime dunes or inland sandy places; round coasts of whole BI, very local inland mostly in En.

11. C. disticha Huds. - *Brown Sedge.* Rhizomatous, with stems borne singly or in 812
pairs, to 1m, with acute angles, rough; lowest bract usually bristle-like, sometimes leaf-like and exceeding inflorescence; utricles 4-5.5(7)mm with beak 1-1.5(3)mm; 2n=62. Native; marshes, fens and wet meadows; frequent throughout most of BI.

12. C. chordorrhiza L. f. - *String Sedge.* Rhizomatous, with stems borne singly, to RR
40cm, ± terete, smooth; utricles 3.5-4.5mm, with beak 0.5-1mm; (2n=60, 62, 70). 812
Native; very wet acid bogs; very rare in W Sutherland and Easterness.

13. C. divisa Huds. - *Divided Sedge.* Rhizomatous but stems often clustered, to R
80cm, with rounded angles, rough distally; lowest bract bristle- or leaf-like, usually 812
just exceeding inflorescence; utricles 3.2-4mm, with beak 0.5-0.8mm; (2n=60). Native; damp, usually brackish grassy places, in marshes, pastures and ditches; locally frequent around coasts of S & C Br, scattered N to Cheviot and Pembs (formerly S Lancs), Jersey, Co Wexford, formerly very rare in Sc.

14. C. maritima Gunnerus - *Curved Sedge.* Long-rhizomatous, with stems single or R
loosely grouped, to 18cm, terete, smooth; utricles 3.5-4.5mm, with beak 0.5-1mm; 812
(2n=60). Native; sand-dunes and damp sandy places; local on coasts of N Br S to Cheviot and Westmorland (formerly S Lancs), mainly N Sc.

15. C. remota L. - *Remote Sedge.* Densely tufted; stems to 75cm, with acute angles 812
(or compressed distally), rough distally; lowest bract leaf-like, far exceeding inflorescence; utricles 2.5-3.8mm, with beak 0.5-0.8mm; 2n=60, 62. Native; woods, hedgerows, shady banks and ditchsides; frequent to common throughout most of BI.

16. C. ovalis Gooden. - *Oval Sedge.* Densely tufted; stems to (40)90cm, with acute 812
angles, rough distally; lowest bract bristle-like, sometimes as long as inflorescence; utricles 3.8-5mm, with beak 1-1.5mm, winged; 2n=64. Native; damp or dry grassy places; common throughout BI.

17. C. echinata Murray - *Star Sedge.* Densely tufted; stems to 40cm, with rounded 812
angles or subterete, rough distally; lowest bract glume-like or bristle-like, rarely as long as inflorescence; utricles 2.8-4mm, with beak 1-1.5mm; (2n=58). Native; wide range of acid to basic bogs and marshes; throughout BI, common in W & N, very scattered in C & E En.

17 x 18. C. echinata x C. dioica = C. x gaudiniana Guthnick has been found in Denbs and W Mayo.

17 x 22. C. echinata x C. curta = C. x biharica Simonk. occurs on mountains in C Sc.

18. C. dioica L. - *Dioecious Sedge.* Shortly rhizomatous, with stems loosely tufted, 812
to (20)30cm, terete, smooth; lowest bract 0; usually dioecious but sometimes with both sexes in the spike; utricles 2.5-3.5mm, with beak 0.7-1.3mm; 2n=52. Native; base-rich bogs and flushes; throughout most of Br, common in N, very scattered in C and absent from most of S, scattered in Ir.

19. C. davalliana Sm. - *Davall's Sedge.* Densely tufted; stems to 40cm, with RR
rounded angles, rough distally; usually dioecious, but sometimes with both sexes in 812
the spike; lowest bract 0; utricles 3.5-4.5mm, with beak 1-1.7mm; (2n=46). Possibly formerly native, now extinct; calcareous fen in N Somerset until c.1845.

20. C. elongata L. - *Elongated Sedge.* Densely tufted; stems to 80cm, with acute R
angles, rough; utricles 2.5-4mm, with beak 0.5-0.8mm; (2n=560. Native; damp 812
places in wet meadows and boggy woods, by ditches and streams; scattered in Br N to Dunbarton and in N Ir.

21. C. lachenalii Schkuhr - *Hare's-foot Sedge.* Shortly rhizomatous, with stems RR
tufted, to 20(30)cm, with rounded angles, rough distally; utricles 2.5-3.3mm, with 812
beak 0.5-0.8mm; (2n=58, 62, 64). Native; wet acid places on mountains ≥750m, usually where snow lies late; very local in C Sc.

21 x 22. C. lachenalii x C. curta = C. x helvola Blytt ex Fr. occurs with the parents in 2 places in S Aberdeen.

22. C. curta Gooden. - *White Sedge.* Shortly rhizomatous, with stems loosely 812
tufted, to 50cm, with acute angles, rough distally; utricles 2-3mm, with beak 0.5-0.7mm; (2n=56, 62). Native; wet acid places on heaths, in bogs and boggy woods and mountainsides; throughout most of Br and Ir, common in N 1/2 of Br, N Ir and Wa, very scattered elsewhere.

Subgenus* 2 - *CAREX (spp. 23-71). Spikes ≥2, usually the apical ≥1 male and the basal ≥1 female and differing obviously in appearance, sometimes with some male flowers on female spikes and vice versa; lowest bract leaf-like unless otherwise stated; spikes sessile to long-stalked, sometimes the lower ones pendent and remote from upper ones, the lateral ones usually with a scale between bract and lowest glume (often close to and ± hidden by former); stigmas usually 3 and nut with 3 rounded angles but often 2 and nut biconvex.

In *C. atrata*, *C. norvegica* and *C. buxbaumii*, and sometimes *C. atrofusca*, the terminal spike is bisexual and all spikes are rather similar in appearance, but all these have 3 stigmas, 3-angled nuts and a scale between bract and lowest glume.

23. C. hirta L. - *Hairy Sedge.* Rhizomatous or shortly so, with stems loosely 813
tufted, to 70cm, with rounded angles, ± smooth; lowest bract rarely exceeding inflorescence, with long pubescent sheath; utricles 4.5-7mm, with beak 1.5-2.5mm, pubescent; 2n=112. Native; damp grassy places in many habitats; common throughout BI except N Sc.

23 x 29. C. hirta x C. vesicaria = C. x grossii Fiek occurs in a dune-slack in Co Wicklow.

24. C. lasiocarpa Ehrh. - *Slender Sedge.* Rhizomatous; stems to 1.2m, with 813
rounded angles, smooth; lowest bract often exceeding inflorescence, with short sheath; utricles 3-5mm, with beak 0.5-1mm, pubescent; (2n=56). Native; bogs and fens; scattered throughout Br and Ir but absent from most of S & C En except E Anglia. The very long narrow leaves are distinctive.

24 x 26. C. lasiocarpa x C. riparia = C. x evoluta Hartm. has been found in N Somerset and Cambs.

25. C. acutiformis Ehrh. - *Lesser Pond-sedge.* Rhizomatous; stems tufted, to 1.5m, 813
with acute angles, rough; lowest bract usually exceeding inflorescence, with 0 or short sheath; utricles 3.5-5mm, with beak 0.3-0.8mm; 2n=38. Native; marshes, wet meadows and swamps, by ponds and streams; common throughout lowland Br and Ir, rare in N & C Sc, S & W Ir and SW En. Stigmas usually 3; fertile plants with 2 stigmas might be derivatives of *C. acutiformis* x *C. acuta*. *C. acuta* has shorter, wider utricles with a much shorter beak.

25 x 26. C. acutiformis x C. riparia = C. x sooi Jákucs was found in 1990 in E Suffolk.

25 x 29. C. acutiformis x C. vesicaria = C. x ducellieri Beauverd was found in S Hants in 1986.

25 x 66. C. acutiformis x C. acuta = C. x subgracilis Druce has been found in scattered localities in En and Wa; flowers with 2 and 3 stigmas occur in the same spike.

26. C. riparia Curtis - *Greater Pond-sedge.* Differs from *C. acutiformis* in larger 813
spikes, glumes and utricles; utricles 5-8mm, with beak 1-2mm; female glumes 7-10mm (not 4-5mm); and male spikes 3-6 (not 2-3) with acuminate (not subacute to obtuse) glumes; (2n=72). Native; similar places to *C. acutiformis* and often with (but not hybridizing with) it; common in C & S Br, scattered N to C Sc, in Ir and in CI.

26 x 29. C. riparia x C. vesicaria = C. x csomadensis Simonk. has been found in Co Wicklow, and there are more doubtful records in S En.

27. C. pseudocyperus L. - *Cyperus Sedge*. Shortly rhizomatous, with stems 813
loosely tufted, to 90cm, with acute angles, rough; lowest bract far exceeding inflorescence, with short sheath; utricles 4-5mm, with stalk-like base, with beak 1.5-2.5mm; (2n=66). Native; in marshes and swamps, by ponds, rivers and canals; locally common in C & S Br, very scattered in N Br, Ir and CI.

27 x 28. C. pseudocyperus x C. rostrata = C. x justi-schmidtii Junge was found in 1955 in W Norfolk but has not been seen recently.

28. C. rostrata Stokes - *Bottle Sedge*. Rhizomatous; stems slightly tufted or not, to 813
1m, with rounded angles, rough distally; lowest bract often exceeding inflorescence, with 0 or short sheath; utricles 3.5-6.5mm, with beak 1-1.5mm; (2n=72-74, 76, 80, 82). Native; acid swamps, lake-margins and reed-beds; throughout Br and Ir, common in N & W, absent from parts of C & S En. See *C. vesicaria* for differences.

28 x 29. C. rostrata x C. vesicaria = C. x involuta (Bab.) Syme occurs in scattered places over much of Br and Ir; 1 of the commoner hybrids.

29. C. vesicaria L. - *Bladder-sedge*. Shortly rhizomatous; stems slightly tufted, to 813
1.2m, with rounded angles, rough distally; lowest bract exceeding inflorescence, with short to long, rarely 0, sheath; utricles (4)5-8mm, with beak 1.5-2.5mm; (2n=74, 82). Native; swamps, marshes, lake margins; frequent through BI. Often confused with *C. rostrata*, but has yellowish-green (not glaucous) leaves and a non-swollen (not a swollen) stem-base.

30. C. x grahamii Boott (*C. stenolepis* auct. non Less., *C. ewingii* E.S. Marshall; *C.* 813
saxatilis x ?*C. vesicaria*) - *Mountain Bladder-sedge*. Rhizomatous, with stems in small tufts, to 50cm, with rounded angles, ± rough distally; lowest bract exceeding inflorescence or not, without sheath; utricles 4-5mm, with beak 0.5-0.9mm. Native; mountain flushes >750m, often with *C. saxatilis* but not with *C. vesicaria*; very local in C Sc.

31. C. saxatilis L. - *Russet Sedge*. Differs from *C. x grahamii* in stems to 40cm; R
female spikes ≤2cm (not ≤3cm); and see Key F (couplet 8); (2n=80). Native; wet 813
places, especially where snow lies late on mountains >750m; local in N & W Sc. Stigmas usually 3; fertile plants with 2 stigmas might be derivatives of hybrids with *C. vesicaria* or *C. rostrata*.

32. C. pendula Huds. - *Pendulous Sedge*. Densely tufted; stems to 1.8m, with 813
rounded angles, smooth; lowest bract usually shorter than inflorescence, with long sheath; utricles 3-3.5mm, with beak 0.3-0.6mm; (2n=58, 60). Native; rich heavy soils in woods and damp copses; common in S Br, decreasing to N, scattered in Ir, Sc and CI, absent from highlands and N Sc, also grown for ornament and sometimes natd.

33. C. sylvatica Huds. - *Wood-sedge*. Densely tufted; stems to 70cm, with rounded 813
angles, smooth; lowest bract sometimes longer than inflorescence, with long sheath; utricles 3-5mm, with beak 1-2.5mm; 2n=58. Native; on damp usually heavy soils in woods, hedgerows and scrub; frequent throughout most of BI, common in S Br, rare in N Sc, not in Outer Isles.

34. C. capillaris L. - *Hair Sedge*. Densely tufted; stems to (20)40cm, with rounded R
angles, smooth; lowest bract leaf-like but narrow, sometimes exceeding 813
inflorescence, with long sheath; utricles 2.5-3mm, with beak 0.5-1mm; 2n=c.54. Native; base-rich or calcareous flushes, mineral-rich bogs; local in N En and Sc, rare in Caerns.

35. C. strigosa Huds. - *Thin-spiked Wood-sedge*. Densely tufted; stems to 75cm, 813
with rounded angles or subterete, smooth; lowest bract usually shorter than inflorescence, with long sheath; utricles 3-4mm, with beak 0.3-0.5mm; 2n=66. Native; damp base-rich soils in woods or woodland clearings; locally frequent in Br N to NE Yorks, scattered in N & C Ir.

36. C. flacca Schreb. - *Glaucous Sedge*. Glaucous, rhizomatous; stems loosely 813
tufted, to 60cm, with rounded angles or subterete, smooth; lowest bract sometimes

exceeding inflorescence, with short to long sheath; utricles 2-3mm, with beak <0.3mm, papillose; 2n=76. Native; wet or dry grassland on chalk and limestone, sand-dunes and base-rich clay, mountain flushes; common throughout BI (the most widespread and one of the most variable spp.). See *C. panicea* for differences.

37. C. panicea L. - *Carnation Sedge*. Glaucous, shortly rhizomatous; stems loosely **813**
tufted, to 60cm, with rounded angles or subterete, smooth; lowest bract shorter than inflorescence, with long to medium sheath; utricles 3-4mm, with beak <0.3(0.5)mm; 2n=32. Native; wet, usually acid, heaths and moors, bogs, mountain flushes; throughout BI, common in N & W, local in SE. Vegetatively distinguished from *C. flacca* by the leaf-tips with 3 rounded angles in section (not flat).

38. C. vaginata Tausch - *Sheathed Sedge*. Rhizomatous; stems tufted, to 40cm, **R**
with rounded angles or subterete, smooth; lowest bract much shorter than **813**
inflorescence, with long loose sheath; utricles 4-4.5mm, with beak 0.5-1mm; (2n=32). Native; wet rocky places, damp slopes and flushes above 600m; local in S, C & formerly N mainland Sc.

39. C. depauperata Curtis ex With. - *Starved Wood-sedge*. Shortly rhizomatous; **RRR**
stems to 1m, loosely tufted, with rounded angles or subterete, smooth; lowest bract **813**
often exceeding inflorescence, with long sheath; utricles 7-9mm (longest of our spp.), with beak 2.5-3mm; 2n=44. Native; dry woods and hedgebanks on chalk or limestone; very rare in N Somerset, Surrey and M Cork, probably extinct in Anglesey and W Kent.

40. C. laevigata Sm. - *Smooth-stalked Sedge*. Densely tufted; stems to 1.2m, with **818**
rounded angles, smooth; lowest bract usually shorter than inflorescence, with long sheath; utricles 4-6mm, with beak 1-2mm; 2n=71, 72. Native; damp shady places, especially woods on heavy soils; scattered throughout most of BI, common in S & W Br, absent from most of C & E En and N Sc.

40 x 41. C. laevigata x C. binervis = C. x deserta Merino was found in 1961 in Caerns.

40 x 47. C. laevigata x C. viridula was found in 1970 in Merioneth; endemic. *C. viridula* ssp. *oedocarpa* was involved.

40 x 48. C. laevigata x C. pallescens was found in 1973 in Westerness; endemic.

41. C. binervis Sm. - *Green-ribbed Sedge*. Densely tufted; stems to 1.2m, subterete, **818**
smooth; lowest bract shorter than inflorescence, with long sheath; utricles 3.5-4.5mm, with beak 1-1.5mm; 2n=74. Native; damp heaths, moors, rocky places and mountainsides; frequent throughout most of BI, common in N & W Br.

41 x 43. C. binervis x C. punctata was found in 1954 in Merioneth.

41 x 47. C. binervis x C. viridula = C. x corstorphinei Druce was found in 1915 in Angus; probably endemic. *C. viridula* ssp. *oedocarpa* was involved.

42. C. distans L. - *Distant Sedge*. Densely tufted; stems to 1m, subterete, smooth; **818**
lowest bract much shorter than inflorescence, with long sheath; utricles 3-4.5mm, with beak 0.7-1mm; 2n=74. Native; brackish and fresh-water marshes, wet rocky places, mostly near sea; round most coasts of BI, frequent inland in S Br, rarely so elsewhere.

42 x 44. C. distans x C. extensa = C. x tornabenii Chiov. has been found in Merioneth and W Cornwall.

42 x 45. C. distans x C. hostiana = C. x muelleriana F.W. Schultz has been found in N Hants and Co Dublin.

42 x 47. C. distans x C. viridula = C. x binderi Podp. has been found in Kintyre and doubtful records exist from elsewhere. *C. viridula* ssp. *brachyrrhyncha* was involved.

43. C. punctata Gaudin - *Dotted Sedge*. Differs from *C. distans* in stems with **R**
rounded angles; lowest bract sometimes just exceeding inflorescence; utricles patent **818**
(not erecto-patent), usually more shiny and more strongly ribbed, often purple-dotted; 2n=68. Native; similar places to *C. distans* and often with it, but strictly coastal; local in S, W & N Ir, CI and S & W Br from S Hants to Wigtowns, formerly Berwicks.

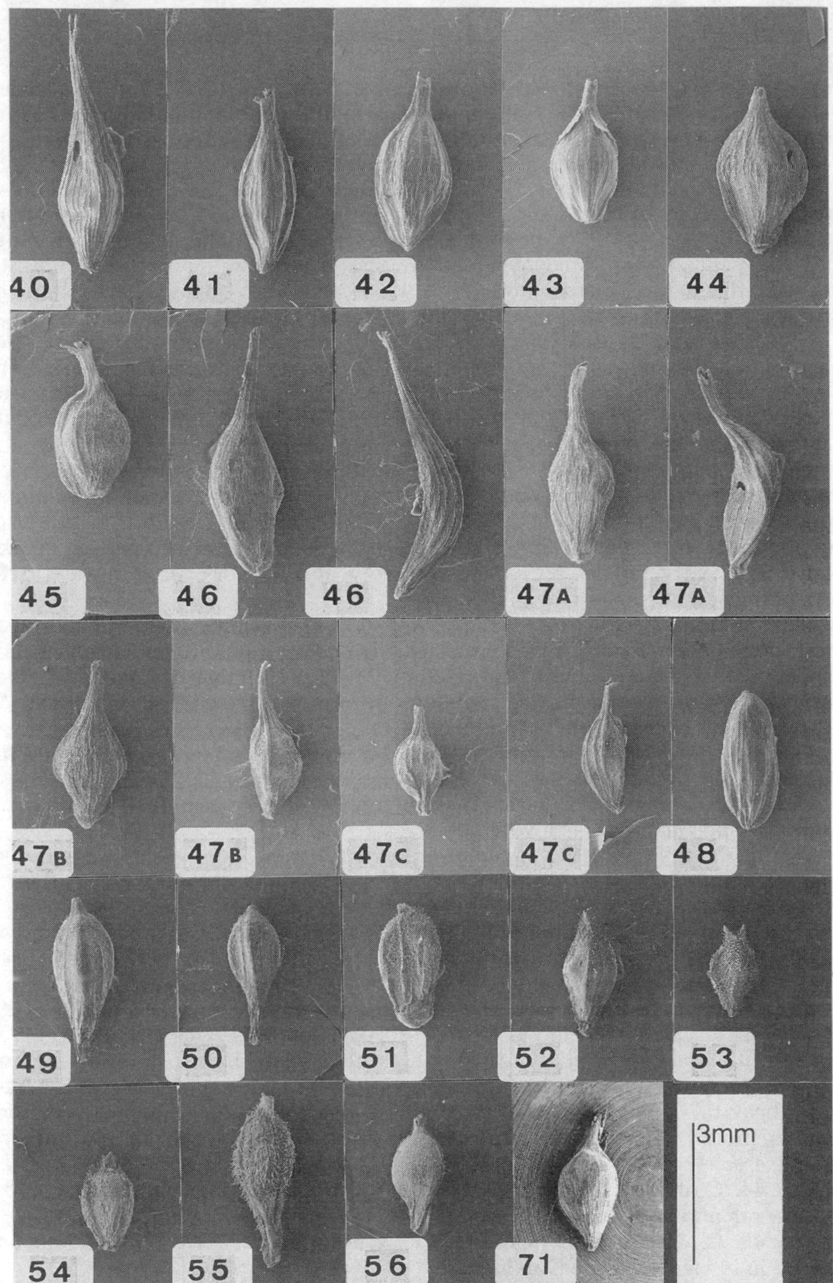

FIG 818 - Utricles of *Carex* subgenus *Carex*. 40, *C. **laevigata***. 41, *C. **binervis***. 42, *C. **distans***. 43, *C. **punctata***. 44, *C. **extensa***. 45, *C. **hostiana***. 46, *C. **flava*** (abaxial and lateral views). 47A, *C. **viridula*** ssp. ***brachyrrhyncha*** (abaxial and lateral views). 47B, *C. **viridula*** ssp. ***oedocarpa*** (abaxial and lateral views). 47C, *C. **viridula*** ssp. ***viridula*** (abaxial and lateral views). 48, *C. **pallescens***. 49, *C. **digitata***. 50, *C. **ornithopoda***. 51, *C. **humilis***. 52, *C. **caryophyllea***. 53, *C. **filiformis***. 54, *C. **ericetorum***. 55, *C. **montana***. 56, *C. **pilulifera***. 71, *C. **buchananii***.

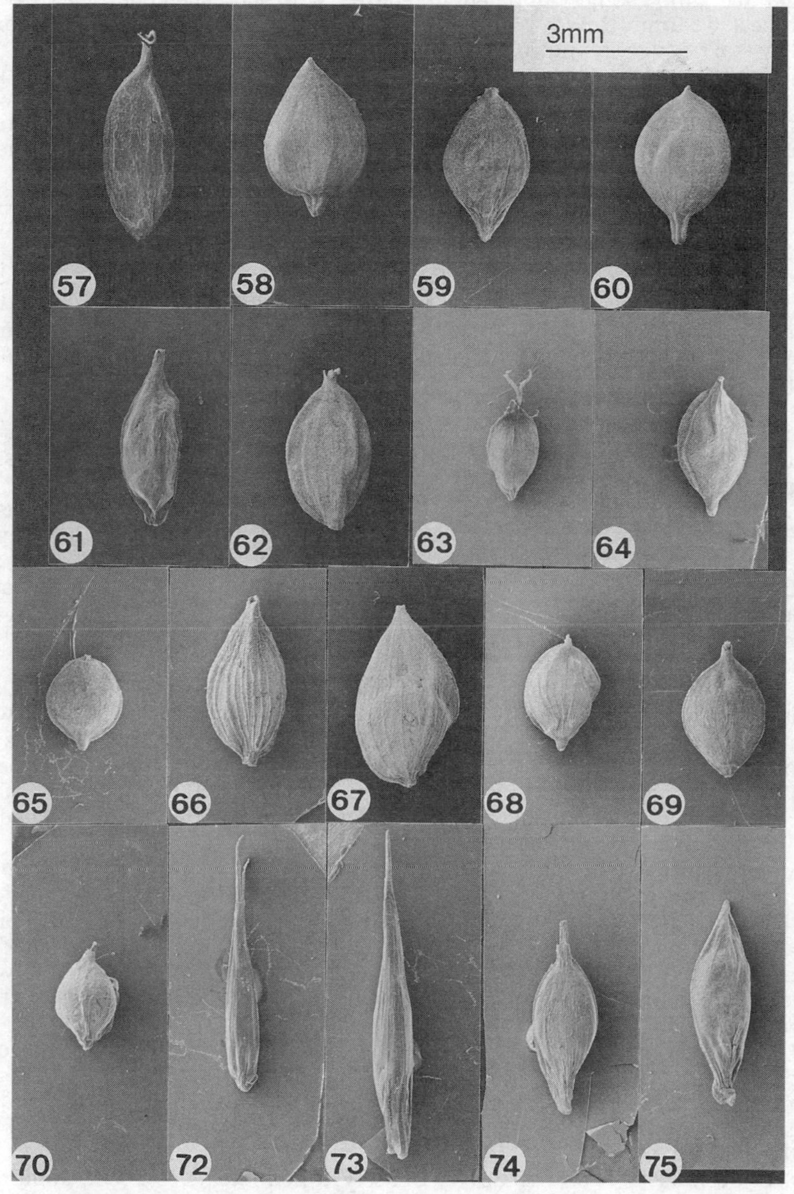

FIG 819 - Utricles of *Carex* subgenera *Carex* and *Primocarex*. 57, *C. atrofusca*. 58, *C. limosa*. 59, *C. rariflora*. 60, *C. magellanica*. 61, *C. atrata*. 62, *C. buxbaumii*. 63, *C. norvegica*. 64, *C. recta*. 65, *C. aquatilis*. 66, *C. acuta*. 67, *C. trinervis*. 68, *C. nigra*. 69, *C. elata*. 70, *C. bigelowii*. 72, *C. microglochin*. 73, *C. pauciflora*. 74, *C. rupestris*. 75, *C. pulicaris*.

44. C. extensa Gooden. - *Long-bracted Sedge.* Glaucous or bluish-greyish-green; 818
densely tufted; stems to 40cm, with rounded angles, smooth; lowest bract much longer than inflorescence but patent to reflexed, with short sheath; utricles 3-4mm, with beak 0.5-1mm; 2n=60. Native; muddy or sandy brackish places in estuaries and by sea; frequent round coasts of most of BI except parts of E coast of Br.

45. C. hostiana DC. - *Tawny Sedge.* Shortly rhizomatous; stems loosely tufted, to 818
65cm, with rounded angles, smooth; lowest bract shorter than inflorescence, with long sheath; utricles 3.5-5mm, with beak 0.8-1.2mm (slightly longer and more abruptly contracted than in *C. distans*); 2n=56. Native; marshes, flushes and fens, inland and maritime; common in W & N Br, locally frequent in Ir and rest of Br.

45 x 47. C. hostiana x C. viridula = C. x fulva Gooden. (*C. x appeliana* Zahn) occurs frequently throughout most of Br and Ir wherever *C. hostiana* meets any of the 3 sspp. of *C. viridula*; probably the commonest hybrid.

46. C. flava L. - *Large Yellow-sedge.* Densely tufted; stems to 70cm, with rounded RR
angles, smooth; lowest bract much longer than inflorescence, often patent or 818
reflexed, with short sheath; utricles 5.5-6.5mm, with reflexed beak 1.5-2.5mm; 2n=60. Native; base-rich fen by lake; now only in Roudsea Wood, Westmorland; formerly occurred in localities now supporting its hybrids and in Cumberland.

46 x 47. C. flava x C. viridula = C. x alsatica Zahn (*C. x pieperiana* Junge, *C. x ruedtii* Kneuck.) occurs (with *C. viridula* ssp. *oedocarpa* as the parent) in the Roudsea Wood locality of *C. flava*, and (with *C. viridula* ssp. *brachyrrhyncha* as the parent) in MW Yorks, Hants and NE Galway.

47. C. viridula Michx. - *Yellow-sedge.* Differs from *C. flava* as in Key F (couplets 16 & 17). Native. Fertile intermediates occur in all 3 combinations between the sspp.

a. Ssp. brachyrrhyncha (Celak.) B. Schmid (*C. lepidocarpa* Tausch, *C. marshallii* 818
A. Benn.). Stems to 70cm, usually straight; leaves usually c.1/2 as long as stems; utricles 3.5-5mm, with very abruptly contracted reflexed beak 1.3-2mm; 2n=68. Base-rich fens, flushes, and lakesides; frequent throughout most of Br and Ir except in acid areas and much of S & C En and Wa.

b. Ssp. oedocarpa (Andersson) B. Schmid (*C. demissa* Hornem.). Stems to 50cm, 818
usually curved; leaves usually nearly as long as stems; utricles 3-4mm, with abruptly contracted not or slightly reflexed beak 0.8-1.3mm (but lower utricles are wholly reflexed); 2n=70. Acid or medium base-rich fens, bogs, flushes, wet fields and by lakes; common over most of BI except parts of C & E En. Commoner than ssp. *brachyrrhyncha* everywhere except on limestone. In dried material unequal drying of utricles can cause beaks to become reflexed.

c. Ssp. viridula (*C. serotina* Mérat, *C. scandinavica* E.W. Davies, *C. bergrothii* 818
Palmgr.). Stems to 25cm (often <10cm), usually straight; leaves usually c. as long as stems; utricles 1.75-3.5mm, with abruptly contracted ± straight beak 0.2-1mm; 2n=70. Acid (mostly) or basic wet places in bogs, marshes and dune-slacks and by lakes; scattered through most of BI, frequent by coast in W & N. Plants with utricles 1.75-2mm with beaks 0.2-0.3mm, mainly in N & W Sc, have been separated as var. **pulchella** (Lönnr.) B. Schmid (*C. scandinavica*).

48. C. pallescens L. - *Pale Sedge.* Densely tufted; stems to 60cm, with acute 818
angles, rough distally; lowest bract exceeding inflorescence, ± without sheath, crimped at base; utricles 2.5-3.5mm, with beak ≤0.1mm; 2n=64. Native; damp grassland, woodland clearings and stream-banks; frequent over most of Br, scattered in Ir.

49. C. digitata L. - *Fingered Sedge.* Densely tufted; stems to 25cm, subterete, R
smooth; lowest bract glume-like, sheathing; utricles 3-4mm, pubescent, with beak 818
0.2-0.5mm; 2n=48, 50. Native; open woodland, scrub and grassy rocky slopes on chalk and limestone; very local in En from N Somerset (formerly Dorset) and Mons to Westmorland and NE Yorks.

50. C. ornithopoda Willd. - *Bird's-foot Sedge.* Differs from *C. digitata* in stems to RR
15(20)cm; utricle beak 0.1-0.3mm; and see Key C (couplet 9); (2n=46, 54). Native; 818
short limestone grassland; very local in Derbys, Westmorland, Cumberland and

MW & NW Yorks.

51. C. humilis Leyss. - *Dwarf Sedge*. Densely tufted; stems to 10(15)cm, subterete, R
smooth; lowest bract glume-like, sheathing and nearly enclosing lowest female 818
spike; utricles 2-3mm, pubescent, with beak ≤0.1mm; 2n=36. Native; short limestone grassland; very locally common in SW En from Dorset to Herefs.

52. C. caryophyllea Latourr. - *Spring-sedge*. Shortly rhizomatous; stems loosely 818
tufted, to 30cm, with acute to rounded angles, smooth; lowest bract leaf- to bristle-like, with sheath 3-5mm; utricles 2-3mm, pubescent, with beak <0.3mm; 2n=68. Native; acid or basic dry to damp short grassland; frequent throughout BI but rare in N Sc.

53. C. filiformis L. (*C. tomentosa* L.) - *Downy-fruited Sedge*. Rhizomatous; stems RR
not or loosely tufted, to 50cm, with rounded angles, smooth to slightly rough 818
distally; lowest bract shorter than to ± as long as inflorescence, with 0 or very short sheath; utricles 2-3mm, pubescent, with beak <0.4mm; 2n=48. Native; damp grassy places in fields and on waysides and in woodland rides; very local in Surrey, N Wilts, Oxon and E Gloucs, formerly W Gloucs and Middlesex.

54. C. ericetorum Pollich - *Rare Spring-sedge*. Shortly rhizomatous; stems loosely R
mat-forming, to 20cm, with rounded angles, smooth; lowest bract glume- to bristle- 818
like, with 0 or very short sheath; utricles 2-3mm, pubescent, with beak ≤0.3mm; 2n=30. Native; dry short calcareous grassland; very local in E & N En from Cambs and W Suffolk to Co Durham and Westmorland.

55. C. montana L. - *Soft-leaved Sedge*. Rhizomatous; stems loosely tufted, to R
40cm, with acute to rounded angles, rough distally; lowest bract glume- to bristle- 818
like, with 0 or very short sheath; utricles 3-4mm, pubescent, with beak <0.3mm; 2n=38. Native; wet or dry grassy places in open or light shade on acid or basic soils, very local in S Br N to Derbys.

56. C. pilulifera L. - *Pill Sedge*. Densely tufted; stems to 40cm, with acute angles, 818
rough distally; lowest bract bristle- to leaf-like, shorter than to rarely slightly longer than inflorescence, with 0 or very short sheath; utricles 2-3.5mm, pubescent, with beak 0.3-0.5mm; 2n=18. Native; grassy or barish places in open or woodland on base-poor sandy or peaty soils; frequent ± throughout BI.

57. C. atrofusca Schkuhr - *Scorched Alpine-sedge*. Shortly rhizomatous; stems RR
loosely tufted, to 35cm, with acute to rounded angles, smooth; lowest bract leaf- to 819
bristle-like, shorter than inflorescence, with short to long sheath; utricles 4-4.5mm, with beak 0.3-0.7mm; (2n=36, 38, 40). Native; mountain flushes 540-1050m; very rare in M Perth, Westerness and Argyll.

58. C. limosa L. - *Bog-sedge*. Rhizomatous; stems not or loosely tufted, to 40cm, 819
with rounded angles, usually rough distally; lowest bract c. as long as spike, with 0 or very short sheath; utricles 3-4mm, with beak ≤0.2mm; (2n=56, 62, 64). Native; very wet blanket- or valley-bogs; locally frequent in N & C Ir and N & W Br, very scattered in and absent from most of S Ir and En.

59. C. rariflora (Wahlenb.) Sm. - *Mountain Bog-sedge*. Shortly rhizomatous; stems RR
loosely tufted or carpet-forming, to 20cm, with rounded angles, usually smooth; 819
lowest bract shorter than inflorescence, with short sheath; utricles 3-4mm, with beak ≤0.2mm; (2n=50, 52, 54). Native; wet peaty mountain slopes or in flushes at 750-1050m, where snow lies late; very local in CE Sc.

60. C. magellanica Lam. (*C. paupercula* Michx.) - *Tall Bog-sedge*. Shortly R
rhizomatous; stems to 40cm, with rounded angles, smooth; lowest bract usually 819
exceeding inflorescence, with 0 or very short sheath; utricles 3-4mm, with beak ≤0.1mm; 2n=58, 60. Native; wet bogs with some water movement, often among *Sphagnum;* scattered in Br from Lake District to NW Sc, N & WC Wa, Co Antrim. Our plant is ssp. **irrigua** (Wahlenb.) Hiitonen.

61. C. atrata L. - *Black Alpine-sedge*. Shortly rhizomatous; stems loosely tufted, to R
55cm, with acute angles, smooth or rough distally; lowest bract exceeding 819
inflorescence, with 0 or very short sheath; utricles 3-4mm, with beak 0.3-0.5mm; 2n=54. Native; wet rock-ledges above 720m; locally frequent in C & NW Sc, very

rare in S Sc, Lake District and Caerns.

62. C. buxbaumii Wahlenb. - *Club Sedge.* Shortly rhizomatous; stems not or loosely tufted, to 70cm, with acute angles, smooth; lowest bract exceeding inflorescence or not, with 0 or very short sheath; utricles 3-4.5mm, with beak 0-0.2mm; (2n=74, 106). Native; wet fens; very rare in Argyll, Westerness and Easterness, formerly Co Antrim. **RR** **819**

63. C. norvegica Retz. - *Close-headed Alpine-sedge.* Densely tufted; stems to 30cm, with rounded angles, smooth or rough distally; lowest bract exceeding inflorescence, with 0 or very short sheath; utricles 1.8-2.5mm, minutely papillose, with beak 0.1-0.3mm; (2n=54, 56, 66). Native; N-facing damp rock-ledges and rocky slopes at 690-990m, where snow lies late; 5 places in C Sc. **RR** **819**

64. C. recta Boott - *Estuarine Sedge.* Rhizomatous; stems tufted, to 1.1m, with acute to rounded angles, smooth; lowest bract exceeding inflorescence, with 0 or very short sheath; utricles 2.5-3mm, with beak ≤0.3mm; 2n=c.74 (c.70, 84). Native; forming extensive patches in wet estuarine areas; very local in NE Sc S to Easterness. **RR** **819**

64 x 65. C. recta x C. aquatilis = C. x grantii A. Benn. occurs on banks of the Wick River, Caithness, where it forms a hybrid swarm.

65. C. aquatilis Wahlenb. - *Water Sedge.* Differs from *C. recta* in leaves green (not glaucous) on upperside; stems with rounded angles or subterete, brittle (not flexible); utricles 2-2.5mm; and female glumes subacute to obtuse and shorter than utricles (not long-pointed and much longer than utricles); 2n=76, 77. Native; swampy areas by lakes and rivers and in marshes; locally frequent in Br N from Lake District and Wa, scattered in Ir. **819**

65 x 68. C. aquatilis x C. nigra = C. x hibernica A. Benn. occurs in scattered places in Br N from C Sc, W Ir; 2n=c.80, c.120.

65 x 70. C. aquatilis x C. bigelowii = C. x limula Fr. occurs at c.800-900m in E highlands of C Sc.

66. C. acuta L. - *Slender Tufted-sedge.* Differs from *C. recta* in stems with acute angles, rough distally; utricles 2-3.5mm, obviously ribbed (not ± smooth); and acuminate female glumes usually longer than utricles; 2n=82, 83, 84. Native; by ponds, ditches, canals and rivers and in marshes; locally frequent throughout BI N to C Sc. **819**

66 x 68. C. acuta x C. nigra (? = *C x elytroides* Fr.) occurs in very scattered places in Br, including some sites in N Br without *C. acuta.*

66 x 69. C. acuta x C. elata = C. x prolixa Fr. has been found in a few places in S En and S Wa.

67. C. trinervis Degl. - *Three-nerved Sedge.* Rhizomatous; stems scarcely tufted, to 40cm, with rounded angles, smooth; lowest bract exceeding inflorescence, with 0 or very short sheath; utricles 3.5-5mm, distinctly veined, with beak ≤0.1mm. Possibly formerly native, now extinct; found in inland area of E Norfolk in 1869 (dune-slack sp. on Continent), probably with or entirely *C. trinervis* x *C. nigra.* **RR** **819**

68. C. nigra (L.) Reichard - *Common Sedge.* Rhizomatous to very shortly so; stems tufted to single, to 70cm, with rounded angles, smooth; lowest bract shorter than to c. as long as inflorescence, without sheath; utricles 2.5-3.5mm, distinctly ribbed, with beak ≤0.2mm; 2n=83, 84, 85. Native; wide range of wet acid to basic places, especially marshes and flushes; common throughout BI. Probably our commonest and most variable sp. See *C. bigelowii* for differences. **819**

68 x 69. C. nigra x C. elata = C. x turfosa Fr. occurs in scattered places in Ir and C & S Br; 2n=79.

68 x 70. C. nigra x C. bigelowii = C. x decolorans Wimm. occurs in the same area as *C. aquatilis* x *C. bigelowii*, often near *C. aquatilis* x *C. nigra* as well.

69. C. elata All. - *Tufted-sedge.* Densely tufted, often tussock-forming; stems to 1m, with acute angles, rough distally; lowest bract shorter than inflorescence, with 0 or very short sheath; utricles 2.5-4mm, distinctly ribbed, with beak ≤0.3mm; 2n=74, 75, 76. Native; bogs, fens, reedswamps and by rivers and lakes; locally **819**

frequent in Ir and C Br, very scattered in S En and N to C Sc.

70. C. bigelowii Torr. ex Schwein. - *Stiff Sedge*. Shortly rhizomatous; stems 819
scarcely tufted but often close, to 30cm, with acute angles, rough; lowest bract shorter than inflorescence, without sheath; utricles 2-3mm, smooth, with beak ≤0.2mm; 2n=68, 69, 70, 71. Native; stony and heathy areas, often where snow lies late, and in flushed gullies, above 600m; common in highlands of Sc, scattered S to N En and N Wa, very scattered in N & W (formerly E) Ir. Usually distinguishable from mountain variants of *C. nigra* by its reddish-brown (not blackish-brown) basal leaf-sheaths.

71. C. buchananii Berggr. - *Silver-spiked Sedge*. Densely tufted, with stems and 818
leaves strongly red-coloured; stems to 60cm, subterete, smooth; lowest bract much longer than inflorescence, with long sheath; utricles 2.5-3mm, smooth, with notched beak 0.5-1mm. Intrd; natd on rough ground; 1 site in Glasgow (Lanarks) since 1990, self-sowing and increasing; New Zealand. The ± completely hyaline female glumes are diagnostic and give the spikes a conspicuous silvery-white colour.

Subgenus* 3 - *PRIMOCAREX Kük. (spp. 72-75). Spike 1, terminal, with male flowers at top and female below; bract 0; stigmas 2 and nut biconvex (*C. pulicaris*) or stigmas 3 and nut with 3 rounded angles (other 3 spp.). An unsatisfactory subgenus, whose members would be better placed in 1 of the other 2 subgenera if only their relationships were clear.

72. C. microglochin Wahlenb. - *Bristle Sedge*. Rhizomatous; stems usually single, RR
to 12cm, with rounded angles or terete, smooth; utricles usually strongly reflexed, 819
3.5-5(6)mm tapered to beak 1-1.5mm, plus bristle exserted 1-2mm; stigmas 3; (2n=50, 58). Native; base-rich flushes on open stony slopes at 600-900m, 1 locality in M Perth, discovered 1923.

73. C. pauciflora Lightf. - *Few-flowered Sedge*. Shortly rhizomatous; stems not or 819
very loosely tufted, to 25cm, with rounded angles, smooth; utricles usually strongly reflexed, 5-7.5mm tapered to beak 1-2mm; stigmas 3; (2n=38, 39, 44, 76). Native; acid blanket bogs; frequent in C & N Sc, scattered S to NE Yorks and Caerns, Co Antrim.

74. C. rupestris All. - *Rock Sedge*. Shortly rhizomatous; stems loosely tufted, to R
20cm, with acute angles, ± smooth; utricles ± erect, 2-3.5mm, with beak 0.2-0.3mm; 819
stigmas 3; (2n=50, 52). Native; on rock ledges or stony ground on limestone or calcareous-flushed sandstone usually above 600m but in N Sc ± down to sea-level; locally frequent in Sc from M Perth and Angus to W Sutherland.

75. C. pulicaris L. - *Flea Sedge*. Densely tufted to shortly rhizomatous; stems to 819
30cm, terete, smooth; utricles usually strongly reflexed, 3.5-6mm, with beak 0.2-0.5mm; stigmas 2; (2n=58, 60). Native; bogs, fens and flushes, usually base-rich; frequent throughout most of BI, but absent from much of C & E En.

157. POACEAE - *Grass family*
(Gramineae)

Annuals or herbaceous perennials (rarely woody perennials - *bamboos*), often with rhizomes or stolons, with usually hollow, cylindrical (rarely flattened or other shapes but not 3-angled) stems; leaves alternate, with long, usually linear, entire, thin (but often rolled up or folded along long axis) blade (*leaf*), with long, stem-sheathing, often cylindrical lower part (*sheath*), usually with *ligule* (a membrane, a fringe of hairs, or a membrane with a distal fringe of hairs) at top of sheath on adaxial side, sometimes with small wing-like extension (*auricle*) on either side at top of sheath. Flowers much reduced, 1-many in discrete units (*spikelets*) very variously arranged in terminal inflorescences, mostly bisexual but often unisexual and bisexual mixed in same spikelet, rarely male and female in different spikelets or parts of plants (dioecious in *Cortaderia*), hypogynous; perianth represented by 2 minute scales (*lodicules*) at base of ovary (rarely fused or 0, 1 or 3); stamens usually 3, rarely 2, 4 or 6; ovary 1-celled, with 1 ovule; styles 2, rarely 1 or 3; stigmas elongated, feathery; fruit a typical *caryopsis*, rarely the wall not fused to the seed inside. Spikelets consisting of a series of bracts; usually 2 (sterile) *glumes* (*lower* and *upper*) at base, rarely 1 or 0, with empty axils; 1-many *florets* above consisting of the bisexual or unisexual flower proper plus 2 (fertile) glumes on either side - *lemma* on abaxial and *palea* on adaxial side (latter sometimes 0); the florets borne on slender axis (*rhachilla*), often 1 or more sterile or even reduced to vestigial scales; lemmas often with horny region (*callus*) at base, this often vestigial but sometimes well developed (often pointed and bristly); lemmas and/or glumes often with short to long dorsal to terminal bristles (*awns*).

Easily distinguished from all other grass-like plants by the distinctive inflorescence and flower structure, and usually from the Cyperaceae by the hollow stems.

Before attempting use of the keys it is essential to dissect a spikelet and to understand thoroughly its structure, the detailed characteristics of all its parts, and the arrangement of spikelets in the inflorescence. The growth-form of the plant is also important; perennials can be distinguished by the presence of sterile leafy shoots (*tillers*) as well as flowering stems (*culms*), and they might have rhizomes and/or stolons as well. Spikelet, glume and lemma lengths exclude awns unless otherwise stated.

General key

1 Ligule a dense fringe of hairs, or membranous but breaking into dense fringe of hairs distally ***Key A***

1 Ligule membranous, sometimes jagged or pubescent but not densely fringed with hairs distally, sometimes 0 2

2 Bamboos - stems woody; leaves with distinct short petiole between blade and sheath ***Key B***

2 Stems not woody; leaves without petiole between blade and sheath 3

3 Maize - female spikelets in simple raceme (cob) low down on plant; male spikelets in terminal panicle or umbel of racemes (tassel) **94. ZEA**

3 Male and female spikelets not in separate inflorescences 4

4 Spikelets arising in groups of 2-7, one fatter and bisexual, the other 1-6 thinner and sterile or male ***Key C***

4 Spikelets all bisexual and similar 5

5 Inflorescence a simple spike, or a simple raceme whose spikelets have pedicels ≤2mm ***Key D***

5 Inflorescence more complex than a simple spike or raceme of spikelets (but often very condensed) 6

6 Inflorescence an umbel or raceme of spikes, or of racemes whose spikelets have pedicels <2mm ***Key E***

6 Inflorescence a panicle, or a raceme whose spikelets have pedicels >3mm 7

7 Spikelets regularly proliferating to form small leafy plantlets (sexual spikelets present or not) 8

7 Spikelets not or only irregularly proliferating (if so, diseased, or because of sterility, or very late in season) 10

8 Lemmas (whether proliferating or not) distinctly keeled on back along midrib **22. POA**

8 Lemmas (whether proliferating or not) rounded on back 9

9 Lemmas with awn arising from dorsal surface, with tuft of hairs arising just below base **38. DESCHAMPSIA**

9 Lemmas with awn 0 or terminal, without tuft of hairs arising just below base **15. FESTUCA**

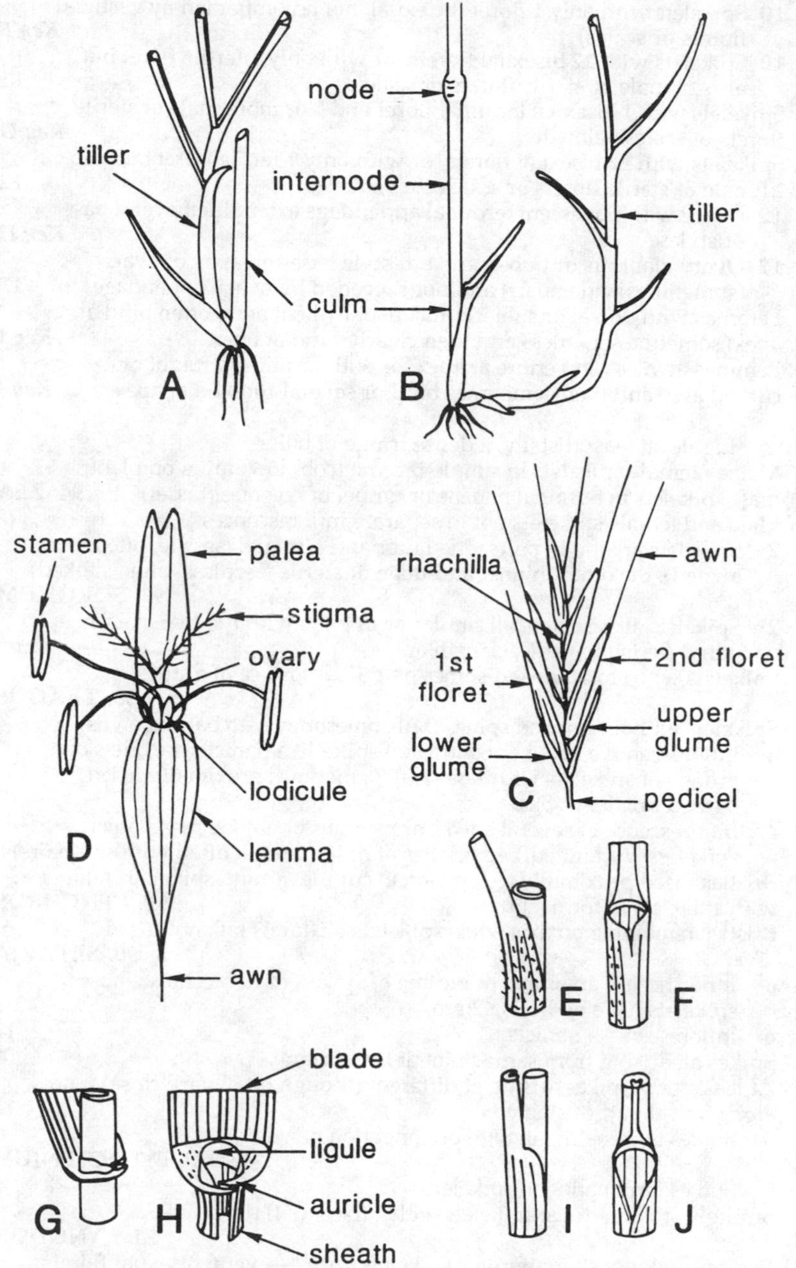

FIG 825 - Poaceae terminology. A, intravaginal innovation shoot. B, extravaginal innovation shoot. C, spikelet. D. floret with lemma pulled back. E-F, innovation leaf-sheath of ***Festuca rubra***. G-H, innovation leaf sheath of ***F. pratensis***. I-J, innovation leaf-sheath of ***F. ovina***. Drawings by S. Ogden.

10 Spikelets with only 1 floret (bisexual, not accompanied by vestigial florets or scales) *Key F*
10 Spikelets with ≥2 bisexual florets, or with only 1 fertile floret but also ≥1 male or sterile florets or scales 11
11 Spikelets with 1 bisexual terminal floret and 1 or more male or sterile florets or scales below it *Key G*
11 Spikelets with ≥2 bisexual florets, or with only 1 fertile floret but also ≥1 male or sterile florets or scales above it 12
12 Ovary with pubescent terminal appendage extending beyond base of styles *Key H*
12 Ovary glabrous or pubescent, but style bases on apex of ovary (sometimes wide apart) and not exceeded by ovary appendage 13
13 Lemmas with dorsal or subterminal, usually bent awn, often bifid at apex, sometimes awnless and then clearly bifid at apex *Key I*
13 Lemmas awnless and entire at apex, or with terminal straight or curved awn and then sometimes bifid or several-toothed at apex *Key J*

Key A - Ligule, at least distally, a dense fringe of hairs
1 Maize - female spikelets in simple raceme (cob) low down on plant; male spikelets in terminal panicle or umbel of racemes (tassel) **94. ZEA**
1 Male and female spikelets not in separate inflorescences 2
2 Spikelets arising in pairs, one fatter and bisexual (sessile, often awned), the other thinner and male or sterile (awnless, often stalked) **93. SORGHUM**
2 Spikelets all bisexual, all similar or in pairs with the two in each pair differing in pedicel length only 3
3 Spikelets with stout hooked spines on dorsal surface of glume **83. TRAGUS**
3 Spikelets without hooked spines (but sometimes with barbed awns) 4
4 Inflorescence a spike or contracted spike-like panicle; spikelets or groups of spikelets with 1-several (sometimes proximally fused) barbed bristles at base 5
4 Inflorescence of several spikes or racemes or obviously a panicle; spikelets without barbed bristles at base (though often with soft hairs) 6
5 Bristles fused proximally to form small cupule around spikelets, falling with spikelets to form a bur **92. CENCHRUS**
5 Bristles remaining on axis when spikelets or florets fall, not fused **90. SETARIA**
6 Inflorescence an umbel or raceme of spikes, or of racemes whose spikelets have pedicels ≤2mm 7
6 Inflorescence a panicle 17
7 Spikes all arising from same point at tip of stem 8
7 At least some spikes arising at different (though often very close) points along apical part of stem 11
8 Spikes terminating in bare prolongation of axis **78. DACTYLOCTENIUM**
8 Spikes terminating in spikelet 9
9 Strongly stoloniferous; spikelets with 1 (bisexual) floret only **81. CYNODON**
9 Plant tufted, not stoloniferous; spikelets with 1-several bisexual florets, if only 1 then with sterile florets or scales distal to it 10
10 Spikelets with ≥3 bisexual florets; lemmas not awned **77. ELEUSINE**
10 Spikelets with 1 bisexual floret plus 1-few sterile ones more distal; lemmas awned **80. CHLORIS**
11 Spikelets in pairs, the two in each a pair differing ± only in pedicel length **(MISCANTHUS)**

11 Spikelets not in pairs 12

12 Spikelets with ≥3 florets **77. ELEUSINE**

12 Spikelets with 1-2 florets 13

13 Spikelets ≥8mm **82. SPARTINA**

13 Spikelets ≤6mm 14

14 Spikelets with small globose swelling at base, immediately below glumes **88. ERIOCHLOA**

14 Spikelets without globose swelling at base 15

15 Spikelets with 2 scales (upper and lower glume) at base <2/3 as long as spikelet, with 1 (bisexual) floret **79. SPOROBOLUS**

15 Spikelets with 0 or 1 scale (lower glume) at base <2/3 as long as spikelet, with 2 florets (the upper bisexual, the lower male or sterile) 16

16 Upper lemma with awn 0.3-1mm (hidden between upper glume and lower lemma); lower lemma and upper glume shortly acuminate **87. UROCHLOA**

16 Upper lemma awnless; lower lemma and upper glume obtuse **86. BRACHIARIA**

17 Spikelets in pairs, the two in each a pair differing ± only in pedicel length **(MISCANTHUS)**

17 Spikelets not in pairs 18

18 Most leaves >1m long, with very rough, cutting edges; plants dioecious **72. CORTADERIA**

18 Leaves <1m, without cutting edges; plants bisexual 19

19 Spikelets consisting of 3 florets, 2 lower empty much larger than 1 upper fertile **10. EHRHARTA**

19 Spikelets rarely consistently with 3 florets, if so then the lowest bisexual 20

20 Spikelets with basal tuft of long silky hairs becoming very conspicuous in fruit **74. PHRAGMITES**

20 Spikelets glabrous to pubescent, but not with basal tuft of long silky hairs 21

21 Spikelets with 1 floret only 22

21 Spikelets with ≥2 florets, sometimes only 1 bisexual and ≥1 male or sterile 25

22 Lemmas conspicuously awned 23

22 Lemmas awnless 24

23 Lemma with single bent awn **12. STIPA**

23 Lemma with trifid straight awn **(ARISTIDA)**

24 Lemmas with 1 vein; spikelets all with 1 floret **79. SPOROBOLUS**

24 Lemmas with 3-5 veins; usually many spikelets with >1 floret **73. MOLINIA**

25 Spikelets with 2 florets, the distal bisexual, the proximal male or sterile **84. PANICUM**

25 Spikelets with ≥2 florets, at least the lowest bisexual 26

26 Lemma entire at apex, not awned, (1-)3(5)-veined; glumes much shorter than rest of spikelet 27

26 Lemma notched or 2-3-lobed at apex, sometimes awned, 5-9-veined; glumes as long as spikelet (excl. awns) or nearly so 28

27 Perennial of wet peaty areas; lemmas ≥3mm **73. MOLINIA**

27 Alien annuals or sometimes perennials; lemmas <3mm **76. ERAGROSTIS**

28 Lemmas with bent awn >5mm **70. RYTIDOSPERMA**

28 Lemmas awnless or with straight awn <1mm 29

29 Lemmas deeply 2-lobed; spikelets falling whole; lemmas <3mm, with very broad hyaline margins; annual **71. SCHISMUS**

29 Lemmas minutely 2-3-toothed; florets falling separately from glumes; lemmas >4mm, with very narrow hyaline margins; perennial **69. DANTHONIA**

Key B - Bamboos (stems woody; leaf-blades with short petiole)

1 Main stems ± square in section **8. CHIMONOBAMBUSA**
1 Main stems cylindrical, or flattened or grooved on one side 2
 2 Main stems flattened or grooved on 1 side at least at upper internodes 3
 2 Main stems ± cylindrical throughout except sometimes just above each node 4
3 Nodes of mid-region of main stems mostly with 2 unequal branches and often a very small 3rd one; stems flattened or grooved on 1 side throughout **(PHYLLOSTACHYS)**
3 Nodes of mid-region of main stems mostly with 3-5 branches; stems often flattened or grooved only at upper internodes **1. SEMIARUNDINARIA**
 4 Nodes of mid-region of main stems mostly with 1(-2) lateral branches 5
 4 Nodes of mid-region of main stems mostly with ≥3 lateral branches 8
5 Leaves pubescent on lowerside **6. SASAELLA**
5 Leaves glabrous (except sometimes on margin) 6
 6 Leaves 4-5x as long as wide, with 5-14 veins on either side of midrib **5. SASA**
 6 Leaves >6x as long as wide, with 2-9 veins on either side of midrib 7
7 Stems 2.5-5m; leaves 15-30cm, 2-4cm wide, with 5-9 veins on either side of midrib **7. PSEUDOSASA**
7 Stems 0.2-2m; leaves 2.5-20cm, 0.3-2.5cm wide, with 2-7 veins on either side of midrib **4. PLEIOBLASTUS**
 8 Stems ≤2.5m **4. PLEIOBLASTUS**
 8 Stems >2.5m 9
9 Leaves 15-25mm wide, mostly with 4-7 veins on either side of midrib **4. PLEIOBLASTUS**
9 Leaves ≤13(20)mm wide, mostly with 2-4 veins on either side of midrib 10
 10 Stems bearing sheaths with distinct falcate auricles and many stiff bristles at apex **2. YUSHANIA**
 10 Stems bearing sheaths without auricles and with no or few bristles at apex 11
11 Sheaths very shortly pubescent when young, without apical bristles; leaf-sheaths with ligules ≤2mm; inflorescence an open panicle with narrow sheaths at base **3. FARGESIA**
11 Sheaths glabrous, with few apical bristles; leaf-sheaths with ligules >2mm; inflorescence short, partly enclosed by large, wide sheaths **(THAMNOCALAMUS)**

Key C - Spikelets arising in groups of 2-7, one fatter and bisexual, the other 1-6 thinner and male or sterile (N.B. *Tragus* (Keys A & D) may have 2-4 fertile and 1-3 sterile spikelets in groups)

1 Inflorescence a spike with 3 spikelets (each with 1 floret) per node, the central one bisexual, the 2 laterals male or sterile **65. HORDEUM**
1 Inflorescence a panicle (sometimes strongly contracted and spike-like), not with regularly 3 spikelets per node and not all spikelets with 1 floret 2
 2 Sterile or male spikelets with ≤3 florets, with glumes at least nearly as long as spikelet 3
 2 Sterile or male spikelets clearly with >5 florets, with glumes much shorter than spikelet 4
3 Spikelets in pairs, 1 bisexual, 1 male or sterile, not falling as a unit **93. SORGHUM**
3 Spikelets in groups of 3-7, 1 bisexual, the rest male or sterile, all falling as a unit **44. PHALARIS**
 4 Bisexual spikelets with 1 floret plus a sterile vestige, accompanied by 2-4 sterile spikelets, all falling as a unit **19. LAMARCKIA**
 4 Bisexual spikelets with (1)2-5 florets, accompanied by 1-few sterile

spikelets, the latter and the glumes of fertile spikelets not falling **18. CYNOSURUS**

Key D - Spikelets all bisexual and similar; inflorescence a simple spike or raceme with pedicels ≤2mm

1 Spikelets with stout hooked spines on dorsal surface of glume **83. TRAGUS**
1 Spikelets without hooked spines (but sometimes with barbed awns) 2
2 Spikelets or groups of spikelets with 1-several (sometimes proximally fused) stiff bristles at base (NB do not confuse with bristle-like glumes) 3
2 Spikelets without stiff bristles at base 4
3 Bristles fused proximally to form small cupule around spikelets, falling with spikelets to form a bur **92. CENCHRUS**
3 Bristles remaining on axis when spikelets or grains fall, not fused **90. SETARIA**
4 One 1-floreted spikelet at each node 5
4 Each node with >1 spikelet or with 1 spikelet with >1 floret 8
5 Densely tufted perennial; lemma awned; glumes 1-2, much shorter than floret **11. NARDUS**
5 Annual; lemma not awned; glumes 1-2, at least as long as floret 6
6 Spikelets not sunk in hollows in axis; glumes obtuse; lemma pubescent **51. MIBORA**
6 Spikelets sunk in hollows in axis; glumes acute to acuminate; lemma glabrous 7
7 Glumes 2 on all spikelets **27. PARAPHOLIS**
7 Glumes 2 on terminal spikelet, 1 on lateral spikelets **28. HAINARDIA**
8 Spikelets 2-3 at each node 9
8 Spikelets 1 at each node 13
9 Spikelets with 1-2 florets; lemmas long-awned 10
9 Spikelets with (2)3-6 florets; lemmas awned or not 12
10 Spikelets 2 per node, with mostly 2 florets (upper reduced and sterile) **(TAENIATHERUM)**
10 Spikelets all or mostly 3 per node, with mostly 1 floret 11
11 Two glumes of each spikelet free to base **65. HORDEUM**
11 Two glumes of each spikelet fused at base **64. HORDELYMUS**
12 Lowest lemma >14mm, awnless; leaves very glaucous; all or ± all nodes with 2 spikelets **63. LEYMUS**
12 Lowest lemma <14mm, awned; leaves not glaucous; upper nodes with 1 spikelet **62 x 65. X ELYTRORDEUM**
13 Lower glume (except in terminal spikelet) 0 or ≤3/4 as long as upper 14
13 Lower glume ≥3/4 as long as upper 17
14 Lemma with a bent awn from dorsal surface **34. GAUDINIA**
14 Lemma awnless or with straight to curved terminal or subterminal awn 15
15 Upper glume with 1-3 veins; lemma very gradually narrowed to long terminal awn **17. VULPIA**
15 Upper glume with 5(-9) veins; lemma awnless or rather abruptly narrowed to long or short terminal or subterminal awn 16
16 Lower glume 0 except in terminal spikelet **16. LOLIUM**
16 Lower glume present on all or most spikelets **15 x 16. X FESTULOLIUM**
17 Perennial; sterile shoots and often rhizomes present 18
17 Annual; sterile shoots and rhizomes 0 22
18 Spikelets scarcely flattened, on distinct pedicels 0.5-2mm **60. BRACHYPODIUM**

18 Spikelets flattened, sessile or with vestigial pedicel <0.5mm 19
19 Inflorescence axis not breaking up at maturity 20
19 Inflorescence axis breaking up at maturity, 1 segment falling off with each spikelet 21
20 Plant densely tufted, without rhizomes; lemmas usually with awns >7mm; spikelets breaking up below each lemma at maturity, leaving glumes on rhachis; anthers ≤3mm **61. ELYMUS**
20 Plant with long rhizomes, not densely tufted; lemmas rarely with awns >7mm; spikelets eventually falling whole, not leaving 2 glumes alone on rhachis; anthers ≥3.5mm **62. ELYTRIGIA**
21 Lemmas awned **62 x 65. X ELYTRORDEUM**
21 Lemmas not awned **62. ELYTRIGIA**
22 Glumes truncate at apex (awned or not) 23
22 Glumes acuminate to obtuse, sometimes shouldered or notched, at apex (awned or not) 24
23 Glumes rounded on back (except near apex) **(AEGILOPS)**
23 Glumes 1-2-keeled **68. TRITICUM**
24 Glumes and lemmas <4mm, not awned **25. CATAPODIUM**
24 Glumes and lemmas >4mm, the lemmas long-awned 25
25 Glumes linear-lanceolate; spikelets with 2(-3) florets (all bisexual) **66. SECALE**
25 Glumes lanceolate to ovate; spikelets with 2-3 bisexual florets plus 1-few distal sterile ones, or with >4 bisexual florets 26
26 Spikelets with >4 bisexual florets, usually flattened narrow-side to inflorescence axis **60. BRACHYPODIUM**
26 Spikelets with 2-3 bisexual plus 1-few distal sterile florets, flattened broad-side to inflorescence axis **67. X TRITICOSECALE**

Key E - Spikelets all bisexual and similar; inflorescence an umbel or raceme of ≥2 spikes or racemes with pedicels ≤2mm

1 Spikes all arising from same point at tip of stem 2
1 At least some spikes arising at different (though often very close) points along apical part of stem 7
2 Lemmas conspicuously awned **80. CHLORIS**
2 Lemmas not awned 3
3 Spikelets with 2-c.10 florets, if with 2 then lower one bisexual 4
3 Spikelets with 1-2 florets, if with 2 then lower one male or sterile 5
4 Spikes terminating in bare prolongation of axis **78. DACTYLOCTENIUM**
4 Spikes terminating in a spikelet **77. ELEUSINE**
5 Fertile floret with 3-4 scales (1-2 glumes, lemma and palea of lower sterile floret) below it; annual **91. DIGITARIA**
5 Fertile floret with 2 scales (2 glumes, or upper glume and lemma representing lower sterile floret) below it; stoloniferous perennial 6
6 Spikelet with 2 lower scales (glumes) 1-veined, shorter than fertile lemma **81. CYNODON**
6 Spikelet with 2 lower scales (glume and sterile lemma) 3-veined, longer than fertile lemma **89. PASPALUM**
7 Spikelets with 2 florets, the upper bisexual, the lower male or sterile 8
7 Spikelets with 1-many florets, at least the lowest bisexual 12
8 Ligule 0 **85. ECHINOCHLOA**
8 Ligule present (membranous or a fringe of hairs) 9
9 Ligule membranous Return to 5
9 Ligule a fringe of hairs or membranous with a fringe of hairs distally 10
10 Spikelets with small bead-like swelling at base; lower glume ± 0 **88. ERIOCHLOA**

10 Spikelets without bead-like swelling at base; lower glume present 11
11 Upper lemma with awn 0.3-1mm (hidden between upper glume and lower lemma) **87. UROCHLOA**
11 Upper (and lower) lemma awnless **86. BRACHIARIA**
12 Spikelets with 1-2 florets 13
12 Spikelets with ≥3 florets 17
13 Spikelets ≥8mm **82. SPARTINA**
13 Spikelets ≤6mm 14
14 Ligule membranous 15
14 Ligule a fringe of hairs 16
15 Spikes 2(-4), 2-7cm; spikelets appressed-pubescent on 1 side, glabrous on other **89. PASPALUM**
15 Spikes usually >4, 0.5-2cm; spikelets glabrous on both sides **54. BECKMANNIA**
16 Two basal scales of spikelet much shorter than rest of spikelet, glabrous **79. SPOROBOLUS**
16 Two basal scales of spikelet longer than rest of spikelet, densely silky-pubescent **88. ERIOCHLOA**
17 All except terminal spikelet in each branch with 1 glume **16. LOLIUM**
17 All spikelets with 2 glumes 18
18 Lemmas 5-veined, >5.5mm **15 x 16. X FESTULOLIUM**
18 Lemmas 1-3-veined (sometimes with 1-3 extra veins very close to midrib, forming thickened keel), <5.5mm, keeled 19
19 Axis of inflorescence much longer than longest spike; tip of lemma bifid (often awned from notch) **75. LEPTOCHLOA**
19 Axis of inflorescence usually shorter (rarely slightly longer) than longest spike; tip of lemma acute to obtuse (often awned from tip) **77. ELEUSINE**

Key F - Spikelets all bisexual and similar; inflorescence a panicle; spikelets with 1 (bisexual) floret and no other scales or sterile florets

1 Spikelets with small bead-like swelling at base **88. ERIOCHLOA**
1 Spikelets without bead-like swelling at base 2
2 Glumes ± 0 (reduced to rims below floret) **9. LEERSIA**
2 Both glumes well developed 3
3 Panicle a soft woolly ovoid dense head; lemmas tapered to 2 apical bristles, with a longer awn **49. LAGURUS**
3 Panicle rarely a soft woolly ovoid dense head, if so then lemmas blunt and awnless; lemmas awned or not but without 2 long apical bristles 4
4 Both glumes notched at apex and with awn from sinus; lemma with dorsal awn 5
4 Glumes usually not awned, often tapered to fine apex but not notched at apex, if awned then lemma awnless 6
5 Awns of glumes (not lemma) >3mm; lemma awned from apex; fertile annual with deciduous spikelets **52. POLYPOGON**
5 Awns of glumes <3mm; lemma awned from below apex; sterile perennial with ± persistent spikelets **45 x 52. X AGROPOGON**
6 Lemmas with long terminal awn with much twisted, stout proximal part **12. STIPA**
6 Lemmas with terminal, dorsal or 0 awn but without twisted proximal part 7
7 Floret with tuft of white hairs at base, the hairs >1/4 as long as lemma 8
7 Floret variously pubescent or glabrous, but without basal tuft of white hairs >1/4 as long as lemma 12
8 Spikelets <9(10)mm; anthers <3mm 9
8 Spikelets >(9)10mm; anthers >3mm 11
9 Lemma keeled, acute to obtuse, awnless; palea almost as long as

lemma **22. POA**

9 Lemma rounded on back, truncate or variously toothed at apex, often awned; palea c.2/3-3/4 as long as lemma 10

10 Tuft of hairs at base of lemma c.0.3-0.6mm **45. AGROSTIS**

10 Tuft of hairs at base of lemma >1mm **46. CALAMAGROSTIS**

11 Panicle very pale, spike-like; spikelets 10-16mm; lemmas with basal hairs <1/2 as long as lemma, with awn 0-1mm; anthers 4-7mm, shedding pollen **47. AMMOPHILA**

11 Panicle usually purplish-green, usually ± lobed; spikelets 9-12mm; lemmas with basal hairs ≥1/2 as long as lemma, with awn 1-2mm; anthers 3-4.5mm, not opening **46 x 47. X CALAMMOPHILA**

12 Both glumes ≤3/4 as long as spikelet, usually 1 or both c.1/2 as long 13

12 Both glumes >3/4 as long as spikelet, usually 1 or both as long or longer 16

13 Lemmas with awns >3mm; annual **50. APERA**

13 Lemmas awnless; normally perennial 14

14 Lemmas ± truncate; ligule membranous **24. CATABROSA**

14 Lemmas obtuse to acute; ligule a fringe of hairs 15

15 Lemmas with 1 vein, <3mm; spikelets all with 1 floret **79. SPOROBOLUS**

15 Lemmas with 3-5 veins, >3mm; usually some spikelets with >1 floret **73. MOLINIA**

16 Glumes with swollen ± hemispherical base and long very tapering distal part **48. GASTRIDIUM**

16 Glumes without swollen base 17

17 Lemma shiny and much harder and tougher than glumes when mature 18

17 Lemma about same texture as glumes or more flimsy and delicate 19

18 Lemmas awnless; ligules 2-10mm **14. MILIUM**

18 Lemmas with deciduous awn 2-5mm; ligules c.1mm **13. ORYZOPSIS**

19 Lemmas with subterminal awn 4-10mm **50. APERA**

19 Lemmas with awn 0 or <4mm, or with awn ≥4mm and arising from lower 1/2 of back of lemma 20

20 Panicle very compact, ovoid or oblong to cylindrical, spike-like, with very short branches 21

20 Panicle diffuse or contracted but with obvious branches 22

21 Lemmas awnless; palea present; florets falling at maturity leaving glumes on panicle **55. PHLEUM**

21 Lemmas usually awned (often shortly); palea absent; spikelets falling as a whole at maturity **53. ALOPECURUS**

22 Lemmas acute to obtuse, strongly keeled **22. POA**

22 Lemmas truncate or variously toothed at apex, rounded on back 23

23 Glumes with short pricklets or at least rough all over back; palea nearly as long as lemma; lemma c.1/2 as long as glumes **52. POLYPOGON**

23 Glumes often with short pricklets on midrib but ± smooth otherwise; palea ≤3/4 as long as lemma; lemma c.2/3-3/4 as long as glumes 24

24 Panicle branches bearing spikelets ± to base, disarticulating (if at all) near base of pedicels; very rare **45 x 52. X AGROPOGON**

24 Panicle branches with clear bare region at base, disarticulating (sometimes not) at base of lemmas; very common **45. AGROSTIS**

Key G - Spikelets all bisexual and similar; inflorescence a panicle; spikelets with 1 bisexual floret plus 1 or more male or sterile florets or scales below it

1 Spikelets or groups of spikelets with 1-several (proximally fused or free) stiff bristles at base 2

1 Spikelets without stiff bristles at base 3

2 Bristles fused proximally to form small cupule around spikelets, falling with spikelets to form bur **92. CENCHRUS**

2 Bristles not fused, retained on axis when spikelets or florets fall **90. SETARIA**

3 Glumes minute, well separated from lowest lemmas by conspicuously pubescent callus **10. EHRHARTA**

3 Glumes overlapping lowest lemmas 4

4 Ligule a fringe of hairs, or membranous with a distal fringe of hairs **84. PANICUM**

4 Ligule membranous or 0 5

5 Bisexual floret with 1 male or sterile floret below it 6

5 Bisexual floret with 2 male or sterile florets below it 8

6 Lower lemma awned from proximal 1/2 of back; perennial **32. ARRHENATHERUM**

6 Lower lemma awnless or with terminal awn; annual 7

7 Ligule membranous; sterile floret a minute scale; glumes equal, keeled **44. PHALARIS**

7 Ligule 0; sterile floret as large as fertile one; glumes very unequal, not keeled **85. ECHINOCHLOA**

8 Lower glume c.1/2 as long as upper; lemmas of lower 2 florets with awns ≥1.5mm **41. ANTHOXANTHUM**

8 Glumes equal or subequal; lemmas of lower 2 florets awnless or with awn <1mm 9

9 Panicle ± diffuse; lower 2 florets male, longer than bisexual floret **42. HIEROCHLOE**

9 Panicle compact; lower 2 florets sterile, shorter than bisexual floret **44. PHALARIS**

Key H - Spikelets all bisexual and similar; inflorescence a panicle or a raceme with pedicels >3mm; spikelets with ≥2 bisexual florets; ovary with pubescent terminal appendage extending beyond base of styles (NB *Brachypodium* and *Gaudinia* have similar ovaries, but the inflorescence is a spike or raceme, not a panicle - Key D).

1 At least lowest lemma with conspicuously bent awn arising dorsally from lower 2/3 of its length, or awnless with glumes longer than rest of spikelet 2

1 Lemmas with straight or slightly bent awns arising from apex or subapically, or awnless with glumes much shorter than rest of spikelet 4

2 Easily uprooted annual without non-flowering shoots; upper glume >(15)20mm **33. AVENA**

2 Firmly rooted perennial with non-flowering shoots; upper glume <15(20)mm 3

3 Spikelets usually <11mm (excl. awns); awn of lowest lemma arising c.1/3 from base; upper glume 7-11mm; lowest floret usually male (sometimes bisexual) **32. ARRHENATHERUM**

3 Spikelets usually >11mm (excl. awns); all lemmas with awn arising near middle or above it; upper glume 10-20mm; lowest floret bisexual **31. HELICTOTRICHON**

4 Lemmas strongly keeled on back **59. CERATOCHLOA**

4 Lemmas rounded on back or keeled near apex 5

5 Perennial, with sterile shoots at flowering time, often with rhizomes 6

5 Annual (or biennial), without sterile shoots at flowering time, without rhizomes 7

6 Spikelets <15mm (excl. awns), narrowed to apex; lemmas ≤8mm (excl. awns) **15. FESTUCA**

6 Spikelets ≥15mm (excl. awns), ± parallel-sided almost to apex; lemmas ≥8mm (excl. awns) **57. BROMOPSIS**

7 Spikelets ovate to lanceolate, slightly compressed, markedly narrowed towards top; lemmas 5-11mm, with awn c. as long or shorter; lower

glume 3-7-veined; upper glume 5-9-veined **56. BROMUS**

7 Spikelets ± straight-sided, widening distally; lemmas 9-36mm, with awn c. as long or longer, lower glume 1(-3)-veined; upper glume 3(-5)-veined **58. ANISANTHA**

Key I - Spikelets all bisexual and similar; inflorescence a panicle or a raceme with pedicels >3mm; spikelets with ≥2 bisexual florets, or with only 1 and that basal and with male or sterile florets distal to it; ovary glabrous or pubescent but without pubescent terminal appendage extending beyond base of styles; lemmas with subterminal to dorsal, usually bent awn, often bifid at apex, sometimes unawned and then always bifid at apex

1 Lemmas >6mm, usually with awn ≥10mm 2

1 Lemmas <6mm, never with awn >10mm 4

2 Easily uprooted annual without non-flowering shoots; upper glume >(15)20mm **33. AVENA**

2 Firmly rooted perennial with non-flowering shoots; upper glume <15(20)mm 3

3 Spikelets usually ≤11mm (excl. awns); awn of lowest lemma arising c.1/3 from base; upper glume 7-11mm; upper lemma(s) often ± unawned **32. ARRHENATHERUM**

3 Spikelets usually ≥11mm (excl. awns); all lemmas with awns arising near middle or above it; upper glume 10-20mm **31. HELICTOTRICHON**

4 Easily uprooted annual without non-flowering shoots 5

4 Firmly rooted (unless in sand) perennial with non-flowering shoots 6

5 Spikelets with 2 florets; lemmas with bent awn arising from below 1/2 way **41. AIRA**

5 Spikelets with 3-5 florets; lemmas with usually ± straight subterminal awn **37. ROSTRARIA**

6 Awns slightly but distinctly widened towards apex (club-shaped) **40. CORYNEPHORUS**

6 Awns parallel-sided or tapered to apex 7

7 Glumes pubescent at least along midrib, falling with florets at maturity; lower lemma unawned; upper floret usually male **39. HOLCUS**

7 Glumes glabrous, remaining on plant when florets fall; lowest lemma awned; all florets usually bisexual 8

8 Spikelets with 2 florets; upper glume as long as spikelet or almost so; lemmas truncate to very blunt and usually jagged at apex **38. DESCHAMPSIA**

8 Spikelets with 2-4 florets; upper glume distinctly shorter than spikelet; lemmas acute and finely bifid at apex **35. TRISETUM**

Key J - Spikelets all bisexual and similar; inflorescence a panicle or a raceme with pedicels >3mm; spikelets with ≥2 bisexual florets, or with only 1 and that basal and with male or sterile florets distal to it; ovary glabrous or pubescent but without pubescent terminal appendage extending beyond base of styles; lemmas unawned and entire at apex, or with terminal straight to curved awn and then sometimes bifid or several-toothed at apex

1 Leaf-sheaths fused almost to apex to form tube round stem 2

1 Leaf-sheaths with free, usually overlapping margins 4

2 Lemmas finely pointed or awned, 5-veined **15. FESTUCA**

2 Lemmas acute to rounded or ± 3-lobed at apex, 7(-9)-veined 3

3 Fertile florets 1-3, with a club-shaped group of sterile florets beyond **30. MELICA**

3 Fertile florets 4-c.16, with 0 or 1 reduced sterile floret beyond **29. GLYCERIA**

4 Lemmas with 3-5 short very pointed or shortly awned teeth at ± truncate apex 5

4 Lemmas 1-pointed to rounded or 2-lobed at apex, awned or not 6

5 Ligule a fringe of hairs; glumes 3-5-veined (lateral veins often short); lemmas 7-9-veined, the veins ending short of the apical points **69. DANTHONIA**

5 Ligule membranous; glumes 1-veined; lemmas 3-5-veined, the veins running into the apical points **26. SESLERIA**

6 Lemmas 2-lobed or -toothed at apex, awned or not from the sinus 7

6 Lemmas 1-pointed (sometimes very minutely notched) to rounded at apex, awned or not from tip 9

7 Lemmas with bent awn >5mm **70. RYTIDOSPERMA**

7 Lemmas awnless or with straight awn <1mm 8

8 Lemmas deeply 2-lobed; spikelets falling whole; annual **71. SCHISMUS**

8 Lemmas minutely 2-toothed; florets falling separately from glumes; perennial **69. DANTHONIA**

9 Ligule a fringe of hairs 10

9 Ligule membranous 11

10 Perennial of wet peaty areas; lemmas ≥3mm **73. MOLINIA**

10 Alien annuals or sometimes perennials; lemmas <3mm **76. ERAGROSTIS**

11 Lemma strongly keeled throughout length 12

11 Lemma rounded on back or keeled only distally 14

12 Lemmas 3-veined, with wide membranous ± shiny margins, rendering panicle silvery **36. KOELERIA**

12 Lemmas 5-veined, often with membranous margins but scarcely shiny and panicle not silvery 13

13 Spikelets borne in dense 1-sided clusters; lemmas long-acuminate to shortly awned **23. DACTYLIS**

13 Spikelets not aggregated into dense 1-sided clusters; lemmas obtuse to acute **22. POA**

14 Lemmas awned 15

14 Lemmas not awned 18

15 Easily uprooted annual without non-flowering shoots **17. VULPIA**

15 Firmly rooted perennial with non-flowering shoots 16

16 Lower glume mostly >3/4 as long as upper; anthers dehiscent **15. FESTUCA**

16 Lower glume mostly <3/4 as long as upper; anthers indehiscent 17

17 Pointed auricles present at junction of leaf-sheath and -blade; lemmas rather abruptly narrowed to awn; upper glume with 5-9 veins **15 x 16. X FESTULOLIUM**

17 Auricles 0; lemmas very gradually narrowed to awn; upper glume with 1-3 veins **15 x 17. X FESTULPIA**

18 Lemmas acute to acuminate at apex 19

18 Lemmas subacute to rounded at apex 20

19 Both glumes >1/2 as long as spikelet; lemmas 3-veined **36. KOELERIA**

19 At least lower glume, and usually both glumes, <1/2 as long as spikelet; lemmas 5-veined **15. FESTUCA**

20 Lemmas strongly cordate at base, wider than long; spikelets pendent at maturity **21. BRIZA**

20 Lemmas not or scarcely cordate at base, longer than wide; spikelets not pendent 21

21 Lemmas 3-veined; spikelets with 2-3 florets; leaf-sheaths compressed, keeled on the back **24. CATABROSA**

21 Lemmas 5-veined; spikelets usually with >4 florets; leaf-sheaths rounded

on the back 22
22 Pointed auricles present at junction of leaf-sheath and -blade; lower glume mostly <3/5 as long as upper **15 x 16. X FESTULOLIUM**
22 Auricles 0; lower glume ≥3/5 as long as upper 23
23 Lemmas minutely pubescent at base, with conspicuous membranous tips and margins; usually perennial **20. PUCCINELLIA**
23 Lemmas glabrous, with very narrow membranous margins; annual **25. CATAPODIUM**

Other genera - The following are all covered in the generic keys.

Of the many extra genera occurring as rare casuals, especially as birdseed-aliens or wool-aliens, mention should be made of **TAENIATHERUM** Nevski and **AEGILOPS** L. (Triticeae, Key D), from Mediterranean. **T. caput-medusae** (L.) Nevski is an annual with flexuous awns up to 12cm. Several annual *Aegilops* spp. (especially **A. ventricosa** Tausch, **A. cylindrica** Host and **A. triuncialis** L.) have been recorded; they have short spikes swollen at base and with long awns.

ARISTIDA L. (Aristideae, subfam. Arundinoideae, Key A), from Mediterranean, is a tufted perennial (annual with us) with long-awned panicles. **A. adscensionis** L. is often mistaken for a *Vulpia* but the 1-floreted spikelets and trifid awns easily separate it; it is a rare wool-alien.

MISCANTHUS Andersson (Andropogoneae, Key A) is a large graceful ornamental with large inflorescences of rather irregularly arranged racemes occasionally persisting in the wild (spikelets are wind-dispersed via the plumed rhachis) and has been used experimentally in biomass production; **M. sinensis** Andersson, from E Asia, is the main sp. used.

PHYLLOSTACHYS Siebold & Zucc. (Bambuseae, Key B), represented mainly by **P. bambusoides** Siebold & Zucc., from China, has stems to 12m and 5-12cm thick. It resembles *Semiarundinaria* in having culm internodes flattened or grooved on 1 side and in being densely clump-forming. **THAMNOCALAMUS** Munro includes **T. spathiflorus** (Trin.) Munro, from Nepal and India; it is similar in general aspect to *Yushania anceps* and *Fargesia spathacea* but differs as in Key B.

SUBFAMILY 1 - BAMBUSOIDEAE (tribes 1-3; genera 1-10). Woody bamboos or herbaceous perennials; leaves often with short false-petiole separating them from sheath, often with conspicuous cross-veins as well as longitudinal veins; ligule membranous, rarely a membrane fringed with hairs; sheaths not fused, the lower ones often without or with very reduced leaves; inflorescence a panicle; spikelets with ≥3 florets, with all florets (except often the distal ones) bisexual, or with 1 bisexual floret with or without 2 large sterile ones below it; glumes (0-)2(3); lemmas firm, 5-many-veined, awnless or with terminal awn; paleas with 1-many veins, 1 on mid-line; stamens 3-6; stigmas 2 or 3; lodicules 2 or 3; embryo bambusoid or oryzoid; photosynthesis C3-type alone, with non-Kranz leaf anatomy; fusoid cells often present; microhairs present; chromosome base-number 12.

TRIBE 1 - BAMBUSEAE (Arundinarieae) (genera 1-8). Woody bamboos; leaves with short false-petiole and with cross-veins as well as longitudinal veins; ligule membranous; lower sheaths usually without or with very reduced leaves; spikelets with ≥3 florets, with all (except often the distal) florets bisexual; glumes (0-)2(3); lemmas 5-many-veined, awnless; paleas with ≥4 veins; stamens 3-6; stigmas 2 or 3; lodicules 2 or 3.

Sheaths refers to those on the main stems and not bearing leaf-blades, as opposed to *leaf-sheaths*.

Flowering in all bamboos is very spasmodic and some have never been known to flower. Many of our spp. flowered for the first time in living memory in the 1960s-1980s, and some still are producing flowers, but flower production is too irregular and uncommon to be of much use in determination. Our spp. are grown in parks, wild gardens and estates, and become natd after neglect because of their persistence as thickets suppressing other vegetation. Most of the 11 spp. treated here have a strong rhizome system (*Semiarundinaria* is the main exception); other spp. forming compact clumps are no less well established but persist rather than spread.

1. SEMIARUNDINARIA Nakai - *Narihira Bamboo.*
Stems >3m, densely clump-forming, flattened or grooved on 1 side (alternating in successive internodes, at least at upper internodes); nodes with 3 lateral branches or 3 larger ones plus some smaller ones; leaves 1.5-4cm wide, glabrous, mostly with 4-6 veins on either side of midrib; stamens 3; stigmas 3.

1. S. fastuosa (Lat.-Marl. ex Mitford) Makino ex Nakai (*Arundinaria fastuosa* (Lat.-Marl. ex Mitford) Houz.) - *Narihira Bamboo*. Stems 3-6(8)m, 1.5-4(8)cm thick, green suffused with purple; leaves 12-24 x 1.5-2.5(4)cm, often purplish; sheaths glabrous except very shortly pubescent at base; (2n=48). Intrd; relic in old gardens, natd on bank of River Thames in Surrey; Japan.

2. YUSHANIA Keng f. - *Fountain-bamboos*
Stems mostly >3m, terete except just above nodes; nodes mostly with ≥3 lateral branches; leaves 5-12mm wide, ± glabrous, mostly with 2-4 veins on either side of midrib; stamens 3; stigmas 2-3.

1. Y. anceps (Mitford) W.C. Lin (*Sinarundinaria anceps* (Mitford) C.S. Chao & Renvoize, *Arundinaria anceps* Mitford, *A. jaunsarensis* Gamble) - *Indian Fountain-bamboo*. Stems 3-6m, 0.7-2cm thick, purplish-green; leaves 6-12(16) x 0.5-1.2(2)cm; sheaths glabrous; (2n=48). Intrd; natd in scattered places in W & S Br, MW Yorks, Ir and CI; India.

3. FARGESIA Franch. (*Sinarundinaria* Nakai) - *Fountain-bamboos*
Stems mostly >3m, terete except just above nodes; nodes with ≥3 lateral branches; leaves 5-15mm wide, glabrous, mostly with 3-4 veins on either side of midrib; stamens 3; stigmas 2-3.

Other spp. - F. murieliae (Gamble) T.P. Yi (*Arundinaria murieliae* Gamble, *Sinarundinaria murieliae* (Gamble) Nakai) (*Umbrella Bamboo*), from W China, persists in some neglected parks and gardens; it has thinner, lighter coloured culms with more pubescent sheaths and long-acuminate (not acute to shortly acuminate) leaf-apices. The last character also distinguishes it from *Yushania anceps*. There has been much confusion regarding the application of the epithets *spathacea, nitida* and *murieliae*, and concerning the synonymy of the genera in which they are placed; recent flowering of the spp. in Br and Man has shown the correct affinities. Formerly *murieliae* and *spathacea* were considered synonyms (or very close) but *nitida* a separate sp., and *Fargesia* was considered a synonym of *Thamnocalamus* but *Sinarundinaria* a separate genus.

1. F. spathacea Franch. (*F. nitida* (Stapf) Keng f., *Sinarundinaria nitida* (Stapf) Nakai, *Arundinaria nitida* Stapf, *A. spathacea* (Franch.) D.C. McClint., *Thamnocalamus spathaceus* (Franch.) Soderstr.) - *Chinese Fountain-bamboo*. Stems 4-6m, 1-1.5cm thick, purplish-green; leaves 3.5-7(10) x 0.5-1.3cm; sheaths pubescent; (2n=48). Intrd; natd in woods, parks and commons scattered over En; W China.

4. PLEIOBLASTUS Nakai - *Bamboos*
Stems 0.2-5(8)m, terete except just above nodes; nodes with 1-many lateral branches; leaves 0.3-3cm wide, glabrous, mostly with 2-7 veins on either side of

midrib; stamens 3; stigmas 3.

1 Stems to 0.75(1.2)m, <5mm thick; branches 1-2 per node; all leaves <10cm **1. P. pygmaeus**
1 Stems 1.2-5(8)m, >5mm thick; branches 2-many per node; most leaves >10cm 2
2 Stems 1.2-2(3)m, 1-1.5cm thick; leaves uniformly green on lowerside **2. P. chino**
2 Stems (2.5)3-5(8)m, 2-3cm thick; leaves on lowerside green on 1 side and greyish-green on other side **3. P. simonii**

Other spp. - **P. humilis** (Mitford) Nakai (*Arundinaria humilis* Mitford), another low-growing sp. from Japan, is sometimes found in neglected parks and estates; it is 0.6-2m tall with stems 0.3-1cm thick, has leaves 8-20 x 1-2.5cm and differs from *P. chino* in its leaves conspicuously bicoloured on lowerside.

1. P. pygmaeus (Miq.) Nakai (*Arundinaria pygmaea* (Miq.) Asch. & Graebn.) - *Dwarf Bamboo*. Stems (0.2)0.4-0.75(1.2)m, 0.1-0.3cm thick; leaves 2.5-8 x 0.3-1.5cm, with 2-3 veins on either side of midrib, uniformly glaucous-green on lowerside; sheaths glabrous except on margins; (2n=48, 54). Intrd; natd rarely but widely in BI, best site in Guernsey; Japan. Our plant is var. **distichus** (Mitf.) Nakai.

2. P. chino (Franch. & Sav.) Makino (*Arundinaria chino* (Franch. & Sav.) Makino) - *Maximowicz's Bamboo*. Stems 1.2-2(3)m, 1-1.5cm thick; leaves (8)12-20(25) x 0.8-3cm, uniformly green on lowerside; sheaths glabrous; (2n=48). Intrd; scattered sites in Br and Ir; Japan.

3. P. simonii (Carrière) Nakai (*Arundinaria simonii* (Carrière) C. Rivière & Rivière) - *Simon's Bamboo*. Stems (2.5)3-5(8)m, 2-3cm thick; leaves 15-25 x 1-2.5(3.5)cm, conspicuously bicoloured on lowerside; sheaths glabrous or very shortly pubescent only at base; (2n=48). Intrd; few sites in SW En, SW Ir and Guernsey; Japan.

5. SASA Makino & Shibata - *Bamboos*

Stems 0.5-3m, terete except just above nodes; nodes mostly with 1 lateral branch; leaves 2.5-9cm wide, glabrous except often with marginal hairs or bristles, mostly with 5-13 veins on either side of midrib; stamens 6; stigmas 3.

1. S. palmata (Burb.) E.G. Camus - *Broad-leaved Bamboo*. Stems 2-3m, 7-10mm thick; leaves 12-30(40) x 3.5-9cm, with 8-14 veins on either side of midrib; false-petioles usually green; sheaths glabrous; (2n=48). Intrd; natd widely in BI; Japan. The widest-leaved of our bamboos; flowered very abundantly in the 1960s.

2. S. veitchii (Carrière) Rehder - *Veitch's Bamboo*. Stems 0.5-1.5m, 5-7mm thick; leaves 10-25 x 2.5-6cm, with 5-9 veins on either side of midrib; false-petioles often purplish; sheaths pubescent especially near base when young; (2n=48). Intrd; natd widely but rarely in BI, mostly S En; Japan.

6. SASAELLA Makino - *Hairy Bamboo*

Stems 0.5-1.5m, terete except just above nodes; nodes mostly with 1 lateral branch; leaves 1-3cm wide, sparsely pubescent on upperside, pubescent to densely so on lowerside, mostly with 3-5 veins either side of midrib; stamens 6; stigmas 3.

1. S. ramosa (Makino) Makino (*Sasa ramosa* (Makino) Makino & Shibata, *Arundinaria vagans* Gamble) - *Hairy Bamboo*. Stems 0.5-1.5m, 3-8mm thick; leaves 8-20 x 1-3cm. Intrd; natd widely but rarely in Br and Ir, especially in woods; Japan.

7. PSEUDOSASA Makino ex Nakai - *Arrow Bamboo*

Stems to 5m, terete except just above nodes; nodes mostly with 1 lateral branch; leaves 2-5cm wide, glabrous, mostly with 5-9 veins either side of midrib; stamens

3; stigmas 3.

1. P. japonica (Siebold & Zucc. ex Steud.) Makino ex Nakai (*Arundinaria japonica* Siebold & Zucc. ex Steud.) - *Arrow Bamboo.* Stems 2.5-5m, 1-2cm thick; leaves 15-30 x 2-4cm; (2n=48). Intrd; frequently natd throughout much of BI, easily the commonest grown and natd bamboo, frequently flowers; Japan and Korea. **841**

8. CHIMONOBAMBUSA Makino - *Square-stemmed Bamboo*
Stems to 8m, square in section with rough edges and faces; nodes mostly with 3 lateral branches; leaves 1-3cm wide, minutely pubescent when young, mostly with 8-14 veins on either side of midrib; stamens 3; stigmas 2.

1. C. quadrangularis (Franceschi) Makino (*Arundinaria quadrangularis* (Franceschi) Makino) - *Square-stemmed Bamboo.* Stems 5-8m, 2-4cm thick; leaves 10-20 x 1-3cm; (2n=48). Intrd; natd in SW En and SW Ir; China.

TRIBE 2 - ORYZEAE (genus 9). Herbaceous rhizomatous perennials; leaves without false-petiole, without cross-veins; ligule membranous; spikelets with 1 floret; glumes 0 or vestigial; lemmas 5-veined, awnless; palea 3-veined; stamens 3; lodicules 2.

9. LEERSIA Sw. - *Cut-grass*
1. L. oryzoides (L.) Sw. - *Cut-grass.* Culms to 1.2m, the sheaths and leaves rough with downward-pointed bristles; ligule 0.5-1.5mm; panicle diffuse or remaining 1/2-enclosed in sheaths (and then cleistogamous); florets 4-6mm; (2n=48, 60). Native; wet meadows, ditches, canal-banks and riversides; very locally frequent from S Somerset to Surrey and W Sussex, decreasing. **RR**

TRIBE 3 - EHRHARTEAE (genus 10). Herbaceous annuals or rhizomatous perennials; leaves without false-petiole, without cross-veins; ligule a membrane fringed with hairs; spikelets with 1 small bisexual floret with awnless 3-veined lemma and 1-veined palea, and two much larger sterile florets below each consisting of 1 long-awned 5-veined lemma only, the former obscured by the latter two; glumes 2, very small and separated from florets by stalk-like callus; stamens 4; lodicules 2.

10. EHRHARTA Thunb. (*Microlaena* R.Br.) - *Weeping-grass*

Other spp. - E. longiflora Sm. (*Annual Veldtgrass*), from S Africa, is a rarer wool-alien; it differs in its annual habit, smooth or softly pubescent leaves, glumes 2-6mm, ligule 1-2mm and sterile lemmas (10)17-22mm with short awn.

1. E. stipoides Labill. (*Microlaena stipoides* (Labill.) R. Br.) - *Weeping-grass.* Erect to procumbent perennial to 70cm, the leaves rough with upward-directed bristles; ligule c.0.5mm; panicle sparsely branched, narrow; glumes <1mm; sterile lemmas 8-16mm plus awn up to 20mm; (2n=40). Intrd; rather infrequent wool-alien; scattered over En; Australia and New Zealand. **841**

SUBFAMILY 2 - POOIDEAE (Festucoideae) (tribes 4-14, genera 11-68). Annual to perennial herbs; leaves without false-petiole or cross-veins; ligule membranous, sometimes a membrane fringed with hairs; sheaths usually not fused, sometimes fused; inflorescence a spike, raceme or panicle; spikelets with 1-many florets, with all florets (except often the distal ones) bisexual or some variously male or sterile; glumes (0-)2; lemmas firm to membranous, 3-9-veined, awnless or with terminal or dorsal awn; palea normally 2(-4)-veined without vein on mid-line, rarely with midrib or 0; stamens 1-3; stigmas 1-2; lodicules 0 or 2(-3), or 2 fused laterally;

embryo pooid; photosynthesis C3-type alone, with non-Kranz leaf anatomy; fusoid cells 0; microhairs usually 0, present in *Nardus*; chromosome base-number usually 7, rarely higher or lower.

TRIBE 4 - NARDEAE (genus 11). Perennials; ligule membranous; sheaths not fused; inflorescence a ± 1-sided spike with 1 spikelet at each node; spikelets with 1 bisexual floret; glumes very short, upper often 0; lemma with 2-3 keels, 3-veined, with terminal awn; stamens 3; stigma 1; lodicules 0; ovary glabrous.

11. NARDUS L. - *Mat-grass*

1. N. stricta L. - *Mat-grass.* Densely tufted; culms to 40(60)cm, wiry; leaves tightly inrolled; ligules 0.5-2mm; lemmas 6-9mm plus awn 1-3mm; (2n=26). Native; acid heaths, moors and mountain slopes, often dominant over large areas; throughout BI in suitable places, but not in much of C & E En or C Ir.

TRIBE 5 - STIPEAE (Milieae) (genera 12-14). Annuals or perennials; ligule membranous or sometimes a membrane with a fringe of hairs; sheaths not fused; inflorescence a ± diffuse panicle; spikelets with 1 bisexual floret; glumes equal or ± so, longer than body of lemma; lemma rounded on back, 3-5-veined, awnless or with long terminal awn, becoming hard and tightly wrapped round caryopsis; stamens 3; stigmas 2; lodicules 2-3; ovary glabrous.

12. STIPA L. (*Nassella* (Trin.) E. Desv.) - *Needle-grasses*
Annuals or perennials; ligule a membrane with a fringe of hairs; lemma with long basal callus pointed and with apically directed bristles, with long terminal awn with stouter spirally twisted proximal part separated from narrow distal part by conspicuous bend; lodicules 2-3.

The genus **Nassella**, with veinless (not 2-veined) palea <3x (not >3x) as long as lemma, is sometimes segregated.

Other spp. - >25 other spp. have been recorded as casuals, mainly wool-aliens due to the very penetrating propagules (grain tightly enclosed by lemma with its pointed callus and long awn). The 3 least rare are probably **S. aristiglumis** F. Muell., from Australia, and **S. formicarum** Delile (*Nassella formicarum* (Delile) Barkworth) and **S. trichotoma** Nees (*Nassella trichotoma* (Nees) Hack. ex Arechav.), from S America. They are wool-aliens with slenderer and shorter awns (≤4cm) than the 2 spp. treated fully. *S. aristiglumis* has lemmas c.6mm with densely appressed hairs and awns ≤2cm; *S. formicarum* has lemmas c.5mm with a cupule as in *S. neesiana* and glabrous apart from basal hair-tuft, and awns 2-4cm; *S. trichotoma* has lemmas c.2mm without a membranous cupule and glabrous apart from basal hair-tuft, and awns 3.5-4cm. Next in abundance are **S. ambigua** Speg. from S America and **S. variabilis** Hughes and **S. verticillata** Nees ex Spreng. from Australia.

1. S. neesiana Trin. & Rupr. (*Nassella neesiana* (Trin. & Rupr.) Barkworth) - 841
American Needle-grass. Perennial; culms to 60cm; glumes c.2cm incl. long fine 3-pronged apex, hyaline at margins and apex; lemma c.10mm incl. callus, glabrous except for dense tuft of hairs at base and on callus, with distinct membranous cupule at apex; awn 5-9cm; (2n=28). Intrd; wool-alien, sometimes ± natd in SE En; scattered in En; S America.

2. S. capensis Thunb. - *Mediterranean Needle-grass.* Annual; culms to 60cm; glumes c.1.5-2cm incl. long fine 1-pointed apex, wholly hyaline; lemma 6-8mm incl. callus, pubescent, with cupule of hairs at apex; awn 7-10cm; (2n=36). Intrd; wool-alien; scattered in En; Mediterranean.

13. ORYZOPSIS Michx. (*Piptatherum* P. Beauv.) - *Smilo-grass*
Perennials; ligule membranous; lemma shiny, with short smooth basal callus, with

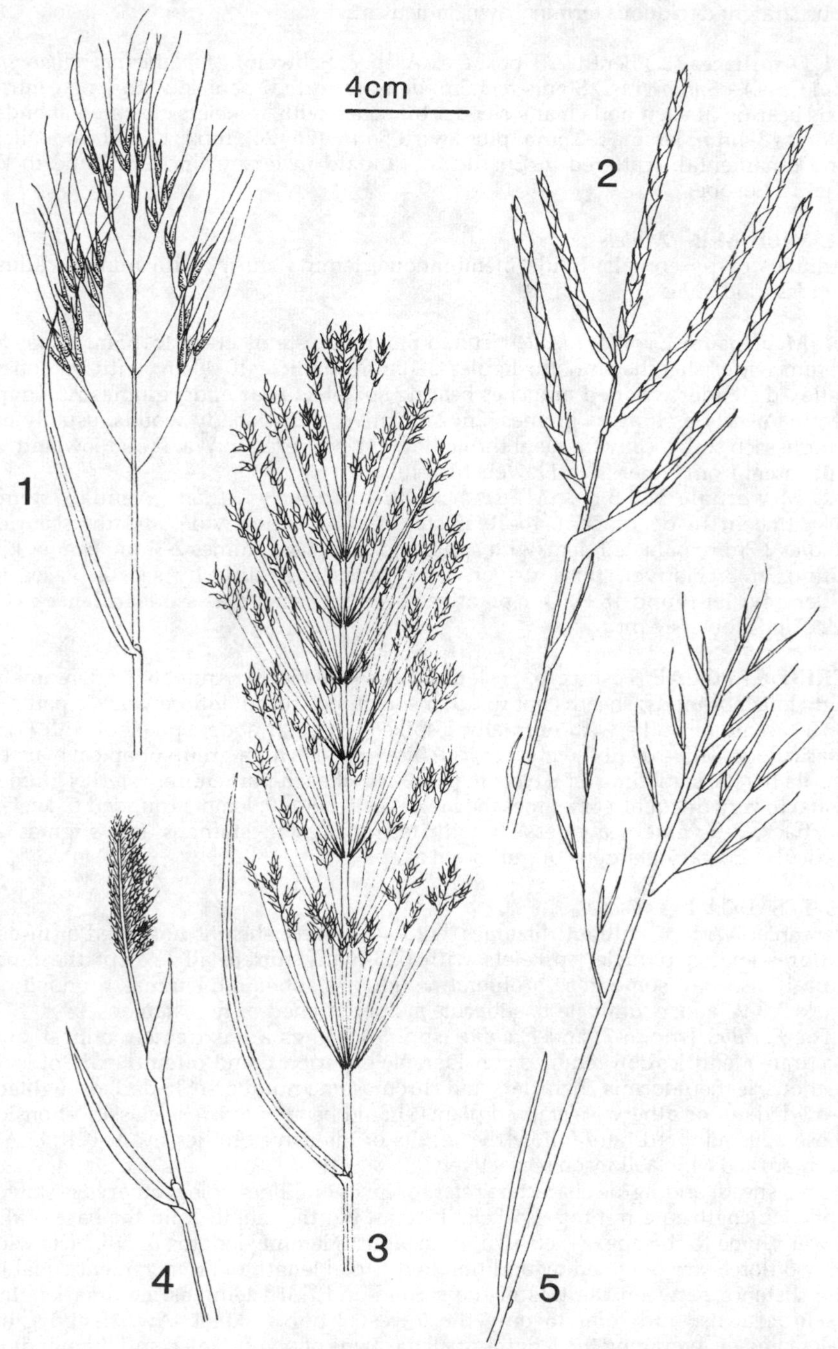

FIG 841 - Poaceae. 1, *Stipa neesiana*. 2, *Pseudosasa japonica*. 3, *Oryzopsis miliacea*. 4, *Lamarckia aurea*. 5, *Ehrharta stipoides*.

long straight deciduous terminal awn; lodicules 2.

1. O. miliacea (L.) Benth. & Hook. f. ex Asch. & Schweinf. (*Piptatherum miliaceum* (L.) Coss.) - *Smilo-grass*. Stems to 1.5m, densely tufted; panicle with long central axis bearing at each node many slender branches with spikelets clustered at ends; glumes 3-4mm; lemma 2-2.5mm plus awn 3-5mm; (2n=24). Intrd; casual wool-alien and ornamental; scattered in En and S Wa, natd in Jersey since 1931 and in W Kent; S Europe. **841** **881**

14. MILIUM L. - *Millets*

Annuals or perennials; ligule membranous; lemma shiny, with minute callus, awnless; lodicules 2.

1. M. effusum L. - *Wood Millet*. Tufted perennial; stems erect, to 1.5m; leaves 5-15mm wide; sheaths smooth; ligules 3-10mm; panicle 10-40cm, with patent to reflexed slender whorled branches bearing spikelets near ends; glumes 2.5-4mm; lemma nearly as long as glumes; 2n=28. Native; moist shady woods, usually on humus-rich soil; locally frequent throughout En, scattered in Wa, Ir and lowland Sc but absent from Outer Isles. Flowers May-Jul.

2. M. vernale M. Bieb. (*M. scabrum* Rich.) - *Early Millet*. Annual; stems procumbent to decumbent, to 10(15)cm; leaves 1-3mm wide; sheaths scabrid; ligules 2-3mm; panicle 1-4cm, with appressed branches; glumes 2-3mm; lemma 1.5-2mm; 2n=8. Native; short turf on sand-dunes and cliffs by sea; 2 places in Guernsey, 1st found 1899. Our plant has been segregated as ssp. **sarniense** D.C. McClint. Flowers Apr. **RR**

TRIBE 6 - POEAE (Festuceae, Seslerieae) (genera 15-26). Annuals or perennials; ligule membranous; sheaths not fused or sometimes fused; inflorescence a panicle, or a raceme or spike with normally 1 spikelet at each node; spikelets with (1)3-many florets all (except the most apical) bisexual or with a group of apical male or sterile florets, sometimes spikelets in pairs, one of each pair entirely sterile; glumes equal to very unequal, sometimes the lower vestigial or 0; lemma rounded or keeled on back, 3-9-veined, awnless or with terminal awn; stamens 1-3; stigmas 2; lodicules 2; ovary glabrous or pubescent at apex.

15. FESTUCA L. - *Fescues*

Perennials with or without rhizomes, without stolons; sheaths not fused or fused; inflorescence a panicle; spikelets with (2)3-many florets all (except the most apical) bisexual, sometimes proliferating; glumes subequal; lemmas rounded on back, 3-5-veined, acuminate to subacute at apex, awned or not; stamens 3.

The *F. rubra* (spp. 6-7) and *F. ovina* (spp. 8-15) aggs are extremely critical, and accurate identification requires considerable experience and often the aid of leaf-sections, leaf epidermis characters and chromosome numbers. Shaded, droughted, crowded-out or otherwise starved plants should be avoided. The classifications of these aggs adopted here follow the results of intensive studies by A.-K.K.A. Al-Bermani and M.J. Wilkinson respectively.

Leaf, sheath and ligule characters refer to leaves on tillers unless otherwise stated. Spikelet lengths are not total spikelet lengths, but the length from the base of the lower glume to the apex (excl. awn) of the fourth lemma; lengths of spikelets with only 3 florets are obtained by addition to the total length of an increment equal to the distance between the tips of the second and third lemmas. Lemma lengths exclude awns, and refer to only the lowest 2 per spikelet. Awn lengths are calculated by averaging the lengths of all the awns of one spikelet, and then finding the mean of this value for 5-10 spikelets. All measurements given are a range of means of 5-10 measurements per plant (not a range of extremes). The distinction between *extravaginal* and *intravaginal* tillers is fundamental. The latter arise parallel to the parent shoot and remain enclosed within the parental leaf-sheath for some distance. The former arise ± at right angles to the parent shoot and break through the parental leaf-sheath at its base. Rhizomes always start off extravaginally; hence the presence of rhizomes indicates the existence of extravaginal shoots, but the absence of rhizomes does not necessarily mean that all tillers are intravaginal. Leaf-sheath fusion should be observed on the next to most apical sheath of a tiller by stripping off all the more mature sheaths below it.

1 Base of leaf on each side extended into pointed auricles clasping stem at level of and on opposite side to ligule 2
1 Leaves without auricles, or with short rounded auricles not clasping stem 4
  2 Lemmas with awns longer than body; exposed nodes of culms dark violet-purple **3. F. gigantea**
  2 Lemmas with awns 0 or much shorter than body; exposed nodes of culms green, sometimes tinged purplish 3
3 Auricles glabrous; lowest 2 panicle nodes with two unequal branches, the shorter with 1-2(3) spikelets **1. F. pratensis**
3 Auricles usually fringed with minute hairs (often few, wearing off with age); lowest 2 panicle nodes with two subequal branches, the shorter with (3)4-many spikelets **2. F. arundinacea**
  4 Leaves of culms and tillers flat (or folded longitudinally when dry), >4mm wide (or >2mm from midrib to edge) 5
  4 Leaves of tillers and usually culms folded longitudinally or the edges also inrolled, ≤4mm wide (or ≤2mm from midrib to edge) 6
5 Leaves 4-14mm wide; ligules >1mm; lemmas 3-veined, awnless; rhizomes 0; ovary with pubescent apex **4. F. altissima**
5 Leaves ≤5mm wide; ligules <1mm; lemmas 5-veined, awned; rhizomes present; ovary with glabrous apex **8. F. rubra**
  6 Ovary obovoid, free from palea, with pubescent apex; most of lemma width translucent, not green; leaves all folded, with sharp apices; ligules 0.5-1mm **5. F. gautieri**
  6 Ovary ellipsoid to oblong, adherent to palea, usually glabrous, if with pubescent apex then culm leaves flat; most of lemma width green, not translucent; leaves without sharp apices; ligules ≤0.5mm 7
7 Young leaves with sheaths fused almost up to top; some or all tillers extravaginal 8
7 Young leaves with sheaths not fused near apex but with overlapping margins; all tillers intravaginal (*F. ovina* agg.) 10
  8 Ovary (and caryopsis) with pubescent apex; leaves with 3(-5) veins; leaves of culms and tillers markedly different, the former flat and 2-4mm wide, the latter folded and ≤0.6mm from midrib to edge **6. F. heterophylla**
  8 Ovary (and caryopsis) glabrous; leaves with 5-9(11) veins; leaves of culms and tillers similar to obviously different 9
9 Leaves with densely pubescent adaxial ribs, rounded on midrib abaxially, with abaxial sclerenchyma usually continuous or semi-continuous, with distinct sclerenchyma bundles in adaxial ribs; always coastal **7. F. arenaria**
9 Leaves with scabrid or sparsely pubescent adaxial ribs, obtuse to keeled on midrib abaxially, with abaxial sclerenchyma in discrete bundles, often without or with very sparse sclerenchyma in adaxial ribs; coastal or inland **8. F. rubra**
  10 Leaves with 5-9 veins, with 4(-6) adaxial grooves; lemmas with awns usually >1.2mm and often >1.6mm 11
  10 Leaves with 5-7 veins, with 2(-4) adaxial grooves; lemmas with awns <1.6mm and often <1.2mm 12
11 Pedicels 1.2-2.8mm; spikelets 6.1-8.5mm; lemmas with awns 1.2-2.6mm;

sheaths often sparsely pubescent; leaves not or slightly glaucous, with abaxial sclerenchyma in 3 main islets (at midrib and edges) **16. F. brevipila**

11 Pedicels 0.5-1.8mm; spikelets 5.4-7mm; lemmas with awns 0.5-1.5mm; sheaths glabrous; leaves usually very glaucous, with abaxial sclerenchyma in ≥5 main islets (at midrib, edges and variously in between) **15. F. longifolia**

12 Spikelets 4.7-7.5mm; lemmas with awns usually 0-1mm; panicles mostly <8cm, with <26 spikelets and lowest 2 nodes <2cm apart; leaves usually <0.57mm midrib to edge 13

12 Spikelets often >7mm, rarely <6mm; lemmas with awns usually >1mm; panicles ≤13cm, with ≤40 spikelets and lowest 2 nodes ≤3.7cm apart; leaves >0.57mm midrib to edge 15

13 Spikelets all or mostly proliferating; sexual florets (if present) with lemmas 3.4-4.2mm and awns 0-0.2mm **10. F. vivipara**

13 Spikelets not or rarely some proliferating; sexual florets with lemmas 2.5-4.9mm and awns 0-1mm 14

14 Spikelets ≤5.5mm; lemmas ≤3.5mm, with awns 0-0.6mm; leaves glabrous, 0.3-0.45(0.53)mm from midrib to edge **11. F. filiformis**

14 Spikelets ≥5.3mm; lemmas ≥3.3mm, with awns 0-1.6mm; leaves often pubescent at base, 0.3-0.75mm from midrib to edge **9. F. ovina**

15 Culms 19-66cm; panicles 3.7-8.6cm; pedicels 0.6-2.5mm; widespread **14. F. lemanii**

15 Culms 11-35cm; panicles 2.3-5.9cm; pedicels 0.8-1.5mm; coastal in CI only 16

16 Panicles well exserted from sheath at anthesis; leaves usually not glaucous; culms erect; usually on dunes **12. F. armoricana**

16 Panicles not completely or only just exserted from sheath at anthesis; leaves often slightly glaucous; culms erect to procumbent; usually on cliffs **13. F. huonii**

Other spp. - Much mis-application of names has occurred in this genus, and records of other spp. are mainly attributable to that. **F. glauca** Vill. is often grown for ornament but is not natd; records of it refer to *F. longifolia*, most of whose records refer in turn to *F. brevipila* and/or *F. lemanii*. Recent records of **F. guestfalica** Boenn. ex Rchb. refer to *F. brevipila*. Records of **F. arvernensis** Auquier, Kerguélen & Markgr.-Dann. and **F. indigesta** Boiss. are also errors.

1. F. pratensis Huds. (*F. elatior* auct. non L.) - *Meadow Fescue*. Culms to 825
80(120)cm; rhizomes 0; sheaths not fused; leaves flat, 3-8mm wide, with glabrous pointed auricles; ligules c.1mm; panicle erect to pendent; spikelets 9-11mm; lemmas 6-7mm, awnless; 2n=14. Native; meadows, hedgerows, waysides, by ditches and rivers, usually on rich moist soil; frequent throughout most of BI.

1 x 2. F. pratensis x F. arundinacea = F. x aschersoniana Dörfl. resembles *F. arundinacea* but shows some influence of *F. pratensis*, has (2n=28), and is sterile; it has been reported only from Warks and Argyll but is probably widespread.

1 x 3. F. pratensis x F. gigantea = F. x schlickumii Grantzow resembles *F. gigantea* but shows some influence of *F. pratensis*, has (2n=28), and is sterile; it has been reported only from W Norfolk.

2. F. arundinacea Schreb. (*F. elatior* L.) - *Tall Fescue*. Culms to 120(200)cm; rhizomes 0; sheaths not fused; leaves usually flat, (1)3-12mm wide, with pointed auricles usually minutely pubescent at margins; ligules ≤2mm; panicle erect or pendent; spikelets (8)9-12mm; lemmas 6-7.5(9)mm with awn 0-4mm; 2n=42. Native; grassy places, rough and marginal ground on wide range of soils; common throughout BI, generally under-recorded. Very variable, probably represented by several ecotypes of different habitats, but the situation is obscured by use of the sp. as a hay-grass and its frequent naturalization from this source.

2 x 3. F. arundinacea x F. gigantea = F. x fleischeri Rohlena (*F. x gigas* Holmb.) resembles *F. x aschersoniana* but has hairs on the auricles, has (2n=42), but is sterile; scattered in Br N from Hunts and Brecs, probably elsewhere.

3. F. gigantea (L.) Vill. - *Giant Fescue.* Culms to 100(150)cm; rhizomes 0; sheaths not fused; leaves flat, 6-18mm wide, with glabrous pointed auricles; ligules $\leq$2.5mm; panicle pendent; spikelets 8-13mm; lemmas 6-9mm with usually wavy awn 10-18mm; 2n=42. Native; woods, hedgerows and other shady places: common in Br and Ir except N & NW Sc. Sometimes confused with *Bromopsis ramosa*, with which it very often grows, but the latter has pubescent (not glabrous) leaves and sheaths, and awns shorter than body of lemma.

4. F. altissima All. - *Wood Fescue.* Culms to 120(150)cm; rhizomes 0; sheaths not fused; leaves flat, 4-14mm wide, without auricles; ligules 1-5mm; panicle ± pendent; spikelets 5-8mm; lemmas 4-6mm, awnless; 2n=14. Native; moist stony slopes and ravines in woods and copses; scattered in Br from S Wa to N mainland Sc, E Sussex, very scattered in Ir. The base of the culms characteristically bears persistent, short, pointed, blade-less leaf-sheaths.

5. F. gautieri (Hack.) K. Richt. - *Spiky Fescue.* Culms to 45cm, arising from mat-forming vegetative shoots; rhizomes 0; sheaths not fused; leaves folded, 0.3-0.6mm from midrib to edge, without auricles, with 7 abaxial sclerenchyma islets sometimes partially joining up; ligules c.1mm; spikelets 6-11.5mm; lemmas 4.3-5.5mm, with awn 0.2-0.6mm; 2n=14. Intrd; natd in limestone quarry in NW Yorks probably since c.1920, also briefly on a bank in S Devon; Pyrenees. Has a superficial resemblance to *F. ovina* agg., but differs in many details: ligules and anthers (2.9-4.1mm, not $\leq$3mm) are longer and the sharp-pointed leaves and translucent lemmas are diagnostic. Our plant is ssp. **scoparia** (A. Kern. & Hack.) Kerguélen.

6. F. heterophylla Lam. - *Various-leaved Fescue.* Culms to 100(120)cm, densely **847** tufted; rhizomes 0; sheaths fused ± to apex; leaves folded, 0.3-0.6mm from midrib to edge, without auricles, with 5 small abaxial sclerenchyma islets; ligules <0.5mm; spikelets 8-11.5mm; lemmas (4.7)5-6.5(8)mm, with awn to 4.5(6)mm; (2n=28, 42). Intrd; grown for ornament and appearing as contaminant of other grass-seed, natd in woods and wood-borders on light soils; scattered in Br, mainly S En, Co Limerick; Europe.

7. F. arenaria Osbeck (*F. juncifolia* St.-Amans, *F. rubra* ssp. *arenaria* (Osbeck) F. **R** Aresch.) - *Rush-leaved Fescue.* Culms to 75(90)cm, scattered; rhizomes very long, **847** sheaths fused ± to apex; leaves folded, 0.5-1.9mm from midrib to edge, without auricles, usually with continuous or subcontinuous abaxial sclerenchyma, often with islets with 'tails', sometimes with 5-13 discrete islets; ligules <0.5mm; spikelets 8-14.2mm; upper glume 3.5-9.1mm; lemmas 6-10.1mm, with awns 0.5-2.6mm; 2n=56. Native; mobile sand-dunes and sandy shingle by sea; frequent on coasts of Br N to E Ross and S Wa, rare to W Ross and in E Ir, CI.

8. F. rubra L. - *Red Fescue.* Culms to 75(100)cm, often much less; sheaths fused ± **825** to apex; leaves flat or folded, 0.5-1.4(2.5)mm from midrib to edge, without auricles, with abaxial sclerenchyma in 5-9 discrete islets; ligules <0.5mm. Extremely variable; just worthy of division into 7 (perhaps more) sspp., some of which are often recognized as spp. 6 of the sspp. represent ± distinct combinations of characters, but all are closely approached by variants of ssp. *rubra*. Several of the sspp. are sold as grass-seed and become natd on road-verges, etc. In upland or damp areas examples with proliferating spikelets are not rare; they are often misnamed *F. vivipara.*

1 Rhizomes 0 or very few and very short ... **d. ssp. commutata**
1 Rhizomes well developed (plants densely tufted or not) ... 2
  2 Leaves 0.8-1.4(2.5)mm from midrib to edge, with distinct islets of sclerenchyma in adaxial ribs; lemmas usually 6-8mm ... 3
  2 Leaves 0.5-1.2mm from midrib to edge, usually without or with very sparse sclerenchyma in adaxial rihs (except in ssp. *juncea*); lemmas 4.2-6.5mm ... 4

3 Leaves of culms and often tillers flat; culms up to 1m; panicle diffuse; widespread **g. ssp. megastachys**
3 Leaves of culms and tillers folded; culms up to 70cm; panicle compact with ± appressed branches; N Br only **f. ssp. scotica**
  4 Lemmas 4.2-6mm, with awns 0.1-1.1mm, glaucous, usually with dense white hairs (sometimes glabrous but usually some pubescent plants nearby); N Wa and northwards only **e. ssp. arctica**
  4 Lemmas 4.4-8mm, with awns 0.5-2.8mm, glaucous or not, the hairs (if present) not dense and white; widespread 5
5 Spikelets 8.7-11.2mm; lemmas 5.7-8 x ≥2mm, with awns 1.1-2.8mm; saline sand or mud, often forming dense mats **c. ssp. litoralis**
5 Spikelets 6.8-10.2mm; lemmas 4.4-6.7 x 1.5-2.4mm, with awns 0.5-2.2mm; rarely in saline sand or mud but often maritime 6
  6 Rhizomes short, forming dense tufts, often some or all plants in population with very glaucous leaves **b. ssp. juncea**
  6 Rhizomes medium to long, forming loose patches; plants rarely glaucous **a. ssp. rubra**

a. Ssp. rubra. Rhizomes well developed, usually forming loose patches; culms to **847**
75cm; leaves folded, rarely flat, 0.6-1.3mm from midrib to edge; spikelets 6.9-10.2mm; upper glume 2.6-5.3mm; lemmas 4.6-6.7mm with awns 0.8-2.1mm; 2n=42, 56. Native; all kinds of grassy places; common throughout BI.

b. Ssp. juncea (Hack.) K. Richt. (ssp. *pruinosa* (Hack.) Piper). Rhizomes short, **847**
forming dense tufts; culms to 75cm; leaves folded, 0.5-1.2mm from midrib to edge; spikelets 6.8-9.9mm; upper glume 3.2-5.3mm; lemmas 4.4-6.5mm, with awns 0.5-2.2mm; 2n=42. Native; maritime cliffs and inland grassy rocky places; round coasts of BI and in hilly areas of N Br, rarely inland elsewhere. Separation of ssp. *pruinosa* from ssp. *juncea* is often impossible in practice and the 2 are amalgamated here. The former is nearly always coastal and often has markedly pruinose leaves, but even on the coast wholly non-pruinose populations occur. It is also usually a shorter plant, but this might be a dwarfing effect of the coastal environment.

c. Ssp. litoralis (G. Mey.) Auquier. Rhizomes rather short, often forming dense **847**
mats; culms to 55cm; leaves folded, 0.6-1mm from midrib to edge; spikelets 8.7-11.2mm; upper glume 4.3-6.2mm; lemmas 5.7-8mm, with awns 1.1-2.8mm; 2n=42. Native; salt-marshes and other muddy or sandy saline areas; probably in suitable places round coasts of BI. Has characteristically short culms, large lemmas and narrow, short leaves, but the distinctive habitat is diagnostic.

d. Ssp. commutata Gaudin (ssp. *caespitosa* Hack., var. *fallax* sensu Tutin, *F.* **847**
nigrescens Lam.). Rhizomes 0 or very few and short, forming dense tufts; culms to 75cm; leaves folded, 0.6-1.1mm from midrib to edge; spikelets 7.2-9.1mm; upper glume 3.2-5mm; lemmas 4.4-6.3mm, with awns 1.1-2.6mm; 2n=42. Native; grassy places and rough ground, usually in well-drained soils; probably throughout Br due to extensive use as grass-seed (*Chewing's Fescue*).

e. Ssp. arctica (Hack.) Govor. (*F. richardsonii* Hook.). Rhizomes well developed, **R**
forming loose patches; culms to 50cm; leaves folded, 0.5-1.2mm from midrib to **847**
edge; spikelets 6.6-8.8mm; upper glume 3.1-5.5mm; lemmas 4.2-6mm, with awns 0.1-1.1mm; 2n=42, 63. Native; wet mountain slopes and gulleys, rock-crevices and flushes down to sea-level, often on serpentine; scattered from C Sc to Shetland, Caerns, Westmorland, Cumberland, probably under-recorded. The densely white-pubescent, glaucous spikelets with short awns are diagnostic, but in many populations some plants are glabrous.

f. Ssp. scotica S. Cunn. ex Al-Bermani. Rhizomes well developed, forming loose **R**
patches; culms to 70cm; leaves folded, 0.8-1.4mm from midrib to edge; spikelets **847**
9.4-11.8mm; upper glume 4.6-6.5mm; lemmas 6.1-7.9mm, with awns 1.2-2.3mm; 2n=56, 63, 70. Native; grassy and rocky places from near sea-level to >800m; Sc from Argyll to Shetland and Outer Hebrides, Cumberland, probably under-recorded. The long spikelets, lemmas and awns and erect panicle-branches are

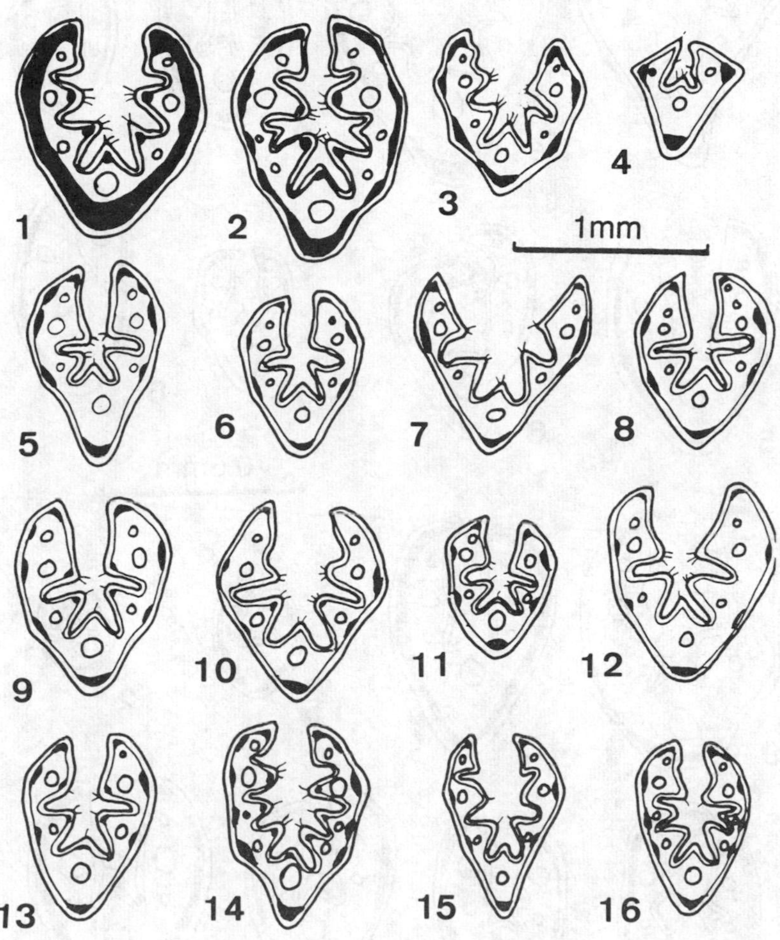

FIG 847 - Transverse sections of innovation leaves of *Festuca rubra* agg. 1-3, ***F. arenaria***. 4, ***F. heterophylla***. 5-16, ***F. rubra***. 5, ssp. ***rubra***. 6, ssp. ***littoralis***. 7-8, ssp. ***arctica***. 9-10, ssp. ***juncea***. 11-12, ssp. ***scotica***. 13-14, ssp. ***megastachys***. 15-16, ssp. ***commutata***. Sclerenchyma in black. Redrawn by C.A. Stace from photographs by A.-K. Al-Bermani.

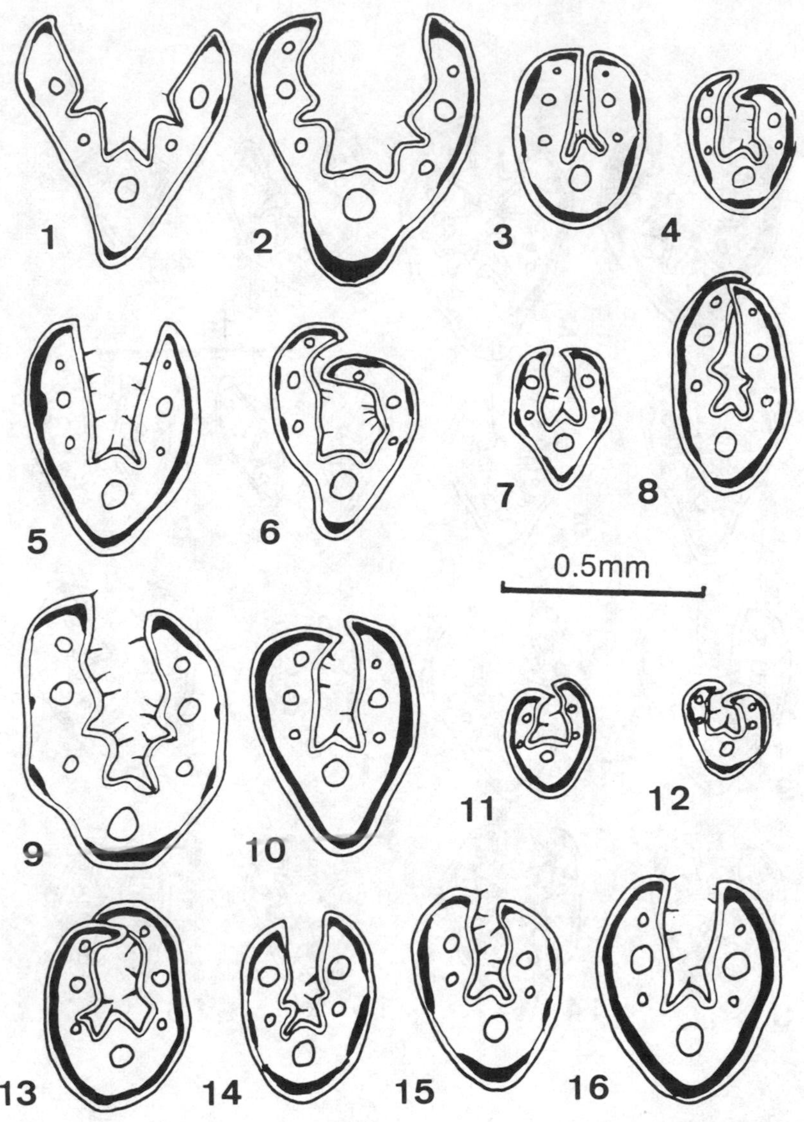

FIG 848 - Transverse sections of innovation leaves of ***Festuca ovina*** agg.
1-2, ***F. brevipila***. 3, ***F. ovina*** ssp. ***ovina***. 4, ***F. ovina*** ssp. ***hirtula***. 5-6, ***F. lemanii***.
7, ***F. vivipara***. 8, ***F. ovina*** ssp. ***ophioliticola***. 9-10, ***F. longifolia***.
11-12, ***F. filiformis***. 13-14, ***F. armoricana***. 15-16, ***F. huonii***. Sclerenchyma in black.
Redrawn by C.A. Stace from drawings by M.J. Wilkinson.

diagnostic.

g. Ssp. megastachys Gaudin (ssp. *fallax* (Thuill.) Nyman, ssp. *multiflora* Piper, 847
var. *planifolia* Hack., *F. diffusa* Dumort., *F. heteromalla* Pourr., non *F. rubra* var. *fallax* sensu Tutin). Rhizomes well developed, forming diffuse patches; culms to 100cm; leaves often flat, sometimes folded, 1-1.4(2.5)mm from midrib to edge; spikelets 7-11.2mm; upper glume 3.5-6.2mm; lemmas 4.7-7.9mm, with awns 0.5-3mm; 2n=56, ?70. Probably intrd; grassy places, especially on waysides where it is much planted; scattered throughout Br; Europe.

9-16. F. ovina L. agg. Plants densely tufted, without rhizomes; sheaths not fused; tiller and culm leaves folded, 0.3-0.95mm from midrib to edge, with very small rounded auricles, with continuous or discontinuous abaxial sclerenchyma; ligules <0.5mm.

9. F. ovina L. - *Sheep's-fescue.* Culms to 50cm; leaves not glaucous or sometimes 825
slightly so, 0.4-0.7mm from midrib to edge, with 5-7 veins and 2(-4) adaxial grooves, with abaxial sclerenchyma in thin broken or sometimes continuous band; panicles 1.5-7.8cm; pedicels 0.8-2.8mm; spikelets 5.3-7.2mm; lemmas 3.3-4.8mm, with awns 0-1mm.

1 Spikelets 5.5-7.5mm; lemmas 3.6-4.9mm; leaves with (5-)7 veins **c. ssp. ophioliticola**
1 Spikelets 5.3-6.3mm; lemmas 3.1-4.2mm; leaves with 5-7 veins 2
  2 Awns 0-0.8mm; leaves usually pubescent at base; lemmas usually pubescent; stomata mostly >31.5 microns **b. ssp. hirtula**
  2 Awns 0-1.2mm; leaves and lemmas usually scabrid; stomata mostly <31.5 microns **a. ssp. ovina**

a. Ssp. ovina. Culms 10-35cm; leaves 0.33-0.67mm from midrib to edge; panicles 848
2.2-7.3cm; pedicels 0.8-2.2mm; lemmas with awns 0-1.1mm; 2n=14. Native; grassy places on well-drained, usually acid soils; common in N, C & SW Br, very sparse in SC & SE En, ?Ir.

b. Ssp. hirtula (Hack. ex Travis) M.J. Wilk. (*F. tenuifolia* var. *hirtula* (Hack. ex 848
Travis) Howarth, *F. ophioliticola* ssp. *hirtula* (Hack. ex Travis) Auquier). Culms 6-45cm; leaves 0.35-0.67mm from midrib to edge; panicles 1.5-6.6cm; pedicels 0.9-2.7mm; lemmas with awns 0-0.8mm; 2n=28. Native; grassy places on well-drained, usually acid soils; common throughout BI. Small plants in S En and CI are scarcely separable from ssp. *ovina*.

c. Ssp. ophioliticola (Kerguélen) M.J. Wilk. (*F. ophioliticola* Kerguélen). Culms 20- 848
50cm; leaves 0.39-0.75mm from midrib to edge; panicles 2.8-8cm; pedicels 1.3-3.6mm; lemmas with awns 0-1.6mm; 2n=28. Native; grassy places on well-drained, often calcareous or serpentine soils; locally common throughout Br and Ir.

10. F. vivipara (L.) Sm. - *Viviparous Sheep's-fescue.* Culms to 44(50)cm; differs 848
from *F. ovina* in all or most spikelets proliferating; completely fertile spikelets (if present) 5.5-6.2mm, with lemmas 3.4-4.2mm with awns 0-0.2mm; 2n=21, 28. Native; grassy places in hilly districts, usually on rocky ground; common in C & N Sc, local in S Sc, N En, Wa and Ir.

Triploid (often proliferating) *Sheep's-fescues* (2n=21) have also been claimed as hybrids between *F. ovina*, *F. filiformis* and *F. vivipara* in various combinations; possibly all 3 hybrids occur, but they need careful study.

11. F. filiformis Pourr. (*F. tenuifolia* Sibth., *F. ovina* ssp. *tenuifolia* (Sibth.) Dumort.) 848
- *Fine-leaved Sheep's-fescue.* Culms to 35cm; leaves not glaucous, 0.3-0.53mm from midrib to edge, with (4)5(-7) veins and 2(-3) adaxial grooves, with abaxial sclerenchyma in thin broken or sometimes continuous band; panicles 2.1-7.2cm; pedicels 0.6-2.3mm; spikelets 4.5-5.5mm; lemmas 2.5-3.5mm, with awns 0-0.6mm; 2n=14. Native; grassy places on usually acid sandy soils; frequent throughout BI.

12. F. armoricana Kerguélen (*F. ophioliticola* ssp. *armoricana* (Kerguélen) Auquier) - RR
Breton Fescue. Culms to 40cm; leaves not glaucous, 0.5-0.88mm from midrib to edge, 848

with 7 veins and 2-4 adaxial grooves, with abaxial sclerenchyma in thin broken or sometimes continuous band; panicles 3.3-5.9cm; pedicels 0.8-1.5mm; spikelets 6-8mm; lemmas 3.9-5.1mm, with awns 0.8-1.6mm; 2n=28. Native; fixed dunes on W coast of Jersey, perhaps overlooked elsewhere.

13. F. huonii Auquier - *Huon's Fescue.* Culms to 25cm, often procumbent; leaves sometimes ± glaucous, often strongly curved, 0.4-0.75mm from midrib to edge, with 5-7 veins and 2-4 adaxial grooves, with abaxial sclerenchyma in thin broken or sometimes continuous band; panicles 2.3-5.4cm; pedicels 0.7-1.8mm; spikelets 5.5-7.5mm; lemmas 3.5-4.9mm, with awns 0.9-1.8mm; 2n=42. Native; grassy cliff-tops and -bases; Jersey, Guernsey and other islands in CI. Reported in 1992 from S Devon but needs confirming. **RR** **848**

14. F. lemanii Bastard (*F. bastardii* Kerguélen & Plonka, *F. longifolia pro parte* sensu C.E. Hubb. et al. non Thuill.) - *Confused Fescue.* Culms to 66cm; leaves often ± glaucous, 0.43-0.95mm from midrib to edge, with (5-)7 veins and 2-4 adaxial grooves, with abaxial sclerenchyma in continuous or subcontinuous (sometimes broken) band; panicles 3.7-8.6cm; pedicels 0.6-2.5mm; spikelets 6.1-8.5mm; lemmas 4-5.3mm, with awns 0.3-1.9mm; 2n=42. Probably native; grassy places on well-drained, acid or calcareous soils, often with *F. ovina;* very scattered in Br, probably under-recorded. **R** **848**

15. F. longifolia Thuill. (*F. caesia* Sm., *F. glauca* var. *caesia* (Sm.) Howarth, *F. glauca* auct. non Vill.) - *Blue Fescue.* Culms to 40cm; leaves usually strongly glaucous, 0.55-0.9mm from midrib to edge, with 7(-9) veins and 4(-6) adaxial grooves, with abaxial sclerenchyma in thick broken or sometimes continuous band; panicles 2.4-5.6cm; pedicels 0.5-1.8mm; spikelets 5.4-7mm; lemmas 3.5-4.8mm, with awns 0.5-1.5mm; 2n=14. Native; very dry acid heaths and maritime cliff-tops; in the Breckland of W Suffolk and a few similar sites in N Lincs, and in S Devon, Jersey, Guernsey and other islands in CI. **RR** **848**

16. F. brevipila R. Tracey (*F. trachyphylla* (Hack.) Krajina non Hack. ex Druce, *F. longifolia pro parte* sensu C.E. Hubb. et al. non Thuill.) - *Hard Fescue.* Culms to 70cm; leaves usually somewhat glaucous, 0.6-1mm from midrib to edge, with (5)7-9 veins and 4(-6) adaxial grooves, with abaxial sclerenchyma usually as 3 islets with tails at midrib and edges, often with smaller islets opposite other veins, rarely subcontinuous; panicles 3.5-9.5cm; pedicels 1.2-2.8mm; spikelets 6.1-8.5mm; lemmas 3.9-5.5mm, with awns 1.2-2.6mm; 2n=42. Intrd; much grown from grass-seed mixtures, natd on roadsides, commons and rough ground especially on acid well-drained soils; frequent in SE En and E Anglia, scattered N to SW Yorks, Jersey, greatly under-recorded; Europe. **848**

15 x 16. FESTUCA x LOLIUM = X FESTULOLIUM Asch. & Graebn.
Inflorescences variously intermediate, at 1 extreme a simple raceme (rarely a spike as in *Lolium*), at the other a branching panicle, often a raceme with a few racemose branches near base; most spikelets with 2 glumes, but lower much shorter than upper; anthers ± indehiscent with ± empty pollen grains, but some degree of fertility exists and some backcrossing may occur.

L. perenne and *L. multiflorum* hybridise with *F. pratensis*, *F. arundinacea* and *F. gigantea*, and 5 of the 6 possible combinations have been found. A single report of *F. rubra* x *L. perenne* = **X F. fredericii** Cugnac & A. Camus was an error, but all combinations are probably under-recorded.

15/1 x 16/1. F. pratensis x L. perenne = X F. loliaceum (Huds.) P. Fourn. (*Hybrid Fescue*) has glabrous auricles and awnless lemmas; inflorescence usually a simple or little-branched raceme; 2n=14, 21. Native; pastures, meadows, riversides and roadsides, often on damp rich soils; throughout most of BI, commonest in S En.

15/1 x 16/2. F. pratensis x L. multiflorum = X F. braunii (K. Richt.) A. Camus differs from X *F. loliaceum* in its awned lemmas; (2n=14). Native and probably intrd in grass-seed; grassy places, rough ground, waysides; scattered in En, Denbs.

15/2 x 16/1. F. arundinacea x L. perenne = X F. holmbergii (Dörfl.) P. Fourn. has a well-branched inflorescence but often with subsessile spikelets; auricles minutely pubescent at margins; (2n=28). Native; similar places to X *F. loliaceum*; very scattered in S En N to Warks, Dunbarton.

15/2 x 16/2. F. arundinacea x L. multiflorum differs from X *F. holmbergii* in its awned lemmas; (2n=28). Native and probably intrd in grass-seed; similar places to X *F. braunii*; very scattered in S En N to Warks.

15/3 x 16/1. F. gigantea x L. perenne = X F. brinkmannii (A. Braun) Asch. & Graebn. resembles either *F. pratensis* x *L. multiflorum* or *F. arundinacea* x *L. multiflorum* in inflorescence shape; differs from both in its longer lemmas and wider leaves. Native; grassy places near the parents; Cambs, Pembs and Merioneth.

F. gigantea x *L. multiflorum* = **X F. nilssonii** Cugnac & A. Camus might occur but would be very difficult to distinguish from X *F. brinkmannii*.

15 x 17. FESTUCA X VULPIA = X FESTULPIA Melderis ex Stace & R. Cotton

Plants perennial and vegetatively close to the *Festuca* parent, but with fewer and shorter (or 0) rhizomes and some overlapping sheaths; panicles narrower and less branched than in *Festuca*, with markedly longer awns; lower glume c.1/2 as long as upper; anthers indehiscent with ± empty pollen grains, but some degree of fertility may exist.

F. arenaria and *F. rubra* hybridise with *V. fasciculata*, *V. bromoides* and *V. myuros*, and 4 of the 6 possible combinations have been found.

15/7 x 17/1. F. arenaria x V. fasciculata = X F. melderisii Stace & R. Cotton has lower glume 5.2-8mm; upper glume 8-11.5mm (incl. awn); lemmas 9.5-10.5mm plus awn 3.5-5mm; anthers 3, 1.5-2mm; 2n=42. Native; on open sand-dunes with parents; very local in S En and S Wa.

15/8 x 17/1. F. rubra x V. fasciculata = X F. hubbardii Stace & R. Cotton has **860**
lower glume 2.4-4.4mm; upper glume (incl. awn) 3.5-7.2mm; lemmas 6-9.5mm plus awn 2-5.5mm; anthers 3, 1.5-2mm; 2n=35. Native; on open sand-dunes with parents; probably frequent in CI and Br N to Westmorland.

15/8 x 17/2. F. rubra x V. bromoides has lower glume 2-3.4mm; upper glume 3.4-5.9mm; lemmas 4.5-7mm plus awn 3.2-6mm; anthers 3, 0.8-1.7mm. Differs from X *F. hubbardii* in not having markedly distally thickened pedicels, in having only 1 (not 2-several) ovary-less floret at apex of spikelets, and in the awnless upper glumes. Native; waste and rough ground with the parents; scattered in Br N to SE Yorks, Guernsey. Probably both ssp. *rubra* and ssp. *commutata* of *F. rubra* have produced this hybrid.

15/8 x 17/3. F. rubra x V. myuros has lower glume 1.5-3.3mm; upper glume 3.2-5mm; lemmas 4.5-6.2mm plus awn 3-6mm; anthers 3, 0.6-1.5mm; 2n=42. Differs from *F. rubra* x *V. bromoides* in its long narrow panicle, showing the influence of *V. myuros*. Native; waste and rough ground with the parents; scattered in Br N to S Lancs. Probably ssp. *rubra*, ssp. *commutata* and ssp. *litoralis* of *F. rubra* have produced this hybrid.

16. LOLIUM L. - *Rye-grasses*

Annuals to perennials without rhizomes or stolons; sheaths not fused; inflorescence normally a spike with spikelets compressed and orientated with backs of lemmas adjacent to spike axes; spikelets partly lying in concavities in spike axis, with 2-many florets all (except the most apical) bisexual; glumes 2 in terminal spikelet, 1 (the upper, abaxial) in lateral spikelets; lemmas rounded on back, 5-9-veined, obtuse to subacute or minutely bifid at apex, with or without subterminal awn; stamens 3.

Abnormal plants with branched inflorescences are not rare. Sp. limits are very unsatisfactory; all spp. may have awned or unawned lemmas, and all except *L. perenne* have rolled (not folded) young leaves. All the spp. are diploids and produce fertile hybrids. Apart from the 1 hybrid covered, *L. multiflorum* x *L. temulentum*, *L. multiflorum* x *L. rigidum* and *L. rigidum* x *L. temulentum* have all been recorded as rare casuals with grain, wool and distillery refuse, imported as hybrid caryopses. Their determination is very difficult and must be considered tentative.

1 Lemmas ovate to elliptic, ≤3x as long as wide, becoming thick and hard at base in fruit; caryopsis ≤3x as long as wide 2
1 Lemmas narrowly oblong-ovate, >3x as long as wide, not becoming thick or hard; caryopsis >3x as long as wide 3
  2 Lowest 2 lemmas (4.6)5-8.5mm; glume (7)10-30mm **4. L. temulentum**
  2 Lowest 2 lemmas 3.5-5(5.5)mm; glume 5-12(15)mm **5. L. remotum**
3 Perennial, with tillers at flowering and fruiting time; leaves folded along midrib when young; lemmas usually unawned **1. L. perenne**
3 Annual or biennial, without tillers at flowering and fruiting time; leaves rolled along long axis when young; lemmas awned or not 4
  4 Lemmas nearly always awned; spikelets usually with ≥11 florets **2. L. multiflorum**
  4 Lemmas usually unawned; spikelets usually with ≤11 florets **3. L. rigidum**

Other spp. - **L. persicum** Boiss. & Hohen. ex Boiss., from SW Asia, is a rather rare grain-alien; it would key to couplet 4 but has 4-7(9) florets per spikelet, lowest 2 lemmas 6.5-11mm with awn 5-21mm, and glume >2/3 as long as spikelet.

1. L. perenne L. - *Perennial Rye-grass.* Perennial to 50(90)cm; spikelets with 4-14 florets; glume 1/3 to >lx as long as rest of spikelet; lowest 2 lemmas 3.5-9mm; awns very rare (≤8mm); 2n=14. Native; grassy places, waste and rough ground, also a common escape from lawns, roadside plantings and pastures; abundant throughout BI.

1 x 2. L. perenne x L. multiflorum = L. x boucheanum Kunth (*L. x hybridum* Hausskn.) occurs occasionally in Br and Ir as a natural product and now commonly in lowland Br and CI as an escape from its cultivation as a valuable pasture or meadow grass; it is annual to perennial, and has rolled young leaves and intermediate spikelet structure (always shortly awned).

2. L. multiflorum Lam. - *Italian Rye-grass.* Annual or biennial to 100(120)cm; spikelets with (5)11-22 florets; glume 1/4-3/4 as long as rest of spikelet; lowest 2 lemmas 4-8mm; awns nearly always present (≤15mm); (2n=14). Intrd; rough and waste ground, field borders, waysides; scattered throughout BI, common in lowland Br and CI; S Europe.

3. L. rigidum Gaudin (*L. loliaceum* (Bory & Chaub.) Hand.-Mazz.) - *Mediterranean* 870
Rye-grass. Annual to 70cm; spikelets with 4-8(11) florets; glume 3/4 to >1x as long as rest of spikelet; lowest 2 lemmas 3.5-8.5mm; awns very rare (≤10mm); (2n=14). Intrd; rather frequent alien from grain, wool and other sources in waste ground and on tips; scattered in Br; Mediterranean. Spikelets usually very narrow, well sunk in concavities of rhachis and largely concealed by glume; easily confused with poorly-grown *L. perenne* if vegetative characters are ignored.

4. L. temulentum L. - *Darnel.* Annual to 75(100)cm; spikelets with 4-10(15) florets; glume 3/4-1.5x as long as rest of spikelet; lowest 2 lemmas (4.6)5-8.5mm; awns present (often ≥10mm) or less often 0; (2n=14). Intrd; formerly common in cornfields, now casual on tips and waste places from many sources especially grain; scattered in BI; Mediterranean.

5. L. remotum Schrank - *Flaxfield Rye-grass.* Annual to 75cm; spikelets with 4-10 florets; glume 2/3-1.5x as long as rest of spikelet; lowest 2 lemmas 3.5-5(5.5)mm; awn present (≤10mm) or more often 0; (2n=14). Intrd; formerly a typical flaxfield alien, now a very occasional alien from grain and other sources; scattered and sporadic in En; E Europe.

17. VULPIA C.C. Gmel. (*Nardurus* Rchb.) - *Fescues*
Annuals; sheaths not fused; inflorescence a sparsely branched narrow panicle or a raceme; spikelets with 2-many florets all (except the most apical) bisexual or with a group of sterile florets at apex; glumes 2, very unequal, the lower at most 3/4 as long as upper; lemmas rounded on back, 3-5-veined, acute or acuminate with long terminal awn; stamens 1-3. Glume lengths and ratios must be measured in spikelets that are not apical on the inflorescence or its branches; in apical spikelets the lower glume is often much longer than normal.

1 Lemmas with basal pointed minutely scabrid callus; ovary and caryopsis with minute apical hairy appendage; lemmas 8-18mm excl. awn; upper glume 10-30mm incl. awn **1. V. fasciculata**

1 Lemma with basal rounded glabrous callus; ovary and caryopsis glabrous; lemmas 3-7.5mm excl. awn; upper glume 1.5-9mm incl. awn if present 2

2 Anthers 3, 0.7-1.3(1.9)mm, well exserted at anthesis; lemmas 3-5mm excl. awn; inflorescence ± always a raceme **5. V. unilateralis**

2 Anthers 1(-3), 0.4-0.8(1.8)mm, usually not exserted at anthesis; lemmas 4-7.5mm; inflorescence a panicle except in starved plants 3

3 Spikelets with 1-3 bisexual and 3-7 distal sterile (but scarcely smaller) florets; lemma of fertile florets 3(-5)-veined **4. V. ciliata**

3 Spikelets with 2-5 bisexual and 1-2 distal much reduced sterile florets; lemma of fertile florets 5-veined 4

4 Lower glume 2.5-5mm, 1/2-3/4 as long as upper; lemmas usually >1.3mm wide when flattened; inflorescence normally well exserted from uppermost sheath at maturity **2. V. bromoides**

4 Lower glume 0.4-2.5mm, 1/10-2/5 as long as upper; lemmas usually <1.3mm wide when flattened; inflorescence normally not fully exserted from uppermost sheath at maturity **3. V. myuros**

Other spp. - Several spp. are found as rare casuals, especially as wool-aliens. The least rare are **V. muralis** (Kunth) Nees (*V. broteri* Boiss. & Reut.), from Europe, which resembles *V. bromoides* in its inflorescence shape but has narrow lemmas as in *V. myuros*, and lower glume 1-3mm and 1/4-1/2 as long as upper and is perhaps overlooked; and **V. geniculata** (L.) Link, from W Mediterranean, which is distinct from all above in its broad lax inflorescence and 3 exserted anthers >2mm.

1. V. fasciculata (Forssk.) Fritsch (*V. membranacea* auct. non (L.) Dumort.) - *Dune* R
Fescue. Culms to 50cm; inflorescence a panicle or raceme 3-11cm; spikelets 10-18mm excl. awns, with distal group of 3-6 reduced sterile florets; lower glume 0.1-2.6mm, <1/6 as long as upper; upper glume 10-30mm incl. awn 3-12mm; fertile lemmas 8-18mm excl. awn; 2n=28. Native; open parts of sand-dunes; locally frequent on coasts of BI N to Cheviot, Man and Co Louth, commoner in W.

2. V. bromoides (L.) Gray - *Squirreltail Fescue*. Culms to 50cm; inflorescence a panicle or sometimes a raceme 1-11cm; spikelets 6.5-11.5mm excl. awns, with 1-2 distal sterile reduced florets; lower glume 2.5-5mm, 1/2-3/4 as long as upper; upper glume 4.5-9mm incl. awn 0-2mm; fertile lemmas 4.5-7.5mm excl. awn; 2n=14. Native; open grassy places on well drained soils, rough and waste ground; frequent over most of BI, common in S Br.

3. V. myuros (L.) C.C. Gmel. (*V. megalura* (Nutt.) Rydb.) - *Rat's-tail Fescue*. Culms to 65cm; inflorescence a panicle or raceme 5-35cm; spikelets 6-10.5mm excl. awns, with 1-2 distal sterile reduced florets; lower glume 0.4-2.5mm, 1/10-2/5 as long as upper; upper glume 2.5-6.5mm incl. awn 0-1mm; fertile lemmas 4.5-7.5mm excl. awn; 2n=42. Probably native; open ground, walls, rough or waste ground, roadsides and by railways, also frequent grain- and wool-alien; increased in recent years, now throughout BI except C & N Sc and most of N Ir, common in S Br. Lemmas are usually scabrid, but may be pubescent at edges (f. **megalura** (Nutt.) Stace & R. Cotton) or pubescent dorsally (f. **hirsuta** (Hack.) Blom) in intrd material.

4. V. ciliata Dumort. - *Bearded Fescue.* Culms to 45cm; inflorescence a panicle or R
raceme 3-20cm; spikelets mostly 5-7mm excl. awns, with distal group of 3-7 sterile (but scarcely smaller) florets; lower glume 0.1-1mm, <1/4 as long as upper; upper glume 1.5-4mm, never awned; fertile lemmas 4-5mm excl. awn, minutely scabrid; sterile lemmas ≤6mm excl. awn, minutely scabrid; 2n=28. Native; on maritime or submaritime sand or shingle; local in S Br N to N Lincs and Merioneth, CI, perhaps increasing. Our plant is ssp. **ambigua** (Le Gall) Stace & Auquier (*V. ambigua* (Le Gall) More). Ssp. **ciliata**, from S & C Europe, was formerly natd in railway sidings at Ardingly, E Sussex; it differs in spikelets mostly 7-10.5mm excl. awns; fertile lemmas 5-6.5mm excl. awn, pubescent on dorsal midline and margins; sterile lemmas ≤8mm excl. awn, densely pubescent at edges. Both sspp. are rather rare casuals from wool and grain.

5. V. unilateralis (L.) Stace (*Nardurus maritimus* (L.) Murb.) - *Mat-grass Fescue.* R
Culms to 40cm; inflorescence usually a raceme 1-10cm, rarely slightly branched below; spikelets 4-7(8)mm excl. awns, with 1-2 distal sterile reduced florets; lower glume 1.5-3.5mm, 1/2-3/4 as long as upper; upper glume 3-5mm, unawned; fertile lemmas (2)3-4(5)mm excl. awn; 2n=14. Native; open grassy places on chalk, also in waste places and waysides; very scattered in Br N to E Gloucs and W Norfolk, formerly to Derbys.

18. CYNOSURUS L. - *Dog's-tails*

Annuals or perennials without rhizomes or stolons; sheaths not fused; inflorescence a compact spike-like panicle; spikelets of 2 kinds - fertile with (1)2-5 florets all (except the most apical) bisexual, and sterile consisting of numerous very narrow acuminate lemmas in herring-bone arrangement, normally 1 fertile and 1 sterile together; glumes 2, subequal; fertile lemmas rounded on back, 5-veined, acute to obtuse or minutely bifid, awnless or with long terminal or subterminal awn; stamens 3.

1. C. cristatus L. - *Crested Dog's-tail.* Tufted perennial to 75cm; leaves 1-4mm wide; ligules 0.5-1.5mm, ± truncate; panicle linear-oblong, 1-10(14) x 0.4-1cm; fertile lemmas 3-4mm plus awn 0-1mm; 2n=14. Native; grassy places on a great range of soils; common throughout BI.

2. C. echinatus L. - *Rough Dog's-tail.* Tufted annual to 75(100)cm; leaves 2-10mm wide; ligules 2-10mm, acute to obtuse; panicle asymmetrically ovoid, 1-4(8) x 0.7-2cm; fertile lemmas 4.5-7mm plus awn 6-16mm; (2n=14). Intrd; casual on waste and rough open ground scattered in BI N to C Sc, natd in sunny places on sandy or rocky ground on coasts of S En and CI; Europe.

19. LAMARCKIA Moench - *Golden Dog's-tail*

Annuals; sheaths not fused; inflorescence a rather compact panicle; spikelets of 2 kinds - fertile with 1 bisexual and 1 vestigial floret, and sterile consisting of numerous flat, overlapping obtuse lemmas, normally 3 sterile and 2 fertile together, but 1 of the 2 fertile often reduced and not producing a caryopsis; glumes 2, subequal; lemmas rounded on back, 5-veined, minutely bifid, with long awn from sinus; stamens 3.

1. L. aurea (L.) Moench - *Golden Dog's-tail.* Culms tufted, to 20(30)cm; leaves 2- 841
6mm wide; ligules 5-10mm, obtuse to jagged at apex; panicle 3-9 x 2-3cm, oblong; 881
fertile lemmas 2-3.5mm, with awn 5-10mm; vestigial lemma also long-awned but lemmas of sterile spikelets unawned; (2n=14). Intrd; casual occasionally persisting for few years in waste or rough ground and on tips, mainly as wool-alien; very

scattered in Br, mainly S & C En and S Wa; Mediterranean.

20. PUCCINELLIA Parl. - *Saltmarsh-grasses*
Annuals to perennials without rhizomes, with or without stolons; sheaths not fused; inflorescence a panicle; spikelets with 2-many florets all (except the most apical) bisexual; glumes 2, slightly unequal; lemmas rounded on back, 5-veined, subacute to rounded at apex, unawned; stamens 3.

1 Lemmas 2.8-4.6mm ... 2
1 Lemmas 1.8-2.5(2.8)mm ... 3
  2 Perennial with many tillers and usually rooting stolons at flowering; anthers 1.3-2.5mm ... **1. P. maritima**
  2 Annual or biennial with 0 or few tillers and 0 stolons at flowering; anthers 0.75-1mm ... **4. P. rupestris**
3 At least some of panicle branches at lower nodes bearing spikelets ± to base; lemmas subacute to obtuse, with midrib reaching apex; anthers mostly <0.75mm ... **3. P. fasciculata**
3 Panicle branches at lower nodes ± all with conspicuous basal region bare of spikelets; lemmas broadly obtuse to rounded, with midrib falling short of apex; anthers mostly >0.75mm ... **2. P. distans**

1. P. maritima (Huds.) Parl. - *Common Saltmarsh-grass*. Perennial, usually with long stolons forming large patches; culms to 80cm; spikelets 5-13mm, with 3-10 florets; lemmas 2.8-4.6mm, subacute to obtuse, with midrib usually just reaching apex; anthers 1.3-2.5mm; 2n=56 (and 49?, 63?, 70?, 77?). Native; bare or semi-bare mud in salt-marshes and estuaries, rarely saline areas and by salted roads inland; common round coasts of BI, few places in C En.

1 x 2. P. maritima x P. distans = P. x hybrida Holmb. occurs rarely on coasts of En from W Sussex to Durham, and has been found by a salted road in S Northumb and (involving *P. distans* ssp. *borealis*) in Outer Hebrides; it often has stolons, has intermediate panicle-shape and lemma size, and is sterile; 2n=49, 51.

1 x 4. P. maritima x P. rupestris = P. x krusemaniana Jansen & Wacht. was recorded from W Sussex in 1920 and S Hants in 1977; it is said to be intermediate and sterile.

2. P. distans (Jacq.) Parl. - *Reflexed Saltmarsh-grass*. Tufted perennial; culms to 60cm; spikelets 3-9mm, with 2-9 florets; lemmas broadly obtuse to rounded, the midrib not reaching apex; anthers 0.5-1.2mm. Native.

a. Ssp. distans. Culms to 60cm; leaves usually flat, 1.5-4mm wide; lower panicle branches strongly reflexed at maturity; lemmas 2-2.5mm; 2n=42. Semi-bare mud, rough and waste ground in estuaries, upper edges of salt-marshes, inland saline areas and by salted main roads; round coasts of BI N to CE Sc, rare in W, common in E, now common inland in C & E En.

b. Ssp. borealis (Holmb.) W.E. Hughes (*P. capillaris* (Lilj.) Jansen). Culms to 40cm; leaves usually folded along midrib, 1-2mm wide; lower panicle branches patent (sometimes weakly reflexed) to suberect at maturity; lemmas (1.8)2.2-2.8mm; 2n=28. Stony or rocky, sometimes sandy places and on sea-walls; coasts of N & E Sc (incl. Orkney and Shetland) from N Outer Hebrides to E Lothian and Midlothian.

2 x 3. P. distans x P. fasciculata occurs rarely on coasts of En from W Sussex to E Norfolk; it has intermediate panicle- and lemma-shape and is sterile; 2n=35.

2 x 4. P. distans x P. rupestris = P. x pannonica (Hack.) Holmb. occurs rarely on coasts of En from S Devon to W Norfolk; it has intermediate panicle-shape and lemma-length and is sterile; 2n=42.

3. P. fasciculata (Torr.) E.P. Bicknell (*P. pseudodistans* (Crép.) Jansen & Wacht.) - **R**
Borrer's Saltmarsh-grass. Tufted perennial; culms to 50cm; spikelets 3-6mm, with 3-8 florets; lemmas 1.8-2.3mm, the midrib reaching apex; anthers 0.4-0.8mm; 2n=28.

Native; in barish places, on sea-walls and banks and by dykes; locally frequent on coasts of S Br from Carms to W Norfolk, by salted roads in E & W Kent. **P. pseudodistans** appears to some extent intermediate between *P. distans* and *P. fasciculata*, especially in its panicle shape, but some of its branches bear spikelets to the base, its lemma midrib reaches the apex, and it has 2n=28; it is best considered a var. of *P. fasciculata* and has a similar distribution in Br.

4. P. rupestris (With.) Fernald & Weath. - *Stiff Saltmarsh-grass*. Tufted annual or **R** biennial; culms to 40cm; spikelets 5-9mm, with 3-6 florets; lemmas 2.8-4mm, the midrib usually reaching apex; anthers 0.75-1mm; 2n=42. Native; in bare places on mud and clay and among rocks and stones; coasts of S & E Br from Pembs to Yorks (formerly Cheviot), by salted road in W Kent.

21. BRIZA L. - *Quaking-grasses*

Annuals, or perennials with short rhizomes; sheaths overlapping; inflorescence a panicle or sometimes a raceme with pedicels >5mm; spikelets with 4-many florets all (except most apical) bisexual, characteristically flattened, broadly ovate and pendent; glumes 2, subequal; lemmas rounded on back, cordate at base, rounded to very obtuse at apex, 7-9-veined, the veins not reaching apex, unawned; stamens 3.

1 Perennial, with tillers at flowering; ligule 0.5-1.5mm; leaves 2-4mm wide **1. B. media**

1 Annual, without tillers at flowering; ligule 2-6mm; leaves 2-10mm wide 2

2 Spikelets 2.5-5mm, >20 per panicle **2. B. minor**

2 Spikelets 8-25mm, <15 per panicle **3. B. maxima**

1. B. media L. - *Quaking-grass*. Perennial; culms to 75cm; spikelets 4-7mm, >20 per panicle; 2n=14. Native; grassland on light to heavy, acid to calcareous, very dry to damp, but usually base-rich soils; locally common throughout BI except N & NW Sc.

2. B. minor L. - *Lesser Quaking-grass*. Annual; culms to 60cm; spikelets 2.5-5mm, **R** >20 per panicle; (2n=10). Possibly native; arable fields, bulb-fields, waste places; locally frequent in SW & CS En and CI, rare casual elsewhere; Mediterranean.

3. B. maxima L. - *Greater Quaking-grass*. Annual; culms to 75cm; spikelets 8-25mm, <15 per panicle; (2n=14). Intrd; natd in dry open places and on banks and field-margins; distribution as for *B. minor*, usually as garden escape; Mediterranean.

22. POA L. (*Parodiochloa* C.E. Hubb.) - *Meadow-grasses*

Annuals or perennials with or without stolons or rhizomes; sheaths overlapping; inflorescence a panicle; spikelets with 1-many florets all (or all except the most apical) bisexual, sometimes proliferating; glumes 2, subequal; lemmas keeled on back, 5-veined, awnless or rarely with short terminal awn, its callus often with tuft of cottony hairs; stamens 3.

1 At least some spikelets proliferating 2

1 Spikelets not proliferating 4

2 Base of culms swollen, bulb-like; plant green **14. P. bulbosa**

2 Base of culms not swollen; plant usually glaucous 3

3 Leaves 2-4.5mm wide when flattened, parallel-sided and ± abruptly narrowed to apex, the uppermost usually arising below 1/2 way up culm **15. P. alpina**

3 Leaves 1-2mm wide when flattened, gradually tapered to apex, the uppermost usually arising above 1/2 way up culm **9. P. x jemtlandica**

4 Plant with distinct, often far-creeping rhizomes 5

4 Plant without rhizomes, sometimes with stolons 8

5 Culms strongly compressed, 4-6(9)-noded, usually slightly bent at each

node **10. P. compressa**

5 Culms terete to somewhat compressed, 2-4-noded, usually straight except near base 6

6 Glumes subequal, both usually 3-veined and distinctly acuminate; culms usually all solitary; sheaths usually with some hairs at junction with blade **4. P. humilis**

6 Glumes distinctly unequal, the lower often 1-veined, acute; culms usually in small or dense clusters; sheaths glabrous at junction with blade 7

7 Tiller leaves 0.5-2mm wide; lemmas 2-3mm; culms usually in dense clusters **6. P. angustifolia**

7 Tiller leaves 2-4(5)mm wide; lowest lemmas 3-4mm; culms usually in small clusters **5. P. pratensis**

8 All or some leaves >5mm wide 9

8 Leaves ≤5mm wide 11

9 Panicle-branches erect; lowest lemma usually >4.5mm, with awn 1.2-2.4mm **16. P. flabellata**

9 Panicle-branches patent; lowest lemma ≤4.5mm, awnless 10

10 Leaves ≤6mm wide; lowest lemma ≤3mm; ligules often >2mm; uppermost culm-leaf usually as long or longer than its sheath **11. P. palustris**

10 Leaves often >6mm wide; lowest lemma >3mm; ligules <2mm; uppermost culm-leaf much shorter than its sheath **7. P. chaixii**

11 Base of culms swollen, bulb-like **14. P. bulbosa**

11 Base of culms not swollen 12

12 Base of culms surrounded by dense mass of dead leaf-sheaths; mountains 13

12 Base of culms with few or no persistent dead leaf-sheaths; widespread 14

13 Leaves 2-4.5mm wide when flattened, parallel-sided and ± abruptly narrowed to apex, the uppermost usually arising below 1/2 way up culm **15. P. alpina**

13 Leaves 1-2mm wide when flattened, gradually tapered to apex, the uppermost usually arising above 1/2 way up culm **8. P. flexuosa**

14 Annual, or perennial due to procumbent stems rooting; culms usually procumbent to ascending; anthers ≤1(1.3)mm; usually some leaves transversely wrinkled 15

14 Perennial; culms usually erect; anthers ≥1.3mm; leaves not transversely wrinkled 16

15 Anthers 0.6-0.8(1.3)mm, 2-3x as long as wide; panicle-branches usually patent to reflexed at fruiting **2. P. annua**

15 Anthers 0.2-0.5mm, 1-1.5x as long as wide; panicle-branches usually erect to suberect at fruiting **1. P. infirma**

16 Ligule of uppermost culm-leaf 4-10mm, acute; sheaths rough **3. P. trivialis**

16 Ligule of uppermost culm-leaf ≤5mm, obtuse to rounded; sheaths smooth 17

17 Ligule of uppermost culm-leaf 0.2-0.5mm, usually truncate **13. P. nemoralis**

17 Ligule of uppermost culm-leaf 0.8-4(5)mm, usually obtuse 18

18 Lowland plant c.30-150cm; lowest panicle node with (3)4-6(8) branches; ligule of uppermost culm-leaf 2-4(5)mm **11. P. palustris**

18 Mountain plant 10-40cm; lowest panicle node with 1-2(4) branches; ligule of uppermost culm-leaf 1-2.5(3)mm **12. P. glauca**

Other spp. - P. labillardieri Steud., from Australia, is a rare wool-alien; it is a tufted perennial with a short ligule and lemmas pubescent in lower 1/2.

1. P. infirma Kunth - *Early Meadow-grass.* Annual; culms erect to procumbent, to 10(25)cm; leaves 1-3(4)mm wide; ligules 1-3.5mm; lowest panicle-node with 1-2 branches; 2n=14. Native; rough ground, waysides and on paths, usually near sea; CI, Scillies, E & W Cornwall scattered E to S Hants, common in Scillies and CI, apparently spreading in S En. RR

2. P. annua L. - *Annual Meadow-grass.* Annual, or perennial due to procumbent stems rooting; culms erect to procumbent, to 20(30)cm; leaves 1-4(5)mm wide; ligules 1-5mm; lowest panicle-node with 1-2(3) branches; 2n=28. Native; rough, waste and cultivated ground, waysides, on paths, in lawns and other close-cut turf; abundant throughout BI.

3. P. trivialis L. - *Rough Meadow-grass.* Perennial with many, often procumbent tillers, some becoming stolons; culms erect, to 70(90)cm; leaves 1.5-5mm wide; ligules 4-10mm; lowest panicle-node with 3-7 branches; 2n=14, 28. Native; open woods, marshes, ditches, riversides, damp grassland, by ponds and lakes, cultivated and rough ground; very common throughout BI.

4. P. humilis Ehrh. ex Hoffm. (*P. subcaerulea* Sm., P. *pratensis* ssp. *irrigata* (Lindm.) H. Lindb.) - *Spreading Meadow-grass.* Perennial with extensive rhizomes; culms erect, to 30(40)cm; leaves 1.5-4mm wide; ligules 0.5-2mm; lowest panicle-node with 2-5 branches; (2n=54-109). Native; grassland, roadsides, on old walls, usually on sandy soil but often near water; throughout BI, but greatly under-recorded.

5. P. pratensis L. - *Smooth Meadow-grass.* Perennial with strong but short rhizomes; culms erect, to 75(100)cm; leaves 2-4(5)mm wide; ligules 1-3mm; lowest panicle-node with 3-5 branches; 2n=84 (42-98). Native; meadows, pastures, waysides, rough and waste ground; very common throughout BI.

6. P. angustifolia L. (*P. pratensis* ssp. *angustifolia* (L.) Dumort.) - *Narrow-leaved Meadow-grass.* Perennial with rhizomes; culms erect, to 70cm; leaves 0.5-2(3)mm wide; ligules 0.5-2mm; lowest panicle-node with 3-5 branches; 2n=c.51, c.62 (42-72). Native; grassy places, rough ground, on and by walls, banks, on well-drained soil; probably frequent throughout BI but greatly under-recorded.

7. P. chaixii Vill. - *Broad-leaved Meadow-grass.* Densely tufted perennial; culms erect, to 1.3m; leaves (4)6-10mm wide; ligules 0.5-2mm; lowest panicle-node with 4-7 branches; (2n=14). Intrd; grown for ornament, natd in woods and copses; scattered throughout Br, Co Down; Europe.

8. P. flexuosa Sm. - *Wavy Meadow-grass.* Tufted perennial; culms erect, to 25cm; leaves 1-2mm wide; ligules 1-3.5mm; lowest panicle-node with 2-5 branches; 2n=42. Native; mountain screes and ledges at 800-1100m; very local in highlands of C Sc. RR

9. P. x jemtlandica (Almq.) K. Richt. (*P. flexuosa* x *P. alpina*) - *Swedish Meadow-grass.* Differs from *P. flexuosa* in spikelets all or nearly all proliferating, with very little external suggestion of *P. alpina* except leaves and stems usually glaucous; 2n=c.36 (37). Native; similar habitat and range to *P. flexuosa,* often (?always) with it and *P. alpina*. RR

10. P. compressa L. - *Flattened Meadow-grass.* Rhizomatous perennial; culms erect to ascending, to 60cm; leaves 1-3(4)mm wide; ligules 0.5-2mm; lowest panicle-node with 2-3 short branches; 2n=42. Native; walls, paths, waysides, stony ground and banks on well-drained soils; rather scattered throughout BI except N Sc, perhaps under-recorded.

11. P. palustris L. - *Swamp Meadow-grass.* Tufted perennial; culms erect, to 1(1.5)m; leaves 2-6mm wide; ligules 2-4(5)mm; lowest panicle-node with 3-8 branches; (2n=14, 28, 42). Intrd; marshes, fens, ditches and damp grassland, natd from previous use as fodder-grass, also rare wool- and grain-alien; scattered in BI N to C Sc (mainly S & C En), often sporadic; Europe. No evidence of its being native as sometimes claimed.

12. P. glauca Vahl (*P. balfourii* Parn.) - *Glaucous Meadow-grass.* Tufted perennial; culms erect, to 40cm; leaves usually glaucous, 2-4mm wide; ligules 1-2.5(3)mm; R lowest panicle-node with 1-2(4) branches; 2n=42, 56, 70. Native; damp mountain rock-ledges and -crevices and rocky slopes at 610-910m; very local in C & N Sc (not Outer Isles), Lake District, Caerns. *P. balfourii* has been shown to be a shade variant of *P. glauca* with laxer habit and ± non-glaucous leaves.

13. P. nemoralis L. - *Wood Meadow-grass*. Tufted perennial; culms erect, to 75(90)cm; leaves 1-3mm wide; ligules 0.2-0.5mm; lowest panicle-node with 3-6 branches; 2n=42, 56. Native; woods, hedgebanks, walls and other shady places; frequent to common in most of BI, but probably intrd in much of Ir and NW Br.

14. P. bulbosa L. - *Bulbous Meadow-grass*. Tufted perennial; culms erect to **R** decumbent, to 40cm, with swollen bulb-like base; leaves 0.5-2mm wide; ligules 1-3(4)mm; lowest panicle-node with 2-3 branches; 2n=28, 45. Native; barish places in short grassland and open ground on sandy soil, shingle or limestone near sea, very rare inland; coasts of S Br from Glam to N Lincs, CI. Spikelets rarely (?never) proliferating except in Glam (all plants) and Surrey and Jersey (both sorts present).

15. P. alpina L. - *Alpine Meadow-grass*. Tufted perennial; culms erect to ± pendent, **R** to 40cm; leaves 2-4.5mm wide; ligules 2-5mm; lowest panicle-node with 1-3 branches; 2n=c.32, 35, 39, c.84. Native; damp mountain rock-ledges and -crevices and rocky slopes at 300-1200m; local in C & N Sc, NW En, Caerns, S Kerry, Co Sligo. Spikelets usually proliferating; sexual plants (sometimes alone) confined to a few places in Sc and NW En.

16. P. flabellata (Lam.) Raspail (*Parodiochloa flabellata* (Lam.) C.E. Hubb.) - **860** *Tussac-grass*. Very densely tufted perennial forming large tussocks up to 1m x 1m (excl. culms); culms erect, to 80cm; leaves (3)5-10mm wide; ligules 4-10mm; lowest panicle-node with 2-4 branches; 2n=42. Intrd; very persistent where planted in yards and on walls in Shetland; extreme S S America.

23. DACTYLIS L. - *Cock's-foot*

Perennials without stolons or rhizomes, with strongly compressed tillers; sheaths overlapping; inflorescence a ± 1-sided panicle simply lobed or formed of stalked dense clusters of spikelets; spikelets with 2-5 florets all (except the most apical) bisexual; glumes 2, unequal; lemma keeled on back, 5-veined, without or with short terminal awn; stamens 3.

Other spp. - D. hispanica Roth (*D. glomerata* ssp. *hispanica* (Roth) Nyman), from SW Europe, was reported from W Cork in 1963; it is a short, densely tillering tetraploid with compact lobed panicles, and differs from similar coastal variants of *D. glomerata* in its split lemma. It was possibly intrd, perhaps casual, but its presence in SW Ir as a native needs investigating.

1. D. glomerata L. - *Cock's-foot*. Culms ± densely tufted, to 1.4m; leaves and sheaths very rough, ± glaucous; lemmas with hairs or prickles on keel, with awn 1.5-2mm; 2n=28. Native; grassland, open woodland and rough, waste and cultivated ground; common throughout BI. Formerly much grown for hay and pasture; the more robust plants of artifical habitats are probably of intrd stock.

2. D. polygama Horv. (*D. glomerata* ssp. *aschersoniana* (Graebn.) Thell., ssp. *lobata* **870** (Drejer) H. Lindb.) - *Slender Cock's-foot*. Differs from *D. glomerata* in leaves and sheaths not or only slightly rough, green; inflorescence slenderer, with smaller clusters of spikelets; lemmas glabrous to obscurely prickly on keel, with awn 0 or <0.5mm; (2n=14). Intrd; grown for ornament and natd in woods; scattered in S En from Bucks and Surrey to Dorset; Europe.

24. CATABROSA P. Beauv. - *Whorl-grass*

Perennials with stolons, without rhizomes; sheaths overlapping; inflorescence a diffuse panicle; spikelets with 1-3 florets all or all except the most apical bisexual; glumes 2, unequal; lemma rounded on back, 3-veined, truncate, awnless; stamens 3.

FIG 860 - Poaceae: Pooideae. 1, *Hainardia cylindrica*. 2, *Helictotrichon neesii*. 3, *Poa flabellata*. 4, *Gaudinia fragilis*. 5, *X Festulpia hubbardii*.

1. C. aquatica (L.) P. Beauv. - *Whorl-grass.* Culms erect to decumbent, to 75cm, often purple-tinged, glabrous or ± so; 2n=20. Native; wet meadows, marshes, ditches, by ponds and streams, often on barish mud; scattered throughout most of lowland BI, rare and decreasing in S. Plants from coastal parts of NW Br are short and ± consistently have spikelets with 1-2 florets; they are best recognized as var. **uniflora** Gray (ssp. *minor* (Bab.) F.H. Perring & P.D.Sell).

25. CATAPODIUM Link - *Fern-grasses*
Annuals; sheaths not fused, inflorescence a stiff raceme or little-branched panicle; spikelets with 3(5)-14 florets all (except the most apical) bisexual; glumes 2, subequal; lemma rounded on back or keeled distally, 5-veined, acute to obtuse or emarginate at apex, awnless; stamens 3.

1. C. rigidum (L.) C.E. Hubb. (*Desmazeria rigida* (L.) Tutin) - *Fern-grass.* Culms erect to procumbent, to 15(60)cm; inflorescence a spike-like raceme to little-branched panicle; lower glume (1)1.3-2mm; upper glume (1.4)1.5-2.3(2.5)mm; lemmas 2-2.6(3)mm; 2n=14. Native; dry, barish places on banks, walls, sand, shingle, chalk and stony ground, especially near sea; locally common in BI N to C Sc. Some coastal plants in SW Br, S Ir and CI have branched panicles with the branches spreading in 3 dimensions, not racemes or panicles spreading only in 1 plane; they are best recognized as var. **majus** (C. Presl) Laínz (ssp. *majus* (C. Presl) F.H. Perring & P.D. Sell).

1 x 2. C. rigidum x C. marinum (1 sterile plant) was found in 1960 in Merioneth; endemic.

2. C. marinum (L.) C.E. Hubb. (*Desmazeria marina* (L.) Druce) - *Sea Fern-grass.* Culms erect to procumbent, to 25cm; inflorescence a spike-like raceme, rarely with few simple branches below; lower glume 2-3mm; upper glume 2.3-3.3mm; lemmas 2.2-3mm; 2n=14, 28. Native; dry barish places by sea on walls, banks, sand and shingle, often with *C. rigidum.* Native; locally common round coasts of BI except parts of E & N Br.

26. SESLERIA Scop. - *Blue Moor-grass*
Tufted perennials with short rhizomes, without stolons; sheaths not fused; inflorescence a small, very compact panicle; spikelets with 2(-3) florets all bisexual; glumes 2, subequal; lemma rounded on back, 3-5-veined, 3-5-toothed at apex, each tooth with awn 0-1.5mm; stamens 3.

1. S. caerulea (L.) Ard. (*S. albicans* Kit. ex Schult.) - *Blue Moor-grass.* Culms to 45cm, usually bare of leaves in distal 1/2; ligule extremely short; panicle 1-3cm, ovoid, usually bluish-violet tinged; 2n=28. Native; barish grassland, rock-crevices and -ledges, screes and 'pavement' on limestone and (in Sc) calcareous mica-schists; locally common in N En S to MW Yorks and in W Ir, very local in C Ir and C Sc, doubtful records for Wa, 1 site in Derbys (found 1989). **R**

TRIBE 7 - HAINARDIEAE (Monermeae auct. non C. E. Hubb.) (genera 27-28). Annuals; ligule membranous; sheaths not fused; inflorescence a very slender cylindrical spike with alternating spikelets partly sunk in cavities in rhachis which breaks up into 1-spikeleted segments at fruiting; spikelets with 1 (bisexual) floret; glumes 1-2, strongly veined and ± horny; lemma delicate, 3-veined with very short lateral veins, acute, unawned; stamens 3; stigmas 2; ovary with rounded glabrous appendage beyond style-bases; lodicules 2; ovary glabrous.

27. PARAPHOLIS C.E. Hubb. - *Hard-grasses*
Glumes 2, inserted side-by-side and together covering rhachis-cavity except at anthesis; lemma with its side towards rhachis.

1. P. strigosa (Dumort.) C.E. Hubb. - *Hard-grass.* Culms usually erect, sometimes ascending or curved, rarely procumbent, to 25(40)cm, very slender; spike fully exserted from uppermost leaf-sheath or not, 2-20cm; anthers 1.5-3(3.5)mm; 2n=28. Native; sparsely-grassed ground on salt-affected soil by salt-marshes and creeks and on rough and waste ground; frequent on coasts of BI N to C Sc.

2. P. incurva (L.) C.E. Hubb. - *Curved Hard-grass.* Culms decumbent to ascending, R to 10(20)cm; spike usually not fully exserted from uppermost leaf-sheath, 1-8(15)cm; anthers 0.5-1(1.5)mm; (2n=36, 38, 42). Native; similar places to *P. strigosa* but also drier spots on cliff-tops and banks, also rare wool- and ballast-alien; local on coasts of Br N to Caerns and S Northumb, CI, very local in SE Ir.

28. HAINARDIA Greuter (*Monerma* auct. non P. Beauv.) - *One-glumed Hard-grass*
Glumes 2 in terminal spikelet, 1 (the upper) in all others and inserted so as to cover the rhachis-cavity except at anthesis; lemma with its back towards rhachis.

1. H. cylindrica (Willd.) Greuter (*Monerma cylindrica* (Willd.) Coss. & Durieu) - **860** *One-glumed Hard-grass.* Culms erect to ascending, straight or curved, to 30(45)cm; spike fully exserted from uppermost leaf-sheath or not, 2-25cm; anthers 1.5-3.5mm; (2n=26). Intrd; fairly frequent casual from birdseed and other sources on tips and waste ground; scattered in S En; S Europe.

TRIBE 8 - MELICEAE (Glycerieae) (genera 29-30). Perennials with rhizomes or stolons; ligule membranous; sheaths fused into tube (often splitting later); inflorescence a little- or much-branched panicle or a raceme with pedicels >3mm; spikelets with 1-many bisexual florets, if with <4 florets then with group of sterile ones beyond; glumes 2, subequal or equal; lemmas rounded on back, 7-9-veined, subacute to rounded, awnless; stamens 3; stigmas 2; lodicules fused laterally into single scale shorter than wide; ovary glabrous.

29. GLYCERIA R. Br. - *Sweet-grasses*
Aquatic or marsh grasses with rhizomes and/or stolons; panicles much- to rather little-branched; ligules acute or acuminate to rounded, 2-10(15)mm; spikelets with 4-16 florets all (except the most apical) bisexual and each falling separately when fruit ripe; glumes 1(-3)-veined; lemmas 7-veined.

1 Spikelets 5-12mm, with 4-10 florets; paleas not winged on keels; culms erect and self-supporting, usually >1m **1. G. maxima**
1 Spikelets 10-35mm, with 6-17 florets; paleas winged on keels distally; culms decumbent to ascending, if erect not self-supporting, rarely >1m 2
  2 Anthers remaining indehiscent; pollen grains all or mostly empty and shrunken; spikelets remaining intact after flowering, not forming fruits **3. G. x pedicellata**
  2 Anthers dehiscent; pollen grains full and turgid; spikelets breaking up between florets when fruit ripe 3
3 Lemmas 5.5-6.5(7.5)mm; anthers 1.5-2.5(3)mm **2. G. fluitans**
3 Lemmas 3.5-5mm; anthers 0.6-1.3mm 4
  4 Lemmas distinctly 3(-5)-toothed at apex, exceeded by 2 sharply pointed apical teeth of palea **4. G. declinata**
  4 Lemmas not or scarcely toothed at apex, not exceeded by 2 (very short) apical teeth of palea **5. G. notata**

1. G. maxima (Hartm.) Holmb. - *Reed Sweet-grass.* Culms erect, to 2.5m; panicles much-branched; lemmas 3-4mm, entire; palea scarcely toothed at apex; anthers (1)1.2-1.8(2)mm; 2n=60. Native; in and by rivers, canals, ponds and lakes, usually in deeper water than other spp.; common in most of En except N, scattered in Wa, Ir and Sc, 1 record in Guernsey, not in N or NW Sc.

2. G. fluitans (L.) R. Br. - *Floating Sweet-grass*. Culms decumbent to ascending (or erect), to 1m; panicles sparsely branched, narrow; lemmas 5.5-6.5(7.5)mm, entire; palea sharply toothed at apex but shorter than lemma; anthers 1.5-2.5(3)mm; 2n=40. Native; on mud or in shallow water by ponds, rivers and canals and in marshes, ditches and wet meadows; common throughout BI.

2 x 4. G. fluitans x G. declinata is a sterile hybrid differing from *G.* x *pedicellata* in its obscurely 3-toothed lemmas and 2n=30; it occurs rarely with the parents very scattered in En and E Cork.

3. G. x pedicellata F. Towns. (*G. fluitans* x *G. notata*) - *Hybrid Sweet-grass*. Differs from *G. fluitans* in lemmas (4)5-5.5(6)mm; anthers 1-1.8mm; and see key (couplet 2); 2n=40. Native; similar places to *G. fluitans*, with 1, both or neither parents, often forming large patches; scattered over most of BI, frequent in En, sometimes in areas (e.g. Jersey, NW Sc) where *G. notata* does not occur.

4. G. declinata Bréb. - *Small Sweet-grass*. Differs from *G. fluitans* in culms to 60cm; lemmas 4-5mm, distinctly 3(-5)-toothed at apex; palea sharply pointed, the points exceeding lemma; anthers 0.6-1.3mm; 2n=20. Native; similar places to *G. fluitans*; scattered throughout BI, probably under-recorded.

5. G. notata Chevall. (*G. plicata* (Fr.) Fr.) - *Plicate Sweet-grass*. Differs from *G. fluitans* in panicles more branched and less narrow; lemmas 3.5-5mm, much blunter at apex; palea with very short teeth; anthers 0.7-1.3mm; 2n=40. Native; similar places to *G. fluitans;* frequent throughout most of BI except most of N & NW Sc and most of CI, perhaps under-recorded.

30. MELICA L. - *Melicks*

Woodland or mountain grasses with short rhizomes; inflorescence a raceme or a sparsely branched panicle; ligules truncate, <2mm; spikelets with 1-3 bisexual florets plus distal ± club-shaped cluster of sterile vestiges, all the florets falling as a unit when fruit ripe; glumes 3-5-veined; lemmas 7-9-veined.

1. M. nutans L. - *Mountain Melick*. Culms erect or pendent distally, to 60cm; sheaths without apical bristle; inflorescence usually simple raceme; spikelets pendent, with 2-3 fertile florets; lower glume 5-veined; 2n=18. Native; woods, scrub and shady rock-crevices on limestone; scattered in N & W Br S to Mons and Northants.

2. M. uniflora Retz. - *Wood Melick*. Culms erect or pendent distally, to 60cm; sheaths with long bristle at apex on side opposite ligule; inflorescence a sparsely branched panicle; spikelets erect, with 1 fertile floret; lower glume 3-veined; 2n=18. Native; woods and shady hedgebanks; scattered and locally common throughout BI except N Sc and CI.

TRIBE 9 - AVENEAE (genera 31-41). Annuals or perennials with or without rhizomes, without stolons; ligule membranous; sheaths not fused or rarely fused to form tube; inflorescence a panicle, sometimes densely contracted, rarely a spike; spikelets with 2-6(11) florets, sometimes 1 or more floret (basal or apical) male, sometimes proliferating; glumes 2, equal to unequal, often nearly as long to longer than rest of spikelet, often with wide ± shiny hyaline margins; lemmas 5-9-veined, rounded on back, acute or obtuse to bifid or variously toothed at apex, awnless or with short terminal awn or more often with dorsal awn with conspicuous bend; stamens 3; stigmas 2; lodicules 2; ovary glabrous or pubescent at apex or all over.

31. HELICTOTRICHON Besser ex Schult. & Schult. f. (*Avenula* (Dumort.) Dumort., *Avenochloa* Holub, *Amphibromus* Nees) - *Oat-grasses*

Tufted perennials with short or 0 rhizomes; inflorescence a rather sparsely branched, ± diffuse panicle; spikelets with 2-7 florets all except the most apical bisexual; glumes clearly unequal, the lower 1-3-veined, the upper 3-5-veined; lemmas 5-7-veined, variously shortly toothed at apex, with long, bent, dorsal awn;

rhachilla-segments pubescent, the hairs in a longer tuft at apex of each segment around lemma-base; ovary glabrous or pubescent at apex.

The genus *Amphibromus*, including *H. neesii*, is sometimes segregated; it differs in its glabrous (not pubescent) ovary. *Avenula* is separable only on anatomical characters and similarly does not merit generic status.

1 Ovary and caryopsis glabrous; lower glume 4-5mm; upper glume 5-7mm; lemmas 5-8mm (excl. awn), papillose **3. H. neesii**
1 Ovary and caryopsis with pubescent apex; lower glume 7-15mm; upper glume 10-20mm; lemmas 9-17mm (excl. awn), smooth to scabrid 2
 2 Lower culm-sheaths softly pubescent; spikelets with 2-3(4) florets; rhachilla hair-tuft 3-6(7)mm; palea with smooth keels **1. H. pubescens**
 2 Culm-sheaths glabrous; spikelets with 3-6(8) florets; rhachilla hair-tuft 1-3mm; palea with scabrid keels **2. H. pratense**

1. H. pubescens (Huds.) Pilg. (*Avenula pubescens* (Huds.) Dumort., *Avenochloa pubescens* (Huds.) Holub) - *Downy Oat-grass*. Culms erect, to 1m; lower culm-sheaths and usually leaves softly pubescent; spikelets 10-17mm, with 2-3(4) florets; lower glume 7-13mm; upper glume 10-18mm; lemmas 9-14mm with awn to 22mm; 2n=14. Native; grassland usually on base-rich soils; ± throughout Bl, common on chalk and limestone in En.

2. H. pratense (L.) Besser (*Avenula pratensis* (L.) Dumort., *Avenochloa pratensis* (L.) Holub) - *Meadow Oat-grass*. Culms usually erect, to 80cm; leaves and sheaths glabrous; spikelets 11-28mm, with 3-6(8) florets; lower glume 10-15mm; upper glume 12-20mm; lemmas 10-17mm with awn to 22(27)mm; 2n=126. Native; similar places to *H. pubescens* and often with it, but usually in shorter turf and commoner in mountains; ± throughout Br except in Outer Isles, C Wa and extreme SW En, common on chalk and limestone.

3. H. neesii (Steud.) Stace (*Amphibromus neesii* Steud.) - *Swamp Wallaby-grass*. **860**
Culms erect, to 1m; leaves and sheaths glabrous; spikelets 9-15mm, with (2)4-7 florets; awn up to 18mm; rhachilla hair-tuft 1-2mm. Intrd; rather characteristic wool-alien on tips, waste ground and fields; scattered in En; Australia.

32. ARRHENATHERUM P. Beauv. - *False Oat-grass*

Loosely tufted perennials; inflorescence a fairly well-branched ± diffuse panicle; spikelets with 2(-5) florets, the lower (lowest) male, the upper bisexual, rarely both bisexual; glumes unequal, the lower 1-veined, the upper 3-veined; lemmas 7-veined, bifid at apex; lemma of male floret with long bent dorsal awn, that of bisexual floret(s) awnless or with short terminal awn or rarely with dorsal bent awn; rhachilla-segments with apical hair-tuft 1-2mm; ovary pubescent at apex.

1. A. elatius (L.) P. Beauv. ex J. & C. Presl - *False Oat-grass*. Culms usually erect, to 1.8m; leaves and sheaths glabrous to sparsely pubescent; spikelets 7-11mm; lower glume 4-6mm; upper glume 7-10mm; lowest lemma 7-10mm with awn up to 20mm; 2n=28. Native; coarse grassy places, waysides, hedgerows, maritime sand and shingle, rough and waste ground; abundant throughout Bl. Var. **bulbosum** (Willd.) St-Amans (ssp. *bulbosum* (Willd.) Hyl., *A. tuberosum* (Gilib.) F.W. Schultz) has the basal, very short culm internodes swollen and corm-like; they are effective propagules in arable land (*Onion Couch*). Variants with >1 bisexual floret or >1 floret with a long awn are easily confused with *Helictotrichon*, but in the latter the spikelets are longer and the lowest awn arises from higher up the lemma.

33. AVENA L. - *Oats*

Annuals; inflorescence a diffuse panicle; spikelets with 2-3 florets all bisexual or the distal 1 or 2 reduced and male or sterile; glumes subequal, 7-11-veined; lemmas 7-9-veined, bifid or with 2 bristles at apex, with or without long, bent, dorsal awn;

rhachilla-segments with or without hair-tuft; ovary pubescent at apex or all over.

A difficult genus, in which general appearance and size of parts are often of little value. For accurate determination spikelets with fully ripe fruits are needed.

1 Lemmas bifid, the 2 apical points (1)3-9mm and each with 1 or more veins entering from main body of lemma and reaching apex 2
1 Lemmas bifid, the 2 apical points 0.5-2mm and without veins or with vein(s) not reaching apex 3
2 Rhachilla disarticulating between florets at maturity, releasing 1-fruited disseminules each with elliptic basal scar; lemmas with dense long hairs on lower 1/2 **1. A. barbata**
2 Rhachilla not disarticulating at maturity, whole spikelets acting as disseminules, or the florets breaking away irregularly without basal scar; lemmas glabrous or sparsely pubescent on lower 1/2 **2. A. strigosa**
3 Rhachilla not disarticulating at maturity, whole spikelets acting as disseminules, or the florets breaking away irregularly without basal scar; lemmas usually unawned, if awned then awn nearly straight, usually glabrous **5. A. sativa**
3 Rhachilla disarticulating at maturity at least above glumes, often also between florets, hence at least lowest floret with basal scar; lemmas with long, strongly bent awns, usually pubescent 4
4 Rhachilla disarticulating at maturity above glumes only, releasing 2-3-fruited disseminules, hence only lowest floret with (ovate) basal scar; longer glume mostly 25-30mm **4. A. sterilis**
4 Rhachilla disarticulating at maturity between florets, releasing 1-fruited disseminules each with ovate basal scar; longer glume mostly 18-25mm **3. A. fatua**

Other spp. - A. byzantina K. Koch (*Algerian Oat*) is a minor Mediterranean crop occasionally occurring as a grain contaminant; it differs from *A. sativa* (in which it is often included) in that the rhachilla eventually breaks just above (not just below) each floret and hence remains attached to the next floret above (not to the next floret below).

1. A. barbata Pott ex Link - *Slender Oat*. Stems to 1m; spikelets with 2-3 florets, each with basal scar; longer glume 15-30mm; lowest lemma 12-18mm, pubescent in lower 1/2, plus 2 apical points 3-5mm; (2n=28). Intrd; rare grain-alien, natd in Guernsey since 1970, perhaps overlooked for *A. strigosa*; Mediterranean.

2. A. strigosa Schreb. - *Bristle Oat*. Stems to 1.2m; spikelets with 2-3 florets, none with basal scar; longer glume 15-26mm; lowest lemma 10-17mm, usually glabrous in lower 1/2, plus 2 apical points 3-9mm and 2 smaller fine bristles alongside; (2n=14, 28). Intrd; formerly (rarely still) grown as minor crop in Wa, Sc and Ir and then a frequent (now very local) cornfield weed sometimes natd, also infrequent casual grain-alien; Spain.

3. A. fatua L. - *Wild-oat*. Stems to 1.5m; spikelets with 2-3 florets, each with basal scar; longer glume (15)18-25mm; lowest lemma 14-20mm, pubescent in lower 1/2, plus 2 apical points ≤0.5mm; 2n=42. Intrd; weed of arable, waste and rough ground; common in most of En, very scattered elsewhere in BI; Europe.

3 x 5. A. fatua x A. sativa occurs rarely in Br in and around fields of *A. sativa* infested by *A. fatua*; it resembles *A. sativa* but shows the influence of *A. fatua* in its longer awns and tardily disarticulating spikelets, and has low fertility.

4. A. sterilis L. (*A. ludoviciana* Durieu) - *Winter Wild-oat*. Stems to 1.5m; spikelets with 2-3 florets, of which only the lowest has basal scar; longer glume (20)25-30(32)mm; lowest lemma 15-25mm, pubescent in lower 1/2, plus 2 apical points ≤1.5mm; (2n=42). Intrd; similar places to *A. fatua* but usually on heavy soils and replacing it there; formerly frequent, now scattered in CS & SE En, also rare but increasing wool- and grain-alien elsewhere; S Europe. Our plant is ssp. **ludoviciana** (Durieu) Gillet & Magne. Ssp. **sterilis** is a rare alien from various sources; it differs in its larger parts (spikelets with 3-5 florets; longer glume 32-45mm; lemmas 25-33mm; ligule >5mm (not <5mm)).

5. A. sativa L. - *Oat*. Stems to 1(1.5)m; spikelets with 2-3 florets, none with basal scar; longer glume 17-30mm; lowest lemma 12-20(25)mm, usually glabrous, plus 2 apical points <1mm; (2n=42). Intrd; locally common crop, frequent on tips and waysides and as grain-alien; throughout BI; W Mediterranean.

34. GAUDINIA P. Beauv. - *French Oat-grass*
Annuals sometimes lasting a few years; inflorescence a spike whose axis breaks giving 1-spikeleted segments at fruiting; spikelets with 4-11 florets all (except most apical) bisexual; lower glume c.1/2 as long as upper, 3-5- and 5-11-veined respectively; lemmas 5-9-veined, minutely bifid at apex, with long, bent, dorsal awn; rhachilla segments ± glabrous; ovary with distinct hairy apex remaining conspicuous as projection on fruit.

1. G. fragilis (L.) P. Beauv. - *French Oat-grass*. Stems to 45(100)cm; leaves and **860**
sheaths pubescent; spikelets 9-20mm; lowest lemma 7-11mm with awn 5-13mm; **881**
(2n=14). Intrd; natd in grassy fields, rough ground and waysides on a wide range of soils; natd locally in SC En, SW Ir and CI, infrequent casual grain-alien elsewhere; S Europe.

35. TRISETUM Pers. - *Yellow Oat-grass*
Perennials without rhizomes or stolons; inflorescence a well-branched panicle; spikelets with 2-4 florets all (except the most apical) bisexual; glumes unequal, the lower 1-, the upper 3-veined; lemmas 5-veined, bifid with 2 short bristle-points at apex, with long, bent dorsal awn; rhachilla-segments pubescent, the hairs <1mm at apex; ovary glabrous.

1. T. flavescens (L.) P. Beauv. - *Yellow Oat-grass*. Stems loosely tufted, to 80cm; lower leaves and sheaths pubescent; spikelets 5-7.5mm (excl. awns); lowest lemma 4-5.5mm with awn 4.5-9mm; 2n=28. Native; meadows, pastures, grassy waysides, especially on base-rich soil, throughout most of lowland BI, common in En and SE Sc, rather scattered elsewhere.

36. KOELERIA Pers. - *Hair-grasses*
Tufted perennials without rhizomes or stolons; inflorescence a spike-like panicle with very short branches; spikelets with 2-3(5) florets all (except most apical) bisexual; glumes unequal, the lower 1-, the upper 3-veined; lemmas 3-veined, acute or with extremely short apical awn, otherwise awnless; rhachilla-segments shortly pubescent; ovary glabrous.

1. K. vallesiana (Honck.) Gaudin - *Somerset Hair-grass*. Stems erect to **RR**
procumbent, to 40cm; stems and sheaths very shortly pubescent; lower sheaths very persistent and rotting to form reticulated network of fibres round swollen culm-bases; panicles 1.5-7 x 0.6-1.2cm; spikelets 4-6mm, shortly appressed-pubescent; 2n= mostly 42, some 14, 28, 49, 63. Native; short limestone grassland; 7 sites in Mendip Hills, N Somerset.

1 x 2. K. vallesiana x K. macrantha occurs in most of the *K. vallesiana* populations; it is sterile and has intermediate leaf-sheath characters, but chromosome number, 2n=35, is the main diagnostic character; endemic.

2. K. macrantha (Ledeb.) Schult. (*K. cristata* auct. non (L.) Pers., *K. gracilis* Pers. nom. illeg., *K. albescens* auct. non DC., *K. britannica* (Domin ex Druce) Ujhelyi, *K. glauca* auct. non (Schrad.) DC.) - *Crested Hair-grass*. Stems erect to ascending, to 60cm; stems and sheaths variously pubescent to glabrous; lower sheaths not very persistent, not forming reticulated fibres and culm-base not swollen; panicles 1-10 x 0.5-2cm; spikelets 3.5-6mm, glabrous to pubescent; 2n=28. Native; short limestone or sandy base-rich grassland, dunes, less often on inland sandy soils; throughout most of BI, mostly on calcareous soils in S, mostly coastal in N. Very variable in stature, pubescence, leaf rigidity and inrolling, and colour, shape and denseness of panicle. Several segregates have been recognized, variously said to occur throughout Br or be confined to coasts of S & SW, but no thorough studies of our plants have been made. Until they have, and until the identity of our plants with the types of names used for them is demonstrated, acceptance of any of these segregates is premature. Despite several claims, the diploid *K. glauca* (Schrad.) DC. has not been confirmed from BI.

37. ROSTRARIA Trin. (*Lophochloa* Rchb.) - *Mediterranean Hair-grass*
Annuals; inflorescence a spike-like panicle with very short branches; spikelets with 3-5(11) florets all (except apical 1-few) bisexual; glumes unequal, the lower 1-, the upper 3-veined; lemmas 5-veined, shortly bifid, with short subterminal awn; rhachilla-segments pubescent; ovary glabrous.

1. R. cristata (L.) Tzvelev (*Lophochloa cristata* (L.) Hyl., *Koeleria phleoides* (Vill.) 870
Pers.) - *Mediterranean Hair-grass.* Stems erect, to 20(60)cm; stems and sheaths 881
glabrous to pubescent; panicles 1-10 x 0.4-1cm; spikelets 3-8mm; awns 1-3mm; (2n=26). Intrd; fairly characteristic wool-alien, rarely from other sources; scattered in Br; Mediterranean.

38. DESCHAMPSIA P. Beauv. - *Hair-grasses*
Densely tufted perennials usually without rhizomes or stolons; inflorescence a very diffuse panicle with fine branches; spikelets with 2 florets both bisexual or sometimes proliferating; glumes unequal, the lower 1-, the upper (1-)3-veined; lemmas 4-5-veined, rounded, obtuse or jagged-toothed at apex, with dorsal (rarely subterminal) straight or bent awn; rhachilla-segments pubescent with longer hair-tuft at base of each; ovary glabrous.

1 Spikelets proliferating **1. D. cespitosa**
1 Spikelets not proliferating 2
2 Leaves >1mm even if rolled up; awns of lemmas not or scarcely exceeding glumes **1. D. cespitosa**
2 Leaves <1mm even if opened out; awns of lemmas conspicuously exceeding glumes 3
3 Lemmas 2-3mm, toothed at apex with marginal teeth the longest; ligule 2-8mm, very acute; palea bifid **2. D. setacea**
3 Lemmas 3-5.5mm, subacute to minutely toothed at apex with marginal teeth not longer than inner ones; ligule 0.5-3mm, obtuse; palea entire at apex **3. D. flexuosa**

1. D. cespitosa (L.) P. Beauv. - *Tufted Hair-grass.* Stems to 1.5(2)m; leaves ≥2mm wide when flattened out, with very rough edges cutting flesh; spikelets proliferating, or sexual and 2.5-6mm. Native.
1 At least some spikelets proliferating 2
1 Spikelets not proliferating 3
2 Leaves distinctly hooded at apex; panicle-branches and pedicels smooth, the main branches usually reflexed; awn arising from middle or upper 1/2 of lemma **c. ssp. alpina**
2 Leaves scarcely or not hooded at apex; panicle-branches and pedicels with minute (sometimes very sparse) pricklets, the main branches rarely reflexed; awn arising from lower 1/2 of lemma **a. ssp. cespitosa**
3 Spikelets 2-3(3.5)mm; hair-tuft at base of lower lemma not reaching

apex of rhachilla-segment above; lowland woodland **b. ssp. parviflora**

3 Spikelets (3)3.5-5(6)mm; hair-tuft at base of lower lemma reaching apex of rhachilla-segment above; lowland meadows and uplands **a. ssp. cespitosa**

a. Ssp. cespitosa. Leaves scarcely or not hooded at apex; panicle-branches and pedicels with minute pricklets; awn arising from lower 1/2 of lemma; spikelets (3)3.5-5(6)mm, or proliferating. Common throughout BI, a tetraploid, 2n=52, in lowland damp meadows, waysides and ditches, and a diploid, 2n=26, triploid or near triploid, 2n=31-44, or tetraploid, 2n=52, in similar habitats and hilly country in N.

b. Ssp. parviflora (Thuill.) Dumort. Differs from ssp. *cespitosa* as in key; diploid, 2n=26. Woods and shady hedgerows; common in lowland Br, ?Ir.

c. Ssp. alpina (L.) Hook. f. (*D. alpina* (L.) Roem. & Schult.). Leaves distinctly hooded at apex; panicle-branches and pedicels smooth; awn arising from middle or upper 1/2 of lemma; spikelets usually all proliferating; near triploid to near tetraploid, 2n=34-56. Damp grassy places on mountains; frequent in W & C Highlands of Sc, very local in Caerns and Lake District, S (formerly N) Kerry and W Mayo.

2. D. setacea (Huds.) Hack. - *Bog Hair-grass.* Stems to 70cm; leaves <1mm wide, usually bluish-green, slightly scabrid; culm-sheaths smooth; spikelets 3-5mm; awn arising from lower 1/2 of lemma; 2n=14. Native; bogs and boggy pools and ditches; very local and scattered in Br and Ir, mostly near coasts of W & N Sc and SC En. **R**

3. D. flexuosa (L.) Trin. - *Wavy Hair-grass.* Stems to 60(100)cm; leaves <1mm wide, usually mid to dark green, slightly scabrid; culm-sheaths slightly scabrid; spikelets 4-6(7)mm; awn arising from lower 1/2 of lemma; 2n=14, 28. Native; acid heaths, moors and open woods, drier parts of bogs; throughout BI, but absent from much of C En and Ir without suitable soils.

39. HOLCUS L. - *Soft-grasses*

Densely tufted or rhizomatous perennials; inflorescence a rather compact panicle; spikelets with 2 florets, the lower bisexual, the upper male; glumes subequal in length but unequal in width, the lower 1-, the upper 3-veined; lemmas 5-veined, ± rounded at apex, the lower awnless, the upper with dorsal awn arising from upper 1/2; rhachilla-segments ± glabrous but lemmas with basal tuft of hairs; ovary glabrous.

1. H. lanatus L. - *Yorkshire-fog.* Stems densely tufted, to 1m; rhizomes 0; leaves and sheaths softly and (sparsely to) densely patent-pubescent; glumes obtuse to subacute at apex; awn of upper lemma recurved to backwardly-hooked, included in glumes; 2n=14. Native; rough grassland, lawns, arable, rough and waste ground, open woods; common throughout BI.

1 x 2. H. lanatus x H. mollis = H. x hybridus Wein is very scattered in Br and Ir but probably under-recorded; it is sterile and resembles *H. mollis*, but has blunter glumes, less exserted awns and more pubescent culms; 2n=21, 22.

2. H. mollis L. - *Creeping Soft-grass.* Stems loosely tufted, to 1m; plant rhizomatous; leaves and sheaths variably pubescent (tiller sheaths often ± densely and softly pubescent) but uppermost culm-sheath glabrous with conspicuously patent-pubescent node below it; glumes acute to acuminate at apex; awn of upper lemma slightly bent, well exserted from glumes; 2n=28, 35, 49. Native; woods, hedgerows, less often open grassland, mostly on acid soils; common throughout most of BI but absent from areas of calcareous or base-rich soil.

40. CORYNEPHORUS P. Beauv. - *Grey Hair-grass*

Densely tufted perennial without rhizomes; inflorescence a rather compact panicle; spikelets with 2 florets, both bisexual; glumes subequal, 1-veined or the upper also with 2 very short laterals; lemmas with 1 central and often 2 pairs of very short lateral veins; obtuse to very shortly bifid at apex, with dorsal bent awn twisted proximally and with club-shaped apex; rhachilla-segments pubescent and with tuft of hairs at base of each lemma; ovary glabrous.

1. C. canescens (L.) P. Beauv. - *Grey Hair-grass*. Stems erect to decumbent, to 35cm; leaves very glaucous, <1mm wide and tightly inrolled; ligules 2-4mm, very acute; spikelets 3-4mm; awns scarcely exceeding glumes; 2n=14. Native; secondarily open sand on leached fixed dunes and inland sandy heathland on acid soils; very local on and near coasts of Jersey, W Suffolk and E & W Norfolk, rare inland in E Suffolk, probably intrd on coasts of S Lancs, Moray, Westerness and E Lothian, natd in Staffs. **RR**

41. AIRA L. - *Hair-grasses*

Annuals; inflorescence a compact to diffuse panicle; spikelets with 2 florets, both bisexual; glumes subequal, 1-3-veined; lemmas with 5 short veins, shortly bifid, with dorsal, slightly bent awn; rhachilla-segments extremely short; lemma with short hair-tuft at base; ovary glabrous.

Other spp. - A. elegantissima Schur, from S Europe, is a wool- or seed-alien rather rarely found in rough ground; it differs from *A. caryophyllea* in its more diffuse panicle with pedicels 2-5x (not 1-2x) as long as spikelets and more abruptly thickened distally.

1. A. caryophyllea L. (*A. multiculmis* Dumort., *A. armoricana* F. Albers) - *Silver Hair-grass*. Stems erect to decumbent, to 25(50)cm; sheaths slightly rough; panicle diffuse, with conspicuous suberect to erecto-patent branches; (2n=28). Native; dry sandy, gravelly or rocky ground, on walls, heaths and dunes; frequent throughout BI. 3 segregates are sometimes recognised but are not consistently separable on external characters and are doubtfully worth maintaining even as sspp. Ssp. **caryophyllea** has spikelets 2.5-3mm; anthers 0.3-0.45mm; caryopsis 1.2-1.6mm; and some pedicels >5mm. Ssp. **multiculmis** (Dumort.) Bonnier & Layens has spikelets 2.2-2.6mm; anthers 0.3-0.5mm; caryopsis 1.1-1.5mm; and all pedicels usually <5mm. Ssp. **armoricana** (F. Albers) Kerguélen, recorded only from W Cornwall, has spikelets 3-3.5mm; anthers 0.3-0.5mm; caryopsis 1.5-1.9mm; and all pedicels usually <5mm. The relative distributions of sspp. *caryophyllea* and *multiculmis* are disputed.

2. A. praecox L. - *Early Hair-grass*. Stems erect to procumbent, to 10(15)cm; sheaths smooth; panicle compact, with short erect branches largely obscured; spikelets 2.5-3.5mm; anthers 0.3-0.3mm; caryopsis 1.4-1.9mm; 2n=14. Native; similar places to *A. caryophyllea*; common throughout BI.

TRIBE 10 - PHALARIDEAE (genera 42-44). Annuals or perennials with or without rhizomes, without stolons; ligule membranous; sheaths not fused; inflorescence a panicle, usually contracted; spikelets with (2-)3 florets, the lower (1-)2 male or sterile and often much reduced, the upper 1 bisexual, rarely the spikelets in groups consisting of 1 central bisexual and 4-6 surrounding sterile (or male) spikelets; glumes 2, equal or unequal, 1 or both nearly as long as to longer than rest of spikelet; lemmas of bisexual florets 3-7-veined, keeled or rounded on back, awnless; palea with 1 vein; lower 2 lemmas minute to longer than upper 1, 0- or 4-5-veined, awnless or with dorsal, bent awn; stamens 2 or 3; stigmas 2; lodicules 0 or 2; ovary glabrous.

Anthoxanthum and *Hierochloe* may be better placed in Aveneae.

42. HIEROCHLOE R. Br. - *Holy-grass*

Rhizomatous perennials; inflorescence a diffuse panicle; lower 2 florets with 3 stamens, with 5-veined lemma slightly longer than lemma of bisexual floret;

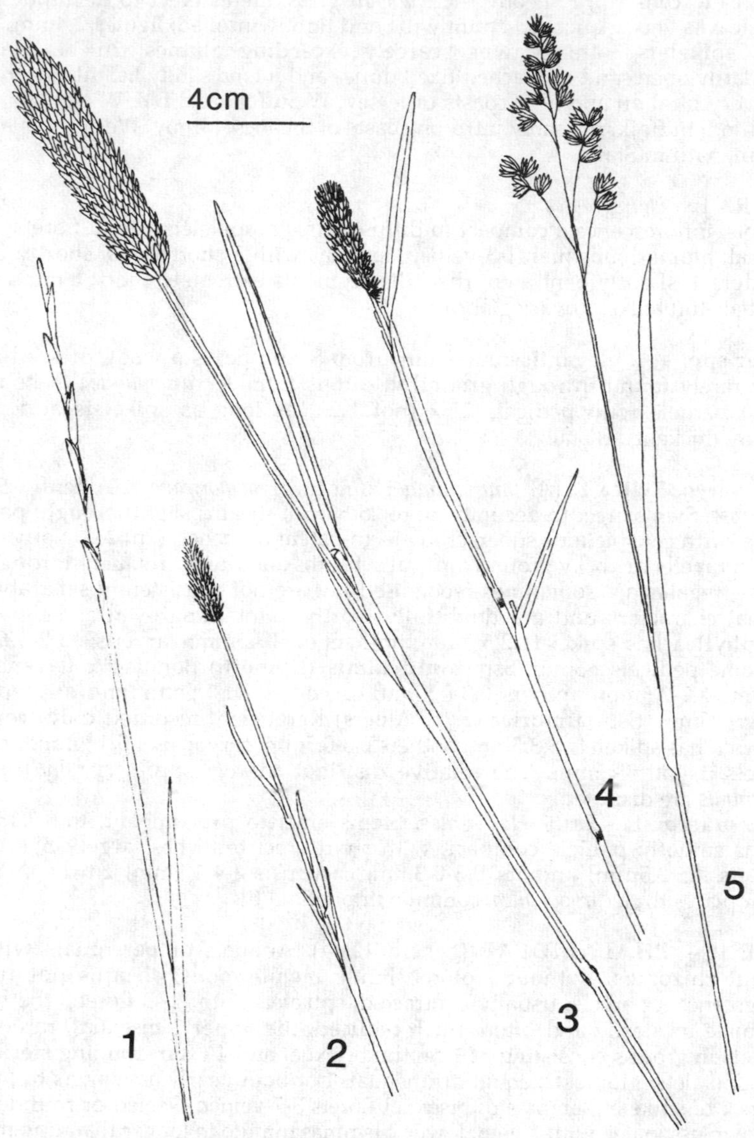

FIG 870 - Poaceae: Pooideae. 1, *Lolium rigidum.* 2, *Rostraria cristata.* 3, *Phalaris aquatica.* 4, *P. paradoxa.* 5, *Dactylis polygama.*

terminal floret with 2 stamens, with 5-veined lemma, with 2 lodicules; glumes subequal, keeled, slightly shorter than rest of spikelet, 1-3-veined. Crushed or dried plant smells strongly of new-mown hay.

1. H. odorata (L.) P. Beauv. - *Holy-grass.* Stems to 60cm; panicles with patent to erecto-patent thin branches; spikelets 3.5-5mm, greenish-purple, becoming characteristic golden-brown; 2n=28. Native; banks of rivers and lakes, wet meadows, flushed cliff-bases by sea; very local on and near coast in Sc, Co Antrim. **RR**

43. ANTHOXANTHUM L. - *Vernal-grasses*
Annuals or tufted perennials; inflorescence a contracted panicle; lower 2 florets sterile, with 4-5-veined lemma slightly longer than lemma of bisexual floret and with long dorsal awn; terminal floret with 2 stamens, 5-veined awnless lemma and 1-veined palea; lodicules 0; glumes very unequal, lower 1-veined, upper 3-veined and longer than rest of spikelet. Crushed or dried plant smells strongly of new-mown hay.

1. A. odoratum L. (*A. alpinum* auct. non Á. & D. Löve) - *Sweet Vernal-grass.* Tufted perennial; culms unbranched, to 50(100)cm; ligules 1-5mm; spikelets 6-10mm; glumes usually pubescent; awns not or only slightly exceeding glumes; awn of lower lemma 2-4mm, of 2nd lemma 6-9mm; anthers 3-4.5mm; 2n=20. Native; in all kinds of grassy places, acid and calcareous, heavy or light soils, lowland or montane; abundant throughout BI. Records of the diploid *A. alpinum* Á. & D. Löve from upland Br were errors for subglabrous plants of *A odoratum.*

2. A. aristatum Boiss. (*A. puelii* Lecoq & Lamotte) - *Annual Vernal-grass.* Tufted annual; culms usually well branched, to 40cm; ligules 0.6-2mm; spikelets 5-7.5mm; glumes glabrous; awns much exceeding glumes; awn of lower lemma 4-5mm, of 2nd lemma 7-10mm; anthers 2.5-3.5mm; (2n=10). Intrd; formerly natd in sandy cultivated or rough ground in Surrey, E Suffolk and perhaps elsewhere but not seen since 1970, now only a rare casual; scattered records in En, S Wa and CI; S Europe. Our plant is ssp. **puelii** (Lecoq & Lamotte) P. Silva.

44. PHALARIS L. - *Canary-grasses*
Annuals or rhizomatous perennials; inflorescence a contracted (often spike-like) panicle; lower 2 florets reduced to scales or rarely only 1 present; terminal floret with 3 stamens, with 5-veined awnless lemma, with 2 lodicules; glumes equal, sharply keeled, longer than rest of spikelet.

1 Perennial with very short to long rhizomes and tillers at flowering time 2
1 Annual without rhizomes or tillers 3
2 Panicle at least distinctly lobed, usually with conspicuous branches; glumes strongly keeled but not winged **1. P. arundinacea**
2 Panicle not lobed, oblong to lanceolate in outline, without visible branches; glumes with distinct wing on keel **2. P. aquatica**
3 Spikelets in groups of 3-7, one bisexual the rest sterile, falling as a group when fruits ripe **6. P. paradoxa**
3 Spikelets all bisexual, the (2-)3 florets of each falling at maturity leaving glumes on panicle 4
4 At least 1 glume with its keel-wing minutely toothed on at least some spikelets **5. P. minor**
4 Wings of glume-keels entire 5
5 Sterile florets 2, ≥1/2 as long as fertile floret **3. P. canariensis**
5 Sterile florets 1-2, <1/3 as long as fertile floret **4. P. brachystachys**

Other spp. - P. angusta Nees ex Trin., from N & S America, and **P. coerulescens** Desf., from Mediterranean, are rarely found on tips and waste ground; *P. angusta* differs from *P. canariensis*, *P. minor* and *P. brachystachys* in spikelets 3-5.5mm, wing of glume-keel minutely serrate and 2 sterile florets; *P. coerulescens* resembles *P. paradoxa* in spikelet structure but is perennial, has glabrous (not pubescent) pedicels and a several-toothed wing of glume-keel.

1. P. arundinacea L. - *Reed Canary-grass*. Perennial to 2m, with long rhizomes; panicle distinctly branched, rather *Dactylis*-like; spikelets 4.5-6.5mm, all bisexual; glumes not winged on keel; sterile florets 1/4-1/2 as long as bisexual floret; 2n=28. Native; by lakes and rivers, in ditches, wet meadows and marshes, also rough and waste ground; common throughout most of BI.

2. P. aquatica L. (*P. tuberosa* L.) - *Bulbous Canary-grass*. Perennial to 1.5m, with short rhizomes; panicle spike-like; spikelets 4.5-6mm, all bisexual; glumes with entire wings on keels; sterile florets ≤1/4 as long as bisexual floret; (2n=28). Intrd; grown as game-cover and -food, sometimes for grazing or silage, natd in fields and rough ground, also casual wool-alien; increasingly frequent in C & S En, more widely scattered as casual; S Europe. **870**

3. P. canariensis L. - *Canary-grass*. Annual to 1.2m; panicle spike-like, usually ovoid; spikelets (4)6-10mm, all bisexual; glumes with entire wings on keels; sterile florets ≥1/2 as long as bisexual floret; 2n=12. Intrd; usually casual birdseed-alien on tips and waste ground, sometimes ± natd, also grown as birdseed crop in S & C En; frequent throughout BI; NW Africa and Canaries.

4. P. brachystachys Link - *Confused Canary-grass*. Annual to 60cm; differs from *P. canariensis* in panicles usually slenderer (narrowly ovoid); and see key; (2n=12). Intrd; rather frequent wool-alien on tips and waste ground, also from other sources; scattered in En, perhaps overlooked for *P. canariensis*; Mediterranean.

5. P. minor Retz. - *Lesser Canary-grass*. Annual to 60cm; differs from *P. canariensis* in panicles usually slenderer (narrowly ovoid) and sterile floret(s) ≤1/3 as long as fertile floret; and from *P. canariensis* and *P. brachystachys* as in key (couplet 4); (2n=28). Intrd; rather frequent casual wool-alien in Br, rarely from other sources, natd in sandy places in Guernsey since at least 1791 and in Scillies; SW & S Europe.

6. P. paradoxa L. - *Awned Canary-grass*. Annual to 1.6m; panicle spike-like, narrowly elliptic to cylindrical; spikelets in groups of 3-7 (1 bisexual, others sterile), the fertile ones 3-6(8)mm with glumes narrowed to short awns and with an apically 1-toothed wing on the keel; the sterile ones all or mostly with club-shaped glumes; sterile florets <1/4 as long as fertile floret; (2n=14). Intrd; natd weed of arable fields, increasingly common in C & S En and S Wa, also a more widespread casual in Br from wool, grain and other sources; SW & S Europe. **870**

TRIBE 11 - AGROSTIDEAE (genera 45-55). Annuals or perennials with or without rhizomes and/or stolons; ligule membranous; sheaths not fused; inflorescence a panicle, contracted or diffuse, often spike-like, rarely a raceme or spike; spikelets with 1 (bisexual) floret (in *Beckmannia* with 2 bisexual or 1 bisexual plus 1 male or sterile florets); glumes 2, usually equal or nearly equal, usually both at least as long as rest of spikelet; lemmas 3-7-veined, rounded on back, awnless or with terminal, subterminal or dorsal awn; stamens 3; stigmas 2; lodicules 0 or 2; ovary glabrous.

45. AGROSTIS L. - *Bents*

Annuals or perennials with or without rhizomes and/or stolons; inflorescence a slightly contracted to very diffuse panicle with obvious branches; glumes 2, equal or nearly so, 1-3-veined, longer than rest of spikelet; lemmas 3-5-veined, awnless or with subterminal or dorsal awn, with or without hair-tuft on callus; palea vestigial to nearly as long as lemma, (usually weakly) 2-veined or veinless; disarticulation at maturity at base of lemma.

A difficult genus due to plasticity and genetic variation, hybridization, and overuse in the past of unreliable characters, notably presence of awns. The palea length and presence of hair-tufts on the lemma-callus are important (though minute) characters (x≥10 lens essential). Presence of rhizomes (as opposed to stolons) is also very reliable but their absence is not, as they are often not developed until late on (in fruit) and in some habitats never.

1 Palea minute, <2/5 as long as lemma 2

1 Palea >2/5 as long as lemma (c.1/2 as long to nearly as long) 6

2 Anthers 0.2-0.6mm; main panicle-branches bare of spikelets for proximal ≥2/3 of length; alien (often casual) of waste ground 3

2 Anthers 1-2mm; main panicle-branches bare of spikelets for proximal ≤1/2 of length; natives 4

3 Spikelets ≥2mm; lemma 1.5-1.7mm, distinctly exceeding caryopsis; leaves 1-3mm wide **10. A. scabra**

3 Spikelets <2mm; lemma 1-1.2mm, not or scarcely exceeding caryopsis; leaves <1mm wide **11. A. hyemalis**

4 Tiller leaves ≤0.3mm wide, bristle-like; panicle always with ± erect branches; rhizomes and stolons 0 **7. A. curtisii**

4 Tiller leaves flat or inrolled, >(0.6)1mm wide even when inrolled; panicle often with patent to erecto-patent branches at or after flowering; rhizomes or stolons usually present 5

5 Stolons 0; rhizomes usually present; ligule on 2nd culm-leaf usually ≤1.5x as long as wide, acute to obtuse **9. A. vinealis**

5 Rhizomes 0; stolons usually present, bearing tufts of leaves or shoots at nodes; ligule on 2nd culm-leaf usually ≥1.5x as long as wide, acute to acuminate **8. A. canina**

6 Anthers 0.2-0.8mm; lemma-callus with tuft of hairs ≥0.3mm 7

6 Anthers 1-1.5mm; lemma-callus glabrous or pubescent 8

7 Lemmas with awn ≥2mm, well exserted from glumes; main panicle-branches bare of spikelets for proximal ≥2/3 of length, patent to erecto-patent after flowering; rhachilla extended above lemma-base, reaching ≥1/2 way up lemma **6. A. avenacea**

7 Lemmas with awn 0-0.5mm, not exserted from glumes; main panicle-branches bare of spikelets for proximal ≤1/2 of length, erect to erecto-patent after flowering; rhachilla not extended above lemma-base **5. A. lachnantha**

8 Lemma-callus with tuft of hairs >0.3mm; awn 0 or, if present, arising from basal 1/3 of lemma, often exceeding glumes **3. A. castellana**

8 Lemma-callus glabrous or with hairs <0.2mm; awn 0 or, if present, arising from apical 1/2 of lemma, rarely exceeding glumes 9

9 Panicle contracted after flowering; rhizomes 0 or short and with ≤3 scale-leaves; ligules of culm-leaves subacute to rounded; stolons usually well developed **4. A. stolonifera**

9 Panicle-branches patent or nearly so after flowering; rhizomes usually present, with >3 scale-leaves; ligules of culm-leaves truncate; stolons 0 or poorly developed 10

10 Ligules of tillers shorter than wide; fruiting panicle-branches with spikelets all well separated; leaves rarely >5mm wide **1. A. capillaris**

10 Ligules of tillers longer than wide; fruiting panicle-branches bearing spikelets in small ± dense clusters at tips; leaves often >5mm wide **2. A. gigantea**

Other spp. - A. exarata Trin. (*Spike Bent*), from N America, has been found once in Bucks as a weed of grass-seed but may be commoner and overlooked as a contaminant of *Highland Bent*. It has a minute palea and anthers <0.6mm, but panicle-branches bearing spikelets ± to the base; the lemma-callus is glabrous and the lemmas awned or unawned. The panicle-branches are ± erect forming a narrow inflorescence superficially resembling that of *Apera interrupta*.

1. A. capillaris L. (*A. tenuis* Sibth.) - *Common Bent*. Rhizomatous perennial to 75cm; ligules truncate or rounded, ≤2mm, those of tillers shorter than wide; panicle diffuse at fruiting, with spikelets all separated; awns 0 or short; 2n=28. Native; all kinds of grassy places and rough ground, especially on acid soils; abundant throughout BI.

1 x 2. A. capillaris x A. gigantea = A. x bjoerkmanii Widén is a vigorous highly sterile pentaploid, 2n=35, with intermediate ligules and panicle; very few scattered records in Br but probably overlooked.

1 x 3. A. capillaris x A. castellana = A. x fouilladei P. Fourn. is partially fertile and variably intermediate between the parents and hence difficult to determine; it has scattered records in En arising from intrd seed-mixture and perhaps in situ, but is likely to become commoner.

1 x 4. A. capillaris x A. stolonifera = A. x murbeckii Fouill. ex P. Fourn. is a vigorous highly sterile tetraploid, 2n=28, with usually both rhizomes and stolons and intermediate ligule- and panicle-shape; it has been recorded from scattered localities in Br but is probably common throughout BI.

1 x 9. A. capillaris x A. vinealis (?=*A. x sanionis* Asch. & Graebn.) is a highly sterile tetraploid, (2n=28), with awns 0 to long and basal, palea c.1/3 as long as lemma, and ligule of tillers c. as long as wide; there are scattered records in Br on poor sandy soils.

2. A. gigantea Roth - *Black Bent*. Rhizomatous perennial to 1(1.2)m; ligules truncate or rounded, ≤6mm, those of tillers longer than wide; panicle diffuse at fruiting, with spikelets in small clusters at branch-tips; awns 0 or short; 2n=42. Native; grassy places, rough, cultivated and waste ground, mostly on disturbed sandy soils; throughout BI, common in S & C En, scattered elsewhere but probably overlooked.

2 x 4. A. gigantea x A. stolonifera is a vigorous highly sterile pentaploid, (2n=35), with intermediate ligule- and panicle-shape and often strong rhizomes and stolons; there are very scattered records in Br.

3. A. castellana Boiss. & Reut. - *Highland Bent*. Rhizomatous perennial to 60cm; ligules subacute to rounded, ≤3mm, those of tillers c. as long to slightly longer than wide; panicles contracted at fruiting; awns 0, short or long; (2n=28, 42). Intrd; lawns, roadsides, amenity and sports areas where sown and escaped; throughout Br, becoming commoner, much under-recorded.

4. A. stolonifera L. - *Creeping Bent*. Stoloniferous perennial to 75cm, with rhizomes 0 or very short; ligules subacute to rounded, ≤7mm, those of tillers longer than wide; panicles contracted at fruiting; awns 0 or short; 2n=28. Native; damp meadows, ditches, marshes, by lakes, ponds, canals and rivers, damp arable and rough ground, dune-slacks; abundant throughout BI.

4 x 9. A. stolonifera x A. vinealis is a highly sterile tetraploid, (2n=28), with awns 0 to long and basal, palea c.1/2 as long as lemma, ligule of tillers longer than wide, and often stolons and rhizomes; it has been recorded only from W Cornwall, but, although rare, is probably overlooked.

5. A. lachnantha Nees - *African Bent*. Loosely tufted annual (with us) or perennial; ligules acute to obtuse, ≤7mm, those of tillers longer than wide; panicles contracted at fruiting; awns 0 or ± so; (2n=28). Intrd; a rather infrequent wool-alien, but perhaps overlooked; scattered in En; C & S Africa.

6. A. avenacea J.F. Gmel. - *Blown-grass*. Tufted perennial (often annual with us) **875**
to 60cm; ligules ≤10mm, acute to rounded, those of tillers longer than wide; panicles very widely spreading at fruiting; awns long; (2n=56). Intrd; frequent wool-alien now natd in waste and rough ground and by roads and railways; scattered in Br; Australia and New Zealand.

7. A. curtisii Kerguélen (*A. setacea* Curtis non Vill.) - *Bristle Bent*. Densely tufted

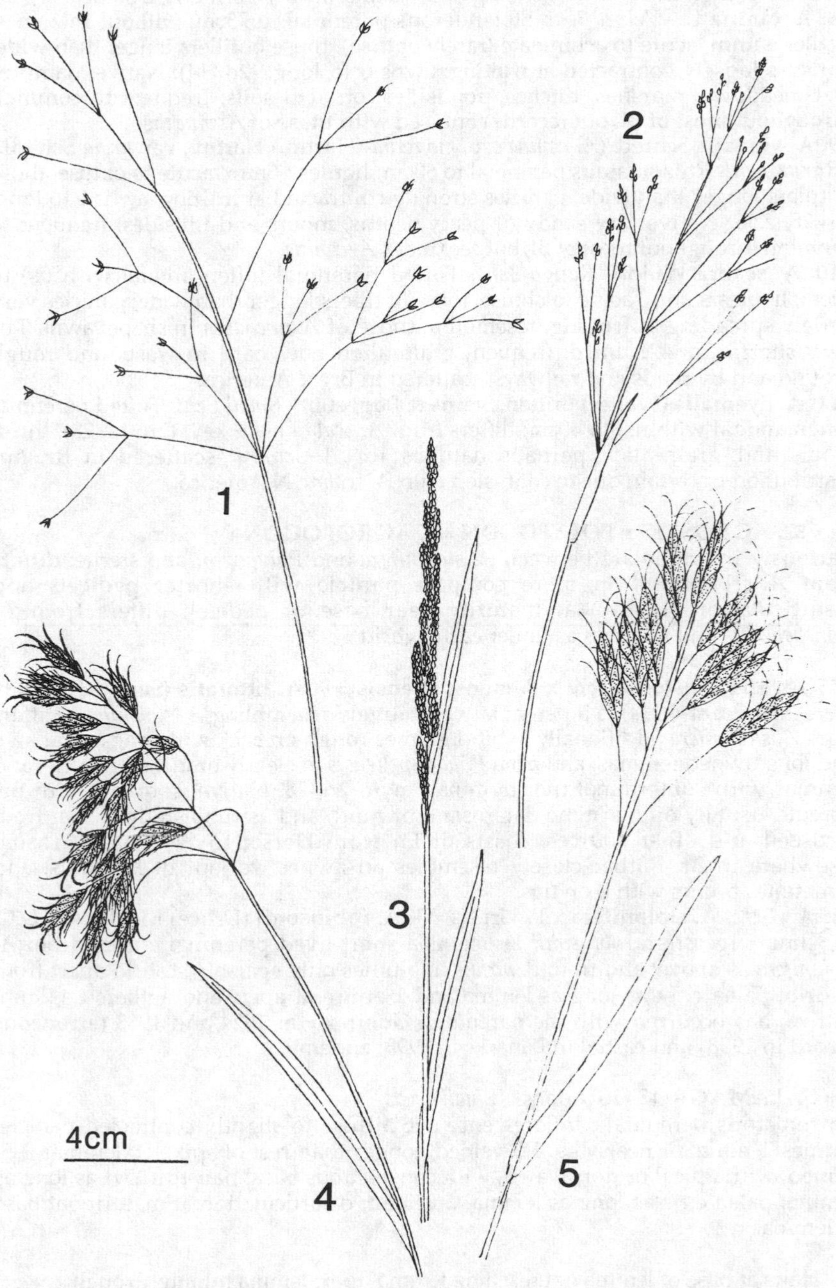

FIG 875 - Poaceae: Pooideae. 1, *Agrostis avenacea.* 2, *A. hyemalis.* 3, *Beckmannia syzigachne.* 4, *Bromus japonicus.* 5, *B. lanceolatus.*

perennial to 60cm; ligules ≤4mm, acute, those of tillers longer than wide; panicles contracted at fruiting; awns long; 2n=14. Native; dry sandy or peaty heaths; locally common in SW En, extending to S Wa and Surrey, formerly E Sussex.

8. A. canina L. - *Velvet Bent.* Stoloniferous perennial to 75cm, without rhizomes; ligules ≤4mm, acute to acuminate (rarely obtuse), those of tillers longer than wide; panicles loosely contracted at fruiting; awns 0 to long; (2n=14). Native; damp or wet meadows, marshes, ditches, pondsides, on acid soils; frequent to common throughout most of BI, but records confused with those of *A. vinealis.*

9. A. vinealis Schreb. (*A. canina* ssp. *montana* (Hartm.) Hartm., var. *arida* Schltdl.) - *Brown Bent.* Rhizomatous perennial to 60cm; ligules ≤5mm, acute to obtuse, those of tillers longer than wide; panicles strongly contracted at fruiting; awns 0 to long; 2n=28, c.56. Native; dry sandy or peaty heaths, moors and hillsides; frequent to common throughout most of BI, but see under *A. canina.*

10. A. scabra Willd. - *Rough Bent.* Tufted perennial (often annual with us) to 60cm; ligules ≤5mm, acute to obtuse, those of tillers longer than wide; panicles very widely spreading at fruiting, resembling those of *A. avenacea* in shape; awns 0 or very short; (2n=42). Intrd; frequent grain-alien now natd in waste and rough ground and by roads and railways; scattered in Br; N America.

11. A. hyemalis (Walter) Britton, Sterns & Poggenb. - *Small Bent.* Tufted perennial **875**
(often annual with us) to 40cm; differs from *A. scabra* as in key; (2n=14, 28). Intrd; wool- and grain-alien perhaps natd as for *A. scabra*; scattered in Br, but distribution uncertain due to confusion with *A. scabra*; N America.

45 x 52. AGROSTIS x POLYPOGON = X AGROPOGON P. Fourn.
Variously intermediate between *A. stolonifera* and *Polypogon* and sterile; differs from *A. stolonifera* in more compact panicle with shorter pedicels and disarticulation (if any) at maturity near base of pedicel; differs from *P. monspeliensis* and *P. viridis* as under each hybrid.

45/4 x 52/1. A. stolonifera x P. monspeliensis = X A. littoralis (Sm.) C.E. Hubb. (*Perennial Beard-grass*) is a perennial vegetatively resembling *A. stolonifera*; differs from *A. stolonifera* additionally in bifid glumes rough on back with long apical awn and long-awned lemmas; and from *P. monspeliensis* in clearly branched panicle, and lemmas with subterminal (not terminal) awn; 2n=28. Native; sporadic with the parents, usually on maritime damp sand or mud, and as rubbish-tip casual from birdseed and other sources; coasts of En from Dorset to W Norfolk, casual elsewhere in Br. Rather closely resembles an awned variant of *P. viridis*, and sometimes occurs with it on tips.

45/4 x 52/2. A. stolonifera x P. viridis = X A. robinsonii (Druce) Melderis & D.C. McClint. (*Agrostis x robinsonii* Druce) is a short-lived perennial; differs from *A. stolonifera* as above; and from *P. viridis* in glumes only sparsely scabrid apart from midrib, palea c.3/4 as long as lemma and ± entire at apex, and anthers c.1.2mm. Native; has occurred with the parents in Guernsey in 1924 and 1953 (erroneous record in 1958) and casual in Lanarks in 1993; endemic.

46. CALAMAGROSTIS Adans. - *Small-reeds*
Rhizomatous perennials; inflorescence a ± diffuse to slightly contracted panicle; glumes 2, equal or nearly so, 1-3-veined, longer than rest of spikelet; lemmas 3-5-veined, with apical or dorsal awn, with conspicuous basal hair-tuft ≥1/2 as long as lemma; palea c.2/3 as long as lemma, 2-veined; disarticulation at maturity at base of lemma.

1 Hairs at base of lemma not reaching lemma-apex; lemma minutely rough, with awn arising 1/4-1/3 way up 2
1 Hairs at base of lemma reaching at least to apex of lemma; lemma smooth, with awn arising from middle, apex or upper 1/2 3

2 Spikelets 3-4(4.5)mm; lower glume acute; culms usually rough just below panicle **4. C. stricta**
2 Spikelets (4)4.5-6mm; lower glume acuminate; culms smooth throughout **5. C. scotica**
3 Culms mostly with 5-8 nodes; ligules 7-10(14)mm; pollen 0; anthers indehiscent **3. C. purpurea**
3 Culms mostly with 2-5 nodes; ligules (1)2-9(12)mm; pollen present; anthers dehiscent 4
4 Leaf uppersides (often sparsely) pubescent; ligules (1)2-4(6)mm; lemmas 3-5-veined, with basal hairs 1.1-1.5x as long as lemma **2. C. canescens**
4 Leaf uppersides scabrid, not pubescent; ligules 4-9(12)mm; lemmas 3-veined, with basal hairs >1.5x as long as lemma **1. C. epigejos**

1. C. epigejos (L.) Roth - *Wood Small-reed.* Culms to 2m, with 2-4 nodes; leaves not pubescent on upperside; ligules 4-9(12)mm; glumes 4-7mm; lemmas with basal hairs c.1.5-2x as long as lemma; 2n=28, (28, 56). Native; damp woods and wood-margins, ditches, fens, dune-slacks; scattered throughout much of Br, common in parts of S & E En, rare in most of Sc, Wa, Ir and CI.

2. C. canescens (F.H. Wigg.) Roth - *Purple Small-reed.* Culms to 1.2m, with 3-5 nodes; differs from C. *epigejos* in less dense panicles, less finely pointed glumes; and see key; 2n=28. Native; fens, marshes and open wet woods; scattered in En (mostly CE), Kircudbrights and Selkirks.

2 x 4. C. canescens x C. stricta = C. x gracilescens (Blytt) Blytt occurs for certain only by a canal in SE Yorks where partially fertile octoploids, 2n=56, intermediate between the parents and sterile tetraploids, 2n=28, closer to C. *stricta* occur along with C. *stricta.* Other populations of C. *stricta* in En and Sc probably have been introgressed by C. *canescens* in the past (leaves slightly pubescent on upperside, ligules >3mm, glumes >4mm, lemma-hairs ± equalling lemma, awn arising 1/3-1/2 way up lemma).

3. C. purpurea (Trin.) Trin. - *Scandinavian Small-reed.* Culms to 1.5m, with 5-8 nodes; leaves pubescent on upperside; ligules 7-10(14)mm; glumes 4-6(7)mm; lemmas with basal hairs 1.1-1.5x as long as lemma; 2n=56. Native; fens, marshes, ditches and lakesides; c.6 localities in Argyll, E Perth, S Aberdeen and Angus, 1 locality in Westmorland, not recognized until 1980. An apomictic taxon probably derived from C. *epigejos* x C. *canescens.* Our plant is ssp. **phragmitoides** (Hartm.) Tzvelev. RR

4. C. stricta (Timm) Koeler - *Narrow Small-reed.* Culms to 1m, with 2-3 nodes; leaves pubescent on upperside; ligules 1-3mm; glumes 3-4(4.5)mm; lemmas with basal hairs 0.5-0.8x as long as lemma; 2n=28. Native; marshes, fens and lakesides; very scattered in N BI S to W Suffolk, Cheshire and Co Antrim (formerly Co Armagh). RR

5. C. scotica (Druce) Druce - *Scottish Small-reed.* Differs from C. *stricta* as in key; perhaps an introgressed variant of it. Native; marshes and fens; 1 locality (formerly 3) in Caithness, possibly 1 in Roxburghs (identity not certain); endemic. RR

46 x 47. CALAMAGROSTIS x AMMOPHILA = X CALAMMOPHILA Brand (X *Ammocalamagrostis* P. Fourn.)

46/1 x 47/1. X C. baltica (Flüggé ex Schrad.) Brand (X *Ammocalamagrostis baltica* (Flüggé ex Schrad.) P. Fourn; *C. epigejos* x *A. arenaria*) (*Purple Marram*) occurs on maritime dunes with *A. arenaria* and near *C. epigejos* in E Suffolk, E Norfolk, Cheviot and W Sutherland, and planted for sand-binding with *A. arenaria* in E Suffolk, E & W Norfolk and S Hants. It is vigorous, sterile and more closely resembles *A. arenaria,* with long rhizomes, ligules 10-25mm and culms to 1.5m with dense, linear-ellipsoid, often purple-tinged panicles; spikelets intermediate, 9-12mm, with lemmas 7-9mm with 3-7 veins, subterminal awn 1-2mm and basal hair-tuft >1/2 as long as lemma; palea 2-4-veined; (2n=28, 42). Our plant is var. **baltica**; on the Continent vars more exactly intermediate and nearer *C. epigejos* respectively, and with different origins, also occur.

47. AMMOPHILA Host - *Marram*
Strongly rhizomatous perennials; inflorescence a compact linear-ellipsoid panicle; glumes 2, subequal, the lower 1-3-, the upper 3-veined, slightly longer than rest of spikelet; lemmas 5-7-veined, awnless or with minute subapical awn <1mm, with basal hair-tuft <1/2 as long as lemma; palea nearly as long as lemma, 2-4-veined; disarticulation at maturity at base of lemma.

1. A. arenaria (L.) Link - *Marram.* Culms to 1.2m; leaves usually tightly inrolled with minutely densely pubescent upperside; ligules 10-30mm; glumes 10-16mm; 2n=28. Native; on mobile sand-dunes, often dominant; common round coasts of BI, rare casual inland.

2. A. breviligulata Fernald - *American Marram.* Differs from *A. arenaria* in ligules 1-3mm; (2n=28). Intrd; planted on sand-dunes at Newborough, Anglesey in late 1950s and well natd in 1 small area, not seen since c.1978 and probably now gone; E N America. Perhaps only a ssp. of *A. arenaria.*

48. GASTRIDIUM P. Beauv. - *Nit-grasses*
Annuals; inflorescence a compact linear-ellipsoid panicle; glumes 2, unequal, 1-veined, much longer than rest of spikelet, linear-lanceolate, with swollen ± hemispherical base; lemmas 5-veined, with dorsal usually bent awn, without basal hair-tuft; palea ± as long as lemma, 2-veined; disarticulation at maturity at base of lemma.

1. G. ventricosum (Gouan) Schinz & Thell. - *Nit-grass.* Stems procumbent to erect, **RR**
to 50(90)cm; panicles 0.5-10(16)cm; spikelets (2)3-5mm; lemmas c.1mm, subglabrous to sparsely pubescent, with awn 0-4mm, rarely exceeding glumes; 2n=14. Native; barish or sparsely grassed, well-drained, calcareous ground, also a frequent wool- and grain-alien; very local as native mostly near coast in SW Br from S Devon and Wight to Glam (plants small and often procumbent to ascending), perhaps CI, casual scattered in En and S Wa, formerly frequent as cornfield weed in SE En (plants often robust and usually erect).

2. G. phleoides (Nees & Meyen) C.E. Hubb. - *Eastern Nit-grass.* Differs from *G. ventricosum* in stems usually erect; spikelets (4)5-8mm; lemma pubescent to densely so, with awn 4-7(8)mm, often exceeding glumes; (2n=28). Intrd; rather infrequent wool-alien; scattered in En; SW Asia. Scarcely distinct from *G. ventricosum.*

49. LAGURUS L. - *Hare's-tail*
Annuals; inflorescence a very compact, ovoid, densely silky-pubescent panicle; glumes 2, equal, 1-veined, linear-lanceolate and tapered to apical awn, longer than rest of spikelet, lemmas 5-veined, with 2 apical bristles reaching c. as far as glume-awns, with dorsal bent awn well exserted from glumes, pubescent but without basal hair-tuft; palea shorter than body of lemma, 2-veined; disarticulation at maturity at base of lemma.

1. L. ovatus L. - *Hare's-tail.* Stems to 60cm, erect; leaves and sheaths softly pubescent; panicle 1-7 x 0.5-2cm; spikelets 7-10mm; awns 8-20mm; (2n=14). Intrd; planted and now often abundant on mobile sand-dunes in Jersey and Guernsey, rare casual elsewhere in En, S Wa and Ir, locally natd in S Devon, W Sussex and Co Wexford; S Europe.

50. APERA Adans. - *Silky-bents*
Annuals; inflorescence a diffuse to ± contracted panicle; glumes 2, unequal, the

lower 1-, the upper 3-veined, ± as long as rest of spikelet; lemmas 5-veined, with subterminal ± straight awn much longer than body, minutely pubescent at base; palea shorter than to ± as long as body of lemma, 2-veined; disarticulation at maturity at base of lemma.

1. A. spica-venti (L.) P. Beauv. - *Loose Silky-bent.* Stems to 1m; leaves 2-10mm wide; ligules 3-10mm; panicles up to 25 x 15cm, very diffuse, the longer branches bare for proximal 1/2; spikelets 2.4-3mm; awns 5-10mm; anthers 1-2mm; (2n=14). Possibly native; frequent in dry sandy arable fields and marginal habitats in E Anglia, parts of SE En and Humber basin, intrd as grain-alien elsewhere in BI but very scattered and usually casual. **R**

2. A. interrupta (L.) P. Beauv. - *Dense Silky-bent.* Stems to 40(70)cm; leaves 0.5-4mm wide; ligules 2-5mm; panicles up to 20 x 1.5cm, loosely contracted, the branches with spikelets nearly to base; spikelets 1.8-2.5mm; awns 4-10mm; anthers 0.3-0.4mm; 2n=14. Possibly native; dry sandy fields and rough ground; locally frequent in E Anglia, scattered N to SE Yorks and S to Berks; S Europe. **R**

51. MIBORA Adans. - *Early Sand-grass*

Annuals; inflorescence a slender 1-sided raceme with pedicels <0.5mm; glumes 2, equal, 1-veined, longer than rest of spikelet; lemmas 5-veined, awnless, pubescent but without basal hair-tuft; palea as long as lemma, 2-veined; disarticulation at maturity at base of lemma.

1. M. minima (L.) Desv. - *Early Sand-grass.* Culms usually densely tufted, very slender, to 8(15)cm, with leaves usually all on basal 1/3; racemes 0.5-2cm; spikelets 1.8-3mm; glumes and lemmas ± truncate; 2n=14. Native; on loose sand of maritime dunes and similar places near sea in Anglesey and CI, possibly native in Glam, natd in few places on S & E coasts of Br N to E Lothian, rare casual elsewhere. **RR**

52. POLYPOGON Desf. - *Beard-grasses*

Annuals or perennials with stolons; inflorescence a contracted or semi-diffuse panicle; glumes 2, equal, 1-veined or the upper 3-veined, much longer than rest of spikelet, awned from apex or unawned; lemmas 5-veined, truncate and finely toothed at apex, without basal hair-tuft, awnless or with short terminal awn; palea nearly as long as lemma, 2-veined; disarticulation at maturity near base or apex of pedicel.

Other spp. - P. maritimus Willd., from S & W Europe, differs from *P. monspeliensis* in its slenderer habit, much more deeply bifid and longer-pubescent glumes, and unawned lemma; it is a very infrequent wool-alien but might be overlooked. **P. elongatus** Kunth, from C America, is another rare wool-alien; it is a perennial with a dense inflorescence and glumes with awns 2-3mm.

1. P. monspeliensis (L.) Desf. (*P. paniceus* (L.) Lag.) - *Annual Beard-grass.* Annual; culms to 80cm; inflorescence densely contracted, ± cylindrical; spikelets disarticulating near apex of pedicel; glumes 2-3mm, notched at apex, each with apical awn 3.5-7mm; lemma with awn (0)1-2mm; anthers 0.4-0.7mm; 2n=28. Native; drier parts of salt-marshes and damp places near sea; near coasts of S & SE En from Dorset to E Norfolk, formerly W Norfolk and Guernsey, also in Br and CI N to C Sc as casual of tips and waste ground, sometimes natd. **R**

2. P. viridis (Gouan) Breistr. (*P. semiverticillatus* (Forssk.) Hyl., *Agrostis semiverticillata* (Forssk.) C. Chr.) - *Water Bent.* Stoloniferous perennial; culms to 60(100)cm; inflorescence semi-diffuse; spikelets disarticulating near base of pedicel; glumes 1.6-2.3mm, obtuse, unawned; lemma unawned; anthers 0.4-0.7mm; (2n=28). Intrd; natd on roadsides and rough ground and by pools in Guernsey and Jersey, scattered casual elsewhere in Br on tips and waste land; S Europe.

53. ALOPECURUS L. - *Foxtails*

Annuals or perennials without rhizomes, sometimes with stolons; inflorescence a very contracted spike-like panicle; glumes 2, equal, 3-veined, sometimes with their margins fused proximally round spikelet, very slightly shorter to very slightly longer than rest of spikelet, keeled, rounded to very acute or apiculate at apex; lemmas 4-veined, obtuse to truncate or notched, without basal hair-tuft, sometimes with margins fused proximally round carpel and stamens, with dorsal long or short awn from lower 1/2; palea 0; disarticulation at maturity near base of pedicel.

1 Lemmas unawned or with awn shorter than body and not exserted from glumes or exserted by ≤0.5mm 2
1 Lemmas with awn longer than body and exserted from glumes by ≥1mm 3
 2 Panicles >3x as long as wide; glumes with hairs <0.5mm; lemma with margins fused proximally for 1/3-1/2 their length; lowlands **4. A. aequalis**
 2 Panicles <3x as long as wide; glumes with hairs >(0.5)1mm; lemma with margins fused proximally for <1/4 their length; mountains **5. A. borealis**
3 Margins of lemma free or fused proximally for <1/4 their length; 2 glumes fused only at extreme base 4
3 Margins of lemma fused proximally for c.1/3-1/2 their length; 2 glumes fused proximally for c.1/4-1/2 their length 5
 4 Glumes acute; basal culm internode swollen, (1)2-4.5(6)mm wider than normal culm width **3. A. bulbosus**
 4 Glumes obtuse; basal culm internode 0-1(1.5)mm wider than normal culm width **2. A. geniculatus**
5 Annual; glumes fused proximally for 1/3-1/2 their length, subglabrous or with hairs <0.5mm on keel, margins and at base, with winged keel **6. A. myosuroides**
5 Perennial; glumes fused proximally for c.1/4 their length, conspicuously pubescent with hairs >0.5mm, with keel unwinged **1. A. pratensis**

1. A. pratensis L. - *Meadow Foxtail.* Perennial; culms usually erect, to 1.2m; panicles 2-12 x 0.5-1.2cm; spikelets 4-6mm; glumes acute, fused proximally for c.1/4 their length, conspicuously pubescent; anthers 2-3.5mm, yellow or purple; 2n=28. Native; grassy places, mostly on damp rich soils; common ± throughout lowland BI.

1 x 2. A. pratensis x A. geniculatus = A. x brachystylus Peterm. (*A. x hybridus* Wimm.) is scattered with the parents throughout Br; it is intermediate in habit, fusion of lemma and glumes, spikelet length (3-4.5mm), and anther length and is usually highly sterile; 2n=28.

2. A. geniculatus L. - *Marsh Foxtail.* Perennial; culms usually decumbent to ascending, often rooting at lower nodes, to 40(50)cm; panicles 1.5-7 x 0.3-0.7cm; spikelets 2-3(3.5)mm; glumes obtuse, fused only at extreme base, conspicuously pubescent; anthers 0.8-2mm, yellow or purple; 2n=28. Native; wet meadows, marshes, ditches, pondsides; frequent to common throughout BI.

2 x 3. A. geniculatus x A. bulbosus = A. x plettkei Mattf. has been found with the parents in S & E En from Dorset to N Lincs, Carms; it is intermediate in culm shape and culm swelling and highly sterile; 2n=21.

2 x 4. A. geniculatus x A. aequalis = A. x haussknechtianus Asch. & Graebn. has been found scattered with the parents from Merioneth and Brecs to W Norfolk and Beds; it is intermediate in spikelet and awn lengths and lemma margin fusion and is highly sterile.

3. A. bulbosus Gouan - *Bulbous Foxtail.* Perennial; culms usually erect to R
ascending, not rooting at nodes, to 30(40)cm; differs from *A. geniculatus* in spikelets 2.5-3.3mm; anthers 1.2-2.2mm; 2n=14; and see key. Native; wet grassy

FIG 881 - Spikelets of **Poaceae**. 1, *Oryzopsis miliacea*. 2, *Lamarckia aurea* (group of 3 sterile and 2 fertile florets). 3, *Schismus barbatus*. 4, *Rostraria cristata*. 5, *Beckmannia syzigachne*. 6, *Rytidosperma racemosum*. 7, *Gaudinia fragilis*. 8, *Cortaderia selloana*. 9, *Eriochloa pseudoacrotricha*. 10-11, *Brachiaria platyphylla* (11, upper glume removed to show upper lemma). 12-13, *Urochloa panicoides* (13, upper glume removed to show upper lemma). 14, *Paspalum distichum*.

places, usually brackish and grazed, near sea or in estuaries; local on coasts of S & E Br from Carms and E Cornwall to N Lincs, Guernsey.

4. A. aequalis Sobol. - *Orange Foxtail.* Annual to perennial; culms usually decumbent to ascending, sometimes rooting at lower nodes, to 40cm; panicles 1-6 x 0.3-0.6cm; spikelets 2-2.5mm; glumes obtuse, fused only at extreme base, conspicuously shortly pubescent; anthers 0.8-1.3mm, orange; 2n=14. Native; similar places to *A. geniculatus*; scattered in C & S Br N to NW Yorks (mainly C & E En), Man, E Cork (found 1992).

5. A. borealis Trin. (*A. alpinus* Sm. non Vill.) - *Alpine Foxtail.* Perennial; culms ± **R**
erect, to 50cm; panicles 1-3 x 0.7-1.2cm; spikelets 3-4.5mm; glumes acute, fused only at extreme base, very conspicuously long-pubescent; anthers 2-2.5mm; 2n=100, c.112, 117. Native; mountain springs and flushes usually dominated by bryophytes, at 600-1200m; very local in N Br from Westmorland to E Ross.

6. A. myosuroides Huds. - *Black-grass.* Annual; culms erect, to 80cm; panicles 2-12 x 0.3-0.6cm; spikelets 4.5-7mm; glumes acute, fused proximally for 1/3-1/2 their length, very inconspicuously pubescent on keels, edges and base only; anthers 2.5-4mm; 2n=14. Native; weed of arable fields and waste ground; frequent (but decreasing) in S, C & E En, very scattered to SW En, CI, Wa, E Ir and C Sc and there mostly casual.

54. BECKMANNIA Host - *Slough-grasses*

Annuals or perennials without stolons, rarely with rhizomes; inflorescence a long raceme of closely packed appressed spikes forming a long narrow inflorescence; glumes 2, equal, 3-veined with connecting veins between, enclosing rest of spikelet except for apiculate tip of lemma(s), strongly keeled and hooded; florets 1 or 2, the 1st bisexual, the second bisexual, male or sterile; lemmas 5-veined, apiculate but not otherwise awned, without basal hair-tuft; palea nearly as long as lemma, 2-veined; disarticulation at maturity below glumes.

Abnormal in Agrostideae in racemose inflorescence and often 2-floreted spikelets with glumes with branched veins, but even more out of place elsewhere.

Other spp. - B. eruciformis (L.) Host (*European Slough-grass*), from E Europe, might occur as more than a rare casual but has been much confused with *B. syzigachne* and recent records of it are errors for the latter; it differs in having 2 regularly bisexual florets per spikelet, usually pubescent lemmas, and anthers 1.5-2mm, and is a longer lived perennial with culms often swollen at base.

1. B. syzigachne (Steud.) Fernald - *American Slough-grass.* Annual or short-lived **875**
perennial to 50(100)cm; inflorescence up to 15(25)cm; spikelets flattened, closely **881**
packed on spikes 5-20mm, 2.2-3.2mm, with 1 floret or sometimes a 2nd male, bisexual or sterile one; glumes and usually lemmas glabrous; anthers 0.7-1.2mm; (2n=14). Intrd; casual of waste ground and tips, mainly from grain or birdseed, scattered in S En and S Wa, formerly persistent in Avonmouth, W Gloucs; N America.

55. PHLEUM L. - *Cat's-tails*

Annuals or perennials, sometimes with rhizomes or stolons; inflorescence a very contracted spike-like panicle; glumes 2, equal, 3-veined, strongly keeled, usually with stiff hairs on keel, apiculate to shortly awned at apex; lemmas 3-7-veined, irregularly truncate to rounded at apex, without basal hair-tuft, unawned; palea as long or nearly as long as lemma, 2-veined; disarticulation at maturity below the lemma.

1 Annual without tillers at flowering; glumes acute to subacute at apex, gradually narrowed to awn; anthers ≤1mm **5. P. arenarium**
1 Perennial with tillers at flowering; glumes obtuse or truncate at apex,

abruptly or very abruptly narrowed to awn; anthers ≥1mm 2

2 Glumes 5-8.5mm incl. awns 2-3mm; panicles (1)2-3(5)x as long as wide; mountains >600m in N Br **3. P. alpinum**

2 Glumes 2-5.5mm incl. awns ≤2mm; panicles (2)3-20(30)x as long as wide; widespread 3

3 Glumes obtuse (and shortly awned) at apex; culms not swollen at base; ligules 0.5-2mm; CE En only **4. P. phleoides**

3 Glumes truncate (and shortly awned) at apex; culms usually swollen at base; ligules 1-9mm; widespread 4

4 Spikelets (3.5)4-5.5mm incl. awns (0.8)1-2mm; panicle 6-10mm wide; leaves 3-9mm wide; ligule usually obtuse **1. P. pratense**

4 Spikelets 2-3.5mm incl. awns 0.2-1(1.2)mm; panicle 3-6mm wide; leaves 2-6mm wide; ligule usually acute **2. P. bertolonii**

Other spp. - P. subulatum (Savi) Asch. & Graebn., from S Europe, is a rare grain-alien; it is an annual differing from *P. arenarium* in semi-elliptic (not linear-lanceolate) glumes, rhachilla prolonged above floret and lemma 2/3 (not 1/3) as long as glumes.

1. P. pratense L. - *Timothy*. Perennial; culms erect, to 1.5m; panicles up to 20(30) x 1cm, cylindrical; glumes (3.5)4-5.5mm incl. awns, truncate at apex with awn 0.8-2mm; 2n=42. Native; grassy places and rough ground; common throughout BI.

2. P. bertolonii DC. (*P. nodosum* auct. non L., *P. hubbardii* D. Kováts, *P. pratense* ssp. *bertolonii* (DC.) Bornm., ssp. *serotinum* (Jord.) Berher) - *Smaller Cat's-tail*. Differs from *P. pratense* in culms to 50(100)cm; panicles up to 8 x 0.5cm; glumes 2-3.5mm incl. awns 0.2-1.2mm; 2n=14; and see key. Native; similar places to *P. pratense* but usually confined to grassland; probably throughout BI but uncertain due to confusion with *P. pratense*. It is possible that *P. bertolonii* refers to a non-British tetraploid (2n=28), in which case **P. serotinum** Jord. would be our plant; ssp. *serotinum* is the correct name at ssp. level in either case.

3. P. alpinum L. (*P. commutatum* Gaudin) - *Alpine Cat's-tail*. Perennial; culms erect, to 50cm; panicles up to 5 x 1.2cm, ovoid to shortly cylindrical; glumes 5-8.5mm incl. awns, truncate at apex with awn 2-3mm; 2n=28. Native; grassy, rocky or mossy wet places on mountains at 600-1200m; very local in N Br from Westmorland to E Ross. Similar distribution to *Alopecurus borealis* and not rarely with it. **R**

4. P. phleoides (L.) H. Karst. - *Purple-stem Cat's-tail*. Perennial; culms erect, to 60cm, the lower sheaths often purplish-tinged; panicles up to 10 x 0.7cm, ± cylindrical but often narrowed at each end; glumes 2.5-3mm incl. awns, obtuse at apex with awn ≤0.5mm; 2n=14. Native; dry sandy and chalky pastures and adjacent rough ground; very local in CE En from Beds and Herts to W Norfolk and E Suffolk. **RR**

5. P. arenarium L. - *Sand Cat's-tail*. Annual; culms erect, to 20(30)cm; panicles up to 5 x 0.7cm, subcylindrical to narrowly ellipsoid; glumes 3-4mm incl. awns, acute or subacute at apex with awn ≤0.7mm; 2n=14. Native; maritime sand-dunes and inland on sandy heaths; frequent on most coasts of BI except N & W Sc, inland in E Anglia.

TRIBE 12 - BROMEAE (genera 56-59). Annuals or perennials without stolons, with or without rhizomes; ligule membranous; sheaths usually fused when young but soon splitting; inflorescence a panicle, rarely slightly contracted; spikelets with several to many bisexual florets, the apical 1 or 2 (sometimes more) often reduced and male or sterile; glumes 2, unequal, 1-9-veined, much shorter than rest of spikelet, unawned; lemmas 5-11-veined, rounded or keeled on back, usually minutely bifid at apex, usually with long subterminal awn; stamens 2-3; stigmas 2; lodicules 2; ovary with pubescent terminal appendage (the styles arising below it).

56. BROMUS L. - *Bromes*

Annuals; spikelets ovoid to narrowly so, terete or slightly compressed; lower glume 3-5(7)-veined; upper glume 5-7(9)-veined; lemmas 7-9(11)-veined, rounded on back, subacute to obtuse or rounded-obtuse, and minutely bifid at apex; stamens 3.

All the spp. are very variable in habit, becoming very small (often with only 1 spikelet) in dry conditions, and in extreme cases spikelet and lemma measurements also vary outside the normal range. Dwarf, starved plants should be avoided; more normal plants are usually nearby. Lemma measurements should be made from the middle or low part of spikelets.

1 Caryopsis thick, with inrolled margins; lemma with margins wrapped around caryopsis when mature, hence lemma margins not overlapping next higher lemma, but rhachilla ± revealed between florets; rhachilla disarticulating tardily 2
1 Caryopsis thin, flat or with weakly inrolled margins; lemma margins not wrapped round caryopsis, overlapping next higher lemma and obscuring rhachilla; rhachilla disarticulating readily 3
2 Spikelets 12-20mm, glabrous or pubescent; lemmas 6.5-9(10)mm; palea equalling lemma; caryopsis 6-9mm; sheaths usually glabrous or sparsely pubescent **8. B. secalinus**
2 Spikelets 8-12mm, glabrous; lemmas 5-6mm; palea shorter than lemma; caryopsis 4-4.5mm; sheaths pubescent **9. B. pseudosecalinus**
3 Palea divided nearly to base; panicle with mostly subsessile spikelets densely clustered in groups of 3; extinct **7. B. interruptus**
3 Palea entire to shortly bifid; panicle various but spikelets not subsessile in groups of 3 4
4 Anthers 3.5-5mm, ≥1/2 as long as lemmas; panicle branches long, forming very open panicle **1. B. arvensis**
4 Anthers 0.2-3mm, <1/2 as long as lemmas; panicles various, often not very open 5
5 Awns curved or bent outwards at maturity, their apices widely diverging 6
5 Awns ± straight to slightly flexuous or curved at maturity, the apices of those of the more apical lemmas ± parallel or even convergent (if ± curved outwards the culms procumbent to ascending) 9
6 Panicle-branches and pedicels much shorter than spikelets; pedicels ≤10mm; spikelets ≤18mm; awn ± cylindrical from base 7
6 At least some panicle-branches and pedicels on well grown plants longer than spikelets; pedicels often >15mm; spikelets ≥18mm; awn strongly flattened at base 8
7 Lemmas 6.5-8.5mm; culms usually <15cm, with usually <10 spikelets; maritime native **4c. B. hordeaceus ssp. ferronii**
7 Lemmas 8-11mm; culms usually >15cm, with usually >10 spikelets; alien **4b. B. hordeaceus ssp. divaricatus**
8 Lemmas 11-18mm; panicles usually rather stiffly erect **11. B. lanceolatus**
8 Lemmas 8-10mm; panicles usually lax with patent to pendent branches **10. B. japonicus**
9 Lemmas (4.5)5.5-6.5mm; caryopsis longer than palea **6. B. lepidus**
9 Lemmas 6.5-11mm; caryopsis shorter than to as long as palea 10
10 Panicle ± lax with at least some pedicels longer than spikelets; lemmas rather coriaceous, with rather obscure veins; anthers 1-3mm 11
10 Panicle ± dense, usually with all pedicels shorter than spikelets; lemmas papery, with prominent veins; anthers 0.5-1.5(2)mm 12
11 Lemmas 6.5-8mm; anthers mostly 1.5-3mm; spikelets 10-16mm; lowest rhachilla-segment mostly 0.7-1mm **3. B. racemosus**

11 Lemmas 8-11mm; anthers mostly 1-1.5mm; spikelets 15-28mm; lowest rhachilla-segment mostly 1.3-1.7mm **2. B. commutatus**
12 Lemmas 8-11mm, usually pubescent **4a. B. hordeaceus ssp. hordeaceus**
12 Lemmas 6.5-8mm, usually glabrous 13
13 Culms to 8(12)cm, procumbent to ascending; caryopsis shorter than palea; maritime dunes **4d. B. hordeaceus ssp. thominei**
13 Culms usually >10cm, usually erect; caryopsis c. as long as palea; widespread **5. B. x pseudothominei**

Other spp. - Several spp. are found as rare casuals on waste ground mainly from seed-mixes. The following 3, from S Europe, would key to couplet 6; they are related to *B. japonicus* and *B. lanceolatus*, with divergent awns strongly flattened at base. **B. squarrosus** L. differs from *B. japonicus* in its simple 1-sided inflorescence, longer spikelets (to 70mm, not to 40mm) and wider lemmas (6-8mm, not 3-5mm). **B. scoparius** L. and **B. alopecuros** Poir. differ from all 3 spp. in their dense panicle with pedicels much shorter than spikelets; *B. scoparius* has lemmas 6.5-11mm with awns 6.5-12mm, *B. alopecuros* has lemmas 12-15mm with awns 12-15mm. **B. pectinatus** Thunb., from S Africa, is very close to *B. japonicus*, but has acute (not obtuse) lobes at lemma-tip and its awns are not recurved. **B. briziformis** Fisch. & C.A. Mey., from SW Asia, is very distinctive in its inflated, pendent *Briza*-like spikelets with awns 0-1mm.

1. B. arvensis L. - *Field Brome*. Culms erect, to 1m; panicle diffuse, ± pendent at maturity, with branches much longer than spikelets; lemmas 7-9mm, ± glabrous, with straight awn 6-10mm; (2n=14). Intrd; usually casual weed of arable and grass fields and waste ground; very scattered in C & S Br (mostly C & E En), decreasing, extremely sporadic elsewhere, natd in E Gloucs; Europe.

2. B. commutatus Schrad. - *Meadow Brome*. Culms erect, to 1m; panicle ± diffuse, drooping to 1 side at maturity, with most branches longer than spikelets; lemmas 8-11mm, usually glabrous (rarely pubescent), with straight awn 3-10mm; (2n=14, 28). Native; grassy places, waysides and rough ground, especially in damp rich meadows; locally frequent in C & S Br, decreasing, very scattered casual elsewhere. Perhaps better as *B. racemosus* ssp. *commutatus* (Schrad.) Syme.

2 x 3. B. commutatus x B. racemosus occurs rather frequently with the parents in C & S Br; it is fertile and backcrosses, forming a spectrum of intermediates between the parents and making them difficult to separate in some areas.

3. B. racemosus L. - *Smooth Brome*. Culms erect, to 1m; panicle fairly open but rather narrow and erect, sometimes drooping to 1 side at maturity, with many branches longer than spikelets; lemmas 6.5-8mm, glabrous, with straight awn 5-9mm; 2n=28. Native; similar places and distribution to *B. commutatus*.

4. B. hordeaceus L. - *Soft-brome*. Culms erect to procumbent, to 80cm; panicle rather to very compact, erect, sometimes drooping to 1 side at maturity, with branches shorter than spikelets; lemmas 6.5-11mm. The 4 sspp. are separated in the key to spp.

a. Ssp. hordeaceus (*B. mollis* L.). Stems to 80cm (often much less), erect; panicle up to 10(16)cm, with few to many spikelets, often drooping to 1 side at maturity; lemmas 8-11mm, pubescent, with ± straight awn 4-11mm; 2n=28. Native; grassy places, waysides, rough ground; frequent throughout lowland BI.

b. Ssp. divaricatus (Bonnier & Layens) Kerguélen (ssp. *molliformis* (Jn. Lloyd) Maire & Weiller, *B. molliformis* Jn. Lloyd). Stems to 60cm, erect; panicle up to 10cm, with many spikelets, stiffly erect, with very short branches and pedicels; lemmas 8-11mm, pubescent, with awn 5-10mm and curved outwards at maturity; (2n=28). Intrd; casual from wool, grass-seed and other sources in waste places and waysides; scattered in En; S Europe.

c. Ssp. ferronii (Mabille) P.M. Sm. (*B. ferronii* Mabille). Stems to 15(20)cm, erect to ascending; panicle up to 5cm, with few spikelets, stiffly erect; lemmas 6.5-8(8.5)mm, pubescent, with awn 2-6mm and curved outwards at maturity; 2n=28. Native; grassy cliff-tops and sandy or shingly ground by sea; locally frequent on coasts of CI and S & SW Br scattered N to Kirkcudbrights and Angus, Man.

d. Ssp. thominei (Hardouin) Braun-Blanq. (*B. thominei* Hardouin). Culms to 8(12)cm, procumbent to ascending; panicle up to 3cm, with few spikelets, erect; lemmas 6.5-7.5mm, glabrous (usually) or pubescent, with straight or slightly curved-out awn 3-7mm; (2n=28). Native; sandy places by sea; coasts of CI and Br probably N to Sc, but much over-recorded for *B. x pseudothominei* and distribution uncertain.

5. B. x pseudothominei P.M. Sm. (*B. thominei* auct. non Hardouin; *B. hordeaceus* x *B. lepidus*) - *Lesser Soft-brome.* Culms erect, to 60cm; panicle ± compact, erect, with all branches shorter than spikelets; lemmas 6.5-8mm, usually glabrous, with straight awn 3-7mm; (2n=28). Probably native; grassland, waysides, rough ground, usually with *B. hordeaceus* and often with *B. lepidus* (and commoner than it), but often without either close by; scattered throughout BI, frequent in C & S Br.

6. B. lepidus Holmb. - *Slender Soft-brome.* Culms erect, to 80cm; panicle compact, erect, with branches shorter than spikelets; lemmas (4.5)5.5-6.5mm, usually glabrous, with straight awn 2-5.5mm; 2n=28. Probably intrd; grassland, waysides and rough ground; scattered throughout BI, frequent in C & S Br; origin uncertain.

7. B. interruptus (Hack.) Druce - *Interrupted Brome.* Culms erect, to 1m; panicle compact, often spaced out below, erect, with very short branches and spikelets often tightly clustered in threes; lemmas 7.5-9mm, pubescent, with straight awn 3-8mm; 2n=28. Native; arable and waste land, especially as weed in *Onobrychis*, *Lolium* or *Trifolium* crops; formerly scattered in SC and SE En, from E Kent and N Somerset to S Lincs, last seen in 1972 in Cambs but retained in cultivation; endemic. **RR**

8. B. secalinus L. - *Rye Brome.* Culms erect, to 1.2m; panicle rather diffuse to ± compact, erect or drooping to 1 side at maturity, with some branches longer than spikelets; lemmas 6.5-9(10)mm, glabrous or (usually) pubescent, with straight awn 0-8mm; (2n=28). Intrd; weed of cereals, marginal and waste ground, much decreased and now an infrequent casual; scattered throughout most of BI except N & W Sc, formerly frequent in C & S En; Europe.

9. B. pseudosecalinus P.M. Sm. - *Smith's Brome.* Culms erect, to 60cm; differs from *B. secalinus* in awn 2-6mm; 2n=14; and see key (couplet 2). Intrd; grassy fields and waysides, probably a grass-seed contaminant; very scattered in Br and Ir; origin uncertain.

10. B. japonicus Thunb. ex Murray - *Thunberg's Brome.* Culms erect, to 80cm; **875**
panicle diffuse, ± pendent at maturity, with branches usually much longer than spikelets; lemmas 8-10mm, glabrous or pubescent, with awns 4-14mm and widely divergent at maturity; (2n=14). Intrd; sporadic casual in waste places from birdseed, wool and other sources; scattered in En; Europe.

11. B. lanceolatus Roth (*B. macrostachys* Desf.) - *Large-headed Brome.* Culms erect, **875**
to 70cm; panicle rather dense, erect, with most branches shorter but some usually longer than spikelets; lemmas 11-18mm, densely pubescent, with awns 6-20mm and widely divergent at maturity; (2n=28). Intrd; sporadic casual in waste places from wool, birdseed and as garden escape; scattered in C & S Br; S Europe.

57. BROMOPSIS (Dumort.) Fourr. (*Zerna* auct. non Panz., *Bromus* sect. *Pnigma* Dumort.) - *Bromes*

Perennials with long to very short rhizomes; spikelets narrowly oblong then tapered to apex, terete or slightly compressed; lower glume 1(-3)-veined; upper glume 3(-5)-veined; lemmas 5-7-veined, rounded or slightly keeled on back, acute to shortly acuminate and minutely bifid at apex; stamens 3.

1 Inflorescence very lax, the branches pendent or all swept to 1 side;

leaf-sheaths with distinct pointed auricles at apex 2
1 Inflorescence dense to fairly lax, the branches erect to erecto-patent; leaf-sheaths without or with short rounded auricles at apex 3
2 Lowest panicle-node with usually >2 branches, some with 1 or very few spikelets, and with small ± glabrous scale; panicle branches swept to 1 side **2. B. benekenii**
2 Lowest panicle-node usually with 2 branches, both long and with >3 spikelets, and with small pubescent scale; panicle branches pendent **1. B. ramosa**
3 Plant densely tufted, with short rhizomes; lemmas with awns (2)3-8mm; leaves of tillers usually folded or inrolled along long axis **3. B. erecta**
3 Plant not densely tufted, with long rhizomes; lemmas awnless or less often with awns up to 3(6)mm; leaves of tillers usually flat **4. B. inermis**

1. B. ramosa (Huds.) Holub (*Bromus ramosus* Huds., *Zerna ramosa* (Huds.) Lindm.) - *Hairy-brome*. Rhizomes very short; culms to 2m; leaf-sheaths all or sometimes all except uppermost with long patent to down-pointed hairs; panicle with pendent branches at maturity; lemmas 10-14mm, with awns 4-8mm; 2n=42. Native; woods, wood-margins and hedgerows; frequent throughout most of lowland BI except Jersey and Outer Isles. See *Festuca gigantea* for differences.

2. B. benekenii (Lange) Holub (*Bromus benekenii* (Lange) Trimen, *Zerna benekenii* R
(Lange) Lindm.) - *Lesser Hairy-brome*. Differs from *B. ramosa* in culms to 1.2m; uppermost leaf-sheath usually glabrous; 2n=28; and see key. Native; similar places to *B. ramosa*, which often accompanies it; very scattered in mainland Br, perhaps overlooked. Distinction from *B. ramosa* needs investigation; ssp. status might be preferable.

3. B. erecta (Huds.) Fourr. (*Bromus erectus* Huds., *Zerna erecta* (Huds.) Gray) - *Upright Brome*. Rhizomes very short; culms to 1.2m; leaf-sheaths glabrous or with patent hairs; panicle ± erect; lemmas 8-15mm, with awn (2)3-8mm; (2n=28, 42, 56, 70, 84. 112). Native; dry grassland and grassy slopes, especially on calcareous soils; common on base-rich soils in C, S & E En, scattered N to Fife and W to W Cornwall, Pembs and Caerns, very scattered in Ir and CI.

4. B. inermis (Leyss.) Holub - *Hungarian Brome*. Rhizomes long; culms to 1.5m; leaf-sheaths glabrous or sometimes with patent hairs; panicle ± erect; lemmas (7)9-13(16)mm. Intrd.

a. Ssp. inermis (*Bromus inermis* Leyss., *Zerna inermis* (Leyss.) Lindm.). Leaf- **889**
sheaths usually glabrous; culm-nodes glabrous or with short hairs just below; lemmas glabrous to scabrid or with sparse hairs on margins, with awn 0(-3)mm; (2n=28, 42, 56, 70). Formerly sown for fodder, now mostly a seed-contaminant, natd (and casual) in rough grassy places, waysides and field-margins; scattered in Br; Europe.

b. Ssp. pumpelliana (Scribn.) W.A. Weber (*B. pumpelliana* (Scribn.) Holub, *Bromus pumpellianus* (Scribn.) Wagnon). Leaf-sheaths usually pubescent; culm-nodes pubescent; lemmas appressed-pubescent on margins, with awn 0-6mm; (2n=42, 56). Natd in 1 site in S Essex: N America. A ± distinct ssp. in N America (where ssp. *inermis* is well natd), but in Br the 2 sspp. merge and probably hybrids rather than true ssp. *pumpelliana* occur.

58. ANISANTHA K. Koch (*Bromus* sect. *Genea* Dumort.) - *Bromes*
Annuals; spikelets ± parallel-sided or widening distally, slightly compressed; lower glume 1(-3)-veined; upper glume 3(-5)-veined; lemmas 7-veined, rounded on back, acute to acuminate and minutely bifid at apex; stamens 2 or 3.

Lemma-lengths should be measured on only the two basal florets.

1 Lemmas 20-36mm 2
1 Lemmas 9-20mm 3

2 Panicle lax, with branches spreading laterally or pendent; callus-scar at base of lemma ovate, rounded **1. A. diandra**

2 Panicle dense, with erect branches; callus-scar at base of lemma elliptic, pointed **2. A. rigida**

3 Panicle lax, with branches spreading laterally or pendent 4

3 Panicle dense, with stiffly erect branches 5

4 Lemmas 9-13mm: inflorescence compound, the larger branches with 3-8 spikelets (except in depauperate plants); spikelets with >3 apical sterile florets **4. A. tectorum**

4 Lemmas 13-20mm; inflorescence simple or the larger branches slightly branched and with up to 3(5) spikelets; spikelets with 1-2 sterile apical florets **3. A. sterilis**

5 Lemmas mostly 12-20mm; spikelets with 1-2(3) sterile apical florets **5. A. madritensis**

5 Lemmas 9-13(15)mm; spikelets with >(2)3 apical sterile florets **6. A. rubens**

1. A. diandra (Roth) Tutin ex Tzvelev (*A. gussonei* (Parl.) Nevski, *Bromus diandrus* Roth) - *Great Brome.* Stems to 80cm; inflorescence lax, with spikelets pendent at maturity; longer branches longer than spikelets, bearing mostly 1-2 spikelets; lemmas 20-36mm, with awns 25-60mm; (2n=56). Intrd; natd in rough and waste ground, waysides and open grassland on warm sandy soils; frequent and well natd in CI and E Anglia, very scattered and usually casual (especially from wool and grain) elsewhere in Br N to C Sc; Europe.

2. A. rigida (Roth) Hyl. (*Bromus rigidus* Roth) - *Ripgut Brome.* Stems to 60cm; inflorescence stiffly erect; all branches shorter than spikelets; lemmas 20-25mm, with awns 25-50mm; (2n=42). Intrd; similar places to *A. diandra;* rather infrequently natd in CI and S Br, rare casual (mainly from same sources) elsewhere; Europe. Despite statements in the literature that *A. diandra* and *A. rigida* differ in anther length most British plants of both have anthers 0.7-1.2mm long. The two taxa are probably better treated as sspp. of 1 sp.

3. A. sterilis (L.) Nevski (*Bromus sterilis* L.) - *Barren Brome.* Stems to 80cm; inflorescence as in *A. diandra*; lemmas 13-20mm, with awns 15-35mm; 2n=14. Native; rough and waste ground, waysides, open grassland, weed of arable land and gardens; throughout lowland BI, common in C & S Br, scattered in Sc and Ir.

4. A. tectorum (L.) Nevski (*Bromus tectorum* L.) - *Drooping Brome.* Stems to 60cm; inflorescence lax, with spikelets pendent at maturity; longer branches longer than spikelets, bearing mostly 4-8 spikelets; spikelets with apical group of c.4-6 sterile florets becoming dispersed with topmost fertile floret; lemmas 9-13mm, with awns 10-18mm; (2n=14). Intrd; similar places to *A. diandra;* natd in W Suffolk and W Norfolk, infrequent casual (mainly from same sources) elsewhere; Europe.

5. A. madritensis (L.) Nevski (*Bromus madritensis* L.) - *Compact Brome.* Stems to 60cm; inflorescence stiffly erect; all or at least many branches shorter than spikelets; lowest panicle node with shortest branch >5mm and longest branch 0-1-branched; lemmas mostly 12-20mm, with awns 12-25mm; (2n=28). Intrd; natd in similar places to *A. diandra;* locally natd in SW En, S Wa and CI, occasional casual (especially from wool) elsewhere in Br N to C Sc and S Ir; Europe.

6. A. rubens (L.) Nevski (*Bromus rubens* L.) - *Foxtail Brome.* Stems to 40cm; 889
inflorescence stiffly erect; all branches shorter than spikelets; lowest panicle node with shortest branch ≤6mm and longest branch 2-5-branched; spikelets with apical group of (3)4-6 sterile florets becoming dispersed with topmost fertile floret; lemma 9-13(15)mm, with awns 8-15mm; (2n=28). Intrd; rather infrequent grain- and wool-alien, casual in waste places; sporadic and very scattered in Br; Europe.

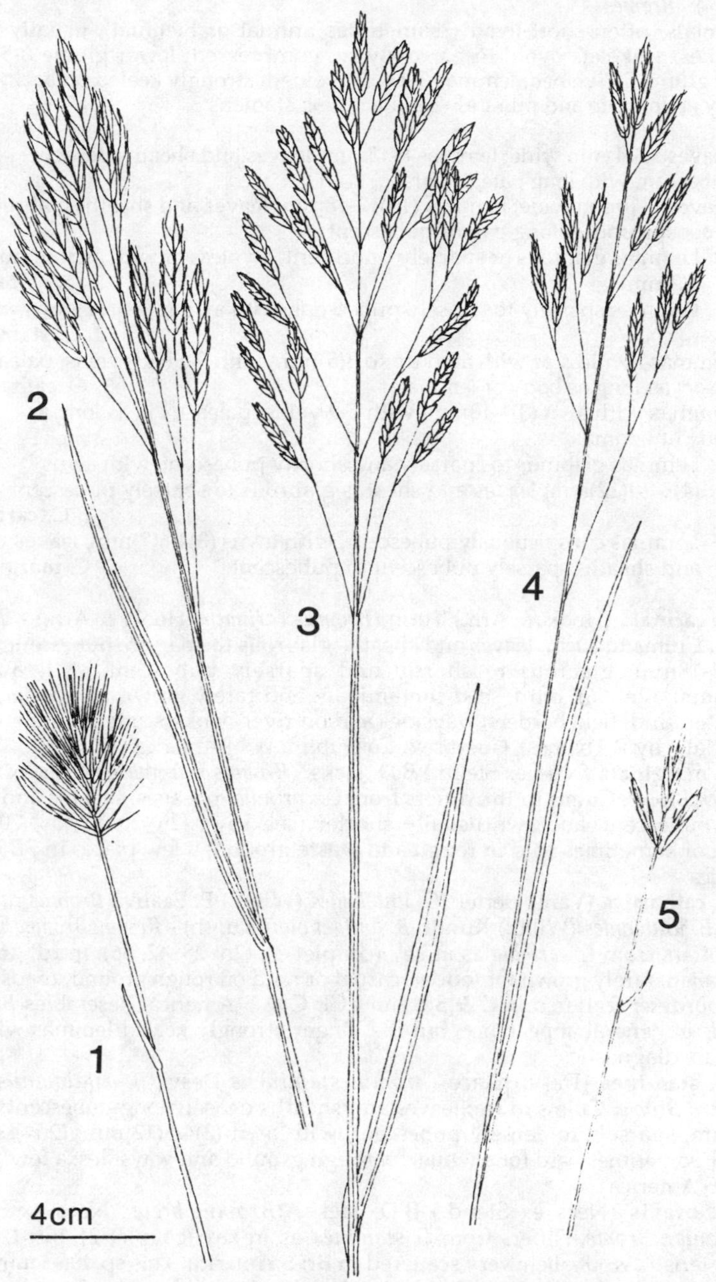

FIG 889 - Poaceae: Pooideae. 1, *Anisantha rubens*. 2, *Ceratochloa carinata*. 3, *Bromopsis inermis* ssp. *inermis*. 4, *Ceratochloa cathartica*. 5, *Brachypodium distachyon*.

59. CERATOCHLOA DC. & P. Beauv. (*Bromus* sect. *Ceratochloa* (DC. & P. Beauv.) Griseb.) - *Bromes*

Perennials, often short-lived (?sometimes annual or biennial), usually without rhizomes; spikelets ovoid to narrowly so, compressed; lower glume 3-5-veined; upper glume 5-7-veined; lemmas 7-11(13)-veined, strongly keeled on back, acute to shortly acuminate and minutely bifid at apex; stamens 3.

1 Leaves <3(4)mm wide; lemmas 8-13mm; leaves and sheaths densely pubescent with long patent hairs 2

1 Leaves 4-10mm wide; lemmas (10)12-18mm; leaves and sheaths glabrous to conspicuously long-patent-pubescent 3

2 Lemmas glabrous or sparsely pubescent, awnless or with awn up to 1(2)mm **5. C. brevis**

2 Lemmas sparsely to densely pubescent, with awn (3)4-8(12)mm **4. C. staminea**

3 Lemmas awnless or with awn up to 3(5)mm, with 9-11(13) veins; palea 1/2-3/4 as long as body of lemma **3. C. cathartica**

3 Lemmas with awn (3)4-10mm, with 7-9 veins; palea 3/4-lx as long as body of lemma 4

4 Lemmas glabrous to sparsely and shortly pubescent, with awns (4)6-10(12)mm; leaves and sheaths glabrous to sparsely pubescent **1. C. carinata**

4 Lemmas conspicuously pubescent, with awns (3)4-6(7)mm; leaves and sheaths sparsely pubescent to pubescent **2. C. marginata**

1. C. carinata (Hook. & Arn.) Tutin (*Bromus carinatus* Hook. & Arn.) - *California* 889
Brome. Culms to 80cm; leaves and sheaths glabrous to sparsely pubescent; lemmas (12)14-18mm, glabrous to shortly and sparsely pubescent, with awn (4)6-10(12)mm; (2n=56). Intrd; seed contaminant and rarely grown for fodder, natd in rough ground, field borders, waysides and on river-banks; scattered in Br, mostly S (especially by R Thames), Guernsey, Co Dublin; W N America.

2. C. marginata (Nees ex Steud.) B.D. Jacks. (*Bromus marginatus* Nees ex Steud.) - *Western Brome*. Culms to 1m; differs from *C. carinata* in leaves, sheaths and lemmas more pubescent, and awn usually shorter (see key); (2n=28, 42, 56, 70). Intrd; casual or sometimes natd in rough and waste ground; a few places in SE En; W N America.

3. C. cathartica (Vahl) Herter (*C. unioloides* (Willd.) P. Beauv., *Bromus catharticus* 889
Vahl, *B. unioloides* (Willd.) Kunth, *B. willdenowii* Kunth) - *Rescue Brome*. Culms to 1m; differs from *C. carinata* as in key (couplet 3); (2n=28, 42, 56). Intrd; grain- and wool-alien, rarely grown for fodder, casual or natd on rough ground, roadsides and field-borders; scattered in C & S Br and CI; C & S America. Resembles *Bromopsis inermis* in general appearance, but the longer strongly keeled lemmas with more veins are diagnostic.

4. C. staminea (Desv.) Stace (*Bromus stamineus* Desv., *B. valdivianus* Phil.) - *Southern Brome*. Culms to 1m; leaves and sheaths densely long-pubescent; lemmas 8-13mm, sparsely to densely pubescent, with awn (3)4-8(12)mm; (2n=42). Intrd; casual, sometimes natd for a while, on rough ground and waysides; a few places in S En; S America.

5. C. brevis (Nees ex Steud.) B.D. Jacks. (*Bromus brevis* Nees ex Steud.) - *Patagonian Brome*. Differs from *C. staminea* as in key (couplet 2). Intrd; a rather characteristic wool-alien; very scattered in Br; S America. This sp. has lemmas with curved keels and hence very closely overlapping (usually awnless) tips, producing a very neat narrowly ovate outline to the spikelet distinct from that of all other spp.

TRIBE 13 - BRACHYPODIEAE (genus 60). Annuals or rhizomatous perennials;

ligule membranous; sheaths not fused; inflorescence a raceme with usually 1 spikelet at each node with pedicels ≤2(2.8)mm; spikelets with many bisexual florets, the apical 1 or 2 reduced and male or sterile; glumes 2, unequal, 3-9-veined, much shorter than rest of spikelet, sometimes shortly awned; lemmas mostly 7-veined, rounded on back, acute to acuminate and usually with short to long awn; stamens 3; stigmas 2; lodicules 2; ovary with pubescent terminal appendage (the styles arising from below it).

60. BRACHYPODIUM P. Beauv. - *False Bromes*

1 Annual; anthers 0.3-1mm; spikelets distinctly compressed **3. B. distachyon**
1 Perennials with rhizomes and tillers; anthers 3-6(8)mm; spikelets subterete 2
  2 Plant weakly rhizomatous, usually densely tufted; culms with 4-8 internodes; raceme usually pendent at apex; leaves usually >6mm wide; lemmas with awns 7-15mm **2. B. sylvaticum**
  2 Plant strongly rhizomatous, usually scarcely tufted; culms with 3-4(5) internodes; raceme usually erect; leaves usually <6mm wide; lemmas with awns 1-5mm **1. B. pinnatum**

Other spp. - B. rupestre (Host) Roem. & Schult. was reported from S & C En and S Wa in 1991 but requires confirmation; it is close to *B. pinnatum*.

1. B. pinnatum (L.) P. Beauv. - *Tor-grass*. Extensively rhizomatous; culms to 1.2m, usually erect; leaves usually 3-6mm wide; sheaths usually glabrous, sometimes pubescent; raceme with (3)6-15 spikelets; lemmas 6-11mm, often glabrous, with awn 1-5mm; 2n=28. Native; grassland, mainly on chalk and limestone; common and often dominant in suitable places in C, S & E En, very scattered in NW & SW En, Wa and Ir, sporadic and usually casual in Sc and CI, apparently increasing.

1 x 2. B. pinnatum x B. sylvaticum = B. x cugnacii A. Camus is probably the identity of some variably sterile intermediates found very scattered in En and Ir with the parents, but they have not been cytologically confirmed and all intermediates examined have the chromosome number of one or other parent. Although the 2 spp. are easily recognized once familiar, no one character can unequivocally separate them. The best single character is awn length, with the other key characters in support; pubescence is of very limited value.

2. B. sylvaticum (Huds.) P. Beauv. - *False Brome*. Shortly (or scarcely) rhizomatous; culms to 1m; usually pendent at tip of raceme; leaves usually 6-10mm wide; sheaths usually pubescent, sometimes glabrous; raceme with (3)5-12 spikelets; lemmas 6-12mm, rarely glabrous, with awn 7-15mm; 2n=18. Native; woods, scrub and shady wood-borders and hedgerows, in open grassland mainly in the N; common throughout BI except much of N Sc.

3. B. distachyon (L.) P. Beauv. - *Stiff Brome*. Stiffly erect annual to 15cm; leaves **889**
mostly 2-4mm wide, usually pubescent; sheaths usually glabrous; raceme with 1-4(6) spikelets; lemmas 7-10mm, glabrous or pubescent, with awn 7-15mm; (2n=10, 20, 30). Intrd; casual in waste places from wool or rarely grain; very scattered in En; Mediterranean.

TRIBE 14 - TRITICEAE (Hordeeae) (genera 61-68). Annuals or perennials with or without rhizomes, without stolons; ligule membranous; sheaths not fused; inflorescence a spike with 1-3 spikelets at each node; spikelets with 1-many florets, often some florets (or some spikelets if spikelets >1 per node) male or sterile; glumes 2, equal or slightly unequal, 1-11-veined, often with long terminal awn; lemmas mostly 5-7-veined, rounded on back, very acute to obtuse, awnless or with terminal awn; stamens 3; stigmas 2; lodicules 2; ovary usually pubescent or pubescent at apex, sometimes with pubescent terminal appendage (the styles arising from below it).

61. ELYMUS L. (*Agropyron* auct. non Gaertn., *Roegneria* K. Koch) - *Couches*
Perennials without rhizomes; spikelets 1 per node, with several to many florets with all but the apical 1 or 2 bisexual, flattened broadside on to rhachis; glumes 2-5-veined, acute or narrowly acute, often awned; lemmas 5-veined, usually long-awned, sometimes awnless or short-awned; spikelets breaking up below each lemma at maturity, leaving glumes on rhachis.

Other spp. - **E. canadensis** L., from N America, was formerly natd in Middlesex; it would key out as *E. caninus* but the glumes both have awns c. as long as the body; it still occurs as a rare casual on waste ground.

1. E. caninus (L.) L. (*Agropyron caninum* (L.) P. Beauv., *A. donianum* F.B. White, *Elymus trachycaulus* (Link) Gould ex Shinners ssp. *donianus* (F.B. White) Á. Löve). - *Bearded Couch*. Culms to 1.2m; leaves mostly flat, 4-12mm wide; spikelets ± contiguous; glumes very acute, finely pointed with awn ≤4mm, reaching >1/2 way up body of adjacent lemma; lemmas gradually narrowed to awn 7-18mm or rarely with awn 1-3(7)mm; 2n=28. Native; woods, hedgerows, shady river-banks and mountain gullies and cliff-ledges; scattered throughout Br and Ir but absent from large parts of C & N Sc and Ir. Var. **donianus** (Buch.-White) Melderis, from mountain rock-ledges and -crevices in C & N Sc, has awns 0-3mm but forms fertile hybrids with var. **caninus** that show every degree of intermediacy.

2. E. scabrus (Labill.) Á. Löve (*Agropyron scabrum* (Labill.) P. Beauv.) - *Australian Couch*. Culms to 1.2m; leaves mostly flat, 2-6mm wide, at least in lower part of spike; spikelets ± distant, each (excl. awns) not reaching as far as next spikelet on opposite side of rhachis; glumes acute, unawned, reaching ≤1/2 way up body of adjacent lemma; lemmas gradually narrowed to awn 10-25mm; (2n=28, 42). Intrd; occasional wool-alien of fields and waste ground; scattered in En; Australia. **896**

62. ELYTRIGIA Desv. (*Agropyron* auct. non Gaertn., *Thinopyrum* Á. Löve) - *Couches*
Perennials with long rhizomes; spikelets 1 per node, with several to many florets with all but the apical 1-2 bisexual, flattened broadside on to rhachis; glumes 3-11-veined, acute to very obtuse, rarely awned; lemmas 5-veined, unawned or with short to rarely long awns; spikelets not breaking up easily at maturity, usually falling whole or rhachis breaking up.

1 Rhachis breaking up between each spikelet at maturity, smooth; ribs on leaf upperside densely minutely pubescent — **3. E. juncea(*)**
1 Rhachis not breaking up between each spikelet at maturity, scabrid; ribs on leaf upperside glabrous to scabrid, sometimes with sparse long hairs — 2
  2 Leaf-sheaths with glabrous margin; leaves usually flat when fresh and turgid, their upperside ribs with rounded tops — **1. E. repens(*)**
  2 At least middle and lower leaf-sheaths with minute (often sparse) fringe of hairs on exposed free margin; leaves usually inrolled even when fresh, their upperside ribs ± flat-topped — **2. E. atherica(*)**

(*)Also hybrids of *E. juncea* with *E. repens* or *E. atherica*.

1. E. repens (L.) Desv. ex Nevski - *Common Couch*. Culms to 1.5m; spikelets contiguous to rather distant; glumes acute, unawned or rarely awned, with 3-7 veins; lemmas obtuse to narrowly acute, awnless or with awn rarely up to 15mm. Native.

a. Ssp. repens (*Agropyron repens* (L.) P. Beauv., *Elymus repens* (L.) Gould). Leaves mostly flat when fresh, green or glaucous, basal ones 3-10mm wide, upper ones 2-8.5mm wide, with fine well-spaced ribs on upperside; spikes up to 20(30)cm; spikelets 10-20mm, with 3-8 florets; glumes 7-12mm, with 3-7 veins, usually awnless; lemmas 8-13mm, usually awnless but sometimes with awn up to 15mm (var. **aristata** ined.); 2n=42. Cultivated, waste and rough ground; common throughout BI, incl. a range of maritime hahitats.

b. Ssp. arenosa (Spenn.) Á. Löve (*Elymus repens* ssp.*arenosus* (Spenn.) Melderis, *Agropyron maritimum* Jansen & Wacht. non (L.) P. Beauv.). Leaves mostly inrolled, ± glaucous, basal ones 3-5mm wide, upper ones 1-3.4mm wide, with thick (0.15-0.2mm wide), close ribs on upperside; spikes up to 9cm; spikelets (7)9-14mm, with (2)3-6 florets; glumes (4.4)6-9(10)mm, with 3(-7) veins, usually with awn (0.2)1-2.3mm; lemmas (5.5)7-10(12)mm, with awn (0.3)1.8-2.8(4.6)mm. Maritime sand, often on dunes; S & E coasts of En from S Somerset to W Norfolk, Guernsey.

1 x 2. E. repens x E. atherica = E. x oliveri (Druce) Kerguélen ex Carreras Mart. (*Agropyron x oliveri* Druce, *Elymus x oliveri* (Druce) Melderis & D.C. McClint.) occurs with the parents in scattered places round the coasts of Br from S En to N Sc and in SW Ir, but is probably not common; it has intermediate leaf-ribs and sheath-margin pubescence, and is sterile, with indehiscent anthers and empty pollen.

1 x 3. E. repens x E. juncea = E. x laxa (Fr.) Kerguélen (*Elymus x laxus* (Fr.) Melderis & D.C. McClint., *Agropyron x laxum* (Fr.) Tutin) occurs rather frequently with the parents on coasts of Br N of Merioneth and SE Yorks, and on the coasts of Ir and CI, but is rare in S 1/2 of Br; it is sterile and has a tardily breaking smooth or slightly scabrid rhachis, distant lower spikelets, and conspicuously scabrid to ± pubescent leaf-ribs; 2n=35. Distinguished from *E. x obtusiuscula* by the glabrous free margin of the sheaths.

2. E. atherica (Link) Kerguélen ex Carreras Mart. (*Elymus athericus* (Link) Kerguélen, *E. pycnanthus* (Godr.) Melderis, *Agropyron pycnanthum* (Godr.) Godr., *A. pungens* auct. non (Pers.) Roem. & Schult.) - *Sea Couch*. Culms to 1.2m; leaves usually inrolled even when fresh, mostly 2-9mm wide, with prominent, rather crowded, ± flat-topped ribs on upperside; spikelets ± contiguous; glumes acute to obtuse, sometimes acuminate, rarely awned, with 4-7 veins; lemmas acute to obtuse, awnless or with awn rarely up to 10mm (var. **setigera** ined.); 2n=42. Native; wet sandy, gravelly or muddy places by sea, often at margins of dunes, creeks or salt-marshes; frequent round coasts of BI N to Co Louth, Dumfriess and NE Yorks. The leaves are often very glaucous, but this is equally true of maritime variants of *E. repens*; the 2 spp. are often very difficult to distinguish, especially late in the season when the leaf-sheath fringe of hairs may wear off.

2 x 3. E. atherica x E. juncea = E. x obtusiuscula (Lange) Hyl., (*Elymus x obtusiusculus* (Lange) Melderis & D.C. McClint., *Agropyron x obtusiusculum* Lange) occurs rather frequently with the parents on coasts of Br N to Durham and Cumberland, and in SW Ir and CI; it closely resembles *E.* x *laxa* but has some minute hairs on free margin of the leaf-sheaths; 2n=35.

3. E. juncea (L.) Nevski (*Agropyron junceum* (L.) P. Beauv., *Elymus farctus* (Viv.) Runemark ex Melderis, non *E. junceus* Fisch.) - *Sand Couch*. Culms to 60(80)cm; leaves usually inrolled, mostly 2-6mm wide, with prominent, crowded, densely and minutely pubescent ribs on upperside; spikelets distant, at least in lower part of spike; glumes subacute to obtuse, unawned, with 7-11 veins; lemmas obtuse, awnless; 2n=28. Native; maritime sand-dunes; common round coasts of BI. Our plant is ssp. **boreoatlantica** (Simonet & Guin.) Hyl. (*Agropyron junceiforme* (Á. & D. Löve) Á. & D. Löve, *Thinopyrum junceiforme* (Á. & D. Löve) Á. Löve, *Elymus farctus* ssp. *boreoatlanticus* (Simonet & Guin.) Melderis).

62 x 65. ELYTRIGIA x HORDEUM = X ELYTRORDEUM Hyl.

62/1 x 65/8. X E. langei (K. Richt.) Hyl. (*X Elyhordeum langei* (K. Richt.) Melderis, 896
X Agrohordeum langei (K. Richt.) E.G. Camus ex A. Camus; *E. repens* x *H. secalinum*)
has been found in wet meadows, fields and roadsides in a few scattered localities

in En from Scillies to Northumb. It exists in two variants: 1 clearly intermediate (known only in 1 place in W Gloucs, 1945-1954), differing from *E. repens* in that the rhachis disarticulates and the glumes and lemmas are awned, and from *H. secalinum* in having rhizomes and mostly 1 spikelet with 2-4 florets per node; the other close to *E. repens* var. *aristata*, from which it differs in that the rhachis disarticulates and the spikelets have 3-5 florets. Both sorts are usually sterile, but the latter sometimes produces some good pollen and caryopses; (2n=49).

63. LEYMUS Hochst. (*Elymus* auct. non L.) - *Lyme-grass*
Perennials with long rhizomes; spikelets 2(-3) per node, with 3-6 florets, with all but the most apical bisexual, flattened broadside on to rhachis; glumes 3-5-veined, finely pointed but not awned; lemmas mostly 7-veined, acute, unawned; spikelets breaking at maturity below each lemma.

1. L. arenarius (L.) Hochst. (*Elymus arenarius* L.) - *Lyme-grass*. Culms to 1.5(2)m, very glaucous; leaves flat, becoming inrolled, 8-20mm wide; spike up to 35cm, dense; spikelets 20-32mm, overlapping; 2n=56. Native; mobile sand on maritime dunes, rarely casual or natd inland; frequent round coasts of most of BI, but absent from large parts of S En and S Ir and from CI.

64. HORDELYMUS (Jess.) Jess. ex Harz - *Wood Barley*
Perennials with very short rhizomes; spikelets (2-)3 per node, with 1(-2) florets, all bisexual; glumes 1-3-veined, finely pointed and with long awn, each pair fused at base; lemmas 5-veined, with very long awn; spikelets breaking at maturity above the glumes, but often not until late autumn.

1. H. europaeus (L.) Jess. ex Harz - *Wood Barley*. Culms to 1.2m; leaves flat, 5- **R**
14mm wide; spike 5-10cm, dense; lemmas 8-10mm with awns 15-25mm; (2n=28). Native; woods and copses; local in Br N to S Northumb, formerly Co Antrim and Berwicks.

65. HORDEUM L. (*Critesion* Raf.) - *Barleys*
Annuals or less often perennials without rhizomes; spikelets 3 per node, each with 1 floret, the central spikelet bisexual, the laterals bisexual, male or sterile; glumes very narrow, 1-3-veined, with long awn; lemmas of bisexual florets 5-veined, with very long awn; rhachis breaking up at each node at maturity, or (in cultivated taxa) below each bisexual lemma.

Most of the spp. are superficially very similar. For accurate identification a *triplet* of spikelets (the 3 spikelets at 1 rhachis node) from near the middle of the spike should be isolated, and the 6 glumes and 3 florets identified. The spikelets, florets and lemmas are here referred to as central or lateral according to their position in the triplet.

1 Rhachis not breaking up at maturity, the caryopsis-containing florets breaking away from the rest of the spikelet which remains on the rhachis; awns of central lemmas usually >10cm (rarely very short) 2
1 Rhachis breaking up at maturity, the triplet of spikelets forming the dispersal unit; all awns ≤10cm 3
  2 All 3 florets in each triplet producing a caryopsis and with a long-awned lemma **1. H. vulgare**
  2 Only central floret of each triplet producing a caryopsis; lateral lemmas awnless or ± so **2. H. distichon**
3 Glumes of lateral spikelets >3cm, awn-like from base to apex; lateral florets extremely reduced, usually simply an awn-like outgrowth; awn of central lemma usually >5cm **6. H. jubatum**
3 Glumes of lateral spikelets <3cm, if >2cm then at least 1 of each pair

distinctly widened at base; lateral florets male or sterile but with obvious floret construction; awns of central lemma <5cm 4

4 Perennial, with tillers at flowering 5

4 Annual, without tillers 6

5 Proximal part of glumes with very short soft hairs >0.1mm; anthers 0.8-2mm; awns strongly divergent at maturity; upper leaf-blades 1.5-2(3)mm wide **7. H. pubiflorum**

5 Proximal part of glumes with minute rough prickles <0.1mm; anthers 3-4mm; awns stiffly erect at maturity; upper leaf-blades 2-6mm wide **8. H. secalinum**

6 Glumes of central spikelet with conspicuous marginal hairs >0.5mm; leaves usually with well-developed pointed auricles **3. H. murinum**

6 Glumes of central spikelets with only pricklets <0.1mm; leaves usually without or with small rounded auricles 7

7 Lateral florets distinctly stalked, the stalk (above glumes) c.1mm and c. as long as stalk (below glumes) of lateral spikelets; longest awns of triplet usually <1cm 8

7 Lateral florets sessile or nearly so, the stalk (above glumes) 0-0.5mm and much shorter than 1-1.5mm stalk (below glumes) of lateral spikelets; longest awns of triplet usually >1cm 9

8 Lateral lemmas obtuse to acute, 1.7-3.3mm **5. H. euclaston**

8 Lateral lemmas strongly acuminate or with awn ≤2mm, 2.8-6mm incl. awn **4. H. pusillum**

9 Glumes of lateral spikelets slightly heteromorphic, the inner with flattened basal part 0.3-0.7mm wide (c.2x as wide as basal part of outer glume); lower leaf-sheaths pubescent with hairs ≥0.5mm **10. H. geniculatum**

9 Glumes of lateral spikelets strongly heteromorphic, the inner with ± winged basal part 0.7-1.2mm wide (c.3-4x as wide as basal part of outer glume); lower leaf-sheaths glabrous to pubescent with hairs ≤0.25mm **9. H. marinum**

1. H. vulgare L. - *Six-rowed Barley*. Annual to 1m; all 3 spikelets in each triplet bisexual, ± sessile; lemmas usually with awns >10cm (almost awnless cultivars exist); glumes with awn 0-2cm; (2n=14). Intrd; a barley now rarely cultivated, casual as grain-alien and rare relic in waste places, waysides and field-borders; fairly frequent throughout BI; SW Asia. Usually the 3 fertile florets per triplet produce 6 vertical rows of caryopses in the spike, but in some cultivars the 2 lateral florets of triplets on opposite sides of the rhachis are superimposed, producing 4 vertical rows (*Four-rowed Barley*).

2. H. distichon L. - *Two-rowed Barley*. Annual to 75cm; differs from *H. vulgare* in lateral spikelets sterile, much reduced, ± awnless, usually stalked; (2n=14). Intrd; the common cultivated barley, common as a relic in waste places, fields, waysides, etc. throughout BI; SW Asia. On biological grounds better amalgamated with *H. vulgare* as convars. *vulgare* and *distichon* (L.) Alef., but very distinct morphologically.

3. H. murinum L. - *Wall Barley*. Annual to 60cm; central spikelet bisexual, sessile; lateral spikelets male or sterile but not or little reduced, stalked; lemmas with awns 1-5cm; glumes with awns 1-3cm.

1 Lemma-body and palea of central floret longer than those of lateral florets; central floret with stalk (above glumes) <0.6mm **a. ssp. murinum**

1 Lemma-body and palea of central floret shorter than those of lateral florets; central floret with stalk (above glumes) 0.6-1.5mm 2

2 Anthers of central florets usually blackish, <0.6mm, ≤1/3 as long as those of lateral florets; leaves usually glaucous **c. ssp. glaucum**

2 Anthers of central florets usually yellowish, >0.6mm long, 1/2-lx as long as those of lateral florets; leaves not glaucous **b. ssp. leporinum**

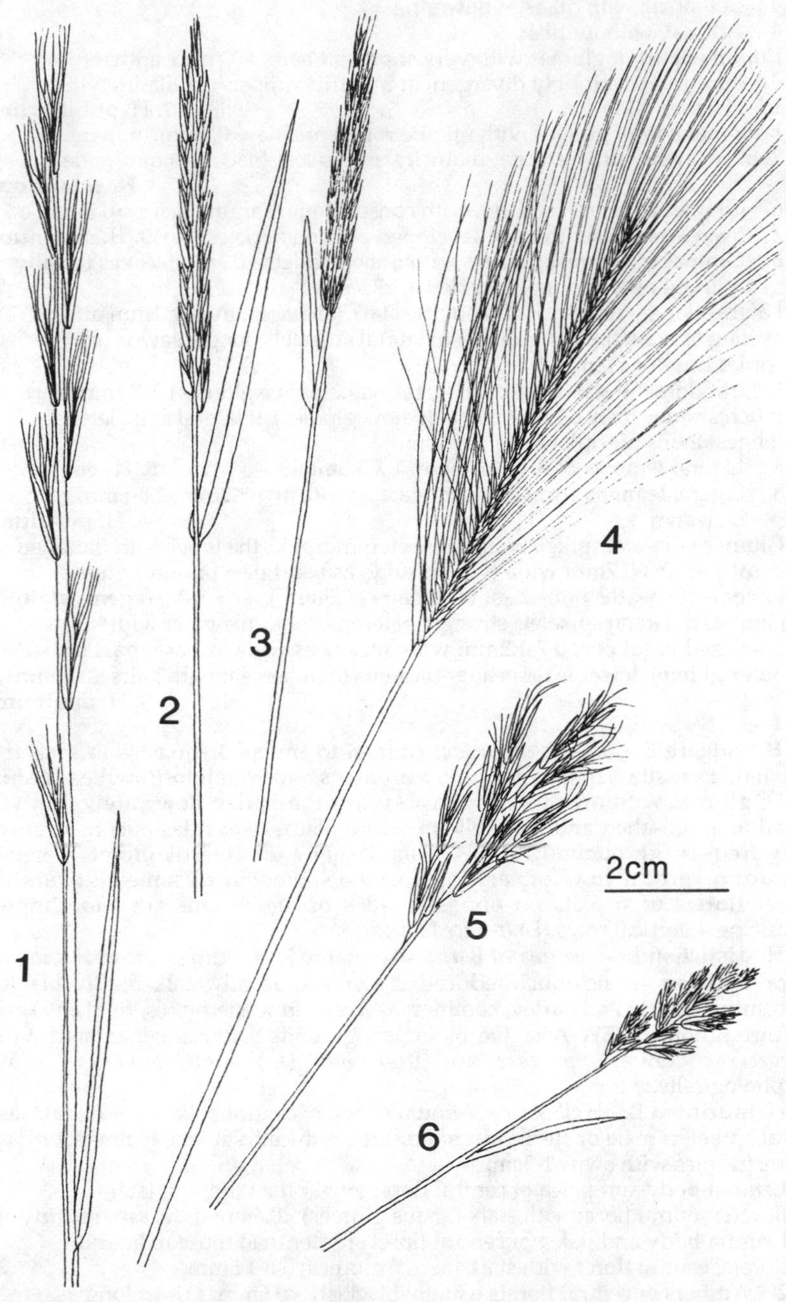

FIG 896 - Poaceae. 1, *Elymus scabrus*. 2, *X Elytrordeum langei*. 3, *Hordeum pusillum*. 4, *H. jubatum*. 5, *Rytidosperma racemosum*. 6, *Schismus barbatus*.

a. Ssp. murinum. Central floret longer than lateral florets, with pedicel <0.6mm; anthers 0.7-1.4mm; 2n=28. Native; weed of waste and rough ground and barish patches in rough grassland; common in C, S & E En and CI, becoming more scattered N & W to NE & SW Sc and N Wa, local in Ir, mainly in S & E.

b. Ssp. leporinum (Link) Arcang. (*H. leporinum* Link). Central floret shorter than lateral florets, with pedicel ≥0.6mm; anthers 0.7-1.4mm; (2n=28, 42). Intrd; rather frequent casual of waste ground, from wool and other sources, sometimes natd for short time; scattered in En and Sc; S Europe.

c. Ssp. glaucum (Steud.) Tzvelev (*H. glaucum* Steud.). Differs from ssp. *leporinum* as in key (anthers of central florets 0.2-0.5mm); (2n=14). Intrd; habitat and distribution as for ssp. *leporinum*, but rarer; Mediterranean.

4. H. pusillum Nutt. - *Little Barley*. Annual to 45cm; central spikelet bisexual, **896**
sessile; lateral spikelets male or sterile, considerably reduced, stalked; central lemma with awn 0.5-1cm; glumes with awns 0.4-0.8cm; (2n=14). Intrd; rather frequent wool-alien on tips and waste ground and in fields; scattered in En; N America. See *H. euclaston* for confusion.

5. H. euclaston Steud. - *Argentine Barley*. Differs from *H. pusillum* as in key (couplet 8); (2n=14). Intrd; source, habitat and distribution as for *H. pusillum*, but relative abundance and distribution uncertain due to confusion; S America.

6. H. jubatum L. - *Foxtail Barley*. Tufted perennial to 60cm; central spikelet **896**
bisexual, sessile; lateral spikelets sterile, greatly reduced (floret often simply a single awn-like lemma), stalked; central lemma with awn (2)4-10cm; glumes awn-like to base, 3-8cm; (2n=28). Intrd; alien from wool, birdseed and grass-seed and garden outcast, casual in waste places, now natd especially along main roads, especially those salted in winter; locally frequent in E Br from E Kent to C Sc, scattered casual elsewhere; N America.

7. H. pubiflorum Hook. f. - *Antarctic Barley*. Tufted perennial to 40cm; central spikelet bisexual, sessile; lateral spikelets sterile, considerably reduced, stalked; central lemma with awn 1-1.6cm; glumes ± awn-like to base, 1.5-2.5cm; (2n=14). Intrd; wool-alien on tips, waste ground and in fields, formerly persistent in W Kent; scattered in En, possibly overlooked for *H. jubatum*; S S America.

8. H. secalinum Schreb. - *Meadow Barley*. Tufted perennial to 80m; central spikelet bisexual, sessile; lateral spikelets male or sterile, slightly reduced, stalked; central lemma with awn 0.6-1.2cm; glumes ± awn-like to base, 1-1.6cm; 2n=14 (14, 28). Native; meadows and pastures, especially on heavy soils; common in C, S & E En, very scattered W & N to Scillies, NW Wa and Durham, formerly Fife, rare and very scattered in CI and Ir.

9. H. marinum Huds. - *Sea Barley*. Annual to 40cm; central spikelet bisexual, **R**
sessile; lateral spikelets sterile, considerably reduced, stalked; central lemma with awn 1.5-2.5cm; glumes ± awn-like to base and 1-2.5cm except inner ones of lateral spikelets which have expanded base; 2n=14. Native; barish or sparsely grassed often salty ground near sea, by saltmarshes, on banks and walls and in rough or waste ground; locally common in S Br N to S Lincs and Flints, formerly very scattered N to CE Sc, rare casual elsewhere.

10. H. geniculatum All. (*H. hystrix* Roth, *H. marinum* ssp. *gussoneanum* (Parl.) Thell.) - *Mediterranean Barley*. Differs from *H. marinum* as in key (couplet 9); (2n=14, 28). Intrd; casual from wool, grain and other sources on tips, waste ground and in fields, natd in Guernsey during 19th Century; scattered in Br; S Europe.

66. SECALE L. - *Rye*

Annuals; spikelets 1 per node, each with 2(-3) bisexual florets; glumes very narrow, acute, 1-veined, awnless or shortly awned; lemmas 5-veined, keeled, acuminate, usually very long-awned; spikelets disarticulating at maturity below each caryopsis, leaving glumes, lemma and palea on rhachis.

1. S. cereale L. - *Rye*. Culms to 1.5m; spikes 5-15cm, usually pendent at maturity;

lemmas with awn c.2-5cm; (2n=14). Intrd; cultivated now on rather small scale, casual as relic and from grain on tips and in waste places; frequent throughout most of BI; SW Asia. Easily confused at a glance with barley (but spikelet structure totally different) or awned wheat (florets >2, only lower fertile; glumes truncate, keeled).

67. X TRITICOSECALE Wittm. ex A. Camus (*SECALE* x *TRITICUM*) - *Triticale*
A new grain-crop, derived by artificial hybridization of *Triticum* and *Secale*, is now being grown on a field scale and is increasingly being found as a relic. It has a pendent spike and long awns and varies greatly in height, sometimes up to 1.8m. It differs from *Triticum* in its obtuse (not truncate) glumes and from *Secale* in its broader glumes and distal sterile florets; (2n=42, 56). Several widely differing cultivars are grown, but there is no valid specific epithet to cover them; they are best referred to as genus-cultivar combinations, e.g. **X Triticosecale** 'Lasko'.

68. TRITICUM L. - *Wheats*
Annuals; spikelets 1 per node, each with 3-7(9) florets, the apical ≥2 sterile and reduced; glumes keeled, truncate to bifid, apiculate to shortly awned; lemmas 5-veined, keeled, truncate to bifid, apiculate to very long-awned; spikelets disarticulating at maturity below each caryopsis, leaving glumes, lemma and palea on rhachis.

Other spp. - T. durum Desf. (*Pasta Wheat*) is a rare casual but perhaps under-recorded for *T. turgidum*, from which it differs in the usually black (not green to yellowish-brown) often longer (to 20cm) awns, flinty (not floury) texture to endosperm when chewed, and usually longer glumes (nearly as long as lowest lemmas, not c.2/3 as long).

1. T. aestivum L. - *Bread Wheat*. Culms to 1.5m; spikes 4-18cm; rhachis glabrous; glumes strongly keeled in upper 1/2, scarcely so in lower 1/2; lemmas awnless or with awn up to 16cm; (2n=42). Intrd; the common cultivated wheat, common as relic in fields and waste ground and on roadsides and tips throughout BI; SW Asia.
2. T. turgidum L. - *Rivet Wheat*. Culms to 1.5m; spikes 7-12cm; rhachis with hair-tufts at each node; glumes strongly keeled throughout; lemmas with awn 8-16cm; (2n=28). Intrd; formerly a crop in N & W Br and Ir, now rarely grown and when so mainly for animal feed, occasional casual as relic or grain-alien; scattered in Br and Ir; SW Asia.

SUBFAMILY 3 - ARUNDINOIDEAE (tribe 15, genera 69-74). Annual or perennial herbs; leaves without false-petiole or cross-veins; ligule a fringe of hairs; sheaths not fused; inflorescence a panicle; spikelets with (1)2-many florets, with all florets (except often the distal ones) bisexual or the lowest male or sterile; glumes 2; lemmas firm, 1-9-veined, sometimes bifid at apex, awnless or with terminal awn; stamens 3; stigmas 2; lodicules 2; embryo arundinoid; photosynthesis C3-type alone, with non-Kranz leaf anatomy; fusoid cells 0; microhairs present; chromosome base-number 9 or 12.

TRIBE 15 - ARUNDINEAE (Cortaderieae, Danthonieae, Molinieae) (genera 69-74).

69. DANTHONIA DC. (*Sieglingia* Bernh.) - *Heath-grass*
Densely tufted perennials; inflorescence a small panicle with rarely >12 spikelets; spikelets with 4-6 florets, all or all except most apical bisexual; glumes ovate, c. or nearly as long as rest of spikelet, 3-7-veined; lemmas 7-9-veined, minutely 3-toothed at apex, awnless, with tuft of short hairs at base and fringe up each side to c.1/2 way.

1. D. decumbens (L.) DC. (*Sieglingia decumbens* (L.) Bernh.) - *Heath-grass*. Culms decumbent to erect, to 40(60)cm; panicle 2-7cm; spikelets 6-12mm; 2n=36. Native; sandy or peaty often damp soil, usually acid but also mountain limestones, mostly on heaths, moors and mountains; common in suitable places throughout BI. Florets often cleistogamous, and sometimes (often whitish) solitary spikelets occur hidden in basal leaf-sheaths.

70. RYTIDOSPERMA Steud. - *Wallaby-grass*
Densely tufted perennials; inflorescence a rather compact to elongated panicle; spikelets with 6-10 florets, lower ones bisexual, upper ones male or sterile and reduced; glumes lanceolate to narrowly ovate, with wide hyaline margins, slightly longer to slightly shorter than rest of spikelet (excl. awns), 5-7-veined; lemmas 7-veined, with 2 long acuminate lobes at apex tipped with straight awns, with long, bent terminal awn from sinus, with dense white silky hairs at base and middle reaching or nearly reaching apex of body of lemma.

1. R. racemosum (R. Br.) Connor & Edgar (*Danthonia racemosa* R. Br.) - *Wallaby-* **881**
grass. Culms erect, to 60cm; panicles 3-5cm; spikelets 7-16mm excl. awns; lemmas **896**
with bent terminal awn 5-15mm, lateral awns 2-8mm; (2n=24). Intrd; occasional wool-alien in fields and waste places and on tips; scattered in En; Australia and New Zealand. Not over-wintering with us.

71. SCHISMUS P. Beauv. - *Kelch-grass*
Tufted annuals; inflorescence a rather compact panicle; spikelets with 4-10 florets, lower ones bisexual, upper ones male or sterile and reduced; glumes lanceolate, with wide hyaline margins, slightly longer to slightly shorter than rest of spikelet, 5-7-veined; lemmas 9-veined, deeply and acutely 2-lobed at apex, awnless or ± so, with short hairs at base and long silky hairs on back not reaching apex.

Other spp. - S. arabicus Nees, from E Mediterranean and W Asia, occurs similarly but is rather rare; it is very similar to *S. barbatus* but has lemma c.3.5mm with apical lobes longer than (not c. as long as) wide, and tip of palea reaching lemma notch to 1/2 way up lobes (not from 1/2 way to near tip of lobes).

1. S. barbatus (L.) Thell. - *Kelch-grass*. Culms erect, to 25cm; panicle 1-4cm; **881**
spikelets 5-6mm; lemmas c.2mm; (2n=12). Intrd; rather characteristic wool-alien in **896**
fields and waste land and on tips; scattered in En; Mediterranean.

72. CORTADERIA Stapf - *Pampas-grasses*
Densely tufted perennials; inflorescence a very large spreading panicle; dioecious or gynodioecious (with female and bisexual plants); spikelets with 2-7 florets; glumes lanceolate, hyaline, slightly unequal, 1-veined, at least upper ± as long to longer than rest of spikelet and with long terminal awn; lemmas 3-5-veined, long-awned at apex, with tuft of long fine hairs at base reaching ± to apex of body of lemma.

1. C. selloana (Schult. & Schult. f.) Asch. & Graebn. - *Pampas-grass*. Plant forming **881**
tussocks often >1m across; leaves 0.9-2.7m, with fiercely cutting serrated edges; culms erect, to 3m; panicles 40-120cm, becoming a silvery-white or red-tinged silky-pubescent mass at fruiting; lemmas bifid, with bristle-like points, pubescent all over; flowers Aug-Nov; dioecious; (2n=72). Intrd; grown as ornament, becoming natd where thrown-out or planted; rough ground, waysides, old gardens, maritime cliffs and dunes; scattered in S En, S Wa and CI, also Dunbarton and Co Wexford; S America.

2. C. richardii (Endl.) Zotov - *Early Pampas-grass*. Differs from *C. selloana* in leaves 0.6-1.2m; panicles 30-60cm, silvery- to yellowish-white; lemmas acuminate, glabrous distally; flowers Jul-Aug; gynodioecious; (2n=90). Intrd; similar places to

C. selloana but less common; reported from several sites in S En and Wa, well natd among *Ulex* on hillside in Cards; New Zealand. Seed-set is abundant on bisexual plants, as in Cards.

73. MOLINIA Schrank - *Purple Moor-grass*
Densely tufted perennials; inflorescence a ± diffuse to ± contracted panicle; spikelets with 1-4 florets, all except most apical bisexual; glumes ovate, slightly unequal, 1-3-veined, much shorter than rest of spikelet; lemmas 3-5-veined, acute to obtuse, awnless, glabrous.

1. M. caerulea (L.) Moench - *Purple Moor-grass.* Plant often forming tussocks; culms to 1.3m, erect; panicles up to 60cm, often purplish; spikelets 3-7.5mm. Native.

a. Ssp. caerulea. Culms usually <65cm; panicle usually <30cm, narrow, with branches mostly <5cm; spikelets 3-5.5mm; lemmas 3-4mm; 2n=36. Heaths, moors, bogs, fens, mountain grassland and cliffs and lake-shores, always on at least seasonally wet ground; common in suitable places throughout BI.

b. Ssp. arundinacea (Schrank) K. Richt. (ssp. *altissima* (Link) Domin, *M. litoralis* Host). Culms mostly 65-125(160)cm; panicles mostly 30-60cm, with very uneven lengthed branches often >10cm, usually spreading at least during flowering; spikelets (3)4-7.5mm; lemmas (3.2)3.5-5.4(5.7)mm; (2n=90). Fens, fen-scrub, fen-type vegetation by rivers and canals; scattered but frequent in suitable places in C & S Br, very scattered N to S & W Sc and in S Ir, Guernsey.

74. PHRAGMITES Adans. - *Common Reed*
Extensively rhizomatous perennials; inflorescence a very large spreading panicle; spikelets with 2-6(more) florets, the lowest male or sterile, the rest bisexual; glumes unequal, narrowly elliptic-ovate, 3-5-veined, much shorter than rest of spikelet; lemmas lanceolate, acute to acuminate at apex, with 1-3 veins, awnless; rhachilla-segments with long, white, silky hairs becoming very conspicuous in fruit.

1. P. australis (Cav.) Trin. ex Steud. (*P. communis* Trin.) - *Common Reed.* Culms to 3.5m but sometimes <1m; panicles 20-60cm, usually purple; spikelets 8-16mm, with rhachilla-hairs up to 10mm; 2n=c.48 (36-96). Native; on mud or in shallow water by lakes, rivers, canals, marshes, fens, bog-margins and edges of salt-marshes and estuaries; common in suitable places throughout BI.

SUBFAMILY 4 - CHLORIDOIDEAE (tribes 16-17, genera 75-83). Annual to perennial herbs; leaves without false-petiole or cross-veins; ligules usually a fringe of hairs, sometimes membranous or a membrane fringed with hairs; sheaths not fused; inflorescence usually an umbel or a raceme of spikes or racemes, sometimes a panicle; spikelets with 1-many florets with all florets (except often the distal ones) bisexual; glumes 2; lemmas membranous to firm, 1-3(9)-veined, awnless or less often awned; stamens 3; stigmas 2; lodicules 0 or 2; embryo chloridoid (or rarely arundinoid); photosynthesis C3- plus C4-type, with Kranz leaf anatomy; fusoid cells 0; microhairs present; chromosome base-number >7.

TRIBE 16 - ERAGROSTIDEAE (Sporoboleae) (genera 75-79). Annuals or perennials with or without rhizomes; ligules membranous or a fringe of hairs; inflorescence a diffuse to contracted panicle, an umbel of spikes, or a raceme of racemes; spikelets with 3-many florets (or with 1 floret in *Sporobolus*), all (except 1-2 most apical) bisexual; glumes 2, much shorter than rest of spikelet; lemmas (1-)3-veined (with extra veins close to midrib in *Eleusine*), usually keeled, awnless (very shortly awned in *Leptochloa*).

75. LEPTOCHLOA P. Beauv. (*Diplachne* P. Beauv.) - *Beetle-grasses*
Rhizomatous perennials (but annual with us); ligule membranous; inflorescence a loose, long raceme of racemes, the spikelets well spaced out, the rhachis ending in a spikelet; spikelets with 6-10(14) florets; glumes unequal, 1-veined; lemmas 3-veined, slightly keeled with prominent midrib and submarginal laterals, with long silky hairs at base and on margins and base of dorsal midline, bifid or shouldered or with 1-few teeth on either side at apex, with very short apical awn; spikelets disarticulating between florets; pericarp adherent to seed.

Other spp. - **L. muelleri** (Benth.) Stace (*Diplachne muelleri* Benth.), from Australia, and **L. uninervia** (C. Presl) Hitchc. & Chase (*D. uninervia* (C. Presl) Parodi), from N & S America, resemble *L. fusca* closely; they have been recorded rarely but may be overlooked. *L. muelleri* differs in the lemmas with rounded subapical teeth and a minute awn shorter than them; *L. uninervia* differs in the lemmas with 1-2 minute teeth or merely a shoulder either side of the shortly apiculate apex, and anthers <0.5mm (not >1mm). The former is perhaps not distinct from *L. fusca*.

1. L. fusca (L.) Kunth (*Diplachne fusca* (L.) P. Beauv.) - *Brown Beetle-grass*. Culms **902**
to 1m; inflorescences up to 40cm, with long, straight erecto-patent unbranched **905**
racemes from main axis; spikelets 8-15mm; lemmas 3-6mm, with awn ≤1.6mm between 2 shorter subapical acute teeth; (2n=20). Intrd; occasional casual wool-alien in fields and waste places and on tips; scattered in En; Australia to tropical Africa. Resembles *Eragrostis plana* in general appearance but has long membranous ligule.

76. ERAGROSTIS Wolf - *Love-grasses*
Annuals or tufted perennials; ligule a fringe of hairs; inflorescence a usually diffuse panicle; spikelets with 3-many florets, often very narrow and parallel-sided; glumes subequal to unequal, (0)1(-3)-veined; lemmas 3-veined, keeled, acute to obtuse, rounded or emarginate; awnless (sometimes apiculate); spikelets disarticulating between florets or the caryopsis falling free leaving persistent rhachilla; pericarp adherent to seed.

Superficially often similar to *Poa*, but the 3-veined lemmas and ligule a ring of hairs distinguish it.

1 Anthers (0.8)1-1.3mm; plant potentially perennial **1. E. curvula**
1 Anthers 0.2-0.6mm; plant annual (at least with us) 2
  2 Plant with minute, sessile, wart-like glands (x≥10 lens) on leaf-margins, sheath-midrib, lemma- and glume-veins, panicle-branches and/or pedicels (if 0 on leaves then always ≥1 on pedicels) 3
  2 Plant without minute sessile glands 4
3 Leaves often >5mm wide; spikelets ≥2mm wide; lemmas (1.7)2-2.8mm; sessile glands prominent on lemma veins, usually not so on pedicels **2. E. cilianensis**
3 Leaves <5mm wide; spikelets ≤2mm wide; lemmas 1.5-2mm; 1-2 sessile glands prominent near pedicel-apex, usually none on lemmas **3. E. minor**
  4 Caryopsis 1-1.3mm; upper glume 1.7-3mm; lemmas 2-2.7mm **6. E. tef**
  4 Caryopsis 0.5-0.8mm; upper glume 0.7-1.4mm; lemmas 1.2-1.8mm 5
5 Panicle-branches erecto-patent at maturity, with spikelets borne nearly to base; lemmas greenish-grey, purple-tinged at apex **4. E. pilosa**
5 Panicle-branches patent at maturity, bare of spikelets for lowest c.1/4-1/3; lemmas dark grey, not purple-tinged **5. E. parviflora**

Other spp. - >40 other spp. have been found on tips and docksides, etc., as rare casuals from wool, grain and other commodities. The following 12 are the least

FIG 902 - Spikelets of **Poaceae: Chloridoideae**. 1-6, ***Eragrostis***. 1, ***E. minor***. 2, ***E. parviflora***. 3, ***E. cilianensis***. 4, ***E. tef***. 5, ***E. curvula***. 6, ***E. pilosa***. 7, ***Leptochloa fusca***. 8, ***Eleusine indica***. 9, ***Chloris truncata***. 10, ***Sporobolus africanus***. 11, ***Dactyloctenium radulans***. 12, ***Cynodon dactylon***.

rare: **E. lugens** Nees, **E. neomexicana** Vasey ex L.H. Dewey, **E. pectinacea** (Michx.) Nees and **E. virescens** Nees from America; **E. lehmanniana** Nees and **E. plana** Nees (see note under *Leptochloa fusca*) from Africa; **E. dielsii** Pilg., **E. lacunaria** F. Muell. ex Benth., **E. leptostachya** (R. Br.) Steud. and **E. trachycarpa** (Benth.) Domin from Australia; and **E. barrelieri** Daveau from the W Mediterranean.

1. E. curvula (Schrad.) Nees (*E. chloromelas* Steud.) - *African Love-grass*. Tufted **902**
perennial but often annual with us; culms erect, to 1.2m; leaves usually <2mm wide **905**
and inrolled; panicle up to 30 x 15cm, very diffuse; spikelets linear, ≤2mm wide, greenish-grey; lemmas 2-3mm; caryopsis ≤0.8mm; (2n=20, 40, 50, 54, 60, 70, 80). Intrd; occasional wool-alien on tips and waste ground; casual scattered in En and Wa, ± natd in Southampton (S Hants); tropical Africa.

2. E. cilianensis (All.) Vignolo ex Janch. (*E. megastachya* (Koeler) Link) - *Stink-* **902**
grass. Annual; culms erect or ascending, to 75cm; leaves usually >5mm wide, flat; **905**
panicle up to 20 x 8cm, fairly dense; spikelets ovate to oblong-ovate, 2-3(4)mm wide, greenish-grey; lemmas (1.7)2-2.8mm; caryopsis ≤0.8mm; (2n=20, 40). Intrd; rather frequent casual wool- and grain-alien on tips and waste ground; scattered in En and Wa; S Europe.

3. E. minor Host (*E. pooides* P. Beauv.) - *Small Love-grass*. Annual; culms erect or **902**
ascending, to 50cm; leaves 2-5mm wide, flat; panicle up to 20 x 8cm, fairly dense; spikelets linear to narrowly ovate-elliptic, 1.3-2mm wide, yellowish-green often purplish- or grey-tinged; lemmas 1.5-2mm; caryopsis ≤0.8mm; (2n=20, 30, 40, 44, 60, 80). Intrd; fairly frequent casual wool- and grain-alien on tips and waste ground; scattered in C & S Br; S Europe.

4. E. pilosa (L.) P. Beauv. (*E. multicaulis* Steud.) - *Jersey Love-grass*. Annual; culms **902**
erect or ascending, to 70cm; leaves 1-3.5mm wide, flat; panicle up to 25 x 15cm, **905**
diffuse; spikelets linear to linear-lanceolate, 1-1.5mm wide, greenish-grey, tinged purple; lemmas 1.2-1.8mm; caryopsis 0.5-0.8mm; (2n=20, 30, 36, 40, 50, 60, 72). Intrd; wool-, birdseed- and grain-alien natd in Jersey since 1961; Mediterranean.

5. E. parviflora (R.Br.) Trin. - *Weeping Love-grass*. Differs from *E. pilosa* in **902**
sometimes perennial in native area; culms to 60cm; panicle up to 30 x 20cm, very diffuse but spikelets often lying ± appressed to panicle-branches; and see key. Intrd; fairly frequent casual wool-alien on tips and waste ground; scattered in En; Australia.

6. E. tef (Zucc.) Trotter - *Teff*. Annual; culms erect, to 1m; leaves flat, 2-4mm **902**
wide; panicle up to 30 x 15cm, diffuse or contracted; spikelets linear to linear-lanceolate, 1-1.5(2)mm wide, straw-coloured or green, often tinged reddish; lemmas 2-2.7mm; caryopsis 1-1.3mm; (2n=40). Intrd; minor grain-crop in warm countries, casual wool- or birdseed-alien on tips and waste ground; scattered in En; tropical Africa.

77. ELEUSINE Gaertn. - *Yard-grasses*

Annuals or tufted perennials; ligule membranous with a sparse or dense fringe of hairs; inflorescence an umbel or very short raceme of spikes with very crowded spikelets, the rhachis ending in a spikelet; spikelets with 3-6 florets; glumes subequal to unequal, 1-7-veined; lemmas with 3 main veins and 2-4 extra veins close to midrib, keeled, acute to apiculate, not awned; spikelets disarticulating between florets; pericarp not adherent to seed, which eventually falls out separately.

1 Lemmas and glumes obtuse, with hooded tip; perennial **2. E. tristachya**
1 Lemmas and glumes pointed (either acute, or subacute to obtuse and apiculate), without hooded tip; annual 2
  2 Spikes (3.5)5-15cm, all or most in terminal umbel; lemmas <1mm wide from keel to edge **1. E. indica**

2 Spikes 1-3cm, in very short terminal raceme; lemmas c.1.5mm wide from keel to edge **3. E. multiflora**

1. E. indica (L.) Gaertn. - *Yard-grass.* Annual; culms to 90cm; spikes (1)2-c.10, all 902
or most in umbel, (3.5)5-15 x <1cm. Intrd; casual birdseed-, grain-, wool-, cotton- 905
or pulse-alien, on tips and waste ground; scattered in En. Relative abundance and distribution in En of the 2 sspp. are unknown due to confusion, except that ssp. *africana* is much the commoner (or the only) wool-alien, and ssp. *indica* is the commoner grain-, pulse- and cotton-alien.

a. Ssp. indica. Ligule sparsely and minutely pubescent at apex; lower glume 1-veined, 1.1-2.3mm; upper glume 1.8-3mm; lemmas 2.4-4mm; seeds 1-1.3mm, with very fine close striations between and at right-angles to main ridges (x≥20 lens); (2n=18). Asia, but now world-wide.

b. Ssp. africana (Kenn.-O'Byrne) S.M. Phillips (*E. africana* Kenn.-O'Byrne). Ligule with strong pubescent fringe at apex; lower glume (1)2-3-veined, 2-3.5mm; upper glume 3-4.7mm; lemmas 3.7-5mm; seeds 1.2-1.6mm, with granulations between main ridges (x≥20 lens); (2n=36). Africa, but now more widespread.

2. E. tristachya (Lam.) Lam. - *American Yard-grass.* Perennial (but not over-wintering with us); culms to 40cm; spikes (1)2-4, all in umbel (rarely 1 below), 1-4 x ≤1cm; (2n=18). Intrd; fairly frequent wool- and birdseed-alien on tips and waste ground; scattered in En; S America.

3. E. multiflora A. Rich - *Fat-spiked Yard-grass.* Annual; culms to 45cm; spikes 2-8, in short raceme but congested at stem-apex, 1-3 x >1cm; (2n=16, 18). Intrd; fairly frequent wool-alien on tips and waste ground; scattered in En; tropical Africa.

78. DACTYLOCTENIUM Willd. - *Button-grass*

Annuals but sometimes rooting at lower nodes; ligule membranous, sometimes slightly fringed at apex; inflorescence an umbel of spikes with very crowded spikelets, the rhachis ending in a short projection; spikelets with 3-5 florets; glumes unequal, 1-veined, the upper with a long awn-like point; lemmas 3-veined, keeled, acuminate to acute-apiculate, not awned; spikelets disarticulating above glumes (not between florets); pericarp not adherent to seed, which eventually falls out separately.

1. D. radulans (R. Br.) P. Beauv. - *Button-grass.* Stems erect to decumbent, to 902
40cm; inflorescence of 4-10 crowded umbellate spikes each (0.5)1-2cm; spikelets 3- 905
5mm. Intrd; fairly frequent wool-alien on tips and waste ground; scattered in En; Australia.

79. SPOROBOLUS R. Br. - *Dropseeds*

Tufted perennials; ligule a fringe of hairs; inflorescence a narrow panicle with short to long, closely appressed, erect branches (the ultimate branches with closely borne small spikelets resembling a spikelet with many florets at a casual glance); spikelets with 1 floret; glumes unequal, 0-1-veined, much shorter than rest of spikelet; lemma 1-3-veined, rounded on back, acute to acuminate, awnless, inrolled and ± cylindrical but tapering at apex; spikelet disarticulating below lemma; pericarp not adherent to seed, which eventually falls out separately.

Other spp. - S. indicus (L.) R. Br., from tropical America, and **S. elongatus** R. Br., from Australia, closely resemble *S. africanus*; they have both been recorded rarely but may be overlooked. Both differ from *S. africanus* in having lemmas 1.3-1.9mm; *S. elongatus* differs from the other 2 spp. in having longer (≥3cm), not appressed lower panicle-branches and 2 (not 3) stamens; *S. indicus* has the caryopsis nearly as long as lemma.

1. S. africanus (Poir.) A. Robyns & Tournay - *African Dropseed.* Culms erect to 902

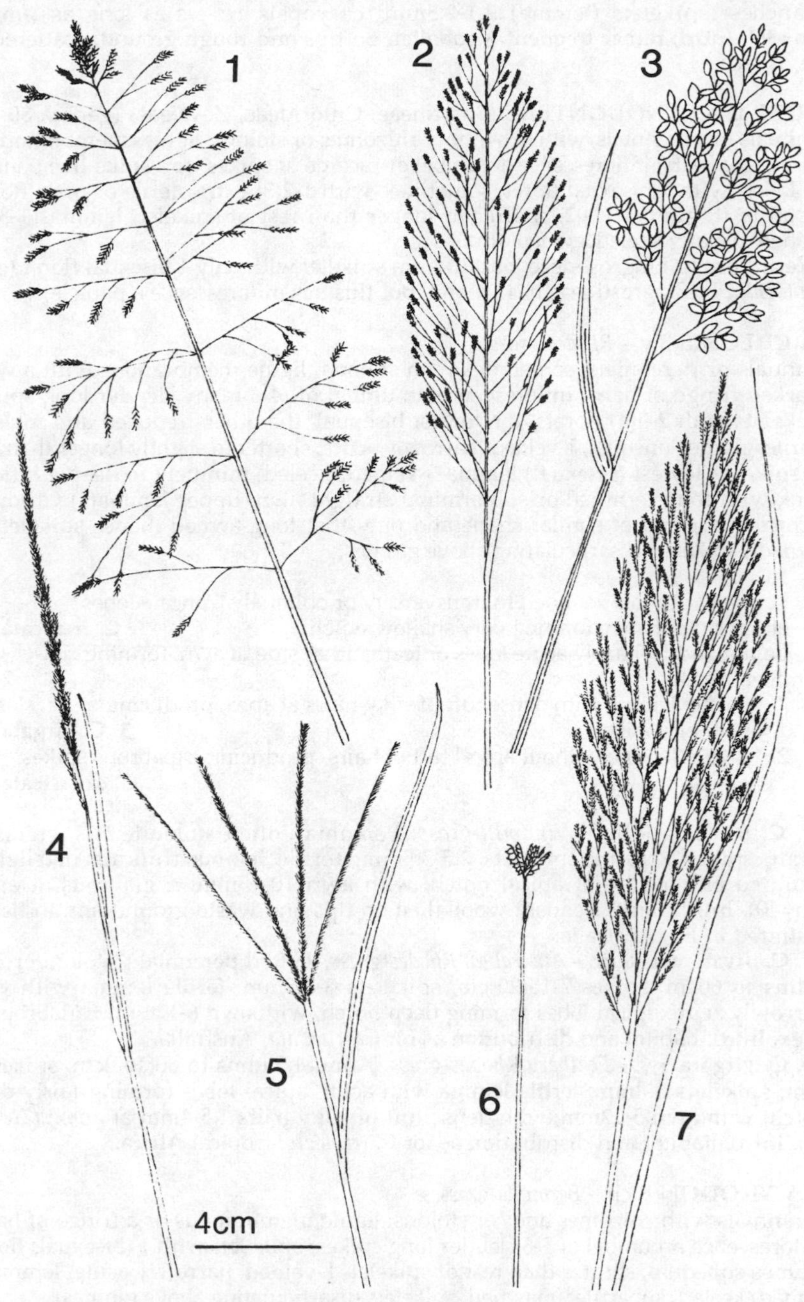

FIG 905 - Poaceae: Chloridoideae. 1, *Eragrostis curvula*. 2, *E. pilosa*. 3, *E. cilianensis*. 4, *Sporobolus africanus*. 5, *Eleusine indica*. 6, *Dactyloctenium radulans*. 7, *Leptochloa fusca*.

ascending, to 1m; panicle very narrow (<1cm), up to 35cm, with short (≤2cm) erect **905**
branches; spikelets (lemma) 2.1-2.5mm; caryopsis 1/2-2/3 as long as lemma; (2n=36). Intrd; rather frequent wool-alien on tips and rough ground; scattered in En; Africa.

TRIBE 17 - CYNODONTEAE (Spartineae, Chlorideae, Zoysieae) (genera 80-83). Annuals or perennials, with or without rhizomes or stolons; ligules membranous or a fringe of hairs; inflorescence an umbel or raceme of spikes, or a spike-like panicle; spikelets with 1 bisexual floret, sometimes with 1-2(3) extra sterile or male florets distal to it; glumes (1-)2, shorter to longer than rest of spikelet; lemmas 1-3(9)-veined, usually keeled, awned or not.

Very close to Eragrostideae, but differ in spikelet with only 1 bisexual floret (only *Leptochloa* of Eragrostideae has 1 floret, but this has inflorescence a panicle).

80. CHLORIS Sw. - *Rhodes-grasses*
Annuals or perennials, sometimes with stolons; ligule membranous with a well-marked fringe of hairs; inflorescence an umbel of (4)6-many slender long spikes; spikelets with 2-3(4) florets, the lowest bisexual, the others reduced and male or sterile; glumes unequal, 1-veined, narrowly acute, shorter to slightly longer than rest of spikelet; lowest (bisexual) lemma 3-veined, keeled, minutely to deeply bifid at apex with long terminal or subterminal straight awn; upper lemma(s) variously reduced, but 2nd of similar shape and only that long-awned (hence spikelets 2-awned); spikelets disarticulating above glumes.

1 Lemma with low rounded to transversely or obliquely truncate lobes either side of awn, forming very shallow notch **1. C. truncata**
1 Lemma with sharply acute lobes or teeth either side of awn, forming deep notch 2
 2 Fertile lemma with dense tuft of silky hairs at apex, producing feathery spikes **3. C. virgata**
 2 Fertile lemma without apical tuft of hairs, producing ± glabrous spikes **2. C. divaricata**

1. C. truncata R.Br. - *Windmill-grass.* Perennial, often stoloniferous; culms to **902**
45cm; spikes 7-15cm; spikelets 2.5-3.5mm; fertile lemma truncate or slightly **909**
rounded either side of apical notch, with awn 10-15mm, ± glabrous at apex; (2n=40). Intrd; fairly frequent wool-alien on tips and waste ground and in fields; scattered in En; Australia.

2. C. divaricata R. Br. - *Australian Rhodes-grass.* Tufted perennial (?stoloniferous); culms to 60cm; spikes 7-15(20)cm; spikelets 3-4.5mm; fertile lemma with very narrowly acute apical lobes forming deep notch, with awn 6-12mm, ± glabrous at apex. Intrd; habitat and distribution as for *C. truncata*; Australia.

3. C. virgata Sw. - *Feathery Rhodes-grass.* Annual; culms to 60(100)cm; spikes 2-8cm; spikelets 3-4mm; fertile lemma with acute apical lobes forming fairly deep notch, with awn 5-12mm, with dense tuft of silky hairs 1.5-4mm at apex; (2n=20, 40). Intrd; habitat and distribution as for *C. truncata*; tropical Africa.

81. CYNODON Rich. - *Bermuda-grasses*
Perennials with rhizomes and/or stolons; ligule membranous or a fringe of hairs; inflorescence an umbel of 3-6 slender long spikes; spikelets with 1 (bisexual) floret; glumes subequal, shorter than rest of spikelet, 1-veined, narrowly acute; lemma 3-veined, keeled, subacute, unawned; spikelets disarticulating above glumes.

1. C. dactylon (L.) Pers. - *Bermuda-grass.* Extensively rhizomatous and **RR**
stoloniferous; culms erect, to 30cm; ligule a fringe of short hairs <0.5mm with longer **902**
tuft at each edge; spikes 2-5cm; spikelets 2-3mm; rhachilla of spikelet continued

beyond base of floret as fine projection between upper glume and floret >1/2 as long as floret; 2n=36. Probably intrd; natd in rough sandy ground, waysides and short grassland near sea; local in SW En, S Wa and CI, casual from wool and other sources scattered elsewhere in En and Wa, sometimes ± natd for short while; world-wide in warm areas.

2. C. incompletus Nees - *African Bermuda-grass*. Differs from *C. dactylon* in rhizomes 0; ligule membranous, 0.4-1mm, with sparse hairs at apex and longer tuft at base of each edge; rhachilla of spikelet not extended beyond base of floret. Intrd; fairly frequent wool-alien on tips and rough ground and in fields; casual or ± natd for short while in scattered places in En; S Africa.

82. SPARTINA Schreb. - *Cord-grasses*

Strongly rhizomatous perennials; ligule a dense fringe of hairs; inflorescence a raceme of (1)2-12(30) long ± erect spikes; spikelets with 1 (bisexual) floret; glumes unequal, the upper as long as or longer than rest of spikelet, 1-9-veined, narrowly acute, awned or not; lemma 1-3-veined, keeled, acute or minutely notched, unawned; spikelets falling entire at maturity.

1 Upper glume very scabrid on keel with rigid pricklets ≥0.3mm, with awn 3-8mm; spikes with 2 rows of spikelets each crowded 4-10 per cm; inland **5. S. pectinata**
1 Upper glume glabrous or with soft hairs <0.3mm on keel, awnless; spikes with 2 rows of spikelets each spaced out 1-3 per cm; coastal 2
  2 Glumes glabrous or with hairs on keel only, sometimes very sparse on body also **4. S. alterniflora**
  2 Glumes softly pubescent on keel and body 3
3 Ligules 1.8-3mm at longest point (beware damaged ones); anthers (5)7-10(13)mm, with full pollen >45 microns across **3. S. anglica**
3 Ligules 0.2-1.8mm at longest point; anthers 4-8(10)mm, if >7mm then indehiscent with empty pollen; pollen <45 microns across 4
  4 Ligules 0.2-0.6mm; anthers 4-6.5mm, dehiscent, with full pollen **1. S. maritima**
  4 Ligules 1-1.8mm; anthers 5-7(10)mm, indehiscent, with empty pollen **2. S. x townsendii**

1. S. maritima (Curtis) Fernald - *Small Cord-grass*. Culms to 50(80)cm; ligules 0.2-0.6mm; spikes (1)2-3(5), 3-8cm; spikelets 11-15mm; glumes softly appressed-pubescent, awnless; anthers 4-6.5mm, with full pollen <45 microns across; 2n=60. Native; tidal sandy or muddy bare places by sea or in estuaries; local in S & E En from Wight (formerly S Devon) to N Lincs, intrd in Co Dublin. R

2. S. x townsendii H. & J. Groves (*S. maritima* x *S. alterniflora*) - *Townsend's Cord-grass*. Culms to 130cm; ligules 1-1.8mm; spikes 2-8, 6-24cm; spikelets (12)14-18(20)mm; glumes softly appressed-pubescent, awnless; anthers 5-7(10)mm, with empty pollen <45 microns across; 2n=49-66, 76. Native; tidal mud-flats; arose prior to 1870 in Southampton Water (S Hants) and still there with the parents, now spread Dorset to W Sussex, but also scattered elsewhere in Br and E Ir, either intrd with *S. anglica* or derived from it, in absence of 1 or both parents (but not of *S. anglica*).

3. S. anglica C.E. Hubb. (*S. x townsendii* auct. non H. & J. Groves) - *Common Cord-grass*. Culms to 130cm; ligules 1.8-3mm; spikes 2-12, 7-23cm; spikelets (15)17-21(26)mm; glumes softly appressed-pubescent, awnless; anthers (5)7-10(13)mm, with full pollen >45 microns across; 2n=120-124. Native; tidal mud-flats; arose c.1890 in Southampton Water as amphidiploid of *S. x townsendii* and spread with it naturally from Dorset to W Sussex, planted extensively elsewhere in BI N to C Sc and now dominant over large areas.

4. S. alterniflora Loisel. (*S. glabra* Muhl. ex Bigelow) - *Smooth Cord-grass*. Culms to 120cm; ligules 1-1.8mm; spikes 3-13, (3)5-15cm; spikelets 10-18mm; glumes glabrous or sparsely pubescent (usually on keel only), awnless; anthers 5-7mm, sometimes indehiscent, with full or sometimes empty pollen <45 microns across; 2n=62. Intrd; planted in 3 sites in S Hants from early 1800s but now extinct in all but 1 despite earlier spread, also planted in E Ross from 1920, S Essex in 1935 and Dorset in 1963; N America. Var. **glabra** (Muhl. ex Bigelow) Fernald differs in its larger leaf-blades and more tightly overlapping lemmas; the S Essex and E Ross plants and some from S Hants belong to this.

5. S. pectinata Bosc ex Link - *Prairie Cord-grass.* Culms to 180cm; ligules 0.5- 909
3mm; spikes 5-20(30), 2-8(10)cm; spikelets 8-12mm; glumes with strongly scabrid keel, glabrous to sparsely pubescent elsewhere, the upper with awn 3-8mm; anthers 5-7mm, with full pollen <45 microns across; (2n=40). Intrd; grown for ornament, persistent and spreading where neglected by fresh-water lake in W Galway since 1967, by lake in S Northumb since 1970 and by quarry in N Hants since 1986, used recently in biomass production trials; N America.

83. TRAGUS Haller - *Bur-grasses*

Annuals; ligule a dense fringe of hairs; inflorescence a spike or spike-like, with 2-5 spikelets on extremely short branch at each node; spikelets with 1 (bisexual) floret; glumes very unequal, the lower 0 or vestigial, the upper at least as long as floret, 5-7-veined, each vein with line of strong hooked spines, acute, unawned; lemma 3-veined, not keeled, acute; each nodal group of spikelets falling as a bur at maturity, the spikelets facing the centre of the bur with the glume-hooks outermost.

Distinctive vegetatively in the strong curved spines on proximal part of leaf-margins.

1 Upper glume 2-3mm **3. T. berteronianus**
1 Upper glume 3.5-4.5mm 2
 2 Spikelets 2 at all or almost all nodes; upper glume with 5 veins (and rows of spines) **2. T. australianus**
 2 Spikelets 3-5 at all or almost all nodes; upper glume with 7 veins (and rows of spines) (one pair of veins sometimes thinner and with smaller spines) **1. T. racemosus**

1. T. racemosus (L.) All. - *European Bur-grass.* Culms to 40cm; inflorescence 2-10 x 0.6-1cm; spikelets 3-5 on common stalk at each node, 1-3 of them often reduced, each 3.5-4.5mm; upper glumes with 7 veins, each bearing row of hooked spines; (2n=40). Intrd; rather frequent wool-alien on tips and rough-ground and in fields; scattered in En; S Europe.

2. T. australianus S.T. Blake - *Australian Bur-grass.* Differs from *T. racemosus* as in 909
key; the 2 spikelets at each node (both bisexual) are closely facing with the glume- 912
hooks outermost, the pair resembling a fruit of *Anthriscus caucalis*. Intrd; habitat and distribution as for *T. racemosus;* Australia.

3. T. berteronianus Schult. - *African Bur-grass.* Culms to 60cm; inflorescence 2-15 x 0.4-0.6cm, spikelets 2 at all or almost all nodes, 2-3mm; upper glumes with 5 veins, each bearing row of hooked spines; (2n=20). Intrd; habitat and distribution as for *T. racemosus*; tropical Africa.

SUBFAMILY 5 - PANICOIDEAE (tribes 18-19, genera 84-94). Annual to perennial herbs; leaves without false-petiole or cross-veins; ligule 0 or a fringe of hairs or a membrane fringed with hairs, rarely membranous; sheaths not fused; inflorescence a panicle, a spike, or an umbel or raceme of racemes or spikes; spikelets with 2 florets, the upper bisexual, the lower male or sterile (*Zea* is monoecious; *Sorghum* has paired spikelets, 1 bisexual the other male or sterile); glumes 2; lemmas firm to thick, 5-11-veined, awnless or with terminal awn; stamens 3; stigmas 2; lodicules 2 (0 in female *Zea*); embryo panicoid;

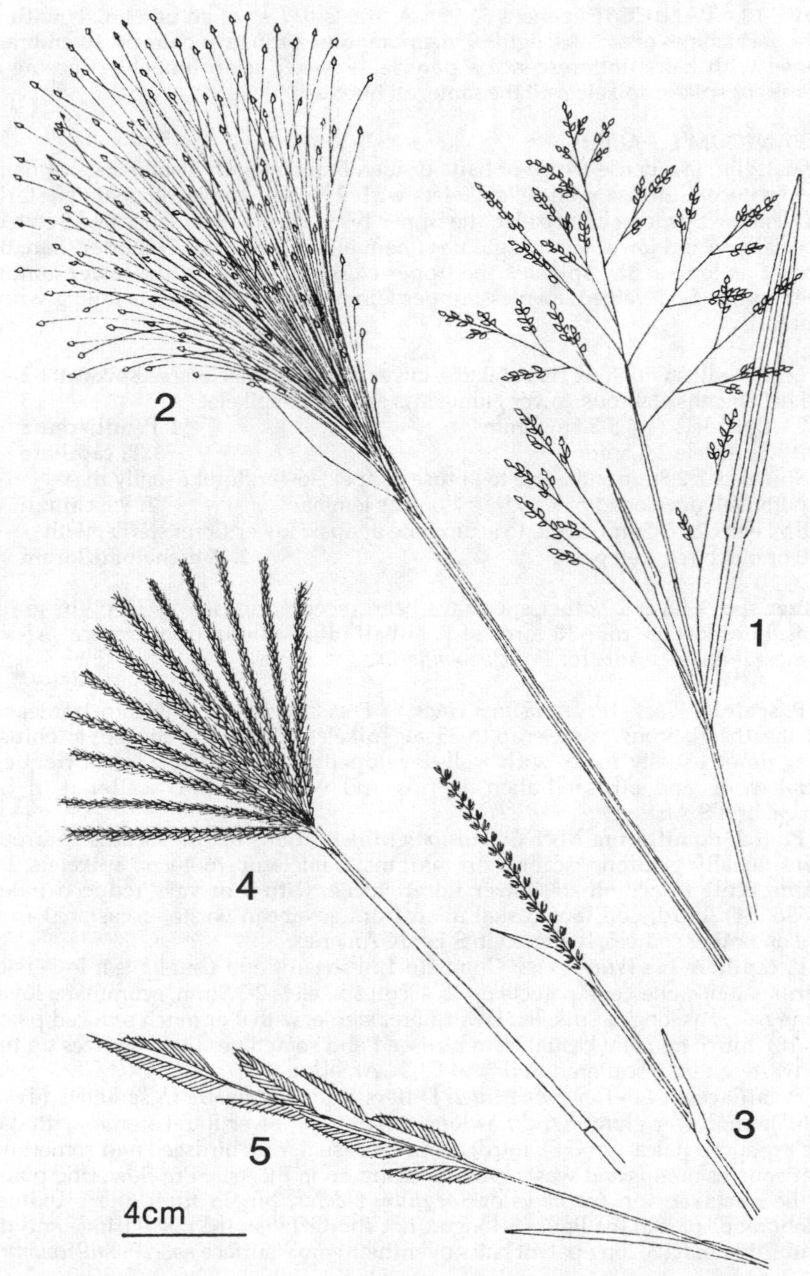

FIG 909 - **Poaceae**. 1, *Panicum dichotomiflorum*. 2, *P. capillare*. 3, *Tragus australianus*. 4, *Chloris truncata*. 5, *Spartina pectinata*.

photosynthesis C3-type alone or C3- plus C4-type, with or without Kranz leaf anatomy; fusoid cells 0; microhairs present; chromosome base-number 5, 9 or 10.

TRIBE 18 - PANICEAE (genera 84-92). Annuals or less often perennials with or without rhizomes or stolons; ligule 0, membranous, a fringe of hairs or a membrane fringed with hairs; inflorescence a panicle, a spike, or an umbel or raceme of racemes or spikes; spikelets all the same, all bisexual.

84. PANICUM L. - *Millets*
Annuals; ligule a dense fringe of hairs or membranous with distal fringe of hairs; inflorescence a diffuse panicle; spikelets with 2 florets, the lower male or sterile with lemma ± as long as spikelet, the upper bisexual, smaller, concealed between upper glume and lower lemma; glumes unequal, the lower much shorter than, the upper ± as long as the spikelet, the upper closely resembling the lower lemma; lower lemma 5-11-veined, awnless; upper lemma awnless; spikelets falling whole at maturity.

1 Leaf-sheaths with long patent hairs; lower glume >1/3 as long as spikelet 2
1 Leaf-sheaths glabrous; lower glume ≤1/3 as long as spikelet 3
  2 Spikelets (4)4.5-5.5(6.5)mm **4. P. miliaceum**
  2 Spikelets 2-3.5mm **3. P. capillare**
3 Spikelets 2-2.8mm, subacute to obtuse at apex; lower floret usually male, with well developed palea >1/2 as long as lemma **1. P. schinzii**
3 Spikelets 2.7-3.5mm, acute to acuminate at apex; lower floret sterile, with 0 or much reduced palea **2. P. dichotomiflorum**

Other spp. - Over 20 other spp. have been recorded mostly as wool- or grain-aliens, but most are rare. Records of **P. subalbidum** Kunth, from tropical Africa, are most if not all errors for *P. dichotomiflorum.*

1. P. schinzii Hack. (*P. laevifolium* Hack.) - *Transvaal Millet.* Culms to 1m; leaves 912
and sheaths glabrous; panicle up to 35cm; spikelets 2-2.8mm, subacute to obtuse; lower floret usually male, with well-developed palea; (2n=18). Intrd; frequent casual wool- and birdseed-alien on tips and waste ground; scattered in En; tropical and S Africa.

2. P. dichotomiflorum Michx. - *Autumn Millet.* Culms to 1m; sheaths glabrous; 909
leaves usually glabrous, scabrid on margins; panicle up to 40cm; spikelets 2.7- 912
3.5mm, acute to acuminate; lower floret sterile, with 0 or very reduced palea; (2n=36, 54). Intrd; constant casual alien from soyabean waste, occasional from wool and other sources; scattered in S En; N America.

3. P. capillare L. - *Witch-grass.* Culms to 1m; sheaths and usually leaf lowerside 909
midrib patent-pubescent; panicle up to 40cm; spikelets 2-3.5mm, acuminate; lower 912
glume 2/5-1/2 as long as spikelet; lower floret sterile, with 0 or much reduced palea; (2n=18). Intrd; frequent casual from birdseed and sometimes other sources on tips and waste ground; scattered in Br and CI; N America.

4. P. miliaceum L. - *Common Millet.* Differs from *P. capillare* in spikelets (4)4.5- 912
5.5(6.5)mm; lower glume 1/2-2/3 as long as spikelet; lower floret sterile, with 0 or much reduced palea; (2n=36). Intrd; common casual from birdseed and sometimes other sources on tips and waste ground; scattered in BI; Asia. Pre-flowering plants can be mistaken for *Zea mays* or *Sorghum bicolor*, but in these the proximal, membranous part of the ligule is longer (not shorter) than the distal fringe and the sheaths do not have long patent hairs over their whole surface as in *P. miliaceum.*

85. ECHINOCHLOA P. Beauv. - *Cockspurs*
Annuals; ligule 0; inflorescence a raceme of ± dense spikes or racemes, or the secondary racemes again racemosely branched, often with long stiff hairs especially

in tufts at branch-points, the spikelets usually in >2 rows; spikelets with 2 florets, the lower male or sterile with lemma ± as long as spikelet, the upper bisexual, smaller, concealed between upper glume and lower lemma; glumes unequal, the lower much shorter than, the upper ± as long as the spikelet, the upper closely resembling the lower lemma; lower lemma 5-7-veined, awned or awnless; upper lemma awnless; spikelets falling whole at maturity.

Sp. limits are very uncertain. *E. crus-galli* and *E. colona* are usually distinct, but their cultivated derivatives *E. esculenta* and *E. frumentacea* are extremely similar, making all 4 into a difficult complex.

1 Inflorescence with lateral spikes or racemes all or mostly clearly separate, obviously branched; lower floret male or sterile, its lemma awned or not 2
1 Inflorescence with fat lateral spikes or racemes close together and forming entirely or for most part a single, lobed, elongate head; lower floret sterile, its lemma not awned 3
  2 Primary branches of inflorescence simple, ≤3cm; spikelets 1.5-3mm; lower floret usually male; lower lemma acute to apiculate, with awn 0-2mm; leaf-blades ≤8mm wide **3. E. colona**
  2 Lower primary branches of inflorescence usually branched again; spikelets 3-4mm; lower floret usually sterile; lower lemma often awned (awn ≤5cm); leaf-blades mostly >1cm wide **1. E. crus-galli**
3 Glumes and lower lemma yellowish-green to straw-coloured; spikelets 2.5-3.5mm; lower lemma acute to subacute, sometimes minutely apiculate **4. E. frumentacea**
3 Glumes and lower lemma bright green usually strongly tinged purplish, sometimes completely purplish; spikelets 3-4mm; lower lemma acuminate **2. E. esculenta**

1. E. crus-galli (L.) P. Beauv. - *Cockspur.* Culms to 1.2m; leaves 1-3(8)cm wide; 912
inflorescence up to 20cm, usually strongly purplish-tinged, the lower branches usually branched again (except in depauperate material), all or most clearly separated; spikelets long-acuminate to long-awned; (2n=36, 42, 48, 54, 72). Intrd; casual mostly from birdseed, also from wool and soyabean and other sources, on tips, waysides and waste ground, also weed of cultivated ground sometimes natd there; scattered throughout most of BI, especially S, natd in S Br and CI; tropics.

2. E. esculenta (A. Braun) H. Scholz (*E. utilis* Ohwi & Yabuno, *E. frumentacea* auct. non Link) - *Japanese Millet.* Differs from *E. crus-galli* in fat congested inflorescence-branches forming lobed single head; spikelets unawned; (2n=54). Intrd; casual birdseed-alien on tips and waste ground; scattered in Br, mainly S; cultivated derivative of *E. crus-galli*, originated in Japan.

3. E. colona (L.) Link - *Shama Millet.* Differs from *E. crus-galli* in main branches 912
usually borne at points separated by >1/2 their length; and see key (couplet 2); (2n=36, 48, 54, 72). Intrd; casual on tips and waste ground from same sources as *E. crus-galli*; occasional in S Br; tropics.

4. E. frumentacea Link - *White Millet.* Habit of *E. esculenta* but differs as in key 912
(couplet 3); (2n=36, 48, 54). Intrd; source, habitat and distribution as for *E. esculenta*; cultivated derivative of *E. colona*, originated in India. The vernacular *Japanese Millet* is often misapplied to this sp.

86. BRACHIARIA (Trin.) Griseb. - *Signal-grasses*

Annuals; ligule a dense fringe of hairs; inflorescence a raceme of racemes with spikelets in 2 rows on 1 side of rhachis; spikelets with 2 florets, the lower male or sterile with lemma ± as long as spikelet, the upper bisexual, smaller, concealed between upper glume and lower lemma; glumes unequal, the lower <1/2 as long as upper, the upper closely resembling the lower lemma; lower lemma 5-7-veined, obtuse, awnless; upper lemma obtuse to rounded, awnless; spikelets falling whole

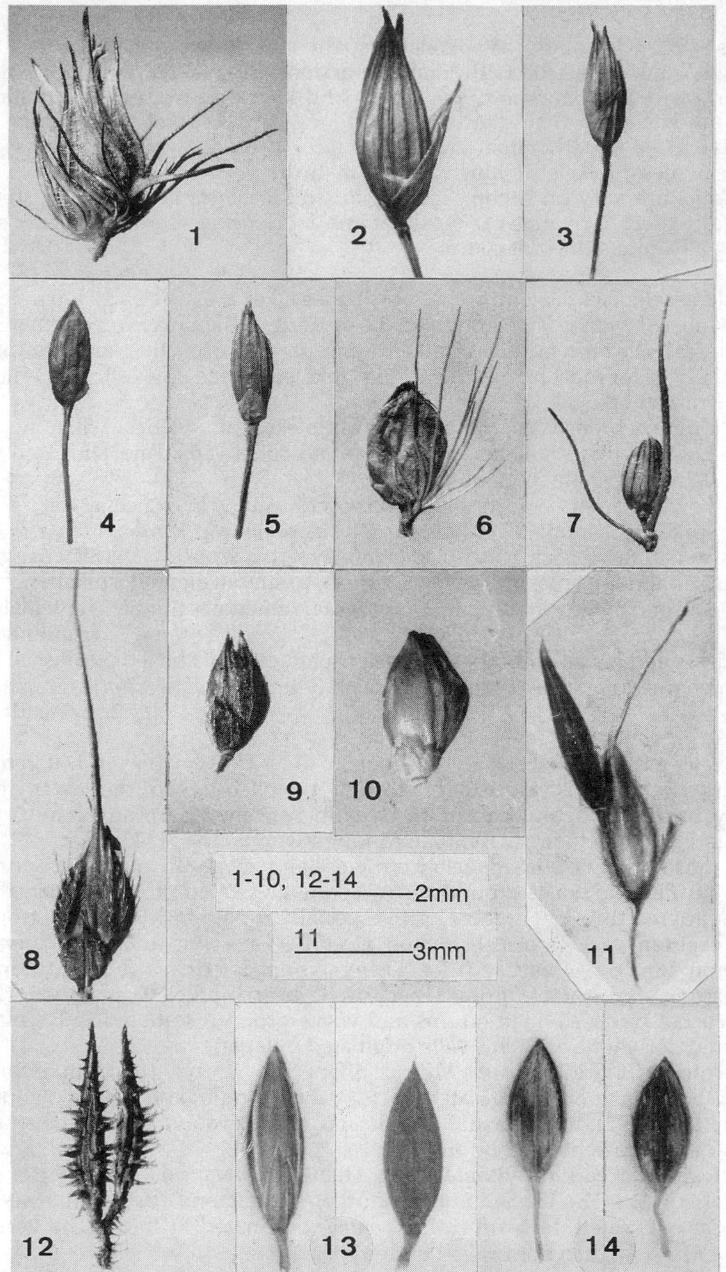

FIG 912 - Spikelets of **Poaceae**. 1, ***Cenchrus echinatus***. 2, ***Panicum miliaceum***. 3, ***P. capillare***. 4, ***P. schinzii***. 5, ***P. dichotomiflorum***. 6, ***Setaria pumila***. 7, ***S. verticillata***. 8, ***Echinochloa crus-galli***. 9, ***E. colona***. 10, ***E. frumentacea***. 11, ***Sorghum halepense*** (group of 1 fertile and 1 sterile spikelets). 12, ***Tragus australianus***. 13-14, ***Digitaria*** (2 views, left-hand view showing upper glume side). 13, ***D. sanguinalis***. 14, ***D. ischaemum***.

at maturity.

Possibly better united with *Urochloa*.

Other spp. - **B. eruciformis** (Sm.) Griseb., from Africa, is a rare birdseed-alien differing from *B. platyphylla* in its much smaller (1.7-2.7mm) pubescent spikelets and 3-14 racemes. It resembles some plants of *Echinochloa colona*, but differs in its more pubescent spikelets borne in only 2 rows, obtuse lower lemma, well developed ligule, and often pubescent (not glabrous) leaf-sheaths.

1. B. platyphylla (Griseb.) Nash - *Broad-leaved Signal-grass*. Culms decumbent to **881**
erect, to 50cm; leaves up to 12 x 1.2cm, glabrous; sheaths glabrous; racemes 2-6, 3- **917**
8cm; spikelets 3.5-4.5mm, glabrous; (2n=18, 36). Intrd; casual from birdseed and soyabean waste; occasional on tips and waste ground; scattered in S En; N America.

87. UROCHLOA P. Beauv. - *Signal-grasses*

Differs from *Brachiaria* in upper glume and lower lemma shortly acuminate, 7-veined; and upper lemma with distinct terminal awn.

1. U. panicoides P. Beauv. - *Sharp-flowered Signal-grass*. Differs from *Brachiaria* **881**
platyphylla in leaves, sheaths and inflorescence-rhachis glabrous or with sparse **917**
long hairs; spikelets 3.5-5mm, glabrous to pubescent; and in generic characters above; (2n=28, 30, 32, 36, 42, 46, 48). Intrd; casual from birdseed on tips and waste ground; scattered in C & S En; Africa and W Asia.

88. ERIOCHLOA Kunth - *Cup-grasses*

Annuals or perennials; ligule a dense fringe of hairs; inflorescence a rather irregular raceme of racemes, the main branches ± appressed to main axis and often slightly branched again, with spikelets scarcely in recognisable rows; spikelets with 2 florets, the lower sterile with lemma almost as long as spikelet and palea 0, the upper bisexual, smaller, concealed between upper glume and lower lemma and with palea, with small bead-like swelling at apex of pedicel; lower glume ± 0; upper glume as long as spikelet (slightly longer than lower lemma but otherwise very similar); lower lemma 5-veined, acuminate to awned to 2mm; upper lemma obtuse to rounded, with awn 0.3-1mm (as in *Urochloa*); spikelets falling whole, disarticulating immediately below bead-like swelling.

The ± absence of a lower glume, and the lower floret being represented by only a glume-like lemma, produces an apparently 2-glumed spikelet with 1 floret.

Other spp. - The spp. found in Br have not been fully investigated; **E. crebra** S.T. Blake, from Australia, and **E. fatmensis** (Hochst. & Steud.) Clayton, from Africa, closely resemble *E. pseudoacrotricha* and records of the last could refer to all 3 and others too.

1. E. pseudoacrotricha (Stapf ex Thell.) S.T. Blake - *Perennial Cup-grass*. **881**
Perennials; culms to 60(100)cm; inflorescence up to 15 x 1cm; spikelets 3.6-6mm **917**
incl. awn ≤2mm, with dense white silky hairs, acuminate; (2n=36). Intrd; rather infrequent wool-alien on tips and in fields and waste places; scattered in En; Australia.

89. PASPALUM L. - *Finger-grasses*

Perennials with or without stolons; ligule membranous; inflorescence a raceme of 2(-4) racemes with spikelets in 2 rows on 1 side of rhachis; spikelets with 2 florets, the lower sterile with lemma as long as spikelet and palea very small, the upper bisexual, smaller, concealed between upper glume and lower lemma and with palea; lower glume ± 0; upper glume very similar to lower lemma; lower lemma 3-5-

veined, acute or slightly apiculate; upper lemma subacute, apiculate; spikelets falling whole.

The ± absence of a lower glume, and the lower floret being represented by little more than a glume-like lemma, produces an apparently 2-glumed spikelet with 1 floret.

Other spp. - **P. dilatatum** Poir., from S Europe (originally S America), is a densely tufted sp. up to 1.5m with dense patent hairs on upper glume; it occurs in waste ground as a seed or grain contaminant.

1. P. distichum L. (*P. paspalodes* (Michx.) Scribn.) - *Water Finger-grass*. Decumbent **881**
stoloniferous perennial; culms to 50cm but usually <20cm high, subglabrous; **917**
racemes 2-7cm; spikelets 2.5-3.5mm, appressed-pubescent on upper glume, glabrous on lemma; (2n=40, 60). Intrd; natd in damp ground by sea at Mousehole (W Cornwall) since 1971 and by canal in E London (Middlesex) since 1984; tropics, natd in Mediterranean.

90. SETARIA P. Beauv. - *Bristle-grasses*
Annuals, or perennials with rhizomes; ligule a dense fringe of hairs; inflorescence a dense spike-like panicle with very short or ± vestigial crowded branches, sometimes interrupted in lower part; spikelets with 2 florets, the lower male or sterile with lemma as long as spikelet, the upper bisexual, slightly smaller, concealed between or protruding from within upper glume and lower lemma, with 1-c.12 strong bristles borne on pedicel and usually awn-like and exceeding spikelets; glumes unequal, the lower ± 0 to c.2/3 as long as spikelet, the upper c.2/3 to as long as spikelet; lower lemma 5-veined, obtuse, awnless; upper lemma obtuse to rounded, awnless; spikelets falling whole at maturity, or (in *S. italica*), the upper floret falling leaving the lower floret and glumes on panicle, the bristles always remaining on panicle.

The number of bristles (1-3, or more) borne on the pedicel below each spikelet is a valuable character but easily over-estimated where spikelets have aborted; in such cases the bristles of aborted spikelets can be wrongly counted in with those of an adjacent well-developed spikelet, but close inspection shows the presence of spikelet-less pedicel(s).

1 Bristles (4)6-8(12) below each spikelet; upper glume scarcely longer than lower glume, c.1/2-2/3 as long as spikelet; lower (sterile) floret with palea almost as long as lemma 2
1 Bristles 1-3 below each spikelet (see note above); upper glume much longer than lower glume, c.2/3-1x as long as spikelet; lower (sterile) floret with palea ≤1/2 as long as lemma 3
　2 Spikes ≥6mm wide when mature (excl. bristles); spikelets (2.5)3-3.3mm **2. S. pumila**
　2 Spikes <5mm wide when mature (excl. bristles); spikelets 2-2.5(3)mm **1. S. parviflora**
3 Main rhachis (often rather sparsely) hispid, with pricklets <0.2mm; bristles usually with backward-directed (rarely with forward-directed) barbs 4
3 Main rhachis densely pubescent with hairs >(0.2)0.5mm; bristles always with forward-directed barbs 5
　4 Spikelets 1.7-2mm; leaf-sheaths glabrous **4. S. adhaerens**
　4 Spikelets 2-2.3mm; leaf-sheaths pubescent on margin **3. S. verticillata**
5 Spikelets disarticulating below upper lemma, leaving glumes and lower lemma on rhachis; panicle often >15 x 1.5cm, the bristles often not or scarcely longer than spikelet-clusters; upper lemma smooth **7. S. italica**
5 Spikelets falling whole, leaving only pedicels and bristles on rhachis;

panicle rarely as much as 15 x 1.5cm, the bristles always much longer than spikelets; upper lemma finely transversely rugose 6

6 Upper glume c.3/4 as long as spikelet; spikelets (2.5)2.7-3mm; leaves pubescent, often sparsely so **6. S. faberi**

6 Upper glume as long or almost as long as spikelet; spikelets (1.8)2-2.5(2.7)mm; leaves glabrous **5. S. viridis**

1. S. parviflora (Poir.) Kerguélen (*S. geniculata* P. Beauv., *Panicum geniculatum* Willd. non Lam.) - *Knotroot Bristle-grass*. Shortly rhizomatous perennial; culms to 75cm; panicles up to 10 x 0.4cm, not interrupted, with densely shortly pubescent rhachis; bristles with forward-directed barbs; spikelets 2-2.5(3)mm; upper lemma rather coarsely transversely rugose; (2n=36, 72). Intrd; casual from wool and birdseed on tips and waste ground; scattered in S En and S Wa; N America. Annual with us and often not developing rhizomes.

2. S. pumila (Poir.) Schult. (*S. lutescens* F.T. Hubb., *S. glauca* auct. non (L.) P. **912**
Beauv.) - *Yellow Bristle-grass*. Annual; culms to 75cm; panicles up to 15 x 1cm, not interrupted, yellowish at maturity, with densely shortly pubescent rhachis; bristles with forward-directed barbs; spikelets (2.5)3-3.3mm; upper lemma coarsely rugose; (2n=36, 72). Intrd; casual weed of cultivated and waste ground and on tips, from wool, birdseed, soyabeans and other sources; occasional in S & C Br and CI, very scattered in N Br and E Ir; warm-temperate Old World.

3. S. verticillata (L.) P. Beauv. (*S. verticilliformis* Dumort., *S. ambigua* (Guss.) **912**
Guss.) - *Rough Bristle-grass*. Annual; culms to 60cm; panicles up to 12 x 1cm, often ± interrupted below; bristles with backward-directed or rarely (var. **ambigua** (Guss.) Parl.) forward-directed barbs; spikelets 2-2.3mm; upper lemma finely transversely rugose; (2n=18, 36, 54). Intrd; casual from wool and birdseed and other sources; habitat and distribution as for *S. glauca*; warm-temperate Old World. *S. verticilliformis* (*S. ambigua*) has been interpreted as the hybrid *S. verticillata* x *S. viridis*, but is in fact a var. of *S. verticillata* with forward-barbed bristles.

4. S. adhaerens (Forssk.) Chiov. (*S. verticillata* ssp. *aparine* (Steud.) T. Durand & Schinz) - *Adherent Bristle-grass*. Differs from *S. verticillata* in bristles always with backward-directed barbs; and see key (couplet 4); (2n=18). Intrd; casual from birdseed and perhaps other sources on tips and waste and cultivated ground; scattered in S En and CI; tropics of Old World. Perhaps better a ssp. or var. of *S. verticillata*.

5. S. viridis (L.) P. Beauv. - *Green Bristle-grass*. Annual; culms to 1m; panicles up to 12(17) x 1.2cm, not interrupted; bristles with forward-directed barbs; spikelets (1.8)2-2.5(2.7)mm; upper lemma finely transversely rugose; (2n=18, 36). Intrd; weed of cultivated and waste ground, sometimes natd, also casual from wool, birdseed and other sources (rarely soyabeans) on tips, etc.; frequent in S & C Br and CI, sporadic elsewhere; warm-temperate Old World.

6. S. faberi Herrm. - *Nodding Bristle-grass*. Differs from *S. viridis* in usually rather larger with markedly curved to pendent (not ± erect) panicles; and see key (couplet 6); (2n=36). Intrd; characteristic casual from soyabean waste, also from grain; scattered in S En, probably overlooked; E Asia, but arriving here from N America (natd).

7. S. italica (L.) P. Beauv. - *Foxtail Bristle-grass*. Annual; culms to 1.5m; panicles up to 30 x 3cm, usually pendent at apex, often interrrupted below; bristles with forward-directed barbs, often shorter than spikelet-clusters; spikelets 2-3mm; upper lemma smooth; (2n=18). Intrd; common casual on tips and waste ground, from birdseed; scattered in Br and CI, mostly S; derived from *S. viridis* as crop probably in China. The well-known cage-birds' 'millet-spray'.

91. DIGITARIA Haller - *Finger-grasses*

Annuals sometimes rooting at lower nodes or rarely perennials; ligule membranous;

inflorescence an umbel of 2-many long narrow racemes or with some racemes borne just below terminal cluster; spikelets with 2 florets, the lower sterile with lemma as long as spikelet, the upper bisexual, as long or slightly shorter, concealed between or protruding from within upper glume and lower lemma; glumes very unequal, the lower very short to ± 0, the upper 1/3-1x as long as spikelet; lower lemma 5-7-veined, obtuse, awnless; upper lemma acute to subacute, awnless; spikelet falling whole.

1 Upper glume and lower lemma ± same length; spikelets 2-2.3(2.5)mm, conspicuously minutely pubescent; leaf-sheaths glabrous except at mouth; pedicel-apex slightly cup-shaped **1. D. ischaemum**
1 Upper glume ≤3/4 as long as lower lemma; spikelets 2.5-3.5(3.8)mm, sparsely appressed-pubescent; leaf-sheaths usually pubescent, sometimes not; pedicels often slightly thicker near apex but simply truncate 2
 2 Lower lemma with smooth veins, often pubescent at edges; upper glume (1/2)2/3-3/4 as long as spikelet, tapering-acute **3. D. ciliaris**
 2 Lower lemma with minutely scabrid veins (x20 lens), rarely pubescent at edges; upper glume 1/3-1/2(2/3) as long as spikelet, rather abruptly acute **2. D. sanguinalis**

Other spp. - **D. brownii** (Roem. & Schult.) Hughes, from Australia, is a rare wool-alien; it differs from the 3 spp. treated here in being perennial and having spikelets with long hairs exceeding the spikelets.

1. D. ischaemum (Schreb. ex Schweigg.) Muhl. - *Smooth Finger-grass*. Culms 912
decumbent to erect, to 35cm, glabrous except at mouth of sheaths; racemes 2-8, 2-8cm; (2n=36, 45). Intrd; weed of cultivated ground, waste places and tips, etc., sometimes ± natd; very scattered in S En, formerly commoner; S Europe.

2. D. sanguinalis (L.) Scop. - *Hairy Finger-grass*. Culms decumbent to erect, to 912
50cm, often pubescent on sheaths and leaves; racemes (2)4-16, 3-16cm; (2n=18, 28, 36-48, 54, 76). Intrd; weed of similar habitats to *D. ischaemum*, sometimes ± natd, but commoner as a birdseed-, wool- or soyabean-alien and less common in cultivated ground; scattered in S Br and CI, once rare but now much commoner than *D. ischaemum*; S Europe.

3. D. ciliaris (Retz.) Koeler (*D. adscendens* (Kunth) Henrard) - *Tropical Finger-grass*. Differs from *D. sanguinalis* in leaves usually glabrous on upperside; and see key; (2n=36, 54, 72). Intrd; casual from similar sources as *D. sanguinalis*, not natd; scattered in S Br; tropics of Old World.

92. CENCHRUS L. - *Sandburs*
Annuals to tufted perennials; ligule a fringe of hairs; inflorescence spike-like, with rhachis bearing groups of 1-few spikelets on very short stalk, the group surrounded and enclosed by spiny bur composed of fused spines and bristles; spikelets with 2 florets, the lower sterile, the upper bisexual; glumes very unequal, the lower often ± 0; lower lemma 5-veined, awnless; spikelets falling as group within bur.

Other spp. - A difficult genus; our spp. have not been worked out, and perhaps several are recorded as *C. echinatus*. The most often reported of the others are **C. incertus** M.A. Curtis and **C. pauciflorus** Benth. (these 2 perhaps the same), from America, and **C. longispinus** (Hack.) Fernald, from Australia. All 3 differ from *C. echinatus* in having bur-spines all similar in several whorls, without finer bristles.

1. C. echinatus L. - *Spiny Sandbur*. Culms to 60cm; panicles to 10cm; burs ± 912
globose, 8-15mm across incl. spines; spines varying from thin bristles at base to 917
flattened plates at apex, all with minute backward-directed barbs; (2n=34, 68). Intrd; infrequent casual on tips and waste ground from wool, birdseed and

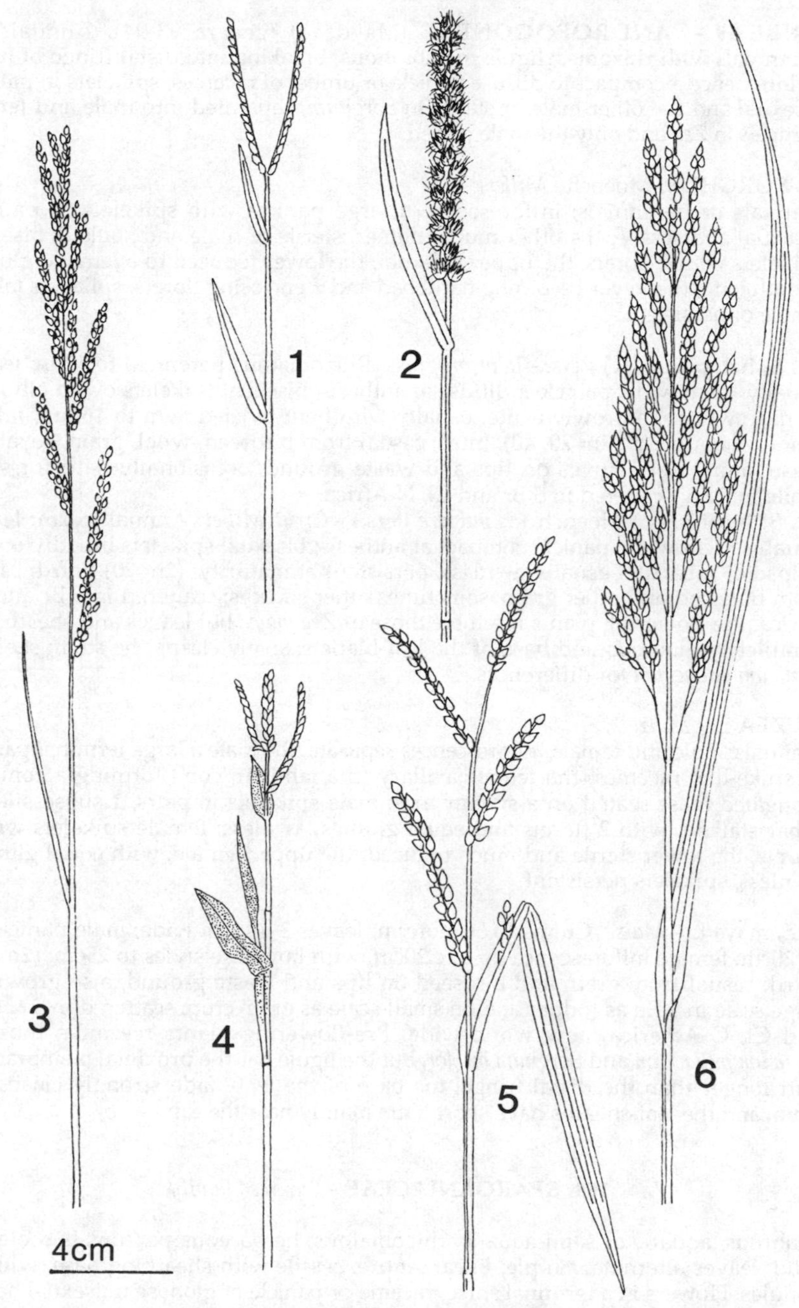

FIG 917 - Poaceae: Panicoideae. 1, *Paspalum distichum*. 2, *Cenchrus echinatus*. 3, *Eriochloa pseudoacrotricha*. 4, *Urochloa panicoides*. 5, *Brachiaria platyphylla*. 6, *Sorghum halepense*.

soyabean waste; scattered in S En; originally America, now pan-tropical.

TRIBE 19 - ANDROPOGONEAE (Maydeae) (genera 93-94). Annuals, or perennials with rhizomes; ligule membranous, breaking into distal fringe of hairs; inflorescence a compact to diffuse panicle or umbel of racemes; spikelets in pairs, 1 bisexual and the other male or sterile in *Sorghum*, separated into male and female panicles in *Zea* and only the male paired.

93. SORGHUM Moench - *Millets*
Annuals or perennials; inflorescence a large panicle with spikelets in pairs, 1 bisexual and sessile, the other much thinner, sterile or male and stalked; bisexual spikelets with 2 florets, the upper bisexual, the lower reduced to a lemma; glumes 2, both long, the lower becoming hardened and ± enclosing florets; spikelets falling whole or persistent.

1. S. halepense (L.) Pers. - *Johnson-grass*. Rhizomatous perennial to 1.5m; leaves **912**
usually <2cm wide; panicle ± diffuse at anthesis; bisexual spikelets ovoid-ellipsoid **917**
or narrowly so, narrowly acute, usually with bent twisted awn to 16mm, falling whole at maturity; (2n=20, 40). Intrd; casual from birdseed, wool, grain, soyabean waste and other sources on tips and waste ground, occasionally natd for short while in S En; scattered in S Br and CI; N Africa.

2. S. bicolor (L.) Moench (*S. vulgare* Pers.) - *Great Millet*. Annual to 2m; leaves usually >2cm wide; panicle compact at anthesis; bisexual spikelets broadly ovoid-ellipsoid, subacute, usually awnless, persistent at maturity; (2n=20). Intrd; casual from birdseed and other grain, sometimes other sources; scattered in S Br and CI; Africa. Pre-flowering plants resemble those of *Zea mays*, but leaves and sheaths are completely glabrous and base of the leaf-blade scarcely clasps the stem; see also *Panicum miliaceum* for differences.

94. ZEA L. - *Maize*
Annuals; male and female inflorescences separate, the male a large terminal panicle of spike-like racemes, the female axillary (the familiar 'cob') forming a compact elongated mass seated on a spongy axis; male spikelets in pairs, 1 subsessile the other stalked, with 2 florets and equal glumes, awnless; female spikelets with 2 florets, the lower sterile and much reduced, the upper female, with equal glumes, awnless; spikelets persistent.

1. Z. mays L. - *Maize*. Culms to 3(-more)m; leaves 3-c.12cm wide; male panicle up to 20cm; female inflorescence up to c.20cm, with immense styles to 25cm; (2n=20). Intrd; casual from grain and birdseed on tips and waste ground, also grown on large scale in S En as fodder and on small-scale as grain crop; scattered in S & C Br and CI; C America, now world-wide. Pre-flowering plants resemble those of *Panicum miliaceum* and *Sorghum bicolor*, but the ligule has the proximal membranous part longer than the distal fringe, the base of the leaf-blade strongly clasps the stem, and the leaf-sheaths have short hairs mainly near the top.

158. SPARGANIACEAE - *Bur-reed family*

Glabrous, aquatic or semi-aquatic, rhizomatous, herbaceous perennials rooted in mud; leaves alternate, simple, linear, entire, sessile with sheathing base, without stipules. Flowers in a terminal spike, raceme or panicle of globose unisexual heads, the more distal heads male, the more proximal female, hypogynous, actinomorphic; perianth of 1-6 inconspicuous scales, often not easy to distinguish from bracts nor to ascribe tepals to 1 or adjacent flowers; male flowers with 3-8 stamens; female flowers with 1-2(3) fused carpels with as many cells, ovules, styles and stigmas;

stigmas linear to subcapitate; fruit a small, dry, spongy drupe with 1-2(3) seeds.

Unmistakable aquatic plants with globose unisexual heads of flowers and fruits.

1. SPARGANIUM L. - *Bur-reeds*

1 Inflorescence branched, the male heads borne at apex of branches as well as of main axis; tepals dark-tipped **1. S. erectum**

1 Inflorescence not branched, the male heads all at apex of main axis; tepals ± translucent, not dark-tipped 2

  2 Stem-leaves not inflated but strongly keeled at base; male heads 3-10, clearly separated **2. S. emersum**

  2 Stem-leaves often inflated but not keeled at base; male heads 1-2(3), if >1 then close together and appearing ± as 1 elongated head 3

3 Bract of lowest female head ≥10cm, ≥2x as long as inflorescence; lowest female peduncle arising from bract axil but fused to main axis for short distance above axil; male heads mostly 2 **3. S. angustifolium**

3 Bract of lowest female head <10cm, barely longer than inflorescence; lowest female peduncle arising directly from bract axil; male head usually 1 **4. S. natans**

1. S. erectum L. - *Branched Bur-reed.* Stems erect, to 1.5m; leaves strongly keeled throughout; inflorescence branched, with >1 head on main branches; (2n=30). Native; by ponds, lakes, slow rivers and canals, in marshy fields and ditches; common throughout most of BI.

1 Fruits distinctly shouldered below beak 2

1 Fruits gradually rounded below beak 3

  2 Fruit with flat top (excl. beak) (3)4-6(7)mm across **a. ssp. erectum**

  2 Fruit with rounded (domed) top (excl. beak) 2.5-4.5mm across **b. ssp. microcarpum**

3 Fruit ellipsoid, gradually tapered to beak, 2-4.5mm across **c. ssp. neglectum**

3 Fruit subglobose, abruptly contracted to beak, 4-7mm across **d. ssp. oocarpum**

a. Ssp. erectum. Fruits often dark brown at apex, with flat top with beak in **920**
centre. Mainly C & S Br, Man, rare in Ir.

b. Ssp. microcarpum (Neuman) Domin. Fruits often dark brown at apex, with **920**
rounded top very abruptly narrowed to beak. Throughout Br, Ir and Man.

c. Ssp. neglectum (Beeby) K. Richt. Fruits pale brown, ellipsoid, gradually **920**
narrowed to beak. Throughout Br and Ir.

d. Ssp. oocarpum (Celak.) Domin. Fruits pale brown, subglobose, abruptly **920**
narrowed to beak. Ir, S & C Br, CI. Usually very few fruits form per head; possibly the hybrid ssp. *erectum* x ssp. *neglectum*.

2. S. emersum Rehmann - *Unbranched Bur-reed.* Stems erect, sometimes ± floating, **920**
to 60cm; aerial leaves strongly keeled at least in lower part, not inflated at base; inflorescence simple, the heads sessile or stalked; fruits ellipsoid, tapered to beak; (2n=30). Native; similar places to *S. erectum* but rarely not in water; frequent throughout BI.

2 x 3. S. emersum x S. angustifolium = S. x diversifolium Graebn. has been found in a few places in W Sc from Wigtowns to W Sutherland; it has the inflated leaf-bases of *S. angustifolium* and the remote male heads of *S. emersum*, and is fertile and backcrosses.

3. S. angustifolium Michx. - *Floating Bur-reed.* Stems usually floating (rarely **920**
erect), to 1m; leaves usually floating, inflated at base, flat distally; inflorescence simple, the heads sessile or stalked; fruits ellipsoid, tapered to beak; (2n=30). Native; in peaty, acid lakes or pools; local in W Wa, NW En, W & N Sc, Ir.

4. S. natans L. (*S. minimum* Wallr.) - *Least Bur-reed.* Stems usually floating (very **920**

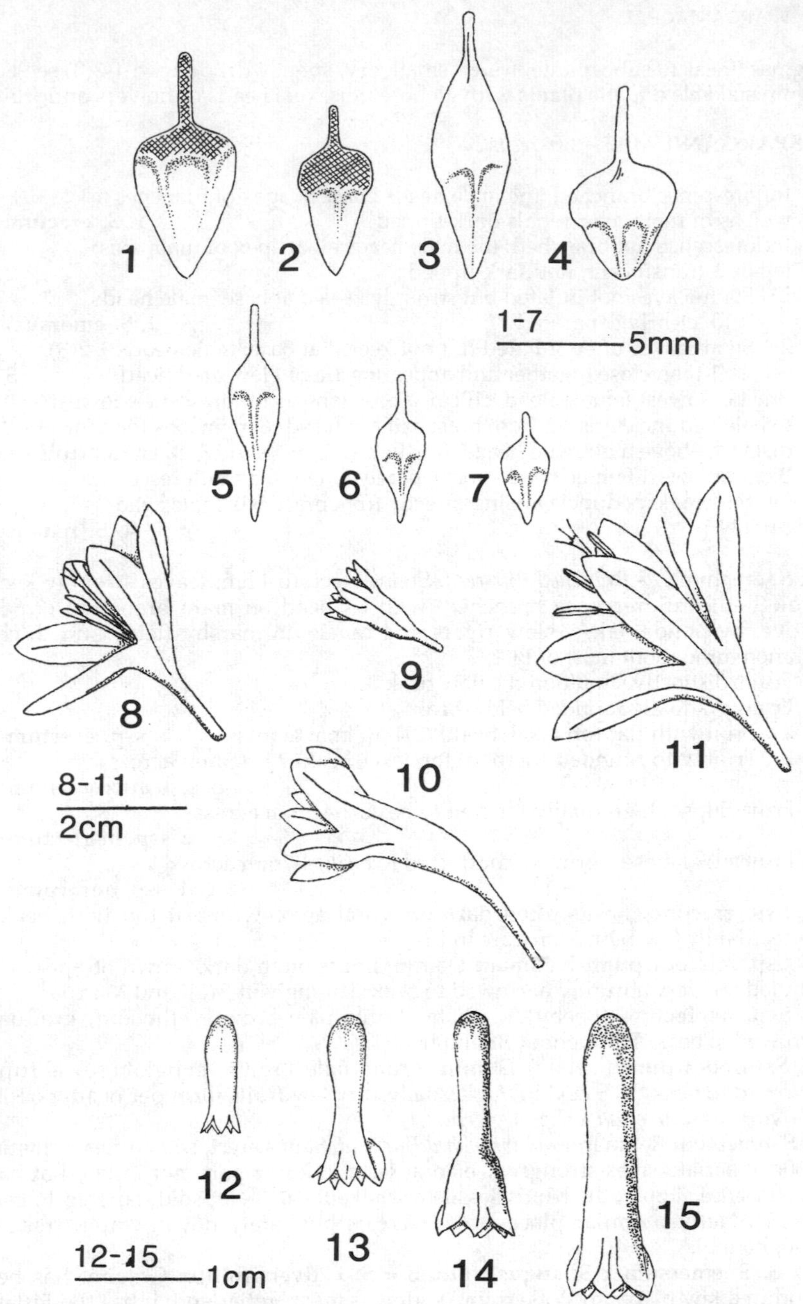

FIG 920 - *Sparganium, Crocosmia, Polygonatum*. 1-7, fruits of *Sparganium*. 1, *S. erectum* ssp. *erectum*. 2, *S. erectum* ssp. *microcarpum*. 3, *S. erectum* ssp. *neglectum*. 4, *S. erectum* ssp. *oocarpum*. 5, *S. emersum*. 6, *S. angustifolium*. 7, *S. natans*. 8-11, flowers (excl. calyx) of *Crocosmia*. 8, *C. x crocosmiiflora*. 9, *C. pottsii*. 10, *C. paniculata*. 11, *C. masoniorum*. 12-15, corollas of *Polygonatum*. 12, *P. verticillatum*. 13, *P. multiflorum*. 14, *P. x hybridum*. 15, *P. odoratum*.

rarely erect), to 50cm; leaves usually floating, barely inflated at base, flat; inflorescence simple, the heads sessile or stalked; fruits ellipsoid-obovoid, tapered to beak; (2n=30). Native; in acid or alkaline lakes, pools or ditches with high organic content; scattered over most of Br and Ir but absent from most of C & S En and S Wa.

159. TYPHACEAE - *Bulrush family*

Glabrous, aquatic or semi-aquatic, rhizomatous, herbaceous perennials rooted in mud; leaves alternate, simple, linear, entire, sessile with sheathing base, without stipules. Flowers very numerous in dense, cylindrical, complex spike, unisexual, the male distal, the female lower down (either contiguous or separated by short bare axis), hypogynous, actinomorphic; perianth of few-many bristles and/or narrow scales; male flowers with 1-5 stamens with usually fused filaments; female flowers with 1 1-celled, stalked ovary with 1 ovule; style 1; stigma clavate to linear; fruit a small 1-seeded capsule.

Unmistakable aquatic plants with conspicuous cylindrical inflorescence, male distally, female proximally.

1. TYPHA L. - *Bulrushes*

1. T. latifolia L. - *Bulrush.* Stems erect, to 3m, usually overtopped by some leaves; leaves 8-24mm wide; male and female parts of spike usually contiguous (rarely ≤2.5cm apart); female part mostly 18-30mm wide, without scales (but with many bristles); 2n=30. Native; reed-swamps, lakes, ponds, slow rivers, ditches; frequent throughout most of BI but absent from much of N & W Sc.

1 x 2. T. latifolia x T. angustifolia = T. x glauca Godr. occurs in scattered places throughout En and Wa and in Stirlings and is probably overlooked; it is variously intermediate in all characters and highly (?completely) sterile.

2. T. angustifolia L. - *Lesser Bulrush.* Differs from *T. latifolia* in leaves 3-6(10)mm wide; male and female parts of inflorescence separated by (0.5)3-8(12)cm; female part mostly 13-25mm wide, with dark scales as well as bristles; (2n=30, 60). Native; similar places to *T. latifolia* but often on more organic soils; scattered throughout most of BI but absent from C & N Sc, Man and most of Ir.

160. BROMELIACEAE - *Rhodostachys family*

Glabrous, glaucous, dome-shaped, evergreen almost woody plants; leaves in close spiral, simple, linear, with strong spines on margins, sessile with sheathing base, without stipules. Flowers ± sessile in dense ± globose terminal heads c.5cm across, bisexual, epigynous, ± actinomorphic; perianth of 3 outer and 3 inner free tepals, coloured; stamens 6; ovary 3-celled, each cell with numerous ovules on axile placenta; style 1; stigmas 3, linear; fruit a berry.

Unmistakable large dome-shaped pineapple-like plants with spiny linear leaves and globose inflorescences.

1 Inflorescences ± sessile in terminal leaf-rosettes; petals blue, with basal scale-like nectaries; leaves flat distally, with marginal spines ± all apically directed **1. FASCICULARIA**

1 Inflorescences arising from terminal leaf-rosettes on distinct stalks ≥10cm; petals pink, without scale-like nectaries; leaves concave throughout, with the lower marginal spines patent to recurved **2. OCHAGAVIA**

1. FASCICULARIA Mez - *Rhodostachys*

Inflorescences sessile; petals blue, with basal scale-like nectaries.

1. F. bicolor (Ruiz & Pav.) Mez (*F. pitcairniifolia* auct. non (Verlot) Mez) - *Rhodostachys*. Plant dome-shaped, to 75cm high and 2m across; leaves up to 35cm, with sharp marginal forward-directed spines, the lower spines patent with forward-directed tips, the most apical leaves turning red at base at flowering. Intrd; well established where planted on maritime dunes or shingle in Scillies and Guernsey; Chile.

2. OCHAGAVIA Phil. - *Tresco Rhodostachys*
Inflorescences stalked; petals pink, without scale-like nectaries.

1. O. carnea (Beer) L.B. Sm. & Looser - *Tresco Rhodostachys*. Very similar to *F. bicolor* in general appearance but leaves shorter (<25cm), with longer, stouter spines of which the lower are patent to recurved, and not or scarcely turning red at base; and see key. Intrd; well established where planted on dunes in Tresco, Scillies; Chile.

161. PONTEDERIACEAE - *Pickerelweed family*

Aquatic, glabrous, rhizomatous, herbaceous perennials rooted in mud; leaves emergent, alternate, simple, entire, petiolate, with sheathing base, without stipules. Flowers sessile in simple terminal emergent spike with bladeless leaf-sheath below it, bisexual, hypogynous, slightly zygomorphic; perianth of 6 violet-blue tepals fused at base; stamens 6; ovary 3-celled with 2 cells empty, or 1-celled; ovule 1; style 1; stigma capitate; fruit dry, 1-seeded, indehiscent.

The only aquatic with ovate-cordate leaves and a spike of blue flowers.

1. PONTEDERIA L. - *Pickerelweed*

1. P. cordata L. - *Pickerelweed*. Semi-submerged, with creeping or floating stems to 1m; leaves mostly basal, 5-25cm, triangular-ovate to narrowly so, mostly cordate at base, with long petioles; spike 3-15cm, dense; flowers 6-10mm across; (2n=16). Intrd; grown for ornament, natd at edges of ponds; scattered in S En, S Lancs; N America.

162. LILIACEAE - *Lily family*

(Alliaceae, Alstroemeriaceae, Amaryllidaceae, Asparagaceae, Asphodelaceae, Colchicaceae, Convallariaceae, Hemerocallidaceae, Hyacinthaceae, Melanthiaceae, Ruscaceae, Trilliaceae)

Usually erect, mostly glabrous, herbaceous perennials or occasionally small evergreen shrubs, rhizomatous or with a corm or bulb or tuberous roots; leaves all basal or alternate, occasionally whorled, simple, entire, sessile or petiolate, with or without sheathing base, without stipules, sometimes reduced to scales and functionally replaced by leaf-like lateral stems (*cladodes*). Flowers solitary (terminal or axillary) or in racemes or umbels, less often in cymes or panicles, bisexual or rarely functionally dioecious, usually conspicuous, hypogynous or epigynous, actinomorphic or slightly zygomorphic; perianth usually of 6 free to fused tepals, often brightly coloured, the 3 outer not or obviously different from the 3 inner, rarely 4 or in 2 whorls of 4-6 each, sometimes with long or short funnel- or collar-shaped *corona* within the rows of tepals; stamens usually 6, rarely 3, 4 or 7-12; ovary 3(-5)-celled with numerous (or fewer, down to 2) ovules on axile placentas; styles 3-4(5) with capitate to linear stigma, or 1 with 1 or 3-4(5) stigmas; fruit a capsule or berry, sometimes a succulent capsule.

A very variable family, usually recognized by conspicuous flowers with 6 petaloid tepals, 6 stamens and 3-celled ovary; exceptions are *Maianthemum*, *Ruscus* and *Paris*. All but *Ruscus* are herbaceous perennials arising from a corm, bulb, rhizome or tuberous roots. See Pontederiaceae, Iridaceae and Agavaceae for differences.

Cronquist's circumscription of the Liliaceae is drawn very widely; some other classifications differ considerably and up to 13 extra families are sometimes recognized. The subfamilial classification adopted here is more or less that of Melchior, but numerous other versions exist.

General key

1 Leaves *Iris*-like, i.e. vertical, with 2 identical surfaces, borne on 2 opposite sides of stem, each with leaf-base sheathing that of next higher leaf 2

1 Leaves not *Iris*-like 3

2 Styles 3; tepals creamy- or greenish-white; filaments glabrous **1. TOFIELDIA**

2 Style 1; tepals yellow; filaments densely pubescent **2. NARTHECIUM**

3 Leaves (3)4(-8), in 1 whorl on stem below single flower **18. PARIS**

3 Leaves not all in 1 whorl on stem below flower; flowers often >1 4

4 Flowers *Crocus*-like, appearing from soil in autumn without leaves or stems; ovary subterranean, emerging with leaves to fruit in spring **7. COLCHICUM**

4 Flowers with stems and leaves; if flower ± *Crocus*-like then ovary above ground at flowering 5

5 Leaves all reduced to small scales, with green ± leaf-like stems (cladodes) arising from main stems in axils 6

5 At least some leaves green and photosynthetic 7

6 Evergreen shrub; cladodes elliptic, very rigid, borne singly **38. RUSCUS**

6 Herb; cladodes linear, soft, borne in clusters of ≥4 **37. ASPARAGUS**

7 Flowers in an umbel with 1-few spathe-like bracts at base, or solitary with spathe-like bract(s) at base, sometimes flowers replaced by bulbils ***Key A***

7 Flowers in a cyme or raceme with or without bracts but without basal spathe-like bract(s), or rarely solitary or in umbel without spathe-like bract(s) at base ***Key B***

Key A - Flowers in an umbel or solitary, with 1-few spathe-like bracts at base (excl. *Colchicum*)

1 Flowers entirely replaced by bulbils 2

1 At least some flowers present 3

2 Stems <4cm; leaves <10cm **9. GAGEA**

2 Stems >5cm; at least some leaves >15cm **25. ALLIUM**

3 Funnel- or collar-like corona present inside perianth 4

3 Corona absent 5

4 Stamens arising from below corona and not fused to it; flowers winter to spring, white to yellow **35. NARCISSUS**

4 Stamens fused to corona at base; flowers summer, white **36. PANCRATIUM**

5 Ovary inferior or semi-inferior 6

5 Ovary superior 11

6 Perianth yellow **32. STERNBERGIA**

6 Perianth white, pink or red, often tinged green 7

7 Flowers pink or red, often tinged green, rarely white but then without green or yellow patches 8

7 Flowers white with green or yellow patches 10

8 Ovary semi-inferior; perianth <2cm, greenish-red, with free tepals **26. NECTAROSCORDUM**

8 Ovary inferior; perianth >4cm, bright pink, with a proximal tube 9

9 Perianth-tube <2cm, funnel-shaped, gradually widened into lobed

part of perianth; spathe herbaceous; flowers appearing before leaves; leaves flat **30. AMARYLLIS**

9 Perianth-tube >3cm, narrowly tubular for most part, abruptly widened into lobed part of perianth; spathe scarious; flowers appearing with leaves; leaves channelled **31. CRINUM**

10 All 6 tepals similar **33. LEUCOJUM**

10 3 inner tepals much shorter and blunter than 3 outer **34. GALANTHUS**

11 Perianth yellow 12

11 Perianth white to various shades of red or blue 13

12 Leaves >12mm wide; bracts (spathes) at base of inflorescence ovate, <15mm; flowers >5 **25. ALLIUM**

12 Leaves ≤12mm wide; bracts (spathes) at base of inflorescence linear, >15mm; flowers 1-5 **9. GAGEA**

13 Perianth with tube >10mm, usually some shade of blue 14

13 Perianth with tepals free or fused for <5mm, very rarely blue or bluish 15

14 Flowers in umbel, held horizontally, slightly zygomorphic, 3-5cm **28. AGAPANTHUS**

14 Flowers solitary, erect, actinomorphic, 2.5-3.5cm **29. TRISTAGMA**

15 Tepals free; style arising from base of ovary; plant with onion-like smell when fresh **25. ALLIUM**

15 Tepals fused at base; style arising from top of ovary; plant without onion-like smell **27. NOTHOSCORDUM**

Key B - Flowers in a cyme or raceme, rarely in an umbel or solitary, without spathe-like bract at base (excl. *Tofieldia, Narthecium, Paris*)

1 Ovary inferior **39. ALSTROEMERIA**

1 Ovary superior 2

2 Tepals and stamens 4; leaves strongly cordate at base **16. MAIANTHEMUM**

2 Tepals and stamens 6(-8); leaves cuneate to rounded at base 3

3 Tepals united into proximal tube >1/5 of their length 4

3 Tepals free or united just at extreme base 11

4 Perianth yellow to orange or red, >3.5cm 5

4 Perianth white to blue, pink or purple, very rarely pale yellow, ≤3.5cm 6

5 Flowers very numerous; perianth <5cm, tubular to narrowly bell-shaped **6. KNIPHOFIA**

5 Flowers <c.20; perianth >5cm, funnel-shaped **5. HEMEROCALLIS**

6 Flowers borne in groups of 1-c.5 in axils of main foliage leaves **15. POLYGONATUM**

6 Flowers terminal, or in terminal inflorescences, with bracts 0 or much reduced from leaves 7

7 Leaves linear to narrowly elliptic, narrowed at base; plant rhizomatous 8

7 Leaves linear, not narrowed at base; plant with bulb 9

8 Flowers pendent, stalked, usually white, with perianth-tube longer than -lobes; leaves with distinct petiole, up to 30 x 10cm incl. petiole **14. CONVALLARIA**

8 Flowers erect to patent, sessile, pink, with perianth-tube shorter than -lobes; leaves only slightly narrowed at base, up to 40 x 2cm **17. REINECKEA**

9 Perianth-tube >2x as long as lobes; corolla contracted at mouth **24. MUSCARI**

9 Perianth-tube <2x as long as lobes; corolla spread open at mouth 10

10 Perianth-tube much shorter than -lobes, the lobes bent outwards at junction with tube **23. CHIONODOXA**

10 Perianth-tube c. as long as -lobes, the lobes gradually curved outwards **22. HYACINTHUS**

11 Inflorescence with few-many flowers, each bractless or with bracts much reduced from leaves; leaves all basal 12
11 Inflorescence with 1-several flowers; stems bearing at least 1 leaf, or if leaves all basal at least lowest bract ± leaf-like or flower 1 16
12 Inflorescence a panicle; filaments densely pubescent **4. SIMETHIS**
12 Inflorescence a raceme; filaments glabrous 13
13 Bracts 2 per flower; tepals fused at extreme base **21. HYACINTHOIDES**
13 Bracts 0 or 1 per flower; tepals free 14
14 Tepals usually blue, sometimes pink or pure white **20. SCILLA**
14 Tepals white with green to reddish-brown stripe on abaxial surface 15
15 Plant with bulb; bracts whitish **19. ORNITHOGALUM**
15 Plant without bulb, with swollen roots; bracts brown **3. ASPHODELUS**
16 Tepals <2cm 17
16 Tepals ≥2cm 18
17 Tepals white with purplish veins **8. LLOYDIA**
17 Tepals yellow with green stripes or tinge on abaxial side **9. GAGEA**
18 Stigmas sessile **11. TULIPA**
18 Stigmas on obvious style 19
19 Stigmas not or scarcely longer than wide; filaments loosely fixed to middle of anther; flowers several **13. LILIUM**
19 Stigmas linear; filaments ± rigidly fixed to base of anthers; flowers mostly 1 20
20 Stem-leaves 2, near stem-base, reddish-blotched; tepals strongly reflexed **10. ERYTHRONIUM**
20 Stem-leaves 3-6(8), most not near stem-base, not blotched; tepals not reflexed **12. FRITILLARIA**

Other genera - Several genera of garden plants are very persistent where thrown out or neglected. **VERATRUM viride** Aiton (*Green False-helleborine*) (Asphodeloideae), from N America, differs from *Asphodelus* in having elliptic leaves on the stems, stems to 2m and green flowers in profuse terminal panicles; 1 clump survived in Dunbarton from at least 1980 until c.1993. **CAMASSIA quamash** (Pursh) Greene (*Quamash*) (Scilloideae), from W N America, resembles a large (to 80cm) robust *Scilla*, but has 3-veined white to blue tepals; a patch was found on a grassy verge in Surrey in 1995.

SUBFAMILY 1 - MELANTHIOIDEAE (tribe Narthecieae) (genera 1-2). Plant rhizomatous; leaves all or mostly basal, *Iris*-like (vertical, flat with 2 identical faces); inflorescence a terminal raceme; tepals yellow to greenish-white, free; styles 1 or 3; ovary superior; fruit a capsule.

1. TOFIELDIA Huds. - *Scottish Asphodel*
Tepals creamy- or greenish-white; filaments glabrous; anthers c. as long as wide, dehiscing inwards; styles 3; capsule splitting where ovary-cells meet; seeds ovoid-curved.

1. T. pusilla (Michx.) Pers. - *Scottish Asphodel.* Stems to 20cm, with 5-10 flowers near apex; leaves up to 8cm x 3mm; tepals 1.5-2.5mm; 2n=30. Native; by streams and in flushes on mountains; very local in N En (Upper Teesdale), locally frequent in C & N Sc, possibly long ago in Leics.

2. NARTHECIUM Huds. - *Bog Asphodel*
Tepals yellow; filaments densely pubescent; anthers >2x as long as wide, dehiscing outwards; style 1; capsule splitting along centre of ovary-cells; seeds with long fine projections at each end.

1. N. ossifragum (L.) Huds. - *Bog Asphodel.* Stems to 45cm, with 6-20 flowers near apex; leaves up to 30cm x 5mm; tepals 6-9mm; 2n=26. Native; bogs and other wet peaty acid places on heaths, moors and mountains; common in Ir and W & N Br, absent from most of C & E En.

SUBFAMILY 2 - ASPHODELOIDEAE (tribes Asphodeleae, Hemerocallideae) (genera 3-6). Plant rhizomatous or with swollen roots; leaves all or nearly all basal, linear; inflorescence a raceme or terminal compound cyme; tepals various colours (not blue); ovary superior; style 1; fruit a capsule splitting along centre of ovary-cells.

3. ASPHODELUS L. - *White Asphodel*
Plant with swollen roots; inflorescence racemose; tepals white with greenish to reddish-purple stripe on outside, erecto-patent, free; filaments glabrous; ovules 2 per cell; flowers actinomorphic.

Other spp. - A. fistulosus L. (*Hollow-leaved Asphodel*), from Mediterranean, a smaller plant with hollow subterete leaves to 35cm, a well-branched inflorescence and tepals 5-12mm, is a rare wool- and grain-alien.

1. A. albus Mill. - *White Asphodel.* Stems to 1m, with dense ± unbranched terminal raceme; leaves up to 60 x 3cm, flat and strongly keeled; tepals 15-20mm; (2n=28, 56). Intrd; natd on grassy bank in Jersey since early 1970s; S Europe.

4. SIMETHIS Kunth - *Kerry Lily*
Rhizomatous; inflorescence cymose; tepals purplish outside, white inside, ± patent, free; filaments densely pubescent; ovules 2 per cell; flowers actinomorphic.

1. S. planifolia (L.) Gren. - *Kerry Lily.* Stems to 40cm, with terminal ± lax panicle; **RRR**
leaves up to 50cm x 7.5mm; tepals 8-11mm; (2n=24, 48). Native; rocky heathland near sea with *Ulex* over c.30 square km near Derrynane, S Kerry, formerly natd on heathland near sea in Dorset and S Hants.

5. HEMEROCALLIS L. - *Day-lilies*
Rhizomatous; inflorescence cymose; tepals yellow to orange, fused to form proximal tube, funnel- to trumpet-shaped; filaments glabrous; ovules many per cell; flowers slightly zygomorphic, at least by upward curvature of stamens and style in laterally-directed flowers.

1. H. fulva (L.) L. - *Orange Day-lily.* Stems to c.1m, with up to c.20 flowers; leaves up to 90 x 2.5cm; flowers dull orange, 7-10cm, ± scentless; (2n=22, 33, 36). Intrd; much grown in gardens and very persistent when neglected or thrown out, often forming dense clumps on rough ground, banks and grassy places; scattered ± throughout Br and CI; garden origin.

2. H. lilioasphodelus L. - *Yellow Day-lily.* Stems to c.80cm, with up to c.12 flowers; leaves up to 65 x 1.5cm; flowers yellow, 7-8cm, sweetly scented; (2n=22). Intrd; natd as for *H. fulva;* scattered in Br, less common than *H. fulva* in En and Wa but more common in Sc; E Asia.

6. KNIPHOFIA Moench - *Red-hot-pokers*
Densely tufted, with short rhizomes; inflorescence a dense raceme; tepals red to yellow or greenish-white, fused to form proximal tube; cylindrical to narrowly bell-shaped; filaments glabrous; ovules many per cell; flowers very slightly zygomorphic due to curvature of perianth and stamens.

Sometimes placed in Aloeaceae.

Other spp. - The garden plants escaping into the wild are not well understood; probably many or most are hybrids in which case attempts to identify them as spp. are futile; **K. ensifolia** Baker might also be present in Alderney.

1. K. uvaria (L.) Oken - *Red-hot-poker*. Stems to 1.2m; leaves up to 80 x 1.8cm, V-shaped in section; bracts 3-9mm, ovate to oblong-ovate, rounded to subacute at apex; raceme up to 12 x 6cm; perianth 2.8-4cm, red at first, becoming yellow and pendent; stamens included or just exserted; (2n=12). Intrd; much grown in gardens, very persistent where thrown out or planted, dunes or waste ground usually near sea; scattered in S & W Br, extending to E Kent and E Suffolk, CI; S Africa.

2. K. x praecox Baker - *Greater Red-hot-poker*. Differs from *K. uvaria* in stems to 2m; leaves up to 200 x 4cm; bracts 8-12mm, lanceolate to linear-oblong, acute to acuminate at apex or rounded at extreme apex; raceme 12-30 x 6-7cm; perianth 2.4-3.5cm; stamens exserted 4-15mm ± as soon as corolla opens; (2n=12). Intrd; grown and natd as for *K. uvaria* and much confused with it; probably similar distribution, E & W Cornwall and CI and N to Caerns and Flints; S Africa. Probably a hybrid complex of *K. uvaria*, **K. bruceae** (Codd) Codd and **K. linearifolia** Baker. Plants from Flints (very well natd on dunes) have been wrongly determined as *K. rufa* Baker.

SUBFAMILY 3 - WURMBAEOIDEAE (tribe Colchiceae) (genus 7). Plant with a corm; leaves ± all on stem, appearing in spring with fruits, linear-oblong; flowers appearing in autumn without leaves, 1-few each arising from ground; tepals pinkish to pale purple, united into long tube proximally; ovary superior, but subterranean at flowering, emerging above ground at stem apex at fruiting; styles 3; fruit a capsule splitting where ovary-cells meet.

Crocus-like, but with 6 (not 3) stamens and 3 simple stigmas.

7. COLCHICUM L. - *Meadow Saffron*

1. C. autumnale L. - *Meadow Saffron*. Tepals with narrow erect tube 5-20cm; capsule(s) produced on stem with sheathing leaves; leaves up to 35 x 5cm; (2n=24, 36, 38, 42). Native; damp meadows and open woods on rich soils; local in C & S Br (common around Severn estuary), very local in S Ir, natd in Jersey, Kirkcudbrights and Co Armagh.

SUBFAMILY 4 - LILIOIDEAE (tribes Tulipeae, Lilieae, Convallarieae, Polygonateae, Parideae) (genera 8-18). Plant rhizomatous or with a bulb; leaves all basal, all on stem, or both, linear to ovate or elliptic; inflorescence a single flower or a spike or raceme; tepals various colours but not blue, free or fused; style 1 or 0; ovary superior; fruit a berry or a capsule splitting along centre of ovary-cells.

8. LLOYDIA Salisb. ex Rchb. - *Snowdon Lily*
Plant with a bulb; leaves mostly basal, some on stem, linear or ± so; flowers 1-2(3) at stem-apex; tepals free, erecto-patent; style 1; fruit a capsule.

1. L. serotina (L.) Rchb. - *Snowdon Lily*. Stems to 15cm, with 2-4 leaves; basal **RRR**
leaves up to 25cm x 2mm; tepals 9-12mm, white with purple veins; 2n=24. Native; cracks in basic mountain rocks; very local in c.5 sites in Caerns.

9. GAGEA Salisb. - *Star-of-Bethlehems*
Plant with 1-2 bulbs; leaves basal (1-4) and on stem (1-4), linear or ± so; flowers 1-5 at stem-apex; tepals free, ± patent; style 1; fruit a capsule.

1. G. lutea (L.) Ker Gawl. - *Yellow Star-of-Bethlehem*. Stems to 25cm, with 1 bulb at base, with 1-5 subumbellate flowers and 2-3 leaf-like bracts; basal leaf usually 1, 15-45cm x 7-15mm; tepals yellow, 10-18mm; 2n=72. Native; damp base-rich woods, hedgerows and rough fields; scattered in Br N to C Sc, rare except in C & N En. Flowers Mar-May.

2. G. bohemica (Zauschn.) Schult. & Schult. f. - *Early Star-of-Bethlehem.* Stems to 4cm, with 2 bulbs at base, with 1(-4) subumbellate flowers and usually 4-6 leaf-like bracts; basal leaves usually 2 per bulb, 4-9cm x c.1mm; tepals yellow, 12-18mm; 2n=60. Native; cracks and ledges of basic rocks; 1 site in Rads, discovered 1965. Flowers Jan-Mar. **RRR**

10. ERYTHRONIUM L. - *Dog's-tooth-violet*
Plant with a bulb; leaves 2, at base of stem, elliptic-oblong, reddish-blotched; flower 1, pendent; tepals free, sharply reflexed; style 1; fruit a capsule.

1. E. dens-canis L. - *Dog's-tooth-violet*. Stems to 30cm; leaves up to 9 x 4cm; tepals 18-30mm, bright pink or sometimes white, reflexed back to expose long-exserted stamens and style; (2n=24). Intrd; planted in woodland, natd in old estates and parks; Midlothian and Fife, formerly NW En; S Europe.

11. TULIPA L. - *Tulips*
Plant with a bulb; leaves basal and on stem, elliptic to linear-elliptic; flowers 1-2 at stem apex, erect at maturity; tepals free, forming cup- or bowl-shaped flower; style 0; fruit a capsule.

1 Flower rounded at base; filaments completely glabrous; buds erect **3. T. gesneriana**
1 Flowers cuneate at extreme base due to narrowing of perianth; filaments pubescent near base; buds pendent 2
  2 Tepals ± uniform yellow **1. T. sylvestris**
  2 Tepals pink to purple with yellow blotch on inside at base **2. T. saxatilis**

1. T. sylvestris L. - *Wild Tulip*. Stems to 50cm; leaves linear-elliptic, up to 30 x 1.8cm; flowers 1(-2); tepals ± uniform yellow, 2-6cm; (2n=24, 48). Intrd; natd in woods, meadows and neglected estates; scattered in En and S & C Sc, formerly much more frequent; S Europe.

2. T. saxatilis Sieber ex Spreng. - *Cretan Tulip*. Stems to 50cm; leaves linear-elliptic to narrowly elliptic, up to 35 x 4.5cm; flowers 1-2(4); tepals pink to purple with basal yellow blotch inside, 3.5-5.5cm; (2n=36). Intrd; natd on stony rough ground, Tresco (Scillies), since 1976; Crete.

3. T. gesneriana L. - *Garden Tulip*. Stems to 60cm; leaves elliptic to narrowly so, up to 35 x 8cm; flower 1; tepals variously yellow to red, pink, purple or white, often with basal blackish blotch inside, 4-8cm; (2n=24, 36, 48). Intrd; the common garden tulip, persistent on tips, waysides and rough ground where thrown out; scattered in C & S En and CI; garden origin.

12. FRITILLARIA L. - *Fritillary*
Plant with a bulb; leaves all or most on stem, linear; flowers 1(-few) at stem apex, pendent even at maturity; tepals free, forming cup-, bowl- or bell-shaped flower; style 1, with 3 linear stigmas; fruit a capsule.

Other spp. - F. pyrenaica L., from SW Europe, has been reported as persisting in S Hants; it differs from *F. meleagris* in its shorter, narrower, bell-shaped perianth which is dark purplish-brown.

1. F. meleagris L. - *Fritillary*. Stems to 30(50)cm, with 3-6(8) leaves up to 20 x 1cm; perianth cup- to bowl-shaped, conspicuously chequered light and dark purple and cream, sometimes white, 3-5cm; (2n=24). Doubtfully native; damp meadows and pastures; local in En N to Staffs and W Norfolk, much less common than **R**

formerly, now frequent only in Suffolk and Thames Valley, planted and natd more widely in En and in Wa.

13. LILIUM L. - *Lilies*
Plant with a bulb; leaves all or most on stem, linear to elliptic; flowers few to many in terminal raceme, with free tepals; flowers pendent with tepals diverging and very strongly rolled back forming 'Turk's-cap' shape in our 2 spp.; flowers erect or facing sideways with slightly recurved tepals in other spp.; style 1, with 3-lobed stigma; fruit a capsule.

Other spp. - Several showy garden taxa may persist for a short while where thrown out or neglected. Most common are **L. candidum** L. (*Madonna Lily*), from Greece, with white tepals and elliptic leaves; **L. regale** E.H. Wilson (*Royal Lily*), from China, with tepals purple outside and white inside and linear leaves; and **L. x hollandicum** Woodcock & Stearn (*L. bulbiferum* L. x *L. maculatum* Thunb.) (*Orange Lily*), of garden origin, with red, orange or yellow tepals and narrowly elliptic leaves. None of these has 'Turk's-cap' flowers; the first two have erect to horizontal funnel-shaped and the third erect cup-shaped flowers. **L. bulbiferum** itself, from Europe, has also been recorded; it differs from *L. x hollandicum* in its more pubescent stem, always orange tepals and 3-7-veined (not 3-veined) leaves.

1. L. martagon L. - *Martagon Lily*. Stems to 1.5m; leaves oblanceolate to elliptic, at least some in whorls; flowers c.4cm across, with tepals up to 3.5cm when straightened, purple with darker spots, sometimes white, sickly scented; (2n=24). Intrd; well natd in woods; scattered in Br N to C Sc, Man; Europe.

2. L. pyrenaicum Gouan - *Pyrenean Lily*. Stems to 1m; leaves linear, spiral; flowers c.5cm across, with tepals up to 6.5cm when straightened, greenish- to orange-yellow with darker spots, strongly sickly scented; (2n=24). Intrd; well natd in woods, hedgerows and field margins; scattered in Br, mostly W & N, Man, Jersey; Pyrenees.

14. CONVALLARIA L. - *Lily-of-the-valley*
Plant rhizomatous; leaves all basal or ± at base of stem, elliptic or narrowly so; flowers 6-20 in long terminal raceme, pendent; tepals fused into tube longer than lobes, forming bell-shaped flower, white; style 1; fruit a red berry.

1. C. majalis L. - *Lily-of-the-valley*. Stems to 25cm; leaves up to 30 x 10cm; flowers 5-10 x 5-10mm, strongly sweetly scented; (2n=32, 36, 38). Native; dry woods, scrub and hedgebanks usually on base-rich soil; scattered through most of Br N to C Sc, much grown in gardens and natd in many places (often more robust plants than native ones).

15. POLYGONATUM Mill. - *Solomon's-seals*
Plant rhizomatous; leaves all on stem, linear-elliptic to elliptic; flowers 1-6 in axillary, stalked, pendent clusters; tepals fused into ± cylindrical tube longer than lobes, white or cream with green markings; style 1; fruit a purple or bluish-black berry.

1 Leaves linear, most in whorls of 3-8 **4. P. verticillatum**
1 Leaves elliptic, all alternate 2
2 Perianths not contracted in middle; filaments glabrous; flowers 1 or 2 per leaf-axil **3. P. odoratum**
2 Perianths slightly contracted in middle; filaments sparsely pubescent; flowers 1-6 per leaf-axil 3
3 Perianths 9-15(20) x 2-4mm; stems terete **1. P. multiflorum**

3 Perianths 15-22(25) x 3-6mm; stems ridged to slightly angled **2. P. x hybridum**

Other spp. - **P. biflorum** (Walter) Elliott, from E N America, resembles a large (to 2m) *P. multiflorum* but has stamens borne c.1/2 way up perianth-tube (not near apex) and glabrous filaments; it has been reported as a garden relic in Surrey.

1. P. multiflorum (L.) All. - *Solomon's-seal.* Stems erect then arching, to 80cm, 920
terete; flowers 1-6 per leaf-axil, 9-15(20) x 2-4mm, contracted in middle; berry bluish-black; 2n=18, 30. Native; woods, mostly on basic soils; locally frequent in S Br, scattered N to N En, rarely natd elsewhere in Br.

2. P. x hybridum Brügger (*P. multiflorum* x *P. odoratum*) - *Garden Solomon's-seal.* 920
Stems erect, then arching, to 1m, ridged to slightly angled; flowers 1-6 per leaf-axil, 15-22(25) x 3-6mm, slightly contracted in middle; berry rarely produced; (2n=28). Intrd; much grown in gardens (by far the commonest taxon there), natd in woods, scrub and rough ground throughout CI and Br, very scattered in Ir; garden origin.

3. P. odoratum (Mill.) Druce - *Angular Solomon's-seal.* Stems erect or erect then R
arching, to 40cm, distinctly angled; flowers 1-2 per leaf-axil, 15-30 x 4-9mm, not 920
contracted; berry bluish-black; 2n=26, 29. Native; in woods on limestone; very local in NW En, Peak District, and around Severn estuary, very scattered in S Wa, rarely natd elsewhere in Br.

4. P. verticillatum (L.) All. - *Whorled Solomon's-seal.* Stems erect, to 80cm, RRR
distinctly angled; flowers 1-4 per leaf-axil, 5-10 x 1.5-3mm, slightly contracted in 920
middle; berry turning from red to purple; 2n=30, 84. Native; mountain woods; very rare in M & E Perth, Angus and S Northumb.

16. MAIANTHEMUM Weber - *May Lilies*
Plant rhizomatous; leaves few from rhizome, 2(-3) on stem, ovate, cordate; flowers numerous in terminal raceme; tepals 4, free, patent, white; stamens 4; style 1; fruit a red berry.

1. M. bifolium (L.) F.W. Schmidt - *May Lily.* Stems to 20cm, pubescent distally; RR
leaves up to 3-6 x 2-5cm; raceme 1-4(5)cm; flowers 2.5-6mm across; 2n=36, c.38, c.42. Native; woods on acid soils; extremely local in Durham, NE Yorks and N Lincs, probably intrd in E Norfolk and W Lothian, formerly elsewhere in N & E En, rarely persistent where planted in woods.

2. M. kamtschaticum (Cham.) Nakai (*M. dilatatum* (Wood) A. Nelson & J.F. Macbr.). - *False Lily-of-the-valley.* Differs from *M. bifolium* in larger size (stems to 35(45)cm; leaves 5-11(20)cm; raceme 2.5-7.5cm); and stems glabrous; (2n=32, 36, 54). Intrd; natd in woodland; S Somerset since 1983 and spreading, MW Yorks; E Asia and W N America.

17. REINECKEA Kunth - *Reineckea*
Plant rhizomatous; leaves all basal, linear; flowers numerous in short terminal spike, erect to patent; tepals fused into tube slightly shorter than patent to reflexed lobes, pink; style 1; fruit a red berry.

1. R. carnea (Haw.) Kunth - *Reineckea.* Leaves in rosette, up to 40 x 2cm; spike 4-9cm on stalk 3-5cm; flowers rather crowded, with tube c.5mm and lobes c.7mm; (2n=38, 42). Intrd; planted as ground-cover, natd in woodland near Lizard, W Cornwall; Japan and China.

18. PARIS L. - *Herb-Paris*
Plant rhizomatous; leaves (3)4(-8), all in 1 whorl at top of stem, elliptic to obovate; inflorescence a single terminal, erect, long-stalked flower; tepals 8(-12), 4(-6) outer lanceolate, 4(-6) inner linear, all green, free, patent; stamens 8(-12); ovary superior,

4(-5)-celled; style 1, short; stigmas 4(-5), linear; fruit a dehiscent black berry.

1. P. quadrifolia L. - *Herb-Paris*. Stems to 40cm; leaves up to 15 x 8cm; tepals 2-3.5cm; 2n=60. Native; in moist woods on calcareous soils; rather local in Br, absent from most of Wa, SW En and N & W Sc.

SUBFAMILY 5 - SCILLOIDEAE (tribe Scilleae) (genera 19-24). Plant with a bulb; leaves all basal, linear or nearly so; inflorescence a terminal raceme; tepals usually white or blue, sometimes pink or brownish, very rarely pale yellow, free or fused; style 1, with 1 (often 3-lobed) stigma; ovary superior; fruit a capsule splitting along centre of ovary-cells.

19. ORNITHOGALUM L. - *Star-of-Bethlehems*
Flowers each with 1 bract; tepals free, white with green stripe(s) on outside; stamens inserted on receptacle, with flattened filaments.

1 Bracts longer than pedicels; filaments with 1 acute lobe at apex on either side of anther **3. O. nutans**
1 Bracts shorter than pedicels (at least on lower flowers); filaments without apical lobes 2
2 Inflorescence corymbose; tepals >14mm **2. O. angustifolium**
2 Inflorescence an elongated raceme; tepals <14mm **1. O. pyrenaicum**

1. O. pyrenaicum L. - *Spiked Star-of-Bethlehem*. Stems to 80(100)cm; inflorescence R
a narrow elongated raceme with >20 flowers; bracts shorter than pedicels; tepals 6-13mm; (2n=16, 32). Native; woods and scrub; very local in SC En N to Hunts, rarely natd elsewhere in En.

2. O. angustifolium Boreau (*O. umbellatum* auct. non L.) - *Star-of-Bethlehem*. Stems to 30cm; inflorescence corymbose, with 4-12 erect flowers; bracts shorter than pedicels; tepals 15-20mm; 2n=27, 28 (triploid). Native; grassy places, rough ground and open woods; scattered throughout Br and CI, but perhaps native only in E En. The closely related **O. umbellatum** L., from Europe, has longer tepals (up to 30mm), more flowers (up to 20) and a bulb producing numerous subspherical (not few elongated) bulblets, and is tetraploid to hexaploid, not diploid to tetraploid. It might be common in gardens and some natd plants could be this; a plant with 2n=54 was reported from Cambs in 1968. Further study of wild plants is needed.

3. O. nutans L. - *Drooping Star-of-Bethlehem*. Stems to 60cm; inflorescence a 1-sided raceme of 2-12 ± pendent flowers; bracts longer than pedicels; tepals 15-30mm; (2n=14-16, 30, 40-42). Intrd; grown in gardens, natd in grassy places; scattered in C & S Br; C Europe.

20. SCILLA L. - *Squills*
Flowers each with 0 or 1 bract; tepals free, blue, rarely white or pink; stamens inserted on base of perianth, with narrow or flattened filaments.

1 Bracts >4mm 2
1 Bracts 0 or <4mm 4
2 Flowers usually >20; lower bracts ≥3cm **7. S. peruviana**
2 Flowers <15; bracts <3cm 3
3 Leaves 2-5mm wide; tepals 5-8mm **5. S. verna**
3 Leaves 10-30mm wide; tepals 8-12mm **6. S. liliohyacinthus**
4 Flowers pendent; tepals 12-16mm **4. S. siberica**
4 Flowers erect to patent; tepals 3-10mm 5
5 Flowering Jul-Sep without leaves; tepals 3-6mm **8. S. autumnalis**
5 Flowering Feb-Apr with leaves; tepals 5-10mm 6

6 Leaves (1)2(-3), sheathing flowering stem 1/4-1/2 way up from base **1. S. bifolia**
6 Leaves 3-7, sheathing flowering stem only at base 7
7 All pedicels <10mm; bracts c.1mm **2. S. messeniaca**
7 Lower pedicels up to 2.5cm; bracts (1.5)2-3mm **3. S. bithynica**

Other spp. - Several other spp. are grown in gardens and some persist locally where neglected or thrown out.

1. S. bifolia L. - *Alpine Squill.* Stems to 20cm, with usually <10 ± erect flowers; 933
bracts 0 or minute; tepals ± patent, 5-10mm; (2n=16, 18, 36, 54). Intrd; grown in gardens, natd and spreading where planted in churchyards and on banks; very scattered in S En, SE Yorks; Europe.

2. S. messeniaca Boiss. - *Greek Squill.* Stems to 15cm, with 7-15(20) ± erect flowers; bracts c.1mm; tepals ± patent, 6-7mm. Intrd; similar status to *S. bifolia* in churchyards and open woods; N Somerset and W Suffolk; Greece.

3. S. bithynica Boiss. - *Turkish Squill.* Stems to 30cm, with (3)7-10(15) flowers; bracts 2-3mm; tepals ± patent, 5-10mm; (2n=12). Intrd; similar status to *S. bifolia* in churchyards and open woods and on banks; very scattered in S En; Turkey and Bulgaria.

4. S. siberica Haw. - *Siberian Squill.* Stems to 20cm, with usually <5 ± pendent 933
flowers; bracts 1-2mm; tepals forming cup- to funnel-shaped flower, 12-16mm; (2n=12, 18). Intrd; much grown in gardens, ± natd where planted and neglected; scattered in Br, mainly SE En; W Russia (not Siberia).

5. S. verna Huds. - *Spring Squill.* Stems to 15cm, with usually <12 suberect flowers; bracts 5-15mm; tepals ± patent, 5-8mm; 2n=20, 22. Native; dry short grassland near sea, especially cliff-tops; locally common on coasts of W Br from S Devon to Shetland, down E coast S to Cheviot, E coast of Ir.

6. S. liliohyacinthus L. - *Pyrenean Squill.* Stems to 40cm, with c.5-15 suberect flowers; bracts 10-25mm; tepals ± patent, 9-12mm; (2n=22). Intrd; natd where planted or neglected in open woodland; scattered from S Somerset and Berks to C Sc; France and Spain.

7. S. peruviana L. - *Portuguese Squill.* Stems to 50cm, with 20-100 suberect flowers; lower bracts 30-80mm; tepals ± patent, (5)8-15mm; (2n=16, 20, 28). Intrd; natd where planted or neglected; very scattered in Br, Man and CI; W Mediterranean.

8. S. autumnalis L. - *Autumn Squill.* Stems to 25cm, with 4-20 suberect flowers; R
bracts 0; tepals ± patent, 3-6mm; 2n=28, 42. Native; short grassland usually near sea; local in CI, SW En scattered E to S Essex and Surrey, formerly Glam.

21. HYACINTHOIDES Heist. ex Fabr. (*Endymion* Dumort.) - *Bluebells*
Flowers each with 2 bracts; tepals free, blue, sometimes white or pink; stamens inserted on or at base of perianth, with narrow filaments.

Doubtfully distinct from *Scilla; H. italica* fits ± equally well into either genus.

1 Tepals ± patent, 5-8mm; all stamens inserted at base of perianth **1. H. italica**
1 Tepals erect to erecto-patent, >10mm; at least 3 outer stamens fused from base to >1/4 way up perianth 2
2 Racemes pendent at apex; perianth tubular for most of length; anthers cream **2. H. non-scripta**
2 Racemes erect; perianth bell-shaped; anthers same colour as tepals **3. H. hispanica**

1. H. italica (L.) Rothm. (*Scilla italica* L.) - *Italian Bluebell.* Stems to 40cm; leaves 933
up to 12mm wide; racemes pyramidal; flowers suberect, scentless; tepals ± patent,

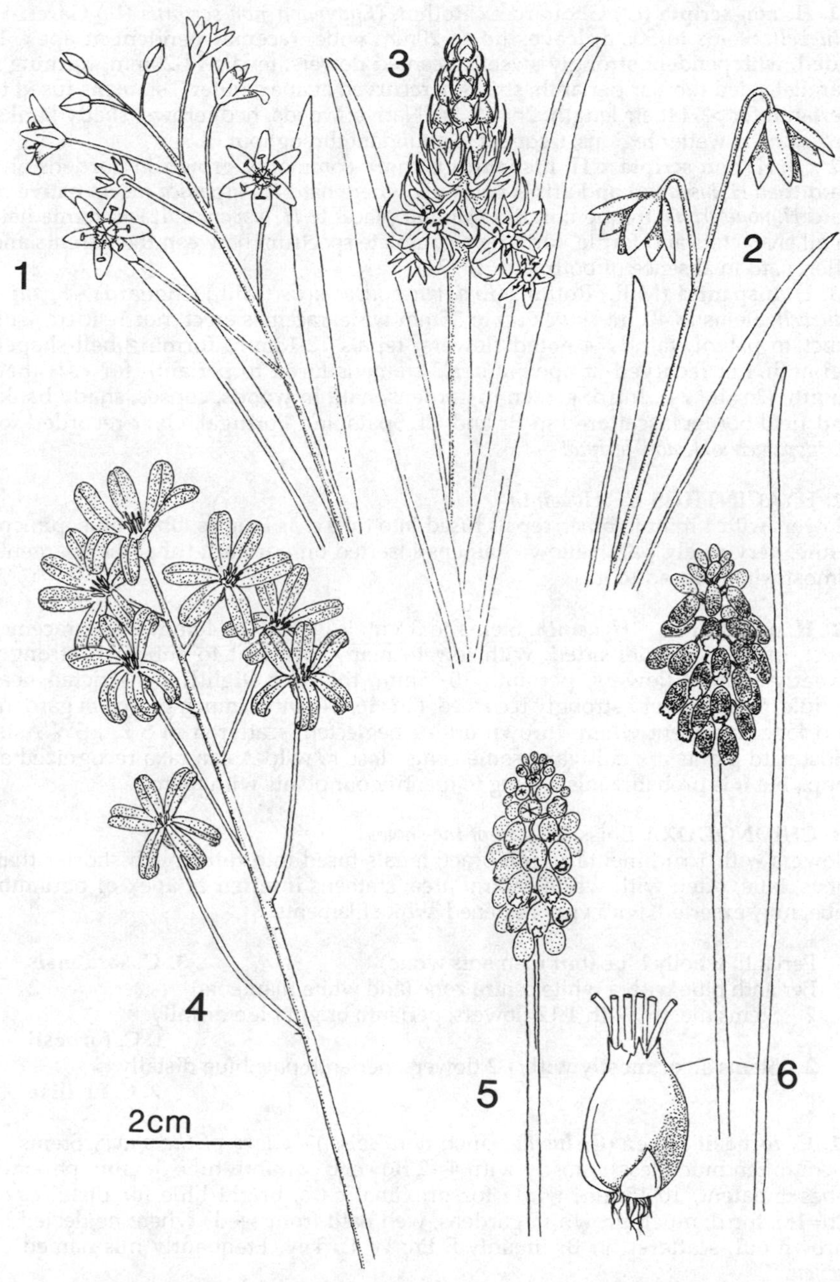

FIG 933 - Liliaceae: Scilloideae. 1, *Scilla bifolia.* 2, *S. siberica.* 3, *Hyacinthoides italica.* 4, *Chionodoxa forbesii.* 5, *Muscari armeniacum.* 6, *M. neglectum.*

5-8mm; (2n=16). Intrd; natd in neglected old woodland in Ayrs, Dorset and S Essex; SW Europe.

2. H. non-scripta (L.) Chouard ex Rothm. (*Endymion non-scriptus* (L.) Garcke) - *Bluebell.* Stems to 50cm; leaves up to 20mm wide; racemes pendent at apex, 1-sided, with pendent strongly sweetly scented flowers; tepals 14-20mm, forming ± parallel-sided tubular perianth, strongly recurved at apex, outer 3 stamens fused to perianth for >3/4 their length; 2n=16, 24. Native; woods, hedgerows, shady banks, grassland in wetter regions; frequent to abundant throughout BI.

2 x 3. H. non-scripta x H. hispanica is more commonly grown in gardens and natd than *H. hispanica* and arises naturally where natd *H. hispanica* meets native or natd *H. non-scripta*; frequent in BI in similar places to *H. hispanica*. It is intermediate in all characters and fertile, forming a complete spectrum between the parents and often natd in absence of both; (2n=16, 24).

3. H. hispanica (Mill.) Rothm. (*Endymion hispanicus* (Mill.) Chouard) - *Spanish Bluebell.* Stems to 40cm; leaves up to 35mm wide; racemes erect, not 1-sided, with erect to patent, faintly scented flowers; tepals 12-18mm, forming bell-shaped perianth, not recurved at apex; outer 3 stamens fused to perianth for <3/4 their length; (2n=16, 24). Intrd; grown in gardens, natd in woods, copses, shady banks and field-borders; scattered in Br and CI; Spain and Portugal. Over-recorded for *H. hispanica* x *H. non-scripta.*

22. HYACINTHUS L. - *Hyacinth*

Flowers with 1 minute bract; tepals fused into tube c. as long as lobes, blue, pink or white, very rarely pale yellow; stamens inserted on perianth-tube, the filaments almost wholly fused to it.

1. H. orientalis L. - *Hyacinth.* Stems to 30cm; leaves up to 40mm wide; racemes erect, often ± parallel-sided, with few to many, pendent to suberect, strongly sweetly scented flowers; perianth 10-35mm, the tube slightly constricted near middle, the lobes very strongly recurved; (2n=16, 24). Intrd; much grown in gardens and long persistent where thrown out or neglected; scattered in S En; SW Asia. Most natd plants are cultivars; some come close to wild Asian taxa recognized as sspp., but it is probably misleading to identify our plants with them.

23. CHIONODOXA Boiss. - *Glory-of-the-snows*

Flowers with 1 rudimentary or 0 bract; tepals fused into tube much shorter than lobes, blue, often with white central area; stamens inserted at apex of perianth-tube, fully exserted, with very flattened, white filaments.

1 Perianth wholly blue (but filaments white) **3. C. sardensis**
1 Perianth blue with a white centre zone (and white filaments) 2
 2 Stems mostly with 4-12 flowers; perianth bright blue distally **1. C. forbesii**
 2 Stems all or mostly with 1-2 flowers; perianth pale blue distally **2. C. luciliae**

1. C. forbesii Baker (*C. luciliae* auct. non Boiss.) - *Glory-of-the-snow.* Stems to 933
30cm (often much less), mostly with 4-12 flowers; perianth-tube 3-5mm; perianth lobes ± patent, 10-15mm, white for proximal c.1/3, bright blue for distal c.2/3; (2n=18). Intrd; much grown in gardens, well natd from seeds where neglected or thrown out; scattered in Br, mainly S En; W Turkey. Frequently mis-named *C. luciliae.*

2. C. luciliae Boiss. - *Boissier's Glory-of the-snow.* Differs from *C. forbesii* in stems mostly with 1-2 flowers; perianth-tube 2.5-4mm; perianth-lobes 12-20mm, pale blue for distal c.2/3; (2n=18, 20). Intrd; natd on grassy slope; Middlesex; W Turkey.

3. C. sardensis Drude - *Lesser Glory-of-the-snow.* Differs from *C. forbesii* in stems mostly with 6-16 flowers; perianth-lobes (5)8-10mm, wholly bright blue; (2n=18). Intrd; natd as for *C. forbesii*, less common though perhaps overlooked; Surrey and Middlesex; Turkey.

24. MUSCARI Mill. - *Grape-hyacinths*
Flowers with 1 minute or 0 bract, the apical group sterile, the lower ones fertile and often of different colour; tepals fused for most of length, blue to blackish-blue, with white lobes, or brownish; stamens inserted c.1/2 way up perianth-tube, included, with narrow filaments.

1 Fertile flowers brownish-buff, on pedicels mostly >5mm; apical sterile flowers bright bluish-violet, some on pedicels >5mm **4. M. comosum**
1 All flowers blue to blackish-blue, on pedicels <5mm 2
  2 Leaves linear with ± spathulate apex; corolla ± spherical with strongly recurved lobes **3. M. botryoides**
  2 Leaves linear to oblanceolate; corolla ellipsoid-ovoid, distinctly longer than wide, with erecto-patent lobes 4
3 Perianth of fertile flowers blackish-blue to dark violet-blue **1. M. neglectum**
3 Perianth of fertile flowers bright blue **2. M. armeniacum**

Other spp. - M. azureum Fenzl, from Turkey, differs from the others in its pale blue corolla not being constricted at the mouth; it was formerly natd in S Somerset.

1. M. neglectum Guss. ex Ten. (*M. atlanticum* Boiss. & Reut., *M. racemosum* Lam. RR
& DC. non (L.) Mill.) - *Grape-hyacinth.* Stems to 30cm; leaves up to 30cm, 2-8mm 933
wide, linear to oblanceolate; racemes dense, 1.5-5cm; fertile flowers 3.5-7.5 x 1.5-3.5mm, dark violet- to blackish-blue with white lobes; sterile flowers smaller, pale blue; (2n=18, 36, 45, 54, 63, 72). Native; dry grassland, hedgebanks and field borders; very local in E & W Suffolk and Cambs, formerly E & W Norfolk, rarely natd elsewhere but over-recorded for *M. armeniacum.*

2. M. armeniacum Leichtlin ex Baker - *Garden Grape-hyacinth.* Differs from *M.* 933
neglectum in fertile flowers 3.5-5.5mm; and see key (couplet 3); (2n=18, 36). Intrd; the common garden *Muscari*, an escape and throwout spreading vegetatively and by seed on rough ground, banks and grassy places, sometimes with *M. neglectum;* scattered in Br, mainly S but N to C Sc, and CI; Balkans to Caucasus.

3. M. botryoides (L.) Mill. - *Compact Grape-hyacinth.* Stems to 25cm; leaves up to 20cm, conspicuously broadened at apex; racemes dense, 1-6cm; fertile flowers 2.5-5 x 2.5-4mm, bright blue with white lobes; sterile flowers smaller and paler; (2n=18, 36). Intrd; garden plant natd where thrown out or neglected; Surrey, but now gone; S Europe.

4. M. comosum (L.) Mill. - *Tassel Hyacinth.* Stems to 60cm; leaves up to 40 x 2cm, linear; racemes rather loose, 4-8cm; fertile flowers 5-10 x 3-5mm, brownish-buff with pale lobes; sterile flowers bright bluish-violet, forming conspicuous apical tuft; (2n=18, 28). Intrd; persistent weed of cultivated and rough ground, sometimes dunes and open grassland; local in SW En, S Wa and CI, rare and usually casual elsewhere; Europe.

SUBFAMILY 6 - ALLIOIDEAE (tribes Allieae, Agapantheae) (genera 25-29). Plant with a rhizome or a bulb, mostly smelling of garlic or onion when fresh; leaves usually all basal, sometimes on stem, linear to ± cylindrical, linear-oblong or elliptic; inflorescence a terminal umbel with usually scarious spathe at base, sometimes reduced to 1 flower or some or all flowers replaced by bulbils but still with spathe; tepals various colours, free or fused proximally; style 1, with capitate to 3-lobed stigma; ovary superior or semi-inferior; fruit a capsule splitting along centre of ovary-cells.

25. ALLIUM L. - *Onions*

Plant with bulb(s), smelling of onion or garlic when fresh; leaves linear to ± cylindric, or elliptic; flowers in umbel, some or all often replaced by bulbils; tepals free or ± so, white to greenish, pink, purple or yellow; ovary superior; ovules usually 2 per cell.

General key

1 Inflorescence consisting entirely of bulbils — *Key A*
1 Inflorescence with at least 1 flower — 2
  2 Inflorescence with bulbil(s) and flower(s) — *Key B*
  2 Inflorescence with flowers only — *Key C*

Key A - Inflorescence consisting entirely of bulbils

1 Leaves circular to semi-circular or subcircular in section — 2
1 Leaves obviously bifacial, flat to strongly keeled — 4
  2 Stem hollow, inflated and bulging just below middle; leaves usually >4mm wide — **2. A. cepa**
  2 Stem solid or nearly so, not inflated; leaves <4mm wide — 3
3 Spathe of 2 persistent valves each with apical attenuate part much longer than basal part — **12. A. oleraceum**
3 Spathe of 1 ± deciduous valve with apical attenuate part c. as long as basal part — **19. A. vineale**
  4 Stems triangular in section; leaves all basal — **10. A. paradoxum**
  4 Stems ± circular in section; at least some leaves borne on stem — 5
5 Leaves <4mm wide — **13. A. carinatum**
5 Leaves >5mm wide — 6
  6 Leaves 2-5; main bulb single, with often numerous small bulblets outside its covering — **17. A. scorodoprasum**
  6 Leaves 4-10; main bulb composed of several ± equal bulblets within common cover — **14. A. sativum**

Key B - Inflorescence consisting of both flowers and bulbils

1 Leaves circular to semi-circular or subcircular in section — 2
1 Leaves obviously bifacial, flat to strongly keeled — 5
  2 Stem hollow, inflated and bulging just below middle; leaves usually >4mm wide — **2. A. cepa**
  2 Stem solid or ± so, not inflated; leaves <4mm wide — 3
3 Stamens shorter than tepals; filaments simple — **12. A. oleraceum**
3 Stamens longer than tepals; inner 3 filaments divided distally into 3 points, the middle one anther-bearing — 4
  4 Spathe 1-valved; lateral points of inner 3 filaments >2x as long as central point — **19. A. vineale**
  4 Spathe 2-valved; lateral points of inner 3 filaments <2x as long as central point — **18. A. sphaerocephalon**
5 Stems triangular in section — **10. A. paradoxum**
5 Stems ± circular in section — 6
  6 Tepals yellow — **7. A. moly**
  6 Tepals pink to white, greenish or purplish — 7
7 Filaments simple; leaves often <5mm wide — 8
7 Inner 3 filaments divided distally into 3 points, the middle one anther-bearing; leaves ≥5mm wide — 9
  8 Stamens shorter than tepals; spathe shorter than pedicels — **4. A. roseum**
  8 Stamens longer than tepals; spathe longer than pedicels — **13. A. carinatum**
9 Stamens longer than tepals — 10
9 Stamens shorter than tepals — 11

10 Bulb scarcely swollen at base, without bulblets; style shorter than tepals; spathe persistent at least until flowering **16. A. porrum**

10 Bulb swollen at base, with bulblets around it within common cover; style longer than tepals; spathe usually deciduous before flowering **15. A. ampeloprasum**

11 Leaves 2-5; main bulb single, with often numerous small bulblets outside its cover; common part of inner 3 filaments 2-3x as long as central distal anther-bearing point **17. A. scorodoprasum**

11 Leaves 4-10; main bulb composed of several ± equal bulblets within common cover; common part of inner 3 filaments c. as long as central distal anther-bearing point **14. A. sativum**

Key C - Inflorescence consisting entirely of flowers

1 Leaves circular to semi-circular or subcircular in section 2

1 Leaves obviously bifacial, flat to strongly keeled 5

2 Stem hollow, inflated and bulging just below middle; leaves usually >4mm wide **2. A. cepa**

2 Stem solid or nearly so, not inflated; leaves <4mm wide 3

3 Filaments simple; stamens shorter than tepals **1. A. schoenoprasum**

3 Inner 3 filaments divided distally into 3 points, the middle one anther-bearing; stamens at least as long as tepals 4

4 Spathe 1-valved; lateral points of inner 3 filaments ≥2x as long as central point **19. A. vineale**

4 Spathe 2-valved; lateral points of inner 3 filaments <2x as long as central point **18. A. sphaerocephalon**

5 Tepals yellow **7. A. moly**

5 Tepals white to pink, greenish or purplish 6

6 Leaves with distinct petiole, the blade elliptic to narrowly so **11. A. ursinum**

6 Leaves without petiole, linear to filiform 7

7 Stem triangular in section 8

7 Stem ± circular in section 10

8 Stigma simple; spathe 1-valved **5. A. neapolitanum**

8 Stigma 3-lobed; spathe 2-valved 9

9 Umbel 1-sided, with pendent flowers; tepals never opening >45° **8. A. triquetrum**

9 Umbel not 1-sided, with erect and pendent flowers; tepals opening >45° at first, less so later **9. A. pendulinum**

10 Inner 3 filaments divided distally into 3 points, the middle one anther-bearing 11

10 Filaments simple 12

11 Bulb scarcely swollen at base, without bulblets; style shorter than tepals; spathe persistent at least until flowering **16. A. porrum**

11 Bulb swollen at base, with bulblets around it within common cover; style longer than tepals; spathe usually deciduous before flowering **15. A. ampeloprasum**

12 Leaves conspicuously pubescent at edge **6. A. subhirsutum**

12 Leaves glabrous 13

13 Stamens longer than tepals; leaves ≤3mm wide; spathe with valves much longer than pedicels **13. A. carinatum**

13 Stamens shorter than tepals; leaves most or all >4mm wide; spathe with valves rarely as long as pedicels 14

14 Leaves >2cm wide **20. A. nigrum**

14 Leaves <1.5cm wide 15

15 Covering of bulb minutely pitted; spathe with 1 primary valve, often deeply ≥2-lobed **4. A. roseum**

15 Covering of bulb with undulating or net-like markings; spathe 2-valved **3. A. unifolium**

Other spp. - A. fistulosum L. (*Welsh Onion*) is grown in gardens as a minor vegetable and might occur as a relic or throwout; it resembles variants of *A. cepa* with several bunched narrow bulbs but differs in its cylindrical leaves, stem widest at middle, and longer tepals (7-9mm).

1. A. schoenoprasum L. - *Chives*. Stems to 50cm, terete, hollow; leaves terete, R
hollow, 1-5mm wide; inflorescence of flowers only; tepals 7-14mm, pink to pale purple; stamens shorter than tepals; filaments simple; 2n=16. Native; rocky ground, usually on limestone, also grown as leaf-vegetable; local in SW & N En and S Wa, formerly Berwicks and E Mayo, very scattered relic or throwout elsewhere in Br.

2. A. cepa L. - *Onion*. Stems to 1m, terete, hollow, bulging just below middle; leaves subterete to hemi-cylindrical, hollow, 2-20mm wide; inflorescence of flowers, bulbils or both; tepals 3-4.5mm, greenish-white; stamens longer than tepals; filaments simple except for small basal tooth each side; (2n=16, 24). Intrd; much grown as vegetable, frequent throwout or relic; scattered throughout Br; garden origin. Many variants occur: the commonest sorts have a large single bulb and usually no bulbils in the inflorescence; *Shallot* (*A. ascalonicum* auct. non L.) has a number of much smaller bulbs; *Spring Onion* has a cluster of small elongated bulbs with white (not brown) covering; *Tree Onion* produces mostly or only bulbils in the inflorescence.

3. A. unifolium Kellogg - *American Garlic*. Stems to 40cm, terete; leaves flat, scarcely keeled, 2-8mm wide; inflorescence of flowers only; tepals 10-17mm, bright pink; stamens shorter than tepals; filaments simple; (2n=14, 32). Intrd; garden plant natd in woods in W Kent and formerly Dunbarton; W N America.

4. A. roseum L. - *Rosy Garlic*. Stems to 75cm, terete; leaves flat, scarcely keeled, **939**
4-12mm wide; inflorescence of flowers with or without bulbils; tepals 7-12mm, pink, rarely white; stamens shorter than tepals; filaments simple; (2n=16, 28, 32, 40, 48). Intrd; garden plant and weed well natd in rough or cultivated ground, old dunes, hedgerows and waysides; frequent in SW En, S Wa and CI, scattered elsewhere; Mediterranean. Presence or absence of bulbils is not worth ssp. ranking.

5. A. neapolitanum Cirillo - *Neapolitan Garlic*. Stems to 50cm, triangular in **939**
section with 2 edges much more acute than other; leaves flat, keeled, 5-20mm wide; inflorescence of flowers only; tepals 7-12mm, white; stamens shorter than tepals; filaments simple; (2n=14, 21, 28, 35). Intrd; garden plant natd in rough and cultivated ground, hedgebanks and waysides; frequent in SW En and CI, very scattered elsewhere in En and Man; Mediterranean.

6. A. subhirsutum L. - *Hairy Garlic*. Stems to 45cm, terete; leaves flat, scarcely **939**
keeled, 2-10mm wide; inflorescence of flowers only; tepals 7-9mm, white; stamens shorter than tepals; filaments simple; (2n=14, 28, 32). Intrd; similar habitats and distribution to *A. neapolitanum*; Mediterranean.

7. A. moly L. - *Yellow Garlic*. Stems to 45cm, terete; leaves flat, narrowly elliptic, **939**
8-35mm wide; inflorescence of flowers only; tepals 9-12mm, yellow; stamens shorter than tepals; filaments simple; (2n=14). Intrd; garden plant natd on warm banks and hedgerows; scattered in S En, Jersey, rarely elsewhere; Spain and France.

8. A. triquetrum L. - *Three-cornered Garlic*. Stems to 45cm, triangular in section with very acute angles; leaves 2-5, flat, scarcely keeled, 4-12mm wide; inflorescence of flowers only; tepals 10-18mm, white with strong green line; stamens shorter than tepals; filaments simple; (2n=18, 27). Intrd; weed of rough, waste and cultivated ground, copses, hedgerows and waysides; common in SW En and CI, scattered elsewhere in En, Wa and Ir; W Mediterranean.

9. A. pendulinum Ten. - *Italian Garlic*. Differs from *A. triquetrum* in stems to 25cm; leaves 3-8mm wide, usually 2; (2n=14, 18); and see Key C (couplet 9). Intrd;

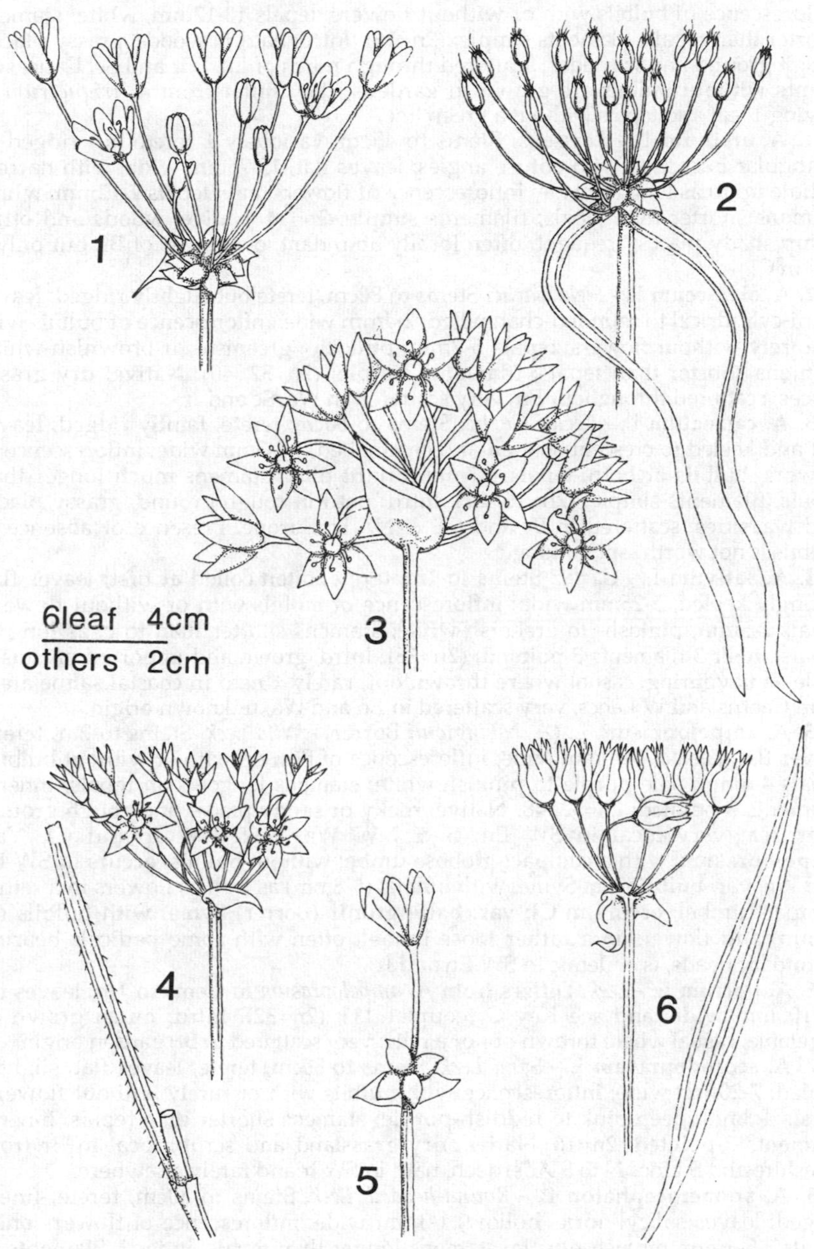

FIG 939 - *Allium*. 1, *A. roseum*. 2, *A. carinatum*. 3, *A. neapolitanum*. 4, *A. subhirsutum*. 5, *A. paradoxum*. 6, *A. moly*.

natd in neglected estates, S Essex and Middlesex; Italy.

10. A. paradoxum (M. Bieb.) G. Don - *Few-flowered Garlic.* Stems to 40cm, triangular in section with acute angles; leaves flat, scarcely keeled, 5-25mm wide; inflorescence of bulbils with or without flowers; tepals 10-12mm, white; stamens shorter than tepals; filaments simple; (2n=16). Intrd; natd in woods, grassy places, rough ground and waysides; scattered through much of En, E Ir and Sc; Caucasus. Plants without bulbils are grown in gardens; they differ from *A. triquetrum* in having 1 leaf and tepals without a green line. **939**

11. A. ursinum L. - *Ramsons.* Stems to 45cm, variously ± terete but ridged to triangular in section with obtuse angles; leaves flat, 15-75mm wide, with narrow petiole up to as long as blade; inflorescence of flowers only; tepals 7-12mm, white; stamens shorter than tepals; filaments simple; 2n=14. Native; woods and other damp shady places; frequent, often locally abundant, over most of BI, but only 1 site in CI.

12. A. oleraceum L. - *Field Garlic.* Stems to 80cm, terete but slightly ridged; leaves hemi-cylindrical to rounded-channelled, 2-4mm wide; inflorescence of bulbils with or rarely without flowers; tepals 5-7mm, pinkish-, greenish- or brownish-white; stamens shorter than tepals; filaments simple; (2n=32, 40). Native; dry grassy places; scattered throughout En, very scattered in Wa, Sc and Ir.

13. A. carinatum L. - *Keeled Garlic.* Stems to 60cm, terete, faintly ridged; leaves flat and keeled to crescent-shaped and channelled, 1.5-3mm wide; inflorescence of flowers, bulbils or both; tepals 4-7mm, bright pink; stamens much longer than tepals; filaments simple; (2n=16, 24). Intrd; natd in rough ground, grassy places and waysides; scattered in Br and CE & NE Ir; Europe. Presence or absence of bulbils is not worth ssp. ranking. **939**

14. A. sativum L. - *Garlic.* Stems to 1m, terete, often coiled at first; leaves flat, strongly keeled, 5-25mm wide; inflorescence of bulbils with or without flowers; tepals 2-5mm, pinkish- to greenish-white; stamens shorter than to c. as long as tepals; inner 3 filaments 3-pointed; (2n=16). Intrd; grown and imported on small-scale as flavouring, casual where thrown out, rarely ± natd in coastal saline areas as in Caerns and W Lancs; very scattered in En and Wa; unknown origin.

15. A. ampeloprasum L. (*A. babingtonii* Borrer) - *Wild Leek.* Stems to 2m, terete; leaves flat, keeled, 5-40mm wide; inflorescence of flowers with or without bulbils; tepals 4-6mm, pale purple to pinkish-white; stamens longer than tepals; inner 3 filaments 3-pointed; 2n=32, 48. Native; rocky or sandy places and rough ground near sea; very local in SW En, S & NW Wa, N & CW Ir and CI. Var. **ampeloprasum**, with a compact globose umbel without bulbils, occurs in SW En and Wa; var. **bulbiferum** Syme, with bulbils (6-8mm) as well as flowers in a rather compact umbel, occurs in CI; var. **babingtonii** (Borrer) Syme, with bulbils (8-15mm) and flowers in a rather loose umbel, often with some pedicels bearing secondary heads, is endemic to SW En and Ir. **RR**

16. A. porrum L. - *Leek.* Differs from *A. ampeloprasum* in stems to 1m; leaves up to 100mm wide; and see Key C (couplet 11); (2n=32). Intrd; much grown as vegetable, casual where thrown out or a relic; very scattered in Br; garden origin.

17. A. scorodoprasum L. - *Sand Leek.* Stems to 80cm, terete; leaves flat, slightly keeled, 7-20mm wide; inflorescence with bulbils with or rarely without flowers; tepals 4-8mm, deep pink to reddish-purple; stamens shorter than tepals; inner 3 filaments 3-pointed; 2n=16. Native; dry grassland and scrub; local in Br from Cheshire and S Lincs N to S Aberdeen, natd in SW Ir and rarely elsewhere.

18. A. sphaerocephalon L. - *Round-headed Leek.* Stems to 80cm, terete, finely ridged; leaves subcylindric, hollow, 1-3mm wide; inflorescence of flowers only; tepals 3.5-6mm, pinkish-purple; stamens longer than tepals; inner 3 filaments 3-pointed; (2n=16, 32). Native; on limestone rocks, W Gloucs, found 1847, and on sandy waste ground by sea, Jersey, found 1836, rarely natd from gardens elsewhere. **RRR**

19. A. vineale L. - *Wild Onion.* Differs from *A. sphaerocephalon* in inflorescence of flowers, bulbils or both; tepals 3-5mm, pink or greenish-white (blue-flowered plants occur in N Kerry); and see Key C (couplet 4); 2n=32. Native; grassy places, rough ground, banks and waysides; common in S En, frequent to scattered in rest of BI except absent in N Sc. By far the commonest narrow-leaved *Allium*.

20. A. nigrum L. - *Broad-leaved Leek*. Stems to 80cm, terete; leaves flat, 30-80mm wide; inflorescence of flowers only; tepals 6-9mm, pink to white with green line; stamens shorter than tepals; filaments simple; (2n=16, 32). Intrd; garden plant natd in rough ground in few places in S En; S Europe.

26. NECTAROSCORDUM Lindl. - *Honey Garlic*

Plant with bulb, smelling of garlic when fresh; leaves linear, strongly keeled; flowers in umbel, sweetly scented; tepals free, greenish-red; ovary semi-inferior; ovules numerous per cell.

1. N. siculum (Ucria) Lindl. (*Allium siculum* Ucria) - *Honey Garlic.* Stems to 1.2m, 957
terete; leaves 1-2cm wide; tepals 12-17mm; stamens shorter than tepals; (2n=16, 18). Intrd; garden plant natd in rough ground; very scattered in En, mostly S; S Europe. Probably both ssp. **siculum** and ssp. **bulgaricum** (Janka) Stearn (tepals greenish-cream tinged with pink, not greenish-red, and usually less acute and forming more broadly bell-shaped perianth; most fruiting pedicels nearly as long as longest, not with a few conspicuously longer than all others) occur, but are scarcely distinct. Intermediates occur in gardens.

27. NOTHOSCORDUM Kunth - *Honeybells*

Plant with bulb, not smelling of garlic or onion; leaves linear, scarcely or not keeled; flowers in umbel, sweetly scented; tepals fused at base, greenish-white with pink midrib; ovary superior; ovules numerous per cell.

1. N. borbonicum Kunth (*N. gracile* auct. non (Aiton) Stearn, *N. inodorum* auct. 957
non (Aiton) G. Nicholson, *N. fragrans* (Vent.) Kunth) - *Honeybells*. Stems to 60cm, terete; leaves 4-15mm wide; tepals 8-14mm; stamens shorter than tepals; (2n=16-20, 24, 32). Intrd; garden plant natd in rough and arable land and neglected estates; scattered in S & SW En, S Ir and CI; S America. Distinguished from white-flowered spp. of *Allium* lacking bulbils by the fused tepals and lack of garlic scent.

28. AGAPANTHUS L'Hér. - *African Lily*

Plant with short tuber-like rhizome, not smelling of garlic or onion; leaves oblong-linear, scarcely keeled; flowers in umbel, slightly zygomorphic; tepals fused in lower 1/2, bright blue, very rarely white; ovary superior; ovules numerous per cell.

1. A. praecox Willd. - *African Lily*. Plant forming dense clump; stems to 1m, terete; leaves 20-70 x 1.5-5.5cm; perianth 30-50mm; stamens c. as long as perianth; (2n=32). Intrd; well natd on sandy soil by sea in Scillies, scarcely so in CI and Surrey; S Africa. Our plant is spp. **orientalis** (F.M. Leight.) F.M. Leight.

29. TRISTAGMA Poepp. (*Ipheion* Raf.) - *Starflowers*

Plant with bulb, smelling of garlic when fresh; leaves linear, slightly keeled; flowers solitary, terminal, with spathe below, sweetly scented; tepals fused in lower 1/2, pale bluish-violet with dark midrib outside; ovary superior; ovules numerous per cell.

1. T. uniflorum (Lindl.) Traub (*Ipheion uniflorum* (Graham) Raf.) - *Spring* 944
Starflower. Stems to 35cm, terete; leaves 4-8mm wide; perianth 25-35mm, the tube 12-16mm, narrow, the lobes patent, acute, forming flower 30-45mm across; stamens as long as perianth-tube; (2n=12). Intrd; weed of cultivated and waste ground; natd in W Cornwall, Scillies and CI, rare casual or relic elsewhere but

perhaps increasing; S America.

SUBFAMILY 7 - AMARYLLIDOIDEAE (genera 30-36). Plant with a bulb; leaves all basal, linear to narrowly elliptic or linear-oblong; inflorescence of 1 flower or a terminal umbel with usually scarious spathe at base; tepals white to yellow or orange, rarely pink, free or fused proximally, sometimes with funnel- or collar-shaped corona within the rows of tepals; style 1, with simple or slightly 3-lobed stigma; ovary inferior; fruit a capsule dehiscing irregularly or along centre of ovary cells, often slightly succulent.

30. AMARYLLIS L. - *Jersey Lily*
Flowers in umbel, erecto-patent, trumpet-shaped, slightly zygomorphic, without corona; tepals fused at base into short tube, pink, all ± similar; flowers appearing in late summer or autumn, before leaves.

1. A. belladonna L. - *Jersey Lily*. Stems very stout, to 60cm; leaves 30-45 x 1.5-3cm, flat; perianth 5-10cm, with tube 1-1.5cm; (2n=20, 22). Intrd; grown for ornament in CI (especially Jersey) and a frequent relic in old fields, rough ground and sandy places, much less so in Scillies and W Cornwall; S Africa. Can be confused with **Nerine sarniensis** (L.) Herb. (*Guernsey Lily*), also from S Africa, grown in CI but not natd, but this has much more open flowers with well exposed stamens and perianth-tube <5mm.

31. CRINUM L. - *Cape-lilies*
Flowers in umbel, erecto-patent, trumpet-shaped, actinomorphic, without corona; tepals fused at base into long tube, pink, all ± similar; flowers appearing in late summer or autumn, with leaves.

1. C. x powellii Baker (*C. bulbispermum* (Burm. f.) Milne-Redh. & Schweick. x *C. moorei* Hook.) - *Powell's Cape-lily*. Stems to 80cm; leaves 60-120 x 4-10cm, channelled; perianth 12-20cm, incl. tube 6-10cm; (2n=22). Intrd; persistent and spreading garden escape or throwout; natd on dunes in Alderney since 1950s; garden origin.

32. STERNBERGIA Waldst. & Kit. - *Winter Daffodil*
Flowers solitary, erect, ± *Crocus*-like, actinomorphic, without a corona; tepals fused into narrow proximal tube, yellow, all ± similar; flowers appearing in autumn, ± with leaves.

Differs from *Crocus* in its ovary aerial at flowering, ± entire stigma, 6 stamens and leaves without central pale stripe.

1. S. lutea (L.) Ker-Gawl. ex Spreng. - *Winter Daffodil*. Stems 2.5-10(20)cm; leaves 7-10 x 0.4-1.5cm; perianth with tube 0.5-2cm, the lobes 3-5.5cm; (2n=12, 16, 18, 20, 22, 24, 44). Intrd; natd on grassy slopes by sea in Jersey since before 1919, formerly Guernsey; Mediterranean.

33. LEUCOJUM L. - *Snowflakes*
Flowers solitary or few in umbel, pendent, bell-shaped or bowl-shaped, actinomorphic, without a corona; tepals free, white with green or yellow patches, all ± similar; flowers appearing in late winter or spring, ± with leaves.

1. L. aestivum L. - *Summer Snowflake*. Forming clumps; stems to 60cm, with (1)2-5(7) flowers; leaves up to 50 x 0.5-1.5cm; tepals 10-22mm, each with green patch near apex; usually flowers Mar-May; seeds black. **944**

a. Ssp. aestivum. Stems with the 2 sharp edges remotely and often inconspicuously denticulate, at least in lower 1/2; flowers (2)3-5(7); spathe mostly **RR**

30-50 x 7-11mm; tepals 13-22mm; (2n=22). Native; wet meadows and willow scrub by rivers; very local in S En N to Oxon (mainly in Thames Valley) and S Ir, natd as garden escape elsewhere in BI but much confused with ssp. *pulchellum*.

b. Ssp. pulchellum (Salisb.) Briq. Stems with the 2 sharp edges entire throughout; flowers (1)2-4; spathe mostly 30-50 x 4-7mm; tepals 10-15mm; (2n=22). Intrd; more common in gardens than ssp. *aestivum*, natd in damp places and rough ground; scattered through BI from CI to Easterness; W Mediterranean. Slenderer and smaller than ssp. *aestivum*.

2. L. vernum L. - *Spring Snowflake*. Differs from *L. aestivum* in stems to 40cm, **RR**
with 1(-2) flowers; leaves up to 30 x 0.5-2.5cm; tepals 15-25mm, with green or **944**
yellow patch near apex; usually flowers Jan-Apr; seeds pale; (2n=20, 22, 24). Possibly native; damp scrub and stream-banks; 2 sites, S Somerset and Dorset, rarely natd elsewhere in En and Fife.

34. GALANTHUS L. - *Snowdrops*

Flowers solitary, pendent, actinomorphic, without a corona; tepals free, the inner white with green patch(es) and forming bell-shaped to bowl-shaped whorl, the outer white and spreading when in full flower; flowers appearing in late winter or spring, ± with leaves or before them, sometimes *flore pleno*.

1 Leaf upperside clear green, not glaucous **6. G. ikariae**
1 Leaf upperside glaucous, at least along central band 2
  2 Leaves with margins folded under at least along most of length, especially when young, glaucous on upper surface **3. G. plicatus**
  2 Leaves flat, or with margins inrolled especially when young, or if weakly folded under then not wholly glaucous on upperside 3
3 Leaves flat or folded under only at base as they unfold, ± linear, ≤1cm wide at flowering 4
3 Leaves inrolled as they unfold, oblanceolate, at least one >1cm wide at flowering 5
  4 Leaf upperside with glaucous central band and dull green lateral bands **2. G. reginae-olgae**
  4 Leaf upperside wholly glaucous **1. G. nivalis**
5 Inner tepals with green patches at apex and base **4. G. elwesii**
5 Inner tepals with green patch at apex only **5. G. caucasicus**

1. G. nivalis L. - *Snowdrop*. Stems to 20cm; leaves 9-15 x 0.5-1cm at flowering, **944**
flat, linear, wholly glaucous; outer tepals 12-25mm; inner tepals 4-10mm, with green patch at apex only; often *flore pleno*; (2n=24-28, 36). Possibly native; woods, damp grassy places, banks and streamsides; scattered throughout Br and CI, usually if not always intrd and rarely seeding, rare escape in Ir.

1 x 3. G. nivalis x G. plicatus. Many plants named *G. plicatus* are probably this, intermediate in leaf-folding and often occurring in absence of 1 or both parents; certainly in SW & SE En. Both sspp. of *G. plicatus* are involved.

1 x 4. G. nivalis x G. elwesii occurs rarely in SE En where both parents are natd together; it is intermediate in size, leaf characters and tepal-patches.

2. G. reginae-olgae Orph. - *Queen Olga's Snowdrop*. Stems to 20cm; leaves 8-15 x 3-6mm at flowering, flat or slightly folded under near base, linear, glaucous except for dull green lateral bands on upperside; outer tepals 15-25mm; inner tepals 9-12mm, with green patch at apex only; (2n=24). Intrd; natd in grassy places; W Kent; E Mediterranean. Differs from all other spp. in usually (but not always) flowering in late autumn or early winter before leaves appear.

3. G. plicatus M. Bieb. - *Pleated Snowdrop*. Stems to 25cm; leaves 5-20 x (0.4)1- **944**
2cm at flowering, with edges folded under, linear to slightly oblanceolate, glaucous with paler central band on upperside; outer tepals 15-30mm; inner tepals 7-12mm. Intrd; natd as for *G. elwesii*; local in S En, but relative distributions of sspp.

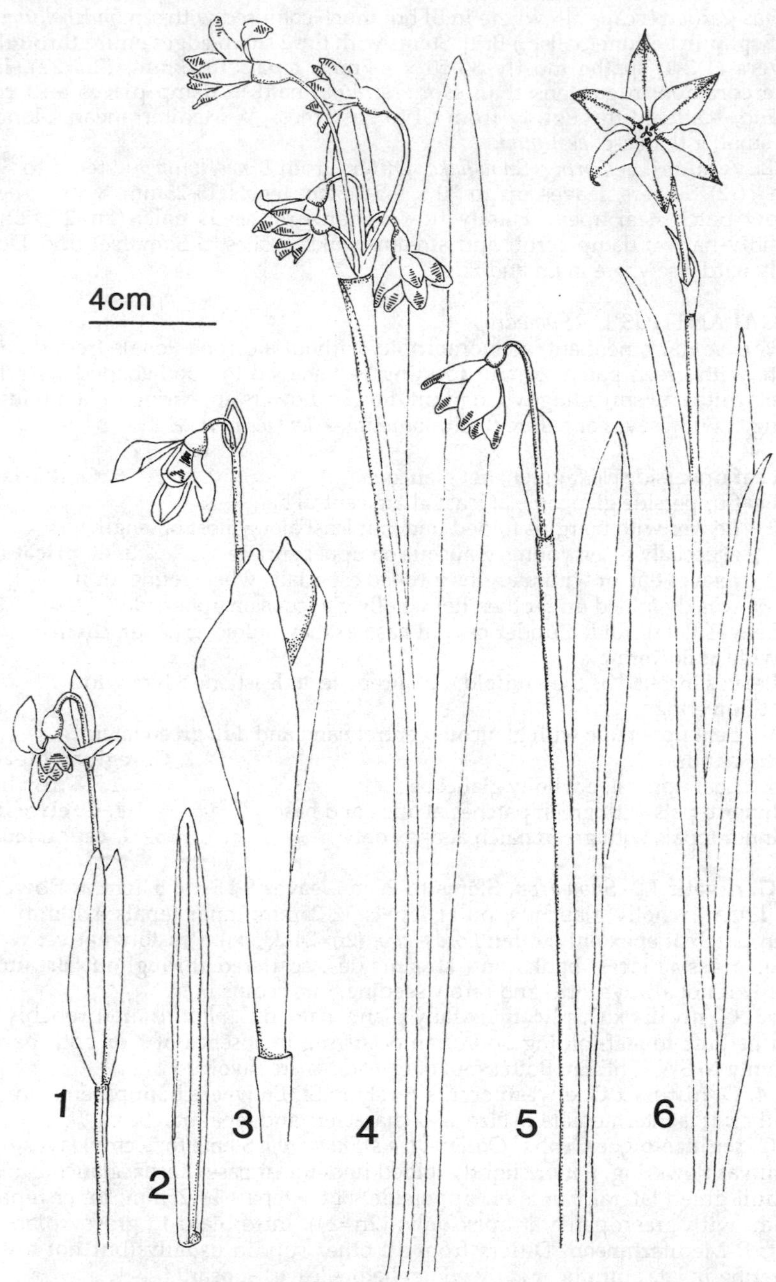

FIG 944 - Liliaceae. 1, *Galanthus nivalis*. 2, *G. plicatus* (leaf only). 3, *G. elwesii*. 4, *Leucojum aestivum*. 5, *L. vernum*. 6, *Tristagma uniflorum*.

unknown.; Romania, Turkey and Crimea.

a. Ssp. plicatus. Inner tepals with green patch at apex only; (2n=24). The commoner ssp., certainly from Cornwall to Kent and further N; Romania and Crimea.

b. Ssp. byzantinus (Baker) D.A. Webb. Inner tepals with green patches at apex and base, sometimes partly joining up; (2n=24). Certainly in SE & SW En; Turkey.

3 x 4. G. plicatus x G. elwesii occurs in Cambs.

4. G. elwesii Hook. f. - *Greater Snowdrop*. Stems to 30cm, leaves 6-30 x 0.6-1.3cm **944**
at flowering, inrolled, hooded at apex, oblanceolate, wholly glaucous; outer tepals 15-30mm; inner tepals 8-13mm, with green patches at base and apex; (2n=24). Intrd; commonly grown in gardens, natd in woods and damp grassland; local in S En, Man; SE Europe.

5. G. caucasicus (Baker) Grossh. - *Caucasian Snowdrop*. Stems to 15cm; leaves 5-11 x 1-2cm at flowering, inrolled, slightly hooded at apex, oblanceolate, wholly glaucous; outer tepals 15-20mm; inner tepals 9-12mm, with green patch at apex only; (2n=24). Intrd; natd as for *G. elwesii* but much less frequently grown; very local in SE En and S Somerset; Caucasus. Differs from *G. elwesii* ± only in smaller parts and absence of basal green patch on inner tepals, and probably better a ssp. of it.

6. G. ikariae Baker - *Green Snowdrop*. Stems to 25cm; leaves 5-16 x 0.5-2cm at flowering, inrolled, hooded at apex, linear-oblanceolate, wholly green; outer tepals 17-26mm; inner tepals 8-13mm, with green patch at apex only; (2n=24). Intrd; natd in grassy and marginal places; scattered in S En, Dunbarton; Aegean.

35. NARCISSUS L. - *Daffodils*

Flowers solitary or few in umbel, pendent to erecto-patent, actinomorphic, with a corona; tepals and corona fused to form hypanthial tube between base of tepals and apex of ovary; tepals white to yellow; corona white to yellow or orange; flowers appearing in spring, with leaves, sometimes *flore pleno*.

An extremely popular garden genus with numerous interspecific hybrids and thousands of cultivars, many with uncertain parentage. Many occur natd in fields, waysides, woods, rough ground, banks, etc., and are very difficult to classify. The descriptions mainly apply to the commonest variants of each taxon, but others (especially colour variants) occur and are then best assigned to the recognized Divisions of the *International Daffodil Checklist*:

Division 1 - Flower 1, with corona at least as long as tepals
Division 2 - Flower 1, with corona >1/3 as long as but shorter than tepals
Division 3 - Flower 1, with corona ≤1/3 as long as tepals
Division 4 - *Flore pleno* variants of garden origin of any affinity
Division 5 - Derivatives of *N. triandrus*, with characteristics of that species evident
Division 6 - Derivatives of *N. cyclamineus*, with characteristics of that species evident
Division 7 - Derivatives of *N. jonquilla*, with characteristics of that species evident
Division 8 - Derivatives of *N. tazetta*, with characteristics of that species evident
Division 9 - Derivatives of *N. poeticus*, with characteristics of that species evident

1 Tepals reflexed back through 180° or almost so 2
1 Tepals patent to erecto-patent 3
  2 Hypanthial tube 10-20mm; corona c. as long as wide; upper 3 stamens and stigma exserted from corona; leaves subcylindrical, 1.5-3mm wide **7. N. triandrus**
  2 Hypanthial tube 2-3mm; corona distinctly longer than wide; all stamens and stigma included in corona; leaves flat or grooved on upperside, 3-6mm wide **11. N. cyclamineus**
3 Hypanthial tube parallel-sided, sometimes abruptly expanded at apex;

corona <10mm, wider than long; stamens of 2 distinct lengths 4

3 Hypanthial tube distinctly widening towards apex; corona usually >10mm, if <10mm then longer than tepals, c. as long as to much longer than wide; stamens all of same length or ± so 7

4 Flower 1; corona yellow with sharply contrasting narrow red rim **4. N. poeticus**

4 Flowers (1)2-8(20); corona white or yellow all over 5

5 Flowers (1)2(-3); hypanthial tube ≥20mm; pollen sterile **3. N. x medioluteus**

5 Flowers (2)3-8(20); hypanthial tube ≤20mm; pollen fertile 6

6 Corona yellow; flowers ≤8(15) **1. N. tazetta**

6 Corona white; flowers ≤20 **2. N. papyraceus**

7 Tepals linear to very narrowly triangular or lanceolate, ≤5mm wide **8. N. bulbocodium**

7 Tepals ovate or triangular-ovate to suborbicular, ≥1cm wide 8

8 Corona c. as long as tepals; hypanthial tube <2x as long as greatest width 9

8 Corona distinctly shorter than tepals; hypanthial tube >2x as long as greatest width 10

9 Hypanthial tube 15-25mm **9. N. pseudonarcissus**

9 Hypanthial tube 9-15mm **10. N. minor**

10 Flower 1; corona usually conspicuously deeper in colour than tepals; leaves ± glaucous, ≥8mm wide; stem distinctly 2-edged **5. N. x incomparabilis**

10 Flowers (1)2-4; tepals and corona the same colour; leaves green, ≤8mm wide; stem subterete **6. N. x odorus**

Other spp. - Several other spp., hybrids and groups of cultivars are grown and may occasionally persist in the wild. **N. jonquilla** L., from Spain and Portugal, differs from its hybrid *N. x odorus* in its ± terete stem, cylindrical hypanthial tube and 2-5 smaller flowers (tepals 10-15mm, corona 3-5mm); **N. x intermedius** Loisel. (*N. tazetta* x *N. jonquilla*), garden origin, differs from *N. x odorus* in its cylindrical hypanthial tube and smaller flowers (tepals 10-18mm, corona 3-4mm), and from *N. jonquilla* in its corona darker in colour than the tepals and wider (5-8mm) leaves; **N. bicolor** L. (*N. abscissus* (Haw.) Schult. & Schult. f.), from SW Europe, differs from *N. pseudonarcissus* in its hypanthial tube 8-12mm and slightly larger tepals and corona.

1. N. tazetta L. - *Bunch-flowered Daffodil.* Leaves flat, 5-15mm wide; flowers (2)3- 947
8(15); hypanthial tube 12-18mm; tepals 8-22mm, white; corona 3-6mm, yellow to deep yellow; (2n=14, 20-24, 28, 30-32). Intrd; rather rare relic in SW En and CI; W & C Mediterranean.

2. N. papyraceus Ker Gawl. - *Paper-white Daffodil.* Differs from *N. tazetta* in flowers ≤20; hypanthial tube 10-20mm; tepals 8-18mm; corona white; (2n=22). Intrd; natd as for *N. tazetta;* W & C Mediterranean.

3. N. x medioluteus Mill. (*N. x biflorus* Curtis; *N. tazetta* x *N. poeticus*) - *Primrose-peerless*. Leaves flat, 7-10mm wide; flowers (1)2(-3); hypanthial tube 20-25mm; tepals 18-22mm, usually white; corona 3-5mm, usually yellow; (2n=17, 24). Intrd; rather frequent relic in most of BI, mainly S; garden origin. This belongs to Division 8; *Primrose-peerless* itself often closely resembles *N. tazetta* (see key for differences), but other cultivars may have tepals and corona the same colour (white or yellow).

4. N. poeticus L. - *Pheasant's-eye Daffodil.* Leaves flat, 5-13mm wide; flower 1; hypanthial tube 20-30mm; tepals 15-30mm, white; corona 1-3mm, yellow with sharply contrasting narrow red rim. Intrd; S Europe. Cultivars of this belong to Division 9.

a. Ssp. poeticus (*N. majalis* Curtis). Leaves 6-10(12)mm wide; tepals usually 20- 947
25mm, obovate to suborbicular, strongly overlapping; corona c.14mm across, ±

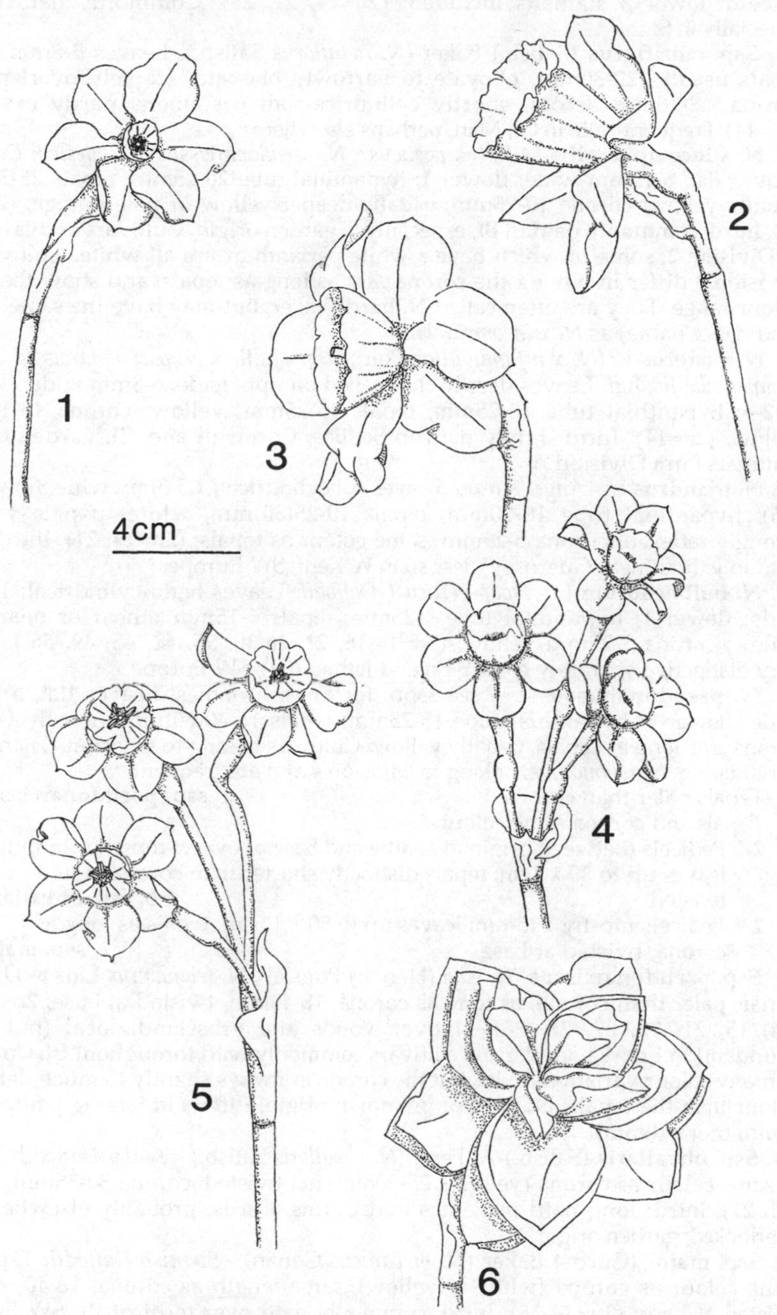

FIG 947 - *Narcissus*. 1, *N. poeticus* (Division 9). 2, *N. pseudonarcissus* (Division 1). 3, *N. x incomparabilis* (Division 2). 4, *N. x odorus* (Division 7). 5, *N. tazetta*. 6, *N. pseudonarcissus* (Division 4).

discoid; lower 3 stamens included; (2n=14, 21, 28). Commonly natd in BI, especially in S.

b. Ssp. radiiflorus (Salisb.) Baker (*N. radiiflorus* Salisb.). Leaves 5-8mm wide; tepals usually 22-30mm, obovate to narrowly obovate, scarcely overlapping; corona c.8-10mm across, shortly cylindrical; all 6 stamens partly exserted; (2n=14). Frequent relic in CI, Man, perhaps elsewhere.

5. N. x incomparabilis Mill. (*N. poeticus* x *N. pseudonarcissus*) - *Nonesuch Daffodil.* **947**
Leaves flat, 8-12mm wide; flower 1; hypanthial tube 20-25mm; tepals 25-35mm, usually yellow; corona 10-25mm, usually deeper yellow or pale orange; (2n=14, 21). Intrd; commonly natd in BI, especially S; garden origin. Cultivars of this belong to Division 2, some of which have a white perianth or are all white. Cultivars of Division 3 differ in having the corona $\leq$1/3 as long as tepals, and show the same colour range. They are often called **N. barrii** Baker, but may have the same origin (and hence name) as *N. x incomparabilis.*

6. N. x odorus L. (*N. x infundibulum* Poir.; *N. jonquilla* x *N. pseudonarcissus* agg.) - **947**
Campernelle Jonquil. Leaves deeply channelled on upperside, 6-8mm wide; flowers (1)2-4; hypanthial tube 15-25mm; tepals 15-25mm, yellow; corona 13-18mm, yellow; (2n=14). Intrd; rarely natd in Scillies, Cornwall and CI; garden origin. Cultivars form Division 7.

7. N. triandrus L. - *Angel's-tears.* Leaves subcylindrical, 1.5-3mm wide; flowers 1-3(6); hypanthial tube 10-20mm; tepals 10-25(30)mm, white to pale yellow, strongly reflexed; corona 5-25mm, same colour as tepals; (2n=14, 21). Intrd; well natd in Jethou (near Guernsey), less so in W Kent; SW Europe.

8. N. bulbocodium L. - *Hoop-petticoat Daffodil.* Leaves hemi-cylindrical, 1-2mm wide; flower 1; hypanthial tube 4-25mm; tepals 6-15mm, linear or nearly so, yellow; corona 7-25mm yellow; (2n=14-16, 21, 26-30, 35, 42, 45, 49, 56)). Intrd; very distinctive sp. rarely natd; S En and Jethou (CI); SW Europe.

9. N. pseudonarcissus L. - see sspp. for English names. Leaves flat, 5-15mm wide; flower 1; hypanthial tube 15-25mm; tepals 18-40(60)mm, usually yellow; corona ± as long as tepals, usually yellow. Cultivars belong to Division 1; *flore pleno* variants, e.g. *N. telamonius*, belong to Division 4 and are frequent.

1 Tepals paler than corona **a. ssp. pseudonarcissus**
1 Tepals and corona same colour 2
  2 Pedicels (between origin of spathe and base of ovary) mostly <15mm; leaves up to 30 x 1cm; tepals distinctly shorter than corona, not twisted **b. ssp. obvallaris**
  2 Pedicels mostly >15mm; leaves up to 50 x 1.5cm; tepals as long as corona, twisted at base **c. ssp. major**

a. Ssp. pseudonarcissus (*N. gayi* (Hénon) Pugsley, *N. telamonius* Link) - *Daffodil.* **947**
Tepals paler than corona, as long as corona, 18-40mm, twisted at base; 2n=14, 43 (14, 15, 21, 22, 27, 28, 30). Native; woods and grassland; local (but often abundant) in En, Wa and Jersey, cultivars commonly natd throughout BI. Cultivars of many colour variations exist, but the corona is always slightly to much darker in colour than the tepals. **N. gayi**, of unknown origin, differs in having pedicels 10-35mm (not <10mm).

b. Ssp. obvallaris (Salisb.) A. Fern. (*N. obvallaris* Salisb.) - *Tenby Daffodil.* Tepals of same colour as corona (yellow), 25-30mm, not twisted; corona 30-35mm; 2n=21 (14, 21). Intrd; long natd in Pembs and Carms, Cards, probably elsewhere but overlooked; garden origin.

c. Ssp. major (Curtis) Baker (*N. hispanicus* Gouan) - *Spanish Daffodil.* Tepals of same colour as corona (white to yellow), same length as corona, 18-40(60)mm, twisted at base; (2n=14-18). Intrd; commonly natd over most of BI; SW Europe. Cultivars of many colour variations exist.

9 x 11. N. pseudonarcissus x N. cyclamineus is the parentage of many garden cultivars including the much-planted **'February Gold'**; these form Division 6. They resemble small *N pseudonarcissus* but have the characteristic reflexed tepals of *N.*

cyclamineus to varying degrees. Reported natd in W Kent.

10. N. minor L. - *Lesser Daffodil*. Like a small *N. pseudonarcissus*, differing in leaves 3-10 mm wide; hypanthial tube 9-15mm; tepals 16-22mm, same colour or paler than corona; (2n=14). Intrd; well natd in Man, less so in few places in S En; SW Europe.

11. N. cyclamineus DC. - *Cyclamen-flowered Daffodil*. Leaves ± flat or shallowly grooved on upperside, 3-6mm wide; flower 1; hypanthial tube 2-3mm; tepals 15-22mm, linear or nearly so, yellow, reflexed through 180°; corona 15-22mm, yellow; (2n=14, 28). Intrd; very distinctive sp. rarely natd; W Kent and Surrey; Spain and Portugal.

36. PANCRATIUM L. - *Sea Daffodil*

Flowers several in umbel, erect to erecto-patent, actinomorphic, with a corona with toothed margin; tepals and corona white, fused to form a long slender greenish hypanthial tube above ovary; stamens fused to corona at base, exserted; flowers appearing in summer, with leaves.

1. P. maritimum L. - *Sea Daffodil*. Stems to 60cm; leaves up to 50 x 2cm, very glaucous; flowers very fragrant; hypanthial tube 6-8cm; tepals 3-5cm; corona 2-3cm; (2n=20, 22, 28). Intrd; 5 plants natd on sand-dunes in S Devon since c.1990, origin unknown; Mediterranean and SW Europe.

SUBFAMILY 8 - ASPARAGOIDEAE (tribes Asparageae, Rusceae) (genera 37-38). Plant rhizomatous; leaves all on stems, reduced to small scales, replaced functionally by stems (cladodes) arising from their axils; flowers dioecious or ± so; inflorescence an inconspicuous cluster of 1-few flowers borne in scale-leaf axil on main stem or on cladode; tepals greenish to yellowish-white, free or fused just at base, without corona; style 1 or ± 0, with capitate or 3-lobed stigma; ovary superior; fruit a spherical red berry.

37. ASPARAGUS L. - *Asparagus*

Stems herbaceous; cladodes borne in cluster of 4-10(more), cylindrical to slightly flattened; flowers 1-2(3) in axils of scale-leaves on main stems; tepals fused at base; berry 5-10mm.

1. A. officinalis L. - see sspp. for English names.

a. Ssp. prostratus (Dumort.) Corb. - *Wild Asparagus*. Stems procumbent to decumbent, to 30(60)cm; cladodes on main lateral branches 4-10(15)mm, rigid, usually glaucous; pedicels 2-6(8)mm; seeds usually <5; 2n=40, 80. Native; grassy sea-cliffs; very local in E & W Cornwall, Pembs, SE Ir and CI, formerly elsewhere in SW En and S Wa N to Anglesey, decreasing (many colonies now with only 1 sex). **RR**

b. Ssp. officinalis - *Garden Asparagus*. Stems erect, to 1.5(2)m; cladodes on main lateral branches (5)10-20(25)mm, flexible, usually green; pedicels 6-10(15)mm; seeds usually 5-6; 2n=20. Intrd; grown as vegetable, very well natd in dry sandy soils among sparse grass, especially on maritime dunes and E Anglia heathland; scattered throughout BI N to C Sc; Europe.

38. RUSCUS L. - *Butcher's-brooms*

Evergreen shrubs; cladodes borne singly, flattened and leaf-like; flowers 1-2 in axils of scale-leaves borne in centre of adaxial side of cladode; tepals free; berry 8-13mm.

1. R. aculeatus L. - *Butcher's-broom*. Stems erect, to 75(100)cm, much branched, cladodes 1-3(4) x 0.4-1cm, with sharp spine at apex; bract of inflorescence ≤3 x 1mm, scale-like; 2n=40. Native; woods, hedgerows, rocky places on dry soils; rather local in CI and Br N to Norfolk and N Wa, natd further N and in native areas.

2. R. hypoglossum L. - *Spineless Butcher's-broom.* Stems oblique, to 40cm, simple; cladodes 3-10 x 1-3.3cm, spineless; bract of inflorescence 10-30 x 3.5-13mm, green; (2n=40). Intrd; grown as ornament, natd in few shady places in W En, formerly Midlothian; SE Europe.

SUBFAMILY 9 - ALSTROEMERIOIDEAE (genus 39). Plant with tuberous roots; leaves all on stems, lanceolate; inflorescence a terminal simple or compound umbel without a spathe; tepals orange with darker markings, free, without corona; flowers slightly zygomorphic, with curved stamens; style 1, with 3-lobed stigma; ovary inferior; fruit a capsule splitting along centre of ovary cells.

39. ALSTROEMERIA L. - *Peruvian Lily*

1. A. aurea Graham (*A. aurantiaca* D. Don) - *Peruvian Lily.* Stems erect, to 1m; leaves 7-10cm; tepals 4-6cm, orange with red spots and streaks; stamens shorter than tepals; (2n=16). Intrd; much grown in gardens, natd in grassy places, rough ground and old garden sites; scattered in Br N to Man and C Sc, mainly in N; Chile.

163. IRIDACEAE - *Iris family*

Usually erect, mostly glabrous, herbaceous perennials, rhizomatous or with a corm or rarely a bulb or swollen roots; leaves all or mostly basal, those on stems alternate and usually smaller and few, simple, entire, sessile, usually with sheathing base, without stipules. Flowers solitary or in terminal spikes or panicles usually with sheathing (often spathe-like) bracts, bisexual, conspicuous, epigynous, actinomorphic or less often zygomorphic; perianth of 6 tepals usually fused into tube proximally, sometimes free, petaloid, the 3 outer not or obviously different from 3 inner; stamens 3; ovary 3-celled, with numerous ovules on axile placentas, rarely 1-celled with 3 parietal placentas; style usually with 3 branches, the branches divided or not, with stigmas at tips, rarely simple with 3 stigmas, the branches sometimes ± petaloid; fruit a capsule splitting along centre of ovary-cells.

Easily distinguished from Liliaceae by the 3 stamens (only the distinctive *Ruscus* in the latter family has 3 stamens).

1 Style-branches broad and petaloid; flowers *Iris*-like 2
1 Style-branches not petaloid, narrow; flowers not *Iris*-like 3
2 Plant with rhizome or bulb; roots not tuberous; ovary 3-celled **5. IRIS**
2 Plant without rhizome or bulb; roots tuberous; ovary 1-celled **4. HERMODACTYLUS**
3 Flowers 1-few, erect, arising direct from ground or on very short stems, *Crocus*-like 4
3 Flowers few-many, erect or laterally-directed, usually arising from aerial green stems in spikes or panicles, not *Crocus*-like 5
4 Leaves subterete, without white line; perianth-tube <1cm, sheathed by green bract **7. ROMULEA**
4 Leaves flat, channelled and with central whitish line on upperside; perianth-tube >1.5cm, sheathed by white or brown bract **8. CROCUS**
5 Perianth actinomorphic, with radially symmetrical lobes and straight tube; plant with or without a corm 6
5 Perianth zygomorphic, often with bilaterally symmetrical lobes but sometimes only so due to curved tube; plant with a corm 11
6 Perianth-lobes fused proximally into tube >5mm 7
6 Perianth-lobes completely free or fused proximally into tube <5mm 8
7 Perianth-tube >3cm, the lobes shorter; bracts <2cm, 3-toothed at apex,

without dark streaks **10. IXIA**

7 Perianth-tube <2cm, the lobes >2cm; bracts >2cm, deeply and jaggedly toothed at apex, with irregular dark longitudinal streaks **11. SPARAXIS**

8 Inner tepals c.2x as long as outer **1. LIBERTIA**

8 Inner and outer tepals same or nearly same length 9

9 Stem terete, arising from a corm; flowers sessile **10. IXIA**

9 Stem flattened, narrowly winged, arising from rhizome or fibrous roots; flowers stalked 10

10 Tepals twisting spirally after flowering; filaments free, arising from top of short perianth-tube **3. ARISTEA**

10 Tepals not twisting after flowering; filaments fused either just at base or for most of length, arising from base of perianth **2. SISYRINCHIUM**

11 Style 3-branched, each branch bifid 12

11 Style with 3 simple branches, or unbranched with 3-lobed stigma 13

12 Bracts >2cm; spike erect, with flowers on 2 sides of axis; leaves tough; seeds winged **6. WATSONIA**

12 Bracts <1.5cm; spike bent horizontally near lowest flower, with flowers on 1 side; leaves soft; seeds not winged **12. FREESIA**

13 Uppermost perianth-lobe >2x as long as rest **14. CHASMANTHE**

13 Uppermost perianth-lobe slightly longer to slightly shorter than rest 14

14 Perianth-lobes strongly narrowed at base; style-branches greatly widened distally **9. GLADIOLUS**

14 Perianth-lobes not narrowed at base; style-branches filiform with minutely capitate stigmas **13. CROCOSMIA**

Other genera - **HOMERIA** Vent., from S Africa, would key out as *Sisyrinchium* because of its actinomorphic flowers, free tepals and united filaments, but has a corm; **H. collina** (Thunb.) Vent. (*H. breyniana* G.J. Lewis) (*Cape-tulip*) has pale yellow or pink fragrant flowers 3-3.5cm and occurs self-sown in large gardens in Scillies and Guernsey, and might escape.

1. LIBERTIA Spreng. - *Chilean-irises*
Plants with short rhizomes; leaves *Iris*-like; inflorescence a small terminal panicle; flowers actinomorphic; tepals white, free, the inner c.2x as long as outer but of similar shape; filaments slightly fused at base; style with 3 entire linear branches.

1. L. formosa Graham (*L. chilensis* (Molina) Klotzsch ex Baker nom. nud.) - *Chilean-iris*. Densely tufted; stems erect, to 1.2m, unbranched; leaves dark green, up to 75 x 1.2cm; flowers with pedicels shorter than and obscured by bracts, c.25mm across; inner tepals 12-18mm; (2n=114). Intrd; garden plant natd on rough ground, waysides and rocky lakeshore and coasts; Scillies, W Cornwall, Man, CW Sc, S Kerry; Chile.

2. L. elegans Poepp. - *Lesser Chilean-iris*. Differs from *L. formosa* in more delicate habit; inflorescence branched; flowers with pedicels exceeding bracts by c.5-10mm, c.13mm across; inner tepals 6-9mm. Intrd; rare garden plant natd in roadside ditches on Gigha, Kintyre, formerly on railway embankment in Dunbarton; Chile.

2. SISYRINCHIUM L. - *Blue-eyed-grasses*
Plants with fibrous roots and short or 0 rhizomes; leaves *Iris*-like; inflorescence a terminal cyme or a panicle of terminal and lateral cymes; flowers actinomorphic; tepals pale to bright yellow or blue, nearly free, ± equal; filaments slightly fused at base to fused for most of length; style with 3 entire linear branches.

1 Tepals blue 2

1 Tepals cream to yellow for most part 3

2 Stem unbranched, with 1 terminal inflorescence; perianth 25-35mm

across, violet-blue; pedicels erect in fruit **2. S. montanum**

2 At least some stems branched, each branch with 1 terminal inflorescence; perianth 15-20mm across, pale blue; pedicels arched to pendent in fruit **1. S. bermudiana**

3 Stem unbranched, with terminal and several lateral cymes; leaves >1cm wide **5. S. striatum**

3 Stem branched or unbranched, with 1 terminal cyme on each branch; leaves ≤1cm wide 4

4 Tepals bright yellow, stem unbranched **3. S. californicum**

4 Tepals cream to pale yellow; stem branched **4. S. laxum**

1. S. bermudiana L. (*S. graminoides* E.P. Bicknell, *S. hibernicum* Á. & D. Löve) - *Blue-eyed-grass.* Stems to 50cm, usually branched; leaves up to 5mm wide; tepals blue, 6-10mm; 2n=64, 88, 96. Probably native; wet meadows and stony ground by lakes; very local in W Ir from W Cork to W Donegal, known only since 1845. There is argument as to whether our plant is a native endemic or an American introduction, and in latter case as to its correct name. Records of natd plants from Br need checking. **R**

2. S. montanum Greene (*S. bermudiana* auct. non L.) - *American Blue-eyed-grass.* Differs from *S. bermudiana* in tepals 10-18mm; and see key; (2n=32, 96). Intrd; natd in grassy places, rough ground and waysides; scattered in Br N to Easterness, CI; N America. Much confused with *S. bermudiana*; identity of natd plants in Ir needs checking.

3. S. californicum (Ker Gawl.) W.T. Aiton (*S. boreale* (E.P. Bicknell) J.K. Henry) - *Yellow-eyed-grass.* Stems to 60cm, unbranched; leaves up to 6mm wide; tepals bright yellow, 12-18mm; 2n=34. Intrd; grown in gardens, natd in damp grassy places near sea; Mons, Pembs, Co Wexford, W Galway; W N America. **S. boreale** may be a different sp., to which our plant belongs.

4. S. laxum Otto ex Sims (*S. iridifolium* Kunth ssp. *valdivianum* (Phil.) Ravenna) - *Veined Yellow-eyed-grass.* Stems to 45cm, branched; leaves up to 10mm wide; tepals whitish to pale yellow with purple veins, 12-15mm. Intrd; garden plant natd on gravelly paths in Jersey, formerly N Hants; S America.

5. S. striatum Sm. - *Pale Yellow-eyed-grass.* Stems to 75cm, unbranched; leaves up to 20mm wide; tepals pale yellow, 15-18mm; (2n=18). Intrd; grown in gardens, natd (often short-lived) on tips, waste ground, banks and waysides; scattered in S En from Surrey to Scillies; S America.

3. ARISTEA Aiton - *Blue Corn-lily*

Plants with rhizomes; leaves *Iris*-like; inflorescence a loose terminal panicle of few-flowered clusters; flowers actinomorphic; tepals united proximally into tube <5mm, blue, ± equal; filaments free, arising from top of perianth-tube; style very slender, with 3-lobed stigma.

1. A. ecklonii Baker - *Blue Corn-lily.* Stems to 60cm, flattened, bearing reduced leaves; leaves up to 60 x 1.2cm, linear; perianth 8-15mm; (2n=32, 64, 65). Intrd; natd in rough ground as garden escape, Tresco (Scillies); S Africa. Often misdetermined as various other spp.

4. HERMODACTYLUS Mill. - *Snake's-head Iris*

Plants with tuberous roots (no rhizomes, bulbs or corms); leaves subterete, very long, 4-angled; flowers solitary, terminal, actinomorphic, *Iris*-like; tepals united proximally into tube; tepals, stamens and styles ± like those of *Iris*.

1. H. tuberosus (L.) Mill. - *Snake's-head Iris.* Stems to 40cm; leaves longer than stems, up to 50cm x 3mm; outer tepals yellowish-green; inner tepals yellowish-green on claw, purplish-brown to blackish on blade; (2n=20). Intrd; garden plant natd in grassy places and hedgerows; N Somerset, N Devon and W Cornwall; Mediterranean.

5. IRIS L. - *Irises*

Plants with rhizomes or rarely bulbs; leaves *Iris*-like (vertical, flat, with 2 identical faces) or subterete or 4-angled; inflorescence terminal, rather simple, cymose; flowers actinomorphic; tepals united proximally into perianth-tube; outer tepals usually longer and wider than the inner, patent, recurved or reflexed, with a narrow proximal part (*claw*) and expanded distal part (*blade*); inner tepals usually erect, less differentiated into blade and claw; filaments free, borne at base of outer tepals; style with 3 long, broad, petaloid branches each with 2 lobes at apex beyond stigma, each covering a stamen.

1 Leaves subterete to slightly flattened, angled or channelled; plant with a bulb 2

1 Leaves flat, not channelled or angled, vertical, with 2 identical faces; plant with rhizome 4

2 Perianth-tube >10mm **13. I. x hollandica**

2 Perianth-tube <10mm 3

3 Leaves evergreen; claw of outer tepals ≤10mm wide, 1.5-2x as long as blade **12. I. xiphium**

3 Leaves dying down in winter; claw of outer tepals >20mm wide, no longer than blade **11. I. latifolia**

4 Outer tepals bearded, i.e. with mass of stout multicellular hairs on inner face **1. I. germanica**

4 Outer tepals not bearded, sometimes softly pubescent with unicellular hairs 5

5 Tepals predominantly yellow or yellow and white, without blue, purple, mauve or violet or only small spots or veining of it 6

5 Tepals predominantly of some shade of blue, purple, mauve or violet 8

6 Leaves evergreen, dark green, with stinking smell when crushed; seeds bright orange; stems distinctly compressed **9. I. foetidissima**

6 Leaves dying in winter, mid- to pale-green, not stinking; seeds brownish; stems subterete 7

7 Inner tepals white; outer tepals white with large yellow patch on blade; petaloid style-lobes subentire **8. I. orientalis**

7 Tepals yellow all over, the outer often with brownish or purple spots or veins; petaloid style-lobes deeply serrate **3. I. pseudacorus**

8 Flowering stems 0 or very short; perianth-tube 6-28cm; style-branches with yellow glands near margins **10. I. unguicularis**

8 Flowering stems well developed; perianth-tube ≤2cm; style-branches without yellow glands 9

9 Leaves evergreen, dark green, with stinking smell when crushed; seeds bright orange **9. I. foetidissima**

9 Leaves dying in winter, mid- to pale-green, not stinking; seeds brownish 10

10 Stems hollow; perianth-tube 4-7mm; bracts brown and papery at flowering **2. I. sibirica**

10 Stems solid; perianth-tube 7-20mm; bracts at least partly green at flowering 11

11 Upper part of ovary sterile, narrower than ovary below and perianth-tube above, forming acuminate beak on capsule ≥5mm 12

11 Ovary without sterile apical part; capsule with 0 or short beak <5mm 13

12 Capsule with beak 5-8mm, with 1 rib where 2 ovary-cells meet; leaves mostly <10mm wide; flowers mostly >8cm across **6. I. ensata**

12 Capsule with beak 8-16mm, with 2 ridges where 2 ovary-cells meet; leaves mostly >10mm wide; flowers mostly <8cm across **7. I. spuria**

13 Outer tepals glabrous on central patch; capsules setting many seeds **4. I. versicolor**

13 Outer tepals pubescent on central patch; capsules setting 0-few seeds **5. I. x robusta**

1. I. germanica L. - *Bearded Iris*. Rhizomatous; leaves flat, (20)30-60mm wide; stems to 90cm, usually branched; flowers 8-15cm across, usually blue to violet or purple with yellow bearded region on outer tepals; (2n=24, 28, 36, 44, 48). Intrd; much grown in gardens, natd on banks, rough and waste ground, waysides, old planted areas; frequent in C & S Br and CI; garden origin. The true *I. germanica* is only 1 of a group of hybrid origin, for which no overall name exists; flower colour varies from white or yellow to blue, violet or purple, with many variations in size and shape.

2. I. sibirica L. - *Siberian Iris*. Rhizomatous; leaves flat, (2)4-10mm wide; stems to 1.2m, usually branched; flowers 6-7cm across, blue or bluish-violet with white patch on outer tepals; (2n=28). Intrd; garden plant natd in rough, often wet or shaded, ground; scattered throughout Br N to Easterness; C Europe to S Asia.

3. I. pseudacorus L. - *Yellow Iris*. Rhizomatous; leaves flat, 10-30mm wide; stems to 1.5m, usually branched; flowers 7-10cm across, yellow with deeper yellow patch on outer tepals; 2n=34. Native; wet meadows, fens and ditches, by lakes and rivers; common throughout BI.

4. I. versicolor L. - *Purple Iris*. Rhizomatous; leaves flat, 8-25mm wide; stems to 1m, usually branched; flowers 6-8cm across, purple to violet with greenish-yellow patch surrounded by whitish area on outer tepals; 2n=108. Intrd; natd by lakes and rivers and in reed-swamps; scattered in En and N to C Sc; E N America.

5. I. x robusta E.S. Anderson (*I. versicolor* x *I. virginica* L.) - *Windermere Iris*. Differs from *I. versicolor* as in key (couplet 13); 2n=89. Intrd; natd in reed-swamp and rough pasture by Lake Windermere, Westmorland, without either parent; E N America and garden origin.

6. I. ensata Thunb. (*I. kaempferi* Siebold ex Lem.) - *Japanese Iris*. Rhizomatous; leaves flat, 4-12mm wide; stems to 90cm, usually unbranched; flowers 8-15cm across, purple with yellow claws and base of blade of outer tepals; (2n=24, 40, 80). Intrd; natd in swamp in W Kent; E Asia.

7. I. spuria L. - *Blue Iris*. Rhizomatous; leaves flat, 6-20mm wide; stems to 90cm, usually unbranched; flowers 6-9cm across, bluish-violet, sometimes with small yellow or white area at base of blade of outer tepals; (2n=40, 42, 44). Intrd; natd in Dorset and N Somerset and by fen ditches in N Lincs since 1836; Europe.

8. I. orientalis Mill. (*I. ochroleuca* L., *I. spuria* ssp. *ochroleuca* (L.) Dykes) - *Turkish Iris*. Rhizomatous; leaves flat, 10-25mm wide; stems to 1.2m, usually little-branched; flowers 8-10cm across, white with yellow patch on outer tepals; (2n=28, 40). Intrd; natd in fields in Dorset since 1960, in limestone scrub in N Somerset since at least 1950 and on banks in W Kent since 1984; E Mediterranean.

9. I. foetidissima L. - *Stinking Iris*. Rhizomatous; leaves flat, 10-25mm wide, evergreen; stems to 80cm, branched; flowers 5-7cm across, dull purplish, rarely pale yellow; 2n=40. Native; dry places in woods, hedges, banks and cliffs near sea, mostly on calcareous soils; locally frequent in CI and Br N to N Wa and Notts, natd elsewhere in Br and Ir.

10. I. unguicularis Poir. - *Algerian Iris*. Rhizomatous; leaves flat, 1-5(10)mm wide, evergreen; stems 0 or very short; flowers 6-9cm across, purple, lilac and whitish, with yellow band on outer tepals; (2n=28, 40, 50). Intrd; garden escape or throwout natd by lane; S Somerset since 1989; N Africa and E Mediterranean.

11. I. latifolia (Mill.) Voss (*I. xiphioides* Ehrh.) - *English Iris*. Bulbous; leaves channelled, whitish in channel, 5-8mm wide; stems to 50cm, unbranched; flowers 8-12cm across, bluish-violet with yellow patch on outer tepals; (2n=42). Intrd; grown in gardens, natd in grassy places in Shetland and W Kent; Pyrenees.

12. I. xiphium L. - *Spanish Iris*. Differs from *I. latifolia* in leaves often <5mm wide,

not or scarcely whitish in channel; perianth-tube 1-3mm (not 3-5mm); (2n=28, 34); and see key (couplet 3). Intrd; grown in gardens and bulb-fields, natd as relic in old fields, rough ground and waste places in Scillies and CI; SW Europe.
13. I. x hollandica hort. (*I. filifolia* Boiss. x *I. tingitana* Boiss. & Reut.) - *Dutch Iris*. Bulbous; leaves 0.5-3mm wide, channelled, evergreen; stems to 50cm, unbranched; flowers 8-12cm across, white, yellow or blue to purple. Intrd; grown and natd as for *I. xiphium*; frequent in CI; garden origin.

6. WATSONIA Mill. - *Bugle-lily*
Plants with corms; leaves *Iris*-like; inflorescence a spike; flowers slightly zygomorphic, with curved perianth-tube; tepals united into tube longer than the lobes, white, the lobes ± equal; filaments free, borne in perianth-tube; style very slender, with 3 bifid stigmas.

1. W. borbonica (Pourr.) Goldblatt (*W. ardernei* Sander) - *Bugle-lily*. Stems to 1m; leaves up to 60 x 4cm; perianth-tube 2.5-3.5cm; perianth-lobes 2.5-3.5cm; (2n=16). Intrd; grown in gardens, natd in rough ground; Tresco (Scillies); S Africa. Our plant is the white-flowered variant of ssp. **ardernei** (Sander) Goldblatt.

7. ROMULEA Maratti - *Sand Crocuses*
Plants with corms; leaves subterete, 4-grooved; flowers 1-several on very short stem, *Crocus*-like, actinomorphic; tepals united into tube much shorter than lobes, white or mauve, the lobes equal; filaments free, borne in perianth-tube; style very slender, with 3 bifid stigmas; ovary above ground at flowering.

1. R. columnae Sebast. & Mauri - *Sand Crocus*. Corm obliquely narrowed at base; **RRR** leaves 5-10cm x 0.6-1mm, recurved; perianth 7-15mm incl. tube 2.5-5.5mm, usually mauve, sometimes white, pale yellow inside at base; (2n=c.48). Native; maritime sandy turf; very local near Dawlish, S Devon, common in all of CI, formerly E Cornwall.
2. R. rosea (L.) Eckl. - *Oniongrass*. Corm rounded at base; leaves 15-25cm x 1-2.5mm, erect or ± so; perianth 15-45mm incl. tube 2.5-8mm, white with yellow inside at base; (2n=18). Intrd; natd at wall-base and in sparsely grassy area on gravel in Guernsey since at least 1969; S Africa. Our plant is var. **australis** (Ewart) M.P. de Vos.

8. CROCUS L. - *Crocuses*
Plants with corms; leaves linear, ± flattened, with central whitish channel; flowers erect, 1-few on short underground pedicels that elongate at fruiting, actinomorphic; tepals united into long narrow tube, various colours, the lobes equal; filaments free, borne at apex of perianth-tube; style slender, with 3 (or more) branches near apex, each with variously divided stigmas; ovary subterranean at flowering.
Below the flower and sheathing the ovary and part of the perianth-tube are 1-2 bracts, borne immediately below the ovary, and 0-1 spathe, borne at the base of the pedicel. See *Colchicum* and *Sternbergia* (Liliaceae) for differences.

1 Flowers appearing in autumn (Sep-Dec), often without leaves, never predominantly yellow 2
1 Flowers appearing in spring (Jan-Apr), usually with leaves, often predominantly yellow 7
  2 Anthers white to cream-coloured 3
  2 Anthers yellow 4
3 Throat of corolla uniformly deep yellow; filaments densely pubescent; style with many yellow or orange main branches **12. C. pulchellus**
3 Throat of corolla whitish with yellow blotches; filaments glabrous to

minutely pubescent; style with 3 cream to yellow main branches **13. C. kotschyanus**

4 Style with 3 main branches; throat of corolla uniformly yellow **8. C. longiflorus**

4 Style with many main branches; throat of corolla whitish to pale yellow 5

5 Corm with covering splitting into rings at base, not becoming fibrous; perianth conspicuously darker-veined outside **11. C. speciosus**

5 Corm with covering becoming fibrous, not splitting into rings at base; perianth usually not darker-veined outside 6

6 Leaves developing well after flowering, 3-4 per shoot, 2-4mm wide; throat of corolla white to pale purple **9. C. nudiflorus**

6 Leaves emerging during flowering, 5-7 per shoot, 0.5-2mm wide; throat of corolla white to pale yellow **10. C. serotinus**

7 Perianth predominantly pale to deep yellow, sometimes tinged or striped dark purple 8

7 Perianth predominantly white or pale mauve to dark purple, sometimes yellow on throat 10

8 Corm with covering splitting horizontally into rings at base, scarcely vertically and not becoming fibrous or reticulated; ground colour of flowers creamy-white to yellow **5. C. chrysanthus**

8 Corm with covering splitting vertically, not horizontally, becoming fibrous or reticulated; ground colour of flowers yellow to deep yellow 9

9 Perianth-lobes wholly bright yellow; leaves 0.5-1mm wide **4. C. ancyrensis**

9 Perianth-lobes suffused or striped purplish-brown on outside, or if uniformly yellow then leaves 1-4mm wide **7. C. x stellaris**

10 Throat of corolla yellow; spathe 0; bracts 2, white 11

10 Throat of corolla white to mauve or purple; spathe 1, papery; bract 1, white 12

11 Corm with covering splitting horizontally into rings at base, scarcely vertically and not becoming fibrous or reticulated; corolla usually darker-veined or -striped on outside **6. C. biflorus**

11 Corm with covering splitting vertically, not horizontally, becoming fibrous and reticulated; corolla not darker-veined or -striped on outside **3. C. sieberi**

12 Leaves mostly 4-8mm wide; flowers white to deep purple, often with dark stripes outside; perianth-tube usually mauve to purple, white only if rest of perianth is white **1. C. vernus**

12 Leaves mostly 2-3mm wide; flowers mauve to pale purple with white perianth-tube **2. C. tommasinianus**

Other spp. - C. sativus L. (*Saffron Crocus*), of unknown origin, was formerly grown for its styles (the spice) and was natd in a few places; it flowers in autumn with the leaves, and has yellow anthers, a strongly striped perianth not yellow inside at base, and 3 simple style-branches >1/2 as long as perianth-lobes.

1. C. vernus (L.) Hill (*C. purpureus* Weston) - *Spring Crocus.* Corm-covering **957**
fibrous; leaves mostly 4-8mm wide; perianth white to deep purple, often strongly more darkly striped outside, with tube not paler than lobes, with white to purple, glabrous to pubescent throat. Intrd; the most commonly grown sp., frequently natd in grassy places, meadows, banks, churchyards; scattered in BI; S Europe.

a. Ssp. vernus. Style usually equalling or exceeding stamens; tepal-lobes (2.5)3-5.5 x 0.9-2cm, usually coloured; (2n=8, 10, 12, 16, 18, 19, 20, 22, 23). Italy eastwards.

b. Ssp. albiflorus (Kit. ex Schult.) Asch. & Graebn. Style usually distinctly shorter than stamens; tepal-lobes 1.5-3.5(5) x 0.4-1.2cm, very often white; (2n=8). Albania

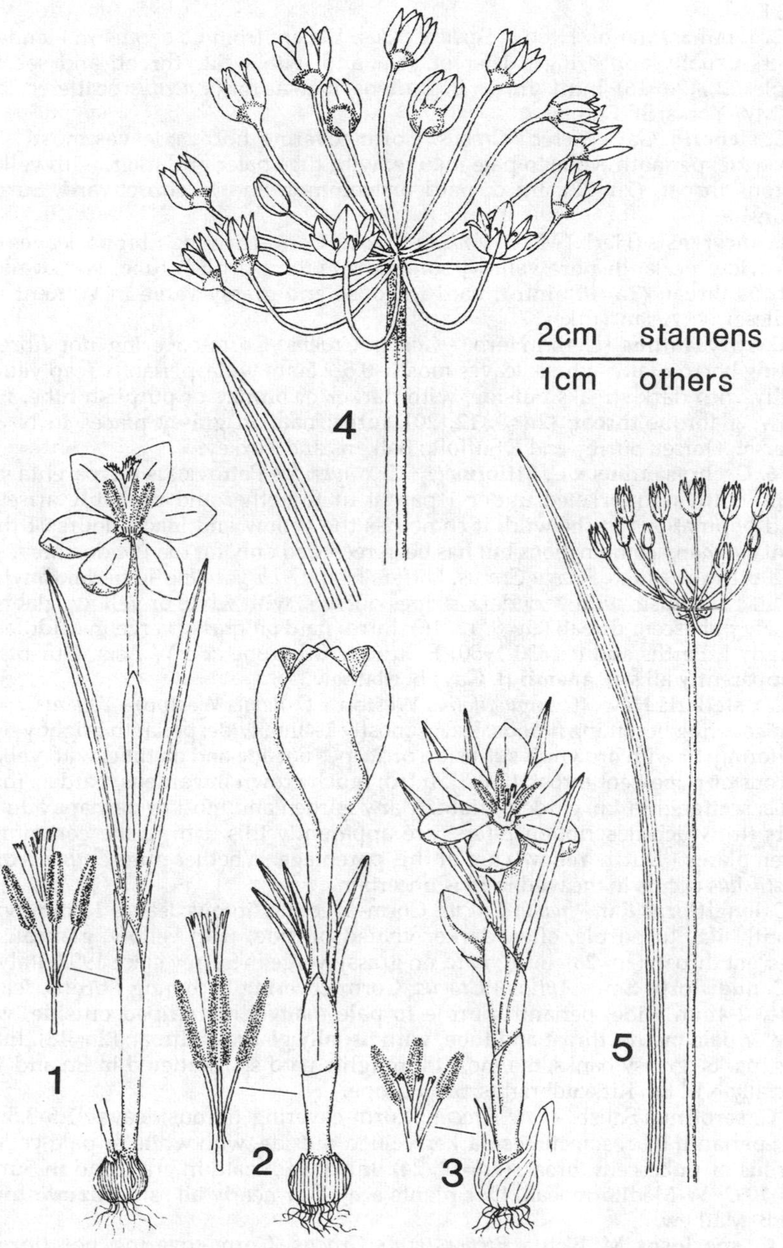

FIG 957 - *Crocus*, **Allioideae.** 1-3, *Crocus*. 1, *C. tommasinianus*. 2, *C. vernus*. 3, *C. x stellaris*. 4, *Nectaroscordum siculum*. 5, *Nothoscordum borbonicum*.

westwards. Smaller and less handsome than ssp. *vernus* and less often grown, but natd plants have been reported.

1 x 2. C. vernus x C. tommasinianus occurs in mixed populations of the 2 spp. in SE En.

2. C. tommasinianus Herb. - *Early Crocus.* Differs from *C. vernus* in slenderer flowers usually appearing earlier in year and with white throat; and see key (couplet 12); (2n=16). Intrd; grown in gardens, natd as for *C. vernus;* scattered in Br N to MW Yorks; SE Europe. **957**

3. C. sieberi J. Gay - *Sieber's Crocus.* Corm-covering fibrous; leaves mostly 1.5-2mm wide; perianth white to pale mauve, with tube paler or darker, with yellow, glabrous throat; (2n=22). Intrd; natd on common and in churchyard, Surrey; Balkans.

4. C. ancyrensis (Herb.) Maw - *Ankara Crocus.* Corm-covering fibrous; leaves 0.5-1mm wide; perianth pure yellow, sometimes with purplish tube, with yellow, glabrous throat; (2n=10). Intrd; natd in copse and grassy verge in W Kent and Middlesex; W Asian Turkey.

5. C. chrysanthus (Herb.) Herb. - *Golden Crocus.* Corm-covering not fibrous, splitting horizontally at base; leaves mostly 0.5-2.5mm wide; perianth deep yellow, usually with dark streaks outside, with darker or bronzy or purplish tube, with yellow, glabrous throat; (2n=8, 12, 20). Intrd; natd in grassy places in N & S Somerset, Dorset, Surrey and E Suffolk; Balkans and Turkey.

5 x 6. C. chrysanthus x C. biflorus (?= C. *x hybridus* Petrovic) is the parentage of several cultivars marketed under 1 parent or the other and probably arises in mixed populations in the wild; it combines the yellow and lilac colours of the 2 parents in many combinations but has been recorded only for the London area.

6. C. biflorus Mill. - *Silvery Crocus.* Differs from *C. chrysanthus* in perianth white, pale lilac or bluish, with very dark stripes outside, with white or yellow, glabrous or finely pubescent throat; (2n=8, 12, 16). Intrd; natd on grassy verge in Middlesex, formerly E Perths and (c.1830-1950) E Suffolk; S Europe and W Asia. Our plants are apparently all ssp. **adamii** (J. Gay) B. Mathew.

7. C. x stellaris Haw. (C. *angustifolius* Weston x C. *flavus* Weston) - *Yellow Crocus.* **957**
Corm-covering becoming fibrous; leaves mostly 1-4mm wide; perianth bright yellow ± uniformly or with brownish suffusion or stripes outside and on tube, with yellow, glabrous or pubescent throat; (2n=8). Intrd; much grown in gardens, natd as for C. *vernus;* scattered in En; garden origin. Many garden and most or perhaps all natd plants (to which description refers) are apparently this hybrid; the commonest garden plants ('Dutch Yellow') are of this parentage. Whether pure C. *flavus* or C. *angustifolius* occurs in the wild in BI is uncertain.

8. C. longiflorus Raf. - *Italian Crocus.* Corm-covering fibrous; leaves 1-3mm wide; perianth lilac to purple, often darker-veined outside, with yellow, glabrous or pubescent throat; (2n=28). Intrd; natd on grassy verge in Surrey since 1992; Italy.

9. C. nudiflorus Sm. - *Autumn Crocus.* Corm-covering becoming fibrous; leaves mostly 2-4mm wide; perianth purple to pale mauve, not striped outside, with white or pale mauve throat and tube, with usually glabrous throat; (2n=48). Intrd; fields, parks, grassy banks, the most thoroughly natd sp.; scattered in En and Wa, especially NW En, Kirkcudbrights; SW Europe.

10. C. serotinus Salisb. - *Late Crocus.* Corm-covering fibrous; leaves 1.5-3.5mm wide; perianth lilac, sometimes darker-veined outside, with white or pale yellow, glabrous or pubescent throat; (2n=22, 24). Intrd; well natd in grassland in Surrey since 1992; W Mediterranean. Our plants are all or nearly all ssp. **salzmannii** (J. Gay) B. Mathew.

11. C. speciosus M. Bieb. - *Bieberstein's Crocus.* Corm-covering not fibrous, splitting horizontally at base; leaves mostly 3-5mm wide; perianth pale mauve to purple with conspicuous dark veining outside, with white or nearly white tube, with whitish, glabrous throat; (2n=18). Intrd; natd in churchyards and on waysides in Surrey, Berks and E Suffolk; SW Asia.

12. C. pulchellus Herb. - *Hairy Crocus*. Differs from *C. speciosus* in throat yellow, glabrous or slightly pubescent; anthers white to cream (not yellow); filaments pubescent (not glabrous); (2n=12). Intrd; natd in churchyard in W Suffolk since 1983 but perhaps now gone; Balkans and W Turkey.

13. C. kotschyanus K. Koch - *Kotschy's Crocus*. Corm-covering fibrous; leaves mostly 1.5-4mm wide; perianth pale mauve to purple with conspicuous dark veining outside, with pubescent, whitish throat with yellow blotches, with white or nearly white tube; (2n=8, 10). Intrd; natd in meadows and on grassy tracksides; Surrey and E & W Suffolk since 1981; Turkey.

9. GLADIOLUS L. - *Gladioluses*

Plants with corms; leaves *Iris*-like; inflorescence a spike; flowers zygomorphic, with curved perianth-tube; tepals united into tube shorter than lobes, pinkish- to purplish-red, the lobes unequal, much narrowed at base; filaments free, borne on perianth-tube; style slender, with 3 short terminal lobes greatly widened distally.

Other spp. - Various garden and florists' Gladioluses rarely occur on tips, but do not persist. **G. italicus** Mill. (*G. segetum* Ker Gawl.), from S Europe, used to be natd in parts of S En, but modern records seem to be errors for *G. communis* from which it differs in the anthers longer than the filaments and the unwinged (not winged) seeds.

1. G. illyricus W.D.J. Koch - *Wild Gladiolus*. Stems to 50(90)cm, unbranched; **RRR** leaves up to 40 x 1cm; flowers 3-8(10); perianth 3.5-5cm, the lobes 2.5-4 x 0.6-1.6cm; 2n=60. Native; among bracken in scrub in New Forest, S Hants, formerly Wight; records of natd plants elsewhere may be other spp.

2. G. communis L. (*G. byzantinus* Mill.) - *Eastern Gladiolus*. Stems to 1m, often branched; leaves up to 70 x 2.5cm; flowers mostly 10-20; perianth 4-5.5cm, the lobes 3-4.5 x 1.5-2.5cm; (2n=90, 120). Intrd; persistent relic of cultivation in old bulb-fields, field-margins, roadsides and rough ground; scattered in extreme S En, frequent in CI and Scillies; Mediterranean. Our plant is ssp. **byzantinus** (Mill.) R.C.V. Douin.

10. IXIA L. - *Corn-lilies*

Plants with corms; leaves *Iris*-like; inflorescence a spike or a raceme of spikes; flowers actinomorphic; tepals united proximally into short or long tube, variously white or yellow to red, ± equal; filaments free, arising from top of or within perianth-tube; style very slender, with 3-lobed stigma.

1. I. campanulata Houtt. (*I. speciosa* J. Kenn.) - *Red Corn-lily*. Stems to 15(40)cm; leaves up to 10(25) x 0.5(1.5)cm, perianth-tube 2-3mm; perianth-lobes 1.2-2.5cm, mainly red but with white and/or yellow stripes. Intrd; grown in gardens and bulb-fields, natd in old fields or rough ground; Scillies; S Africa.

2. I. paniculata D. Delaroche - *Tubular Corn-lily*. Stems to 1m; leaves up to 60 x 1.2cm; perianth-tube 3-7cm, very slender; perianth-lobes 1.2-2.5cm, cream to pale yellow tinged with red. Intrd; grown and natd as for *I. campanulata*; Scillies; S Africa.

11. SPARAXIS Ker Gawl. - *Harlequinflowers*

Plants with corms; leaves *Iris*-like; inflorescence a spike; flowers ± actinomorphic, with brown, jaggedly-toothed, dark-streaked bracts; tepals united into straight tube shorter than lobes, mostly red and white, the lobes ± equal; filaments free, borne in perianth-tube, slightly asymmetrically arranged; style very slender, with 3 linear branches.

1. S. grandiflora (D. Delaroche) Ker Gawl. - *Plain Harlequinflower*. Stems to 45cm;

leaves up to 30 x 1.3cm; perianth-tube 0.8-1.4cm; perianth-lobes 2-3cm, red or red-and-white striped, often yellow near base; (2n=20). Intrd; grown in bulb-fields, natd in and by old fields; Scillies; S Africa.

12. FREESIA Eckl. ex Klatt - *Freesia*
Plants with corms; leaves *Iris*-like; inflorescence a spike; flowers slightly zygomorphic, with curved perianth-tube, sweetly scented; tepals united into tube longer than lobes, white, yellow, orange, pink, purple or mauve, the lobes ± equal; filaments free, borne in perianth-tube; style very slender, with 3 bifid stigmas.

1. F. x hybrida L.H. Bailey (*F. refracta* auct. non (Jacq.) Eckl. ex Klatt) - *Freesia*. Stems to 40cm; leaves up to 30 x 1cm; perianth-tube 1.5-3cm; perianth-lobes 0.8-1.5cm; (2n=22, 33, 44). Intrd; grown in bulb-fields, relic sometimes ± natd in and by old fields; Guernsey and Scillies; S Africa.

13. CROCOSMIA Planch. (*Curtonus* N.E. Br.) - *Montbretias*
Plants with corms that produce rhizomes; leaves *Iris*-like; inflorescence an often branched spike; flowers zygomorphic, with curved perianth-tube; tepals united into tube longer or shorter than lobes, orange to brick-red, the lobes rather unequal; filaments free, borne asymmetrically in perianth-tube; style slender, with 3 short branches.

1 Leaves ribbed and pleated at least when young, at least some >3cm wide; perianth >4.5cm 2
1 Leaves ribbed but not pleated, <3cm wide; perianth <4(5)cm 3
  2 Perianth-lobes c.1/2 as long as -tube or less, erecto-patent; stamens shorter than perianth **1. C. paniculata**
  2 Perianth-lobes c. as long as -tube, widely spreading; stamens slightly exceeding perianth **2. C. masoniorum**
3 Perianth-lobes c.1/2 as long as -tube, ± erect; perianth-tube very narrow at base, abruptly widened distally **3. C. pottsii**
3 Perianth-lobes c. as long as -tube, ± patent; perianth-tube gradually expanded distally **4. C. x crocosmiiflora**

1. C. paniculata (Klatt) Goldblatt (*Curtonus paniculatus* (Klatt) N.E. Br.) - *Aunt-* 920
Eliza. Stems to 1.2m; leaves up to 90 x 8cm; perianth 4.5-6cm, with tube narrow at base then rather abruptly expanded, with erecto-patent lobes 1-2cm; (2n=22). Intrd; grown in gardens, natd in marginal habitats, rough and waste ground; local in W Sc, W Ir, NW En, W Wa and S Br; S Africa.

2. C. masoniorum (L. Bolus) N.E. Br. - *Giant Montbretia*. Differs from *C. paniculata* 920
in perianth with tube rather gradually widened, with spreading lobes 2.4-3cm; and see key; (2n=22). Intrd; grown and natd as for *C. paniculata*; E Cornwall and N Somerset, formerly Dunbarton; S Africa.

3. C. pottsii (Macnab ex Baker) N.E. Br. - *Potts' Montbretia*. Stems to 80cm; leaves 920
up to 80 x 2.5cm; perianth 2-3cm, with tube narrow at base then abruptly expanded, with ± erect lobes 0.5-1cm; stamens slightly exceeding perianth-tube; (2n=22). Intrd; rarely grown in gardens but natd by roads, lakes and rivers in CW & SW Sc, Man, Caerns, E Cornwall and SW, MW & NW Ir; S Africa.

4. C. x crocosmiiflora (Lemoine) N.E. Br. (*Tritonia x crocosmiiflora* (Lemoine) G. 920
Nicholson; *C. pottsii* x *C. aurea* (Hook.) Planch.) - *Montbretia*. Differs from *C. pottsii* in stems to 60cm; perianth 2.5-4(5)cm, with tube gradually widened distally, with spreading lobes 1.2-2.6cm; stamens nearly as long as perianth; (2n=22-24, 33). Intrd; much grown in gardens, very well natd in hedgerows, woods, by lakes and rivers, and on waste ground; scattered throughout BI, common in Ir, W Br and CI.

14. CHASMANTHE N.E. Br. - *Chasmanthe*
Plants with corms; leaves *Iris*-like; inflorescence a spike; flowers strongly zygomorphic, with curved perianth-tube; tepals united, variously red to orange, the lobes extremely unequal, the uppermost at least as long as tube and continuing its curvature, the 2 adjacent parallel to it but shorter, the 3 lower much shorter and slightly down-turned; filaments free, borne on perianth-tube; stamens conspicuously exserted; style long-exserted, very narrow, with 3 terminal thin lobes.

Other spp. - The reported presence of **C. aethiopica** (L.) N.E. Br. natd in the Scillies needs investigating; it differs in having leaves to 2.5cm wide, lateral perianth-lobes (7)10-15mm and patent or recurved (not 5-8mm and suberect), and style-branches 4-5mm (not 3-3.5mm).

1. C. bicolor (Gasp. ex Ten.) N.E. Br. - *Chasmanthe*. Stems to 1.3m; leaves to 80 x 3.5cm, with strong midrib; perianth-tube 3-3.5cm, yellow on lowerside, orange-red on upperside; uppermost perianth-lobe orange-red, 2-3.5cm. Intrd; grown for ornament, natd in damp shady places nearby; Scillies, mainly Tresco; S Africa.

164. AGAVACEAE - *Centuryplant family*

Perennials with thick, woody, sparsely branched stems with leaves tufted at branch-ends, or with stemless giant rosette of leaves; leaves in rosettes, tough, often succulent, simple, sessile, entire or distantly toothed, with ± sheathing base, without stipules. Flowers in large terminal panicles, bisexual, hypogynous or epigynous, actinomorphic or slightly zygomorphic; perianth of 6 lobes fused into tube proximally; stamens 6, borne on perianth-tube; ovary 3-celled, each cell with numerous ovules on axile placentas; style 1, short and thick to long and slender, with 3-lobed stigma; fruit a dehiscent or indehiscent capsule or a berry.

Flower structure as in Liliaceae, but the huge rosettes of tough leaves, on the ground or at the ends of branches, separate Agavaceae at a glance.

1	Leaf-rosettes sessile on ground or ± so; perianth greenish- or brownish-yellow	2
1	Leaf-rosettes at ends of woody branches; perianth whitish	3
2	Leaves with extremely strong spines at margins and apex	**2. AGAVE**
2	Leaves spineless	**4. PHORMIUM**
3	Leaves spine-tipped, strongly recurved; perianth >4cm	**1. YUCCA**
3	Leaves often sharply pointed but not spine-tipped, not recurved; perianth <1cm	**3. CORDYLINE**

1. YUCCA L. - *Spanish-daggers*
Stems usually branched, with leaf-rosettes at ends of branches; leaves with sharp spine at apex, entire or with few inconspicuous teeth; flowers hypogynous; perianth actinomorphic, bell-shaped, with lobes much longer than tube; fruit an indehiscent capsule.

1. Y. recurvifolia Salisb. - *Curved-leaved Spanish-dagger*. Plant to 2(5)m; leaves up to 100 x 5cm, mostly strongly recurved; perianth creamy- or greenish-white, 5-8cm; (2n=60). Intrd; natd on sand-dunes and in gravel-pits; Glam since 1982, S Devon and Worcs; SE USA.

2. AGAVE L. - *Centuryplant*
Stems ± 0, the leaf-rosette sessile on ground; leaves with extremely sharp spine at apex and many more along margins; flowers epigynous; perianth ± actinomorphic,

tubular, with tube much longer than lobes; fruit a dehiscent capsule.

1. A. americana L. - *Centuryplant*. Rosettes mostly 2-3m across, with massive succulent, tough, very spiny leaves 1-2m x 15-30cm; flowering stem rarely produced, to 7(12)m; perianth greenish-yellow, 7-10cm; (2n=20, 60, 120, 180, 240). Intrd; very persistent where planted, surviving from suckers when main rosette dies after the single, rare flowering; CI; Mexico.

3. CORDYLINE Comm. ex A. Juss. - *Cabbage-palm*
Stems well-developed, simple or branched, with leaf-rosettes at ends of branches; leaves entire, sharply pointed but without a spine; flowers hypogynous; perianth actinomorphic, with short tube and wide-spreading much longer lobes; fruit a berry, becoming dry with age.

1. C. australis (G. Forst.) Endl. - *Cabbage-palm*. Stems to 20m, becoming branched after first flowering; leaves up to 100 x 6cm; perianth white, 5-6mm, c.1cm across; (2n=38, 120). Intrd; much planted in W Br, Man, Ir and CI, persistent in SW En, S Ir, Man and CI, producing seedlings in Man, CI and W Cornwall; New Zealand.

4. PHORMIUM J.R. & G. Forst. - *New Zealand Flaxes*
Stems ± 0, the leaf-rosette sessile on ground; leaves entire, not spiny, folded proximally, nearly flat distally; flowers hypogynous; perianth slightly zygomorphic, with short tube and longer lobes, ± tubular; fruit a dehiscent capsule.

1. P. tenax J.R. & G. Forst. - *New Zealand Flax*. Rosettes 1-2m across, with suberect, extremely tough, fibrous leaves up to 3m x 12cm; flowering stem to 4m; perianth 3-5cm, with outer lobes brownish-red, with inner lobes greenish-yellow with not or slightly recurved tips; (2n=32). Intrd; very persistent where planted on cliffs or rocky places by sea; natd in W Cornwall, Scillies, W Cork, Man and CI, self-sown mainly in Scillies; New Zealand.

2. P. cookianum Le Jol. (*P. colensoi* Hook. f.) - *Lesser New Zealand Flax*. Differs from *P. tenax* in smaller size; leaves up to 2m x 7cm; flowering stem to 2m: perianth 2.5-4cm, with greenish-yellow lobes tinged with red, ≥1 of the inner lobes strongly recurved at tip; (2n=32). Intrd; planted and natd as for *P. tenax*; Scillies since 1920, often self-sown; New Zealand.

165. DIOSCOREACEAE - *Black Bryony family*

Glabrous, twining, herbaceous perennials with subterranean tuber; leaves alternate, simple, entire, petiolate, stipulate. Flowers in axillary, simple or branched racemes, dioecious, inconspicuous, epigynous, actinomorphic; perianth of 6 tepals united only at base, sepaloid, all ± similar; stamens 6, vestigial in female flowers; ovary 3-celled, each cell with 2 ovules on axile placentas; style 1; stigmas 3, recurved, each 2-lobed, vestigial in male flowers; fruit a berry with ≤6 seeds.

The only herbaceous twiner with dioecious inconspicuous flowers and red berries except *Bryonia* (Cucurbitaceae), which has pubescent, palmately lobed leaves and stem-tendrils.

1. TAMUS L. - *Black Bryony*

1. T. communis L. - *Black Bryony*. Stems to 5m; leaves c.5-15 x 4-11cm, broadly ovate, strongly cordate; perianth yellowish-green, 3-6mm across; berry 10-13mm across, bright red; 2n=48. Native; scrambling over hedges, shrubs and wood-margins; local in CI, common in Br N to Cumberland and Durham, rarely intrd further N and in Man and Ir.

166. ORCHIDACEAE - *Orchid family*
(Cypripediaceae)

Erect, herbaceous perennials, sometimes ± chlorophyll-less saprophytes, with succulent roots, subterranean tubers or rhizomes; leaves alternate, sometimes mostly basal, those on stem often reduced, simple, entire, sessile, usually clasping stem, without stipules. Flowers in terminal raceme or spike with bract to each flower (rarely only 1 flower), very rarely single and terminal, bisexual, usually conspicuous, epigynous, zygomorphic; perianth of 6 free tepals in 2 whorls of 3, usually all petaloid (though sometimes greenish or brownish), the 3 outer ones ('sepals') similar, 2 of the inner ones ('petals') similar and often ± similar to 3 outer, the other (usually apparently the lowest but actually the uppermost due to twisting of the flower through 180°) usually strongly different and forming a lip (*labellum*) which is usually the largest and most conspicuous part of the flower, and often is extended from its base behind the flower into a hollow spur; stamens and stigmas borne on a special structure (*column*) in the centre of the flower; stamens 2 (*Cypripedium*) or 1 (others), each with sessile anther containing pollen as many single grains (*Cypripedium*) or as 2-4 masses (*pollinia*) (others): pollinia often stalked, often provided with a sticky pad (*viscidium*) (at base of stalk when present); ovary 1-celled, with extremely numerous ovules on 3 parietal placentae; style 0; stigmas 3, either all receptive (*Cypripedium*) or 2 receptive and the third a sterile protrusion (*rostellum*) or ± 0 (others); fruit a capsule.

Many of the spp. are prone to produce plants with unusual labellum-shapes or flower-colours (e.g. albinos), but these usually occur in populations of normal plants.

The distinctive flowers could be confused only by the very inexperienced with a few petaloid monocotyledons or dicotyledons (e.g. *Impatiens*, *Orobanche*, *Pinguicula*): the inferior ovary and 1(-2) stamens on a column will dispel the confusion.

Hybrids occur frequently within the tribe Orchideae, especially within *Dactylorhiza*, and should be looked for whenever ≥2 spp. of the tribe occur close together. Hybrids between 2 spp. with the same chromosome number are usually fertile, and even those between spp. with different chromosome numbers are not always completely sterile. 7 intergeneric hybrid combinations also occur, 6 of them between genera 14-18. Other intergeneric hybrid combinations have also been recorded from BI, several of them (e.g. *Platanthera* x *Coeloglossum*, *Platanthera* x *Dactylorhiza*) known from the Continent, but are of uncertain or erroneous identity.

1 Plants saprophytic, without green leaves **2**
1 Plants holophytic (not parasitic or saprophytic), with green leaves **4**
  2 Flowers not twisted upside down, hence labellum and spur directed ± upwards; spur >5mm; very rare **4. EPIPOGIUM**
  2 Flowers twisted upside down (as normal in orchids), hence labellum directed downwards; spur 0 **3**
3 Labellum 8-12mm, c.2x as long as other tepals, brown; flowers usually >20 **5. NEOTTIA**
3 Labellum ≤6mm, c. as long as other tepals, whitish-cream with reddish markings; flowers ≤12 **11. CORALLORRHIZA**
  4 Spur present, sometimes very short **5**
  4 Spur 0 **14**
5 Labellum with 3 lobes, the central 3-6cm, linear, ribbon-like, in a loose spiral **22. HIMANTOGLOSSUM**
5 Labellum without a central ribbon-like lobe ≥3cm **6**
  6 Spur <3mm **7**
  6 Spur >3mm **10**
7 Flowers greenish-brown; labellum parallel-sided, with 3 short apical

lobes **17. COELOGLOSSUM**

7 Flowers white, cream or greenish-white, sometimes with red or pink markings; labellum not parallel-sided, with 1 or more laterally protruding lobes on either side 8

8 Central lobe of labellum entire; flowers never with pink or red tinge or markings **15. PSEUDORCHIS**

8 Central lobe of labellum conspicuously 2-3-lobed at apex; flowers often with pink or red tinge or markings 9

9 Flowers white to pinkish; labellum <5mm, scarcely longer than outer tepals; leaves usually spotted; Ir and Man only **19. NEOTINEA**

9 Flowers white and dark purple, the latter predominating in unopened flowers hence at top of spike; labellum >5mm, c.2x as long as outer tepals; leaves not spotted; Br only **20. ORCHIS**

10 Labellum linear, entire; flowers always white (or green-tinged) **13. PLATANTHERA**

10 Labellum not linear, lobed (if scarcely so then very wide); flowers pure white only in rare albinos 11

11 Spur >(8)11mm, filiform, >6x as long as widest point 12

11 Spur ≤11mm, not filiform, <6x as long as widest point 13

12 Labellum plane, not raised into plates; spike ± cylindrical; flowers strongly scented; each pollinium becoming detached separately, each with a small stalk and basal sticky pad **16. GYMNADENIA**

12 Labellum with 2 raised plates near its base; spike pyramidal; flowers not scented; each pollinium with its own stalk but becoming detached together on a common sticky pad **14. ANACAMPTIS**

13 Lower bracts herbaceous, green, often suffused purplish; labellum terminating in a single pointed to rounded tooth or lobe much smaller than or rarely nearly as large as the portions on either side **18. DACTYLORHIZA**

13 Bracts membranous, brown, often suffused purplish; labellum terminating in a forked lobe (often with a tooth in the notch) larger than lobes on either side, or in a small truncate or notched lobe smaller than portions on either side **20. ORCHIS**

14 Labellum yellow, c.3cm, concavely bowl-shaped; other tepals maroon, 3-5cm; very rare **1. CYPRIPEDIUM**

14 Labellum not concavely bowl-shaped; other tepals<2cm 15

15 Labellum velvety in texture, resembling insect's abdomen **23. OPHRYS**

15 Labellum not velvety, not resembling insect's abdomen 16

16 At least some flowers not twisted upside down, hence labellum directed ± upwards 17

16 Flowers all twisted upside down, hence labellum directed downwards 18

17 Labellum entire and flat at margin; leaves ≤2cm, with minute tubercles near apex **10. HAMMARBYA**

17 Labellum crenate or crisped at margin; leaves >2cm, without tubercles **9. LIPARIS**

18 Labellum with 2 lobes at its apex exceeding all others 19

18 Labellum with 1 lobe at its apex exceeding all others, sometimes this slightly notched 20

19 Leaves 2, on stem; labellum with 0 or 2 small lateral lobes much shorter than 2 apical lobes **6. LISTERA**

19 Leaves usually >2, the largest ones basal; labellum with 2 lateral lobes longer than 2 apical lobes **21. ACERAS**

20 Flowers white to greenish- or yellowish-white, in 1-3 distinct spirals in spike, or if forming a strictly 1-sided spike then leaves all in basal rosette 21

20 Flowers white or whitish only in rare albinos, not in distinct spirals, if forming a strictly 1-sided spike then leaves mainly on stems 22
21 Shortly rhizomatous; leaves conspicuously net-veined, in basal rosettes; labellum not frilly at edge **8. GOODYERA**
21 Tufted, with swollen roots; leaves not or inconspicuously net-veined, if in basal rosette then labellum frilly at edge **7. SPIRANTHES**
22 Main leaves 2(-3), basal; labellum with apical lobe and 2 shorter laterals, not constricted **12. HERMINIUM**
22 Main leaves (2)3-many, on stems (sometimes near base); labellum constricted at base of apical lobe, the proximal part often with 2 lateral lobes 23
23 Apical lobe of labellum pendent, flat, entire **(SERAPIAS)**
23 Apical lobe of labellum variously held, variously crenate, frilly, undulate or toothed 24
24 Flowers erect to erecto-patent, sessile; proximal part of labellum partly wrapped round column **2. CEPHALANTHERA**
24 Flowers patent to pendent, stalked; proximal part of labellum not wrapped round column **3. EPIPACTIS**

Other genera - **SERAPIAS** L. (*Tongue-orchids*), from the Mediterranean, can be identified in the above key. 2 plants of **S. parviflora** Parl. were found in E Cornwall in 1989, probably the result of deliberate introduction: stems to 30cm; leaves linear-elliptic; labellum 15-20mm, dark brownish-red. 3 plants appeared in 1990 and 5 in 1991, but none since. Similarly 1 plant of the larger **S. lingua** L. occurred in Guernsey in 1991, but not since.

TRIBE 1 - CYPRIPEDIEAE (genus 1). Stamens 2 plus a large sterile projection (staminode); pollen dispersed as separate grains; receptive stigmas 3; labellum a deeply concave bowl; spur 0; flower single (very rarely 2), terminal.

1. CYPRIPEDIUM L. - *Lady's-slipper*

1. C. calceolus L. - *Lady's-slipper*. Stems to 30cm, rather pubescent; leaves 3-4, all on stem, ovate, pubescent; labellum mainly yellow, c.3cm; other tepals maroon, 3-5cm, patent; (2n=20). Native; north-facing grassy slope on limestone; 1 locality in MW Yorks, formerly widespread on limestone in N En, planted in a few sites in N En. **RRR**

TRIBE 2 - NEOTTIEAE (genera 2-8). Fertile stamen 1; pollen dispersed as 2 often rather friable (or 2 each with 2 halves) pollinia; receptive stigmas 2; third stigma 0 or a sterile bulge (rostellum); stamen borne at back of column; pollinia sessile or with an apical stalk, with or without a sticky pad (pollinia with basal stalk and sticky pad in *Epipogium*); labellum of various shapes, often without well-marked lobes, often constricted near middle so delimiting proximal and distal parts; spur 0 (present in *Epipogium*).

2. CEPHALANTHERA Rich. - *Helleborines*

Shortly rhizomatous; leaves several, all on stem; flowers sessile or ± so, borne spirally, white or purplish-pink; spur 0; labellum constricted c.1/2 way into proximal and distal parts, neither markedly lobed, the proximal part partly wrapped round column; rostellum 0; pollinia without stalks.

1 Flowers purplish-pink; ovaries with glandular hairs; labellum acute **3. C. rubra**
1 Flowers white with yellow or orange marks on labellum; ovaries glabrous; labellum obtuse 2
2 Lower leaves ovate to rather narrowly so; bracts longer than ovaries;

sepals obtuse **1. C. damasonium**

2 Lower leaves lanceolate to narrowly elliptic-oblong; bracts shorter than ovaries; sepals acute **2. C. longifolia**

1. C. damasonium (Mill.) Druce - *White Helleborine.* Stems to 60cm; lower leaves ovate to rather narrowly so; bracts longer than ovaries; flowers 3-11(16), white, 15-25mm; sepals obtuse; (2n=32, 36, 54). Native; shady woods, commonly of *Fagus* with little ground cover, on chalk and limestone; locally frequent in S En N to Northants and Herefs.

1 x 2. C. damasonium x C. longifolia = C. x schulzei E.G. Camus, Bergon & A. Camus was found in 1974 and 1975 in S Hants in woodland with both parents; it is intermediate in leaf and flower characters.

2. C. longifolia (L.) Fritsch - *Narrow-leaved Helleborine.* Stems to 60cm; lower leaves lanceolate to narrowly elliptic-oblong; bracts shorter than ovaries; flowers 3-15(20), white, 10-16mm; sepals acute; (2n=32, 34). Native; woods and shady places on calcareous soils; much less common but much more widespread than C. *damasonium*, scattered in Br and Ir N to W Sutherland, decreasing. **R**

3. C. rubra (L.) Rich. - *Red Helleborine.* Stems to 60cm; lower leaves lanceolate to narrowly elliptic-oblong; bracts longer than ovaries; flowers 3-8(15), purplish-pink, 15-25mm; sepals acute; (2n=36, 44, 48). Native; *Fagus* woods on chalk or limestone, rarely with other vegetation; very rare in N Hants, Bucks and E Gloucs, formerly elsewhere in S En. **RRR**

3. EPIPACTIS Zinn - *Helleborines*

Rhizomatous, mostly shortly so, leaves several, all on stem; flowers distinctly pedicellate, borne spirally or ± on 1 side of stem, various dull colours; spur 0; labellum usually differentiated c.1/2 way into proximal and distal parts, neither markedly lobed, the proximal part ± cup-shaped and not wrapped round column; rostellum obvious and secreting a white sticky cap (*viscidium*), or minute and with 0 or vestigial viscidium; pollinia without stalks.

E. phyllanthes, *E. leptochila* and *E. youngiana* form a problematical complex of self-pollinated plants in which sp. limits are uncertain and disputed. They can be distinguished from the other spp. by having a rostellum which secretes little or no viscidium that usually withers soon after the flower opens. In the other (cross-pollinated) spp. the rostellum is obvious and the viscidium remains in the open flower until removed along with the pollinia by visiting insects. This can be effected with a match-stick or similar object; when the viscidium is touched and the object pulled away the pollinia are drawn out with it. In the self-pollinated spp. the pollinia crumble apart and cannot easily be pulled out whole.

1 Rhizome long; labellum strongly constricted separating proximal and distal portions, the proximal with erect triangular lobe on each side **1. E. palustris**

1 Rhizome short or ± 0; labellum not or slightly constricted between proximal and distal portions, the proximal with 0 or obscure lateral lobes 2

2 Inflorescence-axis glabrous or nearly so; flowers pendent as soon as they open; leaves often shorter than internodes **7. E. phyllanthes**

2 Inflorescence-axis pubescent to densely so; at least younger flowers usually patent to erecto-patent; leaves usually longer than internodes 3

3 Ovary pubescent to densely so; perianth usually reddish-purple all over **2. E. atrorubens**

3 Ovary glabrous to sparsely pubescent; perianth usually greenish often marked or tinged with pink, purple or violet, but not so coloured all over 4

4 Upper leaves usually spirally arranged; rostellum secreting obvious, white, persistent viscidium; pollinia becoming detached as integral units 5

4 Upper leaves usually obviously 2-ranked; rostellum without or with sparse, soon disappearing viscidium; pollinia crumbling apart 6

5 Leaves dark green, the lowest wider than long or almost so; distal part of labellum wider than long, with 2 usually rough brownish bosses near base **4. E. helleborine**

5 Leaves greyish-green, often tinged violet, the lowest considerably longer than wide; distal part of labellum at least as long as wide, with 2 smoothly pleated pinkish bosses near base **3. E. purpurata**

6 Rostellum >1/2 as long as anthers; stigma with 2 basal bosses, with the rostellum forming a 3-horned shape; ovary usually glabrous; petals pinkish; distal part of labellum wider than long **5. E. youngiana**

6 Rostellum ≤1/2 as long as anthers; stigma without marked basal bosses hence not 3-horned; ovary usually pubescent; petals pale green; distal part of labellum longer than wide or wider than long **6. E. leptochila**

1. E. palustris (L.) Crantz - *Marsh Helleborine.* Stems to 45(60)cm, pubescent; leaves spiral, mostly >2x as long as wide; perianth predominantly white, with red and yellow markings; viscidium well developed; ovary pubescent; (2n=40, 44, 46, 48). Native; fens, base-rich marshy fields, dune-slacks; locally frequent in BI N to C Sc, extinct in many inland sites.

2. E. atrorubens (Hoffm.) Besser - *Dark-red Helleborine.* Stems to 30(60)cm, densely whitish-pubescent; leaves 2-ranked, mostly >2x as long as wide; perianth usually reddish-purple; viscidium well developed; ovary pubescent; (2n=38, 40, 60). Native; limestone scrub, grassland, scree and rocky places; very locally frequent in CW Ir, N Sc, N Wa, N En S to Derbys, formerly Brecs. R

2 x 4. E. atrorubens x E. helleborine = E. x schmalhausenii K. Richt. has been reported from N Wa, N En and N Sc with both parents, but due to its fertility is very difficult to determine certainly; the most convincing specimens are from Arnside Knott, Westmorland.

3. E. purpurata Sm. - *Violet Helleborine.* Stems to 60(80)cm, often densely clumped, pubescent above; leaves spiral, mostly >2x as long as wide; perianth predominantly green with white, pink-tinged labellum; viscidium well developed; ovary shortly pubescent; (2n=40). Native; woods (often dense) on calcareous or sandy soil; frequent in SE & SC En N to Salop and Leics.

3 x 4. E. purpurata x E. helleborine = E. x schulzei P. Fourn. has been reported frequently from within the area of *E. purpurata*, but is difficult to determine certainly due to its fertility.

4. E. helleborine (L.) Crantz - *Broad-leaved Helleborine.* Stems to 80(100)cm, pubescent above; upper leaves spiral, <2x as long as wide, the lowest usually wider than long; perianth predominantly green, tinged pink; viscidium well developed; ovary usually sparsely pubescent; (2n=36, 38, 40, 44). Native; woods, scrub and hedgerows; frequent in Br and Ir except rare in N Sc.

5. E. youngiana A.J. Richards & A.F. Porter - *Young's Helleborine.* Stems to 60cm, pubescent above; upper leaves ± 2-ranked, distinctly longer than wide; perianth predominantly green and pink; rostellum well developed; ovary ± glabrous. Native; woodland on heavy, often heavy-metal-polluted, soils; S Northumb and Lanarks; endemic. Described in 1982 but of uncertain origin; perhaps a hybrid derivative of *E. helleborine* and *E. phyllanthes.* RRR

6. E. leptochila (Godfery) Godfery (*E. cleistogama* C.A. Thomas, *E. dunensis* (T. & T.A. Stephenson) Godfery, *E. muelleri* Godfery) - *Narrow-lipped Helleborine.* Stems to 60cm, pubescent at least above; leaves 2-ranked, mostly >2x as long as wide; perianth predominantly yellowish-green, the labellum mottled or tinged with red or pink; viscidium minute or 0; ovary pubescent; (2n=36, 40). Native; woods mostly on calcareous or heavy-metal-polluted soils, river-gravels and dunes; locally scattered in Br N to Lanarks. Var. **dunensis** T. & T. A. Stephenson occurs on R dunes in C & N Br but intermediate plants occur inland in N En; it differs from var. **leptochila** in distal part of the labellum wider than long and reflexed (not longer than wide and non-reflexed). **E. cleistogama** has green ± cleistogamous flowers; it occurs in 1 area of W Gloucs and is probably best considered another var. of *E. leptochila*. **E. muelleri** has been recorded in E Sussex but seems scarcely different from var. *dunensis*.

7. E. phyllanthes G.E. Sm. (*E. cambrensis* C.A. Thomas, *E. vectensis* (T. & T.A. Stephenson) Brooke & F. Rose, *E. pendula* C.A. Thomas non A.A. Eaton, *E. confusa* D.P. Young) - *Green-flowered Helleborine*. Stems to 40cm, glabrous or ± so; leaves 2-ranked, mostly distinctly longer than wide; perianth predominantly green; viscidium minute; ovary glabrous; (2n=36). Native; woods on calcareous or sandy soils sometimes heavy-metal-polluted, and on dunes; scattered in Br N to Cheviot and Westmorland, Lanarks, very scattered in Ir. Very variable from area to area, as in many self-pollinating plants. The above 4 synonyms and other taxa are probably best recognized as vars.; some of these are cleistogamous. The type var. has an undifferentiated green labellum resembling the 2 petals. R

4. EPIPOGIUM J.G. Gmel. ex Borkh. - *Ghost Orchid*

Saprophytic, chlorophyll-less, with coral-like rhizome producing thin creeping rhizomes; leaves few, small and scale-like; flowers shortly pedicellate, not twisted upside down so spur and labellum point upwards, pale pink; spur present, c. as long as labellum; labellum with 2 short rounded lateral lobes at base; rostellum well developed; pollinia with basal stalk ending in viscidium.

1. E. aphyllum Sw. - *Ghost Orchid*. Stems to 25cm, pinkish; flowers 1-2(4), c.15-20mm vertically across, patent to slightly pendent; (2n=68). Native; in deep shade of *Fagus* or *Quercus* woods on leaf-litter or rotten stumps; very rare in 1 site each in Herefs, Oxon and Bucks, formerly Salop. RRR

5. NEOTTIA Guett. - *Bird's-nest Orchid*

Saprophytic, chlorophyll-less, with very short rhizome wrapped with succulent roots; leaves few, small and scale-like; flowers shortly pedicellate, pale brown; spur 0; labellum divided apically into 2 lobes and with 2 small lateral teeth near base; other 5 tepals convergent to form loose hood; rostellum well developed; pollinia not stalked.

1. N. nidus-avis (L.) Rich. - *Bird's-nest Orchid*. Stems to 50cm, brown; flowers numerous, crowded, c.15-20mm vertically across, patent, with labellum c.10-12mm; (2n=36). Native; on leaf-litter in shady woods, often of *Fagus* on calcareous soils; scattered throughout most of Br and Ir, locally frequent in S En.

6. LISTERA R. Br. - *Twayblades*

Shortly rhizomatous; leaves normally 2, in opposite pair on stem (on non-flowering stems the leaves are at the apex); flowers shortly pedicellate, yellowish-green to dull reddish; spur 0; labellum deeply divided apically into 2 lobes, sometimes with short tooth between them; rostellum well developed; pollinia not stalked.

1. L. ovata (L.) R. Br. - *Common Twayblade*. Stems 20-60(75)cm; leaves ovate-elliptic, 5-20cm, with 3-5 prominent longitudinal veins; labellum yellowish-green, 7-15mm; other 5 tepals ± convergent to form loose hood; flowers usually >15(≤c.100); 2n=34-36. Native; woods, hedgerows, grassy fields, dune-slacks, sometimes on *Calluna*-moors; frequent throughout BI.

2. L. cordata (L.) R.Br. - *Lesser Twayblade*. Stems to 10(25)cm; leaves triangular-ovate, 1-2.5cm, with prominent midrib; labellum dull reddish, 3.5-4.5mm; other 5 tepals ± patent, 2-2.5mm; flowers c.3-15; (2n=36, 38, 40, 42, 44). Native; upland woods and moors in usually wet, acid places, often among *Sphagnum* or other moss and under *Calluna* or other moorland shrubs; frequent in Sc, scattered in Ir and in Br S to Derbys and N Devon, not in C, SC or SE En except 1 site in E Sussex.

7. SPIRANTHES Rich. - *Lady's-tresses*
With tuberous roots; leaves several, basal and on stem or ± all basal; flowers sessile, white with green markings, usually borne spirally in tight spike; spur 0; labellum ± unlobed, with slightly to markedly frilly distal edge, ± appressed to other tepals to form tubular or trumpet-shaped perianth; rostellum well developed; pollinia not stalked.

1 Leaves at flowering time obovate-elliptic, all in tight rosette adjacent to base of flowering stem which bears only reduced scale-leaves **1. S. spiralis**
1 Leaves at flowering time linear-lanceolate to -oblanceolate, around base of flowering stem and short way up it 2
 2 Flowers in 1 spiral row in spike, 6-8mm excl. ovary; bracts 6-9mm; leaves subacute to obtuse **2. S. aestivalis**
 2 Flowers in 3 spiral rows in spike, 10-14mm excl. ovary; bracts 10-20(30)mm; leaves acute **3. S. romanzoffiana**

1. S. spiralis (L.) Chevall. - *Autumn Lady's-tresses*. Stems to 15(20)cm, with flowers in single spiral or 1-sided spike; leaves all in basal rosette, patent; flowers 4-6mm excl. ovary; 2n=30. Native; short permanent grassland and grassy dunes; locally frequent in BI N to NE Yorks, Man and Co Sligo, extinct in many inland sites.

2. S. aestivalis (Poir.) Rich. - *Summer Lady's-tresses*. Stems to 40cm, with flowers in single spiral; leaves basal and on stem, ± erect; flowers 6-8mm excl. ovary; (2n=30). Native; marshy ground in Hants until 1959, bog in Guernsey until 1914, by pond in Jersey until 1926; extinct. RR

3. S. romanzoffiana Cham. - *Irish Lady's-tresses*. Stems to 30cm, with flowers in 3 close spirals; leaves basal and on stem, ± erect; flowers 10-14mm excl. ovary; 2n=60. Native; marshy meadows near streams, rivers or lakes; extremely local in SW, W & NE Ir, CW & NW Sc, S Devon. R

8. GOODYERA R. Br. - *Creeping Lady's-tresses*
With rhizomes giving rise to sterile leaf-rosettes and flowering stems with basal leaf-rosette and reduced scale-like stem-leaves; flowers sessile, like those of *Spiranthes* but in weak spiral or 1-sided spike, and labellum with entire distal edge.

1. G. repens (L.) R.Br. - *Creeping Lady's-tresses*. Stems to 20(25)cm; leaves ovate-elliptic, ± patent; flowers 3-5mm excl. ovary; 2n=30. Native; on barish ground under *Pinus* or *Betula* or rarely on moist dunes; local in N Br S to Cumberland (formerly SE Yorks), ?intrd in E & W Norfolk and E Suffolk. R

TRIBE 3 - EPIDENDREAE (genera 9-11). Fertile stamen 1; pollen dispersed as 4 (or 2, each with 2 halves) pollinia; receptive stigmas 2; third stigma a minute sterile bulge (rostellum); stamen borne at apex of column; pollinia sessile, with minute sticky pads; labellum rather small, simple or with 2 short lateral lobes; spur 0 or ± so.

9. LIPARIS Rich. - *Fen Orchid*
Leaves usually 2, on stem, green; stem with 2 basal tubers side-by-side; labellum directed upwards, downwards or any intermediate direction, frilly on margins, scarcely lobed.

1. L. loeselii (L.) Rich. - *Fen Orchid*. Stems to 20cm; leaves 2.5-8cm, elliptic; RRR flowers <20, yellowish-green, c.10mm vertically across, the labellum c. as long as other tepals; (2n=26, 32). Native; wet peaty fens and dune-slacks; very local in N Devon, E & W Norfolk, Glam and Carms, formerly elsewhere in E Anglia and E Kent, greatly decreased.

10. HAMMARBYA Kuntze - *Bog Orchid*
Leaves 2(-4), on stem, green; stem with 2 basal tubers, 1 above the other; labellum directed upwards, entire on margins, not lobed.

1. H. paludosa (L.) Kuntze - *Bog Orchid*. Stems to 8(12)cm; leaves 0.5-2cm, elliptic, with marginal fringe of tiny bulbils; flowers <20, yellowish-green, c.7mm vertically across, the labellum somewhat shorter than other tepals; (2n=28). Native; on wet *Sphagnum* in bogs; formerly scattered throughout most of Br and Ir except C En, now very rare except in CW & NW Sc and locally in S Hants and C Wa, ± extinct in E En. R

11. CORALLORRHIZA Ruppius ex Gagnebin - *Coralroot Orchid*
Yellowish-brown or yellowish-green, saprophytic; leaves all on stem, scale-like; stem with coral-like rhizome; labellum directed downwards, distinctly 3-lobed, entire on margins.

1. C. trifida Châtel. - *Coralroot Orchid*. Stems to 20cm; leaves reduced to few sheaths on stem; flowers ≤12, yellowish-green tinged brown, c.6mm vertically across, the labellum somewhat shorter than other tepals; (2n=40, 42, 84). Native; damp peaty or mossy ground under trees or shrubs in woods, scrub and dune-slacks; scattered in N Br S to MW Yorks. R

TRIBE 4 - ORCHIDEAE (genera 12-23). Fertile stamen 1; pollen dispersed as 2 pollinia; receptive stigmas 2; third stigma a small to large sterile bulge (rostellum); stamen borne in front of column; pollinia on long (short in *Herminium*) stalks each with a sticky pad or the 2 sharing 1 sticky pad; labellum often large and conspicuous, very variably lobed; spur 0 to very long.

12. HERMINIUM L. - *Musk Orchid*
Leaves 2(-4), near base of stem, plus usually 1 reduced leaf higher up; all tepals ± incurved; labellum narrow, with 2 short lateral lobes; the 2 petals rather similar but with shorter lobes; spur 0; plant with 1 ± globose underground tuber.

1. H. monorchis (L.) R. Br. - *Musk Orchid*. Stems to 15(25)cm; leaves elliptic-oblong, 2-7cm; flowers yellowish-green, c.6-8mm vertically across, rather dense in spike; (2n=38, 40). Native; chalk and limestone grassland; local in S Br N to E Gloucs and Beds, formerly to Glam and W Norfolk, decreasing. R

13. PLATANTHERA Rich. - *Butterfly-orchids*
Leaves 2(-3), near base of stem, plus few reduced leaves higher up; upper 3 tepals ± incurved; labellum linear-oblong, entire; 2 lateral sepals spreading; spur long and slender; plant with 2 ellipsoid underground tubers with tapering apices.

1. P. chlorantha (Custer) Rchb. - *Greater Butterfly-orchid*. Stems to 60cm; leaves 5-15cm; flowers pure white to greenish-white, c.18-23mm transversely across; labellum 10-16mm; pollinia 3-4mm (conspicuously yellow against white perianth), divergent downwards along stalks so that viscidia are c.4mm apart; spur 19-28 x c.1mm, often strongly curved; 2n=42. Native; woods and (in N) in open grassland, usually on calcareous soils; locally frequent throughout Br and Ir except Orkney and Shetland, much commoner than *P. bifolia* in S Br.

1 x 2. P. chlorantha x P. bifolia = P. x hybrida Brügger has been reported from several places but never confirmed; the parents sometimes occur together.

2. P. bifolia (L.) Rich. - *Lesser Butterfly-orchid*. Differs from *P. chlorantha* in usually smaller stature and smaller in all parts; flowers usually fewer, c.11-18mm transversely across; labellum 6-10mm; pollinia c.2mm, parallel, the viscidia c. 1mm apart; spur 15-20 x c.1mm, usually slightly curved; 2n=42. Native; similar habitats and distribution to *P. chlorantha*, commoner than it in N, hence more often in open habitats.

13 x 15. PLATANTHERA X PSEUDORCHIS = X PSEUDANTHERA McKean

13/1 x 15/1. X P. breadalbanensis McKean (*Pl. chlorantha* x *Ps. albida*) was described from a specimen found with the supposed parents in M Perth in 1980, but this has been re-identified as an abnomal variant of *Platanthera chlorantha*.

13 x 18. PLATANTHERA X DACTYLORHIZA = X DACTYLANTHERA P.F. Hunt & Summerh.(*X Rhizanthera* P.F. Hunt & Summerh. nom. inval.)

13/2 x 18/2. P. bifolia x D. maculata = X D. chevallieriana (E.G. Camus) ined. (*X Rhizanthera chevallieriana* (E.G. Camus) Soó) has been recorded from M Perth and Outer Hebrides intermittently since the late 19th century, most recently on Ben Lawers (M Perth) in 1979, 1984 and 1996, but the plant in the last locality has been redetermined as abnormal *D. maculata*; there are thus no confirmed records.

14. ANACAMPTIS Rich. - *Pyramidal Orchid*

Leaves several, decreasing in size up stem; upper 3 tepals ± incurved; labellum deeply and nearly equally 3-lobed, with 2 raised plates near its base; 2 lateral sepals spreading; spur long and slender; plant with 2 subglobose underground tubers.

See *Gymnadenia* for differences.

1. A. pyramidalis (L.) Rich. - *Pyramidal Orchid*. Stems to 60cm; leaves lanceolate, the lowest c.8-15cm; flowers in ± pyramidal dense spike, pinkish-purple (rarely white), c.10-12mm vertically across, with spur 12-14 x <1mm; (2n=20, 36, 42, 54, 72). Native; chalk and limestone grassland, calcareous dunes; locally frequent in BI N to S Ebudes and Fife.

14 x 16. ANACAMPTIS x GYMNADENIA = X GYMNANACAMPTIS Asch. & Graebn.

14/1 x 16/1. X G. anacamptis (F.H. Wilms) Asch. & Graebn. (*A. pyramidalis* x *G. conopsea*) has been recorded from Hants, Gloucs and Co Durham; it has the labellum plates of *Anacamptis* and the scent and cylindrical spike of *Gymnadenia*.

15. PSEUDORCHIS Ség. (*Leucorchis* E. Mey.) - *Small-white Orchid*

Leaves several, decreasing in size up stem; upper 5 tepals ± incurved; labellum quite deeply and nearly equally 3-lobed, without raised plates at base; spur short, wide, rounded at apex; plant with cluster of tapering underground tubers.

1. P. albida (L.) Á. & D. Löve (*L. albida* (L.) E. Mey.) - *Small-white Orchid*. Stems to 20(40)cm; leaves oblong-oblanceolate, the lowest 2.5-8cm; flowers in a cylindrical dense spike, creamy-white, c.2-4mm vertically across, with spur 2-3mm; (2n=40, 42). Native; short grassland, usually base-rich and upland; frequent in C, W & N Sc, very scattered elsewhere in N Br, Ir and Wa, formerly S to Derbys and in Sussex, now very rare in En, extinct in many places. Superficially resembles *Neotinea*, but labellum-shape is totally different.

15 x 16. PSEUDORCHIS x GYMNADENIA = X PSEUDADENIA P.F. Hunt

15/1 x 16/1. X P. schweinfurthii (Hegelm. ex A. Kern.) P.F. Hunt (*X Gymleucorchis schweinfurthii* (Hegelm. ex A. Kern.) T. & T.A. Stephenson; *P. albida* x *G. conopsea*)

has been recorded from several places in N Br and is still frequent in NW Sc with both parents; it is intermediate in size, perianth shape (especially spur) and colour (pale pink).

15 x 18. PSEUDORCHIS x DACTYLORHIZA = X PSEUDORHIZA P.F. Hunt

15/1 x 18/2. X P. bruniana (Brügger) P.F. Hunt (*P. albida* x *D. maculata*) was found in Orkney in 1977; it resembles *D. maculata* in stem and leaf characters, and *P. albida* in inflorescence shape, size and colour, but has intermediate floral characters.

16. GYMNADENIA R. Br. - *Fragrant Orchid*

Leaves several, decreasing in size up stem; upper 3 tepals ± incurved; labellum shallowly 3-lobed, without raised plates at base; 2 lateral sepals spreading; spur long and slender; plant with several divided tapering underground tubers.

Often grows with *Anacamptis*, but flowers earlier (little or no overlap); the shape of the tubers, labellum and spike, and the scented flowers, distinguish it.

Other spp. - Old records of the European **G. odoratissima** (L.) Rich., with smaller flowers (lateral sepals 2.5-3mm; labellum 2.3-3mm; spur 4-5mm) have never been confirmed.

1. G. conopsea (L.) R. Br. - *Fragrant Orchid*. Stems to 40(75)cm; leaves linear-lanceolate, the lowest c.6-15cm; flowers in ± cylindrical dense spike, sweetly scented, usually lilac-purple, sometimes other shades of red (rarely white), c.8-12mm vertically across, with spur 8-17 x c.1mm. Native.

1 Lateral sepals mostly 4-5 x c.2mm; labellum obscurely lobed, (3)3.5-4(5)mm wide **c. ssp. borealis**
1 Lateral sepals mostly 5-7 x c.1mm; labellum conspicuously lobed, (4.5)5.5-7(8)mm wide 2
  2 Labellum scarcely wider than long, mostly 5-6 x 5.5-6.5mm; lateral sepals c.5-6mm; spur mostly 12-14mm **a. ssp. conopsea**
  2 Labellum much wider than long, mostly 3.5-4 x 6.5-7mm; lateral sepals c.6-7mm; spur mostly 14-16mm **b. ssp. densiflora**

a. Ssp. conopsea. Flowers (7)10-11(13)mm horizontally across; labellum conspicuously lobed, (4)5-6(6.5) x (4.5)5.5-6.5(7)mm; lateral sepals bent downwards, 5-6 x c.1mm; spur (11)12-14(17)mm; 2n=40. Dry chalk or limestone grassland; frequent in Br N to Co Durham, N Ir.

b. Ssp. densiflora (Wahlenb.) E.G. Camus, Bergon & A. Camus. Flowers (10)11-13(14.5)mm horizontally across; labellum conspicuously lobed, (3)3.5-4(4.5) x (5.5)6.5-7(8)mm; lateral sepals held horizontally, 6-7 x c.1mm; spur (13)14-16(17)mm; (2n=40). Base-rich fens and usually N-facing chalk grassland; scattered and local in Br N to Westmorland, W Ross, scattered through Ir (commonest ssp.).

c. Ssp. borealis (Druce) F. Rose. Flowers (7)8-10(12)mm horizontally across; labellum obscurely lobed, (3.5)4-4.5(5) x (3)3.5-4(5)mm; lateral sepals bent downwards, 4-5 x c.2mm; spur (8)11-14(15)mm; 2n=40. Base-rich to -poor hilly grassland in Sc, W Wa and N & SW En, bogs in S Hants and E Sussex, ?Ir.

16 x 17. GYMNADENIA x COELOGLOSSUM = X GYMNAGLOSSUM Rolfe

16/1 x 17/1. X G. jacksonii (Quirk) Rolfe (*G. conopsea* x *C. viride*) has been recorded sporadically throughout much of Br and Ir; the inflorescences resemble those of *Gymnadenia* but are tinged green and have a much shorter spur, and labellum- and leaf-shapes are intermediate.

16 x 18. GYMNADENIA x DACTYLORHIZA = X DACTYLODENIA Garay & H.R. Sweet (*X Dactylogymnadenia* Soó)

Hybrids of this combination often resemble *Dactylorhiza* in general appearance but have usually faintly spotted leaves and scented flowers with a longer spur; the perianth is variously intermediate in details. Precise parentage is difficult to determine without knowledge of the sp. or spp. of *Dactylorhiza* present nearby.

16/1 x 18/1. X D. st-quintinii (Godfery) J. Duvign. (? *X D. heinzeliana* (Reichardt) Garay & H.R. Sweet, *X Dactylogymnadenia cookei* (Hesl.-Harr.) Soó; *G. conopsea* x *D. fuchsii*) has been found in scattered localities throughout most of Br and Ir.

16/1 x 18/2. X D. legrandiana (E.G. Camus) Peitz (*X Dactylogymnadenia legrandiana* (E.G. Camus) Soó; *G. conopsea* x *D. maculata*) has been found in scattered localities throughout most of Br and Ir.

16/1 x 18/3. X D. vollmannii (M. Schulze) Peitz (*X Dactylogymnadenia vollmannii* (M. Schulze) Soó; *G. conopsea* x *D. incarnata*) was confirmed for W Cornwall in 1984, an earlier record being doubtful.

16/1 x 18/4. X D. wintoni (Druce) Peitz (*X Dactylogymnadenia wintoni* (Druce) Soó; *G. conopsea* x *D. praetermissa*) has been found in S En; ?endemic.

16/1 x 18/5. X D. varia (T. & T.A. Stephenson) Aver. (*X Dactylogymnadenia varia* (T. & T.A. Stephenson) Soó; *G. conopsea* x *D. purpurella*) has been found in Cumberland, various parts of Sc (mostly W) and Co Down; ?endemic.

17. COELOGLOSSUM Hartm. - *Frog Orchid*
Leaves several, decreasing in size up stem; upper 5 tepals incurved; labellum oblong, shallowly 3-lobed near tip with the central lobe the shortest; spur very short, rounded; plant with 2 divided tapering underground tubers.

1. C. viride (L.) Hartm. - *Frog Orchid*. Stems to 20(35)cm; leaves elliptic-oblong, sometimes broadly so, the lowest c.1.5-5cm; flowers in ± cylindrical dense spike, yellowish-green tinged with reddish-brown, c.6-10mm vertically across; labellum 3.5-6mm; spur c.2mm; 2n=40. Native; grassland, especially on base-rich or calcareous soils; locally frequent throughout Br and Ir, extinct in most places in S Sc and C & E En.

17 x 18. COELOGLOSSUM X DACTYLORHIZA = X DACTYLOGLOSSUM P.F. Hunt & Summerh.
Hybrids of this combination usually have flowers of the colour of the *Dactylorhiza* parent variously tinged or overlaid with green; other characters of habit, leaves and perianth shape are variously intermediate. Precise parentage is difficult to determine without knowledge of the sp. or spp. of *Dactylorhiza* present nearby. Hybrids involving *D. praetermissa*, *D. incarnata* and *D. majalis* have been reported, but the evidence is weak.

17/1 x 18/1. X D. mixtum (Asch. & Graebn.) Rauschert (*C. viride* x *D. fuchsii*) has been found widely scattered in Br and in Co Down.

17/1 x 18/2. X D. conigerum (Norman) Rauschert (*X D. dominianum* (E.G. Camus, Bergon & A. Camus) Soó, *X D. drucei* (E.G. Camus) Soó; *C. viride* x *D. maculata*) has very scattered records in Br.

17/1 x 18/5. X D. viridella (Hesl.-Harr. f.) Soó (*C. viride* x *D. purpurella*) has been found in Co Durham, M & N Ebudes and Outer Hebrides; ?endemic.

18. DACTYLORHIZA Necker ex Nevski (*Dactylorchis* (Klinge) Verm.) - *Marsh-orchids*
Leaves several, the lower sheathing stem, the upper transitional to bracts and not sheathing (though often clasping) stem; upper 3 tepals ± incurved; labellum usually shallowly 3-lobed, sometimes more deeply so or ± unlobed, nearly always as wide as or wider than long; 2 lateral sepals spreading, erect or bent down; spur down-pointed, usually <10mm, mostly rather wide; plant with divided tapering underground tubers.

A very difficult genus owing to ready hybridization between any of the spp., and the complex pattern of variation within most spp. whereby considerable differences between populations are often evident. There is much disagreement as to sp. limits; the views of R.H. Roberts are in the main followed here. Except with typical material it is often not possible to identify single specimens; before using the key the population should be surveyed and means of 5-10 non-extreme plants calculated.

Of the 8 spp., *D. fuchsii* and *D. incarnata* are diploid (2n=40) and the others are tetraploid (2n=80). Hybrids within a ploidy level are highly fertile, those between ploidy levels highly but not completely sterile; even in the case of triploid hybrids (2n=60) backcrossing and introgression can occur. Identification of hybrids involves careful examination of the characters of the putative parents in the sites concerned. The hybrids are intermediate in most characters, have low pollen fertility if triploid, and often show marked hybrid vigour.

1 Stem solid; leaves nearly always spotted; usually 2-6 reduced non-sheathing leaves present on stem transitional between main (sheathing) leaves and bracts; lateral sepals spreading horizontally or bent down; spur usually <2mm wide at midpoint 2
1 Stem usually hollow, at least below; leaves often not spotted; usually 0-2 reduced non-sheathing leaves present on stem transitional between main (sheathing) leaves and bracts; lateral sepals ± erect; spur usually >2mm wide at midpoint 3
  2 Labellum lobed c.1/2 way to base with central lobe usually exceeding the two laterals and ≥1/2 as wide as them; leaves mostly subacute to obtuse, with spots usually ± transversely elongated **1. D. fuchsii**
  2 Labellum lobed much <1/2 way to base, with central lobe as long as or shorter than 2 laterals and much ≤1/2 as wide as them; leaves mostly narrowly acute to subacute, with usually ± circular spots **2. D. maculata**
3 Leaves unspotted or with spots on both surfaces, yellowish-green, narrowly hooded at apex; labellum usually with markedly reflexed sides (if slightly reflexed then leaves with spots on both surfaces) hence appearing very narrow from front, usually with 2 distinct dark loop-shaped marks side by side **3. D. incarnata**
3 Leaves unspotted or with spots mostly on upperside, mid-, dark- or greyish-green, not or broadly hooded at apex; labellum usually without markedly reflexed sides, often nearly flat, usually without 2 distinct dark loops 4
  4 Total number of leaves usually <5, the widest <1.5(2)cm wide; labellum usually distinctly 3-lobed, with central lobe distinctly exceeding 2 laterals and usually >1/2(1/3) as long as unlobed basal part 5
  4 Total number of leaves usually ≥5 (except in *D. majalis* ssp. *scotica*), the widest >(1.5)2cm wide; labellum usually not strongly 3-lobed, if so then central lobe usually <1/3 as long as unlobed basal part (except *D. majalis*) 6
5 Leaves with strong dark spots and rings on upperside; bracts predominantly green, suffused reddish-purple at margins and apex, with dark spots and rings on upperside; transitional non-sheathing leaves (0-)2 **8. D. lapponica**
5 Leaves with 0 or faint dark spots or rings on upperside; bracts strongly suffused reddish-purple all over, without dark spots or rings; transitional non-sheathing leaves 0-1(2) **7. D. traunsteineri**
  6 Labellum with well-marked narrow central lobe usually 1/3-1/2 as long as unlobed basal part, with dark spots or lines mostly in central

part but usually some extending almost to margins; leaves unmarked (parts of Ir only) or with strong spots or blotches mostly >2mm across **6. D. majalis**

6 Labellum usually obscurely lobed, with dark spots or lines usually confined to central part; leaves usually unmarked, sometimes with small spots <2mm across or with rings (rarely larger spots) 7

7 Leaves usually broadly hooded at apex, their dark markings (if present) small spots <2mm across or rarely larger; labellum ± rhombic, usually reddish-purple, usually <7.5 x 9.5mm **5. D. purpurella**

7 Leaves flat or slightly hooded at apex, their dark markings (if present) mainly as rings; labellum orbicular to transversely broadly elliptic, usually pinkish- or pale mauvish-purple, usually >7.5 x 9.5mm **4. D. praetermissa**

Other spp. - D. russowii (Klinge) Holub was reported from NE Yorks in 1989 but requires confirmation; it is close to *D. traunsteineri*.

1. D. fuchsii (Druce) Soó (*D. maculata* ssp. *fuchsii* (Druce) Hyl., *D. longebracteata* auct. non (F.W. Schmidt) Holub, *Dactylorchis fuchsii* (Druce) Verm.) - *Common Spotted-orchid*. Stems to 50(70)cm; largest leaves ≤4(5.5)cm wide, usually with dark transversely elongated spots, flat at apex; labellum lobed c.1/2 way to base, the central lobe the longest, with pale pink to white ground-colour; 2n=40. Native; damp woods, banks and meadows, marshes and fens, usually on base-rich soil; ± throughout BI, the commonest orchid in En and Ir. Some populations in Ir, Sc, Man and Cornwall are statistically separable and have received ssp. status, but are worth only var. rank. 977

1 x 2. D. fuchsii x D. maculata = D. x transiens (Druce) Soó has been found scattered throughout BI but is over-recorded for plants difficult to determine; (2n=60).

1 x 3. D. fuchsii x D. incarnata = D. x kerneriorum (Soó) Soó has been found scattered throughout Br and Ir; (2n=40).

1 x 4. D. fuchsii x D. praetermissa = D. x grandis (Druce) P.F. Hunt occurs throughout the range of *D. praetermissa* in Br (and CI?); probably the commonest hybrid orchid in S Br; 2n=60.

1 x 5. D. fuchsii x D. purpurella = D. x venusta (T. & T.A. Stephenson) Soó occurs throughout the range of *D. purpurella* in Br and Ir; it is sometimes partially fertile; 2n=60; ?endemic.

1 x 6. D. fuchsii x D. majalis = D. x braunii (Halácsy) Borsos & Soó has been found in Anglesey, SE Yorks, W Sc and Co Clare populations of *D. majalis*.

1 x 7. D. fuchsii x D. traunsteineri (= *D. x kelleriana* P.F. Hunt nom. inval.) has been found in the Irish, Anglesey and Yorks areas of *D. traunsteineri*.

2. D. maculata (L.) Soó (*Dactylorchis maculata* (L.) Verm.) - *Heath Spotted-orchid*. 977
Stems to 40(50)cm; largest leaves ≤2(2.5)cm wide, usually with dark ± rounded spots, usually flat (sometimes hooded) at apex; labellum lobed much <1/2 way to base, the central lobe shorter than to as long as laterals, with pale pink to ± white ground-colour; 2n=80. Native; damp peaty places in bogs, marshes and ditches; ± throughout BI, the commonest orchid in Sc and Ir. Our plant is ssp. **ericetorum** (E.F. Linton) P.F. Hunt & Summerh.; sspp. **maculata** and/or **elodes** (Griseb.) Soó have also been reported here but the evidence is weak. Ssp. **rhoumensis** (Hesl.-Harr. f.) Soó, from Rhum, is a diploid, 2n=40, variously placed under *D. fuchsii* and *D. maculata* but is of very doubtful ssp. status anyway.

2 x 3. D. maculata x D. incarnata = D. x carnea (E.G. Camus) Soó (*D. x maculatiformis* (Rouy) Borsos & Soó, *D. x claudiopolitana* Soó nom. nud.) has been found scattered in Br and Ir; (2n=60).

2 x 4. D. maculata x D. praetermissa = D. x hallii (Druce) Soó probably occurs throughout the range of *D. praetermissa* in Br (and CI?), but much more rarely than *D. x grandis* due to different habitat preferences.

2 x 5. D. maculata x D. purpurella = D. x formosa (T. & T.A. Stephenson) Soó occurs throughout the range of *D. purpurella* in Br and Ir; probably the commonest hybrid orchid in N Br and Ir.

2 x 6. D. maculata x D. majalis = D. x dinglensis (Wilmott) Soó (*D. x townsendiana* auct. non (Rouy) Soó) has been found in the Irish and Cards populations of *D. majalis*.

2 x 7. D. maculata x D. traunsteineri = D. x jenensis (Brand) Soó has been found in the NW Wa, Yorks and Irish populations of *D. traunsteineri*; (2n=60).

3. D. incarnata (L.) Soó (*Dactylorchis incarnata* (L.) Verm.) - *Early Marsh-orchid*. 977
Stems to 40(80)cm; largest leaves 1-2(3.5)cm wide, narrowly hooded at apex; labellum obscurely lobed or lobed much <1/2 way to base, ± rhombic, the sides usually strongly reflexed. Native; decreasing due to land-drainage. Very variable; some of the main variants occupy distinct habitats or geographical regions and have been recognized as sspp. Of these, the most distinct is ssp. *cruenta*; most of the others are possibly better considered vars. of ssp. *incarnata*, but the present consensus is followed here.

1 At least some plants with leaves with spots on both surfaces **e. ssp. cruenta**
1 Plants with unmarked leaves, rarely some with spots on upperside only 2
  2 Perianth pale yellow or cream; labellum usually >6.5 x 8mm and with well-marked lobes, the lateral ones usually indented; lowest bract usually >30 x 20mm **f. ssp. ochroleuca**
  2 Perianth variously pink to purple, rarely white or cream (if so then labellum usually <6.5 x 8mm and without indented lateral lobes and lowest bract usually <30 x 20mm) 3
3 Ground-colour of perianth pink; bracts usually lacking anthocyanin 4
3 Ground-colour of perianth red to purple; bracts usually strongly suffused with anthocyanin 5
  4 Plants to 40cm with <6 leaves; labella usually <7 x 8.5mm, marked with lines; spurs usually <7.5mm **a. ssp. incarnata**
  4 Plants to50(80)cm with ≥6 leaves; labella usually >7 x 8.5mm, marked with dots; spurs usually >7.5mm **b. ssp. gemmana**
5 Plants mostly >20cm; perianth ground-colour reddish-purple **d. ssp. pulchella**
5 Plants mostly <20cm; perianth ground-colour vivid ruby- or crimson-red **c. ssp. coccinea**

a. Ssp. incarnata (*Orchis strictifolia* Opiz). Plants mostly 20-40cm; leaves not spotted or rarely with few small dots on upperside; perianth ground-colour pale pink; labellum usually <7 x 8.5mm, usually obscurely lobed; 2n=40. Wet meadows, fens and marshes on base-rich or neutral soils; locally frequent in En and Wa, extremely scattered in Sc and Ir.

b. Ssp. gemmana (Pugsley) P.D. Sell (*Dactylorchis incarnata* ssp. *gemmana* (Pugsley) Hesl.-Harr. f.). Plants to 50(80)cm; leaves not spotted or rarely with few small dots on upperside; perianth ground-colour pale pink; labellum usually >7 x 8.5mm, usually obscurely lobed. Base-rich fens and marshes; very local in E Norfolk and W Galway, perhaps elsewhere in C & S En; ?endemic.

c. Ssp. coccinea (Pugsley) Soó (*Dactylorchis incarnata* ssp. *coccinea* (Pugsley) Hesl.-Harr. f.). Plants mostly <20cm; leaves not spotted; perianth ground-colour vivid ruby- or crimson-red; labellum usually <6.5 x 8mm, usually obscurely lobed; 2n=40.

FIG 977 - Labella of ***Dactylorhiza***. 1, ***D. incarnata*** (4th in row, ssp. ***ochroleuca***). 2, ***D. praetermissa***. 3 (1st 3), ***D. purpurella***; 3 (4th in row), ***D. praetermissa*** var. ***junialis***. 4, ***D. majalis*** ssp. ***occidentalis***. 5, ***D. majalis*** ssp. ***cambrensis***. 6, ***D. traunsteineri***. 7, ***D. lapponica***. 8, ***D. fuchsii***. 9, ***D. maculata***. Drawings by R.H. Roberts.

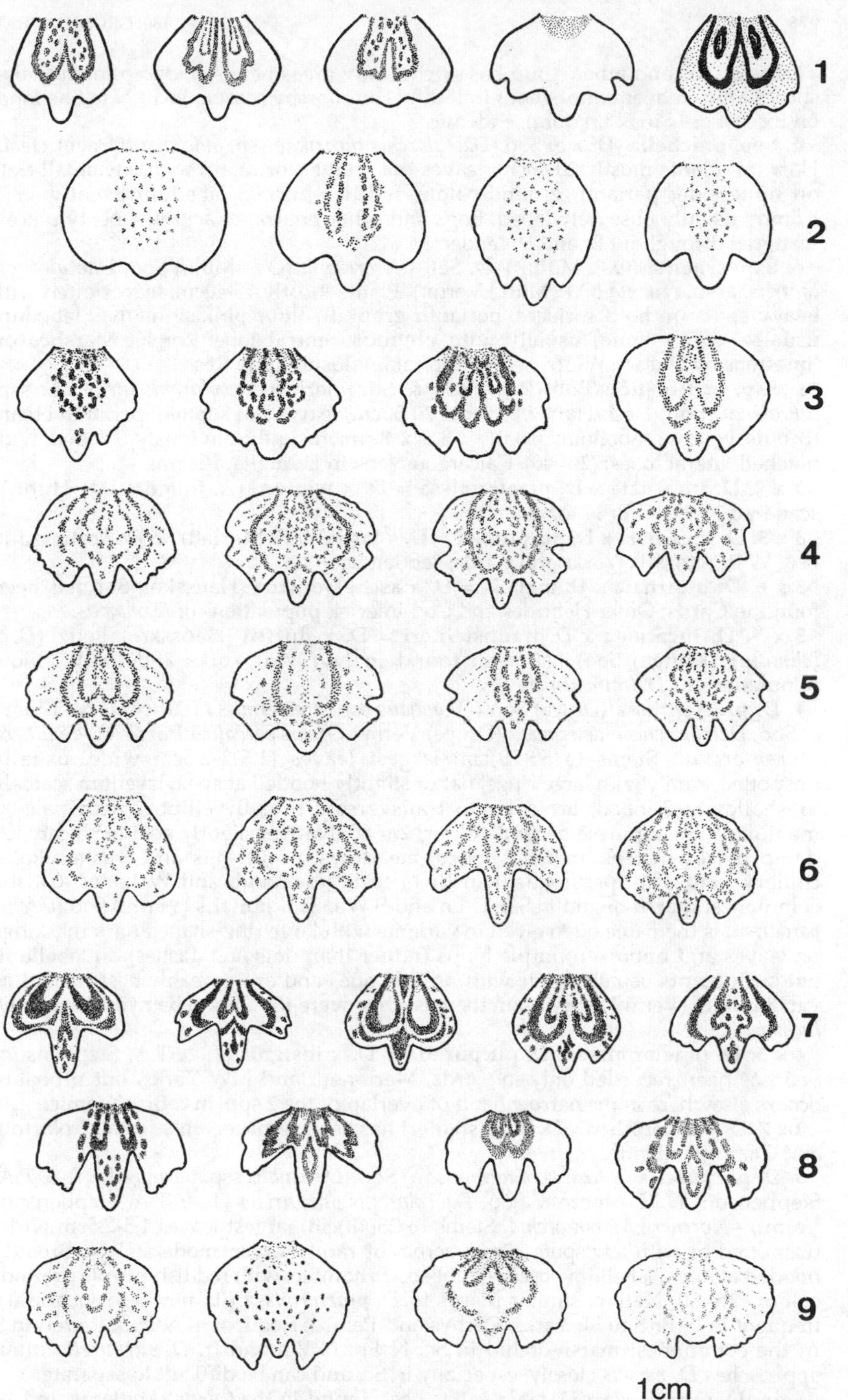

FIG 977 - see caption opposite

Dune-slacks and other damp base-rich sandy areas near sea, damp inland lake-shores in Ir; frequent throughout Ir, locally common by coast in W Br N to Shetland, on E coast only in N En (rare); endemic.

d. Ssp. pulchella (Druce) Soó (*Dactylorchis incarnata* ssp. *pulchella* (Druce) Hesl.-Harr. f.). Plants mostly 20-40cm; leaves not spotted or rarely with few small dots on upperside; perianth ground-colour reddish-purple; labellum usually <7 x 8.5mm, usually obscurely lobed. Bogs and other neutral to acid wet peaty places; scattered throughout Br and Ir; ?endemic.

e. Ssp. cruenta (O.F. Müll.) P.D. Sell (*D. cruenta* (O.F. Müll.) Soó, *Dactylorchis incarnata* ssp. *cruenta* (O.F. Müll.) Verm.). Plants mostly 15-40cm; leaves often with heavy spots on both surfaces; perianth ground-colour pinkish-mauve; labellum usually <7.5 x 8mm, usually with obvious central lobe; 2n=40. Marshes on limestone by lakes in WC Ir, neutral mountain flushes in W Ross.

f. Ssp. ochroleuca (Boll) P.F. Hunt & Summerh. (*Dactylorchis incarnata* ssp. **977**
ochroleuca (Boll) Hesl.-Harr. f.). Plants 20-50cm; leaves not spotted; perianth cream to pale yellow; labellum usually >6.5 x 8mm, usually obviously 3-lobed with notched lateral lobes; (2n=40). Calcareous fens in E Anglia, ?Carms.

3 x 4. D. incarnata x D. praetermissa = D. x wintoni (A. Camus) P.F. Hunt is scattered in S & C Br N to S Lancs.

3 x 5. D. incarnata x D. purpurella = D. x latirella (P.M. Hall) Soó is scattered in N & W Br S to MW Yorks and Cards; ?endemic.

3 x 6. D. incarnata x D. majalis = D. x aschersoniana (Hausskn.) Soó has been found in Cards, Outer Hebrides and Co Limerick populations of *D. majalis*.

3 x 7. D. incarnata x D. traunsteineri = D. x dufftii (Hausskn.) Peitz (*D. x lehmanii* (Klinge) Soó) has been found in NW Wa, Yorks and Co Wicklow populations of *D. traunsteineri*.

4. D. praetermissa (Druce) Soó (*D. majalis* ssp. *praetermissa* (Druce) D.M. Moore **977**
& Soó, *Dactylorchis praetermissa* (Druce) Verm., *Orchis pardalina* Pugsley) - *Southern Marsh-orchid*. Stems to 50(70)cm; largest leaves (1.5)2-2.5cm wide, usually unspotted, rarely with large rings, flat or slightly hooded at apex; labellum scarcely or shallowly 3-lobed, orbicular to transversely broadly elliptic, with pale to medium pinkish-purple ground-colour; 2n=80. Native; slightly acid to calcareous damp places in fens, marshes, bogs, meadows, gravel-pits and waste alkali-, colliery- and ash-tips; frequent in Br N to S Northumb and W Lancs, CI, the commonest marsh-orchid in S & C En and S Wa. Ssp. **junialis** (Verm.) Soó (*Orchis pardalina*) is the name often given to variants with large ring-shaped dark markings on leaves and unbroken purple loops (rather than dots and dashes) on labellum, but such plants usually occur with normal ones and are probably best treated as var. **junialis** (Verm.) Senghas; in the past they were thought to be hybrids with *D. fuchsii*.

4 x 5. D. praetermissa x D. purpurella = D. x insignis (T. & T.A. Stephenson) Soó has been recorded only in Cards, Merioneth and MW Yorks, but probably occurs elsewhere in the narrow band of overlap of the 2 spp. in C Br; endemic.

4 x 7. D. praetermissa x D. traunsteineri has been recorded only from W Norfolk and Cambs; endemic.

5. D. purpurella (T. & T.A. Stephenson) Soó (*D. majalis* ssp. *purpurella* (T. & T.A. **977**
Stephenson) D.M. Moore & Soó, *Dactylorchis purpurella* (T. & T.A. Stephenson) Verm.) - *Northern Marsh-orchid*. Stems to 25(40)cm; largest leaves 1.5-2.5cm wide, unspotted or with few spots <2mm across or rarely larger, moderately to broadly hooded at apex; labellum obscurely lobed, ± rhombic, with reddish-purple ground-colour; 2n=80. Native; similar places to *D. praetermissa* (its northern vicariant); frequent in N Br S to SE Yorks, Derbys and Pembs, frequent in N Ir, scattered in S Ir, the commonest marsh-orchid in Sc, N En, N Wa and Ir. *D. purpurella* often approaches *D. majalis* closely, especially in Sc, and can be difficult to separate.

5 x 6. D. purpurella x D. majalis has been found in the Cards, Anglesey and W Sc populations of *D. majalis*, and probably occurs in Ir too; endemic.

6. D. majalis (Rchb.) P.F. Hunt & Summerh. (*Dactylorchis majalis* (Rchb.) Verm.) - *Western Marsh-orchid*. Stems to 30(35)cm; largest leaves mostly 1.5-2.8cm wide, usually heavily spotted or blotched, rarely unspotted, flat or slightly hooded at apex; labellum usually distinctly lobed with prominent central lobe, subrhombic to transversely broadly elliptic, with light reddish-purple ground-colour. Native; marshes, fens, wet meadows and dune-slacks. A difficult sp. whose variation can perhaps best be summarized by the recognition of 2 sspp., both distinct from the Continental type ssp.

a. Ssp. occidentalis (Pugsley) P.D. Sell (ssp. *kerryensis* (Wilmott) Senghas nom. **R**
illeg., var. *kerryensis* (Wilmott) R.M. Bateman & Denholm, ssp. *scotica* E. Nelson **977**
nom. illeg., var. *ebudensis* Wiefelspütz ex R.M. Bateman & Denholm, *D. kerryensis* (Wilmott) P.F. Hunt & Summerh., *Dactylorchis majalis* ssp. *occidentalis* (Pugsley) Hesl.-Harr. f.). Stems mostly <20cm; largest leaves mostly <10cm x >2cm; spur usually <3.5mm wide at entrance; 2n=80. Scattered over Ir (mostly S & W), N Uist (Outer Hebrides); endemic.

b. Ssp. cambrensis (R.H. Roberts) R.H. Roberts (var. *cambrensis* (R.H. Roberts) **R**
R.M. Bateman & Denholm, *D. purpurella* ssp. *majaliformis* E. Nelson ex Løjtnant, **977**
Dactylorchis majalis ssp. *cambrensis* R.H. Roberts). Stems mostly >20cm; largest leaves mostly >10cm x <2cm; spur usually >3.5mm wide at entrance; 2n=80. NW & MW Wa, SE Yorks, and NW & N Sc; endemic.

More work is needed on Scottish populations to see whether they are worth separate ssp. status (*scotica* in N Uist; *majaliformis* on the mainland).

7. D. traunsteineri (Saut. ex Rchb.) Soó (*D. majalis* ssp. *traunsteinerioides* **R**
(Pugsley) R.M. Bateman & Denholm, var. *eborensis* (Godfery) R.M. Bateman & **977**
Denholm, var. *francis-drucei* (Wilmott) R.M. Bateman & Denholm, *Orchis latifolia* L. var. *eborensis* Godfery, *O. francis-drucei* Wilmott, *Dactylorchis traunsteineri* (Saut. ex Rchb.) Verm., *D. traunsteinerioides* (Pugsley) Verm.) - *Narrow-leaved Marsh-orchid*. Stems to 30(40)cm; widest leaves mostly <1.5cm wide, unspotted or rather faintly spotted, slightly hooded at apex, labellum distinctly 3-lobed but even if not then with a prominent central projection, with light reddish-purple ground-colour and usually many dark markings ± to margins; 2n=80. Native; calcareous fens and other damp base-rich grassy places; local in Ir, W Sc, NW & SW Wa, N En and E Anglia, very scattered in C S En.

8. D. lapponica (Hartm.) Soó (*D. traunsteineri* ssp. *lapponica* (Hartm.) Soó, *D.* **RRR**
pseudocordigera (Neuman) Soó) - *Lapland Marsh-orchid*. Stems to 21cm; widest **977**
leaves 1.1-1.5cm wide, heavily spotted and blotched, not or scarcely hooded at apex; labellum usually distinctly lobed but even if not then with a prominent central projection, with reddish-purple ground-colour and usually heavy dark spots and lines; (2n=80). Native; slightly acidic to base-rich hillside flushes; very local in Westerness, Kintyre, N Ebudes and Outer Hebrides, 1st recognized in Br in 1986.

19. NEOTINEA Rchb. f. - *Dense-flowered Orchid*

Leaves 2-3(4) near base of stem, with reduced leaves up stem; upper 5 tepals incurved; labellum with 2 large lateral lobes and larger terminal lobe usually shallowly subdivided at apex or with small tooth between the divisions; spur short, rounded at apex; plant with 2 ovoid underground tubers.

1. N. maculata (Desf.) Stearn (*N. intacta* (Link) Rchb. f.) - *Dense-flowered Orchid*. **R**
Stems to 30(40)cm; leaves oblong-elliptic, sometimes purple-spotted (in W Ir only) but usually not, 2-6cm; flowers in dense cylindrical spike, creamy-white, sometimes pink-tinged, c.5-6mm vertically across, with labellum 4-5mm, with spur c.2mm; (2n=40, 42). Native; rocky and sandy grassy places and maritime dunes; very local in CW & SW Ir, very rare in W Donegal (found 1983) and Man (found 1966). Superficially resembles *Pseudorchis*, but labellum-shape is totally different.

20. ORCHIS L. - *Orchids*

Leaves several, on stem, with few smaller ones above; upper 3 or upper 5 tepals incurved, the 2 lateral sepals incurved or erect to patent; labellum with 2 lateral lobes and a terminal lobe, the latter often larger than the laterals and usually 2-3-lobed at its apex; spur long or short, rounded to truncate or emarginate at apex; plant with 2 ovoid underground tubers.

1 Upper 3 tepals incurved to form a 'helmet'; 2 lateral sepals erect to patent 2
1 All 5 upper tepals incurved to form a 'helmet' 3
2 Labellum with terminal lobe exceeded by 2 laterals, sometimes ± 0; spur shorter than ovary; bracts 3-veined or the lowest few 5-veined; leaves never spotted **1. O. laxiflora**
2 Labellum with terminal lobe exceeding laterals; spur at least as long as ovary; bracts 1-veined or the lowest few 3-veined; leaves usually dark-spotted **2. O. mascula**
3 Area of central lobe of labellum from smaller than to slightly larger than that of each lateral lobe 4
3 Area of central lobe of labellum at least 2x that of each lateral lobe 5
4 Spur horizontal or directed upwards, ± as long as ovary; labellum mauvish-purple with paler, spotted, central area **3. O. morio**
4 Spur directed downwards, <1/2 as long as ovary; labellum white with reddish-purple spots **4. O. ustulata**
5 Outside of all 3 sepals (forming 'helmet') dark reddish-purple, contrasting strongly with very pale labellum; 2 main sublobes of terminal lobe of labellum wider than long **5. O. purpurea**
5 Outside of sepals pale pinkish-purple, scarcely or not contrasting with labellum; 2 main sublobes of terminal lobe of labellum longer than wide 6
6 Two main sublobes of terminal lobe of labellum oblong, >2x as wide as lateral lobes **6. O. militaris**
6 Two main sublobes of terminal lobe of labellum linear, c. as wide as lateral lobes **7. O. simia**

1. O. laxiflora Lam. - *Loose-flowered Orchid.* Stems to 50(80)cm; leaves lanceolate to linear-oblong, unspotted; flowers rather uniformly purple; labellum with 2 large lateral and 0 or smaller terminal lobe; (2n=36). Native; marshy meadows and by lakes; locally common in Jersey and Guernsey. RR

1 x 3. O. laxiflora x O. morio = O. x alata Fleury occurs sporadically with the parents in Jersey and occurred on Guernsey in 1949.

2. O. mascula (L.) L. - *Early-purple Orchid.* Stems to 40(60)cm; leaves elliptic-oblong to narrowly so, usually spotted; flowers uniformly pinkish-purple to purple; labellum with 3 often ± equal lobes, the terminal one shallowly bilobed; (2n=42). Native; neutral or base-rich grassland, scrub and woods, usually in shade in S but in open in N; frequent to common throughout BI.

2 x 3. O. mascula x O. morio = O. x morioides Brand occurs rarely and sporadically in En and Wa.

3. O. morio L. - *Green-winged Orchid.* Stems to 20(40)cm; leaves narrowly elliptic-oblong, unspotted; flowers rather uniformly mauvish-purple but labellum with pale, darker-spotted central region and upper tepals with green veins; labellum with 3 often ± equal lobes, the terminal one often shallowly bilobed; 2n=36. Native; base-rich to neutral short undisturbed grassland; formerly frequent over most of En, Wa, Ir and CI, now greatly reduced and local, Ayrs.

4. O. ustulata L. - *Burnt Orchid.* Stems to 15(30)cm; leaves elliptic-oblong, unspotted; outside of sepals dark reddish-purple, contrasting strongly with white labellum with reddish-purple spots, the contrast greatest when upper flowers are still unopened; labellum with 3 ± equal lobes, the terminal one shallowly bilobed; (2n=42). Native; short grassland on chalk and limestone; formerly locally frequent R over much of En, now greatly reduced and extremely local, Glam.

5. O. purpurea Huds. - *Lady Orchid.* Stems to 50(100)cm; leaves elliptic-oblong, unspotted; outside of sepals dark reddish-purple, contrasting strongly with white or pale pink labellum with pink to reddish-purple spots, the contrast greatest when upper flowers are still unopened; labellum with 3 lobes, the terminal one much larger than laterals and with 2 broad sublobes with usually small tooth between; (2n=42). Native; woods and scrub, rarely open grassland, on chalk; locally frequent on N Downs in E & W Kent, very scattered (mostly extinct) elsewhere in S En. **R**

6. O. militaris L. - *Military Orchid.* Stems to 45(60)cm; leaves elliptic-oblong, unspotted; flowers pinkish- to reddish-purple; sepals paler on outside than labellum or sepals on inside; labellum with 3 lobes, the terminal one much larger than laterals with 2 oblong sublobes with small tooth between, paler or white with purplish spots in central part; 2n=42. Native; chalk grassland and old chalk-pit with invading trees and shrubs; 1 site each in Bucks and W Suffolk, 2 sites sporadically in Oxon, formerly more widespread in mid and lower Thames valley. **RRR**

6 x 7. O. militaris x O. simia = O. x beyrichii A. Kern. occurred up to mid-19th Century in M Thames valley when the 2 parents co-existed there.

7. O. simia Lam. - *Monkey Orchid.* Stems to 30(40)cm; leaves elliptic-oblong, unspotted; flowers pinkish-purple with a pale or white pinkish-purple-spotted area in centre of labellum; labellum with 3 lobes, the terminal one much larger than laterals with 2 linear sublobes with small tooth between; (2n=42). Native; chalk grassland and open scrub; 2 sites in E Kent, 2 in Oxon, formerly very scattered elsewhere in SE En and 1 site in SE Yorks. **RRR**

20 x 21. ORCHIS x ACERAS = X ORCHIACERAS E.G. Camus

20/7 x 21/1. X O. bergonii (De Nant.) E.G. Camus (*O. simia* x *A. anthropophorum*) was found at an *O. simia* site in E Kent in 1985; it has a labellum shape similar to that of *O. simia* but is intermediate in flower colour and sepal shape and size.

21. ACERAS R. Br. - *Man Orchid*

Leaves several, near stem-base, with few smaller ones above; upper 5 tepals incurved to form 'helmet'; labellum with 3 lobes, the lateral linear, the terminal larger and with 2 linear terminal sublobes; spur 0; plant with 2 ovoid underground tubers.

Differs from *Orchis* only in absence of spur and probably better united with it.

1. A. anthropophorum (L.) W.T. Aiton - *Man Orchid.* Stems to 40(50)cm; leaves narrowly elliptic-oblong, unspotted; flowers greenish-yellow, often tinged reddish-brown; (2n=42). Native; chalk and limestone grassland or scrub; local in SE En (frequent only in E & W Kent), scattered W to Dorset (formerly N Somerset) and N to S Lincs. **R**

22. HIMANTOGLOSSUM W.D.J. Koch - *Lizard Orchid*

Leaves several, decreasing in size up stem; upper 5 tepals incurved to form 'helmet'; labellum very long and narrow, with 2 linear lateral lobes and a long terminal lobe spirally coiled at first and with 2 small sublobes at apex; spur short; plant with 2 ovoid underground tubers.

Close to *Orchis* and perhaps not distinct.

1. H. hircinum (L.) Spreng. - *Lizard Orchid.* Stems to 70(90)cm; leaves elliptic-oblong, purple-mottled or not; flowers greyish-yellowish-green; labellum 4-7cm, the central lobe 3-6cm; (2n=24, 36). Native; on calcareous soils in rough ground, dunes, scrub and marginal places usually among tall grass; scattered places (often sporadic) in S & E En W to N Somerset (formerly N Devon) and N to W Suffolk (formerly NE Yorks), formerly Jersey. **RRR**

23. OPHRYS L. - *Bee-orchids*

Leaves several, decreasing in size up stem; upper 5 tepals all patent, the 2 petals markedly different from the 3 sepals; labellum velvety in texture, subentire or with 2 small lateral lobes, the terminal lobe large and resembling an insect's abdomen; spur 0; plant with 2 ovoid underground tubers.

1 Labellum with distinct lateral lobes; the 2 petals filiform; labellum distinctly longer than wide **1. O. insectifera**
1 Labellum with 0 or obscure lateral lobes; the 2 petals oblong to linear-oblong; labellum not or only just longer than wide 2
 2 Sepals yellowish- to brownish-green; the 2 petals yellowish-green, >1/2 as long as sepals **2. O. sphegodes**
 2 Sepals pink or greenish-pink; the 2 petals pink or greenish-pink, ≤1/2 as long as sepals 3
3 Apex of labellum shortly bilobed, with the short simple projection between directed downwards and backwards (hence ± invisible from front of flower) **3. O. apifera**
3 Apex of labellum shortly bilobed, with the short projection between directed prominently forwards and often 3-toothed **4. O. fuciflora**

Other spp. - O. bertolonii Moretti, from Mediterranean, was found in Dorset in 1976 but was almost certainly planted and has been removed.

1. O. insectifera L. - *Fly Orchid*. Stems to 60cm; sepals yellowish-green; petals purplish-brown, filiform, slightly <1/2 as long as sepals; labellum with well-marked lateral lobes and terminal lobe conspicuously bilobed, purplish-brown with shining blue central area; (2n=36). Native; woods, scrub, grassland, spoil-heaps, fens and lakesides on calcareous soils; scattered throughout Br N to Westmorland and NE Yorks, C Ir, frequent in SE En.

1 x 2. O. insectifera x O. sphegodes = O. x hybrida Pokorny has occurred sporadically in E Kent; it is intermediate in flower shape.

1 x 3. O. insectifera x O. apifera (*O. x pietzschii* Kümpel nom. inval.) has occurred in woodland in N Somerset since 1968; it is intermediate in sepal colour and petal and labellum shape.

2. O. sphegodes Mill. - *Early Spider-orchid*. Stems to 20(35)cm; sepals yellowish- **RRR** to brownish-green; petals yellowish, oblong, >1/2 as long as sepals; labellum subentire or notched (sometimes with small tooth in notch) at apex, dark purplish-brown with variable blue markings; (2n=36). Native; grassland or spoil-heaps on chalk or limestone; very local from E Kent and W Suffolk to Dorset and W Gloucs, intrd and natd in Herts, formerly to W Cornwall and Northants and in Denbs and Jersey.

3. O. apifera Huds. - *Bee Orchid*. Stems to 45(60)cm; sepals pink or greenish-pink, sometimes very pale; petals greenish-pink, oblong, ≤1/2 as long as sepals; labellum with short lateral lobes, shallowly bilobed at apex with backward-directed projection in notch, reddish-brown with various markings of yellow, gold and brown; (2n=36). Native; grassland, scrub, spoil-heaps and sand-dunes on calcareous or base-rich soils; locally frequent in Br N to Cumberland and Durham, CI, scattered in Ir.

3 x 4. O. apifera x O. fuciflora = O. x albertiana E.G. Camus has been found in E Kent.

4. O. fuciflora (Crantz) Moench (*O. holoserica* auct. non (Burm. f.) Greuter) - *Late* **RRR** *Spider-orchid*. Stems to 35(55)cm; differs from *O. apifera* in sepals and petals usually clear pink; and see key; (2n=36). Native; short grassland on chalk; very local in E Kent.

GLOSSARY

For some special terms, used only in 1 or few families, direct reference to the relevant family (families) is made; in those cases the family description and the notes immediately following it should be consulted. Words in the definitions that are given in **bold** are themselves defined elsewhere in the glossary.

abaxial - of a lateral organ, the side away from the axis, normally the **lowerside**
achene - a dry, indehiscent, 1-seeded fruit, ± hard, with papery to leathery wall; **achene-pit**, see 139. Asteraceae
acicle - a slender prickle with scarcely widened base
actinomorphic - of a flower with radial (i.e. >1 plane of) symmetry
acuminate - gradually tapering to a point; Fig 989
acute - with point <90°; Fig 989
adaxial - of a lateral organ, the side towards the axis, normally the **upperside**
adherent - joined or fused
aerial - above-ground or above-water
agamospecies - a group of **apomictic** plants treated at the species level, but usually exhibiting a much narrower range of variation than a sexual species
alien - not **native**, introduced to a region deliberately or accidentally by man
alternate - lateral organs on an axis 1 per **node**, successive ones on opposite sides
anastomosing - dividing up and then joining again, usually applied to veins
androecium - the group of male parts of a flower; all the **stamens**; Fig 992
andromonoecious - having male and **bisexual** flowers on the same plant
annual - completing its life-cycle in ≤12 months (but often not within 1 calendar year)
anther - pollen-bearing part of a **stamen**, usually terminal on a stalk or **filament**; Fig 992
anthesis - flowering time; strictly pollen-shedding time
apiculus - a small, abruptly delimited point; **apiculate**, with an apiculus; Fig 989
apomictic - producing seed wholly female in origin, without fertilization
appendage - small extra protrusion or extension, such as on a **petal, sepal** or seed
appendix - see 151. Araceae
appressed - lying flat against another organ
aril - **succulent** covering around a seed, outside the **testa** (not the **pericarp**)
aristate - extended into a long bristle; Fig 989
ascending - sloping or curving upwards
auricle - basal extension of a leaf-blade, especially in Poaceae; Fig 825
awn - see 157. Poaceae
axil - angle between main and lateral axes; **axillary**, in the axil; see also **subtend**
axile - of a **placenta** formed by central axis of an **ovary** that is connected by **septa** to the wall; Fig 990

beak - a narrow, usually apical, projection
berry - a **succulent** fruit, the seeds usually >1 and without a stony coat
biennial - completing its life-cycle in >1 but <2 years, not flowering in the first year
bifid - divided into two, usually deeply, at apex
bifurcate - dividing into two branches
biotype - a genetically fixed variant of a **taxon** particularly adapted to some (usually environmental) condition

birdseed-alien - **alien** introduced as contaminant of birdseed
bisexual - of a plant or flower, bearing both sexes
blade - main part of a flat organ (e.g. **petal**, leaf); cf. **claw, petiole**; Figs 825, 989
bloom - delicate, waxy, easily removed covering to fruit, leaves etc.; see also **pruinose**
bract - modified, often scale-like, leaf **subtending** a flower, less often a branch; **bracteate**, with bract(s); Fig 991
bracteole - a supplementary or secondary **bract** or a bract once removed; Fig 991
bud-scales - scales enclosing a bud before it expands
bulb - swollen underground organ consisting of condensed stem and **succulent** scale-leaves
bulbil - a small bulb or tuber, usually **axillary**, on an aerial part of the plant
bullate - with the surface raised into blister-like swellings

callus - see 157. Poaceae
calyx (plural **calyces**) - the outer **whorl(s)** of the **perianth**, if different from the inner; all the **sepals**; **calyx-tube**, **calyx-lobes**, the **proximal** fused and **distal** free parts of a calyx in which the **sepals** are partly fused; Fig 992
capillary - hair-like
capitate - head-like, such as a tight inflorescence on a stalk, a knob-like stigma on a style, or a stalked gland
capitulum - see 138. Dipsacaceae, 139. Asteraceae; Fig 991
capsule - a dry, many-seeded dehiscent fruit formed from >1 carpel
carpel - the basic female reproductive unit of Magnoliopsida, 1-many per flower, if >1 then **free** or fused; Fig 992
carpophore - a stalk-like sterile part of a flower between the **receptacle** and **carpels**, as in some Apiaceae and Caryophyllaceae; Fig 992
cartilaginous - cartilage-like in consistency, hard but easily cut with a knife, not green
caryopsis - see 157. Poaceae
casual - an **alien** plant not **naturalized**
catkin - a condensed **spike** of reduced flowers on a long axis, often flexible and wind-pollinated
cell - of an **ovary**, the chambers into which it may be divided (often each one corresponding to a **carpel**); Fig 990
cladode - see 162. Liliaceae
clavate - club-shaped, slender and distally thickened
claw - **proximal**, narrow part of a flat organ such as a **petal**, bearing the **blade distally**
cleistogamous - of flowers, not opening, becoming self-pollinated in the bud stage
column - a stout stalk formed by fusion of various floral parts in, e.g., Orchidaceae, Geraniaceae, *Rosa*; **columnar**, column-like
commissure - see 111. Apiaceae
compound - not **simple**, of a leaf divided right to the **rhachis** into **leaflets**; Fig 990
compressed - flattened
cone - compact body composed of axis with lateral organs bearing spores or seeds, as in Lycopodiopsida, Equisetopsida, Pinopsida; **cone-scales**, the lateral organs of a cone
connective - part of **anther** connecting its 2 halves
contiguous - touching at the edges with no gap between
convergent - of ≥2 organs with apices closer together than their bases
cordate - of the base of a flat organ; see Fig 989
coriaceous - of leathery texture
corm - short, usually erect, swollen underground stem
corolla - the inner **whorls** of the **perianth**, if different from the outer; all the **petals**; **corolla-tube**, **corolla-lobes**, the **proximal** fused and **distal** free parts of a

corolla in which the **petals** are partly fused; Fig 992
corona - see 162. Liliaceae
corymb - a **raceme** in which the lower flowers have longer **pedicels**, producing a ± flat-topped **inflorescence**; **corymbose**, corymb-like though not necessarily strictly a corymb; Fig 991
cotyledon - the first leaves of a plant, 1 in Liliidae, usually 2 in Magnoliidae, 2-several in Pinopsida, usually quite different in appearance from all subsequent leaves
crenate - of the margin of a flat organ; see Fig 989
culm - see 157. Poaceae
cuneate - of the base of a flat organ; see Fig 989
cupule - see 41. Fagaceae
cuspidate - abruptly narrowed to a point; Fig 989
cyme - an **inflorescence** in which each flower terminates the growth of a branch, more **distal** flowers being produced by longer branches lateral to it; **cymose**, in the form of a cyme; Fig 991

decaploid - see **polyploid**
deciduous - not persistent, e.g. leaves falling in autumn or **petals** falling after **anthesis**
decumbent - **procumbent** but with the apex turning up to become **ascending** or **erect**
decurrent - of a lateral organ, having its base prolonged down the main axis
decussate - **opposite**, with successive pairs at right angles to each other
dehiscent - opening naturally
dentate - of the margin of a flat organ; see Fig 989
denticulate - minutely or finely **dentate**
depressed-globose - similar to **globose** but wider than long
dichasium - cyme with 2 lateral branches at each **node**; Fig 991
dimorphic - occurring in 2 forms
dioecious - having the 2 sexes on different plants
diploid - having 2 matching sets of chromosomes, as in **sporophytic** tissue
disc - anything disc-shaped, e.g. top of *Rosa* or *Nuphar* fruit, nectar-secreting ring inside flower at base; **disc flowers**, see 139. Asteraceae
dissected - deeply divided up into segments
distal - at the end away from point of attachment; cf. **proximal**
divaricate - dividing into widely **divergent** branches
divergent - of ≥2 organs with apices further apart than their bases
dorsiventral - with distinct upperside and lowerside
drupe - a **succulent** or spongy fruit, the seeds usually 1 and with a stony coat
dry - not **succulent**

e- - without, e.g. **eglandular, ebracteate**
ellipsoid - a solid shape elliptic in side view
elliptic - a flat shape widest in middle and 1.2-3x as long as wide (if less **broadly** so; if more, **narrowly** so); Fig 989
emarginate - with a pronounced, angled notch at the apex; Fig 989, cf. **retuse**
embryo-sac - see **gametophyte, ovule**
endemic - confined to one particular area, i.e. (in this book) to BI
endosperm - In Magnoliopsida the nutritive tissue for the embryo in the developing seed; it might or might not remain as the food-store in the mature seed
entire - of the margin of a flat organ, not **toothed or lobed**; see Fig 989
epicalyx - organs on the outside of a flower, calyx-like but outside and additional to the **calyx**
epicormic - of new shoots borne direct from the trunk of a **tree**
epigynous - of a flower with an **inferior ovary**; Fig 992

erect - upright
erecto-patent - between **erect** and **patent**
escape - a plant growing outside a garden but having spread vegetatively or by seed from one
evergreen - retaining leaves throughout the year
exceeding - longer than
exserted - protruding from

falcate - sickle- or scythe-shaped
false-fruit - an apparent **fruit** actually formed by tissue (e.g. **receptacle, bracts**) in addition to the real fruit
fascicle - a bunch or bundle, usually with short and indeterminate branching pattern; **fasciculate**, with or in the form of fascicles
fastigiate - a plant with upright branches forming a narrow outline
fibrous roots - a root system in which there is no main axis; cf. **tap-root**
filament - stalk part of a **stamen**; Fig 992
filiform - thread- or wire-like
flexuous - of a stem or hair, wavy
flore pleno - 'double' flower, with many more **petals** than normal, usually due to conversion of **stamens** to petals; in Asteraceae, a 'double' **capitulum**, with **disc flowers** all or many converted to **ray flowers**
floret - see 157. Poaceae; Fig 825
foliaceous - leaf-like, of an organ not normally thus
follicle - a dry, usually many-seeded fruit **dehiscent** along 1 side, formed from 1 **carpel**
free - separate, not fused to another organ or to one another except at point of origin
free-central - of a **placenta** formed by central axis of any **ovary** that is not connected by **septa** to the wall; Fig 992
fruit - the ripe, fertilized **ovary**, containing seeds

gametophyte - the **haploid** generation of a plant that bears the true sex-organs (that produce the gametes), in **pteridophytes** the **prothallus**, in **spermatophytes** the pollen grains (male) and **embryo-sac** (female)
glabrous - hair-less
gland - a secreting structure, usually round or ± so, on the surface of an organ, below the surface, or raised on a stalk
glandular - with the functions of, or bearing, glands
glaucous - bluish-white in colour (rather than green)
globose - spherical
glume - see 156. Cyperaceae, 157. Poaceae; Fig 825
grain-alien - alien introduced as contaminant of grain
granulose - with a fine sand-like surface texture
gynodioecious - having female and **bisexual** plants
gynoecium - the group of female parts of a flower; all the **carpels**; Fig 992

half-epigynous - of a flower with a **semi-inferior ovary**; Fig992
haploid - having only 1 set of chromosomes, as in **gametophytic** tissue
hastate - of the base of a flat organ; Fig 989
heptaploid - see **polyploid**
herb - a plant dying down to ground-level each year
herbaceous - not woody, dying down each year; leaf-like as opposed to **woody**, horny, **scarious** or spongy
heterophyllous - having leaves of ≥2 distinct forms
heterosporous - having **spores** of 2 sorts (**megaspores**, female; and **microspores**, male), as in all **spermatophytes** and a few **pteridophytes**

heterostylous - having 2 forms (not sexes) of flower on different plants, the 2 sorts with different **styles** and/or **stigmas** (and pollen)
hexaploid - see **polyploid**
hilum - scar on a **seed** where it left its point of attachment
hispid - with harsh hairs or bristles
homosporous - having **spores** all of 1 sort, as in most **pteridophytes**
homostylous - not **heterostylous**
hypanthium - extension of **receptacle** above base of **ovary**, in **perigynous** and **epigynous** flowers; Fig 992
hyaline - thin and ± transparent
hypogynous - of a flower with a **superior ovary**, the **calyx**, **corolla** and **stamens** being inserted at the base of the ovary; Fig 992

imparipinnate - pinnate with an unpaired terminal **leaflet**; Fig 990
included - not exserted
incurved - curved inwards
indehiscent - not **dehiscent**
indusium - small flap or pocket of tissue covering groups of **sporangia** in many Pteropsida
inferior - of an **ovary** that is borne below the point of origin of the **sepals, petals** and **stamens** and is fused with the **receptacle** (**hypanthium**) surrounding it; Fig 992
inflated - of an organ that is dilated, leaving a gap between it and its contents
inflorescence - a group of flowers with their branching system and associated **bracts** and **bracteoles**; Fig 991
insertion - the position and form of the point of attachment of an organ
internode - the stem between adjacent **nodes**; Fig 825
introduced - a plant that owes its existence in this country to importation (deliberate or not) by man
introgression - the acquiring of characteristics by one species from another by hybridization followed by backcrossing
isodiametric - of any shape or organ, ± the same distance across in any plane
isophyllous - not **heterophyllous**

keel - a longitudinal ridge on an organ, like the keel of a boat; see also 79. Fabaceae

labellum - see 166. Orchidaceae
laciniate - irregularly and deeply toothed; Fig 989
laminar - in the form of a flat leaf
Lammas growth - extra, usually abnormal, growth put on in summer by some **trees** (around Lammas Day, 1st Aug)
lanceolate - very narrowly **ovate**, c.6x as long as wide; Fig 989
latex - milky juice
lax - loose or diffuse, not dense
leaflet - a division of a **compound** leaf; Fig 990
leaf-opposed - a lateral organ borne on the stem on opposite side from a leaf, not in a leaf-**axil** as usual
leaf-rosette - a radiating cluster of leaves, often at the base of a stem at soil level
legume - a usually dry, usually many-seeded **fruit dehiscent** along 2 sides, formed from 1 **carpel**
lemma - see 157. Poaceae; Fig 825
lenticular - lens-shaped; can vary from biconvex to biconcave
ligulate - see 139. Asteraceae
ligule - minute **membranous** flap at base of leaf of *Isoetes* or *Selaginella*; see also 139. Asteraceae, 157. Poaceae; Fig 825
limb - **distal** expanded part of a **calyx** or **corolla**, as distinct from the **tube** or

throat

linear - long and narrow with ± parallel margins, i.e. extremely narrowly **oblong**; Fig 989

lip - part of the **distal** region of a **calyx** or **corolla** sharply differentiated from the rest due to fusion or close association of its parts

lobe - a substantial division of a leaf, **calyx** or **corolla**; cf. **tooth**

lodicule - see 157. Poaceae; Fig 825

long-shoot - stem of potentially unlimited growth, especially in **trees** or **shrubs**

lowerside - the under surface of a flat organ

lunate - crescent moon-shaped

mealy - with a floury texture

megasporangium - in a **heterosporous** plant, the **sporangia** bearing **megaspores**

megaspore - in a **heterosporous** plant, the female **spores** that give rise to female **gametophytes**

meiosis - special form of cell-division (in **sporangia**, **pollen-sacs** or **ovules**) in which the chromosome number is halved, producing **haploid spores**

membranous - like a membrane in consistency

mericarp - a 1-seeded portion formed by splitting up of a 2-many-seeded **fruit**, as in Geraniaceae, Apiaceae, Boraginaceae, Malvaceae, etc.; see also **schizocarp**

-merous - divided into or composed of a particular number of parts; hence **trimerous, tetramerous, pentamerous**

micron - a micrometre, i.e. one-thousandth of a millimetre or one-millionth of a metre

microsporangium - in a **heterosporous** plant, the **sporangia** bearing **microspores**

microspore - in a **heterosporous** plant, the male **spores** that give rise to male **gametophytes**

midrib - the central, main **vein**; Fig 989

monocarpic - living for >1 year, flowering and fruiting, and then dying

monochasium - **cyme** with 1 lateral branch at each **node**; Fig 991

monoecious - having separate male and female flowers on the same plant

monomorphic - not **dimorphic** or **trimorphic**, occurring in 1 form only

mucronate - having a very short bristle-like tip; **mucro**, the tip itself; Fig 989

mycorrhiza - fungal cells that live within or intimately around the roots of vascular plants

naked - not enclosed

naturalized - an **alien** plant that has become self-perpetuating in BI

native - opposite of alien, a plant that colonizied BI by natural means, often long ago

nec - nor, nor of

nectariferous - nectar-bearing

nectar-pit - a **nectariferous** pit

nectary - any **nectariferous** organ, usually a small knob or a modified **petal** or **stamen**

nodding - bent over and **pendent** at tip

node - the position on a stem where leaves, flowers or lateral stems arise; Fig 825

non - not, not of

nonaploid - see **polyploid**

nothomorph - one of ≥2 variants of a particular hybrid

nut - a dry, indehiscent 1-seeded **fruit** with a hard **woody** wall, often large

nutlet - a small nut, or a **woody**-walled **mericarp**

ob- - the other way up from normal, usually flattened or widened at the **distal** rather than **proximal** end, e.g. **obovoid, obtrullate**

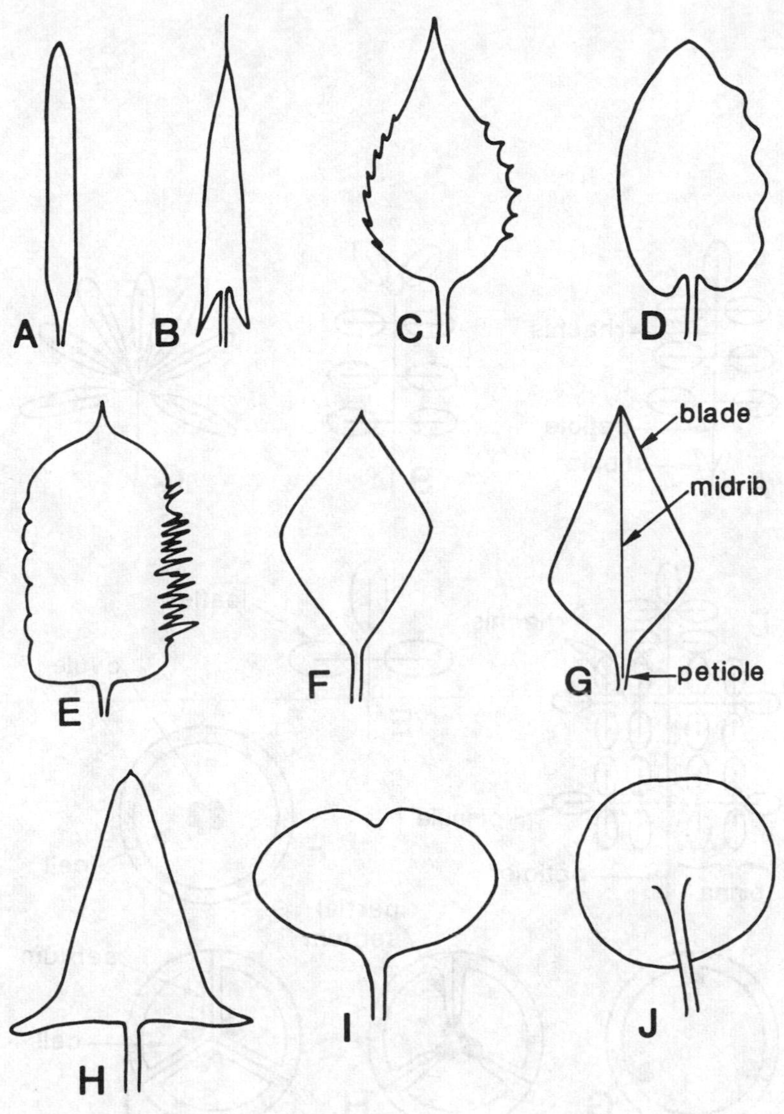

FIG 989 - Simple leaf-shapes. A, **linear**, apex acute, base cuneate, margin entire.
B, **lanceolate**, apex aristate, base sagittate, margin entire.
C, **ovate**, apex acuminate, base rounded, margin serrate (left) and dentate (right).
D, **elliptic**, apex obtuse, base cordate, margin entire (left) and sinuous (right).
E, **oblong**, apex cuspidate, base truncate, margin crenate (left) and laciniate (right).
F, **rhombic**, apex mucronate, base cuneate, margin entire.
G, **trullate**, apex acute, base cuneate, margin entire.
H, **triangular**, apex apiculate, base hastate, margin entire.
I, **transversely elliptic**, apex retuse, base rounded, margin entire.
J, **orbicular** and peltate, margin entire. Drawings by S. Ogden.

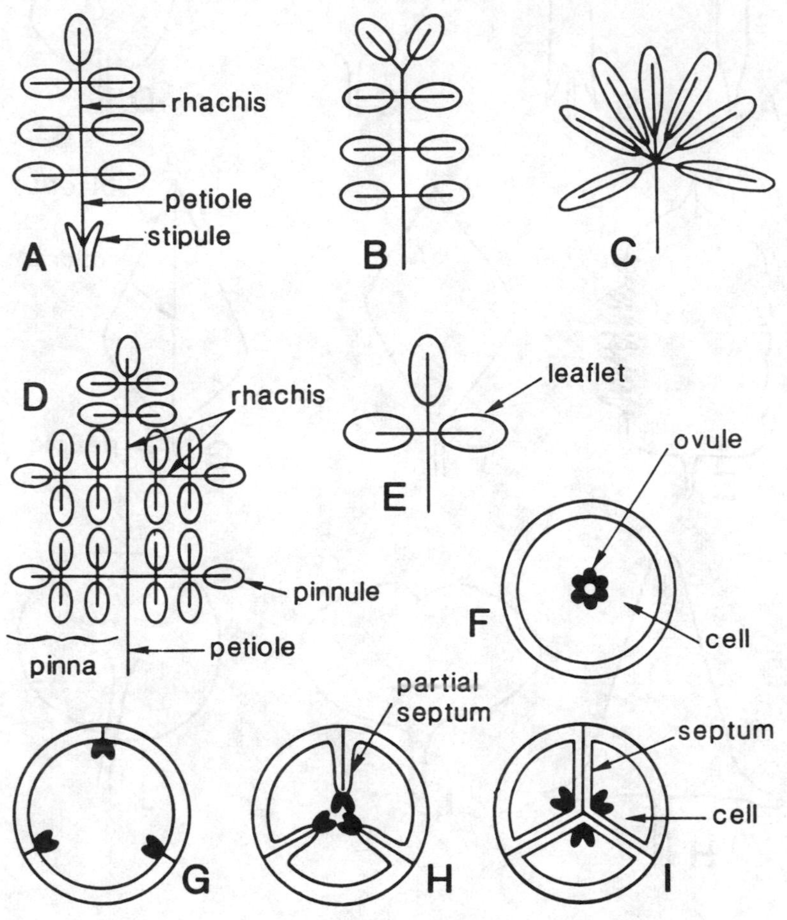

FIG 990 - A-E, Compound leaf-types. A, **pinnate** (imparipinnate, stipulate). B, **pinnate** (paripinnate). C, **palmate**. D, **2-pinnate**. E, **ternate**. F-I, ovaries in transverse section to show septa and placentation. F, 1-celled, **free-central** placentation. G, 1-celled, **parietal** placentation. H, 1-celled, **parietal** placentation. I, 3-celled, **axile** placentation. Drawings by S. Ogden.

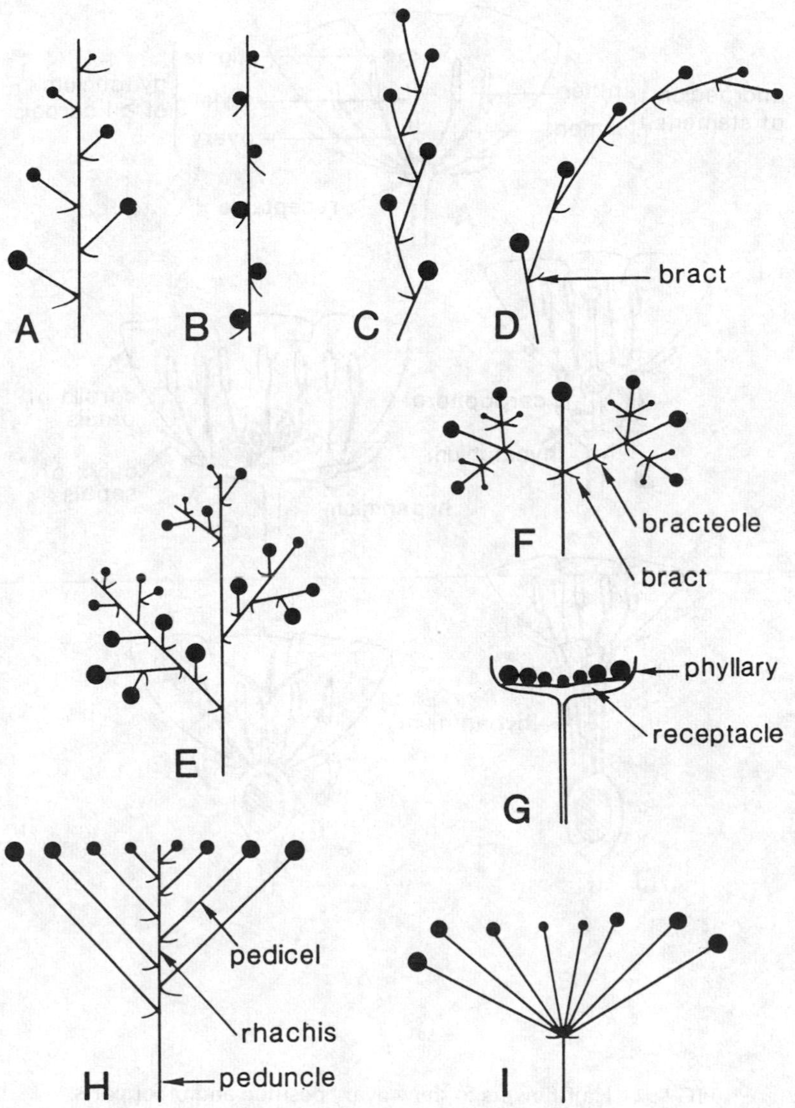

FIG 991 - Inflorescences. A, **raceme**. B, **spike**. C-D, **cymes (monochasial)**. E, **panicle**. F, **cyme (dichasial)**. G, **capitulum**. H, **corymb**. I, **umbel**. Drawings by S. Ogden.

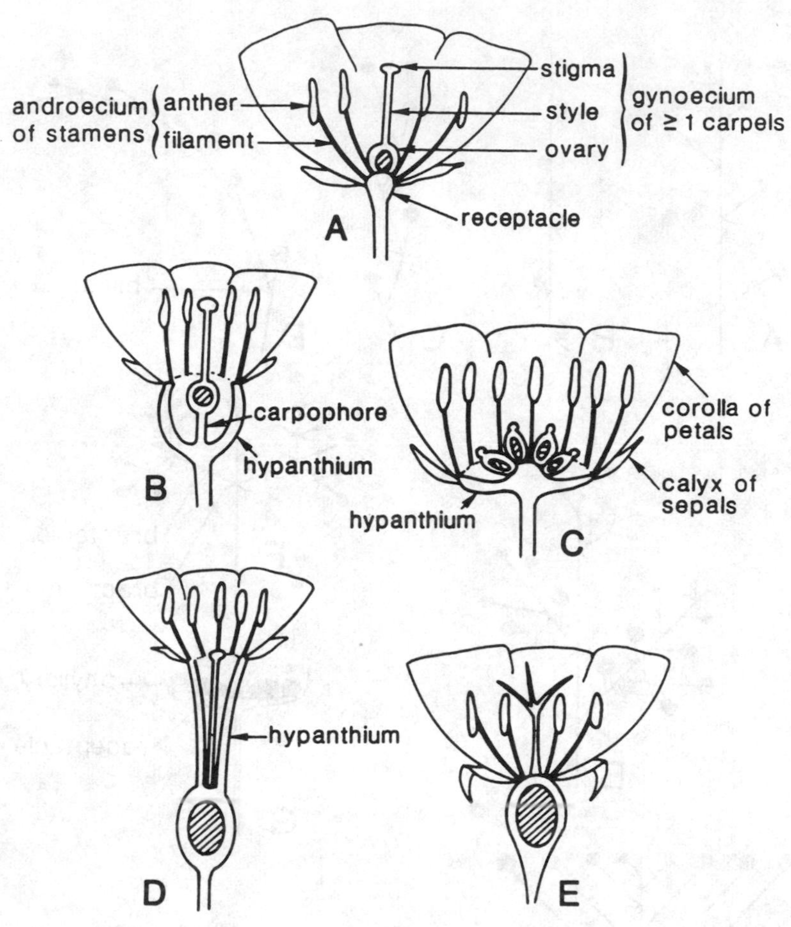

FIG 992 - Half-flowers to show ovary position and other parts.
A, **hypogynous flower**, ovary superior, carpels 1 or >1 and fused.
B, **perigynous flower** with cup-shaped hypanthium, ovary superior, with carpophore, carpels 1 or >1 and fused.
C, **perigynous flower** with flat hypanthium, ovary superior, carpels 4, free.
D, **epigynous flower** with tubular hypanthium, ovary inferior, carpels 1 or >1 and fused.
E, **epigynous flower**, ovary inferior, carpels 3, fused. Drawings by S. Ogden.

oblong - a flat shape with middle part ± parallel-sided, 1.2-3x as long as wide (if less, **broadly** so; if more, **narrowly** so); Fig 989
obtuse - with a point >90°; Fig 989
octoploid - see **polyploid**
oilseed-alien - **alien** introduced as contaminant of oilseed
opposite - of 2 organs arising laterally at 1 **node** on opposite sides of the stem
orbicular - a flat shape circular in outline; Fig 989
ovary - the basal part of the **gynoecium** containing the **ovules**; Figs 825, 990, 992
ovate - a flat shape widest nearer the base and 1.2-3x as long as wide (if less, **broadly** so; if more, **narrowly** so); Fig 989
ovoid - a solid shape **ovate** in side view
ovule - organ (inside the **ovary** in Magnoliopsida, **naked** in Pinopsida) that contains the **embryo-sac** (which in turn contains the egg) developing into the **seed** after fertilization; Fig 992

palea - see 157. Poaceae: Fig 825
palmate - a **compound** leaf, with >3 **leaflets** all arising at 1 point; Fig 990
panicle - a **compound** or much-branched **inflorescence**, either **racemose** or **cymose**; **paniculate**, in the form of a panicle; Fig 991
papilla - small nipple-like projection; **papillose** or **papillate**, covered with papillae
pappus - see 139. Asteraceae
parasite - plant that gets all or some of its nourishment by attachment (often but not always under the ground) to other plants
parietal - of a **placenta** formed by central axis of an **ovary** that is connected by **septa** to the wall; Fig 990
paripinnate - **pinnate** without an unpaired terminal **leaflet**; Fig 990
partial septum - a **septum** that is incomplete; Fig 990
patent - projecting ± at right-angles
pedicel - stalk of a flower; **pedicellate**, having a pedicel; Figs 825, 991
peduncle - stalk of a group of ≥2 flowers; **pedunculate**, having a peduncle; Fig 991
peltate - of a flat shape with its stalk arising from the plane surface, not the edge; Fig 989
pendent - hanging down
pentamerous - divided into or composed of five parts
pentaploid - see **polyploid**
perennial - living >2 years
perianth - the outer non-sexual covering layers of the flower (the **calyx** and **corolla** together), usually used when the calyx and corolla are not or little differentiated; **perianth-lobes**, **perianth-tube**, the **lobes** and **tube** of a partially fused perianth
pericarp - the wall of a **fruit**, originally the **ovary** wall
perigynous - of a flower with a **superior ovary** but with the **calyx**, **corolla** and **stamens** inserted above the base of the ovary on an extension of the **receptacle** (**hypanthium**) that is not fused with the ovary; Fig 992
persistent - remaining attached longer than normal
petal - one of the segments of the inner **whorl(s)** of the **perianth**; **petaloid**, petal-like; Fig 992
petiole - the stalk of a leaf; **petiolate**, with a petiole; Figs 989, 990
phyllary - see 139. Asteraceae; Fig 991
pinna - the primary division of a **≥2-pinnate** leaf; Fig 990
pinnate - a **compound** leaf, with >3 **leaflets** arising in **opposite** pairs along the **rhachis**; **2-**(etc)**pinnate**, pinnate with the **pinnae** pinnate again (etc.); see also **paripinnate**, **imparipinnate**; Fig 990
pinnule - the ultimate division of a **≥2-pinnate** leaf, usually applied only in ferns; Fig 990

placenta - points of origin of **ovules** in an **ovary**; **placentation**, the arrangement of placentae; Fig 990
plastic - varying in form according to environmental conditions, not according to genetic characteristics
pollen-sac - the microsporangium of a **spermatophyte**; one of the chambers in an **anther** in which the pollen is formed
pollinium - see 166. Orchidaceae
polyploid - having >2 sets of chromosomes, e.g. 3(triploid), 4(tetraploid), 5(pentaploid), 6(hexaploid), 7(heptaploid), 8(octoploid), 9(nonaploid), 10(decaploid)
prickle - spiny outgrowth with a broadened base
pricklet - small, weak **prickle**
procumbent - trailing along the ground
proliferating - with **inflorescences** bearing plantlets instead of flowers or **fruits**
pro parte - partly; in part
prothallus - small **gametophyte** generation of a plant bearing the true sex-organs, mostly applied to the free-living gametophytes of **pteridophytes**
proximal - at the end near the point of attachment, cf. **distal**
pruinose - with a **bloom**
pteridophytes - ferns and fern allies, i.e. Lycopodiopsida, Equisetopsida, Pteropsida
pubescent - with hairs; **pubescence**, the hair covering
punctate - marked with dots or transparent spots

raceme - an **inflorescence** with the oldest flowers (or **spikelets** in Poaceae) the most **proximal** lateral ones and a potentially continuously growing apex; **racemose**, in the form of a raceme; Fig 991
radiate - see 139. Asteraceae
rank - a vertical file of lateral organs; **2-ranked**, etc., with 2 (etc.) ranks of lateral organs
ray - anything that radiates outwards, e.g. branches of an **umbel**, **stigma**-ridges in *Papaver* or *Nuphar*; **ray flowers**, see 139. Asteraceae
receptacle - the usually expanded, often cup-shaped or tubular, apical part of a **pedicel** on which the flower parts are inserted; **receptacular scales** or **bristles**, see 139. Asteraceae; Figs 991, 992
recurved - curved down or back
reflexed - bent down or back
reniform - kidney-shaped
resiniferous - producing resin; **resinous**, resin-like; **resin-duct**, microscopic canal producing resin
reticulate - forming or covered with a network
retuse - with a shallow blunt notch at the apex, cf. **emarginate**
revolute - rolled back or down
rhachilla - see 157. Poaceae; Fig 825
rhachis - the axis (not the stalk) of an **inflorescence** or **pinnate** leaf; Figs 990, 991
rhizome - an underground or ground-level, usually horizontal or down-growing stem, often ± swollen; **rhizomatous**, bearing or in the form of a rhizome; cf. **stolon**
rhombic - a flat shape, widest in the middle and ± angled (not rounded) there, 1.2-3x as long as wide (if less, **broadly** so; if more, **narrowly** so); Fig 989
rigid - stiff, not flexible
rostellum - see 166. Orchidaceae
rounded - without a point or angle; Fig 989
rugose - with a wrinkled surface

sagittate - of the base of a flat organ; see Fig 989

saprophyte - a plant deriving its nourishment from decaying organisms, usually leaf-mould
scabrid - rough to the touch, with minute **prickles** or bristly hairs
scale-leaf - a leaf reduced to a small scale
scape - a flowering stem of a plant in which all the leaves are basal, none on the scape
scarious - of thin, papery texture and not green
schizocarp - a **fruit** that breaks into 1-seeded portions or **mericarps**
sclerenchyma - **woody** tissue in a partly or mostly non-woody organ
scorpioid - a **monochasial cyme** that is coiled up like a scorpion's tail when young
scrambler - a plant sprawling over other plants, fences, etc.
seed - a fertilized **ovule**
self-compatible - self-fertile, able to self-fertilize
self-incompatible - self-sterile, not able to self-fertilize
semi-inferior - of an **ovary** of which the lower part is **inferior**, but the upper part is **free** and projects above the **sepals**, etc.
sensu lato - in the broad sense
sensu stricto - in the narrow sense
sepal - one of the segments of the outer **whorl(s)** of the **perianth**; **sepaloid**, sepal-like; Fig 992
septum - a wall or membrane dividing the **ovary** into **cells**; Fig 990
serrate - with a row of ± apically directed **teeth**; **serration**, that sort of toothing; Fig 989
sessile - not stalked
sheath - see 157. Poaceae; Fig 825
short-shoot - a short stem of strictly limited growth, usually lateral on a **long-shoot**, especially on **trees** and **shrubs**
shrub - a **woody** plant that is not a **tree**
simple - not **compound**; Fig 989
sinuous - wavy, either of a hair or stalk, or of the margin of a leaf and then the sinuation in the same plane as the leaf surface; Fig 989
sinus - the space or indentation between 2 **lobes** or **teeth**; **basal sinus**, the sinus at the base of a leaf, either side of the **petiole** if present
solitary - borne singly
sorus - a group of **sporangia** in Pteropsida
soyabean-alien - **alien** introduced as contaminant of soyabeans
spadix - see 151. Araceae
spathe - an ensheathing **bract**, as in Lemnaceae, Araceae, Hydrocharitaceae
spathulate - paddle- or spoon-shaped
spermatophyte - a seed-plant, i.e. Pinopsida and Magnoliopsida
spike - a **racemose inflorescence** in which the flowers (or **spikelets** in Poaceae) have no stalks; **spikiform**, in the form of a spike; Fig 991
spikelet - see 50/1. Plumbaginaceae/*Limonium*, 156. Cyperaceae, 157. Poaceae; Fig 825
spine - a sharp, stiff, straight **woody** outgrowth, usually not greatly widened at base; **spinose**, spine-like; **spinulose**, diminutive of last; spiny, with spines
spiral - lateral organs on an axis 1 per **node**, successive ones not at 180° to each other
sporangium - a body producing **spores** in **pteridophytes**
spore - the **haploid** product of **meiotic** division, produced on the **sporophyte** and developing into the **gametophyte**
sporophyte - the **diploid** generation of a plant that bears the **sporangia** (**ovules** and **pollen-sacs** in Magnoliopsida); the main plant body of all vascular plants
spreading - growing out **divergently**, not straight or **erect**
spur - a protrusion or tubular or pouch-like outgrowth of any part of a flower

stamen - the basic male reproductive unit of Magnoliopsida, 1-many per flower, sometimes fused; Figs 825, 992
staminode - a sterile **stamen**, sometimes modified to perform some other function, e.g. that of a **petal** or **nectary**
standard - see 79. Fabaceae
stellate - star-shaped, with radiating arms
stem-leaves - leaves borne on the stem as opposed to basally
stigma - the apical part of a **gynoecium** that is receptive to pollen, simple to much branched; Figs 825, 992
stipule - one of a (usually) pair of appendages at the base of a leaf or its **petiole**, often but often not **foliaceous**; **stipulate**, with stipules; Fig 990
stolon - an **aerial** or **procumbent** stem, usually not swollen; **stoloniferous**, bearing stolons; cf. **rhizome**
style - the stalk on any **ovary** bearing the **stigma(s)**, sometimes absent; Fig 992
stylopodium - see 111. Apiaceae
sub- - almost, as in subacute, subglabrous, subglobose, subentire, subequal; sometimes under, as in subaquatic
subshrub - a **perennial** with a short **woody** surface stem producing **aerial herbaceous** stems
subtend - of a lateral organ, to have another organ in its **axil**
subulate - tapering ± constantly from a narrow base to a fine point
succulent - fleshy and juicy or pulpy
sucker - a new **aerial** shoot borne (often underground) on the roots of a **tree** or **shrub**
superior - of any **ovary** that is borne above the **calyx, corolla** and **stamens** or, if below or partly below them, then not fused laterally to the **receptacle**; Fig 992
suture - a seam of a union, often splitting open in later development

tap-root - a main descending root bearing laterals; cf. **fibrous roots**
taxon - any taxonomic grouping, such as a genus or species
tendril - a spirally coiled thread-like outgrowth from a stem or leaf, used by the plant for support
tepal - one of the segments of the **perianth**, used when **sepals** and **petals** are not differentiated
terete - rounded in section
terminal - at the very apex
ternate - a **compound** leaf with 3 **leaflets**; **2-**(etc.)**ternate**, ternate with the 3 divisions ternate again; Fig 990
testa - the outer coat of a seed
tetrad - a group of 4 **spores** or pollen grains formed by **meiotic** division
tetramerous - divided into or composed of four parts
tetraploid - see **polyploid**
throat - the opening where the **tube** joins the **limb** of a **corolla** or **calyx**
tiller - see 157. Poaceae; Fig 825
tomentose - a very dense often ± matted hair-covering
tooth - a shallow division of a leaf, **calyx** or **corolla**, or of the apex of a **capsule**; cf. **lobe, valve**
transverse - lying cross-ways; **transversely elliptic** (etc.), **elliptic** (etc.) but with the point of attachment at the side, not at one end; Fig 989
tree - a **woody** plant, usually ≥5m, with a single trunk
triangular - a flat shape widest at the base and 1.2-3x as long as wide (if less, **broadly** so; if more, **narrowly** so); Fig 989
trifid - divided into 3, usually deeply, at apex
trimerous - divided into or composed of three parts
trimorphic - occurring in 3 forms

tripartite - divided into 3 parts
triploid - see **polyploid**
trullate - a flat shape widest nearer the base and ± angled (not rounded) there, 1.2-3x as long as wide (if less, **broadly** so; if more, **narrowly** so); Fig 989
truncate - of the base or apex of a flat organ, straight or flat; Fig 989
tube - narrow, cylindrical, **proximal** part of a **calyx** or **corolla**, as distinct from the **limb, lobes** or **throat**
tuber - swollen roots or subterranean stems; **tuberous**, tuber-like
tubercle - a small ± spherical or **ellipsoid** swelling; **tuberculate**, with a surface texture covered in minute tubercles
tubular - in the form of a hollow cylinder; **tubular flowers**, see 139. Asteraceae
tufted - of elongated organs or stems that are clustered together
twig - ultimate branch of a woody stem

umbel - an **inflorescence** in which all the **pedicels** arise from one point; **compound umbel**, an umbel of umbels; **umbellate**, umbel-like; Fig 991
undulate - wavy at the edge in the plane at right-angles to the surface
unifacial - with only 1 surface, not with a **lowerside** and **upperside**
unisexual - of a plant or flower, bearing only 1 sex
upperside - the upper surface of a flat organ

valve - a deep division or **lobe** of a **capsule** apex; cf. **tooth**
vascular - pertaining to the **veins** or wood (i.e. conducting tissue) of an organ; **vascular bundle**, one anatomically discrete file of vascular tissue; **vascular plant**, a plant with vascular bundles, i.e. **pteridophytes** and **spermatophytes**
vegetative - not reproductive
vein - a strand of **vascular tissue** consisting of ≥1 **vascular bundles**; **venation**, the pattern of veins
verrucose - covered in small wart-like outgrowths
viscid - sticky
viscidium - see 166. Orchidaceae

whorl - a group of lateral organs borne >2 at each **node**; **whorled**, in the form of a whorl
wing - any **membranous** or **foliaceous** extension of an organ, e.g. a stem, **seed** or **fruit**; **winged**, with a wing; see also 79. Fabaceae
woody - hard and wood-like, not quickly dying or withering
wool-alien - an **alien** introduced as a contaminant of raw wool imports
woolly - clothed with shaggy hairs

zygomorphic - of a flower with bilateral (i.e. 1 plane of) symmetry

INDEX

Compiled by R. Gwynn Ellis

Accepted Latin names are in **bold**, synonyms in *italics* and English names in Roman. References are given to the main mention in the text and to illustrations, the latter designated by 'FIG'.

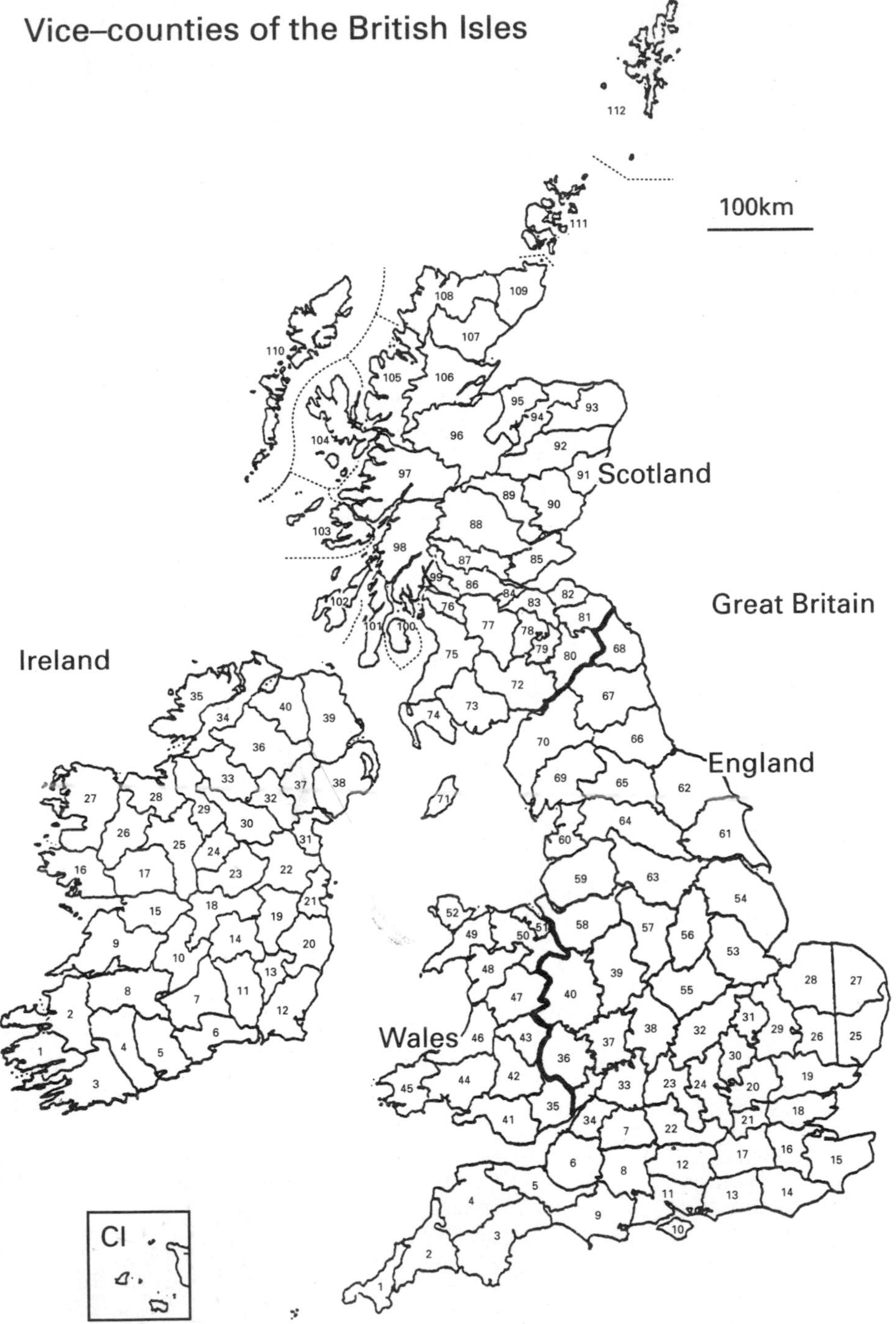

Vice–counties of the British Isles
100km
Scotland
Great Britain
Ireland
England
Wales
CI